Safety Symbols		Hazard	Examples	Precaution	Remedy
Disposal		Special disposal procedures need to be followed.	Certain chemicals, living organisms	Do not dispose of these materials in the sink or trash can.	Dispose of wastes as directed by your teacher.
Biological		Organisms or other biological materials that might be harmful to humans	Bacteria, fungi, blood, unpreserved tissues, plant materials	Avoid skin contact with these materials. Wear mask and gloves.	Notify your teacher if you suspect contact with material. Wash hands thoroughly.
Extreme Temperature		Objects that can burn skin by being too cold or too hot	Boiling liquids, hot plates, dry ice, liquid nitrogen	Use proper protection when handling.	Go to your teacher for first aid.
Sharp Object		Use of tools or glassware that can easily puncture or slice skin	Razor blades, pins, scalpels, pointed tools, dissecting probes, broken glass	Practice common-sense behavior and follow guidelines for use of the tool.	Go to your teacher for first aid.
Fume		Possible danger to respiratory tract from fumes	Ammonia, acetone, nail polish remover, heated sulfur, moth balls	Be sure there is good ventilation. Never smell fumes directly. Wear a mask.	Leave foul area and notify your teacher immediately.
Electrical		Possible danger from electrical shock or burn	Improper grounding, liquid spills, short circuits, exposed wires	Double-check setup with teacher. Check condition of wires and apparatus.	Do not attempt to fix electrical problems. Notify your teacher immediately.
Irritant		Substances that can irritate the skin or mucous membranes of the respiratory tract	Pollen, moth balls, steel wool, fiberglass, potassium permanganate	Wear dust mask and gloves. Practice extra care when handling these materials.	Go to your teacher for first aid.
Chemical		Chemicals that can react with and destroy tissue and other materials	Bleaches such as hydrogen peroxide; acids such as sulfuric acid, hydrochloric acid; bases such as ammonia, sodium hydroxide	Wear goggles, gloves, and an apron	Immediately flush the affected area with water and notify your teacher.
Toxic		Substance may be poisonous if touched, inhaled, or swallowed.	Mercury, many metal compounds, iodine, poinsettia plant parts	Follow your teacher's instructions.	Always wash hands thoroughly after use. Go to your teacher for first aid.
Flammable		Flammable chemicals may be ignited by open flame, spark, or exposed heat.	Alcohol, kerosene, potassium permanganate	Avoid open flames and heat when using flammable chemicals.	Notify your teacher immediately. Use fire safety equipment if applicable.
Open Flame		Open flame in use, may cause fire.	Hair, clothing, paper, synthetic materials	Tie back hair and loose clothing. Follow teacher's instruction on lighting and extinguishing flames.	Notify your teacher immediately. Use fire safety equipment if applicable.

Eye Safety

Proper eye protection should be worn at all times by anyone performing or observing science activities.

Clothing Protection

This symbol appears when substances could stain or burn clothing.

Radioactivity

This symbol appears when radioactive materials are used.

Handwashing

After the lab, wash hands with soap and water before removing goggles.

GLENCOE

PHYSICS

PRINCIPLES & PROBLEMS

Zitzewitz

Haase

Harper

Navid Baraty/Flickr/Getty Images

Mc Graw Hill Education

Bothell, WA • Chicago, IL • Columbus, OH • New York, NY

ABOUT THE COVER

How did he do that?! The stack of rocks on the cover might look impossible, but it's real! Bill Dan is a world-renowned rock balance artist. He has been balancing rocks along the shores of San Francisco for most of his life and has helped to elevate rock stacking to an art form. Learn more about how rock stacking uses the principles of physics by watching Physics TV.

Cover Navid Baraty/Flickr/Getty Images
Image above Courtesy of Bill Dan

The McGraw-Hill Companies

connectED.mcgraw-hill.com

Education

Send all inquiries to:
McGraw-Hill Education
STEM Learning Solutions Center
8787 Orion Place
Columbus, OH 43240-4027

ISBN: 978-0-07-659252-4
MHID: 0-07-659252-9
Printed in the United States of America.

8 9 DOW 16 15 14

McGraw-Hill is committed to providing instructional materials in Science, Technology, Engineering, and Mathematics (STEM) that give all students a solid foundation, one that prepares them for college and careers in the 21st century.

Courtesy Bill Dan

CONTENTS IN BRIEF

Full House Images/Riser/Getty Images

PHYSICS ONLINE

Introducing

iLab Station

At connectED.mcgraw-hill.com, you will find this one-stop online resource for laboratory investigations and their worksheets. Look for the iLab Station button throughout your eStudentEdition for links to these laboratory experiences.

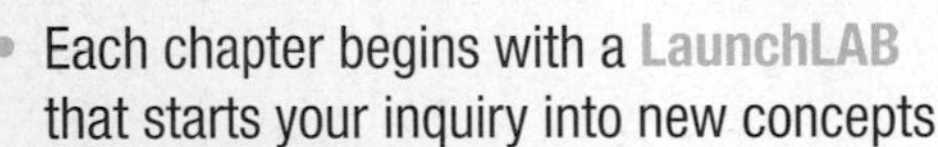

- Each chapter begins with a LaunchLAB that starts your inquiry into new concepts.
- MiniLABs and PhysicsLABs are found throughout each chapter, providing both quick and more in-depth investigations of key physics concepts.
- Online, editable lab worksheets from the iLab Station will let your teacher tailor the investigation to your specific needs.

In the iLab Station at connectED.mcgraw-hill.com, you will find laboratory investigations, procedures, and worksheets:

Date ____ Period ____ Name ____

CHAPTER 14 LaunchLAB

WAVES ON A SPRING

Procedure

1. Read the procedure and the safety information, and complet
2. Stretch out a coiled spiral spring, but do not overstretch it. O end still, while the other person generates a sideways pulse pulse while it travels along the spring and when it hits the h observations.
3. Repeat step 2 with a larger pulse. Record your observation
4. Generate a different pulse by compressing the spring at or your observations.
5. Generate a third type of pulse by twisting one end of the Record your observations.

Analysis

6. What happens to the pulses as they travel through the s the end of the spring? How did the pulse in step 2 comp
7. **Critical Thinking** What are some properties that seem through the spring?

Date ____ Period ____ Name ____

CHAPTER 14 MiniLAB

WAVE INTERACTION

Procedure

1. Read the procedure and the safety information, and complete the lab form.
2. Design an experiment using a **coiled-spring toy** to test what happens when waves from different directions meet.
3. Perform your experiment and record your observations.

Analysis

4. Does the speed of either wave change?
5. Do the waves bounce off each other, or do they pass through each other?

Date ____ Period ____ Name ____

PhysicsLAB

PENDULUM VIBRATIONS

Design Your Own

Background: A pendulum can provide a simple model for the investigation of wave properties. In this investigation, you will design a procedure to use the pendulum to examine amplitude, period, and frequency of a wave. You also will determine the gravitational field strength from the period and length of the pendulum.

Question

How can a pendulum demonstrate the properties of waves?

Objectives

- **Determine** what variables affect a pendulum's period.
- **Investigate** the frequency, period, and amplitude of a pendulum.
- **Measure** *g*, the gravitational field strength, using the period and length of a pendulum.

Procedure

1. Read the procedure and the safety information, and complete the lab form.
2. Design a pendulum using a ring stand, a string with a paper clip, and a sinker attached to the paper clip. Be sure to check with your teacher and have your design approved before you proceed with the lab.

Go online!
connectED.mcgraw-hill.com

"It's EASY to get my assignments online and QUICK to find everything I need."

CHAPTER 4

Forces in One Dimension

BIGIDEA Net forces cause changes in motion.

SECTIONS
1 Force and Motion
2 Weight and Drag Force
3 Newton's Third Law

LaunchLAB iLab Station

FORCES IN OPPOSITE DIRECTIONS
What happens when more than one force acts on an object?

WATCH THIS! Video

APPARENT WEIGHT
Have you ever had the sensation of feeling weightless while on a freefall ride? An elevator acts in much the same way, giving the sensation of being lighter when you're moving down and heavier when you're moving up.

(l)The McGraw-Hill Companies, (r)James Lauritz/Digital Vision/Getty Images

88 Chapter 4 • Forces in One Dimension

Your introduction to the chapter concepts begins with PHYSICS T.V. videos, connecting physics to your life.

Fancy Photography/Veer

A² DEVELOPMENT PROCESS

Why A²?

Accuracy Assurance is central to McGraw-Hill's commitment to high-quality, learner-oriented, real-world, and error-free products. Also at the heart of our A² Development Process is a commitment to make the text **Accessible** and **Approachable** for both students and teachers. A collaboration among authors, content editors, academic advisors, and classroom teachers, the A² Development Process provides opportunities for continual improvement through customer feedback and thorough content review.

The A² Development Process begins with a review of the previous edition and a look forward to state and national standards. The authors for *Physics: Principles and Problems* combine expertise in teacher training and education with a mastery of physics content knowledge. As manuscript is created and edited, consultants review the accuracy of the content while our Teacher Advisory Board members examine the program from the points of view of both teacher and student. Student labs, technology resources, and teacher demonstrations are reviewed for both accuracy of content and safety. As design elements are applied, chapter text is again reviewed, as are photos and diagrams.

Throughout the life of the program, McGraw-Hill continues to troubleshoot and incorporate improvements. Our goal is to deliver to you a program that has been created, refined, tested, and validated as a successful tool for your continued **Academic Achievement.**

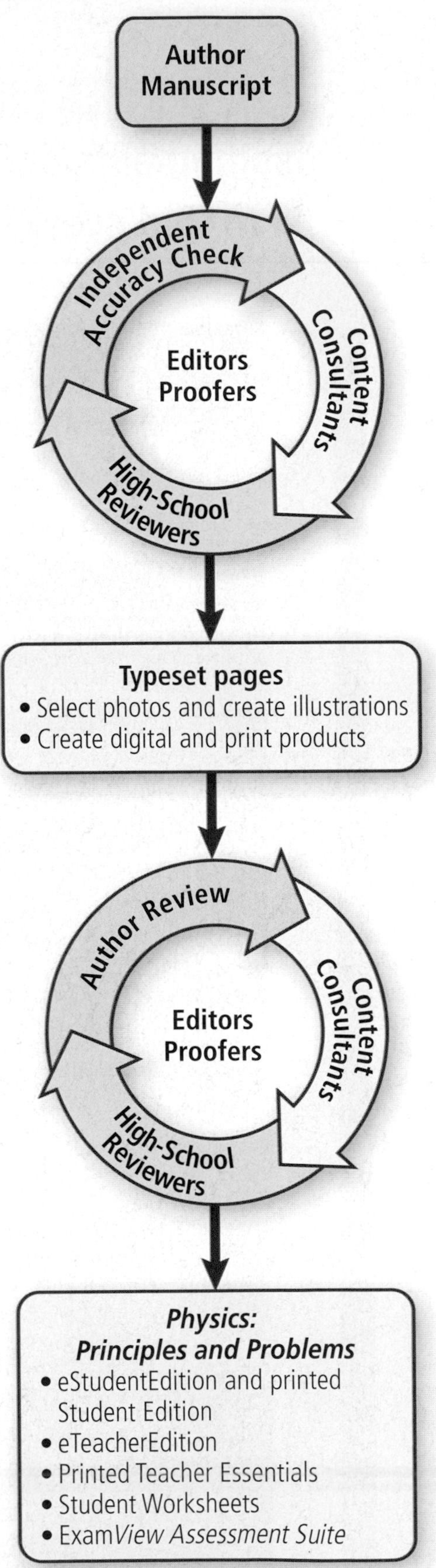

Authors

The authors of *Physics: Principles and Problems* used their physics content knowledge and teaching expertise to craft manuscript that is accessible, accurate, and geared toward student achievement.

Paul W. Zitzewitz, lead author

is an emeritus professor of physics and of science education at the University of Michigan–Dearborn. He received his BA from Carleton College and his MA and PhD from Harvard University, all in physics. Dr. Zitzewitz taught physics to undergraduates at the University of Michigan-Dearborn for 36 years and, as an active experimenter in the field of atomic physics, published more than 50 research papers. He was named a Fellow of the American Physical Society for his contributions to physics and science education for high school and middle school teachers and students. He is now treasurer of the American Association of Physics Teachers and has been the president of the Michigan Section of the American Association of Physics Teachers and chair of the American Physical Society's Forum on Education.

David G. Haase

is an Alumni Distinguished Undergraduate Professor of Physics at North Carolina State University. He earned a BA in physics and mathematics at Rice University and an MA and a PhD in physics at Duke University, where he was a J.B. Duke Fellow. He has been an active researcher in experimental low-temperature and nuclear physics. He teaches undergraduate and graduate physics courses and has worked many years in K–12 teacher training. He was the founding director of The Science House at NC State, a science-math learning center that leads teacher training and student programs across North Carolina. He has co-authored over 100 papers in experimental physics and in science education. He is a Fellow of the American Physical Society. He received the Alexander Holladay Medal for Excellence, NC State University; was awarded the Pegram Medal for Physics Teaching Excellence; and was chosen 1990 Professor of the Year in the state of North Carolina by the Council for the Advancement and Support of Education (CASE).

Kathleen A. Harper

is an auxiliary faculty member in the Engineering Education Innovation Center at The Ohio State University. She received her MS in physics and BS in electrical engineering and applied physics from Case Western Reserve University and her PhD in physics from The Ohio State University. She has taught introductory physics, astronomy, and engineering courses to undergraduates for nearly 20 years and has been instrumental in offering Modeling Instruction workshops to in-service high school teachers, both in Ohio and nationwide. Her research interests include the teaching and learning of problem-solving skills and the development of alternative problem formats. She is active in AAPT, both on the local and national levels, often presenting talks and workshops on teaching problem solving. Additionally, she is co-editor of a selection of articles available through AAPT's Compadre portal entitled, "Getting Started in Physics Education Research."

Teacher Advisory Board

The Teacher Advisory Board gave the editorial staff and design team feedback on the content and design of both the Student Edition and the Teacher Essentials. They were instrumental in providing valuable input toward the development of the 2013 edition of *Physics: Principles and Problems.* We thank these teachers for their hard work and creative suggestions.

Janet Adams
Physics Teacher
Mars Area High School
Mars, PA

Craig Dowler
Physics Teacher
West Genesee High School
Camillus, NY

Chris Foust, M.S.
Physics Teacher
Hermitage High School
Richmond, VA

Ryan Hall
Physics Teacher
Palatine High School
Palatine, IL

Stan Hutto
Physics Teacher
Alamo Heights High School
San Antonio, TX

Richard A. Lines, Jr.
Physics Teacher
Garland High School
Garland, TX

Nikki Malatin
Physics Teacher
West Caldwell High School
Lenoir, NC

Jennifer McDonnell
Coordinator of Math and Science
School District U-46
Elgin, IL

Jeremy Paschke
Physics Teacher
York High School
Elmhurst, IL

Don Pata
Physics Teacher
Grosse Pointe North High School
Grosse Pointe Woods, MI

Charles Payne
Physics Teacher
Northern High School
Durham, NC

Content Consultants

Content consultants each reviewed selected chapters of *Physics: Principles and Problems* for content accuracy and clarity.

Dr. Solomon Bililign, PhD
Professor of Physics
Director: NOAA-ISET Center
North Carolina A&T State University
Greensboro, NC

Ruth Howes
Professor Emerita of Physics and Astronomy
Ball State University
Muncie, IN

Dr. Keith H. Jackson
Professor and Chair of Department of Physics
Morgan State University
Baltimore, MD

Kathleen Johnston
Associate Professor of Physics
Louisiana Tech University
Ruston, LA

Dr. Monika Kress
Associate Professor
Department of Physics & Astronomy
San Jose State University
San Jose, CA

Jorge Lopez
Professor
University of Texas at El Paso
El Paso, TX

Dr. Ramon E. Lopez
Department of Physics
University of Texas at Arlington
Arlington, TX

Albert J. Osei
Professor of Physics
Oakwood University
Huntsville, AL

Charles Ruggiero
Visiting Assistant Professor
Allegheny College
Meadeville, PA

Toni Sauncy
Associate Professor of Physics
Angelo State University
San Angelo, TX

Sally Seidel
Professor of Physics
University of New Mexico
Albuquerque, NM

Contributing Writers

Additional science writers added feature content, teacher materials, assessment, and laboratory investigations.

Molly Wetterschneider
Austin, TX

Steve Whitt
Columbus, OH

Safety Consultant

The safety consultant reviewed labs and lab materials for safety and implementation.

Kenneth Russell Roy, PhD
Director of Science and Safety
Glastonbury Public Schools
Glastonbury, CT

Teacher Reviewers

Each teacher reviewed selected chapters of *Physics: Principles and Problems* and provided feedback and suggestions regarding the effectiveness of the instruction.

Joseph S. Bonanno
Physics Teacher
Red Creek Central School
Red Creek, NY

Beverly Trina Cannon
Physics Teacher
Highland Park High School
Dallas TX

Katharine Chole
Physics Teacher
Villa Duchesne School
St. Louis, MO

Cheryl Rawlins Cowley
Lead Physics Teacher
Sherman High School
Sherman, TX

B. Wayne Davis
Physics Teacher (retired)
Henrico County Public Schools
Richmond, VA

Nina Morley Daye
Physics Teacher
Orange High School
Hillsborough, NC

David Eberst
Physics Teacher
Bishop Watterson High School
Columbus, OH

Terry Elmer
Physics Teacher
Red Creek Central School
Red Creek, NY

Michael Fetsko
Physics Teacher
Mills E. Godwin High School
Henrico, VA

Chris Foust, M.S.
Physics Teacher
Hermitage High School
Richmond, VA

Elaine Gwinn
Physics Teacher
Shenandoah High School
Middletown, IN

Janie Head
Physics Teacher
Foster High School
Richmond, TX

Stan Hutto
Physics Teacher
Alamo Heights High School
San Antonio, TX

Emily James
Physics Teacher
Brewster Academy
Wolfeboro, NH

Dr. Christopher D. Jones
School of Applied and Engineering Physics
Cornell University
Ithaca, NY

Dr. Mike Papadimitriou
Headmaster
Academy for Science and Health Professions
Conroe, TX

Julia Quaintance
Physics Teacher
Morgan Local High School
McConnelsville, OH

Stephen Rea
University of Michigan, Dearborn
Plymouth, MI

Patricia Rollison
Physics Teacher
St. Gertrude High School
Richmond, VA

Patrick Slattery
Physics Teacher
South Elgin High School
South Elgin, IL

James Stankevitz
Physics Teacher
Wheaton Warrenville South High School
Wheaton, IL

Jason Sterlace
Physics Teacher
J.R. Tucker High School
Henrico, VA

Christopher White
Physics Teacher
Seneca High School
Seneca, SC

Michael Young, MS
ACT, Inc.
Iowa City, IA

Tom Young
Physics Teacher
Whitehouse High School
Whitehouse, TX

ONLINE PROBLEMS, HELP, & INVESTIGATIONS

Example Problems throughout each chapter give you step-by-step instructions and hints on how to solve each type of physics problem. Each Example Problem is followed by a number of Practice Problems on the same topic, allowing you to practice the skill.

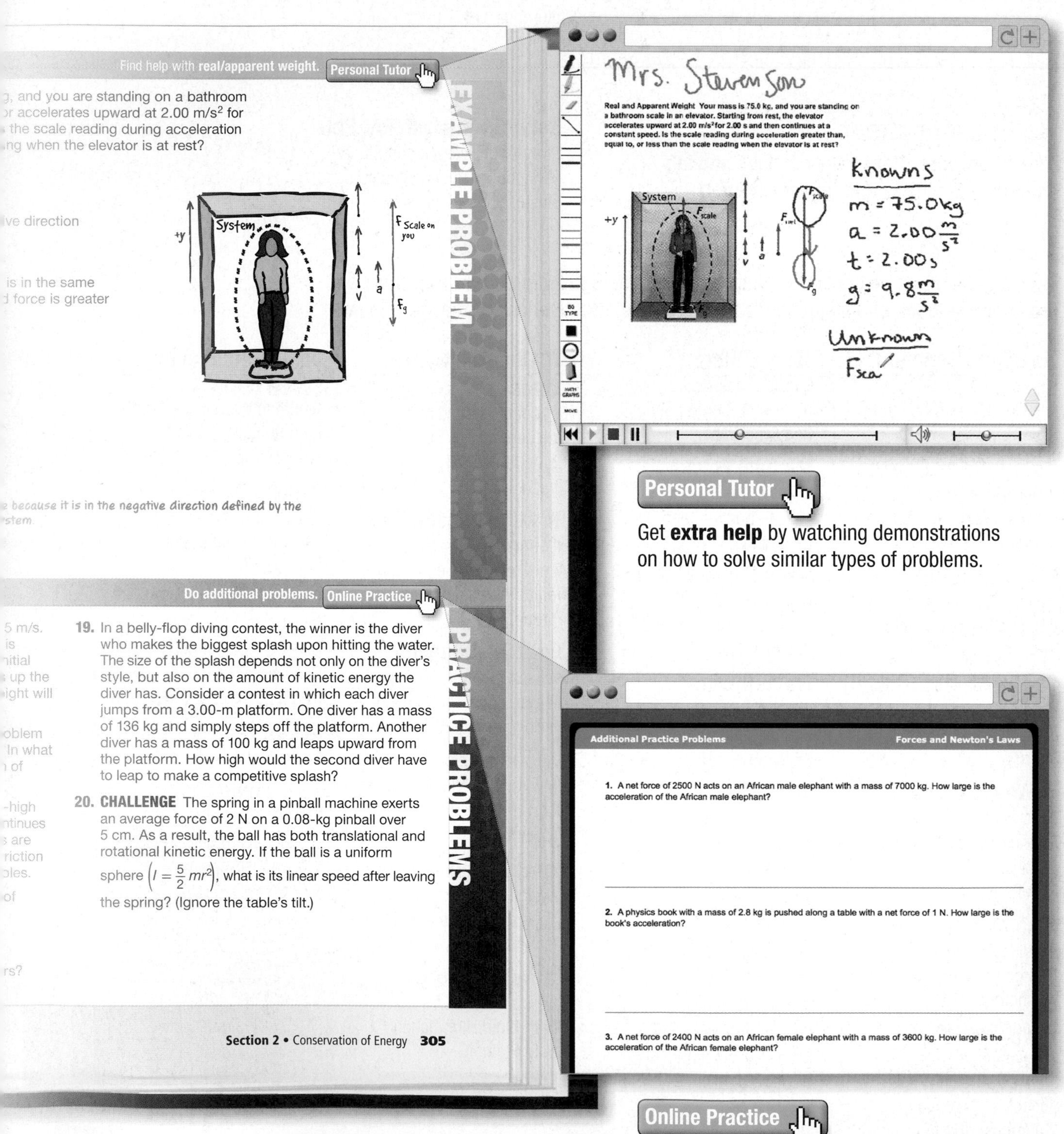

Find help with real/apparent weight. Personal Tutor
EXAMPLE PROBLEM
System
+y
F Scale on you
a
v
F_g
Mrs. Stevenson
Real and Apparent Weight Your mass is 75.0 kg, and you are standing on a bathroom scale in an elevator. Starting from rest, the elevator accelerates upward at 2.00 m/s² for 2.00 s and then continues at a constant speed. Is the scale reading during acceleration greater than, equal to, or less than the scale reading when the elevator is at rest?
Knowns
m = 75.0kg
a = 2.00 m/s²
t = 2.00s
g = 9.8 m/s²
Unknown
Fsca
Do additional problems. Online Practice
PRACTICE PROBLEMS
19. In a belly-flop diving contest, the winner is the diver who makes the biggest splash upon hitting the water. The size of the splash depends not only on the diver's style, but also on the amount of kinetic energy the diver has. Consider a contest in which each diver jumps from a 3.00-m platform. One diver has a mass of 136 kg and simply steps off the platform. Another diver has a mass of 100 kg and leaps upward from the platform. How high would the second diver have to leap to make a competitive splash?
20. CHALLENGE The spring in a pinball machine exerts an average force of 2 N on a 0.08-kg pinball over 5 cm. As a result, the ball has both translational and rotational kinetic energy. If the ball is a uniform sphere ($I = \frac{5}{2}mr^2$), what is its linear speed after leaving the spring? (Ignore the table's tilt.)
Section 2 • Conservation of Energy 305
Additional Practice Problems
Forces and Newton's Laws
1. A net force of 2500 N acts on an African male elephant with a mass of 7000 kg. How large is the acceleration of the African male elephant?
2. A physics book with a mass of 2.8 kg is pushed along a table with a net force of 1 N. How large is the book's acceleration?
3. A net force of 2400 N acts on an African female elephant with a mass of 3600 kg. How large is the acceleration of the African female elephant?

Personal Tutor

Get **extra help** by watching demonstrations on how to solve similar types of problems.

Online Practice

Additional problems provide opportunities to practice the skills and concepts associated with each Example Problem.

Go **online!**
connectED.mcgraw-hill.com

"The e-book includes VIDEOS, ANIMATIONS, and tools to help me LEARN."

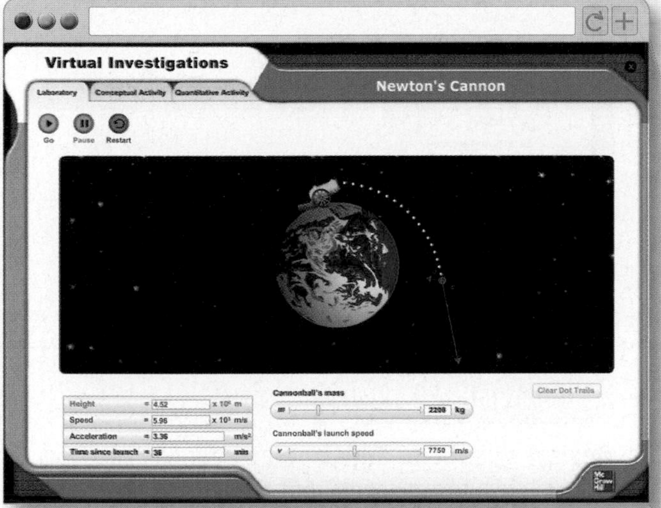

Virtual Investigations

Take labs to the next level, launching rockets, firing cannons, exploring electrons, and more in an **online laboratory.**

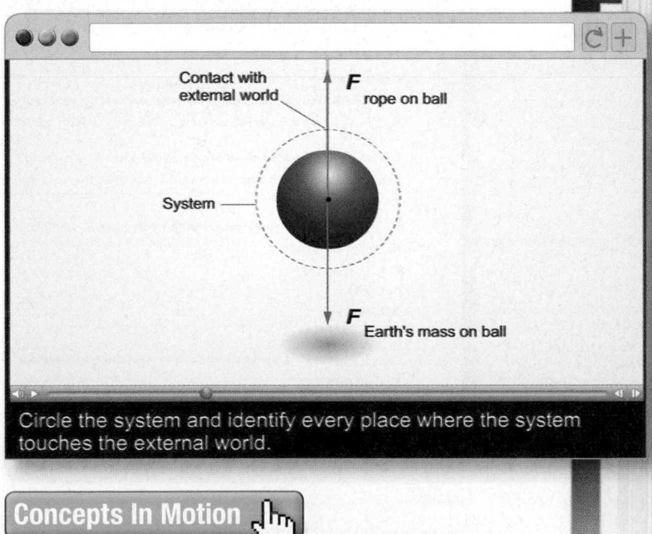

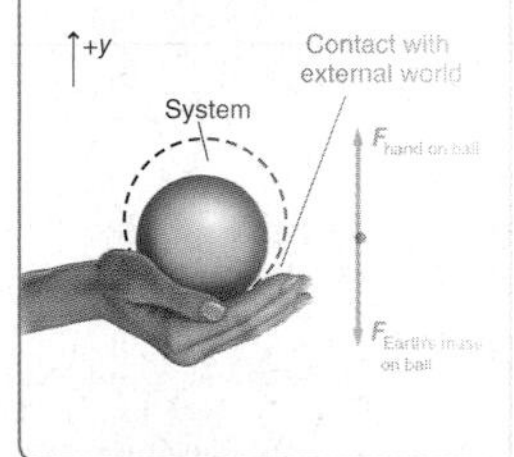

Figure 3 The drawings of the ball in the hand and the ball hanging from the string are both pictorial models. The free-body diagram for each situation is shown next to each pictorial model.

View an **animation of free body diagrams.**

Concepts In Motion

Concepts In Motion

See physics content come to life in **animated figures** and moving diagrams.

Ocean/CORBIS

ONLINE STUDY TOOLS

Quizzes, assessments, and study tools provide opportunities for self-assessment, review, and additional practice.

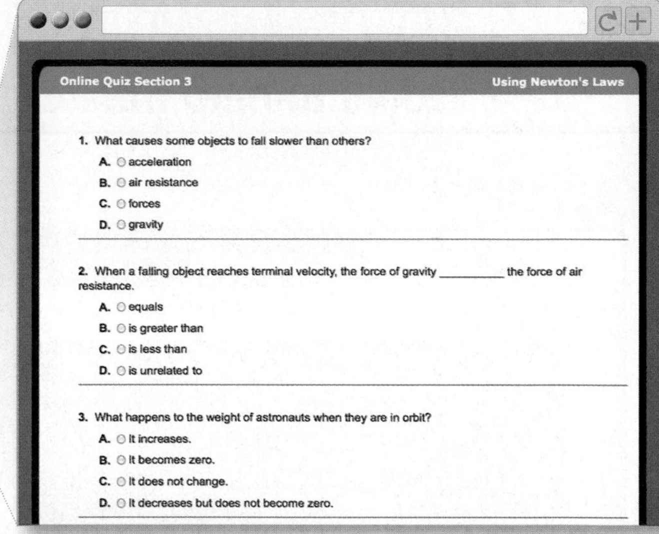

Online Quiz Section 3 — Using Newton's Laws

1. What causes some objects to fall slower than others?
   - A. acceleration
   - B. air resistance
   - C. forces
   - D. gravity

2. When a falling object reaches terminal velocity, the force of gravity ________ the force of air resistance.
   - A. equals
   - B. is greater than
   - C. is less than
   - D. is unrelated to

3. What happens to the weight of astronauts when they are in orbit?
   - A. It increases.
   - B. It becomes zero.
   - C. It does not change.
   - D. It decreases but does not become zero.

Section Self-Check

Review questions and problems for each section help you identify concepts that require additional study.

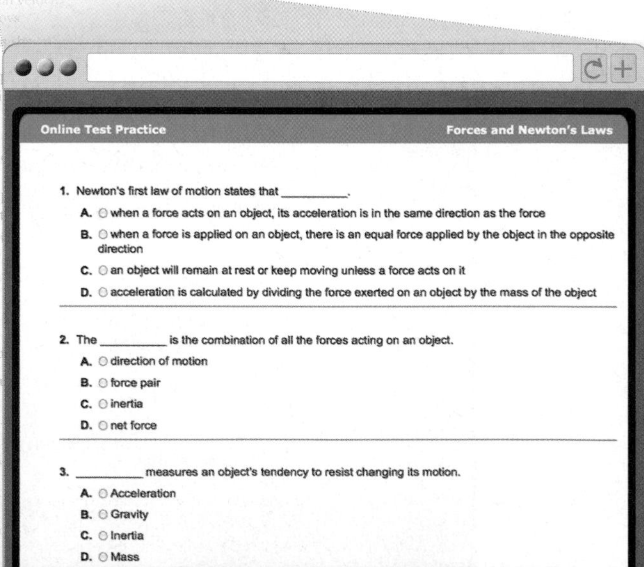

Online Test Practice — Forces and Newton's Laws

1. Newton's first law of motion states that ________.
   - A. when a force acts on an object, its acceleration is in the same direction as the force
   - B. when a force is applied on an object, there is an equal force applied by the object in the opposite direction
   - C. an object will remain at rest or keep moving unless a force acts on it
   - D. acceleration is calculated by dividing the force exerted on an object by the mass of the object

2. The ________ is the combination of all the forces acting on an object.
   - A. direction of motion
   - B. force pair
   - C. inertia
   - D. net force

3. ________ measures an object's tendency to resist changing its motion.
   - A. Acceleration
   - B. Gravity
   - C. Inertia
   - D. Mass

Chapter Self-Check

Chapter-level **study tools** provide more opportunities for review and practice.

Go **online!**
connectED.mcgraw-hill.com

"I have the online STUDY TOOLS I need to help me SUCCEED."

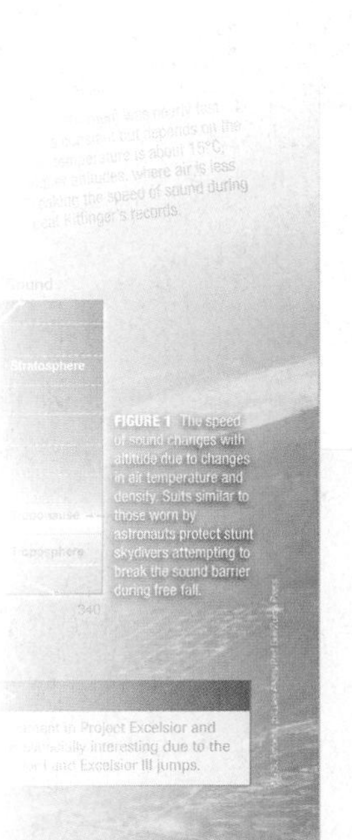

- weight (p. 102)
- gravitational field (p. 100)
- apparent weight (p. 102)
- weightlessness (p. 102)
- drag force (p. 104)
- terminal velocity (p. 105)

VOCABULARY
- **interaction pair** (p. 106)
- **Newton's third law** (p. 106)
- **tension** (p. 109)
- **normal force** (p. 111)

Games and Multilingual eGlossary

Vocabulary Practice

Vocabulary Practice

The **multilingual e-Glossary** and vocabulary study tools drive home important concepts.

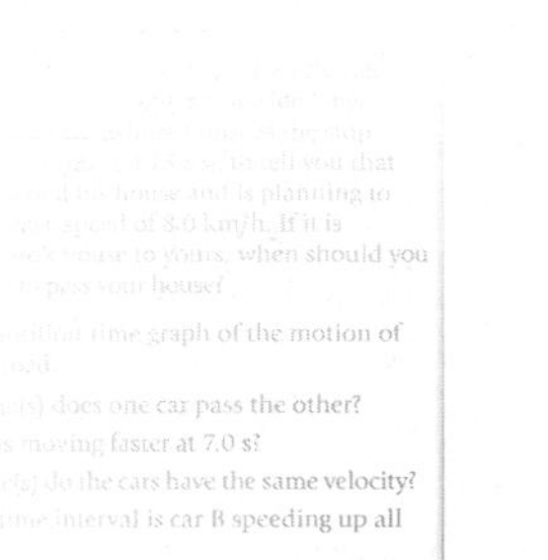

3. If a motorcycle starts from rest and maintains a constant acceleration of 3 m/s^2, what will its velocity be after 10 s?
   a. 10 m/s c. 90 m/s
   b. 30 m/s d. 100 m/s
4. How does an object's acceleration change if the net force on the object is doubled?
   a. The acceleration is cut in half.
   b. The acceleration does not change.
   c. The acceleration is doubled.
   d. The acceleration is multiplied by four.
5. What is the weight of a 225-kg space probe on the Moon? The gravitational field on the Moon is 1.62 N/kg.
   a. 139 N c. 1.35×10^3 N
   b. 364 N d. 2.21×10^3 N

Online Test Practice

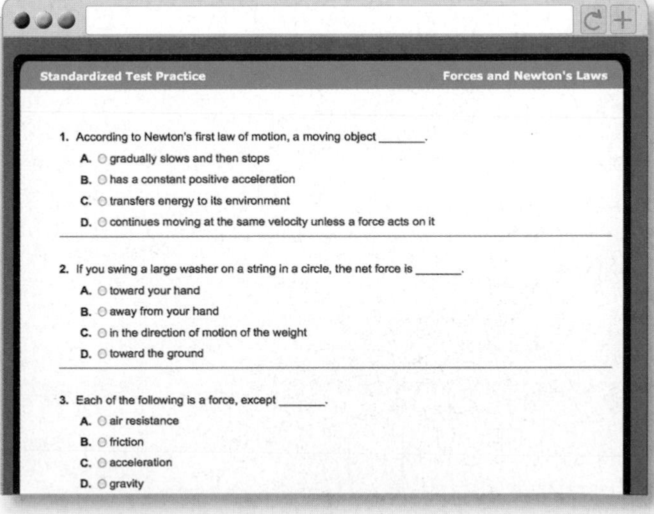

Online Test Practice

Practice taking **standardized tests** as you also review chapter content.

PhotoDisc/SuperStock

TABLE OF CONTENTS

Your eStudentEdition, found at connectED.mcgraw-hill.com, includes interactive features to improve your understanding of the physics content presented in each chapter.

Concepts in Motion include animations and interactive diagrams to help explain topics and eliminate misconceptions. BrainPOP Videos connect physics to your world. Virtual Investigations are online laboratory experiences, and Personal Tutors provide video demonstrations of solutions to various example problems.

Throughout each chapter, you will see references to laboratory experiences from the iLabStation, your online resource for lab activities and worksheets.

Each chapter begins with a LaunchLAB, an introductory laboratory investigation designed to introduce the concepts in that chapter. MiniLABs are short investigations that can improve your understanding of physics content. You will also find one or more PhysicsLABs in each chapter, providing opportunities for more in-depth investigations.

CHAPTER 1

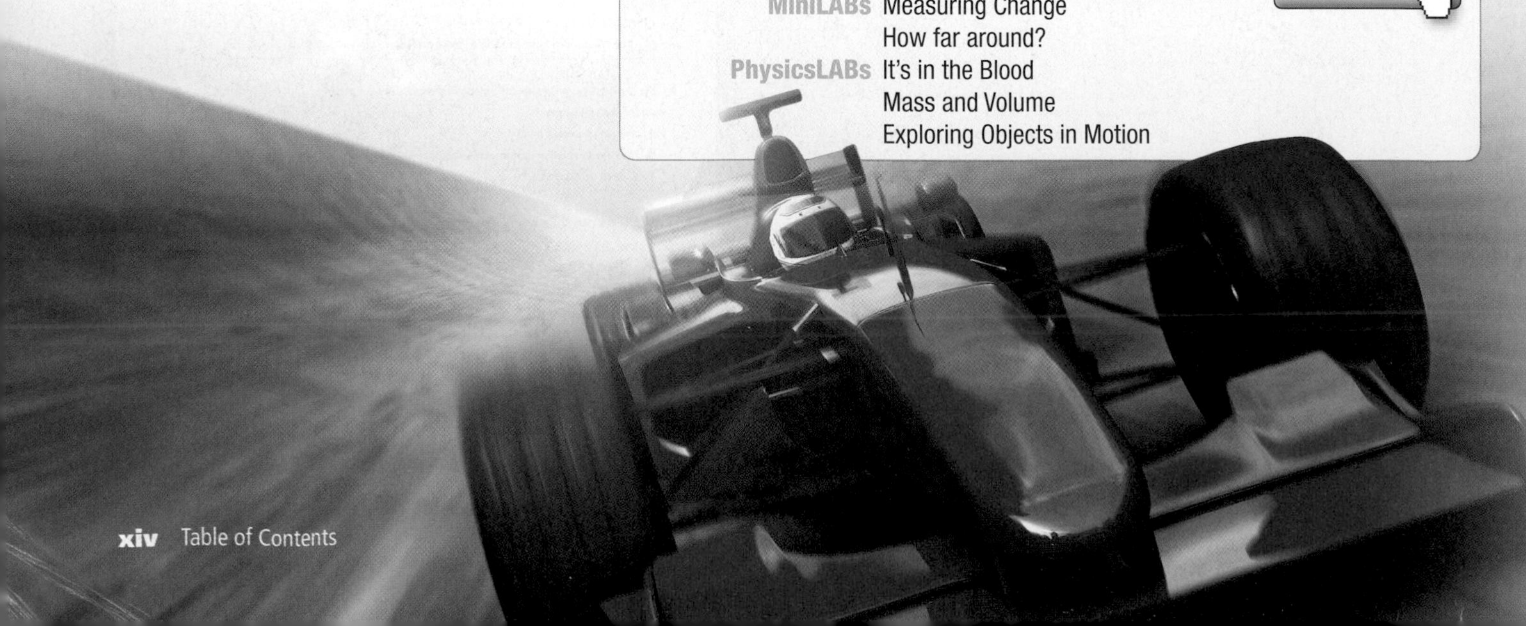

Colin Anderson/Photographer's Choice/Getty Images

Mechanics

CHAPTER 2

Representing Motion 32

Go online!

Concepts in Motions: Motion Diagrams v. Particle Motion; Position-Time Graphs
Personal Tutor: Average Velocity and Average Speed

iLab Station

LaunchLAB: Toy Car Race
MiniLABs: Vector Models; Velocity Vectors
PhysicsLABs: Motion Diagrams; Constant Speed; Measure Velocity

CHAPTER 3

Accelerated Motion 58

Go online!

Concepts in Motion: Velocity v. Time Graphs
Virtual Investigation: Investigate Accelerated Motion
BrainPOP: Acceleration
Personal Tutor: Displacement

iLab Station

LaunchLAB: Graphing Motion
MiniLABs: Steel Ball Race; Free Fall
PhysicsLABs: Acceleration; Tossed-Ball Motion; Measuring Acceleration; Free-Fall Acceleration

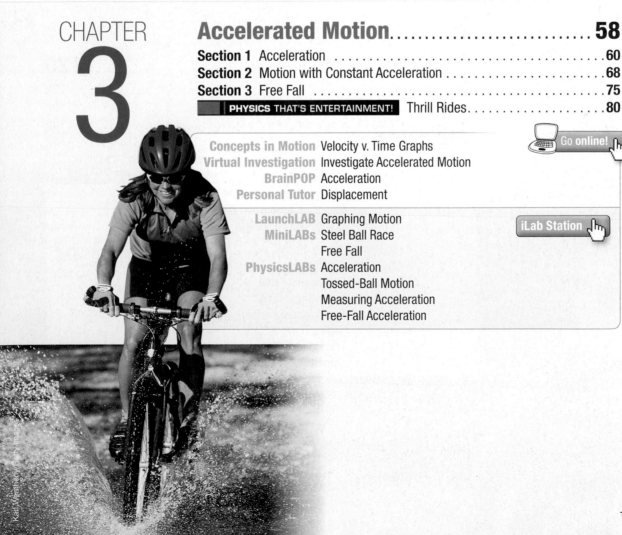

Karl Weatherly/Getty Images

Mechanics

CHAPTER 4

Forces in One Dimension 88

CHAPTER 5

Displacement and Force in Two Dimensions 120

David Young-Wolff/PhotoEdit

Go online! connectED.mcgraw-hill.com

Mechanics

CHAPTER 6

Go online!

Concepts in Motion Circular Motion and Centripetal Acceleration
Virtual Investigation Soccer Kick
Personal Tutor Flight of a Ball

iLab Station

LaunchLAB Projectile Motion
MiniLABs Over the Edge
Projectile Path
PhysicsLABs Launch an Investigation
On Target
Centripetal Force
Moving Reference Frame

CHAPTER 7

Go online!

Concepts in Motions Kepler's First Law
Kepler's Second Law
Kepler's Third Law
Law of Universal Gravitation
Cavendish Experiment
Virtual Investigations Universal Gravitation
Newton's Cannon
Personal Tutor Orbital Speed

iLab Station

LaunchLAB Model Mercury's Motion
MiniLABs Weightless Water
Weight in Free Fall
PhysicsLABs Modeling Orbits
How can you measure mass?
Inertial Mass and Gravitational Mass

Mechanics

CHAPTER 8

CHAPTER 9

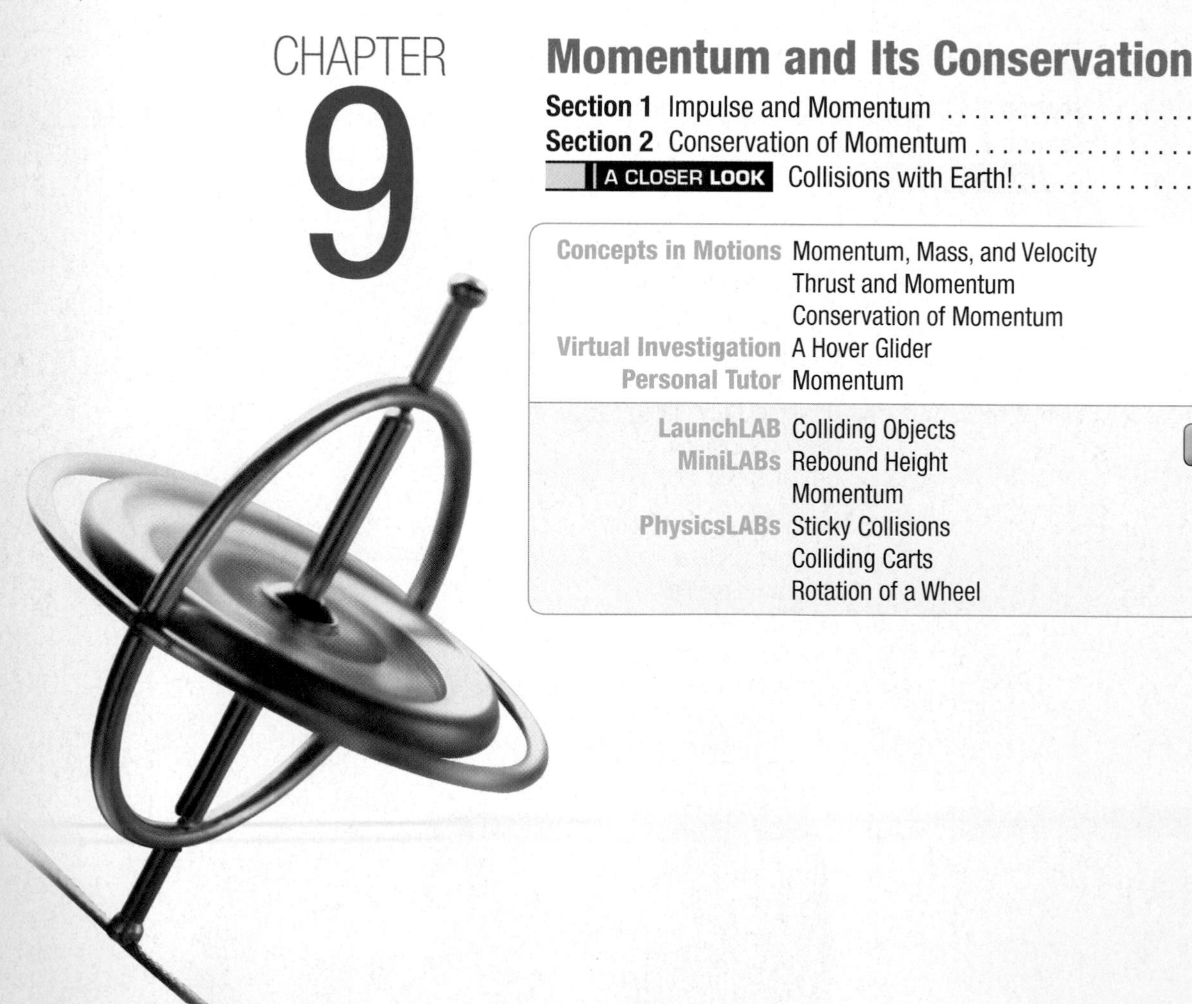

Pete Gardner/Digital Vision/Getty Images

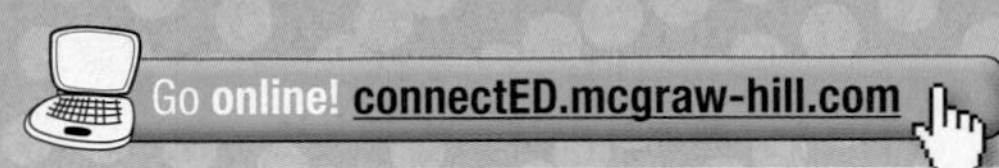

Energy

CHAPTER 10

CHAPTER 11

Fuse/Jupiterimages

Energy

CHAPTER 12

CHAPTER 13

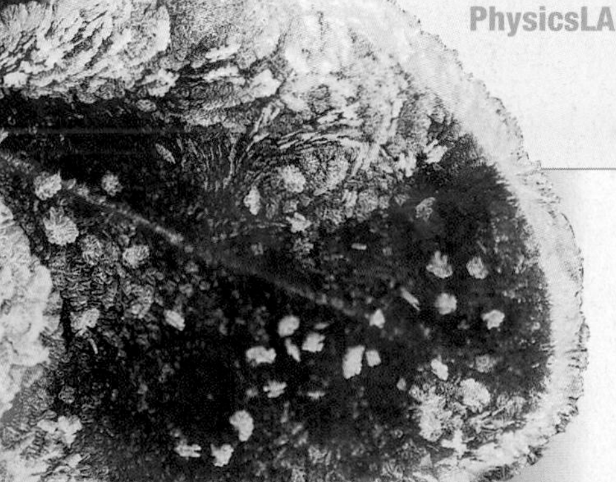

Ingram Publishing

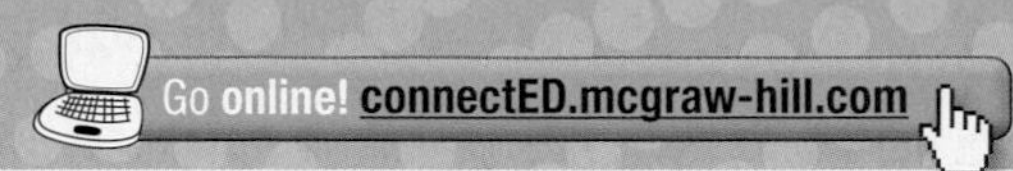

Waves and Light

CHAPTER 14

CHAPTER 15

Don Farrall/Photodisc/Getty Images

Waves and Light

CHAPTER 16

CHAPTER 17

Richard Hutchings/Digital Light Source

CHAPTER 18

Go online!

Concepts in Motions: Refraction; Total Internal Reflection; Drawing a Ray Diagram; Virtual Optics Lab; Compound Microscope; Gravitational Lensing
Virtual Investigations: Snell's Law; Ray Tracing and Lenses
BrainPOP: The Eye
Personal Tutor: Angle of Refraction

iLab Station

LaunchLAB: Broken Straw Appearance
MiniLABs: Personal Rainbow; Lens Masking Effect; Water Lenses
PhysicsLABs: How does light bend?; A Whole Spectrum of Possibilities; Convex Lens Images; Convex Lenses and Focal Length

CHAPTER 19

Go online!

Concepts in Motions: Monochromatic Light; Single-Slit Diffraction; First Dark Band
Virtual Investigation: Double-Slit Interference
Personal Tutor: Diffraction Grating

iLab Station

LaunchLAB: Patterns of Light
MiniLABs: Soap Film; Diffraction Gratings; Retinal Projection Screen
PhysicsLABs: What is the wavelength?; Double-Slit Interference; Holograms

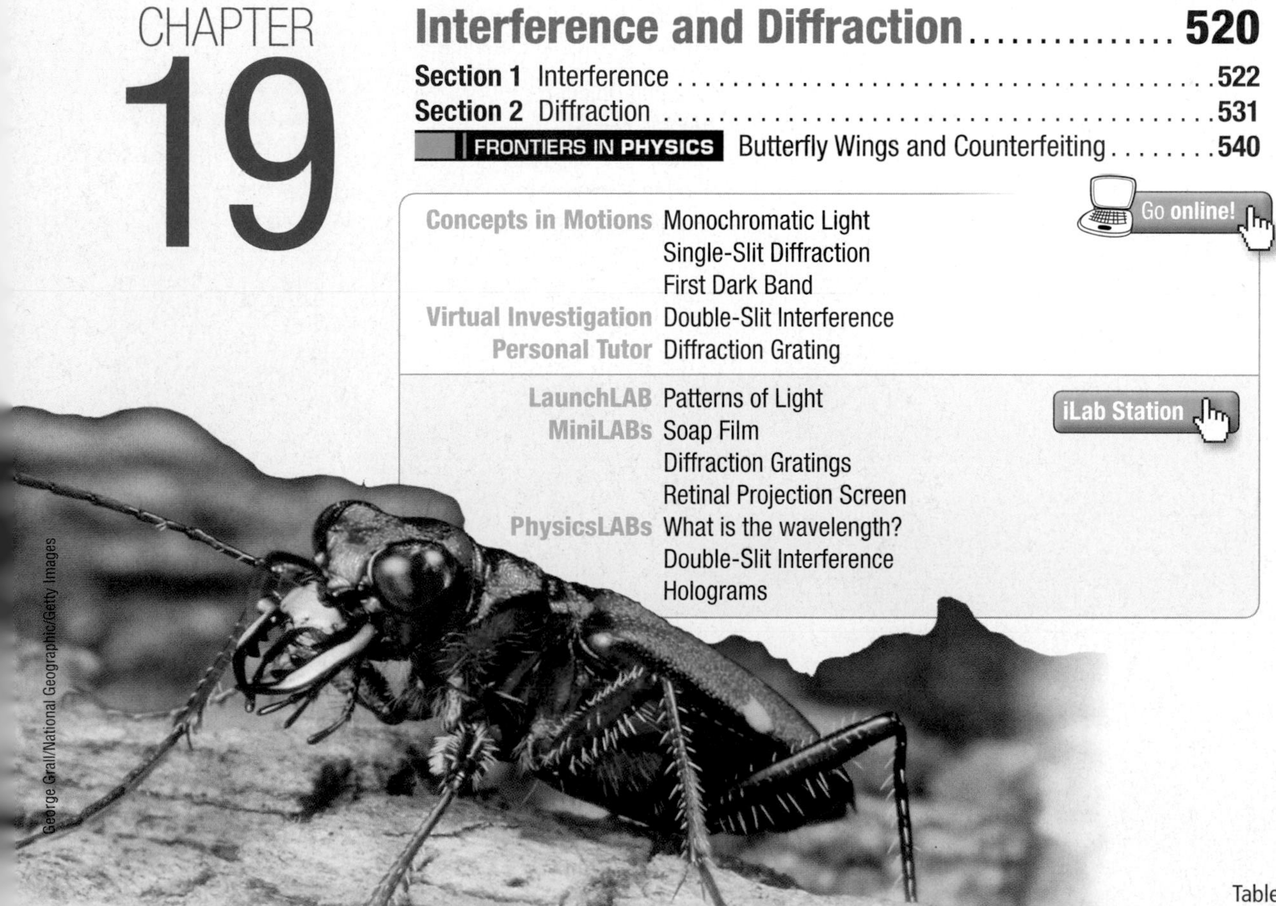

George Grall/National Geographic/Getty Images

Electricity and Magnetism

CHAPTER 20

CHAPTER 21

Roger Ressmeyer/Encyclopedia/CORBIS

Electricity and Magnetism

CHAPTER 22

Electric Current ... 596

Go online!

Concepts in Motion Current and Circuits
Electric Circuits
Virtual Investigations Circuits and Ohm's Law
BrainPOP Current Electricity
Personal Tutor Electric Heat

iLab Station

LaunchLAB Light the Bulb
MiniLABs Current Affairs
Making Electrical Energy
PhysicsLABs Voltage, Current, and Resistance
Variables in Circuits
Energy Conservation

CHAPTER 23

Series and Parallel Circuits ... 622

Go online!

Concepts in Motion Circuit Lab
Virtual Investigation Electric Circuits
Video Electric Circuits
Personal Tutor Resistance and Current

LaunchLAB Fuses and Electric Circuits
MiniLABs Parallel Resistance
Parallel Circuit Measurements
PhysicsLABs Resistors in Series Circuits
Circuit Configurations
How do parallel resistors work?
Series and Parallel Circuits

Photodisc Collection/Getty Images

Electricity and Magnetism

Subatomic Physics

CHAPTER 27

Go online!

Concepts in Motions Photoelectric Effect
Virtual Investigation The Photoelectric Effect
Personal Tutor Units of hc and Photon Energy

iLab Station

LaunchLAB Lightbulb Spectrum
MiniLABs Glows in the Dark
Modeling the Quantum
PhysicsLAB Modeling the Photoelectric Effect
Relating Color and LED Voltage Drop

CHAPTER 28

Concepts in Motions Rutherford's Experiment
Spectra and Quantized Energy
Energy Levels
Bohr Model of the Atom
Hydrogen's Emission Spectrum
Spontaneous and Stimulated Emission
BrainPOP Atomic Model
Personal Tutor Finding the Energy of the Hydrogen Atom

LaunchLAB Identifying Samples
MiniLABs Bright Line Spectra
Diffraction of Laser Light
PhysicsLABs Emission Spectra
Electron Transitions

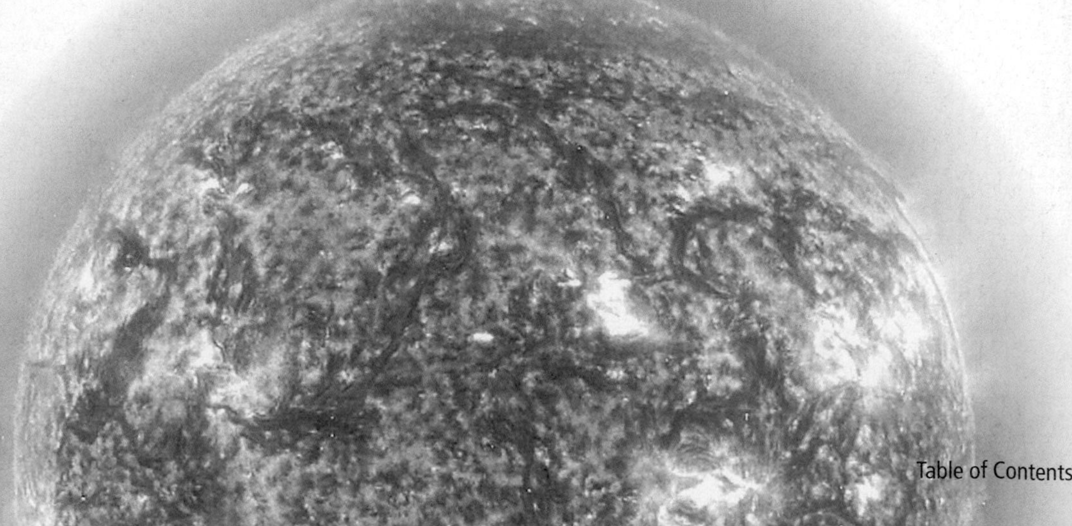

SOHO/ESA/NASA

Go online! connectED.mcgraw-hill.com

Subatomic Physics

CHAPTER 29

Concepts in Motion Diodes
Transistors
Personal Tutor Doped Semiconductors

iLab Station

LaunchLAB Conduction in a Diode
MiniLABs Optoisolator
Red Light
PhysicsLABs Diode Current and Potential Difference
Computer Logic

CHAPTER 30

Go online!

Concepts in Motions Radioactivity
Alpha, Beta, and Gamma Radiation
Radioactive Decay
Chain Reaction
Nuclear Fusion
Linear Accelerator
Virtual Investigations Half-Life of an Element
Videos Nuclear Medicine
Nuclear Power Plant
BrainPOPs Isotopes
Nuclear Energy
Personal Tutors Joules and Electron Volts
Energy Equivalent of Mass
Calculating Half-Life

iLab Station

LaunchLAB A Nuclear Model
MiniLABs Modeling Radioactive Decay
What's in the background?
PhysicsLABs Exploring Radiation
Blocking Radiation

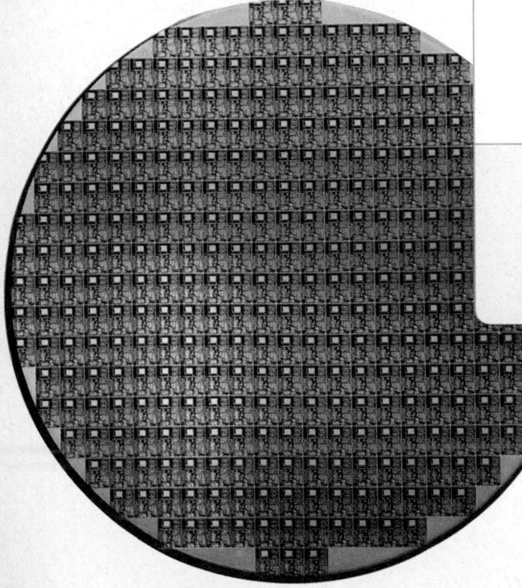

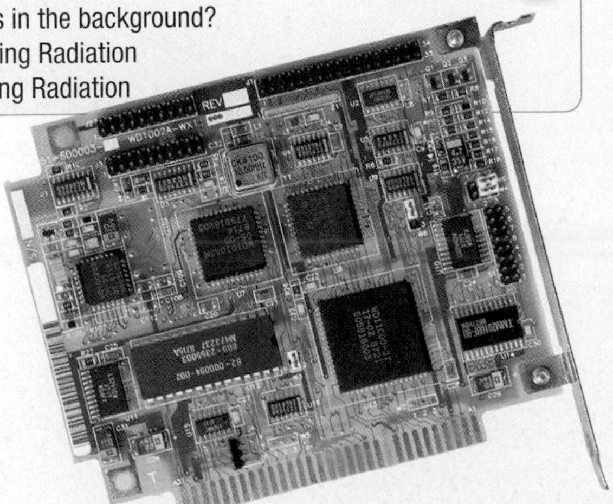

(l)CORBIS, (r)Comstock Images/Alamy

STUDENT RESOURCES

Math Skill Handbook 835

Selected Solutions 860

Reference Tables 885

Glossary ▪ Glosario 894

Index 920

Rubberball/Getty Images

REAL-WORLD PHYSICS

Physics: Principles and Problems **makes physics real. Throughout the text, find personal physics connections, surprising examples of physics in careers, and how physicists are engaged in cutting-edge science research.**

SECTION 1

Force and Motion

PHYSICS 4 YOU

To start a skateboard moving on a level surface, you must push on the ground with your foot. If the skateboard is on a ramp, however, gravity will pull you down the slope. In both cases unbalanced forces change the skateboard's motion.

MAINIDEA

A force is a push or a pull.

Essential Questions

- What is a force?
- What is the relationship between force and acceleration?
- How does motion change when the net force is zero?

Review Vocabulary

acceleration the rate at which the velocity of an object changes

New Vocabulary

Force

Consider a textbook resting on a table. To cause it to move, you could either push or pull on it. In physics, a push or a pull is called a **force.** If you push or pull harder on an object, you exert a greater force on the object. In other words, you increase the magnitude of the applied force. The direction in which the force is exerted also matters—if you push the resting book to the right, the book will start moving to the right. If you push the book to the left, it will start moving to the left. Because forces have both magnitude and direction, forces are vectors. The symbol $\boldsymbol{F}$ is vector notation that represents the size and direction of a force, while F represents only the magnitude. The magnitude of a force is measured in units called newtons (N).

Unbalanced forces change motion Recall that motion diagrams describe the positions of an object at equal time intervals. For example, the motion diagram for the book in **Figure 1** shows the distance between the dots increasing. This means the speed of the book is increasing. At $t = 0$, it is at rest, but after 2 seconds it is moving at 1.5 m/s. This change in speed means it is accelerating. What is the cause of this acceleration? The book was at rest until you pushed it, so the cause of the acceleration is the force exerted by your hand. In fact, all accelerations are the result of an unbalanced force acting on an object. What is the relationship between force and acceleration?

PHYSICS 4 YOU at the beginning of each section tells you how the physics you are about to learn relates to your life.

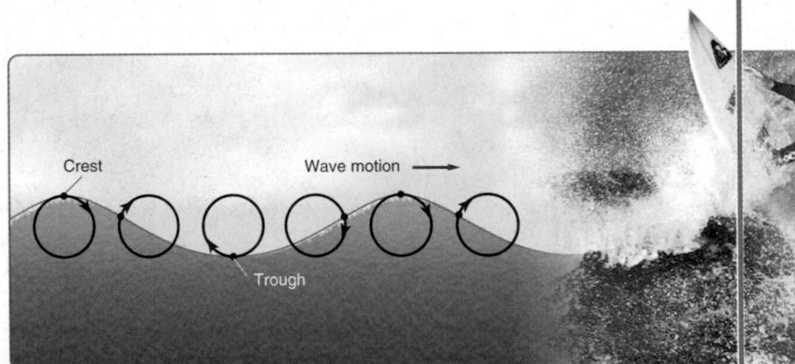

Figure 8 Surface waves in water cause movement both parallel and perpendicular to the direction of wave motion. When these waves interact with the shore, the regular circular motion is disrupted and the waves break on the beach.

Surface waves Waves that are deep in a lake or an ocean are longitudinal. In a **surface wave,** however, the medium's particles follow a circular path that is at times parallel to the direction of travel and at other times perpendicular to the direction of wave travel, as shown in **Figure 8.** Surface waves set particles in the medium, in this case water, moving in a circular pattern. At the top and bottom of the circular path, particles are moving parallel to the direction of the wave's travel. This is similar to a longitudinal wave. At the left and right sides of each circle, particles are moving up or down. This up-and-down motion is perpendicular to the wave's direction, similar to a transverse wave.

Wave Properties

Many types of waves share a common set of wave properties. Some wave properties depend on how the wave is produced, whereas others depend on the medium through which the wave travels.

Amplitude How does the pulse generated by gently shaking a rope differ from the pulse produced by a violent shake? The difference is similar to the difference between a ripple in a pond and an ocean breaker—they have different amplitudes. You read earlier that the amplitude of periodic motion is the greatest distance from equilibrium. Similarly, as shown in **Figure 9,** a transverse wave's amplitude is the maximum distance of the wave from equilibrium.

REAL-WORLD PHYSICS

Tsunamis On March 11, 2011, a wall of water estimated to be ten meters high hit areas on the East coast of Japan —tsunami! A tsunami is a series of ocean waves that can have wavelengths over 100 km, periods of one hour, and wave speeds of 500–1000 km/h.

Throughout the book, **REAL-WORLD PHYSICS** demonstrates how the physics you are learning applies to the world around you.

End-of-chapter features highlight physics in careers, how it connects to the real world, and what today's physicists are doing to learn more about our universe.

PHYSICS THAT'S ENTERTAINMENT!

How does physics apply to entertainment? You might be surprised! Explore the physics of special effects, 3-D movies, theater acoustics, and more!

A CLOSER LOOK

Take "A Closer Look" at a number of physics topics and discover the story behind some of the most interesting physics applications!

ON THE JOB

You might be surprised to discover the physics that can be found in many different careers. Explore jobs that unexpectedly rely on an understanding of physics.

FRONTIERS IN PHYSICS

What is being discovered in today's physics research? Explore the work being done by today's physicists.

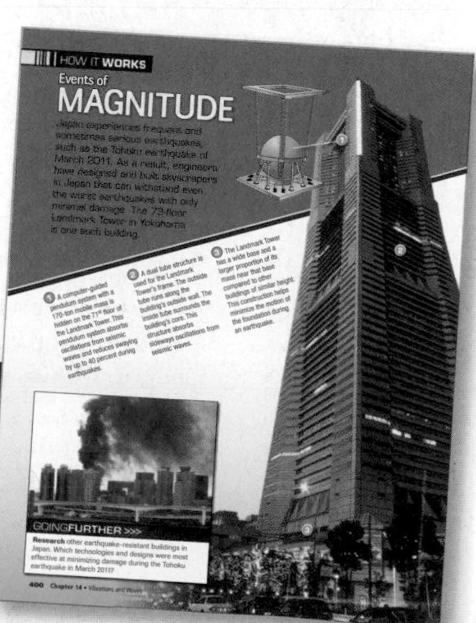

HOW IT WORKS

Explore the physics of everyday objects or natural phenomena by discovering how they "work."

UNDERSTANDING PHYSICS

At the start of each chapter, you will see the BIGIDEA that will help you understand how what you are about to investigate fits into the big picture of science.

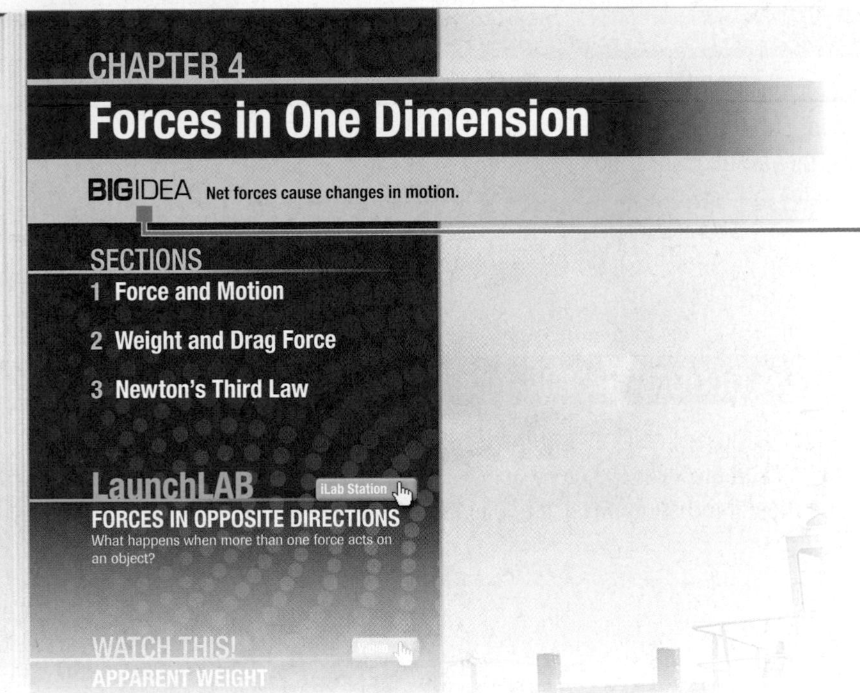

CHAPTER 4
Forces in One Dimension
BIGIDEA Net forces cause changes in motion.
SECTIONS
1 Force and Motion
2 Weight and Drag Force
3 Newton's Third Law
LaunchLAB
iLab Station
FORCES IN OPPOSITE DIRECTIONS
What happens when more than one force acts on an object?
WATCH THIS!
APPARENT WEIGHT

The **BIG**IDEA is the focus of the chapter. By reading the text, doing labs and answering Practice Problems, Section Reviews, and Chapter Assessments, you will build an in-depth understanding of this idea.

CHAPTER 4 ASSESSMENT

SECTION 1 **Force and Motion**

Mastering Concepts

39. BIGIDEA You kick a soccer ball across a field. It slows down and comes to a stop. You ask your younger brother to explain what happened to the ball. He says, "The force of your foot was transferred to the ball, which made it move. When that force ran out, the ball stopped." Would Newton agree with that explanation? If not, explain how Newton's laws would describe it.

40. Cycling Imagine riding a single-speed bicycle. Why do you have to push harder on the pedals to start the bicycle moving than to keep it moving at a constant velocity?

Mastering Problems

41. What is the net force acting on a 1.0-kg ball in free-fall? Ignore air resistance.

42. Skating Joyce and Efua are skating. Joyce pushes Efua, whose mass is 40.0 kg, with a force of 5.0 N. What is Efua's resulting acceleration?

43. A 2300-kg car slows down at a rate of 3.0 m/s^2 when approaching a stop sign. What is the magnitude of the net force causing it to slow down?

44. Breaking the Wishbone After Thanksgiving, Kevin and Gamal use the turkey's wishbone to make a wish. If Kevin pulls on it with a force 0.17 N larger than the force Gamal pulls with in the opposite direction and the wishbone has a mass of 13 g, what is the wishbone's initial acceleration?

SECTION 2 Weight and Drag Force

48. Before a skydiver opens her parachute, she m... falling at a velocity higher than the termin... that she will have after the parachute deplo...

a. Describe what happens to her velocity as sh... opens the parachute.

b. Describe the sky diver's velocity from whe... parachute has been open for a time until sh... about to land.

49. Three objects are dropped simultaneously fr... top of a tall building: a shot put, an air-filled ... loon, and a basketball.

a. Rank the objects in the order in which the... reach terminal velocity, from first to last.

b. Rank the objects according to the order in... they will reach the ground, from first to la...

c. What is the relationship between your an... parts a and b?

Mastering Problems

50. What is your weight in newtons? Show you...

51. A rescue helicopter lifts two people using a wi... and a rescue ring as shown in **Figure 21**.

a. The winch is capable of exerting a 2000 N... What is the maximum mass it can lift?

b. If the winch applies a force of 1200 N, wha... rescuer and victims' acceleration? Draw a fr... body diagram for the people being lifted.

c. Using the acceleration from part b, how lon... it take to pull the people up to the helicop... Assume the people are initially at rest.

In the Chapter Assessment, you will find a question or problem that will help you evaluate your understanding of the **BIG**IDEA.

At the start of each section, you will find a Reading Preview that summarizes what you will learn while exploring the section.

The **MAIN** IDEA is the core concept covered in the section. Together, the Main Ideas from all the sections in the chapter support the chapter's Big Idea.

Essential Questions reflect the important objectives of the section. Together, an understanding of these questions will lead toward understanding the section's Main Idea.

In the Section Review, you will find a question that will help you to assess your understanding of the section's **MAIN** IDEA.

The remaining questions assess your understanding of the **Essential Questions**

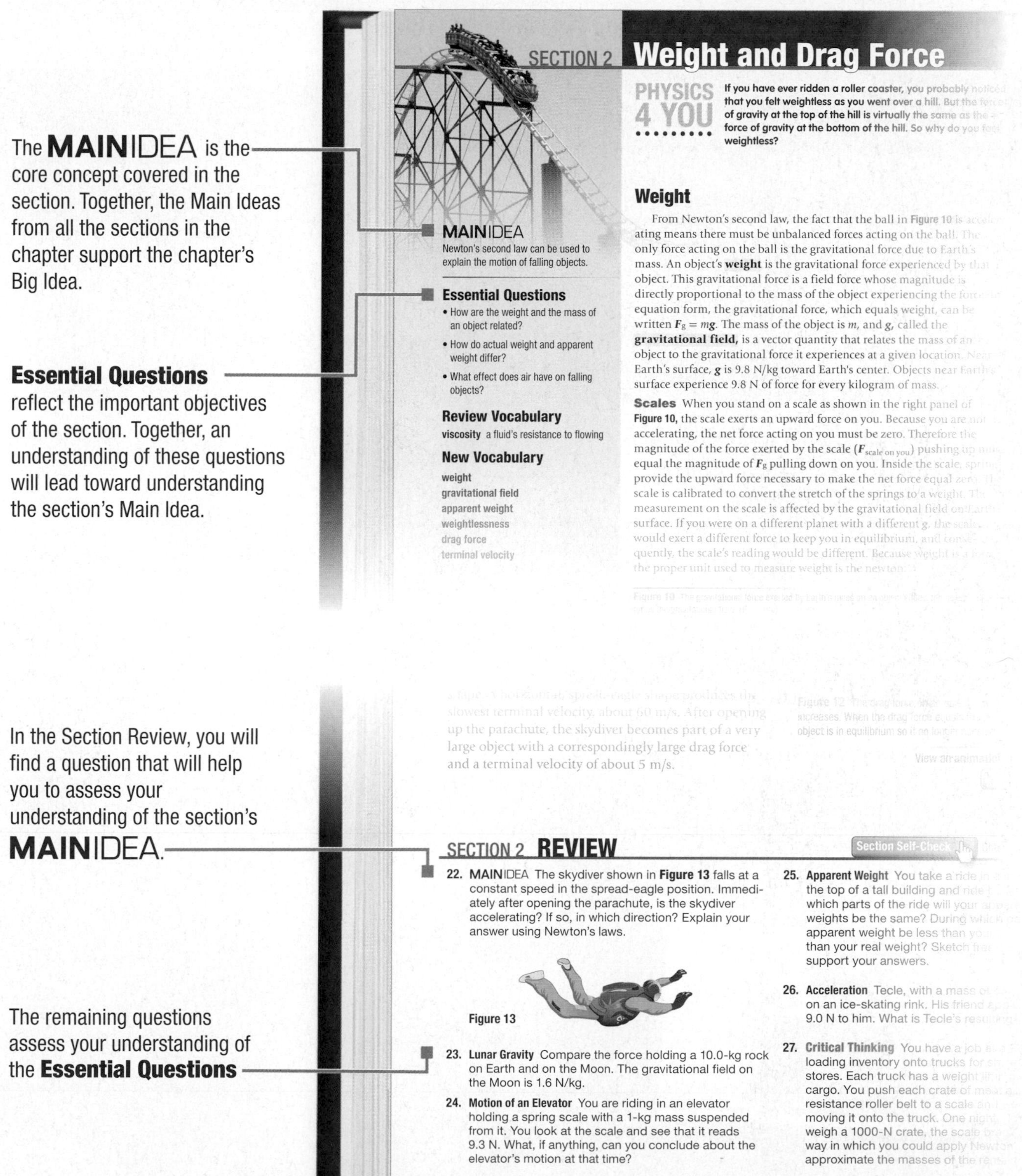

SECTION 2 Weight and Drag Force
PHYSICS 4 YOU
If you have ever ridden a roller coaster, you probably noticed that you felt weightless as you went over a hill. But the force of gravity at the top of the hill is virtually the same as the force of gravity at the bottom of the hill. So why do you feel weightless?
MAIN IDEA
Newton's second law can be used to explain the motion of falling objects.
Essential Questions
• How are the weight and the mass of an object related?
• How do actual weight and apparent weight differ?
• What effect does air have on falling objects?
Review Vocabulary
viscosity a fluid's resistance to flowing
New Vocabulary
weight
gravitational field
apparent weight
weightlessness
drag force
terminal velocity
Weight
From Newton's second law, the fact that the ball in Figure 10 is accelerating means there must be unbalanced forces acting on the ball. The only force acting on the ball is the gravitational force due to Earth's mass. An object's weight is the gravitational force experienced by that object. This gravitational force is a field force whose magnitude is directly proportional to the mass of the object experiencing the force. In equation form, the gravitational force, which equals weight, can be written $F_g = mg$. The mass of the object is m, and g, called the gravitational field, is a vector quantity that relates the mass of an object to the gravitational force it experiences at a given location. Near Earth's surface, g is 9.8 N/kg toward Earth's center. Objects near Earth's surface experience 9.8 N of force for every kilogram of mass.
Scales When you stand on a scale as shown in the right panel of Figure 10, the scale exerts an upward force on you. Because you are not accelerating, the net force acting on you must be zero. Therefore the magnitude of the force exerted by the scale ($F_{scale\ on\ you}$) pushing up must equal the magnitude of F_g pulling down on you. Inside the scale, springs provide the upward force necessary to make the net force equal zero. The scale is calibrated to convert the stretch of the springs to a weight. The measurement on the scale is affected by the gravitational field on Earth's surface. If you were on a different planet with a different g, the scale would exert a different force to keep you in equilibrium, and consequently, the scale's reading would be different. Because weight is a force, the proper unit used to measure weight is the newton.
slowest terminal velocity, about 60 m/s. After opening up the parachute, the skydiver becomes part of a very large object with a correspondingly large drag force and a terminal velocity of about 5 m/s.
SECTION 2 REVIEW
Section Self-Check
22. MAIN IDEA The skydiver shown in Figure 13 falls at a constant speed in the spread-eagle position. Immediately after opening the parachute, is the skydiver accelerating? If so, in which direction? Explain your answer using Newton's laws.
Figure 13
23. Lunar Gravity Compare the force holding a 10.0-kg rock on Earth and on the Moon. The gravitational field on the Moon is 1.6 N/kg.
24. Motion of an Elevator You are riding in an elevator holding a spring scale with a 1-kg mass suspended from it. You look at the scale and see that it reads 9.3 N. What, if anything, can you conclude about the elevator's motion at that time?
25. Apparent Weight You take a ride in a ... the top of a tall building and ride ... which parts of the ride will your ... weights be the same? During which ... apparent weight be less than your ... than your real weight? Sketch ... support your answers.
26. Acceleration Tecle, with a mass of ... on an ice-skating rink. His friend ... 9.0 N to him. What is Tecle's resulting ...
27. Critical Thinking You have a job ... loading inventory onto trucks for ... stores. Each truck has a weight ... cargo. You push each crate of ... resistance roller belt to a scale ... moving it onto the truck. One night ... weigh a 1000-N crate, the scale ... way in which you could apply Newton's ... approximate the masses of the re...
Section 2 • Weight ...

CHAPTER 1

A Physics Toolkit

BIGIDEA **Physicists use scientific methods to investigate energy and matter.**

SECTIONS

1 **Methods of Science**

2 **Mathematics and Physics**

3 **Measurement**

4 **Graphing Data**

LaunchLAB

MASS AND FALLING OBJECTS

Does mass affect the rate at which an object falls?

WATCH THIS!

ROCK STACKING

How do you get a stack of rocks to remain upright and balanced? It's all about understanding the physics! Explore the science behind the art of rock stacking.

(l)Jose Luis Stephens/Getty Images, (r)Brandi Simons/Stringer/Getty Images News/Getty Images

Go online!
connectED.mcgraw-hill.com

SECTION 1 # Methods of Science

PHYSICS 4 YOU **Think about what the world would be like if we still thought Earth was flat or if we didn't have indoor plumbing or electricity. Science helps us learn about the natural world and improve our lives.**

MAIN IDEA

Scientific investigations do not always proceed with identical steps but do contain similar methods.

Essential Questions

- What are the characteristics of scientific methods?
- Why do scientists use models?
- What is the difference between a scientific theory and a scientific law?
- What are some limitations of science?

Review Vocabulary

control the standard by which test results in an experiment can be compared

New Vocabulary

physics
scientific methods
hypothesis
model
scientific theory
scientific law

What is physics?

Science is not just a subject in school. It is a method for studying the natural world. After all, science comes from the Latin word *scientia,* which means "knowledge." Science is a process based on inquiry that helps develop explanations about events in nature. **Physics** is a branch of science that involves the study of the physical world: energy, matter, and how they are related.

When you see the word *physics* you might picture a chalkboard full of formulas and mathematics: $E = mc^2$, $I = \frac{V}{R}$, $x = \left(\frac{1}{2}\right)at^2 + v_0t + x_0$. Maybe you picture scientists in white lab coats or well-known figures such as Marie Curie and Albert Einstein. Alternatively, you might think of the many modern technologies created with physics, such as weather satellites, laptop computers, or lasers. Physicists investigate the motions of electrons and rockets, the energy in sound waves and electric circuits, the structure of the proton and of the universe. The goal of this course is to help you better understand the physical world.

People who study physics go on to many different careers. Some become scientists at universities and colleges, at industries, or in research institutes. Others go into related fields, such as engineering, computer science, teaching, medicine, or astronomy, as shown in **Figure 1.** Still others use the problem-solving skills of physics to work in finance, construction, or other very different disciplines. In the last 50 years, research in the field of physics has led to many new technologies, including satellite-based communications and high-speed microscanners used to detect disease.

Figure 1 Physicists may choose from a variety of careers.

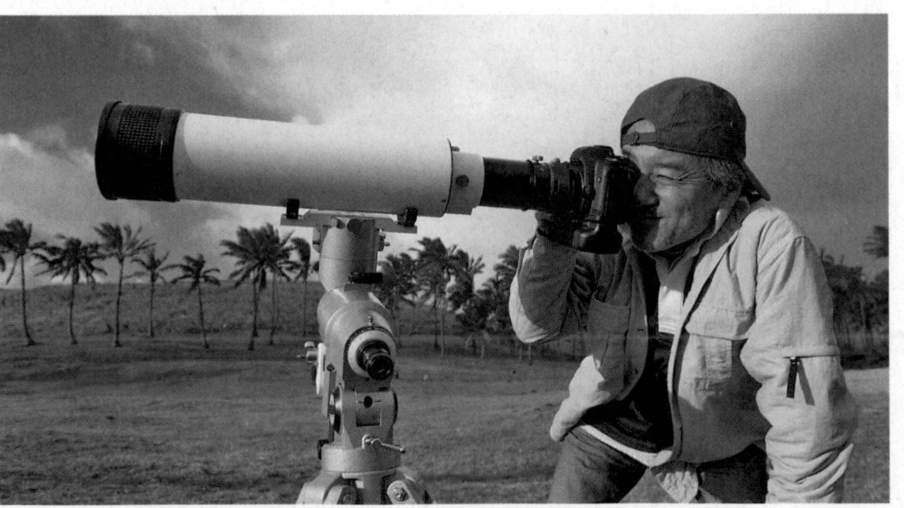

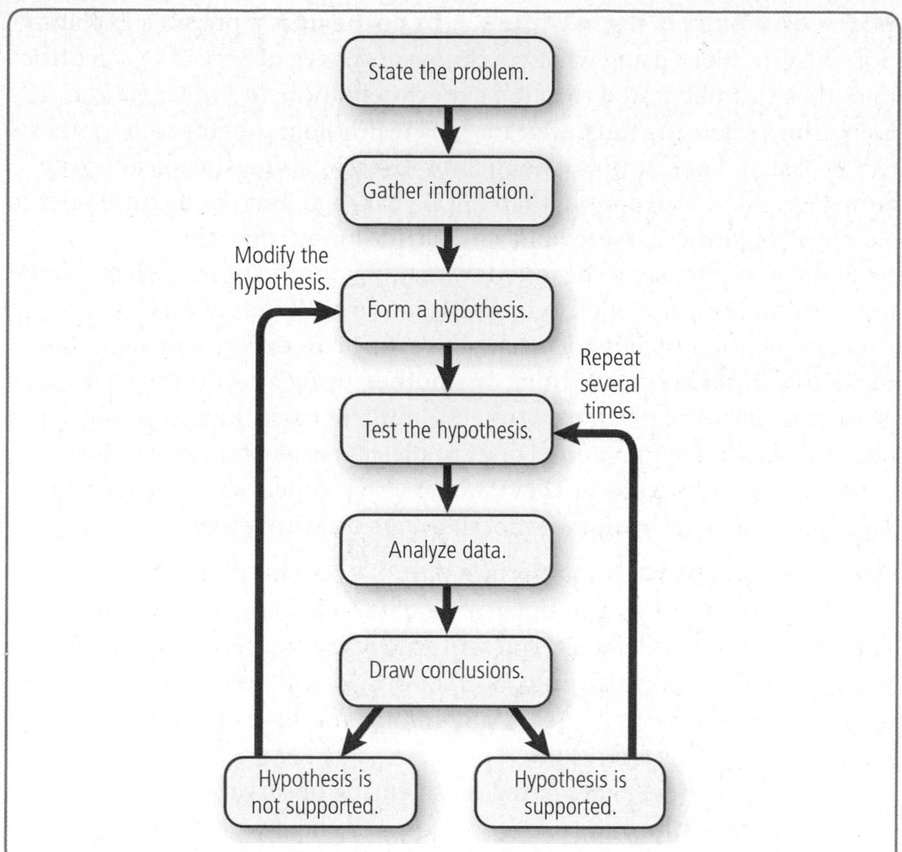

Figure 2 The series of procedures shown here is one way to use scientific methods to solve a problem.

Scientific Methods

Although physicists do not always follow a rigid set of steps, investigations often follow similar patterns. These patterns of investigation procedures are called **scientific methods.** Common steps found in scientific methods are shown in **Figure 2.** Depending on the particular investigation, a scientist might add new steps, repeat some steps, or skip steps altogether.

State the problem When you begin an investigation, you should state what you are going to investigate. Many investigations begin when someone observes an event in nature and wonders why or how it occurs. The question of "why" or "how" is the problem.

Scientists once posed questions about why objects fall to Earth, what causes day and night, and how to generate electricity for daily use. Many times a statement of a problem arises when an investigation is complete and its results lead to new questions. For example, once scientists understood why we experience day and night, they wanted to know why Earth rotates.

Sometimes a new question is posed during the course of an investigation. In the 1940s, researcher Percy Spencer was trying to answer the question of how to mass-produce the magnetron tubes used in radar systems. When he stood in front of an operating magnetron, which produces microwaves, a candy bar in his pocket melted. The new question of how the magnetron was cooking food was then asked.

Research and gather information Before beginning an investigation, it is useful to research what is already known about the problem. Making and examining observations and interpretations from reliable sources fine-tune the question and form it into a hypothesis.

View a **BrainPOP video on scientific methods.**

Form and test a hypothesis A **hypothesis** is a possible explanation for a problem using what you know and have observed. A scientific hypothesis can be tested through experimentation and observation. Sometimes scientists must wait for new technologies before a hypothesis can be tested. For example, the first hypotheses about the existence of atoms were developed more than 2300 years ago, but the technologies to test these hypotheses were not available for many centuries.

Some hypotheses can be tested by making observations. Others can be tested by building a model and relating it to real-life situations. One common way to test a hypothesis is to perform an experiment. An experiment tests the effect of one thing on another, using a control. Sometimes it is not possible to perform experiments; in these cases, investigations become descriptive in nature. For example, physicists cannot conduct experiments in deep space. They can, however, collect and analyze valuable data to help us learn more about events occurring there.

MiniLAB

MEASURING CHANGE

How does increasing mass affect the length of a spring?

Analyze the data An important part of every investigation includes recording observations and organizing data into easy-to-read tables and graphs. Later in this chapter, you will study ways to display data. When you are making and recording observations, you should include all results, even unexpected ones. Many important discoveries have been made from unexpected results.

Scientific inferences are based on scientific observations. All possible scientific explanations must be considered. If the data are not organized in a logical manner, incorrect conclusions can be drawn. When a scientist communicates and shares data, other scientists will examine those data, how the data were analyzed, and compare the data to the work of others. Scientists, such as the physicist in **Figure 3,** share their data and analyses through reports and conferences.

Draw conclusions Based on the analysis of the data, the next step is to decide whether the hypothesis is supported. For the hypothesis to be considered valid and widely accepted, the results of the experiment must be the same every time it is repeated. If the experiment does not support the hypothesis, the hypothesis must be reconsidered. Perhaps the hypothesis needs to be revised, or maybe the experimenter's procedure needs to be refined.

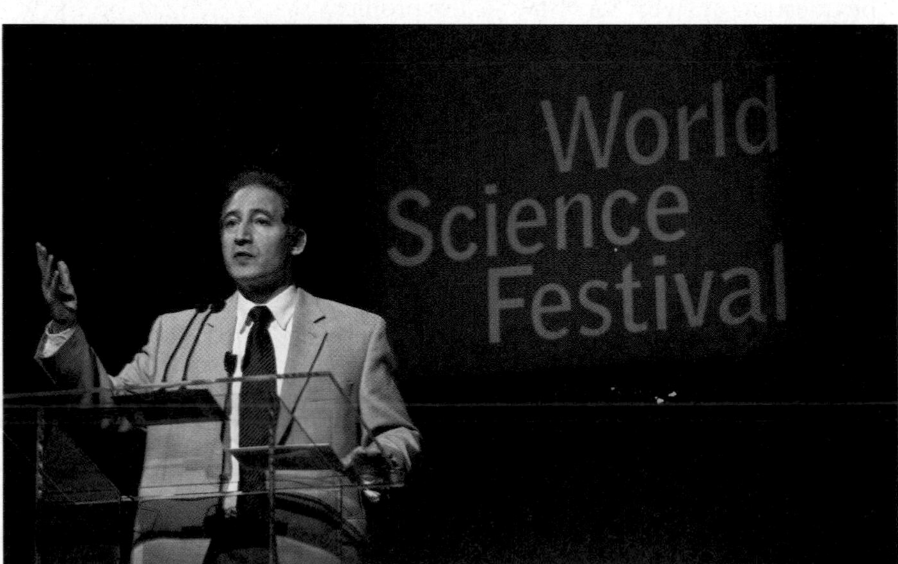

Lynn Goldsmith/CORBIS News/CORBIS

Figure 3 An important part of scientific methods is to share data and results with other scientists. This physicist is giving a presentation at the World Science Festival.

Peer review Before it is made public, science-based information is reviewed by scientists' peers—scientists who are in the same field of study. Peer review is a process by which the procedures and results of an experiment are evaluated by peer scientists of those who conducted the research. Reviewing other scientists' work is a responsibility that many scientists have.

Being objective One also should be careful to reduce bias in scientific investigations. Bias can occur when the scientist's expectations affect how the results are analyzed or the conclusions are made. This might cause a scientist to select a result from one trial over those from other trials. Bias might also be found if the advantages of a product being tested are used in a promotion and the drawbacks are not presented. Scientists can lessen bias by running as many trials as possible and by keeping accurate notes of each observation made.

Models

Sometimes, scientists cannot see everything they are testing. They might be observing an object that is too large or too small, a process that takes too much time to see completely, or a material that is hazardous. In these cases, scientists use models. A **model** is a representation of an idea, event, structure, or object that helps people better understand it.

Models in history Models have been used throughout history. In the early 1900s, British physicist J.J. Thomson created a model of the atom that consisted of electrons embedded in a ball of positive charge. Several years later, physicist Ernest Rutherford created a model of the atom based on new research. Later in the twentieth century, scientists discovered the nucleus is not a solid ball but is made of protons and neutrons. The present-day model of the atom is a nucleus made of protons and neutrons surrounded by an electron cloud. All three of these models are shown in **Figure 4.** Scientists use models of atoms to represent their current understanding because of the small size of an atom.

Figure 4 Throughout history, scientists have made models of the atom.

Infer Why have models of the atom changed over the years?

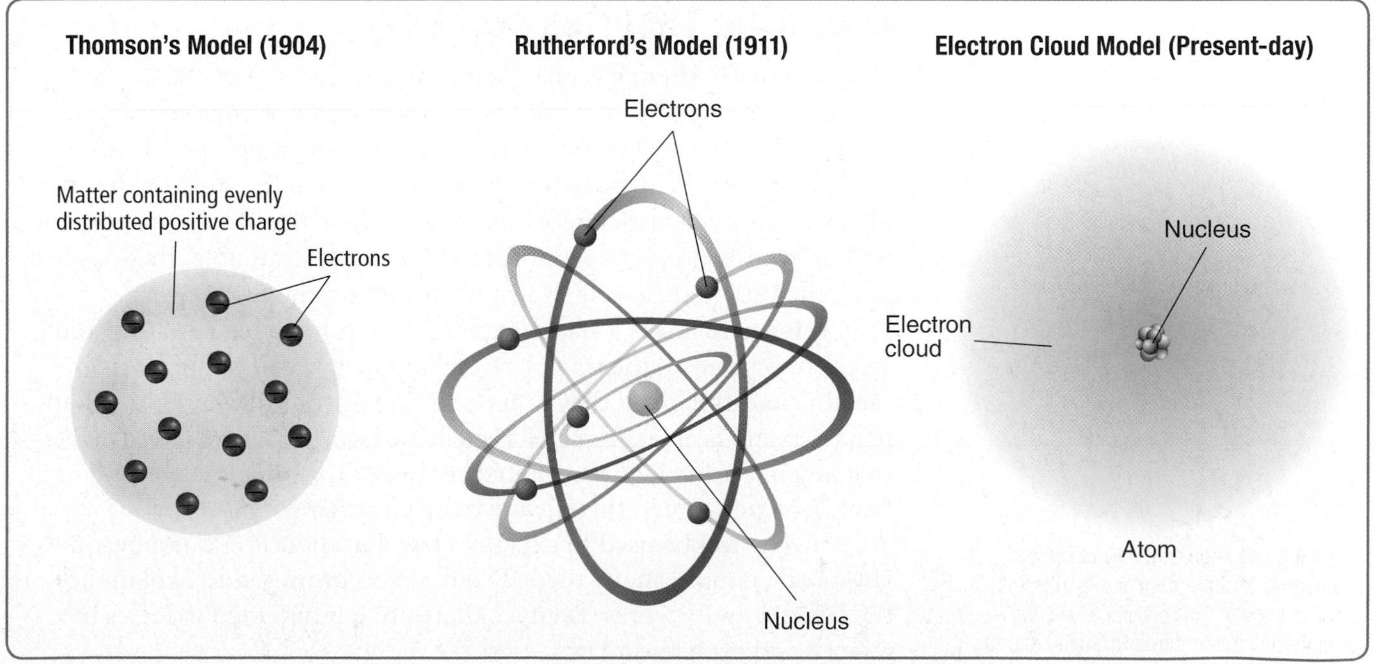

Figure 5 This is a computer simulation of an aircraft landing on a runway. The image on the screen in front of the pilot mimics what he would see if he were landing a real plane.

Identify other models around your classroom.

Michael Dunning/Photographer's Choice/Getty Images

High-tech models Scientific models are not always something you can touch. Another type of model is a computer simulation. A computer simulation uses a computer to test a process or procedure and to collect data. Computer software is designed to mimic the processes under study. For instance, it is not possible for astronomers to observe how our solar system was formed, but when models of the process are proposed, they can be tested with computers.

Computer simulations also enable pilots, such as the ones shown in **Figure 5,** to practice all aspects of flight without ever leaving the ground. In addition, the computer simulation can simulate harsh weather conditions or other potentially dangerous in-flight challenges that pilots might face.

☑ **READING CHECK** **Identify** two advantages of using computer simulations.

Scientific Theories and Laws

A **scientific theory** is an explanation of things or events based on knowledge gained from many observations and investigations.

It is not a guess. If scientists repeat an investigation and the results always support the hypothesis, the hypothesis can be called a theory. Just because a scientific theory has data supporting it does not mean it will never change. As new information becomes available, theories can be refined or modified, as shown in **Figure 6** on the next page.

A **scientific law** is a statement about what happens in nature and seems to be true all the time. Laws tell you what will happen under certain conditions, but they don't explain why or how something happens. Gravity is an example of a scientific law. The law of gravity states that any one mass will attract another mass. To date, no experiments have been performed that disprove the law of gravity.

A theory can be used to explain a law, but theories do not become laws. For example, many theories have been proposed to explain how the law of gravity works. Even so, there are few accepted theories in science and even fewer laws.

Greek philosophers proposed that objects fall because they seek their natural places. The more massive the object, the faster it falls.

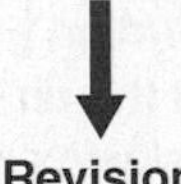

Revision

Galileo showed that the speed at which an object falls depends on the amount of time for which that object has fallen and not on the object's mass.

Revision

Newton provided an explanation for why objects fall. Newton proposed that objects fall because the object and Earth are attracted by a force. Newton also stated that there is a force of attraction between any two objects with mass.

Revision

Einstein suggested that the force of attraction between two objects is due to mass causing the space around it to curve.

Figure 6 If experiments provide new insight and evidence about a theory, the theory is modified accordingly. The theory describing the behavior of falling objects has undergone many revisions based on new evidence.

The Limitations of Science

Science can help you explain many things about the world, but science cannot explain or solve everything. Although it is the scientist's job to make guesses, the scientist also has to make sure his or her guesses can be tested and verified.

Questions about opinions, values, or emotions are not scientific because they cannot be tested. For example, some people may find a particular piece of art beautiful while others do not. Some people may think that certain foods, such as pizza, taste delicious while others do not. Or, some people might think that the best color is blue, while others think it is green. You might take a survey to gather opinions about such questions, but that would not prove the opinions are true for everyone.

SECTION 1 REVIEW

Section Self-Check Check your understanding.

1. **MAIN**IDEA Summarize the steps you might use to carry out an investigation using scientific methods.
2. **Define** the term hypothesis and identify three ways in which a hypothesis can be tested.
3. **Describe** why it is important for scientists to avoid bias.
4. **Explain** why scientists use models. Give an example of a scientific model not mentioned in this section.
5. **Explain** why a scientific theory cannot become a scientific law.
6. **Analyze** Your friend conducts a survey, asking students in your school about lunches provided by the cafeteria. She finds that 90 percent of students surveyed like pizza. She concludes that this scientifically proves that everyone likes pizza. How would you respond to her conclusion?
7. **Critical Thinking** An accepted value for free-fall acceleration is 9.8 m/s^2. In an experiment with pendulums, you calculate that the value is 9.4 m/s^2. Should the accepted value be tossed out to accommodate your new finding? Explain.

SECTION 2

Mathematics and Physics

PHYSICS 4 YOU **If you were to toss a tennis ball straight up into the air, how could you determine how far the ball would rise or how long it would stay in the air? How could you determine the velocity of the skydiver in the photo? Physicists use mathematics to help find the answers to these and other questions about motion, forces, energy, and matter.**

MAIN IDEA

We use math to express concepts in physics.

Essential Questions

- Why do scientists use the metric system?
- How can dimensional analysis help evaluate answers?
- What are significant figures?

Review Vocabulary

SI Système International d'Unités–the improved, universally accepted version of the metric system that is based on multiples of ten; also called the International System of Units

New Vocabulary

dimensional analysis
significant figures

Mathematics in Physics

Physicists often use the language of mathematics. In physics, equations are important tools for modeling observations and for making predictions. Equations are one way of representing relationships between measurements. Physicists rely on theories and experiments with numerical results to support their conclusions. For example, you can predict that if you drop a penny, it will fall, but can you predict how fast it will be going when it strikes the ground below? Different models of falling objects give different answers to how the speed of the object changes as it falls or on what the speed depends. By measuring how an object falls, you can compare the experimental data with the results predicted by different models. This tests the models, allowing you to pick the best one or to develop a new model.

SI Units

To communicate results, it is helpful to use units that everyone understands. The worldwide scientific community currently uses an adaptation of the metric system for measurements. **Table 1** shows that the Système International d'Unités, or SI, uses seven base quantities. Other units, called derived units, are created by combining the base units in various ways. Velocity is measured in meters per second (m/s). Often, derived units are given special names. For example, electric charge is measured in ampere-seconds (A•s), which are also called coulombs (C).

The base quantities were originally defined in terms of direct measurements. Scientific institutions have since been created to define and regulate measurements. SI is regulated by the International Bureau of Weights and Measures in Sèvres, France.

Table 1 SI Base Units

Base Quantity	Base Unit	Symbol
Length	meter	m
Mass	kilogram	kg
Time	second	s
Temperature	kelvin	K
Amount of a substance	mole	mol
Electric current	ampere	A
Luminous intensity	candela	cd

Digital Vision/Getty Images

This bureau and the National Institute of Science and Technology (NIST) in Gaithersburg, Maryland, keep the standards of length, time, and mass against which our metersticks, clocks, and balances are calibrated. The standard for a kilogram is shown in **Figure 7.**

You probably learned in math class that it is much easier to convert meters to kilometers than feet to miles. The ease of switching between units is another feature of SI. To convert between units, multiply or divide by the appropriate power of 10. Prefixes are used to change SI base units by powers of 10, as shown in **Table 2.** You often will encounter these prefixes in daily life, as in, for example, milligrams, nanoseconds, and centimeters.

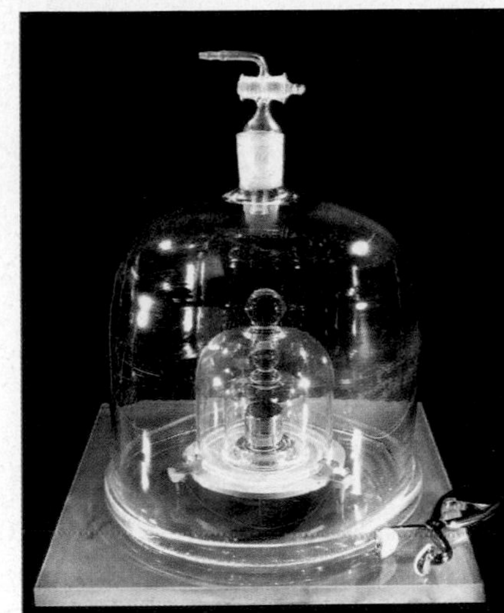

Figure 7 The International Prototype Kilogram, the standard for the mass of a kilogram, is a mixture of platinum and iridium. It is kept in a vacuum so it does not lose mass. Scientists are working to redefine the standard for a kilogram, using a perfect sphere made of silicon.

Describe Why is it important to have standards for measurements?

Table 2 Prefixes Used with SI Units

Prefix	Symbol	Multiplier	Scientific Notation	Example
femto–	f	0.000000000000001	10^{-15}	femtosecond (fs)
pico–	p	0.000000000001	10^{-12}	picometer (pm)
nano–	n	0.000000001	10^{-9}	nanometer (nm)
micro–	μ	0.000001	10^{-6}	microgram (μg)
milli–	m	0.001	10^{-3}	milliamps (mA)
centi–	c	0.01	10^{-2}	centimeter (cm)
deci–	d	0.1	10^{-1}	deciliter (dL)
kilo–	k	1000	10^{3}	kilometer (km)
mega–	M	1,000,000	10^{6}	megagram (Mg)
giga–	G	1,000,000,000	10^{9}	gigameter (Gm)
tera–	T	1,000,000,000,000	10^{12}	terahertz (THz)

☑ **READING CHECK Identify** the prefix that would be used to express 2,000,000,000 bytes of computer memory.

Dimensional Analysis

You can use units to check your work. You often will need to manipulate a formula, or use a string of formulas, to solve a physics problem. One way to check whether you have set up a problem correctly is to write out the equation or set of equations you plan to use. Before performing calculations, check that the answer will be in the expected units. For example, if you are finding a car's speed and you see that your answer will be measured in s/m or m/s^2, you have made an error in setting up the problem. This method of treating the units as algebraic quantities that can be cancelled is called **dimensional analysis.** Knowing that your answer will be in the correct units is not a guarantee that your answer is right, but if you find that your answer will have the wrong units, you can be sure that you have made an error. Dimensional analysis also is used in choosing conversion factors. A conversion factor is a multiplier equal to 1.

Find help with **dimensional analysis.**

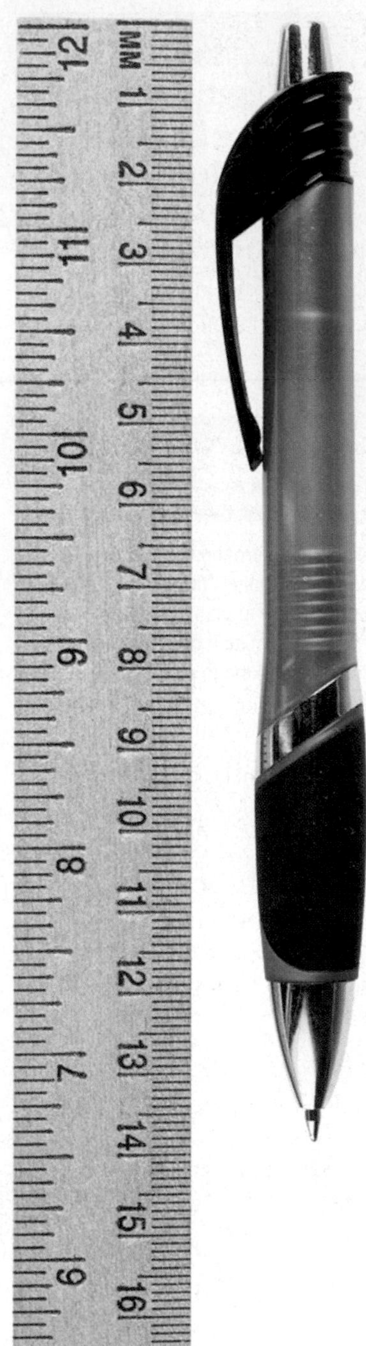

Figure 8 The student measuring this pen recorded the length as 138.1 mm.

Infer Why is the last digit in this measurement uncertain?

For example, because 1 kg = 1000 g, you can construct the following conversion factors:

$$1 = \frac{1\ \text{kg}}{1000\ \text{g}} \qquad 1 = \frac{1000\ \text{g}}{1\ \text{kg}}$$

Choose a conversion factor that will make the initial units cancel, leaving the answer in the desired units. For example, to convert 1.34 kg of iron ore to grams, do as shown below.

$$1.34\ \cancel{\text{kg}}\left(\frac{1000\ \text{g}}{1\ \cancel{\text{kg}}}\right) = 1340\ \text{g}$$

You also might need to do a series of conversions. To convert 43 km/h to m/s, do the following:

$$\left(\frac{43\ \cancel{\text{km}}}{1\ \cancel{\text{h}}}\right)\left(\frac{1000\ \text{m}}{1\ \cancel{\text{km}}}\right)\left(\frac{1\ \cancel{\text{h}}}{60\ \cancel{\text{min}}}\right)\left(\frac{1\ \cancel{\text{min}}}{60\ \text{s}}\right) = 12\ \text{m/s}$$

Significant Figures

Suppose you use a metric ruler to measure a pen and you find that the end of the pen is just past 138 mm, as shown in **Figure 8.** You estimate that the pen is one-tenth of a millimeter past the last tic mark on the ruler and record the pen as being 138.1 mm long. This measurement has four valid digits: the first three digits are certain, and the last one is uncertain. The valid digits in a measurement are referred to as **significant figures.** The last digit given for any measurement is the uncertain digit. All nonzero digits in a measurement are significant.

Are all zeros significant? No. For example, in the measurement 0.0860 m, the first two zeros serve only to locate the decimal point and are not significant. The last zero, however, is the estimated digit and is significant. The measurement 172,000 m could have 3, 4, 5, or 6 significant figures. This ambiguity is one reason to use scientific notation. It is clear that the measurement 1.7200×10^5 m has five significant figures.

Arithmetic with significant figures When you perform any arithmetic operation, it is important to remember that the result never can be more precise than the least-precise measurement.

To add or subtract measurements, first perform the operation, then round off the result to correspond to the least-precise value involved. For example, 3.86 m + 2.4 m = 6.3 m because the least-precise measure is to one-tenth of a meter.

To multiply or divide measurements, perform the calculation and then round to the same number of significant figures as the least-precise measurement. For example, $\frac{409.2\ \text{km}}{11.4\ \text{L}} = 35.9\ \text{km/L}$, because the least-precise measurement has three significant figures. Some calculators display several additional digits, while others round at different points. Be sure to record your answers with the correct number of digits.

Find help with **significant figures** and **scientific notation.**

Solving Problems

As you continue this course, you will complete practice problems. Most problems will be complex and require a strategy to solve. This textbook includes many example problems, each of which is solved using a three-step process. Read Example Problem 1 and follow the steps to calculate a car's average speed using distance and time.

EXAMPLE PROBLEM 1

Get help with **example problems.** Personal Tutor

USING DISTANCE AND TIME TO FIND SPEED When a car travels 434 km in 4.5 h, what is the car's average speed?

1 ANALYZE AND SKETCH THE PROBLEM

The car's speed is unknown. The known values include the distance the car traveled and time. Use the relationship among speed, distance, and time to solve for the car's speed.

KNOWN	**UNKNOWN**
distance = 434 km	speed = ? km/h
time = 4.5 h	

2 SOLVE FOR THE UNKNOWN

$\text{distance} = \textbf{speed} \times \text{time}$ ◀ *State the relationship as an equation.*

$\textbf{speed} = \frac{\text{distance}}{\text{time}}$ ◀ *Solve the equation for speed.*

$\textbf{speed} = \frac{434\ \text{km}}{4.5\ \text{h}}$ ◀ *Substitute distance = 434 km and time = 4.5 h.*

$\textbf{speed} = \textbf{96.4 km/h}$ ◀ *Divide, and calculate units.*

3 EVALUATE THE ANSWER

Check your answer by using it to calculate the distance the car traveled.

$\text{distance} = \text{speed} \times \text{time} = 96.4\ \text{km}/\cancel{\text{h}} \times 4.5\ \cancel{\text{h}} = 434\ \text{km}$

The calculated distance matches the distance stated in the problem. This means the average speed is correct.

THE PROBLEM

1. Read the problem carefully.
2. Be sure you understand what is being asked.

ANALYZE AND SKETCH THE PROBLEM

1. Read the problem again.
2. Identify what you are given, and list the known data. If needed, gather information from graphs, tables, or figures.
3. Identify and list the unknowns.
4. Determine whether you need a sketch to help solve the problem.
5. Plan the steps you will follow to find the answer.

SOLVE FOR THE UNKNOWN

1. If the solution is mathematical, write the equation and isolate the unknown factor.
2. Substitute the known quantities into the equation.
3. Solve the equation.
4. Continue the solution process until you solve the problem.

EVALUATE THE ANSWER

1. Reread the problem. Is the answer reasonable?
2. Check your math. Are the units and the significant figures correct?

EXAMPLE PROBLEM

SECTION 2 REVIEW

Section Self-Check — Check your understanding.

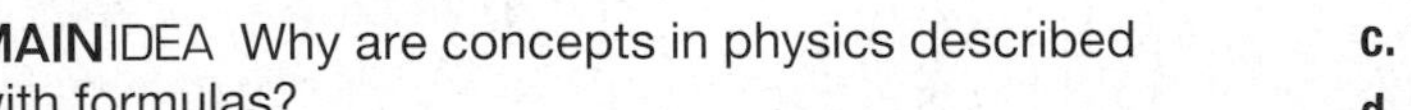

8. **MAINIDEA** Why are concepts in physics described with formulas?
9. **SI Units** What is one advantage to using SI Units in science?
10. **Dimensional Analysis** How many kilohertz are 750 megahertz?
11. **Dimensional Analysis** How many seconds are in a leap year?
12. **Significant Figures** Solve the following problems, using the correct number of significant figures each time.
    - **a.** 10.8 g − 8.264 g
    - **b.** 4.75 m − 0.4168 m
    - **c.** 139 cm × 2.3 cm
    - **d.** 13.78 g / 11.3 mL
    - **e.** 6.201 cm + 7.4 cm + 0.68 m + 12.0 cm
    - **f.** 1.6 km + 1.62 m + 1200 cm
13. **Solving Problems** Rewrite $F = Bqv$ to find v in terms of F, q, and B.
14. **Critical Thinking** Using values given in a problem and the equation of distance = speed × time, you calculate a car's speed to be 290 km/h. Is this a reasonable answer? Why or why not? Under what circumstances might this be a reasonable answer?

Measurement

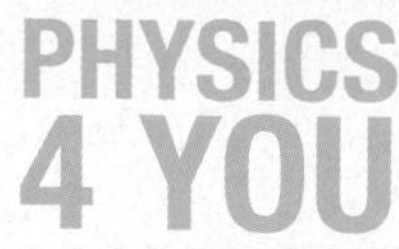

There are many devices that you often use to make measurements. Clocks measure time, rulers measure distance, and speedometers measure speed. What other measuring devices have you used?

MAIN IDEA

Making careful measurements allows scientists to repeat experiments and compare results.

Essential Questions

- Why are the results of measurements often reported with an uncertainty?
- What is the difference between precision and accuracy?
- What is a common source of error when making a measurement?

Review Vocabulary

parallax the apparent shift in the position of an object when it is viewed from different angles

New Vocabulary

measurement
precision
accuracy

What is measurement?

When you visit the doctor for a checkup, many measurements are taken: your height, weight, blood pressure, and heart rate. Even your vision is measured and assigned numbers. Blood might be drawn so measurements can be made of blood cells or cholesterol levels. Measurements quantify our observations: a person's blood pressure isn't just "pretty good," it's $\frac{110}{60}$, the low end of the good range.

A **measurement** is a comparison between an unknown quantity and a standard. For example, if you measure the mass of a rolling cart used in an experiment, the unknown quantity is the mass of the cart and the standard is the gram, as defined by the balance or the spring scale you use. In the MiniLab in Section 1, the length of the spring was the unknown and the centimeter was the standard.

Comparing Results

As you learned in Section 1, scientists share their results. Before new data are fully accepted, other scientists examine the experiment, look for possible sources of error, and try to reproduce the results. Results often are reported with an uncertainty. A new measurement that is within the margin of uncertainty is in agreement with the old measurement.

For example, archaeologists use radiocarbon dating to find the age of cave paintings, such as those from the Lascaux cave, in **Figure 9,** and the Chauvet cave. Each radiocarbon date is reported with an uncertainty. Three radiocarbon ages from a panel in the Chauvet cave are 30,940 $\pm$ 610 years, 30,790 $\pm$ 600 years, and 30,230 $\pm$ 530 years. While none of the measurements exactly matches, the uncertainties in all three overlap, and the measurements agree with each other.

Figure 9 These drawings are from the Lascaux cave in France. Scientists estimate that the drawings were made 17,000 years ago.

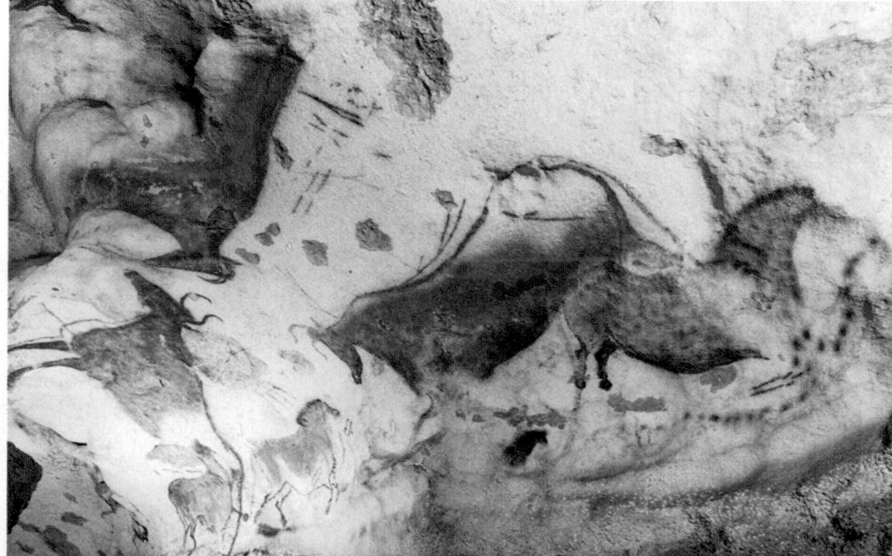

Suppose three students performed the MiniLab from Section 1 several times, starting with springs of the same length. With two washers on the spring, student 1 made repeated measurements, which ranged from 14.4 cm to 14.8 cm. The average of student 1's measurements was 14.6 cm, as shown in **Figure 10.** This result was reported as (14.6 ± 0.2) cm. Student 2 reported finding the spring's length to be (14.8 ± 0.3) cm. Student 3 reported a length of (14.0 ± 0.1) cm.

Could you conclude that the three measurements are in agreement? Is student 1's result reproducible? The ranges of the results of students 1 and 2 overlap between 14.5 cm and 14.8 cm. However, there is no overlap and, therefore, no agreement, between their results and the result of student 3.

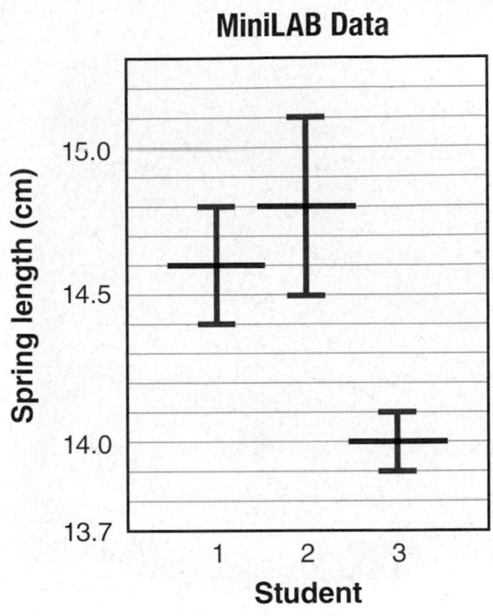

Figure 10 Three students took multiple measurements. The red bars show the uncertainty of each measurement.

Explain Are the measurements in agreement? Is student 3's result reproducible? Why or why not?

Precision Versus Accuracy

Both precision and accuracy are characteristics of measured values as shown in **Figure 11.** How precise and accurate are the measurements of the three students above? The degree of exactness of a measurement is called its **precision.** In the example above, student 3's measurements are the most precise, within ± 0.1 cm. Both the measurements of student 1 and student 2 are less precise because they have a larger uncertainty (student 1 = ± 0.2 cm, student 2 = ± 0.3 cm).

Precision depends on the instrument and technique used to make the measurement. Generally, the device that has the finest division on its scale produces the most precise measurement. The precision of a measurement is one-half the smallest division of the instrument. For example, suppose a graduated cylinder has divisions of 1 mL. You could measure an object to within 0.5 mL with this device. However, if the smallest division on a beaker is 50 mL, how precise would your measurements be compared to those taken with the graduated cylinder?

The significant figures in an answer show its precision. A measure of 67.100 g is precise to the nearest thousandth of a gram. Recall from Section 2 the rules for performing operations with measurements given to different levels of precision. If you add 1.2 mL of acid to a beaker containing 2.4×10^2 mL of water, you cannot say you now have 2.412×10^2 mL of fluid because the volume of water was not measured to the nearest tenth of a milliliter, but to the nearest 10 mL.

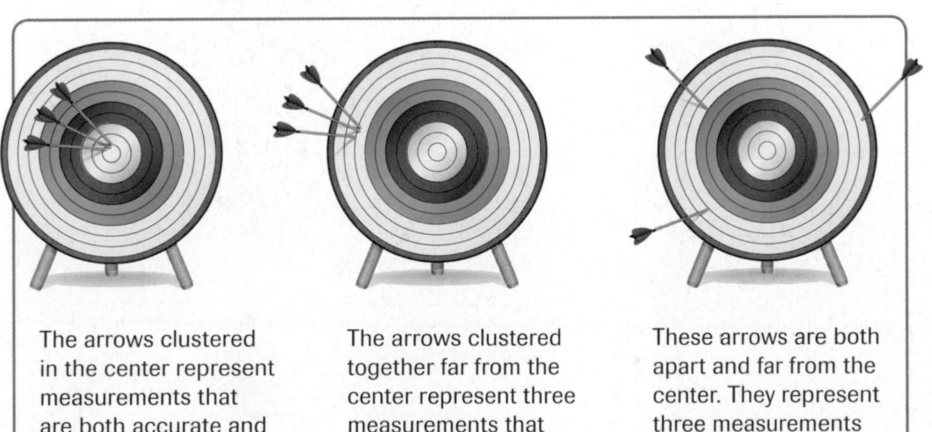

The arrows clustered in the center represent measurements that are both accurate and precise.
The arrows clustered together far from the center represent three measurements that are precise but not accurate.
These arrows are both apart and far from the center. They represent three measurements that are inaccurate and imprecise.

Figure 11 The yellow area in the center of each target represents an accepted value for a particular measurement. The arrows represent measurements taken by a scientist during an experiment.

View an **animation of precision and accuracy.**

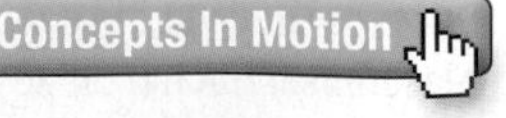

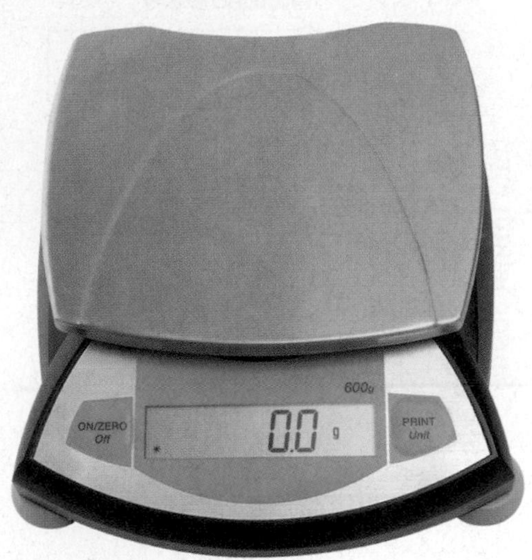

Figure 12 Accuracy is checked by zeroing an instrument before measuring.

Infer Is this instrument accurate? Why or why not?

Accuracy describes how well the results of a measurement agree with the "real" value; that is, the accepted value as measured by competent experimenters, as shown in **Figure 11.** If the length of the spring that the three students measured had been 14.8 cm, then student 2 would have been most accurate and student 3 least accurate. What might have led someone to make inaccurate measurements? How could you check the accuracy of measurements?

A common method for checking the accuracy of an instrument is called the two-point calibration. First, does the instrument read zero when it should, as shown in **Figure 12?** Second, does it give the correct reading when it is measuring an accepted standard? Regular checks for accuracy are performed on critical measuring instruments, such as the radiation output of the machines used to treat cancer.

☑ READING CHECK **Compare and contrast** precision and accuracy.

Techniques of Good Measurement

To assure accuracy and precision, instruments also have to be used correctly. Measurements have to be made carefully if they are to be as precise as the instrument allows. One common source of error comes from the angle at which an instrument is read. Scales should be read with one's eye directly in front of the measure, as shown in **Figure 13.** If the scale is read from an angle, also shown in **Figure 13,** a different, less accurate, value will be obtained. The difference in the readings is caused by parallax, which is the apparent shift in the position of an object when it is viewed from different angles. To experiment with parallax, place your pen on a ruler and read the scale with your eye directly over the tip, then read the scale with your head shifted far to one side.

View a **BrainPOP video on measuring matter.**

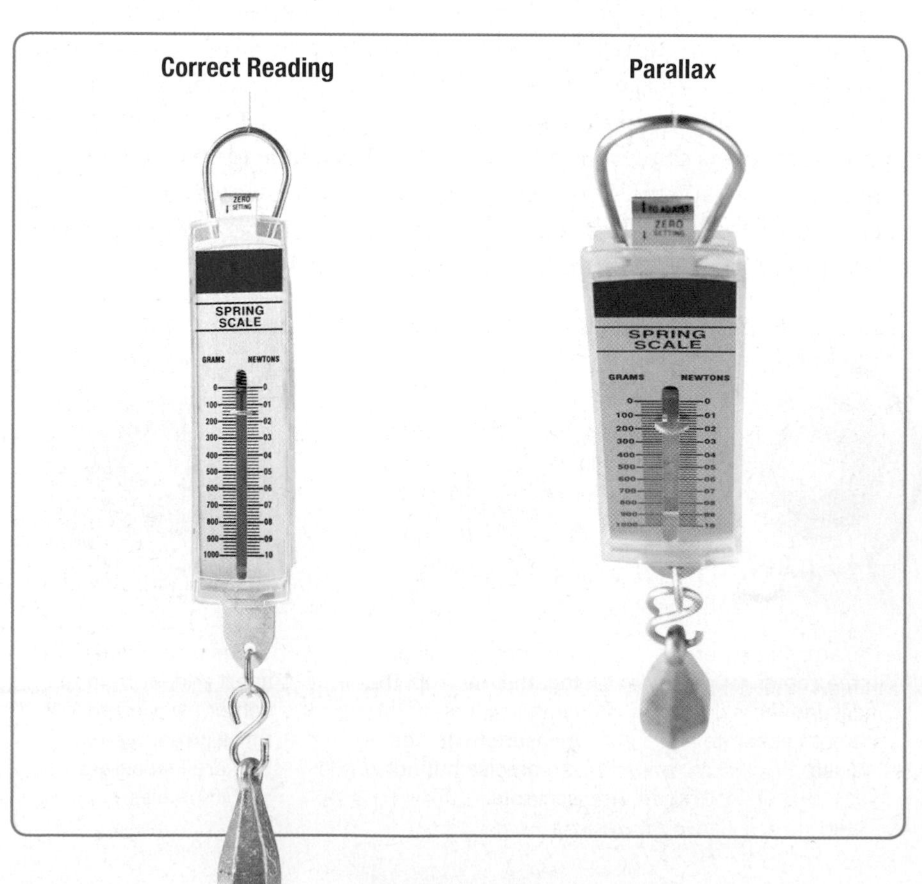

Figure 13 By positioning the scale head on (left), your results will be more accurate than if you read your measurements at an angle (right).

Identify How far did parallax shift the measurement on the right?

GPS The Global Positioning System, or GPS, offers an illustration of accuracy and precision in measurement. The GPS consists of 24 satellites with transmitters in orbit and numerous receivers on Earth. The satellites send signals with the time, measured by highly accurate atomic clocks. The receiver uses the information from at least four satellites to determine latitude, longitude, and elevation. (The clocks in the receivers are not as accurate as those on the satellites.)

Receivers have different levels of precision. A device in an automobile might give your position to within a few meters. Devices used by geophysicists, as in **Figure 14,** can measure movements of millimeters in Earth's crust.

The GPS was developed by the United States Department of Defense. It uses atomic clocks, which were developed to test Einstein's theories of relativity and gravity. The GPS eventually was made available for civilian use. GPS signals now are provided worldwide free of charge and are used in navigation on land, at sea, and in the air, for mapping and surveying, by telecommunications and satellite networks, and for scientific research into earthquakes and plate tectonics.

Figure 14 This scientist is setting up a highly accurate GPS receiver in order to record and analyze the movements of continental plates.

Ty Milford/Aurora/Getty Images

PhysicsLAB

MASS AND VOLUME

How does mass depend on volume?

SECTION 3 REVIEW

Section Self-Check — Check your understanding.

15. **MAIN IDEA** You find a micrometer (a tool used to measure objects to the nearest 0.001 mm) that has been badly bent. How would it compare to a new, high-quality meterstick in terms of its precision? Its accuracy?

16. **Accuracy** Some wooden rulers do not start with 0 at the edge, but have it set in a few millimeters. How could this improve the accuracy of the ruler?

17. **Parallax** Does parallax affect the precision of a measurement that you make? Explain.

18. **Uncertainty** Your friend tells you that his height is 182 cm. In your own words, explain the range of heights implied by this statement.

19. **Precision** A box has a length of 18.1 cm and a width of 19.2 cm, and it is 20.3 cm tall.
    - **a.** What is its volume?
    - **b.** How precise is the measurement of length? Of volume?
    - **c.** How tall is a stack of 12 of these boxes?
    - **d.** How precise is the measurement of the height of one box? Of 12 boxes?

20. **Critical Thinking** Your friend states in a report that the average time required for a car to circle a 1.5-mi track was 65.414 s. This was measured by timing 7 laps using a clock with a precision of 0.1 s. How much confidence do you have in the results of the report? Explain.

SECTION 4

Graphing Data

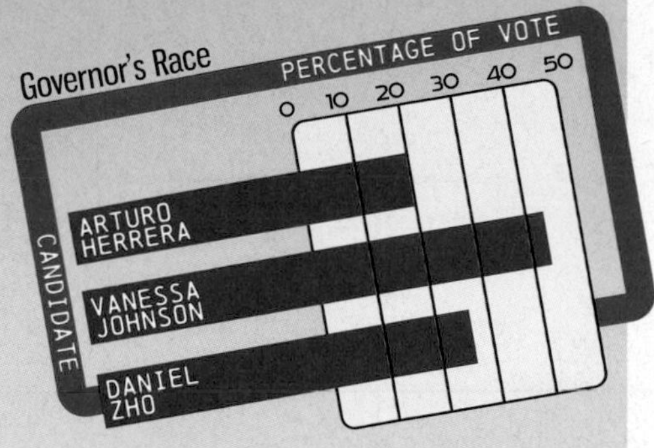

PHYSICS 4 YOU Graphs are often used in news stories after elections. Bar and circle graphs are used to show the number or percentage of votes various candidates received. Other graphs are used to show increases and decreases in population or resources over years.

MAIN IDEA

Graphs make it easier to interpret data, identify trends, and show relationships among a set of variables.

Essential Questions

- What can be learned from graphs?
- What are some common relationships in graphs?
- How do scientists make predictions?

Review Vocabulary

slope on a graph, the ratio of vertical change to horizontal change

New Vocabulary

independent variable
dependent variable
line of best fit
linear relationship
quadratic relationship
inverse relationship

Identifying Variables

When you perform an experiment, it is important to change only one factor at a time. For example, **Table 3** gives the length of a spring with different masses attached. Only the mass varies; if different masses were hung from different types of springs, you wouldn't know how much of the difference between two data pairs was due to the different masses and how much was due to the different springs.

Independent and dependent variables A variable is any factor that might affect the behavior of an experimental setup. The factor that is manipulated during an investigation is the **independent variable.** In this investigation, the mass was the independent variable. The factor that depends on the independent variable is the **dependent variable.** In this investigation, the amount the spring stretched depended on the mass, so the amount of stretch was the dependent variable. A scientist might also look at how radiation varies with time or how the strength of a magnetic field depends on the distance from a magnet.

Line of best fit A line graph shows how the dependent variable changes with the independent variable. The data from **Table 3** are graphed in **Figure 15.** The line in blue, drawn as close to all the data points as possible, is called a **line of best fit.** The line of best fit is a better model for predictions than any one point along the line. **Figure 15** gives detailed instructions on how to construct a graph, plot data, and sketch a line of best fit.

A well-designed graph allows patterns that are not immediately evident in a list of numbers to be seen quickly and simply. The graph in **Figure 15** shows that the length of the spring increases as the mass suspended from the spring increases.

Table 3 Length of a Spring for Different Masses

Mass Attached to Spring (g)	Length of Spring (cm)
0	13.7
5	14.1
10	14.5
15	14.9
20	15.3
25	15.7
30	16.0
35	16.4

■ Plotting Line Graphs

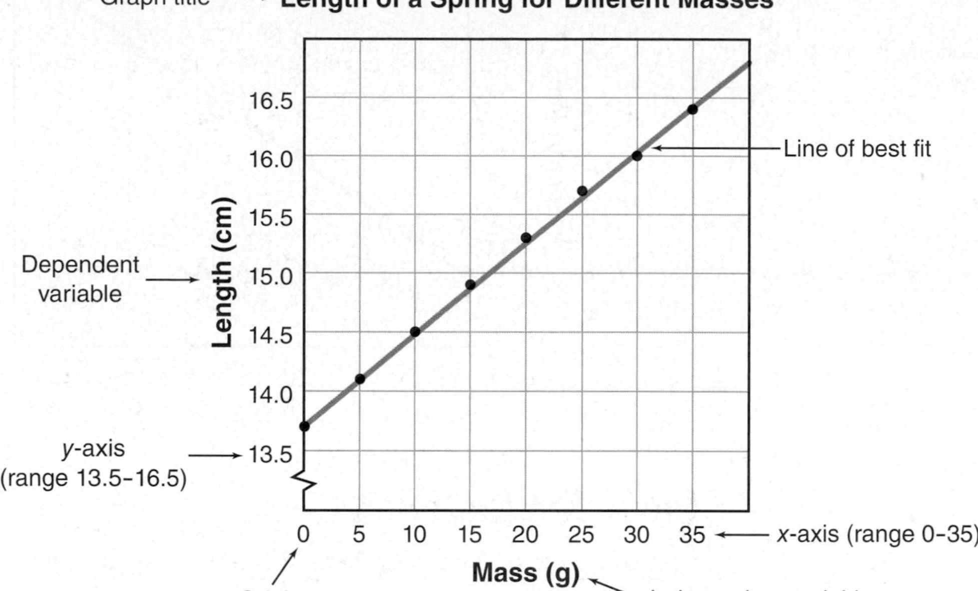

1. Identify the independent variable and dependent variable in your data. In this example, the independent variable is mass (g) and the dependent variable is length (cm). The independent variable is plotted on the horizontal axis, the *x*-axis. The dependent variable is plotted on the vertical axis, the *y*-axis.
2. Determine the range of the independent variable to be plotted. In this case the range is 0-35.
3. Decide whether the origin (0,0) is a valid data point.
4. Spread the data out as much as possible. Let each division on the graph paper stand for a convenient unit. This usually means units that are multiples of 2, 5, or 10.
5. Number and label the horizontal axis. The label should include the units, such as Mass (g).
6. Repeat steps 2-5 for the dependent variable.
7. Plot the data points on the graph.
8. Draw the best-fit straight line or smooth curve that passes through as many data points as possible. This is sometimes called *eyeballing.* Do not use a series of straight-line segments that connect the dots. The line that looks like the best fit to you may not be exactly the same as someone else's. There is a formal procedure, which many graphing calculators use, called the least-squares technique, that produces a unique best-fit line, but that is beyond the scope of this textbook.
9. Give the graph a title that clearly tells what the graph represents.

Figure 15 Use the steps above to plot line graphs from data tables.

View an **animation of graphing data.**

Concepts In Motion

Figure 16 In a linear relationship, the dependent variable—in this case, length—varies linearly with the independent variable. The independent variable in this experiment is mass.

Describe What happens to the length of the spring as mass decreases?

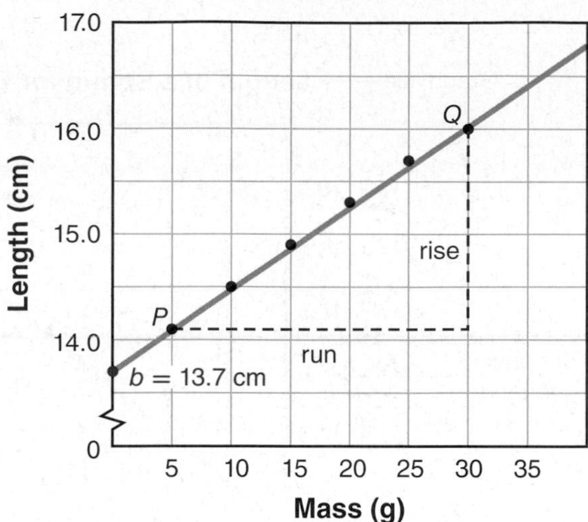

Linear Relationships

Scatter plots of data take many different shapes, suggesting different relationships. Three of the most common relationships include linear relationships, quadratic relationships, and inverse relationships. You probably are familiar with them from math class.

When the line of best fit is a straight line, as in **Figure 15,** there is a linear relationship between the variables. In a **linear relationship,** the dependent variable varies linearly with the independent variable. The relationship can be written as the following equation.

MiniLAB

HOW FAR AROUND?

What is the relationship between circumference and diameter?

LINEAR RELATIONSHIP BETWEEN TWO VARIABLES $y = mx + b$

Find the y-intercept (b) and the slope (m) as illustrated in **Figure 16.** Use points on the line—they may or may not be data points. The slope is the ratio of the vertical change to the horizontal change. To find the slope, select two points, P and Q, far apart on the line. The vertical change, or rise (Δy), is the difference between the vertical values of P and Q. The horizontal change, or run (Δx), is the difference between the horizontal values of P and Q.

Get help with **determining slope.**

SLOPE

The slope of a line is equal to the rise divided by the run, which also can be expressed as the vertical change divided by the horizontal change.

$$m = \frac{rise}{run} = \frac{\Delta y}{\Delta x}$$

In **Figure 16:** $m = \frac{(16.0 \text{ cm} - 14.1 \text{ cm})}{(30 \text{ g} - 5 \text{ g})} = 0.08 \text{ cm/g}$

If y gets smaller as x gets larger, then $\frac{\Delta y}{\Delta x}$ is negative, and the line slopes downward from left to right. The y-intercept (b) is the point at which the line crosses the vertical axis, or the y-value when the value of x is zero. In this example, $b = 13.7$ cm. This means that when no mass is suspended by the spring, it has a length of 13.7 cm. When $b = 0$, or $y = mx$, the quantity y is said to vary directly with x. In physics, the slope of the line and the y-intercept always contain information about the physical system that is described by the graph.

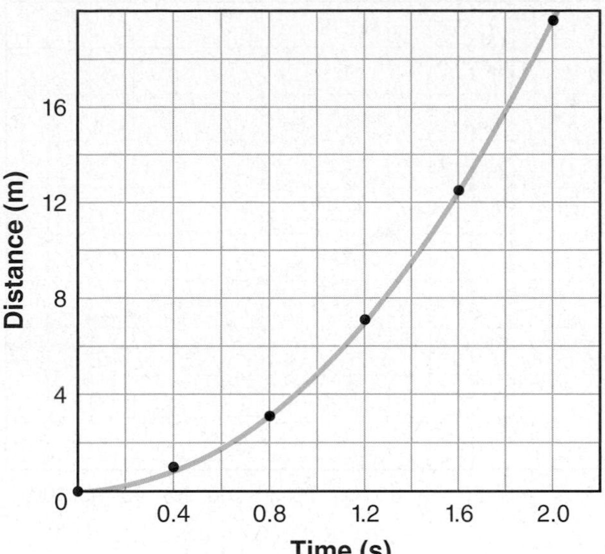

Figure 17 The quadratic, or parabolic, relationship shown here is an example of a nonlinear relationship.

Nonlinear Relationships

Figure 17 shows the distance a brass ball falls versus time. Note that the graph is not a straight line, meaning the relationship is not linear. There are many types of nonlinear relationships in science. Two of the most common are the quadratic and inverse relationships.

Quadratic relationship The graph in **Figure 17** is a quadratic relationship, represented by the equation below. A **quadratic relationship** exists when one variable depends on the square of another.

Get help with **quadratic graphs** and **quadratic equations.**

QUADRATIC RELATIONSHIP BETWEEN TWO VARIABLES

$$y = ax^2 + bx + c$$

A computer program or graphing calculator easily can find the values of the constants *a*, *b*, and *c* in this equation. In **Figure 17,** the equation is $d = 5t^2$. See the Math Handbook in the back of this book or online for more on making and using line graphs.

READING CHECK Explain how two variables related to each other in a quadratic relationship.

PHYSICS CHALLENGE

An object is suspended from spring 1, and the spring's elongation (the distance it stretches) is x_1. Then the same object is removed from the first spring and suspended from a second spring. The elongation of spring 2 is x_2. x_2 is greater than x_1.

1. On the same axes, sketch the graphs of the mass versus elongation for both springs.
2. Should the origin be included in the graph? Why or why not?
3. Which slope is steeper?
4. At a given mass, $x_2 = 1.6x_1$. If $x_2 = 5.3$ cm, what is x_1?

Spring 1 Spring 2

x_1 x_2

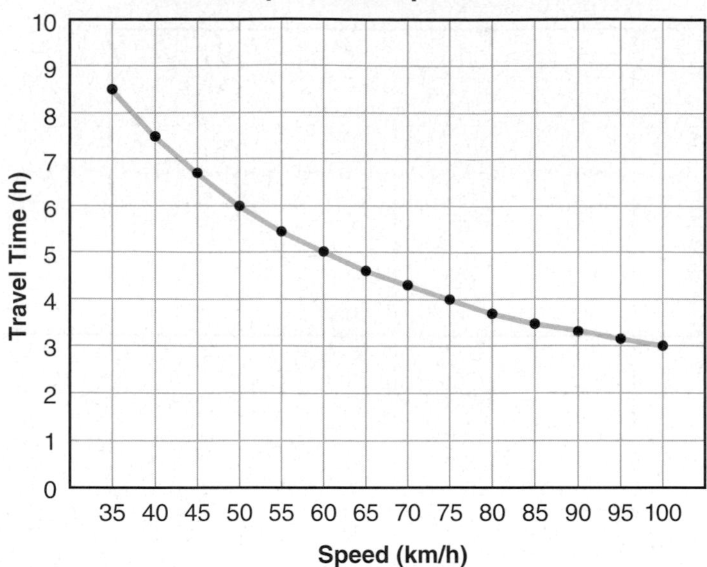

Figure 18 This graph shows the inverse relationship between speed and travel time.

Describe How does travel time change as speed increases?

Inverse relationship The graph in **Figure 18** shows how the time it takes to travel 300 km varies as a car's speed increases. This is an example of an inverse relationship, represented by the equation below. An **inverse relationship** is a hyperbolic relationship in which one variable depends on the inverse of the other variable.

INVERSE RELATIONSHIP BETWEEN TWO VARIABLES $y = \frac{a}{x}$

The three relationships you have learned about are a sample of the relations you will most likely investigate in this course. Many other mathematical models are used. Important examples include sinusoids, used to model cyclical phenomena, and exponential growth and decay, used to study radioactivity. Combinations of different mathematical models represent even more complex phenomena.

PhysicsLAB

IT'S IN THE BLOOD

FORENSICS LAB How can blood spatter provide clues?

iLab Station

☑ **READING CHECK Explain** how two variables are related to each other in an inverse relationship.

PRACTICE PROBLEMS

Do additional problems. Online Practice 

21. The mass values of specified volumes of pure gold nuggets are given in **Table 4.**

a. Plot mass versus volume from the values given in the table and draw the curve that best fits all points.

b. Describe the resulting curve.

c. According to the graph, what type of relationship exists between the mass of the pure gold nuggets and their volume?

d. What is the value of the slope of this graph? Include the proper units.

e. Write the equation showing mass as a function of volume for gold.

f. Write a word interpretation for the slope of the line.

Table 4 Mass of Pure Gold Nuggets

Volume (cm^3)	Mass (g)
1.0	19.4
2.0	38.6
3.0	58.1
4.0	77.4
5.0	96.5

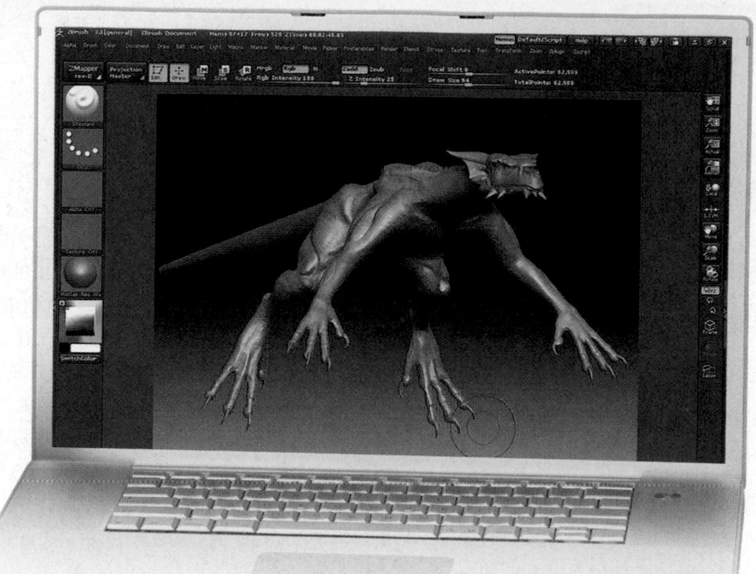

Figure 19 In order to create a realistic animation, computer animators use mathematical models of the real world to create a convincing fictional world. This computer model of a dragon Is In development on an animator's laptop.

Predicting Values

When scientists discover relationships like the ones shown in the graphs in this section, they use them to make predictions. For example, the equation for the linear graph in **Figure 16** is as follows:

$$y = (0.08\ \text{cm/g})x + 13.7\ \text{cm}$$

Relationships, either learned as formulas or developed from graphs, can be used to predict values you haven't measured directly. How far would the spring in **Table 3** stretch with 49 g of mass?

$$y = (0.08\ \text{cm/g})(49\ \text{g}) + 13.7\ \text{cm}$$
$$= 18\ \text{cm}$$

It is important to decide how far you can extrapolate from the data you have. For example, 90 g is a value far outside the ones measured, and the spring might break rather than stretch that far.

Physicists use models to accurately predict how systems will behave: what circumstances might lead to a solar flare (an immense outburst of material from the Sun's surface into space), how changes to a grandfather clock's pendulum will change its ability to keep accurate time, or how magnetic fields will affect a medical instrument. People in all walks of life use models in many ways. One example is shown in **Figure 19.** With the tools you have learned in this chapter, you can answer questions and produce models for the physics questions you will encounter in the rest of this textbook.

PhysicsLAB

EXPLORING OBJECTS IN MOTION

INTERNET LAB How can you determine the speed of a vehicle?

SECTION 4 REVIEW

Check your understanding.

22. MAINIDEA Graph the following data. Time is the independent variable.

Time (s)	0	5	10	15	20	25	30	35
Speed (m/s)	12	10	8	6	4	2	2	2

23. **Interpret a Graph** What would be the meaning of a nonzero *y*-intercept in a graph of total mass versus volume?

24. **Predict** Use the relationship illustrated in **Figure 16** to determine the mass required to stretch the spring 15 cm.

25. **Predict** Use the relationship shown in **Figure 18** to predict the travel time when speed is 110 km/h.

26. **Critical Thinking** Look again at the graph in **Figure 16.** In your own words, explain how the spring would be different if the line in the graph were shallower or had a smaller slope.

Coming to Life

The Physics Behind Animation

If you were asked to name careers that use physics, animator would probably not be the first that comes to your mind. Three-dimensional (3-D) computer animation has replaced traditional two-dimensional, hand-drawn animation as the preferred medium for big-screen animated features. Knowing the physics involved in movement and light interaction is important for would-be animators who aim to create physically accurate models.

Modeling movement Initially, 3-D models are either sculpted by hand or modeled directly in the computer. Internal control points are connected to a larger grid with fewer external control points called a cage, shown in **Figure 1.** Linear geometric equations linking the cage to animation variables allow animators to produce complex, physically accurate movement without needing to move each individual control point.

Computer power The computer power required to render all of these equations is substantial. For example, the rendering equation needed for global illumination—the simulation of light bouncing around an environment—typically involves 10 million points, each with its own equation. Each frame of the animation, representing 0.04 s of screen time, generally takes about six hours to render.

Realistic characters In the past, proponents of math-based animation avoided using complicated characters, such as human beings, who appeared jarringly unrealistic compared to their nonhuman counterparts. In these cases, many animation studios preferred the technique of motion capture. Improvements in the last decade have led to increasingly complex virtual environments, however, such as oceans, and more compelling "purely animated" human characters.

FIGURE 1 Each point on the numerous triangles that make up the character grid are linked by geometric equations.

GOING FURTHER >>>

Research There is a debate that motion capture is a technique that takes the art out of animation. Compare the benefits and drawbacks of math-based animation with those of motion-capture animation.

STUDY GUIDE

BIGIDEA Physicists use scientific methods to investigate energy and matter.

VOCABULARY
- **physics** *(p. 4)*
- **scientific methods** *(p. 5)*
- **hypothesis** *(p. 6)*
- **model** *(p. 7)*
- **scientific theory** *(p. 8)*
- **scientific law** *(p. 8)*

SECTION 1 Methods of Science

MAINIDEA Scientific investigations do not always proceed with identical steps but do contain similar methods.

- Scientific methods include making observations and asking questions about the natural world.
- Scientists use models to represent things that may be too small or too large, processes that take too much time to see completely, or a material that is hazardous.
- A scientific theory is an explanation of things or events based on knowledge gained from observations and investigations. A scientific law is a statement about what happens in nature, which seems to be true all the time.
- Science can't explain or solve everything. Questions about opinions or values can't be tested.

VOCABULARY
- **dimensional analysis** *(p. 11)*
- **significant figures** *(p. 12)*

SECTION 2 Mathematics and Physics

MAINIDEA We use math to express concepts in physics.

- Using the metric system helps scientists around the world communicate more easily.
- Dimensional analysis is used to check that an answer will be in the correct units.
- Significant figures are the valid digits in a measurement.

VOCABULARY
- **measurement** *(p. 14)*
- **precision** *(p. 15)*
- **accuracy** *(p. 16)*

SECTION 3 Measurement

MAINIDEA Making careful measurements allows scientists to repeat experiments and compare results.

- Measurements are reported with uncertainty because a new measurement that is within the margin of uncertainty confirms the old measurement.
- Precision is the degree of exactness with which a quantity is measured. Accuracy is the extent to which a measurement matches the true value.
- A common source of error that occurs when making a measurement is the angle at which an instrument is read. If the scale of an instrument is read at an angle, as opposed to at eye level, the measurement will be less accurate.

VOCABULARY
- **independent variable** *(p. 18)*
- **dependent variable** *(p. 18)*
- **line of best fit** *(p. 18)*
- **linear relationship** *(p. 20)*
- **quadratic relationship** *(p. 21)*
- **inverse relationship** *(p. 22)*

SECTION 4 Graphing Data

MAINIDEA Graphs make it easier to interpret data, identify trends, and show relationships among a set of variables.

- Graphs contain information about the relationships among variables. Patterns that are not immediately evident in a list of numbers are seen more easily when the data are graphed.
- Common relationships shown in graphs include linear relationships, quadratic relationships, and inverse relationships. In a linear relationship the dependent variable varies linearly with the independent variable. A quadratic relationship occurs when one variable depends on the square of another. In an inverse relationship, one variable depends on the inverse of the other variable.
- Scientists use models and relationships between variables to make predictions.

Games and Multilingual eGlossary

Vocabulary Practice

CHAPTER 1 ASSESSMENT

Chapter Self-Check

SECTION 1 Methods of Science

Mastering Concepts

27. Describe a scientific method.

28. Explain why scientists might use each of the models listed below.

a. physical model of the solar system

b. computer model of airplane aerodynamics

c. mathematical model of the force of attraction between two objects

SECTION 2 Mathematics and Physics

Mastering Concepts

29. Why is mathematics important to science?

30. What is the SI system?

31. How are base units and derived units related?

32. Suppose your lab partner recorded a measurement as 100 g.

a. Why is it difficult to tell the number of significant figures in this measurement?

b. How can the number of significant figures in such a number be made clear?

33. Give the name for each of the following multiples of the meter.

a. $\frac{1}{100}$ m

b. $\frac{1}{1000}$ m

c. 1000 m

34. To convert 1.8 h to minutes, by what conversion factor should you multiply?

35. Solve each problem. Give the correct number of significant figures in the answers.

a. 4.667×10^{4} g + 3.02×10^{5} g

b. $(1.70\times10^{2}$ J$)$ ÷ $(5.922\times10^{-4}$ cm$^{3})$

Mastering Problems

36. Convert each of the following measurements to meters.

a. 42.3 cm

b. 6.2 pm

c. 21 km

d. 0.023 mm

e. 214 μm

f. 57 nm

37. Add or subtract as indicated.

a. 5.80×10^{9} s + 3.20×10^{8} s

b. 4.87×10^{-6} m − 1.93×10^{-6} m

c. 3.14×10^{-5} kg + 9.36×10^{-5} kg

d. 8.12×10^{7} g − 6.20×10^{6} g

38. Ranking Task Rank the following numbers according to the number of significant figures they have, from most to least: 1.234, 0.13, 0.250, 7.603, 0.08. Specifically indicate any ties.

39. State the number of significant figures in each of the following measurements.

a. 0.00003 m

b. 64.01 fm

c. 80.001 m

d. 6×10^{8} kg

e. 4.07×10^{16} m

40. Add or subtract as indicated.

a. 16.2 m + 5.008 m + 13.48 m

b. 5.006 m + 12.0077 m + 8.0084 m

c. 78.05 cm^{2} − 32.046 cm^{2}

d. 15.07 kg − 12.0 kg

41. Multiply or divide as indicated.

a. $(6.2\times10^{18}$ m$)(4.7\times10^{-10}$ m$)$

b. $\frac{(5.6\times10^{-7}\text{ m})}{(2.8\times10^{-12}\text{ s})}$

c. $(8.1\times10^{-4}$ km$)(1.6\times10^{-3}$ km$)$

d. $\frac{(6.5\times10^{5}\text{ kg})}{(3.4\times10^{3}\text{ m}^{3})}$

42. Gravity The force due to gravity is $F = mg$ where $g = 9.8$ N/kg.

a. Find the force due to gravity on a 41.63-kg object.

b. The force due to gravity on an object is 632 N. What is its mass?

43. Dimensional Analysis Pressure is measured in pascals, where 1 Pa = 1 kg/(m•s^{2}). Will the following expression give a pressure in the correct units?

$$\frac{(0.55\text{ kg})(2.1\text{ m/s})}{9.8\text{ m/s}^{2}}$$

SECTION 3 Measurement

Mastering Concepts

44. What determines the precision of a measurement?

45. How does the last digit differ from the other digits in a measurement?

Chapter Self-Check

Mastering Problems

46. A water tank has a mass of 3.64 kg when it is empty and a mass of 51.8 kg when it is filled to a certain level. What is the mass of the water in the tank?

47. The length of a room is 16.40 m, its width is 4.5 m, and its height is 3.26 m. What volume does the room enclose?

48. The sides of a quadrangular plot of land are 132.68 m, 48.3 m, 132.736 m, and 48.37 m. What is the perimeter of the plot?

49. How precise a measurement could you make with the scale shown in **Figure 20?**

Figure 20

50. Give the measurement shown on the meter in **Figure 21** as precisely as you can. Include the uncertainty in your answer.

Figure 21

51. Estimate the height of the nearest door frame in centimeters. Then measure it. How accurate was your estimate? How precise was your estimate? How precise was your measurement? Why are the two precisions different?

52. Temperature The temperature drops linearly from 24°C to 10°C in 12 hours.

a. Find the average temperature change per hour.

b. Predict the temperature in 2 more hours if the trend continues.

c. Could you accurately predict the temperature in 24 hours? Explain why or why not.

SECTION 4 Graphing Data

Mastering Concepts

53. How do you find the slope of a linear graph?

54. When driving, the distance traveled between seeing a stoplight and stepping on the brakes is called the reaction distance. Reaction distance for a given driver and vehicle depends linearly on speed.

a. Would the graph of reaction distance versus speed have a positive or a negative slope?

b. A driver who is distracted takes a longer time to step on the brake than a driver who is not. Would the graph of reaction distance versus speed for a distracted driver have a larger or smaller slope than for a normal driver? Explain.

55. During a laboratory experiment, the temperature of the gas in a balloon is varied and the volume of the balloon is measured. Identify the independent variable and the dependent variable.

56. What type of relationship is shown in **Figure 22?** Give the general equation for this type of relation.

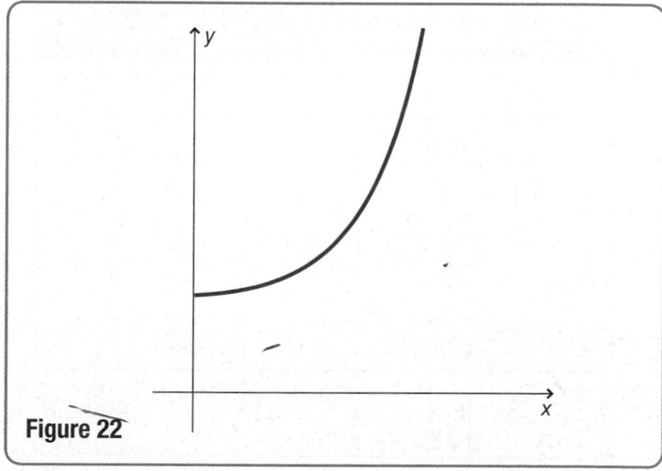

Figure 22

57. Given the equation $F = \frac{mv^2}{R}$, what kind of relationship exists between each of the following?

a. F and R

b. F and m

c. F and v

Chapter Self-Check

Mastering Problems

58. **Figure 23** shows the masses of three substances for volumes between 0 and 60 cm^3.

a. What is the mass of 30 cm^3 of each substance?

b. If you had 100 g of each substance, what would be each of their volumes?

c. In one or two sentences, describe the meaning of the slopes of the lines in this graph.

d. Explain the meaning of each line's y-intercept.

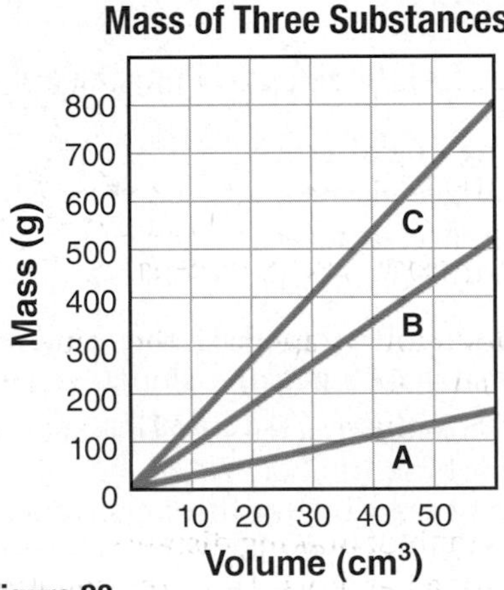

Figure 23

59. Suppose a mass is placed on a horizontal table that is nearly frictionless. Various horizontal forces are applied to the mass. The distance the mass traveled in 5 seconds for each force applied is measured. The results of the experiment are shown in **Table 5.**

Table 5 Distance Traveled with Different Forces

Force (N)	Distance (cm)
5.0	24
10.0	49
15.0	75
20.0	99
25.0	120
30.0	145

a. Plot the values given in the table and draw the curve that best fits all points.

b. Describe the resulting curve.

c. Use the graph to write an equation relating the distance to the force.

d. What is the constant in the equation? Find its units.

e. Predict the distance traveled when a 22.0-N force is exerted on the object for 5 s.

60. Suppose the procedure from the previous problem changed. The mass was varied while the force was kept constant. Time and distance were measured, and the acceleration of each mass was calculated. The results of the experiment are shown in **Table 6.**

Table 6 Acceleration of Different Masses

Mass (kg)	Acceleration (m/s^2)
1.0	12.0
2.0	5.9
3.0	4.1
4.0	3.0
5.0	2.5
6.0	2.0

a. Plot the values given in the table and draw the curve that best fits all points.

b. Describe the resulting curve.

c. Write the equation relating acceleration to mass given by the data in the graph.

d. Find the units of the constant in the equation.

e. Predict the acceleration of an 8.0-kg mass.

61. During an experiment, a student measured the mass of 10.0 cm^3 of alcohol. The student then measured the mass of 20.0 cm^3 of alcohol. In this way, the data in **Table 7** were collected.

Table 7
The Mass Values of Specific Volumes of Alcohol

Volume (cm^3)	Mass (g)
10.0	7.9
20.0	15.8
30.0	23.7
40.0	31.6
50.0	39.6

a. Plot the values given in the table and draw the curve that best fits all the points.

b. Describe the resulting curve.

c. Use the graph to write an equation relating the volume to the mass of the alcohol.

d. Find the units of the slope of the graph. What is the name given to this quantity?

e. What is the mass of 32.5 cm^3 of alcohol?

Chapter Self-Check

Applying Concepts

62. Is a scientific method one set of clearly defined steps? Support your answer.

63. Explain the difference between a scientific theory and a scientific law.

64. **Figure 24** gives the height above the ground of a ball that is thrown upward from the roof of a building, for the first 1.5 s of its trajectory. What is the ball's height at $t = 0$? Predict the ball's height at $t = 2$ s and at $t = 5$ s.

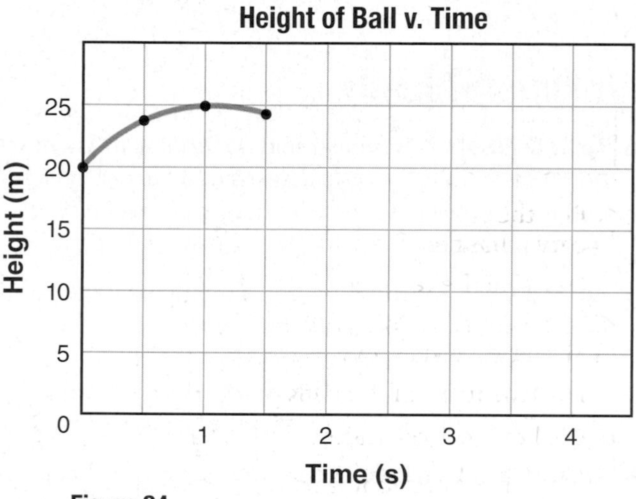

Figure 24

65. Density The density of a substance is its mass divided by its volume.

a. Give the metric unit for density.

b. Is the unit for density a base unit or a derived unit?

66. What metric unit would you use to measure each of the following?

a. the width of your hand

b. the thickness of a book cover

c. the height of your classroom

d. the distance from your home to your classroom

67. Size Make a chart of sizes of objects. Lengths should range from less than 1 mm to several kilometers. Samples might include the size of a cell, the distance light travels in 1 s, and the height of a room.

68. Time Make a chart of time intervals. Sample intervals might include the time between heartbeats, the time between presidential elections, the average lifetime of a human, and the age of the United States. In your chart, include several examples of very short and very long time intervals.

69. Speed of Light Two scientists measure the speed of light. One obtains $(3.001 \pm 0.001)\times10^8$ m/s; the other obtains $(2.999 \pm 0.006)\times10^8$ m/s.

a. Which is more precise?

b. Which is more accurate? (You can find the speed of light in the back of this textbook.)

70. You measure the dimensions of a desk as 132 cm, 83 cm, and 76 cm. The sum of these measures is 291 cm, while the product is 8.3×10^5 cm^3. Explain how the significant figures were determined in each case.

71. Money Suppose you receive \$15.00 at the beginning of a week and spend \$2.50 each day for lunch. You prepare a graph of the amount you have left at the end of each day for one week. Would the slope of this graph be positive, zero, or negative? Why?

72. Data are plotted on a graph, and the value on the *y*-axis is the same for each value of the independent variable. What is the slope? Why? How does *y* depend on *x*?

73. Driving The graph of braking distance versus car speed is part of a parabola. Thus, the equation is written $d = av^2 + bv + c$. The distance (d) has units in meters, and velocity (v) has units in meters/second. How could you find the units of a, b, and c? What would they be?

74. How long is the leaf in **Figure 25?** Include the uncertainty in your measurement.

Figure 25

75. Explain the difference between a hypothesis and a scientific theory.

76. Give an example of a scientific law.

77. What reason might the ancient Greeks have had not to question the (incorrect) hypothesis that heavier objects fall faster than lighter objects? *Hint: Did you ever question which falls faster?*

Laura Sifferlin

Chapter Self-Check

78. A graduated cylinder is marked every mL. How precise a measurement can you make with this instrument?

79. Reverse Problem Write a problem with real-life objects for which the graph in **Figure 26** could be part of the solution.

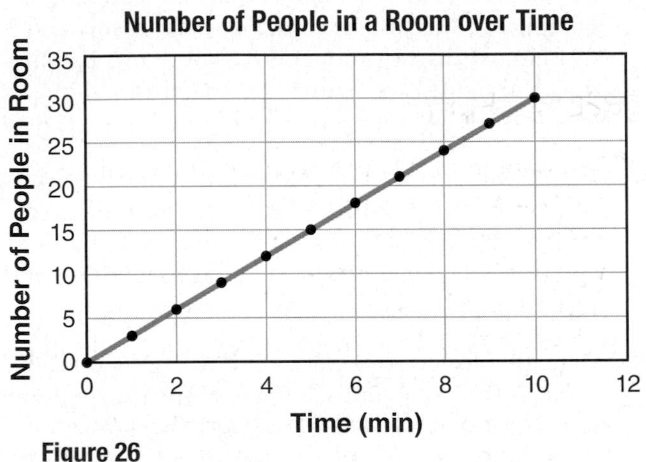

Figure 26

Mixed Review

80. Arrange the following numbers from most precise to least precise: 0.0034 m, 45.6 m, 1234 m.

81. Figure 27 shows an engine of a jet plane. Explain why a width of 80 m would be an unreasonable value for the diameter of the engine. What would be a reasonable value?

Figure 27

82. You are cracking a code and have discovered the following conversion factors: 1.23 longs = 23.0 mediums, and 74.5 mediums = 645 shorts. How many shorts are equal to one long?

83. You are given the following measurements of a rectangular bar: length = 2.347 m, thickness = 3.452 cm, height = 2.31 mm, mass = 1659 g. Determine the volume, in cubic meters, and density, in g/cm^3, of the beam.

84. A drop of water contains 1.7×10^{21} molecules. If the water evaporated at the rate of one million molecules per second, how many years would it take for the drop to completely evaporate?

85. A 17.6-gram sample of metal is placed in a graduated cylinder containing 10.0 cm^3 of water. If the water level rises to 12.20 cm^3, what is the density of the metal?

Thinking Critically

86. Apply Concepts It has been said that fools can ask more questions than the wise can answer. In science, it is frequently the case that one wise person is needed to ask the right question rather than to answer it. Explain.

87. Apply Concepts Find the approximate mass of water in kilograms needed to fill a container that is 1.40 m long and 0.600 m wide to a depth of 34.0 cm. Report your result to one significant figure. (Use a reference source to find the density of water.)

88. Analyze and Conclude A container of gas with a pressure of 101 kPa has a volume of 324 cm^3 and a mass of 4.00 g. If the pressure is increased to 404 kPa, what is the density of the gas? Pressure and volume are inversely proportional.

89. BIGIDEA **Design an Experiment** How high can you throw a ball? What variables might affect the answer to this question?

90. Problem Posing Complete this problem so that the final answer will have 3 significant figures: "A home remedy used to prevent swimmer's ear calls for equal parts vinegar and rubbing alcohol. You measure 45.62 mL of vinegar . . . ."

Writing in Physics

91. Research and describe a topic in the history of physics. Explain how ideas about the topic changed over time. Be sure to include the contributions of scientists and to evaluate the impact of their contributions on scientific thought and the world outside the laboratory.

92. Explain how improved precision in measuring time would have led to more accurate predictions about how an object falls.

STANDARDIZED TEST PRACTICE

MULTIPLE CHOICE

1. Two laboratories use radiocarbon dating to measure the age of two wooden spear handles found in the same grave. Lab A finds an age of 2250 ± 40 years for the first object; lab B finds an age of 2215 ± 50 years for the second object. Which is true?
   - **A.** Lab A's reading is more accurate than lab B's.
   - **B.** Lab A's reading is less accurate than lab B's.
   - **C.** Lab A's reading is more precise than lab B's.
   - **D.** Lab A's reading is less precise than lab B's.

2. Which of the following is equal to 86.2 cm?
   - **A.** 8.62 m
   - **B.** 0.862 mm
   - **C.** 8.62×10^{-4} km
   - **D.** 862 dm

3. Jario has a problem to do involving time, distance, and velocity, but he has forgotten the formula. The question asks him for a measurement in seconds, and the numbers that are given have units of m/s and km. What could Jario do to get the answer in seconds?
   - **A.** Multiply the km by the m/s, then multiply by 1000.
   - **B.** Divide the km by the m/s, then multiply by 1000.
   - **C.** Divide the km by the m/s, then divide by 1000.
   - **D.** Multiply the km by the m/s, then divide by 1000.

4. What is the slope of the graph?
   - **A.** $0.25\ \text{m/s}^2$
   - **B.** $0.4\ \text{m/s}^2$
   - **C.** $2.5\ \text{m/s}^2$
   - **D.** $4.0\ \text{m/s}^2$

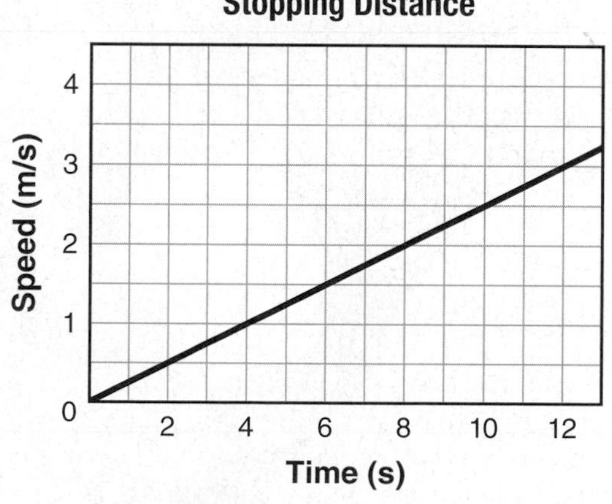

5. Which formula is equivalent to $D = \frac{m}{V}$?
   - **A.** $V = \frac{m}{D}$
   - **B.** $V = Dm$
   - **C.** $V = \frac{mD}{V}$
   - **D.** $V = \frac{D}{m}$

6. A computer simulation is an example of what?
   - **A.** a hypothesis
   - **B.** a model
   - **C.** a scientific law
   - **D.** a scientific theory

FREE RESPONSE

7. You want to calculate an acceleration, in units of m/s^2, given a force, in N, and the mass, in g, on which the force acts. ($1\ \text{N} = 1\ \text{kg}\cdot\text{m/s}^2$)
   - **a.** Rewrite the equation $F = ma$ so a is in terms of m and F.
   - **b.** What conversion factor will you need to multiply by to convert grams to kilograms?
   - **c.** A force of 2.7 N acts on a 350-g mass. Write the equation you will use, including the conversion factor, to find the acceleration.

8. Find an equation for a line of best fit for the data shown below.

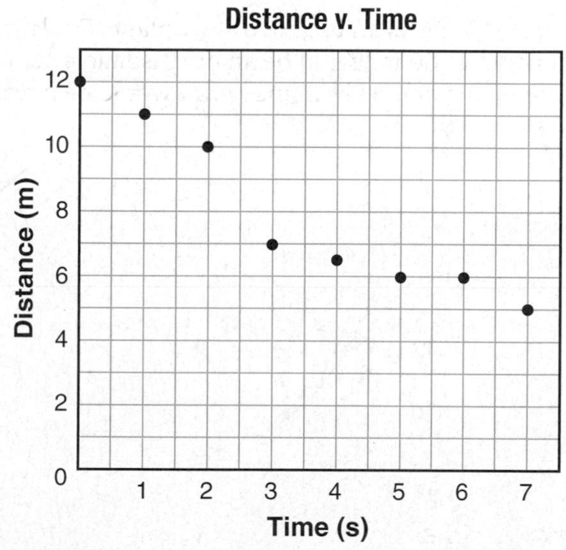

NEED EXTRA HELP?

If You Missed Question	1	2	3	4	5	6	7	8
Review Section	3	2	2	4	2	1	2	4

Online Test Practice

CHAPTER 2

Representing Motion

BIGIDEA **You can use displacement and velocity to describe an object's motion.**

SECTIONS

1 **Picturing Motion**

2 **Where and When?**

3 **Position-Time Graphs**

4 **How Fast?**

LaunchLAB

TOY CAR RACE

What factors determine an object's speed?

WATCH THIS!

Video

MEASURING SPEED

Have you ever been passed by another car on the freeway? If you know a few important details, it's possible to determine how fast that car is going. It's physics in action on the freeway.

(l)Steve Allen/Brand X Pictures, (r)Photodisc/Getty Images

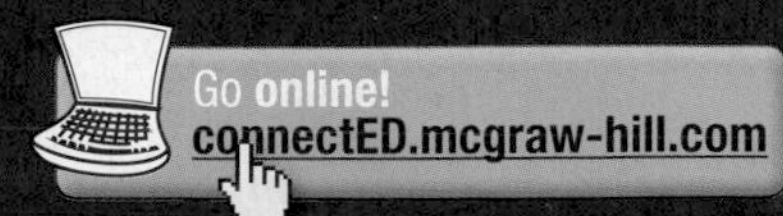
Go online!
connectED.mcgraw-hill.com

SECTION 1 # Picturing Motion

PHYSICS 4 YOU **Look at this multiple-exposure photograph of a bird's movement. Physicists can use these photographs to evaluate changes in position and velocity.**

MAIN IDEA

You can use motion diagrams to show how an object's position changes over time.

Essential Questions

- How do motion diagrams represent motion?
- How can you use a particle model to represent a moving object?

Review Vocabulary

model a representation of an idea, event, structure, or object to help people better understand it

New Vocabulary

motion diagram
particle model

All Kinds of Motion

You have learned about scientific processes that will be useful in your study of physics. You will now begin to use these tools to analyze motion. In subsequent chapters, you will apply these processes to many kinds of motion. You will use words, sketches, diagrams, graphs, and equations. These concepts will help you determine how fast and how far an object moves, in which direction that object is moving, whether that object is speeding up or slowing down, and whether that object is standing still or moving at a constant speed.

Changes in position What comes to your mind when you hear the word *motion?* A spinning ride at an amusement park? A baseball soaring over a fence for a home run? Motion is all around you—from fast trains and speedy skiers to slow breezes and lazy clouds. Objects move in many different ways, such as the straight-line path of a bowling ball in a bowling lane's gutter, the curved path of a car rounding a turn, the spiral of a falling kite, and swirls of water circling a drain. When an object is in motion, such as the subway train in **Figure 1,** its position changes.

Some types of motion are more complicated than others. When beginning a new area of study, it is generally a good idea to begin with the least complicated situation, learn as much as possible about it, and then gradually add more complexity to that simple model. In the case of motion, you will begin your study with movement along a straight line.

Figure 1 The subway train appears blurry in the photograph because its position changed during the time the camera shutter was open.

Describe how the picture would be different if the train were sitting still.

(t)Creativity+ Timothy K. Hamilton/Flickr/Getty Images, (b)Gavin Hellier/Robert Harding World Imagery/Getty Images

Movement along a straight line In general, an object can move along many different kinds of paths, but straight-line motion follows a path directly between two points without turning left or right. For example, you might describe an object's motion as forward and backward, up and down, or north and south. In each of these cases, the object moves along a straight line.

Suppose you are reading this textbook at home. As you start to read, you glance over at your pet hamster and see that it is sitting in a corner of the cage. Sometime later you look over again, and you see that it now is sitting next to the food dish in the opposite corner of the cage. You can infer that your hamster has moved from one place to another in the time between your observations. What factors helped you make this inference about the hamster's movement?

The description of motion is a description of place and time. You must answer the questions of where an object is located and when it is at that position in order to clearly describe its motion. Next, you will look at some tools that help determine when an object is at a particular place.

☑ READING CHECK **Identify** two factors you must know in order to describe the motion of an object along a straight line.

Motion Diagrams

Consider the following example of straight-line motion: a runner jogs along a straight path. One way of representing the runner's motion is to create a series of images showing the runner's position at equal time intervals. You can do this by photographing the runner in motion to obtain a sequence of pictures. Each photograph will show the runner at a point that is farther along the straight path.

Consecutive images Suppose you point a camera in a direction and a runner crosses the camera's field of view. Then you take a series of photographs of the runner at equal time intervals, without moving the camera. **Figure 2** shows what a series of consecutive images for a runner might look like. Notice that the runner is in a different position in each image, but everything in the background remains in the same position. This indicates that, relative to the camera and the ground, only the runner is in motion.

☑ READING CHECK **Decide** whether the spaces between a moving object's position must be equal if photographs are taken of the object at equal time intervals. Explain.

Figure 2 You can tell that the jogger is in motion because her position changes relative to the tree and the ground.

Figure 3 Combining the images from **Figure 2** produces this motion diagram of the jogger's movement. The series of dots at the bottom of the figure is a particle model that corresponds to the motion diagram.

Explain how the particle model shows that the jogger's speed is not changing.

View an **animation of motion diagrams v. particle motion.**

PhysicsLAB

MOTION DIAGRAMS

How do the motion diagrams of a fast toy car and a slow toy car differ?

Combining images Suppose that you layered the four images of the runner from **Figure 2** one on top of the other. **Figure 3** shows what such a layered image might look like. You see more than one image of the moving runner, but you see only a single image of the tree and other motionless objects in the background. A series of images showing the positions of a moving object at equal time intervals is called a **motion diagram.**

Particle Models

Keeping track of the runner's motion is easier if you disregard the movement of her arms and her legs and instead concentrate on a single point at the center of her body. In effect, you can disregard the fact that the runner has size and imagine that she is a very small object located precisely at that central point. In a **particle model,** you replace the object or objects of interest with single points. Use of the particle model is common throughout the study of physics.

To use the particle model, the object's size must be much less than the distance it moves. The object's internal motions, such as the waving of the runner's arms or the movement of her legs, are ignored in the particle model. In the photographic motion diagram, you could identify one central point on the runner, such as a point centered at her waistline, and draw a dot at its position at different times. The bottom of **Figure 3** shows the particle model for the runner's motion. In the next section, you will learn how to create and use a motion diagram that shows how far an object moved and how much time it took to move that far.

SECTION 1 REVIEW

Section Self-Check Check your understanding.

1. **MAIN**IDEA How does a motion diagram represent an object's motion?
2. **Motion Diagram of a Bike Rider** Draw a particle model motion diagram for a bike rider moving at a constant pace along a straight path.
3. **Motion Diagram of a Car** Draw a particle model motion diagram corresponding to the motion diagram in **Figure 4** for a car coming to a stop at a stop sign. What point on the car did you use to represent the car?

Figure 4

4. **Motion Diagram of a Bird** Draw a particle model motion diagram corresponding to the motion diagram in **Figure 5** for a flying bird. What point on the bird did you choose to represent the bird?

Figure 5

5. **Critical Thinking** Draw particle model motion diagrams for two runners during a race in which the first runner crosses the finish line as the other runner is three-fourths of the way to the finish line.

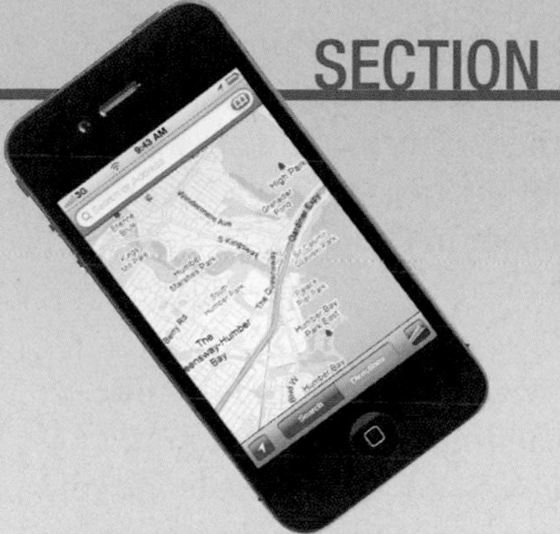

SECTION 2

Where and When?

PHYSICS 4 YOU

Have you ever used an electronic map for directions? These useful devices display the distances and directions you need to go. Many even show the time for different parts of the trip. To find your way to a place, you need clear directions for getting there.

MAIN IDEA

A coordinate system is helpful when you are describing motion.

Essential Questions

- What is a coordinate system?
- How does the chosen coordinate system affect the sign of objects' positions?
- How are time intervals measured?
- What is displacement?
- How are motion diagrams helpful in answering questions about an object's position or displacement?

Review Vocabulary

dimension extension in a given direction; one dimension is along a straight line; three dimensions are height, width, and length

New Vocabulary

coordinate system
origin
position
distance
magnitude
vector
scalar
time interval
displacement
resultant

Oleksiy Maksymenko/age fotostock

Coordinate Systems

Is it possible to measure distance and time on a motion diagram? Before photographing a runner, you could place a long measuring tape on the ground to show where the runner is in each image. A stopwatch within the camera's view could show the time. But where should you place the end of the measuring tape? When should you start the stopwatch?

Position and distance It is useful to identify a system in which you have chosen where to place the zero point of the measuring tape and when to start the stopwatch. A **coordinate system** gives the location of the zero point of the variable you are studying and the direction in which the values of the variable increase. The **origin** is the point at which all variables in a coordinate system have the value zero. In the example of the runner, the origin, which is the zero point of the measuring tape, could be 6 m to the left of the tree. Because the motion is in a straight line, your measuring tape should lie along this line. The straight line is an axis of the coordinate system.

You can indicate how far the runner in **Figure 6** is from the origin at a certain time on the motion diagram by drawing an arrow from the origin to the point that represents the runner, shown at the bottom of the figure. This arrow represents the runner's **position,** the distance and direction from the origin to the object. In general, **distance** is the entire length of an object's path, even if the object moves in many directions. Because the motion in **Figure 6** is in one direction, the arrow lengths represent distance.

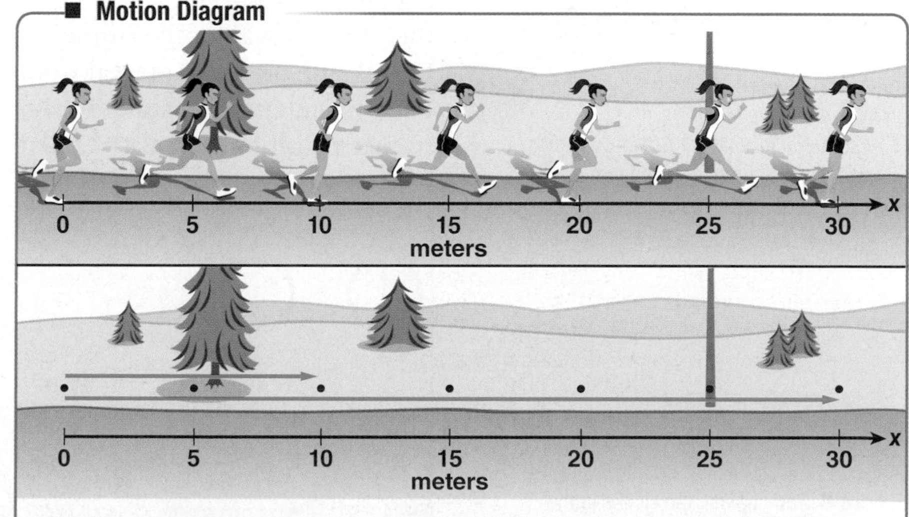

Figure 6 A simplified motion diagram uses dots to represent a moving object and arrows to indicate positions.

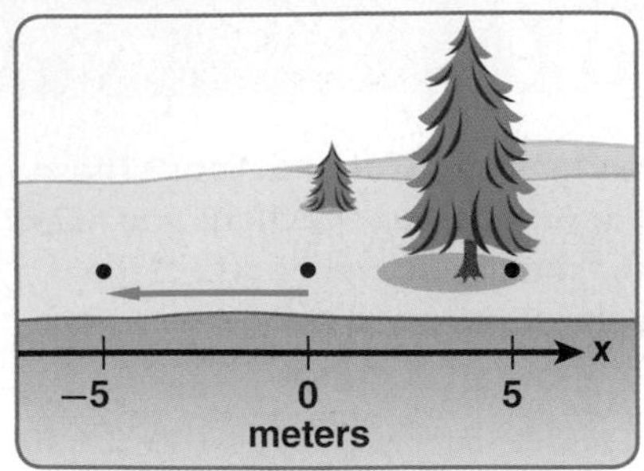

Figure 7 The green arrow indicates a negative position of –5 m if the direction right of the origin is chosen as positive.

Infer What position would the arrow indicate if you chose the direction left of the origin as positive?

Negative position Is there such a thing as a negative position? Suppose you chose the coordinate system just described but this time placed the origin 4 m left of the tree with the *x*-axis extending in a positive direction to the right. A position 9 m left of the tree, or 5 m left of the origin, would be a negative position, as shown in **Figure 7.**

Vectors and Scalars

Many quantities in physics have both size, also called **magnitude,** and direction. A quantity that has both magnitude and direction is called a **vector.** You can represent a vector with an arrow. The length of the arrow represents the magnitude of the vector, and the direction of the arrow represents the direction of the vector. A quantity that is just a number without any direction, such as distance, time, or temperature, is called a **scalar.** In this textbook, we will use boldface letters to represent vector quantities and regular letters to represent scalars.

Time intervals are scalars. When analyzing the runner's motion, you might want to know how long it took her to travel from the tree to the lamppost. You can obtain this value by finding the difference between the stopwatch reading at the tree and the stopwatch reading at the lamppost. **Figure 8** shows these stopwatch readings. The difference between two times is called a **time interval.**

A common symbol for a time interval is Δt, where the Greek letter delta (Δ) is used to represent a change in a quantity. Let t_i represent the initial (starting) time, when the runner was at the tree. Let t_f represent the final (ending) time of the interval, when the runner was at the lamppost. We define a time interval mathematically as follows.

TIME INTERVAL

The time interval is equal to the change in time from the initial time to the final time.

$$\Delta t = t_f - t_i$$

The subscripts i and f represent the initial and final times, but they can be the initial and final times of any time interval you choose. In the example of the runner, the time it takes for her to go from the tree to the lamppost is $t_f - t_i = 5.0\ s - 1.0\ s = 4.0\ s$. You could instead describe the time interval for the runner to go from the origin to the lamppost. In this case the time interval would be $t_f - t_i = 5.0\ s - 0.0\ s = 5.0\ s$. The time interval is a scalar because it has no direction. What about the runner's position? Is it also a scalar?

VOCABULARY
Science Usage v. Common Usage

Magnitude
- Science usage
a measure of size
When drawing vectors, the magnitude of a vector is proportional to that vector's length.
- Common usage
great size or extent
The magnitude of the Grand Canyon is difficult to capture in photographs.

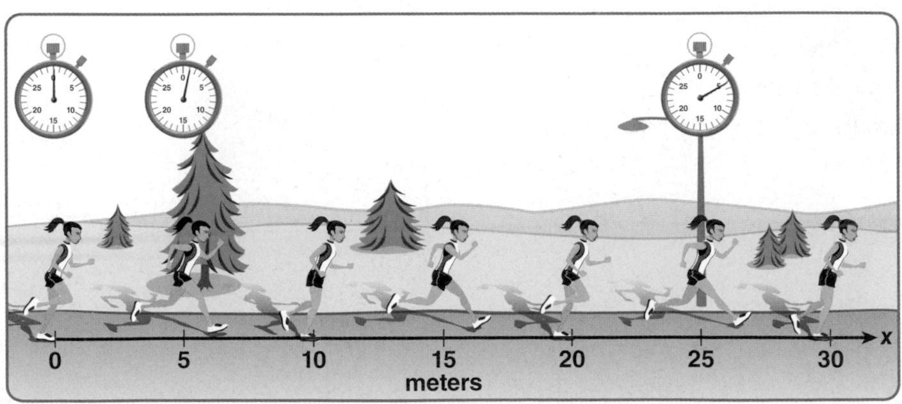

Figure 8 You can use the clocks in the figure to calculate the time interval (Δt) for the runner's movement from one position to another.

Positions and displacements are vectors. You have already seen how a position can be described as negative or positive in order to indicate whether that position is to the left or the right of a coordinate system's origin. This suggests that position is a vector because position has direction—either right or left in this case.

Figure 9 shows the position of the runner at both the tree and the lamppost. Notice that you can draw an arrow from the origin to the location of the runner in each case. These arrows have magnitude and direction. In common speech, a position refers to a certain place, but in physics, the definition of a position is more precise. A position is a vector with the arrow's tail at the origin of a coordinate system and the arrow's tip at the place.

You can use the symbol $\boldsymbol{x}$ to represent position vectors mathematically. In **Figure 9,** the symbol $\boldsymbol{x}_i$ represents the position at the tree, and the symbol $\boldsymbol{x}_f$ represents the position at the lamppost. The symbol $\Delta\boldsymbol{x}$ represents the change in position from the tree to the lamppost. Because a change in position is described and analyzed so often in physics, it has a special name. In physics, a change in position is called a **displacement.** Because displacement has direction, it is a vector.

READING CHECK **Contrast** the distance an object moves and the object's displacement for straight-line motion.

What was the runner's displacement when she ran from the tree to the lamppost? By looking at **Figure 9,** you can see that this displacement is 20 m to the right. Notice also, that the displacement from the tree to the lamppost ($\Delta\boldsymbol{x}$) equals the position at the lamppost ($\boldsymbol{x}_f$) minus the position at the tree ($\boldsymbol{x}_i$). This is true in general; displacement equals final position minus initial position.

DISPLACEMENT

Displacement is the change in position from initial position to final position.

$$\Delta\boldsymbol{x} = \boldsymbol{x}_f - \boldsymbol{x}_i$$

Remember that the initial and final positions are the start and the end of any interval you choose. Although position is a vector, sometimes the magnitude of a position is described without the boldface. In this case, a plus or minus sign might be used to indicate direction.

READING CHECK **Describe** what the direction and length of a displacement arrow indicate.

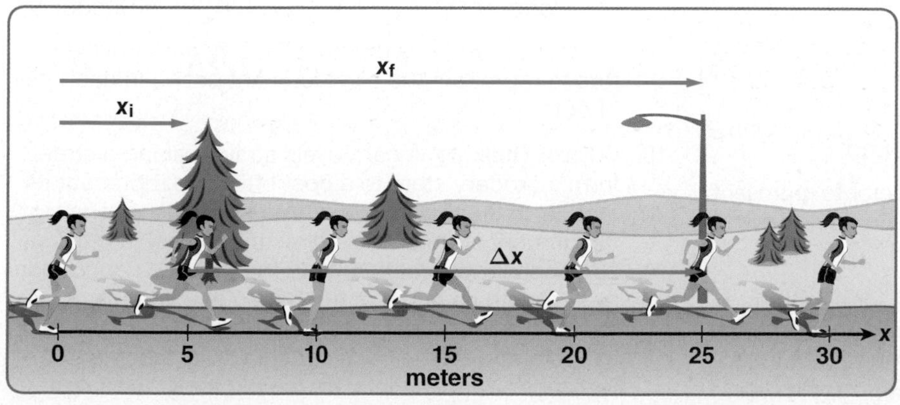

Figure 9 The vectors $\boldsymbol{x}_i$ and $\boldsymbol{x}_f$ represent positions. The vector $\Delta\boldsymbol{x}$ represents displacement from $\boldsymbol{x}_i$ to $\boldsymbol{x}_f$.

Describe the displacement from the lamppost to the tree.

MiniLAB

VECTOR MODELS

How can you model vector addition using construction toys?

Vector addition and subtraction You will learn about many different types of vectors in physics, including velocity, acceleration, and momentum. Often, you will need to find the sum of two vectors or the difference between two vectors. A vector that represents the sum of two other vectors is called a **resultant.** **Figure 10** shows how to add and subtract vectors in one dimension. In a later chapter, you will learn how to add and subtract vectors in two dimensions.

Figure 10 You can use a diagram or an equation to combine vectors.

Analyze What is the sum of a vector 12 m north and a vector 8 m north?

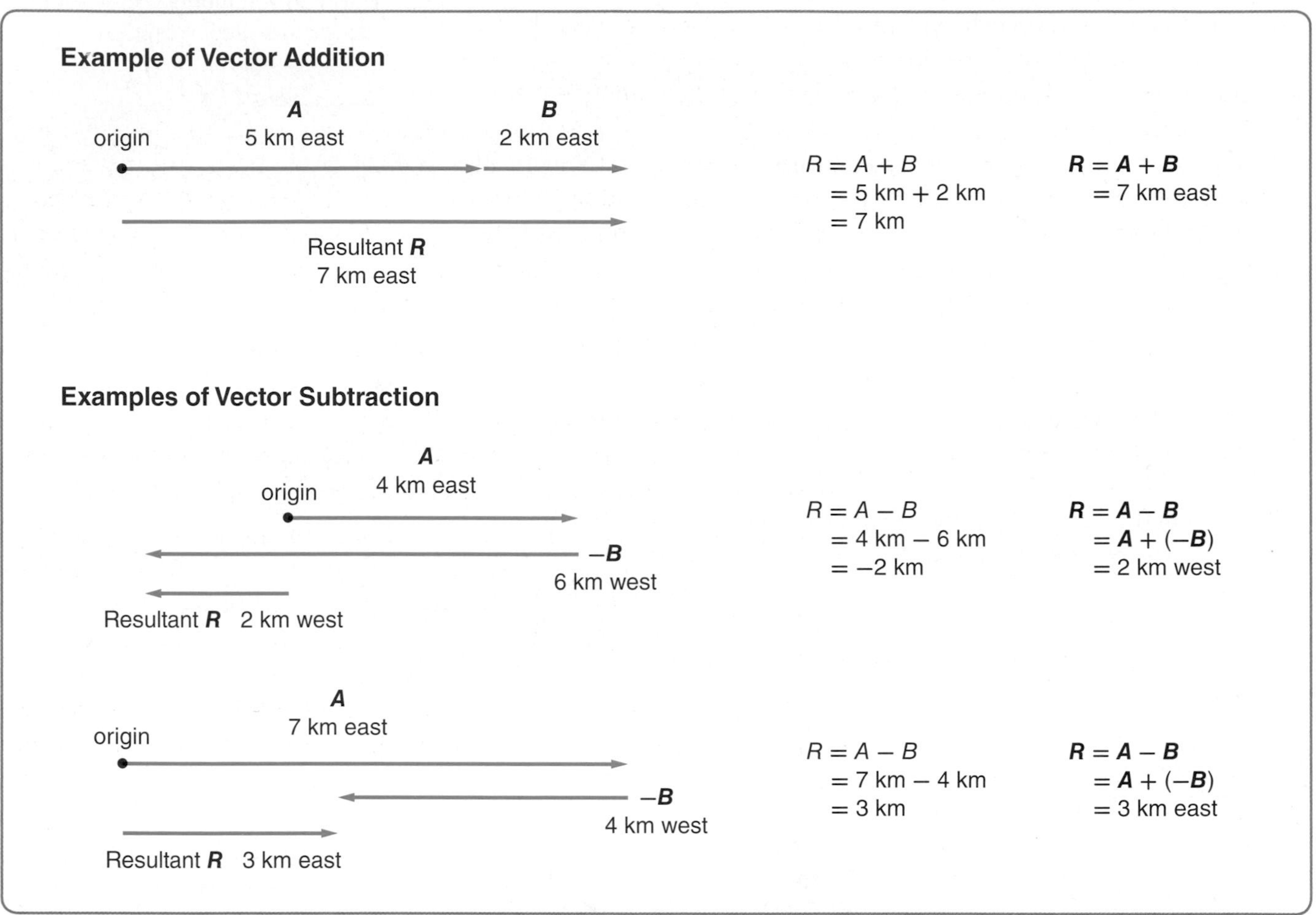

SECTION 2 REVIEW

Section Self-Check Check your understanding.

6. **MAIN IDEA** Identify a coordinate system you could use to describe the motion of a girl swimming across a rectangular pool.

7. **Displacement** The motion diagram for a car traveling on an interstate highway is shown below. The starting and ending points are indicated.

   Start • • • • • • End

   Make a copy of the diagram. Draw a vector to represent the car's displacement from the starting time to the end of the third time interval.

8. **Position** Two students added a vector for a moving object's position at $t = 2$ s to a motion diagram. When they compared their diagrams, they found that their vectors did not point in the same direction. Explain.

9. **Displacement** The motion diagram for a boy walking to school is shown below.

   Home • • • • • • • • • • School

   Make a copy of this motion diagram, and draw vectors to represent the displacement between each pair of dots.

10. **Critical Thinking** A car travels straight along a street from a grocery store to a post office. To represent its motion, you use a coordinate system with its origin at the grocery store and the direction the car is moving as the positive direction. Your friend uses a coordinate system with its origin at the post office and the opposite direction as the positive direction. Would the two of you agree on the car's position? Displacement? Distance? The time interval the trip took? Explain.

SECTION 3

Position-Time Graphs

Many graphs show trends over time. For example, a graph might show the price of gasoline through the course of several months or years. Similarly, a position-time graph can show how a rower's position changes through time. Rowers can use graphs to analyze their performances.

MAIN IDEA

You can use position-time graphs to determine an object's position at a certain time.

Essential Questions

- What information do position-time graphs provide?
- How can you use a position-time graph to interpret an object's position or displacement?
- What are the purposes of equivalent representations of an object's motion?

Review Vocabulary

intersection a point where lines meet and cross

New Vocabulary

position-time graph
instantaneous position

Finding Positions

When analyzing complex motion, it often is useful to represent the motion in a variety of ways. A motion diagram contains information about an object's position at various times. Tables and graphs can also show this same information. Review the motion diagrams in **Figure 8** and **Figure 9.** You can use these diagrams to organize the times and corresponding positions of the runner, as in **Table 1.**

Plotting data The data listed in **Table 1** can be presented on a **position-time graph,** in which the time data is plotted on a horizontal axis and the position data is plotted on a vertical axis. The graph of the runner's motion is shown in **Figure 11.** To draw this graph, first plot the runner's positions. Then, draw a line that best fits the points.

Estimating time and position Notice that the graph is not a picture of the runner's path—the graphed line is sloped, but the runner's path was horizontal. Instead, the line represents the most likely positions of the runner at the times between the recorded data points. Even though there is no data point exactly when the runner was 12.0 m beyond her starting point or where she was at $t = 4.5$ s, you can use the graph to estimate the time or her position. The example problem on the next page shows how.

Table 1 Position v. Time

Time (s)	Position (m)
0.0	0.0
1.0	5.0
2.0	10.0
3.0	15.0
4.0	20.0
5.0	25.0

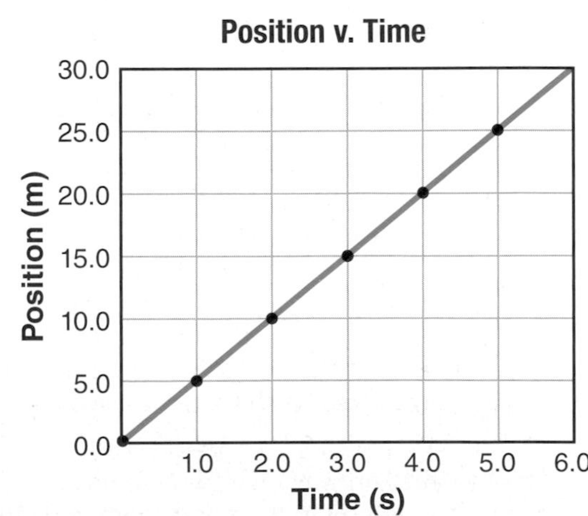

Figure 11 You can create a position-time graph by plotting the positions and times from the table. By drawing a best-fit line, you can estimate other times and positions.

Explain Why is the line on the graph sloped even though it describes motion along a flat path?

View an **animation of position-time graphs.**

EXAMPLE PROBLEM 1

Find help with **interpolating and extrapolating.** Math Handbook

ANALYZE A POSITION-TIME GRAPH When did the runner whose motion is described in **Figure 11** reach 12.0 m beyond the starting point? Where was she after 4.5 s?

1 ANALYZE THE PROBLEM

Restate the questions.

Question 1: At what time was the magnitude of the runner's position (x) equal to 12.0 m?

Question 2: What was the runner's position at time $t = 4.5$ s?

2 SOLVE FOR THE UNKNOWN

Question 1

Examine the graph to find the intersection of the best-fit line with a horizontal line at the 12.0 m mark. Next, find where a vertical line from that point crosses the time axis. The value of t there is 2.4 s.

Question 2

Find the intersection of the graph with a vertical line at 4.5 s (halfway between 4.0 s and 5.0 s on this graph). Next, find where a horizontal line from that point crosses the position axis. The value of x is approximately 22.5 m.

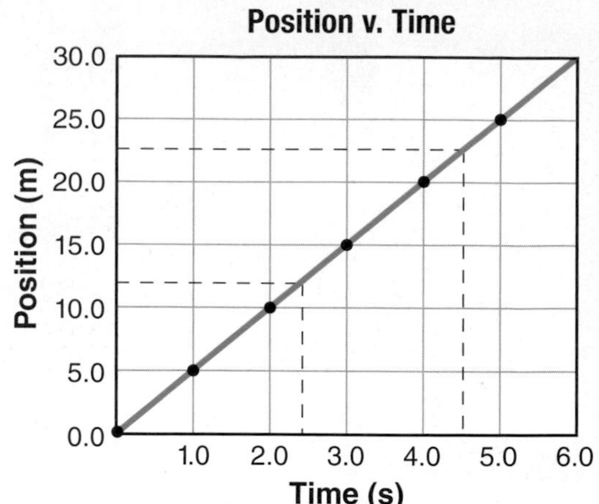

PRACTICE PROBLEMS

Do additional problems. Online Practice

*For problems 11–13, refer to **Figure 12.***

11. The graph in **Figure 12** represents the motion of a car moving along a straight highway. Describe in words the car's motion.

12. Draw a particle model motion diagram that corresponds to the graph.

13. Answer the following questions about the car's motion. Assume that the positive x-direction is east of the origin and the negative x-direction is west of the origin.

a. At what time was the car's position 25.0 m east of the origin?

b. Where was the car at time $t = 1.0$ s?

c. What was the displacement of the car between times $t = 1.0$ s and $t = 3.0$ s?

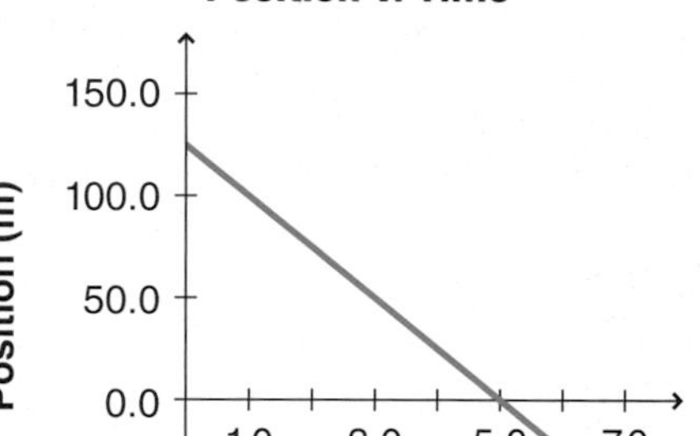

Figure 12

14. The graph in **Figure 13** represents the motion of two pedestrians who are walking along a straight sidewalk in a city. Describe in words the motion of the pedestrians. Assume that the positive direction is east of the origin.

15. CHALLENGE Ari walked down the hall at school from the cafeteria to the band room, a distance of 100.0 m. A class of physics students recorded and graphed his position every 2.0 s, noting that he moved 2.6 m every 2.0 s. When was Ari at the following positions?

a. 25.0 m from the cafeteria

b. 25.0 m from the band room

c. Create a graph showing Ari's motion.

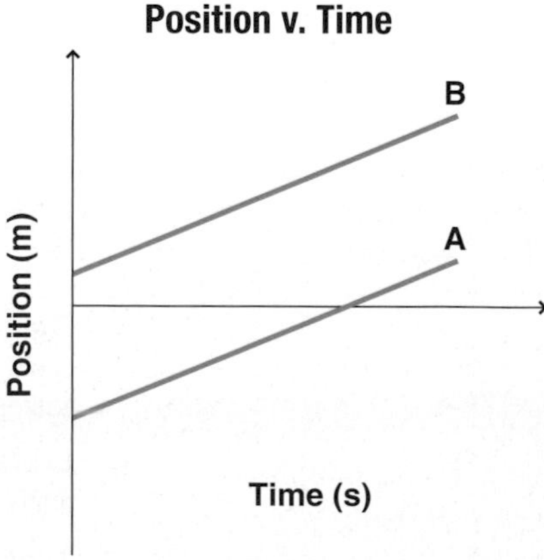

Figure 13

Time (s)	Position (m)
0.0	0.0
1.0	5.0
2.0	10.0
3.0	15.0
4.0	20.0
5.0	25.0

Table 1 Position v. Time

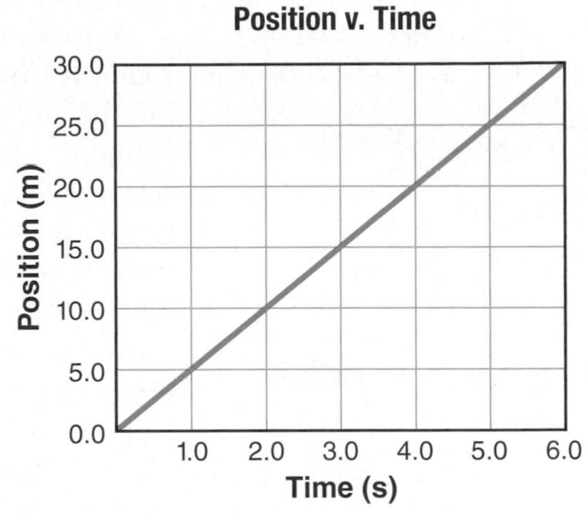

Motion Diagram

Begin • • • • • • • End

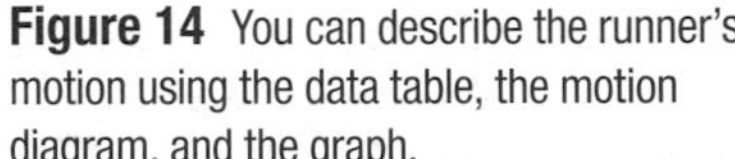

Figure 14 You can describe the runner's motion using the data table, the motion diagram, and the graph.

Identify one benefit the table has over the graph.

Instantaneous position How long did the runner spend at any location? Each position has been linked to a time, but how long did that time last? You could say "an instant," but how long is that? If an instant lasts for any finite amount of time, then the runner would have stayed at the same position during that time, and she would not have been moving. An instant is not a finite period of time, however. It lasts zero seconds. The symbol x represents the runner's **instantaneous position**—the position at a particular instant. Instantaneous position is usually simply called position.

☑ **READING CHECK** **Explain** what is meant by the instantaneous position of a runner.

Equivalent representations As shown in **Figure 14,** you now have several different ways to describe motion. You might describe motion using words, pictures (or pictorial representations), motion diagrams, data tables, or position-time graphs. All of these representations contain the same information about the runner's motion. However, depending on what you want to learn about an object's motion, some types of representations will be more useful than others. In the pages that follow, you will practice constructing these equivalent representations and learn which ones are most useful for solving different kinds of problems.

PHYSICS CHALLENGE

POSITION-TIME GRAPHS Natana, Olivia, and Phil all enjoy exercising and often go to a path along the river for this purpose. Natana bicycles at a very consistent 40.25 km/h, Olivia runs south at a constant speed of 16.0 km/h, and Phil walks south at a brisk 6.5 km/h. Natana starts biking north at noon from the waterfalls. Olivia and Phil both start at 11:30 A.M. at the canoe dock, 20.0 km north of the falls.

1. Draw position-time graphs for each person.
2. At what time will the three exercise enthusiasts be located within the smallest distance interval from each other?
3. What is the length of that distance interval?

Karl Weatherly/Getty Images

Multiple Objects on a Position-Time Graph

A position-time graph for two different runners is shown in Example Problem 2 below. Notice that runner A is ahead of runner B at time $t = 0$, but the motion of each runner is different. When and where does one runner pass the other? First, you should restate this question in physics terms: At what time are the two runners at the same position? What is their position at this time? You can evaluate these questions by identifying the point on the position-time graph at which the lines representing the two runners' motions intersect.

The intersection of two lines on a position-time graph tells you when objects have the same position, but does this mean that they will collide? Not necessarily. For example, if the two objects are runners and if they are in different lanes, they will not collide, even though they might be the same distance from the starting point.

READING CHECK **Explain** what the intersection of two lines on a position-time graph means.

What else can you learn from a position-time graph? Notice in Example Problem 2 that the lines on the graph have different slopes. What does the slope of the line on a position-time graph tell you? In the next section, you will use the slope of a line on a position-time graph to determine the velocity of an object. When you study accelerated motion, you will draw other motion graphs and learn to interpret the areas under the plotted lines. In later studies, you will continue to refine your skills with creating and interpreting different types of motion graphs.

EXAMPLE PROBLEM 2

Find help with **interpolating and extrapolating.** Math Handbook

INTERPRETING A GRAPH The graph to the right describes the motion of two runners moving along a straight path. The lines representing their motion are labeled A and B. When and where does runner B pass runner A?

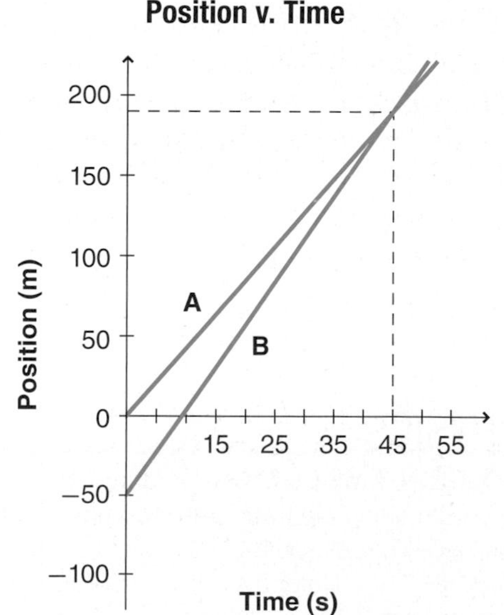

1 ANALYZE THE PROBLEM

Restate the questions.

Question 1: At what time are runner A and runner B at the same position?

Question 2: What is the position of runner A and runner B at this time?

2 SOLVE FOR THE UNKNOWN

Question 1

Examine the graph to find the intersection of the line representing the motion of runner A with the line representing the motion of runner B. These lines intersect at time 45 s.

Question 2

Examine the graph to determine the position when the lines representing the motion of the runners intersect. The position of both runners is about 190 m from the origin.

Runner B passes runner A about 190 m beyond the origin, 45 s after A has passed the origin.

PRACTICE PROBLEMS

Do additional problems. Online Practice

For problems 16–19, refer to the figure in Example Problem 2 on the previous page.

16. Where was runner A located at $t = 0$ s?

17. Which runner was ahead at $t = 48.0$ s?

18. When runner A was at 0.0 m, where was runner B?

19. How far apart were runners A and B at $t = 20.0$ s?

20. CHALLENGE Juanita goes for a walk. Later her friend Heather starts to walk after her. Their motions are represented by the position-time graph in **Figure 15.**

a. How long had Juanita been walking when Heather started her walk?

b. Will Heather catch up to Juanita? How can you tell?

c. What was Juanita's position at $t = 0.2$ h?

d. At what time was Heather 5.0 km from the start?

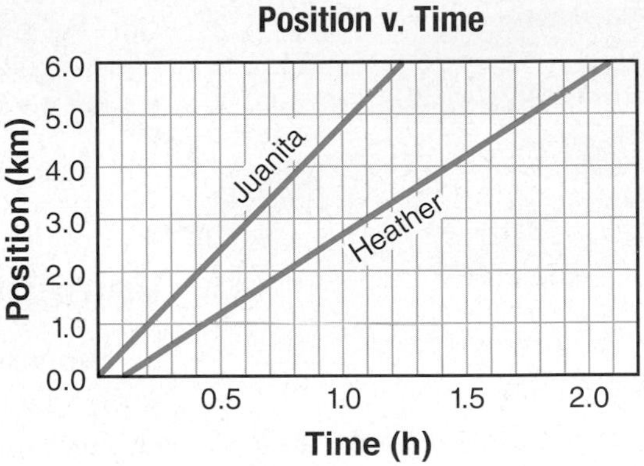

Figure 15

PRACTICE PROBLEMS

SECTION 3 REVIEW

Section Self-Check Check your understanding.

21. MAINIDEA Using the particle model motion diagram in **Figure 16** of a baby crawling across a kitchen floor, plot a position-time graph to represent the baby's motion. The time interval between successive dots on the diagram is 1 s.

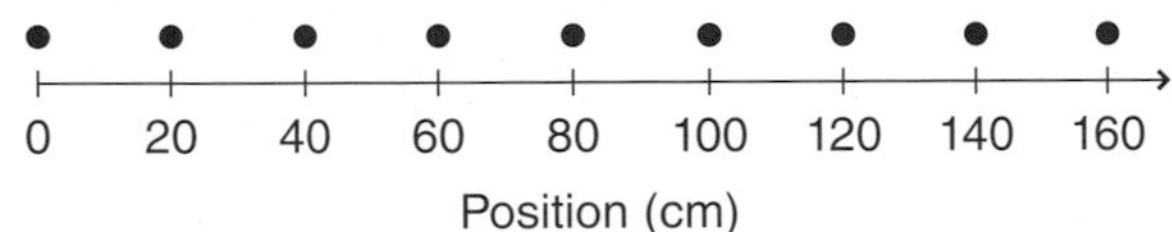

Figure 16

*For problems 22–25, refer to **Figure 17.***

22. Particle Model Create a particle model motion diagram from the position-time graph in **Figure 17** of a hockey puck gliding across a frozen pond.

23. Time Use the hockey puck's position-time graph to determine the time when the puck was 10.0 m beyond the origin.

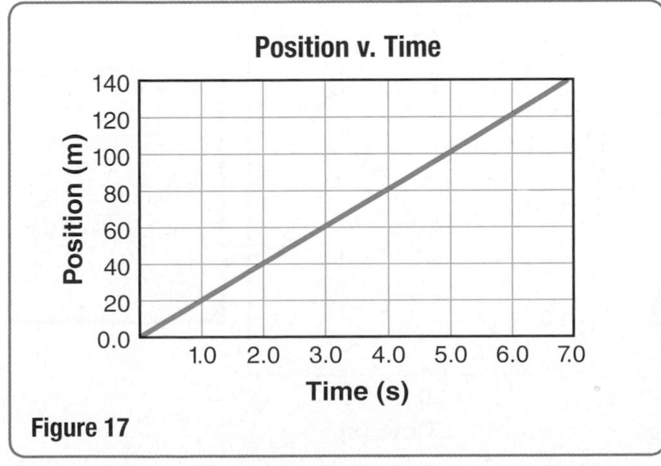

Figure 17

24. Distance Use the position-time graph in **Figure 17** to determine how far the hockey puck moved between times 0.0 s and 5.0 s.

25. Time Interval Use the position-time graph for the hockey puck to determine how much time it took for the puck to go from 40 m beyond the origin to 80 m beyond the origin.

26. Critical Thinking Look at the particle model diagram and the position-time graph shown in **Figure 18.** Do they describe the same motion? How do you know? Do not confuse the position coordinate system in the particle model with the horizontal axis in the position-time graph. The time intervals in the particle model diagram are 2 s.

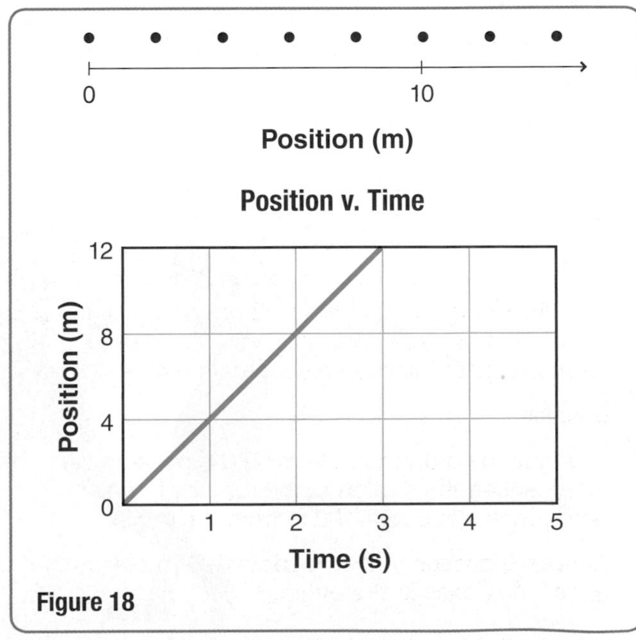

Figure 18

SECTION 4 How Fast?

PHYSICS 4 YOU **Snails move much slower than cheetahs. You can see this by observing how far the animals travel during a given time period. For example, a cheetah can travel 30 m in a second, but a snail might move only 1 cm in that time interval.**

MAIN IDEA

An object's velocity is the rate of change in its position.

Essential Questions

- What is velocity?
- What is the difference between speed and velocity?
- How can you determine an object's average velocity from a position-time graph?
- How can you represent motion with pictorial, physical, and mathematical models?

Review Vocabulary

absolute value magnitude of a number, regardless of sign

New Vocabulary

average velocity
average speed
instantaneous velocity

Velocity and Speed

Suppose you recorded the motion of two joggers on one diagram, as shown by the graph in **Figure 19.** The position of the jogger wearing red changes more than that of the jogger wearing blue. For a fixed time interval, the magnitude of the displacement ($\Delta \boldsymbol{x}$) is greater for the jogger in red because she is moving faster. Now, suppose that each jogger travels 100 m. The time interval (Δt) for the 100 m would be smaller for the jogger in red than for the one in blue.

Slope on a position-time graph Compare the lines representing the joggers in the graph in **Figure 19.** The slope of the red jogger's line is steeper, indicating a greater change in position during each time interval. Recall that you find the slope of a line by first choosing two points on the line. Next, you subtract the vertical coordinate (x in this case) of the first point from the vertical coordinate of the second point to obtain the rise of the line. After that, you subtract the horizontal coordinate (t in this case) of the first point from the horizontal coordinate of the second point to obtain the run. The rise divided by the run is the slope.

Figure 19 A greater slope shows that the red jogger traveled faster.

Analyze How much farther did the red jogger travel than the blue jogger in the 3 s interval described by the graph?

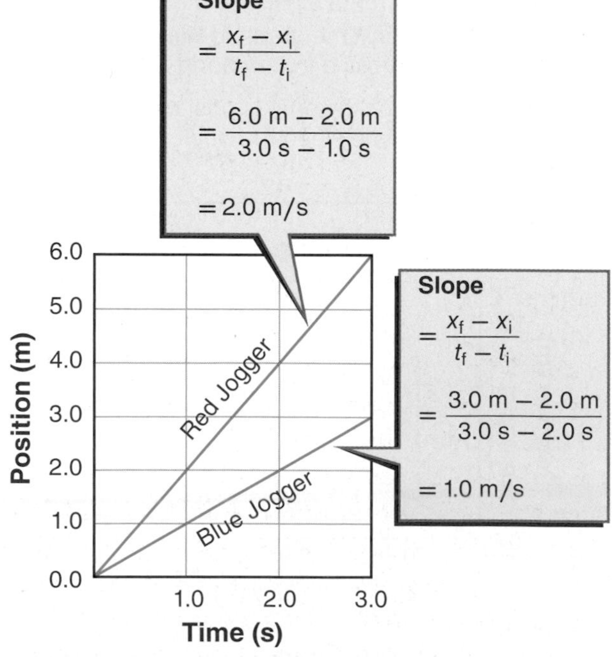

Average velocity Notice that the slope of the faster runner's line in **Figure 19** is a greater number. A greater slope indicates a faster speed. Also notice that the slope's units are meters per second. Looking at how the slope is calculated, you can see that slope is the change in the magnitude of the position divided by the time interval during which that change took place: $\frac{x_f - x_i}{t_f - t_i}$, or $\frac{\Delta x}{\Delta t}$. When Δx gets larger, the slope gets larger; when Δt gets larger, the slope gets smaller. This agrees with the interpretation given on the previous page of the speeds of the red and blue joggers. **Average velocity** is the ratio of an object's change in position to the time interval during which the change occurred. If the object is in uniform motion, so that its speed does not change, then its average velocity is the slope of its position-time graph.

AVERAGE VELOCITY

Average velocity is defined as the change in position divided by the time during which the change occurred.

$$\bar{v} \equiv \frac{\Delta x}{\Delta t} = \frac{x_f - x_i}{t_f - t_i}$$

The symbol ≡ means that the left-hand side of the equation is defined by the right-hand side.

Interpreting slope The position-time graph's slope in **Figure 20** is −5.0 m/s. Notice that the slope of the graph indicates both magnitude and direction. By calculating the slope from the rise divided by the run between two points, you find that the object whose motion is represented by the graph has an average velocity of −5.0 m/s. The object started out at a positive position and moves toward the origin. After 4 s, it passes the origin and continues moving in the negative direction at a rate of 5.0 m/s.

READING CHECK Explain the meaning of a position-time graph slope that is upward or downward, and above or below the *x*-axis.

Average speed The slope's absolute value is the object's **average speed,** 5.0 m/s, which is the distance traveled divided by the time taken to travel that distance. For uniform motion, average speed is the absolute value of the slope of the object's position-time graph. The combination of an object's average speed ($\bar{v}$) and the direction in which it is moving is the average velocity ($\bar{\boldsymbol{v}}$). Remember that if an object moves in the negative direction, its change in position is negative. This means that an object's displacement and velocity are both always in the same direction.

PhysicsLABs

CONSTANT SPEED

How can you determine average speed by measuring distance and time?

MEASURE VELOCITY

PROBEWARE LAB How can you measure velocity with a motion detector?

Figure 20 The downward slope of this position-time graph shows that the motion is in the negative direction.

Analyze What would the graph look like if the motion were at the same speed, but in the positive direction?

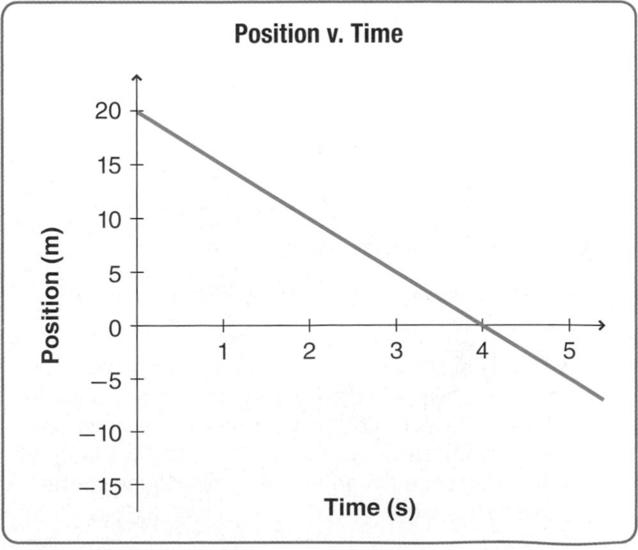

EXAMPLE PROBLEM

EXAMPLE PROBLEM 3

Get help with **average velocity and average speed.** Personal Tutor

AVERAGE VELOCITY The graph at the right describes the straight-line motion of a student riding her skateboard along a smooth, pedestrian-free sidewalk. What is her average velocity? What is her average speed?

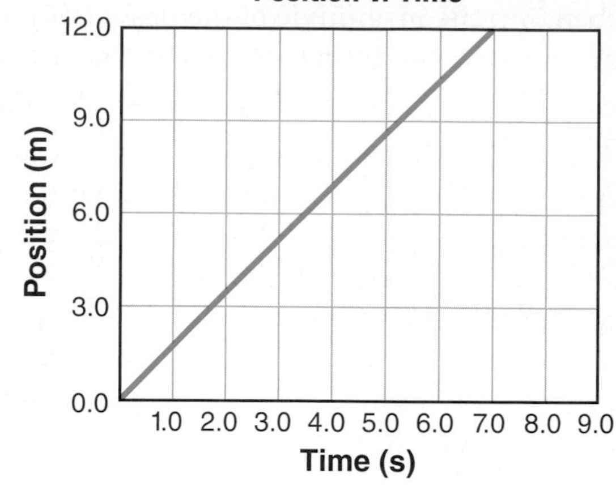

1 ANALYZE AND SKETCH THE PROBLEM

Identify the graph's coordinate system.

UNKNOWN

$\bar{\mathbf{v}} = ?$ $\bar{v} = ?$

2 SOLVE FOR THE UNKNOWN

Find the average velocity using two points on the line.

$$\bar{v} = \frac{\Delta x}{\Delta t}$$

$$= \frac{x_f - x_i}{t_f - t_i}$$

$$= \frac{12.0\text{ m} - 0.0\text{ m}}{7.0\text{ s} - 0.0\text{ s}}$$

◀ Substitute $x_2 = 12.0$ m, $x_1 = 0.0$ m, $t_2 = 7.0$ s, $t_1 = 0.0$ s.

$\bar{\mathbf{v}} = 1.7$ m/s in the positive direction

The average speed ($\bar{v}$) is the absolute value of the average velocity, or 1.7 m/s.

3 EVALUATE THE ANSWER

- **Are the units correct?** The units for both velocity and speed are meters per second.
- **Do the signs make sense?** The positive sign for the velocity agrees with the coordinate system. No direction is associated with speed.

PRACTICE PROBLEMS

PRACTICE PROBLEMS

Do additional problems. Online Practice

27. The graph in **Figure 21** describes the motion of a cruise ship drifting slowly through calm waters. The positive *x*-direction (along the vertical axis) is defined to be south.

a. What is the ship's average speed?

b. What is its average velocity?

28. Describe, in words, the cruise ship's motion in the previous problem.

29. What is the average velocity of an object that moves from 6.5 cm to 3.7 cm relative to the origin in 2.3 s?

30. The graph in **Figure 22** represents the motion of a bicycle.

a. What is the bicycle's average speed?

b. What is its average velocity?

31. Describe, in words, the bicycle's motion in the previous problem.

32. CHALLENGE When Marshall takes his pet dog for a walk, the dog walks at a very consistent pace of 0.55 m/s. Draw a motion diagram and a position-time graph to represent Marshall's dog walking the 19.8-m distance from in front of his house to the nearest stop sign.

Position v. Time

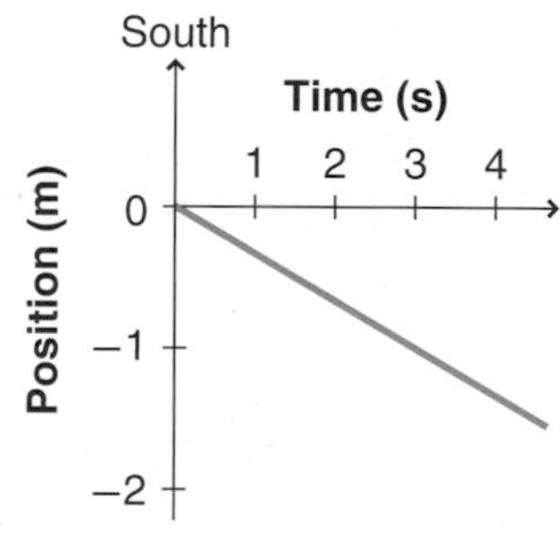

Figure 21

Position v. Time

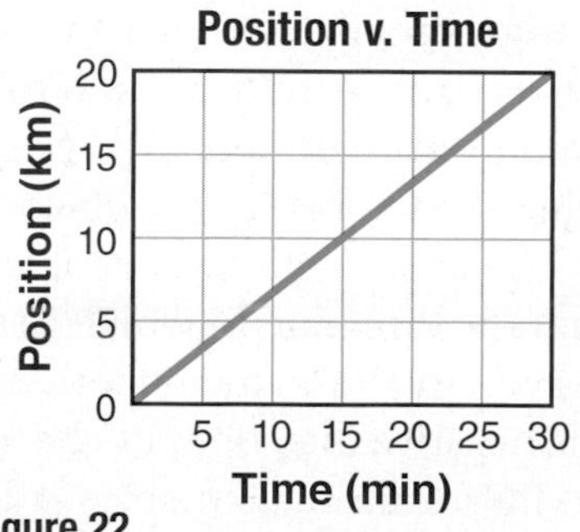

Figure 22

Instantaneous velocity Why do we call the quantity $\frac{\Delta x}{\Delta t}$ average velocity? Why don't we just call it velocity? A motion diagram shows the position of a moving object at the beginning and end of a time interval. It does not, however, indicate what happened within that time interval. During the time interval, the object's speed could have remained the same, increased, or decreased. The object may have stopped or even changed direction. You can find the average velocity for each time interval in the motion diagram, but you cannot find the speed and the direction of the object at any specific instant. The speed and the direction of an object at a particular instant is called the **instantaneous velocity.** In this textbook, the term velocity will refer to instantaneous velocity, represented by the symbol $\boldsymbol{v}$.

☑ READING CHECK **Explain** how average velocity is different from velocity.

Average velocity on motion diagrams When an object moves between two points, its average velocity is in the same direction as its displacement. The two quantities are also proportional—when displacement is greater during a given time interval, so is average velocity. A motion diagram indicates the average velocity's direction and magnitude.

Imagine two cars driving down the road at different speeds. A video camera records the motion of the cars at the rate of one frame every second. Imagine that each car has a paintbrush attached to it that automatically descends and paints a red line on the ground for half a second every second. The faster car would paint a longer line on the ground. The vectors you draw on a motion diagram to represent the velocity are like the lines that the paintbrushes make on the ground below the cars. In this book, we use red to indicate velocity vectors on motion diagrams. **Figure 23** shows motion diagrams with velocity vectors for two cars. One is moving to the right, and the other is moving to the left.

☑ READING CHECK **Identify** what the lengths of velocity vectors mean.

Equation of Motion

Often it is more efficient to use an equation, rather than a graph, to solve problems. Any time you graph a straight line, you can find an equation to describe it. Take another look at the graph in **Figure 20** for the object moving with a constant velocity of −5.0 m/s. Recall that you can represent any straight line with the equation $y = mx + b$, where y is the quantity plotted on the vertical axis, m is the line's slope, x is the quantity plotted on the horizontal axis, and b is the line's y-intercept.

For the graph in **Figure 20,** the quantity plotted on the vertical axis is position, represented by the variable $\boldsymbol{x}$. The line's slope is −5.0 m/s, which is the object's average velocity ($\bar{\boldsymbol{v}}$). The quantity plotted on the horizontal axis is time (t). The y-intercept is 20.0 m. What does this 20.0 m represent? This shows that the object was at a position of 20.0 m when $t = 0.0$ s. This is called the initial position of the object and it is designated $\boldsymbol{x}_i$.

REAL-WORLD PHYSICS

SPEED RECORDS The world record for the men's 100-m dash is 9.58 s, established in 2009 by Usain Bolt. The world record for the women's 100-m dash is 10.49 s, established in 1988 by Florence Griffith-Joyner.

MiniLAB

VELOCITY VECTORS

How can velocity vectors represent the motion of a mass on a string?

Figure 23 The length of each velocity vector is proportional to the magnitude of the velocity that it represents.

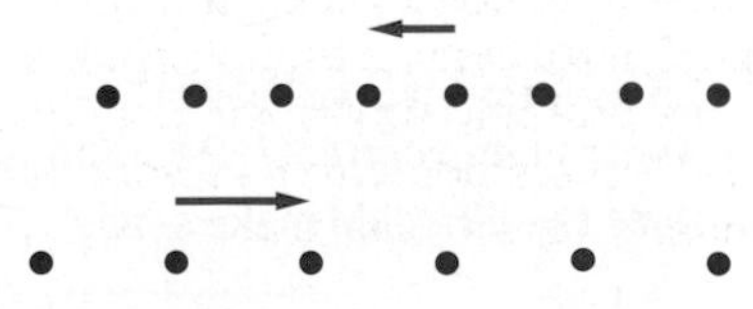

Paul Gilham/Getty Images Sport/Getty Images

CONNECTING MATH TO PHYSICS

Lines and Graphs Symbols used in the point-slope equation of a line relate to symbols used for motion variables on a position-time graph.

General Variable	Specific Motion Variable	Value in Figure 20
y	$\boldsymbol{x}$	
m	$\bar{\boldsymbol{v}}$	−5.0 m/s
x	t	
b	$\boldsymbol{x}_i$	20.0 m

A summary is given to the left of how the general variables in the straight-line formula are changed to the specific variables you have been using to describe motion. The table also shows the numerical values for the average velocity and initial position. Consider the graph shown in **Figure 20.** The mathematical equation for the line graphed is as follows:

$$y = (-5.0 \text{ m/s})x + 20.0 \text{ m}$$

You can rewrite this equation, using $\boldsymbol{x}$ for position and t for time.

$$\boldsymbol{x} = (-5.0 \text{ m/s})t + 20.0 \text{ m}$$

It might be confusing to use y and x in math but use $\boldsymbol{x}$ and t in physics. You do this because there are many types of graphs in physics, including position v. time graphs, velocity v. time graphs, and force v. position graphs. For a position v. time graph, the math equation $y = mx + b$ can be rewritten as follows:

POSITION

An object's position is equal to the average velocity multiplied by time plus the initial position.

$$\boldsymbol{x} = \bar{\boldsymbol{v}}t + \boldsymbol{x}_i$$

This equation gives you another way to represent motion. Note that a graph of x v. t would be a straight line.

EXAMPLE PROBLEM 4

Find help with **solving equations.** Math Handbook

POSITION The figure shows a motorcyclist traveling east along a straight road. After passing point *B*, the cyclist continues to travel at an average velocity of 12 m/s east and arrives at point *C* 3.0 s later. What is the position of point *C*?

A B C

x_i = 46 m east

x = ?

1 ANALYZE THE PROBLEM

Choose a coordinate system with the origin at *A*.

KNOWN

$\bar{\boldsymbol{v}}$ = 12 m/s east

$\boldsymbol{x}_i$ = 46 m east

t = 3.0 s

UNKNOWN

$\boldsymbol{x}$ = ?

2 SOLVE FOR THE UNKNOWN

$x = \bar{v}t + x_i$ ◀ Use magnitudes for the calculations.

$= (12 \text{ m/s})(3.0 \text{ s}) + 46 \text{ m}$ ◀ Substitute $\bar{v}$ = 12 m/s, t = 3.0 s, and x_i = 46 m.

$= 82 \text{ m}$

$\boldsymbol{x}$ = 82 m east

3 EVALUATE THE ANSWER

- **Are the units correct?** Position is measured in meters.
- **Does the direction make sense?** The motorcyclist is traveling east the entire time.

EXAMPLE PROBLEM

PRACTICE PROBLEMS

Do additional problems. Online Practice

For problems 33–36, refer to ***Figure 24.***

33. The diagram at the right shows the path of a ship that sails at a constant velocity of 42 km/h east. What is the ship's position when it reaches point *C*, relative to the starting point, *A*, if it sails from point *B* to point *C* in exactly 1.5 h?

34. Another ship starts at the same time from point *B*, but its average velocity is 58 km/h east. What is its position, relative to *A*, after 1.5 h?

35. What would a ship's position be if that ship started at point *B* and traveled at an average velocity of 35 km/h west to point *D* in a time period of 1.2 h?

36. CHALLENGE Suppose two ships start from point *B* and travel west. One ship travels at an average velocity of 35 km/h for 2.2 h. Another ship travels at an average velocity of 26 km/h for 2.5 h. What is the final position of each ship?

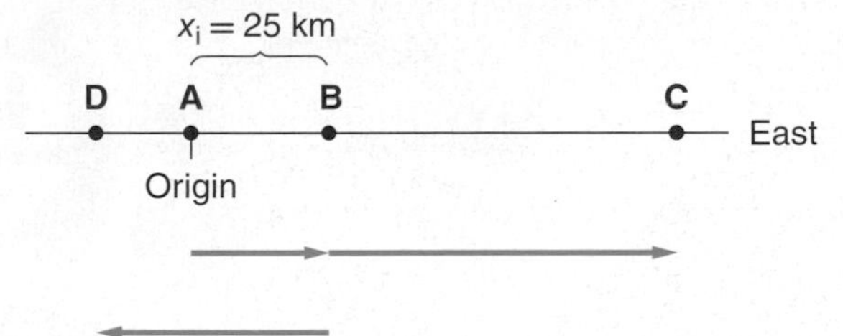

Figure 24

PRACTICE PROBLEMS

SECTION 4 REVIEW

Section Self-Check Check your understanding.

37. MAINIDEA How is an object's velocity related to its position?

For problems 38–40, refer to ***Figure 25.***

38. Ranking Task Rank the position-time graphs according to the average speed, from greatest average speed to least average speed. Specifically indicate any ties.

39. Contrast Average Velocities Describe differences in the average velocities shown on the graph for objects A and B. Describe differences in the average velocities shown on the graph for objects C and D.

40. Ranking Task Rank the graphs in **Figure 25** according to each object's initial position, from most positive position to most negative position. Specifically indicate any ties. Would your ranking be different if you ranked according to initial distance from the origin?

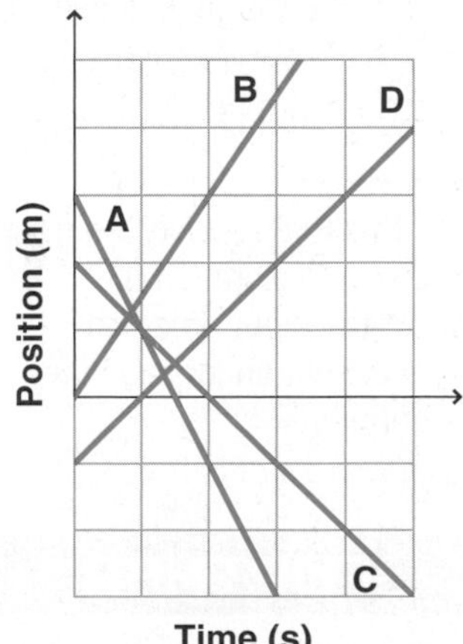

Figure 25

41. Average Speed and Average Velocity Explain how average speed and average velocity are related to each other for an object in uniform motion.

42. Position Two cars are traveling along a straight road, as shown in **Figure 26.** They pass each other at point B and then continue in opposite directions. The red car travels for 0.25 h from point B to point C at a constant velocity of 32 km/h east. The blue car travels for 0.25 h from point B to point D at a constant velocity of 48 km/h west. How far has each car traveled from point B? What is the position of each car relative to the origin, point A?

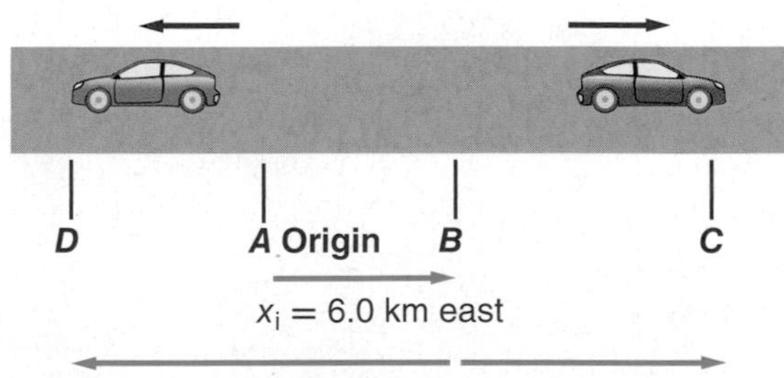

Figure 26

43. Position A car travels north along a straight highway at an average speed of 85 km/h. After driving 2.0 km, the car passes a gas station and continues along the highway. What is the car's position relative to the start of its trip 0.25 h after it passes the gas station?

44. Critical Thinking In solving a physics problem, why is it important to create pictorial and physical models before trying to solve an equation?

Got the time?

What is time? If one hour of time passes for you, does one hour of time also pass for your friend? You might think that the answer is yes, but it is actually no. Time passes at different rates depending on your point of view.

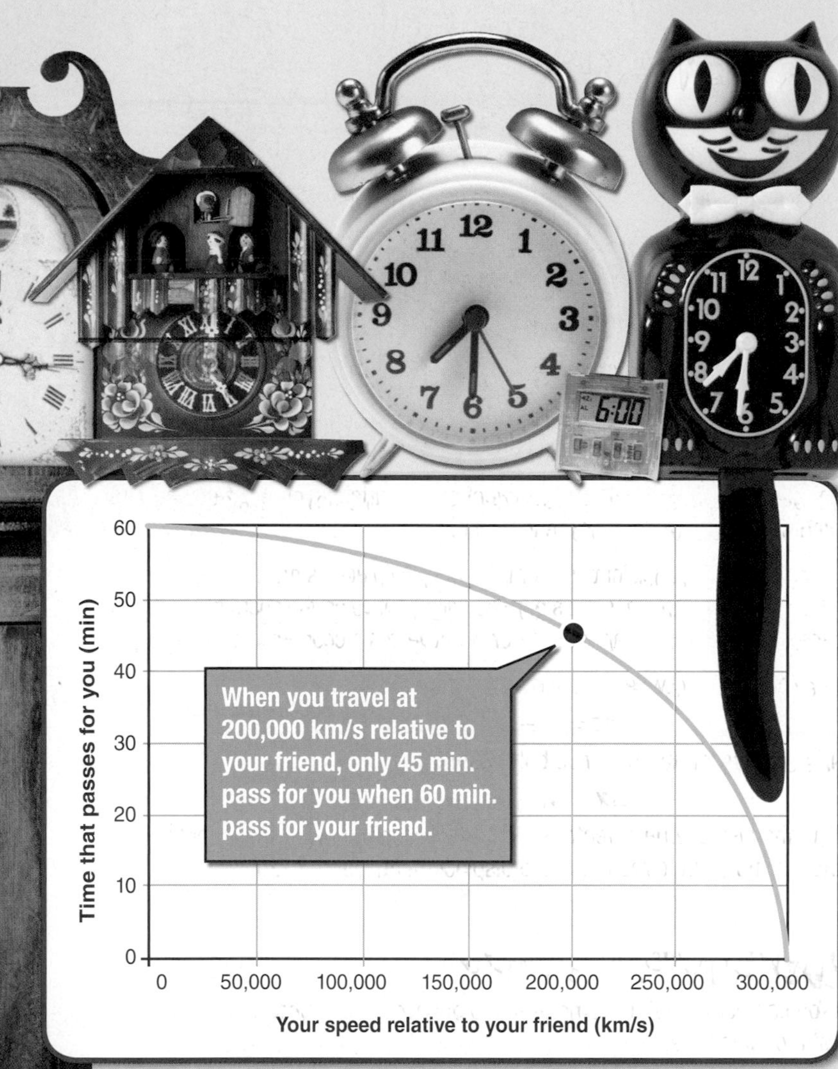

FIGURE 1 In this graph, 60 minutes always passes for your friend, but other amounts of time pass for you.

Speed and time Think about how wrong that last sentence seems. For example, suppose that you tell your friend to meet you at the mall in one hour. You both assume that when one hour passes for you, one hour also passes for your friend.

This is because you and your friend move very slowly relative to each other. At slow speeds, one hour for you is almost exactly the same as one hour for your friend. As you move faster relative to your friend, however, the difference between your time and your friend's time increases.

How fast? You would need to travel very fast relative to your friend in order for any difference to be noticeable. If you travel at 100,000 km/s, then only 57 minutes passes for you when one hour passes for your friend. At 200,000 km/s, only 45 minutes passes for you during your friend's hour. **Figure 1** shows how your time compares to one hour of your friend's time as you travel faster and faster relative to your friend.

Real–World Application All of this might seem rather pointless. After all, even the fastest spacecraft travel at less than 100 km/s. Have you ever used a GPS receiver, such as the one shown in **Figure 2?** At 4 km/s, a GPS satellite travels fast enough for time differences to affect the accuracy of the GPS receiver. The effect is small—approximately 10 μs in one day. It is enough, however, that the GPS would become completely useless within one month if engineers did not account for it.

FIGURE 2 This GPS receiver would be completely inaccurate if the designers of the Global Positioning System did not understand the relativity of time.

GOING FURTHER >>>

Research Gravity also affects time. Research how gravity affects time on Earth and on a GPS satellite.

CHAPTER 2 STUDY GUIDE

BIGIDEA You can use displacement and velocity to describe an object's motion.

VOCABULARY
- **motion diagram** *(p. 36)*
- **particle model** *(p. 36)*

SECTION 1 Picturing Motion

MAINIDEA You can use motion diagrams to show how an object's position changes over time.

- A motion diagram shows the position of an object at successive equal time intervals.
- In a particle model motion diagram, an object's position at successive times is represented by a series of dots. The spacing between dots indicates whether the object is moving faster or slower.

VOCABULARY
- **coordinate system** *(p. 37)*
- **origin** *(p. 37)*
- **position** *(p. 37)*
- **distance** *(p. 37)*
- **magnitude** *(p. 38)*
- **vector** *(p. 38)*
- **scalar** *(p. 38)*
- **time interval** *(p. 38)*
- **displacement** *(p. 39)*
- **resultant** *(p. 40)*

SECTION 2 Where and When?

MAINIDEA A coordinate system is helpful when you are describing motion.

- A coordinate system gives the location of the zero point of the variable you are studying and the direction in which the values of the variable increase.
- A vector drawn from the origin of a coordinate system to an object indicates the object's position in that coordinate system. The directions chosen as positive and negative on the coordinate system determine whether the objects' positions are positive or negative in the coordinate system.
- A time interval is the difference between two times.

$$\Delta t = t_f - t_i$$

- Change in position is displacement, which has both magnitude and direction.

$$\Delta \boldsymbol{x} = \boldsymbol{x}_f - \boldsymbol{x}_i$$

- On a motion diagram, the displacement vector's length represents how far the object was displaced. The vector points in the direction of the displacement, from $\boldsymbol{x}_i$ to $\boldsymbol{x}_f$.

VOCABULARY
- **position-time graph** *(p. 41)*
- **instantaneous position** *(p. 43)*

SECTION 3 Position-Time Graphs

MAINIDEA You can use a position-time graph to determine an object's position at a certain time.

- Position-time graphs provide information about the motion of objects. They also might indicate where and when two objects meet.
- The line on a position-time graph describes an object's position at each time.
- Motion can be described using words, motion diagrams, data tables, or graphs.

VOCABULARY
- **average velocity** *(p. 47)*
- **average speed** *(p. 47)*
- **instantaneous velocity** *(p. 49)*

SECTION 4 How Fast?

MAINIDEA An object's velocity is the rate of change in its position.

- An object's velocity tells how fast it is moving and in what direction it is moving.
- Speed is the magnitude of velocity.
- Slope on a position-time graph describes the average velocity of the object.

$$\bar{\boldsymbol{v}} \equiv \frac{\Delta \boldsymbol{x}}{\Delta t} = \frac{\boldsymbol{x}_f - \boldsymbol{x}_i}{t_f - t_i}$$

- You can represent motion with pictures and physical models. A simple equation relates an object's initial position ($\boldsymbol{x}_i$), its constant average velocity ($\bar{\boldsymbol{v}}$), its position ($\boldsymbol{x}$), and the time (t) since the object was at its initial position.

$$\boldsymbol{x} = \bar{\boldsymbol{v}}t + \boldsymbol{x}_i$$

Games and Multilingual eGlossary

CHAPTER 2 ASSESSMENT

Chapter Self-Check

SECTION 1 Picturing Motion

Mastering Concepts

45. What is the purpose of drawing a motion diagram?

46. Under what circumstances is it legitimate to treat an object as a particle when solving motion problems?

SECTION 2 Where and When?

Mastering Concepts

47. The following quantities describe location or its change: position, distance, and displacement. Briefly describe the differences among them.

48. How can you use a clock to find a time interval?

SECTION 3 Position-Time Graphs

Mastering Concepts

49. In-line Skating How can you use the position-time graphs for two in-line skaters to determine if and when one in-line skater will pass the other one?

SECTION 4 How Fast?

Mastering Concepts

50. BIGIDEA Which equation describes how the average velocity of a moving object relates to its displacement?

51. Walking Versus Running A walker and a runner leave your front door at the same time. They move in the same direction at different constant velocities. Describe the position-time graphs of each.

52. What does the slope of a position-time graph measure?

53. If you know the time it took an object to travel between two points and the positions of the object at the points, can you determine the object's instantaneous velocity? Its average velocity? Explain.

Mastering Problems

54. You ride a bike at a constant speed of 4.0 m/s for 5.0 s. How far do you travel?

55. Astronomy Light from the Sun reaches Earth in about 8.3 min. The speed of light is 3.00×10^{8} m/s. What is the distance from the Sun to Earth?

56. Problem Posing Complete this problem so that someone must solve it using the concept of average speed: "A butterfly travels 15 m from one flower to another …."

57. Nora jogs several times a week and always keeps track of how much time she runs each time she goes out. One day she forgets to take her stopwatch with her and wonders if there is a way she can still have some idea of her time. As she passes a particular bank building, she remembers that it is 4.3 km from her house. She knows from her previous training that she has a consistent pace of 4.0 m/s. How long has Nora been jogging when she reaches the bank?

58. Driving You and a friend each drive 50.0 km. You travel at 90.0 km/h; your friend travels at 95.0 km/h. How much sooner will your friend finish the trip?

Applying Concepts

59. Ranking Task The position-time graph in **Figure 27** shows the motion of four cows walking from the pasture back to the barn. Rank the cows according to their average velocity, from slowest to fastest.

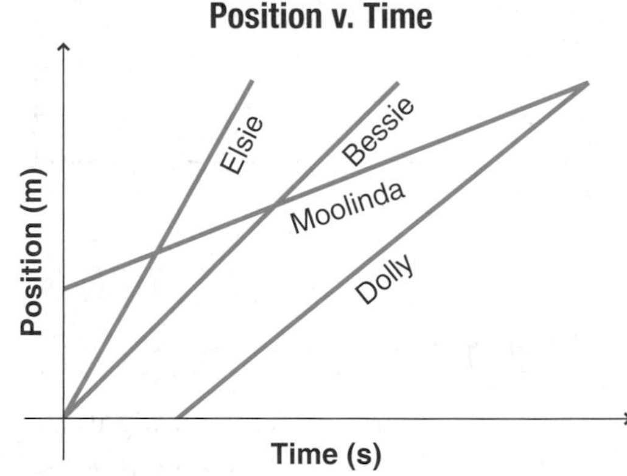

Figure 27

60. Figure 28 is a position-time graph for a rabbit running away from a dog. How would the graph differ if the rabbit ran twice as fast? How would it differ if the rabbit ran in the opposite direction?

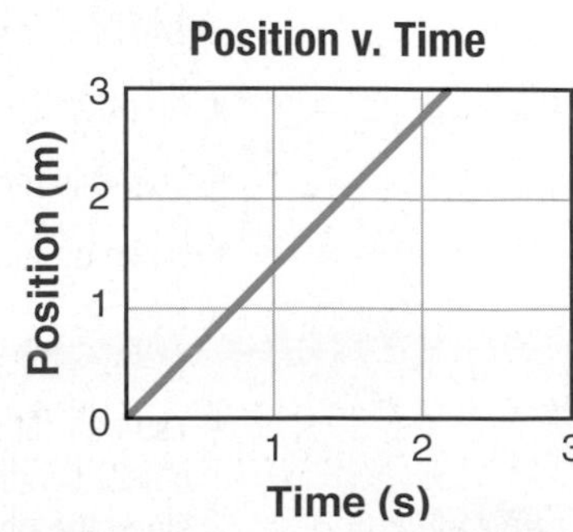

Figure 28

61. Test the following combinations and explain why each does not have the properties needed to describe the concept of velocity: $\Delta x + \Delta t$, $\Delta x - \Delta t$, $\Delta x \times \Delta t$, $\frac{\Delta t}{\Delta x}$.

62. Football When solving physics problems, what must be true about the motion of a football in order for you to treat the football as if it were a particle?

63. Figure 29 is a graph of two people running.

a. Describe the position of runner A relative to runner B at the y-intercept.

b. Which runner is faster?

c. What occurs at point P and beyond?

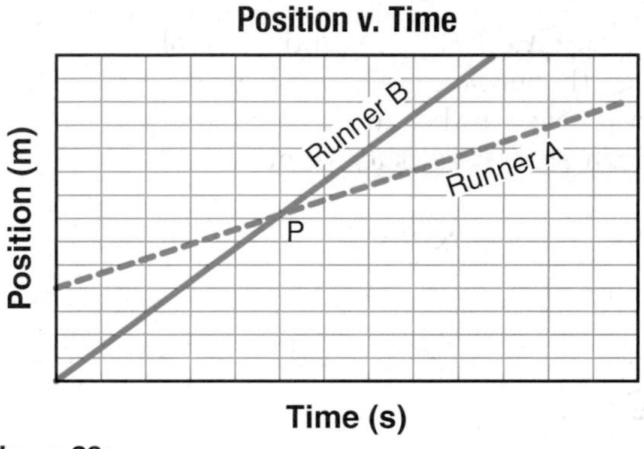

Figure 29

Mixed Review

64. Cycling A cyclist traveling along a straight path maintains a constant velocity of 5.0 m/s west. At time $t = 0.0$ s, the cyclist is 250 m west of point A.

a. Plot a position-time graph of the cyclist's location from point A at 10.0-s intervals for a total time of 60.0 s.

b. What is the cyclist's position from point A at 60.0 s?

c. What is the displacement from the starting position at 60.0 s?

65. Figure 30 is a particle model diagram for a chicken casually walking across a road. Draw the corresponding position-time graph, and write an equation to describe the chicken's motion.

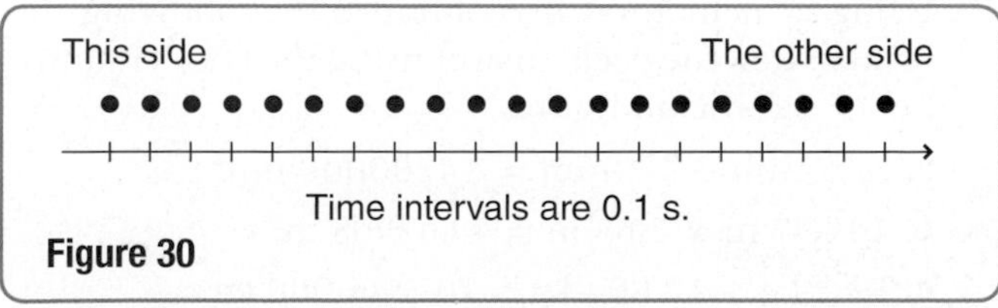

Figure 30

66. Figure 31 shows position-time graphs for Joszi and Heike paddling canoes in a local river.

a. At what time(s) are Joszi and Heike in the same place?

b. How long does Joszi paddle before passing Heike?

c. Where on the river does it appear that there might be a swift current?

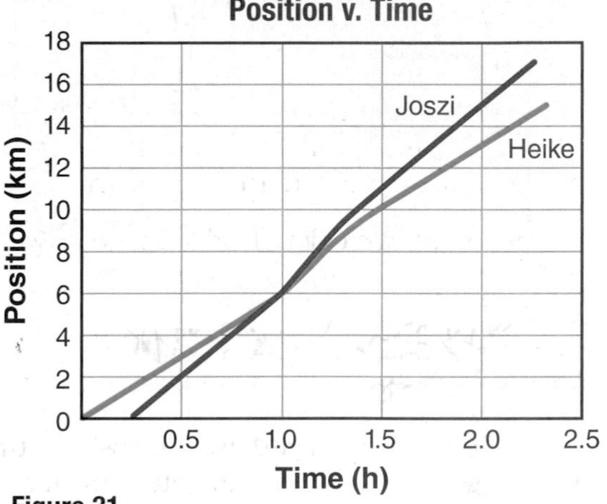

Figure 31

67. Driving Both car A and car B leave school when a stopwatch reads zero. Car A travels at a constant 75 km/h, and car B travels at a constant 85 km/h.

a. Draw a position-time graph showing the motion of both cars over 3 hours. How far are the two cars from school when the stopwatch reads 2.0 h? Calculate the distances and show them on your graph.

b. Both cars passed a gas station 120 km from the school. When did each car pass the gas station? Calculate the times and show them on your graph.

68. Draw a position-time graph for two cars traveling to a beach that is 50 km from school. At noon, car A leaves a store that is 10 km closer to the beach than the school is and moves at 40 km/h. Car B starts from school at 12:30 P.M. and moves at 100 km/h. When does each car get to the beach?

69. Two cars travel along a straight road. When a stopwatch reads $t = 0.00$ h, car A is at $x_A = 48.0$ km moving at a constant speed of 36.0 km/h. Later, when the watch reads $t = 0.50$ h, car B is at $x_B = 0.00$ km moving at 48.0 km/h. Answer the following questions, first graphically by creating a position-time graph and then algebraically by writing equations for the positions x_A and x_B as a function of the stopwatch time (t).

a. What will the watch read when car B passes car A?

b. At what position will car B pass car A?

c. When the cars pass, how long will it have been since car A was at the reference point?

Chapter Self-Check

70. The graph in **Figure 32** depicts Jim's movement along a straight path. The origin is at one end of the path.

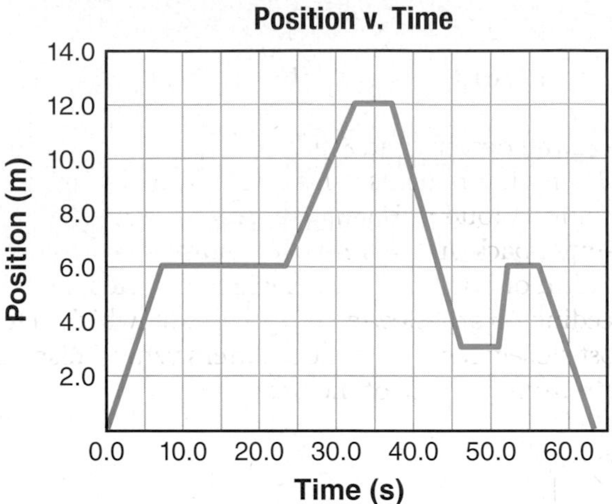

Figure 32

a. Reverse Problem Write a story describing Jim's movements along the path that would correspond to the motion represented by the graph.

b. When is Jim 6.0 m from the origin?

c. How much time passes between when Jim starts moving and when he is 12.0 m from the origin?

d. What is Jim's average velocity between 37.0 s and 46.0 s?

Thinking Critically

71. Apply Calculators Members of a physics class stood 25 m apart and used stopwatches to measure the time at which a car traveling on the highway passed each person. **Table 2** shows their data.

Table 2 Position v. Time

Time (s)	Position (m)
0.0	0.0
1.3	25.0
2.7	50.0
3.6	75.0
5.1	100.0
5.9	125.0
7.0	150.0
8.6	175.0
10.3	200.0

Use a graphing calculator to fit a line to a position-time graph of the data and to plot this line. Be sure to set the display range of the graph so that all the data fit on it. Find the line's slope. What was the car's speed?

72. Apply Concepts You want to average 90 km/h on a car trip. You cover the first half of the distance at an average speed of 48 km/h. What average speed must you have for the second half of the trip to meet your goal? Is this reasonable? Note that the velocities are based on half the distance, not half the time.

73. Design an Experiment Every time someone drives a particular red motorcycle past your friend's home, his father becomes angry. He thinks the motorcycle is going too fast for the posted 25 mph (40 km/h) speed limit. Describe a simple experiment you could do to determine whether the motorcycle is speeding the next time it passes your friend's house.

74. Interpret Graphs Is it possible for an object's position-time graph to be a horizontal line? A vertical line? If you answer yes to either situation, describe the associated motion in words.

Writing in Physics

75. Physicists have determined that the speed of light is 3.00×10^8 m/s. How did they arrive at this number? Read about some of the experiments scientists have performed to determine light's speed. Describe how the experimental techniques improved to make the experiments' results more accurate.

76. Some species of animals have good endurance, while others have the ability to move very quickly, but only for a short amount of time. Use reference sources to find two examples of each quality, and describe how it is helpful to that animal.

Cumulative Review

77. Convert each of the following time measurements to its equivalent in seconds:

a. 58 ns

b. 0.046 Gs

c. 9270 ms

d. 12.3 ks

78. State the number of significant figures in the following measurements:

a. 3218 kg

b. 60.080 kg

c. 801 kg

d. 0.000534 kg

79. Using a calculator, Chris obtained the following results. Rewrite each answer using the correct number of significant figures.

a. 5.32 mm + 2.1 mm = 7.4200000 mm

b. 13.597 m × 3.65 m = 49.62905 m^2

c. 83.2 kg − 12.804 kg = 70.3960000 kg

CHAPTER 2 STANDARDIZED TEST PRACTICE

MULTIPLE CHOICE

1. Which statement would be true about the particle model motion diagram for an airplane flying at a constant speed of 850 km/h?

A. The dots would start close together and get farther apart as the plane moved away from the airport.

B. The dots would be far apart at the beginning and get closer together as the plane moved away from the airport.

C. The dots would form an evenly spaced pattern.

D. The dots would start close together, get farther apart, and then get close together again as the airplane traveled away from the airport.

2. Which statement about drawing vectors is true?

A. The vector's length should be proportional to its magnitude.

B. You need a vector diagram to solve all physics problems properly.

C. A vector is a quantity that has a magnitude but no direction.

D. All quantities in physics are vectors.

3. The figure below shows a simplified graph of a bicyclist's motion. (Speeding up and slowing down motion is ignored.) When is the person's velocity greatest?

A. section I　　**C.** point D

B. section III　　**D.** point B

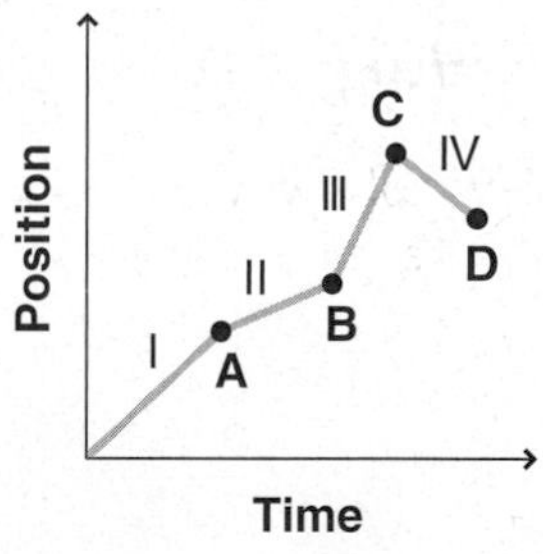

4. What is the average velocity of a train moving along a straight track if its displacement is 192 m east during a time period of 8.0 s?

A. 12 m/s east　　**C.** 48 m/s east

B. 24 m/s east　　**D.** 96 m/s east

5. A squirrel descends an 8-m tree at a constant speed in 1.5 min. It remains still at the base of the tree for 2.3 min. A loud noise then causes the squirrel to scamper back up the tree in 0.1 min to the exact position on the branch from which it started. Ignoring speeding up and slowing down motion, which graph most closely represents the squirrel's vertical displacement from the base of the tree?

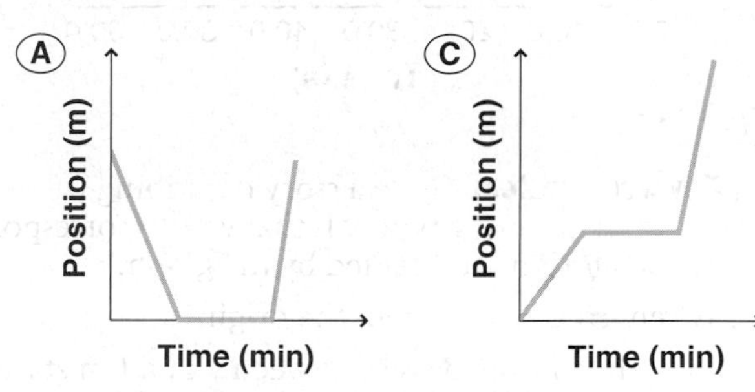

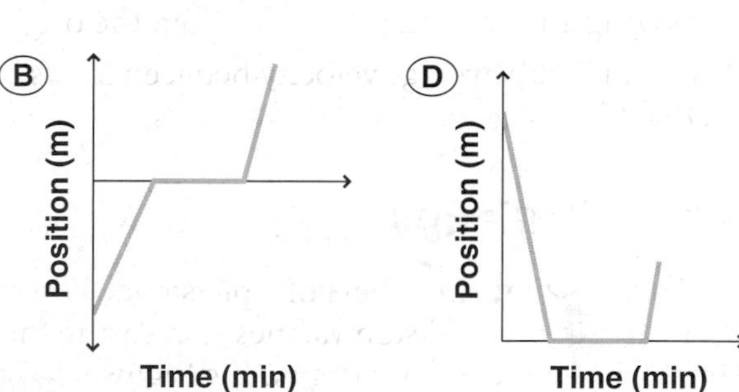

FREE RESPONSE

6. A rat is moving along a straight path. Find the rat's position relative to its starting point if it moves 12.8 cm/s north for 3.10 s.

NEED EXTRA HELP?

If You Missed Question	1	2	3	4	5	6
Review Section	1	2	3	4	4	4

Online Test Practice

CHAPTER 3

Accelerated Motion

BIG IDEA **Acceleration is the rate of change in an object's velocity.**

SECTIONS

LaunchLAB

GRAPHING MOTION

How does a graph showing constant speed compare to a graph of an object that is accelerating?

WATCH THIS!

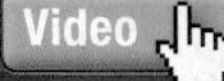

SKATEBOARD PHYSICS

How does a trip to your local skate park involve physics? You might be surprised! Explore acceleration as skateboarders show off their best moves.

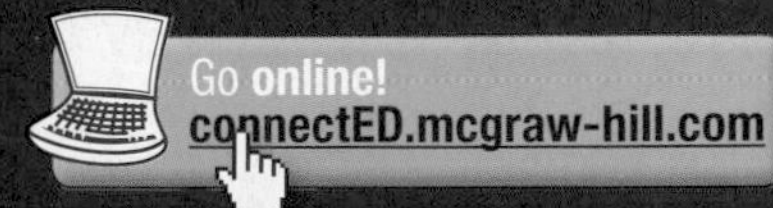
Go online!
connectED.mcgraw-hill.com

SECTION 1 Acceleration

PHYSICS 4 YOU As an airplane takes off, its speed changes from 5 m/s on the runway to nearly 300 m/s once it's in the air. If you've ever ridden on an airplane, you've felt the seat push against your back as the plane rapidly accelerates.

MAIN IDEA
An object accelerates when its velocity changes—that is, when it speeds up, slows down, or changes direction.

Essential Questions
- What is acceleration?
- How is acceleration different from velocity?
- What information can you learn from velocity-time graphs?

Review Vocabulary
vector a quantity that has magnitude and direction

New Vocabulary
acceleration
velocity-time graph
average acceleration
instantaneous acceleration

Nonuniform Motion Diagrams

An object in uniform motion moves along a straight line with an unchanging velocity, but few objects move this way all the time. More common is nonuniform motion, in which velocity is changing. In this chapter, you will study nonuniform motion along a straight line. Examples include balls rolling down hills, cars braking to a stop, and falling objects. In later chapters you will analyze nonuniform motion that is not confined to a straight line, such as motion along a circular path and the motion of thrown objects, such as baseballs.

Describing nonuniform motion You can feel a difference between uniform and nonuniform motion. Uniform motion feels smooth. If you close your eyes, it feels as if you are not moving at all. In contrast, when you move around a curve or up and down a roller coaster hill, you feel pushed or pulled.

How would you describe the motion of the person in **Figure 1**? In the first diagram, the person is motionless, but in the others, her position is changing in different ways. What information do the diagrams contain that could be used to distinguish the different types of motion? Notice the distances between successive positions. Because there is only one image of the person in the first diagram, you can conclude that she is at rest. The distances between images in the second diagram are the same because the jogger is in uniform motion; she moves at a constant velocity. In the remaining two diagrams, the distance between successive positions changes. The change in distance increases if the jogger speeds up. The change decreases if the jogger slows down.

Motion Diagram

a. The person is motionless.

b. Equally spaced images show her moving at a constant speed.

c. She is speeding up.

d. She is slowing down.

Figure 1 The distance the jogger moves in each time interval indicates the type of motion.

Joe Drivas/Photographer's Choice/Getty Images

Particle model diagram What does a particle model motion diagram look like for an object with changing velocity? **Figure 2** shows particle model motion diagrams below the motion diagrams of the jogger when she is speeding up and slowing down. There are two major indicators of the change in velocity in this form of the motion diagram. The change in the spacing of the dots and the differences in the lengths of the velocity vectors indicate the changes in velocity. If an object speeds up, each subsequent velocity vector is longer, and the spacing between dots increases. If the object slows down, each vector is shorter than the previous one, and the spacing between dots decreases. Both types of motion diagrams indicate how an object's velocity is changing.

READING CHECK Analyze What do increasing and decreasing lengths of velocity vectors indicate on a motion diagram?

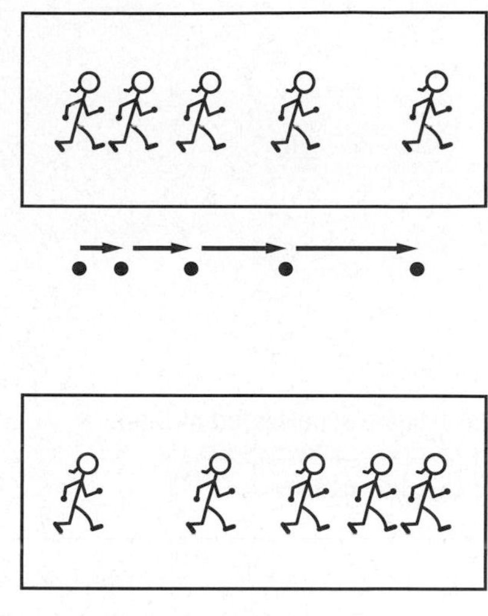

Figure 2 The change in length of the velocity vectors on these motion diagrams indicates whether the jogger is speeding up or slowing down.

Displaying acceleration on a motion diagram For a motion diagram to give a full picture of an object's movement, it should contain information about the rate at which the object's velocity is changing. The rate at which an object's velocity changes is called the **acceleration** of the object. By including acceleration vectors on a motion diagram, you can indicate the rate of change for the velocity.

Figure 3 shows a particle motion diagram for an object with increasing velocity. Notice that the lengths of the red velocity vectors get longer from left to right along the diagram. The figure also describes how to use the diagram to draw an acceleration vector for the motion. The acceleration vector that describes the increasing velocity is shown in violet on the diagram.

Notice in the figure that if the object's acceleration is constant, you can determine the length and direction of an acceleration vector by subtracting two consecutive velocity vectors and dividing by the time interval. That is, first find the change in velocity, $\Delta \boldsymbol{v} = \boldsymbol{v}_f - \boldsymbol{v}_i = \boldsymbol{v}_f + (-\boldsymbol{v}_i)$, where $\boldsymbol{v}_i$ and $\boldsymbol{v}_f$ refer to the velocities at the beginning and the end of the chosen time interval. Then divide by the time interval (Δt). The time interval between each dot in **Figure 3** is 1 s. You can draw the acceleration vector from the tail of the final velocity vector to the tip of the initial velocity vector.

Finding Acceleration Vectors

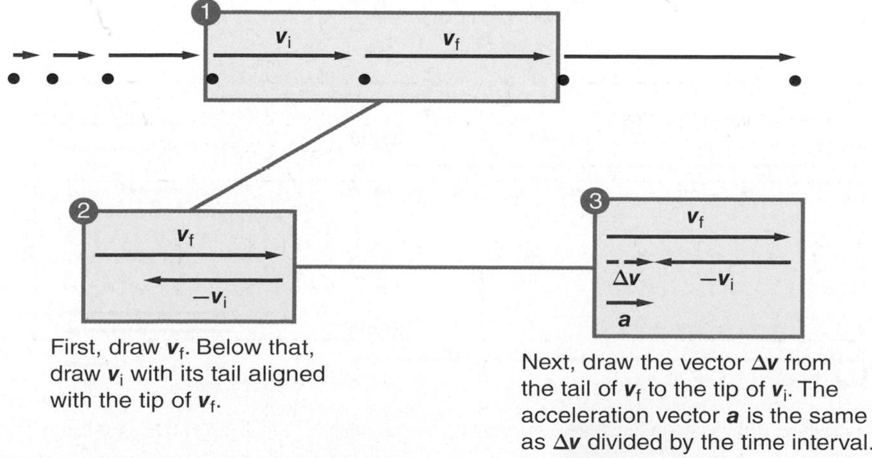
1
v_i
v_f
2
v_f
$-v_i$
3
v_f
Δv
$-v_i$
a
First, draw $\boldsymbol{v}_f$. Below that, draw $\boldsymbol{v}_i$ with its tail aligned with the tip of $\boldsymbol{v}_f$.
Next, draw the vector $\Delta \boldsymbol{v}$ from the tail of $\boldsymbol{v}_f$ to the tip of $\boldsymbol{v}_i$. The acceleration vector $\boldsymbol{a}$ is the same as $\Delta \boldsymbol{v}$ divided by the time interval.

Figure 3 For constant acceleration, an acceleration vector on a particle model diagram is the difference in the two velocity vectors divided by the time interval: $\boldsymbol{a} = \frac{\Delta \boldsymbol{v}}{\Delta t}$.

Analyze Can you draw an acceleration vector for two successive velocity vectors that are the same length and direction? Explain.

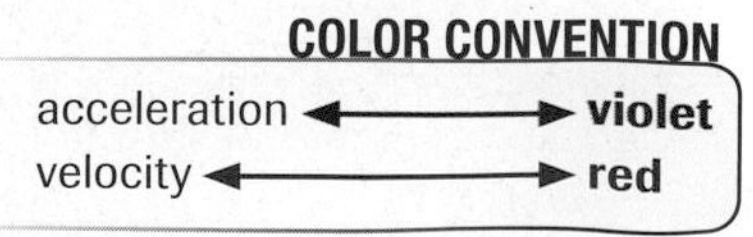
COLOR CONVENTION
acceleration
violet
velocity
red

Direction of Acceleration

Consider the four situations shown in **Figure 4** in which an object can accelerate by changing speed. The first motion diagram shows the car moving in the positive direction and speeding up. The second motion diagram shows the car moving in the positive direction and slowing down. The third shows the car speeding up in the negative direction, and the fourth shows the car slowing down as it moves in the negative direction. The figure also shows the velocity vectors for the second time interval of each diagram, along with the corresponding acceleration vectors. Note that Δt is equal to 1 s.

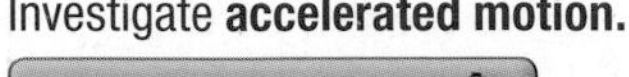

Investigate **accelerated motion.**

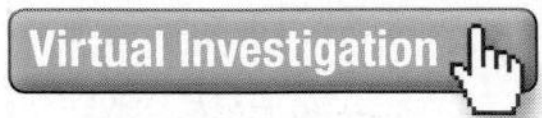

In the first and third situations, when the car is speeding up, the velocity and acceleration vectors point in the same direction. In the other two situations, in which the acceleration vector is in the opposite direction from the velocity vectors, the car is slowing down. In other words, when the car's acceleration is in the same direction as its velocity, the car's speed increases. When they are in opposite directions, the speed of the car decreases.

Both the direction of an object's velocity and its direction of acceleration are needed to determine whether it is speeding up or slowing down. An object has a positive acceleration when the acceleration vector points in the positive direction and a negative acceleration when the acceleration vector points in the negative direction. It is important to notice that the sign of acceleration alone does not indicate whether the object is speeding up or slowing down.

READING CHECK **Describe** the motion of an object if its velocity and acceleration vectors have opposite signs.

Figure 4 You need to know the direction of both the velocity and acceleration vectors in order to determine whether an object is speeding up or slowing down.

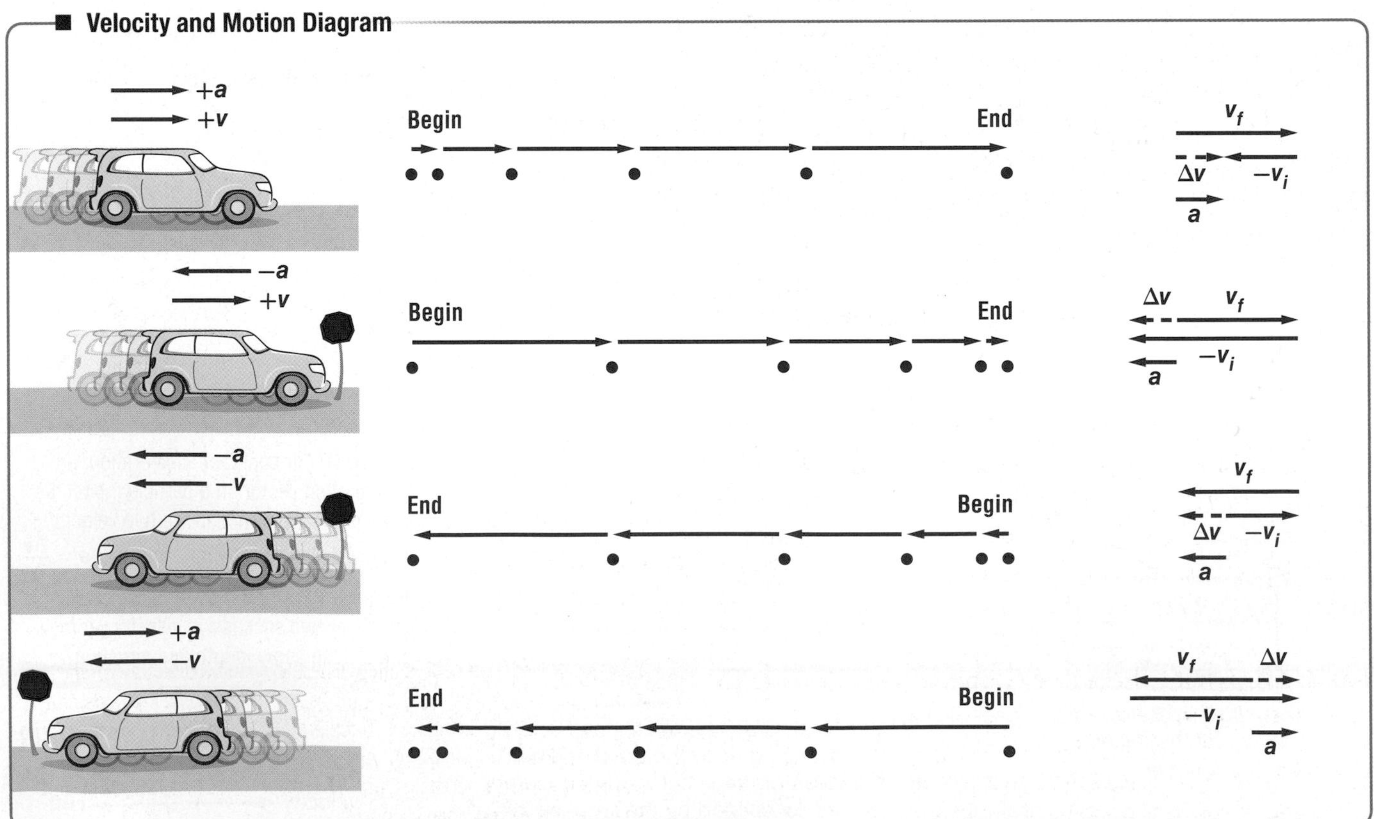

Velocity-Time Graphs

Just as it was useful to graph position versus time, it also is useful to plot velocity versus time. On a **velocity-time graph,** or *v-t* graph, velocity is plotted on the vertical axis and time is plotted on the horizontal axis.

Slope The velocity-time graph for a car that started at rest and sped up along a straight stretch of road is shown in **Figure 5.** The positive direction has been chosen to be the same as that of the car's motion. Notice that the graph is a straight line. This means the car sped up at a constant rate. The rate at which the car's velocity changed can be found by calculating the slope of the velocity-time graph.

The graph shows that the slope is 5.00 (m/s)/s, which is commonly written as 5 m/s^2. Consider the time interval between 4.00 s and 5.00 s. At 4.00 s, the car's velocity was 20.0 m/s in the positive direction. At 5.00 s, the car was traveling at 25.0 m/s in the same direction. Thus, in 1.00 s, the car's velocity increased by 5.0 m/s in the positive direction. When the velocity of an object changes at a constant rate, it has a constant acceleration.

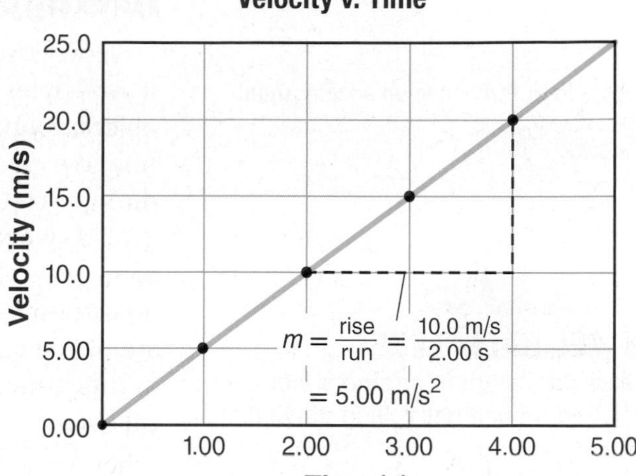

Figure 5 You can determine acceleration from a velocity-time graph by calculating the slope of the data. The slope is the rise divided by the run using any two points on the line.

View an **animation on velocity v. time graphs.**

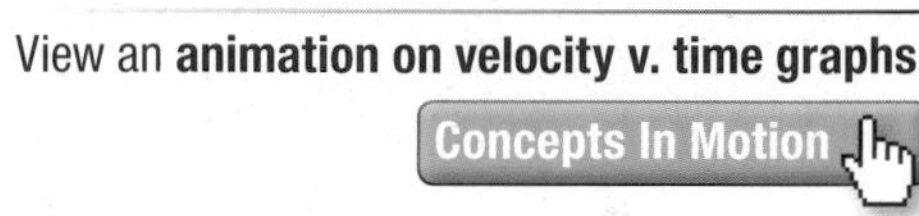

Reading velocity-time graphs The motions of five runners are shown in **Figure 6.** Assume that the positive direction is east. The slopes of Graphs A and E are zero. Thus, the accelerations are zero. Both graphs show motion at a constant velocity—Graph A to the east and Graph E to the west. Graph B shows motion with a positive velocity eastward. Its slope indicates a constant, positive acceleration. You can infer that the speed increases because velocity and acceleration are positive. Graph C has a negative slope. It shows motion that begins with a positive velocity, slows down, and then stops. This means the acceleration and the velocity are in opposite directions. The point at which Graphs C and B cross shows that the runners' velocities are equal at that time. It does not, however, identify their positions.

Graph D indicates motion that starts out toward the west, slows down, for an instant has zero velocity, and then moves east with increasing speed. The slope of Graph D is positive. Because velocity and acceleration are initially in opposite directions, the speed decreases to zero at the time the graph crosses the *x*-axis. After that time, velocity and acceleration are in the same direction, and the speed increases.

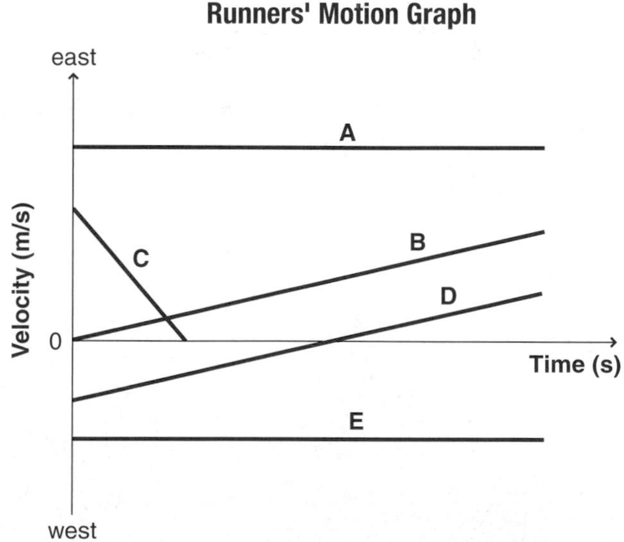

Figure 6 Because east is chosen as the positive direction on the graph, velocity is positive if the line is above the horizontal axis and negative if the line is below it. Acceleration is positive if the line is slanted upward on the graph. Acceleration is negative if the line is slanted downward on the graph. A horizontal line indicates constant velocity and zero acceleration.

READING CHECK Describe the meaning of a line crossing the *x*-axis in a velocity-time graph.

View a **BrainPOP video on acceleration.**

MiniLAB

STEEL BALL RACE

Does the height of a ramp affect the motion of a ball rolling down it?

PhysicsLABs

ACCELERATION

How can you use motion measurements to calculate the acceleration of a rolling ball?

TOSSED-BALL MOTION

PROBEWARE LAB What does the graph of ball tossed upward look like?

Average and Instantaneous Acceleration

How does it feel differently if the car you ride in accelerates a little or if it accelerates a lot? As with velocity, the acceleration of most moving objects continually changes. If you want to describe an object's acceleration, it is often more convenient to describe the overall change in velocity during a certain time interval rather than describing the continual change.

The **average acceleration** of an object is its change in velocity during some measurable time interval divided by that time interval. Average acceleration is measured in meters per second per second (m/s/s), or simply meters per second squared (m/s^2). A car might accelerate quickly at times and more slowly at times. Just as average velocity depends only on the starting and ending displacement, average acceleration depends only on the starting and ending velocity during a time interval. **Figure 7** shows a graph of motion in which the acceleration is changing. The average acceleration during a certain time interval is determined just as it is in **Figure 5** for constant acceleration. Notice, however, that because the line is curved, the average acceleration in this graph varies depending on the time interval that you choose.

The change in an object's velocity at an instant of time is called **instantaneous acceleration.** You can determine the instantaneous acceleration of an object by drawing a tangent line on the velocity-time graph at the point of time in which you are interested. The slope of this line is equal to the instantaneous acceleration. Most of the situations considered in this textbook assume an ideal case of constant acceleration. When the acceleration is the same at all points during a time interval, the average acceleration and the instantaneous accelerations are equal.

☑ **READING CHECK Contrast** How is instantaneous acceleration different from average acceleration?

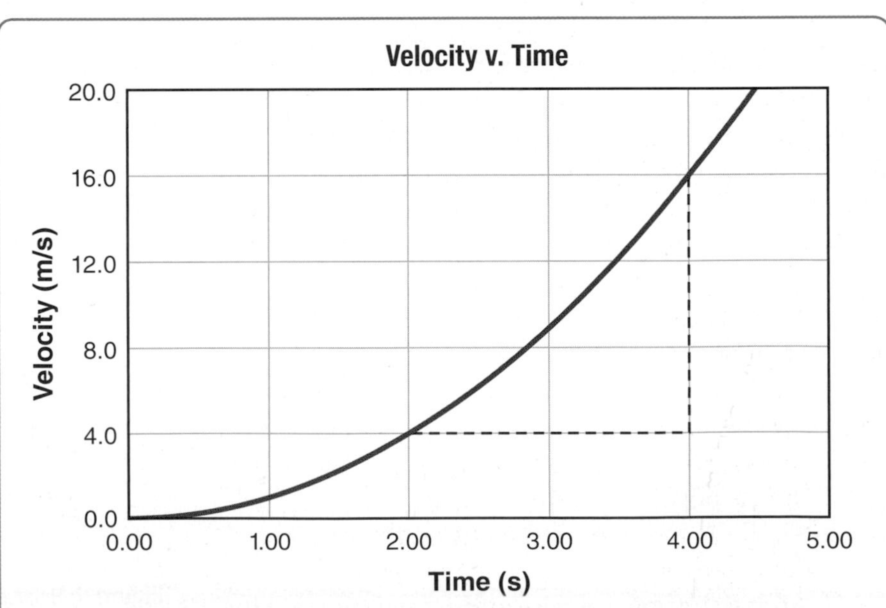

Figure 7 A curved line on a velocity-time graph shows that the acceleration is changing. The slope indicates the average acceleration during a time interval that you choose.

Calculate How large is the average acceleration between 0.00 s and 2.00 s?

Calculating Acceleration

How can you describe the acceleration of an object mathematically? Recall that the acceleration of an object is the slope of that object's velocity v. time graph. On a velocity v. time graph, slope equals $\Delta \boldsymbol{v}/\Delta t$.

AVERAGE ACCELERATION
Average acceleration is defined as the change in velocity divided by the time it takes to make that change.

$$\bar{\boldsymbol{a}} \equiv \frac{\Delta \boldsymbol{v}}{\Delta t} = \frac{\boldsymbol{v}_f - \boldsymbol{v}_i}{t_f - t_i}$$

Suppose you run wind sprints back and forth across the gym. You first run at a speed of 4.0 m/s toward the wall. Then, 10.0 s later, your speed is 4.0 m/s as you run away from the wall. What is your average acceleration if the positive direction is toward the wall?

$$\bar{\boldsymbol{a}} \equiv \frac{\Delta \boldsymbol{v}}{\Delta t} = \frac{\boldsymbol{v}_f - \boldsymbol{v}_i}{t_f - t_i}$$

$$= \frac{-4.0 \text{ m/s} - 4.0 \text{ m/s}}{10.0 \text{ s}} = -0.80 \text{ m/s}^2$$

EXAMPLE PROBLEM 1

Find help with **slope.** 

VELOCITY AND ACCELERATION How would you describe the sprinter's velocity and acceleration as shown on the graph?

1 ANALYZE AND SKETCH THE PROBLEM
From the graph, note that the magnitude of the sprinter's velocity starts at zero, increases rapidly for the first few seconds, and then, after reaching about 10.0 m/s, remains almost constant.

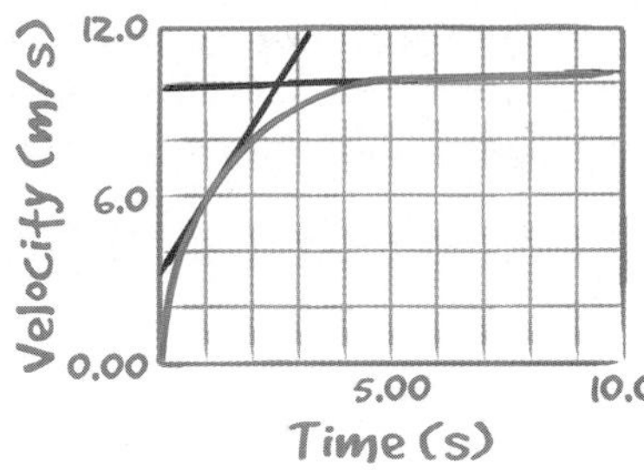

KNOWN	UNKNOWN
v = varies	a = ?

2 SOLVE FOR THE UNKNOWN
Draw tangents to the curve at two points. Choose $t = 1.00$ s and $t = 5.00$ s.
Solve for magnitude of the acceleration at 1.00 s:

$$a = \frac{\text{rise}}{\text{run}}$$

◀ *The slope of the line at 1.00 s is equal to the acceleration at that time.*

$$= \frac{10.0 \text{ m/s} - 6.0 \text{ m/s}}{2.4 \text{ s} - 1.00 \text{ s}}$$

$$= 2.9 \text{ m/s/s} = 2.9 \text{ m/s}^2$$

Solve for the magnitude of the instantaneous acceleration at 5.0 s:

$$a = \frac{\text{rise}}{\text{run}}$$

◀ *The slope of the line at 5.0 s is equal to the acceleration at that time.*

$$= \frac{10.3 \text{ m/s} - 10.0 \text{ m/s}}{10.0 \text{ s} - 0.00 \text{ s}}$$

$$= 0.030 \text{ m/s/s} = 0.030 \text{ m/s}^2$$

The acceleration is not constant because its magnitude changes from 2.9 m/s^2 at 1.0 s to 0.030 m/s^2 at 5.0 s.

The acceleration is in the direction chosen to be positive because both values are positive.

3 EVALUATE THE ANSWER
Are the units correct? Acceleration is measured in m/s^2.

EXAMPLE PROBLEM

PRACTICE PROBLEMS

Do additional problems. Online Practice

1. The velocity-time graph in **Figure 8** describes Steven's motion as he walks along the midway at the state fair. Sketch the corresponding motion diagram. Include velocity vectors in your diagram.
2. Use the *v-t* graph of the toy train in **Figure 9** to answer these questions.
   a. When is the train's speed constant?
   b. During which time interval is the train's acceleration positive?
   c. When is the train's acceleration most negative?
3. Refer to **Figure 9** to find the average acceleration of the train during the following time intervals.
   a. 0.0 s to 5.0 s **b.** 15.0 s to 20.0 s **c.** 0.0 s to 40.0 s
4. **CHALLENGE** Plot a *v-t* graph representing the following motion: An elevator starts at rest from the ground floor of a three-story shopping mall. It accelerates upward for 2.0 s at a rate of 0.5 m/s^2, continues up at a constant velocity of 1.0 m/s for 12.0 s, and then slows down with a constant downward acceleration of 0.25 m/s^2 for 4.0 s as it reaches the third floor.

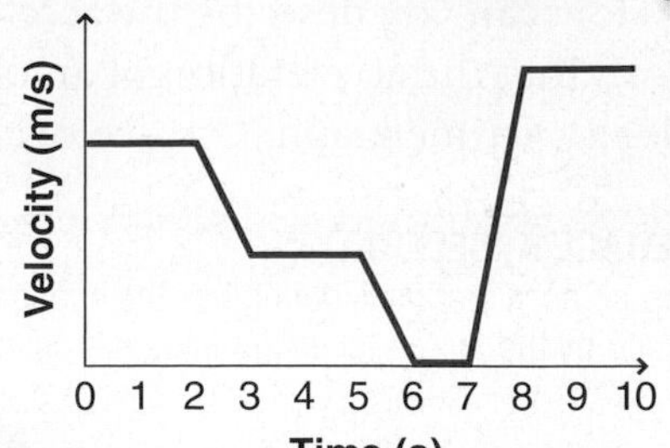

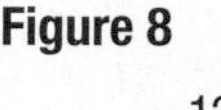
Figure 8

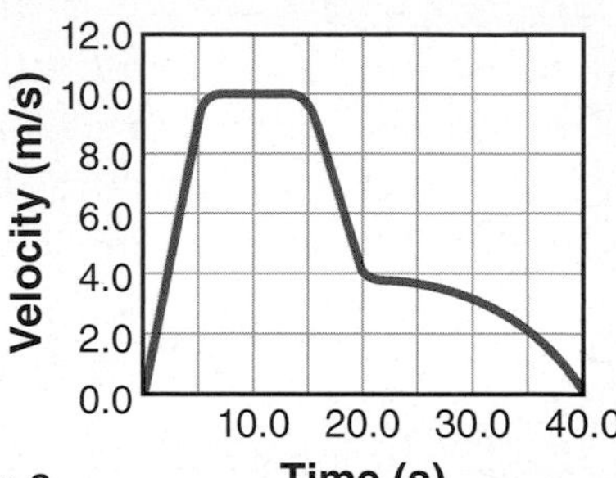

Figure 9

EXAMPLE PROBLEM 2

Find help with **significant figures.** Math Handbook

ACCELERATION Describe a ball's motion as it rolls up a slanted driveway. It starts at 2.50 m/s, slows down for 5.00 s, stops for an instant, and then rolls back down. The positive direction is chosen to be up the driveway. The origin is where the motion begins. What are the sign and the magnitude of the ball's acceleration as it rolls up the driveway?

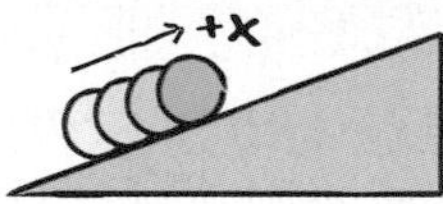

Begin

End

a

1 ANALYZE AND SKETCH THE PROBLEM

- Sketch the situation.
- Draw the coordinate system based on the motion diagram.

KNOWN	UNKNOWN
$\mathbf{v}_i = +2.50$ m/s	$\mathbf{a} = ?$
$\mathbf{v}_f = 0.00$ m/s at $t = 5.00$ s	

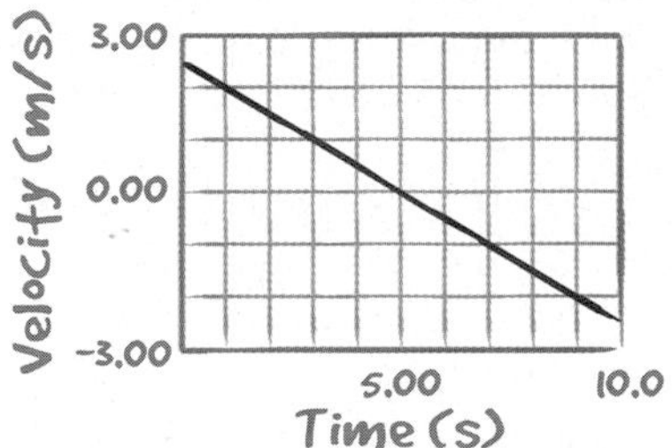

2 SOLVE FOR THE UNKNOWN

Find the acceleration from the slope of the graph.

Solve for the change in velocity and the time taken to make that change.

$\Delta \mathbf{v} = \mathbf{v}_f - \mathbf{v}_i$

$= 0.00 \text{ m/s} - 2.50 \text{ m/s} = -2.50 \text{ m/s}$ ◀ Substitute $v_f = 0.00$ m/s at $t_f = 5.00$ s, $v_i = 2.50$ m/s at $t_i = 0.00$ s.

$\Delta t = t_f - t_i$

$= 5.00 \text{ s} - 0.00 \text{ s} = 5.00 \text{ s}$ ◀ Substitute $t_f = 5.00$ s, $t_i = 0.00$ s.

Solve for the acceleration.

$\bar{\mathbf{a}} \equiv \frac{\Delta \mathbf{v}}{\Delta t} = (-2.50 \text{ m/s}) / 5.00 \text{ s}$ ◀ Substitute $\Delta v = -2.50$ m/s, $\Delta t = 5.00$ s.

$= -0.500 \text{ m/s}^2$ or 0.500 m/s^2 down the driveway

3 EVALUATE THE ANSWER

- **Are the units correct?** Acceleration is measured in m/s^2.
- **Do the directions make sense?** As the ball slows down, the direction of acceleration is opposite that of velocity.

PRACTICE PROBLEMS

Do additional problems. Online Practice 

5. A race car's forward velocity increases from 4.0 m/s to 36 m/s over a 4.0-s time interval. What is its average acceleration?

6. The race car in the previous problem slows from 36 m/s to 15 m/s over 3.0 s. What is its average acceleration?

7. A bus is moving west at 25 m/s when the driver steps on the brakes and brings the bus to a stop in 3.0 s.

a. What is the average acceleration of the bus while braking?

b. If the bus took twice as long to stop, how would the acceleration compare with what you found in part **a?**

8. A car is coasting backward downhill at a speed of 3.0 m/s when the driver gets the engine started. After 2.5 s, the car is moving uphill at 4.5 m/s. If uphill is chosen as the positive direction, what is the car's average acceleration?

9. Rohith has been jogging east toward the bus stop at 3.5 m/s when he looks at his watch and sees that he has plenty of time before the bus arrives. Over the next 10.0 s, he slows his pace to a leisurely 0.75 m/s. What was his average acceleration during this 10.0 s?

10. CHALLENGE If the rate of continental drift were to abruptly slow from 1.0 cm/y to 0.5 cm/y over the time interval of a year, what would be the average acceleration?

Acceleration with Constant Speed

Think again about running wind sprints across the gym. Notice that your speed is the same as you move toward the wall of the gym and as you move away from it. In both cases, you are running at a speed of 4.0 m/s. How is it possible for you to be accelerating?

Acceleration can occur even when speed is constant. The average acceleration for the entire trip you make toward the wall of the gym and back again is $-0.80\ \mathrm{m/s^2}$. The negative sign indicates that the direction of your acceleration is away from the wall because the positive direction was chosen as toward the wall. The velocity changes from positive to negative when the direction of motion changes. A change in velocity results in acceleration. Thus, acceleration can also be associated with a change in the direction of motion.

☑ **READING CHECK** **Explain** how it is possible for an object to accelerate when the object is traveling at a constant speed.

SECTION 1 REVIEW

Section Self-Check Check your understanding.

11. **MAIN**IDEA What are three ways an object can accelerate?

12. **Position-Time and Velocity-Time Graphs** Two joggers run at a constant velocity of 7.5 m/s east. **Figure 10** shows the positions of both joggers at time $t = 0$.

a. What would be the difference(s) in the position-time graphs of their motion?

b. What would be the difference(s) in their velocity-time graphs?

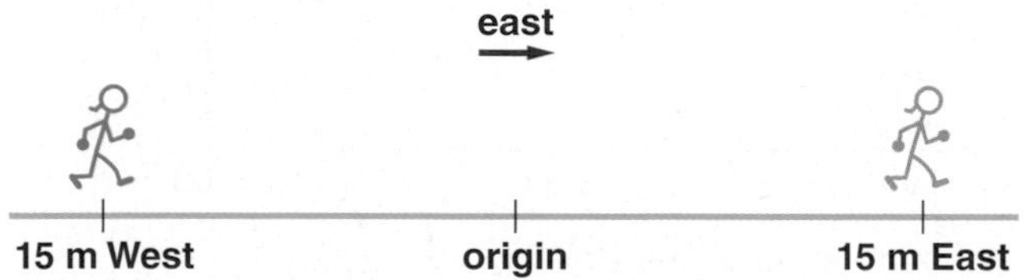

Figure 10

13. **Velocity-Time Graph** Sketch a velocity-time graph for a car that goes east at 25 m/s for 100 s, then west at 25 m/s for another 100 s.

14. **Average Velocity and Average Acceleration** A canoeist paddles upstream at a velocity of 2.0 m/s for 4.0 s and then floats downstream at 4.0 m/s for 4.0 s.

a. What is the average velocity of the canoe during the 8.0-s time interval?

b. What is the average acceleration of the canoe during the 8.0-s time interval?

15. **Critical Thinking** A police officer clocked a driver going 32 km/h over the speed limit just as the driver passed a slower car. When the officer stopped the car, the driver argued that the other driver should get a ticket as well. The driver said that the cars must have been going the same speed because they were observed next to each other. Is the driver correct? Explain with a sketch and a motion diagram.

SECTION 2 Motion with Constant Acceleration

PHYSICS 4 YOU **Suppose a car is moving along a road and suddenly the driver sees a fallen tree blocking the way ahead. Will the driver be able to stop in time? It all depends on how effectively the car's brakes can cause the car to accelerate in the direction opposite its motion.**

MAIN IDEA

For an object with constant acceleration, the relationships among position, velocity, acceleration, and time can be described by graphs and equations.

Essential Questions

- What do a position-time graph and a velocity-time graph look like for motion with constant acceleration?
- How can you determine the displacement of a moving object from its velocity-time graph?
- What are the relationships among position, velocity, acceleration, and time?

Review Vocabulary

displacement change in position having both magnitude and direction; it is equal to the final position minus the initial position

Position with Constant Acceleration

If an object experiences constant acceleration, its velocity changes at a constant rate. How does its position change? The positions at different times of a car with constant acceleration are graphed in **Figure 11.** The graph shows that the car's motion is not uniform. The displacements for equal time intervals on the graph get larger and larger. As a result, the slope of the line in **Figure 11** gets steeper as time goes on. For an object with constant acceleration, the position-time graph is a parabola.

The slopes from the position-time graph in **Figure 11** have been used to create the velocity-time graph on the left in **Figure 12.** For an object with constant acceleration, the velocity-time graph is a straight line.

A unique position-time graph cannot be created using a velocity-time graph because it does not contain information about position. It does, however, contain information about displacement. Recall that for an object moving at a constant velocity, the velocity is the displacement divided by the time interval. The displacement is then the product of the velocity and the time interval. On the right graph in **Figure 12** on the next page, $\boldsymbol{v}$ is the height of the plotted line above the horizontal axis, and Δt is the width of the shaded triangle. The area is $\left(\frac{1}{2}\right)\boldsymbol{v}\Delta t$, or $\Delta\boldsymbol{x}$. Thus, the area under the v-t graph equals the displacement.

READING CHECK Identify What is the shape of a position-time graph of an object traveling with constant acceleration?

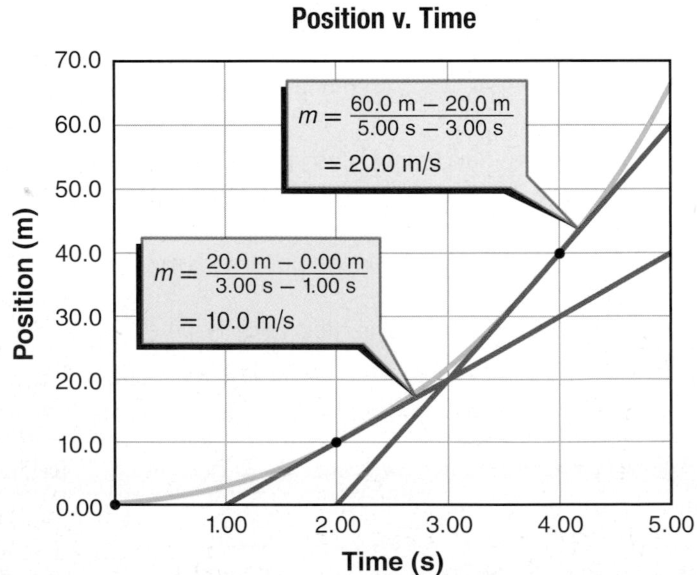

Figure 11 The slope of a position-time graph changes with time for an object with constant acceleration.

MARTIN OESER/DDP/AFP/Getty Images

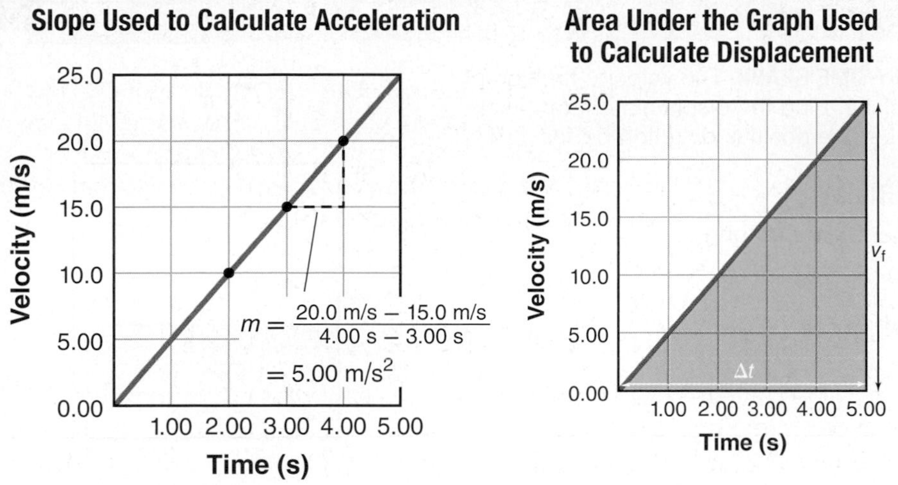

Slope Used to Calculate Acceleration

$$m = \frac{20.0 \text{ m/s} - 15.0 \text{ m/s}}{4.00 \text{ s} - 3.00 \text{ s}} = 5.00 \text{ m/s}^2$$

Area Under the Graph Used to Calculate Displacement

Figure 12 The slopes of the position-time graph in **Figure 11** are shown in these velocity-time graphs. The rise divided by the run gives the acceleration on the left. The area under the curve gives the displacement on the right.

Calculate What is the slope of the velocity-time graph on the left between $t = 2.00$ s and $t = 5.00$ s?

Velocity with Average Acceleration

You have read that the equation for average velocity can be algebraically rearranged to show the new position after a period of time, given the initial position and the average velocity. The definition of average acceleration can be manipulated similarly to show the new velocity after a period of time, given the initial velocity and the average acceleration.

If you know an object's average acceleration during a time interval, you can use it to determine how much the velocity changed during that time. You can rewrite the definition of average acceleration ($\overline{\boldsymbol{a}} \equiv \frac{\Delta \boldsymbol{v}}{\Delta t}$) as follows:

$$\Delta \boldsymbol{v} = \overline{\boldsymbol{a}}\Delta t$$
$$\boldsymbol{v}_{\mathrm{f}} - \boldsymbol{v}_{\mathrm{i}} = \overline{\boldsymbol{a}}\Delta t$$

The equation for final velocity with average acceleration can be written:

FINAL VELOCITY WITH AVERAGE ACCELERATION

The final velocity is equal to the initial velocity plus the product of the average acceleration and the time interval.

$$\boldsymbol{v}_{\mathrm{f}} = \boldsymbol{v}_{\mathrm{i}} + \overline{\boldsymbol{a}}\Delta t$$

In cases when the acceleration is constant, the average acceleration ($\overline{\boldsymbol{a}}$) is the same as the instantaneous acceleration ($\boldsymbol{a}$). This equation can be rearranged to find the time at which an object with constant acceleration has a given velocity. You can also use it to calculate the initial velocity of an object when both a velocity and the time at which it occurred are given.

PhysicsLAB

MEASURING ACCELERATION

PROBEWARE LAB What does a graph show about the motion of a cart?

PRACTICE PROBLEMS

Do additional problems.

16. A golf ball rolls up a hill toward a miniature-golf hole. Assume the direction toward the hole is positive.

a. If the golf ball starts with a speed of 2.0 m/s and slows at a constant rate of 0.50 m/s², what is its velocity after 2.0 s?

b. What is the golf ball's velocity if the constant acceleration continues for 6.0 s?

c. Describe the motion of the golf ball in words and with a motion diagram.

17. A bus traveling 30.0 km/h east has a constant increase in speed of 1.5 m/s². What is its velocity 6.8 s later?

18. If a car accelerates from rest at a constant rate of 5.5 m/s² north, how long will it take for the car to reach a velocity of 28 m/s north?

19. CHALLENGE A car slows from 22 m/s to 3.0 m/s at a constant rate of 2.1 m/s². How many seconds are required before the car is traveling at a forward velocity of 3.0 m/s?

EXAMPLE PROBLEM 3

Find help with **significant figures.** Math Handbook

EXAMPLE PROBLEM

FINDING DISPLACEMENT FROM A VELOCITY-TIME GRAPH The velocity-time graph at the right shows the motion of an airplane. Find the displacement of the airplane for $\Delta t = 1.0$ s and for $\Delta t = 2.0$ s. Let the positive direction be forward.

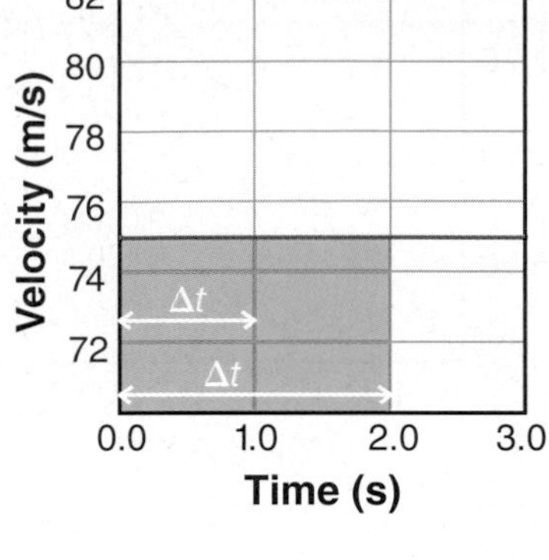

1 ANALYZE AND SKETCH THE PROBLEM

- The displacement is the area under the *v-t* graph.
- The time intervals begin at $t = 0.0$ s.

KNOWN	UNKNOWN
$\mathbf{v} = +75$ m/s	$\Delta \mathbf{x} = ?$
$\Delta t = 1.0$ s	
$\Delta t = 2.0$ s	

2 SOLVE FOR THE UNKNOWN

Use the relationship among displacement, velocity, and time interval to find $\Delta \mathbf{x}$ during $\Delta t = 1.0$ s.

$\Delta \mathbf{x} = \mathbf{v}\Delta t$

$= (+75 \text{ m/s})(1.0 \text{ s})$ ◀ Substitute $v = +75$ m/s, $\Delta t = 1.0$ s.

$= +75$ m

Use the same relationship to find $\Delta \mathbf{x}$ during $\Delta t = 2.0$ s.

$\Delta \mathbf{x} = \mathbf{v}\Delta t$

$= (+75 \text{ m/s})(2.0 \text{ s})$ ◀ Substitute $v = +75$ m/s, $\Delta t = 2.0$ s.

$= +150$ m

3 EVALUATE THE ANSWER

- **Are the units correct?** Displacement is measured in meters.
- **Do the signs make sense?** The positive sign agrees with the graph.
- **Is the magnitude realistic?** Moving a distance of about one football field per second is reasonable for an airplane.

PRACTICE PROBLEMS

Do additional problems. Online Practice

PRACTICE PROBLEMS

20. The graph in **Figure 13** describes the motion of two bicyclists, Akiko and Brian, who start from rest and travel north, increasing their speed with a constant acceleration. What was the total displacement of each bicyclist during the time shown for each? *Hint: Use the area of a triangle: area =* $\left(\frac{1}{2}\right)$*(base)(height).*

21. The motion of two people, Carlos and Diana, moving south along a straight path is described by the graph in **Figure 14.** What is the total displacement of each person during the 4.0-s interval shown on the graph?

22. CHALLENGE A car, just pulling onto a straight stretch of highway, has a constant acceleration from 0 m/s to 25 m/s west in 12 s.

a. Draw a *v-t* graph of the car's motion.

b. Use the graph to determine the car's displacement during the 12.0-s time interval.

c. Another car is traveling along the same stretch of highway. It travels the same distance in the same time as the first car, but its velocity is constant. Draw a *v-t* graph for this car's motion.

d. Explain how you knew this car's velocity.

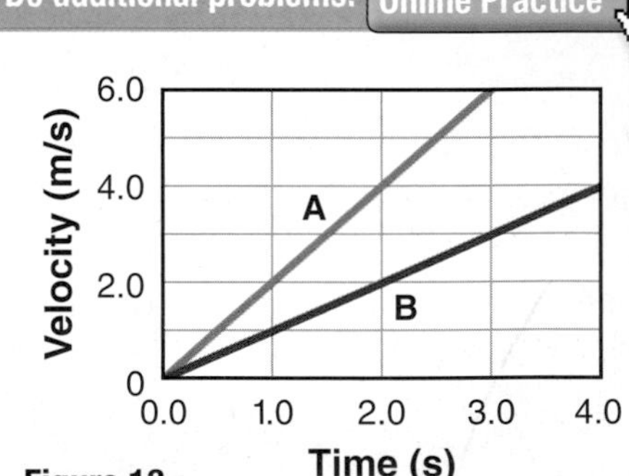

Figure 13

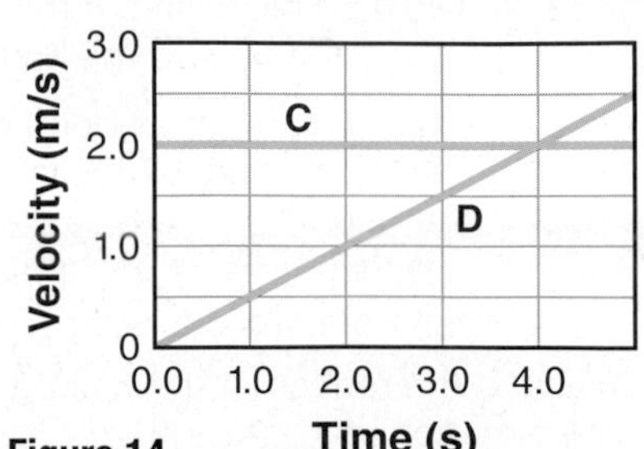

Figure 14

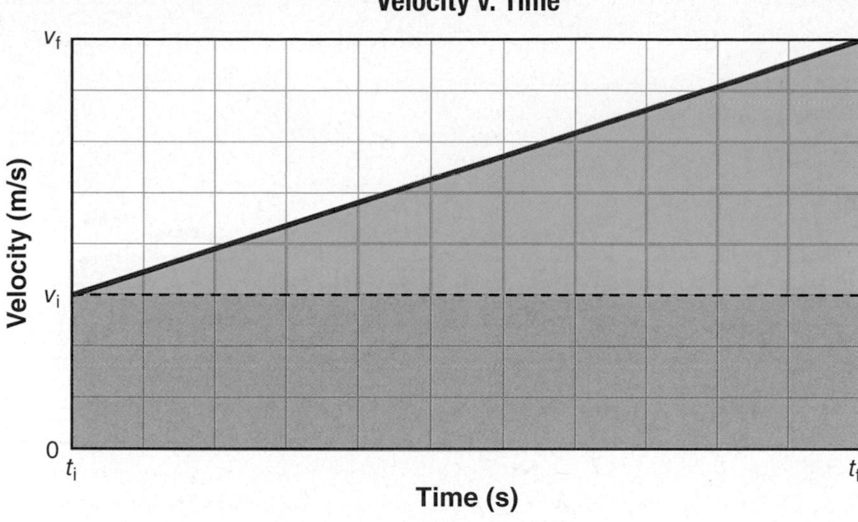

Figure 15 For motion with constant acceleration, if the initial velocity on a velocity-time graph is not zero, the area under the graph is the sum of a rectangular area and a triangular area.

Motion with an initial nonzero velocity The graph in **Figure 15** describes constant acceleration that started with an initial velocity of $\boldsymbol{v}_i$. To determine the displacement, you can divide the area under the graph into a rectangle and a triangle. The total area is then:

$$\Delta \boldsymbol{x} = \Delta \boldsymbol{x}_{\text{rectangle}} + \Delta \boldsymbol{x}_{\text{triangle}} = \boldsymbol{v}_i(\Delta t) + \left(\frac{1}{2}\right)\Delta \boldsymbol{v}\Delta t$$

Substituting $a\Delta t$ for the change in velocity in the equation yields:

$$\Delta \boldsymbol{x} = \Delta \boldsymbol{x}_{\text{rectangle}} + \Delta \boldsymbol{x}_{\text{triangle}} = \boldsymbol{v}_i(\Delta t) + \left(\frac{1}{2}\right)\boldsymbol{a}(\Delta t)^2$$

When the initial or final position of the object is known, the equation can be written as follows:

$$\boldsymbol{x}_f - \boldsymbol{x}_i = \boldsymbol{v}_i(\Delta t) + \left(\frac{1}{2}\right)\boldsymbol{a}(\Delta t)^2 \text{ or } \boldsymbol{x}_f = \boldsymbol{x}_i + \boldsymbol{v}_i(\Delta t) + \left(\frac{1}{2}\right)\boldsymbol{a}(\Delta t)^2$$

If the initial time is $t_i = 0$, the equation then becomes the following.

POSITION WITH AVERAGE ACCELERATION

An object's final position is equal to the sum of its initial position, the product of the initial velocity and the final time, and half the product of the acceleration and the square of the final time.

$$\boldsymbol{x}_f = \boldsymbol{x}_i + \boldsymbol{v}_i t_f + \left(\frac{1}{2}\right)\boldsymbol{a} t_f^2$$

REAL-WORLD PHYSICS

DRAG RACING A dragster driver tries to obtain maximum acceleration over a course. The fastest U.S. National Hot Rod Association time on record for the 402-m course is 3.771 s. The highest final speed on record is 145.3 m/s (324.98 mph).

An Alternative Equation

Often, it is useful to relate position, velocity, and constant acceleration without including time. Rearrange the equation $\boldsymbol{v}_f = \boldsymbol{v}_i + \boldsymbol{a}t_f$ to solve for time: $t_f = \frac{\boldsymbol{v}_f - \boldsymbol{v}_i}{\boldsymbol{a}}$.

You can then rewrite the position with average acceleration equation by substituting t_f to obtain the following:

$$\boldsymbol{x}_f = \boldsymbol{x}_i + \boldsymbol{v}_i\left(\frac{\boldsymbol{v}_f - \boldsymbol{v}_i}{\boldsymbol{a}}\right) + \left(\frac{1}{2}\right)\boldsymbol{a}\left(\frac{\boldsymbol{v}_f - \boldsymbol{v}_i}{\boldsymbol{a}}\right)^2$$

This equation can be solved for the velocity ($\boldsymbol{v}_f$) at any position ($\boldsymbol{x}_f$).

VELOCITY WITH CONSTANT ACCELERATION

The square of the final velocity equals the sum of the square of the initial velocity and twice the product of the acceleration and the displacement since the initial time.

$$v_f^2 = v_i^2 + 2a(x_f - x_i)$$

EXAMPLE PROBLEM 4

Get help with **displacement.** Personal Tutor

EXAMPLE PROBLEM

DISPLACEMENT An automobile starts at rest and accelerates at 3.5 m/s² after a traffic light turns green. How far will it have gone when it is traveling at 25 m/s?

1 ANALYZE AND SKETCH THE PROBLEM

- Sketch the situation.
- Establish coordinate axes. Let the positive direction be to the right.
- Draw a motion diagram.

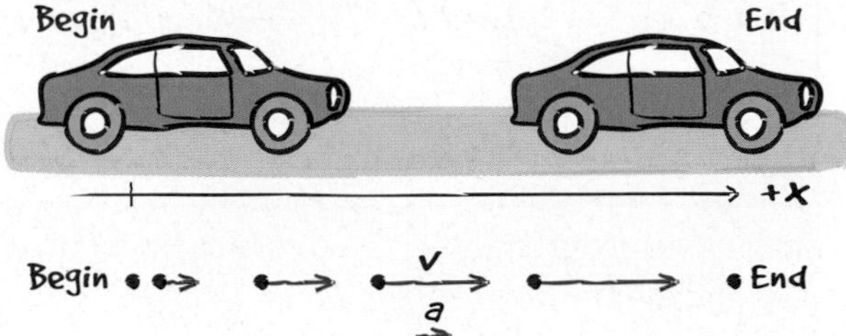

KNOWN	UNKNOWN
$x_i = 0.00$ m	$x_f = ?$
$v_i = 0.00$ m/s	
$v_f = +25$ m/s	
$\bar{a} = a = +3.5$ m/s²	

2 SOLVE FOR THE UNKNOWN

Use the relationship among velocity, acceleration, and displacement to find x_f.

$$v_f^2 = v_i^2 + 2a(x_f - x_i)$$

$$x_f = x_i + \frac{v_f^2 - v_i^2}{2a}$$

$$= 0.00 \text{ m} + \frac{(+25 \text{ m/s})^2 - (0.00 \text{ m/s})^2}{2(+3.5 \text{ m/s}^2)}$$

◀ Substitute $x_i = 0.00$ m, $v_f = +25$ m/s, $v_i = 0.00$ m/s, $a = +3.5$ m/s².

$$= +89 \text{ m}$$

3 EVALUATE THE ANSWER

- **Are the units correct?** Position is measured in meters.
- **Does the sign make sense?** The positive sign agrees with both the pictorial and physical models.
- **Is the magnitude realistic?** The displacement is almost the length of a football field. The result is reasonable because 25 m/s (about 55 mph) is fast.

PRACTICE PROBLEMS

Do additional problems. Online Practice

PRACTICE PROBLEMS

23. A skateboarder is moving at a constant speed of 1.75 m/s when she starts up an incline that causes her to slow down with a constant acceleration of −0.20 m/s². How much time passes from when she begins to slow down until she begins to move back down the incline?

24. A race car travels on a straight racetrack with a forward velocity of 44 m/s and slows at a constant rate to a velocity of 22 m/s over 11 s. How far does it move during this time?

25. A car accelerates at a constant rate from 15 m/s to 25 m/s while it travels a distance of 125 m. How long does it take to achieve the final speed?

26. A bike rider pedals with constant acceleration to reach a velocity of 7.5 m/s north over a time of 4.5 s. During the period of acceleration, the bike's displacement is 19 m north. What was the initial velocity of the bike?

27. CHALLENGE The car in **Figure 16** travels west with a forward acceleration of 0.22 m/s². What was the car's velocity ($\mathbf{v}_i$) at point $\mathbf{x}_i$ if it travels a distance of 350 m in 18.4 s?

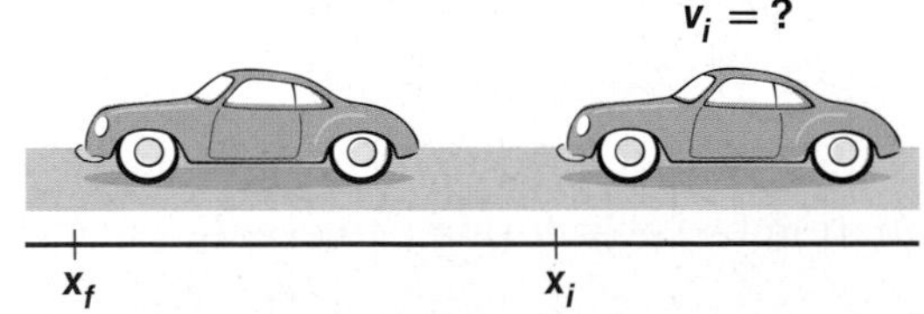

Figure 16

EXAMPLE PROBLEM 5

Find help with **isolating a variable.** Math Handbook

EXAMPLE PROBLEM

TWO-PART MOTION You are driving a car, traveling at a constant velocity of 25 m/s along a straight road, when you see a child suddenly run onto the road. It takes 0.45 s for you to react and apply the brakes. As a result, the car slows with a steady acceleration of 8.5 m/s^2 in the direction opposite your motion and comes to a stop. What is the total displacement of the car before it stops?

1 ANALYZE AND SKETCH THE PROBLEM

- Sketch the situation.
- Choose a coordinate system with the motion of the car in the positive direction.
- Draw the motion diagram, and label ***v*** and ***a***.

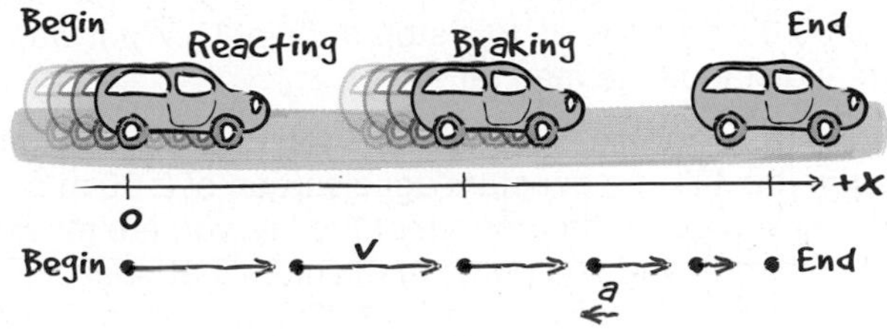

KNOWN	UNKNOWN
$\mathbf{v}_{reacting} = +25$ m/s	$\mathbf{x}_{reacting} = ?$
$t_{reacting} = 0.45$ s	$\mathbf{x}_{braking} = ?$
$\mathbf{a}_{braking} = -8.5$ m/s^2	$\mathbf{x}_{total} = ?$
$\mathbf{v}_{i,\ braking} = +25$ m/s	
$\mathbf{v}_{f,\ braking} = 0.00$ m/s	

2 SOLVE FOR THE UNKNOWN

Reacting:

Use the relationship among displacement, velocity, and time interval to find the displacement of the car as it travels at a constant speed.

$$\mathbf{x}_{reacting} = \mathbf{v}_{reacting}\, t_{reacting}$$

$$\mathbf{x}_{reacting} = (+25\text{ m/s})(0.45\text{ s}) = +11\text{ m}$$

◀ Substitute $v_{reacting} = +25$ m/s, $t_{reacting} = 0.45$ s.

Braking:

Use the relationship among velocity, acceleration, and displacement to find the displacement of the car while it is braking.

$$v_{f,\ braking}{}^2 = v_{reacting}{}^2 + 2a_{braking}(x_{braking})$$

Solve for $x_{braking}$.

$$x_{braking} = \frac{v_{f,\ braking}{}^2 - v_{reacting}{}^2}{2a_{braking}}$$

$$= \frac{(0.00\text{ m/s})^2 - (+25\text{ m/s})^2}{2(-8.5\text{ m/s}^2)}$$

◀ Substitute $v_{f,\ braking} = 0.00$ m/s, $v_{reacting} = +25$ m/s, $a_{braking} = -8.5$ m/s^2.

$$= +37\text{ m}$$

The total displacement is the sum of the reaction displacement and the braking displacement.

Solve for $\mathbf{x}_{total}$.

$$\mathbf{x}_{total} = \mathbf{x}_{reacting} + \mathbf{x}_{braking}$$

$$= +11\text{ m} + 37\text{ m}$$

◀ Substitute $x_{reacting} = +11$ m, $x_{braking} = +37$ m.

$$= +48\text{ m}$$

3 EVALUATE THE ANSWER

- **Are the units correct?** Displacement is measured in meters.
- **Do the signs make sense?** Both $\mathbf{x}_{reacting}$ and $\mathbf{x}_{braking}$ are positive, as they should be.
- **Is the magnitude realistic?** The braking displacement is small because the magnitude of the acceleration is large.

PRACTICE PROBLEMS

Do additional problems. Online Practice

28. A car with an initial velocity of 24.5 m/s east has an acceleration of 4.2 m/s^2 west. What is its displacement at the moment that its velocity is 18.3 m/s east?

29. A man runs along the path shown in **Figure 17.** From point A to point B, he runs at a forward velocity of 4.5 m/s for 15.0 min. From point B to point C, he runs up a hill. He slows down at a constant rate of 0.050 m/s^2 for 90.0 s and comes to a stop at point C. What was the total distance the man ran?

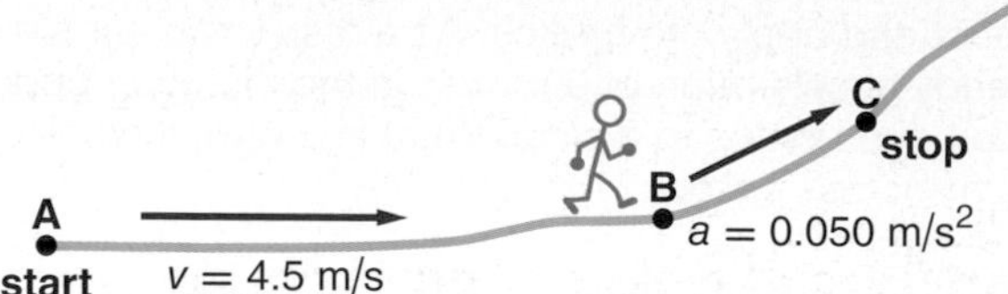

Figure 17

30. You start your bicycle ride at the top of a hill. You coast down the hill at a constant acceleration of 2.00 m/s^2. When you get to the bottom of the hill, you are moving at 18.0 m/s, and you pedal to maintain that speed. If you continue at this speed for 1.00 min, how far will you have gone from the time you left the hilltop?

31. Sunee is training for a 5.0-km race. She starts out her training run by moving at a constant pace of 4.3 m/s for 19 min. Then she accelerates at a constant rate until she crosses the finish line 19.4 s later. What is her acceleration during the last portion of the training run?

32. **CHALLENGE** Sekazi is learning to ride a bike without training wheels. His father pushes him with a constant acceleration of 0.50 m/s^2 east for 6.0 s. Sekazi then travels at 3.0 m/s east for another 6.0 s before falling. What is Sekazi's displacement? Solve this problem by constructing a velocity-time graph for Sekazi's motion and computing the area underneath the graphed line.

SECTION 2 REVIEW

Section Self-Check Check your understanding.

33. **MAIN IDEA** If you were given initial and final velocities and the constant acceleration of an object, and you were asked to find the displacement, what mathematical relationship would you use?

34. **Acceleration** A woman driving west along a straight road at a speed of 23 m/s sees a deer on the road ahead. She applies the brakes when she is 210 m from the deer. If the deer does not move and the car stops right before it hits the deer, what is the acceleration provided by the car's brakes?

35. **Distance** The airplane in **Figure 18** starts from rest and accelerates east at a constant 3.00 m/s^2 for 30.0 s before leaving the ground.

    a. What was the plane's displacement ($\Delta \boldsymbol{x}$)?

    b. How fast was the airplane going when it took off?

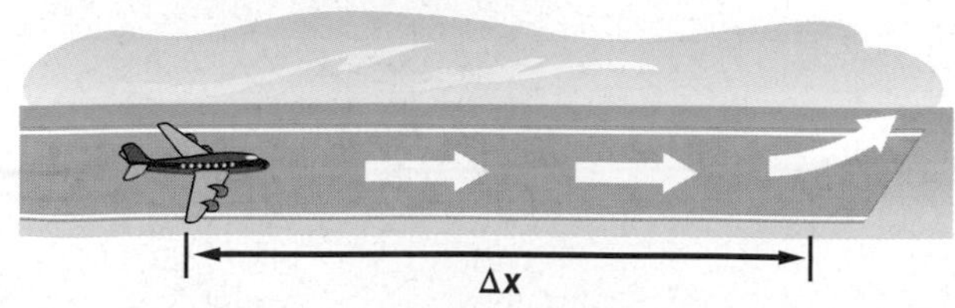

Figure 18

36. **Distance** An in-line skater first accelerates from 0.0 m/s to 5.0 m/s in 4.5 s, then continues at this constant speed for another 4.5 s. What is the total distance traveled by the in-line skater?

37. **Final Velocity** A plane travels a distance of 5.0×10^2 m north while being accelerated uniformly from rest at the rate of 5.0 m/s^2. What final velocity does it attain?

38. **Final Velocity** An airplane accelerated uniformly from rest at the rate of 5.0 m/s^2 south for 14 s. What final velocity did it attain?

39. **Graphs** A sprinter walks up to the starting blocks at a constant speed and positions herself for the start of the race. She waits until she hears the starting pistol go off and then accelerates rapidly until she attains a constant velocity. She maintains this velocity until she crosses the finish line, and then she slows to a walk, taking more time to slow down than she did to speed up at the beginning of the race. Sketch a velocity-time and a position-time graph to represent her motion. Draw them one above the other using the same time scale. Indicate on your position-time graph where the starting blocks and finish line are.

40. **Critical Thinking** Describe how you could calculate the acceleration of an automobile. Specify the measuring instruments and the procedures you would use.

SECTION 3 Free Fall

PHYSICS 4 YOU

Before their parachutes open, skydivers sometimes join hands to form a ring as they fall toward Earth. What happens if the skydivers have different masses? Do they fall at the same rate or different rates?

MAIN IDEA

The acceleration of an object in free fall is due to gravity alone.

Essential Questions

- What is free-fall acceleration?
- How do objects in free fall move?

Review Vocabulary

origin the point at which both variables in a coordinate system have the value zero

New Vocabulary

free fall
free-fall acceleration

Galileo's Discovery

Which falls with more acceleration, a piece of paper or your physics book? If you hold one in each hand and release them, the book hits the ground first. Do heavier objects accelerate more as they fall? Try dropping them again, but first place the paper flat on the book. Without air pushing against it, the paper falls as fast as the book. For a lightweight object such as paper, collisions with particles of air have a greater effect than they do on a heavy book.

To understand falling objects, first consider the case in which air does not have an appreciable effect on motion. Recall that gravity is an attraction between objects. **Free fall** is the motion of an object when gravity is the only significant force acting on it.

About 400 years ago, Galileo Galilei discovered that, neglecting the effect of the air, all objects in free fall have the same acceleration. It doesn't matter what they are made of or how much they weigh. The acceleration of an object due only to the effect of gravity is known as **free-fall acceleration.** **Figure 19** depicts the results of a 1971 free-fall experiment on the Moon in which astronauts verified Galileo's results.

Near Earth's surface, free-fall acceleration is about 9.8 m/s^2 downward (which is equal to about 22 mph/s downward). Think about the skydivers above. Each second the skydivers fall, their downward velocity increases by 9.8 m/s. When analyzing free fall, whether you treat the acceleration as positive or negative depends on the coordinate system you use. If you define upward as the positive direction, then the free-fall acceleration is negative. If you decide that downward is the positive direction, then free-fall acceleration is positive.

Figure 19 In 1971 astronaut David Scott dropped a hammer and a feather at the same time from the same height above the Moon's surface. The hammer's mass was greater, but both objects hit the ground at the same time because the Moon has gravity but no air.

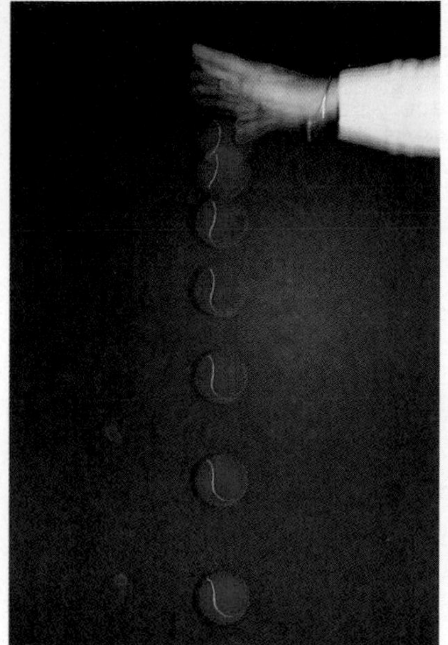

Figure 20 Because of free-fall acceleration, the speed of this falling ball increases 9.8 m/s each second.

MiniLAB

FREE FALL

How can you use the motion of a falling object to estimate free-fall acceleration?

iLab Station

Free-Fall Acceleration

Galileo's discovery explains why parachutists can form a ring in midair. Regardless of their masses, they fall with the same acceleration. To understand the acceleration that occurs during free fall, look at the multiflash photo of a dropped ball in **Figure 20.** The time interval between the images is 0.06 s. The distance between each pair of images increases, so the speed is increasing. If the upward direction is positive, then the velocity is becoming more and more negative.

Ball thrown upward Instead of a dropped ball, could this photo also illustrate a ball thrown upward? Suppose you throw a ball upward with a speed of 20.0 m/s. If you choose upward to be positive, then the ball starts at the bottom of the photo with a positive velocity. The acceleration is $\boldsymbol{a} = -9.8 \text{ m/s}^2$. Because velocity and acceleration are in opposite directions, the speed of the ball decreases. If you think of the bottom of the photo as the start, this agrees with the multiflash photo.

Rising and falling motion After 1 s, the ball's velocity is reduced by 9.8 m/s, so it now is traveling at +10.2 m/s. After 2 s, the velocity is +0.4 m/s, and the ball still is moving upward. What happens during the next second? The ball's velocity is reduced by another 9.8 m/s and equals −9.4 m/s. The ball now is moving downward. After 4 s, the velocity is −19.2 m/s, meaning the ball is falling even faster.

Velocity-time graph The v-t graph for the ball as it goes up and down is shown in **Figure 21.** The straight line sloping downward does not mean that the speed is always decreasing. The speed decreases as the ball rises and increases as it falls. At around 2 s, the velocity changes smoothly from positive to negative. As the ball falls, its speed increases in the negative direction. The figure also shows a closer view of the v-t graph. At an instant of time, near 2.04 s, the velocity is zero.

Figure 21 The velocity-time graph describes the change in the ball's speed as it rises and falls. The graph on the right gives a close-up view of the change in velocity at the top of the ball's trajectory.

Analyze What would the graph look like if downward were chosen as the positive direction?

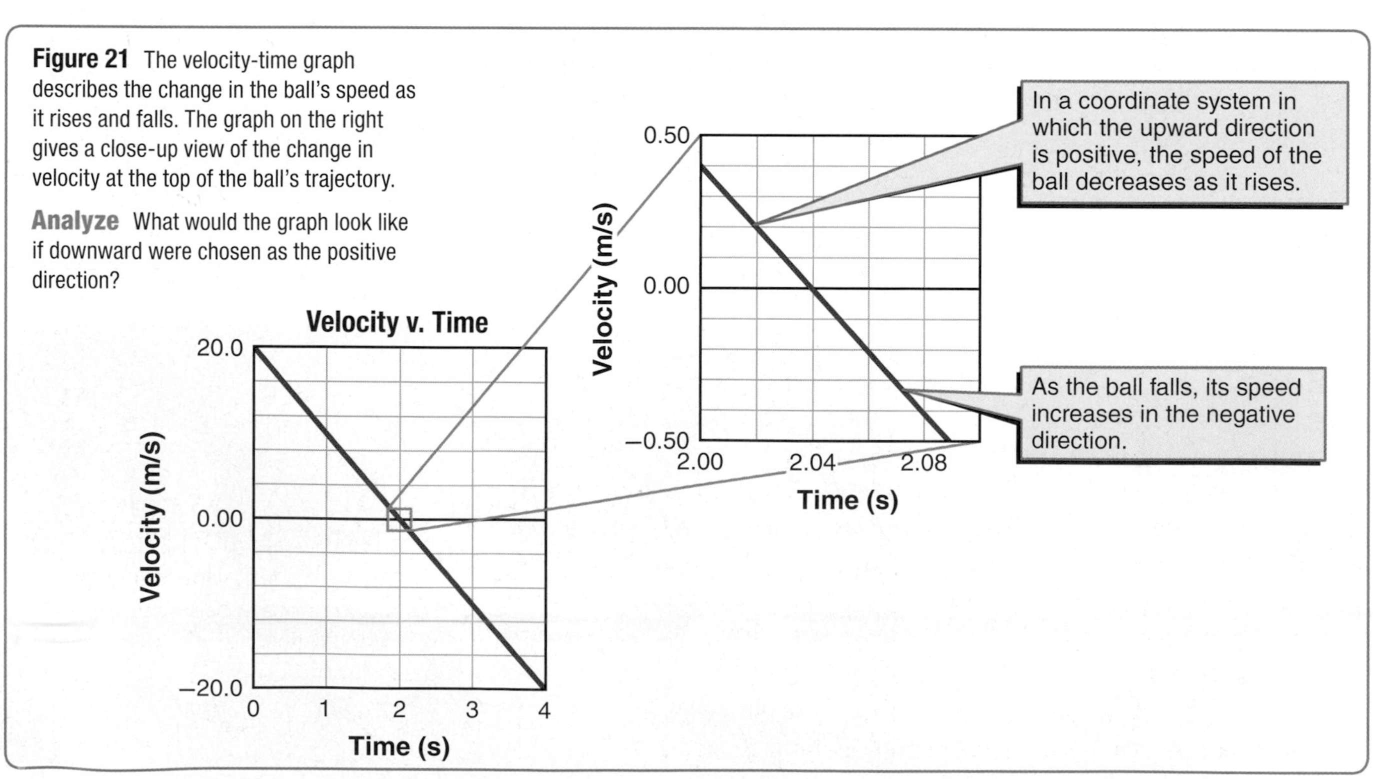

Velocity v. Time
20.0
0.00
−20.0
Velocity (m/s)
0
1
2
3
4
Time (s)
0.50
0.00
−0.50
Velocity (m/s)
2.00
2.04
2.08
Time (s)
In a coordinate system in which the upward direction is positive, the speed of the ball decreases as it rises.
As the ball falls, its speed increases in the negative direction.

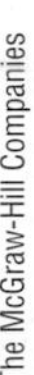

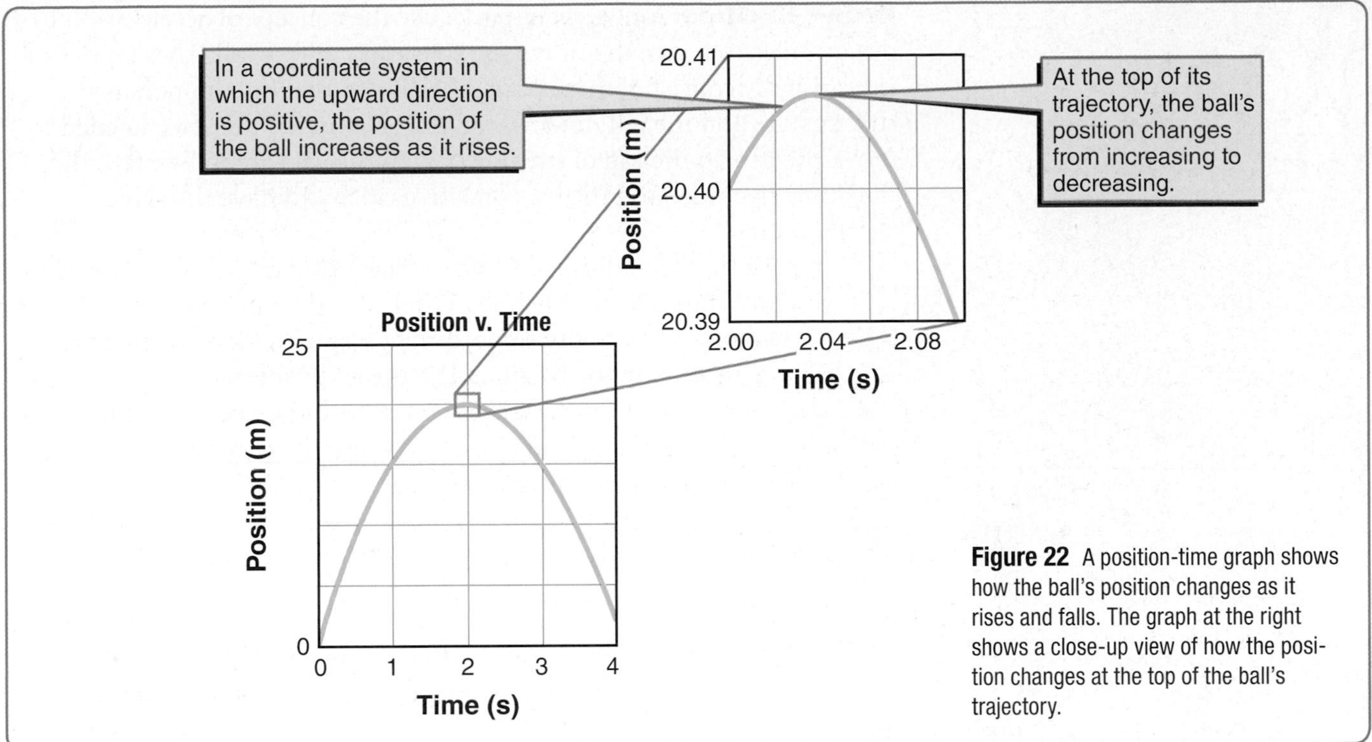
In a coordinate system in which the upward direction is positive, the position of the ball increases as it rises.
At the top of its trajectory, the ball's position changes from increasing to decreasing.
Position v. Time
Position (m)
Time (s)
25
0
0
1
2
3
4
20.41
20.40
20.39
2.00
2.04
2.08

Figure 22 A position-time graph shows how the ball's position changes as it rises and falls. The graph at the right shows a close-up view of how the position changes at the top of the ball's trajectory.

Position-time graph Look at the position-time graphs in **Figure 22.** These graphs show how the ball's height changes as it rises and falls. If an object is moving with constant acceleration, its position-time graph forms a parabola. Because the ball is rising and falling, its graph is an inverted parabola. The shape of the graph shows the progression of time. It does not mean that the ball's path was in the shape of a parabola. The close-up graph on the right shows that at about 2.04 s, the ball reaches its maximum height.

Maximum height Compare the close-up graphs in **Figure 21** and **Figure 22.** Just before the ball reaches its maximum height, its velocity is decreasing in the negative direction. At the instant of time when its height is maximum, its velocity is zero. Just after it reaches its maximum height, the ball's velocity is increasing in the negative direction.

Acceleration The slope of the line on the velocity-time graph in **Figure 21** is constant at -9.8 m/s^2. This shows that the ball's free-fall acceleration is 9.8 m/s^2 in the downward direction the entire time the ball is rising and falling.

It may seem that the acceleration should be zero at the top of the trajectory, but this is not the case. At the top of the flight, the ball's velocity is 0 m/s. If its acceleration were also zero, the ball's velocity would not change and would remain at 0 m/s. The ball would not gain any downward velocity and would simply hover in the air. Have you ever seen that happen? Objects tossed in the air on Earth always fall, so you know the acceleration of an object at the top of its flight must not be zero. Further, because the object falls down, you know the acceleration must be downward.

READING CHECK Analyze If you throw a ball straight up, what are its velocity and acceleration at the uppermost point of its path?

VOCABULARY

Science Usage v. Common Usage

Free fall

• Science usage

motion of a body when air resistance is negligible and the acceleration can be considered due to gravity alone

Acceleration during free fall is 9.8 m/s² downward.

• Common usage

a rapid and continuing drop or decline

The stock market's free fall in 1929 marked the beginning of the Great Depression.

PhysicsLAB

FREE-FALL ACCELERATION

INTERNET LAB How can you use motion data to calculate free-fall acceleration?

iLab Station

Free-fall rides Amusement parks use the concept of acceleration to design rides that give the riders the sensation of free fall. These types of rides usually consist of three parts: the ride to the top, momentary suspension, and the fall downward. Motors provide the force needed to move the cars to the top of the ride. When the cars are in free fall, the most massive rider and the least massive rider will have the same acceleration.

Suppose the free-fall ride shown in **Figure 23** starts from the top at rest and is in free fall for 1.5 s. What would be its velocity at the end of 1.5 s? Choose a coordinate system with a positive axis upward and the origin at the initial position of the car.

Because the car starts at rest, $\boldsymbol{v}_i$ would be equal to 0.0 m/s. To calculate the final velocity, use the equation for velocity with constant acceleration.

$$\begin{aligned}\boldsymbol{v}_f &= \boldsymbol{v}_i + \bar{\boldsymbol{a}}t_f \\ &= 0.0\ \text{m/s} + (-9.8\ \text{m/s}^2)(1.5\ \text{s}) \\ &= -15\ \text{m/s}\end{aligned}$$

How far do people on the ride fall during this time? Use the equation for displacement when time and constant acceleration are known.

$$\begin{aligned}\boldsymbol{x}_f &= \boldsymbol{x}_i + \boldsymbol{v}_i t_f + \left(\frac{1}{2}\right)\bar{\boldsymbol{a}}t_f^2 \\ &= 0.0\ \text{m} + (0.0\ \text{m/s})(1.5\ \text{s}) + \left(\frac{1}{2}\right)(-9.8\ \text{m/s}^2)(1.5\ \text{s})^2 \\ &= -11\ \text{m}\end{aligned}$$

Figure 23 The people on this amusement-park ride experience free-fall acceleration.

PRACTICE PROBLEMS

Do additional problems. 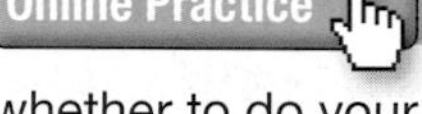

41. A construction worker accidentally drops a brick from a high scaffold.

a. What is the velocity of the brick after 4.0 s?

b. How far does the brick fall during this time?

42. Suppose for the previous problem you choose your coordinate system so that the opposite direction is positive.

a. What is the brick's velocity after 4.0 s?

b. How far does the brick fall during this time?

43. A student drops a ball from a window 3.5 m above the sidewalk. How fast is it moving when it hits the sidewalk?

44. A tennis ball is thrown straight up with an initial speed of 22.5 m/s. It is caught at the same distance above the ground.

a. How high does the ball rise?

b. How long does the ball remain in the air? *Hint: The time it takes the ball to rise equals the time it takes to fall.*

45. You decide to flip a coin to determine whether to do your physics or English homework first. The coin is flipped straight up.

a. What are the velocity and acceleration of the coin at the top of its trajectory?

b. If the coin reaches a high point of 0.25 m above where you released it, what was its initial speed?

c. If you catch it at the same height as you released it, how much time did it spend in the air?

46. CHALLENGE A basketball player is holding a ball in her hands at a height of 1.5 m above the ground. She drops the ball, and it bounces several times. After the first bounce, the ball only returns to a height of 0.75 m. After the second bounce, the ball only returns to a height of 0.25 m.

a. Suppose downward is the positive direction. What would the shape of a velocity-time graph look like for the first two bounces?

b. What would be the shape of a position-time graph for the first two bounces?

Herbert Kehrer/Bridge/CORBIS

Variations in Free Fall

When astronaut David Scott performed his free-fall experiment on the Moon, the hammer and the feather did not fall with an acceleration of magnitude 9.8 m/s^2. The value 9.8 m/s^2 is free-fall acceleration only near Earth's surface. The magnitude of free-fall acceleration on the Moon is approximately 1.6 m/s^2, which is about one-sixth its value on Earth.

When you study force and motion, you will learn about factors that affect the value of free-fall acceleration. One factor is the mass of the object, such as Earth or the Moon, that is responsible for the acceleration. Free-fall acceleration is not as great near the Moon as near Earth because the Moon has much less mass.

Free-fall acceleration also depends on the distance from the object responsible for it. The rings drawn around Earth in **Figure 24** show how free-fall acceleration decreases with distance from Earth. It is important to understand, however, that variations in free-fall acceleration at different locations on Earth's surface are very small, even with great variations in elevation. In New York City, for example, the magnitude of free-fall acceleration is about 9.81 m/s^2. In Denver, Colorado, it is about 9.79 m/s^2, despite a change in elevation of almost 1600 m greater. For calculations in this book, a value of 9.8 m/s^2 will be used for free-fall acceleration.

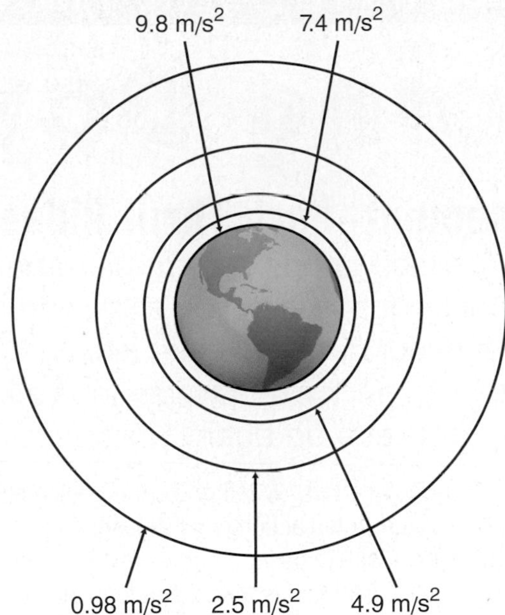

Figure 24 As the distance from Earth increases, the effect of free-fall acceleration decreases.

Analyze According to the diagram, what is the magnitude of free-fall acceleration a distance above Earth's surface equal to Earth's radius?

SECTION 3 REVIEW

Check your understanding.

47. **MAINIDEA** Suppose you hold a book in one hand and a flat sheet of paper in another hand. You drop them both, and they fall to the ground. Explain why the falling book is a good example of free fall, but the paper is not.

48. **Final Velocity** Your sister drops your house keys down to you from the second-floor window, as shown in **Figure 25.** What is the velocity of the keys when you catch them?

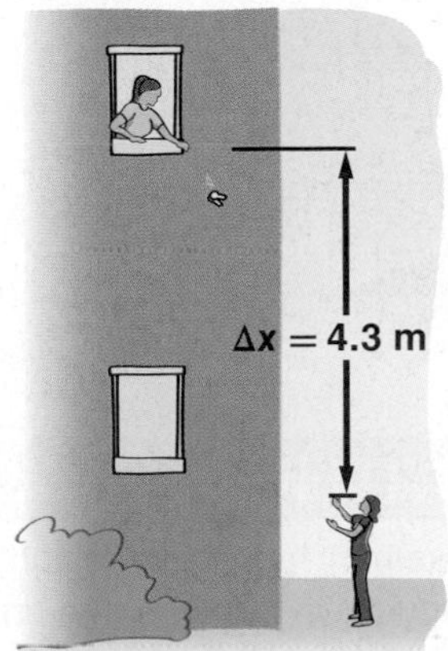

Figure 25

49. **Free-Fall Ride** Suppose a free-fall ride at an amusement park starts at rest and is in free fall. What is the velocity of the ride after 2.3 s? How far do people on the ride fall during the 2.3-s time period?

50. **Maximum Height and Flight Time** The free-fall acceleration on Mars is about one-third that on Earth. Suppose you throw a ball upward with the same velocity on Mars as on Earth.

    a. How would the ball's maximum height compare to that on Earth?

    b. How would its flight time compare?

51. **Velocity and Acceleration** Suppose you throw a ball straight up into the air. Describe the changes in the velocity of the ball. Describe the changes in the acceleration of the ball.

52. **Critical Thinking** A ball thrown vertically upward continues upward until it reaches a certain position, and then falls downward. The ball's velocity is instantaneously zero at that highest point. Is the ball accelerating at that point? Devise an experiment to prove or disprove your answer.

Going Down?

Amusement–Park Thrill Rides

Your stomach jumps as you plummet 100 m downward. Just when you think you might smash into the ground, the breaks kick in and the amusement-park ride brings you to a slow and safe stop. "Let's go again!" cries your friend.

It's all about acceleration. Whether it's the 100-m plunge of a drop tower or the gentle up-and-down action of a carousel ride, the thrills of many amusement-park rides are based on the same principle—changes in velocity are exciting. Thrill rides take advantage of accelerations due to both changes in speed and changes in direction. A roller coaster is exciting because of accelerations produced by hills, loops, and banked turns, as shown in **Figure 1.**

Rides that use free-fall acceleration Many rides use free-fall acceleration to generate thrills. Drop-tower rides let passengers experience free fall and then carefully slow their descent just before they reach the ground. Pendulum rides, such as the one in **Figure 2,** act like giant swings. Passengers experience a stomach-churning moment as the pendulum reaches the top of its swing and begins to move downward. Passengers on a roller coaster experience a similar moment at the top of a hill.

FIGURE 1 Passengers accelerate as they move around a banked turn of a roller coaster. Acceleration is also part of what provides the excitement in a series of loops.

FIGURE 2 Passengers sit in the pendulum of this pirate ship–themed ride, which swings back and forth like the pendulum on a grandfather clock.

DROP TO

GOINGFURTHER >>>

Research amusement-park ride design online. Design your own ride and present your design to the class. Identify points during the ride where the passengers accelerate.

CHAPTER 3

STUDY GUIDE

BIG IDEA Acceleration is the rate of change in an object's velocity.

VOCABULARY

- **acceleration** *(p. 61)*
- **velocity-time graph** *(p. 63)*
- **average acceleration** *(p. 64)*
- **instantaneous acceleration** *(p. 64)*

SECTION 1 Acceleration

MAIN IDEA An object accelerates when its velocity changes—that is, when it speeds up, slows down, or changes direction.

- Acceleration is the rate at which an object's velocity changes.
- Velocity and acceleration are not the same thing. An object moving with constant velocity has zero acceleration. When the velocity and the acceleration of an object are in the same direction, the object speeds up; when they are in opposite directions, the object slows down.
- You can use a velocity-time graph to find the velocity and the acceleration of an object. The average acceleration of an object is the slope of its velocity-time graph.

$$\bar{\mathbf{a}} \equiv \frac{\Delta \mathbf{v}}{\Delta t} = \frac{\mathbf{v}_f - \mathbf{v}_i}{t_f - t_i}$$

SECTION 2 Motion with Constant Acceleration

MAIN IDEA For an object with constant acceleration, the relationships among position, velocity, acceleration, and time can be described by graphs and equations.

- If an object is moving with constant acceleration, its position-time graph is a parabola, and its velocity-time graph is a straight line.
- The area under an object's velocity-time graph is its displacement.
- In motion with constant acceleration, position, velocity, acceleration, and time are related:

$$\mathbf{v}_f = \mathbf{v}_i + \bar{\mathbf{a}}\Delta t$$

$$\mathbf{x}_f = \mathbf{x}_i + \mathbf{v}_i t_f + \frac{1}{2}\bar{\mathbf{a}} t_f^2$$

$$v_f^2 = v_i^2 + 2\bar{a}(x_f - x_i)$$

VOCABULARY

- **free fall** *(p. 75)*
- **free-fall acceleration** *(p. 75)*

SECTION 3 Free Fall

MAIN IDEA The acceleration of an object in free fall is due to gravity alone.

- Free-fall acceleration on Earth is about 9.8 m/s^2 downward. The sign associated with free-fall acceleration in equations depends on the choice of the coordinate system.
- When an object is in free fall, gravity is the only force acting on it. Equations for motion with constant acceleration can be used to solve problems involving objects in free fall.

Games and Multilingual eGlossary

Vocabulary Practice

CHAPTER 3 ASSESSMENT

Chapter Self-Check

SECTION 1 Acceleration

Mastering Concepts

53. BIGIDEA How are velocity and acceleration related?

54. Give an example of each of the following:

a. an object that is slowing down but has a positive acceleration

b. an object that is speeding up but has a negative acceleration

c. an object that is moving at a constant speed but has an acceleration

55. Figure 26 shows the velocity-time graph for an automobile on a test track. Describe how the velocity changes with time.

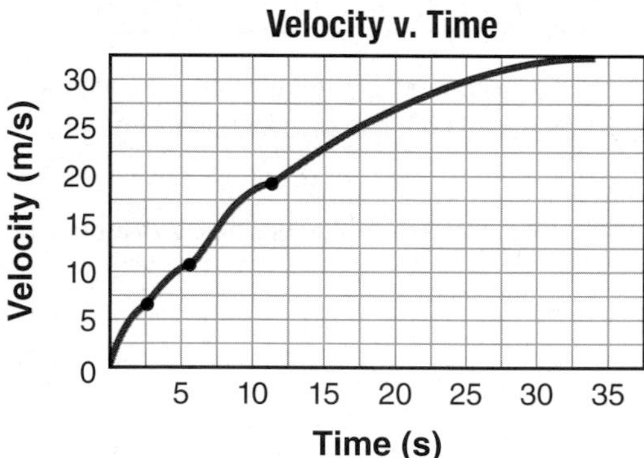

Figure 26

56. If the velocity-time graph of an object moving on a straight path is a line parallel to the horizontal axis, what can you conclude about its acceleration?

Mastering Problems

57. Ranking Task Rank the following objects according to the magnitude of the acceleration, from least to greatest. Specifically indicate any ties.

A. A falling acorn accelerates from 0.50 m/s to 10.3 m/s in 1.0 s.

B. A car accelerates from 20 m/s to rest in 1.0 s.

C. A centipede accelerates from 0.40 cm/s to 2.0 cm/s in 0.50 s.

D. While being hit, a golf ball accelerates from rest to 4.3 m/s in 0.40 s.

E. A jogger accelerates from 2.0 m/s to 1.0 m/s in 8.3 s.

58. Problem Posing Complete this problem so that it can be solved using the concept listed: "Angela is playing basketball . . ."

a. acceleration

b. speed

59. The graph in **Figure 27** describes the motion of an object moving east along a straight path. Find the acceleration of the object at each of these times:

a. during the first 5.0 min of travel

b. between 5.0 min and 10.0 min

c. between 10.0 min and 15.0 min

d. between 20.0 min and 25.0 min

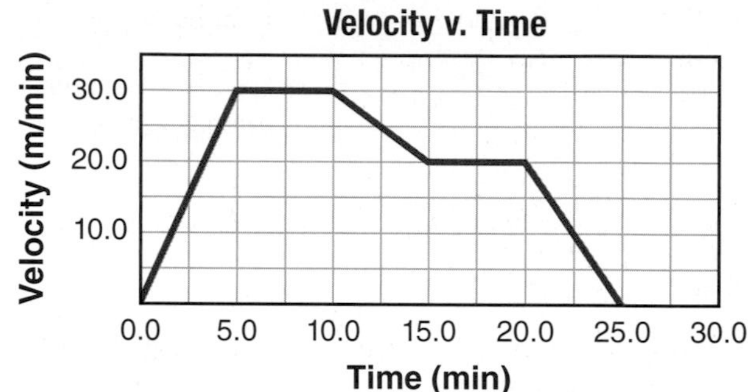

Figure 27

60. Plot a velocity-time graph using the information in **Table 1,** and answer the following questions:

a. During what time interval is the object speeding up? Slowing down?

b. At what time does the object reverse direction?

c. How does the average acceleration of the object between 0.0 s and 2.0 s differ from the average acceleration between 7.0 s and 12.0 s?

Table 1 Velocity v. Time

Time (s)	Velocity (m/s)
0.00	4.00
1.00	8.00
2.00	12.0
3.00	14.0
4.00	16.0
5.00	16.0
6.00	14.0
7.00	12.0
8.00	8.00
9.00	4.00
10.0	0.00
11.0	−4.00
12.0	−8.00

61. Determine the final velocity of a proton that has an initial forward velocity of 2.35×10^5 m/s and then is accelerated uniformly in an electric field at the rate of -1.10×10^{12} m/s^2 for 1.50×10^{-7} s.

62. Ranking Task Marco wants to buy the used sports car with the greatest acceleration. Car A can go from 0 m/s to 17.9 m/s in 4.0 s. Car B can accelerate from 0 m/s to 22.4 m/s in 3.5 s. Car C can go from 0 to 26.8 m/s in 6.0 s. Rank the three cars from greatest acceleration to least. Indicate if any are the same.

SECTION 2
Motion with Constant Acceleration

Mastering Concepts

63. What quantity does the area under a velocity-time graph represent?

64. Reverse Problem Write a physics problem with real-life objects for which the graph in **Figure 28** would be part of the solution.

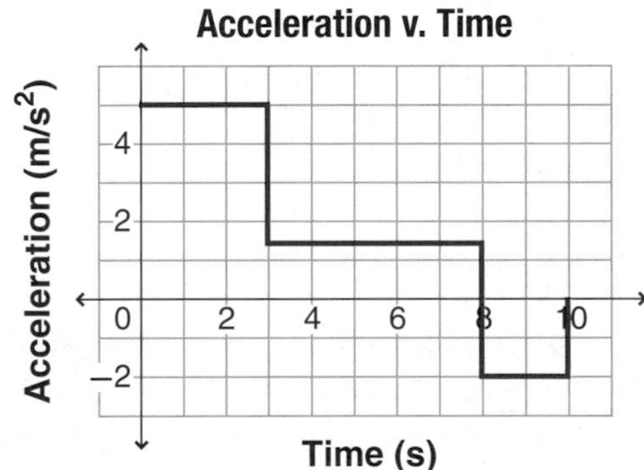

Figure 28

Mastering Problems

65. A car moves forward up a hill at 12 m/s with a uniform backward acceleration of 1.6 m/s^2.

a. What is its displacement after 6.0 s?

b. What is its displacement after 9.0 s?

66. Airplane Determine the displacement of a plane that experiences uniform acceleration from 66 m/s north to 88 m/s north in 12 s.

67. Race Car A race car is slowed with a constant acceleration of 11 m/s^2 opposite the direction of motion.

a. If the car is going 55 m/s, how many meters will it travel before it stops?

b. How many meters will it take to stop a car going twice as fast?

68. Refer to **Figure 29** to find the magnitude of the displacement during the following time intervals. Round answers to the nearest meter.

a. $t = 5.0$ min and $t = 10.0$ min

b. $t = 10.0$ min and $t = 15.0$ min

c. $t = 25.0$ min and $t = 30.0$ min

d. $t = 0.0$ min and $t = 25.0$ min

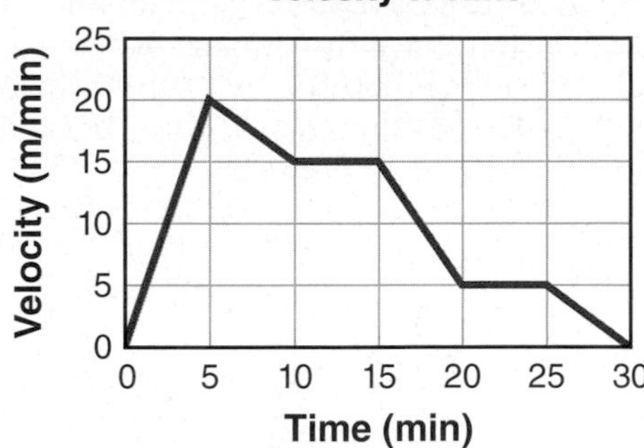

Figure 29

SECTION 3 **Free Fall**

Mastering Concepts

69. Explain why an aluminum ball and a steel ball of similar size and shape, dropped from the same height, reach the ground at the same time.

70. Give some examples of falling objects for which air resistance can and cannot be ignored.

Mastering Problems

71. Suppose an astronaut drops a feather from a height of 1.2 m above the surface of the Moon. If the free-fall acceleration on the Moon is 1.62 m/s^2 downward, how long does it take the feather to hit the Moon's surface?

72. A stone that starts at rest is in free fall for 8.0 s.

a. Calculate the stone's velocity after 8.0 s.

b. What is the stone's displacement during this time?

73. A bag is dropped from a hovering helicopter. The bag has fallen for 2.0 s. What is the bag's velocity? How far has the bag fallen? Ignore air resistance.

74. You throw a ball downward from a window at a speed of 2.0 m/s. How fast will it be moving when it hits the sidewalk 2.5 m below?

75. If you throw the ball in the previous problem up instead of down, how fast will it be moving when it hits the sidewalk?

Chapter Self-Check

76. Beanbag You throw a beanbag in the air and catch it 2.2 s later at the same place at which you threw it.

a. How high did it go?

b. What was its initial velocity?

Applying Concepts

77. Croquet A croquet ball, after being hit by a mallet, slows down and stops. Do the velocity and the acceleration of the ball have the same signs?

78. Explain how you would walk to produce each of the position-time graphs in **Figure 30.**

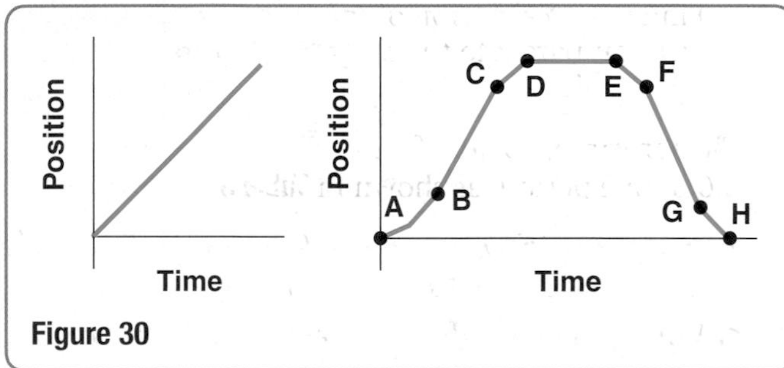

Figure 30

79. If you were given a table of velocities of an object at various times, how would you determine whether the acceleration was constant?

80. Look back at the graph in **Figure 26.** The three notches in the graph occur where the driver changed gears. Describe the changes in velocity and acceleration of the car while in first gear. Is the acceleration just before a gear change larger or smaller than the acceleration just after the change? Explain your answer.

81. An object shot straight up rises for 7.0 s before it reaches its maximum height. A second object falling from rest takes 7.0 s to reach the ground. Compare the displacements of the two objects during this time interval.

82. Draw a velocity-time graph for each of the graphs in **Figure 31.**

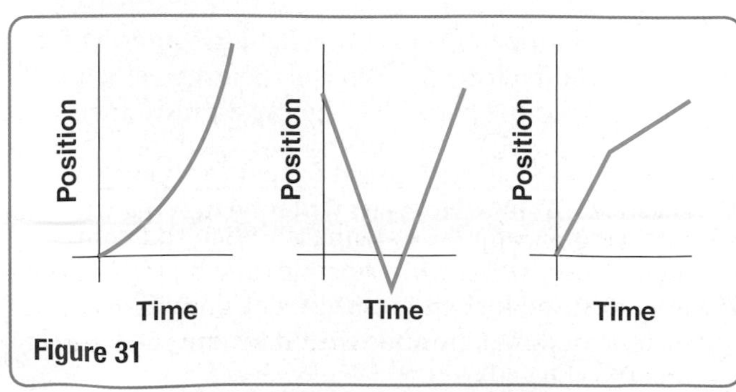

Figure 31

83. The Moon The value of free-fall acceleration on the Moon is about one-sixth of its value on Earth.

a. Would a ball dropped by an astronaut hit the surface of the Moon with a greater, equal, or lesser speed than that of a ball dropped from the same height to Earth?

b. Would it take the ball more, less, or equal time to fall?

84. Jupiter An object on the planet Jupiter has about three times the free-fall acceleration as on Earth. Suppose a ball could be thrown vertically upward with the same initial velocity on Earth and on Jupiter. Neglect the effects of Jupiter's atmospheric resistance, and assume that gravity is the only force on the ball.

a. How would the maximum height reached by the ball on Jupiter compare to the maximum height reached on Earth?

b. If the ball on Jupiter were thrown with an initial velocity that is three times greater, how would this affect your answer to part **a?**

85. Rock A is dropped from a cliff, and rock B is thrown upward from the same position.

a. When they reach the ground at the bottom of the cliff, which rock has a greater velocity?

b. Which has a greater acceleration?

c. Which arrives first?

Mixed Review

86. Suppose a spaceship far from any star or planet had a uniform forward acceleration from 65.0 m/s to 162.0 m/s in 10.0 s. How far would the spaceship move?

87. Figure 32 is a multiflash photo of a horizontally moving ball. What information about the photo would you need and what measurements would you make to estimate the acceleration?

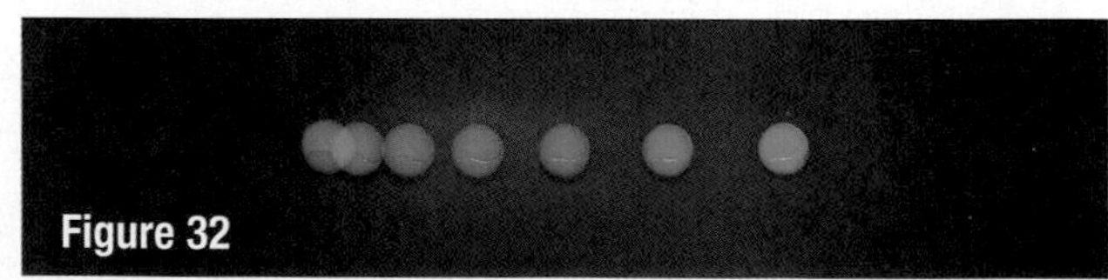

Figure 32

88. Bicycle A bicycle accelerates from 0.0 m/s to 4.0 m/s in 4.0 s. What distance does it travel?

89. A weather balloon is floating at a constant height above Earth when it releases a pack of instruments.

a. If the pack hits the ground with a downward velocity of 73.5 m/s, how far did the pack fall?

b. How long did it take for the pack to fall?

90. The total distance a steel ball rolls down an incline at various times is given in **Table 2.**

a. Draw a position-time graph of the motion of the ball. When setting up the axes, use five divisions for each 10 m of travel on the *x*-axis. Use five divisions for 1 s of time on the *t*-axis.

b. Calculate the distance the ball has rolled at the end of 2.2 s.

Table 2 Position v. Time

Time (s)	Position (m)
0.0	0.0
1.0	2.0
2.0	8.0
3.0	18.0
4.0	32.0
5.0	50.0

91. Engineers are developing new types of guns that might someday be used to launch satellites as if they were bullets. One such gun can give a small object a forward velocity of 3.5 km/s while moving it through a distance of only 2.0 cm.

a. What acceleration does the gun give this object?

b. Over what time interval does the acceleration take place?

92. Safety Barriers Highway safety engineers build soft barriers, such as the one shown in **Figure 33,** so that cars hitting them will slow down at a safe rate. Suppose a car traveling at 110 km/h hits the barrier, and the barrier decreases the car's velocity at a rate of 32 m/s^2. What distance would the car travel along the barrier before coming to a stop?

Figure 33

CGIBackgrounds.com/Collection Mix: Subjects/Getty Images

93. Baseball A baseball pitcher throws a fastball at a speed of 44 m/s. The ball has constant acceleration as the pitcher holds it in his hand and moves it through an almost straight-line distance of 3.5 m. Calculate the acceleration. Compare this acceleration to the free-fall acceleration on Earth.

94. Sleds Rocket-powered sleds are used to test the responses of humans to acceleration. Starting from rest, one sled can reach a speed of 444 m/s in 1.80 s and can be brought to a stop again in 2.15 s.

a. Calculate the acceleration of the sled when starting, and compare it to the magnitude of free-fall acceleration, 9.8 m/s^2.

b. Find the acceleration of the sled as it is braking, and compare it to the magnitude of free-fall acceleration.

95. The forward velocity of a car changes over an 8.0-s time period, as shown in **Table 3.**

a. Plot the velocity-time graph of the motion.

b. What is the car's displacement in the first 2.0 s?

c. What is the car's displacement in the first 4.0 s?

d. What is the displacement of the car during the entire 8.0 s?

e. Find the slope of the line between $t = 0.0$ s and $t = 4.0$ s. What does this slope represent?

f. Find the slope of the line between $t = 5.0$ s and $t = 7.0$ s. What does this slope indicate?

Table 3 Velocity v. Time

Time (s)	Velocity (m/s)
0.0	0.0
1.0	4.0
2.0	8.0
3.0	12.0
4.0	16.0
5.0	20.0
6.0	20.0
7.0	20.0
8.0	20.0

96. A truck is stopped at a stoplight. When the light turns green, the truck accelerates at 2.5 m/s^2. At the same instant, a car passes the truck going at a constant 15 m/s. Where and when does the truck catch up with the car?

Chapter Self-Check

97. Karate The position-time and velocity-time graphs of George's fist breaking a wooden board during karate practice are shown in **Figure 34.**

a. Use the velocity-time graph to describe the motion of George's fist during the first 10 ms.

b. Estimate the slope of the velocity-time graph to determine the acceleration of his fist when it suddenly stops.

c. Express the acceleration as a multiple of the magnitude of free-fall acceleration, 9.8 m/s^2.

d. Determine the area under the velocity-time curve to find the displacement of the fist in the first 6 ms. Compare this with the position-time graph.

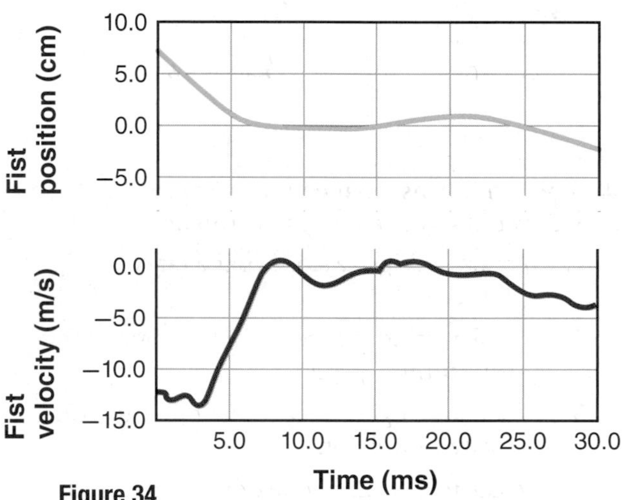

Figure 34

98. Cargo A helicopter is rising at 5.0 m/s when a bag of its cargo is dropped. The bag falls for 2.0 s.

a. What is the bag's velocity?

b. How far has the bag fallen?

c. How far below the helicopter is the bag?

Thinking Critically

99. Probeware Design a probeware lab to measure the distance an accelerated object moves over time. Use equal time intervals so that you can plot velocity over time as well as distance. A pulley at the edge of a table with a mass attached is a good way to achieve uniform acceleration. Suggested materials include a motion detector, lab cart, string, pulley, C-clamp, and masses. Generate position-time and velocity-time graphs using different masses on the pulley. How does the change in mass affect your graphs?

100. Analyze and Conclude Which (if either) has the greater acceleration: a car that increases its speed from 50 km/h to 60 km/h or a bike that goes from 0 km/h to 10 km/h in the same time? Explain.

101. Analyze and Conclude An express train traveling at 36.0 m/s is accidentally sidetracked onto a local train track. The express engineer spots a local train exactly 1.00×10^2 m ahead on the same track and traveling in the same direction. The local engineer is unaware of the situation. The express engineer jams on the brakes and slows the express train at a constant rate of 3.00 m/s^2. If the speed of the local train is 11.0 m/s, will the express train be able to stop in time, or will there be a collision? To solve this problem, take the position of the express train when the engineer first sights the local train as a point of origin. Next, keeping in mind that the local train has exactly a 1.00×10^2 m lead, calculate how far each train is from the origin at the end of the 12.0 s it would take the express train to stop (accelerate at −3.00 m/s^2 from 36 m/s to 0 m/s).

a. On the basis of your calculations, would you conclude that a collision will occur?

b. To check the calculations from part **a** and to verify your conclusion, take the position of the express train when the engineer first sights the local train as the point of origin and calculate the position of each train at the end of each second after the sighting. Make a table showing the distance of each train from the origin at the end of each second. Plot these positions on the same graph and draw two lines. Compare your graph to your answer to part **a**.

Writing in Physics

102. Research and describe Galileo's contributions to physics.

103. Research the maximum acceleration a human body can withstand without blacking out. Discuss how this impacts the design of three common entertainment or transportation devices.

Cumulative Review

104. Solve the following problems. Express your answers in scientific notation.

a. 6.2×10^{-4} m + 5.7×10^{-3} m

b. 8.7×10^{8} km − 3.4×10^{7} km

c. $(9.21\times10^{-5}$ cm$)(1.83\times10^{8}$ cm$)$

d. $\dfrac{2.63\times10^{-6}\text{ m}}{4.08\times10^{6}\text{ s}}$

105. The equation below describes an object's motion. Create the corresponding position-time graph and motion diagram. Then write a physics problem that could be solved using that equation. Be creative.

$$x = (35.0\text{ m/s})\,t - 5.0\text{ m}$$

CHAPTER 3 STANDARDIZED TEST PRACTICE

MULTIPLE CHOICE

Use the following information to answer the first two questions.

A ball rolls down a hill with a constant acceleration of 2.0 m/s^2. The ball starts at rest and travels for 4.0 s before it reaches the bottom of the hill.

1. How far did the ball travel during this time?

A. 8.0 m **C.** 16 m
B. 12 m **D.** 20 m

2. What was the ball's speed at the bottom of the hill?

A. 2.0 m/s **C.** 12 m/s
B. 8.0 m/s **D.** 16 m/s

3. A driver of a car enters a new 110-km/h speed zone on the highway. The driver begins to accelerate immediately and reaches 110 km/h after driving 500 m. If the original speed was 80 km/h, what was the driver's forward acceleration?

A. 0.44 m/s^2 **C.** 8.4 m/s^2
B. 0.60 m/s^2 **D.** 9.8 m/s^2

4. A flowerpot falls off a balcony 85 m above the street. How long does it take to hit the ground?

A. 4.2 s **C.** 8.7 s
B. 8.3 s **D.** 17 s

5. A rock climber's shoe loosens a rock, and her climbing buddy at the bottom of the cliff notices that the rock takes 3.20 s to fall to the ground. How high up the cliff is the rock climber?

A. 15.0 m **C.** 50.0 m
B. 31.0 m **D.** 1.00×10^2 m

6. A car traveling at 91.0 km/h approaches the turnoff for a restaurant 30.0 m aheadxd. If the driver slams on the brakes with an acceleration of -6.40 m/s^2, what will be her stopping distance?

A. 14.0 m **C.** 50.0 m
B. 29.0 m **D.** 100.0 m

7. What is the correct formula manipulation to find acceleration when using the equation $v_f^2 = v_i^2 + 2ax$?

A. $\frac{v_f^2 - v_i^2}{x}$ **C.** $\frac{(v_f + v_i)^2}{2x}$
B. $\frac{v_f^2 + v_i^2}{2x}$ **D.** $\frac{v_f^2 - v_i^2}{2x}$

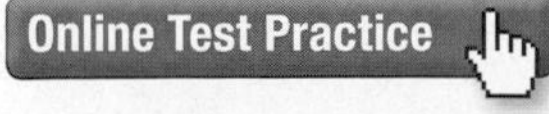

8. The graph below shows the motion of a farmer's truck. What is the truck's total displacement? Assume north is the positive direction.

A. 150 m south **C.** 300 m north
B. 125 m north **D.** 600 m south

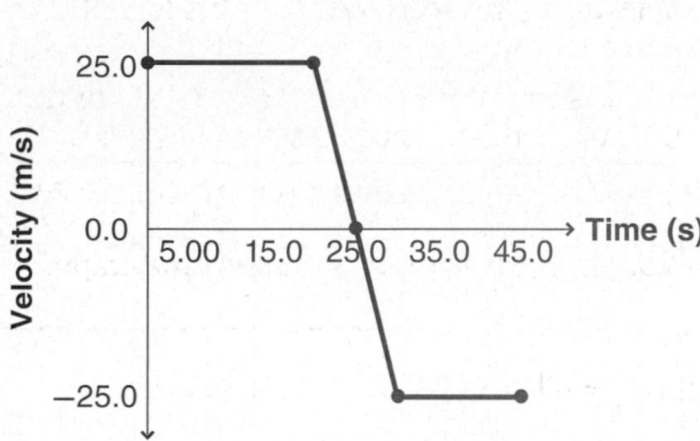

9. How can the instantaneous acceleration of an object with varying acceleration be calculated?

A. by calculating the slope of the tangent on a position-time graph
B. by calculating the area under the graph on a position-time graph
C. by calculating the area under the graph on a velocity-time graph
D. by calculating the slope of the tangent on a velocity-time graph

FREE RESPONSE

10. Graph the following data, and then show calculations for acceleration and displacement after 12.0 s on the graph.

Time (s)	Velocity (m/s)
0.00	8.10
6.00	36.9
9.00	51.3
12.00	65.7

NEED EXTRA HELP?

If you Missed Question	1	2	3	4	5	6	7	8	9	10
Review Section	2	2	2	3	3	2	2	2	1	2

CHAPTER 4

Forces in One Dimension

BIGIDEA **Net forces cause changes in motion.**

SECTIONS

1 **Force and Motion**

2 **Weight and Drag Force**

3 **Newton's Third Law**

LaunchLAB

iLab Station

FORCES IN OPPOSITE DIRECTIONS

What happens when more than one force acts on an object?

WATCH THIS!

Video

APPARENT WEIGHT

Have you ever had the sensation of feeling weightless while on a freefall ride? An elevator acts in much the same way, giving the sensation of being lighter when you're accelerating down and heavier when you're moving up.

(l)The McGraw-Hill Companies, (r)James Lauritz/Digital Vision/Getty Images

Go online!
connectED.mcgraw-hill.com

SECTION 1 Force and Motion

PHYSICS 4 YOU **To start a skateboard moving on a level surface, you must push on the ground with your foot. If the skateboard is on a ramp, however, gravity will pull you down the slope. In both cases unbalanced forces change the skateboard's motion.**

MAIN IDEA
A force is a push or a pull.

Essential Questions

- What is a force?
- What is the relationship between force and acceleration?
- How does motion change when the net force is zero?

Review Vocabulary

acceleration the rate at which the velocity of an object changes

New Vocabulary

force
system
free-body diagram
net force
Newton's second law
Newton's first law
inertia
equilibrium

Force

Consider a textbook resting on a table. To cause it to move, you could either push or pull on it. In physics, a push or a pull is called a **force.** If you push or pull harder on an object, you exert a greater force on the object. In other words, you increase the magnitude of the applied force. The direction in which the force is exerted also matters—if you push the resting book to the right, the book will start moving to the right. If you push the book to the left, it will start moving to the left. Because forces have both magnitude and direction, forces are vectors. The symbol $\boldsymbol{F}$ is vector notation that represents the size and direction of a force, while F represents only the magnitude. The magnitude of a force is measured in units called newtons (N).

Unbalanced forces change motion Recall that motion diagrams describe the positions of an object at equal time intervals. For example, the motion diagram for the book in **Figure 1** shows the distance between the dots increasing. This means the speed of the book is increasing. At $t = 0$, it is at rest, but after 2 seconds it is moving at 1.5 m/s. This change in speed means it is accelerating. What is the cause of this acceleration? The book was at rest until you pushed it, so the cause of the acceleration is the force exerted by your hand. In fact, all accelerations are the result of an unbalanced force acting on an object. What is the relationship between force and acceleration? By the end of this section, you will be able to answer that question as well as apply laws of motion to solve many different types of problems.

☑ READING CHECK **Identify** the cause of all accelerations.

Figure 1 The hand pushing on the book exerts a force that causes the book to accelerate in the direction of the unbalanced force.

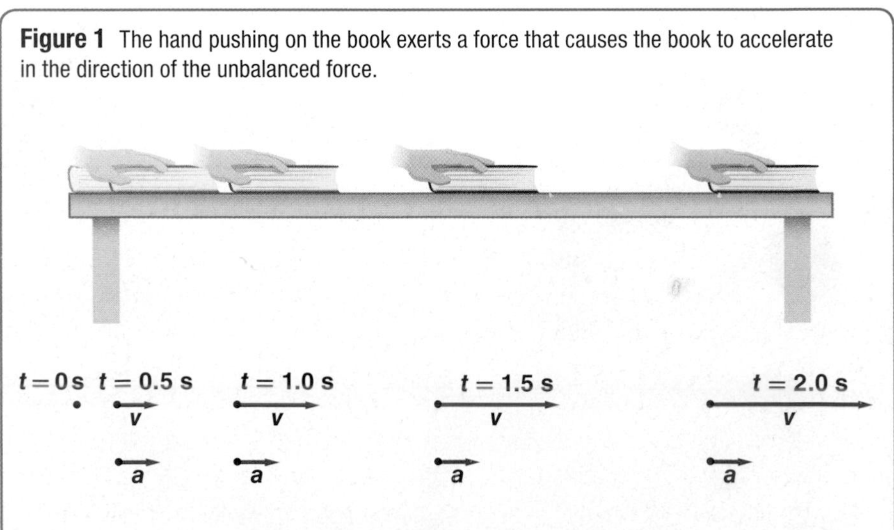

Royalty-Free/CORBIS

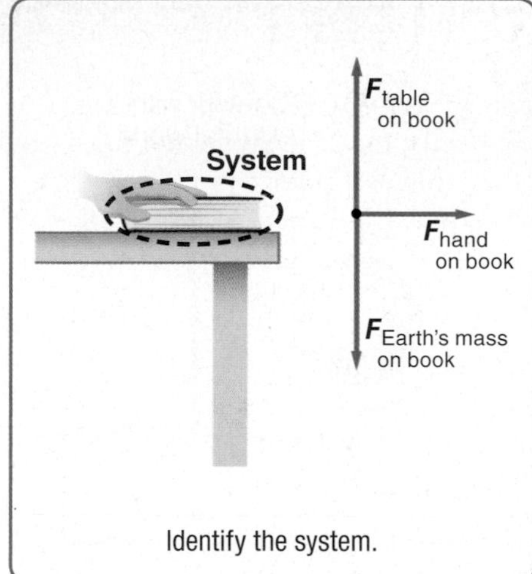

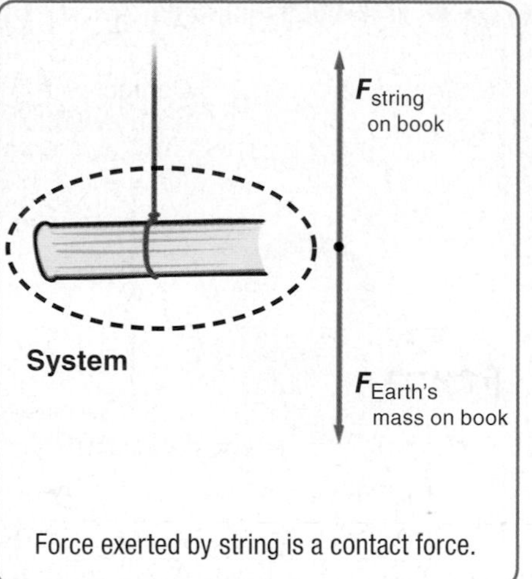

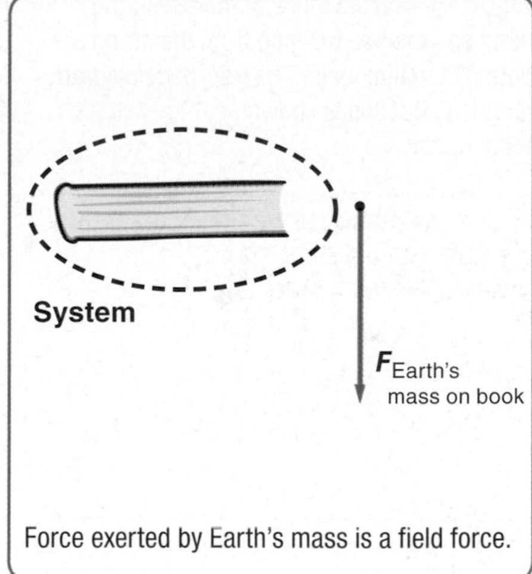

Figure 2 The book is the system in each of these situations.

Classify each force in the first panel as either a contact force or a field force.

Systems and external world When considering how a force affects motion, it is important to identify the object or objects of interest, called the **system.** Everything around the system with which the system can interact is called the external world. In **Figure 2,** the book is the system. Your hand, Earth, string and the table are parts of the external world that interact with the book by pushing or pulling on it.

View an **animation on systems.**

Contact forces Again, think about the different ways in which you could move a textbook. You could push or pull it by directly touching it, or you could tie a string around it and pull on the string. These are examples of contact forces. A contact force exists when an object from the external world touches a system, exerting a force on it. If you are holding this physics textbook right now, your hands are exerting a contact force on it. If you place the book on a table, you are no longer exerting a contact force on the book. The table, however, is exerting a contact force because the table and the book are in contact.

VOCABULARY
Science Use v. Common Use

Force
• Science usage
a push or a pull exerted on an object
The force of gravity exerted by the Sun on Earth pulls Earth into orbit around the Sun.
• Common usage
to compel by physical, moral, or intellectual means
Emi forced her younger brother to wash the dishes.

Field forces There are other ways in which the motion of the textbook can change. You could drop it, and as you learned in a previous chapter, it would accelerate as it falls to the ground. The gravitational force of Earth acting on the book causes this acceleration. This force affects the book whether or not Earth is actually touching it. Gravitational force is an example of a field force. Field forces are exerted without contact. Can you think of other kinds of field forces? If you have ever investigated magnets, you know that they exert forces without touching. You will investigate magnetism and other field forces in future chapters. For now, the only field force you need to consider is the gravitational force.

Agents Forces result from interactions; every contact and field force has a specific and identifiable cause, called the agent. You should be able to name the agent exerting each force as well as the system upon which the force is exerted. For example, when you push your textbook, your hand (the agent) exerts a force on the textbook (the system). If there are not both an agent and a system, a force does not exist. What about the gravitational force? The agent is the mass of Earth exerting a field force on the book. The labels on the forces in **Figure 2** are good examples of how to identify a force's agent and the system upon which the force acts.

Figure 3 The drawings of the ball in the hand and the ball hanging from the string are both pictorial models. The free-body diagram for each situation is shown next to each pictorial model.

View an **animation of free body diagrams.**

Concepts In Motion

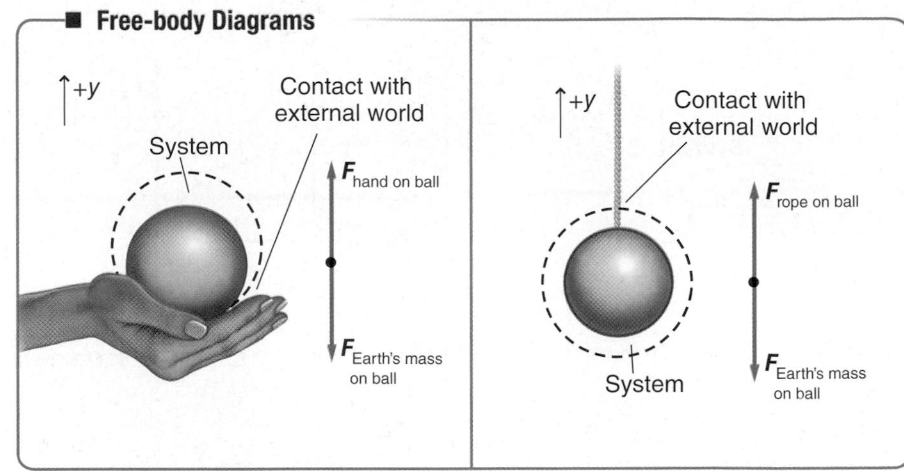

Free-body diagrams Just as pictorial representations and motion diagrams are useful in solving problems about motion, similar representations will help you analyze how forces affect motion. The first step is to sketch the situation, as shown in **Figure 3.** Circle the system, and identify every place where the system touches the external world. It is at these places that an agent exerts a contact force. Then identify any field forces on the system. This gives you the pictorial representation.

A **free-body diagram** is a physical representation that shows the forces acting on a system. Follow these guidelines when drawing a free-body diagram:

- The free-body diagram is drawn separately from the sketch of the problem situation.
- Apply the particle model, and represent the object with a dot.
- Represent each force with an arrow that points in the direction the force is applied. Always draw the force vectors pointing away from the particle, even when the force is a push.
- Make the length of each arrow proportional to the size of the force. Often you will draw these diagrams before you know the magnitudes of all the forces. In such cases, make your best estimate.
- Label each force. Use the symbol $\boldsymbol{F}$ with a subscript label to identify both the agent and the object on which the force is exerted.
- Choose a direction to be positive, and indicate this on the diagram.

Using free-body diagrams and motion diagrams Recall that all accelerations are the result of unbalanced forces. If a motion diagram shows that an object is accelerating, a free-body diagram of that object should have an unbalanced force in the same direction as the acceleration.

☑ READING CHECK **Compare** the direction of an object's acceleration with the direction of the unbalanced force exerted on the object.

Combining Forces

What happens if you and a friend each push a table and exert 100 N of force on it? When you push together in the same direction, you give the table twice the acceleration that it would have if just one of you applied 100 N of force. When you push on the table in opposite directions with the same amount of force, as in **Figure 4,** there is no unbalanced force, so the table does not accelerate but remains at rest.

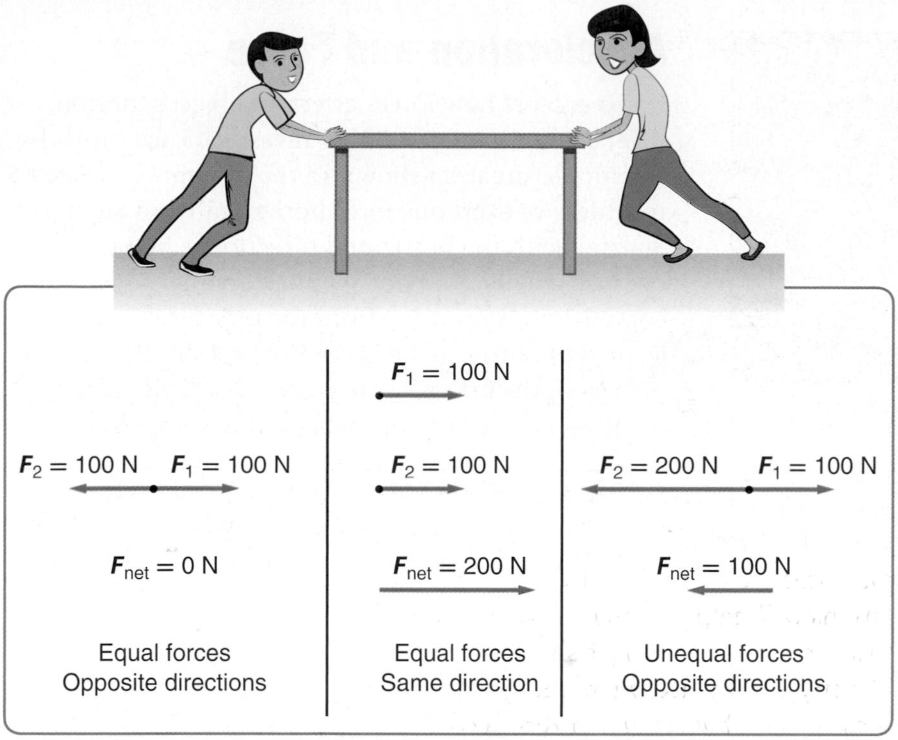

Figure 4 The net force acting on the table is the vector sum of all the forces acting on the table. This case only considers the horizontal forces acting on the table.

$F_2 = 100$ N $F_1 = 100$ N	$F_1 = 100$ N; $F_2 = 100$ N	$F_2 = 200$ N $F_1 = 100$ N
$F_{net} = 0$ N	$F_{net} = 200$ N	$F_{net} = 100$ N
Equal forces Opposite directions	Equal forces Same direction	Unequal forces Opposite directions

Net force The bottom portion of **Figure 4** shows free-body diagrams for these two situations. The third diagram in **Figure 4** shows the free-body diagram for a third situation in which your friend pushes on the table twice as hard as you in the opposite direction. Below each free-body diagram is a vector representing the resultant of the two forces. When the force vectors are in the same direction, they can be replaced by one vector with a length equal to their combined length. When the forces are in opposite directions, the resultant is the length of the difference between the two vectors. Another term for the vector sum of all the forces on an object is the **net force.**

You also can analyze the situation mathematically. Call the positive direction the direction in which you are pushing the table with a 100 N force. In the first case, your friend is pushing with a negative force of 100 N. Adding them together gives a total force of 0 N, which means there is no acceleration. In the second case, your friend's force is 100 N, so the total force is 200 N in the positive direction and the table accelerates in the positive direction. In the third case, your friend's force is −200 N, so the total force is −100 N and the table accelerates in the negative direction.

PRACTICE PROBLEMS

Do additional problems. Online Practice

For each of the following situations, specify the system and draw a motion diagram and a free-body diagram. Label all forces with their agents, and indicate the direction of the acceleration and of the net force. Draw vectors of appropriate lengths. Ignore air resistance unless otherwise indicated.

1. A skydiver falls downward through the air at constant velocity. (The air exerts an upward force on the person.)
2. You hold a softball in the palm of your hand and toss it up. Draw the diagrams while the ball is still touching your hand.
3. After the softball leaves your hand, it rises, slowing down.
4. After the softball reaches its maximum height, it falls down, speeding up.
5. **CHALLENGE** You catch the ball in your hand and bring it to rest.

Applying a Constant Force

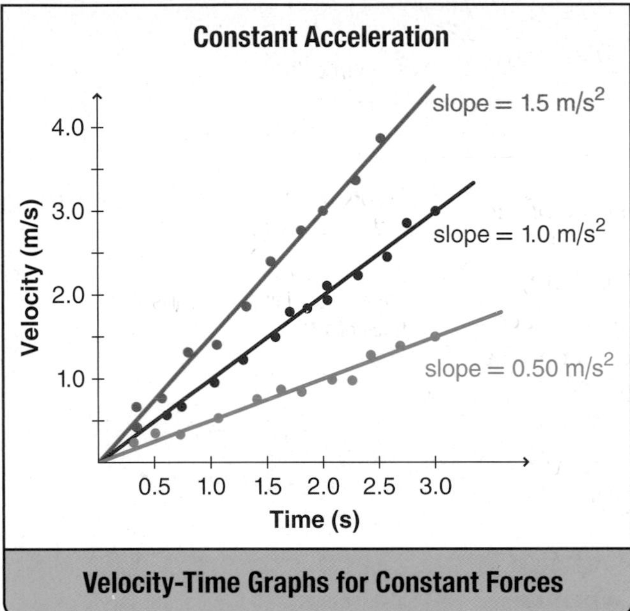

Velocity-Time Graphs for Constant Forces

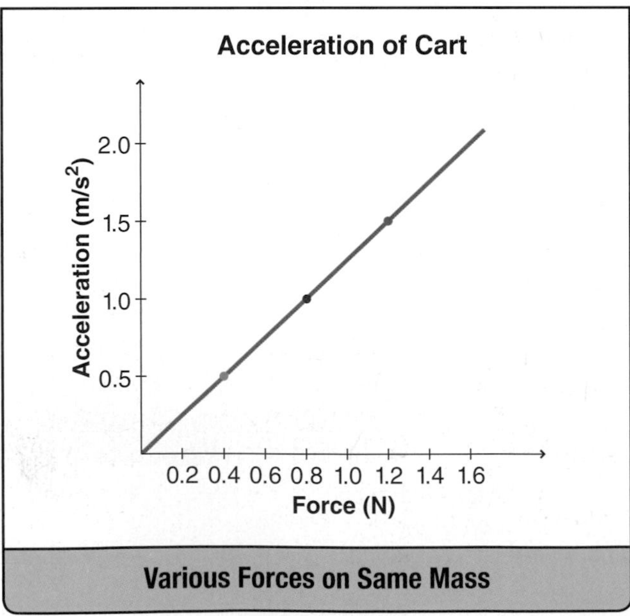

Various Forces on Same Mass

Figure 5 A spring scale exerts a constant unbalanced force on the cart. Repeating the investigation with different forces produces velocity-time graphs with different slopes.

View an **animation of force and acceleration.**

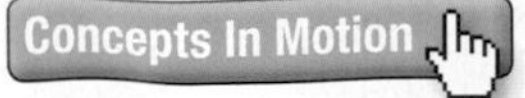

Acceleration and Force

To explore how forces affect an object's motion, think about doing a series of investigations. Consider the simple situation shown in the top photo of **Figure 5** in which we exert one force horizontally on an object. Starting with the horizontal direction is helpful because gravity does not act horizontally. To reduce complications resulting from the object rubbing against the surface, the investigations should be done on a smooth surface, such as a well-polished table. We'll also use a cart with wheels that spin easily.

Apply constant force How can you exert a constant unbalanced force? One way is to use a device called a spring scale. Inside the scale is a spring that stretches proportionally to the magnitude of the applied force. The front of the scale is calibrated to read the force in newtons. If you pull on the scale so that the reading on the front stays constant, the applied force is constant. The top photo in **Figure 5** shows a spring scale pulling a low-resistance cart with a constant unbalanced force.

If you perform this investigation and measure the cart's velocity for a period of time, you could construct a graph like the green line shown in the velocity-time graphs for constant forces in the middle panel of **Figure 5.** The constant slope of the line in the velocity-time graph indicates the cart's velocity increases at a constant rate. The constant rate of change of velocity means the acceleration is constant. This constant acceleration is a result of the constant unbalanced force applied by the spring scale to the cart.

How does the acceleration depend on the force? Repeat the investigation with a larger constant force. Then repeat it again with an even greater force. For each force, plot a velocity-time graph like the red and blue lines in the middle panel of **Figure 5.** Recall that the line's slope is the cart's acceleration. Calculate the slope of each line and plot the results for each force to make an acceleration-force graph, as shown in the bottom panel of **Figure 5.**

The graph indicates the relationship between force and acceleration is linear. Because the relationship is linear, you can apply the equation for a straight line:

$$y = kx + b$$

The y-intercept is 0, so the linear equation simplifies to $y = kx$. The y-variable is acceleration and the x-variable is force, so acceleration equals the slope of the line times the applied net force.

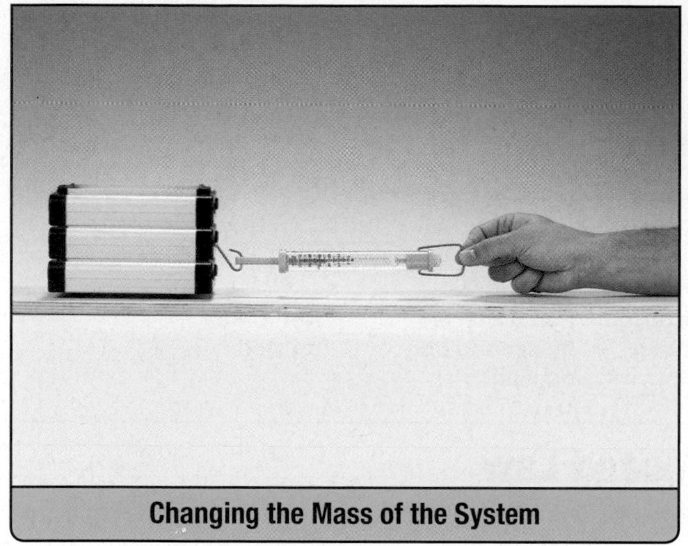
Changing the Mass of the System

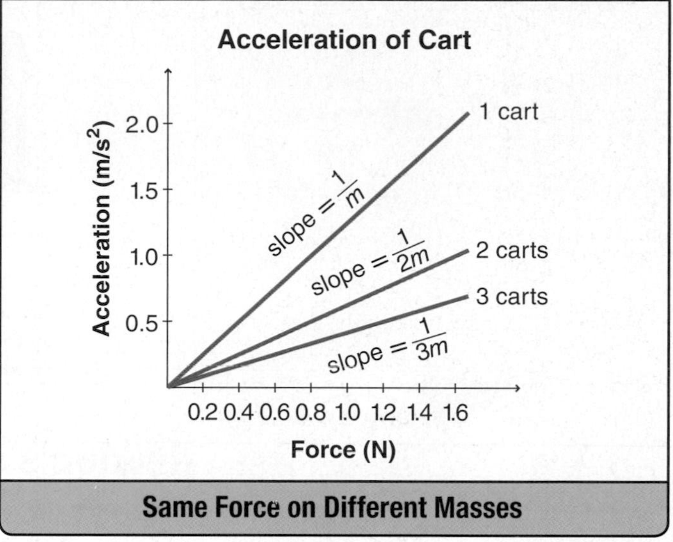

Same Force on Different Masses

Figure 6 Changing an object's mass affects that object's acceleration.

Compare the acceleration of one cart to the acceleration of two carts for an applied force of 1 N.

Interpreting slope What is the physical meaning of the slope of the acceleration-force graph? Does it describe something about the object that is accelerating? To see, change the object. Suppose that a second, identical cart is placed on top of the first, and then a third cart is added as in **Figure 6.** The spring scale would be pulling two carts and then three. A plot of the force versus acceleration for one, two, and three carts is shown in the graph in **Figure 6.**

The graph shows that if the same force is applied in each case, the acceleration of two carts is $\frac{1}{2}$ the acceleration of one cart, and the acceleration of three carts is $\frac{1}{3}$ the acceleration of one cart. This means that as the number of carts increases, the acceleration decreases. In other words, a greater force is needed to produce the same acceleration. The slopes of the lines in **Figure 6** depend upon the number of carts; that is, the slope depends on the total mass of the carts. In fact, the slope is the reciprocal of the mass (slope $= \frac{1}{\text{mass}}$). Using this value for slope, the mathematical equation $y = kx$ becomes the physics equation $a = \frac{F_{\text{net}}}{m}$.

What information is contained in the equation $a = \frac{F_{\text{net}}}{m}$? It tells you that a net force applied to an object causes that object to experience a change in motion—the force causes the object to accelerate. It also tells you that for the same object, if you double the force, you will double the object's acceleration. Lastly, if you apply the same force to objects with different masses, the one with the most mass will have the smallest acceleration and the one with the least mass will have the greatest acceleration.

READING CHECK Determine how the force exerted on an object must be changed to reduce the object's acceleration by half.

Recall that forces are measured in units called newtons. Because $F_{\text{net}} = ma$, a newton has the units of mass times the units of acceleration. So one newton is equal to one kg•m/s^2. To get an approximate idea of the size of 1 N, think about the downward force you feel when you hold an apple in your hand. The force exerted by the apple on your hand is approximately one newton. **Table 1** shows the magnitudes of some other common forces.

Table 1 Common Forces

Description	*F* (N)
Force of gravity on a coin (nickel)	0.05
Force of gravity on a 0.45-kg bag of sugar	4.5
Force of gravity on a 70-kg person	686
Force exerted by road on an accelerating car	3000
Force of a rocket engine	5,000,000

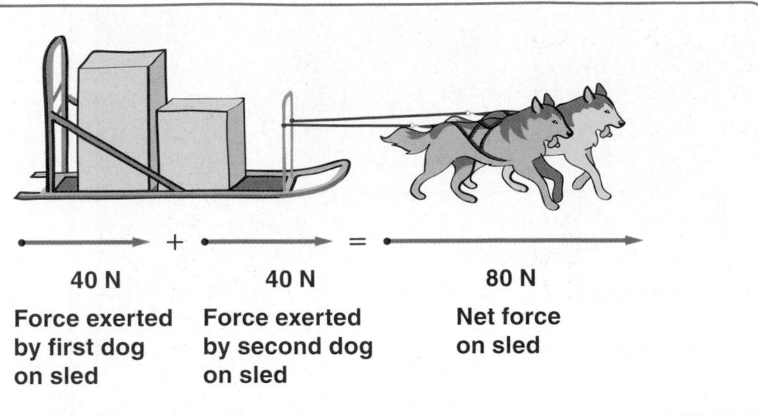

Figure 7 The net force acting on an object is the vector sum of all the forces acting on that object.

Investigate **Newton's second law.**

Newton's Second Law

Figure 7 shows two dogs pulling a sled. Each dog pulls with a force of 40 N. From the cart and spring-scale investigations, you know that the sled accelerates as a result of the unbalanced force acting it. Would the acceleration change if instead of two dogs each exerting a 40-N force, there was one bigger, stronger dog exerting a single 80-N force on the sled? When considering forces and acceleration, it is important to find the sum of all forces, called the net force, acting on a system.

Newton's second law states that the acceleration of an object is proportional to the net force and inversely proportional to the mass of the object being accelerated. This law is based on observations of how forces affect masses and is represented by the following equation.

NEWTON'S SECOND LAW

The acceleration of an object is equal to the sum of the forces acting on the object divided by the mass of the object.

$$\boldsymbol{a} = \frac{\boldsymbol{F}_{\text{net}}}{m}$$

Solving problems using Newton's second law One of the most important steps in correctly applying Newton's second law is determining the net force acting on the object. Often, more than one force acts on an object, so you must add the force vectors to determine the net force. Draw a free-body diagram showing the direction and relative strength of each force acting on the system. Then, add the force vectors to find the net force. Next, use Newton's second law to calculate the acceleration. Finally, if necessary, you can use what you know about accelerated motion to find the velocity or position of the object.

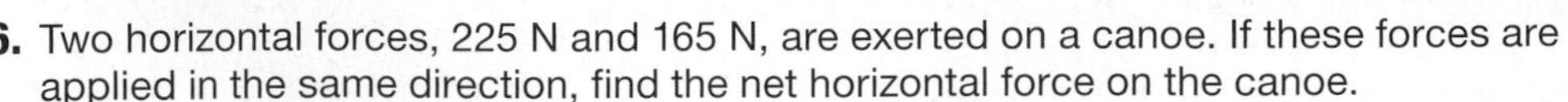

PRACTICE PROBLEMS Do additional problems. Online Practice

6. Two horizontal forces, 225 N and 165 N, are exerted on a canoe. If these forces are applied in the same direction, find the net horizontal force on the canoe.

7. If the same two forces as in the previous problem are exerted on the canoe in opposite directions, what is the net horizontal force on the canoe? Be sure to indicate the direction of the net force.

8. CHALLENGE Three confused sled dogs are trying to pull a sled across the Alaskan snow. Alutia pulls east with a force of 35 N, Seward also pulls east but with a force of 42 N, and big Kodiak pulls west with a force of 53 N. What is the net force on the sled?

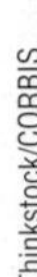

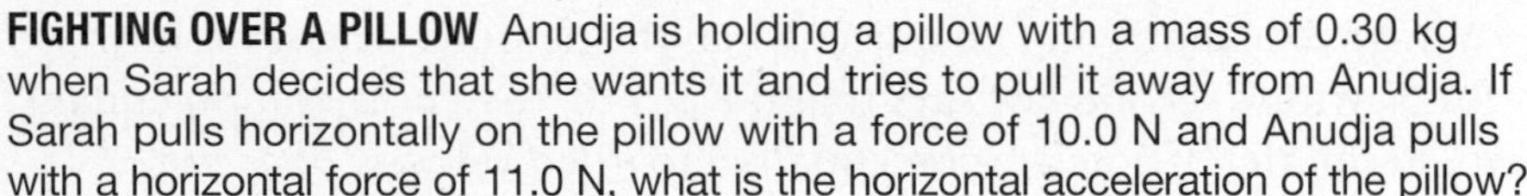

EXAMPLE PROBLEM 1

Find help with **operations with significant digits.** Math Handbook

FIGHTING OVER A PILLOW Anudja is holding a pillow with a mass of 0.30 kg when Sarah decides that she wants it and tries to pull it away from Anudja. If Sarah pulls horizontally on the pillow with a force of 10.0 N and Anudja pulls with a horizontal force of 11.0 N, what is the horizontal acceleration of the pillow?

1 ANALYZE AND SKETCH THE PROBLEM

- Sketch the situation.
- Identify the pillow as the system, and the direction in which Anudja pulls as positive.
- Draw the free-body diagram. Label the forces.

KNOWN	UNKNOWN
$m = 0.30$ kg	$a = ?$
$F_{\text{Anudja on pillow}} = 11.0$ N	
$F_{\text{Sarah on pillow}} = 10.0$ N	

2 SOLVE FOR THE UNKNOWN

$$F_{\text{net}} = F_{\text{Anudja on pillow}} + (-F_{\text{Sarah on pillow}})$$

Use Newton's second law.

$$a = \frac{F_{\text{net}}}{m}$$

$$= \frac{F_{\text{Anudja on pillow}} + (-F_{\text{Sarah on pillow}})}{m}$$

$$= \frac{11.0\text{ N} - 10.0\text{ N}}{0.30\text{ kg}}$$

◀ *Substitute $F_{\text{Anudja on pillow}} = 11.0$ N, $F_{\text{Sarah on pillow}} = 10.0$ N, $m = 0.30$ kg.*

$$= 3.3\text{ m/s}^2$$

$$\boldsymbol{a} = 3.3\text{ m/s}^2 \text{ toward Anudja}$$

3 EVALUATE THE ANSWER

- **Are the units correct?** m/s^2 is the correct unit for acceleration.
- **Does the sign make sense?** The acceleration is toward Anudja because Anudja is pulling toward herself with a greater force than Sarah is pulling in the opposite direction.
- **Is the magnitude realistic?** The net force is 1 N and the mass is 0.3 kg, so the acceleration is realistic.

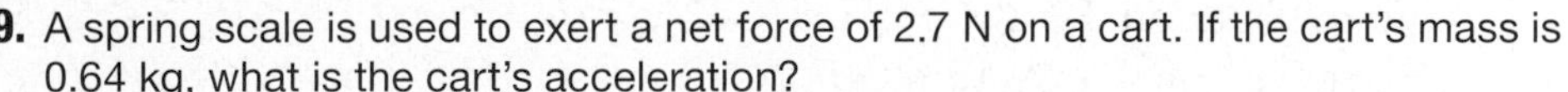

PRACTICE PROBLEMS

Do additional problems. Online Practice

9. A spring scale is used to exert a net force of 2.7 N on a cart. If the cart's mass is 0.64 kg, what is the cart's acceleration?

10. Kamaria is learning how to ice skate. She wants her mother to pull her along so that she has an acceleration of 0.80 m/s^2. If Kamaria's mass is 27.2 kg, with what force does her mother need to pull her? (Neglect any resistance between the ice and Kamaria's skates.)

11. CHALLENGE Two horizontal forces are exerted on a large crate. The first force is 317 N to the right. The second force is 173 N to the left.

a. Draw a force diagram for the horizontal forces acting on the crate.

b. What is the net force acting on the crate?

c. The box is initially at rest. Five seconds later, its velocity is 6.5 m/s to the right. What is the crate's mass?

EXAMPLE PROBLEM

Newton's First Law

What is the motion of an object when the net force acting on it is zero? Newton's second law says that if $F_{net} = 0$, then acceleration equals zero. Recall that if acceleration equals zero, then velocity does not change. Thus a stationary object with no net force acting on it will remain at rest. What about a moving object, such as a ball rolling on a surface? How long will the ball continue to roll? It will depend on the surface. If the ball is rolled on a thick carpet that exerts a force on the ball, it will come to rest quickly. If it is rolled on a hard, smooth surface that exerts very little force, such as a bowling alley, the ball will roll for a long time with little change in velocity. Galileo did many experiments and he concluded that if he could remove all forces opposing motion, horizontal motion would never stop. Galileo was the first to recognize that the general principles of motion could be found by extrapolating experimental results to an ideal case.

In the absence of a net force, the velocity of the moving ball and the lack of motion of the stationary object do not change. Newton recognized this and generalized Galileo's results into a single statement. Newton's statement, "an object that is at rest will remain at rest, and an object that is moving will continue to move in a straight line with constant speed, if and only if the net force acting on that object is zero," is called **Newton's first law.**

Inertia Newton's first law is sometimes called the law of inertia because **inertia** is the tendency of an object to resist changes in velocity. The car and the red block in **Figure 8** demonstrate the law of inertia. In the left panel, both objects are moving to the right. In the right panel, the wooden box applies a force to the car, causing it to stop. The red block does not experience the force applied by the wooden box. It continues to move to the right with the same velocity as in the left panel.

Is inertia a force? No. Forces are results of interactions between two objects; they are not properties of single objects, so inertia cannot be a force. Remember that because velocity includes both the speed and direction of motion, a net force is required to change either the speed or the direction of motion. If the net force is zero, Newton's first law means the object will continue with the same speed and direction.

Figure 8 The car and the block approach the wooden box at the same speed. After the collision, the block continues on with the same horizontal speed.

Matt Meadows

Figure 9 An object is in equilibrium if its velocity isn't changing. In both cases pictured here, velocity isn't changing, so the net force must be zero.

Equilibrium According to Newton's first law, a net force causes the velocity of an object to change. If the net force on an object is zero, then the object is in **equilibrium.** An object is in equilibrium if it is moving at a constant velocity. Note that being at rest is simply a special case of the state of constant velocity, $\boldsymbol{v} = 0$. Newton's first law identifies a net force as something that disturbs a state of equilibrium. Thus, if there is no net force acting on the object, then the object does not experience a change in speed or direction and is in equilibrium. As **Figure 9** indicates, at least in terms of net forces, there is no difference between sitting in a chair and falling at a constant velocity while skydiving—velocity isn't changing, so the net force is zero.

When analyzing forces and motion, it is important to keep in mind that the real world is full of forces that resist motion, called frictional forces. Newton's ideal, friction-free world is not easy to obtain. If you analyze a situation and find that the result is different from a similar experience that you have had, ask yourself whether this is because of the presence of frictional forces. For example, if you shove a textbook across a table, it will quickly come to a stop. At first thought, it might seem Newton's first law is violated in this case because the book's velocity changes even though there is no apparent force acting on the book. The net force acting on the book, however, is a frictional force between the table and the book in the direction opposite motion.

VOCABULARY
Science Use v. Common Use

Equilibrium

• **Science usage**
the condition in which the net force on an object is zero
When an object is in equilibrium, the object's velocity will be constant.

• **Common usage**
a state of balance
The pollution disturbed the ecosystem's equilibrium.

SECTION 1 REVIEW

Check your understanding.

12. MAINIDEA Identify each of the following as either **a, b,** or **c**: mass, inertia, the push of a hand, friction, air resistance, spring force, gravity, and acceleration.
    - **a.** contact force
    - **b.** a field force
    - **c.** not a force
13. **Free-Body Diagram** Draw a free-body diagram of a bag of sugar being lifted by your hand at an increasing speed. Specifically identify the system. Use subscripts to label all forces with their agents. Remember to make the arrows the correct lengths.
14. **Free-Body Diagram** Draw a free-body diagram of a water bucket being lifted by a rope at a decreasing speed. Specifically identify the system. Label all forces with their agents and make the arrows the correct lengths.
15. **Critical Thinking** A force of 1 N is the only horizontal force exerted on a block, and the horizontal acceleration of the block is measured. When the same horizontal force is the only force exerted on a second block, the horizontal acceleration is three times as large. What can you conclude about the masses of the two blocks?

SECTION 2

Weight and Drag Force

PHYSICS 4 YOU

If you have ever ridden a roller coaster, you probably noticed that you felt weightless as you went over a hill. But the force of gravity at the top of the hill is virtually the same as the force of gravity at the bottom of the hill. So why do you feel weightless?

MAIN IDEA

Newton's second law can be used to explain the motion of falling objects.

Essential Questions

- How are the weight and the mass of an object related?
- How do actual weight and apparent weight differ?
- What effect does air have on falling objects?

Review Vocabulary

viscosity a fluid's resistance to flowing

New Vocabulary

weight
gravitational field
apparent weight
weightlessness
drag force
terminal velocity

Weight

From Newton's second law, the fact that the ball in **Figure 10** is accelerating means there must be unbalanced forces acting on the ball. The only force acting on the ball is the gravitational force due to Earth's mass. An object's **weight** is the gravitational force experienced by that object. This gravitational force is a field force whose magnitude is directly proportional to the mass of the object experiencing the force. In equation form, the gravitational force, which equals weight, can be written $\boldsymbol{F}_g = m\boldsymbol{g}$. The mass of the object is m, and $\boldsymbol{g}$, called the **gravitational field,** is a vector quantity that relates the mass of an object to the gravitational force it experiences at a given location. Near Earth's surface, $\boldsymbol{g}$ is 9.8 N/kg toward Earth's center. Objects near Earth's surface experience 9.8 N of force for every kilogram of mass.

Scales When you stand on a scale as shown in the right panel of **Figure 10,** the scale exerts an upward force on you. Because you are not accelerating, the net force acting on you must be zero. Therefore the magnitude of the force exerted by the scale ($\boldsymbol{F}_{\text{scale on you}}$) pushing up must equal the magnitude of $\boldsymbol{F}_g$ pulling down on you. Inside the scale, springs provide the upward force necessary to make the net force equal zero. The scale is calibrated to convert the stretch of the springs to a weight. The measurement on the scale is affected by the gravitational field on Earth's surface. If you were on a different planet with a different $\boldsymbol{g}$, the scale would exert a different force to keep you in equilibrium, and consequently, the scale's reading would be different. Because weight is a force, the proper unit used to measure weight is the newton.

Figure 10 The gravitational force exerted by Earth's mass on an object equals the object's mass times the gravitational field, ($\boldsymbol{F}_g = m\boldsymbol{g}$).

Identify the forces acting on you when you are in equilibrium while standing on a scale.

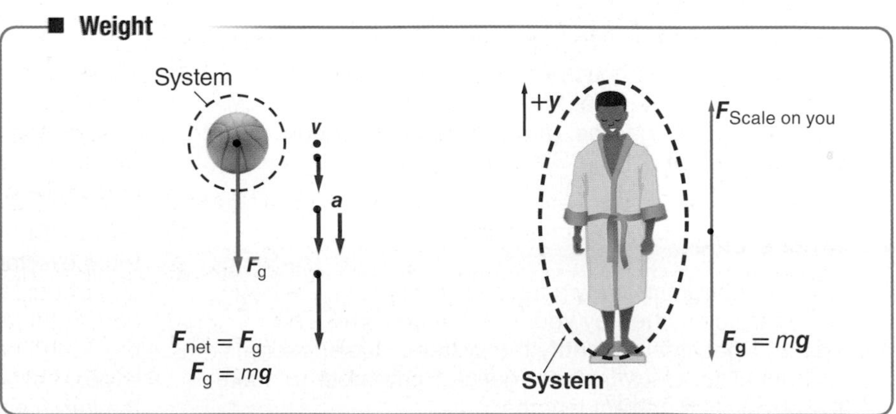

Dynamic Graphics/SuperStock

EXAMPLE PROBLEM 2

Find help with **operations with significant digits.** Math Handbook

COMPARING WEIGHTS Kiran holds a brass cylinder in each hand. Cylinder A has a mass of 100.0 g and cylinder B has a mass of 300.0 g. What upward forces do his two hands exert to keep the cylinders at rest? If he then drops the two, with what acceleration do they fall? (Ignore air resistance.)

1 ANALYZE AND SKETCH THE PROBLEM

- Sketch the situation.
- Identify the two cylinders as the systems, and choose the upward direction as positive.
- Draw the free-body diagrams. Label the forces.

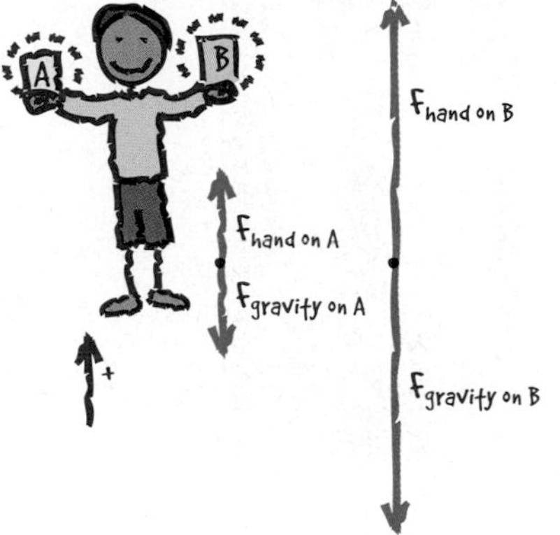

KNOWN	UNKNOWNS
$m_A = 0.1000$ kg	$F_{Hand\ on\ A} = ?$
$m_B = 0.3000$ kg	$F_{Hand\ on\ B} = ?$
$g = -9.8$ N/kg	$a_A = ?$ $a_B = ?$

2 SOLVE FOR THE UNKNOWNS

For cylinder A:

$$F_{Net\ on\ A} = F_{Hand\ on\ A} + F_{Gravity\ on\ A}$$
$$0 = F_{Hand\ on\ A} + F_{Gravity\ on\ A}$$
$$F_{Hand\ on\ A} = -F_{Gravity\ on\ A}$$
$$F_{Hand\ on\ A} = -m_A g$$
$$= -(0.1000\ \text{kg})(-9.8\ \text{N/kg})$$
$$= 0.98\ \text{N up}$$

For cylinder B:

$$F_{Net\ on\ B} = F_{Hand\ on\ B} + F_{Gravity\ on\ B}$$
$$0 = F_{Hand\ on\ B} + F_{Gravity\ on\ B}$$
$$F_{Hand\ on\ B} = -F_{Gravity\ on\ B}$$
$$F_{Hand\ on\ B} = -m_B g$$
$$= -(0.3000\ \text{kg})(-9.8\ \text{N/kg})$$
$$= 2.9\ \text{N up}$$

After the cylinders are dropped, the only force on each is the force of gravity. Use Newton's second law.

$$a_A = \frac{F_{Net\ on\ A}}{m_A} \qquad a_B = \frac{F_{Net\ on\ B}}{m_B}$$

$$a_A = \frac{m_A g}{m_A} = g \qquad a_B = \frac{m_B g}{m_B} = g$$ ◀ *Substitute $F_{Net\ on\ A} = m_A g$ and $F_{Net\ on\ B} = m_B g$.*

$$= -9.8\ \text{m/s}^2 \qquad = -9.8\ \text{m/s}^2$$ ◀ *Substitute $g = -9.8$ N/kg $= -9.8$ m/s².*

3 EVALUATE THE ANSWER

- **Are the units correct?** N is the correct unit for force. m/s^2 is the correct unit for acceleration.
- **Does the sign make sense?** The direction of the fall is downward, the negative direction, and the object is speeding up, so the acceleration should be negative.
- **Is the magnitude realistic?** Forces are 1–5 N, typical of that exerted by objects that have a mass of one kg or less. The accelerations are both equal to free fall acceleration.

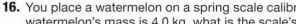

PRACTICE PROBLEMS

Do additional problems. 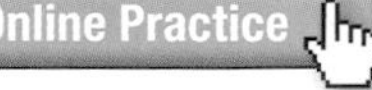

16. You place a watermelon on a spring scale calibrated to measure in newtons. If the watermelon's mass is 4.0 kg, what is the scale's reading?

17. You place a 22.50-kg television on a spring scale. If the scale reads 235.2 N, what is the gravitational field at that location?

18. A 0.50-kg guinea pig is lifted up from the ground. What is the smallest force needed to lift it? Describe the particular motion resulting from this minimum force.

19. CHALLENGE A grocery sack can withstand a maximum of 230 N before it rips. Will a bag holding 15 kg of groceries that is lifted from the checkout counter at an acceleration of 7.0 m/s^2 hold?

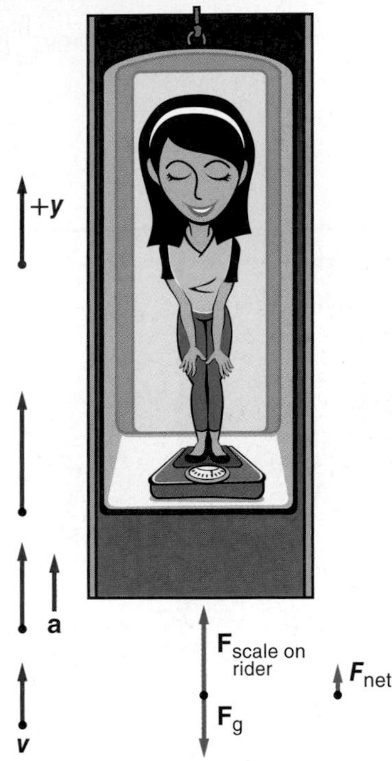

Figure 11 If you are accelerating upward, the net force acting on you must be upward. The scale must exert an upward force greater than the downward force of your weight.

View an **animation of apparent weight.**

Apparent weight What is weight? Because the weight force is defined as $\boldsymbol{F}_g = m\boldsymbol{g}$, $\boldsymbol{F}_g$ changes when $\boldsymbol{g}$ varies. On or near the surface of Earth, $\boldsymbol{g}$ is approximately constant, so an object's weight does not change appreciably as it moves around near Earth's surface. If a bathroom scale provides the only upward force on you, then it reads your weight. What would it read if you stood with one foot on the scale and one foot on the floor? What if a friend pushed down on your shoulders or lifted up on your elbows? Then there would be other contact forces on you, and the scale would not read your weight.

What happens if you stand on a scale in an elevator? As long as you are not accelerating, the scale will read your weight. What would the scale read if the elevator accelerated upward? **Figure 11** shows the pictorial and physical representations for this situation. You are the system, and upward is the positive direction. Because the acceleration of the system is upward, the net force must be upward. The upward force of the scale must be greater than the downward force of your weight. Therefore, the scale reading is greater than your weight.

If you ride in an elevator accelerating upward, you feel as if you are heavier because the floor presses harder on your feet. On the other hand, if the acceleration is downward, then you feel lighter, and the scale reads less than your weight. The force exerted by the scale is an example of **apparent weight,** which is the support force exerted on an object.

☑ **READING CHECK** **Describe** the reading on the scale as the elevator accelerates upward from rest, reaches a constant speed, then comes to a stop.

Imagine that the cable holding the elevator breaks. What would the scale read then? The scale and you would both accelerate at $a = g$. According to this formula, the scale would read zero and your apparent weight would be zero. That is, you would be weightless. However, **weightlessness** does not mean that an object's weight is actually zero; rather, it means that there are no contact forces acting to support the object, and the object's *apparent weight* is zero. Similar to the falling elevator, astronauts experience weightlessness in orbit because they and their spacecraft are in free fall. You will study gravity and weightlessness in greater detail in a later chapter.

PhysicsLAB

FORCES IN AN ELEVATOR

INTERNET LAB How does apparent weight change when riding in an elevator?

MiniLAB

MASS AND WEIGHT

How are mass and weight related?

PROBLEM-SOLVING STRATEGIES

FORCE AND MOTION

When solving force and motion problems, use the following strategies.

1. Read the problem carefully, and sketch a pictorial model.
2. Circle the system and choose a coordinate system.
3. Determine which quantities are known and which are unknown.
4. Create a physical model by drawing a motion diagram showing the direction of the acceleration.
5. Create a free-body diagram showing all the forces acting on the object.
6. Use Newton's laws to link acceleration and net force.
7. Rearrange the equation to solve for the unknown quantity.
8. Substitute known quantities with their units into the equation and solve.
9. Check your results to see whether they are reasonable.

EXAMPLE PROBLEM 3

Find help with **real/apparent weight.** Personal Tutor

EXAMPLE PROBLEM

REAL AND APPARENT WEIGHT Your mass is 75.0 kg, and you are standing on a bathroom scale in an elevator. Starting from rest, the elevator accelerates upward at 2.00 m/s^2 for 2.00 s and then continues at a constant speed. Is the scale reading during acceleration greater than, equal to, or less than the scale reading when the elevator is at rest?

1 ANALYZE AND SKETCH THE PROBLEM

- Sketch the situation.
- Choose a coordinate system with the positive direction as upward.
- Draw the motion diagram. Label *v* and *a.*
- Draw the free-body diagram. The net force is in the same direction as the acceleration, so the upward force is greater than the downward force.

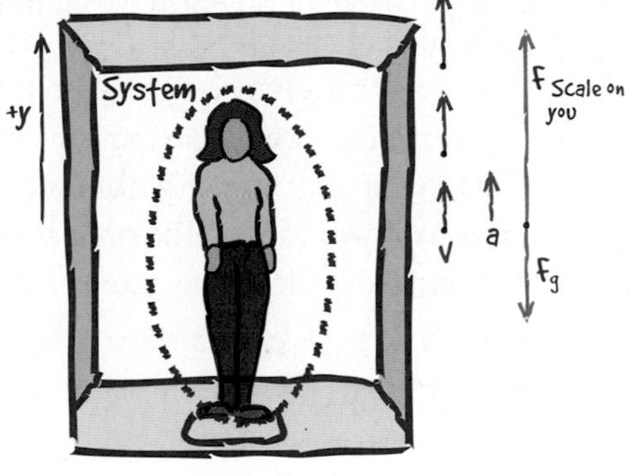

KNOWN

$m = 75.0$ kg

$a = 2.00$ m/s^2

$t = 2.00$ s

$g = 9.8$ N/kg

UNKNOWN

$F_{scale} = ?$

2 SOLVE FOR THE UNKNOWN

$F_{net} = ma$

$F_{net} = F_{scale} + (-F_g)$ ◀ F_g is negative because it is in the negative direction defined by the coordinate system.

Solve for F_{scale}.

$F_{scale} = F_{net} + F_g$

Elevator at rest:

$F_{scale} = F_{net} + F_g$ ◀ The elevator is not accelerating. Thus, $F_{net} = 0.00$ N.

$= F_g$ ◀ Substitute $F_{net} = 0.00$ N.

$= mg$ ◀ Substitute $F_g = mg$.

$= (75.0\text{ kg})(9.8\text{ N/kg})$ ◀ Substitute $m = 75.0$ kg, $g = 9.8$ N/kg.

$= 735$ N

Elevator accelerating upward:

$F_{scale} = F_{net} + F_g$

$= ma + mg$ ◀ Substitute $F_{net} = ma$, $F_g = mg$

$= (75.0\text{ kg})(2.00\text{ m/s}^2) + (75.0\text{ kg})(9.8\text{ N/kg})$ ◀ Substitute $m = 75.0$ kg, $a = 2.00$ m/s^2, $g = 9.8$ N/kg

$= 885$ N

The scale reading when the elevator is accelerating (885 N) is larger than the scale reading when the elevator is at rest (735 N).

3 EVALUATE THE ANSWER

- **Are the units correct?** kg•m/s^2 is the force unit, N.
- **Does the sign make sense?** The positive sign agrees with the coordinate system.
- **Is the magnitude realistic?** $F_{scale} = 885$ N is larger than it would be at rest when F_{scale} would be 735 N. The increase is 150 N, which is about 20 percent of the rest weight. The upward acceleration is about 20 percent of that due to gravity, so the magnitude is reasonable.

PRACTICE PROBLEMS

Do additional problems. Online Practice 

20. On Earth, a scale shows that you weigh 585 N.

a. What is your mass?

b. What would the scale read on the Moon ($g = 1.60$ N/kg)?

21. **CHALLENGE** Use the results from Example Problem 3 to answer questions about a scale in an elevator on Earth. What force would be exerted by the scale on a person in the following situations?

a. The elevator moves upward at constant speed.

b. It slows at 2.0 m/s^2 while moving downward.

c. It speeds up at 2.0 m/s^2 while moving downward.

d. It moves downward at constant speed.

e. In what direction is the net force as the elevator slows to a stop as it is moving down?

PhysicsLAB

TERMINAL VELOCITY

PROBEWARE LAB How does air resistance affect objects in free fall?

MiniLAB

UPSIDE-DOWN PARACHUTE

How does terminal velocity depend on mass?

Drag Force

It is true that the particles in the air around an object exert forces on that object. Air actually exerts huge forces, but in most cases, it exerts balanced forces on all sides, and therefore it has no net effect. Can you think of any experiences that help to prove that air can exert a force? One such example would be holding a piece of paper at arm's length and blowing hard at the paper. The paper accelerates in response to the air striking it, meaning the air exerts a net force on the paper.

So far, you have neglected the force of air on an object moving through the air. In actuality, when an object moves through any fluid, such as air or water, the fluid exerts a force on the moving object in the direction opposite the object's motion. A **drag force** is the force exerted by a fluid on an object opposing motion through the fluid. This force is dependent on the motion of the object, the properties of the object, and the properties of the fluid that the object is moving through. For example, as the speed of the object increases, so does the magnitude of the drag force. The size and shape of the object also affect the drag force. The fluid's properties, such as its density and viscosity, also affect the drag force.

PHYSICS CHALLENGE

A 415-kg container of food and water is dropped from an airplane at an altitude of 300 m. First, consider the situation ignoring air resistance. Then calculate the more realistic situation involving a drag force provided by a parachute.

1. If you ignore air resistance, how long will it take the container to fall 300 m to the ground?
2. Again, ignoring air resistance, what is the speed of the container just before it hits the ground?
3. The container is attached to a parachute designed to produce a drag force that allows the container to reach a constant downward velocity of 6 m/s. What is the magnitude of the drag force when the container is falling at a constant 6 m/s down?

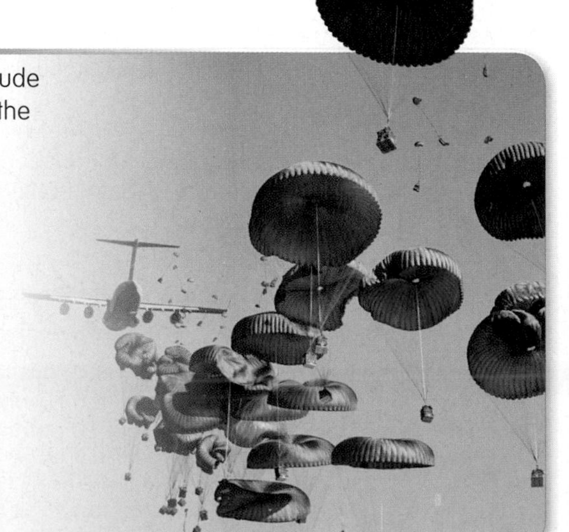

U.S. Air Force photo by Tech. Sgt. James L. Harper Jr.

Terminal velocity If you drop a tennis ball, as in **Figure 12,** it has very little velocity at the start and thus only a small drag force. The downward force of gravity is much stronger than the upward drag force, so there is a downward acceleration. As the ball's velocity increases, so does the drag force. Soon the drag force equals the force of gravity. When this happens, there is no net force, and so there is no acceleration. The constant velocity that is reached when the drag force equals the force of gravity is called the **terminal velocity.**

When light objects with large surface areas are falling, the drag force has a substantial effect on their motion, and they quickly reach terminal velocity. Heavier, more-compact objects are not affected as much by the drag force. For example, the terminal velocity of a table-tennis ball in air is 9 m/s, that of a basketball is 20 m/s, and that of a baseball is 42 m/s. Competitive skiers increase their terminal velocities by decreasing the drag force on them. They hold their bodies in an egg shape and wear smooth clothing and streamlined helmets.

Skydivers can increase or decrease their terminal velocity by changing their body orientation and shape. A horizontal, spread-eagle shape produces the slowest terminal velocity, about 60 m/s. After opening up the parachute, the skydiver becomes part of a very large object with a correspondingly large drag force and a terminal velocity of about 5 m/s.

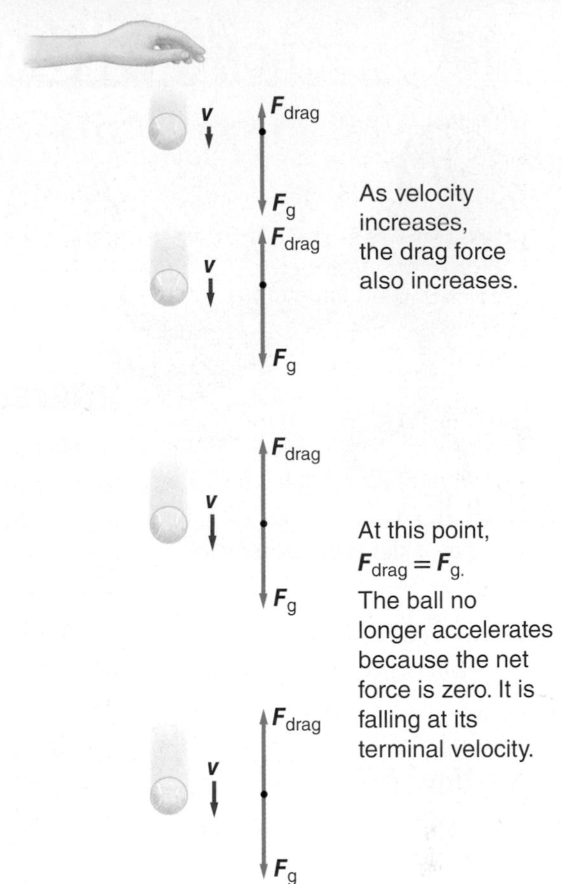

Figure 12 The drag force on an object increases as its velocity increases. When the drag force equals the gravitational force, the object is in equilibrium so it no longer accelerates.

View an **animation of terminal velocity.**

Concepts In Motion

SECTION 2 REVIEW

Section Self-Check Check your understanding.

22. **MAIN**IDEA The skydiver shown in **Figure 13** falls at a constant speed in the spread-eagle position. Immediately after opening the parachute, is the skydiver accelerating? If so, in which direction? Explain your answer using Newton's laws.

Figure 13

23. **Lunar Gravity** Compare the force holding a 10.0-kg rock on Earth and on the Moon. The gravitational field on the Moon is 1.6 N/kg.
24. **Motion of an Elevator** You are riding in an elevator holding a spring scale with a 1-kg mass suspended from it. You look at the scale and see that it reads 9.3 N. What, if anything, can you conclude about the elevator's motion at that time?
25. **Apparent Weight** You take a ride in a fast elevator to the top of a tall building and ride back down. During which parts of the ride will your apparent and real weights be the same? During which parts will your apparent weight be less than your real weight? More than your real weight? Sketch free-body diagrams to support your answers.
26. **Acceleration** Tecle, with a mass of 65.0 kg, is standing on an ice-skating rink. His friend applies a force of 9.0 N to him. What is Tecle's resulting acceleration?
27. **Critical Thinking** You have a job at a meat warehouse loading inventory onto trucks for shipment to grocery stores. Each truck has a weight limit of 10,000 N of cargo. You push each crate of meat along a low-resistance roller belt to a scale and weigh it before moving it onto the truck. One night, right after you weigh a 1000-N crate, the scale breaks. Describe a way in which you could apply Newton's laws to approximate the masses of the remaining crates.

SECTION 3 Newton's Third Law

PHYSICS 4 YOU **If you push against a wall while sitting in a chair that has wheels, you will accelerate across the floor. What applies the unbalanced force that causes your acceleration? Newton's third law helps answer this question.**

MAIN IDEA
All forces occur in interaction pairs.

Essential Questions

- What is Newton's third law?
- What is the normal force?

Review Vocabulary

symmetry correspondence of parts on opposite sides of a dividing line

New Vocabulary

interaction pair
Newton's third law
tension
normal force

Interaction Pairs

Figure 14 illustrates the idea of forces as interaction pairs. There is a force from the boy on the dog's toy, and there is a force from the dog's toy on the boy. Forces always come in pairs similar to this example. Consider the boy (A) as one system and the toy (B) as another. What forces act on each of the two systems? Looking at the force diagrams in **Figure 14,** you can see that each system exerts a force on the other. The two forces, $\boldsymbol{F}_{\text{A on B}}$ and $\boldsymbol{F}_{\text{B on A}}$, are the forces of interaction between the two. Notice the symmetry in the subscripts: A on B and B on A.

The forces $\boldsymbol{F}_{\text{A on B}}$ and $\boldsymbol{F}_{\text{B on A}}$ are an **interaction pair,** which is a set of two forces that are in opposite directions, have equal magnitudes, and act on different objects. Sometimes, an interaction pair is called an action-reaction pair. This might suggest that one causes the other; however, this is not true. For example, the force of the boy pulling on the toy doesn't cause the toy to pull on the boy. The two forces either exist together or not at all.

☑ **READING CHECK** **Predict** the magnitude and direction of the force applied on you if you push against a tree with a force of 15 N directed to the left.

Definition of Newton's third law In **Figure 14,** the force exerted by the boy on the toy is equal in magnitude and opposite in direction to the force exerted by the toy on the boy. Such an interaction pair is an example of **Newton's third law,** which states that all forces come in pairs. The two forces in a pair act on different objects and are equal in strength and opposite in direction.

Figure 14 The force that the toy exerts on the boy and the force that the boy exerts on the toy are an interaction pair.

NEWTON'S THIRD LAW

The force of A on B is equal in magnitude and opposite in direction of the force of B on A.

$$F_{\text{A on B}} = -F_{\text{B on A}}$$

Using Newton's third law Consider the situation of holding a book in your hand. You can draw one free-body diagram for you and one for the book. Are there any interaction pairs? When identifying interaction pairs, keep in mind that they always occur in two different free-body diagrams, and they always will have the symmetry of subscripts noted on the previous page. In this case, the interaction pair is $F_{\text{book on hand}}$ and $F_{\text{hand on book}}$.

The ball in **Figure 15** interacts with the table and with Earth. First, analyze the forces acting on one system, the ball. The table exerts an upward force on the ball, and the mass of Earth exerts a downward gravitational force on the ball. Even though these forces are in opposite directions, they are not an interaction pair because they act on the same object. Now consider the ball and the table together. In addition to the upward force exerted by the table on the ball, the ball exerts a downward force on the table. This is an interaction pair.

Notice also that the ball has a weight. If the ball experiences a force due to Earth's mass, then there must be a force on Earth's mass due to the ball. In other words, they are an interaction pair.

$$F_{\text{Earth's mass on ball}} = -F_{\text{ball on Earth's mass}}$$

An unbalanced force on Earth would cause Earth to accelerate. But acceleration is inversely proportional to mass. Because Earth's mass is so huge in comparison to the masses of other objects that we normally consider, Earth's acceleration is so small that it can be neglected. In other words, Earth can be often treated as part of the external world rather than as a second system. The problem-solving strategies below summarize how to deal with interaction pairs.

☑ **READING CHECK Explain** why Earth's acceleration is usually very small compared to the acceleration of the object that Earth interacts with.

PROBLEM-SOLVING STRATEGIES

Use these strategies to solve problems in which there is an interaction between objects in two different systems.

1. Separate the system or systems from the external world.
2. Draw a pictorial model with coordinate systems for each system.
3. Draw a physical model that includes free-body diagrams for each system.
4. Connect interaction pairs by dashed lines.
5. To calculate your answer, use Newton's second law to relate the net force and acceleration for each system.
6. Use Newton's third law to equate the magnitudes of the interaction pairs and give the relative direction of each force.
7. Solve the problem and check the reasonableness of the answers' units, signs, and magnitudes.

Newton's Third Law

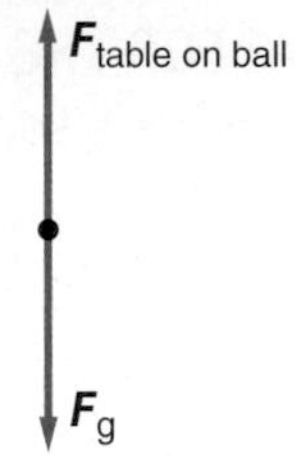

The two forces acting on the ball are $F_{\text{table on ball}}$ and $F_{\text{Earth's mass on ball}}$. These forces are not an interaction pair.

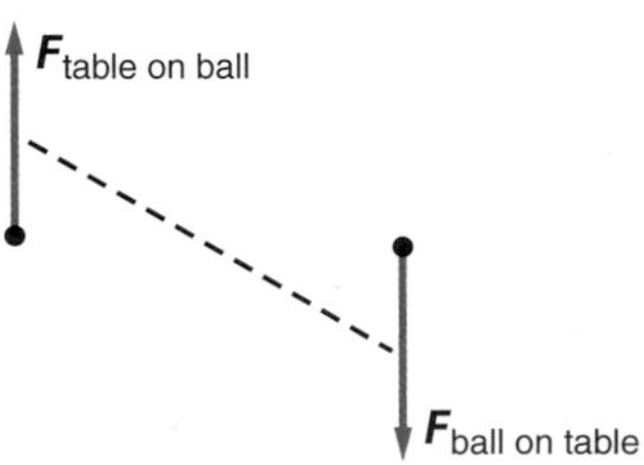

Force interaction pair between ball and table.

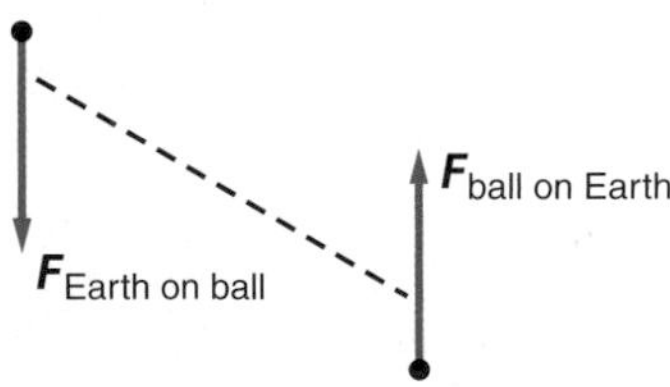

Force interaction pair between ball and Earth.

Figure 15 A ball resting on a table is part of two interaction pairs.

EXAMPLE PROBLEM 4

Find help with **operations with scientific notation.** Math Handbook

EARTH'S ACCELERATION A softball has a mass of 0.18 kg. What is the gravitational force on Earth due to the ball, and what is Earth's resulting acceleration? Earth's mass is 6.0×10^{24} kg.

1 ANALYZE AND SKETCH THE PROBLEM

- Draw free-body diagrams for the two systems: the ball and Earth.
- Connect the interaction pair by a dashed line.

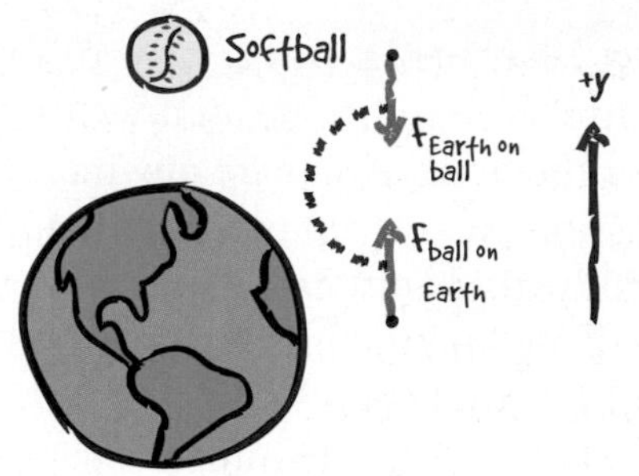

KNOWN	UNKNOWN
$m_{\text{ball}} = 0.18$ kg	$\boldsymbol{F}_{\text{Earth on ball}} = ?$
$m_{\text{Earth}} = 6.0\times10^{24}$ kg	$\boldsymbol{a}_{\text{Earth}} = ?$
$\boldsymbol{g} = 9.8$ N/kg	

2 SOLVE FOR THE UNKNOWN

Use Newton's second law to find the weight of the ball.

$$\boldsymbol{F}_{\text{Earth on ball}} = m_{\text{ball}}\boldsymbol{g}$$
$$= (0.18 \text{ kg})(-9.8 \text{ N/kg}) \quad \text{◀ Substitute } m_{\text{ball}} = 0.18 \text{ kg}, g = -9.8 \text{ N/kg.}$$
$$= -1.8 \text{ N}$$

Use Newton's third law to find $\boldsymbol{F}_{\text{ball on Earth}}$.

$$\boldsymbol{F}_{\text{ball on Earth}} = -\boldsymbol{F}_{\text{Earth on ball}}$$
$$= -(-1.8 \text{ N}) \quad \text{◀ Substitute } F_{\text{Earth on ball}} = -1.8 \text{ N.}$$
$$= +1.8 \text{ N}$$

Use Newton's second law to find $\boldsymbol{a}_{\text{Earth}}$.

$$\boldsymbol{a}_{\text{Earth}} = \frac{\boldsymbol{F}_{\text{net}}}{m_{\text{Earth}}}$$
$$= \frac{1.8 \text{ N}}{6.0\times10^{24} \text{ kg}} \quad \text{◀ Substitute } F_{\text{net}} = 1.8 \text{ N}, m_{\text{Earth}} = 6.0\times10^{24} \text{ kg.}$$
$$= 2.9\times10^{-25} \text{ m/s}^2 \text{ toward the softball}$$

3 EVALUATE THE ANSWER

- **Are the units correct?** Force is in N and acceleration is in m/s^2.
- **Do the signs make sense?** Force and acceleration should be positive.
- **Is the magnitude realistic?** It makes sense that Earth's acceleration should be so small because Earth is so massive.

PRACTICE PROBLEMS

Do additional problems. Online Practice

28. You lift a relatively light bowling ball with your hand, accelerating it upward. What are the forces on the ball? What forces does the ball exert? What objects are these forces exerted on?

29. A brick falls from a construction scaffold. Identify any forces acting on the brick. Also identify any forces the brick exerts and the objects on which these forces are exerted. (Air resistance may be ignored.)

30. A suitcase sits on a stationary airport luggage cart, as in **Figure 16.** Draw a free-body diagram for each object and specifically indicate any interaction pairs between the two.

31. CHALLENGE You toss a ball up in the air. Draw a free-body diagram for the ball after it has lost contact with your hand but while it is still moving upward. Identify any forces acting on the ball. Also identify any forces that the ball exerts and the objects on which these forces are exerted. Assume that air resistance is negligible.

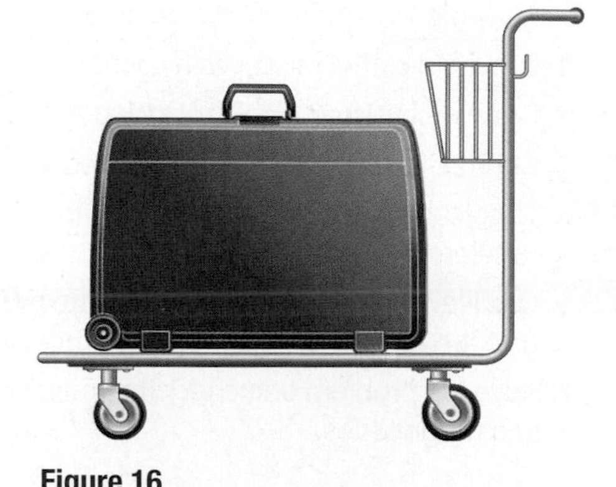

Figure 16

Tension

Tension is simply a specific name for the force that a string or rope exerts. A simplification within this textbook is the assumption that all strings and ropes are massless. In **Figure 17,** the rope is about to break in the middle. If the rope breaks, the bucket will fall; before it breaks, there must be forces holding the rope together. The force that the top part of the rope exerts on the bottom part is $\boldsymbol{F}_{\text{top on bottom}}$. Newton's third law states that this force must be part of an interaction pair. The other member of the pair is the force that the bottom part of the rope exerts on the top, $\boldsymbol{F}_{\text{bottom on top}}$. These forces, equal in magnitude but opposite in direction, also are shown in **Figure 17.**

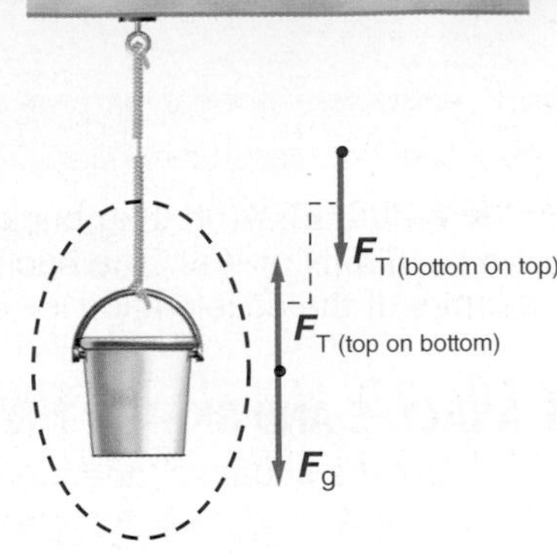

Figure 17 The tension in the rope is equal to the weight of all the objects hanging from it.

Think about this situation in another way. Before the rope breaks, the bucket is in equilibrium. This means that the force of its weight downward must be equal in magnitude but opposite in direction to the tension in the rope upward. Similarly, if you look at the point in the rope just above the bucket, it also is in equilibrium. Therefore, the tension of the rope below it pulling down must be equal to the tension of the rope above it pulling up. You can move up the rope, considering any point in the rope, and see that the tension forces at any point in the rope are pulling equally in both directions. Thus, the tension in the rope equals the weight of all objects below it.

Examine the tension forces shown in **Figure 18.** If team A is exerting a 500-N force and the rope does not accelerate, then team B also must be pulling with a force of 500 N. What is the tension in the rope? If each team pulls with 500 N of force, is the tension 1000 N? To decide, think of the rope as divided into two halves. The left side is not accelerating, so the net force on it is zero. Thus, $F_{\text{A on left side}} = F_{\text{right side on left side}} = 500$ N. Similarly, $F_{\text{B on right side}} = F_{\text{left side on right side}} = 500$ N. But the two tensions, $\boldsymbol{F}_{\text{right side on left side}}$ and $\boldsymbol{F}_{\text{left side on right side}}$, are an interaction pair, so they are equal and opposite. Thus, the tension in the rope equals the force with which each team pulls, or 500 N. To verify this, you could cut the rope in half and tie the ends to a spring scale. The scale would read 500 N.

Figure 18 The rope is not accelerating, so the tension in the rope equals the force with which each team pulls.

David Young-Wolff/PhotoEdit

EXAMPLE PROBLEM 5

Find help with **isolating a variable.** Math Handbook

EXAMPLE PROBLEM

LIFTING A BUCKET A 50.0-kg bucket is being lifted by a rope. The rope will not break if the tension is 525 N or less. The bucket started at rest, and after being lifted 3.0 m, it moves at 3.0 m/s. If the acceleration is constant, is the rope in danger of breaking?

1 ANALYZE AND SKETCH THE PROBLEM

- Draw the situation, and identify the forces on the system.
- Establish a coordinate system with the positive axis upward.
- Draw a motion diagram; include ***v*** and ***a***.
- Draw the free-body diagram, and label the forces.

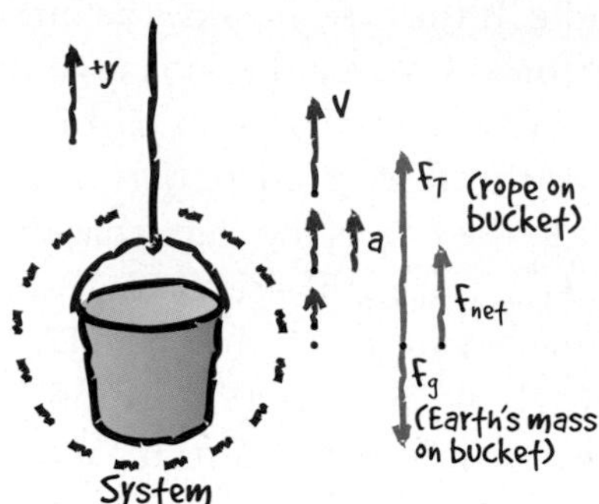

KNOWN		UNKNOWN
$m = 50.0$ kg	$v_f = 3.0$ m/s	$F_T = ?$
$v_i = 0.0$ m/s	$d = 3.0$ m	

2 SOLVE FOR THE UNKNOWN

F_{net} is the sum of the positive force of the rope pulling up (F_T) and the negative weight force ($-F_g$) pulling down as defined by the coordinate system.

$$F_{net} = F_T + (-F_g)$$

$$F_T = F_{net} + F_g$$

$$= ma + mg$$

◀ Substitute $F_{net} = ma$, $F_g = mg$

v_i, v_f, and d are known.

$$v_f^2 = v_i^2 + 2ad$$

$$a = \frac{v_f^2 - v_i^2}{2d}$$

$$= \frac{v_f^2}{2d}$$

◀ Substitute $v_i = 0.0$ m/s^2.

$$F_T = ma + mg$$

$$= m\left(\frac{v_f^2}{2d}\right) + mg$$

◀ Substitute $a = v_f^2 / (2d)$.

$$= (50.0\text{ kg})\left(\frac{(3.0\text{ m/s})^2}{2(3.0\text{ m})}\right) + (50.0\text{ kg})(9.8\text{ N/kg})$$

◀ Substitute $m = 50.0$ kg, $v_f = 3.0$ m/s, $d = 3.0$ m, $g = 9.8$ N/kg.

$$= 560\text{ N}$$

The rope is in danger of breaking because the tension exceeds 525 N.

3 EVALUATE THE ANSWER

- **Are the units correct?** Dimensional analysis verifies kg•m/s^2, which is N.
- **Does the sign make sense?** The upward force should be positive.
- **Is the magnitude realistic?** The magnitude is a little larger than 490 N, which is the weight of the bucket. $F_g = mg = (50.0\text{ kg})(9.8\text{ N/kg}) = 490\text{ N}$

PRACTICE PROBLEMS

Do additional problems. Online Practice

32. Diego and Mika are trying to fix a tire on Diego's car, but they are having trouble getting the tire loose. When they pull together in the same direction, Mika with a force of 23 N and Diego with a force of 31 N, they just barely get the tire to move off the wheel. What is the magnitude of the force between the tire and the wheel?

33. CHALLENGE You are helping to repair a roof by loading equipment into a bucket that workers hoist to the rooftop. If the rope is guaranteed not to break as long as the tension does not exceed 450 N and you fill the bucket until it has a mass of 42 kg, what is the greatest acceleration that the workers can give the bucket as they pull it to the roof?

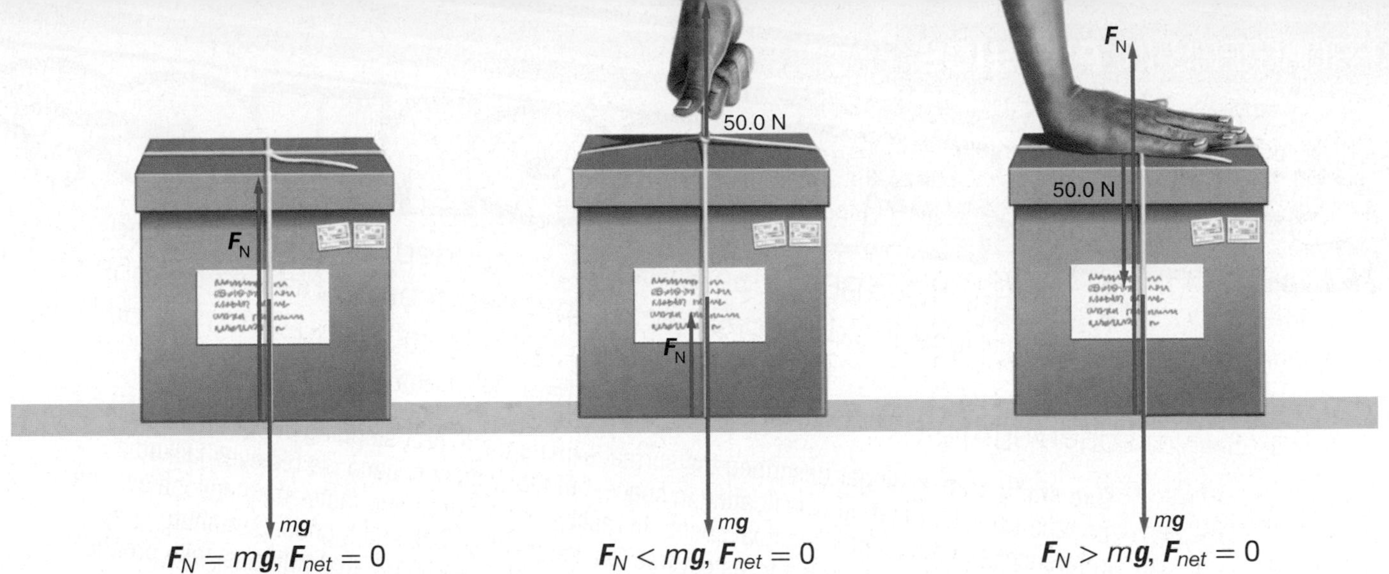

Figure 19 The normal force is not always equal to the object's weight.

The Normal Force

Any time two objects are in contact, they exert a force on each other. Think about a box sitting on a table. There is a downward force on the box due to Earth's gravitational attraction. There also is an upward force that the table exerts on the box. This force must exist because the box is in equilibrium. The **normal force** is the perpendicular contact force that a surface exerts on another surface.

The normal force always is perpendicular to the plane of contact between two objects, but is it always equal to the weight of an object? **Figure 19** shows three situations involving a box with the same weight. What if you tied a string to the box and pulled up on it a little bit, but not enough to accelerate the box, as shown in the middle panel in **Figure 19?** When you apply Newton's second law to the box and the forces acting on the box, you see $F_N + F_{\text{string on box}} - F_g = ma = 0$ N, which can be rearranged to show $F_N = F_g - F_{\text{string on box}}$.

You can see that in this case the normal force that the table exerts on the box is less than the box's weight (F_g). Similarly, if you pushed down on the box on the table as shown in the final panel in **Figure 19,** the normal force would be more than the box's weight. Finding the normal force will be important when you study friction in detail.

PhysicsLAB

NEWTON'S THIRD LAW

What are the interaction pairs between train cars?

SECTION 3 REVIEW

Section Self-Check Check your understanding.

34. MAINIDEA Hold a ball motionless in your hand in the air as in **Figure 20.** Identify each force acting on the ball and its interaction pair.

Figure 20

35. Force Imagine lowering the ball in **Figure 20** at increasing speed. Do any of the forces or their interaction-pair partners change? Draw separate free-body diagrams for the forces acting on the ball and for each set of interaction pairs.

36. Tension A block hangs from the ceiling by a massless rope. A second block is attached to the first block and hangs below it on another piece of massless rope. If each of the two blocks has a mass of 5.0 kg, what is the tension in each rope?

37. Tension A block hangs from the ceiling by a massless rope. A 3.0-kg block is attached to the first block and hangs below it on another piece of massless rope. The tension in the top rope is 63.0 N. Find the tension in the bottom rope and the mass of the top block.

38. Critical Thinking A curtain prevents two tug-of-war teams from seeing each other. One team ties its end of the rope to a tree. If the other team pulls with a 500-N force, what is the tension in the rope? Explain.

SUPERSONIC?

On August 16, 1960, Joe Kittinger ascended 31.3 km into the stratosphere inside a capsule suspended from a helium balloon... and stepped out. He fell for 4 min 36 s, reaching 274 m/s before opening his parachute, landing safely 11 min later.

One small step Kittinger described the surreal experience of first stepping out of the capsule. He had the sensation that he was floating in space, but that was only because his eyes lacked a reference point to gauge his actual speed; in reality, he was dropping faster than anyone ever had before. The jump set records for fastest free fall, longest free fall, and highest parachute jump. Kittinger's jump set the stage for crewed space programs by proving that humans, with proper equipment, can survive the extreme conditions of such high altitudes.

Terminal velocity Terminal velocity is achieved when the upward drag force equals the downward gravitational force. Terminal velocity changes depending on the temperature and density of air at given altitudes. From a normal jump altitude of about 3500 m, the fastest speed a skydiver can achieve is around 90 m/s (nearly 200 mph). Due to the low air density a jumper will experience at extreme altitudes, they could reach speeds of 300 m/s (680 mph) or more.

Going supersonic Kittinger's maximum free-fall speed of 274 m/s (614 mph) was nearly fast enough to break the sound barrier. The speed of sound in air is not a constant but depends on the temperature of the air at a given altitude. At sea level, where the temperature is about 15°C, sound travels at 340 m/s. **Figure 1** shows that at much higher altitudes, where air is less dense and much colder, sound travels more slowly. Breaking the speed of sound during free fall is one of the goals of those attempting to beat Kittinger's records.

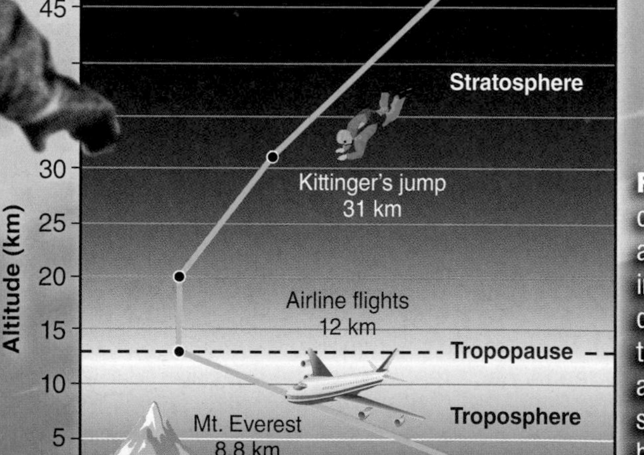

FIGURE 1 The speed of sound changes with altitude due to changes in air temperature and density. Suits similar to those worn by astronauts protect stunt skydivers attempting to break the sound barrier during free fall.

GOING FURTHER >>>

Research Colonel Kittinger's involvement in Project Excelsior and Project Manhigh. Project Excelsior is especially interesting due to the technical problems during the Excelsior I and Excelsior III jumps.

STUDY GUIDE

BIGIDEA Net forces cause changes in motion.

VOCABULARY
- **force** *(p. 90)*
- **system** *(p. 91)*
- **free-body diagram** *(p. 92)*
- **net force** *(p. 93)*
- **Newton's second law** *(p. 96)*
- **Newton's first law** *(p. 98)*
- **inertia** *(p. 98)*
- **equilibrium** *(p. 99)*

SECTION 1 Force and Motion

MAINIDEA A force is a push or a pull.

- A force is a push or a pull. Forces have both direction and magnitude. A force might be either a contact force or a field force.
- Newton's second law states that the acceleration of a system equals the net force acting on it divided by its mass.

$$\boldsymbol{a} = \frac{\boldsymbol{F}_{\text{net}}}{m}$$

- Newton's first law states that an object that is at rest will remain at rest and an object that is moving will continue to move in a straight line with constant speed, if and only if the net force acting on that object is zero. An object with zero net force acting on it is in equilibrium.

VOCABULARY
- **weight** *(p. 100)*
- **gravitational field** *(p. 100)*
- **apparent weight** *(p. 102)*
- **weightlessness** *(p. 102)*
- **drag force** *(p. 104)*
- **terminal velocity** *(p. 105)*

SECTION 2 Weight and Drag Force

MAINIDEA Newton's second law can be used to explain the motion of falling objects.

- The object's weight ($\boldsymbol{F}_g$) depends on the object's mass and the gravitational field at the object's location.

$$\boldsymbol{F}_g = m\boldsymbol{g}$$

- An object's apparent weight is the magnitude of the support force exerted on it. An object with no apparent weight experiences weightlessness.
- A falling object reaches a constant velocity when the drag force is equal to the object's weight. The constant velocity is called the terminal velocity. The drag force on an object is determined by the object's weight, size, and shape as well as the fluid through which it moves.

VOCABULARY
- **interaction pair** *(p. 106)*
- **Newton's third law** *(p. 106)*
- **tension** *(p. 109)*
- **normal force** *(p. 111)*

SECTION 3 Newton's Third Law

MAINIDEA All forces occur in interaction pairs.

- Newton's third law states that the two forces that make up an interaction pair of forces are equal in magnitude, but opposite in direction and act on different objects. In an interaction pair, $\boldsymbol{F}_{\text{A on B}}$ does not cause $\boldsymbol{F}_{\text{B on A}}$. The two forces either exist together or not at all.

$$\boldsymbol{F}_{\text{A on B}} = -\boldsymbol{F}_{\text{B on A}}$$

- The normal force is a support force resulting from the contact between two objects. It is always perpendicular to the plane of contact between the two objects.

Games and Multilingual eGlossary

Vocabulary Practice

Chapter Self-Check

SECTION 1 Force and Motion

Mastering Concepts

39. BIGIDEA You kick a soccer ball across a field. It slows down and comes to a stop. You ask your younger brother to explain what happened to the ball. He says, "The force of your foot was transferred to the ball, which made it move. When that force ran out, the ball stopped." Would Newton agree with that explanation? If not, explain how Newton's laws would describe it.

40. Cycling Imagine riding a single-speed bicycle. Why do you have to push harder on the pedals to start the bicycle moving than to keep it moving at a constant velocity?

Mastering Problems

41. What is the net force acting on a 1.0-kg ball moving at a constant velocity?

42. Skating Joyce and Efua are skating. Joyce pushes Efua, whose mass is 40.0 kg, with a force of 5.0 N. What is Efua's resulting acceleration?

43. A 2300-kg car slows down at a rate of 3.0 m/s^2 when approaching a stop sign. What is the magnitude of the net force causing it to slow down?

44. Breaking the Wishbone After Thanksgiving, Kevin and Gamal use the turkey's wishbone to make a wish. If Kevin pulls on it with a force 0.17 N larger than the force Gamal pulls with in the opposite direction and the wishbone has a mass of 13 g, what is the wishbone's initial acceleration?

SECTION 2 Weight and Drag Force

Mastering Concepts

45. Suppose that the acceleration of an object is zero. Does this mean that there are no forces acting on the object? Give an example using an everyday situation to support your answer.

46. Basketball When a basketball player dribbles a ball, it falls to the floor and bounces up. Is a force required to make it bounce? Why? If a force is needed, what is the agent involved?

47. A cart has a net horizontal force acting on it to the right. Jon says that it must be moving to the right. Joanne says no, it could be moving in either direction. Is either of these two correct? If so, explain and describe the velocity and acceleration (if any) of the cart.

48. Before a skydiver opens her parachute, she might be falling at a velocity higher than the terminal velocity that she will have after the parachute deploys.

- **a.** Describe what happens to her velocity as she opens the parachute.
- **b.** Describe the sky diver's velocity from when her parachute has been open for a time until she is about to land.

49. Three objects are dropped simultaneously from the top of a tall building: a shot put, an air-filled balloon, and a basketball.

- **a.** Rank the objects in the order in which they will reach terminal velocity, from first to last.
- **b.** Rank the objects according to the order in which they will reach the ground, from first to last.
- **c.** What is the relationship between your answers to parts a and b?

Mastering Problems

50. What is your weight in newtons? Show your work.

51. A rescue helicopter lifts two people using a winch and a rescue ring as shown in **Figure 21.**

- **a.** The winch is capable of exerting a 2000-N force. What is the maximum mass it can lift?
- **b.** If the winch applies a force of 1200 N, what is the rescuer's and victim's acceleration? Draw a free-body diagram for the people being lifted.
- **c.** Using the acceleration from part b, how long does it take to pull the people up to the helicopter? Assume the people are initially at rest.

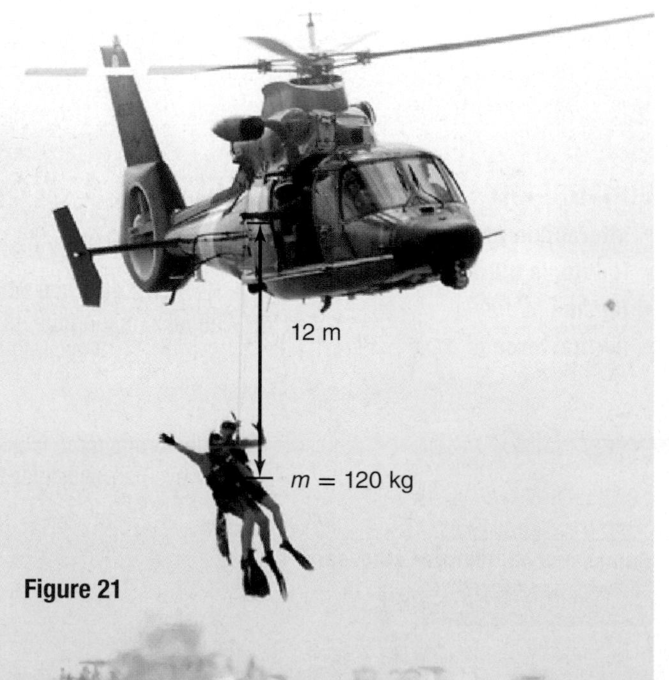

Figure 21

52. What force would a scale in an elevator on Earth exert on a 53-kg person standing on it during the following situations?
   a. The elevator moves up at a constant speed.
   b. It slows at 2.0 m/s^2 while moving upward.
   c. It speeds up at 2.0 m/s^2 while moving downward.
   d. The elevator moves down at a constant speed.
   e. It slows to a stop while moving downward with a constant acceleration of 2.5 m/s^2.

53. **Astronomy** On the surface of Mercury, the gravitational field is 0.38 times its value on Earth.
   a. What would a 6.0-kg mass weigh on Mercury?
   b. If the gravitational field on the surface of Pluto is 0.08 times that of Mercury, what would a 7.0-kg mass weigh on Pluto?

54. A 65-kg diver jumps off of a 10.0-m tower. Assume that air resistance is negligible.
   a. Find the diver's velocity when the diver hits the water.
   b. The diver comes to a stop 2.0 m below the surface. Find the net force exerted by the water.

SECTION 3 Newton's Third Law

Mastering Concepts

55. A rock is dropped from a bridge. Earth pulls on the rock and accelerates it downward. According to Newton's third law, the rock also pulls on Earth, but Earth does not seem to accelerate. Explain.

56. Explain why the tension in a massless rope is constant throughout the rope.

57. Ramon pushes on a bed as shown in **Figure 22.** Draw a free-body diagram for the bed and identify all the forces acting on it. Make a separate list of all the forces that the bed applies to other objects.

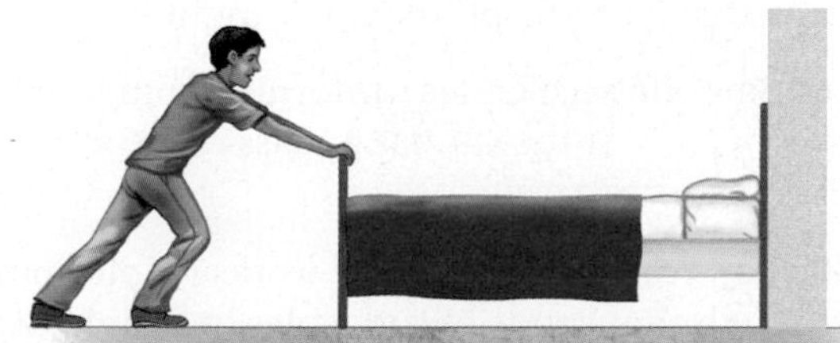

Figure 22

58. **Baseball** A batter swings a bat and hits a baseball. Draw free-body diagrams for the baseball and the bat at the moment of contact. Specifically indicate any interaction pairs between the two diagrams.

59. **Ranking Task** **Figure 23** shows a block in three different situations. Rank them according to the magnitude of the normal force between the block (or spring) and the floor, greatest to least. Specifically indicate any ties.

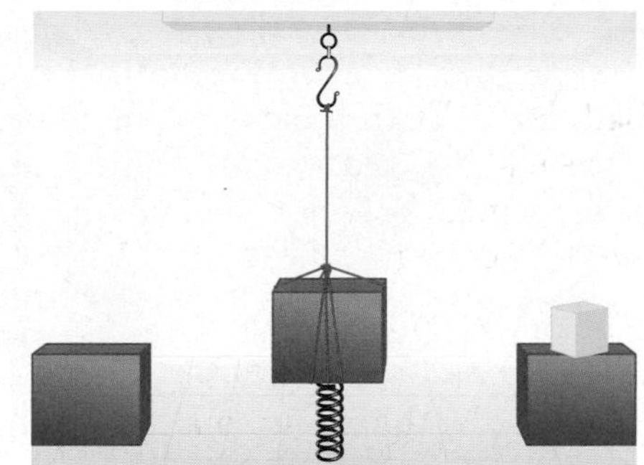

Figure 23

Mastering Problems

60. A 6.0-kg block rests on top of a 7.0-kg block, which rests on a horizontal table.
   a. What is the force (magnitude and direction) exerted by the 7.0-kg block on the 6.0-kg block?
   b. What is the force (magnitude and direction) exerted by the 6.0-kg block on the 7.0-kg block?

61. **Rain** A 2.45-mg raindrop falls to the ground. As it is falling, what magnitude of force does it exert on Earth?

62. Male lions and human sprinters can both accelerate at about 10.0 m/s^2. If a typical lion weighs 170 kg and a typical sprinter weighs 75 kg, what is the difference in the force exerted by the ground during a race between these two species? (Both the forward and normal forces should be calculated.)

63. A 4500-kg helicopter accelerates upward at 2.0 m/s^2. What lift force is exerted by the air on the propellers?

64. Three blocks are stacked on top of one another. The top block has a mass of 4.6 kg, the middle one has a mass of 1.2 kg, and the bottom one has a mass of 3.7 kg. Identify and calculate any normal forces between the objects.

Applying Concepts

65. **Whiplash** If you are in a car that is struck from behind, you can receive a serious neck injury called whiplash.
   a. Using Newton's laws, explain what happens to cause such an injury.
   b. How does a headrest reduce whiplash?

Chapter Self-Check

66. When you look at the label of the product in **Figure 24** to get an idea of how much the box contains, does it tell you its mass, weight, or both? Would you need to make any changes to this label to make it correct for consumption on the Moon?

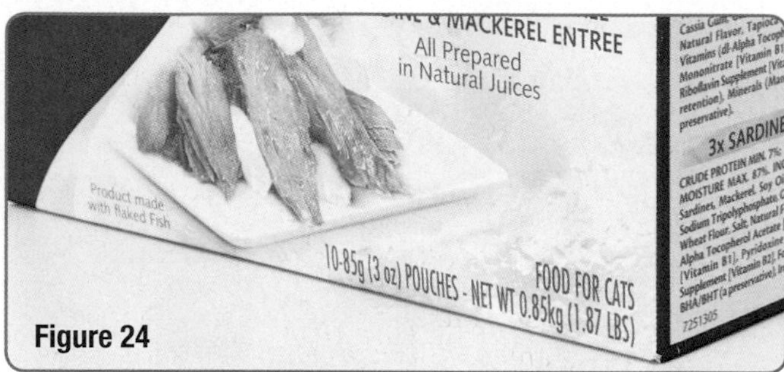

Figure 24

67. From the top of a tall building, you drop two table-tennis balls, one filled with air and the other with water. Both experience air resistance as they fall. Which ball reaches terminal velocity first? Do both hit the ground at the same time?

68. It can be said that 1 kg is equivalent to 2.21 lb. What does this statement mean? What would be the proper way of making the comparison?

69. You toss a ball straight up into the air. Assume that air resistance is negligible.
   - **a.** Draw a free-body diagram for the ball at three points: on the way up, at the very top, and on the way down. Specifically identify the forces and agents acting on the ball.
   - **b.** What is the ball's velocity at the very top of the motion?
   - **c.** What is the ball's acceleration at this point?

70. When receiving a basketball pass, a player doesn't hold his or her hands still but moves them in the direction of the moving ball. Explain in terms of acceleration and Newton's second law why the player moves his or hands in this manner.

Mixed Review

71. A dragster completed a 402.3-m (0.2500-mi) run in 5.023 s. If the car had a constant acceleration, what was its acceleration and final velocity?

72. **Space Station** Pratish weighs 588 N on Earth but is currently weightless in a space station. If she pushes off the wall with a vertical acceleration of 3.00 m/s^2, determine the force exerted by the wall during her push off.

73. **Jet** A 2.75×10^6-N jet plane is ready for takeoff. If the jet's engines supply a constant forward force of 6.35×10^6 N, how much runway will it need to reach its minimum takeoff speed of 285 km/h?

74. **Drag Racing** A 873-kg dragster, starting from rest, attains a speed of 26.3 m/s in 0.59 s.
   - **a.** Find the average acceleration of the dragster.
   - **b.** What is the magnitude of the average net force on the dragster during this time?
   - **c.** What horizontal force does the seat exert on the driver if the driver has a mass of 68 kg?

75. The dragster in the previous problem completed a 402.3-m track in 4.936 s. It crossed the finish line going 126.6 m/s. Does the assumption of constant acceleration hold true? What information is needed to determine whether the acceleration was constant?

76. Suppose a 65-kg boy and a 45-kg girl use a massless rope in a tug-of-war on an icy, resistance-free surface as in **Figure 25.** If the acceleration of the girl toward the boy is 3.0 m/s^2, find the magnitude of the acceleration of the boy toward the girl.

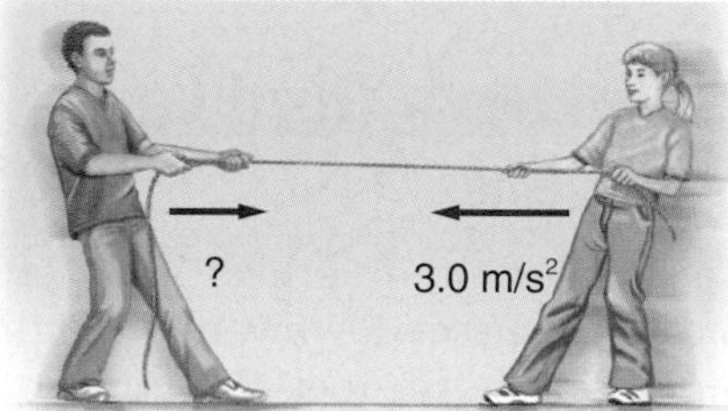

Figure 25

77. **Baseball** As a baseball is being caught, its speed goes from 30.0 m/s to 0.0 m/s in about 0.0050 s. The mass of the baseball is 0.145 kg.
   - **a.** What is the baseball's acceleration?
   - **b.** What are the magnitude and the direction of the force acting on it?
   - **c.** What are the magnitude and the direction of the force acting on the player who caught it?

78. An automobile accelerates uniformly from 0 to 24 m/s in 6.0 s. If the car has a mass of 2.0×10^3 kg, what is the force accelerating it?

79. **Air Hockey** An air-hockey table works by pumping air through thousands of tiny holes in a table to support light pucks. This allows the pucks to move around on cushions of air with very little resistance. One of these pucks has a mass of 0.25 kg and is pushed along by a 12.0-N force for 0.90 s.
   - **a.** What is the puck's acceleration?
   - **b.** What is the puck's final velocity?

80. Weather Balloon The instruments attached to a weather balloon in **Figure 26** have a mass of 8.0 kg. The balloon is released and exerts an upward force of 98 N on the instruments.

a. What is the acceleration of the balloon and the instruments?

b. After the balloon has accelerated for 10.0 s, the instruments are released. What is the velocity of the instruments at the moment of their release?

c. What net force acts on the instruments after their release?

d. When does the direction of the instruments' velocity first become downward?

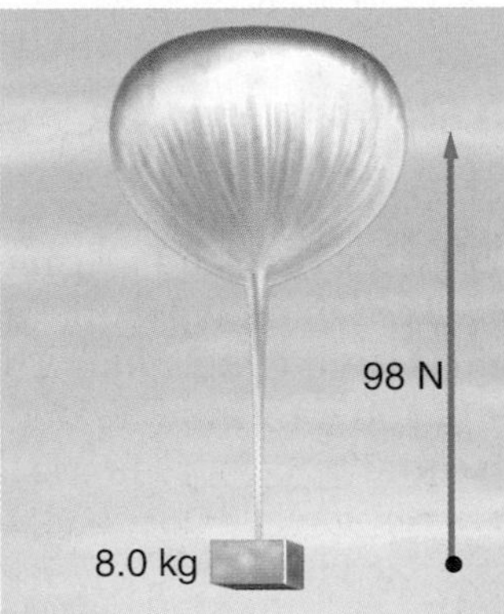

Figure 26

81. When a horizontal force of 4.5 N acts on a block on a resistance-free surface, it produces an acceleration of 2.5 m/s^2. Suppose a second 4.0-kg block is dropped onto the first. What is the magnitude of the acceleration of the combination if the same force continues to act? Assume that the second block does not slide on the first block.

82. **Figure 27** shows two blocks, masses 4.3 kg and 5.4 kg, being pushed across a frictionless surface by a 22.5-N horizontal force applied to the 4.3-kg block.

a. What is the acceleration of the blocks?

b. What is the force of the 4.3-kg block on the 5.4-kg block?

c. What is the force of the 5.4-kg block on the 4.3-kg block?

Figure 27

83. A student stands on a bathroom scale in an elevator at rest on the 64th floor of a building. The scale reads 836 N.

a. As the elevator moves up, the scale reading increases to 936 N. Find the acceleration of the elevator.

b. As the elevator approaches the 74th floor, the scale reading drops to 782 N. What is the acceleration of the elevator?

c. Using your results from parts a and b, explain which change in velocity, starting or stopping, takes the longer time.

84. Two blocks, one of mass 5.0 kg and the other of mass 3.0 kg, are tied together with a massless rope as in **Figure 28.** This rope is strung over a massless, resistance-free pulley. The blocks are released from rest. Find the following:

a. the tension in the rope

b. the acceleration of the blocks

Hint: You will need to solve two simultaneous equations.

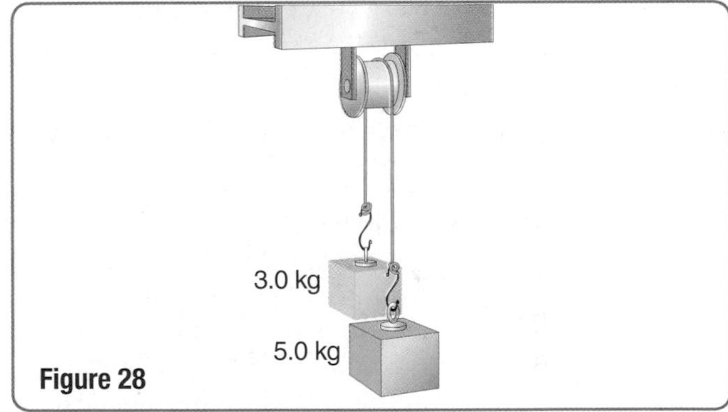

Figure 28

Thinking Critically

85. Reverse Problem Write a physics problem with real-life objects for which the following equation would be part of the solution:

$F = (23\ \text{kg})(1.8\ \text{m/s}^2)$

86. Formulate Models A 2.0-kg mass (m_A) and a 3.0-kg mass (m_B) are connected to a lightweight cord that passes over a frictionless pulley. The pulley only changes the direction of the force exerted by the rope. The hanging masses are free to move. Choose coordinate systems for the two masses with the positive direction being up for m_A and down for m_B.

a. Create a pictorial model.

b. Create a physical model with motion and free-body diagrams.

c. What is the acceleration of the smaller mass?

Chapter Self-Check

87. Use Models Suppose that the masses in the previous problem are now 1.00 kg and 4.00 kg. Find the acceleration of the larger mass.

88. Observe and Infer Three blocks that are connected by massless strings are pulled along a frictionless surface by a horizontal force, as shown in **Figure 29.**

a. What is the acceleration of each block?

b. What are the tension forces in each of the strings? *Hint: Draw a separate free-body diagram for each block.*

Figure 29

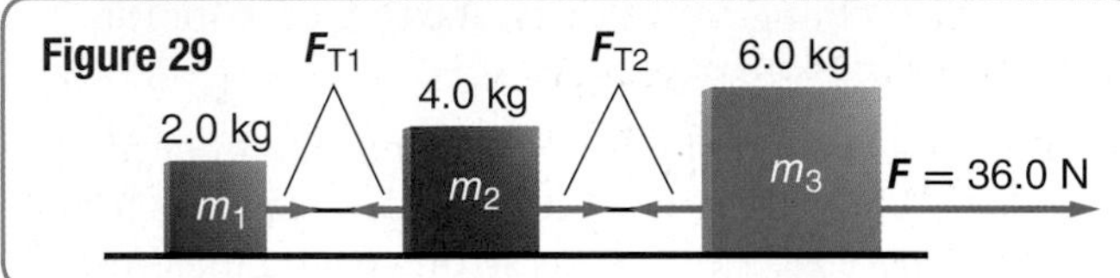

89. Critique Using the Example Problems in this chapter as models, write a solution to the following problem. A 3.46-kg block is suspended from two vertical ropes attached to the ceiling. What is the tension in each rope?

90. Think Critically You are serving as a scientific consultant for a new science-fiction TV series about space exploration. In episode 3, the heroine, Misty Moonglow, has been asked to be the first person to ride in a new interplanetary transport ship. She wants to be sure that the transport actually takes her to the planet she wants to get to, so she needs a device to measure the force of gravity when she arrives. To measure the force of gravity, the script writers would like Misty to perform an experiment involving a scale. It is your job to design a quick experiment Misty can conduct involving a scale to determine which planet she is on. Describe the experiment and include what the results would be for Venus (g = 8.9 N/kg), which is where she is supposed to go, and Mercury (g = 3.7 N/kg), which is where the transport takes her.

91. Apply Concepts Develop a lab that uses a motion detector and either a calculator or a computer program that graphs the distance a free-falling object moves over equal intervals of time. Also graph velocity versus time. Compare and contrast your graphs. Using your velocity graph, determine the gravitational field. Does it equal g?

92. Problem Posing Complete this problem so that it must be solved using the concept listed below: "A worker unloading a truck gives a 10-kg crate of oranges a push across the floor ..."

a. Newton's second law

b. Newton's third law

Writing in Physics

93. Research Newton's contributions to physics and write a one-page summary. Do you think his three laws of motion were his greatest accomplishments? Explain why or why not.

94. Review, analyze, and critique Newton's first law. Can we prove this law? Explain. Be sure to consider the role of resistance.

95. Physicists classify all forces into four fundamental categories: gravitational, electromagnetic, weak nuclear, and strong nuclear. Investigate these forces and describe the situations in which they are found.

Cumulative Review

96. Cross-Country Skiing Your friend is training for a cross-country skiing race, and you and some other friends have agreed to provide him with food and water along his training route. It is a bitterly cold day, so none of you wants to wait outside longer than you have to. Taro, whose house is the stop before yours, calls you at 8:25 A.M. to tell you that the skier just passed his house and is planning to move at an average speed of 8.0 km/h. If it is 5.2 km from Taro's house to yours, when should you expect the skier to pass your house?

97. Figure 30 is a position-time graph of the motion of two cars on a road.

a. At what time(s) does one car pass the other?

b. Which car is moving faster at 7.0 s?

c. At what time(s) do the cars have the same velocity?

d. Over what time interval is car B speeding up all the time?

e. Over what time interval is car B slowing down all the time?

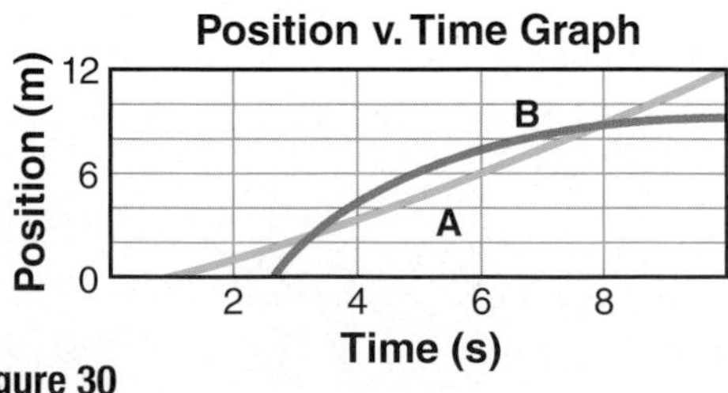

Figure 30

98. Refer to **Figure 30** to find the instantaneous speed for the following:

a. car B at 2.0 s

b. car B at 9.0 s

c. car A at 2.0 s

STANDARDIZED TEST PRACTICE

MULTIPLE CHOICE

1. What is the acceleration of the car described by the graph below?

A. 0.20 m/s^2 **C.** 1.0 m/s^2

B. 0.40 m/s^2 **D.** 2.5 m/s^2

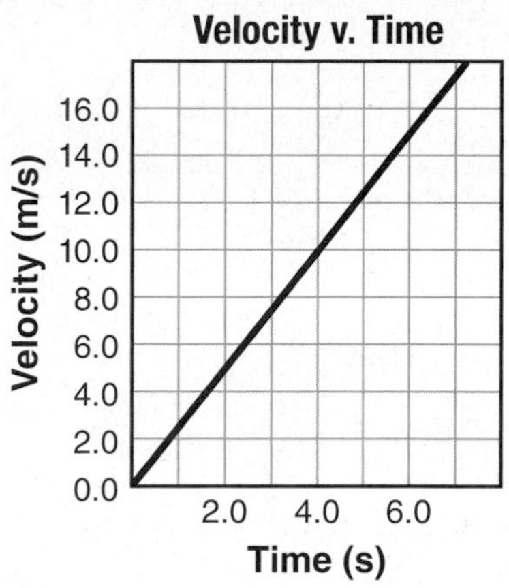

2. What distance will a sprinter travel in 4.0 s if his or her acceleration is 2.5 m/s^2? Assume the sprinter starts from rest.

A. 13 m **C.** 80 m

B. 20 m **D.** 90 m

3. If a motorcycle starts from rest and maintains a constant acceleration of 3 m/s^2, what will its velocity be after 10 s?

A. 10 m/s **C.** 90 m/s

B. 30 m/s **D.** 100 m/s

4. How does an object's acceleration change if the net force on the object is doubled?

A. The acceleration is cut in half.

B. The acceleration does not change.

C. The acceleration is doubled.

D. The acceleration is multiplied by four.

5. What is the weight of a 225-kg space probe on the Moon? The gravitational field on the Moon is 1.62 N/kg.

A. 139 N **C.** 1.35×10^3 N

B. 364 N **D.** 2.21×10^3 N

Online Test Practice

6. A 73-kg woman stands on a scale in an elevator. The scale reads 810 N. What is the magnitude and direction of the elevator's acceleration?

A. 0.23 m/s^2 up **C.** 6.5 m/s^2 down

B. 1.3 m/s^2 up **D.** 11 m/s^2 down

7. A 45-kg child sits on a 3.2-kg tire swing. What is the tension in the rope that hangs from a tree branch?

A. 310 N **C.** 4.5×10^2 N

B. 4.4×10^2 N **D.** 4.7×10^2 N

8. The tree branch in the previous problem sags, and the child's feet rest on the ground. If the tension in the rope is reduced to 220 N, what is the value of the normal force being exerted on the child's feet?

A. 2.2×10^2 N **C.** 4.3×10^2 N

B. 2.5×10^2 N **D.** 6.9×10^2 N

9. In the graph below, what is the force being exerted on the 16-kg cart?

A. 4 N **C.** 16 N

B. 8 N **D.** 32 N

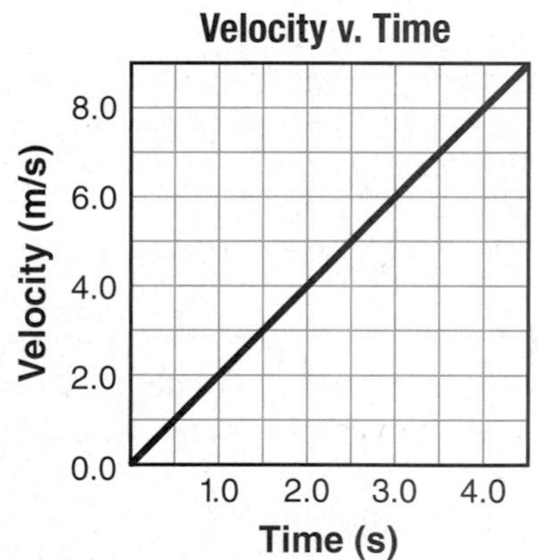

FREE RESPONSE

10. Draw a free-body diagram of a dog sitting on a scale in an elevator. Using words and mathematical formulas, describe what happens to the apparent weight of the dog when the elevator accelerates upward, when the elevator travels at a constant speed downward, and when the elevator falls freely downward.

NEED EXTRA HELP?

If you Missed Question	1	2	3	4	5	6	7	8	9	10
Review Section	1	1	1	1	2	2	3	3	1	2

CHAPTER 5

Displacement and Force in Two Dimensions

BIG IDEA Forces in two dimensions can be described using vector addition and vector resolution.

SECTIONS

1 Vectors

2 Friction

3 Forces in Two Dimensions

LaunchLAB

iLab Station

ADDING VECTORS

How can you add two force vectors with magnitudes of 2 N and find a sum of 2 N?

WATCH THIS!

Video

FORCES AND MOTION

A hurried commuter leaves a full coffee cup on top of his car before leaving for work. As the car starts moving, what will happen to the coffee cup? How long will it stay on the car before falling off?

PHYSICS T.V.

Go online!
connectED.mcgraw-hill.com

SECTION 1 Vectors

PHYSICS 4 YOU Hanging by your fingertips hundreds of meters above the ground may not be your idea of fun, but every year millions of people enjoy the sport of rock climbing. During their ascents, climbers apply forces in many different directions to overcome the force of gravity pulling them down.

MAIN IDEA

All vectors can be broken into *x*- and *y*-components.

Essential Questions

- How are vectors added graphically?
- What are the components of a vector?
- How are vectors added algebraically?

Review Vocabulary

vector a quantity that has magnitude and direction

New Vocabulary

components
vector resolution

Vectors in Two Dimensions

How do climbers cling to a rock wall? Often the climber has more than one support point. This means there are multiple forces acting on him. Because he grips crevices in the rock, the rock pulls back on him. Also the rope secures him to the rock, so there are two contact forces acting on him. Gravity pulls on him as well, so there are a total of three forces acting on the climber.

One aspect of this situation that is different from the ones that you might have studied earlier is that the forces exerted by the rock face on the climber do not push or pull only in the horizontal or vertical direction. You know from previous study that you can pick your coordinate system and orient it in the way that is most useful to analyze a situation. But how can you set up a coordinate system for a net force when you are dealing with more than one dimension? And what happens when the forces are not at right angles to each other?

Vectors revisited Let's begin by reviewing force vectors in one dimension. Consider the case in **Figure 1** in which you and a friend both push on a table together. Suppose that you each exert a 40-N force to the right. The sum of the forces is 80 N to the right, which is what you probably expected. But how do you find the sum of these force vectors? In earlier discussions of adding vectors along a straight line, you learned the resultant vector always points from the tail of the first vector to the tip of the final vector. Does this rule still apply when the vectors do not lie along a straight line?

Figure 1 The sum of the two applied forces is 80 N to the right.

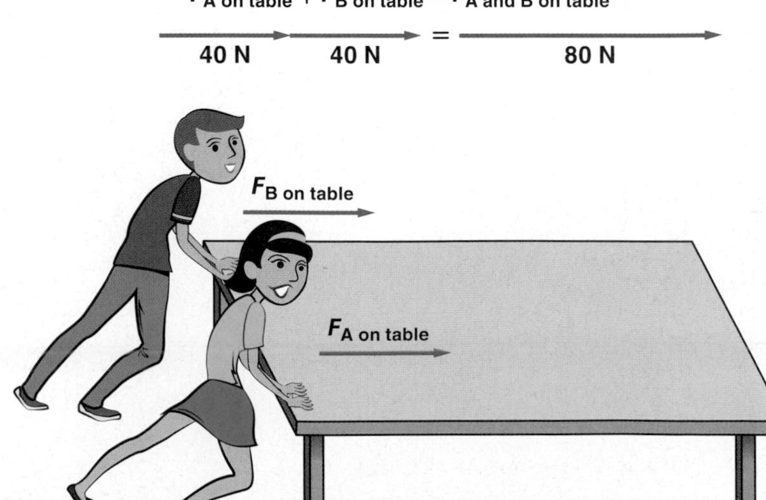

Aaron Black/The Image Bank/Getty Images

Adding vectors in two dimensions Even when vectors do not lie on a straight line, the resultant vector always points from the tail of the first vector to the tip of the final vector. You can use a protractor and a ruler both to draw the vectors at the correct angles and also to measure the magnitude and the direction of the resultant vector. **Figure 2** illustrates how to add vectors graphically in two dimensions. Notice that when vector ***B*** is moved, its magnitude and direction are unchanged. This is always the case—you do not change a vector's length or direction when you move that vector.

☑ **READING CHECK Describe** the process of graphically adding vectors using a protractor and a ruler.

Perpendicular vectors You can also use trigonometry to determine the length and the direction of resultant vectors. Remember that you can use the Pythagorean theorem to find the lengths of a right triangle's sides. If you were adding together two vectors at right angles, such as vector ***A*** pointing east and vector ***B*** pointing north in **Figure 2,** you could use the Pythagorean theorem to find the resultant's magnitude (R).

$$R^2 = A^2 + B^2$$

Angles other than 90° If you are adding two vectors that are at an angle other than 90°, as in **Figure 3,** then you can use the law of sines or the law of cosines. It is best to use the law of sines when you are given two angle measurements and only one vector magnitude. The law of cosines is particularly useful when given two vectors and the angle between the two vectors.

Law of Sines

$$\frac{R}{\sin \theta} = \frac{A}{\sin a} = \frac{B}{\sin b}$$

Law of Cosines

$$R^2 = A^2 + B^2 - 2AB \cos \theta$$

What happens if you apply the law of cosines to a triangle in which $\theta = 90°$? Notice that the first three terms in the law of cosines are the same three terms found in the Pythagorean theorem. The final term in the law of cosines, $-2AB \cos \theta$, equals zero if $\theta = 90°$ because $\cos (90°) = 0$.

$$R^2 = A^2 + B^2 - 2AB \cos 90°$$
$$R^2 = A^2 + B^2 - 2AB\ (0)$$
$$R^2 = A^2 + B^2$$

If $\theta = 90°$, the triangle is a right triangle and the law of cosines reduces to the Pythagorean theorem.

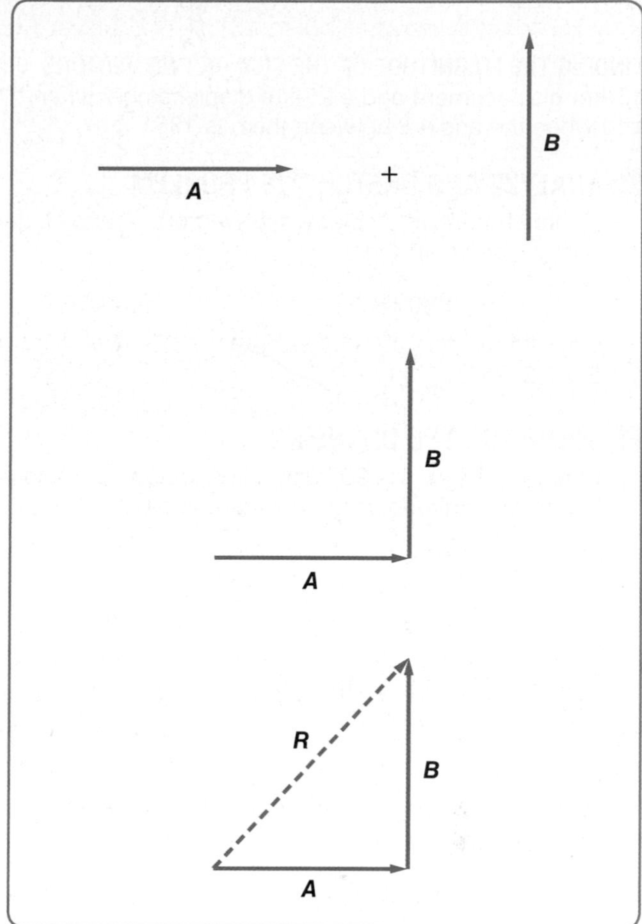

Figure 2 When adding vectors in two dimensions, follow the same process as adding vectors in one dimension: place vectors tip to tail and then connect the tail of the first vector to the tip of the final vector to find the resultant.

Figure 3 If the vectors are not at right angles, the Pythagorean theorem does not apply. Instead, use the law of cosines or the law of sines with the variables as shown below.

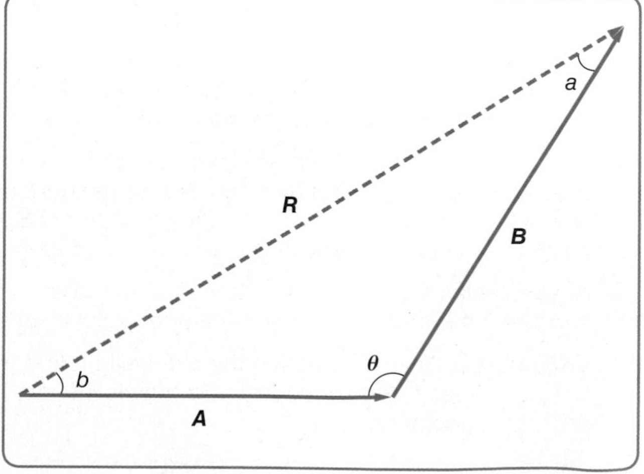

EXAMPLE PROBLEM 1

Find help with **square and cube roots.** Math Handbook

FINDING THE MAGNITUDE OF THE SUM OF TWO VECTORS Find the magnitude of the sum of a 15-km displacement and a 25-km displacement when the angle θ between them is 90° and when the angle θ between them is 135°.

1 ANALYZE AND SKETCH THE PROBLEM

- Sketch the two displacement vectors, A and B, and the angle between them.

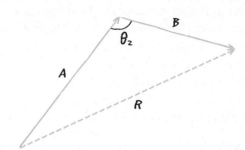

KNOWN		UNKNOWN
$A = 25$ km	$\theta = 90°$ or $\theta_2 = 135°$	$R = ?$
$B = 15$ km		

2 SOLVE FOR THE UNKNOWN

When the angle θ is 90°, use the Pythagorean theorem to find the magnitude of the resultant vector.

$$R^2 = A^2 + B^2$$

$$R = \sqrt{A^2 + B^2}$$

$$= \sqrt{(25\text{ km})^2 + (15\text{ km})^2}$$ ◀ Substitute $A = 25$ km, $B = 15$ km

$$= 29\text{ km}$$

When the angle θ does not equal 90°, use the law of cosines to find the magnitude of the resultant vector.

$$R^2 = A^2 + B^2 - 2AB(\cos\theta_2)$$

$$R = \sqrt{A^2 + B^2 - 2AB(\cos\theta_2)}$$

$$= \sqrt{(25\text{ km})^2 + (15\text{ km})^2 - 2(25\text{ km})(15\text{ km})(\cos 135°)}$$ ◀ Substitute $A = 25$ km, $B = 15$ km, $\theta_2 = 135°$

$$= 37\text{ km}$$

3 EVALUATE THE ANSWER

- **Are the units correct?** Each answer is a length measured in kilometers.
- **Do the signs make sense?** The sums are positive.
- **Are the magnitudes realistic?** From the sketch, you can see the resultant should be longer than either vector.

PRACTICE PROBLEMS

Do additional problems. Online Practice

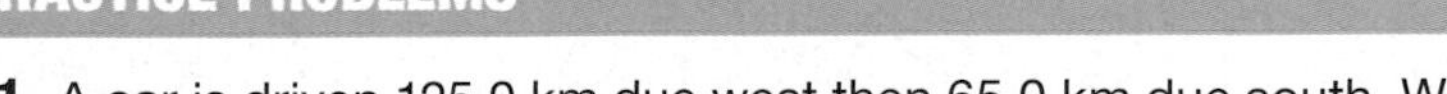

1. A car is driven 125.0 km due west then 65.0 km due south. What is the magnitude of its displacement? Solve this problem both graphically and mathematically, and check your answers against each other.
2. Two shoppers walk from the door of the mall to their car. They walk 250.0 m down a lane of cars, and then turn 90° to the right and walk an additional 60.0 m. How far is the shoppers' car from the mall door? Solve this problem both graphically and mathematically, and check your answers against each other.
3. A hiker walks 4.5 km in one direction then makes a 45° turn to the right and walks another 6.4 km. What is the magnitude of the hiker's displacement?
4. **CHALLENGE** An ant crawls on the sidewalk. It first moves south a distance of 5.0 cm. It then turns southwest and crawls 4.0 cm. What is the magnitude of the ant's displacement?

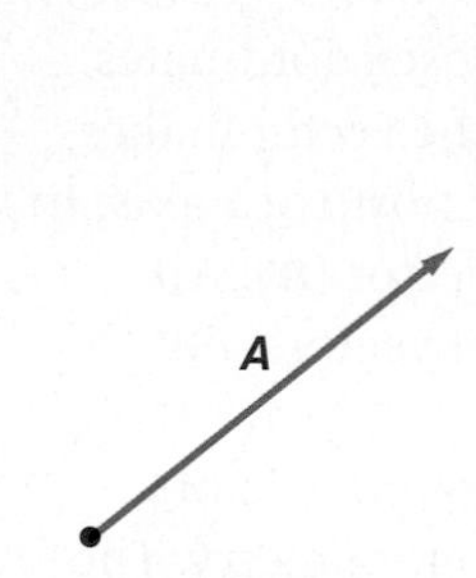

You may place the vector on any coordinate system as long as the vector's direction and magnitude remain unchanged.

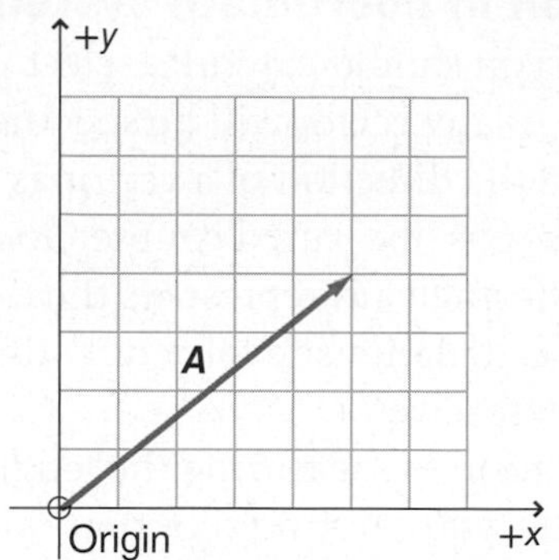

The coordinate system can be oriented to make the problem easier to solve.

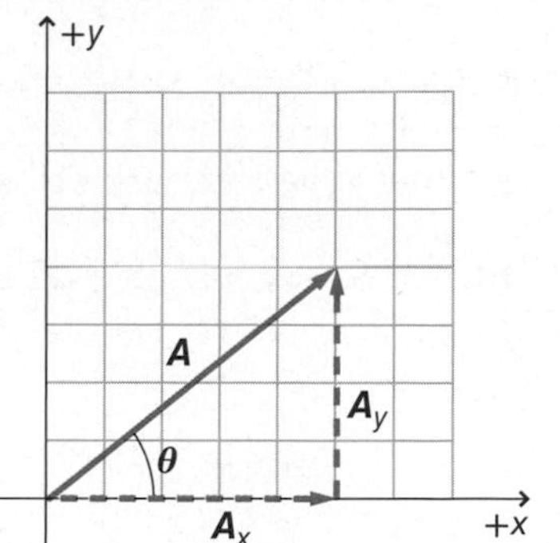

If the components of a vector are added together, they result in the original vector.

Figure 4 Vector ***A*** is placed on a coordinate system. Notice that ***A***'s direction is measured counterclockwise from the positive *x*-axis.

Describe a vector whose *y*-component is zero.

View an **animation of component vectors.**

Vector Components

What is the direction of the vector shown in the left panel in **Figure 4?** The vector isn't pointing directly up or to the right, but somewhere between the two. To determine the exact direction, we need to choose a coordinate system. This coordinate system, such as the one in the center panel of **Figure 4,** is similar to laying a grid drawn on a sheet of transparent plastic on top of a vector problem. You choose where to put the center of the grid (the origin) and establish the directions in which the axes point. Notice that the *x*-axis is drawn through the origin with an arrow pointing in the positive direction. The positive *y*-axis is located 90° from the positive *x*-axis and crosses the *x*-axis at the origin.

How do you choose the direction of the *x*-axis? There is never a single correct answer, but some choices make the problem easier to solve than others. When the motion you are describing is confined to the surface of Earth, it is often convenient to have the *x*-axis point east and the *y*-axis point north. If the motion is on an incline, it's convenient to place the positive *x*-axis in the direction of the motion parallel to the surface of the incline.

Component vectors Defining a coordinate system allows you to describe a vector in a very useful way. Vector **A,** shown in the right panel of **Figure 4,** for example, could now be described as going 5 units in the positive *x*-direction and 4 units in the positive *y*-direction. You can represent this information in the form of two vectors like the ones labeled $\mathbf{A}_x$ and $\mathbf{A}_y$ in the diagram. Notice that $\mathbf{A}_x$ is parallel to the *x*-axis and $\mathbf{A}_y$ is parallel to the *y*-axis. Further, you can see that if you add $\mathbf{A}_x$ and $\mathbf{A}_y$, the resultant is the original vector, **A.** A vector can be broken into its **components,** which are a vector parallel to the *x*-axis and another parallel to the *y*-axis. This can always be done, and the following vector equation is always true.

$$\mathbf{A} = \mathbf{A}_x + \mathbf{A}_y$$

This process of breaking a vector into its components is sometimes called **vector resolution.** Notice that the original vector is the hypotenuse of a right triangle. This means that the magnitude of the original vector will always be greater than or equal to the magnitude of either component vector.

VOCABULARY
Science Usage v. Common Usage

Components
- Science usage
projections of a vector parallel to the *x*- and *y*-axes
A vector is the algebraic sum of its components.
- Common usage
a constituent part or ingredient
Some of the computer's electronic components malfunctioned during the power surge.

Direction in coordinate systems In a two-dimensional force problem, you should carefully select your coordinate system because the direction of any vector will be specified relative to those coordinates. We define the direction of a vector as the angle that the vector makes with the x-axis, measured counterclockwise from the positive x-axis. In **Figure 5,** the angle (θ) represents the direction of the vector (**A**). All algebraic calculations involve only the components of vectors, not the vectors themselves.

In addition to measuring the lengths of the component vectors graphically, you can find the components by using trigonometry. The components are calculated using the equations below.

$$\cos\theta = \frac{\text{adjacent side}}{\text{hypotenuse}} = \frac{A_x}{A}; \text{ therefore, } A_x = A\cos\theta$$

$$\sin\theta = \frac{\text{opposite side}}{\text{hypotenuse}} = \frac{A_y}{A}; \text{ therefore, } A_y = A\sin\theta$$

When the angle that a vector makes with the x-axis is larger than 90°, the sign of one or more components is negative, as shown in the top coordinate system of **Figure 5.**

READING CHECK Explain how you should measure the direction of a vector.

Figure 5 The coordinate plane is divided into four quadrants. A vector's components will be positive or negative depending on the vector's quadrant.

Classify as positive or negative the components of a vector whose angle is 280°. In which quadrant does the vector lie?

■ **Coordinate System**

+y

Second quadrant
$90° < \theta < 180°$

A_x is negative.

A_y is positive.

$\tan\theta$ is negative.

II

First quadrant
$0° < \theta < 90°$

A_x is positive.

A_y is positive.

$\tan\theta$ is positive.

I

+x

III

A_x is negative.

A_y is negative.

$\tan\theta$ is positive.

Third quadrant
$180° < \theta < 270°$

IV

A_x is positive.

A_y is negative.

$\tan\theta$ is negative.

Fourth quadrant
$270° < \theta < 360°$

Example:

+y, A, A_y, $\theta = 130°$, A_x, +x

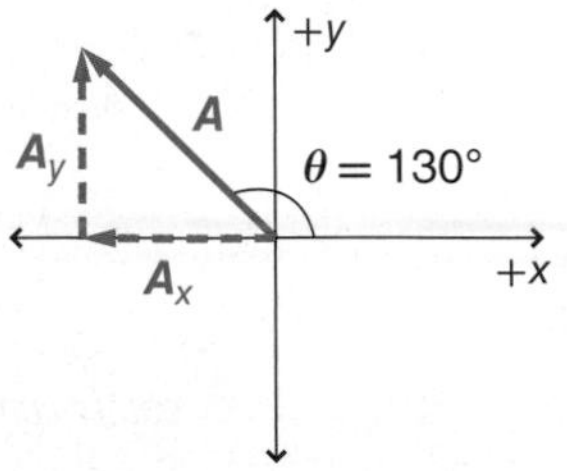

- **A** is in 2nd quadrant.
- Expect A_x to be negative:

$$A_x = A\cos\theta = (5.0\text{ N})\cos 130° = -3.2\text{ N}$$

- Expect A_y to be positive.

$$A_y = A\sin\theta = (5.0\text{ N})\sin 130° = 3.8\text{ N}$$

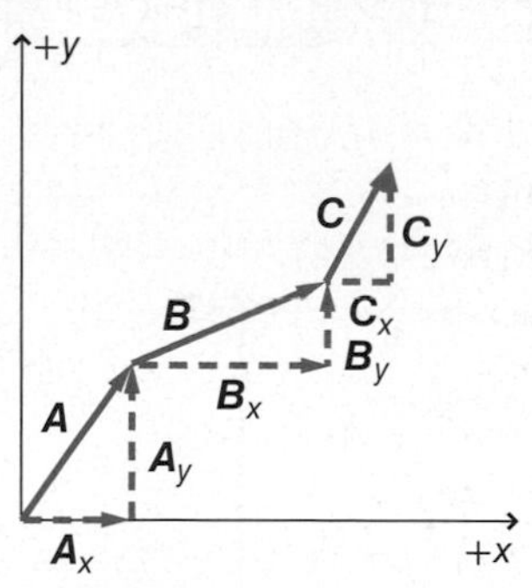

Add the vectors graphically by placing them tip to tail.

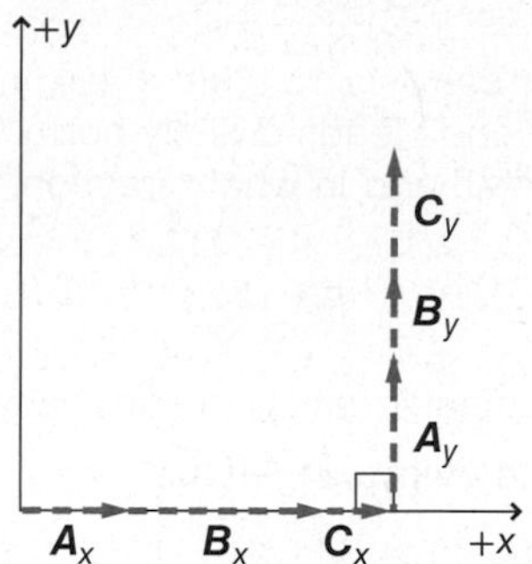

Add the *x*-components together and the *y*-components together.

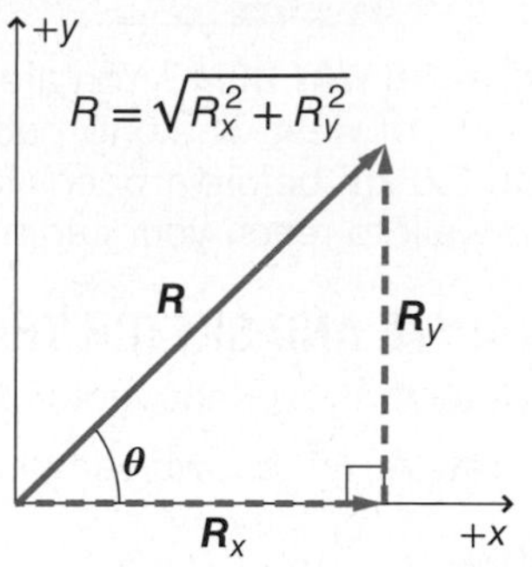

The magnitude of ***R*** can be calculated using the Pythagorean theorem.

Figure 6 The vector sum of ***A, B,*** and ***C*** is the same as the vector sum of $\mathbf{R_x}$ and $\mathbf{R_y}$.

Algebraic Addition of Vectors

You might be wondering why you should resolve vectors into their components. **Figure 6** shows how resolving vectors into components makes adding vectors together much easier. Two or more vectors (***A, B, C,*** etc.) may be added by first resolving each vector into its *x*- and *y*-components. You add the *x*-components to form the *x*-component of the resultant:

$$\mathbf{R_x} = \mathbf{A_x} + \mathbf{B_x} + \mathbf{C_x}.$$

Similarly, you add the *y*-components to form the *y*-component of the resultant:

$$\mathbf{R_y} = \mathbf{A_y} + \mathbf{B_y} + \mathbf{C_y}.$$

Because $\mathbf{R_x}$ and $\mathbf{R_y}$ are at a right angle (90°), you can calculate the magnitude of the resultant vector using the Pythagorean theorem,

$$R^2 = {R_x}^2 + {R_y}^2$$

To find the resultant vector's direction, recall that the angle the vector makes with the *x*-axis is given by the following equation.

$$\theta = \tan^{-1}\left(\frac{R_y}{R_x}\right)$$

You can find the angle by using the $\tan^{-1}$ key on your calculator. Note that when $\tan\theta > 0$, most calculators give the angle between 0° and 90°. When $\tan\theta < 0$, the reported angle will be between 0° and −90°.

PROBLEM-SOLVING STRATEGIES

VECTOR ADDITION

Use the following technique to solve problems for which you need to add or subtract vectors.

1. Choose a coordinate system.
2. Resolve the vectors into their *x*-components using $A_x = A\cos\theta$ and their *y*-components using $A_y = A\sin\theta$, where θ is the angle measured counterclockwise from the positive *x*-axis.
3. Add or subtract the component vectors in the *x*-direction.
4. Add or subtract the component vectors in the *y*-direction.
5. Use the Pythagorean theorem, $R = \sqrt{{R_x}^2 + {R_y}^2}$, to find the magnitude of the resultant vector.
6. Use $\theta = \tan^{-1}\left(\frac{R_y}{R_x}\right)$ to find the angle of the resultant vector.

EXAMPLE PROBLEM 2

Find help with **trigonometry.** Math Handbook

EXAMPLE PROBLEM

FINDING YOUR WAY HOME You are on a hike. Your camp is 15.0 km away, in the direction 40.0° north of west. The only path through the woods leads directly north. If you follow the path 5.0 km before it opens into a field, how far, and in what direction, would you have to walk to reach your camp?

1 ANALYZE AND SKETCH THE PROBLEM

- Draw the resultant vector, ***R,*** from your original location to your camp.
- Draw ***A,*** the known vector, and draw ***B,*** the unknown vector.

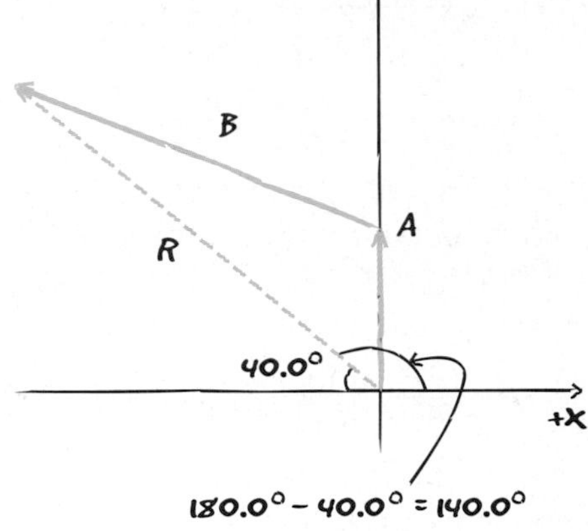

KNOWN	UNKNOWN
$\boldsymbol{A}$ = 5.0 km, due north	$\boldsymbol{B}$ = ?
$\boldsymbol{R}$ = 15.0 km, 40.0° north of west	
θ = 140.0°	

2 SOLVE FOR THE UNKNOWN

Find the components of ***R.***

$R_x = R \cos \theta$

$= (15.0 \text{ km}) \cos 140.0°$ ◀ Substitute R = 15.0 km, θ = 140.0°.

$= -11.5$ km

$R_y = R \sin \theta$

$= (15.0 \text{ km}) \sin 140.0°$ ◀ Substitute R = 15.0 km, θ = 140.0°.

$= 9.64$ km

Because ***A*** is due north, A_x = 0.0 km and $\boldsymbol{A}_y$ = 5.0 km.

Use the components of ***R*** and ***A*** to find the components of ***B.***

$B_x = R_x - A_x$

$= -11.5 \text{ km} - 0.0 \text{ km}$ ◀ Substitute R_x = −11.5 km, A_x = 0.0 km.

$= -11.5$ km ◀ The negative sign means that this component points west.

$B_y = R_y - A_y$

$= 9.64 \text{ km} - 5.0 \text{ km}$ ◀ Substitute R_y = 9.64 km, A_y = 5.0 km.

$= 4.6$ km ◀ This component points north.

Use the components of vector ***B*** to find the magnitude of vector ***B.***

$B = \sqrt{B_x^2 + B_y^2}$

$= \sqrt{(-11.5 \text{ km})^2 + (4.6 \text{ km})^2}$ ◀ Substitute B_x = −11.5 km, B_y = 4.6 km.

$= 12.4$ km

Locate the tail of vector ***B*** at the origin of a coordinate system, and draw the components B_x and B_y. The vector ***B*** is in the second quadrant. Use the tangent to find the direction of vector ***B.***

$\theta = \tan^{-1}\left(\frac{B_y}{B_x}\right)$

$= \tan^{-1}\left(\frac{4.6 \text{ km}}{-11.5 \text{ km}}\right)$ ◀ Substitute B_y = 4.6 km, B_x = −11.5 km.

$= -22°$ or $158°$ ◀ Tangent of an angle is negative in quadrants II and IV, so two answers are possible.

Since ***B*** is in the second quadrant, θ, measured from the positive x-axis, must be 158°. This direction can also be given as 22° north of west. Thus, ***B*** = 12.4 km at 22° north of west.

3 EVALUATE THE ANSWER

- **Are the units correct?** Kilometers and degrees are correct.
- **Do the signs make sense?** They agree with the diagram.
- **Is the magnitude realistic?** The length of ***B*** should be longer than R_x because the angle between ***A*** and ***B*** is greater than 90°.

PRACTICE PROBLEMS

Do additional problems. Online Practice

Solve problems 5–10 algebraically. You may also solve some of them graphically to check your answers.

5. Sudhir walks 0.40 km in a direction 60.0° west of north then goes 0.50 km due west. What is his displacement?

6. You first walk 8.0 km north from home then walk east until your displacement from home is 10.0 km. How far east did you walk?

7. In a coordinate system in which the positive *x*-axis is east, for what range of angles is the *x*-component positive? For what range is it negative?

8. Could a vector ever be shorter than one of its components? Could a vector ever be equal in length to one of its components? Explain.

9. Two ropes tied to a tree branch hold up a child's swing as shown in **Figure 7.** The tension in each rope is 2.28 N. What is the combined force (magnitude and direction) of the two ropes on the swing?

10. CHALLENGE Afua and Chrissy are going to sleep overnight in their tree house and are using some ropes to pull up a 3.20-kg box containing their pillows and blankets. The girls stand on different branches, as shown in **Figure 8,** and pull at the angles with the forces indicated. Find the *x*- and *y*-components of the initial net force on the box. *Hint: Draw a free-body diagram so you do not leave out a force.*

Figure 7

Figure 8

SECTION 1 REVIEW

Check your understanding.

11. MAINIDEA Find the components of vector ***M,*** shown in **Figure 9.**

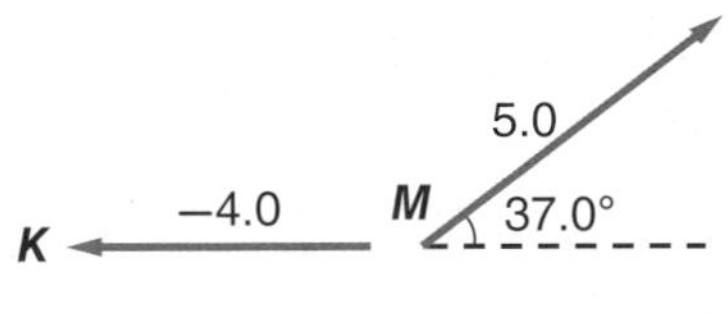

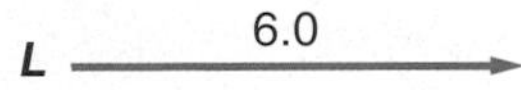

Figure 9

12. Components of Vectors Find the components of vectors ***K*** and ***L*** in **Figure 9.**

13. Vector Sum Find the sum of the three vectors shown in **Figure 9.**

14. Vector Difference Subtract vector ***K*** from vector ***L,*** shown in **Figure 9.**

15. Commutative Operations Mathematicians say that vector addition is commutative because the order in which vectors are added does not matter.

a. Use the vectors from **Figure 9** to show graphically that $\boldsymbol{M} + \boldsymbol{L} = \boldsymbol{L} + \boldsymbol{M}$.

b. Which ordinary arithmetic operations (addition, subtraction, multiplication, and division) are commutative? Which are not? Give an example of each operation to support your conclusion.

16. Distance v. Displacement Is the distance you walk equal to the magnitude of your displacement? Give an example that supports your conclusion.

17. Critical Thinking You move a box through one displacement and then through a second displacement. The magnitudes of the two displacements are unequal. Could the displacements have directions such that the resultant displacement is zero? Suppose you move the box through three displacements of unequal magnitude. Could the resultant displacement be zero? Support your conclusion with a diagram.

SECTION 2 Friction

Imagine trying to play basketball while wearing socks instead of athletic shoes. You would slip and slide all over the basketball court. Shoes help provide the forces necessary to quickly change directions while running up and down the court.

MAIN IDEA

Friction is a type of force between two touching surfaces.

Essential Questions

- What is the friction force?
- How do static and kinetic friction differ?

Review Vocabulary

force push or pull exerted on an object

New Vocabulary

kinetic friction
static friction
coefficient of kinetic friction
coefficient of static friction

Kinetic and Static Friction

Shove your book across a desk. When you stop pushing, the book will quickly come to rest. The friction force of the desk acting on the book accelerated it in the direction opposite to the one in which the book was moving. So far, you have not considered friction in solving problems, but friction is all around you.

Types of friction There are two types of friction. When you moved your book across the desk, the book experienced a type of friction that acts on moving objects. This force is known as **kinetic friction,** and it is exerted on one surface by another when the two surfaces rub against each other because one or both surfaces are moving.

To understand the other type of friction, imagine trying to push a couch across the floor as shown on the left in **Figure 10.** You push on it with a small force, but it does not move. Because it is not accelerating, Newton's laws tell you that the net force on the couch must be zero. There must be a second horizontal force acting on the couch, one that opposes your force and is equal in size. This force is **static friction,** which is the force exerted on one surface by another when there is no motion between the two surfaces. You need to push harder.

If the couch still does not move, the force of static friction must be increasing in response to your applied force. Finally when you push hard enough, the couch will begin to move as in the right side of **Figure 10.** Evidently there is a limit to how large the static friction force can be. Once your force is greater than this maximum static friction, the couch begins moving and kinetic friction begins to act on it.

Figure 10 An applied force is balanced by static friction up to a maximum limit. When this limit is exceeded, the object begins to move.

Identify the type of friction force acting on the couch when it begins to move.

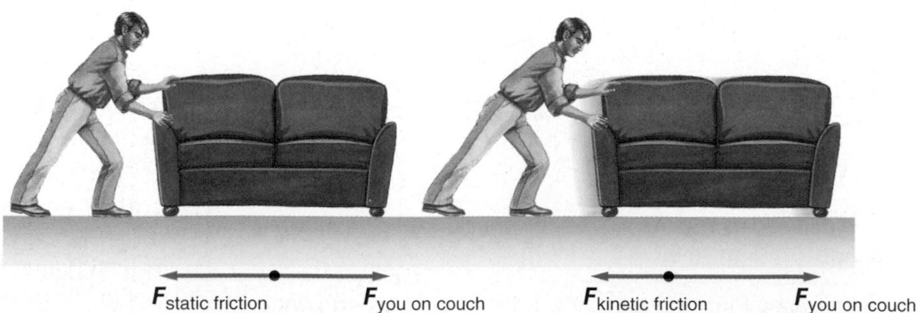

Static friction increases up to a maximum to balance the applied force.

The couch accelerates when the applied force exceeds the maximum static friction force.

Pete Saloutos/Photographer's Choice RF/Getty Images

Mathematical models for friction forces On what does a friction force depend? The materials that the surfaces are made of play a role. For example, there is more friction between your shoes and concrete than there is between your socks and a polished wood floor. It might seem reasonable that the force of friction depends on either the surface area in contact or the speed at which the surfaces move past each other. Experiments have shown that this is not true. The normal force between the two objects does matter, however. The harder one object is pushed against the other, the greater the force of friction that results.

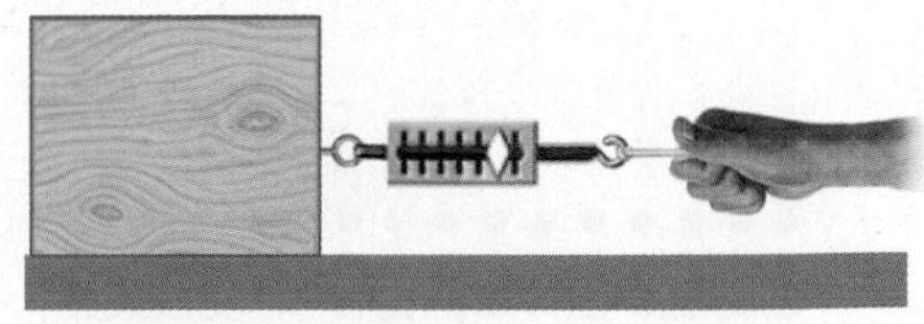

Figure 11 The spring scale applies a constant force on the block.

☑ **READING CHECK** **Identify** the two factors that affect friction forces.

Kinetic friction Imagine pulling a block along a surface at a constant velocity. Since the block is not accelerating, according to Newton's laws, the friction force must be equal and opposite to the force with which you pull. **Figure 11** shows one way in which you can measure the force you exert as you pull a block of known mass along a table at a constant velocity. The spring scale will indicate the force you exert on the block. You can then stack additional blocks on top of the first block to increase the normal force and repeat the measurement. **Table 1** shows the results of such an experiment.

Plotting the data will yield a graph like the one in **Figure 12.** There is a linear relationship between the kinetic friction force and the normal force. The different lines on the graph correspond to dragging the block along different surfaces. Note that the line for the sandpaper surface has a steeper slope than the line for the highly polished table. You would expect it to be much harder to pull the block along sandpaper than along a polished table, so the slope must be related to the magnitude of the resulting friction force. The slope of the line on a kinetic friction force v. normal force graph, designated μ_k, is called the **coefficient of kinetic friction** and relates the friction force to the normal force.

PhysicsLAB

HIT-AND-RUN DRIVER

FORENSICS LAB How can you determine a car's speed by examining the tire marks left at the scene of an accident?

KINETIC FRICTION FORCE

The kinetic friction force equals the product of the coefficient of kinetic friction and the normal force.

$$F_{\text{f, kinetic}} = \mu_k F_N$$

Figure 12 A plot of kinetic friction v. normal force for a block pulled along different surfaces shows a linear relationship between the two forces for each surface. The slope of the line is μ_k.

Compare the coefficient of kinetic friction for the three surfaces shown on the graph.

Table 1 Kinetic Friction v. Normal Force (sandpaper)

Number of blocks	Normal force (N)	Kinetic friction (N)
1	0.98	0.53
2	1.96	0.95
3	2.94	1.4
4	3.92	1.8
5	4.90	2.3
6	5.88	3.1
7	6.86	3.3
8	7.84	4.0

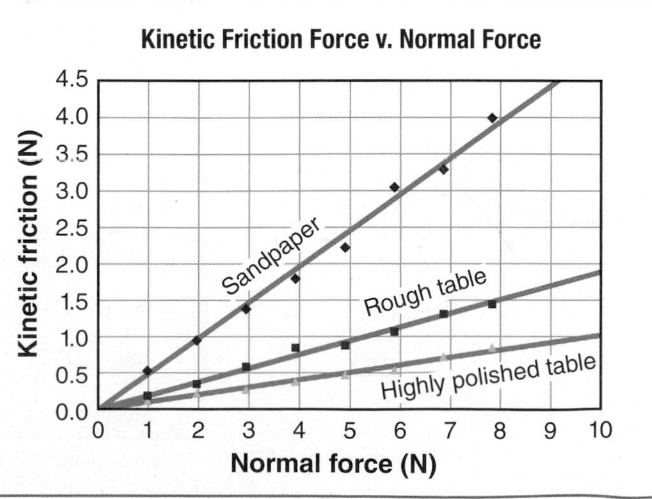

REAL-WORLD PHYSICS

CAUSES OF FRICTION All surfaces, even those that appear to be smooth, are rough at a microscopic level. If you look at a photograph of a graphite crystal magnified by a scanning tunneling microscope, the atomic level surface irregularities of the crystal are revealed. When two surfaces touch, the high points on each are in contact and temporarily bond. This is the origin of both static and kinetic friction. The details of this process are still unknown and are the subject of research in both physics and engineering.

Static friction The maximum static friction force relates to the normal force in a similar way as the kinetic friction force. Remember that the static friction force acts in response to a force trying to cause a stationary object to start moving. If there is no such force acting on an object, the static friction force is zero. If there is a force trying to cause motion, the static friction force will increase up to a maximum value before it is overcome and motion starts.

STATIC FRICTION FORCE

The static friction force is less than or equal to the product of the coefficient of static friction and the normal force.

$$F_{\text{f, static}} \leq \mu_s F_N$$

In the equation for the maximum static friction force, μ_s is the **coefficient of static friction** between the two surfaces. The maximum static friction force that must be overcome before motion can begin is $\mu_s F_N$. In the example of pushing the couch shown in **Figure 10,** the maximum static friction force balances the force of the person pushing on the couch the instant before the couch begins to move.

Note that the equations for the kinetic and maximum static friction forces involve only the magnitudes of the forces. The friction forces (F_f) are always perpendicular to the normal force (F_N).

Measuring coefficients of friction **Table 2** shows coefficients of friction between various surfaces. These coefficients are estimates for each combination of surfaces. Measurements of coefficients of friction are quite sensitive to the conditions of the surfaces. Surface impurities, such as dust or small amounts of oil from experimenters' hands, can significantly affect the results of the measurements. Another important fact regarding **Table 2** is that all the measurements were made on dry surfaces (with the exception of the oiled steel). Wet surfaces behave quite differently than dry surfaces. The linear relationship between the kinetic friction force and the normal force does not always apply to wet surfaces or surfaces treated with oil. All problems in this text assume dry surfaces and a linear relationship between the kinetic friction force and the normal force.

PhysicsLAB

COEFFICIENT OF FRICTION

How are coefficients of static and kinetic friction measured?

Table 2 Typical Coefficients of Friction*

Surfaces	Coefficient of static friction (μ_s)	Coefficient of kinetic friction (μ_k)
Cast iron on cast iron	1.1	0.15
Glass on glass	0.94	0.4
Leather on oak	0.61	0.52
Nonstick coating on steel	0.04	0.04
Oak on oak	0.62	0.48
Steel on steel	0.78	0.42
Steel on steel (with castor oil)	0.15	0.08

*All measurements are for dry surfaces unless otherwise stated.

EXAMPLE PROBLEM 3

Find help with **significant figures.** Math Handbook

BALANCED FRICTION FORCES You push a 25.0-kg wooden box across a wooden floor at a constant speed of 1.0 m/s. The coefficient of kinetic friction is 0.20. How large is the force that you exert on the box?

1 ANALYZE AND SKETCH THE PROBLEM

- Identify the forces, and establish a coordinate system.
- Draw a motion diagram indicating constant v and $a = 0$.
- Draw the free-body diagram.

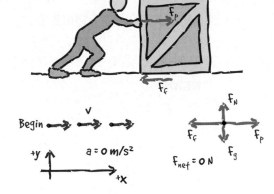

KNOWN		UNKNOWN
$m = 25.0$ kg	$v = 1.0$ m/s	$\mathbf{F}_{\text{Person on box}} = ?$
$a = 0.0$ m/s^2	$\mu_k = 0.20$	

2 SOLVE FOR THE UNKNOWN

The normal force is in the y-direction, and the box does not accelerate in that direction.

$$\mathbf{F}_N = -\mathbf{F}_g$$

$$= -m\mathbf{g}$$ ◀ Substitute $F_g = mg$

$$= -(25.0 \text{ kg})(-9.8 \text{ N/kg})$$ ◀ Substitute $m = 25.0$ kg, $g = -9.8$ N/kg

$$= +245 \text{ N}$$

The pushing force is in the x-direction; v is constant, thus the box does not accelerate.

$$F_{\text{Person on box}} = \mu_k F_N$$

$$= (0.20)(245 \text{ N})$$ ◀ Substitute $\mu_k = 0.20$, $F_N = 245$ N

$$= 49 \text{ N}$$

$$\mathbf{F}_{\text{Person on box}} = 49 \text{ N, to the right}$$

3 EVALUATE THE ANSWER

- **Are the units correct?** Force is measured in newtons.
- **Does the sign make sense?** The positive sign agrees with the sketch.
- **Is the magnitude realistic?** The pushing force is $\frac{1}{5}$ the weight of the box. This corresponds with $\mu_k = 0.20 = \frac{1}{5}$.

PRACTICE PROBLEMS

Do additional problems. Online Practice

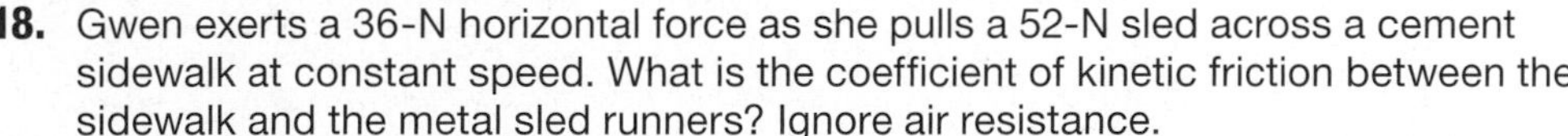

18. Gwen exerts a 36-N horizontal force as she pulls a 52-N sled across a cement sidewalk at constant speed. What is the coefficient of kinetic friction between the sidewalk and the metal sled runners? Ignore air resistance.

19. Mr. Ames is dragging a box full of books from his office to his car. The box and books together have a combined weight of 134 N. If the coefficient of static friction between the pavement and the box is 0.55, how hard must Mr. Ames push horizontally on the box in order to start it moving?

20. Thomas sits on a small rug on a polished wooden floor. The coefficient of kinetic friction between the rug and the slippery wooden floor is only 0.12. If Thomas weighs 650 N, what horizontal force is needed to pull the rug and Thomas across the floor at a constant speed?

21. CHALLENGE You need to move a 105-kg sofa to a different location in the room. It takes a 403-N force to start the sofa moving. What is the coefficient of static friction between the sofa and the carpet?

EXAMPLE PROBLEM 4

Find help with **isolating a variable.** Math Handbook

UNBALANCED FRICTION FORCES Imagine the force you exert on the 25.0-kg box in Example Problem 3 is doubled.

a. What is the resulting acceleration of the box?

b. How far will you push the box if you push it for 3 s?

1 ANALYZE AND SKETCH THE PROBLEM

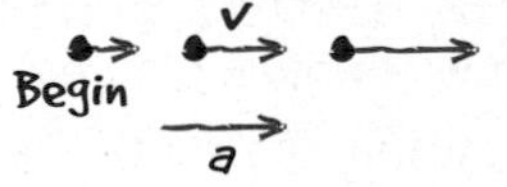

- Draw a motion diagram showing v and a.
- Draw the free-body diagram with a doubled $F_{\text{person on box}}$

KNOWN		UNKNOWN
$m = 25.0$ kg	$\mu_k = 0.20$	$a = ?$
$v = 1.0$ m/s	$F_{\text{person on box}} = 2(49\text{ N}) = 98\text{ N}$	
$t = 3.0$ s		

2 SOLVE FOR THE UNKNOWN

a. The normal force is in the y-direction, and the box does not accelerate in that direction.

$$\mathbf{F}_N = -\mathbf{F}_g$$
$$= -m\mathbf{g}$$

The box does accelerate in the x-direction. So the forces must be unequal.

$$F_{net} = F_{\text{person on box}} - F_f$$

$$ma = F_{\text{person on box}} - F_f$$ ◀ Substitute $F_{net} = ma$.

$$a = \frac{F_{\text{person on box}} - F_f}{m}$$

Find F_f and substitute it into the expression for a.

$$F_f = \mu_k F_N$$

$$= \mu_k mg$$ ◀ Substitute $F_N = mg$.

$$a = \frac{F_{\text{person on box}} - \mu_k mg}{m}$$ ◀ Substitute $F_f = \mu_k mg$.

$$= \frac{98\text{ N} - (0.20)(25.0\text{ kg})(9.8\text{ N/kg})}{25.0\text{ kg}}$$ ◀ Substitute $F_{\text{person on box}} = 98$ N, $\mu_k = 0.20$, $m = 25.0$ kg, $g = 9.8$ N/kg.

$$= 2.0\text{ m/s}^2$$

b. Use the relationship between distance, time, and constant acceleration.

$$x_f = x_i + v_i t_f + \frac{1}{2}a t_f^2$$

$$x_f = v_i t + \left(\frac{1}{2}\right)a t^2$$ ◀ Substitute $x_i = 0$ m.

$$= (1.0\text{ m/s})(3.0\text{ s}) + \left(\frac{1}{2}\right)(2.0\text{ m/s}^2)(3.0\text{ s})^2$$ ◀ Substitute $v_i = 1$ m/s, $t = 3.0$ s, $a = 2.0$ m/s^2.

$$= 12\text{ m}$$

3 EVALUATE THE ANSWER

- **Are the units correct?** a is measured in m/s^2, and x_f is measured in meters.
- **Does the sign make sense?** In this coordinate system, the sign should be positive.
- **Is the magnitude realistic?** 12 m is about the length of two full-sized cars. It is realistic to push a box of this distance in 3 s.

PRACTICE PROBLEMS

Do additional problems. Online Practice

PRACTICE PROBLEMS

22. A 1.4-kg block slides freely across a rough surface such that the block slows down with an acceleration of $-1.25\ m/s^2$. What is the coefficient of kinetic friction between the block and the surface?

23. You want to move a 41-kg bookcase to a different place in the living room. If you push with a force of 65 N and the bookcase accelerates at $0.12\ m/s^2$, what is the coefficient of kinetic friction between the bookcase and the carpet?

24. Consider the force pushing the box in Example Problem 4. How long would it take for the velocity of the box to double to 2.0 m/s?

25. Ke Min is driving at 23 m/s. He sees a tree branch lying across the road. He slams on the brakes when the branch is 60.0 m in front of him. If the coefficient of kinetic friction between the car's locked tires and the road is 0.41, will the car stop before hitting the branch? The car has a mass of 1200 kg.

26. **CHALLENGE** Isabel pushes a shuffleboard disk, accelerating it to a speed of 6.5 m/s before releasing it as indicated in **Figure 13.** If the coefficient of kinetic friction between the disk and the concrete court is 0.31, how far does the disk travel before it comes to a stop? Will Isabel's shot stop in the 10-point section of the board?

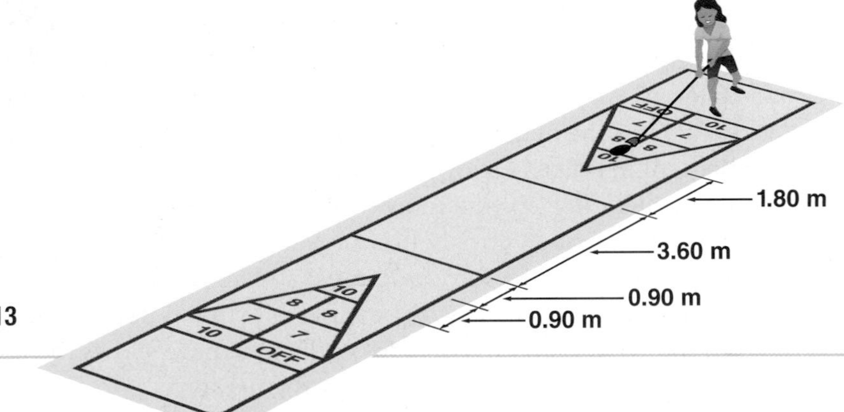

Figure 13

SECTION 2 REVIEW

Section Self-Check Check your understanding.

27. **MAIN**IDEA Compare static friction and kinetic friction. How are the frictional forces similar, and how do the forces differ?

28. **Friction** At a wedding reception, you notice a boy who looks like his mass is about 25 kg running across the dance floor then sliding on his knees until he stops. If the coefficient of kinetic friction between the boy's pants and the floor is 0.15, what is the friction force acting on him as he slides?

29. **Velocity** Dinah is playing cards with her friends, and it is her turn to deal. A card has a mass of 2.3 g, and it slides 0.35 m along the table before it stops. If the coefficient of kinetic friction between the card and the table is 0.24, what was the initial speed of the card as it left Dinah's hand?

30. **Force** The coefficient of static friction between a 40.0-kg picnic table and the ground below that table is 0.43. How large is the greatest horizontal force that could be exerted on the table without moving the table?

31. **Acceleration** You push a 13-kg table in the cafeteria with a horizontal force of 20 N, but the table does not move. You then push the table with a horizontal force of 25 N, and it accelerates at $0.26\ m/s^2$. What, if anything, can you conclude about the coefficients of static and kinetic friction?

32. **Critical Thinking** Rachel is moving to a new apartment and puts a dresser in the back of her pickup truck. When the truck accelerates forward, what force accelerates the dresser? Under what circumstances could the dresser slide? In which direction?

SECTION 3 Forces in Two Dimensions

PHYSICS 4 YOU **The person to the left is riding on a zip line. The tension in the rope provides the upward force necessary to balance the person's weight. If the tension in the rope increases, how would the angle the rope makes with the horizontal change?**

MAIN IDEA

An object is in equilibrium if the net forces in the *x*-direction and in the *y*-direction are zero.

Essential Questions

- How can you find the force required for equilibrium?
- How do you resolve force vector components for motion along an inclined plane?

Review Vocabulary

equilibrium the condition in which the net force on an object is zero

New Vocabulary

equilibrant

Equilibrium Revisited

You have already studied several situations dealing with forces in two dimensions. For example, when friction acts between two surfaces, you must take into account both the friction force that is parallel to the surface and the normal force that is perpendicular to that surface. So far, you have considered only motion along a horizontal surface. Now you will analyze situations in which the forces acting on an object are at angles other than 90°.

Recall that when the net force on an object is zero, the object is in equilibrium. According to Newton's laws, the object will not accelerate because there is no net force acting on it; an object in equilibrium moves with constant velocity. (Remember that staying at rest is a state of constant velocity.) You have already analyzed several equilibrium situations in which two forces act on an object. It is important to realize that equilibrium can also occur if more than two forces act on an object. As long as the net force on the object is zero, the object is in equilibrium.

What is the net force acting on the ring in **Figure 14?** The free-body diagram in **Figure 14** shows the three forces acting on the ring. The ring is not accelerating, so you know the net force must be zero. The free-body diagram, however, does not immediately indicate that the net force is zero. To find the net force, you must add all the vectors together. Remember that vectors may be moved if you do not change their direction (angle) or length. **Figure 15** on the next page shows the process of adding the force vectors to discover the net force.

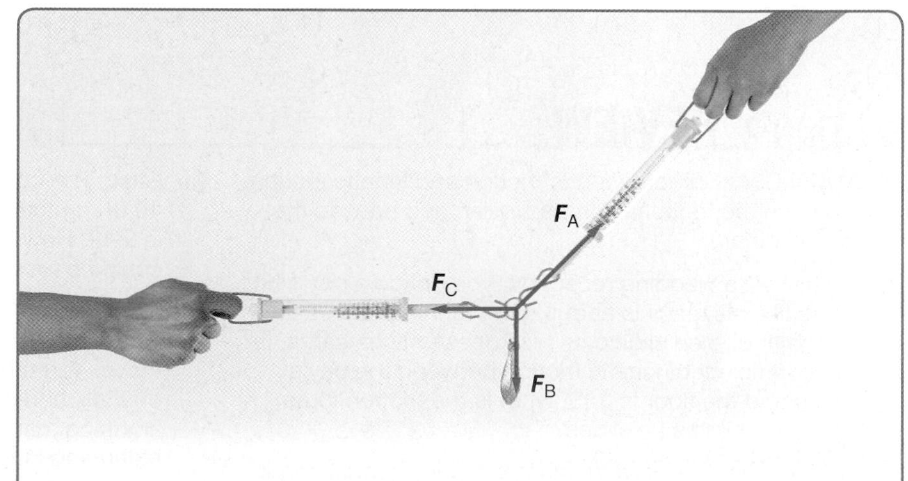

Figure 14 The ring does not accelerate, so the net force acting on it must be zero.

Compare the vertical component of the force pulling up and to the right to the weight of the mass hanging from the ring.

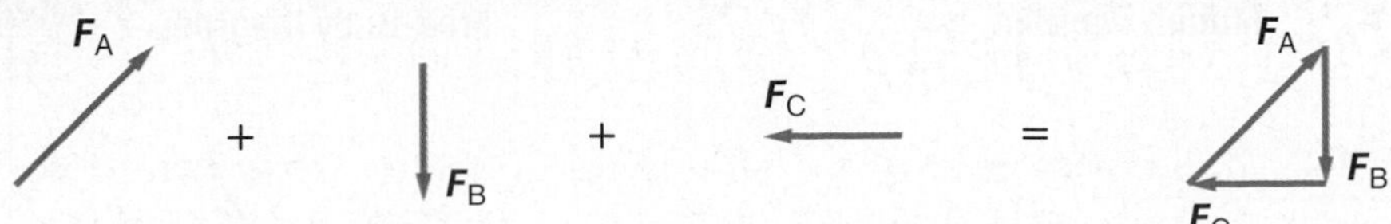

Figure 15 The forces acting on the ring sum to a zero net force.

Equilibrants **Figure 15** shows the addition of the three forces, $\boldsymbol{F}_A$, $\boldsymbol{F}_B$, and $\boldsymbol{F}_C$, acting on the ring. Note that the three vectors form a closed triangle. The sum $\boldsymbol{F}_A + \boldsymbol{F}_B + \boldsymbol{F}_C$ is zero. Now suppose you have a situation where two forces are exerted on an object and the sum is not zero, as shown in the free-body diagram in **Figure 16.** How could you find a third force that, when added to the other two, would add up to zero and put the object in equilibrium? To find this force, first add the forces already being exerted on the object. This single force that produces the same effect as the two or more individual forces together is called the resultant force. In order to put the object in equilibrium, you must add another force, called the **equilibrant,** that has the same magnitude as the resultant force but is in the opposite direction. Notice that in **Figure 15** the force $\boldsymbol{F}_C$ is the equilibrant of $\boldsymbol{F}_A + \boldsymbol{F}_B$. **Figure 16** illustrates the procedure for finding the equilibrant for two vectors. Note that this general procedure for finding the equilibrant works for any number of vectors.

READING CHECK **Identify** the relationship between the equilibrant and the resultant vector.

MiniLAB

EQUILIBRIUM

How can you find the equilibrant of two forces?

Figure 16 The equilibrant is the force required to put an object in equilibrium.

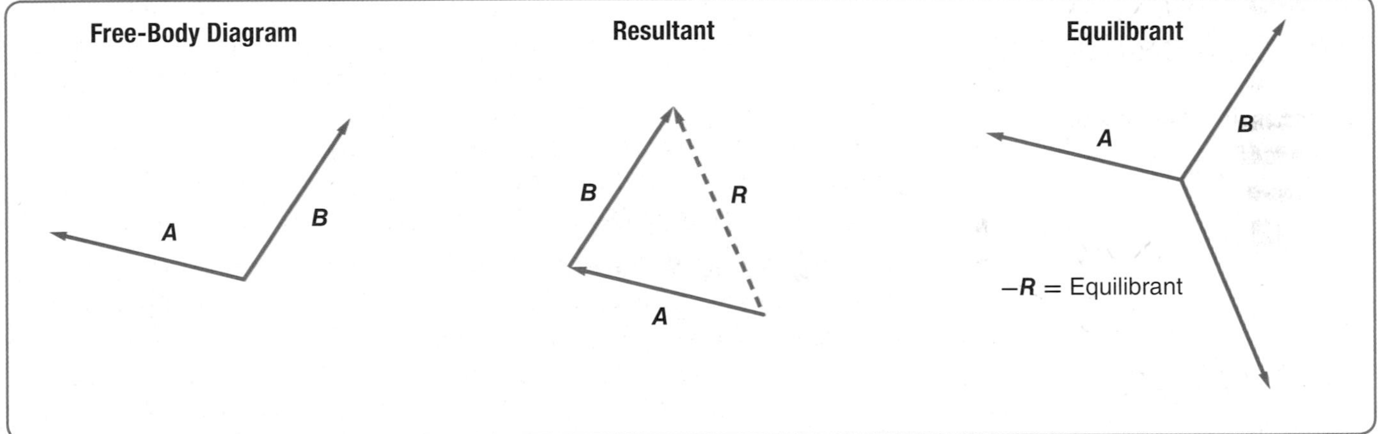

PHYSICS CHALLENGE

Find the equilibrant for the entire set of following forces shown in the figure below.

$\boldsymbol{F}_1 = 61.0$ N at 17.0° north of east
$\boldsymbol{F}_2 = 38.0$ N at 64.0° north of east
$\boldsymbol{F}_3 = 54.0$ N at 8.0° west of north
$\boldsymbol{F}_4 = 93.0$ N at 53.0° west of north
$\boldsymbol{F}_5 = 65.0$ N at 21.0° south of west
$\boldsymbol{F}_6 = 102.0$ N at 15.0° west of south
$\boldsymbol{F}_7 = 26.0$ N south
$\boldsymbol{F}_8 = 77.0$ N at 22.0° east of south
$\boldsymbol{F}_9 = 51.0$ N at 33.0° east of south
$\boldsymbol{F}_{10} = 82.0$ N at 5.0° south of east

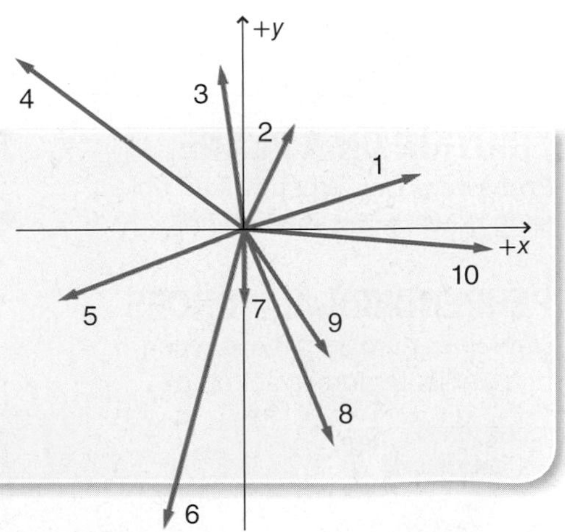

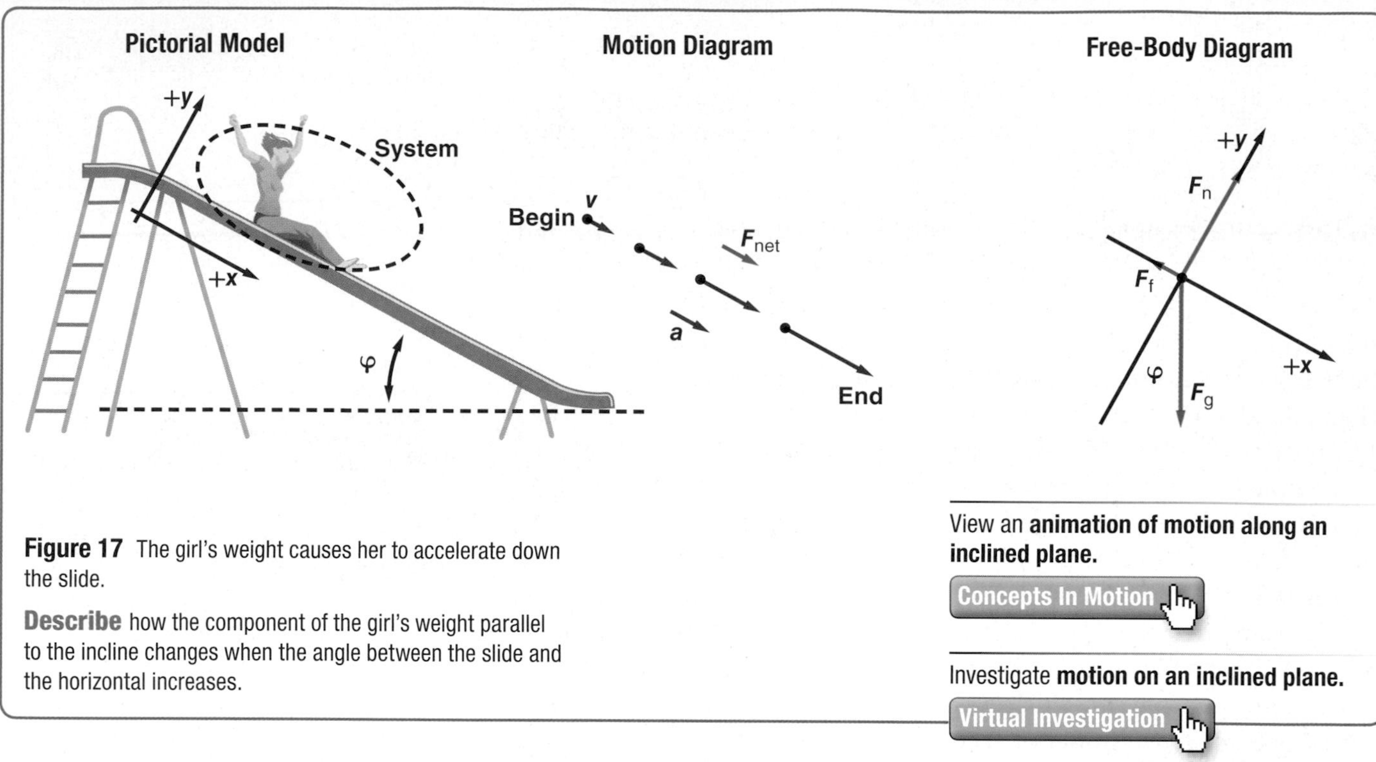

Figure 17 The girl's weight causes her to accelerate down the slide.

Describe how the component of the girl's weight parallel to the incline changes when the angle between the slide and the horizontal increases.

View an **animation of motion along an inclined plane.**

Concepts In Motion

Investigate **motion on an inclined plane.**

Virtual Investigation

MiniLAB

FORCES ON A PLANE

How does an incline's angle change the forces acting on an object on the plane?

PhysicsLABs

SLIDING DOWN A SLOPE

How does an incline's slope affect acceleration?

FRICTION ON A PLANE

Probeware Lab When will an object slide down an inclined plane?

PERPENDICULAR FORCES

How does an object move when perpendicular forces act on it?

Inclined Planes

You have applied Newton's laws to a variety of situations but only to motions that were either horizontal or vertical. How would you apply Newton's laws in a situation such as the one in **Figure 17,** in which you want to find the net force on the girl?

First, identify the forces acting on the system. In this case, the girl is the system. The gravitational force on the girl is downward, toward the center of Earth. The normal force acts perpendicular to the slide, and the kinetic friction force acts parallel to the slide. You can see the resulting free-body diagram in the final panel of **Figure 17.** You know from experience that the girl's acceleration will be along the slide.

Choosing a coordinate system You should next choose your coordinate system carefully. Because the girl's acceleration is parallel to the incline, one axis, the *x*-axis, should be in that direction. The *y*-axis is perpendicular to the *x*-axis and perpendicular to the incline's surface. With this coordinate system, you now have two forces, the normal force and friction force, in the directions of the coordinate axes. The weight, however, has components in both the *x*- and *y*-directions. This means that when you place an object on an inclined plane, the magnitude of the normal force between the object and the plane usually will not equal the object's weight.

☑ **READING CHECK Explain** why you would choose the *x*-direction to be parallel to the slope of an inclined plane.

You will need to apply Newton's laws once in the *x*-direction and once in the *y*-direction. Because the weight does not point in either of these directions, you will need to break this vector into its *x*- and *y*-components before you can sum the forces in these two directions. Example Problem 5 and Example Problem 6 both show this procedure.

EXAMPLE PROBLEM 5

Get help with **components of weight.** Personal Tutor

COMPONENTS OF WEIGHT FOR AN OBJECT ON AN INCLINE A 562-N crate is resting on a plane inclined 30.0° above the horizontal. Find the components of the crate's weight that are parallel and perpendicular to the plane.

1 ANALYZE AND SKETCH THE PROBLEM

- Include a coordinate system with the positive *x*-axis pointing uphill.
- Draw the free-body diagram showing $\boldsymbol{F}_g$, the components $\boldsymbol{F}_{gx}$ and $\boldsymbol{F}_{gy}$, and the angles θ and ϕ.

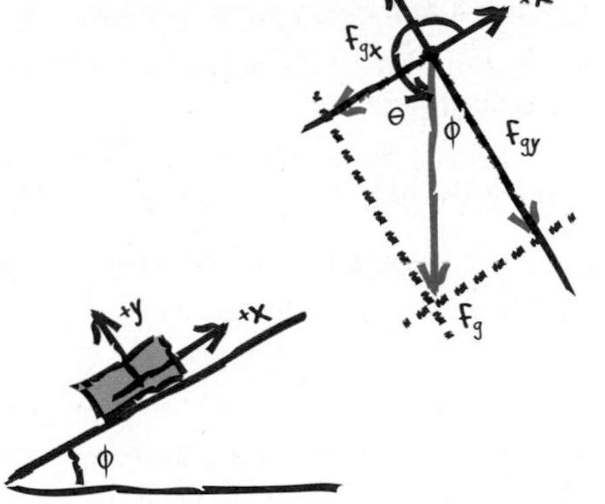

KNOWN	UNKNOWN
$\boldsymbol{F}_g = 562$ N down	$F_{gx} = ?$
$\phi = 30.0°$	$F_{gy} = ?$
	$\theta = ?$

2 SOLVE FOR THE UNKNOWN

$\theta + \phi = 270°$ ◀ *The angle from the positive x-axis to the negative y-axis is 270°.*

$\theta = 270° - 30°$ ◀ *Substitute $\phi = 30°$ and rearrange.*

$= 240°$

$F_{gx} = F_g(\cos\theta)$

$= (562\text{ N})(\cos 240.0°)$ ◀ *Substitute $F_g = 562$ N, $\theta = 240.0°$.*

$= -281\text{ N}$

$F_{gy} = F_g(\sin\theta)$

$= (562\text{ N})(\sin 240.0°)$ ◀ *Substitute $F_g = 562$ N, $\theta = 240.0°$.*

$= -487\text{ N}$

3 EVALUATE THE ANSWER

- **Are the units correct?** Force is measured in newtons.
- **Do the signs make sense?** The components point in directions opposite to the positive axes.
- **Are the magnitudes realistic?** The values are less than F_g, as expected.

PRACTICE PROBLEMS

Do additional problems. Online Practice

33. An ant climbs at a steady speed up the side of its anthill, which is inclined 30.0° from the vertical. Sketch a free-body diagram for the ant.

34. Ryan and Becca are moving a folding table out of the sunlight. A cup of lemonade, with a mass of 0.44 kg, is on the table. Becca lifts her end of the table before Ryan does, and as a result, the table makes an angle of 15.0° with the horizontal. Find the components of the cup's weight that are parallel and perpendicular to the plane of the table.

35. Fernando, who has a mass of 43.0 kg, slides down the banister at his grandparents' house. If the banister makes an angle of 35.0° with the horizontal, what is the normal force between Fernando and the banister?

36. CHALLENGE A suitcase is on an inclined plane as shown in **Figure 18.** At what angle θ will the component of the suitcase's weight parallel to the plane be equal to half the component of its weight perpendicular to the plane?

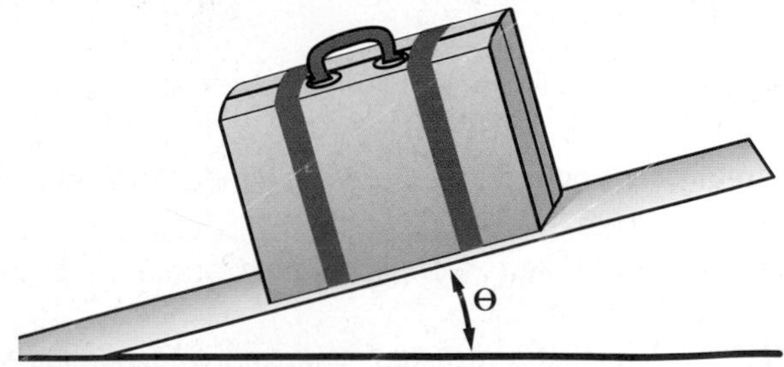

Figure 18

EXAMPLE PROBLEM 6

Find help with **isolating a variable.** Math Handbook

SLIDE Ichiko, who has a mass of 45 kg, is going down a slide sloped at 27°. The coefficient of kinetic friction is 0.23. How fast does she slide 1.0 s after starting from rest?

1 ANALYZE AND SKETCH THE PROBLEM

- Establish a coordinate system.
- Draw a free-body diagram that shows the forces acting on the girl as she travels down the slide.
- The three forces acting on the girl are the normal force, kinetic friction, and the girl's weight.

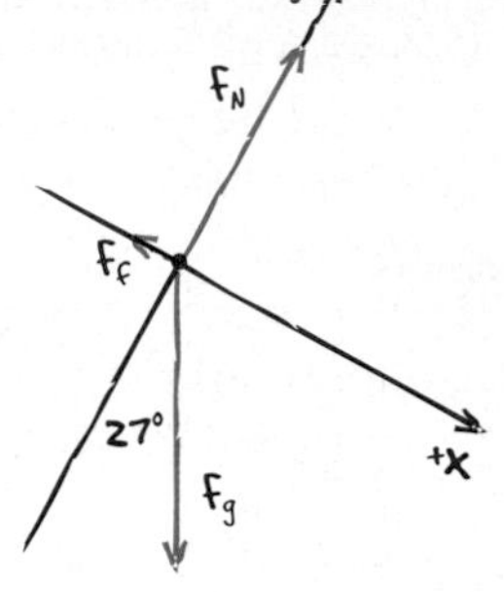

KNOWN		UNKNOWN
$m = 45$ kg	$\phi = 27°$	$a = ?$
$\mu_k = 0.23$	$v_i = 0.0$ m/s	$v_f = ?$
$t = 1.0$ s		$\theta = ?$

2 SOLVE FOR THE UNKNOWN

$\theta = 270° + \phi$ ◀ The angle from the positive x-axis to the negative y-axis is 270°.

$= 270° + 27°$ ◀ Substitute $\phi = 27°$.

$= 297°$

y-direction:

$F_{net,\,y} = ma_y$ ◀ There is no acceleration in the y-direction, so $a_y = 0.0\ m/s^2$.

$= 0.0$ N

Add forces in the y-direction to find F_N.

$F_N + F_{gy} = F_{net,\,y}$

$F_N = -F_{gy}$ ◀ Substitute $F_{net,\,y} = 0.0$ N and rearrange.

$= -mg(\sin\theta)$ ◀ Substitute $F_{gy} = mg\sin\theta$.

x-direction:

Use the net force in the x-direction and Newton's second law to solve for a.

$F_{net,\,x} = F_{gx} - F_f$ ◀ F_f is negative because it is in the negative x-direction.

$ma_x = mg(\cos\theta) - \mu_k F_N$ ◀ Substitute $F_{net,\,x} = ma_x$, $F_{gx} = mg\cos\theta$, $F_f = \mu_k F_N$.

$ma = mg(\cos\theta) + \mu_k mg(\sin\theta)$ ◀ Substitute $a_x = a$ (all acceleration is in the x-direction), $F_N = -mg\sin\theta$.

$a = g(\cos\theta + \mu_k\sin\theta)$

$= (9.8\ m/s^2)(\cos 297° + (0.23)\sin 297°)$ ◀ Substitute $g = 9.8\ m/s^2$, $\theta = 297°$, $\mu_k = 0.23$.

$= 2.4\ m/s^2$

Because v_i, a, and t are all known, use the relationship between velocity, acceleration, and time.

$v_f = v_i + at$

$= 0.0\ m/s + (2.4\ m/s^2)(1.0\ s)$ ◀ Substitute $v_i = 0.0$ m/s, $a = 2.4\ m/s^2$, $t = 1.0$ s.

$= 2.4$ m/s

3 EVALUATE THE ANSWER

- **Are the units correct?** Performing dimensional analysis on the units verifies that v_f is in m/s and a is in m/s^2.
- **Do the signs make sense?** Because v_f and a are both in the +x direction, the signs do make sense.
- **Are the magnitudes realistic?** The velocity is similar to a person's running speed, which is realistic for a steep slide and a low coefficient of kinetic friction.

PRACTICE PROBLEMS

Do additional problems. Online Practice

37. Consider the crate on the incline in Example Problem 5. Calculate the magnitude of the acceleration. After 4.00 s, how fast will the crate be moving?

38. Jorge decides to try the slide discussed in Example Problem 6. Jorge's trip down the slide is quite different from Ichiko's. After giving himself a push to get started, Jorge slides at a constant speed. What is the coefficient of kinetic friction between Jorge's pants and the slide?

39. Stacie, who has a mass of 45 kg, starts down a slide that is inclined at an angle of 45° with the horizontal. If the coefficient of kinetic friction between Stacie's shorts and the slide is 0.25, what is her acceleration?

40. CHALLENGE You stack two physics books on top of each other as shown in **Figure 19.** You tilt the bottom book until the top book just begins to slide. You perform five trials and measure the angles given in **Table 3.**

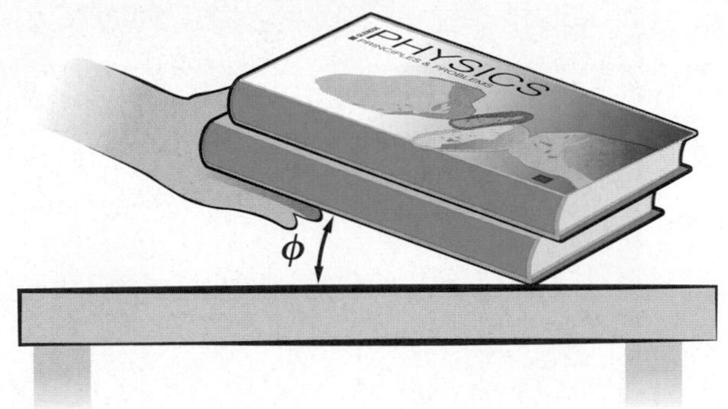

Figure 19

Table 3 Trial Number and Angle of Tilt

Trial	ϕ
1	21°
2	17°
3	21°
4	18°
5	19°

a. What is the average ϕ measured during the five trials?

b. What is the coefficient of static friction between the covers of the two books? Use the average ϕ found in part **a.**

c. You measure the top book's acceleration down the incline to be 1.3 m/s^2. What is the coefficient of kinetic friction? Assume ϕ is the average value found in part **a.**

SECTION 3 REVIEW

Check your understanding.

41. MAINIDEA A rope pulls a 63-kg water skier up a 14.0° incline with a tension of 512 N. The coefficient of kinetic friction between the skier and the ramp is 0.27. What are the magnitude and direction of the skier's acceleration?

42. Forces One way to get a car unstuck is to tie one end of a strong rope to the car and the other end to a tree, then pull the rope at its midpoint at right angles to the rope. Draw a free-body diagram and explain how even a small force on the rope can exert a large force on the car.

43. Mass A large scoreboard is suspended from the ceiling of a sports arena by ten strong cables. Six of the cables make an angle of 8.0° with the verticals while the other four make an angle of 10.0°. If the tension in each cable is 1300 N, what is the scoreboard's mass?

44. Vector Addition What is the sum of three vectors that, when placed tip to tail, form a triangle? If these vectors represent forces on an object, what does this imply about the object? Describe the motion resulting from these three forces acting on the object.

45. Equilibrium You are hanging a painting using two lengths of wire. The wires will break if the force is too great. Should you hang the painting as shown in the top or the bottom image of **Figure 20?** Explain.

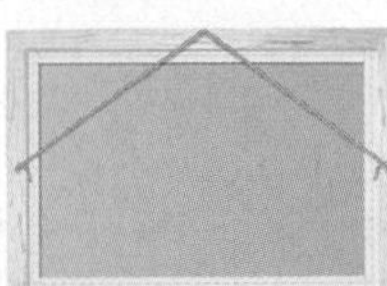

Figure 20

46. Critical Thinking Can the coefficient of friction ever have a value such that a child would be able to slide *up* a slide at a constant velocity? Explain why or why not. Assume that no one pushes or pulls on the child.

ON THE **JOB**

Out on a Limb

Have you ever heard the saying, "What goes up must come down"? Architects rely on a thorough understanding of forces to make sure that buildings and other structures stay up.

Balanced forces You may recall that an object accelerates because of a net force acting on it. But just because an object isn't moving doesn't mean forces aren't acting on it. When an object is at rest, the forces acting on the object are balanced.

For example, at this very moment the gravitational force is pulling you down. At the same time, this gravitational force is balanced by the normal force from your chair pushing up against you. Because these forces are balanced, there is no net force and you do not accelerate. When designing a building, bridge, or other structure, an architect must make sure that the gravitational force is properly balanced by other forces.

Cantilevers Have you ever stood on a balcony such as the one shown in **Figure 1?** With nothing pushing up from below to balance the gravitational force, why doesn't it fall? A cantilever is a beam that is supported on only one end. The weight of the cantilever must be balanced by forces exerted by the rest of the structure.

FIGURE 1 These people are looking out from one of the cantilevers on the Willis Tower in Chicago.

GOING**FURTHER**

Research how forces are distributed in a catenary arch, such as the Gateway Arch in St. Louis, Missouri. Create a poster that shows the forces throughout the arch.

CHAPTER 5

STUDY GUIDE

BIGIDEA Forces in two dimensions can be described using vector addition and vector resolution.

VOCABULARY
- **components** *(p. 125)*
- **vector resolution** *(p. 125)*

SECTION 1 Vectors

MAINIDEA All vectors can be broken into *x*- and *y*-components.

- Vectors are added graphically by placing the tail of the second vector on the tip of the first vector. The resultant is the vector pointing from the tail of the first vector to the tip of the final vector.
- The components of a vector are projections of the component vectors onto axes. Vectors can be summed by separately adding the *x*- and *y*-components.

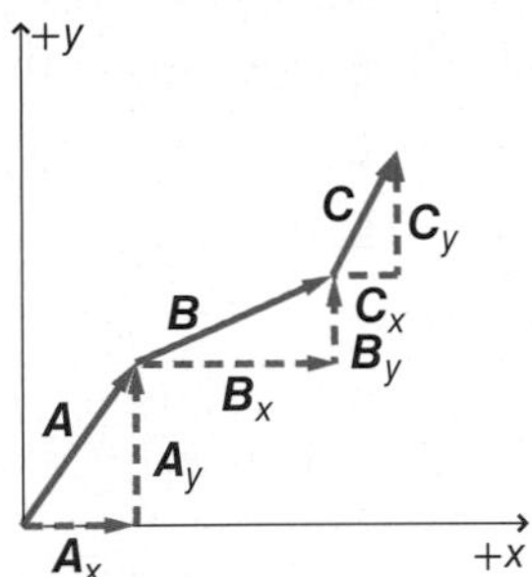

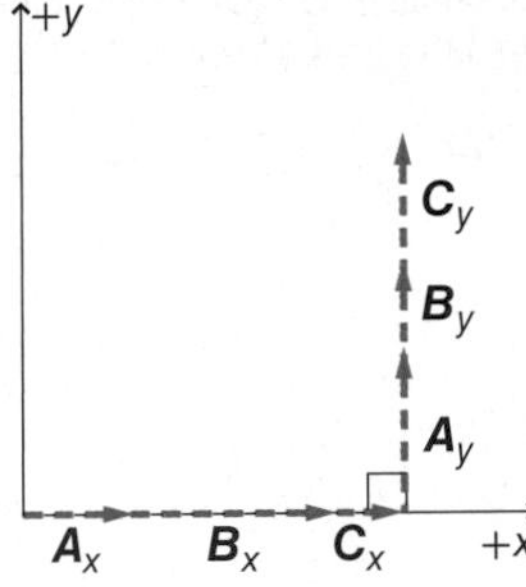

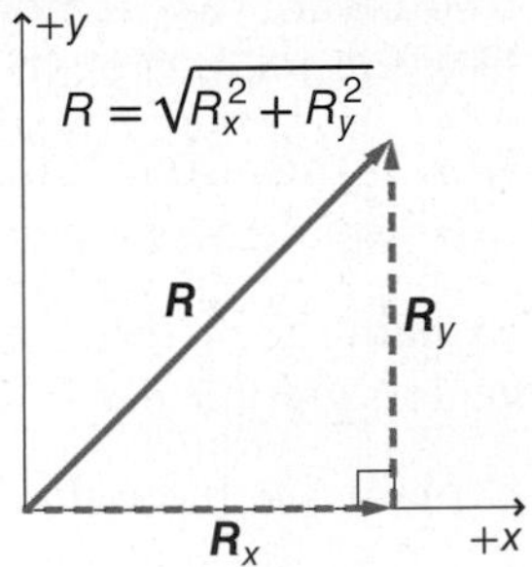

- When two vectors are at right angles, you can use the Pythagorean theorem to determine the magnitude of the resultant vector. The law of cosines and the law of sines can be used to find the resultant of any two vectors.

VOCABULARY
- **kinetic friction** *(p. 130)*
- **static friction** *(p. 130)*
- **coefficient of kinetic friction** *(p. 131)*
- **coefficient of static friction** *(p. 132)*

SECTION 2 Friction

MAINIDEA Friction is a type of force between two touching surfaces.

- Friction is a force that acts parallel to the surfaces when two surfaces touch.
- The kinetic friction force is equal to the coefficient of kinetic friction times the normal force. The static friction force is less than or equal to the coefficient of static friction times the normal force.

$$F_{\text{f,kinetic}} = \mu_k F_N$$

$$F_{\text{f,static}} \leq \mu_s F_N$$

VOCABULARY
- **equilibrant** *(p. 137)*

SECTION 3 Forces in Two Dimensions

MAINIDEA An object is in equilibrium if the net forces in the *x*-direction and in the *y*-direction are zero.

- The equilibrant is a force of the same magnitude but opposite direction as the sum of all the other forces acting on an object.
- Friction forces are parallel to an inclined plane but point up the plane if the motion of the object is down the plane. An object on an inclined plane has a component of the force of gravity parallel to the plane; this component can accelerate the object down the plane.

Games and Multilingual eGlossary

Vocabulary Practice

CHAPTER 5 ASSESSMENT

Chapter Self-Check

SECTION 1 Vectors

Mastering Concepts

47. BIGIDEA How would you add two vectors graphically?

48. Which of the following actions is permissible when you graphically add one vector to another: moving the vector, rotating the vector, or changing the vector's length?

49. In your own words, write a clear definition of the resultant of two or more vectors. Do not explain how to find it; explain what it represents.

50. How is the resultant displacement affected when two displacement vectors are added in a different order?

51. Explain the method you would use to subtract two vectors graphically.

52. Explain the difference between ***A*** and *A*.

53. The Pythagorean theorem usually is written $c^2 = a^2 + b^2$. If this relationship is used in vector addition, what do *a*, *b*, and *c* represent?

54. When using a coordinate system, how is the angle or direction of a vector determined with respect to the axes of the coordinate system?

Mastering Problems

55. Cars A car moves 65 km due east then 45 km due west. What is its total displacement?

56. Find the horizontal and vertical components of the following vectors shown in **Figure 21.** In all cases, assume that up and right are positive directions.

a. E

b. F

c. A

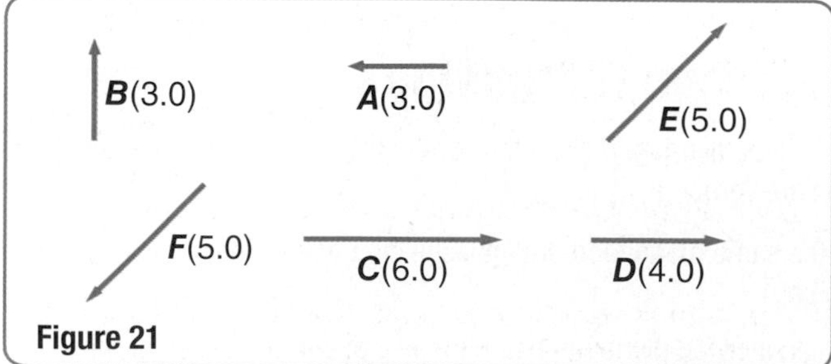

Figure 21

57. Graphically find the sum of the following pairs of vectors, shown in **Figure 21.**

a. D and A

b. C and D

c. C and A

d. E and F

58. Graphically add the following sets of vectors shown in **Figure 21.**

a. A, C, and D

b. A, B, and E

c. B, D, and F

59. Ranking Task Rank the following according to the magnitude of the net force, from least to greatest. Specifically indicate any ties.

A. 20 N up + 10 N down

B. 20 N up + 10 N left

C. 20 N up + 10 N up

D. 20 N up + 10 N 20° below the horizontal

E. 20 N up

60. You walk 30 m south and 30 m east. Find the magnitude and direction of the resultant displacement both graphically and algebraically.

61. A hiker's trip consists of three segments. Path A is 8.0 km long heading 60.0° north of east. Path B is 7.0 km long in a direction due east. Path C is 4.0 km long heading 315° counterclockwise from east.

a. Graphically add the hiker's displacements in the order A, B, C.

b. Graphically add the hiker's displacements in the order C, B, A.

c. What can you conclude about the resulting displacements?

62. Two forces are acting on the ring in **Figure 22.** What is the net force acting on the ring?

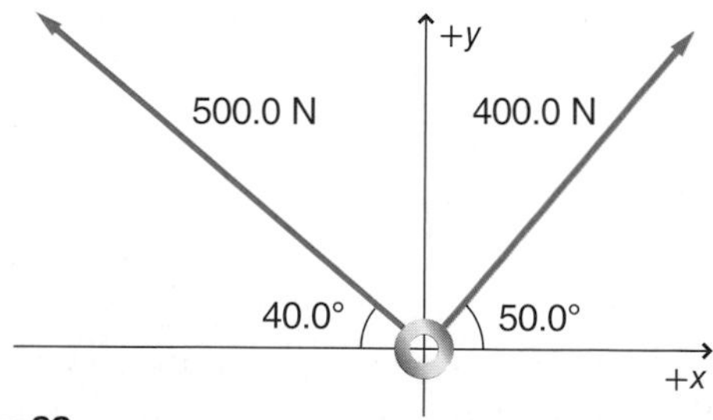

Figure 22

63. Space Exploration A descent vehicle landing on Mars has a vertical velocity toward the surface of Mars of 5.5 m/s. At the same time, it has a horizontal velocity of 3.5 m/s.

a. At what speed does the vehicle move along its descent path?

b. At what angle with the vertical is this path?

c. If the vehicle is 230 m above the surface, how long until it reaches the surface?

64. Three forces are acting on the ring in **Figure 23.** What is the net force acting on the ring?

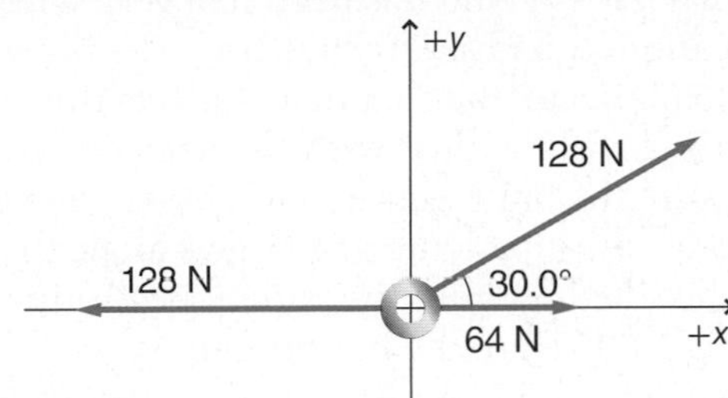

Figure 23

65. A Ship at Sea A ship at sea is due into a port 500.0 km due south in two days. However, a severe storm comes in and blows it 100.0 km due east from its original position. How far is the ship from its destination? In what direction must it travel to reach its destination?

66. Navigation Alfredo leaves camp and, using a compass, walks 4 km E, then 6 km S, 3 km E, 5 km N, 10 km W, 8 km N, and, finally, 3 km S. At the end of two days, he is planning his trip back. By drawing a diagram, compute how far Alfredo is from camp and which direction he should take to get back to camp.

SECTION 2 Friction

Mastering Concepts

67. What is the meaning of a coefficient of friction that is greater than 1.0? How would you measure it?

68. Cars Using the model of friction described in this textbook, would the friction between a tire and the road be increased by a wide rather than a narrow tire? Explain. Assume the tires have the same mass .

Mastering Problems

69. If you use a horizontal force of 30.0 N to slide a 12.0-kg wooden crate across a floor at a constant velocity, what is the coefficient of kinetic friction between the crate and the floor?

70. A 225-kg crate is pushed horizontally with a force of 710 N. If the coefficient of friction is 0.20, calculate the acceleration of the crate.

71. A force of 40.0 N accelerates a 5.0-kg block at 6.0 m/s^2 along a horizontal surface.

a. What would the block's acceleration be if the surface were frictionless?

b. How large is the kinetic friction force?

c. What is the coefficient of kinetic friction?

72. Moving Appliances Your family just had a new refrigerator delivered. The delivery man has left and you realize that the refrigerator is not quite in the right position, so you plan to move it several centimeters. If the refrigerator has a mass of 88 kg, the coefficient of kinetic friction between the bottom of the refrigerator and the floor is 0.13, and the static coefficient of friction between these same surfaces is 0.21, how hard do you have to push horizontally to get the refrigerator to start moving?

73. Stopping at a Red Light You are driving a 1200.0-kg car at a constant speed of 14.0 m/s along a straight, level road. As you approach an intersection, the traffic light turns red. You slam on the brakes. The car's wheels lock, the tires begin skidding, and the car slides to a halt in a distance of 25.0 m. What is the coefficient of kinetic friction between your tires and the road?

SECTION 3 Forces in Two Dimensions

Mastering Concepts

74. Describe a coordinate system that would be suitable for dealing with a problem in which a ball is thrown up into the air.

75. If a coordinate system is set up such that the positive *x*-axis points in a direction 30° above the horizontal, what should be the angle between the *x*-axis and the *y*-axis? What should be the direction of the positive *y*-axis?

76. Explain how you would set up a coordinate system for motion on a hill.

77. If your textbook is in equilibrium, what can you say about the forces acting on it?

78. Can an object that is in equilibrium be moving? Explain.

79. You are asked to analyze the motion of a book placed on a sloping table.

a. Describe the best coordinate system for analyzing the motion.

b. How are the components of the weight of the book related to the angle of the table?

80. For a book on a sloping table, describe what happens to the component of the weight force parallel to the table and the force of friction on the book as you increase the angle the table makes with the horizontal.

a. Which components of force(s) increase when the angle increases?

b. Which components of force(s) decrease?

Chapter Self-Check

Mastering Problems

81. An object in equilibrium has three forces exerted on it. A 33.0-N force acts at 90.0° from the *x*-axis, and a 44.0-N force acts at 60.0° from the *x*-axis. Both angles are measured counterclockwise from the positive *x*-axis. What are the magnitude and the direction of the third force?

82. Five forces act on the object in **Figure 24:** (1) 60.0 N at 90.0°, (2) 40.0 N at 0.0°, (3) 80.0 N at 270.0°, (4) 40.0 N at 180.0°, and (5) 50.0 N at 60.0°. What are the magnitude and the direction of a sixth force that would produce equilibrium?

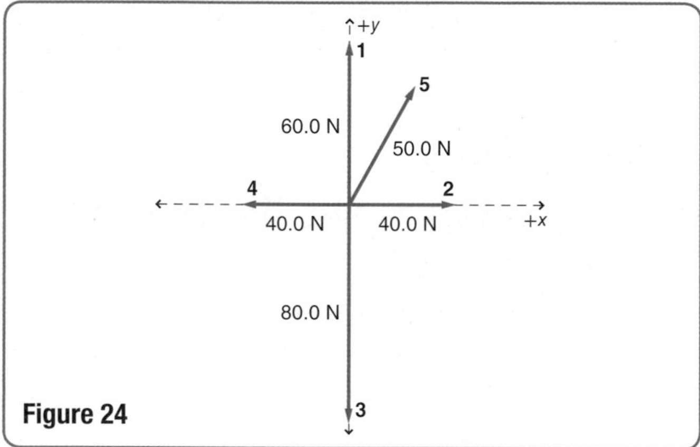

Figure 24

83. Advertising Joe wishes to hang a sign weighing 7.50×10^2 N so that cable A, attached to the store, makes a 30.0° angle, as shown in **Figure 25.** Cable B is horizontal and attached to an adjoining building. What is the tension in cable B?

Figure 25

84. A street lamp weighs 150 N. It is supported by two wires that form an angle of 120.0° with each other. The tensions in the wires are equal.

a. What is the tension in each wire?

b. If the angle between the wires supporting the street lamp is reduced to 90.0°, what is the tension in each wire?

85. A 215-N box is placed on an incline that makes a 35.0° angle with the horizontal. Find the component of the weight parallel to the incline.

86. Emergency Room You are job-shadowing a nurse in the emergency room of a local hospital. An orderly wheels in a patient who has been in a very serious accident and has had severe bleeding. The nurse quickly explains to you that in a case like this, the patient's bed will be tilted with the head downward to make sure the brain gets enough blood. She tells you that, for most patients, the largest angle that the bed can be tilted without the patient beginning to slide off is 45.0° from the horizontal.

a. On what factor or factors does this angle of tilting depend?

b. Find the coefficient of static friction between a typical patient and the bed's sheets.

87. Two blocks are connected by a string over a frictionless, massless pulley such that one is resting on an inclined plane and the other is hanging over the top edge of the plane, as shown in **Figure 26.** The hanging block has a mass of 16.0 kg, and the one on the plane has a mass of 8.0 kg. The coefficient of kinetic friction between the block and the inclined plane is 0.23. The blocks are released from rest.

a. What is the acceleration of the blocks?

b. What is the tension in the string connecting the blocks?

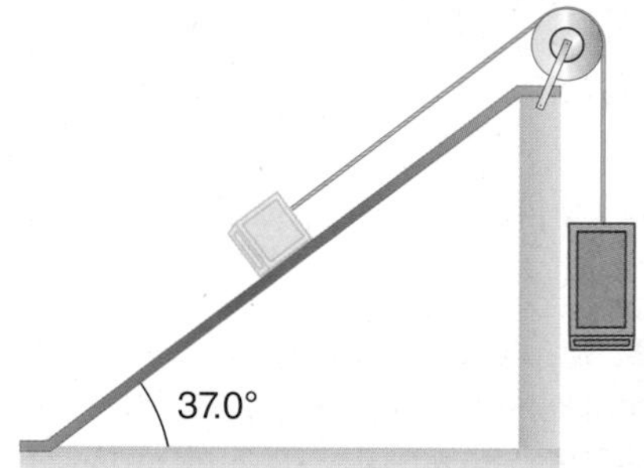

Figure 26

88. In **Figure 27,** a block of mass *M* is pushed with a force (*F*) such that the smaller block of mass *m* does not slide down the front of it. There is no friction between the larger block and the surface below it, but the coefficient of static friction between the two blocks is μ_s. Find an expression for *F* in terms of *M*, *m*, μ_s, and *g*.

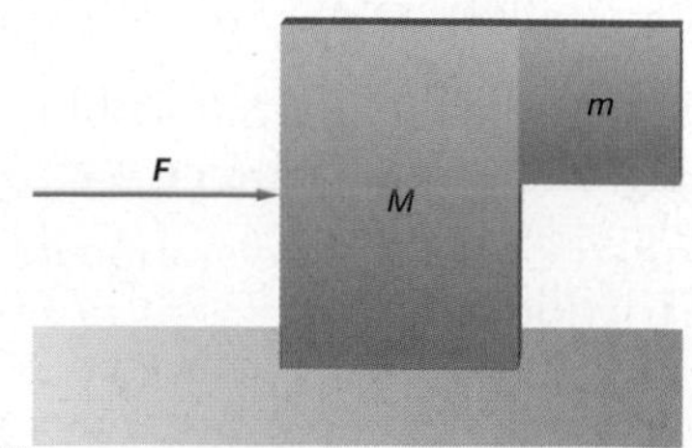

Figure 27

Chapter Self-Check

Applying Concepts

89. A vector that is 1 cm long represents a displacement of 5 km. How many kilometers are represented by a 3-cm vector drawn to the same scale?

90. A vector drawn 15 mm long represents a velocity of 30 m/s. How long should you draw a vector to represent a velocity of 20 m/s?

91. What is the largest possible net displacement resulting from two displacements with magnitudes 3 m and 4 m? What is the smallest possible resultant? Draw sketches to demonstrate your answers.

92. How does the resultant displacement change as the angle between two vectors increases from 0° to 180°?

93. *A* and *B* are two sides of a right triangle, where $\tan \theta = \frac{A}{B}$.

a. Which side of the triangle is longer if $\tan \theta$ is greater than 1.0?

b. Which side is longer if $\tan \theta$ is less than 1.0?

c. What does it mean if $\tan \theta$ is equal to 1.0?

94. Traveling by Car A car has a velocity of 50 km/h in a direction 60° north of east. A coordinate system with the positive *x*-axis pointing east and a positive *y*-axis pointing north is chosen. Which component of the velocity vector is larger, *x* or *y*?

95. Under what conditions can the Pythagorean theorem, rather than the law of cosines, be used to find the magnitude of a resultant vector?

96. A problem involves a car moving up a hill, so a coordinate system is chosen with the positive *x*-axis parallel to the surface of the hill. The problem also involves a stone that is dropped onto the car. Sketch the problem and show the components of the velocity vector of the stone.

97. Pulling a Cart According to legend, a horse learned Newton's laws. When the horse was told to pull a cart, it refused, saying that if it pulled the cart forward, according to Newton's third law, there would be an equal force backwards. Thus, there would be balanced forces, and, according to Newton's second law, the cart would not accelerate. How would you reason with this horse?

98. Tennis When stretching a tennis net between two posts, it is relatively easy to pull one end of the net hard enough to remove most of the slack, but you need a winch to take the last bit of slack out of the net to make the top almost completely horizontal. Why is this true?

99. The weight of a book on an inclined plane can be resolved into two vector components, one along the plane and the other perpendicular to it.

a. At what angle are the components equal?

b. At what angle is the parallel component equal to zero?

100. TV Towers A TV station's transmitting tower is held upright by guy wires that extend from the ground to the top of the tower at an angle of 67° above the horizontal. The force along the guy wires can be resolved into perpendicular and parallel components with respect to the ground. Which one is larger?

Mixed Review

101. The scale in **Figure 28** is being pulled on by three ropes. What net force does the scale read?

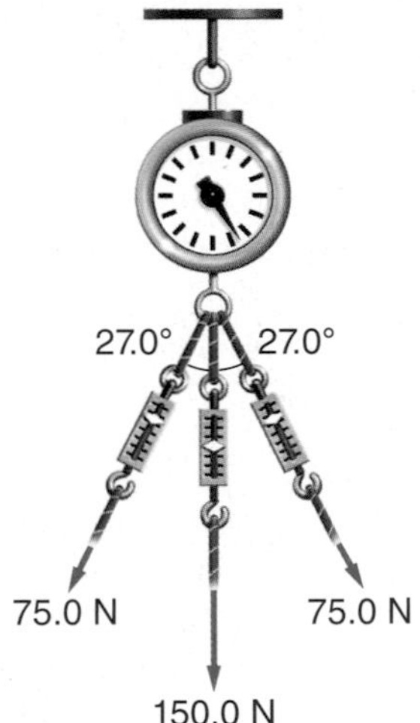

Figure 28

102. Mythology Sisyphus was a character in Greek mythology who was doomed in Hades to push a boulder to the top of a steep mountain. When he reached the top, the boulder would slide back down the mountain and he would have to start all over again. Assume that Sisyphus slides the boulder up the mountain without being able to roll it, even though in most versions of the myth, he rolled it.

a. If the coefficient of kinetic friction between the boulder and the mountainside is 0.40, the mass of the boulder is 20.0 kg, and the slope of the mountain is a constant 30.0°, what is the force that Sisyphus must exert on the boulder to move it up the mountain at a constant velocity?

b. Sisyphus pushes the boulder at a velocity of 0.25 m/s and it takes him 8.0 h to reach the top of the mountain, what is the mythical mountain's vertical height?

Chapter Self-Check

103. Landscaping A tree is being transported on a flatbed trailer by a landscaper. If the base of the tree slides on the trailer, the tree will fall over and be damaged. The coefficient of static friction between the tree and the trailer is 0.50. The truck's initial speed is 55 km/h.

a. The truck must come to a stop at a traffic light without the tree sliding forward and falling on the trailer. What is the maximum possible acceleration the truck can experience?

b. What is the truck's minimum stopping distance if the truck accelerates uniformly at the maximum acceleration calculated in part **a?**

Thinking Critically

104. Use Models Using the Example Problems in this chapter as models, write an example problem to solve the following problem. Include the following sections: Analyze and Sketch the Problem, Solve for the Unknown (with a complete strategy), and Evaluate the Answer. A driver of a 975-kg car traveling 25 m/s puts on the brakes. What is the shortest distance it will take for the car to stop if the coefficient of kinetic friction is 0.65? Assume that the force of friction of the road on the tires is constant and the tires do not slip.

105. Analyze and Conclude Margaret Mary, Doug, and Kako are at a local amusement park and see an attraction called the Giant Slide, which is simply a very long and high inclined plane. Visitors at the amusement park climb a long flight of steps to the top of the 27° inclined plane and are given canvas sacks. They sit on the sacks and slide down the 70-m-long plane. At the time when the three friends walk past the slide, a 135-kg man and a 20-kg boy are each at the top preparing to slide down. "I wonder how much less time it will take the man to slide down than it will take the boy," says Margaret Mary. "I think the boy will take less time," says Doug. "You're both wrong," says Kako. "They will reach the bottom at the same time."

a. Perform the appropriate analysis to determine who is correct.

b. If the man and the boy do not take the same amount of time to reach the bottom of the slide, calculate how many seconds of difference there will be between the two times.

106. Problem Posing Complete this problem so it can be solved using two-dimensional vector addition: "Jeff is cleaning his room when he finds his favorite basketball card under his bed ..."

107. Reverse Problem Write a physics problem with real-life objects for which the **Figure 29** would be part of the solution.

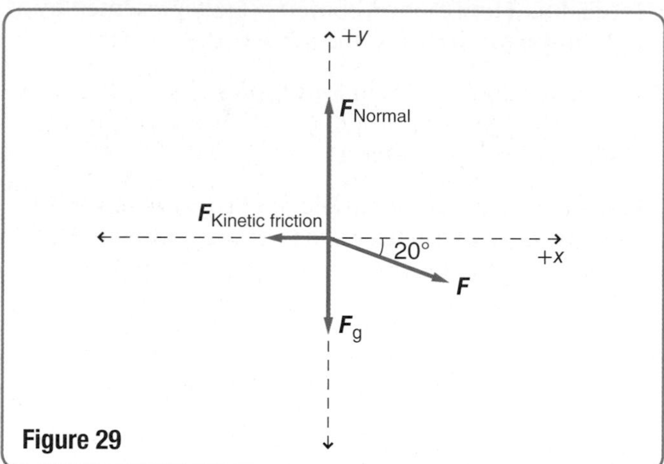

Figure 29

Writing in Physics

108. Investigate some of the techniques used in industry to reduce the friction between various parts of machines. Describe two or three of these techniques and explain the physics of how they work.

109. Olympics In recent years, many Olympic athletes, such as sprinters, swimmers, skiers, and speed skaters, have used modified equipment to reduce the effects of friction and air or water drag. Research a piece of equipment used by one of these types of athletes and the way it has changed over the years. Explain how physics has impacted these changes.

Cumulative Review

110. Perform the indicated addition, subtraction, multiplication, or division and state the answer with the correct number of significant digits.

a. 85.26 g + 4.7 g

b. 1.07 km + 0.608 km

c. 186.4 kg − 57.83 kg

d. (1.26 kg)(9.8 N/kg)

e. $\frac{10\text{ m}}{4.5\text{ s}}$

111. You ride your bike for 1.5 h at an average velocity of 10 km/h, then for 30 min at 15 km/h. What is your average velocity?

112. A 45-N force is exerted in the upward direction on a 2.0-kg briefcase. What is the acceleration of the briefcase?

CHAPTER 5 STANDARDIZED TEST PRACTICE

MULTIPLE CHOICE

1. What is the term used to describe a force that equals the sum of two or more forces?

A. equilibrant
B. normal force
C. resultant force
D. tension force

2. For a winter fair, some students decide to build 30.0-kg wooden pull carts on sled skids. If two 90.0-kg passengers get in, how much force will the puller have to exert to move a pull cart? The coefficient of maximum static friction between the sled skids and the road is 0.15.

A. 1.8×10^2 N
B. 3.1×10^2 N
C. 2.1×10^3 N
D. 1.4×10^4 N

3. It takes a minimum force of 280 N to move a 50.0-kg crate. What is the coefficient of maximum static friction between the crate and the floor?

A. 0.18
B. 0.57
C. 1.8
D. 5.6

4. Two tractors in the figure below pull against a 1.00×10^3-kg log. If the angle of the tractors' chains in relation to each other is 18.0° and each tractor pulls with a force of 8×10^2 N, how large is the force exerted by the tractors on the log?

A. 250 N
B. 7.90×10^2 N
C. 1.58×10^3 N
D. 9.80×10^3 N

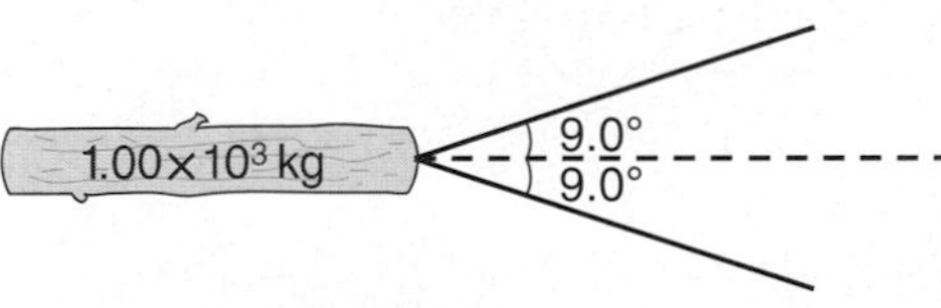

5. An airplane pilot tries to fly directly east with a velocity of 800.0 km/h. If a wind comes from the southwest at 80.0 km/h, what is the relative velocity of the airplane to the surface of Earth?

A. 804 km/h, 5.7° N of E
B. 858 km/h, 3.8° N of E
C. 859 km/h, 4.0° N of E
D. 880 km/h, 45° N of E

Online Test Practice

6. What is the y-component of a 95.3-N force that is exerted at 57.1° to the horizontal?

A. 51.8 N
B. 80.0 N
C. 114 N
D. 175 N

7. As shown in the figure below, a string exerts a force of 18 N on a box at an angle of 34° from the horizontal. What is the horizontal component of the force on the box?

A. 10 N
B. 15 N
C. 21.7 N
D. 32 N

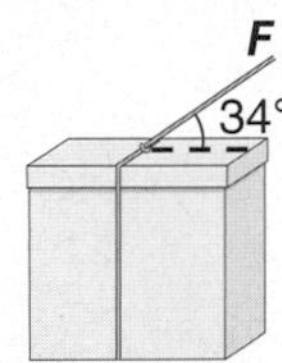

8. Sukey is riding her bicycle on a path when she comes around a corner and sees that a fallen tree is blocking the way 42 m ahead. If the coefficient of friction between her bicycle's tires and the gravel path is 0.36 and she is traveling at 25.0 km/h, how much stopping distance will she require? Sukey and her bicycle, together, have a mass of 95 kg.

A. 3.0 m
B. 4.5 m
C. 6.8 m
D. 9.8 m

FREE RESPONSE

9. A man starts from a position 310 m north of his car and walks for 2.7 min in a westward direction at a constant velocity of 10 km/h. How far is he from his car when he stops?

10. Jeeves is tired of his 41.2-kg son sliding down the banister, so he decides to apply an extremely sticky paste to the top of the banister. The paste increases the coefficient of static friction to 0.72. What will be the magnitude of the static friction force on the boy if the banister is at an angle of 52.4° from the horizontal?

NEED EXTRA HELP?

If you Missed Question	1	2	3	4	5	6	7	8	9	10
Review Section	3	2	2	1	1	1	1	2	1	3

CHAPTER 6

Motion in Two Dimensions

BIGIDEA You can use vectors and Newton's laws to describe projectile motion and circular motion.

SECTIONS

1 **Projectile Motion**

2 **Circular Motion**

3 **Relative Velocity**

LaunchLAB

iLab Station

PROJECTILE MOTION

What does the path of a projectile, such as a ball that is thrown, look like?

WATCH THIS!

Video

PROJECTILE PHYSICS

Have you ever seen a catapult or trebuchet in action? Discover the physics of launching projectiles!

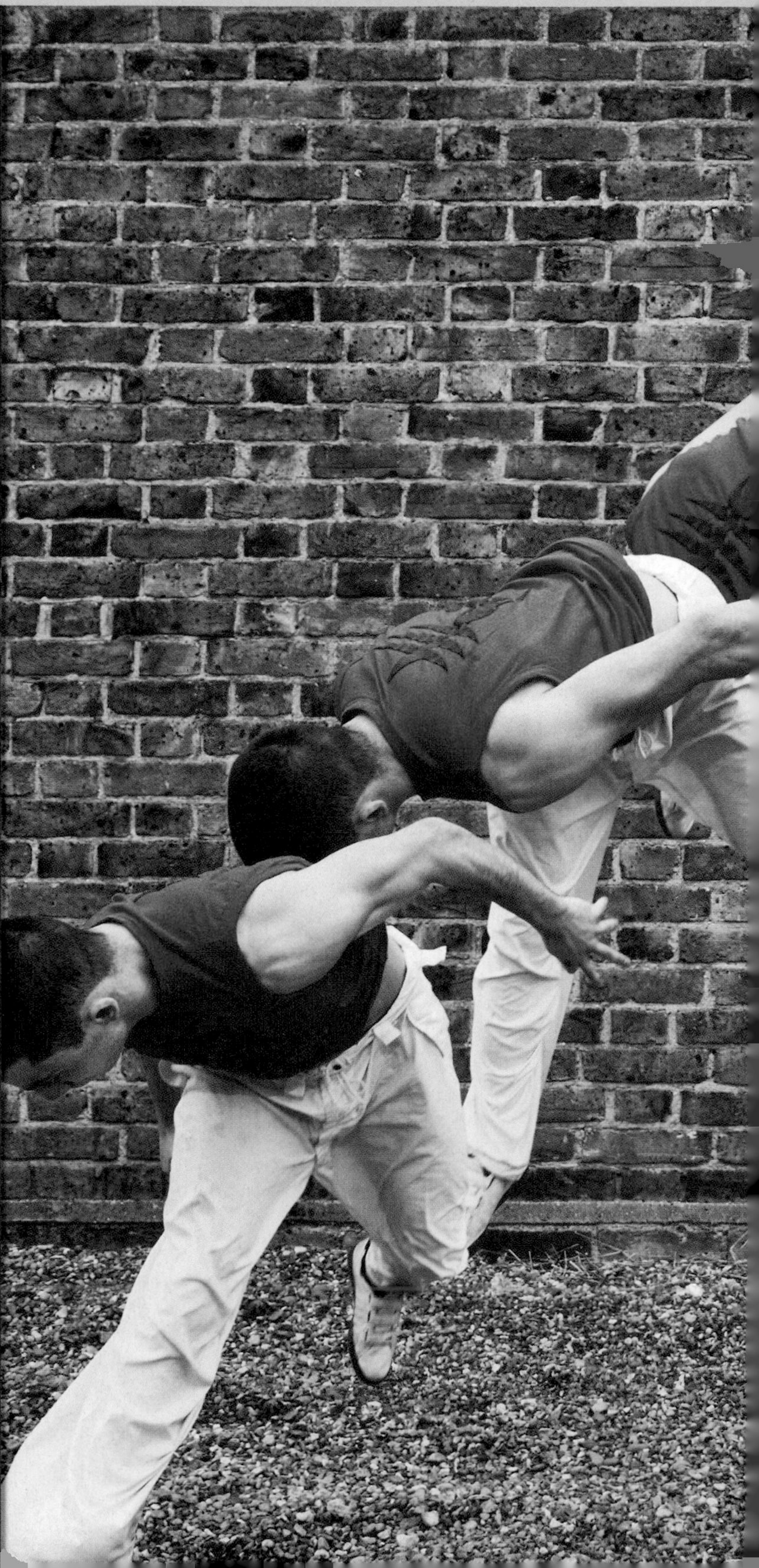

Go online!
connectED.mcgraw-hill.com

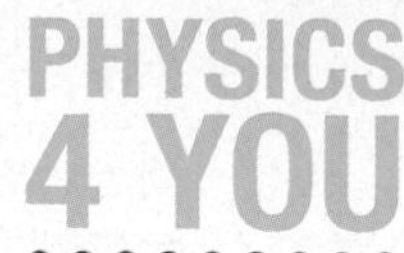

SECTION 1 Projectile Motion

PHYSICS 4 YOU **When you throw a softball or a football, it travels in an arc. These tossed balls are projectiles. The word *projectile* comes from the Latin prefix *pro–*, meaning "forward", and the Latin root *ject* meaning "to throw."**

MAINIDEA

A projectile's horizontal motion is independent of its vertical motion.

Essential Questions

- How are the vertical and horizontal motions of a projectile related?
- What are the relationships between a projectile's height, time in the air, initial velocity, and horizontal distance traveled?

Review Vocabulary

motion diagram a series of images showing the positions of a moving object taken at regular time intervals

New Vocabulary

projectile
trajectory

Path of a Projectile

A hopping frog, a tossed snowball, and an arrow shot from a bow all move along similar paths. Each path rises and then falls, always curving downward along a parabolic path. An object shot through the air is called a **projectile.** You can draw a free-body diagram of a launched projectile and identify the forces acting on it. If you ignore air resistance, after an initial force launches a projectile, the only force on it as it moves through the air is gravity. Gravity causes the object to curve downward. Its path through space is called its **trajectory.** You can determine a projectile's trajectory if you know its initial velocity. In this chapter, you will study two types of projectile motion. The top of **Figure 1** shows water that is launched as a projectile horizontally. The bottom of the figure shows water launched as a projectile at an angle. In both cases, gravity curves the path downward along a parabolic path.

Figure 1 A projectile launched horizontally immediately curves downward, but if it is launched upward at an angle, it rises and then falls, always curving downward.

(t)Visual Ideas/Camilo Morales/Blend Images/Getty Images, (c b)Richard Hutchings/Digital Light Source

Independence of Motion in Two Dimensions

Think about two softball players warming up for a game, tossing high fly balls back and forth. What does the path of the ball through the air look like? Because the ball is a projectile, it has a parabolic path. Imagine you are standing directly behind one of the players and you are watching the softball as it is being tossed. What would the motion of the ball look like? You would see it go up and back down, just like any object that is tossed straight up in the air. If you were watching the softball from a hot-air balloon high above the field, what motion would you see then? You would see the ball move from one player to the other at a constant speed, just like any object that is given an initial horizontal velocity, such as a hockey puck sliding across ice. The motion of projectiles is a combination of these two motions.

Why do projectiles behave in this way? After a softball leaves a player's hand, what forces are exerted on the ball? If you ignore air resistance, there are no contact forces on the ball. There is only the field force of gravity in the downward direction. How does this affect the ball's motion? Gravity causes the ball to have a downward acceleration.

Comparing motion diagrams The trajectories of two balls are shown in **Figure 2.** The red ball was dropped, and the purple ball was given an initial horizontal velocity of 2.0 m/s. What is similar about the two paths? Look at their vertical positions. The horizontal lines indicate the equal vertical distances. At each moment that a picture was taken, the heights of the two balls were the same. Because the change in vertical position was the same for both, their average vertical velocities during each interval were also the same. The increasingly large distance traveled vertically by the balls, from one time interval to the next, shows that they were accelerating downward due to the force of gravity.

Notice that the horizontal motion of the launched ball does not affect its vertical motion. A projectile launched horizontally has initial horizontal velocity, but it has no initial vertical velocity. Therefore, its vertical motion is like that of an object dropped from rest. Just like the red ball, the purple ball has a downward velocity that increases regularly because of the acceleration due to gravity.

☑ **READING CHECK** **Explain** why a dropped object has the same vertical velocity as an object launched horizontally.

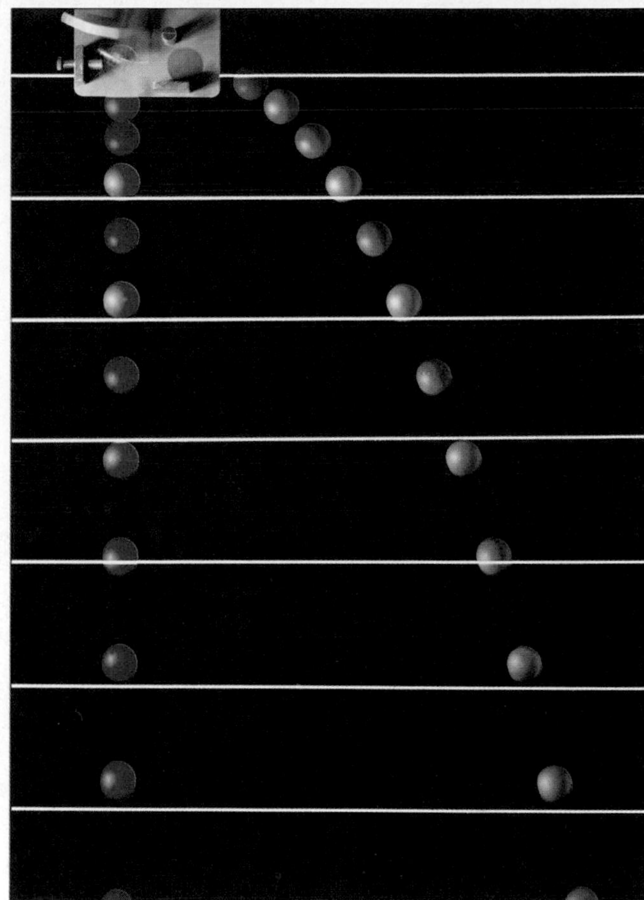

Figure 2 The ball on the left was dropped with no initial velocity. The ball on the right was given an initial horizontal velocity. The balls have the same vertical motion as they fall.

Identify What is the vertical velocity of the balls after falling for 1 s?

Investigate a **soccer kick.**

Virtual Investigation

MiniLABs

OVER THE EDGE

Does mass affect the motion of a projectile?

PROJECTILE PATH

Are the horizontal and vertical motions of a projectile related?

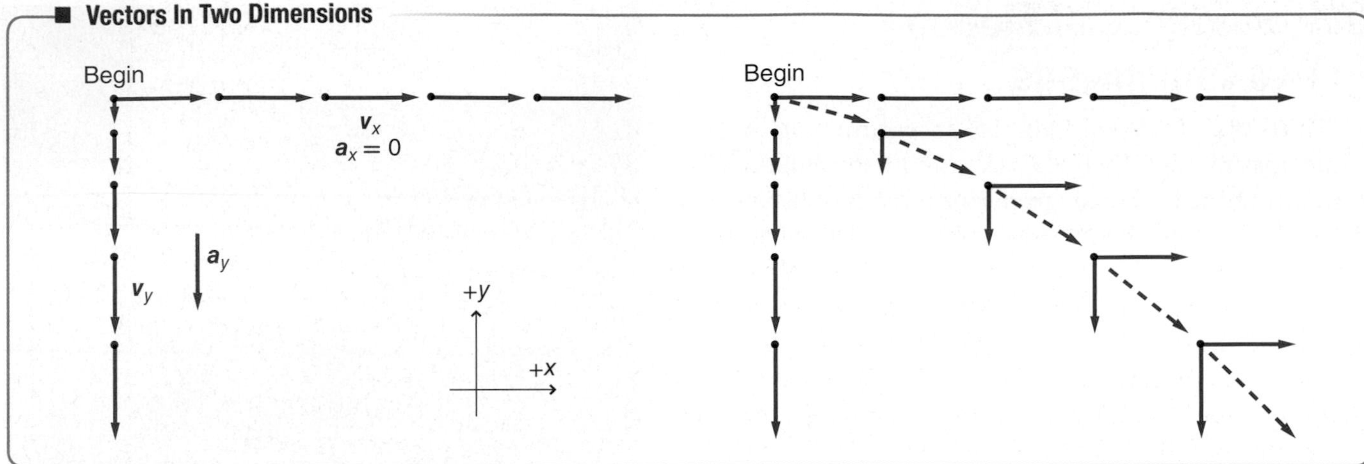

Figure 3 To describe the motion of a horizontally launched projectile, the *x*- and *y*-components can be treated independently. The resultant vectors of the projectile are tangent to a parabola.

Decide What is the value of a_y?

Horizontally Launched Projectiles

Imagine a person standing near the edge of a cliff and kicking a pebble horizontally. Like all horizontally launched projectiles, the pebble will have an initial horizontal velocity, but it will not have an initial vertical velocity. What will happen to the pebble as it falls from the cliff?

Separate motion diagrams Recall that the horizontal motion of a projectile does not affect its vertical motion. It is therefore easier to analyze the horizontal motion and the vertical motion separately. Separate motion diagrams for the *x*-components and *y*-components of a horizontally launched projectile, such as a pebble kicked off a cliff, are shown on the left in **Figure 3.**

Horizontal motion Notice the horizontal vectors in the diagram on the left. Each of the velocity vectors is the same length, which indicates that the object's velocity is not changing. The pebble is not accelerating horizontally. This constant velocity in the horizontal direction is exactly what should be expected, because after the initial kick, there is no horizontal force acting on the pebble. (In reality, the pebble's speed would decrease slightly because of air resistance, but remember that we are ignoring air resistance in this chapter.)

READING CHECK Explain why the horizontal motion of a projectile is constant.

Vertical motion Now look at the vertical velocity vectors in the diagram on the left. Each velocity vector has a slightly longer length than the one above it. The changing length shows that the object's velocity is increasing and accelerating downward. Again, this is what should be expected, because in this case the force of gravity is acting on the pebble.

Parabolic path When the *x*- and *y*-components of the object's motion are treated independently, each path is a straight line. The diagram on the right in **Figure 3** shows the actual parabolic path. The horizontal and vertical components at each moment are added to form the total velocity vector at that moment. You can see how the combination of constant horizontal velocity and uniform vertical acceleration produces a trajectory that has a parabolic shape.

PhysicsLABs

LAUNCH AN INVESTIGATION

FORENSICS LAB How can physics reconstruct a projectile's launch?

ON TARGET

DESIGN YOUR OWN LAB What factors affect projectile motion?

PROBLEM-SOLVING STRATEGIES

MOTION IN TWO DIMENSIONS

When solving projectile problems, use the following strategies.

1. Draw a motion diagram with vectors for the projectile at its initial position and its final position. If the projectile is launched at an angle, also show its maximum height and the initial angle.
2. Consider vertical and horizontal motion independently. List known and unknown variables.
3. For horizontal motion, the acceleration is $a_x = 0.0\ \text{m/s}^2$. If the projectile is launched at an angle, its initial vertical velocity and its vertical velocity when it falls back to that same height have the same magnitude but different direction: $v_{yi} = -v_{yf}$.
4. For vertical motion, $a_y = -9.8\ \text{m/s}^2$ (if you choose up as positive). If the projectile is launched at an angle, its vertical velocity at its highest point is zero: $v_{y,\max} = 0$.
5. Choose the motion equations that will enable you to find the unknown variables. Apply them to vertical and horizontal motion separately. Remember that time is the same for horizontal and vertical motion. Solving for time in one of the dimensions identifies the time for the other dimension.
6. Sometimes it is useful to apply the motion equations to part of the projectile's path. You can choose any initial and final points to use in the equations.

Motion Equations

Horizontal (constant speed)

$$\boldsymbol{x}_f = \boldsymbol{v}t_f + \boldsymbol{x}_i$$

Vertical (constant acceleration)

$$\boldsymbol{v}_f = \boldsymbol{v}_i + \boldsymbol{a}t_f$$

$$\boldsymbol{x}_f = \boldsymbol{x}_i + \boldsymbol{v}_i t_f + \frac{1}{2}\boldsymbol{a}t_f^2$$

$$v_f^2 = v_i^2 + 2a(x_f - x_i)$$

EXAMPLE PROBLEM 1

Find help with **square roots.** Math Handbook

A SLIDING PLATE You are preparing breakfast and slide a plate on the countertop. Unfortunately, you slide it too fast, and it flies off the end of the countertop. If the countertop is 1.05 m above the floor and it leaves the top at 0.74 m/s, how long does it take to fall, and how far from the end of the counter does it land?

1 ANALYZE AND SKETCH THE PROBLEM

Draw horizontal and vertical motion diagrams. Choose the coordinate system so that the origin is at the top of the countertop. Choose the positive *x* direction in the direction of horizontal velocity and the positive *y* direction up.

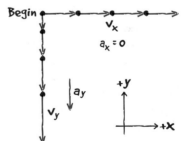

KNOWN		UNKNOWN
$x_i = y_i = 0$ m	$a_x = 0\ \text{m/s}^2$	$t = ?$
$v_{xi} = 0.75$ m/s	$a_y = -9.8\ \text{m/s}^2$	$x_f = ?$
$v_{yi} = 0$ m/s	$y_f = -1.05$ m	

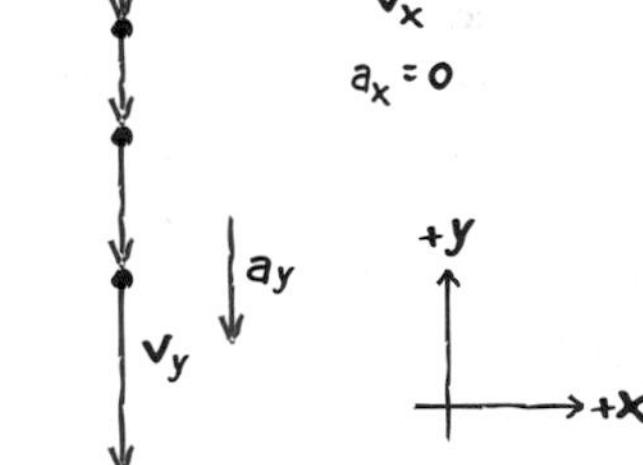

2 SOLVE FOR THE UNKNOWN

Use the equation of motion in the *y* direction to find the time of fall.

$$y_f = y_i + \frac{1}{2}a_y t^2$$

$$t = \sqrt{\frac{2(y_f - y_i)}{a_y}}$$ ◀ Rearrange the equation to solve for time.

$$= \sqrt{\frac{2(-1.05\ \text{m} - 0\ \text{m})}{-9.8\ \text{m/s}^2}} = 0.46\ \text{s}$$ ◀ Substitute $y_f = -1.05$ m, $y_i = 0$ m, $a_y = -9.8\ \text{m/s}^2$.

Use the equation of motion in the *x* direction to find where the plate hits the floor.

$$\boldsymbol{x}_f = \boldsymbol{v}_x t = (0.74\ \text{m/s to the right})(0.46\ \text{s}) = 0.34\ \text{m to the right of the counter}$$

3 EVALUATE THE ANSWER

- **Are the units correct?** Time is measured in seconds. Position in measured in meters.
- **Do the signs make sense?** Both are positive. The position sign agrees with the coordinate choice.
- **Are the magnitudes realistic?** A fall of about a meter takes about 0.5 s. During this time, the horizontal displacement of the plate would be about 0.5 s × 0.74 m/s.

EXAMPLE PROBLEM

PRACTICE PROBLEMS

Do additional problems. Online Practice

1. You throw a stone horizontally at a speed of 5.0 m/s from the top of a cliff that is 78.4 m high.
   - **a.** How long does it take the stone to reach the bottom of the cliff?
   - **b.** How far from the base of the cliff does the stone hit the ground?
   - **c.** What are the horizontal and vertical components of the stone's velocity just before it hits the ground?
2. Lucy and her friend are working at an assembly plant making wooden toy giraffes. At the end of the line, the giraffes go horizontally off the edge of a conveyor belt and fall into a box below. If the box is 0.60 m below the level of the conveyor belt and 0.40 m away from it, what must be the horizontal velocity of giraffes as they leave the conveyor belt?
3. **CHALLENGE** You are visiting a friend from elementary school who now lives in a small town. One local amusement is the ice-cream parlor, where Stan, the short-order cook, slides his completed ice-cream sundaes down the counter at a constant speed of 2.0 m/s to the servers. (The counter is kept very well polished for this purpose.) If the servers catch the sundaes 7.0 cm from the edge of the counter, how far do they fall from the edge of the counter to the point at which the servers catch them?

Figure 4 When a projectile is launched at an upward angle, its parabolic path is upward and then downward. The up-and-down motion is clearly represented in the vertical component of the vector diagram.

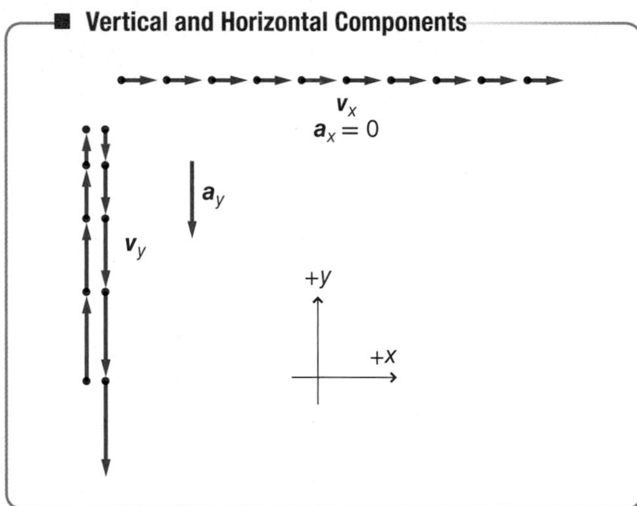

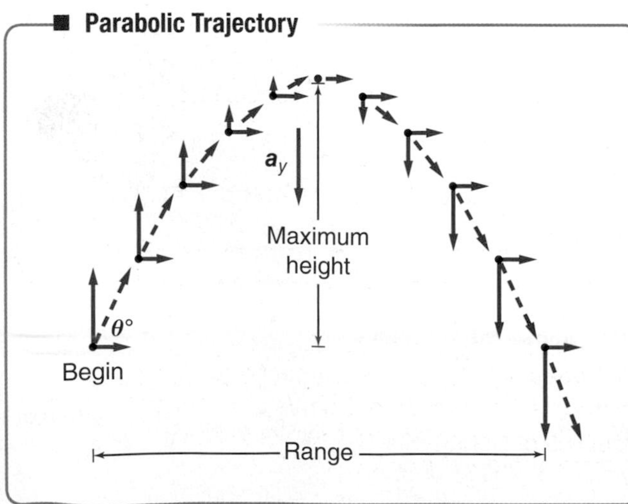

Angled Launches

When a projectile is launched at an angle, the initial velocity has a vertical component as well as a horizontal component. If the object is launched upward, like a ball tossed straight up in the air, it rises with slowing speed, reaches the top of its path where its speed is momentarily zero, and descends with increasing speed.

Separate motion diagrams The upper diagram of **Figure 4** shows the separate vertical- and horizontal-motion diagrams for the trajectory. In the coordinate system, the *x*-axis is horizontal and the *y*-axis is vertical. Note the symmetry. At each point in the vertical direction, the velocity of the object as it is moving upward has the same magnitude as when it is moving downward. The only difference is that the directions of the two velocities are opposite. When solving problems, it is sometimes useful to consider symmetry to determine unknown quantities.

Parabolic path The lower diagram of **Figure 4** defines two quantities associated with the trajectory. One is the maximum height, which is the height of the projectile when the vertical velocity is zero and the projectile has only its horizontal-velocity component. The other quantity depicted is the range (*R*), which is the horizontal distance the projectile travels when the initial and final heights are the same. Not shown is the flight time, which is how much time the projectile is in the air. For football punts, flight time often is called hang time.

☑ **READING CHECK** At what point of a projectile's trajectory is its vertical velocity zero?

EXAMPLE PROBLEM 2

Get help with **the flight of a ball.** Personal Tutor

THE FLIGHT OF A BALL A ball is launched at 4.5 m/s at 66° above the horizontal. It starts and lands at the same distance from the ground. What are the maximum height above its launch level and the flight time of the ball?

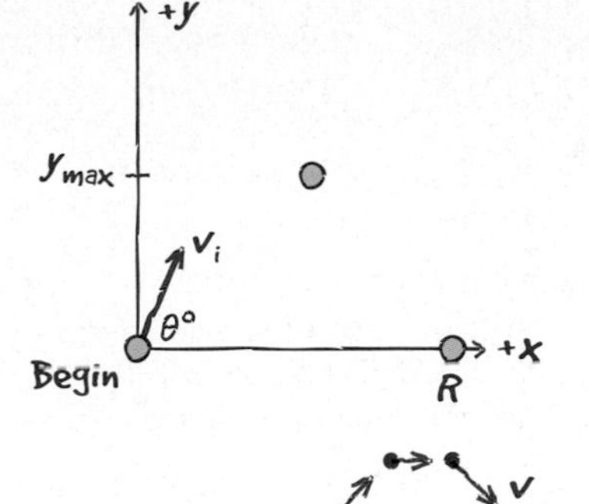

1 ANALYZE AND SKETCH THE PROBLEM

- Establish a coordinate system with the initial position of the ball at the origin.
- Show the positions of the ball at the beginning, at the maximum height, and at the end of the flight. Show the direction of $\boldsymbol{F}_{net}$.
- Draw a motion diagram showing $\boldsymbol{v}$ and $\boldsymbol{a}$.

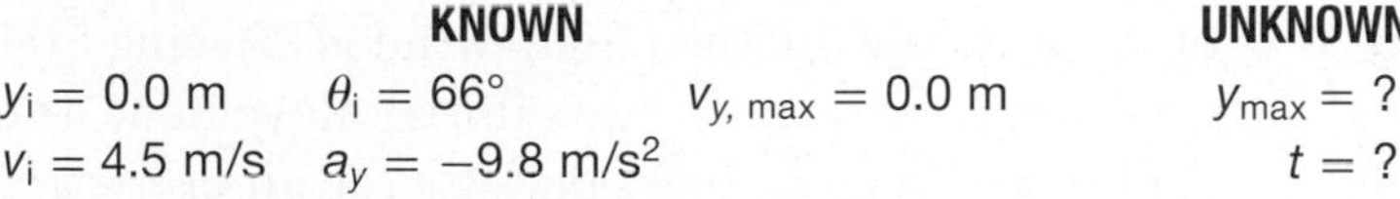

KNOWN			UNKNOWN
$y_i = 0.0$ m	$\theta_i = 66°$	$v_{y,\,max} = 0.0$ m	$y_{max} = ?$
$v_i = 4.5$ m/s	$a_y = -9.8\ \text{m/s}^2$		$t = ?$

$F_g = F_{net}$ v_{yi} v_i v_{xi} v a

2 SOLVE FOR THE UNKNOWN

Find the y-component of v_i.

$$v_{yi} = v_i(\sin \theta_i)$$
$$= (4.5\ \text{m/s})(\sin 66°) = 4.1\ \text{m/s}$$ ◀ Substitute v_i = 4.5 m/s, θ_i = 66°.

Use symmetry to find the y-component of v_f.

$$v_{yf} = -v_{yi} = -4.1\ \text{m/s}$$

Solve for the maximum height.

$$v_{y,\,max}{}^2 = v_{yi}{}^2 + 2a_y(y_{max} - y_i)$$
$$(0.0\ \text{m/s})^2 = v_{yi}{}^2 + 2a_y(y_{max} - 0.0\ \text{m})$$
$$y_{max} = -\frac{v_{yi}{}^2}{2a_y}$$
$$= -\frac{(4.1\ \text{m/s})^2}{2(-9.8\ \text{m/s}^2)} = 0.86\ \text{m}$$ ◀ Substitute v_{yi} = 4.1 m/s; a_y = -9.8 m/s².

Solve for the time to return to the launching height.

$$v_{yf} = v_{yi} + a_y t$$
$$t = \frac{v_{yf} - v_{yi}}{a_y}$$
$$= \frac{-4.1\ \text{m/s} - 4.1\ \text{m/s}}{-9.8\ \text{m/s}^2} = 0.84\ \text{s}$$ ◀ Substitute v_{yf} = -4.1 m/s; v_{yi} = 4.1 m/s; a_y = -9.8 m/s².

3 EVALUATE THE ANSWER

Are the magnitudes realistic? For an object that rises less than 1 m, a time of less than 1 s is reasonable.

PRACTICE PROBLEMS

Do additional problems. Online Practice

4. A player kicks a football from ground level with an initial velocity of 27.0 m/s, 30.0° above the horizontal, as shown in **Figure 5.** Find each of the following. Assume that forces from the air on the ball are negligible.

a. the ball's hang time

b. the ball's maximum height

c. the horizontal distance the ball travels before hitting the ground

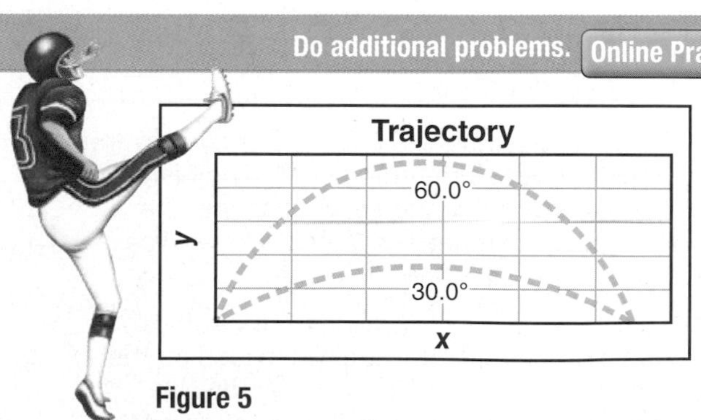

Figure 5

5. The player in the previous problem then kicks the ball with the same speed but at 60.0° from the horizontal. What is the ball's hang time, horizontal distance traveled, and maximum height?

6. CHALLENGE A rock is thrown from a 50.0-m-high cliff with an initial velocity of 7.0 m/s at an angle of 53.0° above the horizontal. Find its velocity when it hits the ground below.

Figure 6 Forces from air can increase or decrease the velocity of a moving object.

No Effect from Air

Force of Air that Increases Velocity

Force of Air that Decreases Velocity

Forces from Air

The effect of forces due to air has been ignored so far in this chapter, but think about why a kite stays in the air or why a parachute helps a skydiver fall safely to the ground. Forces from the air can significantly change the motion of an object.

What happens if there is wind? Moving air can change the motion of a projectile. Consider the three cases shown in **Figure 6.** In the top photo, water is flowing from the hose pipe with almost no effect from air. In the middle photo, wind is blowing in the same direction as the water's initial movement. The path of the water changes because the air exerts a force on the water in the same direction as its motion. The horizontal distance the water travels increases because the force increases the water's horizontal speed. The direction of the wind changes in the bottom photo. The horizontal distance the water travels decreases because the air exerts a force in the direction opposite the water's motion.

What if the direction of wind is at an angle relative to a moving object? The horizontal component of the wind affects only the horizontal motion of an object. The vertical component of the wind affects only the vertical motion of the object. In the case of the water, for example, a strong updraft could decrease the downward speed of the water.

The effects shown in **Figure 6** occur because the air is moving enough to significantly change the motion of the water. Even air that is not moving, however, can have a significant effect on some moving objects. A piece of paper held horizontally and dropped, for example, falls slowly because of air resistance. The air resistance increases as the surface area of the object that faces the moving air increases.

SECTION 1 REVIEW

Check your understanding.

7. **MAIN IDEA** Two baseballs are pitched horizontally from the same height but at different speeds. The faster ball crosses home plate within the strike zone, but the slower ball is below the batter's knees. Why do the balls pass the batter at different heights?

8. **Free-Body Diagram** An ice cube slides without friction across a table at a constant velocity. It slides off the table and lands on the floor. Draw free-body and motion diagrams of the ice cube at two points on the table and at two points in the air.

9. **Projectile Motion** A tennis ball is thrown out a window 28 m above the ground at an initial velocity of 15.0 m/s and 20.0° below the horizontal. How far does the ball move horizontally before it hits the ground?

10. **Projectile Motion** A softball player tosses a ball into the air with an initial velocity of 11.0 m/s, as shown in **Figure 7.** What will be the ball's maximum height?

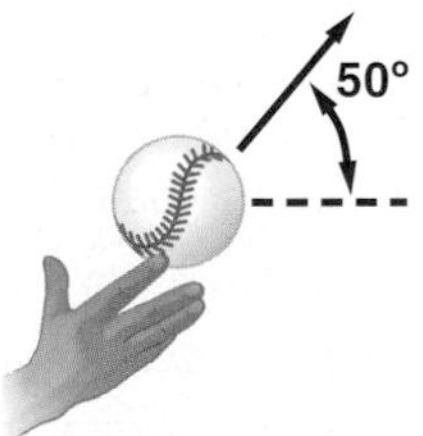

Figure 7

11. **Critical Thinking** Suppose an object is thrown with the same initial velocity and direction on Earth and on the Moon, where the acceleration due to gravity is one-sixth its value on Earth. How will vertical velocity, time of flight, maximum height, and horizontal distance change?

SECTION 2 Circular Motion

PHYSICS 4 YOU Many amusement park and carnival rides spin. When the ride is spinning, forces from the walls or sides of the ride keep the riders moving in a circular path.

MAIN IDEA

An object in circular motion has an acceleration toward the circle's center due to an unbalanced force toward the circle's center.

Essential Questions

- Why is an object moving in a circle at a constant speed accelerating?
- How does centripetal acceleration depend upon the object's speed and the radius of the circle?
- What causes centripetal acceleration?

Review Vocabulary

average velocity the change in position divided by the time during which the change occurred; the slope of an object's position-time graph

New Vocabulary

uniform circular motion
centripetal acceleration
centripetal force

Describing Circular Motion

Consider an object moving in a circle at a constant speed, such as a stone being whirled on the end of a string or a fixed horse on a carousel. Are these objects accelerating? At first, you might think they are not because their speeds do not change. But remember that acceleration is related to the change in velocity, not just the change in speed. Because their directions are changing, the objects must be accelerating.

Uniform circular motion is the movement of an object at a constant speed around a circle with a fixed radius. The position of an object in uniform circular motion, relative to the center of the circle, is given by the position vector $\boldsymbol{r}$. Remember that a position vector is a displacement vector with its tail at the origin. Two position vectors, $\boldsymbol{r}_1$ and $\boldsymbol{r}_2$, at the beginning and end of a time interval are shown on the left in **Figure 8.** As the object moves around the circle, the length of the position vector does not change, but its direction does. The diagram also shows two instantaneous velocity vectors. Notice that each velocity vector is tangent to the circular path, at a right angle to the corresponding position vector.

To determine the object's velocity, you first need to find its displacement vector over a time interval. You know that a moving object's average velocity is defined as $\frac{\Delta \boldsymbol{x}}{\Delta t}$, so for an object in circular motion, $\bar{\boldsymbol{v}} = \frac{\Delta \boldsymbol{r}}{\Delta t}$. The right side of **Figure 8** shows $\Delta \boldsymbol{r}$ drawn as the displacement from $\boldsymbol{r}_1$ to $\boldsymbol{r}_2$ during a time interval. The velocity for this time interval has the same direction as the displacement, but its length would be different because it is divided by Δt.

Position and Velocity Vectors

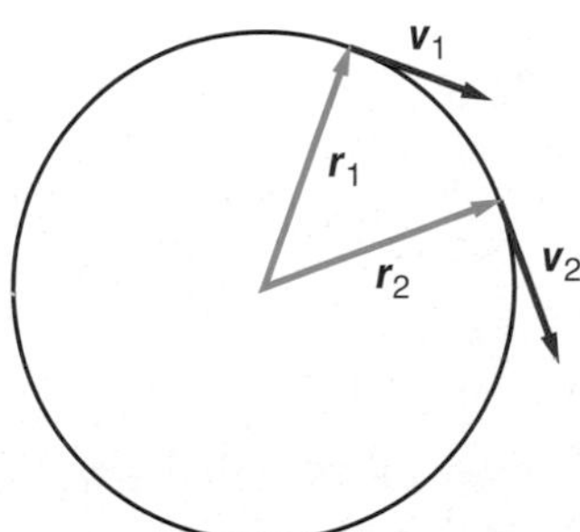

Displacement Vector

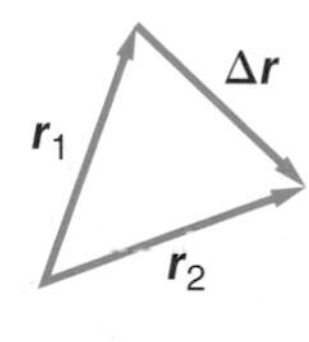

Figure 8 For an object in uniform circular motion, the velocity is tangent to the circle. It is in the same direction as the displacement.

Analyze How can you tell from the diagram that the motion is uniform?

View an **animation of circular motion and centripetal acceleration.**

Andrew Holt/Photographer's Choice/Getty Images

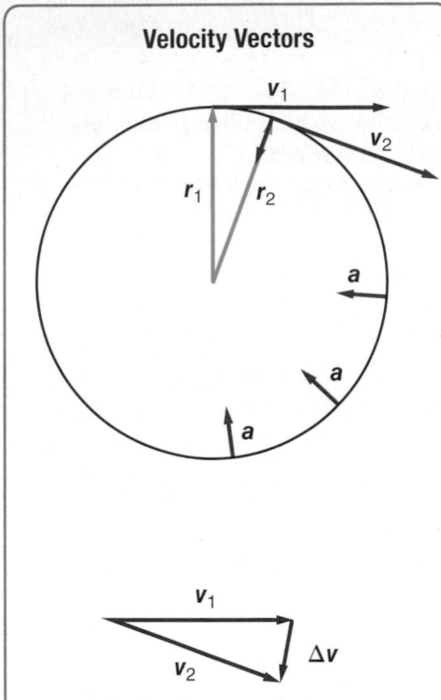

Figure 9 The acceleration of an object in uniform circular motion is the change in velocity divided by the time interval. The direction of centripetal acceleration is always toward the center of the circle.

Centripetal Acceleration

You have read that a velocity vector of an object in uniform circular motion is tangent to the circle. What is the direction of the acceleration? **Figure 9** shows the velocity vectors $\boldsymbol{v}_1$ and $\boldsymbol{v}_2$ at the beginning and end of a time interval. The difference in the two vectors ($\Delta\boldsymbol{v}$) is found by subtracting the vectors, as shown at the bottom of the figure. The average acceleration $\left(\overline{\boldsymbol{a}} = \frac{\Delta\boldsymbol{v}}{\Delta t}\right)$ for this time interval is in the same direction as $\Delta\boldsymbol{v}$. For a very small time interval, $\Delta\boldsymbol{v}$ is so small that $\boldsymbol{a}$ points toward the center of the circle.

Repeat this process for several other time intervals when the object is in different locations on the circle. As the object moves around the circle, the direction of the acceleration vector changes, but it always points toward the center of the circle. For this reason, the acceleration of an object in uniform circular motion is called center-seeking or **centripetal acceleration.**

Magnitude of acceleration What is the magnitude of an object's centripetal acceleration? Look at the starting points of the velocity vectors in the top of **Figure 9.** Notice the triangle the position vectors at those points make with the center of the circle. An identical triangle is formed by the velocity vectors in the bottom of **Figure 9.** The angle between $\boldsymbol{r}_1$ and $\boldsymbol{r}_2$ is the same as that between $\boldsymbol{v}_1$ and $\boldsymbol{v}_2$. Therefore, similar triangles are formed by subtracting the two sets of vectors, and the ratios of the lengths of two corresponding sides are equal. Thus, $\frac{\Delta r}{r} = \frac{\Delta v}{v}$. The equation is not changed if both sides are divided by Δt.

$$\frac{\Delta r}{r\Delta t} = \frac{\Delta v}{v\Delta t}$$

But, $v = \frac{\Delta r}{\Delta t}$ and $a = \frac{\Delta v}{\Delta t}$.

$$\left(\frac{1}{r}\right)\left(\frac{\Delta r}{\Delta t}\right) = \left(\frac{1}{v}\right)\left(\frac{\Delta v}{\Delta t}\right)$$

Substituting $v = \frac{\Delta r}{\Delta t}$ in the left-hand side and $a = \frac{\Delta v}{\Delta t}$ in the right-hand side gives the following equation:

$$\frac{v}{r} = \frac{a}{v}.$$

Solve for acceleration, and use the symbol a_c for centripetal acceleration.

CENTRIPETAL ACCELERATION

Centripetal acceleration always points to the center of the circle. Its magnitude is equal to the square of the speed divided by the radius of motion.

$$a_c = \frac{v^2}{r}$$

Period of revolution One way to describe the speed of an object moving in a circle is to measure its period (T), the time needed for the object to make one complete revolution. During this time, the object travels a distance equal to the circumference of the circle ($2\pi r$). The speed, then, is represented by $v = \frac{2\pi r}{T}$. If you substitute for v in the equation for centripetal acceleration, you obtain the following equation:

$$a_c = \frac{(2\pi r/T)^2}{r} = \frac{4\pi^2 r}{T^2}.$$

Figure 10 As the hammer thrower swings the ball around, tension in the chain is the force that causes the ball to have an inward acceleration.

Predict Neglecting air resistance, how would the horizontal acceleration and velocity of the hammer change if the thrower released the chain?

Centripetal force Because the acceleration of an object moving in a circle is always in the direction of the net force acting on it, there must be a net force toward the center of the circle. This force can be provided by any number of agents. For Earth circling the Sun, the force is the Sun's gravitational force on Earth. When a hammer thrower swings the hammer, as in **Figure 10,** the force is the tension in the chain attached to the massive ball. When an object moves in a circle, the net force toward the center of the circle is called the **centripetal force.** To accurately analyze centripetal acceleration situations, you must identify the agent of the force that causes the acceleration. Then you can apply Newton's second law for the component in the direction of the acceleration in the following way.

NEWTON'S SECOND LAW FOR CIRCULAR MOTION

The net centripetal force on an object moving in a circle is equal to the object's mass times the centripetal acceleration.

$$\boldsymbol{F}_{\text{net}} = m\boldsymbol{a}_{\text{c}}$$

Direction of acceleration When solving problems, you have found it useful to choose a coordinate system with one axis in the direction of the acceleration. For circular motion, the direction of the acceleration is always toward the center of the circle. Rather than labeling this axis *x* or *y*, call it *c*, for centripetal acceleration. The other axis is in the direction of the velocity, tangent to the circle. It is labeled *tang* for tangential. You will apply Newton's second law in these directions, just as you did in the two-dimensional problems you have solved before. Remember that centripetal force is just another name for the net force in the centripetal direction. It is the sum of all the real forces, those for which you can identify agents that act along the centripetal axis.

In the case of the hammer thrower in **Figure 10,** in what direction does the hammer fly when the chain is released? Once the contact force of the chain is gone, there is no force accelerating the hammer toward the center of the circle, so the hammer flies off in the direction of its velocity, which is tangent to the circle. Remember, if you cannot identify the agent of a force, then it does not exist.

PhysicsLAB

CENTRIPETAL FORCE

What keeps an object moving when you swing it in a circle?

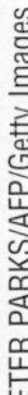

EXAMPLE PROBLEM 3

Find help with **significant figures.** Math Handbook

UNIFORM CIRCULAR MOTION A 13-g rubber stopper is attached to a 0.93-m string. The stopper is swung in a horizontal circle, making one revolution in 1.18 s. Find the magnitude of the tension force exerted by the string on the stopper.

1 ANALYZE AND SKETCH THE PROBLEM

- Draw a free-body diagram for the swinging stopper.
- Include the radius and the direction of motion.
- Establish a coordinate system labeled *tang* and *c*. The directions of a_c and F_T are parallel to *c*.

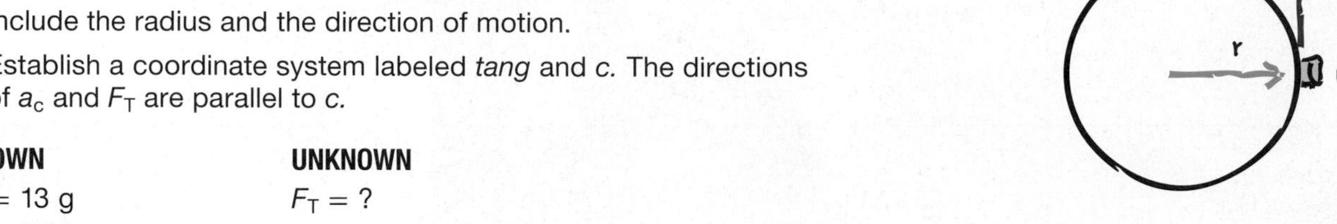

KNOWN	**UNKNOWN**
$m = 13$ g	$F_T = ?$
$r = 0.93$ m	
$T = 1.18$ s	

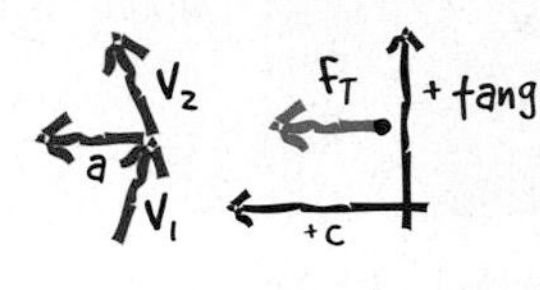

2 SOLVE FOR THE UNKNOWN

Find the magnitude of the centripetal acceleration.

$$a_c = \frac{4\pi^2 r}{T^2}$$

$$= \frac{4\pi^2(0.93\text{ m})}{(1.18\text{ s})^2}$$ ◀ *Substitute $r = 0.93$ m, $T = 1.18$ s.*

$$= 26\text{ m/s}^2$$

Use Newton's second law to find the magnitude of the tension in the string.

$$F_T = ma_c$$

$$= (0.013\text{ kg})(26\text{ m/s}^2)$$ ◀ *Substitute $m = 0.013$ kg, $a_c = 26$ m/s^2.*

$$= 0.34\text{ N}$$

3 EVALUATE THE ANSWER

- **Are the units correct?** Dimensional analysis verifies that a_c is in meters per second squared and F_T is in newtons.
- **Do the signs make sense?** The signs should all be positive.
- **Are the magnitudes realistic?** The force is almost three times the weight of the stopper, and the acceleration is almost three times that of gravity, which is reasonable for such a light object.

PRACTICE PROBLEMS

Do additional problems. Online Practice

12. A runner moving at a speed of 8.8 m/s rounds a bend with a radius of 25 m. What is the centripetal acceleration of the runner, and what agent exerts the centripetal force on the runner?

13. An airplane traveling at 201 m/s makes a turn. What is the smallest radius of the circular path (in kilometers) the pilot can make and keep the centripetal acceleration under 5.0 m/s^2?

14. A 45-kg merry-go-round worker stands on the ride's platform 6.3 m from the center, as shown in **Figure 11.** If her speed (v_{worker}) as she goes around the circle is 4.1 m/s, what is the force of friction (F_f) necessary to keep her from falling off the platform?

15. A 16-g ball at the end of a 1.4-m string is swung in a horizontal circle. It revolves once every 1.09 s. What is the magnitude of the string's tension?

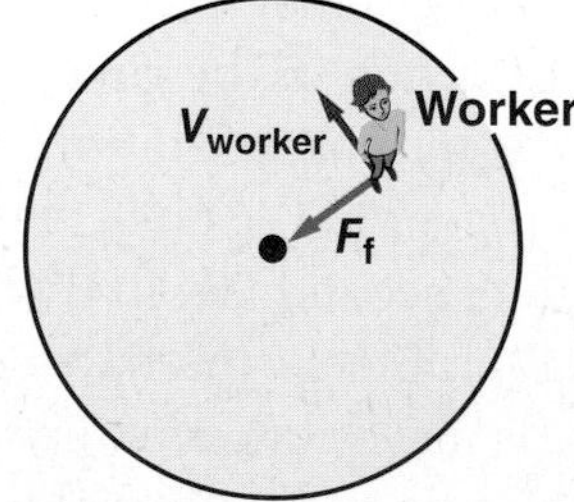

Figure 11

16. CHALLENGE A car racing on a flat track travels at 22 m/s around a curve with a 56-m radius. Find the car's centripetal acceleration. What minimum coefficient of static friction between the tires and the road is necessary for the car to round the curve without slipping?

Centrifugal "Force"

If a car makes a sharp left turn, a passenger on the right side might be thrown against the right door. Is there an outward force on the passenger? Consider a similar situation. If a car in which you are riding stops suddenly, you will be thrown forward into your safety belt. Is there a forward force on you? No, because according to Newton's first law, you will continue moving with the same velocity unless there is a net force acting on you. The safety belt applies the force that accelerates you to a stop.

Figure 12 shows a car turning left as viewed from above. A passenger in the car would continue to move straight ahead if it were not for the force of the door acting in the direction of the acceleration. As the car goes around the curve, the car and the passenger are in circular motion, and the passenger experiences centripetal acceleration. Recall that centripetal acceleration is always directed toward the center of the circle. There is no outward force on the passenger.

If, however, you think about similar situations that you have experienced, you know that it feels as if a force is pushing you outward. The so-called centrifugal, or outward, force is a fictitious, nonexistent force. You feel as if you are being pushed only because you are accelerating relative to your surroundings. There is no real force because there is no agent exerting a force.

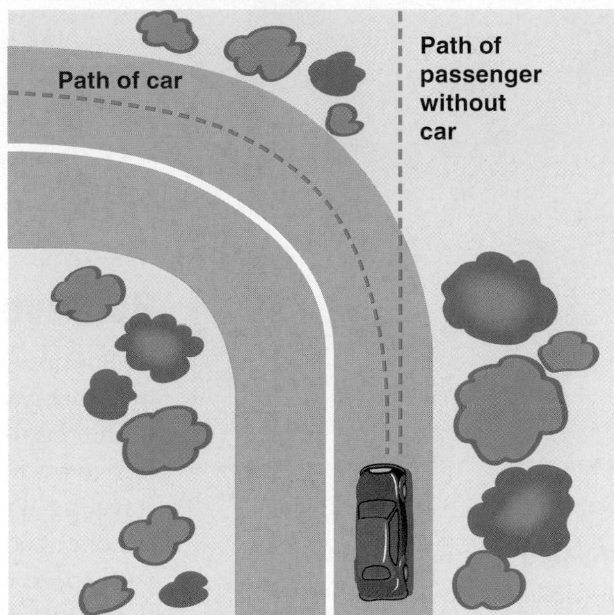

Figure 12 When a car moves around a curve, a passenger feels a fictitious centrifugal force directed outward. In fact, the force on the passenger is the centripetal force, which is directed toward the center of the circle and is exerted by the seat on which the person is sitting.

SECTION 2 REVIEW

Check your understanding.

17. **MAIN**IDEA If you attach a ball to a rope and swing it at a constant speed in a circle above your head, the ball is in uniform circular motion. In which direction does it accelerate? What force causes the acceleration?

18. **Uniform Circular Motion** What is the direction of the force that acts on the clothes in the spin cycle of a top-load washing machine? What exerts the force?

19. **Centripetal Acceleration** A newspaper article states that when turning a corner, a driver must be careful to balance the centripetal and centrifugal forces to keep from skidding. Write a letter to the editor that describes physics errors in this article.

20. **Free-Body Diagram** You are sitting in the back seat of a car going around a curve to the right. Sketch motion and free-body diagrams to answer these questions:
    - **a.** What is the direction of your acceleration?
    - **b.** What is the direction of the net force on you?
    - **c.** What exerts this force?

21. **Centripetal Acceleration** An object swings in a horizontal circle, supported by a 1.8-m string. It completes a revolution in 2.2 s. What is the object's centripetal acceleration?

22. **Centripetal Force** The 40.0-g stone in **Figure 13** is whirled horizontally at a speed of 2.2 m/s. What is the tension in the string?

Figure 13

23. **Amusement-Park Ride** A ride at an amusement park has people stand around a 4.0-m radius circle with their backs to a wall. The ride then spins them with a 1.7-s period of revolution. What are the centripetal acceleration and velocity of the riders?

24. **Centripetal Force** A bowling ball has a mass of 7.3 kg. What force must you exert to move it at a speed of 2.5 m/s around a circle with a radius of 0.75 m?

25. **Critical Thinking** Because of Earth's daily rotation, you always move with uniform circular motion. What is the agent that supplies the force that causes your centripetal acceleration? If you are standing on a scale, how does the circular motion affect the scale's measure of your weight?

SECTION 3
Relative Velocity

PHYSICS 4 YOU **Have you ever noticed that people riding along with you on an escalator don't seem to be moving, while people going in the opposite direction seem to be moving very fast? These people may in fact have the same speed relative to the ground, but their velocities relative to you are very different.**

MAIN IDEA

An object's velocity depends on the reference frame chosen.

Essential Questions

- What is relative velocity?
- How do you find the velocities of an object in different reference frames?

Review Vocabulary

resultant a vector that results from the sum of two other vectors

New Vocabulary

reference frame

Relative Motion in One Dimension

Suppose you are in a school bus that is traveling at a velocity of 8 m/s in a positive direction. You walk with a velocity of 1 m/s toward the front of the bus. If a friend is standing on the side of the road watching the bus go by, how fast would your friend say you are moving? If the bus is traveling at 8 m/s, its speed as measured by your friend in a coordinate system fixed to the road is 8 m/s. When you are standing still on the bus, your speed relative to the road is also 8 m/s, but your speed relative to the bus is zero. How can your speed be different?

Different reference frames In this example, your motion is viewed from different coordinate systems. A coordinate system from which motion is viewed is a **reference frame.** Walking at 1 m/s toward the front of the bus means your velocity is measured in the reference frame of the bus. Your velocity in the road's reference frame is different. You can rephrase the problem as follows: given the velocity of the bus relative to the road and your velocity relative to the bus, what is your velocity relative to the road?

A vector representation of this problem is shown in **Figure 14.** If right is positive, your speed relative to the road is 9 m/s, the sum of 8 m/s and 1 m/s. Suppose that you now walk at the same speed toward the rear of the bus. What would be your velocity relative to the road? **Figure 14** shows that because the two velocities are in opposite directions, the resultant speed is 7 m/s, the difference between 8 m/s and 1 m/s. You can see that when the velocities are along the same line, simple addition or subtraction can be used to determine the relative velocity.

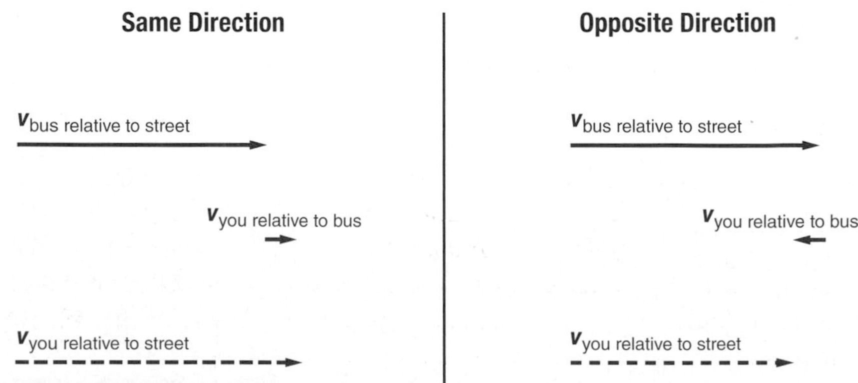

Figure 14 When an object moves in a moving reference frame, you add the velocities if they are in the same direction. You subtract one velocity from the other if they are in opposite directions.

Recall What do the lengths of the velocity vectors indicate?

Brand X Pictures/Punchstock

PHYSICS CHALLENGE

TENSION IN A ROPE Phillipe whirls a stone of mass m on a rope in a horizontal circle above his head such that the stone is at a height h above the ground. The circle has a radius of r, and the magnitude of the tension in the rope is T. Suddenly the rope breaks, and the stone falls to the ground. The stone travels a horizontal distance x from the time the rope breaks until it impacts the ground. Find a mathematical expression for x in terms of T, r, m, and h. Does your expression change if Phillipe is walking 0.50 m/s relative to the ground?

Combining velocity vectors Take a closer look at how the relative velocities in **Figure 14** were obtained. Can you find a mathematical rule to describe how velocities are combined when the motion is in a moving reference frame? For the situation in which you are walking in a bus, you can designate the velocity of the bus relative to the road as $\boldsymbol{v}_{b/r}$. You can designate your velocity relative to the bus as $\boldsymbol{v}_{y/b}$ and the velocity of you relative to the road as $\boldsymbol{v}_{y/r}$. To find the velocity of you relative to the road in both cases, you added the velocity vectors of you relative to the bus and the bus relative to the road. Mathematically, this is represented as $\boldsymbol{v}_{y/b} + \boldsymbol{v}_{b/r} = \boldsymbol{v}_{y/r}$. The more general form of this equation is as follows.

RELATIVE VELOCITY

The relative velocity of object a to object c is the vector sum of object a's velocity relative to object b and object b's velocity relative to object c.

$$\boldsymbol{v}_{a/b} + \boldsymbol{v}_{b/c} = \boldsymbol{v}_{a/c}$$

PhysicsLAB

MOVING REFERENCE FRAME

How can you describe motion in a moving reference frame?

Relative Motion in Two Dimensions

Adding relative velocities also applies to motion in two dimensions. As with one-dimensional motion, you first draw a vector diagram to describe the motion, and then you solve the problem mathematically.

Vector diagrams The method of drawing vector diagrams for relative motion in two dimensions is shown in **Figure 15.** The velocity vectors are drawn tip-to-tail. The reference frame from which you are viewing the motion, often called the ground reference frame, is considered to be at rest. One vector describes the velocity of the second reference frame relative to ground. The second vector describes the motion in that moving reference frame. The resultant shows the relative velocity, which is the velocity relative to the ground reference frame.

READING CHECK Decide Can a moving car be a reference frame?

An example is the relative motion of an airplane. Airline pilots cannot expect to reach their destinations by simply aiming their planes along a compass direction. They must take into account the plane's speed relative to the air, which is given by their airspeed indicators, and their direction of flight relative to the air. They also must consider the velocity of the wind at the altitude they are flying relative to the ground. These two vectors must be combined to obtain the velocity of the airplane relative to the ground. The resultant vector tells the pilot how fast and in what direction the plane must travel relative to the ground to reach its destination. A similar situation occurs for boats that are traveling on water that is flowing.

Figure 15 Vectors are placed tip-to-tail to find the relative velocity vector for two-dimensional motion. The subscript o/g refers to an object relative to ground, o/m refers to an object relative to a moving reference frame, and m/g refers to the moving frame relative to ground.

Analyze How would the resultant vector change if the ground reference frame were considered to be the moving reference frame?

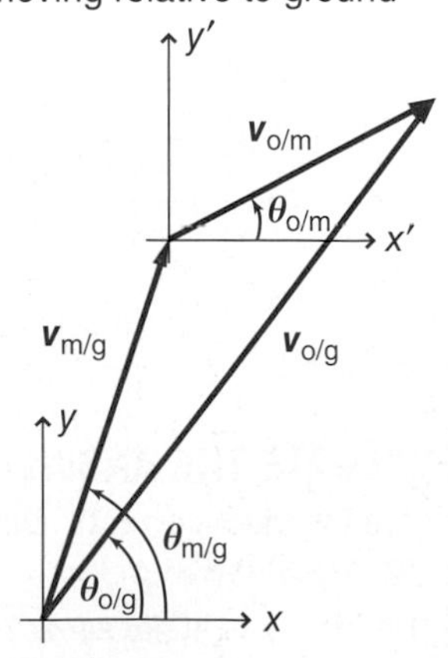

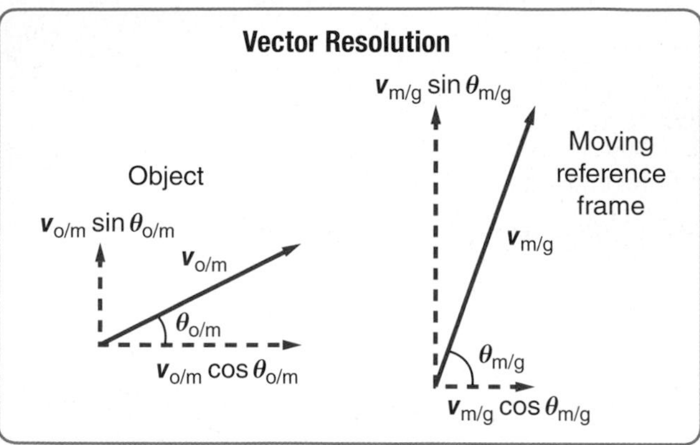

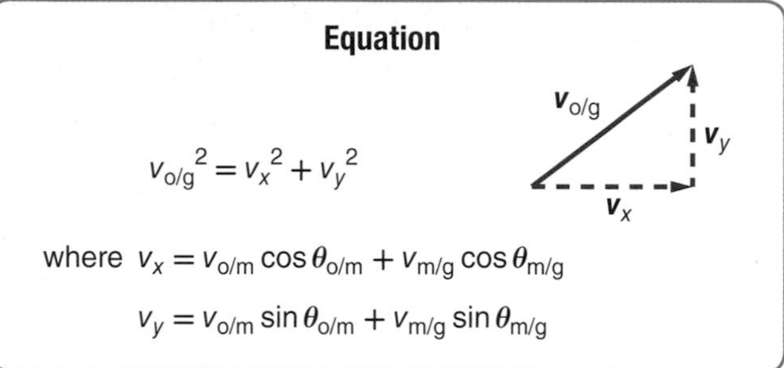

Equation
$v_{o/g}$
v_y
v_x
$v_{o/g}^2 = v_x^2 + v_y^2$
where $v_x = v_{o/m} \cos \theta_{o/m} + v_{m/g} \cos \theta_{m/g}$
$v_y = v_{o/m} \sin \theta_{o/m} + v_{m/g} \sin \theta_{m/g}$

Figure 16 To find the velocity of an object in a moving reference frame, resolve the vectors into *x*-and *y*-components.

Combining velocities You can use the equations in **Figure 16** to solve problems for relative motion in two dimensions. The velocity of a reference frame moving relative to the ground is labeled $\boldsymbol{v}_{m/g}$. The velocity of an object in the moving frame is labeled $\boldsymbol{v}_{o/m}$. The relative velocity equation gives the object's velocity relative to the ground: $\boldsymbol{v}_{m/g} = \boldsymbol{v}_{o/m} + \boldsymbol{v}_{m/g}$.

To determine the magnitude of the object's velocity relative to the ground ($v_{m/g}$), first resolve the velocity vectors of the object and the moving reference frame into *x*- and *y*-components. Then apply the Pythagorean theorem. The general equation is shown in **Figure 16,** but for many problems the equation is simpler because the vectors are along an axis. As shown in the example problem below, you can find the angle of the object's velocity relative to the ground by observing the vector diagram and applying a trigonometric relationship.

READING CHECK Explain How are vectors used to describe relative motion in two dimensions?

EXAMPLE PROBLEM 4

Find help with **inverse tangent.** Math Handbook

RELATIVE VELOCITY OF A MARBLE Ana and Sandra are riding on a ferry boat traveling east at 4.0 m/s. Sandra rolls a marble with a velocity of 0.75 m/s north, straight across the deck of the boat to Ana. What is the velocity of the marble relative to the water?

1 ANALYZE AND SKETCH THE PROBLEM

Establish a coordinate system. Draw vectors for the velocities.

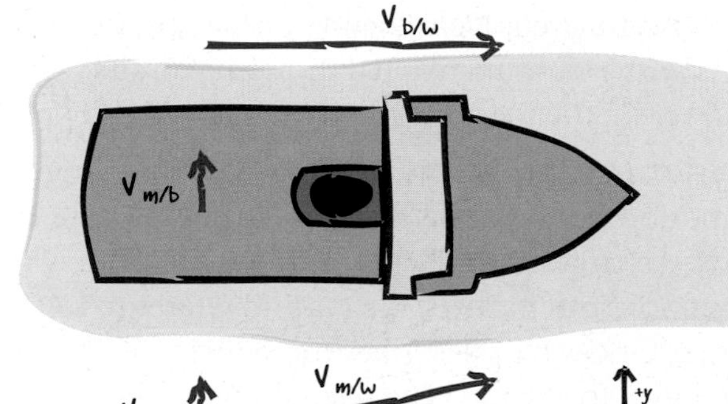

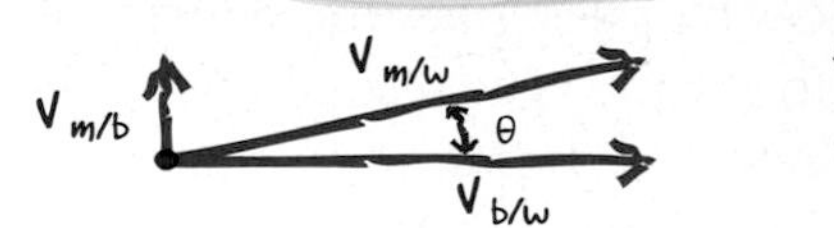

KNOWN		UNKNOWN
$v_{b/w} = 4.0$ m/s	$v_{m/b} = 0.75$ m/s	$v_{m/w} = ?$

2 SOLVE FOR THE UNKNOWN

The velocities are perpendicular, so we can use the Pythagorean theorem.

$$v_{m/w}^2 = v_{b/w}^2 + v_{m/b}^2$$

$$v_{m/w} = \sqrt{v_{b/w}^2 + v_{m/b}^2}$$

$$= \sqrt{(4.0\ \text{m/s})^2 + (0.75\ \text{m/s})^2} = 4.1\ \text{m/s}$$

◀ Substitute $v_{b/w} = 4.0$ m/s, $v_{m/b} = 0.75$ m/s.

Find the angle of the marble's velocity.

$$\theta = \tan^{-1}\left(\frac{v_{m/b}}{v_{b/w}}\right)$$

$$= \tan^{-1}\left(\frac{0.75\ \text{m/s}}{4.0\ \text{m/s}}\right) = 11°\ \text{north of east}$$

The marble travels 4.1 m/s at 11° north of east.

3 EVALUATE THE ANSWER

- **Are the units correct?** Dimensional analysis verifies units of meters per second for velocity.
- **Do the signs make sense?** The signs should all be positive.
- **Are the magnitudes realistic?** The resulting velocity is of the same order of magnitude as the velocities given in the problem and slightly larger than the larger of the two.

PRACTICE PROBLEMS

Do additional problems. Online Practice

26. You are riding in a bus moving slowly through heavy traffic at 2.0 m/s. You hurry to the front of the bus at 4.0 m/s relative to the bus. What is your speed relative to the street?

27. Rafi is pulling a toy wagon through a neighborhood at a speed of 0.75 m/s. A caterpillar in the wagon is crawling toward the rear of the wagon at a rate of 2.0 cm/s. What is the caterpillar's velocity relative to the ground?

28. A boat is rowed directly upriver at a speed of 2.5 m/s relative to the water. Viewers on the shore see that the boat is moving at only 0.5 m/s relative to the shore. What is the speed of the river? Is it moving with or against the boat?

29. A boat is traveling east at a speed of 3.8 m/s. A person walks across the boat with a velocity of 1.3 m/s south.

a. What is the person's speed relative to the water?

b. In what direction, relative to the ground, does the person walk?

30. An airplane flies due north at 150 km/h relative to the air. There is a wind blowing at 75 km/h to the east relative to the ground. What is the plane's speed relative to the ground?

31. CHALLENGE The airplane in **Figure 17** flies at 200.0 km/h relative to the air. What is the velocity of the plane relative to the ground if it flies during the following wind conditions?

a. a 50.0-km/h tailwind

b. a 50.0-km/h headwind

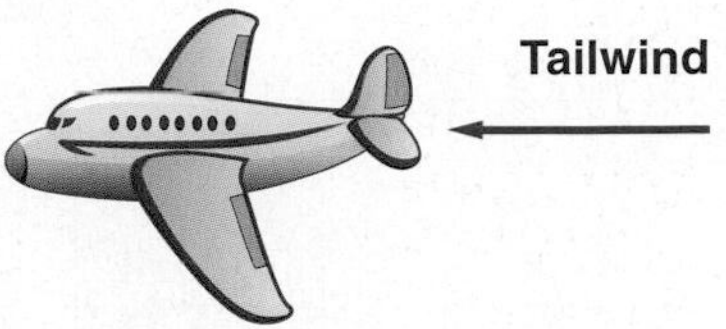

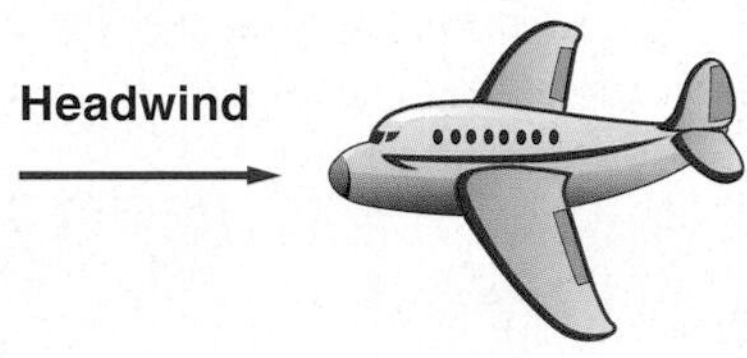

Figure 17

SECTION 3 REVIEW

Section Self-Check — Check your understanding.

32. MAINIDEA A plane has a speed of 285 km/h west relative to the air. A wind blows 25 km/h east relative to the ground. What are the plane's speed and direction relative to the ground?

33. Relative Velocity A fishing boat with a maximum speed of 3 m/s relative to the water is in a river that is flowing at 2 m/s. What is the maximum speed the boat can obtain relative to the shore? The minimum speed? Give the direction of the boat, relative to the river's current, for the maximum speed and the minimum speed relative to the shore.

34. Relative Velocity of a Boat A motorboat heads due west at 13 m/s relative to a river that flows due north at 5.0 m/s. What is the velocity (both magnitude and direction) of the motorboat relative to the shore?

35. Boating You are boating on a river that flows toward the east. Because of your knowledge of physics, you head your boat 53° west of north and have a velocity of 6.0 m/s due north relative to the shore.

a. What is the velocity of the current?

b. What is the speed of your boat relative to the water?

36. Boating Martin is riding on a ferry boat that is traveling east at 3.8 m/s. He walks north across the deck of the boat at 0.62 m/s. What is Martin's velocity relative to the water?

37. Relative Velocity An airplane flies due south at 175 km/h relative to the air. There is a wind blowing at 85 km/h to the east relative to the ground. What are the plane's speed and direction relative to the ground?

38. A Plane's Relative Velocity An airplane flies due north at 235 km/h relative to the air. There is a wind blowing at 65 km/h to the northeast relative to the ground. What are the plane's speed and direction relative to the ground?

39. Critical Thinking You are piloting the boat in **Figure 18** across a fast-moving river. You want to reach a pier directly opposite your starting point. Describe how you would navigate the boat in terms of the components of your velocity relative to the water.

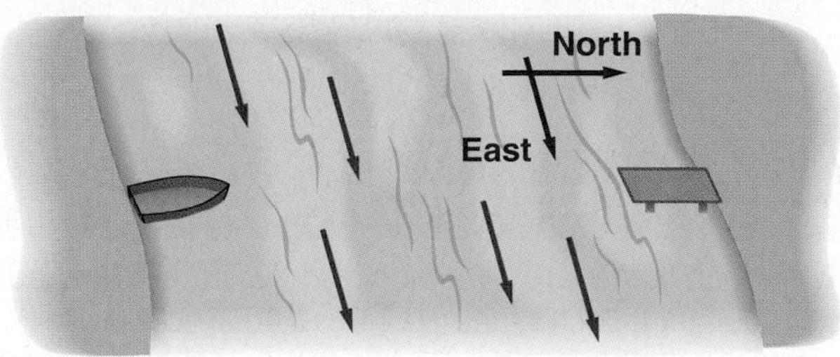

Figure 18

Need for SPEED

Race-Car Driver

The job of a race-car driver is more than just pushing down the gas pedal and following the curve of the track. Managing the extreme forces at work while driving a car at speeds of nearly 320 kilometers per hour takes endurance, strength, and fast reflexes—especially during the turns.

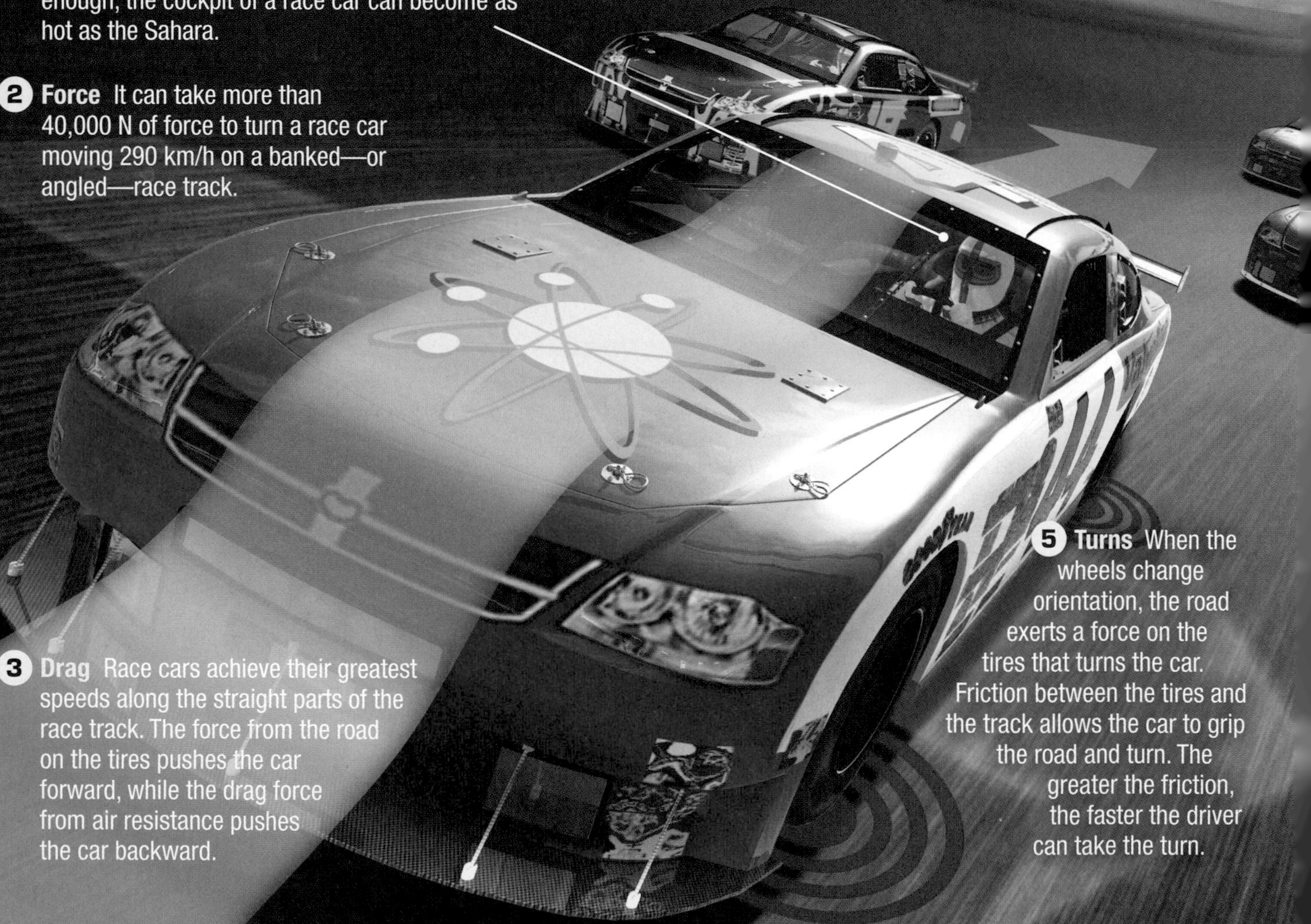

1 Heat A race-car driver wears a helmet to protect the head from impact, a full-body suit that protects against fire, and gloves to improve steering-wheel grip. As if all this gear isn't hot enough, the cockpit of a race car can become as hot as the Sahara.

2 Force It can take more than 40,000 N of force to turn a race car moving 290 km/h on a banked—or angled—race track.

3 Drag Race cars achieve their greatest speeds along the straight parts of the race track. The force from the road on the tires pushes the car forward, while the drag force from air resistance pushes the car backward.

4 Grip The gravitational force pulls the car downward, producing friction between the track and the small area of the tire that touches it, called the contact patch. Air flowing around the car body also produces a downward force on the car, which results in an increased normal force and increased friction at the contact patch.

5 Turns When the wheels change orientation, the road exerts a force on the tires that turns the car. Friction between the tires and the track allows the car to grip the road and turn. The greater the friction, the faster the driver can take the turn.

GOINGFURTHER

Research Compare at least two different kinds of auto racing events, such as drag racing and stock car, in terms of the different forces at work due to the different car styles, track structures, and racing rules.

CHAPTER 6 STUDY GUIDE

BIGIDEA You can use vectors and Newton's laws to describe projectile motion and circular motion.

VOCABULARY
- **projectile** *(p. 152)*
- **trajectory** *(p. 152)*

SECTION 1 Projectile Motion

MAINIDEA A projectile's horizontal motion is independent of its vertical motion.

- The vertical and horizontal motions of a projectile are independent. When there is no air resistance, the horizontal motion component does not experience an acceleration and has constant velocity; the vertical motion component of a projectile experiences a constant acceleration under these same conditions.
- The curved flight path a projectile follows is called a trajectory and is a parabola. The height, time of flight, initial velocity, and horizontal distance of this path are related by the equations of motion. The horizontal distance a projectile travels before returning to its initial height depends on the acceleration due to gravity and on both components of the initial velocity.

VOCABULARY
- **uniform circular motion** *(p. 159)*
- **centripetal acceleration** *(p. 160)*
- **centripetal force** *(p. 161)*

SECTION 2 Circular Motion

MAINIDEA An object in circular motion has an acceleration toward the circle's center due to an unbalanced force toward the circle's center.

- An object moving in a circle at a constant speed has an acceleration toward the center of the circle because the direction of its velocity is constantly changing.
- Acceleration toward the center of the circle is called centripetal acceleration. It depends directly on the square of the object's speed and inversely on the radius of the circle.

$$a_c = \frac{v^2}{r}$$

- A net force must be exerted by external agents toward the circle's center to cause centripetal acceleration.

$$\mathbf{F}_{net} = m\mathbf{a}_c$$

VOCABULARY
- **reference frame** *(p. 164)*

SECTION 3 Relative Velocity

MAINIDEA An object's velocity depends on the reference frame chosen.

- A coordinate system from which you view motion is called a reference frame. Relative velocity is the velocity of an object observed in a different, moving reference frame.
- You can use vector addition to solve motion problems of an object in a moving reference frame.

Games and Multilingual eGlossary

Vocabulary Practice

Chapter Self-Check

SECTION 1 Projectile Motion

Mastering Concepts

40. Some students believe the force that starts the motion of a projectile, such as the kick given a soccer ball, remains with the ball. Is this a correct viewpoint? Present arguments for or against.

41. Consider the trajectory of the cannonball shown in **Figure 19.**

a. Where is the magnitude of the vertical-velocity component largest?

b. Where is the magnitude of the horizontal-velocity component largest?

c. Where is the vertical velocity smallest?

d. Where is the magnitude of the acceleration smallest?

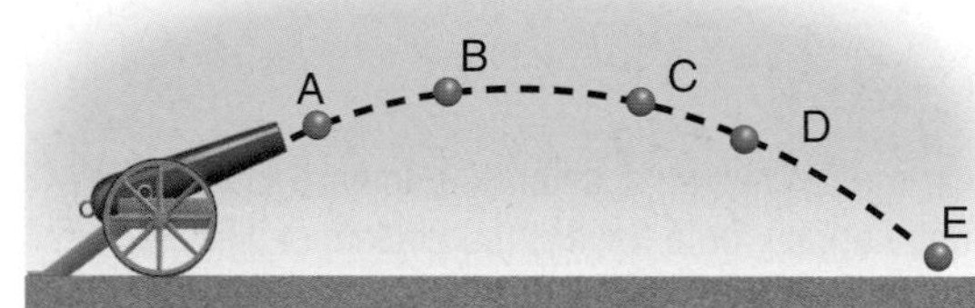

Figure 19

42. Trajectory Describe how forces cause the trajectory of an object launched horizontally to be different from the trajectory of an object launched upward at an angle.

43. Reverse Problem Write a physics problem with real-life objects for which the following equations would be part of the solution. *Hint: The two equations describe the same object.*

$$x = (1.5\ \text{m/s})t \qquad 8.0\ \text{m} = \frac{1}{2}(9.8\ \text{m/s}^2)t^2$$

44. An airplane pilot flying at constant velocity and altitude drops a heavy crate. Ignoring air resistance, where will the plane be relative to the crate when the crate hits the ground? Draw the path of the crate as seen by an observer on the ground.

Mastering Problems

45. You accidentally throw your car keys horizontally at 8.0 m/s from a cliff 64 m high. How far from the base of the cliff should you look for the keys?

46. A dart player throws a dart horizontally at 12.4 m/s. The dart hits the board 0.32 m below the height from which it was thrown. How far away is the player from the board?

47. The toy car in **Figure 20** runs off the edge of a table that is 1.225 m high. The car lands 0.400 m from the base of the table.

a. How long did it take the car to fall?

b. How fast was the car going on the table?

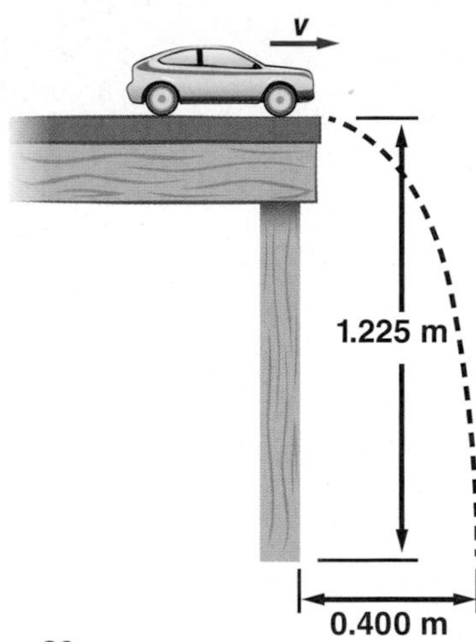

Figure 20

48. Swimming You took a running leap off a high-diving platform. You were running at 2.8 m/s and hit the water 2.6 s later. How high was the platform, and how far from the edge of the platform did you hit the water? Assume your initial velocity is horizontal. Ignore air resistance.

49. Archery An arrow is shot at 30.0° above the horizontal. Its velocity is 49 m/s, and it hits the target.

a. What is the maximum height the arrow will attain?

b. The target is at the height from which the arrow was shot. How far away is it?

50. BIGIDEA A pitched ball is hit by a batter at a 45° angle and just clears the outfield fence, 98 m away. If the top of the fence is at the same height as the pitch, find the velocity of the ball when it left the bat. Ignore air resistance.

51. At-Sea Rescue An airplane traveling 1001 m above the ocean at 125 km/h is going to drop a box of supplies to shipwrecked victims below.

a. How many seconds before the plane is directly overhead should the box be dropped?

b. What is the horizontal distance between the plane and the victims when the box is dropped?

52. Diving Divers in Acapulco dive from a cliff that is 61 m high. What is the minimum horizontal velocity a diver must have to enter the water at least 23 m from the cliff?

Chapter Self-Check

53. Jump Shot A basketball player is trying to make a half-court jump shot and releases the ball at the height of the basket. Assume that the ball is launched at an angle of 51.0° above the horizontal and a horizontal distance of 14.0 m from the basket. What speed must the player give the ball in order to make the shot?

54. The two baseballs in **Figure 21** were hit with the same speed, 25 m/s. Draw separate graphs of y versus t and x versus t for each ball.

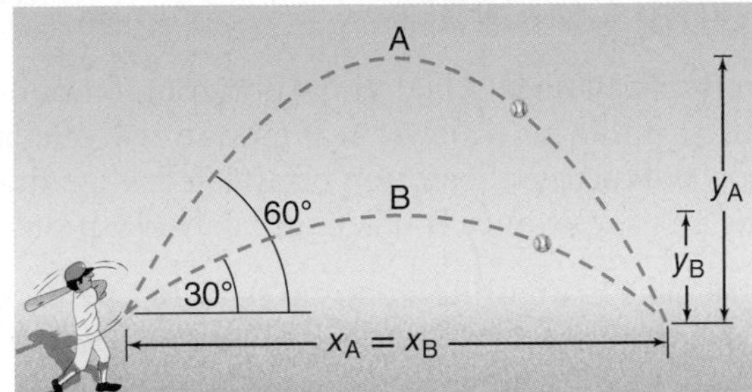

Figure 21

SECTION 2 Circular Motion

Mastering Concepts

55. Can you go around a curve with the following accelerations? Explain.

a. zero acceleration vector

b. constant acceleration vector

56. To obtain uniform circular motion, how must the net force that acts on a moving object depend on the speed of the object?

57. Suppose you whirl a yo-yo about your head in a horizontal circle.

a. In what direction must a force act on the yo-yo?

b. What exerts the force?

c. If you let go of the string on the yo-yo, in which direction would the toy travel? Use Newton's laws in your answer.

Mastering Problems

58. Car Racing A 615-kg racing car completes one lap in a time of 14.3 s around a circular track that has a radius of 50.0 m. Assume the race car moves at a constant speed.

a. What is the acceleration of the car?

b. What force must the track exert on the tires to produce this acceleration?

59. Ranking Task Rank the following objects according to their centripetal accelerations, from least to greatest. Specifically indicate any ties.

A: a 0.50-kg stone moving in a circle of radius 0.6 m at a speed of 2.0 m/s

B: a 0.50-kg stone moving in a circle of radius 1.2 m at a speed of 3.0 m/s

C: a 0.60-kg stone moving in a circle of radius 0.8 m at a speed of 2.4 m/s

D: a 0.75-kg stone moving in a circle of radius 1.2 m at a speed of 3.0 m/s

E: a 0.75-kg stone moving in a circle of radius 0.6 m at a speed of 2.4 m/s

60. Hammer Throw An athlete whirls a 7.00-kg hammer 1.8 m from the axis of rotation in a horizontal circle, as shown in **Figure 22.** If the hammer makes one revolution in 1.0 s, what is the centripetal acceleration of the hammer? What is the tension in the chain?

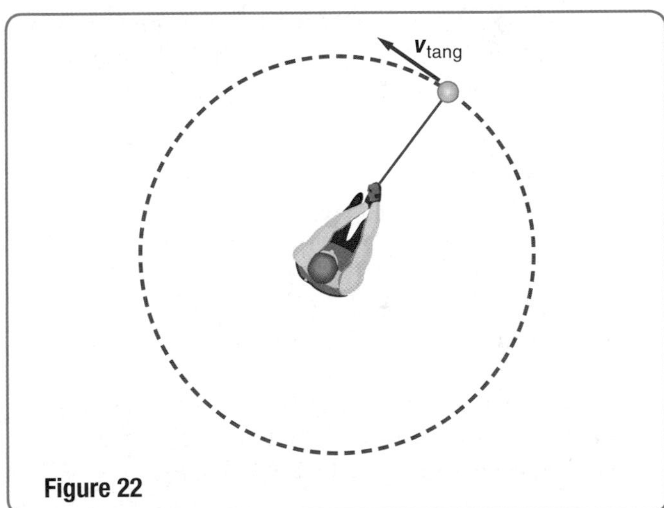

Figure 22

61. A rotating rod that is 15.3 cm long is spun with its axis through one end of the rod. The other end of the rod has a constant speed of 2010 m/s (4500 mph).

a. What is the centripetal acceleration of the end of the rod?

b. If you were to attach a 1.0-g object to the end of the rod, what force would be needed to hold it on the rod?

62. A carnival clown rides a motorcycle down a ramp and then up and around a large, vertical loop. If the loop has a radius of 18 m, what is the slowest speed the rider can have at the top of the loop so that the motorcycle stays in contact with the track and avoids falling? *Hint: At this slowest speed, the track exerts no force on the motorcycle at the top of the loop.*

Chapter Self-Check

63. A 75-kg pilot flies a plane in a loop as shown in **Figure 23.** At the top of the loop, when the plane is completely upside-down for an instant, the pilot hangs freely in the seat and does not push against the seat belt. The airspeed indicator reads 120 m/s. What is the radius of the plane's loop?

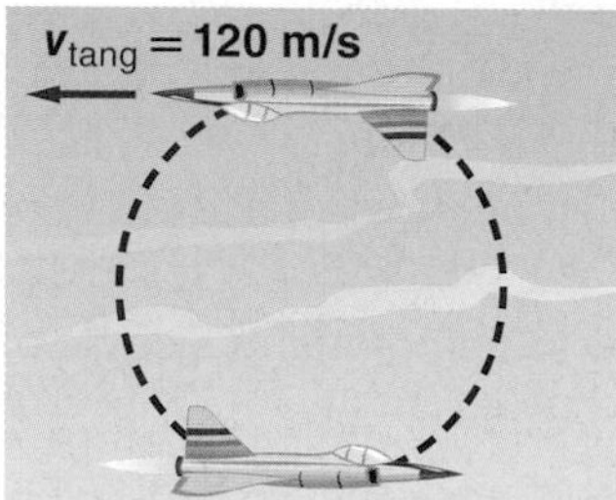

Figure 23

SECTION 3 Relative Velocity

Mastering Concepts

64. Why is it that a car traveling in the opposite direction as the car in which you are riding on the freeway often looks like it is moving faster than the speed limit?

Mastering Problems

65. Odina and LaToya are sitting by a river and decide to have a race. Odina will run down the shore to a dock, 1.5 km away, then turn around and run back. LaToya will also race to the dock and back, but she will row a boat in the river, which has a current of 2.0 m/s. If Odina's running speed is equal to LaToya's rowing speed in still water, which is 4.0 m/s, what will be the outcome of the race? Assume they both turn instantaneously.

66. **Crossing a River** You row a boat, such as the one in **Figure 24,** perpendicular to the shore of a river that flows at 3.0 m/s. The velocity of your boat is 4.0 m/s relative to the water.

    a. What is the velocity of your boat relative to the shore?

    b. What is the component of your velocity parallel to the shore? Perpendicular to it?

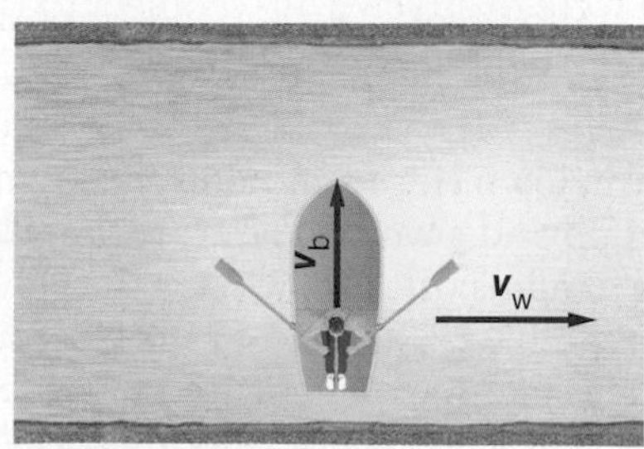

Figure 24

67. **Air Travel** You are piloting a small plane, and you want to reach an airport 450 km due south in 3.0 h. A wind is blowing from the west at 50.0 km/h. What heading and airspeed should you choose to reach your destination in time?

68. **Problem Posing** Complete this problem so that it can be solved using the concept of relative velocity: "Hannah is on the west bank of a 55-m-wide river with a current of 0.7 m/s ...."

Applying Concepts

69. **Projectile Motion** Explain how horizontal motion can be uniform while vertical motion is accelerated. How will projectile motion be affected when drag due to air resistance is taken into consideration?

70. **Baseball** A batter hits a pop-up straight up over home plate at an initial speed of 20 m/s. The ball is caught by the catcher at the same height at which it was hit. At what velocity does the ball land in the catcher's mitt? Neglect air resistance.

71. **Fastball** In baseball, a fastball takes about $\frac{1}{2}$ s to reach the plate. Assuming that such a pitch is thrown horizontally, compare the distance the ball falls in the first $\frac{1}{4}$ s with the distance it falls in the second $\frac{1}{4}$ s.

72. You throw a rock horizontally. In a second horizontal throw, you throw the rock harder and give it even more speed.

    a. How will the time it takes the rock to hit the ground be affected? Ignore air resistance.

    b. How will the increased speed affect the distance from where the rock left your hand to where the rock hits the ground?

73. **Field Biology** A zoologist standing on a cliff aims a tranquilizer gun at a monkey hanging from a tree branch that is in the gun's range. The barrel of the gun is horizontal. Just as the zoologist pulls the trigger, the monkey lets go and begins to fall. Will the dart hit the monkey? Ignore air resistance.

74. **Football** A quarterback throws a football at 24 m/s at a 45° angle. If it takes the ball 3.0 s to reach the top of its path and the ball is caught at the same height at which it is thrown, how long is it in the air? Ignore air resistance.

75. **Track and Field** You are working on improving your performance in the long jump and believe that the information in this chapter can help. Does the height that you reach make any difference to your jump? What influences the length of your jump?

76. Driving on a Freeway Explain why it is that when you pass a car going in the same direction as you on the freeway, it takes a longer time than when you pass a car going in the opposite direction.

77. Imagine you are sitting in a car tossing a ball straight up into the air.

a. If the car is moving at a constant velocity, will the ball land in front of, behind, or in your hand?

b. If the car rounds a curve at a constant speed, where will the ball land?

78. You swing one yo-yo around your head in a horizontal circle. Then you swing another yo-yo with twice the mass of the first one, but you don't change the length of the string or the period. How do the tensions in the strings differ?

79. Car Racing The curves on a race track are banked to make it easier for cars to go around the curves at high speeds. Draw a free-body diagram of a car on a banked curve. From the motion diagram, find the direction of the acceleration.

a. What exerts the force in the direction of the acceleration?

b. Can you have such a force without friction?

Mixed Review

80. Early skeptics of the idea of a rotating Earth said that the fast spin of Earth would throw people at the equator into space. The radius of Earth is about 6.38×10^3 km. Show why this idea is wrong by calculating the following.

a. the speed of a 97-kg person at the equator

b. the force needed to accelerate the person in the circle

c. the weight of the person

d. the normal force of Earth on the person, that is, the person's apparent weight

81. Firing a Missile An airplane moving at 375 m/s relative to the ground fires a missile forward at a speed of 782 m/s relative to the plane. What is the missile's speed relative to the ground?

82. Rocketry A rocket in outer space that is moving at a speed of 1.25 km/s relative to an observer fires its motor. Hot gases are expelled out the back at 2.75 km/s relative to the rocket. What is the speed of the gases relative to the observer?

83. A 1.13-kg ball is swung vertically from a 0.50-m cord in uniform circular motion at a speed of 2.4 m/s. What is the tension in the cord at the bottom of the ball's motion?

84. Two dogs, initially separated by 500.0 m, are running toward each other, each moving with a constant speed of 2.5 m/s. A dragonfly, moving with a constant speed of 3.0 m/s, flies from the nose of one dog to the other, then turns around instantaneously and flies back to the other dog. It continues to fly back and forth until the dogs run into each other. What distance does the dragonfly fly during this time?

85. Banked Roads Curves on roads often are banked to help prevent cars from slipping off the road. If the speed limit for a particular curve of radius 36.0 m is 15.7 m/s (35 mph), at what angle should the road be banked so that cars will stay on a circular path even if there were no friction between the road and the tires? If the speed limit was increased to 20.1 m/s (45 mph), at what angle should the road be banked?

86. The 1.45-kg ball in **Figure 25** is suspended from a 0.80-m string and swung in a horizontal circle at a constant speed.

a. What is the tension in the string?

b. What is the speed of the ball?

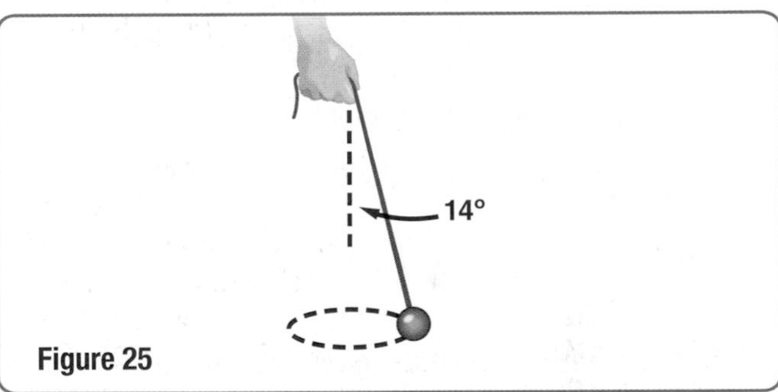

Figure 25

87. A baseball is hit directly in line with an outfielder at an angle of 35.0° above the horizontal with an initial speed of 22.0 m/s. The outfielder starts running as soon as the ball is hit at a constant speed of 2.5 m/s and barely catches the ball. Assuming that the ball is caught at the same height at which it was hit, what was the initial separation between the hitter and the outfielder? *Hint: There are two possible answers.*

88. A Jewel Heist You are a technical consultant for a locally produced cartoon. In one episode, two criminals, Shifty and Crafty, have stolen some jewels. Crafty has the jewels when the police start to chase him. He runs to the top of a 60.0-m tall building in his attempt to escape. Meanwhile, Shifty runs to the convenient hot-air balloon 20.0 m from the base of the building and untethers it, so it begins to rise at a constant speed. Crafty tosses the bag of jewels horizontally with a speed of 7.3 m/s just as the balloon begins its ascent. What must the velocity of the balloon be for Shifty to easily catch the bag?

Chapter Self-Check

Thinking Critically

89. **Apply Concepts** Consider a roller-coaster loop like the one in **Figure 26.** Are the cars traveling through the loop in uniform circular motion? Explain.

Figure 26

90. **Apply Computers and Calculators** A baseball player hits a belt-high (1.0 m) fastball down the left-field line. The player hits the ball with an initial velocity of 42.0 m/s at an angle 26° above the horizontal. The left-field wall is 96.0 m from home plate at the foul pole and is 14 m high. Write the equation for the height of the ball (y) as a function of its distance from home plate (x). Use a computer or graphing calculator to plot the path of the ball. Trace along the path to find how high above the ground the ball is when it is at the wall.
    - **a.** Is the hit a home run?
    - **b.** What is the minimum speed at which the ball could be hit and clear the wall?
    - **c.** If the initial velocity of the ball is 42.0 m/s, for what range of angles will the ball go over the wall?

91. **Analyze** Albert Einstein showed that the rule you learned for the addition of velocities does not work for objects moving near the speed of light. For example, if a rocket moving at speed v_A releases a missile that has speed v_B relative to the rocket, then the speed of the missile relative to an observer that is at rest is given by $v = \frac{v_A + v_B}{1 + \frac{v_A v_B}{c^2}}$, where c is the speed of light, 3.00×10^8 m/s. This formula gives the correct values for objects moving at slow speeds as well. Suppose a rocket moving at 11 km/s shoots a laser beam out in front of it. What speed would an unmoving observer find for the laser light? Suppose that a rocket moves at a speed $\frac{c}{2}$, half the speed of light, and shoots a missile forward at a speed of $\frac{c}{2}$ relative to the rocket. How fast would the missile be moving relative to a fixed observer?

92. **Analyze and Conclude** A ball on a light string moves in a vertical circle. Analyze and describe the motion of this system. Be sure to consider the effects of gravity and tension. Is this system in uniform circular motion? Explain your answer.

Writing In Physics

93. **Roller Coasters** The vertical loops on most roller coasters are not circular in shape. Research and explain the physics behind this design choice.

94. Many amusement-park rides utilize centripetal acceleration to create thrills for the park's customers. Choose two rides other than roller coasters that involve circular motion, and explain how the physics of circular motion creates the sensations for the riders.

Cumulative Review

95. Multiply or divide, as indicated, using significant figures correctly.
    - **a.** $(5\times10^8\ \text{m})(4.2\times10^7\ \text{m})$
    - **b.** $(1.67\times10^{-2}\ \text{km})(8.5\times10^{-6}\ \text{km})$
    - **c.** $\frac{2.6\times10^4\ \text{kg}}{9.4\times10^3\ \text{m}^3}$
    - **d.** $\frac{6.3\times10^{-1}\ \text{m}}{3.8\times10^2\ \text{s}}$

96. Plot the data in **Table 1** on a position-time graph. Find the average speed in the time interval between 0.0 s and 5.0 s.

Table 1 Position v. Time

Clock Reading t (s)	Position x (m)
0.0	30
1.0	30
2.0	35
3.0	45
4.0	60
5.0	70

97. Carlos and his older brother Ricardo are at the grocery store. Carlos, with mass 17.0 kg, likes to hang on the front of the cart while Ricardo pushes it, even though both boys know this is not safe. Ricardo pushes the 12.4-kg cart with his brother on it such that they accelerate at a rate of 0.20 m/s^2.
    - **a.** With what force is Ricardo pushing?
    - **b.** What is the force the cart exerts on Carlos?

MULTIPLE CHOICE

1. A 1.60-m-tall girl throws a football at an angle of 41.0° from the horizontal and at an initial speed of 9.40 m/s. What is the horizontal distance between the girl and the spot when the ball is again at the height above the ground from which the girl threw it?
   A. 4.55 m
   B. 5.90 m
   C. 8.90 m
   D. 10.5 m

2. A dragonfly is sitting on a merry-go-round 2.8 m from the center. If the tangential velocity of the ride is 0.89 m/s, what is the centripetal acceleration of the dragonfly?
   A. 0.11 m/s^2
   B. 0.28 m/s^2
   C. 0.32 m/s^2
   D. 2.2 m/s^2

3. The force exerted by a 2.0-m massless string on a 0.82-kg object being swung in a horizontal circle is 4.0 N. What is the tangential velocity of the object?
   A. 2.8 m/s
   B. 3.1 m/s
   C. 4.9 m/s
   D. 9.8 m/s

4. A 1000-kg car enters an 80.0-m-radius curve at 20.0 m/s. What centripetal force must be supplied by friction so the car does not skid?
   A. 5.0 N
   B. 2.5×10^2 N
   C. 5.0×10^3 N
   D. 1.0×10^3 N

5. A jogger on a riverside path sees a rowing team coming toward him. Relative to the ground, the jogger is running at 10 km/h west and the boat is sailing at 20 km/h east. How quickly does the jogger approach the boat?
   A. 10 km/h
   B. 30 km/h
   C. 20 km/h
   D. 40 km/h

6. What is the maximum height obtained by a 125-g apple that is slung from a slingshot at an angle of 78° from the horizontal with an initial velocity of 18 m/s?
   A. 0.70 m
   B. 16 m
   C. 32 m
   D. 33 m

7. An orange is dropped at the same time and from the same height that a bullet is shot from a gun. Which of the following is true?
   A. The acceleration due to gravity is greater for the orange because the orange is heavier.
   B. Gravity acts less on the bullet than on the orange because the bullet is moving so quickly.
   C. The velocities will be the same.
   D. The two objects will hit the ground at the same time.

FREE RESPONSE

8. A lead cannonball is shot horizontally at a speed of 25 m/s out of the circus cannon, shown in the figure, on the high-wire platform on one side of a circus ring. If the platform is 52 m above the 80-m diameter ring, will the performers need to adjust their cannon so that the ball will land inside the ring instead of past it? Explain.

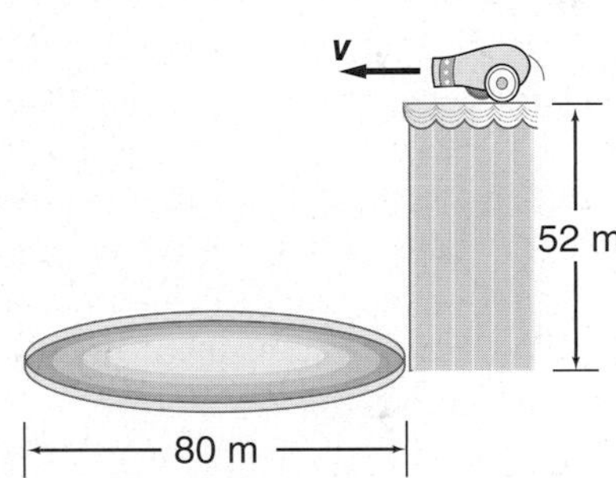

9. A mythical warrior swings a 5.6-kg mace on the end of a magically massless 86-cm chain in a horizontal circle above his head. The mace makes one full revolution in 1.8 s. Find the tension in the magical chain.

Online Test Practice

NEED EXTRA HELP?

If You Missed Question	1	2	3	4	5	6	7	8	9
Review Section	1	2	2	2	3	1	1	1	2

CHAPTER 7

Gravitation

BIG IDEA **Gravity is an attractive field force that acts between objects with mass.**

SECTIONS

1 Planetary Motion and Gravitation

2 Using the Law of Universal Gravitation

LaunchLAB

MODEL MERCURY'S MOTION

How can measurements of angles and distances be used to draw a model of an orbit?

WATCH THIS!

PLANETARY MOTION

What can looking through a telescope tell you about gravitation? Explore the physics of amateur astronomy and find out!

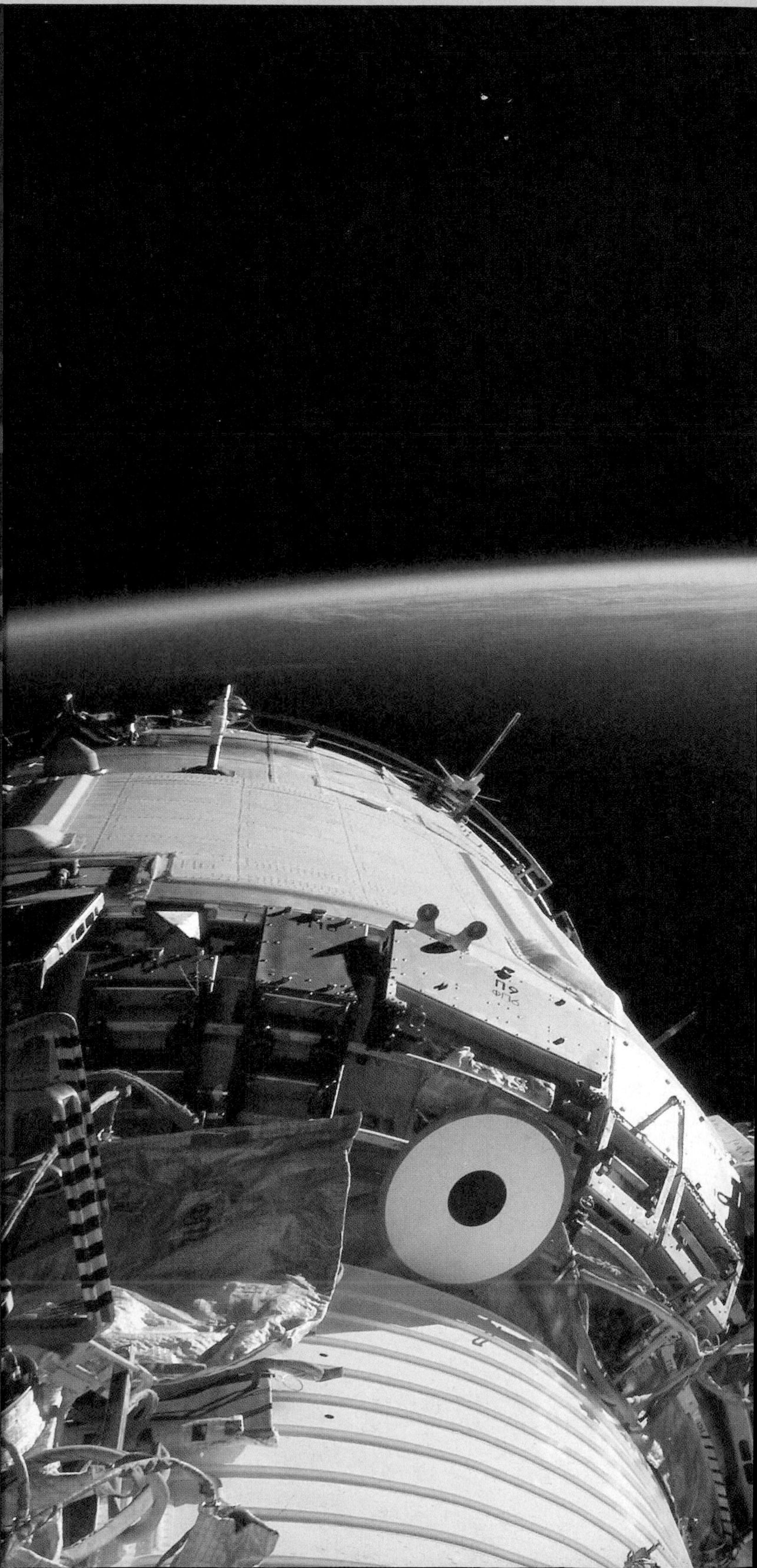

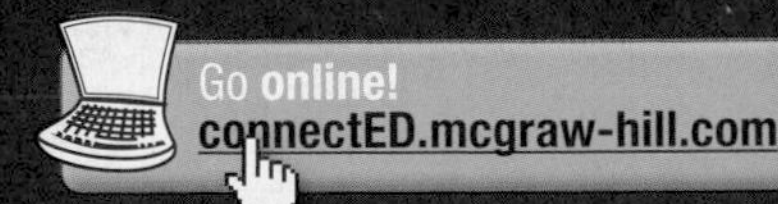
Go online!
connectED.mcgraw-hill.com

Planetary Motion and Gravitation

Our solar system includes the Sun, Earth and seven other major planets, dwarf planets, and interplanetary dust and gas. Various moons orbit the planets. What holds all this together?

MAINIDEA

The gravitational force between two objects is proportional to the product of their masses divided by the square of the distance between them.

Essential Questions

- What is the relationship between a planet's orbital radius and period?
- What is Newton's law of universal gravitation, and how does it relate to Kepler's laws?
- Why was Cavendish's investigation important?

Review Vocabulary

Newton's third law states all forces come in pairs and that the two forces in a pair act on different objects, are equal in strength, and are opposite in direction

New Vocabulary

Kepler's first law
Kepler's second law
Kepler's third law
gravitational force
law of universal gravitation

Early Observations

In ancient times, the Sun, the Moon, the planets, and the stars were assumed to revolve around Earth. Nicholas Copernicus, a Polish astronomer, noticed that the best available observations of the movements of planets did not fully agree with the Earth-centered model.

The results of his many years of work were published in 1543, when Copernicus was on his deathbed. His book showed that the motion of planets is much more easily understood by assuming that Earth and other planets revolve around the Sun. His model helped explain phenomena such as the inner planets Mercury and Venus always appearing near the Sun. Copernicus's view advanced our understanding of planetary motion. He incorrectly assumed, however, that planetary orbits are circular. This assumption did not fit well with observations, and modification of Copernicus's model was necessary to make it accurate.

Tycho Brahe was born a few years after Copernicus died. As a boy of 14 in Denmark, Tycho observed an eclipse of the Sun on August 21, 1560. The fact that it had been predicted inspired him toward a career in astronomy.

As Tycho studied astronomy, he realized that the charts of the time did not accurately predict astronomical events. Tycho recognized that measurements were required from one location over a long period of time. He was granted an estate on the Danish island of Hven and the funding to build an early research institute. Telescopes had not been invented, so to make measurements, Tycho used huge instruments that he designed and built in his own shop, such as those shown in **Figure 1.** Tycho is credited with the most accurate measurements of the time.

Figure 1 Instruments such as these were used by Tycho to measure the positions of planets.

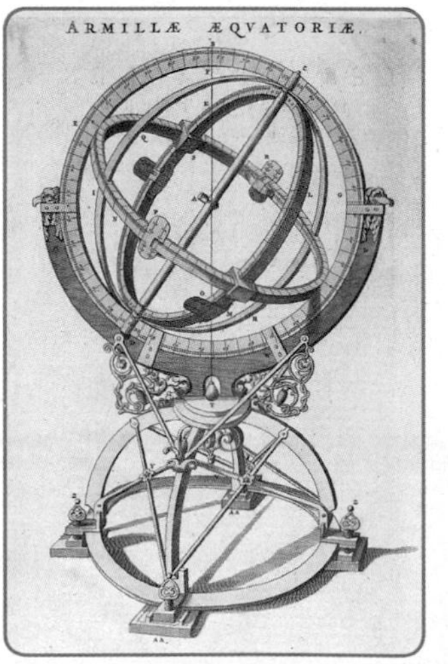

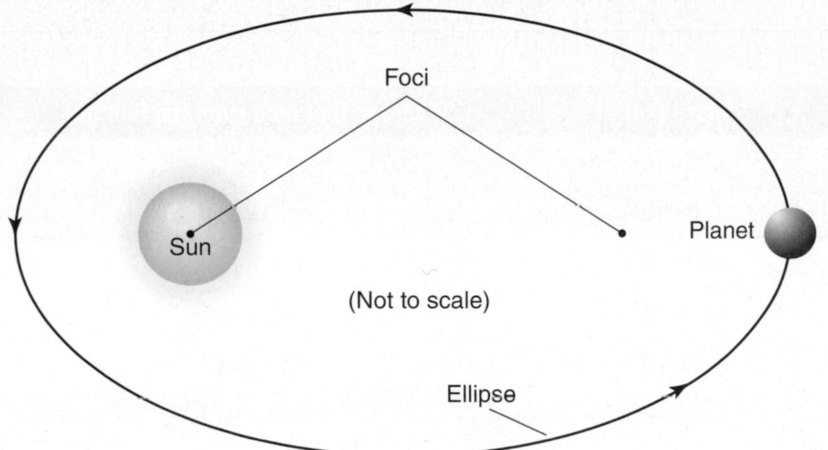

Figure 2 The orbit of each planet is an ellipse, with the Sun at one focus.

Kepler's Laws

In 1600 Tycho moved to Prague where Johannes Kepler, a 29-year-old German, became one of his assistants. Kepler analyzed Tycho's observations. After Tycho's death in 1601, Kepler continued to study Tycho's data and used geometry and mathematics to explain the motion of the planets. After seven years of careful analysis of Tycho's data on Mars, Kepler discovered the laws that describe the motion of every planet and satellite, natural or artificial. Here, the laws are presented in terms of planets.

View an **animation of Kepler's first law.**

Kepler's first law states that the paths of the planets are ellipses, with the Sun at one focus. An ellipse has two foci, as shown in **Figure 2.** Although exaggerated ellipses are used in the diagrams, Earth's actual orbit is very nearly circular. You would not be able to distinguish it from a circle visually.

Kepler found that the planets move faster when they are closer to the Sun and slower when they are farther away from the Sun. **Kepler's second law** states that an imaginary line from the Sun to a planet sweeps out equal areas in equal time intervals, as illustrated in **Figure 3.**

☑ **READING CHECK** **Compare** the distances traveled from point 1 to point 2 and from point 6 to point 7 in **Figure 3.** Through which distance would Earth be traveling fastest?

A period is the time it takes for one revolution of an orbiting body. Kepler also discovered a mathematical relationship between periods of planets and their mean distances away from the Sun.

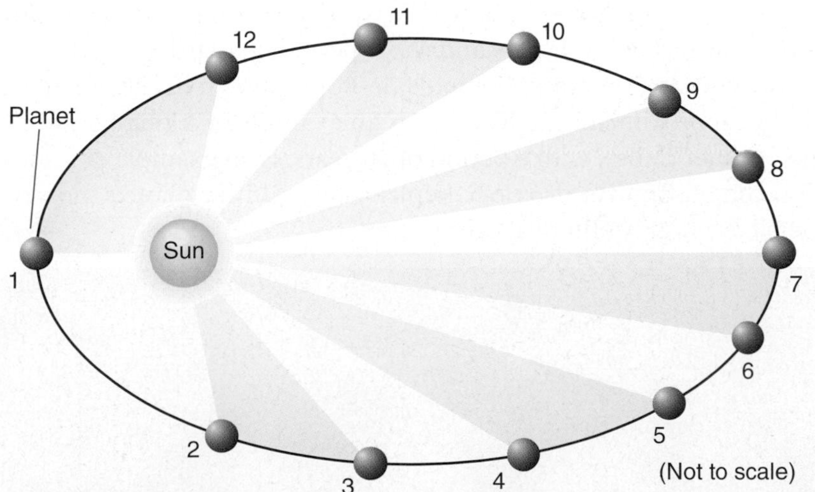

Figure 3 Kepler found that elliptical orbits sweep out equal areas in equal time periods.

Explain why the equal time areas are shaped differently.

View an **animation and a simulation of Kepler's second law.**

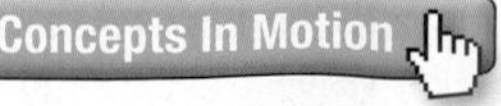

Table 1 Solar System Data

Name	Average Radius (m)	Mass (kg)	Average Distance from the Sun (m)
Sun	6.96×10^{8}	1.99×10^{30}	—
Mercury	2.44×10^{6}	3.30×10^{23}	5.79×10^{10}
Venus	6.05×10^{6}	4.87×10^{24}	1.08×10^{11}
Earth	6.38×10^{6}	5.97×10^{24}	1.50×10^{11}
Mars	3.40×10^{6}	6.42×10^{23}	2.28×10^{11}
Jupiter	7.15×10^{7}	1.90×10^{27}	7.78×10^{11}
Saturn	6.03×10^{7}	5.69×10^{26}	1.43×10^{12}
Uranus	2.56×10^{7}	8.68×10^{25}	2.87×10^{12}
Neptune	2.48×10^{7}	1.02×10^{26}	4.50×10^{12}

Kepler's third law states that the square of the ratio of the periods of any two planets revolving about the Sun is equal to the cube of the ratio of their average distances from the Sun. Thus, if the periods of the planets are T_A and T_B and their average distances from the Sun are r_A and r_B, Kepler's third law can be expressed as follows.

View an **animation and a simulation of Kepler's third law.**

KEPLER'S THIRD LAW

The square of the ratio of the period of planet A to the period of planet B is equal to the cube of the ratio of the distance between the centers of planet A and the Sun to the distance between the centers of planet B and the Sun.

$$\left(\frac{T_A}{T_B}\right)^2 = \left(\frac{r_A}{r_B}\right)^3$$

Note that Kepler's first two laws apply to each planet, moon, and satellite individually. The third law, however, relates the motion of two objects around a single body. For example, it can be used to compare the planets' distances from the Sun, shown in **Table 1,** to their periods around the Sun. It also can be used to compare distances and periods of the Moon and artificial satellites orbiting Earth.

PhysicsLAB

MODELING ORBITS

What is the shape of the orbits of planets and satellites in the solar system?

Comet periods Comets are classified as long-period comets or short-period comets based on orbital periods. Long-period comets have orbital periods longer than 200 years and short-period comets have orbital periods shorter than 200 years. Comet Hale-Bopp, shown in **Figure 4,** with a period of approximately 2400 years, is an example of a long-period comet. Comet Halley, with a period of 76 years, is an example of a short-period comet. Comets also obey Kepler's laws. Unlike planets, however, comets have highly elliptical orbits.

Figure 4 Hale-Bopp is a long-period comet, with a period of 2400 years. This photo was taken in 1997, when Hale-Bopp was highly visible.

Jamie Cooper/SSPL/Getty Images

EXAMPLE PROBLEM 1

Find help with **isolating a variable.** Math Handbook

CALLISTO'S DISTANCE FROM JUPITER Galileo measured the orbital radii of Jupiter's moons using the diameter of Jupiter as a unit of measure. He found that Io, the closest moon to Jupiter, has a period of 1.8 days and is 4.2 units from the center of Jupiter. Callisto, the fourth moon from Jupiter, has a period of 16.7 days. Using the same units that Galileo used, predict Callisto's distance from Jupiter.

1 ANALYZE AND SKETCH THE PROBLEM

- Sketch the orbits of Io and Callisto.
- Label the radii.

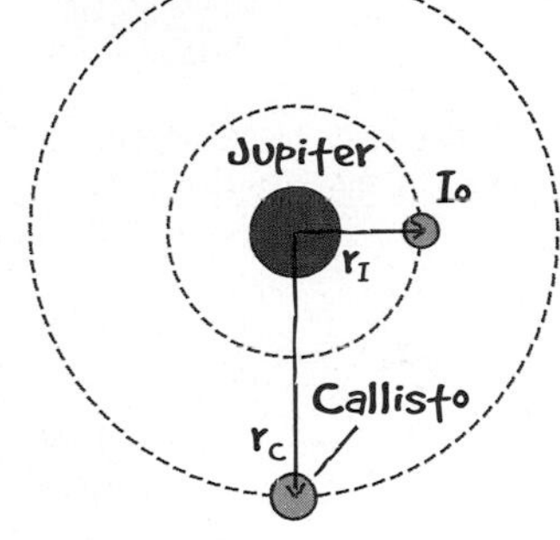

KNOWN	UNKNOWN
$T_C = 16.7$ days	$r_C = ?$
$T_I = 1.8$ days	
$r_I = 4.2$ units	

2 SOLVE FOR CALLISTO'S DISTANCE FROM JUPITER

Solve Kepler's third law for r_C.

$$\left(\frac{T_C}{T_I}\right)^2 = \left(\frac{r_C}{r_I}\right)^3$$

$$r_C{}^3 = r_I{}^3\left(\frac{T_C}{T_I}\right)^2$$

$$r_C = \sqrt[3]{r_I{}^3\left(\frac{T_C}{T_I}\right)^2}$$

◀ Substitute r_I = 4.2 units, T_C = 16.7 days, T_I = 1.8 days

$$= \sqrt[3]{(4.2\text{ units})^3\left(\frac{16.7\text{ days}}{1.8\text{ days}}\right)^2}$$

$$= \sqrt[3]{6.4\times10^3\text{ units}^3}$$

$$= 19\text{ units}$$

3 EVALUATE THE ANSWER

- **Are the units correct?** r_C should be in Galileo's units, like r_I.
- **Is the magnitude realistic?** The period is larger, so the radius should be larger.

PRACTICE PROBLEMS

Do additional problems. Online Practice

1. If Ganymede, one of Jupiter's moons, has a period of 32 days, how many units is its orbital radius? Use the information given in Example Problem 1.
2. An asteroid revolves around the Sun with a mean orbital radius twice that of Earth's. Predict the period of the asteroid in Earth years.
3. Venus has a period of revolution of 225 Earth days. Find the distance between the Sun and Venus as a multiple of Earth's average distance from the Sun.
4. Uranus requires 84 years to circle the Sun. Find Uranus's average distance from the Sun as a multiple of Earth's average distance from the Sun.
5. From **Table 1** you can find that, on average, Mars is 1.52 times as far from the Sun as Earth is. Predict the time required for Mars to orbit the Sun in Earth days.
6. The Moon has a period of 27.3 days and a mean distance of 3.9×10^5 km from its center to the center of Earth.
   a. Use Kepler's laws to find the period of a satellite in orbit 6.70×10^3 km from the center of Earth.
   b. How far above Earth's surface is this satellite?
7. **CHALLENGE** Using the data in the previous problem for the period and radius of revolution of the Moon, predict what the mean distance from Earth's center would be for an artificial satellite that has a period of exactly 1.00 day.

EXAMPLE PROBLEM

PRACTICE PROBLEMS

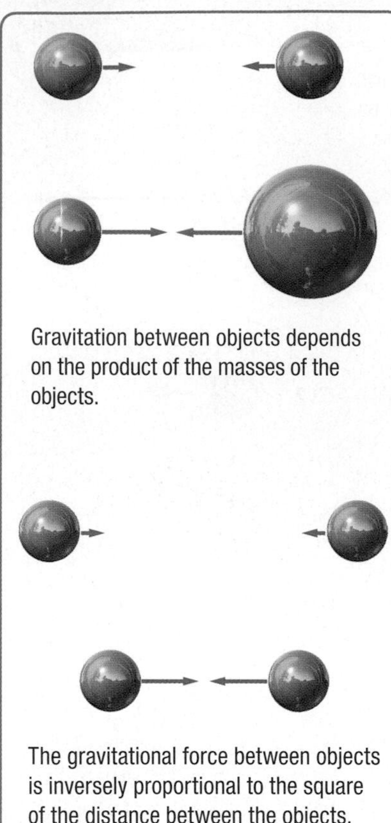

Gravitation between objects depends on the product of the masses of the objects.

The gravitational force between objects is inversely proportional to the square of the distance between the objects.

Figure 5 Mass and distance affect the magnitude of the gravitational force between objects.

View an **animation of the law of universal gravitation.**

Newton's Law of Universal Gravitation

In 1666, Isaac Newton began his studies of planetary motion. It has been said that seeing an apple fall made Newton wonder if the force that caused the apple to fall might extend to the Moon, or even beyond. He found that the magnitude of the force (F_g) on a planet due to the Sun varies inversely with the square of the distance (r) between the centers of the planet and the Sun. That is, F_g is proportional to $\frac{1}{r^2}$. The force ($\mathbf{F}_g$) acts in the direction of the line connecting the centers of the two objects, as shown in **Figure 5.**

Newton found that both the apple's and the Moon's accelerations agree with the $\frac{1}{r^2}$ relationship. According to his own third law, the force Earth exerts on the apple is exactly the same as the force the apple exerts on Earth. Even though these forces are exactly the same, you can easily observe the effect of the force on the apple because it has much lower mass than Earth. The force of attraction between two objects must be proportional to the objects' masses and is known as the **gravitational force.**

Newton was confident that the same force of attraction would act between any two objects anywhere in the universe. He proposed the **law of universal gravitation,** which states that objects attract other objects with a force that is proportional to the product of their masses and inversely proportional to the square of the distance between them as shown below.

LAW OF UNIVERSAL GRAVITATION

The gravitational force is equal to the universal gravitational constant, times the mass of object 1, times the mass of object 2, divided by the distance between the centers of the objects, squared.

$$F_g = \frac{Gm_1m_2}{r^2}$$

According to Newton's equation, F is directly proportional to m_1 and m_2. If the mass of a planet near the Sun doubles, the force of attraction doubles. Use the Connecting Math to Physics feature below to examine how changing one variable affects another. **Figure 6** illustrates the inverse square relationship graphically. The term G is the universal gravitational constant and will be discussed in the next sections.

Figure 6 This is a graphical representation of the inverse square relationship.

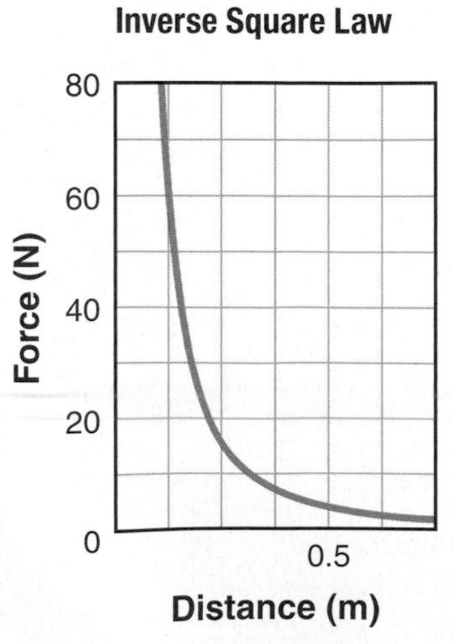

CONNECTING MATH TO PHYSICS

Direct and Inverse Relationships Newton's law of universal gravitation has both direct and inverse relationships.

$F_g \propto m_1m_2$		$F_g \propto \frac{1}{r^2}$	
Change	**Result**	**Change**	**Result**
$(2m_1)m_2$	$2F_g$	$2r$	$\frac{1}{4}F_g$
$(3m_1)m_2$	$3F_g$	$3r$	$\frac{1}{9}F_g$
$(2m_1)(3m_2)$	$6F_g$	$\frac{1}{2}r$	$4F_g$
$\left(\frac{1}{2}\right)m_1m_2$	$\frac{1}{2}F_g$	$\frac{1}{3}r$	$9F_g$

Universal Gravitation and Kepler's Third Law

Newton stated the law of universal gravitation in terms that applied to the motion of planets about the Sun. This agreed with Kepler's third law and confirmed that Newton's law fit the best observations of the day.

Consider a planet orbiting the Sun, as shown in **Figure 7.** Newton's second law of motion, $F_{net} = ma$, can be written as $F_{net} = m_p a_c$, where F_{net} is the magnitude of the gravitational force, m_p is the mass of the planet, and a_c is the centripetal acceleration of the planet. For simplicity, assume circular orbits. Recall from your study of uniform circular motion that for a circular orbit $a_c = \frac{4\pi^2 r}{T^2}$. This means that $F_{net} = m_p a_c$ may now be written $F_{net} = \frac{m_p 4\pi^2 r}{T^2}$. In this equation, T is the time in seconds required for the planet to make one complete revolution about the Sun. If you set the right side of this equation equal to the right side of the law of universal gravitation, you arrive at the following result:

$$\frac{Gm_S m_p}{r^2} = \frac{m_p 4\pi^2 r}{T^2}$$

$$T^2 = \left(\frac{4\pi^2}{Gm_S}\right)r^3$$

$$T = \sqrt{\left(\frac{4\pi^2}{Gm_S}\right)r^3}$$

The period of a planet orbiting the Sun can be expressed as follows.

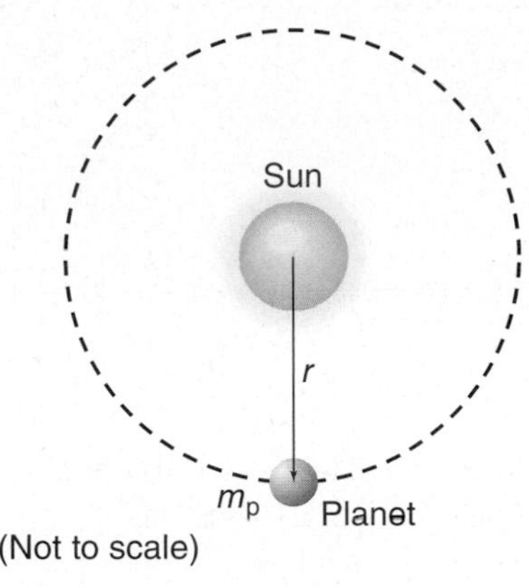

Figure 7 A planet with mass m_p and average distance from the Sun r orbits the Sun. The mass of the Sun is m_S.

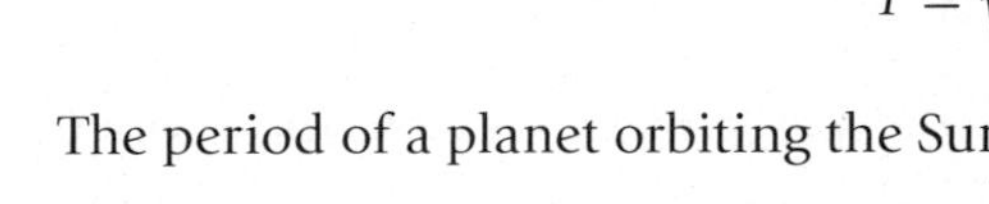

PERIOD OF A PLANET ORBITING THE SUN

The period of a planet orbiting the Sun is equal to 2π times the square root of the average distance from the Sun cubed, divided by the product of the universal gravitational constant and the mass of the Sun.

$$T = 2\pi\sqrt{\frac{r^3}{Gm_S}}$$

Squaring both sides makes it apparent that this equation is Kepler's third law of planetary motion: the square of the period is proportional to the cube of the distance that separates the masses. The factor $\frac{4\pi^2}{Gm_S}$ depends on the mass of the Sun and the universal gravitational constant. Newton found that this factor applied to elliptical orbits as well.

PHYSICS CHALLENGE

Astronomers have detected three planets that orbit the star Upsilon Andromedae. Planet B has an average orbital radius of 0.0595 AU and a period of 4.6171 days. Planet C has an average orbital radius of 0.832 AU and a period of 241.33 days. Planet D has an average orbital radius of 2.53 AU and a period of 1278.1 days. (Distances are given in astronomical units (AU)—Earth's average distance from the Sun. The distance from Earth to the Sun is 1.00 AU.)

1. Do these planets obey Kepler's third law?
2. Find the mass of the star Upsilon Andromedae in units of the Sun's mass. *Hint: compare* $\frac{r^3}{T^2}$ *for these planets with that of Earth in the same units (AU and days).*

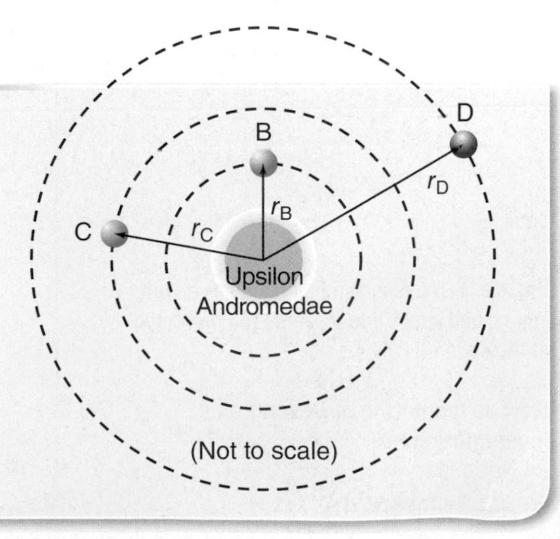

Measuring the Universal Gravitational Constant

How large is the constant G? As you know, the force of gravitational attraction between two objects on Earth is relatively small. The slightest attraction, even between two massive bowling balls, is almost impossible to detect. In fact, it took 100 years from the time of Newton's work for scientists to develop an apparatus that was sensitive enough to measure the force of gravitational attraction.

Cavendish's apparatus In 1798 English scientist Henry Cavendish used equipment similar to the apparatus shown in **Figure 8** to measure the gravitational force between two objects. The apparatus has a horizontal rod with small lead spheres attached to each end. The rod is suspended at its midpoint so that it can rotate. Because the rod is suspended by a thin wire, the rod and spheres are very sensitive to horizontal forces.

To measure G, two large spheres are placed in a fixed position close to each of the two small spheres, as shown in **Figure 8.** The force of attraction between the large and small spheres causes the rod to rotate. When the force required to twist the wire equals the gravitational force between the spheres, the rod stops rotating. By measuring the angle through which the rod turns, the attractive force between the objects can be calculated.

Investigate **universal gravitation.**

READING CHECK Explain why the rod and sphere in Cavendish's apparatus must be sensitive to horizontal forces.

The angle through which the rod turns is measured by using a beam of light that is reflected from the mirror. The distances between the sphere's centers and the force can both be measured. The masses of the spheres are known. By substituting the values for force, mass, and distance into Newton's law of universal gravitation, an experimental value for G is found: when m_1 and m_2 are measured in kilograms, r in meters, and F in newtons, $G = 6.67\times10^{-11}\ \text{N}\cdot\text{m}^2/\text{kg}^2$.

Cavendish Balance

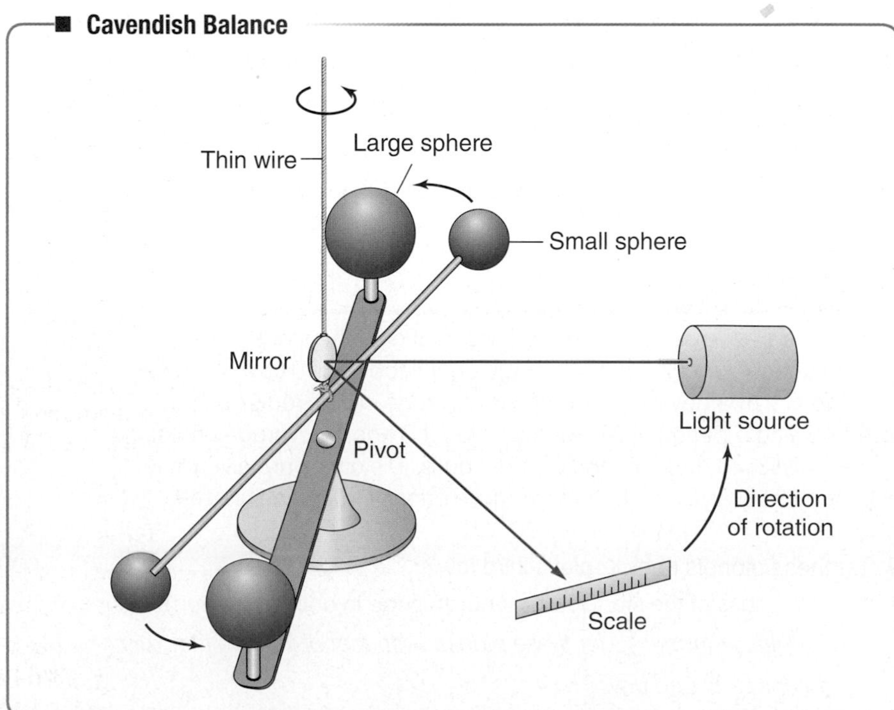

Figure 8 A Cavendish balance uses a light source and a mirror to measure the movement of the spheres.

View an **animation of Cavendish's investigation.**

The importance of G Cavendish's investigation often is called "weighing Earth" because it helped determine Earth's mass. Once the value of G is known, not only the mass of Earth, but also the mass of the Sun can be determined. In addition, the gravitational force between any two objects can be calculated by using Newton's law of universal gravitation. For example, the attractive gravitational force (F_g) between two bowling balls of mass 7.26 kg, with their centers separated by 0.30 m, can be calculated as follows:

$$F_g = \frac{(6.67\times10^{-11}\ \text{N}\cdot\text{m}^2/\text{kg}^2)(7.26\ \text{kg})(7.26\ \text{kg})}{(0.30\ \text{m})^2} = 3.9\times10^{-8}\ \text{N}$$

You know that on Earth's surface, the weight of an object of mass m is a measure of Earth's gravitational attraction: $F_g = mg$. If Earth's mass is represented by m_E and Earth's radius is represented by r_E, the following is true:

$$F_g = \frac{Gm_Em}{{r_E}^2} = mg,\ \text{and so}\ g = \frac{Gm_E}{{r_E}^2}$$

This equation can be rearranged to solve for m_E.

$$m_E = \frac{g{r_E}^2}{G}$$

Using $g = 9.8$ N/kg, $r_E = 6.38\times10^6$ m, and $G = 6.67\times10^{-11}$ N•m²/kg², the following result is obtained for Earth's mass:

$$m_E = \frac{(9.8\ \text{N/kg})(6.38\times10^6\ \text{m})^2}{6.67\times10^{-11}\ \text{N}\cdot\text{m}^2/\text{kg}^2} = 5.98\times10^{24}\ \text{kg}$$

When you compare the mass of Earth to that of a bowling ball, you can see why the gravitational attraction between everyday objects is not easily observed. Cavendish's investigation determined the value of G, confirmed Newton's prediction that a gravitational force exists between any two objects, and helped calculate the mass of Earth **(Figure 9).**

Figure 9 Cavendish's investigations helped calculate the mass of Earth.

NASA/Goddard Space Flight Center Scientific Visualization Studio

SECTION 1 REVIEW

Section Self-Check Check your understanding.

8. **MAIN IDEA** What is the gravitational force between two 15-kg balls whose centers are 35 m apart? What fraction is this of the weight of one ball?

9. **Neptune's Orbital Period** Neptune orbits the Sun at an average distance given in **Figure 10,** which allows gases, such as methane, to condense and form an atmosphere. If the mass of the Sun is 1.99×10^{30} kg, calculate the period of Neptune's orbit.

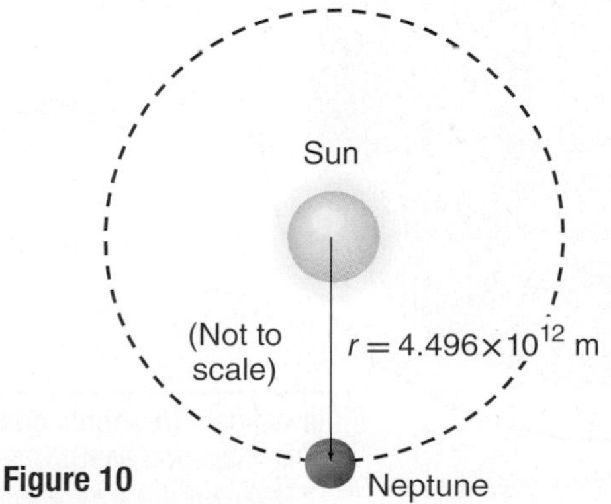

Figure 10

10. **Gravity** If Earth began to shrink, but its mass remained the same, what would happen to the value of g on Earth's surface?

11. **Universal Gravitational Constant** Cavendish did his investigation using lead spheres. Suppose he had replaced the lead spheres with copper spheres of equal mass. Would his value of G be the same or different? Explain.

12. Kepler's three statements and Newton's equation for gravitational attraction are called laws. Were they ever theories? Will they ever become theories?

13. **Critical Thinking** Picking up a rock requires less effort on the Moon than on Earth.
    a. How will the Moon's gravitational force affect the path of the rock if it is thrown horizontally?
    b. If the thrower accidentally drops the rock on her toe, will it hurt more or less than it would on Earth? Explain.

SECTION 2 Using the Law of Universal Gravitation

PHYSICS 4 YOU **Have you ever used a device to locate your position or to map where you want to go? Where does that device get information? The Global Positioning System (GPS) consists of many satellites circling Earth. GPS satellites give accurate position data anywhere on or near Earth.**

MAINIDEA

All objects are surrounded by a gravitational field that affects the motions of other objects.

Essential Questions

- How can you describe orbital motion?
- How are gravitational mass and inertial mass alike, and how are they different?
- How is gravitational force explained, and what did Einstein propose about gravitational force?

Review Vocabulary

centripetal acceleration the center-seeking acceleration of an object moving in a circle at a constant speed

New Vocabulary

inertial mass
gravitational mass

Orbits of Planets and Satellites

The planet Uranus was discovered in 1781. By 1830 it was clear that the law of gravitation didn't correctly predict its orbit. Two astronomers proposed that Uranus was being attracted by the Sun and by an undiscovered planet. They calculated the orbit of such a planet in 1845, and, one year later, astronomers at the Berlin Observatory found the planet now called Neptune. How is it possible for planets, such as Neptune and Uranus, to remain in orbit around the Sun?

Newton used a drawing similar to the one shown in **Figure 11** to illustrate a thought experiment on the motion of satellites. Imagine a cannon, perched high atop a mountain, firing a cannonball horizontally with a given horizontal speed. The cannonball is a projectile, and its motion has both vertical and horizontal components. Like all projectiles on Earth, it would follow a parabolic trajectory and fall back to the ground.

If the cannonball's horizontal speed were increased, it would travel farther across the surface of Earth and still fall back to the ground. If an extremely powerful cannon were used, however, the cannonball would travel all the way around Earth and keep going. It would fall toward Earth at the same rate that Earth's surface curves away. In other words, the curvature of the projectile would continue to just match the curvature of Earth so that the cannonball would never get any closer to or farther away from Earth's curved surface. The cannonball would, therefore, be in orbit.

Figure 11 Newton imagined a projectile launched parallel to Earth. If it has enough speed it will fall toward Earth with a curvature that matches the curvature of Earth's surface.

Identify the factor that is not considered in this example.

Investigate **Newton's cannon.**

Virtual Investigation

Newton's thought experiment ignored air resistance. For the cannonball to be free of air resistance, the mountain on which the cannon is perched would have to be more than 150 km above Earth's surface. By way of comparison, the mountain would have to be much taller than the peak of Mount Everest, the world's tallest mountain, which is only 8.85 km in height. A cannonball launched from a mountain that is 150 km above Earth's surface would encounter little or no air resistance at an altitude of 150 km because the mountain would be above most of the atmosphere. Thus, a cannonball or any object or satellite at or above this altitude could orbit Earth.

A satellite's speed A satellite in an orbit that is always the same height above Earth moves in uniform circular motion. Recall that its centripetal acceleration is given by $a_c = \frac{v^2}{r}$. Newton's second law, $F_{net} = ma_c$, can thus be rewritten $F_{net} = \frac{mv^2}{r}$. If Earth's mass is m_E, then this expression combined with Newton's law of universal gravitation produces the following equation:

$$\frac{Gm_Em}{r^2} = \frac{mv^2}{r}$$

Solving for the speed of a satellite in circular orbit about Earth (v) yields the following.

SPEED OF A SATELLITE ORBITING EARTH

The speed of a satellite orbiting Earth is equal to the square root of the universal gravitational constant times the mass of Earth, divided by the radius of the orbit.

$$v = \sqrt{\frac{Gm_E}{r}}$$

A satellite's orbital period A satellite's orbit around Earth is similar to a planet's orbit about the Sun. Recall that the period of a planet orbiting the Sun is expressed by the following equation:

$$T = 2\pi\sqrt{\frac{r^3}{Gm_S}}$$

Thus, the period for a satellite orbiting Earth is given by the following equation.

PERIOD OF A SATELLITE ORBITING EARTH

The period for a satellite orbiting Earth is equal to 2π times the square root of the radius of the orbit cubed, divided by the product of the universal gravitational constant and the mass of Earth.

$$T = 2\pi\sqrt{\frac{r^3}{Gm_E}}$$

The equations for the speed and period of a satellite can be used for any object in orbit about another. The mass of the central body will replace m_E in the equations, and r will be the distance between the centers of the orbiting body and the central body. Orbital speed (v) and period (T) are independent of the mass of the satellite.

READING CHECK Describe how the mass of a satellite affects that satellite's orbital speed and period.

StockTrek/Getty Images

REAL-WORLD PHYSICS

GEOSYNCHRONOUS ORBIT The GOES weather satellites orbit Earth once a day at an altitude of 35,785 km. The orbital speed of the satellite matches Earth's rate of rotation. Thus, to an observer on Earth, the satellite appears to remain above one spot. Dish antennas on Earth can be directed to one point in the sky and remain in a fixed position as the satellite orbits.

EXAMPLE PROBLEM 2

Get help with **orbital speed.** Personal Tutor

ORBITAL SPEED AND PERIOD Assume that a satellite orbits Earth 225 km above its surface. Given that the mass of Earth is 5.97×10^{24} kg and the radius of Earth is 6.38×10^{6} m, what are the satellite's orbital speed and period?

1 ANALYZE AND SKETCH THE PROBLEM

Sketch the situation showing the height of the satellite's orbit.

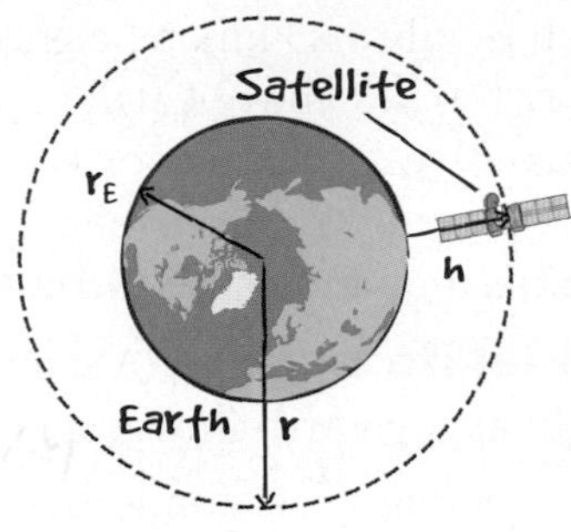

KNOWN	UNKNOWN
$h = 2.25\times10^{5}$ m	$v = ?$
$r_E = 6.38\times10^{6}$ m	$T = ?$
$m_E = 5.97\times10^{24}$ kg	
$G = 6.67\times10^{-11}$ N·m²/kg²	

2 SOLVE FOR ORBITAL SPEED AND PERIOD

Determine the orbital radius by adding the height of the satellite's orbit to Earth's radius.

$$r = h + r_E$$
$$= 2.25\times10^{5}\text{ m} + 6.38\times10^{6}\text{ m} = 6.60\times10^{6}\text{ m}$$

◀ Substitute $h = 2.25\times10^{5}$ m and $r_E = 6.38\times10^{6}$ m.

Solve for the speed.

$$v = \sqrt{\frac{Gm_E}{r}}$$

◀ Substitute $G = 6.67\times10^{-11}$ N·m²/kg², $m_E = 5.97\times10^{24}$ kg, and $r = 6.60\times10^{6}$ m.

$$= \sqrt{\frac{(6.67\times10^{-11}\text{ N·m}^2/\text{kg}^2)(5.97\times10^{24}\text{ kg})}{6.60\times10^{6}\text{ m}}}$$
$$= 7.77\times10^{3}\text{ m/s}$$

Solve for the period.

$$T = 2\pi\sqrt{\frac{r^3}{Gm_E}}$$

◀ Substitute $r = 6.60\times10^{6}$ m, $G = 6.67\times10^{-11}$ N·m²/kg², and $m_E = 5.97\times10^{24}$ kg.

$$= 2\pi\sqrt{\frac{(6.60\times10^{6}\text{ m})^3}{(6.67\times10^{-11}\text{ N·m}^2/\text{kg}^2)(5.97\times10^{24}\text{ kg})}}$$
$$= 5.34\times10^{3}\text{ s}$$

This is approximately 89 min, or 1.5 h.

3 EVALUATE THE ANSWER

Are the units correct? The unit for speed is meters per second, and the unit for period is seconds.

PRACTICE PROBLEMS

Do additional problems. 

For the following problems, assume a circular orbit for all calculations.

14. Suppose that the satellite in Example Problem 2 is moved to an orbit that is 24 km larger in radius than its previous orbit.

a. What is its speed?

b. Is this faster or slower than its previous speed?

c. Why do you think this is so?

15. Uranus has 27 known moons. One of these moons is Miranda, which orbits at a radius of 1.29×10^{8} m. Uranus has a mass of 8.68×10^{25} kg. Find the orbital speed of Miranda. How many Earth days does it take Miranda to complete one orbit?

16. Use Newton's thought experiment on the motion of satellites to solve the following.

a. Calculate the speed that a satellite shot from a cannon must have to orbit Earth 150 km above its surface.

b. How long, in seconds and minutes, would it take for the satellite to complete one orbit and return to the cannon?

17. CHALLENGE Use the data for Mercury in **Table 1** to find the following.

a. the speed of a satellite that is in orbit 260 km above Mercury's surface

b. the period of the satellite

▶ CONNECTION TO EARTH SCIENCE *Landsat 7*, shown in **Figure 12,** is an artificial satellite that provides images of Earth's continental surfaces. *Landsat 7* images have been used to create maps, study land use, and monitor resources and global changes. The *Landsat 7* system enables researchers to monitor small-scale processes, such as deforestation, on a global scale. Satellites, such as *Landsat 7*, are accelerated to the speeds necessary for them to achieve orbit by large rockets, such as shuttle-booster rockets. Because the acceleration of any mass must follow Newton's second law of motion, $F_{net} = ma$, more force is required to launch a more massive satellite into orbit. Thus, the mass of a satellite is limited by the capability of the rocket used to launch it.

Figure 12 *Landsat 7* is capable of providing up to 532 images of Earth per day.

Free-Fall Acceleration

The acceleration of objects due to Earth's gravity can be found by using Newton's law of universal gravitation and his second law of motion. For a free-falling object of mass m, the following is true:

$$F = \frac{Gm_{E}m}{r^2} = ma, \text{ so } a = \frac{Gm_{E}}{r^2}$$

If you set $a = g_E$ and $r = r_E$ on Earth's surface, the following equation can be written:

$$g = \frac{Gm_E}{{r_E}^2}, \text{ thus, } m_E = \frac{g{r_E}^2}{G}$$

You saw above that $a = \frac{Gm_E}{r^2}$ for a free-falling object. Substitution of the above expression for m_E yields the following:

$$a = \frac{G\left(\frac{g{r_E}^2}{G}\right)}{r^2}$$

$$a = g\left(\frac{r_E}{r}\right)^2$$

On the surface of Earth, $r = r_E$ and so $a = g$. But, as you move farther from Earth's center, r becomes larger than r_E, and the free-fall acceleration is reduced according to this inverse square relationship. What happens to your mass as you move farther and farther from Earth's center?

Weight and weightlessness You may have seen photos similar to **Figure 13** in which astronauts are on a spacecraft in an environment often called zero-g or weightlessness. The spacecraft orbits about 400 km above Earth's surface. At that distance, $g = 8.7$ N/kg, only slightly less than that on Earth's surface. Earth's gravitational force is certainly not zero in the shuttle. In fact, gravity causes the shuttle to orbit Earth. Why, then, do the astronauts appear to have no weight?

Remember that you sense weight when something, such as the floor or your chair, exerts a contact force on you. But if you, your chair, and the floor all are accelerating toward Earth together, then no contact forces are exerted on you. Thus, your apparent weight is zero and you experience apparent weightlessness. Similarly, the astronauts experience apparent weightlessness as the shuttle and everything in it falls freely toward Earth.

Figure 13 Astronauts in orbit around Earth are in free fall because their spacecraft and everything in it is accelerating toward Earth along with the astronauts. That is, the floor is constantly falling from beneath their feet.

MiniLABs

WEIGHTLESS WATER

What are the effects of weightlessness in free fall?

WEIGHT IN FREE FALL

What is the effect of free fall on mass?

The Gravitational Field

Recall from studying motion that many common forces are contact forces. Friction is exerted where two objects touch; for example, the floor and your chair or desk push on you when you are in contact with them. Gravity, however, is different. It acts on an apple falling from a tree and on the Moon in orbit. In other words, gravity acts over a distance. It acts between objects that are not touching or that are not close together. Newton was puzzled by this concept. He wondered how the Sun could exert a force on planet Earth, which is hundreds of millions of kilometers away.

Field concept The answer to the puzzle arose from a study of magnetism. In the nineteenth century, Michael Faraday developed the concept of a field to explain how a magnet attracts objects. Later, the field concept was applied to gravity.

Any object with mass is surrounded by a gravitational field, which exerts a force that is directly proportional to the mass of the object and inversely proportional to the square of the distance from the object's center. Another object experiences a force due to the interaction between its mass and the gravitational field ($\boldsymbol{g}$) at its location. The direction of $\boldsymbol{g}$ and the gravitational force is toward the center of the object producing the field. Gravitational field strength is expressed by the following equation.

GRAVITATIONAL FIELD

The gravitational field strength produced by an object is equal to the universal gravitational constant times the object's mass, divided by the square of the distance from the object's center.

$$g = \frac{Gm}{r^2}$$

Suppose the gravitational field is created by the Sun. Then a planet of mass m in the Sun's gravitational field has a force exerted on it that depends on its mass and the magnitude of the gravitational field at its location. That is, $\boldsymbol{F}_g = m\boldsymbol{g}$, toward the Sun. The force is caused by the interaction of the planet's mass with the gravitational field at its location, not with the Sun millions of kilometers away. To find the gravitational field caused by more than one object, calculate all gravitational fields and add them as vectors.

The gravitational field is measured by placing an object with a small mass (m) in the gravitational field and measuring the force ($\boldsymbol{F}_g$) on it. The gravitational field is calculated using $\boldsymbol{g} = \frac{\boldsymbol{F}_g}{m}$. The gravitational field is measured in units of newtons per kilogram (N/kg).

On Earth's surface, the strength of the gravitational field is 9.8 N/kg, and its direction is toward Earth's center. The field can be represented by a vector of length g pointing toward the center of the object producing the field. You can picture the gravitational field produced by Earth as a collection of vectors surrounding Earth and pointing toward it, as shown in **Figure 14.** The strength of Earth's gravitational field varies inversely with the square of the distance from Earth's center. Earth's gravitational field depends on Earth's mass but not on the mass of the object experiencing it.

Figure 14 Earth's gravitational field can be represented by vectors pointing toward Earth's center. The decrease in $\boldsymbol{g}$'s magnitude follows an inverse-square relationship as the distance from Earth's center increases.

Explain why the value of $\boldsymbol{g}$ never reaches zero.

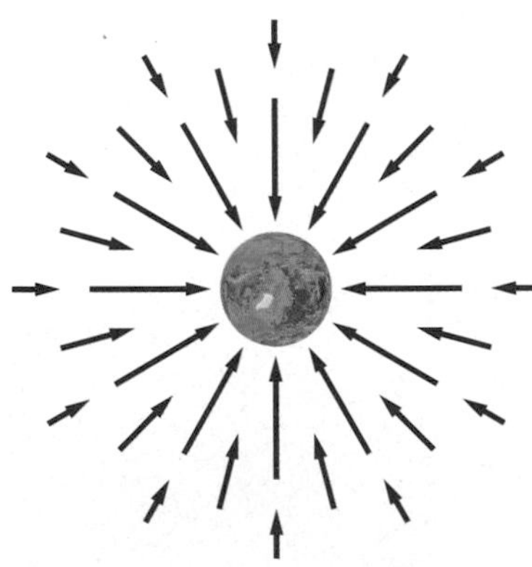

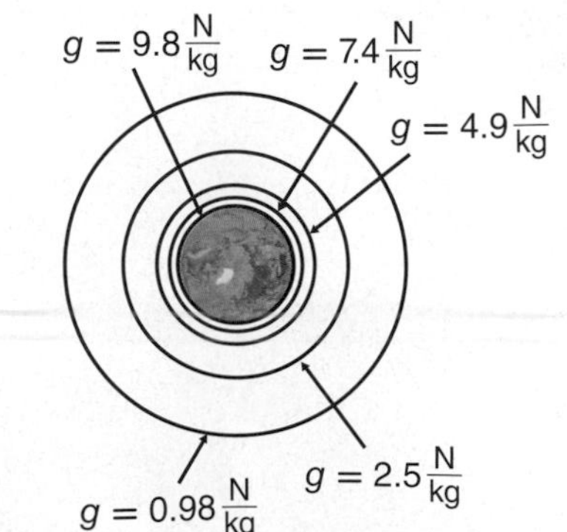

Two Kinds of Mass

You read that mass can be defined as the slope of a graph of force versus acceleration. That is, mass is equal to the net force exerted on an object divided by its acceleration. This kind of mass, related to the inertia of an object, is called inertial mass and is represented by the following equation.

INERTIAL MASS

Inertial mass is equal to the net force exerted on the object divided by the acceleration of the object.

$$m_{\text{inertial}} = \frac{F_{\text{net}}}{a}$$

Inertial mass You know that it is much easier to push an empty cardboard box across the floor than it is to push one that is full of books. The full box has greater inertial mass than the empty one. The **inertial mass** of an object is a measure of the object's resistance to any type of force. Inertial mass of an object is measured by exerting a force on the object and measuring the object's acceleration. The more inertial mass an object has, the less acceleration it undergoes as a result of a net force exerted on it.

Gravitational mass Newton's law of universal gravitation, $F_g = \frac{Gm_1m_2}{r^2}$, also involves mass—but a different kind of mass. Mass as used in the law of universal gravitation is a quantity that measures an object's response to gravitational force and is called **gravitational mass.** It can be measured by using a simple balance, such as the one shown in **Figure 15.** If you measure the magnitude of the attractive force exerted on an object by another object of mass *m*, at a distance *r*, then you can define the gravitational mass in the following way.

GRAVITATIONAL MASS

The gravitational mass of an object is equal to the distance between the centers of the objects squared, times the gravitational force, divided by the product of the universal gravitational constant, times the mass of the other object.

$$m_g = \frac{r^2F_g}{Gm}$$

How different are these two kinds of mass? Suppose you have a watermelon in the trunk of your car. If you accelerate the car forward, the watermelon will roll backward relative to the trunk. This is a result of its inertial mass—its resistance to acceleration. Now, suppose your car climbs a steep hill at a constant speed. The watermelon will again roll backward. But this time, it moves as a result of its gravitational mass. The watermelon is pulled downward toward Earth.

Newton made the claim that inertial mass and gravitational mass are equal in magnitude. This hypothesis is called the principle of equivalence. All investigations conducted so far have yielded data that support this principle. Most of the time we refer simply to the mass of an object. Albert Einstein also was intrigued by the principle of equivalence and made it a central point in his theory of gravity.

PhysicsLABs

HOW CAN YOU MEASURE MASS?

How is an inertial balance used to measure mass?

INERTIAL MASS AND GRAVITATIONAL MASS

How can you determine the relationship between inertial mass and gravitational mass?

Figure 15 A simple balance is used to determine the gravitational mass of an object.

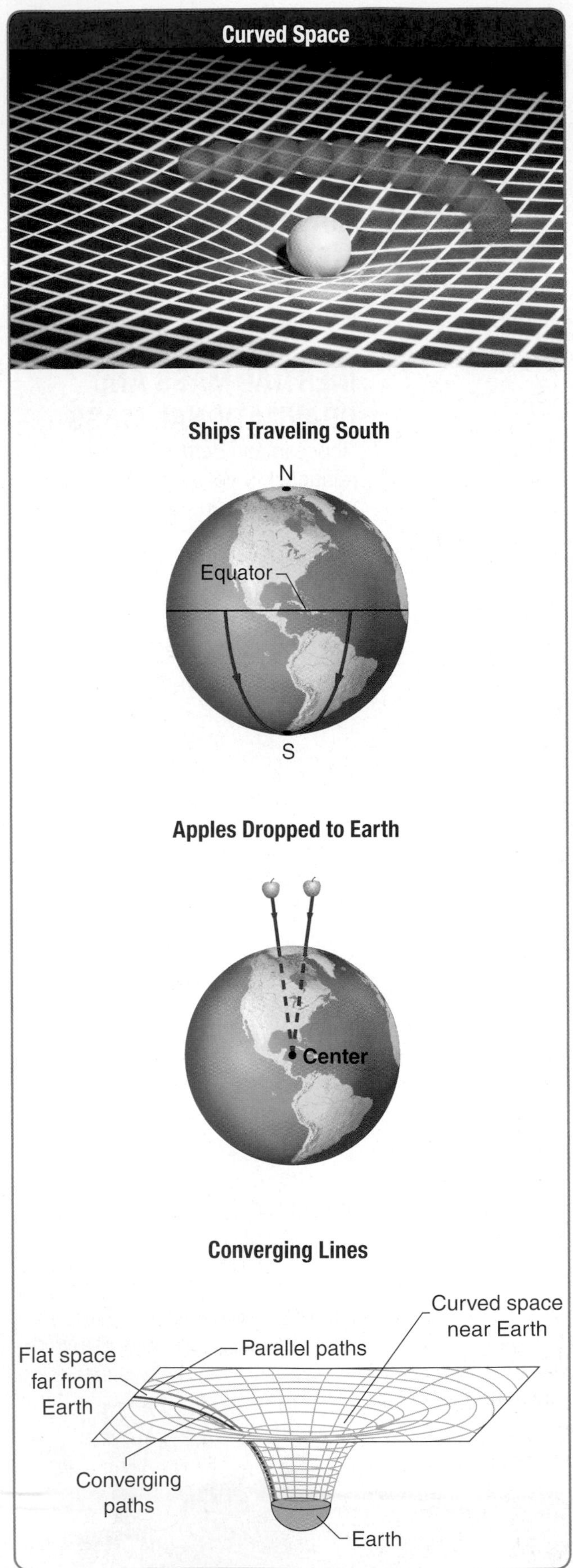

Figure 16 Visualizing how space is curved is difficult. Analogies can help you understand difficult concepts.

Einstein's Theory of Gravity

Newton's law of universal gravitation allows us to calculate the gravitational force that exists between two objects because of their masses. Newton was puzzled, however, as to how two objects could exert forces on each other if those two objects were millions of kilometers away from each other. Albert Einstein proposed that gravity is not a force but rather an effect of space itself. According to Einstein's explanation of gravity, mass changes the space around it. Mass causes space to be curved, and other bodies are accelerated because of the way they follow this curved space.

Curved space One way to picture how mass affects space is to model three-dimensional space as a large, two-dimensional sheet, as shown in the top part of **Figure 16.** The yellow ball on the sheet represents a massive object. The ball forms an indentation on the sheet. A red ball rolling across the sheet simulates the motion of an object in space. If the red ball moves near the sagging region of the sheet, it will be accelerated. In a similar way, Earth and the Sun are attracted to each other because of the way space is distorted by the two objects.

Ships traveling south The following is another analogy that might help you understand the curvature of space. Suppose you watch from space as two ships travel due south from the equator. At the equator, the ships are separated by 4000 km. As they approach the South Pole, the distance decreases to 1 km. To the sailors, their paths are straight lines, but because of Earth's curvature, they travel in a curve, as viewed far from Earth's surface, as in **Figure 16.**

Apples dropped to Earth Consider a similar motion. Two apples are dropped to Earth, initially traveling in parallel paths, as in **Figure 16.** As they approach Earth, they are pulled toward Earth's center. Their paths converge.

Converging lines This convergence can be attributed to the curvature of space near Earth. Far from any massive object, such as a planet or star, space is flat, and parallel lines remain parallel. Then they begin to converge. In flat space, the parallel lines would remain parallel. In curved space, the lines converge.

Einstein's theory or explanation, called the general theory of relativity, makes many predictions about how massive objects affect one another. In every test conducted to date, Einstein's theory has been shown to give the correct results.

Ted Kinsman/Photo Researchers

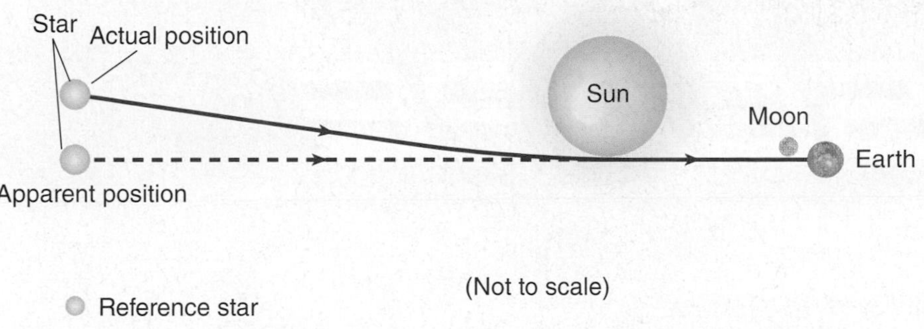

Figure 17 Light is bent around massive objects in space, altering the apparent position of the light's source.

Describe how this effect contradicts your experience of light's behavior.

Deflection of light Einstein's theory predicts that massive objects deflect and bend light. Light follows the curvature of space around the massive object and is deflected, as shown in **Figure 17.** In 1919, during an eclipse of the Sun, astronomers found that light from distant stars that passed near the Sun was deflected an amount that agreed with Einstein's predictions.

Another result of general relativity is the effect of gravity on light from extremely massive objects. If an object is massive and dense enough, the light leaving it is totally bent back to the object. No light ever escapes the object. Objects such as these, called black holes, have been identified as a result of their effects on nearby stars. Black holes have been detected through the radiation produced when matter is pulled into them.

While Einstein's theory provides very accurate predictions of gravity's effects, it is still incomplete. It does not explain the origin of mass or how mass curves space. Physicists are working to understand the deeper meaning of gravity and the origin of mass itself.

SECTION 2 REVIEW

Check your understanding.

18. **MAIN IDEA** The Moon is 3.9×10^5 km from Earth's center and Earth is 14.96×10^7 km from the Sun's center. The masses of Earth and the Sun are 5.97×10^{24} kg and 1.99×10^{30} kg, respectively. During a full moon, the Sun, Earth, and the Moon are in line with each other, as shown in **Figure 18.**
    - **a.** Find the ratio of the gravitational fields due to Earth and the Sun at the center of the Moon.
    - **b.** What is the net gravitational field due to the Sun and Earth at the center of the Moon?

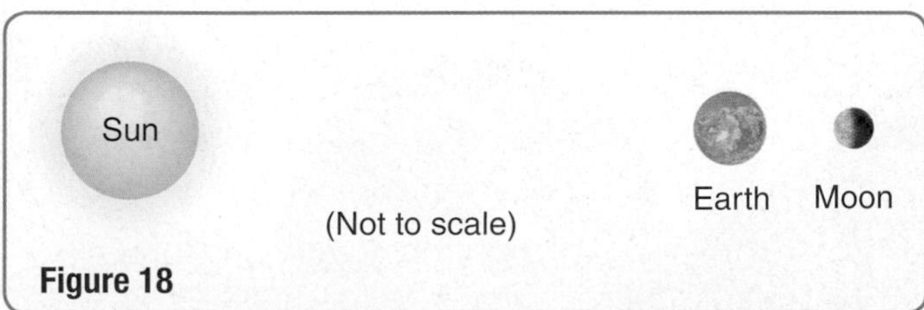

Figure 18

19. **Apparent Weightlessness** Chairs in an orbiting spacecraft are weightless. If you were on board such a spacecraft and you were barefoot, would you stub your toe if you kicked a chair? Explain.
20. **Gravitational Field** The mass of the Moon is 7.3×10^{22} kg and its radius is 1785 km. What is the strength of the gravitational field on the surface of the Moon?
21. **Orbital Period and Speed** Two satellites are in circular orbits about Earth. One is 150 km above the surface, the other is 160 km.
    - **a.** Which satellite has the larger orbital period?
    - **b.** Which has the greater speed?
22. **Theories and Laws** Why is Einstein's description of gravity called a theory, while Newton's is a law?
23. **Astronaut** What would be the strength of Earth's gravitational field at a point where an 80.0-kg astronaut would experience a 25.0 percent reduction in weight?
24. **A Satellite's Mass** When the first artificial satellite was launched into orbit by the former Soviet Union in 1957, U.S. president Dwight D. Eisenhower asked scientists to calculate the mass of the satellite. Would they have been able to make this calculation? Explain.
25. **Critical Thinking** It is easier to launch a satellite from Earth into an orbit that circles eastward than it is to launch one that circles westward. Explain.

NO ESCAPE

Is a BLACK HOLE really a hole?

You may have wondered about black holes. What are they and where do they come from?

A star explodes What happens when a giant star runs low on fuel? The star cannot maintain its temperature, causing it to collapse under its own weight. The resulting explosion, called a supernova, is usually brighter than the entire galaxy it is in.

If what's left of the star is massive enough (more than three times the mass of the Sun), the remnant becomes one of the strangest objects in the universe: a black hole.

Escape from Earth Imagine standing on the surface of Earth and throwing a ball straight up. As the ball moves up, gravitational force robs the ball of its upward velocity. Finally, when the upward velocity reaches zero, the ball begins to fall back down. **Figure 1** illustrates this.

Of course, this happens due to the limitations of your throwing arm. If you could throw the ball fast enough (about 11,000 m/s), it would not fall back to Earth but instead would escape Earth entirely. This speed is the escape velocity from the surface of our planet.

Escape from a black hole? A black hole is a very compact, massive object with an escape velocity so high that nothing, not even light, can escape. The image in **Figure 2** shows a star being swallowed by a black hole. Anything that passes through the boundary of the influence of a black hole is truly lost forever.

Are black holes really holes? NO! They are extremely dense objects in space.

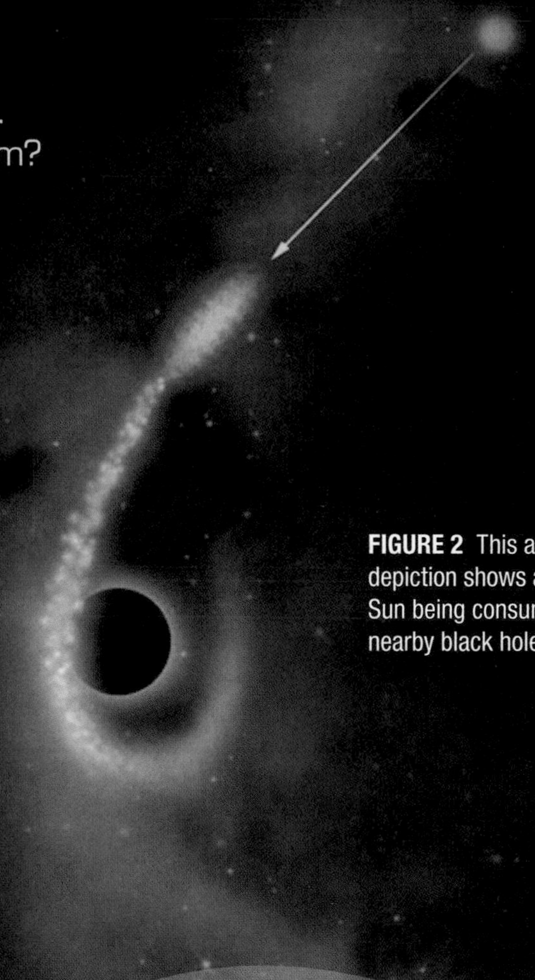

FIGURE 2 This artist's depiction shows a star like our Sun being consumed by a nearby black hole.

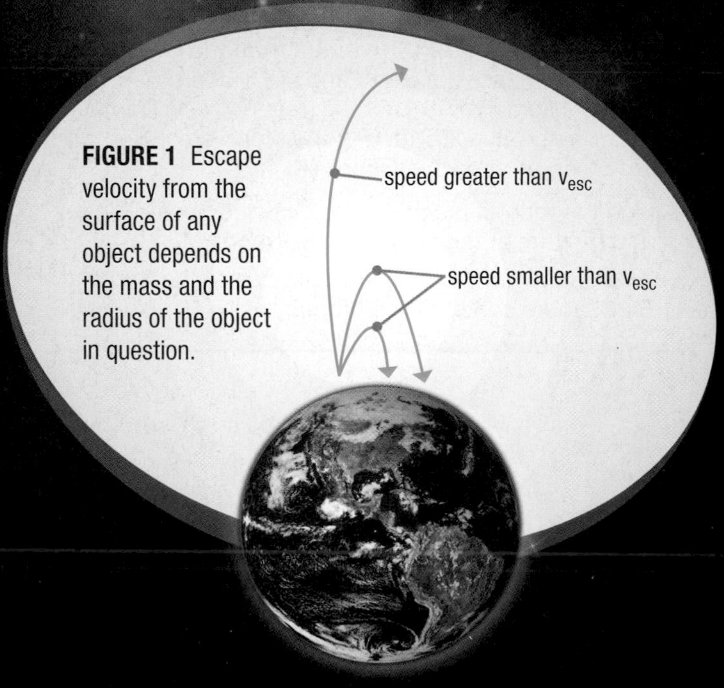

speed greater than v_{esc}

speed smaller than v_{esc}

FIGURE 1 Escape velocity from the surface of any object depends on the mass and the radius of the object in question.

GOING FURTHER >>>

Research Light traveling through a vacuum cannot slow down. So how, then, does a black hole prevent a light beam from escaping? To find out, investigate and write about an effect known as the gravitational red shift.

STUDY GUIDE

BIGIDEA Gravity is an attractive field force that acts between objects with mass.

VOCABULARY
- **Kepler's first law** *(p. 179)*
- **Kepler's second law** *(p. 179)*
- **Kepler's third law** *(p. 180)*
- **gravitational force** *(p. 182)*
- **law of universal gravitation** *(p. 182)*

SECTION 1 Planetary Motion and Gravitation

MAINIDEA The gravitational force between two objects is proportional to the product of their masses divided by the square of the distance between them.

- Kepler's first law states that planets move in elliptical orbits, with the Sun at one focus, and Kepler's second law states that an imaginary line from the Sun to a planet sweeps out equal areas in equal times. Kepler's third law states that the square of the ratio of the periods of any two planets is equal to the cube of the ratio of the distances between the centers of the planets and the center of the Sun.

$$\left(\frac{T_A}{T_B}\right)^2 = \left(\frac{r_A}{r_B}\right)^3$$

- Newton's law of universal gravitation can be used to rewrite Kepler's third law to relate the radius and period of a planet to the mass of the Sun. Newton's law of universal gravitation states that the gravitational force between any two objects is directly proportional to the product of their masses and inversely proportional to the square of the distance between their centers. The force is attractive and along a line connecting the centers of the masses.

$$F = \frac{Gm_1m_2}{r^2}$$

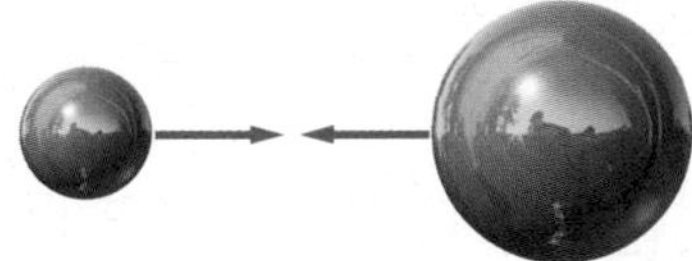

- Cavendish's investigation determined the value of G, confirmed Newton's prediction that a gravitational force exists between two objects, and helped calculate the mass of Earth.

VOCABULARY
- **inertial mass** *(p. 191)*
- **gravitational mass** *(p. 191)*

SECTION 2 Using the Law of Universal Gravitation

MAINIDEA All objects are surrounded by a gravitational field that affects the motions of other objects.

- The speed and period of a satellite in circular orbit describe orbital motion. Orbital speed and period for any object in orbit around another are calculated with Newton's second law.
- Gravitational mass and inertial mass are two essentially different concepts. The gravitational and inertial masses of an object, however, are numerically equal.
- All objects have gravitational fields surrounding them. Any object within a gravitational field experiences a gravitational force exerted on it by the gravitational field. Einstein's general theory of relativity explains gravitational force as a property of space itself.

Games and Multilingual eGlossary

Vocabulary Practice

Chapter Self-Check

SECTION 1
Planetary Motion and Gravitation

Mastering Concepts

26. Problem Posing Complete this problem so that it can be solved using Kepler's third law: "Suppose a new planet was found orbiting the Sun in the region between Jupiter and Saturn . . ."

27. In 1609 Galileo looked through his telescope at Jupiter and saw four moons. The name of one of the moons that he saw is Io. Restate Kepler's first law for Io and Jupiter.

28. Earth moves more slowly in its orbit during summer in the northern hemisphere than it does during winter. Is it closer to the Sun in summer or in winter?

29. Is the area swept out per unit of time by Earth moving around the Sun equal to the area swept out per unit of time by Mars moving around the Sun? Explain your answer.

30. Why did Newton think that a force must act on the Moon?

31. How did Cavendish demonstrate that a gravitational force of attraction exists between two small objects?

32. What happens to the gravitational force between two masses when the distance between the masses is doubled?

33. According to Newton's version of Kepler's third law, how would the ratio $\frac{T^2}{r^3}$ change if the mass of the Sun were doubled?

Mastering Problems

34. Jupiter is 5.2 times farther from the Sun than Earth is. Find the length of Jupiter's orbital period in Earth years.

35. The dwarf planet Pluto has a mean distance from the Sun of 5.87×10^{12} m. What is its orbital period of Pluto around the Sun in years?

36. Use **Table 1** to compute the gravitational force that the Sun exerts on Jupiter.

37. The gravitational force between two electrons that are 1.00 m apart is 5.54×10^{-71} N. Find the mass of an electron.

38. Two bowling balls each have a mass of 6.8 kg. The centers of the bowling balls are located 21.8 cm apart. What gravitational force do the two bowling balls exert on each other?

39. **Figure 19** shows a Cavendish apparatus like the one used to find G. It has a large lead sphere that is 5.9 kg in mass and a small one with a mass of 0.047 kg. Their centers are separated by 0.055 m. Find the force of attraction between them.

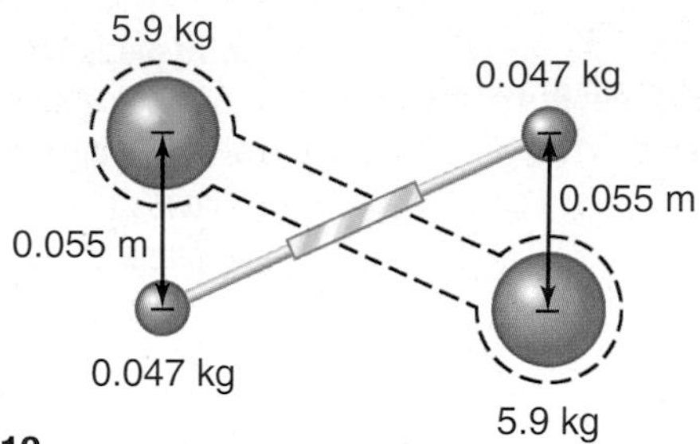

Figure 19

40. Assume that your mass is 50.0 kg. Earth's mass is 5.97×10^{24} kg, and its radius is 6.38×10^{6} m.

a. What is the force of gravitational attraction between you and Earth?

b. What is your weight?

41. A 1.0-kg mass weighs 9.8 N on Earth's surface, and the radius of Earth is roughly 6.4×10^{6} m.

a. Calculate the mass of Earth.

b. Calculate the average density of Earth.

42. Tom's mass is 70.0 kg, and Sally's mass is 50.0 kg. Tom and Sally are standing 20.0 m apart on the dance floor. Sally looks up and sees Tom. She feels an attraction. Supposing that the attraction is gravitational, find its size. Assume that both Tom and Sally can be replaced by spherical masses.

43. BIGIDEA The centers of two balls are 2.0 m apart, as shown in **Figure 20.** One ball has a mass of 8.0 kg. The other has a mass of 6.0 kg. What is the gravitational force between them?

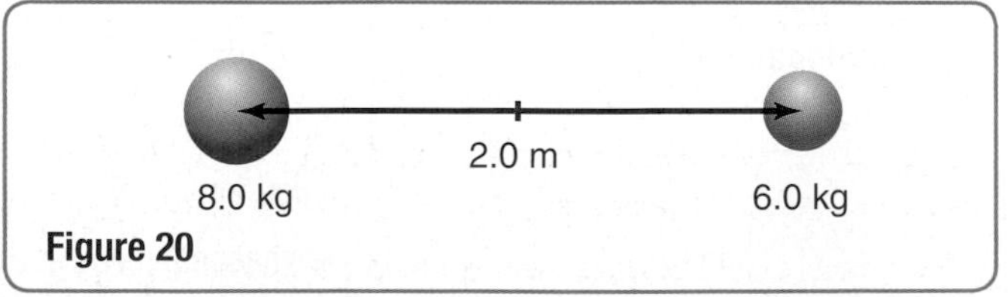

Figure 20

44. The star HD102272 has two planets. Planet A has a period of 127.5 days and a mean orbital radius of 0.615 AU. Planet B has a period of 520 days and a mean orbital radius of 1.57 AU. What is the mass of the star in units of the Sun's mass?

45. If a small planet, D, were located 8.0 times as far from the Sun as Earth is, how many years would it take the planet to orbit the Sun?

46. The Moon's center is 3.9×10^8 m from Earth's center. The Moon is 1.5×10^8 km from the Sun's center. If the mass of the Moon is 7.3×10^{22} kg, find the ratio of the gravitational forces exerted by Earth and the Sun on the Moon.

47. **Ranking Task** Using the solar system data in the reference tables at the end of the book, rank the following pairs of planets according to the gravitational force they exert on each other, from least to greatest. Specifically indicate any ties.

A. Mercury and Venus, when 5.0×10^7 km apart

B. Jupiter and Saturn, when 6.6×10^8 km apart

C. Jupiter and Earth, when 6.3×10^8 km apart

D. Mercury and Earth, when 9.2×10^7 km apart

E. Jupiter and Mercury, when 7.2×10^8 km apart

48. Two spheres are placed so that their centers are 2.6 m apart. The gravitational force between the two spheres is 2.75×10^{-12} N. What is the mass of each sphere if one of the spheres is twice the mass of the other sphere?

49. **Toy Boat** A force of 40.0 N is required to pull a 10.0-kg wooden toy boat at a constant velocity across a smooth glass surface on Earth. What is the force that would be required to pull the same wooden toy boat across the same glass surface on the planet Jupiter?

50. Mimas, one of Saturn's moons, has an orbital radius of 1.87×10^8 m and an orbital period of 23.0 h. Use Newton's version of Kepler's third law to find Saturn's mass.

51. **Halley's Comet** Every 76 years, comet Halley is visible from Earth. Find the average distance of the comet from the Sun in astronomical units. (AU is equal to the Earth's average distance from the Sun. The distance from Earth to the Sun is defined as 1.00 AU.)

52. Area is measured in m^2, so the rate at which area is swept out by a planet or satellite is measured in m^2/s.

a. How quickly is an area swept out by Earth in its orbit about the Sun?

b. How quickly is an area swept out by the Moon in its orbit about Earth? Use 3.9×10^8 m as the average distance between Earth and the Moon and 27.33 days as the period of the Moon.

53. The orbital radius of Earth's Moon is 3.9×10^8 m. Use Newton's version of Kepler's third law to calculate the period of Earth's Moon if the orbital radius were doubled.

SECTION 2

Using the Law of Universal Gravitation

Mastering Concepts

54. How do you answer the question, "What keeps a satellite up?"

55. A satellite is orbiting Earth. On which of the following does its speed depend?

a. mass of the satellite

b. distance from Earth

c. mass of Earth

56. What provides the force that causes the centripetal acceleration of a satellite in orbit?

57. During space flight, astronauts often refer to forces as multiples of the force of gravity on Earth's surface. What does a force of $5g$ mean to an astronaut?

58. Newton assumed that a gravitational force acts directly between Earth and the Moon. How does Einstein's view of the attractive force between the two bodies differ from Newton's view?

59. Show that the units of g, previously given as N/kg, are also m/s^2.

60. If Earth were twice as massive but remained the same size, what would happen to the value of g?

Mastering Problems

61. **Satellite** A geosynchronous satellite is one that appears to remain over one spot on Earth, as shown in **Figure 21.** Assume that a geosynchronous satellite has an orbital radius of 4.23×10^7 m.

a. Calculate its speed in orbit.

b. Calculate its period.

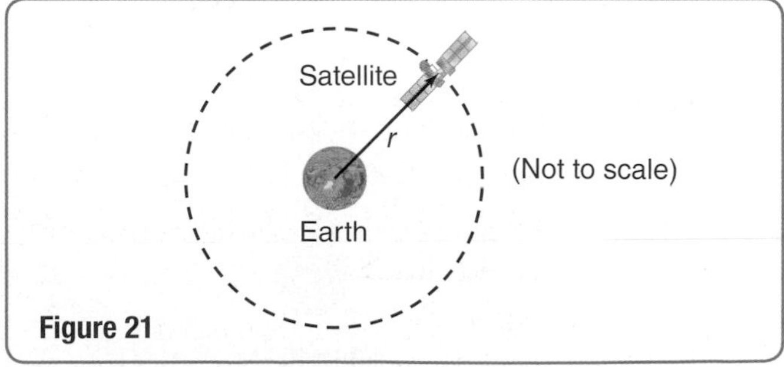

Figure 21

62. **Asteroid** The dwarf planet Ceres has a mass of 7×10^{20} kg and a radius of 500 km.

a. What is g on the surface of Ceres?

b. How much would a 90-kg astronaut weigh on Ceres?

Chapter Self-Check

63. The Moon's mass is 7.34×10^{22} kg, and it has an orbital radius of 3.9×10^{8} m from Earth. Earth's mass is 5.97×10^{24} kg.

a. Calculate the gravitational force of attraction between Earth and the Moon.

b. Find the magnitudes of Earth's gravitational field at the Moon.

64. Reverse Problem Write a physics problem with real-life objects for which the following equation would be part of the solution:

$$8.3\times10^{3}\ \text{m/s} = \sqrt{\frac{(6.67\times10^{-11}\ \text{N}\cdot\text{m}^2/\text{kg}^2)(5.97\times10^{24}\ \text{kg})}{r}}$$

65. The radius of Earth is about 6.38×10^{3} km. A spacecraft with a weight of 7.20×10^{3} N travels away from Earth. What is the weight of the spacecraft at each of the following distances from the surface of Earth?

a. 6.38×10^{3} km

b. 1.28×10^{4} km

c. 2.55×10^{4} km

66. Rocket How high does a rocket have to go above Earth's surface before its weight is half of what it is on Earth?

67. Two satellites of equal mass are put into orbit 30.0 m apart. The gravitational force between them is 2.0×10^{-7} N.

a. What is the mass of each satellite?

b. What is the initial acceleration given to each satellite by gravitational force?

68. Two large spheres are suspended close to each other. Their centers are 4.0 m apart, as shown in **Figure 22.** One sphere weighs 9.8×10^{2} N. The other sphere weighs 1.96×10^{2} N. What is the gravitational force between them?

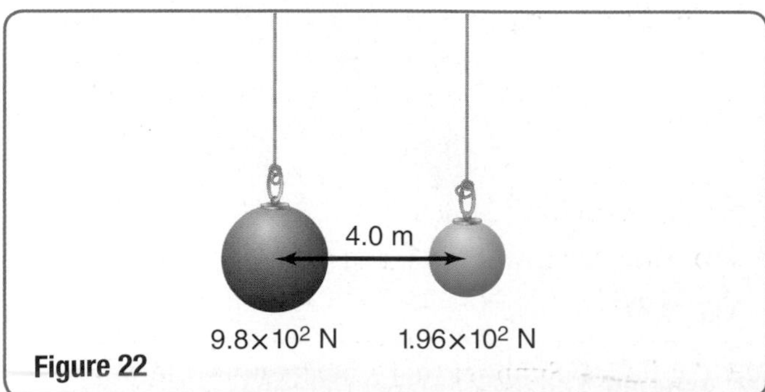

Figure 22

69. Suppose the centers of Earth and the Moon are 3.9×10^{8} m apart, and the gravitational force between them is about 1.9×10^{20} N. What is the approximate mass of the Moon?

70. On the surface of the Moon, a 91.0-kg physics teacher weighs only 145.6 N. What is the value of the Moon's gravitational field at its surface?

71. The mass of an electron is 9.1×10^{-31} kg. The mass of a proton is 1.7×10^{-27} kg. An electron and a proton are about 0.59×10^{-10} m apart in a hydrogen atom. What gravitational force exists between the proton and the electron of a hydrogen atom?

72. Consider two spherical 8.0-kg objects that are 5.0 m apart.

a. What is the gravitational force between the two objects?

b. What is the gravitational force between them when they are 5.0×10^{1} m apart?

73. If you weigh 637 N on Earth's surface, how much would you weigh on the planet Mars? Mars has a mass of 6.42×10^{23} kg and a radius of 3.40×10^{6} m.

74. Find the value of g, the gravitational field at Earth's surface, in the following situations.

a. Earth's mass is triple its actual value, but its radius remains the same.

b. Earth's radius is tripled, but its mass remains the same.

c. Both the mass and the radius of Earth are doubled.

Applying Concepts

75. Acceleration The force of gravity acting on an object near Earth's surface is proportional to the mass of the object. **Figure 23** shows a table-tennis ball and a golf ball in free fall. Why does a golf ball not fall faster than a table-tennis ball?

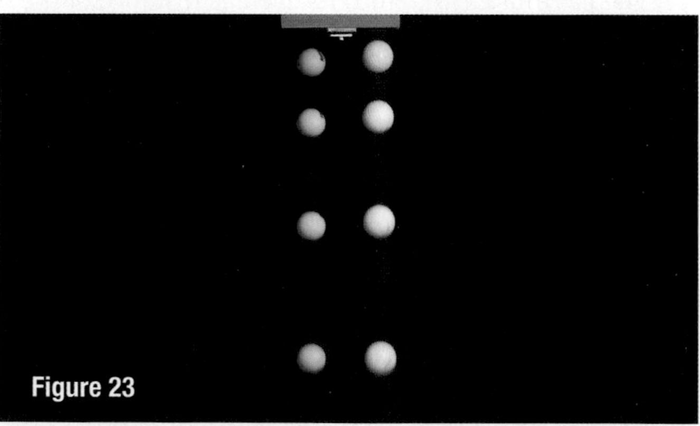

Figure 23

Richard Megna/Fundamental Photographs, NYC

76. What information do you need to find the mass of Jupiter using Newton's version of Kepler's third law?

77. Why was the mass of the dwarf planet Pluto not known until a satellite of Pluto was discovered?

Chapter Self-Check

78. A satellite is one Earth radius above Earth's surface. How does the acceleration due to gravity at that location compare to acceleration at the surface of Earth?

79. What would happen to the value of G if Earth were twice as massive, but remained the same size?

80. An object in Earth's gravitational field doubles in mass. How does the force exerted by the field on the object change?

81. Decide whether each of the orbits shown in **Figure 24** is a possible orbit for a planet.

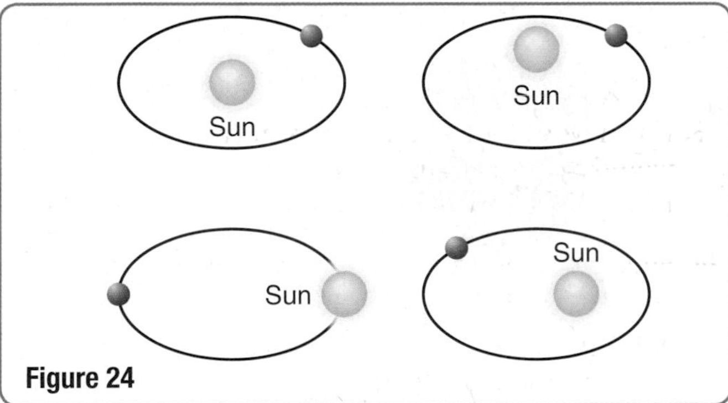

Figure 24

82. The Moon and Earth are attracted to each other by gravitational force. Does the more massive Earth attract the Moon with a greater force than the Moon attracts Earth? Explain.

83. **Figure 25** shows a satellite orbiting Earth. Examine the equation $v = \sqrt{\frac{Gm_E}{r}}$, relating the speed of an orbiting satellite and its distance from the center of Earth. Does a satellite with a large or small orbital radius have the greater velocity?

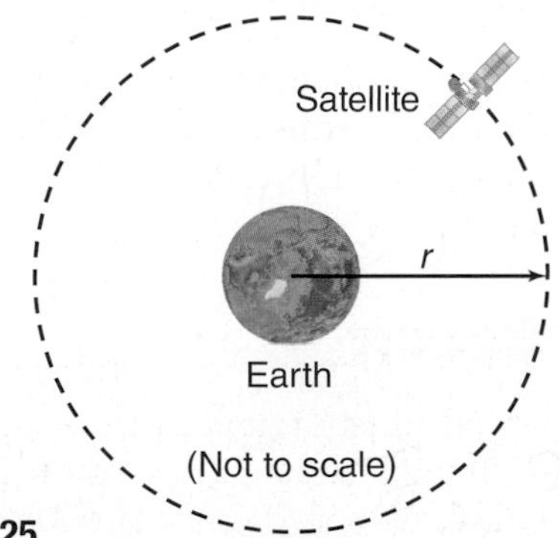

Figure 25

84. **Space Shuttle** If a space shuttle goes into a higher orbit, what happens to the shuttle's period?

85. Jupiter has about 300 times the mass of Earth and about ten times Earth's radius. Estimate the size of *g* on the surface of Jupiter.

86. Mars has about one-ninth the mass of Earth. **Figure 26** shows satellite M, which orbits Mars with the same orbital radius as satellite E, which orbits Earth. Which satellite has a smaller period?

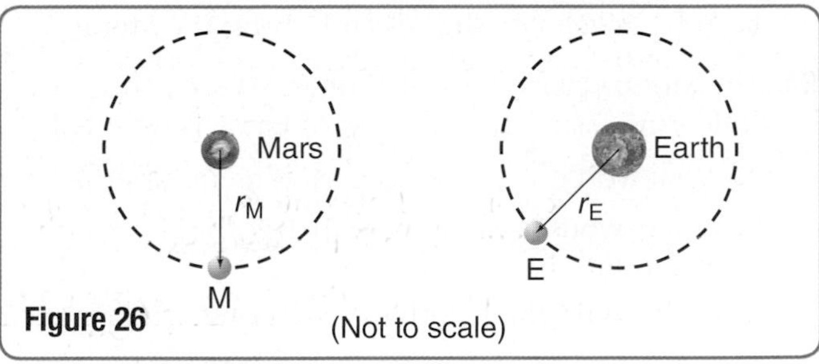

Figure 26

87. **Weight** Suppose that yesterday your body had a mass of 50.0 kg. This morning you stepped on a scale and found that you had gained weight. Assume your location has not changed.
   a. What happened, if anything, to your mass?
   b. What happened, if anything, to the ratio of your weight to your mass?

88. As an astronaut in an orbiting space shuttle, how would you go about "dropping" an object down to Earth?

89. **Weather Satellites** The weather pictures you see on television come from a spacecraft that is in a stationary position relative to Earth, 35,700 km above the equator. Explain how the satellite can stay in exactly the same position. What would happen if it were closer? Farther out? *Hint: Draw a pictorial model.*

Mixed Review

90. Use the information for Earth to calculate the mass of the Sun, using Newton's version of Kepler's third law.

91. The Moon's mass is 7.3×10^{22} kg and its radius is 1738 km. Suppose you perform Newton's thought experiment in which a cannonball is fired horizontally from a very high mountain on the Moon.
   a. How fast would the cannonball have to be fired to remain in orbit?
   b. How long would it take the cannonball to return to the cannon?

92. **Car Races** Suppose that a Martian base has been established. The inhabitants want to hold car races for entertainment. They want to construct a flat, circular race track. If a car can reach speeds of 12 m/s, what is the smallest radius of a track for which the coefficient of friction is 0.50?

Chapter Self-Check

93. ***Apollo 11*** On July 19, 1969, *Apollo 11* was adjusted to orbit the Moon at a height of 111 km. The Moon's radius is 1738 km, and its mass is 7.3×10^{22} kg.

a. What was the period of *Apollo 11* in minutes?

b. At what velocity did *Apollo 11* orbit the Moon?

94. The Moon's period is one month. Answer the following assuming the mass of Earth is doubled.

a. What would the Moon's period be in months?

b. Where would a satellite with an orbital period of one month be located?

c. How would the length of a year on Earth change?

95. Satellite A satellite is in orbit, as in **Figure 27,** with an orbital radius that is half that of the Moon's. Find the satellite's period in units of the Moon's period.

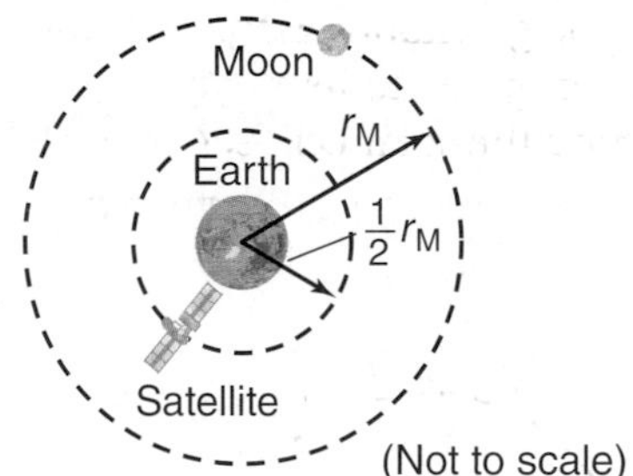

Figure 27

96. How fast would a planet of Earth's mass and size have to spin so that an object at the equator would be weightless? Give the period in minutes.

Thinking Critically

97. Make and Use Graphs Use Newton's law of universal gravitation to find an equation where x is equal to an object's distance from Earth's center and y is its acceleration due to gravity. Use a graphing calculator to graph this equation, using 6400–6600 km as the range for x and 9–10 m/s^2 as the range for y. The equation should be of the form $y = c\left(\frac{1}{x^2}\right)$. Use this graph and find y for these locations: sea level, 6400 km; the top of Mt. Everest, 6410 km; a satellite in typical orbit, 6500 km; a satellite in higher orbit, 6600 km.

98. Suppose the Sun were to disappear—its mass destroyed. If the gravitational force were action at a distance, Earth would experience the loss of the gravitational force of the Sun immediately. But, if the force were caused by a field or Einstein's curvature of space, the information that the Sun was gone would travel at the speed of light. How long would it take this information to reach Earth?

99. Analyze and Conclude The tides on Earth are caused by the pull of the Moon. Is this statement true?

a. Determine the forces (in newtons) that the Moon and the Sun exert on a mass (m) of water on Earth. Answer in terms of m.

b. Which celestial body, the Sun or the Moon, has a greater pull on the waters of Earth?

c. What is the difference in force exerted by the Moon on water at the near surface and water at the far surface (on the opposite side) of Earth, as illustrated in **Figure 28.** Answer in terms of m.

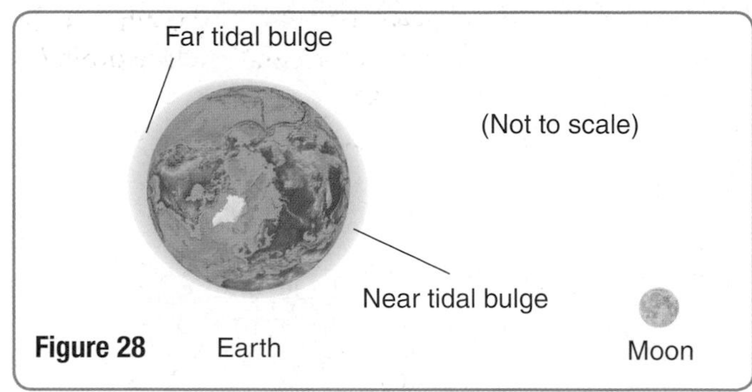

Figure 28

d. Determine the difference in force exerted by the Sun on water at the near surface and on water at the far surface (on the opposite side) of Earth.

e. Which celestial body has a greater difference in pull from one side of Earth to the other?

f. Why is it misleading to say the tides result from the pull of the Moon? Make a correct statement to explain how the Moon causes tides on Earth.

Writing in Physics

100. Research and report the history of how the distance between the Sun and Earth was determined.

101. Explore the discovery of planets around other stars. What methods did the astronomers use? What measurements did they take? How did they use Kepler's third law?

Cumulative Review

102. Airplanes A jet left Pittsburgh at 2:20 P.M. and landed in Washington, DC, at 3:15 P.M. on the same day. If the jet's average speed was 441.0 km/h, what is the distance between the cities?

103. Carolyn wants to weigh her brother Jared. He agrees to stand on a scale, but only if they ride in an elevator. He steps on the scale while the elevator accelerates upward at 1.75 m/s^2. The scale reads 716 N. What is Jared's usual weight?

STANDARDIZED TEST PRACTICE

MULTIPLE CHOICE

1. Two satellites are in orbit around a planet. One satellite has an orbital radius of 8.0×10^{6} m. The period of revolution for this satellite is 1.0×10^{6} s. The other satellite has an orbital radius of 2.0×10^{7} m. What is this satellite's period of revolution?

A. 5.0×10^{5} s **C.** 4.0×10^{6} s
B. 2.5×10^{6} s **D.** 1.3×10^{7} s

2. The illustration below shows a satellite in orbit around a small planet. The satellite's orbital radius is 6.7×10^{4} km and its speed is 2.0×10^{5} m/s. What is the mass of the planet around which the satellite orbits? ($G = 6.67\times10^{-11}$ N•m^2/kg^2)

A. 2.5×10^{18} kg **C.** 2.5×10^{23} kg
B. 4.0×10^{20} kg **D.** 4.0×10^{28} kg

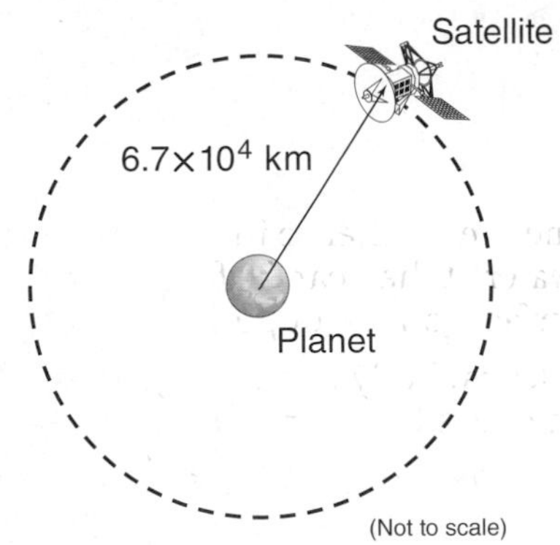

3. Two satellites are in orbit around the same planet. Satellite A has a mass of 1.5×10^{2} kg, and satellite B has a mass of 4.5×10^{3} kg. The mass of the planet is 6.6×10^{24} kg. Both satellites have the same orbital radius of 6.8×10^{6} m. What is the difference in the orbital periods of the satellites?

A. no difference **C.** 2.2×10^{2} s
B. 1.5×10^{2} s **D.** 3.0×10^{2} s

Online Test Practice

4. A moon revolves around a planet with a speed of 9.0×10^{3} m/s. The distance from the moon to the center of the planet is 5.43×10^{6} m. What is the orbital period of the moon?

A. $1.2\pi\times10^{2}$ s **C.** $1.2\pi\times10^{3}$ s
B. $6.0\pi\times10^{2}$ s **D.** $1.2\pi\times10^{9}$ s

5. A moon in orbit around a planet experiences a gravitational force not only from the planet, but also from the Sun. The illustration below shows a moon during a solar eclipse, when the planet, the moon, and the Sun are aligned. The moon has a mass of about 3.9×10^{21} kg. The mass of the planet is 2.4×10^{26} kg, and the mass of the Sun is 1.99×10^{30} kg. The distance from the moon to the center of the planet is 6.0×10^{8} m, and the distance from the moon to the Sun is 1.5×10^{11} m. What is the ratio of the gravitational force on the moon due to the planet, compared to its gravitational force due to the Sun during the solar eclipse?

A. 0.5 **C.** 5.0
B. 2.5 **D.** 7.5

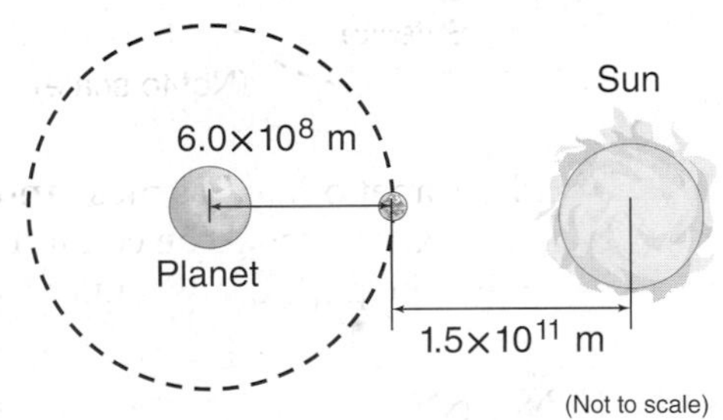

FREE RESPONSE

6. Two satellites are in orbit around a planet. Satellite S_1 takes 20 days to orbit the planet at a distance of 2×10^{5} km from the center of the planet. Satellite S_2 takes 160 days to orbit the planet. What is the distance of satellite S_2 from the center of the planet?

NEED EXTRA HELP?

If you Missed Question	1	2	3	4	5	6
Review Section	1	2	2	2	2	1

CHAPTER 8

Rotational Motion

BIGIDEA **Applying a torque to an object causes a change in that object's angular velocity.**

SECTIONS

1 **Describing Rotational Motion**

2 **Rotational Dynamics**

3 **Equilibrium**

LaunchLAB

iLab Station

ROLLING OBJECTS

How do different objects rotate as they roll?

WATCH THIS!

DANGEROUS CURVE

Have you ever seen a road sign depicting a truck tipping over? You will usually find them near curves. They warn truck and bus drivers to slow down while rounding the curve. What can happen if these drivers do not slow down on the curve?

(l)Photodisc/Getty Images, (r)LMR Group/Alamy

Go online!
connectED.mcgraw-hill.com

SECTION 1

Describing Rotational Motion

Many cars have a tachometer that displays the rate at which the motor's shaft rotates. This speed is measured in thousands of revolutions per minute. Why might a driver need to know this information?

MAIN IDEA

Angular displacement, angular velocity, and angular acceleration all help describe angular motion.

Essential Questions

- What is angular displacement?
- What is average angular velocity?
- What is average angular acceleration, and how is it related to angular velocity?

Review Vocabulary

displacement change in position having both magnitude and direction; it is equal to the final position minus the initial position

New Vocabulary

radian
angular displacement
angular velocity
angular acceleration

Angular Displacement

You probably have observed a spinning object many times. How would you measure such an object's rotation? Find a circular object, such as a DVD. Mark one point on the edge of the DVD so that you can keep track of its position. Rotate the DVD to the left (counterclockwise), and as you do so, watch the location of the mark. When the mark returns to its original position, the DVD has made one complete revolution.

Measuring revolution How can you measure a fraction of one revolution? It can be measured in several different ways, but the two most used are degrees and radians. A degree is $\frac{1}{360}$ of a revolution and is the usual scale marking on a protractor. In mathematics and physics, the radian is related to the ratio of the circumference of a circle to its radius. In one revolution, a point on the edge of a wheel travels a distance equal to 2π times the radius of the wheel. For this reason, the **radian** is defined as $\frac{1}{(2\pi)}$ of a revolution. One complete revolution is an angle of 2π radians. The radian is abbreviated *rad.*

The Greek letter theta (θ) is used to represent the angle of revolution. **Figure 1** shows the angles in radians for several common fractions of a revolution. Note that counterclockwise rotation is designated as positive, while clockwise is negative. As an object rotates, the change in the angle is called the object's **angular displacement.**

Figure 1 A fraction of a revolution can be measured in degrees or radians. Some common angles are shown below measured in radians. Each angle is measured in the counterclockwise direction from $\theta = 0$.

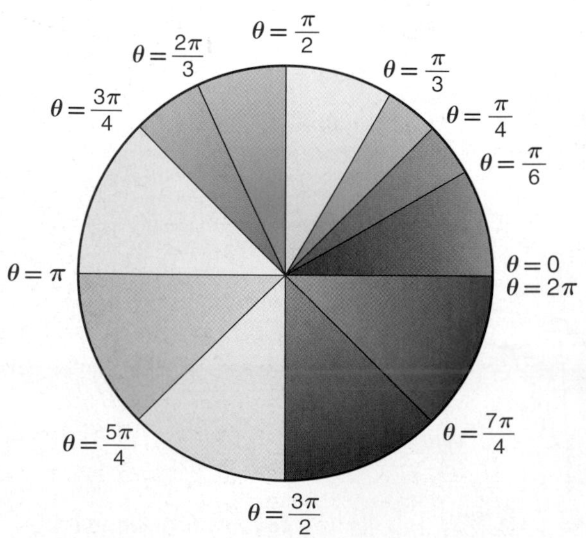

Don Mason/Brand X Pictures/Getty Images

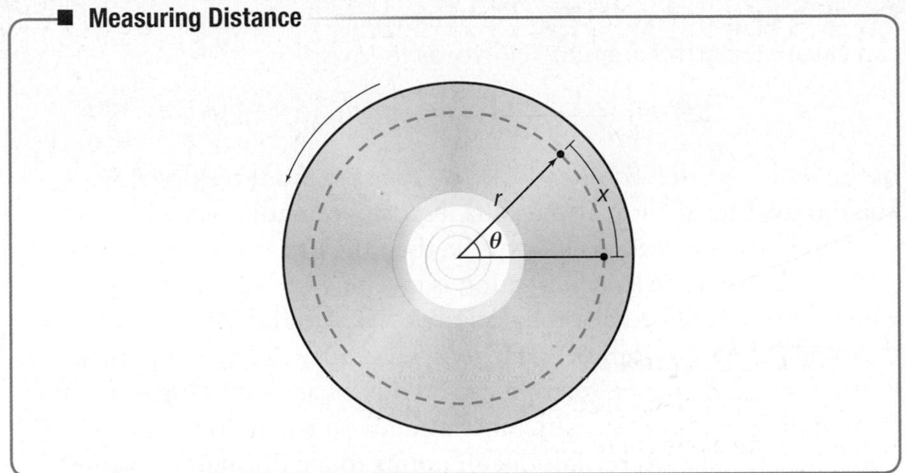

Figure 2 The dashed line shows the path of a point on a DVD as the DVD rotates counterclockwise about its center. The point is located a distance *r* from the center of the DVD and moves a distance *x* as it rotates.

Explain what the variables *r*, *x*, and θ represent.

Earth's revolution As you know, Earth turns one complete revolution, or 2π rad, in 24 h. In 12 h, it rotates through π rad. Through what angle does Earth rotate in 6 h? Because 6 h is one-fourth of a day, Earth rotates through an angle of $\frac{\pi}{2}$ rad during that period. Earth's rotation as seen from the North Pole is positive. Is it positive or negative when viewed from the South Pole?

☑ **READING CHECK** **Identify** the angle that Earth rotates in 48 h.

Measuring distance How far does a point on a rotating object move? You already found that a point on the edge of an object moves 2π times the radius in one revolution. In general, for rotation through an angle (θ), a point at a distance r from the center, as shown in **Figure 2,** moves a distance given by $x = r\theta$. If r is measured in meters, you might think that multiplying it by θ rad would result in x being measured in m•rad. However, this is not the case. Radians indicate the dimensionless ratio between x and r. Thus, x is measured in meters.

Angular Velocity

How fast does a DVD spin? How can you determine its speed of rotation? Recall that velocity is displacement divided by the time taken to make the displacement. Likewise, the **angular velocity** of an object is angular displacement divided by the time taken to make the angular displacement. The angular velocity of an object is given by the following ratio, and is represented by the Greek letter omega (ω).

AVERAGE ANGULAR VELOCITY OF AN OBJECT

The angular velocity equals the angular displacement divided by the time required to make the rotation.

$$\omega = \frac{\Delta\theta}{\Delta t}$$

Recall that if the velocity changes over a time interval, the average velocity is not equal to the instantaneous velocity at any given instant. Similarly, the angular velocity calculated in this way is actually the average angular velocity over a time interval (Δt). Instantaneous angular velocity equals the slope of a graph of angular position versus time.

☑ **READING CHECK** **Define** angular velocity in your own words.

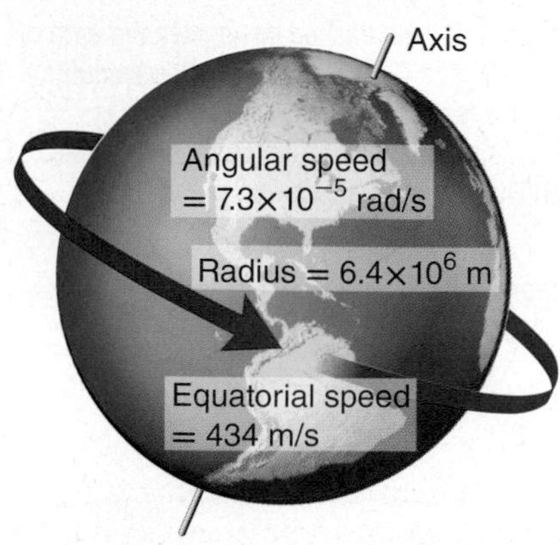

Figure 3 Earth is a rotating, rigid body and all parts rotate at the same rate.

Earth's angular velocity Angular velocity is measured in rad/s. You can calculate Earth's angular velocity as follows:

$$\omega_E = \frac{(2\pi \text{ rad})}{(24.0 \text{ h})(3600 \text{ s/h})} = 7.27\times10^{-5} \text{ rad/s}$$

In the same way that counterclockwise rotation produces positive angular displacement, it also results in positive angular velocity.

If an object's angular velocity is ω, then the linear velocity of a point a distance r from the axis of rotation is given by $v = r\omega$. The speed at which an object on Earth's equator moves as a result of Earth's rotation is given by $v = r\omega = (6.38\times10^6 \text{ m/rad})(7.27\times10^{-5} \text{ rad/s}) = 464 \text{ m/s}$. Earth is an example of a rotating, rigid body, as shown in **Figure 3.** Even though different points at different latitudes on Earth do not move the same distance in each revolution, all points rotate through the same angle. All parts of a rigid body rotate at the same rate. The Sun is not a rigid body. Different parts of the Sun rotate at different angular velocities. Most objects that we will consider in this chapter are rigid bodies.

Angular Acceleration

What if angular velocity is changing? For example, a car could accelerate from 0.0 m/s to 25 m/s in 15 s. In the same 15 s, the angular velocity of the car's 0.64 m diameter wheels would change from 0.0 rad/s to 78 rad/s. The wheels would undergo **angular acceleration,** which is the change in angular velocity divided by the time required to make the change. Angular acceleration (α) is represented by the following equation and is measured in rad/s^2.

AVERAGE ANGULAR ACCELERATION OF AN OBJECT
Angular acceleration is equal to the change in angular velocity divided by the time required to make that change.

$$\alpha = \frac{\Delta\omega}{\Delta t}$$

If the change in angular velocity is positive, then the angular acceleration also is positive. Angular acceleration defined in this way is also the average angular velocity over the time interval Δt. One way to find the instantaneous angular acceleration is to find the slope of a graph of angular velocity as a function of time. The linear acceleration of a point at a distance (r) from the axis of an object with angular acceleration (α) is given by $a = r\alpha$. **Table 1** is a summary of linear and angular relationships discussed previously in this section.

☑ **READING CHECK** **Compare** the angular velocity and angular acceleration of a rotating body.

Table 1 Linear and Angular Measures

Quantity	Linear	Angular	Relationship
Displacement	x (m)	θ (rad)	$x = r\theta$
Velocity	v (m/s)	ω (rad/s)	$v = r\omega$
Acceleration	a (m/s^2)	α (rad/s^2)	$a = r\alpha$

PRACTICE PROBLEMS

Do additional problems. Online Practice

1. What is the angular displacement of each of the following hands of a clock in 1.00 h? State your answer in three significant digits.
   a. the second hand
   b. the minute hand
   c. the hour hand
2. A rotating toy above a baby's crib makes one complete counterclockwise rotation in 1 min.
   a. What is its angular displacement in 3 min?
   b. What is the toy's angular velocity in rad/min?
   c. If the toy is turned off, does it have positive or negative angular acceleration? Explain.
3. If a truck has a linear acceleration of 1.85 m/s^2 and the wheels have an angular acceleration of 5.23 rad/s^2, what is the diameter of the truck's wheels?
4. The truck in the previous problem is towing a trailer with wheels that have a diameter of 48 cm.
   a. How does the linear acceleration of the trailer compare with that of the truck?
   b. How do the angular accelerations of the wheels of the trailer and the wheels of the truck compare?
5. **CHALLENGE** You replace the tires on your car with tires of larger diameter. After you change the tires, how will the angular velocity and number of revolutions be different, for trips at the same speed and over the same distance?

PRACTICE PROBLEMS

Angular frequency An object can revolve many times in a given amount of time. For instance, a spinning wheel may complete several revolutions in 1 min. The number of complete revolutions made by an object in 1 s is called angular frequency. Angular frequency is defined as $f \equiv \frac{\omega}{2\pi}$.

One example of such a rotating object is a computer hard drive. Listen carefully when you start a computer. You often will hear the hard drive spinning. Hard drive frequencies are measured in revolutions per minute (RPM). Inexpensive hard drives rotate at 4800, 5400, and 7200 RPM. More advanced hard drives operate at 10,000 or 15,000 RPM. The faster the hard drive rotates, the quicker the hard drive can access or store information.

SECTION 1 REVIEW

Check your understanding.

6. **MAIN**IDEA The Moon rotates once on its axis in 27.3 days. Its radius is 1.74×10^6 m.
   a. What is the period of the Moon's rotation in seconds?
   b. What is the frequency of the Moon's rotation in rad/s?
   c. A rock sits on the surface at the Moon's equator. What is its linear speed due to the Moon's rotation?
   d. Compare this speed with the speed of a person at Earth's equator due to Earth's rotation.
7. **Angular Displacement** A movie lasts 2 h. During that time, what is the angular displacement of each of the following?
   a. the hour hand
   b. the minute hand
   c. the second hand
8. **Angular Acceleration** In the spin cycle of a clothes washer, the drum turns at 635 rev/min. If the lid of the washer is opened, the motor is turned off. If the drum requires 8.0 s to slow to a stop, what is the angular acceleration of the drum?
9. **Angular Displacement** Do all parts of the minute hand on a watch, shown in **Figure 4,** have the same angular displacement? Do they move the same linear distance? Explain.
10. **Critical Thinking** A CD-ROM has a spiral track that starts 2.7 cm from the center of the disk and ends 5.5 cm from the center. The disk drive must turn the disk so that the linear velocity of the track is a constant 1.4 m/s. Find the following.
    a. the angular velocity of the disk (in rad/s and rev/min) for the start of the track
    b. the disk's angular velocity at the end of the track
    c. the disk's angular acceleration if the disk is played for 76 min

Figure 4

Jill Braaten/The McGraw-Hill Companies, Inc

SECTION 2 Rotational Dynamics

MAIN IDEA

Torques cause changes in angular velocity.

Essential Questions

- What is torque?
- How is the moment of inertia related to rotational motion?
- How are torque, the moment of inertia, and Newton's second law for rotational motion related?

Review Vocabulary

magnitude a measure of size

New Vocabulary

lever arm
torque
moment of inertia
Newton's second law for rotational motion

PHYSICS 4 YOU Have you ever watched a washing machine spin? During the rinse cycle, it starts spinning slowly, but its angular velocity soon increases until the clothes are a blur. The rotational motion causes water in the clothes to move to the drum. The holes in the drum allow the water to drain from the washer.

Force and Angular Velocity

How do you start the rotation of an object? That is, how do you increase its angular velocity? A toy top is a handy round object that is easy to spin. If you wrap a string around it and pull hard, you could make the top spin rapidly. The force of the string is exerted at the outer edge of the top and at right angles to the line from the center of the top to the point where the string leaves the top's surface.

You have learned that a force changes the velocity of a point object. In the case of a toy top, a force that is exerted in a very specific way changes the angular velocity of an extended object. An extended object is an object that has a definite shape and size.

Consider how you open a door: you exert a force. How can you exert the force to open the door most easily? To get the most effect from the least force, you exert the force as far from the axis of rotation as possible, as shown in **Figure 5.** In this case, the axis of rotation is an imaginary vertical line through the hinges. The doorknob is near the outer edge of the door. You exert the force on the doorknob at right angles to the door. Thus, the magnitude of the force, the distance from the axis to the point where the force is exerted, and the direction of the force determine the change in angular velocity.

Figure 5 When opening a door that is free to rotate about its hinges, apply the force farthest from the hinges, at an angle perpendicular to the door.

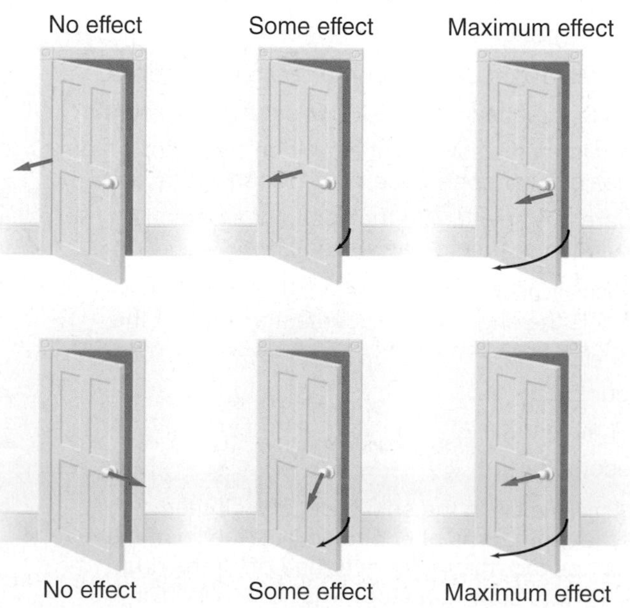

Ryan McVay/Photodisc/Getty Images

Lever Arm Length

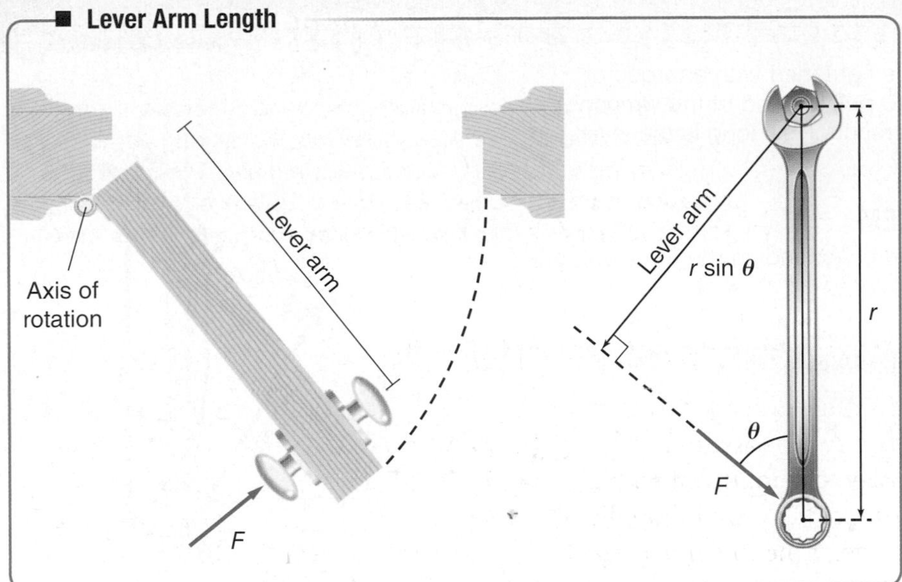

Figure 6 The lever arm is the perpendicular distance from the axis of rotation to the point where the force is exerted. For the door, the lever arm is along the width of the door, from the hinge to the point where the force is exerted. For the wrench, the lever arm is equal to $r \sin \theta$, when the angle (θ) between the force and the radius of rotation is not equal to 90°.

Explain why the formula $r \sin \theta$ is used to find the length of the lever arm.

View a **simulation of a rotating ladder.**

Lever arm For a given applied force, the change in angular velocity depends on the **lever arm,** which is the perpendicular distance from the axis of rotation to the point where the force is exerted. If the force is perpendicular to the radius of rotation, as it was with the toy top, then the lever arm is the distance from the axis (r). For the door example, the lever arm is the distance from the hinges to the point where you exert the force, as illustrated on the left in **Figure 6.** If a force is not exerted perpendicular to the radius, the length of the lever arm is reduced. You must use mathematics to find the length of the lever arm.

To find the lever arm, you must extend the line of the force until it forms a right angle with a line from the center of rotation, as shown in **Figure 6.** The distance between this intersection and the axis is the lever arm. Using trigonometry, the lever arm (L) can be calculated by the equation $L = r \sin \theta$. In this equation, r is the distance from the axis of rotation to the point where the force is exerted, and θ is the angle between the force and the radius from the axis of rotation to the point where the force is applied.

☑ **READING CHECK** **Explain** what each of the variables represent in the equation $L = r \sin \theta$.

Torque The term **torque** describes the combination of force and lever arm that can cause an object to rotate. The magnitude of a torque is the product of the force and the perpendicular lever arm. Because force is measured in newtons and distance is measured in meters, torque is measured in newton-meters (N•m). Torque is represented by the Greek letter tau (τ) and is represented by the equation shown below.

TORQUE

Torque is equal to the force F times the lever arm ($r \sin \theta$).

$$\tau = Fr \sin \theta$$

☑ **READING CHECK** **Identify** what each of the variables in the torque equation—τ, F, r, and θ—represents.

EXAMPLE PROBLEM 1

Find help with **trigonometric ratios.** Math Handbook

LEVER ARM A bolt on a car engine must be tightened with a torque of 35 N•m. You use a 25-cm-long wrench and pull the end of the wrench at an angle of 60.0° to the handle of the wrench. How long is the lever arm, and how much force must you exert?

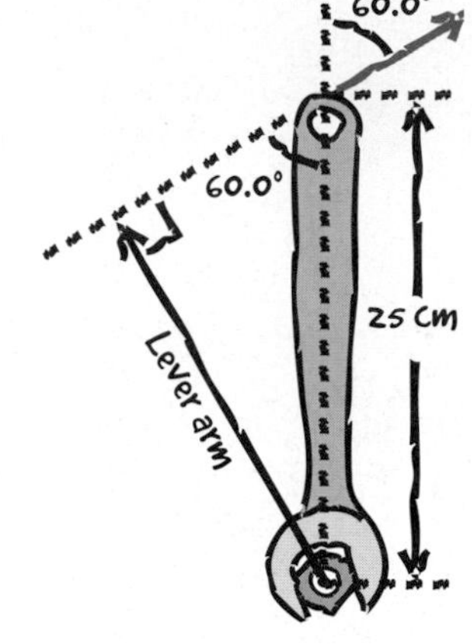

1 ANALYZE AND SKETCH THE PROBLEM

Sketch the situation. Find the lever arm by extending the force vector backward until a line that is perpendicular to it intersects the axis of rotation.

KNOWN		UNKNOWN
$r = 0.25$ m	$\tau = 35$ N•m	$L = ?$
$\theta = 60.0°$		$F = ?$

2 SOLVE FOR THE UNKNOWN

Solve for the length of the lever arm.

$$L = r \sin \theta$$

$$= (0.25 \text{ m})(\sin 60.0°)$$ ◀ Substitute $r = 0.25$ m and $\theta = 60.0°$ into the equation. Then, solve the equation.

$$= 0.22 \text{ m}$$

Solve for the force.

$$\tau = Fr \sin \theta$$

$$F = \frac{\tau}{(r \sin \theta)}$$

$$= \frac{(35 \text{ N}\cdot\text{m})}{(0.25 \text{ m})(\sin 60.0°)}$$ ◀ Substitute $\tau = 35$ N•m, $r = 0.25$ m, and $\theta = 60.0°$ into the equation.

$$= 1.6\times10^2 \text{ N}$$ ◀ Then, solve the equation. Remember to use significant digits.

3 EVALUATE THE ANSWER

- **Are the units correct?** Force is measured in newtons.
- **Does the sign make sense?** Only the magnitude of the force needed to rotate the wrench clockwise is calculated.

PRACTICE PROBLEMS

Do additional problems. Online Practice

11. Consider the wrench in Example Problem 1. What force is needed if it is applied to the wrench pointing perpendicular to the wrench?
12. If a torque of 55.0 N•m is required to turn a bolt and the largest force you can exert is 135 N, how long a lever arm must you use to turn the bolt?
13. You have a 0.234-m-long wrench. A job requires a torque of 32.4 N•m, and you can exert a force of 232 N.
    a. What is the smallest angle, with respect to the handle of the wrench, at which you can pull on the wrench and get the job done?
    b. A friend can exert 275 N. What is the smallest angle she can use to accomplish the job?
14. You stand on a bicycle pedal, as shown in **Figure 7.** Your mass is 65 kg. If the pedal makes an angle of 35° above the horizontal and the pedal is 18 cm from the center of the chain ring, how much torque would you exert?
15. **CHALLENGE** If the pedal in the previous problem is horizontal, how much torque would you exert? How much torque would you exert when the pedal is vertical?

Figure 7

Richard Hutchings/Digital Light Source

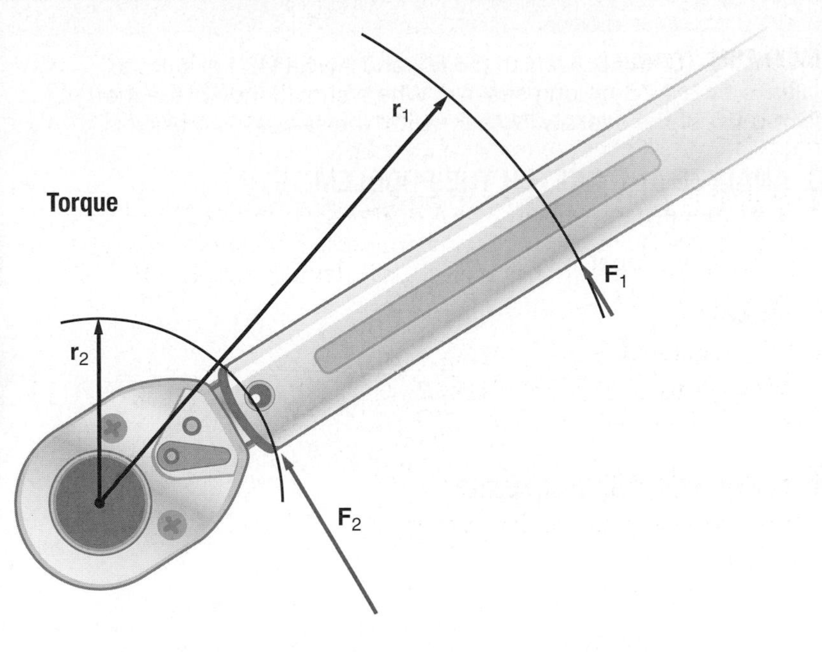

Figure 8 This worker uses a long wrench because it requires him to exert less force to tighten and loosen the nut. The wrench has a long lever arm, and less force is required if the force is applied farther from the axis of rotation (the center of the nut).

Finding Net Torque

Figure 8 shows a practical application of increasing torque to make a task easier. For another example of torque, try the following investigation. Collect two pencils, some coins, and some transparent tape. Tape two identical coins to the ends of the pencil and balance it on the second pencil, as shown in **Figure 9.** Each coin exerts a torque that is equal to the distance from the balance point to the center of the coin (r) times its weight (F_g), as follows:

$$\tau = rF_g$$

But the torques are equal and opposite in direction. Thus, the net torque is zero:

$$\tau_1 - \tau_2 = 0$$
or
$$r_1F_{g1} - r_2F_{g2} = 0$$

How can you make the pencil rotate? You could add a second coin on top of one of the two coins, thereby making the two forces different. You also could slide the balance point toward one end or the other of the pencil, thereby making the two lever arms of different length.

PhysicsLABs

LEVERAGE

FORENSICS LAB Can a person use a simple lever to open a heavy locked door?

TORQUES

Can you measure forces that produce torque?

View an **animation of net torque.**

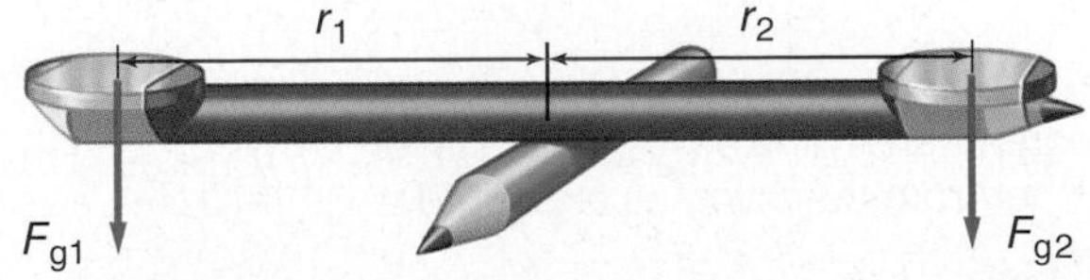

Figure 9 The torque exerted by the first coin ($F_{g1}r_1$) is equal and opposite in direction to the torque exerted by the second coin ($F_{g2}r_2$) when the pencil is balanced.

EXAMPLE PROBLEM 2

Find help with **isolating a variable.** Math Handbook

BALANCING TORQUES Kariann (56 kg) and Aysha (43 kg) want to balance on a 1.75-m-long seesaw. Where should they place the pivot point of the seesaw? Assume that the seesaw is massless.

1 ANALYZE AND SKETCH THE PROBLEM

- Sketch the situation.
- Draw and label the vectors.

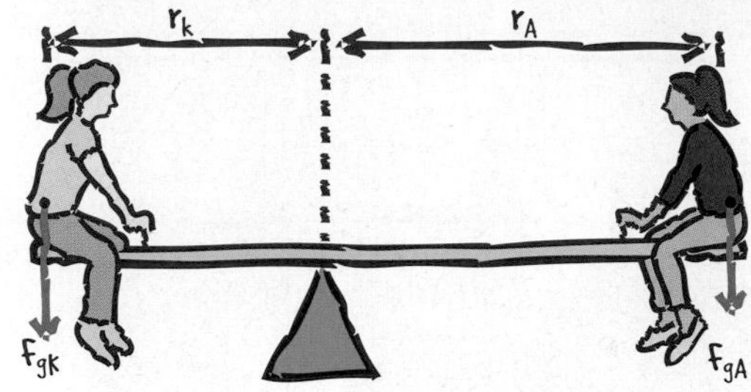

KNOWN	**UNKNOWN**
$m_K = 56$ kg	$r_K = ?$
$m_A = 43$ kg	$r_A = ?$
$r_K + r_A = 1.75$ m	

2 SOLVE FOR THE UNKNOWN

Find the two forces.

Kariann:

$$F_{gK} = m_K g$$
$$= (56\text{ kg})(9.8\text{ N/kg})$$
$$= 5.5{\times}10^2\text{ N}$$

◀ Substitute the known values into the equation: $m_K = 56$ kg, $g = 9.8$ N/kg.

Aysha:

$$F_{gA} = m_A g$$
$$= (43\text{ kg})(9.8\text{ N/kg})$$
$$= 4.2{\times}10^2\text{ N}$$

◀ Substitute the known values into the equation: $m_A = 43$ kg, $g = 9.8$ N/kg.

Define Kariann's distance in terms of the length of the seesaw and Aysha's distance.

$$r_K = 1.75\text{ m} - r_A$$

When there is no rotation, the sum of the torques is zero.

$$F_{gK}r_K = F_{gA}r_A$$
$$F_{gK}r_K - F_{gA}r_A = 0.0\text{ N·m}$$
$$F_{gK}(1.75\text{ m} - r_A) - F_{gA}r_A = 0.0\text{ N·m}$$

◀ Substitute the relationship between Kariann's distance in terms of Aysha's distance in to the equation: $r_K = 1.75$ m $- r_A$.

Solve for r_A.

$$F_{gK}(1.75\text{ m}) - F_{gK}(r_A) - F_{gA}r_A = 0.0\text{ N·m}$$
$$F_{gK}r_A + F_{gA}r_A = F_{gK}(1.75\text{ m})$$
$$(F_{gK} + F_{gA})r_A = F_{gK}(1.75\text{ m})$$
$$r_A = \frac{F_{gK}(1.75\text{ m})}{(F_{gK} + F_{gA})}$$
$$= \frac{(5.5{\times}10^2\text{ N})(1.75\text{ m})}{(5.5{\times}10^2\text{ N} + 4.2{\times}10^2\text{ N})}$$
$$= 0.99\text{ m}$$

◀ Substitute $F_{gK} = 5.5{\times}10^2$ N and $F_{gA} = 4.2{\times}10^2$ N.

3 EVALUATE THE ANSWER

- **Are the units correct?** Distance is measured in meters.
- **Do the signs make sense?** Distances are positive.
- **Is the magnitude realistic?** Aysha is about 1 m from the center, so Kariann is about 0.75 m away from it. Because Kariann's weight is greater than Aysha's weight, the lever arm on Kariann's side should be shorter. Aysha is farther from the pivot, as expected.

PRACTICE PROBLEMS

Do additional problems. 

16. Ashok, whose mass is 43 kg, sits 1.8 m from a pivot at the center of a seesaw. Steve, whose mass is 52 kg, wants to seesaw with Ashok. How far from the center of the seesaw should Steve sit?
17. A bicycle-chain wheel has a radius of 7.70 cm. If the chain exerts a 35.0-N force on the wheel in the clockwise direction, what torque is needed to keep the wheel from turning?
18. Two stationary baskets of fruit hang from strings going around pulleys of different diameters, as shown in **Figure 10.** What is the mass of basket A?

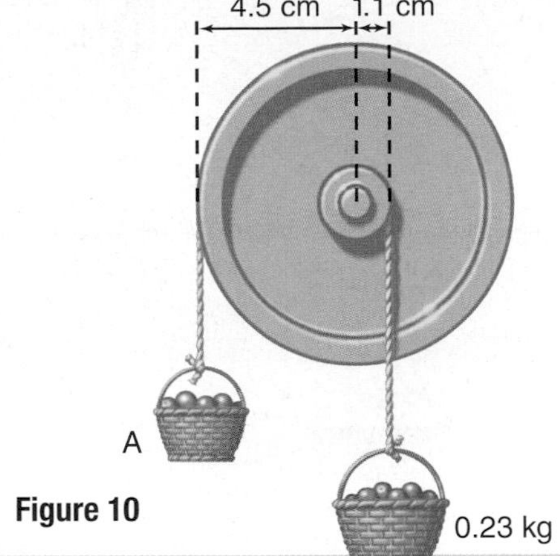

Figure 10

19. Suppose the radius of the larger pulley in problem 18 was increased to 6.0 cm. What is the mass of basket A now?
20. **CHALLENGE** A bicyclist, of mass 65.0 kg, stands on the pedal of a bicycle. The crank, which is 0.170 m long, makes a 45.0° angle with the vertical, as shown in **Figure 11.** The crank is attached to the chain wheel, which has a radius of 9.70 cm. What force must the chain exert to keep the wheel from turning?

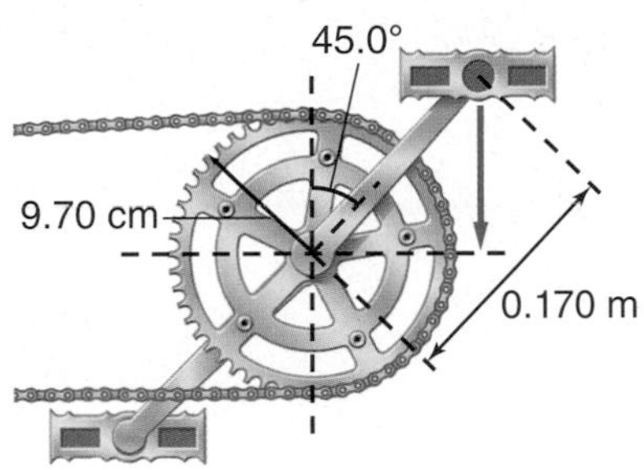

Figure 11

The Moment of Inertia

If you exert a force on a point mass, its acceleration will be inversely proportional to its mass. How does an extended object rotate when a torque is exerted on it? To observe firsthand, recover the pencil, the coins, and the transparent tape that you used earlier in this chapter. First, tape the coins at the ends of the pencil. Hold the pencil between your thumb and forefinger, and wiggle it back and forth. Take note of the forces that your thumb and forefinger exert. These forces create torques that change the angular velocity of the pencil and coins.

Now move the coins so that they are only 1 or 2 cm apart. Wiggle the pencil as before. Did you notice that the pencil is now easier to rotate? The torque that was required was much less this time. The mass of an object is not the only factor that determines how much torque is needed to change its angular velocity; the distribution or location of the mass also is important.

The resistance to rotation is called the **moment of inertia,** which is represented by the symbol I and has units of mass times the square of the distance. For a point object located at a distance (r) from the axis of rotation, the moment of inertia is given by the following equation.

MOMENT OF INERTIA OF A POINT MASS

The moment of inertia of a point mass is equal to the mass of the object times the square of the object's distance from the axis of rotation.

$$I = mr^2$$

MiniLAB

BALANCING TORQUES

Can you find the equilibrium point on a beam?

iLab Station

Table 2 Moments of Inertia for Various Objects

Object	Location of Axis	Diagram	Moment of Inertia
Thin hoop of radius r	through central diameter	Axis, r	mr^2
Solid, uniform cylinder of radius r	through center	Axis, r	$\left(\frac{1}{2}\right)mr^2$
Uniform sphere of radius r	through center	Axis, r	$\left(\frac{2}{5}\right)mr^2$
Long, uniform rod of length l	through center	Axis, l	$\left(\frac{1}{12}\right)ml^2$
Long, uniform rod of length l	through end	Axis, l	$\left(\frac{1}{3}\right)ml^2$
Thin, rectangular plate of length l and width w	through center	Axis, l, w	$\left(\frac{1}{12}\right)m(l^2 + w^2)$

Moment of inertia and mass As you have seen, the moments of inertia for extended objects, such as the pencil and coins, depend on how far the masses are from the axis of rotation. A bicycle wheel, for example, has almost all of its mass in the rim and tire. Its moment of inertia about its axle is almost exactly equal to mr^2, where r is the radius of the wheel. For most objects, however, the mass is distributed closer to the axis so the moment of inertia is less than mr^2. For example, as shown in **Table 2,** for a solid cylinder of radius r, $I = \left(\frac{1}{2}\right)mr^2$, while for a solid sphere, $I = \left(\frac{2}{5}\right)mr^2$.

☑ **READING CHECK** **Write** the equation for the moment of inertia of a hoop.

Moment of inertia and rotational axis The moment of inertia also depends on the location and direction of the rotational axis, as illustrated in **Figure 12.** To observe this firsthand, hold a book in the upright position by placing your hands at the bottom of the book. Feel the torque needed to rock the book toward you and then away from you. Now put your hands in the middle of the book and feel the torque needed to rock the book toward you and then away from you. Note that much less torque is needed when your hands are placed in the middle of the book because the average distance of the book's mass from the rotational axis is much less in this case.

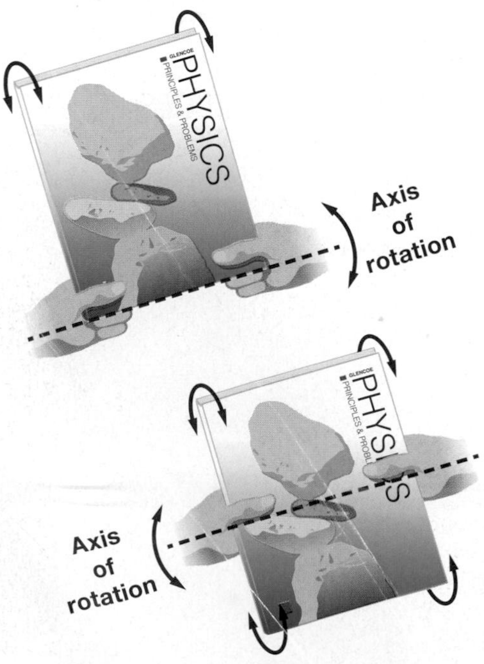

Figure 12 The moment of inertia of a book depends on the axis of rotation. The moment of inertia of the book on the top is larger than the moment of inertia of the book on the bottom because the average distance of the book's mass from the rotational axis is larger.

Identify which book requires more torque to rotate it and why.

EXAMPLE PROBLEM 3

Find help with **significant figures.** Math Handbook

MOMENT OF INERTIA A simplified model of a twirling baton is a thin rod with two round objects at each end. The length of the baton is 0.66 m, and the mass of each object is 0.30 kg. Find the moment of inertia of the baton as it is rotated about an axis at the midpoint between the round objects and perpendicular to the rod. What is the moment of inertia of the baton if the axis is moved to one end of the rod? Which is greater? The mass of the rod is negligible compared to the masses of the objects at the ends.

1 ANALYZE AND SKETCH THE PROBLEM

Sketch the situation. Show the baton with the two different axes of rotation and the distances from the axes of rotation to the masses.

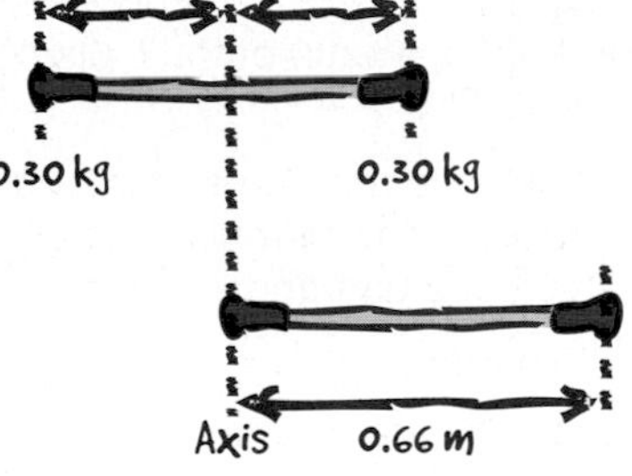

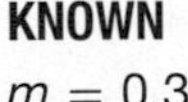

KNOWN	**UNKNOWN**
$m = 0.30$ kg	$I = ?$
$l = 0.66$ m	

2 SOLVE FOR THE UNKNOWN

Calculate the moment of inertia of each mass separately.

Rotating about the center of the rod:

$$r = \left(\frac{1}{2}\right)l$$

$$= \left(\frac{1}{2}\right)(0.66\text{ m})$$ ◀ Substitute the known value, $l = 0.66$ m, into the equation.

$$= 0.33\text{ m}$$

$$I_{\text{single mass}} = mr^2$$

$$= (0.30\text{ kg})(0.33\text{ m})^2$$ ◀ Substitute $m = 0.30$ kg and $r = 0.33$ m into the equation.

$$= 0.033\text{ kg}\cdot\text{m}^2$$

Find the moment of inertia of the baton.

$$I = 2I_{\text{single mass}}$$

$$= 2(0.033\text{ kg}\cdot\text{m}^2)$$ ◀ Substitute $I_{\text{single mass}} = 0.033$ kg•m² into the equation.

$$= 0.066\text{ kg}\cdot\text{m}^2$$

Rotating about one end of the rod:

$$I_{\text{single mass}} = mr^2$$

$$= (0.30\text{ kg})(0.66\text{ m})^2$$ ◀ Substitute $m = 0.30$ kg and $r = 0.66$ m into the equation.

$$= 0.13\text{ kg}\cdot\text{m}^2$$

The $I_{\text{single mass}} = mr^2$

$$= (0.30\text{ kg})(0.0\text{ m})^2$$

$= 0$ for the other mass.

Find the moment of inertia of the baton.

$$I = I_{\text{single mass}} + 0$$

$$= 0.13\text{ kg}\cdot\text{m}^2$$

The moment of inertia is greater when the baton is swung around one end.

3 EVALUATE THE ANSWER

- **Are the units correct?** Moment of inertia is measured in kg•m^2.
- **Is the magnitude realistic?** Masses and distances are small, and so are the moments of inertia. Doubling the distance increases the moment of inertia by a factor of 4. Thus, doubling the distance increases the moment of inertia more than having only one mass decreases the moment of inertia.

PRACTICE PROBLEMS

Do additional problems. Online Practice

PRACTICE PROBLEMS

21. Two children of equal masses sit 0.3 m from the center of a seesaw. Assuming that their masses are much greater than that of the seesaw, by how much is the moment of inertia increased when they sit 0.6 m from the center? Ignore the moment of inertia for the seesaw.

22. Suppose there are two balls with equal diameters and masses. One is solid, and the other is hollow, with all its mass distributed at its surface. Are the moments of inertia of the balls equal? If not, which is greater?

23. Calculate the moments of inertia for each object below using the formulas in **Table 2.** Each object has a radius of 2.0 m and a mass of 1.0 kg.

a. a thin hoop

b. a solid, uniform cylinder

c. a solid, uniform sphere

24. CHALLENGE Figure 13 shows three equal-mass spheres on a rod of very small mass. Consider the moment of inertia of the system, first when it is rotated about sphere A and then when it is rotated about sphere C.

a. Are the moments of inertia the same or different? Explain. If the moments of inertia are different, in which case is the moment of inertia greater?

b. Each sphere has a mass of 0.10 kg. The distance between spheres A and C is 0.20 m. Find the moment of inertia in the following instances: rotation about sphere A, rotation about sphere C.

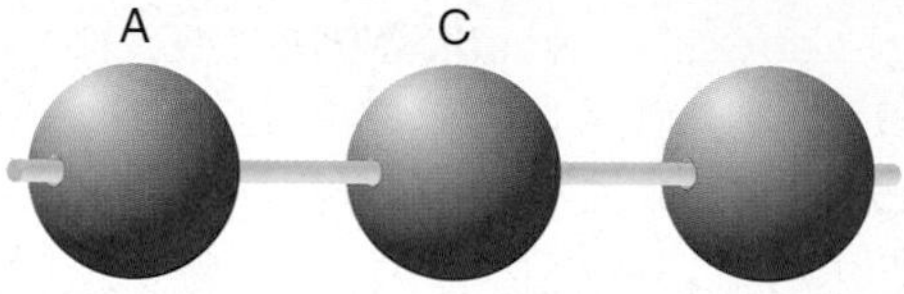

Figure 13

Newton's Second Law for Rotational Motion

Newton's second law for linear motion is expressed as $a = \frac{F_{net}}{m}$. If you rewrite this equation to represent rotational motion, acceleration is replaced by angular acceleration (α) force is replaced by net torque (τ_{net}) and mass is replaced by moment of inertia (I). Angular acceleration is directly proportional to the net torque and inversely proportional to the moment of inertia as stated in **Newton's second law for rotational motion.** This law is expressed by the following equation.

NEWTON'S SECOND LAW FOR ROTATIONAL MOTION

The angular acceleration of an object about a particular axis equals the net torque on the object divided by the moment of inertia.

$$\alpha = \frac{\tau_{net}}{I}$$

If the torque on an object and the angular velocity of that object are in the same direction, then the angular velocity of the object increases. If the torque and angular velocity are in different directions, then the angular velocity decreases.

PHYSICS CHALLENGE

Moments of Inertia Rank the objects shown in the diagram from least to greatest according to their moments of inertia about the indicated axes. All spheres have equal masses and all separations are the same. Assume that the rod's mass is negligible.

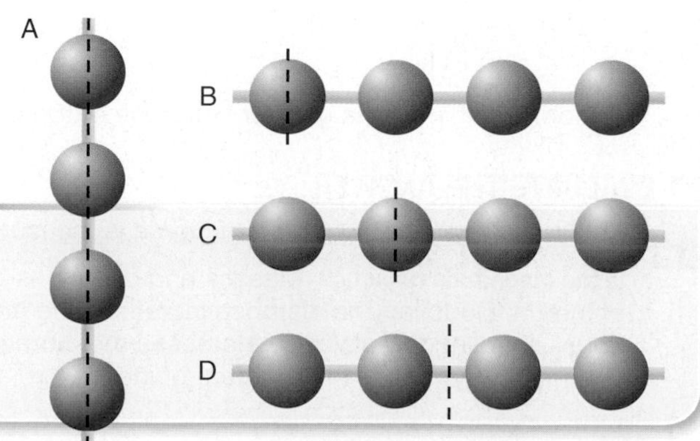

EXAMPLE PROBLEM 4

Get help with **torque.** Personal Tutor

TORQUE A solid steel wheel is free to rotate about a motionless central axis. It has a mass of 15 kg and a diameter of 0.44 m and starts at rest. You want to increase this wheel's rotation about its central axis to 8.0 rev/s in 15 s.

a. What torque must be applied to the wheel?

b. If you apply the torque by wrapping a strap around the outside of the wheel, how much force should you exert on the strap?

1 ANALYZE AND SKETCH THE PROBLEM

Sketch the situation. The torque must be applied in a counterclockwise direction; force must be exerted as shown.

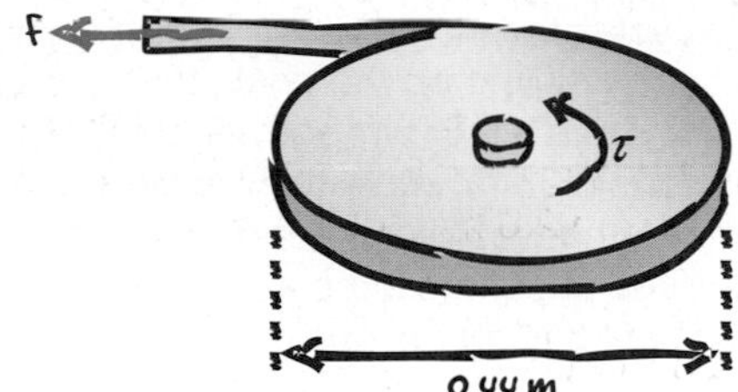

KNOWN	UNKNOWN
$m = 15$ kg	$\alpha = ?$
$r = \left(\frac{1}{2}\right)(0.44\text{ m}) = 0.22$ m	$I = ?$
$\omega_i = 0.0$ rad/s	$\tau = ?$
$\omega_f = 2\pi(8.0\text{ rev/s})$	$F = ?$
$t = 15$ s	

2 SOLVE FOR THE UNKNOWN

a. Solve for angular acceleration.

$$\alpha = \frac{\Delta\omega}{\Delta t}$$

$$= \frac{(16\pi\text{ rad/s} - (0.0\text{ rad/s}))}{15\text{ s}}$$ ◀ Substitute $\omega_f = 16\pi$ rad/s and $\omega_i = 0.0$ rad/s into the equation.

$$= 3.4\text{ rad/s}^2$$

Solve for the moment of inertia.

$$I = \left(\frac{1}{2}\right)mr^2$$

$$= \left(\frac{1}{2}\right)(15\text{ kg})(0.22\text{ m})^2$$ ◀ Substitute $m = 15$ kg and $r = 0.22$ m into the equation.

$$= 0.36\text{ kg·m}^2$$

Solve for torque.

$$\tau = I\alpha$$

$$= (0.36\text{ kg·m}^2)(3.4\text{ rad/s}^2)$$ ◀ Substitute $I = 0.36$ kg·m^2 and $\alpha = 3.4$ rad/s^2 into the equation.

$$= 1.2\text{ kg·m}^2\text{/s}^2$$

$$= 1.2\text{ N·m}$$

b. Solve for force.

$$\tau = Fr$$

$$F = \frac{\tau}{r}$$

$$= \frac{(1.2\text{ N·m})}{(0.22\text{ m})}$$ ◀ Substitute $\tau = 1.2$ N·m and $r = 0.22$ m into the equation.

$$= 5.5\text{ N}$$

3 EVALUATE THE ANSWER

- **Are the units correct?** Torque is measured in N·m and force is measured in N.
- **Is the magnitude realistic?** Despite its large mass, the small size of the wheel makes it relatively easy to spin.

EXAMPLE PROBLEM

PRACTICE PROBLEMS

Do additional problems. Online Practice

PRACTICE PROBLEMS

25. Consider the wheel in Example Problem 4. If the force on the strap were twice as great, what would be the angular frequency of the wheel after 15 s?

26. A solid wheel accelerates at 3.25 rad/s² when a force of 4.5 N exerts a torque on it. If the wheel is replaced by a wheel which has all of its mass on the rim, the moment of inertia is given by $I = mr^2$. If the same angular velocity were desired, what force should be exerted on the strap?

27. A bicycle wheel on a repair bench can be accelerated either by pulling on the chain that is on the gear or by pulling on a string wrapped around the tire. The tire's radius is 0.38 m, while the radius of the gear is 0.14 m. What force would you need to pull on the string to produce the same acceleration you obtained with a force of 15 N on the chain?

28. The bicycle wheel in the previous problem is used with a smaller gear whose radius is 0.11 m. The wheel can be accelerated either by pulling on the chain that is on the gear or by pulling a string that is wrapped around the tire. If you obtained the needed acceleration with a force of 15 N on the chain, what force would you need to exert on the string?

29. A chain is wrapped around a pulley and pulled with a force of 16.0 N. The pulley has a radius of 0.20 m. The pulley's rotational speed increases from 0.0 to 17.0 rev/min in 5.00 s. What is the moment of inertia of the pulley?

30. CHALLENGE A disk with a moment of inertia of 0.26 kg•m² is attached to a smaller disk mounted on the same axle. The smaller disk has a diameter of 0.180 m and a mass of 2.5 kg. A strap is wrapped around the smaller disk, as shown in **Figure 14.** Find the force needed to give this system an angular acceleration of 2.57 rad/s².

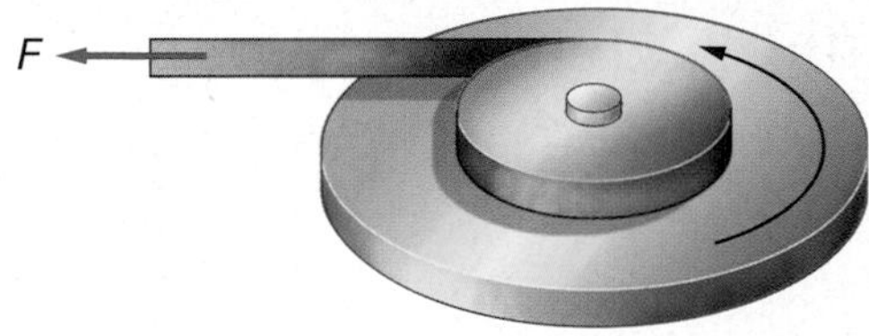

Figure 14

SECTION 2 REVIEW

Section Self-Check Check your understanding.

31. MAINIDEA Vijesh enters a revolving door that is not moving. Explain where and how Vijesh should push to produce a torque with the least amount of force.

32. Lever Arm You open a door by pushing at a right angle to the door. Your friend pushes at the same place, but at an angle of 55° from the perpendicular. If both you and your friend exert the same torque on the door, how do the forces you and your friend applied compare?

33. The solid wheel, shown in **Figure 15,** has a mass of 5.2 kg and a diameter of 0.55 m. It is at rest, and you need it to rotate at 12 rev/s in 35 s.

a. What torque do you need to apply to the wheel?

b. If a nylon strap is wrapped around the outside of the wheel, how much force do you need to exert on the strap?

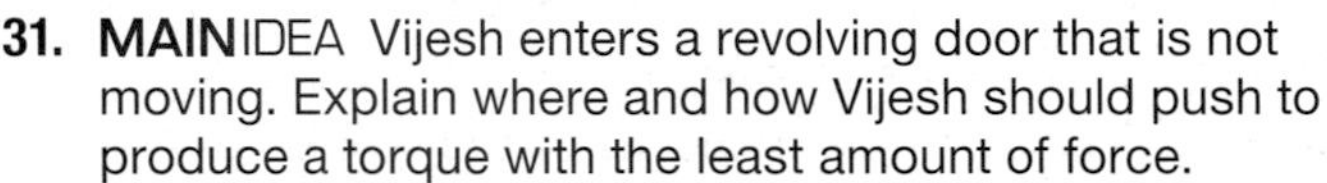

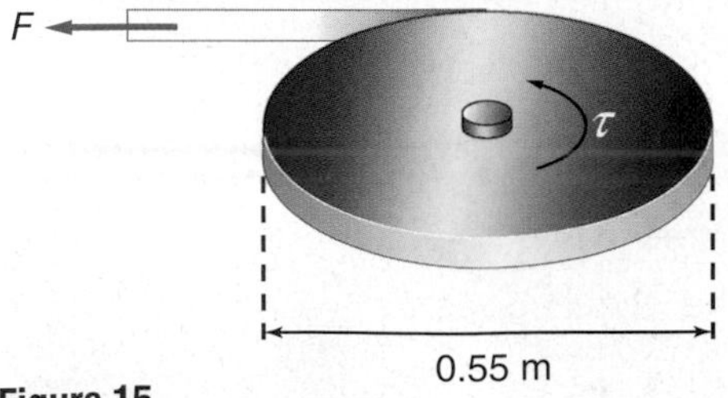

Figure 15

34. Net Torque Two people are pulling on ropes wrapped around the edge of a large wheel. The wheel has a mass of 12 kg and a diameter of 2.4 m. One person pulls in a clockwise direction with a 43-N force, while the other pulls in a counterclockwise direction with a 67-N force. What is the net torque on the wheel?

35. Moment of Inertia Refer to **Table 2,** and rank the moments of inertia from least to greatest of the following objects: a sphere, a wheel with almost all of its mass at the rim, and a solid disk. All have equal masses and diameters. Explain the advantage of using the object with the least moment of inertia.

36. Newton's Second Law for Rotational Motion A rope is wrapped around a pulley and pulled with a force of 13.0 N. The pulley's radius is 0.150 m. The pulley's rotational speed increases from 0.0 to 14.0 rev/min in 4.50 s. What is the moment of inertia of the pulley?

37. Critical Thinking A ball on an extremely low-friction, tilted surface will slide downhill without rotating. If the surface is rough, however, the ball will roll. Explain why, using a free-body diagram.

SECTION 3 Equilibrium

PHYSICS 4 YOU

This gymnast moves and shifts his body during his routine to control his movements. Here, he has increased his moment of inertia to help establish balance. At other times, he decreases his moment of inertia to facilitate rapid changes in position.

MAIN IDEA

An object in static equilibrium experiences a net force of zero and a net torque of zero.

Essential Questions

- What is center of mass?
- How does the location of the center of mass affect the stability of an object?
- What are the conditions for equilibrium?
- How do rotating frames of reference give rise to apparent forces?

Review Vocabulary

torque a measure of how effectively a force causes rotation; the magnitude is equal to the force times the lever arm

New Vocabulary

center of mass
centrifugal "force"
Coriolis "force"

The Center of Mass

Why are some vehicles more likely than others to roll over when involved in an accident? What causes a vehicle to roll over? The answers are important to the engineers who design safe vehicles. In this section, you will learn some of the factors that cause an object to tip over.

How does a freely moving object rotate around its center of mass? A wrench may spin about its handle or end-over-end. Does any single point on the wrench follow a straight path? **Figure 16** shows the path of the wrench. You can see that there is a single point whose path traces a straight line, as if the wrench could be replaced by a point particle at that location. The white dot in the photo represents this point. The point on the object that moves in the same way that a point particle would move is the **center of mass** of an object.

Locating the center of mass How can you locate the center of mass of an object? First, suspend the object from any point. When the object stops swinging, the center of mass is somewhere along the vertical line drawn from the suspension point. Draw the line. Then, suspend the object from another point. Again, the center of mass must be directly below this point. Draw a second vertical line. The center of mass is at the point where the two lines cross. The wrench and all other objects that are freely moving through space rotate about an axis that goes through their center of mass. Where would you think the center of mass of a person is located?

☑ **READING CHECK** **Paraphrase** the definition of the center of mass.

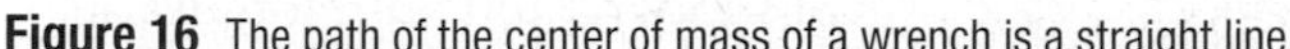

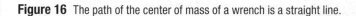

Figure 16 The path of the center of mass of a wrench is a straight line.

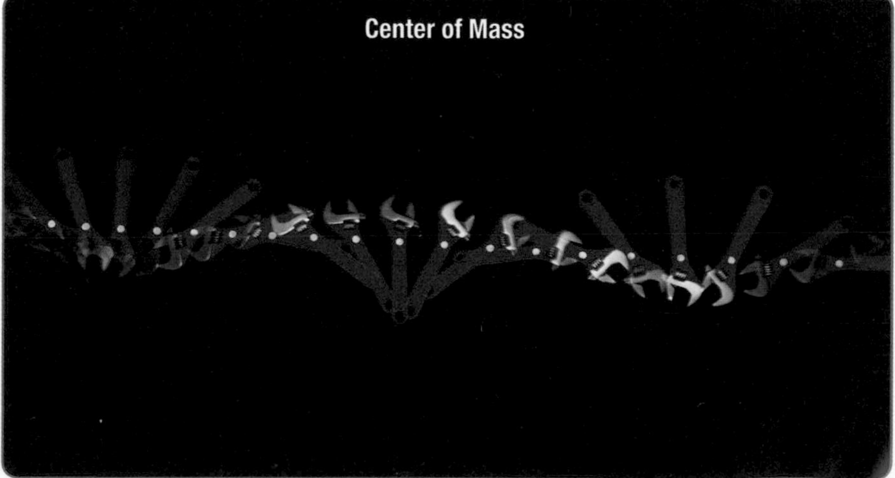

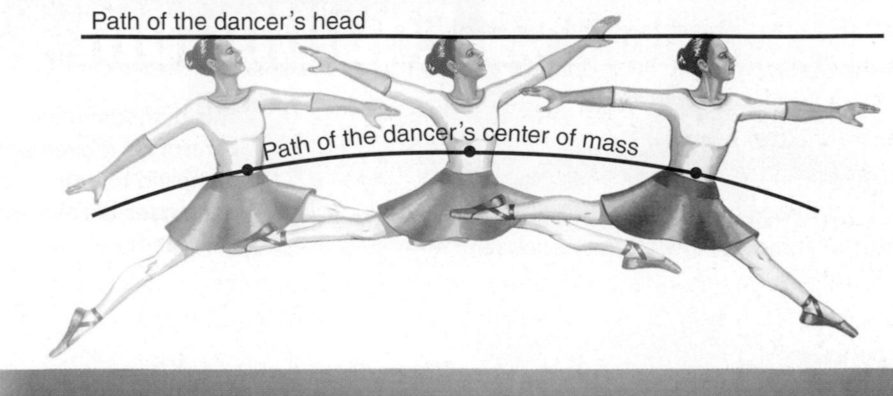

Figure 17 The upward motion of the ballet dancer's head is less than the upward motion of the center of mass. Thus, the head and torso move in a nearly horizontal path. This creates an illusion of floating.

The human body's center of mass For a person who is standing with her arms hanging straight down, the center of mass is a few centimeters below the navel, midway between the front and back of the person's body. The center of mass is farther below the navel for women than men, which often results in better balance for women than men. Because the human body is flexible, however, its center of mass is not fixed. If you raise your hands above your head, your center of mass rises 6 to 10 cm. A ballet dancer, for example, can appear to be floating on air by changing her center of mass in a leap. By raising her arms and legs while in the air, as shown in **Figure 17,** the dancer moves her center of mass up. The path of the center of mass is a parabola, but the dancer's head stays at almost the same height for a surprisingly long time.

REAL-WORLD PHYSICS

THE FOSBURY FLOP In high jumping, a technique called the Fosbury flop allows a high jumper to clear the bar when it is placed at the highest position. This is possible because the athlete's center of mass passes below the bar as he or she somersaults over the bar, with his or her back toward it.

Center of Mass and Stability

What factors determine whether a vehicle is stable or prone to roll over in an accident? To understand the problem, think about tipping over a box. A tall, narrow box standing on end tips more easily than a low, broad box. Why? To tip a box, as shown in **Figure 18,** you must rotate it about a corner. You pull at the top with a force ($\boldsymbol{F}$) applying a torque (τ_F). The weight of the box, acting on the center of mass ($\boldsymbol{F}_g$) applies an opposing torque (τ_w). When the center of mass is directly above the point of support, τ_w is zero. The only torque is the one applied by you. As the box rotates farther, its center of mass is no longer above its base of support, and both torques act in the same direction. At this point, the box tips over rapidly.

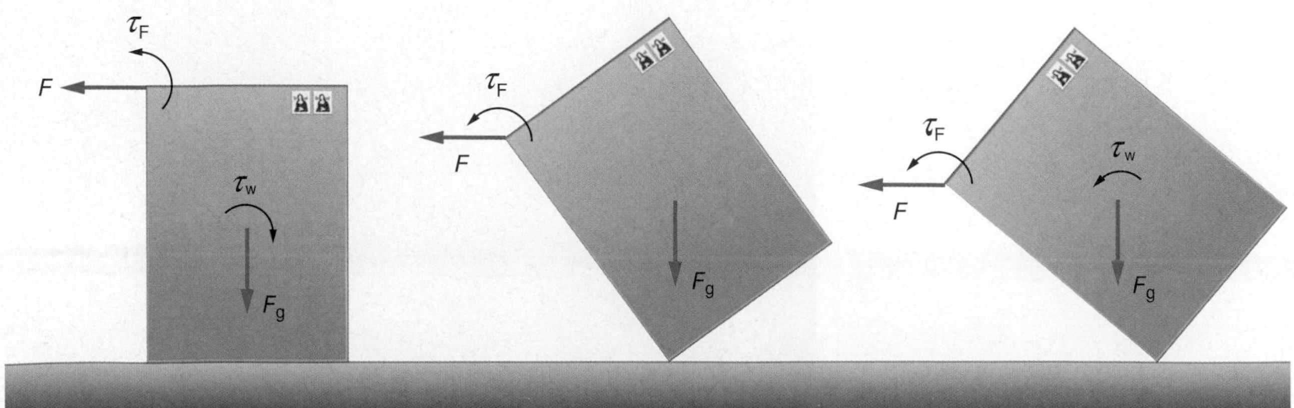

Figure 18 The curved arrows show the direction of the torque produced by the force exerted to tip over a box.

Stability An object is said to be stable if a large external force is required to tip it. The box in **Figure 18** is stable as long as the direction of the torque due to its weight (τ_w) tends to keep it upright. This occurs as long as the box's center of mass lies above its base. To tip the box over, you must rotate its center of mass around the axis of rotation until it is no longer above the base of the box. To rotate the box, you must lift its center of mass. The broader the base, the more stable the object is. Passengers on a city bus, for example, often stand with their feet spread apart to avoid toppling over as the bus starts and stops and weaves through traffic.

Why do vehicles roll over? **Figure 19** shows two vehicles about to roll over. Note that the one with the higher center of mass does not have to be tilted very far for its center of mass to be outside its base—its center of mass does not have to be raised as much as the other vehicle's center of mass. As demonstrated by the vehicles, the lower the location of an object's center of mass, the greater its stability.

You are stable when you stand flat on your feet. When you stand on tiptoe, however, your center of mass moves forward directly above the balls of your feet, and you have very little stability. In judo, aikido, and other martial arts, the fighter uses torque to rotate the opponent into an unstable position, where the opponent's center of mass does not lie above his or her feet. A small person can use torque, rather than force, to defend himself or herself against a stronger person.

In summary, if the center of mass is not located above the base of an object, it is unstable and will roll over without additional torque. If the center of mass is above the base of the object, it is stable. If the base of the object is very narrow and the center of mass is high, then the object might be stable, but the slightest force will cause it to tip over.

☑ **READING CHECK Describe** when an object is the most stable.

Conditions for Equilibrium

If your pen is at rest, what is needed to keep it at rest? You could either hold it up or place it on a desk or some other surface. An upward force must be exerted on the pen to balance the downward force of gravity. You must also hold the pen so that it will not rotate. An object is said to be in static equilibrium if both its velocity and angular velocity are zero or constant. Thus, for an object to be in static equilibrium, it must meet two conditions. First, it must be in translational equilibrium; that is, the net force exerted on the object must be zero. Second, it must be in rotational equilibrium; that is, the net torque exerted on the object must be zero.

MiniLAB

SPINNING TOPS

Where is a spinning object's center of mass?

View an **animation about stability.**

PhysicsLAB

EQUILIBRIUM

Can you make the net torque on scaffolding zero so that it will not rotate?

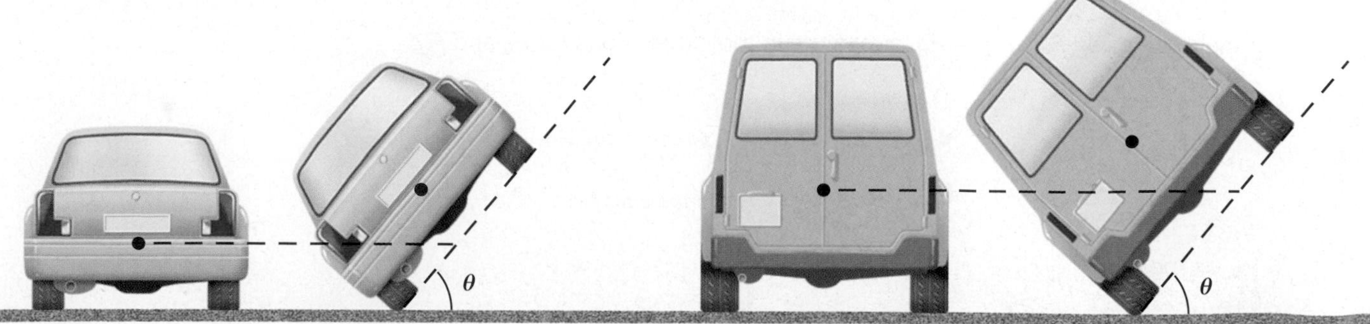

Figure 19 Larger vehicles have a higher center of mass than smaller ones. The higher the center of mass, the smaller the tilt needed to cause the vehicle's center of mass to move outside its base and cause the vehicle to roll over.

EXAMPLE PROBLEM 5

Find help with **isolating a variable.** Math Handbook

STATIC EQUILIBRIUM A 5.8-kg ladder, 1.80 m long, rests on two sawhorses. Sawhorse A is 0.60 m from one end of the ladder, and sawhorse B is 0.15 m from the other end of the ladder. What force does each sawhorse exert on the ladder?

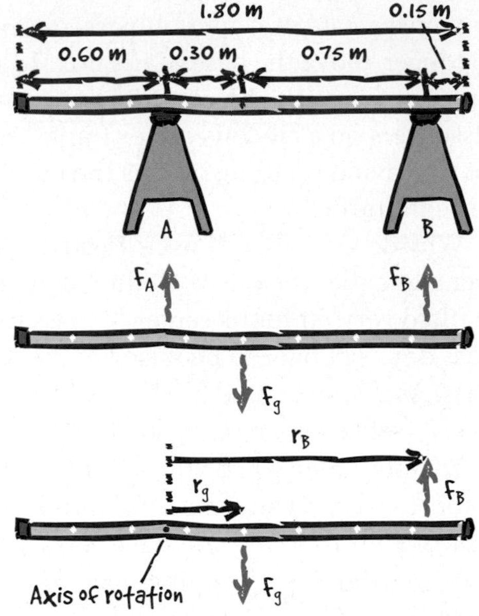

1 ANALYZE AND SKETCH THE PROBLEM

- Sketch the situation.
- Choose the axis of rotation at the point where F_A acts on the ladder. Thus, the torque due to F_A is zero.

KNOWN	UNKNOWN
$m = 5.8$ kg	$F_A = ?$
$l = 1.80$ m	$F_B = ?$
$l_A = 0.60$ m	
$l_B = 0.15$ m	

2 SOLVE FOR THE UNKNOWN

For a ladder that has a constant density, the center of mass is at the center rung.

The net force is the sum of all forces on the ladder.

The ladder is in translational equilibrium, so the net force exerted on it is zero.

$$F_{net} = F_A + F_B + (-F_g)$$

$$0.0 \text{ N} = F_A + F_B + (-F_g)$$ ◀ The ladder is in translational equilibrium, so the net force exerted on it is zero.

Solve for F_A.

$$F_A = F_g - F_B$$

Find the torques due to F_g and F_B.

$$\tau_g = -F_g r_g$$ ◀ τ_g is in the clockwise direction.

$$\tau_B = +F_B r_B$$ ◀ τ_B is in the counterclockwise direction.

The net torque is the sum of all torques on the object.

$$\tau_{net} = \tau_B + \tau_g$$

$$0.0 \text{ N·m} = \tau_B + \tau_g$$ ◀ The ladder is in rotational equilibrium, so $\tau_{net} = 0.0$ N•m.

$$\tau_B = -\tau_g$$

$$F_B r_B = F_g r_g$$ ◀ Substitute $\tau_B = r_B F_B$ and $\tau_g = -r_g F_g$ into the equation.

Solve for F_B.

$$F_B = \frac{F_g r_g}{r_B}$$

$$= \frac{r_g mg}{r_B}$$ ◀ Substitute $F_g = mg$

Using the expression $F_A = F_g - F_B$, substitute in the expressions for F_B and F_g.

$$F_A = F_g - F_B$$

$$= F_g - \frac{r_g mg}{r_B}$$ ◀ Substitute $F_B = \frac{r_g mg}{r_B}$ into the equation.

$$= mg - \frac{r_g mg}{r_B}$$ ◀ Substitute $F_g = mg$ into the equation.

$$= mg\left(1 - \frac{r_g}{r_B}\right)$$

EXAMPLE PROBLEM

Solve for r_g.

$$r_g = \left(\frac{1}{2}\right)l - l_A$$
$$= 0.90\text{ m} - 0.60\text{ m}$$
$$= 0.30\text{ m}$$

◀ For a ladder, which has a constant density, the center of mass is at the center rung.

◀ Substitute $\frac{l}{2}$ = 0.90 m and l_A = 0.60 m into the equation.

Solve for r_B.

$$r_B = (0.90\text{ m} - l_B) + (0.90\text{ m} - l_A)$$
$$= (0.90\text{ m} - 0.15\text{ m}) + (0.90\text{ m} - 0.60\text{ m})$$
$$= 0.75\text{ m} + 0.30\text{ m}$$
$$= 1.05\text{ m}$$

◀ Substitute l_B = 0.15 m and l_A = 0.60 m into the equation.

Calculate F_B.

$$F_B = \frac{r_g mg}{r_B}$$
$$= \frac{(0.30\text{ m})(5.8\text{ kg})(9.8\text{ N/kg})}{(1.05\text{ m})}$$
$$= 16\text{ N}$$

◀ Substitute r_g = 0.30 m, m = 5.8 kg, g = 9.8 N/kg, and r_B = 1.05 m into the equation.

Calculate F_A.

$$F_A = mg\left(1 - \frac{r_g}{r_B}\right)$$
$$= (5.8\text{ kg})(9.8\text{ N/kg})\left(1 - \frac{(0.30\text{ m})}{(1.05\text{ m})}\right)$$
$$= 41\text{ N}$$

◀ Substitute r_g = 0.30 m, m = 5.8 kg, g = 9.8 N/kg, and r_B = 1.05 m into the equation.

3 EVALUATE THE ANSWER

- **Are the units correct?** Forces are measured in newtons.
- **Do the signs make sense?** Both forces are upward.
- **Is the magnitude realistic?** The forces add up to the weight of the ladder, and the force exerted by the sawhorse closer to the center of mass is greater, which makes sense.

PRACTICE PROBLEMS

Do additional problems. Online Practice 

38. What would be the forces exerted by the two sawhorses if the ladder in Example Problem 5 had a mass of 11.4 kg?

39. A 7.3-kg ladder, 1.92 m long, rests on two sawhorses, as shown in **Figure 20.** Sawhorse A, on the left, is located 0.30 m from the end, and sawhorse B, on the right, is located 0.45 m from the other end. Choose the axis of rotation to be the center of mass of the ladder.

a. What are the torques acting on the ladder?

b. Write the equation for rotational equilibrium.

c. Solve the equation for F_A in terms of F_g.

d. How would the forces exerted by the two sawhorses change if A were moved very close to, but not directly under, the center of mass?

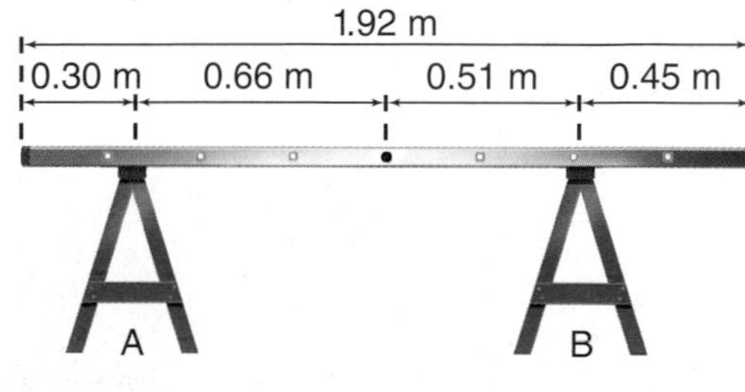

Figure 20

40. A 4.5-m-long wooden plank with a 24-kg mass is supported in two places. One support is directly under the center of the board, and the other is at one end. What are the forces exerted by the two supports?

41. **CHALLENGE** A 85-kg diver walks to the end of a diving board. The board, which is 3.5 m long with a mass of 14 kg, is supported at the center of mass of the board and at one end. What are the forces on the two supports?

PRACTICE PROBLEMS

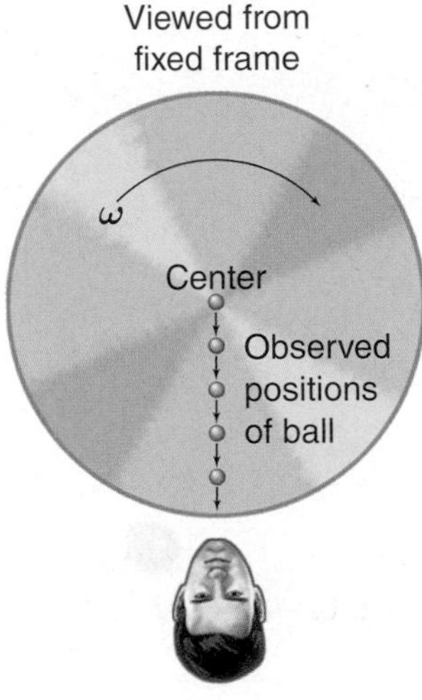

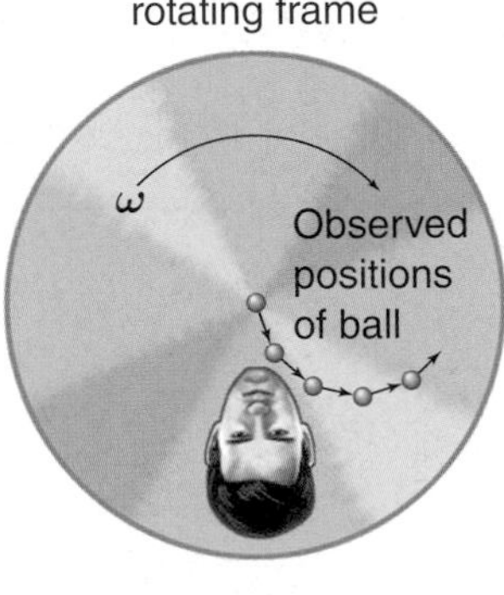

Figure 21 The Coriolis "force" is not a real force. It exists only in rotating reference frames.

Rotating Frames of Reference

When you sit on a spinning amusement-park ride, it feels as if a strong force is pushing you to the outside. A pebble on the floor of the ride would accelerate outward without a horizontal force being exerted on it in the same direction. The pebble would not move in a straight line relative to the floor of the ride. In other words, Newton's laws of motion would not seem to apply. This is because in the ride your point of view, called your frame of reference, is rotating. Newton's laws are valid only in non-rotating or nonaccelerated frames of reference.

Motion in a rotating reference frame is important to us because Earth rotates. The effects of the rotation of Earth are too small to be noticed in the classroom or lab, but they are significant influences on the motion of the atmosphere and, therefore, on climate and weather.

Centrifugal "Force"

Suppose you fasten one end of a spring to the center of a rotating platform. An object lies on the platform and is attached to the other end of the spring. As the platform rotates, an observer on the platform sees the object stretch the spring. The observer might think that some force toward the outside of the platform is pulling on the object. This apparent force seems to pull on a moving object but does not exert a physical outward push on it. This apparent force, which seems to push an object outward, is observed only in rotating frames of reference and is called the **centrifugal "force."** It is not a real force because there is no physical outward push on the object.

As the platform rotates, an observer on the ground would see things differently. This observer sees the object moving in a circle, and it accelerates toward the center because of the force of the spring. As you know, this centripetal acceleration is given by $a_c = \frac{v^2}{r}$. It also can be written in terms of angular velocity, as $a_c = \omega^2 r$. Centripetal acceleration is proportional to the distance from the axis of rotation and depends on the square of the angular velocity. Thus, if you double the rotational frequency, the angular acceleration increases by a factor of four.

☑ **READING CHECK** **Define** centrifugal force in your own words.

The Coriolis "Force"

A second effect of rotation is shown in **Figure 21.** Suppose a person standing at the center of a rotating disk throws a ball toward the edge of the disk. Consider the horizontal motion of the ball as seen by two observers, and ignore the vertical motion of the ball as it falls. An observer standing outside the disk sees the ball travel in a straight line at a constant speed toward the edge of the disk. However, the other observer, who is stationed on the disk and rotating with it, sees the ball follow a curved path at a constant speed. The apparent force that seems to deflect a moving object from its path and is observed only in rotating frames of reference is called the **Coriolis "force."** Like the centrifugal "force," the Coriolis "force" is not a real force. It seems to exist because we observe a deflection in horizontal motion when we are in a rotating frame of reference.

Coriolis "force" due to Earth's rotation

Suppose a cannon is fired from a point on the equator toward a target due north of it. If the projectile were fired directly northward, it would also have an eastward velocity component because of the rotation of Earth. Recall that Earth is actually rotating beneath the projectile.

This eastward speed is greater at the equator than at any other latitude. Thus, as the projectile moves northward, it also moves eastward faster than points on Earth below it do. The result is that the projectile lands east of the target as shown in **Figure 22.**

READING CHECK **Describe** how you would aim a projectile to compensate for the rotation of Earth.

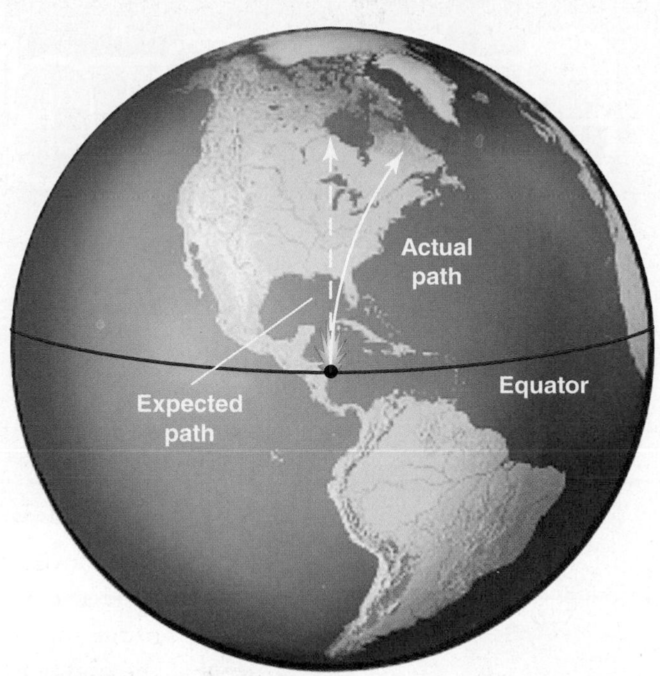

Figure 22 An observer on Earth sees the Coriolis "force" cause a projectile fired due north to deflect to the right of the intended target.

While an observer in space would see Earth's rotation, an observer on Earth could claim that the projectile missed the target because of the Coriolis "force" on the rocket. Note that for objects moving toward the equator, the direction of the apparent force is westward. A projectile will land west of the target when fired due south.

CONNECTION TO EARTH SCIENCE The direction of winds around high- and low-pressure areas results from the Coriolis "force." Winds flow from areas of high to low pressure. Because of the Coriolis "force" in the northern hemisphere, winds from the south go to the east of low-pressure areas. Winds from the north end up west of low-pressure areas. Therefore, winds rotate counter-clockwise around low-pressure areas in the northern hemisphere. In the southern hemisphere, however, winds rotate clockwise around low-pressure areas. This is why tropical cyclones, or hurricanes as they are also called, rotate clockwise in the southern hemisphere and counterclockwise in the northern hemisphere.

SECTION 3 REVIEW

42. **MAIN**IDEA Give an example of an object for each of the following conditions.

- **a.** rotational equilibrium, but not translational equilibrium
- **b.** translational equilibrium, but not rotational equilibrium

43. **Center of Mass** Can the center of mass of an object be located in an area where the object has no mass? Explain.

44. **Stability of an Object** Why is a modified vehicle with its body raised high on risers less stable than a similar vehicle with its body at normal height?

45. **Center of Mass** Where is the center of mass on a roll of masking tape?

46. **Locating the Center of Mass** Describe how you would find the center of mass of this textbook.

47. **Rotating Frames of Reference** A penny is placed on a rotating, old-fashioned record turntable. At the highest speed, the penny starts sliding outward. What are the forces acting on the penny?

48. **Critical Thinking** You have read about how the spin of Earth on its axis affects the winds. Predict the direction of the flow of surface ocean currents in the northern and southern hemispheres.

Spin-Cycle

CENTRIFUGES

Powerful physics is at work in devices known as laboratory centrifuges.

One of the main uses of laboratory centrifuges is separating blood into its components for analysis. Some labs also use centrifuges to separate DNA and proteins from the samples.

Centrifuges must be balanced with a sample on one side offset by a sample or a blank on the other. Much like a washing machine, an out-of-balance centrifuge can quickly malfunction.

Laboratory centrifuges can spin very quickly, up to thousands of revolutions per minute (rpm), producing accelerations of hundreds or thousands of times the free-fall acceleration.

Centrifuges use a principle called sedimentation. The most dense parts of a mixture move toward the bottom of the sample tube, while the least dense portions remain at the top.

GOING FURTHER >>>

Research Interview a technician at a pathology laboratory or a student at a local college or university to learn how centrifuges are used in separating samples to be analyzed.

CHAPTER 8

STUDY GUIDE

BIGIDEA Applying a torque to an object causes a change in that object's angular velocity.

VOCABULARY
- **radian** *(p. 204)*
- **angular displacement** *(p. 204)*
- **angular velocity** *(p. 205)*
- **angular acceleration** *(p. 206)*

SECTION 1 Describing Rotational Motion

MAINIDEA Angular displacement, angular velocity, and angular acceleration all help describe angular motion.

- Angular displacement is the change in the angle (θ) as an object rotates. It is usually measured in degrees or radians.
- Average angular velocity is the object's angular displacement divided by the time taken to make the angular displacement. Average angular velocity is represented by the Greek letter omega (ω) and is determined by the following equation:

$$\omega = \frac{\Delta\theta}{\Delta t}$$

- Average angular acceleration is the change in angular velocity divided by the time required to make the change.

$$\alpha = \frac{\Delta\omega}{\Delta t}$$

VOCABULARY
- **lever arm** *(p. 209)*
- **torque** *(p. 209)*
- **moment of inertia** *(p. 213)*
- **Newton's second law for rotational motion** *(p. 216)*

SECTION 2 Rotational Dynamics

MAINIDEA Torques cause changes in angular velocity.

- Torque describes the combination of a force and a lever arm that can cause an object to rotate. Torque is represented by the Greek letter tau (τ) and is determined by the following equation:

$$\tau = Fr \sin \theta$$

- The moment of inertia is a point object's resistance to changes in angular velocity. The moment of inertia is represented by the letter I and for a point mass, it is represented by the following equation:

$$I = mr^2$$

- Newton's second law for rotational motion states that angular acceleration is directly proportional to the net torque and inversely proportional to the moment of inertia.

$$\alpha = \frac{\tau_{net}}{I}$$

VOCABULARY
- **center of mass** *(p. 219)*
- **centrifugal "force"** *(p. 224)*
- **Coriolis "force"** *(p. 224)*

SECTION 3 Equilibrium

MAINIDEA An object in static equilibrium experiences a net force of zero and a net torque of zero.

- The center of mass of an object is the point on the object that moves in the same way that a point particle would move.
- An object is stable against rollover if its center of mass is above its base.
- An object in equilibrium has no net force exerted on it and there is no net torque acting on it.
- Centrifugal "force" and the Coriolis "force" are two apparent, but nonexistent, forces that seem to exist when a object is analyzed from a rotating frame of reference.

Games and Multilingual eGlossary

Vocabulary Practice

Chapter Self-Check

SECTION 1 Describing Rotational Motion

Mastering Concepts

49. BIGIDEA A bicycle wheel rotates at a constant 25 rev/min. Is its angular velocity decreasing, increasing, or constant?

50. A toy rotates at a constant 5 rev/min. Is its angular acceleration positive, negative, or zero?

51. Do all parts of Earth move at the same rate? Explain.

52. A unicycle wheel rotates at a constant 14 rev/min. Is the total acceleration of a point on the tire inward, outward, tangential, or zero?

Mastering Problems

53. On a test stand a bicycle wheel is being rotated about its axle so that a point on the edge moves through 0.210 m. The radius of the wheel is 0.350 m, as shown in **Figure 23.** Through what angle (in radians) is the wheel rotated?

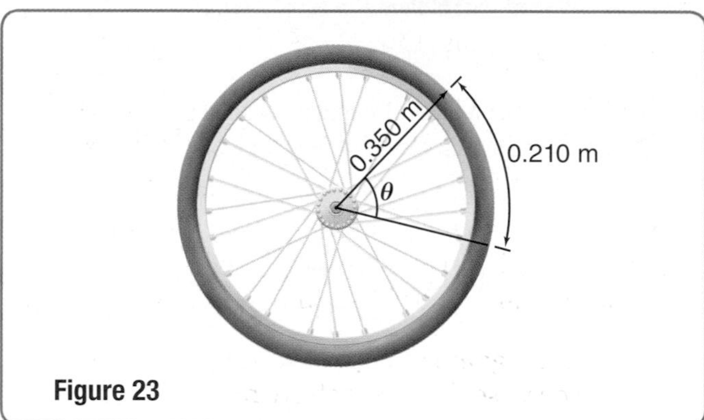

Figure 23

54. The outer edge of a truck tire that has a radius of 45 cm has a velocity of 23 m/s. What is the angular velocity of the tire in rad/s?

55. A steering wheel is rotated through 128°, as shown in **Figure 24.** Its radius is 22 cm. How far would a point on the steering wheel's edge move?

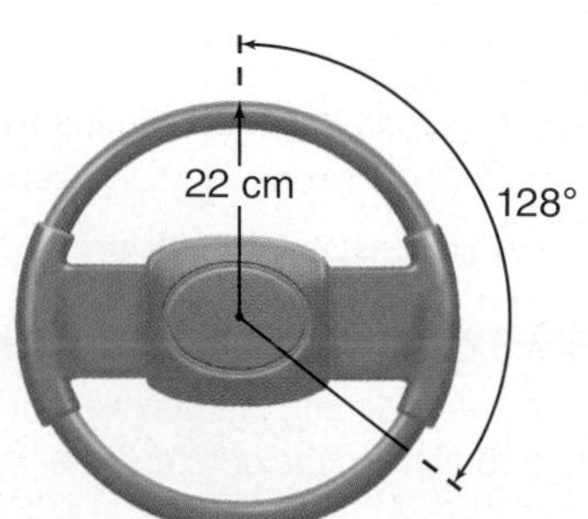

Figure 24

56. Propeller A propeller spins at 1880 rev/min.

a. What is its angular velocity in rad/s?

b. What is the angular displacement of the propeller in 2.50 s?

57. The propeller in the previous problem slows from 475 rev/min to 187 rev/min in 4.00 s. What is its angular acceleration?

58. The automobile wheel shown in **Figure 25** rotates at 2.50 rad/s. How fast does a point 7.00 cm from the center travel?

Figure 25

59. Washing Machine A washing machine's two spin cycles are 328 rev/min and 542 rev/min. The diameter of the drum is 0.43 m.

a. What is the ratio of the centripetal accelerations for the fast and slow spin cycles? Recall that $a_c = \frac{v^2}{r}$ and $v = r\omega$.

b. What is the ratio of the linear velocity of an object at the surface of the drum for the fast and slow spin cycles?

c. Find the maximum centripetal acceleration in terms of g for the washing machine.

60. A laboratory ultracentrifuge is designed to produce a centripetal acceleration of $0.35\times10^6\ g$ at a distance of 2.50 cm from the axis. What angular velocity in revolutions per minute is required?

SECTION 2 Rotational Dynamics

Mastering Concepts

61. Think about some possible rotations of your textbook. Are the moments of inertia about these three axes the same or different? Explain.

62. An auto repair manual specifies the torque needed to tighten bolts on an engine. Why does it not mention force?

63. Ranking Task Rank the torques on the five doors shown in **Figure 26** from least to greatest. Note that the magnitude of all the forces is the same.

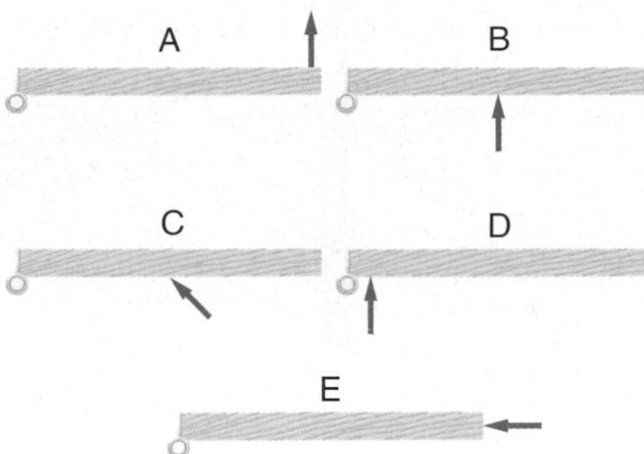

Figure 26

Mastering Problems

64. Wrench A bolt is to be tightened with a torque of 8.0 N•m. If you have a wrench that is 0.35 m long, what is the least amount of force you must exert?

65. What is the torque on a bolt produced by a 15-N force exerted perpendicular to a wrench that is 25 cm long, as shown in **Figure 27?**

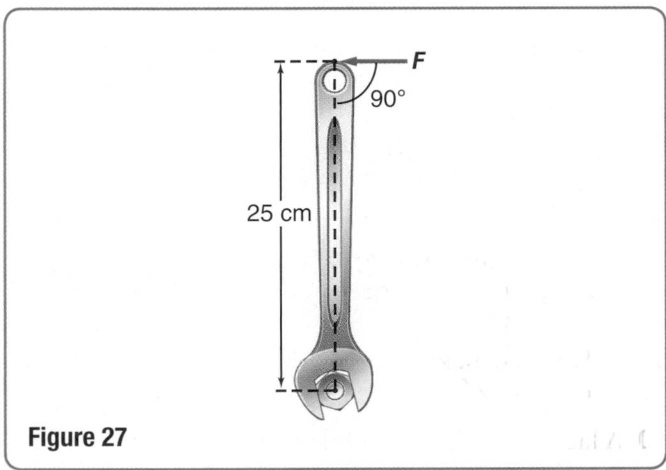

Figure 27

66. A bicycle wheel with a radius of 38 cm is given an angular acceleration of 2.67 rad/s^2 by applying a force of 0.35 N on the edge of the wheel. What is the wheel's moment of inertia?

67. Toy Top A toy top consists of a rod with a diameter of 8.0 mm and a disk of mass 0.0125 kg and a diameter of 3.5 cm. The moment of inertia of the rod can be neglected. The top is spun by wrapping a string around the rod and pulling it with a velocity that increases from zero to 3.0 m/s over 0.50 s.

a. What is the resulting angular velocity of the top?

b. What force was exerted on the string?

68. A toy consisting of two balls, each 0.45 kg, at the ends of a 0.46-m-long, thin, lightweight rod is shown in **Figure 28.** Find the moment of inertia of the toy. The moment of inertia is to be found about the center of the rod.

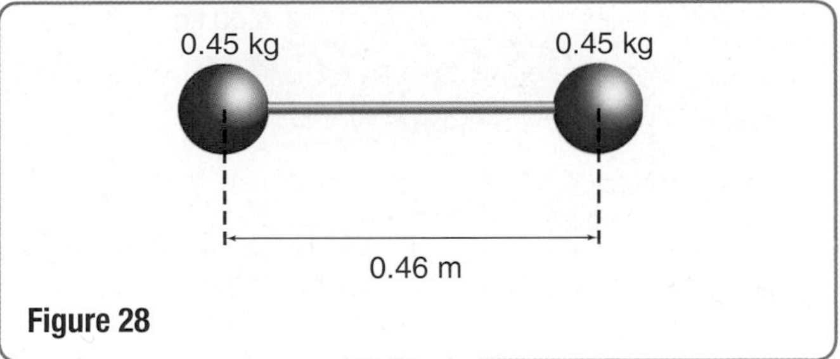

Figure 28

SECTION 3 Equilibrium

Mastering Concepts

69. To balance a car's wheel, it is placed on a vertical shaft and weights are added to make the wheel horizontal. Why is this equivalent to moving the center of mass until it is at the center of the wheel?

70. A stunt driver maneuvers a monster truck so that it is traveling on only two wheels. Where is the center of mass of the truck?

71. Suppose you stand flat-footed, then you rise and balance on tiptoe. If you stand with your toes touching a wall, you cannot balance on tiptoe. Explain.

72. Why does a gymnast appear to be floating on air when she raises her arms above her head in a leap?

73. Why is a vehicle with wheels that have a large diameter more likely to roll over than a vehicle with wheels that have a smaller diameter?

Mastering Problems

74. A 12.5-kg board, 4.00 m long, is being held up on one end by Ahmed. He calls for help in lifting the board, and Judi responds.

a. What is the least force that Judi could exert to lift the board to the horizontal position? What part of the board should she lift to exert this force?

b. What is the greatest force that Judi could exert to lift the board to the horizontal position? What part of the board should she lift to exert this force?

75. A car's specifications state that its weight distribution is 53 percent on the front tires and 47 percent on the rear tires. The wheel base is 2.46 m. Where is the car's center of mass?

Chapter Self-Check

76. Two people are holding up the ends of a 4.25-kg wooden board that is 1.75 m long. A 6.00-kg box sits on the board, 0.50 m from one end, as shown in **Figure 29.** What forces do the two people exert?

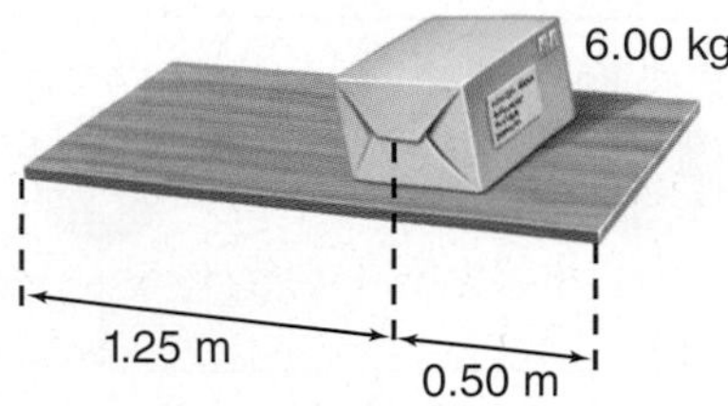

Figure 29

Applying Concepts

77. Two gears are in contact and rotating. One is larger than the other, as shown in **Figure 30.** Compare their angular velocities. Also compare the linear velocities of two teeth that are in contact.

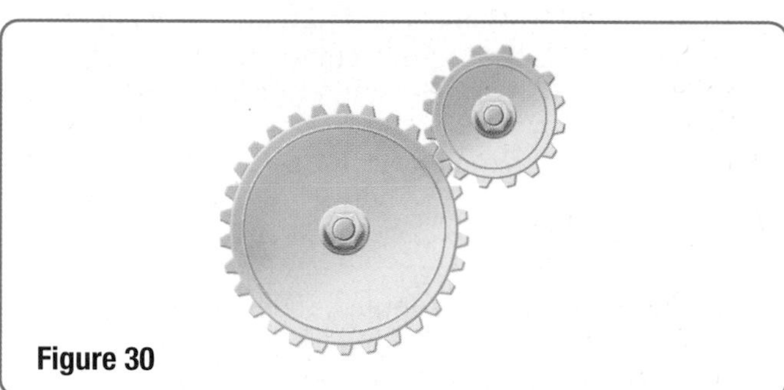
Figure 30

78. How can you experimentally find the moment of inertia of an object?

79. Bicycle Wheels Three bicycle wheels have masses that are distributed in three different ways: mostly at the rim, uniformly, and mostly at the hub. The wheels all have the same mass. If equal torques are applied to them, which one will have the greatest angular acceleration? Which one will have the least?

80. Bowling Ball When a bowling ball leaves a bowler's hand, it does not spin. After it has gone about half the length of the lane, however, it does spin. Explain how its rotation rate increased.

81. Flat Tire Suppose your car has a flat tire. You find a lug wrench to remove the nuts from the bolt studs. You cannot turn the nuts. Your friend suggests ways you might produce enough torque to turn them. What three ways might your friend suggest?

82. Why can you ignore forces that act on the axis of rotation of an object in static equilibrium when determining the net torque?

83. Tightrope Walkers Tightrope walkers often carry long poles that sag so that the ends are lower than the center as shown in **Figure 31.** How does a pole increase the tightrope walker's stability? *Hint: Consider both center of mass and moment of inertia.*

Figure 31

84. Merry-Go-Round While riding a merry-go-round, you toss a key to a friend standing on the ground. For your friend to be able to catch the key, should you toss it a second or two before you reach the spot where your friend is standing or wait until your friend is directly beside you? Explain.

85. In solving problems about static equilibrium, why is the axis of rotation often placed at a point where one or more forces are acting on the object?

86. Ranking Task Rank the objects in **Figure 32** according to their moments of inertia, from least to greatest. Each has a radius of (R) and total mass (M), which is uniformly distributed throughout the shaded region. Specifically indicate any ties.

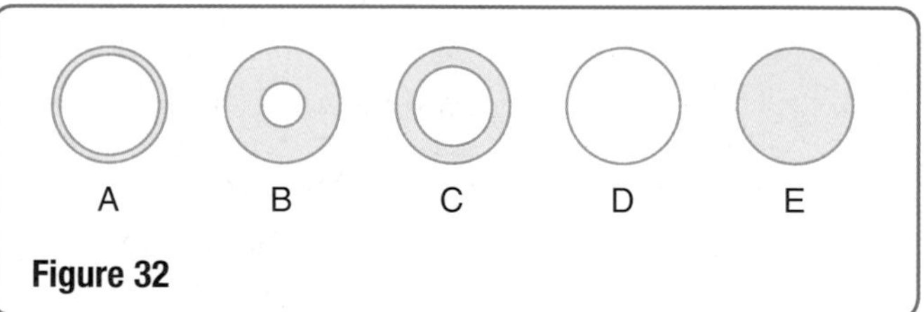

Figure 32

Mixed Review

87. A wooden door of mass m and length l is held horizontally by Dan and Ajit. Dan suddenly drops his end.

a. What is the angular acceleration of the door just after Dan lets go?

b. Is the acceleration constant? Explain.

88. Hard Drive A hard drive on a computer spins at 7200 rpm (revolutions per minute). If the drive is designed to start from rest and reach operating speed in 1.5 s, what is the angular acceleration of the disk?

Chapter Self-Check

89. Topsoil Ten bags of topsoil, each weighing 75 N, are placed on a 2.43-m-long sheet of wood. They are stacked 0.50 m from one end of the sheet of wood, as shown in **Figure 33.** Two people lift the sheet of wood, one at each end. Ignoring the weight of the wood, how much force must each person exert?

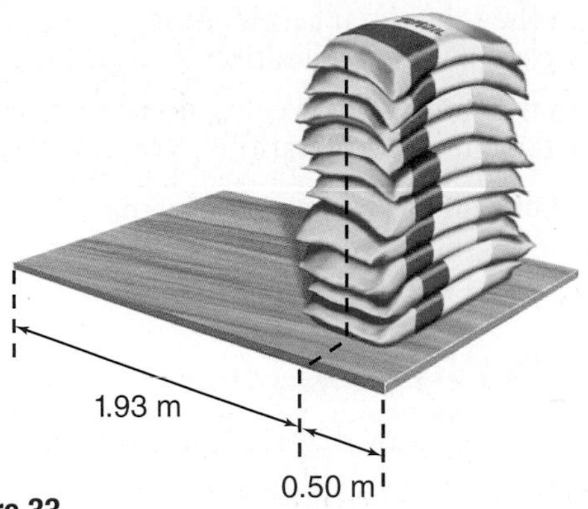

Figure 33

90. A steel beam that is 6.50 m long weighs 325 N. It rests on two supports, 3.00 m apart, with equal amounts of the beam extending from each end. Suki, who weighs 575 N, stands on the beam in the center and then walks toward one end. How close to the end can she walk before the beam begins to tip?

91. The second hand on a watch is 12 mm long. What is the velocity of its tip?

92. A cylinder with a 50-cm diameter, as shown in **Figure 34,** is at rest on a surface. A rope is wrapped around the cylinder and pulled. The cylinder rolls without slipping.

a. After the rope has been pulled a distance of 2.50 m at a constant speed, how far has the center of mass of the cylinder moved?

b. If the rope was pulled a distance of 2.50 m in 1.25 s, how fast was the center of mass of the cylinder moving?

c. What is the angular velocity of the cylinder?

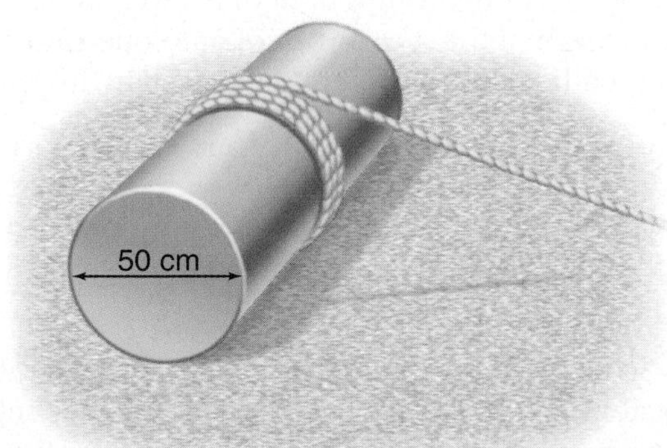

Figure 34

93. Basketball A basketball is rolled down the court. A regulation basketball has a diameter of 24.1 cm, a mass of 0.60 kg, and a moment of inertia of 5.8×10^{-3} kg•m^2. The basketball's initial velocity is 2.5 m/s.

a. What is its initial angular velocity?

b. The ball rolls a total of 12 m. How many revolutions does it make?

c. What is its total angular displacement?

94. The basketball in the previous problem stops rolling after traveling 12 m.

a. If its acceleration was constant, what was its angular acceleration?

b. What torque was acting on it as it was slowing down?

95. Speedometers Most speedometers in automobiles measure the angular velocity of the transmission and convert it to speed. How will increasing the diameter of the tires affect the reading of the speedometer?

96. A box is dragged across the floor using a rope that is a distance h above the floor. The coefficient of friction is 0.35. The box is 0.50 m high and 0.25 m wide. Find the force that just tips the box.

97. Lumber You buy a 2.44-m-long piece of 10 cm × 10 cm lumber. Your friend buys a piece of the same size and cuts it into two lengths, each 1.22 m long, as shown in **Figure 35.** You each carry your lumber on your shoulders.

a. Which load is easier to lift? Why?

b. Both you and your friend apply a torque with your hands to keep the lumber from rotating. Which load is easier to keep from rotating? Why?

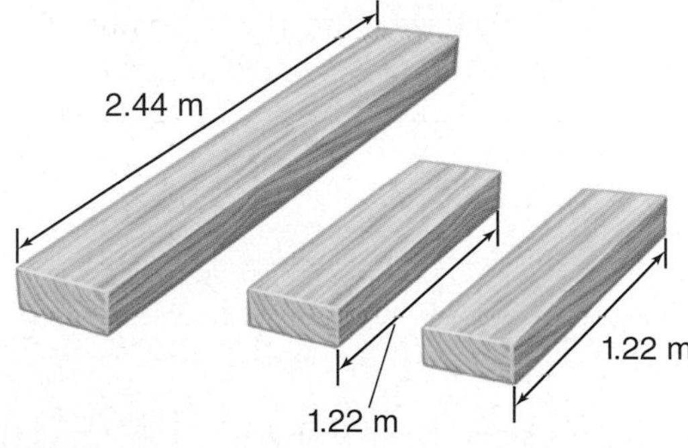

Figure 35

98. Surfboard Harris and Paul carry a surfboard that is 2.43 m long and weighs 143 N. Paul lifts one end with a force of 57 N.

a. What force must Harris exert?

b. What part of the board should Harris lift?

Chapter Self-Check

Thinking Critically

99. Analyze and Conclude A banner is suspended from a horizontal, pivoted pole, as shown in **Figure 36.** The pole is 2.10 m long and weighs 175 N. The banner, which weighs 105 N, is suspended 1.80 m from the pivot point or axis of rotation. What is the tension in the cable supporting the pole?

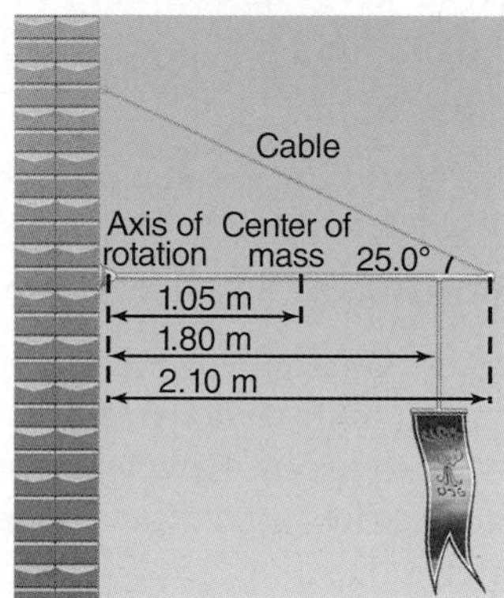

Figure 36

100. Analyze and Conclude A pivoted lamp pole is shown in **Figure 37.** The pole weighs 27 N, and the lamp weighs 64 N.

a. What is the torque caused by each force?

b. Determine the tension in the rope supporting the lamp pole.

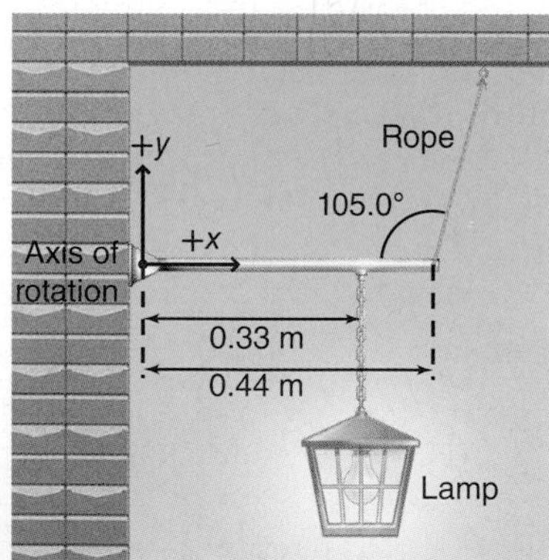

Figure 37

101. Reverse Problem Write a physics problem with real-life objects for which the following equation would be part of the solution:

$$\Delta\theta = \left(20\ \frac{\text{rad}}{\text{s}}\right)(4\ \text{s}) - \frac{1}{2}\left(3.5\ \frac{\text{rad}}{\text{s}^2}\right)(4\ \text{s})^2$$

102. Problem Posing Complete this problem so that it can be solved using the concept of torque: "A painter carries a 3.0-m, 12-kg ladder...."

103. Apply Concepts Consider a point on the edge of a wheel rotating about its axis.

a. Under what conditions can the centripetal acceleration be zero?

b. Under what conditions can the tangential (linear) acceleration be zero?

c. Can the tangential acceleration be nonzero while the centripetal acceleration is zero? Explain.

d. Can the centripetal acceleration be nonzero while the tangential acceleration is zero? Explain.

104. Apply Concepts When you apply the brakes in a car, the front end dips. Why?

Writing in Physics

105. Astronomers know that if a natural satellite is too close to a planet, it will be torn apart by tidal forces. The difference in the gravitational force on the part of the satellite nearest the planet and the part farthest from the planet is stronger than the forces holding the satellite together. Research the Roche limit, and determine how close the Moon would have to orbit Earth to be at the Roche limit.

106. Automobile engines are rated by the torque they produce. Research and explain why torque is an important quantity to measure.

Cumulative Review

107. Two blocks, one of mass 2.0 kg and the other of mass 3.0 kg, are tied together with a massless rope. This rope is strung over a massless, resistance-free pulley. The blocks are released from rest. Find the following:

a. the tension in the rope

b. the acceleration of the blocks

108. Eric sits on a seesaw. At what angle, relative to the vertical, will the component of his weight parallel to the length of the seesaw be equal to one-third the perpendicular component of his weight?

109. The pilot of a plane wants to reach an airport 325 km due north in 2.75 h. A wind is blowing from the west at 30.0 km/h. What heading and airspeed should be chosen to reach the destination on time?

110. A 60.0-kg speed skater with a velocity of 18.0 m/s skates into a curve of 20.0-m radius. How much friction must be exerted between the skates and the ice for her to negotiate the curve?

MULTIPLE CHOICE

1. The illustration below shows two boxes on opposite ends of a massless board that is 3.0 m long. The board is supported in the middle by a fulcrum. The box on the left has a mass (m_1) of 25 kg, and the box on the right has a mass (m_2) of 15 kg. How far should the fulcrum be positioned from the left side of the board in order to balance the masses horizontally?

A. 0.38 m **C.** 1.1 m

B. 0.60 m **D.** 1.9 m

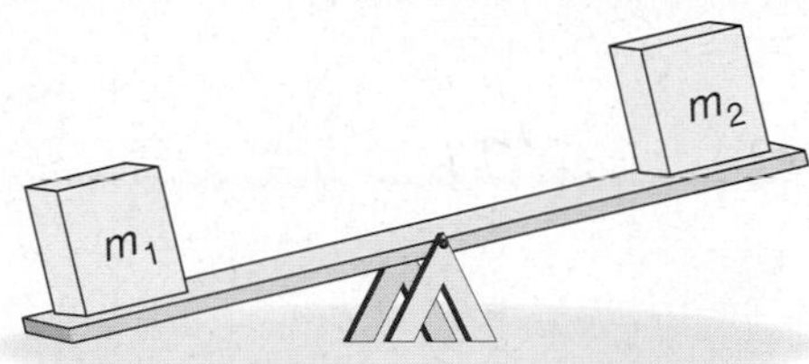

2. A force of 60 N is exerted on one end of a 1.0-m-long lever. The other end of the lever is attached to a rotating rod that is perpendicular to the lever. By pushing down on the end of the lever, you can rotate the rod. If the force on the lever is exerted at an angle of 30° to the perpendicular to the lever, what torque is exerted on the rod? (sin 30° = 0.5; cos 30° = 0.87; tan 30° = 0.58)

A. 30 N **C.** 60 N

B. 52 N **D.** 69 N

3. A child attempts to use a wrench to remove a nut on a bicycle. Removing the nut requires a torque of 10 N•m. The maximum force the child is capable of exerting at a 90° angle is 50 N. What is the length of the wrench the child must use to remove the nut?

A. 0.1 m **C.** 0.2 m

B. 0.15 m **D.** 0.25 m

4. A car moves a distance of 420 m. Each tire on the car has a diameter of 42 cm. Which shows how many revolutions each tire makes as they move that distance?

A. $\left(\frac{(5.0\times10^1)}{\pi}\right)$ rev **C.** $\left(\frac{(1.5\times10^2)}{\pi}\right)$ rev

B. $\left(\frac{(1.0\times10^2)}{\pi}\right)$ rev **D.** $\left(\frac{(1.0\times10^3)}{\pi}\right)$ rev

5. A thin hoop with a mass of 5.0 kg rotates about a perpendicular axis through its center. A force of 25 N is exerted tangentially to the hoop. If the hoop's radius is 2.0 m, what is its angular acceleration?

A. 1.3 rad/s **C.** 5.0 rad/s

B. 2.5 rad/s **D.** 6.3 rad/s

6. Two of the tires on a farmer's tractor have diameters of 1.5 m. If the farmer drives the tractor at a linear velocity of 3.0 m/s, what is the angular velocity of each tire?

A. 2.0 rad/s **C.** 4.0 rad/s

B. 2.3 rad/s **D.** 4.5 rad/s

FREE RESPONSE

7. You use a 25-cm long wrench to remove the lug nuts on a car wheel, as shown in the illustration below. If you pull up on the end of the wrench with a force of 2.0×10^2 N at an angle of 30°, what is the torque on the wrench? (sin 30° = 0.5, cos 30° = 0.87)

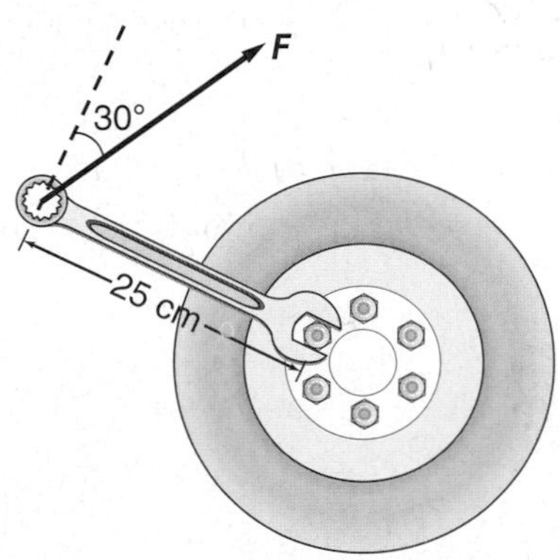

Online Test Practice

NEED EXTRA HELP?

If You Missed Question	1	2	3	4	5	6	7
Review Section	2	2	2	1	1	1	2

CHAPTER 9

Momentum and Its Conservation

BIGIDEA If the net force on a closed system is zero, the total momentum of that system is conserved.

SECTIONS

LaunchLAB

COLLIDING OBJECTS

What factors determine the speed and direction of objects after a collision?

WATCH THIS!

CRASH!

From fender benders to bumper cars, there is a lot of physics involved in collisions. How do forces shape what happens when two objects collide?

PHYSICS T.V.

(l)Mikael Karlsson/Alamy, (r)MIKE CLARKE/AFP/Getty Images

Go online!
connectED.mcgraw-hill.com

SECTION 1 Impulse and Momentum

PHYSICS 4 YOU Lacrosse players wear helmets and padding to protect themselves from flying balls. Lacrosse balls are not very massive (about 145 g), but players might hurl them at speeds over 40 m/s. Why are lacrosse balls so dangerous?

MAIN IDEA
An object's momentum is equal to its mass multiplied by its velocity.

Essential Questions
- What is impulse?
- What is momentum?
- What is angular momentum?

Review Vocabulary
angular velocity the angular displacement of an object divided by the time needed to make the displacement

New Vocabulary
impulse
momentum
impulse-momentum theorem
angular momentum
angular impulse-angular momentum theorem

Impulse-Momentum Theorem

It can be exciting to watch a baseball player hit a home run. The pitcher hurls the baseball toward the plate. The batter swings, and the ball recoils from the impact. Before the collision, the baseball moves toward the bat. During the collision, the ball is squashed against the bat. After the collision, the ball moves at a high velocity away from the bat, and the bat continues along its path but at a slower velocity.

According to Newton's second law of motion, the force from the bat changed the ball's velocity. This force changes over time, as shown in **Figure 1.** Just after contact, the ball is squeezed, and the force increases to a maximum more than 10,000 times the weight of the ball. The ball then recovers its shape and rebounds from the bat. The force on the ball rapidly returns to zero. This whole event takes place within about 3.0 ms. How can you calculate the change in velocity of the baseball?

Impulse Newton's second law of motion ($\boldsymbol{F} = m\boldsymbol{a}$) can be rewritten in terms of the change in velocity divided by the time for that change.

$$\boldsymbol{F} = m\boldsymbol{a} = m\left(\frac{\Delta \boldsymbol{v}}{\Delta t}\right)$$

Multiplying both sides by the time interval (Δt) results in this equation.

$$\boldsymbol{F}\Delta t = m\Delta \boldsymbol{v} = m(\boldsymbol{v}_{\mathrm{f}} - \boldsymbol{v}_{\mathrm{i}})$$

The left side of the equation ($\boldsymbol{F}\Delta t$) is defined as the impulse. The **impulse** on an object is the product of the average force on an object and the time interval over which it acts. Impulse is measured in newton-seconds. If the force $\boldsymbol{F}$ varies with time, the magnitude of the impulse equals the area under the curve of a force-time graph, as in **Figure 1.**

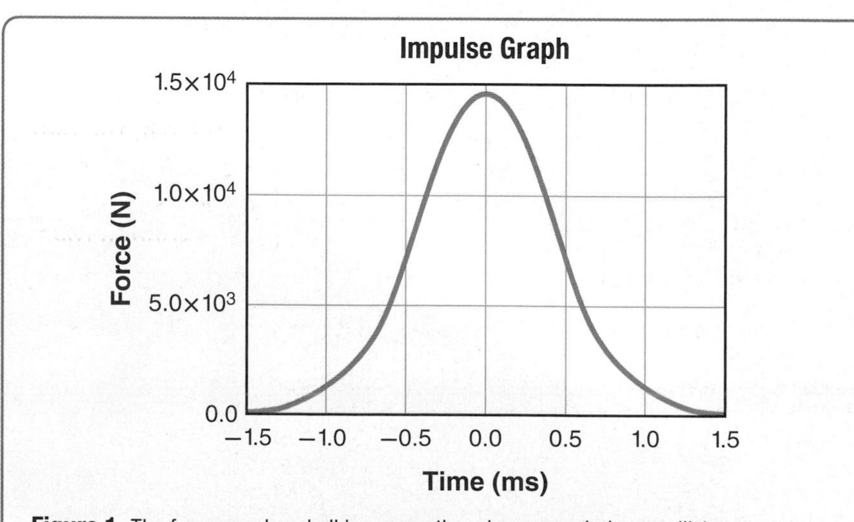

Figure 1 The force on a baseball increases then decreases during a collision.

Rubberball/Getty Images

Momentum The right side of the equation ($m\Delta\boldsymbol{v}$) involves the change in velocity, where $m\Delta\boldsymbol{v} = m\boldsymbol{v}_f - m\boldsymbol{v}_i$. The product of the object's mass (m) and the object's velocity ($\boldsymbol{v}$) is defined as the **momentum** of the object. Momentum is measured in kg•m/s. An object's momentum, also known as linear momentum, is represented by the following equation.

MOMENTUM

The momentum of an object equals the mass of the object times the object's velocity.

$$\boldsymbol{p} = m\boldsymbol{v}$$

Now, you can rewrite the impulse as $\boldsymbol{F}\Delta t = m\Delta\boldsymbol{v} = m\boldsymbol{v}_f - m\boldsymbol{v}_i = \boldsymbol{p}_f - \boldsymbol{p}_i$. Thus, the impulse on an object is equal to the change in its momentum, which is called the **impulse-momentum theorem.** The following relationship expresses the impulse-momentum theorem.

IMPULSE-MOMENTUM THEOREM

An impulse acting on an object is equal to the object's final momentum minus the object's initial momentum.

$$\boldsymbol{F}\Delta t = \boldsymbol{p}_f - \boldsymbol{p}_i$$

If the force on an object is constant, the impulse is the product of the force multiplied by the time interval over which it acts. Generally, the force is not constant; however, you can find the impulse using an average force multiplied by the time interval over which it acts, or by finding the area under a force-time graph.

Because velocity is a vector, momentum also is a vector. Similarly, impulse is a vector because force is a vector. As with any other vector quantity, make sure you use signs consistently to indicate direction.

Using the impulse-momentum theorem According to the impulse-momentum theorem, the impulse from a bat changes a baseball's momentum. You can calculate the impulse using a force-time graph. In **Figure 1,** the area under the curve is about 15 N•s. The direction of the impulse is in the direction of the force. Therefore, the change in momentum of the ball also is 15 N•s. Because 1 N•s is equal to 1 kg•m/s, the momentum gained by the ball is 15 kg•m/s in the direction of the force acting on it.

Suppose a batter hits a fastball. Assume the positive direction is toward the pitcher. Before the collision of the ball and the bat, the ball, with a mass of 0.145 kg, has a velocity of −47 m/s. Therefore, the baseball's momentum is $\boldsymbol{p}_i = (0.145\ \text{kg})(-47\ \text{m/s}) = -6.8\ \text{kg·m/s}$.
The momentum of the ball after the collision is found by solving the impulse-momentum theorem for the final momentum: $\boldsymbol{p}_f = \boldsymbol{p}_i + \boldsymbol{F}\Delta t$.

$$\boldsymbol{p}_f = \boldsymbol{p}_i + \boldsymbol{F}\Delta t = -6.8\ \text{kg·m/s} + 15\ \text{kg·m/s} = +8.2\ \text{kg·m/s}$$

Because $\boldsymbol{p}_f = m\boldsymbol{v}_f$, solving for $\boldsymbol{v}_f$ yields the ball's final velocity:

$$\boldsymbol{v}_f = \frac{\boldsymbol{p}_f}{m} = \frac{+8.2\ \text{kg·m/s}}{+0.145\ \text{kg}} = +57\ \text{m/s}$$

A speed of 52 m/s is fast enough to clear most outfield fences if the batter hits the baseball in the correct direction.

VOCABULARY

Science Usage v. Common Usage

Impulse

• Science usage

the product of average net force and the time interval over which the force acts

The impulse on the rocket from its exhaust lifted the astronauts into orbit.

• Common usage

a sudden tendency toward some action

After seeing the commercial, Amad had an impulse to eat a candy bar.

Momentum

• Science usage

product of object's mass and object's velocity

A moving truck usually has much more momentum than a moving bicycle.

• Common usage

strength gained through the development of events

The team hoped to continue the momentum of the regular season into the play-offs.

PhysicsLAB

STICKY COLLISIONS

What happens to the momentum of an object during a collision?

EXAMPLE PROBLEM 1

Find help with **significant figures.** Math Handbook

EXAMPLE PROBLEM

AVERAGE FORCE A 2200-kg vehicle traveling at 94 km/h (26 m/s) can be stopped in 21 s by gently applying the brakes. It can be stopped in 3.8 s if the driver slams on the brakes, or in 0.22 s if it hits a concrete wall. What is the impulse exerted on the vehicle? What average force is exerted on the vehicle in each of these stops?

1 ANALYZE AND SKETCH THE PROBLEM

- Sketch the system.
- Include a coordinate axis. Select the car's direction as positive.
- Draw a vector diagram for momentum and impulse.

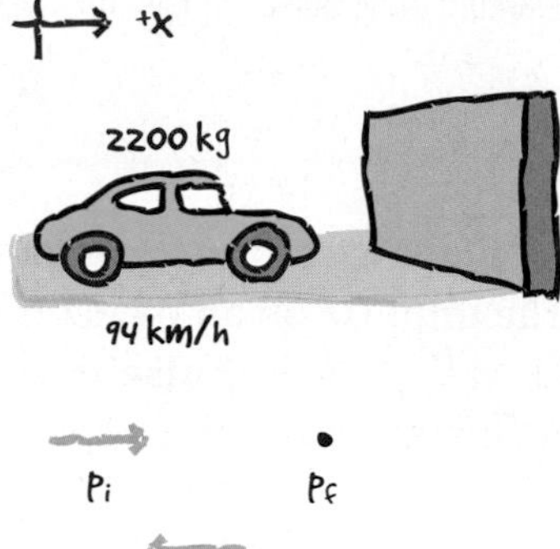

KNOWN		UNKNOWN
$m = 2200$ kg	$\Delta t_{\text{gentle braking}} = 21$ s	$\boldsymbol{F}\Delta t = ?$
$\boldsymbol{v}_i = +26$ m/s	$\Delta t_{\text{hard braking}} = 3.8$ s	$\boldsymbol{F}_{\text{gentle braking}} = ?$
$\boldsymbol{v}_f = 0.0$ m/s	$\Delta t_{\text{hitting a wall}} = 0.22$ s	$\boldsymbol{F}_{\text{hard braking}} = ?$
		$\boldsymbol{F}_{\text{hitting a wall}} = ?$

2 SOLVE FOR THE UNKNOWN

Determine the initial momentum ($\boldsymbol{p}_i$).

$\boldsymbol{p}_i = m\boldsymbol{v}_i$

$= (2200 \text{ kg})(+26 \text{ m/s})$ ◀ Substitute $m = 2200$ kg, $v_i = +26$ m/s.

$= +5.7\times10^4$ kg•m/s

Determine the final momentum ($\boldsymbol{p}_f$).

$\boldsymbol{p}_f = m\boldsymbol{v}_f$

$= (2200 \text{ kg})(0.0 \text{ m/s})$ ◀ Substitute $m = 2200$ kg, $v_f = 0.0$ m/s.

$= 0.0$ kg•m/s

Apply the impulse-momentum theorem to determine the impulse and the force needed to stop the vehicle.

$\boldsymbol{F}\Delta t = \boldsymbol{p}_f - \boldsymbol{p}_i$

$\boldsymbol{F}\Delta t = (0.0 \text{ kg•m/s}) - (5.7\times10^4 \text{ kg•m/s})$ ◀ Substitute $p_f = 0.0$ kg•m/s, $p_i = 5.7\times10^4$ kg•m/s.

$= -5.7\times10^4$ kg•m/s

$$\boldsymbol{F} = \frac{-5.7\times10^4 \text{ kg•m/s}}{\Delta t}$$

$$\boldsymbol{F}_{\text{gentle braking}} = \frac{-5.7\times10^4 \text{ kg•m/s}}{21 \text{ s}}$$ ◀ Substitute $\Delta t_{\text{gentle braking}} = 21$ s.

$= -2.7\times10^3$ N

$$\boldsymbol{F}_{\text{hard braking}} = \frac{-5.7\times10^4 \text{ kg•m/s}}{3.8 \text{ s}}$$ ◀ Substitute $\Delta t_{\text{hard braking}} = 3.8$ s.

$= -1.5\times10^4$ N

$$\boldsymbol{F}_{\text{hitting a wall}} = \frac{-5.7\times10^4 \text{ kg•m/s}}{0.22 \text{ s}}$$ ◀ Substitute $\Delta t_{\text{hitting a wall}} = 0.22$ s.

$= -2.6\times10^5$ N

3 EVALUATE THE ANSWER

- **Are the units correct?** Impulse is in kg•m/s. Force is measured in newtons.
- **Does the direction make sense?** Force is exerted in the direction opposite to the velocity of the car, which means the force is negative.
- **Is the magnitude realistic?** It is reasonable that the force needed to stop a car is thousands or hundreds of thousands of newtons. The impulse is the same for all stops. Thus, as the stopping time decreases by about a factor of 10, the force increases by about a factor of 10.

PRACTICE PROBLEMS

Do additional problems. Online Practice

1. A compact car, with mass 725 kg, is moving at 115 km/h toward the east. Sketch the moving car.

a. Find the magnitude and direction of its momentum. Draw an arrow on your sketch showing the momentum.

b. A second car, with a mass of 2175 kg, has the same momentum. What is its velocity?

2. The driver of the compact car in the previous problem suddenly applies the brakes hard for 2.0 s. As a result, an average force of 5.0×10^3 N is exerted on the car to slow it down.

a. What is the change in momentum, or equivalently, what is the magnitude and direction of the impulse on the car?

b. Complete the "before" and "after" sketches, and determine the momentum and the velocity of the car now.

3. A 7.0-kg object, moving at 2.0 m/s, receives two impulses (one after the other) along the direction of its motion. Both of these impulses are illustrated in **Figure 2.** Find the resulting speed and direction of motion of the object after each impulse.

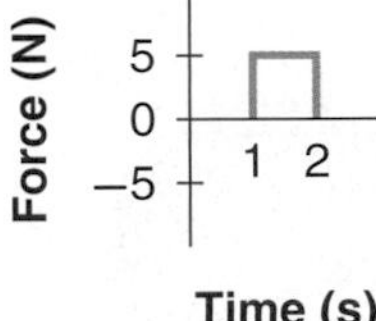

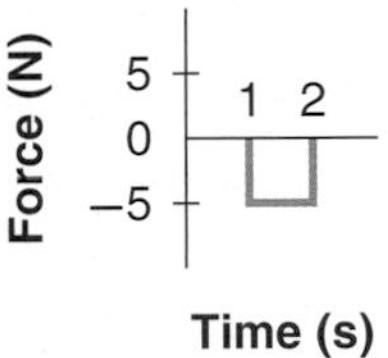

Figure 2

4. The driver accelerates a 240.0-kg snowmobile, which results in a force being exerted that speeds up the snowmobile from 6.00 m/s to 28.0 m/s over a time interval of 60.0 s.

a. Sketch the event, showing the initial and final situations.

b. What is the snowmobile's change in momentum? What is the impulse on the snowmobile?

c. What is the magnitude of the average force that is exerted on the snowmobile?

5. CHALLENGE Suppose a 60.0-kg person was in the vehicle that hit the concrete wall in Example Problem 1. The velocity of the person equals that of the car both before and after the crash, and the velocity changes in 0.20 s. Sketch the problem.

a. What is the average force exerted on the person?

b. Some people think they can stop their bodies from lurching forward in a vehicle that is suddenly braking by putting their hands on the dashboard. Find the mass of an object that has a weight equal to the force you just calculated. Could you lift such a mass? Are you strong enough to stop your body with your arms?

Using the impulse-momentum theorem to save lives The impulse-momentum theorem shows that a large impulse causes a large change in momentum. This large impulse could result either from a large force acting over a short period of time or from a smaller force acting over a longer period of time. A passenger in a car accident might experience a large impulse, but it is the force on the passenger that causes injuries.

What happens to the driver when a crash suddenly stops a car? An impulse, either from the dashboard or from an air bag, brings the driver's momentum to zero. According to the impulse-momentum equation, $\boldsymbol{F}\Delta t = \boldsymbol{p}_f - \boldsymbol{p}_i$. The final momentum ($\boldsymbol{p}_f$) is zero. The initial momentum ($\boldsymbol{p}_i$) is the same with or without an air bag. Thus, the impulse ($\boldsymbol{F}\Delta t$) also is the same. An air bag, such as the one shown in **Figure 3,** increases the time interval during which the force acts on the passenger. Therefore the required force is less. The air bag also spreads the force over a larger area of the person's body, thereby reducing the likelihood of injuries.

Figure 3 An air bag reduces injuries by making the force on a passenger less and by spreading that force over a larger area.

Don Johnston/Stone/Getty Images

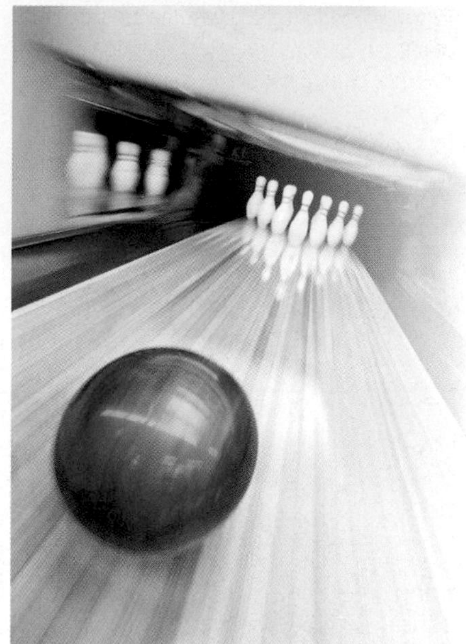

Figure 4 The ball's linear momentum is the product of its mass and its velocity. The ball's angular momentum as it rolls down the lane is the product of its moment of inertia and its angular velocity.

Angular Momentum

The impulse-momentum theorem is useful if the momentum of an object is linear, but how can you describe momentum that is angular, as it is for the rotating bowling ball in **Figure 4?** Recall from your study of rotational motion that the angular velocity of a rotating object changes if torque is applied to it. This is a statement of Newton's law for rotational motion, $\tau = \frac{I\Delta\omega}{\Delta t}$. You can rearrange this relationship, just as Newton's second law of motion was, to produce $\tau\Delta t = I\Delta\omega$.

The left side of this equation ($\tau\Delta t$) is the rotating object's angular impulse. You can rewrite the right side as $I\Delta\omega = I\omega_f - I\omega_i$. The product of a rotating object's moment of inertia and angular velocity is called **angular momentum,** which is represented by the symbol L. The following relationship describes an object's angular momentum.

ANGULAR MOMENTUM

The angular momentum of an object is defined as the product of the object's moment of inertia and the object's angular velocity.

$$L = I\omega$$

Angular momentum is measured in kg•m^2/s. Just as an object's linear momentum changes when an impulse acts on it, the object's angular momentum changes when an angular impulse acts on it. The object's angular impulse is equal to the change in the object's angular momentum, as stated by the **angular impulse-angular momentum theorem.** This theorem can be represented by the following relationship.

ANGULAR IMPULSE-ANGULAR MOMENTUM THEOREM

The angular impulse on an object is equal to the object's final angular momentum minus the object's initial angular momentum.

$$\tau\Delta t = L_f - L_i$$

REAL-WORLD PHYSICS

PADDLE SPORTS A table-tennis player controls the flight of the ball by putting spin on the ball. The player hits the ball at an angle so that the paddle exerts both an impulse to change the momentum and an angular impulse to make the ball spin. The interaction with the air will cause a forward-spinning ball to drop rapidly, making it more difficult for the opponent to return.

If the net force on an object is zero, its linear momentum is constant. If the net torque acting on an object is zero, its angular momentum is also constant, but the two situations are slightly different. Because an object's mass cannot change, if its momentum is constant, then its velocity is also constant. In the case of angular momentum, however, the object's angular velocity can change if the shape of the object changes. This is because the moment of inertia depends on the object's mass and the way it is distributed about the axis of rotation or revolution. Thus, the angular velocity of an object can change even if no torques act on it.

▶ CONNECTION TO ASTRONOMY Astronomers have discovered many examples of two stars that orbit each other. Together the stars are called a binary star system. The torque on the binary system is zero because the gravitational force acts only directly between the stars. Therefore, the binary system's angular momentum is constant. But sometimes the stars are so close that the strong gravitational force rips some material from one star and deposits it on the other star. This movement of matter changes the moment of inertia of the binary system. As a result, the angular velocity of the system will change, even though the angular momentum is unchanged.

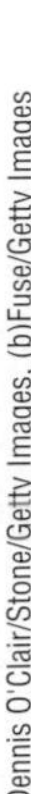

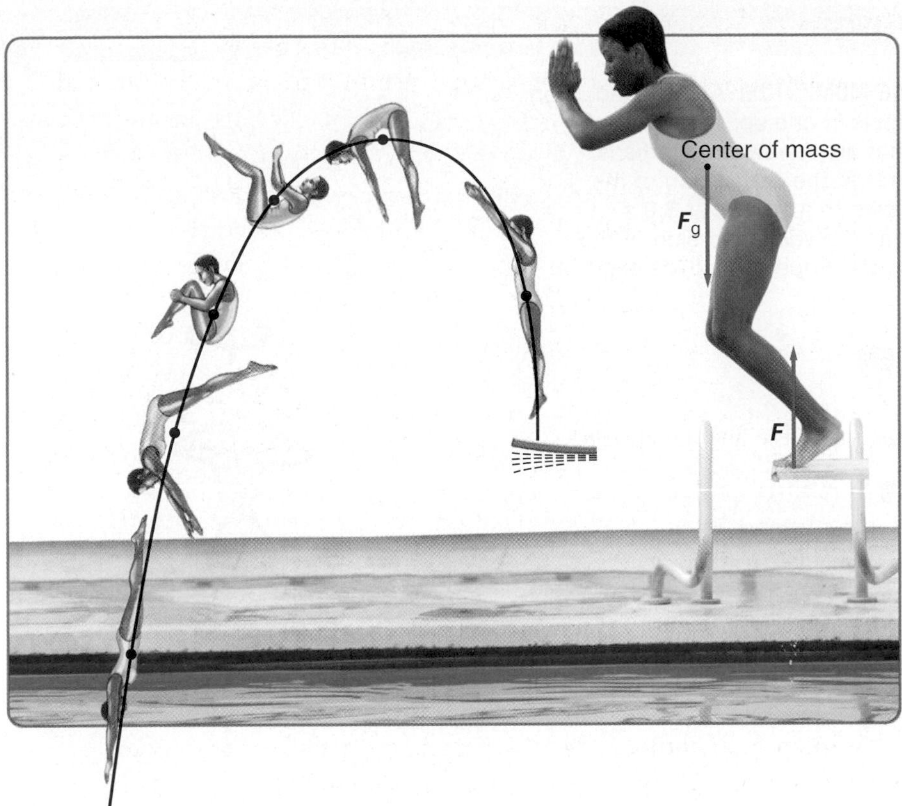

Figure 5 Before she dives, the girl's center of mass is in front of her feet. During her dive, the path of her center of mass forms a parabola. She changes her moment of inertia by moving her arms and legs.

Analyze How does extending her arms affect the diver's motion?

COLOR CONVENTION

force ⟷ blue

Diving into a pool Consider the diver in **Figure 5.** How does she start rotating her body? She uses the diving board to apply an external force to her body. Then, she moves her center of mass in front of her feet and uses the board to give a final upward push to her feet. This provides a torque that acts over time (Δt) and increases the angular momentum of the diver.

Before the diver reaches the water, she can change her angular velocity by changing her moment of inertia. She may go into a tuck position, grabbing her knees with her hands. By moving her mass closer to the axis of rotation, the diver decreases her moment of inertia and increases her angular velocity. When she nears the water, she stretches her body straight, thereby increasing the moment of inertia and reducing the angular velocity. As a result, she goes straight into the water.

☑ **READING CHECK Explain** why going into a tuck position increases a diver's angular velocity.

Ice-skating An ice-skater uses a similar method to spin. To begin rotating on one foot, the ice-skater causes the ice to exert a force on her body by pushing a portion of one skate into the ice. If the skater in **Figure 6** pushes on the ice in one direction, the ice will exert a force on her in the opposite direction. The force results in a torque if the force is exerted some distance away from the pivot point and in a direction that is not toward it. The greatest torque for a given force will result if the push is perpendicular to the lever arm.

The ice-skater then can control her angular velocity by changing her moment of inertia. Both arms and one leg can be extended from the body to slow the rotation or can be pulled in close to the axis of rotation to speed it up. To stop spinning, another torque must be exerted by using the free skate to create a way for the ice to exert the needed force.

Figure 6 When the skater pushes off the ice, the ice exerts a torque on the skater.

EXAMPLE PROBLEM 2

Find help with **significant digits.** Math Handbook

EXAMPLE PROBLEM

ANGULAR IMPULSE AND CHANGE IN ANGULAR MOMENTUM At the conclusion of his competition routine, an ice-skater spins in one spot at a rate of 2.5 rotations per second. If you assume that his moment of inertia is 0.50 kg•m^2, what is his angular momentum? At the conclusion of the spin, the skater slowly and dramatically slows to a stop over 3.0 s by dragging the toe on one skate. What was the value and sign of the angular impulse on the skater? What average torque did the ice exert on the skater?

1 ANALYZE AND SKETCH THE PROBLEM

- Sketch the system.
- Draw the axis of rotation and the direction of the angular impulse.

KNOWN	UNKNOWN
$\omega_i = 2.5$ rotations/s	$L_i = ?$
$\omega_f = 0$ rad/s	angular impulse = ?
$I = 0.50$ kg•m^2	$\tau_{avg} = ?$
$\Delta t = 3.0$ s	
$L_f = 0.0$ kg•m^2/s	

2 SOLVE FOR THE UNKNOWNS

Determine the initial angular momentum.

$L_i = I\omega_i$

$= (0.50 \text{ kg}\cdot\text{m}^2)(2.5 \text{ rotations/s})(2\pi \text{ rad/rotation})$ ◀ *Substitute I = 0.50 kg•m^2 and ω_i = 2.5 rotations/s. Convert from rotations to radians.*

$= 7.9 \text{ kg}\cdot\text{m}^2\text{/s}$ ◀ *Radians is a dimensionless unit and is removed.*

Determine the angular impulse that the ice exerted on the skater.

angular impulse $= \tau_{avg}\,\Delta t = L_f - L_i$ ◀ *The angular impulse equals the change in angular momentum.*

$= 0.0 \text{ kg}\cdot\text{m}^2\text{/s} - 7.9 \text{ kg}\cdot\text{m}^2\text{/s}$ ◀ *Substitute L_f = 0.0 kg•m^2/s, L_i = 7.9 kg•m^2/s.*

$= -7.9 \text{ kg}\cdot\text{m}^2\text{/s}$

Determine the average torque that the ice exerted on the skater. From the angular impulse, you can rearrange the quantities to find the average torque.

$\tau_{avg} = \frac{L_f - L_i}{\Delta t}$ ◀ *Solve for the average torque.*

$= \frac{0.0 \text{ kg}\cdot\text{m}^2\text{/s} - 7.9 \text{ kg}\cdot\text{m}^2\text{/s}}{3.0 \text{ s}}$ ◀ *Substitute L_f = 0.0 kg•m^2/s, L_i = 7.9 kg•m^2/s, and Δt = 3.0 s.*

$= -2.6 \text{ N}\cdot\text{m}$ ◀ *The unit kg•m^2/s^2 is equivalent to N•m.*

3 EVALUATE THE ANSWER

- **Are the units correct?** Yes, a kg•m^2/s^2 is the equivalent of a N•m.
- **Do the signs make sense?** The angular impulse and the torque are each negative, indicating that the angular momentum decreases.
- **Are the magnitudes realistic?** Yes, the value of the torque is small, but the product of the torque and time equals the value of the angular momentum.

PRACTICE PROBLEMS

Do additional problems. Online Practice

6. A 0.25-m-diameter circular saw blade in a workshop rotates at 5.0×10^3 rpm, as shown in **Figure 7.** After the electrical power to the saw is turned off, it takes several seconds for the blade to slow to a complete stop. The moment of inertia of the blade is 8.0×10^{-3} kg•m^2. Friction in the axle provides an average torque of 2.3×10^{-1} N•m to slow the blade. How many seconds does it take for the blade to stop?

7. A baseball pitcher can throw a 132 km/h (82 mph) curve ball that rotates about 6.0×10^2 rpm. What is the angular velocity of the thrown ball? The pitcher's throwing motion lasts about 0.15 s, and the moment of inertia of the ball is 8.0×10^{-5} kg•m^2. What average torque did the pitcher exert on the ball?

8. As a bowler releases the ball onto the alley, the ball does not roll but slides. Slowly the friction of the alley surface causes the ball to roll and have a final angular velocity of 7.00×10^1 rad/s. The moment of inertia of the ball is 0.0350 kg•m^2, and the ball moves down the alley in 2.40 s. What are the angular impulse and the average torque that the alley surface exerts on the bowling ball?

9. A bicycle is clamped upside down on a workbench for the bicycle repair woman to repair a front wheel axle. She gives the front wheel a spin with her hand, and the wheel rotates at 5.0 rev/s. What is the angular velocity of the wheel? If the moment of inertia of the wheel is 0.060 kg•m^2, what angular impulse did the repair woman give the wheel?

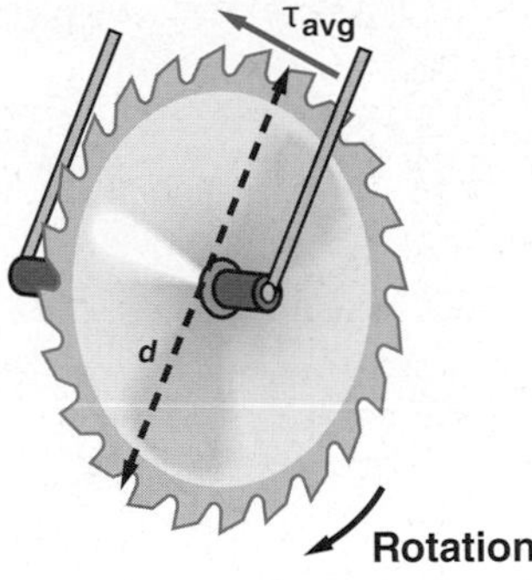

Figure 7

PRACTICE PROBLEMS

SECTION 1 REVIEW

Section Self-Check Check your understanding.

10. **MAIN**IDEA Which has more momentum, a truck that is parked or a falling raindrop? Explain.

11. **Momentum** Is the momentum of a car traveling south different from that of the same car when it travels north at the same speed? Draw the momentum vectors to support your answer.

12. **Impulse and Momentum** When you jump from a height to the ground, you let your legs bend at the knees as your feet hit the floor. Explain why you do this in terms of the physics concepts introduced in this chapter.

13. **Impulse and Momentum** A 0.174-kg softball is pitched horizontally at 26.0 m/s. The ball moves in the opposite direction at 38.0 m/s after it is hit by the bat.
    - **a.** Draw arrows showing the ball's momentum before and after the bat hits it.
    - **b.** What is the change in momentum of the ball?
    - **c.** What is the impulse delivered by the bat?
    - **d.** If the bat and ball are in contact for 0.80 ms, what is the average force the bat exerts on the ball?

14. **Momentum** The speed of a basketball as it is dribbled is the same just before and just after the ball hits the floor. Is the impulse on and the change in momentum of the basketball equal to zero when the basketball hits the floor? If not, in which direction is the change in momentum? Draw the basketball's momentum vectors before and after it hits the floor.

Figure 8

15. **Angular Momentum** The ice-skater in **Figure 8** spins with his arms outstretched. When he pulls his arms in and raises them above his head, he spins much faster than before. Did a torque act on the ice-skater? What caused his angular velocity to increase?

16. **Critical Thinking** An archer shoots arrows at a target. Some of the arrows stick in the target, while others bounce off. Assuming that the masses of the arrows and the velocities of the arrows are the same, which arrows produce a bigger impulse on the target? *Hint: Draw a diagram to show the momentum of the arrows before and after hitting the target for the two instances.*

SECTION 2

Conservation of Momentum

PHYSICS 4 YOU **When a game of billiards is started, the balls are usually arranged in a triangle. A player then shoots the cue ball at them, causing the balls to spread out in all directions. How does the motion of the cue ball before the break affect the motions of the balls after the break?**

MAIN IDEA

In a closed, isolated system, linear momentum and angular momentum are conserved.

Essential Questions

- How does Newton's third law relate to conservation of momentum?
- Under which conditions is momentum conserved?
- How can the law of conservation of momentum and the law of conservation of angular momentum help explain the motion of objects?

Review Vocabulary

momentum the product of an object's mass and the object's velocity

New Vocabulary

closed system
isolated system
law of conservation of momentum
law of conservation of angular momentum

Two-Particle Collisions

In the first section of this chapter, you learned how a force applied during a time interval changes the momentum of a baseball. In the discussion of Newton's third law of motion, you learned that forces are the result of interactions between two objects. The force of a bat on a ball is accompanied by an equal and opposite force of the ball on the bat. Does the momentum of the bat, therefore, also change?

The bat, the hand and arm of the batter, and the ground on which the batter is standing are all objects that interact when a batter hits the ball. Thus, the bat cannot be considered as a single object. In contrast to this complex system, examine for a moment the much simpler system shown in **Figure 9,** the collision of two balls.

Force and impulse During the collision of the two balls, each one briefly exerts a force on the other. Despite the differences in sizes and velocities of the balls, the forces they exert on each other are equal and opposite, according to Newton's third law of motion. These forces are represented by $\boldsymbol{F}_{\text{red on blue}} = -\boldsymbol{F}_{\text{blue on red}}$.

How do the impulses imparted by both balls compare? Because the time intervals over which the forces are exerted are the same, the impulses the balls exert on each other must also be equal in magnitude but opposite in direction: $(\boldsymbol{F}\Delta t)_{\text{red on blue}} = -(\boldsymbol{F}\Delta t)_{\text{blue on red}}$.

Notice that no mention has been made of the masses of the balls. Even though the balls have different sizes and approach each other with different velocities, and even though they may have different masses, the forces and the impulses they exert on each other have equal strength but opposite directions.

Figure 9 The balls have different sizes, masses, and velocities, but if they interact the forces and impulses they exert on each other are equal and opposite.

Momentum How did the momentums of the two balls the girl and the boy rolled toward each other in **Figure 9** change as a result of the collision? Compare the changes shown in **Figure 10.**

$$\text{For ball C: } \boldsymbol{p}_{\text{Cf}} - \boldsymbol{p}_{\text{Ci}} = \boldsymbol{F}_{\text{D on C}}\Delta t$$
$$\text{For ball D: } \boldsymbol{p}_{\text{Df}} - \boldsymbol{p}_{\text{Di}} = \boldsymbol{F}_{\text{C on D}}\Delta t$$

According to the impulse-momentum theorem, the change in momentum is equal to the impulse. Because the impulses are equal in magnitude but opposite in direction, we know the following:

$$\boldsymbol{p}_{\text{Cf}} - \boldsymbol{p}_{\text{Ci}} = -(\boldsymbol{p}_{\text{Df}} - \boldsymbol{p}_{\text{Di}})$$
$$\boldsymbol{p}_{\text{Cf}} + \boldsymbol{p}_{\text{Df}} = \boldsymbol{p}_{\text{Ci}} + \boldsymbol{p}_{\text{Di}}$$

This equation states that the sum of the momentums of the balls in **Figure 10** is the same before and after the collision. That is, the momentum gained by ball D is equal to the momentum lost by ball C. If the system is defined as the two balls, the sum of the gain and loss in momentum is zero, and therefore, momentum is conserved for the system.

Momentum in a Closed, Isolated System

Under what conditions is the momentum of the system of two balls conserved? The first and most obvious condition is that no balls are lost and no balls are gained. Such a system, which does not gain or lose mass, is said to be a **closed system.**

The second condition required to conserve the momentum of a system is that the only forces that are involved are internal forces; that is, there are no unbalanced forces acting on the system from objects outside of it. When the net external force exerted on a closed system is zero, the system is described as an **isolated system.** No system on Earth is absolutely isolated, however, because there will always be some interactions between a system and its surroundings. Often, these interactions are small enough to be ignored when solving physics problems.

Systems can contain any number of objects, and the objects that make up a system can stick together or they can come apart during a collision. But even under these conditions, momentum is conserved. The **law of conservation of momentum** states that the momentum of any closed, isolated system does not change. This law enables you to make a connection between conditions before and after an interaction without knowing any details of the interaction.

☑ **READING CHECK** **Explain** the difference between a closed system and an isolated system.

Figure 10 When two balls of different mass and velocity collide, the momentum of each ball is changed. If the system is isolated, however, the sum of their momentums before the collision equals the sum of their momentums after the collision.

Analyze How can you tell from the diagram that the impulses exerted by the balls have the same magnitude?

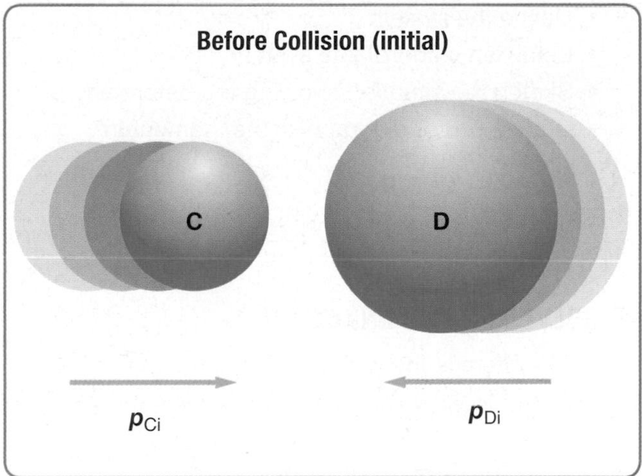

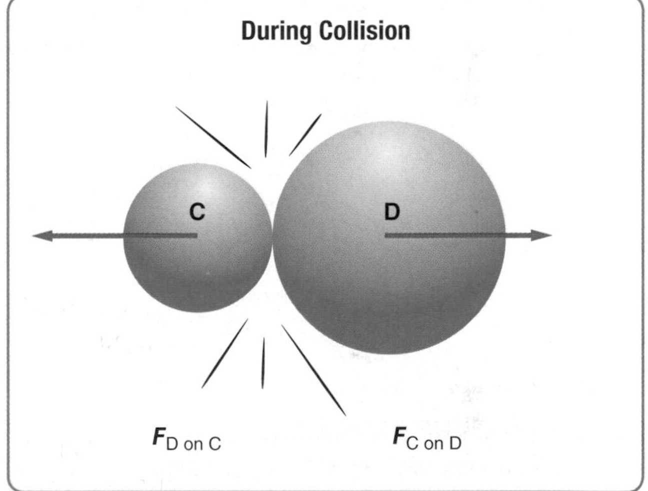

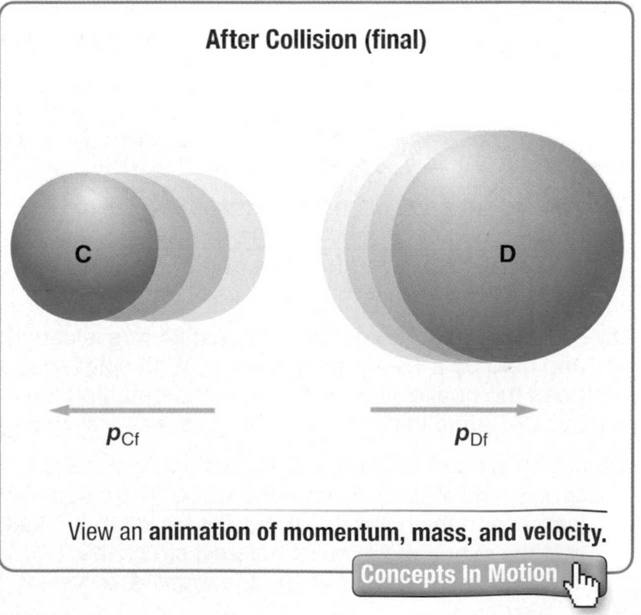

View an **animation of momentum, mass, and velocity.**

Concepts In Motion

EXAMPLE PROBLEM 3

Get help with **order of operations.** Math Handbook

SPEED A 1875-kg car going 23 m/s rear-ends a 1025-kg compact car going 17 m/s on ice in the same direction. The two cars stick together. How fast do the two attached cars move immediately after the collision?

1 ANALYZE AND SKETCH THE PROBLEM

- Define the system.
- Establish a coordinate system.
- Sketch the situation showing the before and after states.
- Draw a vector diagram for the momentum.

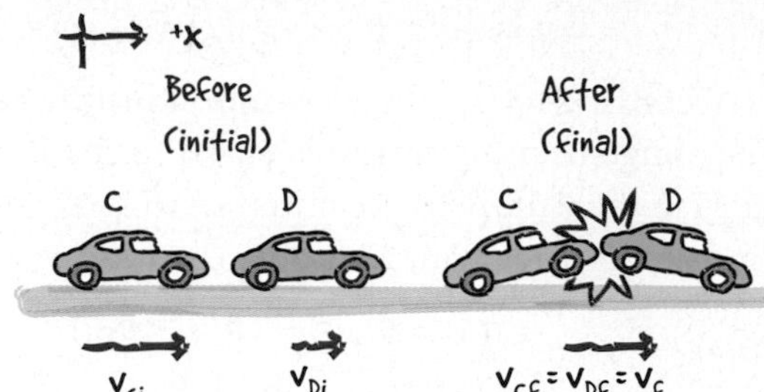

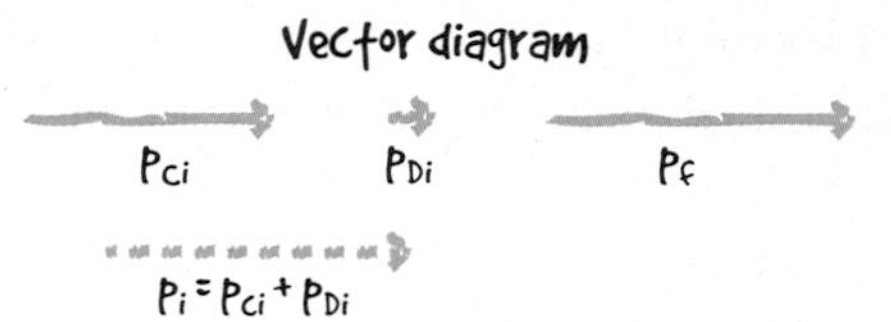

KNOWN		UNKNOWN
$m_C = 1875$ kg	$m_D = 1025$ kg	$\mathbf{v}_f = ?$
$\mathbf{v}_{Ci} = +23$ m/s	$\mathbf{v}_{Di} = +17$ m/s	

2 SOLVE FOR THE UNKNOWN

The system is the two cars. Momentum is conserved because the ice makes the total external force on the cars nearly zero.

$$\mathbf{p}_i = \mathbf{p}_f$$

$$\mathbf{p}_{Ci} + \mathbf{p}_{Di} = \mathbf{p}_{Cf} + \mathbf{p}_{Df}$$

$$m_C\mathbf{v}_{Ci} + m_D\mathbf{v}_{Di} = m_C\mathbf{v}_{Cf} + m_D\mathbf{v}_{Df}$$

Because the two cars stick together, their velocities after the collision, denoted as $\mathbf{v}_f$, are equal.

$$\mathbf{v}_{Cf} = \mathbf{v}_{Df} = \mathbf{v}_f$$

$$m_C\mathbf{v}_{Ci} + m_D\mathbf{v}_{Di} = (m_C + m_D)\mathbf{v}_f$$

Solve for $\mathbf{v}_f$.

$$\mathbf{v}_f = \frac{m_C\mathbf{v}_{Ci} + m_D\mathbf{v}_{Di}}{m_C + m_D}$$

$$= \frac{(1875\text{ kg})(+23\text{ m/s}) + (1025\text{ kg})(+17\text{ m/s})}{1875\text{ kg} + 1025\text{ kg}}$$

◀ Substitute $m_C = 1875$ kg, $v_{Ci} = +23$ m/s, $m_D = 1025$ kg, $v_{Di} = +17$ m/s.

$$= +21\text{ m/s}$$

3 EVALUATE THE ANSWER

- **Are the units correct?** Velocity is measured in meters per second.
- **Does the direction make sense?** $\mathbf{v}_{Ci}$ and $\mathbf{v}_{Di}$ are in the positive direction; therefore, $\mathbf{v}_f$ should be positive.
- **Is the magnitude realistic?** The magnitude of $\mathbf{v}_f$ is between the initial speeds of the two cars, but closer to the speed of the more massive one, so it is reasonable.

PRACTICE PROBLEMS

Do additional problems.

17. Two freight cars, each with a mass of 3.0×10^5 kg, collide and stick together. One was initially moving at 2.2 m/s, and the other was at rest. What is their final speed? Define the system as the two cars.

18. A 0.105-kg hockey puck moving at 24 m/s is caught and held by a 75-kg goalie at rest. With what speed does the goalie slide on the ice after catching the puck? Define the puck and the goalie as a system.

19. A 35.0-g bullet strikes a 5.0-kg stationary piece of lumber and embeds itself in the wood. The piece of lumber and the bullet fly off together at 8.6 m/s. What was the speed of the bullet before it struck the lumber? Define the bullet and the wood as a system.

20. A 35.0-g bullet moving at 475 m/s strikes a 2.5-kg bag of flour at rest on ice. The bullet passes through the bag and exits it at 275 m/s. How fast is the bag moving when the bullet exits?

21. The bullet in the previous problem strikes a 2.5-kg steel ball that is at rest. After the collision, the bullet bounces backward at 5.0 m/s. How fast is the ball moving when the bullet bounces backward?

22. CHALLENGE A 0.50-kg ball traveling at 6.0 m/s collides head-on with a 1.00-kg ball moving in the opposite direction at 12.0 m/s. After colliding, the 0.50-kg ball bounces backward at 14 m/s. Find the other ball's speed and direction after the collision.

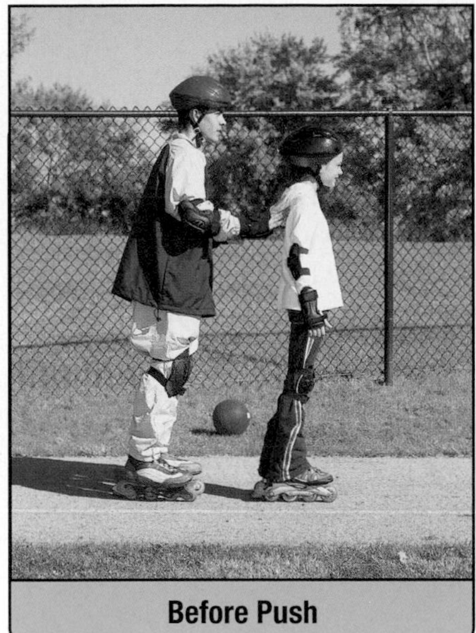

Figure 11 The boy exerts a force, causing the girl to move to the right. The boy's position relative to the ball shows that the equal but opposite force from the girl caused him to recoil to the left.

Infer How can you tell from the motion that the boy's mass is greater than the girl's mass?

Recoil

It is very important to define a system carefully. The momentum of a baseball changes when the external force of a bat is exerted on it. The baseball, therefore, is not an isolated system. On the other hand, the total momentum of two colliding balls within an isolated system does not change because all forces are between the objects within the system.

Can you find the final velocities of the in-line skaters in **Figure 11?** Assume they are skating on a smooth surface with no external forces. They both start at rest, one behind the other.

Skater C, the boy, gives skater D, the girl, a push. Now, both skaters are moving in opposite directions. Because the push was an internal force, you can use the law of conservation of momentum to find the skaters' relative velocities. The total momentum of the system was zero before the push. Therefore, it must be zero after the push.

Before		**After**
$\boldsymbol{p}_{Ci} + \boldsymbol{p}_{Di}$	$=$	$\boldsymbol{p}_{Cf} + \boldsymbol{p}_{Df}$
0	$=$	$\boldsymbol{p}_{Cf} + \boldsymbol{p}_{Df}$
$\boldsymbol{p}_{Cf}$	$=$	$-\boldsymbol{p}_{Df}$
$m_C\boldsymbol{v}_{Cf}$	$=$	$-m_D\boldsymbol{v}_{Df}$

The coordinate system can be chosen so that the positive direction is to the right. The momentums of the skaters after the push are equal in magnitude but opposite in direction. The backward motion of skater C is an example of recoil. Are the skaters' velocities equal and opposite? The last equation can be written to solve for the velocity of skater C.

$$\boldsymbol{v}_{Cf} = \left(\frac{-m_D}{m_C}\right)\boldsymbol{v}_{Df}$$

The velocities depend on the skaters' relative masses. If skater C has a mass of 68.0 kg and skater D's mass is 45.4 kg, then the ratio of their velocities will be 68.0 : 45.4, or 1.50. The less massive skater moves at the greater velocity. Without more information about how hard skater C pushed skater D, however, you cannot find the velocity of each skater.

Laura Sifferlin

PhysicsLAB

COLLIDING CARTS

PROBEWARE LAB What happens to the momentums of two carts when they collide?

MiniLAB

REBOUND HEIGHT

How do mass and velocity affect the momentum of a bouncing ball?

Investigate **a hover glider.**

Virtual Investigation

Propulsion in Space

View an **animation of thrust and momentum.**

Concepts In Motion

How does a rocket in space change velocity? The rocket carries both fuel and oxidizer. When the fuel and oxidizer combine in the rocket motor, the resulting hot gases leave the exhaust nozzle at high speed. The rocket and the chemicals form a closed system. The forces that expel the gases are internal forces, so the system is also an isolated system. Thus, objects in space can accelerate by using the law of conservation of momentum and Newton's third law of motion.

A newer type of space propulsion uses the recoil that results from the force of ions to move a spaceship. An ion thruster accelerates ions in an electric or magnetic field. As the ions are expelled in one direction, the conservation of momentum results in the movement of the spaceship in the opposite direction. The force exerted by an ion engine might only be a fraction of a newton, but running the engine for long periods of time can significantly change the velocitiy of a spaceship.

EXAMPLE PROBLEM 4

Find help with **isolating a variable.** Math Handbook

SPEED An astronaut at rest in space fires a thruster pistol that expels a quick burst of 35 g of hot gas at 875 m/s. The combined mass of the astronaut and pistol is 84 kg. How fast and in what direction is the astronaut moving after firing the pistol?

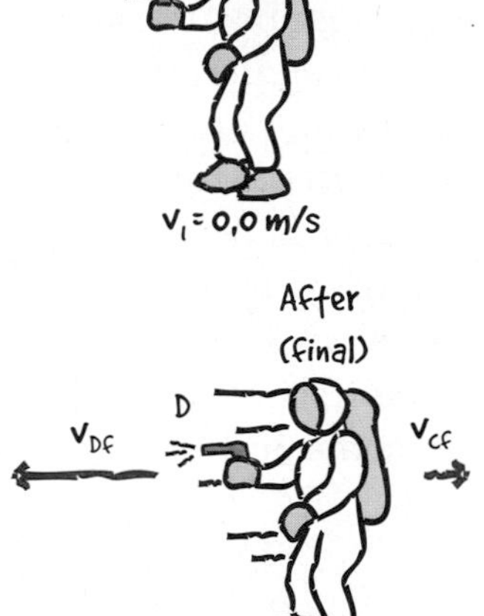

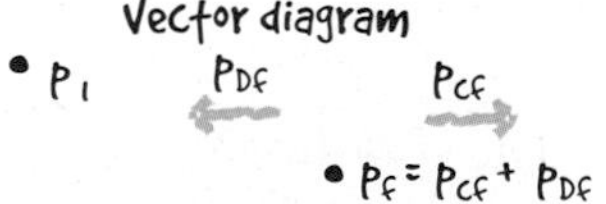

1 ANALYZE AND SKETCH THE PROBLEM

- Define the system and establish a coordinate axis.
- Sketch before and after conditions. Draw a vector diagram.

KNOWN		UNKNOWN
$m_C = 84$ kg	$\mathbf{v}_{Ci} = \mathbf{v}_{Di} = 0.0$ m/s	$\mathbf{v}_{Cf} = ?$
$m_D = 0.035$ kg	$\mathbf{v}_{Df} = -875$ m/s	

2 SOLVE FOR THE UNKNOWN

The system is the astronaut, the gun, and the gas.

$$\mathbf{p}_i = \mathbf{p}_{Ci} + \mathbf{p}_{Di} = 0.0 \text{ kg·m/s}$$

◀ Before the pistol is fired, the system is at rest; initial momentum is zero.

Use the law of conservation of momentum to find $\mathbf{p}_{Cf}$.

$$\mathbf{p}_i = \mathbf{p}_f$$

$$0.0 \text{ kg·m/s} = \mathbf{p}_{Cf} + \mathbf{p}_{Df}$$

$$\mathbf{p}_{Cf} = -\mathbf{p}_{Df}$$

◀ The momentum of the astronaut is equal in magnitude, but opposite in direction, to the momentum of the gas leaving the pistol.

Solve for the final velocity of the astronaut ($\mathbf{v}_{Cf}$).

$$m_C\mathbf{v}_{Cf} = -m_D\mathbf{v}_{Df}$$

$$\mathbf{v}_{Cf} = \frac{-m_D\mathbf{v}_{Df}}{m_C}$$

$$= \frac{-(0.035 \text{ kg})(-875 \text{ m/s})}{84 \text{ kg}}$$

◀ Substitute $m_D = 0.035$ kg, $v_{Df} = -875$ m/s, $m_C = 84$ kg.

$$= +0.36 \text{ m/s}$$

3 EVALUATE THE ANSWER

- **Are the units correct?** The velocity is measured in meters per second.
- **Does the direction make sense?** The astronaut's direction is opposite that of the gas.
- **Is the magnitude realistic?** The astronaut's mass is much larger than that of the gas, so the velocity of the astronaut is much less than that of the expelled gas.

PRACTICE PROBLEMS

Do additional problems. Online Practice 

23. A 4.00-kg model rocket is launched, expelling burned fuel with a mass of 50.0 g at a speed of 625 m/s. What is the velocity of the rocket after the fuel has burned? *Hint: Ignore the external forces of gravity and air resistance.*

24. A thread connects a 1.5-kg cart and a 4.5-kg cart. After the thread is burned, a compressed spring pushes the carts apart, giving the 1.5-kg cart a velocity of 27 cm/s to the left. What is the velocity of the 4.5-kg cart?

25. CHALLENGE Carmen and Judi row their canoe alongside a dock. They stop the canoe, but they do not secure it. The canoe can still move freely. Carmen, who has a mass of 80.0 kg, then steps out of the canoe onto the dock. As she leaves the canoe, Carmen moves forward at a speed of 4.0 m/s, causing the canoe, with Judi still in it, to move also. At what speed and in what direction do the canoe and Judi move if their combined mass is 115 kg?

Two-Dimensional Collisions

Until now you have considered momentum in only one dimension. The law of conservation of momentum holds for all closed systems with no external forces. It is valid regardless of the directions of the particles before or after they interact. But what happens in two or three dimensions? **Figure 12** shows a collision of two billiard balls. Consider the billiard balls to be the system. The original momentum of the moving ball is $\boldsymbol{p}_{\mathrm{Ci}}$, and the momentum of the stationary ball is zero. Therefore, the momentum of the system before the collision is $\boldsymbol{p}_{\mathrm{Ci}}$.

After the collision, both balls are moving and have momentums. Ignoring friction with the tabletop, the system is closed and isolated. Thus, the law of conservation of momentum can be used. The initial momentum equals the sum of the final momentums, so $\boldsymbol{p}_{\mathrm{Ci}} = \boldsymbol{p}_{\mathrm{Cf}} + \boldsymbol{p}_{\mathrm{Df}}$.

The sum of the components of the vectors before and after the collision must also be equal. If the *x*-axis is defined in the direction of the initial momentum, then the *y*-component of the initial momentum is zero. The sum of the final *y*-components also must be zero.

$$\boldsymbol{p}_{\mathrm{Cf},\,y} + \boldsymbol{p}_{\mathrm{Df},\,y} = 0$$

The *y*-components are equal in magnitude but are in the opposite direction and, thus, have opposite signs. The sums of the horizontal components before and after the collision also are equal.

$$\boldsymbol{p}_{\mathrm{Ci}} = \boldsymbol{p}_{\mathrm{Cf},\,x} + \boldsymbol{p}_{\mathrm{Df},\,x}$$

View an **animation of conservation of momentum.**

MiniLAB

MOMENTUM

How can you use the law of conservation of momentum to determine an object's velocity after a collision?

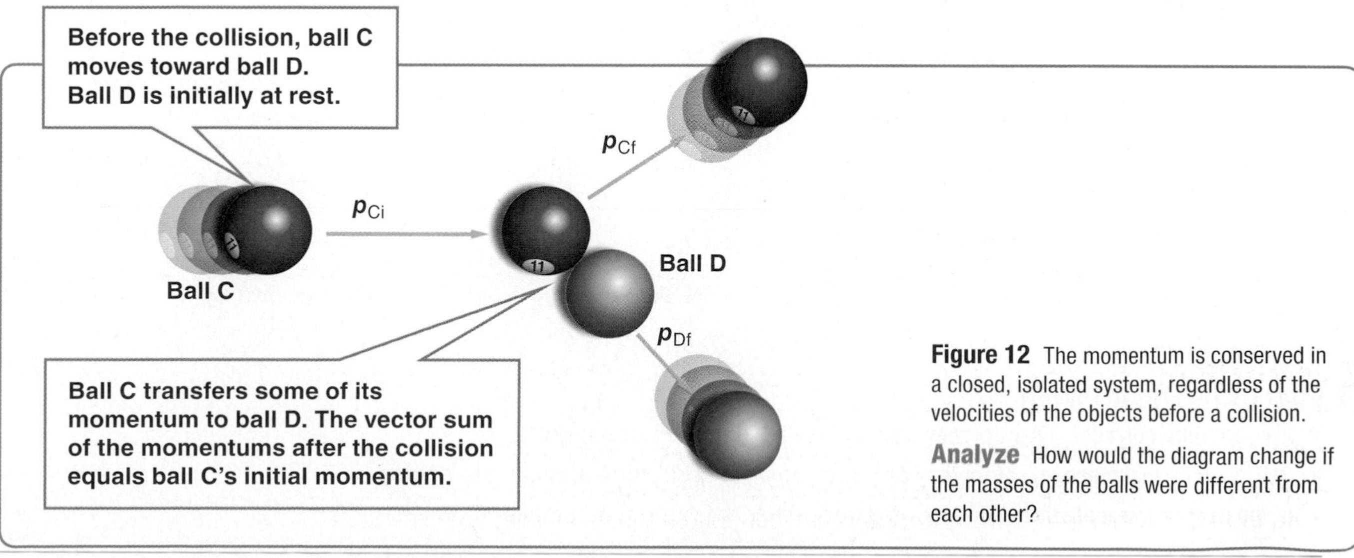

Before the collision, ball C moves toward ball D. Ball D is initially at rest.
p_{Ci}
p_{Cf}
Ball C
Ball D
p_{Df}
Ball C transfers some of its momentum to ball D. The vector sum of the momentums after the collision equals ball C's initial momentum.

Figure 12 The momentum is conserved in a closed, isolated system, regardless of the velocities of the objects before a collision.

Analyze How would the diagram change if the masses of the balls were different from each other?

EXAMPLE PROBLEM 5

Get help with **momentum problems.** Personal Tutor

EXAMPLE PROBLEM

SPEED A 1325-kg car, C, moving north at 27.0 m/s, collides with a 2165-kg car, D, moving east at 11.0 m/s. The collision causes the two cars to stick together. In what direction and with what speed do they move after the collision?

Before (initial)

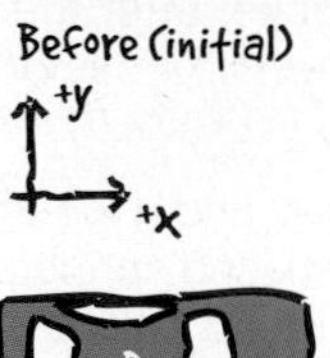

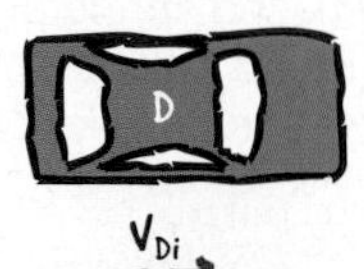

After (initial)

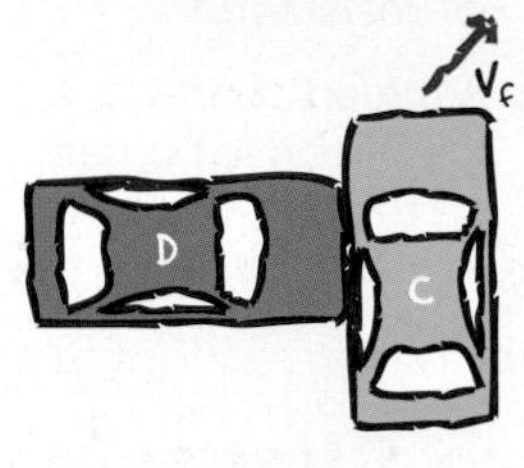

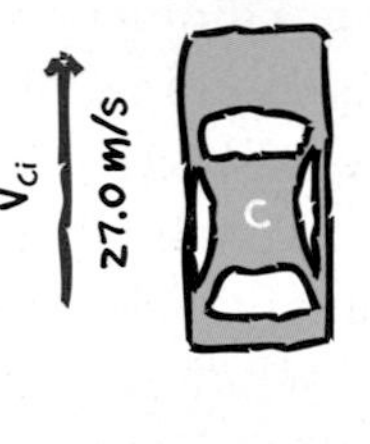

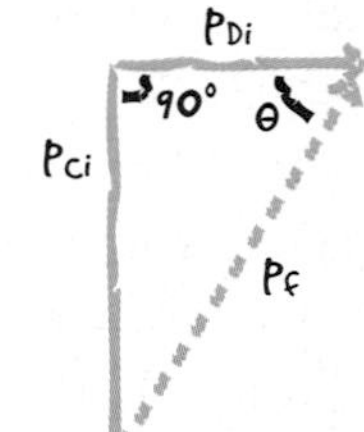

1 ANALYZE AND SKETCH THE PROBLEM

- Define the system.
- Sketch the before and after states.
- Establish the coordinate axis with the *y*-axis north and the *x*-axis east.
- Draw a momentum-vector diagram.

KNOWN	UNKNOWN
$m_C = 1325$ kg	$v_f = ?$
$m_D = 2165$ kg	$\theta = ?$
$v_{Ci,\,y} = 27.0$ m/s	
$v_{Di,\,x} = 11.0$ m/s	

2 SOLVE FOR THE UNKNOWN

Define the system as the two cars. Determine the magnitudes of the initial momenta of the cars and the momentum of the system.

$$p_{Ci} = m_C v_{Ci,\,y}$$
$$= (1325\text{ kg})(27.0\text{ m/s})$$
$$= 3.58\times10^4\text{ kg·m/s (north)}$$

◀ Substitute $m_C = 1325$ kg, $v_{Ci,\,y} = 27.0$ m/s.

$$p_{Di} = m_D v_{Di,\,x}$$
$$= (2165\text{ kg})(11.0\text{ m/s})$$
$$= 2.38\times10^4\text{ kg·m/s (east)}$$

◀ Substitute $m_D = 2165$ kg, $v_{Di,\,x} = 11.0$ m/s.

Use the law of conservation of momentum to find p_f.

$$p_{f,\,x} = p_{i,\,x} = 2.38\times10^4\text{ kg·m/s}$$

◀ Substitute $p_{i,\,x} = p_{Di} = 2.38\times10^4$ kg·m/s.

$$p_{f,\,y} = p_{i,\,y} = 3.58\times10^4\text{ kg·m/s}$$

◀ Substitute $p_{i,\,y} = p_{Ci} = 3.58\times10^4$ kg·m/s.

Use the diagram to set up equations for $p_{f,\,x}$ and $p_{f,\,y}$.

$$p_f = \sqrt{(p_{f,\,x})^2 + (p_{f,\,y})^2}$$
$$= \sqrt{(2.38\times10^4\text{ kg·m/s})^2 + (3.58\times10^4\text{ kg·m/s})^2}$$
$$= 4.30\times10^4\text{ kg·m/s}$$

◀ Substitute $p_{f,\,x} = 2.38\times10^4$ kg·m/s, $p_{f,\,y} = 3.58\times10^4$ kg·m/s.

Solve for θ.

$$\theta = \tan^{-1}\left(\frac{p_{f,\,y}}{p_{f,\,x}}\right)$$
$$= \tan^{-1}\left(\frac{3.58\times10^4\text{ kg·m/s}}{2.38\times10^4\text{ kg·m/s}}\right)$$
$$= 56.4°$$

◀ Substitute $p_{f,\,y} = 3.58\times10^4$ kg·m/s, $p_{f,\,x} = 2.38\times10^4$ kg·m/s.

Determine the final speed.

$$v_f = \frac{p_f}{m_C + m_D}$$
$$= \frac{4.30\times10^4\text{ kg·m/s}}{1325\text{ kg} + 2165\text{ kg}}$$
$$= 12.3\text{ m/s}$$

◀ Substitute $p_f = 4.30\times10^4$ kg·m/s, $m_C = 1325$ kg, $m_D = 2165$ kg.

3 EVALUATE THE ANSWER

- **Are the units correct?** The correct unit for speed is meters per second.
- **Do the signs make sense?** Answers are both positive and at the appropriate angles.
- **Is the magnitude realistic?** The cars stick together, so v_f must be smaller than v_{Ci}.

PRACTICE PROBLEMS

Do additional problems. Online Practice

26. A 925-kg car moving north at 20.1 m/s collides with a 1865-kg car moving west at 13.4 m/s. After the collision, the two cars are stuck together. In what direction and at what speed do they move after the collision? Define the system as the two cars.

27. A 1383-kg car moving south at 11.2 m/s is struck by a 1732-kg car moving east at 31.3 m/s. After the collision, the cars are stuck together. How fast and in what direction do they move immediately after the collision? Define the system as the two cars.

28. A 1345-kg car moving east at 15.7 m/s is struck by a 1923-kg car moving north. They stick together and move with a velocity of 14.5 m/s at $\theta = 63.5°$. Was the north-moving car exceeding the 20.1 m/s speed limit?

29. CHALLENGE A stationary billiard ball with mass 0.17 kg is struck by an identical ball moving 4.0 m/s. Afterwards, the second ball moves 60.0° to the left of its original direction. The stationary ball moves 30.0° to the right of the moving ball's original direction. What is the velocity of each ball after the collision?

Conservation of Angular Momentum

Like linear momentum, angular momentum can be conserved. The **law of conservation of angular momentum** states that if no net external torque acts on a closed system, then its angular momentum does not change, as represented by the following equation.

LAW OF CONSERVATION OF ANGULAR MOMENTUM

An isolated system's initial angular momentum is equal to its final angular momentum.

$$L_i = L_f$$

The spinning ice-skater in **Figure 13** demonstrates conservation of angular momentum. When he pulls in his arms, he spins faster. Without an external torque, his angular momentum does not change and $L = I\omega$ is constant. The increased angular velocity must be accompanied by a decreased moment of inertia. By pulling in his arms, the skater brings more mass closer to the axis of rotation, decreasing the radius of rotation and decreasing his moment of inertia. You can calculate changes in angular velocity using the law of conservation of angular momentum.

$$L_i = L_f$$

$$\text{thus, } I_i\omega_i = I_f\omega_f$$

$$\frac{\omega_f}{\omega_i} = \frac{I_i}{I_f}$$

Figure 13 When an ice-skater tucks his arms, his moment of inertia decreases. Because angular momentum is conserved, his angular velocity increases.

Figure 14 If you spin a top at a tilt, Earth's gravity exerts a torque on it, causing the upper end to precess.

Decide Why does the top precess only when it is tilted?

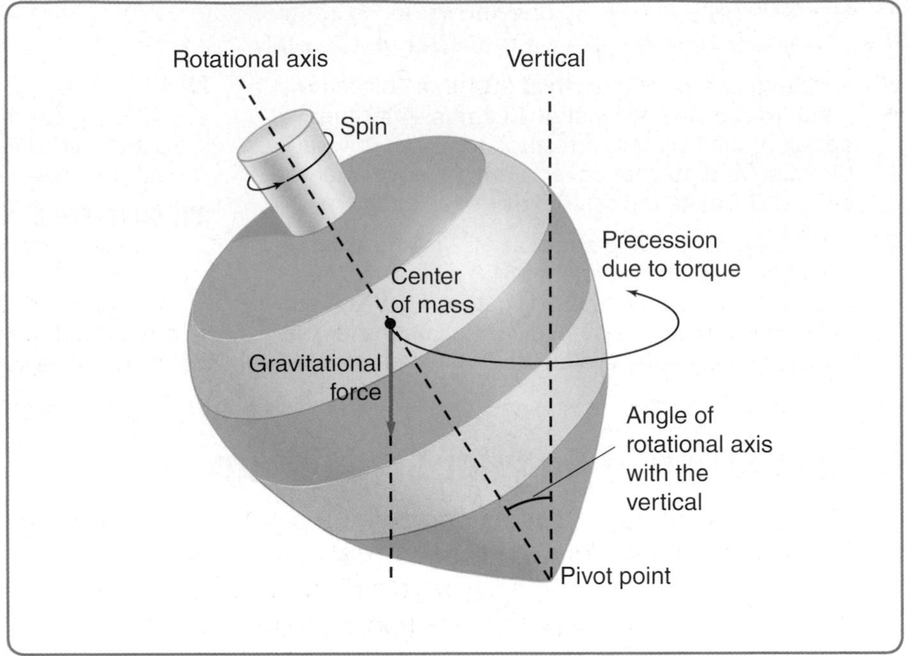

Tops and Gyroscopes

Because of the conservation of angular momentum, the direction of rotation of a spinning object can be changed only by applying a torque. If you played with a top as a child, you may have spun it by twisting its axle between your fingers or by pulling the string wrapped around its axle. When a top is vertical, the force of gravity on the center of mass of the top points through the pivot point. As a result, there is no torque on the top, and the axis of its rotation does not change. If the top is tipped, as shown in **Figure 14,** a torque tends to rotate it downward. Rather than tipping over, however, the axis of the top slowly precesses, which means it revolves slightly away from the vertical line.

Earth's rotation is another example of precession of a spinning object. Earth is not a perfect sphere. Because it has an equatorial bulge, the gravitational pull of the Sun exerts a torque on it, causing it to precess. It takes about 26,000 years for Earth's rotational axis to go through one cycle of precession.

READING CHECK Analyze Why does Earth precess as it rotates?

PhysicsLAB

ROTATION OF A WHEEL

What can set you spinning?

PHYSICS CHALLENGE

COLLIDING CARS Your friend was driving her 1265-kg car north on Oak Street when she was hit by a 925-kg compact car going west on Maple Street. The cars stuck together and slid 23.1 m at 42° north of west. The speed limit on both streets is 22 m/s (50 mph). Define the two cars as a system. Assume that momentum was conserved during the collision and that acceleration was constant during the skid. The coefficient of kinetic friction between the tires and the pavement is 0.65.

1. Your friend claims she was not speeding but the driver of the other car was. How fast was your friend driving before the crash?
2. How fast was the other car moving before the crash? Can you support your friend's case in court?

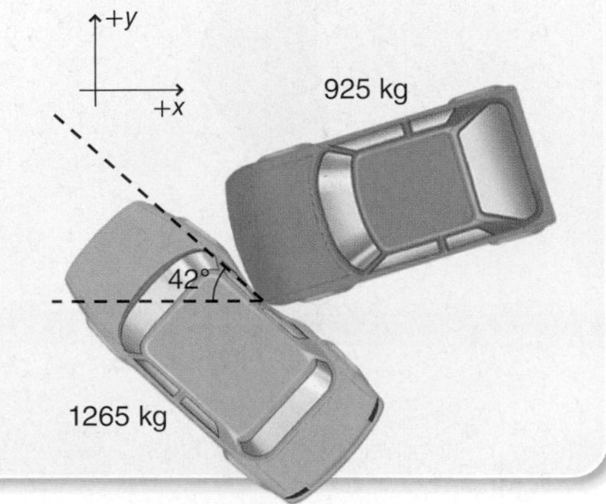

A gyroscope, such as the one shown in **Figure 15,** is a wheel or disk that spins rapidly around one axis while being free to rotate around one or two other axes. The direction of its large angular momentum can be changed only by applying an appropriate torque. Without such a torque, the direction of the axis of rotation does not change.

Gyroscopes are used in airplanes, submarines, and spacecraft to keep an unchanging reference direction. Giant gyroscopes are used in cruise ships to detect changes in orientation so that the ship can reduce its motion in rough water. Gyroscopic compasses, unlike magnetic compasses, maintain direction even when they are not on a level surface.

A football quarterback uses the gyroscope effect to make an accurate forward pass. A spiral is a pass in which the football spins around its longer axis. If the quarterback throws the ball in the direction of its spin axis of rotation, the ball keeps its pointed end forward, thereby reducing air resistance. Thus, the ball can be thrown far and accurately. If its spin direction is slightly off, the ball wobbles. If the ball is not spun, it tumbles end over end.

Spin also stabilizes the flight of a flying disk. A well-spun plastic disk can fly many meters through the air without wobbling. Some people are able to perform tricks with a yo-yo because the yo-yo's fast rotational speed keeps it rotating in one plane.

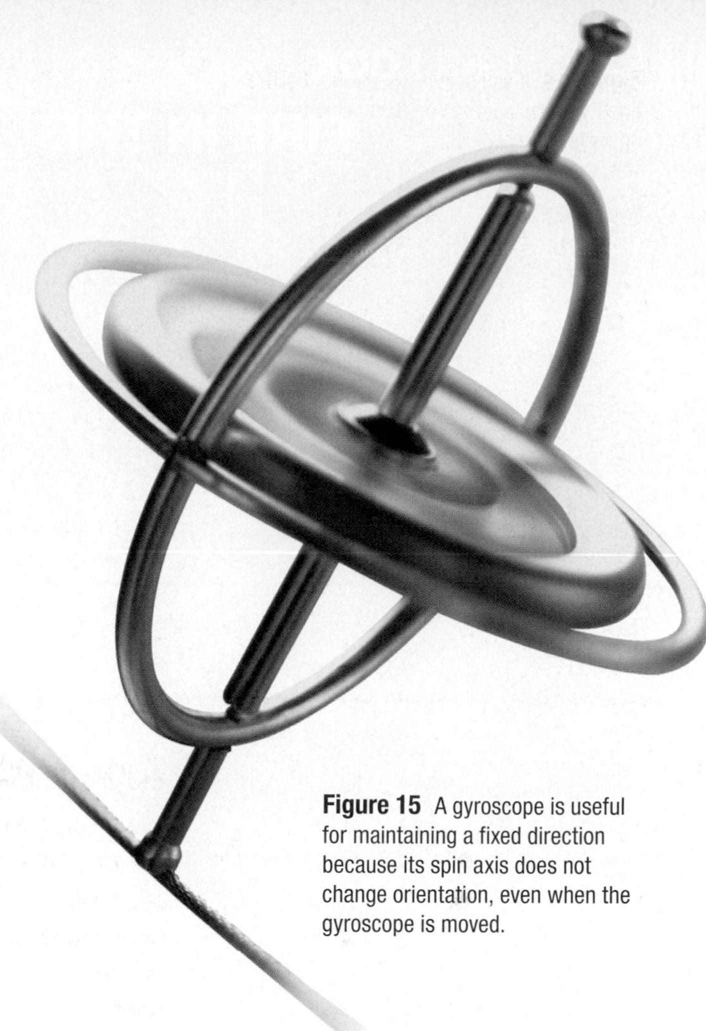

Figure 15 A gyroscope is useful for maintaining a fixed direction because its spin axis does not change orientation, even when the gyroscope is moved.

SECTION 2 REVIEW

Section Self-Check Check your understanding.

30. MAIN IDEA The outer rim of a plastic disk is thick and heavy. Besides making it easier to catch, how does this affect the rotational properties of the plastic disk?

31. Speed A cart, weighing 24.5 N, is released from rest on a 1.00-m ramp, inclined at an angle of 30.0° as shown in **Figure 16.** The cart rolls down the incline and strikes a second cart weighing 36.8 N.

a. Define the two carts as the system. Calculate the speed of the first cart at the bottom of the incline.

b. If the two carts stick together, with what initial speed will they move along?

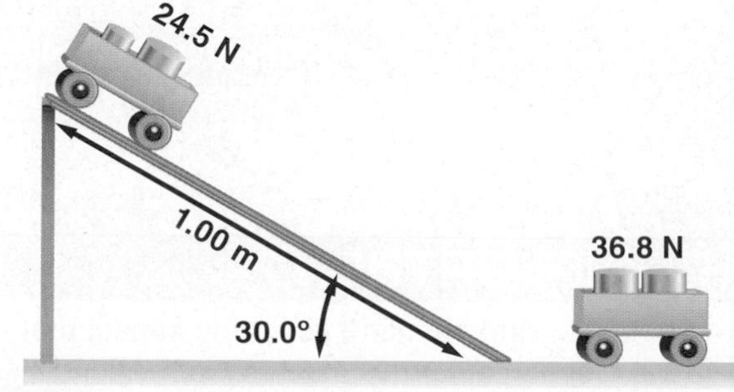

Figure 16

32. Conservation of Momentum During a tennis serve, the racket of a tennis player continues forward after it hits the ball. Is momentum conserved in the collision between the tennis racket and the ball? Explain, making sure you define the system.

33. Momentum A pole-vaulter runs toward the launch point with horizontal momentum. Where does the vertical momentum come from as the athlete vaults over the crossbar?

34. Initial Momentum During a soccer game, two players come from opposite directions and collide when trying to head the ball. The players come to rest in midair and fall to the ground. Describe their initial momentums.

35. Critical Thinking You catch a heavy ball while you are standing on a skateboard, and then you roll backward. If you were standing on the ground, however, you would be able to avoid moving while catching the ball.

a. Identify the system you use in each case.

b. Explain both situations using the law of conservation of momentum.

FIRE IN THE SKY

Imagine a giant asteroid the size of a small city crashing into the planet. Its momentum would certainly move matter on Earth.

1. Scientists think such a collision did occur 65 million years ago, and its effects may have caused the extinction of the dinosaurs. This impact left the crater at Chicxulub, Mexico.

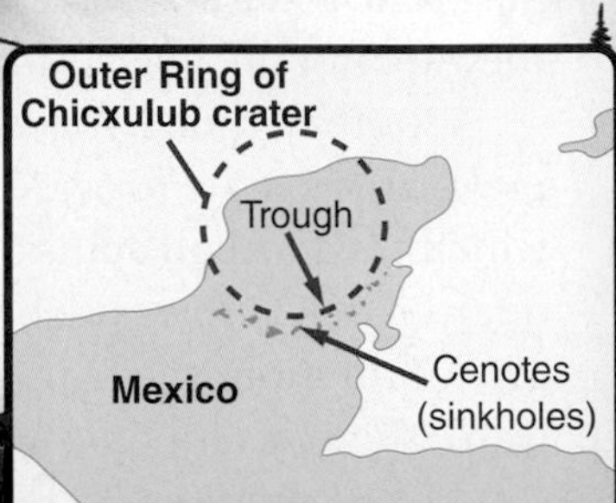

2. Astronomers are tracking more than 1,000 Near Earth Objects (NEOs) that are greater than 1 km in diameter—big enough to devastate the area near where they land. So far, none are predicted to collide with Earth.

3. The impact of a giant asteroid would send up debris that might darken the sky for months. The impulse from its crash might also set in motion giant ocean waves that would flood shores around the globe.

4. Should scientists detect an NEO that is a threat, world leaders hope to find a way to send spacecraft to meet it and alter its course.

5. Some worry that a dangerous NEO might have too much momentum for human technology to significantly change that NEO's path. Scientists hope that early detection will give people plenty of time to prepare for any natural disasters that could be caused by a collision.

GOING FURTHER >>>

Research Find out more about the impact that resulted in the creation of Earth's moon. What evidence is there that such an impact occurred?

CHAPTER 9 STUDY GUIDE

BIGIDEA If the net force on a closed system is zero, the total momentum of that system is conserved.

VOCABULARY
- **impulse** *(p. 236)*
- **momentum** *(p. 237)*
- **impulse-momentum theorem** *(p. 237)*
- **angular momentum** *(p. 240)*
- **angular impulse-angular momentum theorem** *(p. 240)*

SECTION 1 Impulse and Momentum

MAINIDEA An object's momentum is equal to its mass multiplied by its velocity.

- The impulse on an object is the average net force exerted on the object multiplied by the time interval over which the force acts.

$$\textbf{\textit{impulse}} = \mathbf{F}\Delta t$$

- The momentum of an object is the product of its mass and velocity and is a vector quantity.

$$\mathbf{p} = m\mathbf{v}$$

When solving a momentum problem, first define the objects in the system and examine their momentum before and after the event. The impulse on an object is equal to the change in momentum of the object.

$$\mathbf{F}\Delta t = \mathbf{p}_f - \mathbf{p}_i$$

- The angular momentum of a rotating object is the product of its moment of inertia and its angular velocity.

$$L = I\omega$$

The angular impulse-angular momentum theorem states that the angular impulse on an object is equal to the change in the object's angular momentum.

$$\tau\Delta t = L_f - L_i$$

VOCABULARY
- **closed system** *(p. 245)*
- **isolated system** *(p. 245)*
- **law of conservation of momentum** *(p. 245)*
- **law of conservation of angular momentum** *(p. 251)*

SECTION 2 Conservation of Momentum

MAINIDEA In a closed, isolated system, linear momentum and angular momentum are conserved.

- According to Newton's third law of motion and the law of conservation of momentum, the forces exerted by colliding objects on each other are equal in magnitude and opposite in direction.
- Momentum is conserved in a closed, isolated system.

$$\mathbf{p}_f = \mathbf{p}_i$$

- The law of conservation of momentum relates the momentums of objects before and after a collision. Use vector analysis to solve momentum-conservation problems in two dimensions. The law of conservation of angular momentum states that if there are no external torques acting on a system, then the angular momentum is conserved.

$$L_f = L_i$$

Because angular momentum is conserved, the direction of rotation of a spinning object can be changed only by applying a torque.

Games and Multilingual eGlossary

Vocabulary Practice

CHAPTER 9 ASSESSMENT

SECTION 1 Impulse and Momentum

Mastering Concepts

36. Can a bullet have the same momentum as a truck? Explain.

37. During a baseball game, a pitcher throws a curve ball to the catcher. Assume that the speed of the ball does not change in flight.

a. Which player exerts the larger impulse on the ball?

b. Which player exerts the larger force on the ball?

38. Newton's second law of motion states that if no net force is exerted on a system, no acceleration is possible. Does it follow that no change in momentum can occur?

39. Why are cars made with bumpers that can be pushed in during a crash?

40. An ice-skater is doing a spin.

a. How can the skater's angular momentum be changed?

b. How can the skater's angular velocity be changed without changing the angular momentum?

Mastering Problems

41. Golf Rocío strikes a 0.058-kg golf ball with a force of 272 N and gives it a velocity of 62.0 m/s. How long was Rocío's club in contact with the ball?

42. A 0.145-kg baseball is pitched at 42 m/s. The batter then hits the ball horizontally toward the pitcher at 58 m/s.

a. Find the change in momentum of the ball.

b. If the ball and the bat are in contact for 4.6×10^{-4} s, what is the average force during contact?

43. A 0.150-kg ball, moving in the positive direction at 12 m/s, is acted on by the impulse illustrated in the graph in **Figure 17.** What is the ball's speed at 4.0 s?

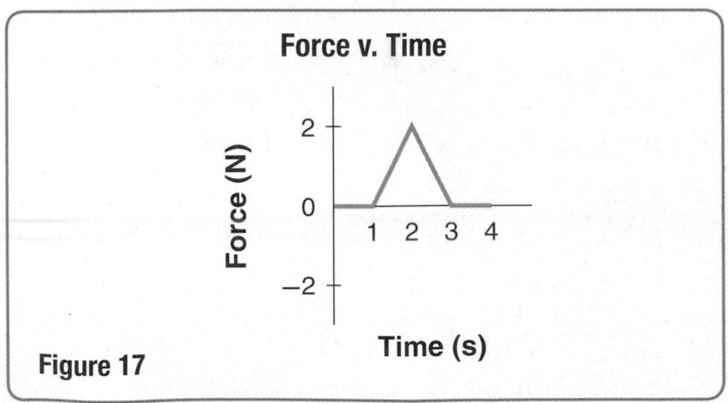

Figure 17

44. Bowling A force of 186 N acts on a 7.3-kg bowling ball for 0.40 s. What is the ball's change in momentum? What is its change in velocity?

45. Hockey A hockey puck has a mass of 0.115 kg and strikes the pole of the net at 37 m/s. It bounces off in the opposite direction at 25 m/s, as shown in **Figure 18.**

a. What is the impulse on the puck?

b. If the collision takes 5.0×10^{-4} s, what is the average force on the puck?

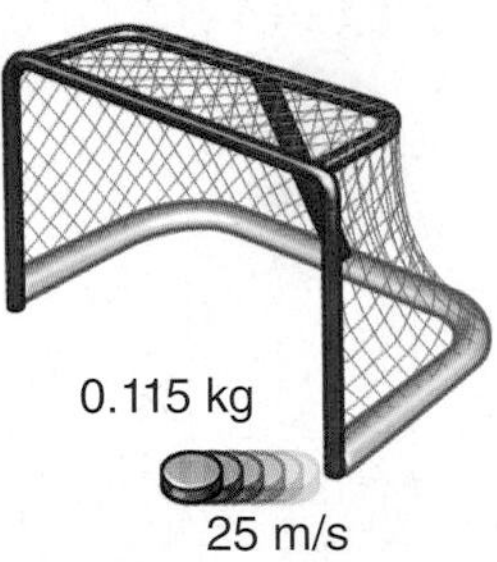

Figure 18

46. A 5500-kg freight truck accelerates from 4.2 m/s to 7.8 m/s in 15.0 s by the application of a constant force.

a. What change in momentum occurs?

b. How large a force is exerted?

47. In a ballistics test at the police department, Officer Rios fires a 6.0-g bullet at 350 m/s into a container that stops it in 1.8 ms. What is the average force that stops the bullet?

48. Volleyball A 0.24-kg volleyball approaches Tina with a velocity of 3.8 m/s. Tina bumps the ball, giving it a speed of 2.4 m/s but in the opposite direction. What average force did she apply if the interaction time between her hands and the ball was 0.025 s?

49. Before a collision, a 25-kg object was moving at 112 m/s. Find the impulse that acted on the object if, after the collision, it moved at the following velocities.

a. +8.0 m/s

b. −8.0 m/s

50. Baseball A 0.145-kg baseball is moving at 35 m/s when it is caught by a player.

a. Find the change in momentum of the ball.

b. If the ball is caught with the mitt held in a stationary position so that the ball stops in 0.050 s, what is the average force exerted on the ball?

c. If, instead, the mitt is moving backward so that the ball takes 0.500 s to stop, what is the average force exerted by the mitt on the ball?

51. Ranking Task Rank the following objects according to the amount of momentum they have, from least to greatest. Specifically indicate any ties.

Object A: mass 2.5 kg, velocity 1.0 m/s east

Object B: mass 3.0 kg, velocity 0.90 m/s west

Object C: mass 3.0 kg, velocity 1.2 m/s west

Object D: mass 4.0 kg, velocity 0.50 m/s north

Object E: mass 4.0 kg, velocity 0.90 m/s east

52. Hockey A hockey player makes a slap shot, exerting a constant force of 30.0 N on the puck for 0.16 s. What is the magnitude of the impulse on the puck?

53. Skateboarding Your brother's mass is 35.6 kg, and he has a 1.3-kg skateboard. What is the combined momentum of your brother and his skateboard if they are moving at 9.50 m/s?

54. A hockey puck has a mass of 0.115 kg and is at rest. A hockey player makes a shot, exerting a constant force of 30.0 N on the puck for 0.16 s. With what speed does it head toward the goal?

55. A nitrogen molecule with a mass of 4.7×10^{-26} kg, moving at 550 m/s, strikes the wall of a container and bounces back at the same speed.

a. What is the molecule's impulse on the wall?

b. If there are 1.5×10^{23} of these collisions each second, what is the average force on the wall?

56. Rockets Small rockets are used to slightly adjust the speeds of spacecraft. A rocket with a thrust of 35 N is fired to change a 72,000-kg spacecraft's speed by 63 cm/s. For how long should it be fired?

57. An animal rescue plane flying due east at 36.0 m/s drops a 175-N bale of hay from an altitude of 60.0 m, as shown in **Figure 19.** What is the momentum of the bale the moment before it strikes the ground? Give both magnitude and direction.

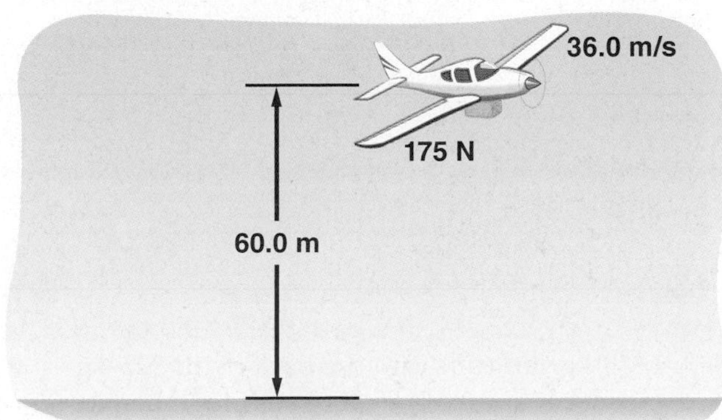

Figure 19

58. Accident A car moving at 10.0 m/s crashes into a barrier and stops in 0.050 s. There is a 20.0-kg child in the car. Assume that the child's velocity is changed by the same amount as that of the car, and in the same time period.

a. What is the impulse needed to stop the child?

b. What is the average force on the child?

c. What is the approximate mass of an object whose weight equals the force in part **b?**

d. Could you lift such a weight with your arm?

e. Why is it advisable to use a proper restraining seat rather than hold a child on your lap?

59. Reverse Problem Write a physics problem with real-life objects for which the following equation would be part of the solution:

$$F=\frac{(1.3\text{ kg})(20.0\text{ cm/s} - 0.0\text{ cm/s})}{0.55\text{ s}}$$

SECTION 2 Conservation of Momentum

Mastering Concepts

60. What is meant by "an isolated system"?

61. A spacecraft in outer space increases its velocity by firing its rockets. How can hot gases escaping from its rocket engine change the velocity of the craft when there is nothing in space for the gases to push against?

62. A cue ball travels across a pool table and collides with the stationary eight ball. The two balls have equal masses. After the collision, the cue ball is at rest. What must be true regarding the speed of the eight ball?

63. BIGIDEA Consider a ball falling toward Earth.

a. Why is the momentum of the ball not conserved?

b. In what system that includes the falling ball is the momentum conserved?

64. A falling basketball hits the floor. Just before it hits, the momentum is in the downward direction, and after it hits the floor, the momentum is in the upward direction.

a. Why isn't the momentum of the basketball conserved even though the bounce is a collision?

b. In what system is the momentum conserved?

65. Only an external force can change the momentum of a system. Explain how the internal force of a car's brakes brings the car to a stop.

66. Children's playgrounds often have circular-motion rides. How could a child change the angular momentum of such a ride as it is turning?

Chapter Self-Check

67. Problem Posing Complete this problem so that it can be solved using conservation of momentum: "Armando, mass 60.0 kg, is at the ice-skating rink . . ."

Mastering Problems

68. A 12.0-g rubber bullet travels at a forward velocity of 150 m/s, hits a stationary 8.5-kg concrete block resting on a frictionless surface, and ricochets in the opposite direction with a velocity of −110 m/s, as shown in **Figure 20.** How fast will the concrete block be moving?

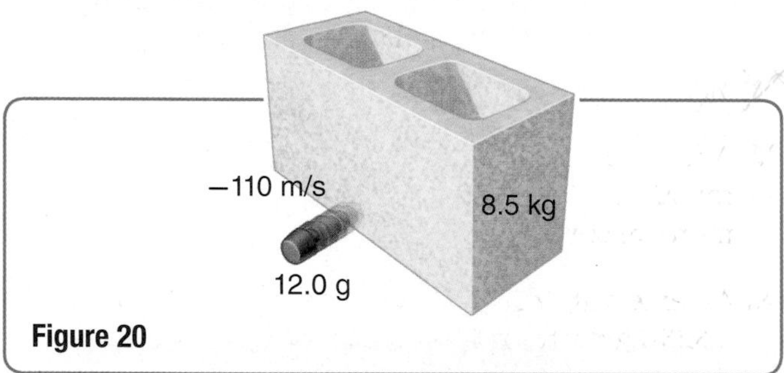

Figure 20

69. Football A 95-kg fullback, running at 8.2 m/s, collides in midair with a 128-kg defensive tackle moving in the opposite direction. Both players end up with zero speed.

a. Identify the before and after situations, and draw a diagram of both.

b. What was the fullback's momentum before the collision?

c. What was the change in the fullback's momentum?

d. What was the change in the defensive tackle's momentum?

e. What was the defensive tackle's original momentum?

f. How fast was the defensive tackle moving originally?

70. Marble C, with mass 5.0 g, moves at a speed of 20.0 cm/s. It collides with a second marble, D, with mass 10.0 g, moving at 10.0 cm/s in the same direction. After the collision, marble C continues with a speed of 8.0 cm/s in the same direction.

a. Sketch the situation, and identify the system. Identify the before and after situations, and set up a coordinate system.

b. Calculate the marbles' momentums before the collision.

c. Calculate the momentum of marble C after the collision.

d. Calculate the momentum of marble D after the collision.

e. What is the speed of marble D after the collision?

71. Two lab carts are pushed together with a spring mechanism compressed between them. Upon being released, the 5.0-kg cart repels with a velocity of 0.12 m/s in one direction, while the 2.0-kg cart goes in the opposite direction. What is the velocity of the 2.0-kg cart?

72. A 50.0-g projectile is launched with a horizontal velocity of 647 m/s from a 4.65-kg launcher moving in the same direction at 2.00 m/s. What is the launcher's velocity after the launch?

73. Skateboarding Kofi, with mass 42.00 kg, is riding a skateboard with a mass of 2.00 kg and traveling at 1.20 m/s. Kofi jumps off, and the skateboard stops dead in its tracks. In what direction and with what velocity did he jump?

74. In-line Skating Diego and Keshia are on in-line skates. They stand face-to-face and then push each other away with their hands. Diego has a mass of 90.0 kg, and Keshia has a mass of 60.0 kg.

a. Sketch the event, identifying the before and after situations, and set up a coordinate axis.

b. Find the ratio of the skaters' velocities just after their hands lose contact.

c. Which skater has the greater speed?

d. Which skater pushed harder?

75. Billiards A cue ball, with mass 0.16 kg, rolling at 4.0 m/s, hits a stationary eight ball of similar mass. If the cue ball travels 45° to the left of its original path and the eight ball travels 45° in the opposite direction, as shown in **Figure 21,** what is the velocity of each ball after the collision?

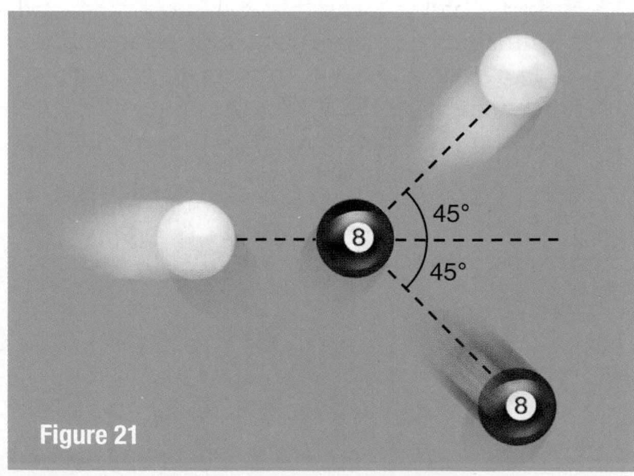

Figure 21

76. A 2575-kg van runs into the back of an 825-kg compact car at rest. They move off together at 8.5 m/s. Assuming that the friction with the road is negligible, calculate the initial speed of the van.

77. A 0.200-kg plastic ball has a forward velocity of 0.30 m/s. It collides with a second plastic ball of mass 0.100 kg, which is moving along the same line at a speed of 0.10 m/s. After the collision, both balls continue moving in the same, original direction. The speed of the 0.100-kg ball is 0.26 m/s. What is the new velocity of the 0.200-kg ball?

Applying Concepts

78. Explain the concept of impulse using physical ideas rather than mathematics.

79. An object initially at rest experiences the impulses described by the graph in **Figure 22.** Describe the object's motion after impulses A, B, and C.

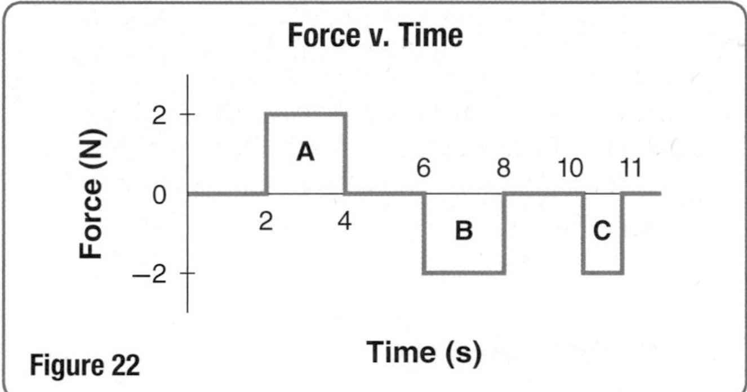

Figure 22

80. Is it possible for an object to obtain a larger impulse from a smaller force than it does from a larger force? Explain.

81. **Foul Ball** You are sitting at a baseball game when a foul ball comes in your direction. You prepare to catch it bare-handed. To catch it safely, should you move your hands toward the ball, hold them still, or move them in the same direction as the moving ball? Explain.

82. A 0.11-g bullet leaves a pistol at 323 m/s, while a similar bullet leaves a rifle at 396 m/s. Explain the difference in exit speeds of the two bullets, assuming that the forces exerted on the bullets by the expanding gases have the same magnitude.

83. During a space walk, the tether connecting an astronaut to the spaceship breaks. Using a gas pistol, the astronaut manages to get back to the ship. Use the language of the impulse-momentum theorem and a diagram to explain why this method was effective.

84. **Tennis Ball** As a tennis ball bounces off a wall, its momentum is reversed. Explain this action in terms of the law of conservation of momentum. Define the system and draw a diagram as a part of your explanation.

85. Two trucks that appear to be identical collide on an icy road. One was originally at rest. The trucks are stuck together and move at more than half the original speed of the moving truck. What can you conclude about the contents of the two trucks?

86. **Bullets** Two bullets of equal mass are shot at equal speeds at equal blocks of wood on a smooth ice rink. One bullet, made of rubber, bounces off the wood. The other bullet, made of aluminum, burrows into the wood. In which case does the block of wood move faster? Explain.

Mixed Review

87. A constant force of 6.00 N acts on a 3.00-kg object for 10.0 s. What are the changes in the object's momentum and velocity?

88. An external, constant force changes the speed of a 625-kg car from 10.0 m/s to 44.0 m/s in 68.0 s.

a. What is the car's change in momentum?

b. What is the magnitude of the force?

89. **Gymnastics Figure 23** shows a gymnast performing a routine. First, she does giant swings on the upper bar, holding her body straight and pivoting around her hands. Then, she lets go of the high bar and grabs her knees with her hands in the tuck position. Finally, she straightens up and lands on her feet.

a. In the second and final parts of the gymnast's routine, around what axis does she spin?

b. Rank in order, from greatest to least, her moments of inertia for the three positions.

c. Rank in order, from greatest to least, her angular velocities in the three positions.

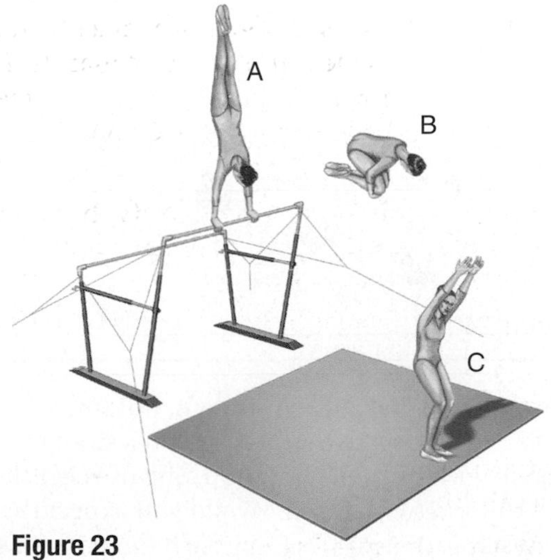

Figure 23

Chapter Self-Check

90. Dragster An 845-kg dragster accelerates on a race track from rest to 100.0 km/h in 0.90 s.

a. What is the change in momentum of the dragster?

b. What is the average force exerted on the dragster?

c. What exerts that force?

91. Ice Hockey A 0.115-kg hockey puck, moving at 35.0 m/s, strikes a 0.365-kg jacket that is thrown onto the ice by a fan of a certain hockey team. The puck and jacket slide off together. Find their velocity.

92. A 50.0-kg woman, riding on a 10.0-kg cart, is moving east at 5.0 m/s. The woman jumps off the front of the cart and lands on the ground at 7.0 m/s eastward, relative to the ground.

a. Sketch the before and after situations, and assign a coordinate axis to them.

b. Find the cart's velocity after the woman jumps off.

93. A 60.0-kg dancer leaps 0.32 m high.

a. With what momentum does he reach the ground?

b. What impulse is needed to stop the dancer?

c. As the dancer lands, his knees bend, lengthening the stopping time to 0.050 s. Find the average force exerted on the dancer's body.

d. Compare the stopping force with his weight.

Thinking Critically

94. Analyze and Conclude Two balls during a collision are shown in **Figure 24,** which is drawn to scale. The balls enter from the left of the diagram, collide, and then bounce away. The heavier ball, at the bottom of the diagram, has a mass of 0.600 kg, and the other has a mass of 0.400 kg. Using a vector diagram, determine whether momentum is conserved in this collision. Explain any difference in the momentum of the system before and after the collision.

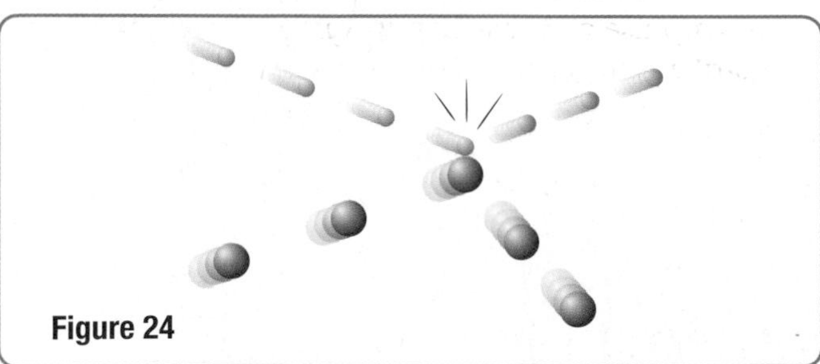

Figure 24

95. Analyze and Conclude A student, holding a bicycle wheel with its axis vertical, sits on a stool that can rotate with negligible friction. She uses her hand to get the wheel spinning. Would you expect the student and stool to turn? If so, in which direction? Explain.

96. Apply Concepts A 92-kg fullback, running at a speed of 5.0 m/s, attempts to dive directly across the goal line for a touchdown. Just as he reaches the line, he is met head-on in midair by two 75-kg linebackers, both moving in the direction opposite the fullback. One is moving at 2.0 m/s, and the other at 4.0 m/s. They all become entangled as one mass.

a. Sketch the before and after situations.

b. What is the players' velocity after the collision?

c. Does the fullback score a touchdown?

Writing in Physics

97. How can highway barriers be designed to be more effective in saving people's lives? Research this issue and describe how impulse and change in momentum can be used to analyze barrier designs.

98. While air bags save many lives, they also have caused injuries and even death. Research the arguments and responses of automobile makers to this statement. Determine whether the problems involve impulse and momentum or other issues.

Cumulative Review

99. The 0.72-kg ball in **Figure 25** is swung vertically from a 0.60-m string in uniform circular motion at a speed of 3.3 m/s. What is the tension in the cord at the top of the ball's motion?

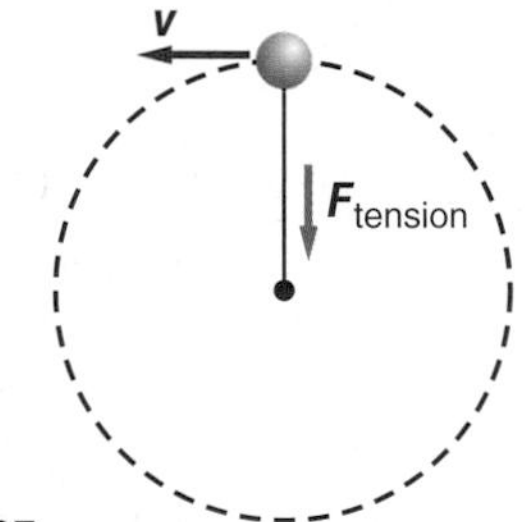

Figure 25

100. You wish to launch a satellite that will remain above the same spot on Earth's surface. This means the satellite must have a period of exactly one day. Calculate the radius of the circular orbit this satellite must have. *Hint: The Moon also circles Earth, and both the Moon and the satellite will obey Kepler's third law. The Moon is 3.9×10^8 m from Earth, and its period is 27.33 days.*

101. A rope is wrapped around a drum that is 0.600 m in diameter. A machine pulls the rope for 2.00 s with a constant 40.0-N force. In that time, 5.00 m of rope unwinds. Find α, ω at 2.00 s, and I.

MULTIPLE CHOICE

1. When a star that is much larger than the Sun nears the end of its lifetime, it begins to collapse but continues to rotate. Which of the following describes the conditions of the collapsing star's moment of inertia (I), angular momentum (L), and angular velocity (ω)?
   - **A.** I increases, L stays constant, ω decreases.
   - **B.** I decreases, L stays constant, ω increases.
   - **C.** I increases, L increases, ω increases.
   - **D.** I increases, L increases, ω stays constant.

2. A 40.0-kg ice-skater glides with a speed of 2.0 m/s toward a 10.0-kg sled at rest on the ice. The ice-skater reaches the sled and holds on to it. The ice-skater and the sled then continue sliding in the same direction in which the ice-skater was originally skating. What is the speed of the ice-skater and the sled after they collide?
   - **A.** 0.4 m/s
   - **B.** 0.8 m/s
   - **C.** 1.6 m/s
   - **D.** 3.2 m/s

3. A bicyclist applies the brakes and slows the motion of the wheels. The angular momentum of each wheel then decreases from 7.0 kg•m^2/s to 3.5 kg•m^2/s over a period of 5.0 s. What is the angular impulse on each wheel?
   - **A.** −0.7 kg•m^2/s
   - **B.** −1.4 kg•m^2/s
   - **C.** −2.1 kg•m^2/s
   - **D.** −3.5 kg•m^2/s

4. A 45.0-kg ice-skater stands at rest on the ice. A friend tosses the skater a 5.0-kg ball. The skater and the ball then move backward across the ice with a speed of 0.50 m/s. What was the speed of the ball at the moment just before the skater caught it?
   - **A.** 2.5 m/s
   - **B.** 3.0 m/s
   - **C.** 4.0 m/s
   - **D.** 5.0 m/s

5. What is the difference in momentum between a 50.0-kg runner moving at a speed of 3.00 m/s and a 3.00×10^3-kg truck moving at a speed of only 1.00 m/s?
   - **A.** 1275 kg•m/s
   - **B.** 2550 kg•m/s
   - **C.** 2850 kg•m/s
   - **D.** 2950 kg•m/s

6. When the large gear in the diagram below rotates, it turns the small gear in the opposite direction at the same linear speed. The larger gear has twice the radius and four times the mass of the smaller gear. What is the angular momentum of the larger gear as a function of the angular momentum of the smaller gear? *Hint: The moment of inertia for a disk is* $\left(\frac{1}{2}\right) mr^2$, *where m is mass and r is the radius of the disk.*
   - **A.** $-2L_{small}$
   - **B.** $-4L_{small}$
   - **C.** $-8L_{small}$
   - **D.** $-16L_{small}$

7. A force of 16 N exerted against a rock with an impulse of 0.8 kg•m/s causes the rock to fly off the ground with a speed of 4.0 m/s. What is the mass of the rock?
   - **A.** 0.2 kg
   - **B.** 0.8 kg
   - **C.** 1.6 kg
   - **D.** 4.0 kg

8. An 82-kg hockey goalie, standing at rest, catches a 0.105-kg hockey puck that is moving at a speed of 46 m/s. With what speed does the goalie slide on the ice?
   - **A.** 0.059 m/s
   - **B.** 0.56 m/s
   - **C.** 1.2 m/s
   - **D.** 5.3 m/s

FREE RESPONSE

9. A 12.0-kg rock falls from a cliff to the ground directly below. Assuming the rock does not bounce, what is the impulse on the rock if its velocity at the moment it strikes the ground is 20.0 m/s downward?

NEED EXTRA HELP?

If you Missed Question	1	2	3	4	5	6	7	8	9
Review Section	2	2	1	2	1	1	1	2	1

Online Test Practice

CHAPTER 10

Work, Energy, and Machines

BIGIDEA **Doing work on a system changes the system's energy.**

SECTIONS

1 **Work and Energy**

2 **Machines**

LaunchLAB

iLab Station

ENERGY AND FALLING

What factors affect the size of the crater a meteor leaves?

WATCH THIS!

MACHINES

What happens when simple machines are combined to do work? Explore the physics of everyday gadgets and learn more about work, energy, and machines.

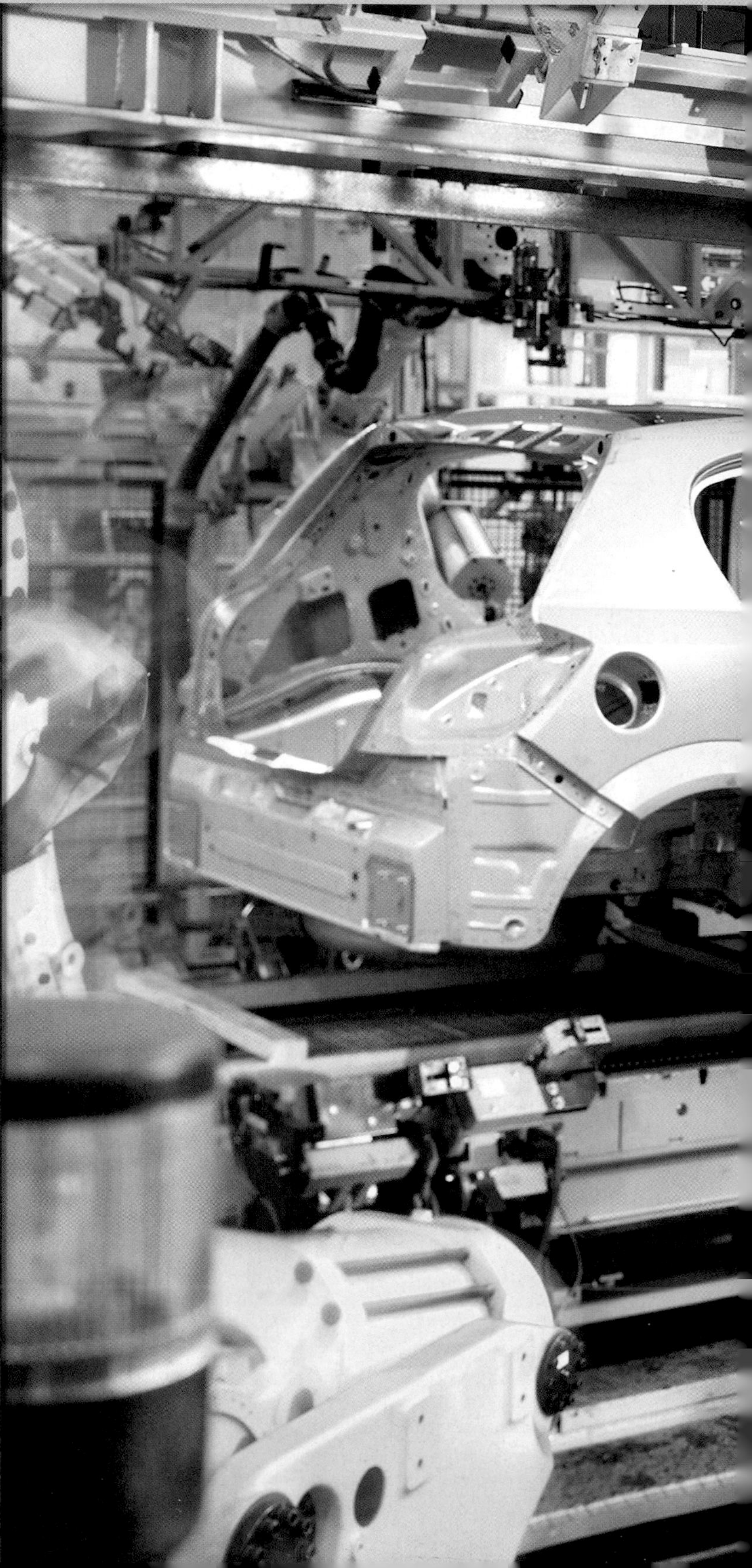

Go online!
connectED.mcgraw-hill.com

SECTION 1 Work and Energy

PHYSICS 4 YOU **Exercising to stay healthy is sometimes called working out. Job related activities are referred to as work. How do scientists define the term *work?***

MAINIDEA

Work is the transfer of energy that occurs when a force is applied through a displacement.

Essential Questions

- What is work?
- What is energy?
- How are work and energy related?
- What is power, and how is it related to work and energy?

Review Vocabulary

law of conservation of momentum states that the momentum of any closed, isolated system does not change

New Vocabulary

work
joule
energy
work-energy theorem
kinetic energy
translational kinetic energy
power
watt

Work

Think about two cars colliding head-on and coming to an immediate stop. You know that momentum is conserved. But the cars were moving prior to the crash and come to a stop after the crash. Thus, it seems that there must be some other quantity that changed as a result of the force acting on each car.

Consider a force exerted on an object while the object moves a certain distance, such as the bookbag in **Figure 1.** There is a net force, so the object is accelerated, $a = \frac{F}{m}$, and its velocity changes. Recall from your study of motion that acceleration, velocity, and distance are related by the equation $v_f^2 = v_i^2 + 2ad$. This can be rewritten as $2ad = v_f^2 - v_i^2$. If you replace a with $\frac{F}{m}$ you obtain $2\left(\frac{F}{m}\right)d = v_f^2 - v_i^2$. Multiplying both sides by $\frac{m}{2}$ gives $Fd = \frac{1}{2}mv_f^2 - \frac{1}{2}mv_i^2$.

The left side of the equation describes an action that was done to the system by the external world. Recall that a system is the object or objects of interest and the external world is everything else. A force (F) was exerted on a system while the point of contact moved. When a force is applied through a displacement, **work** (W) is done on the system.

The SI unit of work is called a **joule** (J). One joule is equal to 1 N•m. One joule of work is done when a force of 1 N acts on a system over a displacement of 1 m. An apple weighs about 1 N, so it takes roughly 1 N of force to lift the apple at a constant velocity. Thus, when you lift an apple 1 m at a constant velocity, you do 1 J of work on it.

Figure 1 Work is done when a force is applied though a displacement.

Identify another example of when a force does work on an object.

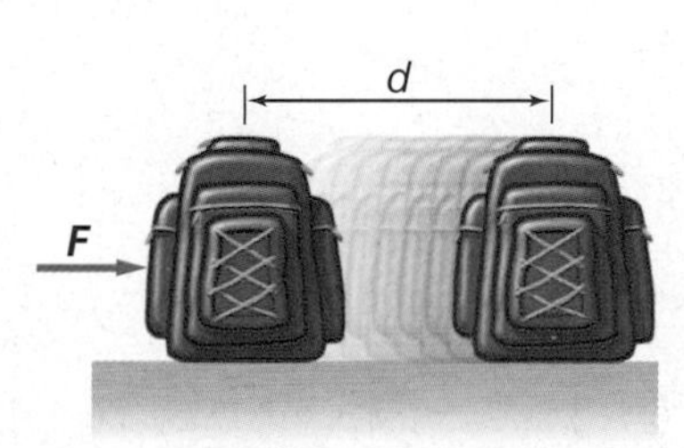

When a force (F) acts through a displacement (d) in the direction of an object's motion, work (W) is done.

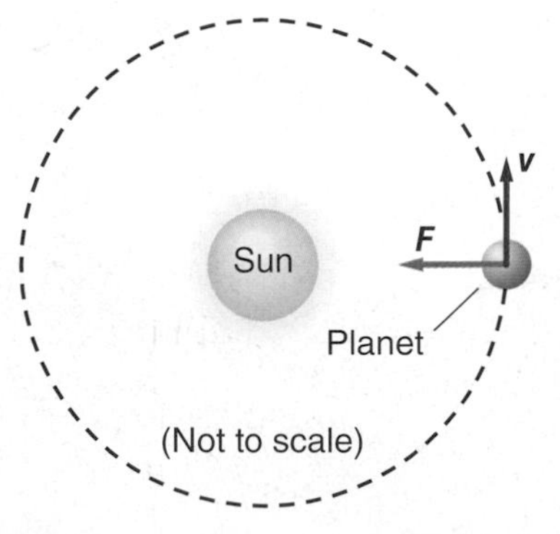

The gravitational force does no work on a planet in a circular orbit because the force is perpendicular to the direction of motion.

Work done by a constant force In the book bag example, F is a constant force exerted in the direction in which the object is moving. In this case, work (W) is the product of the force and the magnitude of the system's displacement. That is, $W = Fd$.

What happens if the exerted force is perpendicular to the direction of motion? For example, for a planet in circular orbit, the force is always perpendicular to the direction of motion, as shown in **Figure 1.** Recall from your study of Newton's laws that a perpendicular force does not change the speed of a system, only its direction. The speed of the planet doesn't change and so the right side of the equation, $\frac{1}{2}mv_f^2 - \frac{1}{2}mv_i^2$, is zero. Therefore, the work done is also zero.

Constant force exerted at an angle What work does a force exerted at an angle do? For example, what work does the person pushing the car in **Figure 2** do? Recall that any force can be replaced by its components. If you use the coordinate system shown in **Figure 2,** the 125-N force (F) exerted in the direction of the person's arm has two components.

The magnitude of the horizontal component (F_x) is related to the magnitude of the applied force (F) by a cosine function: $\cos 25.0° = \frac{F_x}{F}$. By solving for F_x, you obtain

$$F_x = F \cos 25.0° = (125 \text{ N})(\cos 25.0°) = 113 \text{ N}.$$

Using the same method, the vertical component is

$$F_y = -F \sin 25.0° = -(125 \text{ N})(\sin 25.0°) = -52.8 \text{ N}.$$

The negative sign shows that the force is downward. Because the displacement is in the x direction, only the x-component does work. The y-component does no work. The work you do when you exert a force on a system at an angle to the direction of motion is equal to the component of the force in the direction of the displacement multiplied by the displacement. The magnitude of the component (F_x) force acting in the direction of displacement is found by multiplying the magnitude of force (F) by the cosine of the angle (θ) between the force and the direction of the displacement: $F_x = F \cos \theta$. Thus, the work done is represented by the following equation.

WORK

Work is equal to the product of the magnitude of the force and magnitude of displacement times the cosine of the angle between them.

$$W = Fd \cos \theta$$

READING CHECK Determine the work you do when you exert a force of 3 N at an angle of 45° from the direction of motion for 1 m.

Notice that the equation above agrees with our expectations for constant forces exerted in the direction of displacement and for constant forces perpendicular to the displacement. In the book bag example, $\theta = 0°$ and $\cos 0° = 1$. Thus, $W = Fd(1) = Fd$, just as we found before. In the case of the orbiting planet, $\theta = 90°$ and $\cos 90° = 0$. Thus, $W = Fd(0) = 0$. This agrees with our previous conclusions.

VOCABULARY

Science Usage v. Common Usage

Work

- **Science usage**

a force applied through a displacement

The student did work on the car, pushing it out of the mud.

- **Common usage**

physical or mental effort

Studying physics can require a lot of work.

Figure 2 Only the horizontal component of the force the man exerts on the car does work because the car's displacement is horizontal.

View an **animation about work.**

Concepts In Motion

Bjorn Vinter/UpperCut Images/Getty Images

Work done by many forces Suppose you are pushing a box on a frictionless surface while your friend is trying to prevent you from moving it, as shown in **Figure 3.** What forces are acting on the box? You are exerting a force to the right and your friend is exerting a force to the left. Earth's gravity exerts a downward force, and the ground exerts an upward normal force. How much work is done on the box?

When several forces are exerted on a system, calculate the work done by each force, and then add the results. For the box in **Figure 3,** the upward and downward forces (gravity and the normal force) are perpendicular ($\theta = 90°$) to the direction of motion and do no work. For these forces, $\theta = 90°$, which makes $\cos\theta = 0$, and thus, $W = 0$.

The force you exert ($F_{\text{on box by you}}$) is in the direction of the displacement, so the work you do is

$$W = F_{\text{on box by you}}d.$$

Your friend exerts a force (F_{friend}) in the direction opposite the displacement ($\theta = 180°$). Because $\cos 180° = -1$, your friend does negative work:

$$W = -F_{\text{on box by friend}}d$$

The total work done on the box would be

$$W = F_{\text{on box by you}}d - F_{\text{on box by friend}}d.$$

READING CHECK **Explain** why you do positive work on the box and your friend does negative work on the box.

It is also important to consider each distance separately. For example, suppose you push a box 1 m with a force of 3 N. You then pull the box back 1 m to the starting point with a force of 3 N. You might think you did no work because the total displacement is zero.

This would be true if the force you exerted were constant, but your force changed direction. Your push was in the box's direction of motion for the first part, so you did 3 J of work. In the second part, both the force you exerted and the direction of motion reversed. Your pull and the box's motion were in the same direction, and you did 3 J of work on the box. Therefore, you did a total of 6 J of work on the box.

READING CHECK **Describe** another scenario in which you do work on a system, and explain how much work is done on the system.

MiniLAB

FORCE APPLIED AT AN ANGLE

How does the direction of a force affect the amount of work done on an object?

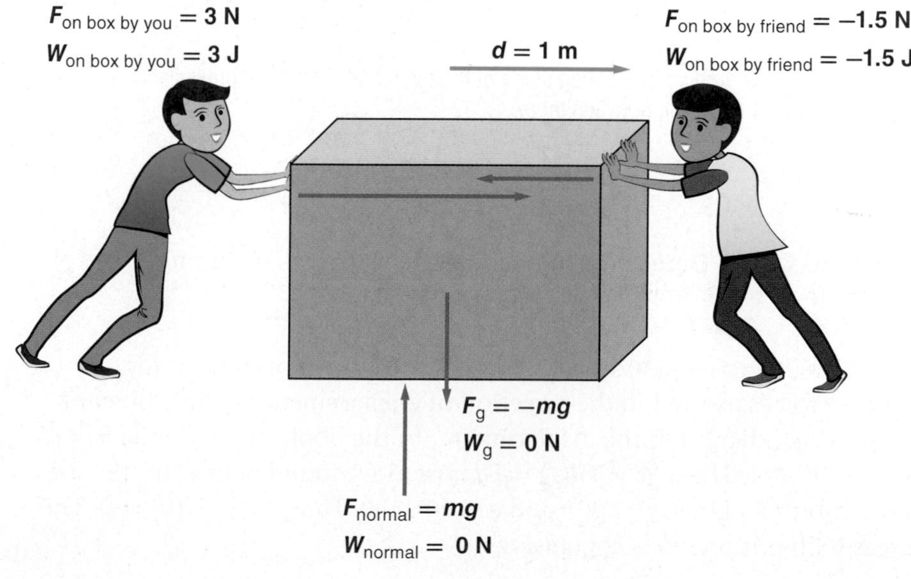

Figure 3 The total work done on a system is the sum of the work done by each agent that exerts a force on the system.

Describe the total work done on the box if your friend exerted a greater force than you did. Be sure to consider the direction of the displacement.

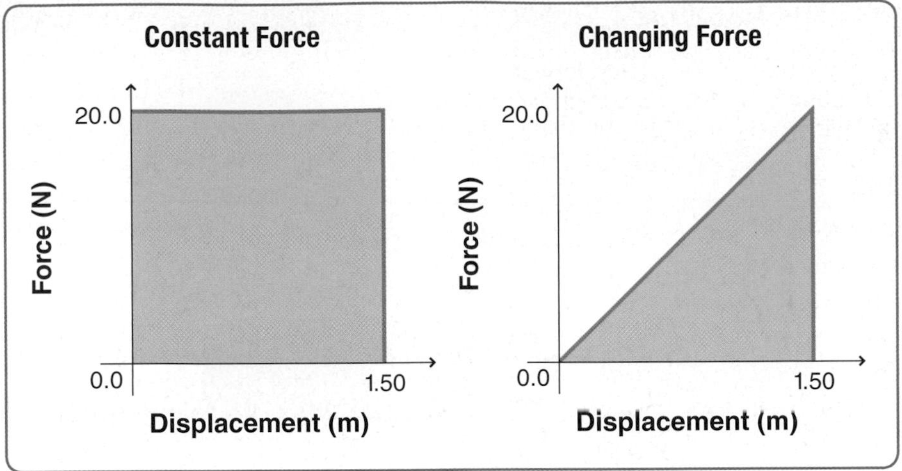

Figure 4 The area under a force-displacement graph is equal to the work.

Finding work done when forces change In the last example, the force changed, but we could determine the work done in each segment. But what if the force changes in a more complicated way? A graph of force versus displacement lets you determine the work done by a force. This graphical method can be used to solve problems in which the force is changing. The left panel of **Figure 4** shows the work done by a constant force of 20.0 N that is exerted to lift an object 1.50 m. The work done by this force is represented by $W = Fd = (20.0\ \text{N})(1.50\ \text{m}) = 30.0\ \text{J}$. Note that the shaded area under the left graph is also equal to (20.0 N)(1.50 m), or 30.0 J. The area under a force-displacement graph is equal to the work done by that force.

This is true even if the force changes. The right panel of **Figure 4** shows the force exerted by a spring, which varies linearly from 0.0 to 20.0 N as it is compressed 1.50 m. The work done by the force that compressed the spring is the area under the graph, which is the area of a triangle, $\left(\frac{1}{2}\right)$(base)(altitude), or $W = \left(\frac{1}{2}\right)(20.0\ \text{N})(1.50\ \text{m}) = 15.0\ \text{J}$. Use the problem-solving strategies below when you solve problems related to work.

PROBLEM-SOLVING STRATEGIES

WORK

When solving work-related problems, use the following strategies:

1. Identify and sketch the system. Show any forces doing work on the system.
2. Establish a coordinate system. Draw the displacement vectors of the system and each force vector doing work on the system.
3. Find the angle (θ) between each force and displacement.
4. Calculate the work done by each force using $W = Fd \cos \theta$.
5. Calculate the net work done.

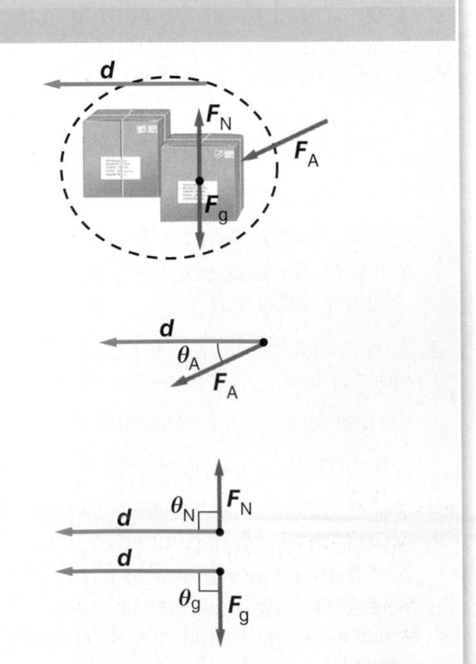

EXAMPLE PROBLEM 1

Find help with **significant figures.** Math Handbook

WORK A hockey player uses a stick to exert a constant 4.50-N force forward to a 105-g puck sliding on ice over a displacement of 0.150 m forward. How much work does the stick do on the puck? Assume friction is negligible.

1 ANALYZE AND SKETCH THE PROBLEM

- Identify the system and the force doing work on it.
- Sketch the situation showing initial conditions.
- Establish a coordinate system with $+x$ to the right.
- Draw a vector diagram.

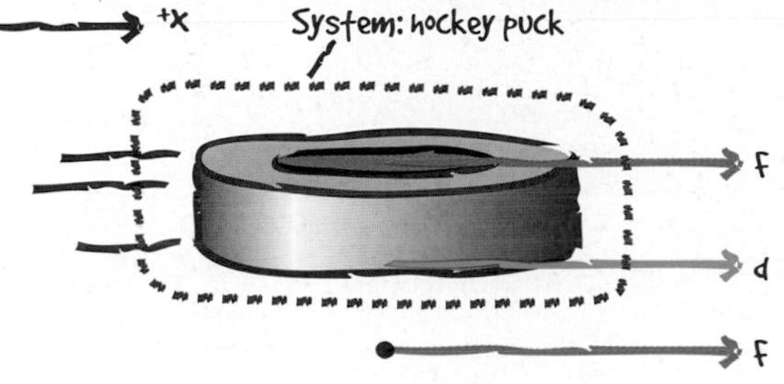

Player does work on hockey puck.

KNOWN	UNKNOWN
$m = 105$ g	$W = ?$
$F = 4.50$ N	
$d = 0.150$ m	
$\theta = 0°$	

2 SOLVE FOR THE UNKNOWN

Use the definition for work.

$$W = Fd\cos\theta$$

$$= (4.50\text{ N})(0.150\text{ m})(1)$$ ◀ Substitute $F = 4.50$ N, $d = 0.150$ m, $\cos 0° = 1$.

$$= 0.675\text{ N·m}$$

$$= 0.675\text{ J}$$ ◀ $1\text{ J} = 1\text{ N·m}$

3 EVALUATE THE ANSWER

- **Are the units correct?** Work is measured in joules.
- **Does the sign make sense?** The stick (external world) does work on the puck (the system), so the sign of work should be positive.

PRACTICE PROBLEMS

Do additional problems. Online Practice

1. Refer to Example Problem 1 to solve the following problem.

a. If the hockey player exerted twice as much force (9.00 N) on the puck over the same distance, how would the amount of work the stick did on the puck be affected?

b. If the player exerted a 9.00-N force, but the stick was in contact with the puck for only half the distance (0.075 m), how much work does the stick do on the puck?

2. Together, two students exert a force of 825 N in pushing a car a distance of 35 m.

a. How much work do the students do on the car?

b. If their force is doubled, how much work must they do on the car to push it the same distance?

3. A rock climber wears a 7.5-kg backpack while scaling a cliff. After 30.0 min, the climber is 8.2 m above the starting point.

a. How much work does the climber do on the backpack?

b. If the climber weighs 645 N, how much work does she do lifting herself and the backpack?

4. CHALLENGE Marisol pushes a 3.0-kg box 7.0 m across the floor with a force of 12 N. She then lifts the box to a shelf 1 m above the ground. How much work does Marisol do on the box?

EXAMPLE PROBLEM 2

Get help with **forces and work.** Personal Tutor

FORCE AND DISPLACEMENT AT AN ANGLE A sailor pulls a boat a distance of 30.0 m along a dock using a rope that makes a 25.0° angle with the horizontal. How much work does the rope do on the boat if its tension is 255 N?

1 ANALYZE AND SKETCH THE PROBLEM

- Identify the system and the force doing work on it.
- Establish coordinate axes.
- Sketch the situation showing the boat with initial conditions.
- Draw vectors showing the displacement, the force, and its component in the direction of the displacement.

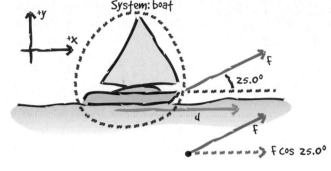

KNOWN		UNKNOWN
$F = 255$ N	$\theta = 25.0°$	$W = ?$
$d = 30.0$ m		

Sailor does work on the boat.

2 SOLVE FOR THE UNKNOWN

Use the definition of work.

$$W = Fd \cos \theta$$

$$= (255 \text{ N})(30.0 \text{ m})(\cos 25.0°)$$ ◀ Substitute $F = 255$ N, $d = 30.0$ m, $\theta = 25.0°$.

$$= 6.93 \times 10^3 \text{ J}$$

3 EVALUATE THE ANSWER

- **Are the units correct?** Work is measured in joules.
- **Does the sign make sense?** The rope does work on the boat, which agrees with a positive sign for work.

PRACTICE PROBLEMS

Do additional problems. Online Practice

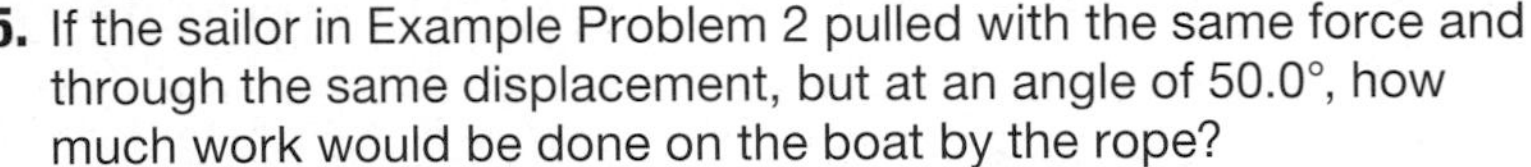

5. If the sailor in Example Problem 2 pulled with the same force and through the same displacement, but at an angle of 50.0°, how much work would be done on the boat by the rope?

6. Two people lift a heavy box a distance of 15 m. They use ropes, each of which makes an angle of 15° with the vertical. Each person exerts a force of 225 N. How much work do the ropes do?

7. An airplane passenger carries a 215-N suitcase up the stairs, a displacement of 4.20 m vertically and 4.60 m horizontally.

a. How much work does the passenger do on the suitcase?

b. The same passenger carries the same suitcase back down the same set of stairs. How much work does the passenger do on the suitcase to carry it down the stairs?

8. A rope is used to pull a metal box a distance of 15.0 m across the floor. The rope is held at an angle of 46.0° with the floor, and a force of 628 N is applied to the rope. How much work does the rope do on the box?

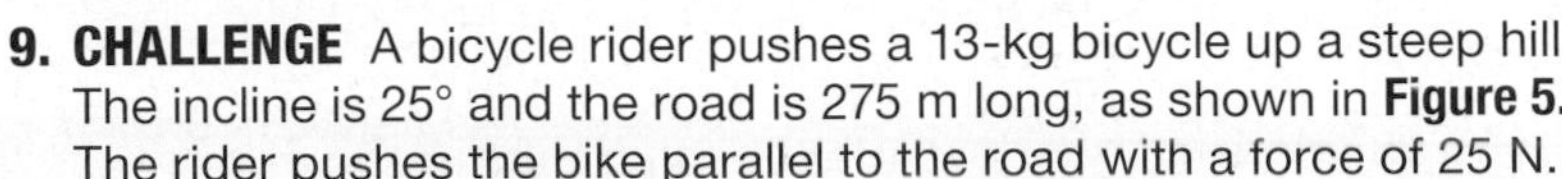

9. CHALLENGE A bicycle rider pushes a 13-kg bicycle up a steep hill. The incline is 25° and the road is 275 m long, as shown in **Figure 5.** The rider pushes the bike parallel to the road with a force of 25 N.

a. How much work does the rider do on the bike?

b. How much work is done by the force of gravity on the bike?

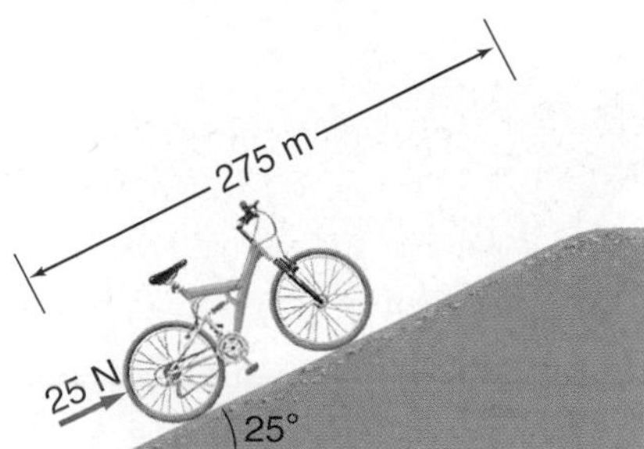

Figure 5

VOCABULARY

Science Usage v. Common Usage

Energy

- Science usage

the ability of a system to produce a change in itself or the world around it

The kinetic energy of the soccer ball decreased as it slowed down.

- Common usage

the capacity of acting or being active

The young students had a lot of energy during recess.

Watch a **BrainPOP video about kinetic energy.**

Energy

Look again at the equation $W = \frac{1}{2}mv_f^2 - \frac{1}{2}mv_i^2$. What property of a system does $\frac{1}{2}mv^2$ describe? A massive, fast-moving vehicle can do damage to objects around it, and a baseball hit at high speed can rise high into the air. That is, a system with this property can produce a change in itself or the world around it. This ability of a system to produce a change in itself or the world around it is called **energy** and is represented by the symbol E.

The right side of the equation, $\frac{1}{2}mv_f^2 - \frac{1}{2}mv_i^2$, indicates a change in a specific kind of energy. That is, work causes a change in energy. This is the **work-energy theorem,** which states that when work is done on a system, the result is a change in the system's energy. The work-energy theorem can be represented by the following equation.

WORK-ENERGY THEOREM

Work done on a system is equal to the change in the system's energy.

$$W = \Delta E$$

Since work is measured in joules, energy must also be measured in joules. In fact, the unit gets its name from the nineteenth-century physicist James Prescott Joule, who established the relationship between work done and the change in energy. Recall that 1 joule equals 1 N•m and that 1 N equals 1 kg•m/s^2. Therefore, 1 joule equals 1 kg•m^2/s^2. These are the same units $\frac{1}{2}mv^2$ has.

Through the process of doing work, energy can move between the external world and the system. The direction of energy transfer can be either way. If the external world does work on a system, then W is positive and the energy of the system increases. If a system does work on the external world, then W is negative and the energy of the system decreases. In summary, work is the transfer of energy that occurs when a force is applied through a displacement.

Changing kinetic energy So far, we have discussed the energy associated with a system's motion. For example, the bobsledders in **Figure 6** do work on their sled to get it moving at the beginning of a race. The energy associated with motion is called **kinetic energy** (KE).

Figure 6 The bobsledders do work on the bobsled when they push it. The result is a change in the bobsled's kinetic energy.

In the examples we have considered, the object was changing position and its energy ($\frac{1}{2}mv^2$) was due to this motion. Energy due to changing position is called **translational kinetic energy** and can be represented by the following equation.

TRANSLATIONAL KINETIC ENERGY

A system's translational kinetic energy is equal to $\frac{1}{2}$ times the system's mass multiplied by the system's speed squared.

$$KE_{trans} = \frac{1}{2}mv^2$$

In the case of the bobsled, work resulted in a change in the object's translational kinetic energy. There are, however, many other forms of energy. Work can cause a change in these other forms, as well. Some of these forms, such as potential energy and thermal energy, will be explored in subsequent chapters.

Doug Pensinger/Getty Images Sport/Getty Images

Power

Suppose you had a stack of books to move from the floor to a shelf. You could lift the entire stack at once, or you could move the books one at a time. How would the amount of work compare between the two cases? Since the total force applied and the displacement are the same in both cases, the work is the same. However, the time needed is different. Recall that work causes a change in energy. The rate at which energy is transformed is **power.** Power is equal to the change in energy divided by the time required for the change.

VOCABULARY
Science Usage v. Common Usage

Power
• **Science usage**
the rate of energy transformation
The elevator's motor must provide enough power to lift 1500 kg to the top of the building in 25 seconds.
• **Common usage**
legal or official authority
The U.S. Congress has the power to declare war.

POWER
Power is equal to the change in energy divided by the time required for the change.

$$P = \frac{\Delta E}{t}$$

When work causes the change in energy, power is equal to the work done divided by the time taken to do the work.

$$P = \frac{W}{t}$$

Consider the two forklifts in **Figure 7.** The left forklift raises the load in 5 seconds, and the right forklift raises the load in 10 seconds. The left forklift is more powerful than the right forklift. Even though the same work is accomplished by both, the left forklift accomplishes it in less time and thus develops more power.

Power is measured in watts (W). One **watt** is 1 J of energy transformed in 1 s. That is, 1 W = 1 J/s. A watt is a relatively small unit of power. For example, a glass of water weighs about 2 N. If you lift it 0.5 m to your mouth at a constant speed, you do 1 J of work. If you lift the glass in 1 s, you are doing work at the rate of 1 W. Because a watt is such a small unit, power often is measured in kilowatts (kW). One kilowatt is equal to 1000 W.

PhysicsLAB

STAIR CLIMBING AND POWER

What is the maximum power you can develop while climbing stairs?

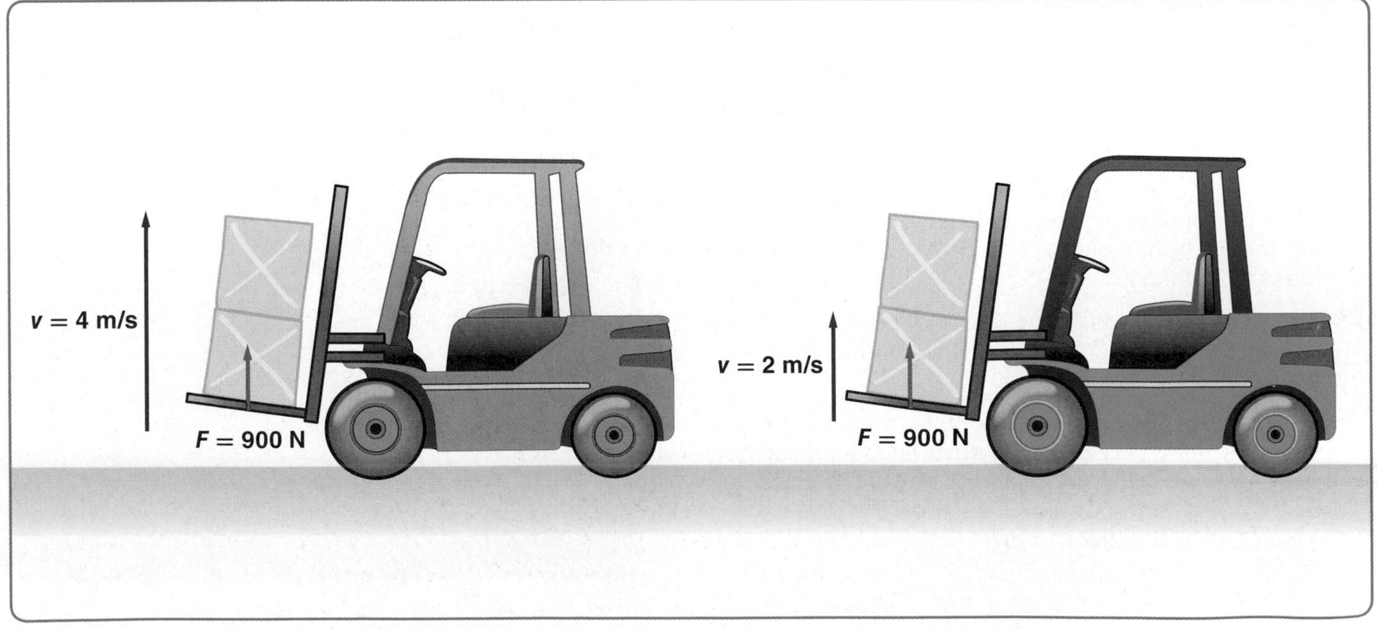

Figure 7 The forklift on the left develops more power than the forklift on the right. It lifts the load at a faster rate.

EXAMPLE PROBLEM 3

Find help with **scientific notation.** Math Handbook

EXAMPLE PROBLEM

POWER An electric motor lifts an elevator 9.00 m in 15.0 s by exerting an upward force of 1.20×10^4 N. What power does the motor produce in kW?

1 ANALYZE AND SKETCH THE PROBLEM

- Sketch the situation showing the system as the elevator with its initial conditions.
- Establish a coordinate system with up as positive.
- Draw a vector diagram for the force and displacement.

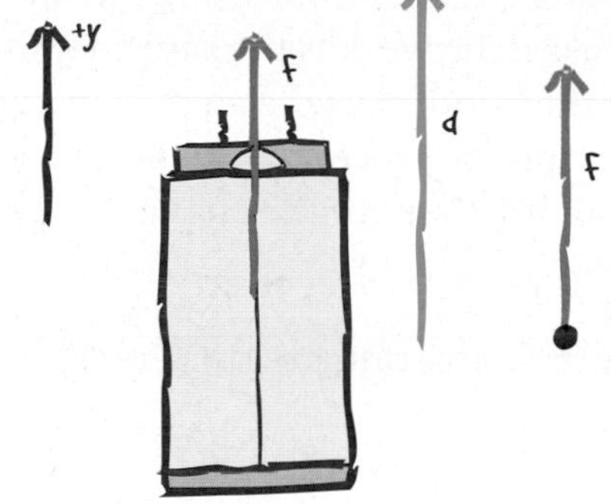

KNOWN	UNKNOWN
$d = 9.00$ m	$P = ?$
$t = 15.0$ s	
$F = 1.20\times10^4$ N	

2 SOLVE FOR THE UNKNOWN

Use the definition of power.

$$P = \frac{W}{t}$$

$$= \frac{Fd}{t}$$ ◀ Substitute $W = Fd \cos 0° = Fd$.

$$= \frac{(1.20\times10^4 \text{ N})(9.00 \text{ m})}{(15.0 \text{ s})}$$ ◀ Substitute $F = 1.20\times10^4$ N, $d = 9.00$ m, $t = 15.0$ s.

$$= 7.20 \text{ kW}$$

3 EVALUATE THE ANSWER

- **Are the units correct?** Power is measured in joules per second or watts.
- **Does the sign make sense?** The positive sign agrees with the upward direction of the force.

PRACTICE PROBLEMS

Do additional problems. Online Practice

10. A 575-N box is lifted straight up a distance of 20.0 m by a cable attached to a motor. The box moves with a constant velocity and the job is done in 10.0 s. What power is developed by the motor in W and kW?

11. You push a wheelbarrow a distance of 60.0 m at a constant speed for 25.0 s by exerting a 145-N force horizontally.

a. What power do you develop?

b. If you move the wheelbarrow twice as fast, how much power is developed?

12. What power does a pump develop to lift 35 L of water per minute from a depth of 110 m? (1 L of water has a mass of 1.00 kg.)

13. An electric motor develops 65 kW of power as it lifts a loaded elevator 17.5 m in 35 s. How much force does the motor exert?

14. CHALLENGE A winch designed to be mounted on a truck, as shown in **Figure 8,** is advertised as being able to exert a 6.8×10^3-N force and to develop a power of 0.30 kW. How long would it take the truck and the winch to pull an object 15 m?

Figure 8

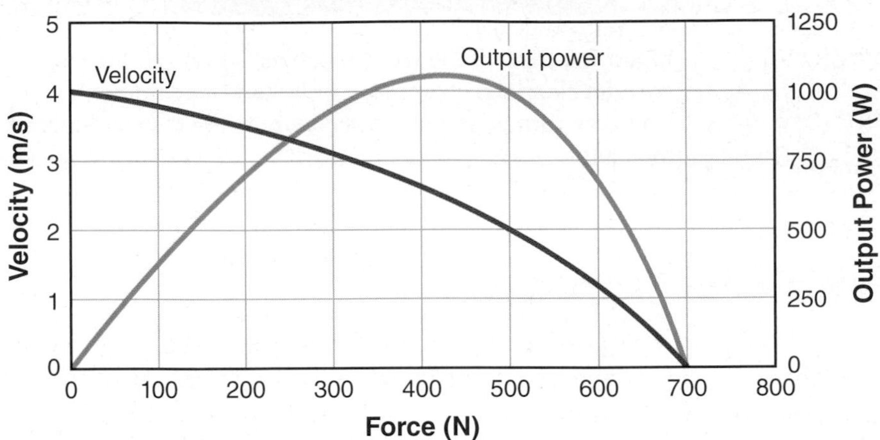

Figure 9 When riding a multispeed bicycle, the output power depends on the force you exert and your speed.

Power and speed You might have noticed in Example Problem 3 that when the force has a component (F_x) in the same direction as the displacement, $P = \frac{F_x d}{t}$. However, because $v = \frac{d}{t}$, power also can be calculated using $P = F_x v$.

When you are riding a multispeed bicycle, how do you choose the correct gear? You want to get your body to deliver the largest amount of power. By considering the equation $P = Fv$, you can see that either zero force or zero speed results in no power delivered. The muscles cannot exert extremely large forces nor can they move very fast. Thus, some combination of moderate force and moderate speed will produce the largest amount of power. **Figure 9** shows that in this particular situation, the maximum power output is over 1000 W when the force is about 400 N and speed is about 2.6 m/s.

All engines—not just humans—have these limitations. Simple machines often are designed to match the force and the speed that the engine can deliver to the needs of the job. You will learn more about simple machines in the next section.

REAL-WORLD PHYSICS

TOUR DE FRANCE A bicyclist in the Tour de France rides at an average speed of about 8.94 m/s for more than 6 h a day. The power output of the racer is about 1 kW. One-fourth of that power goes into moving the bike against the resistance of the air, gears, and tires. Three-fourths of the power is used to cool the racer's body.

SECTION 1 REVIEW

Check your understanding.

15. **MAIN IDEA** If the work done on an object doubles its kinetic energy, does it double its speed? If not, by what ratio does it change the speed?

16. **Work** Murimi pushes a 20-kg mass 10 m across a floor with a horizontal force of 80 N. Calculate the amount of work done by Murimi on the mass.

17. **Work** Suppose you are pushing a stalled car. As the car gets going, you need less and less force to keep it going. For the first 15 m, your force decreases at a constant rate from 210.0 N to 40.0 N. How much work did you do on the car? Draw a force-displacement graph to represent the work done during this period.

18. **Work** A mover loads a 185-kg refrigerator into a moving van by pushing it at a constant speed up a 10.0-m, friction-free ramp at an angle of inclination of 11°. How much work is done by the mover on the refrigerator?

19. **Work** A 0.180-kg ball falls 2.5 m. How much work does the force of gravity do on the ball?

20. **Work and Power** Does the work required to lift a book to a high shelf depend on how fast you raise it? Does the power required to lift the book depend on how fast you raise it? Explain.

21. **Power** An elevator lifts a total mass of 1.1×10^3 kg a distance of 40.0 m in 12.5 s. How much power does the elevator deliver?

22. **Mass** A forklift raises a box 1.2 m and does 7.0 kJ of work on it. What is the mass of the box?

23. **Work** You and a friend each carry identical boxes from the first floor of a building to a room located on the second floor, farther down the hall. You choose to carry the box first up the stairs, and then down the hall to the room. Your friend carries it down the hall on the first floor, then up a different stairwell to the second floor. How do the amounts of work done by the two of you on your boxes compare?

24. **Critical Thinking** Explain how to find the change in energy of a system if three agents exert forces on the system at once.

SECTION 2 Machines

PHYSICS 4 YOU When you think of the word *machine,* you might think of vacuum cleaners, computers, or industrial equipment. However, ramps, screws, and crowbars are also considered machines.

MAINIDEA

Machines make tasks easier by changing the magnitude or the direction of the force exerted.

Essential Questions

- What is a machine, and how does it make tasks easier?
- How are mechanical advantage, the effort force, and the resistance force related?
- What is a machine's ideal mechanical advantage?
- What does the term efficiency mean?

Review Vocabulary

work a force applied through a distance

New Vocabulary

machine
effort force
resistance force
mechanical advantage
ideal mechanical advantage
efficiency
compound machine

Benefits of Machines

Machines, whether powered by engines or people, make tasks easier. A **machine** is a device that makes tasks easier by changing either the magnitude or the direction of the applied force. Consider the bottle opener in **Figure 10.** When you lift the handle, you do work on the opener. The opener lifts the cap, doing work on it. The work you do is called the input work (W_i). The work the machine does is called the output work (W_o).

Recall that work transfers energy. When you do work on the bottle opener, you transfer energy to the opener. The opener does work on the cap and transfers energy to it. The opener is not a source of energy. So, the cap cannot receive more energy than what you put into the opener. Thus, the output work can never be greater than the input work. The machine only aids in the transfer of energy from you to the bottle cap.

Mechanical advantage The force exerted by a user on a machine is called the **effort force** (F_e). The force exerted by the machine is called the **resistance force** (F_r). For the bottle opener in **Figure 10,** F_e is the upward force exerted by the person using the bottle opener and F_r is the upward force exerted by the bottle opener. The ratio of resistance force to effort force $\left(\frac{F_r}{F_e}\right)$ is called the **mechanical advantage** (MA) of the machine.

MECHANICAL ADVANTAGE

The mechanical advantage of a machine is equal to the resistance force divided by the effort force.

$$MA = \frac{F_r}{F_e}$$

Figure 10 This bottle opener makes opening a bottle easier. However, it does not lessen the energy required to do so.

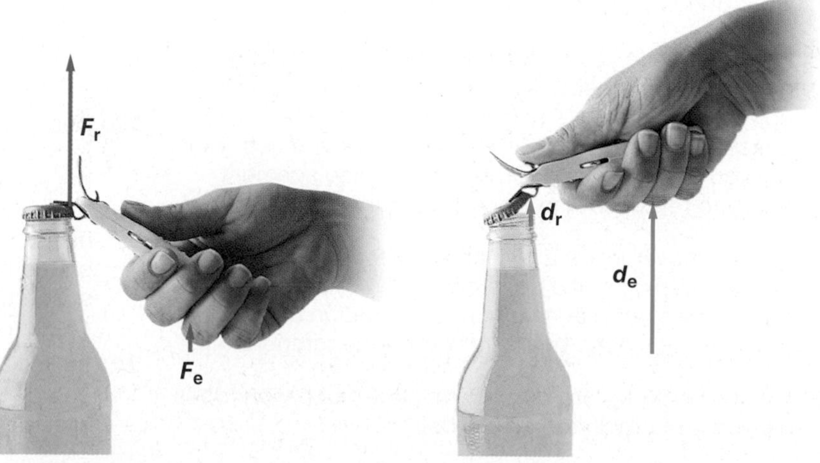

(t)Brand X Pictures/Getty Images, (bl br)Hutchings Photography

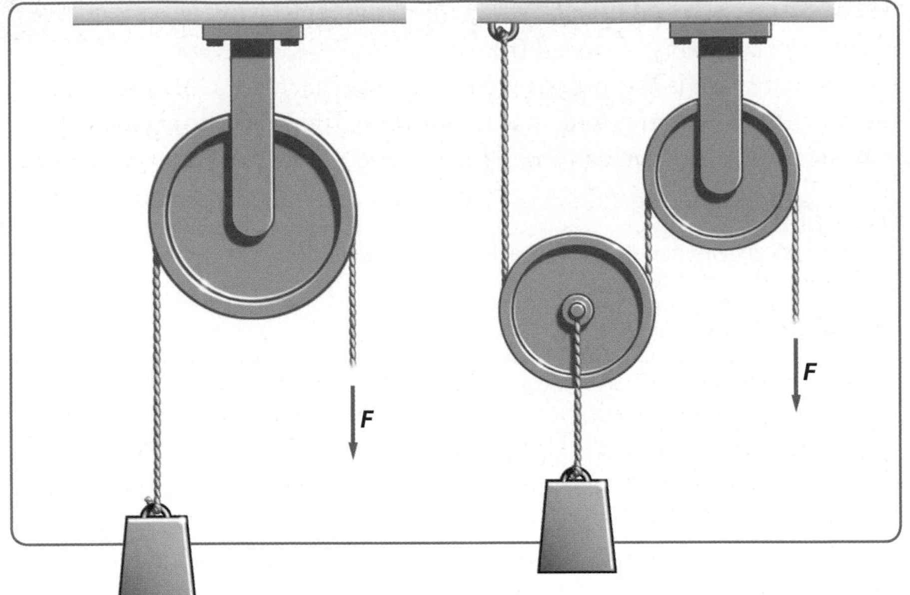

Figure 11 A fixed pulley has a mechanical advantage equal to 1 but is useful because it changes the direction of the force. A pulley system with a movable pulley has a mechanical advantage greater than 1, making the force applied to the object larger than the force originally exerted.

Watch a **BrainPOP video about pulleys.**

For a fixed pulley, such as the one shown on the left in **Figure 11,** the effort force (F_e) and the resistance force (F_r) are equal. Thus, *MA* is 1. What is the advantage of this machine? The fixed pulley is useful, not because the effort force is lessened, but because the direction of the effort force is changed. Many machines, such as the bottle opener shown in **Figure 10** and the pulley system shown on the right in **Figure 11,** have a mechanical advantage greater than 1. When the mechanical advantage is greater than 1, the machine increases the force applied by a person.

READING CHECK **Calculate** A machine has a mechanical advantage of 3. If the input force is 2 N what is the output force?

PhysicsLAB

LIFTING WITH PULLEYS

How does the arrangement of a pulley system affect its ideal mechanical advantage.

Ideal mechanical advantage A machine can increase force, but it cannot increase energy. An ideal machine transfers all the energy, so the output work equals the input work: $W_o = W_i$. The input work is the product of the effort force and the displacement the effort force acts through: $W_i = F_e d_e$. The output work is the product of the resistance force and the displacement the resistance force acts through: $W_o = F_r d_r$. Substituting these expressions into $W_o = W_i$ gives $F_r d_r = F_e d_e$. This equation can be rewritten $\frac{F_r}{F_e} = \frac{d_e}{d_r}$.

Recall that mechanical advantage is given by $MA = \frac{F_r}{F_e}$. Thus, for an ideal machine, **ideal mechanical advantage** (*IMA*) is equal to the displacement of the effort force divided by the displacement of the resistance force. *IMA* can be represented as follows.

IDEAL MECHANICAL ADVANTAGE

The ideal mechanical advantage of a machine is equal to the displacement of the effort force divided by the displacement of the resistance force.

$$IMA = \frac{d_e}{d_r}$$

Notice that you measure the displacements to calculate the ideal mechanical advantage, but you measure the forces exerted to find the actual mechanical advantage.

View an **animation on the benefits of machines.**

VOCABULARY
Science Usage v. Common Usage

Efficiency

• **Science usage**
the ratio of output work to input work
The efficiency of the system of pulleys is 86 percent.

• **Common usage**
production without waste
The new, high-efficiency washing machine uses less water and electricity than an older machine.

Efficiency In a real machine, not all of the input work is available as output work. Energy removed from the system through heat or sound means that there is less output work from the machine. Consequently, the machine is less efficient at accomplishing the task. The **efficiency** of a machine (e) is defined as the ratio of output work to input work.

EFFICIENCY
The efficiency of a machine (in %) is equal to the output work, divided by the input work, multiplied by 100.

$$e = \frac{W_o}{W_i} \times 100$$

An ideal machine has equal output and input work, $\frac{W_o}{W_i} = 1$, and its efficiency is 100 percent. All real machines have efficiencies of less than 100 percent.

Efficiency can be expressed in terms of the mechanical advantage and ideal mechanical advantage. Efficiency, $e = \frac{W_o}{W_i}$, can be rewritten as follows:

$$e = \frac{W_o}{W_i} = \frac{F_r d_r}{F_e d_e}$$

Because $MA = \frac{F_r}{F_e}$ and $IMA = \frac{d_e}{d_r}$, the following expression can be written for efficiency.

EFFICIENCY
The efficiency of a machine (in %) is equal to its mechanical advantage, divided by the ideal mechanical advantage, multiplied by 100.

$$e = \left(\frac{MA}{IMA}\right) \times 100$$

A machine's design determines its ideal mechanical advantage. An efficient machine has an *MA* almost equal to its *IMA*. A less-efficient machine has a small *MA* relative to its *IMA*. To obtain the same resistance force, a greater force must be exerted in a machine of lower efficiency than in a machine of higher efficiency.

READING CHECK Determine If a machine has an efficiency of 50 percent and an *MA* of 3, what is its *IMA?*

PHYSICS CHALLENGE

An electric pump pulls water at a rate of 0.25 m^3/s from a well that is 25 m deep. The water leaves the pump at a speed of 8.5 m/s.

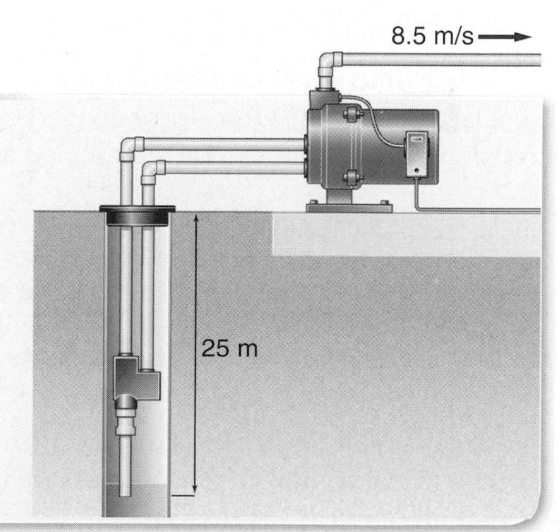

1. What power is needed to lift the water to the surface?
2. What power is needed to increase the water's kinetic energy?
3. If the pump's efficiency is 80 percent, how much power must be delivered to the pump?

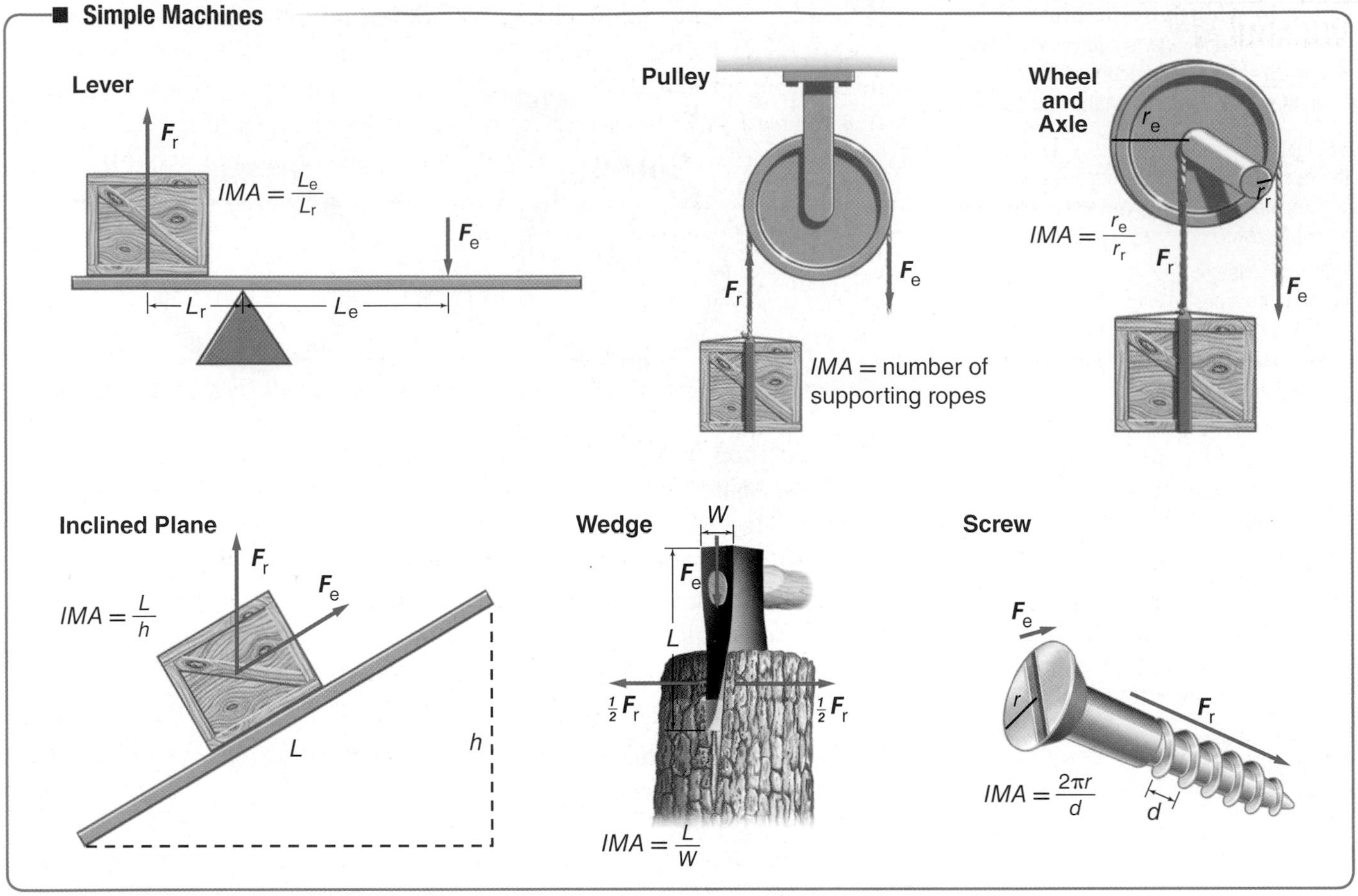

Figure 12 Each type of simple machine makes work easier by changing the direction or magnitude of the force. Notice that the *IMA* of each machine is related to the properties of the machine. For example, the *IMA* of a lever is equal to the length of the effort arm (L_e) divided by the length of the resistance arm (L_r). Similarly, the *IMA* of a wedge is equal to its length (L) divided by its width (W).

Identify an everyday example of each simple machine.

Compound Machines

Most machines, no matter how complex, are combinations of one or more of the six simple machines: the lever, pulley, wheel and axle, inclined plane, wedge, and screw. These machines are shown in **Figure 12.**

The *IMA* of each machine shown in **Figure 12** is the ratio of the displacement of the effort force to the displacement of the resistance force. For machines such as the lever and the wheel and axle, this ratio can be replaced by the ratio of the displacements between the place where the force is applied and the pivot point. A common version of the wheel and axle is a steering wheel, such as the one shown in **Figure 13.** The *IMA* is the ratio of the radii of the wheel and axle.

A machine consisting of two or more simple machines linked in such a way that the resistance force of one machine becomes the effort force of the second is called a **compound machine.** Some examples of compound machines are scissors (wedges and levers) and wheelbarrows (lever and wheel and axle).

Figure 13 A steering wheel is an example of a wheel and axle. Its *IMA* is $\frac{r_e}{r_r}$.

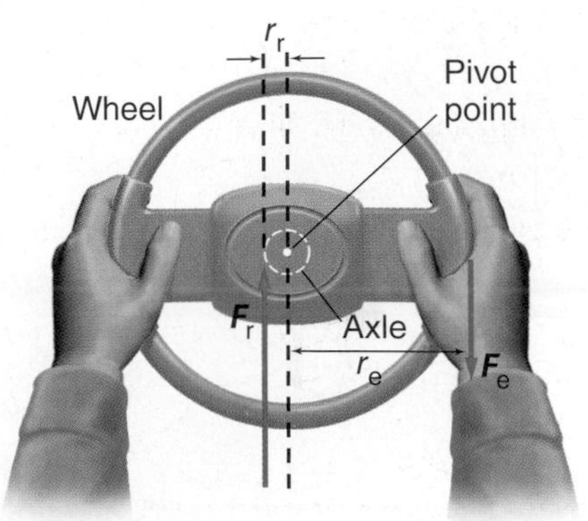

READING CHECK **Compare and contrast** simple machines and compound machines.

Figure 14 A bicycle is a compound machine. The rider exerts a force on the pedals, which exert a force on the chain. The chain then exerts a force on the rear gear.

MiniLAB

WHEEL AND AXLE

The gear mechanism on your bicycle multiplies the distance that you travel, but what does it do to the force you exert?

Mechanical advantage and bicycles A bicycle, such as the one shown in **Figure 14,** is a compound machine, consisting of two wheel-and-axle systems. The first is the pedal and front gear. Here, the effort force is the force that the rider exerts on the pedal, $F_{\text{rider on pedal}}$. The force of your foot is most effective when the force is exerted perpendicular to the arm of the pedal; that is, when the torque is largest. Therefore, we will assume that $F_{\text{rider on pedal}}$ is applied perpendicular to the pedal arm. The resistance is the force that the front gear exerts on the chain, $F_{\text{gear on chain}}$.

READING CHECK Identify the effort force and the resistance force for the pedal and front gear.

The rear gear and the rear wheel act like another wheel and axle. The chain exerts an effort force on the rear gear, $F_{\text{chain on gear}}$. This force is equal to the force of the front gear on the chain. That is $F_{\text{gear on chain}} = F_{\text{chain on gear}}$. The resistance force is the force that the wheel exerts on the road, $F_{\text{wheel on road}}$.

The *MA* of a compound machine is the product of the *MA*s of the simple machines from which it is made. Therefore, the bicycle's mechanical advantage is given by the following equation:

$$MA_{\text{total}} = MA_{\text{pedal gear}} \times MA_{\text{rear wheel}}$$

$$MA_{\text{total}} = \left(\frac{F_{\text{gear on chain}}}{F_{\text{rider on pedal}}}\right)\left(\frac{F_{\text{wheel on road}}}{F_{\text{chain on gear}}}\right) = \frac{F_{\text{wheel on road}}}{F_{\text{rider on pedal}}}$$

Recall that $F_{\text{gear on chain}} = F_{\text{chain on gear}}$. Therefore, they cancel in the equation shown above.

Similarly, the *IMA* of the bicycle is $IMA = IMA_{\text{pedal gear}} \times IMA_{\text{rear wheel}}$. For each wheel and axle, the IMA is the ratio of the wheel to the axle.

For the pedal gear, $IMA = \frac{\text{pedal radius}}{\text{front gear radius}}$

For the rear wheel, $IMA = \frac{\text{rear gear radius}}{\text{wheel radius}}$

For the bicycle, then,

$$IMA = \left(\frac{\text{pedal radius}}{\text{front gear radius}}\right)\left(\frac{\text{rear gear radius}}{\text{wheel radius}}\right)$$

$$= \left(\frac{\text{rear gear radius}}{\text{front gear radius}}\right)\left(\frac{\text{pedal radius}}{\text{wheel radius}}\right)$$

READING CHECK Explain What are the units of *MA* and *IMA* for the bicycle?

EXAMPLE PROBLEM 4

Find help with **isolating a variable.** Math Handbook

MECHANICAL ADVANTAGE You examine the rear wheel on your bicycle. It has a radius of 35.6 cm and has a gear with a radius of 4.00 cm. When the chain is pulled with a force of 155 N, the wheel rim moves 14.0 cm. The efficiency of this part of the bicycle is 95.0 percent.

a. What is the *IMA* of the wheel and gear?

b. What is the *MA* of the wheel and gear?

c. What is the resistance force?

d. How far was the chain pulled to move the rim 14.0 cm?

1 ANALYZE AND SKETCH THE PROBLEM

- Sketch the wheel and axle.
- Sketch the force vectors.

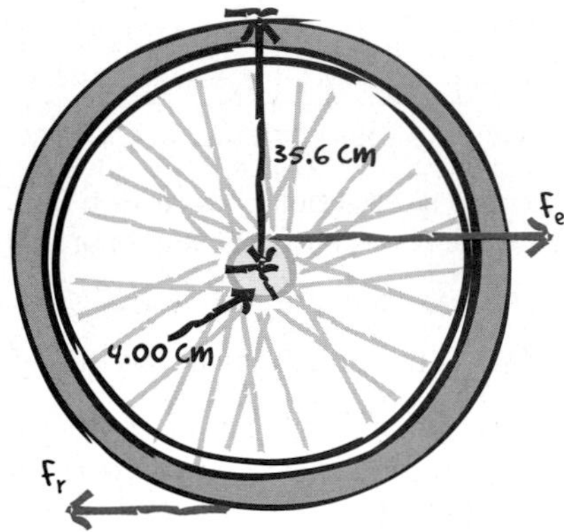

KNOWN		UNKNOWN	
$r_e = 4.00$ cm	$e = 95.0\%$	$IMA = ?$	$F_r = ?$
$r_r = 35.6$ cm	$d_r = 14.0$ cm	$MA = ?$	$d_e = ?$
$F_e = 155$ N			

2 SOLVE FOR THE UNKNOWN

a. Solve for *IMA.*

$$IMA = \frac{r_e}{r_r}$$

◀ For a wheel-and-axle machine, *IMA* is equal to the ratio of radii.

$$= \frac{4.00 \text{ cm}}{35.6 \text{ cm}} = 0.112$$

◀ Substitute $r_e = 4.00$ cm, $r_r = 35.6$ cm.

b. Solve for *MA.*

$$e = \frac{MA}{IMA} \times 100$$

$$MA = \left(\frac{e}{100}\right) \times IMA$$

$$= \left(\frac{95.0}{100}\right) \times 0.112 = 0.106$$

◀ Substitute $e = 95.0\%$, $IMA = 0.112$.

c. Solve for force.

$$MA = \frac{F_r}{F_e}$$

$$F_r = (MA)(F_e)$$

$$= (0.106)(155 \text{ N}) = 16.4 \text{ N}$$

◀ Substitute $MA = 0.106$, $F_e = 155$ N.

d. Solve for distance.

$$IMA = \frac{d_e}{d_r}$$

$$d_e = (IMA)(d_r)$$

$$= (0.112)(14.0 \text{ cm}) = 1.57 \text{ cm}$$

◀ Substitute $IMA = 0.112$, $d_r = 14.0$ cm.

3 EVALUATE THE ANSWER

- **Are the units correct?** Force is measured in newtons, and distance in centimeters.
- **Is the magnitude realistic?** *IMA* is low for a bicycle because a greater F_e is traded for a greater d_r. *MA* is always smaller than *IMA.* Because *MA* is low, F_r also will be low. The small distance the axle moves results in a large distance covered by the wheel. Thus, d_e should be very small.

EXAMPLE PROBLEM

PRACTICE PROBLEMS

Do additional problems. Online Practice

25. If the gear radius of the bicycle in Example Problem 4 is doubled while the force exerted on the chain and the distance the wheel rim moves remain the same, what quantities change, and by how much?

26. A sledgehammer is used to drive a wedge into a log to split it. When the wedge is driven 0.20 m into the log, the log is separated a distance of 5.0 cm. A force of 1.7×10^4 N is needed to split the log, and the sledgehammer exerts a force of 1.1×10^4 N.

a. What is the *IMA* of the wedge?

b. What is the *MA* of the wedge?

c. Calculate the efficiency of the wedge as a machine.

27. A worker uses a pulley to raise a 24.0-kg carton 16.5 m, as shown in **Figure 15.** A force of 129 N is exerted, and the rope is pulled 33.0 m.

a. What is the *MA* of the pulley?

b. What is the efficiency of the pulley?

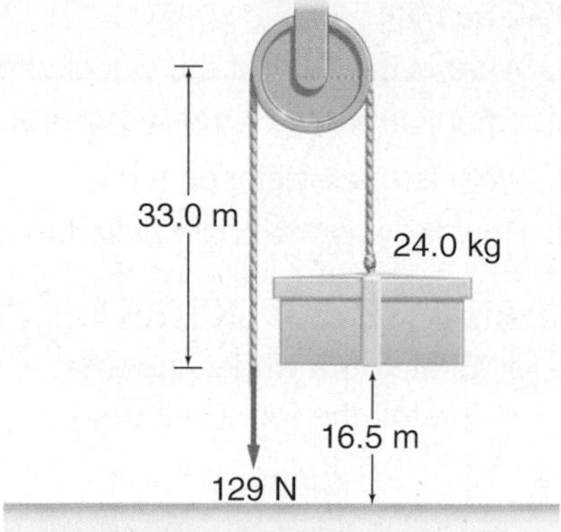

Figure 15

28. A winch has a crank with a 45-cm radius. A rope is wrapped around a drum with a 7.5-cm radius. One revolution of the crank turns the drum one revolution.

a. What is the ideal mechanical advantage of this machine?

b. If, due to friction, the machine is only 75 percent efficient, how much force would have to be exerted on the handle of the crank to exert 750 N of force on the rope?

29. CHALLENGE You exert a force of 225 N on a lever to raise a 1.25×10^3-N rock a distance of 13 cm. If the efficiency of the lever is 88.7 percent, how far did you move your end of the lever?

Multi-gear bicycle Shifting gears on a bicycle is a way of adjusting the ratio of gear radii to obtain the desired *IMA*. On a multi-gear bicycle, the rider can change the *IMA* of the machine by choosing the size of one or both gears. **Figure 16** shows a rear gear with five different gear sizes. When accelerating or climbing a hill, the rider increases the ideal mechanical advantage to increase the force that the wheel exerts on the road.

To increase the *IMA*, the rider needs to make the rear gear radius large compared to the front gear radius (refer to the *IMA* equation earlier in the section). For the same force exerted by the rider, a larger force is exerted by the wheel on the road. However, the rider must rotate the pedals through more turns for each revolution of the wheel.

On the other hand, less force is needed to ride the bicycle at high speed on a level road. The rider needs a small rear gear and a large front gear, resulting in a smaller *IMA*. Thus, for the same force exerted by the rider, a smaller force is exerted by the wheel on the road. However, in return, the rider does not have to move the pedals as far for each revolution of the wheel.

An automobile transmission works in the same way. To accelerate a car from rest, large forces are needed and the transmission increases the *IMA* by increasing the gear ratio. At high speeds, however, the transmission reduces the gear ratio and the *IMA* because smaller forces are needed. Even though the speedometer shows a high speed, the tachometer indicates the engine's low angular speed.

Figure 16 A rider can change the *IMA* of the bicycle by shifting gears.

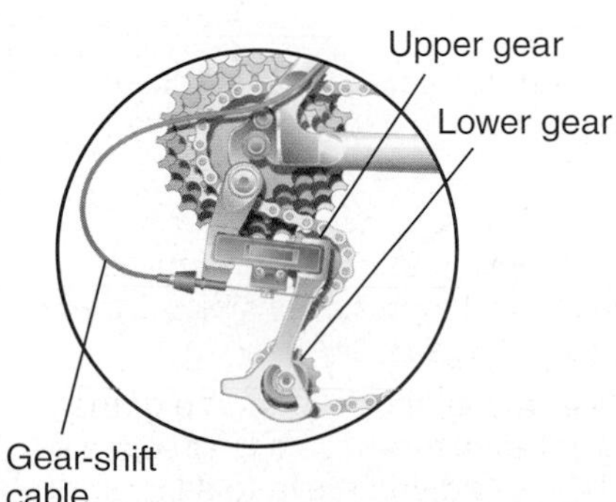

☑ **READING CHECK Explain** why your car needs multiple gears.

The Human Walking Machine

▸ **CONNECTION TO BIOLOGY** Movement of the human body is explained by the same principles of force and work that describe all motion. Simple machines, in the form of levers, give humans the ability to walk, run, and perform many other activities. The lever systems of the human body are complex. However, each system has the following four basic parts:

1. a rigid bar (bone)
2. a source of force (muscle contraction)
3. a fulcrum or pivot (movable joints between bones)
4. a resistance (the weight of the body or an object being lifted or moved)

Figure 17 shows these parts in the lever system in a human leg.

Lever systems of the body are not very efficient, and mechanical advantages are low. This is why walking and jogging require energy (burn calories) and help people lose weight.

When a person walks, the hip acts as a fulcrum and moves through the arc of a circle, centered on the foot. The center of mass of the body moves as a resistance around the fulcrum in the same arc. The length of the radius of the circle is the length of the lever formed by the bones of the leg. Athletes in walking races increase their velocity by swinging their hips upward to increase this radius.

A tall person's body has lever systems with less mechanical advantage than a short person's does. Although tall people usually can walk faster than short people can, a tall person must apply a greater force to move the longer lever formed by the leg bones. How would a tall person do in a walking race? What are the factors that affect a tall person's performance? Walking races are usually 20 or 50 km long. Because of the inefficiency of their lever systems and the length of a walking race, very tall people rarely have the stamina to win.

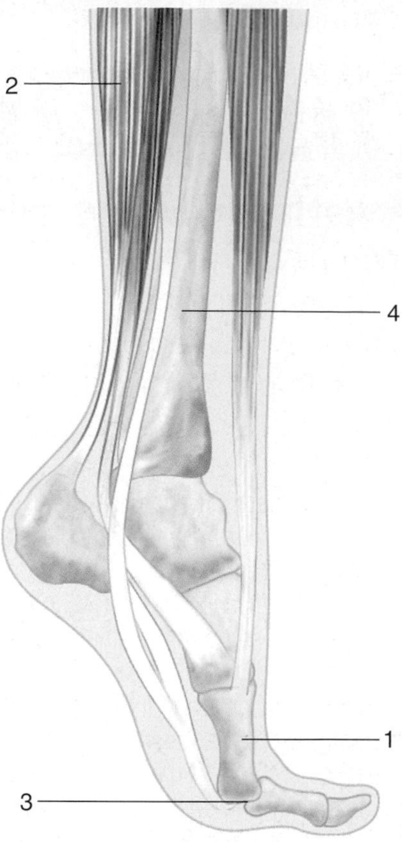

Figure 17 The human leg and foot function as a compound machine.

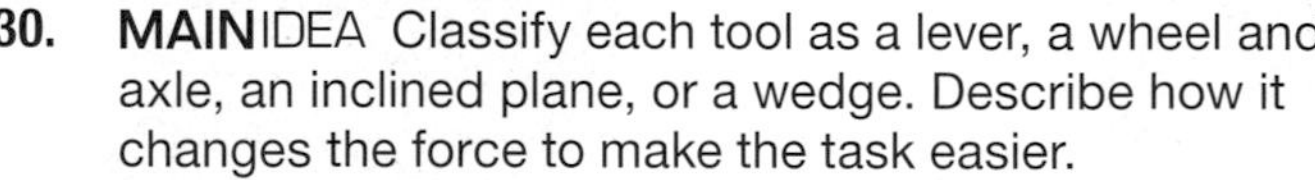

SECTION 2 REVIEW

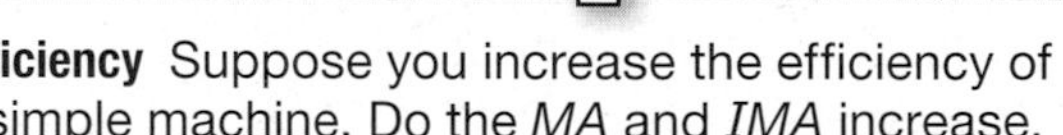

Check your understanding.

30. **MAINIDEA** Classify each tool as a lever, a wheel and axle, an inclined plane, or a wedge. Describe how it changes the force to make the task easier.

a. screwdriver
b. pliers
c. chisel
d. nail puller

31. **IMA** A worker is testing a multiple pulley system to estimate the heaviest object that he could lift. The largest downward force he can exert is equal to his weight, 875 N. When the worker moves the rope 1.5 m, the object moves 0.25 m. What is the heaviest object that he could lift?

32. **Compound Machines** A winch has a crank on a 45-cm arm that turns a drum with a 7.5-cm radius through a set of gears. It takes three revolutions of the crank to rotate the drum through one revolution. What is the *IMA* of this compound machine?

33. **Efficiency** Suppose you increase the efficiency of a simple machine. Do the *MA* and *IMA* increase, decrease, or remain the same?

34. **Critical Thinking** The mechanical advantage of a multi-gear bicycle is changed by moving the chain to a suitable rear gear.

a. To start out, you must accelerate the bicycle, so you want to have the bicycle exert the greatest possible force. Should you choose a small or large gear?

b. As you reach your traveling speed, you want to rotate the pedals as few times as possible. Should you choose a small or large gear?

c. Many bicycles also let you choose the size of the front gear. If you want even more force to accelerate while climbing a hill, would you move to a larger or smaller front gear?

disADVANTAGE?

Prosthetic Limbs In the past, losing both legs from the knee down might mean that an amputee could never again walk, let alone run. However, with modern running prosthetics—artificial legs with curved carbon-fiber blades—amputees can compete in world-class racing events.

Energy Runners with natural legs use their thighs, knees, calves, and ankles to absorb energy as each foot hits the ground and pushes off, as in **Figure 1.** With a prosthetic leg, some of the kinetic energy of the swinging blade is stored as the blade compresses when it hits the ground. The compressed blade acts like a spring, transforming the stored energy back into kinetic energy and pushing the runner forward with each stride, as shown in **Figure 2.**

Is running with prosthetics easier? Many athletes think that prosthetics might even give amputees, who can wear technology that has been engineered to maximize their performance, an advantage over runners who have their natural legs. Scientists test prosthetics to determine how their performance measures up against natural legs and feet.

When it comes to running, prosthetic legs have disadvantages as well. At the starting block, runners with prosthetic legs lose more time launching into a sprint than do runners with natural legs. Furthermore, there is evidence that runners with prosthetics experience a disadvantage in force production, which could reduce their running speed.

FIGURE 1 The energy transformations for a natural leg are less efficient than those for a prosthetic leg.

FIGURE 2 When the foot of the prosthetic hits the ground, it compresses, storing energy from the impact.

Hannah Johnston/Getty Image Sport/Getty Images

GOINGFURTHER >>>

Research the controversy surrounding athletic events in which some participants wear prosthetics. Write an opinion piece about the fairness of races in which runners that have both natural legs compete with double and single amputees.

CHAPTER 10 STUDY GUIDE

BIGIDEA Doing work on a system changes the system's energy.

VOCABULARY
- **work** *(p. 264)*
- **joule** *(p. 264)*
- **energy** *(p. 270)*
- **work-energy theorem** *(p. 270)*
- **kinetic energy** *(p. 270)*
- **translational kinetic energy** *(p. 270)*
- **power** *(p. 271)*
- **watt** *(p. 271)*

SECTION 1 Work and Energy

MAINIDEA Work is the transfer of energy that occurs when a force is applied through a displacement.

- Work is done when a force is applied through a displacement. Work is the product of the force exerted on a system and the component of the distance through which the system moves that is parallel to the force.

$$W = Fd\cos\theta$$

The work done can be determined by calculating the area under a force-displacement graph.

- Energy is the ability of a system to produce a change in itself or its environment. A moving object has kinetic energy. Objects that are changing position have translational kinetic energy.

$$KE_{trans} = \frac{1}{2}mv^2$$

- The work done on a system is equal to the change in energy of the system. This is called the work-energy theorem.

$$W = \Delta E$$

- Power is the rate at which energy is transformed. When work causes the change in energy, power is equal to the rate of work done.

$$P = \frac{\Delta E}{t} = \frac{W}{t}$$

VOCABULARY
- **machine** *(p. 274)*
- **effort force** *(p. 274)*
- **resistance force** *(p. 274)*
- **mechanical advantage** *(p. 274)*
- **ideal mechanical advantage** *(p. 275)*
- **efficiency** *(p. 276)*
- **compound machine** *(p. 277)*

SECTION 2 Machines

MAINIDEA Machines make tasks easier by changing the magnitude or the direction of the force exerted.

- Machines, whether powered by engines or humans, do not change the amount of work done, but they do make the task easier by changing the magnitude or direction of the effort force.
- The mechanical advantage (*MA*) is the ratio of resistance force to effort force.

$$MA = \frac{F_r}{F_e}$$

- The ideal mechanical advantage (IMA) is the ratio of the distances moved.

$$IMA = \frac{d_e}{d_r}$$

- The efficiency of a machine is the ratio of output work to input work.

$$e = \left(\frac{W_o}{W_i}\right) \times 100$$

The efficiency of a machine can be found from the real and ideal mechanical advantages. In all real machines, *MA* is less than *IMA,* and *e* is less than 100 percent.

$$e = \left(\frac{MA}{IMA}\right) \times 100$$

Games and Multilingual eGlossary

Vocabulary Practice

Chapter Self-Check

SECTION 1 Work and Energy

Mastering Concepts

35. In what units is work measured?

36. A satellite orbits Earth in a circular orbit. Does Earth's gravity do work on the satellite? Explain.

37. An object slides at constant speed on a frictionless surface. What forces act on the object? What is the work done by each force on the object?

38. Define *work* and *power.*

39. What is a watt equivalent to in terms of kilograms, meters, and seconds?

Mastering Problems

40. The third floor of a house is 8 m above street level. How much work must a pulley system do to lift a 150-kg oven at a constant speed to the third floor?

41. Haloke does 176 J of work lifting himself 0.300 m at a constant speed. What is Haloke's mass?

42. Tug-of-War During a tug-of-war, team A does 2.20×10^5 J of work in pulling team B 8.00 m. What average force did team A exert?

43. To travel at a constant velocity, a car exerts a 551-N force to balance air resistance. How much work does the car do on the air as it travels 161 km from Columbus to Cincinnati?

44. Cycling A cyclist exerts a 15.0-N force while riding 251 m in 30.0 s. What power does the cyclist develop?

45. A student librarian lifts a 2.2-kg book from the floor to a height of 1.25 m. He carries the book 8.0 m to the stacks and places the book on a shelf that is 0.35 m above the floor. How much work does he do on the book?

46. A horizontal force of 300.0 N is used to push a 145-kg mass 30.0 m horizontally in 3.00 s.

a. Calculate the work done on the mass.

b. Calculate the power developed.

47. Wagon A wagon is pulled by a force of 38.0 N exerted on the handle at an angle of 42.0° with the horizontal. If the wagon is pulled for 157 m, how much work is done on the wagon?

48. Lawn Mower To mow the yard, Shani pushes a lawn mower 1.2 km with a horizontal force of 66.0 N. Does all of the applied force do work on the mower, and how much work does Shani do on the mower?

49. A 17.0-kg crate is to be pulled a distance of 20.0 m, requiring 1210 J of work to be done on the crate. The job is done by attaching a rope and pulling with a force of 75.0 N. At what angle is the rope held?

50. Lawn Tractor The lawn tractor in **Figure 18** goes up a hill at a constant velocity. Calculate the power that is developed by the tractor.

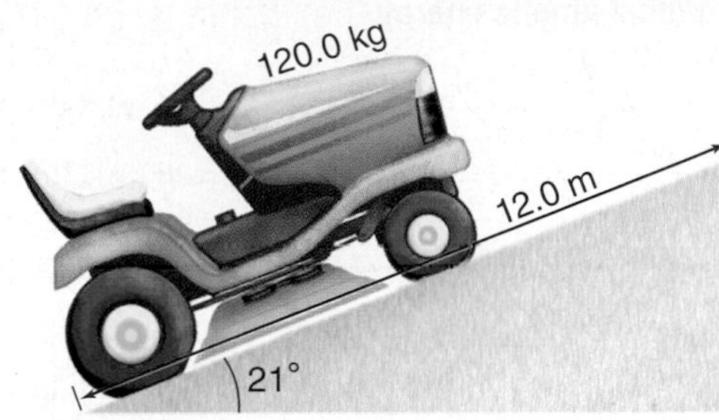

Figure 18

51. You slide a crate up a ramp at an angle of 30.0° to a vertical height of 1.15 m. You exert a 225-N force parallel to the ramp, and the crate moves at a constant speed. The coefficient of friction is 0.28. How much work do you do on the crate?

52. A 4.2×10^3-N piano is wheeled up a 3.5-m ramp at a constant speed. The ramp makes an angle of 30.0° with the horizontal. Find the work done by a man wheeling the piano up the ramp.

53. Sled Diego pulls a sled across level snow as shown in **Figure 19.** If the sled moves a distance of 65.3 m, how much work does Diego do on the sled?

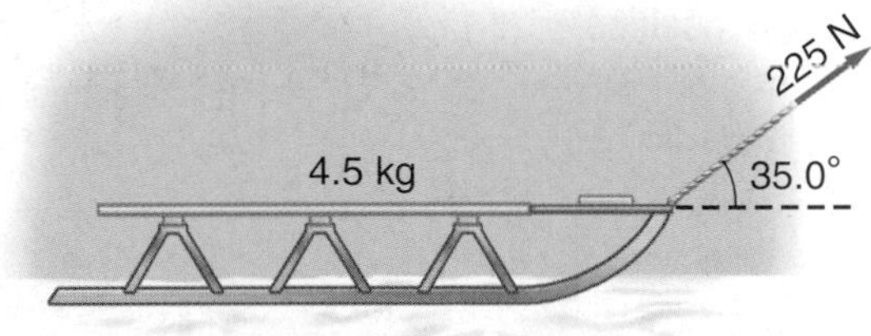

Figure 19

54. Escalator Sau-Lan's mass is 52 kg. She rides up the escalator at Ocean Park in Hong Kong. This is the world's longest escalator, with a length of 227 m and an average inclination of 31°.

a. How much work does the escalator do on Sau-Lan?

b. Sue's mass is 65 kg and she rides the escalator, too. How much work does the escalator do on Sue?

55. Lawn Roller A lawn roller is pushed across a lawn by a force of 115 N along the direction of the handle, which is 22.5° above the horizontal. If 64.6 W of power is developed for 90.0 s, what distance is the roller pushed?

56. Boat Engine An engine moves a boat through the water at a constant speed of 15 m/s. The engine must exert a force of 6.0 kN to balance the force that the water exerts against the hull. What power does the engine develop?

57. Maricruz slides a crate up an inclined ramp that is attached to a platform as shown in **Figure 20.** A 400.0-N force, parallel to the ramp, is needed to slide the crate up the ramp at a constant speed.

a. How much work does Maricruz do in sliding the crate up the ramp?

b. How much work would be done on the crate if Maricruz simply lifted the crate straight up from the floor to the platform at a constant speed?

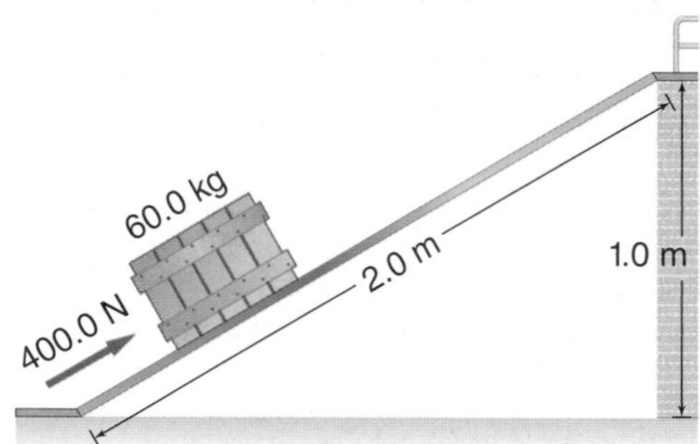

Figure 20

58. A worker pushes a 93-N crate up an inclined plane at a constant speed. As shown in **Figure 21,** the worker pushes parallel to the ground with a force of 85 N.

a. How much work does the worker do on the crate?

b. How much work is done by gravity on the crate? (Be careful with the signs you use.)

c. The coefficient of friction is $\mu = 0.20$. How much energy is transformed by friction? (Be careful with the signs you use.)

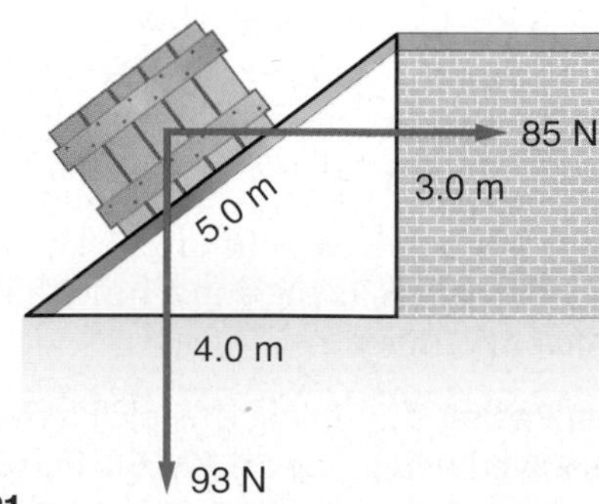

Figure 21

59. In **Figure 22,** the magnitude of the force necessary to stretch a spring is plotted against the distance the spring is stretched.

a. Calculate the slope of the graph (k), and show that $F = kd$, where $k = 25$ N/m.

b. Use the graph to find the work done in stretching the spring from 0.00 m to 0.20 m.

c. Show that the answer to part **b** can be calculated using the formula $W = \left(\frac{1}{2}\right)kd^2$, where W is the work, $k = 25$ N/m (the slope of the graph), and d is the distance the spring is stretched (0.20 m).

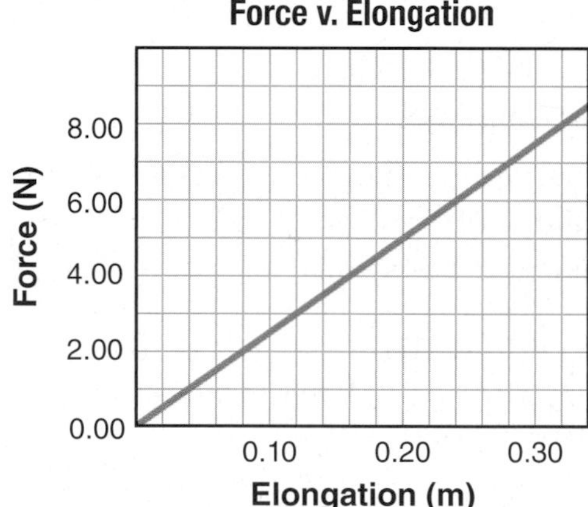

Figure 22

60. Use the graph in **Figure 22** to find the required work to stretch the spring from 0.12 m to 0.28 m.

61. The graph in **Figure 23** shows the force exerted on and displacement of an object being pulled.

a. Find the work done to pull the object 7.0 m.

b. Calculate the power that would be developed if the work was done in 2.0 s.

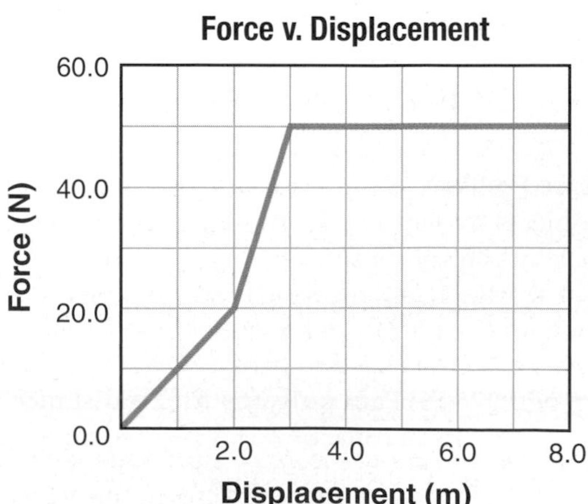

Figure 23

Chapter Self-Check

62. Oil Pump In 35.0 s, a pump delivers 0.550 m^3 of oil into barrels on a platform 25.0 m above the intake pipe. The oil's density is 0.820 g/cm^3.

a. Calculate the work done by the pump on the oil.

b. Calculate the power produced by the pump.

63. Conveyor Belt A 12.0-m-long conveyor belt, inclined at 30.0°, is used to transport bundles of newspapers from the mail room up to the cargo bay to be loaded onto delivery trucks. The mass of a newspaper is 1.0 kg, and each bundle has 25 newspapers. Find the power the conveyor develops if it delivers 15 bundles per minute.

SECTION 2 Machines

Mastering Concepts

64. Is it possible to get more work out of a machine than you put into it?

65. Explain how bicycle pedals are a simple machine.

Mastering Problems

66. Piano Takeshi raises a 1200-N piano a distance of 5.00 m using a set of pulleys. He pulls in 20.0 m of rope.

a. How much effort force would Takeshi apply if this were an ideal machine?

b. What force is used to balance the friction force if the actual effort is 340 N?

c. What is the output work?

d. What is the input work?

e. What is the mechanical advantage?

67. Because there is very little friction, the lever is an extremely efficient simple machine. Using a 90.0-percent-efficient lever, what input work is needed to lift an 18.0-kg mass a distance of 0.50 m?

68. A student exerts a force of 250 N on a lever through a distance of 1.6 m as he lifts a 150-kg crate. If the efficiency of the lever is 90.0 percent, how far is the crate lifted?

69. Reverse Problem Write a physics problem with real-life objects for which the following equation would be part of the solution:

$$(12.5\ \text{N})d = \frac{1}{2}(6.0\ \text{kg})(1.10\ \text{m/s})^2 - \frac{1}{2}(6.0\ \text{kg})(0.05\ \text{m/s})^2$$

70. A pulley system is used to lift a 1345-N weight a distance of 0.975 m. Paul pulls the rope a distance of 3.90 m, exerting a force of 375 N.

a. What is the *IMA* of the system?

b. What is the mechanical advantage?

c. How efficient is the system?

71. A force of 1.4 N is exerted on a rope in a pulley system. The force is exerted through a distance of 40.0 cm, lifting a 0.50-kg mass 10.0 cm. Calculate the following:

a. the *MA*

b. the *IMA*

c. the efficiency

72. Use **Figure 24** to answer the following questions.

a. What force, parallel to the ramp (F_A), is required to slide a 25-kg box at constant speed to the top of the ramp? Ignore friction.

b. What is the *IMA* of the ramp?

c. What are the actual *MA* and the efficiency of the ramp if a 75-N parallel force is needed?

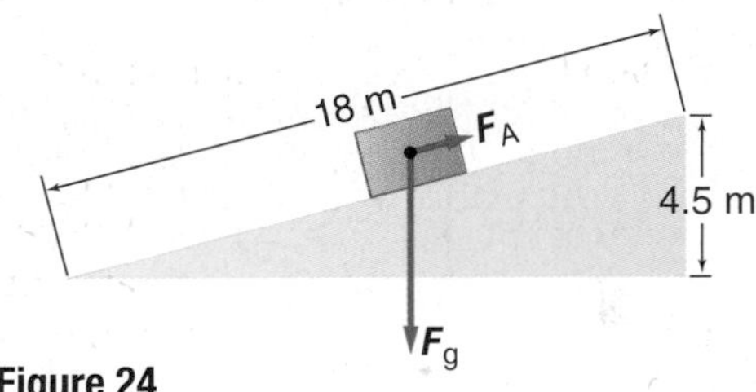

Figure 24

73. Bicycle Luisa pedals a bicycle with the wheel shown in **Figure 25.** If the wheel revolves once, what is the length of the chain that was used?

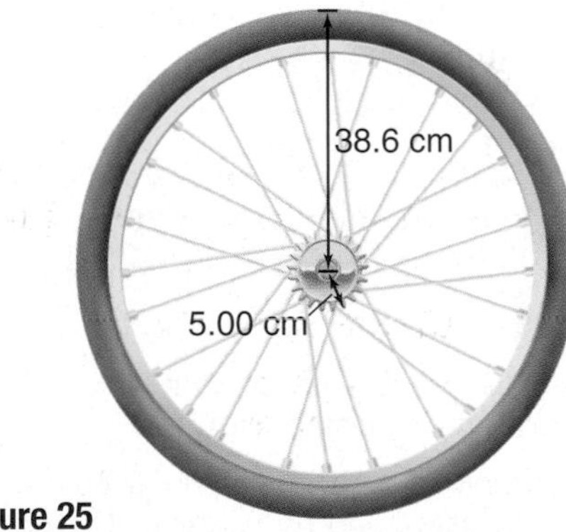

Figure 25

74. A motor with an efficiency of 88 percent runs a crane with an efficiency of 42 percent. The power supplied to the motor is 5.5 kW. At what constant speed does the crane lift a 410-kg crate?

75. What work is required to lift a 215-kg mass a distance of 5.65 m, using a machine that is 72.5 percent efficient?

76. Problem Posing Complete this problem so that it can be solved using power: "While rearranging furniture, Opa needs to move a 50-kg sofa ..."

Chapter Self-Check

77. A compound machine is made by attaching a lever to a pulley system. Consider an ideal compound machine consisting of a lever with an *IMA* of 3.0 and a pulley system with an *IMA* of 2.0.

a. Show that the compound machine's *IMA* is 6.0.

b. If the compound machine is 60.0 percent efficient, how much effort must be applied to the lever to lift a 540-N box?

c. If you move the effort side of the lever 12.0 cm, how far is the box lifted?

Applying Concepts

78. Which requires more work—carrying a 420-N backpack up a 200-m-high hill or carrying a 210-N backpack up a 400-m-high hill? Why?

79. Lifting You slowly lift a box of books from the floor and put it on a table. Earth's gravity exerts a force, magnitude mg, downward, and you exert a force, magnitude mg, upward. The two forces have equal magnitudes and opposite directions. It appears that no work is done, but you know that you did work. Explain what work was done.

80. You have an after-school job carrying cartons of new copy paper up a flight of stairs and then carrying recycled paper back down the stairs. The mass of the paper is the same in both cases. Your physics teacher says that you did no work, so you should not be paid. In what sense is the physics teacher correct? What arrangement of payments might you make to ensure that you are properly compensated?

81. Once downstairs, you carry the cartons of paper along a 15-m-long hallway. Are you doing work by carrying the boxes down the hall? Explain.

82. Climbing Stairs Two people of the same mass climb the same flight of stairs. The first person climbs the stairs in 25 s; the second person does so in 35 s.

a. Which person does more work? Explain.

b. Which person produces more power? Explain.

83. Show that power can be written as $P = Fv \cos \theta$.

84. How can you increase the ideal mechanical advantage of a machine?

85. BIGIDEA **Orbits** Explain why a planet orbiting the Sun does not violate the work-energy theorem.

86. Claw Hammer A standard claw hammer is used to pull a nail from a piece of wood. Where should you place your hand on the handle and where should the nail be located in the claw to make the effort force as small as possible?

87. Wedge How can you increase the mechanical advantage of a wedge without changing its ideal mechanical advantage?

Mixed Review

88. Ramps Isra has to get a piano onto a 2.0-m-high platform. She can use a 3.0-m-long frictionless ramp or a 4.0-m-long frictionless ramp. Which ramp should Isra use if she wants to do the least amount of work?

89. Brutus, a champion weight lifter, raises 240 kg of weights a distance of 2.35 m at a constant speed.

a. Find the work Brutus does on the weights.

b. How much work is done by Brutus on the weights holding the weights above his head?

c. How much work is done by Brutus on the weights lowering them back to the ground?

d. Does Brutus do work if he lets go of the weights and they fall back to the ground?

90. A 805-N horizontal force is needed to drag a crate across a horizontal floor at a constant speed. You drag the crate using a rope held at a 32° angle.

a. What force do you exert on the rope?

b. How much work do you do on the crate if you move it 22 m?

c. If you complete the job in 8.0 s, what power is developed?

91. Dolly and Ramp A dolly is used to move a 115-kg refrigerator up a ramp into a house. The ramp is 2.10 m long and rises 0.850 m. The mover pulls the dolly with a force of 496 N parallel to the ramp. The dolly and ramp constitute a machine.

a. What work does the mover do on the dolly?

b. What is the work done on the refrigerator by the machine?

c. What is the efficiency of the machine?

92. Sally does 11.4 kJ of work dragging a wooden crate 25.0 m across a floor at a constant speed. The rope she uses to pull the crate makes an angle of 48.0° with the horizontal.

a. What force does the rope exert on the crate?

b. Find the force of friction acting on the crate.

c. How much energy is transformed by the force of friction between the floor and the crate?

93. Sledding An 845-N sled is pulled a distance of 185 m. The task requires 1.20×10^4 J of work and is done by pulling on a rope with a force of 125 N. At what angle is the rope held?

Chapter Self-Check

94. An electric winch pulls an 875-N crate up a 15° incline at 0.25 m/s. The coefficient of friction between the crate and incline is 0.45.

a. What power does the winch develop?

b. How much electrical power must be delivered to the winch if it is 85 percent efficient?

Thinking Critically

95. Apply Concepts A 75-kg sprinter runs the 50.0-m dash in 8.50 s. Assume the sprinter's acceleration is constant throughout the race.

a. Find the sprinter's average power for the race.

b. What is the maximum power the sprinter develops?

96. Apply Concepts The sprinter in the previous problem runs the 50.0-m dash again in 8.50 s. This time, however, the sprinter accelerates in the first second and runs the rest of the race at a constant velocity.

a. Calculate the average power produced for that first second.

b. What is the maximum power the sprinter now generates?

97. Analyze and Conclude You are carrying boxes to a storage loft that is 12 m above the ground. You need to move 30 boxes with a total mass of 150 kg as quickly as possible. You could carry more than one up at a time, but if you try to move too many at once, you will go very slowly and rest often. If you carry only one box at a time, most of the energy will go into raising your own body. The power that your body can develop over a long time depends on the mass that you carry, as shown in **Figure 26.** Find the number of boxes to carry on each trip that would minimize the time required. What time would you spend doing the job? Ignore the time needed to go back down the stairs and to lift and lower each box.

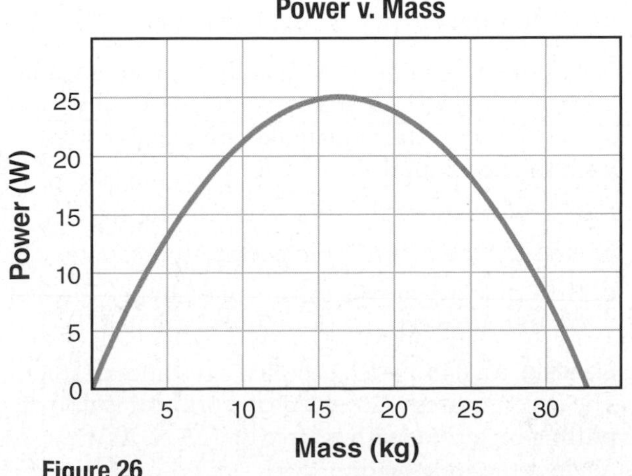

Figure 26

98. Ranking Task A 20-kg boy interacts with a bench as shown in **Figure 27.** Rank each interaction according to the work the boy does on the bench, from least to greatest. Clearly indicate any ties.

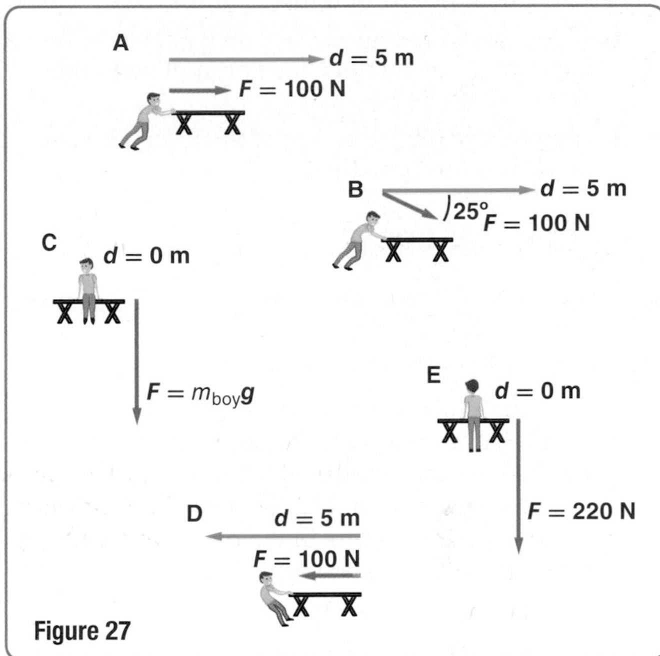

Figure 27

Writing in Physics

99. Just as a bicycle is a compound machine, so is an automobile. Find the efficiencies of the component parts of the power train (engine, transmission, wheels, and tires). Explore possible improvements in each of these efficiencies.

100. The terms *force, work, power,* and *energy* are often used as synonyms in everyday use. Obtain examples from radio, television, print media, and advertisements that illustrate meanings for these terms that differ from those used in physics.

Cumulative Review

101. You are gardening and fill a garbage can with soil and weeds. The 24-kg can is too heavy to lift so you push it across the yard. The coefficient of kinetic friction between the can and the muddy grass is 0.27, and the coefficient of static friction is 0.35. How hard must you push horizontally to get the can to just start moving?

102. Baseball A major league pitcher throws a fastball horizontally at a speed of 40.3 m/s (90 mph). How far has it dropped by the time it crosses home plate 18.4 m (60 ft, 6 in.) away?

MULTIPLE CHOICE

1. A 4-N soccer ball sits motionless on a field. A player's foot exerts a force of 5 N on the ball for a distance of 0.1 m, and the ball rolls a distance of 10 m. How much kinetic energy does the ball gain from the player?

A. 0.5 J **C.** 9 J

B. 0.9 J **D.** 50 J

2. A pulley system consists of two fixed pulleys and two movable pulleys that lift a rock that has a 300-N weight at a constant speed. If the effort force used to lift the rock is 100 N, what is the mechanical advantage of the system?

A. $\frac{1}{3}$ **C.** 3

B. $\frac{3}{4}$ **D.** 6

3. A compound machine used to raise heavy boxes consists of a ramp and a pulley. The efficiency of pulling a 100-kg box up the ramp is 50 percent. If the efficiency of the pulley is 90 percent, what is the overall efficiency of the compound machine?

A. 40 percent **C.** 50 percent

B. 45 percent **D.** 70 percent

4. A 20.0-N block is attached to the end of a rope, and the rope is looped around a pulley system. If you pull the opposite end of the rope a distance of 2.00 m, the pulley system raises the block a distance of 0.40 m. What is the pulley system's ideal mechanical advantage?

A. 2.5 **C.** 5.0

B. 4.0 **D.** 10.0

5. Two people carry identical 40.0-N boxes up a ramp. The ramp is 2.00 m long and rests on a platform that is 1.00 m high. One person walks up the ramp in 2.00 s, and the other person walks up the ramp in 4.00 s. What is the difference in power the two people use to carry the boxes up the ramp?

A. 5.00 W **C.** 20.0 W

B. 10.0 W **D.** 40.0 W

Online Test Practice

6. A skater with a mass of 50.0 kg slides across an icy pond with negligible friction. As he approaches a friend, both he and his friend hold out their hands, and the friend exerts a force in the direction opposite to the skater's movement, which slows the skater's speed from 2.0 m/s to 1.0 m/s. What is the change in the skater's kinetic energy?

A. −25 J **C.** −100 J

B. −75 J **D.** −150 J

7. The box in the diagram is being pushed up the ramp with a force of 100.0 N. What is the work done on the box?
($\sin 30° = 0.50$, $\cos 30° = 0.87$, $\tan 30° = 0.58$)

A. 150 J **C.** 450 J

B. 260 J **D.** 600 J

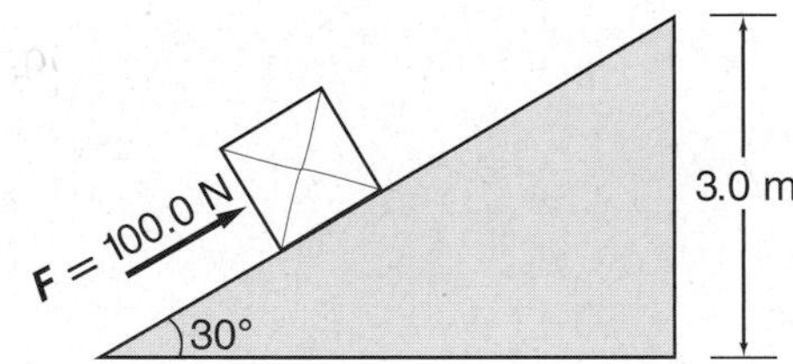

FREE RESPONSE

8. The diagram shows a box being pulled along a horizontal surface with a force of 200.0 N. Calculate the work done on the box and the power required to pull it a distance of 5.0 m in 10.0 s.
($\sin 45° = \cos 45° = 0.71$)

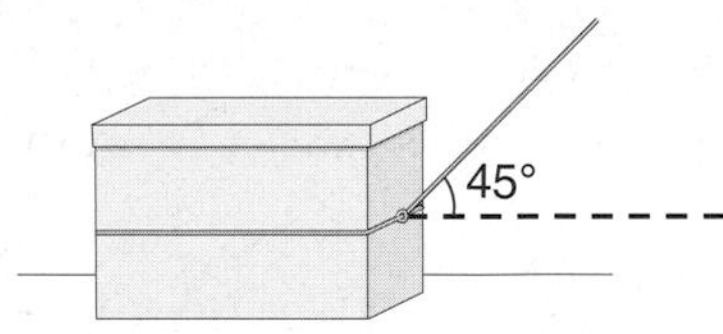

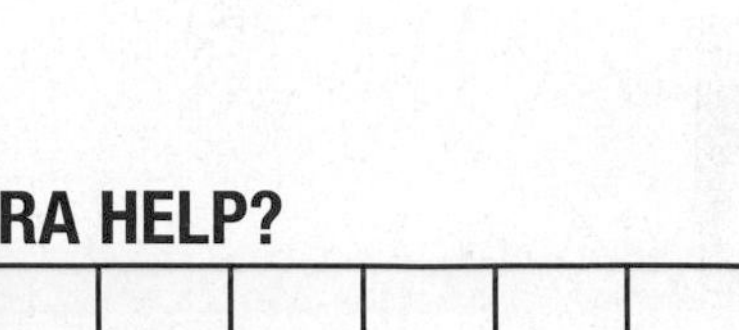

NEED EXTRA HELP?

If You Missed Question	1	2	3	4	5	6	7	8
Review Section	1	2	2	2	1	1	1	1

CHAPTER 11

Energy and Its Conservation

BIGIDEA **Within a closed, isolated system, energy can change form, but the total energy is constant.**

SECTIONS

1 The Many Forms of Energy

2 Conservation of Energy

LaunchLAB

iLab Station

ENERGY OF A BOUNCING BALL

What determines how high a basketball bounces?

WATCH THIS!

FUEL EFFICIENCY

The phrase *fuel-efficient* is often used as a selling point when describing the features of a car. But what does that phrase really mean? How is fuel-efficiency calculated? And how does it relate to the conservation of energy?

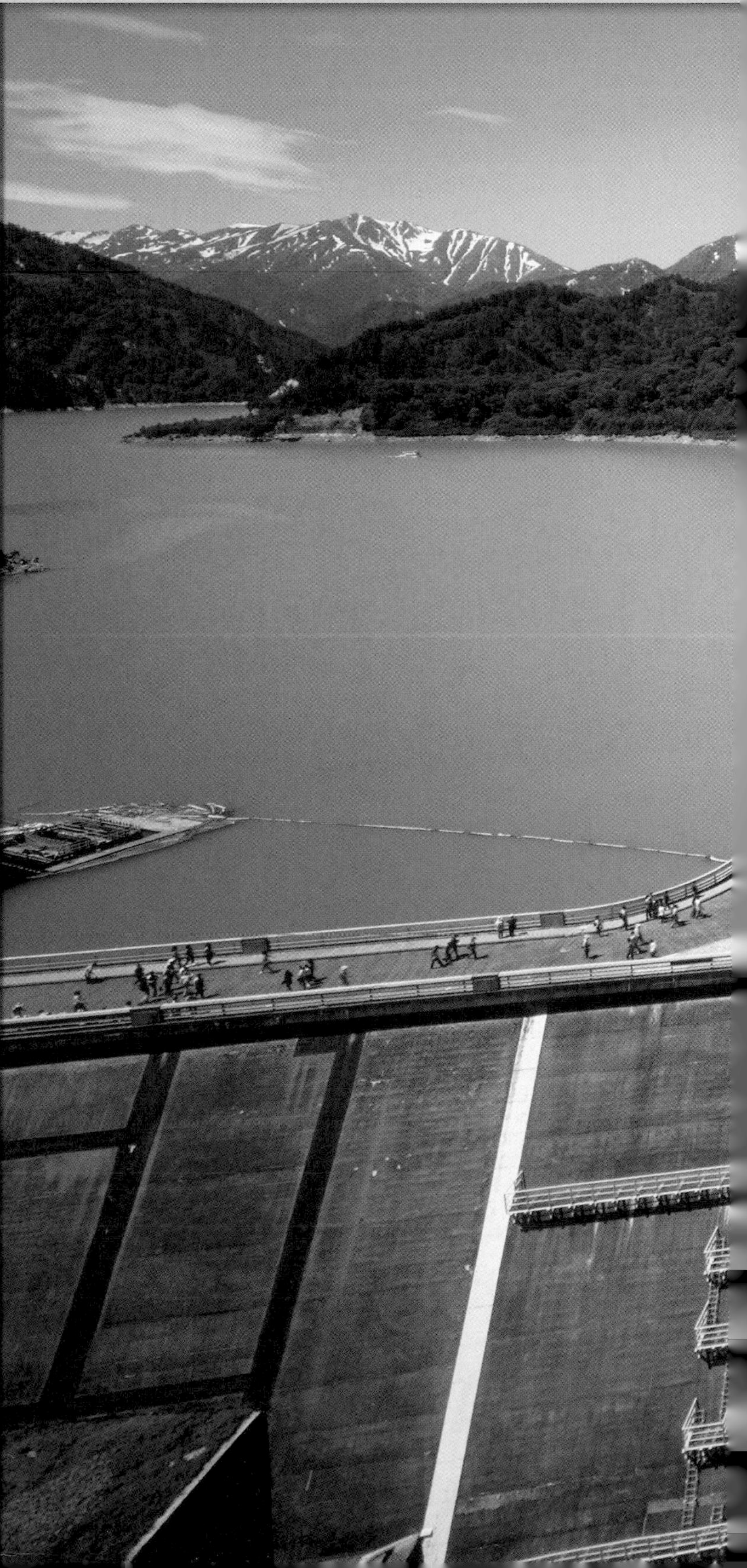

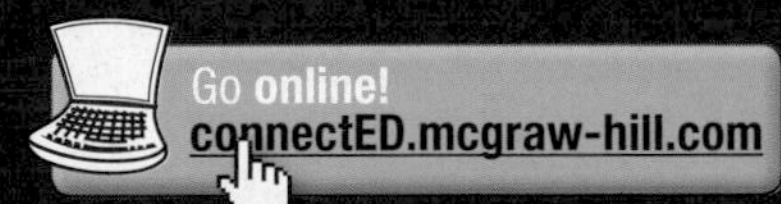
Go online!
connectED.mcgraw-hill.com

SECTION 1

The Many Forms of Energy

PHYSICS 4 YOU **Water has several forms. It forms ice and water vapor. In a similar way, energy has different forms, such as energy due to motion or energy due to object interactions.**

MAIN IDEA

Kinetic energy is energy due to an object's motion, and potential energy is energy stored due to the interactions of two or more objects.

Essential Questions

- How is a system's motion related to its kinetic energy?
- What is gravitational potential energy?
- What is elastic potential energy?
- How are mass and energy related?

Review Vocabulary

system the object or objects of interest that can interact with each other and with the outside world

New Vocabulary

rotational kinetic energy
potential energy
gravitational potential energy
reference level
elastic potential energy
thermal energy

A Model of the Work-Energy Theorem

The word *energy* is used in many different ways in everyday speech. Some fruit-and-cereal bars are advertised as energy sources. Athletes use energy in sports. Companies that supply your home with electricity, natural gas, or heating fuel are called energy companies.

Scientists and engineers use the term *energy* more precisely. Recall that the work-energy theorem states that doing work on a system causes a change in the energy of that system. That is, work is the process that transfers energy between a system and the external world. Recall that a system is the object or objects of interest. When an agent performs work on a system, the system's energy increases. When the system does work on its surroundings, the system's energy decreases.

Modeling transformations In some ways, energy is like ice cream—it comes in different varieties. You can have vanilla, chocolate, or peach ice cream. These are different varieties, but they are all ice cream. However, unlike ice cream, energy can be changed from one form to another. A system of objects can possess energy in a variety of forms. In this chapter, you will learn how energy is transformed from one variety (or form) to another and how to keep track of the changes.

Keeping track of energy changes is much like keeping track of your savings account. Bar diagrams, such as those shown in **Figure 1,** can be used to track money or energy. The amount of money you have changes when you earn more or spend it. Similarly, energy can be stored or it can be used. In each case, the behavior of a system is affected.

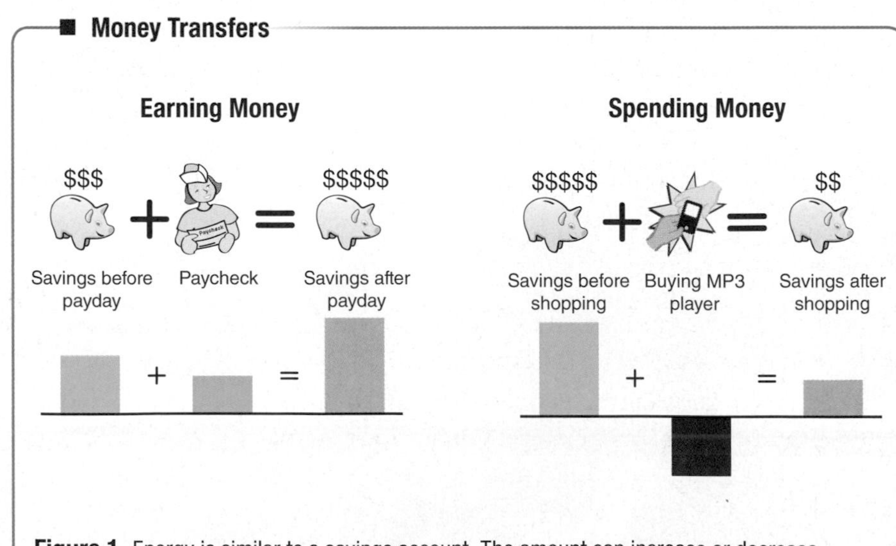

Figure 1 Energy is similar to a savings account. The amount can increase or decrease.

Richard Hutchings/Digital Light Source

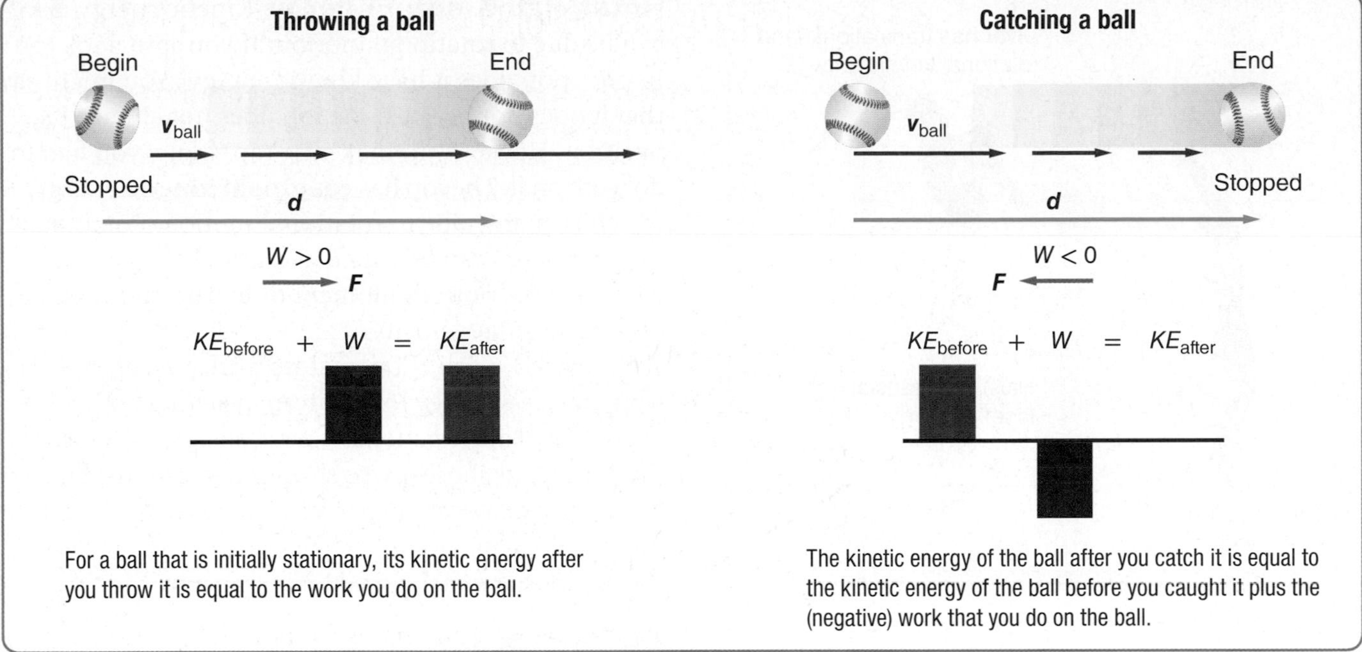

Throwing a ball
Begin
End
v_{ball}
Stopped
d
W > 0
F
KE_{before} + W = KE_{after}
For a ball that is initially stationary, its kinetic energy after you throw it is equal to the work you do on the ball.
Catching a ball
Begin
End
v_{ball}
Stopped
d
W < 0
F
KE_{before} + W = KE_{after}
The kinetic energy of the ball after you catch it is equal to the kinetic energy of the ball before you caught it plus the (negative) work that you do on the ball.

Figure 2 When you do work on an object, such as a ball, you change its energy.

Identify two other events in which work causes a change in energy.

Throwing a ball Gaining and losing energy can be illustrated by throwing and catching a ball. Recall that when you exert a constant force (F) on an object through a distance (d) in the direction of the force, you do an amount of work, represented by $W = Fd$. The work is positive because the force and the motion are in the same direction. The energy of the object increases by an amount equal to the work (W). Suppose the object is a stationary ball, and you exert a force to throw the ball. As a result of the work you do, the ball gains kinetic energy. This process is shown on the left in **Figure 2.** You can represent the process using an energy bar diagram. The height of the bar represents the amount of work (in joules). The kinetic energy after the work is done is equal to the sum of the initial kinetic energy plus the work done on the ball.

Catching a ball A similar process occurs when you catch the ball. Before the moving ball is caught, it has kinetic energy. In catching it, you exert a force on the ball in the direction *opposite* to its motion. Therefore, you do negative work on it, causing it to stop. Now that the ball is not moving, it has no kinetic energy. This process is represented in the energy bar diagram on the right in **Figure 2.** Kinetic energy is always positive, so the initial kinetic energy of the ball is positive. The work done on the ball is negative and the final kinetic energy is zero. Again, the kinetic energy after the ball has stopped equals the sum of the initial kinetic energy and the work done on the ball.

Kinetic Energy

Recall that translational kinetic energy is due to an object's change in position. It is represented by the equation $KE_{trans} = \frac{1}{2}mv^2$, where m is the object's mass and v is the magnitude of the object's velocity or speed. Kinetic energy is proportional to the object's mass. For example, a 7.26-kg bowling ball thrown through the air has more kinetic energy than a 0.148-kg baseball with the same speed. An object's kinetic energy is also proportional to the square of the object's speed. A car moving at 20 m/s has four times the kinetic energy of the same car moving at 10 m/s.

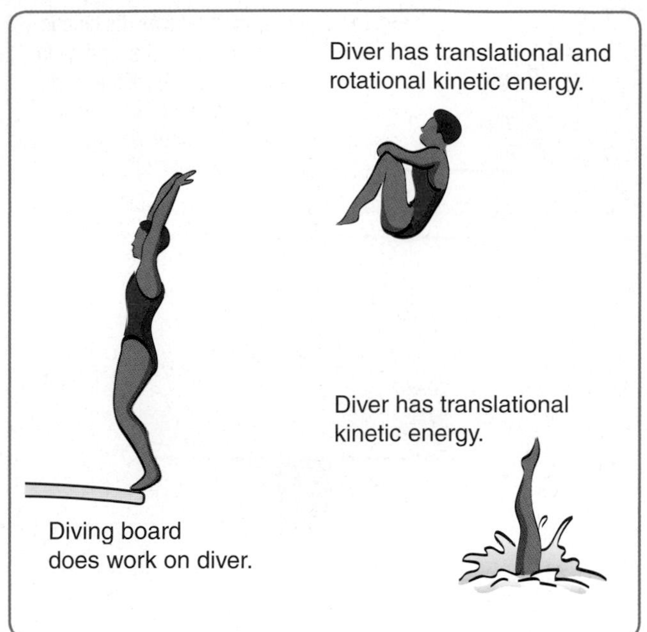
Diver has translational and
rotational kinetic energy.
Diver has translational
kinetic energy.
Diving board
does work on diver.

Figure 3 The diving board does work on the diver. This work increases the diver's kinetic energy.

Watch a **BrainPOP video about potential energy.**

Rotational Kinetic Energy Kinetic energy also can be due to rotational motion. If you spin a toy top in one spot, does it have kinetic energy? You might say that it does not because the top does not change its position. However, to make the top rotate, you had to do work on it. The top has **rotational kinetic energy,** which is energy due to rotational motion. Rotational kinetic energy can be calculated using $KE_{\text{rot}} = \frac{1}{2}I\omega^2$, where I is the object's moment of inertia and ω is the object's angular velocity.

In **Figure 3,** a diving board does work on a diver. This work transfers energy to the diver, including both translational and rotational kinetic energy. Her center of mass moves as she leaps, so she has translational kinetic energy. She rotates about her center of mass, so she has rotational kinetic energy. When she slices into the water, she has mostly translational kinetic energy.

Potential Energy

The money model that was discussed earlier also illustrates the transformation of energy from one form to another. Both money and energy can be in different forms. You can have one five-dollar bill, 20 quarters, or 500 pennies. In all of these forms, you still have five dollars. In the same way, energy can be kinetic energy, stored energy, or another form.

For an example of stored energy, consider a small group of boulders high on a mountain. These boulders have been moved away from Earth's center by geological processes against the gravitational force; thus, the system that includes both Earth and the boulders has stored energy. Energy that is stored due to interactions between objects in a system is called **potential energy.**

Not all potential energy is due to gravity. A spring-loaded toy, such as a jack-in-the-box, is a system that has potential energy, but the energy is stored due to a compressed spring, not gravity. Can you think of other ways that a system could have potential energy?

PRACTICE PROBLEMS

Do additional problems. Online Practice

1. A 52.0-kg skater moves at 2.5 m/s and glides to a stop over a distance of 24.0 m. Find the skater's initial kinetic energy. How much of her kinetic energy is transformed into other forms of energy by friction as she stops? How much work must she do to speed up to 2.5 m/s again?
2. An 875.0-kg car speeds up from 22.0 m/s to 44.0 m/s. What are the initial and final kinetic energies of the car? How much work is done on the car to increase its speed?
3. A comet with a mass of 7.85×10^{11} kg strikes Earth at a speed of 25.0 km/s. Find the kinetic energy of the comet in joules, and compare the work that is done by Earth in stopping the comet to the 4.2×10^{15} J of energy that was released by the largest nuclear weapon ever exploded.
4. **CHALLENGE** A 2-kg wheel rolls down the road with a linear speed of 15 m/s. Find its translational and rotational kinetic energies. (*Hint:* $I = mr^2$)

Gravitational Potential Energy

Look at the three oranges being juggled in **Figure 4.** If you consider the system to be only one orange, then it has several external forces acting on it. The force of the juggler's hand does work, giving the orange its initial kinetic energy. After the orange leaves the juggler's hand, only the gravitational force acts on it, assuming that air resistance is negligible. How much work does gravity do on the orange as its height changes?

Work done by gravity Let h represent the orange's height measured from the juggler's hand. On the way up, its displacement is upward, but the gravitational force on the orange (F_g) is downward, so the work done by gravity is negative: $W_g = -mgh$. On the way back down, the force and displacement are in the same direction, so the work done by gravity is positive: $W_g = mgh$. Thus, while the orange is moving upward, gravity does negative work, slowing the orange to a stop. On the way back down, gravity does positive work, increasing the orange's speed and thereby increasing its kinetic energy. The orange recovers all of its original kinetic energy when it returns to the same height as the juggler's hand, as shown in the energy bar diagrams at the bottom of the page. Notice in the diagram that the orange has equal amounts of gravitational potential energy and kinetic energy at $\frac{1}{2}h$.

READING CHECK Describe the work done by gravity as the orange rises from the jugglers hand.

Recall that according to Newton's law of universal gravitation, there is a gravitational attraction between any pair of objects. This gravitational attraction between the objects is a force that performs work when one of the objects moves. Usually, we are most concerned with the gravitational potential energy between an object and Earth. For this particular example, we choose to discuss the object and Earth together as a system. If the object moves away from Earth, the system stores energy due to the gravitational force between the object and Earth. The stored energy due to gravity is called **gravitational potential energy,** represented by the symbol *GPE*. The height to which the object has risen is determined by using a **reference level,** the position where *GPE* is defined to be zero. In the example of the juggler, the reference level is the juggler's hand.

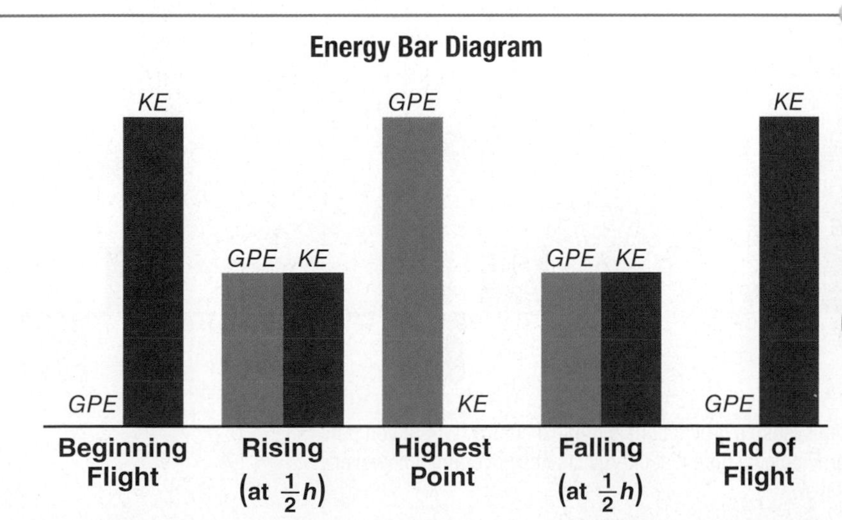

Figure 4 As the orange rises, its kinetic energy transforms into potential energy in the system composed of the orange and Earth. As it falls, the potential energy is transformed back into kinetic energy. In this process, Earth also gains and loses very small amounts of kinetic energy. The energy bar diagram at the bottom of the page illustrates how the energy changes as the orange moves along its path.

View an **animation about gravitational potential energy.**

Concepts In Motion

Full House Images/Riser/Getty Images

Calculating gravitational potential energy

For an object with mass (m) that has risen to a height (h) above the reference level, the gravitational potential energy of the object-Earth system is represented by the following equation.

GRAVITATIONAL POTENTIAL ENERGY

The gravitational potential energy of an object-Earth system is equal to the product of the object's mass, the gravitational field strength, and the object's distance above the reference level.

$$GPE = mgh$$

In this equation, g is the gravitational field strength. Gravitational potential energy, like kinetic energy, is measured in joules.

Energy transformations Consider the energy of a system consisting of Earth and an orange being juggled. The energy of the system is in two forms: kinetic energy and gravitational potential energy. At the beginning of the orange's flight, all the energy is in the form of kinetic energy, as shown on the left in **Figure 5.** As the orange rises, it slows down and the system's energy changes from kinetic energy to gravitational potential energy. At the highest point of the orange's flight, the velocity is zero and all the energy is in the form of gravitational potential energy.

As the orange falls, gravitational potential energy changes back into kinetic energy. The sum of kinetic energy and gravitational potential energy is constant at all times because no work is done on the system by forces outside of the system.

Reference levels So far, we have set the reference level to be the juggler's hand. That is, the height of the orange is measured from where it left contact with the juggler's hand. Thus, at the juggler's hand, $h = 0$ m and $GPE = 0$ J. However, you can set the reference level at any height that is convenient for solving a given problem.

Suppose the reference level is set at the highest point of the orange's flight. Then, $h = 0$ m and the system's $GPE = 0$ J at that point, as illustrated on the right in **Figure 5.** The system's potential energy is negative at the beginning of the orange's flight, zero at the highest point, and negative at the end of the orange's flight. If you were to calculate the total energy of the system represented on the left in **Figure 5,** it would be different from the total energy of the system represented on the right in **Figure 5.** This is because the reference levels are different in each case. However, the total energy of the system in each situation is constant at all times during the flight of the orange.

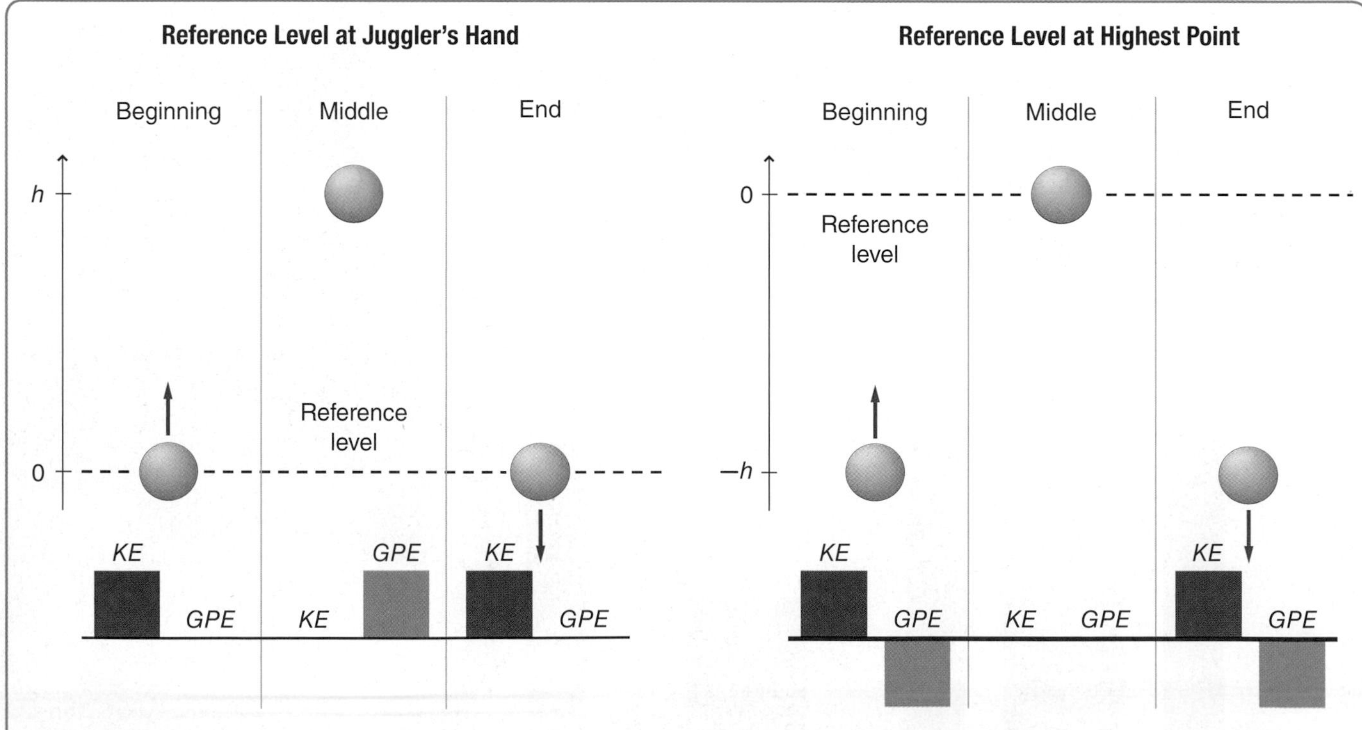

Figure 5 Where you put the reference level affects the system's gravitational potential energy. Where you put the reference level does not affect how the system's gravitational potential energy changes over time, however.
Analyze Where else might you place the reference level?

EXAMPLE PROBLEM 1

Find help with **order of operations.** Math Handbook

GRAVITATIONAL POTENTIAL ENERGY You lift a 7.30-kg bowling ball from the storage rack and hold it up to your shoulder. The storage rack is 0.61 m above the floor and your shoulder is 1.12 m above the floor.

a. When the bowling ball is at your shoulder, what is the ball-Earth system's gravitational potential energy relative to the floor?

b. When the bowling ball is at your shoulder, what is the ball-Earth system's gravitational potential energy relative to the rack?

c. How much work was done by gravity as you lifted the ball from the rack to shoulder level?

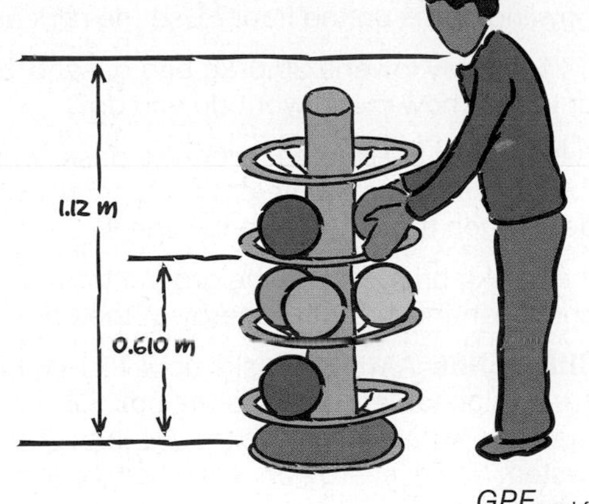

1 ANALYZE AND SKETCH THE PROBLEM

- Sketch the situation.
- Choose a reference level.
- Draw an energy bar diagram showing the gravitational potential energy with the floor as the reference level.

$GPE_{r\ rel\ f}$ + $W_{by\ you}$ = $GPE_{s\ rel\ f}$

KNOWN		UNKNOWN
$m = 7.30$ kg	$g = 9.8$ N/kg	$GPE_{s\ rel\ f} = ?$
$h_r = 0.61$ m (rack relative to the floor)		$GPE_{s\ rel\ r} = ?$
$h_s = 1.12$ m (shoulder relative to the floor)		$W = ?$

2 SOLVE FOR THE UNKNOWN

a. Set the reference level to be at the floor.

Determine the gravitational potential energy of the system when the ball is at shoulder level.

$GPE_{s\ rel\ f} = mgh_s = (7.30\text{ kg})(9.8\text{ N/kg})(1.12\text{ m})$ ◀ Substitute $m = 7.30$ kg, $g = 9.8$ N/kg, $h_s = 1.12$ m

$= 8.0\times10^1$ J

b. Set the reference level to be at the rack height.

Determine the height of your shoulder relative to the rack.

$h = h_s - h_r$

Determine the gravitational potential energy of the system when the ball is at shoulder level.

$GPE_{s\ rel\ r} = mgh = mg(h_s - h_r)$ ◀ Substitute $h = h_s - h_r$

$= (7.30\text{ kg})(9.8\text{ N/kg})(1.12\text{ m} - 0.61\text{ m})$ ◀ Substitute $m = 7.30$ kg, $g = 9.8$ N/kg, $h_s = 1.12$ m, $h_r = 0.61$ m

$= 36$ J ◀ This also is equal to the work done by you as you lifted the ball.

c. The work done by gravity is the weight of the ball times the distance the ball was lifted.

$W = Fd = -(mg)h = -(mg)(h_s - h_r)$ ◀ The weight opposes the motion of lifting, so the work is negative.

$= -(7.30\text{ kg})(9.8\text{ N/kg})(1.12\text{ m} - 0.61\text{ m})$ ◀ Substitute $m = 7.30$ kg, $g = 9.8$ N/kg, $h_s = 1.12$ m, $h_r = 0.61$ m

$= -36$ J

3 EVALUATE THE ANSWER

- **Are the units correct?** Both potential energy and work are measured in joules.
- **Does the answer make sense?** The system should have a greater *GPE* measured relative to the floor than relative to the rack because the ball's distance above the floor level is greater than the ball's distance above the rack.

PRACTICE PROBLEMS

Do additional problems. Online Practice

5. In Example Problem 1, what is the potential energy of the ball-Earth system when the bowling ball is on the floor? Use the rack as your reference level.
6. If you slowly lower a 20.0-kg bag of sand 1.20 m from the trunk of a car to the driveway, how much work do you do?
7. A boy lifts a 2.2-kg book from his desk, which is 0.80 m high, to a bookshelf that is 2.10 m high. What is the potential energy of the book-Earth system relative to the desk when the book is on the shelf?
8. If a 1.8-kg brick falls to the ground from a chimney that is 6.7 m high, what is the change in the potential energy of the brick-Earth system?
9. **CHALLENGE** A worker picks up a 10.1-kg box from the floor and sets it on a 1.1-m-high table. He slides the box 5.0 m along the table and then lowers it back to the floor. What were the changes in the box-Earth system's energy, and how did the system's total energy change? (Ignore friction.)

Elastic Potential Energy

When the string on the bow shown in **Figure 6** is pulled, work is done on the bow string and energy is transferred to the bow. If you identify the system as the bow, the arrow, and Earth, the energy of the system increases. When the string and the arrow are released, the stored energy is changed into kinetic energy. The stored energy due to the pulled string is elastic potential energy. **Elastic potential energy** is stored energy due to an object's change in shape. Systems that include springs, rubber bands, and trampolines often have elastic potential energy.

☑ **READING CHECK** **Define** the term *elastic potential energy.*

You can also store elastic potential energy in a system when you bend an elastic object in that system. For example, when stiff metal or bamboo poles were used in pole-vaulting, the poles did not bend easily. Little work was done on the poles, and the poles did not store much potential energy. Since flexible fiberglass poles were introduced, however, record pole-vaulting heights have soared.

Figure 6 The archer-bow-arrow system has maximum elastic potential energy before the string is released, as shown on the left. When the arrow and string disengage, the elastic potential energy is completely transformed into kinetic energy, as shown on the right.

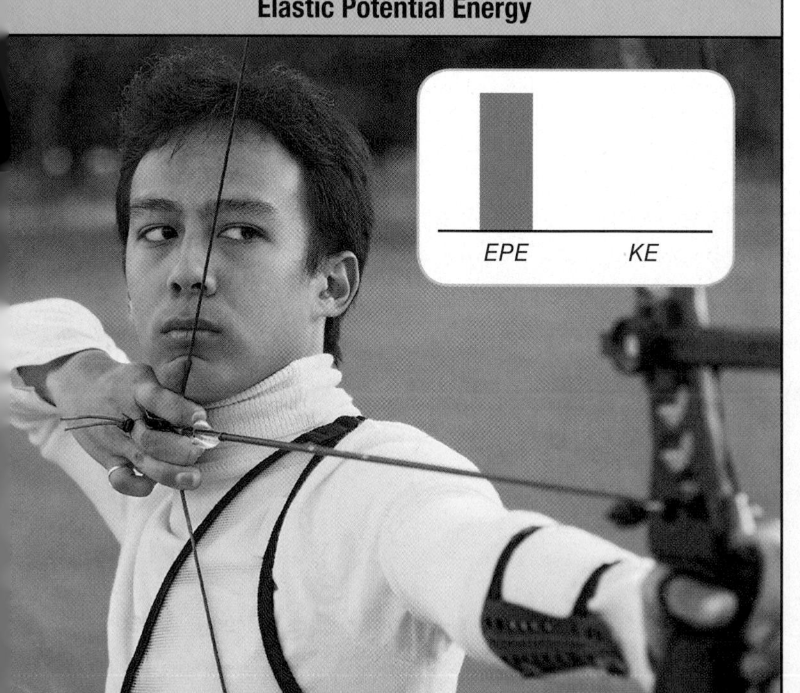

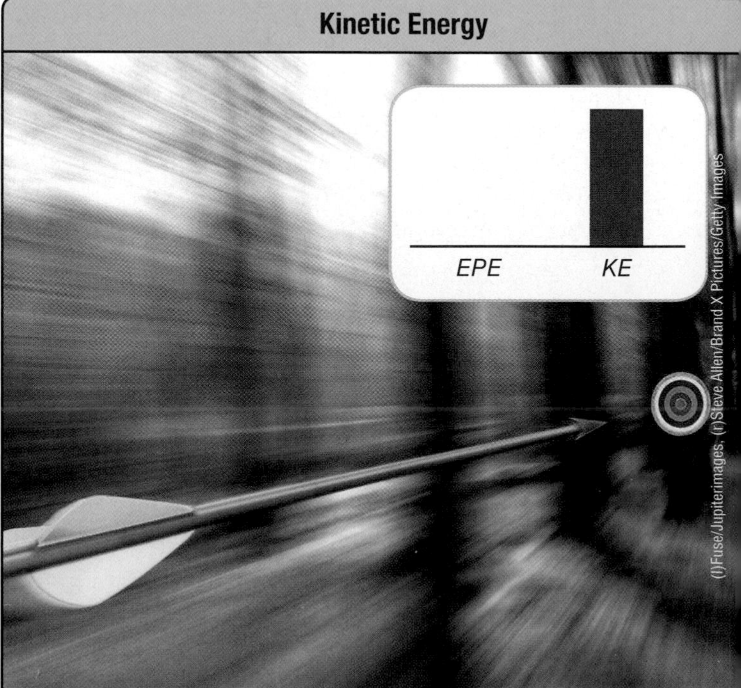

A pole-vaulter runs with a flexible pole and plants its end into a socket in the ground. When the pole-vaulter bends the pole, as shown in **Figure 7,** some of the pole-vaulter's kinetic energy is transformed to elastic potential energy. When the pole straightens, the elastic potential energy is transformed to gravitational potential energy and kinetic energy as the pole-vaulter is lifted as high as 6 m above the ground. Unlike stiff metal poles or bamboo poles, fiberglass poles have an increased capacity for storing elastic potential energy. Thus, pole-vaulters can clear bars that are set very high.

Figure 7 The pole-vaulter stores elastic potential energy in the vaulter-pole-track system. The elastic potential energy of this system then changes into kinetic energy and gravitational potential energy.

Mass

Albert Einstein recognized yet another form of potential energy that is proportional to the object's mass. He demonstrated that mass represents a form of energy. This energy is called the rest energy (E_0) and can be calculated using the following formula.

REST ENERGY
An object's rest energy is equal to its mass times the speed of light squared.

$$E_0 = mc^2$$

According to this famous formula, mass is a form of energy that can be transformed into other forms of energy. Because the speed of light is very fast (300,000,000 m/s), even a small amount of mass is equivalent to a large amount of energy. For example, 1 g of mass is equivalent to 90 trillion J of energy. As a result, mass-energy equivalence is only apparent when large amounts of energy are involved, such as during nuclear explosions and particle physics experiments.

Other Forms of Energy

Think about all the forms and sources of energy you encounter every day. Gasoline provides energy to run your car. You eat food to obtain energy. Power plants harness the energy of wind, water, fossil fuels, and atoms and transform it into electrical energy.

Chemical and nuclear energy Recall that fossil fuels release chemical energy when they are burned. A similar process occurs when you digest food. Your body uses the energy from the chemical bonds in your food as an energy source. The bonds inside an atom's nucleus also store energy. This energy is called nuclear energy and is released when the structure of an atom's nucleus changes. You will learn more about nuclear energy in later chapters.

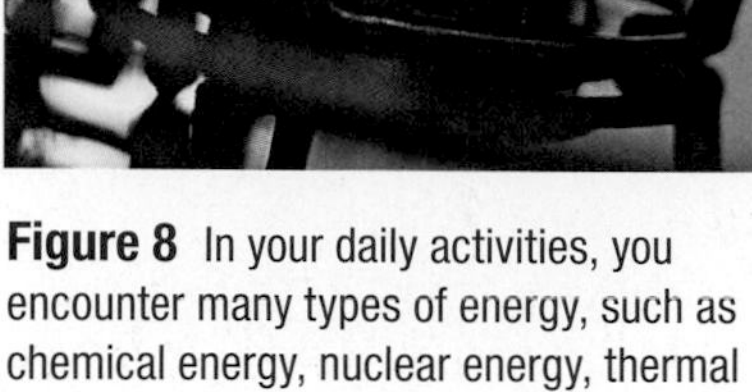

Figure 8 In your daily activities, you encounter many types of energy, such as chemical energy, nuclear energy, thermal energy, and electrical energy.

Thermal, electrical, and radiant energy Thermal energy, which is associated with temperature, is another form of energy. **Thermal energy** is the sum of the kinetic energy and potential energy of the particles in a system. When you warm your hands by rubbing them together, you transform kinetic energy into thermal energy. Thermal energy can also be transferred. For example, the stove in **Figure 8** transfers thermal energy to the pan.

Power plants, such as the one shown in **Figure 8,** transform various types of energy into electrical energy, which is energy associated with charged particles. Electrical energy can be transmitted through wires, such as those shown in **Figure 8,** to your home. Appliances transform this electrical energy into other forms of energy, such as radiant energy in lightbulbs or into kinetic energy of moving fan blades. Radiant energy is energy transferred by electromagnetic waves. You will learn more about these forms of energy in later chapters.

SECTION 1 REVIEW

Check your understanding.

10. MAINIDEA How can you apply the work-energy theorem to lifting a bowling ball from a storage rack to your shoulder?

11. **Elastic Potential Energy** You get a spring-loaded jumping toy ready by compressing the spring. The toy then flies straight up. Draw bar graphs that describe the forms of energy present in the following instances. Assume the system includes the spring toy and Earth.
    a. The toy is pushed down thereby compressing the spring.
    b. The spring expands and the toy jumps.
    c. The toy reaches the top of its flight.

12. **Potential Energy** A 25.0-kg shell is shot from a cannon at Earth's surface. The reference level is Earth's surface.
    a. What is the shell-Earth system's gravitational potential energy when the shell's height is 425 m?
    b. What is the change in the system's potential energy when the shell falls to a height of 225 m?

13. **Potential Energy** A 90.0-kg rock climber climbs 45.0 m upward, then descends 85.0 m. The initial height is the reference level. Find the potential energy of the climber-Earth system at the top and at the bottom. Draw bar graphs for both situations.

14. **Rotational Kinetic Energy** On a playground, some children push a merry-go-round so that it turns twice as fast as it did before they pushed it. What are the relative changes in angular momentum and rotational kinetic energy of the merry-go-round?

15. **Critical Thinking** Karl uses an air hose to exert a constant horizontal force on a puck, which is on a frictionless air table. The force is constant as the puck moves a fixed distance.
    a. Explain what happens in terms of work and energy. Draw bar graphs.
    b. Suppose Karl uses a different puck with half the first one's mass. All other conditions remain the same. How should the kinetic energy and work done differ from those in the first situation?
    c. Describe what happened in parts a and b in terms of impulse and momentum.

SECTION 2
Conservation of Energy

PHYSICS 4 YOU **If you count your money and find that you are $3 short, you would not assume that the money just disappeared. You might search for it. Just as money does not vanish into thin air, energy does not vanish.**

MAIN IDEA
In a collision in a closed, isolated system, the total energy is conserved, but kinetic energy might not be conserved.

Essential Questions

- Under what conditions is energy conserved?
- What is mechanical energy, and when is it conserved?
- How are momentum and kinetic energy conserved or changed in a collision?

Review Vocabulary

closed system a system that does not gain or lose mass

New Vocabulary

law of conservation of energy
mechanical energy
elastic collision
inelastic collision

The Law of Conservation of Energy

Scientists look for energy that seems to be missing by determining where and how the energy was transferred. We certainly expect that the total amount of energy in a system remains constant as long as the system is closed and isolated from external forces.

The **law of conservation of energy** states that in a closed, isolated system, energy can neither be created nor destroyed; rather, energy is conserved. Under these conditions, energy can change form but the system's total energy in all of its forms remains constant.

Mechanical energy For many situations you focus on the energy that comes from the motions of and interactions between objects. The sum of the kinetic energy and potential energy of the objects in a system is the system's **mechanical energy** (*ME*). The kinetic energy includes both the translational and rotational kinetic energies of the objects in the system. Potential energy includes the gravitational and elastic potential energies of the system. In any system, mechanical energy is represented by the following equation.

MECHANICAL ENERGY OF A SYSTEM

The mechanical energy of a system is equal to the sum of the kinetic energy and potential energy of the system's objects.

$$ME = KE + PE$$

Imagine a system consisting of a 10.0-N bowling ball falling to Earth, as shown in **Figure 9.** For this case, the system's mechanical energy is 20 J.

Figure 9 When a bowling ball is dropped, mechanical energy is conserved.

Predict Suppose the ball started with an upward velocity. What would its energy graph look like?

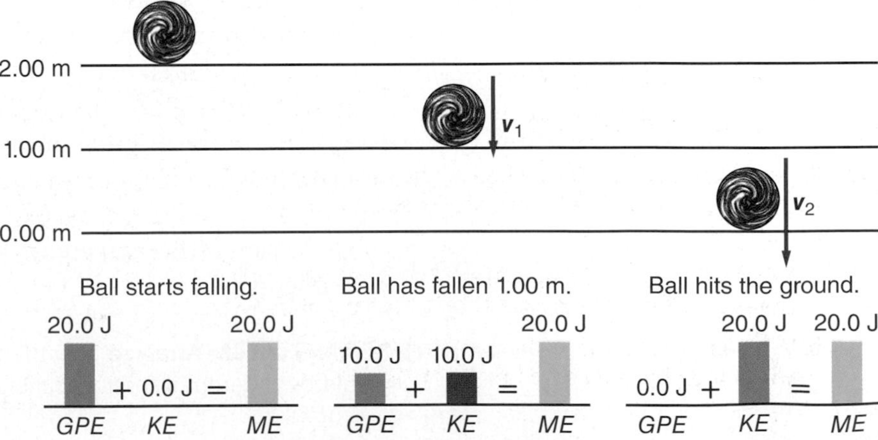

D. Hurst/Alamy

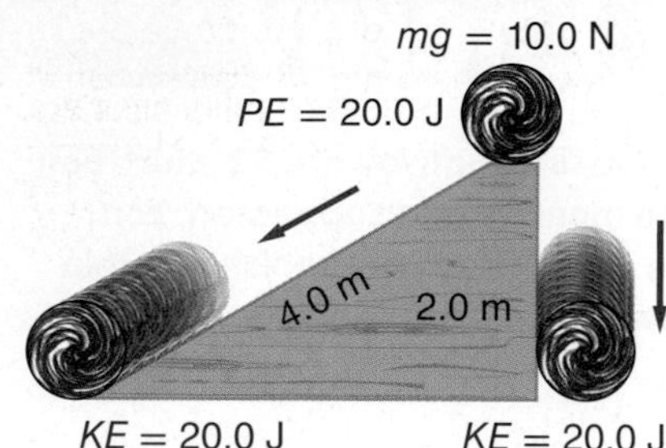

Figure 10 If friction does no work on the ball, the ball's final kinetic energy is equal to its initial gravitational potential energy regardless of which path it follows.

View an **animation of conservation of mechanical energy.**

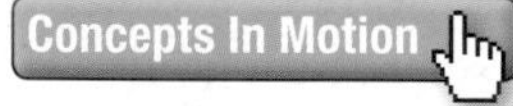

Conservation of mechanical energy Suppose that you let go of the bowling ball. As it falls, the gravitational potential energy of the ball-Earth system is transformed into kinetic energy. When the ball is 1.00 m above Earth's surface: $GPE = mgh = F_g h =$(10.0 N)(1.00 m) = 10.0 J. What is the system's kinetic energy when the ball is at a height of 1.00 m? The system consisting of the ball and Earth is closed and isolated because no external forces are doing work upon it. Therefore, the total mechanical energy of the system remains constant at 20.0 J.

$$ME = KE + PE$$
$$KE = ME - PE$$
$$KE = 20.0\text{ J} - 10.0\text{ J} = 10.0\text{ J}$$

When the ball reaches ground level, the system's potential energy is zero, and the system's kinetic energy is 20.0 J. The equation that describes conservation of mechanical energy can be written as follows.

CONSERVATION OF MECHANICAL ENERGY

When mechanical energy is conserved, the sum of the system's kinetic energy and potential energy before an event is equal to the sum of the system's kinetic energy and potential energy after that event.

$$KE_i + PE_i = KE_f + PE_f$$

What happens if the ball does not fall straight down but rolls down a ramp, as shown in **Figure 10?** If friction does not affect the system, then the system of the ball and Earth remains closed and isolated. The ball still moves down a vertical distance of 2.00 m, so the decrease in potential energy is 20.0 J. Therefore, the increase in kinetic energy is 20.0 J. As long as friction doesn't affect the system's energy, the path the ball takes does not matter. However, in the case with the ramp, the kinetic energy is split between translational kinetic energy (it is moving forward) and rotational kinetic energy (it is rolling). Three situations involving conservation of mechanical energy are shown in **Figure 11** on the next page.

Conservation and other forms of energy A swinging pendulum eventually stops, a bouncing ball comes to rest, and the heights of roller-coaster hills get lower and lower as the ride progresses. Where does the mechanical energy in such systems go? Objects moving through the air experience the forces of air resistance. In a roller coaster, there are also frictional forces between the wheels and the tracks. Each of these forces transforms mechanical energy into other forms of energy.

When a ball bounces from a surface, it compresses. Most of the kinetic energy transforms into elastic potential energy. Some, but not all, of this elastic potential energy transforms back into kinetic energy during the bounce. But as in the cases of the pendulum and the roller coaster, some of the original mechanical energy in the system transforms into another form of energy within the system or transmits to energy outside the system, as in air resistance. The bouncing ball becomes warmer (thermal energy), and we hear the sound of the bounce (sound energy).

READING CHECK **Analyze** how the ball's final kinetic energy in **Figure 10** would be different if friction transformed some of the system's energy.

MiniLABs

ENERGY EXCHANGE

How does energy transform when a ball is shot upward?

INTERRUPTED PENDULUM

How is the mechanical energy of a pendulum affected by an obstruction?

Figure 11 The conservation of mechanical energy is an important consideration in designing roller coasters, ski slopes, and the pendulums for grandfather clocks.

Investigate **roller coasters.**

Virtual Investigation

Roller Coasters The roller-coaster car is nearly at rest at the top of the first hill, and the total mechanical energy in the Earth-roller coaster car system is the system's gravitational potential energy at that point. If a hill farther along the track were higher than the first one, the roller-coaster car would not be able to climb the higher hill because the energy required to do so would be greater than the total mechanical energy of the system.

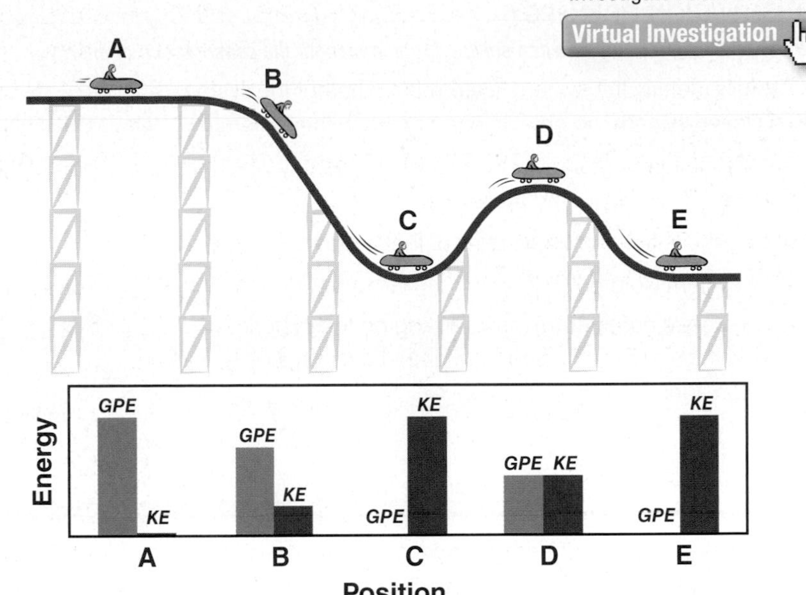

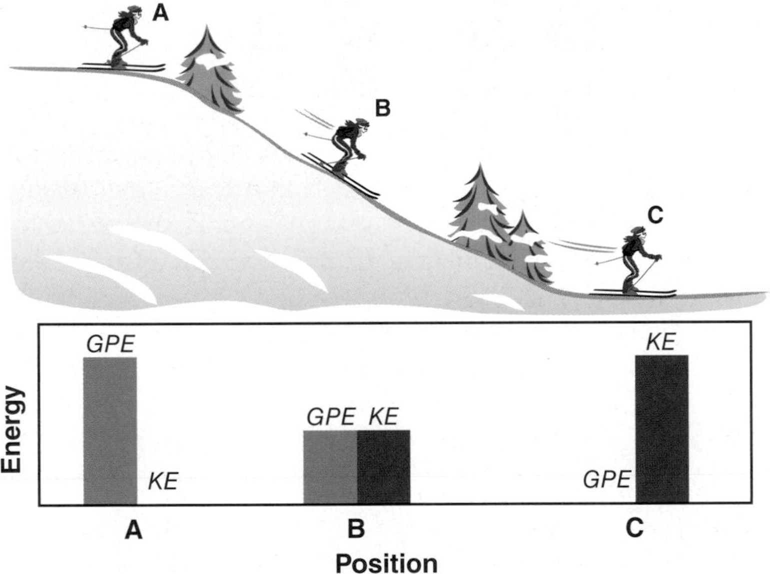

Skiing When you ski down a steep slope, you begin from rest at the top of the slope and the total mechanical energy is equal to the gravitational potential energy. Once you start skiing downhill, the gravitational potential energy is transformed to kinetic energy. As you ski down the slope, your speed increases as more potential energy is transformed to kinetic energy.

Pendulums The simple oscillation of a pendulum also demonstrates conservation of mechanical energy. The system is the pendulum bob and Earth. The reference level is chosen to be the height of the bob at the lowest point. When an external force pulls the bob to one side, the force does work that adds mechanical energy to the system. At the instant the bob is released, all the energy is in the form of potential energy, but as the bob swings downward, potential energy is transformed to kinetic energy. When the bob is at the lowest point, the gravitational potential energy is zero, and the kinetic energy is equal to the total mechanical energy.

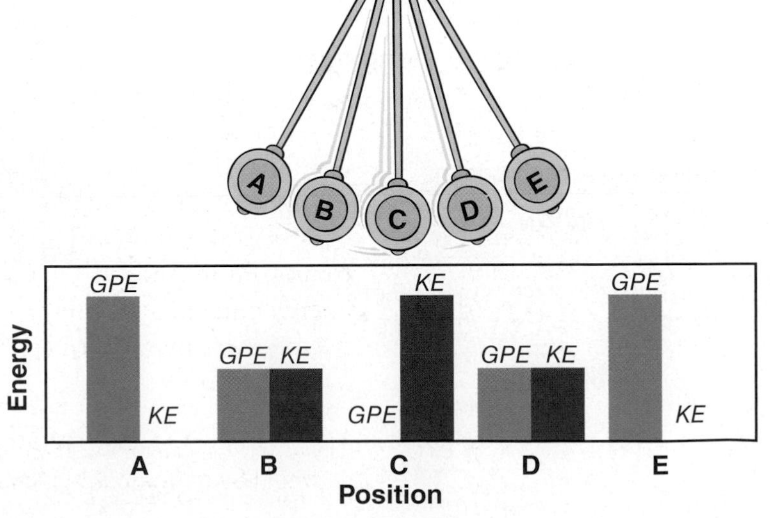

PROBLEM-SOLVING STRATEGIES

CONSERVATION OF ENERGY

Use the following strategies when solving problems about the conservation of energy.

1. Carefully identify the system. Determine whether the system is closed. In a closed system, no objects enter or leave the system.
2. Identify the forms of energy in the system. Identify which forms are part of the mechanical energy of the system.
3. Identify the initial and final states of the system.
4. Is the system isolated?
   a. If there are no external forces acting on the system, then the system is isolated and the total energy of the system is constant.
   $$E_{initial} = E_{final}$$
   b. If there are external forces, then the final energy is the sum of the initial energy and the work done on the system. Remember that work can be negative.
   $$E_{initial} + W = E_{final}$$
5. For an isolated system, identify the types of energy in the system before and after. If the only forms of energy are potential and kinetic, mechanical energy is conserved.
   $$KE_i + PE_i = KE_f + PE_f$$

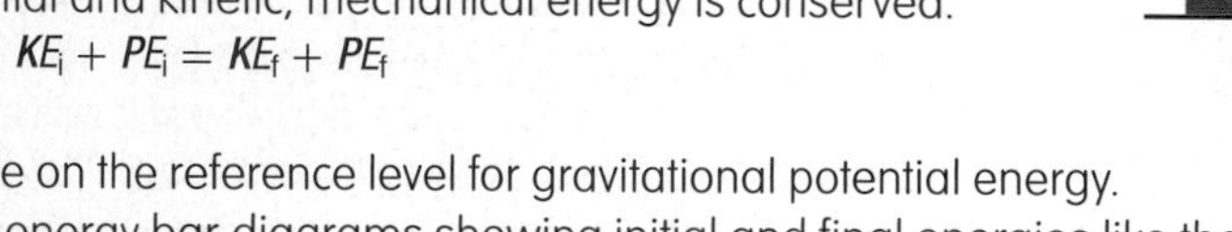

Decide on the reference level for gravitational potential energy. Draw energy bar diagrams showing initial and final energies like the diagram shown to the right.

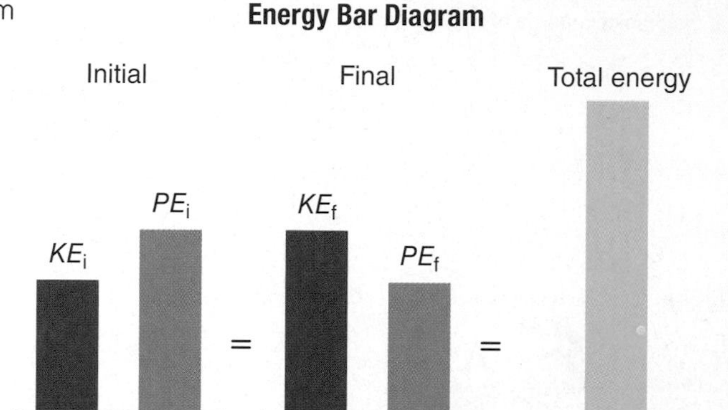

EXAMPLE PROBLEM 2

Find help with **square and cube roots.** Math Handbook

CONSERVATION OF MECHANICAL ENERGY A 22.0-kg tree limb is 13.3 m above the ground. During a hurricane, it falls on a roof that is 6.0 m above the ground.

a. Find the kinetic energy of the limb when it reaches the roof. Assume that the air does no work on the tree limb.

b. What is the limb's speed when it reaches the roof?

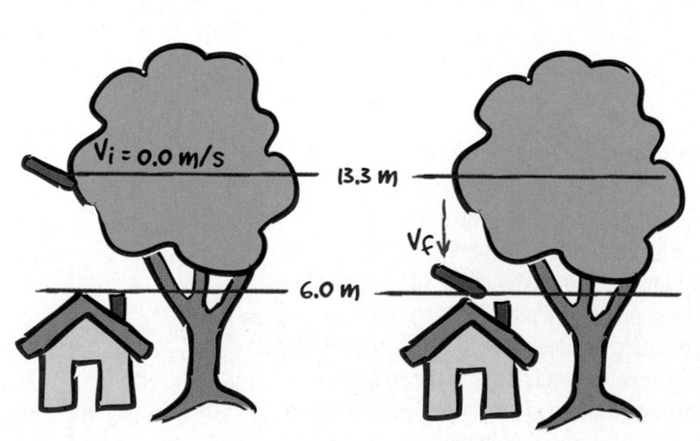

1 ANALYZE AND SKETCH THE PROBLEM

- Sketch the initial and final conditions.
- Choose a reference level.
- Draw an energy bar diagram.

KNOWN

$m = 22.0$ kg	$g = 9.8$ N/kg
$h_{limb} = 13.3$ m	$v_i = 0.0$ m/s
$h_{roof} = 6.0$ m	$KE_i = 0.0$ J

UNKNOWN

$GPE_i = ?$	$KE_f = ?$
$GPE_f = ?$	$v_f = ?$

EXAMPLE PROBLEM

2 SOLVE FOR THE UNKNOWN

a. Set the reference level as the height of the roof.

Find the initial height of the limb relative to the roof.

$h = h_{limb} - h_{roof} = 13.3\text{ m} - 6.0\text{ m} = 7.3\text{ m}$ ◀ Substitute h_{limb} = 13.3 m, h_{roof} = 6.0 m

Determine the initial potential energy of the limb-Earth system.

$GPE_i = mgh = (22.0\text{ kg})(9.8\text{ N/kg})(7.3\text{ m})$
$= 1.6{\times}10^3\text{ J}$ ◀ Substitute m = 22.0 kg, g = 9.8 N/kg, h = 7.3 m

Identify the initial kinetic energy of the system.

$KE_i = 0.0\text{ J}$ ◀ The tree limb is initially at rest.

Identify the final potential energy of the system.

$GPE_f = 0.0\text{ J}$ ◀ h = 0.0 m at the roof.

Use the law of conservation of energy to find the KE_f.

$$KE_f + GPE_f = KE_i + GPE_i$$
$$KE_f = KE_i + GPE_i - GPE_f$$
$$= 0.0\text{ J} + 1.6{\times}10^3\text{ J} - 0.0\text{ J}$$
$$= 1.6{\times}10^3\text{ J}$$

◀ Substitute KE_i = 0.0 J, GPE_i = 1.6×10³ J, and GPE_i = 0 J.

b. Determine the speed of the limb.

$$KE_f = \frac{1}{2}mv_f^2$$
$$v_f = \sqrt{\frac{2KE_f}{m}} = \sqrt{\frac{2(1.6{\times}10^3\text{ J})}{22.0\text{ kg}}}$$
$$= 12\text{ m/s}$$

◀ Substitute KE_f = 1.6×10³ J, m = 22.0 kg

3 EVALUATE THE ANSWER

- **Are the units correct?** The magnitude of velocity is measured in m/s, and energy is measured in kg•m²/s² = J.
- **Do the signs make sense?** *KE* and speed are both positive in this scenario.

PRACTICE PROBLEMS

Do additional problems. Online Practice 

16. A bike rider approaches a hill at a speed of 8.5 m/s. The combined mass of the bike and the rider is 85.0 kg. Choose a suitable system. Find the initial kinetic energy of the system. The rider coasts up the hill. Assuming friction is negligible, at what height will the bike come to rest?

17. Suppose that the bike rider in the previous problem pedaled up the hill and never came to a stop. In what system is energy conserved? From what form of energy did the bike gain mechanical energy?

18. A skier starts from rest at the top of a 45.0-m-high hill, skis down a 30° incline into a valley, and continues up a 40.0-m-high hill. The heights of both hills are measured from the valley floor. Assume that friction is negligible and ignore the effect of the ski poles.

a. How fast is the skier moving at the bottom of the valley?

b. What is the skier's speed at the top of the second hill?

c. Do the angles of the hills affect your answers?

19. In a belly-flop diving contest, the winner is the diver who makes the biggest splash upon hitting the water. The size of the splash depends not only on the diver's style, but also on the amount of kinetic energy the diver has. Consider a contest in which each diver jumps from a 3.00-m platform. One diver has a mass of 136 kg and simply steps off the platform. Another diver has a mass of 100 kg and leaps upward from the platform. How high would the second diver have to leap to make a competitive splash?

20. CHALLENGE The spring in a pinball machine exerts an average force of 2 N on a 0.08-kg pinball over 5 cm. As a result, the ball has both translational and rotational kinetic energy. If the ball is a uniform sphere $\left(I = \frac{5}{2}mr^2\right)$, what is its linear speed after leaving the spring? (Ignore the table's tilt.)

Figure 12 In a car crash, the kinetic energy decreases.

Explain What happened to the energy?

Analyzing Collisions

A collision between two objects, whether the objects include an automobile such as the one shown in **Figure 12,** hockey players, or subatomic particles, is one of the most common situations analyzed in physics. Because the details of a collision can be very complex during the collision itself, the strategy is to find the motion of the objects just before and just after the collision. What conservation laws can be used to analyze such a system? If the system is closed and isolated, then momentum and energy are conserved. However, the potential energy or thermal energy in the system might decrease, remain the same, or increase. Therefore, you cannot predict whether kinetic energy is conserved. **Figure 13** on the next page shows four different kinds of collisions. Look at the calculations for kinetic energy for each of the cases. In all the cases except case 1, there is a change in the amount of kinetic energy.

In case 1, the momentum of the system before and after the collision is represented by the following:

$$p_{\mathrm{i}} = p_{\mathrm{Ai}} + p_{\mathrm{Bi}} = (1.00\ \mathrm{kg})(1.00\ \mathrm{m/s}) + (1.00\ \mathrm{kg})(0.00\ \mathrm{m/s}) = 1.00\ \mathrm{kg{\cdot}m/s}$$

$$p_{\mathrm{f}} = p_{\mathrm{Af}} + p_{\mathrm{Bf}} = (1.00\ \mathrm{kg})(0.00\ \mathrm{m/s}) + (1.00\ \mathrm{kg})(1.00\ \mathrm{m/s}) = 1.00\ \mathrm{kg{\cdot}m/s}$$

Thus, in case 1, the momentum is conserved.

READING CHECK Calculate the initial and final momentums for the remaining cases in **Figure 13.** Verify that momentum is conserved for each case.

Elastic and inelastic collisions Next, consider again the kinetic energy of the system in each of these cases. In case 1 the kinetic energy of the system before and after the collision was the same. A collision in which the kinetic energy does not change is called an **elastic collision.** Collisions between hard objects, such as those made of steel, glass, or hard plastic, often are called nearly elastic collisions.

Kinetic energy decreased in cases 2 and 3. Some of it may have been transformed to thermal energy. A collision in which kinetic energy decreases is called an **inelastic collision.** Objects made of soft, sticky materials, such as clay, act in this way. In case 2, the objects stuck together after colliding. A collision in which colliding objects stick together, such as in case 2, is called a perfectly inelastic collision.

In case 4, the kinetic energy of the system increased. If energy in the system is conserved, then one or more other forms of energy must have decreased. Perhaps potential energy from a compressed spring was released during the collision, adding kinetic energy to the system. This kind of collision is called a superelastic or explosive collision.

The four kinds of collisions can be represented using energy bar diagrams, also shown in **Figure 13.** The kinetic energies before and after the collisions can be calculated, but we do not always have detailed information about the other forms of energy involved. In automobile collisions, kinetic energy can be transformed into thermal energy and sound, but they are difficult to measure. We infer their amounts from the amount of kinetic energy that is transformed.

PhysicsLABs

CONSERVATION OF ENERGY

Is energy conserved when a ball rolls down a ramp?

IS ENERGY CONSERVED?

How does friction affect the energy of a system?

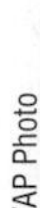

Figure 13 In a closed, isolated system, momentum is conserved when objects collide. Whether kinetic energy is conserved depends on whether the collision is elastic, inelastic, or superelastic.

Case 1: Elastic Collision

$KE_f = KE_i$

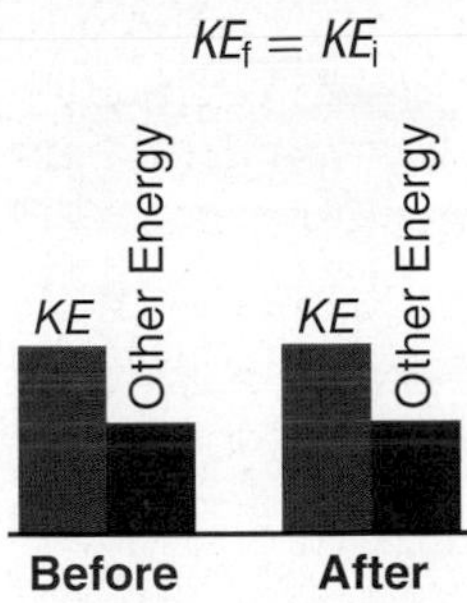

Before collision

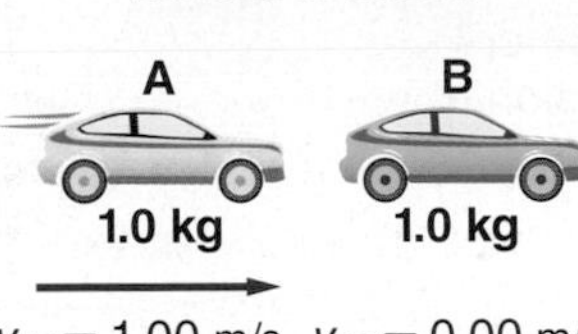

$v_{Ai} = 1.00$ m/s $v_{Bi} = 0.00$ m/s

$$KE_i = KE_{Ai} + KE_{Bi}$$
$$= \frac{1}{2}(1.00\ \text{kg})(1.00\ \text{m/s})^2 + \frac{1}{2}(1.00\ \text{kg})(0.00\ \text{m/s})^2$$
$$= 0.50\ \text{J}$$

After collision

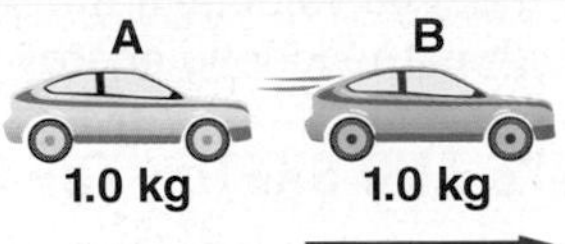

$v_{Af} = 0.00$ m/s $v_{Bf} = 1.00$ m/s

$$KE_i = KE_{Ai} + KE_{Bi}$$
$$= \frac{1}{2}(1.00\ \text{kg})(0.00\ \text{m/s})^2 + \frac{1}{2}(1.00\ \text{kg})(1.00\ \text{m/s})^2$$
$$= 0.50\ \text{J}$$

Case 2: Perfectly Inelastic Collision

$KE_f < KE_i$

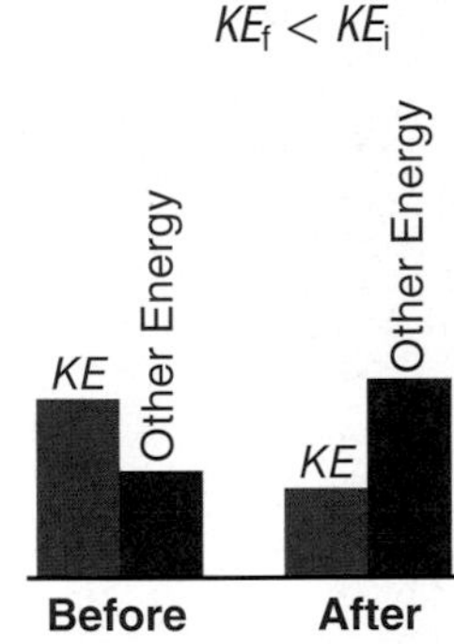

Before collision

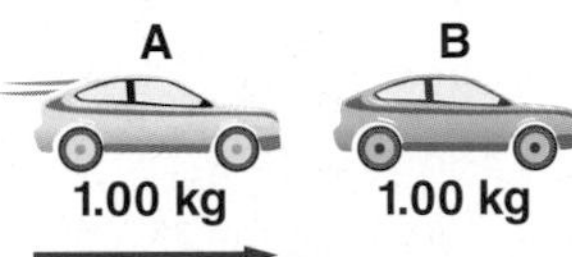

$v_{Ai} = 1.00$ m/s $v_{Bi} = 0.00$ m/s

$$KE_i = KE_{Ai} + KE_{Bi}$$
$$= \frac{1}{2}(1.00\ \text{kg})(1.00\ \text{m/s})^2 + \frac{1}{2}(1.00\ \text{kg})(0.00\ \text{m/s})^2$$
$$= 0.50\ \text{J}$$

After collision

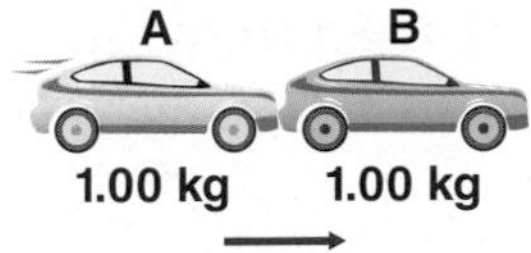

$v_{Af} = v_{Bf} = 0.50$ m/s

$$KE_f = KE_{Af} + KE_{Bf}$$
$$= \frac{1}{2}(1.00\ \text{kg})(0.50\ \text{m/s})^2 + \frac{1}{2}(1.00\ \text{kg})(0.50\ \text{m/s})^2$$
$$= 0.25\ \text{J}$$

Case 3: Inelastic Collision

$KE_f < KE_i$

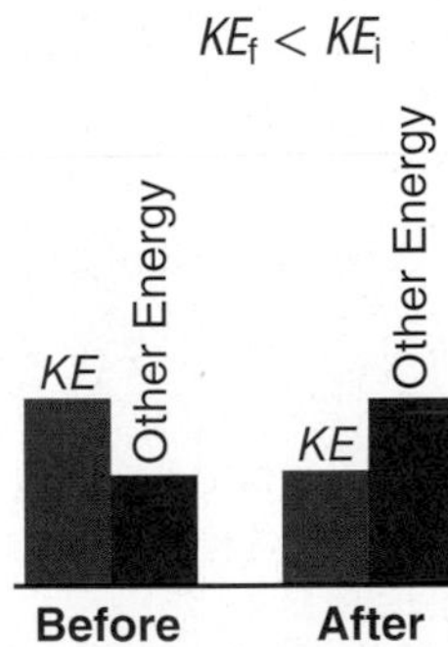

Before collision

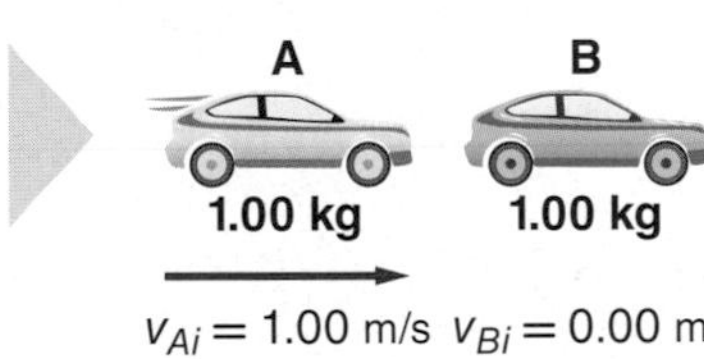

$v_{Ai} = 1.00$ m/s $v_{Bi} = 0.00$ m/s

$$KE_i = KE_{Ai} + KE_{Bi}$$
$$= \frac{1}{2}(1.00\ \text{kg})(1.00\ \text{m/s})^2 + \frac{1}{2}(1.00\ \text{kg})(0.00\ \text{m/s})^2$$
$$= 0.50\ \text{J}$$

After collision

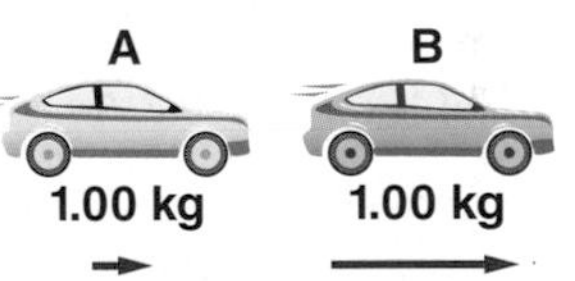

$v_{Af} = 0.25$ m/s $v_{Bf} = 0.75$ m/s

$$KE_f = KE_{Af} + KE_{Bf}$$
$$= \frac{1}{2}(1.00\ \text{kg})(0.25\ \text{m/s})^2 + \frac{1}{2}(1.00\ \text{kg})(0.75\ \text{m/s})^2$$
$$= 0.31\ \text{J}$$

Case 4: Superelastic Collision

$KE_f > KE_i$

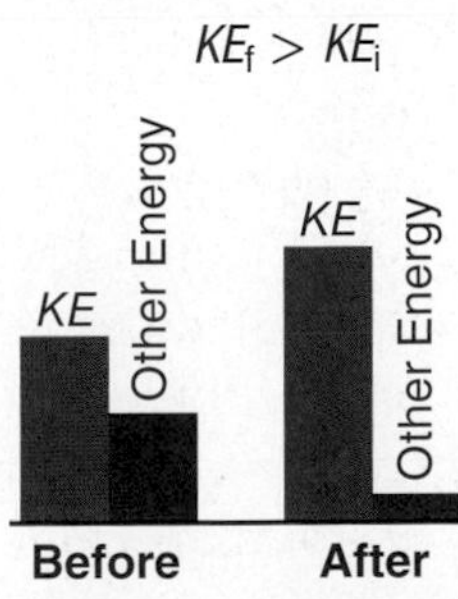

Before collision

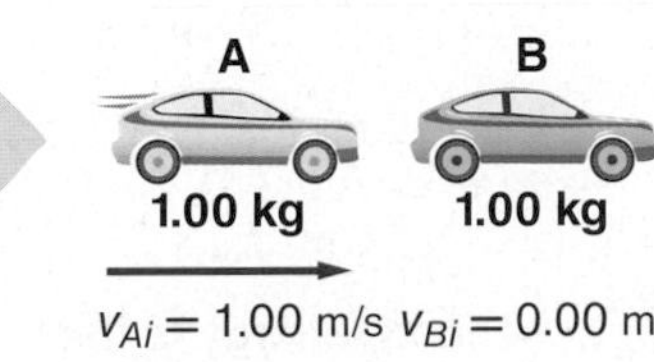

$v_{Ai} = 1.00$ m/s $v_{Bi} = 0.00$ m/s

$$KE_i = KE_{Ai} + KE_{Bi}$$
$$= \frac{1}{2}(1.00\ \text{kg})(1.00\ \text{m/s})^2 + \frac{1}{2}(1.00\ \text{kg})(0.00\ \text{m/s})^2$$
$$= 0.50\ \text{J}$$

After collision

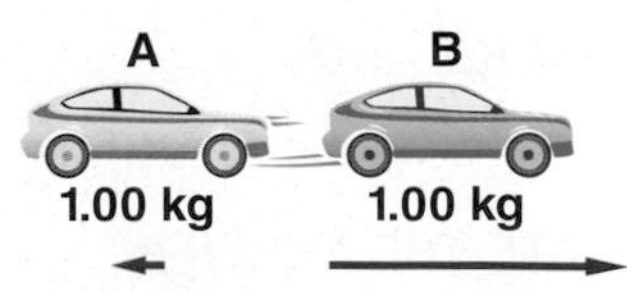

$v_{Af} = -0.20$ m/s $v_{Bf} = 1.20$ m/s

$$KE_f = KE_{Af} + KE_{Bf}$$
$$= \frac{1}{2}(1.00\ \text{kg})(-0.20\ \text{m/s})^2 + \frac{1}{2}(1.00\ \text{kg})(0.00\ \text{m/s})^2$$
$$= 0.74\ \text{J}$$

EXAMPLE PROBLEM 3

Get help with **inelastic collisions.** Personal Tutor

INELASTIC COLLISION In an accident on a slippery road, a compact car with a mass of 1150 kg moving at 15.0 m/s smashes into the rear end of a car with mass of 1575 kg moving at 5.0 m/s in the same direction.

a. What is the final velocity if the wrecked cars lock together?

b. How much is the kinetic energy decreased in the collision?

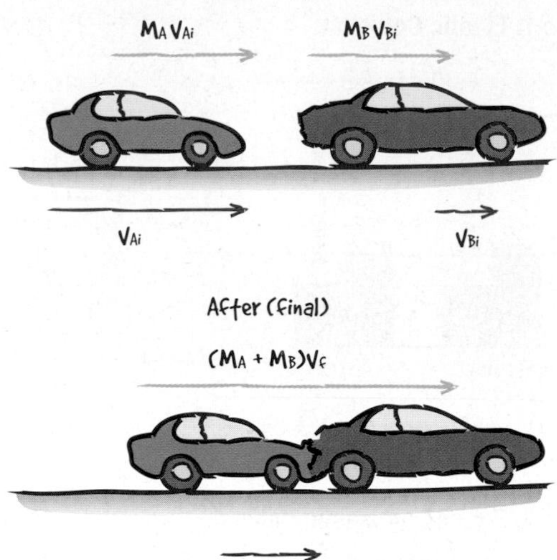

1 ANALYZE AND SKETCH THE PROBLEM

- Sketch the initial and final conditions.
- Sketch the momentum diagram.

KNOWN	UNKNOWN
$m_A = 1150$ kg	$v_f = ?$
$m_B = 1575$ kg	$\Delta KE = KE_f - KE_i = ?$
$v_{Ai} = 15.0$ m/s	
$v_{Bi} = 5.0$ m/s	
$v_{Af} = v_{Bf} = v_f$	

2 SOLVE FOR THE UNKNOWN

a. Use conservation of momentum to find the final velocity.

$$p_{Ai} + p_{Bi} = p_{Af} + p_{Bf}$$

$$m_A v_{Ai} + m_B v_{Bi} = (m_A + m_B)v_f$$

$$v_f = \frac{m_A v_{Ai} + m_B v_{Bi}}{m_A + m_B}$$

$$= \frac{(1150\ \text{kg})(15.0\ \text{m/s}) + (1575\ \text{kg})(5.0\ \text{m/s})}{(1150\ \text{kg} + 1575\ \text{kg})}$$

◀ Substitute $m_A = 1150$ kg, $v_{Ai} = 15.0$ m/s, $m_B = 1575$ kg, $v_{Bi} = 5.0$ m/s

$= 9.2$ m/s, in the direction of the motion before the collision

b. To determine the change in kinetic energy of the system, KE_f and KE_i are needed.

$$KE_f = \frac{1}{2}mv_f^2 = \frac{1}{2}(m_A + m_B)v_f^2$$

◀ Substitute $m = m_A + m_B$

$$= \frac{1}{2}(1150\ \text{kg} + 1575\ \text{kg})(9.2\ \text{m/s})^2$$

◀ Substitute $m_A = 1150$ kg, $m_B = 1575$ kg, $v_f = 9.2$ m/s

$$= 1.2\times10^5\ \text{J}$$

$$KE_i = KE_{Ai} + KE_{Bi}$$

$$= \frac{1}{2}(m_A v_{Ai}^2) + \frac{1}{2}(m_B v_{Bi}^2)$$

◀ Substitute $KE_{Ai} = \frac{1}{2}(m_A v_{Ai}^2)$, $KE_{Bi} = \frac{1}{2}(m_B v_{Bi}^2)$

$$= \frac{1}{2}(1150\ \text{kg})(15.0\ \text{m/s})^2 + \frac{1}{2}(1575\ \text{kg})(5.0\ \text{m/s})^2$$

◀ Substitute $m_A = 1150$ kg, $m_B = 1575$ kg, $v_{Ai} = 15.0$ m/s, $v_{Bi} = 5.0$ m/s

$$= 1.5\times10^5\ \text{J}$$

Solve for the change in kinetic energy of the system.

$$\Delta KE = KE_f - KE_i$$

$$= 1.2\times10^5\ \text{J} - 1.5\times10^5\ \text{J} = -3\times10^4\ \text{J}$$

◀ Substitute $KE_f = 1.2\times10^6$ J, $KE_i = 1.5\times10^6$ J

3 EVALUATE THE ANSWER

- **Are the units correct?** Velocity is measured in meters per second; energy is measured in joules.
- **Does the sign make sense?** The change in kinetic energy is negative, meaning that kinetic energy decreased as expected in an inelastic collision.

PRACTICE PROBLEMS

Do additional problems. Online Practice

21. An 8.00-g bullet is fired horizontally into a 9.00-kg block of wood on an air table and is embedded in it. After the collision, the block and bullet slide along the frictionless surface together with a speed of 10.0 cm/s. What was the initial speed of the bullet?

22. A 91.0-kg hockey player is skating on ice at 5.50 m/s. Another hockey player of equal mass, moving at 8.1 m/s in the same direction, hits him from behind. They slide off together.

a. What are the total mechanical energy and momentum of the system before the collision?

b. What is the velocity of the two hockey players after the collision?

c. How much was the system's kinetic energy decreased in the collision?

23. CHALLENGE A 0.73-kg magnetic target is suspended on a string. A 0.025-kg magnetic dart, shot horizontally, strikes the target head-on. The dart and the target together, acting like a pendulum, swing 12.0 cm above the initial level before instantaneously coming to rest.

a. Sketch the situation and choose a system.

b. Decide what is conserved in each step of the process and explain your decision.

c. What was the initial velocity of the dart?

PRACTICE PROBLEMS

PHYSICS CHALLENGE

A bullet of mass m, moving at speed v_1, goes through a motionless wooden block and exits with speed v_2. After the collision, the block, which has mass m_B, is moving.

1. What is the final speed (v_B) of the block?
2. What was the change in the bullet's mechanical energy?
3. How much energy was lost to friction inside the block?

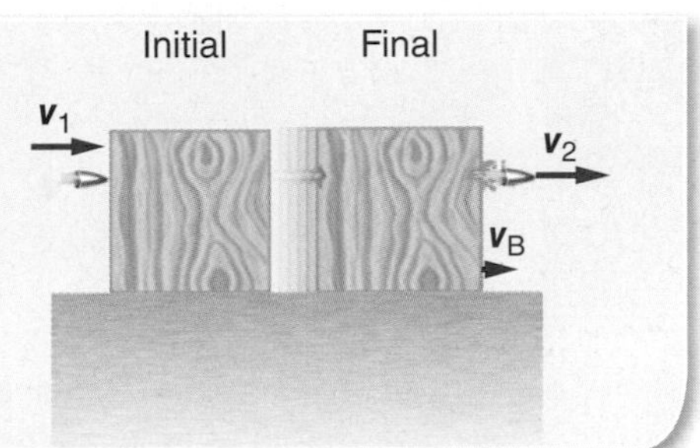

SECTION 2 REVIEW

Check your understanding.

24. MAINIDEA A child jumps on a trampoline. Draw energy bar diagrams to show the forms of energy present in the following situations.

a. The child is at the highest point.

b. The child is at the lowest point.

25. Closed Systems Is Earth a closed, isolated system? Support your answer.

26. Kinetic Energy Suppose a glob of chewing gum and a small, rubber ball collide head-on in midair and then rebound apart. Would you expect kinetic energy to be conserved? If not, what happens to the energy?

27. Kinetic Energy In table tennis, a very light but hard ball is hit with a hard rubber or wooden paddle. In tennis, a much softer ball is hit with a racket. Why are the two sets of equipment designed in this way? Can you think of other ball-paddle pairs in sports? How are they designed?

28. Potential Energy A rubber ball is dropped from a height of 8.0 m onto a hard concrete floor. The ball hits the floor and bounces repeatedly. Each time the ball hits the floor, the ball-Earth system loses 1/5 of its mechanical energy. How many times will the ball bounce before it bounces back up to a height of only about 4 m?

29. Energy As shown in **Figure 14,** a child slides down a playground slide. At the bottom of the slide, she is moving at 3.0 m/s. How much energy was transformed by friction as she slid down the slide?

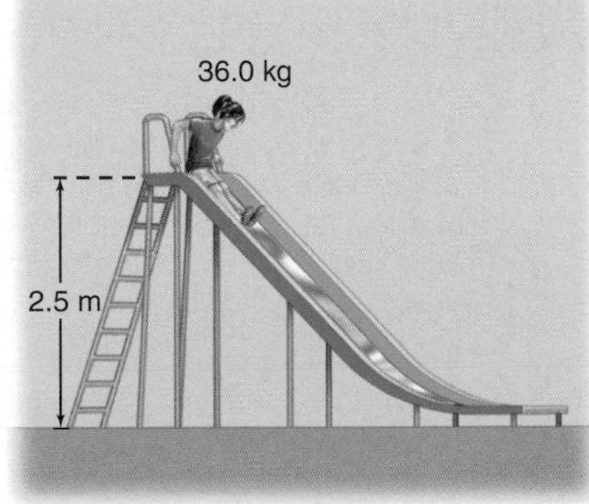

Figure 14

30. Critical Thinking A ball drops 20 m. When it has fallen half the distance, or 10 m, half of the energy is potential and half is kinetic. When the ball has fallen for half the amount of time it takes to fall, will more, less, or exactly half of the energy be potential energy?

What's the Alternative?

Are you familiar with the terms *energy crisis* and *alternative energy*? An energy crisis can occur when the cost of energy rises too high or the supply of an energy resource dips too low. Scientists and engineers are searching for ways to provide alternative energy that is plentiful, less expensive, and green. Most of these sources have one thing in common—they can be used to turn a turbine attached to an electric generator that converts mechanical energy to electrical energy.

Energy from the wind
As wind spins the blades of wind turbines, some of the wind's kinetic energy turns a shaft connected to an electric generator. A group of wind turbines is called a wind farm.

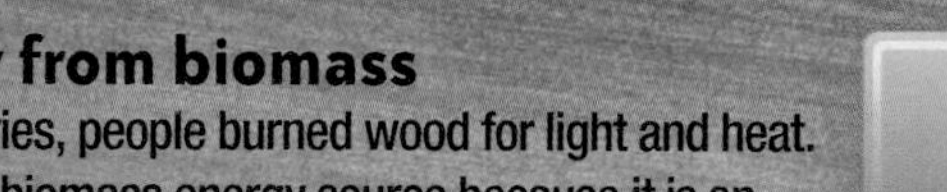

Energy from biomass
For centuries, people burned wood for light and heat. Wood is a biomass energy source because it is an organic material that was living or recently living. Today, wood, garbage, waste products, landfill gases, and alcohol fuels are burned as biomass energy sources or produce other energy sources.

Energy from water
In hydroelectric plants, the kinetic energy of falling water turns the turbines of electric generators. These plants are highly efficient and produce little pollution.

Energy from the Sun
A photovoltaic cell is a device that uses radiant energy to knock electrons loose from atoms. The flow of these electrons in an electric circuit is electricity. Solar panels use photovoltaic cells to convert sunlight directly into electricity.

Energy from Earth
Geothermal plants harness steam produced in Earth's hot interior to turn turbines and generate electricity. In addition, hot springs water and ground-source heat pumps can be used to heat buildings.

Energy from nuclear reactions
Two nuclear reactions generate large amounts of thermal energy that can be used to heat water and produce steam that spins turbines—fission and fusion. Within the core of a nuclear power plant, fission splits uranium nuclei. Research is ongoing into fusion, the process of joining small nuclei to form a larger nucleus.

GOINGFURTHER >>>

Class Debate Many factors must be considered when deciding which type of energy to use. Brainstorm a list of important factors, such as cost, availability, and efficiency. Research the pros and cons of various alternative energies, then debate the future of each.

CHAPTER 11 STUDY GUIDE

BIGIDEA Within a closed, isolated system, energy can change form, but the total energy is constant.

VOCABULARY
- **rotational kinetic energy** *(p. 294)*
- **potential energy** *(p. 294)*
- **gravitational potential energy** *(p. 295)*
- **reference level** *(p. 295)*
- **elastic potential energy** *(p. 298)*
- **thermal energy** *(p. 300)*

SECTION 1 The Many Forms of Energy

MAINIDEA Kinetic energy is energy due to an object's motion, and potential energy is energy stored due to the interactions of two or more objects.

- The translational kinetic energy of an object is proportional to its mass and the square of its velocity. The rotational kinetic energy of an object is proportional to the object's moment of inertia and the square of its angular velocity.
- The gravitational potential energy of an object-Earth system depends on the object's weight and that object's distance from the reference level. The reference level is the position where the gravitational potential energy is defined to be zero ($h = 0$).

$$GPE = mgh$$

- Elastic potential energy is energy stored due to stretching, compressing, or bending an object.
- Albert Einstein recognized that mass itself is a form of energy. This energy is called rest energy.

$$E_0 = mc^2$$

VOCABULARY
- **law of conservation of energy** *(p. 301)*
- **mechanical energy** *(p. 301)*
- **elastic collision** *(p. 306)*
- **inelastic collision** *(p. 306)*

SECTION 2 Conservation of Energy

MAINIDEA In a collision in a closed, isolated system, the total energy is conserved, but kinetic energy might not be conserved.

- The total energy of a closed, isolated system is constant. Within the system, energy can change form, but the total amount of energy does not change. Thus, energy is conserved.
- The sum of kinetic and potential energy is called mechanical energy.

$$ME = KE + PE$$

In a closed, isolated system where the only forms of energy are kinetic energy and potential energy, mechanical energy is conserved. The mechanical energy before an event is the same as the mechanical energy after the event.

$$KE_{before} + PE_{before} = KE_{after} + PE_{after}$$

- Momentum is conserved in collisions if the external force is zero. The kinetic energy may be unchanged or decreased by the collision, depending on whether the collision is elastic or inelastic. The type of collision in which the kinetic energy before and after the collision is the same is called an elastic collision. The type of collision in which the kinetic energy after the collision is less than the kinetic energy before the collision is called an inelastic collision.

Games and Multilingual eGlossary

Vocabulary Practice

Chapter Self-Check

SECTION 1 The Many Forms of Energy

Mastering Concepts

Unless otherwise noted, air resistance does no work.

31. Explain how work and energy change are related.

32. Explain how force and energy change are related.

33. What form of energy does a system that contains a wound-up spring toy have? What form of energy does the toy have when it is going? When the toy runs down, what has happened to the energy?

34. A ball is dropped from the top of a building. You choose the top of the building to be the reference level, while your friend chooses the bottom. Explain whether the energy calculated using these two reference levels is the same or different for the following situations:

- **a.** the ball-Earth system's potential energy
- **b.** the change in the system's potential energy as a result of the fall
- **c.** the kinetic energy of the system at any point

35. Can a baseball's kinetic energy ever be negative?

36. Can a baseball-Earth system ever have a negative gravitational potential energy? Explain without using a formula.

37. If a sprinter's velocity increases to three times the original velocity, by what factor does the kinetic energy increase?

38. What energy transformations take place when an athlete is pole-vaulting?

39. The sport of pole-vaulting was drastically changed when the stiff, wooden poles were replaced by flexible fiberglass poles. Explain why.

Mastering Problems

Unless otherwise noted, air resistance does no work.

40. A 1600-kg car travels at a speed of 12.5 m/s. What is its kinetic energy?

41. Shawn and his bike have a combined mass of 45.0 kg. Shawn rides his bike 1.80 km in 10.0 min at a constant velocity. What is the system's kinetic energy?

42. Tony has a mass of 45 kg and a speed of 10.0 m/s.

- **a.** Find Tony's kinetic energy.
- **b.** Tony's speed decreases to 5.0 m/s. Now what is his kinetic energy?
- **c.** Find the ratio of the kinetic energies in parts a and b. Explain how this ratio relates to the change in speed.

43. A racing car has a mass of 1525 kg. What is its kinetic energy if it has a speed of 108 km/h?

44. Katia and Angela each have a mass of 45 kg and are moving together with a speed of 10.0 m/s.

- **a.** What is their combined kinetic energy?
- **b.** What is the ratio of their combined mass to Katia's mass?
- **c.** What is the ratio of their combined kinetic energy to Katia's kinetic energy? Explain how this ratio relates to the ratio of their masses.

45. Train In the 1950s an experimental train with a mass of 2.50×10^4 kg was powered along 509 m of level track by a jet engine that produced a thrust of 5.00×10^5 N. Assume friction is negligible.

- **a.** Find the work done on the train by the jet engine.
- **b.** Find the change in kinetic energy.
- **c.** Find the final kinetic energy of the train if it started from rest.
- **d.** Find the final speed of the train.

46. Car Brakes The driver of the car in **Figure 15** suddenly applies the brakes, and the car slides to a stop. The average force between the tires and the road is 7100 N. How far will the car slide after the brakes are applied?

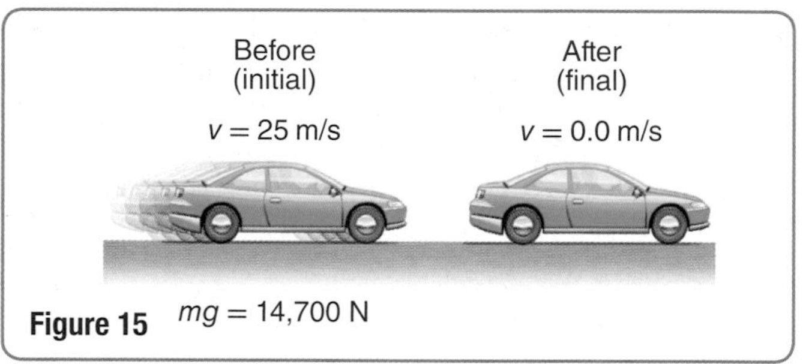

Figure 15

47. A 15.0-kg cart moves down a level hallway with a velocity of 7.50 m/s. A constant 10-N force acts on the cart, slowing its velocity to 3.20 m/s.

- **a.** Find the change in the cart's kinetic energy.
- **b.** How much work was done on the cart?
- **c.** How far did the cart move while the force acted?

48. DeAnna, with a mass of 60.0 kg, climbs 3.5 m up a gymnasium rope. How much energy does a system containing DeAnna and Earth gain from this climb?

49. A 6.4-kg bowling ball is lifted 2.1 m to a shelf. Find the increase in the ball-Earth system's energy.

50. Mary weighs 505 N. She walks down a 5.50-m-high flight of stairs. What is the change in the potential energy of the Mary-Earth system?

51. Weight Lifting A weight lifter raises a 180-kg barbell to a height of 1.95 m. What is the increase in the potential energy of the barbell-Earth system?

52. A science museum display about energy has a small engine that pulls on a rope to lift a block 1.00 m. The display indicates that 1.00 J of work is done. What is the mass of the block?

53. Antwan raised a 12.0-N book from a table 75 cm above the floor to a shelf 2.15 m above the floor. Find the change in the system's potential energy.

54. Tennis A professional tennis player serves a ball. The 0.060-kg ball is in contact with the racket strings, as shown in **Figure 16,** for 0.030 s. If the ball starts at rest, what is its kinetic energy as it leaves the racket?

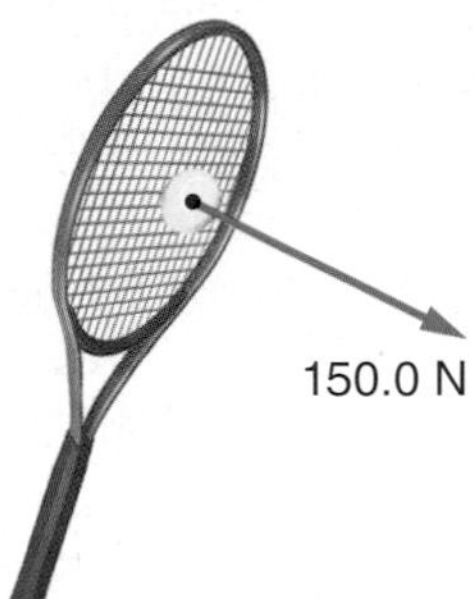

Figure 16

55. Pam has a mass of 45 kg. Her rocket pack supplies a constant force for 22.0 m, and Pam acquires a speed of 62.0 m/s as she moves on frictionless ice.

a. What is Pam's final kinetic energy?

b. What is the magnitude of the force?

56. Collision A 2.00×10^3-kg car has a speed of 12.0 m/s when it hits a tree. The tree doesn't move, and the car comes to rest, as shown in **Figure 17.**

a. Find the change in kinetic energy of the car.

b. Find the amount of work done by the tree on the car as the front of the car crashes into the tree.

c. Find the magnitude of the force that pushed in the front of the car by 50.0 cm.

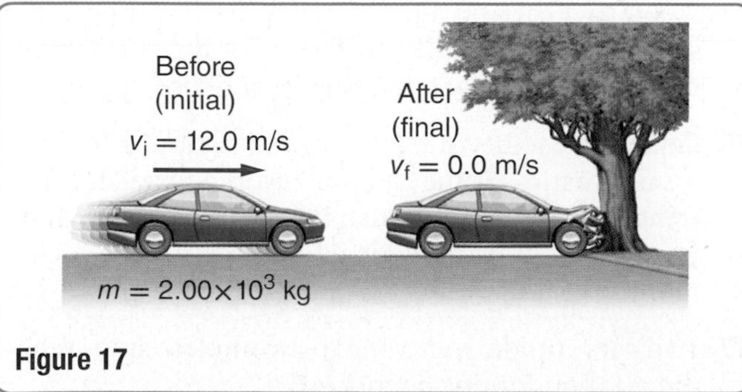

Figure 17

Chapter Self-Check

Section 2 Conservation of Energy

Mastering Concepts

Unless otherwise noted, air resistance does no work.

57. You throw a clay ball at a hockey puck on ice. The smashed clay ball and the hockey puck stick together and move slowly.

a. Is momentum conserved in the collision? Explain.

b. Is kinetic energy conserved? Explain.

58. Draw energy bar diagrams for the following processes.

a. An ice cube, initially at rest, slides down a frictionless slope.

b. An ice cube, initially moving, slides up a frictionless slope and comes momentarily to rest.

59. BIGIDEA Describe the transformations from kinetic energy to potential energy and vice versa for a roller-coaster ride.

60. Describe how the kinetic energy and elastic potential energy of a bouncing rubber ball decreases. What happens to the ball's motion?

Mastering Problems

Unless otherwise noted, friction is negligible and air resistance does no work.

61. A 10.0-kg test rocket is fired vertically. When the engine stops firing, the rocket's kinetic energy is 1960 J. After the fuel is burned, to what additional height will the rocket rise?

62. A constant net force of 410 N is applied upward to a stone that weighs 32 N. The upward force is applied through a distance of 2.0 m, and the stone is then released. To what height, from the point of release, will the stone rise?

63. A 98.0-N sack of grain is hoisted to a storage room 50.0 m above the ground floor of a grain elevator.

a. How much work was done?

b. What is the increase in potential energy of a system containing the sack of grain and Earth?

c. The rope being used to lift the sack of grain breaks just as the sack reaches the storage room. What kinetic energy does the sack have just before it strikes the ground floor?

64. A 2.0-kg rock is initially at rest. The rock falls, and the potential energy of the rock-Earth system decreases by 407 J. How much kinetic energy does the system gain as the rock falls?

Chapter Self-Check

65. A rock sits on the edge of a cliff, as shown in **Figure 18.**

a. What potential energy does the rock-Earth system possess relative to the base of the cliff?

b. The rock falls without rolling from the cliff. What is its kinetic energy just before it strikes the ground?

c. What is the rock's speed as it hits the ground?

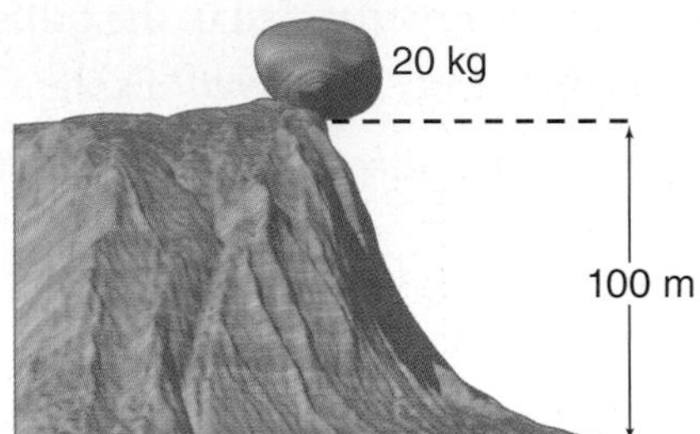

Figure 18

66. Archery An archer fits a 0.30-kg arrow to the bowstring. He exerts an average force of 201 N to draw the string back 0.5 m.

a. If all the energy goes into the arrow, with what speed does the arrow leave the bow?

b. The arrow is shot straight up. What is the arrow's speed when it reaches a height of 10 m above the bow?

67. Railroad Car A 5.0×10^5-kg railroad car collides with a stationary railroad car of equal mass. After the collision, the two cars lock together and move off, as shown in **Figure 19.**

a. Before the collision, the first railroad car was moving at 8.0 m/s. What was its momentum?

b. What was the total momentum of the two cars after the collision?

c. What were the kinetic energies of the two cars before and after the collision?

d. Account for the change of kinetic energy.

$m_{\text{each car}} = 5.0\times10^5$ kg
$v_f = 4.0$ m/s

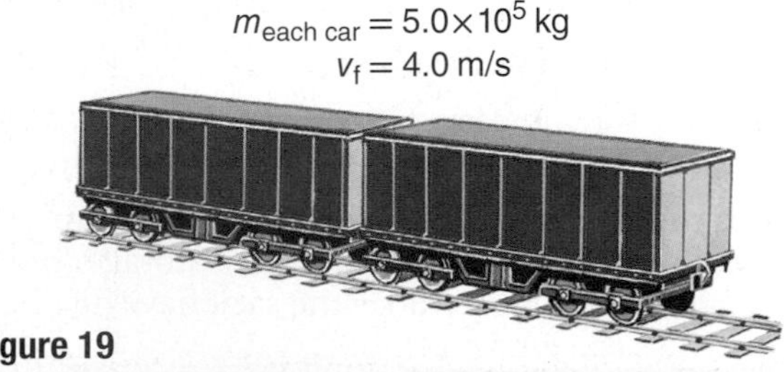

Figure 19

68. Slide Lorena's mass is 28 kg. She climbs the 4.8-m ladder of a slide and reaches a velocity of 3.2 m/s at the bottom of the slide. How much mechanical energy did friction transform to other forms?

69. From what height would a compact car have to be dropped to have the same kinetic energy that it has when being driven at 1.00×10^2 km/h?

70. Problem Posing Complete this problem so that it can be solved using each concept listed below: "A cartoon character is holding a 50-kg anvil at the edge of a cliff..."

a. conservation of mechanical energy

b. Newton's second law

71. Kelli weighs 420 N, and she sits on a playground swing that hangs 0.40 m above the ground. Her mom pulls the swing back and releases it when the seat is 1.00 m above the ground.

a. How fast is Kelli moving when the swing passes through its lowest position?

b. If Kelli moves through the lowest point at 2.0 m/s, how much work was done on the swing by friction?

72. Hakeem throws a 10.0-g ball straight down from a height of 2.0 m. The ball strikes the floor at a speed of 7.5 m/s. Find the ball's initial speed.

73. A 635-N person climbs a ladder to a height of 5.0 m. Use the person and Earth as the system.

a. Draw energy bar diagrams of the system before the person starts to climb and after the person stops at the top. Has the mechanical energy changed? If so, by how much?

b. Where did this energy come from?

Applying Concepts

74. The driver of a speeding car applies the brakes and the car comes to a stop. The system includes the car but not the road. Apply the work-energy theorem to the following situations to describe the changes in energy of the system.

a. The car's wheels do not skid.

b. The brakes lock and the car's wheels skid.

75. A compact car and a trailer truck are traveling at the same velocity. Did the car engine or the truck engine do more work in accelerating its vehicle?

76. Catapults Medieval warriors used catapults to assault castles. Some catapults worked by using a tightly wound rope to turn the catapult arm. What forms of energy are involved in catapulting a rock into the castle wall?

77. Two cars collide and come to a complete stop. Where did all their kinetic energy go?

Chapter Self-Check

78. **Skating** Two skaters of unequal mass have the same speed and are moving in the same direction. If the ice exerts the same frictional force on each skater, how will the stopping distances of their bodies compare?

79. You swing a mass on the end of a string around your head in a horizontal circle at constant speed, as shown from above in **Figure 20.**
   a. How much work is done on the mass by the tension of the string in one revolution?
   b. Is your answer to part a in agreement with the work-energy theorem? Explain.

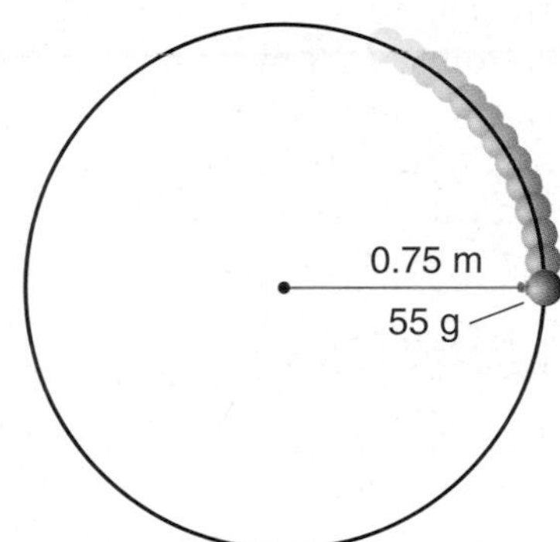

Figure 20

80. **Roller Coaster** You have been hired to make a roller coaster more exciting. The owners want the speed at the bottom of the first hill doubled. How much higher must the first hill be built?

81. Two identical balls are thrown from the top of a cliff, each with the same initial speed. One is thrown straight up, the other straight down. How do the kinetic energies and speeds of the balls compare as they strike the ground?

82. Give specific examples that illustrate the following processes.
   a. Work is done on a system, increasing kinetic energy with no change in potential energy.
   b. Potential energy is changed to kinetic energy with no work done on the system.
   c. Work is done on a system, increasing potential energy with no change in kinetic energy.
   d. Kinetic energy is reduced, but potential energy is unchanged. Work is done by the system.

Mixed Review

83. Suppose a chimpanzee swings through the jungle on vines. If it swings from a tree on a 13-m-long vine that starts at an angle of 45°, what is the chimp's speed when it reaches the ground?

84. An 0.80-kg cart rolls down a frictionless hill of height 0.32 m. At the bottom of the hill, the cart rolls on a flat surface, which exerts a frictional force of 2.0 N on the cart. How far does the cart roll on the flat surface before it comes to a stop?

85. A stuntwoman finds that she can safely break her fall from a one-story building by landing in a box filled to a 1-m depth with foam peanuts. In her next movie, the script calls for her to jump from a five-story building. How deep a box of foam peanuts should she prepare?

86. **Football** A 110-kg football linebacker collides head-on with a 150-kg defensive end. After they collide, they come to a complete stop. Before the collision, which player had the greater kinetic energy? Which player had the greater momentum?

87. A 2.0-kg lab cart and a 1.0-kg lab cart are held together by a compressed spring. The lab carts move at 2.1 m/s in one direction. The spring suddenly becomes uncompressed and pushes the two lab carts apart. The 2-kg lab cart comes to a stop, and the 1.0-kg lab cart moves ahead. How much energy did the spring add to the lab carts?

88. A 55.0-kg scientist roping through the treetops in the jungle sees a lion about to attack an antelope. She swings down from her 12.0-m-high perch and grabs the antelope (21.0 kg) as she swings. They barely swing back up to a tree limb out of the lion's reach. How high is the tree limb?

89. A cart travels down a hill as shown in **Figure 21.** The distance the cart must roll to the bottom of the hill is 0.50 m/sin 30.0° = 1.0 m. The surface of the hill exerts a frictional force of 5.0 N on the cart. Does the cart reach the bottom of the hill?

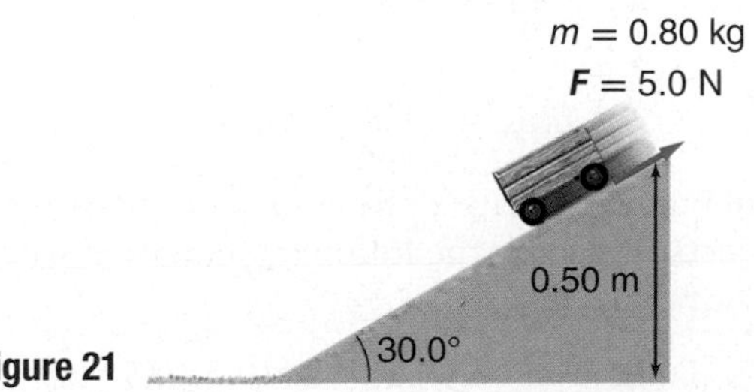

Figure 21

90. Object A, sliding on a frictionless surface at 3.2 m/s, hits a 2.0-kg object, B, which is motionless. The collision of A and B is completely elastic. After the collision, A and B move away from each other in opposite directions at equal speeds. What is the mass of object A?

Chapter Self-Check

Thinking Critically

91. Apply Concepts A golf ball with a mass of 0.046 kg rests on a tee. It is struck by a golf club with an effective mass of 0.220 kg and a speed of 44 m/s. Assuming that the collision is elastic, find the speed of the ball when it leaves the tee.

92. Apply Concepts A fly hitting the windshield of a moving pickup truck is an example of a collision in which the mass of one of the objects is many times larger than the other. On the other hand, the collision of two billiard balls is one in which the masses of both objects are the same. How is energy transferred in these collisions? Consider an elastic collision in which billiard ball m_1 has velocity v_1 and ball m_2 is motionless.

a. If $m_1 = m_2$, what fraction of the initial energy is transferred to m_2?

b. If $m_1 \gg m_2$, what fraction of the initial energy is transferred to m_2?

c. In a nuclear reactor, neutrons must be slowed down by causing them to collide with atoms. (A neutron is about as massive as a proton.) Would hydrogen, carbon, or iron atoms be more desirable to use for this purpose?

93. Reverse Problem Write a physics problem with real-life objects for which the energy bar diagram in **Figure 22** would be part of the solution:

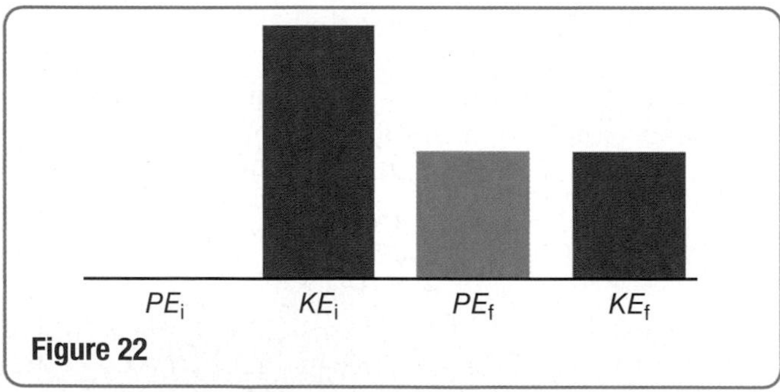

Figure 22

94. Reverse Problem Write a physics problem with real-life objects for which the following equation would be part of the solution:

$$\frac{1}{2}(6.0\text{ kg})v^2 = (6.0\text{ kg})(9.8\text{ N/kg})(0.92\text{ m})$$

95. Analyze and Conclude In a perfectly elastic collision, both momentum and kinetic energy are conserved. Two balls, with masses m_A and m_B, are moving toward each other with speeds v_A and v_B, respectively. Solve the appropriate equations to find the speeds of the two balls after the collision.

96. Ranking Task Five clay pots are dropped or thrown from the same rooftop as described below. Rank them according to their speed when they strike the ground, from least to greatest. Specifically indicate any ties.

A. 1 kg, dropped from rest

B. 1 kg, thrown downward with $v_i = 2$ m/s

C. 1 kg, thrown upward with $v_i = 2$ m/s

D. 1 kg, thrown horizontally with $v_i = 2$ m/s

E. 2 kg, dropped from rest

97. Analyze and Conclude A 25-g ball is fired with an initial speed of v_1 toward a 125-g ball that is hanging motionless from a 1.25-m string. The balls have a perfectly elastic collision. As a result, the 125-g ball swings out until the string makes an angle of 37.0° with the vertical. What is v_1?

Writing in Physics

98. Most energy comes from the Sun and allows us to live and to operate our society. In what forms does this solar energy come to us? Research how the Sun's energy is turned into a form that we can use. After we use the Sun's energy, where does it go? Explain.

99. All forms of energy can be classified as either kinetic or potential energy. How would you describe nuclear, electrical, chemical, biological, solar, and light energy, and why? For each of these types of energy, research what objects are moving and how energy is stored due to the interactions between those objects.

Cumulative Review

100. A satellite is in a circular orbit with a radius of 1.0×10^7 m and a period of 9.9×10^3 s. Calculate the mass of Earth. *Hint: Gravity is the net force on such a satellite. Scientists have actually measured Earth's mass this way.*

101. A 5.00-g bullet is fired with a velocity of 100.0 m/s toward a 10.00-kg stationary solid block resting on a frictionless surface.

a. What is the change in momentum of the bullet if it is embedded in the block?

b. What is the change in momentum of the bullet if it ricochets in the opposite direction with a speed of 99 m/s?

c. In which case does the block end up with a greater speed?

CHAPTER 11 STANDARDIZED TEST PRACTICE

MULTIPLE CHOICE

1. You lift a 4.5-kg box from the floor and place it on a shelf that is 1.5 m above the ground. How much energy did you use in lifting the box?

A. 9.0 J
B. 49 J
C. 11 J
D. 66 J

2. A bicyclist increases her speed from 4.0 m/s to 6.0 m/s. The combined mass of the bicyclist and the bicycle is 55 kg. How much work did the bicyclist do in increasing her speed?

A. 11 J
B. 28 J
C. 55 J
D. 550 J

3. You move a 2.5-kg book from a shelf that is 1.2 m above the ground to a shelf that is 2.6 m above the ground. What is the change in potential energy of a system containing the book and Earth?

A. 1.4 J
B. 25 J
C. 3.5 J
D. 34 J

4. The illustration below shows a ball swinging on a rope. The mass of the ball is 4.0 kg. Assume that the mass of the rope is negligible. Assume that friction is negligible. What is the maximum speed of the ball as it swings back and forth?

A. 0.14 m/s
B. 21 m/s
C. 7.0 m/s
D. 49 m/s

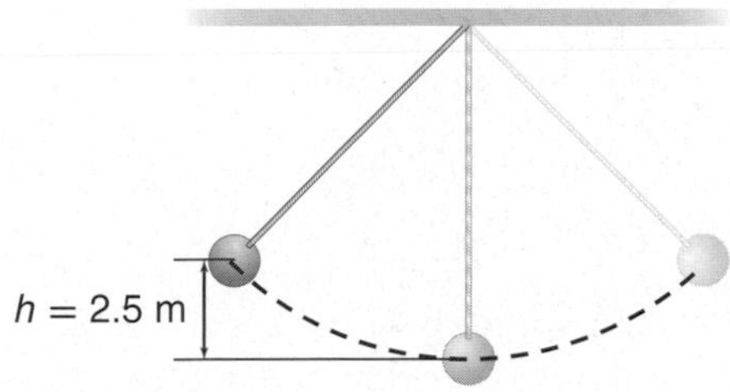

5. A hockey puck of mass m slides along the ice at a speed of v_1. It strikes a wall and bounces back in the opposite direction. The energy of the puck after striking the wall is half its initial energy. Assuming friction is negligible, which expression gives the puck's new speed as a function of its initial speed?

A. $\frac{1}{2}v_1$
B. $\sqrt{2}(v_1)$
C. $\frac{\sqrt{2}}{2}(v_1)$
D. $2v_1$

Online Test Practice

6. The illustration below shows a box on a curved, frictionless track. The box starts with zero velocity at the top of the track. It then slides from the top of the track to the horizontal part at the ground. Its velocity just at the moment it reaches the ground is 14 m/s. What is the height (h) from the ground to the top of the track?

A. 7 m
B. 14 m
C. 10 m
D. 20 m

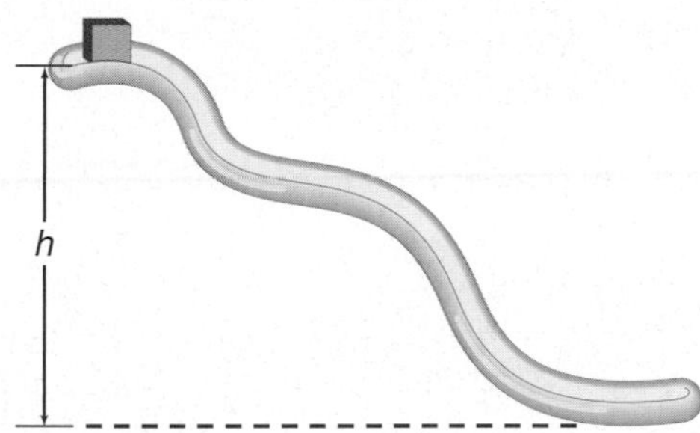

7. You drop a 6.0×10^{-2}-kg ball from a height of 1.0 m above a hard, flat surface. The ball strikes the surface and its energy decreases by 0.14 J. It then bounces back upward. How much kinetic energy does the ball have just after it bounces from the flat surface?

A. 0.20 J
B. 0.59 J
C. 0.45 J
D. 0.73 J

FREE RESPONSE

8. A box sits on a platform supported by a compressed spring. The box has a mass of 1.0 kg. When the spring is released, it does 4.9 J of work on the box, and the box flies upward. What will be the maximum height above the platform reached by the box before it begins to fall?

9. A 90.0-kg hockey player moves at 5.0 m/s and collides head-on with a 110-kg player moving at 3.0 m/s in the opposite direction. After the collision, they move off together at 1.0 m/s. Find the decrease in kinetic energy from the collision for the system containing both hockey players.

NEED EXTRA HELP?

If You Missed Question	1	2	3	4	5	6	7	8	9
Review Section	1	1	1	2	2	2	2	2	2

CHAPTER 12

Thermal Energy

BIGIDEA **Thermal energy is related to the motion of an object's particles and can be transferred and transformed.**

SECTIONS

1 **Temperature, Heat, and Thermal Energy**

2 **Changes of State and Thermodynamics**

LaunchLAB

THERMAL ENERGY TRANSFER

How is thermal energy transferred between your hands and a glass of water?

WATCH THIS!

WARM-BLOODED V. COLD-BLOODED

You might remember from biology that some organisms generate body heat internally, while others modify their behavior to regulate their body heat. What is the physics of body heat?

(l)Ingram Publishing, (r)Jetta Productions/Dana Neely/Getty Images

Go online!
connectED.mcgraw-hill.com

SECTION 1 Temperature, Heat, and Thermal Energy

PHYSICS 4 YOU **Have you seen a balloon outside on a cold day? It was probably shrunken. But if you took it into a warm house, it would return to its normal size. Why does temperature affect the balloon's size?**

MAIN IDEA

Heat is a transfer of thermal energy that occurs spontaneously from a warmer object to a cooler object.

Essential Questions

- How are temperature and thermal energy related?
- How are thermal equilibrium and temperature related?
- How is thermal energy transferred?
- What is specific heat?

Review Vocabulary

thermal energy the sum of the kinetic and potential energies of the particles that make up an object

New Vocabulary

thermal conduction
thermal equilibrium
heat
convection
radiation
specific heat

Thermal Energy

You have studied how objects collide and trade kinetic energies. Every material is made of microscopic particles. The many particles present in a gas have linear and rotational kinetic energies. The particles also might have potential energy due to their internal bonds and interactions with each other. As gas particles collide with each other and with the walls of the container, they transfer energy. There are numerous molecules that make up the gas, resulting in many collisions. The energies of the particles become randomly distributed.

Thus, it is convenient to discuss the total energy of the particles that compose the gas and the average energy per particle in the gas. Recall that the sum of the particles' energies is the object's thermal energy. The average kinetic energy per particle is related to the temperature of the gas. The relationship between the particles' random motions and the bulk property of the material is described by kinetic theory.

Hot objects and cold objects What makes an object hot? Consider a helium-filled balloon. The balloon is kept inflated by the repeated pounding from helium atoms on the balloon wall. Each of the approximately 10^{22} helium atoms in the balloon collides with the balloon wall, bounces back, and hits the balloon wall again somewhere else. The size and the temperature of the balloon are affected by the average kinetic energy of the helium atoms, as shown in **Figure 1.**

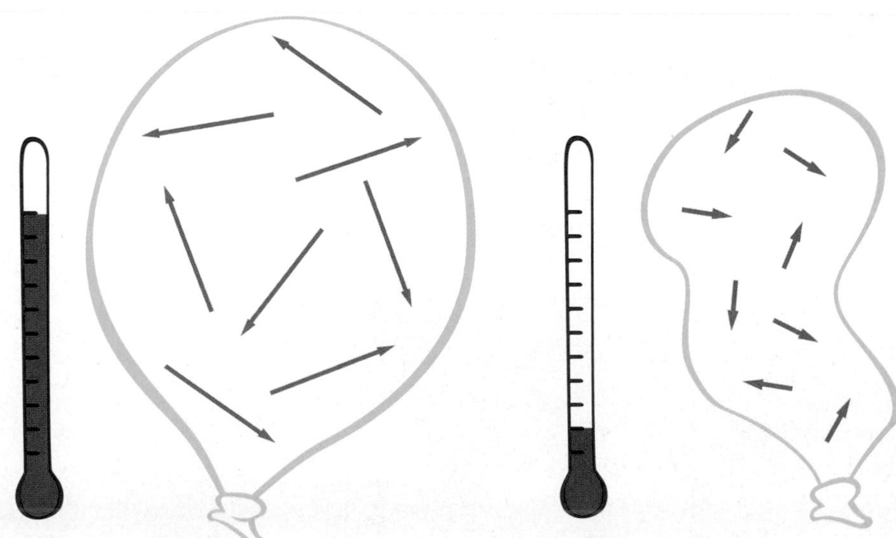

Figure 1 The temperature of an object is related to the average kinetic energy of its particles. The average kinetic energy of the particles that make up a hot object is greater than the average kinetic energy of particles that make up a cold object.

If you put a balloon in sunlight, energy absorbed from the sunlight makes each of the helium atoms move faster in random directions and bounce off the rubber walls of the balloon more often. Each atomic collision with the balloon wall puts a greater force on the balloon and stretches the rubber. Thus, the warmed balloon expands. On the other hand, if you refrigerate a balloon, you will find that it shrinks. It must do so because the particles are moving more slowly. The refrigeration has removed some of their thermal energy.

Thermal energy in solids The atoms or molecules in solids also have kinetic energy, but they are unable to move everywhere as gas atoms do. One way to illustrate the structure of a solid is to picture a number of atoms that are held in place next to each other by atomic forces that act like springs. The atoms cannot move freely, but they do bounce back and forth, with some bouncing more than others. Each atom has some kinetic energy and some potential energy. If a solid has *N* atoms, then the total thermal energy in the solid is equal to the average kinetic energy plus potential energy per atom times *N*.

Thermal Energy and Temperature

You have just seen that, on average, a particle in a hot object has more kinetic energy than a particle in a cold object. This does not mean that each of the particles that compose an object has the same amount of energy; they have a wide range of energies. The average kinetic energy of the particles that compose a hot object, however, is greater than the average kinetic energy of the particles that compose a cold object.

To understand this, consider the heights of students in a 9th-grade class and the heights of students in a 12th-grade class. The students' heights vary, as shown in **Figure 2,** but you can calculate the average height for each class. The average height of the 12th-grade class is greater than the average of height of the 9th-grade class, even though some 9th-grade students might be taller than some 12th-grade students.

Temperature depends only on the average kinetic energy of the particles in the object. It does not depend on the number of particles that compose the object. For example, consider the two muffins shown in **Figure 3.** They are at the same temperature, but the large muffin has ten times as many particles as the small muffin. Thus, the large muffin has ten times the thermal energy of the small muffin. The thermal energy of an object depends on both its temperature and the number of particles that make up that object.

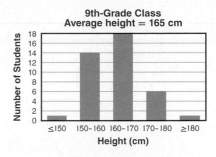

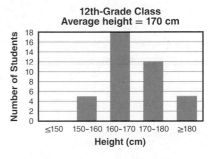

9th-Grade Class
Average height = 165 cm

12th-Grade Class
Average height = 170 cm

Figure 2 The average height in a 9th-grade class is less than the average height in a 12th-grade class. Similarly, the average kinetic energy of a hot object's particles is greater than the average kinetic energy of a cold object's particles.

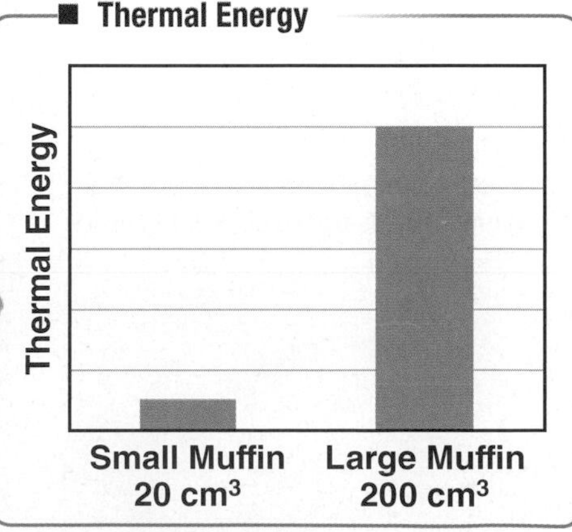

Figure 3 Two muffins at the same temperature can have different thermal energies.

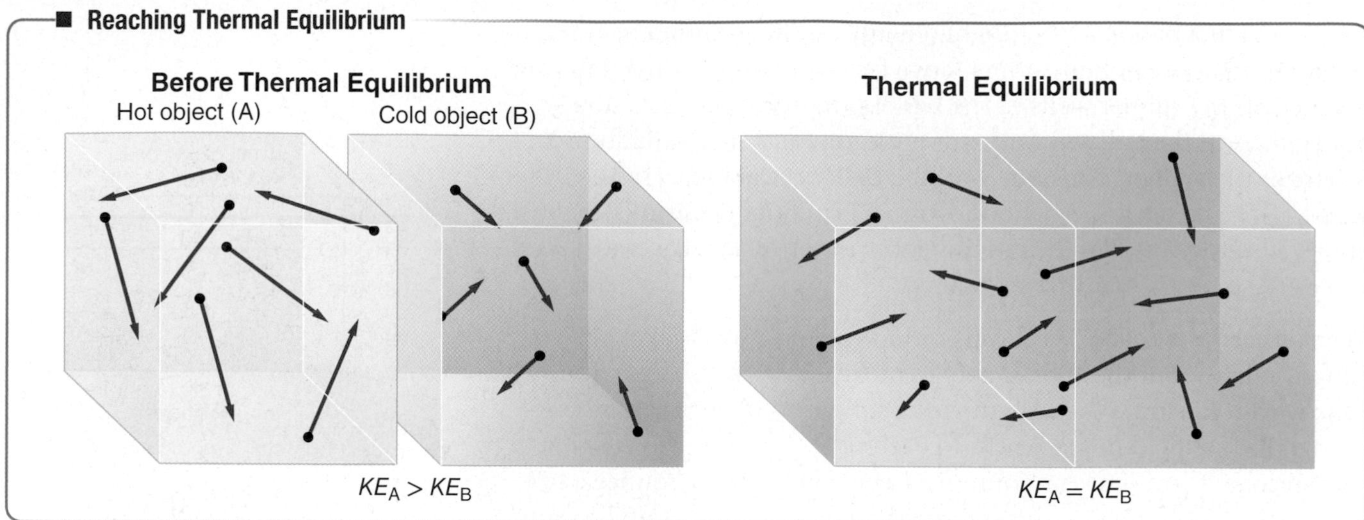

Figure 4 When a hot object and a cold object are in contact, there is a net transfer of thermal energy from the hot object to the cold object. When the two objects reach thermal equilibrium, the transfer of energy between the objects is equal and the objects are at the same temperature.

Equilibrium and Thermometers

How do you measure your body temperature? You might place a thermometer in your mouth and wait for a moment before checking the reading. Measuring your temperature involves random collisions and energy transfers between the particles of the thermometer and the particles of your body. Your body is hotter than the thermometer. That is, the average kinetic energy of the particles that compose your body is greater than the average kinetic energy of the thermometer's particles.

When the cool thermometer touches your skin, the particles of your skin collide with the particles of the thermometer. On average, the more energetic skin particles will transfer energy to the less energetic particles of the thermometer. **Thermal conduction** is the transfer of thermal energy that occurs when particles collide. As a result of these collisions, the thermal energy of the thermometer's particles increases. At the same time, the thermal energy of your skin's particles decreases.

Thermal equilibrium The thermometer's particles also transfer energy to your body's particles. As the thermometer's particles gain more energy, the amount of energy they give back to the skin increases. At some point, the rate of energy transfer from the thermometer to your body is equal to the rate of transfer in the other direction. At this point, your body and the thermometer have reached thermal equilibrium. **Thermal equilibrium** is the state in which the rates of thermal energy transfer between two objects are equal and the objects are at the same temperature. **Figure 4** shows two blocks reaching equilibrium.

READING CHECK **Identify** a situation where two objects are in thermal equilibrium and a situation where two objects are not in thermal equilibrium.

Thermometers Every thermometer has some useful property that changes with temperature. Household thermometers often contain colored alcohol that expands when heated. The hotter the thermometer, the more the alcohol expands and the higher it rises in the tube. The liquid crystal thermometer in **Figure 5** uses a variety of long molecules that rearrange and cause a color change at specific temperatures. Medical thermometers and the thermometers that monitor automobile engines use very small, temperature-sensitive electronic circuits to take rapid measurements.

Figure 5 Liquid crystal thermometers change color with temperature.

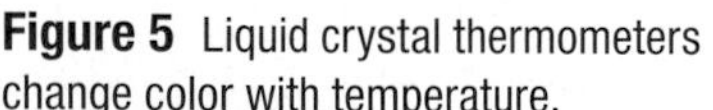

Summarize the process that occurs when this thermometer is placed on your forehead.

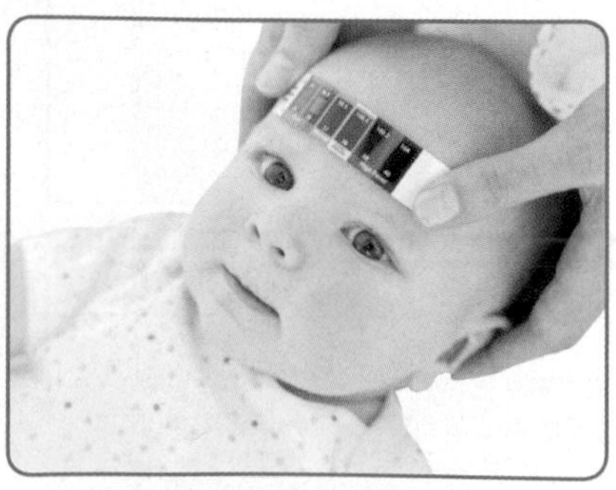

IAN HOOTON/Science Photo Library/Getty Images

Temperature limits You might say that a fire is hot and a freezer is cold. But the temperatures of everyday objects are only a small part of the wide range of temperatures present in the universe, as shown in **Figure 6.** Temperatures do not appear to have an upper limit. The interior of the Sun is at least 1.5×10^{7}°C. Supernova cores are even hotter. On the other hand, liquefied gases can be very cold. For example, helium liquefies at −269°C. Even colder temperatures can be reached by making use of special properties of solids, helium isotopes, and atoms and lasers.

Temperatures do, however, have a lower limit. Generally, materials contract as they cool. If an ideal atomic gas in a balloon were cooled to −273.15°C, it would contract in such a way that it occupied a volume that is only the size of the atoms, and the atoms would become motionless. At this temperature, all the thermal energy that could be removed has been removed from the gas, and the temperature cannot be reduced any further. Therefore, there can be no temperature less than −273.15°C, which is called absolute zero.

☑ **READING CHECK** **Explain** why the term *absolute zero* is appropriate for the coldest temperature possible.

Figure 6 Temperatures in the universe range from just above absolute zero to more than 10^{10} K.

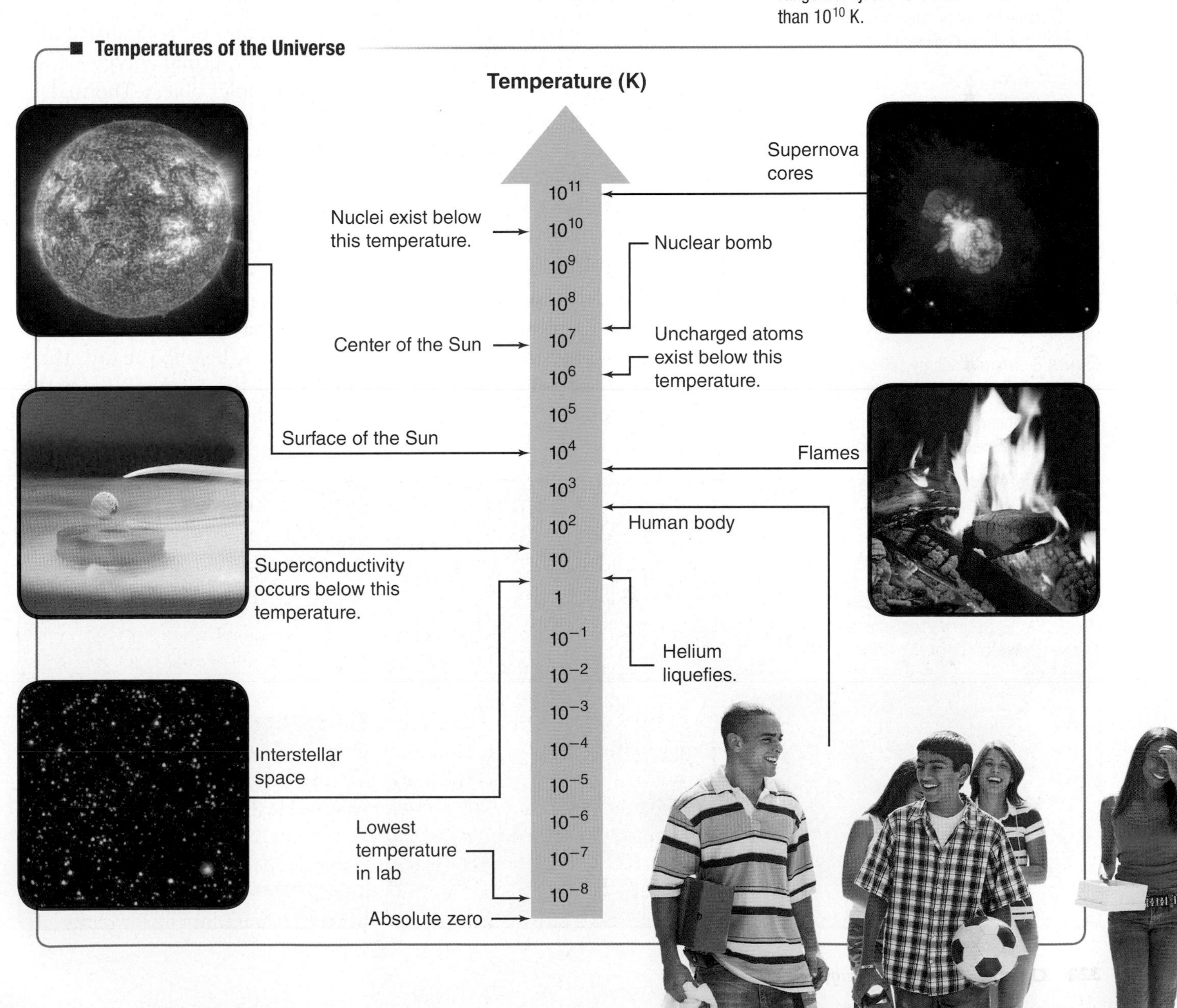

Comparing Temperature Scales

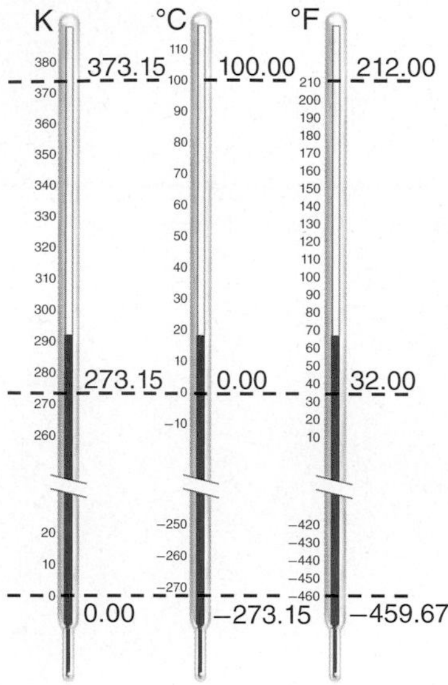

Figure 7 The Kelvin and Celsius scales are used by scientists. In the United States, the Fahrenheit scale is often used for weather reports and cooking.

Temperature scales In the United States, weather agencies report temperatures in degrees Fahrenheit. Scientists, however, use the Celsius and Kelvin scales. The Celsius scale is based on the properties of water and was devised in 1741 by Swedish physicist Anders Celsius. On the Celsius scale, the freezing point of pure water at sea level is defined to be 0°C. The boiling point of pure water at sea level is defined to be 100°C. The Celsius scale is useful for day-to-day measurements of temperature.

The Kelvin scale On the Celsius scale, temperatures can be negative. Negative temperatures suggest a particle could have negative kinetic energy. Because temperature represents average kinetic energy of the object's particles, it makes more sense to use a temperature scale whose zero temperature is where the particles' kinetic energy is also zero. Therefore the zero point of the Kelvin scale is defined to be absolute zero. On the Kelvin scale, the freezing point of water (0°C) is about 273 K and the boiling point of water is about 373 K. Each interval on this scale, called a kelvin, is equal to 1°C. Thus, $T_C + 273 = T_K$. **Figure 7** compares the Fahrenheit, Celsius, and Kelvin scales.

Heat and Thermal Energy Transfer

When two objects come in contact with each other, they redistribute their thermal energies. **Heat** (Q) is the transfer of thermal energy, which occurs spontaneously from a hotter object to a cooler object. Thermal energy cannot be transferred from a colder object to a hotter object without work being done. Like work and energy, heat is measured in joules. In the thermometer example, thermal energy was transferred from the warm skin to the cold thermometer because of the collisions of particles. If thermal energy has been absorbed by an object, Q is positive. If thermal energy is transferred from an object, Q is negative.

Conduction, convection, and radiation **Figure 8** shows the three types of heat—conduction, convection, and radiation. If you place one end of a metal rod in a flame, the hot gas conducts heat to the rod. The other end of the rod also becomes warm because the particles that make up the rod conduct thermal energy to their neighbors.

Figure 8 Thermal energy can be transferred by conduction, convection, or radiation.

Identify other common occurrences of conduction, convection, and radiation.

Have you ever noticed the motion on the surface of a pot of water just about to boil? The water at the bottom of the pot is heated by conduction, expands, and floats to the top, while the colder, denser water at the top sinks to the bottom. The motion of the hot water rapidly carries heat from the bottom of the pot to the top surface of the water. Heating caused by the motion of fluid in a liquid or gas due to temperature differences is called **convection.** Atmospheric turbulence is caused by convection of gases in the atmosphere. Thunderstorms and hurricanes are excellent examples of large-scale atmospheric convection. Convection also contributes to ocean currents that move water and materials over large distances.

The third method of thermal transfer, unlike the first two, does not depend on the presence of matter. The Sun warms Earth from over 150 million km away via **radiation,** which is the transfer of energy by electromagnetic waves. These waves carry the energy from the hot Sun through the vacuum of space to the much cooler Earth.

PhysicsLAB

SOLAR COLLECTORS

PROBEWARE LAB How efficient are solar collectors at collecting radiant energy from the Sun?

Specific Heat

Some objects are easier to warm up than others. On a bright summer day, the Sun radiates thermal energy to the sand on a beach and to the ocean water. The sand on the beach becomes quite hot, while the ocean water stays relatively cool. When an object is heated, its thermal energy increases and its temperature can increase. The amount of the increase in temperature depends on the size of the object and its composition.

The **specific heat** of a material is the amount of energy that must be added to a unit mass of the material to raise its temperature by one temperature unit. In SI units, specific heat (C) is measured in J/(kg·K).

READING CHECK Define the term *specific heat.*

Table 1 provides values of specific heat for some common substances. For example, 897 J must be added to 1 kg of aluminum to raise its temperature by 1 K. The specific heat of aluminum is therefore 897 J/(kg·K). Materials with different specific heats are used for different purposes. Metals, such as those used to make the pans in **Figure 9,** have low specific heats and are good thermal conductors. Notice that liquid water has a high specific heat compared to other substances. Ice and water vapor also have relatively high specific heats. These high specific heats have had significant effects on our climate and our bodies.

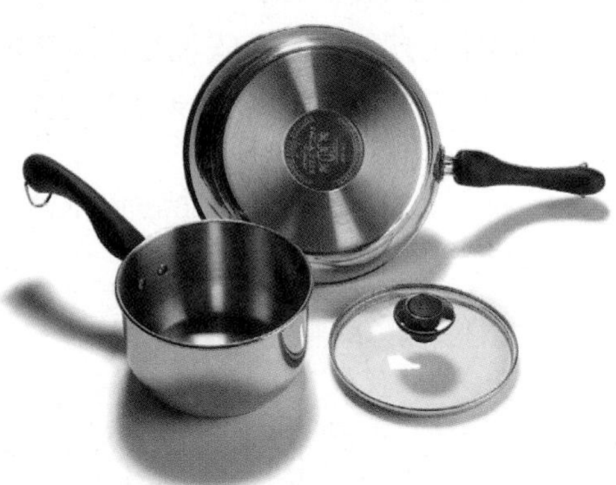

Figure 9 These pans are made of stainless steel and have copper bottoms and plastic handles.

Explain how the selection of these materials is related to their specific heats.

Table 1 Specific Heat of Common Substances

Material	Specific Heat (J/(kg·K))	Material	Specific Heat (J/(kg·K))
Aluminum	897	Lead	130
Brass	376	Methanol	2450
Carbon	710	Silver	235
Copper	385	Water Vapor	2020
Glass	840	Water	4180
Ice	2060	Zinc	388
Iron	450		

PhysicsLAB

HEATING AND COOLING

How does a constant supply of thermal energy affect the temperature of water?

Measuring heat When a substance is heated, the substance's temperature can change. The change of temperature (ΔT) depends on heat (Q), the mass of the substance, and the specific heat of the substance. By using the following equation, you can calculate the heat (Q) required to change the temperature of an object.

HEAT

Heat is equal to the mass of an object times the specific heat of the object times the difference between the final and initial temperatures.

$$Q = mC\Delta T = mC(T_f - T_i)$$

For example, when the temperature of 10.0 kg of water is increased from 80 K to 85 K, the heat is $Q = (10.0\ \text{kg})(4180\ \text{J/(kg·K)})(85\ \text{K} - 80\ \text{K}) = 2.1\times10^5$ J. Remember that the temperature interval for the Kelvin scale is the same as that for the Celsius scale. For this reason, you can calculate ΔT on the Kelvin scale or on the Celsius scale.

EXAMPLE PROBLEM 1

Find help with **order of operations.** Math Handbook

HEAT TRANSFER A 5.10-kg cast-iron skillet is heated on the stove from 295 K to 373 K. How much thermal energy had to be transferred to the iron?

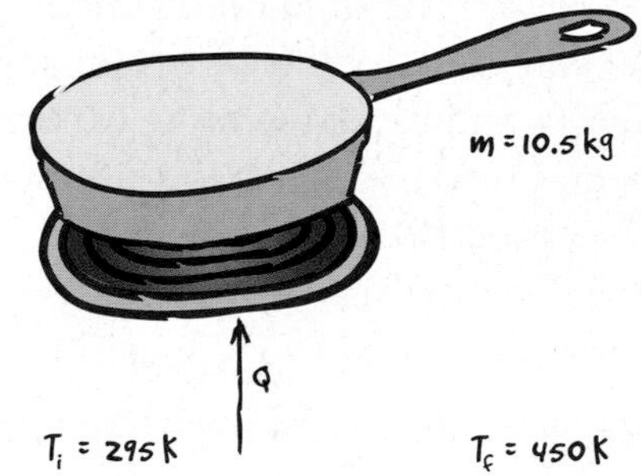

1 ANALYZE AND SKETCH THE PROBLEM

Sketch the thermal energy transfer into the skillet from the stove top.

KNOWN		UNKNOWN
$m = 5.10$ kg	$C = 450$ J/(kg·K)	$Q = ?$
$T_i = 295$ K	$T_f = 373$ K	

2 SOLVE FOR THE UNKNOWN

$Q = mC(T_i - T_f)$

$= (5.10\ \text{kg})(450\ \text{J/(kg·K)})(373\ \text{K} - 295\ \text{K})$ ◀ Substitute $m = 5.10$ kg, $C = 450$ J/(kg·K), $T_f = 373$ K, $T_i = 295$ K.

$= 1.8\times10^5$ J

3 EVALUATE THE ANSWER

- **Are the units correct?** Heat is measured in joules.
- **Does the sign make sense?** Temperature increased, so Q is positive.

PRACTICE PROBLEMS

Do additional problems. Online Practice

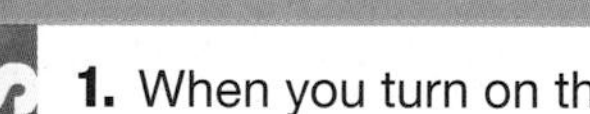

1. When you turn on the hot water to wash dishes, the water pipes heat up. How much thermal energy is absorbed by a copper water pipe with a mass of 2.3 kg when its temperature is raised from 20.0°C to 80.0°C?
2. Electrical power companies sell electrical energy by the kilowatt-hour, where 1 kWh $= 3.6\times10^6$ J. Suppose that it costs $0.15 per kWh to run your electric water heater. How much does it cost to heat 75 kg of water from 15°C to 43°C to fill a bathtub?
3. **CHALLENGE** A car engine's cooling system contains 20.0 L of water (1 L of water has a mass of 1 kg).
   a. What is the change in the temperature of the water if 836.0 kJ of thermal energy is added?
   b. Suppose that it is winter, and the car's cooling system is filled with methanol. The density of methanol is 0.80 g/cm^3. What would be the increase in temperature of the methanol if it absorbed 836.0 kJ of thermal energy?
   c. Which coolant, water or methanol, would better remove thermal energy from a car's engine? Explain.

Measuring Specific Heat

A calorimeter, such as the simple one shown in **Figure 10,** is a device that measures changes in thermal energy. A calorimeter is carefully insulated so that thermal energy transfer to the external world is kept to a minimum. A measured mass of a substance that has been heated to a high temperature (T_A) is placed in the calorimeter. The calorimeter also contains a known mass of cold water at a measured temperature (T_B). Thermal energy is transferred from the warmer substance to the cooler water until they come to an equilibrium temperature (T_f). By measuring these three temperatures, the specific heat of the unknown substance can be calculated.

Energy conservation The operation of a calorimeter depends on the conservation of energy in an isolated, closed system composed of the water and the substance being measured. Energy can neither enter nor leave this system but can be transferred from one part of the system to another. Therefore, if the thermal energy of the test substance changes by an amount (ΔE_A) then the change in thermal energy of the water (ΔE_B) must be related by the equation $\Delta E_A + \Delta E_B = 0$. This can be rearranged to form the equation:

$$\Delta E_A = -\Delta E_B$$

The change in energy of the cold water is positive, and the change in energy of the hot test substance is negative. A positive change in energy indicates a rise in temperature, and a negative change in energy indicates a fall in temperature.

In an isolated, closed system, no work is done, so the change in thermal energy for each substance is equal to the heat and can be expressed by the following equation:

$$\Delta E = Q = mC\Delta T = mC(T_f - T_i)$$

Combining this equation with $\Delta E_A = -\Delta E_B$ gives:

$$m_A C_A(T_f - T_A) = -m_B C_B(T_f - T_B)$$

The final temperatures of the two substances are equal because they are in thermal equilibrium. Solving for the unknown specific heat (C_A) gives the equation:

$$C_A = \frac{-m_B C_B \Delta T_B}{m_A \Delta T_A}$$

Calorimeter

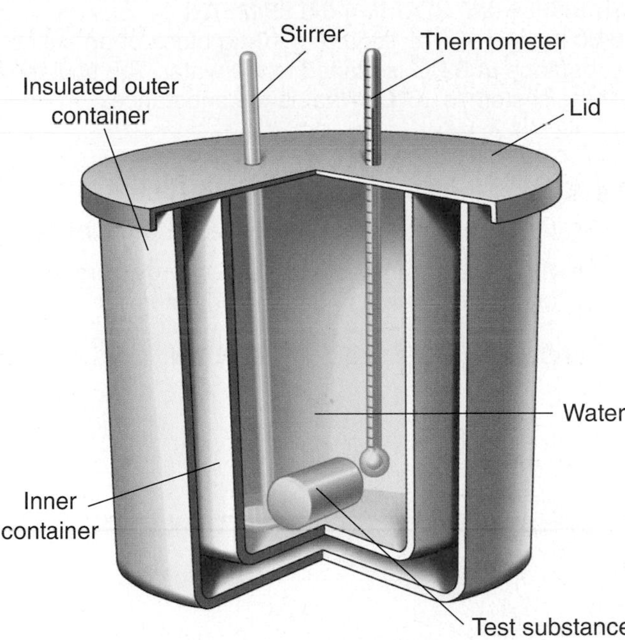

Figure 10 In a simple calorimeter, a hot test substance and a known volume of cold water are placed in an isolated system and allowed to come to thermal equilibrium. The ideal calorimeter has perfect insulation and does not transfer thermal energy to or from the outside. More sophisticated types of calorimeters are used to measure chemical reactions and the energy content of various foods.

View an **animation about using calorimetry to determine specific heat.**

PhysicsLAB

HOW MANY CALORIES ARE THERE?

FORENSIC LAB How can a calorimeter be used to determine energy transfers?

EXAMPLE PROBLEM 2

Find help with **operations with significant figures.** Math Handbook

TRANSFERRING HEAT IN A CALORIMETER A calorimeter contains 0.50 kg of water at 15°C. A 0.10-kg block of an unknown substance at 62°C is placed in the water. The final temperature of the system is 16°C. What is the substance?

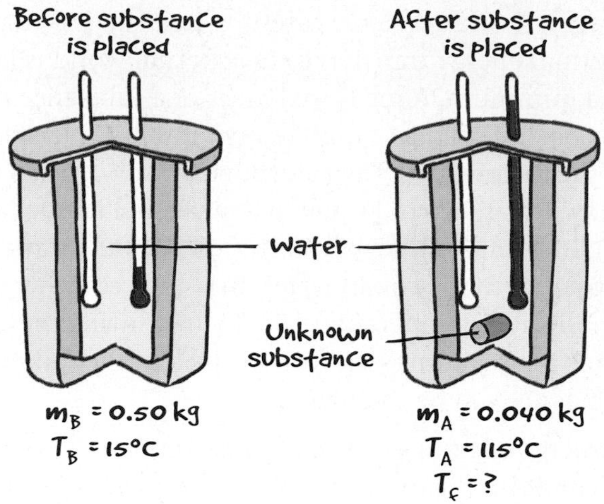

1 ANALYZE AND SKETCH THE PROBLEM

- Let the unknown be sample A and water be sample B.
- Sketch the transfer of thermal energy from the hotter unknown sample to the cooler water.

KNOWN	UNKNOWN
$m_A = 0.10$ kg	$C_A = ?$
$T_A = 62°C$	
$m_B = 0.50$ kg	
$C_B = 4180$ J/(kg·K)	
$T_B = 15°C$	
$T_f = 16°C$	

2 SOLVE FOR THE UNKNOWN

Determine the final temperature using the following equation. Beware of the minus signs.

$$C_A = \frac{-m_B C_B \Delta T_B}{m_A \Delta T_A}$$

◀ Substitute m_A = 0.10 kg, T_A = 62°C, m_B = 0.50 kg, C_B = 4180 J/(kg·K), T_B = 15°C, T_f = 16°C.

$$= \frac{-(0.50 \text{ kg})(4180 \text{ J/(kg·K)})(16°C - 15°C)}{(0.10 \text{ kg})(16°C - 62°C)}$$

$$= 450 \text{ J/(kg·K)}$$

According to **Table 1,** the specific heat of the unknown substance equals that of iron.

3 EVALUATE THE ANSWER

- **Are the units correct?** Specific heat is measured in J/(kg·K).
- **Is the magnitude realistic?** The answer is of the same magnitude as most metals listed in **Table 1.**

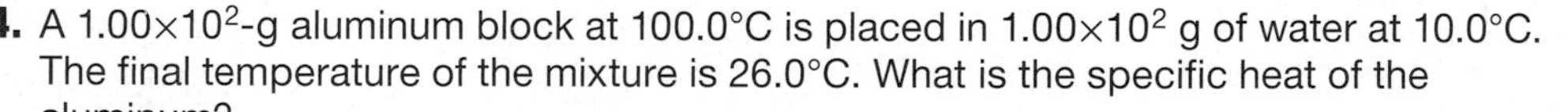

PRACTICE PROBLEMS

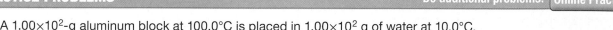

Do additional problems. Online Practice

4. A 1.00×10^2-g aluminum block at 100.0°C is placed in 1.00×10^2 g of water at 10.0°C. The final temperature of the mixture is 26.0°C. What is the specific heat of the aluminum?

5. Three metal fishing weights, each with a mass of 1.00×10^2 g and at a temperature of 100.0°C, are placed in 1.00×10^2 g of water at 35.0°C. The final temperature of the mixture is 45.0°C. What is the specific heat of the metal in the weights?

6. A 2.00×10^2-g sample of water at 80.0°C is mixed with 2.00×10^2 g of water at 10.0°C in a calorimeter. What is the final temperature of the mixture?

7. A 1.50×10^2-g piece of glass at a temperature of 70.0°C is placed in a container with 1.00×10^2 g of water initially at a temperature of 16.0°C. What is the equilibrium temperature of the water?

8. CHALLENGE A 4.00×10^2-g sample of water at 15.0°C is mixed with 4.00×10^2 g of water at 85.0°C. After the system reaches thermal equilibrium, 4.00×10^2 g of methanol at 15°C is added. Assume there is no thermal energy lost to the surroundings. What is the final temperature of the mixture?

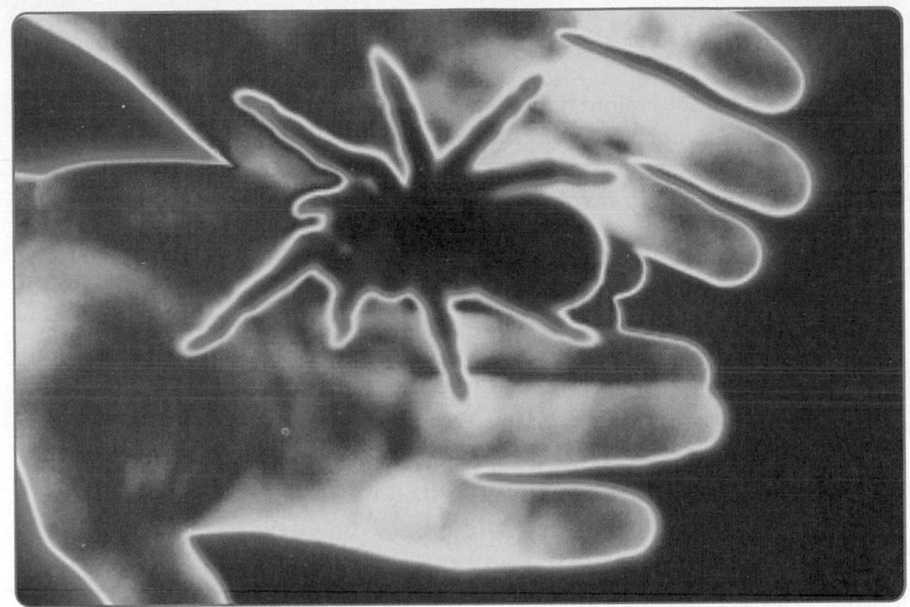

Figure 11 Cold-blooded animals depend on external sources of thermal energy to maintain their body temperatures. In contrast, warm-blooded animals maintain their body temperatures internally. In this thermal image, the spider is at the same temperature as the surrounding air, but the person's hands are significantly warmer than the surrounding air.

Animals and Thermal Energy

CONNECTION TO BIOLOGY Animals can be divided into two groups based on how they control their body temperatures. Most, such as the spider in **Figure 11,** are cold-blooded animals. Their body temperatures depend on the environment. A cold-blooded animal regulates the transfer of thermal energy by its behavior, such as hiding under a rock to keep cool or sunning itself to keep warm.

The others are warm-blooded animals whose body temperatures are controlled internally. That is, a warm-blooded animal's body temperature remains stable regardless of the temperature of the environment. For example, humans are warm-blooded and have body temperatures close 37°C. To regulate its body temperature, a warm-blooded animal relies on bodily responses initiated by the brain, such as shivering and sweating, to counteract a rise or fall in body temperature.

SECTION 1 REVIEW

Section Self-Check Check your understanding.

9. MAINIDEA The hard tile floor of a bathroom always feels cold to bare feet even though the rest of the room is warm. Is the floor colder than the rest of the room?
10. **Temperature** Make the following conversions:
    - **a.** 5°C to kelvins
    - **b.** 34 K to degrees Celsius
    - **c.** 212°C to kelvins
    - **d.** 316 K to degrees Celsius
11. **Units** Are the units the same for heat (Q) and specific heat (C)? Explain.
12. **Types of Energy** Describe the mechanical energy and the thermal energy of a bouncing basketball.
13. **Thermal Energy** Could the thermal energy of a bowl of hot water equal that of a bowl of cold water? Explain your answer.
14. **Cooling** On a dinner plate, a baked potato always stays hot longer than any other food. Why?
15. **Heat and Food** It takes much longer to bake a whole potato than potatoes that have been cut into pieces. Why?
16. **Cooking** Stovetop pans are made from metals such as copper, iron, and aluminum. Why are these materials used?
17. **Specific Heat** If you take a plastic spoon out of a cup of hot cocoa and put it in your mouth, you are not likely to burn your tongue. However, you could very easily burn your tongue if you put the hot cocoa in your mouth. Why?
18. **Critical Thinking** As water heats in a pot on a stove, it might produce some mist above its surface right before the water begins to roll. What is happening?

SECTION 2

Changes of State and Thermodynamics

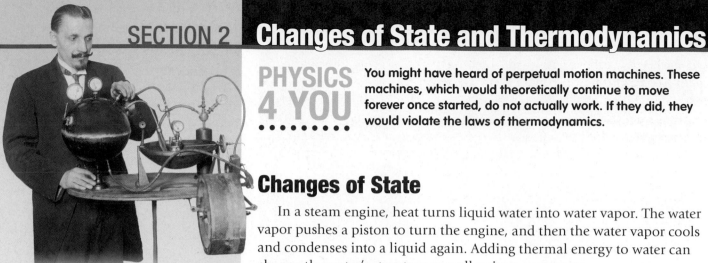

PHYSICS 4 YOU **You might have heard of perpetual motion machines. These machines, which would theoretically continue to move forever once started, do not actually work. If they did, they would violate the laws of thermodynamics.**

When thermal energy is transferred, energy is conserved and the total entropy of the universe will increase.

Essential Questions

- How are the heats of fusion and vaporization related to changes in state?
- What is the first law of thermodynamics?
- How do engines, heat pumps, and refrigerators demonstrate the first law of thermodynamics?
- What is the second law of thermodynamics?

Review Vocabulary

joule (J) the SI unit of work and energy; 1 J of work is done when a force of 1 N acts on an object over a displacement of 1 m

New Vocabulary

heat of fusion
heat of vaporization
first law of thermodynamics
heat engine
second law of thermodynamics
entropy

Changes of State

In a steam engine, heat turns liquid water into water vapor. The water vapor pushes a piston to turn the engine, and then the water vapor cools and condenses into a liquid again. Adding thermal energy to water can change the water's structure as well as its temperature.

The three most common states of matter on Earth are solid, liquid, and gas. As the temperature of a solid rises, that solid usually changes to a liquid. At even higher temperatures, it becomes a gas. If the gas cools, it returns to the liquid state. If the cooling continues, the liquid will return to the solid state. How can these changes be explained? Recall that when the thermal energy of a material changes, the motion of its particles also changes, as does the temperature.

Figure 12 diagrams the changes of state as thermal energy is added to 1.0 kg of water starting at 243 K (ice) and continuing until that water reaches 473 K (water vapor). Between points A and B, the ice is warmed to 273 K. At this point, the added thermal energy gives the water molecules enough energy to partially overcome the forces holding them together. The particles are still touching each other, but they have more freedom of movement. Eventually, the molecules become free enough to slide past each other.

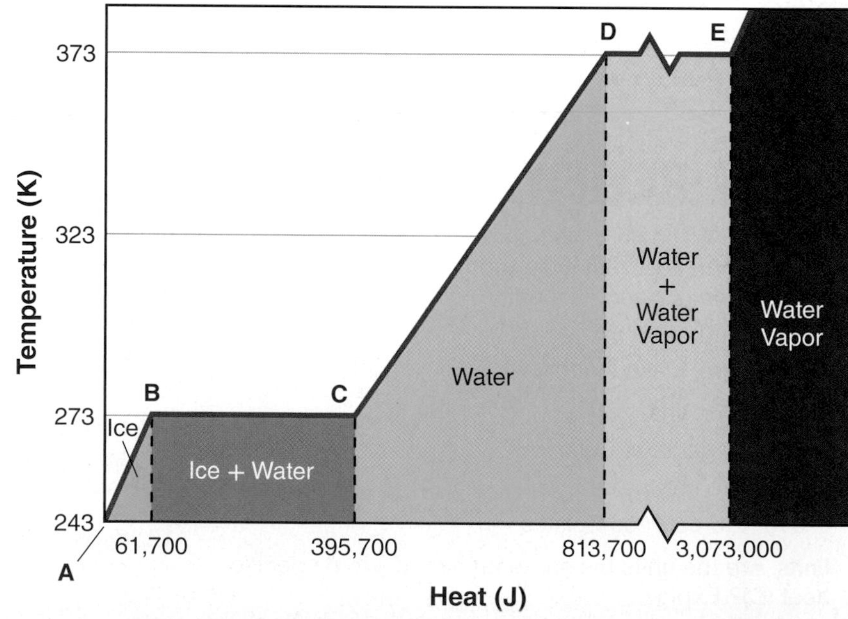

Figure 12 Thermal energy added to a substance can raise the temperature or cause a change in state. Note that the scale is broken between points D and E.

View an **animation of changes of state.**

Melting point At this point, water changes from a solid to a liquid, just as the snowman in **Figure 13** does. The temperature at which this change occurs is the melting point of a substance. When a substance melts, the addition of thermal energy allows the particles to move, rotate, and vibrate in ways not available in the solid. Each of these new motions might add new modes of kinetic or potential energy. The added thermal energy does not change the temperature of the material. This can be observed between points B and C in **Figure 12,** where the added thermal energy melts all the ice at a constant 273 K.

Boiling point Once the ice is completely melted, adding more thermal energy increases the motion of the water molecules again. The temperature rises as shown between points C and D in **Figure 12.** As the temperature increases further, some of the particles that make up the liquid acquire enough energy to break free from the other particles.

At a specific temperature, known as the boiling point, adding more energy to a substance causes it to undergo another change of state. All the added thermal energy converts it from the liquid state to the gas state. As in melting, the temperature does not rise while a liquid boils, as shown between points D and E in **Figure 12.** After the water is entirely converted to gas, any added thermal energy again increases the motion of the molecules, and the temperature rises. After point E, the water vapor is heated to temperatures greater than 373 K.

Heat of fusion and heat of vaporization The amount of thermal energy needed to melt 1 kg of a substance is called the substance's **heat of fusion** (H_f). For ice the heat of fusion is 3.34×10^5 J/kg. If 1 kg of ice at its melting point, 273 K, absorbs 3.34×10^5 J, the ice becomes 1 kg of water at the same temperature, 273 K. The added energy causes a change in state but not in temperature. The horizontal distance in **Figure 12** from point B to point C represents the heat of fusion.

The thermal energy needed to vaporize 1 kg of a liquid is called the **heat of vaporization** (H_v). Water's heat of vaporization is 2.26×10^6 J/kg. The horizontal distance from point D to point E in **Figure 12** represents the heat of vaporization. Every material has a characteristic heat of fusion and heat of vaporization. The values of some heats of fusion (H_f) and heats of vaporization (H_v) are shown in **Table 2.**

Figure 13 Thermal energy is transferred from the warmer air to the snowman, causing the snowman to melt.

Watch a **BrainPOP video on how matter changes states.**

MiniLAB

MELTING

How does heating affect the temperature and state of water?

Table 2 Heats of Fusion and Vaporization of Common Substances

Material	Heat of Fusion H_f (J/kg)	Heat of Vaporization H_v (J/kg)
Copper	2.05×10^5	5.07×10^6
Mercury	1.15×10^4	2.72×10^5
Gold	6.30×10^4	1.64×10^6
Methanol	1.09×10^5	8.78×10^5
Iron	2.66×10^5	6.29×10^6
Silver	1.04×10^5	2.36×10^6
Lead	2.04×10^4	8.64×10^5
Water (ice)	3.34×10^5	2.26×10^6

Andreas Kuehn/The Image Bank/Getty Images

Figure 14 One way to measure the absorption of energy by a material is to add energy from a constant source of thermal energy and measure the change in temperature over time. The plot of temperature v. time is called a heating curve. For this figure, a beaker of cold water was placed on a hot plate. The resulting heating curve is graphed.

Explain why thermal energy must be added at a constant rate to calculate the specific heat of water from this graph.

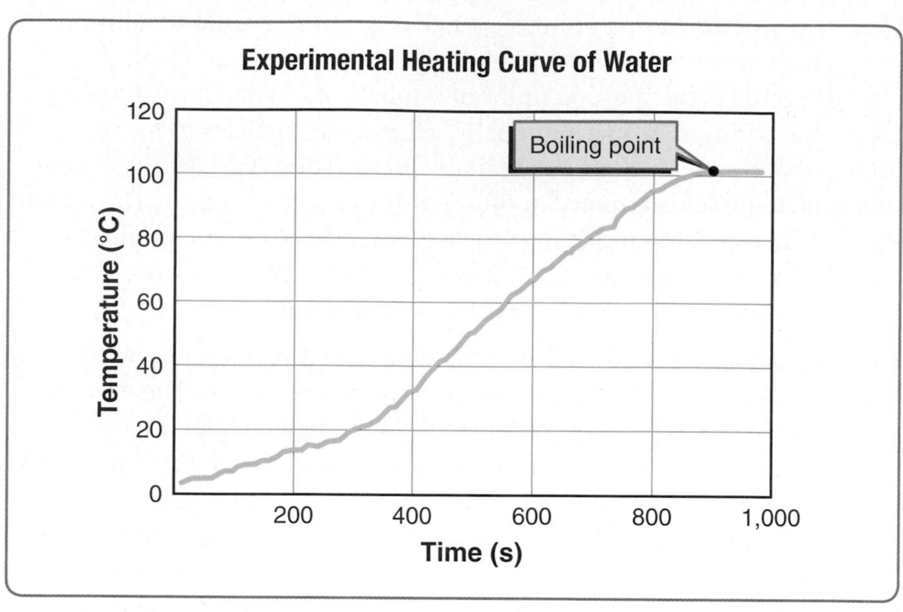

Investigate **phase changes.**

PhysicsLAB

HEAT OF FUSION

How can you measure the heat of fusion of ice?

Energy and changes of state There is a definite slope to the graph in **Figure 14** between about 300 s and 800 s. Heat is added at a constant rate, so this slope is proportional to the reciprocal of the specific heat of water. The slope between points A and B in **Figure 12** is proportional to the reciprocal of the specific heat of ice, and the slope above point E is proportional to the reciprocal of the specific heat of water vapor. The slope for water is less than those of both ice and water vapor. This is because water has a greater specific heat than does ice or water vapor. The heat (Q) required to melt a solid of mass (m) is given by the following equation.

HEAT REQUIRED TO MELT A SOLID

The heat required to melt a solid is equal to the mass of the solid times the heat of fusion of the solid.

$$Q = mH_f$$

Similarly, the heat (Q) required to vaporize a mass (m) of liquid is given by the following equation.

HEAT REQUIRED TO VAPORIZE A LIQUID

The heat required to vaporize a liquid is equal to the mass of the liquid times the heat of vaporization of the liquid.

$$Q = mH_v$$

When a liquid freezes, an amount of thermal energy ($Q = -mH_f$) must be removed from the liquid to turn it into a solid. The negative sign indicates that the thermal energy is transferred from the sample to the external world. In the same way, when a vapor condenses to a liquid, an amount of thermal energy ($Q = -mH_v$) must be removed from the vapor.

Water absorbs significant amounts of thermal energy when it melts or evaporates. Every day you use the large heats of fusion and vaporization of water. Each gram of sweat that evaporates from your skin carries off about 2.3 kJ of thermal energy. This is one cooling process that many warm-blooded animals use to regulate their body temperatures. Similarly the melting of a 24-g cube of ice absorbs enough thermal energy, 8 kJ, to cool a glass of water by about 30°C.

EXAMPLE PROBLEM 3

Get help with **solving for amount of heat.** Personal Tutor

HEAT Suppose that you are camping in the mountains. You need to melt 1.50 kg of snow at 0.0°C and heat it to 70.0°C to make hot cocoa. How much heat will you need?

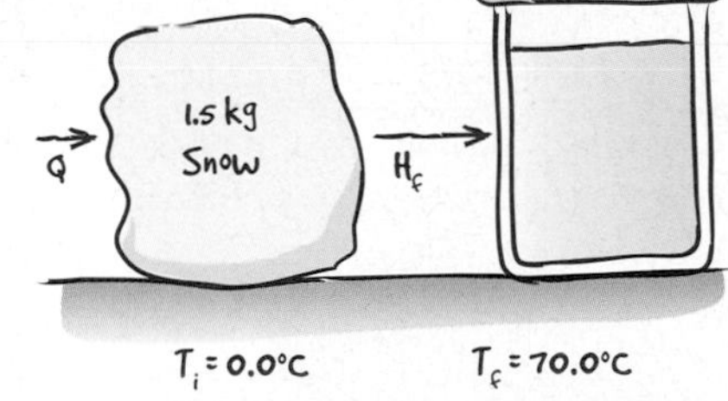

1 ANALYZE AND SKETCH THE PROBLEM

- Sketch the relationship between heat and water in its solid and liquid states.
- Sketch the transfer of heat as the temperature of the water increases.

KNOWN		UNKNOWN
$m = 1.50$ kg	$H_f = 3.34\times10^5$ J/kg	$Q_{\text{melt ice}} = ?$
$T_i = 0.0$°C	$T_f = 70.0$°C	$Q_{\text{heat liquid}} = ?$
$C = 4180$ J/(kg·K)		$Q_{\text{total}} = ?$

2 SOLVE FOR THE UNKNOWN

Calculate the heat needed to melt ice.

$$Q_{\text{melt ice}} = mH_f$$
$$= (1.50\text{ kg})(3.34\times10^5\text{ J/kg})$$
$$= 5.01\times10^5\text{ J} = 5.01\times10^2\text{ kJ}$$

◀ Substitute $m = 1.50$ kg, $H_f = 3.34\times10^5$ J/kg.

Calculate the temperature change.

$$\Delta T = T_f - T_i$$
$$= 70.0°\text{C} - 0.0°\text{C} = 70.0°\text{C}$$

◀ Substitute $T_f = 70.0$°C, $T_i = 0.0$°C.

Calculate the heat needed to raise the water temperature.

$$Q_{\text{heat liquid}} = mC\Delta T$$
$$= (1.50\text{ kg})(4180\text{ J/(kg·K)})(70.0°\text{C})$$
$$= 4.39\times10^5\text{ J} = 4.39\times10^2\text{ kJ}$$

◀ Substitute $m = 1.50$ kg, $C = 4180$ J/(kg·K), $\Delta T = 70.0$°C.

Calculate the total amount of heat needed.

$$Q_{\text{total}} = Q_{\text{melt ice}} + Q_{\text{heat liquid}}$$
$$= 5.01\times10^2\text{ kJ} + 4.39\times10^2\text{ kJ}$$
$$= 9.40\times10^2\text{ kJ}$$

◀ Substitute $Q_{\text{melt ice}} = 5.01\times10^2$ kJ, $Q_{\text{heat liquid}} = 4.39\times10^2$ kJ.

3 EVALUATE THE ANSWER

- **Are the units correct?** Energy units are in joules.
- **Does the sign make sense?** Q is positive when thermal energy is absorbed.
- **Is the magnitude realistic?** To check the magnitude, perform a quick estimation:

$$Q = (1.5\text{ kg})(300{,}000\text{ J/kg}) + (1.5\text{ kg})(4000\text{ J/(kg·K)})(70\text{ K}) = 9\times10^2\text{ kJ}.$$

PRACTICE PROBLEMS

Do additional problems. Online Practice

19. How much thermal energy is absorbed by 1.00×10^2 g of ice at −20.0°C to become water at 0.0°C?

20. A 2.00×10^2-g sample of water at 60.0°C is heated to water vapor at 140.0°C. How much thermal energy is absorbed?

21. Use the graph in **Figure 15** to calculate the heat of fusion and heat of vaporization of water in joules per kilogram.

22. A steel plant operator wishes to change 100 kg of 25°C iron into molten iron (melting point = 1538°C). How much thermal energy must be added?

23. CHALLENGE How much thermal energy is needed to change 3.00×10^2 g of ice at −30.0°C to water vapor at 130.0°C?

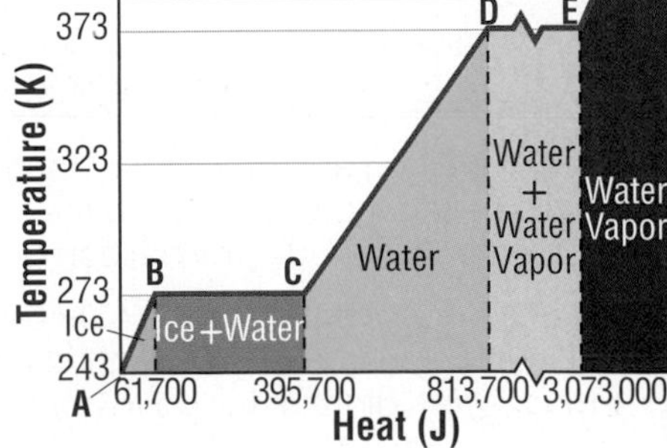

Figure 15

Figure 16 Steam engines change thermal energy into useful mechanical energy.

The First Law of Thermodynamics

The first steam engines were built in the eighteenth century and used to power trains and factories. Steam engines, such as the one in **Figure 16,** change thermal energy into mechanical energy. The invention of the steam engine contributed greatly to the Industrial Revolution and to the study of how heat is related to work. The study of how thermal energy is transformed into other forms of energy is called thermodynamics.

Before 1900, scientists did not realize that the concepts of thermodynamics were linked to the motions of particles in an object and they considered thermodynamics to be a separate topic from mechanics. Today, engineers routinely apply the concepts of thermodynamics to develop higher performance refrigerators, automobile engines, aircraft engines, and numerous other machines.

The first law developed for thermodynamics was a statement about what thermal energy is and where it can go. As you know, you can raise the temperature of a glass of cold water by placing it on a hot plate by stirring it. That is, you can increase the water's thermal energy by heating or by doing work on it. If we consider the system to be the water, the work the system does on you is equal to the negative of the work you do on the system.

The **first law of thermodynamics** states that the change in thermal energy (ΔU) of an object is equal to the heat (Q) that is added to the object minus the work (W) done by the object. Note that ΔU, Q, and W are all measured in joules, the unit of energy.

THE FIRST LAW OF THERMODYNAMICS

The change in thermal energy of an object is equal to the heat added to the object minus the work done by the object.

$$\Delta U = Q - W$$

MiniLAB

CONVERT ENERGY

How are work and thermal energy related?

The first law of thermodynamics is merely a restatement of the law of conservation of energy, which states that energy is neither created nor destroyed but can be changed into other forms.

Another example of changing the amount of thermal energy in a system is a hand pump used to inflate a bicycle tire. As a person pumps, the air and the hand pump become warm. The mechanical energy in the moving piston is converted into thermal energy of the gas. Similarly, other forms of energy, such as light, sound, and electrical energy, can be changed into thermal energy. For example, a toaster converts electrical energy into radiant energy when it toasts bread. You can probably think of many more examples from your everyday life.

PRACTICE PROBLEMS

Do additional problems. Online Practice

24. A gas balloon absorbs 75 J of thermal energy. The balloon expands but stays at the same temperature. How much work did the balloon do in expanding?

25. A drill bores a small hole in a 0.40-kg block of aluminum and heats the aluminum by 5.0°C. How much work did the drill do in boring the hole?

26. How many times would you have to drop a 0.50-kg bag of lead shot from a height of 1.5 m to heat the shot by 1.0°C?

27. When you stir a cup of tea, you do about 0.050 J of work each time you circle the spoon in the cup. How many times would you have to stir the spoon to heat a 0.15-kg cup of tea by 2.0°C?

28. CHALLENGE An expansion valve does work on 100 g of water. The system is isolated, and all of the work is used to convert the 90°C water into water vapor at 110°C. How much work does the expansion valve do on the water?

Heat engines When you rub your hands together to warm them, you convert mechanical energy into thermal energy. This energy conversion is easy to produce. However, the reverse process, the conversion of thermal energy into mechanical energy, is more difficult. A **heat engine** is a device that is able to continuously convert thermal energy to mechanical energy. A heat engine requires a high-temperature source, a low-temperature receptacle, called a sink, and a way to convert the thermal energy into work. **Figure 17** shows that some thermal energy from the source is used to do work and some is transferred to the sink.

Internal combustion engines An automobile's internal combustion engine, such as the one shown in **Figure 18,** is one example of a heat engine. In the engine, input heat (Q_H) is transferred from a high-temperature flame to a mixture of air and gas vapor in the cylinder. The hot air expands and pushes on a piston, thereby changing thermal energy into mechanical energy. The heated air is expelled, the piston returns to the top of the cylinder, and the cycle repeats. Car engines repeat this cycle many times each minute. The thermal energy from the flame is converted into mechanical energy, which propels the car.

Waste heat Not all thermal energy from the flame is converted into mechanical energy. When the engine is running, the gases and the engine parts become hot. The exhaust comes in contact with outside air and heats the air. In addition, thermal energy from the engine is transferred to a radiator. Outside air passes through the radiator and becomes warmer. All of this energy (Q_C) transferred out of the automobile engine is called waste heat. When the engine is working continuously, its internal energy does not change. That is, $\Delta U = 0 = Q - W$. The net heat going into the engine is $Q = Q_H - Q_C$. Thus, the work done by the engine is $W = Q_H - Q_C$. All heat engines generate waste heat, and therefore no engine can ever convert all the energy into useful kinetic energy.

Energy Diagram of a Heat Engine

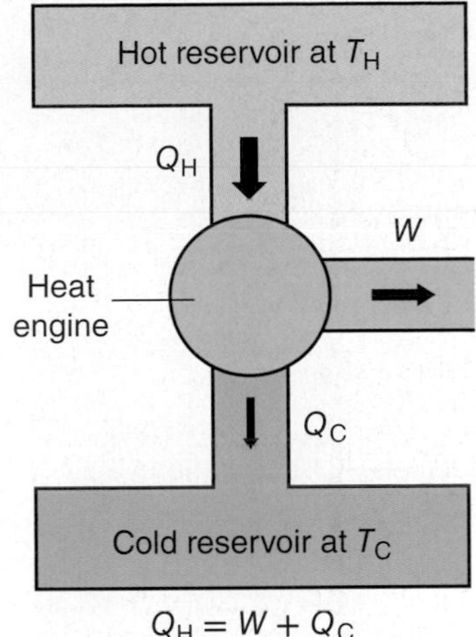
Hot reservoir at T_H
Q_H
W
Heat engine
Q_C
Cold reservoir at T_C
$Q_H = W + Q_C$

Figure 17 Heat engines transform thermal energy into mechanical energy and waste heat. This schematic shows the energy transfers and transformations.

Watch an **animation of a heat engine.**

Concepts In Motion

Figure 18 Internal combustion engines are one type of heat engine. They are used in automobiles.

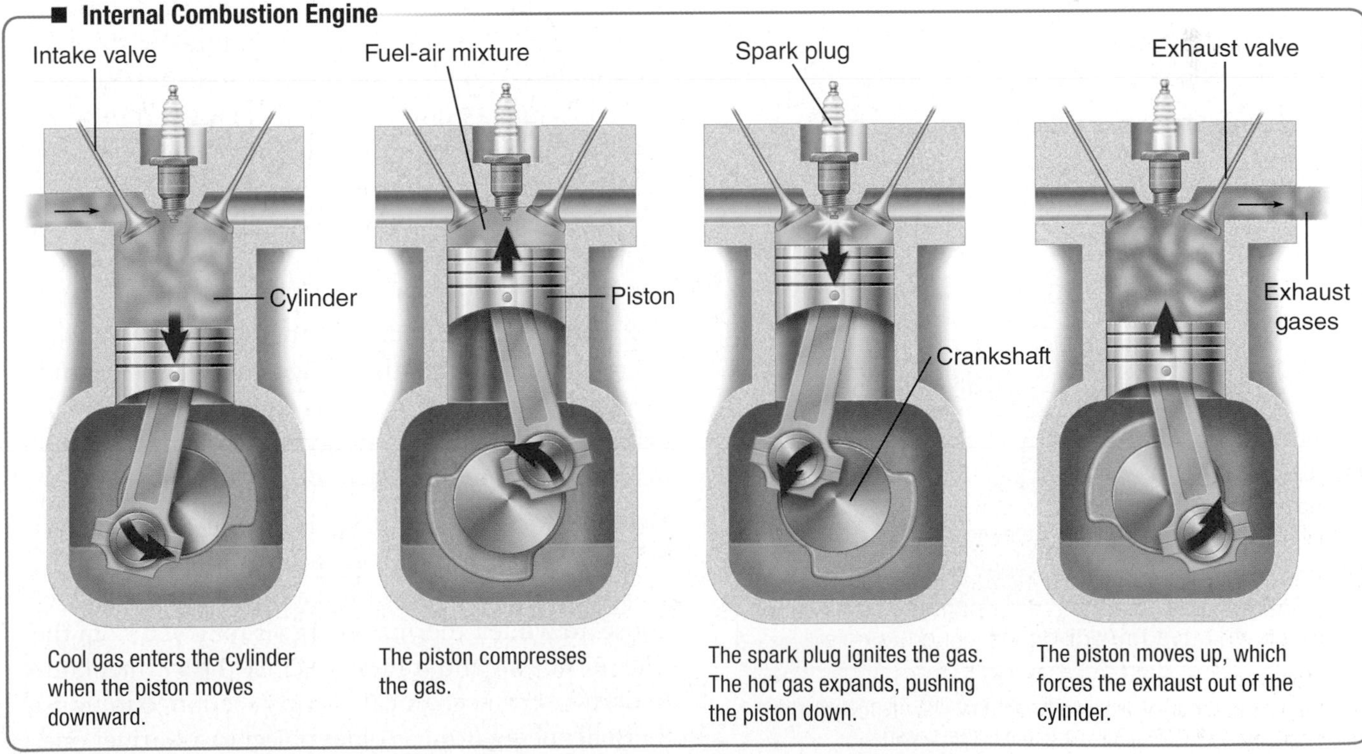
Internal Combustion Engine
Intake valve
Fuel-air mixture
Spark plug
Exhaust valve
Cylinder
Piston
Exhaust gases
Crankshaft
Cool gas enters the cylinder when the piston moves downward.
The piston compresses the gas.
The spark plug ignites the gas. The hot gas expands, pushing the piston down.
The piston moves up, which forces the exhaust out of the cylinder.

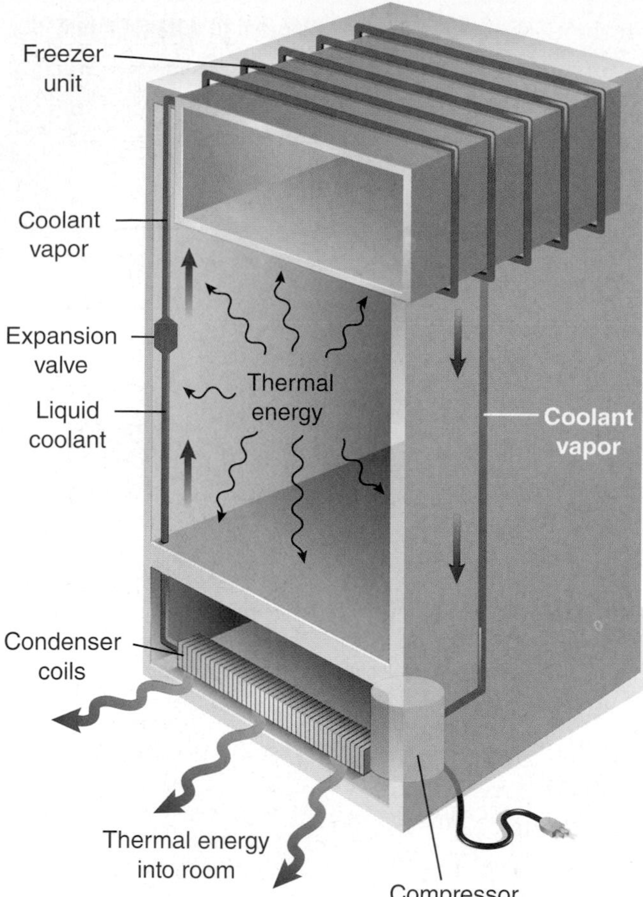

Figure 19 Liquid coolant is pumped into an expansion valve, where it absorbs energy from its surroundings and becomes a gas. The gas then heats up as it absorbs thermal energy from inside the refrigerator. A compressor does work on the gas to cool it to a liquid, and the cycle begins again.

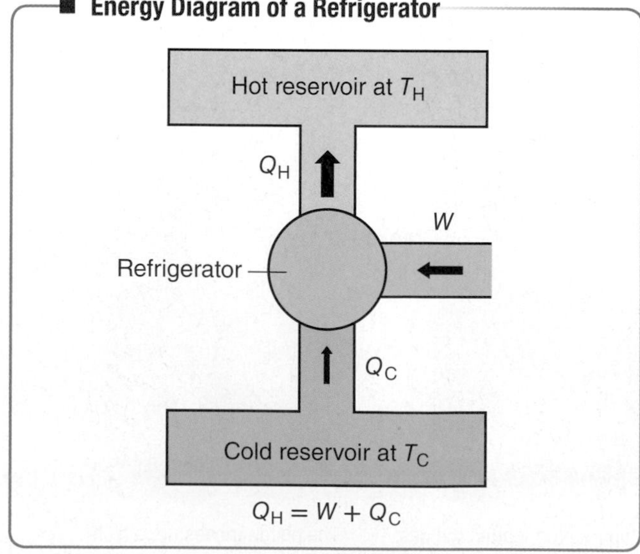

Figure 20 When work is done on the refrigerator, thermal energy is transferred from the cold reservoir to the hot reservoir.

Efficiency Engineers and car salespeople often discuss the fuel efficiencies of automobile engines. They are referring to the amount of the input heat (Q_H) that is turned into useful work (W). The actual efficiency of an engine is given by the ratio W/Q_H. If all the input heat could be turned into useful work by the engine, the engine would have an efficiency of 100 percent. Because there is always waste heat (Q_C), even the most efficient engines fall short of 100 percent efficiency.

☑ **READING CHECK** **Infer** Would a more efficient engine burn more fuel or less fuel for the same amount of work than a less efficient engine?

In fact, most heat engines are significantly less than 100 percent efficient. For example, even the most efficient automobile gasoline engines have an efficiency of less than 40 percent. A typical gasoline engine in an automobile has an efficiency that is closer to 20 percent.

A considerable amount of thermal energy transfers from a hot automobile engine to the cooler surroundings. Does any device transfer thermal energy from a cold object to warmer surroundings?

Refrigerators Thermal energy is transferred from a warm object to a cold object spontaneously. But it is also possible to remove thermal energy from a colder object and add it to a warmer object if work is done. A refrigerator, such as the one in **Figure 19,** is a common example of a device that accomplishes this transfer. Electrical energy runs a motor that does work on a gas and compresses it.

The gas draws thermal energy from the interior of the refrigerator, passes from the compressor through the condenser coils on the outside of the refrigerator, and cools into a liquid. Thermal energy is transferred into the air in the room. The liquid reenters the interior, vaporizes, and absorbs thermal energy from its surroundings. The gas returns to the compressor, and the process is repeated. The overall change in the thermal energy of the gas is zero. Thus, according to the first law of thermodynamics, the sum of the thermal energy removed from the refrigerator's contents and the work done by the motor is equal to the thermal energy expelled. These energy transfers and transformations are shown in **Figure 20.**

Heat pumps A heat pump is a refrigerator that can be run in two directions. In summer, the pump removes thermal energy from a house and cools the house. In winter, thermal energy is removed from the cold outside air and transferred into the warmer house. In both cases, mechanical energy is required to transfer thermal energy from a colder object to a warmer one.

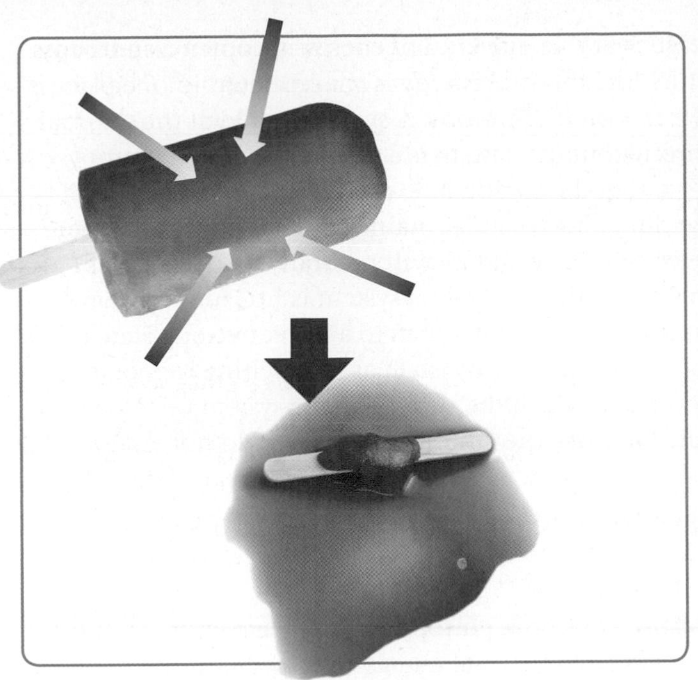

Figure 21 According to the second law of thermodynamics, thermal energy always spreads out. The red arrows represent thermal energy flow. Thermal energy spontaneously flows from a warmer object to a colder object.

(tl)Photodisc/Getty Images, (tr)Robbie Augspurger/Flickr/Getty Images, (bl)Martin Poole/Digital Vision/Getty Images, (br)Barry Wong/Photodisc/Getty Images

The Second Law of Thermodynamics

We take for granted that many daily events occur spontaneously only in one direction. You would be shocked if the reverse of the same events occurred spontaneously. For example, you are not surprised when a metal spoon, heated at one end, soon becomes uniformly hot. Consider your reaction, however, if a spoon lying on a table suddenly, on its own, became red hot at one end and icy cold at the other. This reverse process would not violate the first law of thermodynamics—the thermal energy of the spoon would remain the same. Many processes that are consistent with the first law of thermodynamics have never been observed to occur spontaneously. However, there is more to modeling thermal events than making sure energy is conserved.

Energy spreads out Examine the melting ice pop and the cooling pizza in **Figure 21.** The first law of thermodynamics does not prohibit net thermal energy transfers from the cold ice pop to the air or from the air to the hot pizza. This does not occur, however, because of the second law of thermodynamics. When a hot object is placed in contact with cooler surroundings, the thermal energy in the hot object has the opportunity to disperse, or spread out more. Some of the thermal energy moves into the cold object, warming it and therefore cooling the originally hotter object. The **second law of thermodynamics** states that whenever there is an opportunity for energy dispersal, the energy always spreads out.

Consider the cooling pizza. The particles in the pizza have a greater average kinetic energy than the particles in the air. Some of the pizza's original thermal energy disperses into the air. As a result, the pizza's temperature decreases and the air temperature increases a small amount. When the pizza and the air reach the same temperature, the average kinetic energy of the particles in the pizza and the air will be the same. That is, the energy spreads out among the particles. Similarly, if you leave the ice pop sitting on the counter, thermal energy from the air will be dispersed to the ice pop. The ice pop will heat up and melt, while the air will experience a small temperature decrease.

Investigate **kinetic theory and how energy spreads out.**

Virtual Investigation

Entropy The measure of this dispersal of energy is known as **entropy** (S). A system in which the thermal energy is concentrated in one place is referred to as a system with low entropy. A system in which the thermal energy is spread throughout the system is a system with high entropy.

Another way of stating the second law of thermodynamics is that natural processes go in a direction that maintains or increases the total entropy of the universe. That is, energy will naturally disperse unless some action is taken to localize it. Once a system is in a high-entropy state, it is highly unlikely that it will return to a lower entropy state on its own. Events that occur spontaneously, such as the melting ice pop or the cooling pizza, are events in which the entropy of the system increases. Processes that would decrease the entropy of a system do not tend to occur spontaneously but require work done by an external agent.

☑ **READING CHECK** **State** the second law of thermodynamics using the term *entropy.*

Entropy and heat engines How does entropy relate to heat engines? If heat engines completely converted thermal energy into mechanical energy with no waste heat, energy would still be conserved, and so the first law of thermodynamics would be obeyed. However, waste heat is always generated, dispersing thermal energy beyond the engine. In the nineteenth century, French engineer Sadi Carnot studied the ability of engines to convert thermal energy into mechanical energy. He developed a logical proof that even an ideal engine would generate some waste heat. Carnot's result was one of the first formal analyses leading to the development of the concept of entropy.

Changes in entropy Like energy, entropy is a property of a system. If thermal energy is added to a system, the entropy increases. If thermal energy is removed from a system, its entropy decreases. If a system does work on its surroundings without any transfer of thermal energy, the entropy does not change. For a reversible process, the change in entropy (ΔS) is expressed by the following equation, in which entropy has units of J/K and the temperature is constant and measured in kelvins.

CHANGE IN ENTROPY

For a reversible process, the change in entropy of a system is equal to the heat added to the system divided by the temperature of the system in kelvins.

$$\Delta S = \frac{Q}{T}$$

PHYSICS CHALLENGE

Entropy has some interesting properties. Calculate the change in entropy for the following situations. Explain how and why these changes in entropy are different from each other. For these small temperature changes, you can use the original temperature to find the change in entropy.

1. Heating 1.0 kg of water from 273 K to 274 K.

2. Heating 1.0 kg of water from 353 K to 354 K.

3. Heating 1.0 kg of lead from 273 K to 274 K.

4. Completely melting 1.0 kg of ice at 273 K.

Entropy and the energy crisis The second law of thermodynamics and the increase in entropy also give new meaning to what has been commonly called the energy crisis. The energy crisis refers to the continued use of limited resources such as natural gas and petroleum. When you use a resource, you do not use up the energy in the resource. For example, when you drive a car, such as those in **Figure 22,** the gas ignites and the chemical energy contained in the molecules of the gas is converted into the kinetic energy that runs the car and the thermal energy that heats the engine. Even if friction converts the car's kinetic energy into thermal energy, the energy is not lost or used up.

The entropy, however, has increased. The chemical energy of the unburned gas has dispersed into many more objects contained within a much larger volume. While it is mathematically possible for all of the dispersed energy to be brought back together in some one object, the probability of this happening is very near zero. For this reason, entropy often is used as a measure of the unavailability of useful energy. The energy in the warmed air in a home is not as available to do mechanical work or to transfer thermal energy to other objects as the energy in the original gas molecules was. The lack of usable energy is actually a surplus of entropy.

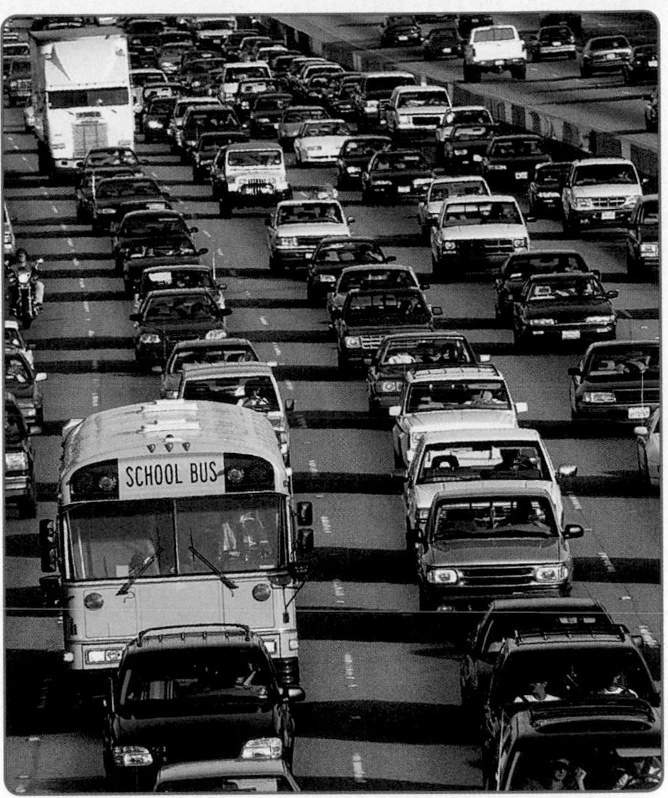

Figure 22 Burning gasoline uses up natural resources and increases entropy but does not use up energy. The energy is no longer in a useful form.

SECTION 2 REVIEW

Check your understanding.

29. MAINIDEA Describe the energy transformations and transfers made by a heat engine, and explain why operating a heat engine causes an increase in entropy.

30. Heat of Vaporization Old heating systems sent water vapor into radiators in each room of a house. In the radiators, the water vapor condensed to water. Analyze this process and explain how it heated a room.

31. Heat of Fusion How much thermal energy is needed to change 50.0 g of ice at −20.0°C to water at 10.0°C?

32. Heat of Vaporization How much energy is needed to heat 1.0 kg of mercury metal from 10.0°C to its boiling point (357°C) and vaporize it completely? For mercury, $C = 140$ J/kg•°C and $H_v = 3.06\times10^5$ J/kg.

33. Mechanical Energy and Thermal Energy A man uses a 320-kg hammer moving at 5.0 m/s to smash a 3.0-kg block of lead against a 450-kg rock. When he measured the temperature of the lead block he found that it had increased by 5.0°C. Explain how this happened.

34. Mechanical Energy and Thermal Energy James Joule carefully measured the difference in temperature of water at the top and the bottom of a waterfall. Why did he expect a difference?

35. Mechanical Energy and Thermal Energy For the waterfall in **Figure 23,** calculate the temperature difference between the water at the top and the bottom of the fall. Assume that the potential energy of the water is all converted to thermal energy.

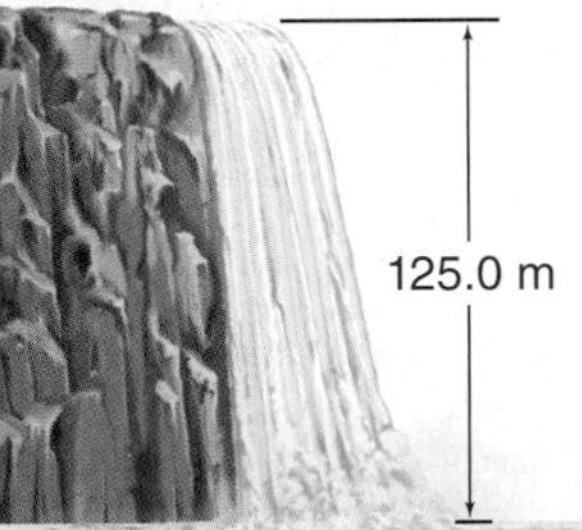

Figure 23

36. Entropy Evaluate why heating a home with natural gas results in increased entropy.

37. Critical Thinking Many outdoor amusement parks and zoos have systems that spray a fine mist of water, which evaporates quickly. Explain why this process cools the surrounding air.

2's Company, 3's a CROWD...

PHYSICS MODELS & PEDESTRIAN DYNAMICS

Have you ever been part of an extremely dense crowd at a concert or sports event? The forces can change direction quickly, pushing you along. You might feel helpless to move in any particular direction. Uncontrolled crowding leads to thousands of deaths every year, notably in sports stadiums and the annual Hajj pilgrimages to Mecca. In an effort to prevent such accidents, stadium managers, public safety officials, and architects are working with physicists to understand how crowds move. Understanding how crowds behave can lead to improvements in building design and crowd safety.

People as particles Physicists noticed that crowds often behave like many-particle systems. When the crowd's density is low, pedestrians can move freely and their motion resembles the behavior of gases. At medium and high densities, crowds resemble liquids. In addition, crowds pass through narrow openings in much the same way that granular materials, such as sand and salt, do. People bunched up near exits have an archlike structure characteristic of granular materials passing though an opening. By applying the same equations used to describe particle movement in homogenous gases, physicist found that they could describe and predict crowd movement.

Social forces Physicists realized, however, that a realistic model for pedestrians must include interactions not governed by Newton's laws. These interactions involve people's internal motivations and are called social forces. Even though human behavior often seems chaotic and unpredictable, behavioral conventions exist. For example, an individual's desire for personal space acts as a repulsive force. People also try to move at the same speed as the crowd and tend to form lanes. Including social forces can greatly improve a pedestrian model.

Future applications Currently, researchers in Germany are developing an evacuation assistant that uses live crowd data. This computerized pedestrian model can help stadium managers plan efficient and safe evacuations routes.

GOING FURTHER

Think Critically Study your school's evacuation plan. Suggest any ways it can be improved.

CHAPTER 12 STUDY GUIDE

BIGIDEA Thermal energy is related to the motion of an object's particles and can be transferred and transformed.

VOCABULARY
- **thermal conduction** *(p. 322)*
- **thermal equilibrium** *(p. 322)*
- **heat** *(p. 324)*
- **convection** *(p. 325)*
- **radiation** *(p. 325)*
- **specific heat** *(p. 325)*

SECTION 1 Temperature, Heat, and Thermal Energy

MAINIDEA Heat is a transfer of thermal energy that occurs spontaneously from a warmer object to a cooler object.

- Thermal energy is the sum of the kinetic and potential energies of an object's particles. An object's temperature is a measure of the average kinetic energy of its particles.
- When two objects are in thermal equilibrium, there is no net transfer of thermal energy between the objects and the two objects are at the same temperature. A thermometer measures temperature by reaching thermal equilibrium with its surroundings. When an object's temperature is at absolute zero, the average kinetic energy of its particles is zero and that object cannot transfer thermal energy.
- Heat is the transfer of thermal energy. Thermal energy is spontaneously transferred from a warm object to a cool object. Thermal energy is transferred by three processes: conduction, convection, and radiation.
- Substances heat differently, based on their specific heats. Specific heat (C) is the heat required to raise the temperature of 1 kg of a substance 1 K.

$$Q = mC\Delta T = mC(T_f - T_i)$$

A calorimeter is a closed system used to measure changes in thermal energies. Specific heat is calculated by using measurements from a calorimeter.

VOCABULARY
- **heat of fusion** *(p. 331)*
- **heat of vaporization** *(p. 331)*
- **first law of thermodynamics** *(p. 334)*
- **heat engine** *(p. 335)*
- **second law of thermodynamics** *(p. 337)*
- **entropy** *(p. 338)*

SECTION 2 Changes of State and Thermodynamics

MAINIDEA When thermal energy is transferred, energy is conserved and the total entropy of the universe will increase.

- Thermal energy transferred during a change of state does not change the temperature of a substance. The heat of fusion is the quantity of heat needed to change 1 kg of a substance from a solid state to a liquid state at its melting point.

$$Q = mH_f$$

The heat of vaporization is the quantity of heat needed to change 1 kg of a substance from a liquid state to a gaseous state at its boiling point.

$$Q = mH_v$$

- The first law of thermodynamics states that the change in the thermal energy of an object is equal to the heat added to the object minus the work done by the object.

$$\Delta U = Q - W$$

- A heat engine converts thermal energy to mechanical energy. A heat pump and a refrigerator use mechanical energy to transfer thermal energy from a region of lower temperature to one of higher temperature.
- The second law of thermodynamics states that whenever there is an opportunity for energy dispersal, the energy always spreads out. Entropy (S) is a measure of the energy dispersal of a system. The second law of thermodynamics indicates that natural processes go in a direction that maintains or increases the total entropy of the universe. The change in entropy of an object is defined as the heat added to the object divided by the object's temperature.

Games and Multilingual eGlossary

Vocabulary Practice

Chapter Self-Check

SECTION 1
Temperature, Heat, and Thermal Energy

Mastering Concepts

38. BIGIDEA Explain the differences among the mechanical energy of a ball, its thermal energy, and its temperature.

39. Can temperature be assigned to a vacuum? Explain.

40. Do all the molecules or atoms in a liquid have the same speed?

41. Is your body a good judge of temperature? On a cold winter day, a metal doorknob feels much colder to your hand than a wooden door does. Explain why this is true.

42. When thermal energy is transferred from a warmer object to a colder object it is in contact with, do the two have the same temperature changes?

Mastering Problems

43. How much thermal energy is needed to raise the temperature of 50.0 g of water from 4.5°C to 83.0°C?

44. Coffee Cup A glass coffee cup is at room temperature. It is then plunged into hot dishwater, as shown in **Figure 24.** If the temperature of the cup reaches that of the dishwater, how much heat does the cup absorb? Assume that the mass of the dishwater is large enough so that its temperature does not change appreciably.

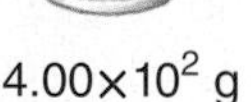

Figure 24

45. A 1.00×10^2-g mass of tungsten at 100.0°C is placed in 2.00×10^2 g of water at 20.0°C. The mixture reaches equilibrium at 21.2°C. Calculate the specific heat of tungsten.

46. A 6.0×10^2-g sample of water at 90.0°C is mixed with 4.00×10^2 g of water at 22.0°C. Assume that there is no thermal energy lost to the container or surroundings. What is the final temperature of the mixture?

47. A 5.00×10^2-g block of metal absorbs 5016 J of thermal energy when its temperature changes from 20.0°C to 30.0°C. Calculate the specific heat of the metal.

48. The kinetic energy of a compact car moving at 100 km/h is 2.9×10^5 J. To get an idea of the amount of energy needed to heat water, how many liters of water would 2.9×10^5 J of energy warm from room temperature (20.0°C) to boiling (100.0°C)?

49. Car Engine A 2.50×10^2-kg cast-iron car engine contains water as a coolant. Suppose that the engine's temperature is 35.0°C when it is shut off, and the air temperature is 10.0°C. The thermal energy given off by the engine and water in it as they cool to air temperature is 4.40×10^6 J. What mass of water is used to cool the engine?

50. Water Heater An electric immersion heater is used to heat a cup of water, as shown in **Figure 25.** The cup is made of glass and contains 250 g of water at 15°C. How much time is needed to bring the water to the boiling point? Assume that the temperature of the cup is the same as the temperature of the water at all times and that no thermal energy is lost to the air.

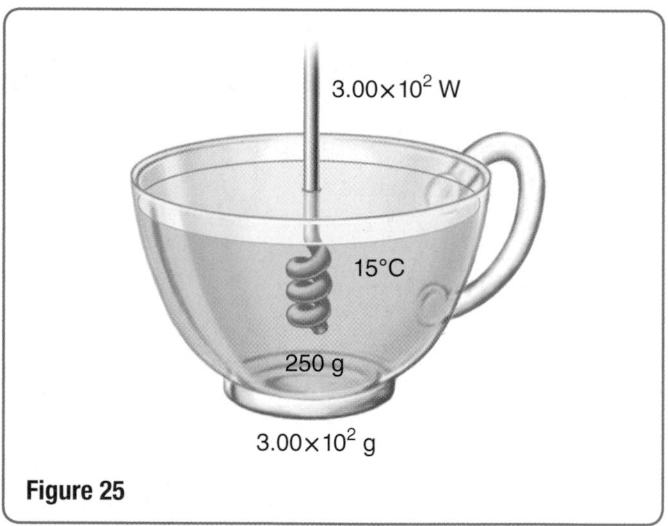

Figure 25

51. Ranking Task The following materials are each placed in identical containers holding equal amounts of room-temperature methanol. Rank the materials according to the amount of thermal energy they transfer to the methanol, from least to greatest. Specifically indicate any ties.

A. 50 g of aluminum at 30°C

B. 60 g of aluminum at 40°C

C. 50 g of glass at 30°C

D. 50 g of silver at 30°C

E. 50 g of zinc at 30°C

52. A piece of zinc at 71.0°C is placed in a container of water, as shown in **Figure 26.** What is the final temperature of the water and the zinc?

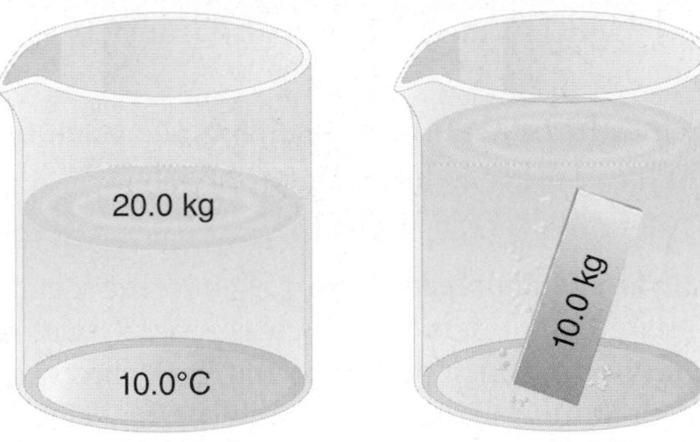

Figure 26

SECTION 2
Changes of State and Thermodynamics

Mastering Concepts

53. Can you add thermal energy to an object without increasing its temperature? Explain.

54. When wax freezes, does it absorb or release energy?

55. Explain why water in a canteen that is surrounded by dry air stays cooler if it has a canvas cover that is kept wet.

56. Which process occurs at the coils of a running air conditioner inside a house, vaporization or condensation? Explain.

Mastering Problems

57. Years ago, a block of ice with a mass of about 20.0 kg was used daily in a home icebox. The temperature of the ice was 0.0°C when it was delivered. As it melted, how much thermal energy did the ice absorb?

58. A 40.0-g sample of chloroform is condensed from a vapor to a liquid at 61.6°C. It releases 9870 J of thermal energy. What is the heat of vaporization of chloroform?

59. A 750-kg car moving at 23 m/s brakes to a stop. Assume that all the kinetic energy is transformed into thermal energy. The brakes contain about 15 kg of iron, which absorbs the energy. What is the increase in temperature of the brakes?

60. How much thermal energy is added to 10.0 g of ice at −20.0°C to convert it to water vapor at 120.0°C?

61. A 4.2-g lead bullet moving at 275 m/s strikes a steel plate and comes to a stop. If all its kinetic energy is converted to thermal energy and none leaves the bullet, what is its temperature change?

62. **Soft Drink** A soft drink from Australia is labeled *Low-Joule Cola.* The label says "100 mL yields 1.7 kJ." The can contains 375 mL of cola. Chandra drinks the cola and then wants to offset this input of food energy by climbing stairs. How high must Chandra climb if her mass is 65.0 kg?

Applying Concepts

63. **Cooking** Sally is cooking pasta in a pot of boiling water. Will the pasta cook faster if the water is boiling vigorously or if it is boiling gently?

64. Which liquid would an ice cube cool faster, water or methanol? Explain.

65. Equal masses of aluminum and lead are heated to the same temperature. The pieces of metal are placed on a block of ice. Which metal melts more ice? Explain.

66. Why do easily vaporized liquids, such as acetone and methanol, feel cool to the skin?

67. Explain why fruit growers spray their trees with water to protect the fruit from freezing when frost is expected.

68. Two blocks of lead have the same temperature. Block A has twice the mass of block B. They are dropped into identical cups of water of equal temperatures. Will the two cups of water have equal temperatures after equilibrium is achieved? Explain.

69. **Windows** Often architects design most windows of a house on the north side. How does this affect the heating and cooling of the house?

Mixed Review

70. What is the efficiency of an engine that outputs 1800 J/s while burning gasoline to produce 5300 J/s? How much waste heat does the engine produce per second?

71. **Stamping Press** A metal stamping machine in a factory does 2100 J of work each time it stamps out a piece of metal. Assume that the work changes only the metal's thermal energy. Each stamped piece is then dipped in a 32.0-kg vat of water for cooling. By how many degrees does the vat heat up each time a piece of stamped metal is dipped into it?

72. **Problem Posing** Complete this problem so that it must be solved using the concepts listed below: "A beaker of water has a temperature of 35°C … ."

a. specific heat

b. entropy

Chapter Self-Check

73. A 1500-kg automobile comes to a stop from 25 m/s. All the energy of the automobile is deposited in the brakes. Assuming that the brakes are about 45 kg of aluminum, what is the change in temperature of the brakes?

74. Iced Tea To make iced tea, you brew the tea with hot water and then add ice. If you start with 1.0 L of 90°C tea, how much ice is needed to cool it to 0°C? Would it be better to let the tea cool to room temperature before adding the ice?

75. A block of copper comes in contact with a block of aluminum, and they come to thermal equilibrium, as shown in **Figure 27.** What are the relative masses of the blocks?

100.0°C Copper	20.0°C Aluminum
60.0°C Copper	60.0°C Aluminum

Figure 27

76. Two copper blocks, each with a mass of 0.35 kg, slide toward each other at the same speed and collide. The two blocks come to a stop together after the collision. Their temperatures increase by 0.20°C as a result of the collision. Assume that all kinetic energy is transformed into thermal energy. What was their speed before the collision?

77. A 2.2-kg block of ice slides across a rough floor. Its initial velocity is 2.5 m/s, and its final velocity is 0.50 m/s. How much of the ice block melted as a result of the work done by friction?

Thinking Critically

78. Analyze and Conclude Chemists use calorimeters to measure the heat produced by chemical reactions. Suppose a chemist dissolves 1.0×10^{22} molecules of a powdered substance into a calorimeter containing 0.50 kg of water. The molecules break up and release their binding energy to the water. The water temperature increases by 2.3°C. What is the binding energy per molecule for this substance?

79. Reverse Problem Write a physics problem with real-life objects for which the following equation would be part of the solution:

$$75\ \text{J/K} = \frac{m(4180\ \text{J/(kg·K)})\ (260\ \text{K} - 250\ \text{K})}{250\ \text{K}}$$

80. Analyze and Conclude A certain heat engine removes 50.0 J of thermal energy from a hot reservoir at temperature $T_H = 545$ K and expels 40.0 J of thermal energy to a colder reservoir at temperature $T_C = 325$ K. In the process, it also transfers entropy from one reservoir to the other.

a. Find the total entropy change of the reservoirs.

b. What would be the total entropy change in the reservoirs if $T_C = 205$ K?

81. Analyze and Conclude During a game, the metabolism of basketball players often increases by as much as 30.0 W. How much perspiration must a player vaporize per hour to dissipate this extra thermal energy?

82. Apply Concepts All energy on Earth comes from the Sun. The Sun's surface temperature is approximately 10^4 K. What would be the effect on Earth if the Sun's surface temperature were 10^3 K?

Writing in Physics

83. Our understanding of the relationship between heat and energy was influenced by a brewer named James Prescott Joule and a soldier named Benjamin Thompson, Count Rumford. Investigate what experiments they did and evaluate whether it is fair that the unit of energy is called the joule and not the thompson.

84. Water has an unusually large specific heat and large heats of fusion and vaporization. The weather and ecosystems depend upon water in all three states. How would the world be different if water's thermodynamic properties were like other materials, such as methanol?

Cumulative Review

85. A rope is wound around a drum with a radius of 0.250 m and a moment of inertia of 2.25 kg·m^2. The rope is connected to a 4.00-kg block. Find the linear acceleration of the block. Find the angular acceleration of the drum. Find the tension (F_T) in the rope. Find the angular velocity of the drum after the block has fallen 5.00 m.

86. A weight lifter raises a 180-kg barbell to a height of 1.95 m. How much work is done by the weight lifter in lifting the barbell?

87. In a Greek myth, Sisyphus is fated to forever roll a huge rock up a hill. Each time he reaches the top, the rock rolls back to the bottom. If the rock has a mass of 215 kg, the hill is 33 m in height, and Sisyphus produces an average power of 0.2 kW, how many times in 1 h can he roll the rock up the hill?

MULTIPLE CHOICE

Use the following information as needed.

$C_{ice} = 2060$ J/(kg•K)
$C_{water} = 4180$ J/(kg•K)
$C_{water\ vapor} = 2020$ J/(kg•K)
$H_{f\ water} = 3.34\times10^5$ J/kg
$H_{v\ water} = 2.26\times10^6$ J/kg

1. Which temperature conversion is incorrect?

A. $-273°C = 0$ K
B. $273°C = 546$ K
C. 298 K $= 571°C$
D. 88 K $= -185°C$

2. What are the units of entropy?

A. J/K
B. K/J
C. J
D. kJ

3. Which statement about two objects in thermal equilibrium is false?

A. Energy exchange between the objects continues to occur.
B. The net flow of energy between the objects is zero.
C. The objects are at the same temperature.
D. There is a net flow of energy from one object to the other.

4. How much heat is needed to warm 363 mL of water from 24°C to 38°C?

A. 21 kJ
B. 36 kJ
C. 121 kJ
D. 820 kJ

5. In the figure below, 81 g of ice melts and warms to 10°C. How much thermal energy is absorbed from the surroundings when this occurs?

A. 0.34 kJ
B. 27 kJ
C. 30 kJ
D. 190 kJ

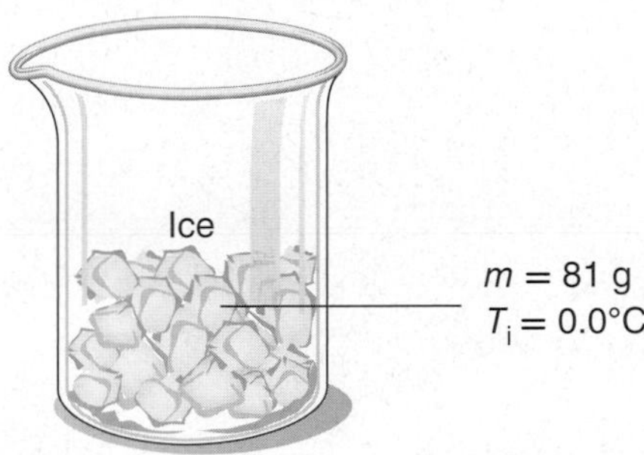

Online Test Practice

6. How much heat is required to heat 87 g of ice at 14 K to water vapor at 140°C?

A. 45 kJ
B. 58 kJ
C. 315 kJ
D. 280 kJ

7. You do 0.050 J of work on the coffee in your cup each time you stir it. What is the increase in entropy in 125 mL of coffee at 65°C when you stir it 85 times?

A. 0.013 J/K
B. 0.050 J
C. 0.095 J/K
D. 4.2 J

8. Why is there always some waste heat in a heat engine?

A. The entropy decreases at each stage.
B. The engine is not as efficient as it could be.
C. The entropy increases at each stage.
D. The energy is being used up.

9. Which statement about entropy and energy is true?

A. When ice freezes, it gives off energy and its entropy increases.
B. When ice freezes, it gives off energy and its entropy decreases.
C. When ice freezes, it absorbs energy and its entropy increases.
D. When ice freezes, it absorbs energy and its entropy decreases.

FREE RESPONSE

10. What is the difference in heat required to melt 454 g of ice at 0.00°C and to turn 454 g of water at 100.0°C into water vapor? Is the amount of this difference greater or less than the amount of energy required to heat the 454 g of water from 0.00°C to 100.0°C?

NEED EXTRA HELP?

If You Missed Question	1	2	3	4	5	6	7	8	9	10
Review Section	1	2	1	1	2	2	2	2	2	2

CHAPTER 13

States of Matter

BIGIDEA The thermal energy of a material and the forces between that material's particles determine its properties.

SECTIONS

1 Properties of Fluids

2 Forces Within Liquids

3 Fluids at Rest and in Motion

4 Solids

LaunchLAB

iLab Station

MEASURING BUOYANCY

How does the density of an object affect how it floats?

WATCH THIS!

PLASMA

You've probably explored the three main states of matter, but how much experience do you have with plasma? You might be surprised!

CORBIS

Go online!
connectED.mcgraw-hill.com

SECTION 1

Properties of Fluids

You might not notice it very often, but gases in the atmosphere are exerting pressure on your body. If you've ever ridden in an elevator in a tall building, been to the top of a mountain, or flown in an airplane, you might have had your ears pop. Your ears pop to help balance the changing pressure between the inside and the outside of your ear.

MAIN IDEA

Fluids flow, have no definite shape, and include liquids, gases, and plasmas.

Essential Questions

- What is a fluid?
- What are the relationships among the pressure, volume, and temperature of a gas?
- What is the ideal gas law?
- What is plasma?

Review Vocabulary

linear relationship a relationship in which the dependent variable varies linearly with the independent variable

New Vocabulary

fluid
pressure
pascal
combined gas law
ideal gas law
thermal expansion
plasma

Liquids and Gases

Water and air are two of the most common substances in the everyday lives of people. We feel their effects when we drink, when we bathe, and literally with every breath we take. Although you might not think about it every day, water and air have a great deal in common. Both water and air flow, and unlike solids, neither one of them has a definite shape. Gases and liquids are two states of matter in which atoms and molecules have great freedom to move. In this chapter, you will learn about the principles that explain how liquids and gases respond to changes in temperature and pressure, how hydraulic systems can multiply applied forces, and how huge metallic ships can float on water.

Fluids Study the ice chunks in the lake in **Figure 1.** Like ice cubes, these chunks have a certain mass and a certain shape, and neither of these quantities depends on the size or shape of the lake basin. What happens, however, when the ice melts? Its mass remains the same, but its shape changes. The water flows to take the shape of the basin and forms a definite, flat, upper surface. As evaporation occurs, the liquid water changes into a gas in the form of water vapor. Like liquid water, water vapor flows and does not have any definite shape. Both liquids and gases are fluids. **Fluids** are materials that can flow and have no definite shape of their own.

Figure 1 The chunks of ice in this lake, which are solids, have definite shapes. However, the liquid water, a fluid, takes the shape of the lake basin.

Identify What fluid is filling the space above the water?

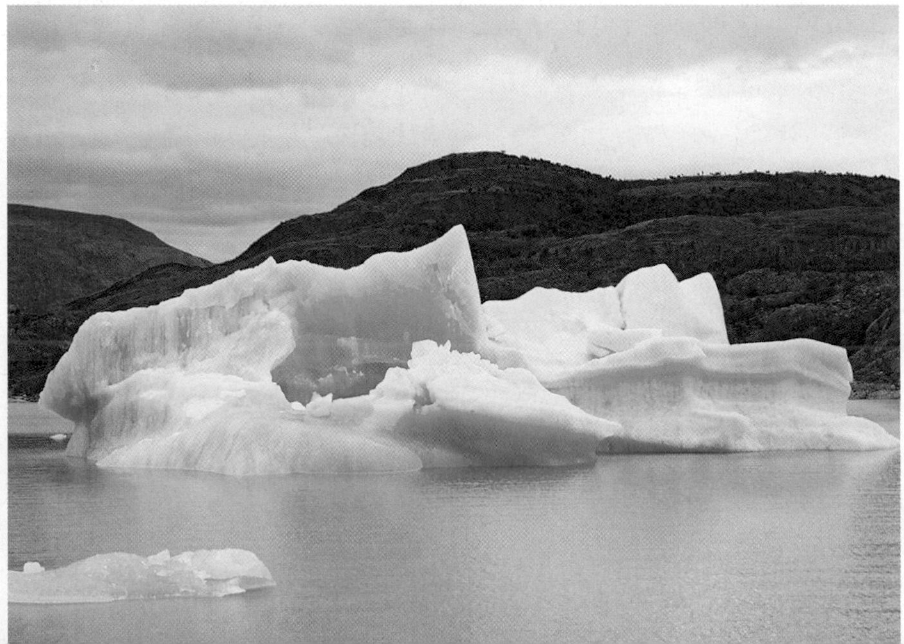

(t)Joaquin Palting/Getty Images, (b)Glow Images/Superstock

Pressure

When considering fluids (as well as solids), it is often useful to think about pressure as well as force. You have probably heard people talk about water pressure and air pressure, and you might already have a sense that pressure and force are related. Pressure and force are not the same, however. **Pressure** is the perpendicular component of a force on a surface divided by the area of the surface. Since pressure is force exerted on a surface, anything that exerts pressure is capable of producing change and doing work. In **Figure 2,** both the astronaut and the legs of the landing module are exerting pressure on the Moon's surface.

PRESSURE

Pressure equals the perpendicular component of the force divided by the surface area to which it is applied.

$$P = \frac{F}{A}$$

Pressure is a scalar. In the SI system, the unit of pressure is the **pascal** (Pa), which is 1 N/m^2. One pascal is a very small amount of pressure, about equal to the pressure that a flat dollar bill exerts on a tabletop. Thus the kilopascal (kPa), equal to 1000 Pa, is usually used. **Table 1** shows pressures in various locations.

Figure 2 The landing module and the astronaut both exert pressure on the surface of the Moon.

Calculate If the lunar module weighed approximately 12,000 N and rested on four pads that were each 91 cm in diameter, what pressure did it exert on the Moon's surface? How could you estimate the pressure exerted by the astronaut?

Table 1 Some Typical Pressures

Location	Pressure (kPa)
The center of Earth	4×10^{8}
The deepest ocean trench	1.1×10^{5}
Standard atmosphere	1.01325×10^{2}
Blood pressure	1.6×10^{1}
Air pressure on top of Mt. Everest	3×10^{1}
The best vacuum	1×10^{-10}

Solids, liquids, and pressure Imagine you are standing on the surface of a frozen lake. The forces your feet exert on the ice are spread over the area of your shoes, resulting in pressure on the ice. Ice is a solid that is made up of vibrating water molecules, and the forces that hold the water molecules in place cause the ice to exert upward forces on your feet that equal your weight. If the ice melted, most of the bonds between the water molecules would be weakened. Although the molecules would continue to vibrate and remain close to each other, they also would slide past one another, and you would break through the surface. The moving water molecules would continue to exert forces on your body.

Gases and pressure The pressure exerted by a gas can be understood by applying the kinetic-molecular theory of gases, which explains the properties of an ideal gas. In this model, particles are treated as taking up no space and having no intermolecular attractive forces. In spite of the fact that particles of a real gas take up space and exert attractive forces, an ideal gas is an accurate model of a real gas under most conditions.

NASA's Goddard Space Flight Center

According to the kinetic-molecular theory, the particles in a gas are in random motion at high speeds and colliding elastically with each other. When a gas particle hits a container's surface, it rebounds, which changes its momentum. The impulses exerted by these collisions result in gas pressure on the surface.

Atmospheric pressure At sea level, gases of the atmosphere exert a force in all directions of approximately 10 N, about the weight of a 1-kg object, on each centimeter of surface area. This atmospheric pressure on your body is so well balanced by your body's outward forces that you seldom notice it. You probably become aware of this pressure only when your ears pop as the result of pressure changes, as when you ride an elevator in a tall building or fly in an airplane. Atmospheric pressure is about 10 N per 1 cm^2 (10^{-4} m^2), which is about 1.0×10^5 N/m^2, or 100 kPa. Other planets in our solar system also have atmospheres that exert pressure. For example, the pressure at the surface of Venus is about 92 times that at Earth's surface.

EXAMPLE PROBLEM 1

Find help with **dimensional calculations.** Math Handbook

EXAMPLE PROBLEM

CALCULATING PRESSURE A child weighs 364 N and sits on a three-legged stool, which weighs 41 N. The bottoms of the stool's legs touch the ground over a total area of 19.3 cm^2.

a. What is the average pressure that the child and the stool exert on the ground?

b. How does the pressure change when the child leans over so that only two legs of the stool touch the floor?

1 ANALYZE AND SKETCH THE PROBLEM

- Sketch the child and the stool, labeling the total force that they exert on the ground.
- List the variables, including the force that the child and the stool exert on the ground and the areas for parts **a** and **b**.

KNOWN

$F_{g\ child} = 364$ N

$F_{g\ stool} = 41$ N

$F_{g\ total} = F_{g\ child} + F_{g\ stool}$
$= 364\text{ N} + 41\text{ N}$
$= 405\text{ N}$

$A_a = 19.3\text{ cm}^2$

$A_b = \frac{2}{3} \times 19.3\text{ cm}^2$
$= 12.9\text{ cm}^2$

UNKNOWN

$P_a = ?$

$P_b = ?$

2 SOLVE FOR THE UNKNOWN

Find each pressure.

$$P = \frac{F}{A}$$

a. $P_a = \left(\frac{405\text{ N}}{19.3\text{ cm}^2}\right)\left(\frac{(100\text{ cm})^2}{(1\text{ m})^2}\right)$ ◀ Substitute $F = F_{g\ total} = 405$ N, $A = A_A = 19.3$ cm^2

$= 2.10\times10^2$ kPa

b. $P_b = \left(\frac{405\text{ N}}{12.9\text{ cm}^2}\right)\left(\frac{(100\text{ cm})^2}{(1\text{ m})^2}\right)$ ◀ Substitute $F = F_{g\ total} = 405$ N, $A = A_B = 12.9$ cm^2

$= 3.14\times10^2$ kPa

3 EVALUATE THE ANSWER

Are the units correct? The units for pressure should be Pa, and 1 N/m^2 = 1 Pa.

PRACTICE PROBLEMS

Do additional problems. Online Practice

1. The atmospheric pressure at sea level is about 1.0×10^5 Pa. What is the force at sea level that air exerts on the top of a desk that is 152 cm long and 76 cm wide?
2. A car tire makes contact with the ground on a rectangular area of 12 cm by 18 cm. If the car's mass is 925 kg, what pressure does the car exert on the ground as it rests on all four tires?
3. A lead brick, 5.0 cm × 10.0 cm × 20.0 cm, rests on the ground on its smallest face. Lead has a density of 11.8 g/cm³. What pressure does the brick exert on the ground?
4. Suppose that during a storm, the atmospheric pressure suddenly drops by 15 percent outside. What net force would be exerted on a front door to a house that is 195 cm high and 91 cm wide? In what direction would this force be exerted?
5. **CHALLENGE** Large pieces of industrial equipment are placed on wide steel plates that spread the weight of the equipment over larger areas. If an engineer plans to install a 454-kg device on a floor that is rated to withstand additional pressure of 5.0×10^4 Pa, how large should the steel support plate be?

The Gas Laws

Think about a container of gas that is held at a constant temperature. If you reduced the volume, what would happen to the pressure of the gas? There would be more collisions between the particles and the container's walls, and so the pressure would increase. Similarly, if you increased the volume, there would be fewer collisions, decreasing the pressure. This inverse relationship was found by seventeenth-century chemist and physicist Robert Boyle. Because the product of inversely related variables is a constant, Boyle's law can be written $PV =$ constant, or $P_1V_1 = P_2V_2$. The subscripts that you see in the gas laws help keep track of different variables, such as pressure and volume, as they change throughout a problem. The relationship between the pressure and the volume of a gas is critical to the scuba diver in **Figure 3.**

About 100 years after Boyle's work, Jacques Charles cooled a gas and found that the volume shrank by $\frac{1}{273}$ of its original volume for every degree cooled, which is a linear relationship. At the time, Charles could not cool gases to the extremely low temperatures achieved in modern laboratories. In order to see what lower limits might be possible, he extended, or extrapolated, the graph of his data to these temperatures. This extrapolation suggested that if the temperature were reduced to –273°C, a gas would have zero volume. The temperature at which a gas would have zero volume is now called absolute zero, which is represented by the zero of the Kelvin temperature scale.

These experiments indicated that under constant pressure, the volume of a sample of gas varies directly with its Kelvin temperature, a result that is now called Charles's law. Charles's law can be written $\frac{V}{T} =$ constant, or $\frac{V_1}{T_1} = \frac{V_2}{T_2}$.

Combined gas law Combining Boyle's law and Charles's law relates the pressure, temperature, and volume of a fixed amount of ideal gas, which leads to the **combined gas law.**

Watch a **video on Charles's law.**

COMBINED GAS LAW

For a fixed amount of an ideal gas, the pressure times the volume, divided by the Kelvin temperature equals a constant.

$$\frac{P_1V_1}{T_1} = \frac{P_2V_2}{T_2} = \text{constant}$$

Figure 3 The gas in the tank on the diver's back is at high pressure. The regulator in the diver's mouth reduces the pressure, making the pressure of the gas the diver breathes equal to that of the water pressure. Bubbles are emitted from the regulator as the diver exhales.

Figure 4 The combined gas law shows the relationship among pressure, temperature, and volume of a fixed amount of an ideal gas. Boyle's law and Charles's law can each be derived from the combined gas law under certain conditions.

Explain What happens if you hold volume constant?

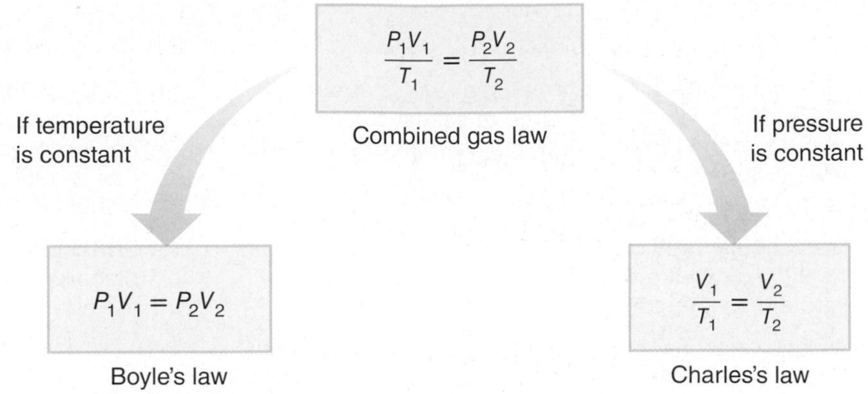

$$\frac{P_1V_1}{T_1} = \frac{P_2V_2}{T_2}$$

Combined gas law

If temperature is constant

$$P_1V_1 = P_2V_2$$

Boyle's law

If pressure is constant

$$\frac{V_1}{T_1} = \frac{V_2}{T_2}$$

Charles's law

As shown in **Figure 4,** the combined gas law reduces to Boyle's law under conditions of constant temperature and to Charles's law under conditions of constant pressure.

View an **animation of the ideal gas law.**

The ideal gas law What might the constant in the combined gas law depend on? Suppose the volume and temperature of an ideal gas are held constant while the number of particles (N) is increased. What happens to the pressure? The particles will have more collisions with the sides of the container, thus increasing the pressure. Removing particles decreases the number of collisions, and thus, decreases the pressure. Therefore, the constant in the combined gas law equation is proportional to N.

$$\frac{PV}{T} = kN$$

The constant (k) is called Boltzmann's constant, and its value is 1.38×10^{-23} Pa•m^3/K. In any practical application, the number of particles (N) is very large. Instead of using N, scientists often use a unit called a mole. One mole (abbreviated mol) is similar to one dozen, except that instead of representing 12 items, one mole represents 6.022×10^{23} particles. This number is called Avogadro's number, after Italian scientist Amedeo Avogadro.

Avogadro's number is numerically equal to the number of particles in a sample of matter whose mass equals the molar mass of the substance. You can use this relationship to convert between mass and n, the number of moles present. Using moles instead of the number of particles, however, changes Boltzmann's constant. This new constant is abbreviated R, and it has the value 8.31 Pa•m^3/(mol•K). Rearranging, you can write the ideal gas law in its most familiar form. The **ideal gas law** states that for an ideal gas, the pressure times the volume is equal to the number of moles multiplied by the constant R and the Kelvin temperature.

MiniLAB

PRESSURE

How much pressure do you exert when standing on one foot?

IDEAL GAS LAW

For an ideal gas, the pressure times the volume is equal to the number of moles times the constant R times the temperature.

$$PV = nRT$$

Note that with the given value of R, volume must be expressed in cubic meters, temperature in kelvin, and pressure in pascals. In practice, the ideal gas law predicts the behavior of gases remarkably well, except under conditions of high pressures or low temperatures.

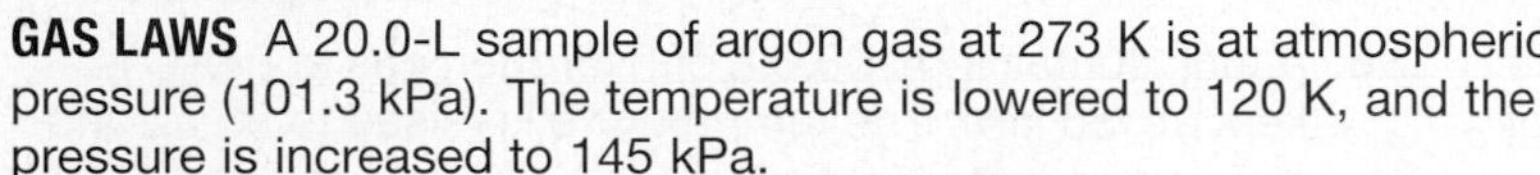

Get help with **gas laws.** Personal Tutor

EXAMPLE PROBLEM

GAS LAWS A 20.0-L sample of argon gas at 273 K is at atmospheric pressure (101.3 kPa). The temperature is lowered to 120 K, and the pressure is increased to 145 kPa.

a. What is the new volume of the argon sample?

b. Find the number of moles of argon atoms in the argon sample.

c. Find the mass of the argon sample. The molar mass (M) of argon is 39.9 g/mol.

1 ANALYZE AND SKETCH THE PROBLEM

- Sketch the situation. Indicate the conditions in the container of argon before and after the change in temperature and pressure.
- List the known and unknown variables.

KNOWN	UNKNOWN
$V_1 = 20.0$ L	$V_2 = ?$
$P_1 = 101.3$ kPa	moles of argon = ?
$T_1 = 273$ K	mass of argon sample = ?
$P_2 = 145$ kPa	
$T_2 = 120$ K	
$R = 8.31$ Pa•m^3/(mol•K)	
$M_{argon} = 39.9$ g/mol	

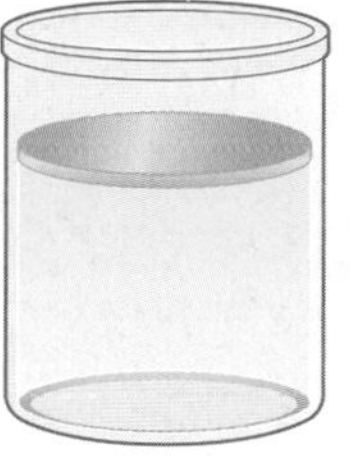
$T_1 = 273$ K
$P_1 = 101.3$ kPa
$V_1 = 20.0$ L

$T_2 = 120$ K
$P_2 = 145$ kPa
$V_2 = ?$

2 SOLVE FOR THE UNKNOWN

a. Use the combined gas law and solve for V_2.

$$\frac{P_1V_1}{T_1} = \frac{P_2V_2}{T_2}$$

$$V_2 = \frac{P_1V_1T_2}{P_2T_1}$$

$$= \frac{(101.3 \text{ kPa})(20.0 \text{ L})(120 \text{ K})}{(145 \text{ kPa})(273 \text{ K})}$$

◀ Substitute $P_1 = 101.3$ kPa, $P_2 = 145$ kPa, $V_1 = 20.0$ L, $T_1 = 273$ K, $T_2 = 120$ K.

$$= 6.1 \text{ L}$$

b. Use the ideal gas law, and solve for n.

$$PV = nRT$$

$$n = \frac{PV}{RT}$$

$$= \frac{(101.3\times10^3 \text{ Pa})(0.0200 \text{ m}^3)}{(8.31 \text{ Pa·m}^3/(\text{mol·K}))(273 \text{ K})}$$

◀ Substitute $P = 101.3\times10^3$ Pa, $V = 0.0200$ m^3, $R = 8.31$ m^3/(mol•K), $T = 273$ K.

$$= 0.893 \text{ mol}$$

c. Use the molar mass to convert from moles of argon in the sample to mass of the sample.

$$m = Mn$$

$$m_{\text{argon sample}} = (39.9 \text{ g/mol})(0.893 \text{ mol})$$

◀ Substitute $M = 39.9$ g/mol, $n = 0.893$ mol.

$$= 35.6 \text{ g}$$

3 EVALUATE THE ANSWER

- **Are the units correct?** The volume (V_2) is in liters, and the mass of the sample is in grams.
- **Is the magnitude realistic?** The change in volume is consistent with an increase in pressure and a decrease in temperature. The calculated mass of the argon sample is reasonable.

PRACTICE PROBLEMS

Do additional problems. Online Practice

6. A tank of helium gas used to inflate toy balloons is at a pressure of 15.5×10^6 Pa and a temperature of 293 K. The tank's volume is 0.020 m^3. How large a balloon would it fill at 1.00 atmosphere and 323 K?
7. What is the mass of the helium gas in the previous problem? The molar mass of helium gas is 4.00 g/mol.
8. A tank containing 200.0 L of hydrogen gas at 0.0°C is kept at 156 kPa. The temperature is raised to 95°C, and the volume is decreased to 175 L. What is the new pressure of the gas?
9. **CHALLENGE** The average molar mass of the components of air (mainly diatomic nitrogen gas and diatomic oxygen gas) is about 29 g/mol. What is the volume of 1.0 kg of air at atmospheric pressure and 20.0°C?

Thermal Expansion

As you applied the combined gas law, you discovered how gases expand as their temperatures increase. **Thermal expansion** is a property of all forms of matter that causes the matter to expand, becoming less dense, when heated. Thermal expansion has many useful applications, such as circulating air in a room.

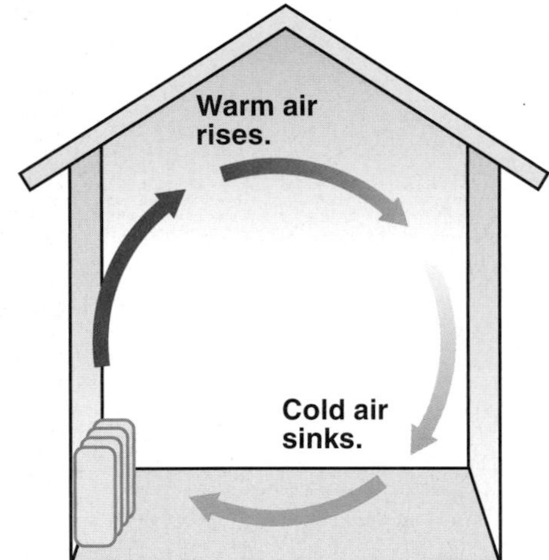

Figure 5 Convection currents occur as warmer, less dense air rises and cooler, denser air sinks.

Convection currents **Figure 5** shows that when the air near the radiator of a room is heated, it becomes less dense and, therefore rises. Gravity pulls the denser, colder air near the ceiling down. The cold air is subsequently warmed by the radiator, and air continues to circulate. This circulation of air within a room is called a convection current. Convection currents also occur in a pot of hot, but not boiling, water on a stove. When the pot is heated from the bottom, the colder and denser water sinks to the bottom where it is warmed and then pushed up by the continuous flow of cooler water from the top.

This thermal expansion occurs in most fluids. A good model for all liquids does not exist, but it is useful to think of a liquid as a finely ground solid. Groups of two, three, or more particles move together as if they were tiny pieces of a solid. When a liquid is heated, particle motion causes these groups to expand in the same way that particles in a solid are pushed apart. The spaces between groups increase. As a result, the whole liquid expands. With an equal change in temperature, liquids expand considerably more than solids but not as much as gases.

☑ **READING CHECK** **Explain** the role of thermal expansion in the formation of a convection current.

Watch a **video on water properties.**

Why ice floats Because matter expands as it is heated, you might predict that ice would be more dense than water, and therefore, it should sink. However, when water is heated from 0°C to 4°C, instead of expanding, it contracts as the forces between particles increase and the ice crystals collapse. These forces between water molecules are strong, and the crystals that make up ice have a very open structure. Even when ice melts, tiny crystals remain. These remaining crystals are melting, and the volume of the water decreases until the temperature reaches 4°C. However, once the temperature of water moves above 4°C, its volume increases because of greater molecular motion. The practical result is that water is most dense at 4°C and ice floats. This unique property of water is very important to our lives and environment. If ice sank, lakes would freeze from the bottom each winter and many would never melt completely in the summer.

Figure 6 Plasma emits light as it conducts electricity. The color produced by glowing plasma depends on the gas inside the tube.

Plasma

If you heat a solid, it melts to form a liquid. Further heating results in a gas. What happens if you increase the temperature still further? Collisions between the particles become violent enough to tear the electrons off the atoms, thereby producing positively charged ions. The gaslike state of negatively charged electrons and positively charged ions is called **plasma.** Plasma is considered to be another state of matter.

The plasma state may seem to be uncommon, but plasma is actually the most common state of matter in the universe. Stars consist mostly of plasma at extremely high temperatures. Much of the matter between stars and galaxies consists of energetic hydrogen that has no electrons. This hydrogen is in the plasma state. The primary difference between gas and plasma is that plasmas can conduct an electric current, whereas gases cannot. Lightning bolts are in the plasma state. Neon signs, such as the one shown in **Figure 6,** contain plasma. The fluorescent bulbs that probably light your school also contain plasma.

SECTION 1 REVIEW

Section Self-Check Check your understanding.

10. MAINIDEA Compare and contrast liquids, gases, and plasmas.

11. Pressure and Force Two boxes are each suspended by thin strings in midair. One is 20 cm × 20 cm × 20 cm. The other is 20 cm × 20 cm × 40 cm.

a. How does the pressure of the air on the outside of the two boxes compare?

b. How does the magnitude of the total force of the air on the two boxes compare?

12. Meteorology A weather balloon used by meteorologists is made of a flexible bag that allows the gas inside to freely expand. If a weather balloon containing 25.0 m^3 of helium gas is released from sea level, what is the volume of gas when the balloon reaches a height of 2100 m, where the pressure is 0.82×10^5 Pa? Assume the temperature is unchanged.

13. Density and Temperature Starting at 0°C, how will the density of water change as it is heated to 4°C? To 8°C?

14. Gas Compression In a certain internal-combustion engine, 0.0021 m^3 of air at atmospheric pressure and 303 K is rapidly compressed to a pressure of 20.1×10^5 Pa and a volume of 0.0003 m^3. What is the final temperature of the compressed gas?

15. The Standard Molar Volume What is the volume of 1.00 mol of a gas at atmospheric pressure and a temperature of 273 K?

16. The Air in a Refrigerator How many moles of air are in a refrigerator with a volume of 0.635 m^3 at a temperature of 2.00°C? If the average molar mass of air is 29 g/mol, what is the mass of the air in the refrigerator?

17. Critical Thinking Compared to the particles that make up carbon dioxide gas, the particles that make up helium gas are very small. What can you conclude about the number of particles in a 2.0-L sample of carbon dioxide gas compared to the number of particles in a 2.0-L sample of helium gas if both samples are at the same temperature and pressure?

SECTION 2

Forces Within Liquids

PHYSICS 4 YOU During exercise or on a hot day, your body perspires to cool itself. As sweat evaporates from the skin's surface, particles with higher-than-average kinetic energy escape from the liquid perspiration. The average kinetic energy of the remaining particles decreases, which leads to a decrease in temperature.

MAIN IDEA

Cohesive forces occur between the particles of a substance, while adhesive forces occur between particles of different substances.

Essential Questions

- What is surface tension?
- What is capillary action?
- How do clouds form?

Review Vocabulary

net force the vector sum of all the forces on an object

New Vocabulary

cohesive forces
adhesive forces

Cohesive Forces

Figure 7 shows a water strider walking across the surface of a pond. This lightweight insect can do this because of surface tension—the tendency of the surface of a liquid to contract to the smallest possible area. Surface tension results from the cohesive forces among the particles of a liquid. **Cohesive forces** are the forces of attraction that like particles exert on one another. Notice that beneath the liquid's surface in **Figure 7,** each particle of the liquid is attracted equally in all directions by neighboring particles. As a result, no net force acts on any of the particles beneath the surface. At the surface, however, the particles are attracted downward and to the sides but not upward. There is a net downward force, which acts on the top layers and causes the surface layer to be slightly compressed. The surface layer acts like a tightly stretched sheet that is strong enough to support the weight of very light objects, such as the water strider.

You might have seen beaded water droplets on a freshly washed and waxed car. Why do these spherical drops form? The force pulling the surface particles into a liquid causes the surface to become as small as possible, and the shape that has the least surface for a given volume is a sphere. The higher the surface tension of the liquid, the more resistant the liquid is to having its surface broken. For example, liquid mercury has much stronger cohesive forces than water does. Liquid mercury forms spherical drops, even when it is placed on a smooth surface. A drop of water flattens out on a smooth surface.

Figure 7 A water strider can walk on water because molecules at the surface experience a net downward force. Below the surface, each particle of liquid is equally attracted in all directions.

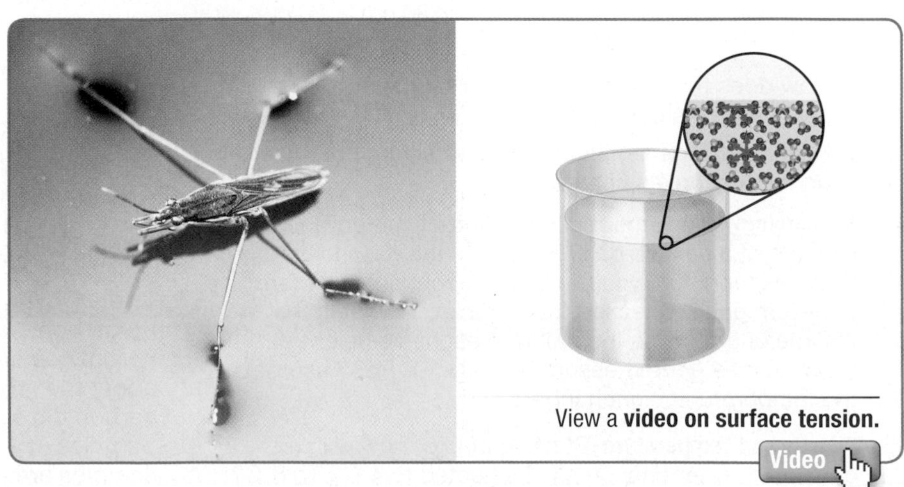

View a **video on surface tension.**

Video

Viscosity In nonideal fluids, the cohesive forces and collisions between fluid molecules cause internal friction that slows the fluid flow and dissipates mechanical energy. The measure of this internal friction is called the viscosity of the liquid. Water is not very viscous, but motor oil is very viscous. As a result of its viscosity, motor oil flows slowly over the parts of an engine to coat the metal and reduce rubbing. Lava, molten rock that flows from a volcano or vent in Earth's surface, is one of the most viscous fluids. There are several types of lava, and the viscosity of each type varies with composition and temperature.

Adhesive Forces

Similar to cohesive forces, **adhesive forces** are attractive forces that act between particles of different substances. When a glass tube is placed in a beaker of water, the surface of the water climbs the outside of the tube, as shown in **Figure 8.** The adhesive forces between the particles that make up the glass and the water molecules are greater than the cohesive forces between the water molecules. In contrast, the cohesive forces between mercury atoms are greater than the adhesive forces between the mercury and the glass, so the liquid does not climb the tube. These forces also cause the mercury's surface to depress, as shown in **Figure 8.**

If a glass tube with a small inner diameter is placed in water, the water rises inside the tube. This happens because the adhesive forces between glass and water molecules are stronger than the cohesive forces between water molecules. The water continues to rise until the weight of the water that is lifted balances the total adhesive force between the glass and water molecules. If the radius of the tube increases, the volume and the weight of the water will increase proportionally faster than the surface area of the tube. Thus, water is lifted higher in a narrow tube than in a wider one. This phenomenon is called capillary action. It causes molten wax to rise in a candle's wick and water to move up through the soil and into the roots of plants.

Evaporation and Condensation

Why does a puddle of water disappear on a hot, dry day? As you have previously read, the particles in a liquid are moving at random speeds. If a fast-moving particle can break through the surface layer, it will escape from the liquid. Because there is a net downward cohesive force at the surface only the most energetic particles escape. This escape of particles is called evaporation.

REAL-WORLD PHYSICS

PLANTS The combination of adhesive and cohesive forces acting on water molecules in plant tissues results in a certain amount of tension. As water molecules evaporate from cells in leaves, this tension pulls adjacent water molecules further into the leaves. Cohesive forces keep water from separating into individual droplets and water is pulled the entire length of a plant, in some cases, as much as 115 m.

Water

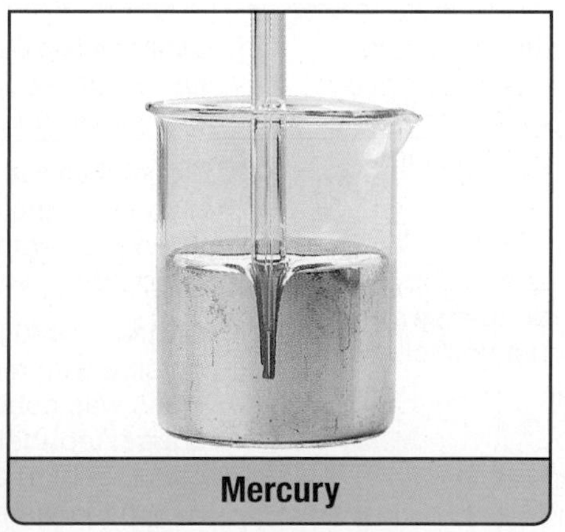

Mercury

Figure 8 Due to adhesive forces, water climbs the outside wall of the glass tube. In the mercury, however, the forces of attraction between mercury atoms are stronger than any adhesive forces between the mercury and the glass. Therefore, the mercury is depressed by the tube.

Figure 9 Warm, moist, surface air rises. Clouds form when the air cools and the water vapor condenses.

Evaporative cooling Evaporation has a cooling effect. On a hot day, your body perspires, and the evaporation of your sweat cools you. In a puddle of water, evaporation causes the remaining liquid to cool. Each time a particle with higher-than-average kinetic energy escapes from the water, the average kinetic energy of the remaining particles decreases. As you learned earlier, a decrease in average kinetic energy is a decrease in temperature. Rubbing alcohol has a noticeable cooling effect when it evaporates from a person's skin. Alcohol molecules evaporate easily because they have weak cohesive forces. A liquid that evaporates quickly is called a volatile liquid.

Have you ever wondered why humid days feel warmer than dry days at the same temperature? On a humid day, the water vapor content of the air is high. Because there are already many water molecules in the air, the water molecules in perspiration are less likely to evaporate from the skin. Evaporation is the body's primary cooling mechanism, so the body is not able to cool itself as effectively on a humid day.

READING CHECK Explain why evaporation has a cooling effect.

Condensation Particles of liquid that have evaporated into the air can also return to the liquid phase if the kinetic energy or temperature decreases, a process called condensation. What happens if you bring a cold glass into a hot, humid area? The outside of the glass soon becomes coated with condensed water. Water molecules moving randomly in the air surrounding the glass strike the cold surface, and if they lose enough energy, the cohesive forces become strong enough to prevent their escape.

The air above any body of water, as shown in **Figure 9,** contains evaporated water vapor, which is water in the form of gas. If the temperature is reduced, the water vapor condenses around tiny dust particles in the air and produces droplets only 0.01 mm in diameter. A cloud of these droplets that forms at Earth's surface is called fog. Fog often forms when moist air is chilled by the cold ground. Fog also occurs briefly when a carbonated drink is opened. The sudden decrease in pressure causes the temperature of the gas in the container to drop, which condenses the water vapor dissolved in that gas.

PhysicsLAB

EVAPORATIVE COOLING

How can you infer the relationship between cohesive forces and evaporation rates?

Watch a **BrainPOP video on humidity.**

SECTION 2 REVIEW

18. **MAIN IDEA** The English language includes the term *adhesive tape* and the phrase *working as a cohesive group.* In these examples, are *adhesive* and *cohesive* being used in the same context as their meanings in physics? Explain your answer.

19. **Surface Tension** A paper clip, which has a density greater than that of water, can be made to stay on the surface of water. What procedures must you follow for this to happen? Explain.

20. **Floating** How can you tell that the paper clip in the previous problem was not floating?

21. **Adhesion and Cohesion** In terms of adhesion and cohesion, explain why alcohol clings to the surface of a glass rod but mercury does not.

22. **Evaporation and Cooling** In the past when a baby had a high fever, the doctor might have suggested gently sponging off the baby with a liquid that evaporates easily. Why would this help?

23. **Critical Thinking** On a hot, humid day, Beth sat outside with a glass of cold water. The outside of the glass was coated with water. Her friend, Sally, suggested that the water had leaked out through the glass. Design an experiment for Beth to show Sally where the water came from.

SECTION 3 Fluids at Rest and in Motion

PHYSICS 4 YOU A 2.5-g penny will sink in a glass of water, but a canoe with several passengers can float on a lake or river. Why does the heavier item float but the lighter item sink? What might happen if the canoe were filled with water?

MAIN IDEA

Hydraulic lifts, floating objects, and carburetors rely on the forces exerted by fluids.

Essential Questions

- What is Pascal's principle?
- How does Archimedes' principle apply to buoyancy?
- What is the role of Bernoulli's principle in airflows?

Review Vocabulary

pressure the perpendicular component of a force on a surface divided by the area of the surface

New Vocabulary

Pascal's principle
buoyant force
Archimedes' principle
Bernoulli's principle
streamlines

Fluids at Rest

If you have ever dived deep into a swimming pool or a lake, you likely felt pressure on your ears. You might have noticed that the pressure you felt did not depend on whether your head was upright or tilted, but that if you swam deeper, the pressure increased.

Pascal's principle Blaise Pascal, a French physician, found that the pressure at a point in a fluid depends on its depth in the fluid and is unrelated to the shape of the fluid's container. He also noted that any change in pressure applied at any point on a confined fluid is transferred undiminished throughout the fluid, a fact that is now known as **Pascal's principle.** Every time you squeeze an open tube of toothpaste, you demonstrate Pascal's principle. The pressure that your fingers exert at the bottom of the tube is transmitted through the toothpaste and forces the paste out at the top. Likewise, if you squeeze one end of a fluid-filled balloon, the other end of the balloon expands.

One application of Pascal's principal is using fluids in machines to multiply forces. In the hydraulic system shown in **Figure 10,** a fluid is confined to two connecting chambers. Each chamber has a piston that is free to move, and the pistons have different surface areas. Recall that if a force (F_1) is exerted on the first piston with a surface area of A_1, the pressure (P_1) exerted on the fluid is $P_1 = \frac{F_1}{A_1}$. The pressure exerted by the fluid on the second piston, with a surface area A_2, is $P_2 = \frac{F_2}{A_2}$.

Figure 10 As F_1 exerts pressure on the smaller piston (piston 1), the pressure is transmitted throughout the fluid. As a result, a multiplied force (F_2) is exerted on the larger piston (piston 2).

Infer How would F_2 change if F_1 increased? Explain why.

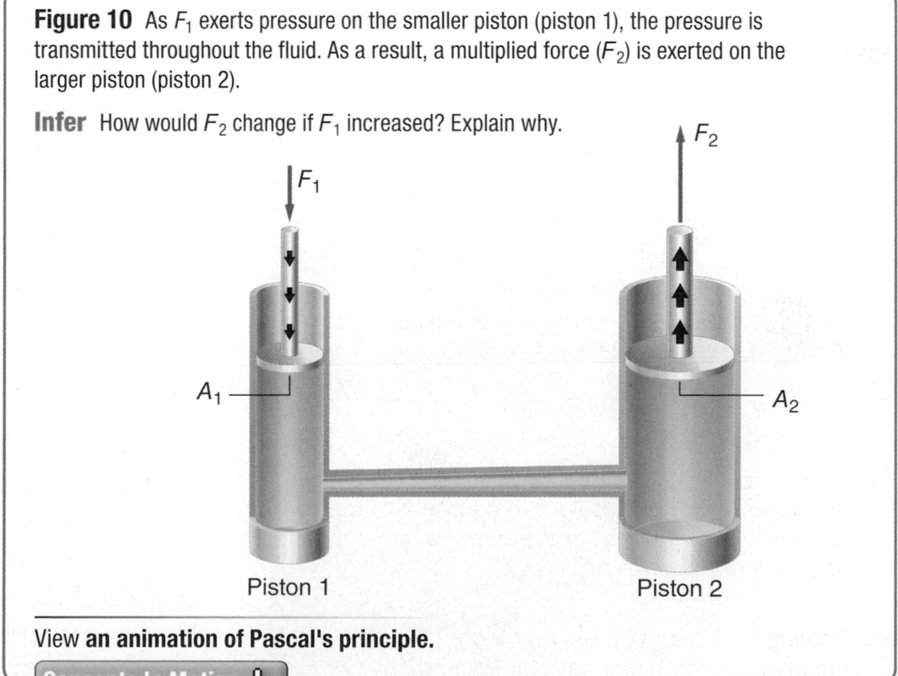

View **an animation of Pascal's principle.**

Concepts In Motion

PRACTICE PROBLEMS

Online Practice

PRACTICE PROBLEMS

24. Dentists' chairs are examples of hydraulic-lift systems. If a chair weighs 1600 N and rests on a piston with a cross-sectional area of 1440 cm^2, what force must be applied to the smaller piston, with a cross-sectional area of 72 cm^2, to lift the chair?

25. A mechanic exerts a force of 55 N on a 0.015 m^2 hydraulic piston to lift a small automobile. The piston the automobile sits on has an area of 2.4 m^2. What is the weight of the automobile?

26. **CHALLENGE** By multiplying a force, a hydraulic system serves the same purpose as a lever or a seesaw. If a 400-N child standing on one piston is balanced by a 1100-N adult standing on another piston, what is the ratio of the areas of their pistons?

PhysicsLAB

UNDER PRESSURE

What causes a scuba diver's ears to hurt?

Figure 11 Submersibles are built to withstand the crushing pressure exerted by the water column.

According to Pascal's principle, pressure is transmitted without change throughout a fluid, so pressure P_2 is equal in value to P_1. You can determine the force exerted by the second piston by setting the pressures equal to each other and solving for F_2.

FORCE EXERTED BY A HYDRAULIC LIFT

The force exerted by the second piston is equal to the force exerted by the first piston multiplied by the ratio of the area of the second piston to the area of the first piston.

$$F_2 = F_1\frac{A_2}{A_1}$$

Swimming Under Pressure

When you are swimming, you feel the pressure of the water increase as you dive deeper. This pressure is a result of gravity; it is related to the weight of the water above you. The deeper you go, the more water there is above you, and the greater the pressure. The pressure of the water is equal to the weight (F_g) of the column of water above you divided by the column's cross-sectional area (A). Even though gravity pulls only in the downward direction, the fluid transmits the pressure in all directions: up, down, and to the sides. As before, the pressure of the water is $P = \frac{F_g}{A}$.

The weight of the column of fluid is $F_g = mg$, and the mass is equal to the density (ρ) of the fluid times its volume, $m = \rho V$. You also know that the volume of the fluid is the area of the base of the column times its height, $V = Ah$. Therefore, $F_g = \rho Ahg$. Substituting ρAhg for F_g gives $P = \frac{F_g}{A} = \frac{\rho Ahg}{A}$. Divide A from the numerator and denominator to arrive at the simplest form of the equation for the pressure exerted by a column of fluid on a submerged body.

PRESSURE OF FLUID ON A BODY

The pressure a column of fluid exerts on a body is equal to the density of the fluid times the height of the column times the free-fall acceleration.

$$P = \rho hg$$

The pressure of a fluid on a body depends on the density of the fluid, its depth, and g. As shown in **Figure 11,** submersibles have explored the deepest ocean trenches and encountered pressures in excess of 1000 times standard air pressure.

CONNECTION TO BIOLOGY Scientists use submersibles to learn more about deep ocean ecosystems. In 1977 the first hydrothermal vents were discovered as the crewed submersible ALVIN cruised over the Pacific Ocean floor. Hydrothermal vents form when superheated water flows up from cracks in the seafloor.

Because these vents are located thousands of meters below the ocean surface, the fluid pressure can be over one hundred times the standard atmospheric pressure. Despite the high pressure and the fact that sunlight does not reach them, hydrothermal vents thrive with life. Giant tube worms harbor bacteria in their tissues. The bacteria use hydrogen sulfide from the vent water to produce sugar, which provides the energy that supports the entire ecosystem. Other organisms that live at hydrothermal vents include fish, mussels, shrimp, clams, and octopuses. Submersibles have been used to explore hydrothermal vents in the Atlantic, Indian, and Arctic Oceans.

Buoyancy What produces the upward force that allows you to swim? The increase in pressure with increasing depth creates an upward force called the **buoyant force.** By comparing the buoyant force on an object with its weight, you can predict whether the object will sink or float.

Suppose that a box is immersed in water. It has a height of l, and its top and bottom each have a surface area of A. Its volume, then, is $V = lA$. Water pressure exerts forces on all sides, as shown in **Figure 12.** Will the box sink or float? As you know, the pressure on the box depends on its depth (h). To find out whether the box will float in water, you will need to analyze the forces acting on it, which are its weight and the forces on each side due to the pressure of the fluid. Compare these two equations:

$$F_{top} = P_{top} A = \rho hgA$$
$$F_{bottom} = P_{bottom} A = \rho(l + h)gA$$

On the four vertical sides, the forces are equal in all directions, so there is no net horizontal force. The upward force on the bottom is larger than the downward force on the top, so there is a net upward force. The buoyant force can now be determined.

$$\begin{aligned} F_{buoyant} &= F_{bottom} - F_{top} \\ &= \rho(l + h)gA - \rho hgA \\ &= \rho lgA = \rho Vg \end{aligned}$$

These calculations show the net upward force to be proportional to the volume of the box. This volume equals the volume of the fluid displaced, or pushed out of the way, by the box. Therefore, the magnitude of the buoyant force (ρVg) equals the weight of the fluid displaced by the object.

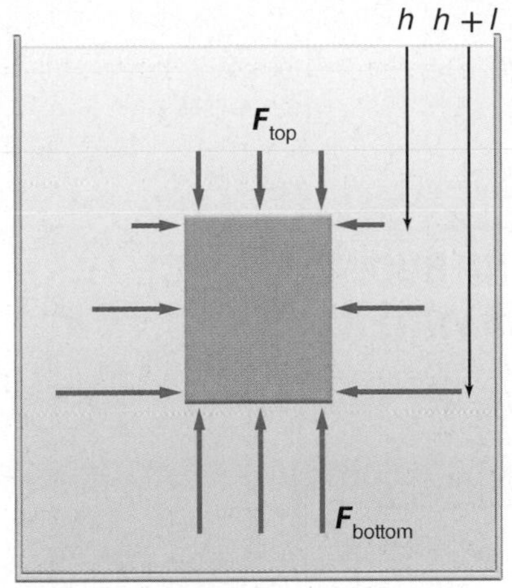

Figure 12 A fluid exerts a greater upward force on the bottom of an immersed object than the downward force on the top of the object. The net upward force is called the buoyant force.

View an **animation of buoyancy.**

BUOYANT FORCE

The buoyant force on an object is equal to the weight of the fluid displaced by the object, which is equal to the density of the fluid in which the object is immersed multiplied by the object's volume and the free-fall acceleration.

$$F_{buoyant} = \rho_{fluid} Vg$$

Archimedes' principle The relationship between buoyant force and the weight of the fluid displaced by an object was discovered in the third century B.C. by Greek scientist and mathematician Archimedes. **Archimedes' principle** states that an object immersed in a fluid has an upward force on it that is equal to the weight of the fluid displaced by the object. The force does not depend on the weight of the object, only on the weight of the displaced fluid.

Sink or float? If you want to know whether an object sinks or floats, you have to take into account all of the forces acting on the object. The buoyant force pushes up, but the weight of the object pulls it down. The difference between the buoyant force and the object's weight determines whether an object sinks or floats.

Suppose you submerge three objects in a tank of water ($\rho_{water} = 1.00\times10^3$ kg/m³). Each object has a volume of 400 cm³, or 4.00×10^{-4} m³. The first object is a steel block with a mass of 3.60 kg. The second is a can of soda with a mass of 0.40 kg. The third is an ice block with a mass of 0.36 kg. How will each item move when it is immersed in water and released? Since the three objects have the same volume, they displace the same volume of water, and the upward force on all three objects is the same, as shown in **Figure 13.** This buoyant force can be calculated as follows.

$$\begin{aligned} F_{buoyant} &= \rho_{water}Vg \\ &= (1.00\times10^3\ \text{kg/m}^3)(4.00\times10^{-4}\ \text{m}^3)(9.8\ \text{N/kg}) \\ &= 3.9\ \text{N} \end{aligned}$$

PhysicsLAB

THE BUOYANT FORCE OF WATER

Why does a rock feel lighter in water?

Figure 13 All of the forces on an object must be accounted for to determine whether an object will sink or float.

Describe the circumstances under which an object will float.

Buoyant Force

Sinking The weight of the block of steel is 8.8 N, much greater than the buoyant force. There is a net downward force, so the block will sink to the bottom of the tank. The net downward force is less than the object's weight. All objects in a liquid, even those that sink, experience a net force that is less than the net force when the object is in air.

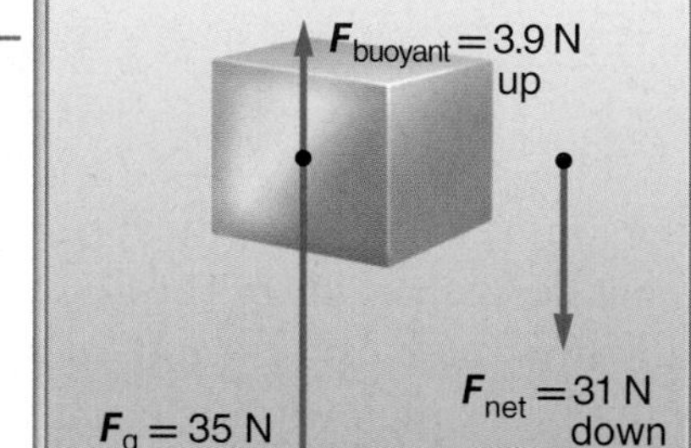

Neutral The weight of the soda can is 0.98 N, the same as the weight of the water displaced. There is, therefore, no net force, and the can will remain wherever it is placed in the water. It is said to have neutral buoyancy.

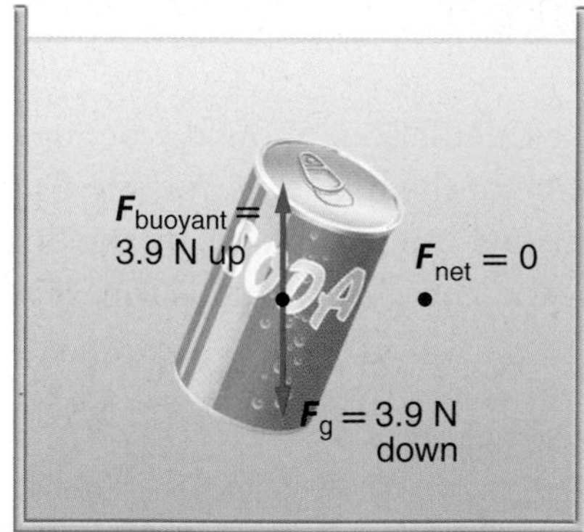

Floating The weight of the ice cube is 0.88 N, less than the buoyant force, so there is a net upward force, and the ice cube will rise. An object will float if its density is less than the density of the fluid in which it is placed.

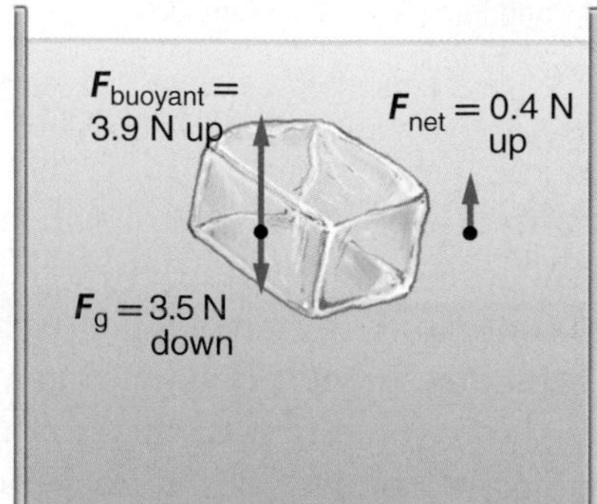

Note: Force vectors are not drawn to scale.

EXAMPLE PROBLEM 3

Get help with **operations with scientific notation.** Math Handbook

ARCHIMEDES' PRINCIPLE A cubic decimeter (1.00×10^{-3} m^3) of a granite building block is submerged in water. The density of granite is 2.70×10^{3} kg/m^3.

a. What is the magnitude of the buoyant force acting on the block?

b. What is the net force on the block?

1 ANALYZE AND SKETCH THE PROBLEM

- Sketch the cubic decimeter of granite immersed in water.
- Show the upward buoyant force and the downward force due to gravity acting on the granite.

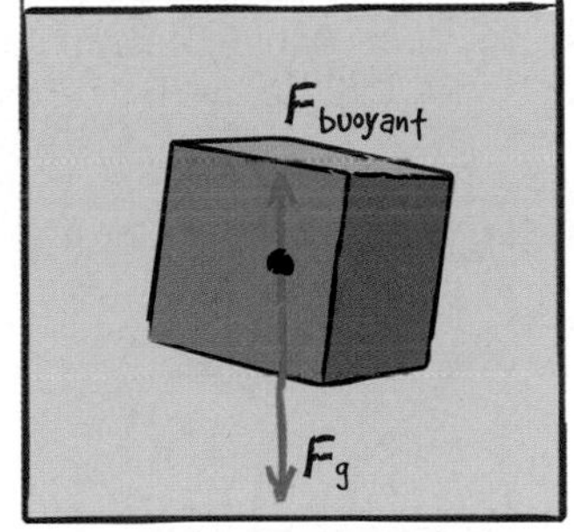

KNOWN	UNKNOWN
$V = 1.00\times10^{-3}\ m^3$	$F_{buoyant} = ?$
$\rho_{granite} = 2.70\times10^{3}\ kg/m^3$	$F_{net} = ?$
$\rho_{water} = 1.00\times10^{3}\ kg/m^3$	

2 SOLVE FOR THE UNKNOWN

a. Calculate the buoyant force on the granite block.

$$F_{buoyant} = \rho_{water}Vg$$
$$= (1.00\times10^{3}\ kg/m^3)(1.00\times10^{-3}\ m^3)(9.8\ N/kg)$$
$$= 9.8\ N$$

◀ Substitute $\rho_{water} = 1.00\times10^{3}$ kg/m^3, $V = 1.00\times10^{-3}$ m^3, $g = 9.8$ N/kg.

b. Calculate the granite's weight, and then find its net force.

$$F_g = mg = \rho_{granite}Vg$$
$$= (2.70\times10^{3}\ kg/m^3)(1.00\times10^{-3}\ m^3)(9.8\ N/kg)$$
$$= 26.5\ N$$

◀ Substitute $\rho_{granite} = 2.70\times10^{3}$ kg/m^3, $V = 1.00\times10^{-3}$ m^3, $g = 9.8$ N/kg.

$$F_{net} = F_g - F_{buoyant}$$
$$= 26.5\ N - 9.8\ N$$
$$= 17\ N$$

◀ Substitute $F_g = 26.5$ N, $F_{buoyant} = 9.8$ N.

3 EVALUATE THE ANSWER

- **Are the units correct?** The forces and the weight are in newtons, as expected.
- **Is the magnitude realistic?** The buoyant force is about one-third the weight of the granite, a sensible answer because the density of water is about one-third that of granite.

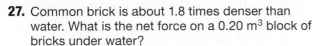

PRACTICE PROBLEMS

Do additional problems. Online Practice

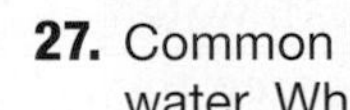

27. Common brick is about 1.8 times denser than water. What is the net force on a 0.20 m^3 block of bricks under water?

28. A girl is floating in a freshwater lake with her head just above the water. If she weighs 610 N, what is the volume of the submerged part of her body?

29. What is the tension in a wire supporting a 1250-N camera submerged in water? The volume of the camera is 16.5×10^{-3} m^3.

30. Plastic foam is about 0.10 times as dense as water. What weight of bricks could you stack on a 1.0-m × 1.0-m × 0.10-m slab of foam so that the slab of foam floats in water and is barely submerged, leaving the bricks dry?

31. CHALLENGE Canoes often have plastic foam blocks mounted under the seats for flotation in case the canoe fills with water. What is the approximate minimum volume of foam needed for flotation for a 480-N canoe?

PhysicsLAB

THE FIRST FORENSIC SCIENTIST

Forensics Lab Was Archimedes the first forensic scientist?

iLab Station

Ships How can ships can be made of steel and still float? You can investigate this by making a small boat out of folded aluminum foil. The boat should float easily. Add a cargo of paper clips or some pennies, and it will ride lower in the water. Crumple the foil into a tight ball, and it will sink. When the boat is hollow and large enough so that its average density is less than the density of water, it floats. As you add cargo, the density increases, and more of the boat is submerged. The crumpled boat has a density greater than the water's and sinks.

Other examples of Archimedes' principle include submarines and fishes. Submarines take advantage of Archimedes' principle as water is pumped into or out of chambers to change the submarine's average density, causing it to rise or sink. Fishes that have swim bladders also use Archimedes' principle to control their depths. To move upward in the water, a fish expands its swim bladder by filling it with gas to displace more water and increase the buoyant force. The fish moves downward by contracting the volume of its swim bladder.

Figure 14 You can demonstrate Bernoulli's principle by narrowing the opening of the hose as water flows out. As the velocity of the water increases, the pressure it exerts decreases.

Richard Hutchings/Digital Light Source

Bernoulli's Principle

Study the flow of water from the hose in **Figure 14.** In the photo on the top, the water flows from the hose unobstructed. In the photo on the bottom, the hose opening has been narrowed by a person's thumb being placed over it. Notice that the stream of water on the bottom looks different than it does on the top. The velocity of the water stream in the bottom photo is greater compared to the velocity of the stream in the top photo. What you can't see from the photographs is that pressure exerted by the water in the bottom photo decreased. The relationship between the velocity and pressure exerted by a moving fluid is named for Swiss scientist Daniel Bernoulli. **Bernoulli's principle** states that as the velocity of a fluid increases, the pressure exerted by that fluid decreases. This principle is a statement of work and energy conservation as applied to fluids.

Another instance in which the velocity of water can change is in a stream. You might have seen the water in a stream speed up as it passed through narrowed sections of the stream bed. As the opening of the hose and the stream channel become wider or narrower, the velocity of the fluid changes to maintain the overall flow of water. The pressure of blood in our circulatory systems depends partly on Bernoulli's principle. Bernoulli's principle also helps explain how smoke is pulled up a fireplace chimney.

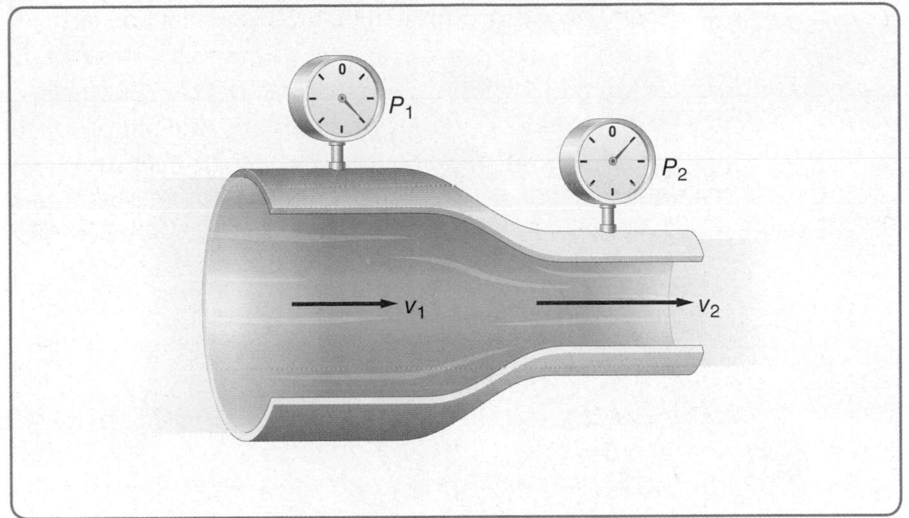

Figure 15 The fluid flowing through this pipe also demonstrates Bernoulli's principle. As the velocity of the fluid increases (v_2 is greater than v_1), the pressure it exerts decreases (P_2 is less than P_1).

Consider a horizontal pipe completely filled with a smoothly flowing ideal fluid. If a certain mass of the fluid enters one end of the pipe, then an equal mass must come out the other end. What happens if the cross section becomes narrower, as shown in **Figure 15?** To keep the same mass of fluid moving through the narrow section in a fixed amount of time, the velocity of the fluid in the tube must increase. As the fluid's velocity increases, so does its kinetic energy. This means that net work has been done on the swifter fluid. This net work comes from the difference between the work that was done to move the mass of fluid into the pipe and the work that was done by the fluid pushing the same mass out of the pipe. The work is proportional to the force on the fluid, which, in turn, depends on the pressure. If the net work is positive, the pressure at the input end of the section, where the velocity is lower, must be larger than the pressure at the output end, where the velocity is higher.

READING CHECK **Describe** the relationship between the velocity of a fluid and the pressure it exerts according to Bernoulli's principle.

Applications of Bernoulli's principle There are many common applications of Bernoulli's principle, such as paint sprayers and sprayers attached to garden hoses to apply fertilizers and pesticides to lawns and gardens. In a hose-end sprayer, a strawlike tube is sunk into the chemical solution in the sprayer. The sprayer is attached to a hose. A trigger on the sprayer allows water from the hose to flow at a high speed, producing an area of low pressure above the tube. The solution is then sucked up through the tube and into the stream of water.

A gasoline engine's carburetor, which is where air and gasoline are mixed, is another common application of Bernoulli's principle. Part of the carburetor is a tube with a constriction, as shown in the diagram in **Figure 16.** The pressure on the gasoline in the fuel supply is the same as the pressure in the thicker part of the tube. Air flowing through the narrow section of the tube, which is attached to the fuel supply, is at a lower pressure, so fuel is forced into the air flow. By regulating the flow of air in the tube, the amount of fuel mixed into the air can be varied. Carburetors are used in motorcycles, stock car race cars, and the motors of small gasoline-powered machines, such as lawn mowers.

Figure 16 In a carburetor, low pressure in the narrow part of the tube draws fuel into the air flow.

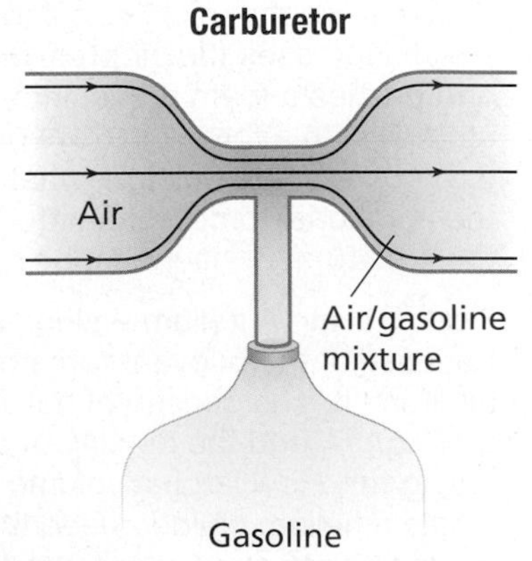

Figure 17 The streamline shows the air flowing above a cyclist pedaling in a wind tunnel.

Andy Sacks/Stone/Getty Images

VOCABULARY
Science Usage v. Common Usage

Streamlines

- **Science Usage**
lines representing the flow of fluids around objects
Simulations of the streamlines around the airplane indicated a design flaw.
- **Common Usage**
to make simpler or more efficient
The new computer system streamlines the registration process.

Streamlines Automobile and aircraft manufacturers spend a great deal of time and money testing new designs in wind tunnels to ensure the greatest efficiency of movement through air. The flow of fluids around objects is represented by **streamlines,** as shown in **Figure 17.** Objects require less energy to move through a smooth streamlined flow.

Streamlines can best be illustrated by a simple demonstration. Imagine carefully squeezing tiny drops of food coloring into a smoothly flowing fluid. If the colored lines that form stay thin and well defined, the flow is said to be streamlined. Notice that if the flow narrows, the streamlines move closer together. Closely spaced streamlines indicate greater velocity and, therefore, reduced pressure. If streamlines swirl and become diffused, the flow of the fluid is said to be turbulent. Bernoulli's principle does not apply to turbulent flow.

SECTION 3 REVIEW

Section Self-Check — Check your understanding.

32. **MAIN IDEA** All soda cans contain the same volume of liquid, 354 mL, and displace the same volume of water. What might be a difference between a can that sinks and one that floats? *Hint: Place a full can of regular soda and a full can of diet soda in water.*

33. **Transmission of Pressure** A toy rocket launcher is designed so that a child stomps on a rubber cylinder, which increases the air pressure in a launching tube and pushes a foam rocket into the sky. If the child stomps with a force of 150 N on a 2.5×10^{-3}-m^2 area piston, what is the additional force transmitted to the 4.0×10^{-4}-m^2 launch tube?

34. **Floating in Air** A helium balloon rises because of the buoyant force of the air lifting it. The density of helium is 0.18 kg/m^3, and the density of air is 1.3 kg/m^3. How large a volume would a helium balloon need to lift the lead brick shown in **Figure 18?**

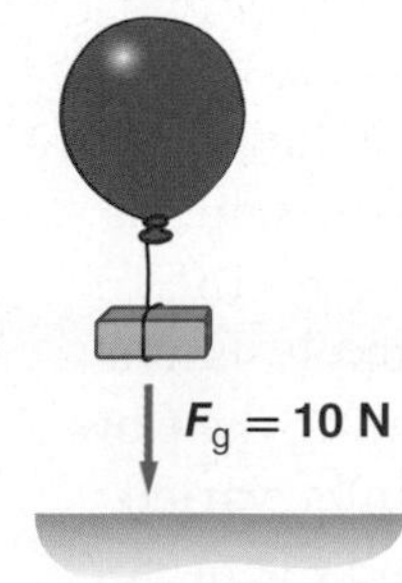

Figure 18

35. **Floating and Density** A fishing bobber made of cork floats with one-tenth of its volume below the water's surface. What is the density of cork?

36. **Pressure and Force** An automobile weighing 2.3×10^4 N is lifted by a hydraulic cylinder with an area of 0.15 m^2.

a. What is the pressure in the hydraulic cylinder?

b. The pressure in the lifting cylinder is produced by pushing on a 0.0082-m^2 cylinder. What force must be exerted on this small cylinder to lift the automobile?

37. Which displaces more water when placed in a pool?

a. a 1.0-kg block of aluminum or a 1.0-kg block of lead

b. a 10-cm^3 block of aluminum or a 10-cm^3 block of lead

38. **Critical Thinking** A tornado passing over a house sometimes makes the house explode from the inside out. How might Bernoulli's principle explain this phenomenon? What could be done to reduce the danger of a door or window exploding outward?

SECTION 4

Solids

If a glass jar with a metal lid will not open, placing it in warm water will often help loosen the lid. This is because, when heated, the metal lid expands more than the glass jar. What would happen if both the lid and the jar were made of the same material?

MAIN IDEA

Solids usually expand when heated.

Essential Questions

- How do a solid's properties relate to that solid's structure?
- Why do solids expand and contract when the temperature changes?
- Why is thermal expansion important?

Review Vocabulary

cohesive force the force of attraction that particles within a substance exert on one another

New Vocabulary

crystal lattice
amorphous solid
coefficient of linear expansion
coefficient of volume expansion

Solid Bodies

How do solids and liquids differ? Solids are stiff, they can be cut in pieces, and they retain their shapes. You can push on solids. Liquids flow and do not retain their shapes. If you push your finger on water, your finger will move through it. Under certain conditions, however, solids and liquids are not easily distinguished. As bottle glass is heated to a molten state, the change from solid to liquid is so gradual that, at times during the process, it is difficult to tell which is which.

When the temperature of a liquid is lowered, the average kinetic energy of the particles decreases. As the particles slow down, the cohesive forces have more effect, and for many solids, the particles become frozen into a fixed pattern called a **crystal lattice,** shown in **Figure 19.** Although the cohesive forces hold the particles in place, the particles in a crystalline solid do not stop moving completely. Rather, they vibrate around their fixed positions. In other materials, such as butter and glass, the particles do not form a fixed crystalline pattern. A substance that has no regular crystal structure but does have a definite volume and shape is called an **amorphous solid.**

Pressure and freezing As a liquid becomes a solid, its particles usually fit more closely together than in the liquid state, making solids more dense than liquids. However, water is an exception because it is most dense at 4°C. Water is also an exception to another general rule. For most liquids, an increase in the pressure on the surface of the liquid increases its freezing point. Because water expands as it freezes, an increase in pressure forces the molecules closer together and opposes the freezing. Therefore, higher pressure lowers the freezing point of water very slightly.

Figure 19 As the temperature of water is lowered and it changes from a liquid to a solid, the particles are frozen in a pattern called a crystal lattice.

It has been hypothesized that the drop in water's freezing point caused by the pressure of an ice-skater's blades produces a thin film of liquid between the ice and the blades. Recent measurements have shown that the friction between the blade and the ice generates enough thermal energy to melt the ice and create a thin layer of water. This explanation is supported by measurements of the spray of ice particles, which are considerably warmer than the ice itself. The same process of melting occurs during snow skiing.

Elasticity of solids External forces applied to a solid object may twist, stretch, or bend it out of shape. The ability of a solid object to return to its original form when the external forces are removed is called the elasticity of the solid. If too much deformation occurs, the object will not return to its original shape because its elastic limit has been exceeded. Elasticity is a property of each substance and depends on the forces holding its particles together. Malleability and ductility are two properties that depend on the structure and elasticity of a substance. Because gold can be flattened and shaped into thin sheets, it is said to be malleable. Copper is a ductile metal because it can be pulled into thin strands of wire.

Thermal Expansion of Solids

It is standard practice for engineers to design small gaps, called expansion joints, into concrete-and-steel highway bridges to allow for the expansion of parts in the heat of summer. Expansion joints are shown in **Figure 20.** Objects expand only a small amount when they are heated, but that small amount could be several centimeters in a 100-m-long bridge. If expansion gaps were not present, the bridge could buckle, or parts of it could break. Some materials, such as the ovenproof glass used for laboratory experiments and cooking, are designed to have the least possible thermal expansion. Large telescope mirrors are made of a ceramic material that is designed to undergo essentially no thermal expansion.

To understand the expansion of heated solids, picture a solid as a collection of particles connected by springs that represent the attractive forces between the particles. When the particles get too close, the springs push them apart. When a solid is heated, the kinetic energy of the particles increases, and they vibrate rapidly and move farther apart, weakening the attractive forces between the particles. As a result, when the particles vibrate more violently with increased temperature, their average separation increases and the solid expands.

Figure 20 Expansion joints are included when bridges, roads, and train tracks are built.

Infer If there were no expansion joints in this road, what might happen during the summer?

Tom Uhlman/Alamy

The change in length of a solid is proportional to the change in temperature, as shown in **Figure 21.** A solid will expand in length twice as much when its temperature is increased by 20°C than when it is increased by 10°C. The expansion also is proportional to its length. This means that a 2-m bar will expand twice as much as a 1-m bar with the same change in temperature. The length (L_2) of a solid at temperature T_2 can be found using the following relationship, where L_1 is the length at temperature T_1 and alpha (α) is the coefficient of linear expansion. The **coefficient of linear expansion** is equal to the change in length divided by the original length and the change in temperature.

$$L_2 = L_1 + \alpha L_1 (T_2 - T_1)$$

With some algebra, you can solve for α.

$$L_2 - L_1 = \alpha L_1 (T_2 - T_1)$$

$$\Delta L = \alpha L_1 \Delta T$$

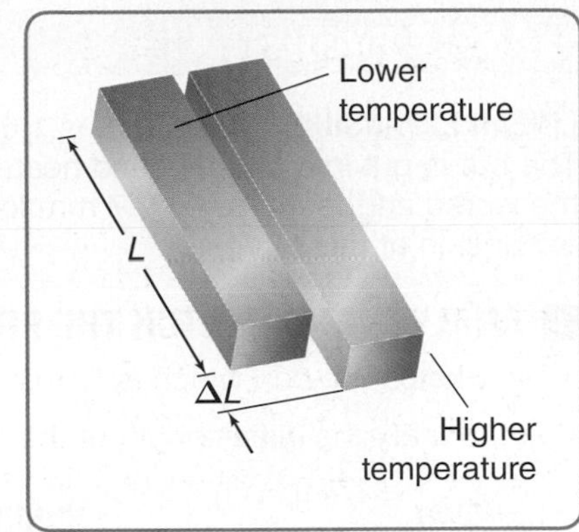

Figure 21 The change in length of a material is proportional to its original length and the change in temperature.

COEFFICIENT OF LINEAR EXPANSION

The coefficient of linear expansion is equal to the change in length divided by the product of the original length and the change in temperature.

$$\alpha = \frac{\Delta L}{L_1 \Delta T}$$

The unit for the coefficient of linear expansion is the reciprocal of degrees Celsius (which can be written as 1/°C or °C^{-1}). The **coefficient of volume expansion** is equal to the change in volume divided by the original volume and the change in temperature. The coefficient of volume expansion (β) is generally about three times the coefficient of linear expansion.

COEFFICIENT OF VOLUME EXPANSION

The coefficient of volume expansion is equal to the change in volume divided by the product of the original volume and the change in temperature.

$$\beta = \frac{\Delta V}{V_1 \Delta T}$$

The unit for β is also 1/°C (°C^{-1}). The two coefficients of thermal expansion for a variety of materials are given in **Table 2.**

Table 2 Coefficients of Thermal Expansion at 20°C

Material	Coefficient of Linear Expansion, α (°C^{-1})	Coefficient of Volume Expansion, β (°C^{-1})
Solids		
Aluminum	23×10^{-6}	69×10^{-6}
Glass (soft)	9×10^{-6}	27×10^{-6}
Glass (ovenproof)	3×10^{-6}	9×10^{-6}
Concrete	12×10^{-6}	36×10^{-6}
Copper	17×10^{-6}	51×10^{-6}
Liquids		
Methanol	Not Applicable	1200×10^{-6}
Gasoline	Not Applicable	950×10^{-6}
Water	Not Applicable	210×10^{-6}

EXAMPLE PROBLEM

EXAMPLE PROBLEM 4

Get help with **operations with scientific notation.** Math Handbook

LINEAR EXPANSION A metal bar is 1.60 m long at room temperature (21°C). The bar is put into an oven and heated to a temperature of 84°C. It is then measured and found to be 1.7 mm longer. What is the coefficient of linear expansion of this material?

1 ANALYZE AND SKETCH THE PROBLEM

- Sketch the bar, which is 1.7 mm longer at 84°C than at 21°C.
- Identify the initial length of the bar (L_1) and the change in length (ΔL).

KNOWN

$L_1 = 1.60$ m

$\Delta L = 1.7\times10^{-3}$ m

$T_1 = 21°C$

$T_2 = 84°C$

UNKNOWN

$\alpha = ?$

2 SOLVE FOR THE UNKNOWN

Calculate the coefficient of linear expansion using the relationship among known length, change in length, and change in temperature.

$$\alpha = \frac{\Delta L}{L_1 \Delta T}$$

$$= \frac{1.7\times10^{-3}\text{ m}}{(1.60\text{ m})(84°C - 21°C)}$$

◀ Substitute $\Delta L = 1.7\times10^{-3}$ m, $L_1 = 1.60$ m, $\Delta T = (T_2 - T_1) = 84°C - 21°C$

$$= 1.7\times10^{-5}\ °C^{-1}$$

3 EVALUATE THE ANSWER

- **Are the units correct?** The units are correctly expressed in $°C^{-1}$.
- **Is the magnitude realistic?** The magnitude of the coefficient is close to the accepted value for copper.

PRACTICE PROBLEMS

PRACTICE PROBLEMS

Do additional problems. Online Practice

39. A piece of aluminum house siding is 3.66 m long on a cold winter day of −28°C. How much longer is it on the hot summer day shown in **Figure 22?**

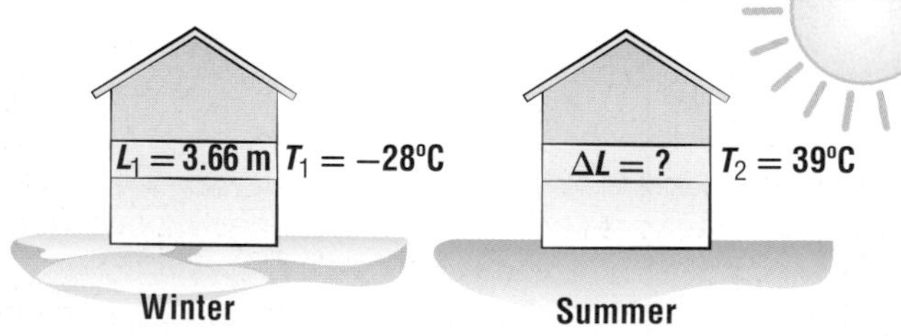

Figure 22

40. A piece of steel is 11.5 cm long at 22°C. It is heated to 1221°C, close to its melting temperature. How long is it?

41. A 400-mL glass beaker at room temperature is filled to the brim with cold water at 4.4°C. When the water warms up to 30.0°C, how much water will spill from the beaker?

42. A tank truck takes on a load of 45,725 L of gasoline in Houston, where the temperature is 28.0°C. The truck delivers its load in Minneapolis, where the temperature is −12.0°C.

a. How many liters of gasoline does the truck deliver?

b. What happened to the gasoline?

43. A hole with a diameter of 0.85 cm is drilled into a steel plate. At 30.0°C, the hole exactly accommodates an aluminum rod of the same diameter. What is the spacing between the plate and the rod when they are cooled to 0.0°C?

44. CHALLENGE A steel ruler is marked in millimeters so that the ruler is absolutely correct at 30.0°C. By what percentage would the ruler be incorrect at −30.0°C?

PHYSICS CHALLENGE

You need to make a 1.00-m-long bar that expands with temperature in the same way as a 1.00-m-long bar of copper would. As shown in the figure at the right, your bar must be made from a bar of iron and a bar of aluminum attached end to end. How long should each of the bars be?

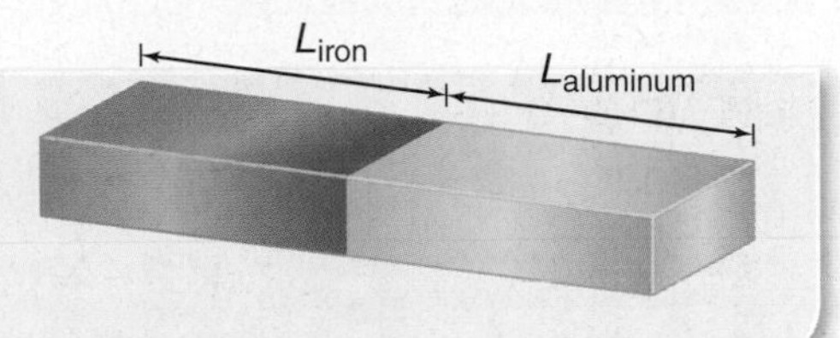

Applications of thermal expansion Engineers take the thermal expansion of materials into consideration as they design structures. You've already read about the expansion joints that are installed on concrete highways and bridges. The regular gaps between slabs of concrete in sidewalks also help keep sidewalks from buckling when the concrete expands during hot weather. Different materials expand at different rates, as indicated by the different coefficients of expansion given in **Table 2.** Engineers also consider different expansion rates when designing systems. Steel bars are often used to reinforce concrete. The steel and concrete must have the same expansion coefficient. Otherwise, the structure could crack on a hot day. Similarly, filling materials used to repair teeth must expand and contract at the same rate as tooth enamel.

Different rates of expansion have useful applications. For example, engineers have taken advantage of these differences to construct a useful device called a bimetallic strip, which is used in thermostats. A bimetallic strip consists of two strips of different metals welded or riveted together. Usually, one strip is brass and the other is iron. When heated, brass expands more than iron does. Thus, when the bimetallic strip of brass and iron is heated, the brass part of the strip becomes longer than the iron part. As a result, the bimetallic strip bends with the brass on the outside of the curve. If the bimetallic strip is cooled, it bends in the opposite direction. The brass is then on the inside of the curve.

In a home thermostat the bimetallic strip is installed so that it bends toward an electric contact as the room cools. When the room cools below the setting on the thermostat, the bimetallic strip bends enough to make electric contact with the switch, which turns on the heater. As the room warms, the bimetallic strip bends in the other direction. When the room's temperature reaches the setting on the thermostat, the electric circuit is broken and the heater switches off.

MiniLAB

JUMPERS

Can heat change the shape of a bimetallic disk?

SECTION 4 REVIEW

Check your understanding.

45. MAINIDEA On a hot day, you are installing an aluminum screen door in a concrete doorframe. You want the door to fit well on a cold winter day. Should you make the door fit tightly in the frame or leave extra room?

46. **Types of Solids** What is the difference between the structure of candle wax and that of ice?

47. **Thermal Expansion** Can you heat a piece of copper enough to double its length?

48. **States of Matter** Does **Table 2** provide a way to distinguish between solids and liquids?

49. **Solids and Liquids** A solid can be defined as a material that can be bent and will resist bending. Explain how these properties relate to the binding of atoms in a solid but do not apply to a liquid.

50. **Critical Thinking** The iron ring in **Figure 23** was made by cutting a small piece from a solid ring. If the ring in the figure is heated, will the gap become wider or narrower? Explain your answer.

Figure 23

BENDING THE RULES!

In a match prior to the 1998 World Cup, Brazilian soccer player Roberto Carlos amazed fans and players by bending a free kick around a wall of defenders by at least a meter and clipping the inside of the post for a goal. The only onlookers who weren't mystified were the physicists in the crowd, as they could easily explain the shot's strange trajectory!

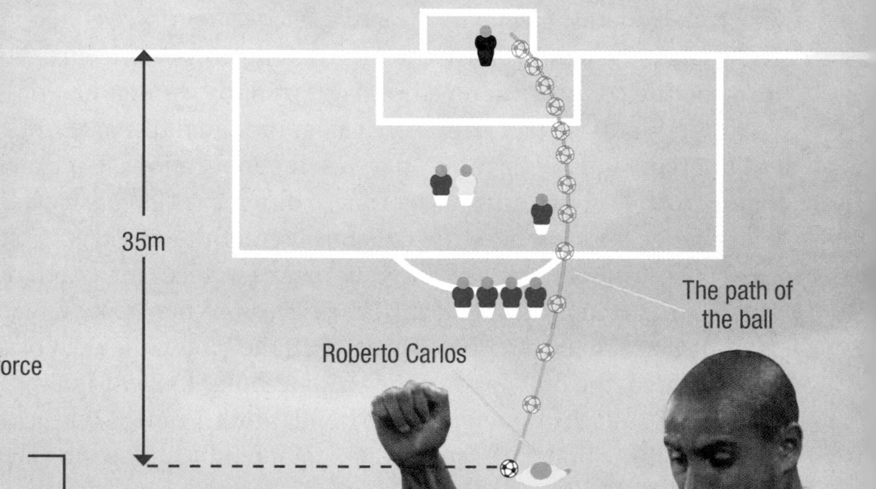

FIGURE 1 A bird's eye view of a soccer ball spinning on its axis perpendicular to the flow of air across it.

35m

The path of the ball

Roberto Carlos

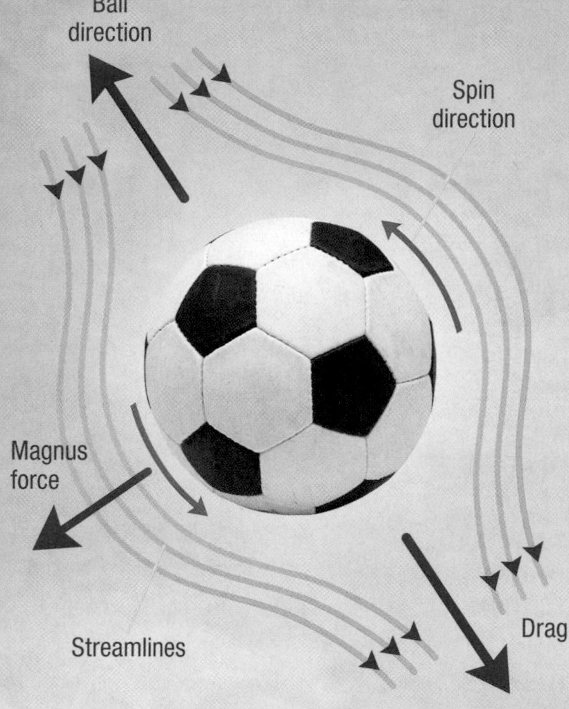

FIGURE 2 The curving path of a rotating ball will spiral exponentially smaller, provided the rotation remains constant.

Bend it like Carlos The path of Carlos's curving shot is shown in **Figure 1.** In a curving shot, the drag force is not the only force that will act on the ball from the air. Another force, called the Magnus force, also acts on the ball. The Magnus force was first explained by Gustav Magnus in 1852. He was trying to determine why spinning artillery shells and bullets deflect to one side. The Magnus force acts on spinning soccer balls too.

In **Figure 2** the left side of the ball is spinning in the same direction as the flow of air around the ball as it moves. As a result, the pressure on the left side of the ball is reduced. Notice that the right side of the ball is spinning in the opposite direction of the airflow. On this side of the ball the drag force is increased. Due to this imbalance of forces, the ball curves to the left. A slow-moving ball with a lot of spin will experience a larger sideways force than a fast-moving ball with the same spin. As the ball slows at the end of its trajectory, the curve becomes more pronounced. Because Carlos had practiced this shot countless times, he knew precisely where the ball would bend in its flight toward the goal.

GOINGFURTHER >>>

Research how different projectiles, such as baseballs, hockey pucks, and flying disks, are designed to travel and how this affects the games in which they are used.

CHAPTER 13 STUDY GUIDE

BIGIDEA The thermal energy of a material and the forces between the material's particles determine its properties.

VOCABULARY
- **fluid** *(p. 348)*
- **pressure** *(p. 349)*
- **pascal** *(p. 349)*
- **combined gas law** *(p. 351)*
- **ideal gas law** *(p. 352)*
- **thermal expansion** *(p. 354)*
- **plasma** *(p. 355)*

SECTION 1 Properties of Fluids

MAINIDEA Fluids flow, have no definite shape, and include liquids, gases, and plasmas.

- Matter in the fluid state flows and has no definite shape of its own.
- The combined gas law represents the relationships among the pressure, volume, and temperature of gases.

$$\frac{P_1V_1}{T_1} = \frac{P_2V_2}{T_2}$$

- The ideal gas law represents the relationship among pressure, volume, temperature, and the number of moles of a gas.
- Plasma is a gaslike state of negatively charged electrons and positively charged ions.

VOCABULARY
- **cohesive forces** *(p. 356)*
- **adhesive forces** *(p. 357)*

SECTION 2 Forces Within Liquids

MAINIDEA Cohesive forces occur between the particles of a substance, while adhesive forces occur between particles of different substances.

- Surface tension is the tendency of the surface of a liquid to contract to the smallest possible area. Surface tension results from the attractive forces that like particles exert on one another.
- Capillary action occurs when a liquid rises in a thin tube because the adhesive forces between the tube and the liquid are stronger than the cohesive forces between liquid's molecules.
- Clouds form when water vapor in the atmosphere cools and condenses, forming droplets around dust particles.

VOCABULARY
- **Pascal's principle** *(p. 359)*
- **buoyant force** *(p. 361)*
- **Archimedes' principle** *(p. 361)*
- **Bernoulli's principle** *(p. 364)*
- **streamlines** *(p. 366)*

SECTION 3 Fluids at Rest and in Motion

MAINIDEA Hydraulic lifts, floating objects, and carburetors rely on the forces exerted by fluids.

- Pascal's principle states that an applied pressure change is transmitted undiminished throughout a fluid.
- According to Archimedes' principle, the buoyant force equals the weight of the fluid displaced by an object.
- Bernoulli's principle states that the pressure exerted by a fluid decreases as its velocity increases.

VOCABULARY
- **crystal lattice** *(p. 367)*
- **amorphous solid** *(p. 367)*
- **coefficient of linear expansion** *(p. 369)*
- **coefficient of volume expansion** *(p. 369)*

SECTION 4 Solids

MAINIDEA Solids usually expand when heated.

- A crystalline solid has a regular pattern of particles, and an amorphous solid has an irregular pattern of particles. Malleability and ductility depend on structure type.
- As the temperature of a solid changes, the kinetic energy of its particles changes accordingly. As the vibrations of the particles change, a solid expands as temperature increases and contracts as temperature decreases.
- Expansion rates of different materials must be considered when designing structures.

Games and Multilingual eGlossary

Chapter Self-Check

SECTION 1 Properties of Fluids

Mastering Concepts

51. How are force and pressure different?

52. A gas is placed in a sealed container, and some liquid is placed in a container of the same size. The gas and liquid both have definite volume. How do they differ?

53. What characteristics do gases and plasmas have in common? How do gases and plasma differ?

54. The Sun is made of plasma. How is this plasma different from the plasmas on Earth?

Mastering Problems

55. Textbooks A 0.85-kg physics book with dimensions of 24.0 cm × 20.0 cm is at rest on a table.

a. What force does the book apply to the table?

b. What pressure does the book apply to the table?

56. Ranking Task Rank the following situations according to the pressure, from least to greatest. Specifically indicate any ties.

A. 20 N exerted over a surface of 0.35 m^2

B. 20 N exerted over a surface of 0.65 m^2

C. 50 N exerted over a surface of 0.05 m^2

D. 50 N exerted over a surface of 0.35 m^2

E. 60 N exerted over a surface of 0.55 m^2

57. As shown in **Figure 24,** a constant-pressure thermometer is made with a cylinder containing a piston that can move freely inside the cylinder. The pressure and the amount of gas enclosed in the cylinder are kept constant. As the temperature increases or decreases, the piston moves up or down in the cylinder. At 0°C, the height of the piston is 20 cm. What is the height of the piston at 100°C?

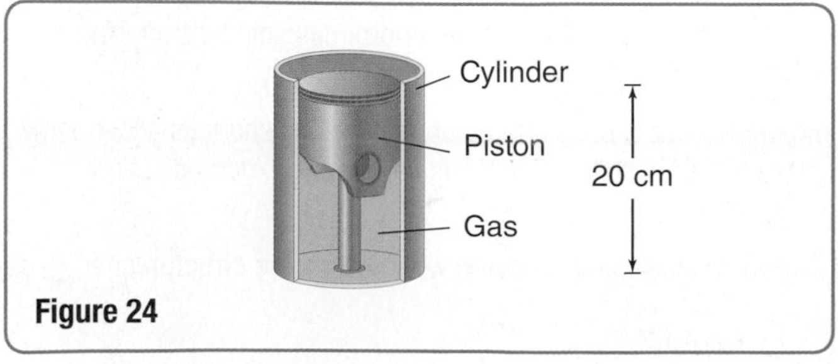

Figure 24

58. Soft Drinks Sodas are made fizzy by the carbon dioxide (CO_2) dissolved in the liquid. An amount of carbon dioxide equal to about 8.0 L of carbon dioxide gas at atmospheric pressure and 300.0 K can be dissolved in a 2-L bottle of soda. The molar mass of CO_2 is 44 g/mol.

a. How many moles of carbon dioxide are in the 2-L bottle? (1 L = 0.001 m^3)

b. What is the mass of the carbon dioxide in the 2-L bottle of soda?

59. A piston with an area of 0.015 m^2 encloses a constant amount of gas in a cylinder with a volume of 0.23 m^3. The initial pressure of the gas equals 1.5×10^5 Pa. A 150-kg mass is placed on the piston, and the piston moves downward to a new position, as shown in **Figure 25.** If the temperature is constant, what is the new volume of the gas in the cylinder?

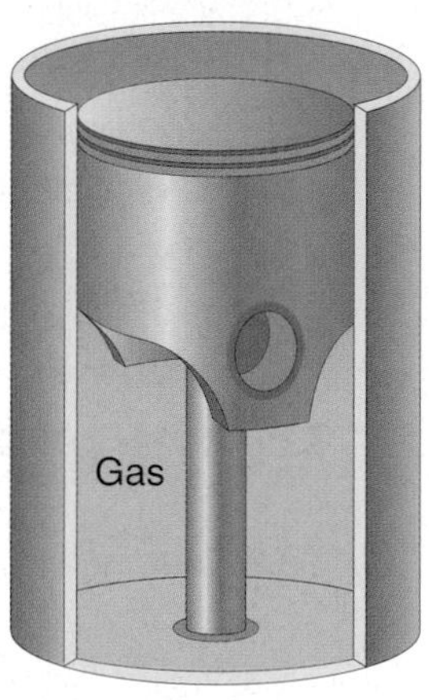

Figure 25

60. Automobiles A certain automobile tire is specified to be used at a gauge pressure of 30.0 psi (psi is pounds per square inch). (One pound per square inch equals 6.90×10^3 Pa.) The term *gauge pressure* means the pressure above atmospheric pressure. Thus, the actual pressure in the tire is 1.01×10^5 Pa + (30.0 psi)(6.90×10^3 Pa/psi) = 3.08×10^5 Pa. As the car is driven, the tire's temperature increases, and the volume and pressure increase. Suppose you filled a car's tire to a volume of 0.55 m^3 at a temperature of 280 K. The initial pressure was 30.0 psi, but during the drive, the tire's temperature increased to 310 K and the tire's volume increased to 0.58 m^3.

a. What is the new pressure in the tire?

b. What is the new gauge pressure?

Chapter Self-Check

SECTION 2 Forces Within Liquids

Mastering Concepts

61. Lakes A frozen lake melts in the spring. What effect does this have on the temperature of the air above the lake?

62. Hiking Canteens used by hikers often are covered with canvas bags. If you wet the canvas bag covering a canteen, the water in the canteen will be cooled. Explain.

SECTION 3 Fluids at Rest and in Motion

Mastering Concepts

63. What do the equilibrium tubes in **Figure 26** tell you about the pressure exerted by a liquid?

Figure 26

64. According to Pascal's principle, what happens to the pressure at the top of a container if the pressure at the bottom is increased?

65. How does the water pressure 1 m below the surface of a small pond compare with the water pressure the same distance below the surface of a lake?

66. Does Archimedes' principle apply to an object inside a flask that is inside a spaceship in orbit?

67. A stream of water goes through a garden hose into a nozzle. As the water speeds up, what happens to the water pressure?

Mastering Problems

68. Reservoirs A reservoir behind a dam is 17 m deep. What is the pressure of the water at the following locations?

a. the base of the dam

b. 4.0 m from the top of the dam

69. A test tube standing vertically in a test-tube rack contains 2.5 cm of oil ($\rho = 0.81$ g/cm^3) and 6.5 cm of water. What is the pressure exerted by the two liquids on the bottom of the test tube?

70. Antiques An antique yellow metal statuette of a bird is suspended from a spring scale. The scale reads 11.81 N when the statuette is suspended in air, and it reads 11.19 N when the statuette is completely submerged in water.

a. Find the volume of the statuette.

b. Is the bird more likely made of gold ($\rho = 19.3\times10^3$ kg/cm^3) or gold-plated aluminum ($\rho = 2.7\times10^3$ kg/cm^3)?

71. During an ecology experiment, an aquarium half-filled with water is placed on a scale. The scale shows a weight of 195 N.

a. A rock weighing 8 N is added to the aquarium. If the rock sinks to the bottom of the aquarium, what will the scale read?

b. The rock is removed from the aquarium, and the amount of water is adjusted until the scale again reads 195 N. A fish weighing 2 N is added to the aquarium. What is the scale reading with the fish in the aquarium?

72. Oceanography As shown in **Figure 27,** a large buoy used to support an oceanographic research instrument is made of a cylindrical, hollow iron tank. The tank is 2.1 m in height and 0.33 m in diameter. The total mass of the buoy and the research instrument is about 120 kg. The buoy must float so that one end is above the water to support a radio transmitter. Assuming that the mass of the buoy is evenly distributed, how much of the buoy will be above the waterline when it is floating?

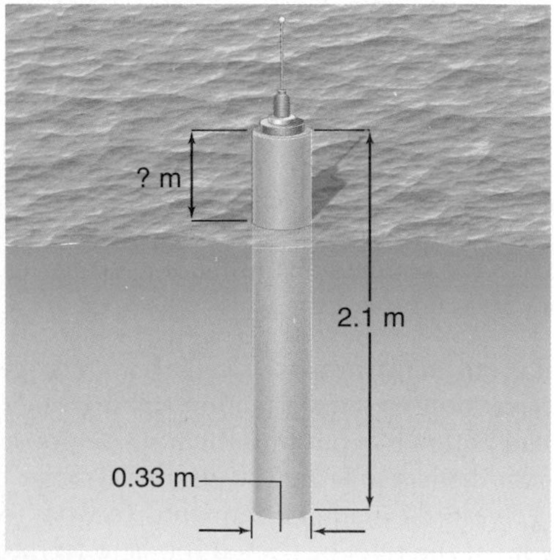

Figure 27

Chapter Self-Check

73. What is the magnitude of the buoyant force on a 26.0-N ball that is floating in fresh water?

74. What is the net force on a rock submerged in water if the rock weighs 45 N in air and has a volume of 2.1×10^{-3} m^3?

75. What is the maximum weight that a balloon filled with 1.00 m^3 of helium can lift in air? Assume that the density of air is 1.20 kg/m^3 and that of helium is 0.177 kg/m^3. Neglect the mass of the balloon.

76. If a rock weighs 54 N in air and experiences a net force of 46 N when submerged in a liquid with a density twice that of water, what will be the net force on it when it is submerged in water?

SECTION 4 Solids

Mastering Concepts

77. How does the arrangement of atoms in a crystalline solid differ from that in an amorphous solid?

78. Does the coefficient of linear expansion depend on the unit of length used? Explain.

Mastering Problems

79. A bar of an unknown metal has a length of 0.975 m at 45°C and a length of 0.972 m at 23°C. What is its coefficient of linear expansion?

80. An inventor constructs a thermometer from an aluminum bar that is 0.500 m in length at 273 K. He measures the temperature by measuring the length of the aluminum bar. If the inventor wants to measure a 1.0-K change in temperature, how precisely must he measure the length of the bar?

81. Bridges How much longer will a 300-m steel bridge be on a 30°C day in August than on a −10°C night in January?

82. What is the change in length of a 2.00-m copper pipe if its temperature is raised from 23°C to 978°C?

83. What is the change in volume of a 1.0-m^3 concrete block if its temperature is raised 45°C?

84. Bridge builders often use rivets that are larger than the rivet hole to make the joint tighter. The rivet is cooled before it is put into the hole. Suppose that a builder drills a hole 1.2230 cm in diameter for a steel rivet 1.2250 cm in diameter. To what temperature must the rivet be cooled if it is to fit into the rivet hole, which is at 20.0°C?

85. A steel tank filled with methanol is 2.000 m in diameter and 5.000 m in height. It is completely filled at 10.0°C. If the temperature rises to 40.0°C, how much methanol (in liters) will flow out of the tank, given that both the tank and the methanol will expand?

86. An aluminum sphere is heated from 11°C to 580°C. If the volume of the sphere is 1.78 cm^3 at 11°C, what is the increase in sphere volume at 580°C?

87. The volume of a copper sphere is 2.56 cm^3 after being heated from 12°C to 984°C. What was the volume of the copper sphere at 12°C?

88. A square iron plate that is 0.3300 m on each side is heated from 0°C to 95°C.

a. What is the change in the length of the sides of the square?

b. What is the change in area of the square?

89. An aluminum cube with a volume of 0.350 m^3 at 350.0 K is cooled to 270.0 K.

a. What is its volume at 270.0 K?

b. What is the length of a side of the cube at 270.0 K?

90. Industry A machinist builds a rectangular mechanical part for a special refrigerator system from two rectangular pieces of steel and two rectangular pieces of aluminum. At 293 K, the part is a perfect square, but at 170 K, the part becomes warped, as shown in **Figure 28.** Which parts were made of steel and which were made of aluminum?

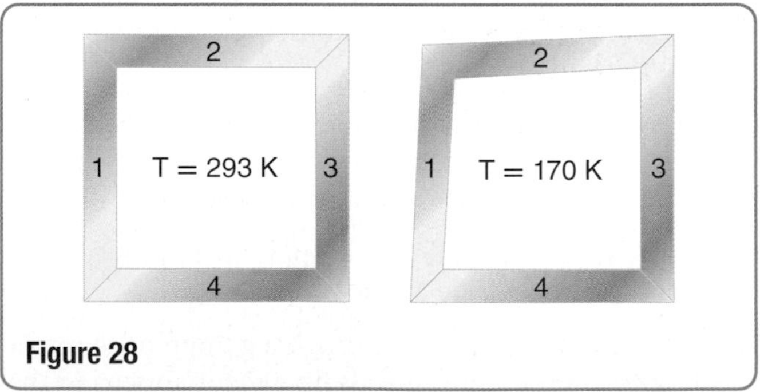

Figure 28

Applying Concepts

91. A rectangular box with its largest surface resting on a table is rotated so that its smallest surface is now on the table. Has the pressure on the table increased, decreased, or remained the same?

92. Show that a pascal is equivalent to a $kg/m{\cdot}s^2$.

Chapter Self-Check

93. Shipping Cargo Compared to an identical empty ship, would a ship filled with table-tennis balls sink deeper into the water or rise in the water? Explain.

94. How deep would a water container have to be to have the same pressure at the bottom as that found at the bottom of a 10.0-cm-deep beaker of mercury, which is 13.55 times as dense as water.

95. Drops of mercury, water, ethanol, and acetone are placed on a smooth, flat surface, as shown in **Figure 29.** From this figure, what can you conclude about the cohesive forces in these liquids?

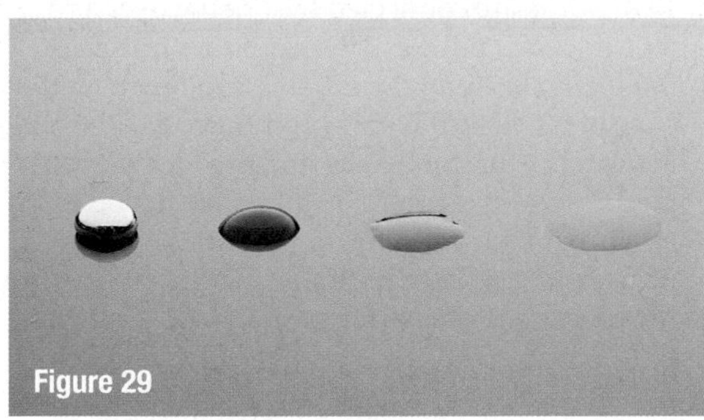
Figure 29

96. Alcohol evaporates more quickly than water does at the same temperature. What does this observation allow you to conclude about the properties of the particles in the two liquids?

97. Five objects with the following densities are put into a tank of water.

a. 85 g/cm^3
b. 0.95 g/cm^3
c. 1.05 g/cm^3
d. 1.15 g/cm^3
e. 1.25 g/cm^3

The density of water is 1.00 g/cm^3. The diagram in **Figure 30** shows six possible positions of these objects. Select a position, from 1 to 6, for each of the five objects. Not all positions need to be selected.

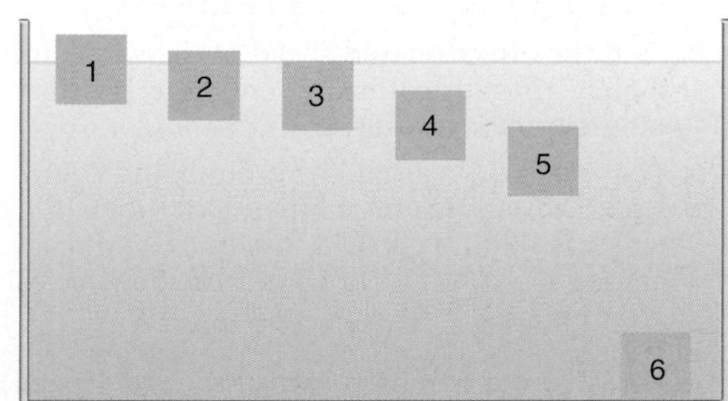

Figure 30

Tim Courlas

98. Equal volumes of water are heated in two tubes that are identical, except that tube A is made of soft glass and tube B is made of ovenproof glass. As the temperature increases, the water level rises higher in tube B than in tube A. Give a possible explanation.

99. A platinum wire easily can be sealed in a glass tube, but a copper wire does not form a tight seal with the glass. Explain.

Mixed Review

100. What is the pressure on the hull of a submarine at a depth of 65 m?

101. Scuba Diving A scuba diver swimming at a depth of 5.0 m under water exhales a 4.2×10^{-6} m^3 bubble of air. What is the volume of that bubble just before it reaches the surface of the water?

102. Reverse Problem Write a physics problem with real-life objects for which the following equation would be part of the solution:

$$T_1 = \frac{(61.2\text{ kPa})(28.0\text{ L})(273\text{ K})}{(77.0\text{ kPa})(25.0\text{ L})}$$

103. BIGIDEA An aluminum bar is floating in a bowl of mercury. When the temperature is increased, does the aluminum float higher or sink deeper into the mercury?

104. There is 100.0 mL of water in an 800.0-mL soft-glass beaker at 15.0°C. How much will the water level have dropped or risen when the beaker and water are heated to 50.0°C?

105. Auto Maintenance A hydraulic jack used to lift cars for repairs is called a 3-ton jack. The large piston is 22 mm in diameter, and the small one is 6.3 mm in diameter. Assume that a force of 3 tons is 3.0×10^4 N.

a. What force must be exerted on the small piston to lift a 3-ton weight?

b. Most jacks use a lever to reduce the force needed on the small piston. If the resistance arm is 3.0 cm, how long must the effort arm of an ideal lever be to reduce the force to 100.0 N?

106. Ballooning A hot-air balloon contains a fixed volume of gas. When the gas is heated, it expands and some gas escapes out the open end. As a result, the mass of the gas in the balloon is reduced. Why would the air in a balloon have to be hotter to lift the same number of people above Vail, Colorado, which has an altitude of 2400 m, than above Norfolk, Virginia, which has an altitude of 3 m?

Chapter Self-Check

Thinking Critically

107. **Problem Posing** Complete this problem so that it can be solved using buoyant force and Newton's second law: "A block of metal has a volume of 2.4 cm^3 and mass of 0.56 kg ...."

108. **Analyze and Conclude** One method of measuring the percentage of body fat is based on the fact that fatty tissue is less dense than muscle tissue. How can a person's average density be assessed with a scale and a swimming pool? What measurements does a physician need to record to find a person's average percentage of body fat?

109. **Analyze and Conclude** A downward force of 700 N is required to fully submerge a plastic foam sphere, as shown in **Figure 31.** The density of the foam is 95 kg/m^3.
   - **a.** What percentage of the sphere would be submerged if the sphere were released to float freely?
   - **b.** What is the weight of the sphere in air?
   - **c.** What is the volume of the sphere?

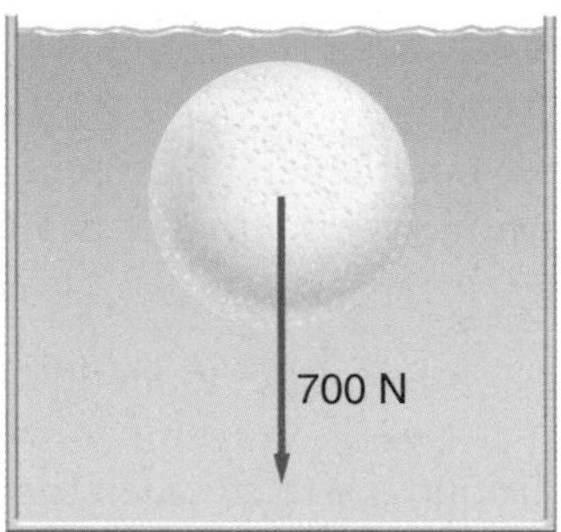

Figure 31

110. **Apply Concepts** Tropical fish for aquariums are often transported home from pet shops in transparent plastic bags filled mostly with water. If you placed a fish in its unopened transport bag in a home aquarium, which of the cases in **Figure 32** best represents what would happen? Explain your reasoning.

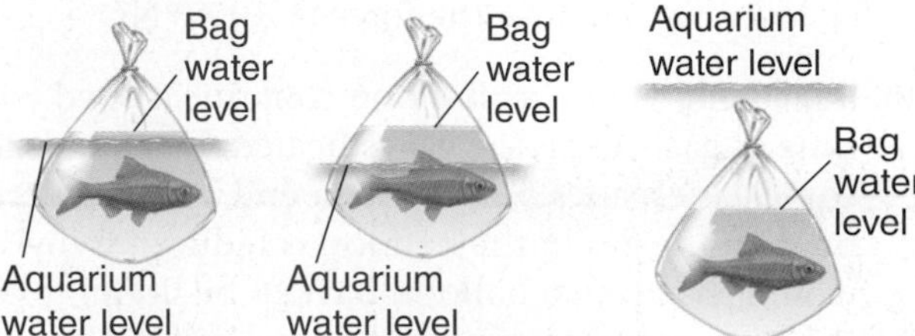

Figure 32

Writing in Physics

111. Some solid materials expand when they are cooled. Water between 4° and 0°C is the most common example, but rubber bands also expand in length when cooled. Research what causes this expansion.

112. Research Joseph Louis Gay-Lussac's contributions to the gas laws. How did Gay-Lussac's work lead to the discovery of the formula for water?

Cumulative Review

113. Two blocks are connected by a string over a frictionless, massless pulley such that one is resting on an inclined plane and the other is hanging over the top edge of the plane, as shown in **Figure 33.** The hanging block has a mass of 3.0 kg and the block on the plane has a mass of 2.0 kg. The coefficient of kinetic friction between the block and the inclined plane is 0.19. Answer the following questions assuming the blocks are released from rest.
   - **a.** What is the acceleration of the blocks?
   - **b.** What is the tension in the string connecting the blocks?

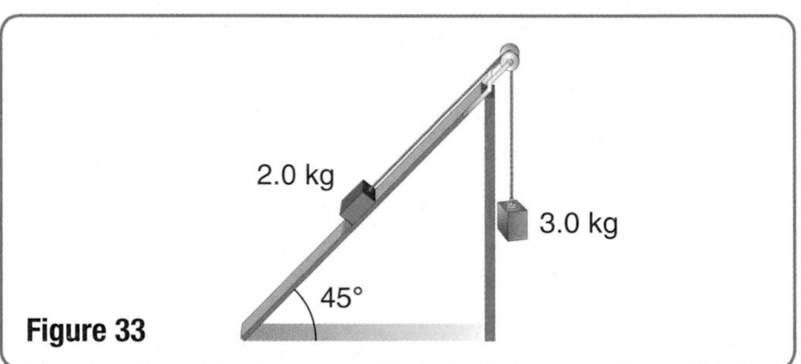

Figure 33

114. A compact car with a mass of 875 kg, moving south at 15 m/s, is struck by a full-sized car with a mass of 1584 kg, moving east at 12 m/s. The two cars stick together, and momentum is conserved.
   - **a.** Sketch the situation, assigning coordinate axes and identifying "before" and "after."
   - **b.** Find the direction and speed of the wreck immediately after the collision, remembering that momentum is a vector quantity.
   - **c.** The wreck skids along the ground and comes to a stop. The coefficient of kinetic friction while the wreck is skidding is 0.55. Assume that the acceleration is constant. How far does the wreck skid after impact?

115. A 188-W motor will lift a load at the rate (speed) of 6.50 cm/s. How great a load can the motor lift at this rate?

STANDARDIZED TEST PRACTICE

MULTIPLE CHOICE

1. Gas with a volume of 10.0 L is trapped in an expandable cylinder. If the pressure is tripled and the temperature is increased by 80.0 percent (as measured on the Kelvin scale), what will be the new volume of the gas?

A. 2.70 L | **C.** 16.7 L
B. 6.00 L | **D.** 54.0 L

2. Nitrogen gas at standard atmospheric pressure, 101.3 kPa, has a volume of 0.080 m^3. If there are 3.6 mol of the gas, what is the temperature?

A. 0.27 K | **C.** 0.27°C
B. 270 K | **D.** 270°C

3. As diagrammed below, an operator applies a force of 200.0 N to the first piston of a hydraulic lift, which has an area of 5.4 cm^2. What is the pressure applied to the hydraulic fluid?

A. 3.7×10^1 Pa | **C.** 3.7×10^3 Pa
B. 2.0×10^3 Pa | **D.** 3.7×10^5 Pa

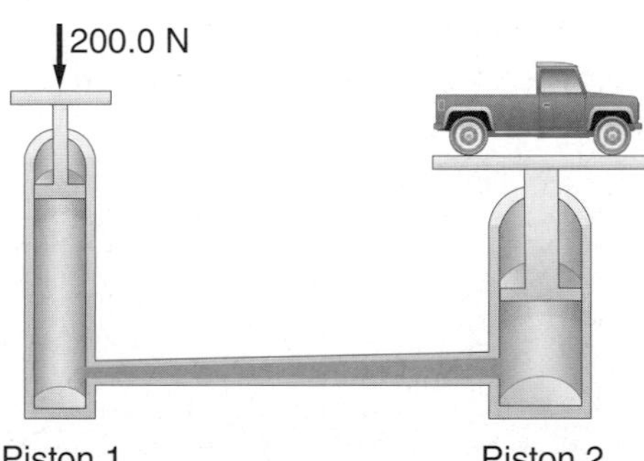

4. If the second piston in the lift diagrammed above exerts a force of 41,000 N, what is the area of the second piston?

A. 0.0049 m^2 | **C.** 0.11 m^2
B. 0.026 m^2 | **D.** 11 m^2

5. The density of cocobolo wood from Costa Rica is 1.10 g/cm^3. What is the net force on a cocobolo wood figurine that displaces 786 mL when submerged in a freshwater lake?

A. 0.770 N | **C.** 7.70 N
B. 0.865 N | **D.** 8.47 N

Online Test Practice

6. What is the buoyant force on a 17-kg object that displaces 85 L of water?

A. 1.7×10^2 N | **C.** 1.7×10^5 N
B. 8.3×10^2 N | **D.** 8.3×10^5 N

7. Which one of the following items does not contain matter in the plasma state?

A. neon lighting
B. stars
C. lightning
D. incandescent lighting

8. Suppose you use a hole punch to make a circular hole in a piece of aluminum foil. If you heat the foil, what will happen to the size of the hole?

A. It will decrease.
B. It will increase.
C. It will decrease, then increase.
D. It will increase, then decrease.

FREE RESPONSE

9. A balloon has a volume of 125 mL of air at standard atmospheric pressure, 101.3 kPa. If the balloon is anchored 1.27 m under the surface of a swimming pool, as illustrated in the diagram below, what is the new volume of the balloon?

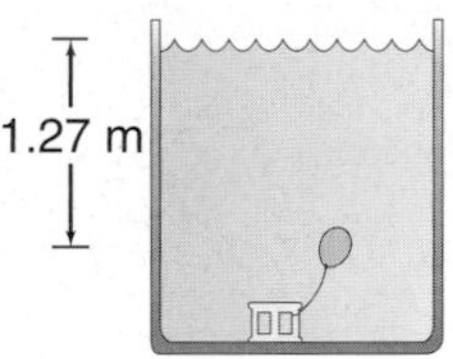

NEED EXTRA HELP?

If you Missed Question	1	2	3	4	5	6	7	8	9
Review Section	1	1	3	3	3	3	1	4	3

CHAPTER 14

Vibrations and Waves

BIGIDEA **Waves and simple harmonic motion are examples of periodic motion.**

SECTIONS

1 **Periodic Motion**

2 **Wave Properties**

3 **Wave Behavior**

LaunchLAB

iLab Station

WAVE INTERACTION

What types of waves can you make on a spring?

WATCH THIS!

Video

PLAYGROUND PHYSICS?

You probably remember playing on swings when you were younger. What did you have to do in order to keep going? How do swings compare to the periodic motion of a pendulum?

(l)Andrew Ward/Life File/Getty Images, (r)JVP photography/Flickr/Getty Images

Go online!
connectED.mcgraw-hill.com

SECTION 1 Periodic Motion

PHYSICS 4 YOU **You might have seen cartoons in which a character is hypnotized using a pocket watch. The watch swings back and forth rhythmically, similar to the pendulum of a grandfather clock. Both the watch and the clock's pendulum undergo periodic motion.**

MAIN IDEA

Periodic motion repeats in a regular cycle.

Essential Questions

- What is simple harmonic motion?
- How much energy is stored in a spring?
- What affects a pendulum's period?

Review Vocabulary

gravitational field (*g*) force per unit mass resulting from the influence of Earth's gravity; near Earth's surface, *g* is 9.8 N/kg toward the center of Earth

New Vocabulary

periodic motion
period
amplitude
simple harmonic motion
Hooke's law
simple pendulum
resonance

Mass on a Spring

The motion of a mass bobbing up and down on a spring is similar to that of a pendulum. The motion repeats, following the same path during the same amount of time. Other similar examples include a vibrating guitar string and a tree swaying in the wind. These motions, which all repeat in a regular cycle, are examples of **periodic motion.**

In each example above, at one position the net force on the object is zero. At that position, the object is in equilibrium. Whenever the object moves away from its equilibrium position, the net force on the system becomes nonzero. This net force acts on the object to bring it back toward equilibrium. The **period** (T) is the time needed for an object to repeat one complete cycle of the motion. The **amplitude** of the motion is the maximum distance the object moves from the equilibrium position.

Simple harmonic motion In **Figure 1,** the force exerted by the spring is directly proportional to the distance the spring is stretched. If you pull the mass down and release it, the mass will bounce up and down through the equilibrium position. Any system in which the force acting to restore an object to its equilibrium position is directly proportional to the displacement of the object shows **simple harmonic motion.**

☑ **READING CHECK** **State** the requirement for simple harmonic motion.

■ **Mass on a Spring**

Figure 1 The force exerted on the mass by the spring is directly proportional to the mass's displacement.

Determine the displacement if the mass is 0.5 *mg*.

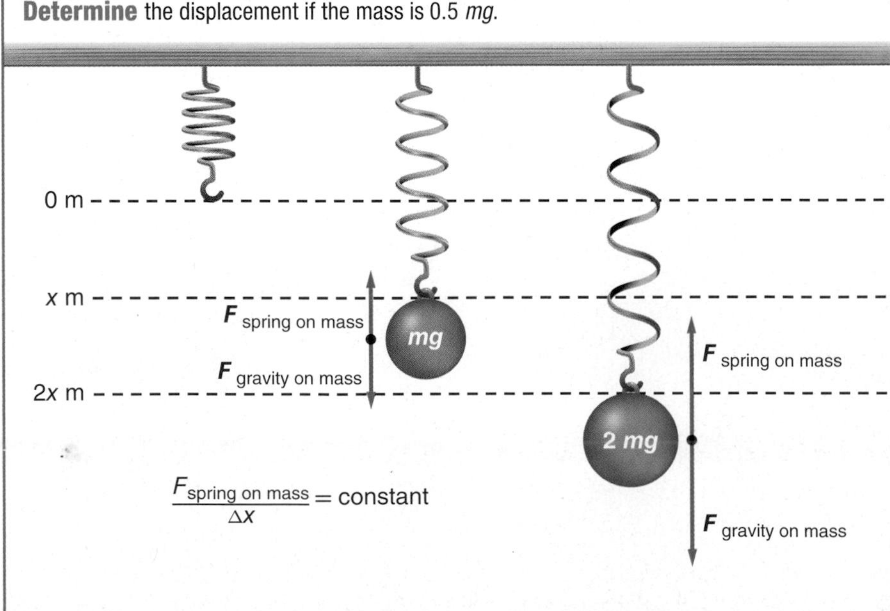

Vstock LLC/Getty Images

Hooke's Law

Table 1
Force Magnitude-Stretch Distance in a Spring

Stretch Distance (m)	Magnitude of Force Exerted by Spring (N)
0.0	0.0
0.030	1.9
0.060	3.7
0.090	6.3
0.12	7.8

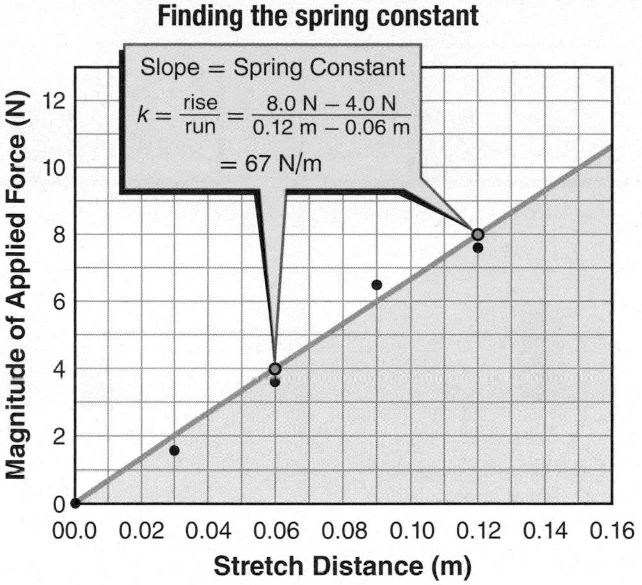

Finding the spring constant

Slope = Spring Constant

$$k = \frac{\text{rise}}{\text{run}} = \frac{8.0\ \text{N} - 4.0\ \text{N}}{0.12\ \text{m} - 0.06\ \text{m}} = 67\ \text{N/m}$$

Magnitude of Applied Force (N): 0, 2, 4, 6, 8, 10, 12

Stretch Distance (m): 00.0, 0.02, 0.04, 0.06, 0.08, 0.10, 0.12, 0.14, 0.16

Figure 2 The spring constant can be determined from the slope of the force magnitude-distance graph. The area under the curve is equal to the potential energy stored in the spring.

Hooke's Law

Table 1 shows the results of an investigation into the relationship between the magnitude of the force exerted by a spring and the distance the spring stretches. **Figure 2** is a graph of the data with the line of best fit. The linear relationship shown in the graph indicates that the magnitude of the force exerted by the spring is directly proportional to the amount the spring is stretched. A spring that exerts a force directly proportional to the distance stretched obeys **Hooke's law.**

HOOKE'S LAW

The magnitude of the force exerted by a spring is equal to the spring constant times the distance the spring is stretched or compressed from its equilibrium position.

$$F = -kx$$

In this equation, k is the spring constant, which depends on the stiffness and other properties of the spring, and x is the distance the spring is stretched from its equilibrium position. Notice that k is the slope of the line in the magnitude of the force v. stretch distance graph. A steeper slope, and thus a larger k, indicates that the spring is harder to stretch. The spring constant has the same units as the slope, newtons/meter (N/m).

The negative sign in Hooke's law indicates that the force is in the direction opposite the stretch or compression direction. The force exerted by the spring on the mass is always directed toward the spring's equilibrium position.

☑ **READING CHECK** **Explain** the negative sign in Hooke's law.

Hooke's law and real springs Not all springs obey Hooke's law. For example, rubber bands do not. Those that do obey Hooke's law are called elastic springs. Even for elastic springs, Hooke's law only applies over a limited range of distances. If a spring is stretched too far, that spring can become so deformed that the force is not proportional to the displacement. We say the spring has exceeded its elastic limit.

Potential energy When you stretch a spring you transfer energy to the spring, giving it elastic potential energy. The work done by an applied force is equal to the area under a force v. distance graph. How much work is done to stretch a spring? In **Figure 2,** the area under the curve represents the work done to stretch the spring. This work is equal to the elastic potential energy stored in the spring. To calculate the potential energy stored in the spring, find the area of the triangle by multiplying one-half the base of the triangle, which is x, by the height of the triangle. According to Hooke's law, the height of the triangle, which is the magnitude of the force, is equal to kx.

POTENTIAL ENERGY IN A SPRING

The potential energy in a spring is equal to one-half times the product of the spring constant and the square of the displacement.

$$PE_{\text{spring}} = \frac{1}{2}kx^2$$

As shown in **Figure 3** on the next page, during completely horizontal simple harmonic motion the spring's elastic potential energy is converted to kinetic energy and then back to potential energy.

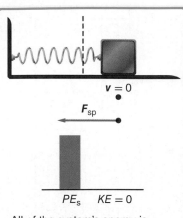

$v = 0$

F_{sp}

PE_s $KE = 0$

All of the system's energy is elastic potential energy. When $t = 0.8$ s, the mass is back to its starting position and repeats the cycle.

v

F_{sp}

PE_s KE

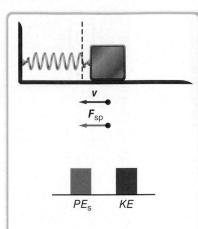

The unbalanced force exerted by the spring accelerates the mass toward equilibrium.

v

$F_{sp} = 0$

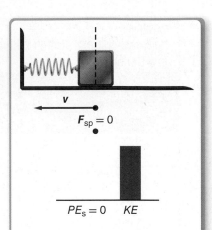

$PE_s = 0$ KE

As the mass passes the equilibrium position, the force is zero, but the velocity and KE are at a maximum.

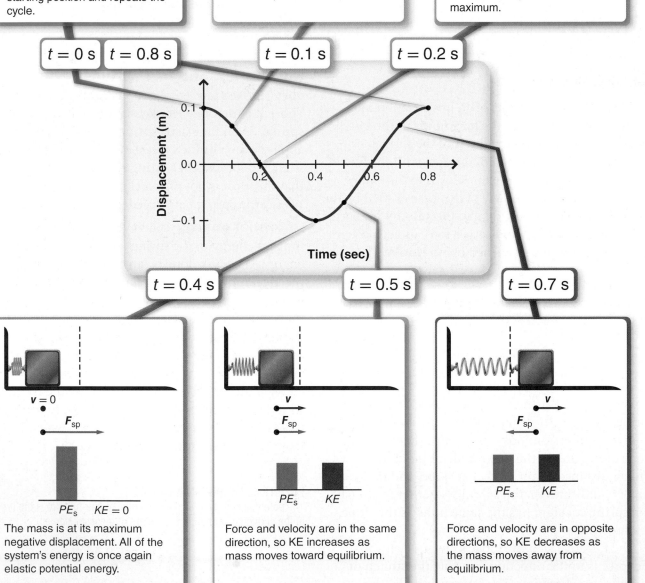

$t = 0$ s $t = 0.8$ s $t = 0.1$ s $t = 0.2$ s

Displacement (m)

0.1

0.0

−0.1

0.2 0.4 0.6 0.8

Time (sec)

$t = 0.4$ s $t = 0.5$ s $t = 0.7$ s

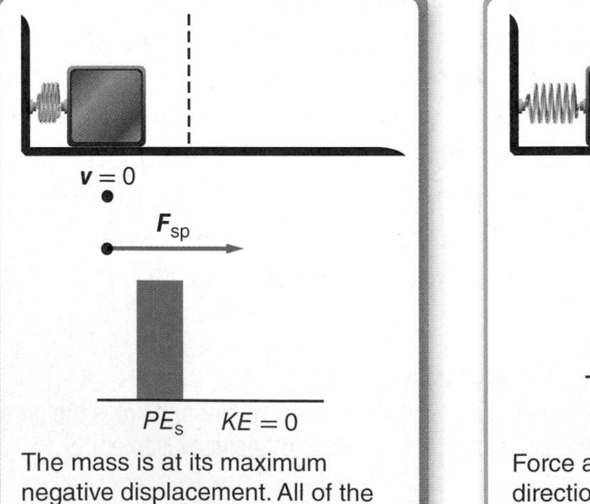

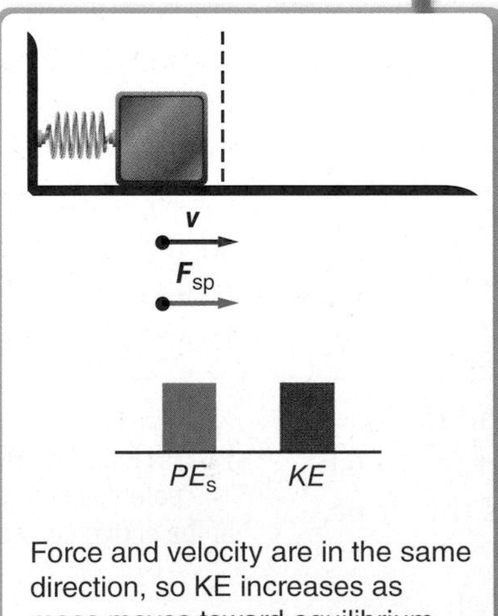

v
F_sp
PE_s KE
Force and velocity are in the same direction, so KE increases as mass moves toward equilibrium.

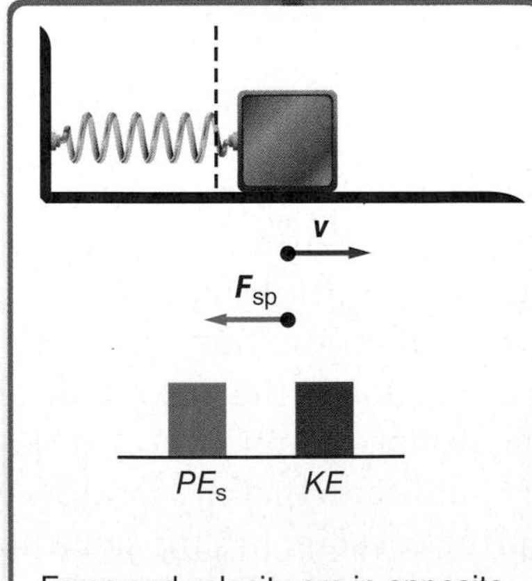

Figure 3 The total mechanical energy of the system is constant throughout the oscillation.

Identify At what position does the mass-and-spring system have the greatest potential energy?

View an **animation of a mass on a spring.**

Concepts In Motion

EXAMPLE PROBLEM 1

Find help with **significant digits.** Math Handbook

THE SPRING CONSTANT AND THE ENERGY OF A SPRING A spring stretches by 18 cm when a bag of potatoes weighing 56 N is suspended from its end.

a. Determine the spring constant.

b. How much elastic potential energy does the spring have when it is stretched this far?

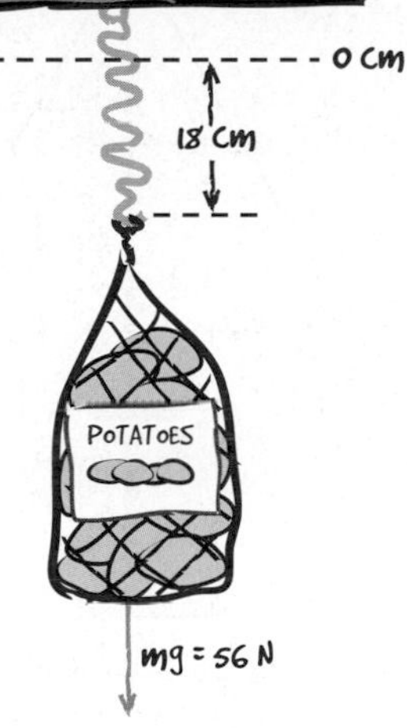

1 ANALYZE AND SKETCH THE PROBLEM

- Sketch the situation.
- Show and label the distance the spring has stretched and its equilibrium position.

KNOWN	UNKNOWN
$x = 18$ cm	$k = ?$
$F = -56$ N	$PE_{sp} = ?$

2 SOLVE FOR THE UNKNOWN

a. Use Hooke's law and isolate k.

$$k = -\frac{F}{x}$$

$$= -\frac{-56\text{ N}}{0.18\text{ m}}$$

◀ Substitute $F = -56$ N, $x = 0.18$ m. The force is negative because it is in the opposite direction of x.

$$= 310\text{ N/m}$$

b. $PE_{sp} = \frac{1}{2}kx^2$

$$= \frac{1}{2}(310\text{ N/m})(0.18\text{ m})^2$$

◀ Substitute $k = 310$ N/m, $x = 0.18$ m.

$$= 5.0\text{ J}$$

3 EVALUATE THE ANSWER

- **Are the units correct?** N/m is the correct unit for the spring constant. $(\text{N/m})(\text{m}^2) = \text{N} \cdot \text{m} = \text{J}$, which is the correct unit for energy.
- **Is the magnitude realistic?** The average magnitude of the force the spring exerts is the average of 0 and 56 N. The work done is $W = Fx = (28\text{ N})(0.18\text{ m}) = 5.0\text{ J}$.

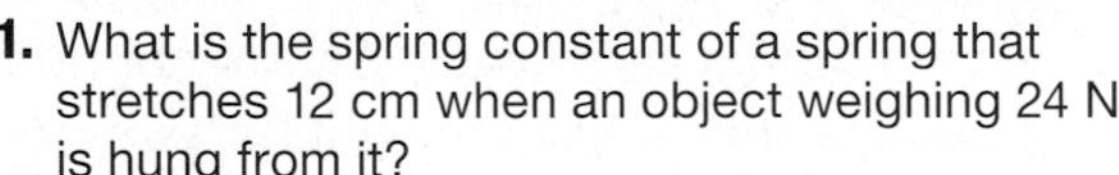

PRACTICE PROBLEMS

Do additional problems. Online Practice

1. What is the spring constant of a spring that stretches 12 cm when an object weighing 24 N is hung from it?
2. A spring with $k = 144$ N/m is compressed by 16.5 cm. What is the spring's elastic potential energy?
3. A spring has a spring constant of 56 N/m. How far will it stretch when a block weighing 18 N is hung from its end?
4. **CHALLENGE** A spring has a spring constant of 256 N/m. How far must it be stretched to give it an elastic potential energy of 48 J?

PHYSICS CHALLENGE

A car of mass m rests at the top of a hill of height h before rolling without friction into a crash barrier located at the bottom of the hill. The crash barrier contains a spring with a spring constant k, which is designed to bring the car to rest with minimum damage.

1. Determine, in terms of m, h, k, and g, the maximum distance (x) the spring will be compressed when the car hits it.
2. If the car rolls down a hill that is twice as high, by what factor will the spring compression increase?
3. What will happen after the car has been brought to rest?

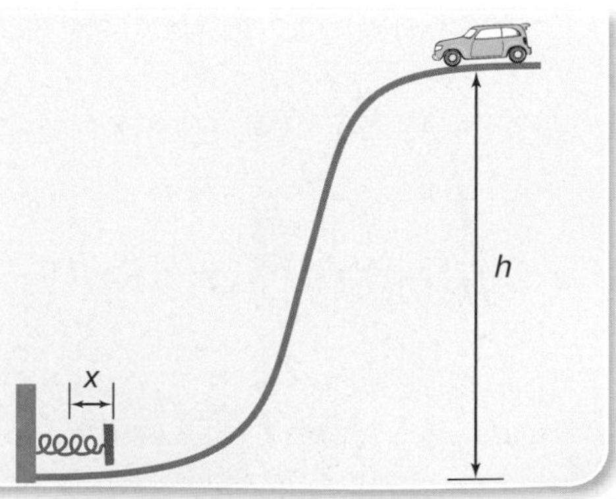

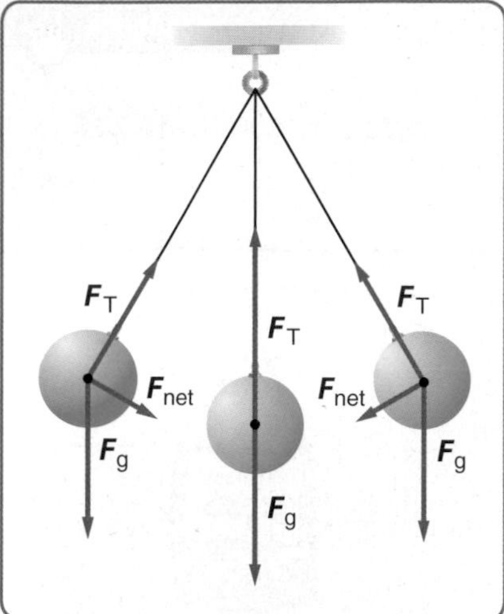

Figure 4 The pendulum's motion is an example of simple harmonic motion because the restoring force is directly proportional to the displacement from equilibrium.

Pendulums

Simple harmonic motion also occurs in the swing of a pendulum. A **simple pendulum** consists of a massive object, called the bob, suspended by a string or a light rod of length ℓ. The bob swings back and forth, as shown in **Figure 4.** The string or rod exerts a tension force ($\mathbf{F_T}$), and gravity exerts a force ($\mathbf{F_g}$) on the bob. Throughout the pendulum's path, the component of the gravitational force in the direction of the pendulum's circular path is a restoring force. At the left and right positions, the restoring force is at a maximum and the velocity is zero. At the equilibrium position, the restoring force is zero and the velocity is maximum.

For small angles (less than about 15°), the restoring force is proportional to the displacement from equilibrium. Similar to the motion of the mass on a spring discussed earlier, the motion of the pendulum is simple harmonic motion. The period of a pendulum is given by the following equation.

PERIOD OF A PENDULUM

The period of a pendulum is equal to 2π times the square root of the length of the pendulum divided by the gravitational field.

$$T = 2\pi\sqrt{\frac{\ell}{g}}$$

Notice that the period depends only on the length of the pendulum and the gravitational field, not on the mass of the bob or the amplitude of oscillation. One practical use of the pendulum is to measure $\mathbf{g}$, which can vary slightly at different locations on Earth.

PhysicsLAB

PENDULUM VIBRATIONS

DESIGN YOUR OWN LAB How can a pendulum demonstrate the properties of waves?

☑ **READING CHECK Describe** the relationship between the period of a simple pendulum and the mass of the pendulum's bob.

Resonance

To get a playground swing going, you can "pump" it by leaning back and pulling the chains at the same point in each swing. Another option is to have a friend give you repeated pushes at just the right times. **Resonance** occurs when forces are applied to a vibrating or oscillating object at time intervals equal to the period of oscillation. As a result, the amplitude of the vibration increases. Other familiar examples of resonance include rocking a car to free it from a snow bank and jumping rhythmically on a trampoline or a diving board to go higher.

Resonance in simple harmonic motion systems causes a larger and larger displacement as energy is added in small increments. The large-amplitude oscillations caused by resonance can produce both useful and catastrophic results. Resonance is used in musical instruments to amplify sounds and in clocks to increase accuracy.

In a particularly dramatic and famous instance, forces generated by the wind contributed to the destruction of the original Tacoma Narrows Bridge, shown in **Figure 5.** Varying forces from of the wind added energy to the bridge's oscillation. These oscillations grew so large that the bridge collapsed into the water below. Architects and engineers must take great care when designing and constructing buildings to insure resonance will not harm the structure.

Figure 5 Wind around the Tacoma Narrows Bridge helped set the bridge in motion. Once in motion, a complex interaction between forces within the bridge's structure and forces applied by the wind caused the bridge's collapse.

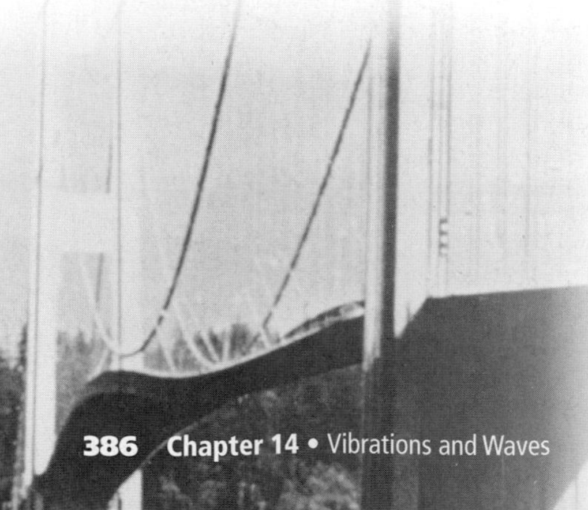

EXAMPLE PROBLEM 2

Find help with **isolating a variable.** Math Handbook

FINDING *g* USING A PENDULUM A pendulum with a length of 36.9 cm has a period of 1.22 s. What is the gravitational field at the pendulum's location?

1 ANALYZE AND SKETCH THE PROBLEM

- Sketch the situation.
- Label the length of the pendulum.

KNOWN	UNKNOWN
$\ell = 36.9$ cm	$g = ?$
$T = 1.22$ s	

36.9 cm

2 SOLVE FOR THE UNKNOWN

$$T = 2\pi\sqrt{\frac{\ell}{g}}$$

◀ Solve for *g*.

$$g = \frac{(2\pi)^2\ell}{T^2}$$

$$= \frac{4\pi^2(0.369\text{ m})}{(1.22\text{ s})^2}$$

◀ Substitute $l = 0.369$ m, $T = 1.22$ s.

$$= 9.78\text{ m/s}^2 = 9.78\text{ N/kg}$$

3 EVALUATE THE ANSWER

- **Are the units correct?** N/kg is the correct unit for gravitational field.
- **Is the magnitude realistic?** The calculated value of *g* is quite close to the accepted value of *g*, 9.8 N/kg. This pendulum could be at a higher elevation above sea level.

EXAMPLE PROBLEM

PRACTICE PROBLEMS

Do additional problems. Online Practice 

5. What is the period on Earth of a pendulum with a length of 1.0 m?

6. How long must a pendulum be on the Moon, where $g = 1.6$ N/kg, to have a period of 2.0 s?

7. CHALLENGE On a planet with an unknown value of *g*, the period of a 0.75-m-long pendulum is 1.8 s. What is *g* for this planet?

SECTION 1 REVIEW

Section Self-Check — Check your understanding.

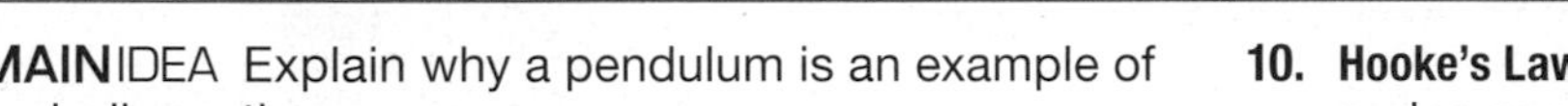

8. **MAIN**IDEA Explain why a pendulum is an example of periodic motion.

9. **Energy of a Spring** The springs shown in **Figure 6** are identical. Contrast the potential energies of the bottom two springs.

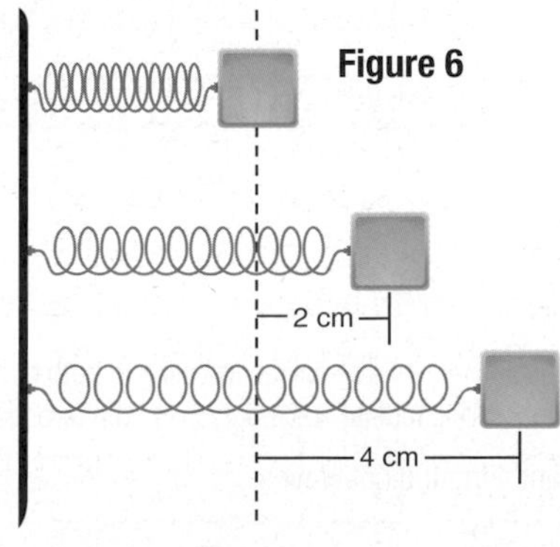

Figure 6

10. **Hooke's Law** Two springs look alike but have different spring constants. How could you determine which one has the larger spring constant?

11. **Hooke's Law** Objects of various weights are hung from a rubber band that is suspended from a hook. The weights of the objects are plotted on a graph against the stretch of the rubber band. How can you tell from the graph whether the rubber band obeys Hooke's law?

12. **Pendulum** How must the length of a pendulum be changed to double its period? How must the length be changed to halve the period?

13. **Resonance** If a car's wheel is out of balance, the car will shake strongly at a specific speed but not at a higher or lower speed. Explain.

14. **Critical Thinking** How is uniform circular motion similar to simple harmonic motion? How are they different?

SECTION 2

Wave Properties

PHYSICS 4 YOU **Imagine tossing a ball to a friend. The ball moves from you and carries kinetic energy. If, however, you and your friend hold the ends of a rope and you give your end a quick shake, the rope remains in your hand. Even though no matter is transferred, the rope carries energy to your friend.**

MAIN IDEA

Waves transfer energy without transferring matter.

Essential Questions

- What are waves?
- How do transverse and longitudinal waves compare?
- What is the relationship between wave speed, wavelength, and frequency?

Review Vocabulary

period in any periodic motion, the amount of time required for an object to repeat one complete cycle of motion

New Vocabulary

wave
wave pulse
transverse wave
periodic wave
longitudinal wave
surface wave
trough
crest
wavelength
frequency

Mechanical Waves

A **wave** is a disturbance that carries energy through matter or space without transferring matter. You have learned how Newton's laws of motion and the law of conservation of energy govern the behavior of particles. These laws also govern the behavior of waves. Water waves, sound waves, and the waves that travel along a rope or a spring are mechanical waves. Mechanical waves travel through a physical medium, such as water, air, or a rope.

Transverse waves A **wave pulse** is a single bump or disturbance that travels through a medium. In the left panel of **Figure 7,** the wave pulse disturbs the rope in the vertical direction, but the pulse travels horizontally. A wave that disturbs the particles in the medium perpendicular to the direction of the wave's travel is called a **transverse wave.** If the disturbances continue at a constant rate, a **periodic wave** is generated.

Longitudinal waves In a coiled spring toy, you can produce another type of wave. If you squeeze together several turns of the coiled spring toy and then suddenly release them, pulses will move away in both directions. The result is called a **longitudinal wave** because the disturbance is parallel to the direction of the wave's travel. Sound waves are longitudinal waves in which the molecules are alternately compressed or decompressed along the path of the wave.

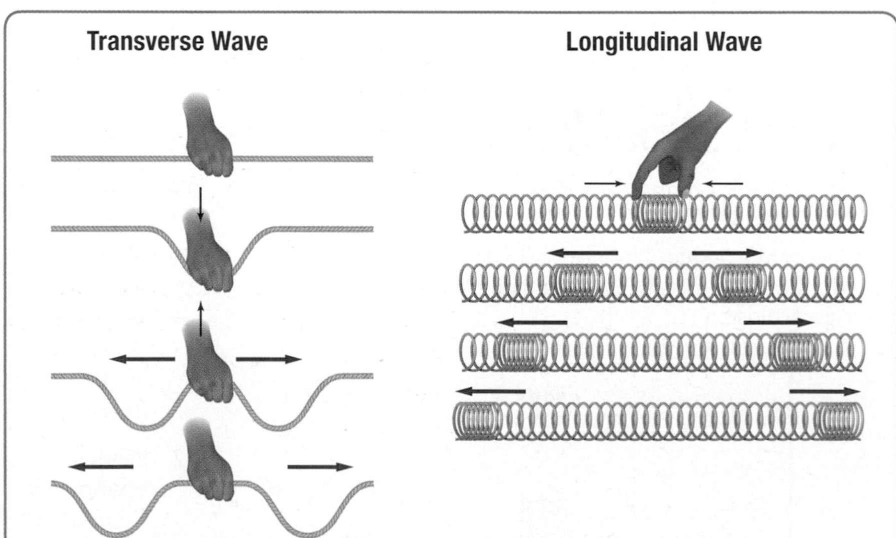

Figure 7 Shaking a rope up and down produces transverse wave pulses traveling in both directions. Squeezing and releasing the coils of a spring produces longitudinal wave pulses in both directions.

Explain the difference between transverse and longitudinal waves.

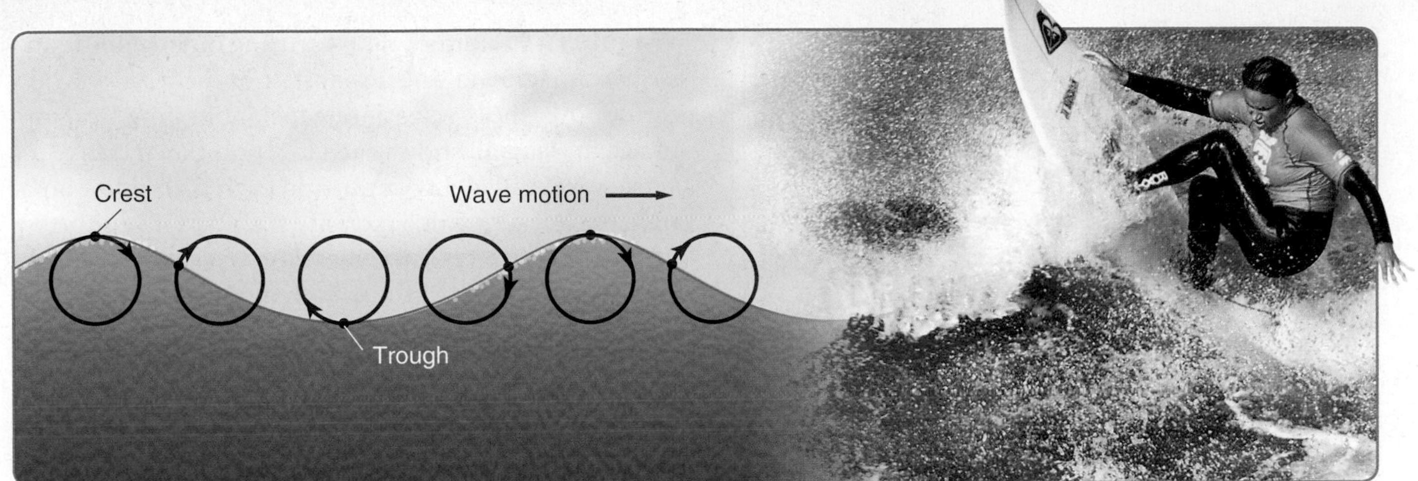

Figure 8 Surface waves in water cause movement both parallel and perpendicular to the direction of wave motion. When these waves interact with the shore, the regular, circular motion is disrupted and the waves break on the beach.

Surface waves Waves that are deep in a lake or an ocean are longitudinal. In a **surface wave,** however, the medium's particles follow a circular path that is at times parallel to the direction of travel and at other times perpendicular to the direction of wave travel, as shown in **Figure 8.** Surface waves set particles in the medium, in this case water, moving in a circular pattern. At the top and bottom of the circular path, particles are moving parallel to the direction of the wave's travel. This is similar to a longitudinal wave. At the left and right sides of each circle, particles are moving up or down. This up-and-down motion is perpendicular to the wave's direction, similar to a transverse wave.

Wave Properties

Many types of waves share a common set of wave properties. Some wave properties depend on how the wave is produced, whereas others depend on the medium through which the wave travels.

Amplitude How does the pulse generated by gently shaking a rope differ from the pulse produced by a violent shake? The difference is similar to the difference between a ripple in a pond and an ocean breaker—they have different amplitudes. You read earlier that the amplitude of periodic motion is the greatest distance from equilibrium. Similarly, as shown in **Figure 9,** a transverse wave's amplitude is the maximum distance of the wave from equilibrium. Since amplitude is a distance, it is always positive. You will learn more about measuring the amplitude of longitudinal waves when you study sound.

Energy of a wave A wave's amplitude depends on how the wave is generated. More energy must be added to the system to generate a wave with a greater amplitude. For example, strong winds produce larger water waves than those formed by gentle breezes. Waves with greater amplitudes transfer more energy. Whereas a wave with a small amplitude might move sand on a beach a few centimeters, a giant wave can uproot and move a tree.

For waves that move at the same speed, the rate at which energy is transferred is proportional to the square of the amplitude. Thus, doubling the amplitude of a wave increases the amount of energy that wave transfers each second by a factor of four.

READING CHECK **Predict** the factor by which the energy transfer per unit time increases if the amplitude of a wave is tripled.

REAL-WORLD PHYSICS

Tsunamis On March 11, 2011, a wall of water estimated to be ten meters high hit areas on the East coast of Japan —tsunami! A tsunami is a series of ocean waves that can have wavelengths over 100 km, periods of one hour, and wave speeds of 500–1000 km/h.

Figure 9 A wave's amplitude is measured from the equilibrium position to the highest or lowest point on the wave.

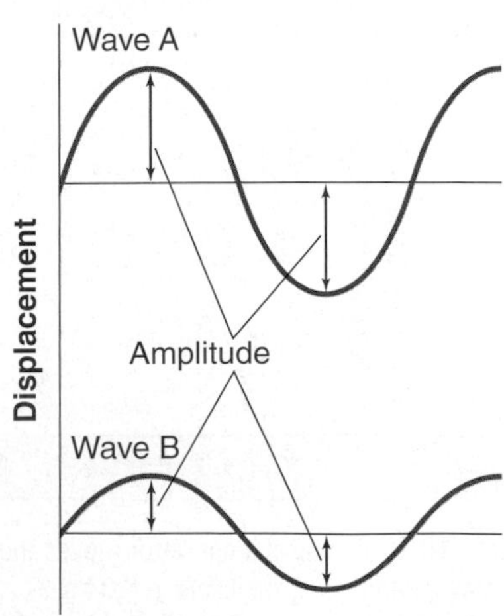

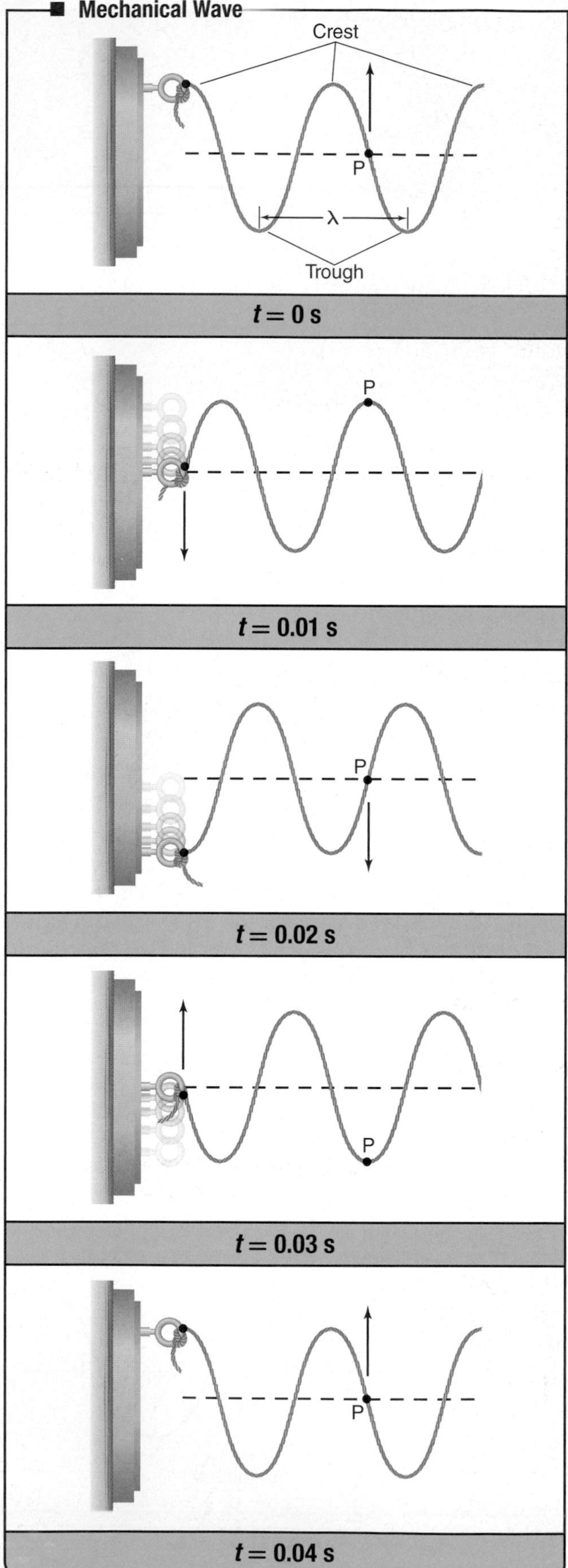

Figure 10 A mechanical oscillator moves the left end of the rope up and down, completing the cycle in 0.04 s.

Wavelength Rather than focusing on one point on a wave, imagine taking a snapshot of the wave so you can see the whole wave at one instant in time. The top image in **Figure 10** shows each low point on a transverse wave, called a **trough,** and each high point on a transverse wave, called a **crest,** of a wave. The shortest distance between points where the wave pattern repeats itself is called the **wavelength.** Crests are spaced by one wavelength. Each trough also is one wavelength from the next. The Greek letter lambda (λ) represents wavelength.

Speed How fast does a wave travel? The speed of a wave pulse can be found in the same way the speed of a moving car is determined. First, measure the displacement of one of the wave's crests or compressions (Δd), then divide this by the time interval (Δt) to find the speed.

$$v = \frac{\Delta d}{\Delta t}$$

For most mechanical waves, both transverse and longitudinal, except water surface waves, wave speed does not depend on amplitude, frequency, or wavelength. Speed depends only on the medium through which the waves move.

READING CHECK Summarize how changing a wave's amplitude, frequency, or wavelength affects the wave's speed.

Phase Any two points on a wave that are one or more whole wavelengths apart are said to be in phase. Particles in the medium are in phase with one another when they have the same displacement from the equilibrium position and the same velocity. Particles with opposite displacements from the equilibrium position and opposite velocities are 180° out of phase. A crest and a trough, for example, are 180° out of phase with each other. Two particles in a wave medium can be anywhere from 0° to 360° out of phase with each other.

Period and frequency Although wave speed and amplitude can describe both wave pulses and periodic waves, period (T) applies only to periodic waves. You have learned that the period of simple harmonic motion, such as the motion of a simple pendulum, is the time it takes for the motion to complete one cycle. Such motion is usually the source, or cause, of a periodic wave. The period of a wave is equal to the period of the source. In **Figure 10** the period (T) equals 0.04 s, which is the time it takes the source to complete one cycle. The same time is taken by P, a point on the rope, to return to its initial position and velocity.

Calculating frequency The **frequency** of a wave (f) is the number of complete oscillations a point on that wave makes each second. Frequency is measured in hertz (Hz). One hertz is one oscillation per second and is equal to 1/s or s^{-1}. The frequency and the period of a wave are related by the following equation.

FREQUENCY OF A WAVE

The frequency of a wave is equal to the reciprocal of the period.

$$f = \frac{1}{T}$$

Both the period and the frequency of a wave depend only on the wave's source. They do not depend on the wave's speed or the medium.

Calculating wavelength You can directly measure a wave's wavelength by measuring the distance between adjacent crests or troughs. You can also calculate it because the wavelength depends on both the frequency of the oscillator and the speed of the wave. In the time interval of one period, a wave moves one wavelength. Therefore, the wavelength of a wave is the speed multiplied by the period, $\lambda = vT$. Using the relation that $f = \frac{1}{T}$, the wavelength equation is very often written in the following way.

View a **BrainPOP video on waves.**

Video

WAVELENGTH

The wavelength of a wave is equal to the velocity divided by the frequency.

$$\lambda = \frac{v}{f}$$

Graphing waves If you took a snapshot of a transverse wave on a coiled spring toy, it might look like one of the waves shown in **Figure 10.** This snapshot could be placed on a graph grid to show more information about the wave, as in the left panel of **Figure 11.** Measuring from peak to peak or trough to trough on such a snapshot provides the wavelength. Now consider recording the motion of a single particle, such as point P in **Figure 10.** That motion can be plotted on a displacement-versus-time graph, as in the right graph in **Figure 11.** Measuring from peak to peak or trough to trough in this graph provides the wave's period.

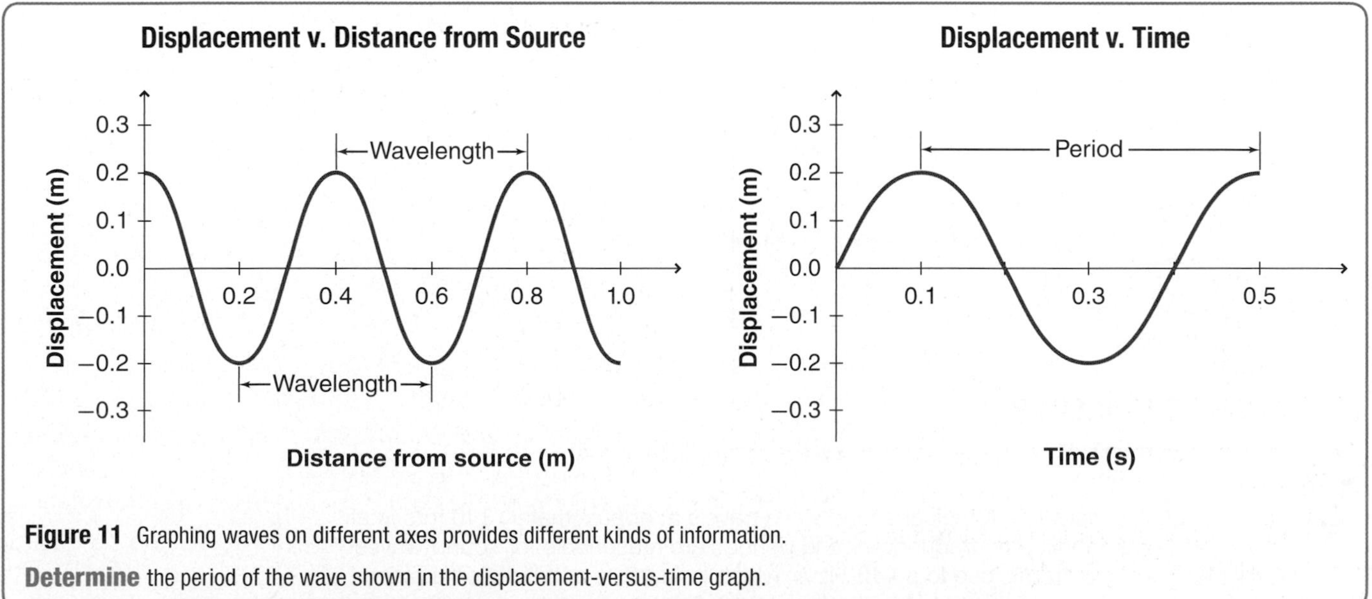

Figure 11 Graphing waves on different axes provides different kinds of information.

Determine the period of the wave shown in the displacement-versus-time graph.

EXAMPLE PROBLEM 3

Get help with **waves.** Personal Tutor

EXAMPLE PROBLEM

CHARACTERISTICS OF A WAVE A sound wave has a frequency of 192 Hz and travels the length of a football field, 91.4 m, in 0.271 s.

a. What is the speed of the wave?

b. What is the wavelength of the wave?

c. What is the period of the wave?

d. If the frequency were changed to 442 Hz, what would be the new wavelength and period?

1 ANALYZE AND SKETCH THE PROBLEM

- Draw a diagram of the wave.
- Draw a velocity vector for the wave.

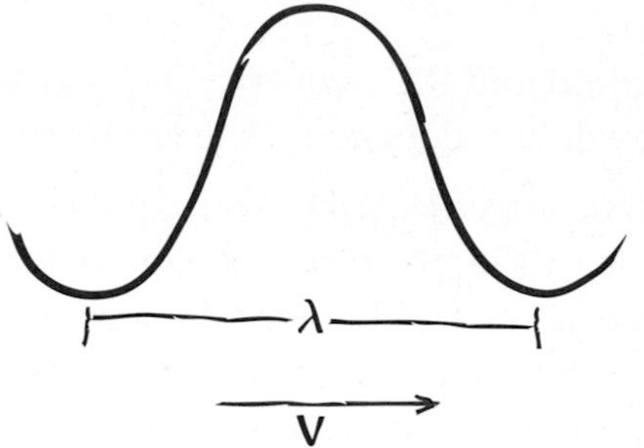

KNOWN	UNKNOWN
$f = 192$ Hz	$v = ?$
$\Delta d = 91.4$ m	$\lambda = ?$
$\Delta t = 0.271$ s	$T = ?$

2 SOLVE FOR THE UNKNOWN

a. Use the definition of velocity.

$$v = \frac{\Delta d}{\Delta t}$$ ◀ Substitute $\Delta d = 91.4$ m, $\Delta t = 0.271$ s

$$= \frac{91.4\text{ m}}{0.271\text{ s}}$$

$$= 337\text{ m/s}$$

b. Use the relationship between wave velocity, wavelength, and frequency.

$$\lambda = \frac{v}{f}$$ ◀ Substitute $v = 337$ m/s, $f = 192$ Hz.

$$= \frac{337\text{ m/s}}{192\text{ Hz}}$$

$$= 1.76\text{ m}$$

c. Use the relationship between period and frequency.

$$T = \frac{1}{f}$$ ◀ Substitute $f = 192$ Hz.

$$= \frac{1}{192\text{ Hz}}$$

$$= 0.00521\text{ s}$$

d. $$\lambda = \frac{v}{f}$$ ◀ Substitute $v = 337$ m/s, $f = 442$ Hz.

$$= \frac{337\text{ m/s}}{442\text{ Hz}}$$

$$= 0.762\text{ m}$$

$$T = \frac{1}{f}$$ ◀ Substitute $f = 442$ Hz.

$$= \frac{1}{442\text{ Hz}}$$

$$= 0.00226\text{ s}$$

3 EVALUATE THE ANSWER

Are the units correct? Hz has the unit s^{-1}, so $(\frac{m}{s})/\text{Hz} = (\frac{m}{s}) \cdot s = m$, which is correct for λ.

Are the magnitudes realistic? A typical sound wave travels at approximately 340 m/s in air, so 337 m/s is reasonable. The frequencies and periods are reasonable for sound waves. The frequency of 442 Hz is close to a 440-Hz A, which is A above middle-C on a piano.

PRACTICE PROBLEMS

Do additional problems. Online Practice

PRACTICE PROBLEMS

15. A sound wave produced by a clock chime is heard 515 m away 1.50 s later.

a. Based on these measurements, what is the speed of sound in air?

b. The sound wave has a frequency of 436 Hz. What is the period of the wave?

c. What is its wavelength?

16. If you want to increase the wavelength of waves in a rope, should you shake it at a higher or lower frequency?

17. What is the speed of a periodic wave disturbance that has a frequency of 3.50 Hz and a wavelength of 0.700 m?

18. How does increasing the wavelength by 50 percent affect the frequency of a wave on a rope?

19. The speed of a transverse wave in a string is 15.0 m/s. If a source produces a disturbance that has a frequency of 6.00 Hz, what is its wavelength?

20. Five wavelengths are generated every 0.100 s in a tank of water. What is the speed of the wave if the wavelength of the surface wave is 1.20 cm?

21. A periodic longitudinal wave that has a frequency of 20.0 Hz travels along a coiled spring toy. If the distance between successive compressions is 0.600 m, what is the speed of the wave?

22. How does the frequency of a wave change when the period of the wave is doubled?

23. Describe the change in the wavelength of a wave when the period is reduced by one-half.

24. If the speed of a wave increases to 1.5 times its original speed while the frequency remains constant, how does the wavelength change?

25. CHALLENGE A hiker shouts toward a vertical cliff as shown in **Figure 12.** The echo is heard 2.75 s later.

a. What is the speed of sound of the hiker's voice in air?

b. The wavelength of the sound is 0.750 m. What is its frequency?

c. What is the period of the wave?

Figure 12

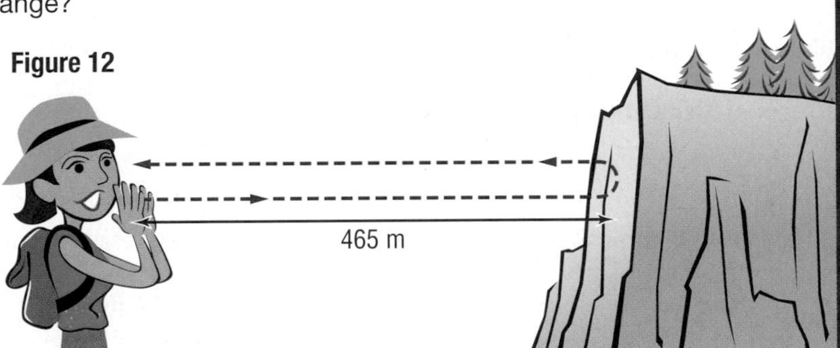

SECTION 2 REVIEW

Section Self-Check Check your understanding.

26. MAINIDEA Suppose you and your lab partner are asked to demonstrate that a transverse wave transports energy without transferring matter. How could you do it?

27. Wave Characteristics You are creating transverse waves on a rope by shaking your hand from side to side. Without changing the distance your hand moves, you begin to shake it faster and faster. What happens to the amplitude, wavelength, frequency, period, and velocity of the wave?

28. Longitudinal Waves Describe longitudinal waves. What types of mediums transmit longitudinal waves?

29. Speeds in Different Mediums If you pull on one end of a coiled spring toy, does the pulse reach the other end instantaneously? What happens if you pull on a rope? What happens if you hit the end of a metal rod? Compare the pulses traveling through these three materials.

30. Critical Thinking If a raindrop falls into a pool, it produces waves with small amplitudes. If a swimmer jumps into a pool, he or she produces waves with large amplitudes. Why doesn't the heavy rain in a thunderstorm produce large waves?

SECTION 3 # Wave Behavior

PHYSICS 4 YOU

You have probably seen the outward ripples that occur when a rock is dropped into a calm lake. But what happens when two rocks are dropped into the lake near each other? How do the ripples from the rocks interact?

MAIN IDEA

Waves can interfere with other waves.

Essential Questions

- How are waves reflected and refracted at boundaries between mediums?
- How does the principle of superposition apply to the phenomenon of interference?

Review Vocabulary

tension the specific name for the force exerted by a rope or a string

New Vocabulary

incident wave
reflected wave
principle of superposition
interference
node
antinode
standing wave
wavefront
ray
normal
law of reflection
refraction

Waves at Boundaries

Recall from Section 2 that the speed of a mechanical wave depends only on the properties of the medium it passes through, not on the wave's amplitude or frequency. For water waves, the depth of the water affects wave speed. For sound waves through air, the temperature affects wave speed. For waves on a spring, the speed depends on the spring's tension and mass per unit length.

Examine what happens when a wave travels across a boundary from one medium into another, such as two springs of different thicknesses joined end to end. **Figure 13** shows a wave pulse traveling from a larger spring into a smaller one. The pulse that strikes the boundary is called the **incident wave.** One pulse from the larger spring continues in the smaller spring, but the speed of the pulse is different in the smaller spring. Note that this transmitted wave pulse remains upward.

Some of the energy of the incident wave's pulse is reflected backward into the larger spring. This returning wave is called the **reflected wave.** Whether the reflected wave is upright or inverted depends on the characteristics of the two springs. For example, if the waves in the smaller spring have a greater speed because the spring is stiffer, then the reflected wave will be inverted.

■ **Wave at a Boundary**

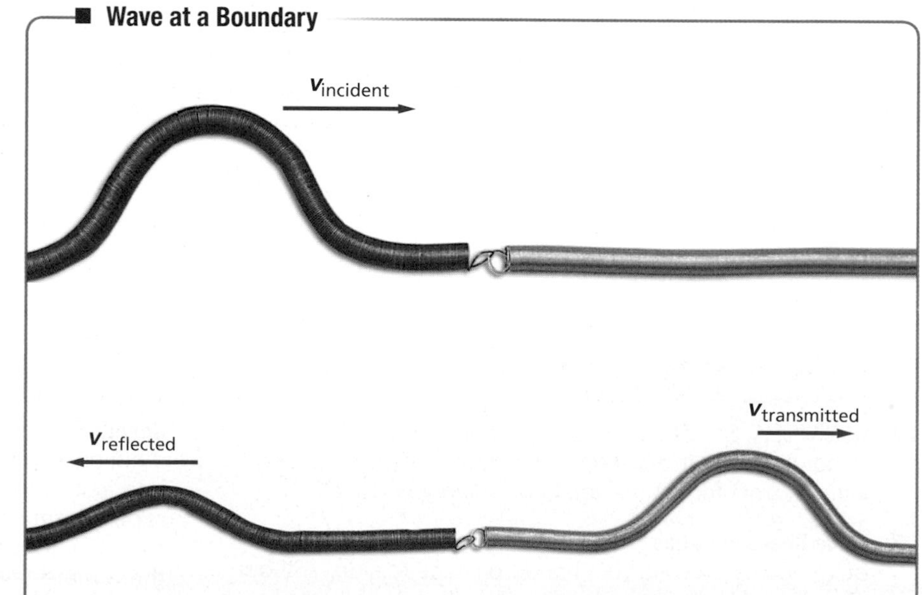

Figure 13 When the wave pulse meets the boundary between the two springs, a transmitted wave pulse and a reflected wave form.

Compare the energy of the incident wave to the energy of the reflected wave.

(tl)Don Farrall/Photodisc/Getty Images (tr br)Tom Pantages

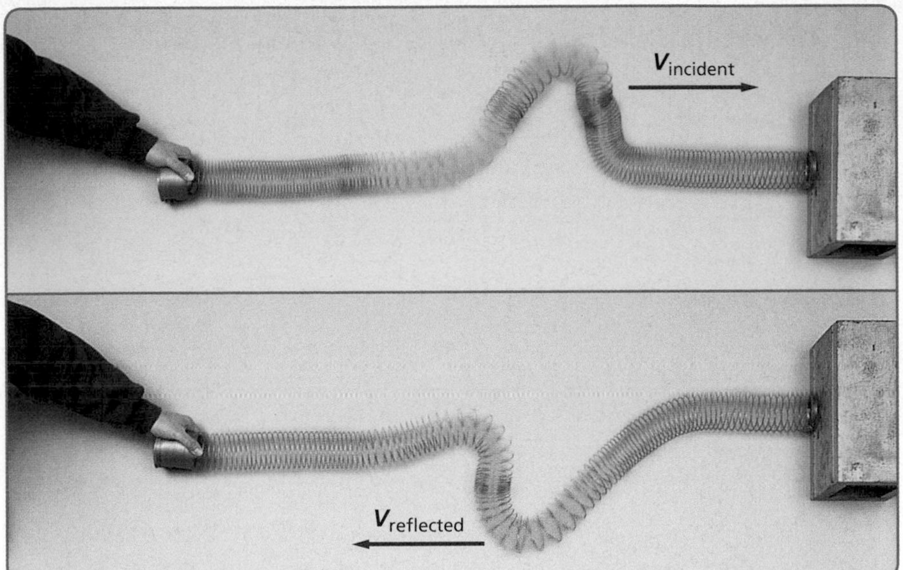

Figure 14 When a wave encounters a rigid boundary, the reflected wave is inverted. Note that the amplitude is not affected by the rigid boundary.

MiniLAB

WAVE REFLECTION

Does reflection change a wave's speed?

Rigid boundaries When a wave pulse hits a rigid boundary, the energy is reflected back, as shown in **Figure 14.** The wall is the boundary of a new medium through which the wave attempts to pass. Instead of passing through, the pulse is reflected from the wall with almost exactly the same amplitude as the pulse of the incident wave. Thus, almost all the wave's energy is reflected back. Very little energy is transmitted into the wall. Also note that the pulse is inverted.

Superposition of Waves

Suppose a pulse traveling along a spring meets a reflected pulse that is coming back from a boundary, as shown in **Figure 15.** In this case, two waves exist in the same place in the medium at the same time. Each wave affects the medium independently. The **principle of superposition** states that the displacement of a medium caused by two or more waves is the algebraic sum of the displacements caused by the individual waves. In other words, two or more waves can combine to form a new wave. If the waves move in the same medium, they can cancel or form a new wave of lesser or greater amplitude. The result of the superposition of two or more waves is called **interference.**

VOCABULARY

Science Usage v. Common Usage

Interference

• **Science usage**

the result of superposition of two or more waves

The amplitude of the interference of several waves was much larger than the amplitude of the individual waves.

• **Common usage**

the act of coming between in a way that hinders or impedes

Ehud was ejected from the game for an interference foul.

■ Interference of Waves

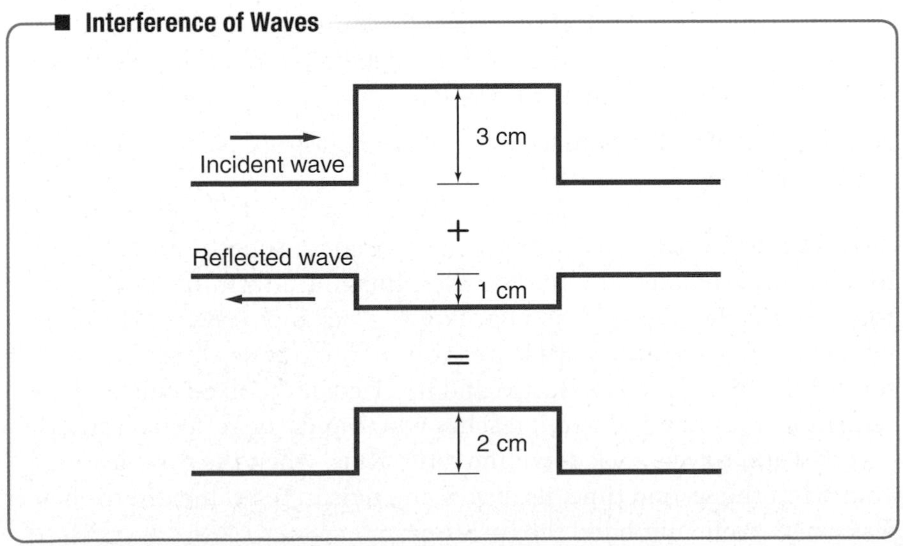

Figure 15 Waves add algebraically during superposition.

Tom Pantages

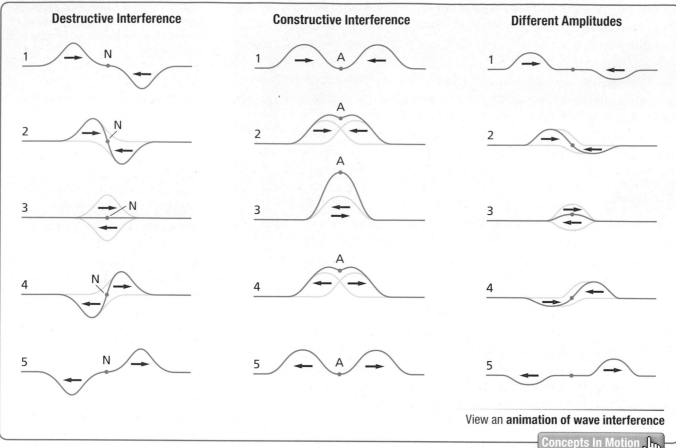

Destructive Interference | **Constructive Interference** | **Different Amplitudes**

View an **animation of wave interference**

Concepts In Motion

Figure 16 When waves add algebraically, the resulting combined waves can be quite different from the individual waves.

Wave interference Wave interference can be either constructive or destructive. The first panel in **Figure 16** shows the superposition of waves with equal but opposite displacements, causing destructive interference. When the pulses meet and are in the same location, the displacement is zero. Point N, which does not move at all, is called a **node.** The pulses travel horizontally and eventually resume their original form.

Constructive interference occurs when wave displacements are in the same direction. The result is a wave that has an amplitude greater than those of the individual waves. A larger pulse appears at point A when the two waves meet. Point A has the largest displacement and is called the **antinode.** The two pulses pass through each other without changing their shapes or sizes. Even if the pulses have unequal amplitudes, the resultant pulse at the overlap is still the algebraic sum of the two pulses, as shown in the final panel of **Figure 16.**

☑ **READING CHECK Compare** the wave medium's displacement at a node and at an antinode.

PhysicsLAB

INTERFERENCE AND DIFFRACTION

How do waves interfere with one another?

Two reflections You can apply the concept of superimposed waves to the control of large-amplitude waves. Imagine attaching one end of a rope to a fixed point, such as a doorknob, a distance L away. When you vibrate the free end, the wave leaves your hand, travels along the rope toward the fixed end, is reflected and inverted at the fixed end, and returns to your hand. When it reaches your hand, the reflected wave is inverted and travels back down the rope. Thus, when the wave leaves your hand the second time, its displacement is in the same direction as it was when it left your hand the first time.

MiniLAB

WAVE INTERACTION

What happens when waves traveling in different directions meet?

Standing waves Suppose you adjust the motion of your hand so that the period of vibration equals the time needed for the wave to make one round-trip from your hand to the door and back. Then, the displacement given by your hand to the rope each time will add to the displacement of the reflected wave. As a result, the amplitude of oscillation of the rope will be much greater than the motion of your hand. This large-amplitude oscillation is an example of mechanical resonance.

The ends of the rope are nodes and an antinode is in the middle, as shown in the top photo in **Figure 17.** Thus, the wave appears to be standing still and is called a **standing wave.** You should note, however, that the standing wave is the interference of waves traveling in opposite directions. If you double the frequency of vibration, you can produce one more node and one more antinode on the rope. Then it appears to vibrate in two segments. When you further increase the vibration frequency, it produces even more nodes and antinodes, as shown in the bottom photo in **Figure 17.**

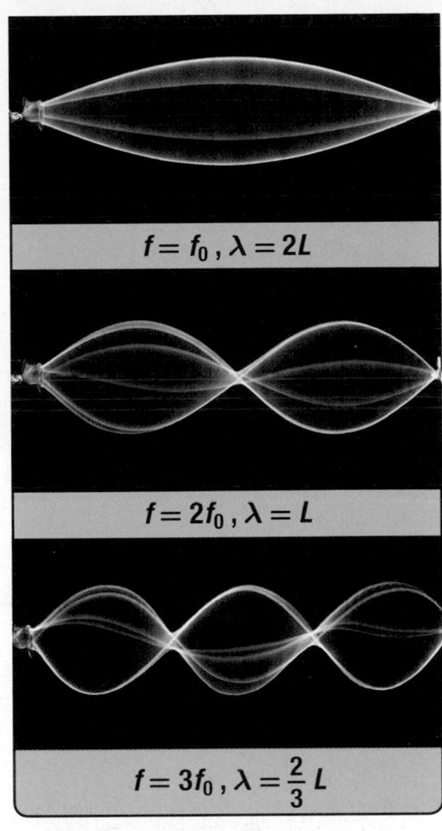

Figure 17 Interference produces standing waves only at certain frequencies.

Predict the wavelength if the frequency is four times the lowest frequency.

View an **animation of standing waves.**

Waves in Two Dimensions

You have studied waves on a rope and on a spring reflecting from rigid supports. During some of these interactions, the amplitude of the waves is forced to be zero by destructive interference. These mechanical waves travel in only one dimension. Waves on the surface of water, however, travel in two dimensions, and sound waves and electromagnetic waves will later be shown to travel in three dimensions. How can two-dimensional waves be represented?

Picturing waves in two dimensions When you throw a small stone into a calm pool of water, you see the circular crests and troughs of the resulting waves spreading out in all directions. You can sketch those waves by drawing circles to represent the wave crests. If you repeatedly dip your finger into water with a constant frequency, the resulting sketch would be a series of concentric circles, called wavefronts, centered on your finger. A **wavefront** is a line that represents the crest of a wave in two dimensions. Wavefronts can be used to show two-dimensional waves of any shape, including circular waves. The photo in **Figure 18** shows circular waves in water. The circles drawn on the diagram show the wavefronts that represent those water waves.

Whatever their shape, two-dimensional waves always travel in a direction that is perpendicular to their wavefronts. That direction can be represented by a **ray,** which is a line drawn at a right angle to the wavefront. When all you want to show is the direction in which a wave is traveling, it is convenient to draw rays instead of wavefronts. The red arrows in **Figure 18** are rays that show the water waves' direction of motion. One advantage of drawing wavefronts is when wavefronts are drawn to scale, they show the wave's wavelengths. In **Figure 18,** the wavelength equals the distance from one circle to the next.

Figure 18 Waves spread out in a circular pattern from the oscillating source.

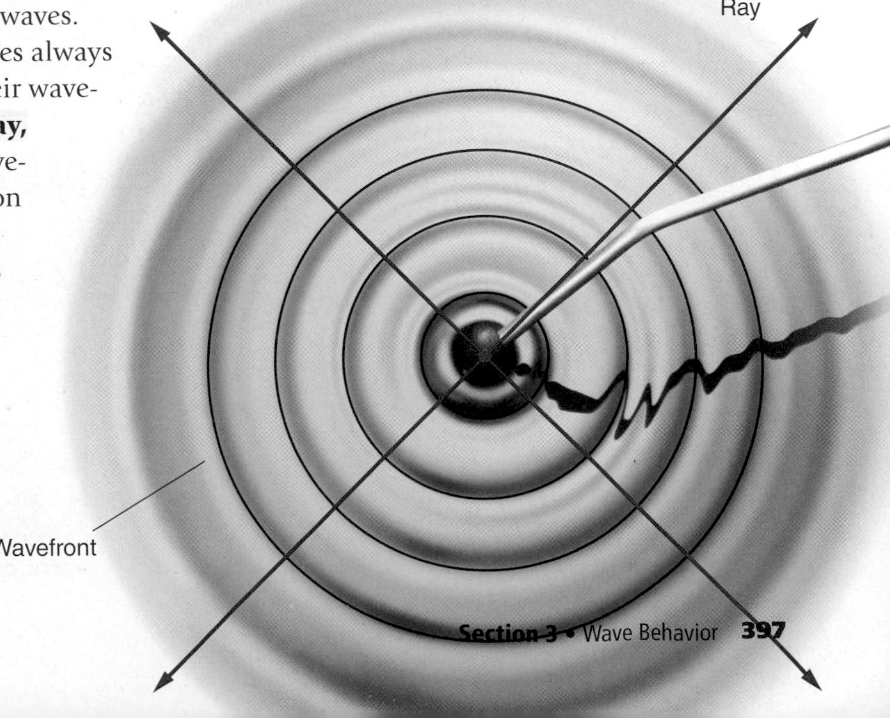

☑ **READING CHECK Identify** the relationship between wavefronts and rays.

VOCABULARY
Science Usage v. Common Usage

Normal
- Science usage
line in a diagram that is drawn perpendicular to a surface
As measured from the normal, angles of incidence and reflection are equal.
- Common usage
conforming to a type, standard, or regular pattern
Such cold temperatures in July are not normal.

Creating two-dimensional waves A ripple tank is a piece of laboratory equipment that is used to investigate the properties of two-dimensional waves. The main portion of the ripple tank shown in **Figure 19** is a shallow tank that contains a thin layer of water. A board attached to a mechanical oscillator produces waves with long, straight wavefronts. A lamp above the tank produces shadows below the tank that show the locations of the crests of the waves. The top photo in **Figure 19** shows a wave traveling through the ripple tank. The direction the wave travels is modeled by a ray diagram. For clarity, the wavefronts are not extended the entire length of the wave.

Reflection of two-dimensional waves The bottom row of pictures in **Figure 19** shows an incident ray encountering a rigid barrier placed at an angle to the ray's path. The orientation of the barrier is shown by a line, called the **normal,** which is drawn perpendicular to the barrier. The angle between the incident ray and the normal is called the angle of incidence and is labeled θ_i in the diagram. The angle between the normal and the reflected ray is called the angle of reflection and is labeled θ_r. The **law of reflection** states that the angle of incidence is equal to the angle of reflection. The law of reflection applies to many different kinds of waves, not just the waves in a ripple tank.

☑ READING CHECK **Explain** how the angle of incidence and the angle of reflection are measured.

Figure 19 The ripple tank produces uniform waves that are useful for modeling wave behavior.

View a **simulation of a ripple tank.**

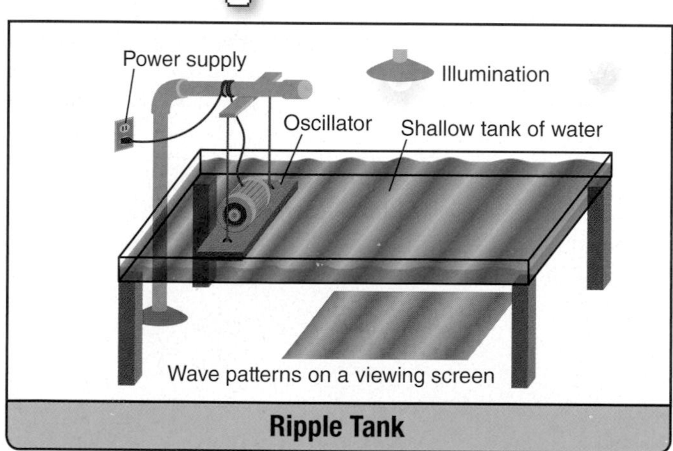

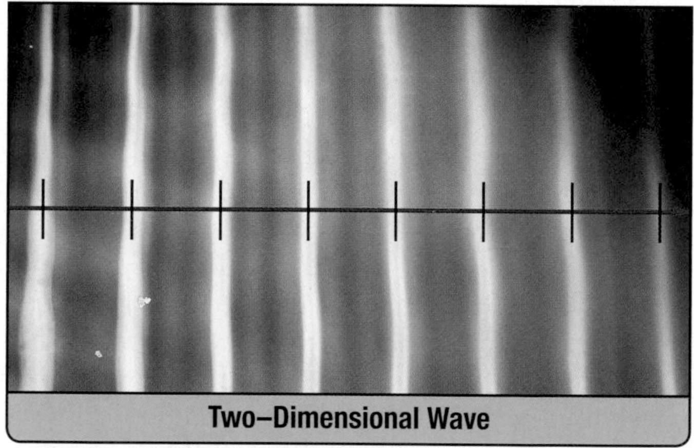

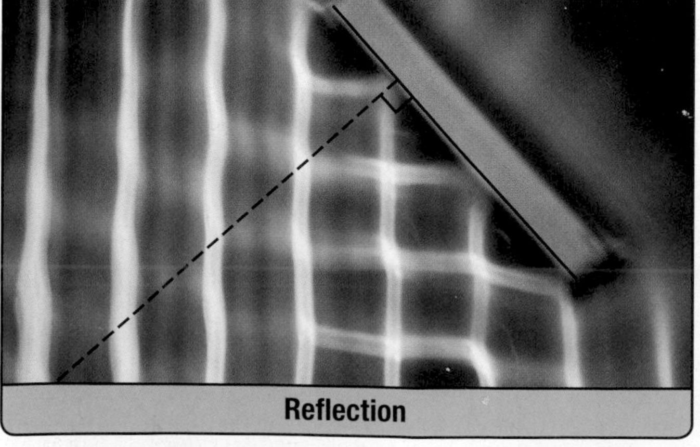

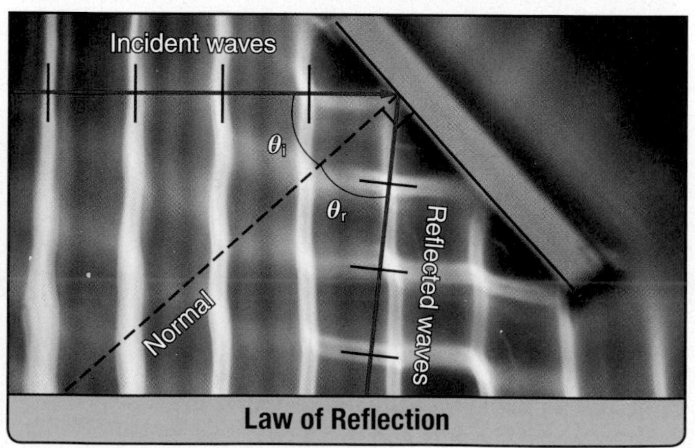

Richard Megna/Fundamental Photographs, NYC

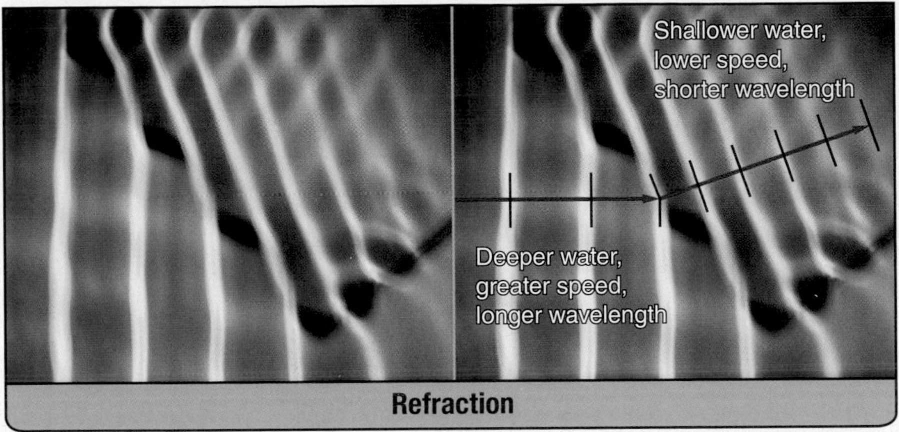

Figure 20 Waves in the ripple tank change direction as they enter shallower water.

Describe how the wavelength changes as the wave travels into the shallow water.

Refraction of waves in two dimensions A ripple tank can also model the behavior of waves as they travel from one medium into another. **Figure 20** shows a glass plate placed under the water in a ripple tank. The water above the plate is shallower than the water in the rest of the tank. As the waves travel from deep to shallow water, their speeds decrease and the direction of the waves changes. Such changes in speed are common when waves travel from one medium to another.

The waves in the shallow water are connected to the waves in the deep water. As a result, the frequency of the waves in the two mediums is the same. Based on the equation $\lambda = \frac{v}{f}$, the decrease in the speed of the waves means the wavelength is shorter in the shallower water. The change in the direction of waves at the boundary between two different mediums is known as **refraction.** **Figure 20** shows a wavefront and ray model of refraction. Part of the wave will refract through the boundary, and part will be reflected from the boundary.

☑ **READING CHECK** **Predict** the factor by which the wavelength changes if the speed of the refracted wave is half that of the incident wave.

Reflection and refraction occur for many different types of waves. Echoes are an example of reflection of sound waves by hard surfaces, such as the walls of a large gymnasium or a distant cliff face. Rainbows are the result of the reflection and refraction of light. As light from the Sun passes through a raindrop, reflection and refraction separate the light into its individual colors, producing a rainbow. The behavior of sound and light is discussed in more detail in other chapters.

PhysicsLAB

REFLECTION AND REFRACTION

How do waves behave at a barrier?

SECTION 3 REVIEW

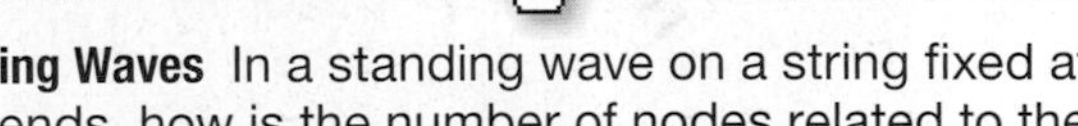

31. **MAIN**IDEA Which characteristics remain unchanged when a wave crosses a boundary into a different medium: frequency, amplitude, wavelength, velocity, and/or direction?
32. **Superposition of Waves** Sketch two wave pulses whose interference produces a pulse with an amplitude greater than either of the individual waves.
33. **Refraction of Waves** In **Figure 20,** the wave changes direction as it passes from one medium to another. Can two-dimensional waves cross a boundary between two mediums without changing direction? Explain.
34. **Standing Waves** In a standing wave on a string fixed at both ends, how is the number of nodes related to the number of antinodes?
35. **Critical Thinking** As another way to understand wave reflection, cover the right-hand side of each drawing in the left panel in **Figure 16** with a piece of paper. The edge of the paper should be at point N, the node. Now, concentrate on the resultant wave, shown in darker blue. Note that it acts as a wave reflected from a boundary. Is the boundary a rigid wall? Repeat this exercise for the middle panel in **Figure 16.**

HOW IT **WORKS**

Events of MAGNITUDE

Japan experiences frequent and sometimes serious earthquakes, such as the Tohoku earthquake of March 2011. As a result, engineers have designed and built skyscrapers in Japan that can withstand even the worst earthquakes with only minimal damage. The 73-floor Landmark Tower in Yokohama is one such building.

1 A computer-guided pendulum system with a 170-ton mobile mass is hidden on the 71st floor of the Landmark Tower. This pendulum system absorbs oscillations from seismic waves and reduces swaying by up to 40 percent during earthquakes.

2 A dual tube structure is used for the Landmark Tower's frame. The outside tube runs along the building's outside wall. The inside tube surrounds the building's core. This building's structure absorbs sideways oscillations from seismic waves.

3 The Landmark Tower has a wide base and a larger proportion of its mass near that base compared to other buildings of similar height. This construction helps minimize the motion of the foundation during an earthquake.

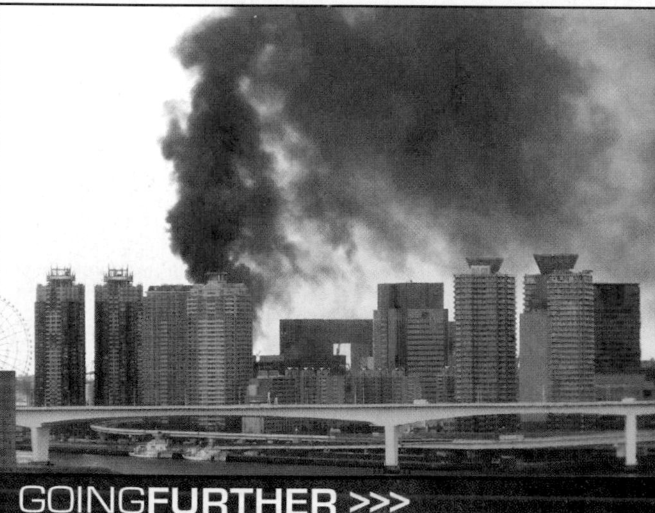

(l)STR/AFP/Getty Images, (r)Rex Butcher/Asia Images/Getty Images

GOINGFURTHER >>>

Research other earthquake-resistant buildings in Japan. Which technologies and designs were most effective at minimizing damage during the Tohoku earthquake in March 2011?

CHAPTER 14

STUDY GUIDE

BIG IDEA Waves and simple harmonic motion are examples of periodic motion.

VOCABULARY
- **periodic motion** *(p. 382)*
- **period** *(p. 382)*
- **amplitude** *(p. 382)*
- **simple harmonic motion** *(p. 382)*
- **Hooke's law** *(p. 383)*
- **simple pendulum** *(p. 386)*
- **resonance** *(p. 386)*

SECTION 1 Periodic Motion

MAIN IDEA Periodic motion repeats in a regular cycle.

- Simple harmonic motion results when the restoring force on an object is directly proportional to the object's displacement from equilibrium.
- The elastic potential energy of a spring that obeys Hooke's law is expressed by the following equation:

$$PE_{sp} = \frac{1}{2}kx^2$$

- The period of a pendulum depends on the pendulum's length and the gravitational field strength at the pendulum's location. The period can be found using the following equation:

$$T = 2\pi\sqrt{\frac{\ell}{g}}$$

VOCABULARY
- **wave** *(p. 388)*
- **wave pulse** *(p. 388)*
- **transverse wave** *(p. 388)*
- **periodic wave** *(p. 388)*
- **longitudinal wave** *(p. 388)*
- **surface wave** *(p. 389)*
- **trough** *(p. 390)*
- **crest** *(p. 390)*
- **wavelength** *(p. 390)*
- **frequency** *(p. 391)*

SECTION 2 Wave Properties

MAIN IDEA Waves transfer energy without transferring matter.

- Waves are disturbances that transfer energy without transferring matter.
- In transverse waves, the displacement of the medium is perpendicular to the direction the wave travels. In longitudinal waves, the displacement is parallel to the direction the wave travels.
- The velocity of a continuous wave is equal to the wave's frequency times its wavelength.

$$v = f\lambda$$

VOCABULARY
- **incident wave** *(p. 394)*
- **reflected wave** *(p. 394)*
- **principle of superposition** *(p. 395)*
- **interference** *(p. 395)*
- **node** *(p. 396)*
- **antinode** *(p. 396)*
- **standing wave** *(p. 397)*
- **wavefront** *(p. 397)*
- **ray** *(p. 397)*
- **normal** *(p. 398)*
- **law of reflection** *(p. 398)*
- **refraction** *(p. 399)*

SECTION 3 Wave Behavior

MAIN IDEA Waves can interfere with other waves.

- When two-dimensional waves are reflected from boundaries, the angles of incidence and reflection are equal. The change in direction of waves at the boundary between two different mediums is called refraction.
- Interference occurs when two or more waves travel through the same medium at the same time. The principle of superposition states that the displacement of a medium resulting from two or more waves is the algebraic sum of the displacements of the individual waves.

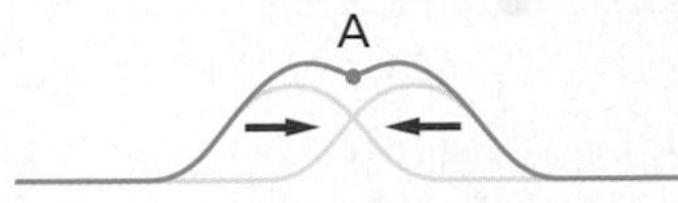

Games and Multilingual eGlossary

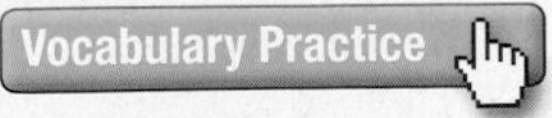

CHAPTER 14 ASSESSMENT

Chapter Self-Check

SECTION 1 Periodic Motion

Mastering Concepts

36. BIGIDEA What is periodic motion? Give three examples of periodic motion.

37. What is the difference between frequency and period? How are they related?

38. What is simple harmonic motion? Give an example of simple harmonic motion.

39. If a spring obeys Hooke's law, how does it behave?

40. How can the spring constant of a spring be determined from a graph of force magnitude versus displacement?

41. How can a spring's potential energy be determined from a graph of force magnitude versus displacement?

42. Does the period of a pendulum depend on the mass of the bob? The length of the string? The amplitude of oscillation? What else does the period depend on?

43. What conditions are necessary for resonance to occur?

Mastering Problems

44. A spring stretches 0.12 m when some apples are suspended from it, as shown in **Figure 21.** What is the spring constant of the spring?

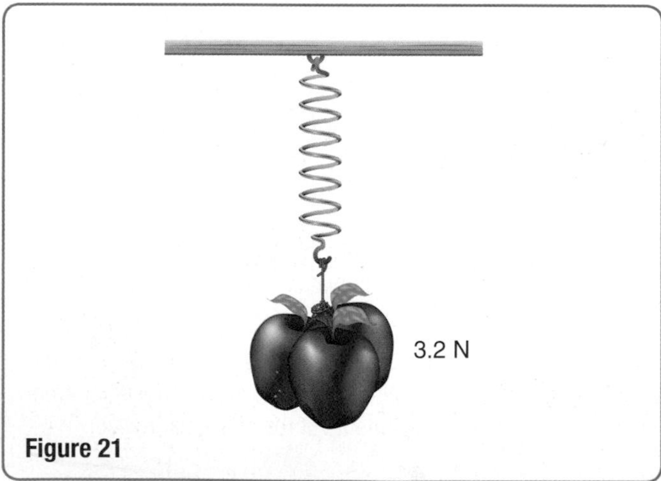

Figure 21

45. **Car Shocks** Each of the coil springs of a car has a spring constant of 25,000 N/m. How much is each spring compressed if it supports one-fourth of the car's 12,000-N weight?

46. A spring with a spring constant of 27 N/m is stretched 16 cm. What is the spring's potential energy?

47. **Rocket Launcher** A toy rocket launcher contains a spring with a spring constant of 35 N/m. How far must the spring be compressed to store 1.5 J of energy?

48. Force magnitude-versus-length data for a spring are plotted on the graph in **Figure 22.**
    a. What is the spring constant of the spring?
    b. What is the spring's potential energy when it is stretched to a length of 0.50 m?

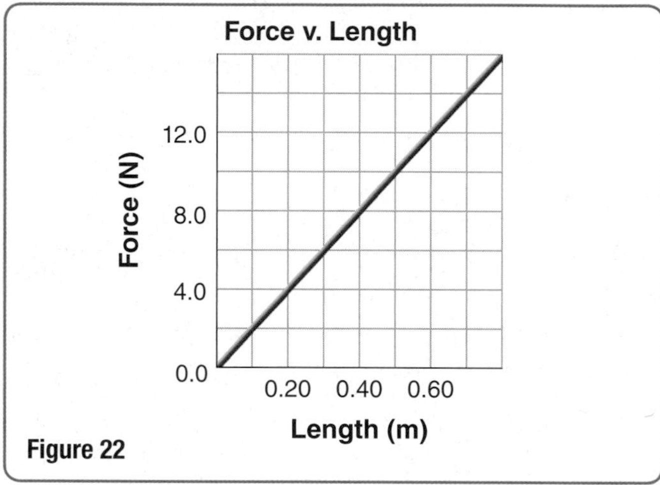

Figure 22

49. How long must a pendulum be to have a period of 2.3 s on the Moon, where $g = 1.6$ N/kg?

50. **Ranking Task** Rank the following pendulums according to period, from least to greatest. Specifically indicate any ties.
    A. 10 cm long, mass = 0.25 kg
    B. 10 cm long, mass = 0.35 kg
    C. 20 cm long, mass = 0.25 kg
    D. 20 cm long, mass = 0.35 kg

SECTION 2 Wave Properties

Mastering Concepts

51. Explain how the process of energy transfer associated with throwing a ball is different from the process of energy transfer associated with a mechanical wave.

52. What are the differences among transverse, longitudinal, and surface waves?

53. Waves are sent along a spring of fixed length.
    a. Can the speed of the waves in the spring be changed? Explain.
    b. Can the frequency of a wave in the spring be changed? Explain.

54. What is the wavelength of a wave?

55. Suppose you send a pulse along a rope. How does the position of a point on the rope before the pulse arrives compare to the point's position after the pulse has passed?

56. What is the difference between a wave pulse and a periodic wave?

57. Describe the difference between wave frequency and wave velocity.

58. Suppose you produce a transverse wave by shaking one end of a spring from side to side. How does the frequency of your hand compare with the frequency of the wave?

59. When are points on a wave in phase with each other? When are they out of phase? Give an example of each.

60. Describe the relationship between the amplitude of a wave and the energy it carries.

Mastering Problems

61. Building Motion The Willis Tower in Chicago sways back and forth in the wind with a frequency of about 0.12 Hz. What is its period of vibration?

62. Ocean Waves An ocean wave has a length of 12.0 m. A wave passes a fixed location every 3.0 s. What is the speed of the wave?

63. The wavelength of water waves in a shallow dish is 6.0 cm. The water moves up and down at a rate of 4.8 oscillations/s.

a. What is the speed of the waves?

b. What is the period of the waves?

64. Water waves in a lake travel 3.4 m in 1.8 s. The period of oscillation is 1.1 s.

a. What is the speed of the water waves?

b. What is their wavelength?

65. Sonar A sonar signal of frequency 1.00×10^{6} Hz has a wavelength of 1.50 mm in water.

a. What is the speed of the signal in water?

b. What is its period in water?

66. The speed of sound in water is 1498 m/s. A sonar signal is sent straight down from a ship at a point just below the water surface, and 1.80 s later the reflected signal is detected. How deep is the water?

67. A sound wave of wavelength 0.60 m and a velocity of 330 m/s is produced for 0.50 s.

a. What is the frequency of the wave?

b. How many complete waves are emitted in this time interval?

c. After 0.50 s, how far is the front of the wave from the source of the sound?

68. Pepe and Alfredo are resting on an offshore platform after a swim. They estimate that 3.0 m separates a trough and an adjacent crest of each surface wave on the lake. They count 12 crests in 20.0 s. Calculate how fast the waves are moving.

69. Earthquakes The velocity of the transverse waves produced by an earthquake is 8.9 km/s, and that of the longitudinal waves is 5.1 km/s. A seismograph records the arrival of the transverse waves 68 s before the arrival of the longitudinal waves. How far away is the earthquake?

SECTION 3 Wave Behavior

Mastering Concepts

70. When a wave crosses a boundary between a thin rope and a thick rope, as shown in **Figure 23,** its wavelength and speed change, but its frequency does not. Explain why the frequency is constant.

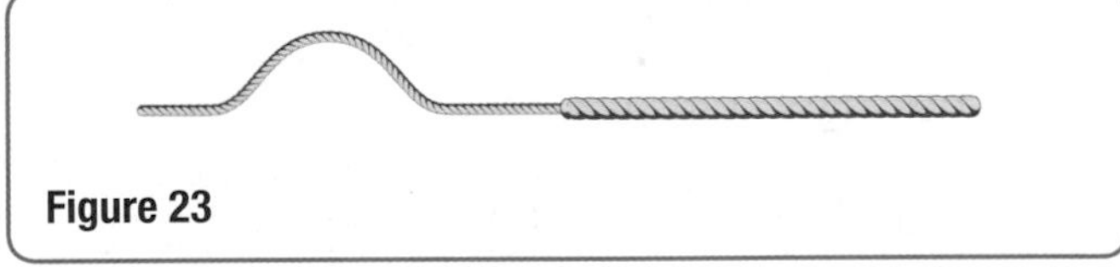

Figure 23

71. How does a wave pulse reflected from a rigid wall differ from the incident pulse?

72. Describe the motion of the particles of a medium located at the nodes of a standing wave.

73. Standing Waves A metal plate is held fixed in the center and sprinkled with sugar. With a violin bow, the plate is stroked along one edge and made to vibrate. The sugar begins to collect in certain areas and move away from others. Describe these regions in terms of standing waves.

74. If a string is vibrating in four parts, there are points where it can be touched without disturbing its motion. Explain. How many of these points exist?

75. Wavefronts pass at an angle from one medium into a second medium, where they travel with a different speed. Describe two changes in the wavefronts. What does not change?

Chapter Self-Check

Mastering Problems

76. Sketch the result for each of the three cases shown in **Figure 24,** when the centers of the two approaching wave pulses lie on the dashed line so that the pulses exactly overlap.

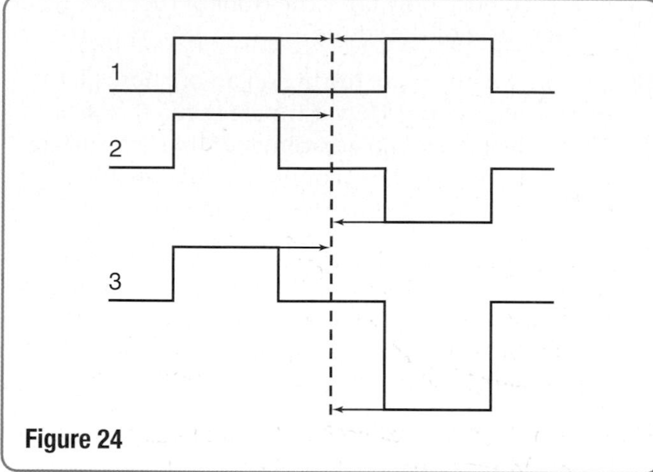

Figure 24

77. Guitars The wave speed in a guitar string is 265 m/s. The length of the string is 63 cm. You pluck the center of the string by pulling it up and letting go. Pulses move in both directions and are reflected off the ends of the string.

a. How long does it take for the pulse to move to the string end and return to the center?

b. When the pulses return, is the string above or below its resting location?

c. If you plucked the string 15 cm from one end of the string, where would the two pulses meet?

78. Standing waves are created in the four strings shown in **Figure 25.** All strings have the same mass per unit length and are under the same tension. The lengths of the strings (*L*) are given. Rank the frequencies of the oscillations, from largest to smallest.

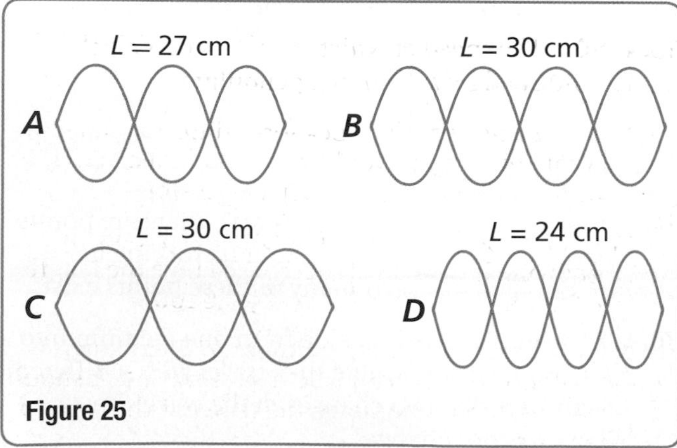

Figure 25

Applying Concepts

79. A ball bounces up and down on the end of a spring. Describe the energy changes that take place during one complete cycle. Does the total mechanical energy change?

80. Can a pendulum clock be used in the orbiting *International Space Station?* Explain.

81. Suppose you hold a 1-m metal bar in your hand and hit its end with a hammer, first, in a direction parallel to its length, and second, in a direction at right angles to its length. Describe the waves produced in the two cases.

82. Suppose you repeatedly dip your finger into a sink full of water to make circular waves. What happens to the wavelength as you move your finger faster?

83. What happens to the period of a wave as the frequency increases?

84. What happens to the wavelength of a wave as the frequency increases?

85. Suppose you make a single pulse on a stretched spring. How much energy is required to make a pulse with twice the amplitude?

86. You can make water slosh back and forth in a shallow pan only if you shake the pan with the correct frequency. Explain.

Mixed Review

87. What is the period of a 1.4-m pendulum?

88. Radio Wave AM-radio signals are broadcast at frequencies between 550 kHz (kilohertz) and 1600 kHz and travel at 3.0×10^8 m/s.

a. What is the signals' range of wavelengths?

b. FM frequencies range between 88 MHz (megahertz) and 108 MHz and travel at the same speed. What is the range of FM wavelengths?

89. The time needed for a water wave to change from the equilibrium level to the crest is 0.18 s.

a. What fraction of a wavelength is this?

b. What is the period of the wave?

c. What is the frequency of the wave?

90. When a 225-g mass is hung from a spring, the spring stretches 9.4 cm. The spring and mass then are pulled 8.0 cm from this new equilibrium position and released. Find the spring constant of the spring.

91. You are floating offshore at the beach. Even though the wave pulses travel toward the beach, you don't move much closer to the beach.

a. What type of wave are you experiencing as you float in the water?

b. Explain why the wave does not move you closer to shore.

c. In the course of 15 s, you count 10 waves that pass you. What is the period of the waves?

d. What is the frequency of the waves?

e. You estimate that the wave crests are 3 m apart. What is the velocity of the waves?

f. After returning to the beach, you learn that the waves are moving at 1.8 m/s. What is the actual wavelength of the waves?

92. Bungee Jumper A 68-kg bungee jumper jumps from a hot-air balloon using a 540-m bungee cord. When the jump is complete and the jumper is suspended unmoving from the cord, it is 1710 m long. What is the spring constant of the bungee cord?

93. You have a mechanical fish scale that is made with a spring that compresses when weight is added to a hook attached below the scale. Unfortunately, the calibrations have completely worn off the scale. However, you have one known mass of 500.0 g that compresses the spring 2.0 cm.

a. What is the spring constant for the spring?

b. If a fish compresses the spring 4.5 cm, what is the mass of the fish?

94. Bridge Swinging In the summer over the New River in West Virginia, several teens swing from bridges with ropes, then drop into the river after a few swings back and forth.

a. If Pam is using a 10.0-m length of rope, how long will it take her to complete one full swing?

b. If Mike has a mass that is 20 kg more than Pam's, how would you expect the period of his swing to differ from Pam's?

c. At what point in the swing is KE at a maximum?

d. At what point in the swing is PE at a maximum?

e. At what point in the swing is KE at a minimum?

f. At what point in the swing is PE at a minimum?

95. Car Springs When you add a 45-kg load to the trunk of a new small car, the two rear springs compress an additional 1.0 cm.

a. What is the spring constant for each of the springs?

b. What is the increase in each of the springs' potential energy after loading the trunk?

96. Amusement Ride You notice that your favorite amusement-park ride seems bigger. The ride consists of a carriage that is attached to a structure so it swings like a pendulum. You remember that the carriage used to swing from one position to another and back again eight times in exactly 1 min. Now it only swings six times in 1 min. Give your answers to the following questions to two significant digits.

a. What was the original period of the ride?

b. What is the new period of the ride?

c. What is the new frequency?

d. How much longer is the arm supporting the carriage on the larger ride?

e. If the park owners wanted to double the period of the ride, what percentage increase would need to be made to the length of the pendulum?

97. Sketch the result for each of the four cases shown in **Figure 26,** when the centers of each of the two wave pulses lie on the dashed line so that the pulses exactly overlap.

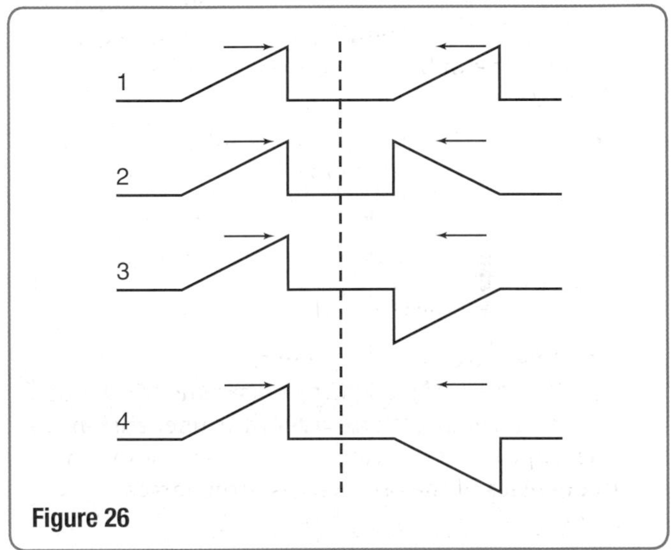

Figure 26

98. Clocks The speed at which a grandfather clock runs is controlled by a swinging pendulum.

a. If you find that the clock loses time each day, what adjustment would you need to make to the pendulum so it would keep better time?

b. If the pendulum currently is 15.0 cm long, by how much would you need to change the length to make the period shorter by 0.0400 s?

c. Another clock has a pendulum 77.5 cm long. Over one day it loses 5.00 min. How much should its length be changed so that it keeps the correct time?

Chapter Self-Check

99. The velocity of a wave on a string depends on how tightly the string is stretched and on the mass per unit length of the string. If F_T is the tension in the string, and μ is the mass/unit length, then the velocity (v) can be determined by the following equation.

$$v = \sqrt{\frac{F_T}{\mu}}$$

A piece of string 5.30 m long has a mass of 15.0 g. What must the tension in the string be to make the wavelength of a 125-Hz wave 120.0 cm?

Thinking Critically

100. Reverse Problem Write a physics problem with real-life objects for which the following equation would be part of the solution:

$$(1.65\ \text{kg})(9.8\ \text{N/kg}) = k\ (0.15\ \text{m})$$

101. Problem Posing Complete this problem so that it must be solved using the concept listed below: "A block is suspended vertically from a spring with a spring constant of 200 N/m..."

a. conservation of energy

b. Newton's second law

c. frequency

102. Make and Use Graphs Several weights were suspended from a spring, and the resulting extensions of the spring were measured. **Table 2** shows the collected data.

Table 2 Weights on a Spring

Force magnitude, F (N)	Extension, x (m)
2.5	0.12
5.0	0.26
7.5	0.35
10.0	0.50
12.5	0.60
15.0	0.71

a. Make a graph of the magnitude of the force applied to the spring versus the spring length. Plot the force on the y-axis.

b. Determine the spring constant from the graph.

c. Using the graph, find the elastic potential energy stored in the spring when it is stretched to 0.50 m.

103. Analyze and Conclude A 20-N force is required to stretch a spring by 0.5 m.

a. What is the spring constant?

b. How much energy does the spring store?

c. Why isn't the work done to stretch the spring equal to the force times the distance, or 10 J?

104. Apply Concepts Gravel roads often develop regularly spaced ridges that are perpendicular to the road, as shown in **Figure 27.** This effect, called washboarding, occurs because most cars travel at about the same speed, and the springs that connect the wheels to the cars oscillate at about the same frequency. If the ridges on a road are 1.5 m apart and cars travel on it at about 5 m/s, what is the frequency of the springs' oscillation?

Figure 27

Writing In Physics

105. Research Christiaan Huygens' work on waves and the controversy between him and Newton over the nature of light. Compare and contrast their explanations of such phenomena as reflection and refraction. Whose model would you choose as the best explanation? Explain why.

Cumulative Review

106. A 1400-kg drag racer automobile can complete a 402-m course in 9.8 s. The final speed of the automobile is 112 m/s.

a. What is the final kinetic energy of the automobile?

b. What is the minimum amount of work that was done by its engine? Why can't you calculate the total amount of work done?

c. What was the average acceleration of the automobile?

107. How much water would a steam engine have to evaporate in 1 s to produce 1 kW of power? Assume that the engine is 20 percent efficient.

STANDARDIZED TEST PRACTICE

MULTIPLE CHOICE

1. What is the value of the spring constant of a spring with a potential energy of 8.67 J when it's stretched 247 mm?

 A. 70.2 N/m **C.** 142 N/m
 B. 71.1 N/m **D.** 284 N/m

2. What is the magnitude of the force acting on a spring with a spring constant of 275 N/m that is stretched 14.3 cm?

 A. 2.81 N **C.** 39.3 N
 B. 19.2 N **D.** 3.93×10^{30} N

3. A mass stretches a spring as it hangs from the spring as shown in the figure below. What is the spring constant?

 A. 0.25 N/m **C.** 26 N/m
 B. 0.35 N/m **D.** 3.5×10^2 N/m

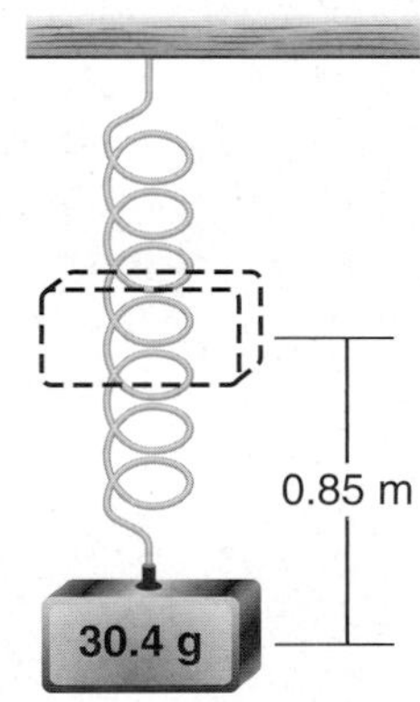

4. A spring with a spring constant of 350 N/m pulls a door closed. How much work is done as the spring pulls the door at a constant velocity from an 85.0-cm stretch to a 5.0-cm stretch?

 A. 110 N·m **C.** 220 N·m
 B. 130 J **D.** 1.1×10^3 J

5. What is the length of a pendulum that has a period of 4.89 s?

 A. 5.94 m **C.** 24.0 m
 B. 11.9 m **D.** 37.3 m

6. What is the frequency of a wave with a period of 3 s?

 A. 0.3 Hz **C.** $\frac{\pi}{3}$ Hz
 B. $\frac{3}{\pi}$ Hz **D.** 3 Hz

Online Test Practice

7. What is the correct rearrangement of the formula for the period of a pendulum to find the length of the pendulum?

 A. $\ell = \frac{4\pi^2 g}{T^2}$ **C.** $\ell = \frac{T^2 g}{(2\pi)^2}$
 B. $\ell = \frac{gT}{4\pi^2}$ **D.** $\ell = \frac{Tg}{2\pi}$

8. Which option describes a standing wave?

	Waves	Direction	Medium
a	Identical	Same	Same
b	Nonidentical	Opposite	Different
c	Identical	Opposite	Same
d	Nonidentical	Same	Different

9. What is the name given to the wave behavior in which a wave changes direction as it moves from one medium to another medium?

 A. interference **C.** reflection
 B. rarefaction **D.** refraction

10. The wave shown in the figure below travels 11.2 m to a wall and back again in 4 s. What is the wave's frequency?

 A. 0.2 Hz **C.** 5 Hz
 B. 2 Hz **D.** 9 Hz

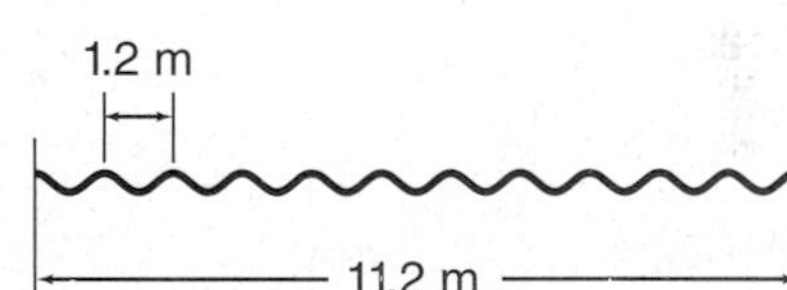

FREE RESPONSE

11. Use dimensional analysis of the equation $kx = mg$ to derive the units of k.

NEED EXTRA HELP?

If You Missed Question	1	2	3	4	5	6	7	8	9	10	11
Review Section	1	1	1	1	1	2	1	3	3	2	1

CHAPTER 15

Sound

BIGIDEA **Sound waves are pressure variations, and many can be detected by the human ear.**

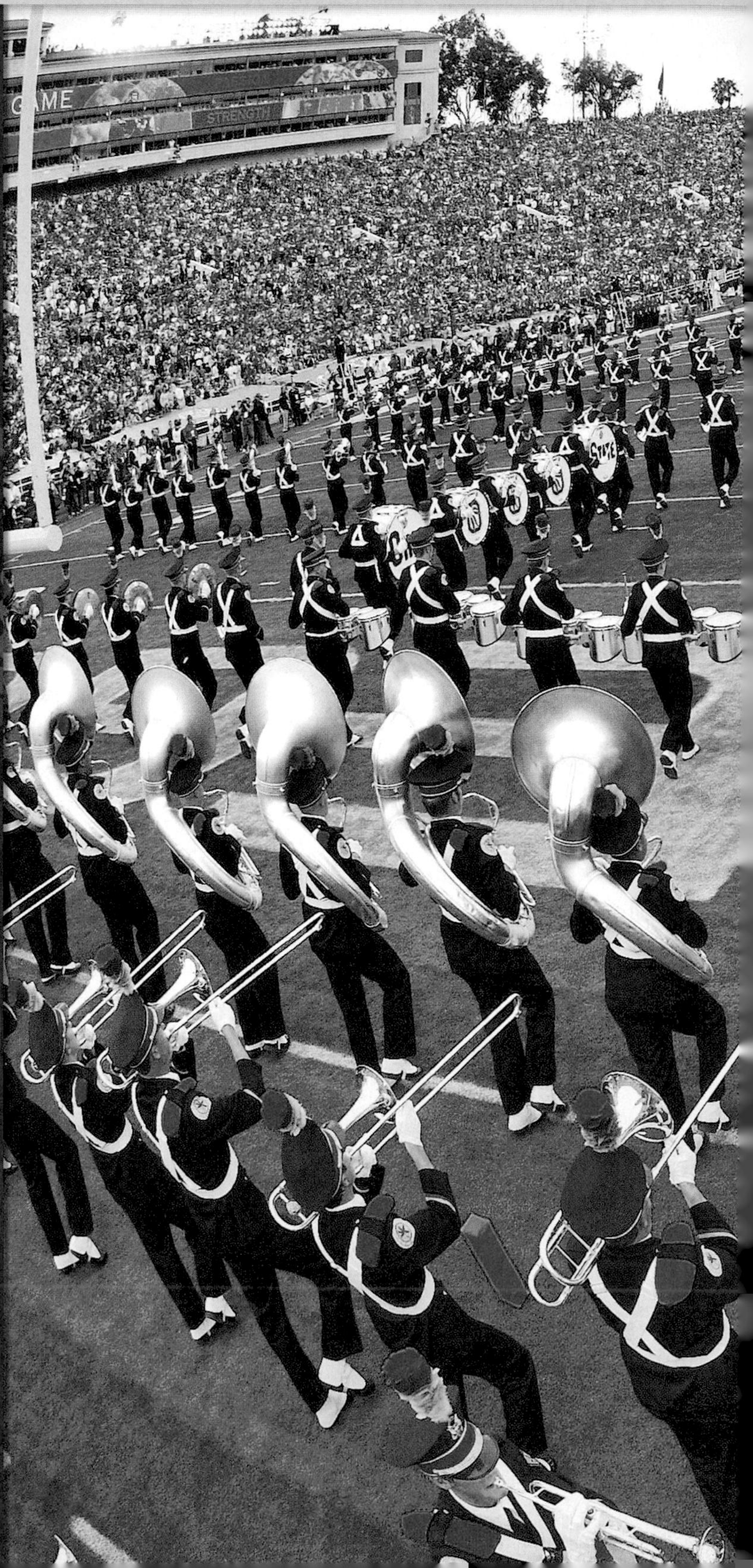

SECTIONS

LaunchLAB

iLab Station

PRODUCING MUSICAL NOTES

How can musical notes be produced using stemmed drinking glasses?

WATCH THIS!

Video

MAKING MUSIC

From woodwinds to brass to percussion, every instrument makes a different kind of sound. What is the physics of making music?

PHYSICS T.V.

(l)Mary-Ella Keith/Alamy, (r)Robert Beck/Sports Illustrated/Getty Images

OHIO STATE UNIVERSITY BAND

SECTION 1

Properties and Detection of Sound

PHYSICS 4 YOU **You have probably noticed that an approaching truck sounds different than the same truck moving away from you. This effect, called the Doppler effect, was first studied in 1845. Dutch scientist C.H.D. Buys Ballot observed the changes in sound as a train carrying horn players went by.**

MAIN IDEA

Our perception of a sound wave depends on that wave's physical properties.

Essential Questions

- What properties does sound share with other waves?
- How do the physical properties of sound waves relate to our perception of sound?
- What is the Doppler effect?
- What are some applications of the Doppler effect?

Review Vocabulary

wave a disturbance that carries energy through matter or space and transfers energy without transferring matter

New Vocabulary

sound wave
pitch
loudness
sound level
decibel
Doppler effect

Sound Waves

Sound is an important part of existence for many living things. For humans, the sound of a siren can heighten our awareness of our surroundings, while the sound of music can relax us. You already are familiar with several of the characteristics of sound, including volume, tone, and pitch, from your everyday experiences. Without thinking about it, you can use these, and other characteristics, to categorize many of the sounds that you hear. For example, some sound patterns are characteristic of speech, while others are characteristic of a musical group. In this chapter, you will study the physical principles of sound, which is a type of wave. You will use your knowledge of waves to describe some of sound's properties and interactions.

Pressure variations Put your fingers against your throat as you hum or speak. Can you feel the vibrations? **Figure 1** shows a vibrating bell that can represent your vocal cords, a loudspeaker, or any other sound source. As it moves back and forth, the edge of the bell strikes the particles in the air. When the edge moves forward, air particles are driven forward; that is, the air particles bounce off the bell with a greater velocity than they would if the bell were not moving. When the edge moves backward, air particles bounce off the bell with a lower velocity.

Figure 1 When the bell is at rest (top) the surrounding air is at average pressure. When the bell is struck (bottom) the vibrating edge creates regions of high and low pressure. For better understanding, the diagram shows the pressure regions moving in one direction. In reality, the waves move out from the bell in all directions.

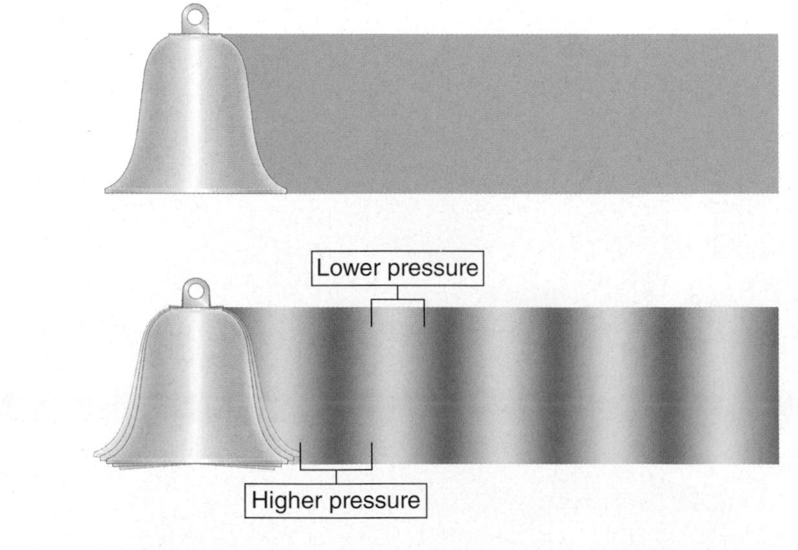

Reed Kaestner/CORBIS

Coiled Spring

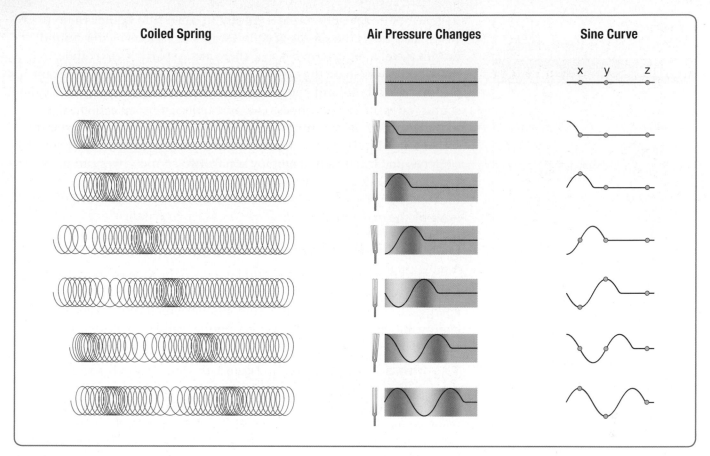

Air Pressure Changes

Sine Curve

x y z

Figure 2 A coiled spring models the oscillations created by a sound wave. As the sound wave travels through the air, the air pressure rises and falls. The changes in the sine curves correspond to the changes in air pressure. Note that the positions of *x*, *y*, and *z* show that the wave, not the matter, travels forward.

The result of these velocity changes is that the forward motion of the bell produces a region where the air pressure is slightly higher than average, as shown in **Figure 1.** The backward motion produces slightly below-average pressure. Collisions of the air particles cause the pressure variations to move away from the bell in all directions. If you were to focus on one spot, you would see the value of the air pressure rise and fall. In this way, the pressure variations are transmitted through matter.

Describing sound A pressure oscillation that is transmitted through matter is a **sound wave.** Sound waves travel through air because a vibrating source produces regular variations, or oscillations, in air pressure. The air particles collide, transmitting the pressure variations away from the source of the sound. The pressure of the air oscillates about the mean air pressure, as shown in **Figure 2.** The frequency of the wave is the number of oscillations in pressure each second. The wavelength is the distance between successive regions of high or low pressure. Because the motion of the particles in air is parallel to the direction of the wave's motion, sound is a longitudinal wave.

Just like any other wave, the speed of sound depends on the medium through which it travels. In air, the speed depends on the temperature, increasing by about 0.6 m/s for each 1°C increase in air temperature. At room temperature (20°C) and sea level, the speed of sound is 343 m/s. For the problems in this book, you may assume these conditions unless otherwise stated.

☑ **READING CHECK Estimate** the speed of sound through air at sea level if the temperature is 25°C.

PhysicsLAB

SOUND TRAVELS THROUGH AIR

PROBEWARE LAB How fast does sound move in air?

Table 1 Speed of Sound in Various Media	
Medium	**m/s**
Air (0°)	331
Air (20°)	343
Helium (0°)	965
Water (25°)	1497
Seawater (25°)	1535
Copper (20°)	4760
Iron (20°)	4994

In general, the speed of sound is greater in solids and liquids than in gases. **Table 1** lists the speeds of sound waves in various media. Sound cannot travel in a vacuum because there are no particles to collide.

Sound waves share the general properties of other waves. For example, they reflect off hard objects, such as the walls of a room. Reflected sound waves are called echoes. The time required for an echo to return to the source of the sound can be used to find the distance between the source and the reflective object. This principle is used by bats, by some cameras, and by ships that employ sonar. Two sound waves can interfere, causing dead spots at nodes where little sound can be heard. Recall that the frequency and wavelength of a wave are related to the speed of the wave by the equation $\lambda = v/f$.

Detection of Pressure Waves

Sound detectors transform sound energy—the kinetic energy of the vibrating particles of the transmitting medium—into another form of energy. A common detector is a microphone, which transforms sound energy into electrical energy. A microphone consists of a thin disk that vibrates in response to sound waves and produces an electrical signal.

The human ear As shown in **Figure 3,** the human ear is a sound detector that receives pressure waves and converts them to electrical impulses. The tympanic membrane, also called the eardrum, vibrates when sound waves enter the auditory canal. Three tiny bones in the middle ear then transfer these vibrations to fluid in the cochlea. Tiny hairs lining the spiral-shaped cochlea detect certain frequencies in the vibrating fluid. These hairs stimulate nerve cells, which send impulses to the brain and produce the sensation of sound.

The ear detects sound waves over a wide range of frequencies and is sensitive to an enormous range of amplitudes. In addition, human hearing can distinguish many different qualities of sound. Knowledge of both physics and biology is required to understand the complexities of the ear. The interpretation of sounds by the brain is even more complex, and is not totally understood.

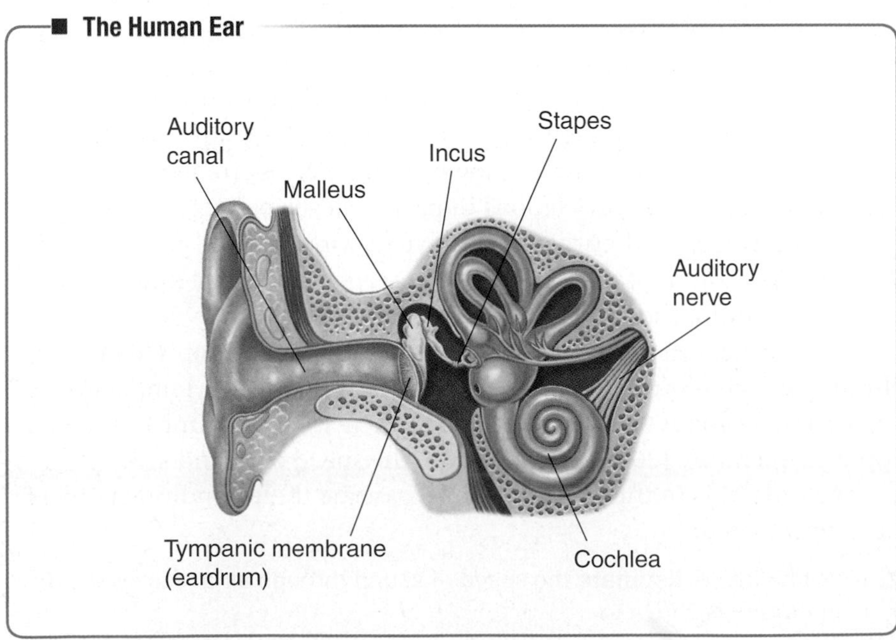

Figure 3 The human ear is a sense organ that translates sound vibrations from the external environment into nerve impulses that are sent to the brain for interpretation. The eardrum vibrates when sound waves enter the auditory canal. The bones in the middle ear—the malleus, the incus, and the stapes—move as a result of the vibrations. The vibrations are then transmitted to the inner ear, where they trigger nerve impulses to the brain.

Perceiving Sound

How humans perceive sound depends partly on the physical characteristics of sound waves, such as frequency and amplitude.

Pitch Marin Mersenne and Galileo first determined that the pitch we hear depends on the frequency of vibration. **Pitch** is the highness or lowness of a sound, and it can be given a name on the musical scale. For instance, the note known as middle C has a frequency of 262 Hz. The highest note on a piano has a frequency of 4186 Hz. The human ear is not equally sensitive to all frequencies. Most people cannot hear sounds with frequencies below 20 Hz or above 16,000 Hz. Many animals, such as dogs, cats, elephants, and bats, are capable of hearing sounds at frequencies that humans cannot hear.

Loudness Frequency and wavelength are two physical characteristics of sound waves. Another physical characteristic of sound waves is amplitude. Amplitude is the measure of the variation in pressure in a wave. The **loudness** of a sound is the intensity of the sound as perceived by the ear and interpreted by the brain. This intensity depends primarily on the amplitude of the pressure wave.

The human ear is extremely sensitive to variations in the intensity of sound waves. Recall that 1 atmosphere of pressure equals 1.01×10^{5} Pa. The ear can detect pressure-wave amplitudes of less than one-billionth of an atmosphere, or 2×10^{-5} Pa. At the other end of the audible range, pressure variations of approximately 20 Pa or greater cause pain. It is important to remember that the ear detects pressure variations only at certain frequencies. Driving over a mountain pass changes the pressure on your ears by thousands of pascals, but this change does not take place at audible frequencies.

Because humans can detect a wide range of intensities, it is convenient to measure these intensities on a logarithmic scale called the **sound level.** The most common unit of measurement for sound level is the **decibel** (dB). The sound level depends on the ratio of the intensity of a given sound wave to that of the most faintly heard sound. This faintest sound is measured at 0 dB. A sound that is ten times more intense registers 20 dB. A sound that is another ten times more intense is 40 dB. Most people perceive a 10-dB increase in sound level as about twice as loud as the original level. **Figure 4** shows the sound level for a variety of sounds. In addition to intensity, pressure variations and the power of sound waves can be described by decibel scales.

VOCABULARY
Science Usage v. Common Usage

Pitch

• Science usage
the highness or lowness of a sound, which depends on the frequency of vibration
The flute and the tuba produce very different pitches.

• Common usage
the delivery of a ball by a pitcher to a batter
Sharon hit the pitch over the fence for a home run.

MiniLAB
SOUND CHARACTERISTICS
What physical factors determine sound?

Figure 4 This decibel scale shows the sound level for a variety of sounds.

Infer About how many times louder does an alarm clock sound than heavy traffic?

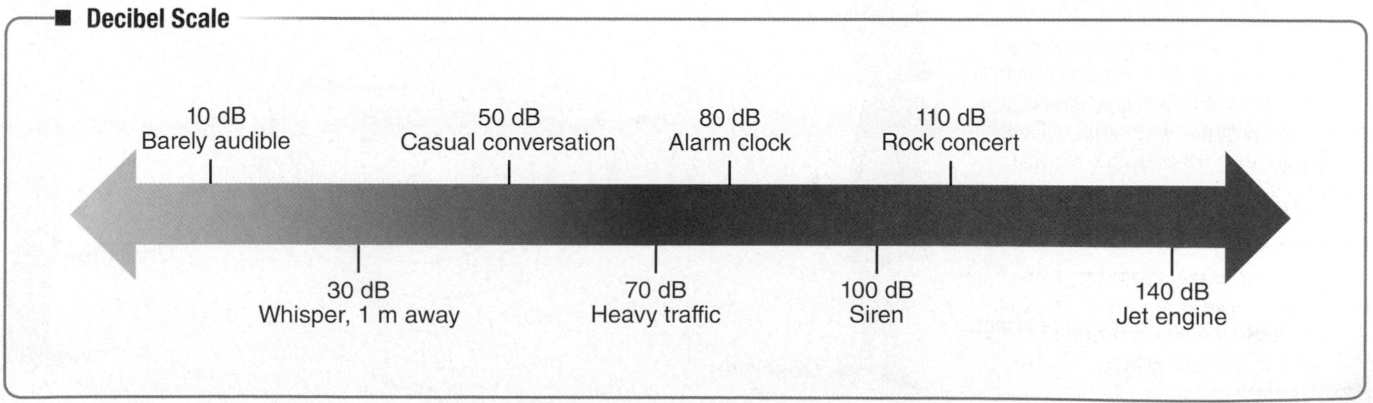

Figure 5 Hearing loss can occur with continuous exposure to loud sounds. Workers in many occupations, such as construction, wear ear protection. The jackhammer this worker is operating has a sound level of 130 dB.

The ear can lose its sensitivity, especially to high frequencies, after exposure to loud sounds in the form of noise or music. The longer a person is exposed to loud sounds, the greater the effect. A person can recover from short-term exposure in a period of hours, but the effects of long-term exposure can last for days or weeks. Long exposure to 100-dB or greater sound levels can produce permanent damage. Hearing loss also can result from loud music being transmitted to stereo headphones from personal radios and CD players. In some cases, the listeners are unaware of just how high the sound levels really are. Cotton earplugs reduce the sound level only by about 10 dB. Special ear inserts can provide a 25-dB reduction. Specifically designed earmuffs and inserts, as shown in **Figure 5,** can reduce the sound level by up to 45 dB.

The Doppler Effect

Have you ever noticed that the pitch of a fast car changed as the vehicle sped past you? The pitch was higher when the vehicle was moving toward you, then it dropped to a lower pitch as the vehicle moved away. The change in frequency of sound caused by the movement of either the source, the detector, or both is called the **Doppler effect.** The Doppler effect is illustrated in **Figure 6.** The sound source (*S*) is moving to the right with a speed of v_s. The waves that the source emits spread in circles centered on the source at the time it produced the waves. As the source moves toward the sound detector, observer A in **Figure 6,** more waves are crowded into the space between them. The wavelength is shortened to λ_A. Because the speed of sound is not changed, more crests reach the ear each second, which means that the frequency of the detected sound increases. When the source is moving away from the detector, observer B in **Figure 6,** the wavelength is lengthened to λ_B, fewer crests reach the ear each second, and the detected frequency is lower.

A Doppler shift also occurs if the detector is moving and the source is stationary. As the detector approaches a stationary source, it encounters more wave crests each second than if it were still, and a higher frequency is detected. If the detector recedes from the source, fewer crests reach it each second, resulting in a lower detected frequency.

PhysicsLAB

WHAT IS A DECIBEL?

How can you measure sound level?

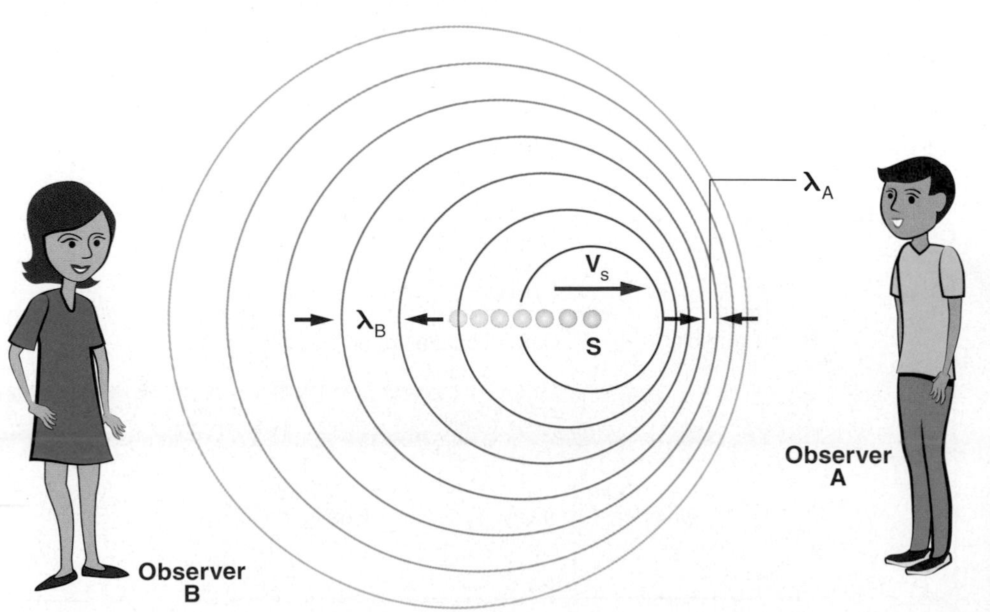

Figure 6 As a sound-producing source moves toward observer A, the wavelength is shortened to λ_A. As the source moves away from observer B, the wavelength is lengthened to λ_B.

Describe What is the relative difference in the frequency of the detected sound for each observer?

View an **animation on the Doppler effect.**

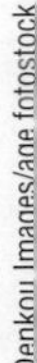

For any combination of moving source and moving observer, the frequency that the observer hears can be found using the relationship below.

DOPPLER EFFECT

The frequency perceived by a detector is equal to the velocity of the detector relative to the velocity of the wave, divided by the velocity of the source relative to the velocity of the wave, multiplied by the wave's frequency.

$$f_d = f_s \frac{v - v_d}{v - v_s}$$

Investigate the **Doppler effect.**

Virtual Investigation

In the Doppler effect equation, v is the velocity of the sound wave, v_s is the velocity of the sound's source, and v_d is the velocity of the observer of interest, who is detecting the sound. The subscript d is used instead of the letter o to avoid confusion with the number zero. The same subscripts are used to denote the corresponding frequencies.

Defining the coordinate system As you solve problems using the above equation, be sure to define the coordinate system so that the positive direction is from the source to the detector. The sound waves will be approaching the detector from the source, so the velocity of sound is always positive. Try drawing diagrams to confirm that the term $\frac{v - v_d}{v - v_s}$ behaves as you would predict based on what you have learned about the Doppler effect. Notice that for a source moving toward the detector (positive direction, which results in a smaller denominator compared to a stationary source) and for a detector moving toward the source (negative direction and increased numerator compared to a stationary detector), the detected frequency (f_d) increases. Similarly, if the source moves away from the detector or if the detector moves away from the source, then f_d decreases. Read the Connecting Math to Physics feature below to see how the Doppler effect equation reduces when the source or observer is stationary.

CONNECTING MATH TO PHYSICS

Reducing Equations When an element in an equation is equal to zero, the equation might reduce to a form that is easier to use.

Stationary detector, source in motion: $v_d = 0$	Stationary source, detector in motion: $v_s = 0$
$f_d = f_s \frac{v - v_d}{v - v_s}$	$f_d = f_s \frac{v - v_d}{v - v_s}$
$= f_s \frac{v}{v - v_s}$	$= f_s \frac{v - v_d}{v}$
$= f_s \frac{\frac{v}{v}}{\frac{v}{v} - \frac{v_s}{v}}$	$= f_s \frac{\frac{v}{v} - \frac{v_d}{v}}{\frac{v}{v}}$
$= f_s \frac{1}{1 - \frac{v_s}{v}}$	$= f_s \frac{1 - \frac{v_d}{v}}{1}$
	$= f_s \left(1 - \frac{v_d}{v}\right)$

EXAMPLE PROBLEM 1

Find help with **fractions.** Math Handbook

THE DOPPLER EFFECT A guitar player sounds C above middle C (523 Hz) while traveling in a convertible at 24.6 m/s. If the car is coming toward you, what frequency would you hear? Assume that the temperature is 20°C.

1 ANALYZE AND SKETCH THE PROBLEM

- Sketch the situation.
- Establish a coordinate axis. Make sure that the positive direction is from the source to the detector.
- Show the velocities of the source and detector.

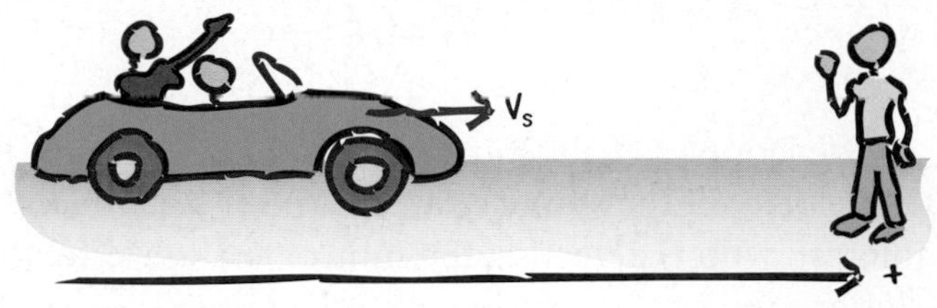

KNOWN	UNKNOWN
$v = +343$ m/s	$f_d = ?$
$v_s = +24.6$ m/s	
$v_d = 0$ m/s	
$f_s = 523$ Hz	

2 SOLVE FOR THE UNKNOWN

Use $f_d = f_s \frac{v - v_d}{v - v_s}$ with $v_d = 0$ m/s.

$$f_d = f_s \frac{1}{1 - \frac{v_s}{v}}$$

$$= 523\text{ Hz} \left(\frac{1}{1 - \frac{24.6\text{ m/s}}{343\text{ m/s}}} \right)$$

◀ Substitute $v = +343$ m/s, $v_s = +24.6$ m/s, and $f_s = 523$ Hz.

$$= 564\text{ Hz}$$

3 EVALUATE THE ANSWER

- **Are the units correct?** Frequency is measured in hertz.
- **Is the magnitude realistic?** The source is moving toward you, so the frequency should be increased.

PRACTICE PROBLEMS

Do additional problems.

1. Repeat Example Problem 1, but with the car moving away from you. What frequency would you hear?
2. You are in an automobile, like the one in **Figure 7**, traveling toward a pole-mounted warning siren. If the siren's frequency is 365 Hz, what frequency do you hear? Use 343 m/s as the speed of sound.

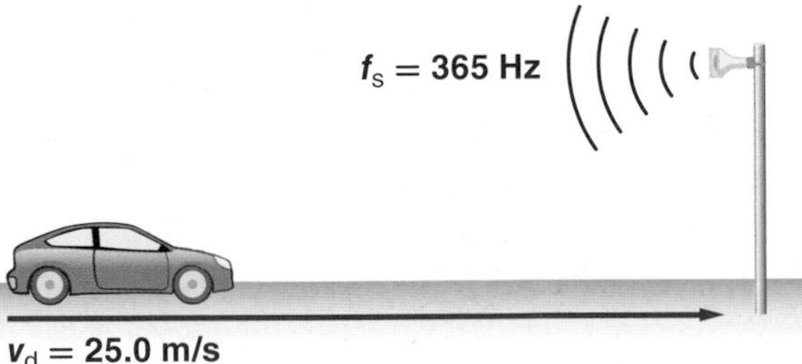

Figure 7

3. You are in an automobile traveling at 55 mph (24.6 m/s). A second automobile is moving toward you at the same speed. Its horn is sounding at 475 Hz. What frequency do you hear? Use 343 m/s as the speed of sound.
4. A submarine is moving toward another submarine at 9.20 m/s. It emits a 3.50-MHz ultrasound. What frequency would the second sub, at rest, detect? The speed of sound in water at the depth the submarines are moving is 1482 m/s.
5. **CHALLENGE** A trumpet plays middle C (262 Hz). How fast would it have to be moving to raise the pitch to C sharp (277 Hz)? Use 343 m/s as the speed of sound.

Figure 8 Bats use the Doppler effect to locate and catch flying insects, such as the moth shown here. As the bat catches up to the moth, the frequency of reflected sound waves increases.

Applications of the Doppler effect The Doppler effect occurs in all wave motion, both mechanical and electromagnetic. It has many applications. Radar detectors use the Doppler effect to measure the speed of baseballs and automobiles. Astronomers observe light from distant galaxies and use the Doppler effect to measure their speeds. Physicians can detect the speed of the moving heart wall in a fetus by means of the Doppler effect in ultrasound.

CONNECTION TO BIOLOGY Bats use the Doppler effect to detect and catch flying insects. When an insect is flying faster than a bat, the reflected frequency is lower, but when the bat is catching up to the insect, as in **Figure 8,** the reflected frequency is higher. Not only do bats use sound waves to navigate and locate their prey, but they often must do so in the presence of other bats. This means they must discriminate their own calls and reflections against a background of many other sounds of many frequencies. Scientists continue to study bats and their amazing abilities to use sound waves.

SECTION 1 REVIEW

Section Self-Check 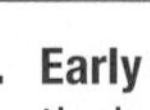 Check your understanding.

6. **MAIN IDEA** What physical characteristic of a sound wave should be changed to alter the pitch of the sound? To alter the loudness?
7. **Graph** The eardrum moves back and forth in response to the pressure variations of a sound wave. Sketch a graph of the displacement of the eardrum versus time for two cycles of a 1.0-kHz tone and for two cycles of a 2.0-kHz tone.
8. **Effect of Medium** List two sound characteristics that are affected by the medium through which the sound passes and two characteristics that are not affected.
9. **Decibel Scale** How many times greater is the sound pressure level of a typical rock concert (110 dB) than a normal conversation (50 dB)?
10. **Early Detection** In the nineteenth century, people put their ears to a railroad track to get an early warning of an approaching train. Why did this work?
11. **Bats** A bat emits short pulses of high-frequency sound and detects the echoes.
    a. In what way would the echoes from large and small insects compare if they were the same distance from the bat?
    b. In what way would the echo from an insect flying toward the bat differ from that of an insect flying away from the bat?
12. **Critical Thinking** Can a trooper using a radar detector at the side of the road determine the speed of a car at the instant the car passes the trooper? Explain.

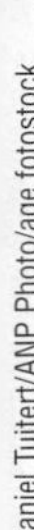

SECTION 2 The Physics of Music

PHYSICS 4 YOU Think about all the musical instruments you have seen or heard. There are violins, flutes, drum kits, and many other instruments. How do you think the size, shape, and materials of an instrument affect the sounds it makes?

MAIN IDEA

Music consists of complex sound waves produced by vibrating objects.

Essential Questions

- What is the origin of sound?
- What are the characteristics of resonance in air columns?
- What are the characteristics of resonance on strings?
- Why are there variations in sound quality among instruments?
- How are beats produced?

Review Vocabulary

frequency the number of complete oscillations that a wave makes each second; is measured in hertz

New Vocabulary

closed-pipe resonator
open-pipe resonator
fundamental
harmonics
dissonance
consonance
beat

Sources of Sound

In the middle of the nineteenth century, German physicist Hermann Helmholtz studied sound production in musical instruments and the human voice. In the twentieth century, scientists and engineers developed electronic equipment that permits not only a detailed study of sound, but also the creation of electronic musical instruments and recording devices that allow us to listen to music whenever and wherever we wish.

Recall that sound is produced by a vibrating object. The vibrations of the object create particle motions that cause pressure oscillations in the air. A loudspeaker has a cone that is made to vibrate by electrical currents. The surface of the cone creates the sound waves that travel to your ear and allow you to hear music. Musical instruments such as gongs, cymbals, and drums are other examples of vibrating surfaces that are sources of sound.

The human voice is produced by vibrations of the vocal cords, which are two membranes located in the throat. Air from the lungs rushing through the throat starts the vocal cords vibrating. The frequency of vibration is controlled by the muscular tension placed on the vocal cords. The more tension on the vocal cords, the more rapidly they vibrate, resulting in a higher pitch sound. If the vocal cords are more relaxed, they vibrate more slowly and produce lower-pitched sounds.

In brass instruments, such as the trumpet, the tuba, and the bugle, the lips of the performer vibrate, as shown in **Figure 9.** Reed instruments, such as the clarinet and the saxophone, have a thin wooden strip called a reed that vibrates as a result of air blown across it, as shown in **Figure 9.** In flutes and organ pipes, air is forced across an opening in a pipe. Air moving past the opening sets the column of air in the instrument into vibration.

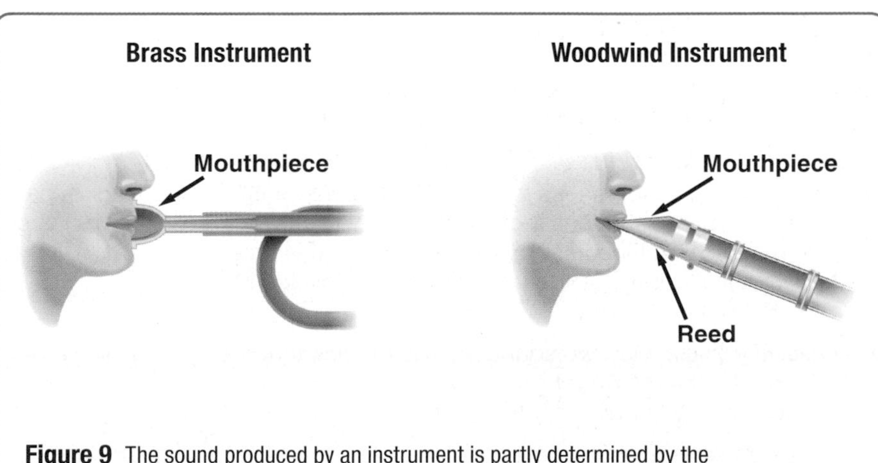

Figure 9 The sound produced by an instrument is partly determined by the structure of the mouthpiece.

Stringed instruments, such as the piano, the guitar, and the violin, work by setting wires or strings into vibration. In the piano, the wires are struck; for the guitar, they are plucked; and for the violin, the friction of the bow causes the strings to vibrate. The strings are attached to a sounding board that vibrates with the strings. The vibrations of the sounding board cause the pressure oscillations in the air that we hear as sound. Electric guitars use electronic devices to detect and amplify the vibrations of the guitar strings.

Resonance in Air Columns

If you have ever used just the mouthpiece of a brass or wind instrument, you know that while the vibration of your lips or the reed alone makes a sound, it is difficult to control the pitch. The long tube that makes up the instrument must be attached if music is to result. When the instrument is played, the air within this tube vibrates at the same frequency, or in resonance, with a particular vibration of the lips or reed. Remember that resonance increases the amplitude of a vibration by repeatedly applying a small external force to the vibrating air particles at the natural frequency of the air column. The length of the air column determines the frequencies of the vibrating air that will resonate. For wind and brass instruments, such as flutes, trumpets, and trombones, changing the length of the column of vibrating air varies the pitch of the instrument. The mouthpiece simply creates a mixture of different frequencies, and the resonating air column acts on a particular set of frequencies to amplify a single note, turning noise into music.

A tuning fork above a hollow tube can provide resonance in an air column, as shown in **Figure 10.** The tube is placed in water so that the bottom end of the tube is below the water surface. A resonating tube with one end closed to air is called a **closed-pipe resonator.** The length of the air column is changed by adjusting the height of the top of the tube above the water. If the tuning fork is struck with a rubber hammer, the sound alternately becomes louder and softer as the length of the air column is varied by moving the tube up and down in the water. The sound is loud when the air column is in resonance with the tuning fork because the resonating air column intensifies the sound of the tuning fork.

PhysicsLAB

SPEED OF SOUND

How can you determine the speed of sound?

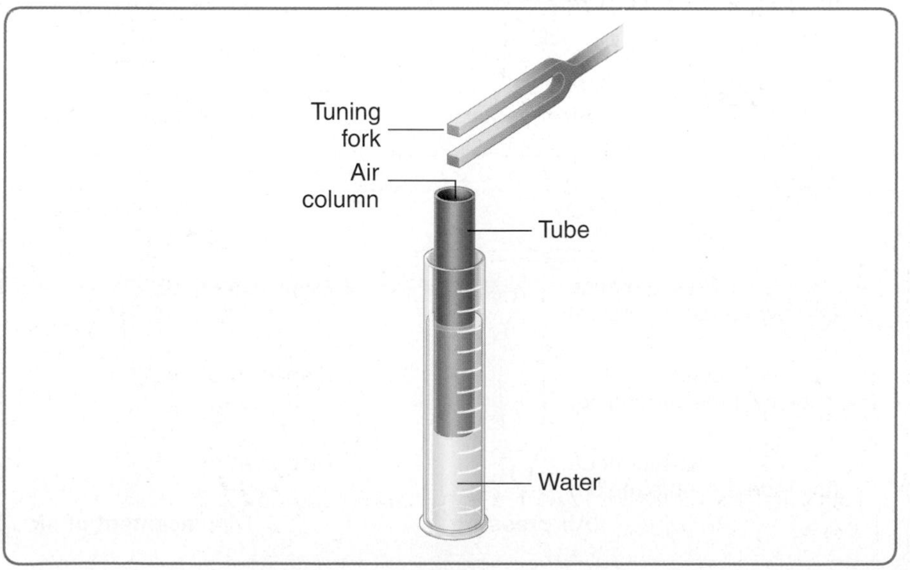

Figure 10 As the tube is raised or lowered, the length of the air column changes, which causes the sound's volume to change.

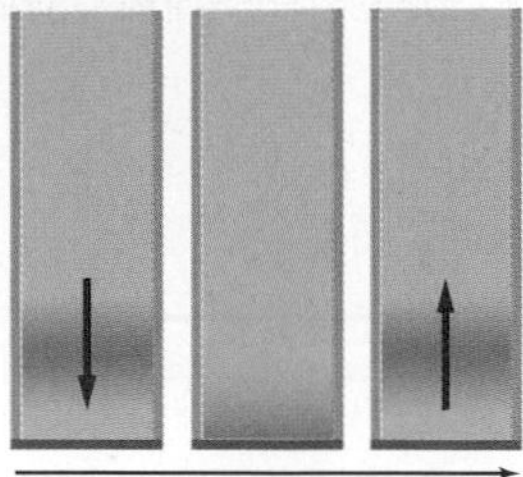

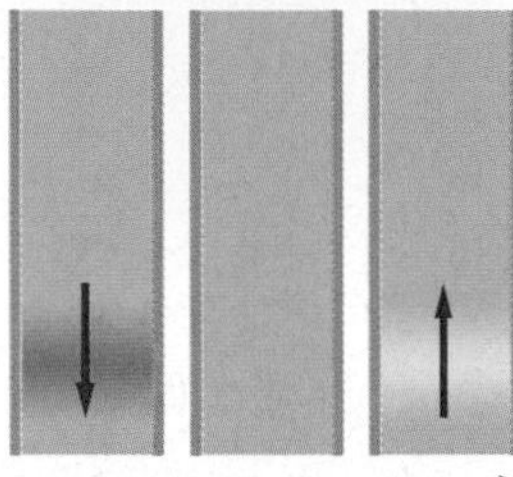

Figure 11 In closed pipes, the sound wave reflects off the closed end. High pressure waves reflect as high pressure. In open pipes, the sound wave reflects off an open end. High pressure waves are reflected as low pressure.

Standing pressure wave How does resonance occur? The vibrating tuning fork produces a sound wave. This wave of alternate high- and low-pressure variations moves down the air column. When the wave hits the water surface, it is reflected back up to the tuning fork, as shown in **Figure 11.** If the reflected high-pressure wave reaches the tuning fork at the same moment that the fork produces another high-pressure wave, then the emitted and returning waves reinforce each other. This reinforcement of waves creates a standing wave, and resonance occurs.

An **open-pipe resonator** is a resonating tube with both ends open that will resonate with a sound source. In this case, the sound wave does not reflect off a closed end, but rather off an open end. If the high-pressure part of the wave strikes the open end, the rebounding wave will be low-pressure at that point, as shown in **Figure 11.**

Resonance lengths **Figure 12** shows a standing sound wave in a pipe represented by a sine wave. Sine waves can represent either the air pressure or the displacement of the air particles. Recall that standing waves have nodes and antinodes. A node is the stationary point where two equal wave pulses meet and are in the same location. An antinode is the place of largest displacement when two wave pulses meet. In the pressure graphs, the nodes are regions of mean atmospheric pressure. At the antinodes, the pressure oscillates between its maximum and minimum values. In the case of the displacement graph, the antinodes are regions of high displacement and the nodes are regions of low displacement. In both cases, two adjacent antinodes (or two nodes) are separated by one-half wavelength.

READING CHECK Explain the difference between a node and an antinode on a displacement graph.

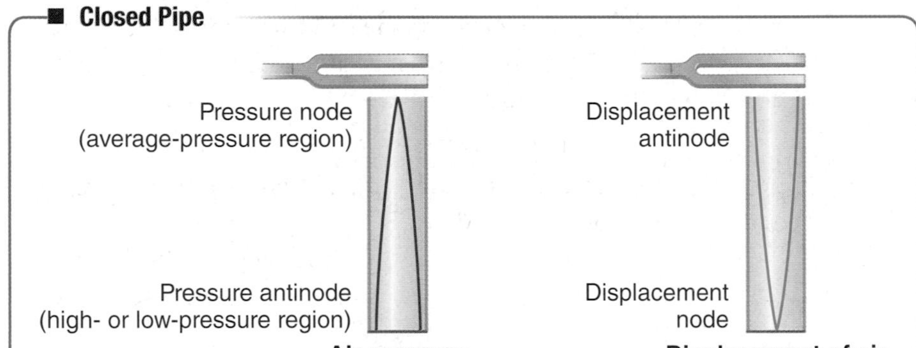

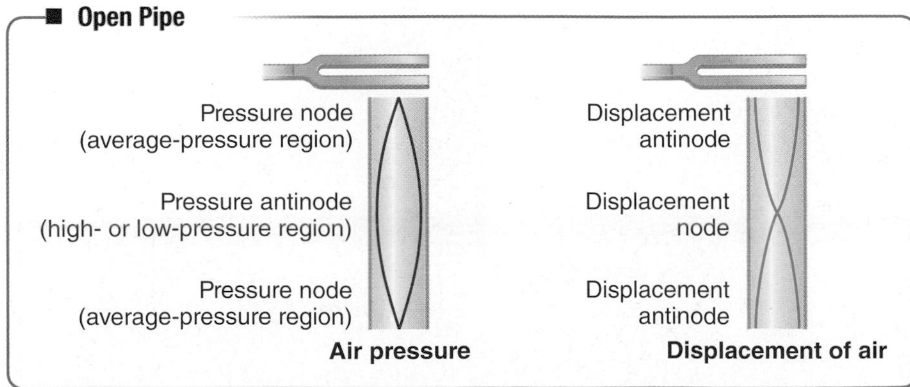

Figure 12 Standing waves in pipes can be represented by sine waves.

Identify Which are the areas of mean atmospheric pressure in the air pressure graphs?

View an **animation on resonance.**

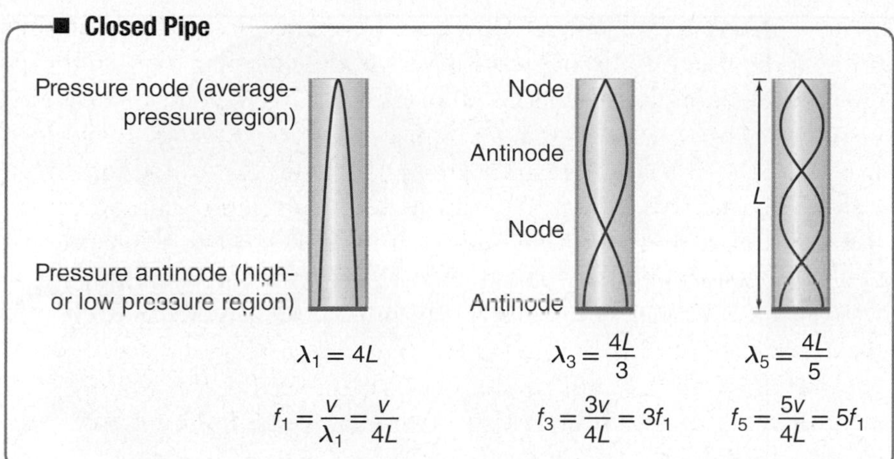

Figure 13 A closed pipe resonates when its length is an odd number of quarter wavelengths.

Resonance frequencies in a closed pipe If a closed end must act as a node, and an open end must act as an antinode, what is the shortest column of air that will resonate in a closed pipe? **Figure 13** shows that it must be one-fourth of a wavelength. As the frequency is increased, additional resonance lengths are found at half-wavelength intervals. Thus, columns of length $\frac{\lambda}{4}, \frac{3\lambda}{4}, \frac{5\lambda}{4}, \frac{7\lambda}{4}$, and so on will all be in resonance with a tuning fork that produces sound of wavelength λ.

In practice, the first resonance length is slightly longer than one-fourth of a wavelength. This is because the pressure variations do not drop to zero exactly at the open end of the pipe. Actually, the node is approximately 0.4 pipe diameters beyond the end. Additional resonance lengths, however, are spaced by exactly one-half of a wavelength. Measurements of the spacing between resonances can be used to find the velocity of sound in air, as in Example Problem 2.

Resonance frequencies in an open pipe The shortest column of air that can have nodes at both ends is one-half of a wavelength long, as shown in **Figure 14.** As the frequency is increased, additional resonance lengths are found at half-wavelength intervals. Thus, columns of length $\frac{\lambda}{2}, \lambda, \frac{3\lambda}{2}, 2\lambda$, and so on will be in resonance with a tuning fork.

If open and closed pipes of the same length are used as resonators, the wavelength of the resonant sound for the open pipe will be half as long as that for the closed pipe. Therefore, the frequency will be twice as high for the open pipe as for the closed pipe. For both pipes, resonance lengths are spaced by half-wavelength intervals.

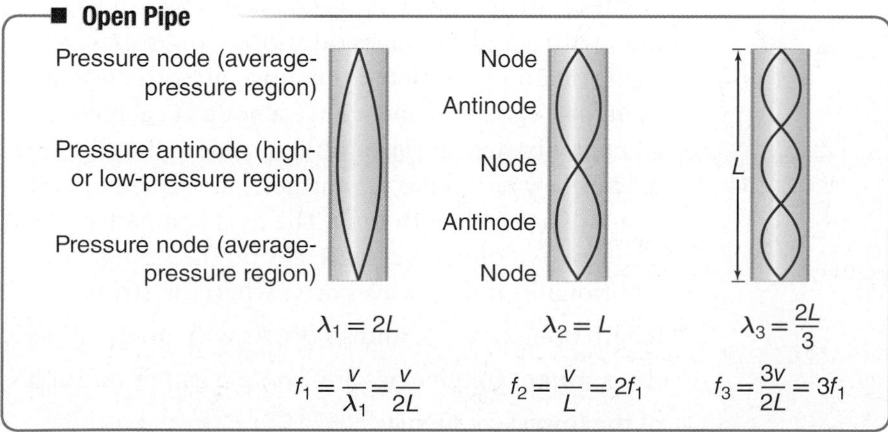

Figure 14 An open pipe resonates when its length is an even number of quarter wavelengths.

Explain How does the length at which an open pipe resonates differ from the length at which a closed pipe resonates?

REAL-WORLD PHYSICS

HEARING AND FREQUENCY The human auditory canal acts as a closed-pipe resonator that increases the ear's sensitivity for frequencies between 2000 and 5000 Hz, but the full range of frequencies that people hear extends from 20 to 20,000 Hz. A dog's hearing extends to frequencies as high as 45,000 Hz, and a cat's extends to frequencies as high as 100,000 Hz.

PhysicsLAB

HOW FAST DOES SOUND TRAVEL?

How can you measure sound speed using an open-pipe resonator?

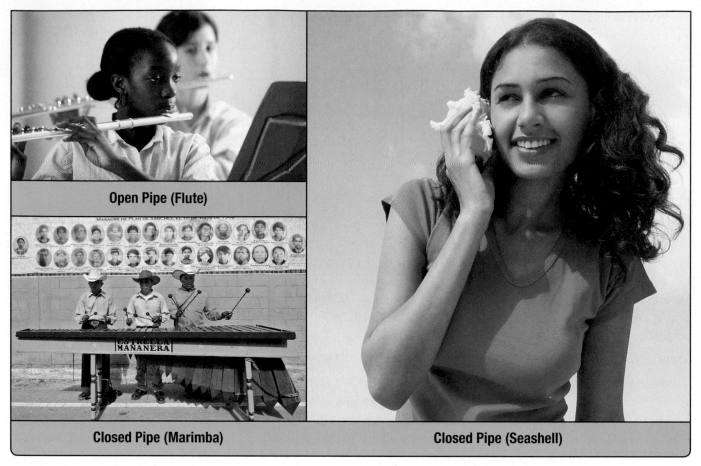

Open Pipe (Flute)

Closed Pipe (Marimba)

Closed Pipe (Seashell)

Figure 15 A flute is an example of an open-pipe resonator. The hanging pipes of a marimba and seashells are examples of closed-pipe resonators.

Hearing resonance Musical instruments use resonance to increase the loudness of particular notes. Open-pipe resonators include flutes, shown in **Figure 15.** Clarinets and the hanging pipes under marimbas and xylophones are examples of closed-pipe resonators. If you shout into a long tunnel, the booming sound you hear is the tunnel acting as a resonator. The seashell in **Figure 15** also acts as a closed-pipe resonator.

Figure 16 A string resonates with standing waves when its length is a whole number of half wavelengths.

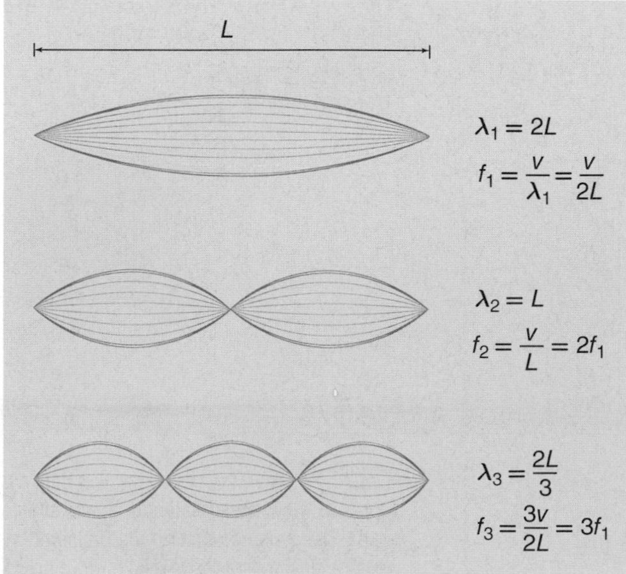

L
$\lambda_1 = 2L$
$f_1 = \frac{v}{\lambda_1} = \frac{v}{2L}$
$\lambda_2 = L$
$f_2 = \frac{v}{L} = 2f_1$
$\lambda_3 = \frac{2L}{3}$
$f_3 = \frac{3v}{2L} = 3f_1$

Resonance on Strings

Although plucking, bowing, or striking strings produces variation in waveforms, waveforms on vibrating strings have many characteristics in common with standing waves on springs and ropes. A string on an instrument is clamped at both ends, and therefore, the string must have a node at each end when it vibrates. In **Figure 16,** you can see that the first mode of vibration has an antinode at the center and is one-half a wavelength long. The next resonance occurs when one wavelength fits on the string, and additional standing waves arise when the string length is $\frac{3\lambda}{2}$, 2λ, $\frac{5\lambda}{2}$, and so on. As with an open pipe, the resonant frequencies are whole-number multiples of the lowest frequency.

Recall that the speed of a wave depends on the medium. For a string, it depends on the tension of the string, as well as its mass per unit length. The tighter the string, the faster the wave moves along it, and therefore, the higher the frequency of its standing waves. This makes it possible to tune a stringed instrument by changing the tension of its strings.

Because strings are so small in cross-sectional area, they move very little air when they vibrate. This makes it necessary to attach them to a sounding board, which transfers their vibrations to the air and produces a stronger sound wave. Unlike the strings themselves, the sounding board should not resonate at any single frequency. Its purpose is to convey the vibrations of all the strings to the air, and therefore it should vibrate well at all frequencies produced by the instrument. Because of the complicated interactions among the strings, the sounding board, and the air, the design and construction of stringed instruments are complex processes, considered by many to be as much an art as a science.

READING CHECK **Describe** the relationship between the tension of a string and the speed of a wave as it travels along the string.

Sound Quality

A tuning fork produces a soft and uniform sound. This is because its tines vibrate like simple harmonic oscillators and produce the simple sine wave shown in the top graph in **Figure 17.** Sounds made by the human voice and musical instruments are much more complex, like the wave in the bottom graph in **Figure 17.** Both waves have the same frequency, or pitch, but they sound very different. The complex sound wave is actually a blend of several of different frequencies. The shape of the wave depends on the relative amplitudes of these frequencies. Different sources provide different combinations of frequencies. In musical terms, the difference between the waves from different instruments is called timbre (TAM bur), tone quality, or color.

MiniLAB

SOUNDS GOOD

How can you determine whether an instrument is an open-pipe resonator or a closed-pipe resonator?

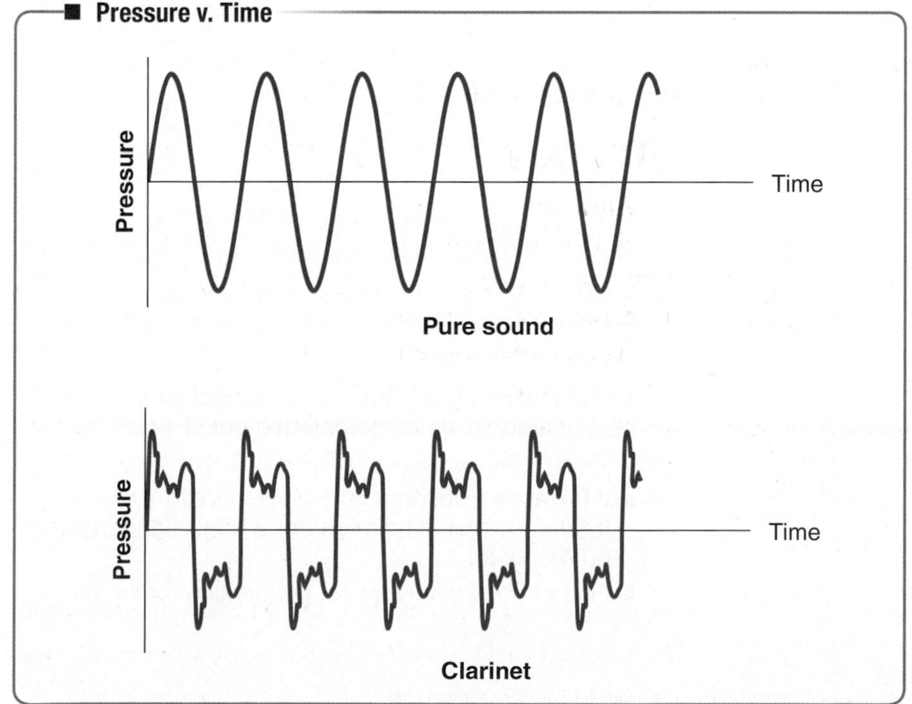

Figure 17 The pure sound produced by a tuning fork is represented by a simple sine wave. The more complex sound produced by a clarinet is represented in the bottom graph.

EXAMPLE PROBLEM 2

Get help with **closed-pipe resonators.** Personal Tutor

EXAMPLE PROBLEM

FINDING THE SPEED OF SOUND USING RESONANCE When a tuning fork with a frequency of 392 Hz is used with a closed-pipe resonator, the loudest sound is heard when the column is 21.0 cm and 65.3 cm long. What is the speed of sound in this case? Is the temperature warmer or cooler than normal room temperature, which is 20°C? Explain your answer.

1 ANALYZE AND SKETCH THE PROBLEM

- Sketch the closed-pipe resonator.
- Mark the resonance lengths.

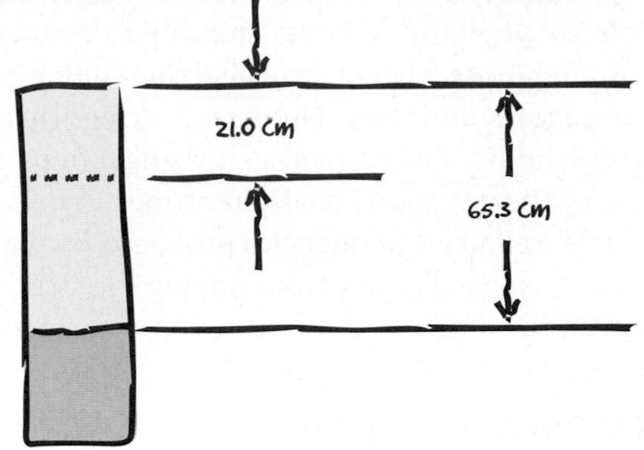

KNOWN	UNKNOWN
$f = 392$ Hz	$v = ?$
$L_A = 21.0$ cm	
$L_B = 65.3$ cm	

2 SOLVE FOR THE UNKNOWN

Solve for the length of the wave using the length-wavelength relationship for a closed pipe.

$$L_B - L_A = \frac{1}{2}\lambda$$

$$\lambda = 2(L_B - L_A)$$ ◀ Rearrange the equation for λ.

$$= 2(0.653 \text{ m} - 0.210 \text{ m})$$ ◀ Substitute $L_B = 0.653$ m, $L_A = 0.210$ m.

$$= 0.886 \text{ m}$$

Use $\lambda = \frac{v}{f}$

$$v = f \cdot \lambda$$ ◀ Rearrange the equation for v.

$$= (392 \text{ Hz})(0.886 \text{ m})$$ ◀ Substitute $f = 392$ Hz, $\lambda = 0.886$ m.

$$= 347 \text{ m/s}$$

The speed is slightly greater than the speed of sound at 20°C, indicating that the temperature is slightly higher than normal room temperature.

3 EVALUATE THE ANSWER

- **Are the units correct?** $(\text{Hz})(\text{m}) = \left(\frac{1}{\text{s}}\right)(\text{m}) = \text{m/s}$. The answer's units are correct.
- **Is the magnitude realistic?** The speed is slightly greater than 343 m/s, which is the speed of sound at 20°C.

PRACTICE PROBLEMS

Do additional problems. Online Practice

13. A 440-Hz tuning fork is used with a resonating column to determine the velocity of sound in helium gas. If the spacing between resonances is 110 cm, what is the velocity of sound in helium gas?

14. The frequency of a tuning fork is unknown. A student uses an air column at 27°C and finds resonances spaced by 20.2 cm. What is the frequency of the tuning fork? Use the speed calculated in Example Problem 2 for the speed of sound in air at 27°C.

15. A 440-Hz tuning fork is held above a closed pipe. Find the spacing between the resonances when the air temperature is 20°C.

16. CHALLENGE A bugle can be thought of as an open pipe. If a bugle were straightened out, it would be 2.65-m long.

a. If the speed of sound is 343 m/s, find the lowest frequency that is resonant for a bugle (ignoring end corrections).

b. Find the next two resonant frequencies for the bugle.

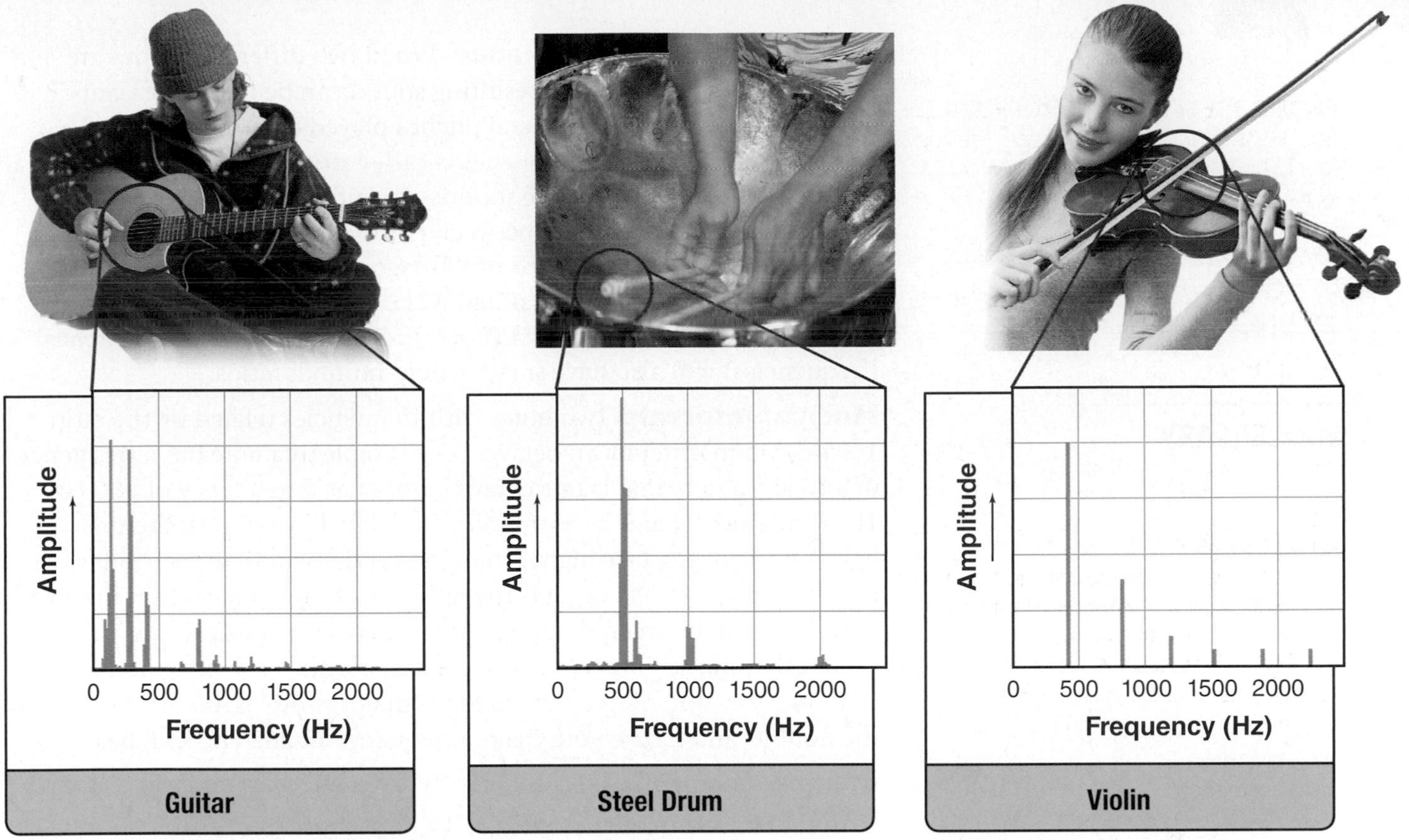

Amplitude → 0 500 1000 1500 2000 Frequency (Hz) **Guitar**

Amplitude → 0 500 1000 1500 2000 Frequency (Hz) **Steel Drum**

Amplitude → 0 500 1000 1500 2000 Frequency (Hz) **Violin**

Figure 18 A guitar, a steel drum, and a violin produce characteristic sound spectra. Each spectrum is unique, as is the timbre of the instrument.

The sound spectrum: fundamental and harmonics The complex sound wave in **Figure 17** was made by a clarinet. Why does the clarinet produce such a sound wave? The air column in a clarinet acts as a closed pipe. Look back at **Figure 13,** which shows three resonant frequencies for a closed pipe. The clarinet acts as a closed pipe, so for a clarinet of length L the lowest frequency (f_1) that will be resonant is $\frac{v}{4L}$. For a musical instrument, the lowest frequency of sound that resonates is called the **fundamental.** A closed pipe also will resonate at $3f_1$, $5f_1$, and so on. These higher frequencies, which are whole-number multiples of the fundamental frequency, are called **harmonics.** It is the addition of these harmonics that gives a clarinet its distinctive timbre.

Some instruments, such as a flute, act as open-pipe resonators. Their fundamental frequency, which is also the first harmonic, is $f_1 = \frac{v}{2L}$ with subsequent harmonics at $2f_1$, $3f_1$, $4f_1$, and so on. Different combinations of these harmonics give each instrument its own unique timbre. Each harmonic on the instrument can have a different amplitude as well. A graph of the amplitude of a wave versus its frequency is called a sound spectrum. The spectra of three instruments are shown in **Figure 18.**

PHYSICS CHALLENGE

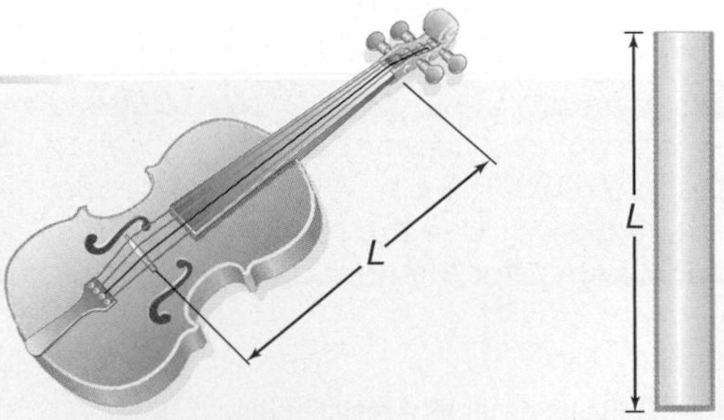

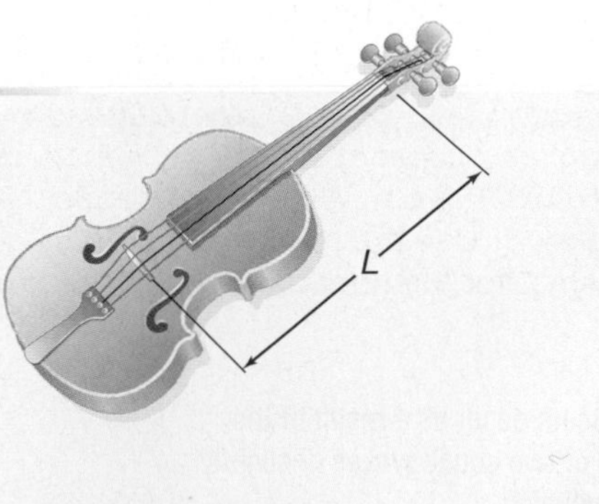

1. Determine the tension, F_T, in a violin string of mass m and length L that will play the fundamental note at the same frequency as a closed pipe also of length L. Express your answer in terms of m, L, and the speed of sound in air, v. The equation for the speed of a wave on a string is $v_{string} = \sqrt{\frac{F_T}{\mu}}$, where F_T is the tension in the string and μ is the mass per unit length of the string.
2. What is the tension in a string of mass 1.0 g and 40.0 cm long that plays the same note as a closed pipe of the same length?

(l)RubberBall Productions, (c)John & Lisa Merrill/The Image Bank/Getty Images, (r)Rayman/Getty Images

Consonance and dissonance When two different pitches are played at the same time, the resulting sound can be either pleasant or jarring. In musical terms, several pitches played together are called a chord. An unpleasant set of pitches is called **dissonance.** If the combination of pitches is pleasant, the sounds are said to be in **consonance.** What sounds pleasing varies between cultures, but most Western music is based upon the observations of Pythagoras of ancient Greece. He noted that pleasing sounds resulted when strings had lengths in small, whole-number ratios, such as 1:2, 2:3, or 3:4. This means their pitches (frequencies) will also have small, whole-number ratios.

VOCABULARY
Science Usage v. Common Usage

Beat
- Science usage
oscillation of wave amplitude that results from the superposition of two sound waves with almost identical frequencies
When the piano tuner no longer heard beats, she knew the piano was tuned properly.
- Common usage
to strike repeatedly
Richard beat the drums while John played the guitar.

Musical intervals Two notes with frequencies related by the ratio 1:2 are said to differ by an octave. For example, if a note has a frequency of 440 Hz, a note that is one octave higher has a frequency of 880 Hz. The fundamental and its harmonics are related by octaves; the first harmonic is one octave higher than the fundamental, the second is two octaves higher, and so on. It is the ratio of two frequencies, not the size of the interval between them, that determines the musical interval.

In other musical intervals, two pitches may be close together. For example, the ratio of frequencies for a "major third" is 4:5. An example is the notes C and E. The note C has a frequency of 262 Hz, so E has a frequency of $\left(\frac{5}{4}\right)(262 \text{ Hz}) = 327 \text{ Hz}$. In the same way, notes in a "fourth" (C and F) have a frequency ratio of 3:4, and those in a "fifth" (C and G) have a ratio of 2:3. More than two notes sounded together also can produce consonance. The three notes called do, mi, and sol make a major chord. For at least 2500 years, western music has recognized this as the sweetest of the three-note chords; it has the frequency ratio of 4:5:6.

Beats

You have seen that consonance is defined in terms of the ratio of frequencies. When the ratio becomes nearly 1:1, the frequencies become very close. Two frequencies that are nearly identical interfere to produce oscillating high and low sound levels called a **beat.** This phenomenon is illustrated in **Figure 19.** The frequency of a beat is the magnitude of difference between the frequencies of the two waves, $f_{\text{beat}} = |f_A - f_B|$. When the difference is less than 7 Hz, the ear detects this as a pulsation of loudness. Musical instruments often are tuned by sounding one against another and adjusting the frequency of one until the beat disappears.

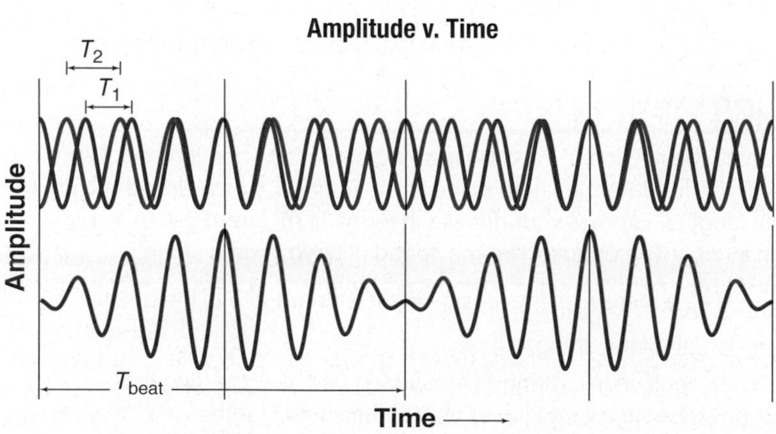

Figure 19 Beats occur as a result of the superposition of two sound waves of slightly different frequencies.

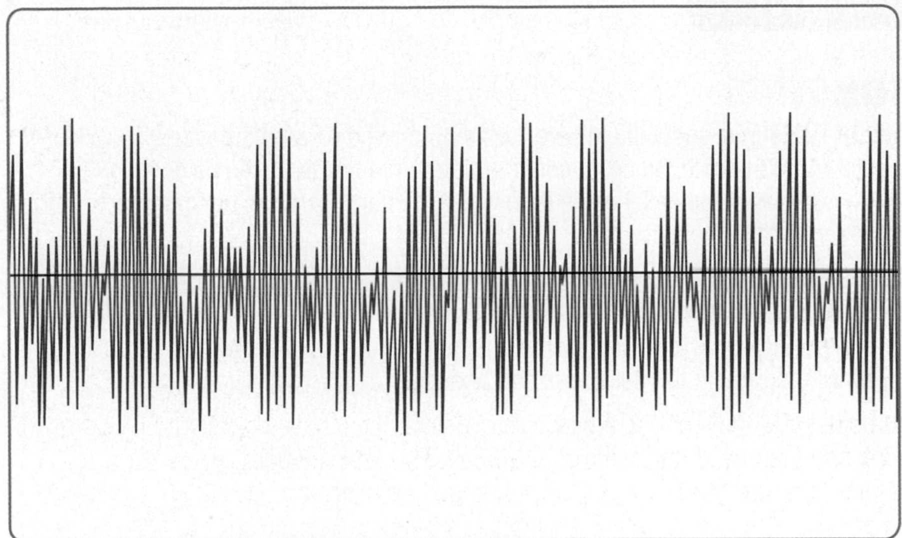

Figure 20 A noise wave consists of many different frequencies, all with about the same amplitude.

Sound Reproduction and Noise

When you listen to a live band or hear your school band practicing, you are hearing music produced directly by a human voice or musical instruments. You may want to hear live music every time you choose to listen to music. Most of the time, however, you likely listen to music that has been recorded and is played via electronic systems. To reproduce the sound faithfully, the system must accommodate all frequencies equally. A good stereo system keeps the amplitudes of all frequencies between 20 and 20,000 Hz to within a range of 3 dB.

A telephone system, on the other hand, needs only to transmit the information in spoken language. Frequencies between 300 and 3000 Hz are sufficient. Reducing the number of frequencies present helps reduce the noise. A noise wave is shown in **Figure 20.** Many frequencies are present with approximately the same amplitude. While noise is not helpful in a telephone system, some people claim that listening to white noise has a calming effect. For this reason, some dentists use noise to help their patients relax.

☑ **READING CHECK Describe** a noise wave.

SECTION 2 REVIEW

Section Self-Check Check your understanding.

17. MAINIDEA What is the vibrating object that produces sounds in each of the following?

a. a human voice

b. a clarinet

c. a tuba

d. a violin

18. Resonance in Air Columns Why is the tube from which a tuba is made much longer than that of a cornet?

19. Resonance in Open Tubes How must the length of an open tube compare to the wavelength of the sound to produce the strongest resonance?

20. Resonance on Strings A violin sounds a note of F sharp, with a pitch of 370 Hz. What are the frequencies of the next three harmonics produced with this note?

21. Resonance in Closed Pipes One closed organ pipe has a length of 2.40 m.

a. What is the frequency of the note played by this pipe?

b. When a second pipe is played at the same time, a 1.40-Hz beat note is heard. By how much is the second pipe too long?

22. Timbre Why do various instruments sound different even when they play the same note?

23. Beats A tuning fork produces three beats per second with a second, 392-Hz tuning fork. What is the frequency of the first tuning fork?

24. Critical Thinking Strike a tuning fork with a rubber hammer and hold it at arm's length. Then press its handle against a desk, a door, a filing cabinet, and other objects. What do you hear? Why?

SOUNDS GOOD!

Theater Acoustics

The ancient Greek theater at Epidaurus, shown in **Figure 1,** has 55 curved rows of limestone benches that can seat up to 14,000 people. Since its construction in the fourth century B.C., a great mystery has intrigued acoustic scientists: how is it possible to hear the voices of actors so clearly in the back-row seats of the theater without the aid of speakers?

It is all in the seats. Some hypothesized that music and voices were carried farther than in other theaters due to the local prevailing winds. Others thought that the Greeks used special masks that helped project the voices of actors. Many even suspected that the slope of the theater seats somehow contributed to the theater's amazing acoustics.

Modern Research It turns out that the seats do a play a role. Researchers have discovered that the shape, placement, and material of the seats all have beneficial acoustic properties. The seats in this theater absorb sound frequencies typical of audience murmurs and other background noises such as wind. At the same time, the front-row seats reflect the sounds of the actors' voices toward the back-row seats. This combination of subtracting unwanted noise and projecting desired noise means that you can enjoy a play even in the "cheap" seats.

Figure 1 Audience members in the farthest back rows of the theater at Epidaurus can still hear the voices of actors on stage—even without microphones—due to the theater's special acoustics.

Purestock/Getty Images

GOING FURTHER >>>

Research home theater acoustic systems and make a poster that explains the recommended placement of speakers and acoustic treatments to maximize listener enjoyment.

CHAPTER 15

STUDY GUIDE

BIGIDEA Sound waves are pressure variations, and many can be detected by the human ear.

SECTION 1 Properties and Detection of Sound

VOCABULARY
- **sound wave** *(p. 411)*
- **pitch** *(p. 413)*
- **loudness** *(p. 413)*
- **sound level** *(p. 413)*
- **decibel** *(p. 413)*
- **Doppler effect** *(p. 414)*

MAINIDEA Our perception of a sound wave depends on that wave's physical properties.

- Sound is a pressure variation transmitted through matter as a longitudinal wave. A sound wave has frequency, wavelength, speed, and amplitude. Sound waves reflect and interfere.
- Sound detectors convert the energy carried by a sound wave into another form of energy. The human ear is a highly efficient and sensitive detector of sound waves. The frequency of a sound wave is heard as its pitch. The loudness of sound as perceived by the ear and brain depends mainly on its amplitude. The pressure amplitude of a sound wave can be measured in decibels (dB).

Decibel Scale

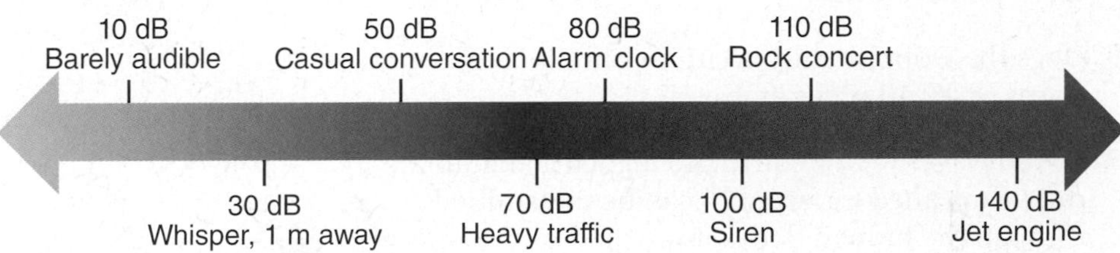

- The Doppler effect is the change in frequency of sound caused by the motion of either the source or the detector.
- The Doppler effect is used in radar detectors, in medical ultrasound machines, and by astronomers. Bats also use the Doppler effect to detect and catch flying insects.

SECTION 2 The Physics of Music

VOCABULARY
- **closed-pipe resonator** *(p. 419)*
- **open-pipe resonator** *(p. 420)*
- **fundamental** *(p. 425)*
- **harmonics** *(p. 425)*
- **dissonance** *(p. 426)*
- **consonance** *(p. 426)*
- **beat** *(p. 426)*

MAINIDEA Music consists of complex sound waves produced by vibrating objects.

- Sound is produced by a vibrating object in a medium.
- An air column can resonate with a sound source, thereby increasing the amplitude of its resonant frequency. A closed pipe resonates when its length is $\frac{\lambda}{4}, \frac{3\lambda}{4}, \frac{5\lambda}{4}$, and so on. Its resonant frequencies are odd-numbered multiples of the fundamental. An open pipe resonates when its length is $\frac{\lambda}{2}, \frac{2\lambda}{2}, \frac{3\lambda}{2}$, and so on. Its resonant frequencies are whole-number multiples of the fundamental.
- A clamped string has a node at each end and resonates when its length is $\frac{\lambda}{2}, \frac{2\lambda}{2}, \frac{3\lambda}{2}$, and so on, just as with an open pipe. The string's resonant frequencies are also whole-number multiples of the fundamental.
- The frequencies and intensities of the complex waves produced by a musical instrument determine the timbre that is characteristic of that instrument.
- Two waves with almost the same frequency interfere to produce beats.

Games and Multilingual eGlossary

Vocabulary Practice

CHAPTER 15 ASSESSMENT

Chapter Self-Check

SECTION 1
Properties and Detection of Sound

Mastering Concepts

25. What are the physical characteristics of sound waves?

26. When timing the 100-m run, officials at the finish line are instructed to start their stopwatches at the sight of smoke from the starter's pistol and not at the sound of its firing. Explain. What would happen to the times for the runners if the timing started when sound was heard?

27. BIGIDEA Name two types of perception of sound and the physical characteristics of sound waves that correspond to them.

28. Does the Doppler shift occur for only some types of waves or for all types of waves?

29. Sound waves with frequencies higher than humans can hear, called ultrasound, can be transmitted through the human body. How could ultrasound be used to measure the speed of blood flowing in veins or arteries? Explain how the waves change to make this measurement possible.

Mastering Problems

30. You hear the sound of the firing of a distant cannon 5.0 s after seeing the flash. How far are you from the cannon?

31. If you shout across a canyon and hear an echo 3.0 s later, how wide is the canyon?

32. A sound wave has a frequency of 4700 Hz and travels along a steel rod. If the distance between compressions, or regions of high pressure, is 1.1 m, what is the speed of the wave?

33. Bats The sound emitted by bats has a wavelength of 3.5 mm. What is the sound's frequency in air?

34. Sound with a frequency of 261.6 Hz travels through water at 25°C. Find the sound's wavelength in water. Do not confuse sound waves moving through water with surface waves moving through water.

35. If the wavelength of a 4.40×10^{2}-Hz sound in freshwater is 3.30 m, what is the speed of sound in freshwater?

36. A fire truck is moving at 35 m/s, and a car in front of the truck is moving in the same direction at 15 m/s. If a 327-Hz siren blares from the truck, what frequency is heard by the driver of the car?

37. Photography As shown in **Figure 21,** some cameras determine the distance to the subject by sending out a sound wave and measuring the time needed for the echo to return to the camera. How long would it take the sound wave to return to such a camera if the subject were 3.00 m away?

Figure 21

38. Sound with a frequency of 442 Hz travels through an iron beam. Find the wavelength of the sound in iron.

39. Aircraft Adam, an airport employee, is working near a jet plane taking off. He experiences a sound level of 150 dB.

a. Adam wears ear protectors that reduce the sound level to that of a typical rock concert. What decrease in dB is provided?

b. Adam then hears something that sounds like a barely audible whisper. What will a person not wearing the ear protectors hear?

40. Music A band plays at an 80-dB sound level. How many times greater is the sound pressure from another band playing at each of the following sound levels?

a. 100 dB

b. 120 dB

41. A coiled-spring toy is shaken at a frequency of 4.0 Hz such that standing waves are observed with a wavelength of 0.50 m. What is the speed of propagation of the wave?

42. Speed of Sound A baseball fan on a warm summer day (30°C) sits in the bleachers 152 m away from home plate.

a. What is the speed of sound in air at 30°C?

b. How long after seeing the ball hit the bat does the fan hear the crack of the bat?

43. On a day when the temperature is 15°C, a person stands some distance (d), as shown in **Figure 22,** from a cliff and claps his hands. The echo returns in 2.0 s. How far away is the cliff?

Figure 22 (Not to scale)

44. Medical Imaging Ultrasound with a frequency of 4.25 MHz can be used to produce images of the human body. If the speed of sound in the body is the same as in salt water, 1.50 km/s, what is the length of a 4.25-MHz pressure wave in the body?

45. Sonar A ship surveying the ocean bottom sends sonar waves straight down into the seawater from the surface. As illustrated in **Figure 23,** the first reflection, off the mud at the seafloor, is received 1.74 s after it was sent. The second reflection, from the bedrock beneath the mud, returns after 2.36 s. The seawater is at a temperature of 25°C, and the speed of sound in mud is 1875 m/s.

a. How deep is the water?

b. How thick is the mud?

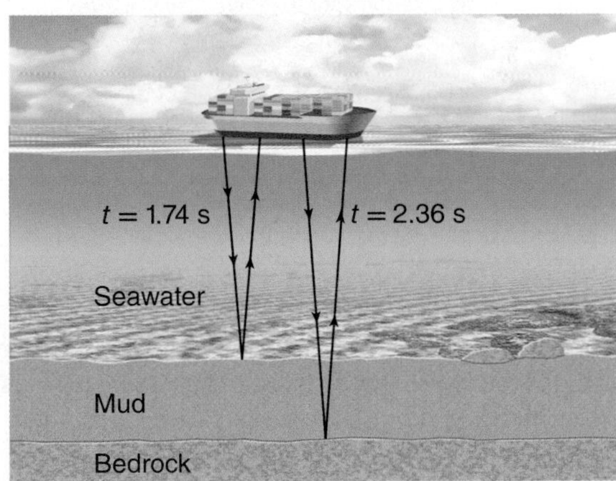

Figure 23 (Not to scale)

46. Determine the increase in sound pressure of a conversation being held at a sound level of 60 dB compared to the softest audible sound.

47. A train moving toward a sound detector at 31.0 m/s blows a 305-Hz whistle. What frequency is detected on each of the following?

a. a stationary train

b. a train moving toward the first train at 21.0 m/s

48. The train in the previous problem now moves away from the detector. What frequency is now detected on each of the following?

a. a stationary train

b. a train moving away from the first train at a speed of 21.0 m/s

SECTION 2 The Physics of Music

Mastering Concepts

49. What is necessary for the production and transmission of sound?

50. Singing How can a certain note sung by an opera singer cause a crystal glass to shatter?

51. Marching In the military, as marching soldiers approach a bridge, the command "route step" is given. The soldiers then walk out-of-step with each other as they cross the bridge. Explain why this is done.

52. Musical Instruments Why don't most musical instruments sound like tuning forks?

53. Musical Instruments What property distinguishes notes played on a trumpet and a clarinet from each other if they have the same pitch and loudness?

54. Trombones Explain how the slide of a trombone, shown in **Figure 24,** changes the pitch of the sound in terms of a trombone being a resonance tube.

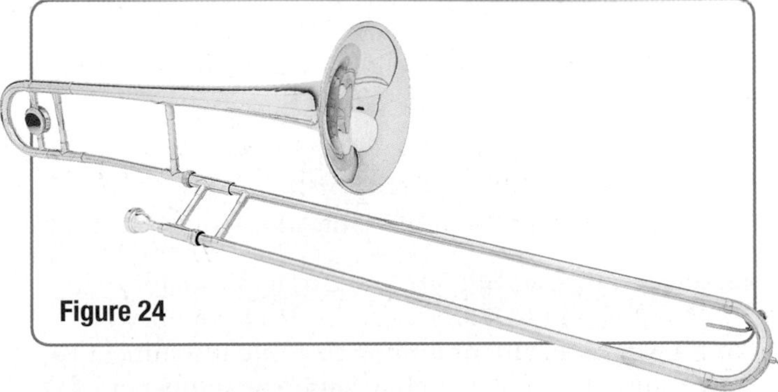

Figure 24

Mastering Problems

55. One tuning fork has a 445-Hz pitch. When a second fork is struck, beat notes occur with a frequency of 3 Hz. What are the two possible frequencies of the second fork?

Tim Fuller

Chapter Self-Check

56. A vertical tube with a tap at the base is filled with water. A tuning fork vibrates over its mouth. As the water level is lowered in the tube, resonance is heard when the water level has dropped 17 cm, and again after 49 cm of distance exists from the water to the top of the tube. What is the frequency of the tuning fork?

57. Human Hearing The auditory canal leading to the eardrum is a closed pipe that is 3.0 cm long. Find the approximate value (ignoring end correction) of the lowest resonance frequency.

58. If you hold a 1.2-m aluminum rod in the center and hit one end with a hammer, it will oscillate like an open pipe. Antinodes of pressure correspond to nodes of molecular motion so there is a pressure antinode in the center of the bar. The speed of sound in aluminum is 6420 m/s. What would be the bar's lowest frequency of oscillation?

59. Ranking Task Rank the following tones according to wavelength, from least to greatest. Specifically indicate any ties.

A. 35 Hz
B. 50 Hz
C. 83 Hz
D. 100 Hz
E. 140 Hz

60. Clarinets The clarinet acts as a closed pipe. If a clarinet sounds a note with a pitch of 370 Hz, what are the frequencies of the lowest three harmonics produced by this instrument?

61. String Instruments A guitar string is 65.0 cm long and is tuned to produce a lowest frequency of 196 Hz.

a. What is the speed of the wave on the string?

b. What are the next two higher resonant frequencies for this string?

62. Musical Instruments The lowest note on an organ is 16.4 Hz.

a. What is the shortest open organ pipe that will resonate at this frequency?

b. What is the pitch if the same organ pipe is closed?

63. Musical Instruments Two instruments are playing the note A (440.0 Hz). Beats with a frequency of 2.5 Hz are heard. Assuming that one instrument is playing the correct pitch, what is the frequency of the pitch played by the second instrument?

64. Musical Instruments One open organ pipe has a length of 836 mm. A second open pipe should have a pitch that is one major third higher. How long should the second pipe be?

65. A flexible, corrugated, plastic tube, shown in **Figure 25,** is 0.85 m long. When it is swung around, it creates a tone that is the lowest pitch for an open pipe of this length. What is the frequency?

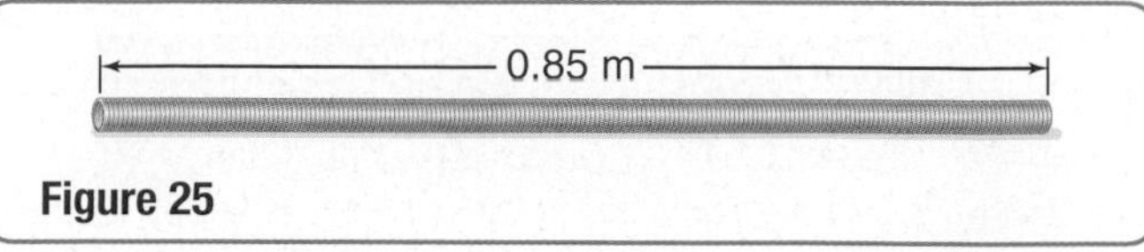

Figure 25

66. The tube from the previous problem is swung faster, producing a higher pitch. What is the new frequency?

67. During normal conversation, the amplitude of a pressure wave is 0.020 Pa.

a. If the area of an eardrum is 0.52 cm^2, what is the force on the eardrum?

b. The mechanical advantage of the three bones in the middle ear is 1.5. If the force in part a is transmitted undiminished to the bones, what force do the bones exert on the oval window, the membrane to which the third bone is attached?

c. The area of the oval window is 0.026 cm^2. What is the pressure increase transmitted to the liquid in the cochlea?

68. As shown in **Figure 26,** a music box contains a set of steel fingers clamped at one end and plucked on the other end by pins on a rotating drum. What is the speed of a wave on a finger that is 2.4 cm long and plays a note of 1760 Hz?

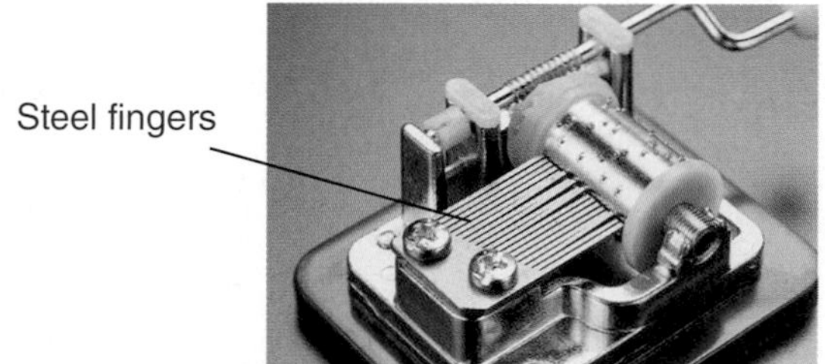

Figure 26

Applying Concepts

69. The speed of sound increases by about 0.6 m/s for each degree Celsius when the air temperature rises. For a given sound, as the temperature increases, what happens to the following?

a. frequency

b. wavelength

70. Does a sound of 40 dB have a factor of 100 times greater pressure variation than the threshold of hearing, or a factor of 40 times greater?

Chapter Self-Check

71. Estimation To estimate the distance in kilometers between you and a lightning flash, count the seconds between the flash and the thunder and divide by 3. Explain how this rule works. Devise a similar rule for miles.

72. Movies In a science-fiction movie, a satellite blows up. The crew of a nearby ship immediately hears and sees the explosion. If you had been hired as an advisor, what two physics errors would you have noticed and corrected?

73. The Redshift Astronomers have observed that the light coming from distant galaxies appears redder than light coming from nearer galaxies. With the help of **Figure 27,** which shows the visible spectrum, explain why astronomers conclude that distant galaxies are moving away from Earth.

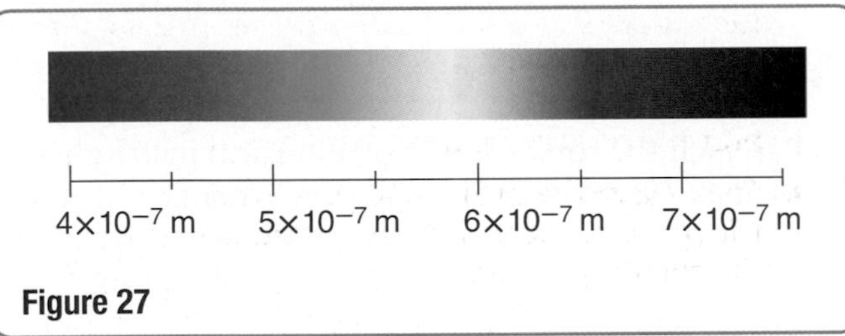

Figure 27

74. Reverse Problem Write a physics problem with real-life objects for which the following equation would be part of the solution:

$$440 \text{ Hz} = \frac{442 \text{ Hz } (343 \text{ m/s} - v_d)}{343 \text{ m/s}}$$

75. If the pitch of a sound is increased, how, if at all, do the following change?

a. frequency

b. wavelength

c. wave velocity

d. amplitude

76. Temperature Changes The speed of sound increases with temperature. Would the pitch of a closed pipe increase or decrease when the temperature of the air rises? Assume that the length of the pipe does not change.

77. Marching Bands Two flutists are tuning up. If the conductor hears the beat frequency increasing, are the two flute frequencies getting closer together or farther apart?

78. Musical Instruments A covered organ pipe plays a certain note. If the cover is removed to make it an open pipe, is the pitch of the sound increased or decreased?

79. Stringed Instruments On a harp, **Figure 28,** long strings produce low notes and short strings produce high notes. On a guitar, **Figure 28,** the strings are all the same length. How can they produce notes of different pitches?

Figure 28

Mixed Review

80. Problem Posing Complete this problem so that it can be solved using the Doppler effect: "A car's horn is stuck and is emitting a pitch of 500 Hz ...."

81. If you drop a stone into a well that is 122.5 m deep, as illustrated in **Figure 29,** how soon after you drop the stone will you hear it hit the bottom of the well?

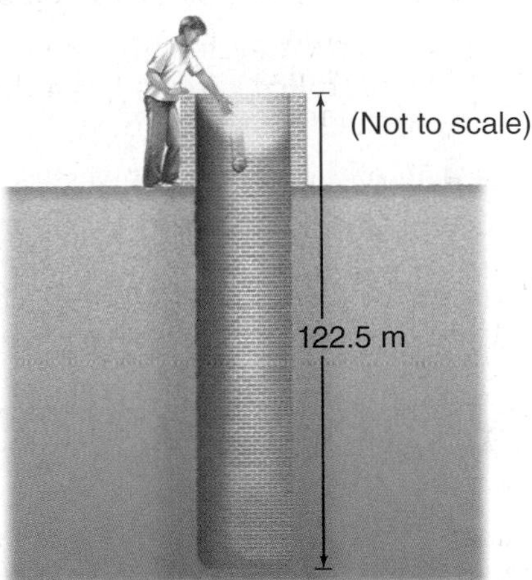

Figure 29

82. In North America, one of the hottest outdoor temperatures ever recorded is 57°C and one of the coldest is −62°C. What are the speeds of sound at those two temperatures?

83. A ship's sonar uses a frequency of 22.5 kHz. The speed of sound in seawater is 1533 m/s. What is the frequency received on the ship that was reflected from a whale traveling at 4.15 m/s away from the ship? Assume that the ship is at rest.

Chapter Self-Check

84. In 1845 Dutch scientist Christoph Buys Ballot tested the Doppler effect. A trumpet player sounded an A (440 Hz) while riding on a flatcar pulled by a train. At the same time, a stationary trumpeter played the same note. Buys Ballot heard 3.0 beats per second. How fast was the train moving toward him?

85. You try to repeat Buys Ballot's experiment from the previous problem. You plan to have a trumpet played in a rapidly moving car. Rather than listening for beat notes, however, you want to have the car move fast enough so that the moving trumpet sounds one major third above a stationary trumpet.

a. How fast would the car have to move?

b. Should you try the experiment? Explain.

86. Guitar Strings The equation for the speed of a wave on a string is $v = \sqrt{\frac{F_T}{\mu}}$, where F_T is the tension in the string and μ is the mass per unit length of the string. A guitar string has a mass of 3.2 g and is 65 cm long. What must be the tension in the string to produce a note whose fundamental frequency is 147 Hz?

Thinking Critically

87. The wavelengths of sound waves produced by a set of tuning forks are shown in **Table 2** below.

a. Plot a graph of the wavelength versus the frequency (controlled variable). What type of relationship does the graph show?

b. Plot a graph of the wavelength versus the inverse of the frequency $\left(\frac{1}{f}\right)$. What kind of graph is this? Determine the speed of sound from this graph.

Table 2 Tuning Forks

Frequency (Hz)	Wavelength (m)
262	1.31
294	1.17
330	1.04
349	0.98
392	0.87
440	0.78

88. Suppose that the frequency of a car horn is 300 Hz when it is stationary. Make a rough sketch of what a graph of frequency versus time would look like as the car approached and then moved past you?

89. Analyze and Conclude Describe how you could use a stopwatch to estimate the speed of sound if you were near the green on a 200-m golf hole as another group of golfers hit their tee shots. Would your estimate of the speed of sound be too large or too small?

90. Apply Concepts A light wave coming from a point on the left edge of the Sun is found by astronomers to have a slightly higher frequency than light from the right edge. What do these measurements tell you about the Sun's motion?

91. Design an Experiment Design an experiment that could test the formula for the speed of a wave on a string. Explain what measurements you would make, how you would make them, and how you would use them to test the formula.

Writing in Physics

92. Research the construction of a musical instrument, such as a violin or a French horn. What factors must be considered besides the length of the strings or the tube? What is the difference between a quality instrument and a cheaper one? How are they tested for tone quality?

93. Research the use of the Doppler effect in the study of astronomy. What is its role in the Big Bang theory? How is it used to detect planets around other stars? To study the motions of galaxies?

Cumulative Review

94. Ball A, rolling west at 3.0 m/s, has a mass of 1.0 kg. Ball B has a mass of 2.0 kg and is stationary. After colliding with ball B, ball A moves south at 2.0 m/s.

a. Sketch the system, showing the velocities and momenta before and after the collision.

b. Calculate the momentum and velocity of ball B after the collision.

95. Chris carries a 10-N carton of milk at a constant velocity along a level hall to the kitchen, a distance of 3.5 m. How much work does Chris do on the carton?

96. A movie stunt person jumps from a five-story building (22 m high) onto a large pillow at ground level. The pillow cushions her fall so that she feels an acceleration of no more than 3.0 m/s^2.

a. If she weighs 480 N, how much energy does the pillow have to absorb?

b. How much force does the pillow exert on her?

STANDARDIZED TEST PRACTICE

MULTIPLE CHOICE

1. How does sound travel from its source to your ear?
   A. by changes in air pressure
   B. by vibrations in wires or strings
   C. by electromagnetic waves
   D. by infrared waves

2. Paulo is listening to classical music in the speakers installed in his swimming pool. A note with a frequency of 330 Hz reaches his ears while he is under water. What is the wavelength of the sound that reaches Paulo's ears? Use 1493 m/s for the speed of sound in water.
   A. 2.19 nm
   B. 4.88×10^{-5} m
   C. 2.19×10^{-1} m
   D. 4.52 m

3. The sound from a trumpet travels at 351 m/s in air. If the frequency of the note is 294 Hz, what is the wavelength of the sound wave?
   A. 9.93×10^{-4} m
   B. 0.849 m
   C. 1.19 m
   D. 1.05×10^{5} m

4. The horn of a car attracts the attention of a stationary observer. If the car is approaching the observer at 60.0 km/h and the horn has a frequency of 512 Hz, what is the frequency of the sound perceived by the observer? Use 343 m/s for the speed of sound in air.
   A. 488 Hz
   B. 512 Hz
   C. 538 Hz
   D. 600 Hz

5. As shown in the diagram below, a car is receding at 72 km/h from a stationary siren. If the siren is wailing at 657 Hz, what is the frequency of the sound perceived by the driver? Use 343 m/s for the speed of sound.
   A. 543 Hz
   B. 620 Hz
   C. 647 Hz
   D. 698 Hz

6. Reba hears 20 beats in 5.0 s when she plays two notes on her piano. She is certain that one note has a frequency of 262 Hz. What are the possible frequencies of the second note?
   A. 242 Hz or 282 Hz
   B. 258 Hz or 266 Hz
   C. 260 Hz or 264 Hz
   D. 270 Hz or 278 Hz

7. Which of the following pairs of instruments have resonant frequencies at each whole-number multiple of the lowest frequency?
   A. a clamped string and a closed pipe
   B. a clamped string and an open pipe
   C. an open pipe and a closed pipe
   D. an open pipe and a reed instrument

FREE RESPONSE

8. The figure below shows the first resonance length of a closed air column. If the frequency of the sound is 488 Hz, what is the speed of the sound?

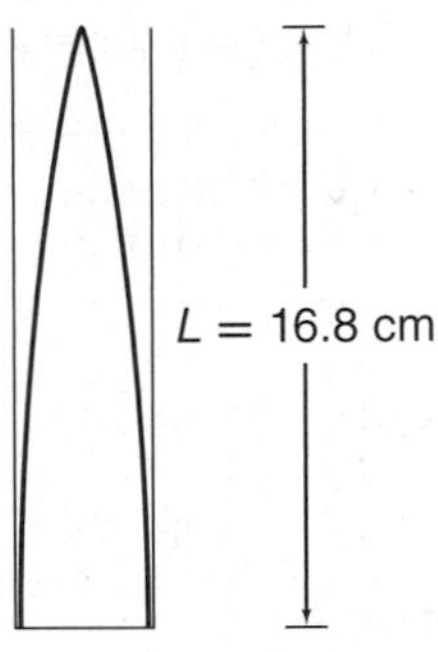

Online Test Practice

NEED EXTRA HELP?

If you Missed Question	1	2	3	4	5	6	7	8
Review Section	1	1	1	1	1	2	2	2

CHAPTER 16

Fundamentals of Light

BIGIDEA Light behaves like a wave and can be detected by the human eye.

SECTIONS

LaunchLAB

iLab Station

LIGHT'S PATH

Can light's path through air be predicted? Use simple everyday materials to find out the answer.

WATCH THIS!

Video

MIXING COLORS

Everyone knows that yellow and blue make green, right? But is that true of mixing paint colors, or combining colored light? Or both? Discover the physics of how colors combine.

(l)Glowimages/Getty Images, (r)Iconotec/Alamy

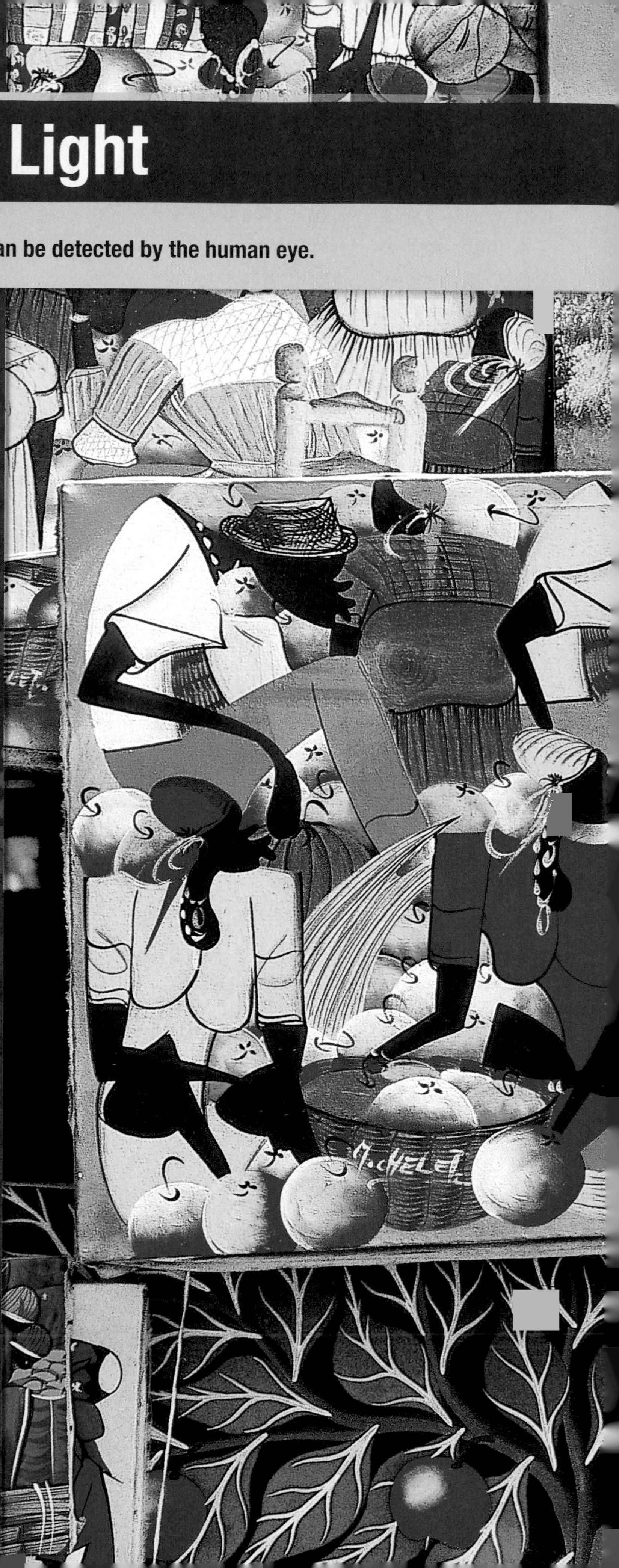

Go online!
connectED.mcgraw-hill.com

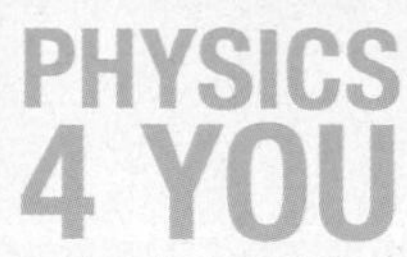

SECTION 1

Illumination

PHYSICS 4 YOU **Sunlight does not reach the deepest parts of the sea. However, many deep-sea fish are bioluminescent, which means they produce their own light. Angler fish, for example, have what appears to be a lamp on a fishing pole attached to their heads to lure prey.**

MAIN IDEA

Light rays travel in straight lines and can change direction only when they encounter a boundary.

Essential Questions

- What is the ray model of light?
- How are distance and illumination related?
- How was the speed of light determined?

Review Vocabulary

precision a characteristic of a measured value describing the degree of exactness of a measurement

New Vocabulary

ray model of light
luminous source
opaque
translucent
transparent
luminous flux
illuminance

Light's Path

How does your body receive information? Many people respond to this question with the five senses, starting with sight and hearing, because light and sound provide a great deal of information about our surroundings. Of the two, light seems to provide the greater variety of information. The human eye can detect tiny changes in the size, position, brightness, and color of objects. You can usually distinguish shadows from solid objects and sometimes can distinguish reflections of objects from the objects themselves. In this section, you will learn where light comes from and how it illuminates your surroundings.

Straight line How does light travel? Think of how a narrow beam of light, such as that of a flashlight or sunlight streaming through a small window, is made visible by dust particles in the air. You see the path of the light as a straight line. When your body blocks sunlight, you see your outline in a shadow, a result of light's straight path. **Figure 1** depicts light's straight path.

When you locate an object with your eyes and walk toward it, you most likely walk in a straight path. This is because the light reflected from the object travels in a straight line. For objects to be visible to you, light must travel from them in a straight line to your eyes. Based on this knowledge of how light travels, models have been developed that describe how light behaves.

☑ READING CHECK **Describe** the evidence in your everyday experiences for light's straight line path.

Figure 1 Light rays traveling in a straight line are evident in many situations.

(t)NaturePL/SuperStock, (b)Brand X Pictures/PunchStock

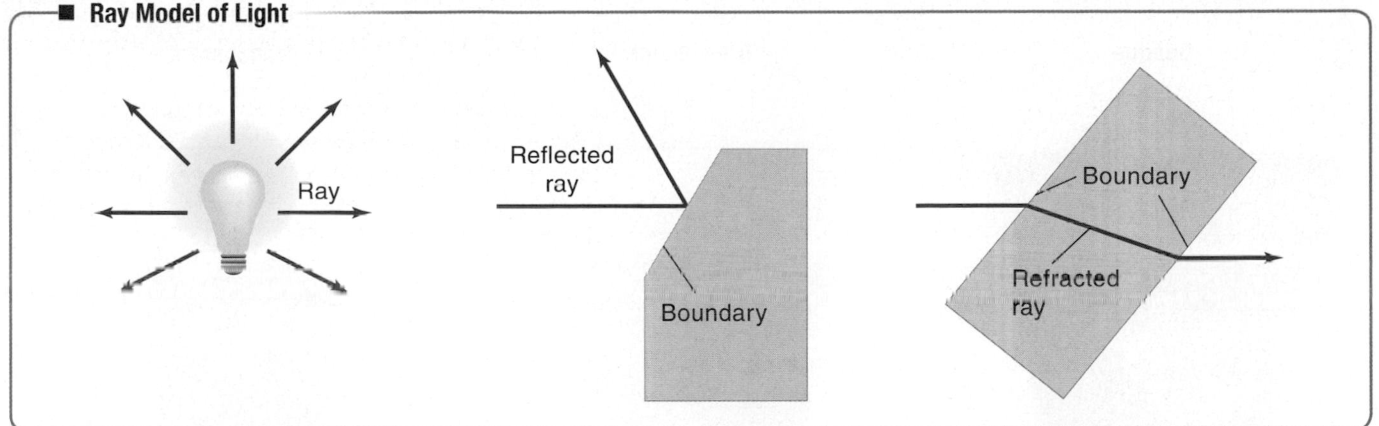

Figure 2 Light's straight line path is demonstrated by the ray model. Light rays change direction when they are reflected or refracted by matter. In either case, light rays continue in a straight path.

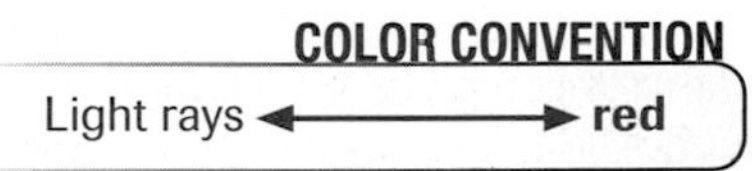

Ray Model of Light

Imagine being in an empty, dark room when a small lightbulb is turned on in the center of the room. You can see around the room, and you can look at the bulb and see it. This must mean the bulb sends light in all directions. You could visualize the light coming from the bulb as an infinite number of arrows traveling straight away from the bulb in all directions. Each arrow represents a ray of light, which travels in a straight path until it reaches a boundary, as shown in **Figure 2.**

After interacting at a boundary, the ray still moves in a straight line, but its direction is changed. These basic principles—that light travels in straight lines and that its direction can be changed by encountering a boundary—constitute the **ray model of light.** The study of light interacting with matter is called ray optics or geometric optics.

Sources of light What is the difference between sunlight and moonlight? For one, sunlight is much brighter. Another important fundamental difference is that the Sun produces and emits its own light, while the Moon is only visible because it reflects the Sun's light. Everything you see fits into one of these two categories. Objects such as the Sun that emit their own light are **luminous sources,** while those that you see due to light reflecting from them are illuminated sources.

Luminous sources include natural sources such as flames and fireflies and human-made devices such as television screens, computer monitors, lasers, and tiny, light-emitting diodes. In **Figure 3,** the luminous source, fluorescent bulbs, produces light from electrical energy. The other objects in the room are illuminated when the light from the bulbs is reflected off of them. In a room with no light, it would be impossible to see anything.

Figure 3 Objects in the room are visible because of reflected light.

Recognize What, if anything, would be visible if the room had no luminous source?

Figure 4 Light is transmitted through the transparent and translucent candle holders. The candle cannot be seen clearly in the opaque holder because the light is absorbed.

Light and matter Objects can absorb, reflect, or transmit light. Objects that reflect and absorb light but do not transmit it are **opaque.** Many common objects—such as books, people, and backpacks—are opaque. Mediums that transmit and reflect light but do not allow objects to be seen clearly through them are **translucent** mediums. The frosted glass in **Figure 4** is a translucent medium. A **transparent** medium, such as air or glass, transmits most of the light that reaches it. **Figure 4** illustrates objects that are opaque, translucent, and transparent. Transparent mediums transmit light, but they also often reflect some light. For example, you can see the glass of the transparent candle holder. This is possible because light is reflected off the glass.

All three types of objects also absorb some light. There are various factors that determine how much light will be absorbed, but opaque objects usually absorb a greater portion of light than translucent or transparent objects.

☑ READING CHECK **Explain** why it is possible to see a fish through a glass fishbowl and also see the glass of the bowl.

Quantity of Light

If you were to have a flashlight shone at you from across the room, what factors would determine how bright that light would appear to you? Three main factors determine the brightness: the quantity of light the flashlight produces, the distance between the lightbulb and your eye, and the angle at which the light rays hit your eye. In this section, you will read about the first two of these factors.

Figure 5 Luminous flux is the rate at which light rays are emitted from a luminous source.

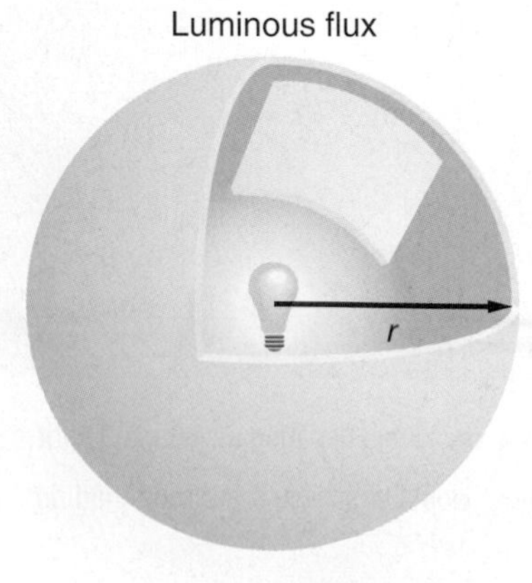

Luminous flux With the ray model of light, a source that is brighter produces more light rays than a less bright source. Imagine again a single lightbulb sending rays in nearly all directions. How could you capture all the light it emits? You would need to construct a surface that completely encloses the bulb, as in **Figure 5.** The rate at which the bulb, a luminous source, produces light energy is called the **luminous flux** (P) and is measured in lumens (lm). The total amount of light that strikes the surface in a given unit of time depends only on the luminous flux of the source.

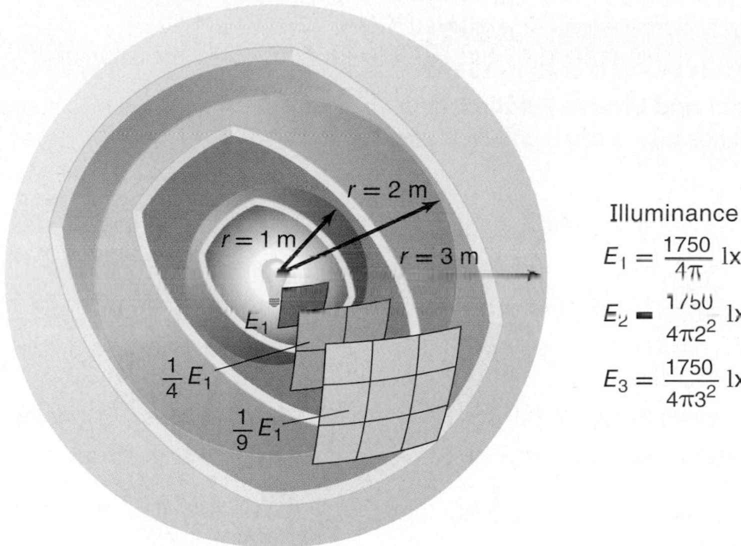

Figure 6 Illuminance (E) is the quantity of light that strikes a surface. As the distance from the luminous source (r) increases, E decreases. E depends on the inverse of r squared.

View an **animation of the inverse square law.**

Illuminance Once you know the quantity of light being emitted by a luminous source, you can determine the amount of illumination the source provides to an object, such as a book. The luminous flux falling on a given surface area at any instant is called **illuminance** (E). It is measured in lux (lx), which is equivalent to lumens per square meter (lm/m^2). In this chapter, we assume for the purpose of simplification that all light sources are point sources.

Consider the setup shown in **Figure 6.** The luminous flux of the source is 1750 lm (typical of a 26-W compact fluorescent bulb). What is the illuminance of the sphere's inside surface at $r = 1$ m? Because all the bulb's luminous flux strikes the surface, divide the luminous flux by the surface area of the sphere, $4\pi r^2$. The surface area is $4\pi(1.00\text{ m})^2 = 4\pi\text{ m}^2$, so the illuminance is $\frac{1750\text{ lm}}{4\pi\text{ m}^2} = 139$ lx. This mathematical relationship means that at a distance of 1.00 m from the bulb, 139 lm strikes each square meter.

Inverse-square relationship What if the sphere surrounding the lamp were larger? If the sphere's radius were 2.00 m, the luminous flux still would total 1750 lm because it only depends on the bulb. With a radius of 2.00 m, however, the area of the sphere would now be equal to $4\pi(2.00\text{ m})^2 = 16.0\pi\text{ m}^2$. The new area is four times larger than that of the 1.00-m sphere, as shown in **Figure 6.** The illuminance of the inside of the 2.00-m sphere is $\frac{1750\text{ lm}}{(16.0\pi\text{ m}^2)} = 34.8$ lx, so 34.8 lm strikes each square meter.

The illuminance on the inside surface of the 2.00-m sphere (E_2) is one-fourth the illuminance on the inside of the 1.00-m sphere. In the same way, the inside of a sphere with a 3.00-m radius has an illuminance only one-ninth $\left[\left(\frac{1}{3}\right)^2\right]$ as large as that of the 1.00-m sphere. **Figure 6** shows that the illuminance produced by a point source is proportional to $\frac{1}{r^2}$, an inverse- square relationship. In the case of the 3.00-m radius, only 15.5 lm strike each square meter inside the sphere. As the light rays spread out in straight lines in all directions from a point source, the number of light rays that illuminate a unit of area decreases as the square of the distance from the point source.

PhysicsLAB

MEASURING ULTRAVIOLET LIGHT

PROBEWARE LAB How effective are various materials at blocking UV light?

CONNECTING MATH TO PHYSICS

Direct and Inverse Relationships The illuminance provided by a source of light has both a direct and an inverse relationship.

Math	Physics
$y = \frac{x}{az^2}$	$E = \frac{P}{4\pi r^2}$
If z is constant, then y is directly proportional to x. • When x increases, y increases. • When x decreases, y decreases.	If r is constant, then E is directly proportional to P. • When P increases, E increases. • When P decreases, E decreases.
If x is constant, then y is inversely proportional to z^2. • When z^2 increases, y decreases. • When z^2 decreases, y increases.	If P is constant, then E is inversely proportional to r^2. • When r^2 increases, E decreases. • When r^2 decreases, E increases.

Luminous intensity Some luminous sources are specified in candelas (cd). A candela is not a measure of luminous flux but of luminous intensity. The luminous intensity of a point source is the luminous flux that falls on 1 m^2 of the inside of a 1-m-radius sphere, so luminous intensity is luminous flux divided by 4π. A bulb with 1750 lm of flux has an intensity of 1750 lm/4π = 139 cd.

In **Figure 7,** the lightbulb is twice as far away from the screen as the candle. For the bulb to provide the same illuminance on its side of the screen as the candle does on the candle side of the screen, the bulb would have to be four times brighter than the candle. The lightbulb's luminous intensity, therefore, would have to be four times the candle's luminous intensity. If both sources in **Figure 7** had the same luminous intensity, the source at $2r$ would only provide one-quarter the illuminance to the screen. This is consistent with the inverse-square relationship we just developed.

READING CHECK Describe what luminous intensity is a measure of and what its relationship is to illuminance.

PhysicsLAB

LIGHT INTENSITY AND DISTANCE

PROBEWARE LAB How does light intensity depend on distance?

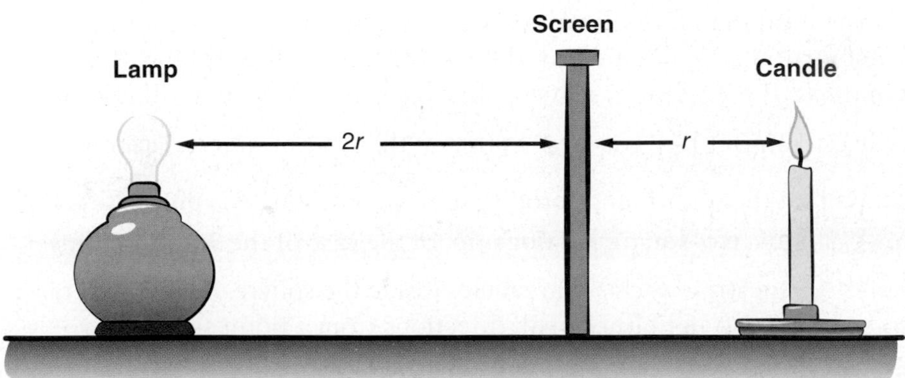

Figure 7 For the lightbulb and the candle to provide the same illuminance to the screen, the luminous intensity of the lightbulb is four times that of the candle.

Surface Illumination

Think again about the scenario in which a flashlight is shining at you from across the room. If the bulb has a small luminous intensity, the light will not be very bright. To increase the brightness, you could use a brighter bulb, thereby increasing the luminous flux, or you could move so that your eyes are closer to the light, decreasing the distance between the light source and your eyes. Following the simplification that we are treating all light sources as point sources, the illuminance and distance will follow the inverse-square relationship. In this case, and in all the cases we will deal with in this book, the illuminance caused by a point light source is represented by the following equation.

REAL-WORLD PHYSICS

ILLUMINATED MINDS When deciding how to achieve the correct illuminance on students' desktops in a classroom, architects must consider the luminous flux of the lights as well as the distance of the lights above the desktops. In addition, the efficiencies of the light sources are an important economic factor.

POINT-SOURCE ILLUMINANCE

If an object is illuminated by a point source of light, then the illuminance at the object is equal to the luminous flux of the light source divided by the surface area of the sphere whose radius is equal to the distance the object is from the light source.

$$E = \frac{P}{4\pi r^2}$$

Remember that the luminous flux of the light source is spreading out in all directions, so only some fraction of the luminous flux is available to illuminate the object. Use of this equation is valid only if the light from the luminous source strikes perpendicular to the surface it is illuminating. It is also only valid if the luminous source is small enough or far enough away to be considered a point source. Thus, the equation does not give accurate values of illuminance for long fluorescent lamps or lightbulbs that are close to the surfaces they illuminate.

Engineers who design lighting systems must understand how the light will be used. If an even illumination is needed to prevent dark areas, the common practice is to evenly space normal lights over the area to be illuminated, as was most likely done with the lights in your classroom. Because such light sources do not produce truly uniform light, however, engineers also design special light sources that control the spread of the light, such that they produce even illuminations over large surface areas. For safety reasons, this is extremely important for automobile headlights, as in **Figure 8.** Automobile engineers must consider these factors when designing headlights.

Figure 8 Dark areas can occur if headlights on cars are not set at the correct angles to adequately illuminate the road.

EXAMPLE PROBLEM

EXAMPLE PROBLEM 1

View a **personal tutor of Illumination.** Personal Tutor

ILLUMINATION OF A SURFACE What is the illuminance on your desktop if it is lit by a 1750-lm lamp that is 2.50 m above your desk?

1 ANALYZE AND SKETCH THE PROBLEM

- Assume the lightbulb is the point source.
- Diagram the position of the bulb and the desktop. Label P and r.

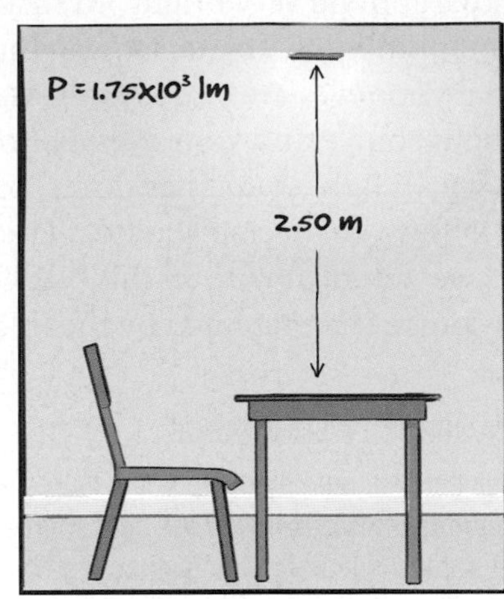

KNOWN	UNKNOWN
$P = 1.75\times10^3$ lm	$E = ?$
$r = 2.50$ m	

2 SOLVE FOR THE UNKNOWN

The surface is perpendicular to the direction in which the light ray is traveling, so you can use the point-source illuminance equation.

$$E = \frac{P}{4\pi r^2}$$

$$= \frac{1.75\times10^3 \text{ lm}}{4\pi(2.50 \text{ m})^2}$$ ◀ Substitute $P = 1.75\times10^3$ lm, $r = 2.50$ m

$$= 22.3 \text{ lm/m}^2$$

$$= 22.3 \text{ lx}$$

3 EVALUATE THE ANSWER

- **Are the units correct?** The units of luminance are $\text{lm/m}^2 = \text{lx}$, which the answer agrees with.
- **Do the signs make sense?** All quantities are positive, as they should be.
- **Is the magnitude realistic?** Illuminance from an 1800-lm lamp at a distance of 2 m is about 20 lx.

PRACTICE PROBLEMS

PRACTICE PROBLEMS

Do additional problems. Online Practice

1. A lamp is moved from 30 cm to 90 cm above the pages of a book. Compare the illumination on the book before and after the lamp is moved.
2. Draw a graph of the illuminance produced by a lamp with a luminous flux of 2275 lm at distances from 0.50 m and 5.0 m.
3. A 64-cd point source of light is 3.0 m away from a painting. What is the illumination on the painting in lux?
4. A screen is placed between two lamps so that they illuminate the screen equally, as shown in **Figure 9.** The first lamp emits a luminous flux of 1445 lm and is 2.5 m from the screen. What is the distance of the second lamp from the screen if the luminous flux is 2375 lm?

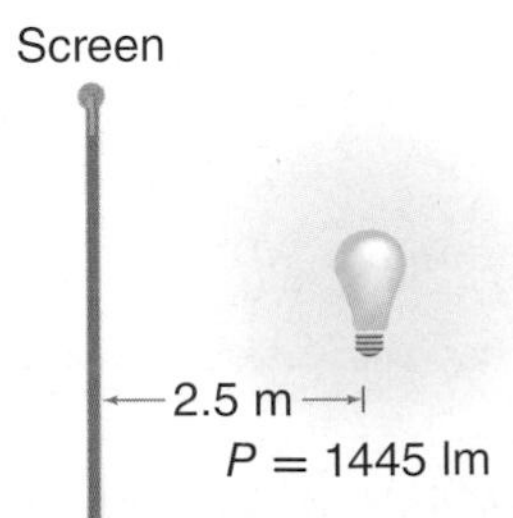

Figure 9

5. What is the illumination on a surface that is 3.0 m below a 150-W incandescent lamp that emits a luminous flux of 2275 lm?
6. A public school law requires a minimum illuminance of 160 lx at the surface of each student's desk. An architect's specifications call for classroom lights to be located 2.0 m above the desks. What is the minimum luminous flux that the lights must produce?
7. **CHALLENGE** Your local public library is planning to remodel the computer lab. The contractors have purchased fluorescent lamps with a rated luminous flux of 1750 lm. The desired illumination on the keyboard surfaces is 175 lx. Assume a single lamp illuminates each keyboard. What distance above the surface should the lights be placed to achieve the desired illumination? If the contractors had also already purchased fixtures to hold the lights that when installed would be 1.5 m above the keyboard surface, would the desired illuminance be achieved? If not, would the illuminance be greater or less than desired? What change in the lamp's luminous flux would be required to achieve the desired illuminance?

The Speed of Light

Arguments that light must travel at a finite speed have existed for more than 2400 years. By the seventeenth century, several scientists had performed experiments that supported the view that light travels at a finite speed, but that this speed is much faster than the speed of sound.

Actually measuring the speed of light was not an easy task in the seventeenth century. As you know from studying motion, if you can measure the time light takes to travel a certain distance, you can calculate the speed of light. However, the time that it takes light to travel between objects on Earth is much shorter than a human's reaction time. How could a seventeenth-century scientist solve this problem?

Clues from Io Danish astronomer Ole Roemer was the first to measure the time it took for light to travel between two points with any success. Between 1668 and 1674, Roemer made 70 measurements of the 1.8-day orbital period of Io, one of Jupiter's moons. He recorded the times when Io emerged from Jupiter's shadow, as shown in **Figure 10.** He made his measurements as part of a project to improve maps by calculating the longitude of locations on Earth. This is an early example of the needs of technology driving scientific advances.

After making many measurements, Roemer was able to predict when the next eclipse of Io would occur. He compared his predictions with the actual measured times and found that Io's observed orbital period increased on average by about 13 s per orbit when Earth was moving away from Jupiter and decreased on average by about 13 s per orbit when Earth was approaching Jupiter. Roemer believed that Jupiter's moons were just as regular in their orbits as Earth's moon; thus, he wondered what might cause this discrepancy in the measurement of Io's orbital period. He considered another variable within the system, the movement and position of Earth relative to Jupiter.

READING CHECK **Explain** why Roemer was making calculations of Io's orbital period.

MiniLAB

THE SPEED OF LIGHT

How can you measure the speed of light?

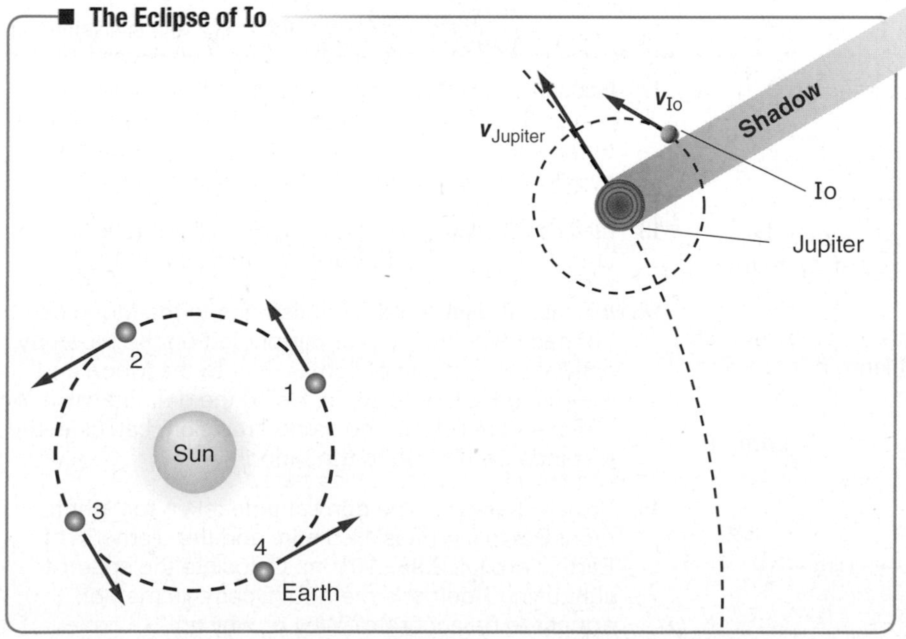

Figure 10 As Earth approaches Jupiter, the light reflected from Io takes less time to reach Earth than when Earth moves away from Jupiter. (Illustration is not to scale.)

Figure 11 Io, one of Jupiter's moons, is the most volcanically active object in the solar system. The dark spots are active volcanoes.

Measuring the speed of light Roemer concluded that as Earth moved away from Jupiter, the light from each new appearance of Io took longer to reach Earth because it traveled farther. As Earth moved toward Jupiter, Io's orbital period seemed to decrease. During the 182.5 days it took for Earth to travel from position 1 to position 3, shown in **Figure 10,** there were (182.5 days)(1 Io eclipse/1.8 days) = 1.0×10^2 Io eclipses. Thus, for light to travel the diameter of Earth's orbit, he calculated that it takes (1.0×10^2 eclipses)(13 s/eclipse) = 1.3×10^3 s, or 22 min. Because distances in space are so great, it was much easier for Roemer to measure differences in the time it took light to travel than it was for scientists conducting Earth-based measurements of the speed of light. At times Io, shown in **Figure 11,** is as much as 9.3×10^{11} m from Earth.

Using the presently known value of the diameter of Earth's orbit (2.93×10^{11} m), Roemer's value of 22 min gives a value for the speed of light of 2.9×10^{11} m/((22 min)(60 s/min)) = 2.2×10^8 m/s. Today, the speed of light is known to be closer to 3.0×10^8 m/s, so light takes 16.5 min, not 22 min, to cross Earth's orbit. Nevertheless, Roemer had successfully proved that light travels at a finite speed.

Michelson's measurements Although many measurements of the speed of light have been made, the most notable were performed by American physicist Albert A. Michelson. Between 1880 and the 1920s, he developed Earth-based techniques to measure the speed of light. In 1926 Michelson measured the time required for light to make a round-trip between two California mountains 35 km apart. Michelson used a set of rotating mirrors to measure such small time intervals. Michelson's best result was $(2.99796 \pm 0.00004)\times10^8$ m/s. For this work, he became the first American to receive a Nobel Prize in science.

The speed of light in a vacuum is a very important and universal value; thus, it has its own special symbol, c. The International Committee on Weights and Measurements has measured and defined the speed of light in a vacuum to be c = 299,792,458 m/s. For many calculations, the value c = 3.00×10^8 m/s is precise enough. At this speed, light travels 9.46×10^{12} km in a year. This distance is called a light-year.

SECTION 1 REVIEW

Section Self-Check — Check your understanding.

8. **MAIN IDEA** What evidence have you observed that light travels in a straight line?

9. **Use of Material Light Properties** Why might you choose a window shade that is translucent? Opaque?

10. **Illuminance** Does one lightbulb provide more or less illuminance than two identical lightbulbs at twice the distance? Explain.

11. **Luminous Intensity** Two lamps illuminate a screen equally from distances shown in **Figure 12.** If lamp A is rated 75 cd, what is lamp B rated?

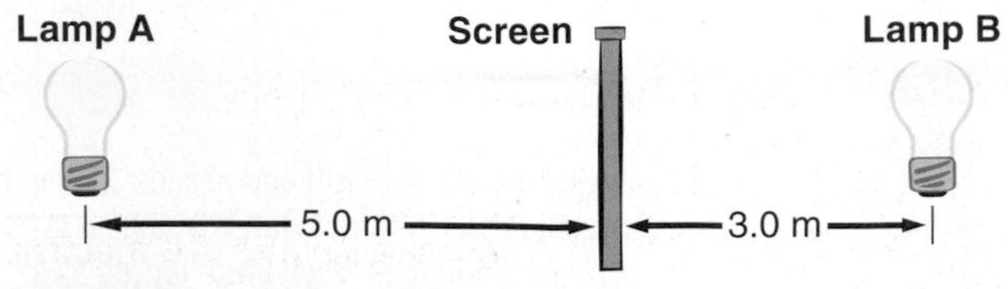

Figure 12

12. **Distance of a Light Source** A lightbulb illuminating your computer keyboard provides only half the illuminance that it should. If it is currently 1.0 m away, how far should it be to provide the correct illuminance?

13. **Distance of Light Travel** How far does light travel in the time it takes sound to travel 1 cm in air at 20°C?

14. **Distance of Light Travel** The distance to the Moon can be found with the help of mirrors left on the Moon by astronauts. A pulse of light is sent to the Moon and returns to Earth in 2.562 s. Using the defined value for the speed of light to the same precision, calculate the distance from Earth to the Moon.

15. **Critical Thinking** The correct time taken for light to cross Earth's orbit is 16.5 min, and the diameter of Earth's orbit is 2.98×10^{11} m. Calculate the speed of light using Roemer's method. Does this method appear to be accurate? Why or why not?

SECTION 2

The Wave Nature of Light

PHYSICS 4 YOU **Take a close look at your shadow. You should notice that the edges of the shadow are not distinct. This blurring happens because light is a wave and it bends around objects—in this case, your body.**

MAIN IDEA

Like all waves, light diffracts around objects, has a wavelength and frequency, and can be Doppler shifted.

Essential Questions

- How does diffraction demonstrate that light has wave properties?
- What are the effects of combining colors of light and mixing pigments?
- How do phenomena such as polarization and the Doppler effect occur?

Review Vocabulary

wavelength the shortest distance between points of a wave where the wave pattern repeats itself, such as from crest to crest or trough to trough

New Vocabulary

diffraction
primary color
secondary color
complementary color
primary pigment
secondary pigment
polarization
Malus's law

Diffraction and the Wave Model

When you are around the corner from a band room, you can hear the music because the sound bends. You can't see the musicians because the light reflecting from them does not reach your eyes. Does the light not bend at all, or does it not bend as much as the sound?

In 1665 Italian scientist Francesco Maria Grimaldi observed that the edges of shadows are not perfectly sharp. He introduced a narrow beam of light into a dark room and held a rod in front of the light such that it cast a shadow on a white surface. The shadow cast by the rod was wider than the shadow should have been if light traveled in a straight line past the edges of the rod. Grimaldi also noted that the shadow was bordered by colored bands. He determined that both of these observations could be explained if light bent slightly. He called the bending of light as it passes the edge of a barrier **diffraction.**

Huygens' principle In 1678 Dutch scientist Christiaan Huygens used a wave model to explain diffraction. According to Huygens' principle, all the points of a wavefront of light can be thought of as new sources of smaller waves. These smaller waves, or wavelets, expand in every direction and are in step with one another. A flat, or plane, wavefront of light consists of an infinite number of point sources in a line. As this wavefront passes by an edge, the edge cuts the wavefront such that each circular wavelet generated by each Huygens' point will propagate as a circular wave in the region where the original wavefront was bent, as shown in **Figure 13.** This wave model explained the diffraction Grimaldi saw.

■ **Huygens' Principle**

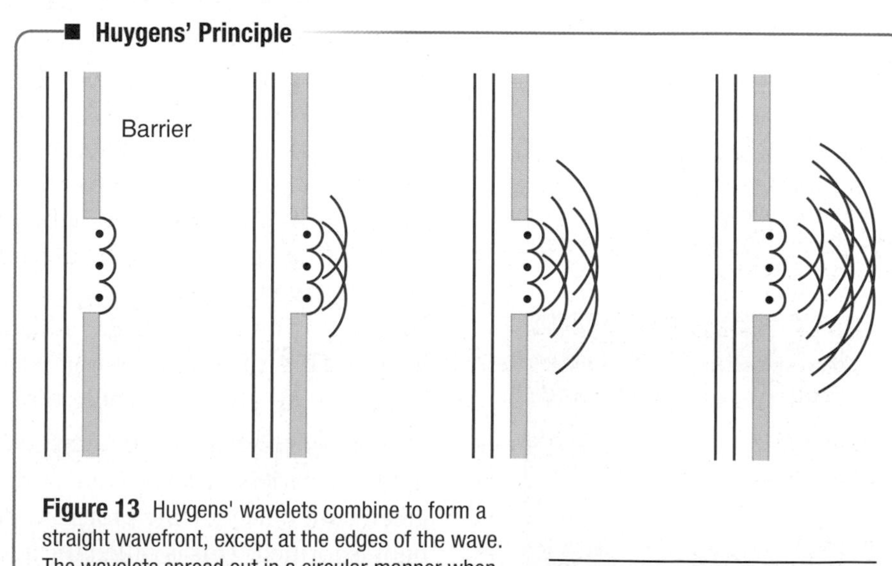

Figure 13 Huygens' wavelets combine to form a straight wavefront, except at the edges of the wave. The wavelets spread out in a circular manner when a barrier creates an edge.

View an **animation of diffraction.**
Concepts In Motion

Figure 14 White light is separated into bands of color by a prism. Each color has a different wavelength.

PhysicsLAB

A WHOLE SPECTRUM OF POSSIBILITIES

FORENSICS LAB How can a spectrum from broken glass at an accident scene help identify a suspect vehicle?

Color

In 1666 Newton performed experiments on the colors produced when a narrow beam of sunlight passed through a glass prism, as shown in **Figure 14.** Newton called the ordered arrangement of colors a spectrum. Using his later-disproved corpuscle (or particle) model of light, he thought that particles of light were interacting with some unevenness in the glass to produce the spectrum.

To test this assumption, Newton allowed the spectrum from one prism to fall on a second prism. If the spectrum was caused by irregularities in the glass, he reasoned that the second prism would increase the spread in colors. Instead, the second prism reversed the spreading of colors and recombined them to form white light. After more experiments, Newton concluded that white light is composed of colors and that a property of the glass other than unevenness caused the light to separate into colors.

Different wavelengths Can the wave model of light explain Newton's observations? For light to be a wave, it must have wavelength and frequency. The work of Grimaldi, Huygens, Newton, and others suggested that the color of light is related to wavelength. Visible light falls within the range of wavelengths from about 400 nm (4.00×10^{-7} m) to 700 nm (7.00×10^{-7} m), as shown in **Figure 14.** The longest visible wavelengths are seen as red light and the shortest as violet.

As white light crosses the boundary from air into glass and back into air in **Figure 14,** its wave nature causes each different color of light to be bent at a different angle. The shorter the wavelength, the more the light is bent. This unequal bending of the different colors causes the white light to be spread into a spectrum.

View an **animation of the primary colors of light.**

Color by addition of light White light can be formed from colored light in a variety of ways. For example, when the correct intensities of red, green, and blue light are projected onto a white screen, as in **Figure 15,** white light is formed. This is called the additive color process, which is used in many television screens. A television screen uses three colors—red, green, and blue. Combinations of these produce the colors you see.

Figure 15 Red, green, and blue light, the primary colors, combine in pairs to produce yellow, cyan, or magenta light. The region where these three colors overlap on the screen appears white.

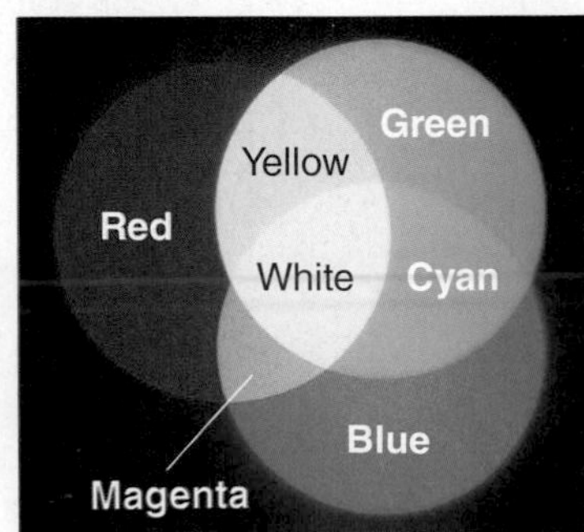

The correct intensities of red, green, and blue light appear on a screen as white light when combined. Because of this phenomenon, they are called the **primary colors** of light. The primary colors can be mixed in pairs to form three additional colors, as shown in **Figure 15.** Red and green light together produce yellow light, blue and green light produce cyan, and red and blue light produce magenta. The colors yellow, cyan, and magenta are called secondary colors. A **secondary color** is a combination of two primary colors. Note that these are slightly different from the primary and secondary colors you might have learned in art class; the reasons for this will be explained shortly.

As shown in **Figure 15,** yellow light can be made from red light and green light. If yellow light and blue light are projected onto a white screen with the correct intensities, the surface will appear to be white. **Complementary colors** are two colors of light that can be combined to produce white light. Thus, yellow is a complementary color of blue, and vice versa, because the two colors of light combine to make white light. In the same way, cyan and red are complementary colors. Magenta and green are the other pair of complementary colors. A practical application of this is that yellowish laundry can be whitened with a bluing agent added to detergent.

View a **BrainPOP video on color.**

Color by subtraction of light As you learned in the first section of this chapter, objects can reflect and transmit light. They also can absorb light. The color of an object depends on the wavelengths present in the light that illuminates the object and also on which wavelengths are absorbed by the object and which wavelengths are reflected. The natural existence or artificial placement of dyes in the material of an object, or pigments on its surface, gives the object color.

Dyes You are probably familiar with dyes that are used to color cloth. Dyes can be made from plant or insect extracts. For example, purple dye can be extracted from the berries of a black mulberry tree. The saffron crocus is a source of yellow dye. One type of red dye is extracted from an insect called a cochineal. A dye is a molecule that absorbs certain wavelengths of light and transmits or reflects others. When light is absorbed, its energy is transferred to the object that it strikes and is transformed into other forms of energy. A red shirt is red because the dyes in it reflect mostly red light to our eyes. When white light falls on the red object shown in **Figure 16,** the dye molecules in the object absorb most of the blue and green light and reflect mostly red light. When only blue light falls on the red object, very little light is reflected and the object appears to be almost black.

Pigments The difference between a dye and a pigment is that pigments usually are made of crushed minerals rather than plant or insect extracts. For example, hematite produces a red pigment, and blue pigment can be obtained from azurite. Pigment particles can be seen with a microscope. A pigment that absorbs only one primary color and reflects two from white light is called a **primary pigment.** Yellow pigment absorbs blue light and reflects red and green light. Yellow, cyan, and magenta are the colors of primary pigments.

READING CHECK Distinguish the difference between color by subtraction and color by addition.

Figure 16 The colors of objects we see are determined by which wavelengths of light are absorbed and which are reflected.

Explain why the die that is yellow in the white light appears red in red light.

White light illuminates these dice.

Blue light illuminates these dice.

Red light illuminates these dice.

Tom Pantages

Figure 17 Magenta, cyan, and yellow are the primary pigments. Secondary pigments, red, green, and blue, are produced from mixing the primary pigments in pairs.

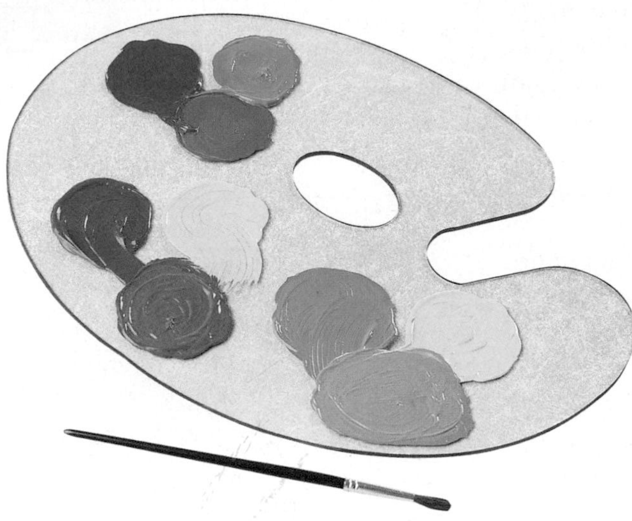

A pigment that absorbs two primary colors and reflects one color is called a **secondary pigment.** The colors of secondary pigments are red (which absorbs green and blue light), green (which absorbs red and blue light), and blue (which absorbs red and green light). Note that the primary pigment colors are the secondary colors of light. In the same way, the secondary pigment colors are light's primary colors.

The primary and secondary pigments are shown in **Figure 17.** When the primary pigments yellow and cyan are mixed, the yellow absorbs blue light and the cyan absorbs red light. Thus, **Figure 17** shows yellow and cyan combining to make green pigment. When yellow pigment is mixed with the secondary pigment blue, which absorbs green and red light, all the primary colors are absorbed, and the result is black. Thus, yellow and blue are complementary pigments. Cyan and red, as well as magenta and green, are also complementary pigments.

CONNECTION TO CHEMISTRY A color printer uses yellow, magenta, and cyan dots of pigment to make a color image on paper. Often, pigments that are used are finely ground compounds, such as titanium(IV) oxide (white), chromium(III) oxide (green), and cadmium sulfide (yellow). Pigments mix to form suspensions rather than solutions. Their chemical form is not changed in a mixture, so they still absorb and reflect the same wavelengths.

CONNECTION TO BIOLOGY You can now begin to understand the colors that you see in **Figure 18.** The plants on the mountain look green because of the chlorophyll in them. One type of chlorophyll absorbs mostly red light and the other absorbs mostly blue light, but they both reflect green light. The energy in the red and blue light that is absorbed is used by the plants during photosynthesis to make food.

MiniLABs

COLOR BY TEMPERATURE

How do colors relate to temperature?

POLARIZATION OF REFLECTED LIGHT

How can light polarized by reflection be observed?

Figure 18 Chlorophyll in green leaves reflects mostly green light, giving the leaves their color.

Explain why leaves are various shades of green.

(t)Laura Sifferlin, (b)Stockbyte/Getty Images

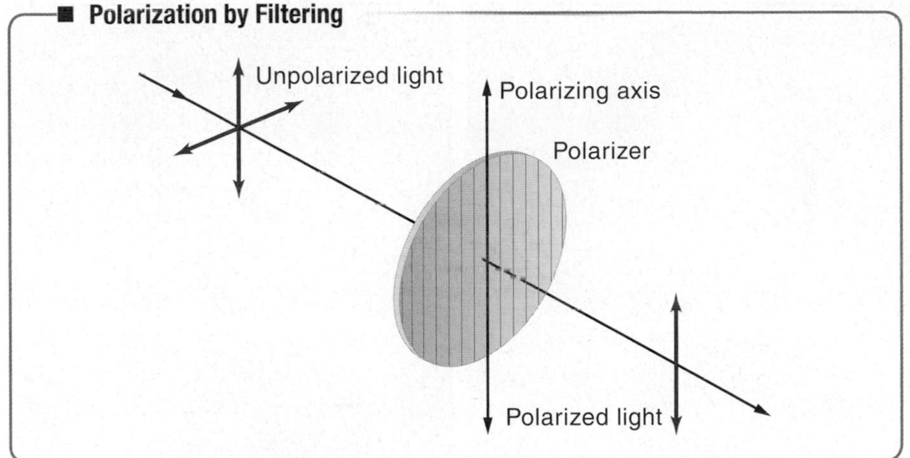

Figure 19 Nonpolarized light rays vibrate randomly in every direction perpendicular to the direction they travel. A polarizing medium blocks light that is not parallel to the polarizing axis.

Polarization of Light

Have you ever looked at light reflected off a road through polarizing sunglasses? If you rotate the glasses, the road first appears to be dark, then light, and then dark again. Light from a lamp, however, changes very little as the glasses are rotated. Why is there a difference? Normal lamplight is not polarized. However, the light that is coming from the road is reflected and has become polarized. **Polarization** is the production of light with a specific pattern of oscillation.

Recall that light behaves as a transverse wave. For waves on a rope, the oscillating medium is the rope. For light waves, the oscillating medium is the electric field. When this electric field oscillates in random directions, the light is nonpolarized. How can you filter nonpolarized light so that whatever passes through the filter is polarized light?

Polarization by filtering The lines in the polarizer in **Figure 19** represent a polarizing axis. The light with the portion of the electric field that oscillates parallel to these lines passes through. The light with the portion of the electric field that oscillates perpendicular to these lines is absorbed. If a polarizer is placed in a beam of nonpolarized light, only the components of the waves in the same direction as the polarizing axis can pass through. As a result, half of the total light passes through, reducing the intensity of the light by half.

CONNECTION TO CHEMISTRY Polarizing mediums contain long molecules in which electrons can oscillate, or move back and forth, all in the same direction. As light travels past the molecules, the electrons absorb light waves that oscillate in the same direction as the electrons. This allows light waves vibrating in one direction to pass, while the waves vibrating in the other direction are absorbed. The direction of a polarizing medium perpendicular to the long molecules is called the polarizing axis. Only waves oscillating parallel to that axis can pass through.

Polarization by reflection When you look through a polarizing filter at the light reflected by a sheet of glass and rotate the filter, you will see the light brighten and dim. The light is partially polarized parallel to the plane of the glass when it is reflected. Polarized reflected light causes glare. Polarizing sunglasses reduce glare from the polarized light reflected off roads. Photographers can use polarizing filters over camera lenses to block reflected light. This result is shown in **Figure 20.**

PhysicsLABs

POLARIZATION
What types of luminous and illuminated sources produce polarized light?

REDUCING GLARE
How do polarizing filters work to reduce glare?

Figure 20 Glare is light that has been polarized by reflection. Photographers use polarizing filters to reduce glare.

Figure 21 Polarizing filters with their axes parallel will allow the light with the same orientation to pass. With the polarizing axes at a 45° angle, the filters allow some light to pass. If the axes of the filters are perpendicular, the second filter will block the light that has passed through the first filter.

Malus's law Suppose you produce polarized light with a polarizing filter. What would happen if you place a second polarizing filter in the path of the polarized light? If the polarizing axis of the second filter is parallel to that of the first, the light will pass through. If the polarizing axis of the second filter is perpendicular to that of the first, no light will pass through, as shown in **Figure 21.**

If the light intensity after the first polarizing filter is I_1 and the intensity after the second filter is I_2, how can you control I_2? I_2 depends only on I_1 and the angle between the axes of the filters, θ. If θ is 0, I_2 equals I_1; if θ is 90°, all of the light is blocked, resulting in I_2 being 0. This indicates that the intensity might depend on the cosine of θ. The actual relationship is a that of a cosine squared. The law that explains the reduction of light intensity as light passes through a second polarizing filter is **Malus's law.**

MALUS'S LAW

The intensity of light coming out of a second polarizing filter is equal to the intensity of polarized light coming out of a first polarizing filter multiplied by the cosine, squared, of the angle between the polarizing axes of the two filters.

$$I_2 = I_1 \cos^2 \theta$$

PHYSICS CHALLENGE

You place an analyzer filter between the two cross-polarized filters, such that its polarizing axis is not parallel to either of the two filters, as shown in the figure to the right.

1. You observe that some light passes through filter 2, though no light passed through filter 2 before you inserted the analyzer filter. Why does this happen?
2. The analyzer filter is placed at an angle of θ relative to the polarizing axis of filter 1. Derive an equation for the intensity of light coming out of filter 2 compared to the intensity of light coming out of filter 1.

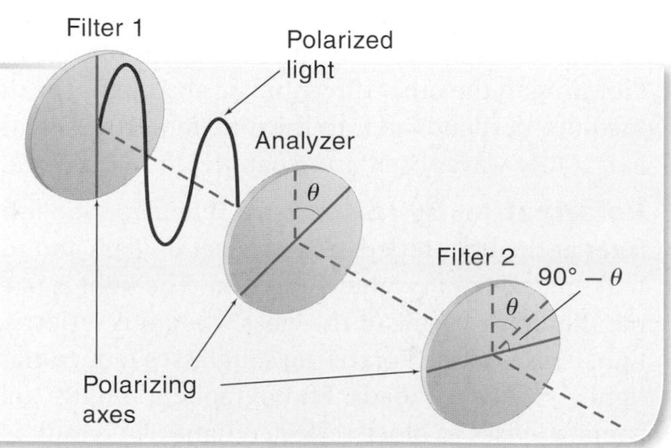

Richard Hutchings/Digital Light Source

Speed, Wavelength, and Frequency of Light

As you have learned, the source of a wave determines that wave's frequency (f), and the medium and the frequency together determine the wavelength (λ) of a wave. Because light has wave properties, the same mathematical models used to describe waves in general can be used to describe light. For light of a given frequency traveling through a vacuum, wavelength is a function of the speed of light (c), which can be written as $\lambda_0 = c/f$. The development of the laser in the 1960s provided new ways to measure the speed of light. The frequency of light can be measured with extreme precision using lasers and the time standard of atomic clocks. Measurements of wavelengths of light, however, are much less precise.

All colors of light travel at c in a vacuum, though the wavelengths are different. Since $\lambda_0 = c/f$, once the frequency of a light wave in a vacuum is measured, the wavelength can be determined.

Relative motion and light What happens if a light source travels toward you or you move toward the source? You have learned that the frequency of a sound heard by a listener changes if either the source or the listener of the sound is moving. The same is true for light. However, when you consider the velocities of a sound source and the observer, you are really considering each one's velocity relative to the medium through which the sound travels. This is not the case for light.

The Doppler effect The nature of light waves is such that they are not vibrations of the particles of a medium. The Doppler effect for light can involve only the relative velocity between the source and the observer. Remember that the only factors in the Doppler effect are the velocity components along the axis between the source and the observer, as shown in **Figure 22.**

Doppler effect for light problems can be simplified by considering axial relative speeds that are much less than the speed of light ($v \ll c$). This simplification is used to develop the equation for the observed light frequency (f_{obs}), shown on the next page.

View an **animation of relative motion and light.**

Figure 22 The Doppler effect describes how light frequency changes if an observer and a light source are moving toward or away from each other.

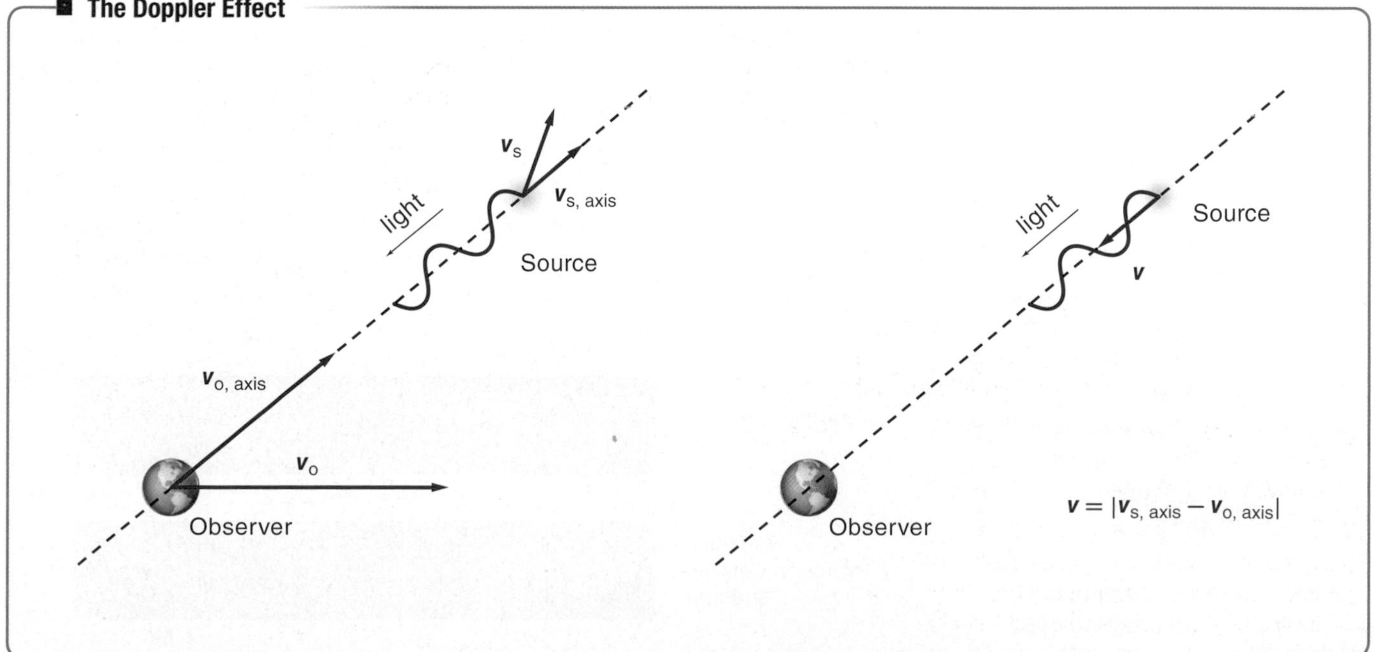

Observed Light Frequency

The observed frequency of light from a source is equal to the actual frequency of the light generated by the source, times the quantity 1 plus the relative speed along the axis, divided by the speed of light, between the source and the observer if they are moving toward each other, or 1 minus the relative speed, divided by the speed of light, if they are moving away from each other.

$$f_{obs} = f\left(1 \pm \frac{v}{c}\right)$$

Applications Most applications of the Doppler effect for light are in astronomy, where phenomena are discussed more in terms of wavelength. Using the relationship $\lambda = c/f$ and the $v \ll c$ simplification, the following equation describes the Doppler shift ($\Delta\lambda$), the difference between the observed and the actual wavelengths.

Doppler Shift

The difference between the observed wavelength of light and the actual wavelength of light generated by a source is equal to the actual wavelength of light generated by the source, times the relative speed of the source and observer, divided by the speed of light.

$$(\lambda_{obs} - \lambda) = \Delta\lambda = \pm\left(\frac{v}{c}\right)\lambda$$

A positive change in wavelength occurs when the relative velocity of the source is away from the observer. In this case, the observed wavelength is longer than the original wavelength. The light appears closer to the red end of the spectrum than it normally would. We say this light is redshifted. A negative change in wavelength occurs when the relative velocity of the source is in a direction toward the observer. In this case, the observed wavelength is shorter than the original wavelength. This is known as a blueshift.

Because the speed of light is constant, when the wavelength is redshifted, the observed frequency is lower than the original due to the inverse relationship between the two variables. When light is blueshifted, the observed frequency is higher.

▶ CONNECTION TO ASTRONOMY Astronomers can determine how objects, such as galaxies, are moving relative to Earth by observing the Doppler shift of their light. This is done by observing the spectrum of light coming from stars in the galaxy using a spectrometer, as shown in **Figure 23.** The same elements that are present in the stars of galaxies emit light of specific wavelengths in labs on Earth. By comparing wavelength, astronomers can learn the velocities of objects toward or away from Earth.

View an **animation of expansion and redshift.**

Concepts In Motion

Figure 23 The bottom hydrogen emission spectrum is redshifted compared to the laboratory spectrum, indicating the light source is moving away from Earth.

PRACTICE PROBLEMS

Do additional problems. Online Practice

16. Oxygen can be made to produce light with a wavelength of 513 nm. What is the frequency of this light?

17. A hydrogen atom in a galaxy moving with a speed of 6.55×10^6 m/s away from Earth emits light with a frequency of 6.16×10^{14} Hz. What frequency of light from that hydrogen atom would be observed by an astronomer on Earth?

18. A hydrogen atom in a galaxy moving with a speed of 6.55×10^6 m/s away from Earth emits light with a wavelength of 486 nm. What wavelength would be observed on Earth from that hydrogen atom?

19. CHALLENGE An astronomer is looking at the spectrum of a galaxy and finds that it has an oxygen spectral line of 525 nm, while the laboratory value is measured at 513 nm. Calculate how fast the galaxy would be moving relative to Earth. Explain whether the galaxy is moving toward or away from Earth and how you know.

In 1929 Edwin Hubble analyzed the light from many galaxies. He observed that the light produced by familiar elements were at longer wavelengths than he had expected them to be. The light was shifted toward the red end of the spectrum. No matter what area of the sky he observed, almost all the galaxies were sending redshifted light to Earth. What do you think caused the spectral lines to be redshifted? Hubble concluded that galaxies are moving away from Earth and suggested that the universe is expanding. Additional studies since then have supported this conclusion. As galaxies move, they sometimes collide and merge. The results of the collision depend on the size and speed of the galaxies.

You have learned that some characteristics of light can be explained with a simple ray model of light, whereas others require a wave model of light. You can use both of these models to study how light interacts with mirrors and lenses. There are some aspects of light that can be understood only through the use of the wave model of light.

SECTION 2 REVIEW

Section Self-Check Check your understanding.

20. **MAIN**IDEA Describe the relative motions of objects when light is redshifted and when light is blueshifted. Answer using the term *Doppler effect*.

21. Addition of Light Colors What color of light must be combined with blue light to obtain white light?

22. Light and Pigment Interaction What color will a yellow banana appear to be when illuminated by each of the following?

a. white light

b. green and red light

c. blue light

23. What are the secondary pigment colors and why do they give objects the appearance of those colors?

24. Combination of Pigments What primary pigment colors must be mixed to produce red? Explain your answer in terms of color subtraction for pigment colors.

25. Polarization Describe a simple experiment you could do to determine whether sunglasses in a store are polarizing.

26. Polarizing Sunglasses Use **Figure 24** to determine the direction the polarizing axis of polarizing sunglasses should be oriented to reduce glare from the surface of a road: vertically or horizontally? Explain.

Figure 24

27. The speed of red light is slower in air and water than in a vacuum. The frequency, however, does not change when red light enters water. Does the wavelength change? If so, how?

28. Critical Thinking Astronomers have determined that our galaxy, the Milky Way, is moving toward Andromeda, a neighboring galaxy. Explain how they determined this. Can you think of a possible reason why the Milky Way is moving toward Andromeda?

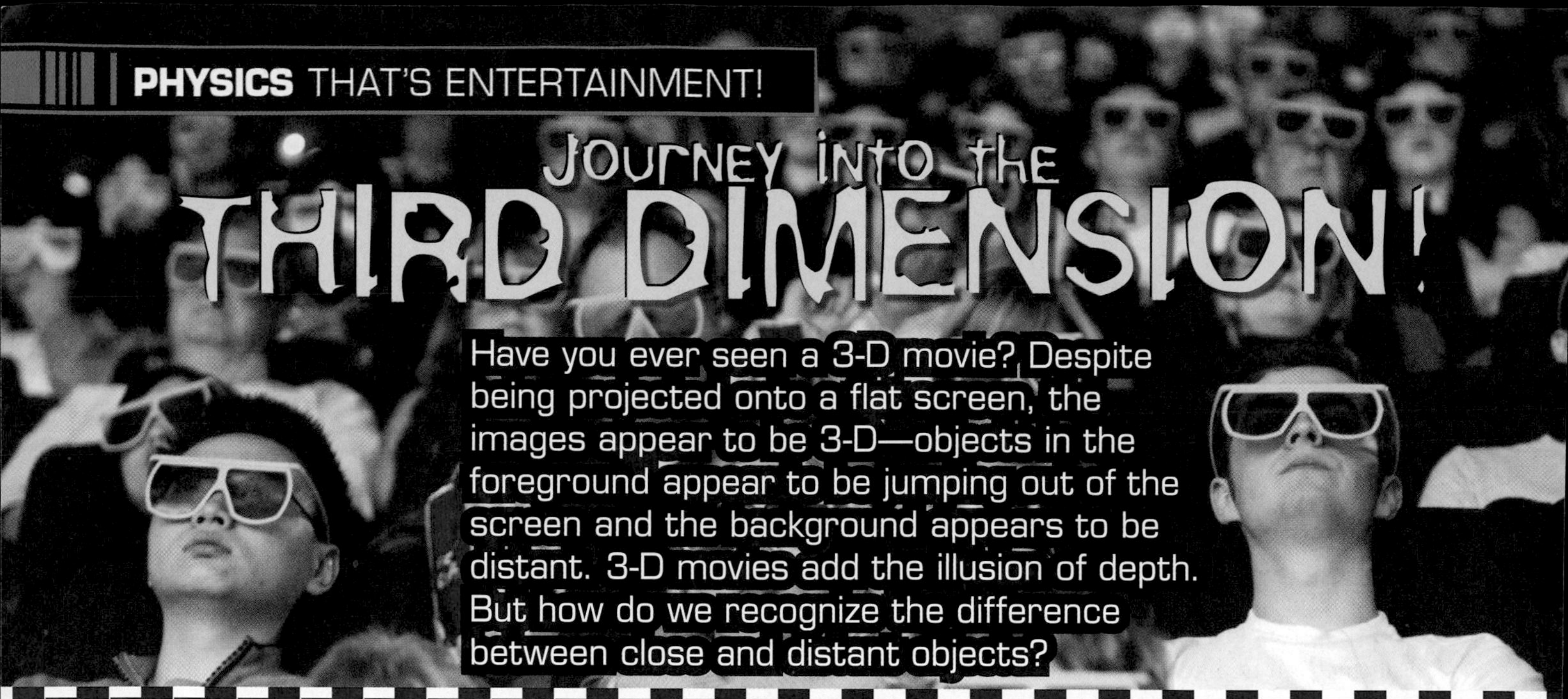

Journey into the THIRD DIMENSION!

Have you ever seen a 3-D movie? Despite being projected onto a flat screen, the images appear to be 3-D—objects in the foreground appear to be jumping out of the screen and the background appears to be distant. 3-D movies add the illusion of depth. But how do we recognize the difference between close and distant objects?

Close one eye and look at an object in the distance. Now hold up your thumb to block it from your view. Without moving your thumb, switch which eye is closed and which is open. What happened?

Two eyes, one image Humans have binocular vision—our brains combine the separate images seen by each eye. The differences between those images are more pronounced with objects that are very close to you and nearly imperceptible with distant objects. That's why your thumb, which is close, appears to move when you switch eyes, but the distant object you were covering with your thumb stays in the same place. This simple concept—one that was figured out nearly 150 years ago—is the basis for the technology of 3-D movies.

Polarizing filters By using two side-by-side cameras to film a scene, filmmakers produce the same effect as binocular vision. The two films are then projected onto the same screen through two different polarizing filters.

A polarizing filter allows light waves through that are vibrating in one particular direction. A pair of 3-D glasses contains two polarizing filters that are perpendicular to each other. As shown in **Figure 1,** the two filters each block one of the two images being projected on the screen, sending a different image to each eye.

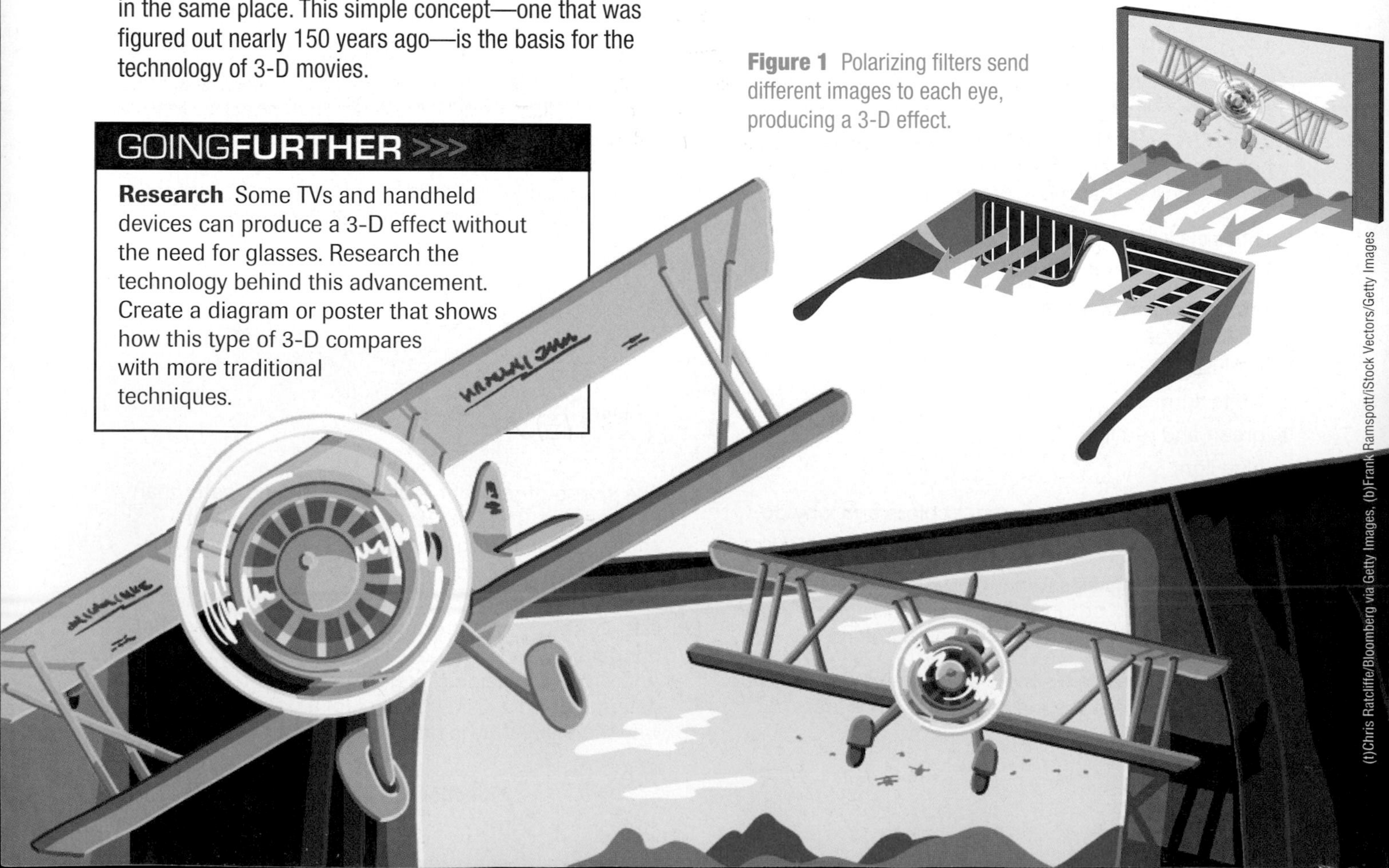

Figure 1 Polarizing filters send different images to each eye, producing a 3-D effect.

GOING FURTHER >>>

Research Some TVs and handheld devices can produce a 3-D effect without the need for glasses. Research the technology behind this advancement. Create a diagram or poster that shows how this type of 3-D compares with more traditional techniques.

CHAPTER 16

STUDY GUIDE

BIGIDEA Light behaves like a wave and can be detected by the human eye.

VOCABULARY

- **ray model of light** *(p. 439)*
- **luminous source** *(p. 439)*
- **opaque** *(p. 440)*
- **translucent** *(p. 440)*
- **transparent** *(p. 440)*
- **luminous flux** *(p. 440)*
- **illuminance** *(p. 441)*

SECTION 1 Illumination

MAINIDEA Light rays travel in straight lines and can only change direction when they encounter a boundary.

- Light can be modeled as a ray that travels in a straight path until it encounters a boundary. Mediums can be characterized as being transparent, translucent, or opaque, depending on how light interacts with them.
- The luminous flux of a light source is the rate at which light is emitted. It is measured in lumens (lm). Illuminance is the luminous flux per unit area. Illuminance is measured in lux (lx), or lumens per square meter (lm/m^2). For a point source, illuminance follows an inverse-square relationship with distance and a direct relationship with luminous flux.

$$E = \frac{P}{4\pi r^2}$$

- Early measurements of the speed of light involved measurement of the time it takes for light to reach Earth from Jupiter's moon Io. Michelson used a land-based technique that involved the distance between two mountains and a set of rotating mirrors. In a vacuum, light has a constant speed of $c = 3.00\times10^8$ m/s.

VOCABULARY

- **diffraction** *(p. 447)*
- **primary color** *(p. 449)*
- **secondary color** *(p. 449)*
- **complementary color** *(p. 449)*
- **primary pigment** *(p. 449)*
- **secondary pigment** *(p. 450)*
- **polarization** *(p. 451)*
- **Malus's law** *(p. 452)*

SECTION 2 The Wave Nature of Light

MAINIDEA Like all waves, light diffracts around objects, has a wavelength and frequency, and can be Doppler shifted.

- In the wave model of light, all the points in a wavefront can be thought of as sources of smaller waves. As light travels past an edge, the wavefront is cut and each new wavelet generates a new circular wave.
- Visible light can have wavelengths between 400 and 700 nm. White light is a combination of the spectrum of colors, each color having a different wavelength. Combining the primary colors—red, blue, and green—forms white light. Combinations of two primary colors form the secondary colors, yellow, cyan, and magenta.

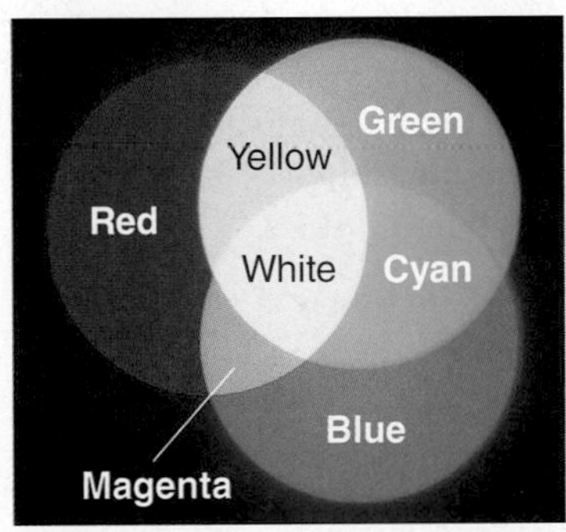

The primary pigments, cyan, magenta, and yellow, are used in combinations of two to produce the secondary pigments, red, blue, and green.

- Polarized light consists of waves whose electric fields oscillate with a specific pattern. Often, the oscillation is in a single plane. Light can be polarized with a polarizing filter or by reflection. Light waves traveling through a vacuum can be characterized in terms of frequency, wavelength, and the speed of light. Light waves are Doppler shifted based on the relative speed of the observer and light source along the axis of the observer and the light source.

Matt Meadows

Games and Multilingual eGlossary

Vocabulary Practice

Chapter Self-Check

SECTION 1 Illumination

Mastering Concepts

29. What are some useful applications of measuring light?

30. Distinguish between a luminous source and an illuminated object.

31. Look carefully at an ordinary, frosted, incandescent bulb. Is it a luminous source or an illuminated object?

32. Explain how you can see ordinary, nonluminous classroom objects.

33. Distinguish among transparent, translucent, and opaque objects.

34. What is the illumination of a surface by a light source directly proportional to? What is it inversely proportional to?

35. What was Ole Roemer the first to measure regarding the propagation of light?

Mastering Problems

36. Find the illumination 4.0 m below a 405-lm lamp.

37. Light takes 1.28 s to travel from the Moon to Earth. What is the distance between them?

38. A three-way bulb uses 50, 100, or 150 W of electrical power to deliver 665, 1620, or 2285 lm in its three settings. The bulb is placed 80 cm above a sheet of paper. If an illumination of at least 175 lx is needed on the paper, what is the minimum setting that should be used?

39. Earth's Speed Ole Roemer found that the average increased delay in the disappearance of Io from one orbit around Jupiter to the next is 13 s.

a. How far does light travel in 13 s?

b. Each orbit of Io takes 42.5 h. Earth travels the distance calculated in part a in 42.5 h. Find the speed of Earth in kilometers per second.

c. Check to make sure your answer for part b is reasonable. Calculate Earth's speed in orbit using its orbital radius, 1.5×10^8 km, and the period, 1.0 y.

40. A student wants to compare the luminous flux of a lightbulb with that of a 1750-lm lamp. The lightbulb and the lamp illuminate a sheet of paper equally. The 1750-lm lamp is 1.25 m away from the sheet of paper; the lightbulb is 1.08 m away. What is the lightbulb's luminous flux?

41. Reverse Problem Write a physics problem with real-life objects for which the following equation would be part of the solution: $95\ \text{lx} = 1100\ \text{lm}/(4\pi r^2)$.

42. Suppose you try to measure the speed of light by putting a mirror on a distant mountain, setting off a camera flash, and measuring the time it takes the flash to reflect off the mirror and return to you, as shown in **Figure 25.** Without instruments, a person can detect a time interval of about 0.10 s. How many kilometers away would the mirror have to be? Compare this distance to some known distances.

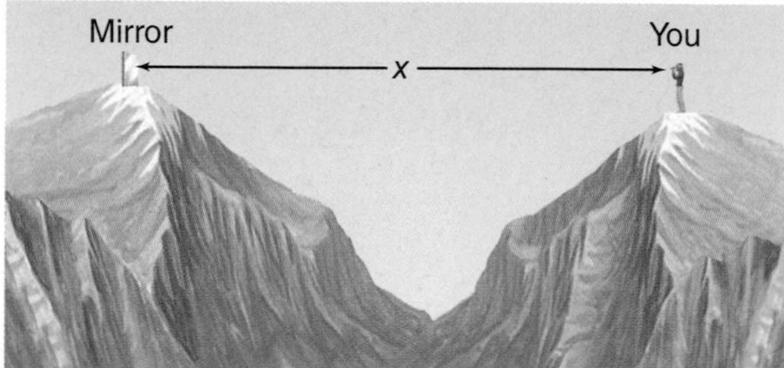

Figure 25

SECTION 2 The Wave Nature of Light

Mastering Concepts

43. Why is the diffraction of sound waves more familiar in everyday experience than is the diffraction of light waves?

44. BIGIDEA What color of light has the shortest wavelength?

45. Ranking Task Rank the following colors of light according to their frequency, from least to greatest: blue, green, red, violet, yellow. Specifically indicate any ties.

46. What is the range of the wavelengths of visible light, from shortest to longest?

47. Ranking Task Rank the following colors of light according to how much they will diffract around the edge of a piece of paper, from bending the most to bending the least: blue, green, red, violet, yellow. Specifically indicate any ties. *Hint: Longer wavelengths diffract more.*

48. Of what colors does white light consist?

49. Can longitudinal waves be polarized? Explain.

50. If a distant galaxy were to emit light in the green region of the light spectrum, would the observed wavelength on Earth shift toward red light or toward blue light? Explain.

Chapter Self-Check

Mastering Problems

51. Convert 700 nm, the wavelength of red light, to meters.

52. **Galactic Motion** How fast is a galaxy moving relative to Earth if light from hydrogen's spectrum of 486 nm is redshifted to 491 nm?

53. Suppose you are facing due east at sunrise. Sunlight is reflected off the surface of a lake, as shown in **Figure 26.** Is the reflected light polarized? If so, in what direction?

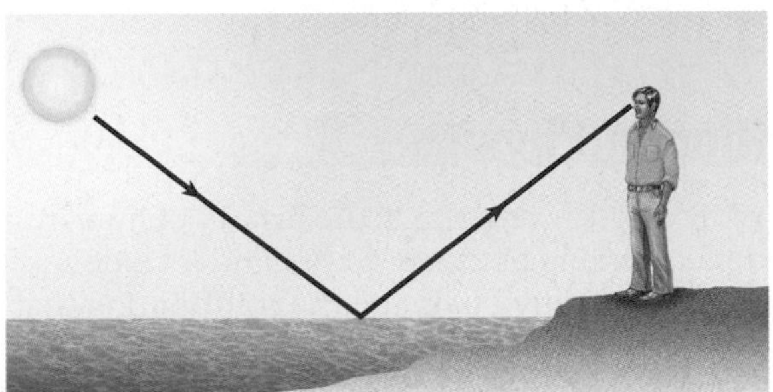

Figure 26

54. **Galactic Motion** Light from hydrogen that is known to be 434 nm is redshifted by 6.50 percent in light coming from a distant galaxy. How fast is the galaxy moving away from Earth?

55. **Problem Posing** Complete this problem so that it must be solved using the speed of light: "A laser produces light of wavelength 655 nm…".

56. Give an example of an unrealistic Doppler shift in terms of wavelength for light from a galaxy moving away from Earth. Why is it unreasonable?

Applying Concepts

57. A point source of light is 2.0 m from screen A and 4.0 m from screen B, as shown in **Figure 27.** How does the illuminance at screen B compare with the illuminance at screen A?

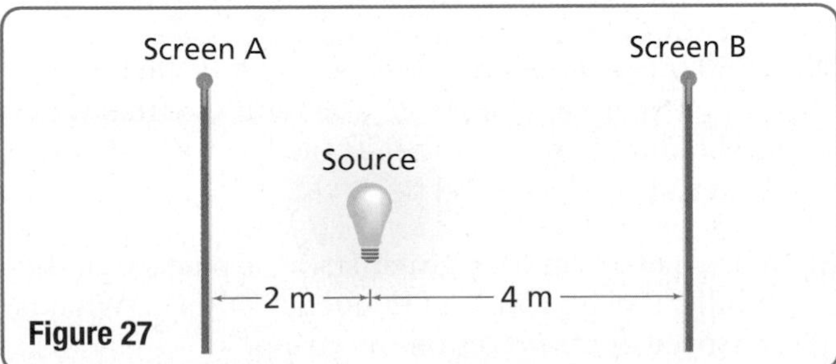

Figure 27

58. Why are the insides of binoculars and cameras painted black?

59. **Reading Lamp** You have a small reading lamp that is 35 cm from the pages of a book. You decide to double the distance.
    a. Is the illuminance at the book the same?
    b. If not, how much more or less is it?

60. **Eye Sensitivity** The eye is most sensitive to yellow-green light. Its sensitivity to red and blue light is less than 10 percent as great. Based on this knowledge, what color would you recommend that fire trucks and ambulances be painted? Why?

61. **Streetlight Color** Some very efficient streetlights contain sodium vapor under high pressure. They produce light that is mainly yellow with some red. Should a community that has these lights buy dark blue police cars? Why or why not?

*Refer to **Figure 28** for problems 62 and 63.*

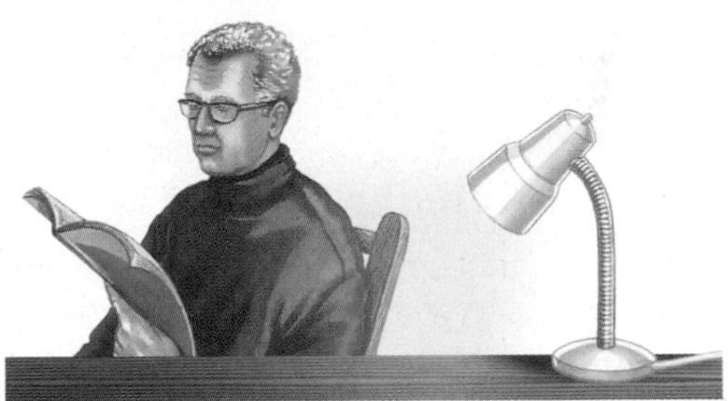

Figure 28

62. What, if anything, happens to the illuminance at a book as the lamp is moved farther away from the book?

63. What, if anything, happens to the luminous intensity of a lamp as it is moved farther away from a book?

64. **Polarized Pictures** Photographers often put polarizing filters over camera lenses to make clouds in the sky more visible. The clouds remain white, while the sky looks darker. Explain this based on your knowledge of polarized light.

65. An apple is red because it reflects red light and absorbs blue and green light.
    a. Why does red cellophane look red in reflected light?
    b. Why does red cellophane make a white lightbulb look red when you hold the cellophane between your eye and the lightbulb?
    c. What happens to the blue and green light?

66. If you have yellow, cyan, and magenta pigments, how can you make a blue pigment? Explain.

Chapter Self-Check

67. You put a piece of red cellophane over one flashlight and a piece of green cellophane over another. You shine the light beams on a white wall. What color will you see where the two flashlight beams overlap?

68. Put both the red and green cellophane pieces over one of the flashlights in the previous problem. If you shine the flashlight beam on a white wall, what color will you see? Explain.

69. Traffic Violation Suppose that you are a traffic officer and you stop a driver for going through a red light. Further suppose that the driver draws a picture for you (**Figure 29**) and explains that the light looked green because of the Doppler effect when he went through it. Explain to him using the Doppler shift equation how fast he would have had to be going for the red light ($\lambda = 645$ nm) to appear green ($\lambda = 545$ nm).

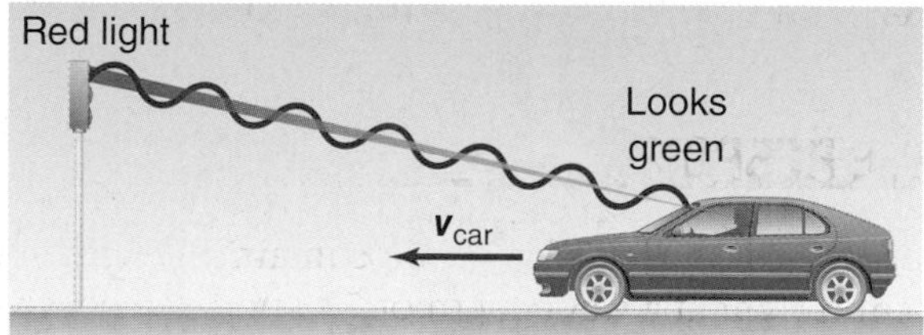

Figure 29

Mixed Review

70. Streetlight Illumination A streetlight contains two identical bulbs that are 3.3 m above the ground. If the community wants to save electrical energy by removing one bulb, how far from the ground should the streetlight be positioned to have the same illumination on the ground under the lamp?

71. An octave in music is a doubling of frequency. Compare the number of octaves that correspond to the human hearing range to the number of octaves in the human vision range. *Hint: There are more than seven octaves on a piano.*

72. Thunder and Lightning Explain why it takes 5 s to hear thunder when lightning is 1.6 km away.

73. Solar Rotation Because the Sun rotates on its axis, one edge of the Sun moves toward Earth and the other moves away. The Sun rotates approximately once every 25 days, and the diameter of the Sun is 1.4×10^{9} m. Hydrogen normally emits light of frequency 6.16×10^{14} Hz. What wavelengths of hydrogen are observed from the receding edge? The approaching edge?

Thinking Critically

74. Make and Use Graphs A 110-cd light source is initially 1.0 m from a screen and then slowly moved away. Determine the illumination on the screen originally and for every meter of increasing distance up to 7.0 m. Graph the data.

a. What is the shape of the graph?

b. What is the relationship between illuminance and distance shown by the graph?

75. Research Why did Galileo's method for measuring the speed of light not work?

Writing in Physics

76. Write an essay describing the history of human understanding of the speed of light. Include significant individuals and the contribution that each individual made.

77. Look up information on the SI unit candela (cd) and explain in your own words the standard that is used to set the value of 1 cd.

Cumulative Review

78. A 2.0-kg object is attached to a 1.5-m long string and swung in a vertical circle at a constant speed of 12 m/s.

a. What is the tension in the string when the object is at the bottom of its path?

b. What is the tension in the string when the object is at the top of its path?

79. A space probe with a mass of 7.600×10^{3} kg is traveling through space at 125 m/s. Mission control decides that a course correction of 30.0° is needed and instructs the probe to fire rockets perpendicular to its present direction of motion. If the gas expelled by the rockets has a speed of 3.200 km/s, what mass of gas should be released?

80. If 250 mL of water in a glass beaker is heated from room temperature to 65°C, what will the final water level be? If the water is then cooled to 5°C, what will the resulting water level be?

81. When a 60.0-cm-long guitar string is plucked in the middle, it plays a note of frequency 440 Hz. What is the speed of the waves on the string?

82. What is the wavelength of a sound wave with a frequency of 17,000 Hz in water at 25°C?

MULTIPLE CHOICE

1. In 1987 a supernova was observed in a neighboring galaxy. Scientists believed the galaxy was 1.66×10^{21} m away. How many years prior to the observation did the supernova explosion actually occur?

A. 5.53×10^{3} y
B. 1.75×10^{5} y
C. 5.53×10^{12} y
D. 1.74×10^{20} y

2. A galaxy is moving away at 5.8×10^{6} m/s. Its light appears to observers to have a frequency of 5.6×10^{14} Hz. What is the emitted frequency of the light?

A. 1.1×10^{13} Hz
B. 5.5×10^{14} Hz
C. 5.7×10^{14} Hz
D. 6.2×10^{14} Hz

3. The illuminance due to a 60.0-W lightbulb at 3.0 m is 9.35 lx. What is the total luminous flux of the bulb?

A. 8.3×10^{22} lm
B. 7.4×10^{21} lm
C. 1.2×10^{2} lm
D. 1.1×10^{3} lm

4. Light from the Sun takes about 8.0 min to reach Earth. About how far away is the Sun?

A. 2.4×10^{9} m
B. 1.4×10^{10} m
C. 1.4×10^{8} km
D. 2.4×10^{9} km

5. What is meant by the phrase *color by subtraction of light*?

A. Adding green, red, and blue light produces white light.
B. Exciting phosphors with electrons in a television produces color.
C. Paint color is changed by subtracting certain colors, such as producing blue paint from green by removing yellow.
D. The color that an object appears to be is a result of the material absorbing specific light wavelengths and reflecting the rest.

Online Test Practice

6. What is the frequency of 404 nm of light in a vacuum?

A. 2.48×10^{23} Hz
B. 7.43×10^{5} Hz
C. 2.48×10^{6} Hz
D. 7.43×10^{14} Hz

7. Which light color combination is incorrect?

A. Red plus green produces yellow.
B. Red plus yellow produces magenta.
C. Blue plus green produces cyan.
D. Blue plus yellow produces white.

8. The illuminance of direct sunlight on Earth is about 1×10^{5} lx. A light on a stage has an intensity in a certain direction of 5×10^{6} cd. At what distance from the stage does a member of the audience experience an illuminance equal to that of sunlight?

A. 1.4×10^{-1} m
B. 7 m
C. 10 m
D. 5×10^{1} m

FREE RESPONSE

9. A celestial object is known to contain an element that emits light at a wavelength of 525 nm. The observed spectral line for this element is at 473 nm. Is the object approaching or receding, and at what speed?

10. Nonpolarized light of intensity I_0 is incident on a polarizing filter, and the emerging light strikes a second polarizing filter, as shown in the figure. What is the light intensity emerging from the second polarizing filter?

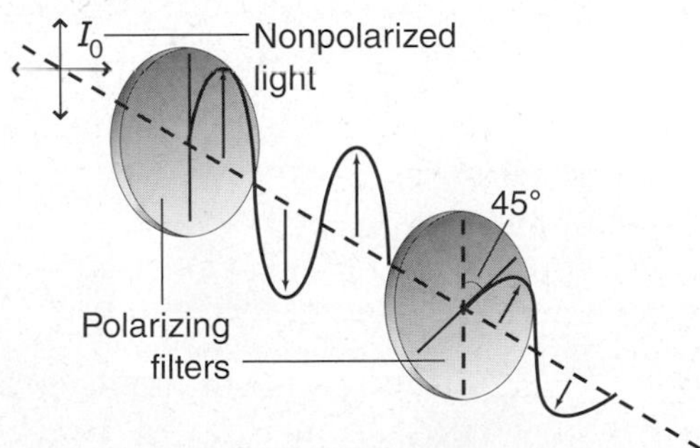

NEED EXTRA HELP?

If You Missed Question	1	2	3	4	5	6	7	8	9	10
Review Section	2	2	1	1	2	2	2	1	2	2

CHAPTER 17

Reflection and Mirrors

BIG IDEA **All surfaces reflect light, but smooth surfaces can produce images.**

SECTIONS

1 Plane Mirrors

2 Curved Mirrors

LaunchLAB

iLab Station

IMAGES FROM MIRRORS

Under what conditions does a mirror produce a clear image on a screen?

WATCH THIS!

Video

SEEING AROUND CORNERS

Have you ever seen an outwardly curved mirror in a store, on a building, or even on a car's side mirror? What's the point of a mirror like that? Explore the physics of reflection.

(l)Comstock/PictureQuest, (r)Alan Copson/Photographer's Choice/Getty Images

Go online!
connectED.mcgraw-hill.com

SECTION 1

Plane Mirrors

PHYSICS 4 YOU

When you think of reflections, you probably think of mirrors. Mirrors are usually glass with a thin layer of aluminum backing. You can, however, also see yourself in windows, calm water, and shiny metals.

MAIN IDEA

The angle of incidence of a light ray is equal to the angle of reflection.

Essential Questions

- What is the law of reflection?
- What is the difference between specular and diffuse reflection?
- How can the images formed by plane mirrors be located?

Review Vocabulary

normal a line that is perpendicular to a barrier or the surface of a mirror

New Vocabulary

specular reflection
diffuse reflection
plane mirror
object
image
virtual image

Reflected Images

Undoubtedly, humans have seen their faces reflected in the quiet water of lakes and ponds for thousands of years. When you look at the surface of a body of water, however, you don't always see a clear reflection as in **Figure 1.** Sometimes, the wind causes ripples in the water, or passing boats produce waves. Disturbances on the water's surface prevent the light from reflecting in a manner such that a clear image is visible.

Almost 4000 years ago, Egyptians understood that the type of reflection you see from a still pond requires a smooth surface. They used polished metal mirrors to view their images. Artisans in sixteenth-century Venice created mirrors by coating the back of a flat piece of glass with a thin sheet of metal. Sharp, well-defined, reflected images were not possible until 1857, however, when Jean Foucault, a French scientist, developed a method of coating glass with silver.

We use ever-increasing precision to make modern mirrors. Today, we don't only use mirrors to view our own reflections. Mirrors are also key components in lasers, telescopes, and other precise optical systems. Today's mirrors are made by evaporating aluminum or silver onto highly polished glass. The same basic physics principles, however, govern reflection from the smooth surface of a pond, from a sixteenth-century mirror, and from a tiny mirror inside a laser.

Figure 1 Disturbances on the surface of a pond or lake produce a distorted reflected image.

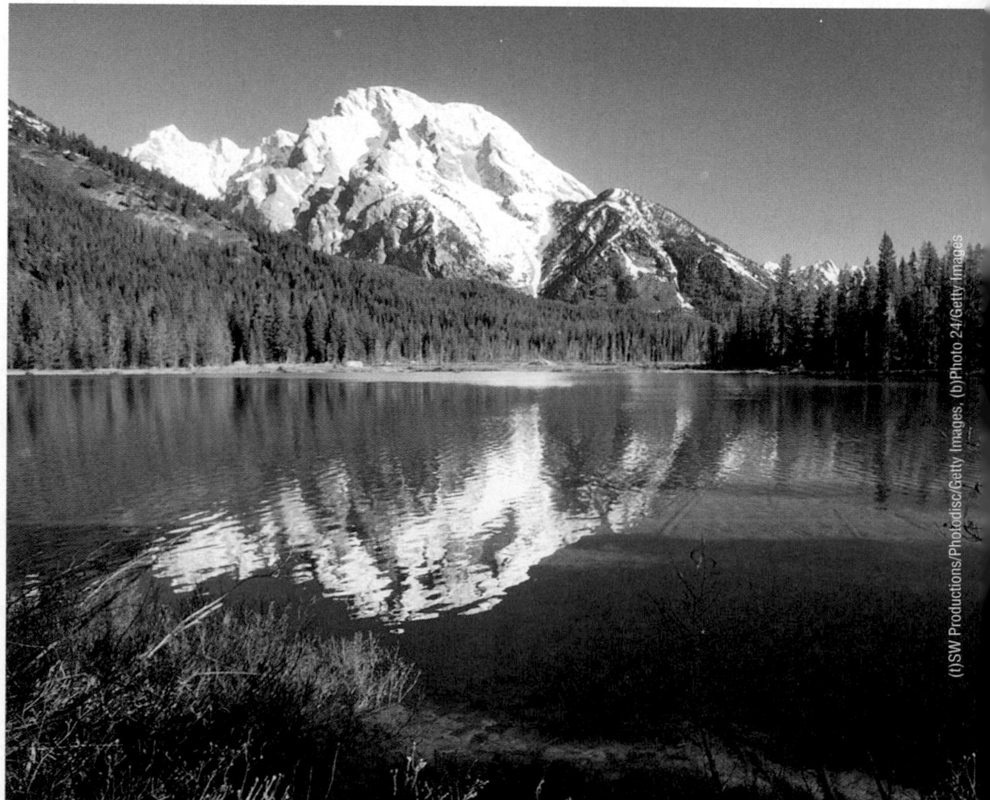

The Law of Reflection

Before you can understand mirrors, you must understand reflection in general. What happens to the light that strikes a book? If you hold a book up to the light, you will see that no light passes through it. Recall that an object like this is called opaque. An opaque object reflects part of the light and absorbs part of the light. The energy of the absorbed light warms the book. The behavior of the reflected light depends on the type of surface.

Consider what happens when you bounce-pass a basketball. As viewed from above, the ball bounces in a straight line to the other player. Recall from studying waves that when a wave traveling in two dimensions encounters a smooth barrier, it reflects with an angle of reflection equal to the angle of incidence. The same two-dimensional reflection relationship applies to light waves.

Light reflects in the same way as a water wave or a basketball. **Figure 2** shows a light ray striking a reflecting surface. The normal is an imaginary line that is perpendicular to a surface at the location where light strikes the surface. The incident ray, the reflected ray, and the normal to the surface always will be in the same plane, which is perpendicular to the surface. Although the light travels in three dimensions, the reflection of the light is planar (two-dimensional). The planar and angle relationships are known together as the law of reflection.

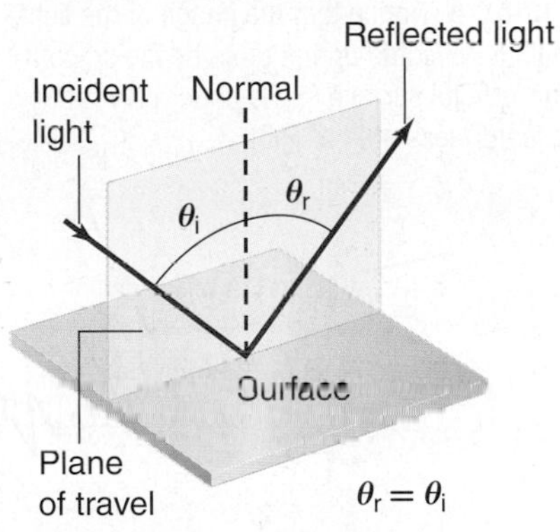

Figure 2 Reflecting light waves have an angle of reflection in the same plane as and equal to the angle of incidence.

View an **animation of the law of reflection.**

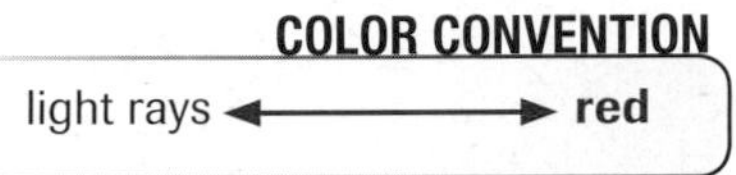

LAW OF REFLECTION

The angle that a reflected ray makes as measured from the normal to a reflective surface equals the angle that the incident ray makes as measured from the same normal.

$$\theta_r = \theta_i$$

Wave model We can describe this law in terms of the wave model of light. **Figure 3** shows a wavefront of light approaching a reflective surface. The wavefront is perpendicular to the light ray. As each point along the wavefront reaches the surface, it reflects off that surface. Because all points are traveling at the same speed, they all travel the same total distance in the same time. Thus, the wavefront as a whole leaves the surface at an angle equal to its incident angle. Note that the wavelength of the light does not affect this process. The surface reflects red, green, and blue light all in the same direction.

Figure 3 All parts of a wavefront reflect from a surface at the same angle. The angle of incidence equals the angle of reflection.

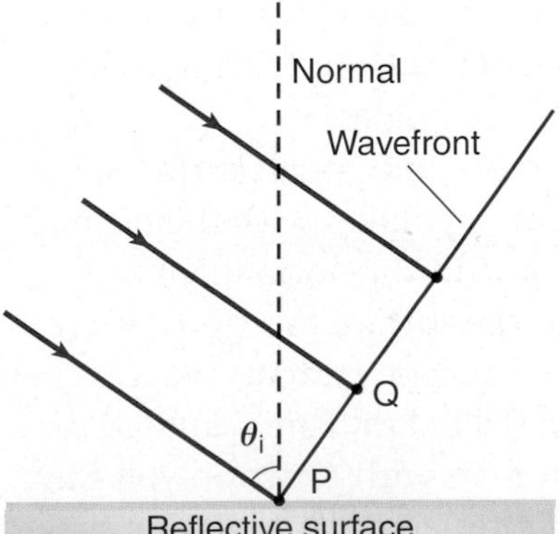

Point P is the point of the wavefront that strikes the surface first.

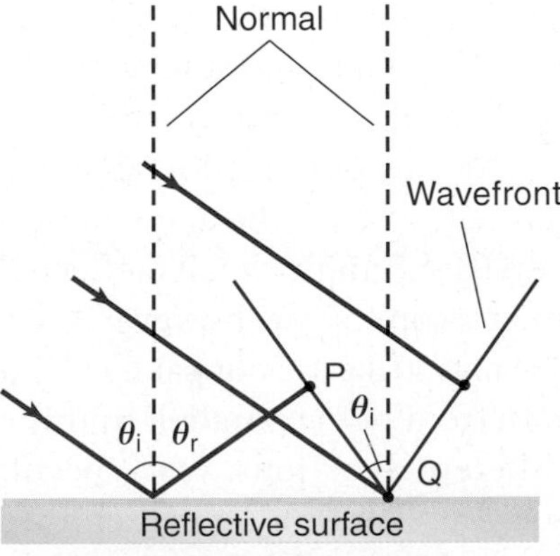

A reflected wavefront forms as the light rays reflect off the surface.

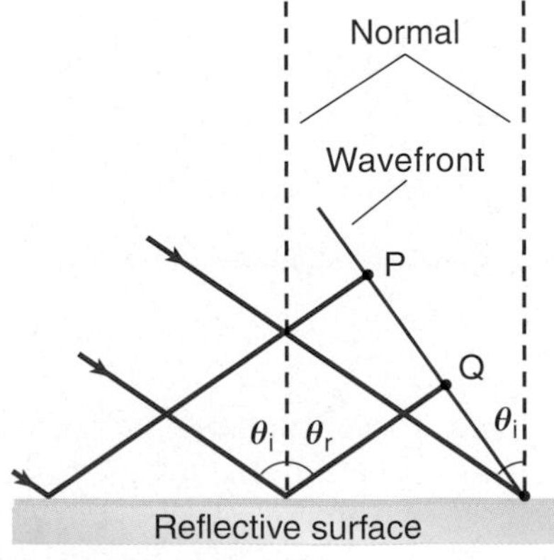

All points of the wavefront reflect at the same angle.

Figure 4 Notice that the image of the lightbulb is reflected on the table by the smooth mirror. The surface of the paper only reflects a featureless area of light.

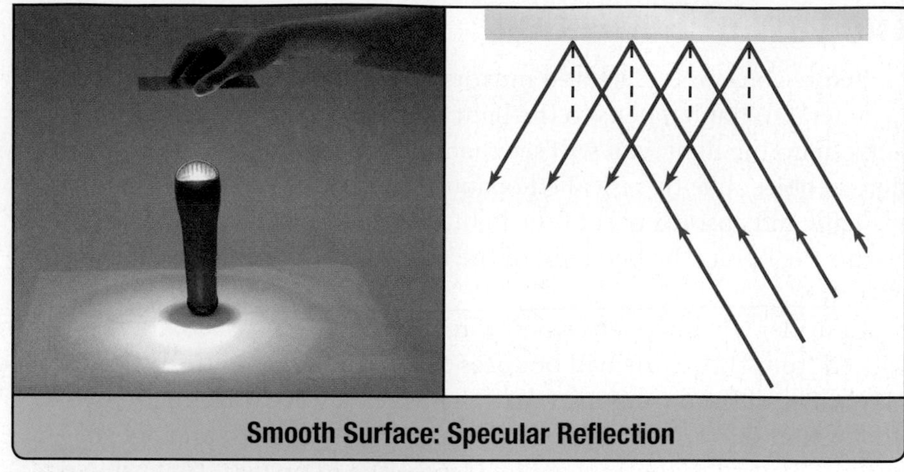

Smooth Surface: Specular Reflection

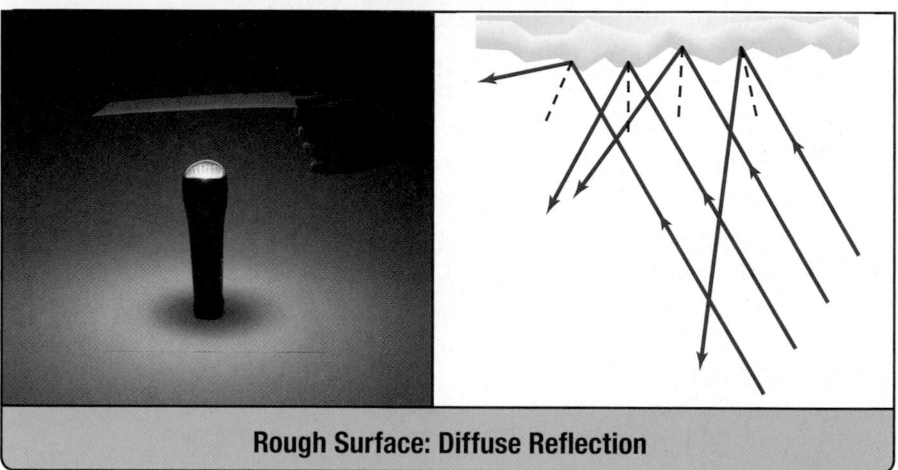

Rough Surface: Diffuse Reflection

Richard Hutchings/Digital Light Source

Smooth surfaces Consider the light rays shown in the top panel of **Figure 4.** All of the rays reflect off the surface parallel to one another as shown. This occurs only if the reflecting surface is not rough on the scale of the wavelength of the light. We consider such a surface to be smooth. A smooth surface, such as a mirror, produces **specular reflection,** in which parallel light rays reflect in parallel.

Rough surfaces What happens when light strikes a sheet of paper? That paper might appear smooth, but on the scale of the wavelength of light, the paper is actually quite rough. Is light reflected? How could you demonstrate this? The bottom panel of **Figure 4** shows light rays reflecting off a sheet of paper. All of the light rays are parallel before they strike the surface, but the reflected rays are not parallel, as shown. This scattering of light off a rough surface is called **diffuse reflection.** Diffuse reflection enables you to read this page from various angles.

The law of reflection applies to both smooth and rough surfaces. For a rough surface, the angle that each incident ray makes with the normal equals the angle that its reflected ray makes with the normal. On a microscopic scale, however, the normals to the surface locations where the rays strike are not parallel. Thus, the rough surface prevents the reflected rays from being parallel. In this case, the reflected rays are scattered in different directions. With specular reflection, as with a mirror, you can see your face. But no matter how much light reflects off a wall or a sheet of paper, you will never be able to use them as mirrors. Recall that you can only see an object, such as a sheet of paper, if light reflects off that object.

EXAMPLE PROBLEM 1

Find help with **operations with significant figures.** Math Handbook

CHANGING THE ANGLE OF INCIDENCE A light ray strikes a plane mirror at an angle of 52.0° to the normal. The mirror then rotates 35.0° around the point where the ray strikes the mirror so that the angle of incidence of the light ray decreases. The axis of rotation is perpendicular to the plane of the incident and the reflected rays. What is the angle between the initial and final reflected ray?

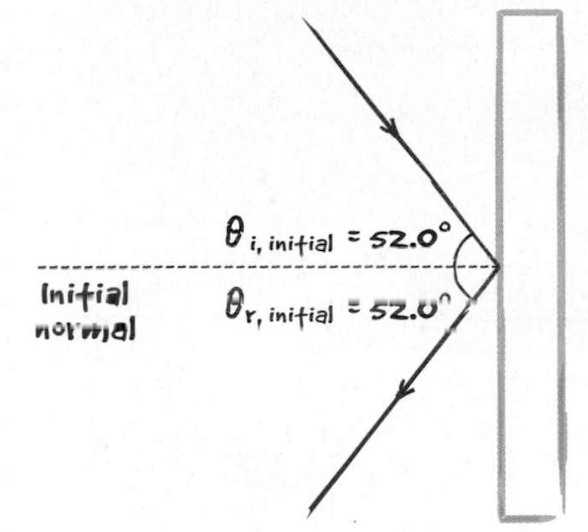

1 ANALYZE AND SKETCH THE PROBLEM

- Sketch the situation before the rotation of the mirror.
- Draw another sketch with the angle of rotation applied to the mirror.
- Draw a third sketch of the reflected rays.

KNOWN	UNKNOWN
$\theta_{i,\ initial} = 52.0°$	$\Delta\theta_r = ?$
$\Delta\theta_{mirror} = 35.0°$	

2 SOLVE FOR THE ANGLE DIFFERENCE

For the angle of incidence to reduce, rotate clockwise.

$\theta_{i,\ final} = \theta_{i,\ initial} - \Delta\theta_{mirror}$

$= 52.0° - 35.0°$ ◀ *Substitute $\theta_{i,\ initial} = 52.0°$, $\Delta\theta_{mirror} = 35.0°$*

$= 17.0°$ clockwise from the new normal

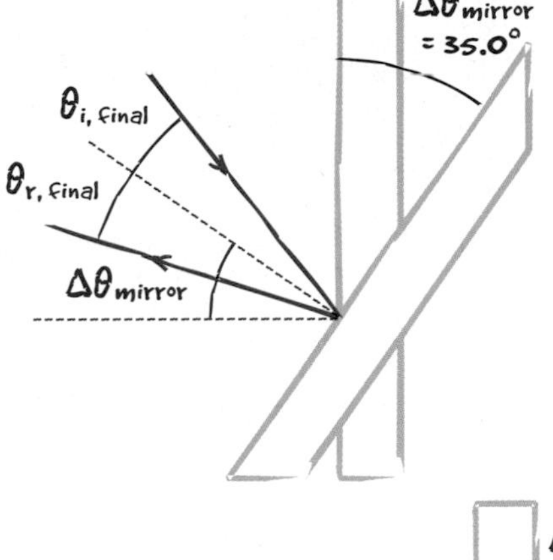

Apply the law of reflection.

$\theta_{r,\ final} = \theta_{i,\ final}$

$= 17.0°$ counterclockwise from the new normal

▲ *Substitute $\theta_{i,\ final} = 17.0°$*

Use the sketches to help determine the angle through which the reflected ray has rotated.

$\Delta\theta_r = \theta_{r,\ initial} + \Delta\theta_{mirror} - \theta_{r,\ final} = 52.0° + 35.0° - 17.0°$

$= 70.0°$ clockwise from the original angle

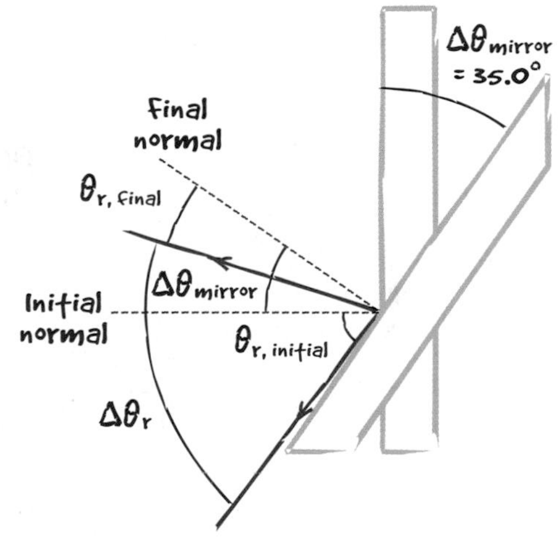

3 EVALUATE THE ANSWER

Is the magnitude realistic? Comparing the final sketch with the initial sketch shows that the angle the light ray makes with the normal decreases as the mirror rotates clockwise. It makes sense, then, that the reflected ray also rotates clockwise and rotates through an angle twice as large as that of the mirror rotation.

PRACTICE PROBLEMS

Do additional problems. Online Practice

1. Explain why the reflection of light off ground glass changes from diffuse to specular if you spill water on it.
2. What is the angle of incidence of a light ray reflected off a plane mirror at an angle of 35° to the normal?
3. Suppose the angle of incidence of a light ray is 42°.
   a. What is the angle of reflection?
   b. What is the angle the incident ray makes with the mirror?
   c. What is the angle between the incident ray and the reflected ray?
4. Light from a laser strikes a plane mirror at an angle of 38° to the normal. If the angle of incidence increases by 13°, what is the new angle of reflection?
5. You position two plane mirrors at right angles to each other. A light ray strikes one mirror at an angle of 60° to the normal and reflects toward the second mirror. What is its angle of reflection off the second mirror?
6. **CHALLENGE** You are asked to design a retroreflector using two mirrors that will reflect a laser beam by 180° independent of the incident direction of the beam. What should be the angle between the two mirrors?

Figure 5 Light from the lamp is reflected from the boy to the mirror. This is diffuse reflection. The mirror reflects light to the boy's eye, enabling him to see an image of himself. Reflection from the mirror is specular reflection.

Describe how the boy is illuminated.

Objects and Plane-Mirror Images

If you looked at yourself in a mirror this morning you saw your reflection in a plane mirror. A **plane mirror** is a flat, smooth surface from which light is reflected by specular reflection. To understand the reflection you saw, you must consider the source of the light and how it was reflected.

When studying illumination, the word *object* is used to refer to a light source. An **object** is either a luminous source of light rays, such as a lightbulb, or an illuminated source of light rays, such as a boy. For most of the light sources that you will study in this chapter, light spreads out from that source in all directions. A mirrored surface can reflect these light rays so an image is visible, as shown in **Figure 5.**

In **Figure 6** the bird is the object. Light reflects diffusely from all parts of the bird. Consider a point on the breast of the bird. What happens to the light reflected from this point? Some of the light rays travel from the bird to the mirror and reflect. What does the girl see? She sees the light that reflects into her eye. Because her brain processes this information as if the rays travel in a straight path, it seems to the girl as if the light follows the dashed lines. The light seems to have come from a point behind the mirror. Just as the rays diverge from the object, they also diverge from the image.

We have only considered one point on the bird, but the girl in **Figure 6** sees rays of light that come from many points on the bird. The combination of light rays reflected from the bird forms the **image** of the bird. It is a **virtual image,** which is a type of image formed by diverging light rays. A virtual image is always on the opposite side of the mirror from the object. The image is virtual because there are no light rays at the image location. Plane mirrors only produce virtual images.

Properties of Plane-Mirror Images

Looking at yourself in a mirror, you can see that your image appears to be the same distance behind the mirror as you are in front of the mirror. How could you test this? Place a ruler between you and the mirror. Where does the image touch the ruler? You also see that your image is vertically oriented as you are, and it matches your size. This is where the expression mirror image originates.

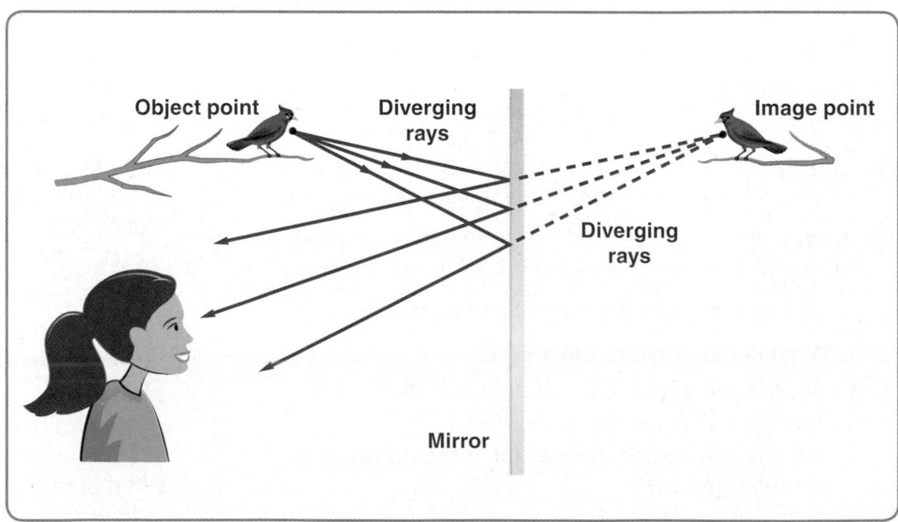

Figure 6 Rays that reflect from the bird will disperse in many directions. Only a few that travel toward the mirror are shown. The image is located where multiple light rays from a point on an object seem to converge.

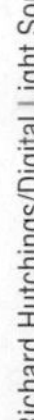

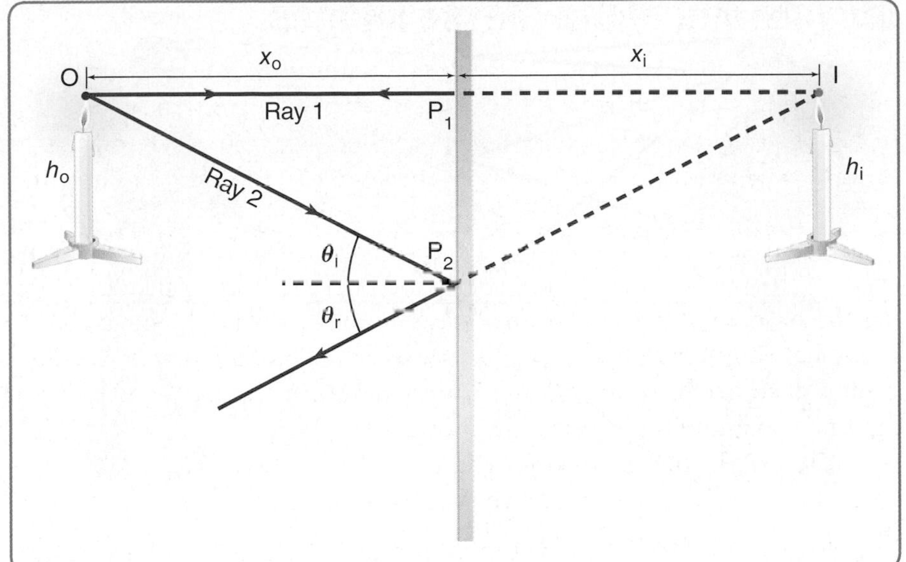

Figure 7 Reflected light rays from the candle (two rays are shown) strike the mirror. Some of those rays reach the viewer's eye. Sight lines (dashed lines) are drawn from where the rays reflect from the mirror to where they converge. The image is located where the sight lines converge.

Explain why $x_i = -x_o$.

Image position and height The geometric model in **Figure 7** demonstrates why the distances are the same. Two rays from point O at the tip of the candle strike the mirror at point P_1 and point P_2, respectively. The mirror reflects both rays according to the law of reflection. We can extend the reflected rays behind the mirror as sight lines. These sight lines converge at point I, which is the image of point O. Ray 1, which strikes the mirror at an angle of incidence of 0°, reflects back on itself, so the sight line is at 90° to the mirror, just as ray 1. Ray 2 also reflects at an angle with respect to the normal equal to the angle of incidence, so the sight line is at the same angle to the mirror as ray 2.

This geometric model reveals that line segments OP_1 and IP_1 are corresponding sides of two congruent triangles, ΔOP_1P_2 and ΔIP_1P_2. The object's position with respect to the mirror (x_o) has a magnitude equal to the length of $\overline{OP_1}$. The apparent position of the image with respect to the mirror, which is called the image position (x_i), has a magnitude equal to the length of $\overline{IP_1}$. From the convention that image position is negative to indicate that the image is virtual, we find the following is true.

PLANE-MIRROR IMAGE POSITION

With a plane mirror, the image position is equal to the negative of the object position. The negative sign indicates that the image is behind the mirror (and therefore virtual).

$$x_i = -x_o$$

You can draw more light rays from the object to the mirror to determine the size of the image. The sight lines of two rays originating from the bottom of the candle in **Figure 7** will converge at the bottom of the image. From the law of reflection and congruent-triangle geometry, we find the following is true of the object height (h_o) and image height (h_i) and any other dimension of the object and image.

PLANE-MIRROR IMAGE HEIGHT

With a plane mirror, image height is equal to object height.

$$h_i = h_o$$

PhysicsLABs

POSITION OF MIRROR REFLECTION

Can you measure the location of an image from a plane mirror?

A LITTLE TIME TO REFLECT

FORENSICS LAB How can you determine the path of light rays incident on a reflective surface, such as glass or a mirror?

MiniLAB

VIRTUAL IMAGE POSITION

Can you focus on a virtual image with a digital camera?

Figure 8 Viewed in a mirror, your height and distance from the mirror appear the same. There is, however, a difference. You are facing the opposite direction.

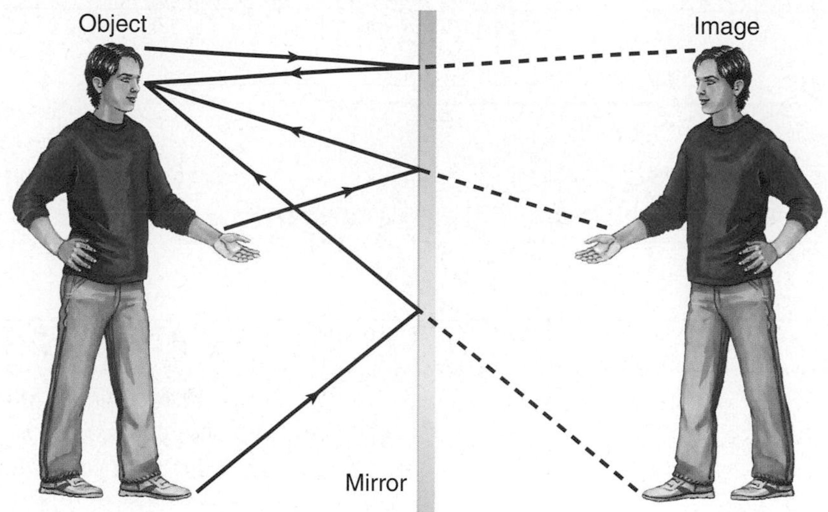

Image orientation A plane mirror produces an image with the same orientation as the object. If you stand on your feet, a plane mirror produces an image of you standing on your feet. If you do a headstand, the mirror shows you doing a headstand. There is, however, a difference between you and your image in a mirror. The mirror gives a front-to-back reversal. You are looking face-to-face at the image of yourself.

Follow the sight lines in **Figure 8.** The mirror reflects an image that is behind the mirror such that the boy appears to be facing himself. When the boy extends his left hand, it appears as though the right hand of the image is extended, but from the boy's perspective, the hand that is extended is on his left-hand side. This happens because a plane mirror does not really reverse left and right. The mirror in **Figure 8** only reverses the boy's image such that it is facing in the opposite direction as the boy, or, in other words, it produces a front-to-back reversal.

Consider the image of the mountain in **Figure 1.** The image of the mountain can be described as upside down, but it is actually a front-to-back reversal of your view of the actual mountain. The lake surface acts like a mirror and is horizontal rather than vertical, so your perspective makes the image look upside down. Turn **Figure 8** 90° clockwise. Now, the actual boy is facing down and his image is facing up. The only thing that has changed is your perspective.

SECTION 1 REVIEW

Section Self-Check Check your understanding.

7. **MAIN**IDEA A light ray strikes a flat, smooth, reflecting surface at an angle of 80° to the normal. What angle does the reflected ray make with the surface of the mirror?
8. **Image Properties** A dog looks at its image, as shown in **Figure 9.** What is the image position, height, and type?

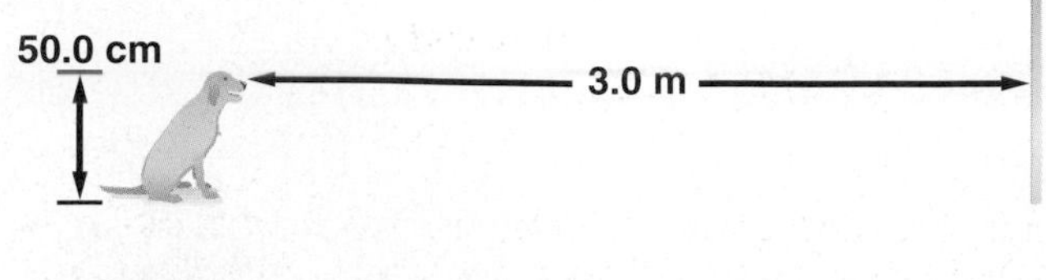

Figure 9

9. **Law of Reflection** Explain how the law of reflection applies to diffuse reflection.
10. **Reflecting Surfaces** Categorize each of the following as a specular or a diffuse reflecting surface: paper, polished metal, window glass, rough metal, plastic milk jug, smooth water surface, and ground glass.
11. **Image Diagram** A car is following another car along a straight road. The first car has a rear window tilted at 45° to the horizontal. Draw a ray diagram showing the position of the Sun that would cause sunlight to reflect into the eyes of the driver of the second car.
12. **Critical Thinking** Explain how diffuse reflection of light off an object enables you to see that object from any angle.

SECTION 2 **Curved Mirrors**

PHYSICS 4 YOU If you have ever looked at your reflection in a metal spoon, you know that your image is usually inverted on the front and upright on the back. Why does this occur?

MAIN IDEA

Curved mirrors can produce real and virtual images and can magnify or reduce the image size.

Essential Questions

- What are some properties and uses of spherical concave mirrors?
- How are ray diagrams used to describe images produced by curved mirrors?
- How are convex mirrors and combinations of mirrors used?
- How can you calculate properties of images produced by curved mirrors?

Review Vocabulary

ray a line that shows the direction a wave is traveling and is drawn at a right angle to the wavefront

New Vocabulary

concave mirror
principal axis
focal point
focal length
real image
spherical aberration
convex mirror
magnification

Properties of Curved Mirrors

A metal spoon acts as a curved mirror, with one side curved inward and the other side curved outward. Move the spoon toward and away from your face. Look at your image on both sides of the spoon. Your image may appear larger or smaller, or it may even be inverted. The properties of curved mirrors and the images that they form depend on the shape of the mirror and on the object's position.

Concave mirrors The inside surface of a shiny metal spoon, the side that holds food, acts as a concave mirror. A **concave mirror,** such as the one illustrated in **Figure 10,** has an inwardly curving reflective surface, the edges of which curve toward the observer. Concave mirrors are placed behind the lightbulbs in theater spotlights and in car headlights. It is a concave mirror that allows the light from a spotlight to be focused into a spot. What affects the behavior of a concave mirror?

The way that a concave mirror reflects light depends on how that mirror is curved. **Figure 10** shows a diagram of a *spherical* concave mirror including some of the important points and distances for understanding how it forms images. A spherical concave mirror is shaped as if it were a section of a hollow sphere with an inner reflective surface. The mirror has the same geometric center (C) as a sphere of radius *r*. For a concave mirror, the distance *r* is also called the radius of curvature. The straight line that includes line segment CM is called the principal axis. The **principal axis** is the line perpendicular to the mirror's surface that divides the mirror in half. Incident light rays that are parallel to the principal axis are reflected and intersect the principal axis at a point halfway between C and M.

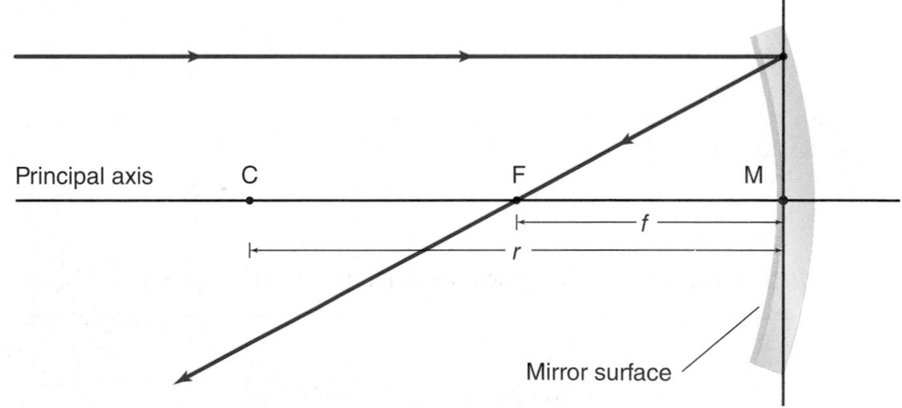

Figure 10 For a concave mirror, the distance from M to the focal point (F) is half the distance from M to C. Light rays that enter parallel to the principal axis converge at the focal point after reflecting off a concave mirror.

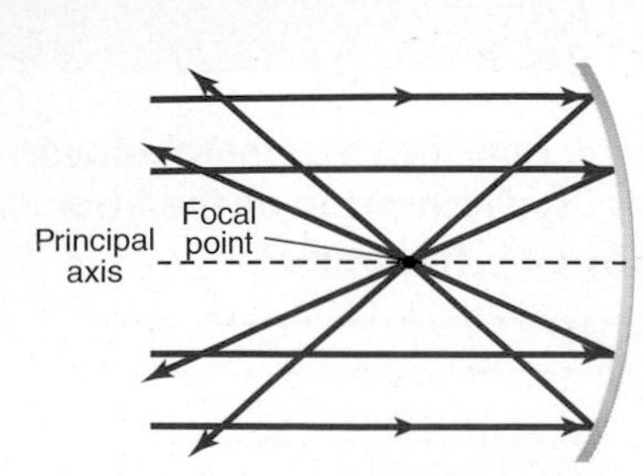

Light rays parallel to the principal axis reflect off a concave mirror and converge at the focal point.

A lightbulb is placed at the focal point of a concave mirror. The parallel reflected rays are evident in this flashlight beam.

Figure 11 If rays are traced from the focal point to a concave mirror, they reflect off the surface as parallel rays.

Focal point In most of the figures in this section, you can see a point marked F. What is point F? To understand the answer to this question, consider what happens when a concave mirror reflects light from the Sun. Because the Sun is so far away, light rays from the Sun to Earth are almost parallel. When you point the principal axis of a concave mirror toward the Sun, all the rays reflect through a very small point. You can locate this point by moving a sheet of paper toward and away from the mirror until the smallest and sharpest spot of sunlight is focused on the paper. This spot is the mirror's **focal point,** which is the point where incident light rays that are parallel to the principal axis converge after reflecting from the mirror. This is point F.

Figure 11 shows that a ray parallel to the principal axis is reflected and crosses the principal axis at point F, the focal point. Note that the distance from M to F is half the distance from M to C. The **focal length** (f) is the distance between the mirror and the focal point and can be expressed as $f = \frac{r}{2}$. The focal length is positive for a concave mirror because it is a point of converging rays.

If you place a light source at the focal point of a concave mirror, the rays will reflect off the mirror parallel to each other and produce a beam. You can see this effect with the flashlight shown in **Figure 11.**

Ray Diagrams for Concave Mirrors

You have already drawn rays to follow the paths of light rays that reflect from plane mirrors. You have also already seen rays shown for the concave mirrors illustrated in **Figure 10** and **Figure 11.** Now you will learn how these rays are shown for concave mirrors. You can use ray diagrams to find the locations, heights, and orientations of images that concave mirrors produce.

Figure 12 shows reflected rays converging at the point I where an image is located. The converging light rays form a **real image** that is inverted and larger than the object. You can see the image floating in space if you place your eye so that the rays that form the image fall on your eye, as in **Figure 12.** You must face a direction that allows the light rays coming from the image to enter your eyes, however. You cannot look at the image from behind. If you were to place a piece of paper or a movie screen at this point, the image would appear on the screen, as shown in **Figure 12.** You cannot project virtual images on a screen, since they are not formed by converging light rays.

Figure 12 Ray diagrams can be used to locate an image reflected from a curved mirror.

View an **animation of the graphical method of finding the image.**

Concepts In Motion

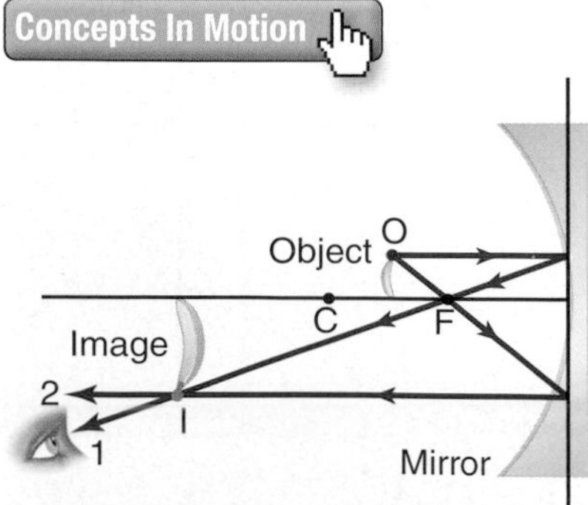

The eye is positioned so that the rays that form the real image strike the eye, allowing the image to be seen.

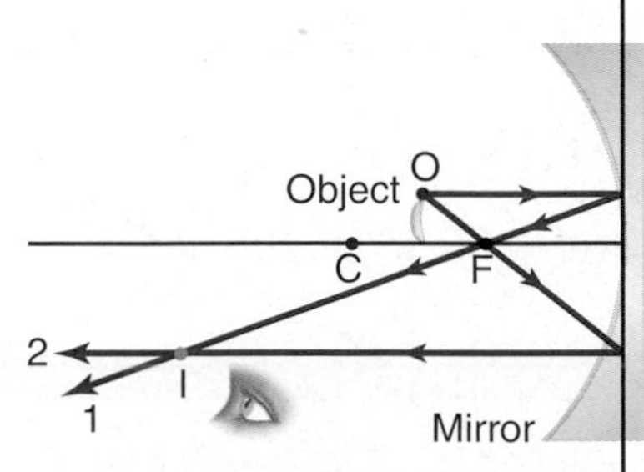

Rays from the object do not reach the eye, and so the image cannot be seen from this position.

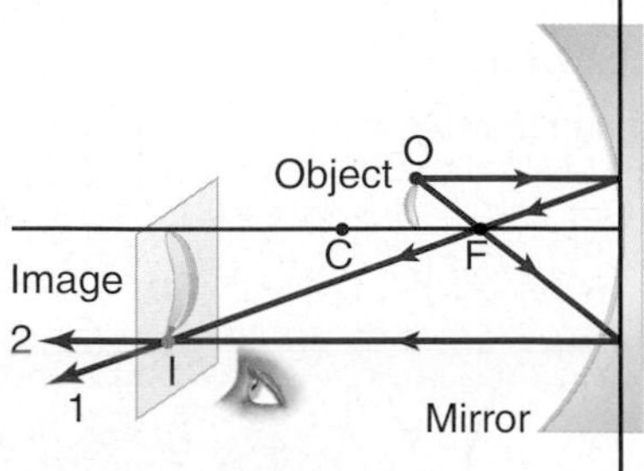

The image can be seen when projected on a white opaque screen.

Nicholas Eveleigh/Getty Images

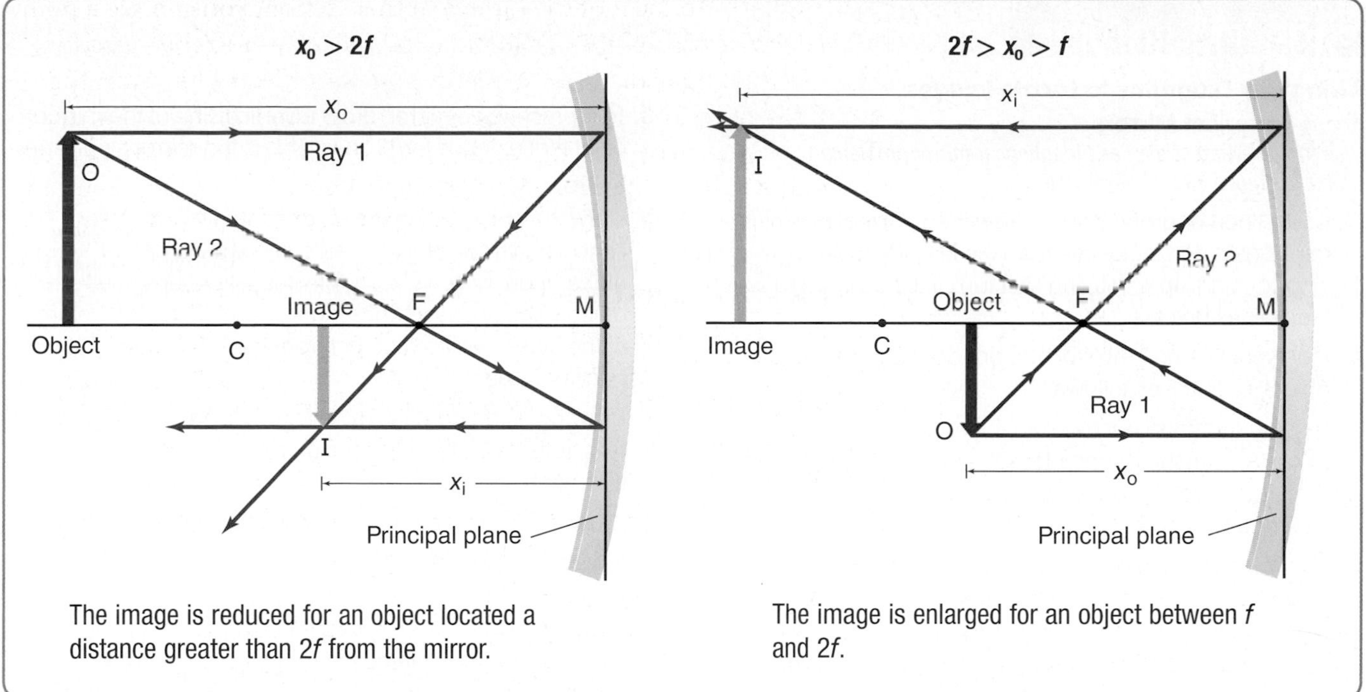

Figure 13 The type of image that results depends on the object's distance from the mirror. *A real, inverted image is formed in both of these situations. Remember that f, the focal length, is the distance from M to F.*

Diagram how other points on the image are formed.

COLOR CONVENTION

object	**purple**
image	**lilac**

Real Images with Concave Mirrors

You can often simplify ray diagrams by using simple, one-dimensional objects, such as the arrow shown in **Figure 13.** A spherical concave mirror produces an inverted real image if the object position (x_o) is greater than twice the focal length (f). The object is then beyond the center of curvature (C). If the object is placed between the center of curvature and the focal point, as shown on the right in **Figure 13,** the image is again real and inverted. The size of the image is now greater than the size of the object. Ray diagrams give you a visual representation of how a concave mirror can enlarge or reduce an image, depending on where the object is placed in relation to the focal point.

CONNECTION TO ASTRONOMY How can the inverted real image produced by a concave mirror be turned right-side up? In 1663, Scottish astronomer James Gregory developed the Gregorian telescope, shown in **Figure 14,** to resolve this problem. It is composed of large and small concave mirrors arranged such that the smaller mirror is outside of the focal point of the larger mirror. Parallel rays of light from distant objects strike the larger mirror and form an image at the focal point. This image is in turn the object for the smaller mirror. The rays then reflect off the smaller mirror and form a real image that is oriented as the object is.

Investigate **ray tracing for mirrors**

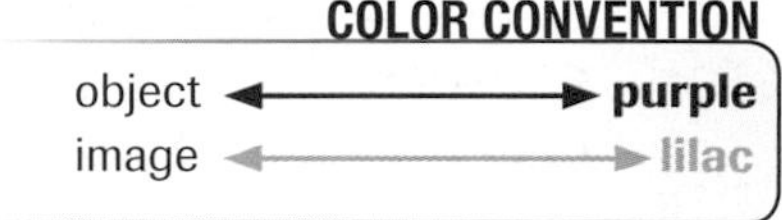

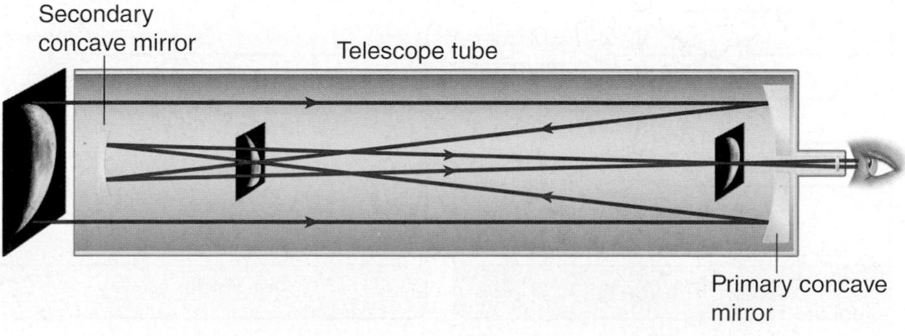

Figure 14 A Gregorian telescope forms a real, upright image for the viewer.

PROBLEM-SOLVING STRATEGIES

Using Ray Diagrams to Locate Images from Spherical Mirrors

*Use the following strategies for spherical-mirror problems. Refer to **Figure 13** and **Figure 16.***

1. Using lined or graph paper, draw the principal axis of the mirror as a horizontal line from the left side to the right side of your paper, leaving six blank lines above and six blank lines below.
2. Place points and labels on the principal axis for the object, C, and F as follows:
   a. If the mirror is concave and the object is beyond C away from the mirror, place the mirror at the right side of the page, the object at the left side of the page, and C and F to scale.
   b. If the mirror is a concave mirror and the object is between C and F, place the mirror at the right side of the page, C at the center of the paper, F halfway between the mirror and C, and the object to scale.
   c. For any other situation, place the mirror in the center of the page, the object or F (whichever is the greater distance from the mirror) at the left side of the page, and the other to scale.
3. To represent the mirror, draw a vertical line at the mirror point that extends for 12 lines. This is the mirror's principal plane.
4. Draw the object as an arrow, and label its top O. For concave mirrors, objects inside of C should not be higher than three lines. For all other situations, the objects should be six lines high. The scale for the height of the object will be different from the scale along the principal axis.
5. Draw ray 1, the parallel ray. It is parallel to the principal axis, reflects off the principal plane, and passes through F.
6. Draw ray 2, the focus ray. It passes through F, reflects off the principal plane, and is reflected parallel to the principal axis.
7. If reflected rays 1 and 2 diverge from each other, then extend sight lines behind the mirror as dotted lines.
8. The image is located where rays 1 and 2 or their sight lines cross after reflection. Label the point of intersection I. Draw the image as a perpendicular arrow from the principal axis to I.

Defects in concave mirrors In diagramming rays, you used a vertical line to represent the mirror. In reality, light rays reflect off the mirror itself, as shown in **Figure 15.** Only parallel rays that are close to the principal axis reflect through the focal point. Other rays converge at points closer to the mirror. This defect, called **spherical aberration,** occurs because the light rays do not converge at a focal point, which makes an image look fuzzy.

A mirror ground to the shape of a parabola, as in **Figure 15,** suffers no spherical aberration. Large, parabolic glass mirrors are expensive to manufacture; so, many telescopes use spherical mirrors and smaller, specially-designed secondary mirrors or lenses to correct for spherical aberration. We can also reduce spherical aberration by reducing the ratio of the mirror's diameter to its radius of curvature.

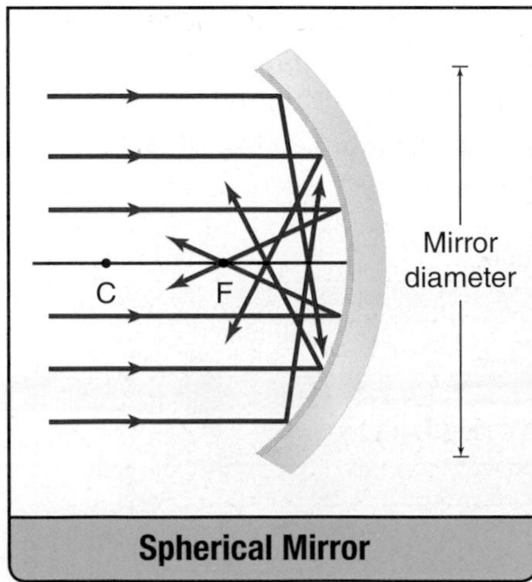

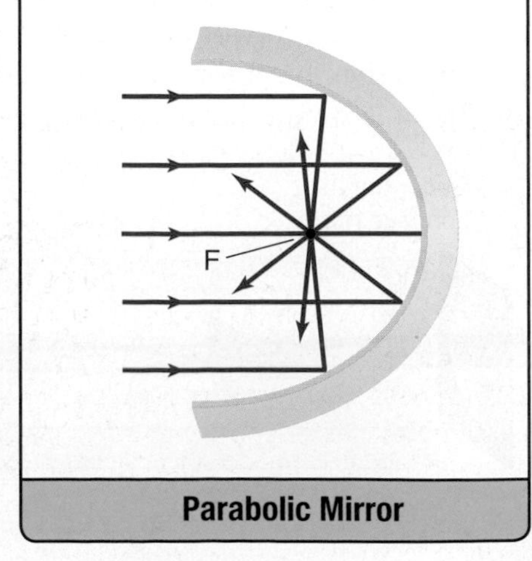

Figure 15 Spherical aberration occurs for spherical mirrors but does not occur for parabolic mirrors.

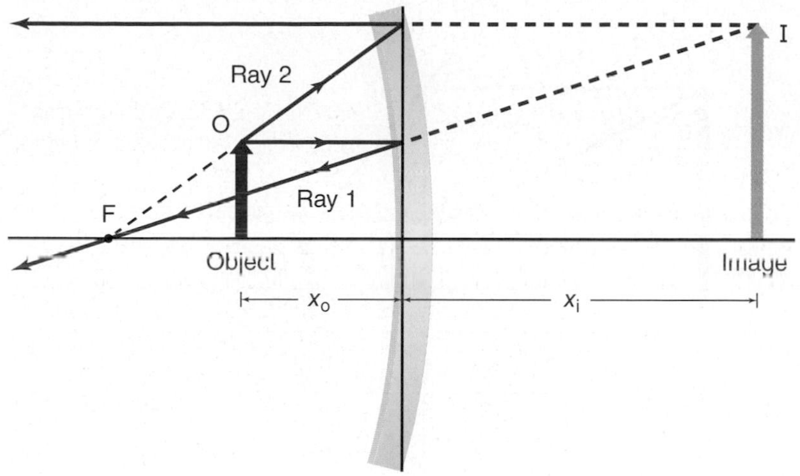

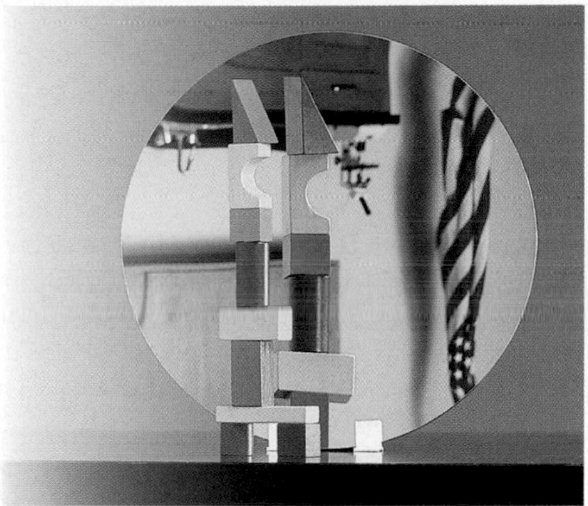

Figure 16 A virtual, upright, enlarged image is formed when an object, such as the block tower, is placed between the focal point and the surface of a concave mirror.

Consider the cause of the appearance of the other images in the mirror.

Virtual Images with Concave Mirrors

You have seen that as an object approaches the focal point (F) of a concave mirror, the image moves farther away from the mirror. If the object is at the focal point, all reflected rays are parallel. They never meet, therefore, and the image is said to be at infinity, so the image could never be seen. What happens if you move the object even closer to the mirror?

What do you see when you move your face close to a concave mirror? The image of your face is upright and behind the mirror. A concave mirror produces a virtual image if the object is located between the mirror and the focal point. This situation is shown in the ray diagram in **Figure 16.**

You draw two rays to locate the image of a point on an object. As before, you draw ray 1 parallel to the principal axis and then reflected through the focal point. Ray 2 is drawn as a line from the point on the object to the mirror, along a line defined by the focal point and the point on the object. At the mirror, ray 2 is reflected parallel to the principal axis. Note in **Figure 16** that ray 1 and ray 2 diverge as they leave the mirror, so there cannot be a real image. However, sight lines extended behind the mirror converge, showing that a virtual image forms behind the mirror. From the position of the point of convergence, you can see that the image is upright and larger than the object.

READING CHECK Explain how you can determine whether an image is virtual.

PhysicsLAB

CONCAVE MIRROR IMAGES

What are the conditions needed to produce real and virtual images with a concave mirror?

PHYSICS CHALLENGE

An object of height h_o is located a distance x_o from a concave mirror with focal length f.

1. Draw and label a ray diagram showing the focal length and location of the object for the case where the image is located twice as far from the mirror as the object. Supply distances in centimeters for the locations of the image, the object, and the focal point.
2. Draw and label a ray diagram showing the location of the object if the image is located twice as far from the mirror as the focal point. Supply distances in centimeters for the locations of the image, the object, and the focal point.
3. Determine where you should place the object so that no image is formed.

Figure 17 Convex mirrors produce images that are always virtual, upright, and smaller than the object.

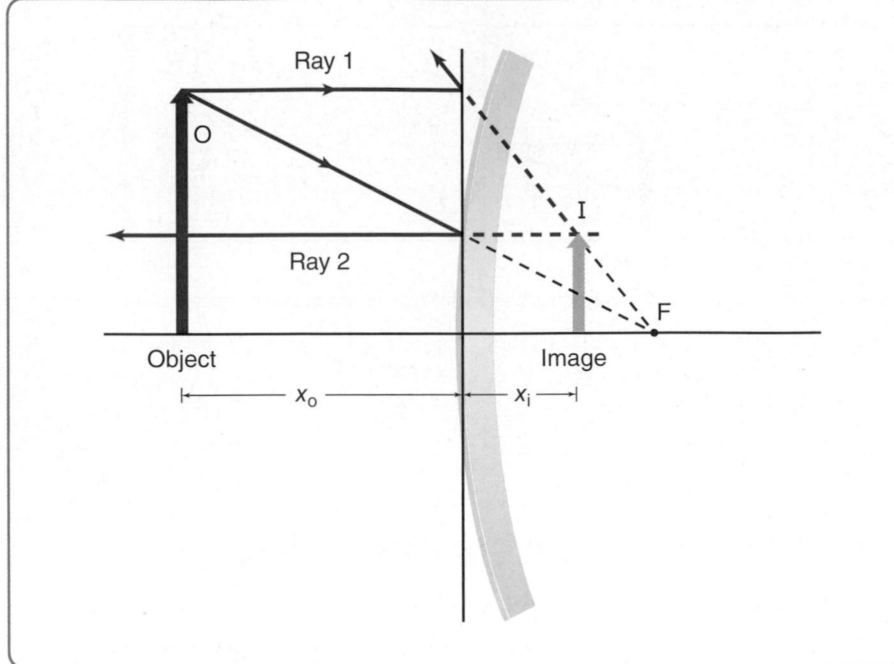

MiniLAB

FINDING THE FOCAL POINT

How can you observe the focal point of curved mirrors?

View a **simulation of a virtual optics lab.**

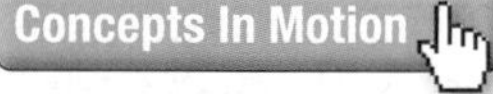

Convex Mirrors

In the first part of this chapter, you learned that the inner surface of a shiny spoon acts as a concave mirror. If you turn the spoon around, the outer surface acts as a **convex mirror,** an outwardly curving reflective surface with edges that curve away from the observer. Store security mirrors and many side-view mirrors on cars are convex mirrors. What do you see when you look at the back of a spoon? You see an upright, but smaller, image of yourself.

Properties of a spherical convex mirror are shown in **Figure 17.** Rays reflected from a convex mirror always diverge. Thus, convex mirrors form only virtual images. The focal point (F) and the geometric center of the mirror (C) are behind the mirror.

The ray diagram in **Figure 17** shows how a spherical convex mirror forms an image. The figure shows two rays, but remember that there are an infinite number of rays reflecting from the object. Ray 1 approaches the mirror parallel to the principal axis. Ray 1 reflects off the mirror following the line between F and the point where ray 1 strikes the principal plane. A sight line extends from the point where ray 1 strikes the principle plane through F. Ray 2 approaches the mirror on a path that, if extended behind the mirror, would pass through F. The reflected part of ray 2 and its sight line are parallel to the principal axis. The two reflected rays diverge. However, the sight lines intersect behind the mirror. The point where they intersect is the location of the image. An image produced by a convex mirror is a virtual image that is upright and smaller than the object.

Field of view It might seem that convex mirrors would have little use because the images they form are smaller than the objects. This property of convex mirrors, however, does have practical uses. By forming smaller images, convex mirrors enlarge the area that an observer sees, which is called the field of view. Also, the image is visible from a large range of angles; thus, the field of view is visible from a wide perspective.

Figure 18 Images from convex mirrors are smaller than the object. This increases the field of view and decreases the driver's blind spot.

Explain why a warning stating that objects in a convex side-view mirror are closer than they appear may be useful.

Because of the increased field of view, convex mirrors are used in cars as passenger-side rearview mirrors, as shown in **Figure 18.** The convex mirrors provide a larger field of view. But, because the reduced image size makes the objects appear farther away than they really are, there is often a warning on passenger-side rearview mirrors of cars stating that objects may be closer than they appear.

Calculating Image Position

We can use the spherical mirror model to develop a simple equation for spherical mirrors. You must use the paraxial ray approximation, which states that only rays that are close to and almost parallel with the principal axis are used to form an image. Using this, in combination with the law of reflection, leads to the mirror equation, relating the focal length (f), object position (x_o), and image position (x_i) of a spherical mirror.

MIRROR EQUATION

The reciprocal of the focal length of a spherical mirror is equal to the sum of the reciprocals of the image position and the object position.

$$\frac{1}{f} = \frac{1}{x_i} + \frac{1}{x_o}$$

Negative values When virtual images are formed, the image position (x_i) has a negative value, indicating that it is located behind the mirror. Recall that virtual images are formed where sight lines intersect. For concave mirrors, a virtual image only forms when the object is between a concave mirror and the focal point. The focal point is in front of the mirror and the focal length has a positive value. For convex mirrors, the focal point is always behind the mirror, and the focal length has a negative value.

When using the mirror equation to solve problems, it is important to remember that it is only approximately correct. The mirror equation does not predict spherical aberration because it uses the paraxial ray approximation. In reality, light coming from an object toward a mirror is diverging, so not all of the light is close to or parallel to the axis. When the mirror diameter is large relative to the radius of curvature to minimize spherical aberration, this equation predicts image properties more precisely.

REAL-WORLD PHYSICS

HUBBLE TROUBLE In 1990, NASA launched the *Hubble Space Telescope* into orbit around Earth. Scientists expected *Hubble* to provide clear images without atmospheric distortions. Soon after it was deployed, however, scientists found that *Hubble* had a spherical aberration. In 1993, astronauts installed corrective optics called COSTAR on *Hubble,* enabling *Hubble* to produce clear images. For over two decades, *Hubble* provided the world with stunning images and enabled scientists to make important discoveries about the universe.

Magnification Another property of a spherical mirror is **magnification** (m), which is the ratio of the image height to the object height. By using similar-triangle geometry, you can also rewrite this ratio in terms of image and object positions.

MAGNIFICATION

The magnification of an object by a spherical mirror, defined as the image height divided by the object height, is equal to the negative of the image position, divided by the object position.

$$m \equiv \frac{h_i}{h_o} = -\frac{x_i}{x_o}$$

The sign of the magnification tells you whether the image is upright or inverted. For virtual images, x_i is negative. This means m is positive. Note that virtual images are always upright, so the height is always positive.

For a real image, the position is positive. The negative sign on the fraction results in a negative value for the magnification, which means that the image is inverted compared to the object. When solving for height, the negative value also means that the image is inverted. If the object is beyond point C, the absolute value of the magnification for the real image is less than 1. This means that the image is smaller than the object. If the object is placed between point C and point F, the absolute value of the magnification is greater than 1, which means the image is larger than the object.

CONNECTING MATH TO PHYSICS

Adding and Subtracting Fractions When using the mirror equation, you first use math to move the fraction that contains the quantity you are seeking to the left-hand side of the equation and everything else to the right. Then you combine the two fractions on the right-hand side by using a common denominator that results from multiplying the denominators.

Math	Physics
$\frac{1}{x} = \frac{1}{y} + \frac{1}{z}$	$\frac{1}{f} = \frac{1}{x_i} + \frac{1}{x_o}$
$\frac{1}{y} = \frac{1}{x} - \frac{1}{z}$	$\frac{1}{x_i} = \frac{1}{f} - \frac{1}{x_o}$
$\frac{1}{y} = \left(\frac{1}{x}\right)\left(\frac{z}{z}\right) - \left(\frac{1}{z}\right)\left(\frac{x}{x}\right)$	$\frac{1}{x_i} = \left(\frac{1}{f}\right)\left(\frac{x_o}{x_o}\right) - \left(\frac{1}{x_o}\right)\left(\frac{f}{f}\right)$
$\frac{1}{y} = \frac{z - x}{xz}$	$\frac{1}{x_i} = \frac{x_o - f}{fx_o}$
$y = \frac{xz}{z - x}$	$x_i = \frac{fx_o}{x_o - f}$

Using this approach, the following relationships can be derived for image position, object position, and focal length:

$$x_i = \frac{fx_o}{x_o - f} \qquad x_o = \frac{fx_i}{x_i - f} \qquad f = \frac{x_i x_o}{x_i + x_o}$$

EXAMPLE PROBLEM 2

Get help with **image position.** Personal Tutor

REAL IMAGE FORMATION BY A CONCAVE MIRROR A concave mirror has a radius of curvature of 20.0 cm. You place a 2.0-cm-tall object 30.0 cm from the mirror. What are the image position and image height?

1 ANALYZE AND SKETCH THE PROBLEM

- Draw a diagram with the object and the mirror.
- Draw two principal rays to locate the image in the diagram

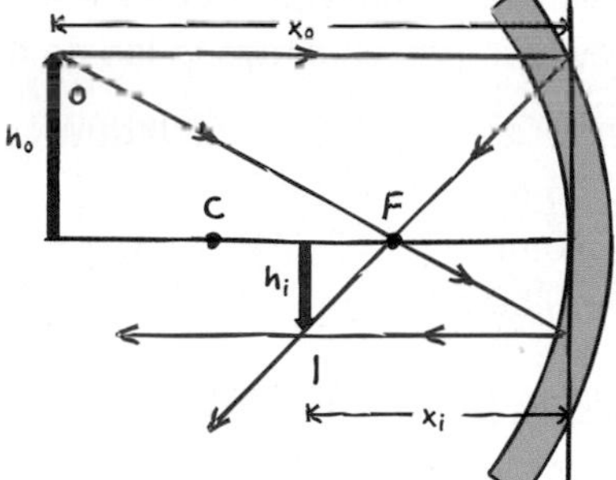

KNOWN	UNKNOWN
$h_o = 2.0$ cm	$x_i = ?$
$x_o = 30.0$ cm	$h_i = ?$
$r = 20.0$ cm	

2 SOLVE FOR THE UNKNOWN

Focal length is half the radius of curvature.

$$f = \frac{r}{2}$$

$$= \frac{20.0\text{ cm}}{2}$$ ◀ *Substitute $r = 20.0$ cm*

$$= 10.0\text{ cm}$$

Use the relationship between the focal length and object position to solve for image position.

$$\frac{1}{f} = \frac{1}{x_i} + \frac{1}{x_o}$$

$$x_i = \frac{fx_o}{x_o - f}$$

$$= \frac{(10.0\text{ cm})(30.0\text{ cm})}{30.0\text{ cm} - 10.0\text{ cm}}$$ ◀ *Substitute $f = 10.0$ cm, $x_o = 30.0$ cm*

$$= 15.0\text{ cm (real image, in front of the mirror)}$$

Use the relationship between object height and object and image position to solve for image height.

$$m \equiv \frac{h_i}{h_o} = \frac{-x_i}{x_o}$$

$$h_i = \frac{-x_i h_o}{x_o}$$

$$= -\frac{(15.0\text{ cm})(2.0\text{ cm})}{30.0\text{ cm}}$$ ◀ *Substitute $x_i = 15.0$ cm, $h_o = 2.0$ cm, $x_o = 30.0$ cm*

$$= -1.0\text{ cm (inverted, smaller image)}$$

3 EVALUATE THE ANSWER

- **Are the units correct?** All positions and heights are in centimeters.
- **Do the signs make sense?** Positive position and negative height agree with the drawing.

EXAMPLE PROBLEM

PRACTICE PROBLEMS

Do additional problems. Online Practice 

13. Use a ray diagram, drawn to scale, to solve Example Problem 2.

14. You place an object 36.0 cm in front of a concave mirror with a 16.0-cm focal length. Determine the image position.

15. You place a 3.0-cm-tall object 20.0 cm from a 16.0-cm-radius concave mirror. Determine the image position and image height.

16. A concave mirror has a 7.0-cm focal length. You place a 2.4-cm-tall object 16.0 cm from the mirror. Determine the image height.

17. CHALLENGE You place an object near a concave mirror with a 10.0-cm focal length. The image is 3.0 cm tall, inverted, and 16.0 cm from the mirror. What are the object position and object height?

EXAMPLE PROBLEM 3

Find help with **isolating a variable.** Math Handbook

IMAGE IN A SECURITY MIRROR A convex security mirror in a warehouse has a –0.50-m focal length. A 2.0-m-tall forklift is 5.0 m from the mirror. What are the image position and image height?

1 ANALYZE AND SKETCH THE PROBLEM

- Draw a diagram with the mirror and the object.
- Draw two principal rays to locate the image in the diagram.

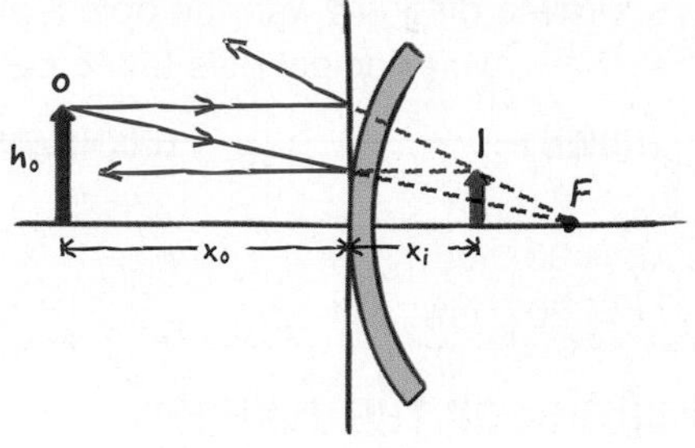

KNOWN	**UNKNOWN**
$h_o = 2.0$ m	$x_i = ?$
$x_o = 5.0$ m	$h_i = ?$
$f = -0.50$ m	

2 SOLVE FOR THE UNKNOWN

Use the relationship between focal length and object position to solve for image position.

$$\frac{1}{f} = \frac{1}{x_i} + \frac{1}{x_o}$$

$$x_i = \frac{fx_o}{x_o - f}$$

$$= \frac{(-0.50\text{ m})(5.0\text{ m})}{5.0\text{ m} - (-0.50\text{ m})}$$ ◀ *Substitute $f = -0.50$ m, $x_o = 5.0$ m*

$$= -0.45\text{ m (virtual image, behind the mirror)}$$

Use the relationship between object height and object and image position to solve for image height.

$$m \equiv \frac{h_i}{h_o} = \frac{-x_i}{x_o}$$

$$h_i = \frac{-x_i h_o}{x_o}$$

$$= \frac{-(-0.45\text{ m})(2.0\text{ m})}{5.0\text{ m}}$$ ◀ *Substitute $x_i = -0.45$ m, $h_o = 2.0$ m, $x_o = 5.0$ m*

$$= 0.18\text{ m (upright, smaller image)}$$

3 EVALUATE THE ANSWER

- **Are the units correct?** All positions and heights are in meters.
- **Do the signs make sense?** A negative position indicates a virtual image; a positive height indicates an image that is upright. These agree with the diagram.

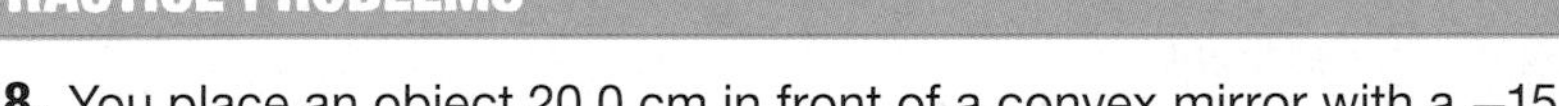

PRACTICE PROBLEMS

Do additional problems. Online Practice

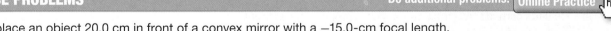

18. You place an object 20.0 cm in front of a convex mirror with a –15.0-cm focal length. Find the image position using both a scale diagram and the mirror equation.

19. A convex mirror has a focal length of –13.0 cm. You place a 6.0-cm diameter lightbulb 60.0 cm from that mirror. What is the lightbulb's image position and diameter?

20. A 7.6-cm-diameter ball is located 22.0 cm from a convex mirror with a radius of curvature of 60.0 cm. What are the ball's image position and diameter?

21. A 1.8-m-tall girl stands 2.4 m from a store's security mirror. Her image appears to be 0.36 m tall.

a. What is the image's distance?

b. What is the focal length of the mirror?

22. CHALLENGE A convex mirror is needed to produce an image that is three-fourths the size of an object and located 24 cm behind the mirror.

a. What is the object's distance?

b. What focal length should be specified?

Table 1 Single-Mirror System Properties

Mirror Type	f	x_o	x_i	m	Image
Plane	∞	$x_o > 0$	$\lvert x_i \rvert = x_o$ (negative)	same size	virtual
Concave	+	$x_o > r$	$r > x_i > f$	reduced, inverted	real
		$r > x_o > f$	$x_i > r$	enlarged, inverted	real
		$f > x_o > 0$	$\lvert x_i \rvert > x_o$ (negative)	enlarged	virtual
Convex	–	$x_o > 0$	$\lvert f \rvert > \lvert x_i \rvert > 0$ (negative)	reduced	virtual

Mirror Comparison

How do the various types of mirrors compare? **Table 1** summarizes the properties of single-mirror systems with objects that are located on the principal axis of the mirror. Virtual images are always behind the mirror, and the image position of a virtual image is always negative. When the absolute value of a magnification is between zero and one, the image is smaller than the object. A negative magnification means the image is inverted relative to the object.

Notice that the single plane mirror and convex mirror produce only virtual images. A concave mirror produces real images when the object is father than the focal distance. A concave mirror produces virtual images when the object is closer than the focal distance. Plane mirrors give reflections on scale with the objects, and convex mirrors provide reduced images, expanding the field of view. A concave mirror acts as a magnifier when an object is within the focal length of the mirror.

A concave mirror enlarges and inverts the image when the object is between the focal length and the radius of curvature. Beyond the radius of curvature, a concave mirror produces a reduced, inverted image.

SECTION 2 REVIEW

Section Self-Check Check your understanding.

23. **MAIN**IDEA If you know the focal length of a concave mirror, where should you place an object so that its image is upright and larger compared to the object? Will this produce a real or virtual image?

24. **Magnification** You place an object 20.0 cm in front of a concave mirror with a focal length of 9.0 cm. What is the magnification of the image?

25. **Object Position** The placement of an object in front of a concave mirror with a focal length of 12.0 cm forms a real image that is 22.3 cm from the mirror. What is the object position?

26. **Image Position and Height** You place a 3.0-cm-tall object 22.0 cm in front of a concave mirror that has a focal length of 12.0 cm. Find the image position and height by drawing a ray diagram to scale. Verify your answer using the mirror and magnification equations.

27. **Ray Diagram** You place a 4.0-cm-tall object 14.0 cm from a convex mirror with a focal length of –12.0 cm. Draw a scale ray diagram showing the image position and height. Verify your answer using the mirror and magnification equations.

28. **Radius of Curvature** You place a 6.0-cm-tall object 16.4 cm from a convex mirror. If the image of the object is 2.8 cm tall, what is the mirror's radius of curvature?

29. **Focal Length** A convex mirror is used to produce an image that is two-thirds the size of an object and located 12 cm behind the mirror. What is the focal length of the mirror?

30. **Critical Thinking** Would spherical aberration be less for a mirror whose height, compared to its radius of curvature, is small or large? Explain.

DISTANT EARTHS

The Hunt for Exoplanets

Scientists have long searched for solar systems beyond ours. But, imagine staring into stadium lights and trying to make out the shape of a tiny mosquito. Astronomers have similar trouble searching for planets outside our solar system—objects called exoplanets.

Hard to find
One reason it is so tricky to find an exoplanet is that they are so very far away. The distance to the nearest star besides the Sun is about 4.3 light years. And, if an exoplanet is like Earth, it is only about a hundredth the diameter of the star it orbits. Detecting the relatively dim light reflected directly from an exoplanet is too difficult for even powerful instruments.

Detecting Exoplanets One way astronomers can find exoplanets is by monitoring the brightness of the light coming from stars. A star's light can appear slightly dimmer when an exoplanet passes in front of it. A light meter called a photometer measures changes in the brightness of light. Such changes can be plotted in a graph, such as the one shown in **Figure 1.**

Kepler Mission NASA's Kepler Mission uses a light meter to scan an area of our galaxy in search of exoplanets. The primary mirror of the light meter is spherical and requires correction for spherical aberration.

Habitable Zone Kepler looks specifically for planets that are within the habitable zones of stars. **Figure 2** shows the habitable zone of our solar system. An exoplanet that has liquid water on it is more likely to exist in this "sweet spot" around a star. Perhaps such a planet, like Earth, could be a home to living things.

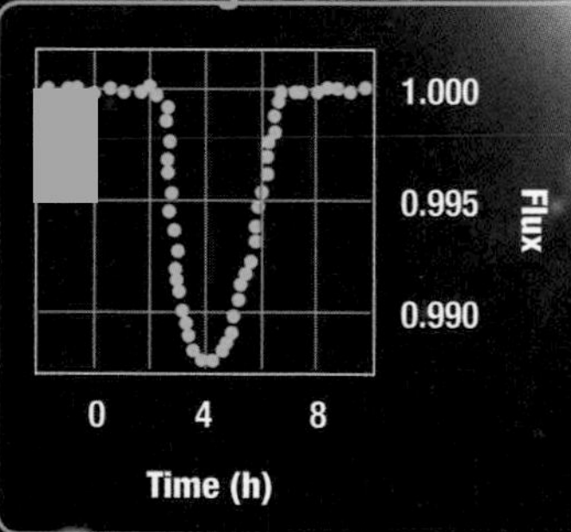

FIGURE 1 This plot represents the brightness of light coming from a star. The curve in the plotted line indicates a dimming of starlight, which could mean a planet has passed in front of the star.

FIGURE 2 In the habitable zone, conditions are right for the existence of liquid water.

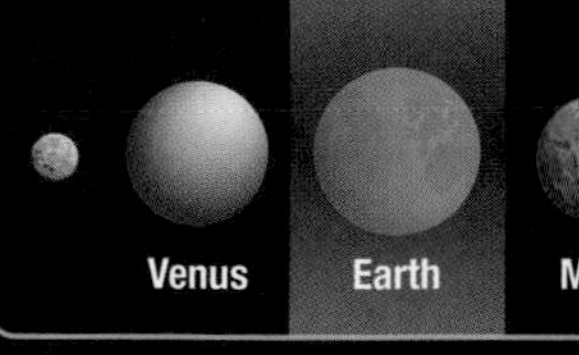

GOING**FURTHER** >>>

Research Visit the NASA Kepler mission Web site to research a recently discovered exoplanet. Summarize the characteristics of that planet, and summarize what characteristics an organism on that planet might need to survive.

CHAPTER 17 STUDY GUIDE

BIG IDEA All surfaces reflect light, but smooth surfaces can produce images.

VOCABULARY
- **specular reflection** *(p. 466)*
- **diffuse reflection** *(p. 466)*
- **plane mirror** *(p. 468)*
- **object** *(p. 468)*
- **image** *(p. 468)*
- **virtual image** *(p. 468)*

SECTION 1 Plane Mirrors

MAIN IDEA The angle of incidence of a light ray is equal to the angle of reflection.

- According to the law of reflection, the angle that an incident ray makes with the normal equals the angle that the reflected ray makes with the normal.

$$\theta_r = \theta_i$$

- The law of reflection applies to both smooth and rough surfaces. A rough surface has normals that are not parallel; therefore, parallel incident rays are not reflected in parallel. A rough surface produces diffuse reflection. A smooth surface has parallel normals; therefore, parallel incident rays are reflected in parallel. A smooth surface produces specular reflection. Specular reflection results in the formation of images that appear to be behind plane mirrors.
- An image produced by a plane mirror is always virtual, is the same size as the object, has the same orientation, and is the same distance from the mirror as the object.

$$x_i = -x_o \qquad h_i = h_o$$

VOCABULARY
- **concave mirror** *(p. 471)*
- **principal axis** *(p. 471)*
- **focal point** *(p. 472)*
- **focal length** *(p. 472)*
- **real image** *(p. 472)*
- **spherical aberration** *(p. 474)*
- **convex mirror** *(p. 476)*
- **magnification** *(p. 478)*

SECTION 2 Curved Mirrors

MAIN IDEA Curved mirrors can produce real and virtual images and can magnify or reduce the image size.

- A spherical concave mirror is shaped as if it were a section of a hollow sphere with the same geometric center (C) and radius of curvature (r) as a sphere of radius r. The focal point (F) of a spherical concave mirror is the point where rays parallel to the principal axis of the mirror converge after reflection. Concave mirrors are used in flashlights, spotlights, and telescopes.
- You can locate the image created by a spherical mirror by drawing two rays from a point on the object to the mirror. The intersection of the two reflected rays or sight lines of the two reflected rays is the location of the image of the object point. A concave mirror produces a real image that is inverted when the object position is greater than the focal length. A concave mirror produces a virtual image that is upright when the object position is less than the focal length. A convex mirror always produces a virtual image that is upright and smaller compared to the object.
- By forming images smaller than the objects, convex mirrors make images seem farther away and produce a wide field of view, which is useful for rearview mirrors and security mirrors. Mirrors can be used in combinations to produce images of any size, orientation, and location desired. The most common use of combinations of mirrors is in telescopes.
- The mirror equation gives the relationship among image position, object position, and focal length of a spherical mirror.

$$\frac{1}{f} = \frac{1}{x_i} + \frac{1}{x_o}$$

The magnification of a mirror image is given by equations relating either the positions or the heights of the image and the object.

$$m \equiv \frac{h_i}{h_o} = \frac{-x_i}{x_o}$$

Games and Multilingual eGlossary

Vocabulary Practice

SECTION 1 Plane Mirrors

Mastering Concepts

31. How does specular reflection differ from diffuse reflection?

32. What is meant by the term *normal to the surface?*

33. Where is the image produced by a plane mirror located?

34. Describe the properties of a plane mirror.

35. A student believes that very sensitive photographic film can detect a virtual image. The student puts photographic film at the location of a virtual image. Does this attempt succeed? Explain.

36. How can you prove to someone that an image is a real image?

37. Is your image as seen in a plane mirror two dimensional (like a photograph) or three dimensional? How can you tell?

Mastering Problems

38. A light ray incident upon a mirror makes an angle of 36° with the mirror. What is the angle between the incident ray and the reflected ray?

39. A light ray strikes a mirror at an angle of 38° to the normal. What is the angle that the reflected angle makes with the normal?

40. Two adjacent plane mirrors form a right angle, as shown in **Figure 19.** A light ray is incident upon one of the mirrors at an angle of 30° to the normal.

a. What is the angle at which the light ray is reflected from the other mirror?

b. The mirrors in **Figure 19** are a retroreflector. A retroreflector is a device that reflects incoming light rays back in a direction opposite to that of the incident rays. Copy the diagram below, and draw the reflected light rays to show that this mirror system acts as a retroreflector.

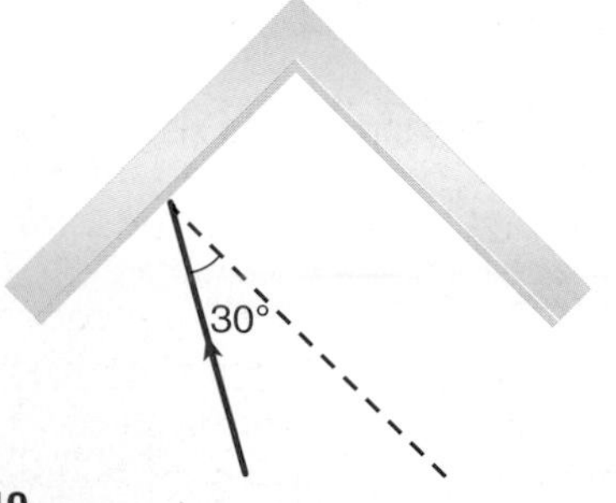

Figure 19

41. A light ray strikes a mirror at an angle of 53° to the normal. What is the angle of reflection? What is the angle between the incident ray and the reflected ray?

42. Draw a ray diagram of a plane mirror to show that if you want to see yourself from your feet to the top of your head, the mirror must be at least half your height.

43. You have a small plane mirror that will be mounted on a wall. If you want to see an image of your knee, where should you place the mirror?

44. Picture in a Mirror Penny wishes to take a picture of her image in a plane mirror, as shown in **Figure 20.** If the camera is 1.2 m in front of the mirror, at what distance should the camera lens be focused?

Figure 20

45. A light ray strikes a mirror at an angle of 60° to the normal. The mirror is then rotated 18° clockwise, as shown in **Figure 21.** What is the angle that the reflected ray makes with the mirror?

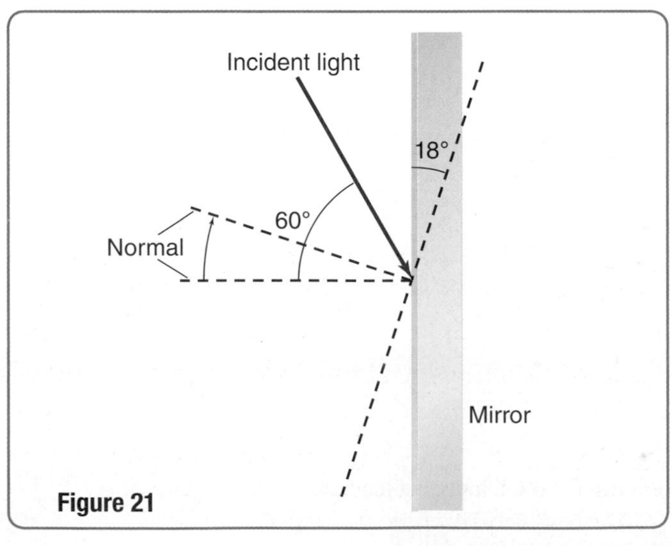

Figure 21

Hutchings Photography

Chapter Self-Check

46. Two plane mirrors are connected at their sides so that they form a 45° angle between them. A light ray strikes one mirror at an angle of 30° to the normal and then reflects off the second mirror. Calculate the angle of reflection of the light ray off the second mirror.

SECTION 2 Curved Mirrors

Mastering Concepts

47. An object produces a virtual image in a concave mirror. Where is the object located?

48. What is the defect that all concave spherical mirrors have, and what causes it?

49. What is the equation relating the focal point, object position, and image position?

50. What is the relationship between the radius of curvature and the focal length of a concave mirror?

51. If you know the image position and object position relative to a curved mirror, how can you determine the mirror's magnification?

52. Why are convex mirrors used as rearview mirrors?

53. Why is it impossible for a convex mirror to form a real image?

Mastering Problems

54. Fun House A boy is standing near a convex mirror in a fun house at a fair. He notices that his image appears to be 0.60 m tall. If the magnification of the mirror is $\frac{1}{3}$, what is the boy's height?

55. You place an object 30.0 cm from a concave mirror of 15.0-cm focal length. The object is 1.8 cm tall. Use the mirror equation to find the image position. What is the image height?

56. A concave mirror has a focal length of 10.0 cm. What is its radius of curvature?

57. Rearview Mirror How far does the image of a car appear behind a convex mirror, with a focal length of 26.0 m, when the car is 10.0 m from the mirror?

58. Star Image Light from a star is collected by a concave mirror. How far from the mirror is the image of the star if the radius of curvature is 150 cm?

59. An object located 18 cm from a convex mirror produces a virtual image 9 cm from the mirror. What is the magnification of the image?

60. Describe the image produced by the object in **Figure 22** as real or virtual, inverted or upright, and smaller or larger than the object.

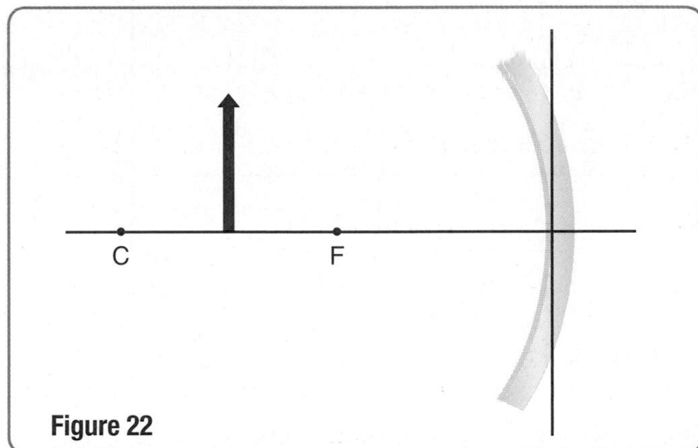

Figure 22

61. Reverse Problem Write a physics problem with real-life objects for which the following equation would be part of the solution: $\frac{1}{0.40\text{ m}} = \frac{1}{0.75\text{ m}} + \frac{1}{x_o}$.

62. Ranking Task Several object and image heights for various mirrors are listed. Rank them according to magnification, from greatest to least. Specifically indicate any ties.

A. object 1.0 cm, image 0.5 cm

B. object 2.0 cm, image 0.5 cm

C. object 2.0 cm, image 1.0 cm

D. object 5.0 cm, image 7.0 cm

E. object 3.0 cm, image 2.0 cm

63. Dental Mirror A dentist uses a small mirror with a radius of curvature of 40 mm to locate a cavity in a patient's tooth. If the mirror is concave and is held 16 mm from the tooth, what is the magnification of the image?

64. An object that is 24 mm tall is placed 12.0 cm from a concave mirror. Sunlight falls on the concave mirror and forms an image that is 3.0 cm from the mirror. Sketch the ray diagram to show the location of the image. Use the mirror equation to calculate the image position. How tall is the image?

65. You place a 3.0-cm-tall object 22.4 cm from a concave mirror. If the mirror has a radius of curvature of 34.0 cm, what are the image position and height?

66. Shiny spheres that are placed on pedestals on a lawn are convex mirrors. One such sphere has a diameter of 40.0 cm. A 12-cm-tall robin sits in a tree that is 1.5 m from the sphere. Where is the image of the robin, and how tall is the image?

Chapter Self-Check

67. Find the image position and height for the object shown in **Figure 23.**

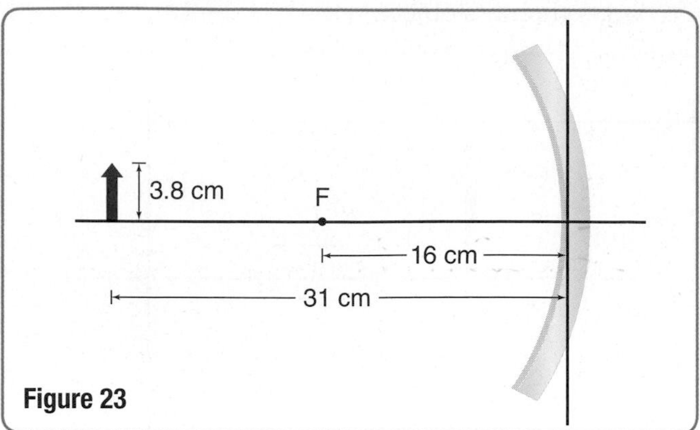

Figure 23

68. Jeweler's Mirror A jeweler inspects a watch with a diameter of 3.0 cm by placing it 8.0 cm in front of a concave mirror of 12.0-cm focal length. Where is the watch's image? What is the image's diameter?

Applying Concepts

69. BIGIDEA A dry road is more of a diffuse reflector than a wet road. Based on **Figure 24,** explain why a wet road appears blacker than a dry road does.

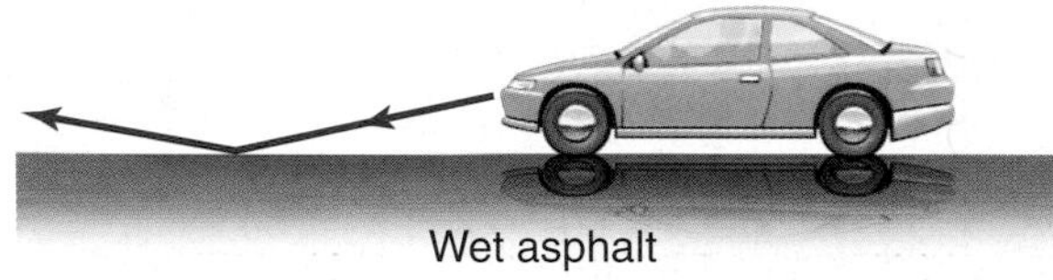

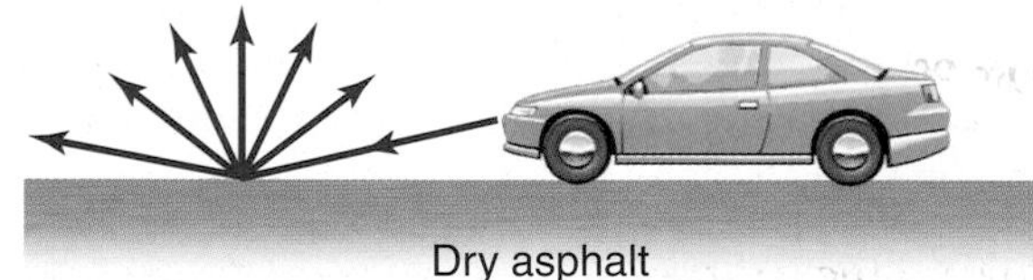

Figure 24

70. Book Pages Why is it desirable that the pages of a book be rough rather than smooth and glossy?

71. Locate and describe the physical properties of the image produced by a concave mirror when the object is located at the center of curvature.

72. An object is located beyond the center of curvature of a spherical concave mirror. Locate and describe the physical properties of the image.

73. Telescope You have to order a large concave mirror for a high-quality telescope. Should you order a spherical mirror or a parabolic mirror? Explain.

74. List all possible arrangements in which you could use a spherical mirror, either concave or convex, to form a real image.

75. Rearview Mirrors The outside rearview mirrors of cars often carry the warning "Objects in the mirror are closer than they appear." What kind of mirrors are these, and what advantage do they have?

76. Describe the properties of the image seen in the single convex mirror in **Figure 25.**

Figure 25

Mixed Review

77. Problem Posing Complete this problem so that the resulting image will be virtual: "An object is placed 32 cm from a mirror ...."

78. An object is located 4.4 cm in front of a concave mirror with a 24.0-cm radius. Locate the image using the mirror equation.

79. A light ray strikes a plane mirror at an angle of 28° to the normal. If the light source is moved so that the angle of incidence increases by 34°, what is the new angle of reflection?

80. A concave mirror has a radius of curvature of 26.0 cm. An object that is 2.4 cm tall is placed 30.0 cm from the mirror. Where is the image position? What is the image height?

81. A convex mirror is needed to produce an image one-half the size of an object and located 36 cm behind the mirror. What focal length should the mirror have?

82. What is the radius of curvature of a concave mirror that magnifies an object by a factor of 13.2 when the object is placed 20.0 cm from the mirror?

Chapter Self-Check

83. A ball is positioned 22 cm in front of a spherical mirror and forms a virtual image. If the spherical mirror is replaced with a plane mirror, the image appears 12 cm closer to the mirror. What kind of spherical mirror was used?

84. Magic Trick A magician uses a concave mirror with a focal length of 8.0 m to make a 3.0-m-tall hidden object, located 18.0 m from the mirror, appear as a real image that is seen by his audience. Draw a scale ray diagram to find the height and location of the image.

85. Copy **Figure 26** on a sheet of paper. Draw rays on the diagram to determine the height and location of the image.

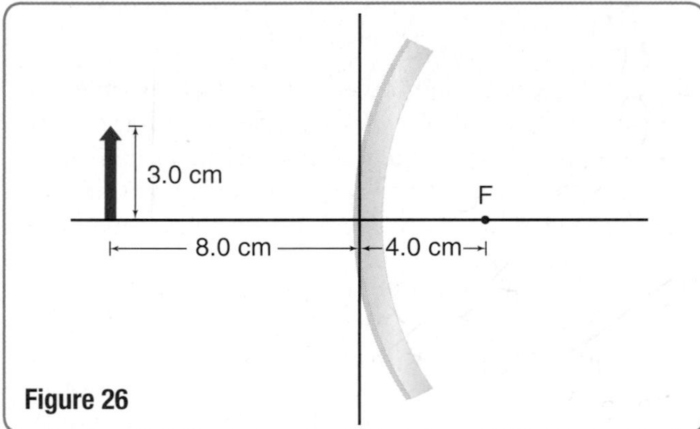

Figure 26

86. A 4.0-cm-tall object is placed 12.0 cm from a convex mirror. If the image of the object is 2.0 cm tall and the image is located at 26.0 cm, what is the focal length of the mirror? Draw a ray diagram to answer the question. Use the mirror equation and the magnification equation to verify your answer.

87. Surveillance Mirror A convenience store uses a surveillance mirror to monitor the store's aisles. Each mirror has a radius of curvature of 3.8 m.

a. What is the image position of a customer who stands 6.5 m in front of the mirror?

b. What is the image height of a customer who is 1.7 m tall?

88. Inspection Mirror A production-line inspector wants a mirror that produces an image that is upright with a magnification of 7.5 when it is located 14.0 mm from a machine part.

a. What kind of mirror would do this job?

b. What is its radius of curvature?

89. A 1.6-m-tall girl stands 3.2 m from a convex mirror. What is the focal length of the mirror if her image appears to be 0.28 m tall?

90. The object in **Figure 27** moves from position 1 to position 2. Copy the diagram onto a sheet of paper. Draw rays showing how the image changes.

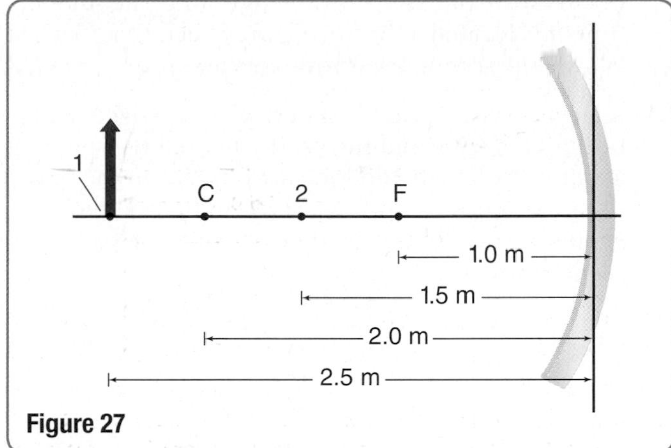

Figure 27

Thinking Critically

91. Analyze and Conclude The object in **Figure 28** is located 22 cm from a concave mirror. What is the focal length of the mirror?

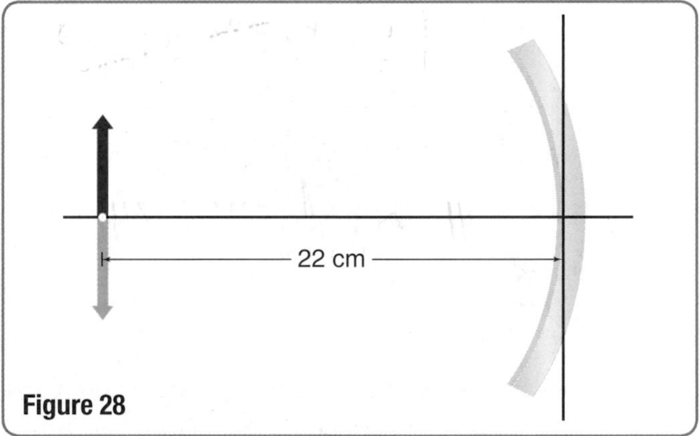

Figure 28

92. Apply Concepts The ball in **Figure 29** slowly rolls toward the concave mirror on the right. Describe how the size of the ball's image changes as it rolls along.

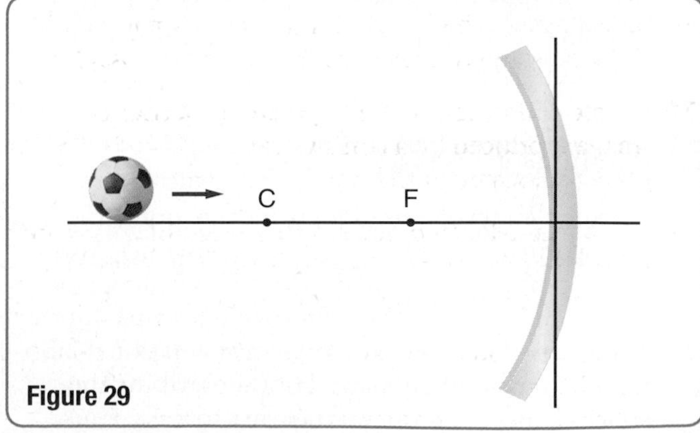

Figure 29

Chapter Self-Check

93. **Analyze and Conclude** The layout of the two-mirror system shown in **Figure 14** is that of a Gregorian telescope. The larger mirror is concave and has a radius of curvature of r and a focal length of f. The smaller mirror is located a distance x away such that $f < x < r$. Why is the secondary mirror concave?

94. A large serving spoon is spherical and reflective on both the outside and inside. If you hold the spoon at a distance of 3 cm and look at your face in the part that holds food, your image is 9 cm behind the spoon. If you flip the spoon over, where is your image located?

95. The image of the primary mirror in a Cassegrain telescope, shown in **Figure 30,** is the object of the secondary mirror. The focal length of the concave primary mirror is 1.0 m and that of the convex secondary mirror −0.50 m. The secondary mirror is located 0.25 m from the focal point of the primary.
    a. What is the object position? The image position?
    b. What is the magnification of the convex mirror? Does it invert the image?

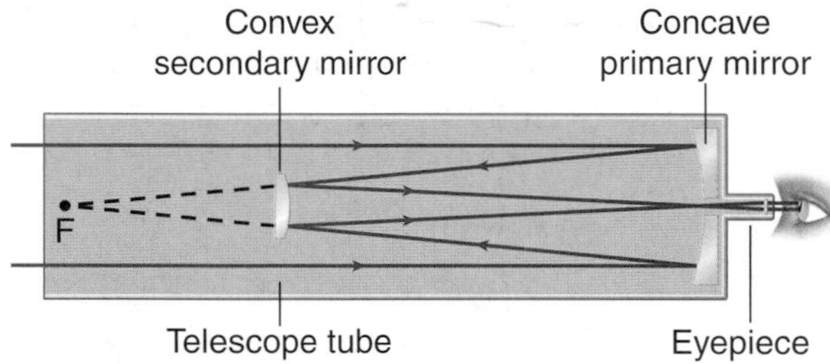

Figure 30

96. **Analyze and Conclude** An optical arrangement used in some telescopes is the Cassegrain system, shown in **Figure 30.** This telescope uses a convex secondary mirror that is positioned between the primary mirror and the focal point of the primary mirror. A single convex mirror produces only virtual images. Explain how the convex mirror in this telescope produces a real image that could be projected onto a CCD detector.

Writing in Physics

97. Research a method used for grinding, polishing, and testing mirrors used in reflecting telescopes. You may report either on methods used by amateur astronomers who make their own telescope optics or on a method used by a project at a national laboratory. Prepare a one-page report describing the method, and present it to the class.

98. Mirrors reflect light because of their metallic coating. Research and write a summary of one of the following:
    a. the different types of coatings used and the advantages and disadvantages of each
    b. the precision optical polishing of aluminum to such a degree of smoothness that no glass is needed in the process of making a mirror

Cumulative Review

99. A child runs down the school hallway and then slides on the newly waxed floor. He was running at 4.7 m/s before he started sliding, and he slid 6.2 m before stopping. What was the coefficient of friction of the waxed floor?

100. A 1.0-g potato bug is walking around the outer rim of an upside down flying disk that has a diameter of 17.2 cm. If the potato bug moves at a rate of 0.63 cm/s, what is the centripetal force acting on the bug? What is the agent that provides this force?

101. A 1.0-g piece of copper falls from a height of 1.0×10^3 m from a hovering helicopter to the ground. Because of air resistance the piece of copper reaches the ground moving at a velocity of 70.0 m/s. Assuming that half of the decrease in the copper's mechanical energy was transformed into thermal energy in the metal, what was its increase in temperature during the fall?

102. It is possible to lift a person who is sitting on a pillow made from a large sealed plastic garbage bag by blowing air into the bag through a soda straw. Suppose that the cross-sectional area of the person sitting on the bag is 0.25 m^2 and the person's weight is 600 N. The soda straw has a cross-sectional area of 2×10^{-5} m^2. With what pressure must you blow into the straw to lift the person that is sitting on the sealed garbage bag?

103. What would be the period of a 2.0-m-long pendulum swinging on the Moon's surface? The Moon's mass is 7.34×10^{22} kg, and its radius is 1.74×10^6 m. What is the period of this pendulum on Earth?

104. **Organ pipes** An organ builder must design a pipe organ that will fit into a small space.
    a. Should he design the instrument to have open pipes or closed pipes? Explain.
    b. Will an organ constructed with open pipes sound the same as one constructed with closed pipes? Explain.

MULTIPLE CHOICE

1. Where is the object located if the image that is produced by a concave mirror is smaller than the object?
   - A. at the mirror's focal point
   - B. between the mirror and the focal point
   - C. between the focal point and center of curvature
   - D. past the center of curvature

2. A concave mirror with a focal length of 16.0 cm produces an image located 38.6 cm from the mirror. What is the distance of the object from the front of the mirror?
   - A. 2.4 cm
   - B. 11.3 cm
   - C. 22.6 cm
   - D. 27.3 cm

3. A convex mirror is used to produce an image that is three-fourths the size of an object and located 8.4 cm behind the mirror. What is the focal length of the mirror?
   - A. −34 cm
   - B. −11 cm
   - C. −6.3 cm
   - D. −4.8 cm

4. A concave mirror produces an inverted image that is 8.5 cm tall, located 34.5 cm in front of the mirror. If the focal point of the mirror is 24.0 cm, then what is the height of the object?
   - A. 2.3 cm
   - B. 3.5 cm
   - C. 14 cm
   - D. 19 cm

5. A cup sits 17 cm from a concave mirror. The image of the cup appears 34 cm in front of the mirror. What are the magnification and orientation of the cup's image?
   - A. 0.5, inverted
   - B. 0.5, upright
   - C. 2.0, inverted
   - D. 2.0, upright

6. What is the focal length of a concave mirror that magnifies, by a factor of +3.2, an object that is placed 30 cm from the mirror?
   - A. 23 cm
   - B. 32 cm
   - C. 44 cm
   - D. 46 cm

7. An object is placed 21 cm in front of a concave mirror with a focal length of 14 cm. What is the image position?
   - A. 242 cm
   - B. 28.4 cm
   - C. 8.4 cm
   - D. 42 cm

8. A light ray strikes a plane mirror at an angle of 23° to the normal. What is the angle between the reflected ray and the mirror?
   - A. 23°
   - B. 46°
   - C. 67°
   - D. 134°

9. The light rays in the illustration below do not properly focus at the focal point. This problem occurs with
   - A. all spherical mirrors.
   - B. all parabolic mirrors.
   - C. only defective spherical mirrors.
   - D. only defective parabolic mirrors.

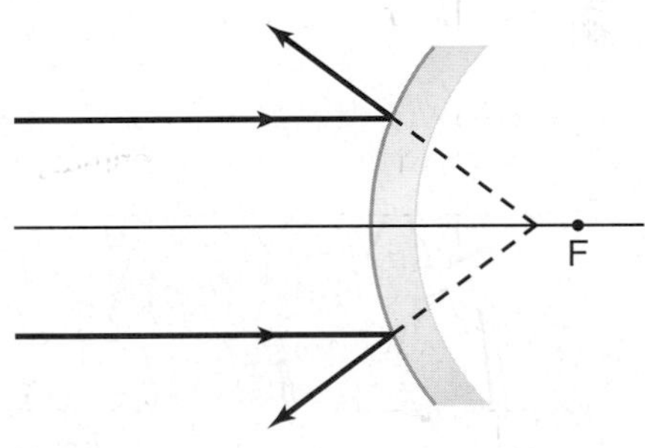

FREE RESPONSE

10. A 5.0-cm-tall object is located 20.0 cm from a convex mirror with a focal length of −14.0 cm. Draw a scale-ray diagram showing the image height.

Online Test Practice

NEED EXTRA HELP?

If you Missed Question	1	2	3	4	5	6	7	8	9	10
Review Section	2	2	2	2	2	2	2	1	2	2

CHAPTER 18

Refraction and Lenses

BIGIDEA Lenses refract light and create images.

SECTIONS

1 Refraction of Light

2 Convex and Concave Lenses

3 Applications of Lenses

LaunchLAB

iLab Station

BROKEN STRAW APPEARANCE

How could you explain the broken appearance of a drinking straw placed in a glass of water?

WATCH THIS!

Video

LIGHTHOUSE LENSES

How does a lighthouse produce a beam of bright, focused light? How are lighthouses similar to spotlights in a theater? It's all about the lens.

PHYSICS T.V.

Go online!
connectED.mcgraw-hill.com

SECTION 1 Refraction of Light

PHYSICS 4 YOU

Have you ever noticed a rainbow when the sky clears after it rains? Rainbows are a colorful example of the interaction of light with a medium. What conditions are required for a rainbow to be visible? Can you ever get close enough to a rainbow to touch it?

MAIN IDEA

The amount of refraction at a boundary depends on the indices of refraction of the two mediums and the angle of incidence.

Essential Questions

- What is Snell's law of refraction?
- What is the meaning of the index of refraction?
- How does total internal reflection occur?
- How does refraction cause various optical effects?

Review Vocabulary

refraction the change in direction of waves at the boundary between two different mediums

New Vocabulary

index of refraction
critical angle
total internal reflection
dispersion

Light and Boundaries

What happens to light as it travels through a window? When light encounters a transparent or translucent medium, some light is reflected from the surface of the medium and some is transmitted through the medium. Recall that when light crosses a boundary between two mediums, it bends.

You might be familiar with this phenomenon already. Have you ever looked down through the water of a swimming pool and thought the water looked shallower than it was? Have you ever looked at the side of a glass of clear soda with a straw in it and the straw appeared broken at the surface? Both of these strange appearances are due to the bending of light at the boundaries between two mediums. This bending of light when it passes from one medium to another is known as refraction.

Refraction Observe the light rays in **Figure 1.** Identical rays of light start in air and pass into three different mediums: water, glass, and diamond. The light hits the surface of each medium at the same angle. What do you notice about the light rays after they cross the boundaries between the mediums? The light rays bend as they cross the boundaries.

What difference do you notice between the three mediums shown? The light rays bend more when traveling from air to diamond than from air to water or air to glass. This phenomenon depends on properties of the mediums that the light rays are traveling from and into.

The rays of light in **Figure 1** all travel from air and enter another medium at the same angle. What do you think the relationship is between the angle of the light as it crosses the boundary between mediums and refraction?

Figure 1 Light refracts as it crosses a boundary. The amount of refraction depends on the properties of the mediums. (Angles are not to scale.)

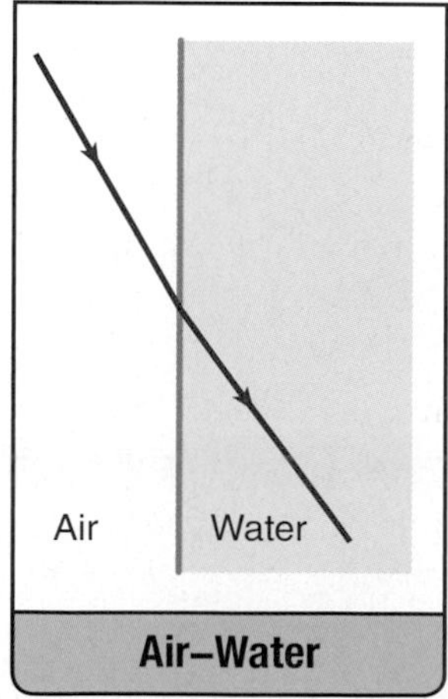

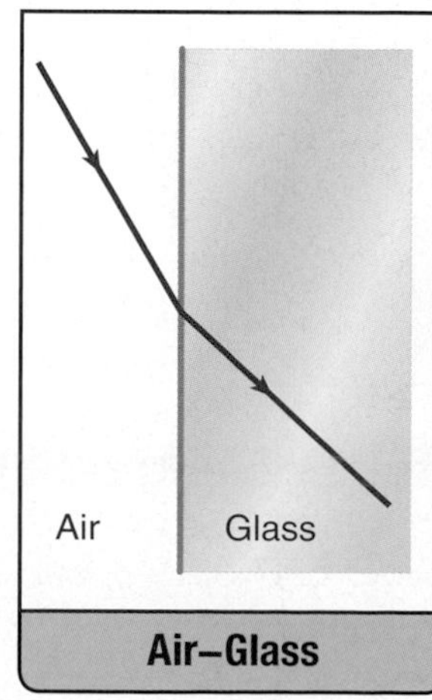

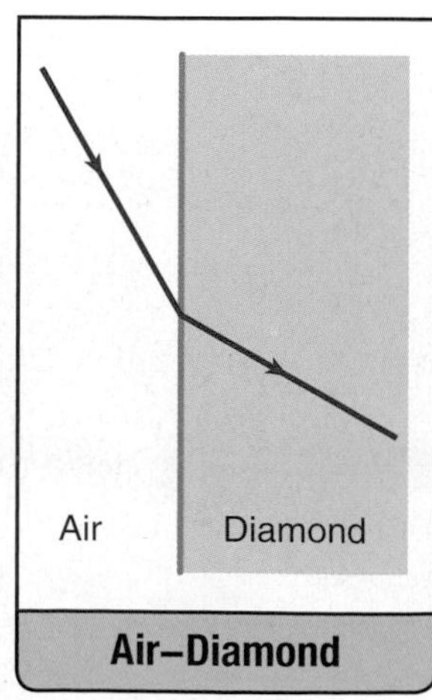

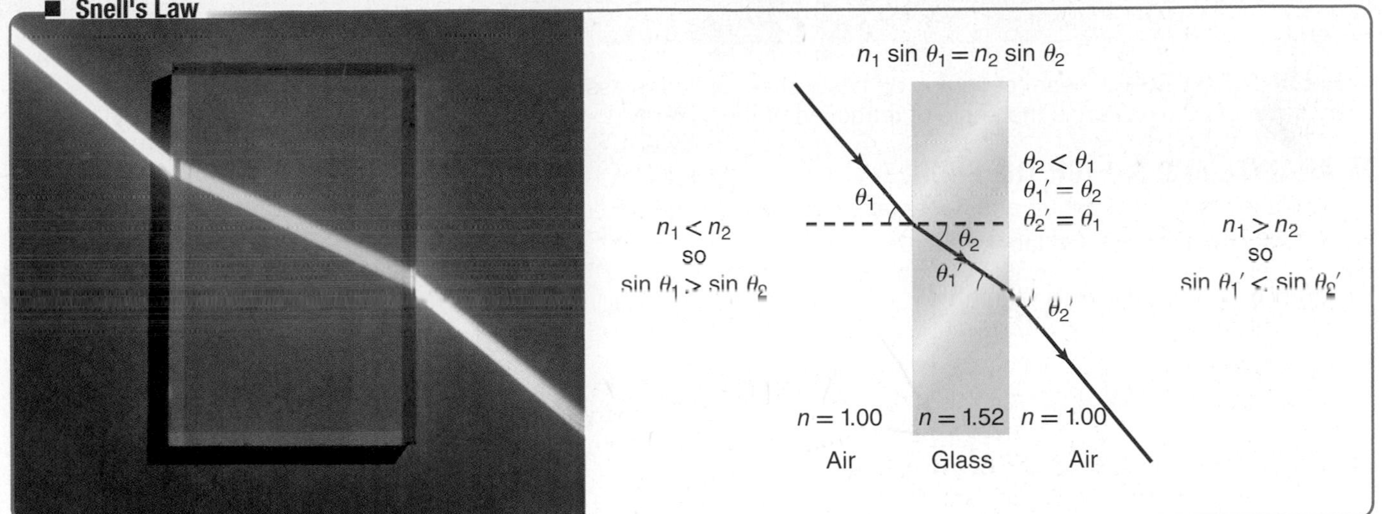

Figure 2 When light travels from air through glass and back to air, it refracts toward and then away from the normal.

Andrew Lambert/Leslie Garland Picture Library/Alamy

Snell's Law of Refraction

René Descartes and Willebrord Snell first studied refraction in the seventeenth century by shining a narrow beam of light onto a transparent medium, such as the glass shown in **Figure 2.** They measured two angles—the angle of incidence and the angle of refraction. The angle of incidence (θ_1) is the angle at which the light ray strikes the surface. The angle of refraction (θ_2) is the angle at which the transmitted light leaves the surface. Both angles are measured with respect to the normal.

Index of refraction Snell found that when light passed from air into a transparent medium, the sines of the angles were related. This **index of refraction** (n) determines the angle of refraction of light as it crosses the boundary between two mediums. Properties of the mediums light is traveling through determine indices of refraction. Values of n for several mediums are listed in **Table 1.**

By doing experiments with other combinations of mediums, Snell developed a law of refraction that is valid when light goes across a boundary between any two mediums.

SNELL'S LAW OF REFRACTION

The product of the index of refraction of the first medium and the sine of the angle of incidence is equal to the product of the index of refraction of the second medium and the sine of the angle of refraction.

$$n_1 \sin \theta_1 = n_2 \sin \theta_2$$

Use **Figure 2** and **Table 1** to show how Snell's law applies when light travels through a piece of glass with parallel surfaces, such as a windowpane. The light is refracted when it enters the glass and again when it leaves the glass. When light goes from air into glass, it travels from a medium with a lower n of 1.00 to one with a higher n of 1.52. The light bends toward the normal.

Traveling from glass to air, light moves from a medium with a higher n (1.52) to one with a lower n (1.00). The light is bent away from the normal. The relative values of n determine whether the light will bend toward or away from the normal. Note the direction of the ray when it leaves the glass. It is the same as before it struck the glass, $\theta_1 = \theta_2'$. This is because the boundaries are between the same two mediums.

View an **animation of refraction.**

Concepts In Motion

Investigate **Snell's law.**

Table 1

Indices of Refraction for Yellow Light (λ = 589 nm in vacuum)

Medium	n
Vacuum	1.00
Air	1.0003*
Water	1.33
Ethanol	1.36
Float glass	1.52
Quartz	1.54
Flint glass	1.62
Diamond	2.42

*The value given for air contains additional significant figures to distinguish it from that for a vacuum. Use a value for n of 1.00 in problems.

EXAMPLE PROBLEM 1

Get help with **finding the angle of refraction.** Personal Tutor

ANGLE OF REFRACTION A beam of light in air hits a sheet of float glass at an angle of 30.0°. What is the angle of refraction of the light ray?

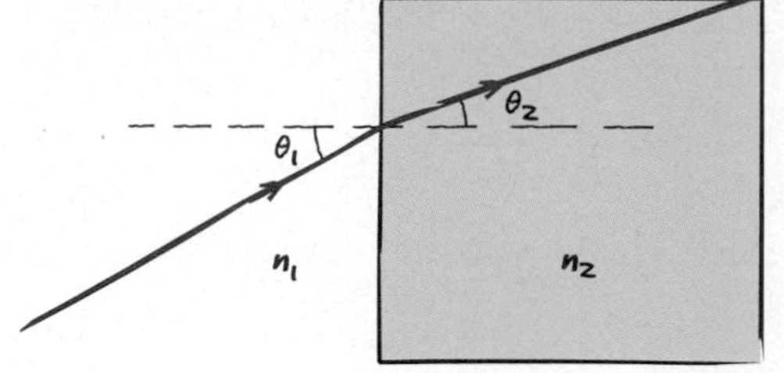

1 ANALYZE AND SKETCH THE PROBLEM

- Make a sketch of the air and float glass boundary.
- Draw the light ray and label θ_1, θ_2, n_1, and n_2.

KNOWN:	**UNKNOWN:**
$\theta_1 = 30.0°$	$\theta_2 = ?$
$n_1 = 1.00$	
$n_2 = 1.52$	

2 SOLVE FOR THE ANGLE OF REFRACTION

Use Snell's law to solve for the sine of the angle of refraction.

$$n_1 \sin \theta_1 = n_2 \sin \theta_2$$

$$\sin \theta_2 = \left(\frac{n_1}{n_2}\right) \sin \theta_1$$

$$\theta_2 = \sin^{-1}\left(\left(\frac{n_1}{n_2}\right) \sin \theta_1\right)$$

$$= \sin^{-1}\left(\left(\frac{1.00}{1.52}\right) \sin 30.0°\right)$$ ◀ *Substitute $n_1 = 1.00$, $n_2 = 1.52$, $\theta_1 = 30.0°$.*

$$= 19.2°$$

3 EVALUATE THE ANSWER

Does the answer make sense? The light beam travels into a medium with a higher n. Therefore, θ_2 should be less than θ_1.

PRACTICE PROBLEMS

Do additional problems. Online Practice

1. A laser beam in air enters ethanol at an angle of incidence of 37.0°. What is the angle of refraction?
2. As light travels from air into water, the angle of refraction is 25.0° to the normal. Find the angle of incidence.
3. Light in air enters a diamond facet at 45.0°. What is the angle of refraction?
4. A block of unknown material is submerged in water. Light in the water enters the block at an angle of incidence of 31°. The angle of refraction of the light in the block is 27°. What is the index of refraction of the material of the block?
5. **CHALLENGE** Light travels from air into another medium. The angle of incidence is 45.0° and the angle of refraction is 27.7°. What is the other medium?

▶ **CONNECTION TO ASTRONOMY** Refraction is responsible for the Moon appearing red during a lunar eclipse, as in **Figure 3.** A lunar eclipse occurs when Earth blocks sunlight from reaching the Moon. As a result, you might expect the Moon to be completely dark. Instead, as sunlight travels between the boundaries of space and Earth's atmosphere, it refracts around Earth toward the Moon. Earth's atmosphere scatters most of the shorter-wavelength blue and green light from the Sun. Thus, longer wavelengths of light pass through Earth's atmosphere. Those light waves are refracted as they travel from space into Earth's atmosphere and then into space. As a result, red light illuminates the Moon.

Figure 3 The Moon appears red during a lunar eclipse due to refraction.

PureStock/Getty Images

The Meaning of the Index of Refraction

The index of refraction describes how much light bends as it enters a medium, but does it have additional significance? Looking at light as a wave yields an interesting connection between the propagation of light through a medium and its index of refraction. Snell's work was done prior to the development of the wave model of light; after the development of the wave model, it was understood that light interacts with atoms in such a way that it moves more slowly through a medium than it does through a vacuum.

Remember that for light traveling through a vacuum, $\lambda_0 = \frac{c}{f}$. This can be written more generally as $\lambda = \frac{v}{f}$, where v is the speed of light in any medium and λ is the wavelength in that medium. Frequency (f) is the number of oscillations a wave makes per second. It does not change at the boundary. Because f stays the same and $\lambda = \frac{v}{f}$, the wavelength decreases when light slows down as it crosses a boundary. And because the speed of light in any medium is slower than the speed of light in a vacuum, the wavelength in any medium is shorter than it is in a vacuum.

Wave model What are the consequences of this change in wavelength? **Figure 4** shows a beam of light made up of a series of parallel, straight wavefronts. Each wavefront represents the crest of a wave and is perpendicular to the beam's direction. The beam's angle of incidence is θ_1. Consider $\triangle$PQR. Because the wavefronts are perpendicular to the beam's direction, $\angle$PQR is a right angle. $\angle$QRP is equal to θ_1. Therefore, $\sin \theta_1$ is equal to $\overline{PQ}$ divided by $\overline{PR}$.

$$\sin \theta_1 = \frac{\overline{PQ}}{\overline{PR}}$$

The angle of refraction (θ_2) can be related in a similar way to $\triangle$PSR. In this case

$$\sin \theta_2 = \frac{\overline{RS}}{\overline{PR}}$$

By taking the ratio of the sines of the two angles, $\overline{PR}$ is canceled, leaving the following equation:

$$\frac{\sin \theta_2}{\sin \theta_1} = \frac{\overline{RS}}{\overline{PQ}}$$

In **Figure 4** $\overline{PQ}$ is the length of three wavelengths of light in air, which can be written as $\overline{PQ} = 3\lambda_1$. In a similar way, $\overline{RS} = 3\lambda_2$. Substituting these values into the previous equation and canceling the common factor of 3 yields an equation relating the angles of incidence and refraction to the wavelength of light in each medium.

$$\frac{\sin \theta_2}{\sin \theta_1} = \frac{3\lambda_2}{3\lambda_1} = \frac{\lambda_2}{\lambda_1}$$

Using $\lambda = \frac{v}{f}$ and canceling the common factor f, because frequency remains constant, the equation can be rewritten:

$$\frac{\sin \theta_2}{\sin \theta_1} = \frac{v_2}{v_1}$$

Wavefronts

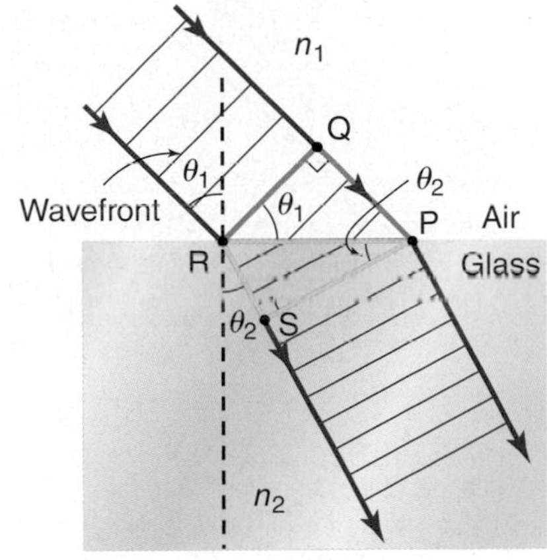

Figure 4 Each wavefront passes the boundary from air to glass at an angle. Part of the wavefront slows, and the ray bends. Since the wave slows and the frequency stays constant, for $\lambda = \frac{v}{f}$ to be true, the wavelength must decrease.

Infer Which medium has a higher index of refraction?

Find help with **sine and trigonometric ratios.**

PhysicsLAB

HOW DOES LIGHT BEND?

Is the index of refraction constant for a medium?

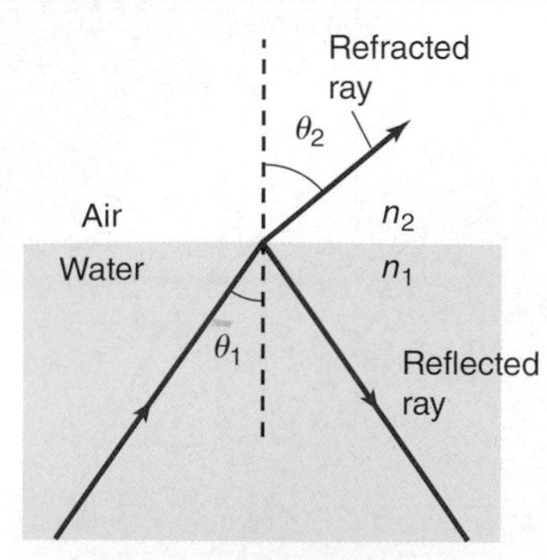

At an angle of incidence less than the critical angle, light is partially refracted and partially reflected.

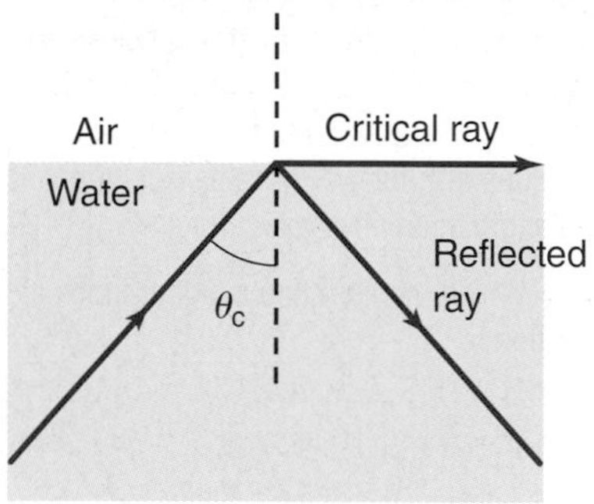

A ray refracted along the boundary of the medium forms the critical angle.

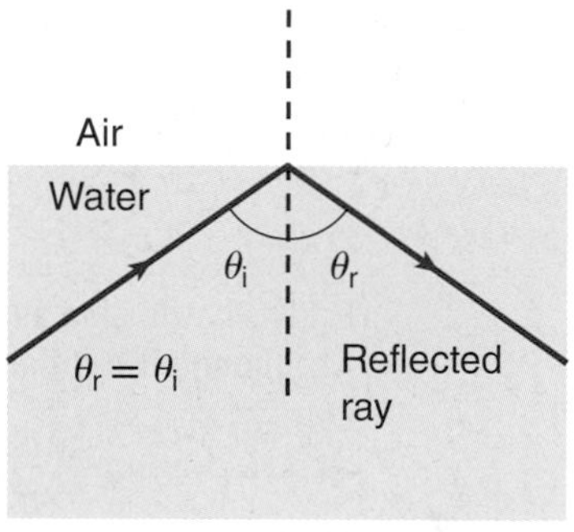

An angle of incidence greater than the critical angle results in total internal reflection, which follows the law of reflection.

Figure 5 Refraction and reflection of light traveling between mediums depend on the angle of incidence θ_1.

View an animation of **total internal reflection.**

Snell's law also can be written as a ratio of the sines of the angles of incidence and refraction.

$$\frac{\sin \theta_2}{\sin \theta_1} = \frac{n_1}{n_2}$$

Using the transitive property of equality, the previous two equations lead to the following equation:

$$\frac{n_1}{n_2} = \frac{v_2}{v_1}$$

In a vacuum, $n = 1.00$ and $v = c$. If either medium is a vacuum, then the equation is simplified to an equation that relates the index of refraction to the speed of light in a medium.

INDEX OF REFRACTION

The index of refraction of a medium is equal to the speed of light in a vacuum divided by the speed of light in the medium.

$$n = \frac{c}{v}$$

This definition of the index of refraction can be used to find the wavelength of light in a medium. In a medium with an index of refraction n, the speed of light is given by $v = \frac{c}{n}$. The wavelength of the light in a vacuum is $\lambda_0 = \frac{c}{f}$. Solve for frequency, and substitute $f = \frac{c}{\lambda_0}$ and $v = \frac{c}{n}$ into $\lambda = \frac{v}{f}$.

$$\lambda = \frac{(c/n)}{(c/\lambda_0)} = \frac{\lambda_0}{n}$$

Note that, as was stated earlier, the wavelength of light in a medium is smaller than its wavelength in a vacuum.

☑ **READING CHECK** **Describe** the relationship between the index of refraction and the speed of light in a medium.

Total Internal Reflection

Recall that when light strikes a transparent boundary, some of the light is transmitted and some is reflected. As light travels from a medium of higher n to one of lower n, the angle of refraction is larger than the angle of incidence, as shown in **Figure 5.** This leads to an interesting phenomenon. As the angle of incidence increases, the angle of refraction increases. At a certain angle of incidence known as the **critical angle** (θ_c), the refracted light ray lies along the boundary of the two mediums.

Total internal reflection occurs when light traveling from a region of a higher index of refraction to a region of a lower index of refraction strikes the boundary at an angle greater than the critical angle such that all light reflects back into the region of the higher index of refraction. This is shown in the bottom diagram of **Figure 5.** To construct an equation for the critical angle of any boundary, you can use Snell's law and substitute $\theta_1 = \theta_c$, and $\sin \theta_2 = \sin 90.0° = 1$.

$$\sin \theta_c = \frac{n_2}{n_1}$$

Total internal reflection causes some curious effects. Suppose you are looking up at the surface from under water in a calm pool. You might see an upside-down reflection of another nearby object that is also under water. You might also see a reflection of the bottom of the pool itself. The surface of the water acts like a mirror, reflecting the image back into the water.

Mirages

On a hot summer day, you sometimes can see the mirage effect shown in **Figure 6.** As you drive down a road, you see what appears to be the reflection of an oncoming car in a pool of water. The pool, however, disappears as you approach it. The mirage is the result of the Sun heating the road. The speed of light, and therefore, the index of refraction, for a gaseous medium can change slightly with temperature. The hot road heats the air above it and produces a thermal layering of air, causing light traveling toward the road from the car to gradually bend upward. This makes the light appear to be coming from a reflection in a pool.

Figure 6 shows how this occurs. As light from a distant object travels downward toward the road, the index of refraction of the air decreases as the air gets hotter, but the temperature change is gradual. Recall that light wavefronts are comprised of Huygens' wavelets. In the case of a mirage, the Huygens' wavelets closer to the ground travel faster than those higher up, causing the wavefronts to gradually turn upward.

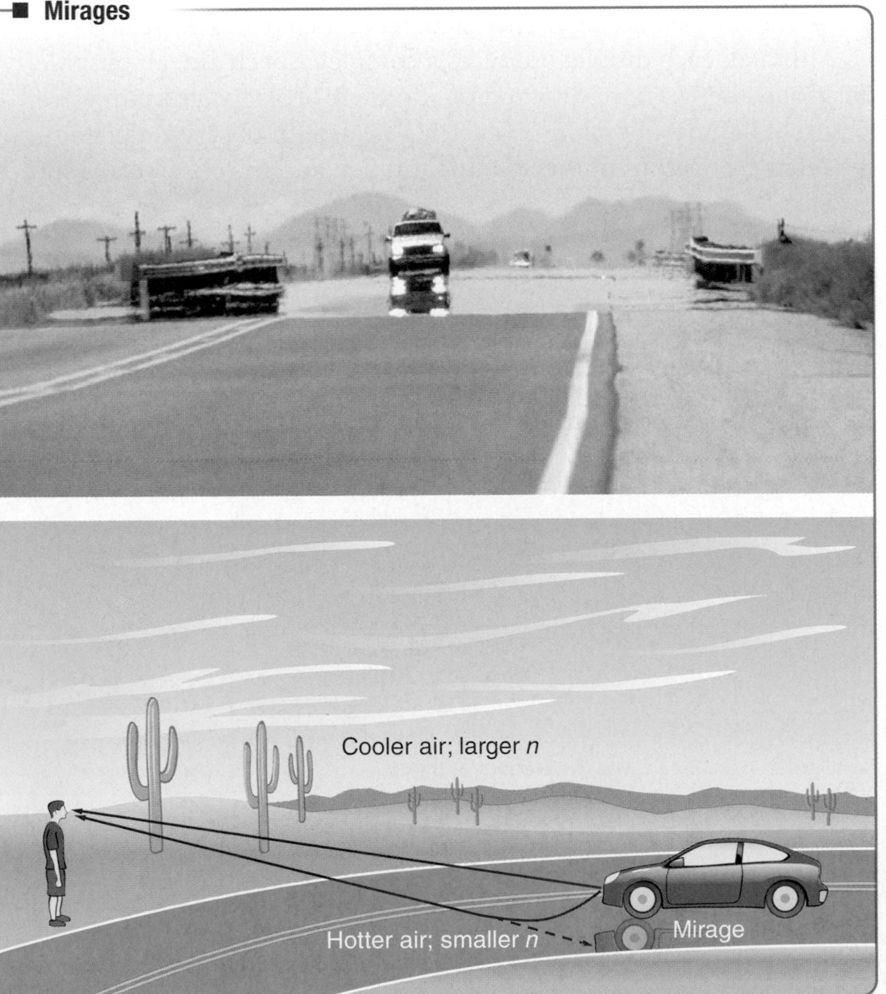

(t)Lawrence Lawry/Photodisc/Getty Images, (b)Kent Wood/Photo Researchers, Inc.

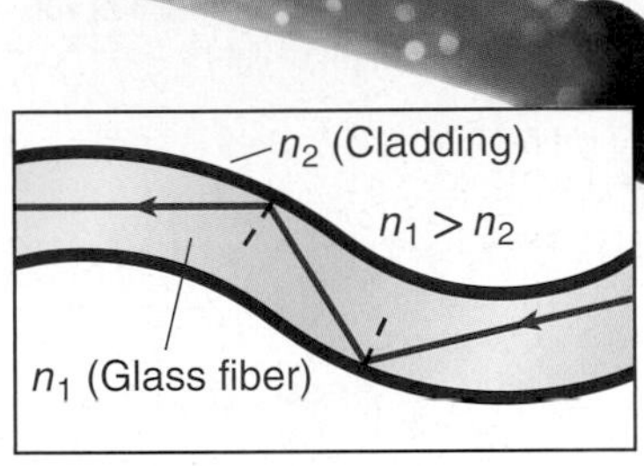

REAL-WORLD PHYSICS

FIBER OPTICS Total internal reflection is used in communication via optical fibers. The light traveling through the transparent fiber always hits the internal boundary of the optical fiber at an angle greater than the critical angle. All the light is reflected, and none of the light is transmitted through the boundary. Light pulses in fiber optic cables carry larger amounts of information over longer distances than other forms of communication. The outer material of each fiber is called cladding.

Figure 6 Light waves reflected from the car on the road are refracted when the air near the surface of the road is hotter than the air above it.

Apply Are the light waves traveling faster near the surface of the road or near the top of the truck?

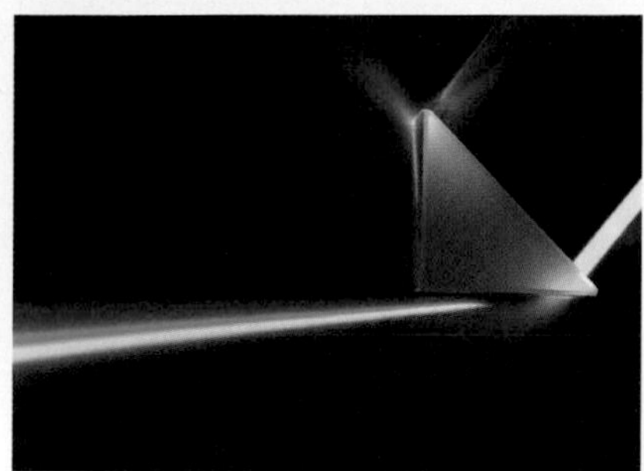

Figure 7 Dispersion through a prism occurs because the index of refraction varies with the wavelength of light.

Dispersion of Light

You might have seen the phenomenon illustrated in **Figure 7** as white light passes through a prism. If you look carefully at the spectrum formed by light passing through the prism, you will notice that violet light is refracted more than red light. This occurs because the speed of violet light through glass is less than the speed of red light through glass. Violet light has a higher frequency than red light, which causes it to interact differently with the atoms of the glass. This results in glass having a slightly higher index of refraction for violet light than it has for red light. This separation of white light into a spectrum of colors is called **dispersion.**

The speed of light in a medium is determined by interactions between the light and the atoms that make up the medium. The speed of light and the index of refraction for a solid or liquid medium vary for different wavelengths of light. The speed of light and index of refraction also change slightly with temperature for a gaseous medium. This is due to the effects of temperature on the energy of particles at the atomic level.

Rainbows A prism is not the only means of dispersing light. A rainbow is a spectrum formed when sunlight is dispersed by water droplets in the atmosphere. Sunlight that falls on a water droplet is refracted. Since each color has a different wavelength, it is refracted at a slightly different angle, as shown in **Figure 8,** resulting in dispersion. At the back surface of the droplet, some of the light undergoes internal reflection. On the way out of the droplet, the light once again is refracted and dispersed.

Although each droplet produces a complete spectrum, an observer positioned between the Sun and the rain will see only a certain wavelength of light from each droplet. The wavelength observed depends on the relative positions of the Sun, the droplet, and the observer, as shown in **Figure 8.**

MiniLAB

PERSONAL RAINBOW

What conditions are needed to make a rainbow?

iLab Station

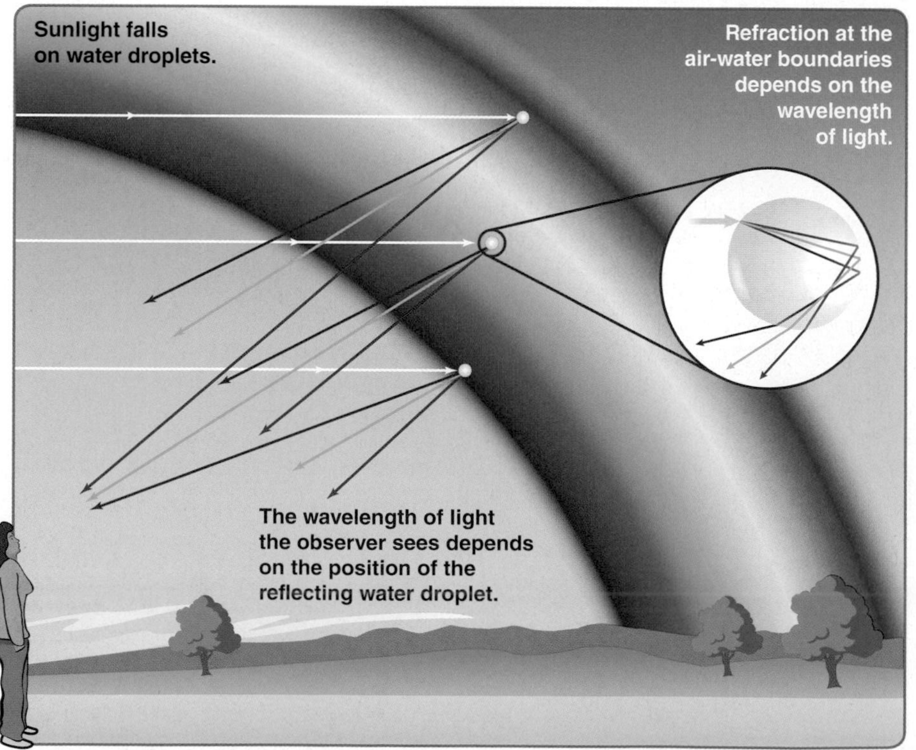
Sunlight falls on water droplets.
Refraction at the air-water boundaries depends on the wavelength of light.
The wavelength of light the observer sees depends on the position of the reflecting water droplet.

Figure 8 Light from the Sun is dispersed by water droplets to form rainbows. Because there are many droplets at various positions in the sky, a complete spectrum is visible.

Consider whether it is possible for a rainbow to be close enough to touch.

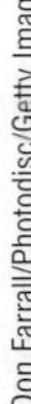
Don Farrall/Photodisc/Getty Images

Figure 9 Light reflected within raindrops can cause the appearance of a secondary rainbow.

PhysicsLAB

A WHOLE SPECTRUM OF POSSIBILITIES

FORENSICS LAB How can a spectrum from broken glass at an accident scene help identify a suspect vehicle?

Second-order rainbow Sometimes, you can see a faint second-order rainbow like the one shown in **Figure 9.** The second rainbow is outside of the first. It is also fainter and has the order of the colors reversed. Light rays that are reflected twice inside water droplets produce this effect.

Very rarely, a third-order rainbow is visible outside the second. A third-order rainbow is even fainter than a second-order rainbow. What is your prediction about how many times light is reflected in the water droplets? What would you predict about the order of appearance of the colors for the third-order rainbow?

SECTION 1 REVIEW

Section Self-Check Check your understanding.

6. **MAIN**IDEA You notice that when a light ray enters a certain liquid from water, it is bent toward the normal, but when it enters the same liquid from float glass, it is bent away from the normal. What can you conclude about the liquid's index of refraction?

7. **Index of Refraction** A ray of light in air has an angle of incidence of 30.0° on a block of unknown material and an angle of refraction of 20.0°, as shown in **Figure 10.** What is the index of refraction of the material?

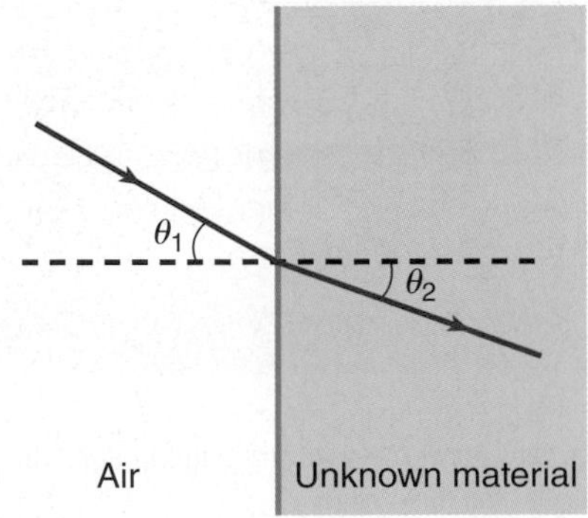

Figure 10

8. **Angle of Refraction** A beam of light passes from water into polyethylene with $n = 1.50$. If $\theta_1 = 57.5°$, what is the angle of refraction in the polyethylene?

9. **Speed of Light** What is the speed of light in chloroform ($n = 1.51$)?

10. **Critical Angle** Is there a critical angle for light traveling from glass to water? From water to glass? Explain your answer.

11. **Total Internal Reflection** If you were to use quartz and float glass to make an optical fiber, which would you use for the cladding layer? Why?

12. **Sunsets** Why can you see the image of the Sun just above the horizon when the Sun itself has already set?

13. **Speed of Light** Could an index of refraction ever be less than 1? What would this imply about the speed of light in that medium?

14. **Critical Thinking** In what direction would you have to be looking to see a rainbow on a rainy late afternoon? Explain.

SECTION 2

Convex and Concave Lenses

PHYSICS 4 YOU **You might have heard that you can light a fire with a magnifying lens. The magnifying lens focuses all the light into a single point. Could you light a fire with any lens, or does the lens need to have certain properties?**

MAIN IDEA

Lenses can be used to enlarge and reduce images.

Essential Questions

- How are real and virtual images formed by single convex and concave lenses?
- How can images formed by lenses be located and described with ray diagrams and equations?
- How can chromatic aberration be reduced?

Review Vocabulary

transparent a property of a medium that allows that medium to transmit light and reflect a fraction of the light, allowing objects to be seen clearly through it

New Vocabulary

lens
convex lens
concave lens
thin lens equation
chromatic aberration
achromatic lens

Types of Lenses

The refraction of light in nature that forms rainbows and red lunar eclipses is beautiful, but refraction also is useful. In 1303, French physician Bernard of Gordon wrote of the use of lenses to correct eyesight. Around 1610, Galileo discovered the moons of Jupiter using a telescope made with two lenses. Since Galileo's time, lenses have been used in many instruments, such as microscopes and cameras.

A **lens** is a piece of transparent material, such as glass or plastic, that is used to focus light and form an image. Each of a lens's two faces might be either curved or flat. A lens that is thicker at the center than at the edges is called a **convex lens,** as shown on the left in **Figure 11.** When it is surrounded by material with a lower index of refraction, such as air, it refracts parallel light rays so that they all pass through a common point, called the focal point, after going through the lens. For this reason, a convex lens is often called a converging lens.

A lens that is thinner in the middle than at the edges is called a **concave lens,** as shown on the right in **Figure 11.** When surrounded by material with a lower index of refraction, rays passing through it spread out, so it is often called a diverging lens.

When light passes through a lens, refraction occurs at both lens surfaces. Using Snell's law and geometry, you can determine the paths of rays passing through lenses. To simplify such problems, you may assume that all refraction occurs at a plane, called the principal plane, that passes through the center of the lens. This approximation, called the thin lens model, applies to all the lenses discussed in this chapter.

Convex Lens

Concave Lens

Figure 11 Convex lenses refract light so that the rays converge after passing through. The light passing through a concave lens does not meet at the focal point.

(t)Photodisc, (bl br)Don Farrall/Photographer's Choice/Getty Images

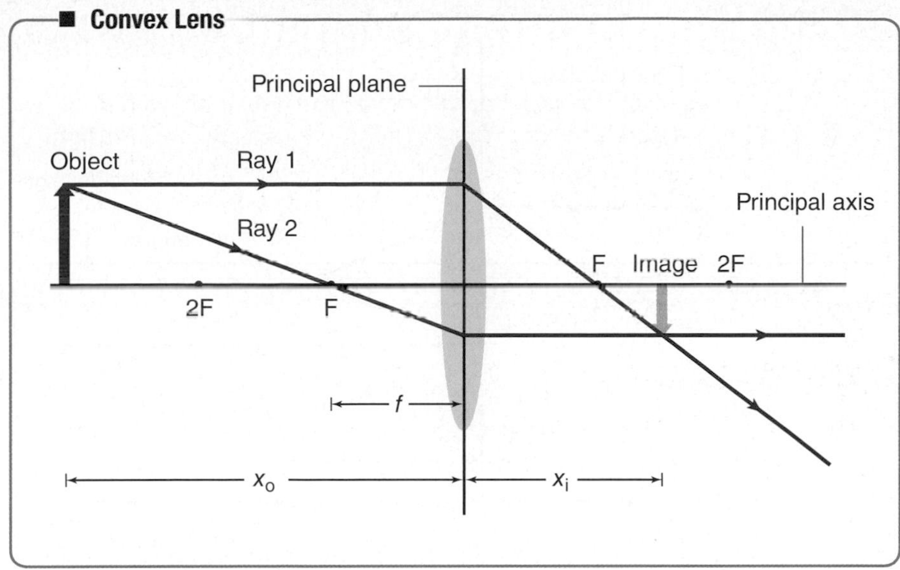

Figure 12 An object placed at a distance greater than twice the focal length from the lens will produce an image that is real, reduced in size, and inverted.

View an **animation of drawing a ray diagram.**

COLOR CONVENTION

light rays ⟷ **red**
lenses ⟷ **light blue**
object ⟷ **purple**
image ⟷ **lilac**

Convex Lenses

If you know the location of the object being viewed and the type and strength of lens used, you can find the image's location. Ray diagrams are a useful representational tool. In a ray diagram, you represent a few important rays to find out how a lens affects the light that passes through it. You can use any two rays to locate the image; with experience, you will learn that some rays are easier to draw and use than others. In **Figure 12,** ray 1 reflects from the top of the object and is parallel to the principal axis. For convex lenses, it passes through the focal point (F) after passing through the lens. In the case shown, with the object placed a distance greater that the focal length (f) from the lens, ray 2 also reflects from the top of the object. It passes through the focal point on the same side of the lens as the object.

Ray diagrams are given for convex lenses with objects placed at different distances from the lens. In these diagrams, x_o is the distance of the object from the lens and x_i is the distance of the image from the lens. In all ray diagrams in this chapter, the thin lens model is used. In this model, light refracts at the center of the lens rather than at the boundaries between air and the surface of the lens.

$x_o \geq 2f$ In **Figure 12,** rays are traced from an object located far from a convex lens. For the purpose of locating the image, you will use two rays. Ray 1 is parallel to the principal axis. It refracts and passes through the focal point (F) on the other side of the lens. Ray 2 passes through F on its way to the lens. After refraction, its path is parallel to the principal axis. The two rays intersect at a point beyond F and locate the image. Rays selected from other points on the object converge at corresponding points to form the complete image. Note that this is a real image that is inverted and smaller compared to the object.

If the object is placed at twice the focal length from the lens at the point 2F, the diagram is similar. The image is real and is found at 2F, but it is no longer reduced in size. Because of symmetry, the image and the object have the same size. Thus, you can conclude that if an object is more than twice the focal length from the lens, the image is smaller than the object.

MiniLAB

LENS MASKING EFFECTS

What effect does covering part of a lens have on the image produced?

Figure 13 An object placed at a distance less than twice the focal length but greater than one focal length from the lens will produce an image that is real, enlarged, and inverted.

$2f > x_o > f$ You can use **Figure 13** to locate the image of an object that is between F and 2F viewed through a convex lens. This is similar to the ray diagram for the object located at a distance greater than twice the focal length with the image and object interchanged. The direction of the rays would be reversed. In this case the image is also real and inverted. For an object placed between F and 2F, the image is enlarged.

When an object is placed at the focal point of a convex lens, ray diagrams cannot be drawn. The refracted rays will emerge in a parallel beam and no image will be seen.

$f > x_o > 0$ **Figure 14** shows how a convex lens forms a virtual image. The object is located between F and the lens. Ray 1, as usual, approaches the lens parallel to the principal axis and is refracted through the focal point, F. Ray 2 travels from the tip of the object in the direction it would have if it had started at F on the object side of the lens. The dashed line from F to the object shows you how to draw ray 2. Ray 2 leaves the lens parallel to the principal axis. Rays 1 and 2 diverge as they leave the lens.

The reflection appears to an observer to come from a spot on the same side of the lens as the object. This is a virtual image that is upright and larger compared to the object. No real image is possible. Drawing sight lines for the two rays back to their apparent intersection locates the virtual image. Note that the actual image is formed by light that passes through the lens, but you can still determine the location of the image by drawing rays that do not have to pass through the lens.

PhysicsLABs

CONVEX LENS IMAGES

What are the characteristics of images formed by a convex lens?

CONVEX LENSES AND FOCAL LENGTH

Does focal length change with image position for a convex lens?

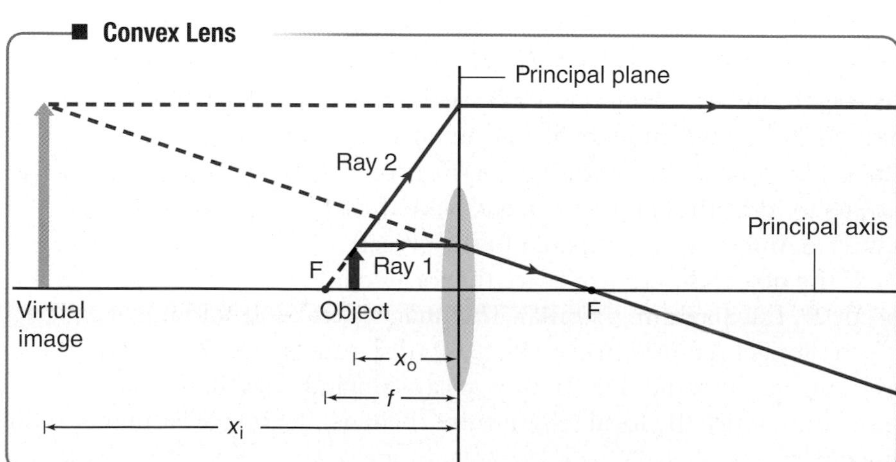

Figure 14 An object placed at a distance less than the focal length from the lens will produce an image that is virtual and enlarged.

Classify an image as virtual or real based on the side of the lens it is on.

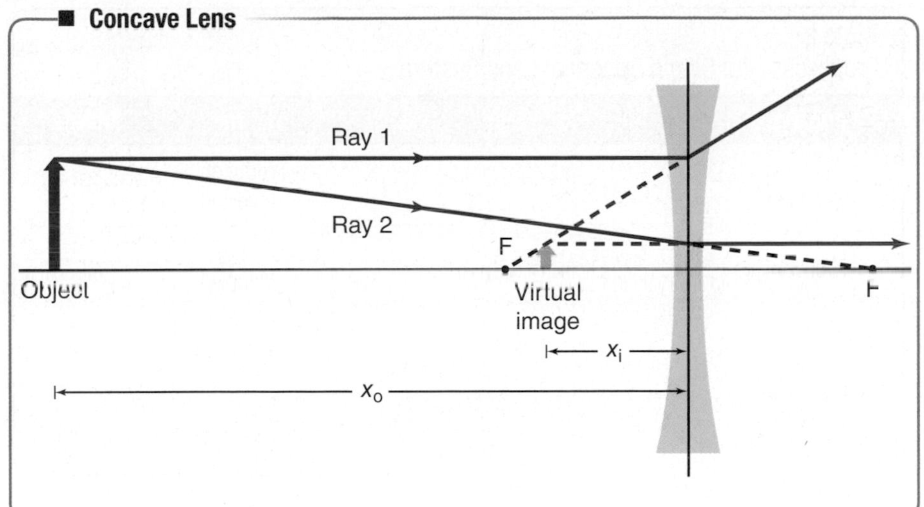

Figure 15 An object placed any distance from a concave lens will always produce an image that is virtual and reduced.

View a **simulation of a virtual optics lab.**

Concepts In Motion

Concave Lenses

A concave lens causes all rays to diverge. **Figure 15** shows how a concave lens forms a virtual image. Ray 1 approaches the lens parallel to the principal axis. It leaves the lens along a line that extends back through the focal point on the object side of the lens. The focal point of a concave lens is found on the same side of the lens as the incoming light. Ray 2 approaches the lens as if it is going to pass through the focal point on the opposite side and leaves the lens parallel to the principal axis.

The sight lines of rays 1 and 2 intersect on the same side of the lens as the object. The image is located at the point from where the two rays appear to intersect, creating a virtual image. The image is upright and smaller than the object. This is true no matter how far from the lens the object is located. The focal length for a diverging lens is negative.

READING CHECK **Describe** why a concave lens always produces a virtual image.

MiniLAB

WATER LENSES

How does the curved surface of water act like a lens?

iLab Station

Lens Equations

Lenses can be constructed with a variety of shapes, but in this book, you will only consider thin lenses, with faces having spherical curvatures. Based on this model of thin spherical lenses, an equation has been developed that looks exactly like the one for spherical mirrors. For lenses, virtual images are always on the same side of the lens as the object, which means the image position is negative. Notice that a concave lens produces only virtual images, whereas a convex lens can produce real images or virtual images.

Thin lens equation The **thin lens equation** relates the focal length of a spherical thin lens, the object position, and the image position.

Investigate **ray tracing and lenses.**

Virtual Investigation

THIN LENS EQUATION

The inverse of the focal length of a spherical lens is equal to the sum of the inverses of the image position and the object position.

$$\frac{1}{f} = \frac{1}{x_i} + \frac{1}{x_o}$$

Table 2
Properties of a Single Spherical Lens System

Lens Type	f	x_o	x_i	m	Image
Convex	+	$x_o > 2f$	$2f > x_i > f$	reduced, inverted	real
		$2f > x_o > f$	$x_i > 2f$	enlarged, inverted	real
		$f > x_o > 0$	$\lvert x_i \rvert > x_o$ (negative)	enlarged	virtual
Concave	–	$x_o > 0$	$\lvert f \rvert > \lvert x_i \rvert > 0$ (negative)	reduced	virtual

When solving problems for concave lenses using the thin lens equation, you should remember that the sign convention for focal length is different from that of convex lenses. This is because a concave lens is a divergent lens. If the focal point for a concave lens is 24 cm from the lens, you should use the value $f = -24$ cm in the thin lens equation. All images for a concave lens are virtual. Thus, if an image distance is given as 20 cm from the lens, then you should use $x_i = -20$ cm. The object position always will be positive.

Magnification A property of spherical thin lenses that measures how much larger or smaller the image is than the object is magnification. The magnification equation for spherical mirrors also can be used for spherical thin lenses. It is used to determine the height and the orientation of the image formed by a spherical thin lens.

MAGNIFICATION
The magnification of an object by a spherical lens, defined as the image height divided by the object height, is equal to the negative of the image position divided by the object position.

$$m \equiv \frac{h_i}{h_o} = -\frac{x_i}{x_o}$$

Magnification gives information about the size and orientation of the image relative to the object. When the absolute value of a magnification is between zero and one, the image is smaller than the object. Magnifications with absolute values greater than one represent images that are larger than the objects. A negative magnification means the image is inverted compared to the object.

Using the equations for lenses It is important that you use the proper sign conventions when using these equations. **Table 2** shows a comparison of the image position, magnification, and type of image formed by single convex and concave lenses when an object is placed at various object positions (x_o) relative to the lens. For convex lenses, the object position relative to the focal point influences the image type.

Notice the similarity of this table to the table for mirrors. As with mirrors, the distance from the principal plane of a lens to its focal point is the focal length (f). The focal length depends on the shape of the lens and the index of refraction of the lens material. Focal lengths, image positions, and image heights can be negative.

EXAMPLE PROBLEM 2

Find help with **operations with significant figures.** Math Handbook

AN IMAGE FORMED BY A CONVEX LENS An object is placed 32.0 cm from a convex lens that has a focal length of 8.0 cm.

a. Where is the image?

b. If the object is 3.0 cm high, how tall is the image?

c. What is the orientation of the image?

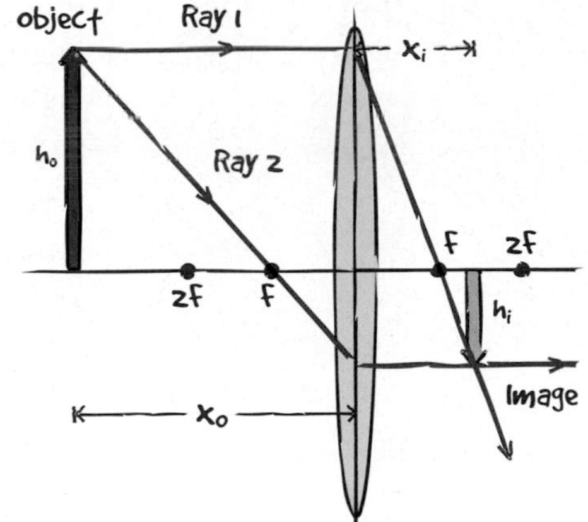

1 ANALYZE AND SKETCH THE PROBLEM

- Sketch the situation, locating the object and the lens.
- Draw the two principal rays.

KNOWN	UNKNOWN
$x_o = 32.0$ cm	$x_i = ?$
$h_o = 3.0$ cm	$h_i = ?$
$f = 8.0$ cm	

2 SOLVE FOR THE IMAGE POSITION AND HEIGHT

a. Use the thin lens equation to determine x_i.

$$\frac{1}{f} = \frac{1}{x_i} + \frac{1}{x_o}$$

$$x_i = \frac{fx_o}{x_o - f}$$

$$= \frac{(8.0 \text{ cm})(32.0 \text{ cm})}{32.0 \text{ cm} - 8.0 \text{ cm}}$$ ◀ Substitute $f = 8.0$ cm, $x_o = 32.0$ cm

$= 11$ cm (11 cm away from the lens on the side opposite the object)

b. Use the magnification equation, and solve for image height.

$$m \equiv \frac{h_i}{h_o} = -\frac{x_i}{x_o}$$

$$h_i = -\frac{x_i h_o}{x_o}$$

$$= -\frac{(11 \text{ cm})(3.0 \text{ cm})}{32.0 \text{ cm}}$$ ◀ Substitute $x_i = 11$ cm, $h_o = 3.0$ cm, $x_o = 32.0$ cm

$= -1.0$ cm (1.0 cm tall)

c. The negative sign for the height in part **b** means the image is inverted.

3 EVALUATE THE ANSWER

- **Are the units correct?** All are in centimeters.
- **Do the signs make sense?** Image position is positive (real image) and image height is negative (inverted compared to the object), which make sense for a convex lens.

PRACTICE PROBLEMS

Do additional problems.   Online Practice

15. A 2.25-cm-tall object is 8.5 cm to the left of a convex lens of 5.5-cm focal length. Find the image position and height.

16. An object near a convex lens produces a 1.8-cm-tall real image that is 10.4 cm from the lens and inverted. If the focal length of the lens is 6.8 cm, what are the object position and height?

17. An object is placed to the left of a convex lens with a 25-mm focal length so that its image is the same size as the object. What are the image and object positions?

18. Calculate the image position and height of a 2.0-cm-tall object located 25 cm from a convex lens with a focal length of 5.0 cm. What is the orientation of the image?

19. Use a scale ray diagram to find the image position of an object that is 30 cm to the left of a convex lens with a 10-cm focal length.

20. CHALLENGE A magnifier with a focal length of 30 cm is used to view a 1-cm-tall object. Use a ray diagram to determine the location and size of the image when the magnifier is positioned 10 cm from the object.

Figure 16 Spherical aberration from the camera lens blurs the edges of this image.

Defects of Spherical Lenses

Throughout this section, you have studied lenses that produce perfect images at specific positions. In reality, spherical lenses, just like spherical mirrors, have intrinsic defects that cause problems with the focus and color of images. Spherical lenses exhibit an aberration associated with their spherical design, just as mirrors do. In addition, the dispersion of light through a spherical lens causes an aberration that mirrors do not exhibit.

Spherical aberration The model you have used for drawing rays through spherical lenses suggests that all parallel rays focus at the same position. However, this is only an approximation. In reality, parallel rays that pass through the edges of a spherical lens focus at positions different from those of parallel rays that pass through the center. This inability of a spherical lens to focus all parallel rays to a single point is called spherical aberration. The effects are shown in **Figure 16.** In reality, most lenses have a slightly different shape to address this, but the spherical approximation works well enough for our purposes. In high-precision instruments, many lenses, often five or more, are used to form sharp, well-defined images.

Chromatic aberration Lenses have a second defect that mirrors do not have. Because the index of refraction of a medium depends on wavelength, different wavelengths of light are refracted at slightly different angles, as you can see in **Figure 17.** Light that passes through a lens is slightly dispersed, especially near the edges, causing an effect called **chromatic aberration.** This is seen as an apparent ring of color around an object viewed through a lens.

Chromatic aberration is always present when a single lens is used. However, this defect can be greatly reduced by an **achromatic lens,** which is a system of two or more lenses, such as a convex lens with a concave lens, that have different indices of refraction. Such a combination of lenses is shown in **Figure 17.** Both lenses in the figure disperse light, but the dispersion caused by the convex lens is almost canceled by the dispersion caused by the concave lens. The index of refraction of the convex lens is chosen so that the combination of lenses still converges the light at the desired location.

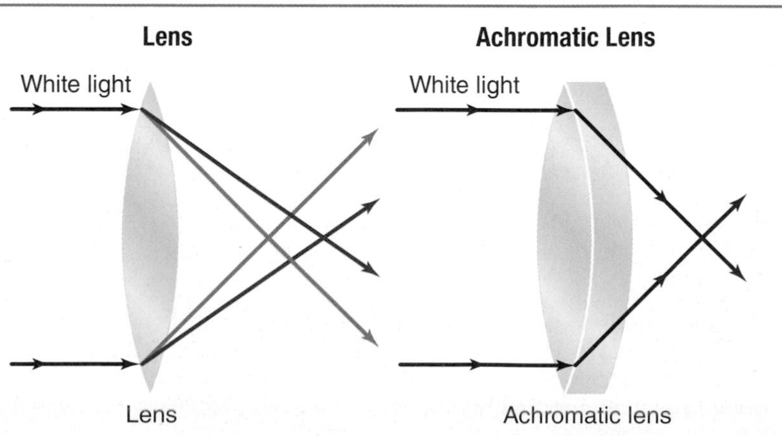

Figure 17 Simple lenses, such as the one shown in the photo on the left, exhibit a rainbow fringing effect called chromatic aberration. Achromatic lenses reduce chromatic aberration.

Explain why the index of refraction is important for achromatic lenses.

SECTION 2 REVIEW

21. **MAIN**IDEA Magnifying lenses normally are used to produce images that are larger than the related objects, but they also can produce images that are smaller than the related objects. Explain.

22. **Types of Lenses** The cross sections of four different thin lenses are shown in **Figure 18.**

    a. Which of these lenses, if any, are convex, or converging, lenses?

    b. Which of these lenses, if any, are concave, or diverging, lenses?

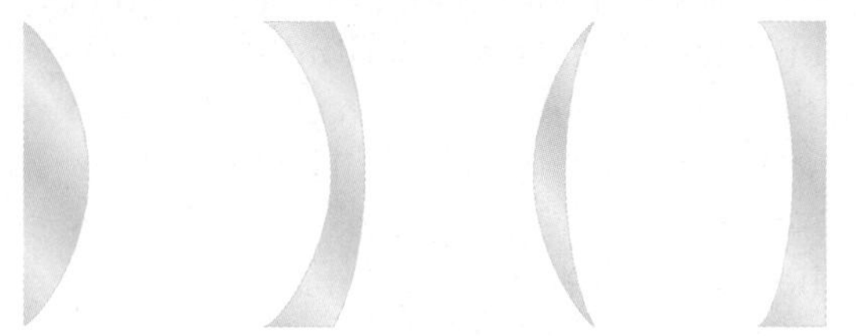

Figure 18

23. **Image Position and Height** Redraw the ray diagram in **Figure 19** and use it to determine the location and size of the image. Use the thin lens equation and the magnification equation to verify your answer.

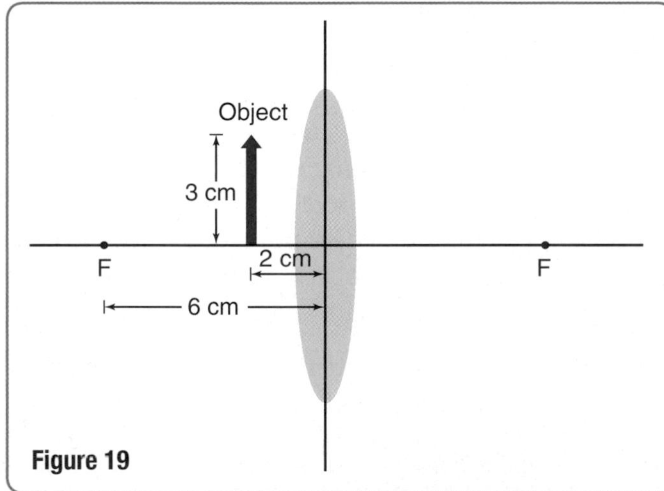

Figure 19

24. **Image Position and Height** An object is placed 1.5 m from a convex lens with a focal length of 1.0 m. Use the thin lens equation to determine the distance of the image from the lens. If the object height is 2.0 m, what is the image height? Is the image real or virtual? Is the image inverted or upright?

25. What do the terms *diverging lens* and *converging lens* mean? What type of lens does each of these terms refer to?

26. Calculations in this chapter use a thin lens approximation. What does this mean? Why is a thin lens approximation used?

27. **Chromatic Aberration** All simple lenses have chromatic aberration. Infer why you do not see this effect when you look through a microscope.

28. **Type of Image** Use the ray diagram in **Figure 20** to determine whether the image for object 1 will be reduced or enlarged, inverted or upright, and real or virtual. Do the same for object 2.

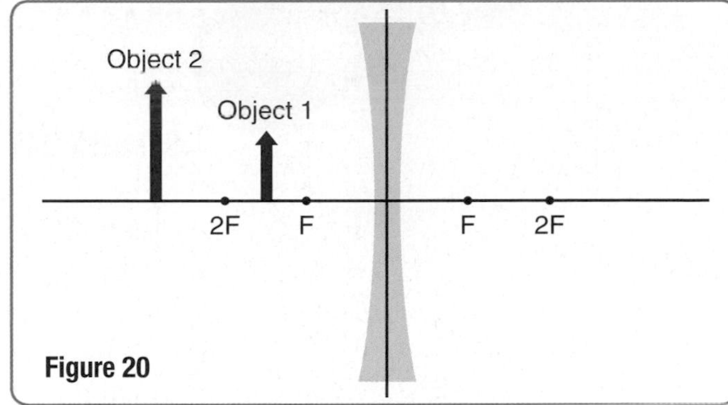

Figure 20

29. **Image Position and Height** A 6.0-cm-tall object is placed 5.0 cm from a convex lens with a focal length of 4.0 cm. Draw a ray diagram to determine the image location and size. Confirm your results using the thin lens equation and the magnification equation.

30. **Diverging Lens** A 6.5-cm tall salt shaker is viewed through a diverging lens with a focal length of 5.0 cm.

    a. If the shaker is 6.0 cm from the lens, what is the image distance from the lens? Is the image virtual or real?

    b. What is the magnification? Is the image smaller or larger than the object?

    c. If the salt shaker is moved to 4 cm from the lens, what is the distance of the image from the lens? What is the magnification? Is the image now smaller or larger than the object?

31. **Chromatic Aberration** You shine white light through a convex lens onto a screen and adjust the distance of the screen from the lens to focus the red light. Which direction should you move the screen to focus the blue light?

32. **Critical Thinking** An air lens constructed of two watch glasses is placed in a tank of water. Copy **Figure 21** and draw the effect of this lens on parallel light rays incident on the lens.

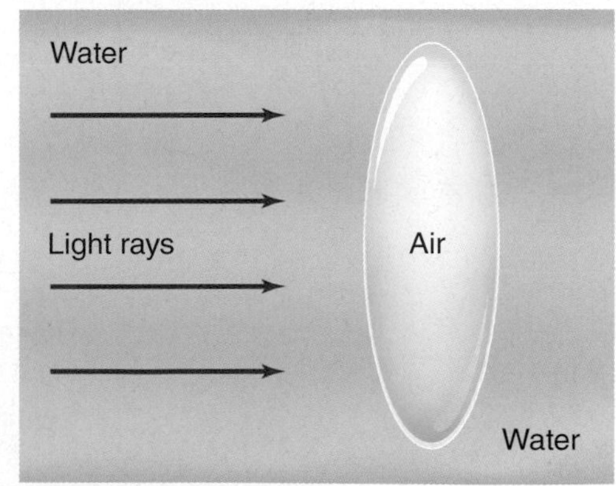

Figure 21

SECTION 3 # Applications of Lenses

PHYSICS 4 YOU **Sailors once used spyglasses—small hand-held telescopes—to see distant objects. Spyglasses and many other telescopes rely on lenses to enlarge images of faraway objects.**

MAIN IDEA

People see objects that they could not otherwise see by using lenses.

Essential Questions

- How does the eye focus light to form an image?
- What are nearsightedness and farsightedness, and how can eyeglass lenses correct these defects?
- What are the characteristics of the optical systems in some common optical instruments?

Review Vocabulary

index of refraction for a medium, the ratio of the speed of light in a vacuum to the speed of light in that medium

New Vocabulary

nearsightedness
farsightedness

Lenses in Eyes

The concepts that you have learned for the refraction of light through lenses apply to almost every optical device. The eye is a remarkable optical device. As shown in **Figure 22,** the eye is a fluid-filled, almost spherical vessel. Light that is emitted from or reflected off an object travels into the eye through the cornea and pupil. The light then passes through the lens and focuses onto the retina that is at the back of the eye. Specialized cells on the retina absorb this light and send information about the image along the optic nerve to the brain.

CONNECTION TO BIOLOGY Because of its name, you might assume that the lens of an eye is responsible for focusing light onto the retina. In fact, light entering the eye is primarily focused by the cornea, because the air-cornea boundary has the greater difference in indices of refraction. The lens is responsible for the fine focus that allows you to clearly see both distant and nearby objects.

READING CHECK Describe the roles of the cornea and the lens in your eye.

The lens uses a process called accommodation. Muscles surrounding the lens can contract or relax, thereby changing the shape of the lens. This, in turn, changes the focal length of the eye. For a healthy eye, when the muscles are relaxed, the image of distant objects is focused on the retina. When the muscles contract, the focal length is shortened, and this allows images of closer objects to be focused on the retina.

The Eye

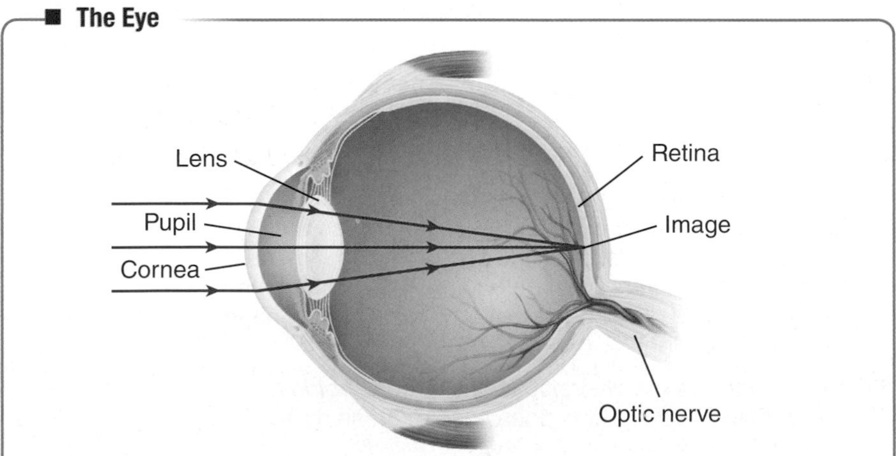

Figure 22 The cornea and the lens of your eye refract light that is reflected off every object you see.

Summarize why most of the refraction occurs at the boundary of the cornea rather than that of the lens.

View a **BrainPOP video on the eye.** Video

Exactostock/SuperStock

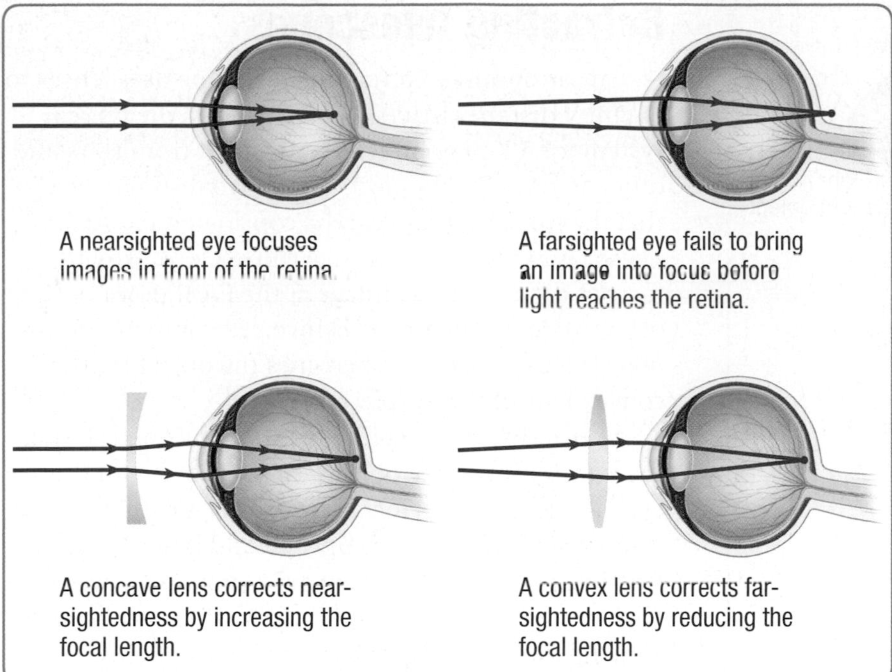

A nearsighted eye focuses images in front of the retina.
A farsighted eye fails to bring an image into focus before light reaches the retina.
A concave lens corrects nearsightedness by increasing the focal length.
A convex lens corrects farsightedness by reducing the focal length.

Figure 23 Objects at a distance are blurred for a nearsighted person. For a farsighted person, objects nearby are blurred.

Explain how bifocal lenses might be made.

Nearsightedness The eyes of many people do not focus sharp images on the retina. Instead, images are focused either in front of the retina or behind it. External lenses, in the form of eyeglasses or contact lenses, adjust the focal length and move images to the retina. **Figure 23** shows the condition of **nearsightedness,** also called myopia. The focal length of the eye is too short to focus light on the retina. Images are formed in front of the retina. As shown in **Figure 23,** concave lenses correct this by diverging light, thereby increasing images' distances from the lens, and forming images on the retina.

Farsightedness You also can see in **Figure 23** that **farsightedness,** also called hyperopia, is the condition in which the focal length of the eye is too long. Images are therefore not formed on the retina. A similar result is caused by the increasing rigidity of the lenses in the eyes of people who are more than about 45 years old. Their muscles cannot shorten the focal length enough to focus images of close objects on the retina. For either defect, convex lenses produce virtual images farther from the eye than the associated objects, as shown in **Figure 23.** The images then become the objects for the eye lens and can be focused on the retina, thereby correcting the defect.

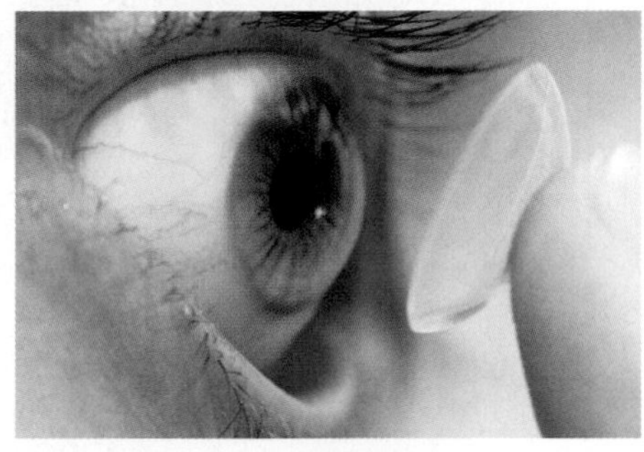

REAL-WORLD PHYSICS

Contacts Contact lenses produce the same results as eyeglasses. These small, thin lenses are placed directly on the corneas. A thin layer of tears between the cornea and lens keeps the lens in place. Most of the refraction occurs at the air-lens boundary, where the difference in indices of refraction is greatest.

PHYSICS CHALLENGE

As light enters the eye, it first encounters the air-cornea boundary. Consider a ray of light that strikes the interface between the air and a person's cornea at an angle of 30.0° to the normal. The index of refraction of the cornea is approximately 1.4.

1. Use Snell's law to calculate the angle of refraction.
2. What would the angle of refraction be if the person were swimming under water with his or her eyes open?
3. Is the refraction greater in air or in water? Does this mean objects under water seem closer or more distant than they would in air?
4. If you want the angle of refraction for the light ray in water to be the same as it is for air, what should the new angle of incidence be?

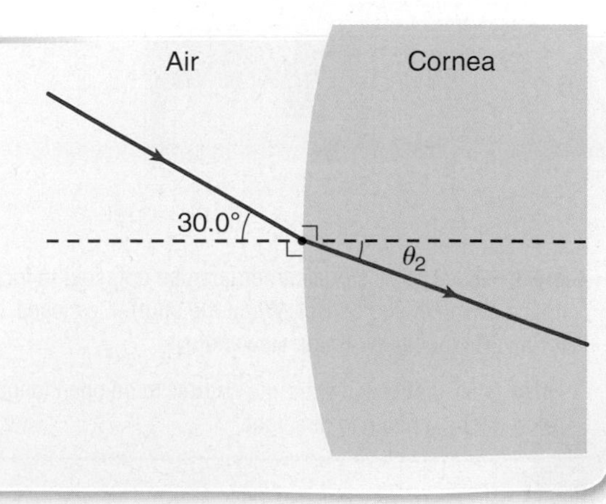

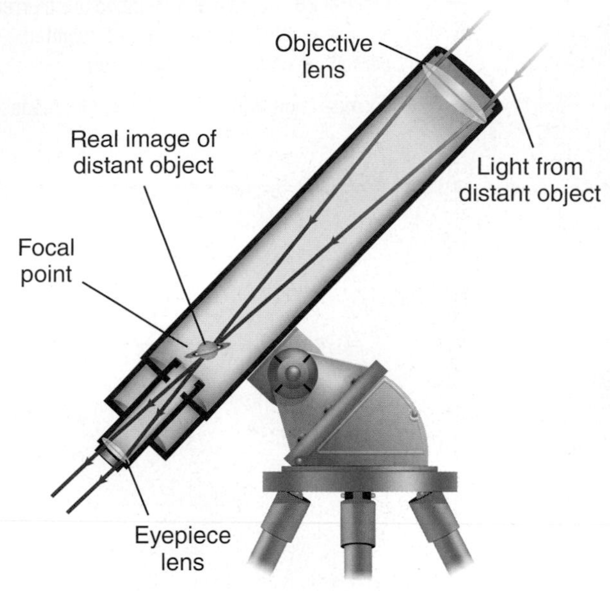

Figure 24 Light from distant objects is gathered by the objective lens and the eyepiece of a refracting telescope.

Evaluate Why is an inverted image acceptable for viewing astronomical objects?

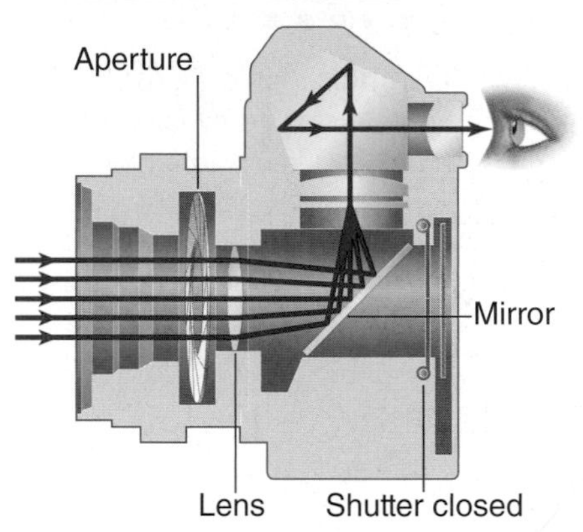

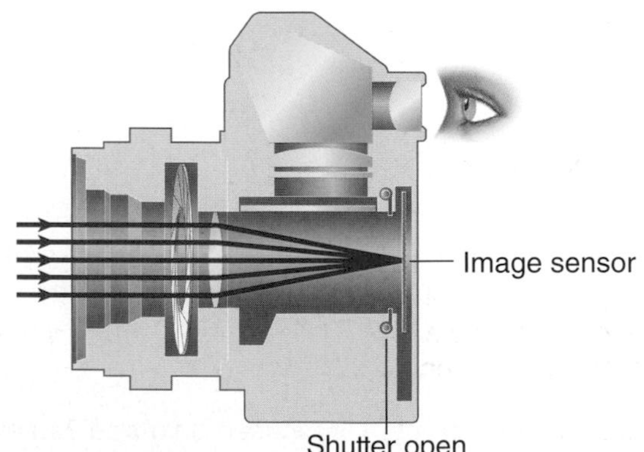

Figure 25 The lens of a camera can be adjusted to focus an image on the image sensor. When the shutter is closed, a mirror diverts the image to the viewer's eye.

Infer Why would you want the shutter to be open longer when taking a photo in dim light?

Refracting Telescopes

An astronomical refracting telescope uses lenses to magnify distant objects. **Figure 24** shows the optical system for a Keplerian telescope. Light from stars and other astronomical objects comes from so far away that the incoming rays can be considered parallel. These rays enter the objective convex lens, which focuses them as a real image at the focal point of the objective lens. The image is inverted compared to the object. This image then becomes the object for the convex lens of the eyepiece.

Notice that the eyepiece lens is positioned so that the focal point of the objective lens is between the eyepiece lens and its focal point. This means a virtual image is produced that is upright and larger than the first image. However, because the first image was already inverted, the final image is still inverted. For viewing astronomical objects, an image that is inverted is acceptable.

In a telescope, the convex lens of the eyepiece is almost always an achromatic lens. Recall that an achromatic lens is a combination of lenses that functions as one lens. The combination of lenses greatly reduces the peripheral colors that can form on images due to chromatic aberration.

☑ **READING CHECK** **Explain** how a refracting telescope works.

Cameras

Figure 25 shows the optical system used in a single-lens reflex camera. As light enters the camera through the aperture, it passes through an achromatic lens. This lens system refracts the light much like a single convex lens would, forming an image that is inverted on the reflex mirror. The image is reflected upward to a prism that inverts and redirects the light to the viewfinder.

When the person holding the camera takes a photograph, he or she presses the shutter-release button, which briefly raises the mirror, as shown in **Figure 25.** The light, instead of being diverted upward to the prism, then travels along a straight path to form an image on the sensor.

The image sensor, which is often a charge-coupled device (CCD), captures a two-dimensional image that corresponds to the image projected onto it. The image information is collected on the photoactive region of the CCD as electric charge proportional to the light intensity. The CCD then processes the charge and transfers the information to a storage device.

Microscopes

Like a telescope, a microscope has both an objective convex lens and a convex eyepiece lens. Microscopes, however, are used to magnify small objects. Images can be hundreds of times larger than the objects they magnify.

The lenses of a microscope have relatively small focal lengths. **Figure 26** is a diagram of the optical system used in a compound microscope.

When viewing a sample with a compound microscope, the object is located between one and two focal lengths from the objective lens. A real image is produced that is inverted and larger than the object. As with a telescope, this image then becomes the object for the eyepiece. This image is between the eyepiece and its focal point. A virtual image is produced that is upright and larger than the image of the objective lens. Thus, the viewer sees an image that is inverted and much larger than the original object.

Figure 26 The objective lens and the eyepiece in this compound microscope produce an image that is inverted and larger compared to the object.

View an **animation of a compound microscope.**

Binoculars

Have you ever used binoculars to view a sporting event or to watch birds? Binoculars, like telescopes, produce magnified images of faraway objects. However, binoculars produce a three-dimensional view. **Figure 27** shows a typical binocular design. Each side of the binoculars is like a small telescope.

Light enters a convex objective lens, which inverts the image. The light then travels through two prisms that use total internal reflection to invert the image again so that the viewer sees an image that is upright compared to the object. The prisms also extend the path along which the light travels and direct it toward the eyepiece of the binoculars.

Just as the separation of your two eyes gives you a sense of three dimensions and depth, the prisms allow a greater separation of the objective lenses, thereby improving the three-dimensional view of a distant object.

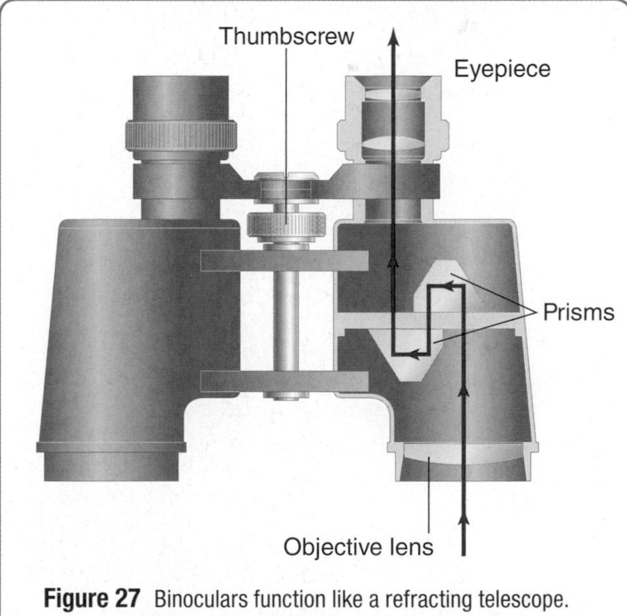

Figure 27 Binoculars function like a refracting telescope.

SECTION 3 REVIEW

Section Self-Check Check your understanding.

33. MAIN IDEA Which type of lens, convex or concave, should a nearsighted person use? Which type should a farsighted person use? See **Figure 28.** Explain.

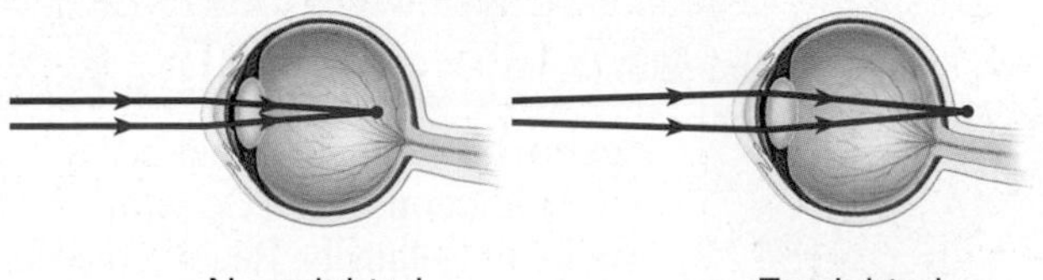

Figure 28

34. Refraction Explain why the cornea is the primary focusing element in the eye.

35. Camera Your camera is focused on a tree 2 m away. You want to focus on a tree that is farther away. Will the lens move closer to the sensor or farther away?

36. Critical Thinking When you use the highest magnification on a microscope, the image is much darker than it is at lower magnifications. What are some possible reasons for the darker image? What could you do to obtain a brighter image?

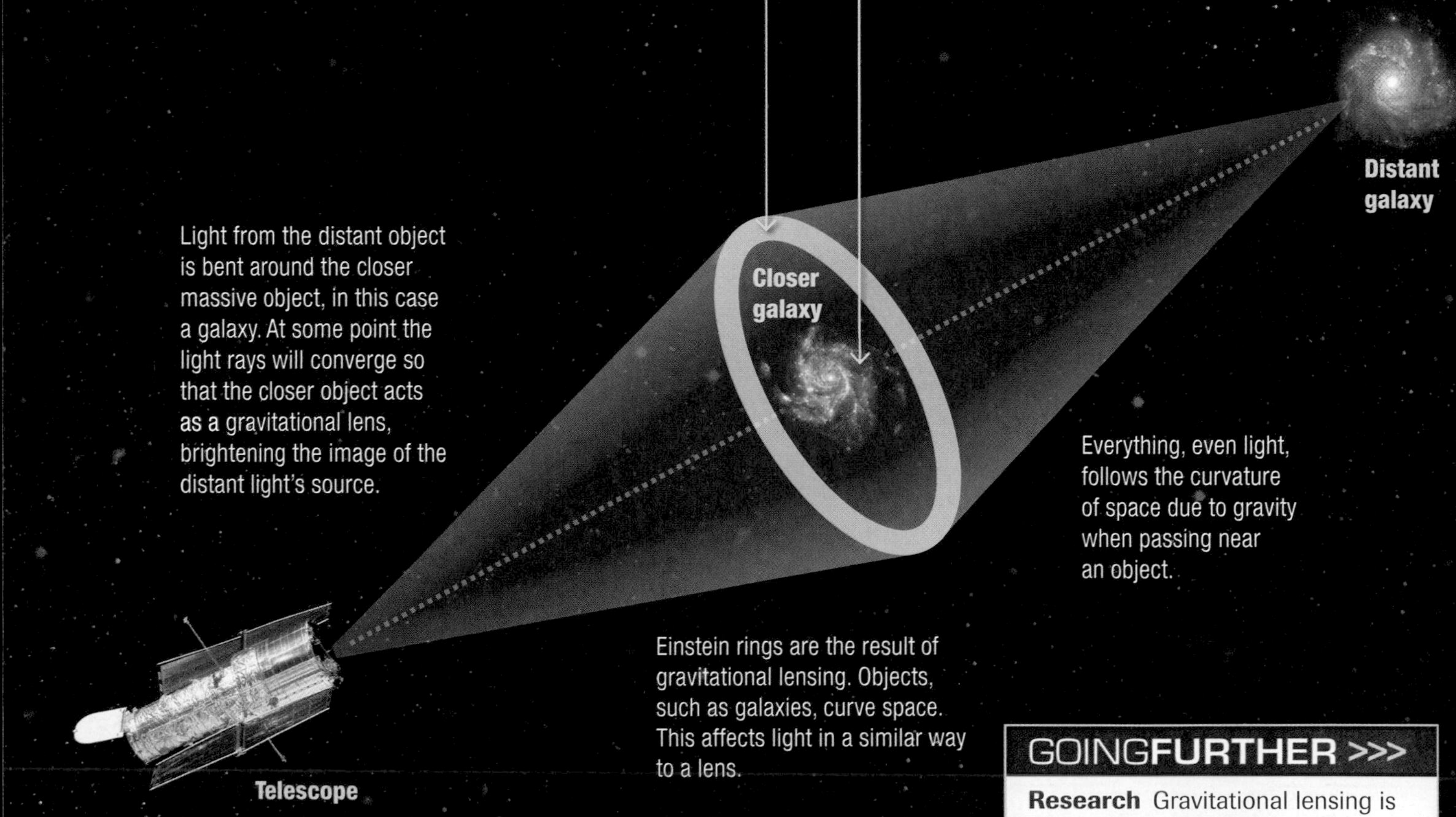

Distant galaxy

Light from the distant object is bent around the closer massive object, in this case a galaxy. At some point the light rays will converge so that the closer object acts as a gravitational lens, brightening the image of the distant light's source.

Closer galaxy

Everything, even light, follows the curvature of space due to gravity when passing near an object.

Einstein rings are the result of gravitational lensing. Objects, such as galaxies, curve space. This affects light in a similar way to a lens.

Telescope

View a **simulation of gravitational lensing.**

Concepts In Motion

GOING**FURTHER** >>>

Research Gravitational lensing is one prediction of the General Theory of Relativity. Research other predictions of the theory. How have these predictions been tested?

CHAPTER 18 STUDY GUIDE

BIGIDEA Lenses refract light and create images.

VOCABULARY
- **index of refraction** *(p. 493)*
- **critical angle** *(p. 496)*
- **total internal reflection** *(p. 496)*
- **dispersion** *(p. 498)*

SECTION 1 Refraction of Light

MAINIDEA The amount of refraction at a boundary depends on the indices of refraction of the two mediums and the angle of incidence.

- A beam of light refracts when it travels across a boundary from one medium with an index of refraction (n_1) into a medium with a different index of refraction (n_2). Refraction is described by Snell's law of refraction.

$$n_1 \sin \theta_1 = n_2 \sin \theta_2$$

- The speed of light in a medium is slower than the speed of light in a vacuum. The ratio of the speed of light in a vacuum (c) to the speed of light in a medium (v) is the index of refraction (n) of the medium.
- When light traveling through a medium hits a boundary with a medium of a smaller index of refraction, if the angle of incidence exceeds the critical angle (θ_c) the light will be reflected back into the original medium by total internal reflection. The indices of refraction for the mediums determine the critical angle.
- Optical effects such as mirages and rainbows are the result of refraction. Mirages occur due to the effect of temperature on n and rainbows occur because refracted white light is dispersed.

VOCABULARY
- **lens** *(p. 500)*
- **convex lens** *(p. 500)*
- **concave lens** *(p. 500)*
- **thin lens equation** *(p. 503)*
- **chromatic aberration** *(p. 506)*
- **achromatic lens** *(p. 506)*

SECTION 2 Convex and Concave Lenses

MAINIDEA Lenses can be used to enlarge and reduce images.

- A single convex lens produces a real image, formed by converging light rays, when the object is farther from the lens than its focal point. A single convex lens produces a virtual image, formed by diverging light rays, when the object is between the lens and the focal point. A single concave lens always produces a virtual image, formed by diverging light rays.
- Ray diagrams use two rays to determine the position, magnification, and orientation of an image formed by a lens. The thin lens equation provides the relationship between focal length (f), object position (x_o), and image position (x_i).

$$\frac{1}{f} = \frac{1}{x_i} + \frac{1}{x_o}$$

The magnification (m) of an image by a lens is defined by the magnification equation.

- All simple lenses have chromatic aberration. Chromatic aberration is reduced by using a combination of lenses with different indices of refraction.

VOCABULARY
- **nearsightedness** *(p. 509)*
- **farsightedness** *(p. 509)*

SECTION 3 Applications of Lenses

MAINIDEA People see objects that they could not otherwise see by using lenses.

- Differences in indices of refraction between air and the cornea are primarily responsible for focusing light in the eye.
- Nearsightedness is the inability to focus clearly on distant objects. A concave lens corrects nearsightedness. Farsightedness is the inability to focus clearly on nearby objects. A convex lens corrects farsightedness.
- Optical instruments use combinations of lenses to obtain clear images of small or distant objects.

Games and Multilingual eGlossary

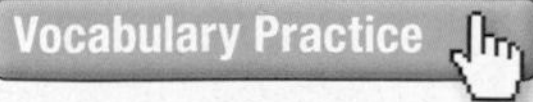

Chapter Self-Check

SECTION 1 Refraction of Light

Mastering Concepts

37. How does the angle of incidence compare with the angle of refraction when a light ray passes from air into glass at a nonzero angle?

38. How does the angle of incidence compare with the angle of refraction when a light ray leaves glass and enters air at a nonzero angle?

39. Regarding refraction, what is the critical angle?

40. Ranking Task Figure 29 depicts a ray of light traveling from air into several mediums. Rank the mediums according to index of refraction from greatest to least. Specifically indicate any ties.

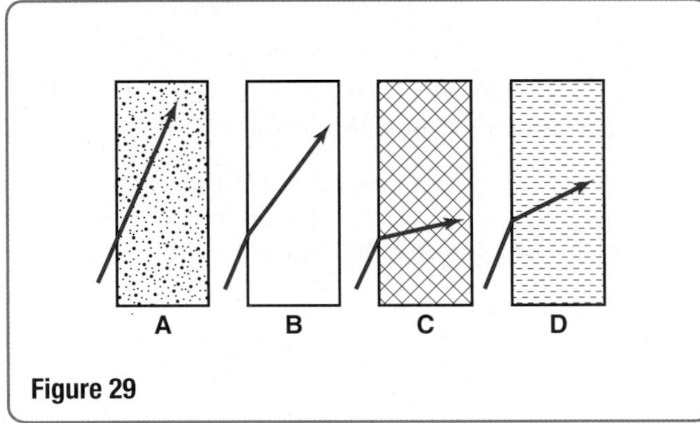

Figure 29

41. Although the light coming from the Sun is refracted while passing through Earth's atmosphere, the light is not separated into its spectrum. What does this indicate about the speeds of different colors of light traveling through air?

42. Explain why the Moon looks red during a lunar eclipse.

Mastering Problems

43. Refer to **Table 1.** Use the index of refraction of diamond to calculate the speed of light in diamond.

44. Light travels from flint glass into ethanol. The angle of refraction in the ethanol is 25.0°. What is the angle of incidence in the glass?

45. Refer to **Table 1.** Find the critical angle for a diamond in air.

46. A beam of light strikes the flat, glass side of a water-filled aquarium at an angle of 40.0° to the normal. For glass, $n = 1.50$.

a. At what angle does the beam enter the glass?

b. At what angle does the beam enter the water?

47. A ray of light travels from air into a liquid, as shown in **Figure 30.** The ray enters the liquid at an angle of 30.0°. The angle of refraction is 22.0°. Using Snell's law, calculate the index of refraction of the liquid. Compare the calculated index of refraction to those in **Table 1.** What might the liquid be?

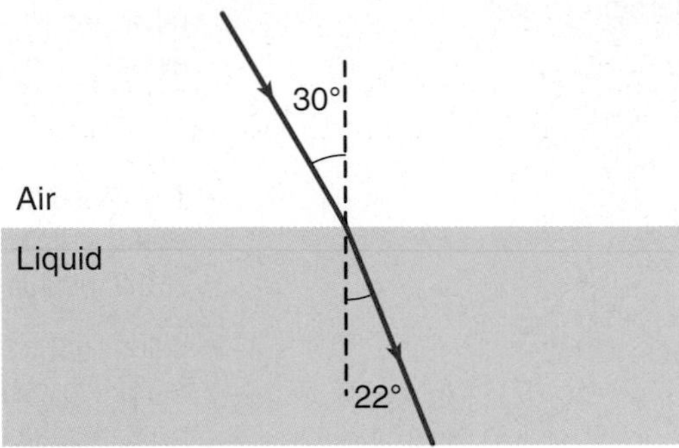

Figure 30

48. Aquarium Tank A thick sheet of plastic, $n = 1.500$, is used as the side of an aquarium tank. Light reflected from a fish in the water has an angle of incidence of 35.0°. At what angle does the light enter the air?

49. Swimming Pool Lights A light source is located 2.0 m below the surface of a swimming pool and 1.5 m from one edge of the pool, as shown in **Figure 31.** The pool is filled to the top with water. At what angle does the light reaching the edge of the pool leave the water? Does the light viewed from this angle appear deeper or shallower than it actually is?

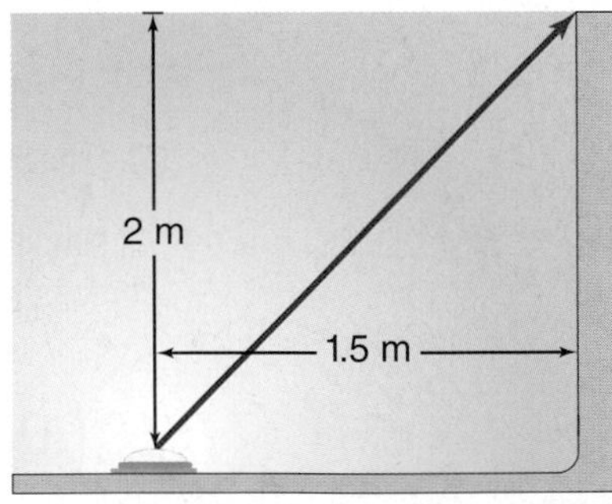

Figure 31 (Not to Scale)

50. The index of refraction of float glass is 1.53 for violet light, and it is 1.51 for red light. What is the speed of violet light in float glass? What is the speed of red light in float glass?

51. The critical angle for a special glass in air is 41.0°. What is the critical angle if the glass is immersed in water?

52. A diamond's index of refraction for red light, 656 nm, is 2.410, while that for blue light, 434 nm, is 2.450. Suppose that white light enters the diamond at 30.0°. Find the angles of refraction for red and blue light.

53. A ray of light in water has an angle of incidence of 55.0°. What is the angle of refraction in air?

54. The ray of light shown in **Figure 32** enters a 60°—60°—60° glass prism, $n = 1.5$.

a. Using Snell's law of refraction, determine the angle θ_2 to the nearest degree.

b. Using elementary geometry, determine the value of θ_1'.

c. Determine θ_2'.

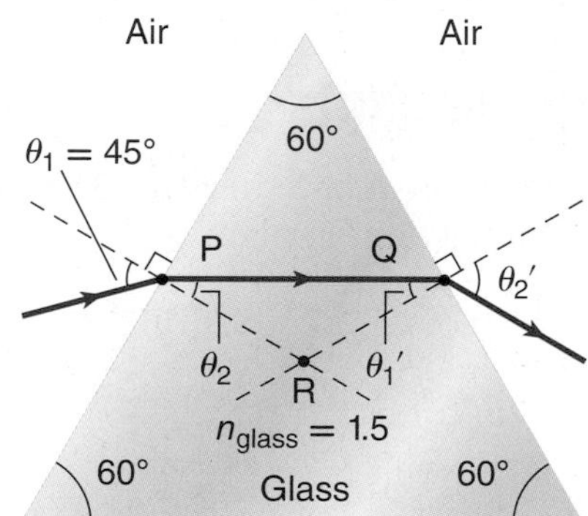

Figure 32

55. The speed of light in clear plastic is 1.90×10^8 m/s. A ray of light strikes the plastic at an angle of 22.0°. At what angle is the ray refracted?

56. A light ray enters a block of float glass, as illustrated in **Figure 33.** Use a ray diagram to trace the path of the ray until it leaves the glass.

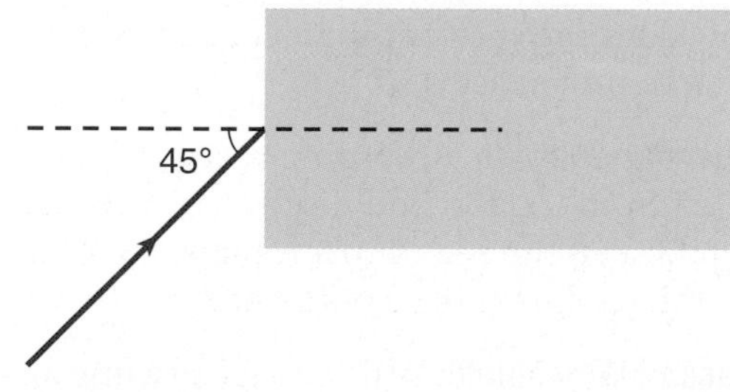

Figure 33

SECTION 2 Convex and Concave Lenses

Mastering Concepts

57. How do the shapes of convex and concave lenses differ?

58. What factor, other than the curvature of the surfaces of a lens, determines the location of the focal point of the lens?

59. Describe why precision optical instruments use achromatic lenses.

60. Locate and describe the properties of the image produced by a convex lens when an object is placed some distance beyond 2F.

61. To project an image from a movie projector onto a screen, the film is placed between F and 2F of a converging lens. This arrangement produces an image that is inverted. Why does the filmed scene appear to be upright when the film is viewed?

Mastering Problems

62. The focal length of a convex lens is 17 cm. A candle is placed 34 cm in front of the lens. Make a ray diagram to locate the image.

63. A cup on the table is placed 72.5 cm from a converging lens with a focal length of 25.5 cm. At what distance from the lens will the image be?

64. You are using a device with a convex lens to reduce the size of a drawing 0.75 times. If you place the drawing 24 cm from the lens to get this result, what is the focal length of the lens?

65. A 2.4-cm-tall piece of candy is located 14.0 cm from a convex lens that has a focal length of 6.0 cm. The candy is 2.4 cm tall.

a. Draw a ray diagram to determine the location, size, and orientation of the image.

b. Solve the problem mathematically.

66. An 8.0-cm-tall bottle of nail polish is placed 15.0 cm in front of a converging lens. A real image is formed 10.0 cm from the lens.

a. What is the focal length of the lens?

b. If the original lens is replaced with a lens having twice the focal length, what are the image position, size, and orientation?

67. A diverging lens has a focal length of 15.0 cm. A game piece placed near it forms a 2.0-cm-high image at a distance of 5.0 cm from the lens.

a. What are the position and the height of the game piece?

b. The diverging lens is now replaced by a converging lens with the same focal length. What are the image position, height, and orientation? Is it a virtual image or a real image?

68. **Reverse Problem** Write a physics problem with real-life objects for which the following equation would be part of the solution: $\frac{1}{0.06\text{ m}} + \frac{1}{x_i} = \frac{1}{0.04\text{ m}}$.

Chapter Self-Check

SECTION 3 Applications of Lenses

Mastering Concepts

69. BIGIDEA Describe how the eye focuses light.

70. What is the condition in which the focal length of the eye is too short to focus light on the retina?

71. What type of image is produced by the objective lens in a refracting telescope? How do you know this?

72. The prisms in binoculars increase the distance between the objective lenses. Why is this useful?

73. What is the purpose of a camera's reflex mirror?

Mastering Problems

74. **Camera Lenses** Camera lenses are described in terms of their focal length. A 50.0-mm lens has a focal length of 50.0 mm.
   a. A camera with a 50.0-mm lens is focused on an object 3.0 m away. What is the image position?
   b. A 1000.0-mm lens is focused on an object 125 m away. What is the image position?

75. **Eyeglasses** To clearly read a book 25 cm away, a farsighted girl needs the image to be 45 cm from her eyes. What focal length is needed for the lenses in her eyeglasses?

76. **Camera** A camera lens with a focal length of 35 mm is used to photograph a distant object. How far from the lens is the real image of the object? Explain.

77. **Copy Machine** The convex lens of a copy machine has a focal length of 25.0 cm. A letter to be copied is placed 40.0 cm from the lens.
   a. How far from the lens is the copy paper?
   b. How much larger will the copy be?

78. **Microscope** A slide of an onion cell is placed 12 mm from the objective lens of a microscope. The focal length of the objective lens is 10.0 mm.
   a. How far from the lens is the image formed?
   b. What is the magnification of this image?
   c. The real image formed is located 10.0 mm beneath the eyepiece lens. If the focal length of the eyepiece is 20.0 mm, where does the final image appear?
   d. What is the final magnification of this compound system?

79. **Telescope** The optical system of a toy refracting telescope consists of a converging objective lens with a focal length of 20.0 cm, located 25.0 cm from a converging eyepiece lens with a focal length of 4.05 cm. It is used to view a 10.0-cm-high object, located 425 cm from the objective lens.
   a. What are the image position, height, and orientation as formed by the objective lens? Is this a real or virtual image?
   b. The objective lens image becomes the object for the eyepiece lens. What are the image position, height, and orientation that a person sees when looking into the telescope? Is this a real image or a virtual image?
   c. What is the magnification of the telescope?

Applying Concepts

80. Which medium, A or B, in **Figure 34** has a larger index of refraction? Explain.

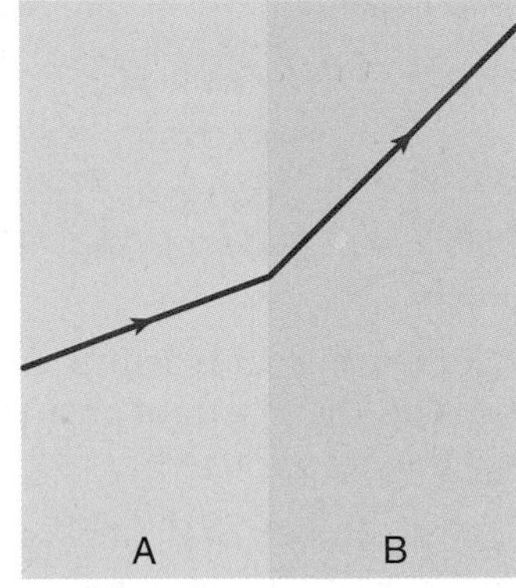

Figure 34

81. How does the speed of light change as the index of refraction increases?

82. **Legendary Mirage** According to legend, Eric the Red sailed from Iceland and discovered Greenland after he had seen the island in a mirage. Describe how the mirage might have occurred.

83. **Cracked Windshield** You see a silvery line along the crack of a cracked windshield. The glass has separated at the crack, and there is air in the crack. The silvery line indicates that light is reflecting off the crack. Draw a ray diagram to explain this. What phenomenon does this illustrate?

84. **Rainbows** Why would you never see a rainbow in the southern sky if you were in the northern hemisphere? In which direction should you look to see rainbows in the late afternoon if you are in the southern hemisphere?

Chapter Self-Check

85. A prism bends blue light more than it bends red light. Explain.

86. How does the size of the critical angle change as the index of refraction of the medium the ray is traveling into increases?

87. Why is chromatic aberration a defect for lenses but not for mirrors?

88. Which pair of mediums—air and water or air and float glass—has the smaller critical angle?

89. Suppose that **Figure 13** is redrawn with a lens of the same focal length but a larger diameter. Explain why the location of the image does not change. Would the image be affected in any way?

90. While under water, a swimmer uses a magnifying lens to observe a small object on the bottom of a swimming pool. She discovers that the magnifying lens does not magnify the object very well. Explain why the magnifying lens is not functioning as it would in air.

91. When subjected to bright sunlight, the pupils of your eyes are smaller than when they are subjected to dimmer light. Explain why your eyes can focus better in bright light.

92. **Binoculars** The objective lenses in binoculars form real upright images compared to their objects. Where are the images located relative to the eyepiece lenses?

Mixed Review

93. A 3.0-cm-tall object is placed 20 cm in front of a converging lens. A real image is formed 10 cm from the lens. What is the focal length of the lens?

94. A critical angle of 45.0° is found for light traveling through a block of float glass into another medium. What is the index of refraction for the medium the light is traveling into?

95. Find the speed of light in antimony trioxide if it has an index of refraction of 2.35.

96. **Astronomy** How many more minutes would it take light from the Sun to reach Earth if the space between them were filled with water rather than a vacuum? The Sun is 1.5×10^8 km from Earth.

97. **Bank Teller Window** A 25-mm-thick sheet of plastic, $n = 1.5$, is used in a bank teller's window. A ray of light strikes the sheet at an angle of 45°. The ray leaves the sheet at 45°, but at a different location. Use a ray diagram to find the distance between the ray that leaves and the one that would have left if the plastic were not there.

98. Derive $n = \frac{\sin \theta_1}{\sin \theta_2}$ from the general form of Snell's law of refraction, $n_1 \sin \theta_1 = n_2 \sin \theta_2$. State any assumptions and restrictions.

99. It is impossible to see through adjacent sides of a square block of glass with an index of refraction of 1.5. The side adjacent to the side that an observer is looking through acts as a mirror. **Figure 35** shows the limiting case for the adjacent side to not act like a mirror. Use your knowledge of geometry and critical angles to show that this ray configuration is not achievable when $n_{glass} = 1.5$.

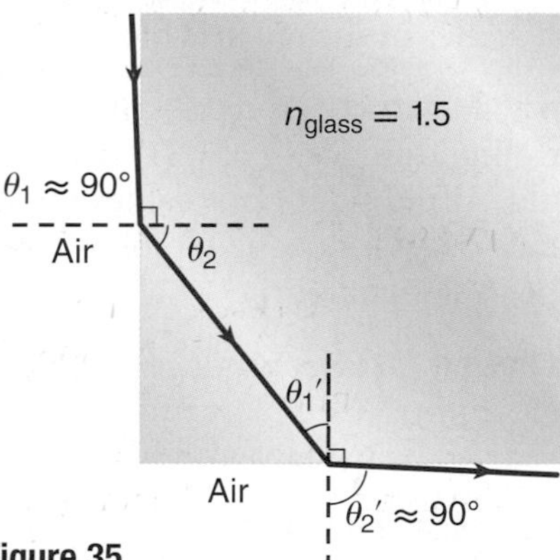

Figure 35

100. **Apparent Depth** Sunlight reflects diffusively off the bottom of an aquarium. **Figure 36** shows two of the many light rays that would reflect diffusively from a point off the bottom of the tank and travel to the surface. The light rays refract into the air as shown. The red dashed line extending back from the refracted light ray is a sight line that intersects with the vertical ray at the location where an observer would see the image of the bottom of the tank. Compute the direction that the refracted ray will travel above the surface of the water. At what depth does the bottom of the tank appear to be if you look into the water? Divide this apparent depth into the true depth and compare it to the index of refraction.

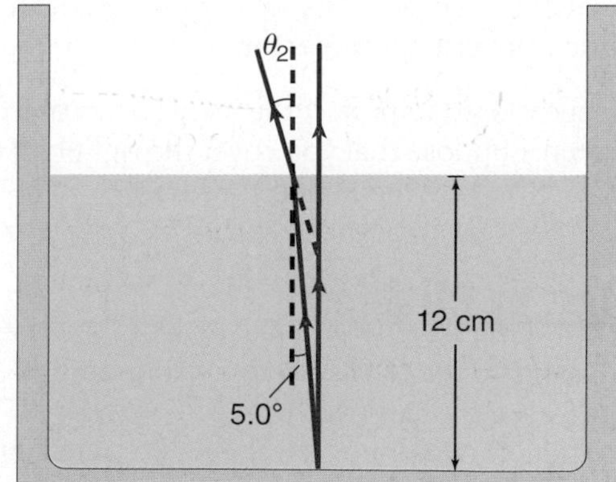

Figure 36

Chapter Self-Check

101. What is the focal length of the cornea in your eyes when you read a book that is held 35.0 cm from your corneas? The distance from each cornea to the retina is 19 mm.

Thinking Critically

102. **Recognize Spatial Relationships** White light traveling through air ($n = 1.00$) enters a slab of glass, incident at exactly 45°. For dense flint glass, $n = 1.7708$ for blue light ($\lambda = 435.8$ nm) and $n = 1.7273$ for red light ($\lambda = 643.8$ nm). What is the difference in the refracted angles of the red and blue light?

103. **Compare and Contrast** Find the critical angle for light traveling from ice ($n = 1.31$) to air. In a very cold world, would fiber-optic cables made of ice or those made of glass do a better job of keeping light inside the cable? Explain.

104. **Problem Posing** "A laser produces light of wavelength 453 nm...." Complete this problem so that it must be solved by using Snell's law.

105. **Recognize Cause and Effect** Your lab partner used a convex lens to produce an image with $x_i = 25$ cm and $h_i = 4.0$ cm. You are examining a concave lens with a focal length of −15 cm. You place the concave lens between the convex lens and the original image, 10 cm from the image. To your surprise, you see a real image on the wall that is larger than the object. You are told that the image from the convex lens is now the object for the concave lens, and because it is on the opposite side of the concave lens, it is a virtual object. Use these hints to find the new image position and image height and to predict whether the concave lens changed the orientation of the original image.

106. **Define Operationally** Name and describe the effect that causes the rainbow-colored fringe commonly seen at the edges of a spot of white light from a slide or overhead projector.

107. A lens is used to project the image of an object onto a screen. Suppose that you cover the right half of the lens. What will happen to the image?

Writing in Physics

108. Investigate the lens system used in an optical device such as an overhead projector or a particular camera or telescope. Prepare a graphics display for the class explaining how the device forms images.

109. The process of accommodation, whereby muscles surrounding the lens in the eye contract or relax to enable the eye to focus on close or distant objects, varies for different species. Investigate this effect for different animals. Prepare a report for the class showing how this fine focusing is accomplished for different eye mechanisms.

Cumulative Review

110. If you drop a 2.0-kg bag of lead shot from a height of 1.5 m, as shown in **Figure 37,** you could assume that half the potential energy will be converted into thermal energy in the lead. The other half would go to thermal energy in the floor. How many times would you have to drop the bag to heat it by 10°C?

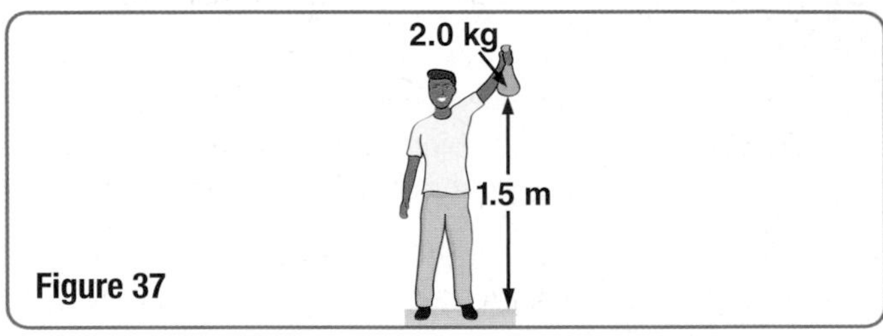

Figure 37

111. A car sounds its horn as it approaches a pedestrian in a crosswalk. Describe the sound of the car horn to the pedestrian as the car brakes to allow him to cross the street?

112. Suppose you could stand on the surface of the Sun and weigh yourself. Also suppose that you could measure the illuminance on your hand from the Sun's visible spectrum produced at that position. Next, imagine yourself traveling to a position 1000 times farther away from the center of the Sun as you were when standing on its surface.
   a. How would the force of gravity on you from the Sun compare to what it was at the surface?
   b. How would the illuminance on your hand from the Sun at the new position compare to what it was when you were standing on its surface? (For simplicity, assume that the Sun is a point source at both positions.)
   c. Compare the effect of distance upon the gravitational force and illuminance.

113. **Beautician's Mirror** The nose of a customer who is trying some face powder is 3.00-cm high and is located 6.00 cm in front of a concave mirror having a 14.0-cm focal length. Find the image position and height of the customer's nose by means of the following:
   a. a ray diagram drawn to scale
   b. the mirror and magnification equations

MULTIPLE CHOICE

1. A flashlight beam is directed at a swimming pool in the dark at an angle of 46° with respect to the normal to the surface of the water. What is the angle of refraction of the beam in the water? (The refractive index for water is 1.33.)

A. 18° **C.** 33°

B. 30° **D.** 44°

2. The speed of light in diamond is 1.24×10^{8} m/s. What is the index of refraction of diamond?

A. 0.0422 **C.** 1.24

B. 0.413 **D.** 2.42

3. Which of the phenomena below is not involved in the formation of rainbows?

A. diffraction **C.** reflection

B. dispersion **D.** refraction

4. George's picture is being taken by Cami, as shown in the figure below, using a camera that has a convex lens with a focal length of 0.0470 m. Determine George's image position.

A. 1.86 cm **C.** 4.82 cm

B. 4.70 cm **D.** 20.7 cm

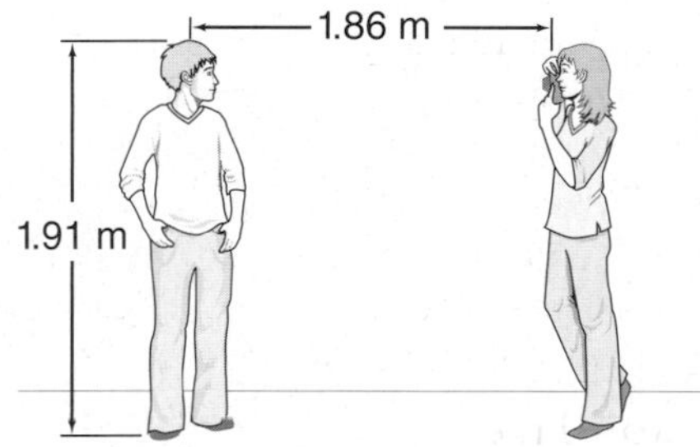

5. What is the magnification of an object that is 4.15 m in front of a camera that has an image position of 5.0 cm?

A. −0.83 **C.** 0.83

B. −0.012 **D.** 1.22

6. Which of the phenomena below is not involved in the formation of mirages?

A. heating of air near the ground

B. Huygens' wavelets

C. reflection

D. refraction

Online Test Practice

7. What is the image position for the situation shown in the figure below?

A. −6.00 m **C.** 0.167 m

B. −1.20 m **D.** 0.833 m

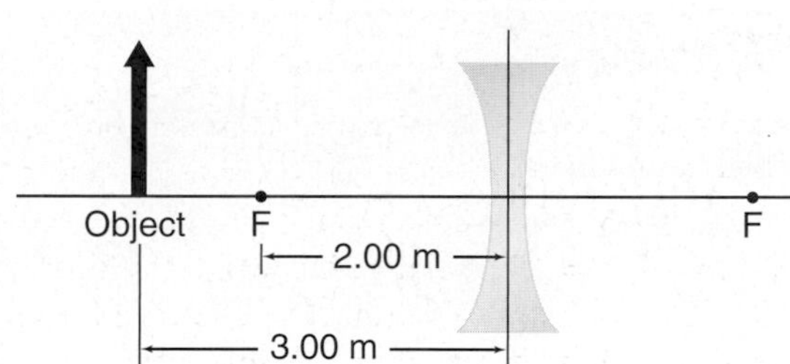

8. What is the critical angle for total internal reflection when light travels from glass (n = 1.52) to water (n = 1.33)?

A. 29.0° **C.** 48.8°

B. 41.2° **D.** 61.0°

9. What happens to the image formed by a convex lens when half the lens is covered?

A. Half the image disappears.

B. The image dims.

C. The image gets blurry.

D. The image inverts.

FREE RESPONSE

10. The critical angle for total internal reflection at a diamond-air boundary is 24.4°. What is the angle of refraction in the air if light enters the boundary at an angle of 20.0°?

11. An object that is 6.98 cm from a lens produces an image that is 2.95 cm from the lens on the same side of the lens. Determine the type of lens that is producing the image and explain how you know.

NEED EXTRA HELP?

If You Missed Question	1	2	3	4	5	6	7	8	9	10	11
Review Section	1	1	1	3	2	1	2	1	2	1	2

CHAPTER 19

Interference and Diffraction

BIG IDEA **Light waves can diffract and interfere with each other.**

SECTIONS

1 Interference

2 Diffraction

LaunchLAB

iLab Station

PATTERNS OF LIGHT

What patterns of light do you observe on a screen when you reflect different colored lights and white light off a compact disc?

WATCH THIS!

CD "RAINBOWS"

Why do DVDs and CDs reflect rainbows? What about oil slicks and insect wings? Investigate the physics of iridescence.

(l)Don Farrall/Getty Images, (r)Fuse/PunchStock

Go online!
connectED.mcgraw-hill.com

SECTION 1

Interference

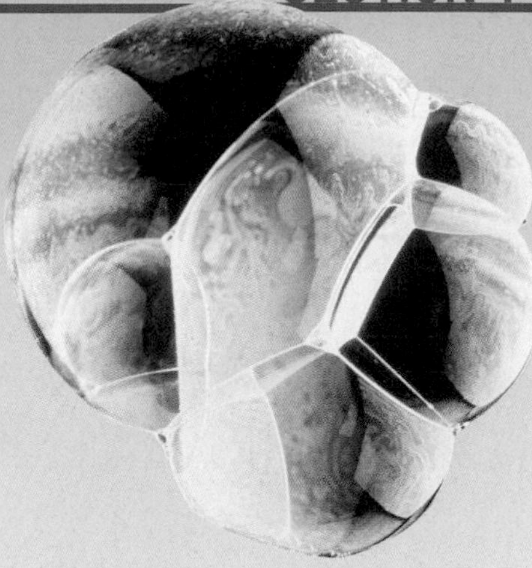

PHYSICS 4 YOU Have you ever seen a rainbow of swirling colors in soap bubbles or in soapy water? This is the result of a phenomenon called thin-film interference. How do light and matter interact to produce these patterns?

MAIN IDEA

Light can interfere when passing through narrow slits or reflecting from a thin film.

Essential Questions

- How does light falling on two slits produce an interference pattern?
- How can you use an interference pattern to find the wavelength of light?
- How can modeling techniques be applied to thin-film interference?

Review Vocabulary

interference results from the superposition of two or more waves

New Vocabulary

incoherent light
coherent light
interference fringes
monochromatic light
thin-film interference

Incoherent and Coherent Light

As you know, light has properties of a wave. Light diffracts as it passes an edge. When studying mirrors and lenses, you learned that reflection and refraction can be explained when light is modeled as a wave. What led scientists to believe that light has wave properties? They discovered that light could be made to interfere, which results from the superposition of waves.

Incoherent light When you look at objects that are illuminated by a white light source such as a nearby lightbulb, you are seeing **incoherent light,** which is light whose waves are not in phase. The effect of incoherence in waves can be seen in the example of heavy rain falling on still water. The surface of the water is choppy and does not have a regular pattern of waves, as shown in **Figure 1.** Because light waves have such a high frequency, incoherent light does not appear choppy to you. Instead, as light from an incoherent white light source illuminates an object, you see the combination of the incoherent light waves as an even, white light.

Coherent light Light made up of waves of the same wavelength that are in phase with each other is **coherent light.** A regular wavefront, which is made of coherent light, can be created by a single point source, as shown in **Figure 1.** A regular wavefront also can be created by multiple point sources when all point sources are in phase. This type of coherent light is produced by a laser.

Figure 1 Choppy, irregular wave patterns model incoherent light. Regular wave patterns model coherent light.

Incoherent Wave

Coherent Wave

Interference of Coherent Light

Between 1801 and 1803, English physician Thomas Young, **Figure 2,** performed a number of investigations establishing the wave properties of light. In the crucial investigation, attributed to Young, light from a small source was passed through two closely spaced slits and produced an interference pattern.

Young selected light from a tiny region of a source and made it coherent by passing through a narrow, single slit. The light was then passed through two closely spaced, narrow slits in a barrier. The overlapping light from the two slits fell on an observing screen. The overlap created a pattern of bright and dark bands called **interference fringes.** Young explained that the bands resulted from constructive and destructive interference of light waves from the two slits in the barrier.

Consider **monochromatic light,** which is light of only one wavelength. In a double-slit interference investigation that uses monochromatic light, constructive interference produces a bright central band of the given color on the screen, as well as other bright bands of near-equal spacing and near-equal width on either side, as shown in **Figure 3.** The intensity of the bright bands decreases the farther the band is from the central band, as you can see. Between the bright bands are dark areas where destructive interference occurs. The positions of the constructive and destructive interference bands depend on the light's wavelength.

When white light is used in a double-slit investigation, however, interference causes the appearance of colored spectra, as shown on the right in **Figure 3.** The various bands of color from the visible spectrum overlap on the screen. All these colors have constructive interference, and the central band is white. Because the positions of the other bright bands of constructive interference depend on wavelength, each color's band is at a different position, resulting in spectra of color.

Figure 2 Thomas Young (1773–1829) is famous for his contributions in many different subject areas. Along with his role in establishing the wave nature of light, he is known for his work deciphering Egyptian hieroglyphics.

PhysicsLAB

HOLOGRAMS

How are the interference of coherent light and holograms related?

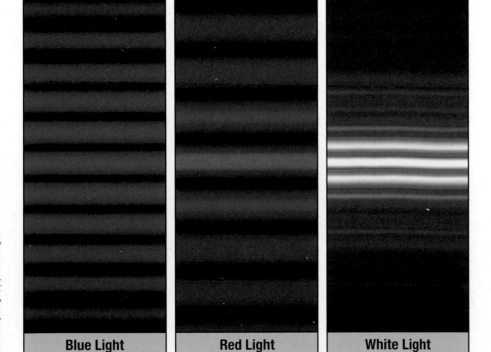

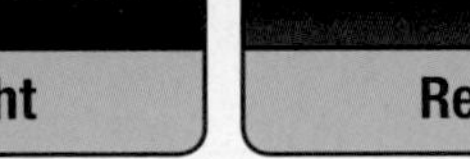

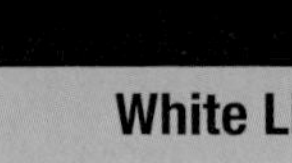

Figure 3 Double-slit interference patterns show a bright central band with a pattern of dark and bright bands on either side.

Figure 4 Nearly cylindrical wavefronts are generated as light passes through the slits.

Consider why the width of the slits is tens to hundreds of light wavelengths.

Investigate **double-slit interference.**

Virtual Investigation

Generation of coherent light Light from a monochromatic source produces incoherent light. Placing a light barrier with a narrow slit in front of the monochromatic light produces coherent light. Because the width of the slit is very small, only light from a tiny region of the source passes through the slit. Diffraction by the slit produces nearly cylindrical wavefronts, as shown in **Figure 4.** The second barrier has two very small slits. Because a cylinder is symmetrical, the two portions of the wavefront arriving at the second barrier are in phase. The two slits at the second barrier produce nearly cylindrical wavefronts. These two wavefronts can then interfere, as shown in **Figure 4.** Depending on their phase relationship, the two waves undergo constructive or destructive interference, as shown in **Figure 5.** If the interference is constructive when the light hits a screen, you will see a bright band. If it is destructive, you will see a dark band.

Figure 5 Coherent waves can experience constructive or destructive interference.

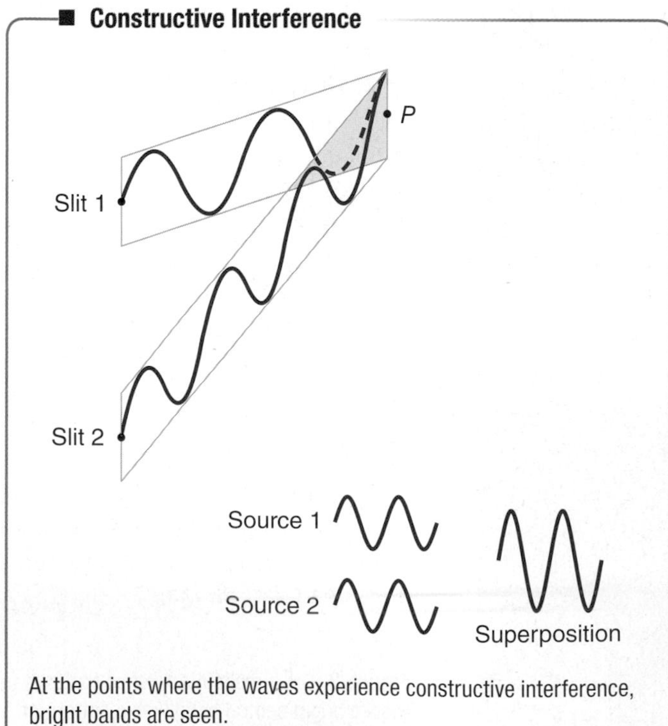

Constructive Interference
P
Slit 1
Slit 2
Source 1
Source 2
Superposition
At the points where the waves experience constructive interference, bright bands are seen.

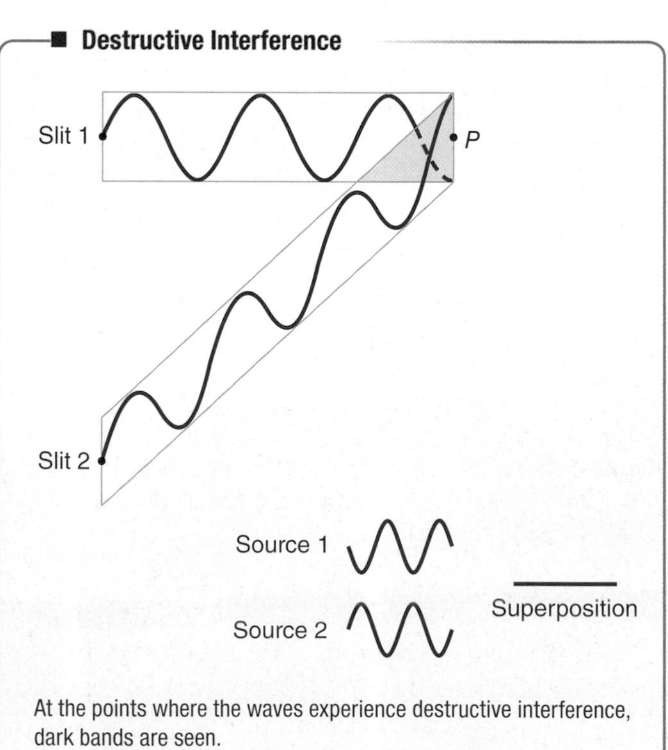

Destructive Interference
Slit 1
P
Slit 2
Source 1
Source 2
Superposition
At the points where the waves experience destructive interference, dark bands are seen.

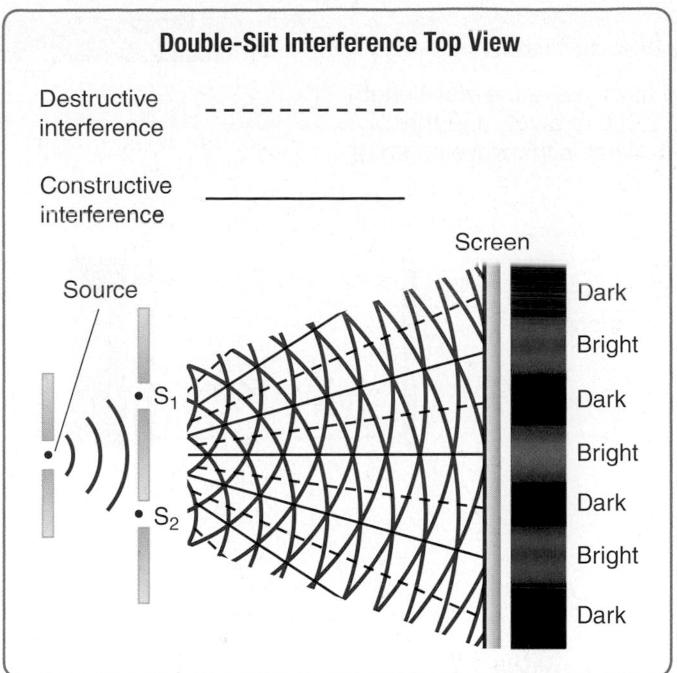

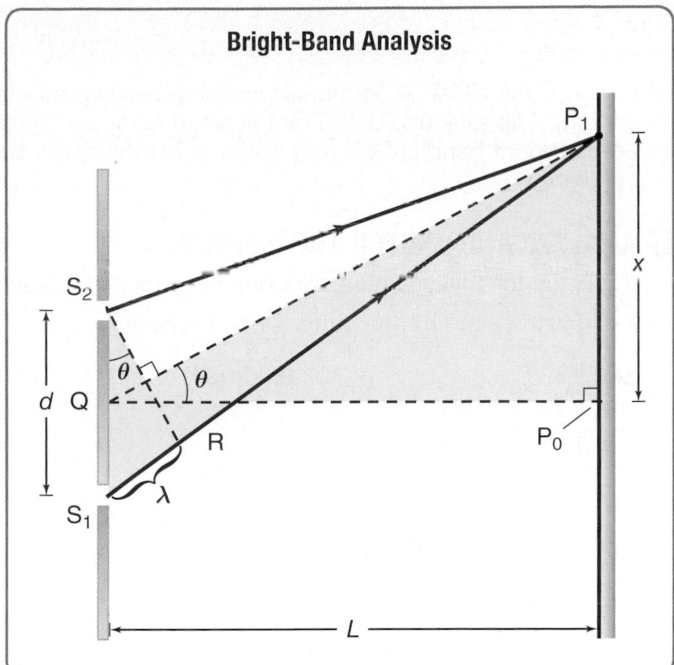

Figure 6 Double-slit interference can be used to determine the wavelength of light. Because L is much larger than d and the angle θ is small, the equation for the wavelength is simplified.

View an **animation of monochromatic light.**

Measuring the wavelength of light A top view of the double-slit investigation is shown in **Figure 6.** The wavefronts interfere constructively and destructively to form a pattern of light and dark bands. The diagram in **Figure 6** shows that light that reaches point P_0 travels the same distance from each slit. Because the waves are in phase, they interfere constructively on the screen to create the central bright band at P_0. There is also constructive interference at the first bright band (P_1) on either side of the central band because line segment P_1S_1 is one wavelength (λ) longer than the line segment P_1S_2. Thus, the waves arrive at P_1 in phase.

There are two triangles shaded in the figure. The larger triangle is a right triangle, so $\tan \theta = x/L$. In the smaller triangle RS_1S_2, the side $\overline{S_1R}$ is the length difference of the two light paths, which is one wavelength. There are now two simplifications for wavelength calculations.

1. If L is much larger than d, then line segments S_1P_1 and S_2P_1 are nearly parallel to each other and to line segment QP_1, and ΔRS_1S_2 is very nearly a right triangle. Thus, $\sin \theta \approx \lambda/d$.

2. If the angle θ is small, then $\sin \theta$ is very nearly equal to $\tan \theta$.

With the above simplifications, the relationships $\tan \theta = x/L$, $\sin \theta \approx \lambda/d$, and $\sin \theta \approx \tan \theta$ combine to form the equation $x/L = \lambda/d$. Solving for λ gives the following.

WAVELENGTH FROM DOUBLE-SLIT INVESTIGATION

The wavelength of light, as measured by a double slit, is equal to the distance on the screen from the central bright band to the first bright band, multiplied by the distance between the slits, divided by the distance from the slits to the screen.

$$\lambda = \frac{xd}{L}$$

Constructive interference occurs at locations x_m on either side of the central bright band, which are determined by the equation $m\lambda = x_m d/L$, where $m = 0, 1, 2$, etc. The central bright band occurs at $m = 0$. The band located where $m = 1$ is often called the first-order band, and so on.

PhysicsLABs

WHAT IS THE WAVELENGTH?

How are double-slit interference patterns related to diffraction and wavelength?

DOUBLE-SLIT INTERFERENCE

How can a double-slit interference pattern of light be used to measure the light's wavelength?

EXAMPLE PROBLEM 1

Find help with **operations with scientific notation.** Math Handbook

EXAMPLE PROBLEM

WAVELENGTH OF LIGHT A double-slit investigation is performed to measure the wavelength of red light. The slits are 0.0190 mm apart. A screen is placed 0.600 m away, and the first-order bright band is 21.1 mm from the central bright band. What is the wavelength of the red light?

1 ANALYZE AND SKETCH THE PROBLEM

- Sketch the investigation, showing the slits and the screen.
- Draw the interference pattern with bands in appropriate locations.

KNOWN	UNKNOWN
$d = 1.90\times10^{-5}$ m	$\lambda = ?$
$x = 2.11\times10^{-2}$ m	
$L = 0.600$ m	

2 SOLVE FOR THE UNKNOWN

$$\lambda = \frac{xd}{L}$$

$$= \frac{(2.11\times10^{-2}\text{ m})(1.90\times10^{-5}\text{ m})}{(0.600\text{ m})}$$ ◀ Substitute $x = 2.11\times10^{-2}$ m, $d = 1.90\times10^{-5}$ m, $L = 0.600$ m.

$$= 6.68\times10^{-7}\text{ m} = 668\text{ nm}$$

3 EVALUATE THE ANSWER

- **Are the units correct?** The answer is in units of length, which is correct for wavelength.
- **Is the magnitude realistic?** The wavelength of red light is near 700 nm, and that of blue is near 400 nm. Thus, the answer is reasonable for red light.

PRACTICE PROBLEMS

Do additional problems. Online Practice 

1. Violet light falls on two slits separated by 1.90×10^{-5} m. A first-order bright band appears 13.2 mm from the central bright band on a screen 0.600 m from the slits. What is λ?
2. Yellow-orange light from a sodium lamp of wavelength 596 nm is aimed at two slits that are separated by 1.90×10^{-5} m. What is the distance from the central band to the first-order yellow band if the screen is 0.600 m from the slits?
3. In a double-slit investigation, physics students use a laser with $\lambda = 632.8$ nm. A student places the screen 1.000 m from the slits and finds the first-order bright band 65.5 mm from the central line. What is the slit separation?
4. **CHALLENGE** Yellow-orange light with a wavelength of 596 nm passes through two slits that are separated by 2.25×10^{-5} m and makes an interference pattern on a screen. If the distance from the central line to the first-order yellow band is 2.00×10^{-2} m, how far is the screen from the slits?

Young presented his findings in 1803, but faced opposition from most physicists because they supported Newton's particle model of light. Young's conclusions gained acceptance after 1820, when Jean Fresnel proposed a mathematical solution for the wave nature of light in a competition. One of the judges, Siméon Denis Poisson, showed that if Fresnel was correct, the shadow of a circular object illuminated with coherent light would have a bright spot at its center. This had never been seen. Another judge, Jean Arago, did the investigation and saw the spot. The prediction and observation of the "Poisson Spot" convinced Poisson and Arago and many others of the wave nature of light.

Thin-Film Interference

Have you ever seen a spectrum of colors produced by a soap bubble or by the oily film on a water puddle in a parking lot, as in **Figure 7?** These colors are not the result of separation of white light by a prism or of absorption by a pigment. The colors are a result of the constructive and destructive interference when light waves reflect from separate surfaces of a thin film, a phenomenon called **thin-film interference.**

If a soap film is held vertically, as in **Figure 8,** its weight makes it thicker at the bottom than at the top. The thickness increases gradually from top to bottom. When a light wave strikes the front surface of the film, it is partially reflected, as shown by ray 1, and partially transmitted. The reflected and transmitted waves have the same frequency as the original. The transmitted wave travels through the film to the back surface where, again, part is reflected, as shown by ray 2. The splitting continues as the light makes many passes through the film. Because matched sets of waves came from the same source, they are coherent.

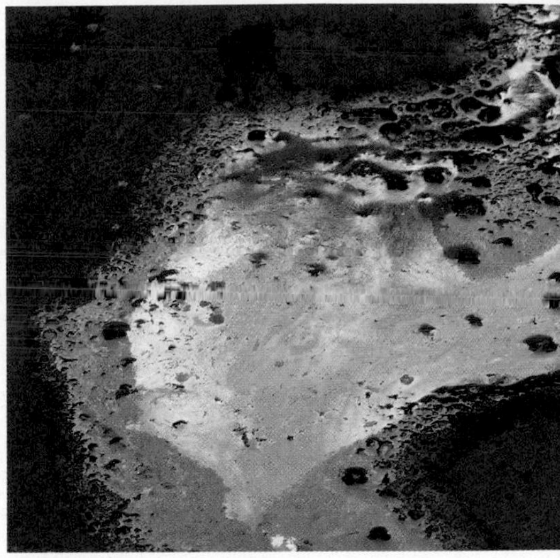

Figure 7 The swirl of colors seen in an oily film is the result of thin-film interference.

Color reinforcement How is the reflection of one color enhanced? This happens when the two reflected waves are in phase for a given wavelength. If the thickness of the soap film in **Figure 8** is one-fourth the wavelength of the light in the film ($\lambda/4$), then the round-trip path length in the film is $\lambda/2$. In this case, you might expect that ray 2 would return to the front surface one-half wavelength out of phase with ray 1 and that the two waves would cancel each other based on the superposition principle.

But when a transverse wave is reflected from a medium in which its speed is slower, the wave is inverted. With light, this happens at the boundary of a medium with a larger index of refraction. As a result, ray 1 is inverted on reflection; whereas ray 2 is reflected from a medium with a smaller index of refraction (air) and is not inverted. Thus, ray 1 and ray 2 are in phase.

If the film thickness (d) satisfies the requirement $d = \lambda/4$, then the color of light with that wavelength will be most strongly reflected. Note that because the wavelength of light in the film is shorter than the wavelength in air, $d = \lambda_{film}/4$, or, in terms of the wavelength in air, $d = \lambda_{vacuum}/4n_{film}$. The two waves reinforce each other as they leave the film. Light with other wavelengths undergoes destructive interference.

MiniLAB

SOAP FILM

What does thin-film interference caused by soap film look like?

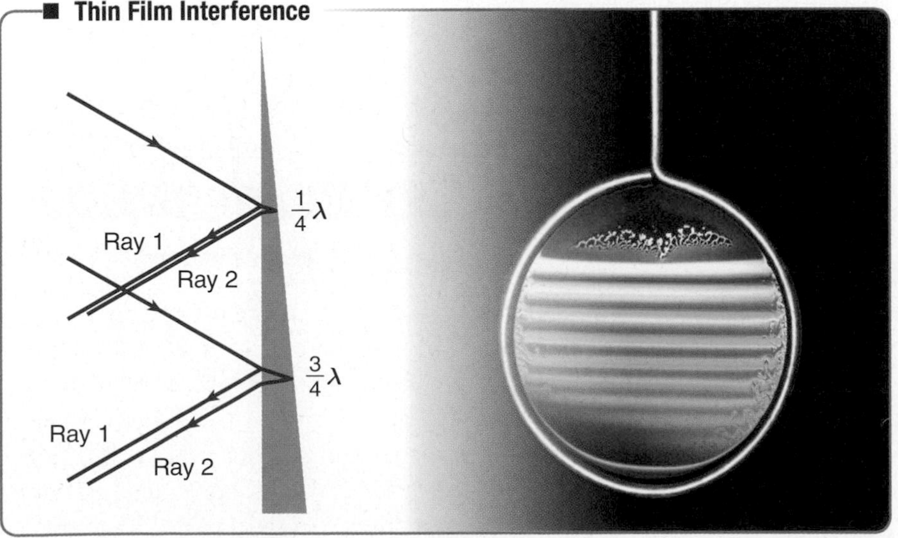

Figure 8 At soap film thicknesses of $\lambda/4$, $3\lambda/4$, $5\lambda/4$, etc., light with wavelength λ is in phase. Bands of that color light are visible at those thicknesses.

(t)Joel Sartore/National Geographic/Getty Images, (b)Tom Pantages

NONREFLECTIVE EYEGLASS LENSES A thin film can be placed on the lenses of eyeglasses to keep them from reflecting wavelengths of light that are highly visible to the human eye. This reduces the glare of reflected light and permits more light to be transmitted.

As you know, different colors of light have different wavelengths. For a film of varying thickness, the wavelength requirement will be met at different thicknesses for different colors. The result is a rainbow of color. Where the film is too thin to produce constructive interference for any wavelength of visible light, the film appears to be black. Notice in **Figure 8** that the pattern of colors that appear on the film repeats. When the thickness of the film is $3\lambda/4$, the round-trip distance is $3\lambda/2$, and constructive interference occurs for light with a wavelength λ again. Any thickness equal to $1\lambda/4$, $3\lambda/4$, $5\lambda/4$, and so on satisfies the conditions for constructive interference for a given wavelength.

Applications of thin-film interference The example of a film of soapy water in air involves constructive interference with one of two waves inverted upon reflection. In the example of a bubble solution or thin film of oil on a puddle of water, as the thickness of the film or the angle the light makes with the film changes, the wavelength undergoing constructive interference changes. This creates a shifting color on the surface of the film when it is under white light.

In other examples of thin-film interference, neither wave or both waves might be inverted. Whether a wave is inverted depends on the refractive indexes of the mediums involved. If both waves are traveling from a lower to a higher index of refraction, they will both be inverted. In this case, the film thicknesses for constructive interference are $1\lambda/2$, λ, $3\lambda/2$, 2λ, $5\lambda/2$, and so on. You can develop a solution for any problem involving thin-film interference by using the following strategy.

PROBLEM-SOLVING STRATEGIES

THIN-FILM INTERFERENCE

When solving thin-film interference problems, construct an equation that is specific to the problem by using the following strategies.

1. Make a sketch of the thin film and the two coherent waves. For simplicity, draw the waves as rays.
2. Read the problem. Is the reflected light of this wavelength brightened or dimmed? When it is bright, the two reflected waves undergo constructive interference. When the reflected light is dimmed, the waves undergo destructive interference.
3. Are either or both waves inverted on reflection? If the index of refraction changes from a lower to a higher value, then the wave is inverted. If it changes from a higher to a lower value, there is no inversion.
4. Find the extra distance the second wave must travel through the thin film to create the needed interference.
   - **a.** If you need constructive interference and one wave is inverted OR you need destructive interference and either both waves or none are inverted, then the difference in distance is an odd number of half wavelengths: $(m + 1/2)\lambda_{film}$, where $m = 0, 1, 2$, etc.
   - **b.** If you need constructive interference and either both waves or none are inverted OR you need destructive interference and one wave is inverted, then the difference is an integer number of wavelengths: $m\lambda_{film}$, where $m = 1, 2, 3$, etc.
5. Set the extra distance traveled by the second ray to twice the film thickness, $2d$.
6. Recall from studying refraction that $\lambda_{film} = \lambda_{vacuum}/n_{film}$.

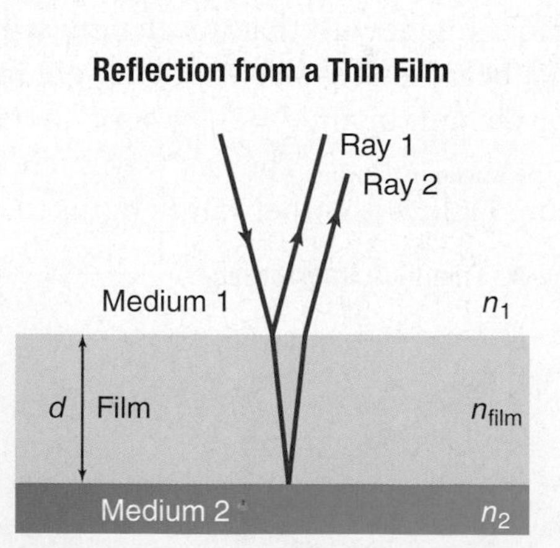

EXAMPLE PROBLEM 2

Find help with **operations with significant figures.** Math Handbook

OIL AND WATER You observe colored rings on a puddle and conclude that there must be an oil slick on the water. You look directly down at the puddle and see a yellow-green ($\lambda = 555$ nm) region. If the refractive index of oil is 1.45 and that of water is 1.33, what is the minimum thickness of oil that could cause this color?

1 ANALYZE AND SKETCH THE PROBLEM

- Sketch the thin film and layers above and below it.
- Draw rays showing reflection off the top of the film as well as the bottom.

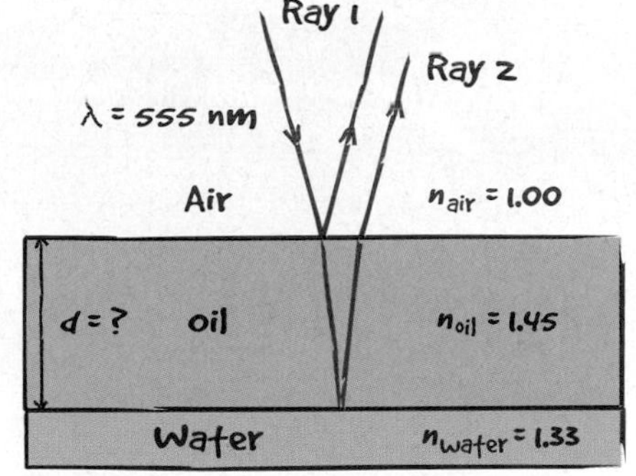

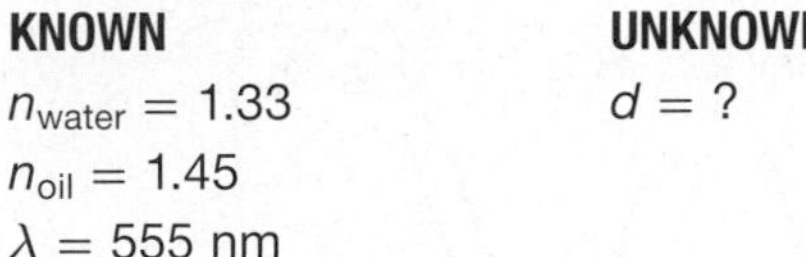

KNOWN	UNKNOWN
$n_{water} = 1.33$	$d = ?$
$n_{oil} = 1.45$	
$\lambda = 555$ nm	

2 SOLVE FOR THE UNKNOWN

Because $n_{oil} > n_{air}$, the wave is inverted on the first reflection. Because $n_{water} < n_{oil}$, there is no inversion on the second reflection. Thus, there is one wave inversion. The wavelength in oil is less than it is in air.

Follow the problem-solving strategy to construct the equation.

$$2d = \left(m + \frac{1}{2}\right)\left(\frac{\lambda}{n_{oil}}\right)$$

Because you want the minimum thickness, $m = 0$.

$$d = \frac{\lambda}{4n_{oil}}$$ ◀ *Substitute $m = 0$.*

$$= \frac{555 \text{ nm}}{(4)(1.45)}$$ ◀ *Substitute $\lambda = 555$ nm, $n_{oil} = 1.45$.*

$$= 95.7 \text{ nm}$$

3 EVALUATE THE ANSWER

- **Are the units correct?** The answer is in nm, which is correct for thickness.
- **Is the magnitude realistic?** The minimum thickness is smaller than one wavelength, which is what it should be.

PRACTICE PROBLEMS

Do additional problems. Online Practice

5. In the situation in Example Problem 2, what would be the thinnest film that would create a reflected red ($\lambda = 635$ nm) band?
6. A glass lens has a nonreflective coating of magnesium fluoride placed on it. How thick should the nonreflective layer be to keep yellow-green light with a wavelength of 555 nm from being reflected? See the sketch in **Figure 9.**
7. You can observe thin-film interference by dipping a bubble wand into some bubble solution and holding the wand in the air. What is the thickness of the thinnest soap film at which you would see a black stripe if the light illuminating the film has a wavelength of 521 nm? Use $n = 1.33$ for the bubble solution.
8. What is the thinnest soap film ($n = 1.33$) for which light of wavelength 521 nm will constructively interfere with itself?
9. **CHALLENGE** A silicon solar cell has a nonreflective coating placed on it. If a film of silicon monoxide, $n = 1.45$, is placed on the silicon, $n = 3.5$, how thick should the layer be to keep yellow-green light ($\lambda = 555$ nm) from being reflected?

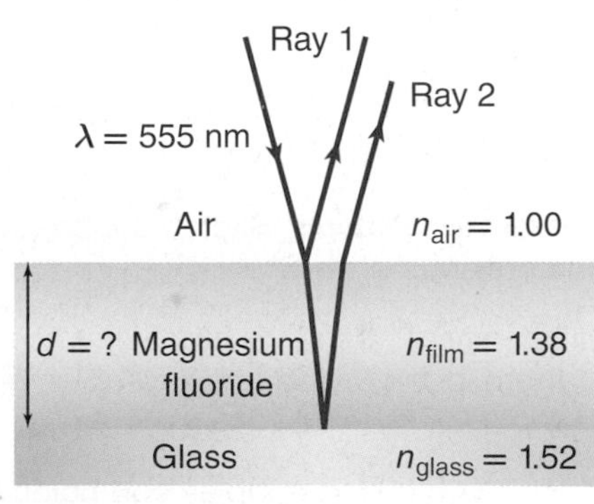

Figure 9

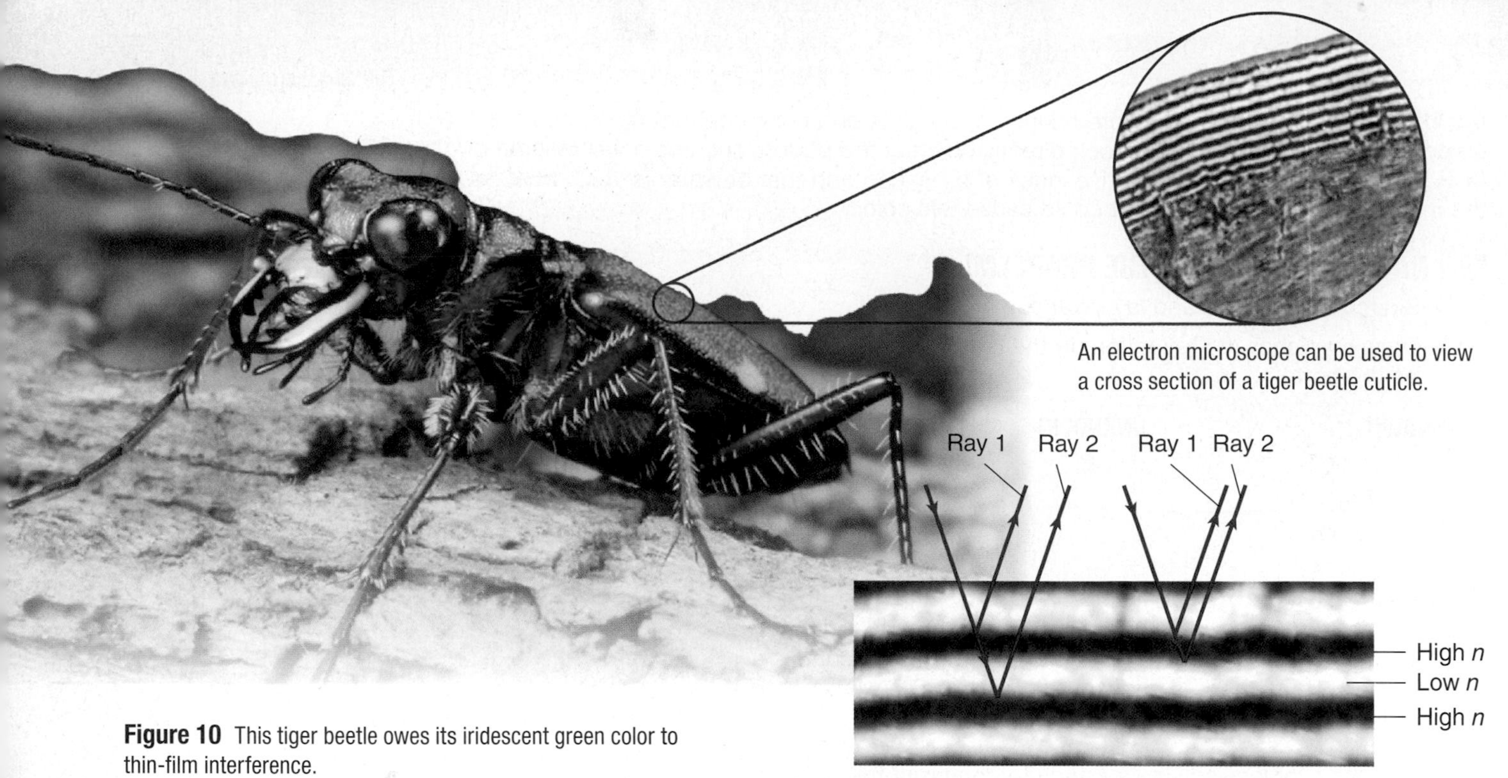

An electron microscope can be used to view a cross section of a tiger beetle cuticle.

Ray 1 Ray 2 Ray 1 Ray 2

High n
Low n
High n

Figure 10 This tiger beetle owes its iridescent green color to thin-film interference.

Interference can occur from individual structures and from multiple structures.

Light interference also occurs naturally in the outer layer of the shells of many beetles, as shown in **Figure 10.** The shimmering green of the tiger beetle is the result of reflection from thin, parallel layers of chitin and sometimes other materials that differ in refractive index. Electron micrograph images show these parallel layers. A diagram showing how these multilayer reflectors work is shown in **Figure 10.** The many layers of the exoskeleton reflect light such that the result is constructive interference of green light. A shimmering appearance results. Many other beetles and butterflies, as well as the gemstone opal, shimmer due to light interference.

SECTION 1 REVIEW

10. MAINIDEA Two very narrow slits are cut close to each other in a large piece of cardboard. They are illuminated by monochromatic red light. A sheet of white paper is placed far from the slits, and a pattern of bright and dark bands is seen on the paper. Describe how a wave behaves when it encounters a slit, and explain why some regions are bright while others are dark.

11. Interference Patterns Sketch the pattern described in the previous problem.

12. Interference Patterns Sketch what happens to the pattern in the previous two problems when the red light is replaced by blue light.

13. Lucien is blowing bubbles and holds the bubble wand with a soap film ($n = 1.33$) in it vertically.

a. What is the second thinnest width of the soap film at which he could see a bright stripe if the light illuminating the film has a wavelength of 575 nm?

b. What other widths produce a bright stripe at 575 nm?

14. Light of wavelength 542 nm falls on a double slit. Use the values from **Figure 11** to determine how far apart the slits are.

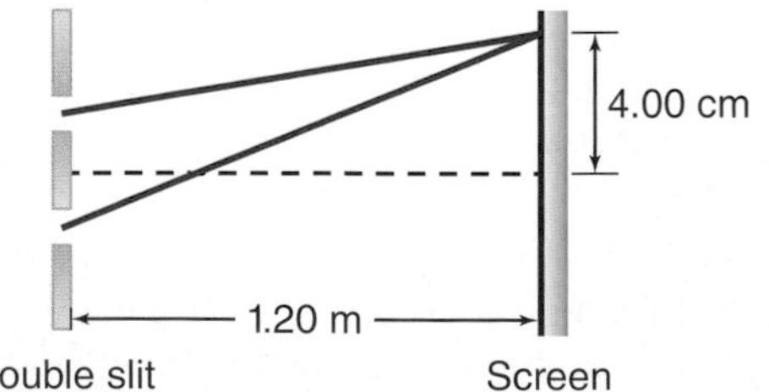

Figure 11

15. Critical Thinking The equation for wavelength from a double-slit investigation uses the simplification that θ is small so that $\sin \theta \approx \tan \theta$. Up to what angle is this a good approximation when your data has two significant figures? Would the maximum angle for a valid approximation increase or decrease as you increase the precision of your angle measurement?

(l)George Grall/National Geographic/Getty Images, (r)Ainsley E Seago/2008 The Royal Society

SECTION 2

Diffraction

PHYSICS 4 YOU **Microscopic pits in compact discs and DVDs can produce a spectrum of reflected light through diffraction. Scientists can also use a knowledge of diffraction to find the wavelengths of light and to study molecular structures, such as DNA.**

MAIN IDEA

Light waves diffract when they pass through a single slit and diffract and interfere when they encounter a diffraction grating.

Essential Questions

- What affects the width of the bright central band in a single-slit diffraction pattern?
- How do diffraction gratings form diffraction patterns?
- How are diffraction gratings used in diffraction grating spectrometers?
- How does diffraction limit the ability to distinguish two closely spaced objects with a lens?

Review Vocabulary

diffraction the bending of light around a barrier

New Vocabulary

diffraction pattern
diffraction grating
Rayleigh criterion

Single-Slit Diffraction

When studying light, you learned that wavefronts of light diffract when they pass around an edge. Diffraction can be explained by using Huygens' principle that a wavefront is made up of many small point-source wavelets. When light passes through a slit that has two closely spaced edges, a pattern is produced on a screen. This pattern, called a **diffraction pattern,** results from constructive and destructive interference of Huygens' wavelets.

When coherent, blue light passes through a single, small opening that is between about 10 and 100 light wavelengths, the light is diffracted by both edges, and a series of bright and dark bands appears on a distant screen, as shown in **Figure 12.** Instead of the nearly equally spaced bands produced by two coherent sources in Young's double-slit investigation, this pattern has a wide, bright central band with dimmer, narrower bands on either side. When using red light instead of blue, the width of the bright central band increases. With white light, the pattern is a combination of patterns of all the colors of the spectrum.

Figure 12 Single-slit diffraction produces one wide dark central band and narrower, dimmer bands on either side.

Compare the colors and widths of the central bands produced by the different colors of light.

(tl)PhotoDisc/Getty Images, (bl bc br)Tom Pantages

Figure 13 To illustrate diffraction patterns with Huygens' wavelets, a pair of points is chosen such that the separation between the points is $\frac{w}{2}$.

View an **animation of single-slit diffraction.**

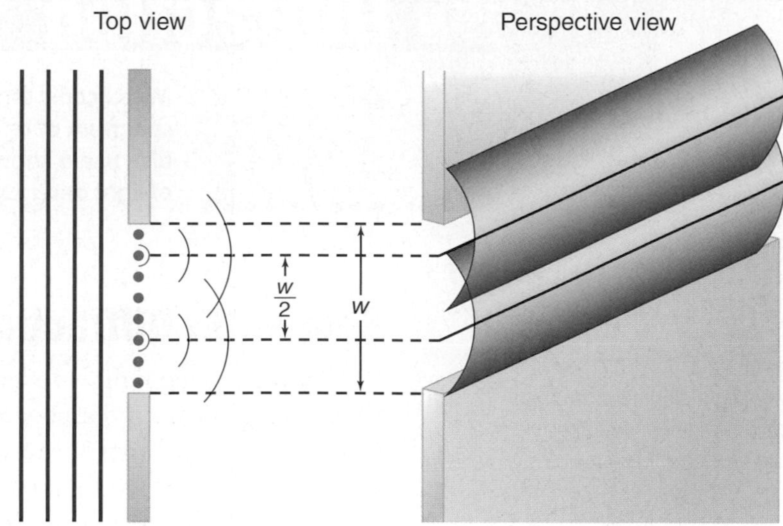

Huygens' wavelets To see how Huygens' wavelets produce the diffraction pattern, imagine a slit of width w as being divided into an even number of Huygens' points, as shown in **Figure 13.** Each point acts as a source of Huygens' wavelets. Divide the slit into two equal halves, and choose one source from each part so that the pair is separated by a distance $w/2$. This pair of sources produces coherent, cylindrical waves that will interfere.

For any Huygens' wavelet produced in the top half, there will be another Huygens' wavelet in the bottom half, a distance $w/2$ away, that it will interfere with destructively to create a dark band on the screen. All similar pairings of Huygens' wavelets interfere destructively at dark bands. Conversely, a bright band on the screen is where pairings of Huygens' wavelets interfere constructively. In the dim regions between bright and dark bands, partial destructive interference occurs.

☑ **READING CHECK Identify** the type of interference of Huygens' wavelets that creates a dark band on the screen.

Diffraction pattern When the single slit is illuminated, a central bright band appears at location P_0 on the screen, as shown in **Figure 14.** The first dark band is at position P_1. At this location, the path lengths r_1 and r_2 of the two Huygens' wavelets differ by one-half wavelength, producing destructive interference. This model is mathematically similar to that of double-slit interference. A comparison of a single-slit diffraction pattern with a double-slit interference pattern using slits of the same width reveals that the entire single-slit diffraction pattern is covered by the narrow dark and light interference bands. Double-slit patterns result from the interference of the light from each single slit.

We will now develop an equation for the diffraction pattern produced by a single slit. We use the same simplifications that were used for double-slit interference, assuming that the distance to the screen is much larger than w. As described above, the separation distance between the sources of the two interfering waves is $w/2$. To find the distance measured on the screen to the first dark band (x_1), note that the path length difference is $\lambda/2$ because at the dark band there is destructive interference. As a result, $x_1/L = \lambda/w$.

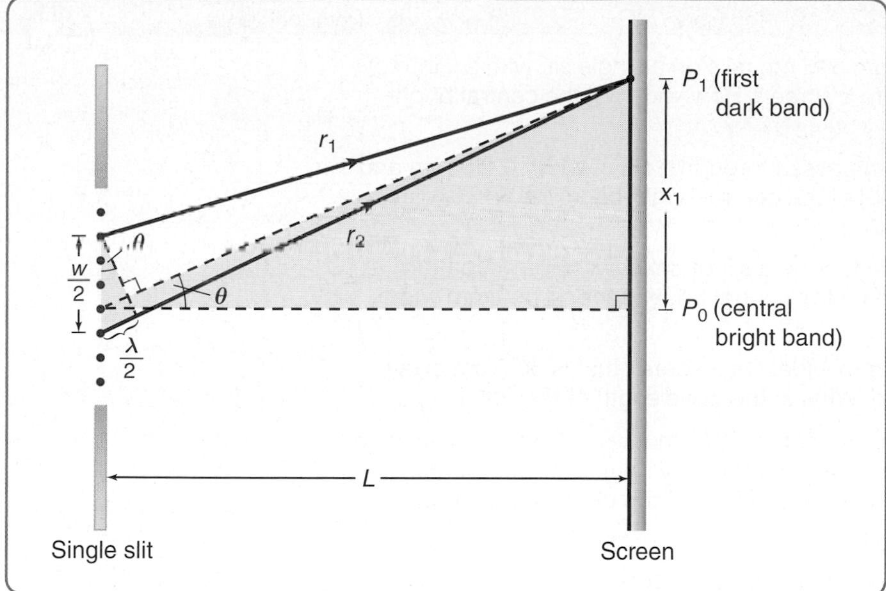

Figure 14 The width of the bright band in single-slit diffraction is related to the wavelength of light, the distance from the slit to the screen, and the width of the slit.

View an **animation describing the first dark band.**

Concepts In Motion

Instead of measuring the distance to the first dark band from the center of the central bright band (x_1), it is better to determine the width of the central bright band ($2x_1$), as in the following equation.

WIDTH OF BRIGHT BAND IN SINGLE-SLIT DIFFRACTION

The width of the central bright band is equal to the product of twice the wavelength times the distance to the screen, divided by the width of the slit.

$$2x_1 = \frac{2\lambda L}{w}$$

Single-slit diffraction patterns make the wave nature of light noticeable when the slits are 10–100 times the wavelength of the light. Larger openings cast sharp shadows, as Isaac Newton first observed. While the single-slit pattern depends on the wavelength of light, it is not very useful for measuring wavelength. A large number of slits put together provides a more useful tool for measuring wavelength.

READING CHECK **Describe** the assumption that is made concerning w and L for the single-slit diffraction equation.

PHYSICS CHALLENGE

You have several unknown substances and wish to use a single-slit diffraction apparatus to determine what each one is. You decide to place a sample of an unknown substance in the region between the slit and the screen and use the data you obtain to determine the identity of each substance by calculating its index of refraction.

1. Come up with a general formula for the index of refraction of an unknown substance in terms of the wavelength of the light (λ_{vacuum}), the width of the slit (w), the distance from the slit to the screen (L), and the distance between the central bright band and the first dark band (x_1).
2. If the source you used had a wavelength of 634 nm, the slit width was 0.10 mm, the distance from the slit to the screen was 1.15 m, and you immersed the apparatus in water ($n_{substance} = 1.33$), then what would you expect the width of the center band to be?

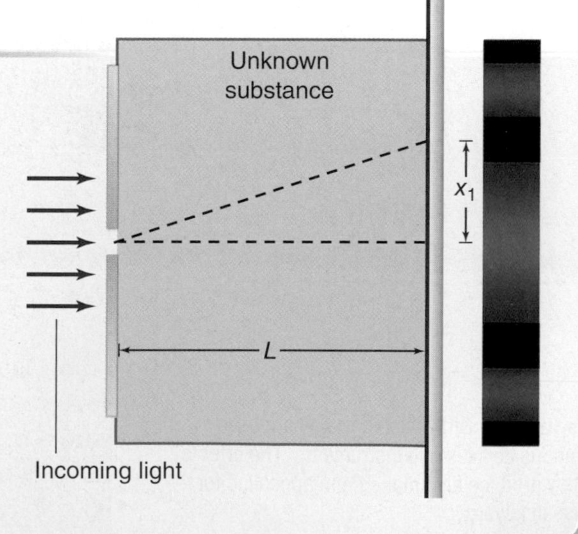

PRACTICE PROBLEMS

Do additional problems. Online Practice

PRACTICE PROBLEMS

16. Monochromatic green light of wavelength 546 nm falls on a single slit with a width of 0.095 mm. The slit is located 75 cm from a screen. How wide will the central bright band be?

17. Yellow light with a wavelength of 589 nm passes through a slit of width 0.110 mm and makes a pattern on a screen. If the width of the central bright band is 2.60×10^{-2} m, how far is it from the slits to the screen?

18. Light from a He-Ne laser ($\lambda = 632.8$ nm) falls on a slit of unknown width. A pattern is formed on a screen 1.15 m away, on which the central bright band is 15.0 mm wide. How wide is the slit?

19. Yellow light falls on a single slit 0.0295 mm wide. On a screen that is 60.0 cm away, the central bright band is 24.0 mm wide. What is the wavelength of the light?

20. **CHALLENGE** White light falls on a single slit that is 0.050 mm wide. A screen is placed 1.00 m away. A student first puts a blue-violet filter ($\lambda = 441$ nm) over the slit, then a red filter ($\lambda = 622$ nm). The student measures the width of the central bright band.

a. Which filter produced the wider band?

b. Calculate the width of the central bright band for both filters.

Diffraction Gratings

Diffraction gratings, such as those shown in **Figure 15,** are often used to make precise measurements of wavelength. A **diffraction grating** is a device that is made up of many small slits that diffract light and form a pattern that is an overlap of single-slit diffraction patterns. This pattern is similar to that of a two-slit interference pattern, but with much narrower and brighter bands. Diffraction gratings can have as many as 10,000 slits per centimeter, which means the spacing between the slits can be as small as 10^{-6} m. A diffraction grating is a useful tool for the study of light and objects that emit or absorb light.

One type of diffraction grating is called a transmission grating. A transmission grating can be made by scratching very fine lines with a diamond point on glass that transmits light. The spaces between the scratched lines act like slits.

Diffraction gratings can be used to enhance the appearance of diamonds. The gratings are etched into certain surfaces of the diamond to improve the dispersion of light and make the gems appear more brilliant.

MiniLAB

DIFFRACTION GRATINGS

What is the effect of wavelength on diffraction patterns from a diffraction grating?

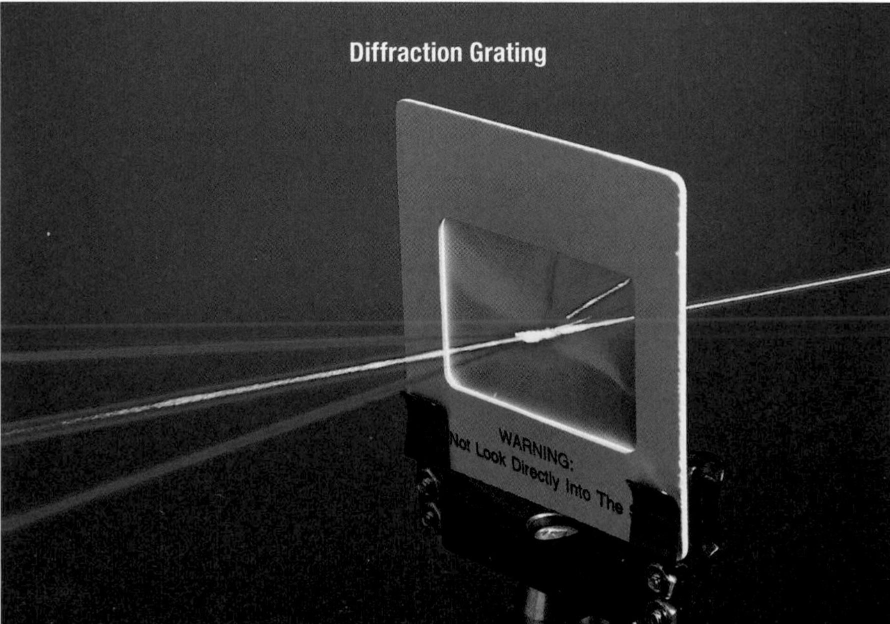

GIPhotoStock X/Alamy

Figure 15 Diffraction gratings are used in various devices and instruments. The effects they produce also make them appealing for use in jewelry.

Holographic diffraction gratings produce the brightest spectra. They are made by using a laser and mirrors to create a diffraction pattern consisting of parallel bright and dark lines. The pattern is projected on a piece of metal that is coated with a light-sensitive material. The light from the laser produces a chemical reaction that hardens the material. The metal is then placed in acid, which attacks the metal wherever it is not protected by the hardened material. The result is a series of hills and valleys in the metal identical to the original diffraction pattern. The metal itself can be used as a reflection grating. In some cases, a plastic film is placed on the heated metal, producing the hills and valleys in the plastic. Because of the sinusoidal shape of the hills and valleys, the diffraction patterns are very bright.

Reflection grating You might have seen light that reflects off CDs or DVDs create a spectrum diffraction pattern, as in **Figure 16.** A type of diffraction grating made by inscribing fine lines on metallic or reflective glass surfaces is called a reflection grating. CDs and DVDs are examples of reflection gratings, producing the color spectra you see when white light reflects off their surfaces. If you were to shine monochromatic light on a DVD, the reflected light would produce a reflection pattern on a screen. Transmission and reflection gratings produce similar patterns, which can be analyzed in the same manner.

Figure 16 A CD is a reflection grating, producing a light spectrum. A magnified view of the surface of a CD shows the arrangement of pits and lands.

CDs, DVDs, and Blu-ray DVDs Why is a music CD or a DVD a diffraction grating? CDs and DVDs are traditional grooved gratings. Their surfaces are actually covered with lines of microscopic indentations called pits separated by flat areas called lands, arranged in a spiral, as shown in **Figure 16.** The turns of the spiral act as a diffraction grating, separating colors by interference. The fact that CDs and DVDs are diffraction gratings is not important to their function, but the way they interact with different wavelengths of light is important.

Storing and reading information A laser is used to "read" the pattern of pits and lands on the CD or DVD. This is similar to the way a blind person reads braille. The light from the laser is reflected from the surface of the disc into a light detector. The laser is focused so that when it reflects from the lands, its bright spot falls on the detector. When it reflects from the pits, it is spread out and dimmer.

The spot size is limited by diffraction, so if shorter laser wavelength is used, the spot size can be reduced and the pits can be closer together, allowing more information to be stored. As laser technology has advanced, shorter wavelength lasers have been used, permitting more information to be put on the disc. Music CDs use infrared light with a wavelength of 780 nm. A CD can hold about 700 megabytes of information. DVDs use red lasers (650 nm), permitting more than 4 gigabytes to be recorded. Blu-ray discs use violet lasers (405 nm). The light appears blue, giving the disc its name. Single-layer blu-ray discs can hold up to 25 gigabytes of information.

☑ **READING CHECK** **Explain** how the amount of information a DVD can store and the wavelength of light used to read it are related.

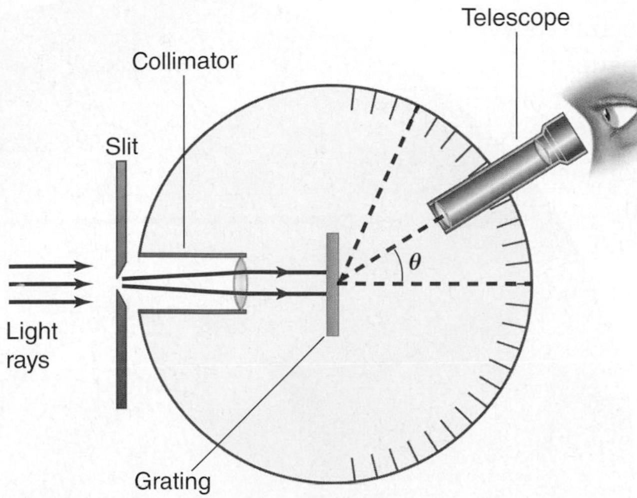

Figure 17 A grating spectroscope is used to accurately measure the wavelength of light.

State the simplification used for double-slit wavelength calculations that does not apply for gratings.

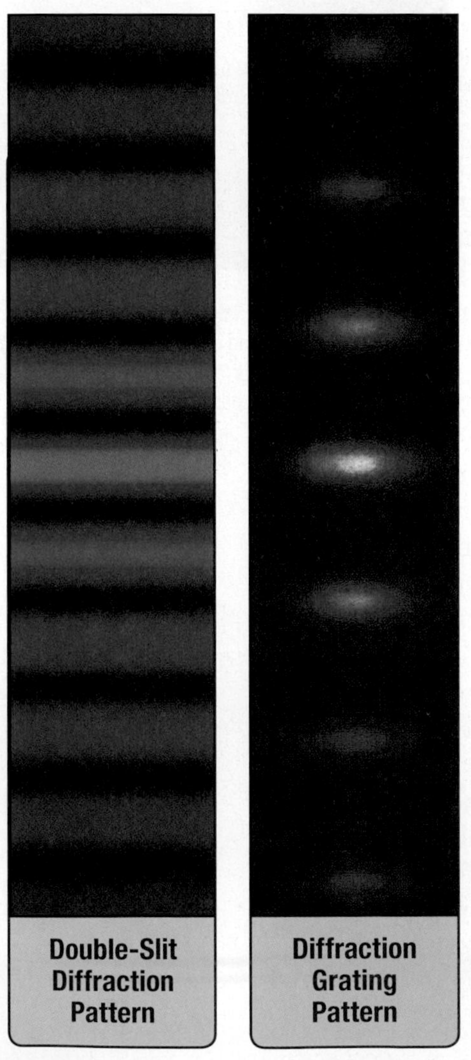

Figure 18 Compare the two diffraction patterns for red light. The diffraction grating pattern provides a more precise measurement.

Measuring wavelength An instrument used to measure light wavelengths using a diffraction grating is called a grating spectroscope, as shown in the diagram in **Figure 17.** The source to be analyzed emits light that is directed through a slit and a collimator and then to a diffraction grating. The grating produces a diffraction pattern that is viewed through a telescope.

If the source of light is a single color, the diffraction pattern produced by a grating has narrow, equally spaced, bright lines, as shown in **Figure 18.** The larger the number of slits per unit length of the grating, the narrower the lines in the diffraction pattern. The narrower the lines, the more precisely the distance between the bright lines can be measured.

Earlier in this chapter, you read that the diffraction pattern produced by a double slit could be used to calculate wavelength. An equation for the diffraction grating can be developed in the same way as for the double slit. However, with a diffraction grating, θ could be large, so the small angle simplification does not apply. Wavelength can be found by measuring the angle (θ) between the central bright line and the first-order bright line.

WAVELENGTH FROM A DIFFRACTION GRATING

The wavelength of light is equal to the slit separation distance times the sine of the angle at which the first-order bright line occurs.

$$\lambda = d \sin \theta$$

Constructive interference from a diffraction grating occurs at angles on either side of the central bright line given by the equation $m\lambda = d \sin \theta$, where $m = 0, 1, 2,$ etc. The central bright line occurs at $m = 0$. Spectroscopists often use the $m = 2$ or 3 lines because measurements of the spacing between bright lines can be made more precisely. Notice that the diffraction grating pattern contains more dark space than the double-slit pattern in **Figure 18.** This is because there is more destructive interference in a diffraction grating than in a double-slit. This results in narrower lines, which also improves the precision of the measurements.

Diffraction gratings are incorporated into spectroscopes used to analyze gemstones. Experienced gemologists recognize the patterns of bands produced by white light passing through different stones. For example, three bright bands of green, yellow, and orange are a strong indication that cobalt is present. This likely means that a blue stone is not an expensive gem such as sapphire or topaz, but rather a cheap piece of glass that has been tinted blue.

EXAMPLE PROBLEM 3

Get help with **diffraction gratings.** Personal Tutor

USING A DVD AS A DIFFRACTION GRATING A student noticed the beautiful spectrum reflected off a rented DVD. She directed a beam from her teacher's green laser pointer at the DVD and found three bright spots reflected on the wall. The label on the pointer indicated that the wavelength was 532 nm. The student found that the spacing between the spots was 1.29 m on the wall, which was 1.25 m away. What is the spacing between the rows on the DVD?

1 ANALYZE AND SKETCH THE PROBLEM

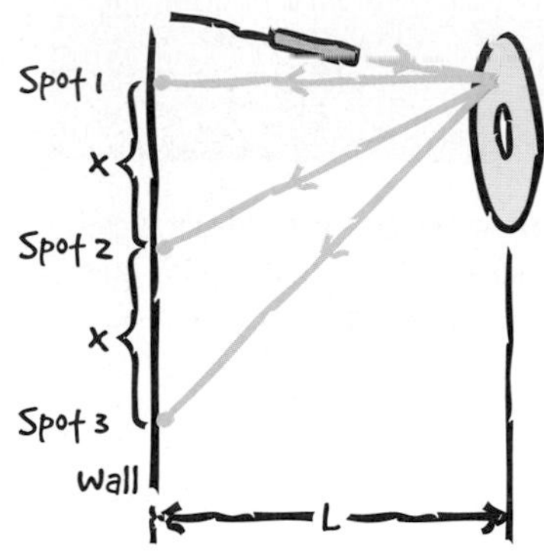

- Sketch the investigation, showing the DVD as a grating and the spots on the wall.
- Identify and label the knowns.

KNOWN	UNKNOWN
$x = 1.29$ m	$d = ?$
$L = 1.25$ m	
$\lambda = 532$ nm	

2 SOLVE FOR THE UNKNOWN

Find the angle between the central bright spot and the one next to it using $\tan \theta = \frac{x}{L}$.

$$\theta = \tan^{-1}\left(\frac{x}{L}\right)$$

$$= \tan^{-1}\left(\frac{1.29\text{ m}}{1.25\text{ m}}\right)$$ ◀ Substitute $x = 1.29$ m, $L = 1.25$ m.

$$= 45.9°$$

Use the diffraction grating wavelength and solve for d.

$$\lambda = d \sin \theta$$

$$d = \frac{\lambda}{\sin \theta}$$

$$= \frac{532\times10^{-9}\text{ m}}{\sin 45.9°}$$ ◀ Substitute $\lambda = 532\times10^{-9}$ m, $\theta = 45.9°$.

$$= 7.41\times10^{-7}\text{ m} = 741\text{ nm}$$

3 EVALUATE THE ANSWER

- **Are the units correct?** The answer is in meters, which is correct for separation.
- **Is the magnitude realistic?** With x and L almost the same size, d is close to λ.

EXAMPLE PROBLEM

PRACTICE PROBLEMS

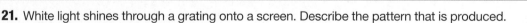

Do additional problems. Online Practice

21. White light shines through a grating onto a screen. Describe the pattern that is produced.

22. If blue light of wavelength 434 nm shines on a diffraction grating and the spacing of the resulting lines on a screen that is 1.05 m away is 0.55 m, what is the spacing between the slits in the grating?

23. A diffraction grating with slits separated by 8.60×10^{-7} m is illuminated by violet light with a wavelength of 421 nm. If the screen is 80.0 cm from the grating, what is the separation of the lines in the diffraction pattern?

24. Blue light shines on the DVD in Example Problem 3. If the dots produced on a wall that is 0.65 m away are separated by 58.0 cm, what is the wavelength of the light?

25. CHALLENGE Light of wavelength 632 nm passes through a diffraction grating and creates a pattern on a screen that is 0.55 m away. If the first bright band is 5.6 cm from the central bright band, how many slits per centimeter does the grating have?

Figure 19 An aperture diffracts light to create a diffraction pattern with a central bright spot with dark and bright rings around it.

Resolving Power of Lenses

The circular lens of a telescope, a microscope, and even your eye acts as a hole, called an aperture, through which light passes. An aperture diffracts light, just as a single slit does. Alternating bright and dark rings occur with a circular aperture, as shown in **Figure 19.** The equation for determining the size of an aperture is similar to that for a single slit. An aperture, however, has a circular edge rather than the two edges of a slit, so slit width (w) is replaced by aperture diameter (D), and a geometric factor of 1.22 is needed, resulting in $x_1 = \frac{1.22\lambda L}{D}$.

▶ CONNECTION TO ASTRONOMY When light from a distant star is viewed through the aperture of a telescope, the image is spread out due to diffraction. If two stars are close together, their images may blur together. In 1879, Lord Rayleigh, a British physicist, mathematician, and Nobel Prize winner, established a criterion for determining whether there are one or two stars in such an image. The **Rayleigh criterion** states that if the center of the bright spot of one source's image falls on the first dark ring of the second, the two images are at the limit of resolution. If the images of two stars are at the limit of resolution, a viewer can tell that there are two stars rather than only one.

If two objects are at the limit of resolution, how can you find the distance between the objects (x_{obj})? From the Rayleigh criterion, the distance between the centers of the bright spots of the two images is x_1. **Figure 20** shows that similar triangles can be used to find that $\frac{x_{obj}}{L_{obj}} = \frac{x_1}{L}$. You can combine this equation with the equation for aperture size $\left(\frac{x_1}{L} = \frac{1.22\lambda}{D}\right)$ and solve for the distance between the objects (x_{obj}).

RAYLEIGH CRITERION

The separation distance between objects that are at the limit of resolution is equal to 1.22, times the wavelength of light, times the distance from the circular aperture to the objects, divided by the diameter of the circular aperture.

$$x_{obj} = \frac{1.22\lambda L_{obj}}{D}$$

MiniLAB

RETINAL PROJECTION SCREEN

How can the retina of your eye be used as a screen?

■ **Telescope Aperture**

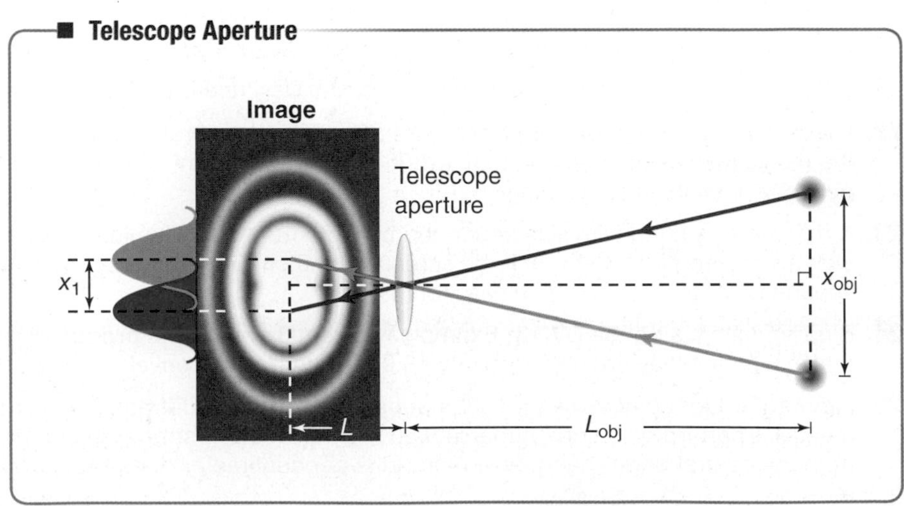

Figure 20 The separation distance of objects can be calculated with similar-triangle geometry. The blue and red colors are only for illustration. (Illustration is not to scale.)

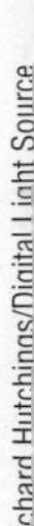

Diffraction in the eye In bright light, the eye's pupil is about 3 mm in diameter. The eye is most sensitive to yellow-green light where $\lambda = 550$ nm. So the Rayleigh criterion applied to the eye gives $x_{obj} = (2\times10^{-4})L_{obj}$. The distance between the pupil and the retina is about 2 cm, so by using $x_1 = 1.22\lambda L/D$, the centers of the bright spots of two barely resolved point sources would be separated by about 4 μm on the retina. The spacing between the cones, which are the light detectors in the retina, in the most sensitive part of the retina, the fovea, is about 2 μm. Thus, in the ideal case, three adjacent cones would record light, dark, and light, as illustrated in **Figure 21.** The distance between the centers of the bright spots from two point sources must be at least the distance between two light-recording cones to be resolved. It seems that the eye is ideally constructed.

Applying the Rayleigh criterion to find the ability of the eye to separate two distance sources shows that the eye could separate two automobile headlights (1.5 m apart) at a distance of 7 km. In practice, however, the eye is not limited by diffraction. Imperfections in the lens and the liquid that fills the eye reduce the eye's resolution to about five times that set by the Rayleigh criterion. The vision processing centers in the human brain also limit detection of small point objects.

Many telescope manufacturers advertise that their instruments are diffraction limited. This means that their telescopes can separate two point sources at the Rayleigh criterion. To reach this limit they must grind the mirrors and lenses to an accuracy of one-tenth of a wavelength (about 55 nm). The larger the diameter of the mirror, the greater the resolution of the telescope. Interactions of light with Earth's atmosphere keep telescopes on Earth from reaching their diffraction limit. Resolution of images from telescopes above Earth's atmosphere is much better than that of larger telescopes on Earth's surface.

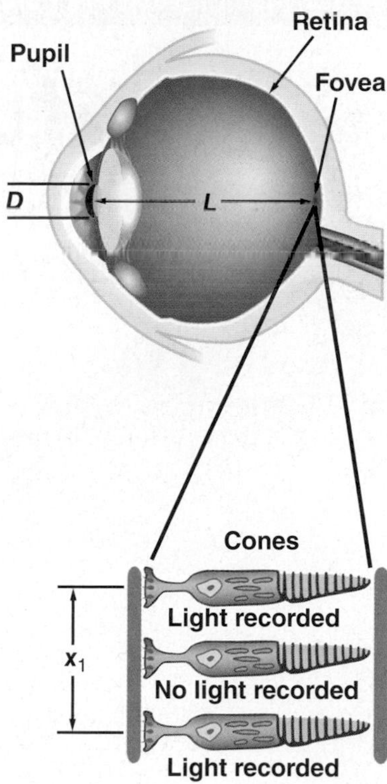

Figure 21 The eye's pupil is an aperture that diffracts light.

SECTION 2 REVIEW

Check your understanding.

26. **MAIN IDEA** Many narrow slits are close to each other and equally spaced in a large piece of cardboard. They are illuminated by monochromatic red light. A sheet of white paper is placed far from the slits, and a pattern of bright and dark bands is visible on the paper. Sketch the pattern that would be seen on the screen.

27. **Rayleigh Criterion** The brightest star in the winter sky in the northern hemisphere is Sirius. In reality, Sirius is a system of two stars that orbit each other. If the *Hubble Space Telescope* (diameter 2.4 m) is pointed at the Sirius system, which is 8.44 light-years from Earth, what is the minimum separation there would need to be between the stars in order for the telescope to be able to resolve them? Assume that the light coming from the stars has a wavelength of 550 nm.

28. **Line Spacing** You shine a red laser light through one diffraction grating and form a pattern of red dots on a screen. Then you substitute a second diffraction grating for the first one, forming a different pattern. The dots produced by the first grating are spread out more than those produced by the second. Which grating has more lines per millimeter?

29. **First-Order Dark Bands** Monochromatic green light with a wavelength equal to 546 nm falls on a single slit of width and location from a screen shown in **Figure 22.** What is the separation of the first-order dark bands?

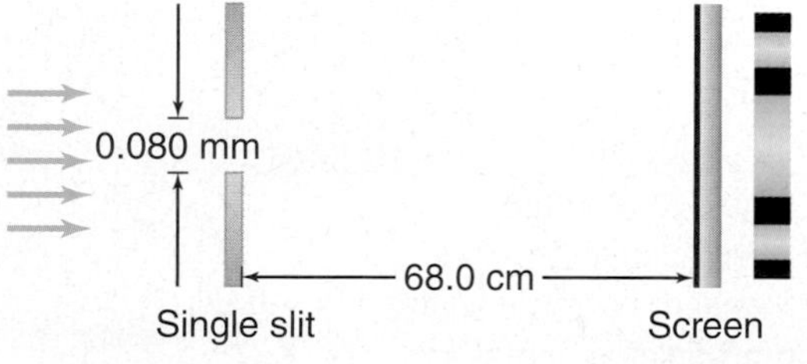

Figure 22

30. **Critical Thinking** You are shown a spectrometer but are not told whether it has been constructed with a prism or a diffraction grating. If you look at a spectrum produced by white light passing through the spectrometer, how could you determine which device produced the spectrum?

Nature's Crime Fighter

IRIDESCENCE IN BUTTERFLIES AND CURRENCY

The shimmering bands of color on the wings of the Indonesian Peacock butterfly show different shades of green and blue as they move. More than just beautiful, this butterfly's iridescent coloration might someday hold the secret to detecting counterfeit money.

Butterfly wings are covered in tiny scales. Viewing these scales under a powerful microscope shows that they have a pattern of tiny, repeating cavities.

Light that hits the center of each cavity reflects directly, but light that hits the edges of the cavity is sent through structures that polarize and reflect light, acting like a diffraction grating. These structures give the wings their bright green or blue-green color, depending on the viewing angle.

Researchers have been able to recreate the microscopic structure of iridescent butterfly scales in an artificial nanomaterial. This material has optical properties similar to those of the butterfly scales it imitates. Its makers foresee applying it as an anti-counterfeiting technology that is much more difficult to forge than current methods.

GOINGFURTHER >>>

Research other examples of color effects in nature that are due to iridescence. Design an informative Web site that illustrates your findings.

STUDY GUIDE

BIGIDEA Light waves can diffract and interfere with each other.

VOCABULARY
- **incoherent light** *(p. 522)*
- **coherent light** *(p. 522)*
- **interference fringes** *(p. 523)*
- **monochromatic light** *(p. 523)*
- **thin-film interference** *(p. 527)*

SECTION 1 Interference

MAINIDEA Light can interfere when passing through narrow slits or reflecting from a thin film.

- The superposition of light waves from coherent light sources can produce an interference pattern. Light passing through two closely spaced, narrow slits produces a pattern of dark and light bands on a screen called interference fringes.
- Interference patterns can be used to measure the wavelength of light.

$$\lambda = \frac{xd}{L}$$

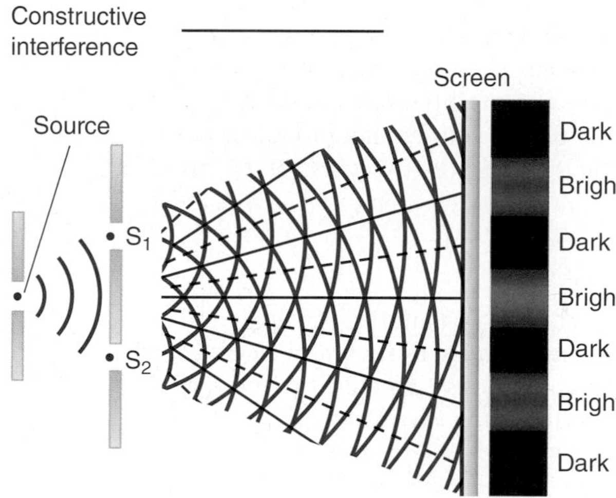

- Interference patterns can result from multiple passes of light through a thin film. Thin-film interference can be modeled by rays reflecting from multiple surfaces of a thin film. The refractive indexes of the mediums the light travels in and the thickness of the film determine how different wavelengths of light will interfere.

VOCABULARY
- **diffraction pattern** *(p. 531)*
- **diffraction grating** *(p. 534)*
- **Rayleigh criterion** *(p. 538)*

SECTION 2 Diffraction

MAINIDEA Light waves diffract when they pass through a single slit and diffract and interfere when they encounter a diffraction grating.

- Light passing through a narrow slit is diffracted, which means spread out from a straight-line path, producing a diffraction pattern on a screen. The width of the bright central band of a single-slit diffraction pattern is related to the wavelength of light used.
- Diffraction gratings consist of large numbers of slits that are very close together and produce narrow spectral lines that result from interference of light diffracted by all the slits.
- Diffraction gratings can be used to measure the wavelength of light precisely or to separate light composed of different wavelengths.

$$\lambda = d \sin \theta$$

- Diffraction limits the ability of an aperture to distinguish two closely spaced objects, because the resulting image contains a diffuse central bright spot. If two bright spots are closer than the limit of resolution, they will overlap and the objects cannot be distinguished.

Games and Multilingual eGlossary

Vocabulary Practice

Chapter Self-Check

SECTION 1 Interference

Mastering Concepts

31. Why is it important that monochromatic light was used to make the interference pattern in the double-slit investigation?

32. Explain why the position of the central bright band of a double-slit interference pattern cannot be used to determine the wavelength of the light.

33. Describe how you could use light of a known wavelength to find the distance between two narrow slits.

34. Describe in your own words what happens in thin-film interference when a colored band is produced by light shining on a soap film suspended in air. Make sure you include in your explanation how the wavelength of the light and the thickness of the film are related.

Mastering Problems

35. Light falls on a pair of slits that are 19.0 μm apart and located 80.0 cm from a screen, as shown in **Figure 23.** The first-order bright band is 1.90 cm from the central bright band. What is the wavelength of the light?

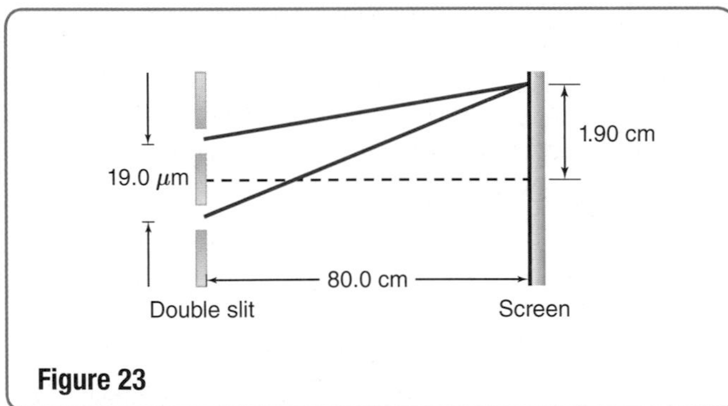

Figure 23

36. Oil Slick After a short spring rain shower, Tom and Ann take their pet Basset Hound for a walk and notice a thin film of oil ($n = 1.45$) on a puddle of water, producing different colors. What is the minimum thickness of a place where the oil creates constructive interference for light with a wavelength equal to 545 nm?

37. Film Thickness A plastic reflecting film ($n = 1.83$) is placed on an auto glass window ($n = 1.52$). What is the thinnest film that will reflect yellow-green light ($\lambda = 555$ nm)? Unfortunately, a film this thin cannot be manufactured. What is the next-thinnest film that will produce the same effect?

38. Insulation Film Winter is approaching and Alejandro is helping to cover the windows in his home with thin sheets of clear plastic ($n = 1.81$) to keep the drafts out. After the plastic is taped up around the windows, the plastic is heated with a hair dryer to shrink-wrap the window, which alters the thickness but not the refractive index of the plastic. Alejandro notices a place on the plastic where there is a blue stripe of color. He realizes that this is created by thin-film interference. What are three possible thicknesses of the plastic where the blue stripe is produced if the wavelength of the light is 445 nm?

39. Ranking Task Five different lasers produce double-slit interference patterns. In each case, the slit separation is 0.035 mm. Rank them according to the wavelength of the lasers, from shortest to longest. Specifically indicate any ties.

A. The screen is 0.95 m from the slits, and adjacent bright spots are separated by 12 mm.

B. The screen is 0.95 m from the slits, and adjacent bright spots are separated by 16 mm.

C. The screen is 1.3 m from the slits, and adjacent bright spots are separated by 20 mm.

D. The screen is 2.8 m from the slits, and adjacent bright spots are separated by 40 mm.

E. The screen is 2.8 m from the slits, and adjacent bright spots are separated by 50 mm.

SECTION 2 Diffraction

Mastering Concepts

40. BIGIDEA White light shines through a diffraction grating. Are the resulting red lines spaced more closely or farther apart than the resulting violet lines? Why?

41. Why do diffraction gratings have large numbers of slits? Why are these slits so close together?

42. Telescopes Why would a telescope with a small diameter be unable to resolve the images of two closely spaced stars?

43. Reverse Problem Write a physics problem with real-life objects for which the following equation would be part of the solution:

$$x_1 = \frac{(2.00 \text{ m})(530 \text{ nm})}{0.20 \text{ nm}}$$

44. Problem Posing Complete this problem so that it can be solved using the Rayleigh criterion: "A telescope currently being designed is to have an aperture that is 8.0 m in diameter...."

45. For a given diffraction grating, which color of visible light produces a bright line closest to the central bright band?

46. When you look at an incandescent lamp through a pair of "fun" glasses you see thin stripes of spectral colors going out from the light in eight directions as in **Figure 24.** You recognize that the glasses act like diffraction gratings. What are the directions of the scratches in these glasses?

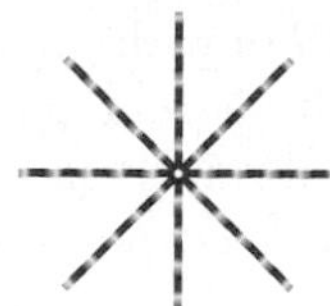

Figure 24

Mastering Problems

47. Monochromatic light passes through a single slit 0.010 cm wide and falls on a screen 100 cm away, as shown in **Figure 25.** If the width of the central band is 1.20 cm, what is the wavelength of the light?

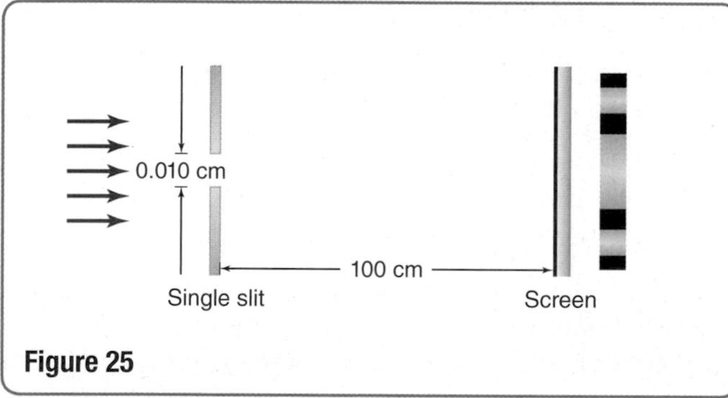

Figure 25

48. A good diffraction grating has 2.5×10^3 lines per cm. What is the distance between two lines?

49. Light with a wavelength of 455 nm passes through a single slit and falls on a screen 100 cm away. If the slit is 0.015 cm wide, what is the distance from the center of the pattern to the first dark band?

50. Kaleidoscope The mirrors have been removed from a kaleidoscope. The diameter of the eyehole at the end is 7.0 mm. If two bluish-purple specks on the other end of the kaleidoscope separated by 40 μm are barely distinguishable, what is the length of the kaleidoscope? Use $\lambda = 650$ nm and assume the resolution is diffraction limited through the eyehole.

51. Monochromatic light with a wavelength of 425 nm passes through a single slit and falls on a screen 75 cm away. If the central bright band is 0.60 cm wide, what is the width of the slit?

52. *Hubble Space Telescope* Suppose the *Hubble Space Telescope,* 2.4 m in diameter, is in orbit 1.0×10^5 m above Earth and is turned to view Earth, as shown in **Figure 26.** If you ignore the effect of the atmosphere, how large an object can the telescope resolve? Use $\lambda = 515$ nm.

Figure 26

53. Spectroscope A spectroscope uses a grating with 12,000 lines/cm. Find the angles at which red light (632 nm) and blue light (421 nm) have first-order bright lines.

Applying Concepts

54. Science Fair At a science fair, one exhibition is a very large soap film that has a fairly consistent thickness. It is illuminated by a light with a wavelength of 432 nm, and nearly the entire surface appears to be a lovely shade of purple. What would you see in the following situations?

a. Film thickness was doubled.

b. The film thickness was increased by half a wavelength of the illuminating light.

c. The film thickness was decreased by one-quarter of a wavelength of the illuminating light.

55. What are the differences in the characteristics of the diffraction patterns formed by diffraction gratings containing 10^4 lines/cm and 10^5 lines/cm?

56. Laser-Pointer Challenge You have two laser pointers, a red one and a green one. Your friends Mark and Carlos disagree about which has the longer wavelength. Mark insists that red light has a longer wavelength, while Carlos is sure that green has the longer wavelength. You have a CD handy. Describe what demonstration you would do with this equipment and how you would explain the results to Carlos and Mark to settle their disagreement.

57. How can you tell whether a pattern is produced by a single slit or a double slit?

Chapter Self-Check

58. **Optical Microscope** Why is blue light used for illumination in an optical microscope?

59. For each of the following examples, indicate whether the color is produced by thin-film interference, refraction, or the presence of pigments.

    a. bubbles
    b. rose petals
    c. oil films
    d. rainbow

60. Describe the changes in a single-slit diffraction pattern as the width of the slit is decreased.

Mixed Review

61. **Record** Marie uses an old $33\frac{1}{3}$ rpm record as a diffraction grating. She shines a laser, $\lambda = 632.8$ nm, on the record, as shown in **Figure 27.** On a screen 4.0 m away, she sees a series of red dots 21 mm apart .

    a. How many grooves are there in a centimeter along the radius of the record?

    b. Marie checks her results by noting that the ridges represent a song that lasts 4.01 min and takes up 16 mm on the record. How many grooves should be in a centimeter?

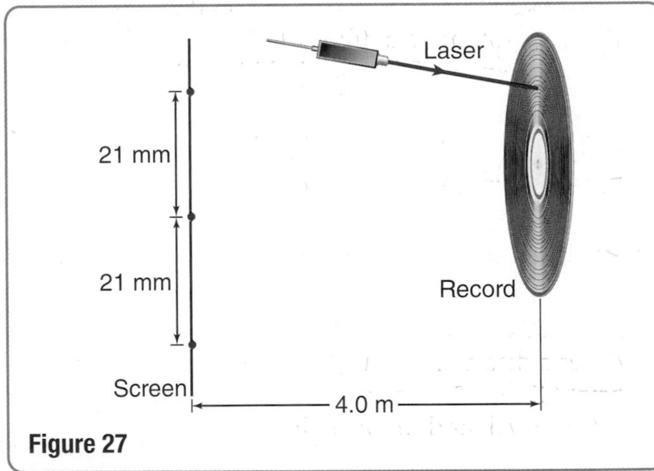

Figure 27

62. **Camera** When a camera with a 50-mm lens is set at $\frac{f}{8}$, its aperture has an opening 6.25 mm in diameter.

    a. A CCD detector is 50.0 mm away and senses light at $\lambda = 550$ nm. What is the resolution of the lens?

    b. The owner of the camera knows that it has 6.3 megapixels in its CCD. The manufacturer says that each pixel is 7.6 μm on a side. Compare the size of the pixel with the width of the central spot calculated in part a.

63. A glass lens has antireflective coating ($n = 1.2$) with a thickness of 125 nm. For which color(s) of light does complete destructive interference occur?

Thinking Critically

64. **Apply Concepts** Yellow light falls on a diffraction grating. On a screen behind the grating, you see three spots: one at zero degrees, where there is no diffraction, and one each at +30° and −30°. You now add a blue light of equal intensity that is in the same direction as the yellow light. What pattern of spots will you now see on the screen?

65. **Apply Concepts** Blue light of wavelength λ passes through a single slit of width w. A diffraction pattern appears on a screen. If you replace the blue light with a green light of wavelength 1.5λ, what slit width will produce the original pattern?

66. **Analyze and Conclude** At night, the pupil of a human eye has an aperture diameter of 8.0 mm. The diameter is smaller in daylight. An automobile's headlights are separated by 1.8 m. How far away can the human eye distinguish the two headlights at night? *Hint: Assume a wavelength of 525 nm.* What besides diffraction might be limiting factors?

Writing in Physics

67. Research and describe Thomas Young's contributions to physics. Evaluate the impact of his research on the scientific thought about light's nature.

68. The gemstone opal has an iridescent sheen. Research and describe how these colors are produced.

69. Many telescopes have adaptive optics that reduce the atmospheric effects that cause stars to twinkle. Research and describe how these systems work.

70. Research and interpret the role of diffraction in medicine and astronomy. Describe at least two applications in each field.

Cumulative Review

71. How much work must be done to push a 0.5-m^3 block of wood to the bottom of a 4-m-deep swimming pool? The density of wood is 500 kg/m^3.

72. What are the wavelengths of microwaves in an oven if their frequency is 2.4 GHz?

73. A concave mirror has a 48.0-cm radius. A 2.0-cm-tall object is placed 12.0 cm from the mirror. Calculate the image position and image height.

74. The focal length of a convex lens is 21.0 cm. A 2.00-cm-tall candle is 7.50 cm from the lens. Use the thin-lens equation to find the image position and image height.

STANDARDIZED TEST PRACTICE

MULTIPLE CHOICE

1. What is the best possible explanation for why the colors of a thin film, such as a soap bubble or oil on water, appear to change and move as you watch?
   - **A.** because convective heat waves in the air next to the thin film distort the light
   - **B.** because the film thickness at any given location changes over time
   - **C.** because the wavelengths in sunlight vary over time
   - **D.** because your vision varies slightly over time

2. Light at 410 nm shines through a slit and falls on a flat screen as shown in the figure below. The width of the slit is 3.8×10^{-6} m. What is the width of the central bright band?
   - **A.** 0.024 m
   - **B.** 0.031 m
   - **C.** 0.048 m
   - **D.** 0.063 m

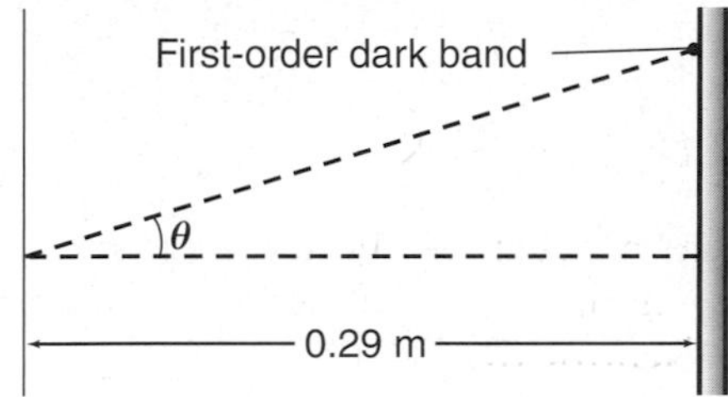

3. For the situation in problem 2, what is the angle (θ) of the first dark band?
   - **A.** 3.1°
   - **B.** 6.2°
   - **C.** 12°
   - **D.** 17°

4. Two stars 6.2×10^4 light-years from Earth are 3.1 light-years apart. What is the smallest diameter telescope that could resolve them using 610 nm light?
   - **A.** 5.0×10^{-5} m
   - **B.** 6.1×10^{-5} m
   - **C.** 1.5×10^{-2} m
   - **D.** 1.5×10^{7} m

5. A grating has slits that are 0.055 mm apart. What is the angle of the first-order bright line for light with a wavelength of 650 nm?
   - **A.** 0.012°
   - **B.** 0.68°
   - **C.** 1.0°
   - **D.** 11°

6. A laser beam at 638 nm illuminates two narrow slits. The third-order band of the resulting pattern is 7.5 cm from the center bright band. The screen is 2.475 m from the slits. How far apart are the slits?
   - **A.** 5.8×10^{-8} m
   - **B.** 6.3×10^{-7} m
   - **C.** 2.1×10^{-5} m
   - **D.** 6.3×10^{-5} m

7. A flat screen is placed 4.200 m from a pair of slits that are illuminated by a beam of monochromatic light. On the screen, the separation between the central bright band and the second-order bright band is 0.082 m. The distance between the slits is 5.3×10^{-5} m. Determine the wavelength of the light.
   - **A.** 2.6×10^{-7} m
   - **B.** 5.2×10^{-7} m
   - **C.** 6.2×10^{-7} m
   - **D.** 1.0×10^{-6} m

8. A clown is blowing soap bubbles, and you notice that the color of one region of a particularly large bubble matches the color of his nose. If the bubble is reflecting 6.5×10^{-7} m red light waves and the index of refraction of the soap film is 1.41, what is the minimum thickness of the soap bubble at the location where it is reflecting red?
   - **A.** 1.2×10^{-7} m
   - **B.** 3.5×10^{-7} m
   - **C.** 9.2×10^{-7} m
   - **D.** 1.9×10^{-6} m

FREE RESPONSE

9. A diffraction grating that has 6000 slits per cm produces a diffraction pattern that has a first-order bright line at 20° from the central bright line. What is the wavelength of the light?

NEED EXTRA HELP?

If you Missed Question	1	2	3	4	5	6	7	8	9
Review Section	1	2	2	2	2	1	1	1	2

Online Test Practice

CHAPTER 20

Static Electricity

BIGIDEA Separated positive and negative charges exert forces on one another.

SECTIONS

1 Electric Charge

2 Electrostatic Force

LaunchLAB

ELECTROSTATIC FORCE: ANOTHER FIELD FORCE

What happens when a charged ruler is brought near some paper scraps?

WATCH THIS!

LIGHTNING

Does your hair ever stand up when you comb or brush it? Have you ever been shocked when petting your cat? Static electricity is everywhere–discover the physics behind the phenomenon.

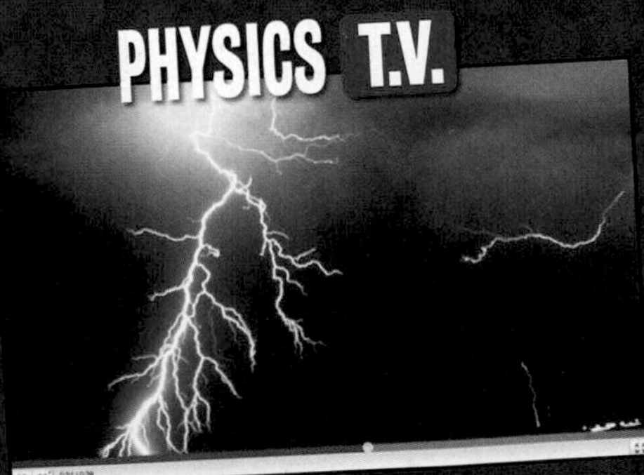

(l)NOAA Photo Library, NOAA Central Library; OAR/ERL/National Severe Storms Laboratory (NSSL), (r)Comstock/JupiterImages

Go online!
connectED.mcgraw-hill.com

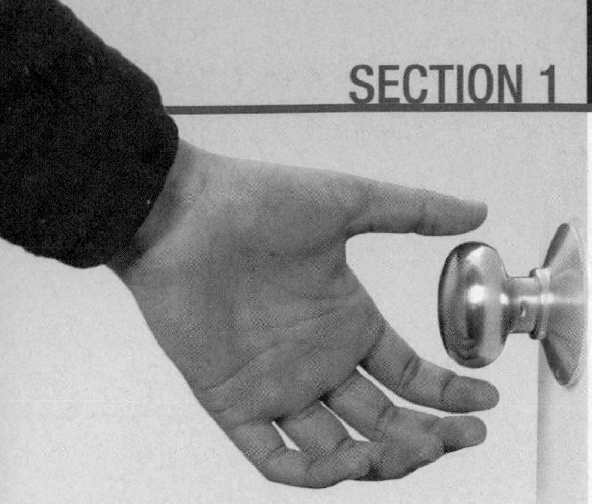

SECTION 1

Electric Charge

PHYSICS 4 YOU **Have you ever touched a metal doorknob on a dry winter day and gotten a shock? You got a shock because electrons were transferred from your hand to the doorknob. How is this shock similar to lightning?**

MAIN IDEA

Like electric charges repel, and unlike electric charges attract.

Essential Questions

- How can you demonstrate that charged objects exert forces, both attractive and repulsive?
- How do we know that charging is the separation, not the creation, of electric charges?
- What are the differences between conductors and insulators?

Review Vocabulary

plasma a gaslike, fluid state of matter made up of negatively charged electrons and positively charged ions that can conduct electric current

New Vocabulary

electrostatics
neutral
insulator
conductor

Evidence of Charge

You might have rubbed your shoes on a carpet and then produced a spark when you touched someone. In 1750 Benjamin Franklin, who made many contributions to the study of electricity, suggested that lightning would have a similar effect on a metal key attached to the string of a kite flown in a thunderstorm. As electricity traveled down the string to the key, loose threads of the string would stand up and repel one another, just as your hair might after rubbing your shoes on a rug. If you touched the key, you would generate a spark and experience a shock. Electric effects produced in this way are called static electricity.

In this chapter, you will investigate **electrostatics,** the study of electric charges that can be collected and held in one place. Electrostatics is the study of static electricity. The effects of static electricity are observable over a vast range of scales, from huge displays of lightning to interactions between electrons on the atomic scale. In later chapters, you will study electric current, which is the net movement of electric charge.

Electrostatic force Have you ever noticed that on a dry day your hair is attracted to your comb after you comb it? Perhaps you have noticed that socks sometimes stick together when you take them out of a clothes dryer. You might also recognize what happens after you rub a balloon on hair, as shown in **Figure 1.** There must be a strong force pulling upward on the hair because it overcomes the gravitational force pulling downward on the hair. The balloon produces an electrostatic force on the hair that is greater than the gravitational force from Earth.

Figure 1 Rubbing a balloon on a cat, especially on a day when the air is dry, can cause the cat's hair to stand up.

Infer Why would the cat's hair be more likely to stand up on a dry day than on a wet day?

The electrostatic force and the gravitational force are both forces, but they act differently. For example, hair is attracted to a balloon only after the balloon has been rubbed; if you wait a while, the attractive property of the balloon diminishes. The gravitational force, on the other hand, does not require rubbing and does not decrease. You will become aware of other differences between the gravitational force and the electrostatic force as you read this chapter.

Like charges You can investigate electric charge using strips of transparent tape. Fold over about 5 mm of the end of a 10-cm strip of tape to use as a handle. Stick the tape strip to a dry, smooth surface, such as a desk, as shown in the top image of **Figure 2.** Attach a second, similar piece of tape next to the first. Quickly pull both strips off the table and bring them near each other. What happens?

Pulling the strips of tape off the table gives them the ability to repel each other, as shown in the center image of **Figure 2.** The strips repel each other because they have accumulated electric charge. The strips were prepared in the same way, so they have the same charge. Two objects with like charges always repel each other.

Unlike charges Now, stick one strip of tape on the desk and place the second strip on top of the first. Use the handle of the bottom strip to slowly pull both strips off the table together. Rub them with the fingers of your other hand until they are no longer attracted to you. By doing this, you discharge the strips. In other words, you remove their charge.

Now, with one hand holding the handle of one strip and the other hand holding the handle of the second strip, quickly pull the two strips apart. You will find that, again, the strips are charged—they are attracted to your hands. Bring them close to each other. Do they still repel each other? No, the strips are now attracted to each other, as shown in the bottom image of **Figure 2.** Because the strips attract each other, you know that they do not have the same charge; like charges repel. This suggests that there is a second type of charge and that unlike charges attract.

Investigating charge You can learn more about charge by doing additional investigations with two pieces of tape. For example, try to determine whether the strength of attraction and repulsion changes with distance. You will find that if you wait a while after charging your tape, especially in humid weather, the tape's electric charge goes away. You can recharge the tape by sticking it to the desk and pulling it off again.

These tape strips become charged when you pull them quickly from the surface of the desk.

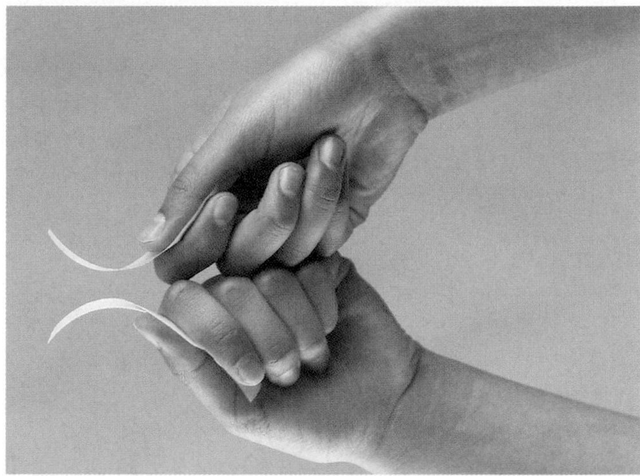

Tape strips repel each other when each has the same charge.

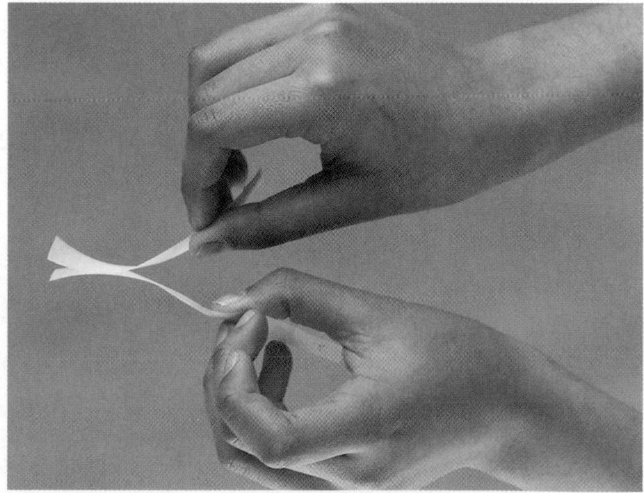

Tape strips attract each other when each has a different charge.

Figure 2 You can use two strips of tape to demonstrate the force between charged objects.

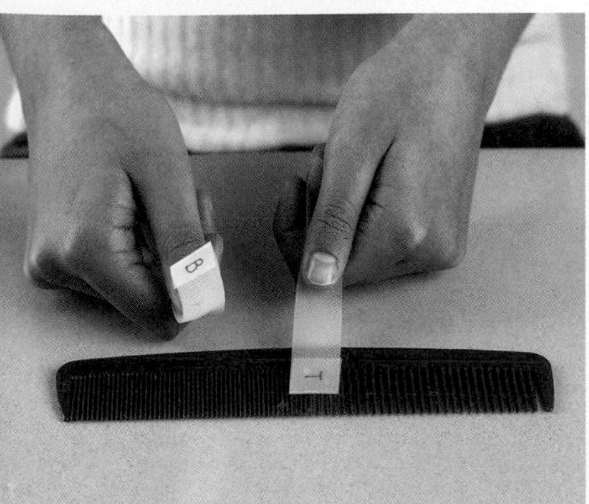

Figure 3 The comb and strip B have the same charge. Like charges repel and unlike charges attract.

PhysicsLAB

STATIC CHARGE

FORENSICS LAB How do positively and negatively charged objects interact with each other and with neutral objects?

Interactions between charged objects Can you use your tape to explore the charge of other objects? Rub a plastic comb on your clothing, and place the comb on a table. Then, prepare your strips of tape as you did before: stick one strip to the table and stick the second strip on top of the first. Label the handle end of the bottom strip B and the handle end of the top strip T. Pull the pair off together. Discharge them, and then pull them apart. Now, holding your two strips of tape by the sticky ends, one in each hand, bring first one and then the other close to the comb, as shown in **Figure 3.** You will find that one strip will be attracted to the comb, and the other will be repelled by it.

☑ READING CHECK **Explain** why one strip of tape is attracted to the comb and the other strip is repelled by the comb.

You can attach your charged tape strips to a desk to explore how they interact with other charged objects. To charge objects, simply rub them with different types of materials. To charge a glass object, rub it with silk, wool, or plastic wrap. To explore how silk or wool behaves, slip a plastic bag over your hand. Rub the cloth and then take your hand out of the bag. Bring both the bag and the cloth near the charged tape.

Did charged objects always attract one strip of tape and repel the other? Did you ever find an object that repels both strips of tape? Did some objects attract both? You quickly found that your fingers attract both strips. You will explore this effect later in this chapter.

Types of charge From your investigations with two strips of tape, you can make two lists. You can label one list B, for objects that have the same charge as the tape you labeled B. And you can label the other list T, for objects that have the same charge as the tape labeled T. There are only two lists because there are only two types of charge. Benjamin Franklin called them positive charges and negative charges.

Materials have varying degrees of ability to acquire charge. Hard rubber and plastic have a tendency to become negatively charged. Glass and wool have a tendency to become positively charged.

Just as you showed that an uncharged pair of tape strips can gain unlike charges, you probably were able to show that if you rubbed plastic with wool, the plastic gained one type of charge and the wool gained the other type of charge. The two kinds of charges did not appear alone, but in pairs. These investigations suggest that matter normally contains both types of charges, positive and negative. Rubbing the materials together separates the two types of charge. To explore this further, consider a microscopic view.

VOCABULARY
Science Usage v. Common Usage

Neutral

- Science Usage
having exact balance between positive and negative charges
Neutral objects do not experience an electric attraction.
- Common Usage
not engaged or supporting either side
Historically, Switzerland has been a neutral country.

A Microscopic View of Charge

Electric charges exist within atoms. In 1897 J.J. Thomson discovered that all materials contain low-mass, negatively charged particles. These particles are called electrons. Between 1909 and 1911, Ernest Rutherford, who had earlier worked as Thomson's assistant, discovered that the atom has a massive, positively charged nucleus surrounded by a cloud of orbiting electrons. You will explore Thomson's and Rutherford's experiments in detail in later chapters. From their experiments and many others, scientists know that atoms are normally neutral. For a **neutral** object, the amount of negative charge exactly balances the amount of positive charge.

Richard Hutchings/Digital Light Source

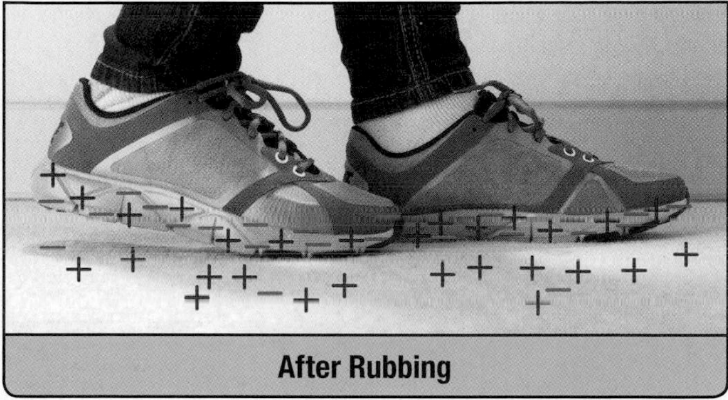

Figure 4 Electrons can be transferred from the wool rug to the rubber shoe.

COLOR CONVENTION

positive charges	+	**red**
negative charges	—	**blue**

Transfer of electrons With the addition of energy, the outer electrons of an atom can be removed from the atom. An atom that is missing electrons has a net positive charge, and, consequently, any matter that has electron-deficient atoms is positively charged. The freed electrons can remain unattached, or they can become attached to other atoms, resulting in atoms with net negative charge. From a microscopic viewpoint, acquiring charge is a process of transferring electrons.

Separation of charge If you rub two neutral objects together, each object can become charged. For instance, when you rub rubber shoes on a wool rug, the energy from the rubbing removes outer electrons from atoms in the wool, and they transfer to the rubber shoe, as shown in **Figure 4.** The extra electrons on the shoe result in a net negative charge on the shoe. The electrons missing from the wool rug result in a net positive charge on the rug. The combined total charge of the two objects remains the same. Charge is conserved, which is one way of saying that individual charges never are created or destroyed. A net positive or negative charge means that electrons have been transferred.

Processes inside a thundercloud can cause the cloud bottom to become negatively charged and the cloud top to become positively charged. Electric charge can be transferred from a road to the car traveling on it. The rubbing of your MP3 player in your pocket can sometimes cause your earbuds to become charged and shock you. In all of these cases, charges are not created, but separated.

View a **BrainPOP video on static electricity.**

Conductors and Insulators

Hold a plastic rod or comb at its midpoint and rub one end on your clothing. You will find that only the rubbed end becomes charged. In other words, the electrons that transferred to the plastic object stayed where they were; they did not move. A material through which a charge will not move easily is called an electric **insulator.** The strips of tape that you charged earlier in this chapter acted as insulators because the charge that accumulated on them did not move; it stayed localized. Glass, dry wood, most plastics, cloth, and dry air are all good insulators.

Suppose you support a metal rod on an insulator so that it is completely surrounded by the insulator. When you touch a charged comb to one end of the metal rod, you will find that the charge spreads very quickly over the entire rod. A material that allows charges to move about easily is called an electric **conductor.** Electrons move and thus conduct electric charge through the metal.

PhysicsLAB

CHARGED OBJECTS

How can you test materials for their abilities to hold positive and negative charges?

Figure 5 Charge spreads out evenly on a conductor. An insulator holds charges where they are placed.

Consider why electrical wires are coated with rubber.

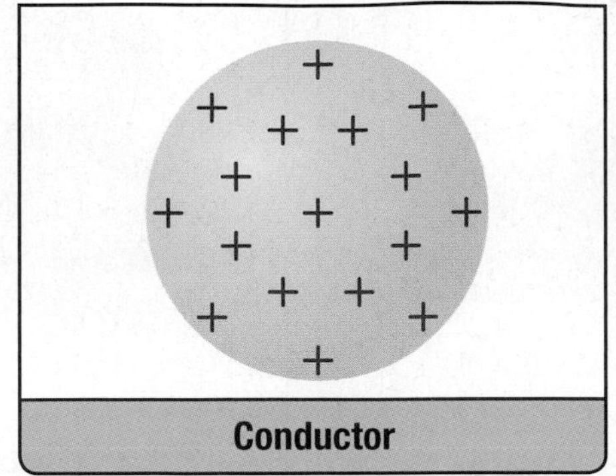

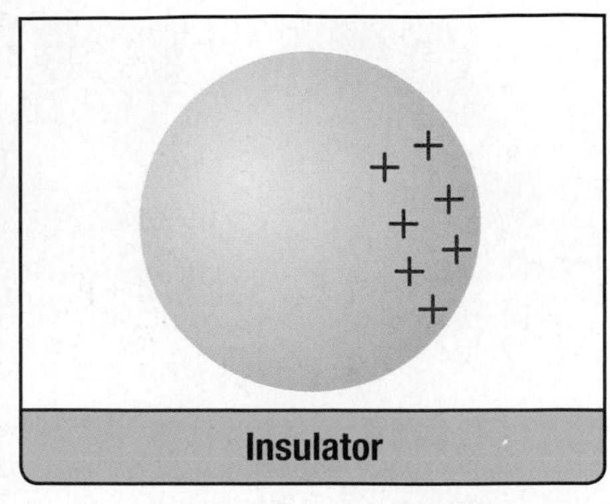

REAL-WORLD PHYSICS

CONDUCTOR OR INSULATOR?

It might be tempting to classify an element as solely a conductor or solely an insulator, but the classification can change depending on various factors, including the form the element takes. For example, carbon in the form of diamond is an insulator, but carbon in the form of graphite conducts charge. This is because of differences in the bonding of the carbon atoms. The carbon bonds in graphite allow some electron movement. As a result, graphite is a much better conductor than diamond, even though both are made of carbon atoms.

Metals From past experience, we know that most metals are good conductors. This is because at least one electron of each atom of a metal can be removed easily. These electrons no longer remain with any particular atom, but move through the metal as a whole. **Figure 5** contrasts how excess charges behave on a conductor and on an insulator. Copper and aluminum are both excellent conductors and are used commercially to carry electric charge. Some nonmetal materials also are good conductors. These include plasma, which consists of negative electrons and positive ions, and graphite.

☑ **READING CHECK** **Explain** why metals are good conductors of electric charge.

Air as a conductor You read that air is an insulator. Under certain conditions, charges can move through air as if it were a conductor. The spark that jumps between your finger and a doorknob after you have rubbed your feet on a carpet discharges you. In other words, you have become neutral because the excess charges have left you. Similarly, lightning discharges a thundercloud. In both of these cases, air becomes a conductor for a brief moment.

Recall that electrons are free to move in conductors. For a spark or lightning to occur, freely moving charged particles must be formed in the normally neutral air. In the case of lightning, excess charges in the cloud and on the ground exert enough force to temporarily remove electrons from the molecules in the air. The electrons and positively charged ions form a plasma, which is a conductor.

SECTION 1 REVIEW

Check your understanding.

1. **MAIN IDEA** In the investigations with tape described in this section, how could you find out which strip of tape, B or T, is positively charged?
2. **Charged Objects** After you rub a comb on a wool sweater, you can use the comb to pick up small pieces of paper. Why does the comb lose this ability after a few minutes?
3. **Types of Charge** A pith ball is a small sphere made of a light material, such as plastic foam, that is often coated with a layer of graphite or aluminum paint. How could you determine whether a pith ball suspended from an insulating thread is neutral, charged positively, or charged negatively?
4. **Charge Separation** You can give a rubber rod a negative charge by rubbing the rod with wool. What happens to the charge of the wool? Why?
5. **Net Charge** An apple contains approximately 10^{26} charged particles. Why don't two apples repel each other when they are brought together?
6. **Charging a Conductor** Suppose you hang a long metal rod from silk threads so that the rod is electrically isolated. You then touch a charged glass rod to one end of the metal rod. Describe the charges on the metal rod.
7. **Charging by Friction** You can charge a rubber rod negatively by rubbing it with wool. What happens when you rub a copper rod with wool?
8. **Critical Thinking** Some scientists once proposed that electric charge is a type of fluid that flows from objects with an excess of the fluid to objects with a deficit. How is the current two-charge model more accurate than the single-fluid model?

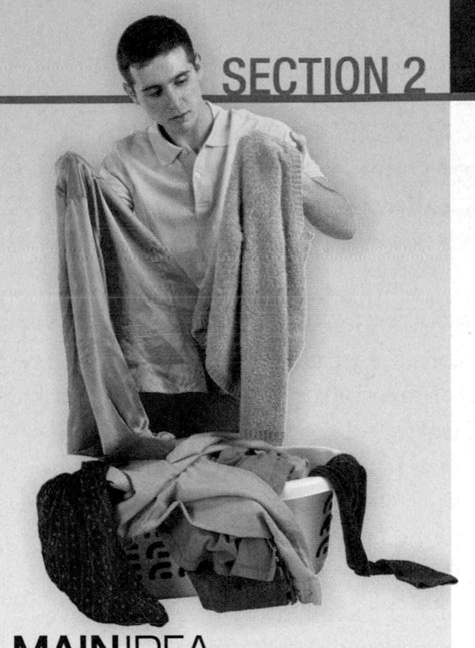

Richard Hutchings/Digital Light Source

SECTION 2 Electrostatic Force

PHYSICS 4 YOU Clothes sometimes come out of the clothes dryer clinging together. This phenomenon is called static cling. You use static cling to affix decals to car windows. An attractive electrostatic force holds the decals in place.

MAIN IDEA

Forces between charged particles are mathematically related to charge and distance.

Essential Questions

- How does the electrostatic force depend on the distance between charges?
- How can you charge objects by conduction and by induction?
- What is Coulomb's law, and how is it used?

Review Vocabulary

force push or pull exerted on an object

New Vocabulary

electroscope
charging by conduction
charging by induction
grounding
Coulomb's law
coulomb
elementary charge

Forces on Charged Objects

In Section 1, you observed that electrostatic forces are typically stronger than gravitational forces, as shown by the charged balloon's lifting of hair. You also observed that electrostatic forces can be either repulsive or attractive, while gravitational forces always are attractive.

Electrostatic forces have been recognized for thousands of years, and over the years many scientists tried to measure them. Daniel Bernoulli, best known for his work with fluids, made some crude measurements in the 1750s. In the 1770s, Henry Cavendish demonstrated a mathematical relationship between forces on charged particles, but he did not publish his work. The discovery of this relationship, described later in this section, is instead credited to the French physicist Charles Coulomb.

Demonstrating forces You demonstrated electrostatic forces using tape strips. You also can demonstrate electrostatic forces by suspending a negatively charged, hard rubber rod so it turns easily, as illustrated in **Figure 6.** If you bring another negatively charged rod near the suspended rod, the suspended rod will turn away. The negative charges on the rods will repel each other even if the rods do not touch. The electrostatic force acts at a distance. If a negatively charged rod is brought near a positively charged rod, however, the two will attract each other. These actions of charged rods can be summarized as follows:

- There are two kinds of electric charge: positive and negative.
- Like charges repel; unlike charges attract.
- Charges exert forces on other charges at a distance.
- The force is stronger when the charges are closer together.

Figure 6 Two charged rods brought near each other will either attract each other or repel each other, depending on their charges.

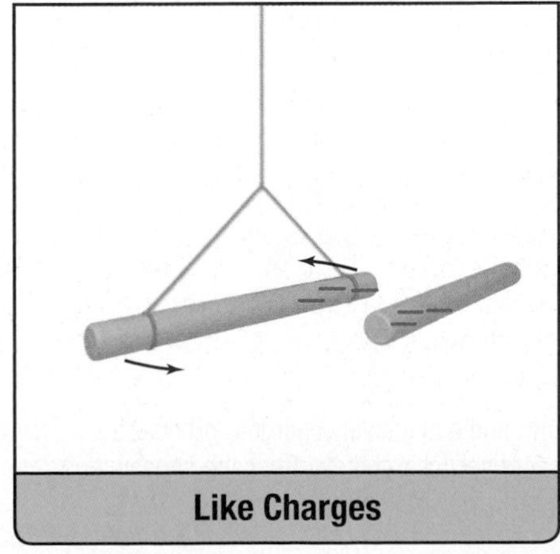

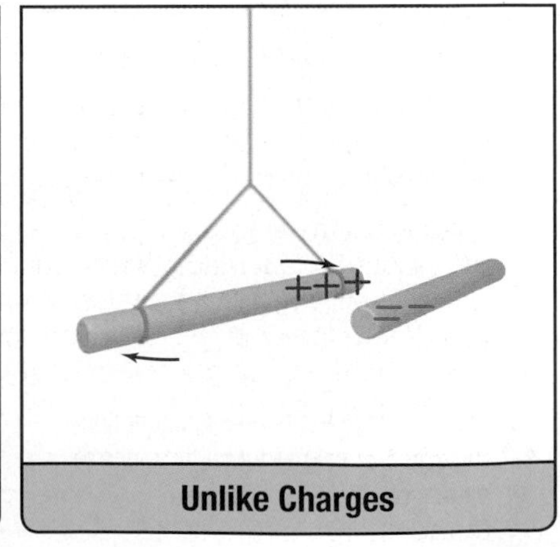

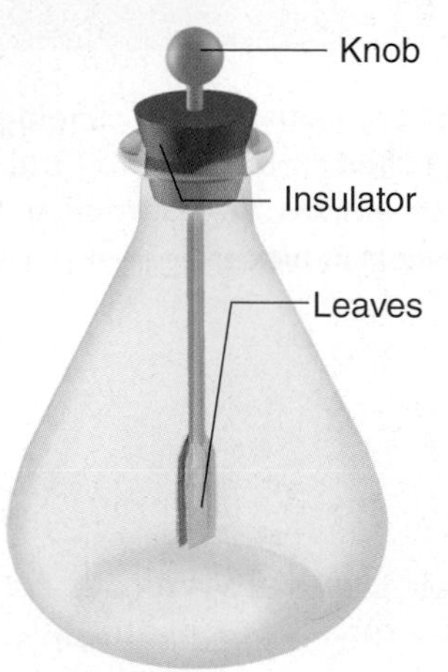

Figure 7 The metal foil leaves on a neutral electroscope hang loosely together.

Determining charge Neither a strip of tape nor a rubber rod that is hanging in the air is a very sensitive or convenient way to determine charge. Instead, you can use a device called an electroscope. An **electroscope** consists of a metal knob connected by a metal stem to two thin, lightweight pieces of metal foil, called leaves, that are enclosed to eliminate air currents. **Figure 7** shows a neutral, uncharged electroscope. Note that the leaves hang loosely.

Charging by conduction When you touch a negatively charged rod to the knob of an electroscope, electrons are transferred to the knob and spread over all the metal surfaces. As shown in the left image of **Figure 8,** the two leaves become negatively charged and repel each other, so they move apart. The leaves also will move apart if you positively charge the electroscope. In either case, you have given the electroscope a net charge. Scientists call charging a neutral object by touching that object with a charged object **charging by conduction.**

If the leaves of an electroscope behave the same no matter the charge, how can you find out whether you have charged an electroscope positively or negatively? You can determine the charge of an electroscope by observing the leaves when you bring a rod of known charge close to but not touching the knob. The leaves will spread farther apart if the rod and the electroscope have the same charge, as shown in the middle image of **Figure 8.** The leaves will fall slightly if the electroscope's charge is opposite that of the rod, as in the far-right image of **Figure 8.**

☑ **READING CHECK Explain** what happens when you bring a positively charged rod near a negatively charged electroscope.

Separation of charge on neutral objects In your investigations with tape, you found that when you held charged tape, the tape was attracted to your fingers. The tape was attracted to your fingers whether the tape was positively charged or negatively charged. Your fingers, however, remained neutral; they had equal amounts of positive and negative charge. Why were the tape strips attracted to them? The tape was attracted to your fingers because the charges in your fingers had separated.

Figure 8 Charging by conduction occurs when charge is transferred between two objects that are touching.

Explain what would happen to the electroscope leaves if the positively charged rod actually touched the knob in the image on the right.

Charging by Conduction

To charge an electroscope by conduction, let a charged metal rod touch the electroscope's knob.

Bringing a negatively charged rod near a negatively charged electroscope causes the leaves to spread apart farther.

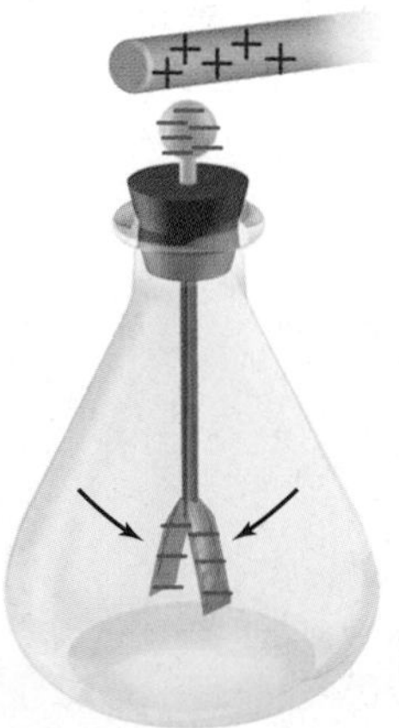

Bringing a positively charged rod near a negatively charged electroscope causes the leaves to fall closer together.

Figure 9 A buildup of negative charge in the bottoms of clouds causes lightning.

Suppose you place a neutral metal rod, or any uncharged conductor, near a positively charged object. The object attracts the electrons in the rod. You know that electrons in conductors move easily. Therefore, some of the rod's electrons move toward the object. The end closest to the object becomes negatively charged, and the other end positively charged. Charges have separated but the rod remains neutral.

The closer charges are to each other, the stronger is the electrostatic force between them; therefore, the separation of charges in the rod results in an attractive force between the rod and the charged object. The force you observed between the tape and your fingers resulted from the same process. Electrons in your fingers moved and charges separated.

Lightning The negative charges at the bottoms of thunderclouds can cause the separation of charges on Earth. Negative charges in the ground below a cloud are repelled from Earth's surface. The forces between the charges in the cloud and those on Earth's surface can break molecules in the air into positively and negatively charged particles. These charged particles are free to move, and they establish a conducting path from the ground to the cloud.

The lightning you observe in a thunderstorm occurs when a bolt of electrons travels down along the conducting path, at speeds close to 14 km/s, and discharges the cloud and Earth, as shown in **Figure 9.**

Charging by induction Suppose that two identical insulated metal spheres are touching, as in **Figure 10.** You can see they have equal charge. If you bring a negatively charged rod close to one sphere, as in the middle image of **Figure 10,** electrons will move from this sphere onto the sphere farther from the rod. This will negatively charge the farther sphere and positively charge the closer sphere. If you separate the spheres while the rod is nearby, each sphere will have a charge, and the charges will be equal but opposite, as shown in the far-right of **Figure 10.** This process of charging a neutral object by bringing a charged object near it is called **charging by induction.**

View an **animation of charging by induction.**

Figure 10 The result of charging by induction is spheres with charges that are equal in magnitude but opposite in sign. No charges were added to the spheres from the rod.

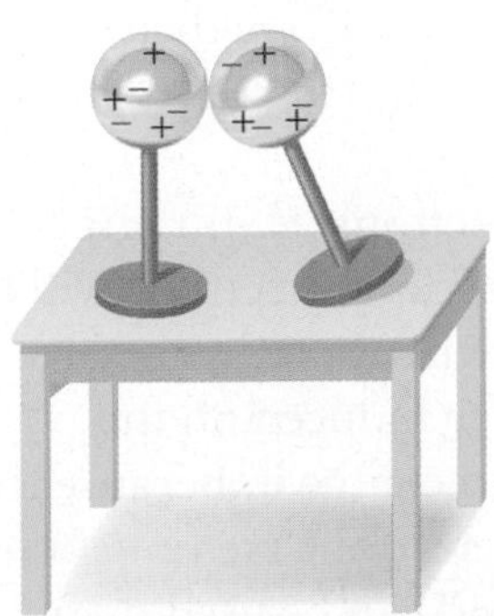

Neutral spheres are touching.

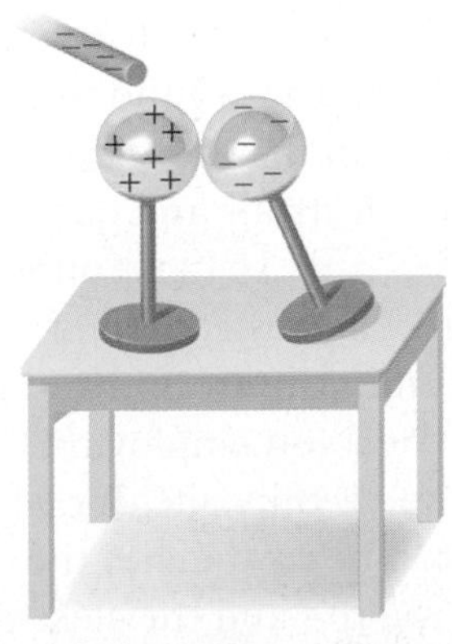

Both spheres are charged by induction.

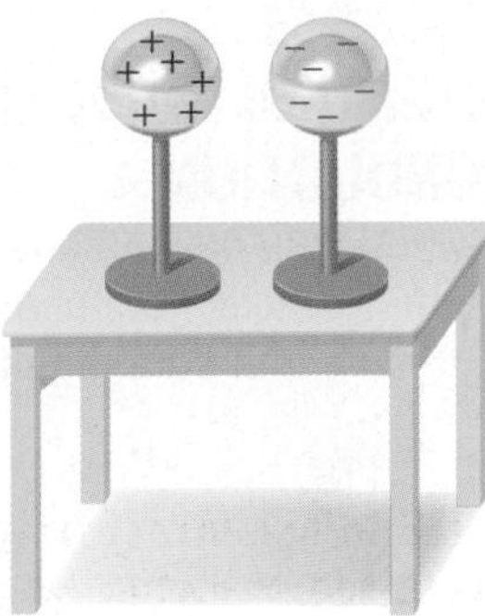

The separated spheres have opposite charges.

■ **Charging by Induction**

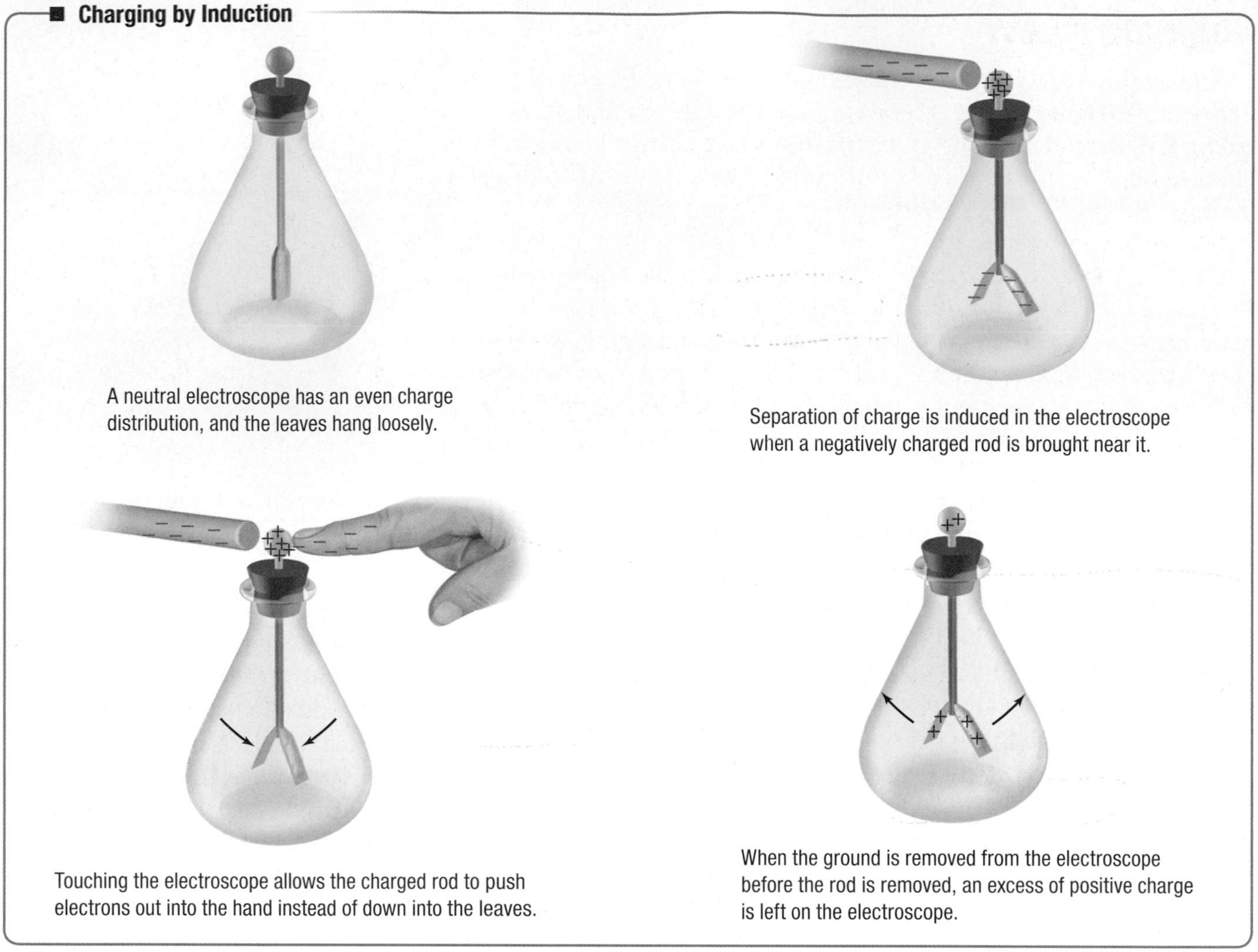

A neutral electroscope has an even charge distribution, and the leaves hang loosely.

Separation of charge is induced in the electroscope when a negatively charged rod is brought near it.

Touching the electroscope allows the charged rod to push electrons out into the hand instead of down into the leaves.

When the ground is removed from the electroscope before the rod is removed, an excess of positive charge is left on the electroscope.

Figure 11 Grounding an electroscope charges the electroscope by induction.

PhysicsLAB

CHARGE IT UP

How can you demonstrate charging by induction and conduction and the separation of charges?

MiniLAB

INVESTIGATING INDUCTION AND CONDUCTION

How can you use a balloon and an electroscope to investigate charging by induction and conduction?

Charge flows to Earth **Figure 11** shows how to charge an electroscope by induction through **grounding,** which is the process of removing excess charge by connecting an object to Earth. Because Earth is very large, it can absorb great amounts of charge without becoming noticeably charged itself. When you ground a charged object, almost any amount of charge can flow to Earth.

If you bring a negatively charged rod close to the knob of the electroscope in **Figure 11,** as in the top-right image, the rod repels electrons onto the leaves. If you ground the knob on the side opposite the charged rod by touching it, electrons will be pushed from the electroscope through your hand and into the ground until the leaves are neutral. If you then remove your hand with the charging rod still in place, the electroscope will have a deficit of electrons and be positively charged.

You also can use the ground as a source of electrons. If you bring a positive rod near the knob of a grounded electroscope, electrons will be attracted from the ground and the electroscope will obtain a negative charge. When you employ this process, the charge induced on the electroscope is opposite that of the object used to charge it. Because the charged rod never touches the electroscope, no electrons are transferred between the rod and the electroscope, and the rod can be used many times to charge objects by induction.

Coulomb's Law

In your investigations with tape, you demonstrated that a force acts between two or more charged objects. You probably found that the strength of the force depended on distance. The closer you brought a charged object to the tape, the stronger the force was. The force also gets stronger as the strength of the charge on the object increases.

How can you vary the quantity of charge in a controlled way? In the late 1770s, Charles Coulomb devised an apparatus that enabled him to determine the amount of charge between two objects. The type of apparatus Coulomb used is shown in **Figure 12.** An insulating rod with a small conducting sphere at one end (A) was suspended from the top by a thin wire that was free to rotate. An equal-sized conducting sphere (B) at the end of another rod was placed in contact with sphere A. When Coulomb touched the rod attached to B with a charged object, the charge spread evenly over the two spheres. Because the two spheres were the same size, they received an equal amount of charge. The symbol for charge is q. We can represent the amount of charge on the spheres as q_A for sphere A and q_B for sphere B; in this case, $q_A = q_B$.

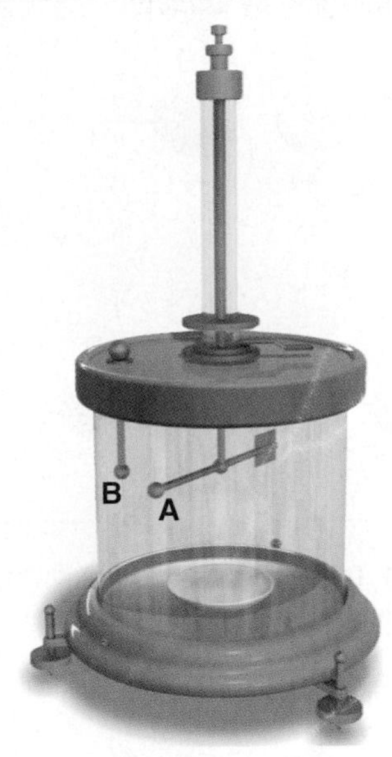

Figure 12 Coulomb used an apparatus similar to this one to measure the electrostatic force between two charged spheres. He observed differences in the amount of deflection of the spheres while varying the distance between them.

Force depends on distance Coulomb found that the force between two charged spheres depends on the distance between them. First, he carefully measured the amount of force needed to twist the suspending wire through a given angle. He then placed equal charges on spheres A and B and varied the distance (r) between them. The force moved A, which twisted the suspending wire. By measuring the deflection of A, Coulomb could calculate the force of repulsion. He showed that the magnitude of the force (F) varies inversely with the square of the distance between the centers of the spheres.

$$F \propto \frac{1}{r^2}$$

Force depends on charge To investigate the way in which the force depends on the amount of charge, Coulomb had to change the charges on the spheres in a measured way. He first charged spheres A and B equally, as before. Then he selected an uncharged sphere (C) of the same size as spheres A and B. When C was placed in contact with B, the spheres shared the charge that had been on B alone. Because the two were the same size, B then had only half of its original charge. Therefore, the charge on B was only one-half the charge on A. After Coulomb adjusted the position of B so that the distance (r) between A and B was the same as before, he found that the force between A and B was half of its former value. That is, he found that the force varies directly with the charge of the bodies.

$$F \propto q_A q_B$$

After many similar measurements, Coulomb summarized the results in a law now known as **Coulomb's law:** the magnitude of the force between point charge q_A and point charge q_B, separated by a distance r, is proportional to the product of the magnitudes of the charges and inversely proportional to the square of the distance between them.

$$F \propto \frac{q_A q_B}{r^2}$$

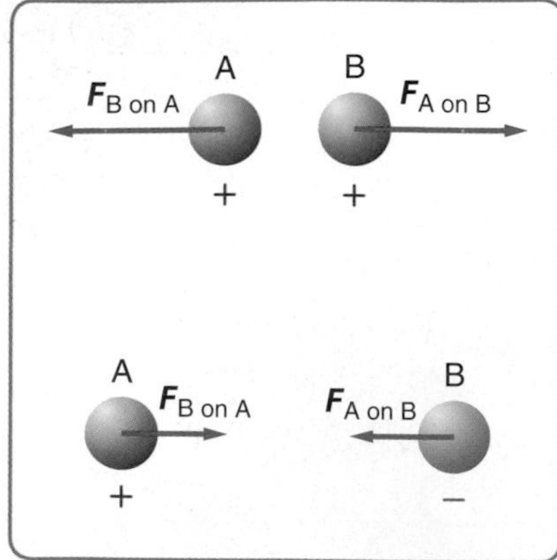

Figure 13 The direction of the force between charged particles depends on the charges. Like charges repel, and unlike charges attract. (Forces are not to scale.)

COLOR CONVENTION

force ⟷ **blue**

The unit of charge: the coulomb The amount of charge an object has is difficult to measure directly. Coulomb's experiments, however, showed that the quantity of charge is related to force. Thus, Coulomb could define a standard quantity of charge in terms of the amount of force that it produces. The SI standard unit of charge is called the **coulomb** (C).

One coulomb is the charge of 6.24×10^{18} electrons or protons. (A proton is an atom's positive charge.) The magnitude of the charge of a single electron or proton is called the **elementary charge.** The elementary charge has a value of 1.602×10^{-19} C. All objects contain charge. In neutral objects the positive and negative charges are balanced, and the net charge is zero. If the negative charge of a book is -1×10^{6} C, and the positive charge of the book is 1×10^{6} C, then the net charge is zero.

In Coulomb's law equation, the magnitude of the force on charge q_A from charge q_B a distance r away from each other can be calculated by incorporating a proportionality constant (K).

COULOMB'S LAW

The force between two charges is equal to a constant times the product of the two charges, divided by the square of the distance between them.

$$F = K\frac{q_A q_B}{r^2}$$

When the charges are measured in coulombs, the distance in meters, and the force in newtons, the constant is 9.0×10^{9} N•m²/C².

View an **animation of Coulomb's law.**

Vector quantity The Coulomb's law equation gives the magnitude of the force that q_A exerts on q_B and also the magnitude of the force that q_B exerts on q_A. These magnitudes are equal. You can observe this example of Newton's third law of motion when you bring two strips of tape with like charges together. Each exerts a force on the other. If you bring a charged comb near either strip, the strip, with its small mass, moves readily. The acceleration of the comb is, of course, much less because of its much greater mass.

The electrostatic force, like all forces, is a vector quantity. Force vectors have both a magnitude and a direction. The Coulomb's law equation gives only the magnitude of the force. To determine direction, you can draw a diagram using force vectors, such as that in **Figure 13.** The direction of force between two objects depends on the objects' charges.

PROBLEM-SOLVING STRATEGIES

ELECTROSTATIC FORCE PROBLEMS

Use these steps to find the magnitude and the direction of the force between charges.

1. Sketch the system, showing all distances and angles to scale.
2. Diagram the vectors of the system.
3. Use Coulomb's law to find the magnitude of the force.
4. Use your diagram, along with trigonometric relations, to find the direction of the force.
5. Perform all algebraic operations on both the numbers and the units. Make sure the units match the variables in question.
6. Consider the magnitude of your answer. Is the magnitude of the net force relatively close to those of the component forces?

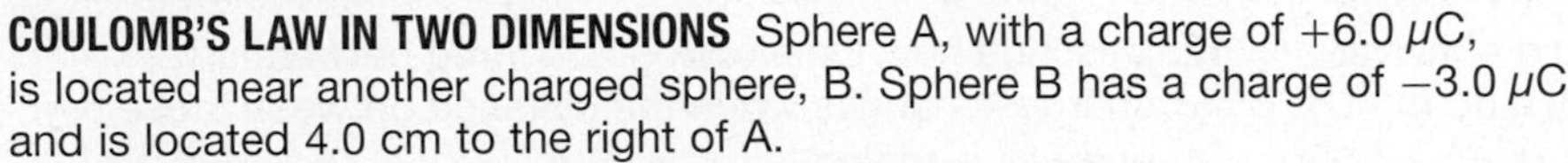

EXAMPLE PROBLEM 1

Get help with **Coulomb's law in two dimensions.** Personal Tutor

COULOMB'S LAW IN TWO DIMENSIONS Sphere A, with a charge of $+6.0\ \mu C$, is located near another charged sphere, B. Sphere B has a charge of $-3.0\ \mu C$ and is located 4.0 cm to the right of A.

a. What is the force of sphere B on sphere A?

b. A third sphere, C, with a $+1.5$-μC charge, is added. If it is located 3.0 cm directly beneath A, what is the new net force on sphere A?

1 ANALYZE AND SKETCH THE PROBLEM

- Establish coordinate axes, and sketch the spheres.
- Show and label the distances between the spheres.
- Diagram and label the force vectors.

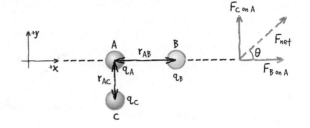

KNOWN		UNKNOWN
$q_A = +6.0\ \mu C$	$r_{AB} = 4.0$ cm	$F_{B\ on\ A} = ?$
$q_B = -3.0\ \mu C$	$r_{AC} = 3.0$ cm	$F_{C\ on\ A} = ?$
$q_C = +1.5\ \mu C$	$K = 9.0\times10^9\ N\cdot m^2/C^2$	$\boldsymbol{F}_{net} = ?$

2 SOLVE FOR THE FORCES ON SPHERE A

a. Find the force of sphere B on sphere A.

$$F_{B\ on\ A} = K\frac{q_A q_B}{r_{AB}^2}$$

$$= (9.0\times10^9\ N\cdot m^2/C^2)\frac{(6.0\times10^{-6}\ C)(3.0\times10^{-6}\ C)}{(4.0\times10^{-2}\ m)^2}$$

◀ *Substitute $q_A = 6.0\ \mu C$, $q_B = 3.0\ \mu C$, $r_{AB} = 4.0$ cm*

$$= 1.0\times10^2\ N$$

Because spheres A and B have unlike charges, the force of B on A is to the right.

b. Find the net force on sphere A.

$$F_{C\ on\ A} = K\frac{q_A q_C}{r_{AC}^2}$$

$$= (9.0\times10^9\ N\cdot m^2/C^2)\frac{(6.0\times10^{-6}\ C)(1.5\times10^{-6}\ C)}{(3.0\times10^{-2}\ m)^2}$$

◀ *Substitute $q_A = 6.0\ \mu C$, $q_C = 1.5\ \mu C$, $r_{AC} = 3.0$ cm*

$$= 9.0\times10^1\ N$$

Spheres A and C have like charges, which repel. The force of C on A is upward.

Find the vector sum of $\boldsymbol{F}_{B\ on\ A}$ and $\boldsymbol{F}_{C\ on\ A}$ to find $\boldsymbol{F}_{net}$ on sphere A.

$$F_{net} = \sqrt{F_{B\ on\ A}^2 + F_{C\ on\ A}^2}$$

$$= \sqrt{(1.0\times10^2\ N)^2 + (9.0\times10^1\ N)^2}$$

◀ *Substitute $F_{B\ on\ A} = 1.0\times10^2$ N, $F_{C\ on\ A} = 9.0\times10^1$ N*

$$= 130\ N$$

$$\tan\theta = \frac{F_{C\ on\ A}}{F_{B\ on\ A}}$$

$$\theta = \tan^{-1}\left(\frac{F_{C\ on\ A}}{F_{B\ on\ A}}\right)$$

$$= \tan^{-1}\left(\frac{9.0\times10^1\ N}{1.0\times10^2\ N}\right)$$

◀ *Substitute $F_{C\ on\ A} = 9.0\times10^1$ N, $F_{B\ on\ A} = 1.0\times10^2$ N*

$$= 42°$$

$\boldsymbol{F}_{net} = 1.3\times10^2$ N, 42° above the x-axis

3 EVALUATE THE ANSWER

- **Are the units correct?** $(N\cdot m^2/C^2)(C)(C)/m^2 = N$. The units work out to be newtons.
- **Does the direction make sense?** Like charges repel; unlike charges attract.
- **Is the magnitude realistic?** The net force is relatively close to the component forces. 130 N is close enough to 90 N and 100 N to be believable.

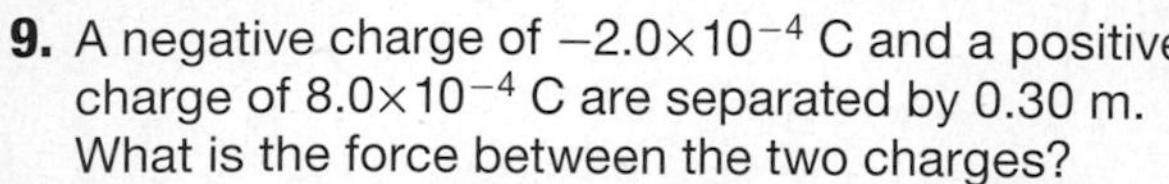

PRACTICE PROBLEMS

Do additional problems. Online Practice

9. A negative charge of -2.0×10^{-4} C and a positive charge of 8.0×10^{-4} C are separated by 0.30 m. What is the force between the two charges?

10. A negative charge of -6.0×10^{-6} C exerts an attractive force of 65 N on a second charge that is 0.050 m away. What is the magnitude of the second charge?

11. Suppose you replace the charge on B in Example Problem 1 with a charge of $+3.00$ μC. Diagram the new situation, and find the net force on A.

12. Describe how the electrostatic force between two charges changes when the distance between those two charges is tripled.

13. Sphere A is located at the origin and has a charge of $+2.0\times10^{-6}$ C. Sphere B is located at $+0.60$ m on the x-axis and has a charge of -3.6×10^{-6} C. Sphere C is located at $+0.80$ m on the x-axis and has a charge of $+4.0\times10^{-6}$ C. Determine the net force on sphere A.

14. CHALLENGE Determine the net force on sphere B in the previous problem.

Using Coulomb's law As you use the Coulomb's law equation, keep in mind that Coulomb's law is valid only for point charges or for charges that are spread uniformly over a sphere. A charged sphere may be treated as if all the charge were located at its center if the charge is spread evenly across its entire surface or throughout its volume.

If the sphere is a conductor and another charge is brought near it, the charges on the sphere will be attracted or repelled and will move, making the sphere no longer uniformly charged. The charge no longer will act as if it were at the sphere's center. Therefore, it is important to consider how large and how far apart two charged spheres are before applying Coulomb's law.

The problems in this textbook assume that charged spheres are small enough and far enough apart to be considered point charges, unless otherwise noted. When you consider shapes such as long wires or flat plates, you must modify Coulomb's law to account for the non-point charge distributions. To do this, you must model the charge distribution as an additive collection of point charges.

MiniLAB

FLYING OBJECTS

What happens to objects charged with a Van de Graaff generator?

iLab Station

Applications of Electrostatic Forces

There are many commercial and industrial applications that take advantage of electrostatic forces. For example, tiny paint droplets can be charged by induction. When sprayed on automobiles and other objects, the droplets repel each other and paint spreads uniformly. Photocopy machines take advantage of static electricity to place black toner on a page so that a precise reproduction of the original document is achieved. Laser printers use static electricity in a similar way. Toner particles are attracted to charged characters on a drum.

PHYSICS CHALLENGE

CHARGES ON SPHERES Two spheres of equal mass (m) and equal positive charge (q) are a distance r apart, as shown in the figure on the right.

1. Derive an expression for the quantity of charge (q) that must be on each sphere so that the spheres are in equilibrium; that is, so that the attractive and repulsive forces between them are balanced.
2. How would doubling the distance between the spheres affect the expression for the value of q you determined in the previous problem? Explain.
3. Assuming that the mass of each sphere is 1.50 kg, determine the charge on each sphere needed to maintain the equilibrium.

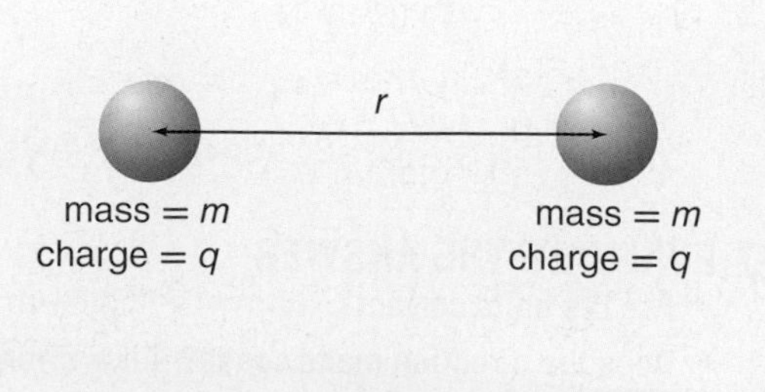

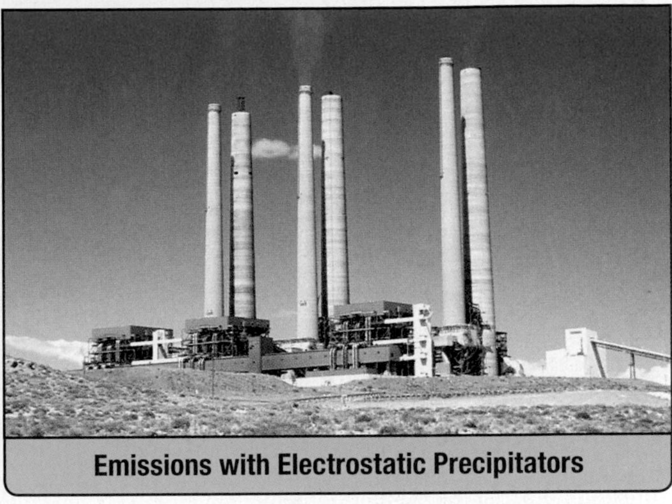

Figure 14 Fly ash from coal combustion is an industrial pollutant. Electrostatic precipitators collect fly ash so it is not released into the atmosphere.

Electrostatic forces also are used to collect emissions in smokestacks. Emissions are charged when they are released from combustion processes. They are then attracted to collectors that have been given an opposite charge. A liquid spray removes the collected emissions safely. Electrostatic precipitators make a huge difference in the amount of pollution released into the air, as shown in **Figure 14.**

In some instances, it is important to avoid the accumulation of static charge. For example, static discharges can damage electronic equipment. Manufacturers pack sensitive electronic components in conductive plastic bags to prevent the build-up of static charge. Computer technicians wear wrist straps connected to the ground to safely conduct charges to Earth.

Electrostatic forces can be dangerous, especially those associated with lightning. Metal lightning rods grounded to Earth protect homes and other structures susceptible to lightning strikes. The rods, placed on the highest part of the roof, safely conduct electric discharge to Earth.

SECTION 2 REVIEW

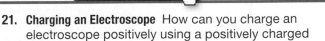

Check your understanding.

15. **MAIN**IDEA Describe the relationship between the magnitude of the electrostatic force, the charge on two objects, and the distance between the objects. What is the equation for this relationship?

16. **Force and Charge** How are electrostatic force and charge related? Describe the force when the charges are like charges and the force when the charges are opposite.

17. **Force and Distance** How are electrostatic force and distance related? How would the force change if the distance between two charges were tripled?

18. **Charging by Induction** In an electroscope being charged by induction, what happens when the charging rod is moved away before the ground is removed from the knob?

19. **Electroscopes** Why do the leaves of a charged electroscope rise to a certain angle and no farther?

20. **Attraction of Neutral Objects** What properties explain how both positively charged objects and negatively charged objects can attract neutral objects?

21. **Charging an Electroscope** How can you charge an electroscope positively using a positively charged rod? Using a negatively charged rod?

22. **Electrostatic Forces** Two charged spheres are held a distance r apart, as shown in **Figure 15.** Compare the force of sphere A on sphere B with the force of sphere B on sphere A.

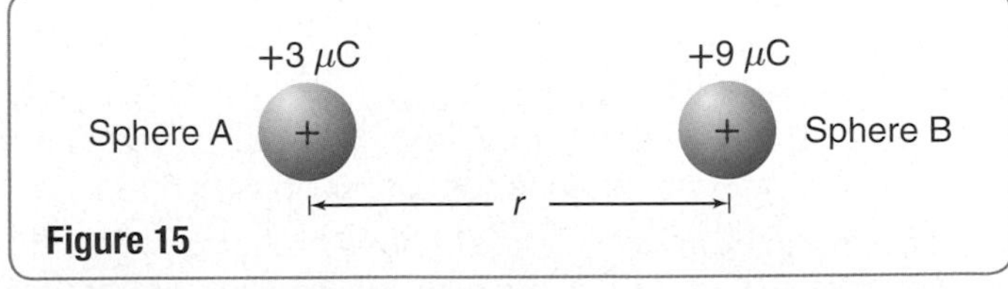

Figure 15

23. **Critical Thinking** Suppose you are testing Coulomb's law using a small, positively charged plastic sphere and a large, positively charged metal sphere. According to Coulomb's law, the force depends on $\frac{1}{r^2}$, where r is the distance between the centers of the spheres. As you bring the spheres close together, the force is smaller than expected from Coulomb's law. Explain.

WHEN SPARKS FLY

GAS PUMP EXPLOSIONS

Quentin felt as if the gas pump was taking forever to refuel his car, so he sat back down in the front seat to wait. When he went back to check, BLAM—an explosion burst from the nozzle.

The culprit: static discharge Most people know of the dangers of refueling a car with the engine running. Explosions such as the one described above can also occur. In Quentin's situation, a spark from static electricity ignited gasoline vapors.

Have you ever experienced an electric shock as you got out of a car? What you felt was the discharge of static electricity. When you brought your charged finger near an electrical conductor, such as the metal of a car door, the charges jumped, producing a spark. Such a spark could ignite gasoline vapors.

What causes static buildup? Static electric charge can build up on a car's driver when fibers in seats and clothing rub together. This is one of the reasons why the person shown in **Figure 1** should not reenter her car until she is completely done pumping gas.

Getting rid of the spark The safest way to avoid an explosion when refueling is to discharge any static electricity as you get out of the car. Always touch something metal before you approach the gas-pump nozzle. Do not get back into your car until you have finished pumping gas.

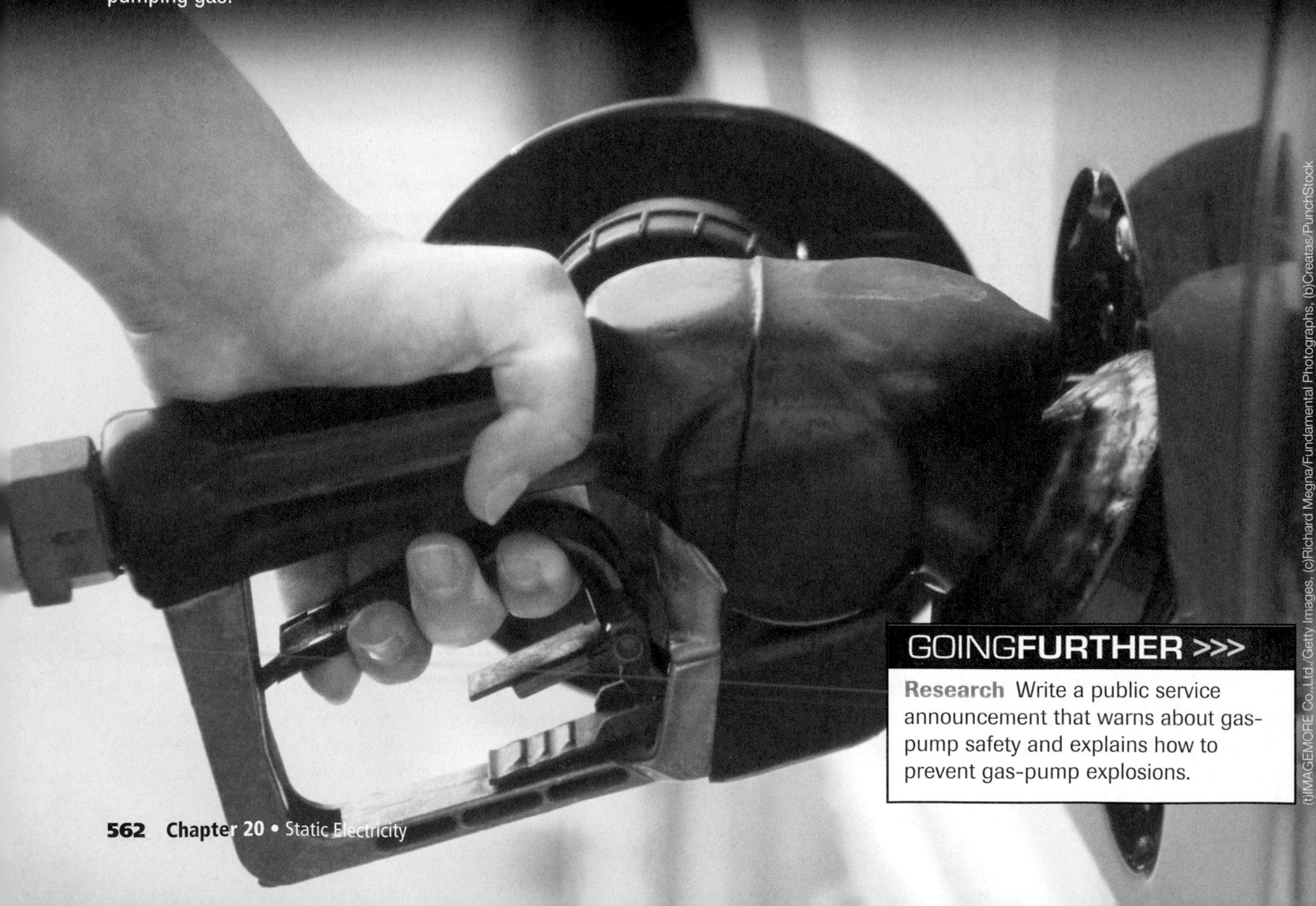

FIGURE 1 You should discharge the unbalanced electric charge from your body before pumping gas. Never get in and out of a vehicle while you are pumping gas.

GOING FURTHER >>>

Research Write a public service announcement that warns about gas-pump safety and explains how to prevent gas-pump explosions.

STUDY GUIDE

BIGIDEA Separated positive and negative charges exert forces on one another.

VOCABULARY
- **electrostatics** *(p. 548)*
- **neutral** *(p. 550)*
- **insulator** *(p. 551)*
- **conductor** *(p. 551)*

SECTION 1 Electric Charge

MAINIDEA Like electric charges repel, and unlike electric charges attract.

- There are two kinds of electric charge, positive and negative. Interactions of these charges explain the attraction and repulsion you observed using strips of tape.
- Electric charge is not created or destroyed; it is conserved. Charging is the separation, not creation, of electric charges. Objects are charged by the transfer of electrons. An area with excess electrons has a net negative charge; an area with a deficit of electrons has a net positive charge.
- Charges added to one part of an insulator remain on that part. Insulators include glass, dry wood, most plastics, and dry air. Charges added to a conductor quickly spread throughout the conductor. Some examples of conductors are graphite, metals, and matter in the plasma state. Under certain conditions, charges can move through a substance that is ordinarily an insulator. Lightning moving through air is one example.

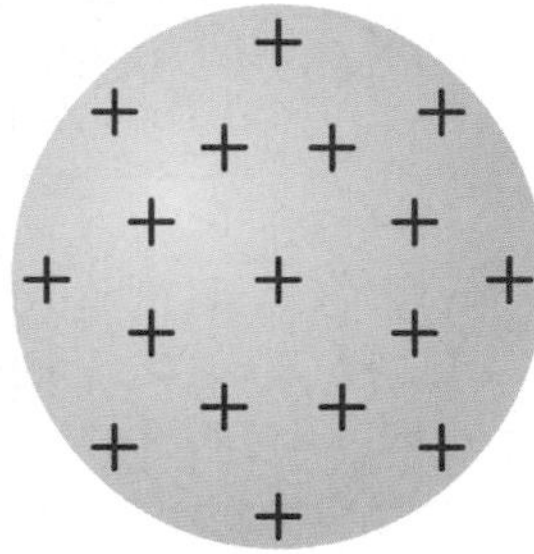

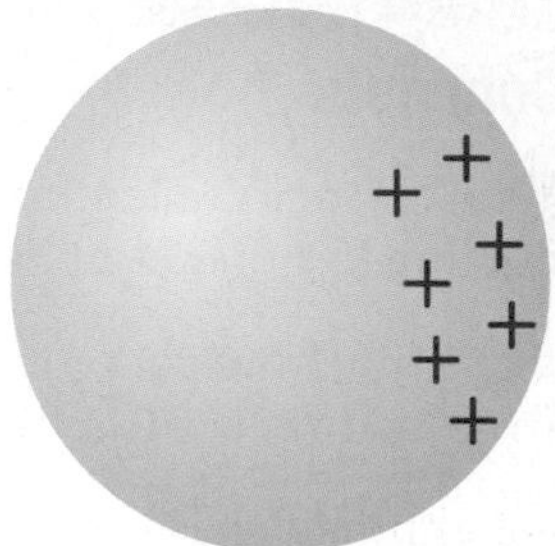

VOCABULARY
- **electroscope** *(p. 554)*
- **charging by conduction** *(p. 554)*
- **charging by induction** *(p. 555)*
- **grounding** *(p. 556)*
- **Coulomb's law** *(p. 557)*
- **coulomb** *(p. 558)*
- **elementary charge** *(p. 558)*

SECTION 2 Electrostatic Force

MAINIDEA Forces between charged particles are mathematically related to charge and distance.

- Electrostatic force acts over a distance and decreases as the distance between charges increases.
- Charging by conduction charges a neutral object by touching that object with a charged object. Conduction requires the transfer of electrons between the charging object and the object being charged. Charging by induction charges a neutral object by bringing a charged object near the neutral object without the two objects touching. Induction makes use of separation of charge or grounding to charge a neutral object. Charging by induction enables the charging object to be used multiple times.
- Coulomb's law states that the force between two charged particles varies directly with the product of their charges and inversely with the square of the distance between them. It is used to find the magnitude of the force two charged particles exert on each other.

$$F = K\frac{q_A q_B}{{r_{AB}}^2}$$

Games and Multilingual eGlossary

Chapter Self-Check

SECTION 1 Electric Charge

Mastering Concepts

24. BIGIDEA If you comb your hair on a dry day, the comb can become negatively charged. Can your hair remain neutral? Explain.

25. If you bring a charged comb near tiny pieces of paper, the pieces will first be attracted to the comb, but after touching they will fly away. Why do they fly away?

26. List some insulators and conductors.

27. What makes metal a good conductor and rubber a good insulator?

SECTION 2 Electrostatic Force

Mastering Concepts

28. Laundry Why do socks taken from a clothes dryer sometimes cling to other clothes?

29. Compact Discs If you wipe a compact disc with a clean cloth, why does the CD then attract dust?

30. Coins The combined charge of all electrons in a nickel is hundreds of thousands of coulombs. Does this imply anything about the net charge on the coin? Explain.

31. How does the distance between two charges impact the force between them? If the distance is decreased while the charges remain the same, what happens to the force?

32. Explain how to charge a conductor negatively if you have only a positively charged rod.

33. Bernoulli's experiments to measure the strength of the electrostatic force used metal disks about 3 cm in diameter. When the disks were close together, would he have found a $\frac{1}{r^2}$ dependence? Explain.

Mastering Problems

34. Atoms Two electrons in an atom are separated by 1.5×10^{-10} m, the typical size of an atom. What is the electrostatic force between them?

35. A positive and a negative charge, each of magnitude 2.5×10^{-5} C, are separated by a distance of 15 cm. Find the force on each of the particles.

36. Two identical positive charges exert a repulsive force of 6.4×10^{-9} N when separated by a distance of 3.8×10^{-10} m. Calculate the charge of each.

37. Lightning A strong lightning bolt transfers about 25 C to Earth. How many electrons are transferred?

38. A positive charge of 3.0 μC is pulled on by two negative charges. As shown in **Figure 16,** one negative charge, −2.0 μC, is 0.050 m to the west, and the other, −4.0 μC, is 0.030 m to the east. What net force is exerted on the positive charge?

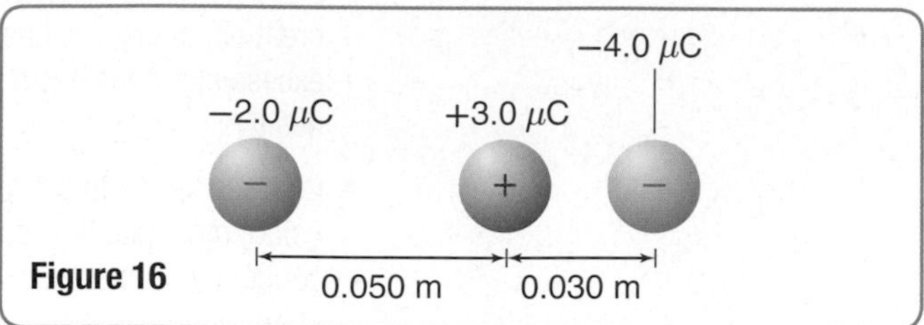

Figure 16

39. Figure 17 shows two positively charged spheres, one with three times the charge of the other. The spheres are 16 cm apart, and the force between them is 0.28 N. What are the charges on the two spheres?

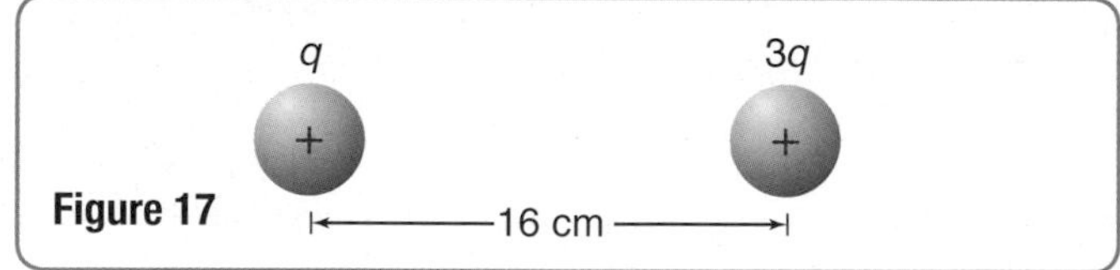

Figure 17

40. Three particles are placed in a line. The left particle has a charge of −55 μC, the middle one has a charge of +45 μC, and the right one has a charge of −78 μC. The middle particle is 72 cm from each of the others, as shown in **Figure 18.**

a. Find the net force on the middle particle.

b. Find the net force on the right particle.

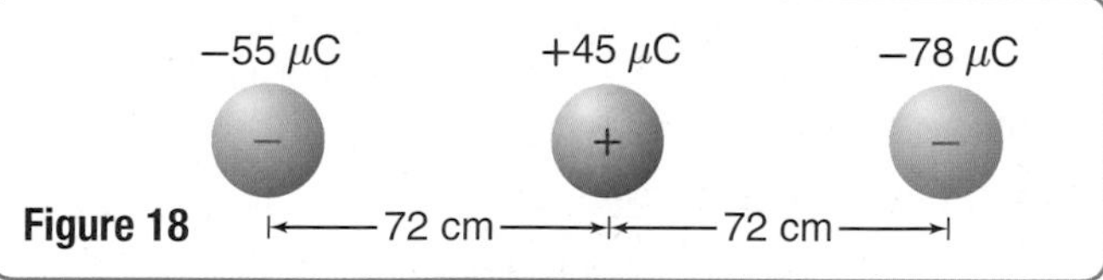

Figure 18

41. Reverse Problem Write a physics problem for which the following equation would be part of the solution:

$$F=\left(9.0\times10^{9}\frac{\text{N}\cdot\text{m}^2}{\text{C}^2}\right)\left(\frac{(3.0\times10^{-6}\text{ C})(2.0\times10^{-6}\text{ C})}{(0.25\text{ m})^2}-\frac{(3.0\times10^{-6}\text{ C})(5.0\times10^{-6}\text{ C})}{(0.45\text{ m})^2}\right)$$

42. Charge in a Coin A nickel has a mass of about 5 g and is 75 percent Cu and 25 percent Ni. On average, each mole of a nickel's atoms will have a mass of about 62 g. Each Cu atom has 29 electrons; each Ni atom has 28 electrons. How many coulombs of charge are on the electrons in a nickel?

Chapter Self-Check

43. **Problem Posing** Complete this problem so that it must be solved using Coulomb's law: "A very small sphere is given a charge of 6.25 μC...."

44. **Ranking Task** Rank the following pairs of point charges according to the magnitude of electrostatic force they exert on each other. Specifically indicate any ties.

A. two 7.0 nC charges separated by 0.20 m

B. two 5.0 nC charges separated by 0.20 m

C. two 2.5 nC charges separated by 0.10 m

D. a 2.5 nC charge and a 5.0 nC charge separated by 0.20 m

E. 1.0 μC charge and a 2.5 μC charge separated by 0.10 m

Applying Concepts

45. Charles Coulomb measured the deflection of sphere A when spheres A and B had equal charges and were a distance r apart. If he made the charge on B one-third the charge on A, how far apart would the two spheres have to be for A to have the same deflection that it had before?

46. Two charged bodies exert a force of 0.145 N on each other. If they are moved so they are one-fourth as far apart, what force is exerted?

47. Electrostatic forces between charges are enormous in comparison to gravitational forces. Yet, you normally do not sense electrostatic forces between yourself and your surroundings, while you do sense gravitational interactions with Earth. Explain.

48. How does the charge of an electron differ from the charge of a proton? How are they similar?

49. Using a charged rod and an electroscope, how can you find whether an object is a conductor?

50. A charged rod is brought near a pile of tiny plastic spheres. Some of the spheres are attracted to the rod, but as soon as the spheres touch the rod, they are flung off in different directions. Explain why this happens.

51. Explain what would happen to the leaves of a positively charged electroscope if a positively charged rod was brought close to, but did not touch, the electroscope. How would the electroscope's leaves behave if the rod were negatively charged?

52. **Lightning** Lightning usually occurs when a negative charge in a cloud is transported to Earth. If Earth is neutral, what provides the attractive force that pulls the electrons toward Earth?

53. The text describes Coulomb's method for charging two spheres, A and B, so that the charge on B was exactly half the charge on A. Suggest a way that Coulomb could have placed a charge on sphere B that was exactly one-third the charge on sphere A.

54. As shown in **Figure 19,** Coulomb's law and Newton's law of universal gravitation appear to be similar. In what ways are the electric and gravitational forces similar? How are they different?

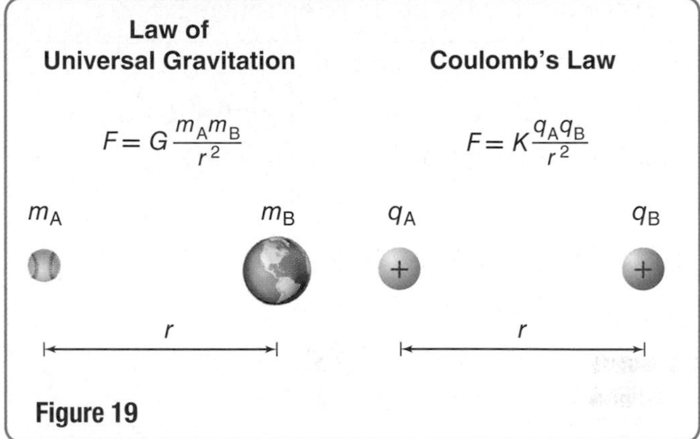

Figure 19

Mixed Review

55. A small metal sphere with charge 1.2×10^{-5} C is touched to an identical neutral sphere and then placed 0.15 m from the second sphere. What is the electrostatic force between the two spheres?

56. **Atoms** What is the electrostatic force between an electron and a proton placed 5.3×10^{-11} m apart, the approximate radius of a hydrogen atom?

57. A small sphere of charge 2.4 μC experiences a force of 0.36 N when a second sphere of unknown charge is placed 5.5 cm from it. What is the charge of the second sphere?

58. Two identically charged spheres placed 12 cm apart have an electrostatic force of 0.28 N between them. What is the charge on each sphere?

59. In an investigation using Coulomb's apparatus, a sphere with a charge of 3.6×10^{-8} C is 1.4 cm from a second sphere of unknown charge. The force between the spheres is 2.7×10^{-2} N. What is the charge of the second sphere?

Thinking Critically

60. **Apply Concepts** Calculate the ratio of the electrostatic force to the gravitational force between the electron and the proton in a hydrogen atom.

Chapter Self-Check

61. Analyze and Conclude Sphere A, with a charge of $+64\ \mu C$, is positioned at the origin. A second sphere, B, with a charge of $-16\ \mu C$ is placed at $+1.00$ m on the x-axis.

a. Where must a third sphere, C, of charge $112\ \mu C$ be placed so there is no net force on it?

b. If the third sphere had a charge of $16\ \mu C$, where should it be placed?

62. Three charged spheres are at the positions shown in **Figure 20.** Find the net force on sphere B.

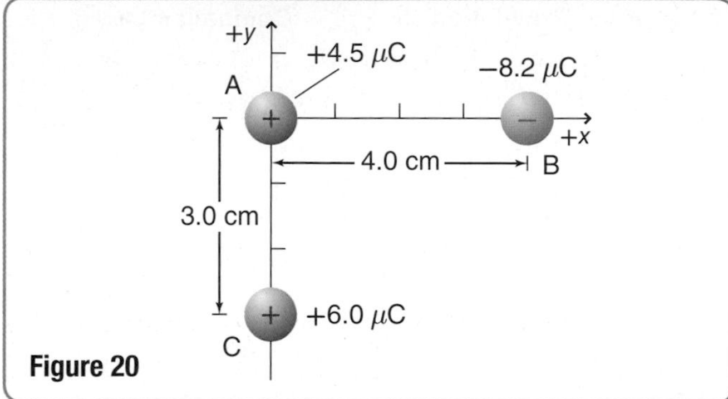

Figure 20

63. The two pith balls in **Figure 21** each have a mass of 1.0 g and an equal charge. One pith ball is suspended by an insulating thread. The other is brought to 3.0 cm from the suspended ball. The suspended ball is now hanging, with the thread forming an angle of 30.0° with the vertical. The ball is in equilibrium with F_E, F_g, and F_T. Calculate each of the following:

a. F_g on the suspended ball

b. F_E

c. the charge on the balls

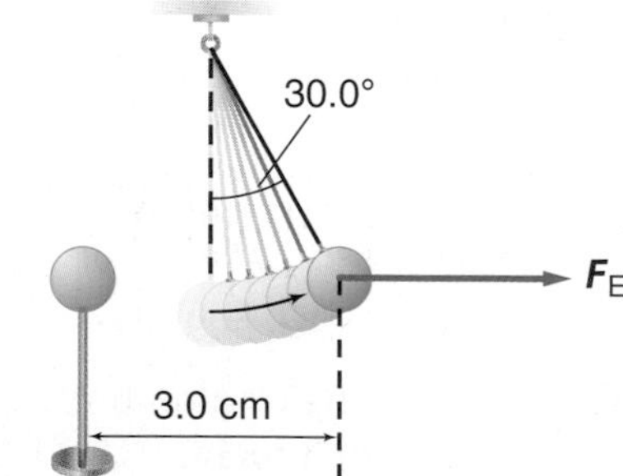

Figure 21

64. A device to trap positively charged ions has four charged rods, equally spaced from the center. The top and bottom rods are charged positively, the left and right rods negatively. Explain why when the ion is at the center of the rods there is no force on it. If the ion moves a small distance up or down, does the net force push it toward or away from the center? What if it moves a small distance to the left or right?

65. Two charges, q_A and q_B, are at rest near a positive charge (q_T) of $7.2\ \mu C$. The first charge (q_A) is a positive charge of $3.6\ \mu C$ located 2.5 cm away from q_T at 35°; q_B is a negative charge of $26.6\ \mu C$ located 6.8 cm away at 125°.

a. Determine the magnitude of each of the forces acting on q_T.

b. Sketch a force diagram.

c. Graphically determine the resultant force on q_T.

Writing in Physics

66. History of Science Research several devices that were used in the seventeenth and eighteenth centuries to study static electricity. Examples you might consider include the Leyden jar and the Wimshurst machine. Discuss how they were constructed and how they were used.

67. When you studied states of matter, you learned that forces exist between water molecules that cause water to be denser as a liquid between 0°C and 4°C than as a solid at 0°C. These forces are electrostatic in nature. Research electrostatic intermolecular forces, such as van der Waals forces and dipole-dipole forces, and describe their effects on matter.

Cumulative Review

68. Explain how a pendulum can be used to determine the acceleration of gravity.

69. Sonar A submarine moving 12.0 m/s sends a sonar ping of frequency 1.50×10^3 Hz toward a seamount directly in front of it. The submarine receives the echo 1.800 s later.

a. How far is the submarine from the seamount?

b. What is the frequency of the sonar wave that strikes the seamount?

c. What is the frequency of the echo?

70. Security Mirror A security mirror is used to produce an image that is three-fourths the size of an object and is located 12.0 cm behind the mirror. What is the focal length of the mirror?

71. A 2.00-cm-tall object is located 20.0 cm away from a diverging lens with a focal length of 24.0 cm. What are the image position, height, and orientation? Is this a real or a virtual image?

72. Spectrometer A spectrometer contains a grating of 11,500 slits/cm. Find the angle at which light of wavelength 527 nm has a first-order bright band.

MULTIPLE CHOICE

1. How many electrons have been removed from a positively charged electroscope if it has a net charge of 7.5×10^{-11} C?

A. 7.5×10^{-11} electrons

B. 2.1×10^{-9} electrons

C. 1.2×10^{8} electrons

D. 4.7×10^{8} electrons

2. The force exerted on a particle with a charge of 5.0×10^{-9} C by a second particle that is 4 cm away is 8.4×10^{-5} N. What is the charge on the second particle?

A. 4.2×10^{-13} C **C.** 3.0×10^{-9} C

B. 2.0×10^{-9} C **D.** 6.0×10^{-5} C

3. Three charges, A, B, and C, are located in a line, as shown below. What is the net force on charge B?

A. 78 N toward A **C.** 130 N toward A

B. 78 N toward C **D.** 210 N toward C

$+8.5\times10^{-6}$ C (A) — 4.2 cm — $+3.1\times10^{-6}$ C (B) — 2.9 cm — $+6.4\times10^{-6}$ C (C)

4. Why is copper a good conductor?

A. Its protons and electrons move readily.

B. Its electrons move readily.

C. It always has a negative charge.

D. It can be insulated.

5. Two equally charged objects exert a force of 90 N on each other. If one object is exchanged for another of the same size but with three times the charge, what is the new force the objects will exert on each other?

A. 10 N **C.** 2.7×10^{2} N

B. 30 N **D.** 8.1×10^{2} N

6. When you touch metal on a dry day, you sometimes receive a shock. What then happens to your charge?

A. You become negatively charged; the metal transferred electrons to you.

B. You become positively charged; you transferred electrons to the metal.

C. You become neutral; the shock discharged you.

D. Nothing; you were merely a conductor.

Online Test Practice

7. What is the charge on an electroscope that has an excess of 4.8×10^{10} electrons?

A. 3.3×10^{-30} C **C.** 7.7×10^{-9} C

B. 4.8×10^{-10} C **D.** 4.8×10^{10} C

8. An alpha particle has a mass of 6.68×10^{-27} kg and a charge of 3.2×10^{-19} C. What is the ratio of the electrostatic force to the gravitational force between two alpha particles?

A. 1 **C.** 2.3×10^{15}

B. 4.8×10^{7} **D.** 3.1×10^{35}

9. Charging a neutral body by touching it with a charged body is called charging by

A. conduction. **C.** grounding.

B. induction. **D.** discharging.

10. Macy rubs a balloon with wool, giving the balloon a charge of -8.9×10^{-14} C. What is the force between the balloon and a metal sphere that is charged to 25 C and is 2 km away?

A. 8.9×10^{-15} N **C.** 2.2×10^{-12} N

B. 5.0×10^{-9} N **D.** 5.6×10^{4} N

FREE RESPONSE

11. According to the diagram, what is the net force exerted by charges A and B on charge C? In your answer, include a diagram showing the force vectors $F_{A\text{ on }C}$, $F_{B\text{ on }C}$, and F_{net}.

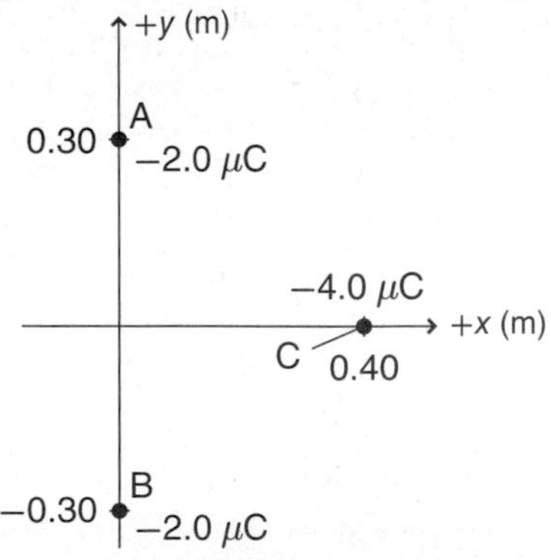

NEED EXTRA HELP?

If You Missed Question	1	2	3	4	5	6	7	8	9	10	11
Review Section	2	2	2	1	2	1	2	2	2	2	2

CHAPTER 21

Electric Fields

BIG IDEA **Electric charges are surrounded by electric fields that exert a force on other charged objects.**

SECTIONS

LaunchLAB

CHARGED OBJECTS AND DISTANCE

How do charged objects interact at a distance?

WATCH THIS!

ELECTRIC SPARKS

Have you ever had an electric spark jump from a doorknob to your hand? What was surrounding the doorknob to produce that spark? Take a trip to your local science museum to find out.

(l)Arthur S. Aubry/Getty Images, (r)Adam Hart-Davis/Photo Researchers

Go online!
connectED.mcgraw-hill.com

SECTION 1

Measuring Electric Fields

PHYSICS 4 YOU

What image do you imagine when you hear the term electric field? If you have seen a Van de Graaff generator in operation, you might imagine someone's hair sticking up. What do a Van de Graaff generator, hair, and an electric field have in common?

MAIN IDEA

An electric field is a property of the space around a charged object that exerts forces on other charged objects.

Essential Questions

- What is an electric field?
- How are charge, electric field, and forces on charged objects related?
- How can you represent electric fields in diagrams and other models?

Review Vocabulary

Coulomb's law states that the force between two point charges varies directly with the product of their charges and inversely with the square of the distance between them

New Vocabulary

electric field
electric field line

Defining the Electric Field

The idea of a force field is probably not new to you if you have watched science-fiction television shows or movies or have read science-fiction books. In these fictional applications, characters use force fields to create invisible barriers to stop invading armies, and to hold prisoners in a cell. As it turns out, force fields really exist but they are not quite like the ones in science fiction.

Recall that according to Coulomb's law, the force between two point charges varies inversely as the distance between those charges changes. How can one object exert a force on another object across empty space? Michael Faraday suggested the idea of a force field. The electric field is one example of a force field. An **electric field** is a property of the space around a charged object that exerts forces on other charged objects. You cannot see an electric field, but there are ways to detect that an electric field is present.

For example, the streams in the plasma sphere shown in **Figure 1** show the effects of the electric field around that plasma sphere's central electrode. A plasma sphere's central electrode is a charged object whose net charge alternates rapidly (120 times per second) between positive net charge and negative net charge. The glass globe contains ions, which are charged particles. The electric field from the central electrode accelerates the ions in the globe. The resulting streams illustrate the paths that the ions travel along and represent the electric field inside the plasma sphere.

Figure 1 The plasma sphere shows how an electric field can interact with matter.

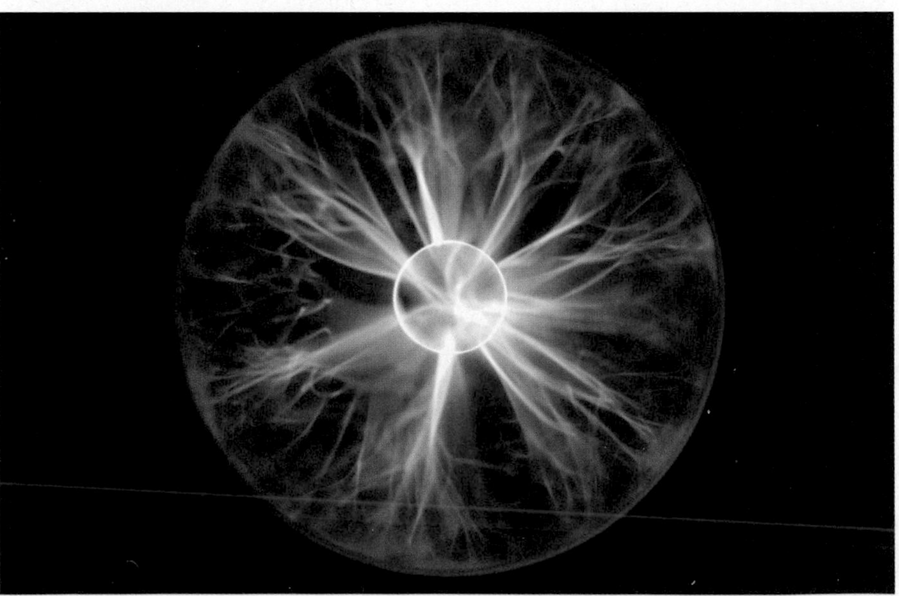

The strength of an electric field You might wonder how you can detect an electric field if you cannot see it. You can envision the force from an electric field by modeling its effects on a small, charged object—a test charge (q')—at some location. If there is an electrostatic force on the object, then there is an electric field at that point. **Figure 2** illustrates a charged object with a net charge of q. Suppose you place the positive test charge at point (A) and measure a force ($\mathbf{F}_A$). According to Coulomb's law, the force is directly proportional to the strength of the test charge (q'). That is, if the charge on the test charge is doubled, so is the force $\left(\frac{2\mathbf{F}}{2q'} = \frac{\mathbf{F}}{q'}\right)$. Therefore, the ratio of the force to the strength of the test charge is a constant. When you divide force ($\mathbf{F}$) by the strength of the test charge (q'), you obtain a vector quantity $\left(\frac{\mathbf{F}}{q'}\right)$. The electric field at point A, the location of q', is represented by the following equation.

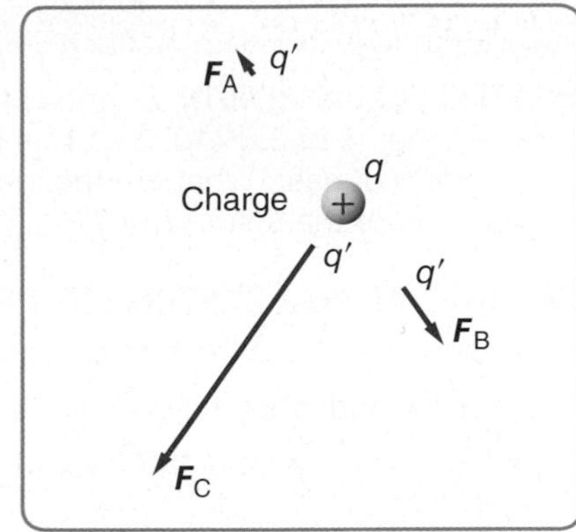

Figure 2 An electric field surrounds particle q. The forces exerted on a positively charged test particle (q') at locations A, B, and C are represented by force vectors. The vectors represent the magnitude and direction of the force from the electric field on the test particle at that point.

ELECTRIC FIELD STRENGTH

The strength of an electric field is equal to the force on a positive test charge divided by the strength of the test charge.

$$\mathbf{E} = \frac{\mathbf{F}_{\text{on } q'}}{q'}$$

The direction of an electric field is the direction of the force on a positive test charge. The magnitude of the electric field is measured in newtons per coulomb, N/C.

READING CHECK **Explain** why some of the variables are bold type in the text and in the equation.

COLOR CONVENTION

- Electric field lines are indigo.
- Positive charges are red.
- Negative charges are blue.

Field vectors You can make a model of an electric field by using arrows to represent the field vectors at various locations, as shown in **Figure 2.** The length of the arrow represents the field strength. The direction of the arrow represents the field direction. To find the field from two charges, add the fields from the individual charges through vector addition. Some typical electric field strengths are shown in **Table 1.**

You also can use a test charge to map the electric field resulting from any collection of test charges. A test charge should be small enough so that its effect on the charge you are testing (q) is negligible. Remember that, according to Newton's third law, the test charge exerts forces back on the charges that produce the electric field. It is important that these forces do not redistribute the charges that you are trying to measure, thereby affecting the electric field that you are trying to map.

Coulomb's law and electric fields If, and only if, the charge q is a point charge or a uniformly charged sphere, you can calculate its electric field from Coulomb's law. Use Coulomb's law in the electric field equation above to find the magnitude of the force exerted on the test charge ($F_{\text{on } q'}$) as shown below.

$$E = \frac{F_{\text{on } q'}}{q'} = F_{\text{on } q'} \times \frac{1}{q'} = \frac{Kqq'}{r^2} \times \frac{1}{q'} = \frac{Kq}{r^2}$$

Table 1 Typical Electric Field Strengths (Approximate)

Field	Value (N/C)
Near a charged, hard-rubber rod	1×10^3
Needed to create a spark in air	3×10^6
At an electron's orbit in a hydrogen atom	5×10^{11}

EXAMPLE PROBLEM 1

Find help with **operations with significant digits.** Math Handbook

ELECTRIC FIELD STRENGTH Suppose that you are measuring an electric field using a positive test charge of 3.0×10^{-6} C. This test charge experiences a force of 0.12 N at an angle of 15° north of east. What are the magnitude and direction of the electric field strength at the location of the test charge?

1 ANALYZE AND SKETCH THE PROBLEM

- Draw and label the test charge, q'.
- Show and label the coordinate system centered on the test charge.
- Diagram and label the force vector at 15° north of east.

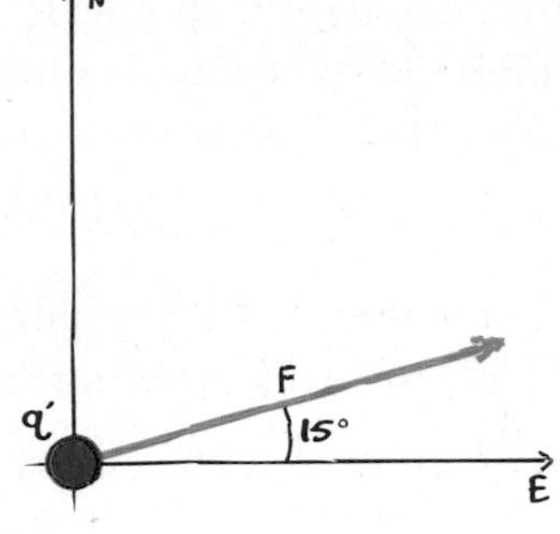

KNOWN	UNKNOWN
$q' = 3.0\times10^{-6}$ C	$E = ?$
$F = 0.12$ N at 15° N of E	

2 SOLVE FOR THE UNKNOWN

$$E = \frac{F}{q'}$$

$$= \frac{0.12\text{ N}}{3.0\times10^{-6}\text{ C}}$$ ◀ Substitute $F = 0.12$ N, $q' = 3.0\times10^{-6}$ C.

$$= 4.0\times10^{4}\text{ N/C}$$

The force on the test charge and the electric field are in the same direction.

$E = 4.0\times10^{4}$ N/C at 15° N of E

3 EVALUATE THE ANSWER

- **Are the units correct?** Electric field strength is correctly measured in N/C.
- **Does the direction make sense?** The field direction is in the direction of the force because the test charge is positive.
- **Is the magnitude realistic?** This field strength is similar to those in **Table 1.**

PRACTICE PROBLEMS

Do additional problems. Online Practice

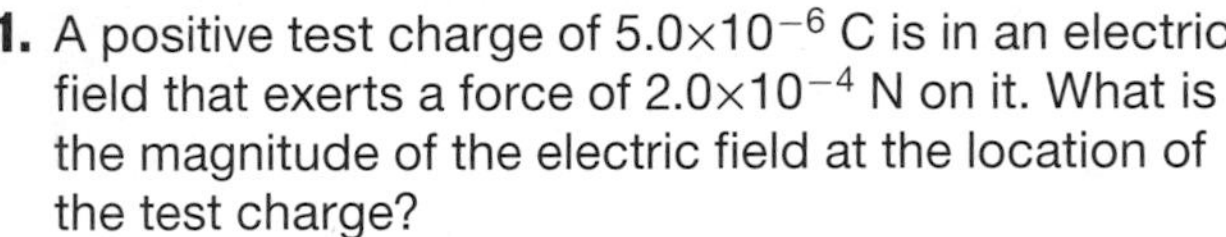

1. A positive test charge of 5.0×10^{-6} C is in an electric field that exerts a force of 2.0×10^{-4} N on it. What is the magnitude of the electric field at the location of the test charge?
2. A negative charge of 2.0×10^{-8} C experiences a force of 0.060 N to the right in an electric field. What are the field's magnitude and direction at that location?
3. Suppose that you place a 2.1×10^{-3}-N pith ball in a 6.5×10^{4} N/C downward electric field. What net charge (magnitude and sign) must you place on the pith ball so that the electrostatic force acting on that pith ball will suspend it against the gravitational force?
4. Complete **Table 2** using your understanding of electric fields.

Table 2 Sample Data

Test Charge Strength (C)	Force Exerted on Test Charge (N)	Electric Field Intensity (N/C)
1.0×10^{-6}	0.30	
2.0×10^{-6}		3.3×10^{5}
	0.45	1.5×10^{5}

5. A positive charge of 3.0×10^{-7} C is located in a field of 27 N/C directed toward the south. What is the force acting on the charge?
6. A negative test charge is placed in an electric field as shown in **Figure 3.** It experiences the force shown. What is the magnitude of the electric field at the location of the charge?

$q' = -5.0\times10^{-6}$ C

q

$F = 0.080$ N toward q

Figure 3

7. **CHALLENGE** You are probing the electric field of a charge of unknown magnitude and sign. You first map the field with a 1.0×10^{-6}-C test charge, then repeat your work with a 2.0×10^{-6}-C test charge.
   a. Would you measure the same forces at the same place with the two test charges? Explain.
   b. Would you find the same field strengths? Explain.

EXAMPLE PROBLEM 2

Find help with **operations with scientific notation.** Math Handbook

ELECTRIC FIELD STRENGTH AND COULOMB'S LAW What is the magnitude of the electric field at a point that is 0.30 m to the right of a small sphere with a net charge of -4.0×10^{-6} C?

1 ANALYZE AND SKETCH THE PROBLEM

- Draw and label the sphere with its net charge (q) as well as the test charge (q'), 0.30 m away.
- Show and label the distance between the charges.
- Diagram and label the force vector acting on q'. Recall that a test charge is usually positive.

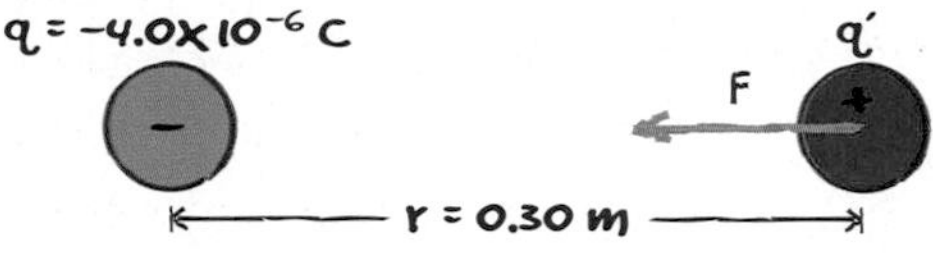

KNOWN	UNKNOWN
$q = -4.0\times10^{-6}$ C	$E = ?$
$r = 0.30$ m	

2 SOLVE FOR THE UNKNOWN

The force and the magnitude of the test charge are unknown, so use Coulomb's law in combination with the electric field strength equation.

$$E = \frac{Kq}{r^2}$$

$$= \frac{(9.0\times10^9\ \text{N}\cdot\text{m}^2/\text{C}^2)(-4.0\times10^{-6}\ \text{C})}{(0.30\ \text{m})^2}$$

◀ Substitue $K = 9.0\times10^9$ N•m²/C², $q = -4.0\times10^{-6}$ C, $r = 0.30$ m.

$$= -4.0\times10^5\ \text{N/C}$$

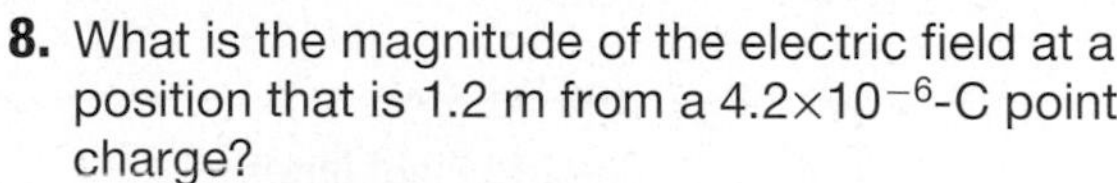

$\boldsymbol{E} = 4.0\times10^5$ N/C toward the sphere, or to the left

3 EVALUATE THE ANSWER

- **Are the units correct?** $\frac{(\text{N}\cdot\text{m}^2/\text{C}^2)(\text{C})}{\text{m}^2}$ = N/C. The units work out to be N/C, which is correct for electric field strength.
- **Does the direction make sense?** The negative sign indicates that the negative charge attracts the positive test charge.
- **Is the magnitude realistic?** This field strength is similar to those in **Table 1.**

PRACTICE PROBLEMS

Do additional problems. Online Practice

8. What is the magnitude of the electric field at a position that is 1.2 m from a 4.2×10^{-6}-C point charge?

9. What is the magnitude of the electric field at a distance twice as far from the point charge in the previous problem?

10. What is the electric field at a position that is 1.6 m east of a point charge of $+7.2\times10^{-6}$ C?

11. The electric field that is 0.25 m from a small sphere is 450 N/C toward the sphere. What is the net charge on the sphere?

12. How far from a point charge of $+2.4\times10^{-6}$ C must you place a test charge in order to measure a field magnitude of 360 N/C?

13. Explain why the strength of the electric field exerted on charge q' by the charged body q is independent of the charge on q'. *Hint: Use mathematics to prove your point.*

14. What is the magnitude of the electric field exerted on the test charge shown in **Figure 4?**

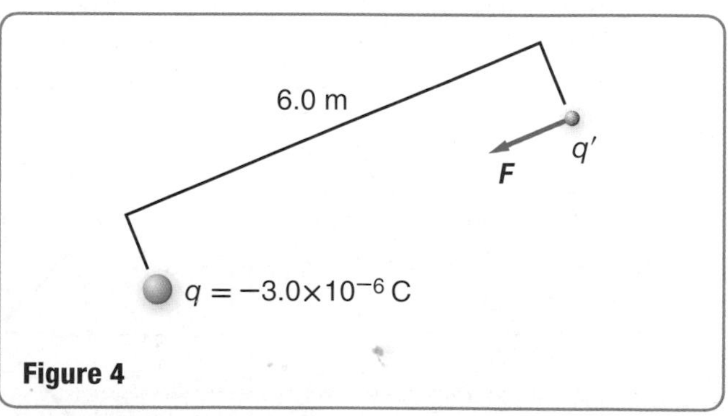

Figure 4

15. CHALLENGE You place a small sphere with a net charge of 5.0×10^{-6} C at one corner of a square that measures 5.0 m on each side. What is the magnitude of the electric field at the opposite corner of the square?

Modeling the Electric Field

An electric field is present even if there is no test charge to measure that field because a test charge is just an imaginary model. Any charge placed in an electric field experiences a force on it resulting from the electric field at that location. So far, you have measured an electric field at a single location. Now, imagine moving the test charge to another location. Measure the force on it again and calculate the electric field again. Repeat this process again and again until you assign to every location in a region a measurement of the electric field. The magnitude of the force depends on the magnitude of the field (E) and the magnitude of the charge (q). Thus, $\mathbf{F} = \mathbf{E}q$. The direction of the force depends on the direction of the field and the sign of the charge.

Electric field lines You have already seen how the streams in a plasma sphere represent an electric field. You also can represent electric fields with electric field lines, such as those shown in **Figure 5.** An **electric field line** indicates the direction of the force due to the electric field on a positive test charge. The spacing between the lines indicates the electric field's strength. The field is stronger where the lines are spaced more closely. The field is weaker where the lines are spaced farther apart.

Because electric field lines indicate the force on a positive test charge, they are directed toward negative charges and away from positive charges. The direction of the electric field ($\mathbf{E}$) at any point is the tangent drawn to a field line at that point, as shown in **Figure 6** on the next page.

Figure 6 shows two ways to represent various electric fields—with electric field lines and with grass seed suspended in oil. An electric field affects electric charges whether or not you represent that field with field lines and whether or not you try to visualize that field with grass seed suspended in oil.

Grass seed can be useful for visualizing electric fields in three dimensions. Both grass seed and electric field lines are only ways of *representing* electric fields, however. Neither is the same as the electric field itself.

Figure 5 Two-dimensional models of electric fields are shown, but remember that electric fields exist in three dimensions.

Explain how you should determine the direction of the arrows for electric field lines.

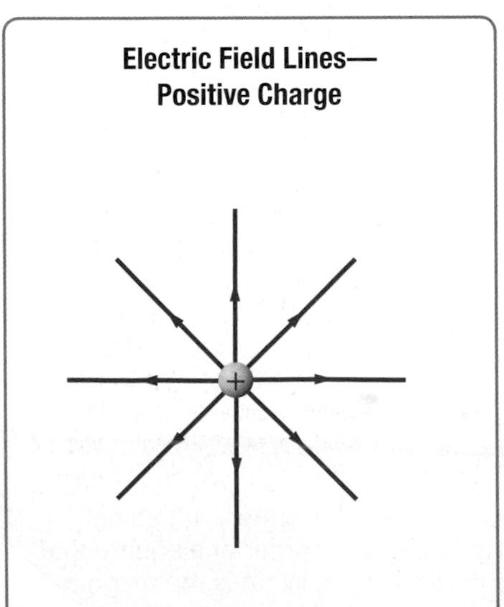

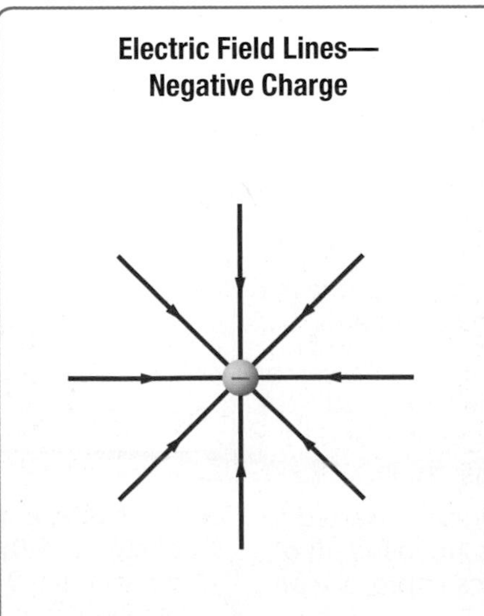

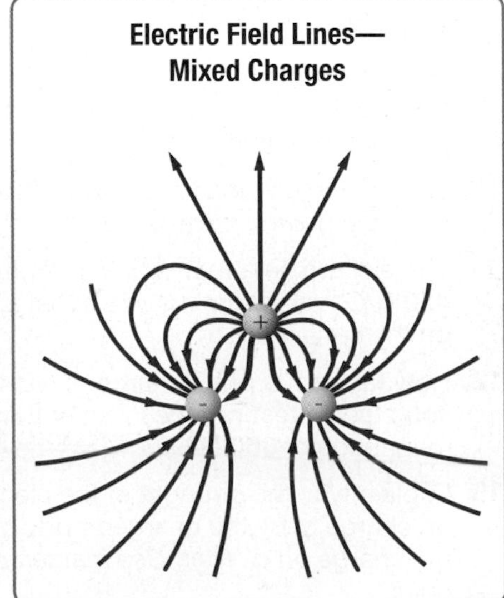

Electrostatic Forces

Figure 6 Models of electric fields around electric charges are shown. Electrostatic forces cause a separation of charge in each long, thin grass seed. The seeds then turn so that they line up along the direction of the electric field. The lines and vectors drawn in the diagrams represent other ways to model electric fields.

Identify the dark circles in each image.

View an animation and an interactive figure of **electric fields and force on a charge.**

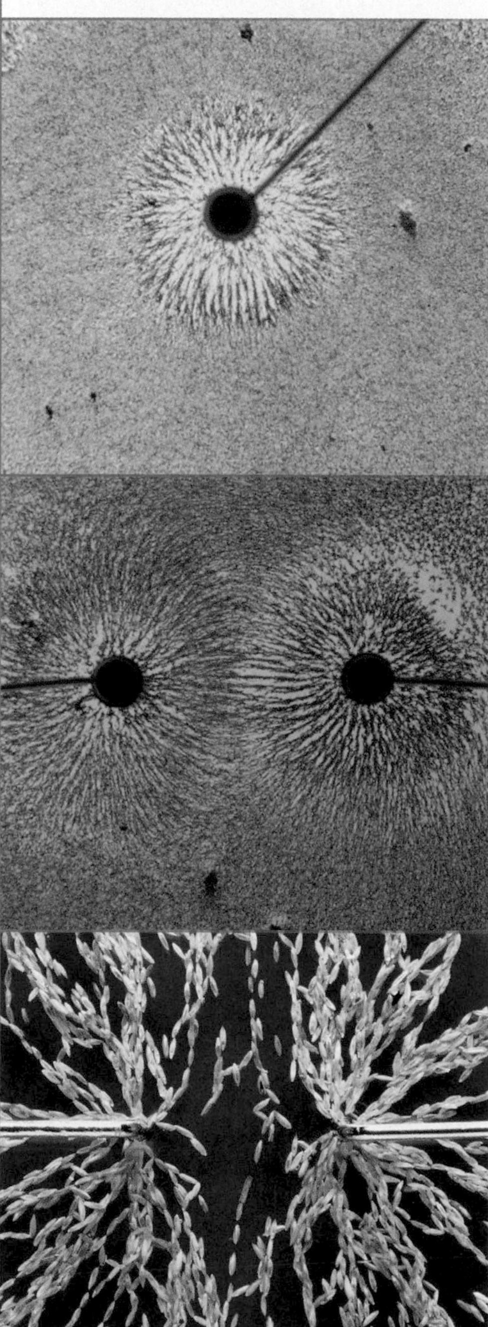

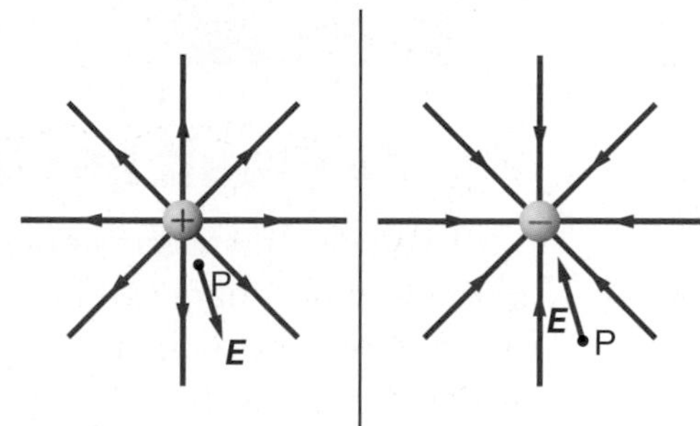

Single Charge For a single charge, the electric field lines radiate from that charge. The field lines radiate out from a positive charge and they radiate inward toward a negative charge. The direction of the electric field (***E***) at any point is the tangent to the field line.

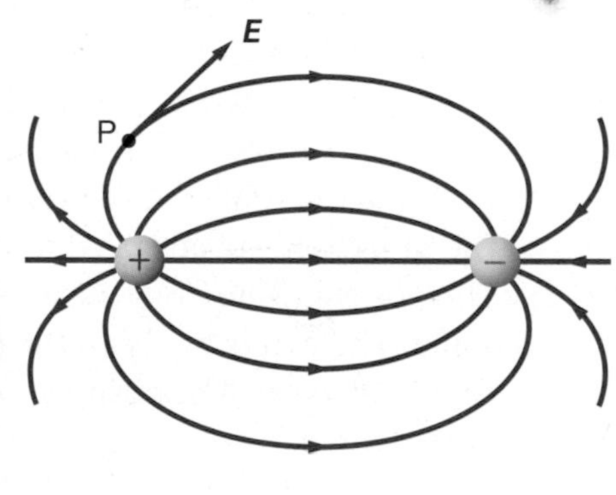

Two Equal and Unlike Charges The electric field lines of two equal and unlike charges form continuous lines from the positive charge to the negative charge. The vector at point P shows the direction of the electric field (***E***) at point P. Where the lines are closest, a charged particle would experience the most force.

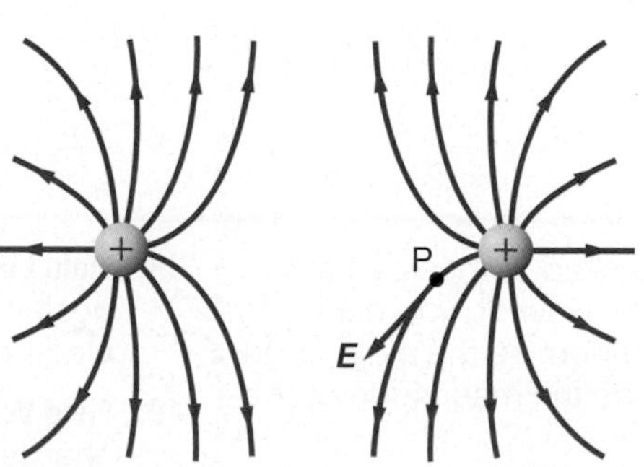

Two Like Charges The electric field lines around two like charges never connect with each other. The vector at point P shows the direction of the electric field (***E***) at point P. A charge would experience no electrostatic force in the center where there are no electric field lines.

Figure 7 Robert Van de Graaff devised his generator in the 1930s to aid in the study of particle acceleration and the study of the structure of the atom. When a person touches the generator, the charges on the person's hairs repel each other, causing the hairs to align with the electric field of the metal dome.

View an animation on electric potential difference.

Van de Graaff Generators

A Van de Graaff generator, shown in **Figure 7,** can build up a large amount of net charge on its metal dome. In a Van de Graaff generator, a belt is driven by two rollers or pulleys. One roller is attached to position A. The other roller is located in the metal dome at position B. When someone turns on the machine, the bottom roller rotates and moves the belt. The roller builds up a strong negative net charge, and the belt builds a weaker positive net charge.

The strong negative electric field on the roller repels electrons in the tips of a brush assembly near position A. The tips of the brushes become positively charged because the electrons move away from them. At the same time, the strong negative electric field of the roller strips electrons from nearby molecules in the air. The stripped electrons are repelled by the roller and are attracted to the positively charged tips of the brush. The positive ions from the air start to move to the negatively charged roller. The belt is between the positive ions and the negatively charged roller, however. The positive ions collect on the belt and are transported to the metal dome.

When a person touches the metal dome, the positively charged dome attracts the electrons in the person's body. The person's hair becomes positively charged, and the individual hairs repel each other. The hairs align with the electric field around the person's head.

SECTION 1 REVIEW

Section Self-Check — Check your understanding.

16. **MAIN**IDEA Suppose you are asked to measure the electric field at a point in space. How do you detect the field at a point? How do you determine the magnitude of the field? How do you choose the magnitude of the test charge?

17. **Field Strength and Direction** A positive test charge of magnitude 2.40×10^{-8} C experiences a force of 1.50×10^{-3} N toward the east. What is the electric field at the position of the test charge?

18. **Field Lines** How can you tell which charges are positive and which are negative by examining the electric field lines?

19. **Field Versus Force** How does the electric field at a test charge differ from the force on that charge?

20. **Critical Thinking** Suppose the top charge in the field on the far right in **Figure 5** is a test charge measuring the field resulting from the two negative charges. Is it small enough to produce an accurate measurement? Explain.

SECTION 2 Applications of Electric Fields

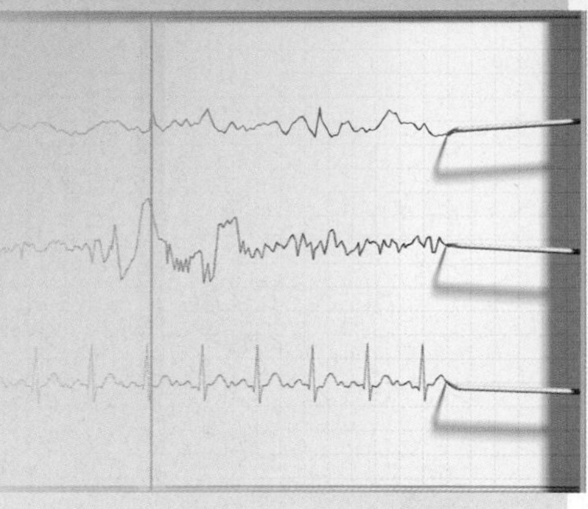

Doctors use several medical procedures to monitor electrical activity in the human body. For example, electrocardiography (ECG or EKG) monitors the electrical activity in the heart, and electroencephalography (EEG) monitors the electrical activity of the brain.

MAIN IDEA

Electric potential (sometimes called voltage) is electric potential energy per unit charge.

Essential Questions

- What is an electric potential difference?
- How is potential difference related to the work required to move a charge?
- What are properties of capacitors?

Review Vocabulary

work the transfer of energy that occurs when a force is applied over a distance; is equal to the product of the object's displacement and the force applied to the object in the direction of displacement

New Vocabulary

electric potential difference
volt
equipotential
capacitor
capacitance

10 - 13

20 - 27

Energy and Electric Potential

As you have read, the concept of energy is extremely useful in mechanics. The law of conservation of energy allows you to predict the motion of objects without knowing the forces in detail. Consider the change in gravitational potential energy of a ball when you lift it, as shown in **Figure 8.** Both the gravitational force (F_g) and the gravitational field $\left(g = \frac{F_g}{m}\right)$ point toward Earth. If you lift the ball against the gravitational force, you do work on the ball. The gravitational potential energy in the ball-Earth system increases.

The same is true in electrical interactions. The work performed moving a charged particle in an electric field can result in the particle gaining electric potential energy or kinetic energy, or both.

The situation is similar with two unlike charges. The charges attract each other, so you must do work to pull one charge away from the other. When you do this work, you increase the electric potential energy of the two-charge system. The larger the charge or the distance moved, the greater the increase in the electric potential energy (ΔPE). In this chapter, only charges at rest and changes in electric potential energy are considered and discussed.

Figure 8 When you lift the ball, you increase the gravitational potential energy of the ball-Earth system. When you separate two unlike charges, you increase the electric potential energy of the two-charge system.

Infer how you would increase the potential energy of a system that contains two like charges.

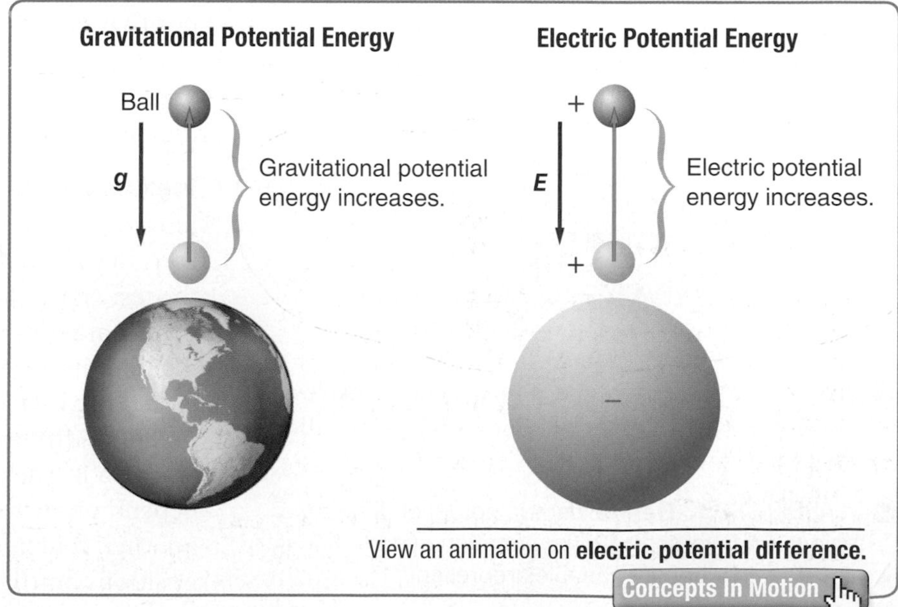

Matthias Kulka/Bridge/CORBIS

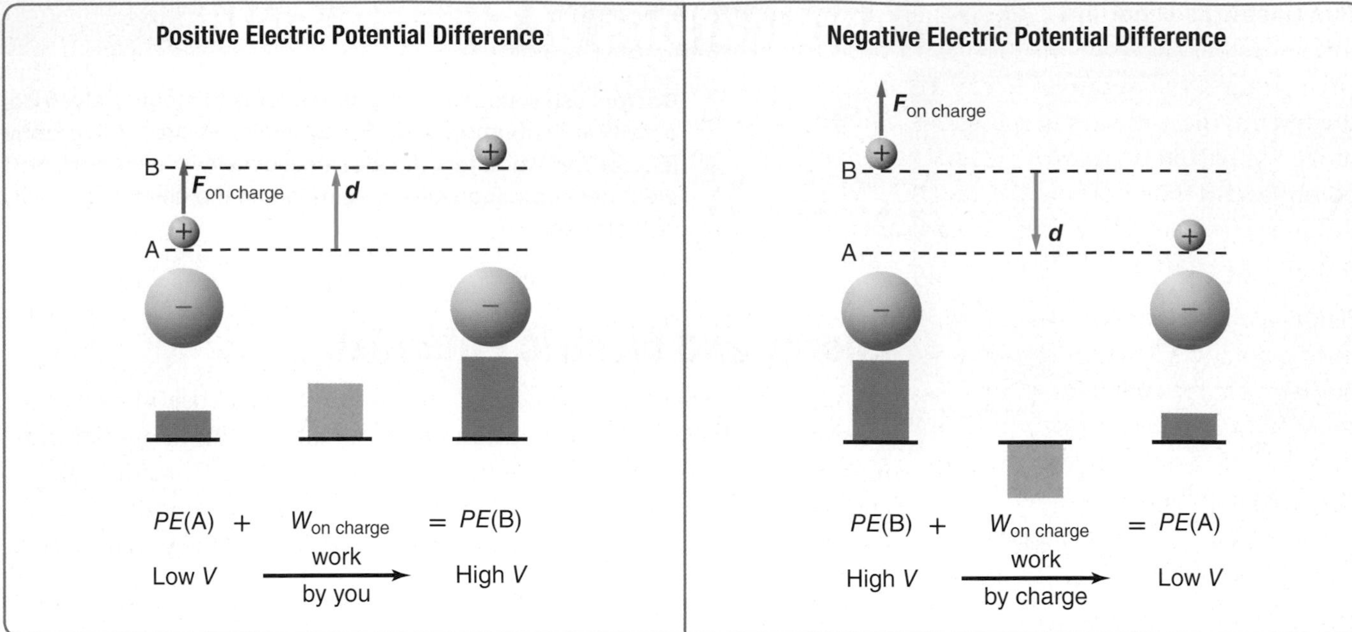

Figure 9 If you move unlike charges apart, then the electric potential difference is positive. If you move unlike charges closer together, then the electric potential difference is negative.

Explain how work and electric potential are related in regard to the test charge.

Electric potential difference You know from your study of mechanics that work done on an object is found using the relationship $W = \boldsymbol{F}d$. This same relationship is used to find work done to move a charge, $W_{\text{on } q'} = \boldsymbol{F}d$.

There are differences, however. The work done on a charge is expressed as work done per unit charge and it is called the electric potential difference. The **electric potential difference** (ΔV), which often is called potential difference, is the work ($W_{\text{on } q}$) needed to move a positive test charge from one point to another, divided by the magnitude of the test charge. You also can think of electric potential difference as the change in electric potential energy (ΔPE) per unit charge.

ELECTRIC POTENTIAL DIFFERENCE

Electric potential difference is the ratio of the work needed to move a charge to the magnitude of that charge.

$$\Delta V \equiv \frac{W_{\text{on } q'}}{q'}$$

Electric potential difference is measured in joules per coulomb (J/C). One joule per coulomb is called a **volt** (V).

READING CHECK Define electric potential difference in your own words, state the equation used to find it, and explain what the variables represent.

Positive electric potential difference The negative charge, in the left panel in **Figure 9,** produces an electric field toward itself. Suppose you place a small positive test charge in the field at position A. That test charge experiences a force in the direction of the field. If you now move the test charge away from the negative charge to position B, you will have to exert a force ($\boldsymbol{F}$) on that test charge. Because the force you exert is in the same direction as the displacement, the work you do on the test charge is positive, and the change in potential energy is positive, as represented by the bar diagram in the figure. Therefore, there is a positive electric potential difference from point A to point B. The potential difference between two points does not depend on the magnitude of the test charge. It depends only on the field and the displacement.

Negative electric potential difference Suppose you now move the test charge back to position A from position B, as in the right panel in **Figure 9.** The work you do on the test charge and the change in electric potential energy are now negative, also shown in the bar diagram in the figure. The electric potential difference is also negative. In fact, it is equal and opposite to the potential difference for the move from position A to position B. The electric potential difference does not depend on the path used to go from one position to another. It depends only on the starting and final positions on that path.

Zero electric potential difference Is there always an electric potential difference between two positions? Suppose you move the test charge in a circle around the negative charge. The force the electric field exerts on the test charge is always perpendicular to the direction in which you moved it, so you do no work on the test charge. Therefore, the electric potential difference is zero in this case. Whenever the electric potential difference between two or more positions is zero, those positions are said to be at **equipotential.**

Electric potential for like charges You have seen that electric potential difference increases as you pull a positive test charge away from a negative charge. What happens when you move a positive test charge away from another positive charge? There is a repulsive force between these two charges. Potential energy decreases as you move the two charges farther apart. Therefore, the electric potential around the positive charge decreases as you move away from that positive charge, as shown in **Figure 10.**

Reference point in a system Recall that the gravitational potential energy of a system can be defined as zero at any reference point. In the same way, the electric potential of any point can be defined as zero. No matter what reference point you choose, the value of the electric potential difference from point A to point B always will be the same.

Voltage versus volts Only differences in gravitational potential energy can be measured. The same is true of electric potential; thus, only differences in electric potential are important. The electric potential difference from point A to point B is defined as $\Delta V = V_B - V_A$. Electric potential differences are measured with a voltmeter. Sometimes, the electric potential difference is simply called the voltage. Do not confuse electric potential difference (ΔV) with the unit for volts (V).

Electric Potential

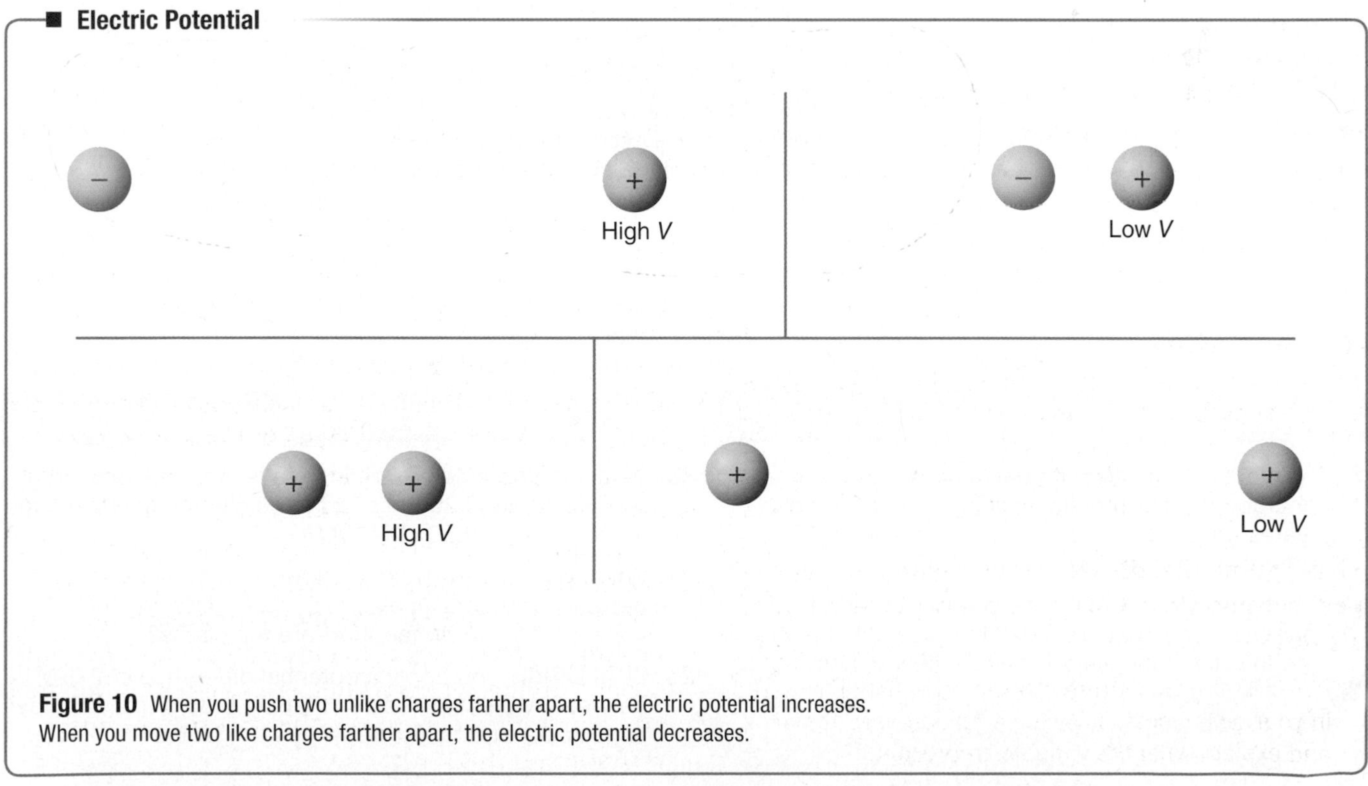

Figure 10 When you push two unlike charges farther apart, the electric potential increases. When you move two like charges farther apart, the electric potential decreases.

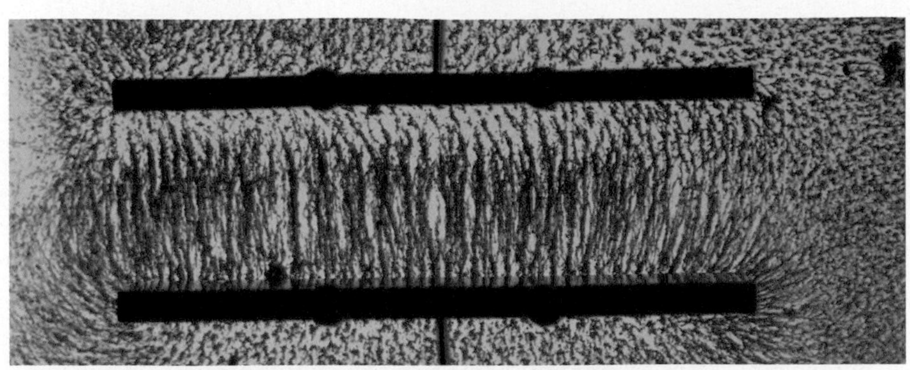

Figure 11 A model of an electric field between two oppositely charged parallel plates is shown. Grass seed is used to model the electric field lines.

Electric Potential in a Uniform Field

You can produce a uniform electric field by placing two large, flat conducting plates parallel to each other. One plate is charged positively, and the other plate is charged negatively. The magnitude and the direction of the electric field are the same at all points between the plates, except at the edges of the plates, and the electric field points from the positive plate to the negative plate. The pattern formed by the grass seeds pictured in **Figure 11** represents the electric field between parallel plates.

Imagine that you move a positive test charge (q') a distance (d) in the direction opposite the electric field direction. Because the field is uniform, the force from that field on the charge is constant. As a result, you can use the relationship $W_{\text{on } q'} = Fd$ to find the work done on the charge. Then, the electric potential difference, the work done per unit charge, is $\Delta V = \frac{Fd}{q'} = \left(\frac{F}{q'}\right)d$. Remember that electric field intensity is force per unit charge $\left(E = \frac{F}{q'}\right)$. Therefore, you can represent the electric potential difference (ΔV) between two points a distance (d) apart in a uniform field ($\boldsymbol{E}$) with the following equation.

ELECTRIC POTENTIAL DIFFERENCE IN A UNIFORM FIELD

The potential difference between two locations in a uniform electric field equals the product of electric field intensity and the distance between the locations parallel to the direction of the field.

$$\Delta V = Ed$$

The electric potential is higher near the positively charged plate and lower near the negatively charged plate.

PRACTICE PROBLEMS

Do additional problems. Online Practice

21. The electric field intensity between two large, charged parallel metal plates is 6000 N/C. The plates are 0.05 m apart. What is the electric potential difference between them?
22. A voltmeter reads 400 V across two charged, parallel plates that are 0.020 m apart. What is the magnitude of the electric field between them?
23. What electric potential difference is between two metal plates that are 0.200 m apart if the electric field between those plates is 2.50×10^{3} N/C?
24. When you apply a potential difference of 125 V between two parallel plates, the field between them is 4.25×10^{3} N/C. How far apart are the plates?
25. **CHALLENGE** You apply a potential difference of 275 V between two parallel plates that are 0.35 cm apart. How large is the electric field between the plates?

EXAMPLE PROBLEM 3

Find help with **operations with scientific notation.** Math Handbook

WORK REQUIRED TO MOVE A PROTON BETWEEN CHARGED PARALLEL PLATES Two charged parallel plates are 1.5 cm apart. The magnitude of the electric field between the plates is 1800 N/C.

a. What is the electric potential difference between the plates?

b. How much work is required to move a proton from the negative plate to the positive plate?

1 ANALYZE AND SKETCH THE PROBLEM

- Draw the plates separated by 1.5 cm.
- Label one plate with positive charges and the other with negative charges.
- Draw uniformly spaced electric field lines from the positive plate to the negative plate.
- Indicate the electric field strength between the plates.
- Place a proton in the electric field.

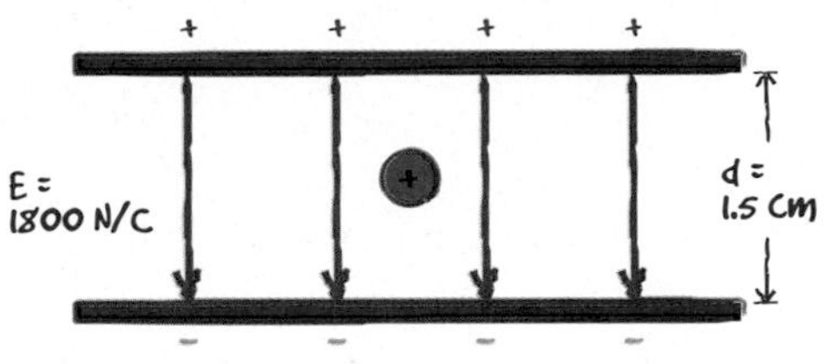

KNOWN	**UNKNOWN**
$E = 1800$ N/C	$\Delta V = ?$
$d = 1.5$ cm	$W = ?$
$q = 1.602\times10^{-19}$ C	

2 SOLVE FOR THE UNKNOWN

a.
$$\begin{aligned}\Delta V &= Ed \\ &= (1800 \text{ N/C})(0.015 \text{ m}) \\ &= 27 \text{ V}\end{aligned}$$

◀ Substitute $E = 1800$ N/C, $d = 0.015$ m.

b.
$$\Delta V = \frac{W}{q}$$
$$\begin{aligned}W &= q\Delta V \\ &= (1.602\times10^{-19} \text{ C})(27 \text{ V}) \\ &= 4.3\times10^{-18} \text{ J}\end{aligned}$$

◀ Substitute $q = 1.602\times10^{-19}$ C, $\Delta V = 27$ V.

3 EVALUATE THE ANSWER

- **Are the units correct?** (N/C)(m) = N•m/C = J/C = V. The units work out to be volts. C•V = C(J/C) = J, the unit for work.
- **Does the sign make sense?** You must do positive work to move a positive charge toward a positive plate.
- **Is the magnitude realistic?** When you move such a small charge through a potential difference of less than 100 volts, the work done is small.

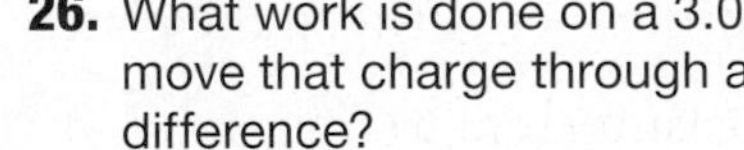

PRACTICE PROBLEMS

Do additional problems. Online Practice

26. What work is done on a 3.0-C charge when you move that charge through a 1.5-V electric potential difference?

27. What is the magnitude of the electric field between the two plates shown in **Figure 12?**

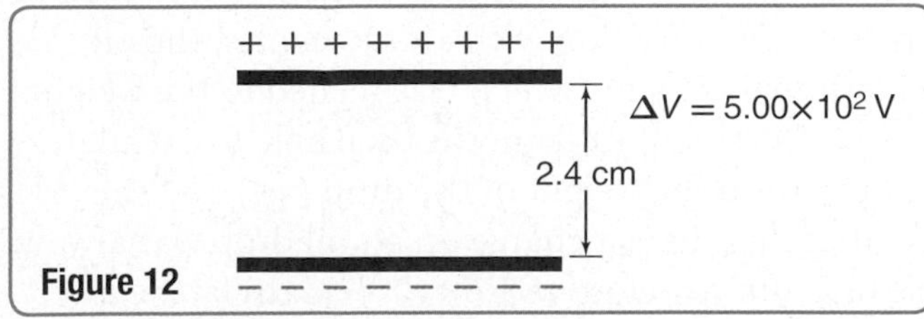

Figure 12

28. An electron in an old television picture tube passes through a potential difference of 18,000 V. How much work is done on the electron as it passes through that potential difference?

29. The electric field in a particle accelerator has a magnitude of 4.5×10^5 N/C. How much work is done to move a proton 25 cm through that field?

30. CHALLENGE A 12-V car battery has 1.44×10^6 C of usable charge on one plate when it is fully energized. How much work can this battery do before it needs to be energized again?

Figure 13 The American physicist Robert A. Millikan used an apparatus similar to this one to determine the charge of a single electron in 1909.

View an **animation of Millikan's oil-drop experiment.**

Concepts In Motion

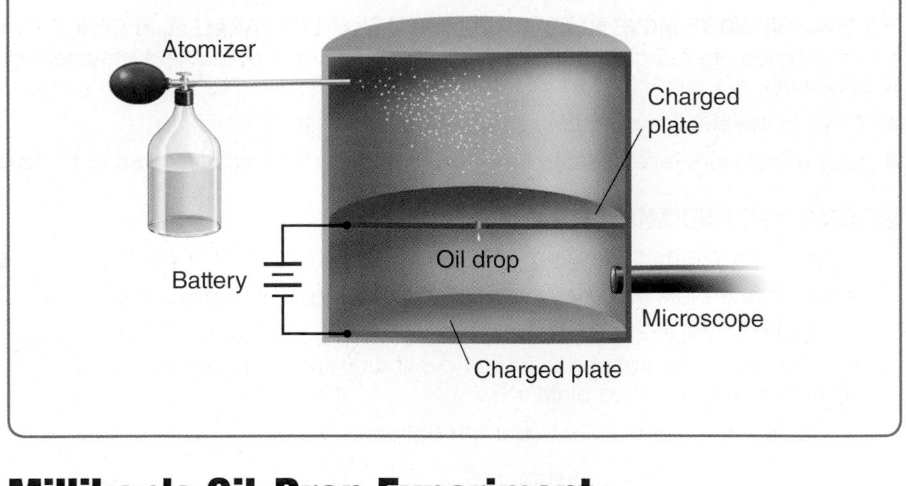

Millikan's Oil-Drop Experiment

You have read that the magnitude of the elementary charge is $e = 1.602\times10^{-19}$ C. The net charge on an object must be some integer multiple of this value. In other words, the net charge on an object can be 2e, −5e, 7e, or even −3157e. The net charge cannot be a fractional charge, such as 0.5e, −23.7e, 6.2e, or −31,524.6e. How do we know?

In 1909, Robert Millikan performed an experiment to test whether charge exists in discrete amounts. Millikan was able to measure the magnitude of the elementary charge with his experiment. A diagram of Millikan's apparatus for this experiment is shown in **Figure 13.** Notice that Millikan used two parallel plates to produce a uniform electric field in his apparatus. How did Millikan perform his experiment?

First, Millikan sprayed fine oil drops from an atomizer into the air. These drops were charged by friction as they were sprayed from the atomizer. Earth's gravitational force pulled the drops downward, and a few of those drops entered the hole in the top plate of Millikan's apparatus. Millikan then adjusted the electric field between the plates until he had suspended a negatively charged drop. At this point, the downward force from Earth's gravitational field and the upward force from the electric field were equal in magnitude.

MiniLAB

ELECTRIC FIELDS

Can you observe the effects of an electric field?

iLab Station

READING CHECK Explain why the following net charges are impossible: 0.66e, 1.554e, and 3.504e.

Calculating charge Millikan calculated the magnitude of the electric field (E) from the electric potential difference between the plates $\left(E = \frac{\Delta V}{d}\right)$. Millikan had to make a second measurement to find the weight (mg) of the tiny drop. To find the weight of a drop, Millikan first suspended it. Then, he turned off the electric field and measured the rate of the fall of the drop. Because of friction with air molecules, the oil drop quickly reached terminal velocity, which was related to the weight of the drop by a complex equation. Millikan was then able to calculate the charge (q) on the drop from the weight of the drop (mg) and the electric field ($\boldsymbol{E}$). He found that the net charge on an oil drop was always some integer multiple of a number close to 1.6×10^{-19} C. In later experiments, others have refined this result to 1.60217653×10^{-19} C.

EXAMPLE PROBLEM 4

Get help with **isolating variables.** Math Handbook

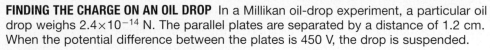

FINDING THE CHARGE ON AN OIL DROP In a Millikan oil-drop experiment, a particular oil drop weighs 2.4×10^{-14} N. The parallel plates are separated by a distance of 1.2 cm. When the potential difference between the plates is 450 V, the drop is suspended.

a. What is the net charge on the oil drop?

b. If the upper plate is positive, how many excess electrons are on the oil drop?

1 ANALYZE AND SKETCH THE PROBLEM

- Draw the plates with the oil drop suspended between them.
- Draw and label vectors representing the forces.
- Indicate the potential difference and the distance between the plates.

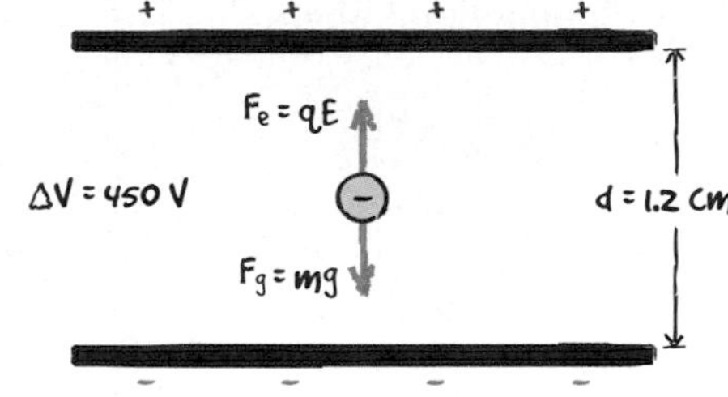

KNOWN	UNKNOWN
$\Delta V = 450$ V	net charge on drop, $q = ?$
$F_g = 2.4\times10^{-14}$ N	number of electrons, $n = ?$
$d = 1.2$ cm	

2 SOLVE FOR THE UNKNOWN

To be suspended, the electric force and gravitational force must be balanced.

$$F_e = F_g$$

$$qE = F_g$$ ◀ Substitute $F_e = qE$.

$$q\left(\frac{\Delta V}{d}\right) = F_g$$ ◀ Substitute $E = \frac{\Delta V}{d}$.

Solve for q.

$$q = \frac{F_g d}{\Delta V}$$

$$= \frac{(2.4\times10^{-14}\ \text{N})(0.012\ \text{m})}{450\ \text{V}}$$ ◀ Substitute $F_g = 2.4\times10^{-14}$ N, $d = 0.012$ m, $\Delta V = 450$ V.

$$= 6.4\times10^{-19}\ \text{C}$$

Solve for the number of electrons on the drop.

$$n = \frac{q}{e}$$

$$= \frac{6.4\times10^{-19}\ \text{C}}{1.602\times10^{-19}\ \text{C}}$$ ◀ Substitute $q = 6.4\times10^{-19}$ C, $e = 1.602\times10^{-19}$ C.

$$= 4.0$$

3 EVALUATE THE ANSWER

- **Are the units correct?** N•m/V = J/(J/C) = C, the unit for charge.
- **Is the magnitude realistic?** This is a small oil drop. It makes sense that the drop would have a small charge and contain only a few electrons.

PRACTICE PROBLEMS

Do additional problems. Online Practice

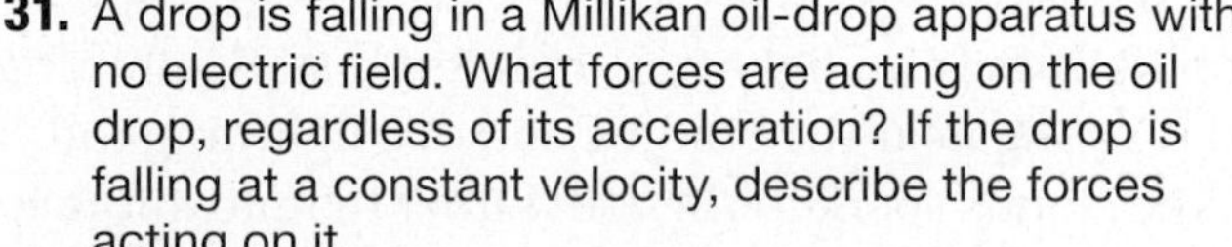

31. A drop is falling in a Millikan oil-drop apparatus with no electric field. What forces are acting on the oil drop, regardless of its acceleration? If the drop is falling at a constant velocity, describe the forces acting on it.

32. An oil drop weighs 1.9×10^{-15} N. You suspend it in an electric field of 6.0×10^{3} N/C. What is the net charge on the drop? How many excess electrons does it carry?

33. An oil drop carries one excess electron and weighs 6.4×10^{-15} N. What electric field strength do you need to suspend the drop so it is motionless?

34. CHALLENGE You suspend a positively charged oil drop that weighs 1.2×10^{-14} N between two parallel plates that are 0.64 cm apart. The potential difference between the plates is 240 V. What is the net charge on the drop? How many electrons is the drop missing?

Figure 14 Charges on a conducting sphere spread far apart to minimize their potential energy.

Explain why the negative charges evenly distribute around the surface of a conductor using what you know about the interactions among negative charges.

Conducting Sphere

On a conducting sphere, the charge is evenly distributed around the surface.

Hollow Sphere

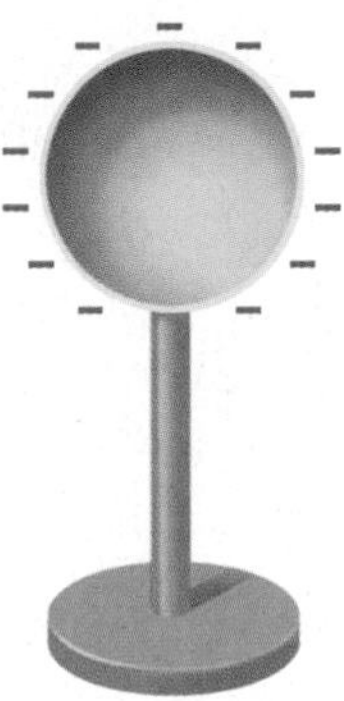

The charges on the hollow sphere are entirely on the outer surface.

Irregular Surface

On an irregular conducting surface, the charges are closest together at sharp points.

Electric Fields Near Conductors

Recall that many of the electrons in a conductor are free to move. Consider the charges on the conducting sphere in **Figure 14.** Because these electrons have like charges, they repel each other. They spread far apart in a way that minimizes their potential energy. The result is these charges come to rest on the surface of the conductor. It does not matter if the conducting sphere is solid or hollow. The excess charges move to the outer surface of the conductor.

Closed metal containers What happens if a closed metal container, such as a box, is charged? You can use a voltmeter to measure the electric potential difference between any two points inside the container. You will find that this potential difference is zero no matter which two points you choose inside the container. What are the consequences of this measurement for the electric field inside of the closed, metal container? Recall that potential difference is equal to the product of the electrical field and distance, or $\Delta V = Ed$. Because the potential difference between any two points inside the container is zero, the equation implies that the field is zero everywhere inside a closed, charged metal container.

Cars are a good example of this scenario. A car is a closed metal box that protects passengers from electric fields generated by lightning. On the outside of the conductor, the electric field often is not zero. The field is always perpendicular to the surface of the conductor. This makes the surface an equipotential; the potential difference between any two locations on the surface is zero.

☑ **READING CHECK** **Summarize** why passengers are protected inside a car if it is struck by lightning.

Irregular surfaces The electric field at the surface does depend on the shape of the conductor, as well as on the electric potential difference between it and other objects. Free charges are closer together at the sharp points of a conductor, as indicated in **Figure 14.** Therefore, the field lines are closer together, and the field is stronger.

This field can become so strong that when electrons are knocked off of atoms, the electrons and resulting ions are accelerated by the field, causing them to strike other atoms, resulting in more ionization of atoms. This chain reaction produces the pink glow seen inside a gas-discharge sphere.

Lightning rods If an electric field is strong enough, when the particles hit other molecules they will produce a stream of ions and electrons that form a plasma, which is a conductor. The result is a spark or, in extreme cases, lightning.

In order to protect buildings from lightning, builders install lightning rods. The electric field is strong near the pointed end of a lightning rod. As a result, charges in the clouds spark to the rod, rather than to another point on the building. From the rod, a conductor takes the charges to the ground. A lightning rod safely diverts lightning into the ground and away from the building.

☑ **READING CHECK** **Describe** how a lightning rod works.

Capacitors

When you lift a book, you increase the gravitational potential energy of the book-Earth system. You can interpret this as storing energy in a gravitational field. In a similar way, you can store energy in an electric field. In 1746 Dutch physician and physicist Pieter Van Musschenbroek invented a small device that could store electrical energy. In honor of the city in which he worked, Musschenbroek called his device a Leyden jar. The modern, much smaller device for storing electrical energy is called a **capacitor.** Manufacturers make capacitors from two conducting plates separated by a thin insulator. They often roll these layers into a cylinder. Engineers design capacitors to have specific capacitances. Capacitors are key components in computers and other electronic devices. You can see an example of capacitors in **Figure 15.**

Figure 15 A computer circuit board uses many capacitors in its operation. The gray cylinders, the disk-shaped objects, and the tiny colored tear-drop shaped objects are all capacitors.

Capacitance Suppose you connect the positive terminal of a 1-V power source to one of the conducting plates of a capacitor and connect the negative terminal of that power source to the other plate of that capacitor. What happens? The power source would produce a potential difference (ΔV) of 1 V between the two plates. This would result in a net positive charge ($+q$) on one plate and a net negative charge of equal magnitude ($-q$) on the other plate.

Now suppose that you increase the voltage of the power source. What happens to the amount of net charge on each plate? **Figure 16** shows the results of an experiment in which the experimenter increases the potential difference across a capacitor's plates from 0 V to 12 V. Notice that the graph of q v. ΔV is a straight line. The slope of the line in a net charge versus potential difference graph is a constant and is called the **capacitance** (C) of the capacitor. You can measure the capacitance of a capacitor in farads (F), where 1 F = 1 C/V. The equation for capacitance is shown below.

MiniLAB

CONSTRUCT A CAPACITOR

Can you make an operational capacitor?

CAPACITANCE

Capacitance is the ratio of the magnitude of the net charge on one plate of the capacitor to the potential difference across the plates.

$$C = \frac{q}{\Delta V}$$

Figure 16 In an experiment, the potential difference results in a charge in each plate. A graph of the data shows the relationship is linear. The slope of this line is the capacitance of the capacitor.

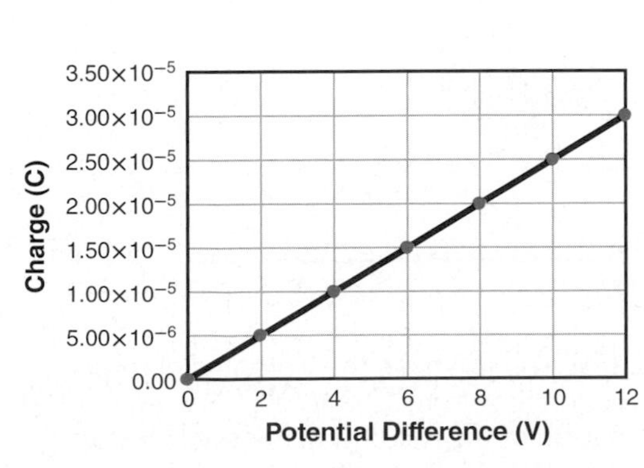

Data Table

Potential Difference (V)	Charge on a Plate (C)
0.0	0.0
2.0	5.2×10^{-6}
4.0	9.7×10^{-6}
6.0	15.0×10^{-6}
8.0	20.3×10^{-6}
10.0	24.7×10^{-6}
12.0	30.1×10^{-6}

Arthur S. Aubry/Getty Images

PhysicsLAB

STORING CHARGE

How can large amounts of electrical energy be stored?

The farad as a unit of measure One farad (F), named after Michael Faraday, is one coulomb per volt (C/V). Just as 1 C is a large amount of net charge, 1 F is also a fairly large capacitance. Most capacitors used in modern electronics have capacitances between 10 picofarads (10×10^{-12} F) and 500 microfarads (500×10^{-6} F). However, the memory capacitors that are used to prevent loss of memory in some computers can have capacitance from 0.5 F to 1.0 F. Note that if the electric potential difference between the plates increases, the net amount of charge on each plate also increases.

EXAMPLE PROBLEM 5

Find help with **operations with significant digits.** Math Handbook

FINDING CAPACITANCE A sphere was connected to the + pole of a 40-V battery while the − pole was connected to Earth. After a period of time, the sphere was charged to 2.4×10^{-6} C. What is the capacitance of the sphere-Earth system?

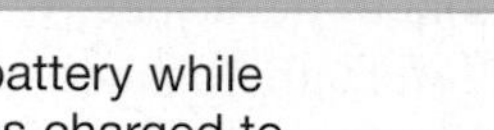

1 ANALYZE AND SKETCH THE PROBLEM

Draw a sphere above Earth, and label the net charge and potential difference.

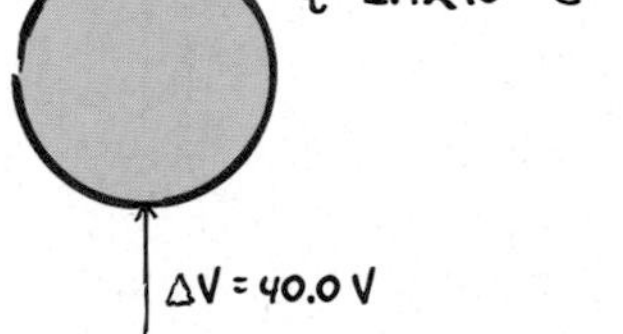

KNOWN

$\Delta V = 40.0$ V

$q = 2.4\times10^{-6}$ C

UNKNOWN

$C = ?$

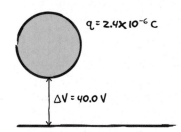

2 SOLVE FOR THE UNKNOWN

$$C = \frac{q}{\Delta V}$$

$$= \frac{2.4\times10^{-6}\text{ C}}{40.0\text{ V}}$$

◀ Substitute $\Delta V = 40.0$ V, $q = 2.4\times10^{-6}$ C.

$$= 6.0\times10^{-8}\text{ F}$$

$$= 0.060\ \mu\text{F}$$

3 EVALUATE THE ANSWER

- **Are the units correct?** C/V = F. The units are farads.
- **Is the magnitude realistic?** The amount of net charge stored on the sphere equals the capacitance multiplied by 40.0 V.

PRACTICE PROBLEMS

Do additional problems.  

35. A 27-μF capacitor has an electric potential difference of 45 V across it. What is the amount the net charge on the positively charged plate of the capacitor?

36. Suppose you connect both a 3.3-μF and a 6.8-μF capacitor across a 24-V electric potential difference. Which capacitor has the greater net charge on its positively charged plate, and what is its magnitude?

37. You later find that the magnitude of net charge on each of the plates for each of the capacitors in the previous problem is 3.5×10^{-4} C. Which capacitor has the larger potential difference across it? What is that potential difference?

38. Suppose that you apply an electric potential difference of 6.0 V across a 2.2-μF capacitor. What does the magnitude of the net charge on one plate need to be to increase the electric potential difference to 15.0 V?

39. A sphere is charged by a 12 V battery and suspended above Earth as shown in **Figure 17.** What is the net charge on the sphere?

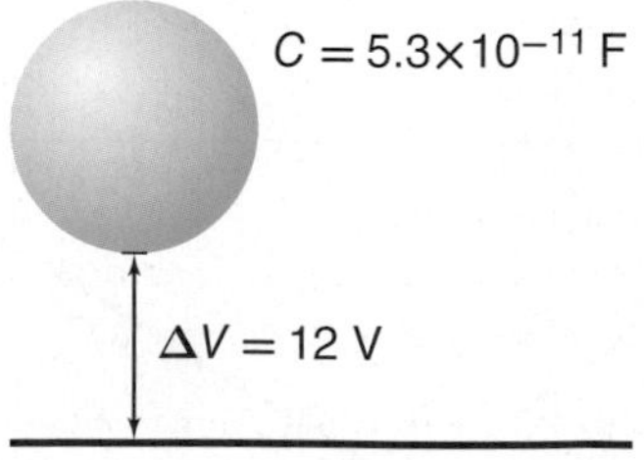

Figure 17

40. CHALLENGE You increase the potential difference across a capacitor from 12.0 V to 14.5 V. As a result, the magnitude of the net charge on each plate increases by 2.5×10^{-5} C. What is the capacitance of the capacitor?

PHYSICS CHALLENGE

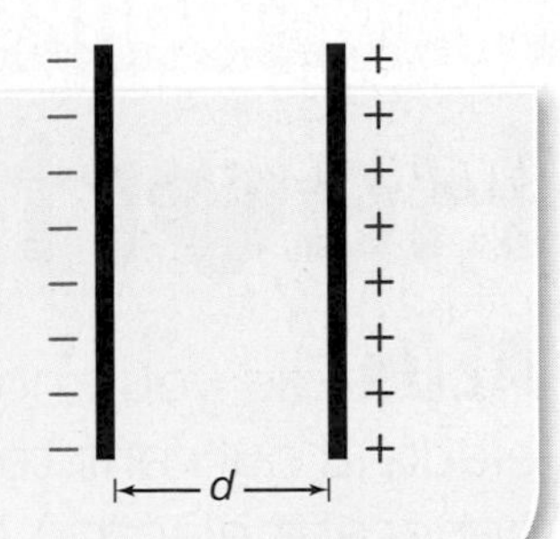

The plates of a capacitor attract each other because they have equal and opposite net charge. A capacitor consisting of two parallel plates that are separated by a distance (d) has capacitance (C).

1. Derive an expression for the force between the two plates when the magnitude of net charge on one of the capacitor's plates is q.
2. The electrostatic force between the plates of a 22-μF capacitor is 2.0 N. What is the magnitude of the net charge on one plate if the separation between those plates is 1.5 mm?

Varieties of capacitors Capacitors have many shapes and sizes. Some are large enough to fill whole rooms and can store enough electrical energy to create artificial lightning or power giant lasers that release thousands of joules of energy in a few billionths of a second. Capacitors are also used in computers, televisions, and digital cameras. These capacitors store much less energy than those used to make artificial lightning, but they can still be dangerous. Charge can remain for hours after the device is turned off. This is why many electronic devices carry warnings not to open the case.

Manufacturers control the capacitance of a capacitor by varying the surface area of the two conductors, or plates, by varying the distance between the plates, and by the nature of the insulating material between the plates. Higher capacitance is obtained by increasing the surface area and decreasing the separation of the plates. Certain insulating materials have the ability to effectively offset some of the charge on the plates and allow more electrical energy to be stored. Capacitors are named for the type of insulating material used to separate the plates and include ceramic, mica, polyester, and paper.

PhysicsLAB

ENERGIZING CAPACITORS

How do the charging times of different capacitors vary with capacitance?

SECTION 2 REVIEW

Section Self-Check Check your understanding.

41. MAINIDEA Suppose a friend asks you to explain how electric potential relates to potential energy. Write a brief explanation that you could use to explain this concept to a friend who does not understand the relationship between the two concepts.
42. **Potential Difference** What is the difference between electric potential energy and electric potential difference?
43. **Electric Field and Potential Difference** Show that a volt per meter is the same as a newton per coulomb.
44. **Millikan Experiment** When the net charge on an oil drop suspended in a Millikan apparatus is changed, the drop begins to fall. How should you adjust the potential difference between the conducting plates to bring the drop back into balance?
45. **Charge and Potential Difference** In the previous problem, if changing the potential difference between the conducting plates has no effect on the falling drop, what does this tell you about the new net charge on the drop?
46. **Capacitance** What is the magnitude of net charge on each conductor plate of a 0.47-μF capacitor when a potential difference of 12 V is applied across that capacitor?
47. **Charge Sharing** If you touch a large, positively charged, conducting sphere with a small, negatively charged, conducting sphere, what can be said about the
    a. potentials of the two spheres;
    b. charges on the two spheres?
48. **Critical Thinking** Explain how the charge in **Figure 18** continues to build up on the metal dome of a Van de Graaff generator. In particular, why isn't charge repelled back onto the belt at point B?

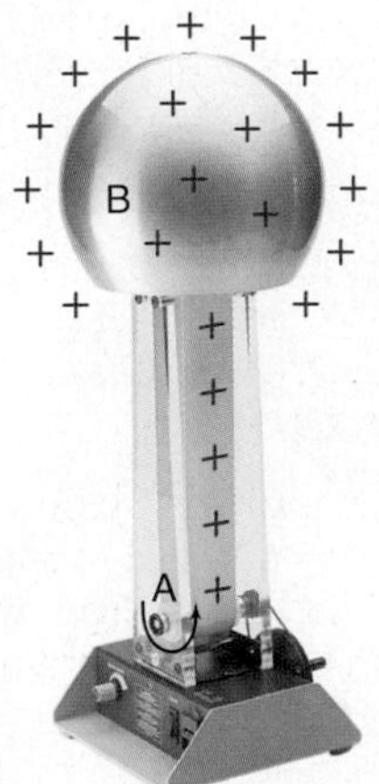

Figure 18

courtesy of PASCO Scientific

Heart-Shaped Box

AEDs Did you know your heart is a pump that needs electric current to work? Cells in the heart muscle generate electric current that triggers the heart muscle to squeeze. If this electric current is disrupted, the heart stops beating—a serious condition called cardiac arrest. Automated external difibrillators (AED) can be used to help restart a heart in cardiac arrest.

1 Call 911 Before using an AED, have someone nearby call 911 for you.

2 Find the nearest AED Many public places have AEDs. Look for a sign that says "AED" or "emergency defibrillation."

3 Read the simple instructions Simple diagrams printed on the AED tell how and where to apply the electrodes on the unconscious person's body.

4 Apply the electrode pads Once placed, the self-stick electrode pads sense whether there is a heartbeat. The electrodes can detect the electric current generated by the heart through the person's skin.

5 Let the AED computer do its job By analyzing the signals from the sensors in the electrode pads, the AED determines whether the heart problem requires an electric shock.

6 Pay attention to the speaker or screen Simple instructions projected from a voice recording and displayed on a digital screen tell the user how and when to deliver the electric shock.

FIGURE 1 AEDs are designed to be used by trained responders as well as untrained bystanders. You can use an AED to deliver an electric shock that could "jumpstart" the pacemaker cells of someone who is suffering from cardiac arrest.

GOING FURTHER >>>

Research where AEDs are thought to be most useful in a community. Then, make a map of locations of AEDs in or around your school or other public locations.

(t) Ashley Cooper/Terra/CORBIS, (b)kpzfoto/Alamy

STUDY GUIDE

BIGIDEA Electric charges are surrounded by electric fields that exert a force on other charged objects.

SECTION 1 Measuring Electric Fields

VOCABULARY
- **electric field** *(p. 570)*
- **electric field line** *(p. 574)*

MAINIDEA An electric field is a property of the space around a charged object that exerts forces on other charged objects.

- An electric field exists around any charged object. The field produces forces on other charged objects.
- Quantities relating to charge, electric fields, and forces are related and can be calculated using these formulas:

$$E = \frac{F}{q'} \qquad E = \frac{Kqq'}{r^2q'} = \frac{Kq}{r^2}$$

- Electric field lines provide a pictorial model of the electric field. They are directed away from positive charges and toward negative charges. They never cross, and their density is related to the strength of the field. The direction of the electric field is the direction of the force on a tiny, positive test charge.

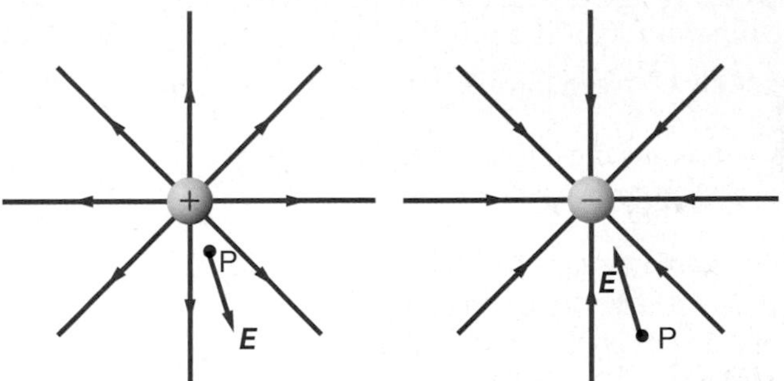

SECTION 2 Applications of Electric Fields

VOCABULARY
- **electric potential difference** *(p. 578)*
- **volt** *(p. 578)*
- **equipotential** *(p. 579)*
- **capacitor** *(p. 585)*
- **capacitance** *(p. 585)*

MAINIDEA Electric potential (sometimes called voltage) is electric potential energy per unit charge.

- Electric potential difference is the change in potential energy per unit charge in an electric field. Electric potential differences are measured in volts.
- Potential difference is related to the work required to move a charge and is represented by the following equation:

$$\Delta V = \frac{W}{q'}$$

- Capacitors are used to store electrical energy. A capacitor consists of two conducting plates separated by an insulator. The capacitance C depends only on the geometry of these plates and insulator. It can be calculated by the following equation.

$$C = \frac{q}{\Delta V}$$

Games and Multilingual eGlossary

Vocabulary Practice

Chapter Self-Check

SECTION 1 Measuring Electric Fields

Mastering Concepts

49. What are the two properties that a test charge must have?

50. How is the direction of an electric field defined?

51. BIGIDEA What are electric field lines?

52. How is the strength of an electric field indicated with electric field lines?

53. Draw some of the electric field lines between each of the following:

a. two like charges of equal magnitude

b. two unlike charges of equal magnitude

c. a positive charge and a negative charge having twice the magnitude of the positive charge

d. two oppositely charged parallel plates

54. In **Figure 19,** where do the electric field lines that leave the positive charge end?

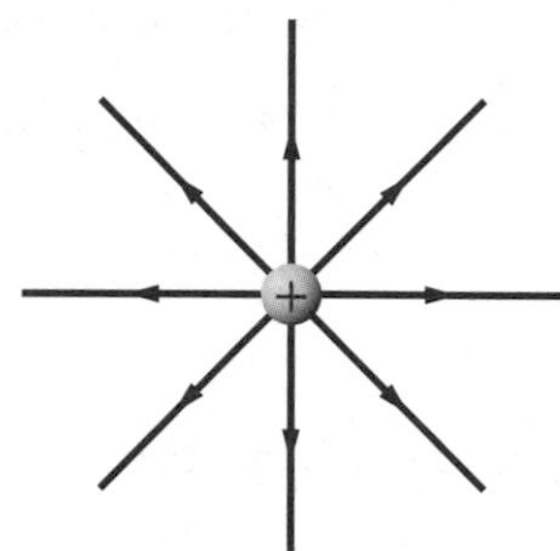

Figure 19

55. What happens to the strength of an electric field when the magnitude of the test charge is halved?

56. You are moving a constant positive charge through an increasing electric field. Does the amount of energy required to move it increase or decrease?

Mastering Problems

The charge of an electron is -1.602×10^{-19} C.

57. What charge exists on a test charge that experiences a force of 1.4×10^{-8} N at a point where the electric field intensity is 5.0×10^{-4} N/C?

58. A test charge experiences a force of 0.30 N on it when it is placed in an electric field intensity of 4.5×10^{5} N/C. What is the magnitude of the charge?

59. What is the electric field strength 20.0 cm from a point charge of 8.0×10^{-7} C?

60. A positive charge of 1.0×10^{-5} C, shown in **Figure 20,** experiences a force of 0.30 N when it is located at a certain point. What is the electric field intensity at that point?

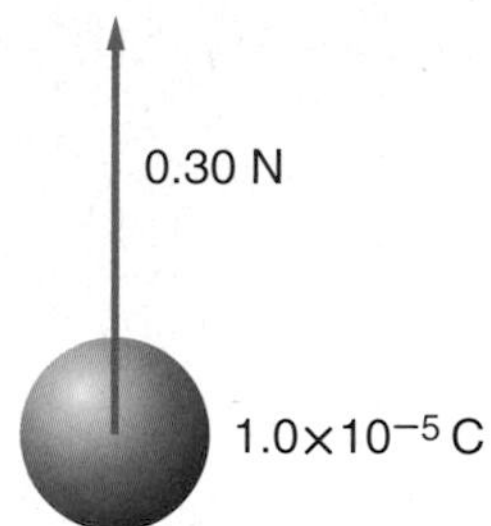

Figure 20

61. The electric field in the atmosphere is about 150 N/C downward.

a. What is the direction of the force on a negatively charged particle?

b. Find the electric force on an electron with charge -1.602×10^{-19} C.

c. Compare the force in part b with the force of gravity on the same electron (mass = 9.1×10^{-31} kg).

62. Carefully sketch each of the following:

a. the electric field produced by a $+1.0$-μC charge

b. the electric field resulting from a $+2.0$-μC charge (Make the number of field lines proportional to the change in charge.)

63. A positive test charge of 6.0×10^{-6} C is placed in an electric field of 50.0-N/C intensity, as in **Figure 21.** What is the strength of the force exerted on the test charge?

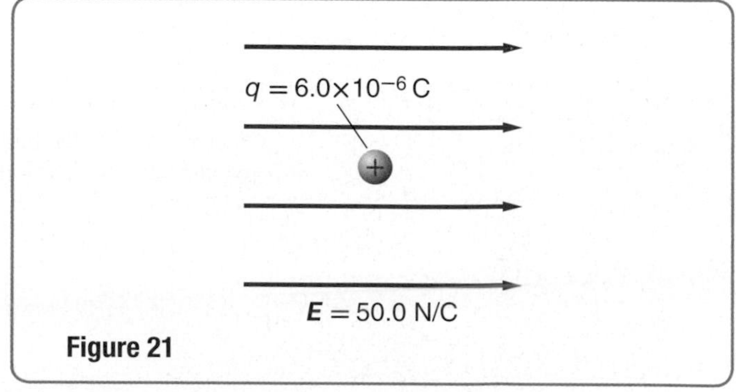

Figure 21

64. A force of 14.005 N exists on a positive test charge (q') that has a charge of 4.005×10^{-19} C. What is the magnitude of the electric field?

Chapter Self-Check

65. Charges X, Y, and Z all are equidistant from each other. X has a +1.0-μC charge. Y has a +2.0-μC charge. Z has a small negative charge.

   a. Draw an arrow representing the force on charge Z.

   b. Charge Z now has a small positive charge on it. Draw an arrow representing the force on it.

66. In an old television picture tube, electrons are accelerated by an electric field having a value of 1.00×10^5 N/C.

   a. Find the force on an electron.

   b. If the field is constant, find the acceleration of the electron (mass 9.11×10^{-31} kg).

67. The nucleus of a lead atom has a charge of 82 protons.

   a. What are the direction and the magnitude of the electric field at 1.0×10^{-10} m from the nucleus?

   b. What are the direction and the magnitude of the force exerted on an electron located at this distance?

SECTION 2

Applications of Electric Fields

Mastering Concepts

68. What SI unit is used to measure electric potential energy? What SI unit is used to measure electric potential difference?

69. Define volt in terms of the change in potential energy of a charge moving in an electric field.

70. Why does a charged object lose its charge when it is touched to the ground?

71. A charged rubber rod that is placed on a table maintains its charge for some time. Why is the charged rod not discharged immediately?

72. **Computers** Delicate parts in electronic equipment, such as those pictured in **Figure 22,** are contained within a metal box inside a plastic case. Why?

Figure 22

Andrew Howe/Photodisc/Getty Images

Mastering Problems

73. If 120 J of work is performed to move 2.4 C of charge from the positive plate to the negative plate shown in **Figure 23,** what potential difference exists between the plates?

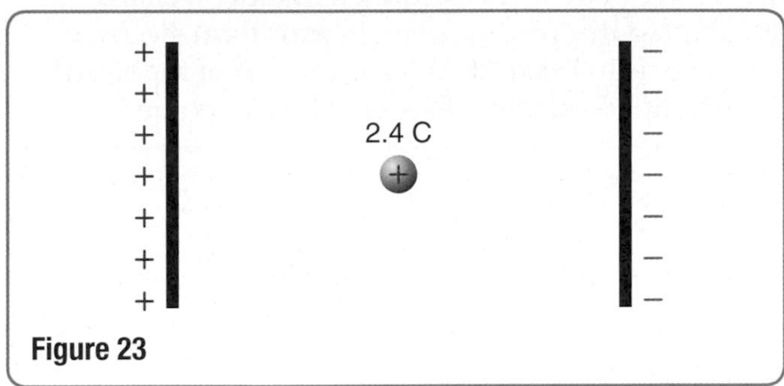

Figure 23

74. How much work is done to transfer 0.15 C of charge through an electric potential difference of 9.0 V?

75. An electron is moved through an electric potential difference of 450 V. How much work is done on the electron?

76. A 12-V battery does 1200 J of work transferring charge. How much charge is transferred?

77. The electric field intensity between two charged plates is 1.5×10^3 N/C. The plates are 0.060 m apart. What is the electric potential difference, in volts, between the plates?

78. A voltmeter indicates that the electric potential difference between two plates is 70.0 V. The plates are 0.020 m apart. What electric field intensity exists between them?

79. A capacitor that is connected to a 45.0-V source has a charge of 90.0 μC. What is the capacitor's capacitance?

80. The oil drop shown in **Figure 24** is negatively charged and weighs 4.5×10^{-15} N. The drop is suspended in an electric field intensity of 5.6×10^3 N/C.

   a. What is the charge on the drop?

   b. How many excess electrons does it carry?

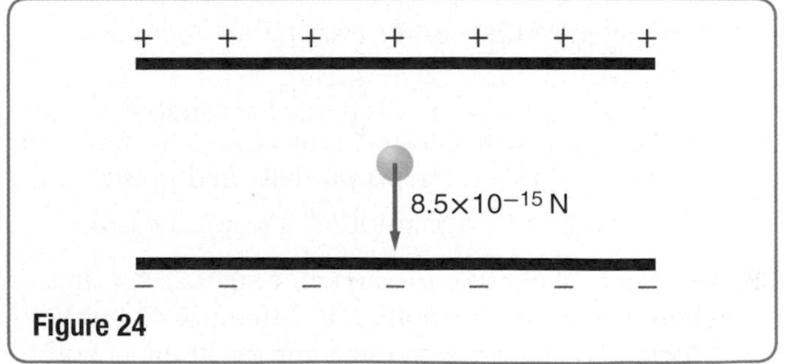

Figure 24

Chapter Self-Check

81. What electric potential difference exists across a 5.4-μF capacitor that has a charge of 8.1×10^{-4} C?

82. What is the charge of a 15.0-pF capacitor when it is connected across a 45.0-V source?

83. A force of 0.065 N is required to move a charge of 37 μC a distance of 25 cm in a uniform electric field, as in **Figure 25.** What is the size of the electric potential difference between the two points?

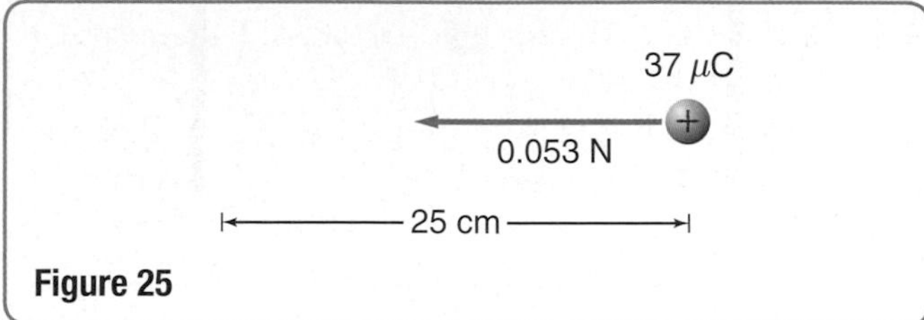

Figure 25

84. Photoflash The potential energy of a capacitor with capacitance (C) and an electric potential difference (ΔV) is represented by $PE = \frac{1}{2}C\Delta V^2$. One application of this is in the electronic photoflash of a strobe light, such the one in **Figure 26.** In such a unit, a capacitor of 10.0 μF has a charge of 3.0×10^{2} V. Find the electrical energy stored.

Figure 26

85. Suppose it took 25 s to energize the capacitor in the previous problem.

a. Find the average power required to energize the capacitor in this time.

b. When the plates are discharged through the strobe lamp, it transfers all its energy in 1.0×10^{-4} s. Find the power delivered to the lamp.

c. How is such a large amount of power possible?

86. The plates of a 0.047-μF capacitor are 0.25 cm apart and are charged to a potential difference of 120 V. How much charge is on one plate of the capacitor?

87. Lasers Lasers are used to try to produce controlled fusion reactions. These lasers require brief pulses of energy that are stored in large rooms filled with capacitors. One such room has a capacitance of 6.1×10^{-2} F and is energized to a potential difference of 10.0 kV.

a. Given that $PE = \frac{1}{2}C\Delta V^2$, find the energy stored in the capacitors.

b. The capacitors' plates are discharged in 10 ns (1.0×10^{-8} s). What power is produced?

c. If the capacitors are energized by a generator with a power capacity of 1.0 kW, how many seconds will be required to energize the capacitors?

Applying Concepts

88. What will happen to the electric potential energy of a charged particle in an electric field when the particle is released and free to move?

89. Figure 27 shows three spheres with charges of equal magnitude, with their signs as shown. Spheres y and z are held in place, but sphere x is free to move. Initially, sphere x is equidistant from spheres y and z. Which path will sphere x begin to follow? Assume that no other forces act on the spheres.

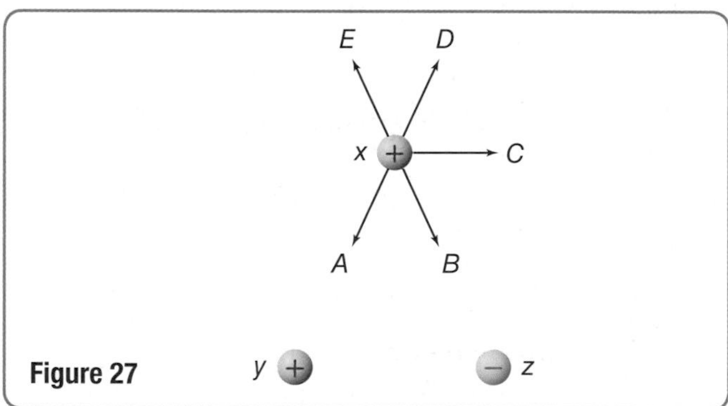

Figure 27

90. Ranking Task Rank the following point charges according to the magnitude of the electric force experienced, from greatest to least. Specifically indicate any ties.

A. a charge of 3 nC at a point where the electric field is 70 N/C

B. a charge of 5 nC at a point where the electric field is 600 N/C

C. a charge of 3 nC at a point where the electric field is 20 N/C

D. a charge of 6 nC at a point where the electric field is 35 N/C

E. a charge of 8 nC at a point where the electric field is 10 N/C

Chapter Self-Check

91. Two unlike-charged oil drops are held motionless in a Millikan oil-drop experiment at the same time.
   a. Can you be sure that the charges are the same?
   b. The ratios of which two properties of the oil drops have to be equal?

92. José and Sue are standing on an insulating platform and holding hands when they are given a charge, as in **Figure 28.** José is larger than Sue. Who has the larger amount of charge from the machine, or do they both have the same amount?

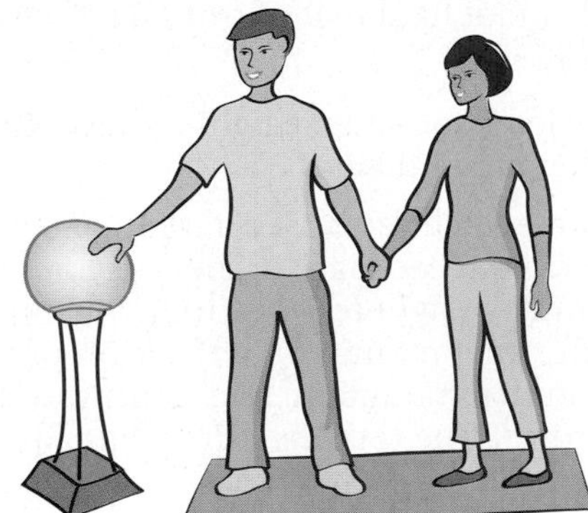

Figure 28

93. **Reverse Problem** Write a physics problem for which the following equation would be part of the solution:
$$E = \frac{(9\ \text{V} - 0\ \text{V})}{0.85\ \text{cm}}$$

94. How can you store different amounts of electrical energy in a capacitor?

Mixed Review

95. How much work does it take to move 0.25 μC between two parallel plates that are 0.40 cm apart if the field between the plates is 6400 N/C?

96. How much charge is on a 0.22-μF parallel plate capacitor if the plates are 1.2 cm apart and the electric field between them is 2400 N/C?

97. Two identical small spheres 25 cm apart carry equal but opposite charges of 0.060 μC, as in **Figure 29.** If the potential difference between them is 300 V, what is the capacitance of the system?

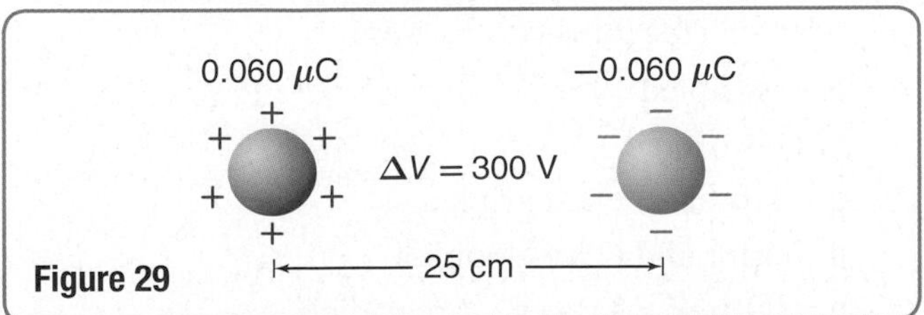

Figure 29

98. **Problem Posing** Complete this problem so that it must be solved using the concept indicated: " A point charge of 4.0 mC is at rest ...."
   a. electric field
   b. electric potential difference

99. The plates of a 0.047-μF capacitor are 0.25 cm apart and are charged to a potential difference of 120 V.
   a. How much charge is on the capacitor?
   b. What is the strength of the electric field between the plates of the capacitor?
   c. An electron is placed between the plates of the capacitor, as in **Figure 30.** What force is exerted on that electron?

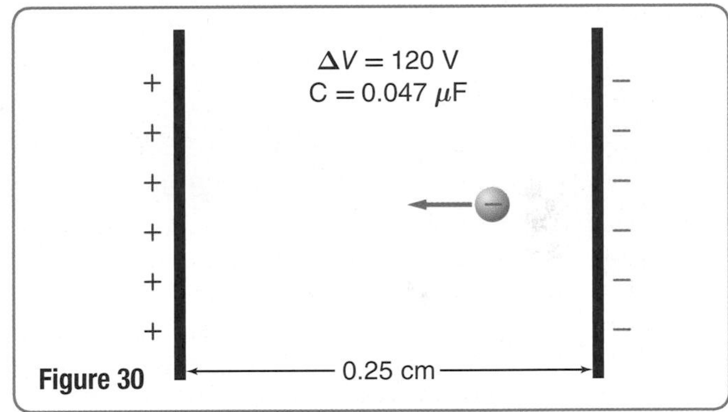

Figure 30

100. How much work would it take to move an additional 0.010 μC between the plates at 120 V in the previous problem?

101. The graph in **Figure 31** represents the amount of charge stored on one plate of a capacitor as a function of the charging potential.
   a. What does the slope of the line represent?
   b. What is the capacitance of the capacitor?
   c. What does the area under the graph line represent?

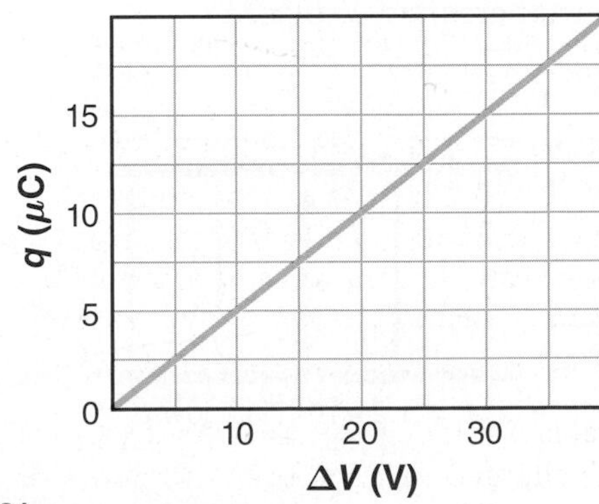

Figure 31

Chapter Self-Check

Thinking Critically

102. Analyze and Conclude Two small spheres, A and B, lie on the *x*-axis, as in **Figure 32.** Sphere A has a charge of $+3.00\times10^{-6}$ C. Sphere B is 0.800 m to the right of sphere A and has a charge of -5.00×10^{-6} C. Find the magnitude and the direction of the electric field strength at a point above the *x*-axis that would form the apex of an equilateral triangle with spheres A and B.

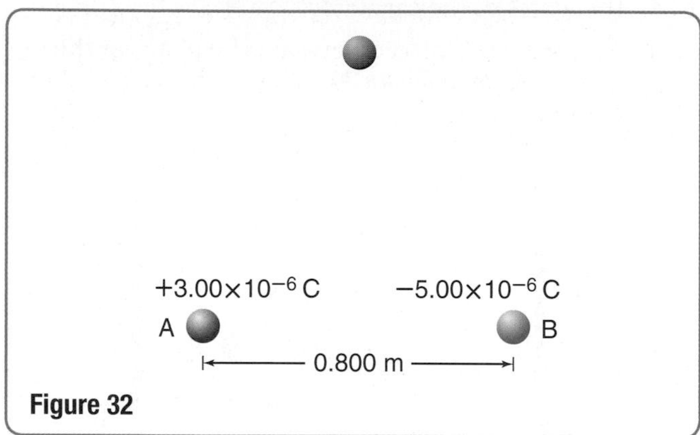

Figure 32

103. Analyze and Conclude In an ink-jet printer, drops of ink are given a certain amount of charge before they move between two large, parallel plates. The plates deflect the charged ink particles as shown in **Figure 33.** The plates have an electric field of $E = 1.2\times10^{6}$ N/C between them and are 1.5 cm long. Drops with a mass $m = 0.10$ ng and a charge $q = 1.0\times10^{-16}$ C are moving horizontally at a speed, $v = 15$ m/s, parallel to the plates. What is the vertical displacement of the drops when they leave the plates? To answer this question, complete the following steps:

a. What is the vertical force on the drops?

b. What is their vertical acceleration?

c. How long are they between the plates?

d. How far are they displaced?

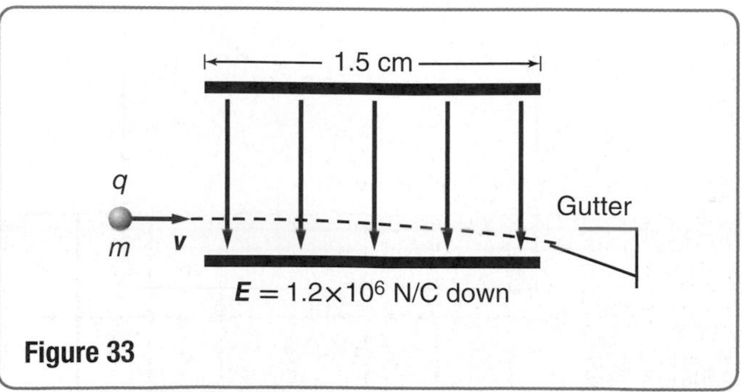

Figure 33

Writing in Physics

104. Choose the name of an electric unit, such as coulomb, volt, or farad, and research the life and work of the scientist for whom it was named. Write a brief essay on this person and include a discussion of the work that justified the honor of having a unit named for him.

Cumulative Review

105. Michelson measured the speed of light by sending a beam of light to a mirror on a mountain 35 km away.

a. How long does it take light to travel the distance to the mountain and back?

b. Assume Michelson used a rotating octagon with a mirror on each face of the octagon. Also assume the light reflects from one mirror travels to the other mountain, reflects off of a fixed mirror on that mountain, and returns to the rotating mirrors. If the rotating mirror has advanced so that when the light returns, it reflects off of the next mirror in the rotation, how fast is the mirror rotating?

c. If each mirror has a mass of 1.0×10^{1} g and rotates in a circle with an average radius of 1.0×10^{1} cm, what is the approximate centripetal force needed to hold the mirror while it is rotating?

106. Mountain Scene You can see an image of a distant mountain in a smooth lake just as you can see a mountain biker next to the lake because light from each strikes the surface of the lake at about the same angle of incidence and is reflected to your eyes. If the lake is about 100 m in diameter, the reflection of the top of the mountain is about in the middle of the lake, the mountain is about 50 km away from the lake, and you are about 2 m tall, then approximately how high above the lake does the top of the mountain reach?

107. A converging lens has a focal length of 38.0 cm. If it is placed 60.0 cm from an object, at what distance from the lens will the image be?

108. A force (F) is measured between two charges (Q and q) separated by a distance (r). What would the new force be for the following?

a. r is tripled

b. Q is tripled

c. both r and Q are tripled

d. both r and Q are doubled

e. all three, r, Q, and q, are tripled

MULTIPLE CHOICE

1. Why is an electric field measured only by a small test charge?

A. so the charge doesn't disturb the field

B. because small charges have small momentum

C. so its size doesn't nudge the charge to be measured aside

D. because an electron always is used as the test charge and electrons are small

2. A positive test charge of 8.7 μC experiences a force of 8.1×10^{-6} N at an angle of 24° N of E. What are the magnitude and direction of the electric field strength at the location of the test charge?

A. 7.0×10^{-8} N/C, 24° N of E

B. 1.7×10^{-6} N/C, 24° S of W

C. 1.1×10^{-3} N/C, 24° W of S

D. 9.3×10^{-1} N/C, 24° N of E

3. What is the potential difference between two plates that are 18 cm apart with a field of 4.8×10^{3} N/C?

A. 27 V

B. 86 V

C. 0.86 kV

D. 27 kV

4. How much work is done on a proton to move it from the negative plate to a positive plate 4.3 cm away if the field is 125 N/C?

A. 5.5×10^{-23} J

B. 8.6×10^{-19} J

C. 1.1×10^{-16} J

D. 5.4 J

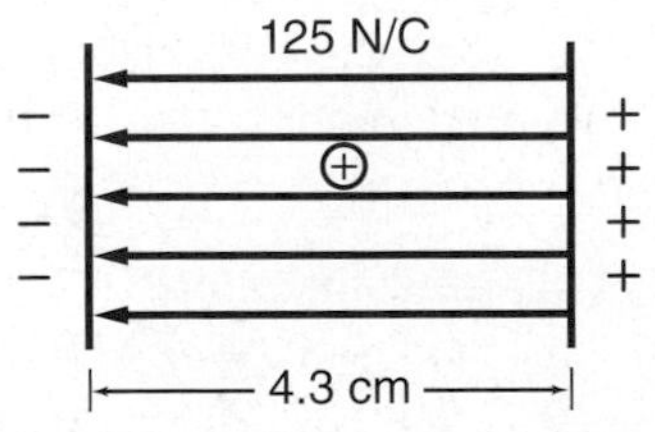

5. How was the magnitude of the field in Millikan's oil-drop experiment determined?

A. using a measurable electromagnet

B. from the electric potential between the plates

C. from the magnitude of the charge

D. by an electrometer

6. In an oil drop experiment, a drop with a weight of 2.0×10^{-14} N was suspended motionless when the potential difference between the plates that were 63 mm apart was 0.78 kV. What was the charge of the drop?

A. -1.6×10^{-18} C

B. -4.0×10^{-16} C

C. -1.2×10^{-15} C

D. -9.3×10^{-13} C

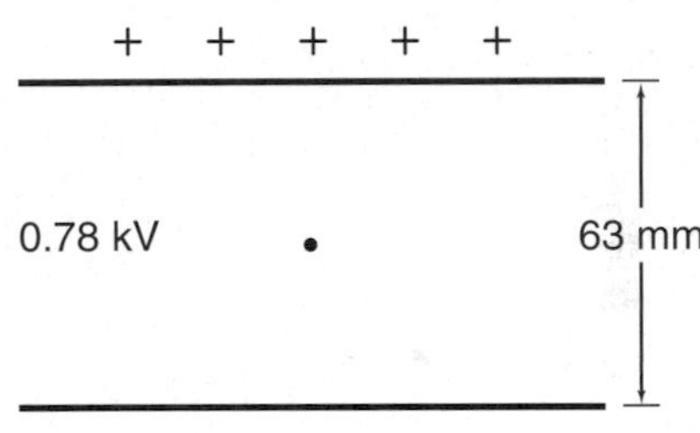

7. A capacitor has a capacitance of 0.093 μF. If the charge of the capacitor is 58 μC, what is the electric potential difference?

A. 5.4×10^{-12} V

B. 1.6×10^{-6} V

C. 6.2×10^{2} V

D. 5.4×10^{3} V

FREE RESPONSE

8. Assume 18 extra electrons are on an oil drop. Calculate the charge of the oil drop, and calculate the potential difference needed to suspend it if it has a weight of 6.12×10^{-14} N and the plates are 14.1 mm apart.

NEED EXTRA HELP?

If you Missed Question	1	2	3	4	5	6	7	8
Review Section	1	1	2	2	2	2	2	2

Online Test Practice

CHAPTER 22

Electric Current

BIGIDEA Electric currents carry electrical energy that can be transformed into other forms of energy.

SECTIONS

1 Current and Circuits

2 Using Electrical Energy

LaunchLAB

iLab Station

LIGHT THE BULB

If someone gave you a few electrical parts, could you make a lightbulb shine?

WATCH THIS!

Video

CIRCUITS

You might be surprised how many simple circuits are around you every day. Explore the physics of electricity and circuits.

PHYSICS T.V.

Go online!
connectED.mcgraw-hill.com

SECTION 1 Current and Circuits

MAIN IDEA

Electric current is the flow of electric charges.

Essential Questions

- What is electric current?
- How does energy change in electric circuits?
- What is Ohm's law?
- How are power, current, potential difference, and resistance mathematically related?

Review Vocabulary

electric potential difference the work done moving a positive test charge between two points in an electric field divided by the magnitude of that test charge

New Vocabulary

electric current
conventional current
battery
electric circuit
ampere
resistance
resistor
parallel connection
series connection

PHYSICS 4 YOU Have you ever seen the circuit board of a computer? There are metal connections between various components, such as semiconductor chips, resistors, and capacitors. Electric charges flow between these components to make the computer work.

Producing Electric Current

You probably can think of many appliances that use electricity, such as hair dryers, MP3 players, stoves, and refrigerators. In this chapter, you will learn more about such devices and electrical energy.

Recall that when two conducting spheres touch, charges flow from the sphere at a higher potential to the one at a lower potential. The flow continues until there is no potential difference between the two spheres. A flow of charged particles is an **electric current.** The diagrams in **Figure 1** show charges flowing through a conductive material. In the diagram on the left, two conductors, A and B, are connected by conductor C. Positive charges flow from the higher potential of B to A through C. The electric current stops when the potential difference between A, B, and C is zero. **Conventional current** is the direction in which a positive test charge moves. However, it usually is the negative charges (electrons) that flow. The flow of electrons and the direction of conventional current are in opposite directions.

Consider **Figure 1** again. You could maintain the electric potential difference between B and A by pumping charged particles from A back to B, as shown on the right in **Figure 1.** Because the pump increases the electric potential energy of the charges, it requires an external energy source. One familiar source, a voltaic or galvanic cell (a common dry cell), transforms chemical energy to electrical energy. Several galvanic cells connected together are called a **battery.**

Figure 1 Positive charges flow from the higher potential at B through the conductive wire C to A, which has a lower potential than B. When the potential difference between B and A is zero, the flow stops. The flow continues in the diagram on the right because a charge pump maintains the potential difference between conductors A and B.

Potential Difference goes to 0

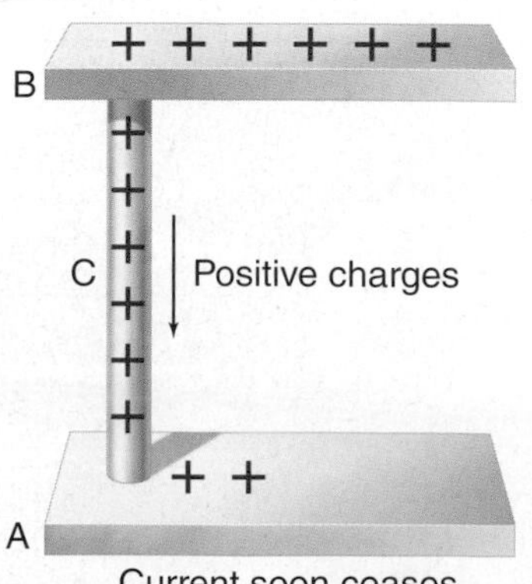

Potential Difference > 0

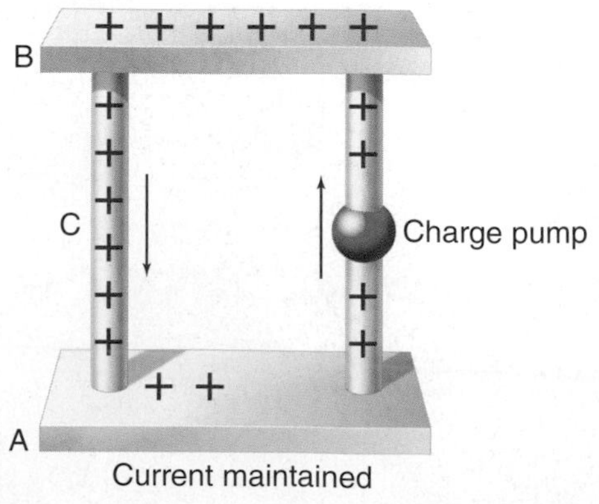

Comstock Images/Alamy

Electric Circuits

The charges in the diagram on the right in **Figure 1** move around a closed loop, cycling from the pump to B, through C, to A, and back to the pump. Any closed loop or conducting path allowing electric charges to flow is called an **electric circuit.** A circuit includes a charge pump, which increases the electric potential energy of the charges flowing from A to B, and a device that reduces the electric potential energy of the charges flowing from B to A. When the charges flow through the device that reduces the electric potential energy ($q\Delta V$), the energy is transformed into some other form of energy. For example, electrical energy is transformed into kinetic energy by a motor, to light energy by a lamp, and to thermal energy by a heater.

Comparing water flow and current A charge pump creates the flow of charged particles that make up a current. Consider a generator driven by a waterwheel, such as the one shown in **Figure 2.** The water falls and rotates the waterwheel and the generator. Thus, the kinetic energy of the water is transformed to electrical energy by the generator. The generator, like the charge pump, increases the electric potential difference (ΔV) between its two wire connections, which gives energy ($q\Delta V$) to each charge (q) flowing through the generator. This energy comes from the transformation of kinetic energy of the water. Not all the water's kinetic energy, however, is transformed to electrical energy, as shown in the flowchart in **Figure 2.**

If the generator attached to the waterwheel is connected to a motor, the charges in the wire flow into the motor. The flow of charges continues through the circuit back to the generator. The motor transforms electrical energy to kinetic energy.

View an **animation about current and circuits.**

Figure 2 The gravitational potential energy of the water is transformed into kinetic energy, then to electrical energy and thermal energy. Energy transformations are not 100 percent efficient. Thermal energy is produced by the splashing water, friction, and electric resistance.

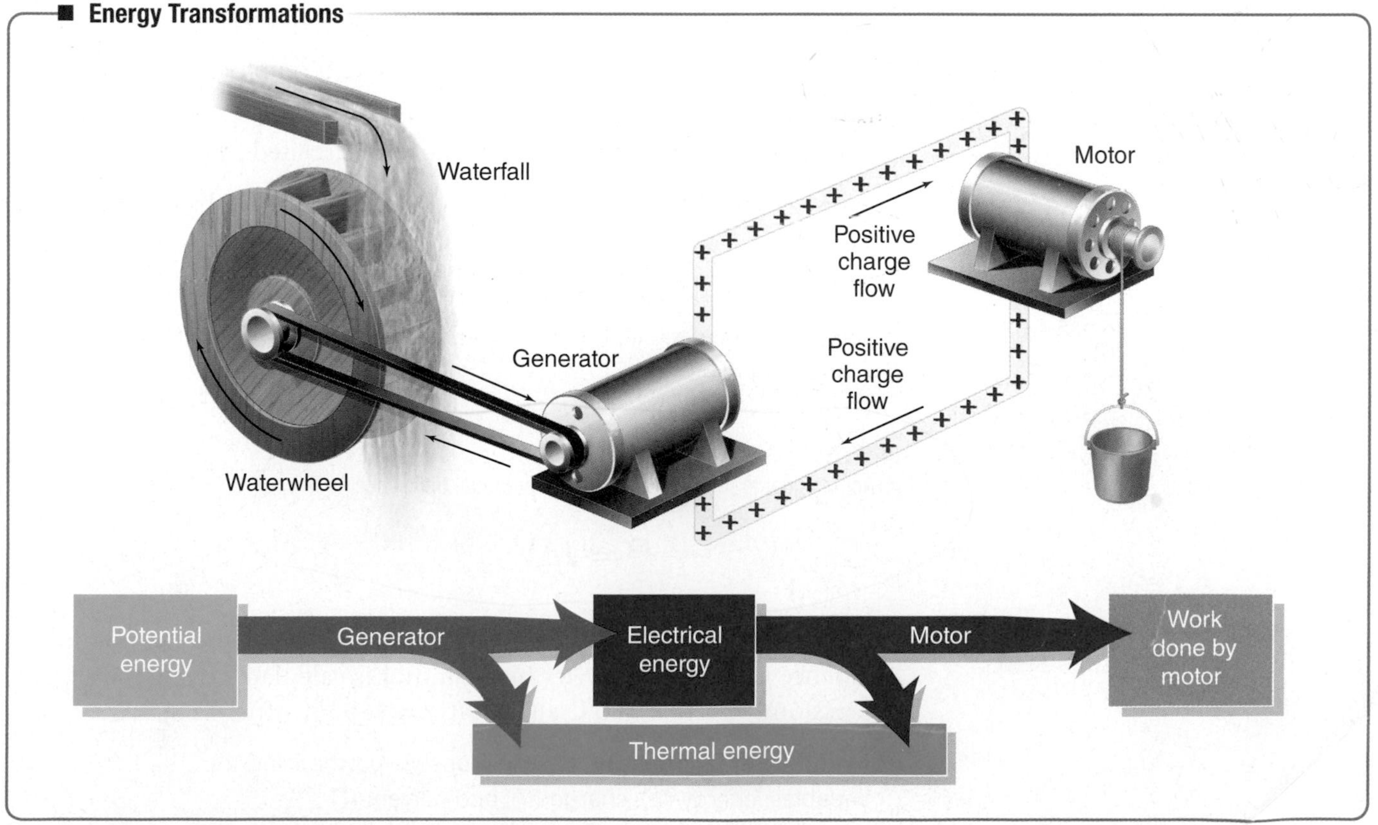

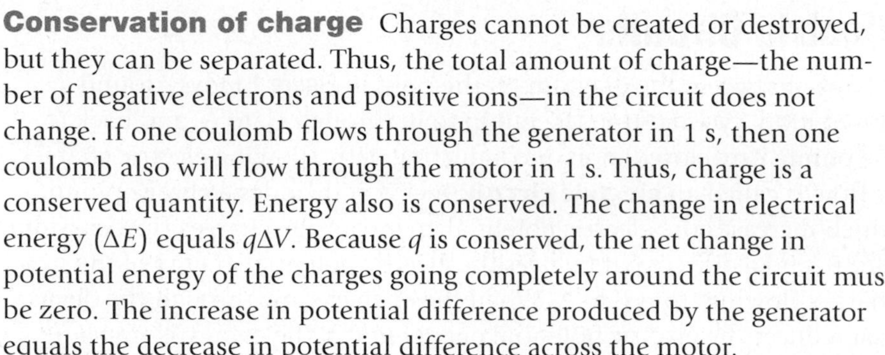

Conservation of charge Charges cannot be created or destroyed, but they can be separated. Thus, the total amount of charge—the number of negative electrons and positive ions—in the circuit does not change. If one coulomb flows through the generator in 1 s, then one coulomb also will flow through the motor in 1 s. Thus, charge is a conserved quantity. Energy also is conserved. The change in electrical energy (ΔE) equals $q\Delta V$. Because q is conserved, the net change in potential energy of the charges going completely around the circuit must be zero. The increase in potential difference produced by the generator equals the decrease in potential difference across the motor.

If the potential difference between two wires is 120 V, the waterwheel and the generator must do 120 J of work on each coulomb of charge that is delivered. Every coulomb of charge moving through the motor delivers 120 J of energy to the motor.

Figure 3 Some appliances, such as hair dryers, are rated in watts. The rating, 1875 W, indicates the rate at which energy changes. In the hair dryer, electrical energy is transformed into thermal energy and kinetic energy.

READING CHECK **Draw** a diagram of an electric circuit, and explain how energy is conserved in it.

Rates of Charge Flow and Energy Transfer

Power, which is measured in watts (W) is the rate at which energy is transferred or transformed. Many appliances, such as the hair dryer shown in **Figure 3,** are rated in watts.

If a generator transfers 1 J of kinetic energy to electrical energy each second, it is transferring energy at the rate of 1 J/s, or 1 W. The energy carried by an electric current depends on the charge transferred (q) and the potential difference across which it moves (ΔV). Thus, $E = q\Delta V$. Recall that the unit for the quantity of electric charge is the coulomb. The rate of flow of electric charge $\left(\frac{q}{t}\right)$ is called electric current and is measured in coulombs per second. Electric current is represented by I, so $I = \frac{q}{t}$. A flow of electric charge equal to one coulomb per second (1 C/s) is called an **ampere** (A).

The energy carried by an electric current is related to the voltage, $E = q\Delta V$. Since current $\left(I = \frac{q}{t}\right)$ is the rate of charge flow, the power $\left(P = \frac{E}{t}\right)$ of an electrical device can be determined by multiplying voltage and current. To derive a useful form of the equation for the power delivered to an electrical device, you can use $P = \frac{E}{t}$ and substitute $E = q\Delta V$ and $q = It$.

POWER

Power is equal to the current times the potential difference.

$$P = I\Delta V$$

If the current through the motor in **Figure 2** is 3.0 A and the potential difference is 120 V, the power in the motor is calculated using the expression $P = I\Delta V = (3.0\ \text{C/s})(120\ \text{J/C}) = 360\ \text{J/s}$, which is 360 W.

READING CHECK **Restate** the equations for power using the variables: energy (E), charge (q), and current (I).

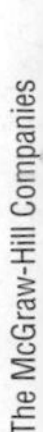

EXAMPLE PROBLEM 1

Find help with **significant figures.** Math Handbook

ELECTRIC POWER AND ENERGY A 6.0-V battery delivers a 0.50-A current to an electric motor connected across its terminals.

a. What power is delivered to the motor?

b. If the motor runs for 5.0 min, how much electrical energy is delivered?

1 ANALYZE AND SKETCH THE PROBLEM

- Draw a circuit showing the positive terminal of a battery connected to a motor and the return wire from the motor connected to the negative terminal of the battery.
- Show the direction of conventional current.

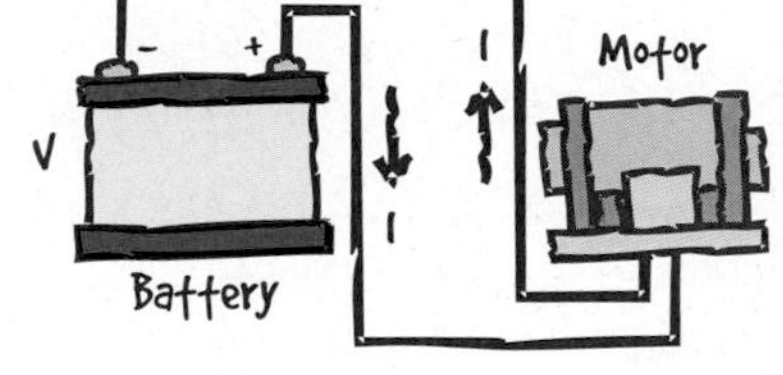

KNOWN	UNKNOWN
$\Delta V = 6.0$ V	$P = ?$
$I = 0.50$ A	$E = ?$
$t = 5.0$ min	

2 SOLVE FOR THE UNKNOWN

a. Use $P = I\Delta V$ to find the power.

$$P = I\Delta V$$

$$P = (0.50\ \text{A})(6.0\ \text{V})$$ ◀ Substitute $I = 0.50$ A, $V = 6.0$ V.

$$= 3.0\ \text{W}$$

b. You learned that $P = \frac{E}{t}$. Solve for E to find the energy.

$$E = Pt$$

$$= (3.0\ \text{W})(5.0\ \text{min})$$ ◀ Substitute $P = 3.0$ W, $t = 5.0$ min.

$$= (3.0\ \text{J/s})(5.0\ \text{min})\left(\frac{60\ \text{s}}{1\ \text{min}}\right)$$

$$= 9.0\times10^{2}\ \text{J}$$

3 EVALUATE THE ANSWER

- **Are the units correct?** Power is measured in watts, and energy is measured in joules.
- **Is the magnitude realistic?** With relatively low voltage and current, a few watts of power is reasonable.

PRACTICE PROBLEMS

Do additional problems. Online Practice

1. A car battery causes a current through a lamp and produces 12 V across it as shown in **Figure 4.** What is the power used by the lamp?

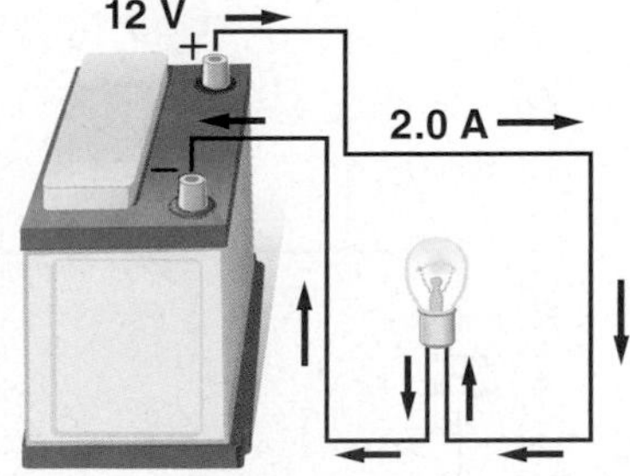

Figure 4

2. What is the current through a 75-W lightbulb that is connected to a 125-V outlet?
3. The current through a lightbulb connected across the terminals of a 125-V outlet is 0.50 A. At what rate does the bulb transform electrical energy to light? (Assume 100 percent efficiency.)
4. The current through the starter motor of a car is 210 A. If the battery maintains 12 V across the motor, how much electrical energy is delivered to the starter in 10.0 s?
5. A 75-V generator supplies 3.0 kW of power. How much current can the generator deliver?
6. A flashlight bulb is rated at 0.90 W. If the lightbulb produces a potential drop of 3.0 V, how much current goes through it?
7. **CHALLENGE** A circuit is changed so the potential difference across a motor doubles and the current through the lightbulb triples. How does this change the motor's power?

MiniLAB

CURRENT AFFAIRS

Does current diminish as it passes through a circuit?

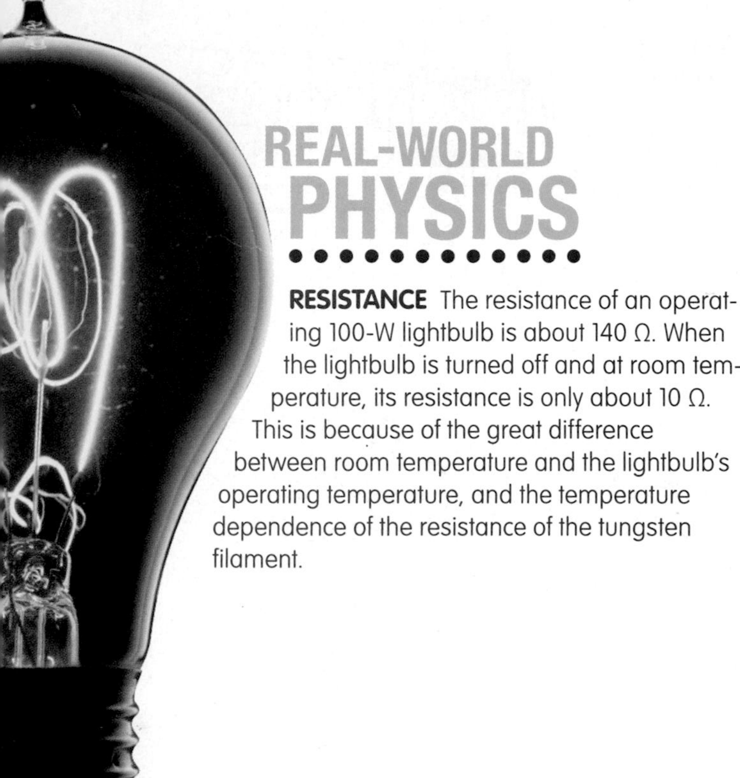

REAL-WORLD PHYSICS

RESISTANCE The resistance of an operating 100-W lightbulb is about 140 Ω. When the lightbulb is turned off and at room temperature, its resistance is only about 10 Ω. This is because of the great difference between room temperature and the lightbulb's operating temperature, and the temperature dependence of the resistance of the tungsten filament.

Diagramming Circuits

A simple circuit can be described in words. It also can be depicted by photographs or artists' drawings of the parts. You might have noticed, however, that circuits are often diagrammed with simple symbols. Such a diagram is called a circuit schematic.

An artist's drawing and a schematic of the same circuit are shown in **Figure 5** at the bottom of this page. Some of the symbols used in circuit schematics are shown in **Figure 6** on the next page. Notice in both the drawing and the schematic that the electric charge is shown flowing out of the positive terminal of the battery. To draw schematic diagrams, use the problem-solving strategy on the next page.

☑ **READING CHECK Explain** how a schematic of an electric circuit compares to an artist's drawing of the same circuit.

An ammeter measures current, and a voltmeter measures potential differences. Often these devices are combined in a single device called a multimeter. Each instrument has two terminals, usually labeled + and −. A voltmeter measures the potential difference between any two points in a circuit. When connecting the voltmeter in a circuit, always connect the + terminal to the end of the circuit component that is closer to the positive terminal of the battery, and connect the − terminal to the other side of the component. The ammeter should be connected so that the current goes *through* the ammeter.

☑ **READING CHECK Identify** what an ammeter and a voltmeter measure.

Figure 5 Simple circuit diagrams can be represented pictorially and schematically.

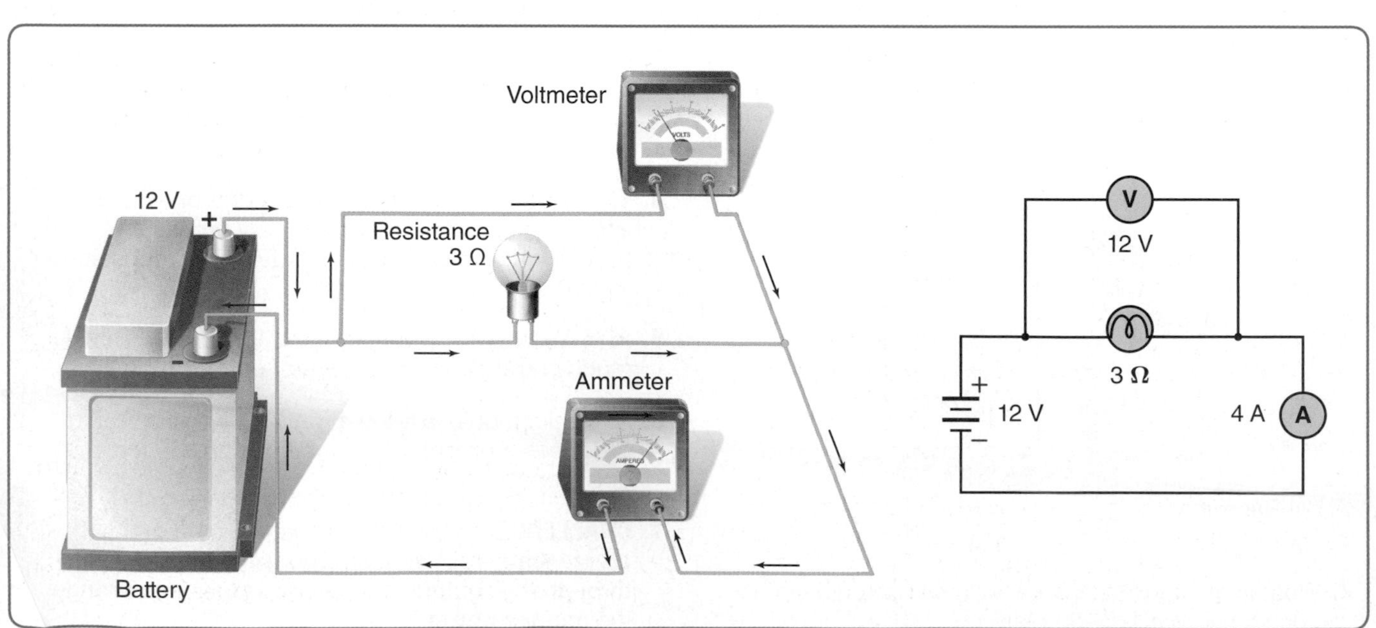

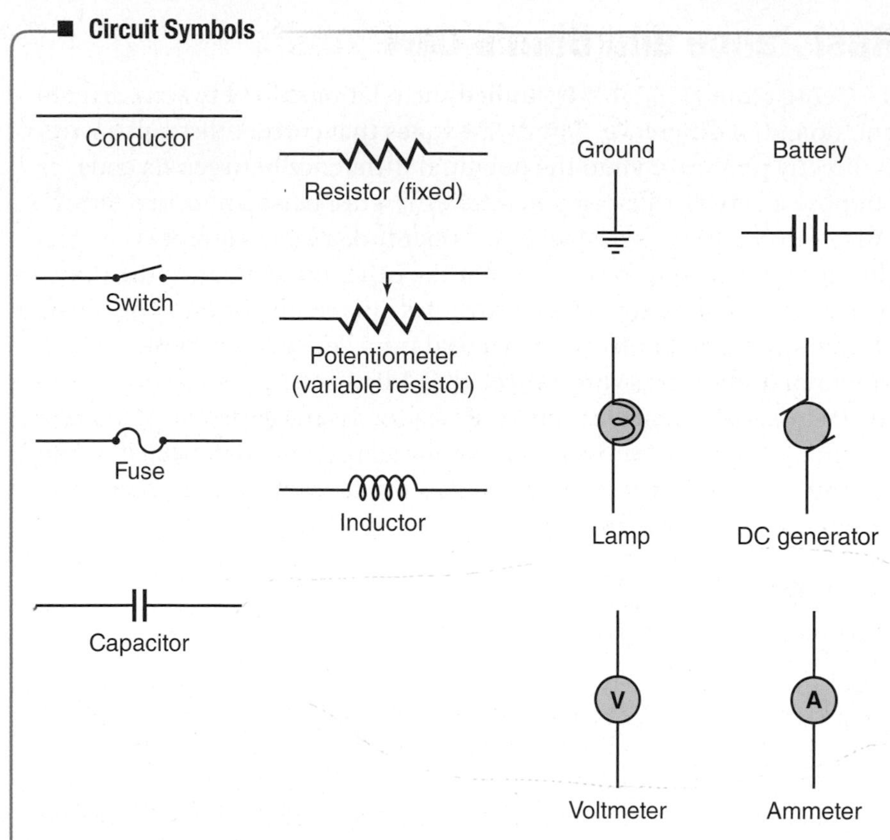

Figure 6 Electric circuit diagrams are commonly drawn using these symbols.

PROBLEM-SOLVING STRATEGIES

DRAWING SCHEMATIC DIAGRAMS

Follow these steps when drawing schematic diagrams.

1. Draw the symbol for the battery or other source of electrical energy, such as a generator, on the left side of the page. Put the positive terminal on top.
2. Draw a wire coming out of the positive terminal. When you reach a resistor or other device, draw the symbol for it.
3. If you reach a point where there are two current paths, such as at a voltmeter, draw a ⊥ in the diagram. Follow one path until the two current paths join again. Then draw the second path.
4. Follow the current path until you reach the negative terminal of the battery.
5. Check your work to make sure that you have included all parts and that there are complete paths for the current to follow.

PRACTICE PROBLEMS

Do additional problems. Online Practice 

8. Draw a circuit diagram to include a 60.0-V battery, an ammeter, and a resistance of 12.5 Ω in series. Draw arrows on your diagram to indicate the direction of the current.

9. Draw a circuit diagram showing a 4.5-V battery, a resistor, and an ammeter that reads 85 mA. Show the direction of the current using conventional rules, and indicate the positive terminal of the battery.

10. Add a voltmeter to measure the potential difference across the resistors in the previous two problems. Label the voltmeters.

11. Draw a circuit using a battery, a lamp, a potentiometer to adjust the lamp's brightness, and an on-off switch.

12. CHALLENGE Repeat the previous problem, adding an ammeter and a voltmeter across the lamp.

Resistance and Ohm's Law

VOCABULARY
Science Use v. Common Use

Resistance
- **Science usage**
the measure of how much an object or a material impedes the flow of electric current induced by a voltage
The resistance of silver is lower than the resistance of copper.
- **Common usage**
the act of opposing
Customer resistance to the menu changes made the chefs reconsider the changes.

Georg Ohm (1787–1854) studied the relationship between current and potential difference. Ohm's law states that current through a wire is directly proportional to the potential difference between its ends. Suppose a battery causes a potential difference between its two terminals. If they are connected with a copper rod, a large current is created through the rod and the battery. On the other hand, connecting the battery terminals with a glass rod allows almost no current. The measure of how strongly an object or a material impedes current produced by a potential difference is **resistance. Table 1** lists some of the factors that impact resistance. Resistance is measured by placing a potential difference across a conductor and dividing the voltage by the current. The resistance (R) is defined as the ratio of electric potential difference (ΔV) to the current (I).

RESISTANCE

Resistance is defined as potential difference divided by current.

$$R \equiv \frac{\Delta V}{I}$$

Investigate **circuits and Ohm's law.**

The resistance of the conductor (R) is measured in ohms. One ohm (1 Ω) is the resistance permitting an electric charge of 1 A to flow when a potential difference of 1 V is applied across the resistance. A simple circuit relating resistance, current, and potential difference is shown in **Figure 7** at the top of the next page. The circuit is completed by a connection to an ammeter, which is a device that measures current.

READING CHECK Define resistance in your own words.

Table 1 Changing Resistance

Factor	How Resistance Changes	Example	
Length	Resistance increases as length increases.	L_1, L_2	$R_{L1} > R_{L2}$
Cross-sectional area	Resistance increases as the cross-sectional area decreases.	A_1, A_2	$R_{A1} > R_{A2}$
Temperature	Resistance usually increases as temperature increases.	T_1, T_2	$R_{T1} > R_{T2}$
Material	Keeping length, cross-sectional area, and temperature constant, resistance varies with the material used.	silver, copper, gold, aluminum, iron, platinum → R increases.	

Ohm's Law Example

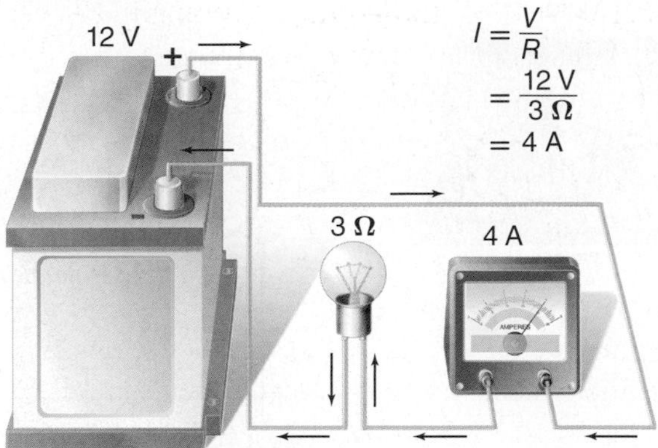

Figure 7 The ammeter and the equation give the same value for the current through the circuit. For a simple circuit with a 3-Ω resistance and a 12-V battery, there is a 4-A current throughout that circuit. The resistance of the ammeter is insignificant.

The ohm The unit for resistance is named for German scientist Georg Ohm, who found that the ratio of potential difference to current is constant for a given conductor. The resistance for most conductors does not vary as the magnitude or the direction of the potential difference applied to it changes. A device having constant resistance independent of the potential difference obeys Ohm's law.

READING CHECK Explain which devices obey Ohm's law.

Resistivity Most metallic conductors obey Ohm's law, at least over a limited range of voltages. Many important devices, however, such as an MP3 player or a pocket calculator, contain transistors and diodes, which do not obey Ohm's law.

Wires used to connect electrical devices have low resistance. A 1-m length of a typical wire used in physics labs has a resistance of about 0.03 Ω. Wires used in home wiring offer as little as 0.004 Ω of resistance for each meter of length. Because wires have so little resistance, there is almost no potential drop across them. To produce greater potential drops, a large resistance concentrated into a small volume is necessary. A **resistor** is a device designed to have a specific resistance. Resistors may be made of carbon, semiconductors, or wires that are long and thin.

Controlling current There are two ways to control the current in a circuit. Because $I = \frac{\Delta V}{R}$, I can be changed by varying ΔV, R, or both. **Figure 8** shows a simple circuit. When ΔV is 6 V and R is 30 Ω, the current is 0.2 A. According to Ohm's law, the greater the voltage placed across a resistor, the larger the current passing through it. If the potential difference across a resistor is halved, the current through that resistor is also halved. In the middle circuit in **Figure 8,** the voltage applied across the resistor is reduced from 6 V to 3 V to reduce the current to 0.1 A. A second way to reduce the current to 0.1 A is to replace the 30-Ω resistor with a 60-Ω resistor, as shown in the bottom circuit in **Figure 8.**

READING CHECK List two ways to control the current flowing through a simple circuit.

Initial Circuit

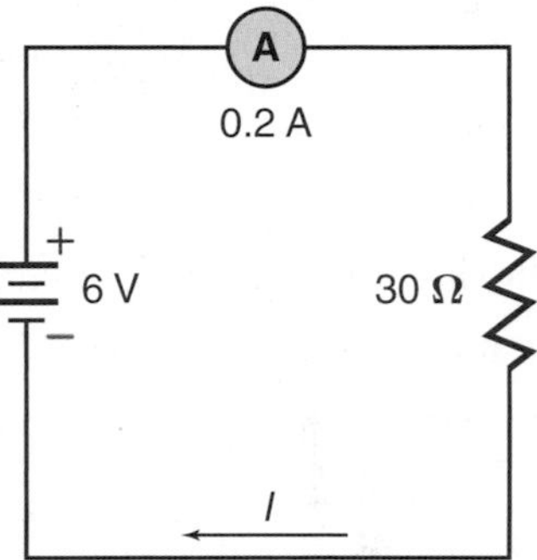

Reduce Voltage

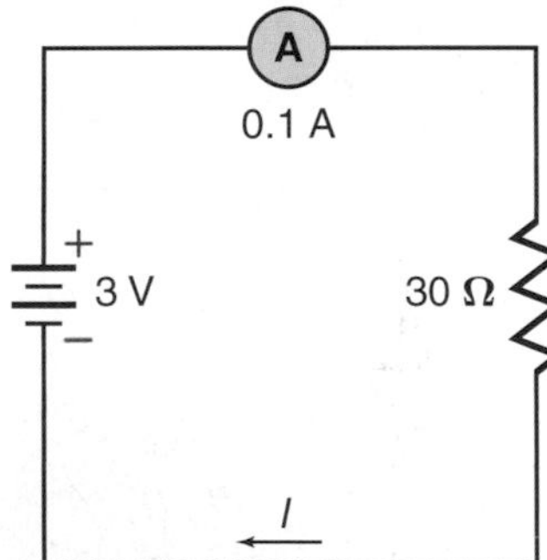

Increase Resistance

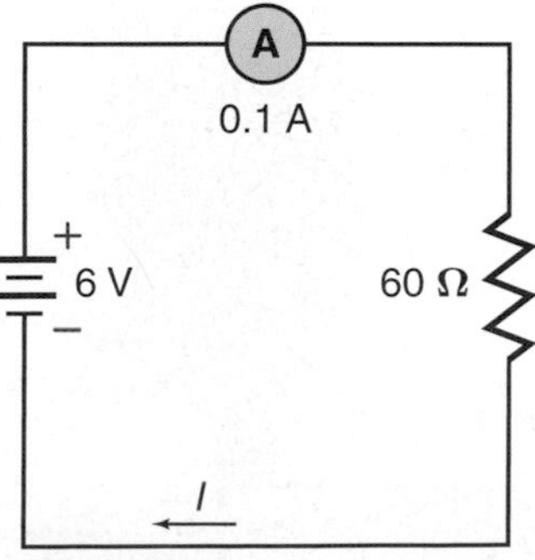

Figure 8 There are two ways to decrease the current through a simple circuit. The voltage can be reduced, or the resistance can be increased.

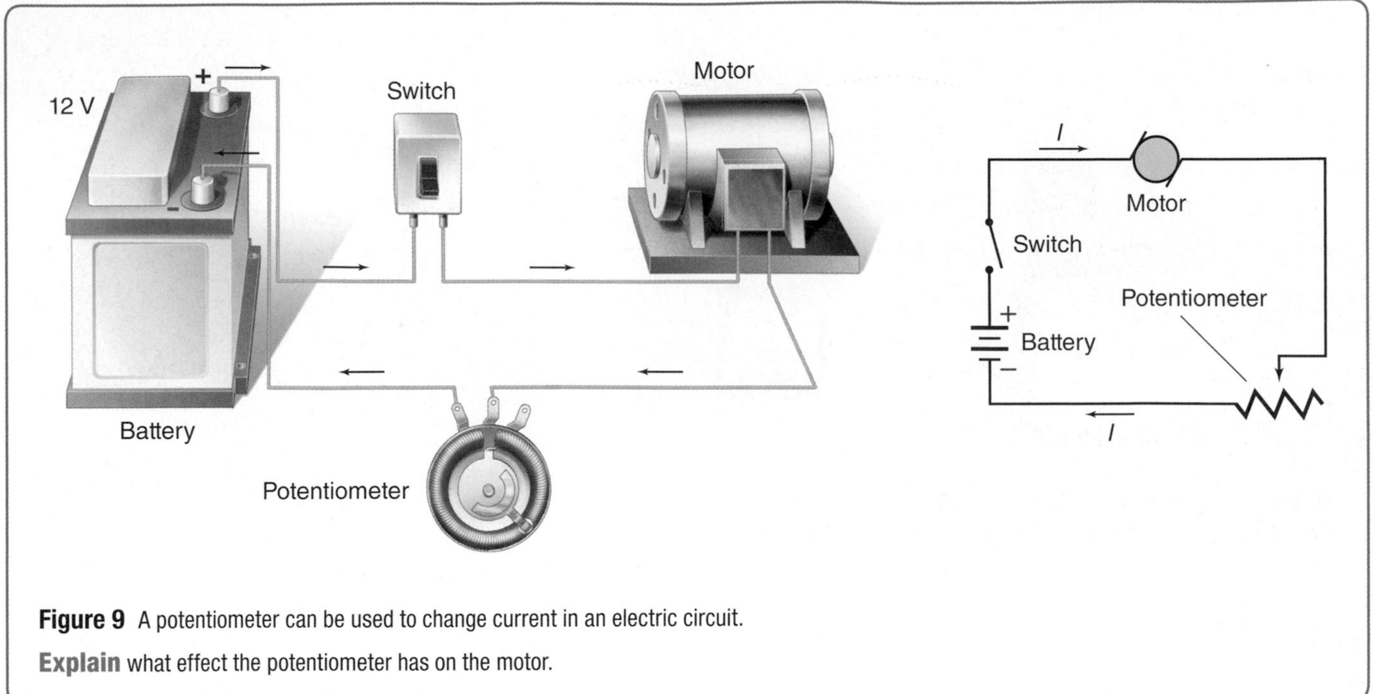

Figure 9 A potentiometer can be used to change current in an electric circuit.

Explain what effect the potentiometer has on the motor.

MiniLAB

MAKING ELECTRICAL ENERGY

Can a series of cells store electrical energy?

Variable resistors Resistors often are used to control the current in circuits or parts of circuits. Sometimes, a smooth, continuous variation of the current is desired. For example, the speed control on some electric motors allows continuous, rather than step-by-step, changes in the rotation of the motor. To achieve this kind of control, a variable resistor, called a potentiometer, is used. A circuit containing a potentiometer is shown in **Figure 9.** Some variable resistors consist of a coil of resistance wire and a sliding contact point. Moving the contact point to various positions along the coil varies the amount of wire in the circuit. As more wire is placed in the circuit, the resistance of the circuit increases; thus, the current changes in accordance with the equation $I = \frac{\Delta V}{R}$. In this way, the speed of a motor can be adjusted from fast, with little wire in the circuit, to slow, with a lot of wire in the circuit. Another example of a potentiometer is shown in **Figure 10.**

▶ **CONNECTION TO HEALTH** The human body acts as a variable resistor. Dry skin's resistance is high enough to keep currents that are produced by small and moderate voltages low. If skin becomes wet, however, its resistance is lowered, and an electric current can rise to dangerous levels. A current as low as 1 mA can be felt as a mild shock, while currents of 15 mA can cause loss of muscle control, and currents of 100 mA can cause death. For safety reasons you should be careful with any electric current, even from a lantern battery.

Figure 10 In a conventional analog joystick, varying the resistance in a potentiometer translates the joystick's position into an electric signal. This signal is used to translate the movement of your hand.

EXAMPLE PROBLEM 2

Find help with **operations with significant figures.** Math Handbook

CURRENT THROUGH A RESISTOR A 30.0-V battery is connected to a 10.0-Ω resistor. What is the current in the circuit?

1 ANALYZE AND SKETCH THE PROBLEM

- Draw a circuit containing a battery, an ammeter, and a resistor.
- Show the direction of the conventional current.

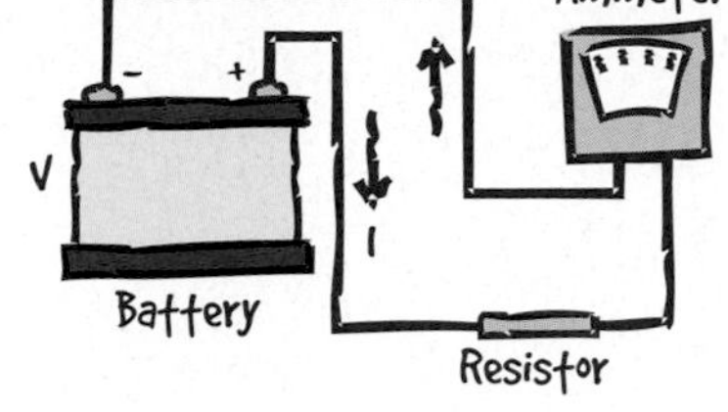

KNOWN	**UNKNOWN**
$\Delta V = 30.0$ V	$I = ?$
$R = 10.0\ \Omega$	

2 SOLVE FOR THE UNKNOWN

Use $I = \frac{\Delta V}{R}$ to determine the current.

$$I = \frac{\Delta V}{R}$$

$$= \frac{30.0\text{ V}}{10.0\ \Omega}$$ ◀ Substitute $\Delta V = 30.0$ V, $R = 10.0$.

$$= 3.00\text{ A}$$

3 EVALUATE THE ANSWER

- **Are the units correct?** Current is measured in amperes.
- **Is the magnitude realistic?** There is a fairly large potential difference and a small resistance, so a current of 3.00 A is reasonable.

EXAMPLE PROBLEM

PRACTICE PROBLEMS

Do additional problems. Online Practice

For all problems, assume the battery voltage and the lamp resistances are constant.

13. An automobile panel lamp with a resistance of 33 Ω is placed across the battery shown in **Figure 11.** What is the current through the circuit?

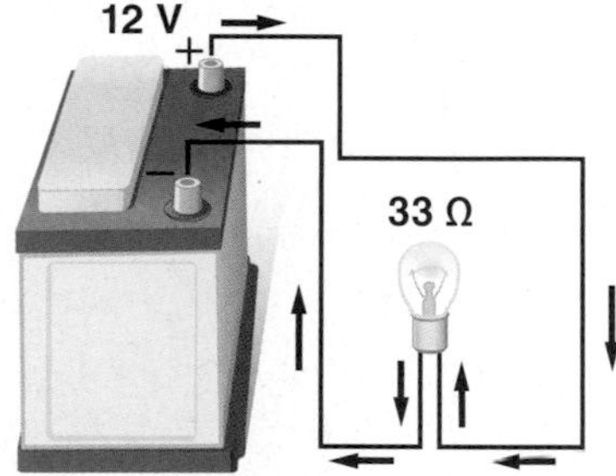

Figure 11

14. A sensor uses 2.0×10^{-4} A of current when it is operated by the battery shown in **Figure 12.** What is the resistance of the sensor circuit?

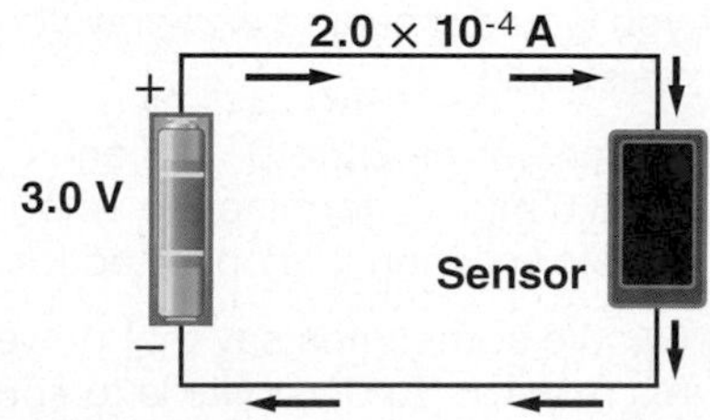

Figure 12

15. A motor with the operating resistance of 32 Ω is connected to a voltage source as shown in **Figure 13.** What is the voltage of the source?

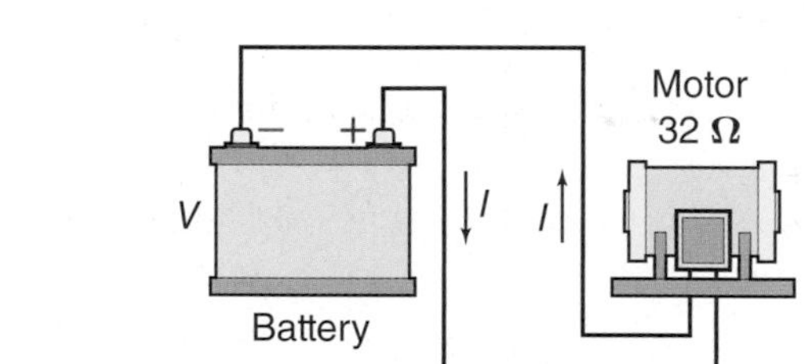

Figure 13

16. A lamp draws a current of 0.50 A when it is connected to a 120-V source.

a. What is the resistance of the lamp?

b. What is the power consumption of the lamp?

17. A 75-W lamp is connected to 125 V.

a. What is the current through the lamp?

b. What is the resistance of the lamp?

18. CHALLENGE A resistor is added to the lamp in the previous problem to reduce the current to half its original value.

a. What is the potential difference across the lamp?

b. How much resistance was added to the circuit?

c. At what rate does the lamp transform electrical energy into radiant and thermal energy?

PRACTICE PROBLEMS

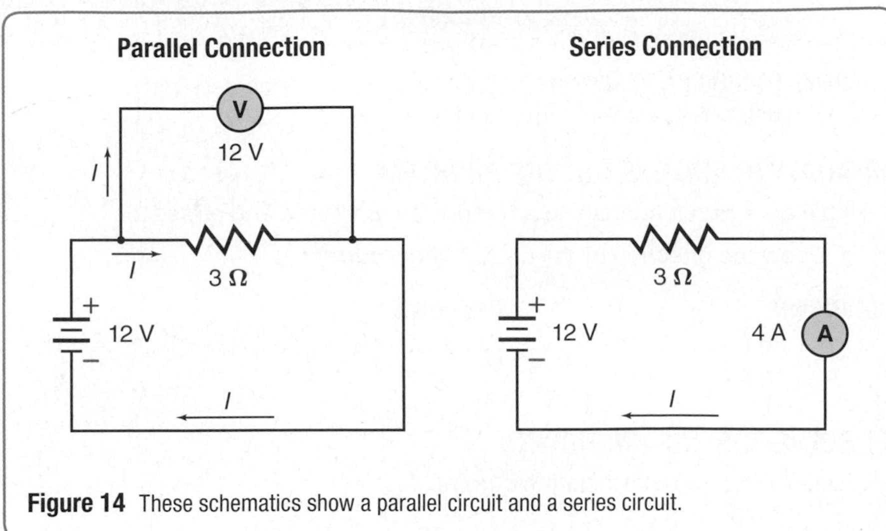

Figure 14 These schematics show a parallel circuit and a series circuit.

PhysicsLAB

VOLTAGE, CURRENT, AND RESISTANCE

What are the mathematical relationships among voltage, current, and resistance?

VARIABLES IN CIRCUITS

PROBEWARE LAB What is the relationship among current, resistance, and potential difference?

Parallel and Series Connections

When a voltmeter is connected across another component, it is called a parallel connection because the circuit component and the voltmeter are aligned parallel to each other in the circuit, as diagrammed in **Figure 14.** Any time the current has two or more paths to follow in a circuit, the connection is a **parallel connection.** The potential difference across the voltmeter is equal to the potential difference across the circuit element. Always associate the words *voltage across* with a parallel connection.

An ammeter measures the current through a circuit component. The same current going through the component must go through the ammeter, so there can be only one current path. A connection with only one current path in a circuit is a **series connection,** which also is shown in **Figure 14.** To add an ammeter to a circuit, the wire connected to the circuit component must be removed and connected to the ammeter instead. Then, another wire is connected from the second terminal of the ammeter to the circuit component. In a series connection, there can be only a single path through the connection. Always associate the words *current through* with a series connection.

SECTION 1 REVIEW

19. MAINIDEA Explain how charged particles are related to electric current, electric circuits, and resistance.

20. Schematic Draw a schematic diagram of a circuit that contains a battery and a lightbulb. Make sure the lightbulb will light in this circuit.

21. Resistance Joe states that because $R = \frac{\Delta V}{I}$, if he increases the voltage, the resistance will increase. Is Joe correct? Explain.

22. Power A circuit has 12 Ω of resistance and is connected to a 12-V battery. Determine the change in power if the resistance decreases to 9.0 Ω.

23. Resistance You want to measure the resistance of a long piece of wire. Show how you would construct a circuit with a battery, a voltmeter, an ammeter, and the wire to be tested to make the measurement. Specify what you would measure and how you would compute the resistance.

24. Energy A circuit transforms 2.2×10^{3} J of energy when it is operated for 3.0 min. Determine the amount of energy it will transform when it is operated for 1 h.

25. Critical Thinking We sometimes say that power is "dissipated" in a resistor. To dissipate is to spread out or disperse. In what sense is something being dispersed when charge flows through a resistor?

SECTION 2 Using Electrical Energy

PHYSICS 4 YOU Think of all the devices in your home that use electrical energy. Identify what type of energy transformations each device must make to operate. For example, lightbulbs transform electrical energy into radiant energy.

MAIN IDEA
Electrical energy can be transformed to radiant energy, thermal energy, and mechanical energy.

Essential Questions

- How is electrical energy transformed into thermal energy?
- How are electrical energy and power related?
- How is electrical energy transmitted with as little thermal energy transformation as possible?

Review Vocabulary

thermal energy the sum of the kinetic and potential energies of the particles in a system

New Vocabulary

superconductor
kilowatt-hour

Electrical Energy, Resistance, and Power

Energy that is supplied to a circuit can be transformed in many useful ways. For example, a lamp changes electrical energy into radiant energy. Unfortunately, not all the energy delivered to a lamp ends up in a useful form. Lightbulbs, especially incandescent lightbulbs, become hot. Some of the electrical energy is transformed into thermal energy.

Heating a resistor Current moving through a resistor increases its thermal energy because flowing electrons bump into the atoms in the resistor. Thus these collisions increase the motion of the atoms and the temperature of the resistor. A space heater, a hot plate, and the heating element in a hair dryer all are designed to transform electrical energy into thermal energy. These and other household appliances, such as those pictured in **Figure 15,** act like resistors when they are in a circuit. When charge (q) moves through a resistor, its potential is reduced by an amount (ΔV). As you have learned, the energy decrease is represented by $q\Delta V$. In practical use, the rate at which energy is changed—the power $\left(P = \frac{E}{t}\right)$—is more important. Earlier, you learned that current is the rate at which charge flows $\left(I = \frac{q}{t}\right)$ and that power transformed in a resistor is represented by $P = I\Delta V$. For a resistor, $\Delta V = IR$. If you know I and R, you can substitute $\Delta V = IR$ into the equation for electrical power to obtain the equation below. Thus, the rate at which energy is transformed in a resistor is proportional both to the square of the current passing through it and to the resistance.

POWER
Power is equal to current squared times resistance.

$$P = I^2R$$

Figure 15 These appliances transform electrical energy into thermal energy.
Explain why the appliances heat up when they are turned on.

Power and thermal energy If you know ΔV and R, but not I, you can substitute $I = \frac{\Delta V}{R}$ into $P = I\Delta V$ to obtain the following equation.

POWER

Power is equal to the potential difference squared divided by the resistance.

$$P = \frac{(\Delta V)^2}{R}$$

Power is the rate at which energy is transformed from one form to another. Energy is changed from electrical to thermal energy, and the temperature of the resistor rises. The resistors in an immersion heater, for example, convert electrical energy to thermal energy fast enough to bring water to a boil in a few minutes.

If power continues to be transformed at a uniform rate, then after time t, the energy transformed to thermal energy will be $E = Pt$. Because $P = I^2R$ and $P = \frac{(\Delta V)^2}{R}$. the total energy to be converted to thermal energy can be written in the following ways.

$$E = Pt \qquad E = I^2Rt \qquad E = \left(\frac{(\Delta V)^2}{R}\right)t$$

EXAMPLE PROBLEM 3

Get help with **electric heat.** Personal Tutor

ELECTRIC HEAT A heater has a resistance of 10.0 Ω. It operates on 120.0 V.

a. What is the power of the heater?

b. What thermal energy is supplied by the heater in 10.0 s?

1 ANALYZE AND SKETCH THE PROBLEM

- Sketch the situation.
- Label the known circuit components, which are a 120.0-V potential difference source and a 10.0-Ω resistor.

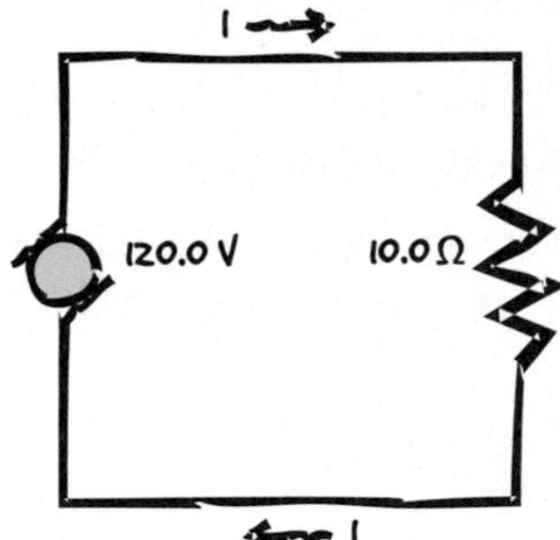

KNOWN	UNKNOWN
$R = 10.0\ \Omega$	$P = ?$
$\Delta V = 120.0\ V$	$E = ?$
$t = 10.0\ s$	

2 SOLVE FOR THE UNKNOWN

a. Because R and ΔV are known, use $P = \frac{(\Delta V)^2}{R}$.

$$P = \frac{(120.0\ V)^2}{10.0\ \Omega}$$ ◀ Substitute $\Delta V = 120.0\ V$, $R = 10.0\ \Omega$.

$= 1440\ W = 1.44\ kW$

b. Solve for the energy.

$E = Pt$

$= (1.44\ kW)(10.0\ s)$ ◀ Substitute $P = 1.44\ kW$, $t = 10.0\ s$.

$= 14.4\ kJ$

3 EVALUATE THE ANSWER

- **Are the units correct?** Power is measured in watts, and energy is measured in joules.
- **Are the magnitudes realistic?** For power, $10^2 \times 10^2 \times 10^{-1} = 10^3$, so kilowatts is reasonable. For energy, $10^3 \times 10^1 = 10^4$, so an order of magnitude of 10,000 joules is reasonable.

PRACTICE PROBLEMS

Do additional problems. Online Practice

26. A 15-Ω electric heater operates on a 120-V outlet.

a. What is the current through the heater?

b. How much energy is used by the heater in 30.0 s?

c. How much thermal energy is liberated in this time?

27. A 39-Ω resistor is connected across a 45-V battery.

a. What is the current in the circuit?

b. How much energy is used by the resistor in 5.0 min?

28. A 100.0-W lightbulb is 22 percent efficient. This means that 22 percent of the electrical energy is transformed to radiant energy.

a. How many joules does the lightbulb transform into radiant energy each minute it is in operation?

b. How many joules of thermal energy does the lightbulb output each minute?

29. The resistance of an electric stove element at operating temperature is 11 Ω.

a. If 220 V are applied across it, what is the current through the stove element?

b. How much energy does the element transform to thermal energy in 30.0 s?

c. The element is used to heat a kettle containing 1.20 kg of water. Assume 65 percent of the thermal energy is absorbed by the water. What is the water's increase in temperature during the 30.0 s?

30. CHALLENGE A 120-V water heater takes 2.2 h to heat a given volume of water to a certain temperature. How long would a 240-V unit operating with the same current take to accomplish the same task?

PRACTICE PROBLEMS

Superconductors A **superconductor** is a material with zero resistance. There is no restriction of current in superconductors, so there is no potential difference (ΔV) across a superconducting wire. Because the rate of energy transformation in a conductor is given by the product $I\Delta V$, a superconductor can conduct electricity without thermal energy transformations. At present, almost all superconductors must be kept at temperatures below 100 K. The practical uses of superconductors today include MRI magnets. Someday superconducting cables may efficiently carry electrical power to cities from distant power plants.

☑ **READING CHECK** **State** the amount of resistance in a superconductor.

PhysicsLAB

ENERGY CONSERVATION

Can you demonstrate the conservation of energy using electrical energy and thermal energy?

iLab Station

Providing Electrical Energy

Hydroelectric facilities, such as the dam on the Columbia River shown in **Figure 16,** are capable of producing a great deal of energy. This hydroelectric energy often is transmitted over long distances to reach homes and industries. How can the transmission occur with as little transformation to thermal energy as possible?

Electrical energy is transformed at a rate represented by $P = I^2R$. Electrical engineers call the resulting unwanted thermal energy the Joule heating loss, or I^2R loss. To reduce this energy transformation, either the current (I) or the resistance (R) must be reduced. All wires have some resistance, even though their resistance is small. The thick wire used to carry electric current into a home has a resistance of 0.20 Ω for a length of 1 km of wire. The total resistance increases with the length of wire.

Suppose that a farmhouse were connected directly to a power plant 3.5 km away. The resistance in the wires needed to carry a current in a circuit to the home and back to the plant is represented by the following equation: $R = 2(3.5\text{ km})(0.20\ \Omega/\text{km}) = 1.4\ \Omega$. An electric stove might cause a 41-A current through the wires. The rate of energy transformation (power) in the wires is represented by the following relationship.

$$P = I^2R = (41\text{ A})^2(1.4\ \Omega) = 2400\text{ W}$$

Figure 16 This hydroelectric dam is one of eleven dams on the Columbia River.

Russell Illig/Getty Images

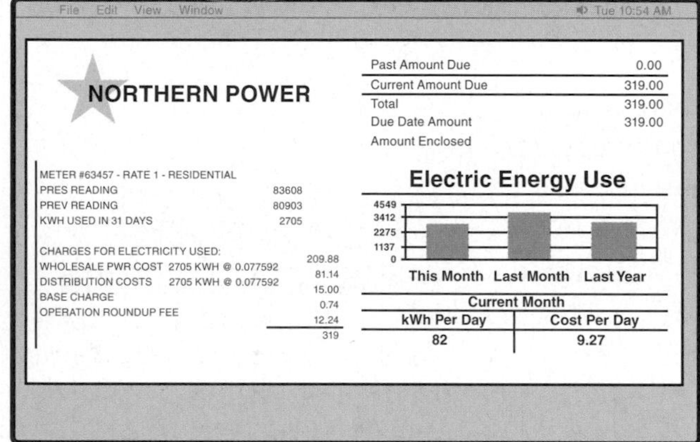

File Edit View Window — Tue 10:54 AM

NORTHERN POWER

Past Amount Due	0.00
Current Amount Due	319.00
Total	319.00
Due Date Amount	319.00
Amount Enclosed	

METER #63457 - RATE 1 - RESIDENTIAL

PRES READING	83608
PREV READING	80903
KWH USED IN 31 DAYS	2705

CHARGES FOR ELECTRICITY USED:	
WHOLESALE PWR COST 2705 KWH @ 0.077592	209.88
DISTRIBUTION COSTS 2705 KWH @ 0.077592	81.14
BASE CHARGE	15.00
OPERATION ROUNDUP FEE	0.74
	12.24
	319

Current Month	
kWh Per Day	Cost Per Day
82	9.27

Figure 17 Smart meters measure the amount of electrical energy used by a consumer. Electric bills often detail the amount of electrical energy used.

View a **simulation of electric circuits.**

Watch a **BrainPOP video about electric currents.**

Electric transmission lines All this electrical energy is transformed to thermal energy and, therefore, is wasted. This waste could be minimized by reducing the resistance. Cables of high conductivity and large diameter (and therefore low resistance) are available, but such cables are expensive and heavy. Because the thermal energy transformation is also proportional to the square of the current in the conductors, it is even more important to minimize the current in the transmission lines.

How can the current in the transmission lines be reduced? The electrical energy per second (power) transferred over a long-distance transmission line is determined by the relationship $P = IV$. The current is reduced without the power being reduced by an increase in the voltage. Some long-distance lines use voltages of more than 500,000 V. The resulting lower current reduces the I^2R loss in the lines by keeping the I^2 factor low. Long-distance transmission lines always operate at voltages much higher than household voltages in order to reduce I^2R loss. The output voltage from the generating plant is reduced upon arrival at electric substations to 2400 V, and again to 240 V or 120 V before being used in homes.

☑ **READING CHECK** **Explain** how current in transmission lines is altered to reduce thermal energy transformations.

The kilowatt-hour While electric companies are called power companies, they actually provide energy rather than power. Power is the rate at which energy is delivered. When consumers pay their home electric bills, an example of which is shown in **Figure 17,** they pay for electrical energy, not power.

The amount of electrical energy used by a device is its rate of energy consumption, in joules per second or watts (W), multiplied by the number of seconds the device is operated. Joules per second times seconds $\left(\frac{\text{J}}{\text{s}}\times\text{s}\right)$ equals the total amount of joules of energy. The joule, also defined as a watt-second, is a relatively small amount of energy, too small for commercial sales use. For this reason, electric companies measure energy sales in a unit of a large number of joules called a kilowatt-hour (kWh). A **kilowatt-hour** is equal to 1000 watts delivered continuously for 3600 s (1 h), or 3.6×10^6 J. Not many household devices other than hot-water heaters, stoves, clothes dryers, microwave ovens, heaters, and hair dryers require more than 1000 W of power.

☑ **READING CHECK** **Define** kilowatt-hour and describe the use of the term.

PRACTICE PROBLEMS

Do additional problems. Online Practice

PRACTICE PROBLEMS

31. An electric space heater draws 15.0 A from a 120-V source. It is operated, on the average, for 5.0 h each day.

a. How much power does the heater use?

b. How much energy in kWh does it consume in 30 days?

c. At $0.12 per kWh, how much does it cost to operate the heater for 30 days?

32. A digital clock has a resistance of 12,000 Ω and is plugged into a 115-V outlet.

a. How much current does it draw?

b. How much power does it use?

c. If the owner of the clock pays $0.12 per kWh, how much does it cost to operate the clock for 30 days?

33. An automotive battery can deliver 55 A at 12 V for 1.0 h and requires 1.3 times as much energy for recharge due to its less-than-perfect efficiency. How long will it take to charge the battery using a current of 7.5 A? Assume the charging voltage is the same as the discharging voltage.

34. CHALLENGE Rework the previous problem by assuming the battery requires the application of 14 V when it is recharging.

PHYSICS CHALLENGE

Use the diagram to the right to help you answer the questions below.

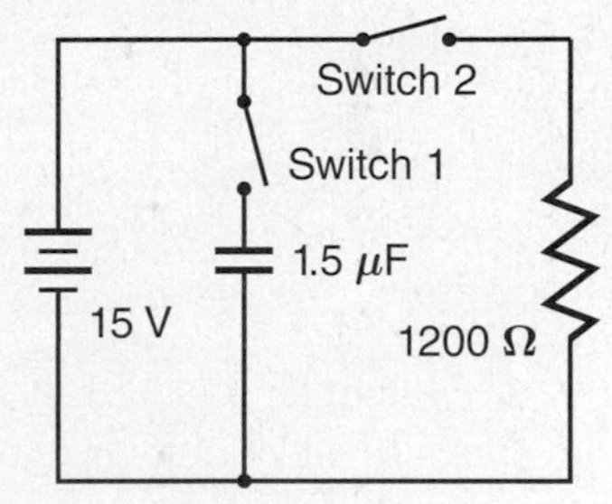

1. Initially, the capacitor is uncharged. Switch 1 is closed, and switch 2 remains open. What is the potential difference across the capacitor?
2. Switch 1 is now opened, and switch 2 remains open. What is the potential difference across the capacitor? Why?
3. Next, switch 2 is closed, while switch 1 remains open. What is the potential difference across the capacitor and the current through the resistor immediately after switch 2 is closed?
4. As time goes on, what happens to the voltage measured across the capacitor and the current through the resistor?

SECTION 2 REVIEW

Check your understanding.

35. MAINIDEA A car engine drives a generator, which transfers electrical energy to the car's battery. The headlights use the energy stored in the car battery to produce light. List the forms of energy in these three operations.

36. Resistance A hair dryer operating from 120 V has two settings, hot and warm. In which setting is the resistance likely to be smaller? Why?

37. Efficiency Evaluate the impact of research to improve power transmission lines on society and the environment.

38. Voltage Why would an electric range and an electric hot-water heater be connected to a 240-V circuit rather than a 120-V circuit?

39. Energy Cost A consumer uses 3098 kWh in 29 days. The utility company charges $0.077592 per kWh for the electricity plus $0.029998 per kWh for the distribution of the electricity. What is the consumer's electric bill for the 29 days?

40. Resistance and Power A toaster is connected to the circuit shown in **Figure 18.**

a. What is the resistance of the toaster?

b. At what rate does the toaster transform energy?

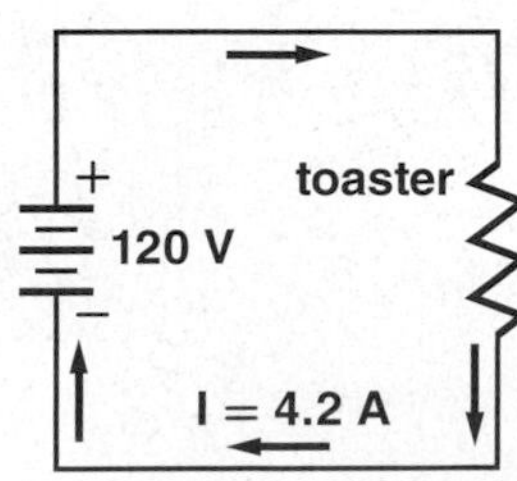

Figure 18

41. Critical Thinking When demand for electric power is high, power companies sometimes reduce the voltage, thereby producing a "brown-out." What is being saved?

A CLOSER **LOOK**

GET SMART

In the early twentieth century, transmission lines and power plants were built throughout the United States to provide homes and businesses with electrical energy. Initially, there were few uses for this technology. Today, the United States is so dependent on electrical energy that all industry and commerce stops when there is an electrical power outage.

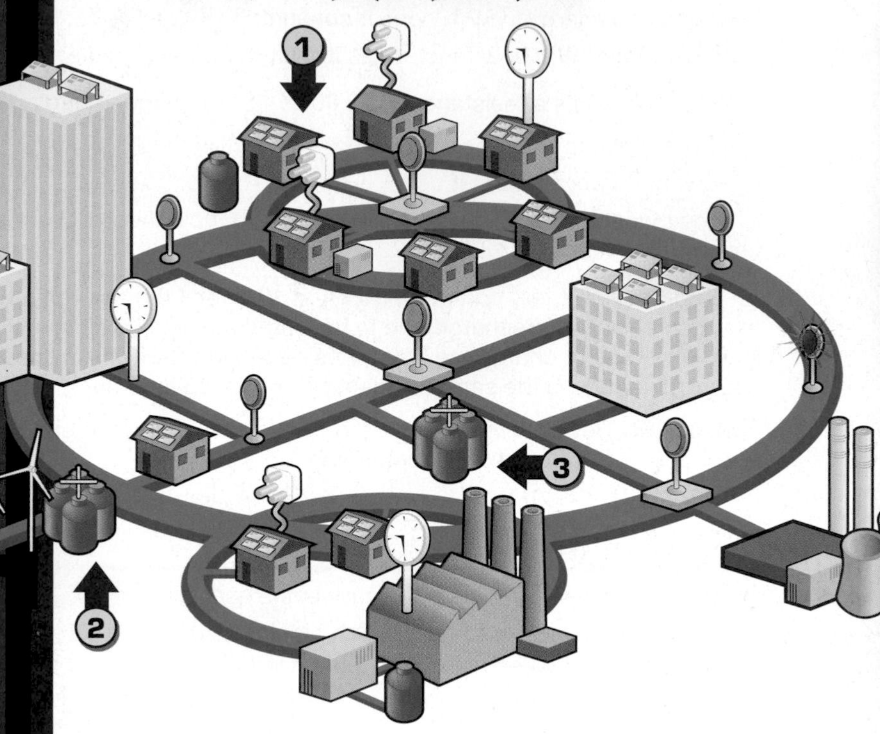

1. The interactive smart grid is efficient by automatically reporting power outages and regulating power as needed.
2. Electrical energy from renewable sources, such as wind and solar, can be easily added to the electrical grid.
3. The control of electrical power can be spread out and decentralized so that it is less vulnerable to natural and human-made disasters.

The electrical grid that was built over the last century needs to be updated so that it can meet the energy demands of the future. The updated grid is called the smart grid because it is more powerful and interactive than the old grid, just like a smart phone is more powerful and interactive than a basic cell phone. The plan is for the smart grid to have the following features:

- **Efficiency** The smart grid must be efficient and it must provide tools for consumers so that they can reduce their overall energy consumption.
- **Reliability** The smart grid must be reliable and power outages must be detected and repaired quickly.
- **Affordability** Improvements to the electric grid keeps electrical energy affordable.
- **Security** The smart grid must be resistant to human-made and natural disasters.
- **Green Friendly** Electrical energy from clean sources, such as solar, wind, and geothermal, must be easily added to the grid for use by all consumers.

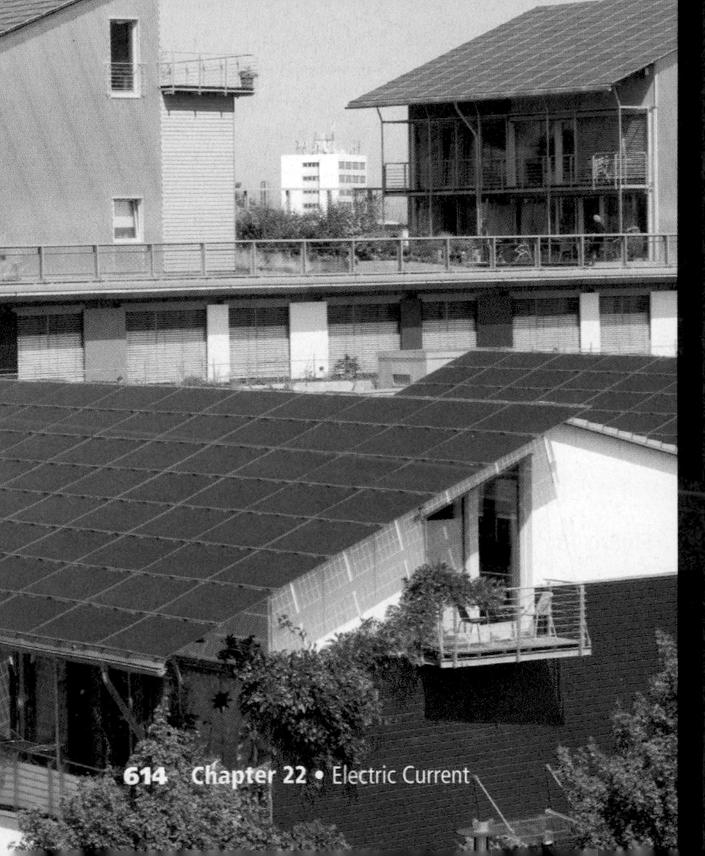

GOING**FURTHER** >>>

Research Talk to a local representative about how improvements to the grid can benefit consumers and energy companies.

STUDY GUIDE

BIGIDEA Electric currents carry electrical energy that can be transformed into other forms of energy.

VOCABULARY

- **electric current** *(p. 598)*
- **conventional current** *(p. 598)*
- **battery** *(p. 598)*
- **electric circuit** *(p. 599)*
- **ampere** *(p. 600)*
- **resistance** *(p. 604)*
- **resistor** *(p. 605)*
- **parallel connection** *(p. 608)*
- **series connection** *(p. 608)*

SECTION 1 Current and Circuits

MAINIDEA Electric current is the flow of electric charges.

- Electric current is a flow of charged particles. By convention, current direction is the direction in which a positive test charge moves.
- A circuit transforms electrical energy to thermal energy, radiant energy, or some other form of energy.
- Ohm's law states that the ratio of potential difference to current is a constant for a given conductor. Any resistance that does not change with potential difference or the direction of charge flow obeys Ohm's law.
- The following equations show how power, current, potential difference, and resistance are mathematically related.

$$P = I\Delta V \qquad R = \frac{\Delta V}{I}$$

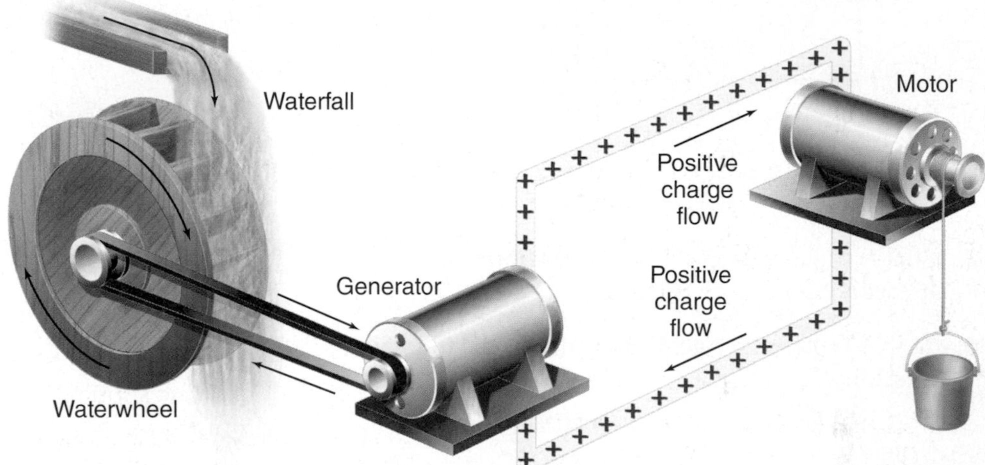

VOCABULARY

- **superconductor** *(p. 611)*
- **kilowatt-hour** *(p. 612)*

SECTION 2 Using Electrical Energy

MAINIDEA Electrical energy can be transformed to radiant energy, thermal energy, and mechanical energy.

- Electrical energy is transformed into thermal energy whenever moving charges transfer energy to other particles.
- If energy is transformed at a uniform rate, the total energy transformed equals power multiplied by time. Power also can be represented by I^2R and $\frac{(\Delta V)^2}{R}$ to give the last two equations.

$$E = Pt$$

$$E = I^2Rt$$

$$E = \left(\frac{(\Delta V)^2}{R}\right)t$$

- The unwanted transformation of electrical energy to thermal energy during transmission is called the joule heating loss, or I^2R loss. The best way to minimize the Joule heating loss is to minimize the current in the transmission wires. Transmitting at higher voltages enables current to be reduced without power being reduced.

Review **vocabulary with translations.**

Games and Multilingual eGlossary

CHAPTER 22 ASSESSMENT

Chapter Self-Check

SECTION 1 Current and Circuits

Mastering Concepts

42. Define the unit of electric current in terms of units of charge and time.

43. How should a voltmeter be connected in **Figure 19** to measure the motor's voltage?

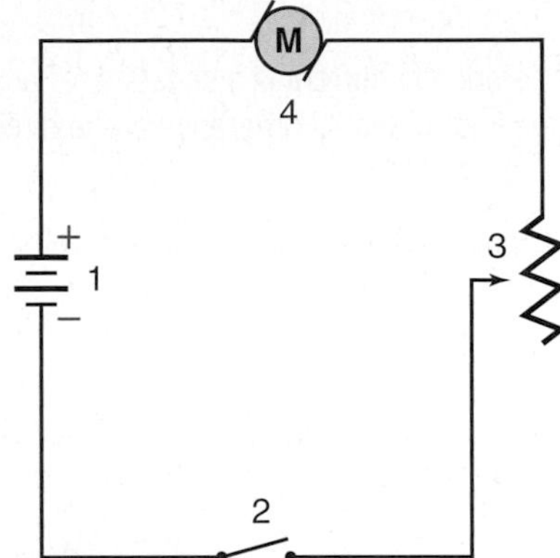

Figure 19

44. How should an ammeter be connected in **Figure 19** to measure the motor's current?

45. What is the direction of the conventional current through the motor in **Figure 19?**

46. BIGIDEA Refer to **Figure 19** to answer the following questions.

a. Which device transforms electrical energy to mechanical energy?

b. Which device transforms chemical energy to electrical energy?

c. Which device turns the circuit on and off?

d. Which device provides a way to adjust speed?

47. Which wire conducts electricity with the least resistance: one with a large cross-sectional diameter or one with a small cross-sectional diameter? Explain.

48. A simple circuit consists of a resistor, a battery, and connecting wires.

a. Draw a circuit schematic of this simple circuit.

b. How must an ammeter be connected in a circuit for the current to be correctly read?

c. How must a voltmeter be connected to a resistor for the potential difference across it to be read?

49. Describe how a potentiometer controls the following devices:

a. electric motors

b. game joystick

Mastering Problems

50. Describe the energy transformations that occur in each of the following devices.

a. an incandescent lightbulb

b. a clothes dryer

c. a digital clock radio

d. a handheld flashlight

51. A motor is connected to a 12-V battery, as shown in **Figure 20.**

a. How much power is delivered to the motor?

b. How much energy is transformed if the motor runs for 15 min?

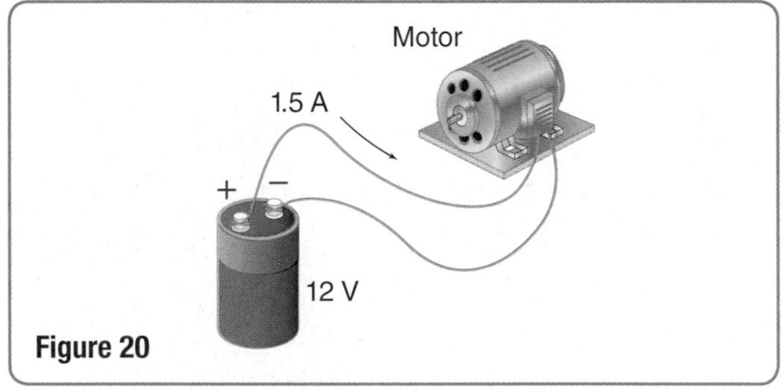

Figure 20

52. Refer to **Figure 21** to answer the following questions.

a. What should the ammeter reading be?

b. What should the voltmeter reading be?

c. How much power is delivered to the resistor?

d. How much energy is delivered to the resistor per hour?

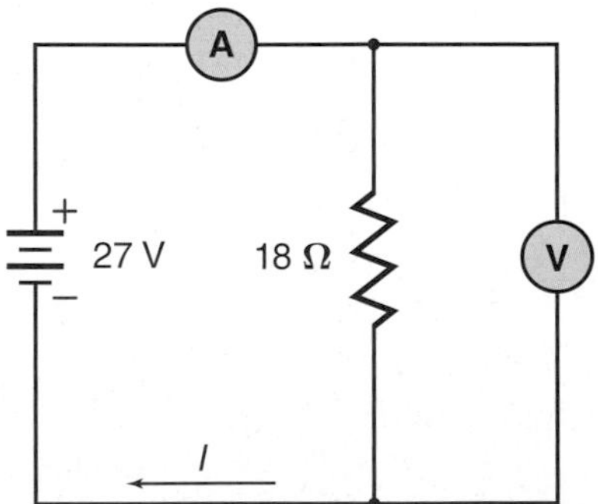

Figure 21

53. Toasters The current through a toaster that is connected to a 120-V source is 8.0 A. What power is dissipated by the toaster?

54. Lightbulbs A current of 1.2 A is measured through a lightbulb when it is connected across a 120-V source. At what rate does the bulb transform energy?

Chapter Self-Check

55. Refer to **Figure 22** to answer the following questions.

a. What should the ammeter reading be?

b. What should the voltmeter reading be?

c. How much power is delivered to the resistor?

d. How much energy is delivered to the resistor per hour?

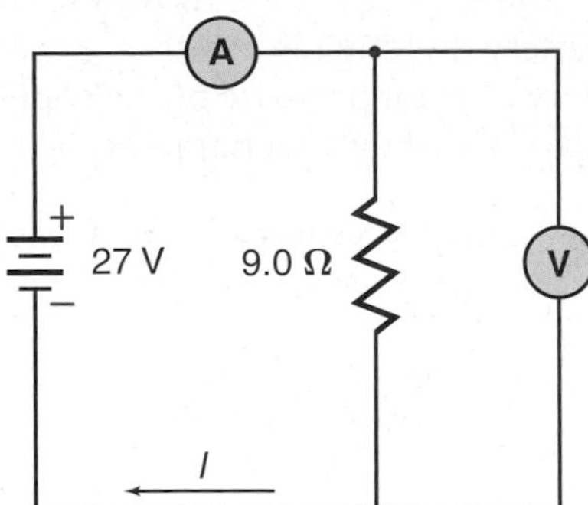

Figure 22

56. Refer to **Figure 23** to answer the following questions.

a. What should the ammeter reading be?

b. What should the voltmeter reading be?

c. How much power is delivered to the resistor?

d. How much energy is delivered to the resistor per hour?

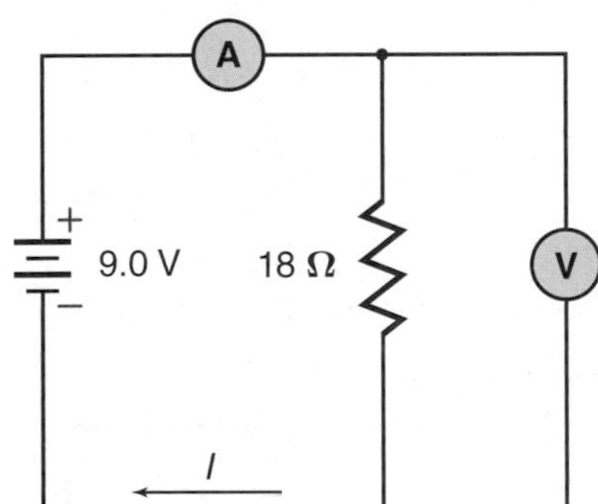

Figure 23

57. A lamp draws 0.50 A from a 120-V generator.

a. How much power is delivered?

b. How much energy is transformed in 5.0 min?

58. A 12-V automobile battery is connected to an electric starter motor. The current through the motor is 210 A.

a. How many joules of energy does the battery deliver to the motor each second?

b. What power, in watts, does the motor use?

59. **Dryers** A 4200-W clothes dryer is connected to a 220-V circuit. How much current does the dryer draw?

60. **Flashlights** A flashlight bulb is connected across a 3.0-V potential difference. The current through the bulb is 1.5 A.

a. What is the power rating of the bulb?

b. How much electrical energy does the bulb transform in 11 min?

61. **Batteries** A resistor of 60.0 V has a current of 0.40 A through it when it is connected to the terminals of a battery. What is the voltage of the battery?

62. What voltage is applied to a 4.0-Ω resistor if the current is 1.5 A?

63. What voltage is placed across a motor with a 15-Ω operating resistance if there is 8.0 A of current?

64. A voltage of 75 V is placed across a 150-Ω resistor. What is the current through the resistor?

65. Some students connected a length of nichrome wire to a variable power supply to produce between 0.00 V and 10.00 V across the wire. They then measured the current through the wire for several voltages. The students recorded the data for the voltages used and the currents measured, as shown in **Table 2.**

a. For each measurement, calculate the resistance.

b. Graph I versus V.

c. Does the nichrome wire obey Ohm's law? If not, for all the potential differences, specify the voltage range for which Ohm's law holds.

Table 2

Potential Difference, ΔV (volts)	Current, I (amps)	Resistance, $R = \frac{\Delta V}{I}$ (Ω)
2.00	0.0140	
4.00	0.0270	
6.00	0.0400	
8.00	0.0520	
10.00	0.0630	
−2.00	−0.0140	
−4.00	−0.0280	
−6.00	−0.0390	
−8.00	−0.0510	
−10.00	−0.0620	

Chapter Self-Check

66. Draw a series circuit diagram to include a 16-Ω resistor, a battery, and an ammeter that reads 1.75 A. Indicate the positive terminal and the voltage of the battery, the positive terminal of the ammeter, and the direction of conventional current.

67. A lamp draws a 66-mA current when connected to a 6.0-V battery. When a 9.0-V battery is used, the lamp draws 75 mA.

a. Does the lamp obey Ohm's law?

b. How much power does the lamp use when it is connected to the 6.0-V battery?

68. Lightbulbs How much energy does a 60.0-W lightbulb use in half an hour? If the lightbulb transforms 12 percent of electrical energy to radiant energy, how much electrical energy is transformed to thermal energy during the half hour?

69. The current through a lamp connected across 120 V is 0.40 A when the lamp is on.

a. What is the lamp's resistance when it is on?

b. When the lamp is cold, its resistance is $\frac{1}{5}$ as great as it is when the lamp is hot. What is the lamp's cold resistance?

c. What is the current through the lamp as it is turned on if it is connected to a potential difference of 120 V?

70. The graph in **Figure 24** shows the current through a device called a silicon diode.

a. A potential difference of +0.70 V is placed across the diode. What is the resistance of the diode?

b. What is the diode's resistance when a +0.60-V potential difference is used?

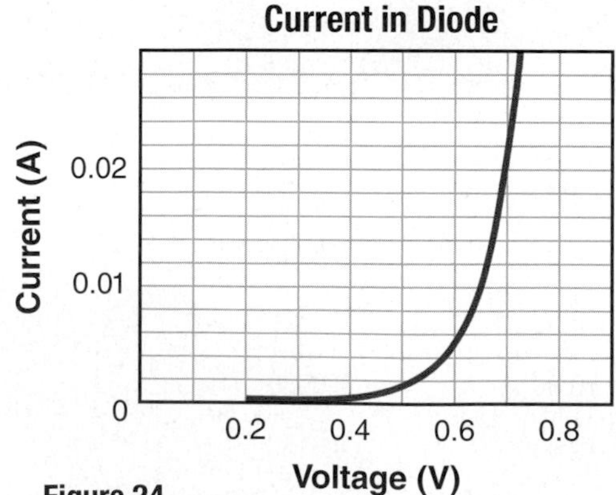

Figure 24

71. Draw a schematic diagram to show a circuit including a 90-V battery, an ammeter, and a resistance of 45 Ω connected in series. What is the ammeter reading? Draw arrows showing the direction of conventional current.

SECTION 2 Using Electrical Energy

Mastering Concepts

72. Why do incandescent lightbulbs burn out more frequently just as they are switched on rather than while they are operating?

73. If a heavy copper wire is used to connect one terminal of a battery directly to the other terminal of that same battery, the temperature of the copper wire rises rapidly. Why does this happen?

74. What electric quantities must be kept small to transmit electrical energy economically over long distances?

75. Define the unit of power in terms of fundamental SI units.

Mastering Problems

76. Batteries A 9.0-V battery costs $3.00 and will deliver 0.0250 A for 26.0 h before it must be replaced. Calculate the cost per kWh.

77. What is the maximum current allowed in a 5.0-W, 220-Ω resistor?

78. A 110-V electric iron draws 3.0 A of current. How much thermal energy does it output in an hour?

79. For the circuit shown in **Figure 25,** the maximum safe power is 5.0×10^1 W. Use the figure to find the following:

a. the maximum safe current

b. the maximum safe voltage

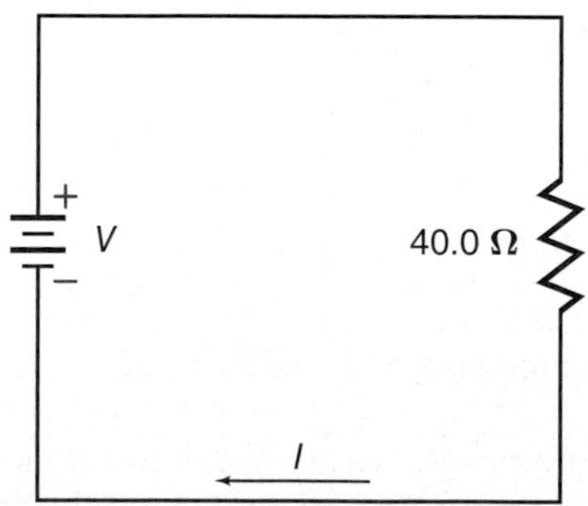

Figure 25

80. Appliances A window air conditioner is estimated to have a cost of operation of $50 per 30 days. This is based on the assumption that the air conditioner will run half of the time and that electricity costs $0.090 per kWh. Determine how much current the air conditioner will take from a 120-V outlet.

81. Utilities **Figure 26** represents an electric furnace. Calculate the monthly (30-day) heating bill if electricity costs $0.10 per kWh and the thermostat is on one-fourth of the time.

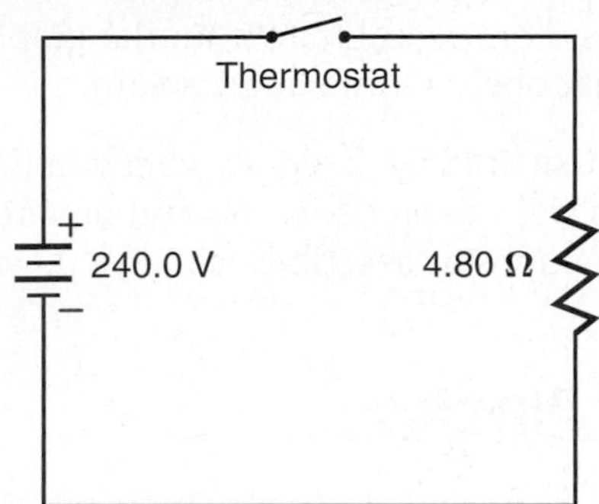

Figure 26

82. Radios A transistor radio operates by means of a 9.0-V battery that supplies it with a 50.0-mA current.

a. If the cost of the battery is $2.49 and it lasts for 300.0 h, what is the cost per kWh to operate the radio in this manner?

b. The same radio, by means of a converter, is plugged into a household circuit by a homeowner who pays $0.12 per kWh. What does it now cost to operate the radio for 300.0 h?

Applying Concepts

83. Batteries When a battery is connected to a complete circuit, charges flow in the circuit almost instantaneously. Explain.

84. Explain why a cow experiences a mild shock when it touches an electric fence.

85. Power Lines Why can birds perch on high-voltage lines without being injured?

86. Describe two ways to increase the current in a circuit.

87. Lightbulbs Two lightbulbs work on a 120-V circuit. One is 50 W and the other is 100 W. Which bulb has a higher resistance? Explain.

88. If the voltage across a circuit is kept constant and the resistance is doubled, what effect does this have on the circuit's current?

89. What is the effect on the current in a circuit if both the voltage and the resistance are doubled? Explain.

90. If the ammeter in the simple circuit in **Figure 8** were moved to the bottom of the diagram, would the ammeter have the same reading? Explain.

91. Ohm's Law Sue finds a device that looks like a resistor. When she connects it to a 1.5-V battery, she measures only 45×10^{-6} A, but when she uses a 3.0-V battery, she measures 25×10^{-3} A. Does the device obey Ohm's law?

92. Two wires can be placed across the terminals of a 6.0-V battery. One has a high resistance, and the other has a low resistance. Which wire will transform energy at a faster rate? Why?

Mixed Review

93. If a person has $5, how long could he or she play a 200-W stereo if electricity costs $0.15 per kWh?

94. A current of 1.2 A is measured through a 50.0-Ω resistor for 5.0 min. How much energy is transformed by the resistor?

95. A 6.0-Ω resistor is connected to a 15-V battery.

a. What is the current in the circuit?

b. How much energy is transformed in 10.0 min?

96. Lightbulbs An incandescent lightbulb with a resistance of 10.0 Ω when it is not lit and a resistance of 40.0 Ω when it is lit has 120 V placed across it.

a. What is the current draw when the bulb is lit?

b. What is the current draw at the instant the bulb is turned on?

c. When does the lightbulb use the most power?

97. A 12-V electric motor's speed is controlled by a potentiometer. At the motor's slowest setting, it uses 0.02 A. At its highest setting, the motor uses 1.2 A. What is the range of the potentiometer?

98. An electric motor operates a pump that irrigates a farmer's crop by pumping 1.0×10^{4} L of water a vertical distance of 8.0 m into a field each hour. The motor has an operating resistance of 22.0 Ω and is connected across a 110-V source.

a. What current does the motor draw?

b. How efficient is the motor?

99. Appliances An electric heater is rated at 500 W.

a. How much energy is delivered to the heater in half an hour?

b. The heater is being used to warm a room containing 50 kg of air. If the specific heat of air is 1.10 kJ/(kg•°C) and 50 percent of the thermal energy warms the air in the room, what is the change in air temperature in half an hour?

c. At $0.08 per kWh, how much does it cost to run the heater 6.0 h per day for 30 days?

Chapter Self-Check

Thinking Critically

100. Apply Concepts An artist's drawing of an electric circuit is shown in **Figure 27.** Draw a schematic of the electric circuit using the correct symbols. Indicate the direction of the conventional current in your drawing.

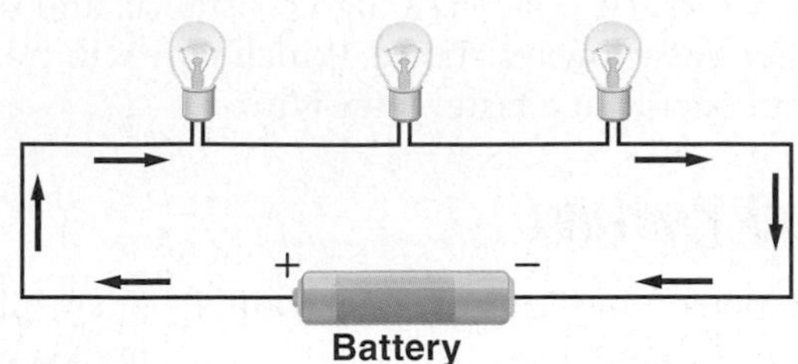

Figure 27

101. Formulate Models The energy needed to increase the potential difference of a charge (q) is represented by $E = q\Delta V$. But in a capacitor, $\Delta V = \frac{q}{C}$. Thus, as charge is added, the potential difference increases. As more charge is added, however, it takes more energy to add the additional charge. Consider a 1.0-F "supercap" used as an energy storage device in a personal computer. Plot a graph of ΔV as the capacitor is charged by adding 5.0 C to it. What is the voltage across the capacitor? The area under the curve is the energy stored in the capacitor. Find the energy in joules. Is it equal to the total charge times the final potential difference? Explain.

102. Ranking Task Rank the following resistors according to the current through each, from greatest to least. Specifically indicate any ties.

A. 10-MΩ resistor with a potential difference across it of 1.5 V

B. 10-MΩ resistor with a potential difference across it of 3.0 V

C. 15-MΩ resistor with a potential difference across it of 1.5 V

D. 15-MΩ resistor with a potential difference across it of 0.75 V

E. 20-kΩ resistor with a potential difference across it of 9.0 V

103. Reverse Problem Write a physics problem with real-life objects for which the following equation would be part of the solution:

$$I = \frac{60\text{ W}}{110\text{ V}}$$

104. Problem Posing Complete this problem so that it must be solved using Ohm's law: "An Ohmic resistor is connected to a 9-V battery . . ."

105. Apply Concepts The sizes of 10-Ω resistors range from a pinhead to a soup can. Why should they be different?

106. Make and Use Graphs The diode graph shown in **Figure 24** is more useful than a similar graph for a resistor that obeys Ohm's law. Explain.

107. Make and Use Graphs Based on what you have learned in this chapter, identify and prepare two parabolic graphs to describe a circuit's power.

Writing in Physics

108. There are three kinds of equations encountered in science: (1) definitions, (2) laws, and (3) derivations. Examples of these are: (1) an ampere is equal to one coulomb per second, (2) force is equal to mass times acceleration, (3) power is equal to voltage squared divided by resistance. Write a one-page explanation of where "resistance is equal to voltage divided by current" fits. Before you begin to write, first research the three categories given above.

109. Recall that you learned that matter expands when it is heated. Research the relationship between thermal expansion and high-voltage transmission lines.

Cumulative Review

110. A person burns energy at the rate of about 8.4×10^{6} J per day. How much does she increase the entropy of the universe in that day? How does this compare to the entropy increase caused by melting 20 kg of ice?

111. When you go up the elevator of a tall building, your ears might pop because of the rapid change in pressure. What is the pressure change caused by riding in an elevator up a 30-story building (150 m)? The density of air is about 1.3 kg/m^3 at sea level.

112. What is the wavelength in air of a 17-kHz sound wave, which is at the upper end of the frequency range of human hearing?

113. Light of wavelength 478 nm falls on a double slit. First-order bright bands appear 3.00 mm from the central bright band. The screen is 0.91 m from the slits. How far apart are the slits?

114. A charge of $+3.0\times10^{-6}$ C is 2.0 m from a second charge of $+6.0\times10^{-5}$ C. What is the magnitude of the force between them?

MULTIPLE CHOICE

1. A 100-W lightbulb is connected to a 120-V electric line. What is the current the lightbulb draws?
   A. 0.8 A **C.** 1.2 A
   B. 1 A **D.** 2 A

2. A 5.0-Ω resistor is connected to a 9.0-V battery. How much energy is transformed in 7.5 min?
   A. 1.2×10^2 J **C.** 3.0×10^3 J
   B. 1.3×10^3 J **D.** 7.3×10^3 J

3. The current in the flashlight shown below is 0.50 A, and the voltage is the sum of the voltages of the individual batteries. What is the power delivered to the bulb of the flashlight?
   A. 0.11 W **C.** 2.3 W
   B. 1.1 W **D.** 4.5 W

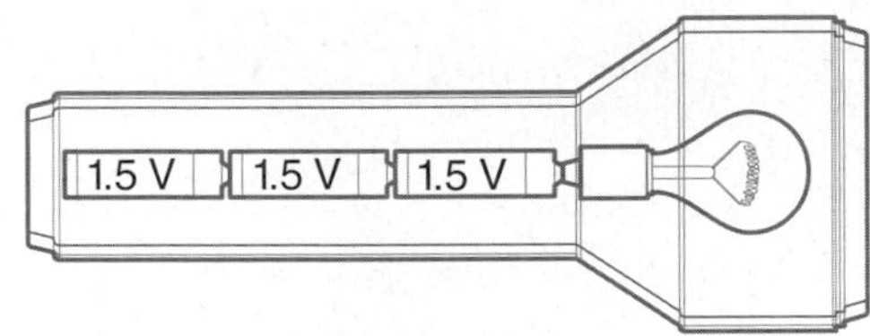

4. If a flashlight with a voltage of 4.5 V and a current of 0.50 A is on for 3.0 min, how much electrical energy is delivered to the bulb?
   A. 6.9 J **C.** 2.0×10^2 J
   B. 14 J **D.** 4.1×10^2 J

5. There is a current of 2.0 A through a circuit containing a motor with a resistance of 12 Ω. How much energy is transformed if the motor runs for one minute?
   A. 4.8×10^1 J **C.** 2.9×10^3 J
   B. 2.0×10^1 J **D.** 1.7×10^5 J

6. What is the effect on the current in a simple circuit if both the voltage and the resistance are reduced by half?
   A. divided by 2 **C.** multiplied by 2
   B. no change **D.** multiplied by 4

7. There is a 5.00-mA current through a circuit with a resistance of 50.0 Ω. What is the power in the circuit?
   A. 1.00×10^{-2} W **C.** 1.25×10^{-3} W
   B. 1.00×10^{-3} W **D.** 2.50×10^{-3} W

8. How much electrical energy is delivered to a 60.0-W lightbulb if the bulb is left on for 2.5 hours?
   A. 4.2×10^{-2} J **C.** 1.5×10^2 J
   B. 2.4×10^1 J **D.** 5.4×10^5 J

FREE RESPONSE

9. The diagram below shows a simple circuit containing a DC generator and a resistor. The table shows the resistances of several small electrical devices. If the resistor in the diagram represents a hair dryer, what is the current in the circuit? How much energy does the hair dryer use if it runs for 2.5 min?

Device	Resistance (Ω)
Hair dryer	8.5
Heater	10.0
Small motor	12.0

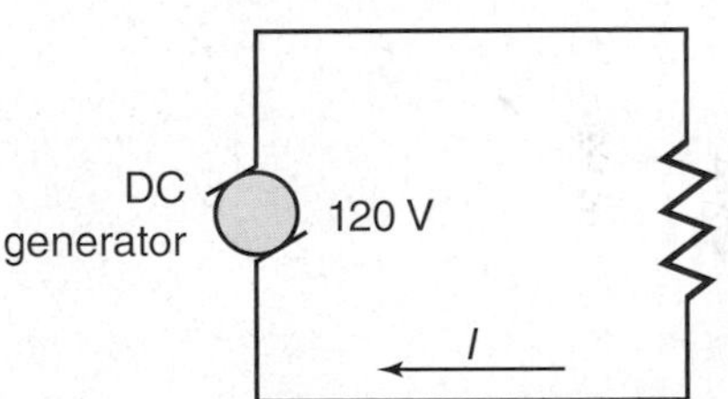

NEED EXTRA HELP?

If You Missed Question	1	2	3	4	5	6	7	8	9
Review Section	1	2	1	2	2	1	1	1	2

Online Test Practice

CHAPTER 23

Series and Parallel Circuits

BIGIDEA **Circuit components can be placed in series, in parallel, or in a combination of series and parallel.**

SECTIONS

1 Simple Circuits

2 Applications of Circuits

LaunchLAB

iLab Station

FUSES AND ELECTRIC CIRCUITS

How do fuses and circuit breakers work?

WATCH THIS!

Video

HOLIDAY LIGHTS

Why do some strings of lights stop working when a lightbulb burns out while others continue working? The answer is physics!

(l)Brand X Pictures/PunchStock, (r)Mira/Alamy

Go online!
connectED.mcgraw-hill.com
SPECIAL SALE

SECTION 1

Simple Circuits

PHYSICS 4 YOU

An electric circuit is a closed loop or pathway that allows electric charge to flow through it. Some electric circuits, such as the circuits inside computers, are extremely complex. Other electric circuits, such as the flashlight you might use when the power goes out, are much simpler.

MAINIDEA

In a series circuit, current follows a single path; in a parallel circuit, current follows more than one path.

Essential Questions

- What are the characteristics of series and parallel circuits?
- How are currents, potential differences, and equivalent resistances in series circuits related?
- How are currents, potential differences, and equivalent resistances in parallel circuits related?

Review Vocabulary

resistance the measure of how much an object or material impedes the current produced by a potential difference; is equal to voltage divided by current

New Vocabulary

series circuit
equivalent resistance
voltage divider
parallel circuit

A River Model

Although the connection might not immediately be clear, you can use a mountain river to model an electric circuit. From its source high in the mountains, the river flows downhill to the plains below. No matter which path the river takes, its change in elevation, from the mountaintop to the plain, is the same. Some rivers flow downhill in a single stream. Other rivers might split into two or more smaller streams as they flow over a waterfall or through a series of rapids. In this case, part of the river follows one path, while other parts of the river follow different paths. No matter how many paths the river takes, however, the total amount of water flowing down the mountain remains unchanged. In other words, the number of paths water takes does not affect the total amount of water flowing downhill.

How does the river shown in **Figure 1** model an electric circuit? The distance the river drops is similar to the potential difference in a circuit. The rate of water flow is similar to current in a circuit. Large rocks and other obstacles produce resistance to water flow and are similar to resistors in a circuit. Which part of the water cycle is similar to a battery or a generator in an electric circuit? The energy source needed to raise water to the top of the mountain is the Sun. Solar energy evaporates water from lakes and seas, leading to the formation of clouds that release rain or snow that falls on the mountaintops. Continue to think about the mountain river model as you read more about electric circuits.

Figure 1 The flow of a mountain stream is similar to electric current because regardless of the path, the amount of water and the amount of electric charges remain the same.

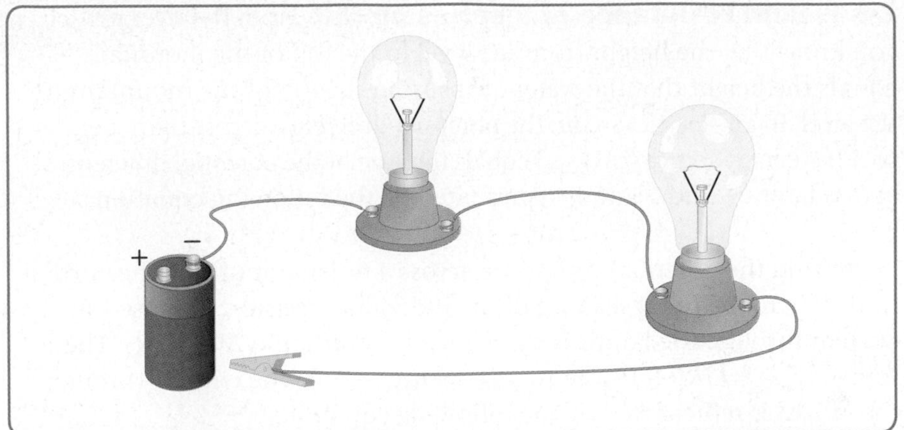

Figure 2 Students constructed this circuit and used it to verify their predictions of the brightness in each lightbulb when the final connection was made.

Watch a **video on electric circuits.**

Series Circuits

Suppose three students connect two identical lamps to a battery, as illustrated in **Figure 2.** Before they connect the battery, their teacher asks them to predict the brightnesses of the two lamps.

The first student predicts that only the lamp close to the positive (+) terminal of the battery will light because all the current will be used up as light energy. The second student predicts that only part of the current will be used up, and the second lamp will glow, but less brightly than the first. The third student predicts that the lamps will be of equal brightness because current is a flow of charge, and the charge leaving the first lamp has nowhere else to go in the circuit except through the second lamp. The third student reasons that because the current will be the same in each lamp, the brightness also will be the same.

Current in a series circuit If you consider the mountain river model for this circuit, you will realize that the third student is correct. Recall that charge cannot be created or destroyed. Because the charge in the circuit has only one path to follow and cannot be destroyed, the same amount of charge that enters the first lamp must also leave that lamp. The current is the same everywhere in the circuit. If you connect three ammeters in the circuit, as shown in **Figure 3,** they all will show the same current. A circuit such as this, in which there is only one path for the current, is called a **series circuit.**

☑ **READING CHECK** **Define** the term *series circuit.*

If current does not decrease when it passes through a lamp, where does the energy to light the lamps come from? The first two students above based their predictions on ideas about energy. The energy to light the lamps must come from somewhere. To these students, it made sense that current would decrease due to a conversion of electrical energy into light energy.

Recall that power, the rate at which electrical energy is converted, can be represented as $P = I\Delta V$. This means that conversions of electrical energy into other forms depend on both current and potential difference. Thus, because current does not change when the lamp converts electrical energy to light energy, there must be a potential difference across the lamp. The resistance of the lamp is defined as $R = \frac{\Delta V}{I}$. Thus, the potential difference across the lamp is $\Delta V = IR$.

Figure 3 Ammeters can be used in a series circuit to show that the current is the same throughout the circuit.

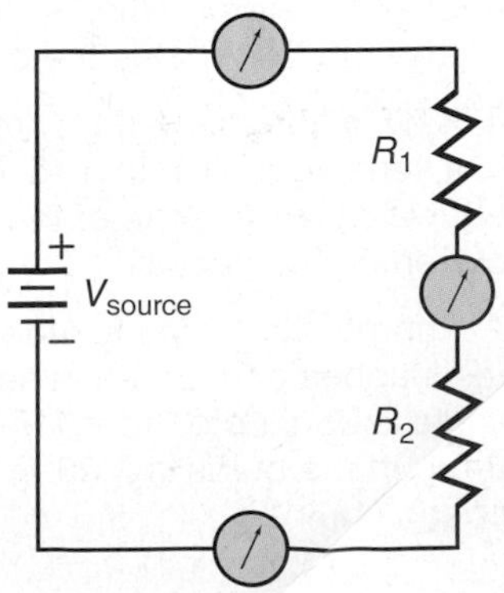

Equivalent resistance of a series circuit From the river model, you know that the height from sea level to the top of the mountain equals the height that the water drops from the top of the mountain to sea level. In an electric circuit, the potential difference across the generator or other energy source (ΔV_{source}) equals the sum of the potential differences across lamps 1 and 2 and is represented by the following equation:

$$\Delta V_{source} = \Delta V_1 + \Delta V_2.$$

To find the potential difference across a resistor, multiply the current in the circuit by the resistance of the individual resistor. Because the current through the lamps is the same, $\Delta V_1 = IR_1$ and $\Delta V_2 = IR_2$. Therefore, $\Delta V_{source} = IR_1 + IR_2$, or $\Delta V_{source} = I(R_1 + R_2)$. The current through the circuit is represented by the following equation:

$$I = \frac{\Delta V_{source}}{(R_1 + R_2)}.$$

The same idea can be extended to any number of resistances in series, not just two. If you replaced all the resistors in a circuit with a single resistor that resulted in the same current, that resistor's value would be the **equivalent resistance** of the circuit. For resistors in series, the same current would exist in the circuit with a single resistor (R) that has a resistance equal to the sum of the individual resistances.

PhysicsLAB

RESISTORS IN SERIES CIRCUITS

PROBEWARE LAB What happens to voltage across a resistor in a series circuit?

EQUIVALENT RESISTANCE FOR RESISTORS IN SERIES

The equivalent resistance of resistors in series equals the sum of the individual resistances of the resistors.

$$R = R_1 + R_2 + \ldots$$

Notice that the equivalent resistance is greater than that of any individual resistor. Therefore, if the battery voltage does not change, adding more devices in series always decreases the current. To find the current through a series circuit, first calculate the equivalent resistance and then use the following equation.

CURRENT

The current through a series circuit is equal to the potential difference across the power source divided by the equivalent resistance.

$$I = \frac{\Delta V_{source}}{R}$$

PRACTICE PROBLEMS

Do additional problems. Online Practice

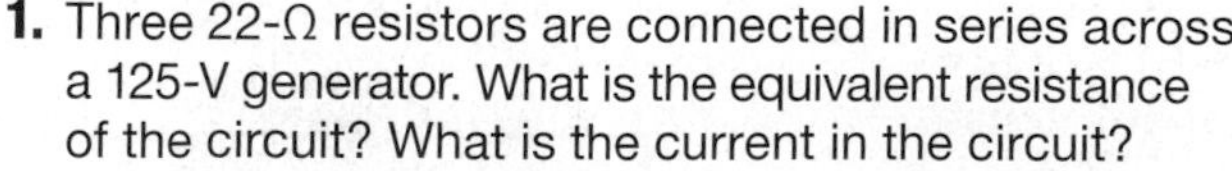

1. Three 22-Ω resistors are connected in series across a 125-V generator. What is the equivalent resistance of the circuit? What is the current in the circuit?
2. A 12-Ω, a 15-Ω, and a 5-Ω resistor are connected in a series circuit with a 75-V battery. What is the equivalent resistance of the circuit? What is the current in the circuit?
3. A string of lights has ten identical bulbs with equal resistances connected in series. When the string of lights is connected to a 117-V outlet, the current through the bulbs is 0.06 A. What is the resistance of each bulb?
4. A 9-V battery is in a circuit with three resistors connected in series.
   - **a.** If the resistance of one of the resistors increases, how will the equivalent resistance change?
   - **b.** What will happen to the current?
   - **c.** Will there be any change in the battery voltage?
5. **CHALLENGE** Calculate the potential differences across three resistors, 12-Ω, 15-Ω, and 5-Ω, that are connected in series with a 75-V battery. Verify that the sum of their potential differences equals the potential difference across the battery.

Voltage dividers Suppose you have a 9-V battery, but you need a potential difference of 5 V for an electric circuit. How can you produce a potential difference of 5 V if your battery produces a potential difference of 9 V? You could use a type of series circuit called a voltage divider. A **voltage divider** produces a source of potential difference that is less than the potential difference across the battery.

Consider the circuit shown in **Figure 4.** Two resistors (R_1 and R_2) are connected in series across a battery with potential difference V. The equivalent resistance of the circuit is

$$R = R_1 + R_2.$$

The current is represented by the following equation:

$$I = \frac{\Delta V}{R} = \frac{\Delta V}{R_1 + R_2}$$

The desired voltage (5 V) is the voltage drop (ΔV_2) across resistor R_2: $\Delta V_2 = IR_2$. Substitute the earlier equation into this equation as shown:

$$\Delta V_2 = IR_2 = \left(\frac{\Delta V}{R_1 + R_2}\right)R_2 = \frac{\Delta V R_2}{R_1 + R_2}$$

By choosing the right resistors, you can produce a potential difference of 5 V across a portion of an electric circuit even if you only have a 9-V battery available.

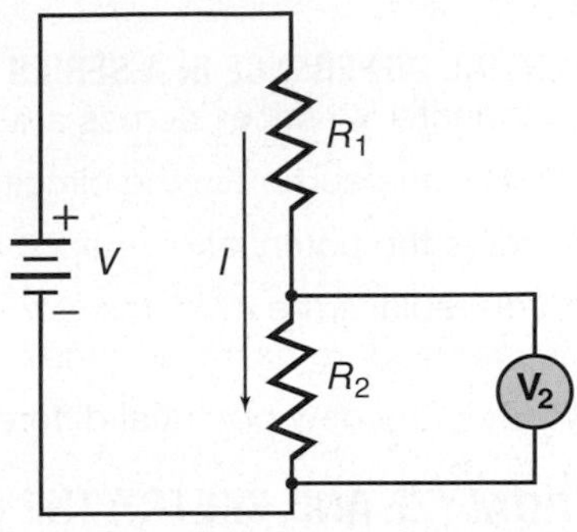

Figure 4 This voltage-divider circuit demonstrates how a voltage of desired magnitude can be achieved by choosing the right combination of resistors.

Explain why the current direction arrow in the diagram is pointing in that direction.

Photoresistors Voltage dividers often are used with sensors, such as photoresistors. A photoresistor's resistance depends on the amount of light that strikes it. A typical photoresistor has a resistance of 400 Ω when light is striking it compared with a resistance of 400,000 Ω when the photoresistor is in the dark. The voltage output of a voltage divider that uses a photoresistor depends upon the amount of light striking that photoresistor at the time.

This circuit can be used as a light meter, such as the one shown in **Figure 5.** In this device, an electronic circuit detects the potential difference and converts it to a measurement of illuminance that can be read on the digital display.

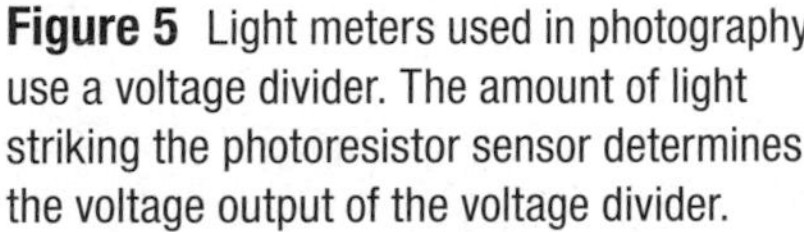

Figure 5 Light meters used in photography use a voltage divider. The amount of light striking the photoresistor sensor determines the voltage output of the voltage divider.

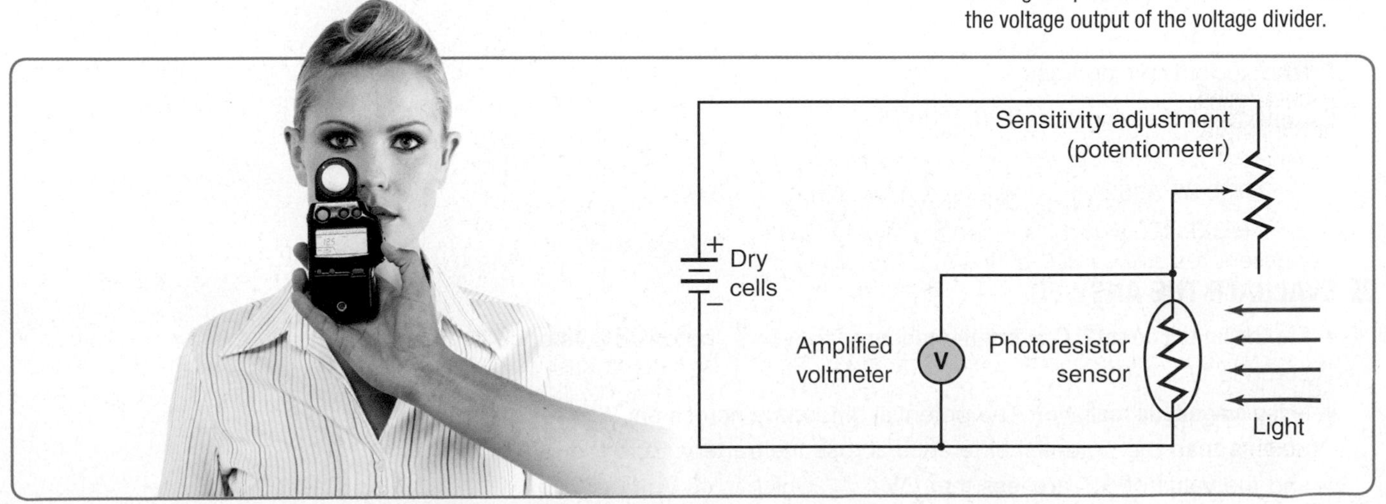

Compassionate Eye Foundation/Matt Hind/OJO Images Ltd/Photodisc/Getty Images

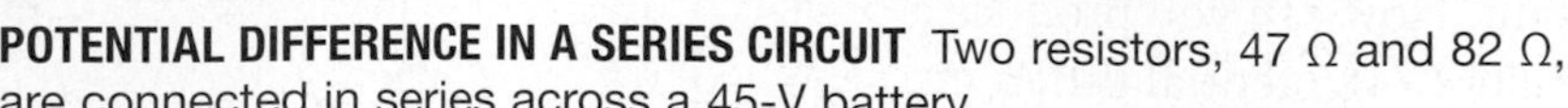

EXAMPLE PROBLEM 1

Find help with **operations with significant figures.** Math Handbook

POTENTIAL DIFFERENCE IN A SERIES CIRCUIT Two resistors, 47 Ω and 82 Ω, are connected in series across a 45-V battery.

a. What is the current in the circuit?

b. What is the potential difference across each resistor?

c. If you replace the 47-Ω resistor with a 39-Ω resistor, will the current increase, decrease, or remain the same?

d. What is the new potential difference across the 82-Ω resistor?

1 ANALYZE AND SKETCH THE PROBLEM

Draw a schematic of the circuit.

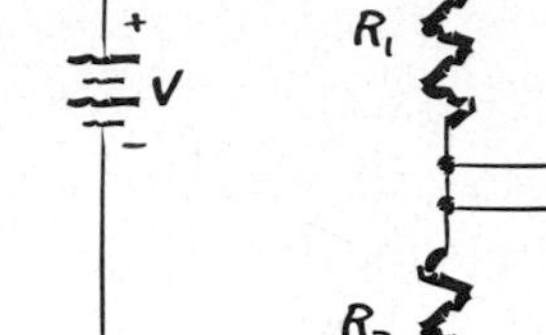

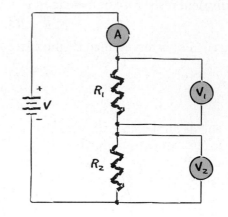

KNOWN	UNKNOWN
$\Delta V_{source} = 45\ V$	$I = ?$
$R_1 = 47\ \Omega$	$\Delta V_1 = ?$
$R_2 = 82\ \Omega$	$\Delta V_2 = ?$

2 SOLVE FOR THE UNKNOWN

a. To determine the current, first find the equivalent resistance.

$$I = \frac{\Delta V_{source}}{R} \text{ and } R = R_1 + R_2$$

$$= \frac{\Delta V_{source}}{R_1 + R_2}$$ ◀ Substitute $R = R_1 + R_2$.

$$= \frac{45\ V}{47\ \Omega + 82\ \Omega}$$ ◀ Substitute $\Delta V_{source} = 45\ \Omega$, $R_1 = 47\ V$, $R_2 = 82\ \Omega$.

$$= 0.35\ A$$

b. Use $\Delta V = IR$ for each resistor.

$$\Delta V_1 = IR_1$$

$$= (0.35\ A)(47\ \Omega)$$ ◀ Substitute $I = 0.35\ A$, $R_1 = 47\ \Omega$.

$$= 16\ V$$

$$\Delta V_2 = IR_2$$

$$= (0.35\ A)(82\ \Omega)$$ ◀ Substitute $I = 0.35\ A$, $R_2 = 82\ \Omega$.

$$= 29\ V$$

c. Calculate current, this time using 39 Ω as R_1.

$$I = \frac{\Delta V_{source}}{R_1 + R_2}$$

$$= \frac{45\ V}{39\ \Omega + 82\ \Omega}$$ ◀ Substitute $\Delta V_{source} = 45\ V$, $R_1 = 39\ \Omega$, $R_2 = 82\ \Omega$.

$$= 0.37\ A$$

The current will increase.

d. Determine the new voltage drop in R_2.

$$\Delta V_2 = IR_2$$

$$= (0.37\ A)(82\ \Omega)$$ ◀ Substitute $I = 0.37\ A$, $R_2 = 82\ \Omega$

$$= 3.0\times10^1\ V$$

3 EVALUATE THE ANSWER

- **Are the units correct?** Current is measured in $A = \frac{V}{\Omega}$; potential difference is measured in $V = A{\cdot}\Omega$.
- **Is the magnitude realistic?** The potential difference across any one resistor must be less than the potential difference across the battery. Both the value of ΔV_2 and the value of ΔV_1 are less than ΔV_{source}, which is 45 V.

EXAMPLE PROBLEM

PRACTICE PROBLEMS

Do additional problems. Online Practice

6. The circuit shown in Example Problem 1 is producing these symptoms: the ammeter reads 0 A, ΔV_1 reads 0 V, and ΔV_2 reads 45 V. What has happened?
7. Suppose the circuit shown in Example Problem 1 has these values: $R_1 = 255\ \Omega$, $R_2 = 290\ \Omega$, and $\Delta V_1 = 17$ V. No other information is available.
   a. What is the current in the circuit?
   b. What is the potential difference across the battery?
   c. What is the total power used in the circuit, and what is the power used in each resistor?
   d. Does the sum of the power used in each resistor in the circuit equal the total power used in the circuit? Explain.
8. Holiday lights often are connected in series and use special lamps that short out when the voltage across a lamp increases to the line voltage. Explain why. Also explain why these light sets might blow their fuses after many bulbs have failed.
9. The circuit in Example Problem 1 has unequal resistors. Explain why the resistor with the lower resistance will operate at a lower temperature.
10. **CHALLENGE** A series circuit is made up of a 12-V battery and three resistors. The potential difference across one resistor is 1.2 V, and the potential difference across another resistor is 3.3 V. What is the voltage across the third resistor?

EXAMPLE PROBLEM 2

Find help with **order of operations.** Math Handbook 

VOLTAGE DIVIDER A 9.0-V battery and two resistors, 390 Ω and 470 Ω, are connected as a voltage divider. What is the potential difference across the 470-Ω resistor?

1 ANALYZE AND SKETCH THE PROBLEM

Draw the battery and resistors in a series circuit.

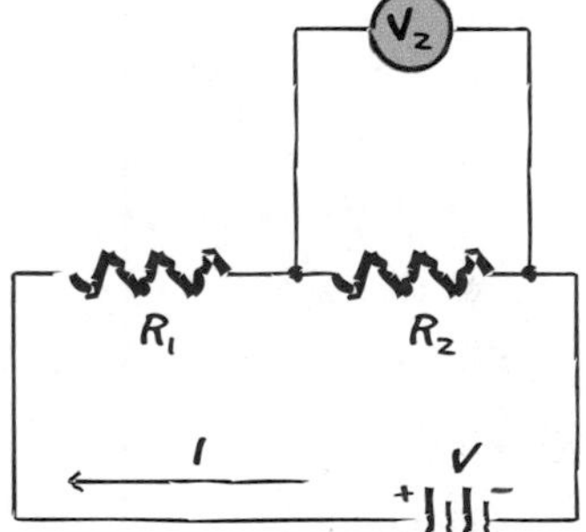

KNOWN

$\Delta V_{source} = 9.0$ V
$R_1 = 390$ V
$R_2 = 470$ V

UNKNOWN

$\Delta V_2 = ?$

2 SOLVE FOR THE UNKNOWN

$$R = R_1 + R_2$$

$$I = \frac{\Delta V_{source}}{R}$$

$$= \frac{\Delta V_{source}}{R_1 + R_2}$$ ◀ Substitute $R = R_1 + R_2$

$$\Delta V_2 = IR_2$$

$$= \frac{\Delta V_{source} R_2}{R_1 + R_2}$$ ◀ Substitute $I = \frac{V_{source}}{R_1 + R_2}$

$$= \frac{(9.0\ \text{V})(470\ \Omega)}{390\ \Omega + 470\ \Omega}$$ ◀ Substitute $\Delta V_{source} = 9.0$ V, $R_1 = 390\ \Omega$, $R_2 = 470\ \Omega$

$$= 4.9\ \text{V}$$

3 EVALUATE THE ANSWER

- **Are the units correct?** Potential difference is measured in $\text{V} = \frac{\text{V}\cdot\Omega}{\Omega}$. The ohms cancel, leaving volts.
- **Is the magnitude realistic?** The potential difference across the resistor is less than the potential difference across the battery. Because 470 Ω is more than half of the equivalent resistance, the potential difference across that resistor is more than half of the potential difference across the battery.

PRACTICE PROBLEMS

Do additional problems. Online Practice

PRACTICE PROBLEMS

11. A 22-Ω resistor and a 33-Ω resistor are connected in series and are connected to a 120-V power source.

a. What is the equivalent resistance of the circuit?

b. What is the current in the circuit?

c. What is the potential difference across each resistor?

12. Three resistors of 3.3 kΩ, 4.7 kΩ, and 3.9 kΩ are connected in series across a 12-V battery.

a. What is the equivalent resistance?

b. What is the current through the resistors?

c. Find the total potential difference across the three resistors.

13. CHALLENGE Select a resistor to be used as part of a voltage divider along with a 1.2-kΩ resistor. The potential difference across the 1.2-kΩ resistor is to be 2.2 V when the supply is 12 V.

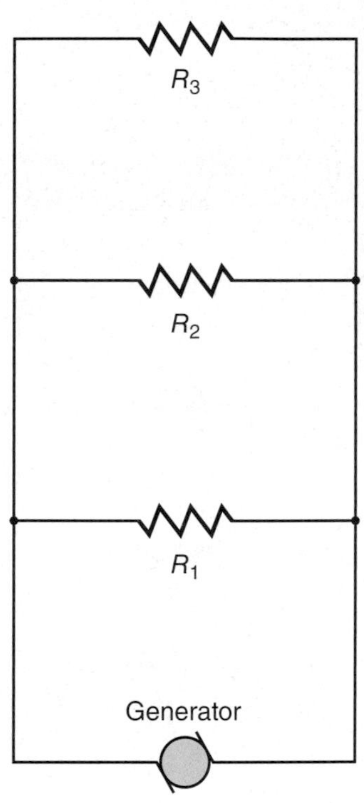

Figure 6 The parallel paths for current in this diagram are analogous to the multiple paths that a river might take down a mountain.

Parallel Circuits

Look at the circuit shown in **Figure 6.** How many current paths are there? The current from the generator can go through any of the three resistors. A circuit in which there are several current paths is called a **parallel circuit.** The three resistors are connected in parallel; both ends of the three paths are connected together. In the mountain river model, such a circuit is illustrated by three paths for the water over a waterfall. Some paths might have a large flow of water, while others might have a small flow. The sum of the water in the flows, however, is equal to the total flow of water over the falls. In addition, regardless of which channel the water flows through, the drop in height is the same. Similarly, in a parallel electric circuit, the total current is the sum of the currents through each path, and the potential difference across each path is the same.

Current in a parallel circuit What is the current through each resistor in a parallel circuit? The answer to this question depends on the individual resistances. For example, in **Figure 7,** the potential difference across each resistor is 120 V. The current through a resistor is given by $I = \frac{\Delta V}{R}$, so you can calculate the current through the 24-Ω resistor as $I = \frac{120\ \text{V}}{24\ \Omega} = 5.0\ \text{A}$. You can calculate the current through the other two resistors in a similar way. The total current through the generator is the sum of the currents through the three paths. In this case, the current through the generator is 38 A.

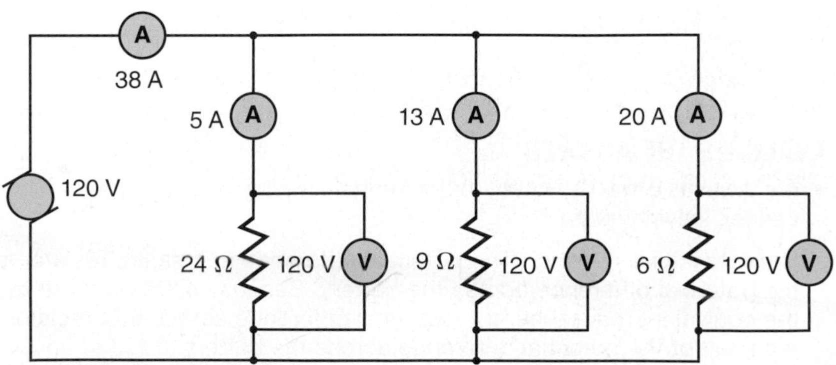

Figure 7 To find the total current in a parallel circuit, find the sum of the currents in each individual path.

State what the symbols V and A mean in the circuit.

Changing the circuit What would happen if you removed the 6-Ω resistor from the circuit in **Figure 7?** Would the current through the 24-Ω resistor change? The current through the 24-Ω resistor depends on the resistance of and the potential difference across that resistor; because neither has changed, the current also has not changed. The same is true for the current through the other resistors. The branches of a parallel circuit are independent of each other. Removing a resistor from the circuit would decrease the total current through the power source, however. For example, if you remove the 6-Ω resistor, the current through the power source would decrease to 18 A.

READING CHECK **Identify** the two factors on which the current through a resistor depends.

Equivalent resistance of a parallel circuit How can you find the equivalent resistance of a parallel circuit? In **Figure 7,** the total current through the power source is 38 A. The value of a single resistor that results in a 38-A current when a 120-V potential difference is placed across it can be calculated by using the following equation:

$$R = \frac{\Delta V}{I}$$
$$= \frac{120\ \text{V}}{38\ \text{A}}$$
$$= 3.2\ \Omega$$

Notice that this resistance is smaller than that of any of the three resistors in parallel. Placing two or more resistors in parallel always decreases the equivalent resistance of a circuit. The resistance decreases because each new resistor provides an additional path for current, thereby increasing the total current while the potential difference remains unchanged.

To calculate the equivalent resistance of a parallel circuit, first note that the total current is the sum of the currents through the branches. If I_1, I_2, and I_3 are the currents through the branches and I is the total current, then

$$I = I_1 + I_2 + I_3.$$

The potential difference across each resistor is the same, so the current through each resistor, for example, R_1, can be found from $I_1 = \frac{\Delta V}{R_1}$. Therefore, the equation for the sum of the currents can be written as:

$$\frac{\Delta V}{R} = \frac{\Delta V}{R_1} + \frac{\Delta V}{R_2} + \frac{\Delta V}{R_3}$$

This equation can be extended for any number of resistors in parallel. Dividing both sides of the equation by V provides an equation for the equivalent resistance of a parallel circuit.

EQUIVALENT RESISTANCE FOR RESISTORS IN PARALLEL

The reciprocal of the equivalent resistance is equal to the sum of the reciprocals of the individual resistances.

$$\frac{1}{R} = \frac{1}{R_1} + \frac{1}{R_2} + \frac{1}{R_3} + \ldots$$

PhysicsLABs

CIRCUIT CONFIGURATIONS

FORENSICS LAB How do different circuit configurations affect the voltages across and currents through the lightbulbs (resistors) in the circuit?

HOW DO PARALLEL RESISTORS WORK?

How is equivalent resistance measured in the lab?

MiniLABs

PARALLEL CIRCUIT MEASUREMENTS

Is voltage constant across parallel components?

PARALLEL RESISTANCE

Can you predict the current in a parallel circuit?

EXAMPLE PROBLEM 3

Get help with **resistance and current.** Personal Tutor

EQUIVALENT RESISTANCE AND CURRENT IN A PARALLEL CIRCUIT Three resistors, 60.0 Ω, 30.0 Ω, and 20.0 Ω, are connected in parallel across a 90.0-V battery.

a. Find the current through each branch of the circuit.

b. Find the equivalent resistance of the circuit.

c. Find the current through the battery.

1 ANALYZE AND SKETCH THE PROBLEM

- Draw a schematic of the circuit.
- Include ammeters to show where you would measure each of the currents.

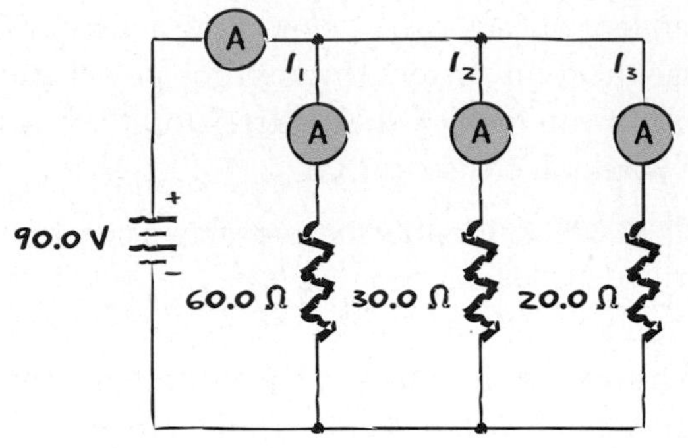

KNOWN	**UNKNOWN**	
$R_1 = 60.0\ \Omega$	$I_1 = ?$	$I = ?$
$R_2 = 30.0\ \Omega$	$I_2 = ?$	$R = ?$
$R_3 = 20.0\ \Omega$	$I_3 = ?$	
$\Delta V = 90.0\text{ V}$		

2 SOLVE FOR THE UNKNOWN

a. Because the voltage across each resistor is the same, use $I = \frac{\Delta V}{R}$ for each branch.

$$I_1 = \frac{\Delta V}{R_1}$$

$$= \frac{90.0\text{ V}}{60.0\ \Omega}$$ ◀ *Substitute $\Delta V = 90.0$ V, $R_1 = 60.0\ \Omega$.*

$$= 1.50\text{ A}$$

$$I_2 = \frac{\Delta V}{R_2}$$

$$= \frac{90.0\text{ V}}{30.0\ \Omega}$$ ◀ *Substitute $\Delta V = 90.0$ V, $R_2 = 30.0\ \Omega$.*

$$= 3.00\text{ A}$$

$$I_3 = \frac{\Delta V}{R_3}$$

$$= \frac{90.0\text{ V}}{20.0\ \Omega}$$ ◀ *Substitute $\Delta V = 90.0$ V, $R_3 = 20.0\ \Omega$.*

$$= 4.50\text{ A}$$

b. Use the equivalent resistance equation for parallel circuits.

$$\frac{1}{R} = \frac{1}{R_1} + \frac{1}{R_2} = \frac{1}{R_3}$$

$$= \frac{1}{60.0\ \Omega} + \frac{1}{30.0\ \Omega} + \frac{1}{20.0\ \Omega}$$ ◀ *Substitute $R_1 = 60.0\ \Omega$, $R_2 = 30.0\ \Omega$, $R_3 = 20.0\ \Omega$.*

$$= 0.100\ \Omega^{-1}$$

$$R = 10.0\ \Omega$$

c. Use $I = \frac{\Delta V}{R}$ to find the total current.

$$I = \frac{\Delta V}{R}$$

$$= \frac{90.0\text{ V}}{10.0\ \Omega}$$ ◀ *Substitute $\Delta V = 90.0$ V, $R = 10.0\ \Omega$.*

$$= 9.00\text{ A}$$

3 EVALUATE THE ANSWER

- **Are the units correct?** Current is measured in amps; resistance is measured in ohms.
- **Is the magnitude realistic?** The equivalent resistance is less than the resistance of any single resistor. The currents through the resistors are inversely proportional to the resistance. The current for the circuit (I) equals the sum of the currents found for each resistor ($I_1 + I_2 + I_3$).

PRACTICE PROBLEMS

Do additional problems. Online Practice

14. You connect three 15.0-Ω resistors in parallel across a 30.0-V battery.

a. What is the equivalent resistance of the parallel circuit?

b. What is the current through the entire circuit?

c. What is the current through each branch of the circuit?

15. Suppose you replace one of the 15.0-Ω resistors in the previous problem with a 10.0-Ω resistor.

a. How does the equivalent resistance change?

b. How does the current through the entire circuit change?

c. How does the current through one of the 15.0-Ω resistors change?

16. You connect a 120.0-Ω resistor, a 60.0-Ω resistor, and a 40.0-Ω resistor in parallel across a 12.0-V battery.

a. What is the equivalent resistance of the parallel circuit?

b. What is the current through the entire circuit?

c. What is the current through each branch of the circuit?

17. CHALLENGE You are trying to reduce the resistance in a branch of a circuit from 150 Ω to 93 Ω. You add a resistor to this branch of the circuit to make this change. What value of resistance should you use, and how should you connect this resistor?

Kirchhoff's Rules

Gustav Robert Kirchhoff was a German physicist who, in 1845, when he was only 21 years old, formulated two rules that govern electric circuits—the loop rule and the junction rule. You can use these two rules to analyze complex electric circuits. Both rules are based on fundamental scientific laws.

The loop rule The loop rule is based on the law of conservation of energy. Imagine you are standing on the side of a hill. You walk around on the side of this hill in a loop, as shown in the top of **Figure 8.** How does your height up the hill before your walk compare to your height up the hill after your walk? Because you begin and end your walk at the same place, your height up the hill after your walk equals your height up the hill before your walk. The sum of increases in height equals the sum of decreases in height during your walk.

A similar situation occurs as an electric charge moves around any loop in an electric circuit. Instead of increases and decreases in height, however, electric charges move through increases and decreases in electric potential. The sum of increases in electric potential around a loop in an electric circuit equals the sum of decreases in electric potential around that loop.

☑ **READING CHECK** **Explain** how Kirchhoff's loop rule and the conservation of energy are related.

For an application of the loop rule, look at the bottom of **Figure 8.** Picture an electric charge traveling clockwise around the red loop. Electric potential increases by 9 V as this charge travels through the battery, and electric potential drops by 5 V as this charge travels through resistor 1. What will be the change in potential as the charge travels through resistor 2? Because the increases in electric potential around a loop must equal the decreases in electric potential around that loop, the drop in electric potential across resistor 2 must be 9 V − 5 V = 4 V.

Notice that resistor 3 does not affect our answer. Why is this? Resistor 3 is not a part of the loop that includes the battery, resistor 1, and resistor 2.

☑ **READING CHECK** **State** the loop rule in your own words.

The Loop Rule

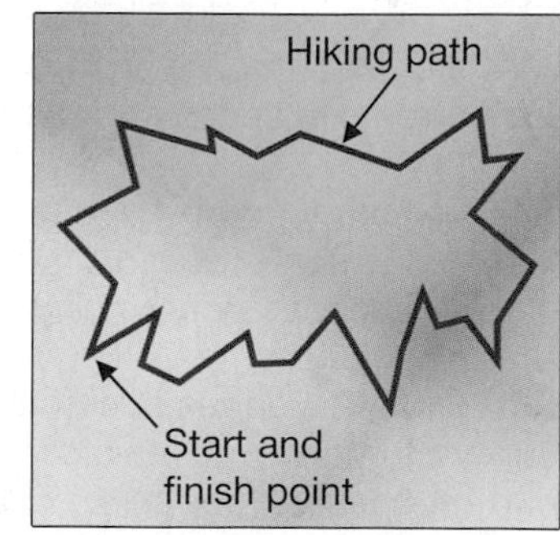

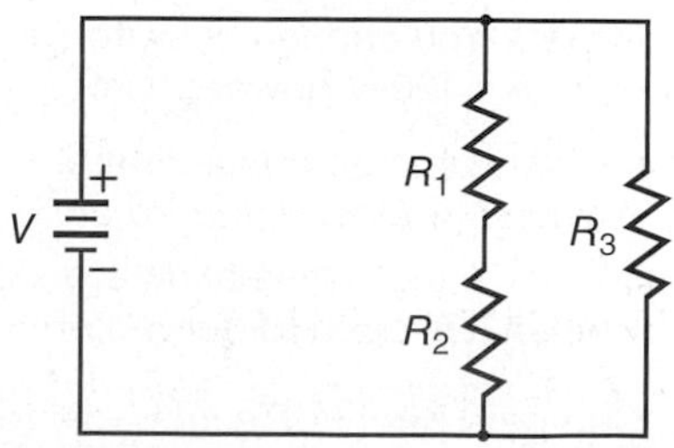

Figure 8 Walking around the side of a hill is similar to electric charges moving around a loop in a circuit.

Figure 9 An application of Kirchhoff's junction rule is shown. At junction A the total current entering (I_1) equals the total current leaving ($I_2 + I_3$). At junction B, the total current entering is $I_2 + I_3$, and the total current leaving is I_1.

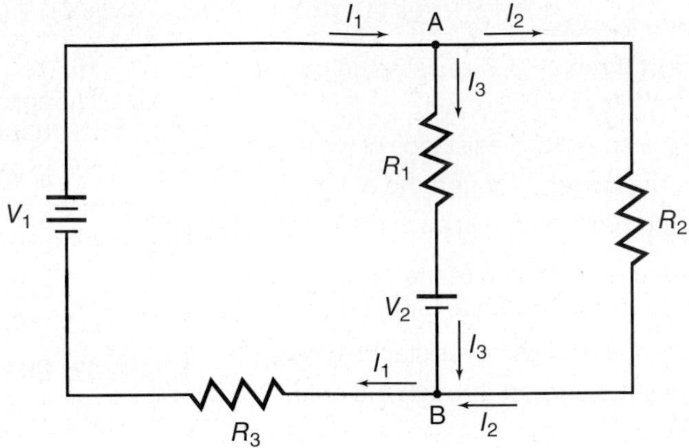

The junction rule The loop rule describes electric potential differences and is based on the law of conservation of energy. The junction rule describes currents and is based on the law of conservation of charge. Recall that the law of conservation of charge states that charge can neither be created nor destroyed. This means that, in an electric circuit, the total current into a section of that circuit must equal the total current out of that same section.

A junction is a location where three or more wires are connected together. The circuit shown in **Figure 9** has two such junctions—one at point A and another at point B. How does the total current into a junction relate to the total current out of that junction? According to the law of conservation of charge, the sum of currents entering a junction is equal to the sum of currents leaving that junction. Otherwise, charge would build up at the junction. This is Kirchhoff's junction rule. In **Figure 9,** $I_1 = I_2 + I_3$ at junction A, and $I_2 + I_3 = I_1$ at junction B. For example, if $I_2 = 0.3$ A and $I_3 = 0.7$ A, then $I_1 = 1.0$ A.

SECTION 1 REVIEW

Check your understanding.

18. MAINIDEA Compare and contrast the voltages and the currents in series and parallel circuits.

19. Total Current A parallel circuit has four branch currents: 120 mA, 250 mA, 380 mA, and 2.1 A. How much current passes through the power source?

20. Total Current A series circuit has four resistors. The current through one resistor is 810 mA. How much current passes through the power source?

21. Circuits You connect a switch in series with a 75-W bulb to a 120-V power source.

a. What is the potential difference across the switch when it is closed (turned on)?

b. What is the potential difference across the switch when it is opened (turned off)?

22. Compare Kirchhoff's loop rule to walking around in a loop on the side of a hill.

23. Explain how Kirchhoff's junction rule relates to the law of conservation of charge.

24. Critical Thinking The circuit in **Figure 10** has four identical resistors. Suppose that a wire is added to connect points A and B. Answer the following questions, and explain your reasoning.

a. What is the current through the wire?

b. What happens to the current through each resistor?

c. What happens to the current through the battery?

d. What happens to the potential difference across each resistor?

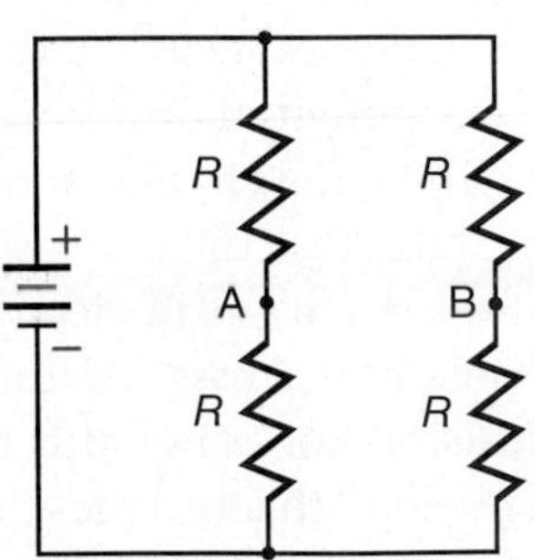

Figure 10

SECTION 2 Applications of Circuits

TRBfoto/Getty Images

PHYSICS 4 YOU You might have circuit breakers, fuses, and ground-fault interrupters (GFIs) in your home. All of these devices are added to circuits to prevent damage. Building codes generally require that all outlets in residential bathrooms be equipped with a ground-fault interrupter with a reset button.

MAIN IDEA

Most circuits are combination series-parallel circuits.

Essential Questions

- How do fuses, circuit breakers, and ground-fault interrupters protect household wiring?
- How can you find currents and potential differences in combined series-parallel circuits?
- How do you use voltmeters and ammeters to measure potential differences and currents in circuits?

Review Vocabulary

electric current a flow of charged particles

New Vocabulary

short circuit
fuse
circuit breaker
ground-fault interrupter
combination series-parallel circuit

Safety Devices

In an electric circuit, fuses and circuit breakers prevent circuit overloads that can occur when too many appliances are turned on at the same time or when a short circuit occurs in one appliance. A **short circuit** occurs when a circuit with very low resistance is formed. When appliances are connected in parallel, each additional appliance placed in operation reduces the equivalent resistance in the circuit and increases the current through the wires. This additional current might produce enough thermal energy to melt the wiring's insulation, cause a short circuit, or even begin a fire.

A **fuse** is a short piece of metal that acts as a safety device by melting and stopping the current when too large a current passes through it. Engineers design fuses to melt before other elements in a circuit are damaged. A **circuit breaker,** shown in **Figure 11,** is an automatic switch that acts as a safety device by stopping the current if the current gets too large and exceeds a threshold value.

Usually, current follows a single path from the power source through an electrical appliance and back to the source. An appliance malfunction or an accidental drop of the appliance into water can result in additional current pathways. A **ground-fault interrupter** (GFI) is a device that contains an electronic circuit that detects small current differences between the two wires in the cord connected to an appliance. An extra current path, such as one through water, could cause this difference. The GFI stops the current when it detects such differences. This often protects a person from electrocution.

Circuit Breaker

Figure 11 When there is too much current through the bimetallic strip in a circuit breaker, the heat that is generated causes the strip to bend and release the latch. The handle moves to the off position, causing the switch to open and the current to stop.

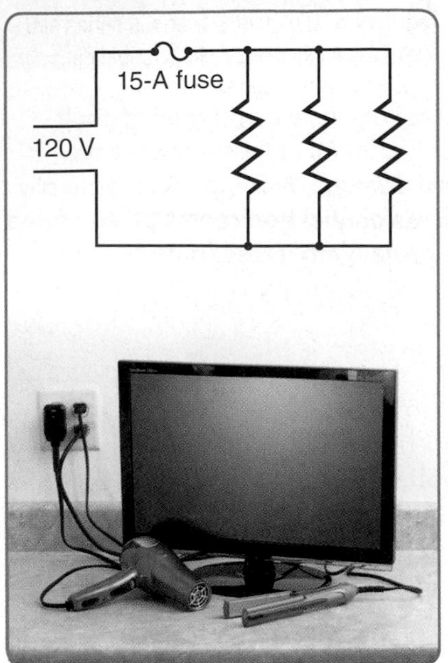

Figure 12 Homes are wired using parallel circuits so that more than one appliance can be used simultaneously.

Identify the current rating of the fuse in the circuit above.

Household circuits Some of the most common uses of fuses, circuit breakers, and GFIs are in household circuits. How are these household circuits designed? Are the lights and outlets in a household circuit wired in series or in parallel? Think about what would happen if all the lights and outlets in a household circuit were wired in series. You would only be able to turn everything on and off at once. This is not what we observe in real houses.

The top panel of **Figure 12** shows a simplified household circuit. Notice that this is a parallel circuit with a 15-A fuse connected in series with the 120-V power source. Each resistor represents a light or an electrical appliance. The current in any one branch of the circuit does not depend on the currents in the other branches. This explains how you can turn off the bathroom light without turning off the television. Everything turns off if the fuse wire melts, however.

Let's consider a specific example. Suppose that the three resistors in **Figure 12** represent a 240-W television, a 720-W straightening iron, and a 1440-W hair dryer. We want to know if the fuse will melt, or "blow," if all three appliances are turned on at the same time.

You can use $I = \frac{P}{\Delta V}$ to find the current through each appliance.

For the television, $I_{TV} = \frac{240\text{ W}}{120\text{ V}} = 2.0\text{ A}$. For the straightening iron,

$I_{iron} = \frac{720\text{ W}}{120\text{ V}} = 6.0\text{ A}$. The current through the hair dryer is

$I_{dryer} = \frac{1440\text{ W}}{120\text{ V}} = 12\text{ A}$.

From Kirchhoff's junction rule, we can see that the current through the fuse equals the sum of the currents through the television, the straightening iron, and the hair dryer.

$$\begin{aligned} I_{fuse} &= I_{TV} + I_{iron} + I_{hair} \\ &= 2.0\text{ A} + 6.0\text{ A} + 12\text{ A} \\ &= 20\text{ A} \end{aligned}$$

Is this enough current to blow the fuse? The fuse in **Figure 12** is rated at 15 A, so the 20-A current from the television, the straightening iron, and the hair dryer would blow the fuse. This would stop current throughout the circuit.

PhysicsLAB

SERIES AND PARALLEL CIRCUITS

How do you determine the equivalent resistance of a series circuit or a parallel circuit?

PHYSICS CHALLENGE

When the galvanometer, a device used to measure very small currents, in this circuit measures zero, the circuit is said to be balanced.

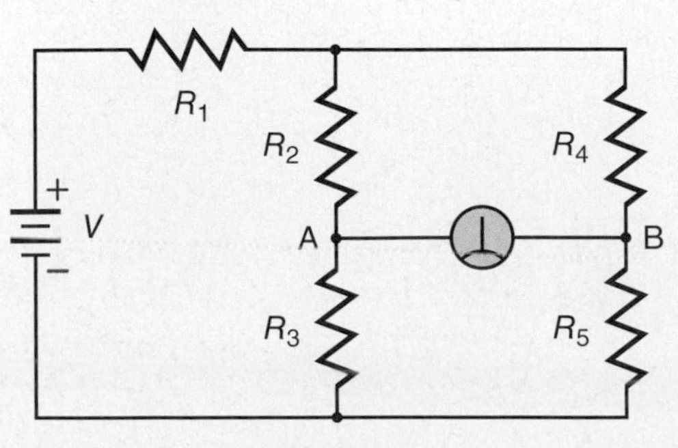

1. Your lab partner states that the only way to balance this circuit is to make all the resistors equal. Will this balance the circuit? Is there more than one way to balance this circuit? Explain.
2. Derive a general equation for a balanced circuit using the given labels. *Hint: Treat the circuit as two voltage dividers.*
3. Which of the resistors can be replaced with a variable resistor and then used to balance the circuit?
4. Which of the resistors can be replaced with a variable resistor and then used as a sensitivity control? Why would this be necessary? How would it be used in practice?

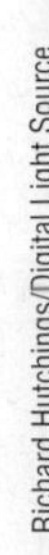

Richard Hutchings/Digital Light Source

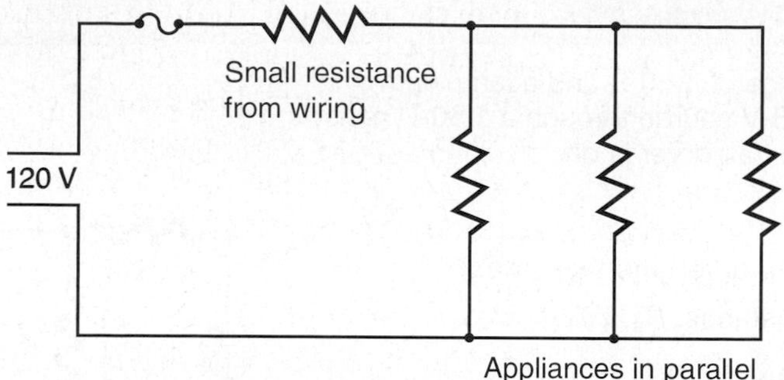

Figure 13 Circuits that contain series and parallel branches are called combination series-parallel circuits.

Explain why the wiring has resistance.

Investigate **electric circuits.**

View a **simulation of a circuit lab.**

Combined Series-Parallel Circuits

Have you ever noticed the light dim when you turned on a hair dryer? The light and the hair dryer are connected in parallel across 120 V. Because of the parallel connection, the current through the light should not have changed when you turned on the hair dryer. Yet the light did dim, so the current must have changed. The dimming occurred because the household wiring has a small resistance. As shown in **Figure 13,** this resistance was in series with the parallel circuit. A circuit that includes series and parallel branches is a **combination series-parallel circuit.** The following are strategies for analyzing such circuits.

PROBLEM-SOLVING STRATEGIES

SERIES-PARALLEL CIRCUITS

When analyzing a combination series-parallel circuit, use the following steps to break down the problem.

1. Draw a schematic diagram of the circuit.
2. Find any parallel resistors. Resistors in parallel have separate current paths. They must have the same potential differences across them. Calculate the single equivalent resistance of a resistor that can replace them. Draw a new schematic using that resistor.
3. Are any resistors (including the equivalent resistor) now in series? Resistors in series have one and only one current path through them. Calculate a single new equivalent resistance that can replace them. Draw a new schematic diagram using that resistor.
4. Repeat steps 2 and 3 until you can reduce the circuit to a single resistor. Find the total circuit current. Then go backward through the circuits to find the currents through and the potential differences across individual resistors.

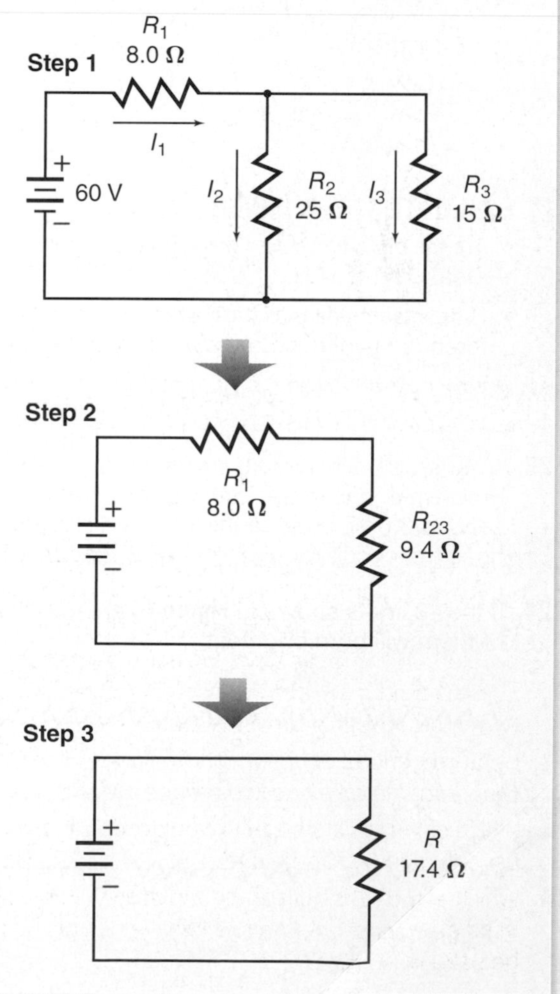

EXAMPLE PROBLEM 4

Find help with **significant digits.** Math Handbook

EXAMPLE PROBLEM

SERIES-PARALLEL CIRCUIT A hair dryer with a resistance of 12.0 Ω and a lamp with a resistance of 125 Ω are connected in parallel to a 125-V source through a 1.50-Ω resistor in series. Find the current through the lamp when the hair dryer is on.

1 ANALYZE AND SKETCH THE PROBLEM

- Draw the series-parallel circuit including the hair dryer and the lamp.
- Replace R_1 and R_2 with a single equivalent resistance, R_p.

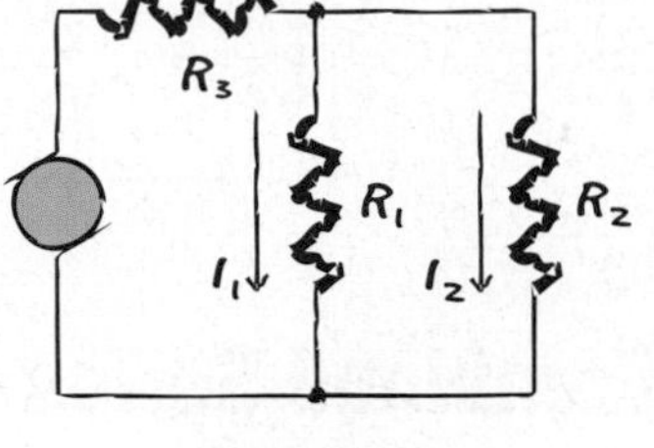

KNOWN		UNKNOWN	
$R_1 = 125\ \Omega$	$R_3 = 1.50\ \Omega$	$I = ?$	$I_1 = ?$
$R_2 = 12.0\ \Omega$	$\Delta V_{source} = 125\ V$	$R = ?$	$R_p = ?$

2 SOLVE FOR THE UNKNOWN

Find the equivalent resistance for the parallel circuit, then find the equivalent resistance for the entire circuit, and then calculate the current.

$$\frac{1}{R_p} = \frac{1}{R_1} + \frac{1}{R_2} = \frac{1}{125\ \Omega} + \frac{1}{12.0\ \Omega}$$ ◀ Substitute $R_1 = 125\ \Omega$, $R_2 = 12.0\ \Omega$.

$$R_p = 10.9\ \Omega$$

$$R = R_3 + R_p = 1.50\ \Omega + 10.9\ \Omega = 12.4\ \Omega$$ ◀ Substitute $R_3 = 1.50\ \Omega$, $R_p = 10.9\ \Omega$.

$$I = \frac{\Delta V_{source}}{R} = \frac{125\ V}{12.4\ \Omega} = 10.1\ A$$ ◀ Substitute $\Delta V_{source} = 125\ V$, $R = 12.4\ \Omega$.

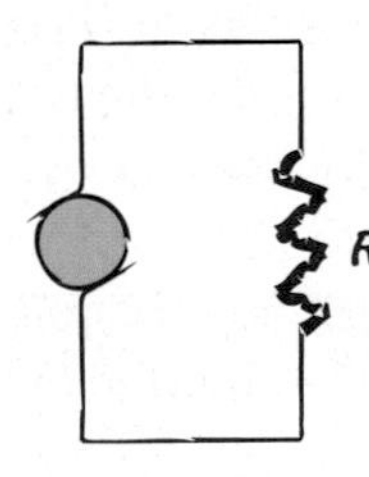

$$\Delta V_3 = IR_3 = (10.1\ A)(1.50\ \Omega) = 15.2\ V$$ ◀ Substitute $I = 10.1\ A$, $R_3 = 1.50\ \Omega$.

$$\Delta V_1 = \Delta V_{source} - \Delta V_3 = 125\ V - 15.2\ V = 1.10\times10^2\ V$$ ◀ Substitute $\Delta V_{source} = 125\ V$, $V_3 = 15.2\ V$.

$$I_1 = \frac{\Delta V_1}{R_1} = \frac{1.10\times10^2\ V}{125\ \Omega} = 0.880\ A$$ ◀ Substitute $\Delta V_1 = 1.10\times10^2\ V$, $R_1 = 125\ \Omega$.

3 EVALUATE THE ANSWER

- **Are the units correct?** Current is measured in amps, and potential differences are measured in volts.
- **Is the magnitude realistic?** The resistance in ohms is greater than the potential difference in volts, so the current should be less than 1 A.

PRACTICE PROBLEMS

Do additional problems. 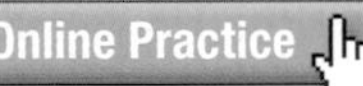

PRACTICE PROBLEMS

25. A series-parallel circuit, similar to the one in Example Problem 4, has three resistors: one uses 2.0 W, the second 3.0 W, and the third 1.5 W. How much current does the circuit require from a 12-V battery?

26. If the 13 lights shown in **Figure 14** are identical, which of them will burn brightest?

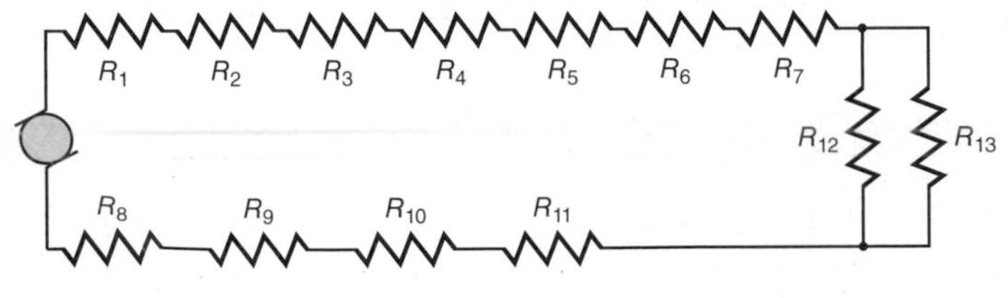

Figure 14

27. CHALLENGE A series-parallel circuit has three appliances on it. A blender and a stand mixer are in parallel, and a toaster is connected in series as shown in **Figure 15.** Find the current through the blender.

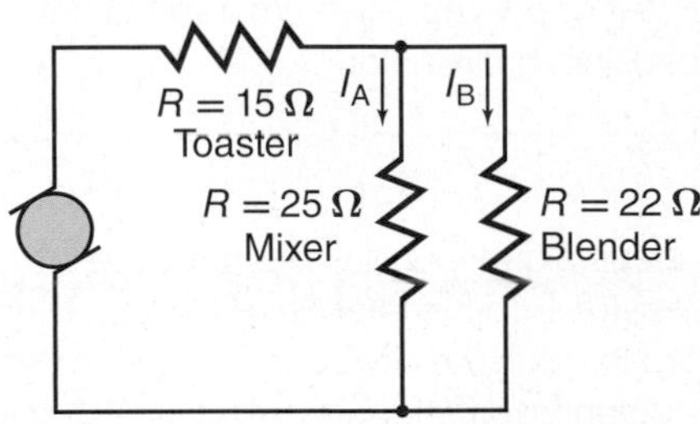

Figure 15

Ammeters and Voltmeters

Recall that an ammeter is a device that is used to measure the current in any branch or part of a circuit. If, for example, you wanted to measure the current through a resistor, you would place an ammeter in series with the resistor. This would require opening the current path and inserting an ammeter. Ideally, the use of an ammeter should not change the current in the circuit. Because the current would decrease if the ammeter increased the resistance in the circuit, the resistance of an ammeter is designed to be as low as possible. **Figure 16** shows an ammeter as a high-resistance meter placed in parallel with a 0.01-Ω resistor. Because the resistance of the ammeter is much less than that of the resistors, the current decrease is negligible.

Another instrument, called a voltmeter, is used to measure the potential difference across a portion of a circuit. To measure the potential difference across a resistor, a voltmeter is connected in parallel with the resistor. Voltmeters are designed to have a very high resistance so as to cause the smallest possible change in currents and voltages in the circuit. Consider the circuit shown in **Figure 16.** A voltmeter is shown as a meter in series with a 10-kΩ resistor. When the voltmeter is connected in parallel with R_2, the equivalent resistance of the combination is smaller than R_2 alone. Thus, the total resistance of the circuit decreases, and the current increases. The value of R_1 has not changed, but the current through it has increased, thereby increasing the potential difference across it. The battery, however, holds the potential difference across R_1 and R_2 constant. Thus, the potential difference across R_2 must decrease. The result of connecting a voltmeter across a resistor is to lower the potential difference across it. If the resistance of the voltmeter is higher, then the change in potential difference across the resistor is lower. Practical meters have resistances of 10 MΩ.

Figure 16 An ammeter is connected in series with two resistors. The small resistance of the ammeter slightly alters the current in the circuit. A voltmeter is connected in parallel with a resistor. The high resistance of the voltmeter results in a negligible change in the circuit current and voltage.

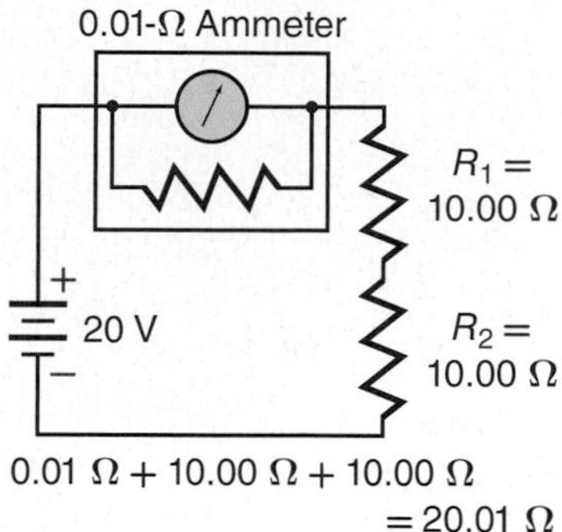

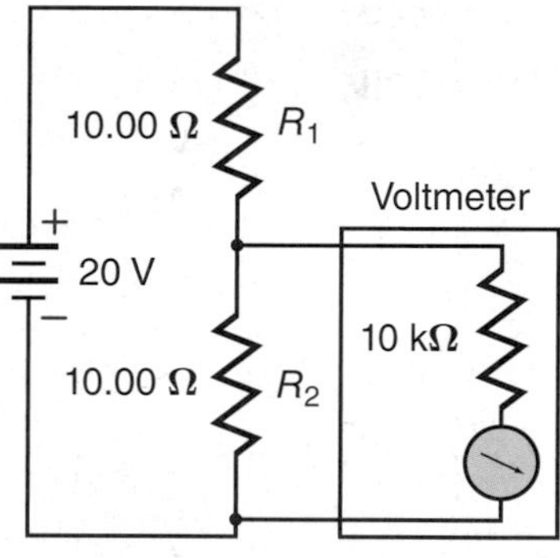

SECTION 2 REVIEW

Check your understanding.

28. MAINIDEA Explain in your own words what a combination series-parallel circuit is.

*Refer to **Figure 17** for questions 29–33 and 35.*

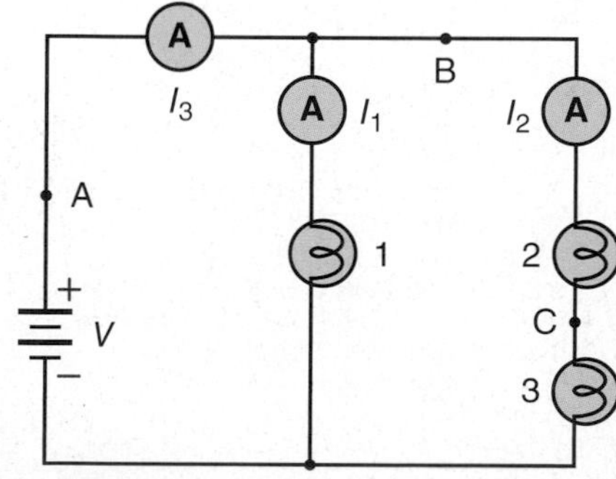

Figure 17

29. Brightness How do the brightness of the bulbs compare?

30. Current If I_3 is 1.7 A and I_1 is 1.1 A, what is the current through bulb 2?

31. Circuits in Series The wire at point C is broken and a small resistor is inserted in series with bulbs 2 and 3. What happens to the brightness of the two bulbs? Explain.

32. Battery Voltage A voltmeter connected across bulb 2 measures 3.8 V, and a voltmeter connected across bulb 3 measures 4.2 V. What is the potential difference across the battery?

33. Circuits Using information from the previous problem, determine whether bulbs 2 and 3 are identical.

34. Circuit Protection Describe three common safety devices associated with household wiring.

35. Critical Thinking How could you rearrange the circuit to make the three bulbs in **Figure 17** burn with equal intensity? Is there more than one way to do this?

Live Wire

ELECTRICIAN

Think about how often you use electricity in a single day. Every time you flip a light switch, open a refrigerator door, or turn on a television, you activate an electric circuit built by an electrician.

Electricians might work in commercial, industrial, or residential settings. In this article we focus on residential applications of electricity.

CIRCUIT BREAKERS Electrical power enters most homes at the circuit breaker box. From this box, it is distributed throughout the rest of the home. The circuit breaker is a sort of switch that opens (breaks the circuit) when too much electric current is drawn through the circuit. Because wires get hot with increased current, it is important that the circuit breaker open before wires get hot enough to burn, break, or short-circuit.

WIRES Wires get hot when there is a current through them. The thinner the wire, the hotter it might become from a current. A wire that is too thin could overheat and start a fire. Electricians must know the current that a wire can safely carry and choose an appropriate wire thickness (or gauge) to handle this current load.

PARALLEL and SERIES Several electrical outlets are usually connected to a single circuit from the circuit breaker box, but these outlets are connected in parallel. Even when some outlets are open circuits (in other words, nothing plugged into the outlets is drawing electric current), other outlets can still be active. In some cases, however, series connections are desired. Electricians must know how and when to use series and parallel circuits when they design home electrical systems.

GOING FURTHER

Research How can you become an electrician? Interview an electrician to find out what education, experience, and special skills are needed for this profession.

CHAPTER 23

STUDY GUIDE

BIGIDEA Circuit components can be placed in series, in parallel, or in a combination of series and parallel.

VOCABULARY
- **series circuit** *(p. 625)*
- **equivalent resistance** *(p. 626)*
- **voltage divider** *(p. 627)*
- **parallel circuit** *(p. 630)*

SECTION 1 Simple Circuits

MAINIDEA In a series circuit, current follows a single path; in a parallel circuit, current follows more than one path.

- The current is the same everywhere in a simple series circuit. If any branch of a parallel circuit is opened, there is no current in that branch. The current in the other branches is unchanged.
- The current in a series circuit is equal to the potential difference divided by the equivalent resistance. The sum of the voltage drops across resistors that are in series is equal to the potential difference applied across the combination. The equivalent resistance of a series circuit is the sum of the resistances of its parts.
- In a parallel circuit, the total current is equal to the sum of the currents in the branches. The voltage drops across all branches of a parallel circuit are the same. The reciprocal of the equivalent resistance of parallel resistors is equal to the sum of the reciprocals of the individual resistances.

VOCABULARY
- **short circuit** *(p. 635)*
- **fuse** *(p. 635)*
- **circuit breaker** *(p. 635)*
- **ground-fault interrupter** *(p. 635)*
- **combination series-parallel circuit** *(p. 637)*

SECTION 2 Applications of Circuits

MAINIDEA Most circuits are combination series-parallel circuits.

- A fuse, a circuit breaker, and a ground-fault interrupter create an open circuit and stop current when currents are dangerously high.

Circuit Breaker

- To find currents and potential differences in complex circuits, a combination of series and parallel branches, any parallel branch first is reduced to a single equivalent resistance. Then, any resistors in series are replaced by a single resistance.
- A voltmeter measures the potential difference (voltage) across any part or combination of parts of a circuit. A voltmeter always has a high resistance and is connected in parallel with the part of the circuit being measured. An ammeter is used to measure the current in a branch or part of a circuit. An ammeter always has a low resistance and is connected in series.

Games and Multilingual eGlossary

Chapter Self-Check

SECTION 1 Simple Circuits

Mastering Concepts

36. Why is it frustrating when one bulb burns out on a string of holiday tree lights connected in series?

37. Why does the equivalent resistance decrease as more resistors are added to a parallel circuit?

38. Several resistors with different values are connected in parallel. How do the values of the individual resistors compare with the equivalent resistance?

39. BIGIDEA Why is household wiring constructed in parallel instead of in series?

40. Why is there a difference in equivalent resistance between three 60-Ω resistors connected in series and three 60-Ω resistors connected in parallel?

41. Compare the amount of current entering a junction in a parallel circuit with that leaving the junction. (A junction is a point where three or more conductors are joined.) Describe which Kirchhoff rule you used in answering the question.

Mastering Problems

42. Ammeter 1 in **Figure 18** reads 0.20 A.

a. What should ammeter 2 indicate?

b. What should ammeter 3 indicate?

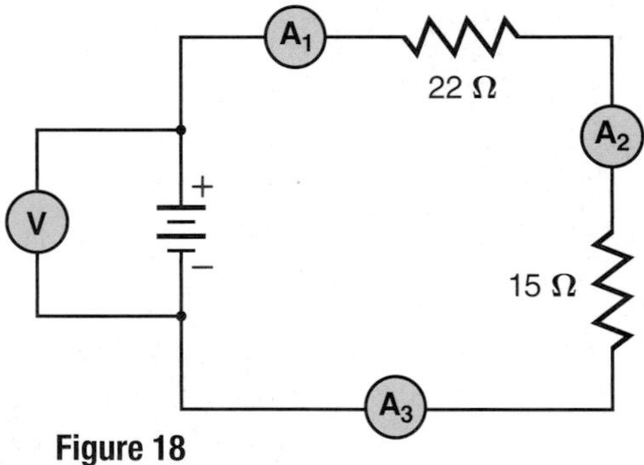

Figure 18

43. Calculate the equivalent resistance of these series-connected resistors: 680 Ω, 1.1 kΩ, and 11 kΩ.

44. Calculate the equivalent resistance of these parallel-connected resistors: 680 Ω, 1.1 kΩ, and 10.2 kΩ.

45. A series circuit has two voltage drops: 5.50 V and 6.90 V. What is the supply voltage?

46. A parallel circuit has two branch currents: 3.45 A and 1.00 A. What is the current through the electric potential source?

47. A flashlight has three batteries, each 1.5 V, and a bulb with a resistance of 15 Ω. But, one of the batteries is put in backward. Use Kirchhoff's loop rule to find the current through the bulb.

48. The change in potential energy of a charge q when its electric potential changes by ΔV is given by $q\Delta V$. Consider the circuit in **Figure 8.**

a. Use Kirchhoff's loop rule to find the change in electric potential energy as charge q goes around the circuit many times. Does the law of conservation of energy hold?

b. Suppose Kirchhoff's loop rule was not true and the potential difference across the battery (ΔV) is larger than the potential difference across the resistors (IR). How would the energy of a charge change as it goes around the loop many times?

49. Ammeter 1 in **Figure 18** reads 0.20 A.

a. What is the total resistance of the circuit?

b. What is the potential difference across the battery?

c. How much power is delivered to the 22-Ω resistor?

d. How much power is supplied by the battery?

50. Ammeter 2 in **Figure 18** reads 0.50 A.

a. Find the potential difference across the 22-Ω resistor.

b. Find the potential difference across the 15-Ω resistor.

c. Find the potential difference across the battery.

51. A 22-Ω lamp and a 4.5-Ω lamp are connected in series and placed across a potential difference of 45 V as shown in **Figure 19.**

a. What is the equivalent resistance of the circuit?

b. What is the current in the circuit?

c. Find the potential difference across each lamp.

d. What is the power used in each lamp?

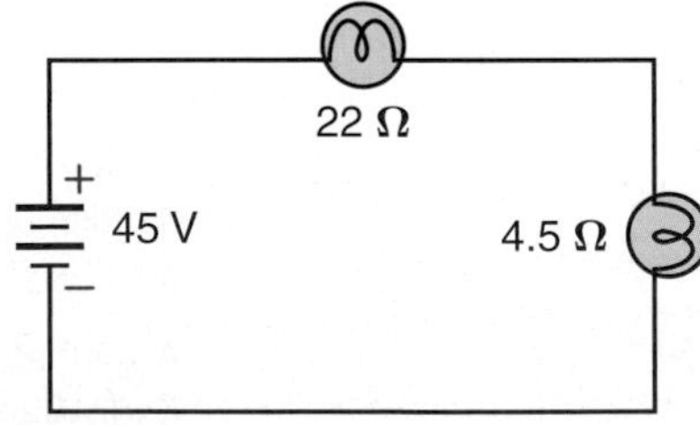

Figure 19

52. A series circuit has two voltage drops: 3.50 V and 4.90 V. What is the supply voltage?

53. A parallel circuit has two branch currents: 1.45 A and 1.00 A. What is the current in the energy source?

54. Refer to **Figure 20** to answer the following questions:

a. What should the ammeter read?

b. What should voltmeter 1 read?

c. What should voltmeter 2 read?

d. How much energy is supplied by the battery per minute?

e. What is the equivalent resistance in the circuit?

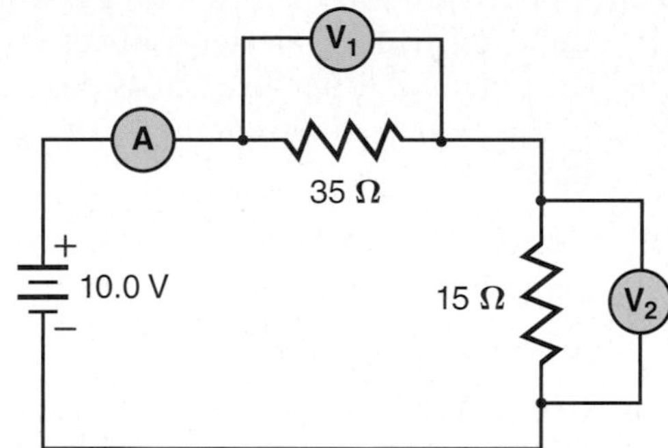

Figure 20

55. For **Figure 21,** the voltmeter reads 70.0 V.

a. Which resistor is the hottest?

b. Which resistor is the coolest?

c. What will the ammeter read?

d. What is the power supplied by the battery?

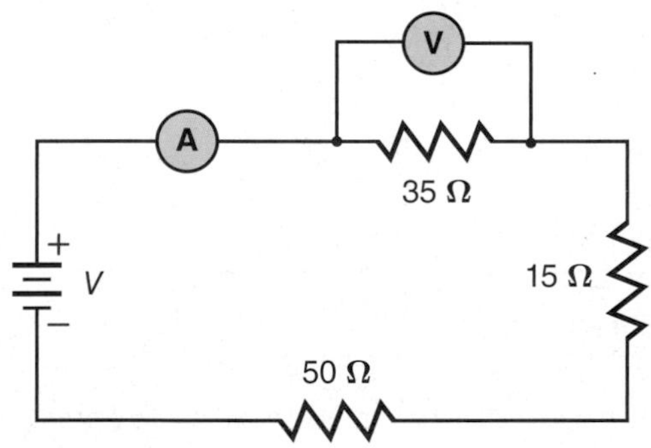

Figure 21

56. The load across a battery consists of two resistors, with values of 15 Ω and 47 Ω, connected in series.

a. What is the total resistance of the load?

b. What is the potential difference across the battery if the current in the circuit is 97 mA?

57. Pete is designing a voltage divider using a 12-V battery and a 82-Ω resistor as R_2. What resistor should be used as R_1 if the potential difference across R_2 is to be 4.0 V?

58. A series-parallel circuit has three resistors, using 5.50 W, 6.90 W, and 1.05 W, respectively. What is the supply power?

59. For **Figure 22,** the battery develops 110 V.

a. Which resistor is the hottest?

b. Which resistor is the coolest?

c. What will ammeter 1 read?

d. What will ammeter 2 read?

e. What will ammeter 3 read?

f. What will ammeter 4 read?

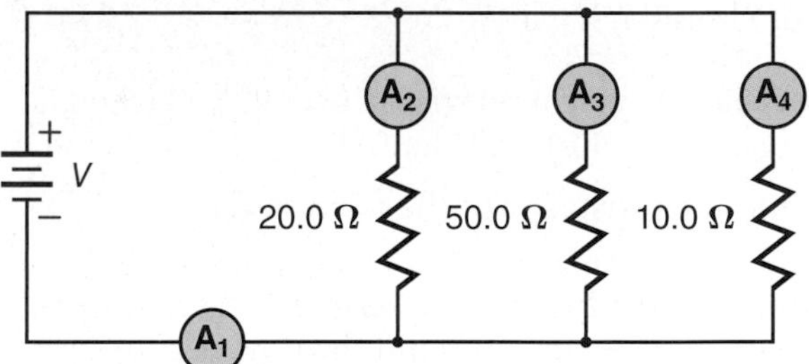

Figure 22

60. For **Figure 22,** ammeter 3 reads 0.40 A.

a. Find the potential difference across the battery.

b. What will ammeter 1 read?

c. What will ammeter 2 read?

d. What will ammeter 4 read?

61. What is the direction of the conventional current in the 50.0-Ω resistor in **Figure 22?**

62. Holiday Lights A string of 18 identical holiday lights is connected in series to a 120-V source. The string uses 64 W.

a. What is the equivalent resistance of the light string?

b. What is the resistance of a single light?

c. What power is used by each light?

63. One of the lights in the previous problem burns out. The light shorts out the bulb filament when it burns out. This drops the resistance of the lamp to zero.

a. What is the resistance of the light string now?

b. Find the power used by the string.

c. Did the power increase or decrease when the bulb burned out?

64. A 16.0-Ω and a 20.0-Ω resistor are connected in parallel. A difference in potential of 40.0 V is applied to the combination.

a. Compute the equivalent resistance of the parallel circuit.

b. What is the total current in the circuit?

c. What is the current in the 16.0-Ω resistor?

Chapter Self-Check

65. A student makes a voltage divider from a 45-V battery, a 475-kΩ resistor, and a 235-kΩ resistor. The output is measured across the smaller resistor. What is the potential difference?

66. Amy needs 5.0 V for an integrated-circuit experiment. She uses a 6.0-V battery and two resistors to make a voltage divider. One resistor is 330 Ω. She decides to make the other resistor smaller. What value should it have?

67. Television A typical television uses 275 W when it is plugged into a 120-V outlet.

a. Find the resistance of the television.

b. The television and 2.5-Ω wires connecting the outlet to the fuse form a series circuit that works like a voltage divider. Find the potential difference across the television.

c. A 12-Ω hair dryer is plugged into the same outlet. Find the equivalent resistance of the two appliances.

d. Find the potential difference across the television and the hair dryer.

SECTION 2 Applications of Circuits

Mastering Concepts

68. Explain how a fuse functions to protect an electric circuit.

69. What is a short circuit? Why is a short circuit dangerous?

70. Why is an ammeter designed to have a very low resistance?

71. Why is a voltmeter designed to have a very high resistance?

72. How does the way in which an ammeter is connected in a circuit differ from the way in which a voltmeter is connected?

Mastering Problems

73. Refer to **Figure 23** and assume that all the resistors are 30.0 Ω. Find the equivalent resistance.

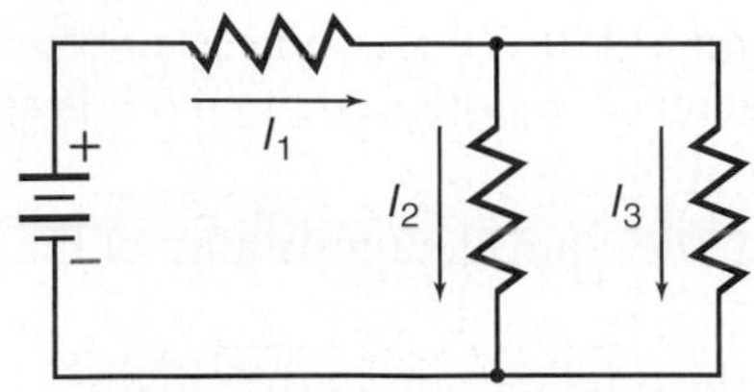

Figure 23

74. Refer to **Figure 23** and assume that each resistor uses 120 mW. Find the total power.

75. Refer to **Figure 23** and assume that $I_1 = 13$ mA and $I_2 = 1.7$ mA. Find I_3.

76. Refer to **Figure 23** and assume that $I_2 = 13$ mA and $I_3 = 1.7$ mA. Find I_1.

77. A circuit contains six 60-W lamps with a resistance of 240 Ω each and a 10.0-Ω heater connected in parallel. The potential difference across the circuit is 120 V. Find the current in the circuit for the following situations:

a. Four lamps are turned on.

b. All the lamps are turned on.

c. Six lamps and the heater are operating.

d. If the circuit has a 12-A fuse, will the fuse melt if all the lamps and the heater are on?

78. Ranking Task Consider the resistors in the circuit in **Figure 24.** Rank them from least to greatest specifically indicating any ties, using the following criteria:

a. the current through each

b. the potential difference across each

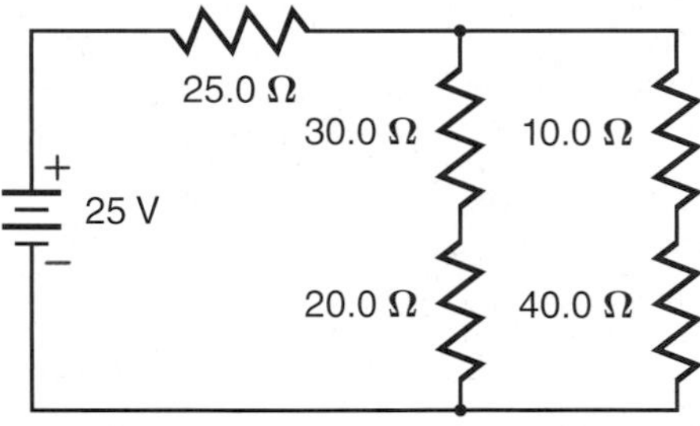

Figure 24

79. During a laboratory exercise, you are supplied with a battery of potential difference ΔV, two heating elements of low resistance that can be placed in water, an ammeter of very small resistance, a voltmeter of extremely high resistance, wires of negligible resistance, a beaker that is well insulated and has negligible heat capacity, and 0.10 kg of water at 25°C. By means of a diagram and standard symbols, show how these components should be connected to heat the water as rapidly as possible.

80. If the voltmeter used in the previous problem holds steady at 45 V and the ammeter reading holds steady at 5.0 A, estimate the time in seconds required to completely vaporize the water in the beaker. Use 4.2 kJ/(kg·°C) as the specific heat of water and 2.3×10^6 J/kg as the heat of vaporization of water.

81. **Home Circuit** A home circuit is shown in **Figure 25.** The wires to the kitchen light each have a resistance of 0.25 Ω. The light has a resistance of 0.24 kΩ. Although the circuit is parallel, the lead lines are in series with each of the components of the circuit.
   a. Compute the equivalent resistance of the circuit consisting of just the light and the lead lines to and from the light.
   b. Find the current through the light.
   c. Find the power used in the light.

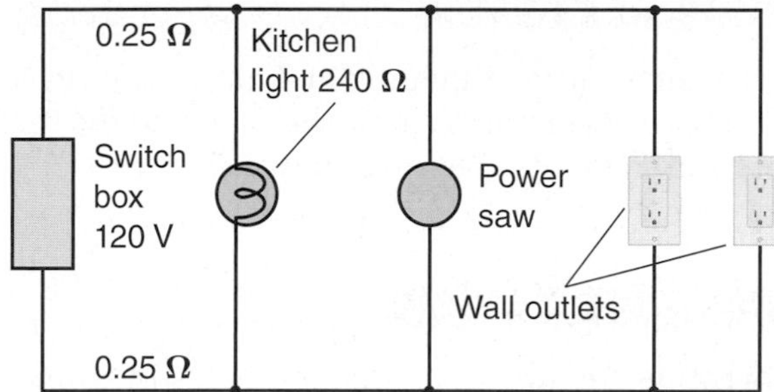

Figure 25

Applying Concepts

82. What happens to the current in the other two lamps if one lamp in a three-lamp series circuit burns out?

83. Suppose the resistor (R_1) in the voltage divider in **Figure 4** is made to be a variable resistor. What happens to the voltage output (ΔV_2) of the voltage divider if the resistance of the variable resistor is increased?

84. Circuit A contains three 60-Ω resistors in series. Circuit B contains three 60-Ω resistors in parallel. How does the current in the second 60-Ω resistor of each circuit change if a switch cuts off the current to the first 60-Ω resistor?

85. What happens to the current in the other two lamps if one lamp in a three-lamp parallel circuit burns out?

86. An engineer needs a 10-Ω resistor and a 15-Ω resistor, but there are only 30-Ω resistors in stock. Must new resistors be purchased? Explain.

87. If you have a 6-V battery and many 1.5-V bulbs, how could you connect them so that they light but do not have more than 1.5 V across each bulb?

88. Two lamps have different resistances, one larger than the other.
   a. If the lamps are connected in parallel, which is brighter (uses more power)?
   b. When the lamps are connected in series, which lamp is brighter?

89. **Household Fuses** Why is it dangerous to replace the 15-A fuse used to protect a household circuit with a fuse that is rated at 30 A?

90. For each of the following, write the form of circuit that applies: series or parallel.
   a. The current is the same everywhere throughout the entire circuit.
   b. The total resistance is equal to the sum of the individual resistances.
   c. The potential difference across each resistor in the circuit is the same.
   d. The potential difference across the battery is proportional to the sum of the resistances of the resistors.
   e. Adding a resistor to the circuit decreases the total resistance.
   f. Adding a resistor to the circuit increases the total resistance.

Mixed Review

91. A voltage divider consists of two 47-kΩ resistors connected in series across a 12-V battery. Determine the measured output across R_2 for the following meters.
   a. an ideal voltmeter
   b. a voltmeter with a resistance of 85 kΩ
   c. a voltmeter with a resistance of 10×10^6 Ω

92. Determine the maximum safe voltage that can be applied across the three series resistors in **Figure 26** if all three are rated at 5.0 W.

Figure 26

93. Determine the maximum safe total power for the circuit in the previous problem.

94. Determine the maximum safe voltage that can be applied across three parallel resistors of 92 Ω, 150 Ω, and 220 Ω, as shown in **Figure 27,** if all three are rated at 5.0 W.

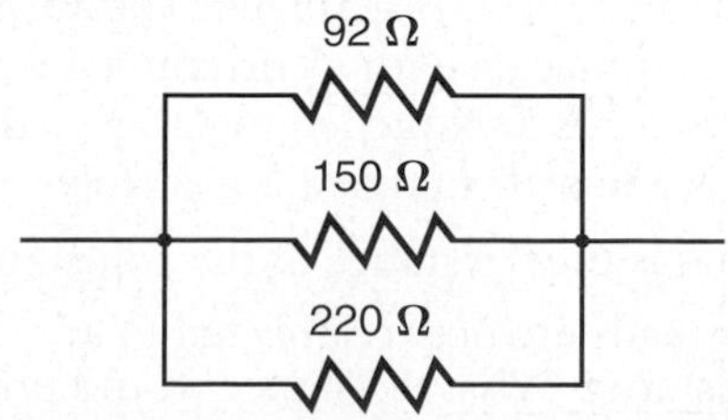

Figure 27

Chapter Self-Check

Thinking Critically

95. Apply Concepts Design a circuit that will light one dozen 12-V bulbs, all to the correct (same) intensity, from a 48-V battery.

a. Design A requires that should one bulb burn out, all other bulbs continue to produce light.

b. Design B requires that should one bulb burn out, those bulbs that continue working must produce the correct intensity.

c. Design C requires that should one bulb burn out, one other bulb also will go out.

d. Design D requires that should one bulb burn out, either two others will go out or no others will go out.

96. Apply Concepts A battery consists of an ideal source of potential difference in series with a small resistance. The electrical energy of the battery is produced by chemical reactions that occur in the battery. However, these reactions also result in a small resistance that, unfortunately, cannot be completely eliminated. A flashlight contains two batteries in series, as shown in **Figure 28.** Each has a potential difference of 1.50 V and an internal resistance of 0.200 Ω. The bulb has a resistance of 22.0 Ω.

a. What is the current through the bulb?

b. How much power does the bulb use?

c. How much greater would the power be if the batteries had no internal resistance?

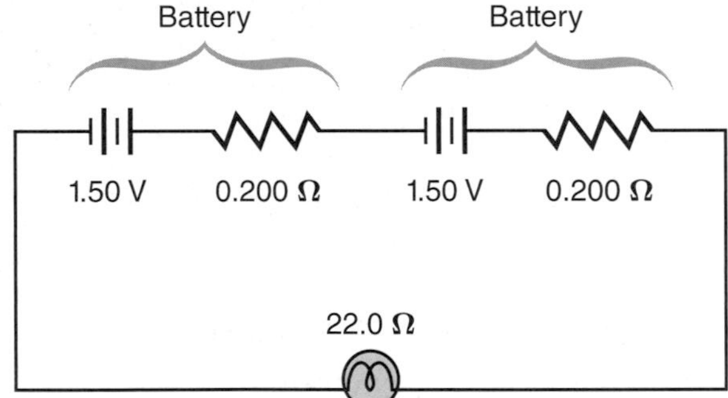

Figure 28

97. Apply Concepts An ohmmeter is made by connecting a 6.0-V battery in series with an adjustable resistor and an ideal ammeter. The ammeter deflects full-scale with a current of 1.0 mA. The two leads are touched together and the resistance is adjusted so that 1.0 mA flows.

a. What is the resistance of the adjustable resistor?

b. The leads are now connected to an unknown resistance. What resistance would produce a current of half-scale (0.50 mA)? Quarter-scale (0.25 mA)? Three-quarters-scale (0.75 mA)?

98. Reverse Problem Write a physics problem with real-life objects for which the following equations would be part of the solution:

$$6.0\ \text{V} = I_1(500\ \Omega)$$

$$6.0\ \text{V} = I_2(100\ \Omega + 200\ \Omega)$$

99. Problem Posing Complete this problem so that it can be solved using Ohm's law: "You are constructing a circuit and have a 4-kΩ resistor. . . ."

Writing in Physics

100. Research Gustav Kirchhoff and his laws. Write a one-page summary of how they apply to the three types of circuits presented in this chapter.

Cumulative Review

101. Airplane An airplane flying through still air produces sound waves. The wave fronts in front of the plane are spaced 0.50 m apart and those behind the plane are spaced 1.50 m apart. The speed of sound is 340 m/s.

a. What would be the wavelength of the sound waves if the airplane were not moving?

b. What is the frequency of the sound waves produced by the airplane?

c. What is the speed of the airplane?

d. What is the frequency detected by an observer located directly in front of the airplane?

e. What is the frequency detected by an observer located directly behind the airplane?

102. An object is located 12.6 cm from a convex mirror with a focal length of 218.0 cm. What is the location of the object's image?

103. The speed of light in a special piece of glass is 1.75×10^{8} m/s. What is its index of refraction?

104. Two charges of 2.0×10^{-5} C and 8.0×10^{-6} C experience a force between them of 9.0 N. How far apart are the two charges?

105. A field strength (E) is measured a distance (d) from a point charge (Q). What would happen to the magnitude of E in the following situations?

a. d is tripled.

b. Q is tripled.

c. Both d and Q are tripled.

d. The test charge q' is tripled.

e. All three, d, Q, and q', are tripled.

STANDARDIZED TEST PRACTICE

MULTIPLE CHOICE

1. Which is the equivalent resistance of the circuit shown below?

A. $\frac{1}{19}$ Ω **C.** 1.5 Ω
B. 1.0 Ω **D.** 19 Ω

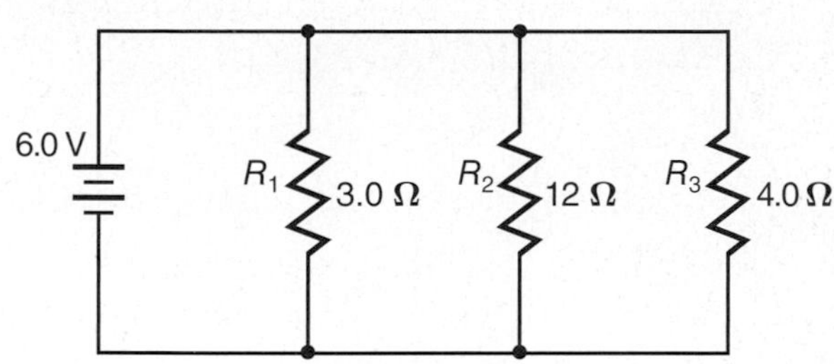

2. Three resistors, 3.0 Ω, 12 Ω, and 4.0 Ω, are connected in parallel across a 6.0-V battery as shown above. What is the current through the battery?

A. 0.32 A **C.** 1.2 A
B. 0.80 A **D.** 4.0 A

3. Three resistors, 3.0 Ω, 5.0 Ω, and 4.0 Ω, are connected in series across a 9.0-V battery. What is the current in the circuit?

A. 0.75 A **C.** 3.0 A
B. 1.8 A **D.** 2.3 A

4. Four resistors, 1.0 Ω, 3.0 Ω, 5.0 Ω, and 4.0 Ω, are connected in series with a 9.0-V battery. What is the equivalent resistance of the circuit?

A. 0.6 Ω **C.** 6.0 Ω
B. 1.8 Ω **D.** 13 Ω

5. Which is the current through the battery in the circuit shown below?

A. 1.15 A **C.** 2.80 A
B. 2.35 A **D.** 5.61 A

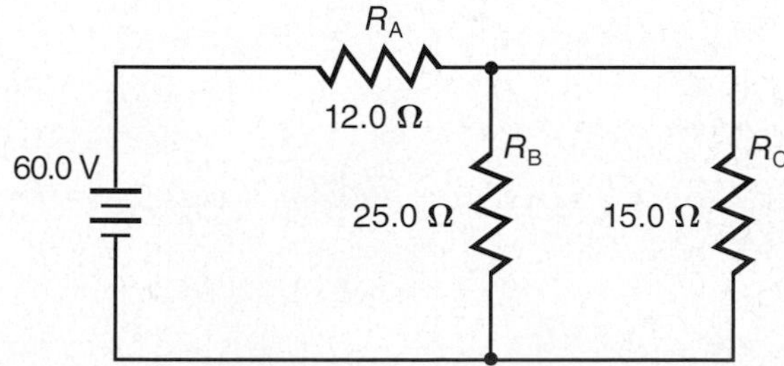

Online Test Practice

6. Which statement is true?

A. The resistance of a typical ammeter is very high.
B. The resistance of a typical voltmeter is very low.
C. Ammeters have zero resistance.
D. A voltmeter causes a small change in current.

7. Nina connects eight 12-Ω lamps in series. What is the total resistance of the circuit?

A. 0.67 Ω **C.** 12 Ω
B. 1.5 Ω **D.** 96 Ω

8. Which device contains an electronic circuit that detects small differences in current caused by an extra current path?

A. fuse
B. circuit breaker
C. ground-fault interrupter
D. short circuit

FREE RESPONSE

9. Chris is throwing a tailgate party before a nighttime football game. To light the tailgate party, he connects 15 large outdoor lamps in series to his 12.0-V car battery. Once connected, the lamps do not glow. An ammeter shows that the current through the lamps is 0.350 A. If the lamps require a 0.500-A current in order to work, how many lamps must Chris remove from the circuit?

10. A series circuit has an 8.0-V battery and four resistors, $R_1 = 4.0$ Ω, $R_2 = 8.0$ Ω, $R_3 = 13.0$ Ω, and $R_4 = 15.0$ Ω. Calculate the current and the power in the circuit.

NEED EXTRA HELP?

If You Missed Question	1	2	3	4	5	6	7	8	9	10
Review Section	1	1	1	1	2	2	1	2	1	1

CHAPTER 24

Magnetic Fields

BIG IDEA **Magnets and moving electric charges are surrounded by magnetic fields that exert forces on magnetic materials and on moving charges.**

SECTIONS

LaunchLAB

DIRECTION OF MAGNETIC FIELDS

Do magnetic fields have direction? Map the fields between two magnets and find out!

WATCH THIS!

MAGNETIC FIELDS

How are magnetic fields used in your life? You might be surprised! Explore the physics of everyday magnets and magnetic fields.

PHYSICS T.V.

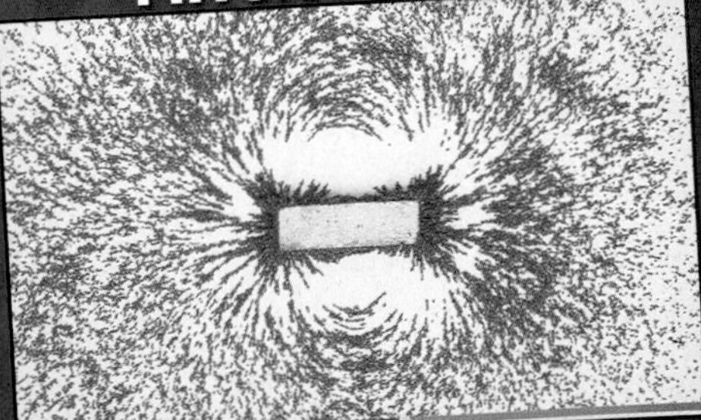

(l)Index Stock Imagery, Inc., (r)Kevin Schafer/The Image Bank/Getty Images

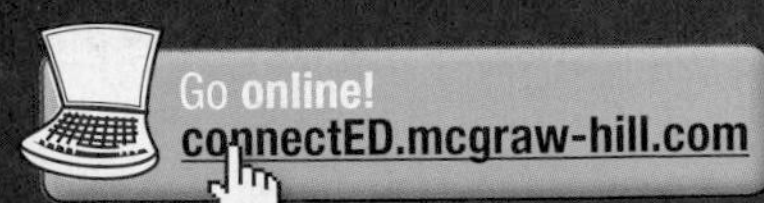
Go online!
connectED.mcgraw-hill.com

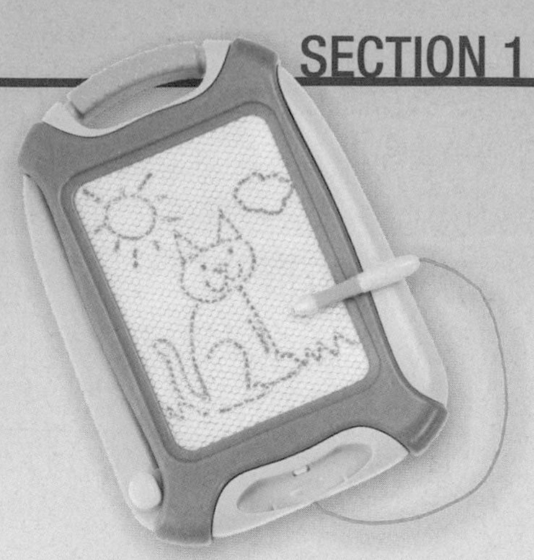

SECTION 1

Understanding Magnetism

PHYSICS 4 YOU **Have you ever played with a toy where you draw pictures by moving tiny iron filings around with a magnet? The magnet you use is a permanent magnet. But some magnets, such as the magnets in scrap yards, are controlled by switches. How can a magnet be turned on and off?**

MAIN IDEA

Magnets and electric currents produce magnetic fields.

Essential Questions

- What are some properties of magnets?
- What causes an object to be magnetic?
- What are the characteristics of magnetic fields?
- What is the relationship between magnetic fields and electric currents?

Review Vocabulary

conventional current the direction in which a positive test charge moves; negative charges move in the opposite direction

New Vocabulary

polarized
domain
magnetic field
magnetic flux
solenoid
electromagnet

Properties of Magnets

Magnets have been known and used for more than 2000 years. Ancient sailors used naturally magnetic rocks, called lodestones, as compasses. Ancient physicians thought lodestones could cure disease. Today, many objects we use every day rely on magnets to work. Electric motors, earbuds, loudspeakers, and computer hard drives all depend on the interaction between magnetic fields and electric currents.

Poles of magnets As a child, you likely did investigations with simple magnets, such as those shown in **Figure 1.** You probably noticed that the two ends of a magnet behave in different ways. That is because magnets are **polarized;** they have two opposite ends, called poles.

Figure 1 All magnets, no matter their size or shape, have two poles.

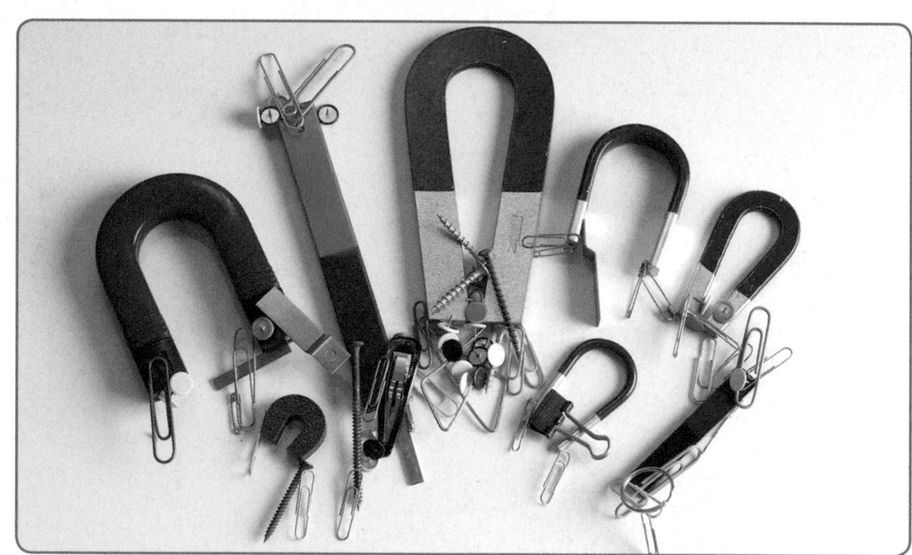

Think about a bar magnet suspended on a string. In what direction do you think it will point when it comes to rest? A magnet that is free to rotate always comes to rest pointing in the north-south direction. The pole pointing north is called the north-seeking pole or, more simply, the north pole. The opposite pole is the south pole. A compass is just a small magnet, mounted so that it is free to rotate.

Earth as a magnet The needle of a compass points in a north-south direction because Earth itself is a giant magnet. A compass's north pole points to Earth's geographic North Pole. As you will read, however, a magnet's north pole is always attracted to a magnetic south pole. Therefore, what we call the North Pole is actually near Earth's magnetic south pole, and the South Pole is near Earth's magnetic north pole.

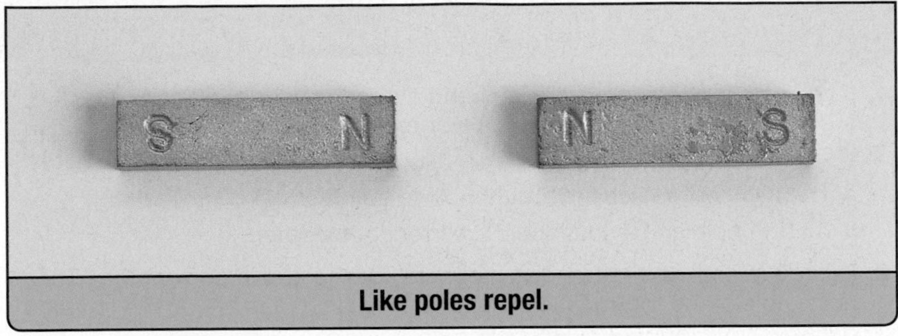

Figure 2 Like poles of two magnets repel each other (top), while unlike poles attract each other (bottom).

View a **BrainPOP video on magnets**.

Opposite poles What's inside a magnet that makes it polarized? You know that when you bring a metal rod near an electric charge, one end of the rod becomes negatively charged and the other end becomes positively charged, polarizing the rod. You might think a magnet is similar, with one half of the magnet positive and the other half negative, but this is not the case. No matter how you cut or break a magnet, a magnet always has two poles. There have been many searches for objects, called monopoles, with only a north pole or only a south pole, but no monopole has ever been found.

VOCABULARY

Science Usage v. Common Usage

Polarized

• **Science usage**
having two opposite ends
All magnets are polarized.

• **Common usage**
broken into opposing factions or groups
Members of Congress were polarized on the issue of Social Security reform.

Poles repel or attract You have likely noticed that forces between two magnets differ depending on how you orient the magnets. When you place the north pole of one magnet next to the north pole of another magnet, the magnets repel each other, as they do in the top of **Figure 2.** The same is true when you bring two south poles together. If, however, you brought the north pole of one magnet next to the south pole of another magnet, the poles would attract each other, as they do in the bottom of **Figure 2.** Like poles repel; unlike poles attract.

Temporary magnets Magnets also attract nails, paper clips, tacks, and other metal objects. These objects have no poles, and both the north and south poles of a magnet attract them. When a magnet touches one of these objects, such as the nail in **Figure 3,** the magnet polarizes the object, making it a temporary magnet. This process is called magnetization by induction.

Magnets only attract some metals. Brass, copper, and aluminum are common metals that are not attracted to magnets. Iron, nickel, and cobalt are strongly attracted. Materials containing these elements, called ferromagnetic materials, can become temporary magnets. A steel nail can become a temporary magnet because it is made of iron with tiny amounts of carbon and other materials. When you remove a nail from a magnet, the nail gradually loses most of its magnetism.

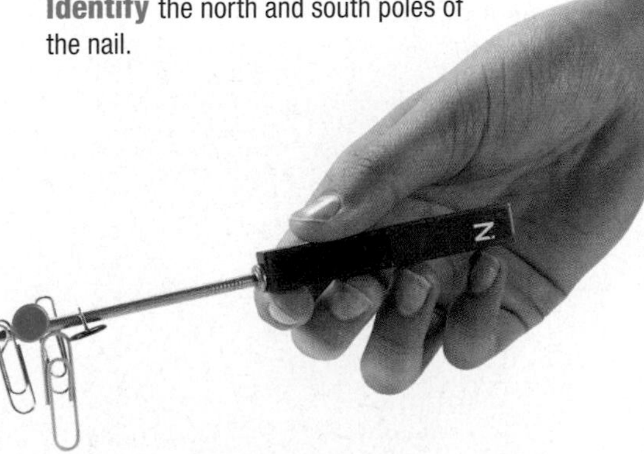

Figure 3 A common nail attached to a magnet becomes a temporary magnet by induction.

Identify the north and south poles of the nail.

Richard Hutchings/Digital Light Source

PRACTICE PROBLEMS

Do additional problems. Online Practice

PRACTICE PROBLEMS

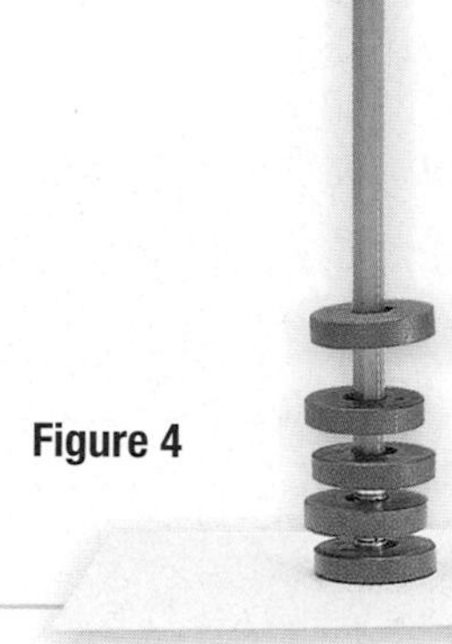

Figure 4

1. If you hold a bar magnet in each hand and bring your hands close together, will the force be attractive or repulsive if the magnets are held in the following ways?
   a. The two north poles are brought close together.
   b. A north pole and a south pole are brought together.
2. **Figure 4** (at left) shows five disk magnets floating above one another. The north pole of the top-most disk faces up. Which poles are on the top side of each of the other magnets?
3. The ends of a compass needle are marked N and S. How would you explain to someone why the pole marked N points north? A complete answer should involve Earth's magnetic poles.
4. **CHALLENGE** When students use magnets and compasses, they often touch the magnets to the compasses. Then they find that the compasses point south. Explain why this might occur.

Magnetic domains What gives a permanent or temporary magnet its magnetic properties? Each atom in a ferromagnetic material acts like a tiny magnet; each has two poles. Each is part of a **domain,** which is a group of neighboring atoms whose poles are aligned. Look at the arrows in **Figure 5.** Each arrow represents a domain. Although domains can contain as many as 10^{20} individual atoms, they are tiny—usually from 10 to 1000 microns across. Even a small sample of a ferromagnetic material contains a huge number of domains.

In a ferromagnetic material that is not magnetized, each domain points in a random direction, as shown in the top panel of **Figure 5.** But if the ferromagnetic material is next to a strong magnet, most of the object's domains preferentially align to point in the same direction as the poles of the external magnet, as shown in the bottom panel of **Figure 5.** When its domains are aligned in the same direction, the material becomes a temporary magnet.

When an external magnet is removed from a temporary magnet, the domains of the temporary magnet return to a random arrangement, and the material loses its magnetization. How long it takes for a temporary magnet to lose its magnetization depends on the interactions between the atoms, which depend on the microscopic structure of the material.

Figure 5 Domains in a nonmagnetized ferromagnetic material point in random directions (top). When a strong magnet is placed near a ferromagnetic material, the domains in that ferromagnetic material align with those of the external magnet (bottom).

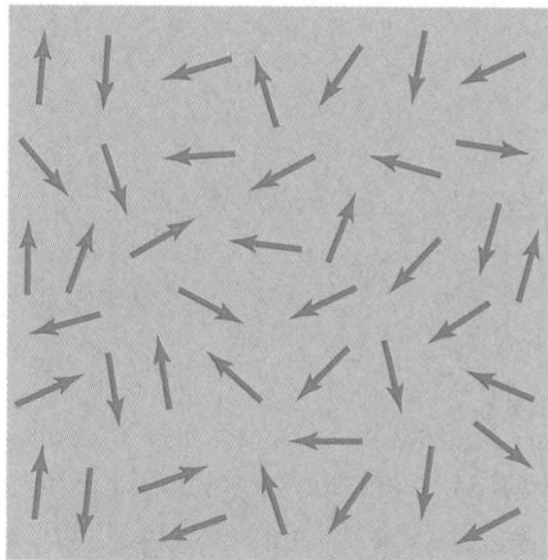

Nonmagnetized Material

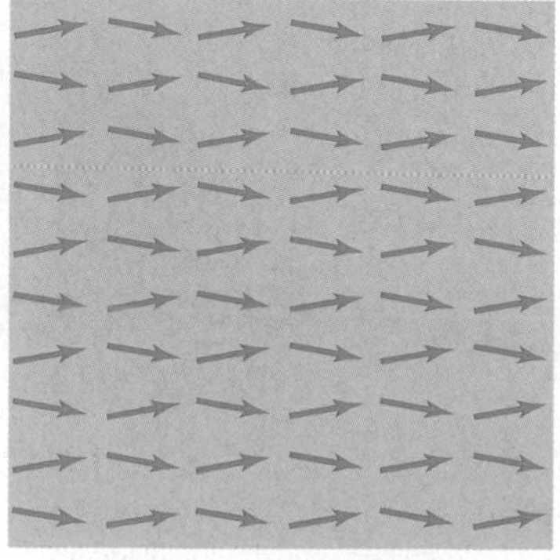

Magnetized Material

Creating permanent magnets The only naturally occurring magnet is the mineral magnetite. The lodestones that ancient sailors used were nothing more than pieces of magnetite. If magnetite is the only naturally occurring magnet, how, then, are commercial permanent magnets made?

When an object containing certain ferromagnetic materials is heated in the presence of a strong magnet, thermal energy frees the atoms in each of the object's domains. The domains can rotate and align with the magnet's poles. The object is then cooled while it is still in the presence of the strong magnet. After cooling, the object's atoms are less free to rotate. Therefore, when the strong magnet is removed from the object, the object remains magnetized. A permanent magnet has been created. If this permanent magnet is later reheated or dropped, however, the atoms can jostle out of alignment, reordering the domains and removing the magnetic properties.

Doug Martin

History of Earth's magnetism Magnetic domains in rocks containing iron record the history of Earth's magnetism. Rocks on the seafloor form when molten rock (magma) pours out of cracks in the bottom of the oceans. As the magma cools into rock, the domains in the iron-containing rocks align in the direction of Earth's magnetic field. These rocks become weak permanent magnets. As more magma pours out of the cracks, the older rocks are pushed away from the cracks. As a result, rocks farther from the cracks are older than those near the cracks.

Scientists who first examined seafloor rocks were surprised to find that the alignment of domains in the iron of rocks of different ages varied. They concluded that Earth's magnetic north and south poles have exchanged places many times during Earth's history.

Magnetic Fields Around Magnets

When you investigate simple magnets, you notice that the forces between magnets are present not only when magnets touch each other, but also when magnets are held apart. Just as the existence of long-range electric and gravitational forces can be interpreted as being the result of electric and gravitational fields, long-range magnetic forces can be interpreted as being the result of magnetic fields.

Magnetic fields are fields that exist in space where magnets would experience a force. They are vector quantities because they have magnitude and direction. The needle of a compass in Earth's magnetic field aligns in the direction of Earth's field. When in a stronger magnetic field, the needle realigns in the direction of the stronger field.

☑ READING CHECK **Compare** How are magnetic fields like electric fields?

Visualizing magnetic fields What do magnetic fields look like? Like electric and gravitational fields, magnetic fields are invisible. But we can visualize them in a few different ways. One way is to place iron filings around a magnet. Each long, thin iron filing around a magnet becomes a temporary magnet by induction. Just like thousands of tiny compass needles, each iron filing rotates until it is parallel to the magnetic field. You can see the results both two-dimensionally and three-dimensionally in **Figure 6.**

MiniLABs

MAGNETIC DOMAINS
What makes a magnet magnetic?

3-D MAGNETIC FIELDS
What does a magnetic field look like?

iLab Station

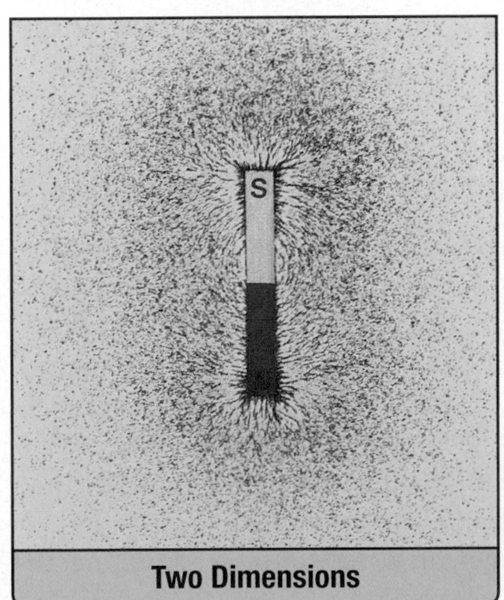

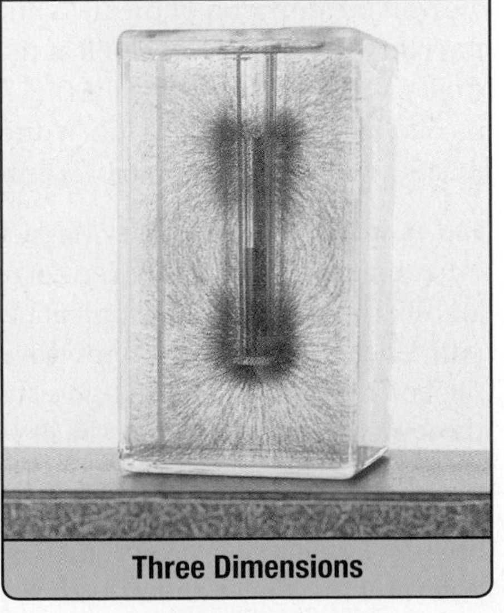

Figure 6 Iron filings illustrate the magnetic field of a bar magnet two-dimensionally (left, on paper) and three-dimensionally (right, in liquid).

Figure 7 Magnetic field lines can be visualized as lines leaving the north pole of a magnet, entering the south pole, and passing through the magnet, forming closed loops. Magnetic fields are traditionally represented by the letter *B*.

COLOR CONVENTION

Magnetic field lines ⟷ green

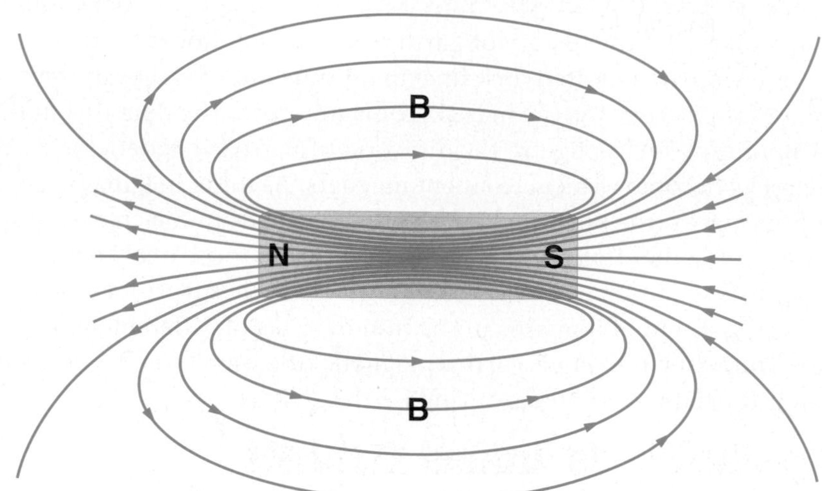

Magnetic field lines Scientists visualize magnetic fields using magnetic field lines, such as those shown in **Figure 7.** Like electric field lines, magnetic field lines are not real. They are used to show the direction as well as the strength of a magnetic field. The number of magnetic field lines passing through a surface perpendicular to the lines is the **magnetic flux.** The flux per unit area is proportional to the strength of the magnetic field. Magnetic flux is most concentrated at magnetic poles, where magnetic field strength is the highest.

The direction of a magnetic field line is defined as the direction in which the north pole of a compass points when placed in a magnetic field. Therefore, field lines emerge from a magnet's north pole and enter at its south pole, as in **Figure 7.** The field lines form closed loops, continuing through a magnet from its south pole to its north pole.

Figure 8 Iron filings can be used to visualize the magnetic field around two like poles (top) and around two unlike poles (bottom). The iron filings help us understand how like poles repel and unlike poles attract.

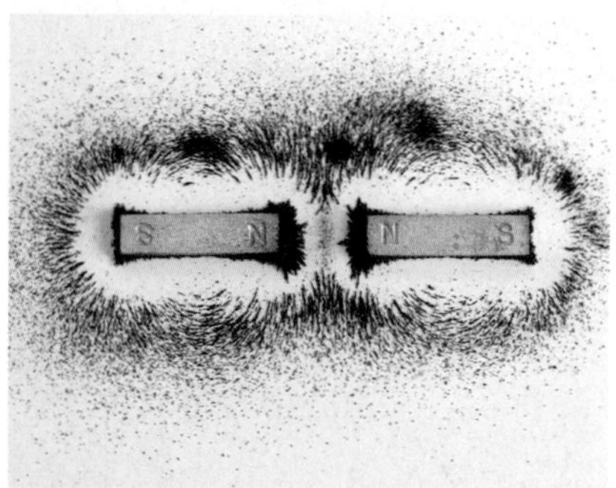

Like poles repel.

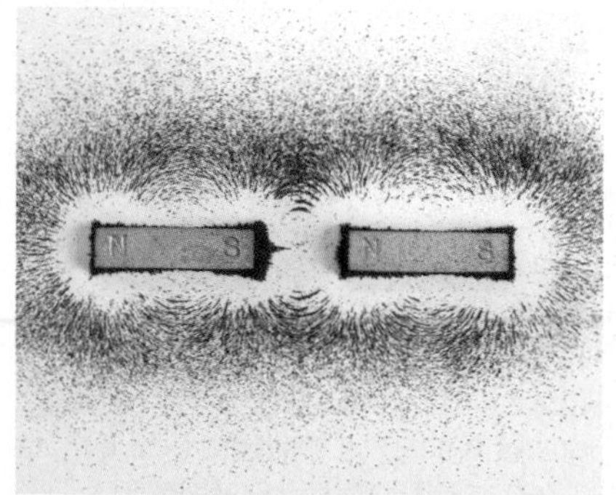

Unlike poles attract.

Forces on objects in magnetic fields

You read earlier that a magnet can polarize ferromagnetic materials. How can this be explained in terms of magnetic fields and forces?

Forces on permanent magnets Magnetic fields exert forces on magnets. When like poles of two magnets are close together, the field produced by the north pole of one magnet pushes the north pole of the second magnet away in the direction of the field lines, as shown by the iron filings in the top panel of **Figure 8.** Now look at the bottom panel of **Figure 8.** The field from the north pole of one magnet now acts on the south pole of the second magnet, attracting it in a direction opposite the field lines. The magnetic field is continuous, forming arcs from one magnet to the other.

Forces on temporary magnets Magnetic fields exert forces not only on other magnets. They also exert forces on ferromagnetic materials. When an object containing a ferromagnetic material is placed in the field of a permanent magnet, field lines leave the magnet's north end and enter the end of the object that is closest to the magnet. The field lines pass through the object and loop back to the magnet's south pole. The domains in the object align their poles along the field lines, making the end of the object closest to the magnet's north pole the object's south pole. The object's new south pole is then attracted to the magnet's north pole, and the object's new north pole is repelled.

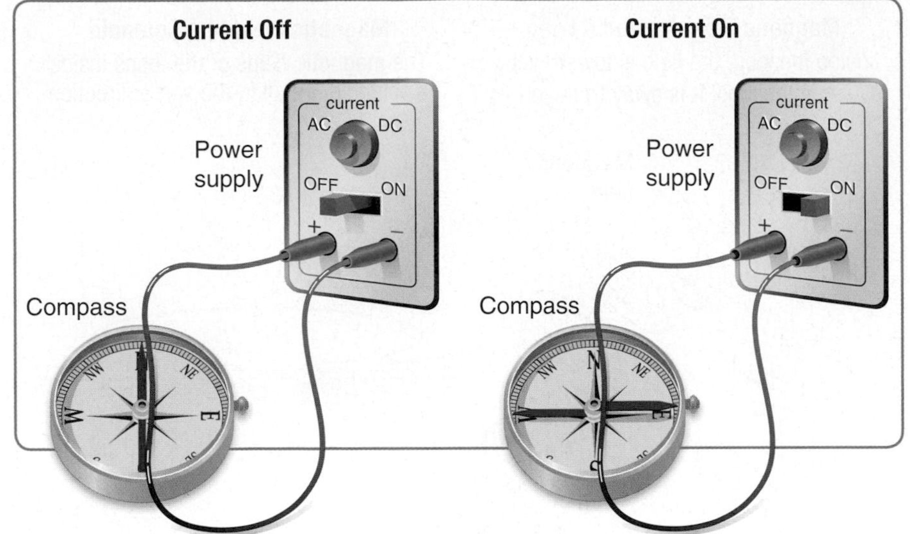

Figure 9 The needle of a compass under a wire and originally parallel to the wire when current is off (left) moves so it is perpendicular to the wire when current is on (right).

View an **animation of a compass and current.**

Electromagnetism

In 1820, while doing a lecture demonstration, Danish physicist Hans Christian Oersted laid a wire across the top of a compass and connected the ends to a battery to complete an electric circuit. The compass was oriented so its needle was parallel to the wire, as shown in the left side of **Figure 9.** When Oersted turned the current on, he was amazed to see that the needle moved so it was perpendicular to the wire, as it is in the right side of **Figure 9.**

When Oersted placed the compass on top of the wire, the needle again became perpendicular to the wire, but it pointed in the other direction. The same thing happened when he reversed the current's direction: the compass needle reversed direction. When he turned off the current, the needle returned to its original position.

Oersted's conclusion—that a current produces a magnetic field—was the first hint that a connection exists between magnetism and electric currents. As you will read, the relationship between magnetism and electric current today underlies the design and operation of many modern devices.

Magnetic fields from current-carrying wires The magnetic field around a current-carrying wire is always perpendicular to that wire. Just as field lines around permanent magnets form closed loops, the field lines around current-carrying wires also form closed loops. The circular pattern of iron filings shown in the top panel of **Figure 10** represents these loops. The strength of the magnetic field around a long, straight wire is proportional to the current in that wire. Magnetic field strength also varies inversely with distance from the wire.

Direction of the magnetic field How can you find the direction of the magnetic field around a current-carrying wire? Scientists use right-hand rules to describe how the directions of electric and magnetic properties relate. In this case, imagine holding a length of wire with your right hand. If your thumb points in the direction of the conventional (positive) current, as it does in the bottom panel of **Figure 10,** the fingers of your hand encircling the wire will point in the direction of the magnetic field.

☑ **READING CHECK Analyze** What happens to the magnetic field around a wire when current changes direction?

PhysicsLAB

CURRENT AND FIELD STRENGTH

How can current produce a strong magnetic field?

Figure 10 The circular patterns formed by iron filings around a current-carrying wire (top) represent the magnetic field around the wire. You can determine the direction of the magnetic field around the wire using a right-hand rule (bottom).

Magnetic Field Around a Wire

Right-Hand Rule

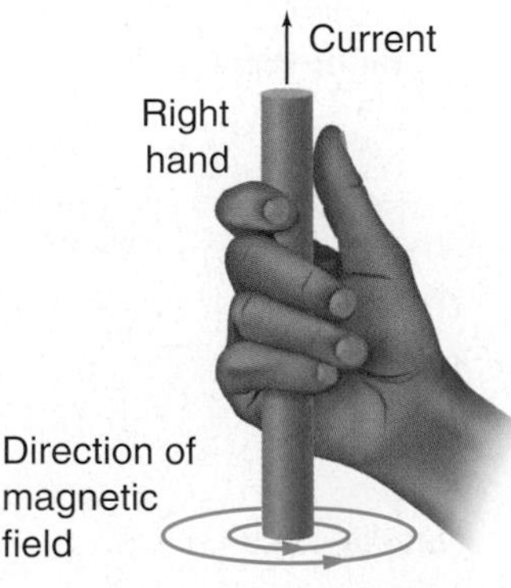

Figure 11 You can model the direction of the magnetic field around a loop of current-carrying wire and around a solenoid.

Assess Is the magnetic field greater inside or outside the solenoid?

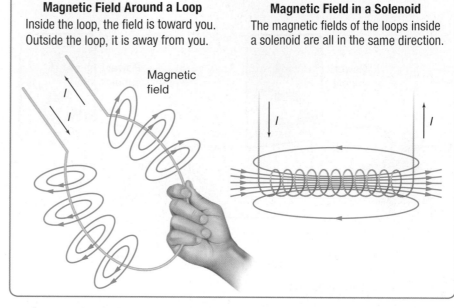

Magnetic Field Around a Loop
Inside the loop, the field is toward you. Outside the loop, it is away from you.

Magnetic field

I

I

Magnetic Field in a Solenoid
The magnetic fields of the loops inside a solenoid are all in the same direction.

I

I

PhysicsLAB

MAKE AN ELECTROMAGNET

DESIGN YOUR OWN What affects the strength of an electromagnet?

Electromagnets You just read that a current in a wire produces a magnetic field encircling that wire. What do you think happens to the magnetic field around a wire formed into a loop? An electric current in a single loop of wire forms a magnetic field all around the loop, as shown in the left panel of **Figure 11.** By applying a right-hand rule to any part of the loop in **Figure 11,** you can see that the direction of the magnetic field inside the loop is always the same.

Now think about a wire with many loops. A wire connected to a circuit and coiled into many spiral loops is a **solenoid.** When current is turned on in a solenoid, each loop produces its own magnetic field. The fields are all in the same direction, as shown in the right panel of **Figure 11,** so the fields add together.

When there is an electric current in a solenoid, the solenoid has a magnetic field similar to the field of a permanent magnet. This kind of magnet is an electromagnet. An **electromagnet** is a magnet whose magnetic field is produced by electric current.

Loops and field strength Solenoids can be exceptionally strong electromagnets, producing magnetic fields much stronger than those around permanent magnets. The strength of the magnetic field in a solenoid is proportional to the current in the solenoid's loops. It is also proportional to the number and spacing of loops. The more loops there are in a solenoid and the closer they are spaced, the greater the solenoid's magnetic field strength.

The magnetic field strength of a solenoid also can be increased by placing an iron-containing rod inside it. An iron rod strengthens the solenoid's magnetism because the solenoid's field produces a temporary magnetic field in the iron, just as a permanent magnet produces a temporary magnet in a ferromagnetic object.

Figure 12 Imagine you are holding the solenoid with your right hand. Your thumb will point toward the solenoid's north pole when you curl your fingers in the direction of the conventional current.

Right-Hand Rule

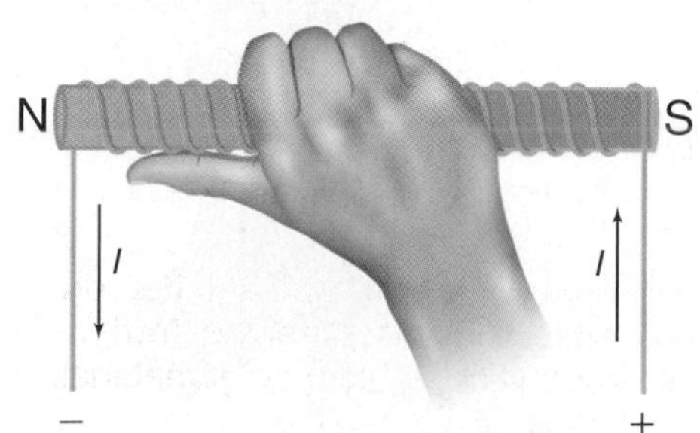

Right-hand rule for a solenoid You can use a right-hand rule to determine the direction of the magnetic field around a solenoid when current is on. Imagine holding a solenoid with your right hand. If you curl your fingers around the solenoid in the direction of the conventional (positive) current, as in **Figure 12,** your thumb will point toward the solenoid's north pole.

PRACTICE PROBLEMS

Do additional problems. Online Practice

5. How does the strength of a magnetic field that is 1 cm from a current-carrying wire compare with each of the following?

a. the strength of the field 2 cm from the wire

b. the strength of the field 3 cm from the wire

6. A long, straight current-carrying wire lies in a north-south direction.

a. The north pole of a compass needle placed above this wire points toward the east. In what direction is the current?

b. If a compass were placed underneath this wire, in which direction would the compass needle point?

7. A student makes a magnet by winding wire around a nail and connecting it to a battery, as shown in **Figure 13.** Which end of the nail—the pointed end or the head—is the north pole?

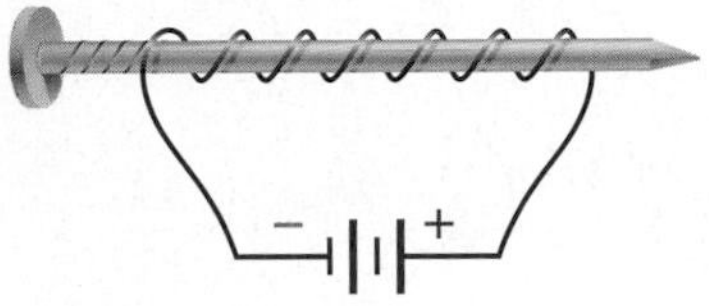

Figure 13

8. You have a battery, a spool of wire, a glass rod, an iron rod, and an aluminum rod. Which rod could you use to make an electromagnet that can pick up steel objects? Explain.

9. CHALLENGE The electromagnet in the previous problem works well, but you would like to make the strength of the electromagnet adjustable by using a potentiometer as a variable resistor. Is this possible? Explain.

PRACTICE PROBLEMS

Electromagnets in computer hard drives A computer uses a tiny solenoid in its hard drive to record information on a spinning disk coated with ferromagnetic material. How does this work? Computer data and software commands are processed digitally in bits. Each bit is either a 1 or a 0. As a solenoid passes over the ferromagnetic material on the spinning disk, the solenoid's magnetic field orientation reverses back and forth, depending on whether the bit is a 1 or a 0. With each pass, the alignment of the domain in the ferromagnetic material directly under it changes.

To read a series of bits, the disk moves under the same wire coil. The coil senses the 1s and 0s and encodes them into number sequences that represent the stored data.

SECTION 1 REVIEW

Section Self-Check Check your understanding.

10. MAINIDEA Explain how to construct an electromagnet.

11. Magnetic Fields What two things about a magnetic field can magnetic field lines represent?

12. Magnetic Forces Identify some magnetic forces around you. How could you demonstrate the effects of those forces?

13. Magnetic Fields Where on a bar magnet is the magnetic field the strongest?

14. Magnetic Fields Two current-carrying wires are close to and parallel to each other and have currents with the same magnitude. If the two currents were in the same direction, how would the magnetic fields of the wires be affected? How would the fields be affected if the two currents were in opposite directions?

15. Direction of Field Describe how to use a right-hand rule to determine the direction of a magnetic field around a straight, current-carrying wire.

16. Electromagnets A glass sheet with iron filings sprinkled on it is placed over an active electromagnet. The iron filings produce a pattern. If this scenario were repeated with the direction of current reversed, what observable differences would result? Explain.

17. Magnetic Domains Explain what happens to the domains of a temporary magnet when the temporary magnet is removed from a magnetic field.

18. Critical Thinking Imagine a toy containing two parallel, horizontal metal rods, one above the other. The top rod is free to move up and down.

a. The top rod floats above the lower rod. When the top rod's direction is reversed, however, it falls down onto the lower rod. Explain how the rods could behave in this way.

b. Assume the toy's top rod was lost and another rod replaced it. The new rod falls on top of the bottom rod no matter its orientation. What type of material is in the replacement rod?

SECTION 2 Applying Magnetic Forces

PHYSICS 4 YOU **Do you use earbuds to listen to your MP3 player when you exercise or ride a bus? An earbud contains a tiny magnet. The magnet's magnetic field creates a force that acts on a current-carrying coil of wire. The coil moves in response to the force and sound waves are produced.**

MAIN IDEA

Many devices, including earbuds and electric motors, rely on forces from magnetic fields in order to work.

Essential Questions

- How is the direction of the force on a current-carrying wire related to the direction of the magnetic field?
- What affects the force on a current-carrying wire in a magnetic field?
- What are the characteristics of the design and operation of an electric motor?
- What affects the force on a charged particle moving in a magnetic field?

Review Vocabulary

resistor a device with a specific resistance

New Vocabulary

galvanometer
electric motor
armature

Forces on Current-Carrying Wires

When you put a magnet in a magnetic field, the magnet can move. What happens when you put a current-carrying wire in a magnetic field? Michael Faraday, who performed many electricity and magnetism experiments during the nineteenth century, discovered that a magnetic field produces a force on a current-carrying wire. The force on the wire is always at right angles to both the direction of the magnetic field and the direction of current, as shown in the left part of **Figure 14.** When current changes direction, so does the force.

Direction of force You can use a right-hand rule to determine the direction of force on a current-carrying wire in a magnetic field. Point the fingers of your right hand in the direction of the magnetic field. Point your thumb in the direction of the wire's conventional (positive) current. The palm of your hand will face in the direction of the force acting on the wire, as shown in the right part of **Figure 14.**

Figure 14 You can use a right-hand rule to determine the direction of force when the current (I) and the magnetic field (B) are known.

Predict what would happen to the force if the current changed direction.

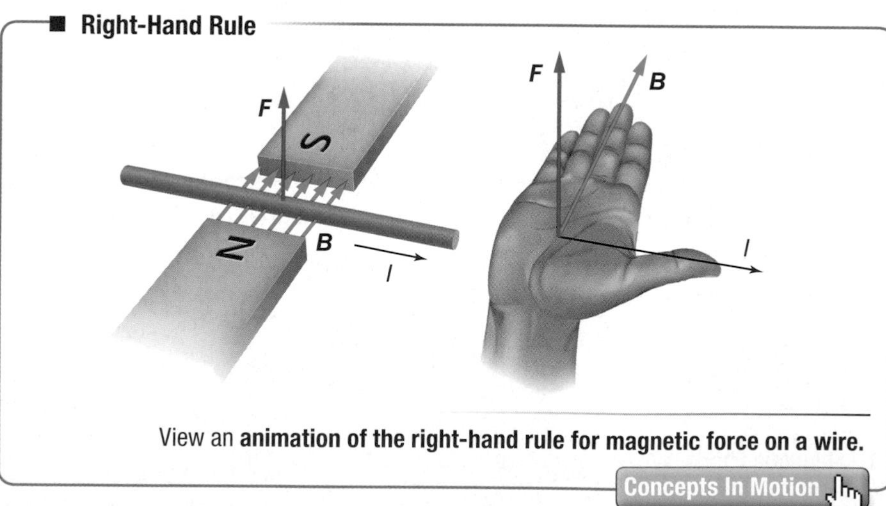

View an **animation of the right-hand rule for magnetic force on a wire.**

Concepts In Motion

Arrows in three dimensions The relationship among magnetic field, electric current, and force is three-dimensional. How do you accurately represent directional arrows in three dimensions on a two-dimensional piece of paper? Imagine an archer shooting an arrow toward you. The arrow looks like a dot. Now imagine the same arrow going away from you. The arrow looks like a cross. You can use dots to represent magnetic fields that go into a piece of paper, and crosses to represent fields that go out of the paper, as shown in **Figure 15** on the opposite page.

Masterfile

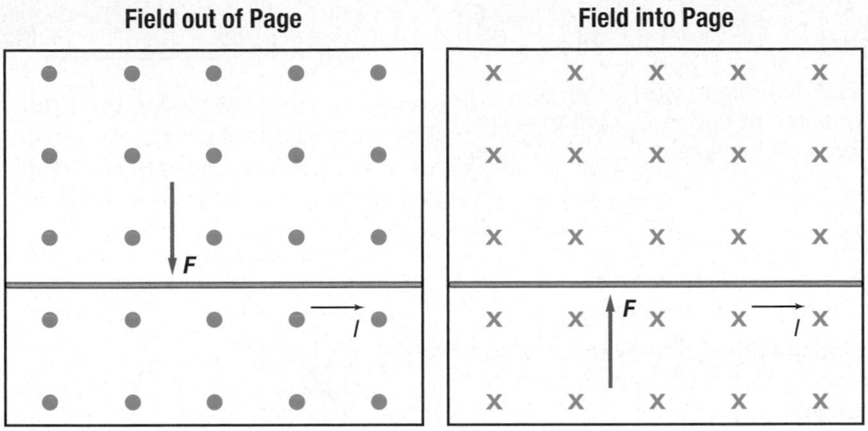

Field out of Page

Field into Page

Figure 15 Dots represent a magnetic field coming out of the page, toward you (left). Crosses represent a magnetic field going into the page, away from you (right). Note that the force on each wire is perpendicular to both the magnetic field and the current.

Magnitude of force You read that you use a right-hand rule to find the direction of the force from a magnetic field on a current-carrying wire. How do you find the magnitude of this force?

Experiments show that the magnitude of the force (F) on a current-carrying wire is proportional to the wire's current (I), the wire's length (L), the strength of the magnetic field (B), and the sine of the angle between the current and the magnetic field ($\sin\theta$). Recall that you measure force in newtons (N) and current in amperes (A). You measure the strength of a magnetic field (B) in teslas (T). One T equals 1 N/(A·m).

FORCE ON A CURRENT-CARRYING WIRE IN A MAGNETIC FIELD

The magnitude of the force on a current-carrying wire in a magnetic field is equal to the product of the current, the length of the wire, the field strength, and the sine of the angle between the current and the magnetic field.

$$F = ILB(\sin\theta)$$

Note that $\sin 0° = 0$, and $\sin 90° = 1$. This means that when the current and the magnetic field are parallel to each other, the force on a current-carrying wire is zero. The force on the wire is greatest when the current and the magnetic field are perpendicular to each other.

Earbuds You might wonder how people apply the relationship among magnetic fields, electric currents, and force in today's technology. You are probably familiar with one example—earbuds.

If you look inside an earbud, such as the one in **Figure 16,** you will find a tiny coil of wire attached to a thin plastic membrane. Beneath the membrane is a permanent magnet. The magnetic field from the permanent magnet is oriented radially so it is perpendicular to both the coil of wire and the direction of motion of the coil.

A music player sends current through an earbud's wires. The current enters the coil, changing direction between 40 and 40,000 times each second, depending on the pitches of the tones it represents. The force from the magnetic field on the coil pushes the coil in and out, depending on the direction of current. This causes the membrane to vibrate, thereby producing sound waves. Each time the current changes direction twice, the membrane vibrates back and forth once.

Most loudspeakers and headphones work in a similar way. A magnetic field exerts a force on a coil of wire mounted on a paper or plastic cone. As the wire moves, it pushes the coil into and out of the field. This motion causes the cone to vibrate and produce sound waves.

Figure 16 An earbud works because a wire in a magnetic field experiences a force. The force makes the wire move, which causes an overlying membrane to vibrate and produce sound waves.

EXAMPLE PROBLEM 1

Find help with **isolating a variable.** Math Handbook

CALCULATE THE STRENGTH OF A MAGNETIC FIELD A straight wire carrying a 5.0-A current is in a uniform magnetic field oriented at right angles to the wire. When 0.10 m of the wire is in the field, the force on the wire is 0.20 N. What is the strength of the magnetic field (B)?

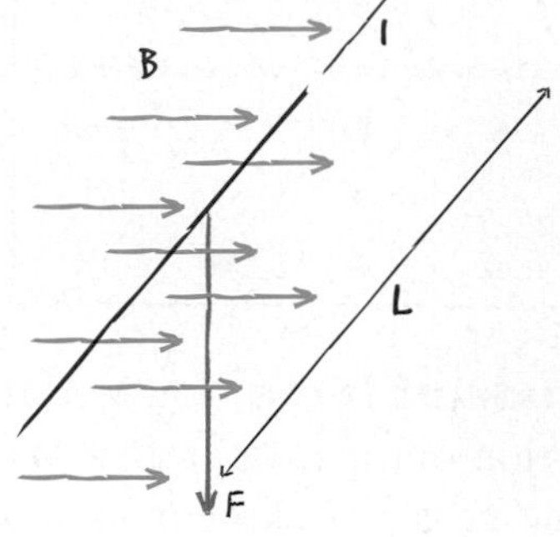

1 ANALYZE AND SKETCH THE PROBLEM

- Sketch the wire and show the direction of the current with an arrow, the magnetic field as B, and the force on the wire as F.
- Determine the direction of the force using the right-hand rule for the force on a current-carrying wire in a magnetic field. The field, the wire, and the force are all at right angles.

KNOWN	UNKNOWN
$I = 5.0$ A	$B = ?$
$L = 0.10$ m	
$F = 0.20$ N	

2 SOLVE FOR THE UNKNOWN

B is uniform, and because B and I are perpendicular to each other, $F = ILB$.

$$F = ILB$$

Solve for B.

$$B = \frac{F}{IL}$$

$$= \frac{0.20\ \text{N}}{(5.0\ \text{A})(0.10\ \text{m})}$$ ◀ Substitute $F = 0.20$ N, $I = 5.0$ A, $L = 0.10$ m.

$$= 0.40\ \text{N/A·m} = 0.40\ \text{T}$$

B is 0.40 T from left to right and perpendicular to I and F.

3 EVALUATE THE ANSWER

- **Are the units correct?** The answer is in teslas, the correct unit for a magnetic field.
- **Is the magnitude realistic?** A magnetic field with a strength under 1 T is realistic.

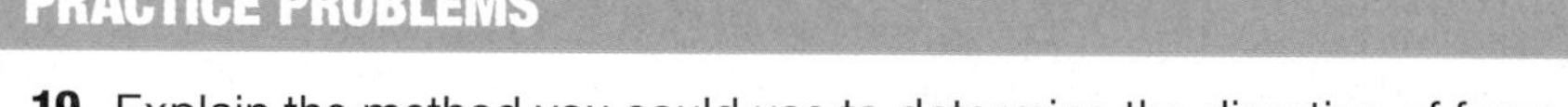

PRACTICE PROBLEMS

Do additional problems. Online Practice

19. Explain the method you could use to determine the direction of force on a current-carrying wire at right angles to a magnetic field. Identify what must be known to use this method.

20. A wire that is 0.50 m long and carrying a current of 8.0 A is at right angles to a 0.40-T magnetic field. How strong is the force that acts on the wire?

21. A wire that is 75 cm long and carrying a current of 6.0 A is at right angles to a uniform magnetic field. The magnitude of the force acting on the wire is 0.60 N. What is the strength of the magnetic field?

22. A 40.0-cm-long copper wire carries a current of 6.0 A and weighs 0.35 N. A certain magnetic field is strong enough to balance the force of gravity on the wire. What is the strength of the magnetic field?

23. How much current would be required to produce a force of 0.38 N on a 10.0-cm length of wire at right angles to a 0.49-T field?

24. CHALLENGE You are making your own loudspeaker. You make a 1-cm-diameter coil with 20 loops of thin wire. You use hot glue to fasten the coil to an aluminum pie plate. The ends of the wire are connected to a plug that goes into the earphone jack on an MP3 music player. You have a bar magnet to produce a magnetic field. How would you orient the magnetic field to make the plate vibrate and produce sound?

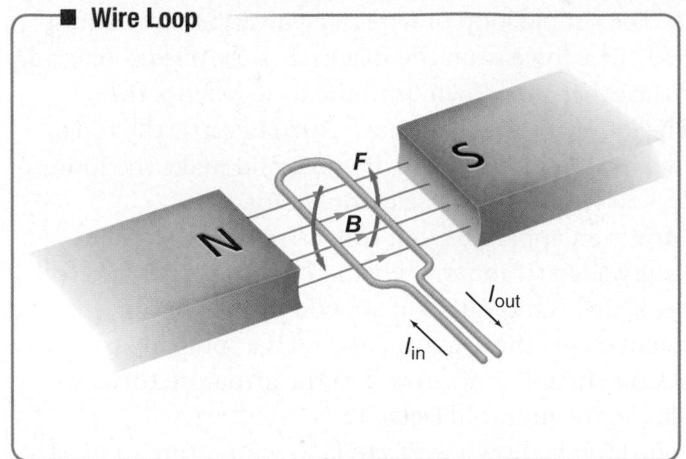

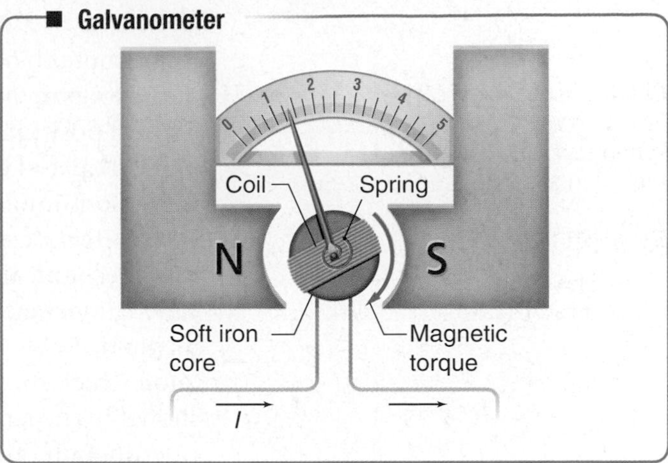

Figure 17 When a wire loop is placed in a magnetic field and current is turned on, the loop rotates (left). The coil in a galvanometer (right) rotates in proportion to the magnitude of the current.

Galvanometers You can use the force that a magnetic field exerts on a loop of current-carrying wire to measure currents. How does this work? Current in the wire loop shown in the left of **Figure 17** passes in one end of the loop and out the other. As it does, the force on the loop pushes one side of the loop down and the other side up. The resulting torque causes the loop to rotate. The magnitude of the torque on the loop is proportional to the magnitude of the current. This is the principle used in a galvanometer.

A **galvanometer** is a device used to measure very small currents. The torque of a small coil spring in this device opposes the torque from the current in the wire loop. The amount of rotation is proportional to the current. The meter is calibrated by determining how much the coil turns when a known current is sent through it, as shown in the right of **Figure 17.** The galvanometer can then be used to measure unknown currents.

Ammeter A galvanometer can produce full-scale deflections with currents as small as 50 μA (50×10^{-6} A). The resistance of the wire loop in a sensitive galvanometer is about 1000 Ω. To measure larger currents, a galvanometer can be converted to an ammeter. To do this, you would place a resistor with small resistance in parallel with the meter, as shown in the top of **Figure 18.** Because current is inversely proportional to resistance, most of the current (I_s) passes through this resistor, called the shunt because it shunts, or bypasses, much of the current around the galvanometer. Only a few microamps (I_m) pass through. The shunt's resistance is chosen to produce the desired meter sensitivity.

Voltmeter You can also connect a galvanometer as a voltmeter. To make a voltmeter, you would place a resistor, called the multiplier, in series with the meter, as in the bottom of **Figure 18.** The galvanometer measures the current through the multiplier. The current is $I = \frac{V}{R}$, where V is the potential difference across the voltmeter, and R is the effective resistance of the galvanometer and the multiplier.

Suppose you want the voltmeter's needle to move across the entire scale when 10 V is placed across it. You would use a resistor so that at 10 V the meter is deflected full-scale by a 50 μA current through the meter and the resistor.

READING CHECK Compare How would you use a resistor to convert a galvanometer first to an ammeter and then to a voltmeter?

Figure 18 A galvanometer can be connected for use as an ammeter (top) and as a voltmeter (bottom).

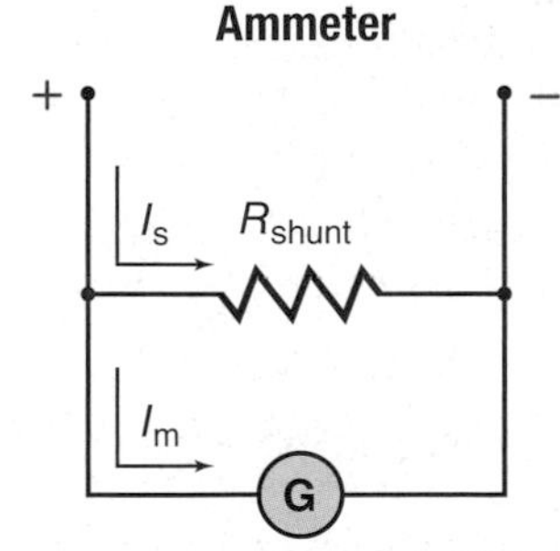

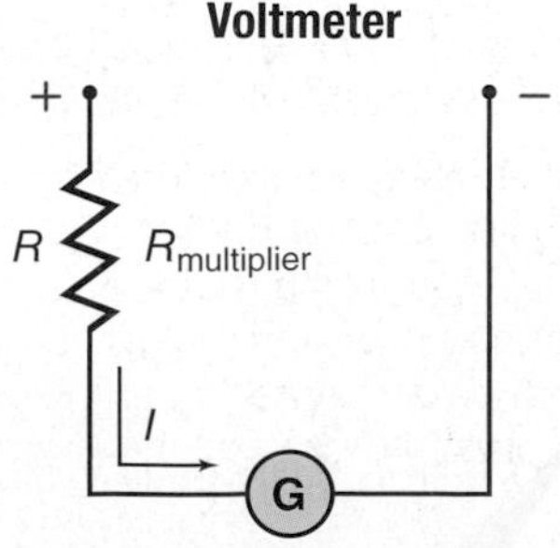

Electric motors The simple loop of wire in a galvanometer cannot rotate more than 180°. The force from the magnetic field pushes one side of the loop up and the other side down until the loop reaches the vertical position. The loop will not continue to turn because the forces are still up and down, parallel to the loop. If you could make the loop rotate continuously, you would have an electric motor.

An **electric motor** is an apparatus that converts electrical energy into mechanical energy. Electric motors rely on a multilooped wire coil called an **armature,** which is mounted on an axle that rotates in a magnetic field. Current enters the armature through a split-ring commutator, which reverses the direction of current as the armature turns, as shown in the simple electric motor in **Figure 19.**

Although only one loop is shown in **Figure 19,** the armature in most electric motors has many loops. The total force acting on the armature is proportional to $nILB$, where n is the total number of turns on the armature (each completing 360°), I is the current, L is the length of wire in each turn that moves perpendicular to the magnetic field, and B is the strength of the magnetic field. The magnetic field is produced either by a permanent magnet or by an electromagnet (called a field coil). The torque on the armature is controlled by varying the current through the motor. The larger the torque, the faster the armature turns.

Forces on Single Charged Particles

You use $F = ILB(\sin \theta)$ to determine the force on a current-carrying wire in a magnetic field. A current is simply a stream of charged particles. How do you determine the force on a single charged particle?

Equation of force The magnetic force on a single charged particle depends on the velocity of the particle, the strength of the magnetic field, and the angle between the directions of the velocity and the field. Consider a single electron moving in a wire of length L that is perpendicular to a magnetic field (B). Current (I) is equal to the charge per unit time entering the wire, $I = \frac{q}{t}$. In this case, q is the charge of the electron and t is the time it takes for the electron to move the distance L.

To find the time required for a particle with speed v to travel distance L, you would use the equation of motion, $x = vt$, or, in this case, $t = \frac{L}{v}$. As a result, you can replace the equation for the current, $I = \frac{q}{t}$, by $I = \frac{qv}{L}$.

Investigate **magnetic fields and the force on a charge.**

FORCE OF A MAGNETIC FIELD ON A MOVING CHARGED PARTICLE

The amount of force from a magnetic field on a particle equals the product of the particle's charge, its speed, the magnetic field strength, and the sine of the angle between the particle's velocity and the magnetic field.

$$F = qvB(\sin \theta)$$

Recall that charge is measured in coulombs (C), velocity in meters per second (m/s), and magnetic field strength in teslas (T). For a particle moving at right angles to a magnetic field, $\sin \theta = 1$, so $F = qvB$.

The direction of the force on a charged particle is perpendicular to that particle's velocity and to the magnetic field. To find the direction of force, you can use the same right-hand rule you use for finding the direction of the force on a current-carrying wire, where the moving charge is the current. If the moving particle is an electron (with a negative charge), the direction of force is reversed.

■ Electric Motor

2. Current reversal Current goes through the brush to the split-ring commutator, which passes it on to the armature. The magnet either repels or attracts the armature, depending on the direction of current. When the armature reaches the vertical position, each half of the commutator changes brushes. This reverses current direction so that the armature can rotate an additional 180°.

3. Continuous rotation Current reverses with each half-turn of the armature; this results in continuous rotation.

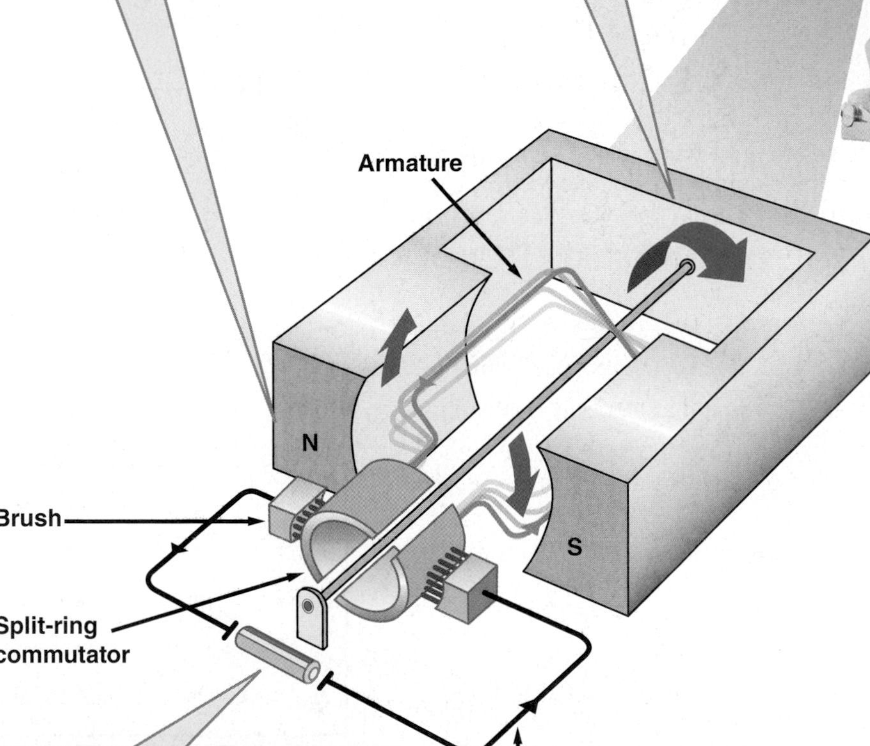

1. Electric connection Current from a battery passes through graphite brushes fixed in position but pressed against a split-ring commutator. The commutator is attached to a wire loop—the armature—along an axle in a magnetic field.

Figure 19 In an electric motor, an armature in a magnetic field rotates 360°. A split-ring commutator changes the direction of the current every 180°, allowing this rotation.

View an **animation of an electric motor.**

Concepts In Motion

Ingram Publishing/Fotosearch

EXAMPLE PROBLEM

EXAMPLE PROBLEM 2

Get help with **force on a charged particle.** Personal Tutor

FORCE ON A CHARGED PARTICLE IN A MAGNETIC FIELD A beam of electrons travels at 3.0×10^6 m/s through a uniform magnetic field of 4.0×10^{-2} T at right angles to the field. How strong is the force acting on each electron?

1 ANALYZE AND SKETCH THE PROBLEM

Draw the beam of electrons and its direction of motion (v). Indicate the magnetic field (B) and the force on the electron beam (F). Note that the direction of force is opposite that given by the right-hand rule because of the electron's negative charge.

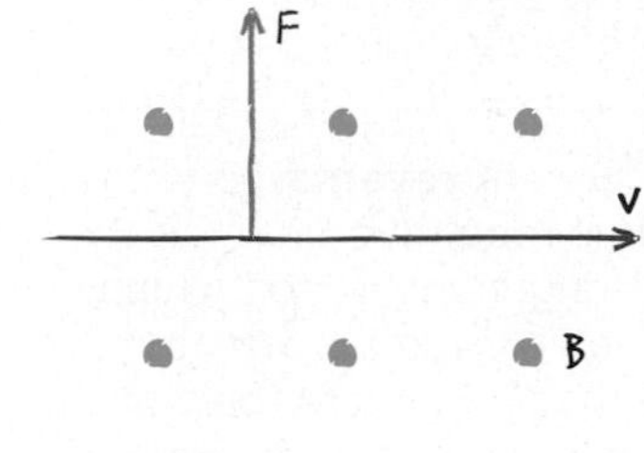

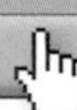

KNOWN	UNKNOWN
$v = 3.0\times10^6$ m/s	$F = ?$
$B = 4.0\times10^{-2}$ T	
$q = -1.602\times10^{-19}$ C	

2 SOLVE FOR THE UNKNOWN

$$F = qvB$$

$$= (-1.602\times10^{-19}\text{ C})(3.0\times10^6\text{ m/s})(4.0\times10^{-2}\text{ T})$$

◀ Substitute $q = -1.602\times10^{-19}$ C, $v = 3.0\times10^6$ m/s, $B = 4.0\times10^{-2}$ T

$$= -1.9\times10^{-14}\text{ N}$$

3 EVALUATE THE ANSWER

- **Are the units correct?** T = N/(A·m) and A = C/s, so T = N·s/(C·m). Thus, (T·C·m)/s = N, the unit for force.
- **Does the direction make sense?** Use the right-hand rule to verify the direction of the force, recalling that the force on the electron is opposite the force given by the right-hand rule due to the electron's negative charge.
- **Is the magnitude realistic?** Forces on electrons and protons are always small fractions of a newton.

PRACTICE PROBLEMS

PRACTICE PROBLEMS

Do additional problems. Online Practice

25. In what direction is the force on an electron if that electron is moving east through a magnetic field that points north?

26. What are the magnitude and direction of the force acting on the proton shown in **Figure 20?**

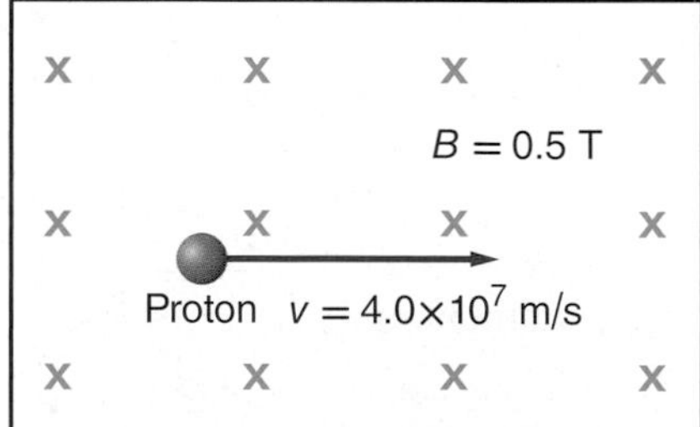

Figure 20

27. A stream of doubly ionized particles (missing two electrons and thus carrying a net positive charge of two elementary charges) moves at a velocity of 3.0×10^4 m/s perpendicular to a magnetic field of 9.0×10^{-2} T. How large is the force acting on each ion?

28. Triply ionized particles in a beam carry a net positive charge of three elementary charge units. The beam enters a magnetic field of 4.0×10^{-2} T. The particles have a speed of 9.0×10^6 m/s and move at right angles to the field. How large is the force acting on each particle?

29. A singly ionized particle experiences a force of 4.1×10^{-13} N when it travels at a right angle through a 0.61-T magnetic field. What is the particle's velocity?

30. CHALLENGE Doubly ionized helium atoms (alpha particles) are traveling at right angles to a magnetic field at a speed of 4.0×10^4 m/s. The force on each particle is 6.4×10^{-16} N. What is the magnetic field strength?

Figure 21 Powerful electromagnets inside the LHC tunnel provide the uniform magnetic field that makes charged particles move in a circular path.

Synchrotrons Because the direction of force is always perpendicular to a charged particle's velocity in a magnetic field, magnets can be used to direct a charged particle's path. For example, accelerating particles in a synchrotron, such as the Large Hadron Collider (LHC), move in a circle as they maintain their velocity at right angles to a uniform magnetic field. You can see several segments of the 27-km-long tunnel housing the LHC in **Figure 21.** As the particles gain speed, the magnetic field in the tunnel is increased to keep the radius of the circle constant. Additional magnets provide horizontal and vertical forces to focus the beam.

Additional segments along the LHC tunnel add fixed amounts of energy that accelerate the particles. Because charged particles go through many accelerators in multiple passes around the synchrotron, the particles can reach extremely high energies. The LHC was designed to give accelerating protons enough energy to travel against a potential difference of 7.2 trillion volts. To reduce electrical power needs, the magnets in the LHC use superconducting wires.

SECTION 2 REVIEW

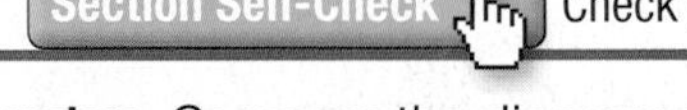

Section Self-Check Check your understanding.

31. MAINIDEA Explain how electric motors use magnets to convert electrical energy to mechanical energy.
32. **Magnetic Forces** Imagine that a current-carrying wire is perpendicular to Earth's magnetic field and runs east-west. If the current is east, in which direction is the force on the wire?
33. **Synchrotrons** In a synchrotron, magnetic fields bend particle beams into segments of a circle, and electric fields accelerate the beams.
    - **a.** A beam of protons circulates in a clockwise direction. In what direction must the magnetic field be oriented? In what direction must the electric fields be oriented?
    - **b.** If a beam of negatively charged antiprotons is to circulate in a counterclockwise direction, must the direction of the magnetic field be changed? Must the direction of the electric fields be changed?
34. **Galvanometers** Compare the diagram of a galvanometer in the left part of **Figure 17** with the electric motor in **Figure 19.** How is the galvanometer similar to an electric motor? How is it different?
35. **Motors** When the plane of an armature in a motor is perpendicular to the magnetic field, the forces do not exert a torque on the coil. Does this mean that the coil does not rotate? Explain.
36. **Resistance** A galvanometer requires 180 μA for full-scale deflection. When it is used as a voltmeter, what total resistance of the meter and the multiplier resistor is needed for a 5.0-V full-scale deflection?
37. **Critical Thinking** Two current-carrying wires move toward each other when they are placed parallel to each other. Compare the directions of the two currents. Explain your reasoning.

Biomagnetic Monitoring

LEAF ME ALONE

Imagine bicycling on the side of a road when a bus accelerates, and a cloud of smoke comes from its exhaust pipe, causing you to cough and rub your eyes. More than coughing and watery eyes, prolonged exposure to the small-particle pollution in exhaust smoke may cause serious health problems.

Measuring air quality Detailed information on pollution levels throughout a city can help city planners design pedestrian and bicycle routes that limit people's exposure to extreme levels of pollution. Air-quality measurements usually rely on a single, centrally located machine to measure pollution levels for a large area. These machines may not provide sufficiently detailed information to study local pollution levels and their effects on human health. Placing more machines in an area would provide more detailed data, but the machines are expensive and difficult to maintain.

Biomagnetic monitoring Environmental scientists around the world are developing a relatively inexpensive process called biomagnetic monitoring to collect detailed air-quality data. The process measures the magnetic properties of tree leaves as an indicator of particulate-pollution levels. In one method of biomagnetic monitoring, researchers briefly expose leaf samples to a strong magnetic field. Researchers then remove the samples from the magnetic field and measure the residual magnetism. Strong residual magnetism indicates iron compounds, which are a part of the particles in automobile exhaust smoke shown in **Figure 1.**

Magnetic leaves indicate pollution. Leaves from trees in high-traffic areas can be up to ten times more magnetic than leaves collected far from busy roads. In addition, leaves collected close to roads used by buses and trucks show the highest magnetism levels.This is because diesel engines, such as those frequently used in buses and trucks, emit the greatest numbers of particles. Areas where vehicles stop for traffic lights or for rush-hour traffic jams are also highly polluted because vehicle engines emit more particles as they accelerate from rest.

FIGURE 1 Automobile exhaust smoke contains particles such as the ones shown below. These particles are harmful to human health when they are inhaled.

Magnification is x8000 at 10 cm.

GOINGFURTHER >>>

Research Create a map showing the location of different tree species around your school, neighborhood, or town. Based on the map, which single species would you use for a biomagnetic monitoring study of traffic exhaust in your area?

STUDY GUIDE

BIGIDEA Magnets and moving electric charges are surrounded by magnetic fields that exert forces on magnetic materials and on moving charges.

VOCABULARY
- **polarized** *(p. 650)*
- **domain** *(p. 652)*
- **magnetic field** *(p. 653)*
- **magnetic flux** *(p. 654)*
- **solenoid** *(p. 656)*
- **electromagnet** *(p. 656)*

SECTION 1 Understanding Magnetism

MAINIDEA Magnets and electric currents produce magnetic fields.

- All magnets have north poles and south poles and are surrounded by magnetic fields.
- Ferromagnetic materials become magnetic when their domains are in alignment with each other.
- Magnetic fields are vector quantities because they have direction and magnitude. They exist in any region in space where a magnet would experience a force. Magnetic fields can be represented by field lines, which exit from a north pole and enter at a south pole, forming closed loops.

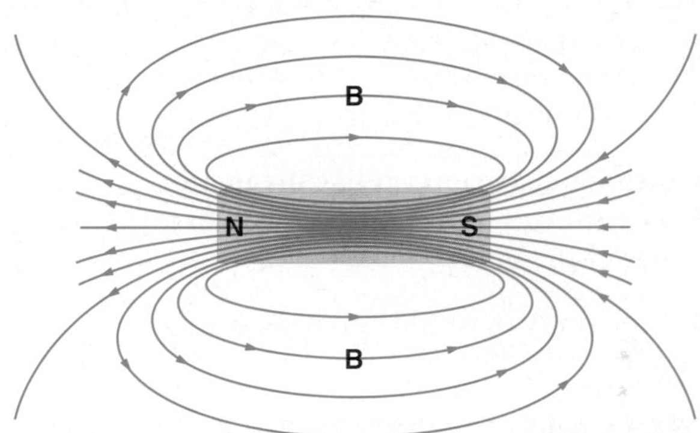

- A magnetic field exists around any current-carrying wire. The magnetic field around a coil of wire is proportional to the number of loops in the coil because the individual fields of the loops are in the same direction.

VOCABULARY
- **galvanometer** *(p. 661)*
- **electric motor** *(p. 662)*
- **armature** *(p. 662)*

SECTION 2 Applying Magnetic Forces

MAINIDEA Many devices, including earbuds and electric motors, rely on forces from magnetic fields in order to work.

- When a current-carrying wire is placed in a magnetic field, a force is exerted on the wire that is perpendicular to both the field and the wire.
- The force on a current-carrying wire in a magnetic field is proportional to the current times the length of the wire times the field strength times the sine of the angle between the current and the magnetic field.

$$F = ILB(\sin \theta)$$

- An electric motor consists of a coil of wire in a magnetic field. When there is current in the coil, the coil rotates as a result of the force on the wire in the magnetic field. Complete 360° rotation is achieved by using a split-ring commutator to switch the direction of the current in the coil as the coil rotates.
- The force that a magnetic field exerts on a moving charged particle depends on the charge of the particle, the velocity of the particle, the strength of the magnetic field, and the angle between the directions of the velocity and the field. The direction of the force is perpendicular to both the field and the particle's velocity.

$$F = qvB(\sin \theta)$$

Games and Multilingual eGlossary

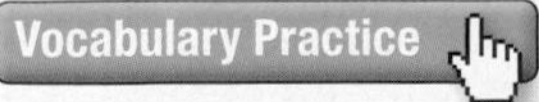

CHAPTER 24 ASSESSMENT

Chapter Self-Check

SECTION 1 Understanding Magnetism

Mastering Concepts

38. State a rule describing magnetic attraction and repulsion.

39. Describe how a temporary magnet differs from a permanent magnet.

40. Name three common ferromagnetic elements.

41. BIGIDEA Draw a small bar magnet with field lines around it. Use arrows to show the direction of the field. Draw a small nail in this magnetic field with the induced north and south poles indicated. The bar magnet's field exerts an attractive force on one pole of the nail and a repulsive force on the other. Why is the nail pulled toward the bar magnet?

42. Draw the magnetic field between two like magnetic poles and then between two unlike magnetic poles. Show the directions of the fields.

43. If you broke a magnet in two, would you have isolated north and south poles? Explain.

44. Describe how to use a right-hand rule to determine the direction of a magnetic field around a straight, current-carrying wire.

45. If a current-carrying wire were bent into a loop, why would the magnetic field inside the loop be stronger than the magnetic field outside?

46. Describe how to use a right-hand rule to determine the north and south ends of an electromagnet.

47. Each atom in a piece of iron is like a tiny magnet. The iron, however, might not be a magnet. Explain.

48. Why will heating a magnet weaken it?

Mastering Problems

49. Refer to **Figure 22** to answer the following questions:

a. Where are the poles?

b. Where is the north pole?

c. Where is the south pole?

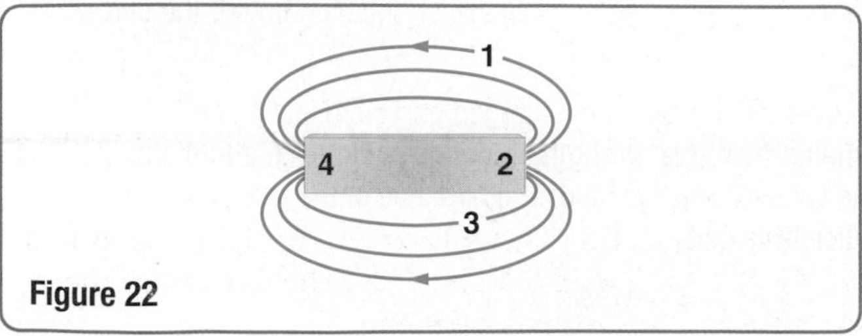

Figure 22

50. As the magnet on the right in **Figure 23** moves toward the magnet suspended on a string, what will the suspended magnet do?

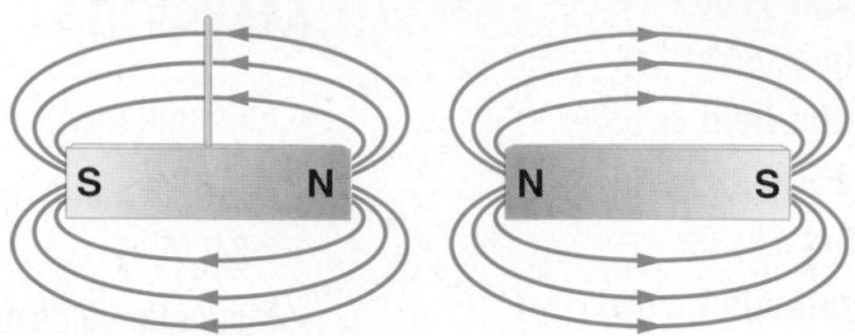

Figure 23

51. As the magnet on the right in **Figure 24** moves toward the magnet suspended on a string, what will the suspended magnet do?

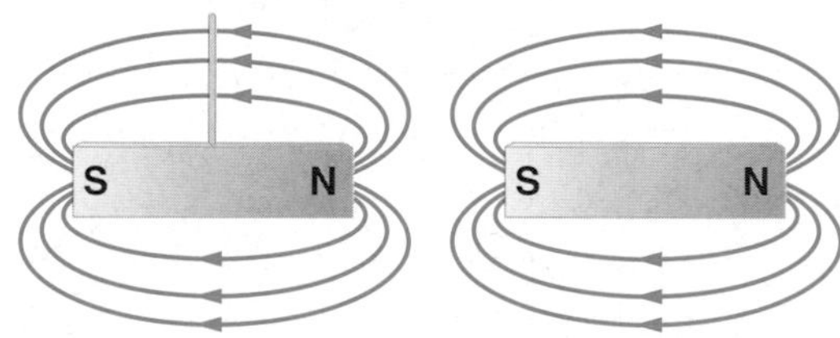

Figure 24

52. **Figure 25** shows the response of a compass in two different positions near a magnet. Where is the south pole of the magnet located?

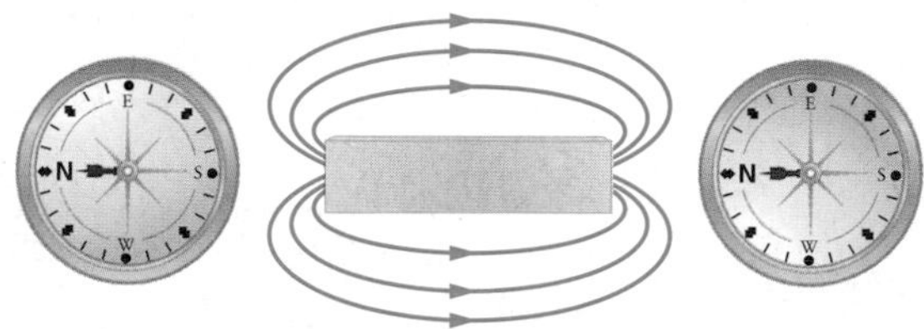

Figure 25

53. Copy the wire segment in **Figure 26** and sketch the magnetic field that its current generates.

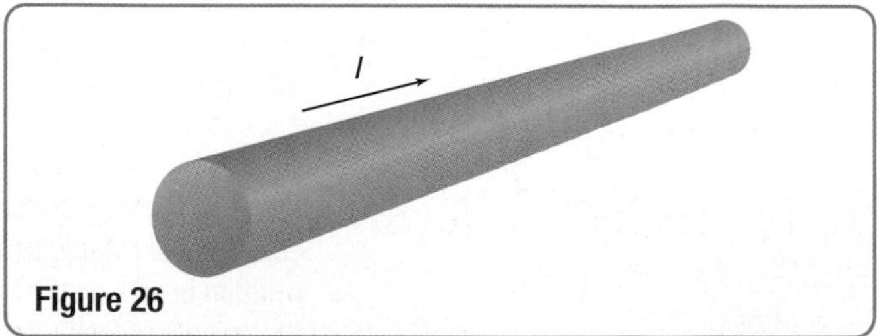

Figure 26

54. The current is coming straight out of the page in the wire in **Figure 27.** Copy the figure and sketch the magnetic field that the current generates.

Figure 27

55. **Figure 28** shows the end view of an electromagnet that is carrying current.

a. What is the direction of the magnetic field inside the loops?

b. What is the direction of the magnetic field outside the loops?

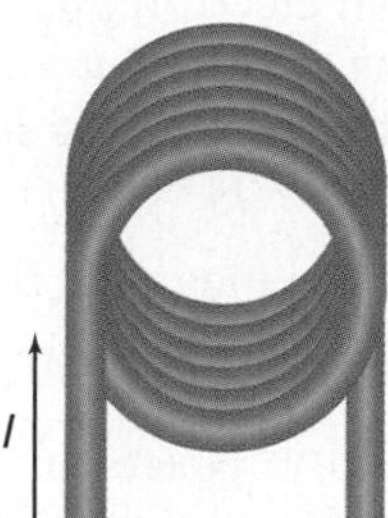

Figure 28

56. The repulsive force between two magnets was measured and found to depend on distance, as given in **Table 1.**

a. Plot the force as a function of distance.

b. Does this force follow an inverse square law?

Table 1 Force v. Distance

Separation, d (cm)	Force, F (N)
1.0	3.93
1.2	0.40
1.4	0.13
1.6	0.057
1.8	0.030
2.0	0.018
2.2	0.011
2.4	0.0076
2.6	0.0053
2.8	0.0038
3.0	0.0028

SECTION 2 Applying Magnetic Forces

Mastering Concepts

57. What kind of meter is created when a shunt is added to a galvanometer?

58. Electric Motor The torque on the armature of an electric motor depends on the length of the wires along the armature's axis of rotation. Are there forces on the wires perpendicular to that axis? If so, why do these forces not contribute to the torque?

59. A strong current suddenly is switched on in a wire. No force, however, acts on the wire. Can you conclude that there is no magnetic field at the location of the wire? Explain.

60. Earbud You are able to listen to music through an earbud because current in a wire coil in the earbud changes direction. What happens to the force on the wire when current changes direction?

61. Synchrotron A synchrotron uses magnets to direct charged particles in a circular path. Explain how the magnetic force on a charged particle makes it move in a circle.

62. The arrangement shown in **Figure 29** is used to convert a galvanometer to what type of device?

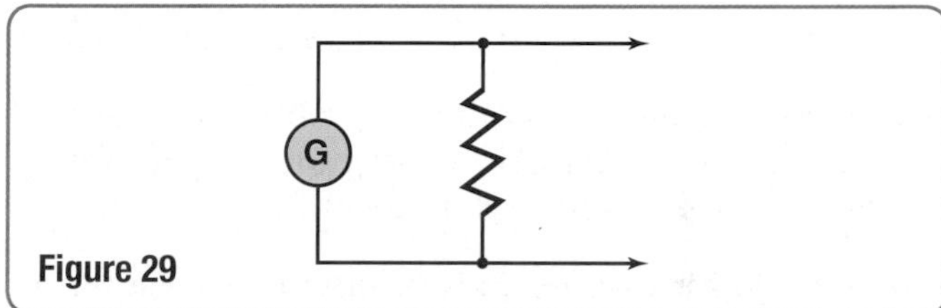

Figure 29

63. What is the resistor shown in **Figure 29** called?

64. The arrangement shown in **Figure 30** is used to convert a galvanometer to what type of device?

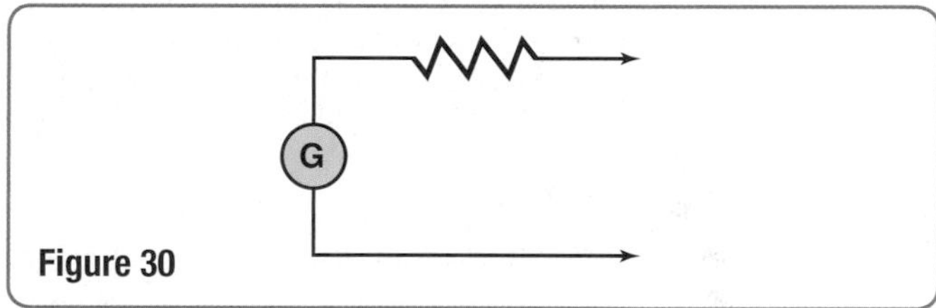

Figure 30

65. What is the resistor shown in **Figure 30** called?

Mastering Problems

66. A wire that is 0.50 m long and carrying a current of 8.0 A is at right angles to a uniform magnetic field. The force on the wire is 0.40 N. What is the strength of the magnetic field?

67. A wire that is 25 cm long is at right angles to a 0.30-T uniform magnetic field. The current through the wire is 6.0 A. What is the magnitude of the force on the wire?

68. A wire that is 35 cm long is parallel to a 0.53-T uniform magnetic field. The current through the wire is 4.5 A. What force acts on the wire?

69. The force acting on a wire that is at right angles to a 0.80-T magnetic field is 3.6 N. The current in the wire is 7.5 A. How long is the wire?

Chapter Self-Check

70. A current-carrying wire is placed between the poles of a magnet, as shown in **Figure 31.** What is the direction of the force on the wire?

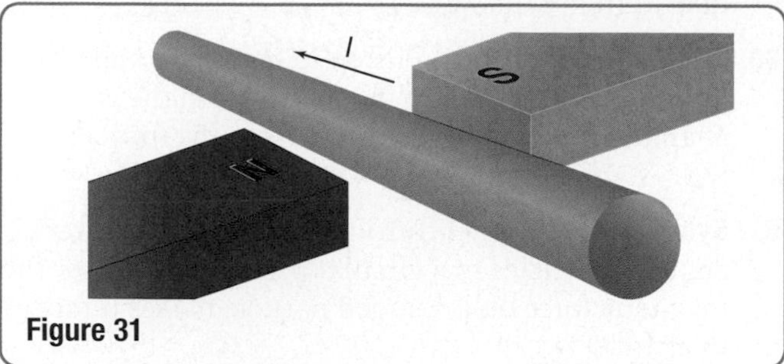

Figure 31

71. The force on a 0.80-m wire that is perpendicular to Earth's magnetic field is 0.12 N. What is the current in the wire? Use 5.0×10^{-5} T for Earth's magnetic field.

72. Problem Posing Complete this problem so that it must be solved using the concept of magnetic force: "A proton has a velocity of 600 m/s ...."

73. A power line carries a 225-A current from east to west, parallel to the surface of Earth.

a. What are the magnitude and direction of the force resulting from Earth's magnetic field acting on each meter of the wire? Use $B_{\text{Earth}} = 5.0\times10^{-5}$ T.

b. In your judgment, would this force be important in designing towers to hold this power line? Explain.

74. Galvanometer A galvanometer deflects full-scale for a 50.0-μA current.

a. What must be the total resistance of the series resistor and the galvanometer to make a voltmeter with 10.0-V full-scale deflection?

b. If the galvanometer has a resistance of 1.0 kΩ, what should be the resistance of the series (multiplier) resistor?

75. The galvanometer in the previous problem is used to make an ammeter that deflects full-scale for 10 mA.

a. What is the potential difference across the galvanometer (1.0 kΩ resistance) when a current of 50 μA passes through it?

b. What is the equivalent resistance of parallel resistors having the potential difference calculated in a circuit with a total current of 10 mA?

c. What resistor should be placed parallel with the galvanometer to make the resistance calculated in part **b**?

76. A beam of electrons moves at right angles to a magnetic field of 6.0×10^{-2} T. The electrons have a velocity of 2.5×10^{6} m/s. What is the magnitude of the force on each electron?

77. A room contains a strong, uniform magnetic field. A loop of fine wire in the room has current in it. The loop is rotated until there is no torque on it as a result of the field. What is the direction of the magnetic field relative to the plane of the coil?

78. Subatomic Particle A muon (a particle with the same charge as an electron but with a mass of 1.88×10^{-28} kg) is traveling at 4.21×10^{7} m/s at right angles to a magnetic field. The muon experiences a force of -5.00×10^{-12} N.

a. How strong is the magnetic field?

b. What acceleration does the muon experience?

79. A force of 5.78×10^{-16} N acts on an unknown particle traveling at a 90° angle through a magnetic field. If the velocity of the particle is 5.65×10^{4} m/s and the magnetic field is 3.20×10^{-2} T, how many elementary charges does the particle carry?

80. Ranking Task Rank the following situations according to the magnitude of magnetic force experienced, from greatest to least. Indicate whether any of them are the same.

A. a stationary charge of 8.0 nC in a magnetic field of 3.0 T

B. a charge of 3.0 nC moving with a speed of 9.0 m/s in a magnetic field of 1.0 T

C. a charge of 1.5 nC moving with a speed of 4.0 m/s in a magnetic field of 1.0 T

D. a charge of 6.0 nC moving with a speed of 9.0 m/s in a magnetic field of 0.5 T

E. a charge of 2.0 nC moving with a speed of 2.0 m/s in a magnetic field of 3.0 T

Applying Concepts

81. A small bar magnet is hidden in a fixed position inside a tennis ball. Describe an experiment you could do to find the location of the magnet's poles.

82. Compass Suppose you are taking a walk in the woods and realize you are lost. You have a compass, but the red paint marking the north pole of the compass needle has worn off. You also have a length of wire and a flashlight with a battery. How could you identify the north pole of the compass?

83. Two parallel wires carry equal currents.

a. If the two currents are in opposite directions, where will the magnetic field from the two wires be larger than the field from either wire alone?

b. Where will the magnetic field from both wires be exactly twice as large as from one wire?

84. A nail is attracted to one pole of a permanent magnet. Describe how you could tell whether the metal is magnetized.

85. Is the magnetic force that Earth exerts on a compass needle less than, equal to, or greater than the force that the needle exerts on Earth? Explain.

86. A magnet attracts a piece of iron that is not a permanent magnet. A charged rubber rod attracts an uncharged insulator. What different microscopic processes produce these similar phenomena?

87. A current-carrying wire runs across a laboratory bench. Describe at least two ways in which you could find the direction of the current.

88. In which direction, in relation to a magnetic field, would you run a current-carrying wire so that the force on it, resulting from the field, is minimized, or even made to be zero?

89. A magnetic field can exert a force on a charged particle. Can the field change the particle's kinetic energy? Explain.

90. As a beam of protons moves from the back to the front of a room, it is deflected upward by a magnetic field. What is the direction of the field?

91. Field lines representing Earth's magnetic field are shown in **Figure 32.** At which location, poles or equator, is the magnetic field strength greatest? Explain.

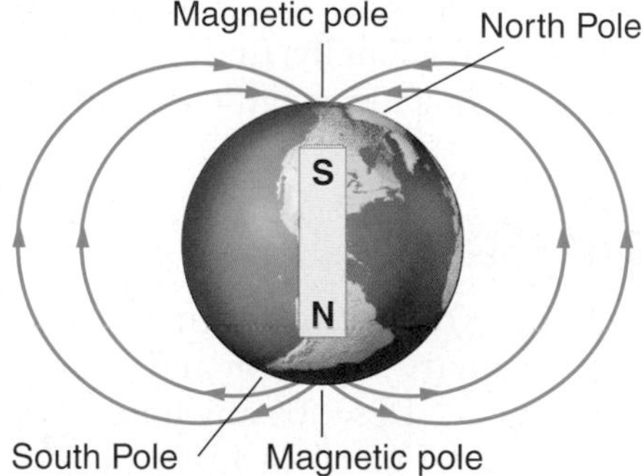

Figure 32

Mixed Review

92. A magnetic field of 16 T acts in a direction due west. An electron is traveling due south at 8.1×10^5 m/s. What are the magnitude and direction of the force acting on the electron?

93. Subatomic Particle A beta particle (high-speed electron) is traveling at right angles to a 0.60-T magnetic field. It has a speed of 2.5×10^7 m/s. What force acts on the particle?

94. A copper wire of insignificant resistance is placed in the center of an air gap between two magnetic poles, as shown in **Figure 33.** The field is confined to the gap and has a strength of 1.9 T.

a. Determine the force on the wire (direction and magnitude) when the switch is open.

b. Determine the force on the wire (direction and magnitude) when the switch is closed.

c. Determine the force on the wire (direction and magnitude) when the switch is closed and the battery is reversed.

d. Determine the force on the wire (direction and magnitude) when the switch is closed and the wire has two 5.5-Ω resistors in series.

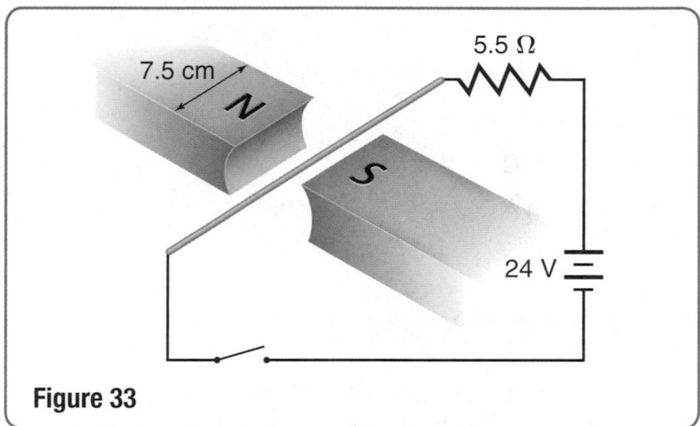

Figure 33

95. Two galvanometers are available. One has 50.0-μA full-scale sensitivity and the other has 500.0-μA full-scale sensitivity. Both have the same coil resistance of 855 Ω. Your challenge is to convert them to measure a current of 100.0 mA, full-scale.

a. Determine the shunt resistor for the 50.0-μA meter.

b. Determine the shunt resistor for the 500.0-μA meter.

c. Determine which of the two is better for actual use. Explain.

96. Loudspeaker The magnetic field in a loudspeaker is 0.15 T. The wire makes 250 loops around a cylindrical form that is 2.5 cm in diameter. The resistance of the wire is 8.0 Ω. Find the force on the wire when 15 V is placed across the wire.

97. A wire carrying 15 A of current has a length of 25 cm in a uniform magnetic field of 0.85 T. Using the equation $F = ILB(\sin\theta)$, find the force on the wire when it makes the following angles with the magnetic field lines.

a. 90°

b. 45°

c. 0°

Chapter Self-Check

98. An electron is accelerated from rest through a potential difference of 2.0×10^4 V, which exists between plates P_1 and P_2, shown in **Figure 34.** The electron then passes through a small opening into a magnetic field of uniform strength. As indicated, the magnetic field is directed into the page.

a. State the direction of the electric field between the plates as either P_1 to P_2 or P_2 to P_1.

b. In terms of the information given, calculate the electron's speed at plate P_2.

c. Describe the motion of the electron through the magnetic field.

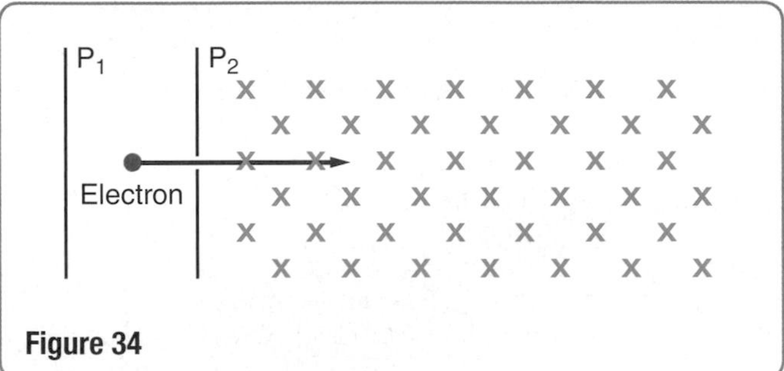

Figure 34

Thinking Critically

99. Synchrotron The LHC synchrotron has a 27-km circumference and 1232 bending magnets. Protons of mass 1.67×10^{-27} kg and charge 1.60×10^{-19} C are all traveling at speeds close to the speed of light, 3.00×10^8 m/s. What magnetic field would be required to bend their paths into a circle? Recall that centripetal force is given by $F = mv^2/r$. The forces come from the particle moving through the magnetic field. The radius of a circle of circumference C is $r = \frac{C}{2\pi}$. Note that due to the effects of special relativity on the protons, the actual magnetic field needed is 8.36 T.

100. Apply Concepts A current is sent through a vertical spring, as shown in **Figure 35.** The end of the spring is in a cup filled with mercury. What will happen? Why?

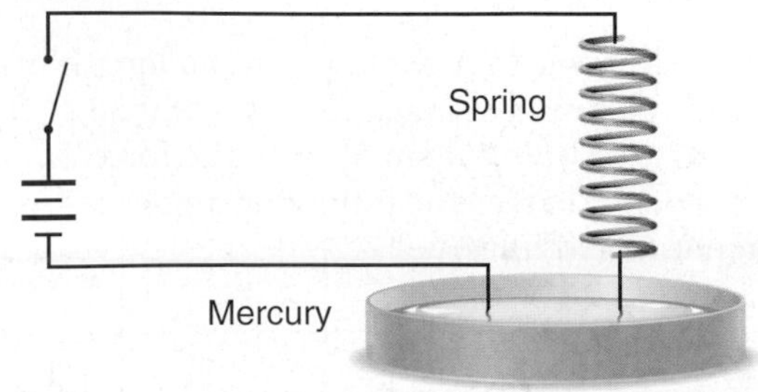

Figure 35

101. Reverse Problem Write a physics problem with real-life objects for which the following equation would be part of the solution:

$$v = \frac{60\ \text{N}}{(1.60\times10^{-19}\ \text{C})\ (2.3\ \text{T})}$$

102. Apply Concepts The magnetic field produced by a long, current-carrying wire is represented by $B = (2\times10^{-7}\ \text{T·m/A})(I/d)$, where B is the field strength in teslas, I is the current in amps, and d is the distance from the wire in meters. Use this equation to estimate the magnetic fields in the following scenarios.

a. The wires in your home seldom carry more than 10 A. How does the magnetic field 0.5 m from one of these wires compare to Earth's magnetic field?

b. High-voltage power transmission lines often carry 200 A with potential differences as high as 765 kV. Estimate the magnetic field on the ground under such a line, assuming that it is about 20 m high. How does this field compare with a magnetic field in your home?

103. Add Vectors In the scenarios described in the previous problem, a second wire carries the same current in the opposite direction. Find the net magnetic field at a distance of 0.10 m from each wire carrying 10 A. The wires are 0.01 m apart. Make a scale drawing of the situation. Calculate the magnitude of the field from each wire and use a right-hand rule to draw vectors showing the directions of the fields. Finally, find the vector sum of the two fields. State its magnitude and direction.

Writing in Physics

104. Research superconducting magnets and write a one-page summary of present and proposed uses for such magnets. Describe any hurdles that stand in the way of their practical application.

Cumulative Review

105. How much work is required to move a charge of 6.40×10^{-3} C through a potential difference of 2500 V?

106. The current in a 120-V circuit increases from 1.3 A to 2.3 A. Calculate the change in power.

107. Determine the total resistance of three 55 Ω resistors connected in parallel and then series-connected to two 55 Ω resistors connected in series.

STANDARDIZED TEST PRACTICE

MULTIPLE CHOICE

1. A straight wire carrying a current of 7.2 A has a field of 8.9×10^{-3} T perpendicular to it. What length of wire in the field will experience a force of 2.1 N?
   - A. 2.6×10^{-3} m
   - B. 3.1×10^{-2} m
   - C. 1.3×10^{-1} m
   - D. 3.3×10^{1} m

2. Assume that a 19-cm length of wire is carrying a current perpendicular to a 4.1-T magnetic field and experiences a force of 7.6 mN. What is the current in the wire?
   - A. 3.4×10^{-7} A
   - B. 9.8×10^{-3} A
   - C. 1.0×10^{-2} A
   - D. 9.8 A

3. A 7.12-μC charge is moving at the speed of light in a magnetic field of 4.02 mT. What is the force on the charge?
   - A. 8.59 N
   - B. 2.90×10^{1} N
   - C. 8.59×10^{12} N
   - D. 1.00×10^{16} N

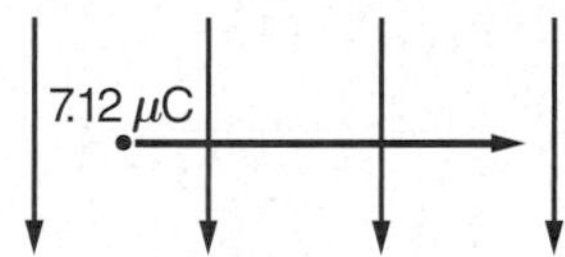

4. An electron is moving at 7.4×10^{5} m/s perpendicular to a magnetic field. It experiences a force of -2.0×10^{-13} N. What is the magnetic field strength?
   - A. 8.2×10^{-15} T
   - B. 1.7×10^{-8} T
   - C. 0.31 T
   - D. 1.7 T

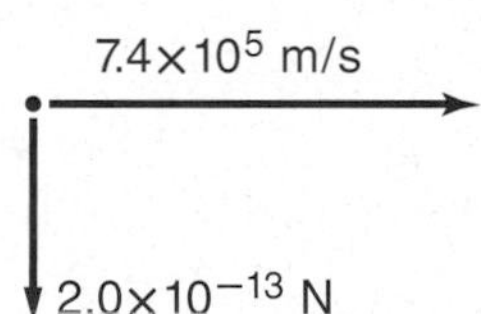

5. Which factor will not affect the strength of a solenoid?
   - A. number of wraps
   - B. strength of current
   - C. thickness of wire
   - D. core type

6. The current through a wire that is 0.80 cm long is 5.0 A. The wire is perpendicular to a 0.60-T magnetic field. What is the magnitude of the force on the wire?
   - A. 2.4×10^{-2} N
   - B. 2.4×10^{-1} N
   - C. 2.4 N
   - D. 24 N

Online Test Practice

7. Which statement about magnetic monopoles is false?
   - A. A monopole is a hypothetical separate north pole.
   - B. Research scientists use them for internal medical testing applications.
   - C. A monopole is a hypothetical separate south pole.
   - D. They don't exist.

8. A uniform magnetic field of 0.25 T points vertically downward. A proton enters the field with a horizontal velocity of 4.0×10^{6} m/s. What are the magnitude and direction of the instantaneous force exerted on the proton as it enters the magnetic field?
   - A. 1.6×10^{-13} N to the left
   - B. 1.6×10^{-13} N downward
   - C. 1.0×10^{6} N upward
   - D. 1.0×10^{6} N to the right

9. How is a temporary magnet different from a permanent magnet?
   - A. A temporary magnet's domains do not align with one another, but a permanent magnet's do.
   - B. A temporary magnet is made from a different type of material than a permanent magnet.
   - C. A temporary magnet has a weaker magnetic field than a permanent magnet.
   - D. A temporary magnet can be turned on and off, but a permanent magnet cannot.

FREE RESPONSE

10. Derive the units of teslas in kilograms, meters, seconds, and coulombs using dimensional analysis and the formulas $F = qvB$ and $F = ILB$.

11. A wire attached to a 5.8-V battery is in a circuit with a resistance of 18 Ω. A 14-cm length of the wire is in a magnetic field of 0.85 T, and the force on the wire is 22 mN. What is the angle of the wire in the field given that the formula for angled wires in fields is $F = ILB(\sin\theta)$?

NEED EXTRA HELP?

If You Missed Question	1	2	3	4	5	6	7	8	9	10	11
Review Section	2	2	2	2	1	2	1	2	1	2	2

CHAPTER 25

Electromagnetic Induction

BIGIDEA **A changing magnetic field can induce current in a conductor.**

SECTIONS

LaunchLAB

CHANGING MAGNETIC FIELDS

What is a changing magnetic field, and how does it affect current in a circuit?

WATCH THIS!

GENERATING ELECTRICITY

Have you ever seen a bicycle headlight that you could power just by pedaling the bicycle? How does such a headlight work? Is the bicycle harder to pedal when the headlight is on?

Go online!
connectED.mcgraw-hill.com

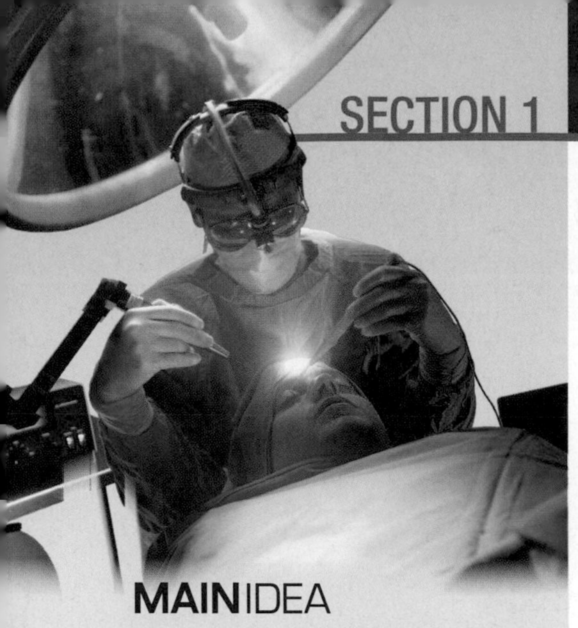

SECTION 1

Inducing Currents

PHYSICS 4 YOU Have you ever watched a hospital television drama where the electricity suddenly goes out? In an instant, with a loud hum, backup generators turn on. Hospitals have backup generators so surgeons can keep operating and live-saving machines can keep functioning.

MAIN IDEA

A changing magnetic field induces an *EMF* in a wire, and the *EMF*, in turn, generates current when the wire is in a circuit.

Essential Questions

- What is induced *EMF?*
- What affects the induced *EMF* and current produced by a changing magnetic field?
- How does a generator produce electrical energy?
- How are the effective current and effective potential difference related to the maximum values of these quantities in an AC circuit?

Review Vocabulary

Potential difference the work needed to move a positive test charge from one point to another divided by the magnitude of the test charge

New Vocabulary

electromagnetic induction
induced electromotive force
electric generator

Changing Magnetic Fields

Following Oersted's discovery that a current produces a magnetic field, Michael Faraday became convinced the reverse was possible: that a magnetic field could produce a current. In 1822 he wrote this goal in his notebook: "Convert magnetism into electricity." After ten years of unsuccessful experiments, he succeeded. He induced a current in a circuit by moving a wire through a magnetic field. In the same year in America, Joseph Henry, a high-school teacher who later became the first secretary of the Smithsonian Institution, made the same discovery.

Figure 1 illustrates a modern version of one of Faraday's and Henry's experiments. A loop of wire that forms a circuit crosses a magnetic field. When the wire is held stationary or moved parallel to the field, nothing happens. When the wire is moved perpendicular to the field, there is a current. When the perpendicular wire is moved in the other direction, the current reverses direction.

Figure 1 The direction of current induced in a wire moving through a magnetic field depends on the direction the wire moves. When motion stops, so does the current.

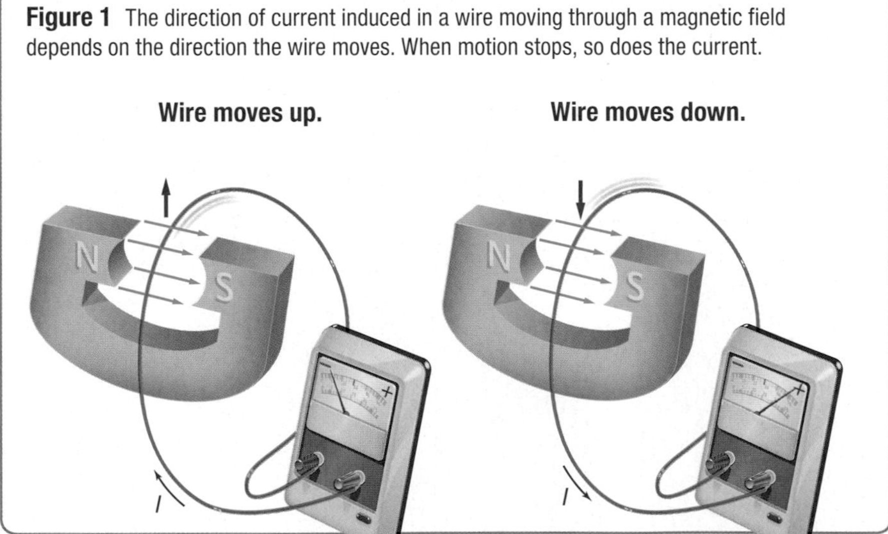

Relative motion An electric current can be generated in a wire in a circuit when at least part of the wire moves through, and cuts, magnetic field lines. Field lines can be cut when a segment of wire moves through a stationary magnetic field, as the wire does in **Figure 1.** Field lines also can be cut when a magnetic field moves past a stationary wire or when the strength of a magnetic field changes around a wire. It is the relative motion between a wire and a magnetic field that can cut field lines and produce current. **Electromagnetic induction** is the process of generating current through a wire in a circuit in a changing magnetic field.

Electromotive force How is current produced by electromagnetic induction? You know that a current requires a source of electrical energy, such as a battery. A battery maintains an electric potential difference in a circuit; charges flow from higher to lower potential. The potential difference across the battery's terminals is the electromotive force, or *EMF.* An *EMF* is a difference in potential, not actually a force. Like many historical terms, it was coined before related principles were well understood.

View **animations of a wire in a changing magnetic field.**

Induced *EMF* You don't need a chemical reaction in a battery to create an *EMF.* When a wire moves perpendicular to a magnetic field, there is a force on the charges in the wire. The force causes negative charges to move to one end of the wire, leaving positive charges at the other end. This separation of charge produces an electric field and therefore a potential difference across the length of the wire. This potential difference is called the **induced electromotive force,** or induced *EMF.*

The magnitude of an induced *EMF* in a wire in a magnetic field depends on the strength of the magnetic field (B), the length of the wire within the field (L), and the component of the velocity of the length of wire that is perpendicular to the field ($v(\sin \theta)$).

INDUCED ELECTROMOTIVE FORCE IN A WIRE

EMF is equal to the strength of the magnetic field times the length of the wire times the component of the velocity of the wire in the field that is perpendicular to the field.

$$EMF = BLv(\sin \theta)$$

If a wire moves perpendicular to a magnetic field, the above equation reduces to $EMF = BLv$, because $\sin 90° = 1$. Note that no *EMF* is induced in a length of wire that moves parallel to a magnetic field because $\sin 0° = 0$.

As in a battery, induced *EMF* is measured in volts (V). Why? Recall that magnetic fields (B) are measured in teslas (T). You can write $B = \frac{F}{IL}$, so the units for B are also N/(A•m). Velocity is measured in meters per second. Using dimensional analysis, you could write the following:

$$\left(\frac{\text{N}}{\text{A}\cdot\text{m}}\right)(\text{m})\left(\frac{\text{m}}{\text{s}}\right) = \frac{(\text{N}\cdot\text{m})}{(\text{A}\cdot\text{s})} = \frac{\text{J}}{\text{C}} = \text{V}$$

Therefore, *EMF* is measured in volts.

Magnitude of current Recall that resistance is equal to potential difference divided by current. Thus, if an *EMF* generates a current in a wire that is part of a circuit with an effective resistance R, and you know the *EMF*, you can determine the current's magnitude by the following equation:

$$I = \frac{EMF}{R}$$

Direction of current How do you determine the direction of an induced current in a wire? As a wire moves through a magnetic field, a force is exerted on the charges in the wire, and they move in the direction of that force, as shown in the top image of **Figure 2.** To find this direction, use the right-hand rule illustrated in the bottom of **Figure 2.** Point your thumb in the direction the wire moves and your fingers in the direction of the magnetic field. Your palm points in the direction of the force on the positive charges and, thus, in the direction of current.

Figure 2 The direction of current in a wire moving through a magnetic field can be found using a right-hand rule. Note that the direction of current is the same as the direction of the force on the charges in the wire.

Direction of Current

F
S
B
v
N
Motion of the wire
I

Right-Hand Rule

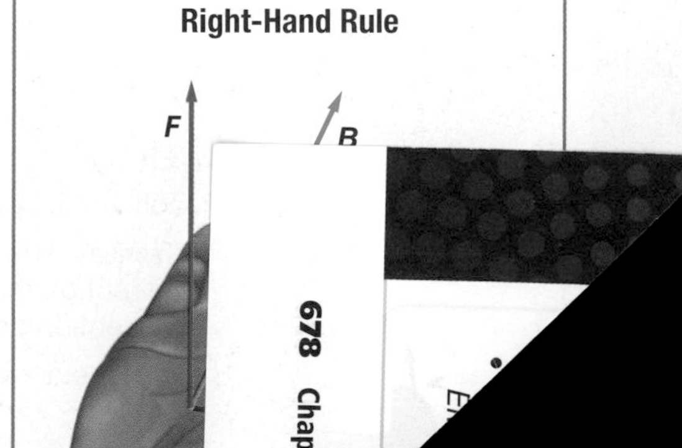

EXAMPLE PROBLEM 1

Get help with **induced *EMF*.** Personal Tutor

EXAMPLE PROBLEM

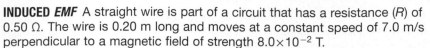

INDUCED *EMF* A straight wire is part of a circuit that has a resistance (R) of 0.50 Ω. The wire is 0.20 m long and moves at a constant speed of 7.0 m/s perpendicular to a magnetic field of strength 8.0×10^{-2} T.

a. What *EMF* is induced in the wire?

b. What is the current through the wire?

c. If a different metal were used for the wire, increasing the circuit's resistance to 0.78 Ω, what would the new current be?

1 ANALYZE AND SKETCH THE PROBLEM

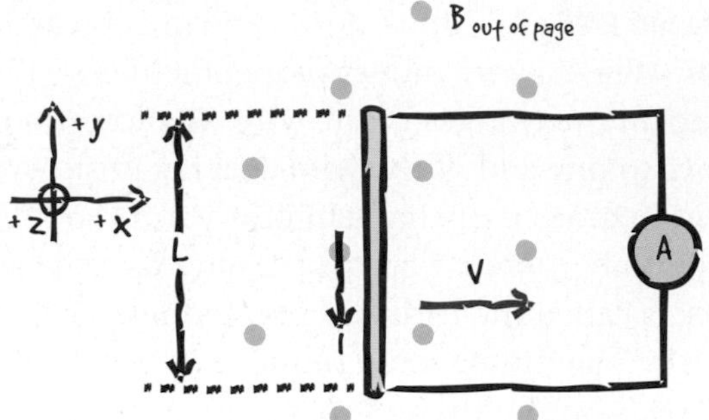

- Draw a straight wire of length L. Connect an ammeter to close the circuit so that there will be a current.
- Show a direction for the magnetic field (B) that is perpendicular to the wire length.
- Show a direction for the velocity (v) that is perpendicular to both the length of the wire and the magnetic field.
- Use the right-hand rule to determine the direction of the current, and show it on the diagram.

KNOWN	UNKNOWN
$v = 7.0$ m/s	$EMF = ?$
$L = 0.20$ m	$I = ?$
$B = 8.0\times10^{-2}$ T	
$R_1 = 0.50\ \Omega$	
$R_2 = 0.78\ \Omega$	

2 SOLVE FOR THE UNKNOWN

a. $EMF = BLv$

$= (8.0\times10^{-2}\text{ T})(0.20\text{ m})(7.0\text{ m/s})$ ◀ Substitute $B = 8.0\times10^{-2}$, $L = 0.20$ m, $v = 7.0$ m/s.

$= 0.11\ \text{T·m}^2\text{/s}$

$= 0.11$ V

b. $I = \frac{EMF}{R}$

$= \frac{0.11\text{ V}}{0.50\ \Omega}$ ◀ Substitute $EMF = 0.11$ V, $R_1 = 0.50\ \Omega$.

$= 0.22$ A

c. $I = \frac{EMF}{R}$

$= \frac{0.11\text{ V}}{0.78\ \Omega}$ ◀ Substitute $EMF = 0.11$ V, $R_2 = 0.78\ \Omega$.

$= 0.14$ A

The current is counterclockwise.

3 EVALUATE THE ANSWER

- **Are the units correct?** Volt is the correct unit for *EMF*. Current is measured in amperes.
- **Does the direction make sense?** The direction obeys a right-hand rule: v is the direction of the thumb, B is the direction the fingers point, and F is the direction the palm faces. Current is in the same direction as the force.
- **Is the magnitude realistic?** The answers are near 10^{-1} A, typical for circuits with small *EMF*s and resistances.

PRACTICE PROBLEMS

Do additional problems. Online Practice

1. You move a straight wire that is 0.5 m long at a speed of 20 m/s vertically through a 0.4-T magnetic field pointed in the horizontal direction.
   a. What *EMF* is induced in the wire?
   b. The wire is part of a circuit with a total resistance of 6.0 Ω. What is the current?
2. A straight wire that is 25 m long is mounted on an airplane flying at 125 m/s. The wire moves in a perpendicular direction through Earth's magnetic field ($B = 5.0\times10^{-5}$ T). What *EMF* is induced in the wire?
3. A straight wire segment in a circuit is 30.0 m long and moves at 2.0 m/s perpendicular to a magnetic field.
   a. A 6.0-V *EMF* is induced. What is the magnetic field?
   b. The total resistance of the circuit is 5.0 Ω. What is the current?
4. **CHALLENGE** A horseshoe magnet is mounted so that the magnetic field lines are vertical. You pass a straight wire between the poles and pull it toward you. The current through the wire is from right to left. Which is the magnet's north pole? Explain.

PRACTICE PROBLEMS

Induced *EMF* in microphones In a previous chapter, you learned how earbuds use magnetic fields to work. Microphones do, too. They are constructed like earbuds, but they work in reverse; they convert sound to electrical energy by electromagnetic induction.

A microphone has a thin aluminum diaphragm attached to a coil in a magnetic field, as shown in **Figure 3.** Sound waves cause the diaphragm to vibrate. This moves the attached coil. The motion of the coil in the magnetic field induces an *EMF* across the coil's ends that generates a current. The induced *EMF* and current vary as the frequency of the sound varies. In this way, the sound waves are converted to electrical signals. The potential differences generated are small, typically 10^{-3} V, but they can be increased, or amplified, by electronic devices.

Figure 3 The microphone converts sound waves from the guitar into electrical signals. The microphone's aluminum diaphragm is attached to a coil in a magnetic field. When sound waves vibrate the diaphragm, the coil moves. The current that is generated is proportional to the sound wave.

Figure 4 An electric current is induced in a rotating wire loop in a circuit that is in a magnetic field. Direction and strength of the induced current change as the loop rotates.

Determine What is the direction of the current when the loop is horizontal to the field, as shown?

View an **animation on electromagnetic induction and generators.**

Electric Generators

Microphones are just one of many technologies that rely on electromagnetic induction. Electromagnetic induction is also the principle governing the operation of electric generators.

An **electric generator** converts mechanical energy to electrical energy. It consists of a number of wire loops in a magnetic field. The wire is wound around an iron core, which concentrates the magnetic flux through the wire. The iron and wire make up the generator's armature, also called a rotor.

Current from a generator Like an armature in a motor, the armature in a generator can rotate freely in a magnetic field. As the armature is turned, its wire loops cut through magnetic field lines. This induces an *EMF* in each loop. Recall that *EMF* depends on wire length, magnetic field strength, and speed. Therefore, increasing the number of loops increases wire length and thus increases the total induced *EMF*. The larger the *EMF*, the stronger the resulting current.

Orientation of loop To understand the steps by which current is produced from the induced *EMF* in a generator, consider the single loop of wire in **Figure 4.** The loop is in a magnetic field. It is also in a closed circuit, so the induced *EMF* produces a current in the loop. As the loop rotates, the strength and the direction of the current change as the orientation of the loop's motion relative to the magnetic field changes.

Recall that only the component of the wire's velocity perpendicular to a magnetic field can induce an *EMF*. Therefore, the current is strongest when the motion of the loop is perpendicular to the field, that is, when the loop is horizontal, as shown in **Figure 4** and in the far left of **Figure 5.**

Current reversals As the loop in **Figure 5** rotates to the vertical position, it moves through the magnetic field lines at an ever-decreasing angle. Thus, it cuts through fewer magnetic field lines per unit of time, and the current decreases. When the loop is in the vertical position, as it is in the middle image of **Figure 5,** current is zero because the wire moves parallel to the magnetic field.

As the loop continues to rotate, the current changes sinusoidally from zero to some maximum value and back to zero during each half-turn. Current reverses each time the loop turns through 180°. A graph of current versus time is shown in the far right of **Figure 5.**

Figure 5 Maximum current is generated when the rotating loop is horizontal to the magnetic field (left). When the loop is vertical to the field (middle), the current is zero. The graph shows that current varies regularly with time as the loop rotates. The numbers on the graph correspond to positions of the loop in the side views. A graph of the variation of *EMF* with time (not shown) would be similar.

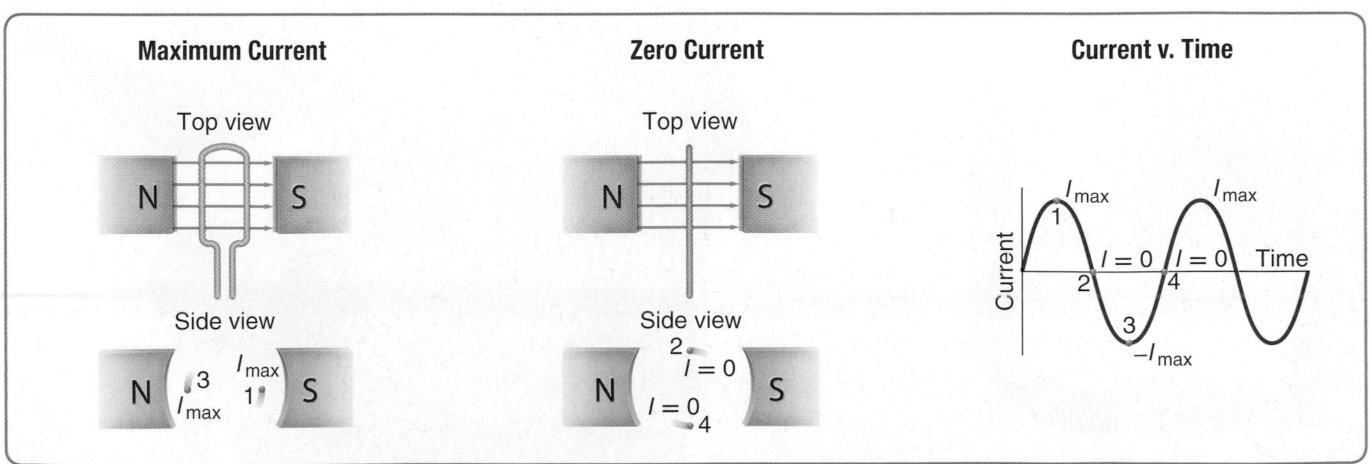

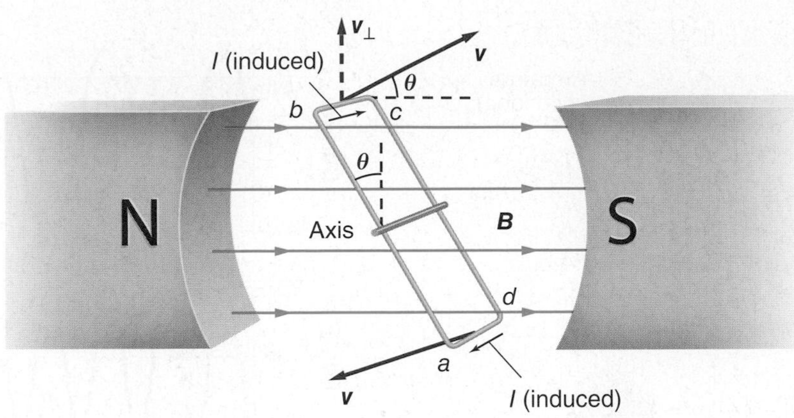

Figure 6 Current is induced in the shorter segments, *bc* and *ad*. No current is induced in the longer segments, *ab* and *cd*. If you pointed your thumb in the direction a long segment is moving and your fingers in the direction of the magnetic field, your thumb would point to the side of the wire. Therefore, no current is induced.

Where is current induced? Not all parts of a loop contribute to an induced *EMF.* Look at **Figure 6.** Only two sides of the wire loop in this figure contribute to an induced *EMF.* When you apply a right-hand rule to segment *ab*, you see that the direction of the induced current would be toward the side of the wire—not along the wire. The same is true of segment *cd*. Thus, no current is induced along either length. In segment *bc*, however, the direction of the induced current is from *b* to *c*, and in segment *ad*, the direction of the current is from *d* to *a*.

Because the conducting loop in **Figure 6** is rotating, the relative angle between a point on the loop and the magnetic field is constantly changing. The *EMF* can be calculated by the equation given earlier, $EMF = BLv(\sin\theta)$, where L is now the combined length of the segments $ad + bc$. The maximum potential difference is induced when a conductor is moving perpendicular to the magnetic field, that is, when $\theta = 90°$.

READING CHECK Apply At what position is the wire loop in a generator when potential difference is greatest?

Sources of mechanical energy Electric generators, such as those in the wind turbines in the image at the start of this chapter, have many wire loops in their armatures, but the total *EMF* induced in them produces current in the same way that an *EMF* produces current in a single loop. Energy from an external source—in this case, the wind—is converted to rotational kinetic energy spinning the turbines. The turbines are attached mechanically to an armature in a magnetic field. As the armature rotates, an *EMF* is induced in each loop. When the generator is connected to an external circuit, the *EMF*s produce current.

Because it is the relative motion between a wire and a magnetic field that induces *EMF*s and current, a current also can be produced in a generator when the armature is stationary and the magnet moves.

Generators and electric motors are almost identical in construction, but they convert energy in opposite directions. A generator converts mechanical energy to electrical energy, while a motor converts electrical energy to mechanical energy. Generators in electrical-generating plants are traditionally powered by steam from water heated by burning fossil fuels or by fissioning uranium. Wind and moving water also can provide the mechanical energy. For smaller generators, people can provide the mechanical energy, as illustrated in **Figure 7.**

Figure 7 Pedaling can provide mechanical energy to turn the armature in a simple generator. The current produced by the generator is enough to power the electric guitar.

Figure 8 This AC generator is similar in construction to an electric motor except it connects to a circuit using a brush-slip-ring device instead of a commutator. As the armature rotates, the direction of the current alternates in time (top right). The power delivered by the generator is always positive (bottom right).

Identify In what position is the armature when current is zero?

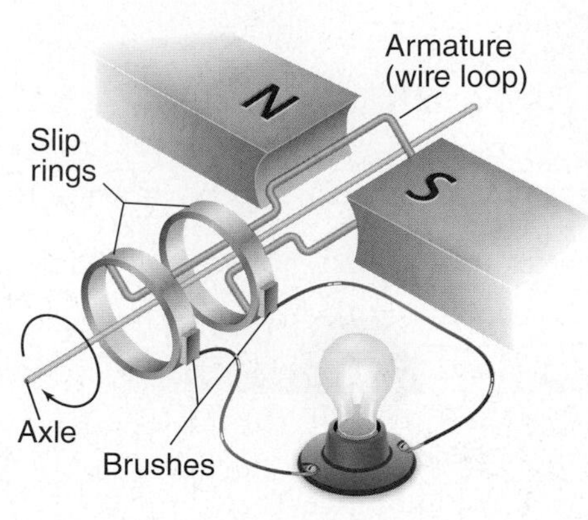

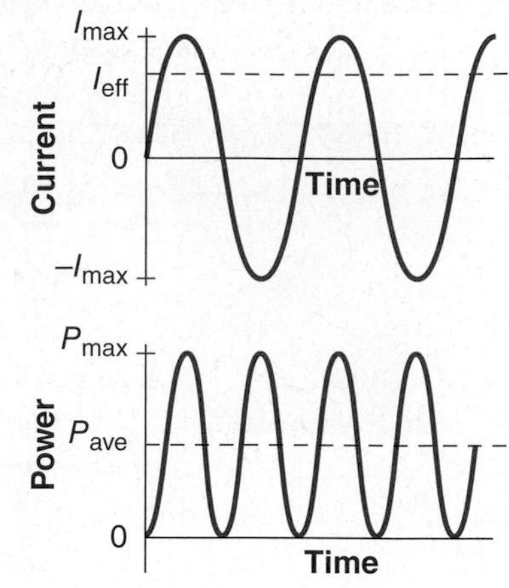

MiniLAB

MOTOR AND GENERATOR

Make a generator and find out what happens when you turn the crank.

Alternating currents and generators There are two main types of generators. In direct current (DC) generators, charges move in a single direction, as they do in batteries. Current is in one direction because the wires of the armature connect to a circuit by means of a commutator.

In contrast, alternating current (AC) generators use a slip-ring device to connect wires to a circuit. In this device, one ring is connected to one end of the armature, and the other ring is connected to the other end of the armature, as shown in the left of **Figure 8.** As the armature rotates through 180°, the induced *EMF* reverses direction. This means the current also reverses direction. The current alternates from positive to negative. In a consistently turned generator, the current alternates at a fixed rate and varies sinusoidally, as shown in the top graph of **Figure 8.**

The frequency of alternating current is different in different countries. In the United States, all electric utilities use a 60-Hz frequency in which current goes from one direction to the other and back to the first 60 times per second. This means that an armature makes 60 complete rotations in one second. Many other countries use a 50-Hz frequency.

Average power The power produced by an AC generator is the product of the current and the potential difference. Because both current and potential difference vary, the power in the circuit varies. The graph at the bottom right of **Figure 8** shows the power versus time produced by an AC generator. Note that power is always positive because I and V are either both positive or both negative. Average power, P_{AC}, is half the maximum power; thus, $P_{AC} = \frac{1}{2} P_{AC\ max}$.

Effective current It is common to describe alternating current and potential difference in terms of effective current and effective potential difference, rather than referring to their maximum values. Recall that $P = I^2R$. Thus, effective current (I_{eff}) can be expressed in terms of the average AC power, as $P_{AC} = I_{eff}{}^2R$. To determine I_{eff} in terms of maximum current (I_{max}), start with the power relationship, $P_{AC} = \frac{1}{2} P_{AC\ max}$, and substitute in I^2R. Then solve for I_{eff}.

EFFECTIVE CURRENT

Effective current is equal to $\frac{\sqrt{2}}{2}$ times the maximum current.

$$I_{eff} = \left(\frac{\sqrt{2}}{2}\right) I_{max} = 0.707 I_{max}$$

Effective potential difference Similarly, because $P = \frac{V^2}{R}$, the effective potential difference is given by the following equation.

EFFECTIVE POTENTIAL DIFFERENCE

Effective potential difference is equal to $\frac{\sqrt{2}}{2}$ times the maximum potential difference.

$$V_{eff} = \left(\frac{\sqrt{2}}{2}\right) V_{max} = 0.707 V_{max}$$

Effective potential difference is commonly referred to as RMS (root mean square) potential difference. In the United States, the RMS potential difference generally available at wall outlets is described as 120 V, where 120 V is the magnitude of the effective potential difference, not the maximum potential difference. In many other countries, the RMS potential difference is 240 V.

War of currents The first electrical distribution system in the United States was developed by Thomas Edison in the late 1800s. Edison's system used direct current. Soon, however, a competing system that used alternating current was developed by Nikola Tesla. Tesla partnered with George Westinghouse to promote AC current, and a "war of the currents" ensued, with Edison pitted against Tesla and Westinghouse. Eventually, AC won. In the next section, you will read why AC is superior to DC for distributing electrical energy.

PRACTICE PROBLEMS

Do additional problems.  

5. A generator develops a maximum potential difference of 170 V.

- **a.** What is the effective potential difference?
- **b.** A 60-W lamp is placed across the generator with an I_{max} of 0.70. What is the effective current through the lamp?
- **c.** What is the resistance of the lamp when it is working?

6. The RMS potential difference of an AC household outlet is 117 V. What is the maximum potential difference across a lamp connected to the outlet? If the RMS current through the lamp is 5.5 A, what is the lamp's maximum current?

7. If the average power used over time by an electric light is 75 W, what is the peak power?

8. CHALLENGE An AC generator delivers a peak potential difference of 425 V.

- **a.** What is the V_{eff} in a circuit connected to the generator?
- **b.** The resistance is 5.0×10^2 Ω. What is the effective current?

SECTION 1 REVIEW

Section Self-Check Check your understanding.

9. MAINIDEA Use the concept of electromagnetic induction to explain how an electric generator works.

10. Generator Could you make a generator by mounting permanent magnets on a rotating shaft and keeping the coil stationary? Explain.

11. Bike Generator A small generator on your bike lights the bike's headlight. What is the source of the energy for the bulb when you ride along a flat road?

12. Microphone Consider the microphone shown in **Figure 3.** What happens when the diaphragm is pushed in?

13. Frequency What changes to an electric generator are required to increase frequency?

14. Output Potential Difference Explain why the output potential difference of an electric gen tor increases when the magnetic field is made str another way to increase the output difference?

15. Critical Thinking A student asks, dissipate any power? The energ when the current is positive is r current is negative. The net cu why this reasoning is wrong.

SECTION 2 Applications of Induced Currents

PHYSICS 4 YOU **Before you fly on an airplane, you must walk through a metal detector. Metal detectors generate magnetic fields that can be used to detect metal objects. What physics principles are behind these detectors?**

MAIN IDEA

The magnetic fields produced by induced currents are crucial to the operation of generators, motors, and transformers.

Essential Questions

- What is Lenz's law and how is it related to induced *EMF*s?
- How do induced *EMF*s affect the operation of motors and generators?
- What is self-inductance, and how does it affect circuits?
- How does the turns ratio in a transformer affect potential difference and current?

Review Vocabulary

law of conservation of energy states that in a closed, isolated system, energy is not created or destroyed

New Vocabulary

Lenz's law
eddy current
self-inductance
transformer
mutual inductance
step-up transformer
step-down transformer

Lenz's Law

Imagine the hand pulling the wires through the magnetic field in **Figure 9** is yours. As you pull, the movement of the wire induces an *EMF*. The wire slides on the wire loop, making a complete circuit, so the *EMF* produces a current. Because the magnetic field (B) is out of the page and the velocity is to the right, the right-hand rule shows that the direction of current in the wire is downward and clockwise in the circuit.

Opposing motion The downward-moving charges in the wire moving through the magnetic field that is out of the page experience a force. The right-hand rule shows that the direction of this force is opposite the movement of the wire. These charges exert a force on the wire that opposes the motion of the wire and makes it harder for you to pull it. The physicist H.F.E. Lenz first demonstrated this force in 1834.

Opposing change While Lenz stated his law in terms of force, there is a second way of interpreting it. The current in the loop produces its own magnetic field. The right-hand rule shows that the direction of this induced field inside the loop is into the paper. That is, it is in the direction that opposes the change in the field that caused it. Thus, **Lenz's law** says that the magnetic field produced by the induced current is in the direction that is opposite the original field.

Figure 9 A wire pulled through a magnetic field generates an *EMF*. The *EMF* produces a current in the circuit (I). Motion of the charges in the wire produces a force (F). The current also produces a magnetic field (blue dots and crosses) that, within the circuit, is in the direction opposite that of the field whose change caused it.

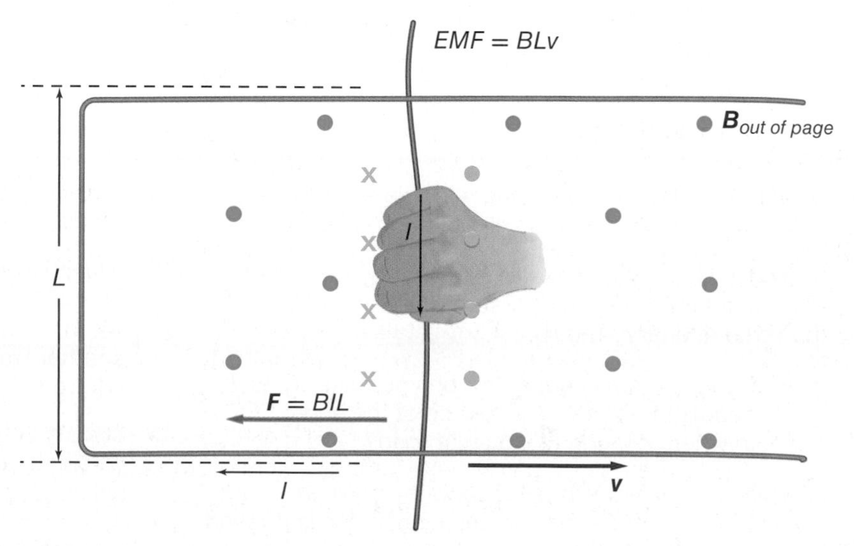

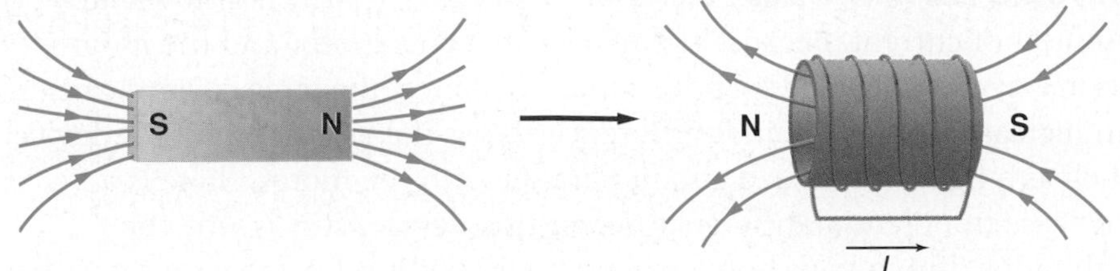

Figure 10 The magnet approaching the coil induces a current in the coil. The right-hand rule shows that the current is in the counter-clockwise direction. The current produces a magnetic field in the direction opposite the change in the field that produced it.

Apply How would the force on the coil change if the magnet was oriented so that its south pole faced the coil?

Lenz's law explains the direction of the current in the coil of wire in **Figure 10.** The approaching magnet causes an increase in the magnetic field inside the coil. The current produced by the induced *EMF* must be in the direction shown in **Figure 10** to produce a magnetic field that opposes the increase in field. The induced field exerts a repelling force on the approaching magnet. By Newton's third law, there must be an equal and opposite force on the coil. This force pushes the coil away from the magnet. If you moved the magnet away from the coil, the field within the coil would decrease. By Lenz's law the induced field would oppose this change by being in the same direction as the field of the magnet and adding to the field. In this case, the current producing the induced field in the coil would be in the clockwise direction.

In either case—moving the magnet toward or away from the coil—an induced magnetic field opposes the change in the field that created it. A decreasing magnetic field induces a field to oppose the decrease; an increasing magnetic field induces a field to oppose the increase.

Generators Lenz's law affects the operation of generators and motors. When a generator is not in a circuit, you can turn its armature, but the induced *EMF* generates no current. Therefore, there is no magnetic field. No force acts on the armature, and the armature is easy to turn.

However, when you turn the armature of a generator in a circuit, the induced *EMF* generates current, and the resulting magnetic field exerts a force on the armature. The force on the armature is opposite the external force that turns the armature, and the armature is difficult to turn. A generator supplying a large current produces a large amount of electric energy. The opposing force on the armature means that a proportionately greater amount of mechanical energy must be supplied to produce it. Note this is consistent with the law of conservation of energy.

Motors When you first turn on a motor, such as when you start a vacuum cleaner, current is large. As soon as the motor begins to turn, however, the motion of the wires across the magnetic field induces an *EMF* that opposes the current. This results in reduced current through the motor.

More load, more energy transfer When a mechanical load is placed on a motor, as when work is being done to lift a weight, the rotation of the motor slows. The slowing of the motor decreases the induced *EMF* opposing the current. Therefore, net current through the motor increases—consistent with the law of conservation of energy. When current increases, so does the rate at which electrical energy is being sent to the motor. This energy is delivered in mechanical form to the load. If the mechanical load is sufficient to stop the motor, the current can be so high that wires overheat.

PhysicsLAB

SWINGING COILS

How does Lenz's law affect the operation of a generator?

MiniLAB

SLOW MOTOR

Observe what happens when you turn a motor on.

Why lights dim when a motor starts When a motor starts, it draws a large amount of current. Because the wires between a motor and the motor's source of electrical energy have some resistance, the large current causes an increase in the potential difference across the wires. If another device, such as a lightbulb, is in a circuit parallel with the motor, there is a decrease in potential difference across this device. This is why the lightbulbs dim in your home when an air conditioner, table saw, or other motor-driven appliance starts operating.

When the current through a motor is interrupted, such as when a switch in the circuit is turned off or when a motor's plug is pulled from an outlet, the magnetic field in the motor drops to zero. Whenever there is a change in a magnetic field around a wire, an *EMF* is induced across the wire that opposes the change that created it. The induced *EMF* can be large enough to cause a spark to jump across the switch or between the plug and the outlet.

Braking effect The effects of Lenz's law can be beneficial. You might have used a laboratory balance to measure masses. A balance takes advantage of induced currents to stop its oscillations when an object is placed on its pan. A piece of metal attached to the balance arm in **Figure 11** rests between the poles of a horseshoe magnet. When an object is placed in the pan, the balance arm swings, and the metal moves through the magnetic field. The induced *EMF* creates currents in the metal, called eddy currents, that slow the metal down.

Eddy currents are currents generated in any piece of metal moving through a magnetic field; the magnetic field they produce opposes the motion that caused the currents. The force on the metal opposes the motion of the metal in either direction, but it does not act when the metal is still. Thus, it does not change measurements of the mass of an object on the pan.

When eddy currents have a braking effect on metal, the effect is called eddy-current damping. Eddy-current damping is commonly used to slow the movement of metal parts. The brakes of some trains and roller coasters are designed to take advantage of eddy-current damping. Often, the damping effect is unwelcome. To reduce eddy-current circulation in the metal parts in motors, the motor cores are constructed from thin metal layers that have insulation added between the layers.

Figure 11 As the metal plate on the end of the balance beam moves through the magnetic field, an eddy current is generated in the metal. The eddy current produces a magnetic field that opposes the motion that caused it. This dampens the beam's oscillations.

View an **animation of Lenz's law in action.**

Concepts In Motion

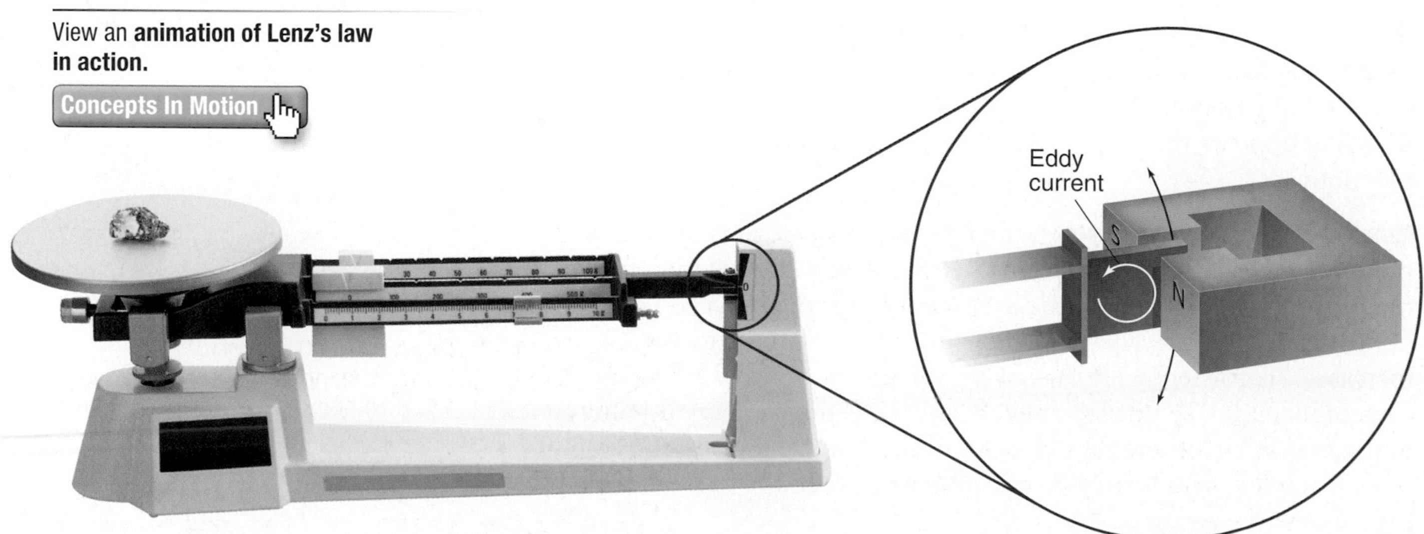

Richard Hutchings

Levitation You have read that Lenz's law dictates that current in a wire in a changing magnetic field is such that the magnetic field it generates opposes the change of the external field. This effect can cause objects to levitate. How does this work?

The AC current in the coil of wire in **Figure 12** generates a changing magnetic field that induces an *EMF* in both of the aluminum rings above it. In the uncut ring on top, the *EMF* produces a current that generates a magnetic field opposing the change in the generating field. Because like poles face each other, the coil and uncut ring repel, and the ring rises.

The lower ring in **Figure 12** does not rise. Because it has been sawed through, it is an incomplete circuit and the *EMF* does not generate a current. Therefore, there is no force on the ring and no levitation.

Levitation caused by repulsion underlies the operation of very fast trains called magnetic levitation (Maglev) trains. As these trains move, they make no physical contact with the tracks.

Figure 12 An *EMF* is induced in both aluminum rings, but current is produced only in the continuous ring because only this ring completes a circuit. Thus, only this ring experiences a force, and only this ring levitates. If the rings were made from a nonconducting material, such as nylon or wood, an *EMF* would not be induced in either one.

Self-inductance An *EMF* can be induced in a wire when the magnetic field in the region of the wire changes. The field can be external, or it can be generated by the current in the wire itself, as in **Figure 13.**

Imagine that the coil of wire in **Figure 13** is suddenly connected to a battery. This causes a change in the potential difference across the coil, producing a changing current through the coil. The changing current generates a changing magnetic field in the coil. This changing field induces an *EMF* in the direction that opposes the change. The induced *EMF* reduces the potential difference across the coil. The result is a net decrease in current.

Therefore, initially the current through the coil is small but increasing. However, as the rate of change of current decreases, so does the opposing *EMF*. When the current reaches a constant final value, the current change is zero, so the induced *EMF* is also zero.

If the potential difference supplied by a battery is decreased, the current is also decreased. So, the induced *EMF* changes to be in a direction to maintain the magnetic field. Therefore, the induced current is in the same direction as the current from the battery. If the coil were suddenly disconnected from the battery, the induced *EMF* could be large enough to produce a spark. The property of a wire, either straight or in a coil, to create an induced *EMF* that opposes the change in the potential difference across the wire is called **self-inductance.**

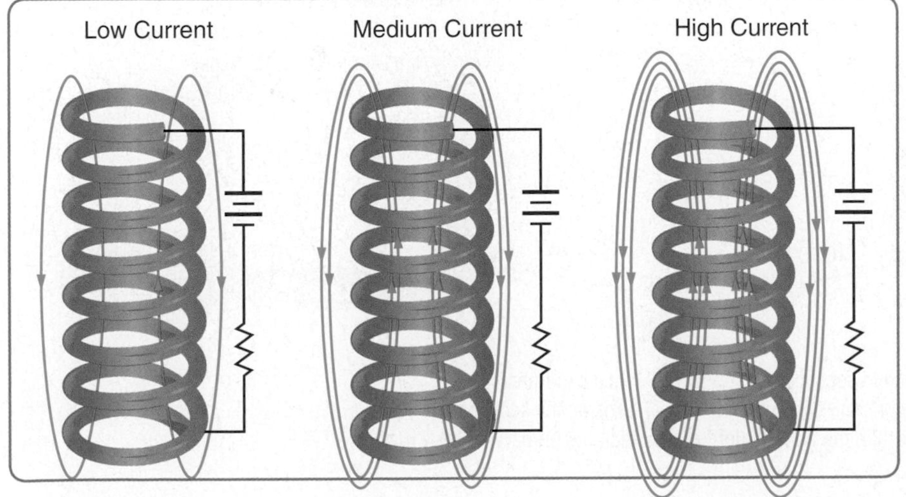

Figure 13 As the current in the coil increases, the magnetic field generated by the current also increases. The increase in the magnetic field induces an *EMF* that opposes the direction of current, and more energy is needed to increase the current further.

Figure 14 Inside this cylinder is a transformer. Transformers reduce high voltages to usable levels before the electrical energy enters your home.

Transformers

You have probably seen metal cylinders attached to utility poles, such as the cylinder on the pole in **Figure 14.** Inside each of these cylinders is a transformer. **Transformers** are devices that increase or decrease potential differences with relatively little waste of energy. Only alternating current (AC) can be sent through a transformer. Direct current (DC) cannot pass through a transformer.

How transformers work You read that self-inductance produces an *EMF* when current changes in a single coil. A transformer has two coils, electrically insulated from each other but wound around the same iron core, as you can see in **Figure 15.** When one coil—the primary coil—is connected to an AC source, the changing current creates a changing magnetic field that is carried through the core to the other coil—the secondary coil. In the secondary coil, the changing magnetic field induces a varying *EMF* and current. An *EMF* and current in one coil due to changing current in another coil is called **mutual inductance.**

The *EMF* induced in the secondary coil, called the secondary potential difference, is proportional to the potential difference provided to the primary coil. The secondary potential difference also depends on what is called the turns ratio. The turns ratio is the number of turns of wire in the secondary coil divided by the number of turns in the primary coil, as shown on the right in the following expressions.

$$\frac{\text{primary potential difference}}{\text{secondary potential difference}} = \frac{\text{number of turns in primary coil}}{\text{number of turns in secondary coil}}$$

$$\frac{V_P}{V_S} = \frac{N_P}{N_S}$$

If the secondary potential difference is larger than the primary potential difference, as it is in the left part of **Figure 15,** the transformer is called a **step-up transformer.** If the secondary potential difference is smaller than the primary potential difference, as in the right part of **Figure 15,** the transformer is called a **step-down transformer.**

READING CHECK Compare How do step-up transformers differ from step-down transformers?

■ Transformers

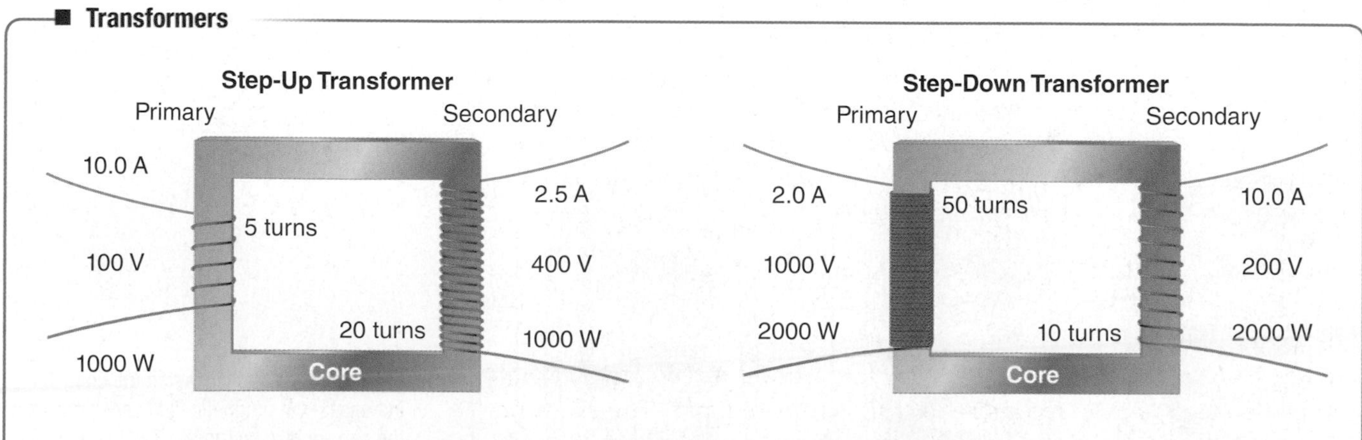

Figure 15 The ratio of primary potential difference to secondary potential difference in a transformer depends on the ratio of the number of turns on the primary coil to the number of turns on the secondary coil. Secondary potential difference can be greater than the primary (left) or less than the primary (right).

The ideal transformer In an ideal transformer, the electrical power delivered to the secondary circuit equals the power supplied to the primary circuit. An ideal transformer dissipates no power itself. Because it is 100 percent efficient, it can be represented by the following equations:

$$P_P = P_S$$
$$V_P I_P = V_S I_S$$

Rearranging the equation to find the ratio $\frac{V_P}{V_S}$ shows that the current in the primary circuit depends on how much current is required by the secondary circuit. This relationship can be combined with the relationship shown earlier (between potential difference and the number of turns) to result in the following equation.

TRANSFORMER EQUATION

The ratio of the current in the secondary coil to the current in the primary coil is equal to the ratio of the potential difference in the primary coil to the potential difference in the secondary coil, which is also equal to the ratio of the number of turns on the primary coil to the number of turns on the secondary coil.

$$\frac{I_S}{I_P} = \frac{V_P}{V_S} = \frac{N_P}{N_S}$$

Changes in potential difference You read that a step-up transformer increases potential difference in the secondary coil. Because transformers cannot increase power output, there must be a corresponding decrease in current that can be supplied to the secondary circuit. Similarly, in a step-down transformer, the current is greater in the secondary circuit than it is in the primary circuit. A decrease in potential difference corresponds to an increase in current, as you can see in the Connecting Math to Physics. Even in real transformers, which are not 100 percent efficient, the current supplied by a step-down transformer is larger than the current in the primary circuit, and a step-up transformer supplies less current than it draws.

Isolation transformers Not all transformers are step-up or step-down. In some transformers, the primary and secondary coils have the same number of turns, so the input and output potential differences are identical. These transformers, called isolation transformers, are often used for safety reasons. They isolate the DC portion of current from sensitive electronic devices where it otherwise might cause interference or electric shock. These sensitive electronic devices include computers, recording instrumentation, and medical tools, such as those used for ultrasound and other diagnostic imaging. Isolation transformers also can be used to reduce electrical noise.

Real transformers Ideal transformers do not convert any electrical energy to thermal energy. In real transformers, however, some of the electrical energy they deliver to the primary circuit is converted into thermal energy, heating the transformer and the air around it. The power that can be delivered by the secondary circuit is thus reduced. The efficiency of a transformer is given by the ratio of the output power to the input power. Typical transformers are 95 to 98 percent efficient. Though the amount of energy lost to thermal energy might seem small, large transformers handle so much power that they can become very hot.

CONNECTING MATH TO PHYSICS

Inequalities Study the following expressions to help you understand the relationships among potential difference (V), current (I), and the number of coils in transformers (N) in primary and secondary circuits.

Step-Up Transformer	Step-Down Transformer
$V_P < V_S$	$V_P > V_S$
$I_P > I_S$	$I_P < I_S$
$N_P < N_S$	$N_P > N_S$

PhysicsLAB

INDUCTION AND TRANSFORMERS

What is the relationship between potential differences across the two coils of a transformer?

EXAMPLE PROBLEM 2

Find help with **significant digits.** Math Handbook

STEP-UP TRANSFORMERS A step-up transformer has a primary coil consisting of 200 turns and a secondary coil consisting of 3000 turns. The primary coil is supplied with an effective AC potential difference of 90.0 V.

a. What is the potential difference in the secondary circuit?

b. The current in the secondary circuit is 2.0 A. What is the current in the primary circuit?

1 ANALYZE AND SKETCH THE PROBLEM

- Draw an iron core that has turns of wire on either side.
- Label the variables I, V, and N.

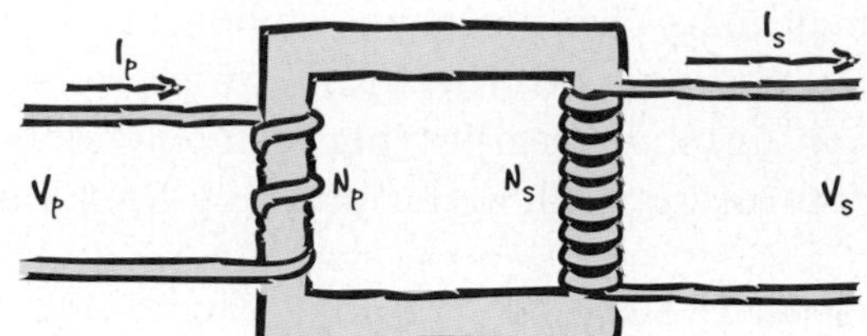

KNOWN		UNKNOWN
$N_p = 200$	$V_p = 90.0$ V	$V_s = ?$
$N_s = 3000$	$I_s = 2.0$ A	$I_p = ?$

2 SOLVE FOR THE UNKNOWN

a. Solve for V_s.

$$\frac{V_s}{V_p} = \frac{N_s}{N_p}$$

$$V_s = \frac{N_s V_p}{N_p}$$

$$= \frac{(3000)(90.0\text{ V})}{200}$$ ◀ *Substitute N_s = 3000, V_p = 90.0 V, N_p = 200.*

$$= 1350\text{ V}$$

b. The power in the primary and secondary circuits is equal assuming 100 percent efficiency.

$$P_p = P_s$$

$$V_p I_p = V_s I_s$$ ◀ *Substitute $P_p = V_p I_p$, $P_s = V_s I_s$.*

Solve for I_p.

$$I_p = \frac{V_s I_s}{V_p}$$

$$= \frac{(1350\text{ V})(2.0\text{ A})}{90.0\text{ V}}$$ ◀ *Substitute V_s = 1350 V, I_s = 2.0 A, V_p = 90.0 V.*

$$= 3.0\times10^1\text{ A}$$

3 EVALUATE THE ANSWER

- **Are the units correct?** Potential difference is in volts, and current is in amps.
- **Is the magnitude realistic?** A large step-up ratio of turns results in a large secondary potential difference yet a smaller secondary current. The answers agree.

PRACTICE PROBLEMS

Do additional problems. Online Practice

For the following problems, effective currents and potential differences are indicated.

16. A step-down transformer has 7500 turns on its primary coil and 125 turns on its secondary coil. The potential difference across the primary circuit is 7.2 kV. What is the potential difference across the secondary circuit? If the current in the secondary circuit is 36 A, what is the current in the primary circuit?

17. CHALLENGE A step-up transformer has 300 turns on its primary coil and 90,000 turns on its secondary coil. The potential difference of the generator to which the primary circuit is attached is 60.0 V. The transformer is 95 percent efficient. What is the potential difference across the secondary circuit? The current in the secondary circuit is 0.50 A. What current is in the primary circuit?

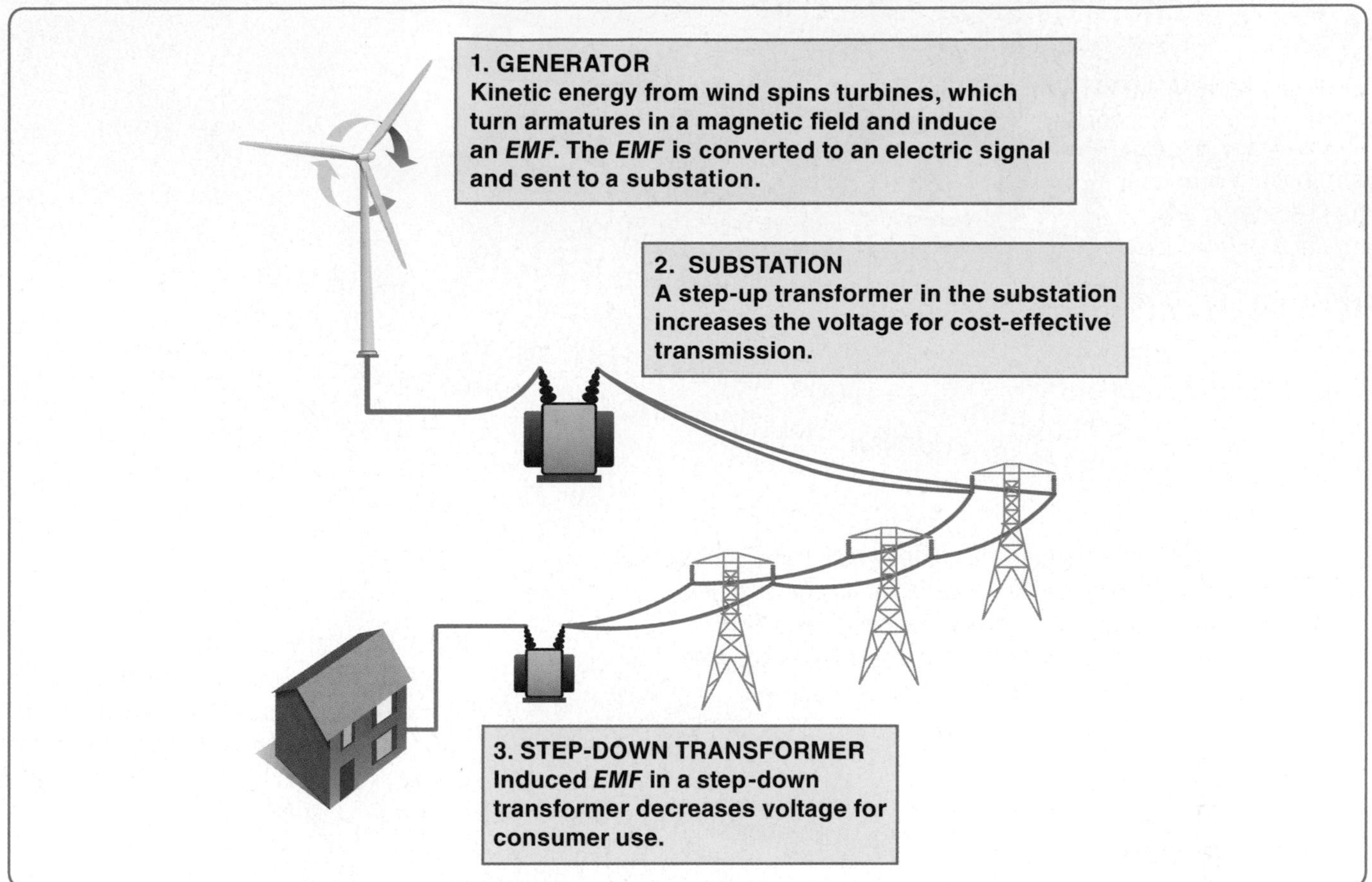

1. GENERATOR
Kinetic energy from wind spins turbines, which turn armatures in a magnetic field and induce an *EMF*. The *EMF* is converted to an electric signal and sent to a substation.
2. SUBSTATION
A step-up transformer in the substation increases the voltage for cost-effective transmission.
3. STEP-DOWN TRANSFORMER
Induced *EMF* in a step-down transformer decreases voltage for consumer use.

Figure 16 Step-up transformers increase potential differences (voltages) in overhead power lines. Step-down transformers decrease potential differences for consumer use.

Everyday uses of transformers You read in a previous chapter that long-distance transmission of electrical energy is economical only if very high potential differences are used. High potential differences reduce the current required in the transmission lines, keeping the wasteful energy transformations low. As shown in **Figure 16,** step-up transformers are used at power sources, where they can develop potential differences as high as 480,000 V. When the energy reaches homes, step-down transformers reduce the potential difference to 120 V.

Transformers in home appliances and other electronic devices further reduce potential differences to usable levels. Game systems, printers, laptop computers, and rechargeable toys have transformers inside their casings or in blocks attached to their cords. The small transformers in these devices reduce potential differences from wall outlets to the 3 V–26 V range.

SECTION 2 REVIEW

Section Self-Check — Check your understanding.

18. MAINIDEA You hang a coil of wire with its ends joined so that it can swing easily. If you now plunge a magnet into the coil, the coil will start to swing. Which way will it swing relative to the magnet and why?

19. **Motors** If you unplugged a running vacuum cleaner from a wall outlet, you would be much more likely to see a spark than you would if you unplugged a lighted lamp from the wall. Why?

20. **Transformers and Current** Explain why a transformer may be operated only on alternating current.

21. **Transformers** Frequently, transformer coils that have only a few turns are made of very thick (low-resistance) wire, while those with many turns are made of thin wire. Why?

22. **Step-Up Transformers** Refer to the step-up transformer shown in **Figure 15.** Explain what would happen to the primary current if the secondary coil were short-circuited.

23. **Critical Thinking** Would permanent magnets make good transformer cores? Explain.

MAGNETIC MONEY

MAGNETIC CARD READERS

You have probably swiped a card in a magnetic card reader at a grocery store, gas station, subway, or library. Some are also used to unlock doors or activate elevators. Have you ever wondered how magnetic card readers work?

1 A card's magnetic stripe (often called a magstripe) consists of linear series of tiny bar magnets arranged in tracks. The polarities of the magnets are oriented in a unique sequence that corresponds to a binary code—a series of 1s and 0s.

2 Why must you swipe the card quickly? Remember Michael Faraday's key discovery—only changing magnetic fields create electric current. If the swipe is too slow, the current is too small to read.

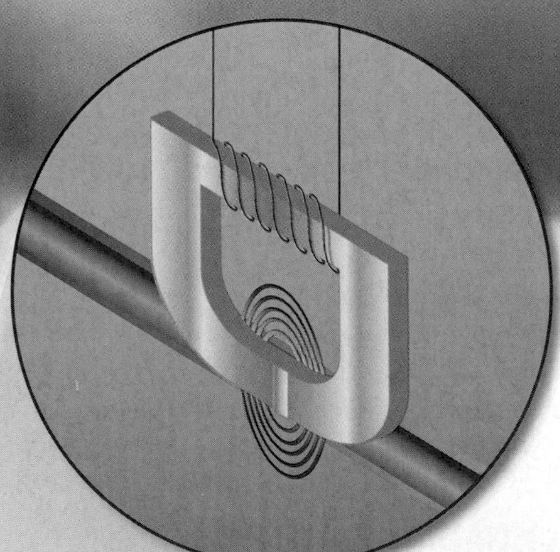

3 When you swipe a card quickly through a card reader, the read head responds to the moving card's sequence of magnetic domains. Each time the read head encounters a domain boundary, it registers a tiny electric current.

4 The read head sends the electric currents to a computer, which reads the sequence of 1s and 0s depending on the type of signal code that is used. The binary code can then be translated into alphanumeric characters.

GOING FURTHER >>>

Research There are many different types of cards and readers. Consider other examples of cards, such as bar code cards, RFID cards, and smart cards. Research and report on how these cards work.

STUDY GUIDE

BIGIDEA A changing magnetic field can induce current in a conductor.

VOCABULARY
- **electromagnetic induction** *(p. 676)*
- **induced electromotive force** *(p. 677)*
- **electric generator** *(p. 680)*

SECTION 1 Inducing Currents

MAINIDEA A changing magnetic field induces an *EMF* in a wire, and the *EMF*, in turn, generates current when the wire is in a circuit.

- An *EMF* is induced in a wire when the magnetic field at the location of the wire changes, as when the wire moves through the magnetic field.
- The *EMF* induced in a length of wire moving through a uniform magnetic field is the product of the magnetic field (B), the length of the wire (L), and the component of the wire's velocity that is perpendicular to the field ($v(\sin\theta)$).

$$EMF = BLv(\sin\theta)$$

- As the armature in a generator is turned by mechanical force, the induced *EMF* creates a current that changes direction each time the armature turns 180°.
- In an AC circuit, the effective current and potential difference are related to their maximum values in the following way:

$$I_{eff} = 0.707\ I_{max}$$
$$V_{eff} = 0.707\ V_{max}$$

VOCABULARY
- **Lenz's law** *(p. 684)*
- **eddy current** *(p. 686)*
- **self-inductance** *(p. 687)*
- **transformer** *(p. 688)*
- **mutual inductance** *(p. 688)*
- **step-up transformer** *(p. 688)*
- **step-down transformer** *(p. 688)*

SECTION 2 Applications of Induced Currents

MAINIDEA The magnetic fields produced by induced currents are crucial to the operation of generators, motors, and transformers.

- Lenz's law states that a current created by an induced *EMF* is in a direction that produces a magnetic field that opposes the change in the magnetic field that generated the current.
- In a generator, an induced *EMF* makes the armature harder to turn when the generator supplies more current. In a motor, induced *EMF* reduces the current through the motor when the motor runs at high speed.
- Self-inductance is a property of a wire or coil that carries a changing current. The faster the current is changing, the greater is the change in the resulting magnetic field and the greater is the induced *EMF* that opposes that change.
- Transformers have two coils close together. An AC current in the primary coil induces an alternating *EMF* in the secondary coil. The turns ratio—the number of turns on the secondary coil divided by the number of turns on the primary coil—is equal to the ratio of the potential difference in the secondary coil to the potential difference in the primary coil. It is also equal to the ratio of the current in the primary coil to the current in the secondary coil.

$$\frac{I_s}{I_p} = \frac{V_p}{V_s} = \frac{N_p}{N_s}$$

Games and Multilingual eGlossary

Vocabulary Practice

CHAPTER 25 ASSESSMENT

Chapter Self-Check

SECTION 1 Inducing Currents

Mastering Concepts

24. BIGIDEA You have a wire connected in a circuit and two magnets. Describe how you could use them to generate potential difference and current.

25. Explain how AC and DC generators differ in the way they connect to electric circuits.

26. Why is iron used in an armature?

For problems 27–28, refer to **Figure 17.**

27. A single conductor moves through a magnetic field and generates a potential difference. In what direction should the wire be moved, relative to the magnetic field, to generate the minimum potential difference?

28. In which direction would the electric field in the wire point if the wire were pulled to the right?

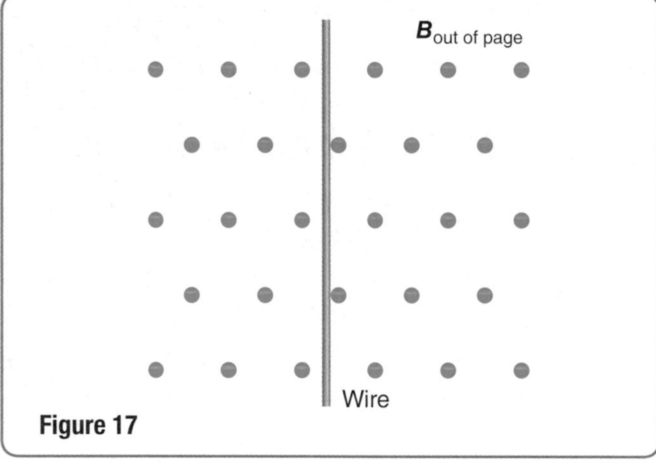

Figure 17

29. What is the effect of increasing the length of the wire of an armature in an electric generator?

30. How were Oersted's and Faraday's results similar? How were they different?

31. What does *EMF* stand for? Why is the name inaccurate?

32. What is the difference between an electric generator and an electric motor?

33. List the major parts of an AC generator.

34. Why is the effective value of an AC current less than its maximum value?

35. Hydroelectricity Water trapped behind a dam turns turbines that rotate generators. List all the forms of energy that take part in the cycle that includes the stored water and the electricity produced.

Mastering Problems

36. A 20.0-m-long wire moves perpendicularly through a magnetic field at 4.0 m/s. An *EMF* of 40 V is induced in the wire. What is the field's strength?

37. Airplanes An airplane traveling at 9.50×10^{2} km/h passes over a region where Earth's magnetic field is 4.5×10^{-5} T and is nearly vertical. What potential difference is induced between the plane's wing tips, which are 75 m apart?

38. A straight wire that is 0.75-m long moves upward through a horizontal 0.30-T magnetic field, as shown in **Figure 18,** at a speed of 16 m/s.

a. What *EMF* is induced in the wire?

b. The wire is part of a circuit with a total resistance of 11 Ω. What is the current?

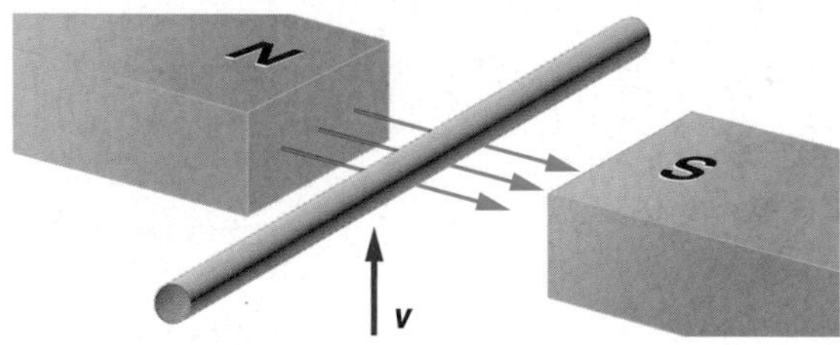

Figure 18

39. At what speed would a 0.20-m length of wire have to move across a 2.5-T magnetic field to induce an *EMF* of 10 V?

40. An AC generator has a maximum potential difference of 565 V. What effective potential difference does the generator deliver to an external circuit?

41. An AC generator develops a maximum potential difference of 150 V. It delivers a maximum current of 30.0 A to an external circuit.

a. What is the effective potential difference of the generator?

b. What effective current does the generator deliver to the external circuit?

c. What is the average power dissipated in the circuit?

42. Stove An electric stove connects to an AC source with an effective potential difference of 240 V.

a. Find the maximum potential difference across one of the stove's elements when it is operating.

b. The resistance of the operating element is 11 Ω. What is the effective current?

Chapter Self-Check

43. You wish to generate an EMF of 4.5 V by moving a wire at 4.0 m/s through a 0.050-T magnetic field. You want to use the shortest length of wire you can. How long must the wire be and what should the angle be between the field and the velocity?

44. You connect both ends of a 0.10-Ω copper wire to the terminals of a 875-Ω galvanometer. You then move a 10.0-cm segment of the wire upward at 1.0 m/s through a 2.0×10^{-2} -T magnetic field. What current will the galvanometer indicate?

45. Refer to Example Problem 1 and **Figure 19** to determine the following.

a. induced potential difference in the conductor

b. current (I)

c. polarity of point A relative to point B

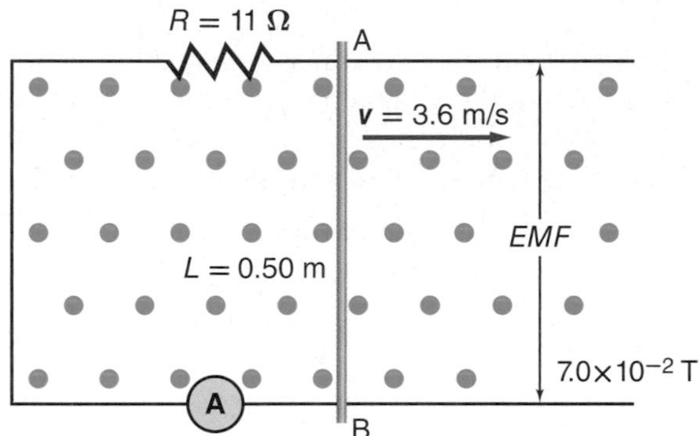

Figure 19

46. The direction of a 0.045-T magnetic field is 60.0° above the horizontal. A wire, 2.5-m long, moves horizontally at 2.4 m/s.

a. What is the vertical component of the magnetic field?

b. What EMF is induced in the wire?

47. **Dams** A generator at a dam can supply 375 MW (375×10^6 W) of electrical power. Assume that the turbine and generator are 85 percent efficient.

a. Find the rate at which falling water must supply energy to the turbine.

b. The energy of the water comes from a change in gravitational potential energy, $GPE = mg\Delta h$. What is the change in GPE needed each second?

c. If the water falls 22 m, what is the mass of the water that must pass through the turbine each second to supply this power?

48. A 20-cm conductor is moving with a constant velocity of 1 m/s in a magnetic field of 4.0 T. What is the induced potential difference when the conductor moves perpendicular to the line of force?

SECTION 2

Applications of Induced Currents

Mastering Concepts

49. State Lenz's law and explain why it is consistent with the law of conservation of energy.

50. Why does the current through a motor decrease when the motor speeds up?

51. **Power Saws** When a power saw is turned on in a garage, why do the lights dim in the house?

52. A 150-W transformer has an input potential difference of 9.0 V and an output current of 5.0 A. What is the ratio of V_{output} to V_{input}?

53. Why is the self-inductance of a coil a major factor when the coil is in an AC circuit but a minor factor when the coil is in a DC circuit?

54. Explain why the word *change* appears so often in this chapter.

55. Upon what does the ratio of the potential difference across the primary coil of a transformer to the potential difference across the secondary coil of the transformer depend?

56. **Trains** The train in **Figure 20** makes no contact with the rails. There are electromagnets in the train but not in the rails. Explain how this train is able to levitate above the rails as long as it is moving.

Figure 20

Mastering Problems

57. **Doorbell** A doorbell requires an effective potential difference of 18 V from a 120-V line.

a. If the primary coil has 475 turns, how many turns does the secondary coil have?

b. The doorbell draws 125-mA of current. What current is in the primary circuit?

Chapter Self-Check

58. Scott connects a transformer to a 24-V source and measures 8.0 V at the secondary circuit. If the primary and secondary circuits were reversed, what would the new output potential difference be?

59. **Hair Dryer** A hair dryer manufactured for use in the United States uses 10 A at 120 V. In a country where the line potential difference is 240 V, the hair dryer is used with a transformer.

 a. What turns ratio should the transformer have?
 b. What current will the hair dryer draw from the 240-V line?

60. A step-up transformer has 80 turns on its primary coil and 1200 turns on its secondary coil. The primary circuit is supplied with an alternating current at 120 V.

 a. What potential difference is being applied across the secondary circuit?
 b. The current in the secondary circuit is 2.0 A. What current is in the primary circuit?
 c. What are the power input and the power output of the transformer?

Applying Concepts

61. A wire is moved horizontally between the poles of a magnet, as shown in **Figure 21.** What is the direction of the induced current?

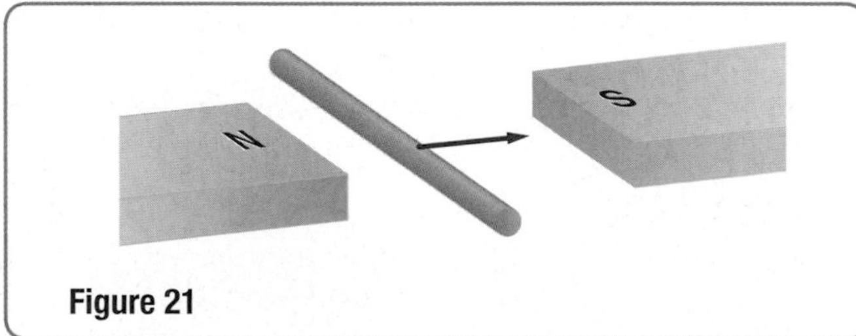

Figure 21

62. You wind wire around a nail, as shown in **Figure 22.** You connect the wire to a battery to make an electromagnet. Is the current larger just after you make the connection or several hundredths of a second after you make the connection? Or, is it always the same? Explain.

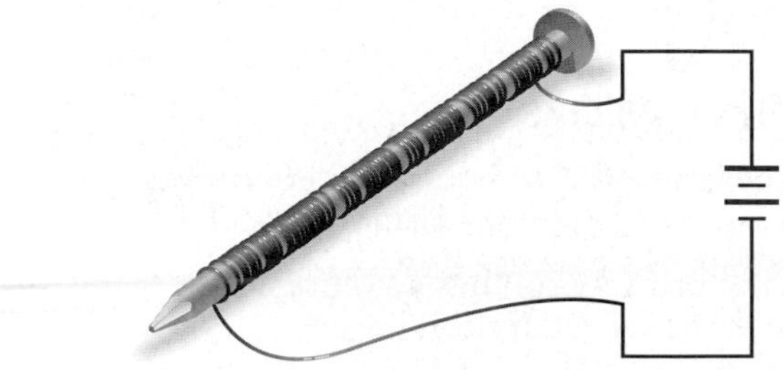

Figure 22

63. You move a length of copper wire down through a magnetic field, as shown in **Figure 23.**

 a. Will the induced current move to the right or to the left in the wire segment?
 b. What will be the direction of the force acting on the wire as a result of the induced current?

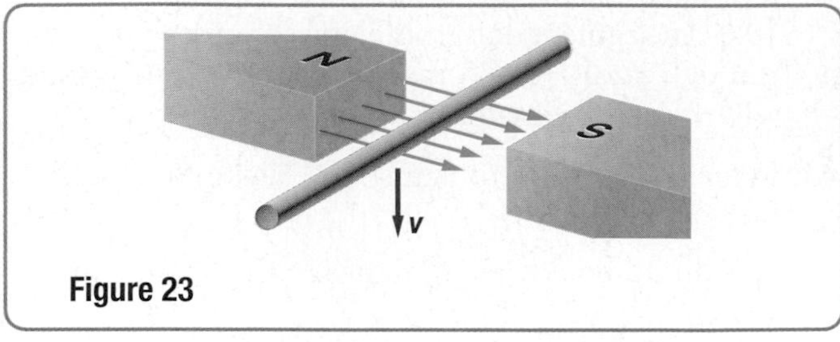

Figure 23

64. Use unit substitution to show that the units of BLv are volts.

65. **Ranking** Consider the transformers described below. Rank them according to the current induced in the secondary coil, from least to greatest. Indicate whether any have equal currents.

 A. primary coil: 50 turns; secondary coil: 25 turns; primary current: 2 A
 B. primary coil: 50 turns; secondary coil: 60 turns; primary current: 2 A
 C. primary coil: 20 turns; secondary coil: 10 turns; primary current: 4 A
 D. primary coil: 20 turns; secondary coil: 40 turns; primary current: 8 A
 E. primary coil: 10 turns; secondary coil: 100 turns; primary current: 1 A

66. When a wire is moved through a magnetic field, does the resistance of the closed circuit affect current only, *EMF* only, both, or neither?

67. A transformer is connected to a battery through a switch, as shown in **Figure 24.** The secondary circuit contains a lightbulb. Will the bulb be lighted as long as the switch is closed, only at the moment the switch is closed, or only at the moment the switch is opened? Explain.

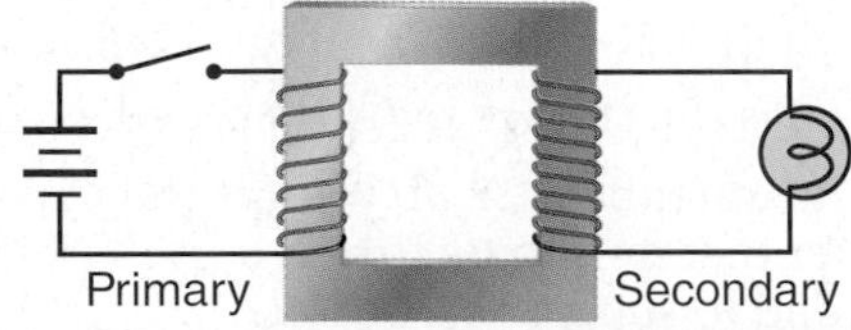

Figure 24

68. Explain why the initial start-up current is so high in an electric motor. Also explain how Lenz's law applies at the instant the motor starts.

Chapter Self-Check

69. A transformer is constructed with a laminated core that is not a superconductor. Because eddy currents cannot be completely eliminated, there is always a small transformation of electrical to thermal energy. This results in a net reduction of power available to the secondary circuit. As long as the core has eddy currents, what fundamental law makes it impossible for the output power to equal the input power?

70. **Earth's Magnetic Field** The direction of Earth's magnetic field in the northern hemisphere is downward and to the north, as shown in **Figure 25.** If an east-west wire moves from north to south, in which direction is the current?

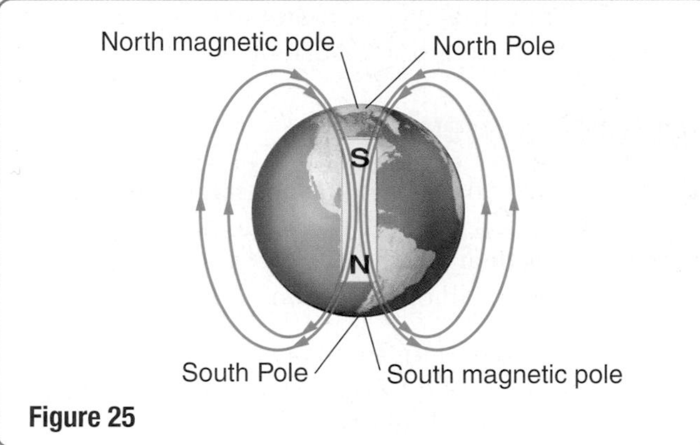

Figure 25

71. A physics instructor drops a magnet through a copper pipe, as illustrated in **Figure 26.** The magnet falls very slowly, and the students in the class conclude that there must be some force opposing gravity.
    - **a.** The magnetic field in the pipe changes. What is induced in the pipe?
    - **b.** How does what is induced in the pipe affect the falling magnet? Explain in terms of energy.
    - **c.** If the instructor used a plastic pipe, would the falling magnet slow down?

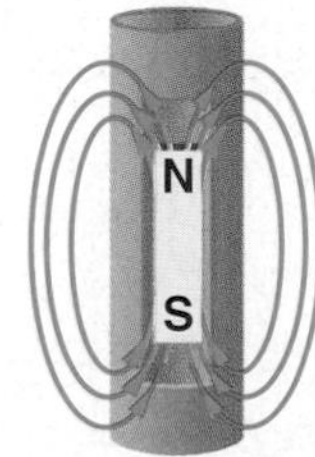

Figure 26

72. **Reverse Problem** Write a physics problem with real-life objects for which the following equation would be part of the solution:

$$v = \frac{6.5\ \text{V}}{(0.15\ \text{T})(0.80\ \text{m})}$$

Mixed Review

73. **Problem Posing** Complete this problem so it can be solved using the concept of electromagnetic inductance: "A copper wire of length 1.25 m...."

74. A step-up transformer's primary coil has 500 turns. Its secondary coil has 15,000 turns. The primary circuit is connected to an AC generator having an *EMF* of 120 V.
    - **a.** Calculate the *EMF* of the secondary circuit.
    - **b.** Find the current in the primary circuit if the current in the secondary circuit is 3.0 A.
    - **c.** What power is drawn by the primary circuit? What power is supplied by the secondary circuit?

75. With what speed must a 0.20-m-long wire cut across a magnetic field for which B is 2.5 T if the wire is to have an *EMF* of 10 V induced across it?

76. At what speed must a wire conductor 50-cm long be moved at right angles to a magnetic field of induction 0.20 T to induce an *EMF* of 1.0 V?

77. A house lighting circuit is rated at 120-V effective potential difference. What is the peak potential difference that can be expected in this circuit?

78. **Toaster** A toaster draws 2.5 A of alternating current. What is the peak current through this toaster?

79. The insulation of a capacitor will break down if the instantaneous potential difference exceeds 575 V. What is the largest effective alternating potential difference that can be applied to the capacitor?

80. **Circuit Breaker** A magnetic circuit breaker will open its circuit if the instantaneous current reaches 21.25 A. What is the largest effective current the circuit will carry?

81. The electricity received at an electrical substation has a potential difference of 240,000 V. What should the turns ratio of the step-down transformer be to have an output of 440 V?

82. An AC generator supplies a 45-kW industrial electric heater. If the effective generator potential difference is 660 V, what peak current is supplied?

83. Hybrid autos use regenerative braking. When the driver applies the brakes, electronic circuits connect the electric motor to the battery in such a way that the battery is charged. How would this connection slow the automobile?

84. Would you expect a real transformer to become hotter if it supplied a small or a large current? Explain.

Chapter Self-Check

85. A step-down transformer has 100 turns on the primary coil and 10 turns on the secondary coil. If a 2.0-kW resistive load is connected to the transformer, what is the effective primary current? Assume the peak secondary potential difference is 60.0 V.

86. A transformer rated at 100 kW has an efficiency of 98 percent.

a. If the connected load consumes 98 kW power, what is the value of the input kW to the transformer?

b. What is the maximum primary current when the transformer delivers its rated power? Assume that $V_P = 600$ V.

87. A 0.40-m-long wire cuts perpendicularly at a velocity of 8.0 m/s across a 2.0-T magnetic field.

a. What *EMF* is induced in the wire?

b. If the wire is in a circuit with a resistance of 6.4 Ω, what is the size of the current in the wire?

88. A 7.50-m-long wire is moved perpendicularly to Earth's magnetic field at 5.50 m/s. What is the size of the current in the wire if the total resistance of the wire is 5.0×10^{-2} Ω? Assume Earth's magnetic field is 5×10^{-5} T.

89. The peak value of the alternating potential difference applied to a 144-V resistor is 1.00×10^{2} Ω. What maximum power can the resistor handle?

90. X-rays A dental X-ray machine uses a step-up transformer to change 240 V to 96,000 V. The secondary side of the transformer has 200,000 turns and an output of 1.0 mA.

a. How many turns does the primary side have?

b. What is the input current?

Thinking Critically

91. Apply Concepts Suppose an "anti-Lenz's law" existed so that an induced current resulted in a force that increased the change in a magnetic field. Thus, when more energy was demanded, the force needed to turn a generator was reduced. What conservation law would be violated by this "anti-law"? Explain.

92. Analyze A step-down transformer that has an efficiency of 92.5 percent is used to obtain 28.0 V from a 125-V household potential difference. The current in the secondary circuit is 25.0 A.

a. Write an expression for transformer efficiency in percent, using power relationships.

b. What is the current in the primary circuit?

c. At what rate is thermal energy generated?

93. Analyze and Conclude A transformer that supplies eight homes has an efficiency of 95 percent. Each home contains an electric oven that draws 35 A from 240-V lines.

a. How much power is supplied to the ovens in the eight homes?

b. How much power is wasted as heat in the transformer?

Writing in Physics

94. Common tools, such as electric drills and circular saws, are typically constructed using universal motors. Using your local library, and other sources, explain how this type of motor may operate on either AC current or DC current.

Cumulative Review

95. Astronomers find that a distant star emits light at a frequency of 4.56×10^{14} Hz. If the star is moving toward Earth at a speed of 2750 km/s, what frequency light will observers on Earth detect?

96. A distant galaxy emits light at a frequency of 7.29×10^{14} Hz. Observers on Earth receive the light at a frequency of 6.14×10^{14} Hz. How fast is the galaxy moving, and in what direction?

97. How much charge is on a 22-μF capacitor with 48 V applied to it?

98. Find the potential difference across a 22-Ω, 5.0-W resistor operating at half of its rating.

99. What is the total resistance of three 85-Ω resistors connected in parallel and then series-connected to two 85-Ω resistors in parallel, as in **Figure 27?**

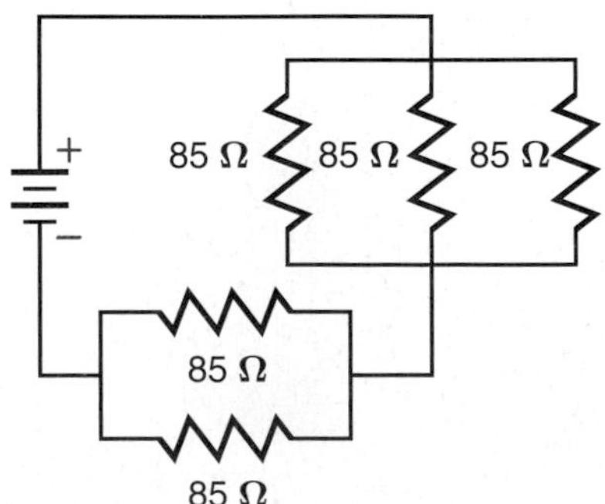

Figure 27

100. An electron with a velocity of 2.1×10^{6} m/s is at right angles to a 0.81-T magnetic field. What is the force on the electron produced by the magnetic field? What is the electron's acceleration? The mass of an electron is 9.11×10^{-31} kg.

MULTIPLE CHOICE

1. Which dimensional analysis is correct for the calculation of *EMF*?

A. (N•A•m)(J)

B. (N/A•m)(m)(m/s)

C. J•C

D. (N•m•A/s)(1/m)(m/s)

2. A 150-W transformer has an input potential difference of 9.0 V and an output current of 5.0 A. What is the ratio of V_{output} to V_{input}?

A. 1 to 3 **C.** 3 to 10

B. 3 to 1 **D.** 10 to 3

3. Which of the following will fail to induce an electric current in the wire?

A.

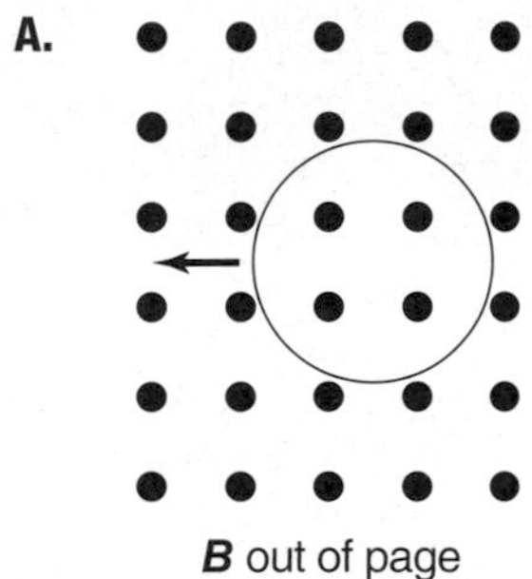

B.

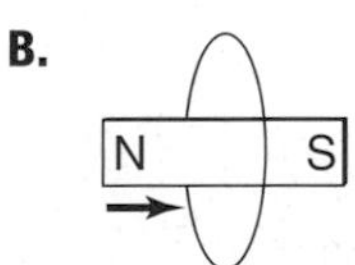

C.

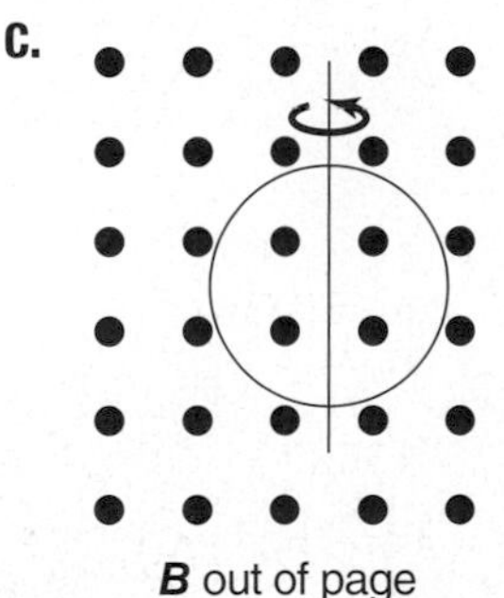

D.

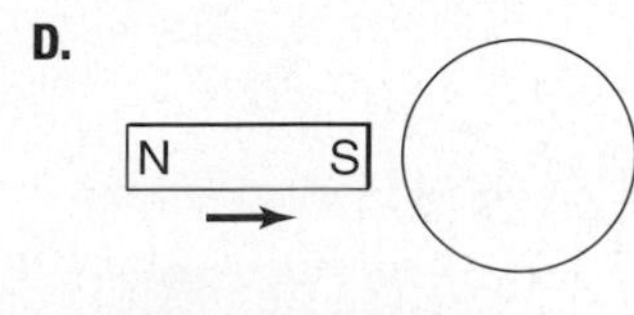

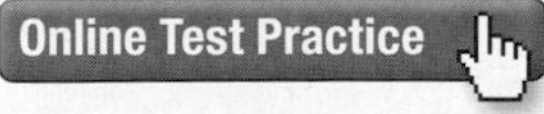

4. A 15-cm length of wire is moving perpendicularly through a magnetic field of strength 1.4 T at the rate of 0.12 m/s. What is the *EMF* induced in the wire?

A. 0 V **C.** 0.025 V

B. 0.018 V **D.** 2.5 V

5. A transformer uses a 91-V supply to operate a 13-V device. If the transformer has 130 turns on the primary coil and the device uses 1.9 A of current from the transformer, what is the current supplied to the primary coil?

A. 0.27 A **C.** 4.8 A

B. 0.70 A **D.** 13.3 A

6. An AC generator that delivers a peak potential difference of 202 V is connected to an electric heater with a resistance of 4.80×10^2 Ω. What is the effective current in the heater?

A. 0.298 A **C.** 2.38 A

B. 1.68 A **D.** 3.37 A

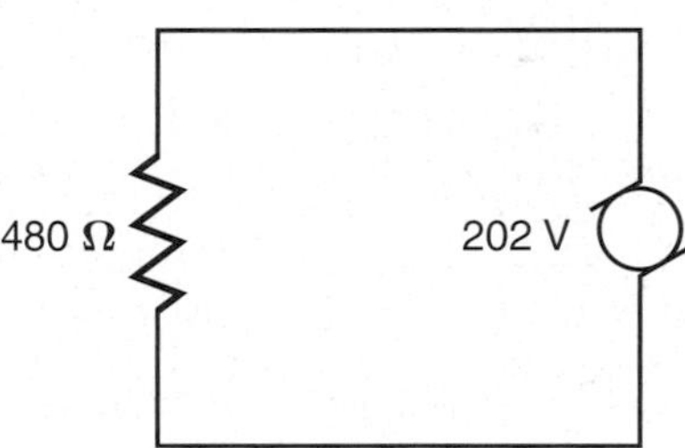

FREE RESPONSE

7. A generating station uses a 440-V transmission line to deliver 80 kW to a neighborhood. The resistance of the line is 0.20 Ω. How much power must the station produce? If the line were replaced by one operated at 12 kV, how much power would be required?

NEED EXTRA HELP?

If You Missed Question	1	2	3	4	5	6	7
Review Section	1	2	1	1	2	1	1

CHAPTER 26

Electromagnetism

BIG IDEA **Electromagnetic waves are coupled, oscillating electric and magnetic fields generated by accelerating electrons.**

SECTIONS

1. **Electric and Magnetic Forces on Particles**
2. **Electric and Magnetic Fields in Space**

LaunchLAB

iLab Station

BROADCASTING RADIO WAVES

What is the relationship between radio signal strength and distance?

WATCH THIS!

Video

MISSION: WI-FI POSSIBLE!

What happens when you need to surf the Web, but you're out of range of your wireless internet signal? If you understand the physics of Wi-Fi, you'll have no problem!

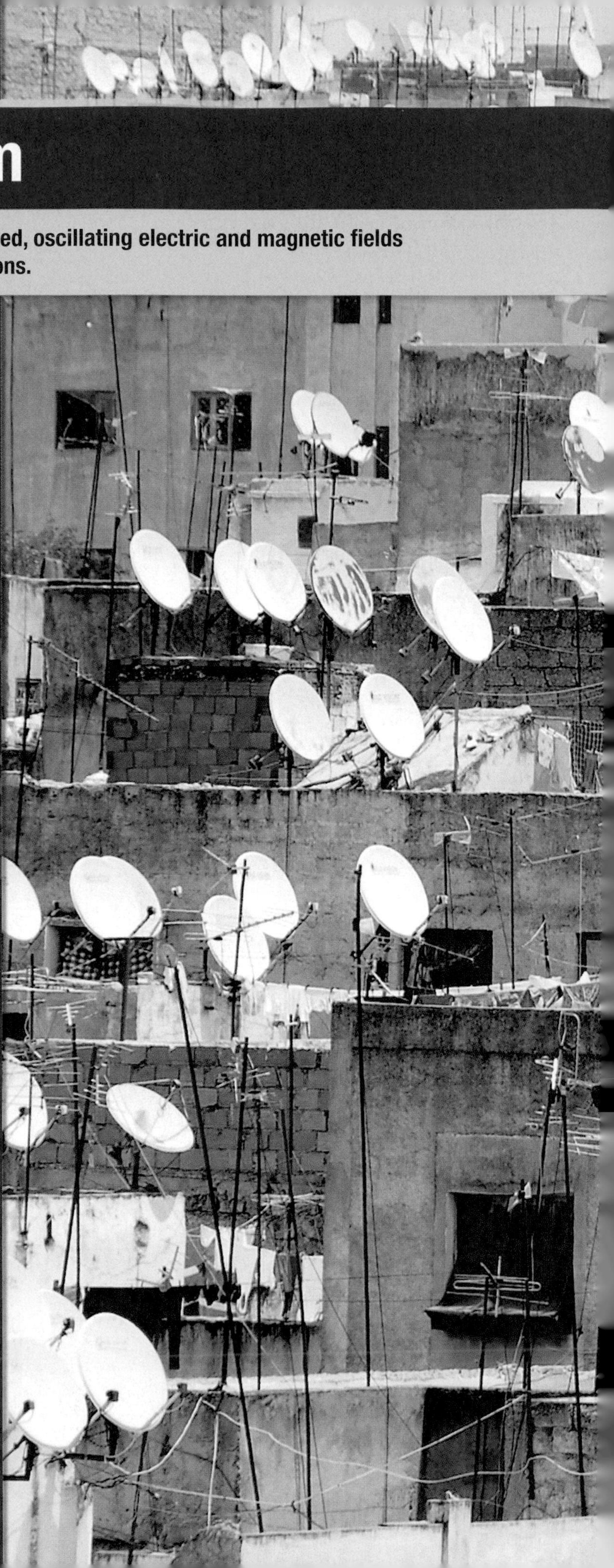

Go online!
connectED.mcgraw-hill.com

SECTION 1 Electric and Magnetic Forces on Particles

PHYSICS 4 YOU Gullies on Mars suggest that water once flowed on Mars's surface. More evidence comes from mass spectrometers aboard Martian rovers. When Mars's soils and gases are subjected to the electric and magnetic fields of a mass spectrometer, telltale signatures of water emerge.

MAIN IDEA

Deflection of moving particles in electric and magnetic fields can be used to find properties of those particles.

Essential Questions

- How did nineteenth century physicists measure the charge-to-mass ratio and the mass of the electron?
- How can you determine velocities of particles in electric and magnetic fields, and how can you find the charge-to-mass ratios of the particles?
- How does a mass spectrometer separate ions of different masses?

Review Vocabulary

Newton's second law states that the acceleration of an object is proportional to the net force and inversely proportional to the object's mass

New Vocabulary

isotope
mass spectrometer

Thomson's Experiments

Electromagnetic waves are an integral part of your everyday life. They bring you messages on your cell phone. They carry music over your radio. They even enable you to see. Before you can understand how electric and magnetic fields combine to make electromagnetic waves, it is necessary to understand the electron. Why? Because electromagnetic waves are produced by accelerating electrons.

Discovery of the electron Throughout the nineteenth century, scientists thought the atom was the smallest unit of matter. But evidence that atoms contained negatively charged particles had been growing, and in 1894 this particle was named the electron. Still, the nature of the electron wasn't known, nor was it known whether the electron could be separated from the atom. In 1897, while doing experiments with a cathode-ray tube, J.J. Thomson was able to extract the negatively charged particles from atoms of various materials. How did he do this?

Thomson evacuated nearly all the air from a cathode-ray tube, similar to the tube in **Figure 1.** A battery connected to the tube produced a large potential difference between the cathode and the anodes. At the end of the tube opposite the cathode, Thomson observed a glowing spot, made by an invisible beam—the cathode ray—that was accelerated from the cathode toward the anodes by the potential difference. As it traveled, the cathode ray passed through slits in the anodes. Thomson realized that the ray consisted of negatively charged particles from the gases in the tiny amount of air remaining in the tube.

Figure 1 In this cathode-ray tube, similar to Thomson's tube, a potential difference between the anodes and the cathode accelerates electrons toward the anodes.

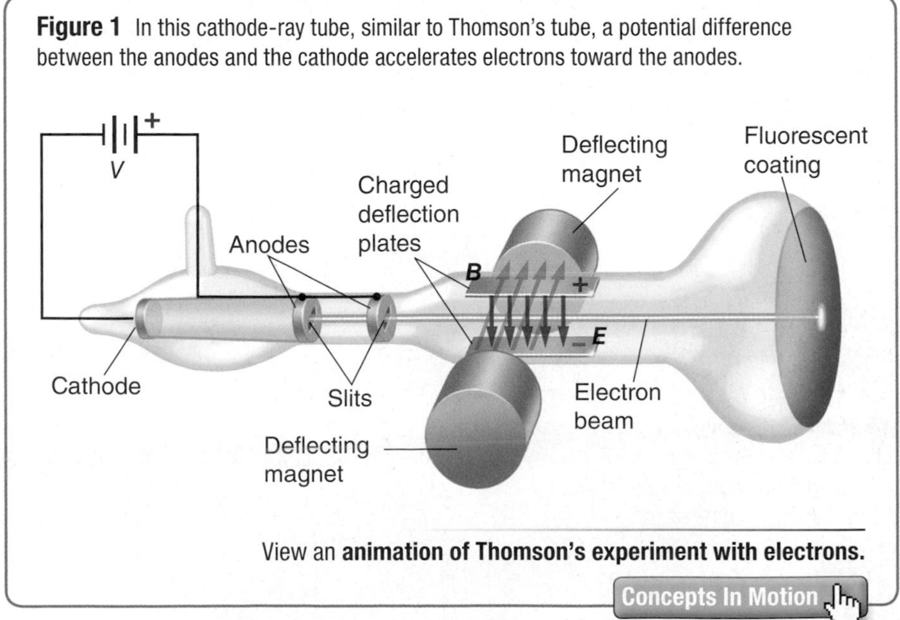

View an **animation of Thomson's experiment with electrons.**

Concepts In Motion

NASA

Although Thomson knew neither the charge (q) nor the mass (m) of the particles that made up the cathode ray, he was able to determine the particles' charge-to-mass ratio. This ratio, he found, was very high, presumably because the particles had very small mass. Thomson further found that the charge-to-mass ratio remained the same no matter what kind of gas he placed in the tube. He concluded that the particles were negative constituents of all atoms—electrons.

Charge-to-mass ratio A closer look at Thomson's experiment reveals how he was able to determine the electron's charge-to-mass ratio. Inside the cathode-ray tube, as in the tube in **Figure 1,** oppositely charged metal plates produced an electric field. Electromagnets outside the tube produced a magnetic field. By manipulating either field, Thomson found he could change the path of the electron beam.

Thomson oriented the electric field perpendicular to the electron beam. This field (of strength E) produced a force (equal to qE) that deflected the beam upward, toward the positive plate. Thomson oriented the magnetic field at right angles to both the beam and the electric field. Recall that the force exerted by a magnetic field is perpendicular both to the field and to the direction of motion of electrons. Thus, the magnetic field in the tube (of strength B) produced a force (equal to Bqv, where v is the electron velocity) that deflected the electron beam downward.

READING CHECK Apply How could you change the magnetic field so the electrons deflected upward?

Thomson adjusted the strengths of the electric and magnetic fields until the electron beam followed a straight, undeflected path. When this occurred, the forces due to the two fields were equal in magnitude and opposite in direction. Mathematically, this can be represented as follows:

$$Bqv = Eq$$

Solving this equation for v yields the following expression:

$$v = \frac{Eq}{Bq} = \frac{E}{B}$$

The above equation shows that the forces are balanced only for electrons that have a specific velocity (v). If the electric field is turned off, only the force due to the magnetic field remains. You know that the direction of motion of an electron is perpendicular to the magnetic force. Therefore, an electron in a uniform magnetic field undergoes centripetal acceleration, having a circular path of radius r. Using Newton's second law of motion for circular motion, the following equation describes the electron's path:

$$Bqv = \frac{mv^2}{r}$$

Solving for $\frac{q}{m}$ results in the following equation.

CHARGE-TO-MASS RATIO OF AN ELECTRON IN A MAGNETIC FIELD

In a magnetic field, the ratio of an electron's charge to its mass is equal to the ratio of the electron's velocity divided by the product of the magnetic field strength and the radius of the electron's circular path.

$$\frac{q}{m} = \frac{v}{Br}$$

MiniLAB

MOVING CHARGED PARTICLES

How do electric and magnetic fields affect moving charged particles?

Mass of an electron To find the value of the charge-to-mass ratio of an electron, Thomson first calculated the straight trajectory velocity (v) of the electron beam's path using measured values of E and B. Next, he turned off the electric field and measured the distance between the fluorescent spot formed by the undeflected beam and the spot formed when the magnetic field alone acted on the beam. Using this distance, he calculated the radius (r) of the electron beam's circular path and obtained q/m as -1.759×10^{11} C/kg.

Calculating the charge-to-mass ratio was the first step in calculating the electron's mass. When Thomson was doing his experiments, the average charge of an electron was known from electrolysis experiments involving many electrons as $q = -1.602\times10^{-19}$ C. Thomson assumed this value could be used for individual electrons. Later, in 1909, Robert Millikan confirmed this value, finding that all electrons have the same charge (e). Therefore, the mass of the electron (m_e) could be calculated.

$$m_e = \frac{e}{q/m} = \frac{-1.602\times10^{-19}\text{ C}}{-1.759\times10^{11}\text{ C/kg}} = 9.107\times10^{-31}\text{ kg}$$

To three significant figures, $m_e = 9.11\times10^{-31}$ kg.

PhysicsLAB

MASS OF AN ELECTRON

How can you determine an electron's mass?

Experiments with positive ions Thomson also used his cathode-ray apparatus to study positive ions. An ion is a charged atom or molecule. Positively charged particles in an electric or magnetic field undergo a deflection opposite to that of electrons, as shown in **Figure 2.**

Mass of a proton To make positive ions, Thomson added a small amount of hydrogen gas to his tube and reversed the direction of the electric field between the cathode and the anodes. The high electric field pulled electrons off the hydrogen atoms, turning the atoms into positive ions. The electric field then accelerated the ions through a tiny slit into the deflection region of the tube. The resulting ion beam passed through electric and magnetic fields on its way toward the fluorescent screen at the end of the tube.

Thomson determined the charge-to-mass ratio of a positive ion of hydrogen (later shown to be a proton) in the same manner as he determined the charge-to-mass ratio of an electron. From this ratio, the mass of a single proton was found to be 1.67×10^{-27} kg. Thomson went on to use his cathode-ray tube to determine the masses of positive ions produced when one or more electrons were stripped from heavier gases, such as helium, neon, and argon.

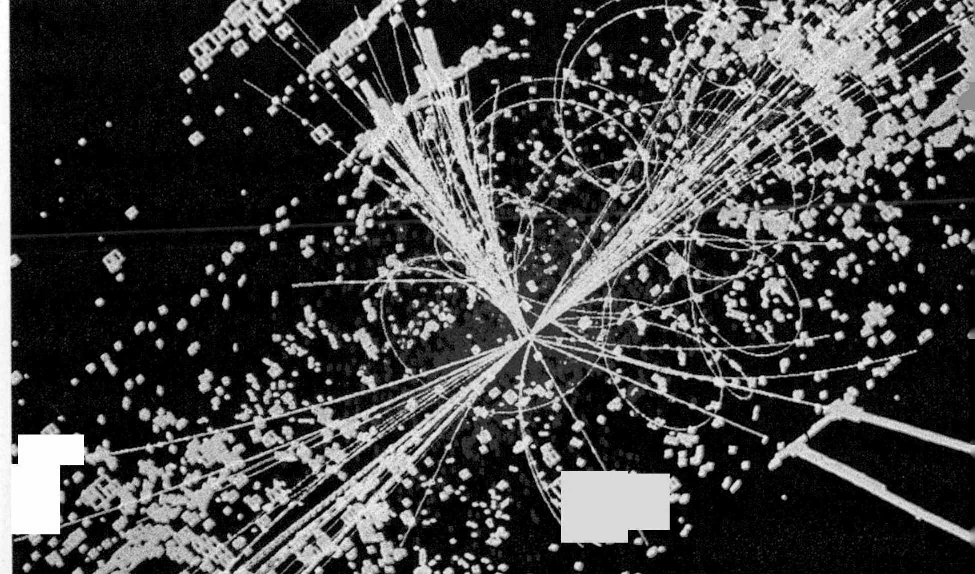

Lucas Taylor/CERN

Figure 2 The tracks of negatively and positively charged particles moving through a magnetic field curve in opposite directions. This computer-generated image simulates a collision event at the Large Hadron Collider.

EXAMPLE PROBLEM 1

Find help with **scientific notation.** Math Handbook

PATH RADIUS An electron (with mass of 9.11×10^{-31} kg) moves through a cathode-ray tube at 2.0×10^{7} m/s perpendicular to a magnetic field of 3.5×10^{-3} T. The electric field is turned off. What is the radius of the electron's circular path?

1 ANALYZE AND SKETCH THE PROBLEM

- Draw the path of the electron and label the velocity (v).
- Sketch the magnetic field perpendicular to the velocity.
- Diagram the force acting on the electron. Add the unknown quantity r of the electron's path.

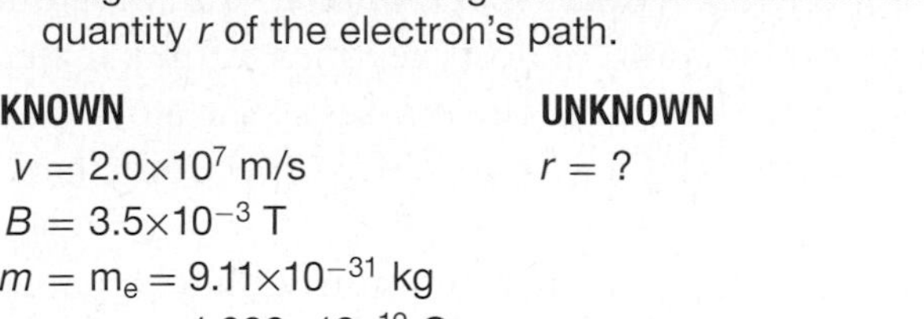

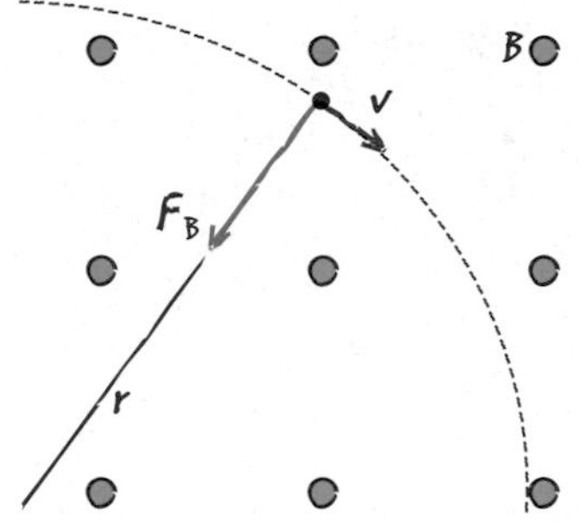

KNOWN	UNKNOWN
$v = 2.0\times10^{7}$ m/s	$r = ?$
$B = 3.5\times10^{-3}$ T	
$m = m_e = 9.11\times10^{-31}$ kg	
$q = e = -1.602\times10^{-19}$ C	

2 SOLVE FOR THE UNKNOWN

Use Newton's second law of motion to describe an electron in a cathode-ray tube subjected to a magnetic field.

$$Bqv = \frac{mv^2}{r}$$

$$r = \frac{mv}{Bq}$$

$$= \frac{(9.11\times10^{-31}\text{ kg})(2.0\times10^{7}\text{ m/s})}{(3.5\times10^{-3}\text{ T})(-1.602\times10^{-19}\text{ C})}$$

◀ Substitute $m = 9.11\times10^{-31}$ kg, $v = 2.0\times10^{7}$ m/s, $B = 3.5\times10^{-3}$ T, $q = -1.602\times10^{-19}$ C.

$$= -3.2\times10^{-2}\text{ m}$$

3 EVALUATE THE ANSWER

- **Are the units correct?** The radius of the circular path is a length measurement, given in units of meters.
- **Are the values reasonable?** Radii of a few centimeters are typical of this kind of apparatus and can easily be measured. The negative sign means that the circle curves in the direction opposite that of a positive ion.

PRACTICE PROBLEMS

Do additional problems. Online Practice

For the following questions, assume all charged particles move perpendicular to a uniform magnetic field.

1. A proton moves at a speed of 7.5×10^{4} m/s as it passes through a magnetic field of 0.080 T. Find the radius of the circular path. Note that the charge carried by the proton is equal to that of the electron but is positive.

2. Electrons move through a magnetic field of 3.0×10^{-3} T balanced by an electric field of 2.4×10^{4} N/C.

a. What is the speed of the electrons?

b. If the electric field were produced by a pair of plates 0.50 cm apart, what would be the potential difference between the two plates?

c. If the electric field were removed, what would be the radius of the circular path that the electrons followed?

3. Protons passing without deflection through a magnetic field of 0.060 T are balanced by an electric field of 9.0×10^{3} V/m. What is the speed of the moving protons?

4. CHALLENGE What trajectory would a positive ion follow moving in a magnetic field that increases linearly with time?

The Mass Spectrometer

An interesting thing happened when Thomson put neon gas into his cathode-ray tube. He observed two glowing dots on the screen instead of just one. Calculating a separate value of q/m for each dot, he concluded that two different atoms of neon could have identical chemical properties but different masses. Each form of the same atom that has the same chemical properties but a different mass is an **isotope.**

Thomson's separation of the isotopes of neon by their mass was the first example of mass spectrometry. A **mass spectrometer** is an instrument that measures the charge-to-mass ratio of positive ions within a material. From this ratio, it is possible to determine the masses of the atomic isotopes that make up the material. One kind of mass spectrometer is illustrated in **Figure 3.**

Material introduced into a mass spectrometer must either be naturally a gas or be heated to the gaseous state. The gas is inserted into a component called the ion source, which you can see on the bottom of **Figure 3.** In the ion source, an energetic beam of electrons collides with the gas's atoms and rips one or more electrons off, producing positive ions. An electric field—the result of a potential difference between two electrodes in the ion source—then accelerates the ions, which enter a magnetic field in the vacuum chamber. The uniform magnetic field causes the ions to move in a circular path before they collide with an electronic detector.

View an **animation of a mass spectrometer.**

Figure 3 Inside this mass spectrometer (blowout on right), a magnet causes the positive ions in a vacuum chamber to be deflected according to their mass. Each particle has a separate mass that is recorded in the vacuum chamber on a detector plate.

Identify the north pole of the magnet.

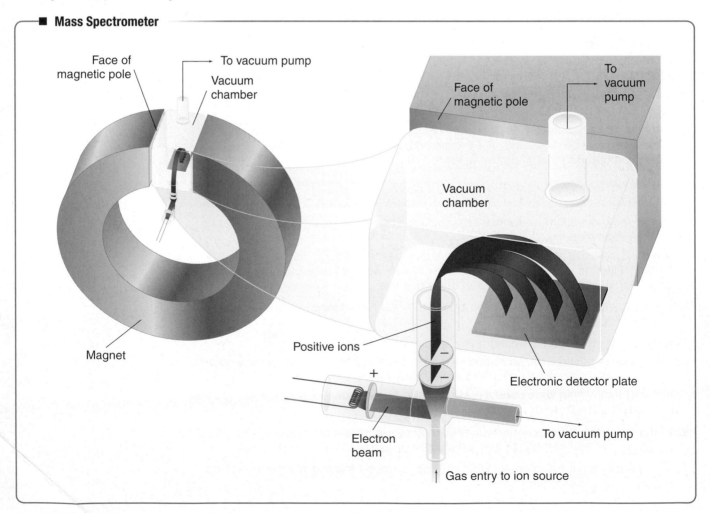

■ Mass Spectrometer

Calculating charge-to-mass ratios Once the ions in a mass spectrometer move into a magnetic field, they follow a circular path. The radius of the path depends on the ion's mass. The lighter the ion, the more it bends and the smaller is the radius of the circle. The radius of the path for each ion can be used to determine the ion's charge-to-mass ratio. The path radius (r) for an ion can be calculated from Newton's second law of motion:

$$Bqv = \frac{mv^2}{r}$$

Solving for r yields the following equation:

$$r = \frac{mv}{qB}$$

The velocity of an accelerated ion can be calculated from the equation for the kinetic energy of ions accelerated from rest through a known potential difference (V_{accel}):

$$KE = \frac{1}{2}mv^2 = qV_{\text{accel}}$$

$$v = \sqrt{\frac{2qV_{\text{accel}}}{m}}$$

Substituting this expression for v in the equation $r = \frac{mv}{qB}$ gives the radius of the ion's circular path:

$$r = \frac{mv}{qB}$$

$$= \frac{m}{qB}\sqrt{\frac{2qV_{\text{accel}}}{m}}$$

$$= \frac{1}{B}\sqrt{\frac{2V_{\text{accel}}m}{q}}$$

Simplifying this equation by multiplying both sides by B yields the following:

$$Br = \sqrt{\frac{2mV_{\text{accel}}}{q}}$$

This equation can be rearranged as shown below.

CHARGE-TO-MASS RATIO OF AN ION IN A MASS SPECTROMETER

The ratio of an ion's charge to its mass is equal to twice the accelerating potential difference divided by the product of the square of the magnetic field strength and the square of the radius of the ion's circular path.

$$\frac{q}{m} = \frac{2V_{\text{accel}}}{B^2r^2}$$

To calculate the mass of an ion, you would divide the charge on the ion by the measured charge-to-mass ratio. The charge of a proton is the same as the charge of an electron except it is positive: 1.602×10^{-19} C. Note that if more than one electron is removed from the atoms of a gas in the ion source, the charge must be multiplied by the number of electrons removed. This number—the number of electrons removed—can be controlled by the mass spectrometer's operator.

☑ **READING CHECK** **Calculate** What is the charge on an ion that has three electrons removed?

EXAMPLE PROBLEM 2

Get help with **finding the mass of an atom.** Personal Tutor

MASS OF A NEON ATOM The operator of a mass spectrometer produces a beam of neon ions from which two electrons have been removed. The ions are thus doubly ionized. The ions first are accelerated by a potential difference of 34 V. When they pass through a magnetic field of 0.050 T, the radius of their path is 53 mm. Determine the mass of the neon atom to the closest whole number of proton masses.

1 ANALYZE AND SKETCH THE PROBLEM

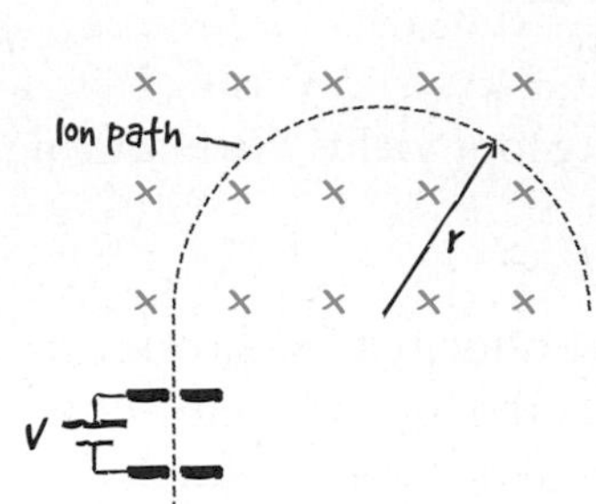

- Draw the circular path of the ions. Label the radius.
- Draw and label the potential difference between the electrodes.

KNOWN		UNKNOWN
$V_{accel} = 34$ V	$m_{proton} = 1.67\times10^{-27}$ kg	$m_{neon} = ?$
$B = 0.050$ T	$q = 2(1.602\times10^{-19}$ C$)$	$N_{proton} = ?$
$r = 0.053$ m	$= 3.20\times10^{-19}$ C	

2 SOLVE FOR THE UNKNOWN

Use the equation for the charge-to-mass ratio of an ion in a mass spectrometer.

$$\frac{q}{m_{neon}} = \frac{2V_{accel}}{B^2r^2}$$

$$m_{neon} = \frac{qB^2r^2}{2V_{accel}}$$

$$= \frac{(3.204\times10^{-19}\text{ C})(0.050\text{ T})^2(0.053\text{ m})^2}{2(34\text{ V})}$$

◀ Substitute $q = 3.204\times10^{-19}$ C, $B = 0.050$ T, $r = 0.053$ m, and $V_{accel} = 34$ V.

$$= 3.3\times10^{-26}\text{ kg}$$

Divide the mass of neon by the mass of a proton to find the number of proton masses.

$$N_{proton} = \frac{m_{neon}}{m_{proton}}$$

$$= \frac{3.3\times10^{-26}\text{ kg}}{1.67\times10^{-27}\text{ kg/proton}}$$

$$\cong 20\text{ protons}$$

3 EVALUATE THE ANSWER

- **Are the units correct?** Mass should be measured in grams or kilograms. The number of protons is not represented by any unit.
- **Is the magnitude realistic?** Neon has two isotopes, with masses of approximately 20 and 22 proton masses.

PRACTICE PROBLEMS

Do additional problems. Online Practice

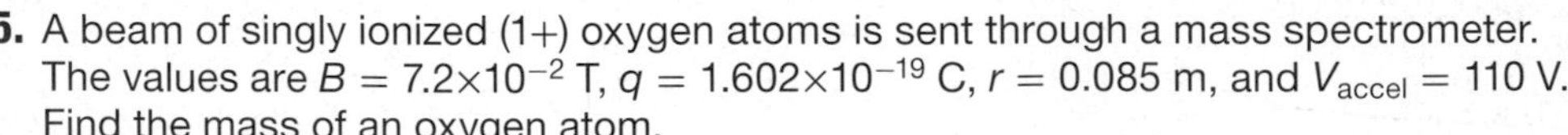

5. A beam of singly ionized (1+) oxygen atoms is sent through a mass spectrometer. The values are $B = 7.2\times10^{-2}$ T, $q = 1.602\times10^{-19}$ C, $r = 0.085$ m, and $V_{accel} = 110$ V. Find the mass of an oxygen atom.
6. A mass spectrometer analyzes and gives data for a beam of doubly ionized (2+) argon atoms. The values are $q = 2(1.602\times10^{-19}$ C$)$, $B = 5.0\times10^{-2}$ T, $r = 0.106$ m, and $V_{accel} = 66.0$ V. Find the mass of an argon atom.
7. A beam of singly ionized (1+) lithium atoms ($m \approx 7m_{proton}$) is accelerated by a potential difference of 320 V. The beam passes through a magnetic field of 1.5×10^{-2} T. What is the radius of curvature of the beam in the magnetic field?
8. **CHALLENGE** Regardless of the energy of the electrons used to produce ions, J.J. Thomson never could remove more than one electron from a hydrogen atom. What could he have concluded about the positive charge of a hydrogen atom?

Isotopic analysis The ions in the mass spectrometer in **Figure 3** strike the detector in different places. Where an ion strikes depends on its mass. The more massive the ion, the larger is the diameter of its curved path. This diameter can be easily measured because it is the distance between the location where the ion strikes and the slit in the electrode. The radius of the path (r) is half this measured distance.

The approximate spacing between ion strikes on a detector for an ionized chromium sample is shown in **Figure 4.** The four distinct marks indicate that a naturally occurring sample of chromium is composed of four isotopes. The abundance of each isotope corresponds to the width of the mark it makes on the detector.

All the chromium ions in **Figure 4** are singly charged because only one electron was removed from the original chromium atoms in the ion source. Charge depends on the number of electrons removed from the original (neutral) atoms. After the first electron is removed, producing a singly ionized (1+) atom, more energy is required to remove the second electron and produce a doubly ionized (2+) atom. This additional energy can come by increasing the electric field, which gives the electron beam more kinetic energy. A beam of higher-energy electrons can produce both singly and doubly charged ions. In this way, the operator of the mass spectrometer chooses the charge on the ion to be studied.

READING CHECK **Describe** how a doubly ionized ion is produced.

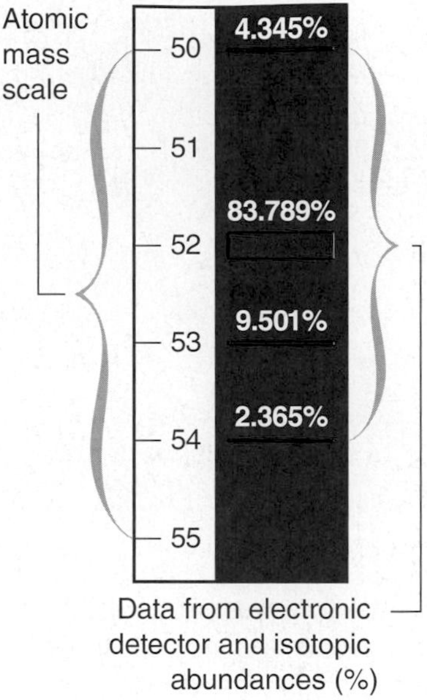

Figure 4 The width of the four isotopes of chromium on an electronic detector provides an indication of abundance. Note that the chromium isotope with a mass number of 52 is the most abundant and that the sum of the percentages of all four equals 100. The weighted average of the masses of an element's isotopes provides the mass of the element shown on the periodic table.

Calculate the average mass of the chromium isotopes.

Mass spectrometry applications Mass spectrometers are so sensitive that they can separate ions with mass differences as small as one ten-thousandth of one percent. Because of this precision, scientists can use mass spectrometers to identify the presence of a single molecule within a 10 billion–molecule sample.

Mass spectrometers have a wide range of applications. As you read in the beginning of this chapter, mass spectrometers are used to analyze the atmospheres and soils of Mars and other objects in the solar system. They also are commonly used in the geological, pharmaceutical, biological, and even the forensic sciences. For example, airports use mass spectrometers to detect traces of molecules found in explosives that might be in the luggage or on the hands or shoes of passengers.

SECTION 1 REVIEW

Section Self-Check Check your understanding.

9. **MAIN IDEA** The radius of the circular path of an ion in a mass spectrometer is given by $r = \frac{1}{B}\sqrt{\frac{2V_{accel}m}{q}}$. Use this equation to explain how a mass spectrometer is able to separate ions of different masses.

10. **Cathode-Ray Tube** Describe how the cathode-ray tube used by J.J. Thomson forms an electron beam.

11. **Magnetic Field** A mass spectrometer can analyze molecules having masses of hundreds of proton masses. If the singly charged ions of these molecules were produced using the same accelerating potential difference as used for smaller ions, how would the mass spectrometer's magnetic field have to be changed for the ions to hit the detector?

12. **Path Radius** A proton moves at a speed of 8.4×10^4 m/s as it passes through a magnetic field of 12.0 mT. Find the radius of the circular path.

13. **Mass** A beam of doubly ionized (2+) oxygen atoms is accelerated by a potential difference of 232 V. The oxygen then enters a magnetic field of 75 mT and follows a curved path with a radius of 8.3 cm. What is the mass of the oxygen atom?

14. **Critical Thinking** In Example Problem 2 (on the previous page), the mass of a neon isotope is determined. Another neon isotope has a mass of 22 proton masses. How far apart on a detector would these two isotopes land?

Electric and Magnetic Fields in Space

PHYSICS 4 YOU Did you know you use microwaves both to cook food and to send text messages on your cell phone? Microwaves carry energy. They can also carry information. How do microwaves and other electromagnetic waves carry information?

MAIN IDEA

Electromagnetic waves are coupled, oscillating electric and magnetic fields that move through space and interact with matter.

Essential Questions

- How do electromagnetic waves propagate through space?
- How does the speed at which electromagnetic waves propagate through different materials vary?
- How do electromagnetic waves transmit information?
- What factors affect an antenna's sensitivity to electromagnetic waves of given wavelengths?

Review Vocabulary

capacitor an electrical device used to store electrical energy that is made up of two conductors separated by an insulator

New Vocabulary

electromagnetic wave
electromagnetic spectrum
electromagnetic radiation
transmitter
antenna
dielectric
carrier wave
piezoelectricity
receiver

What are electromagnetic waves?

You read in Section 1 that accelerating electrons produce electromagnetic waves and that these waves are a combination of electric and magnetic fields. Recall that Hans Christian Oersted was the first to make the connection between electricity and magnetism when he found, in 1820, that moving charges produce magnetic fields. Ten years later, Michael Faraday and Joseph Henry independently discovered the converse—that a wire's magnetic field can induce an electric field.

Moving through space Electric fields can be induced even without a wire. As shown in the left of **Figure 5,** a changing magnetic field alone produces an electric field. The reverse is also true. A changing electric field produces a magnetic field, as shown in the right of **Figure 5.** Scottish physicist James Maxwell quantified the relationship between electric and magnetic fields in 1873, when he summarized the work of Coulomb, Oersted, and Faraday into four equations. These equations predicted that electromagnetic waves could move through space. In 1887, this was confirmed by German physicist Heinrich Hertz. Hertz's name was later given to the unit that measures a wave's oscillation frequency. Recall that 1 Hz equals one cycle per second.

Today, it is known that **electromagnetic waves** are coupled, oscillating electric and magnetic fields that propagate through space and matter. Knowledge of electromagnetic wave properties has led to many technologies that have had a huge impact on society.

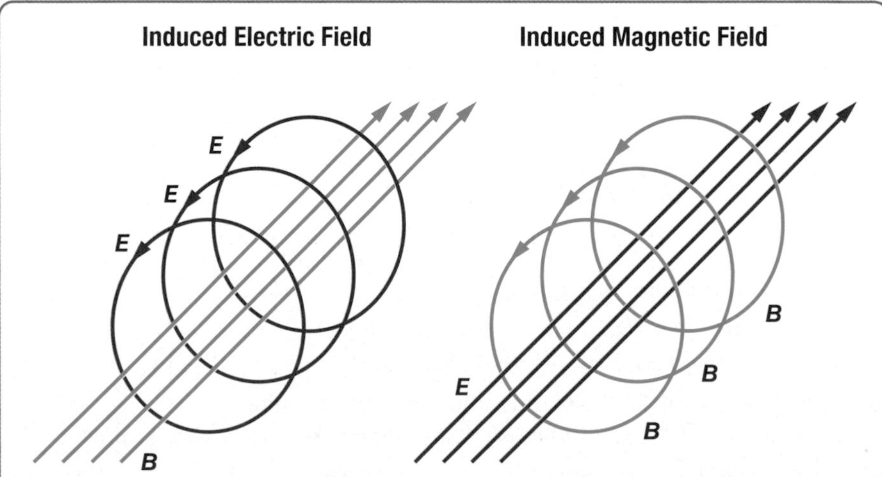

Figure 5 An electric field can be induced by a changing magnetic field (left). Notice that the field lines of the electric field are closed loops. Unlike in an electrostatic field, there are no charges on which the lines begin or end. A magnetic field also can be induced by a changing electric field (right).

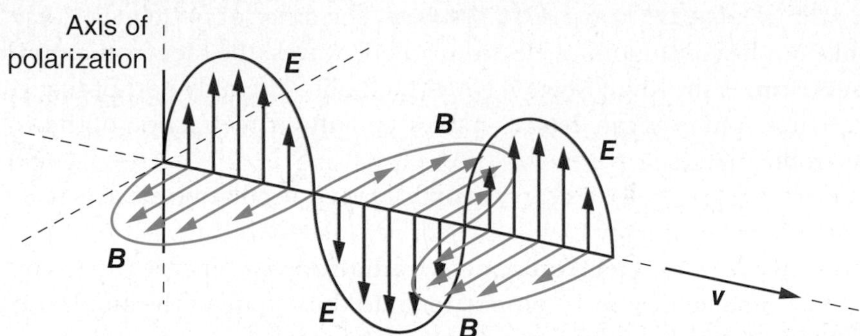

Figure 6 The electric and magnetic fields that make up an electromagnetic wave oscillate at right angles to each other and to the direction of the wave's velocity (v).

Electromagnetic wave properties How are the magnetic and electric fields related in an electromagnetic wave? Accelerating electrons produce the electric field of a wave, and the varying electric field produces the wave's magnetic field. The orientations of these two fields are shown in **Figure 6.** As the wave moves, its electric field oscillates up and down while its magnetic field oscillates at right angles to the electric field. Both the electric field and the magnetic field are at right angles to the wave propagation direction.

Travel in a vacuum All electromagnetic waves move in the same way. Like waves on a rope, electromagnetic waves are transverse waves and can travel through a medium. Unlike other types of waves, electromagnetic waves also can travel in a vacuum. The speed at which electromagnetic waves travel in a vacuum is 299,792,458 m/s, approximated as 3.00×10^8 m/s and denoted as c, the speed of light. The waves travel only slightly slower in air. Recall that the wavelength, frequency, and speed of a wave are related by the following equation.

$$\lambda = \frac{v}{f}$$

In the equation above, the wavelength (λ) is measured in meters, the speed (v) is measured in meters per second, and the frequency (f) is measured in hertz. For an electromagnetic wave traveling in a vacuum, the speed (v) is equal to the speed of light (c). Thus, for an electromagnetic wave, the equation becomes the following:

$$\lambda = \frac{c}{f}$$

Note that in the wavelength-frequency equation, the product of frequency and wavelength is constant—equal to c—for any electromagnetic wave. Thus, as wavelength increases, frequency decreases, and vice versa.

READING CHECK Calculate If the frequency of a wave increases by a factor of 1.25, by what factor does the wavelength change?

PRACTICE PROBLEMS

Do additional problems. Online Practice 

15. What is the wavelength of green light that has a frequency of 5.70×10^{14} Hz?

16. An electromagnetic wave has a frequency of 8.2×10^{14} Hz. What is the wavelength of the wave?

17. What is the frequency of an electromagnetic wave that has a wavelength of 2.2×10^{-2} m?

18. CHALLENGE If an electromagnetic wave is propagating to the right and the electric field is in and out of the page, in what direction is the magnetic field?

Types of electromagnetic waves The range of frequencies that make up the continuum of electromagnetic waves, the **electromagnetic spectrum,** is shown in **Figure 7.** Note that light—the only part of the spectrum your eyes can detect—makes up only a tiny fraction of the electromagnetic spectrum.

Electromagnetic waves carry energy in their oscillating electric and magnetic fields. Energy that is carried, or radiated, by an electromagnetic wave is called **electromagnetic radiation.** The energy carried by an electromagnetic wave is proportional to the square of the amplitude of the electric field and the area the wave crosses.

Some electromagnetic radiation is energetic enough to be harmful to humans. But it is possible to control the energy from electromagnetic waves, even high-energy waves, for communication, industrial, and medical purposes.

Figure 7 The electromagnetic spectrum contains waves that range from long radio wavelengths larger than houses to short gamma wavelengths smaller than atoms. The left column shows examples of radiation sources. Note that wave frequency extends beyond frequencies shown in this image.

Observe Which kind of electromagnetic wave is the size of a proton?

■ **The Electromagnetic Spectrum**

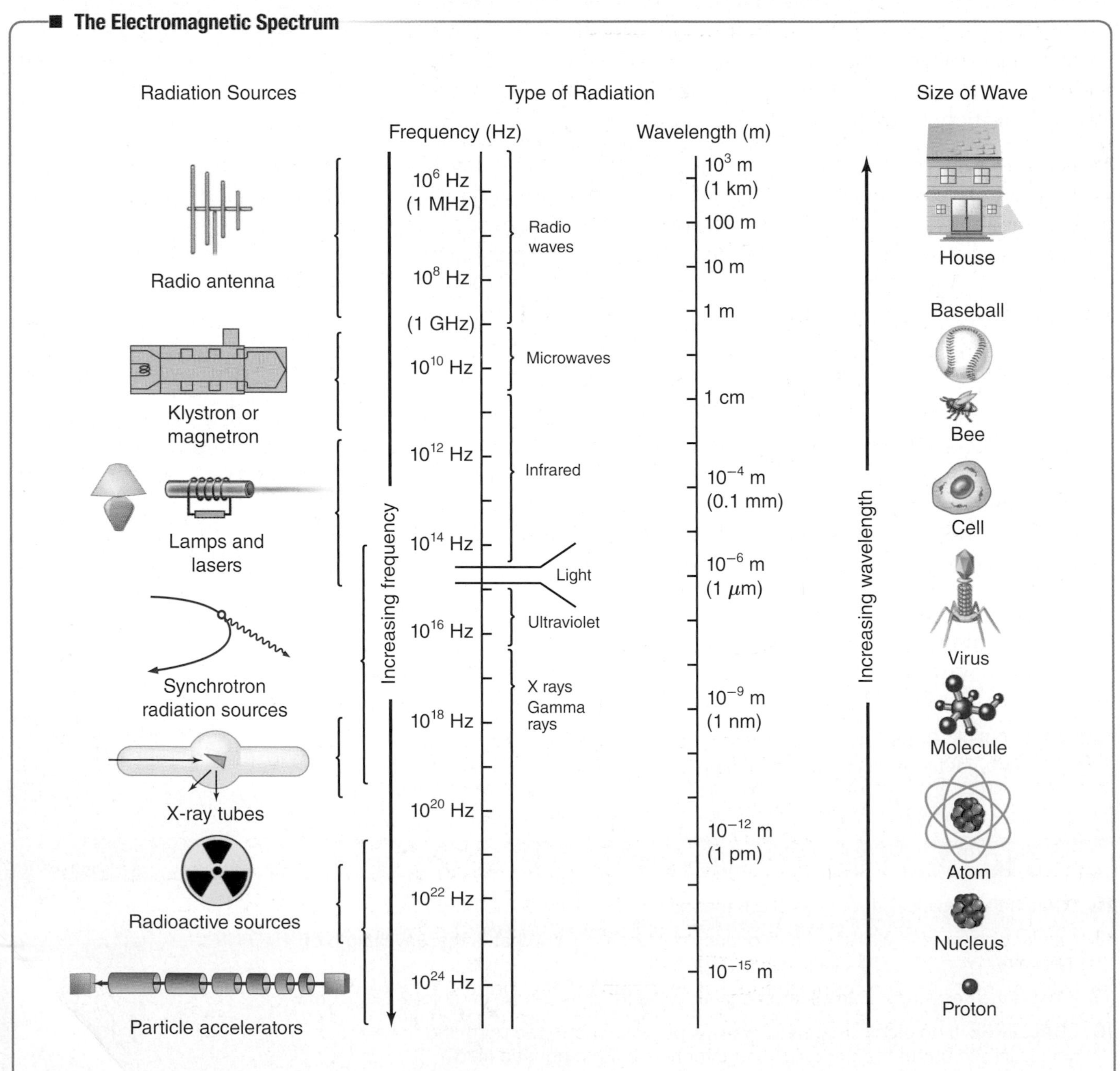

PHYSICS CHALLENGE

WAVELENGTHS OF LIGHT Wavelengths for some of the colors of light are shown in **Table 1.**

1. Which color has the longest wavelength?
2. Which color travels the fastest in a vacuum?
3. Waves with longer wavelengths diffract around objects in their path more than waves with shorter wavelengths. Which color will diffract the most? The least?
4. Calculate the frequency range for each color of light given in **Table 1.**

Table 1
Wavelengths of Light

Color	Wavelength (nm)
Violet-Indigo	390 to 455
Blue	455 to 492
Green	492 to 577
Yellow	577 to 597
Orange	597 to 622
Red	622 to 700

Uses of lower-frequency waves The lowest-frequency waves—radio waves—are used mainly to broadcast information. Radio waves with long wavelengths can be transmitted long distances because they reflect off ions in the atmosphere. Shorter radio waves used for FM radio and television travel in straight paths and so must be relayed from station to station across Earth's curved surface. Cellular phones and the Global Positioning System transmit information using very-short-wavelength radio waves called microwaves. You also use microwaves to cook food. The water and fat in food absorb microwaves, and the waves' energy turns to thermal energy to cook the food.

Infrared waves have shorter wavelengths than microwaves do. Cameras with special sensors to detect infrared radiation can produce images, and infrared night-vision goggles and cameras allow people to see in the dark. Because hot objects emit far infrared waves (with long wavelengths), infrared detectors can measure the temperature of buildings and other objects. Infrared radiation also can be used to heat buildings. Near infrared radiation (with shorter wavelengths and higher frequencies) can carry signals on optical fiber systems or through the air, programmed from remote-control devices.

Ultraviolet (UV) radiation has even higher frequencies. UV radiation can ionize molecules and atoms and cause chemical reactions, such as sunburn. UV radiation also is used in industry to cure polymers and sterilize instruments. In the semiconductor industry, UV radiation is used to etch patterns on silicon wafers in integrated circuits.

☑ **READING CHECK Compare** short and long radio wave propagation.

Uses of higher-frequency waves X-rays are produced when high-energy electrons rip tightly bound electrons off atoms. When the electrons in the atoms rearrange themselves, they emit X-rays. German physicist William Roentgen discovered X-rays in 1895 using a vacuum glass tube similar to that shown in **Figure 8.** In modern X-ray tubes, electrons are accelerated to high speeds by means of potential differences of 20,000 V or more. You are familiar with X-ray pictures of bones and teeth. X-rays are also widely used to kill cancerous cells.

Gamma rays are even higher-frequency electromagnetic radiation. These waves come from the radioactive nuclei of atoms. Gamma rays can be used to detect dangerous substances in shipping containers. In medicine, they are used to treat cancer by destroying cells.

Figure 8 In Roentgen's glass tube, a very high potential difference gave electrons kinetic energies large enough to produce X-rays when they struck the metal anode. The glow they produced on a phosphorescent screen persisted even when Roentgen placed a piece of wood between the tube and the screen. But when Roentgen placed his hand between the tube and photographic film, the bones in his hand blocked the rays.

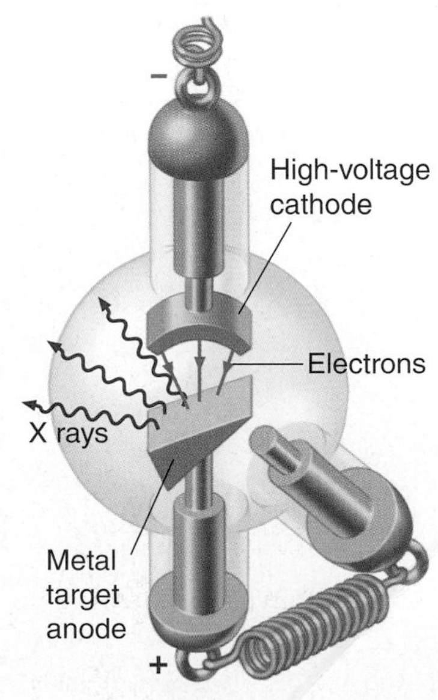

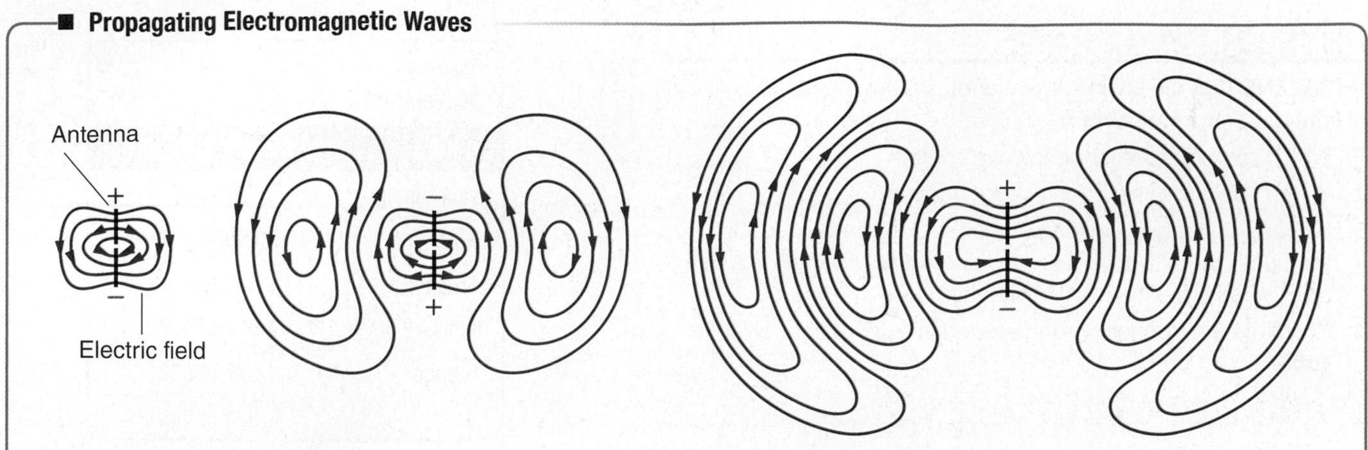

Figure 9 An AC current source connected to an antenna produces an oscillating potential difference across the antenna. This oscillating potential difference accelerates electrons, which generate an oscillating electric field. The oscillating electric field produces a changing magnetic field (not shown), and this magnetic field, in turn, generates a changing electric field. This process continues, and the electromagnetic wave propagates away from the antenna.

Transmitting Electromagnetic Waves

You read that radio waves and microwaves can carry information. How are these waves—and the information they carry—broadcast?

Propagation through space Radio waves and microwaves are broadcast into space by transmitters connected to antennas. A **transmitter** is a device that converts voice, music, pictures, or data to electronic signals, amplifies the signals, and then sends the signals to an antenna. The **antenna** creates the electromagnetic waves that propagate through the air. How does the antenna do this?

The transmitter produces an oscillating potential difference across the metallic antenna that accelerates electrons in the metal. The acceleration generates an oscillating electric field that propagates away from the antenna. You can see how an electric field forms from an antenna in **Figure 9.**

Not shown in **Figure 9** is the changing magnetic field generated by the changing electric field. The magnetic field also propagates away from the antenna at the same speed as the electric field but at right angles to it and to the direction of propagation. Note that an electromagnetic wave produced by an antenna is polarized; that is, its electric field is parallel to the antenna's conductor.

MiniLAB

WAVE SIGNALS

How do remote control devices work?

Propagation through matter Electromagnetic waves can travel through matter other than air. Sunlight shining through a glass of water is an example where light waves travel through three forms of matter: air, glass, and water. These materials are dielectrics. A **dielectric** is a poor conductor of electric current whose electric charges partially align with an electric field. The velocity of an electromagnetic wave in a dielectric is always less than the wave's speed in a vacuum; you can calculate the velocity of a wave through any dielectric using the following equation:

$$v = \frac{c}{\sqrt{k}}$$

In this equation, the wave velocity (v) is measured in meters per second; the speed of light (c) is 3.00×10^8 m/s; and the relative dielectric constant (k) is a dimensionless quantity. In a vacuum, $k = 1.00000$, and the wave velocity is equal to c. In air, $k = 1.00054$, and electromagnetic waves move just slightly slower than c. The dielectric coefficient is the square of the index of refraction, $k = n^2$, so that $\sqrt{k} = n$.

PRACTICE PROBLEMS Do additional problems. Online Practice 

19. What is the speed of an electromagnetic wave traveling through air? Use $c = 299{,}792{,}458$ m/s in your calculation.

20. Water has a dielectric constant of 1.77. What is the speed of light in water?

21. The speed of light traveling through a material is 2.43×10^{8} m/s. What is the dielectric constant of the material?

22. CHALLENGE A radio signal is transmitted from Earth's surface to the Moon's surface, 376,290 km away. What is the shortest time a reply can be expected?

Producing Electromagnetic Waves

Suppose you've just tuned to your favorite radio station. How does that station generate the radio waves you hear as music?

Carrier waves The Federal Communications Commission assigns every commercial radio station in the United States a specific wavelength in the radio part of the electromagnetic spectrum called the station's **carrier wave.** A station broadcasts music or other information by varying its carrier wave, modulating either its frequency or its amplitude. This is done by the station's transmitter.

A transmitter contains three parts. The oscillator generates the carrier wave. The modulator uses music, pictures, words or other data to change the carrier wave's amplitude or frequency. And the amplifier increases the potential difference of the resulting signal.

Setting the oscillation frequency To produce carrier waves up to 400 MHz, an oscillator uses a coil and a capacitor connected in parallel. The oscillator circuit creates a potential difference across the capacitor that produces an electric field and stores charges in the capacitor. When the potential difference is removed, the capacitor discharges and the stored electrons flow through the coil. The current creates a changing magnetic field that induces an *EMF* across the circuit. The *EMF* recharges the capacitor in the reverse direction; the process then repeats in reverse. One complete oscillation cycle is shown in **Figure 10.**

Coil-and-Capacitor Circuit

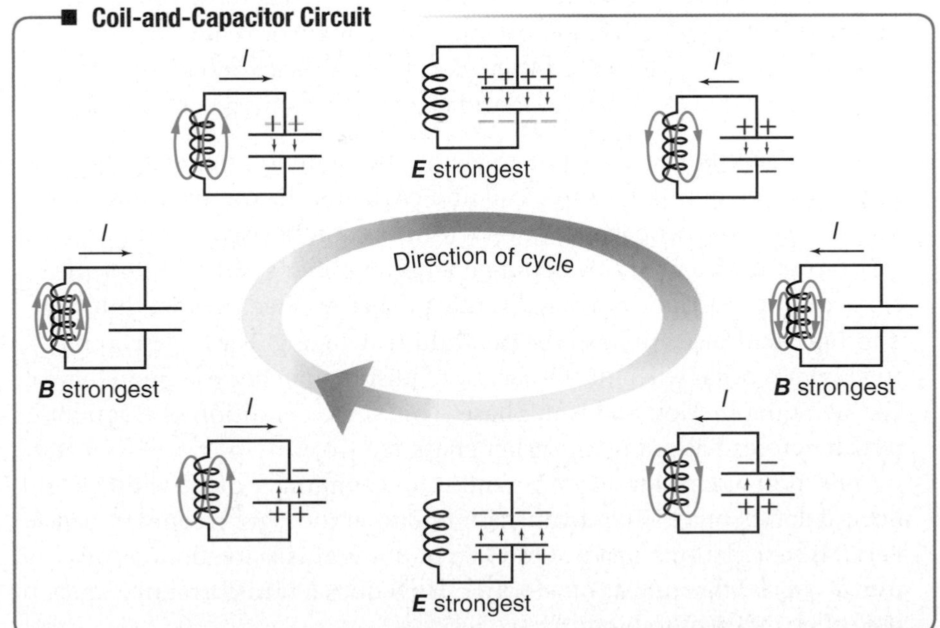

Figure 10 In a complete oscillation cycle of a coil-and-capacitor circuit, the magnetic field is strongest when the electric field is weakest. The number of oscillations per second equals the frequency of the waves produced. If the coil's ability to store magnetic energy or the capacitance changed, so would the frequency of the oscillations.

Figure 11 The electrons in a coil-and-capacitor circuit are like a pendulum's bob. As the bob swings, its displacement changes. The point where the pendulum's motion comes to a stop is analogous to zero current in the circuit.

Determine What is the charge on the capacitor when the current is maximum?

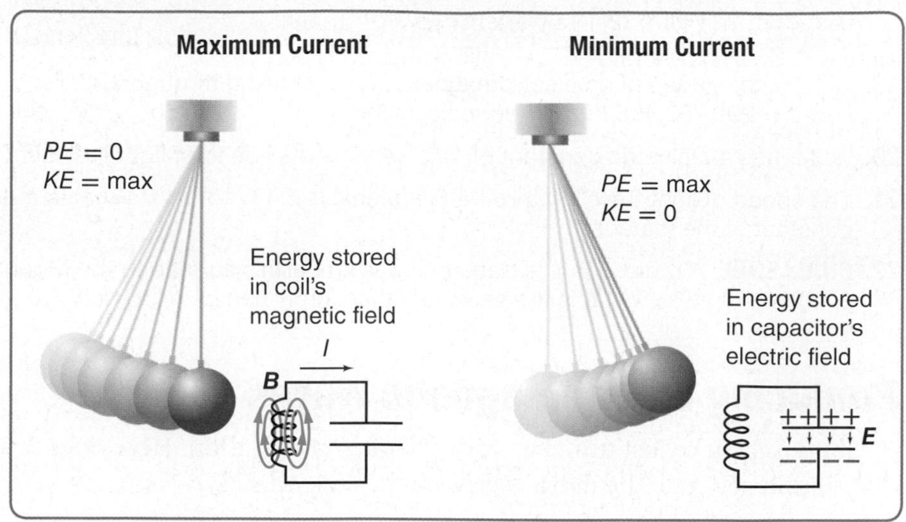

Swinging pendulum analogy You can compare the discharging and recharging process in a coil-and-capacitor circuit to the cyclic oscillations of a swinging pendulum, as shown in **Figure 11.** The displacement of the pendulum's bob from the vertical represents the electrons in the coil and capacitor.

The moving bob has the greatest speed at the bottom of its swing. This point in the pendulum's motion, shown in the left of **Figure 11,** is similar to the point in the circuit where electron flow in the coil is at its maximum and the charge on the capacitor is zero. When the pendulum bob reaches the peak of its swing, shown in the right of **Figure 11,** its displacement is maximum and its velocity is zero. This is similar to the point in the circuit when the capacitor holds the maximum charge and the flow of electrons in the coil is zero.

The pendulum in **Figure 11** has kinetic energy (*KE*) as a result of its motion. It has potential energy (*PE*) due to its displacement. The sum of the *PE* and *KE*—the total energy—is constant throughout the pendulum's motion. The coil-and-capacitor circuit is similar. There is energy both in the circuit's magnetic field, produced by the coil, and in the circuit's electric field, produced by the capacitor. When the current is largest, the energy is all stored in the magnetic field. When the current is zero, the energy is all stored in the electric field. The total energy of the circuit (the sum of the magnetic and electric field energies) is constant.

Steady oscillations Just as friction causes a pendulum to stop swinging if it is left alone, the oscillations in a coil and capacitor die out over time due to the energy being propagated in the electromagnetic wave and due to resistance in the circuit. By adding energy to either system, the oscillations can be made to continue. Gentle pushes applied to a pendulum at appropriate times will keep the pendulum swinging. The largest amplitude swings occur when the frequency of pushes matches the frequency of the swinging motion and is in phase. This is the condition of resonance, which you read about in an earlier chapter.

Just as gentle pushes keep a pendulum swinging, potential differences applied to a coil-and-capacitor circuit at the right frequency can keep the oscillations steady. One way of doing this is to add a second coil to the circuit, forming a transformer. How does a transformer keep the circuit's oscillations going?

As shown in **Figure 12,** the transmitter's oscillator increases the AC potential difference induced in the secondary coil of the transformer. The current is then added back to the coil and capacitor. This enables the circuit to maintain its oscillations.

Frequencies set by a resonant cavity The frequency produced by a coil-and-capacitor circuit can be increased by decreasing the coil's ability to store magnetic energy and the capacitor's capacitance. Above frequencies of about 400 MHz, however, individual coils and capacitors are no longer useful.

Microwaves, with frequencies from 0.4 GHz to 100 GHz, are produced using a resonant cavity, a rectangular metal box that acts both as a coil and as a capacitor. The size of the box determines the frequency of oscillation. Note that in a microwave oven, the size of the oven itself has no effect on wave frequency; it is only the size of the oven's internal resonant cavity that affects frequency.

To produce waves with frequencies higher than 100 GHz, a resonant cavity would have to be reduced to molecular size. Infrared waves, for example, are generated by vibrating nuclei within molecules. Higher-frequency light and ultraviolet waves, as well as X-rays, are generated by electrons within atoms. The highest-frequency waves, gamma rays, are generated by accelerated charges in atomic nuclei.

☑ **READING CHECK Assess** Why can't resonant cavities be used to produce infrared waves?

Frequencies set by piezoelectricity There are other ways of generating oscillating potential differences for transmitters. For example, quartz crystals deform when a potential difference is applied across them, a property known as **piezoelectricity.** The application of an AC potential difference to a cut section of quartz crystal results in sustained oscillations. Just as a piece of metal vibrates at a specific frequency when it is bent and released, so does a quartz crystal. The thinner the crystal, the higher is the vibration frequency.

A crystal's piezoelectric property also generates an *EMF* when the crystal is deformed. This *EMF* is produced at the vibrating frequency of the crystal, so it can be amplified and returned to the crystal to keep it vibrating. Because of their nearly constant vibration frequencies, quartz crystals are commonly used to generate electromagnetic waves in cell phones, television, wireless phones, wireless WiFi routers, and computers.

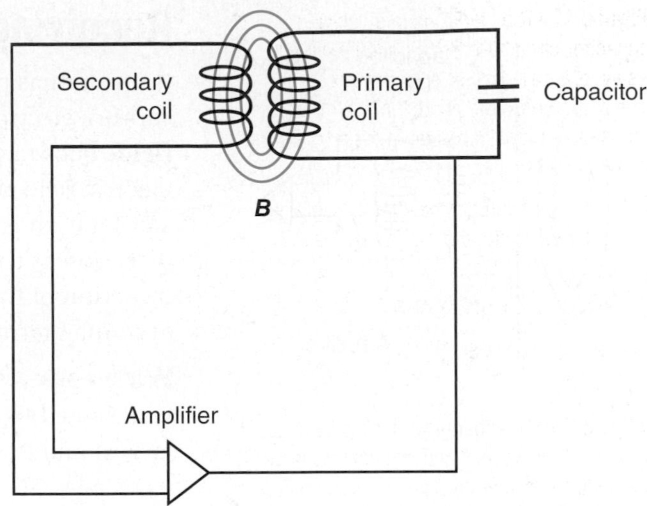

Figure 12 The amplified oscillating current from the secondary coil of the transformer is at the same frequency as the coil-and-capacitor circuit. Therefore, the amplified current keeps the oscillations going.

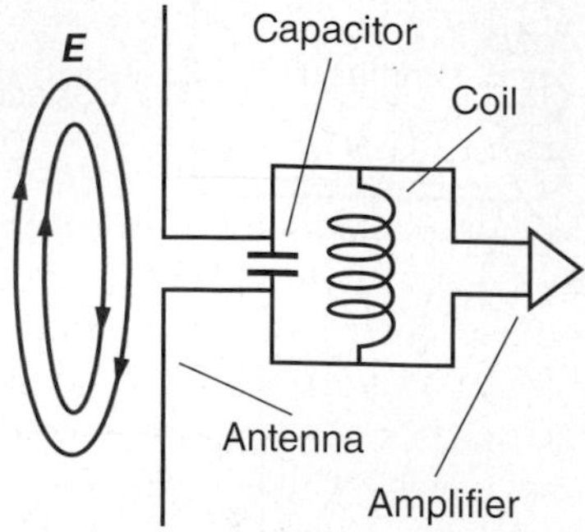

Figure 13 The changing electric fields from a radio station signal cause electrons in an antenna to accelerate. The information carried by the signal can then be decoded and amplified and used to drive a loudspeaker.

Receiving Electromagnetic Waves

Antennas propagate electromagnetic waves into space. Antennas also capture electromagnetic waves, converting the waves' oscillating electric fields back to potential differences. As shown in **Figure 13,** a wave's electric field accelerates electrons in the metal of an antenna. The acceleration is largest when the antenna is positioned in the same direction as the wave polarization; that is, when it is parallel to the direction of the wave's electric field. A potential difference across the antenna's terminals oscillates at the frequency of the wave.

Wire antennas When an antenna is one-half the length of the wave it is designed to detect, the potential difference across its terminals is largest and the antenna is most efficient. Therefore, an antenna designed to receive radio waves is longer than one designed to receive microwaves.

Though antennas that are one-half wavelength long are most efficient, antennas that are one-quarter wavelength long are often used when the connection between the antenna and receiver is at the end rather than the middle of the antenna. Antennas can be made shorter by constructing them from a helical coil or by adding a dielectric material, such as a ceramic, with a dielectric coefficient higher than air.

Cell phones have as many as seven antennas. These phones typically communicate at frequencies near 875, 1850, and 2050 MHz. Further, they receive GPS signals at 1.575 GHz. They send and receive remote earpiece and WiFi signals at 2.45 GHz. Cell phone antennas built on ceramic dielectric blocks using printed conductors are typically only a few millimeters long. Laptop computers also have several antennas to accommodate WiFi and remote devices.

PhysicsLAB
BLOCKING WAVES

Can you block the transmission of radio waves?

Dish antennas All electromagnetic waves, not just light waves, undergo reflection, refraction, and diffraction. Dish antennas, such as the one in **Figure 14,** reflect short-wavelength radio signals, just as parabolic mirrors reflect light waves. A parabolic dish antenna reflects and focuses signals off its surface and into a horn. The horn, supported by a tripod structure over the main dish, contains a short dipole antenna. Like a telescope that shows only a narrow portion of sky, a dish antenna is sensitive only to signals coming from specific directions.

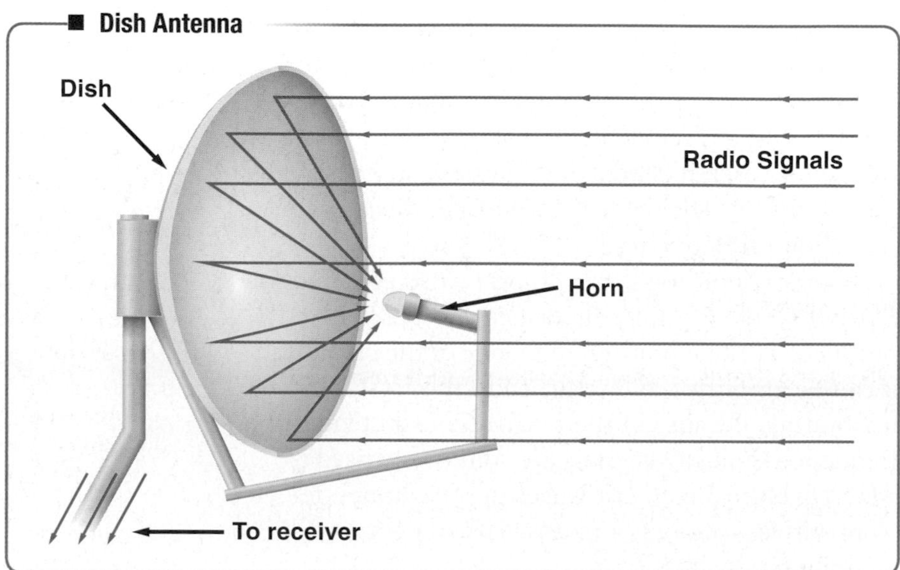

Figure 14 A received signal is reflected off the surface of a dish and focused into the horn, which contains the antenna. The large surface area of the dish collects more electromagnetic wave energy than a wire antenna does, making it well suited for receiving weak radio signals.

Converting waves to information After an antenna converts the electric fields of electromagnetic waves to potential differences, it sends the potential differences to a **receiver,** which converts them to usable information—sound, pictures, or data. Waves of different frequencies strike an antenna and enter the receiver simultaneously. How does the receiver select waves that carry only the desired information?

To select waves of a particular frequency (and reject the others), a receiver uses a tuner. A tuner has a coil-and-capacitor circuit or a resonant cavity. When you turn a radio dial, you select a desired radio station by adjusting the capacitance until the oscillation frequency of the circuit equals the frequency of the desired wave. Then, only waves of the desired frequency can produce oscillating potential differences of significant amplitude in the receiver.

FM and AM signals You read that an electromagnetic wave can carry information only if some property of the wave is changed, or modulated, by the information. A radio station's carrier wave is varied either in amplitude (AM, or Amplitude Modulation) or in frequency (FM, or Frequency Modulation). Commercial AM stations broadcast in the 550–1650 kHz band, while commercial FM stations broadcast between 88 and 108 MHz. FM signals have less noise because most noise sources, such as lightning, create waves of varying amplitude to which FM receivers are insensitive.

☑ **READING CHECK Apply** Why are AM antennas longer than FM antennas?

Digital signals There are many radio services, including commercial, emergency, and two-way walkie talkies, that use other frequencies and either FM or AM modulation. Cell phones, televisions, and computers convert sound, pictures, and data to digital signals. Digital signals are a series of pulses of potential differences. Usually, the pulses are encoded into a binary system of 1s and 0s that vary in duration, not amplitude or frequency. Digital signals can contain more information in the same amount of time as AM or FM and are less affected by noise.

REAL-WORLD PHYSICS

DIGITAL TV On June 13, 2009, all full-power U.S. television stations ceased broadcasting analog signals. Analog TV signals, like FM and AM radio signals, are broadcast by modulating carrier waves. Television stations now broadcast only digital signals, which are encoded into 0s and 1s—the digital code used in computers. Digital TV bandwidth can be compressed to provide four, five, or more channels in the same carrier frequency used by a single channel in analog TV. Digital TV also has higher-fidelity sound and carries about five times more picture information.

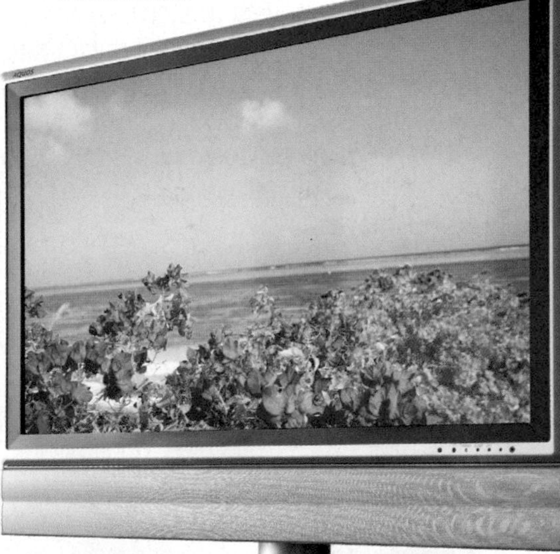

Yasuhide Fumoto/Digital Vision/Getty Images

SECTION 2 REVIEW

Section Self-Check Check your understanding.

23. **MAIN IDEA** Explain how electromagnetic waves propagate through space.

24. **Electromagnetic Waves** What are some primary characteristics of electromagnetic waves? How do electromagnetic waves differ from sound waves and other waves? Explain.

25. **Frequency** An electromagnetic wave has a wavelength of 1.5×10^{-5} m. What is its frequency?

26. **Radio Signals** Radio antennas normally have metal rod elements that are oriented horizontally. From this information, what can you deduce about the directions of the electric fields in radio signals?

27. **Parabolic Receivers** Why is it important for a parabolic dish receiving antenna to be aimed directly toward the transmitter?

28. **Antenna Design** Would an FM antenna designed to be most sensitive to stations near 88 MHz be shorter or longer than one designed to receive stations near 108 MHz? Explain your reasoning.

29. **Dielectric Constant** The speed of light traveling through an unknown material is 1.98×10^{8} m/s. Given that the speed of light in a vacuum is 3.00×10^{8} m/s, what is the dielectric constant of the unknown material?

30. **Critical Thinking** Most of the UV radiation from the Sun is blocked by the ozone layer in Earth's atmosphere. Scientists have found that the ozone layer over Antarctica and the southern hemisphere has thinned. Use what you have learned about electromagnetic waves to explain why some scientists are very concerned about the thinning ozone layer.

A CLOSER LOOK

In The Zone

CELL PHONES

An invisible web surrounds each of us all the time. Cell phones are portals into this web, linking you to people all over the world.

Sending and receiving A cell phone is both a transmitter and a receiver of microwave signals. The cell phone is able to send and receive simultaneously, and to switch frequencies quickly to avoid interference with other signals.

We're all connected A "cell" is the broadcast range of a single cell phone tower. The network of towers that blanket an area communicate with one another and also with the land-based telephone network. When you first turn on your phone, it searches for a signal from the nearest tower. Your phone also transmits its location, so you can be found if someone tries to call you. When a cell phone user moves from one cell to another, the signal switches to the tower in the new cell.

what's up?
1 CELL Each cell has a diameter ranging from about 1,000 m in urban areas up to 8,000 m in rural areas.
2 FREQUENCY As you near a border between two cells, your call is "handed off" from one tower to another.
3 COMMUNICATION Cell towers communicate with one another, with a central switching office, and with land-based telephone networks.
ttyl... :-)

GOINGFURTHER >>>

Research Cell phones and other wireless devices are prohibited on airplanes and in hospitals. Research how these regulations came about and what dangers are posed by the use of cell phones in these locations.

CHAPTER 26 STUDY GUIDE

BIGIDEA Electromagnetic waves are coupled, oscillating electric and magnetic fields generated by accelerating electrons.

SECTION 1 Electric and Magnetic Forces on Particles

VOCABULARY
- **isotope** *(p. 706)*
- **mass spectrometer** *(p. 706)*

MAINIDEA Deflection of moving particles in electric and magnetic fields can be used to find properties of those particles.

- By using electric and magnetic fields to deflect a beam of electrons in a tube nearly evacuated of air—the cathode-ray tube—J.J. Thomson measured the electron's charge-to-mass ratio. R.A. Millikan later measured the charge, allowing the electron's mass to be calculated.
- The charge-to-mass ratio of an electron or positive ion is found by first using crossed electric and magnetic fields to determine the particle's velocity and then using a magnetic field to deflect the particle.
- A mass spectrometer uses both electric and magnetic fields to separate and measure the masses of ionized atoms and molecules. The electric field gives the ion a specific kinetic energy. In the magnetic field, the ion follows a circular path that depends on the ion's mass and charge.

SECTION 2 Electric and Magnetic Fields in Space

VOCABULARY
- **electromagnetic wave** *(p. 710)*
- **electromagnetic spectrum** *(p. 712)*
- **electromagnetic radiation** *(p. 712)*
- **transmitter** *(p. 714)*
- **antenna** *(p. 714)*
- **dielectric** *(p. 714)*
- **carrier wave** *(p. 715)*
- **piezoelectricity** *(p. 717)*
- **receiver** *(p. 719)*

MAINIDEA Electromagnetic waves are coupled, oscillating electric and magnetic fields that move through space and interact with matter.

- The electric and magnetic fields that make up an electromagnetic wave oscillate at right angles to each other and to the direction of the wave's velocity (***v***).

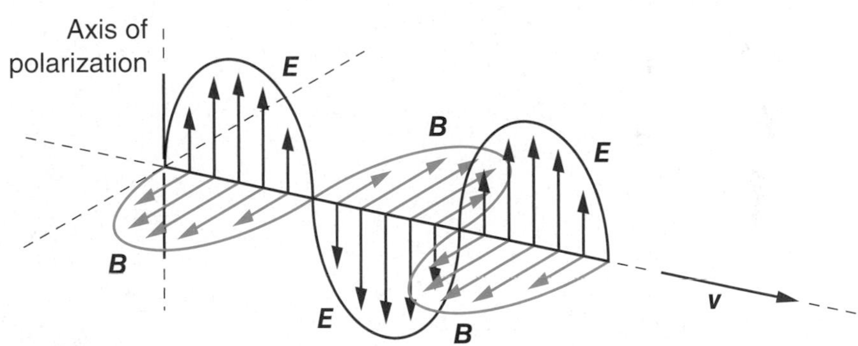

- Wavelength is equal to wave speed divided by wave frequency. For an electromagnetic wave traveling in a vacuum, *v* is equal to the speed of light (c). For an electromagnetic wave traveling through a dielectric, *v* is equal to c divided by the square root of the relative dielectric constant (*k*).
- Electromagnetic waves can carry information if their amplitude or frequency is varied by the data, voice, or video information that is to be transmitted. Waves also can be encoded with digitized information.
- An antenna is most sensitive and most efficient when its length is one-half or one-quarter as long as the wavelength of the electromagnetic wave it is designed to detect.

Games and Multilingual eGlossary

Vocabulary Practice

Chapter Self-Check

SECTION 1 Electric and Magnetic Forces on Particles

Mastering Concepts

31. What are the mass and the charge of an electron?

32. What are isotopes?

33. What must happen to an electron to produce an electromagnetic wave?

34. You have been asked to determine the composition of a sample of soil from a crime scene. How would you use a mass spectrometer to do this?

Mastering Problems

35. Electrons moving at 3.6×10^6 m/s pass through an electric field with a strength of 5.8×10^5 V/m. How large a magnetic field must the electrons experience for their path to be undeflected?

36. A proton moves through a 0.036-T magnetic field, as shown in **Figure 15.** If the proton moves in a circular path with a radius of 0.20 m, what is its speed?

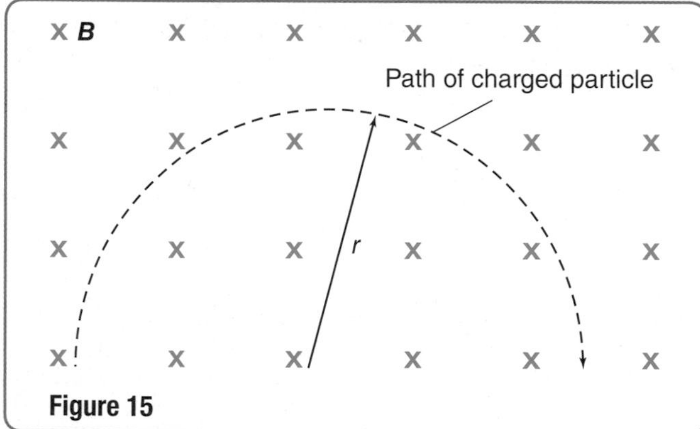

Figure 15

37. A mass spectrometer yields the following data for a beam of doubly ionized (2+) sodium atoms: $B = 8.0\times10^{-2}$ T, $q = 2(1.602\times10^{-19}$ C), $r = 0.077$ m, and $V_{accel} = 156$ V. Calculate the mass of a sodium atom.

38. An alpha particle has a mass of 6.6×10^{-27} kg and a charge of 2+. Such a particle is observed to move through a 0.20-T magnetic field along a path of radius 0.090 m.

a. What potential difference would be required to give the required speed?

b. What is the particle's kinetic energy?

c. What speed does the particle have?

39. Reverse Problem Create a question that involves these quantities: $q = 1.602\times10^{-19}$ C, $m = 12$ $(1.67\times10^{-27}$ kg), $V_{accel} = 515$ V, $B = 75$ mT.

40. Ranking Particles enter regions with magnetic fields as described below. Rank them according to the radius of the circular path they follow, from smallest to largest. Indicate whether any are equal.

A. a proton entering a magnetic field of 0.12 T at a speed of 4.0×10^3 m/s

B. a proton entering a magnetic field of 0.12 T at a speed of 8.0×10^4 m/s

C. a proton entering a magnetic field of 0.24 T at a speed of 4.0×10^3 m/s

D. an electron entering a magnetic field of 0.12 T at a speed of 4.0×10^3 m/s

E. an electron entering a magnetic field of 0.24 T at a speed of 8.0×10^4 m/s

41. In a mass spectrometer, ionized silicon atoms move in curved paths, as shown in **Figure 16.** If the smaller radius corresponds to a mass of 28 proton masses, what is the mass of the other silicon isotope?

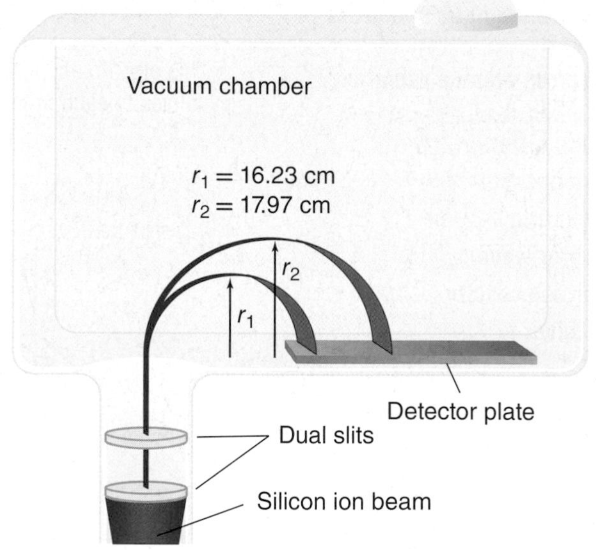

Figure 16

42. An electron is accelerated by a 4.5-kV potential difference. How strong a magnetic field must the electron experience if its path is a circle of radius 5.0 cm?

SECTION 2 Electric and Magnetic Fields in Space

Mastering Concepts

43. BIGIDEA Explain how an antenna is used both to transmit and to receive radio waves.

44. Why is an oscillator used to produce electromagnetic waves? If a battery or a DC generator were used, could it create electromagnetic waves?

45. Sketch the electric and magnetic fields created by a vertical antenna wire as it transmits radio waves.

46. What happens to a quartz crystal when a potential difference is applied across it?

Mastering Problems

47. **Parabolic Dish** The wavelengths of radio waves reflected by a parabolic dish are 2.0 cm long. What is the length of the antenna that detects them?

48. **Bar-Code Scanner** A bar-code scanner uses a laser light source with a wavelength of about 650 nm. Determine the frequency of the laser.

49. **Shortwave radio** A receiving antenna is 4.8 m long. What frequency signal could it best detect?

50. What is the optimum length of an antenna designed to receive a 101.3-MHz FM radio signal?

51. An electromagnetic wave with a frequency of 100 MHz is transmitted through a coaxial cable having a dielectric constant of 2.30. What is the wave's speed?

52. **Cell Phone** A cellular telephone transmitter operates on a carrier frequency of 8.00×10^8 Hz. What is the optimal length of an antenna designed to receive this signal? Note that single-ended antennas, such as those used by cell phones, generate the largest amplitude wave when their length is one-fourth the wavelength of the wave they broadcast or receive.

Applying Concepts

53. **Reverse Problem** Write a physics problem with real-life objects for which the following equation would be part of the solution:

$$q(0.065\text{ T}) = \frac{m(2.8\times10^5\text{ m/s})}{0.045\text{ m}}$$

54. The electrons in the Thomson tube in **Figure 17** travel from left to right. The beam bends upward. Which deflection plate is positively charged?

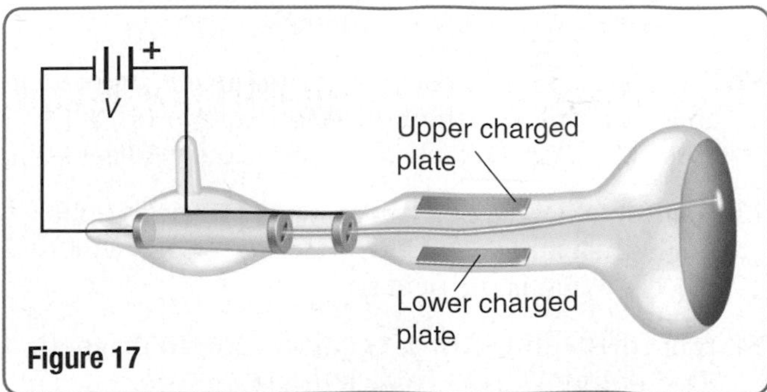

Figure 17

55. Show that the units of E/B are the same as the units for velocity.

56. The vacuum chamber of a mass spectrometer is shown in **Figure 18.** If a sample of ionized neon is being tested in the mass spectrometer, in what direction must the magnetic field be directed to bend the ions into a clockwise semicircle?

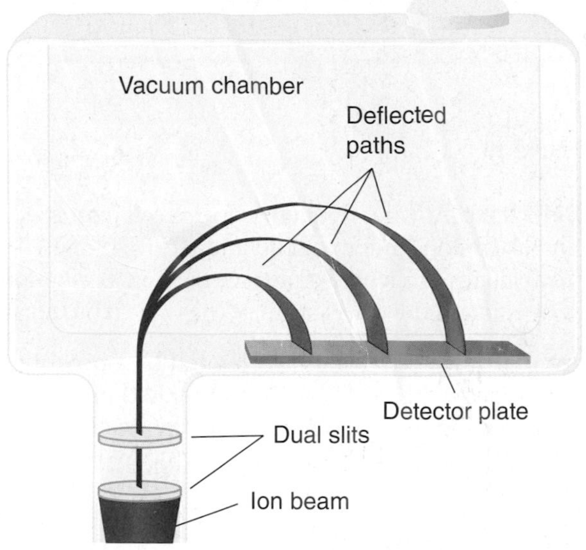

Figure 18

57. If the sign of the charge on the particles in the previous question is changed from positive to negative, do the directions of either or both of the fields have to be changed to keep the particles undeflected? Explain.

58. For each of the following properties, identify whether radio waves, light waves, or X-rays have the largest value.

 a. wavelength

 b. frequency

 c. velocity

59. You are reading a novel about alien beings. The eyes of these aliens are sensitive to microwaves. Would you expect such a being to have larger or smaller eyes than yours? Why?

60. **Cell phone** A cell phone uses a remote earpiece at 2.45 GHz, receives GPS signals at 1.575 GHz, and accesses cell phone bands at 0.90, 1.90, and 2.05 GHz. Rank the optimal antenna lengths for all of these services, from shortest to longest.

61. **Problem Posing** Complete this problem so that it can be solved using both electric and magnetic fields: "A proton has a velocity of 3.7×10^6 m/s east . . . ."

Chapter Self-Check

Mixed Review

62. Radio An FM radio station broadcasts on a frequency of 94.5 MHz using the antenna in **Figure 19.** Assuming the antenna provides the best reception, how long is the part labeled *L*?

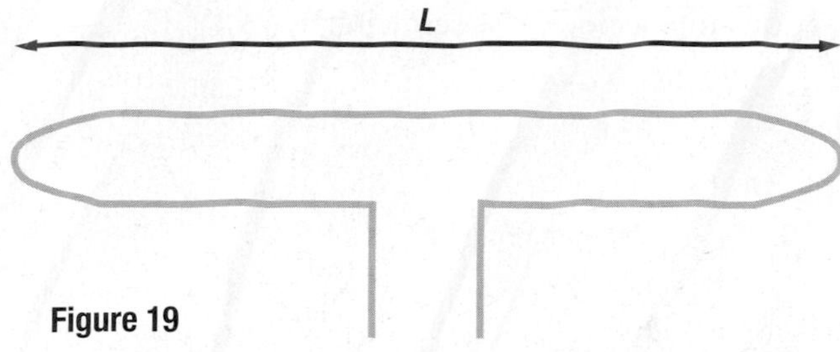

Figure 19

63. Cell Phone At what frequency does a cell phone with an 8.3-cm-long antenna send and receive signals? The ideal length of a single-ended antenna in a cell phone is one-fourth the wavelength of the wave it broadcasts.

64. An unknown particle is accelerated by a potential difference of 1.50×10^{2} V. The particle then enters a magnetic field of 50.0 mT and follows a curved path with a radius of 9.80 cm. What is q/m?

Thinking Critically

65. Apply Concepts Many police departments use radar guns to detect speeding drivers. A radar gun uses a high-frequency electromagnetic signal to measure the speed of a moving object. The frequency of the radar gun's transmitted signal is known. This signal reflects off the moving object and returns to the receiver on the radar gun. Because the object is moving relative to the radar gun, the frequency of the returned signal is different from that of the originally transmitted signal. This phenomenon is known as the Doppler shift. When the object is moving toward the radar gun, the frequency of the returned signal is greater than the frequency of the original signal. If the initial transmitted signal has a frequency of 10.525 GHz and the returned signal shows a Doppler shift of 1850 Hz, what is the speed of the moving object? Use the following equation:

$$v_{\text{target}} = \frac{cf_{\text{Doppler}}}{2f_{\text{transmitted}}}$$

Where,

v_{target} = velocity of target (m/s)

c = speed of light (m/s)

f_{Doppler} = Doppler shift frequency (Hz)

$f_{\text{transmitted}}$ = frequency of transmitted wave (Hz)

66. Apply Concepts H. G. Wells wrote a science-fiction novel called *The Invisible Man* in which a man drinks a potion and becomes invisible, although he retains all his other faculties. Explain why an invisible person would not be able to see.

67. Design an Experiment You are designing a mass spectrometer using the principles discussed in this chapter. You want to distinguish singly ionized (1+) molecules of 175 proton masses from those with 176 proton masses, but the spacing between adjacent cells in your detector is 0.10 mm. The molecules must have been accelerated by a potential difference of at least 500.0 V to be detected. What are some values of *V*, *B*, and *r* that your apparatus should have?

Writing in Physics

68. Portable mass spectrometers are used to detect explosives in airports. Do research on the kind of instruments in use. Briefly describe the instruments that use principles different from those described in this chapter.

Cumulative Review

69. A He–Ne laser (λ = 633 nm) is used to illuminate a slit of unknown width, forming a pattern on a screen located 0.95 m behind the slit. If the first dark band is 8.5 mm from the center of the central bright band, how wide is the slit?

70. The force between two identical metal spheres with the charges shown in **Figure 20** is *F*. If the spheres are touched together and returned to their original positions, what is the new force between them?

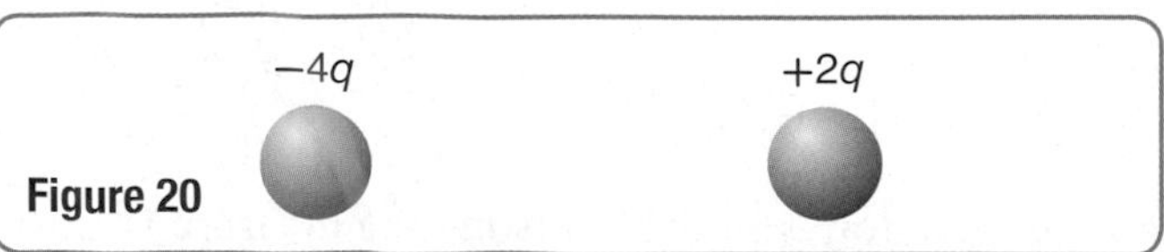

Figure 20

71. What is the electric field strength between two parallel plates spaced 1.2 cm apart if a potential difference of 45 V is applied to them?

72. Calculate the daily cost of operating an air compressor that runs one-fourth of the time and draws 12.0 A from a 245-V circuit when the cost is $0.0950 per kWh.

73. A 440-cm length of wire carrying 7.7 A is at right angles to a magnetic field. The force on the wire is 0.55 N. What is the field strength?

74. A north-south wire moves east through a magnetic field pointing down. Which direction is the current?

MULTIPLE CHOICE

1. For a charged particle moving in a circular trajectory,

A. the magnetic force is parallel to the velocity and is directed toward the center of the circular path.

B. the magnetic force may be perpendicular to the velocity and is directed away from the center of the circular path.

C. the magnetic force always remains parallel to the velocity and is directed away from the center of the circular path.

D. the magnetic force always remains perpendicular to the velocity and is directed toward the center of the circular path.

2. The radius of the circular path traveled by a proton in a uniform 0.10-T magnetic field is 6.6 cm. What is the proton's velocity?

A. 6.3×10^{5} m/s

B. 2.0×10^{6} m/s

C. 6.3×10^{7} m/s

D. 2.0×10^{12} m/s

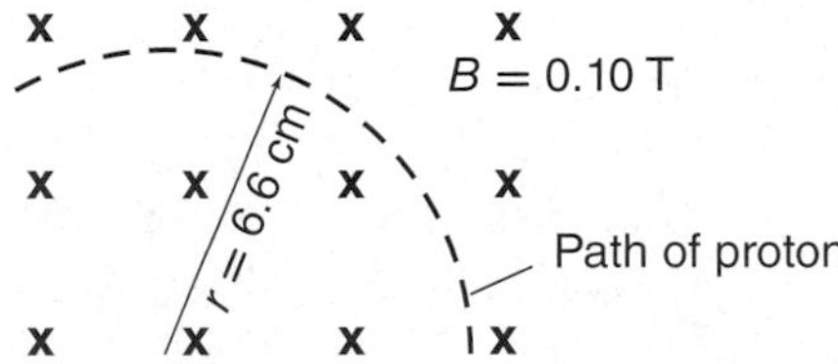

3. The dielectric constant of ruby mica is 5.4. What is the speed of light as it passes through ruby mica?

A. 3.2×10^{3} m/s

B. 9.4×10^{4} m/s

C. 5.6×10^{7} m/s

D. 1.3×10^{8} m/s

4. A certain radio station broadcasts with waves that are 2.87 m long. What is the frequency of the waves?

A. 9.57×10^{-9} Hz

B. 3.48×10^{-1} Hz

C. 1.04×10^{8} Hz

D. 3.00×10^{8} Hz

5. Which of the following situations does not create an electromagnetic wave?

A. A constant potential difference is applied across a piezoelectric quartz crystal.

B. Alternating current passes through a wire contained inside a plastic pipe.

C. A resonant alternating potential difference is applied across a coil-and-capacitor circuit.

D. High-energy electrons strike a metal target in an X-ray tube.

6. The circular path of a proton beam has a radius of 0.52 m as it moves perpendicular to a magnetic field of 0.45 T. If the mass of an individual proton is 1.67×10^{-27} kg, what is the speed of the protons making up the beam?

A. 1.2 m/s

B. 4.7×10^{3} m/s

C. 2.2×10^{7} m/s

D. 5.8×10^{8} m/s

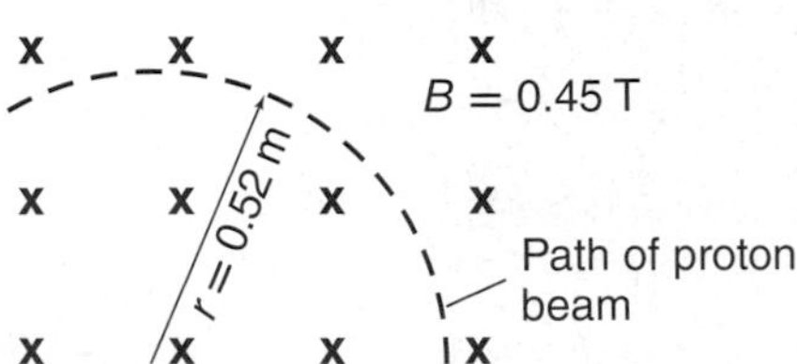

FREE RESPONSE

7. A deuteron (the nucleus of a deuterium atom) has a mass of 3.34×10^{-27} kg and a charge of +e. It has a velocity of 2.88×10^{5} m/s and travels in a magnetic field of 0.150 T. What is the radius of curvature of its path?

Online Test Practice

NEED EXTRA HELP?

If You Missed Question	1	2	3	4	5	6	7
Review Section	1	1	2	2	2	1	1

CHAPTER 27

Quantum Theory

BIGIDEA **Waves can behave like particles, and particles can behave like waves.**

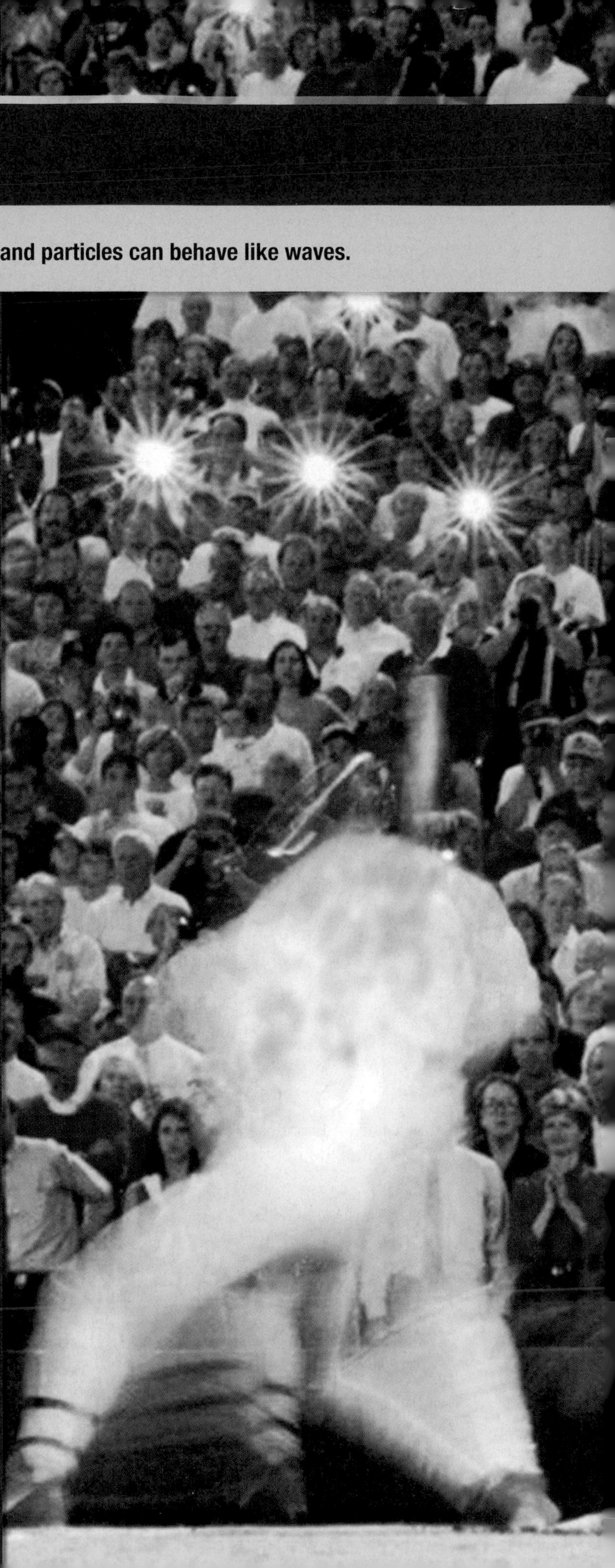

SECTIONS

1 **A Particle Model of Waves**

2 **Matter Waves**

LaunchLAB

iLab Station

LIGHTBULB SPECTRUM

What affects the electromagnetic radiation emitted by an object?

WATCH THIS!

NIGHT-VISION GOGGLES

You've probably seen night-vision gear in spy movies, on television, or in video games, but do you know how they work? Are all night-vision goggles the same? Explore the physics of spy tech.

Go online!
connectED.mcgraw-hill.com

SECTION 1

A Particle Model of Waves

PHYSICS 4 YOU **If you are climbing a flight of stairs, you can stand on the first step, or the second, or the third. But you cannot stand on step 1.381 or step 3.5. In a similar way, energy comes in packets that are integer multiples of the smallest energy.**

MAIN IDEA

Light can behave as massless particles called photons.

Essential Questions

- What are the characteristics of the electromagnetic spectrum emitted by an object?
- What is the photoelectric effect?
- What is the Compton effect?

Review Vocabulary

electromagnetic wave coupled, oscillating electric and magnetic fields that travel through space and matter

New Vocabulary

emission spectrum
quantized
photoelectric effect
threshold frequency
photon
work function
Compton effect

A New Model Based on Packets of Energy

James Maxwell's electromagnetic wave theory was supported by the experiments of Heinrich Hertz. Hertz's experiments firmly established the existence of electromagnetic waves. Maxwell's theory was a great success because it seemed to explain some properties of light, including interference, diffraction, and polarization.

Despite these successes, Maxwell's description of light as a purely electromagnetic wave could not explain several other important phenomena. One such phenomenon was the spectrum of electromagnetic waves emitted by all objects. Calculations based on electromagnetic wave theory predicted that any object, regardless of temperature, would emit an infinite amount of energy in the form of electromagnetic waves. It was also discovered that when a metal surface was exposed to ultraviolet radiation, electrons were emitted from the metal in a curious manner. These two phenomena could be explained once it was understood that electromagnetic waves have particlelike properties as well as wavelike properties.

Electromagnetic radiation from objects The lightbulbs shown in **Figure 1** are connected to a dimmer control that varies the potential difference to the bulb. As the potential difference increases, the temperature of the bulb's glowing filament increases. As a result, the color changes from deep red to orange to yellow and finally to white. This color change occurs because the higher-temperature filament emits higher-frequency radiation. The higher-frequency end of the visible spectrum mixes in blue and violet colors with the red and orange, resulting in the filament appearing whiter.

Figure 1 The color of the glow of the lightbulb depends on the temperature of the filament.

Lower Temperature

Higher Temperature

(t)Chris Tubbs/Red Cover/Getty Images, (bl br)Richard Hutchings/Digital Light Source

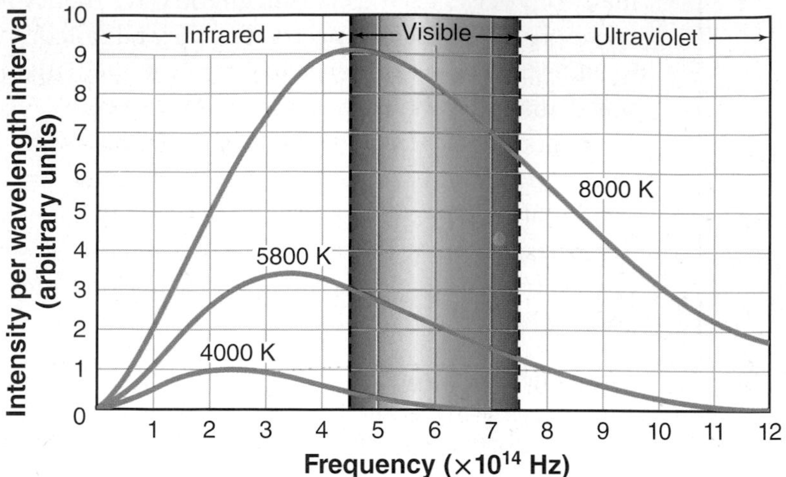

Figure 2 The frequency at which an object's emission spectrum peaks depends on that object's temperature. As the temperature of the object increases, the peak frequency also increases.

As predicted by electromagnetic wave theory, the vibrating charged particles within each bulb filament emit light and infrared radiation. While all objects, from very cold to extremely hot objects, emit a range of electromagnetic waves, the filament glows brightly in the visible light range because it is hot. (This is called incandescence; hence, the name incandescent lightbulb.) The colors you see depend on the relative intensities of the emitted electromagnetic waves of various frequencies and the sensitivity of your eyes to those waves.

Emission spectra What would you expect to see if you viewed the glowing filament through a diffraction grating? When viewed in this way, all the colors of the rainbow would be visible. **Figure 2** shows the visible colors that correspond to the frequencies around 6×10^{14} Hz. The bulb also emits other radiation that you would not see. A graph of the intensity of the radiation emitted from an object over a range of frequencies is known as an **emission spectrum.** Emission spectra of objects at temperatures of 4000 K, 5800 K, and 8000 K are shown in **Figure 2.** Note that at each temperature there is a frequency at which the maximum amount of energy is emitted. If you compare the curves, you will see that as the temperature increases, the frequency at which the maximum amount of energy is emitted also increases.

▶ CONNECTION TO ASTRONOMY The total power emitted by an object also increases with temperature. The power (energy emitted per second) of an electromagnetic wave is proportional to the hot object's kelvin temperature raised to the fourth power (T^4). Thus, hotter objects radiate considerably more energy per second than do cooler objects. Probably the most common example of a hot object radiating a great amount of power is the Sun. The Sun's surface temperature is about 5800 K, and it radiates 4×10^{26} W of power—an enormous quantity. On average, each square meter of Earth's surface receives about 1000 J of energy per second (1000 W) from the Sun. This is enough energy to power ten 100-W lightbulbs.

READING CHECK **Predict** how the power emitted by an object changes if the object's kelvin temperature is doubled.

MiniLAB

GLOWS IN THE DARK

How do different wavelengths of light affect fluorescein?

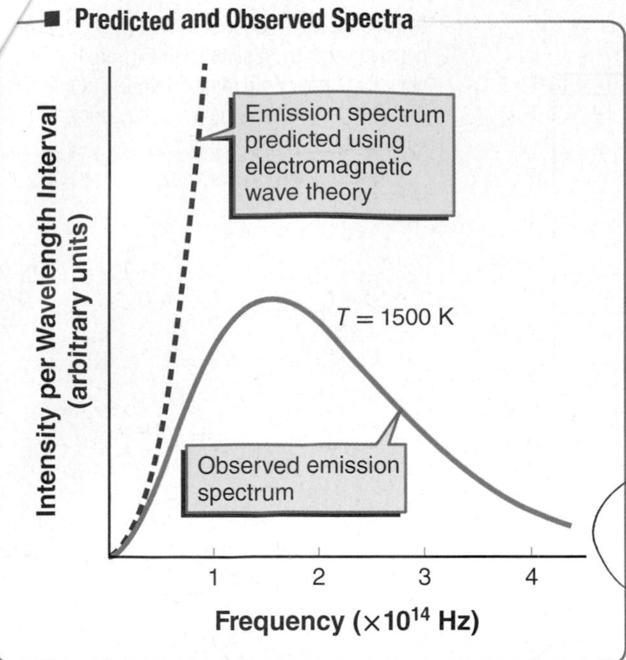

Figure 3 The red, dashed line shows the emission spectrum predicted using the electromagnetic wave theory. The theory predicts any object will radiate an infinite amount of energy. This does not match the observed emission spectrum, shown in blue.

Identify the region of the graph where electromagnetic wave theory is a good model for the observed emission spectrum.

MiniLAB

MODELING THE QUANTUM

How can you determine the smallest unit of a larger sample?

Explaining emission spectra Between 1887 and 1900, many physicists used electromagnetic wave theory to try to predict emission spectra, but they all failed. **Figure 3** shows the difference between the predicted spectrum and the observed spectrum. In 1900, German physicist Max Planck found that he could calculate the correct spectrum if an atom could only emit or absorb particular amounts of energy. Planck hypothesized an atom's energy changes in a solid were proportional to a frequency of vibration times an integer.

ENERGY OF VIBRATION

The energy emitted or absorbed by a vibrating atom is equal to the product of an integer, Planck's constant, and the frequency of vibration.

$$E = nhf$$

In the equation, f is the frequency of vibration of the atom, h is a constant called Planck's constant, which has a value of 6.626×10^{-34} J/Hz, and n is an integer such as 0, 1, 2, 3 . . .

$$n = 0{:}\ E = (0)hf = 0$$
$$n = 1{:}\ E = (1)hf = hf$$
$$n = 2{:}\ E = (2)hf = 2hf$$

The energy of the radiation (E) can have values of hf, 2hf, 3hf, and so on, but it can never have fractional values such as $\frac{2}{3}$ hf or $\frac{3}{4}$ hf. In other words, energy is **quantized**—it exists only in bundles of specific amounts. For calculations, the constant h is usually rounded to 6.63×10^{-34} J/Hz.

Changes in vibrations Electromagnetic wave theory would allow atoms to continuously emit radiation at all times. Instead, Planck proposed that atoms emit radiation only at the particular moments when their vibrational energy changes. For example, if the energy of an atom changes from 3hf to 2hf, according to wave theory, the atom emits radiation only while the energy changes. The energy radiated is equal to the change in energy of the atom, in this case hf. Similarly if the atom absorbed energy hf, its energy could change from 2hf to 3hf.

Planck found that because the constant h has an extremely small value, the energy-changing steps are very small. You would not notice them in the motions of objects in everyday life. Still, the idea of quantized energy troubled physicists, even Planck himself. It hinted that the classical physics of Newton and Maxwell might apply only under certain conditions. Planck's troubling idea was the first step on the way to the development of solar power cells, modern electronics, and computers.

The Photoelectric Effect

Physicists of the early 1900s were also challenged by another experimental result that could not be explained by electromagnetic wave theory. When ultraviolet light shines on a negatively charged zinc plate, the plate can produce an electrical discharge—a stream of electrons jumping from the plate. Even a low-intensity ultraviolet light can produce this effect. When visible light shines on the same charged plate, the plate does not discharge. This is true even when the visible light has a high intensity.

This result is contrary to electromagnetic wave theory, which predicts the plate would emit electrons when any frequency of electromagnetic radiation shines on the plate. Higher-frequency ultraviolet radiation and lower frequency visible light are forms of electromagnetic radiation, so why would one cause the zinc plate to discharge but not the other? The emission of electrons when electromagnetic radiation falls on an object is called the **photoelectric effect.**

The photocell The photoelectric effect can be studied in a photocell, shown in **Figure 4.** The cell contains two metal electrodes sealed in a tube from which the air has been removed. The evacuated tube prevents the metal surfaces from oxidizing and keeps emitted electrons from being slowed or stopped by gas particles. The larger electrode, called the cathode, usually is coated with cesium or another alkali metal. The smaller electrode, called the anode, is made of a thin wire so that it blocks only a very small amount of radiation. The tube allows ultraviolet light to pass through it. A potential difference placed across the electrodes attracts electrons ejected from the cathode to the anode.

When no light falls on the cathode, which is the negative electrode, there is no current in the circuit. However, when light with certain frequencies falls on the cathode, a current is produced, which is measured by the ammeter, as shown in **Figure 4.** Electrons in the cathode absorb the energy from the light. The electrons escape the potential energy that holds them in the cathode and travel to the anode, which is the positive electrode. This flow of photoelectrons forms the current in the circuit.

PhysicsLAB

MODELING THE PHOTOELECTRIC EFFECT

How can you use steel balls to model the photoelectric effect?

Investigate **the photoelectric effect.**

View an **animation of the photoelectric effect.**

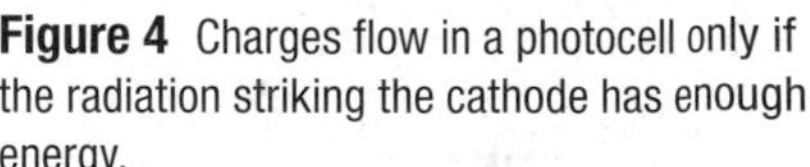

Figure 4 Charges flow in a photocell only if the radiation striking the cathode has enough energy.

Compare the ultraviolet radiation's wavelength to visible light's wavelength. What does the comparison tell you about the relative energies of the two types of radiation?

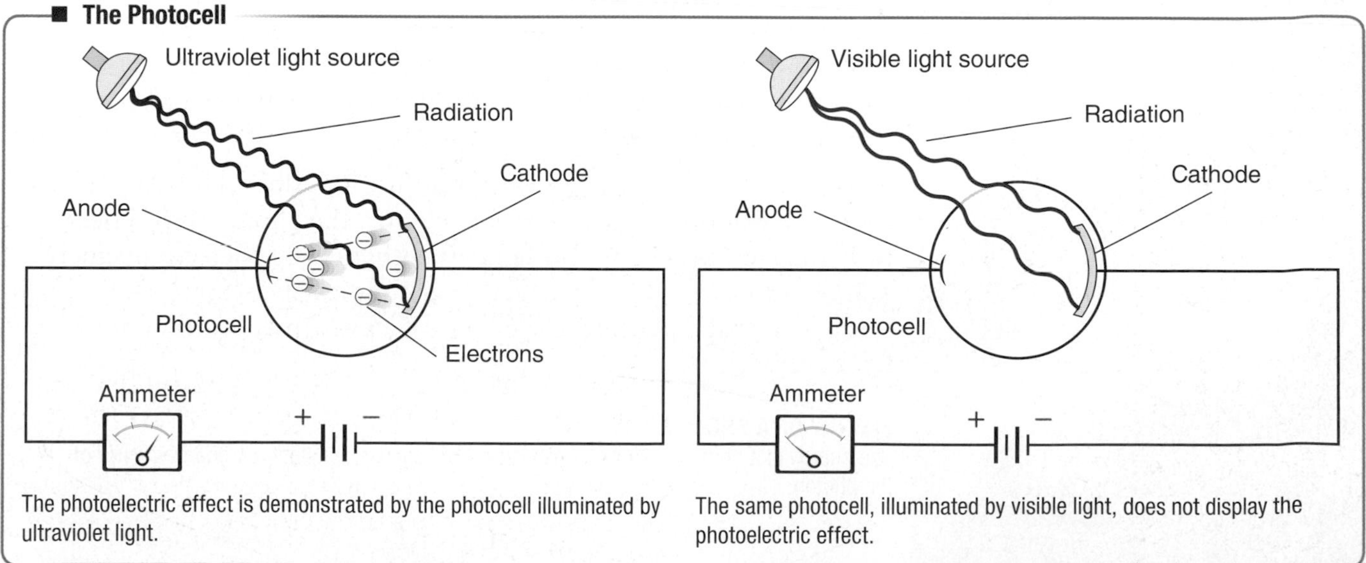

The photoelectric effect is demonstrated by the photocell illuminated by ultraviolet light.

The same photocell, illuminated by visible light, does not display the photoelectric effect.

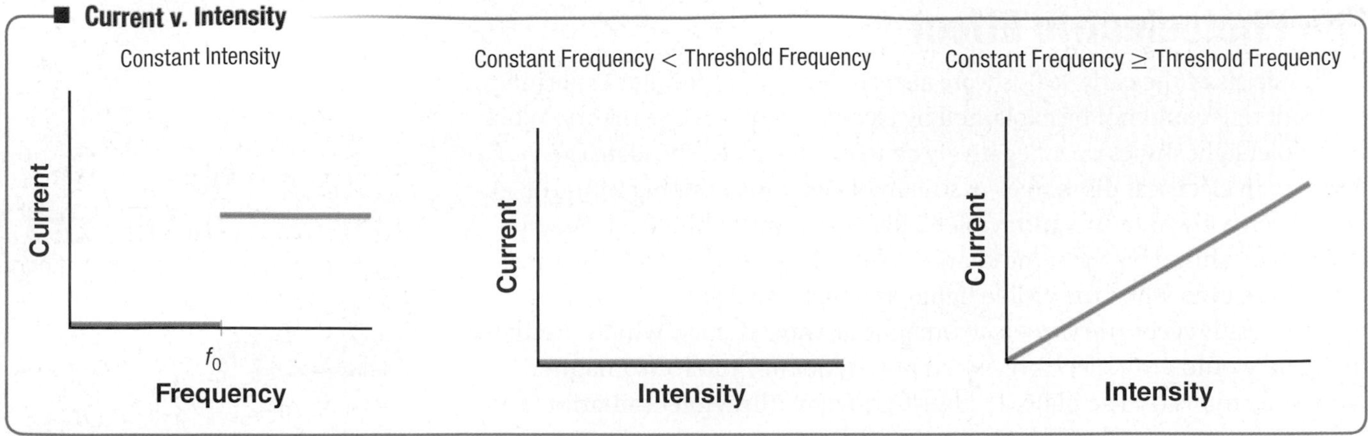

Figure 5 The emission of photoelectrons depends on the incident light's frequency. The threshold frequency must be exceeded in order to produce photoelectrons.

REAL-WORLD PHYSICS

SOLAR PANELS The photoelectric effect is the basis of solar panel technology. A solar panel consists of a grid of solar cells made of semiconducting material. Photons from the Sun that have a certain threshold frequency can cause electrons to break free from atoms of the semiconducting material, resulting in an electric current within the solar panel.

Threshold frequency Not all radiation falling on a photocell causes a current. Electrons eject from the cathode only if the frequency of the incident radiation is greater than a certain minimum value, called the **threshold frequency** (f_0). The graph on the left in **Figure 5** shows there is a current only when the threshold frequency is reached. The value of the threshold frequency varies widely and depends on the type of metal used for the cathode. For example, all frequencies of visible light except low-frequency red eject electrons from cesium. No frequency of visible light ejects electrons from zinc. Higher-frequency ultraviolet radiation ejects electrons from zinc.

The center graph in **Figure 5** shows that even intense radiation with a frequency below f_0 will not cause the emission of electrons. Conversely, the graph on the right shows that even very low-intensity radiation with a frequency at or above f_0 causes the immediate emission of electrons from the cathode. When the incoming radiation, also called the incident radiation, has a frequency equal to or greater than the threshold frequency, increasing the intensity of the radiation increases the current through the photocell.

The electromagnetic wave theory would say that the electric field in the light wave accelerates and ejects the electrons from the metal. The strength and energy of the electric field are related to the intensity of the light (not to its frequency). Thus, it follows that electrons in the metal could absorb energy from a low-intensity, low-frequency light source for a long time and eventually gain enough energy to be ejected. As you just read, however, this is not the case. Observations show that electrons are ejected immediately when even low-intensity radiation at or above the threshold frequency shines on the metal.

Photons and quantized energy In 1905 Albert Einstein published a revolutionary theory that explained the photoelectric effect. According to Einstein, visible light and other forms of electromagnetic radiation consist of discrete, quantized bundles of energy, each of which was later called a **photon.** The energy of a photon depends on its frequency.

ENERGY OF A PHOTON

The energy of a photon is equal to the product of Planck's constant and the frequency of the photon.

$$E = hf$$

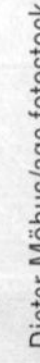

Electron volts In the equation for the energy of a photon, f is frequency, which has units of hertz (Hz), and h is Planck's constant, which has units of joules per hertz (J/Hz). Recall that the unit Hz is defined as $\frac{1}{s}$ or s^{-1}. Therefore, the $\frac{J}{Hz}$ unit of Planck's constant is also equivalent to J•s. Because the joule is too large a unit of energy to use with atomic-sized systems, the more convenient energy unit of the electron volt (eV) is often used. One electron volt is the energy of an electron accelerated across a potential difference of 1 V.

$$1\ \text{eV} = (1.602\times10^{-19}\ \text{C})(1\ \text{V})$$

$$= 1.602\times10^{-19}\ \text{C}\cdot\text{V}$$

$$= 1.602\times10^{-19}\ \text{J}$$

Using the definition of an electron volt allows the photon energy equation to be rewritten in a simplified form, as shown below.

ENERGY OF A PHOTON

The energy of a photon is equal to the constant 1240 eV•nm divided by the wavelength of the photon.

$$E = \frac{hc}{\lambda} = \frac{1240\ \text{eV}\cdot\text{nm}}{\lambda}$$

PhysicsLAB

RELATING COLOR AND LED VOLTAGE DROP

How does the voltage applied to an LED affect the wavelength of the emitted light?

iLab Station

PROBLEM-SOLVING STRATEGIES

Personal Tutor

Units of hc and Photon Energy

Converting the quantity hc to the unit eV•nm results in a simplified equation that you can use to solve problems involving photon wavelength.

1. The energy of a photon of frequency f is given by this equation:

$$E = hf$$

2. Recall that $f = \frac{c}{\lambda}$, where c is the speed of light in a vacuum. The equation above can therefore be written

$$E = \frac{hc}{\lambda}$$

3. When using the equation $E = \frac{hc}{\lambda}$, if the value of hc in eV•nm is divided by λ in nm, you will obtain the energy in eV. Thus, it is useful to know the value of hc in eV•nm. Because h and c are constants, the value of hc is also a constant.

4. The conversion of hc to the unit eV•nm is as follows:

$$hc = (6.626\times10^{-34}\ \text{J/Hz})(2.998\times10^{8}\ \text{m/s})\left(\frac{1\ \text{eV}}{1.602\times10^{-19}\ \text{J}}\right)\left(\frac{10^{9}\ \text{nm}}{1\ \text{m}}\right)$$

$$= 1240\ \text{eV}\cdot\text{nm}$$

5. Substituting hc = 1240 eV•nm into the equation for the energy of a photon yields the following, where λ is in nm and E is in eV:

$$E = \frac{hc}{\lambda} = \frac{1240\ \text{eV}\cdot\text{nm}}{\lambda}$$

6. Use the above equation to solve photon energy problems when energy in eV is desired.

PRACTICE PROBLEMS

Do additional problems. Online Practice

Use E = 1240 eV•nm/λ to solve the following problems.

1. What is a photon's energy if the photon's wavelength is 515 nm?
2. A photon's energy is 2.03 eV. What is the photon's wavelength?
3. Rank the following photons from least to greatest energy.
   A. 4.0 eV
   B. 320 nm
   C. 811 nm
   D. 2.1 eV
4. **CHALLENGE** The diagram in **Figure 6** shows the visible light spectrum. What is the range of energies associated with photons in the visible light spectrum?

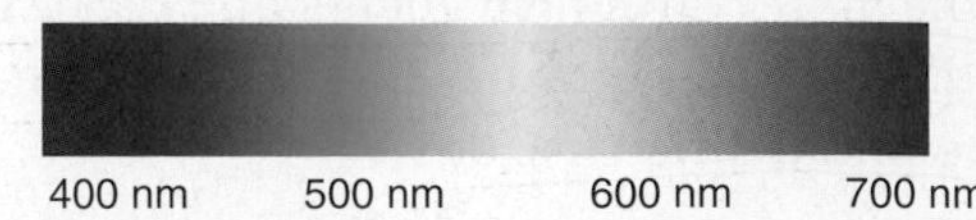

Figure 6

Planck and Einstein

It is important to note that Einstein's photon theory goes further than Planck's hypothesis of radiation from hot objects. While Planck proposed that atoms vibrating at frequency f emit electromagnetic radiation with energy equal to nhf, he did not suggest that light and other forms of electromagnetic radiation act like particles. Einstein's photon theory reinterpreted and extended Planck's hypothesis of radiation emitted by objects.

Einstein's photon theory explains the existence of a threshold frequency and the prompt emission of electrons during the photoelectric effect. A photon with a minimum frequency and energy (hf_0) can eject an electron from metal. If the photon frequency is less than f_0, the photon will not have the energy needed to pry the electron from the metal. Because one photon interacts with one electron, an electron cannot simply store energy from low frequency photons until it collects enough energy to break free. On the other hand, radiation with a frequency greater than f_0 provides more than enough energy to eject an electron. In fact, the excess energy, $hf - hf_0$, becomes the ejected electron's kinetic energy.

KINETIC ENERGY OF AN ELECTRON EJECTED DUE TO THE PHOTOELECTRIC EFFECT

The kinetic energy of an ejected electron is equal to the difference between the incident photon energy (hf) and the energy of a photon with the threshold frequency (hf_0).

$$KE = hf - hf_0$$

Note that hf_0 is the minimum energy needed to free the most loosely held electron in an atom. Not all electrons in a metal have the same energy; some will require more than this minimum energy in order to escape. As a result, the ejected electrons have differing kinetic energies. Thus, it is important to realize that the phrase "kinetic energy of the ejected electrons" refers to the maximum kinetic energy that an ejected electron could have. Some of the ejected electrons will have less.

☑ **READING CHECK Explain** how an electron emitted by the photoelectric effect can have kinetic energy less than the threshold energy.

Testing the photon theory One way to test the photon theory is to measure the electrons' kinetic energy and compare the results to the value predicted by the theory. The device in **Figure 7** can indirectly measure the electrons' kinetic energy. In the device, the potential difference across the tube is adjustable. When the potential difference is adjusted to make the anode negative, the ejected electrons experience a force opposing their motion due to the electric field between the cathode and anode. As a result, the electrons' kinetic energy decreases as they approach the anode. Only electrons ejected with enough kinetic energy reach the anode.

Increasing the opposing potential difference makes the anode more negative, which means the electron's initial kinetic energy must be greater to reach the anode. At a certain potential difference, called the stopping potential, the current stops because no electrons ejected from the cathode have enough kinetic energy to reach the anode.

At the stopping potential, the electrons' kinetic energy at the cathode equals the work done by the electric field to stop them. In equation form this is

$$KE = -e\Delta V_0$$

In this equation, ΔV_0 is the magnitude of the stopping potential difference in volts (J/C), and e is the charge of the electron (-1.602×10^{-19} C). Note that the negative sign in the equation along with the negative value of e yield a positive value for KE. Calculations of electrons' kinetic energy based on this experiment support Einstein's photon theory.

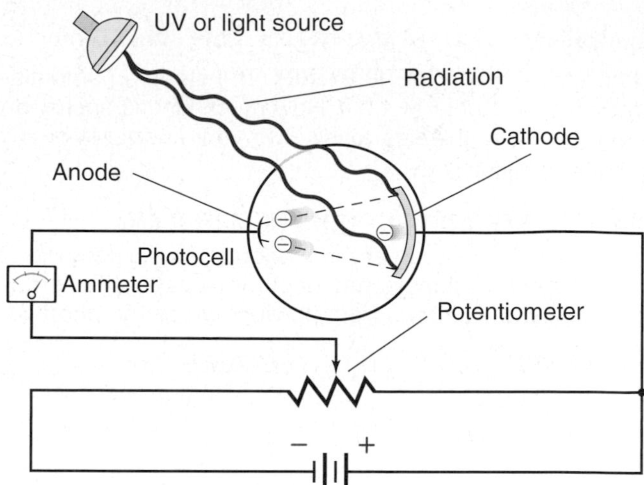

Figure 7 The potentiometer controls the potential difference inside the photocell. By carefully adjusting the potentiometer, you can determine the potential that results in zero current. At the zero current threshold, all electrons fail to reach the anode. Instead, they fall back on the cathode due to the force caused by the electric field between the cathode and the anode.

Application of the photoelectric effect If you have ever used a digital camera, you have used a device that takes advantage of the photoelectric effect. The digital camera shown in **Figure 8** uses a grid of tiny photoelectric detectors—several million of them packed in a few square centimeters. In a typical design, light enters the camera through the lens and passes through a grid of filters placed over the photoelectric detectors. The filters allow only certain wavelengths of light, corresponding to specific colors of light, to pass through to an individual detector. Photons that strike a detector result in photoelectrons being stored in the detector. The amount of charge stored at the detector's location indicates the brightness of light. The information from the entire grid of detectors is then compiled to recreate the image. Because the photoelectrons generate a voltage within the detectors, the process is a variation of the photoelectric effect known as the photovoltaic effect.

Figure 8 A digital camera's image detector relies on the photoelectric effect to generate a voltage.

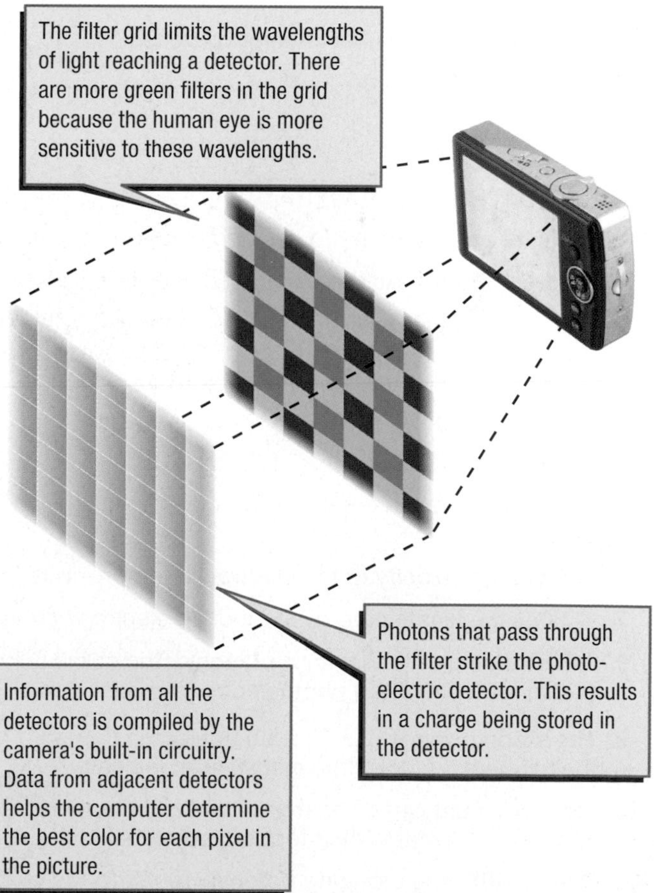

The filter grid limits the wavelengths of light reaching a detector. There are more green filters in the grid because the human eye is more sensitive to these wavelengths.
Photons that pass through the filter strike the photoelectric detector. This results in a charge being stored in the detector.
Information from all the detectors is compiled by the camera's built-in circuitry. Data from adjacent detectors helps the computer determine the best color for each pixel in the picture.

EXAMPLE PROBLEM 1

Find help with **operations with scientific notation.** Math Handbook

PHOTOELECTRON KINETIC ENERGY The stopping potential difference of a certain photocell is 4.0 V. How much kinetic energy does the incident light give the electrons? Give your answer in both joules and electron volts.

1 ANALYZE AND SKETCH THE PROBLEM

Draw the cathode and the anode, the incident radiation, and the direction of the ejected electron. Note that the stopping potential prevents electrons from flowing across the photocell.

KNOWN	UNKNOWN
$\Delta V = 4.0\ \text{V}$	KE (in J and eV) = ?
$e = -1.602\times10^{-19}\ \text{C}$	

2 SOLVE FOR THE UNKNOWN

The electric field does work on the electrons, so the final kinetic energy of an electron equals the initial kinetic energy plus the work done on the electron.

$$KE_{\text{initial}} + W_{\text{electric field on electron}} = KE_{\text{final}}$$

$$KE_{\text{initial}} + W_{\text{electric field on electron}} = 0\ \text{J}$$

◀ The applied potential difference is the stopping potential, so $KE_{\text{final}} = 0$ J.

Solve for KE_{initial}.

$$KE_{\text{initial}} = -W_{\text{electric field on electron}}$$

$$= -e\Delta V_0$$

◀ Use the definition of electric potential difference. $\left(\Delta V = \frac{W_{\text{on }e}}{e}\right)$ to find $W_{\text{on }e} = e\Delta V$.

$$= -(-1.602\times10^{-19}\ \text{C})(4.0\ \text{V})$$

$$= +6.4\times10^{-19}\ \text{J}$$

◀ Substitute $e = -1.602\times10^{-19}$ C, $\Delta V_0 = 4.0$ V

Convert KE from joules to electron volts.

$$KE = (6.4\times10^{-19}\ \text{J})\frac{1\ \text{eV}}{1.602\times10^{-19}\ \text{J}}$$

$$= 4.0\ \text{eV}$$

3 EVALUATE THE ANSWER

- **Are the units correct?** Joules and electron volts are both units of energy.
- **Does the sign make sense?** Kinetic energy is always positive.
- **Is the magnitude realistic?** One electron volt is the energy of an electron accelerated across a potential difference of 1 V. The electron is accelerated by a potential difference of 4 V, so 4 eV is reasonable.

PRACTICE PROBLEMS

Do additional problems. 

5. An electron has an energy of 2.3 eV. What is the kinetic energy of the electron in joules?
6. What is the velocity of the electron in the previous problem?
7. What is the kinetic energy in eV of an electron with a velocity of 6.2×10^{6} m/s?
8. The stopping potential for a photoelectric cell is 5.7 V. Calculate the maximum kinetic energy of the emitted photoelectrons in eV.
9. The stopping potential for a photoelectric cell is 5.1 V. How much kinetic energy does the incident light give the electrons in joules?
10. The maximum kinetic energy of emitted photoelectrons in a photoelectric cell is 7.5×10^{-19} J. What is the stopping potential?
11. **CHALLENGE** The stopping potential required to prevent current through a photocell is 3.2 V. Calculate the maximum kinetic energy in joules of the photoelectrons as they are emitted.

Table 1 Threshold Frequency, Threshold Wavelength, and Work Function for Selected Metals

Metal	Threshold Frequency ($\times10^{14}$ Hz)	Threshold Wavelength (nm)	Work Function (eV)
Cesium	4.70	637	1.95
Magnesium	8.84	339	3.66
Silver	11.1	270	4.6
Sodium	5.70	526	2.36

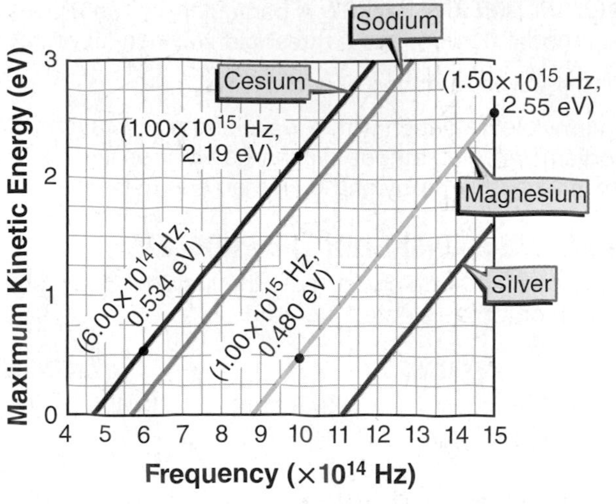

Figure 9 The metal's threshold frequency equals the line's *x*-intercept. Each line's slope equals Planck's constant.

Measuring h A graph of the maximum kinetic energies of the electrons ejected from a metal versus the frequencies of the incident photons is a straight line, as shown in **Figure 9.** All metals have similar graphs with the same slope. This slope equals the rise/run ratio of the line, which is equal to Planck's constant (h).

The graphs of various metals differ only in the threshold frequency needed to free the electrons. In **Figure 9,** the threshold frequency (f_0) is the point at which $KE = 0$. Looking at the graph, we see that f_0 is located at the intersection of the line with the *x*-axis. For example, the *x*-intercept, and therefore the threshold frequency, for cesium is 4.70×10^{14} Hz. A photon with that frequency has just enough energy to eject an electron from the metal. This minimum energy, called the **work function** of a metal, is the energy needed to free the most weakly bound electron from the metal. The magnitude of the work function is equal to hf_0. When a photon of frequency f_0 is incident on a metal, the energy of the photon is sufficient to release an electron but not sufficient to provide the electron with any kinetic energy.

CONNECTING MATH TO PHYSICS

Slope of a Line The lines in **Figure 9** are parallel. That means they all have the same slope. Below are the calculations for cesium and magnesium.

Math	Physics	
	Cesium	Magnesium
(x_1, y_1), (x_2, y_2)	(6.00×10^{14} Hz, 0.534 eV), (1.00×10^{15} Hz, 2.19 eV)	(1.00×10^{15} Hz, 0.480 eV), (1.50×10^{15} Hz, 2.55 eV)
$m = \frac{y_2 - y_1}{x_2 - x_1}$	$m = \frac{2.19\text{ eV} - 0.534\text{ eV}}{1.00\times10^{15}\text{ Hz} - 6.00\times10^{14}\text{ Hz}}$	$m = \frac{2.55\text{ eV} - 0.480\text{ eV}}{1.50\times10^{15}\text{ Hz} - 1.00\times10^{15}\text{ Hz}}$
	$m = 4.14\times10^{-15}$ eV/Hz	$m = 4.14\times10^{-15}$ eV/Hz

▲ All lines in Figure 9 have a slope of 4.14×10^{-15} eV/Hz. When converted to J/Hz, this value becomes the familiar value of Planck's constant:

$$4.14\times10^{-15}\ \frac{\text{eV}}{\text{Hz}}\left(\frac{1.602\times10^{-19}\text{ J}}{1\text{ eV}}\right) = 6.63\times10^{-34}\text{ J/Hz}$$

EXAMPLE PROBLEM 2

Find help with **operations with significant digits.** Math Handbook

WORK FUNCTION AND ENERGY A particular photocell uses a sodium cathode. Sodium has a threshold wavelength of 526 nm.

a. Find the work function of sodium in eV.

b. If ultraviolet radiation with a wavelength of 348 nm falls on sodium, will the cathode discharge electrons? If so, what is the maximum energy of the ejected electrons in eV?

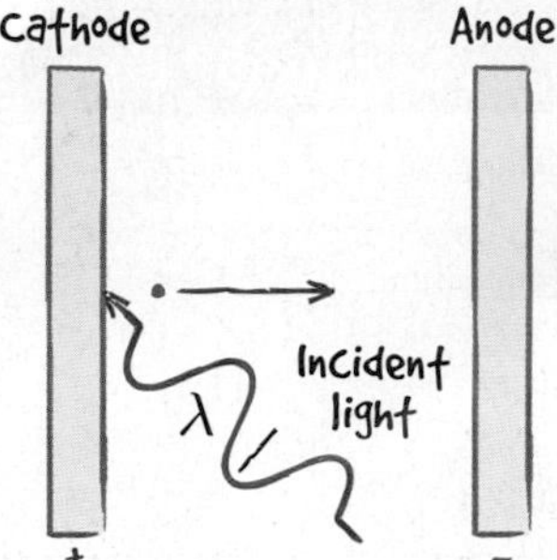

1 ANALYZE AND SKETCH THE PROBLEM

Draw the cathode and anode, the incident radiation, and the direction of the ejected electron.

KNOWN		UNKNOWN
$\lambda_0 = 526$ nm	$\lambda = 348$ nm	$W = ?$ $KE = ?$
$hc = 1240$ eV·nm		

2 SOLVE FOR THE UNKNOWN

a. Find the work function using Planck's constant and the threshold wavelength.

$$W = hf_0 = \frac{hc}{\lambda_0}$$

$$= \frac{1240 \text{ eV·nm}}{526 \text{ nm}}$$ ◀ Substitute $hc = 1240$ eV·nm, $\lambda_0 = 526$ nm

$$= 2.36 \text{ eV}$$

b. Use the relationship between a photon's energy and wavelength to find the photon's energy.

$$E = \frac{1240 \text{ eV·nm}}{\lambda}$$

$$= \frac{1240 \text{ eV·nm}}{348 \text{ nm}}$$ ◀ Substitute $\lambda = 348$ nm

$$= 3.56 \text{ eV}$$

To calculate the kinetic energy of the ejected electron, subtract the work function from the energy of the incident radiation.

$$KE = hf - hf_0 = \frac{hc}{\lambda} - \frac{hc}{\lambda_0}$$

$$= E - W$$ ◀ Substitute $\frac{hc}{\lambda} = E$, $\frac{hc}{\lambda_0} = W$

$$= 3.56 \text{ eV} - 2.36 \text{ eV}$$ ◀ Substitute $E = 3.56$ eV, $W = 2.36$ eV.

$$= 1.20 \text{ eV}$$

3 EVALUATE THE ANSWER

- **Are the units correct?** Performing dimensional analysis on the units verifies that electron volts is the proper unit for kinetic energy.
- **Does the sign make sense?** Kinetic energy is always positive.
- **Are the magnitudes realistic?** Work functions given in **Table 1** are a few electron volts, so the magnitude is realistic.

PRACTICE PROBLEMS

Do additional problems. Online Practice

12. The threshold wavelength of zinc is 310 nm. Find the threshold frequency, in Hz, and the work function, in eV, of zinc.

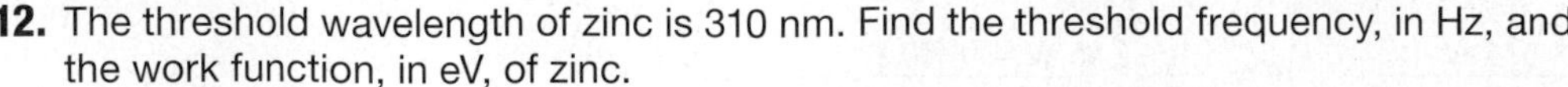

13. The work function for cesium is 1.95 eV. What is the maximum kinetic energy, in eV, of photoelectrons ejected when 425-nm violet light falls on the cesium?

14. When a metal is illuminated with 193-nm ultraviolet radiation, electrons with kinetic energies of 3.5 eV are emitted. What is the work function of the metal?

15. CHALLENGE A researcher illuminates a sample of metal and finds the longest wavelength to eject electrons is 273 nm. Use **Table 1** to identify the most likely metal.

The Compton Effect

The photoelectric effect demonstrates that a photon has kinetic energy even though it has no mass. In 1916 Einstein predicted that the photon should also have another particle property—momentum. He argued that the momentum of a photon should be equal to $\frac{E}{c}$. Because $E = hf$ and $\frac{f}{c} = \frac{1}{\lambda}$, the photon's momentum is given by the following equation.

PHOTON MOMENTUM

The momentum of photon is equal to Planck's constant divided by the photon's wavelength.

$$p = \frac{hf}{c} = \frac{h}{\lambda}$$

Experiments done in 1922 by American physicist Arthur Holly Compton tested Einstein's hypothesis. Compton's results further supported the particle model of light. Compton directed X-rays of a known wavelength at a graphite target, as shown on the left in **Figure 10,** and measured the wavelengths of the X-rays scattered by the target. He observed that some of the X-rays were scattered without change in wavelength, whereas others had a longer wavelength than that of the original radiation. These results are shown on the right panel in **Figure 10.** Note that the peak wavelength for the unscattered X-rays corresponds to the wavelength of the original incident X-rays, whereas the peak wavelength for the scattered X-rays is greater than that of the original incident X-rays.

Recall that the equation for the energy of a photon, $E = hf$, also can be written as $E = \frac{hc}{\lambda}$. This second equation shows that the energy of a photon is inversely proportional to its wavelength. The increase in wavelength that Compton observed meant that the X-ray photons had lost both energy and momentum. The shift in the energy of scattered photons is called the **Compton effect.** This shift in energy, indicated by a shift in wavelength, is very small—only about 10^{-3} nm. It is a measurable effect only when electromagnetic waves have wavelengths less than 10^{-2} nm.

Figure 10 The shift in the peak wavelength of scattered X-rays indicates that the scattered photons lost energy and momentum.

Explain why an increase in wavelength indicates a decrease in energy.

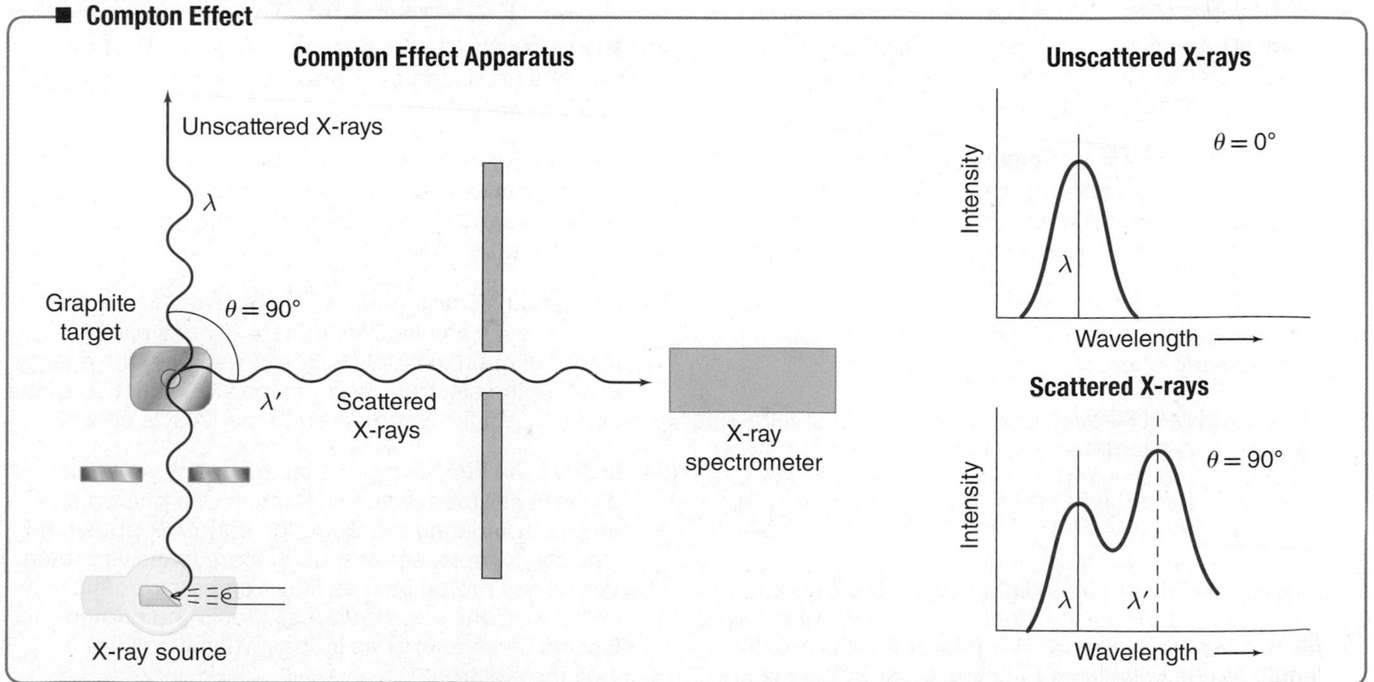

Figure 11 The laws of conservation of energy and momentum apply to collisions between billiard balls and between photons and particles.

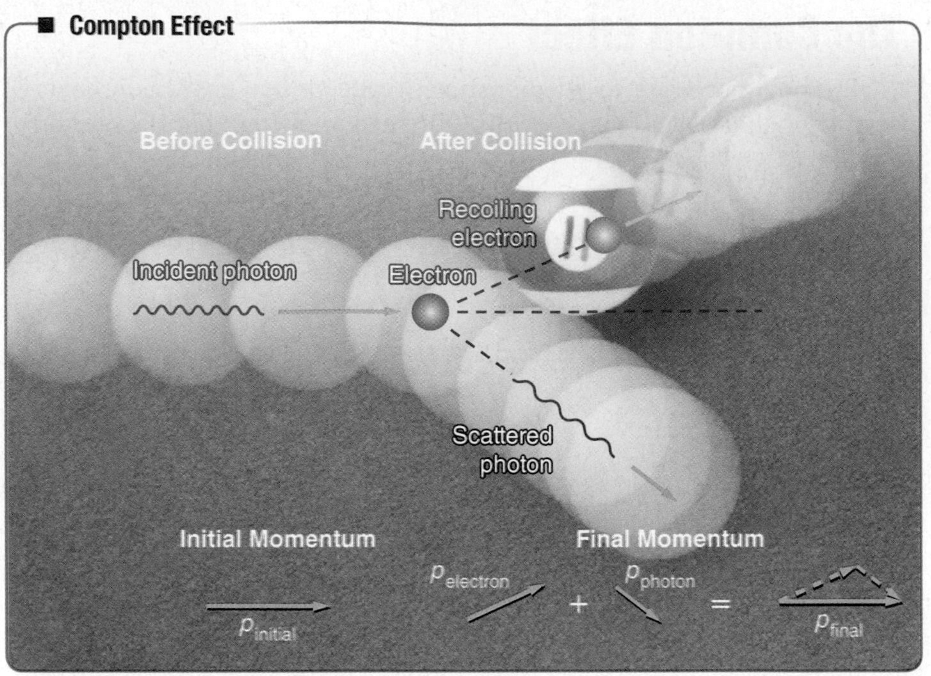

Photons and conservation of energy and momentum In later experiments, Compton observed that electrons were ejected from the graphite block during the experiment. He suggested that the X-ray photons collided with electrons in the graphite target and transferred energy and momentum to them. Compton thought these photon-electron collisions were similar to the elastic collisions experienced by billiard balls, as shown in **Figure 11.** He tested this idea by measuring the energies of the electrons ejected from the graphite block. Compton found that the energy and momentum gained by the ejected electrons equaled the energy and momentum lost by the photons. Thus, photons obey the laws of conservation of energy and momentum when they collide with other particles.

SECTION 1 REVIEW

Check your understanding.

16. **MAIN IDEA** Why is high-intensity, low-frequency light unable to eject electrons from a metal, whereas low-intensity, high-frequency light can?

17. **Frequency and Energy of Hot-Object Radiation** As the temperature of an object is increased, how does the frequency of peak intensity change? How does the total amount of radiated energy from the object change?

18. **Photoelectric and Compton Effects** Distinguish the photoelectric effect from the Compton effect.

19. **Photoelectric and Compton Effects** An experimenter sends an X-ray into a target. An electron, but no other radiation, emerges from the target. Explain whether this event is a result of the photoelectric effect or the Compton effect.

20. **Photoelectric Effect** Green light ($\lambda = 532$ nm) strikes an unknown metal, causing electrons to be ejected. The ejected electrons can be stopped by a potential of 1.44 V. What is the work function, in eV, of the metal?

21. **Energy of a Photon** What is the energy, in eV, of the photons produced by a laser pointer having a 650-nm wavelength?

22. **Compton Effect** An X-ray strikes a bone, collides with an electron, and scatters. How does the wavelength of the scattered X-ray compare to the wavelength of the incident X-ray?

23. **Photoelectric Effect** An X-ray is absorbed in a bone and releases an electron. If the X-ray has a wavelength of approximately 0.02 nm, estimate the energy, in eV, of the electron. Assume the work function of the bone is negligible compared to the X-ray's energy.

24. **Critical Thinking** Imagine that the collision of two billiard balls models the interaction of a photon and an electron during the Compton effect. Suppose the electron is replaced by a much more massive proton. Would this proton gain as much energy from the collision as the electron does? Would the photon lose as much energy as it does when it collides with the electron?

Richard Megna/Fundamental Photographs, NYC

EXAMPLE PROBLEM 3

Find help with **isolating a variable.** Math Handbook

DE BROGLIE WAVELENGTH An electron is accelerated by a potential difference of 75 V. What is its de Broglie wavelength?

1 ANALYZE AND SKETCH THE PROBLEM

Include the positive and the negative plates in your drawing.

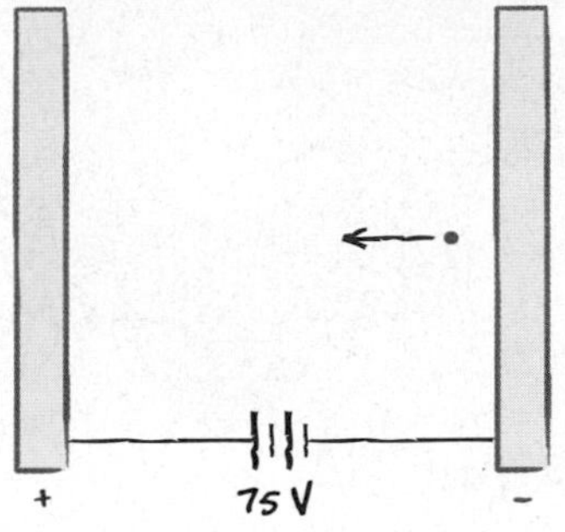

KNOWN		UNKNOWN
$\Delta V = 75\text{ V}$	$m_e = 9.11\times10^{-31}\text{ kg}$	$\lambda = ?$
$e = -1.602\times10^{-19}\text{ C}$	$h = 6.63\times10^{-34}\text{ J·s}$	

2 SOLVE FOR THE UNKNOWN

Write relationships for the kinetic energy of the electron based on potential difference and motion, and use them to calculate the electron's velocity.

$KE = -e\Delta V$, and $KE = \frac{1}{2}mv^2$

$\frac{1}{2}mv^2 = -e\Delta V$ ◀ Equate the expressions for KE.

Solve for v.

$$v = \sqrt{\frac{-2e\Delta V}{m}}$$

$$= \sqrt{\frac{-2(-1.602\times10^{-19}\text{ C})(75\text{ V})}{9.11\times10^{-31}\text{ kg}}}$$ ◀ Substitute $e = -1.602\times10^{-19}$ C, $\Delta V = 75$ V, $m_e = 9.11\times10^{-31}$ kg

$$= 5.1\times10^{6}\text{ m/s}$$

Use the relationship between mass, velocity, and momentum.

$$p = mv$$

$$= (9.11\times10^{-31}\text{ kg})(5.1\times10^{6}\text{ m/s})$$ ◀ Substitute $m_e = 9.11\times10^{-31}$ kg, $v = 5.1\times10^{6}$ m/s

$$= 4.6\times10^{-24}\text{ kg·m/s}$$

Use the relationship between momentum and the de Broglie wavelength.

$$\lambda = \frac{h}{p}$$

$$= \frac{6.63\times10^{-34}\text{ J·s}}{4.6\times10^{-24}\text{ kg·m/s}}$$ ◀ Substitute $h = 6.63\times10^{-34}$ J·s, $p = 4.6\times10^{-24}$ kg·m/s

$$= 1.4\times10^{-10}\text{ m, which equals 0.14 nm}$$

3 EVALUATE THE ANSWER

- **Are the units correct?** Dimensional analysis on the units verifies m/s for v and nm for λ.
- **Does the sign make sense?** Positive values are expected for both v and λ.
- **Are the magnitudes realistic?** The wavelength is close to 0.1 nm, which is in the X-ray region of the electromagnetic spectrum.

PRACTICE PROBLEMS

Do additional problems. Online Practice

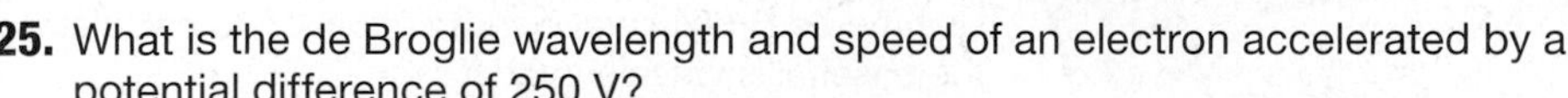

25. What is the de Broglie wavelength and speed of an electron accelerated by a potential difference of 250 V?

26. A 7.0-kg bowling ball rolls with a velocity of 8.5 m/s.

a. What is the de Broglie wavelength of the bowling ball?

b. Why does the bowling ball exhibit no observable wave behavior?

27. What potential difference is needed to accelerate an electron so it has a 0.125-nm wavelength?

28. CHALLENGE The electron in Example Problem 3 has a de Broglie wavelength of 0.14 nm. What is the kinetic energy, in eV, of a proton with the same wavelength?

EXAMPLE PROBLEM 3

Find help with **isolating a variable.** Math Handbook

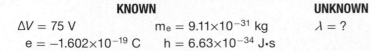

DE BROGLIE WAVELENGTH An electron is accelerated by a potential difference of 75 V. What is its de Broglie wavelength?

1 ANALYZE AND SKETCH THE PROBLEM

Include the positive and the negative plates in your drawing.

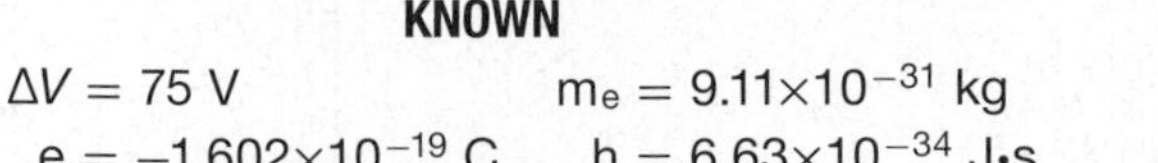

KNOWN		UNKNOWN
$\Delta V = 75\ \text{V}$	$m_e = 9.11\times10^{-31}\ \text{kg}$	$\lambda = ?$
$e = -1.602\times10^{-19}\ \text{C}$	$h = 6.63\times10^{-34}\ \text{J·s}$	

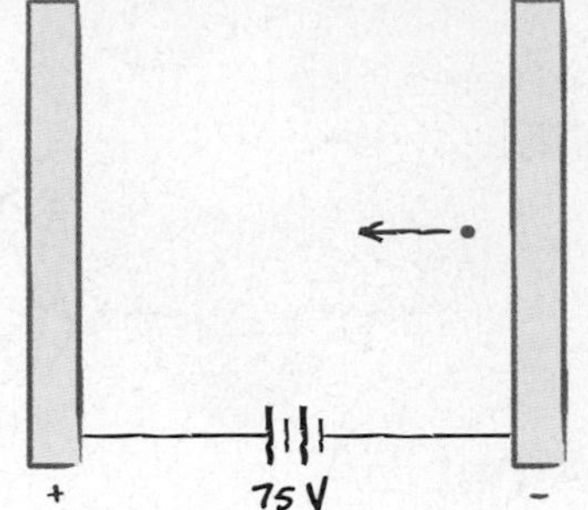

2 SOLVE FOR THE UNKNOWN

Write relationships for the kinetic energy of the electron based on potential difference and motion, and use them to calculate the electron's velocity.

$KE = -e\Delta V$, and $KE = \frac{1}{2}mv^2$

$\frac{1}{2}mv^2 = -e\Delta V$ ◀ Equate the expressions for *KE*.

Solve for *v*.

$$v = \sqrt{\frac{-2e\Delta V}{m}}$$

$$= \sqrt{\frac{-2(-1.602\times10^{-19}\ \text{C})(75\ \text{V})}{9.11\times10^{-31}\ \text{kg}}}$$ ◀ Substitute $e = -1.602\times10^{-19}$ C, $\Delta V = 75$ V, $m_e = 9.11\times10^{-31}$ kg

$$= 5.1\times10^{6}\ \text{m/s}$$

Use the relationship between mass, velocity, and momentum.

$$p = mv$$

$$= (9.11\times10^{-31}\ \text{kg})(5.1\times10^{6}\ \text{m/s})$$ ◀ Substitute $m_e = 9.11\times10^{-31}$ kg, $v = 5.1\times10^{6}$ m/s

$$= 4.6\times10^{-24}\ \text{kg·m/s}$$

Use the relationship between momentum and the de Broglie wavelength.

$$\lambda = \frac{h}{p}$$

$$= \frac{6.63\times10^{-34}\ \text{J·s}}{4.6\times10^{-24}\ \text{kg·m/s}}$$ ◀ Substitute $h = 6.63\times10^{-34}$ J·s, $p = 4.6\times10^{-24}$ kg·m/s

$$= 1.4\times10^{-10}\ \text{m, which equals } 0.14\ \text{nm}$$

3 EVALUATE THE ANSWER

- **Are the units correct?** Dimensional analysis on the units verifies m/s for *v* and nm for λ.
- **Does the sign make sense?** Positive values are expected for both *v* and λ.
- **Are the magnitudes realistic?** The wavelength is close to 0.1 nm, which is in the X-ray region of the electromagnetic spectrum.

PRACTICE PROBLEMS

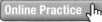

Do additional problems. 

25. What is the de Broglie wavelength and speed of an electron accelerated by a potential difference of 250 V?

26. A 7.0-kg bowling ball rolls with a velocity of 8.5 m/s.

a. What is the de Broglie wavelength of the bowling ball?

b. Why does the bowling ball exhibit no observable wave behavior?

27. What potential difference is needed to accelerate an electron so it has a 0.125-nm wavelength?

28. CHALLENGE The electron in Example Problem 3 has a de Broglie wavelength of 0.14 nm. What is the kinetic energy, in eV, of a proton with the same wavelength?

Location and Momentum

De Broglie's theory of particles' wavelike properties led to a fundamental change in our understanding of atomic-sized objects. In everyday experience, it is logical to think that you can measure any property to any desired level of precision. For example, imagine measuring a ball's position during free fall. You might first try using a meterstick and a stopwatch to make the measurements. If you desire more precise measurements, you might use a video camera and a calculator or a computer. You can imagine using more and more sophisticated equipment to achieve ever greater precision. In fact, classical physics sets no limit on a measurement's precision. de Broglie's revolutionary theory of matter waves forced physicists to redefine the limits of measurement.

Heisenberg uncertainty principle Consider measuring the location of an atomic or subatomic scale particle by shining light on it and then collecting the reflected light in a measuring device. Because of diffraction, the reflected light you use to detect the particle spreads out and makes it impossible to locate the particle exactly. If more precise measurements are needed, you can use shorter-wavelength radiation, which decreases diffraction and decreases the uncertainty in the particle's location.

As a result of the Compton effect, however, when short-wavelength, high-energy radiation strikes a particle, the particle's momentum is changed, as shown in **Figure 13.** Therefore, the act of measuring the location of a particle has the effect of changing the particle's momentum. The more precise the determination of a particle's location, the greater the uncertainty is in its momentum. Similarly, if the momentum of the particle is precisely known, the position of the particle becomes less certain.

☑ **READING CHECK Explain** why measuring a particle's position with high-energy radiation increases the uncertainty of its momentum.

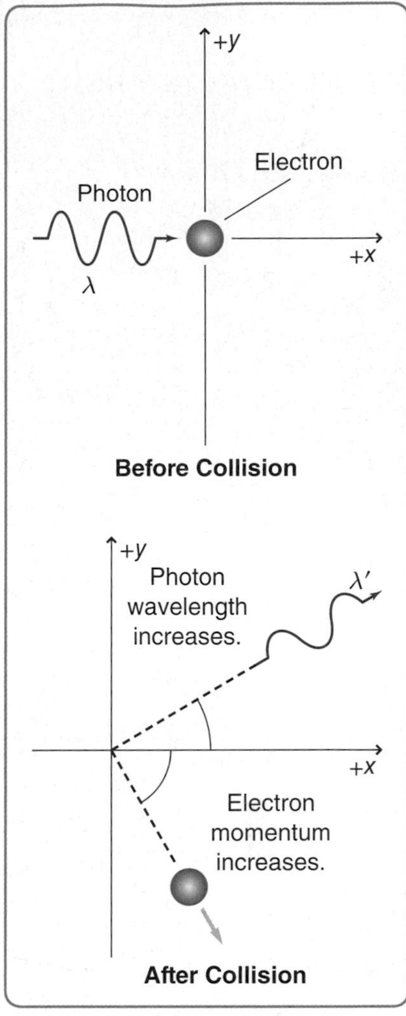

Figure 13 A collision between a photon and the electron is necessary to measure the electron's position. This collision scatters the electron and the photon, changing their momentums.

Compare the photon wavelength before and after the collision. What does this change in wavelength indicate about the photon's energy?

This situation is summarized by the **Heisenberg uncertainty principle,** which states there is a limit to how precisely a particle's position and momentum can simultaneously be measured. This principle, named for German physicist Werner Heisenberg, is the result of light and matter's dual wavelike and particlelike properties. Even though the effects of the limit are only noticeable when measuring atomic-scale particles, Heisenberg's work led to a fundamental change in our understanding of the world. While the classical theories of Newton and Maxwell are successful models of everyday objects, quantum theory and its models of the dual nature of light and matter are required to accurately describe objects on the atomic scale.

SECTION 2 REVIEW

Section Self-Check Check your understanding.

29. **MAIN IDEA** De Broglie's theory of matter waves provides a way to calculate the wavelength of any moving particle. Explain why the wave nature of everyday objects is not obvious.
30. **de Broglie Wavelength** An alpha particle (helium nucleus) has a mass of 6.6×10^{-27} kg and is moving at a speed of 120 m/s. What is this alpha particle's de Broglie wavelength?
31. **de Broglie Wavelength** What is the de Broglie wavelength of an electron accelerated through a potential difference of 125 V?
32. **Critical Thinking** When light or a beam of atoms passes through a double slit, an interference pattern forms even when photons or atoms pass through the slits one at a time. How does the Heisenberg uncertainty principle explain this?

Pushing the limits!

QUANTUM TOUCHSCREENS

If you slam into a brick wall, you know you won't pass right through to the other side. But objects the size of atoms and smaller can tunnel through what might seem like an insurmountable barrier. This bizarre property of matter is called quantum tunneling. It has real applications, both low- and high-tech.

Particlelike and wavelike behavior When an electrician wants to connect two strands of aluminum wire, all she needs to do is twist the wires together. This is odd, as aluminum develops a coating of aluminum oxide that acts as an insulator. If electrons always behaved like particles, they could never pass through this insulation. Because electrons also behave like waves, there is a small chance that the electron will be found outside the wire. If there's another wire nearby (as in the twisted wire example), the electron can tunnel from one wire to the other through what seems like an impenetrable barrier. Since there are lots and lots of electrons, this small chance for each individual electron to travel across the barrier translates into a measurable current across the barrier.

Under pressure A more complex example of quantum tunneling occurs in a new technology for pressure-sensitive touchscreens and keypads. Inside these devices, conductive nanoparticles covered in a nonconductive coating sit far apart. This is the off position. When the device is touched, the particles, around ten nanometers in diameter, move closer together. The greater the applied pressure, the nearer the particles approach one another.

Because the particles are spiky, as shown in **Figure 1,** the spikes from one particle are able to draw very close to the spikes from another particle. The movement of electrons along these spikes is an electric current. But it's not a simple on-off switch, for the greater the pressure applied, the stronger the electric current. This provides a pressure-sensitive control that can be applied in many different ways.

Figure 1 Pressure on the screen or button causes particles to draw closer together, increasing the rate of quantum tunneling along the spines.

New applications This is very different from a traditional touch screen, which is either on (touched) or off (not touched). This new technology adds a third dimension to the screen, almost like the slide on a trombone provides a continuous range of notes. It's the weirdness of quantum mechanics right at your fingertips.

GOING**FURTHER** >>>

Research other types of touchscreen technologies. What are some of the ways that this three-dimensional touchscreen might go beyond what traditional touchscreens can do?

(l) Sheryl Chapman/Photodisc/Getty Images, (r) courtesy Peratech Ltd.

STUDY GUIDE

BIGIDEA Waves can behave like particles, and particles can behave like waves.

VOCABULARY
- **emission spectrum** *(p. 729)*
- **quantized** *(p. 730)*
- **photoelectric effect** *(p. 731)*
- **threshold frequency** *(p. 732)*
- **photon** *(p. 732)*
- **work function** *(p. 737)*
- **Compton effect** *(p. 739)*

SECTION 1 A Particle Model of Waves

MAINIDEA Light can behave as massless particles called photons.

- The spectrum emitted by an object covers a broad range of wavelengths. The spectrum depends on the object's temperature. Planck explained an object's spectrum by supposing that a particle can absorb or emit only certain energies that are integer multiples of a constant, now called Planck's constant.

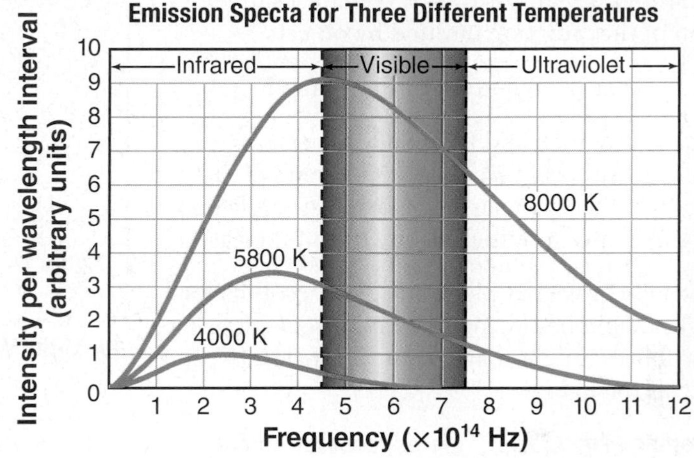

- The photoelectric effect is the emission of electrons by certain metals when they are exposed to electromagnetic radiation. Einstein explained the photoelectric effect by postulating that light exists in bundles of energy called photons. The relationship between a photon's energy, frequency, and wavelength is given below.

$$E = hf = \frac{hc}{\lambda} = \frac{1240\ \text{eV}\cdot\text{nm}}{\lambda}$$

- The Compton effect demonstrates that photons have momentum, as predicted by Einstein. Even though photons, which travel at the speed of light, have zero mass, they do have energy and momentum. A photon's momentum, frequency, and wavelength are related by the following equation:

$$p = \frac{hf}{c} = \frac{h}{\lambda}$$

VOCABULARY
- **de Broglie wavelength** *(p. 741)*
- **Heisenberg uncertainty principle** *(p. 743)*

SECTION 2 Matter Waves

MAINIDEA Moving particles have wavelike properties.

- The wave nature of material particles was suggested by de Broglie and verified experimentally by the diffraction of electrons through crystals. All moving particles have a wavelength, known as the de Broglie wavelength. The following equation is used to calculate a particle's de Broglie wavelength.

$$\lambda = \frac{h}{p} = \frac{h}{mv}$$

- The particle and the wave aspects are complementary parts of the complete nature of both matter and light. The Heisenberg uncertainty principle states that it is not possible to simultaneously measure the precise position and momentum of any particle of light or matter.

Games and Multilingual eGlossary

Vocabulary Practice

Chapter Self-Check

SECTION 1 A Particle Model of Waves

Mastering Concepts

33. Incandescent Light An incandescent lightbulb is controlled by a dimmer. What happens to the color of the light given off by the bulb as the potential difference applied to the bulb decreases?

34. Describe the concept of quantized energy.

35. What quantity is quantized in Max Planck's interpretation of the radiation emitted by objects?

36. BIGIDEA What is a quantum of light called?

37. Light above the threshold frequency shines on the cathode in a photocell. How does Einstein's photoelectric effect theory explain the fact that as the light intensity increases, photoelectron current increases?

38. Describe how Einstein's photon theory accounts for the fact that light below the threshold frequency of a metal produces no photoelectrons, regardless of the intensity of the light.

39. Photographic Film Older cameras recorded images on film. Because certain types of black-and-white film were not sensitive to red light, they could be developed in a darkroom illuminated by red light. How does the photon model of light explain this?

40. How does the Compton effect demonstrate that photons have momentum as well as energy?

Mastering Problems

41. According to Planck's theory, what is the frequency of vibration of an atom if it gives off 5.44×10^{-19} J while changing its value of n by 1?

42. What potential difference is needed to stop electrons with a maximum kinetic energy of 4.8×10^{-19} J?

43. What is the momentum of a photon of violet light that has a wavelength of 4.0×10^{2} nm?

44. The threshold frequency of a certain metal is 3.00×10^{14} Hz. What is the maximum kinetic energy of an ejected photoelectron if the metal is illuminated by light with a wavelength of 6.50×10^{2} nm?

45. The threshold frequency of sodium is 4.4×10^{14} Hz. How much work must be done to free an electron from the surface of sodium?

46. If light with a frequency of 1.00×10^{15} Hz falls on the sodium in the previous problem, what is the maximum kinetic energy of the photoelectrons?

47. The stopping potential of a certain metal is shown in **Figure 14.** What is the maximum kinetic energy of the photoelectrons in the following units?

a. electron volts

b. joules

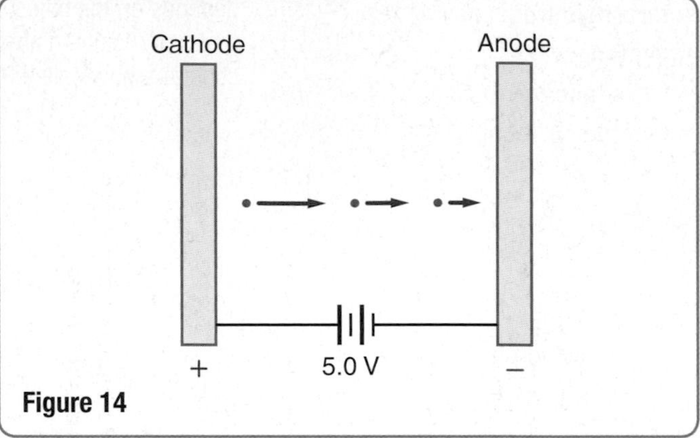

Figure 14

48. Light Meter A photographer's light meter uses a photocell to measure the light falling on the subject to be photographed. What should be the work function of the cathode if the photocell is to be sensitive to red light ($\lambda = 680$ nm) as well as to the other colors of light?

49. Solar Energy A home uses about 4×10^{11} J of energy each year. In many parts of the United States, there are about 3000 h of sunlight each year. On average, each square meter of Earth's surface receives about 1000 J of energy per second (1000 W) from the Sun.

a. How much energy from the Sun falls on one square meter each year?

b. If this solar energy can be converted to useful energy with an efficiency of 20 percent, how large an area of converters would produce the energy needed by the home?

SECTION 2 Matter Waves

Mastering Concepts

50. The momentum (p) of a particle of matter is given by $p = mv$. Can you calculate the momentum of a photon using the same equation? Explain.

51. Describe a method for measuring each of the following electron properties:

a. charge

b. mass

c. wavelength

52. Describe how each of the following photon properties could be measured.
   a. energy
   b. momentum
   c. wavelength

Mastering Problems

53. What is the de Broglie wavelength of an electron moving at 3.0×10^{6} m/s?

54. What velocity would an electron need to have a de Broglie wavelength of 3.0×10^{-10} m?

55. A cathode-ray tube accelerates an electron from rest across a potential difference of 5.0×10^{3} V.
   a. What is the velocity of the electron?
   b. What is the electron's wavelength?

56. The kinetic energy of a hydrogen atom's electron is 13.65 eV.
   a. Find the velocity of the electron.
   b. Calculate the electron's de Broglie wavelength.
   c. Given that a hydrogen atom's radius is 0.519 nm, calculate the circumference of a hydrogen atom and compare it with the de Broglie wavelength for the atom's electron.

57. **Ranking Task** Rank the following particles according to their de Broglie wavelengths, from greatest to least. Specifically indicate any ties.
   A. an electron with a speed of 300 m/s
   B. an electron with a speed of 500 m/s
   C. a proton with a speed of 3 m/s
   D. a proton with a speed of 500 m/s

58. An electron has a de Broglie wavelength of 0.18 nm.
   a. How large a potential difference did it experience if it started from rest?
   b. If a proton has a de Broglie wavelength of 0.18 nm, how large is the potential difference that it experienced if it started from rest?

Applying Concepts

59. Two iron bars are held in a fire. One glows dark red, while the other glows bright orange.
   a. Which bar is hotter?
   b. Which bar is radiating more energy?

60. Will high-frequency light eject a greater number of electrons from a photosensitive surface than low-frequency light, assuming that both frequencies are above the threshold frequency?

61. Potassium can emit photoelectrons when struck by blue light, whereas tungsten requires ultraviolet radiation in order to emit photoelectrons.
   a. Which metal has a higher threshold frequency?
   b. Which metal has a larger work function?

62. Compare the de Broglie wavelength of the baseball shown in **Figure 15** with the diameter of the baseball.

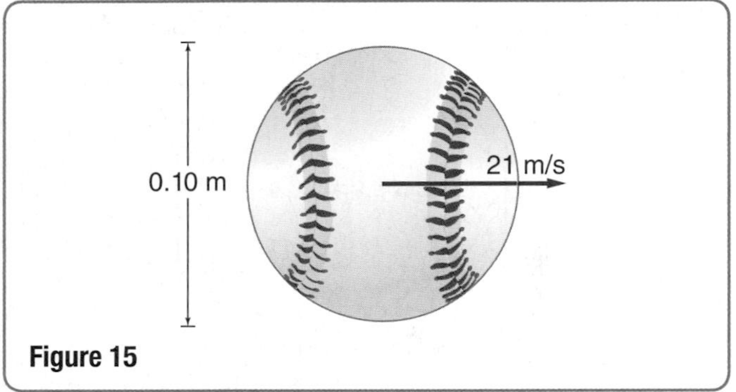

Figure 15

Mixed Review

63. What is the maximum kinetic energy of photoelectrons ejected from a metal that has a stopping potential of 3.8 V?

64. The threshold frequency of a certain metal is 8.0×10^{14} Hz. What is the metal's work function?

65. If light with a frequency of 1.6×10^{15} Hz falls on the metal in the previous problem, what is the maximum kinetic energy of the photoelectrons?

66. Find the de Broglie wavelength of a deuteron (nucleus of ^{2}H isotope) of mass 3.3×10^{-27} kg that moves with a speed of 2.5×10^{4} m/s.

67. The work function for a certain piece of iron is 4.7 eV.
   a. What is the threshold wavelength of iron?
   b. Iron is exposed to radiation of wavelength 150 nm. What is the maximum kinetic energy of the ejected electrons in eV?

68. Barium has a work function of 2.48 eV. What is the longest wavelength of light that will cause electrons to be emitted from barium?

Chapter Self-Check

69. An electron has a de Broglie wavelength of 400.0 nm, the shortest wavelength of visible light.

a. Find the velocity of the electron.

b. Calculate the energy of the electron in eV.

70. Incident radiation falls on tin, as shown in **Figure 16.** The threshold frequency of tin is 1.2×10^{15} Hz.

a. What is the threshold wavelength of tin?

b. What is the work function of tin?

c. The incident electromagnetic radiation has the wavelength indicated in **Figure 16.** What is the kinetic energy of the ejected electrons in eV?

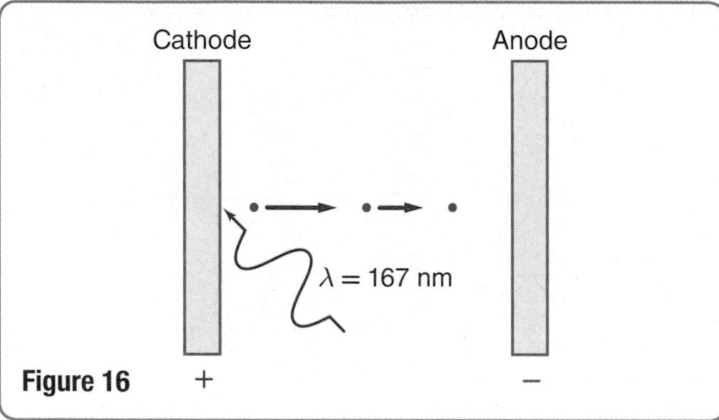

Figure 16

Thinking Critically

71. Apply Concepts Just barely visible light with an intensity of 1.5×10^{-11} W/m^2 enters a person's eye, as shown in **Figure 17.**

a. If this light shines into the person's eye and passes through the person's pupil, what is the power, in watts, that enters the person's eye?

b. Use the given wavelength of the incident light and information provided in **Figure 17** to calculate the number of photons per second entering the eye.

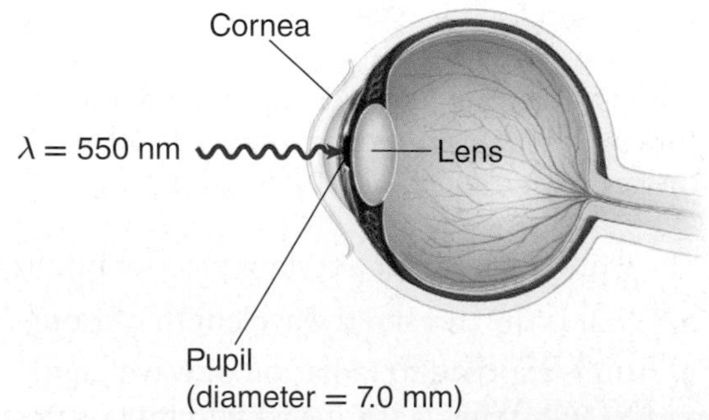

Figure 17

72. Problem Posing Complete this problem so that it must be solved using the work function: "Light of wavelength 443 nm is incident upon an unknown metal ...."

73. Make and Use Graphs A student completed a photoelectric-effect experiment and recorded the stopping potential as a function of wavelength, as shown in **Table 2.** The photocell had a sodium cathode. Plot the stopping potential versus frequency. Use your calculator to draw the best-fit straight line (regression line). From the slope and intercept of the line, find the work function, the threshold wavelength, and the value of h/e from this experiment. Compare the value of h/e to the accepted value.

Table 2 Stopping Potential v. Wavelength, Sodium

λ (nm)	ΔV_0 (V)
200	4.20
300	2.06
400	1.05
500	0.41
600	0.03

74. Reverse Problem Write a physics problem with real objects for which the following equation would be part of the solution:

$$\lambda = \frac{6.63\times10^{-34}\ \text{J·s}}{1.19\times10^{-27}\ \text{kg·m}}$$

Writing in Physics

75. Research the most massive particle for which interference effects have been seen. Describe the experiment and how the interference was created.

Cumulative Review

76. The spring in a pogo stick is compressed 15 cm when a child who weighs 400.0 N stands on it. What is the spring constant of the spring?

77. A marching band sounds flat on a cold day. Why?

78. A charge of 8.0×10^{-7} C experiences a force of 9.0 N when placed 0.02 m from a second charge. What is the magnitude of the second charge?

79. A homeowner buys a dozen identical 120-V light sets. Each light set has 24 bulbs connected in series, and the resistance of each bulb is 6.0 V. Calculate the total load in amperes if the homeowner operates all the sets from a single exterior outlet.

80. The force on a 1.2-m wire is 1.1×10^{-3} N. The wire is perpendicular to Earth's magnetic field. How much current is in the wire?

STANDARDIZED TEST PRACTICE

MULTIPLE CHOICE

1. A helium-neon laser emits photons with a wavelength of 632.8 nm. What is the energy, in joules, of each photon emitted by the laser?

A. 3.135×10^{-19} J **C.** 2.546×10^{8} J
B. 8.231×10^{-17} J **D.** 1.639×10^{34} J

2. What is the work function of a metal?

A. a measure of how much work an electron emitted from the metal can do

B. equal to the threshold frequency

C. the energy needed to free the metal atom's innermost electron

D. the energy needed to free the most weakly bound electron

3. How is the threshold frequency related to the photoelectric effect?

A. It is the minimum frequency of incident radiation needed to cause the ejection of atoms from the anode of a photocell.

B. It is the maximum frequency of incident radiation needed to cause the ejection of atoms from the anode of a photocell.

C. It is the frequency of incident radiation below which electrons will be ejected from an atom.

D. It is the minimum frequency of incident radiation needed to cause the ejection of electrons from an atom.

4. Radiation with an energy of 5.17 eV strikes a photocell, as shown below. If the work function of the photocell is 2.31 eV, what is the energy of the ejected photoelectron?

A. 0.00 eV **C.** 2.86 eV
B. 2.23 eV **D.** 7.48 eV

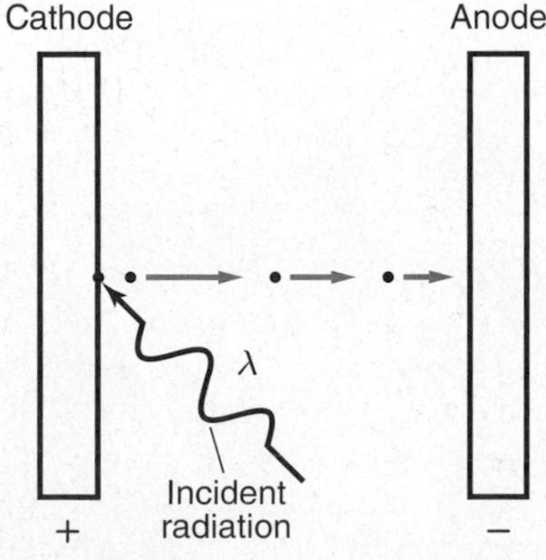

5. A photon has a frequency of 1.14×10^{15} Hz. What is the energy of the photon?

A. 5.82×10^{-49} J **C.** 8.77×10^{-19} J
B. 7.55×10^{-19} J **D.** 1.09×10^{-12} J

6. What is the de Broglie wavelength of an electron moving at 391 km/s? The mass of an electron is 9.11×10^{-31} kg.

A. 3.52×10^{-25} m **C.** 4.27×10^{-15} m
B. 4.79×10^{-15} m **D.** 1.86×10^{-9} m

7. As shown in the diagram below, an electron is accelerated by a potential difference of 95.0 V. What is the de Broglie wavelength of the electron?

A. 5.02×10^{-22} m **C.** 2.52×10^{-10} m
B. 1.26×10^{-10} m **D.** 5.10×10^{6} m

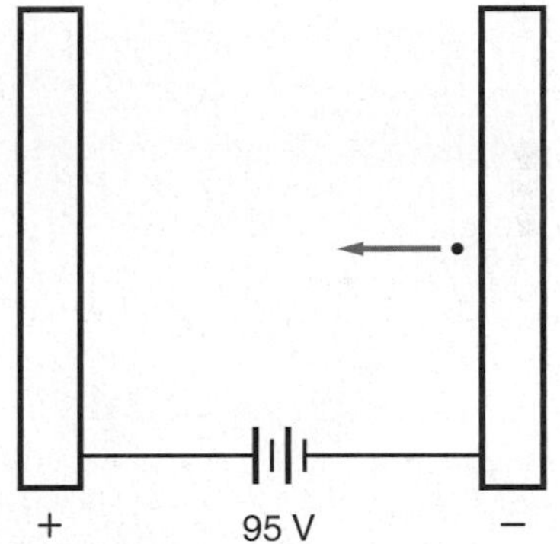

FREE RESPONSE

8. An object has a de Broglie wavelength of 2.3×10^{-34} m when its velocity is 45 m/s. What is the mass, in kilograms of the object?

9. An electron microscope is useful because the de Broglie wavelengths of electrons can be made smaller than the wavelength of visible light. What energy in electron volts has to be given to an electron for it to have a de Broglie wavelength of 20.0 nm?

NEED EXTRA HELP?

If You Missed Question	1	2	3	4	5	6	7	8	9
Review Section	1	1	1	1	1	2	2	2	2

CHAPTER 28

The Atom

BIGIDEA An atom consists of a nucleus of protons and neutrons surrounded by one or more electrons.

SECTIONS

1 Bohr's Model of the Atom

2 The Quantum Model of the Atom

LaunchLAB

IDENTIFYING SAMPLES

How can you use the unique properties of a substance to identify an unknown sample?

WATCH THIS!

LASERS

How are lasers used? A better question might be how *aren't* lasers used? You might be surprised to discover how many applications of lasers are around you every day.

PHYSICS T.V.

(l)Steve Cole/Getty Images, (r)Patrick Bennett/Nomad/CORBIS

Go online!
connectED.mcgraw-hill.com

SECTION 1

Bohr's Model of the Atom

PHYSICS 4 YOU **The red, green, blue, and orange lights often seen in shop and restaurant windows provide evidence of each element's unique atomic structure. Commonly called neon lights, such lights use coated glass tubes, high-voltage electricity, and mixtures of gases to achieve a variety of colors.**

MAIN IDEA

The Bohr model of the atom represents the atom as a massive, central nucleus with orbiting electrons that have quantized energies.

Essential Questions

- What did Rutherford's experiment reveal about the structure of the atom?
- What are emission spectra and absorption spectra?
- How do the radius and energy of electron orbitals depend on the principal quantum number?

Review Vocabulary

photoelectric effect the emission of electrons by certain metals that occurs when they are exposed to electromagnetic radiation

New Vocabulary

alpha particles
nucleus
absorption spectrum
energy level
ground state
excited state
principal quantum number

The Nuclear Model

By the end of the 1800s, most scientists agreed on the existence of atoms. J.J. Thomson's discovery of the electron provided convincing evidence that the atom must be made of even smaller, subatomic particles. Every atom tested contained negatively charged electrons, but these electrons possessed very little mass. Because atoms were known to be much more massive than the total mass of the electrons, scientists had many unanswered questions. Where was the mass located? What was the nature of this yet-to-be-discovered object?

Moreover, atoms were known to be electrically neutral, yet only negatively charged electrons had been identified within the atom. How were the negatively charged electrons arranged in the atom? What was the source of the atom's neutrality? Were positively charged particles also present in the atom? Knowing that their understanding of the atom was far from complete, scientists began searching for answers to many challenging questions.

Testing Thomson's model J.J. Thomson proposed that a massive, positively charged substance filled the atom. He pictured the negatively charged electrons as being distributed throughout this positively charged substance like raisins in a muffin. Ernest Rutherford, along with laboratory collaborators Hans Geiger and Ernest Marsden, however, performed a series of experiments that showed the atom had a very different structure. **Figure 1** shows the equipment Rutherford used to investigate the distribution of mass and charge in the atom.

Rutherford's Gold Foil Experiment

Source of α particles
Beam of α particles
Deflected particles
Circular fluorescent screen
Gold foil

Figure 1 Bombarding metal foil with alpha particles convinced Rutherford's team that most of the atom's mass was concentrated in a small volume in the atom's center.

Ryan McVay/Getty Images

Rutherford's investigation Rutherford made use of substances that emitted penetrating rays. Some of these emissions, later named **alpha particles,** were known to be massive, positively charged particles that move at high speeds. Alpha particles are represented by the symbol α. The α-particles in Rutherford's experiments could be detected by the small flashes of light produced when the particles collided with a zinc-sulfide-coated screen.

As shown in **Figure 1,** Rutherford directed a beam of α-particles at an extremely thin sheet of gold foil. Rutherford was aware of Thomson's model of the atom, and he expected only minor deflections of the α-particles as they passed through the thin gold foil. He hypothesized that the paths of the massive, high-speed α-particles would be only slightly altered as they passed through the evenly distributed positive charge making up each gold atom. The test results amazed him. While most of the α-particles passed through the gold foil either undeflected or only slightly deflected, a few of the particles were scattered through very large angles. Some even rebounded through angles larger than 90°. **Figure 2** shows a diagram of these results. Rutherford compared his amazement to that of firing a 15-inch cannon shell at tissue paper and then having the shell bounce back and hit him.

View an **animation of Rutherford's experiment.**

☑ **READING CHECK** **Explain** why Rutherford was so surprised by the results of the gold leaf experiment.

■ **Deflected Alpha Particles**

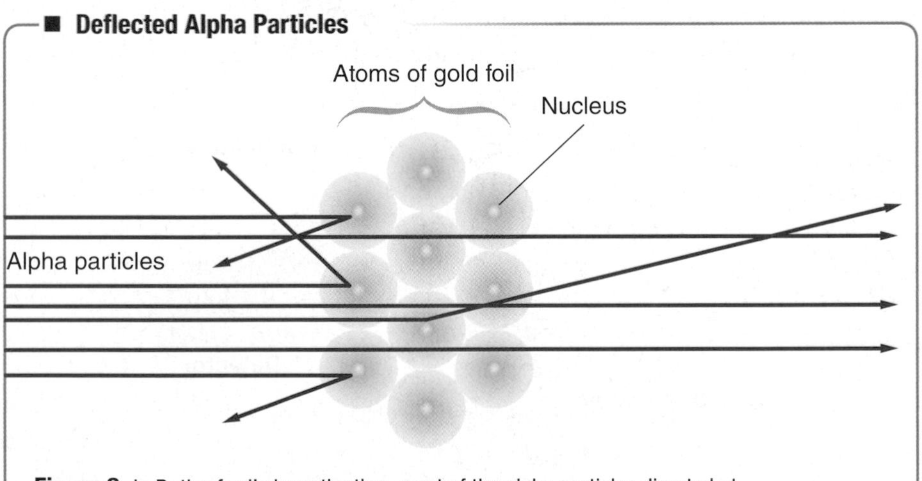

Figure 2 In Rutherford's investigation, most of the alpha particles directed at the thin sheet of gold foil passed through it without deflection. One in 20,000, however, was deflected at a large angle.

Rutherford's revised model of the atom Using Coulomb's law for the force between charged objects and Newton's laws of motion, Rutherford concluded that the results could be explained if all of an atom's positive charge were concentrated in a tiny, massive core, which is now called the **nucleus.** Therefore, Rutherford's model of the atom is called a nuclear model. Researchers have since determined that all the positive charge and more than 99.9 percent of an atom's mass are contained in its nucleus. The electrons do not contribute a significant amount of mass to the atom and are distributed through a large volume that stretches far away from the nucleus. The volume occupied by the electrons defines the overall size, or diameter, of the atom, which is about 50,000 times larger than the diameter of the nucleus. An atom is mostly empty space.

View a **BrainPOP video on the atomic model.**

Emission spectra How are the electrons arranged in the atom? One of the clues scientists used to answer this question came from the study of the light emitted by atoms. As shown in **Figure 3,** atoms of a gaseous sample can be made to emit light in a gas-discharge tube. You probably are familiar with the colorful neon signs used by some businesses. These signs work on the same principles as gas-discharge tubes. A gas-discharge tube consists of a low-pressure gas contained within a glass tube that has electrodes attached to each end. The gas glows when a large potential difference is applied across the tube. What interested scientists the most about this phenomenon was the fact that each different gas glowed with a unique color. The characteristic glows emitted by two gases are shown in **Figure 3.**

When the light emitted by the gas is passed through a diffraction grating or a prism, an emission spectrum is obtained that is a fingerprint of each element in the gas. An emission spectrum can be studied in greater detail using an instrument called a spectroscope. As shown in **Figure 3,** light in a diffraction spectroscope passes through a slit and is diffracted through a diffraction grating. The spectroscope forms an image of the slit at a different position for each wavelength.

As you may have read in an earlier chapter, the spectrum of an incandescent solid, such as a lightbulb filament, is a continuous band of colors from red through violet. The spectrum of a gas, however, is a series of distinct lines of different colors. The bright-line emission spectra for mercury and neon are shown in the bottom portion of **Figure 3.** Each colored line corresponds to a particular wavelength of light emitted by the atoms of that gas.

PhysicsLAB

EMISSION SPECTRA

How do elements' emission spectra differ?

Figure 3 An element's emission spectrum is a unique pattern that can be used to identify an unknown sample.

■ Emission Spectra

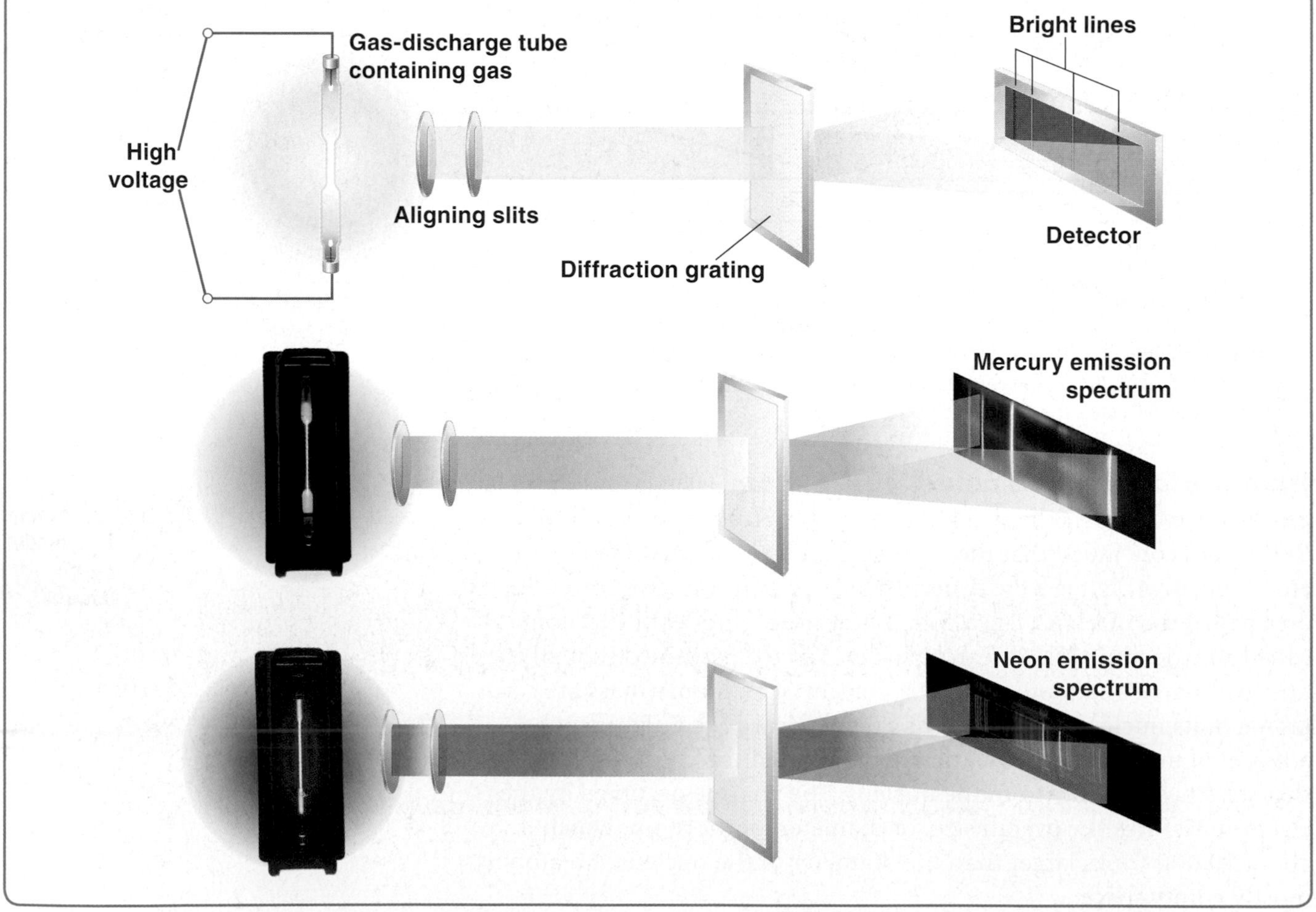

(tl bl)Charles D. Winters/Photo Researchers, Inc., (tr br)Ted Kinsman/Photo Researchers, Inc.

Identifying unknown samples An emission spectrum is a useful analytic tool that can be used to identify an unknown sample of gas. When an unknown gas is placed in a gas-discharge tube, it can be made to emit light. The emitted light includes combinations of wavelengths that are uniquely characteristic of the atoms of that gas. Thus, the unknown gas can be identified by comparing its wavelengths with the wavelengths measured in the spectra of known gas samples.

When the emission spectrum of a combination of elements is photographed, an analysis of the lines on the photograph can indicate the identities and the relative concentrations of the elements present. The elements that are present in the greatest concentrations produce emission lines of greater relative intensities. Through a comparison of the line intensities, the percentage composition of the gaseous material can be determined. This type of analysis is used daily to solve problems in chemistry, engineering, medicine, and environmental science.

Absorption spectra **Figure 4** illustrates another technique used to identify unknown samples. Passing white light through a gas sample and then through a prism results in the normally continuous spectrum of white light, but with dark lines in it corresponding to wavelengths of light absorbed by the gas. The set of wavelengths absorbed by a gas is the **absorption spectrum** of the gas. For a gas, the bright lines of the emission spectrum and the dark lines of the absorption spectrum occur at the same wavelengths, as shown in the bottom portion of **Figure 4.** Thus, cool, gaseous elements absorb the same wavelengths that they emit when excited. The composition of a gas can be determined from the wavelengths of the dark lines of the gas's absorption spectrum.

Figure 4 Passing white light through a sample of gas and a prism produces a nearly continuous spectrum. The dark lines missing from the spectrum are the wavelengths absorbed by the gas. Much like emission spectra, absorption spectra can be used to identify unknown samples.

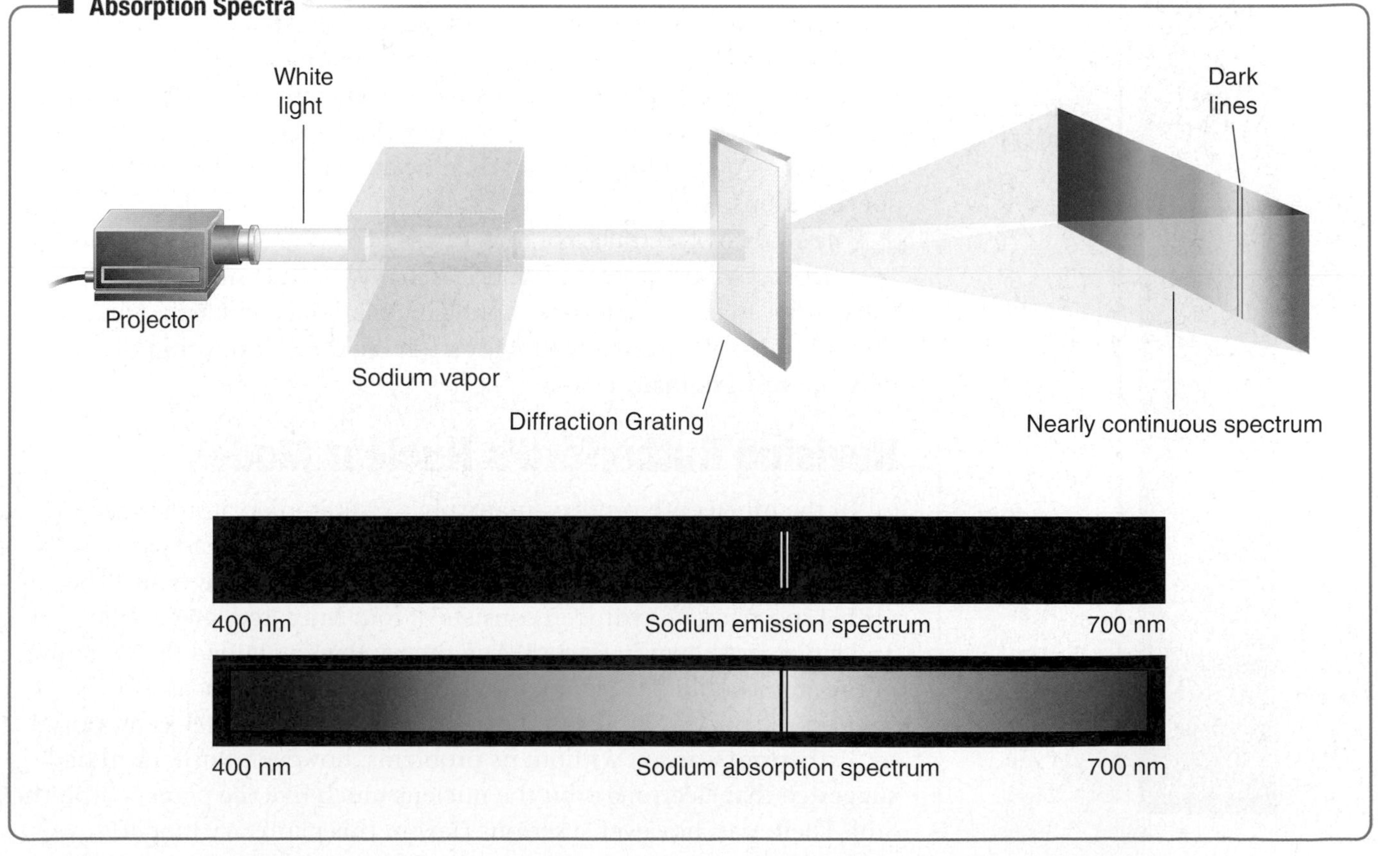

Figure 5 The Sun's absorption spectrum shows characteristic Fraunhofer lines for the elements in the Sun's atmosphere. There are many lines, some of which are faint and others that are very dark, depending on abundance of the elements in the Sun.

▶ CONNECTION TO ASTRONOMY In 1814 Josef von Fraunhofer observed the presence of several dark lines in the spectrum of sunlight. These dark lines, now called Fraunhofer lines, are seen in **Figure 5.** He reasoned that as sunlight passed through the gaseous atmosphere surrounding the Sun, the gases absorbed certain characteristic wavelengths. These absorbed wavelengths produced the dark lines in the observed spectrum. The composition of the Sun's atmosphere was determined by comparing the missing lines in the observed spectrum with the known emission spectra of various elements. The Sun was found to be mostly hydrogen and helium. The compositions of many other stars have been determined using this technique.

☑ **READING CHECK Describe** how scientists are able to determine the composition of stars.

Figure 6 Hydrogen's emission spectrum has four lines in the visible range.

Compare the difference between adjacent wavelengths of light as the energy increases.

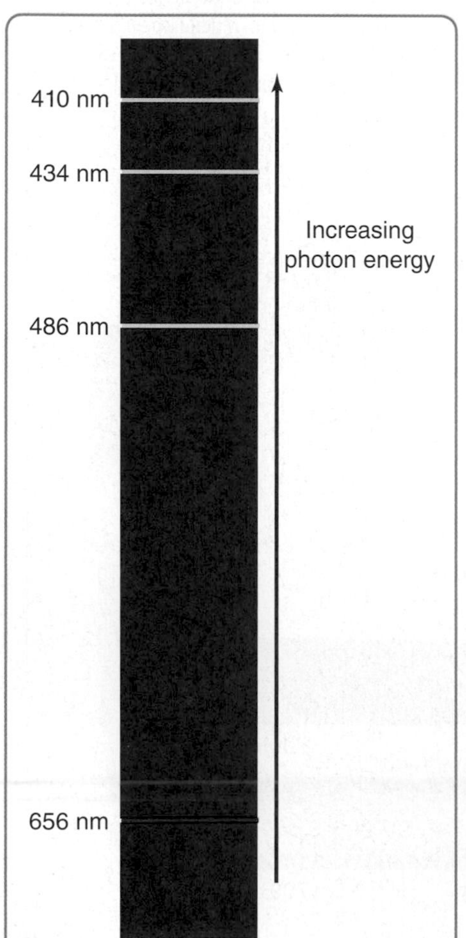

Spectroscopy As a result of the elements' characteristic spectra, scientists are able to analyze, identify, and quantify unknown materials by observing the spectra that they emit or absorb. The emission and absorption spectra of elements are important in industry as well as in scientific research. For example, steel mills reprocess large quantities of scrap iron of varying compositions. The exact composition of a sample of scrap iron can be determined in minutes by spectrographic analysis. The composition of the steel can then be adjusted to suit commercial specifications.

The study of spectra is a branch of science known as spectroscopy. Spectroscopists are employed in research and industrial settings. Spectroscopy is an effective tool for analyzing materials on Earth, and it is the only currently available tool for studying the composition of stars over the vast expanse of space.

Revising Rutherford's Nuclear Model

In the nineteenth century, many physicists studied atomic spectra to determine the structure of the atom. Hydrogen was studied extensively because it is the lightest element and has the simplest spectrum. The visible spectrum of hydrogen consists of four lines: red, green, blue, and violet, as shown in **Figure 6.** Any theory that explained the structure of the atom would have to account for these wavelengths, as well as support the nuclear model of the atom. The nuclear model as proposed by Rutherford was not without its problems, however. Rutherford had suggested that electrons orbit the nucleus much like the planets orbit the Sun. There was, however, a serious flaw in this planetary model.

Images were created by Dr. Donald Mickey, University of Hawaii, Institute for Astronomy, from National Solar Observatory spectra data.

Problems with the planetary model An electron in an orbit constantly accelerates toward the nucleus. From previous study, you know that accelerating electrons radiate energy by emitting electromagnetic waves. At the rate that an orbiting electron would lose energy, it should spiral into the nucleus within a fraction of a second. But we know that atoms are stable, so the planetary model does not agree with the laws of electromagnetism. In addition, the planetary model predicted that the accelerating electrons would radiate energy at all wavelengths. However, as you just read, the light emitted by atoms is radiated only at specific wavelengths.

Danish physicist Niels Bohr went to England in 1911 and joined Rutherford's group to work on determining the structure of the atom. He tried to unite the nuclear model of the atom with Planck's quantized energy levels and Einstein's theory of light. This was a courageous idea, because in 1911, neither of these revolutionary ideas was widely understood or accepted.

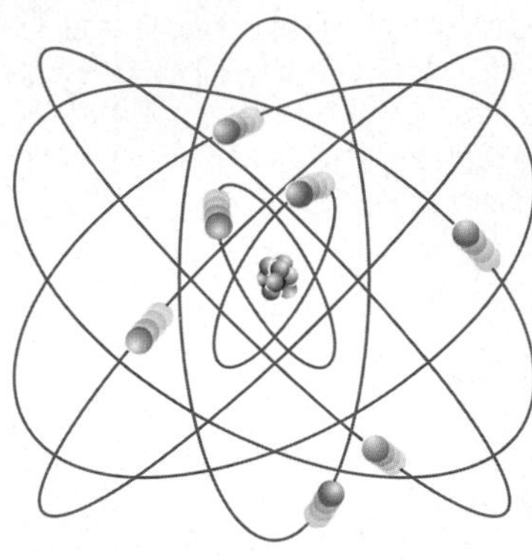

Figure 7 The planetary model of the atom stated electrons move in fixed orbits around the nucleus.

Quantized Energy

Bohr began with the planetary arrangement of electrons, as diagrammed in **Figure 7,** but then he proposed that the laws of electromagnetism do not apply inside the atom. He postulated that an electron in a stable orbit in an atom does not radiate energy, even though it is accelerating. Bohr referred to this stable condition as a stationary state. He went on to assume that each of the stationary states has its own specific amount of energy. In other words, Bohr considered the energy levels in an atom to be quantized.

As shown in **Figure 8,** the quantization of energies of electrons in atoms can be likened to a flight of stairs with decreasing-height steps. To go up the stairs you must move from one step to the next—it is impossible to stop at a midpoint between steps. Instead of steps, the electrons have quantized amounts of energy, each of which is called an **energy level.** Just as you cannot occupy a position between steps, an electron's energy cannot have a value between allowed energy levels. An atom with the smallest allowable amount of energy is said to be in the **ground state.** When such an atom absorbs energy, one or more of its electrons move to higher energy levels. Any energy level above the ground state is called an **excited state.**

View **animations of spectra and quantized energy.**

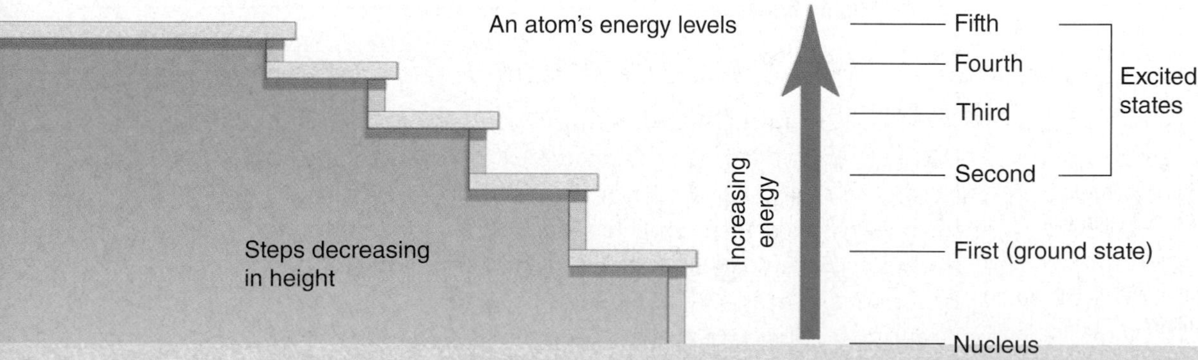

Figure 8 The allowed energy levels in an atom can be compared to a set of stairs. Note how the difference in energy between adjacent energy levels decreases as the energy level increases.

Energy of an atom What determines the amount of energy an atom has? An atom's energy equals the sum of the kinetic energy of the electrons and the potential energy from the attractive force between the electrons and the nucleus. The energy of an atom with electrons in a nearby orbit is less than that of an atom with electrons in a faraway orbit because work must be done to move the electrons away from the nucleus. Thus, atoms in excited, higher-energy states have electrons in larger, more distant, orbits. Because energy is quantized and energy is related to the size of the orbit, the size of the orbit also is quantized. This model of an atom, that of a central nucleus with orbiting electrons having specific quantized energy levels, is the Bohr model of the atom.

If Bohr was correct in hypothesizing that stable atoms do not radiate energy, then what is responsible for an atom's characteristic emission spectrum? To answer this question, Bohr suggested that electromagnetic energy is emitted when the atom changes from one state to another. Incorporating Einstein's photoelectric theory, Bohr knew that the energy of every photon is given by the equation $E_{photon} = hf$. He then postulated that when an atom absorbs a photon, the atom's energy increases by an amount equal to that of the photon. When this excited atom makes a transition to a lower energy level, it emits a photon.

When the atom makes the transition from its initial energy level (E_i) to its final energy level (E_f) the change in energy (ΔE_{atom}) is given by the following equation:

$$\Delta E_{atom} = E_f - E_i.$$

As shown in **Figure 9,** the amount by which the energy of the atom decreases equals the energy of the emitted photon.

$$E_{photon} = -\Delta E_{atom}$$

or

$$E_{photon} = E_i - E_f$$

The following equations summarize the relationships between the change in an atom's energy levels and the emitted photon's energy.

ENERGY OF AN EMITTED PHOTON

The energy of an emitted photon equals the product of Planck's constant and the emitted photon's frequency. The energy of an emitted photon also is equal to the decrease in the atom's energy.

$$E_{photon} = hf = -\Delta E_{atom}$$

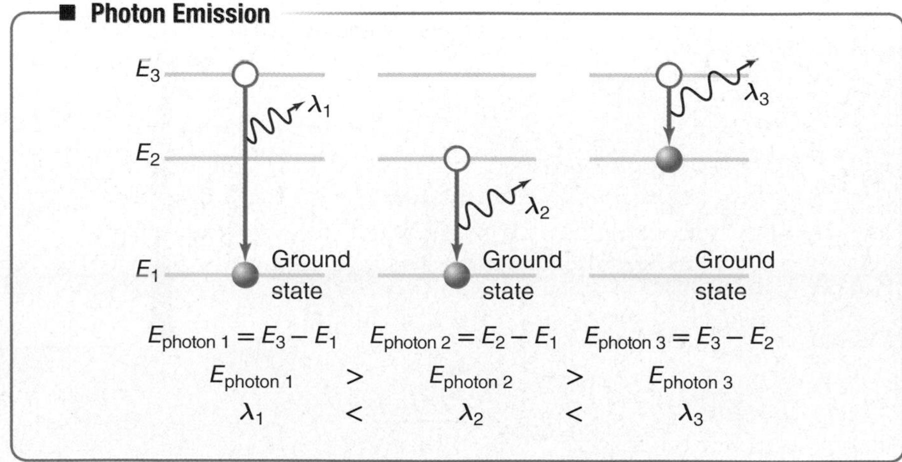

Figure 9 An emitted photon's energy is equal to the amount by which the energy of the atom decreases.

Identify the transition that produces the photon with the greatest energy.

View an **animation of energy levels.**

Predictions of the Bohr Model

A scientific theory must do more than present postulates; it must allow predictions to be made that can be checked against experimental data. A good theory also can be applied to many different problems, and it ultimately provides a simple, unified explanation of some part of the physical world. Bohr used his theory to calculate the wavelengths of light emitted by a hydrogen atom. The calculations were in excellent agreement with the values measured by other scientists. As a result of this agreement between Bohr's predictions and the measured values, Bohr's model was widely accepted.

Unfortunately, the model only worked for the element hydrogen; it could not predict the spectrum of helium, the next-simplest element. In addition, there was not a good explanation as to why the laws of electromagnetism should work everywhere but inside the atom.

MiniLAB

BRIGHT LINE SPECTRA

How do elements' emission spectra differ?

Development of Bohr's model Bohr developed his model by applying Newton's second law of motion ($F_{net} = ma$) to the electron. The net force is described by Coulomb's law for the interaction between an electron of charge −e that is a distance r from a proton of charge +e. That force is given by the following equation:

$$F = \frac{-\mathrm{K}\mathrm{e}^2}{r^2}$$

The acceleration of the electron in a circular orbit about a much more massive proton is given by $a = -\frac{v^2}{r}$, where the negative sign shows that the direction is inward. Thus, Bohr obtained the following relationship:

$$\frac{\mathrm{K}\mathrm{e}^2}{r^2} = \frac{\mathrm{m_e}v^2}{r}$$

Solving this equation for r gives the following relationship:

$$r = \frac{\mathrm{K}\mathrm{e}^2}{\mathrm{m_e}v^2}$$

In the equation, K, the Coulomb's law constant, is 9.0×10^9 N•m²/C².

Next, Bohr considered the angular momentum of the orbiting electron, which is equal to the product of an electron's momentum and the radius of its circular orbit. The angular momentum of the electron is thus mvr. Bohr postulated that angular momentum also is quantized; that is, an electron's angular momentum can have only certain values. Using n to represent an integer, Bohr proposed that the electron's angular momentum equaled an integer multiple of $\frac{\mathrm{h}}{2\pi}$, where h is Planck's constant.

$$\mathrm{m_e}vr = \frac{n\mathrm{h}}{2\pi}$$

Rearranging the above equation to solve for v yields the following:

$$v = \frac{n\mathrm{h}}{2\pi \mathrm{m_e} r}$$

Using this relationship for velocity and the equation for a hydrogen atom's radius given above, Bohr found that the orbital radii of the electrons in a hydrogen atom are given by the following equation.

View a **simulation of the Bohr model of the atom.**

ELECTRON ORBITAL RADIUS IN HYDROGEN

$$r_n = \frac{\mathrm{h}^2 n^2}{4\pi^2 \mathrm{K} \mathrm{m_e} \mathrm{e}^2}$$

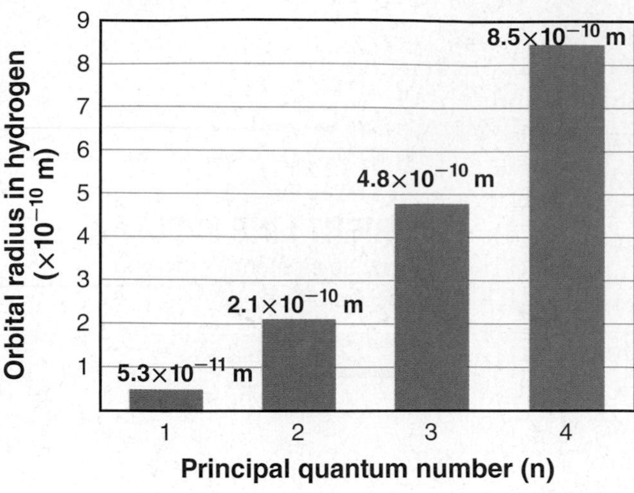

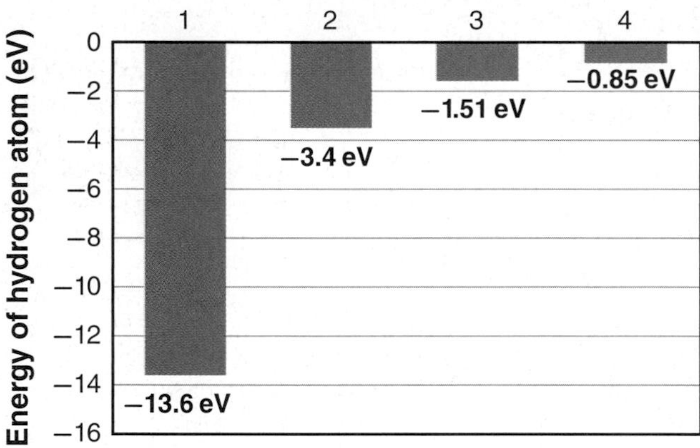

Figure 10 The radius of the hydrogen atom increases as n^2. The energy of the hydrogen atom increases from a minimum of −13.6 eV when $n = 1$ (the ground state) to values closer to 0 eV as n increases.

Get help with **finding the energy of a hydrogen atom.**

Calculating orbital radius You can calculate the radius of the innermost orbit of a hydrogen atom, also known as the Bohr radius, by substituting known values and $n = 1$ into the equation for electron orbital radius on the previous page.

$$r_1 = \frac{(6.626\times10^{-34}\ \text{J}\cdot\text{s})^2(1)^2}{4\pi^2(9.0\times10^9\ \text{N}\cdot\text{m}^2/\text{C}^2)(9.11\times10^{-31}\ \text{kg})(1.60\times10^{-19}\ \text{C})^2}$$

$$= 5.3\times10^{-11}\ \text{J}^2\cdot\text{s}^2/\text{N}\cdot\text{m}^2\cdot\text{kg}$$

$$= 5.3\times10^{-11}\ \text{m, or } 0.053\ \text{nm}$$

By performing a little more algebra, you can show that the total energy of the atom, which is the sum of the kinetic energy of the electron and the potential energy, and is given by $\frac{-\text{K}e^2}{2r}$, is represented by the following equation:

$$E_n = \frac{-2\pi^2\text{K}^2\text{m}_\text{e}\text{e}^4}{\text{h}^2} \times \frac{1}{n^2}$$

By substituting numerical values for the constants, you can calculate the total energy of the atom in joules, which yields the following equation:

$$E_n = -2.17\times10^{-18}\ \text{J} \times \frac{1}{n^2}$$

Converting the relationship to units of electron volts ($1\ \text{eV} = 1.6\times10^{-19}\ \text{J}$) yields the following equation.

ENERGY OF A HYDROGEN ATOM

The total energy of an atom with principal quantum number n is equal to the product of −13.6 eV and the inverse of n^2.

$$E_n = -13.6\ \text{eV} \times \frac{1}{n^2}$$

You previously read that the electron's orbital radius is quantized. The equation above shows that the energy of the atom is quantized as well. The integer (n) that appears in the equations for hydrogen's energy and electron orbital radius is called the **principal quantum number.** The principal quantum number determines the quantized values of r and E.

Figure 10 summarizes these relationships for the first four values of n. The graph at the top of the figure shows that the electron's orbital radius (r) increases as the square of n. The graph at the bottom of the figure shows that the energy (E) of the hydrogen atom depends on $\frac{1}{n^2}$. As the magnitude of the atom's energy decreases, the electron's orbital radius increases.

☑ **READING CHECK Describe** how the orbital radius and the energy depend on the principal quantum number.

Energy Level Transitions

Detail of Balmer Series Transitions

−0.38 eV E_6
−0.54 eV E_5
−0.85 eV E_4
−1.51 eV E_3
−3.40 eV E_2
−13.6 eV E_1

$n = 6$
$n = 5$
$n = 4$
Paschen series (infrared)
$n = 3$
Balmer series (visible)
Lyman series (ultraviolet)
$n = 2$
Ground state
$n = 1$
Nucleus

View an **animation of hydrogen's emission spectrum.**

Concepts In Motion

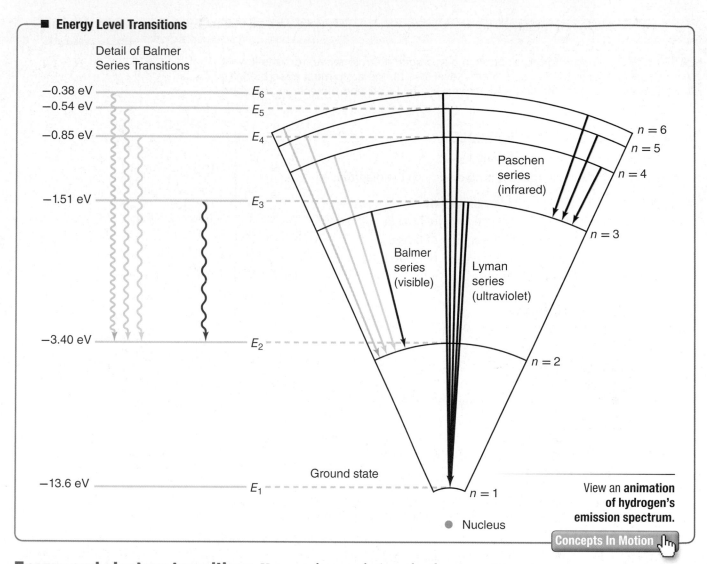

Figure 11 A hydrogen atom emits visible light when electrons transition from higher energy levels to the $n = 2$ energy level. Transitions to other energy levels result in emission of ultraviolet and infrared electromagnetic energy.

Energy and electron transitions You may be wondering why the energy of an atom in the Bohr model has a negative value. Recall from your study of energy that only energy differences affect the motion of objects. The zero energy level can be chosen at will. In this case, zero energy is defined as the energy of the atom when the electron is infinitely far from the nucleus and has no kinetic energy. This condition exists when the atom has been ionized; that is, when the electron has been removed from the atom. Because work has to be done to ionize an atom, the total energy of an atom that has an orbiting electron is less than zero. When an atom makes a transition from a lower to a higher energy level, the total energy becomes less negative, but the overall energy change is positive.

Some of hydrogen's energy levels and the possible energy level transitions that it can undergo are shown in **Figure 11.** Note that an excited hydrogen atom can emit electromagnetic energy in the infrared, visible, or ultraviolet range, depending on the transition that occurs. Ultraviolet light is emitted when the atom drops into its ground state from any excited state. The four visible lines in the hydrogen spectrum are produced when the atom drops from the $n = 3$ or higher energy state into the $n = 2$ energy state.

☑ **READING CHECK** **Describe** the electron transitions that result in the visible lines in the hydrogen emission spectrum.

PhysicsLAB

ELECTRON TRANSITIONS

How do elements' emission spectra differ?

EXAMPLE PROBLEM 1

Find help with **operations with significant figures.** Math Handbook

EXAMPLE PROBLEM

ENERGY LEVELS In order for an electron in a hydrogen atom to move from the lowest energy level ($n = 1$) to the second energy level ($n = 2$), the atom must absorb energy. Determine the energy of the first and second energy levels and the energy absorbed by the atom.

1 ANALYZE AND SKETCH THE PROBLEM

- Diagram the energy levels E_1 and E_2.
- Indicate the direction of increasing energy on the diagram.

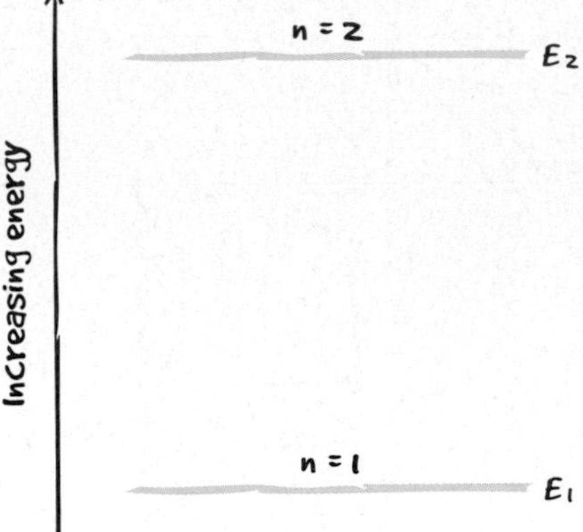

KNOWN	UNKNOWN
quantum number of innermost energy level, $n = 1$	energy of level $E_1 = ?$
quantum number of second energy level, $n = 2$	energy of level $E_2 = ?$
	energy difference, $\Delta E = ?$

2 SOLVE FOR THE UNKNOWN

Use the principal quantum number to calculate the energy of each level.

$$E_n = -13.6 \text{ eV} \times \frac{1}{n^2}$$

$$E_1 = -13.6 \text{ eV} \times \frac{1}{(1)^2} = -13.6 \text{ eV}$$ ◀ Substitute $n = 1$.

$$E_2 = -13.6 \text{ eV} \times \frac{1}{(2)^2} = -3.40 \text{ eV}$$ ◀ Substitute $n = 2$.

The energy absorbed by the atom, ΔE, is equal to the energy difference between the final energy level of the atom, E_f, and the initial energy level of the atom, E_i.

$$\Delta E = E_f - E_i$$

$= E_2 - E_1$ ◀ Substitute $E_f = E_2$, $E_i = E_1$.

$= -3.40 \text{ eV} - (-13.6 \text{ eV})$ ◀ Substitute $E_2 = -3.40$ eV, $E_1 = -13.6$ eV.

$= 10.2 \text{ eV}$ Energy is absorbed.

3 EVALUATE THE ANSWER

- **Are the units correct?** Orbital energy values should be measured in electron volts.
- **Is the sign correct?** The energy difference is positive when electrons move from lower energy levels to higher energy levels.
- **Is the magnitude realistic?** The energy needed to move an electron from the first energy level to the second energy level should be approximately 10 eV, which it is.

PRACTICE PROBLEMS

Do additional problems. Online Practice 

1. Calculate the energies of the second, third, and fourth energy levels in the hydrogen atom.
2. Calculate the energy difference between E_3 and E_2 in the hydrogen atom.
3. Calculate the energy difference between E_4 and E_2 in the hydrogen atom.
4. **CHALLENGE** The text shows the solution of the equation $r_n = \frac{h^2n^2}{4\pi^2Km_ee^2}$ for $n = 1$, the innermost orbital radius of the hydrogen atom. Note that with the exception of n^2, all factors in the equation are constants. The value of r_1 is 5.3×10^{-11} m, or 0.053 nm. Use this information to calculate the radii of the second, third, and fourth energy levels in the hydrogen atom.

The Quantum Model of the Atom

PHYSICS 4 YOU Whether bowling or battling an alien invasion, every time you play a video game you use technology made possible by creative applications of quantum mechanics. For example, the laser used to read the game disc was made possible by physicists' understanding of the quantum model of the atom.

MAIN IDEA

The quantum model of the atom predicts the probability of finding an electron within a specific region.

Essential Questions

- What are the characteristics of the quantum model of the atom?
- How does a laser work?
- What are the properties of laser light?

Review Vocabulary

Heisenberg uncertainty principle states that it is impossible to simultaneously measure a particle's exact position and momentum

New Vocabulary

quantum model
electron cloud
quantum mechanics
stimulated emission
laser

From Orbits to an Electron Cloud

Bohr's model of quantized electron orbits was a great step forward in atomic theory. However, the model could not correctly predict electron energies for atoms with more than one electron. Louis de Broglie's work with the wavelike properties of particles helped lead physicists from Bohr's orbit model to a new understanding of atomic structure.

de Broglie waves in the atom Recall that de Broglie proposed that particles have wave properties, similar to Einstein's theory that light has particle properties. The de Broglie wavelength of a particle with momentum mv is defined as $\lambda = \frac{h}{mv}$. The angular momentum of a particle can be defined as $mvr = \frac{hr}{\lambda}$. Thus, Bohr's required quantized angular-momentum condition, $mvr = \frac{nh}{2\pi}$, can be written in the following way:

$$\frac{hr}{\lambda} = \frac{nh}{2\pi} \text{ or } 2\pi r = n\lambda.$$

Figure 13 illustrates this relationship that the circumference of the Bohr orbit ($2\pi r$) must be equal to a whole number multiple (n) of de Broglie wavelengths (λ).

In 1926 Austrian physicist Erwin Schröedinger extended de Broglie's wave model to create a quantum theory of particles, such as electrons, based on waves. However, the electron itself is not a wave. The wave describes the probability of finding an electron in a particular location or having a particular momentum or energy.

■ **de Broglie waves**

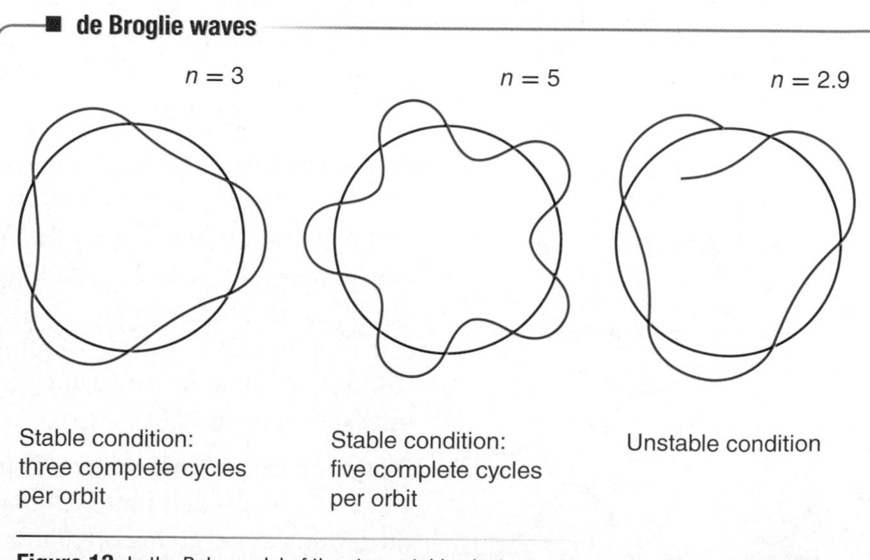

Figure 13 In the Bohr model of the atom, stable electron orbits around the nucleus must have a circumference equal to a whole-number multiple (n) of the de Broglie wavelengths.

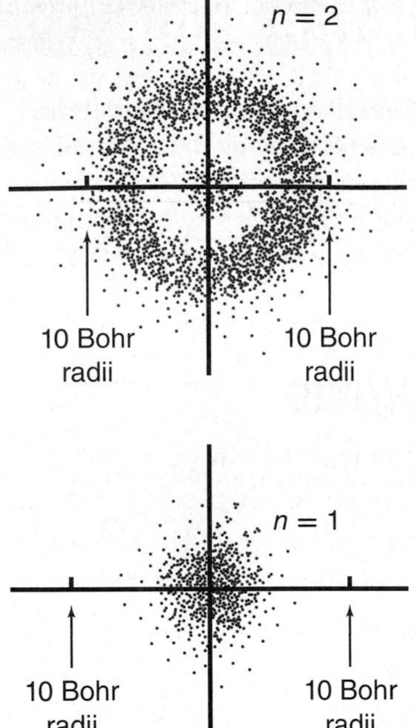

Figure 14 The density of dots in these plots corresponds to the probability of finding the electron in a hydrogen atom in the first and second energy levels.

Figure 15 Coherent light is in phase when each wave is at the same point in its cycle at the same time.

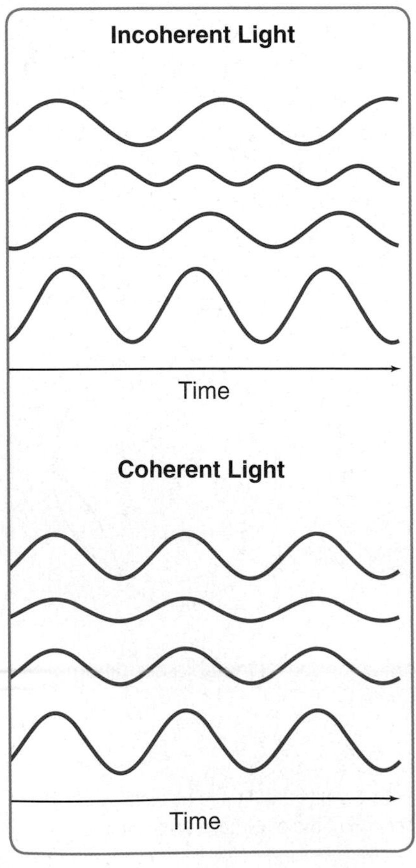

Uncertainty principle applied to atomic theory As you may have read in a previous chapter, Heisenberg proposed that it is impossible to measure exactly both the position and momentum of a particle at the same time. You can only know them approximately. Building on Heisenberg's ideas, the modern **quantum model** of the atom predicts only the probability that an electron is in a specific region. The quantum model predicts that the most probable distance between the electron and the nucleus for a hydrogen atom equals the radius predicted by the Bohr model.

The probability of the electron being at any specific radius can be calculated, and a three-dimensional plot will then show regions of equal probability. The region in which there is a high probability of finding the electron is called the **electron cloud.** **Figure 14** shows a slice through the electron cloud for the two lowest energy states of a hydrogen atom.

☑ **READING CHECK** **Explain** the meaning of the electron cloud.

Even though the quantum model of the atom is difficult to visualize, it was an early success in the development of **quantum mechanics,** the study of the properties of matter using its wave properties. Quantum mechanics is a useful tool for physicists, chemists, biologists, and engineers. It has been used to develop new chemicals, new computer chips, new solar power cells, and new medicines. Quantum mechanics also helped analyze the details of atoms' emission and absorption of light, leading to the development of a new type of light source.

Lasers

Most of the light you see is a jumble of uncoordinated waves having many different wavelengths. It is like music from a choir in which every member is singing a different song. As you know, light emitted by an incandescent source such as a filament lightbulb has a continuum of wavelengths, whereas light produced by an atomic gas consists of only a few distinct wavelengths. Light from each of these sources travels in all directions. The waves are not necessarily all at the same point in their cycle at the same time. Such out-of-step light waves are called incoherent light. Waves that are in phase are said to be coherent. Light waves that are coherent are referred to as coherent light. Both types of light waves are shown in **Figure 15.**

☑ **READING CHECK** **Contrast** coherent and incoherent light.

Light is emitted by atoms that move from a higher energy state to a lower energy state. So far, you have learned two ways in which atoms can be excited to the higher energy state: thermal excitation and electron collision. In a third process, a photon with exactly the right energy can collide with an atom and excite it. In a laser this process can produce intense, coherent light.

Spontaneous and stimulated emission What happens when an atom is in an excited state? After a very short time, it normally returns to the ground state, giving off a photon of the same energy that it absorbed, as shown in the left panel of **Figure 16.** This process is called spontaneous emission.

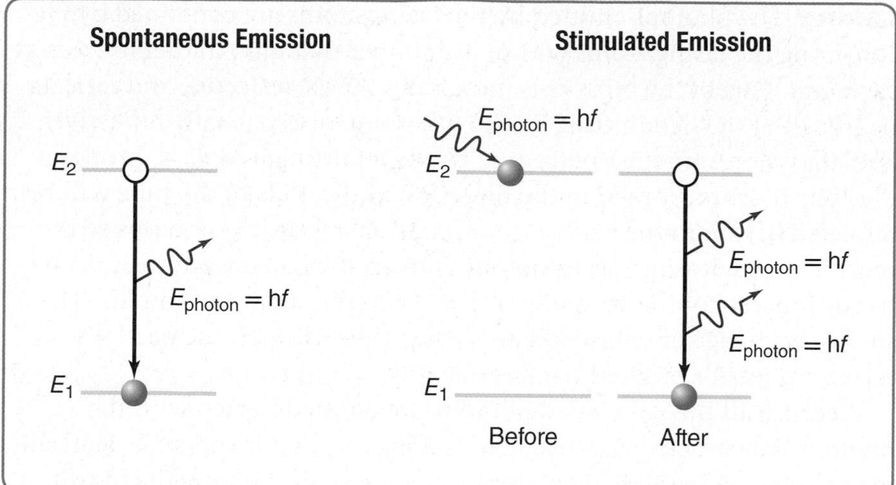

Figure 16 Spontaneous emission occurs when an electron in an atom drops from an excited state to the ground state, emitting a photon. During stimulated emission, an excited atom is struck by a photon with energy $E_2 - E_1$. The atom drops to the ground state and emits two photons. Both of these photons have the same energy: $E_{photon} = E_2 - E_1$.

Compare the wavelengths of the incident and emitted photons.

View an **animation of spontaneous and stimulated emission.**

In 1917 Einstein considered what happens to an atom already in an excited state that is struck by a photon with an energy equal to the energy difference between the excited state and the ground state. He showed that the atom, by a process known as **stimulated emission,** returns to the ground state and emits a photon with an energy equal to the energy difference between the two states. The photon that caused, or stimulated, the emission is not affected. The two photons leaving the atom not only will have the same frequency, but they will be in step, or coherent, as shown in the right panel of **Figure 16.** Either of the two photons can now strike other excited atoms, thereby producing additional photons that are in step with the original photons. This process can continue and produce an avalanche of photons, all of the same wavelength and all in phase.

☑ **READING CHECK** **Compare** the two photons emitted by stimulated emission.

For this process to occur, certain conditions must be met. First, there must be other atoms in the excited state. Second, the atoms must remain in the excited state long enough to be struck by a photon. Third, the photons must be contained so that they are able to strike other excited atoms. In 1960 a device called a **laser** was invented that fulfilled all the conditions needed to produce coherent light. The word laser is an acronym that stands for light amplification by stimulated emission of radiation. An atom that emits light when it is stimulated in a laser is said to lase.

Atom excitation The atoms in a laser can be excited, or pumped. An intense flash of light with a wavelength shorter than that of the laser can be used to pump the atoms. The shorter-wavelength, higher-energy photons produced by the flash collide with and excite the lasing atoms. When one of the excited atoms decays to a lower energy state by emitting a photon, the avalanche of photons begins. This results in the emission of a brief flash, or pulse, of laser light. Alternatively, the lasing atoms can be excited by collisions with other atoms. In the helium-neon lasers often seen in science classrooms, an electric discharge excites the helium atoms. These excited helium atoms collide with the neon atoms, pumping them to an excited state and causing them to lase. The laser light resulting from this process is a continuous beam rather than pulsed flashes.

Lasing The photons emitted by the lasing atoms are contained by confining the lasing atoms within a chamber that has parallel mirrors at each end. One of the mirrors is more than 99.9% reflective and reflects nearly all of the light hitting it. The other mirror is partially reflective and allows only about 1 percent of the light hitting it to pass through. Photons that are emitted in the direction of the ends of the tube will be reflected back into the gas by the mirrors. The reflected photons strike more atoms, releasing more photons with each pass between the mirrors. As the process continues, a high intensity beam of photons builds. The small percentage of photons that exit the tube through the partially reflecting mirror produce the laser beam.

Because all the stimulated photons are emitted in step with the photons that struck the atoms, light leaving the laser is coherent. The light may also be of one wavelength, or monochromatic, because the transition of electrons between only one pair of energy levels in one type of atom is involved. The parallel mirrors used in the laser result in the emitted laser light being highly directional. The beam does not diverge much as it travels, therefore a typical laser beam can have a small diameter, and the light is very intense. It can be focused to a fine point. It is also possible to switch stimulated emission on and off in very short times to produce brief, high-power light pulses. These characteristics of laser light make it particularly well-suited to many different applications.

MiniLAB

DIFFRACTION OF LASER LIGHT

What pattern is produced when you pass laser light through a diffraction grating?

Laser Applications

If you have used a CD or DVD player, then you have used a laser. These lasers, as well as the ones used in laser pointers, are made of semiconducting solids. The laser in a CD player is made of layers of gallium arsenide (GaAs) and gallium aluminum arsenide (GaAlAs). The lasing layer is only 200-nm thick, and each side of the crystal is only 1–2 mm long. The atoms in the semiconducting solid are pumped by an electric current, and the resulting photons are amplified as they bounce between the polished ends of the crystal. Shorter wavelength laser beams, such as those used in the DVD player shown in **Figure 17,** can focus on smaller spots. This allows DVDs to store more data than CDs.

Because laser beams are narrow and highly directional, surveyors use laser beams in applications such as checking the straightness of long tunnels and pipes. When astronauts visited the Moon, they left behind mirrors on the Moon's surface. Scientists on Earth used the mirrors to reflect a laser beam transmitted from Earth to accurately measure the distance from Earth to the Moon. Tracking the Moon's location from different parts of Earth allowed researchers to measure the movement of Earth's tectonic plates.

READING CHECK **Identify** the characteristics of laser light that make it well-suited to use in surveying and measuring large distances.

Figure 17 The laser used in this DVD player emits light with a wavelength of about 650 nm.

Richard Hutchings/Digital Light Source

Because lasers can emit light bursts as brief at 10^{-15} s, they can be used to investigate very fast changes in the structures of atoms or molecules. Other scientists are trying to produce energy from high temperature plasmas using very large lasers that produce petawatt (10^{15} watts) pulses. **Table 1** on the next page summarizes several applications of lasers.

Table 1 Laser Applications

Fiber-Optic Communications

Laser light is commonly used in fiber-optics communications. A fiber optic cable makes use of total internal reflection to transmit light over distances of many kilometers with little loss of signal energy. The laser, typically with a wavelength of 1300–1500 nm, is rapidly switched on and off, transmitting information as a series of pulses through the fiber. All over the world, optical fibers have replaced copper wires for the transmission of telephone calls, computer data, and television signals.

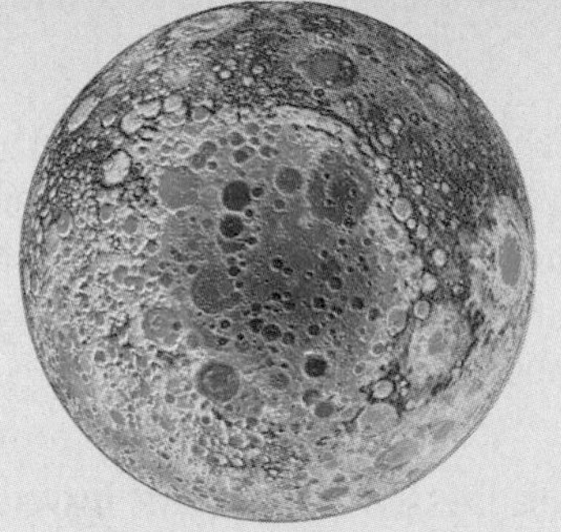

Lunar Research

The Lunar Reconnaissance Orbiter pointed a laser at the surface of the Moon and recorded the time required for the laser's light to bounce off the Moon's surface and return to the satellite. The data collected during this mission produced topographical maps accurate to within 10 cm. In addition to providing clues about the Moon's history, these maps can be used to select safe landing sites for future robotic and manned missions to the Moon.

Laser Printers

1. A series of lenses and mirrors directs the laser beam to scan over the drum. A typical printer laser has a wavelength of about 760 nm.

2. The laser's light changes the electrostatic charge on the photosensitive drum. The precisely guided laser beam forms an image of the printed page on the drum.

3. As the drum rolls past the toner supply, droplets of toner are attracted to the regions of the drum hit by the laser.

4. The drum transfers the charged toner to the oppositely-charged paper. The drum continues to roll past a rubber strip that cleans the drum, making it ready to receive another image from the laser.

5. The fuser heats the paper and toner so the toner melts and sticks permanently to the paper. Fusers can reach temperatures of up to 200° C.

SECTION 2 REVIEW

Section Self-Check Check your understanding.

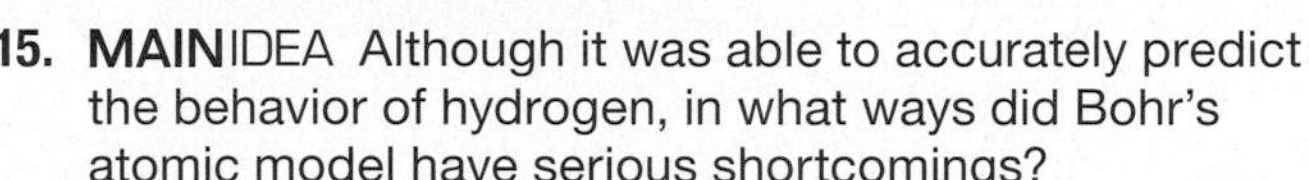

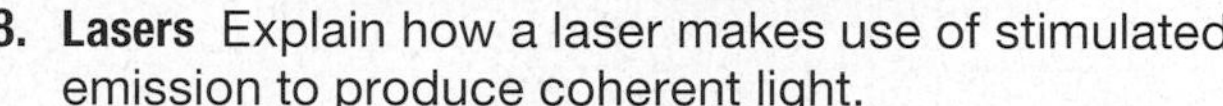

15. **MAIN**IDEA Although it was able to accurately predict the behavior of hydrogen, in what ways did Bohr's atomic model have serious shortcomings?

16. **Quantum Model** Explain why the Bohr model of the atom conflicts with the Heisenberg uncertainty principle, whereas the quantum model does not.

17. **Pumping Atoms** Explain whether green light could be used to pump a red laser. Why could red light not be used to pump a green laser?

18. **Lasers** Explain how a laser makes use of stimulated emission to produce coherent light.

19. **Laser Light** What are four characteristics of laser light that make it useful?

20. **Critical Thinking** Suppose an electron cloud were reduced to almost the size of the nucleus. Use the Heisenberg uncertainty principle to explain why this process would require adding a tremendous amount of energy.

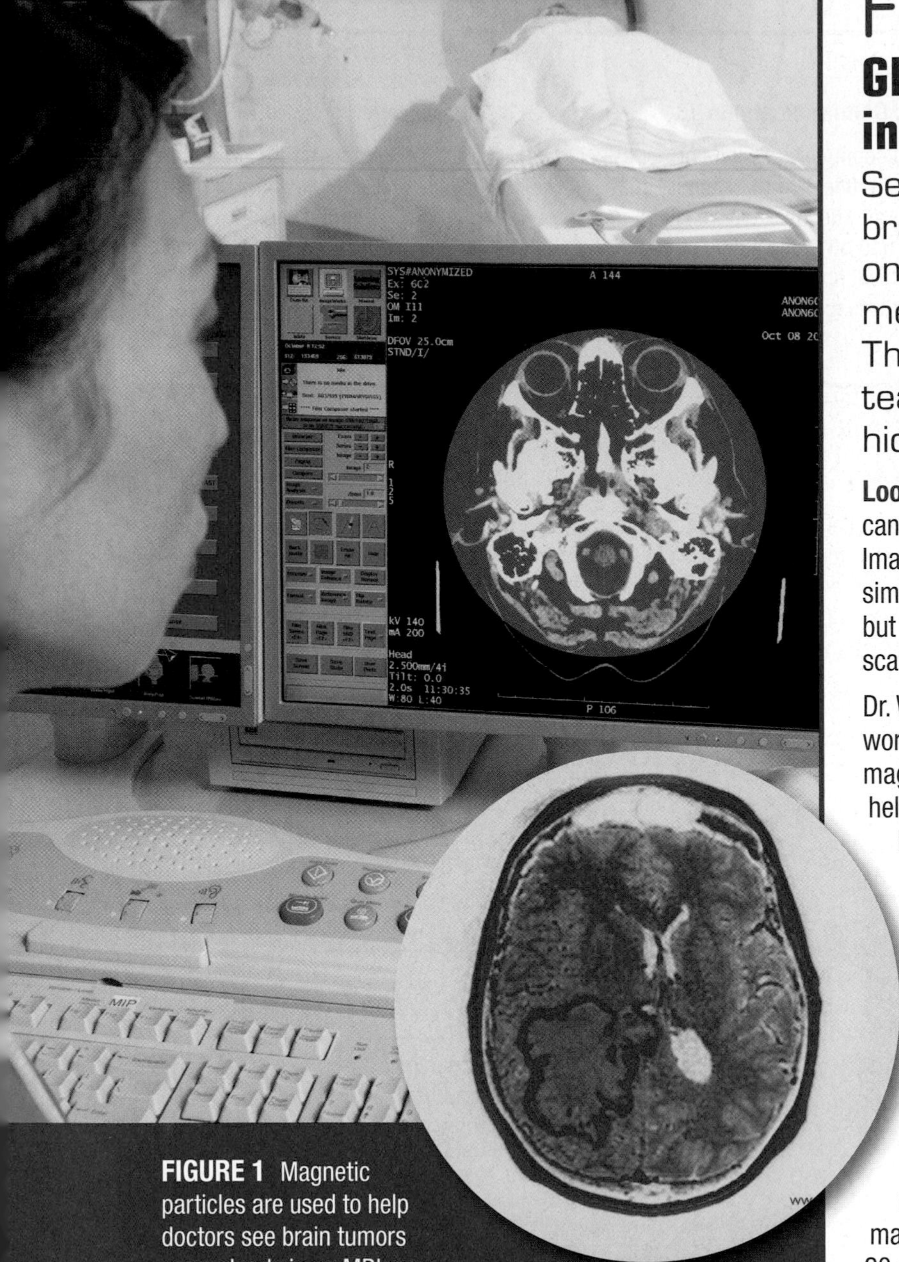

Future Looks Bright

Glowing Nanoparticles in the Brain

Seeing past our skulls and into our brains as they function has long been one of the great challenges of neural medicine. Dr. Jessica Winter of The Ohio State University is part of a team working on new windows into that hidden world.

Looking at tumors A glioblastoma is a deadly brain tumor that can be difficult to remove surgically. Magnetic Resonance Imaging (MRI) is one tool to visualize such tumors. An MRI scan similar to **Figure 1** can reveal a tumor and show its boundaries, but surgeons cannot operate while the MRI machine is scanning the brain.

Dr. Winter and her colleagues at The Ohio State University are working on a solution. They've created particles that are both magnetic and fluorescent. By being magnetic, the particles can help pinpoint the tumor in an MRI. By being fluorescent, as in **Figure 2,** the particles give the surgeon a visible guide, distinguishing the cancerous tissue from the healthy brain during surgery.

Both magnetic and fluorescent The particles are special because magnetism arises from the presence of metal atoms with aligned electron spins. Fluorescence comes from tightly bound outer electrons that absorb and reemit light.

Dr. Winter and colleagues used high temperature to bind two different particles together into a particle called a nanocomposite. The nanocomposite preserved both magnetic and fluorescent properties in particles approximately 20 nm in diameter.

The catch Unfortunately, these magnetic and fluorescent nanocomposites are toxic to the patient. Dr. Winter and her team are working on ways to alter the nanocomposites to make them safe for use in living subjects. If they are successful, surgeons could be guided by gently glowing brains.

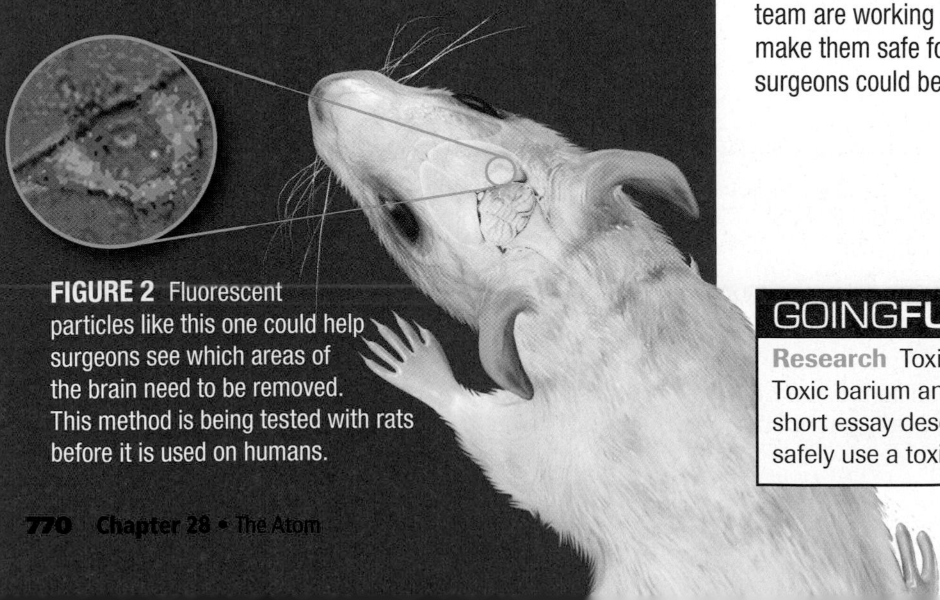

FIGURE 1 Magnetic particles are used to help doctors see brain tumors more clearly in an MRI scan.

FIGURE 2 Fluorescent particles like this one could help surgeons see which areas of the brain need to be removed. This method is being tested with rats before it is used on humans.

GOING FURTHER >>>

Research Toxic materials are often useful in medicine. Toxic barium and gadolinium are two examples. Write a short essay describing ways in which medical professionals safely use a toxic material within a patient's body.

CHAPTER 28 STUDY GUIDE

BIG IDEA An atom consists of a nucleus of protons and neutrons surrounded by one or more electrons.

VOCABULARY

- **alpha particles** *(p. 753)*
- **nucleus** *(p. 753)*
- **absorption spectrum** *(p. 755)*
- **energy level** *(p. 757)*
- **ground state** *(p. 757)*
- **excited state** *(p. 757)*
- **principal quantum number** *(p. 760)*

SECTION 1 Bohr's Model of the Atom

MAIN IDEA The Bohr model of the atom represents the atom as a massive, central nucleus with orbiting electrons that have quantized energies.

- Ernest Rutherford directed positively charged, high-speed alpha particles at thin metal foils. By studying the paths of the reflected particles, he showed that an atom is mostly empty space with a tiny, massive, positively charged nucleus at its center. Electrons move in a large volume around the nucleus but contribute less than 0.1 percent of an atom's mass.
- Atoms make transitions between allowable energy levels, absorbing or emitting energy in the form of photons (electromagnetic waves). Emitted photons form the atom's emission spectrum, and absorbed photons form the absorption spectrum. The photon's energy is equal to the decrease from the initial to the final state of the atom.

$$E_{photon} = -\Delta E$$

- According to the Bohr model, the radius of an electron's orbit can have only certain (quantized) values. The radius of the electron orbit in energy level n of a hydrogen atom is given by the following equation.

$$r_n = \frac{h^2n^2}{4\pi^2Km_ee^2}$$

Niels Bohr's model of the atom correctly showed that the energy of an atom can have only certain values; thus, it is quantized. He showed that the energy of a hydrogen atom in energy level n is equal to the product of −13.6 eV and the inverse of n^2.

$$E_n = -13.6\text{ eV} \times \frac{1}{n^2}$$

VOCABULARY

- **quantum model** *(p. 766)*
- **electron cloud** *(p. 766)*
- **quantum mechanics** *(p. 766)*
- **stimulated emission** *(p. 767)*
- **laser** *(p. 767)*

SECTION 2 The Quantum Model of the Atom

MAIN IDEA The quantum model of the atom predicts the probability of finding an electron within a specific region.

- In the quantum model of the atom, the atom's energy has only specific, quantized values. Only the probability of finding the electron in a specific region can be determined. In the hydrogen atom, the most probable distance of the electron from the nucleus is the same as the electron's orbital radius predicted by the Bohr model.
- Lasers produce coherent light by creating a chain reaction of stimulated emissions. Stimulated emission occurs when an incident photon causes an atom's electron in an excited state to emit a coherent photon as it falls into a lower energy state. The incident photon and the emitted photon can then cause other atoms to emit more coherent photons. The laser cavity is constructed to reflect these coherent photons back and forth through the lasing material, producing even more photons.
- Lasers produce light that is directional, powerful, monochromatic, and coherent. Each of these properties has useful applications.

Games and Multilingual eGlossary

Vocabulary Practice

CHAPTER 28 ASSESSMENT

Chapter Self-Check

SECTION 1 Bohr's Model of the Atom

Mastering Concepts

21. BIGIDEA Describe how Rutherford determined that the positive charge in an atom is concentrated in a tiny region, rather than spread throughout the atom.

22. How does the Bohr model explain why the absorption spectrum of hydrogen contains exactly the same wavelengths as its emission spectrum?

23. Review the planetary model of the atom. What are some of the problems with a planetary model of the atom?

24. Analyze and critique the Bohr model of the atom. What three assumptions did Bohr make in developing his model?

25. Gas-Discharge Tubes Describe how line spectra from gas-discharge tubes are produced.

26. How does the Bohr model account for the spectra emitted by atoms?

27. Explain why line spectra produced by hydrogen gas-discharge tubes are different from those produced by helium gas-discharge tubes.

Mastering Problems

28. A calcium atom drops from 4.68 eV above the ground state to 2.93 eV above the ground state. What is the wavelength of the photon emitted?

29. A calcium atom in an excited state (E_2) has an energy level 1.88 eV above the ground state. A photon of energy 2.03 eV strikes the calcium atom and is absorbed by it. To what energy level is the calcium atom raised? Refer to **Figure 18.**

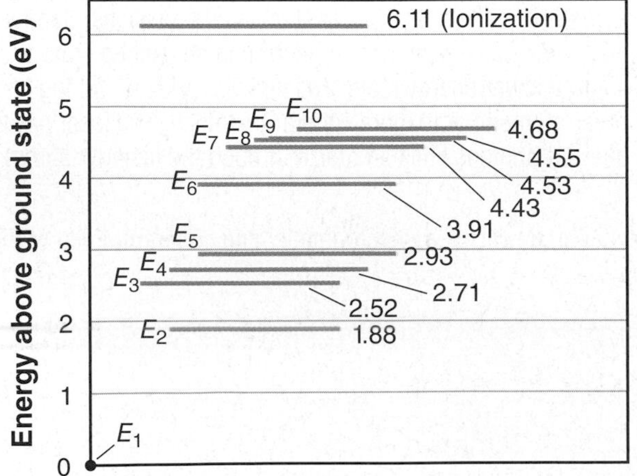

Figure 18

30. A calcium atom is in an excited state at the E_7 energy level. How much energy is released when the atom drops down to the E_3 energy level? Refer to **Figure 18.**

31. A photon of orange light with a wavelength of 6.00×10^2 nm enters a calcium atom in the E_8 excited state and ionizes the atom. What kinetic energy will the electron have as it is ejected from the atom? Refer to **Figure 18.**

32. A mercury atom is in an excited state at the E_6 energy level. Refer to **Figure 19.**

a. How much energy would be needed to ionize the atom?

b. How much energy would be released if the atom dropped down to the E_2 energy level instead?

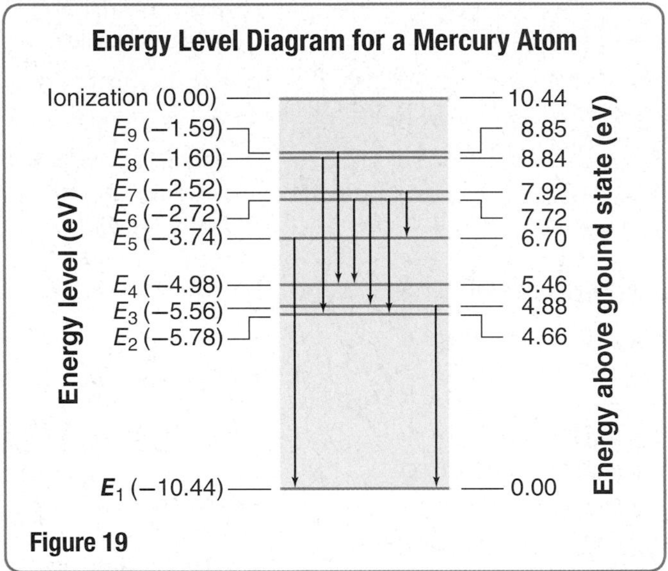

Figure 19

33. A mercury atom in an excited state has an energy of −4.98 eV. It absorbs a photon that raises it to the next-higher energy level. What are the energy and the frequency of the photon that is absorbed? Refer to **Figure 19.**

34. What energies are associated with a hydrogen atom's energy levels of E_2, E_3, E_4, E_5, and E_6?

35. Ranking Task Rank the following energy transitions of hydrogen atoms according to the energy of the released photon, from least to greatest. Specifically indicate any ties.

A. from $n = 5$ to $n = 3$

B. from $n = 5$ to $n = 4$

C. from $n = 4$ to $n = 2$

D. from $n = 3$ to $n = 2$

E. from $n = 2$ to $n = 1$

Chapter Self-Check

36. Reverse Problem Write a physics problem with real-life objects for which the following equation would be part of the solution:

$$E = 13.6\ \text{eV}\left(\frac{1}{1^2} - \frac{1}{3^2}\right)$$

37. For a hydrogen atom with an electron in the $n = 3$ Bohr orbital, find the following.

a. radius of the orbital

b. electrostatic force between the proton and the electron

c. centripetal acceleration of the electron

d. orbital electron speed; compare to speed of light

SECTION 2
The Quantum Model of the Atom

Mastering Concepts

38. Lasers A laboratory laser has a power of only 0.8 mW ($8{\times}10^{-4}$ W). Why is it brighter than the light of a 100-W lamp?

39. What properties of laser light led to its use in light shows?

Mastering Problems

40. DVD Players Lasers used in the DVD players such as the one shown in **Figure 20** typically emit light at 640 nm. What is the difference, measured in eV, between the two lasing energy levels?

Figure 20

41. A laser beam's power equals the photon energy times the number of photons per second that are emitted.

a. If you want a laser at 840 nm to have the same power as one at 427 nm, how many times more photons per second are needed?

b. Find the number of photons per second in a 5.0-mW 840-nm laser.

42. HeNe Lasers The HeNe lasers used in many classrooms can be made to lase at three wavelengths: 632.8 nm, 543.4 nm, and 1152.3 nm.

a. Find the energy difference between the two states involved in the generation of each wavelength.

b. Identify the color of each wavelength.

Applying Concepts

43. Northern Lights The northern lights shown in **Figure 21** are caused by high-energy particles from the Sun striking atoms high in Earth's atmosphere. If you looked at these lights through a spectrometer, would you see a continuous or line spectrum? Explain.

Figure 21

44. If white light were emitted from Earth's surface and observed by someone in space, would its spectrum appear to be continuous? Explain.

45. Is money a good example of quantization? Is water? Explain.

46. A certain atom has four energy levels, with E_4 being the highest and E_1 being the lowest. If the atom can make transitions between any two levels, how many spectral lines can the atom emit? Which transition produces the photon with the highest energy?

47. A photon is emitted when an electron in an excited hydrogen atom drops through energy levels. What is the maximum energy that the photon can have? If this same amount of energy were given to the atom in the ground state, what would happen?

48. Compare the quantum mechanical theory of the atom with the Bohr model.

Mixed Review

49. A 14.0-eV photon enters a hydrogen atom in the ground state and ionizes it. With what kinetic energy will the electron be ejected from the atom?

Chapter Self-Check

50. A hydrogen atom is in the $n = 2$ level.

a. If a photon with a wavelength of 332 nm strikes the atom, show that the atom will be ionized.

b. Assume the electron receives the excess energy from the atom's ionization. What will be the kinetic energy of the electron in joules?

51. A beam of electrons strikes atomic hydrogen gas. What minimum electron energy is needed for the hydrogen atoms to emit the red light produced when the atom goes from the $n = 3$ to the $n = 2$ state?

52. The most precise spectroscopy experiments use "two-photon" techniques. Two photons with identical wavelengths are directed at the target atoms from opposite directions. Each photon has half the energy needed to excite the atoms from the ground state to the desired energy level. What laser wavelength is needed for a precise study of the energy difference between $n = 1$ and $n = 2$ in hydrogen?

Thinking Critically

53. Apply Concepts The result of projecting the spectrum of a high-pressure mercury vapor lamp onto a wall in a dark room is shown in **Figure 22.** What are the differences in energy levels for each of the three visible lines?

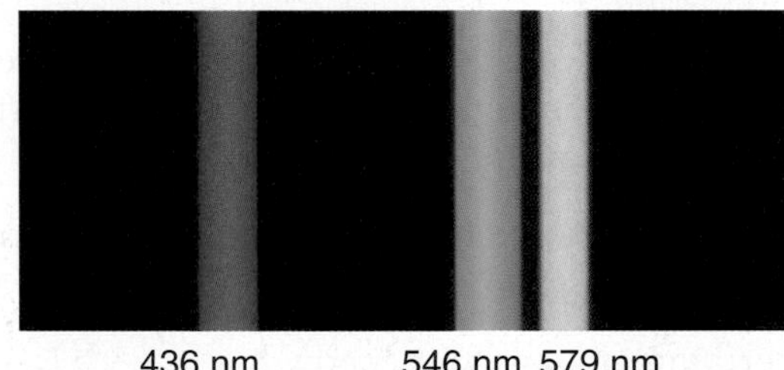

Figure 22

54. Interpret Scientific Illustrations After the emission of the visible photons described in the previous problem, the mercury atom continues to emit photons until it reaches the ground state. From an inspection of **Figure 22,** determine whether any of these photons would be visible. Explain.

55. Analyze and Conclude A positronium atom consists of an electron and a positron bound together. Although the lifetime of this "atom" is very short—on the average, one-seventh of a microsecond—its energy levels can be measured. The Bohr model can be used to calculate energies with the mass of the electron replaced by one-half its mass. Describe how the orbital radii and the energy of each level would be affected. What would be the wavelength of the E_2 to E_1 transition?

56. Problem Posing Complete this problem so that it can be solved using the Bohr model of the hydrogen atom: "A hydrogen atom is in its ground state...."

Writing in Physics

57. Research the history of atomic models. Describe each model. Identify its strengths and weaknesses.

58. Green laser pointers emit light with a wavelength of 532 nm. Research the type of laser used in this type of pointer, and describe its operation. Indicate whether the laser is pulsed or continuous.

Cumulative Review

59. A 1.0-m-long wire shown in **Figure 23** is moving at a speed of 4.0 m/s at right angles to Earth's magnetic field where the magnetic field strength is 5.0×10^{-5} T. What is the *EMF* induced in the wire?

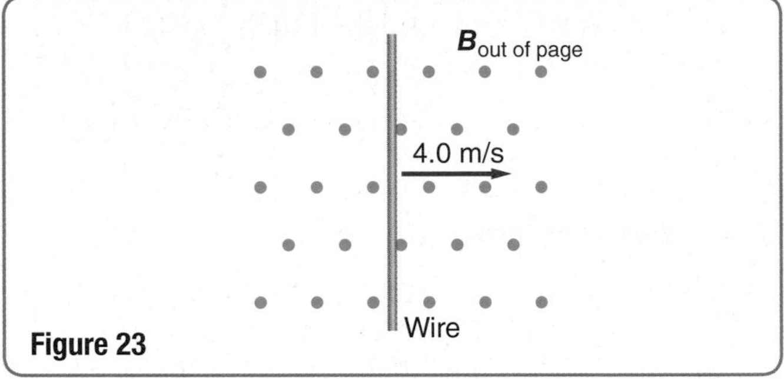

Figure 23

60. The force on a test charge of $+3.00\times10^{-7}$ C is 0.027 N. What is the electric field strength at the position of the test charge?

61. A technician needs a 4-Ω resistor but only has 1-Ω resistors. Is there a way to combine what she has? Explain.

62. The electrons in a beam move at 2.8×10^{8} m/s in an electric field of 1.4×10^{4} N/C. What value must the magnetic field have if the electrons pass through the crossed fields undeflected?

63. Consider modifications needed for J.J. Thomson's cathode-ray tube so that it could accelerate protons rather than electrons. Then answer these questions:

a. To select particles of the same velocity, would the ratio E/B have to be changed? Explain.

b. For the deflection caused by the magnetic field alone to remain the same, would the B field have to be made smaller or larger? Explain.

64. The stopping potential needed to reflect all the electrons ejected from a metal is 7.3 V. What is the electrons' maximum kinetic energy in joules?

MULTIPLE CHOICE

1. Which model of the atom was based on the results of Rutherford's gold foil experiment?

A. Bohr model

B. nuclear model

C. plum pudding model

D. quantum mechanical model

2. A mercury atom emits light with a wavelength of 405 nm. What is the energy difference between the two energy levels involved with this emission?

A. 0.22 eV

B. 2.14 eV

C. 3.06 eV

D. 4.05 eV

3. The diagram below shows the energy levels for a mercury atom. What wavelength of light is emitted when a mercury atom makes a transition from E_7 to E_5?

A. 811 nm

B. 943 nm

C. 984 nm

D. 1020 nm

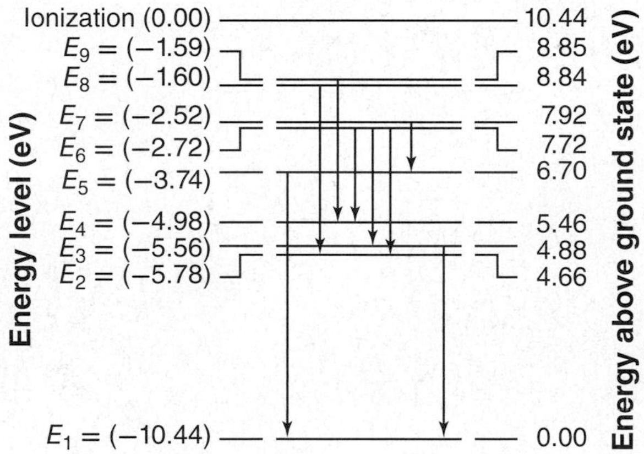

4. Which statement about the quantum model of the atom is false?

A. The possible energy levels of the atom are quantized.

B. The locations of the electrons around the nucleus are known precisely.

C. The electron cloud defines the area where electrons are likely to be located.

D. Stable electron orbits are related to the de Broglie wavelength.

Online Test Practice

Questions 5 and 6 refer to the diagram showing the Balmer series electron transitions in a hydrogen atom.

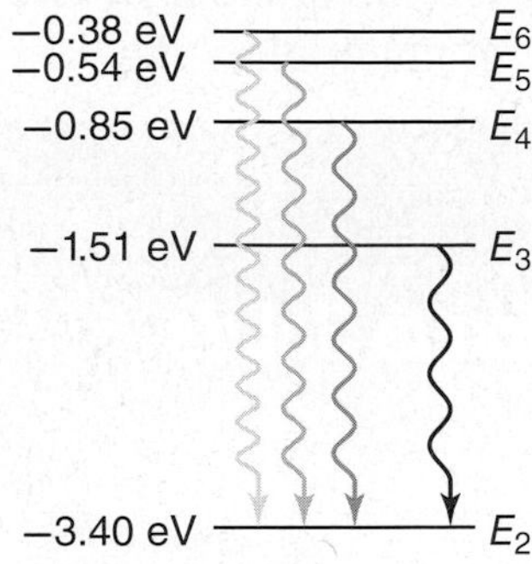

5. Which energy level transition is responsible for the emission of light with the greatest frequency?

A. E_2 to E_5

B. E_3 to E_2

C. E_3 to E_6

D. E_6 to E_2

6. What is the frequency of the Balmer series line related to the energy level transition from E_4 to E_2? (Note that $1\ \text{eV} = 1.60\times10^{-19}$ J.)

A. 2.55×10^{14} Hz

B. 4.32×10^{14} Hz

C. 6.15×10^{14} Hz

D. 1.08×10^{15} Hz

FREE RESPONSE

7. Determine the wavelength of light emitted when a hydrogen atom makes a transition from the $n = 5$ energy level to the $n = 2$ energy level.

NEED EXTRA HELP?

If You Missed Question	1	2	3	4	5	6	7
Review Section	1	1	1	2	1	1	1

CHAPTER 29

Solid-State Electronics

BIGIDEA **We can make a variety of electronic devices by controlling the conductivity of materials.**

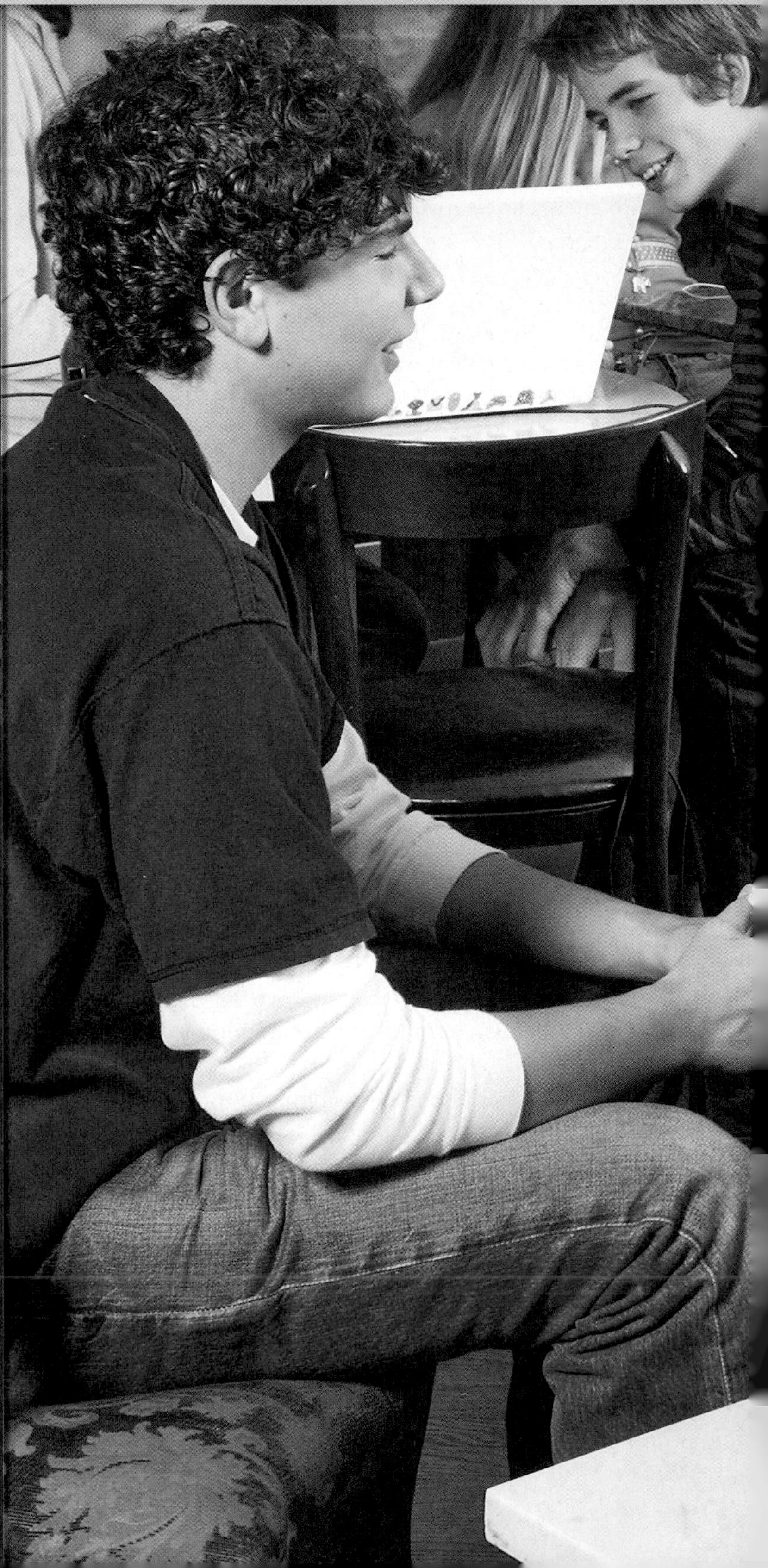

SECTIONS

LaunchLAB

CONDUCTION IN A DIODE

How is a bicolor light-emitting diode (LED) similar to doors labeled *Enter* and *Exit?*

WATCH THIS!

ROBOT PHYSICS

Integrated circuits and microchips are everywhere! Look around you and discover how solid-state electronics allow us to do what used to be impossible.

(l)BananaStock/PunchStock, (r)Marili Forastieri/Getty Images

Go online!
connectED.mcgraw-hill.com
WORLD HISTORY People & Nations • MODERN WORLD

SECTION 1

Conduction in Solids

PHYSICS 4 YOU

You have probably heard of Silicon Valley, a region near San José, California. When the area was named, it was a hub of businesses that developed and made silicon-containing computer chips. Today, it is home to many high-tech industries.

MAIN IDEA

A material's conductivity depends on the energy difference between the valence and conducting bands.

Essential Questions

- How does the band theory of solids explain conduction?
- How do the energy levels in conductors, insulators, and semiconductors vary?
- How can the conductivity of a semiconductor be improved?
- How are *n*-type and *p*-type semiconductors alike? How are they different?

Review Vocabulary

conductor a material, such as copper, through which a charge will move easily

New Vocabulary

semiconductor
band theory
intrinsic semiconductor
dopant
extrinsic semiconductor

Band Theory of Solids

You have learned about electric conductors and insulators. Electric charges can move easily in conductors but not in insulators. A third type of material, called a **semiconductor,** acts as a conductor in certain conditions and as an insulator in others. When you examine these three types of materials at the atomic level, the difference in the way they are able to transmit charges becomes apparent.

In your study of solids, you learned that crystalline solids consist of atoms bound together in regular arrangements. You also know that an atom consists of a dense, positively charged nucleus surrounded by a cloud of negatively charged electrons. These electrons can occupy only certain allowed energy levels. Because the electrons can have only certain energies, any energy changes that occur are quantized; that is, the energy changes occur in specific amounts.

Energy levels Under most conditions, an atom's electrons occupy the lowest possible energy levels. This condition is called the ground state. Consider constructing a solid by bringing atoms together, one by one. You would start with an atom in the ground state. At large interatomic spacings (>0.8 nm), with no very near neighbors, there are two discrete energy levels for the atom. As the solid crystal forms by moving atoms closer to the atom, the electric fields of these other neighboring atoms affect the energy levels of its electrons, as shown in **Figure 1.**

Figure 1 The electric fields of neighboring atoms split the energy levels of an atom. Thus, there is an energy gap between the valence and conduction bands.

Estimate the energy gap of a material with an atomic separation of 0.2 nm.

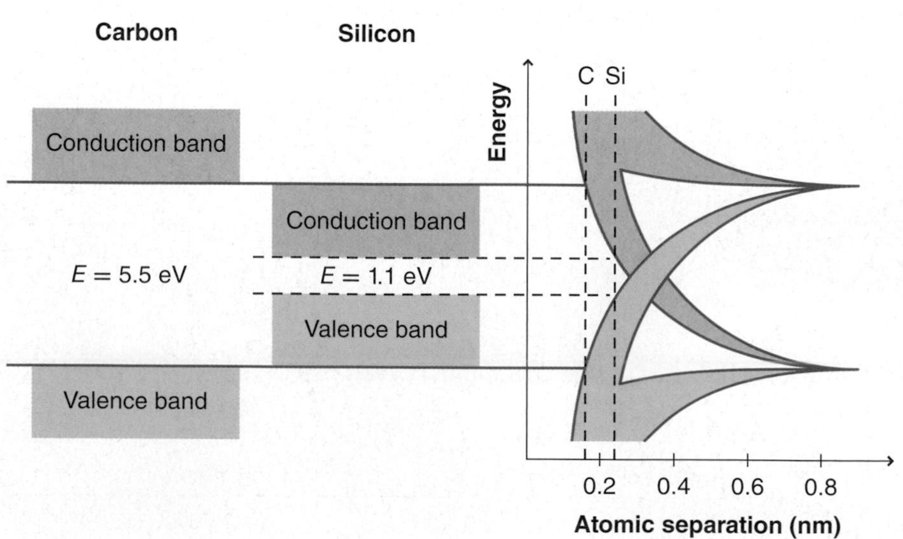

In the solid crystal, the electric fields of the neighboring atoms split the atom's ground-state energy levels into multiple levels. There are so many levels and they are so close together that they no longer appear as distinct levels. Instead, they appear as energy bands, shown in **Figure 1.** The lower energy regions are called valence bands and are occupied by the bonding electrons in the crystal. The higher energy regions, called conduction bands, are available for electrons to move between atoms.

☑ **READING CHECK Explain** What is the difference between the valence band and the conducting band in a solid?

Energy gaps Notice in **Figure 1** that the atomic separations for crystalline silicon and crystalline carbon (diamond) translate to valence bands and conduction bands that are separated by energy gaps. These gaps have no energy levels available for electrons and are called forbidden energy regions. This description of valence and conduction bands, separated by forbidden energy gaps, is known as the **band theory** of solids. Band theory can be used to better understand electric conduction.

Conductivity Consider a sample of silicon. At absolute zero, the valence band of silicon would be completely full and the conduction band would be completely empty. But at room temperature, some of the valence electrons have enough thermal energy to jump the 1.1-eV gap to the conduction band and serve as charge carriers. As the temperature increases and more electrons gain enough energy to jump the gap, the conductivity of silicon increases.

Conductivity also depends on the size of the energy gap. The smaller the energy gap, the less energy required to move electrons from the valence band to the conduction band. Thus, the smaller the band gap, the better conductor a material is, as shown in **Figure 2.** For example, carbon in graphite form is a much better conductor than diamond because graphite has a smaller energy gap. Similarly, crystalline silicon has a smaller energy gap than diamond. It is a better conductor because less energy is required to move electrons to the conduction band.

Germanium has an even smaller energy gap of 0.7 eV. This means that germanium is a better conductor than silicon at any given temperature. However, it also means that germanium is too sensitive to temperature for many electronic applications. Relatively small changes in temperature cause large changes in the conductivity of germanium, making it harder to control and keep stable.

Figure 2 The conductivity of a material increases as the energy gap decreases.

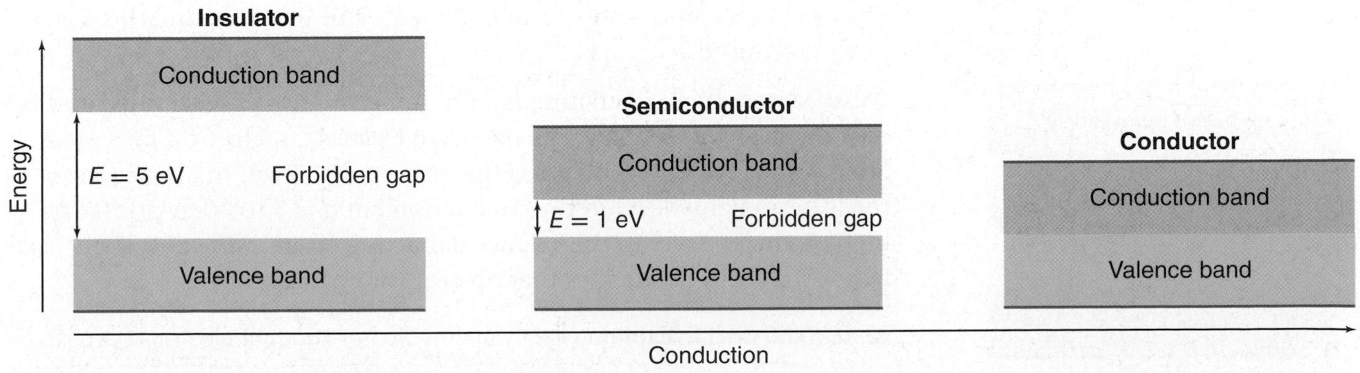

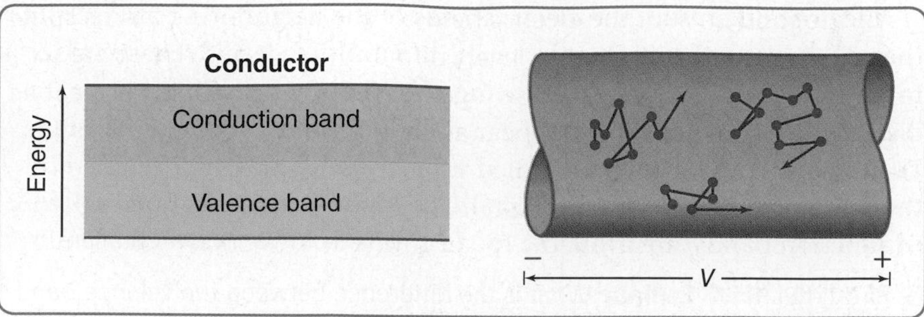

Figure 3 In a conductor, some electrons are in the conduction band. These electrons are free to move between atoms. The electrons move randomly in a conductor, but if a field is applied across the wire, the electrons drift toward one end.

Conductors and Insulators

When a potential difference is placed across a material, the resulting electric field exerts a force on the electrons and does work on them. The electrons accelerate and gain energy. If there are bands within the material that are only partially filled, then there are energy levels available that are only slightly higher than the electrons' ground-state levels. As a result, the electrons that gain energy from the field can move from one atom to the next.

Such movement of electrons from one atom to the next is an electric current, and the entire process is known as electric conduction. The more easily the charges move, the more conductive the material is. Since conductivity is the reciprocal of resistivity, as a material's conductivity is reduced, its resistivity rises.

Conductors Materials with partially filled bands, such as the metals aluminum, lead, and copper, conduct electricity easily. Lead has an interatomic spacing of 0.27 nm. Look back at the graph in **Figure 1.** You will see that this spacing would translate to a conduction band that overlaps the valence band. As one would expect, lead is a good conductor. Materials with overlapping, partially filled bands are conductors, as shown in **Figure 3.**

Free electrons in conductors move about rapidly in a random way and change directions when they collide with the cores of atoms. However, if an electric field is put across a length of wire, there will be a net force pushing the electrons in one direction. Their motion is not greatly affected, but they have a slow overall movement dictated by the electric field, as shown in **Figure 3.** Electrons continue to move within the wire with speeds of 10^6 m/s in random directions. Although the random motion of the electrons is very fast, they might have drift at speeds of just 10^{-3} m/s or slower toward the positive end of the wire. This model of electron motion in conductors is called the electron-gas model. If the temperature of the conductor is increased, the electrons' speeds increase. The electrons collide more frequently with atomic cores, and the conductivity is reduced.

Figure 4 The energy gap in an insulator is large enough that few, if any, electrons have enough thermal energy to reach the conduction band.

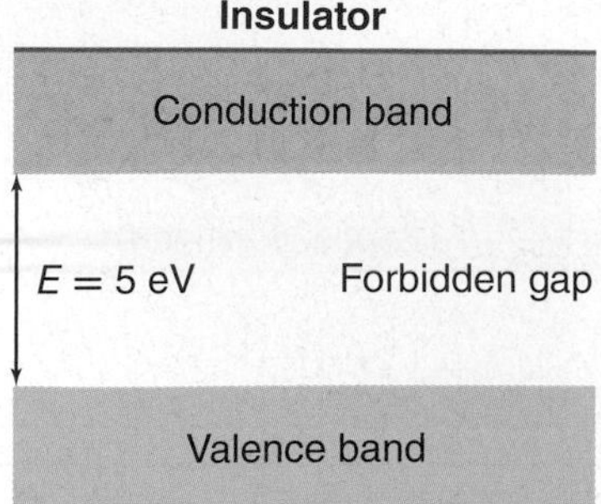

Insulators In an insulating material, the valence band is filled and the conduction band is empty. As shown in **Figure 4,** an electron must gain a large amount of energy to go to the next energy level. In an insulator, the lowest energy level in the conduction band is 5 to 10 eV above the highest energy level in the valence band, as shown in **Figure 4.** There is at least a 5-eV gap of energies that no electrons can occupy.

☑ **READING CHECK** **Explain** why insulators do not conduct electric charges.

Electrons in an insulator have some kinetic energy as a result of their thermal energy. However, the average kinetic energy of these electrons at room temperature is not sufficient for them to jump the forbidden gap. If a small electric field is placed across an insulator, almost no electrons gain enough energy to reach the conduction band. Therefore, there is no current. Electrons in an insulator must be given a large amount of energy to be pulled into the conduction band. As a result, the electrons in an insulator tend to remain in place, and the material does not conduct electricity. Insulators are often used to coat conducting wire to prevent circuits from shorting out.

EXAMPLE PROBLEM 1

Find help with **dimensional calculations.** Math Handbook

EXAMPLE PROBLEM

THE FREE-ELECTRON DENSITY OF A CONDUCTOR How many free electrons exist in a cubic centimeter of copper? Each atom contributes one electron. The density, atomic mass, and number of atoms per mole of copper can be found in the Reference Tables.

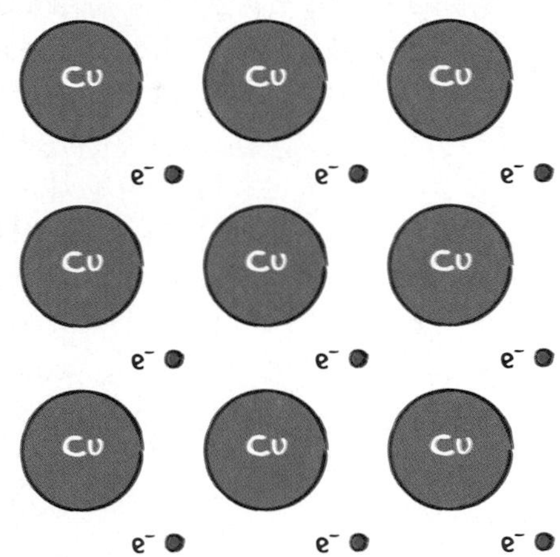

1 ANALYZE THE PROBLEM

Identify the knowns using the Reference Tables.

KNOWN	UNKNOWN
For copper: 1 free e^- per atom	free e^-/cm^3 = ?
$\rho = 8.92\ g/cm^3$	
$M = 63.55\ g/mol$	
$N_A = 6.02\times10^{23}$ atoms/mol	

2 SOLVE FOR THE UNKNOWN

$$\frac{\text{free e}^-}{\text{cm}^3} = \frac{(\text{free e}^-)}{\text{atom}}(N_A)\left(\frac{1}{M}\right)(\rho)$$

$$= \left(\frac{1\ \text{free e}^-}{1\ \text{atom}}\right)\left(\frac{6.02\times10^{23}\ \text{atoms}}{1\ \text{mol}}\right)\left(\frac{1\ \text{mol}}{63.55\ \text{g}}\right)\left(\frac{8.92\ \text{g}}{1\ \text{cm}^3}\right)$$

◀ *Substitute free e^-/atom = 1 free e^-/atom, $N_A = 6.02\times10^{23}$ atoms/mol, $M = 63.55$ g/mol, $\rho = 8.92\ g/cm^3$.*

$$= 8.45\times10^{22}\ \text{free e}^-/\text{cm}^3 \text{ in copper}$$

3 EVALUATE THE ANSWER

- **Are the units correct?** Dimensional analysis on the units confirms the number of free electrons per cubic centimeter.
- **Is the magnitude realistic?** One would expect a large number of electrons in a cubic centimeter.

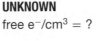

PRACTICE PROBLEMS

Do additional problems. Online Practice

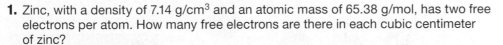

1. Zinc, with a density of 7.14 g/cm^3 and an atomic mass of 65.38 g/mol, has two free electrons per atom. How many free electrons are there in each cubic centimeter of zinc?
2. Silver has one free electron per atom. Use the Reference Tables and determine the number of free electrons in 1 cm^3 of silver.
3. Gold has one free electron per atom. Use the Reference Tables and determine the number of free electrons in 1 cm^3 of gold.
4. Aluminum has three free electrons per atom. Use the Reference Tables and determine the number of free electrons in 1 cm^3 of aluminum.
5. **CHALLENGE** The tip of the Washington Monument was made of 2835 g of aluminum because it was a rare and costly metal in the 1800s. Use problem 4 and determine the number of free electrons in the tip of the Washington Monument.

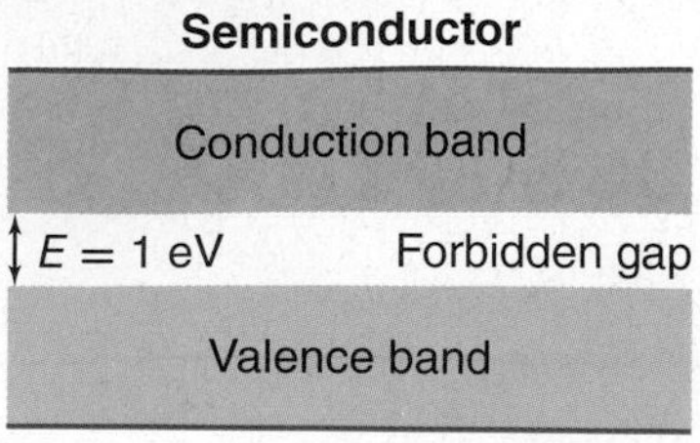

Semiconductor

Conduction band

$E = 1$ eV Forbidden gap

Valence band

Figure 5 In a semiconductor, the energy gap is small enough that some electrons have enough thermal energy to enter the conduction band.

Semiconductors

So far, we have categorized materials as either conductors or insulators. But there are also materials called semiconductors that do not strictly fall into either category. As discussed earlier, electrons move more freely in semiconductors than in insulators but not as easily as in conductors. The energy gap between the valence and conduction bands of a semiconductor is approximately 1 eV, as shown in **Figure 5.** A semiconductor's structure explains its conductive characteristics. Atoms of the most common semiconductors, silicon (Si) and germanium (Ge), have four valence electrons. These four electrons are involved in binding the atoms together into a solid crystal. The valence electrons form a filled band, as in an insulator, but the forbidden gap is much smaller.

Holes and free electrons Not much energy is needed to pull an electron from a silicon atom into the conduction band. The gap is so small that some electrons reach the conduction band as a result of their thermal kinetic energy alone. That is, the random motion of atoms and electrons gives some electrons enough kinetic energy to break free of their home atoms and move around the silicon crystal. Pure semiconductors that conduct as a result of thermally freed electrons are called **intrinsic semiconductors.** There are very few electrons available to carry charge, so intrinsic semiconductors have very low conductivity.

If an electric field is applied to a semiconductor, electrons in the conduction band move through the solid in response. In metals, an increase in temperature reduces the conductivity, but for semiconductors, a higher temperature means that electrons more easily reach the conduction band and the conductivity increases.

READING CHECK Contrast How is the effect of temperature on conductivity different for metals than for semiconductors?

An atom from which an electron has broken free has a net positive charge and contains a hole. As shown in **Figure 6,** a hole is an empty energy level in the valence band. An electron from the conduction band can fill this hole and become bound to the atom. When a hole and a free electron recombine, their opposite charges neutralize each other. The electron, however, left behind a hole at its previous location, which can be filled by a different electron. Thus, negatively charged free electrons move in one direction and positively charged holes move in the opposite direction.

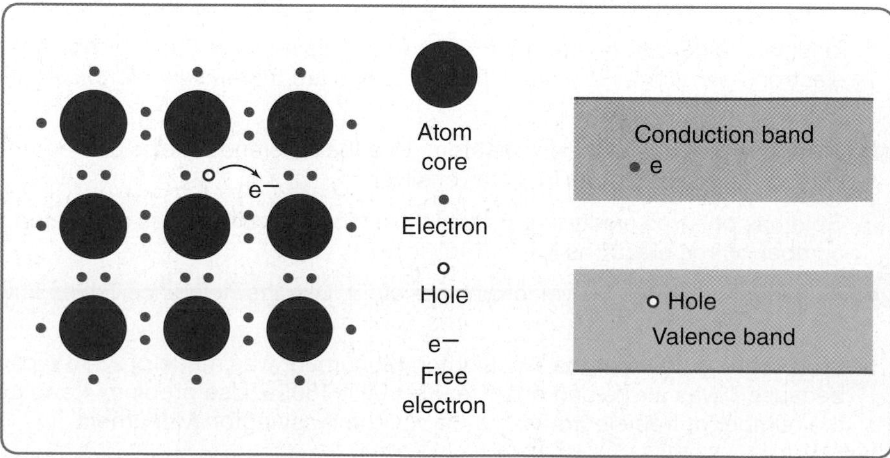

Figure 6 In a semiconductor, some electrons have enough thermal energy to break free and move about the crystal. There are electrons in the conduction band and holes in the valence band.

EXAMPLE PROBLEM 2

Find help with **scientific notation.** Math Handbook

FREE ELECTRON DENSITY IN AN INTRINSIC SEMICONDUCTOR
Because of the thermal kinetic energy of solid silicon at room temperature, there are 1.45×10^{10} free electrons/cm^3. What is the number of free electrons per atom of silicon at room temperature?

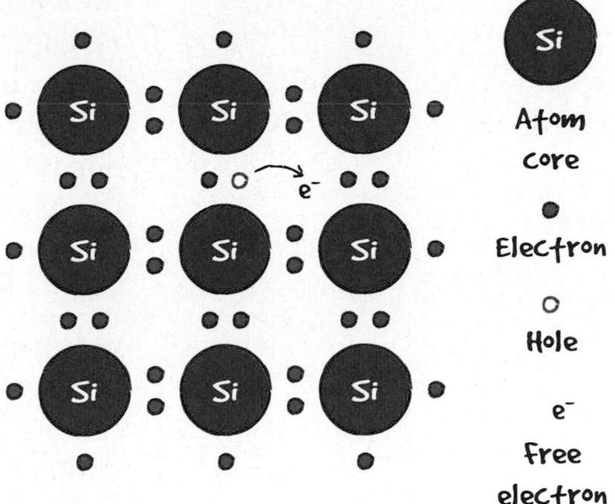

1 ANALYZE THE PROBLEM

Identify the knowns and unknowns.

KNOWN

$\rho = 2.33$ g/cm^3
$M = 28.09$ g/mol
$N_A = 6.02\times10^{23}$ atoms/mol
For Si: 1.45×10^{10} free e$^-$/cm^3

UNKNOWN

free e$^-$/atom of Si = ?

2 SOLVE FOR THE UNKNOWN

$$\frac{\text{free e}^-}{\text{atom}} = \left(\frac{1}{N_A}\right)(M)\left(\frac{1}{\rho}\right)(1.45\times10^{10}\text{ free e}^-/\text{cm}^3\text{ for Si})$$

$$= \left(\frac{1\text{ mol}}{6.02\times10^{23}\text{ atoms}}\right)\left(\frac{28.09\text{ g}}{1\text{ mol}}\right)\left(\frac{1\text{ cm}^3}{2.33\text{ g}}\right)\left(\frac{1.45\times10^{10}\text{ free e}^-}{\text{cm}^3}\right)$$

◀ Substitute $N_A = 6.02\times10^{23}$ atoms/mol, $M = 28.09$ g/mol, $\rho = 2.33$ g/cm^3, free e$^-$/cm^3 Si = 1.45×10^{10} free e$^-$/cm^3.

$$= 2.90\times10^{-13}\text{ free e}^-/\text{atom of Si}$$

3 EVALUATE THE ANSWER

- **Are the units correct?** Using dimensional analysis confirms the correct units.
- **Is the magnitude realistic?** In an intrinsic semiconductor, such as silicon at room temperature, very few atoms have free electrons.

PRACTICE PROBLEMS

Do additional problems. Online Practice

6. In pure germanium, which has a density of 5.32 g/cm^3 and an atomic mass of 72.6 g/mol, there are 2.25×10^{13} free electrons/cm^3 at room temperature. How many free electrons are there per atom?

7. At 200.0 K, germanium has 1.16×10^{10} free electrons/cm^3. How many free electrons are there per atom at this temperature?

8. At 100.0 K, germanium has 3.47 free electrons/cm^3. How many free electrons are there per atom at this temperature?

9. A sample of germanium has 4.81×10^{11} free electrons/cm^3. How many free electrons per atom does the sample have?

10. At 100.0 K, silicon has 9.23×10^{-10} free electrons/cm^3. How many free electrons are there per atom at this temperature? What does this temperature represent on the Celsius scale?

11. At 200.0 K, silicon has 1.89×10^{5} free electrons/cm^3. How many free electrons are there per atom at this temperature? What does this temperature represent on the Celsius scale?

12. CHALLENGE The density of free electrons in a sample of silicon doubles when the temperature is raised 8.00°C. What is the density of free electrons in a silicon sample at 264 K? *(Hint: Use your answer from the previous question.)*

Doped Semiconductors

To make practical devices, the conductivity of intrinsic semiconductors must be increased greatly. That is, the number of holes and free electrons must be increased. This is done by adding atoms of other elements to the intrinsic semiconductor. These electron donor or acceptor atoms that can be added in low concentrations to intrinsic semiconductors are called **dopants.** The doped semiconductors are known as **extrinsic semiconductors.** There are two types of dopants used to increase the conductivity of semiconductors, *n*-type and *p*-type.

***n*-type semiconductors** In some cases, an electron donor with five valence electrons, such as arsenic (As), is used as a dopant for silicon. **Figure 7** shows a silicon crystal where a dopant atom has replaced one of the silicon atoms. Four of the five arsenic valence electrons bind to neighboring silicon. The fifth electron is called the donor electron. The energy of this donor electron is close to the conduction band, as shown in **Figure 7.** Thus, thermal energy can easily move the electron from the dopant atom into the conduction band. Conduction in this kind of semiconductor is increased by the availability of these extra donor electrons to the conduction band. This type of semiconductor is known as an *n*-type semiconductor, since negative begins with *n*.

***p*-type semiconductors** Alternatively, an electron acceptor with three valence electrons, such as gallium (Ga), can dope silicon. When a gallium atom replaces a silicon atom, one binding electron is missing. This creates a hole in the silicon crystal, as shown in **Figure 7.** Electrons in the conduction band can easily drop into these holes, creating new holes. Conduction in such semiconductors is enhanced by the availability of the extra holes, as shown in **Figure 7.** These kinds of semiconductors are called *p*-type, with the *p* standing for positive.

Both *p*-type and *n*-type semiconductors are electrically neutral. Adding dopant atoms of either type does not add any net charge to a semiconductor. Both types of semiconductor use electrons and holes in conduction. Adding only a few dopant atoms per million silicon atoms increases the semiconductor's conductivity by a factor of 1000 or more.

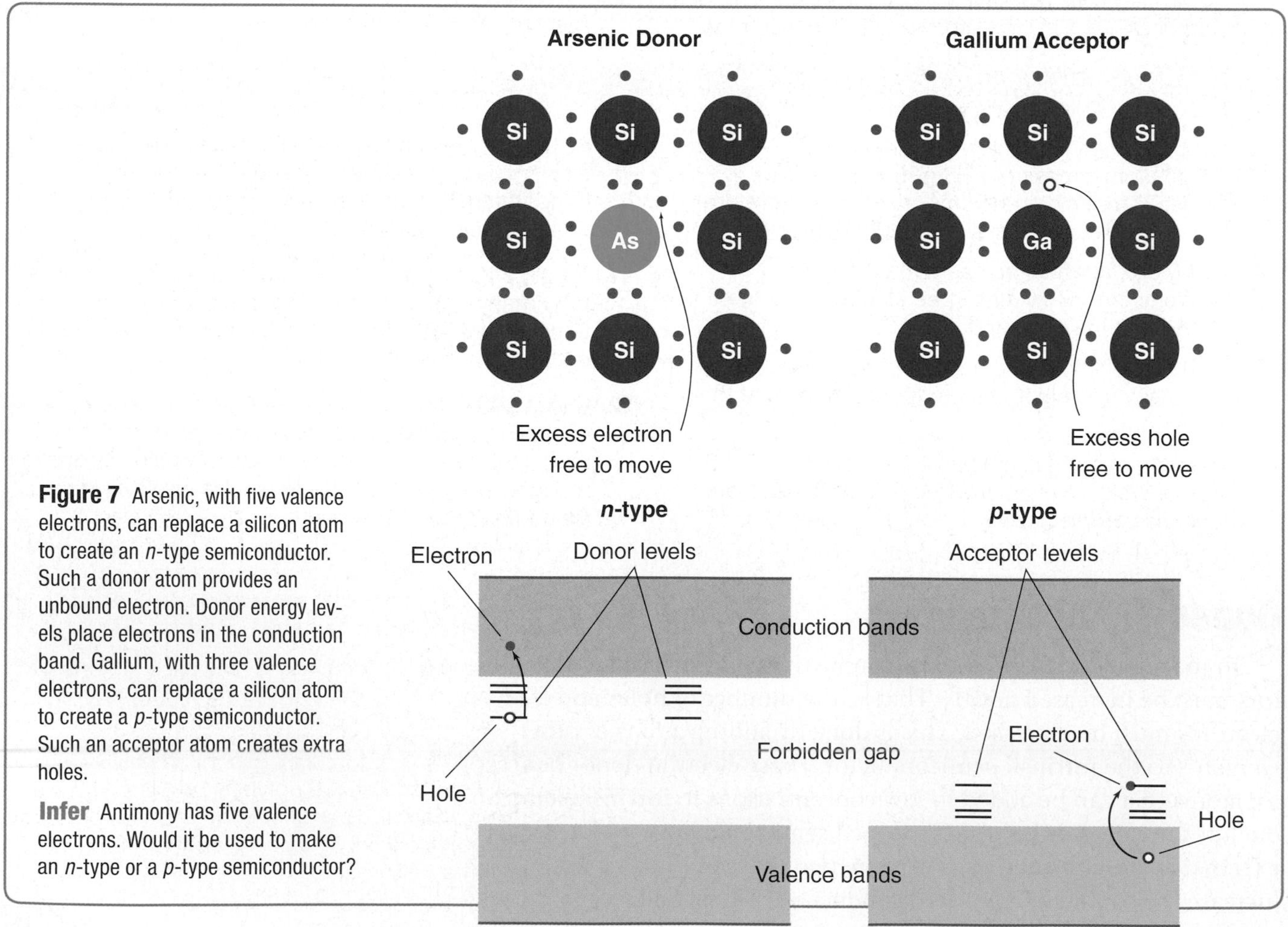

Figure 7 Arsenic, with five valence electrons, can replace a silicon atom to create an *n*-type semiconductor. Such a donor atom provides an unbound electron. Donor energy levels place electrons in the conduction band. Gallium, with three valence electrons, can replace a silicon atom to create a *p*-type semiconductor. Such an acceptor atom creates extra holes.

Infer Antimony has five valence electrons. Would it be used to make an *n*-type or a *p*-type semiconductor?

EXAMPLE PROBLEM 3

Get help with **doped semiconductors.** Personal Tutor

THE CONDUCTIVITY OF DOPED SILICON A scientist dopes silicon with arsenic so that one in every one million silicon atoms is replaced by an arsenic atom. Each arsenic atom donates one electron to the conduction band.

a. What is the density of free electrons?

b. By what ratio is this density greater than that of intrinsic silicon with 1.45×10^{10} free e^-/cm^3?

c. Is conduction mainly by the electrons of the silicon or the arsenic?

Arsenic Donor

Si Si Si Si As Si Si Si Si

Excess electron free to move

1 ANALYZE THE PROBLEM

Identify the knowns and unknowns.

KNOWN	UNKNOWN
1 As atom/10^6 Si atoms	free e^-/cm^3 donated by As = ?
1 free e^-/As atom	ratio of As-donated free e^- to intrinsic free e^- = ?
4.99×10^{22} Si atoms/cm^3	
1.45×10^{10} free e^-/cm^3 in intrinsic Si	

2 SOLVE FOR THE UNKNOWN

a. $$\left(\frac{\text{free } e^-}{cm^3}\text{ from As}\right) = \left(\frac{\text{free } e^-}{\text{As atom}}\right)\left(\frac{\text{As atoms}}{\text{Si atoms}}\right)\left(\frac{\text{Si atoms}}{cm^3}\right)$$

$$\left(\frac{\text{free } e^-}{cm^3}\right) = \left(\frac{1\text{ free } e^-}{1\text{ As atom}}\right)\left(\frac{1\text{ As atom}}{1\times10^{6}\text{ Si atoms}}\right)\left(\frac{4.99\times10^{22}\text{ Si atoms}}{cm^3}\right)$$

$$= 4.99\times10^{16}\text{ free } e^-/cm^3 \text{ from As donor in doped Si}$$

◀ Substitute free e^-/As atom = 1 free e^-/1 As atom, As atoms/Si atoms = 1 As atom/1×10^6 Si atoms, Si atoms/cm^3 = 4.99×10^{22} Si atoms/cm^3.

b. $$\text{Ratio} = \left(\frac{\text{free } e^-/cm^3 \text{ in doped Si}}{\text{free } e^-/cm^3\text{ in intrinsic Si}}\right)$$

$$= \left(\frac{4.99\times10^{16}\text{ free } e^-/cm^3\text{ in doped Si}}{1.45\times10^{10}\text{ free } e^-/cm^3\text{ in intrinsic Si}}\right)$$

$$= 3.44\times10^{6}\text{ As-donated electrons per intrinsic Si electron}$$

◀ Substitute 4.99×10^{16} free e^-/cm^3 in doped Si, 1.45×10^{10} free e^-/cm^3 in intrinsic Si.

c. Because there are over 3 million arsenic-donated electrons for every intrinsic electron, conduction is mainly by the arsenic-donated electrons.

3 EVALUATE THE ANSWER

- **Are the units correct?** Using dimensional analysis confirms the correct units.
- **Is the magnitude realistic?** The ratio is large enough so that intrinsic electrons make almost no contribution to conductivity.

PRACTICE PROBLEMS

Do additional problems. Online Practice

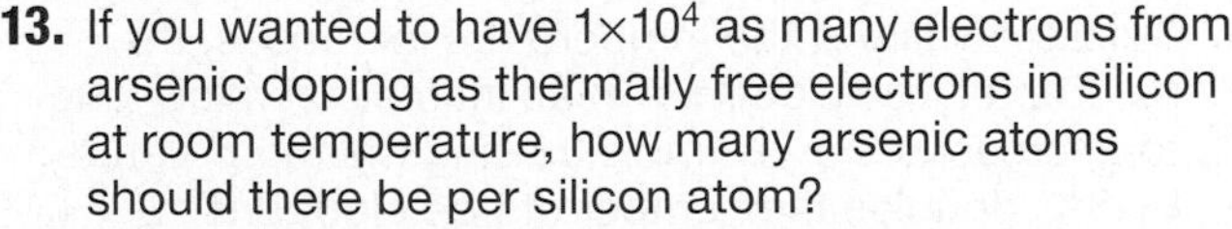

13. If you wanted to have 1×10^4 as many electrons from arsenic doping as thermally free electrons in silicon at room temperature, how many arsenic atoms should there be per silicon atom?

14. Germanium has 5.19×10^{-9} free electrons per atom. If you wanted to have 5×10^3 as many electrons from arsenic doping as thermally free electrons in a germanium semiconductor, how many arsenic atoms should there be per germanium atom?

15. Germanium at 400.0 K has 1.13×10^{15} thermally liberated carriers/cm^3. If it is doped with one As atom per 1 million Ge atoms, what is the ratio of doped carriers to thermal carriers?

16. Germanium at 200.0 K has 1.16×10^{10} free electrons per atom. If you wanted to have 2×10^3 as many electrons from arsenic doping as thermally free electrons in a germanium semiconductor, how many arsenic atoms should there be per germanium atom?

17. Silicon at 400.0 K has 4.54×10^{12} thermally liberated carriers/cm^3. If it is doped with 1 As atom per one million Si atoms, what is the ratio of doped carriers to thermal carriers?

18. CHALLENGE Based on the previous problem, draw a conclusion about the behavior of germanium devices as compared to silicon devices at temperatures in excess of the boiling point of water.

Figure 8 Light meters are made with semiconductors that are sensitive to specific wavelengths of light. Photographers use light meters to assess light levels before taking a picture.

Making and using semiconductors Silicon is doped by putting a pure silicon crystal in a vacuum with a sample of the dopant material. The dopant is heated until it is vaporized, and the atoms condense on the cold silicon. The dopant diffuses into the silicon as it warms. Then, a thin layer of aluminum or gold is evaporated onto the doped crystal. A wire is welded to this metal layer, allowing the user to apply a potential difference across the doped silicon.

Thermistors The electric conductivity of intrinsic and extrinsic semiconductors is sensitive to both temperature and light. In metals, conductivity is reduced when the temperature rises. An increase in the temperature of a semiconductor, however, allows more electrons to reach the conduction band. Thus, conductivity increases and resistance decreases.

One semiconductor device, the thermistor, is designed so that its resistance depends very strongly on temperature. The thermistor can be used as a sensitive thermometer and to compensate for temperature variations of other components in an electric circuit. Thermistors also can be used to detect radio waves, infrared radiation, and other forms of radiation.

☑ READING CHECK **Describe** the relationship between temperature and conductivity in a thermistor.

MiniLAB

OPTOISOLATOR

How can electrical signals be transferred from one circuit to another?

Light meters Other useful applications of semiconductors depend on their light sensitivity. When light falls on a semiconductor, the light can excite electrons from the valence band to the conduction band in the same way that other energy sources excite atoms. Thus, the resistance decreases as the light intensity increases.

Extrinsic semiconductors can be tailored to respond to specific wavelengths of light. These include the infrared and visible regions of the spectrum. Materials such as silicon and cadmium sulfide serve as light-dependent resistors in light meters used by lighting engineers to design the illumination of stores, offices, and homes. They are also used by photographers to adjust their cameras to capture the best images, as shown in **Figure 8.**

SECTION 1 REVIEW

Check your understanding.

19. **MAIN**IDEA Magnesium oxide has a forbidden gap of 8 eV. Is this material a conductor, an insulator, or a semiconductor?
20. **Carrier Mobility** In which type of material—a conductor, a semiconductor, or an insulator—are electrons most likely to remain with the same atom?
21. **Insulator or Conductor?** Silicon dioxide is widely used in the manufacture of solid-state devices. Its energy-band diagram shows a gap of 9 eV between the valence band and the conduction band. Is it more useful as an insulator or a conductor?
22. **Intrinsic and Extrinsic Semiconductors** You are designing an integrated circuit using a single crystal of silicon. You want to have a region with relatively good insulating properties. Should you dope this region or leave it as an intrinsic semiconductor?
23. **Types of Semiconductors** How are *n*-type and *p*-type semiconductors different?
24. **Semiconductors** If the temperature increases, the number of free electrons in an intrinsic semiconductor increases. For example, raising the temperature by 8°C doubles the number of free electrons in silicon. Is it more likely that an intrinsic semiconductor or a doped semiconductor will have a conductivity that depends on temperature? Explain.
25. **Critical Thinking** Silicon produces a doubling of thermally liberated carriers for every 8°C increase in temperature, and germanium produces a doubling of thermally liberated carriers for every 13°C increase. It would seem that germanium would be superior for high-temperature applications, but the opposite is true. Explain.

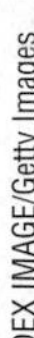

SECTION 2

Electronic Components

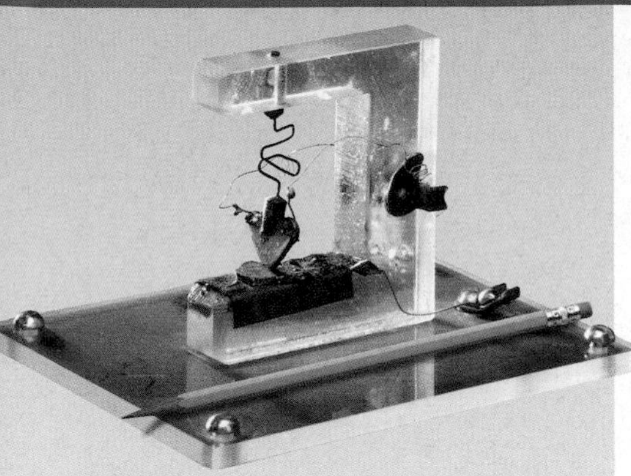

PHYSICS 4 YOU

If you have seen pictures of early computers, you know that they were large enough to take up several rooms. Today, computers can be small enough to fit in your pocket. This size reduction was largely due to the invention of a small electronic component called a transistor, shown to the left. Advances have led to even smaller transistors.

MAIN IDEA

Diodes and transistors are the fundamental components in today's electronic circuits.

Essential Questions

- What is a diode?
- What are some uses of diodes?
- What is a transistor?
- What is the major function of a transistor?

Review Vocabulary

laser a device that produces coherent, directional, monochromatic light that can be used to excite other atoms

New Vocabulary

diode
depletion layer
transistor
microchip

The Birth of Solid-State Electronics

All electronic devices owe their origins to the vacuum tubes of the early 1900s. In vacuum tubes, electron beams flow through space to amplify and control faint electric signals. Vacuum tubes are big, require lots of electric power, and generate considerable thermal energy. They have heated filaments, which require the replacement of the tubes after one to five years.

In the late 1940s, solid-state devices made of semiconducting materials, such as the transistor shown above, that could do the jobs of vacuum tubes were invented. These devices amplify and control very weak electric signals through the movement of electrons within a tiny crystalline space. Very few electrons flow in semiconductor devices, and these devices have no filaments. Thus, they operate with a low power input. They are very small, generate little heat, and are inexpensive to manufacture. Their estimated useful life is 20 years or more.

A decade later, these devices were combined with other electrical components to make the first integrated circuit, as shown in **Figure 9.** Today's electronic instruments, such as smart phones, televisions, videodisc players, and laptops, rely on semiconductor devices that are combined on chips of silicon a few millimeters wide. In these devices, current and potential difference vary in more complex ways than Ohm's law. Semiconductor devices can convert current from AC to DC and amplify potential differences.

Figure 9 The first integrated circuit was built in 1958. The strip of germanium in the center of the circuit is about 1 cm long.

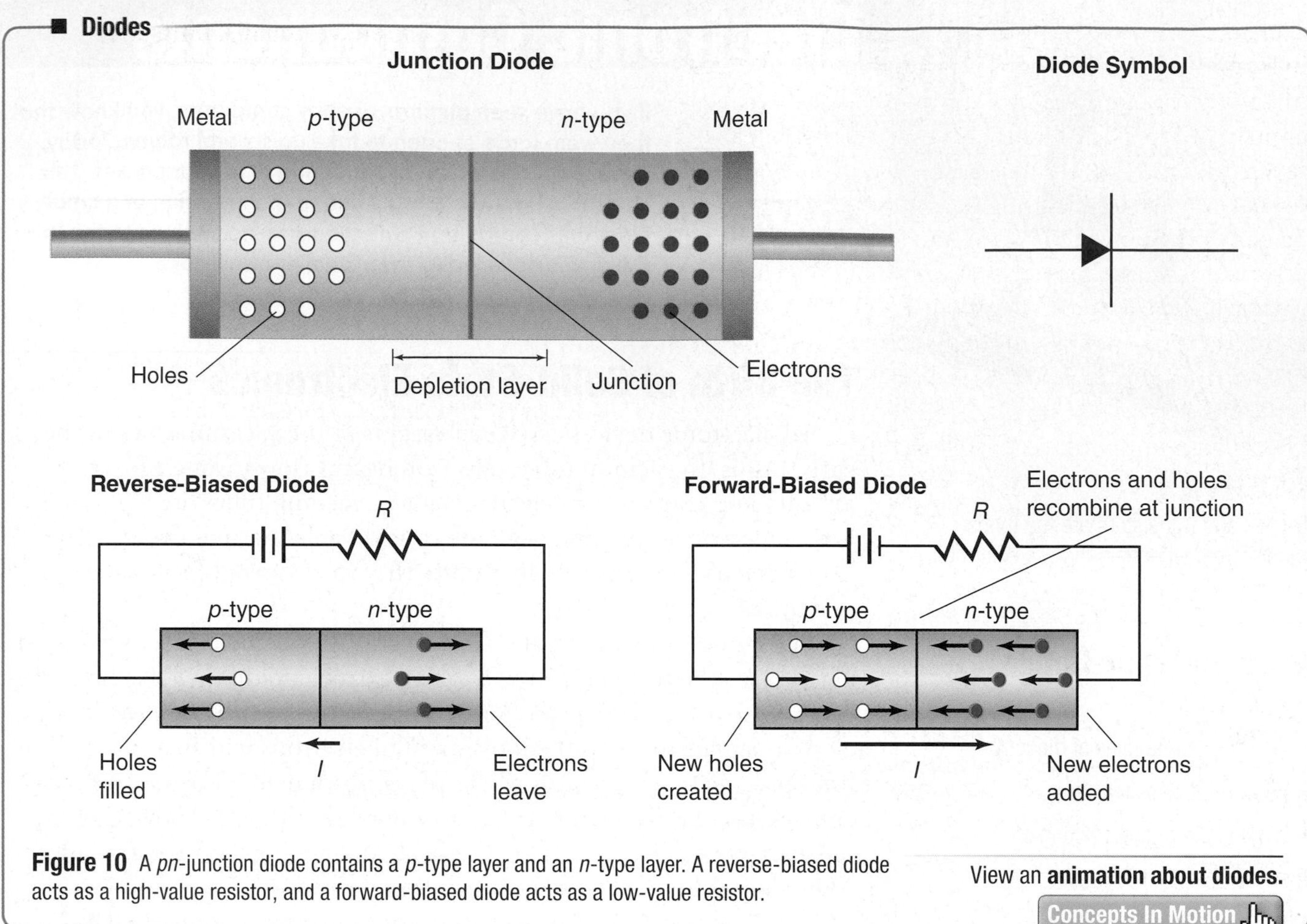

Figure 10 A *pn*-junction diode contains a *p*-type layer and an *n*-type layer. A reverse-biased diode acts as a high-value resistor, and a forward-biased diode acts as a low-value resistor.

View an **animation about diodes.**

Concepts In Motion

Diodes

The simplest semiconductor device is the diode. A **diode** consists of a sandwich of *p*-type and *n*-type semiconductors, as shown in **Figure 10.** To make a diode, a single sample of silicon is treated first with a *p*-dopant and then with an *n*-dopant. The boundary between the *p*-side and the *n*-side is called the junction, so this diode is a *pn*-junction diode. Metal contacts on each end allow wires to be attached.

The free electrons on the *n*-side of the junction are attracted to the positive holes on the *p*-side. The electrons move easily into the *p*-side and recombine with the holes. Similarly, holes from the *p*-side move into the *n*-side and recombine with electrons. As a result of this flow, the *n*-side has a net positive charge, and the *p*-side has a net negative charge.

These charges produce forces in the opposite direction that stop further movement of charge carriers. The region around the junction is left with neither holes nor free electrons and is called the **depletion layer.** It has no charge carriers, so it is a poor conductor of electricity. Thus, a junction diode consists of two relatively good conductors that surround a poor conductor.

Forward and reverse bias When a diode is connected into a circuit in the way shown in the bottom left panel of **Figure 10,** both the free electrons in the *n*-type semiconductor and the holes in the *p*-type semiconductor are attracted toward the battery. The width of the depletion layer is increased, and no charge carriers meet. Almost no current passes through the diode. It acts like a very high-value resistor; it is almost an insulator. A diode oriented in this manner is called a reverse-biased diode.

If the battery is connected in the opposite direction, as shown in the bottom right panel of **Figure 10,** charge carriers move toward the junction. If the battery's potential difference is large enough—0.6 V for a silicon diode—electrons reach the *p*-end and fill the holes. The depletion layer is eliminated, and a current passes through the diode.

The battery continues to supply electrons for the *n*-end. It removes electrons from the *p*-end, which is the same as supplying holes. A diode in this kind of circuit is a forward-biased diode.

The arrow in the diode symbol indicates the direction that the diode is facing. That is, the arrow indicates the direction that conventional current will be allowed through the diode.

Current and potential difference The graph in **Figure 11** shows the current through a silicon diode as a function of potential difference across it. If the applied potential difference is negative, the diode is reverse-biased and acts like a very high-value resistor. Only a tiny current, about 10^{-11} A for a silicon diode, passes. If the potential difference is positive, the diode is forward-biased and acts like a low-value resistor. Increasing the battery's potential difference increases the current in the diode. The diode does not, however, obey Ohm's law. The potential difference across a typical diode is 0.7 V.

One major use of a diode is to convert alternating current (AC) to direct current (DC). That is, it acts like a one-way door. When a diode is used in such a way, it is called a rectifier.

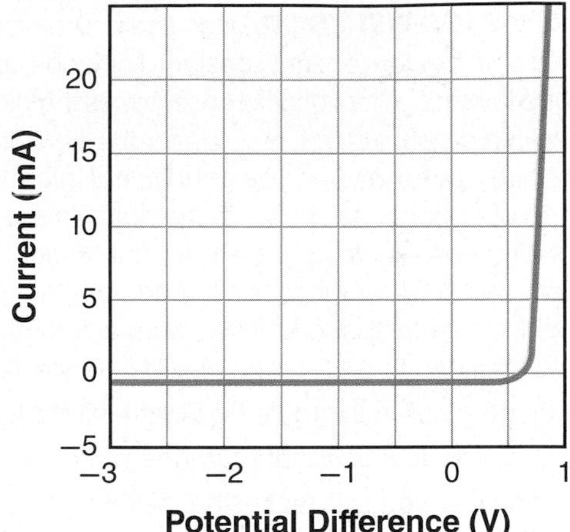

Figure 11 For potential differences below about 0.6 V, the diode is reverse-biased, and there is no current. Above the threshold, the diode is forward-biased, and there is a current.

EXAMPLE PROBLEM 4

Find help with **operations with significant figures.** Math Handbook

A DIODE IN A SIMPLE CIRCUIT A 0.70-V silicon diode, with $\frac{I}{\Delta V}$ characteristics like those shown in **Figure 11,** is connected to a power supply through a 470-Ω resistor. The power supply forward-biases the diode, and its potential difference is adjusted until the diode current is 12 mA. What potential difference does the power supply produce?

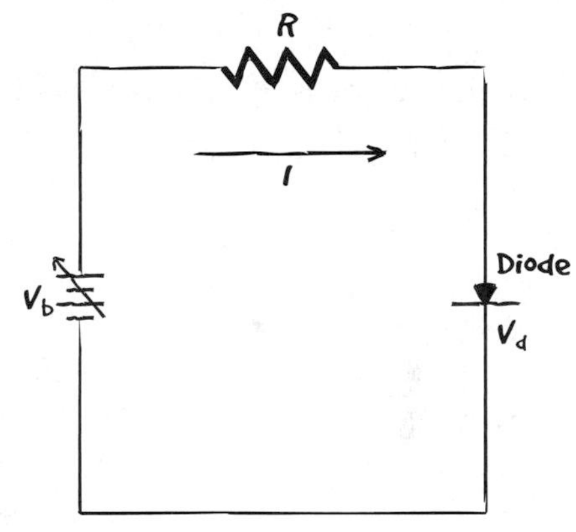

1 ANALYZE AND SKETCH THE PROBLEM

Draw a circuit diagram connecting a diode, a 470-Ω resistor, and a power supply. Indicate the direction of current.

KNOWN	UNKNOWN
$I = 0.012$ A	$\Delta V_b = ?$
$\Delta V_d = 0.70$ V	
$R = 470\ \Omega$	

2 SOLVE FOR THE UNKNOWN

The potential difference across the resistor is known from $\Delta V = IR$. By Kirchhoff's Law, the power supply's potential difference is the sum of the potential differences across the resistor and the diode.

$$\Delta V_b = IR + \Delta V_d$$

$$= (0.012\ \text{A})(470\ \Omega) + 0.70\ \text{V} = 6.3\ \text{V}$$

◀ Substitute $I = 0.012$ A, $R = 470$, $\Delta V_d = 0.70$ V.

3 EVALUATE THE ANSWER

- **Are the units correct?** The power supply's potential difference is in volts.
- **Is the magnitude realistic?** It is in accord with the current and the resistance.

PRACTICE PROBLEMS

Do additional problems. Online Practice

Unless otherwise noted, $\Delta V_{diode} = 0.70$ V.

26. What battery potential difference would be needed to produce a current of 2.5 mA in the diode in Example Problem 4?

27. A 0.40-V germanium diode is wired in series with a 470-Ω resistor. What potential difference would a battery need to have in order to produce a diode current of 12 mA?

28. What battery potential difference would be needed to produce a current of 2.5 mA if two identical diodes are wired in series with a 470-Ω resistor?

29. Describe how the diodes in the previous problem should be connected.

30. CHALLENGE Describe what would happen in the previous two problems if the diodes were connected in series but with improper polarity.

EXAMPLE PROBLEM

PHYSICS CHALLENGE

DIODE CHARACTERISTICS We often use approximations in diode circuits because diode resistance is not constant. For diode circuits, the first approximation ignores the forward potential difference across the diode. The second approximation takes into account a typical value for the potential difference across the diode. A third approximation uses additional information about the diode, often in the form of a graph, as shown to the right. This is the characteristic current-potential difference curve for the diode. The straight line shows current-potential difference conditions for all possible diode potential differences for a 180-Ω resistor, a 1.8-V battery, and a diode, from a zero diode potential difference and 10.0 mA at one end to a 1.8-V potential difference, 0.0 mA at the other end.

Use the diode circuit in Example Problem 4 with $\Delta V_b = 1.8$ V, but with $R = 180$ Ω.

1. Determine the diode current using the first approximation.
2. Determine the diode current using the second approximation and assuming a 0.70-V diode drop.
3. Determine the diode current using the third approximation by using the accompanying diode graph.
4. Estimate the error for all three approximations, ignoring the battery and the resistor. Discuss the impact of greater battery potential differences on the errors.

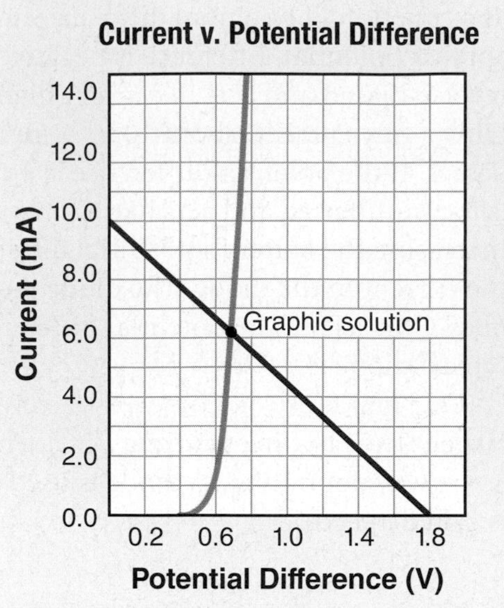

PhysicsLAB

DIODE CURRENT AND POTENTIAL DIFFERENCE

Do different diodes have different current-potential difference characteristics?

Light-emitting diodes Diodes that are made from certain combinations of elements, such as gallium and aluminum with arsenic and phosphorus, emit light when they are forward-biased. When electrons reach the holes in the junction, they recombine and release excess energy as radiant energy. These diodes are called light-emitting diodes (LEDs). Red LEDs typically have potential differences of about 1.9 V. White LEDs typically have potential differences of about 3.5 V.

Basic LEDs are used as indicator lights on computers, to make display signs, in traffic lights, and in some lamps. LEDs are bright, do not generate much heat, and are extremely efficient. Diodes can detect light as well as emit it. Light falling on the junction of a reverse-biased *pn*-junction diode produces electrons and holes, resulting in a current that depends on the light intensity.

Diode laser Some LEDs are configured to emit a narrow beam of coherent, monochromatic laser light. Such diode lasers are inexpensive, compact, powerful light sources. They are used in CD players, laser pointers, and supermarket bar-code scanners, as shown in **Figure 12.** A typical diode laser emits light at approximately 800 nm, which is the near infrared. The beam is output from a small spot on a gallium-aluminum-arsenic chip. When powered by 80 mA, the diode has a forward potential difference of about 2 V. Diode lasers are also used in many video game controllers.

Figure 12 The diode laser in a bar-code scanner emits light and detects light reflected from the bar code.

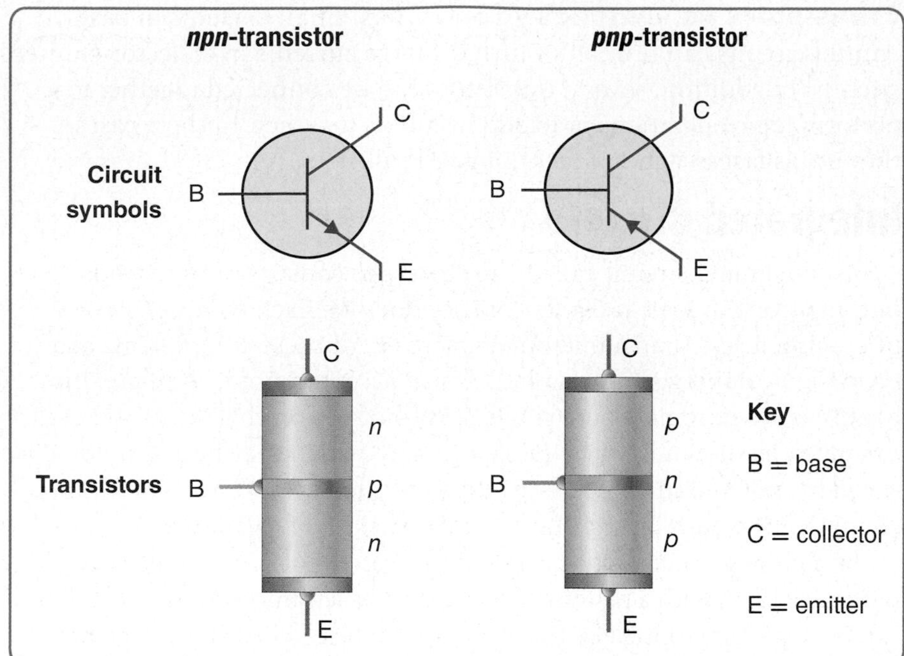

Figure 13 Transistors are made of three layers of semiconductor materials. Both *pnp*-transistors and *npn*-transistors have a collector and an emitter sandwiched around the base layer.

State what the arrow on the emitter means.

View an **animation about transistors.**

Transistors

A **transistor** is a device made of doped semiconductor material that can be used to amplify currents. One type is the *npn*-transistor, which consists of layers of *n*-type semiconductor on either side of a thin *p*-type layer, as shown in **Figure 13.** The *pnp*-transistor is similar. The central layer is called the base, and the regions on either side are the emitter and the collector. The schematic symbols for the two transistor types are also shown **Figure 13.** The arrow on the emitter shows the direction of conventional current.

Amplification An *npn*-transistor can be used to amplify current, as illustrated in **Figure 14.** You can think of the two *pn*-junctions in the transistor as two back-to-back diodes. If only the right battery (V_C) is connected, it keeps the collector more positive than the emitter. The base-collector diode is reverse-biased, with a wide depletion layer, so there is no current from the collector to the base. When the battery on the left (V_B) is connected, the base is more positive than the emitter. That makes the base-emitter diode forward-biased, allowing current (I_B) from the base to the emitter.

The very thin base region is part of both diodes in the transistor. The charges injected by I_B reduce the reverse bias of the base-collector diode, permitting charge to flow from the collector to the emitter. A small change in I_B thus produces a large change in I_C. The collector current causes a potential difference across resistor R_C. If V_B increases, the base current (I_B) increases. The collector current (I_C) will also increase, but many times more (perhaps 100 times or so). Small changes in the base potential (V_B) produce large changes in the collector current. The transistor amplifies small potential differences into much larger ones.

The current gain from the base to the collector $\left(\frac{I_C}{I_B}\right)$ is a useful indicator of a transistor's performance. The current gain ranges from 50 to 300 for general-purpose transistors. A *pnp*-transistor works the same way, except that the potentials of both batteries are reversed.

MiniLAB

RED LIGHT

What are the characteristics of an LED?

Figure 14 Transistors can be used to amplify currents. This circuit uses an *npn*-transistor.

Draw a circuit using a *pnp*-transistor.

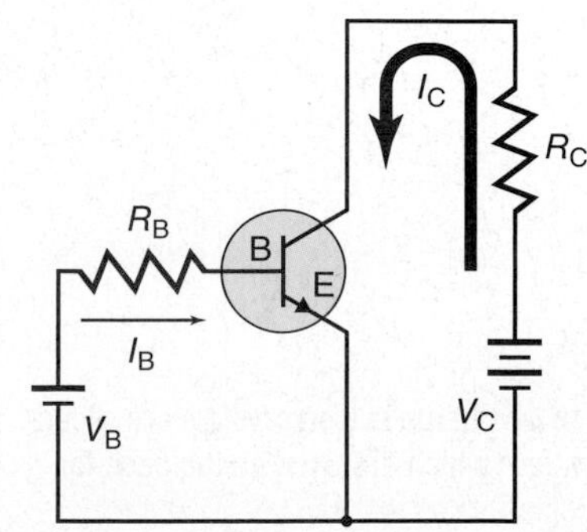

Transistors as switches In computers, small currents in base-emitter circuits can turn on or turn off large currents in collector-emitter circuits. In addition, several transistors can be connected together to perform logic operations or to add numbers together. In these cases, they act as fast switches rather than as amplifiers.

Integrated Circuits

An integrated circuit, called a **microchip,** consists of thousands of transistors, diodes, resistors, and conductors. Each component of a microchip is less than a micrometer across. All these components can be made by doping silicon with donor or acceptor atoms. A microchip begins as an extremely pure single crystal of silicon that is 10–30 cm in diameter, as shown in **Figure 15.** The crystal, which may be 1–2 m long, is sliced by a diamond-coated saw into wafers less than 1 mm thick. The circuit is then built layer by layer on the surface of this wafer.

By a photographic process, most of the wafer's surface is covered by a protective layer, with a pattern of selected areas left uncovered so that they can be doped appropriately. The wafer is then placed in a vacuum chamber. Vapors of a dopant such as arsenic enter the machine, doping the wafer in the unprotected regions. By controlling the amount of exposure, the engineer can control the conductivity of the exposed regions of the chip.

This process creates resistors, as well as one of the two layers of a diode or one of the three layers of a transistor. The protective layer is removed, and another one with a different pattern of exposed areas is applied. Then the wafer is exposed to another dopant, often gallium, producing *pn*-junctions. If a third layer is added, *npn*-transistors can be formed. The wafer also might be exposed to oxygen to produce areas of silicon dioxide insulation. A layer exposed to aluminum vapors can produce a pattern of thin conducting pathways among the resistors, diodes, and transistors. **Figure 15** also shows a wafer after it has completed this process.

☑ **READING CHECK** **Explain** Why is aluminum, instead of gallium or arsenic, used to make conducting pathways on the silicon wafer?

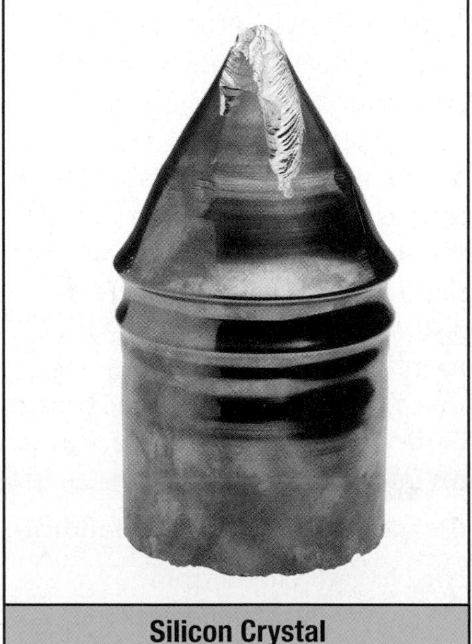

Silicon Crystal

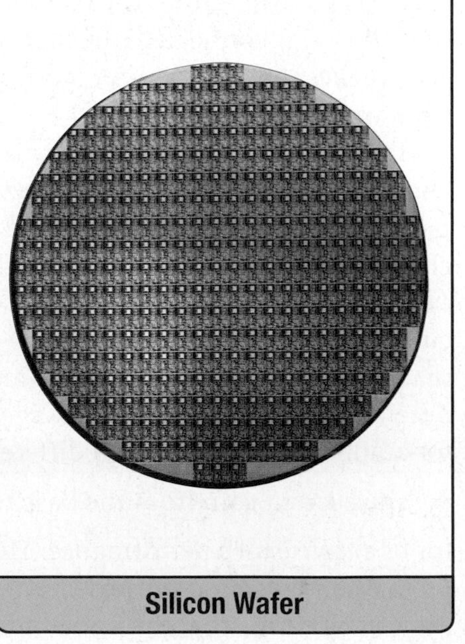

Silicon Wafer

Figure 15 Large silicon crystals are sliced into wafers which are used as the base for microchips.

Figure 16 The central processing unit of a computer is a collection of microchips.

Computing and microchips Thousands of identical circuits, usually called chips, are produced at one time on a single wafer. The chips are then tested, sliced apart, and mounted in a carrier. Wires are attached to the contacts, and the final assembly is then sealed into a protective plastic case. The tiny size of microchips allows the placement of complicated circuits in a small space, as shown in **Figure 16**. Because electronic signals need only travel tiny distances, this miniaturization has increased the processing speed, the functionality, and the reliability of computers. Chips are also used in appliances and automobiles as well as in computers.

Producing semiconductor electronics requires that physicists, chemists, and engineers work together. Physicists contribute their understanding of the motion of electrons and holes in semiconductors. Physicists and chemists together add precisely controlled amounts of dopants to extremely pure silicon to alter its conductivity. Engineers develop the means of mass-producing chips containing thousands of miniaturized diodes and transistors. Together, their efforts have brought our world into this electronic age.

PhysicsLAB

COMPUTER LOGIC

How do computers make decisions?

iLab Station

SECTION 2 REVIEW

Section Self-Check Check your understanding.

31. MAIN IDEA You have a low-current, AC signal. Describe how you could use a diode and a transistor to obtain a larger DC signal.

32. Diode Resistance Compare the resistance provided by a *pn*-junction diode when it is forward-biased and when it is reverse-biased.

33. Diode Polarity In a light-emitting diode, which battery terminal should be connected to the *p*-end to make the diode light?

34. Current Gain The base current in a transistor circuit measures 55 μA, and the collector current measures 6.6 mA. What is the base-to-collector current gain?

35. Transistor Circuit The emitter current in a transistor circuit is always equal to the sum of the base current and the collector current: $I_E = I_B + I_C$. If the current gain from the base to the collector is 95, what is the ratio of emitter current to base current?

36. Diode Circuit A 0.70-V silicon diode is connected to a power supply through a 110-Ω resistor. The power supply forward-biases the diode, and the adjusted current is 12 mA. What is the potential difference?

37. Diode Potential Difference A diode is characterized in **Figure 10.** It is forward-biased by a battery and a series resistor so that there is more than 10 mA of current and the potential difference is always about 0.70 V. The battery potential difference is then increased by 1 V.

a. By how much does the potential difference across the diode increase? Across the resistor?

b. By how much does the current through the resistor increase?

38. Critical Thinking Could you replace an *npn*-transistor with two separate diodes that are connected by their *p*-terminals? Explain.

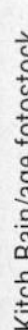

Kitch Bain/age fotostock

A CLOSER LOOK

ROBOTS IN SPACE!

FIGURE 1 The Cassini space probe explored Saturn and some of its moons. This is an artist's depiction of Cassini passing Saturn.

Far from Earth, robotic space probes explore the Sun, planets, and moons, making discoveries that couldn't have been made any other way. They are the farthest-flung objects we have ever produced. Voyager 1 and Voyager 2, launched by NASA in 1997, currently send scientific data from the outer edge of the solar system and will enter interstellar space in 2015.

Early robots Space probes in the 1960s were essentially flying cameras. Limited computing power meant that these spacecraft had highly restricted capabilities. They would fly by a target, such as the Moon, snapping as many photographs as possible.

New tools Later, robotic space probes completed more complex tasks. Increasingly sophisticated instrument packages allowed robotic probes to explore in new ways. Radio, microwave, infrared, and ultraviolet imaging opened new vistas. Orbiting robotic probes, such as the Cassini space probe shown in **Figure 1,** built detailed maps of our planetary neighbors. Robotic rovers, such as the Mars rover shown in **Figure 2,** performed complex chemical analyses right on the surface.

Artificial intelligence? In 1999, Deep Space 1, an experimental robotic probe with artificial intelligence, was tested. When spacecraft operations were turned over to the software, a computer was in charge of making decisions about spacecraft operations for the first time. Later, Deep Space 1 sent data from an asteroid and a comet.

The future As robotic probes perform more and more complex tasks far from home, they will need to become autonomous. Rovers on Mars, for instance, could explore much more of the planet if they could adjust operations on their own instead of waiting for signals delayed by several minutes to arrive from controllers back on Earth.

FIGURE 2 The Mars rover Spirit and its twin Opportunity spent over five years sampling and analysing the composition of Mars' surface.

GOINGFURTHER >>>

Design a robotic probe mission to a star system beyond the Sun, with the goal of exploring any planets there. What are the capabilities your robot must possess? Remember the distance, and therefore the time signal lag, as well as the fact that much about this faraway solar system will remain unknown until the robotic probe arrives.

CHAPTER 29 STUDY GUIDE

BIGIDEA We can make a variety of electronic devices by controlling the conductivity of materials.

VOCABULARY
- **semiconductor** *(p. 778)*
- **band theory** *(p. 779)*
- **intrinsic semiconductor** *(p. 782)*
- **dopant** *(p. 783)*
- **extrinsic semiconductor** *(p. 783)*

SECTION 1 Conduction in Solids

MAINIDEA A material's conductivity depends on the energy difference between the valence and conducting bands.

- The band theory of solids states that, in solids, the electric fields of neighboring atoms spread the allowed energy levels for an atom's outer electrons into broad bands. The valence and conduction bands are separated by forbidden energy gaps. The smaller the forbidden gap, the better the material's conduction.
- In conductors, electrons can move through the solid because the conduction band is partially filled. In insulators, more energy is needed to move electrons into the conduction band than is generally available. In semiconductors, the energy gap is small enough that some electrons can reach the conduction band.
- Conductivity of semiconductors increases with increasing temperature or illumination. Conduction in semiconductors is also enhanced by doping pure crystals with small amounts of other kinds of atoms, called dopants.
- The *n*-type semiconductors are doped with electron donor atoms. They conduct by the response of these donor electrons to applied potential differences. The *p*-type semiconductors are doped with electron acceptor atoms. They conduct by making holes available to electrons in the conduction band.

VOCABULARY
- **diode** *(p. 788)*
- **depletion layer** *(p. 788)*
- **transistor** *(p. 791)*
- **microchip** *(p. 792)*

SECTION 2 Electronic Components

MAINIDEA Diodes and transistors are the fundamental components in today's electronic circuits.

- A *pn*-junction diode consists of a layer of a *p*-type semiconductor joined with a layer of an *n*-type semiconductor. Diodes conduct charges in one direction only. If a potential difference is applied across the diode in the direction of polarity, the diode is forward-biased and conducts charge. If a potential difference is applied opposite its polarity, the diode is reverse-biased and does not conduct charge.
- Diodes can be used to convert AC to DC. Some diodes emit light when a potential difference is applied; these diodes can also detect light.
- A transistor is a sandwich of three layers of semiconductor material, configured as either *npn* or *pnp* layers. The center base layer is very thin compared to the other layers, which are called the emitter and the collector.
- A transistor can amplify a weak signal into a much stronger one. The ratio of the collector-emitter current to the base current is known as the current gain and is a useful measure of transistor amplification.

Games and Multilingual eGlossary.

Vocabulary Practice

Chapter Self-Check

SECTION 1 Conduction in Solids

Mastering Concepts

39. How do the energy levels in a crystal of an element differ from the energy levels in a single atom of that element?

40. BIGIDEA Why does heating a semiconductor increase its conductivity?

41. What is the main current carrier in a *p*-type semiconductor?

Mastering Problems

42. How many free electrons exist in a cubic centimeter of sodium? Its density is 0.971 g/cm^3, its atomic mass is 22.99 g/mol, and there is one free electron per atom.

43. At a temperature of 0°C, thermal energy frees 1.55×10^9 e^-/cm^3 in pure silicon. The density of silicon is 2.33 g/cm^3, and the atomic mass of silicon is 28.09 g/mol. What is the fraction of atoms that have free electrons?

44. Ranking Task Rank the following according to the number of free electrons per atom, from least to greatest. Specifically indicate any ties.

A. aluminum
B. gallium
C. indium
D. platinum
E. zinc

SECTION 2 Electronic Components

Mastering Concepts

45. An ohmmeter is an instrument that places a potential difference across a device to be tested, measures the resulting current, and displays the device's resistance. If you connect an ohmmeter across a diode, will the current you measure depend on which end of the diode is connected to the positive terminal of the ohmmeter? Explain.

46. What is the significance of the arrowhead at the emitter in a transistor circuit symbol?

47. Describe the structure of a forward-biased diode, and explain how it works.

Mastering Problems

48. Problem Posing Complete this problem so that its solution requires the use of a diode: "A 3.0-V alternating source is connected to a 450-W resistor …"

49. Diode A silicon diode with a voltage of 0.70 V is connected to a battery through a 270-Ω resistor. The battery forward-biases the diode, and the diode current is 15 mA. What is the battery's potential difference?

50. LED The potential difference across a glowing LED is about 1.2 V. In **Figure 17,** the potential difference across the resistor is the difference between the battery potential difference and the LED's potential difference. What is the current through each of the following?

a. the LED

b. the resistor

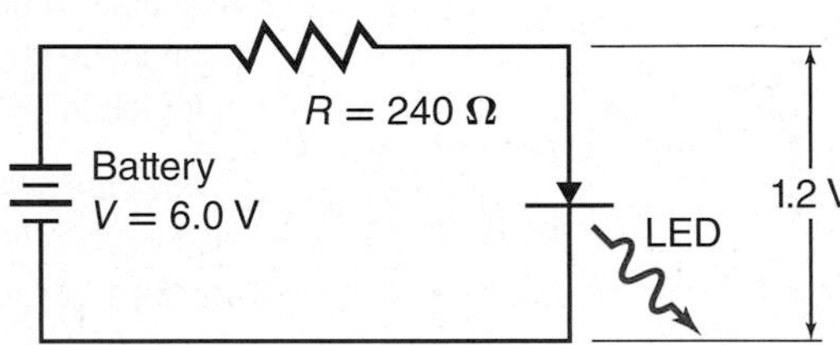

Figure 17

51. Jon wants to raise the current through the LED in **Figure 17** up to 3.0×10^1 mA so that it glows brighter. The potential difference across the LED is still 1.2 V. What resistor should be used?

52. Assume that the switch shown in **Figure 18** is off.

a. Determine the base current.

b. Determine the collector current.

c. Determine the voltmeter reading.

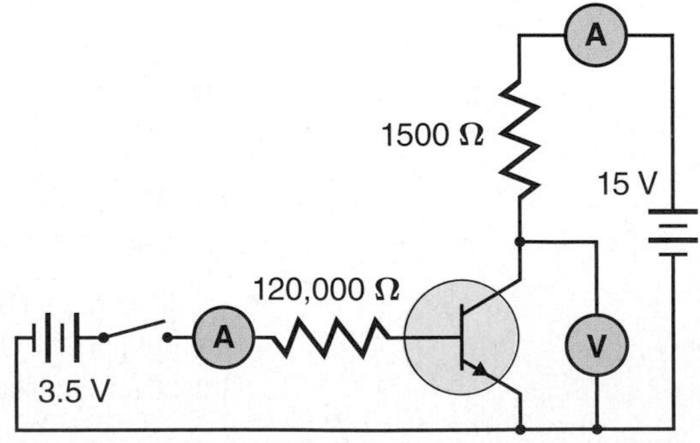

Figure 18

53. When the switch shown in **Figure 18** is on, there is a 0.70-V drop across the base-emitter junction and a current gain from base to collector of 220.

a. Determine the base current.

b. Determine the collector current.

c. Determine the voltmeter reading.

Chapter Self-Check

Applying Concepts

54. For the energy-band diagrams shown in **Figure 19,** which one represents a material with an extremely high resistance?

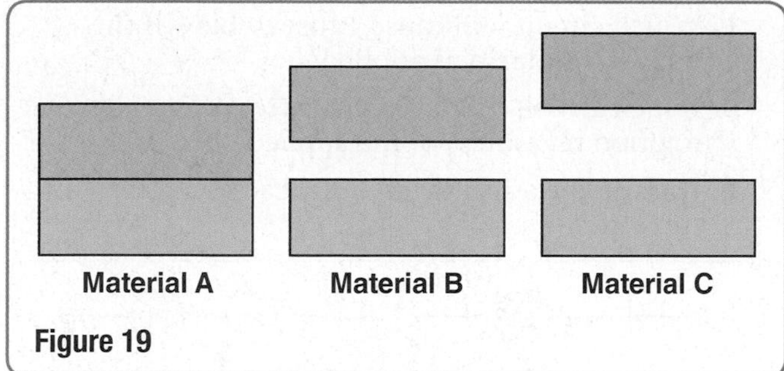

Figure 19

55. For the energy-band diagrams shown in **Figure 19,** which has a half-full conduction band?

56. For the energy-band diagrams shown in **Figure 19,** which one represents semiconductors?

57. Which of the following materials would make the best insulator: one with a forbidden gap 8 eV wide, one with a forbidden gap 3 eV wide, or one with no forbidden gap?

58. Consider atoms of the materials in the previous problem. From which of the three materials would it be most difficult to remove an electron?

59. State whether the bulb in each of the circuits of **Figure 20** is lighted.

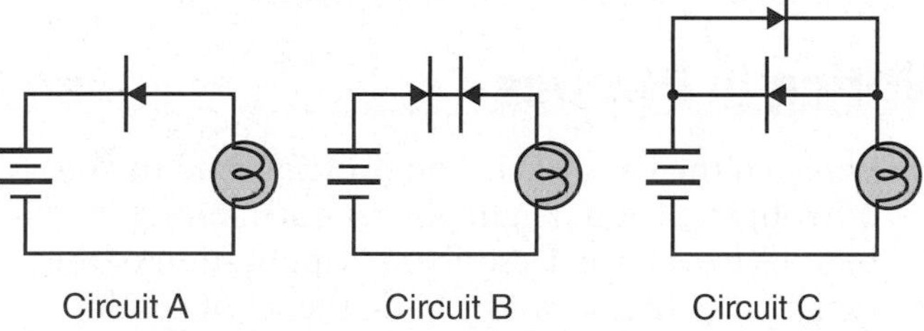

Figure 20

60. In the circuit shown in **Figure 21,** state whether lamp L_1, lamp L_2, both, or neither is lighted.

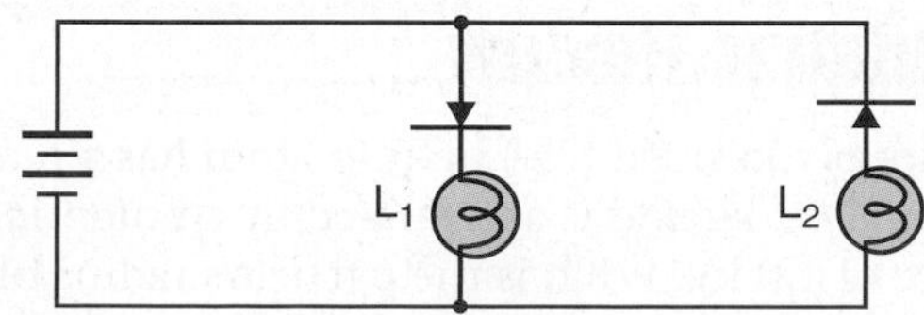

Figure 21

61. If you dope pure germanium with gallium alone, do you produce a resistor, a diode, or a transistor?

62. Use the periodic table to determine which of these elements could be added to germanium to make a *p*-type semiconductor: B, N, P, Si, Al, Ga, As, Sn, Sb.

63. Graphite's resistance decreases as temperature rises. Does graphite conduct electricity more like copper or more like silicon does?

64. Does an ohmmeter show a higher resistance when a diode is forward-biased or reverse-biased?

65. If the ohmmeter in the previous problem shows the lower resistance, is the ohmmeter lead on the arrow side of the diode at a higher or lower potential than the lead connected to the other side?

66. Draw the time-versus-amplitude waveform for point A in the circuit in **Figure 22,** assuming the input AC waveform shown in the figure.

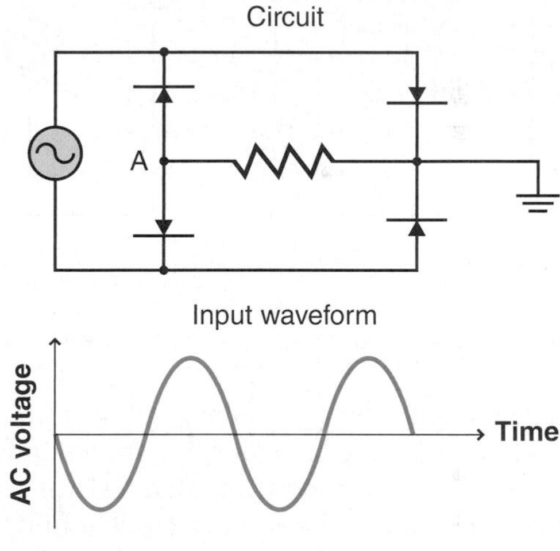

Figure 22

Mixed Review

67. Silicon's forbidden gap is 1.1 eV. Electromagnetic waves striking the silicon cause valence electrons to move into the conduction band. What is the longest wavelength of radiation that could excite an electron in this way? Recall that $E = \frac{1240 \text{ eV}\cdot\text{nm}}{\lambda}$.

68. Si Diode A particular silicon diode at 0°C shows a current of 1.0 nA when it is reverse-biased. What current can be expected if the temperature increases to 104°C? Assume that the reverse-bias potential difference remains constant. (The thermal carrier production of silicon doubles for every 8°C increase in temperature.)

69. LED A light-emitting diode produces green light ($\lambda = 550$ nm) when an electron moves from the conduction band to the valence band. Find the width of the forbidden gap in eV in this diode.

Chapter Self-Check

70. Ge Diode A particular germanium diode at 0°C shows a current of 1.5 μA when it is reverse-biased. What current can be expected if the temperature increases to 104°C? Assume that the reverse-bias potential difference remains constant. (The thermal charge-carrier production of germanium doubles for every 13°C increase in temperature.)

71. Refer to **Figure 23.**

a. Determine the voltmeter reading.

b. Determine the reading of A_1.

c. Determine the reading of A_2.

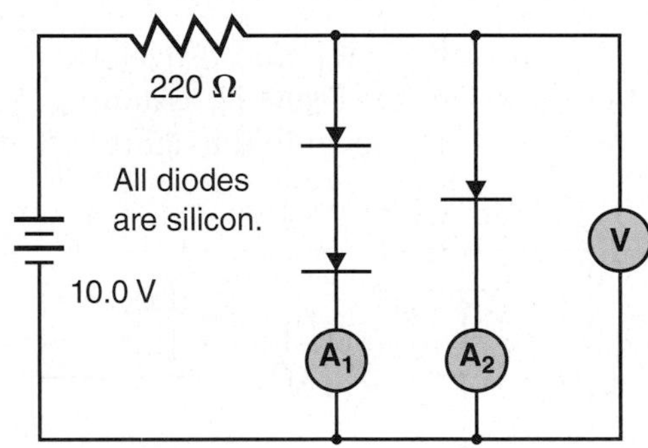

Figure 23

Thinking Critically

72. Apply Concepts The $\frac{I}{\Delta V}$ characteristics of two LEDs that glow with different colors are shown in **Figure 24.** Each is to be connected through a resistor to a 9.0-V battery. If each is to be run at a current of 0.040 A, what resistors should be chosen for each?

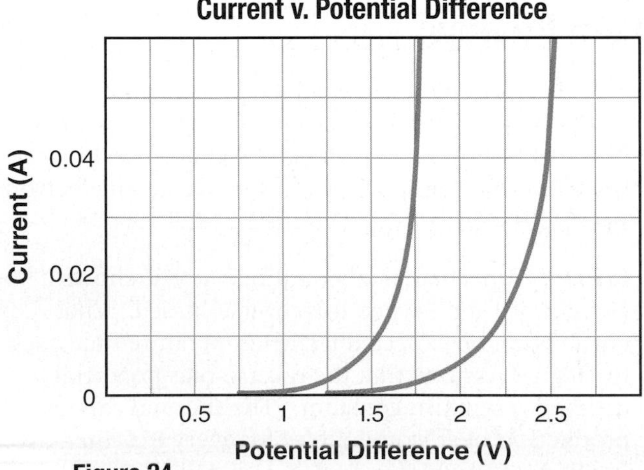

Figure 24

73. Apply Concepts Suppose that the two LEDs in the previous problem are now connected in series. If the same battery is to be used and a current of 0.035 A is desired, what resistor should be used?

74. Apply Concepts The motor in **Figure 25** runs in one direction with a given polarity applied and reverses direction with the opposite polarity.

a. Which circuit will allow the motor to run in only one direction?

b. Which circuit will cause a fuse to blow if the incorrect polarity is applied?

c. Which circuit produces the correct direction of rotation regardless of the applied polarity?

d. Discuss the advantages and disadvantages of all three circuits.

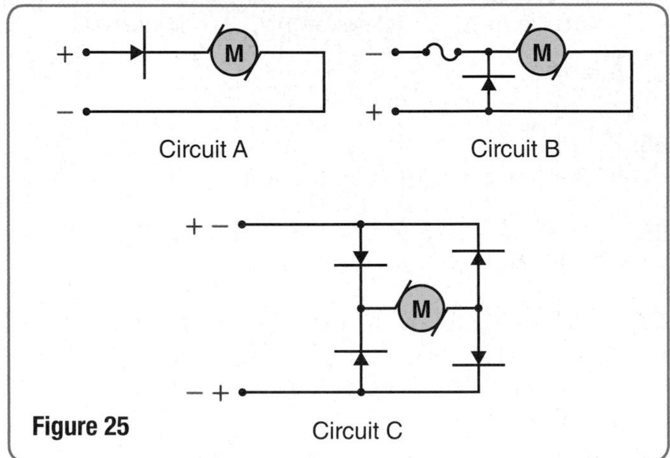

Figure 25

75. Reverse Problem Write a physics problem with real-life objects for which the following expression would be part of the solution:

$$\left(\frac{2e^-}{1 \text{ atom}}\right)\left(\frac{6.02\times10^{23} \text{ atoms}}{1 \text{ mol}}\right)\left(\frac{1 \text{ mol}}{112.411 \text{ g}}\right)\left(\frac{8.645 \text{ g}}{\text{cm}^3}\right).$$

Writing in Physics

76. Research the Pauli exclusion principle and the life of Wolfgang Pauli. Highlight his outstanding contributions to science. Describe the application of the exclusion principle to the band theory of conduction, especially in semiconductors.

77. Write a one-page paper discussing the Fermi energy level as it applies to energy-band diagrams for semiconductors. Include at least one drawing.

Cumulative Review

78. A doubly ionized (2+) helium atom has a mass of 6.7×10^{-27} kg and is accelerated by a potential difference of 1.0 kV. What is the particle's radius of curvature in a uniform magnetic field of 6.5×10^{-2} T?

79. What potential difference is needed to stop photoelectrons that have a maximum kinetic energy of 8.0×10^{-19} J?

CHAPTER 29 STANDARDIZED TEST PRACTICE

MULTIPLE CHOICE

1. Which line in the following table best describes both n- and p-type silicon semiconductors?

	n-type	*p*-type
A.	gallium-doped	added electrons
B.	added electrons	arsenic-doped
C.	arsenic-doped	added holes
D.	added holes	gallium-doped

2. Which statement about diodes is false?

A. Diodes can amplify potential differences.
B. Diodes can detect light.
C. Diodes can emit light.
D. Diodes can rectify AC.

3. The base current in a transistor circuit measures 45 μA, and the collector current measures 8.5 mA. What is the current gain from base to collector?

A. 110 **C.** 205
B. 190 **D.** 240

4. In the previous problem, if the base current is increased by 5 μA, how much is the collector current increased?

A. 5 μA **C.** 10 mA
B. 1 mA **D.** 190 μA

5. A transistor circuit shows a collector current of 4.75 mA, and the base-to-collector current gain is 250. What is the base current?

A. 1.19 μA **C.** 4.75 mA
B. 19.0 μA **D.** 1190 mA

6. Cadmium has two free electrons per atom. How many free electrons are there per cubic centimeter of cadmium? The density of cadmium is 8650 kg/m^3, and its atomic mass is 112.4 g/mol.

A. 1.24×10^{21} **C.** 9.26×10^{24}
B. 9.26×10^{22} **D.** 1.17×10^{27}

7. Thermal electron production in silicon doubles for every 8°C increase in temperature. A silicon diode at 0°C shows a current of 2.0 nA when reverse-biased. What will be the current at 112°C if the reverse-bias potential difference is constant?

A. 11 μA **C.** 44 μA
B. 33 μA **D.** 66 μA

8. Which line in the following table best describes the behavior of intrinsic silicon semiconductors to increasing temperature?

	Effect of Increasing Temperature on Intrinsic Silicon Semiconductors	
	Conductivity	Resistance
A.	increases	increases
B.	increases	decreases
C.	decreases	increases
D.	decreases	decreases

FREE RESPONSE

9. A silicon diode is connected in the forward-biased direction to a power supply though a 485-Ω resistor, as shown below. If the diode potential difference is 0.70 V, what is the power supply potential difference when the diode current is 14 mA?

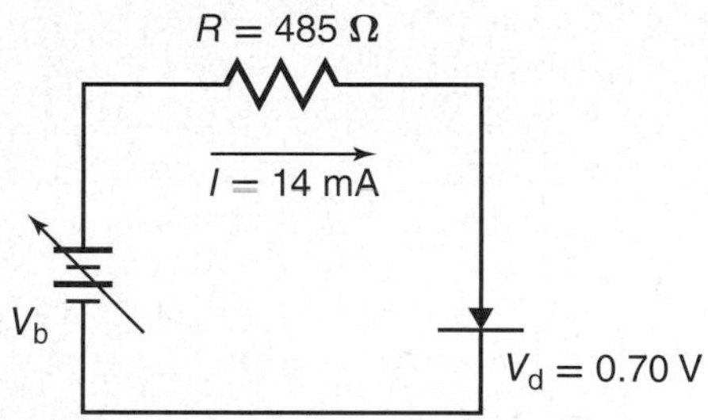

NEED EXTRA HELP?

If You Missed Question	1	2	3	4	5	6	7	8	9
Review Section	1	2	2	2	2	1	2	1	2

CHAPTER 30

Nuclear and Particle Physics

BIGIDEA **Splitting and fusing atomic nuclei release energy.**

SECTIONS

1 **The Nucleus**

2 **Nuclear Decay and Reactions**

3 **The Building Blocks of Matter**

LaunchLAB

A NUCLEAR MODEL

If protons repel each other, how does the nucleus—which is made of protons and neutrons—stay together?

WATCH THIS!

SMOKE DETECTORS

You probably have a smoke detector in your home. Many smoke detectors use radiation in order to function. Where does this radiation come from? And how does the radiation help the smoke detector work?

(l)S. Wanke/PhotoLink/Getty Images, (r)Imagebroker/Photolibrary

Go online!
connectED.mcgraw-hill.com

SECTION 1

The Nucleus

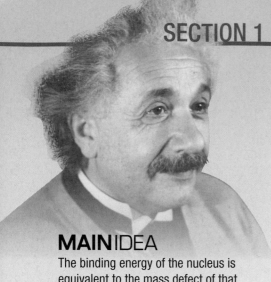

PHYSICS 4 YOU **$E = mc^2$ might be the most famous equation in physics. It shows that energy is the product of mass and the speed of light squared, but what does this mean?**

MAIN IDEA

The binding energy of the nucleus is equivalent to the mass defect of that nucleus.

Essential Questions

- What properties does the nucleus of an atom have?
- In what ways do the nuclei of isotopes differ?
- What force holds the nucleus together?
- What are the binding energy and the mass defect of the nucleus?
- How is the mass defect of a nucleus related to the binding energy?

Review Vocabulary

nucleus the tiny, massive, positively charged central core of an atom

New Vocabulary

nucleon
atomic number
mass number
atomic mass unit
strong nuclear force
mass defect
binding energy

Description of the Nucleus

Recall from your study of atoms that Ernest Rutherford and his team measured the deflection of alpha particles as they hit gold foil. From this investigation, Rutherford concluded that the atom has a small, dense, positively charged nucleus surrounded by nearly massless electrons, as shown in **Figure 1.** However, the investigation gave no information about the structure of the nucleus. Henry Mosely, a member of Rutherford's team, found the magnitude of the nuclear charge during X-ray scattering experiments. The results showed that the positively charged protons accounted for roughly half the mass of the nucleus. What accounted for the other half?

Protons and neutrons One hypothesis was that the nucleus contained protons and enough electrons to reduce the charge to the observed value. A second hypothesis was that the nucleus contained uncharged particles as well as protons. In 1932 English physicist James Chadwick discovered the neutron, a neutral particle with a mass similar to that of a proton. This particle accounted for the missing mass of the nucleus and its charge.

To describe an atom's nucleus, you need to know how many protons and neutrons it has. Protons and neutrons are collectively referred to as **nucleons.** The number of protons is the **atomic number** (Z). To find the charge of the nucleus, multiply this number by the elementary charge (e). For example, a nucleus with five protons has a charge of +5e. The number of nucleons in an atom is the **mass number** (A). Protons and neutrons have almost the same mass—approximately 1.66×10^{-27} kg, or 1 **atomic mass unit** (u). This is more than the mass of 1800 electrons. To find the approximate nuclear mass, multiply the mass number by u.

$$\text{approximate nuclear mass} \cong A \times \text{u}$$

The Nucleus

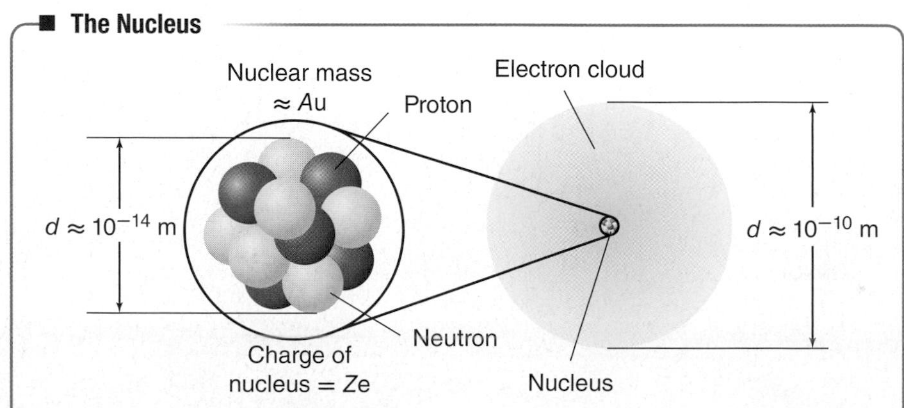

Figure 1 The atom consists of a very dense nucleus surrounded by an electron cloud. The nucleus is made of protons and neutrons.

Sylvia Salmi/Bettman/CORBIS

Size of the nucleus Rutherford's investigations produced the first measurements of the size of the nucleus. He found that nuclei have diameters of about 10^{-14} m. A typical atom might have a radius 10,000 times larger than the diameter of its nucleus. Although the nucleus contains nearly all the mass of an atom, proportionally the nucleus occupies less space in the atom than the Sun does in our solar system. The nucleus is incredibly dense—about 1.4×10^{18} kg/m^3. If a nucleus could be 1 cm^3, it would have a mass of about 1 billion tons. This is roughly 90,000 times more massive than the largest ocean liners.

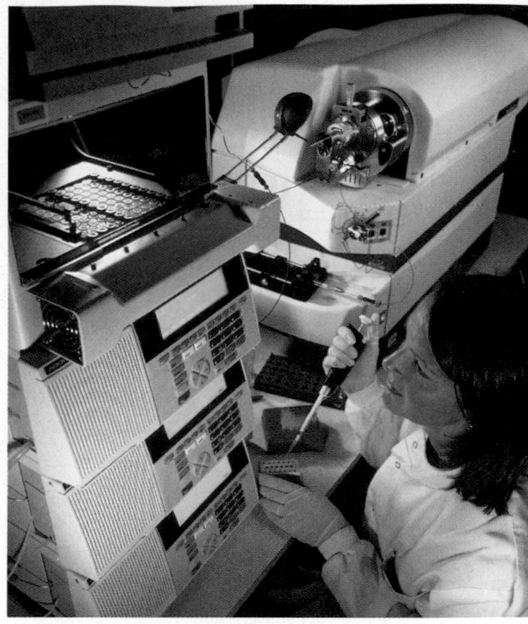

Figure 2 Mass spectrometers are used to measure the mass differences of isotopes.

Isotopes

Look at the periodic table in the back of your book, and examine the atomic masses of the elements. The first four elements have atomic masses near whole numbers of atomic mass units. Because the nucleus contains nearly all the mass and each nucleon has a mass of approximately 1 u, this seems to make sense. Boron, however, has a mass of 10.8 u. How can this be?

Nuclei of isotopes The puzzle of atomic masses that are not whole numbers can be solved by applying electromagnetism. Recall that experiments with mass spectrometers, such as the one in **Figure 2,** show that an element can have atoms with different masses. For example, in an analysis of a pure helium sample, two spots appeared on the detector of the spectrometer. The two spots were produced by helium atoms of different masses. One type of helium atom had a mass of 3 u, while the second type had a mass of 4 u. All neutral helium atoms have 2 protons and 2 electrons. But one kind of helium atom has 1 neutron in its nucleus, while the other has 2 neutrons. The two kinds of atoms are called isotopes of helium.

All nuclei of an element have the same number of protons but have different numbers of neutrons, as shown by the hydrogen and helium nuclei in **Figure 3.** All isotopes of a neutral element have the same number of electrons as protons and behave chemically in the same way.

Watch a **BrainPOP video about isotopes.**

☑ **READING CHECK** **Compare and contrast** the isotopes carbon-12 and carbon-14.

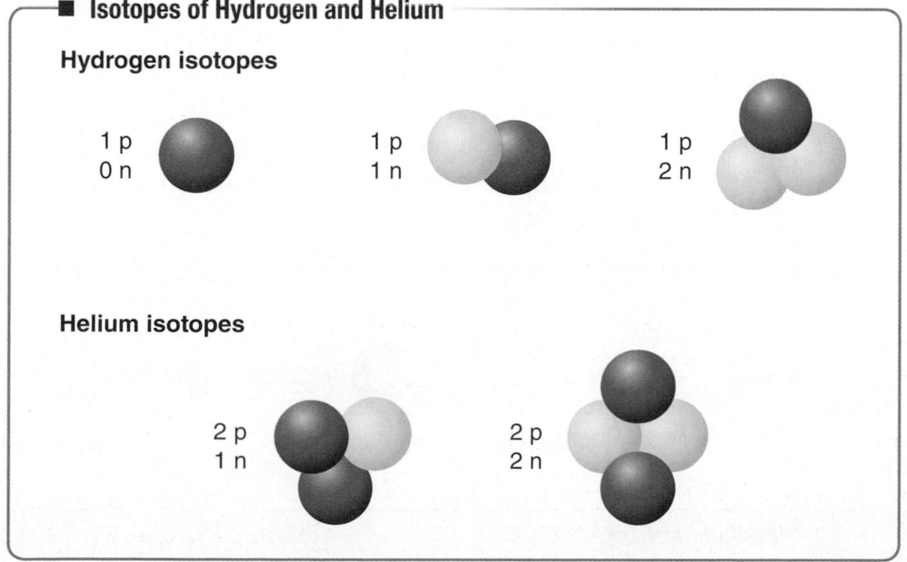

Figure 3 Isotopes of an element have the same number protons but different numbers of neutrons.

Identify the mass number and the electric charge of each nucleus.

Colin Cuthbert/Photo Researchers, Inc.

PRACTICE PROBLEMS

Online Practice

1. Three isotopes of uranium have mass numbers of 234, 235, and 238. The atomic number of uranium is 92. How many neutrons are in the nucleus of each isotope?
2. How many neutrons are in an atom of the mercury isotope $^{200}_{80}Hg$?
3. An isotope of oxygen has a mass number of 15. How many neutrons are in a nucleus of this isotope?
4. **CHALLENGE** Write the symbols for the three isotopes of hydrogen that have zero, one, and two neutrons in the nucleus.

A special notation is used to describe an isotope.

$$^{A}_{Z}X \quad \begin{array}{l} X = \text{element symbol} \\ A = \text{mass number} \\ Z = \text{atomic number} \end{array}$$

For example, hydrogen-2 (deuterium) is $^{2}_{1}H$, and the three naturally occurring isotopes of neon are $^{20}_{10}Ne$, $^{21}_{10}Ne$, and $^{22}_{10}Ne$.

Average mass The measured mass of chlorine is 35.453 u. This figure is now understood to be the average mass of the naturally occurring isotopes of helium. Thus, while the mass of an individual atom is close to a whole number of mass units, the atomic mass determined from an average sample of atoms does not have to be. The atomic mass of chlorine, 35.453 u, is the weighted average of its two isotopes, chlorine-35 and chlorine-36. Most elements have several isotopic forms that occur naturally. The mass of the isotope carbon-12 is now used to define the mass unit. One u is defined to be $\frac{1}{12}$ the mass of the carbon-12 isotope.

The Strong Nuclear Force

The negatively charged electrons around the positively charged nucleus of an atom are held in place by the attractive electromagnetic force. However, the nucleus consists of positively charged protons and neutral neutrons. The protons repel each other. Why don't the protons cause the nucleus to fly apart? We must conclude that an even stronger attractive force exists within the nucleus.

When any pair of nucleons gets within about a proton's radius (1.4×10^{-15} m) of each other, they experience an attraction that is more than 100 times stronger than the electromagnetic force. This **strong nuclear force,** also called the strong force, is an attractive force between nucleons that are close together, as they are in a nucleus. The strong force has the same strength between all nucleon pairs (proton-proton, neutron-neutron, or proton-neutron), as shown in **Figure 4.**

Figure 4 The strong force is the same between all pairs of nucleons.

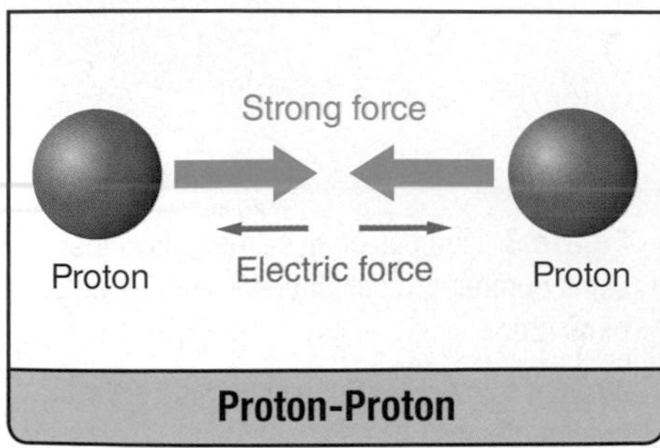

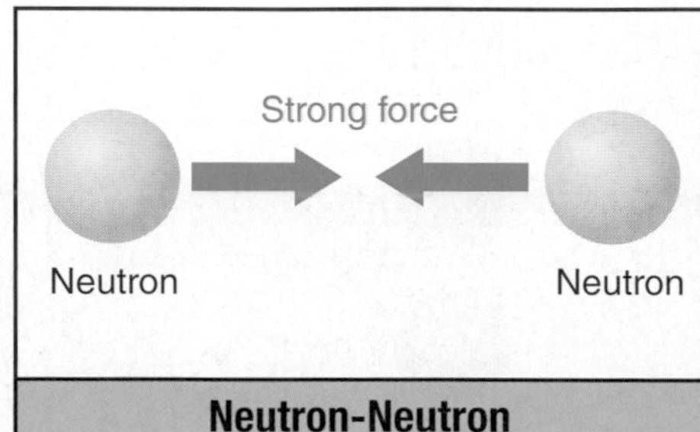

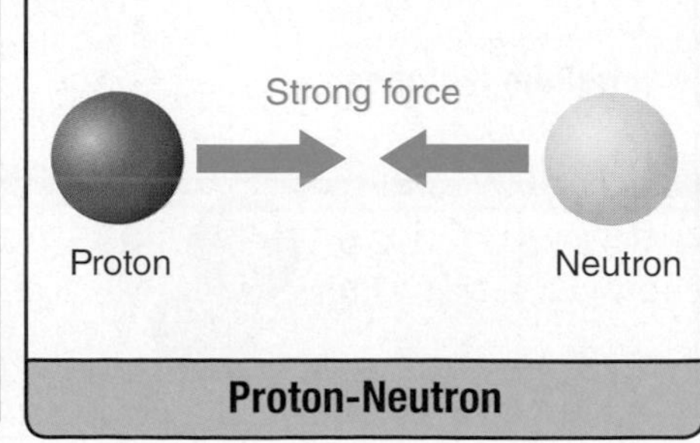

Mass defect and binding energy An initially surprising observation is that the mass of an assembled nucleus is less than the sum of the masses of the protons and neutrons that compose it. For example, the helium nucleus $^{4}_{2}He$ consists of two protons and two neutrons. The mass of a proton is 1.007276 u. The mass of a neutron is 1.008665 u. You might expect the mass of the nucleus to be 4.031882 u, but this is not the case. Careful measurement shows the mass of a helium nucleus is only 4.002603 u. The actual mass of the helium nucleus is less than the mass of its constituent parts by 0.029279 u. This difference between the sum of the masses of the individual nucleons and the actual mass is called the **mass defect.** Why is there a mass defect?

The answer comes from Einstein's work. He showed that mass and energy are the same. To pull a nucleon out of a nucleus, you have to do work to overcome the strong nuclear force. Doing work adds energy to the system. Thus, the system of separated protons and neutrons has more energy than the assembled nucleus. This energy difference between separated nucleons and the assembled nucleus is the **binding energy** of the nucleus. Because the assembled nucleus has less energy, all binding energies are negative, as shown in **Figure 5.** The binding energy can be expressed in the form of an equivalent amount of mass, according to Einstein's famous equation.

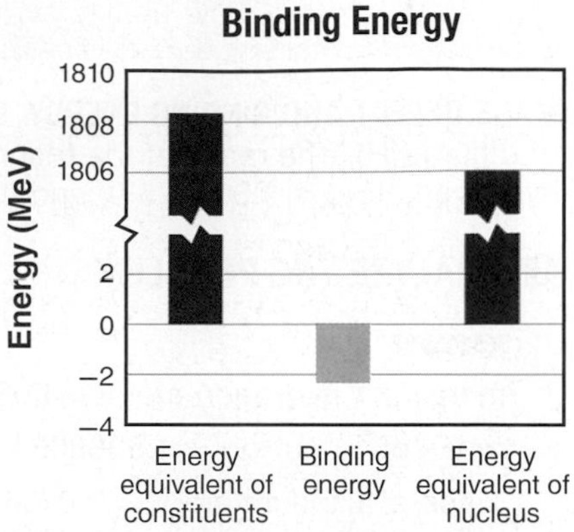

Figure 5 The energy equivalent of the mass of the individual nucleons is equal to the energy equivalent of the mass of the bound nucleus plus the binding energy.

Get help with **joules and electron volts.**

ENERGY EQUIVALENT OF MASS

The energy equivalent to a mass is equal to the mass times the square of the speed of light in a vacuum.

$$E = mc^2$$

Determining the mass defect Mass spectrometers usually measure the masses of isotopes; that is, they measure the mass of the nucleus and the electrons. When calculating the mass defect of a nucleus, one must make sure that the mass of the atom's electrons is accounted for properly. Thus, the mass of hydrogen (one proton and one electron) usually is given in mass-defect problems. Furthermore, masses are usually measured in atomic mass units. It is useful, then, to use Einstein's equation to determine the energy equivalent of 1 u (1.6605×10^{-27} kg). In this calculation, each quantity will be expressed with five significant digits.

$$\begin{aligned} E &= mc^2 \\ &= (1.6605\times10^{-27}\ \text{kg})(2.9979\times10^{8}\ \text{m/s})^2 \\ &= 1.4924\times10^{-10}\ \text{J} \end{aligned}$$

At this scale, the most convenient unit of energy to use is the electron volt (1 eV = 1.60218×10^{-19} J).

$$\begin{aligned} E &= (1.4924\times10^{-10}\ \text{J})\left(\frac{1\ \text{eV}}{1.60218\times10^{-19}\ \text{J}}\right) \\ &= 9.3149\times10^{8}\ \text{eV} = 931.49\ \text{MeV} \end{aligned}$$

Hence, each atomic mass unit of mass is equivalent to 931.49 MeV of energy. This ratio simplifies calculations when determining the binding energy of an isotope.

READING CHECK Describe the meaning of the equation $E = 931.49$ MeV/u, and explain why it is important.

EXAMPLE PROBLEM 1

Find help with **operations with significant figures.** Math Handbook

MASS DEFECT AND BINDING ENERGY Find the mass defect and binding energy of tritium ($^{3}_{1}H$). The mass of the tritium isotope is 3.016049 u, the mass of a hydrogen atom is 1.007825 u, and the mass of a neutron is 1.008665 u.

1 ANALYZE THE PROBLEM

KNOWN	**UNKNOWN**
mass of 1 hydrogen atom = 1.007825 u	total mass of tritium's parts = ?
mass of 1 neutron = 1.008665 u	mass defect = ?
mass of 1 tritium atom = 3.016049 u	binding energy of tritium = ?
energy equivalent of of 1 u = 931.49 MeV	

2 SOLVE FOR THE UNKNOWN

Determine the total mass of tritium's parts.

mass of 1 hydrogen atom:	1.007825 u	◀ *Add the masses of tritium's parts: one hydrogen atom (one proton and one electron) and two neutrons.*
mass of 2 neutrons:	+ 2.017330 u	
total mass of nucleons and electron:	3.025155 u	

Determine the mass defect.

mass of tritium:	3.016049 u	◀ *The mass defect is equal to the mass of the assembled tritium minus the mass of the sum of its parts.*
mass of tritium's pieces:	− 3.025155 u	
mass defect:	− 0.009106 u	

Determine the binding energy of tritium.

E = (mass defect in u)(energy equivalent of 1 u)

◀ *The binding energy is the energy equivalent of the mass defect.*

$E = (-0.009106\ \text{u})(931.49\ \text{MeV/u})$

$= -8.4821\ \text{MeV}$

◀ *Substitute mass defect = -0.009106 u and binding energy per u = 931.49 MeV.*

3 EVALUATE THE ANSWER

- **Are the units correct?** Mass is measured in u, and energy is measured in MeV.
- **Does the sign make sense?** Binding energy should be negative.
- **Is the magnitude realistic?** According to **Figure 6,** binding energies per nucleon in this range are between −2 MeV and −3 MeV, so the answer for three nucleons is reasonable.

PRACTICE PROBLEMS

Do additional problems. Online Practice

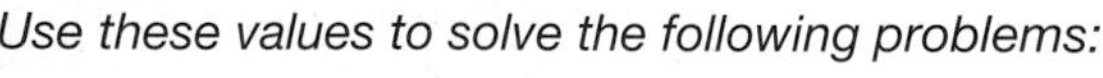

Use these values to solve the following problems:

mass of hydrogen = 1.007825 u

mass of neutron = 1.008665 u

1 u = 931.49 MeV

5. The carbon isotope $^{12}_{6}C$ has a mass of 12.000000 u.

a. Calculate its mass defect.

b. Calculate its binding energy.

6. Deuterium ($^{2}_{1}H$) has a mass of 2.014102 u.

a. Calculate its mass defect.

b. Calculate its binding energy.

7. The nitrogen isotope $^{15}_{7}N$ has a mass defect of −0.113986 u.

a. Calculate the mass of this isotope.

b. Calculate the binding energy of the nucleus.

8. CHALLENGE The oxygen isotope $^{16}_{8}O$ has a mass of 15.994915 u. A second oxygen isotope is $^{18}_{8}O$, which has a mass of 17.999160 u.

a. What is the mass defect of each isotope?

b. What is the binding energy of each nucleus?

c. Why do you think the binding energies are different?

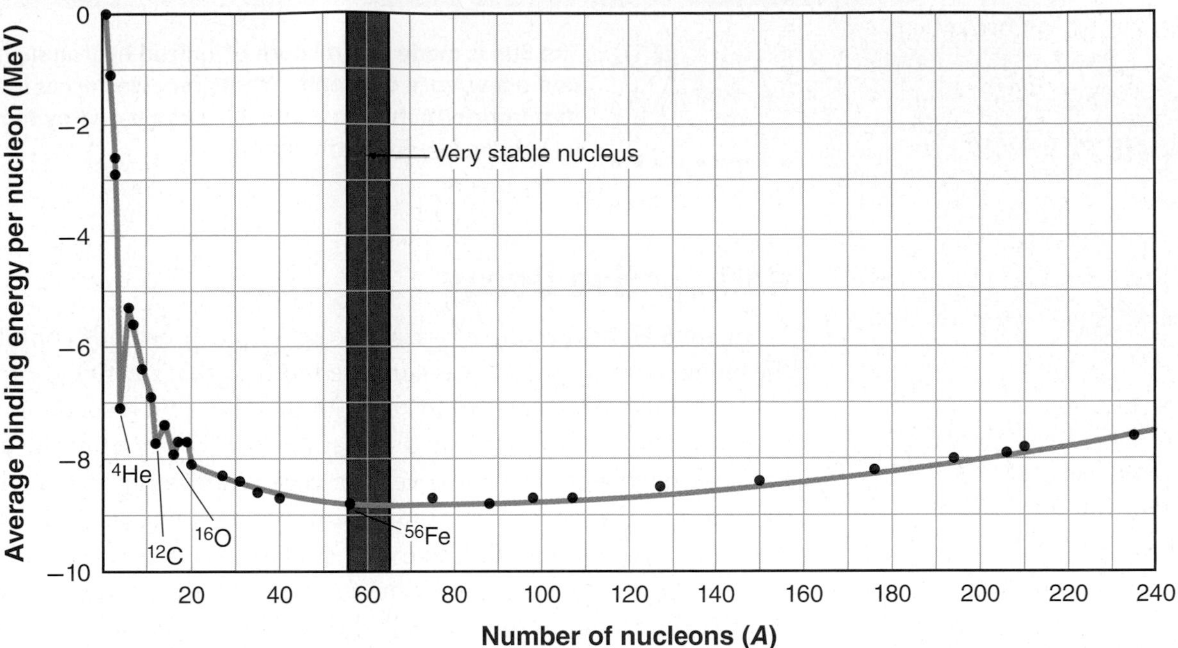

Figure 6 The binding energy per nucleon changes as the number of nucleons (A) varies. Iron-56 has the most tightly bound nucleus and is the most stable isotope known.

Infer Which nucleus, $^{131}_{52}I$ or $^{236}_{94}Pu$, would you expect to have a larger-magnitude average binding energy per nucleon?

Binding energy and nuclear mass **Figure 6** shows that the binding energy per nucleon depends on the mass of the nucleus. Except for a few nuclei, the binding energy per nucleon becomes more negative as the number of nucleons (A) increases to 56, which is that of iron (Fe). A more negative average binding energy indicates tightly bound nucleons and a more stable nucleus. The most stable nucleus is $^{56}_{26}Fe$ because the magnitude of its binding energy is greatest. Smaller nuclei become more stable as the number of nucleons approaches 56. Nuclei with more than 56 nucleons are less strongly bound and therefore less stable.

Large, unstable nuclei will readily decay into smaller, more stable nuclei. For example, all nuclei with atomic numbers greater than 92 decay readily. Under certain conditions, such as very high temperature, nuclei smaller than $^{56}_{26}Fe$ can fuse into larger, more stable nuclei. In the Sun and other stars, for example, hydrogen is converted into helium, carbon, and other heavier elements in a reaction that releases energy, causing the electromagnetic radiation that you experience as visible light.

SECTION 1 REVIEW

Check your understanding.

9. **MAIN IDEA** The radioactive carbon isotope $^{14}_{6}C$ has a mass of 14.003074 u.
   a. What is the mass defect of this isotope?
   b. What is the binding energy of its nucleus?
10. **Nuclei** Consider these two pairs of nuclei: $^{12}_{6}C$ and $^{13}_{6}C$, and $^{11}_{5}B$ and $^{11}_{6}C$. In which way are the two pairs alike? In which way are they different?
11. **Binding Energy** When a tritium nucleus ($^{3}_{1}H$) decays, it emits an electron. The tritium isotope becomes the helium isotope $^{3}_{2}He$. Which of the two nuclei would you expect to have a more negative binding energy? Explain.
12. **Mass Defect** Which of the two nuclei in the previous problem has the larger mass defect?
13. **Strong Nuclear Force** The range of the strong nuclear force is so short that only nucleons that are adjacent to each other are affected by the force. Use this fact to explain why in large nuclei the repulsive electromagnetic force can overcome the strong attractive force and make the nucleus unstable.
14. **Critical Thinking** In stars, small nuclei fuse to form nuclei with smaller binding energy per nucleon, such as helium and carbon. Older stars also produce oxygen ($Z = 8$) and silicon ($Z = 14$). What is the atomic number of the heaviest nucleus that could be formed in this way? Explain.

SECTION 2 Nuclear Decay and Reactions

The Sun is made of a mixture of ionized hydrogen, helium, and a few trace elements. What processes occur in the Sun that transform this mass into the radiant energy that gives us light and keeps Earth warm?

MAINIDEA

Radioactive decay releases particles and energy and can change an atom into another element.

Essential Questions

- What are some characteristics of the main types of radioactive decay?
- What is an isotope's half-life?
- How does nuclear fission release energy? How is this used in a nuclear reactor?
- How does nuclear fusion release energy?

Review Vocabulary

isotope each of the differing forms of the same atom that have different masses but have the same chemical properties

New Vocabulary

radioactive
alpha decay
beta decay
gamma decay
nuclear reaction
half-life
activity
fission
chain reaction
fusion

Radioactive Decay

In 1896 Henri Becquerel was working with compounds containing the element uranium. He was surprised to find that nearby photographic plates that had been covered to keep out light became fogged, or partially exposed. This fogging suggested that some kind of ray from the uranium had passed through the plate coverings. **Figure 7** shows a photographic plate that Becquerel exposed to this radiation. Several materials other than uranium or its compounds also were found to emit these penetrating rays.

In 1909, after years of experiments, Ernest Rutherford was able to identify the radiation that Becquerel had found. This radiation was composed of helium-4 nuclei. The larger uranium nuclei were decaying and emitting these helium nuclei. Nuclei that emit particles and energy are said to be **radioactive.** A material that emits particles is also said to decay. Nuclei decay in order to become more stable. Radioactive decay is a natural process.

In few areas of physics has basic knowledge led to applications as quickly as in the field of nuclear physics. The medical use of the radioactive element radium began within 20 years of its discovery. Proton accelerators were used for medical applications less than one year after being invented. In the case of nuclear fission (splitting nuclei), the military application was under development before the basic physics was even known. Peaceful applications, such as power plants, followed in less than 10 years.

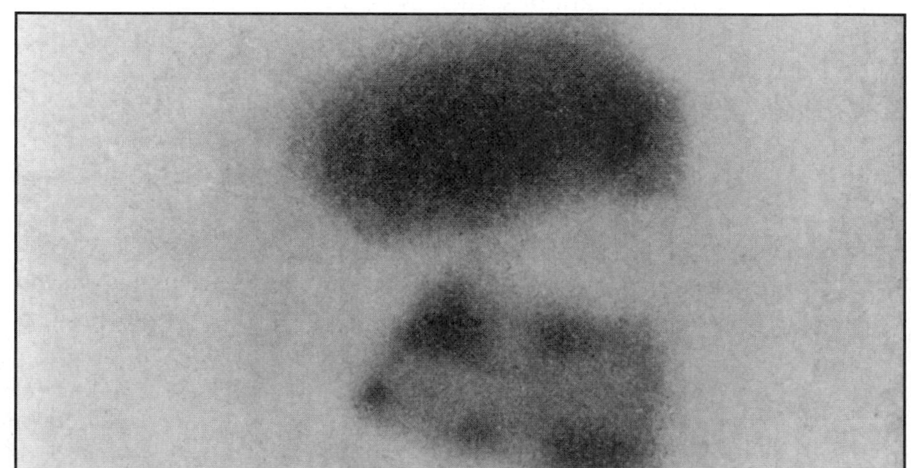

Figure 7 Henri Becquerel found that photographic plates could be exposed by nearby uranium compounds. The dark shadows are the exposed portions of the plates.

View an **animation about radioactivity.**

Types of Radiation

Rutherford's team found that uranium compounds produce three kinds of radiation. He classified the radiation by how easy it was to stop and named them alpha (α), beta (β), and gamma (γ) radiation. A thick sheet of paper can stop alpha radiation, and 6 mm of aluminum can stop most beta radiation. Gamma rays are stopped by several centimeters of lead. We now know that each type of radiation involves a different particle.

Alpha decay Alpha particles are helium nuclei ($^{4}_{2}\text{He}$). The mass number of an α particle is 4, and the atomic number is 2. The emission of an α particle from a nucleus is a process called **alpha decay.** Alpha decay occurs in nuclei with at least 106 nucleons. When a nucleus emits an α particle, its mass number (A) is reduced by 4, and its atomic number (Z) is reduced by 2. The element changes, or transmutes, into a different element. For example, $^{238}_{92}\text{U}$ transmutes into $^{234}_{90}\text{Th}$ by alpha decay.

Beta decay Beta particles are electrons emitted by the nucleus. The nucleus does not contain electrons, so where do these electrons come from? **Beta decay** occurs when a neutron changes to a proton or a proton changes to a neutron within the nucleus. In all reactions, charge is conserved. That is, the total charge before the decay equals the total charge after the decay. In beta decay, when a neutron (charge 0) changes to a proton (charge +1e), an electron (charge −1e) also appears. A nucleus with N neutrons and Z protons ends up as a nucleus of $N - 1$ neutrons and $Z + 1$ protons. Another neutral, nearly massless particle called an antineutrino is also is emitted in beta decay.

Another type of beta decay occurs when a proton changes to a neutron in the nucleus. Because a positron (positively charged electron) is emitted, this type of beta decay is called positron emission.

Gamma decay Gamma rays are high-energy photons emitted by the nucleus during gamma decay. **Gamma decay** occurs when there is a redistribution of energy within the nucleus and a gamma ray is emitted. Neither the mass number nor the atomic number changes in gamma decay. Gamma radiation often accompanies alpha and beta decay. The three main types of radiation are summarized in **Table 1.**

Radioactive elements often go through a series of successive decays until they form a stable nucleus. For example, $^{238}_{92}\text{U}$ undergoes 14 separate decays before the stable lead isotope $^{206}_{82}\text{Pb}$ is produced.

Table 1 Three Types of Radiation

Alpha Decay	Beta Decay	Gamma Decay
α particle ($^{4}_{2}\text{He}$)	β particle ($^{0}_{-1}\text{e}$)	γ particle (photon)
charge +2	charge −1	no charge
least penetration	medium penetration	highest penetration
transmutes nucleus: $A \rightarrow A - 4$ $Z \rightarrow Z - 2$ $N \rightarrow N - 2$	transmutes nucleus: $A \rightarrow A$ $Z \rightarrow Z + 1$ $N \rightarrow N - 1$	changes only energy: $A \rightarrow A$ $Z \rightarrow Z$ $N \rightarrow N$

PhysicsLABs

COMMON SOURCES OF RADIATION

What are some common sources of radiation that are found in your home or school?

EXPLORING RADIATION

How does the radiation level vary with the distance from the source?

Watch **an animation about alpha, beta, and gamma radiation.**

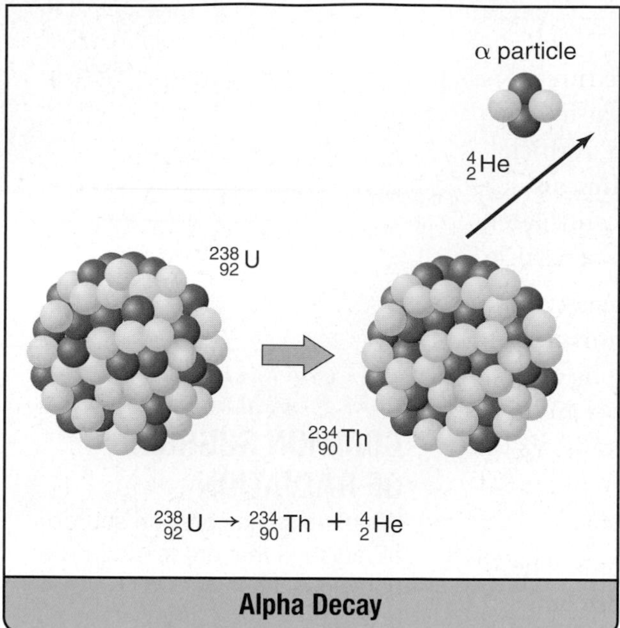

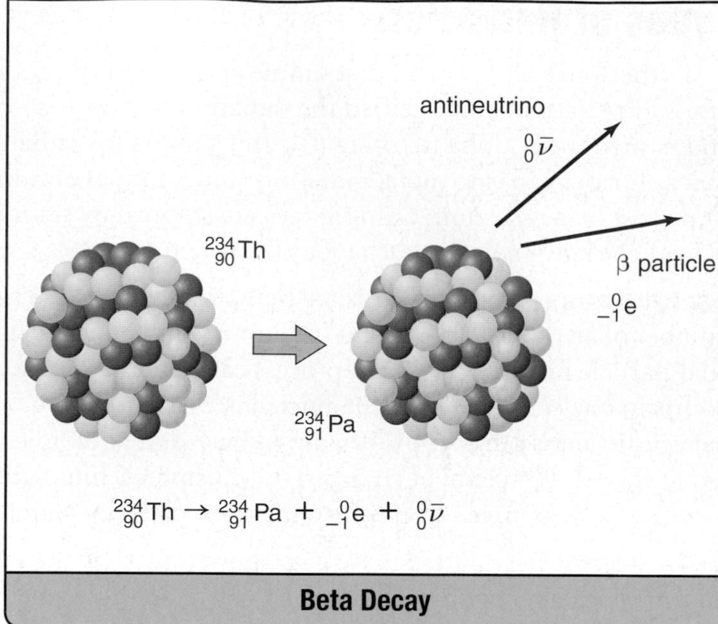

Figure 8 When uranium-238 undergoes alpha decay, an alpha particle is emitted and the resulting nucleus is thorium-234. Thorium-234 can undergo beta decay to produce protactinium-234 and a beta particle.

Nuclear Reactions and Equations

Whenever the energy or the number of neutrons or protons in a nucleus changes, the resulting event is called a **nuclear reaction.** Just as in chemical reactions, some nuclear reactions release energy, while others occur only when energy is added to a nucleus. One form of nuclear reaction is the emission of particles by radioactive nuclei. The reaction releases excess energy in the form of the kinetic energy of the emitted particles. Two such reactions are shown in **Figure 8.**

☑ **READING CHECK** **Explain** why burning methane is not a nuclear reaction, while the alpha decay of uranium is.

Nuclear reactions can be described by words, diagrams, and equations. The symbols in nuclear equations simplify the calculation of atomic number and mass number in nuclear reactions. For example, the alpha decay in **Figure 8** can be expressed by the following equation:

$$^{238}_{92}\text{U} \rightarrow ^{234}_{90}\text{Th} + ^{4}_{2}\text{He}$$

This equation states that uranium-238 decays into thorium-234 by emitting an alpha particle.

The total number of nuclear particles (protons and neutrons) stays the same during the reaction, so the sum of the superscripts on each side of the equation must be equal: $238 = 234 + 4$. The total charge also is conserved during the reaction, so the sum of the subscripts on each side must be equal: $92 = 90 + 2$.

During beta decay, an electron ($^{0}_{-1}\text{e}$) and an antineutrino ($^{0}_{0}\overline{\nu}$) are produced. **Figure 8** shows the transmutation of a thorium-234 nucleus into a protactinium-234 nucleus by the emission of a β particle. The reaction is described by the following equation. Check that charge is conserved in this reaction.

$$^{234}_{90}\text{Th} \rightarrow ^{234}_{91}\text{Pa} + ^{0}_{-1}\text{e} + ^{0}_{0}\overline{\nu}$$

Note that the superscript on the left-hand side of the equation equals the sum of the superscripts on the right-hand side. Equality also exists between the subscripts on both sides of the equation. In alpha and beta decay, one nucleus, shown on the left of the equation, decays to another nucleus and one or more radioactive particles, shown on the right of the equation. Another example of transmutation occurs when a particle collides with the nucleus, often resulting in the emission of other particles. For example, ${}^{12}_{6}C + {}^{1}_{1}H \rightarrow {}^{13}_{7}N$. Such reactions are illustrated in Example Problem 3 and in the discussion of fission later in this section.

MiniLAB

MODELING RADIOACTIVE DECAY

How is probability related to radioactive decay?

iLab Station 

EXAMPLE PROBLEM 2

Find help with **solving equations.** Math Handbook

EXAMPLE PROBLEM

ALPHA AND BETA DECAY Write the nuclear equation for each radioactive process.

a. The radioactive radium isotope ${}^{226}_{88}Ra$ emits an α particle and becomes the radon isotope ${}^{222}_{86}Rn$.

b. The radioactive lead isotope ${}^{209}_{82}Pb$ decays into the bismuth isotope ${}^{209}_{83}Bi$ by the emission of a β particle and an antineutrino.

1 ANALYZE THE PROBLEM

KNOWN	UNKNOWN
a. Initial: ${}^{226}_{88}Ra$ Final: α particle $({}^{4}_{2}He)$, ${}^{226}_{86}Rn$	What is the nuclear equation?
b. Initial: ${}^{209}_{82}Pb$ Final: ${}^{209}_{83}Bi$, β particle $({}^{0}_{-1}e)$, antineutrino $({}^{0}_{0}\bar{\nu})$	What is the nuclear equation?

2 SOLVE FOR THE UNKNOWN

a. ${}^{226}_{88}Ra \rightarrow {}^{4}_{2}He + {}^{222}_{86}Rn$ ◀ *Substitute ${}^{4}_{2}He$ for α particle.*

b. ${}^{209}_{82}Pb \rightarrow {}^{209}_{83}Bi + {}^{0}_{-1}e + {}^{0}_{0}\bar{\nu}$ ◀ *Substitute ${}^{0}_{-1}e$ for β particle and ${}^{0}_{0}\bar{\nu}$ for antineutrino.*

3 EVALUATE THE ANSWER

- **Is the number of nucleons conserved?**
  a. 226 = 4 + 222, so mass number is conserved.
  b. 209 = 209 + 0 + 0, so mass number is conserved.
- **Is charge conserved?**
  a. 88 = 2 + 86, so charge is conserved.
  b. 82 = 83 − 1 + 0, so charge is conserved.

PRACTICE PROBLEMS

Do additional problems. Online Practice

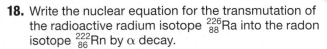

15. Write the nuclear equation for the transmutation of the radioactive uranium isotope ${}^{234}_{92}U$ into the thorium isotope ${}^{230}_{90}Th$ by the emission of an α particle.

16. Write the nuclear equation for the transmutation of the radioactive thorium isotope ${}^{230}_{90}Th$ into the radioactive radium isotope ${}^{226}_{88}Ra$ by the emission of an α particle.

17. Write the nuclear equation for the transmutation of americium-241 to neptunium-237 by the emission of an α particle.

18. Write the nuclear equation for the transmutation of the radioactive radium isotope ${}^{226}_{88}Ra$ into the radon isotope ${}^{222}_{86}Rn$ by α decay.

19. The radioactive lead isotope ${}^{214}_{82}Pb$ can change to the radioactive bismuth isotope ${}^{214}_{83}Bi$ by the emission of a β particle and an antineutrino. Write the nuclear equation.

20. Challenge The radioactive carbon isotope ${}^{14}_{6}C$ undergoes β decay. Determine the product of the decay, and write the nuclear equation.

EXAMPLE PROBLEM

EXAMPLE PROBLEM 3

Find help with **solving equations.** Math Handbook

SOLVE NUCLEAR EQUATIONS When nitrogen gas is bombarded with α particles, high-energy protons are emitted. What new isotope is created? Use the periodic table at the back of the book.

1 ANALYZE THE PROBLEM

KNOWN	UNKNOWN
nitrogen = $^{14}_{7}N$	What isotope belongs on the right side of the equation?
$\alpha = {}^{4}_{2}He$	
proton = $^{1}_{1}H$	

2 SOLVE FOR THE UNKNOWN

Write the equation for the nuclear reaction.

$^{4}_{2}He + {}^{14}_{7}N \rightarrow {}^{1}_{1}H + {}^{A}_{Z}X$

Solve for Z and for A.

$Z = 2 + 7 - 1 = 8$

$A = 4 + 14 - 1 = 17$

The element with $Z = 8$ is oxygen. The isotope must be $^{17}_{8}O$.

3 EVALUATE THE ANSWER

Does the equation balance? The number of nucleons is conserved: $4 + 14 = 1 + 17$. The charge is conserved: $2 + 7 = 1 + 8$.

PRACTICE PROBLEMS

Do additional problems. Online Practice 

21. Use the periodic table at the back of your book to complete the following:

a. $^{33}_{15}P \rightarrow ? + {}^{0}_{-1}e + {}^{0}_{0}\bar{\nu}$

b. $^{55}_{24}Cr \rightarrow ? + {}^{0}_{-1}e + {}^{0}_{0}\bar{\nu}$

22. Write the nuclear equation for the transmutation of the seaborgium isotope $^{263}_{106}Sg$ into the rutherfordium isotope $^{259}_{104}Rf$ by the emission of an alpha particle.

23. A proton collides with the nitrogen isotope $^{15}_{7}N$, forming a new isotope and an alpha particle. What is the isotope? Write the nuclear equation.

24. CHALLENGE Write the nuclear equations for the beta decay of the following isotopes:

a. $^{210}_{82}Pb$

b. $^{210}_{83}Bi$

c. $^{234}_{90}Th$

d. $^{239}_{93}Np$

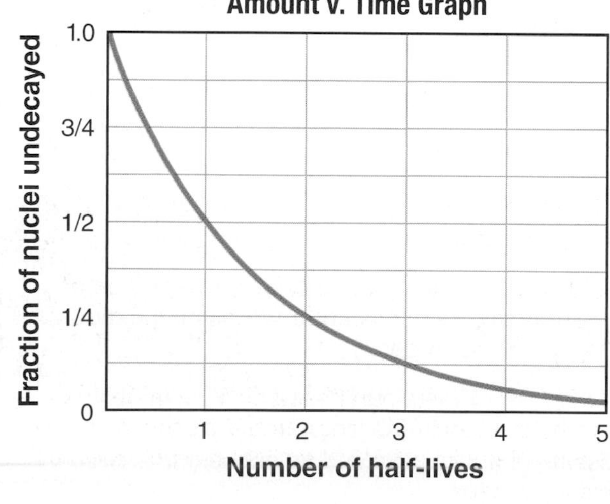

Figure 9 Half of the radioactive nuclei decay each half-life.

View a **simulation of radioactive decay.**

Concepts In Motion

Half-Life

Radioactive isotopes vary greatly in stability. Some exist for only a fraction of a second before decaying, while others may exist for billions of years. Scientists describe this property in terms of half-life. The **half-life** is the time required for half of the atoms in any given quantity of a radioactive isotope to decay. After each half-life, approximately half the nuclei remain undecayed, as shown in **Figure 9.** The half-life is an average found by observing a large number of decays. Each isotope has its own half-life.

For example, the half-life of the radium isotope $^{226}_{88}Ra$ is 1600 years. That is, in 1600 years half a given quantity of $^{226}_{88}Ra$ decays into another element—radon. In another 1600 years, half the remaining radium will decay. In other words, after 3200 years, one-fourth of the original amount will remain.

Table 2 Half-Lives of Selected Isotopes

Element	Isotope	Half-Life	Radiation Produced
Hydrogen	$^{3}_{1}H$	12.3 years	β
Carbon	$^{14}_{6}C$	5730 years	β
Cobalt	$^{60}_{27}Co$	30 years	β, γ
Iodine	$^{131}_{53}I$	8.07 days	β, γ
Lead	$^{212}_{82}Pb$	10.6 hours	β
Polonium	$^{194}_{84}Po$	0.7 seconds	α
Polonium	$^{210}_{84}Po$	138 days	α, γ
Uranium	$^{235}_{92}U$	7.1×10^{8} years	α, γ
Uranium	$^{238}_{92}U$	4.51×10^{9} years	α, γ
Plutonium	$^{236}_{94}Pu$	2.85 years	α
Plutonium	$^{242}_{94}Pu$	3.79×10^{5} years	α, γ

The half-lives of selected isotopes are shown in **Table 2.** If you know the original amount of a substance and its half-life, you can calculate the amount remaining after a given number of half-lives.

Investigate **the half-life of an element.**

Get help with **calculating half-life.**

HALF-LIFE

The amount of a radioactive isotope remaining in a sample equals the original amount times $\frac{1}{2}$ to the *t*, where *t* is the number of half-lives that have passed.

$$\text{remaining} = \text{original}\left(\frac{1}{2}\right)^{t}$$

Half-lives of radioactive isotopes are used to date objects. The age of a sample of organic material can be found by measuring the amount of carbon-14 remaining. The age of Earth was calculated based on the decay of uranium into lead.

The number of decays per second of a radioactive substance is called its **activity.** Activity is proportional to the number of radioactive atoms present. Therefore, the activity of a sample also is reduced by one-half in one half-life. Consider $^{131}_{53}I$, with a half-life of 8.07 days. If the activity of a certain sample of iodine-131 is 8×10^{5} decays/s, then 8.07 days later, its activity will be 4×10^{5} decays/s. After another 8.07 days, its activity will be 2×10^{5} decays/s. A sample's activity is also related to its half-life. The shorter the half-life, the higher the activity. Consequently, if you know the activity of a substance and the amount of that substance, you can determine its half-life. The SI unit for decays per second is a Becquerel (Bq).

PRACTICE PROBLEMS

Do additional problems. Online Practice

Use **Figure 9** *and* **Table 2** *to solve the following problems.*

25. A 1.0-g sample of tritium ($^{3}_{1}H$) is produced. After 24.6 years, what mass of tritium will remain?

26. The isotope $^{238}_{93}Np$ has a half-life of 2.0 days. If 4.0 g of neptunium is produced on Monday, what will be the mass of neptunium remaining on Tuesday of the next week?

27. A sample of $^{210}_{84}Po$ is purchased on September 1. Its activity is 2×10^{6} Bq. The sample is used in an experiment on June 1. What activity can be expected?

28. CHALLENGE Tritium ($^{3}_{1}H$) once was used in watches to produce a fluorescent glow with a brightness proportional to tritium's activity. Compare the watch's brightness after 6 years to its original brightness.

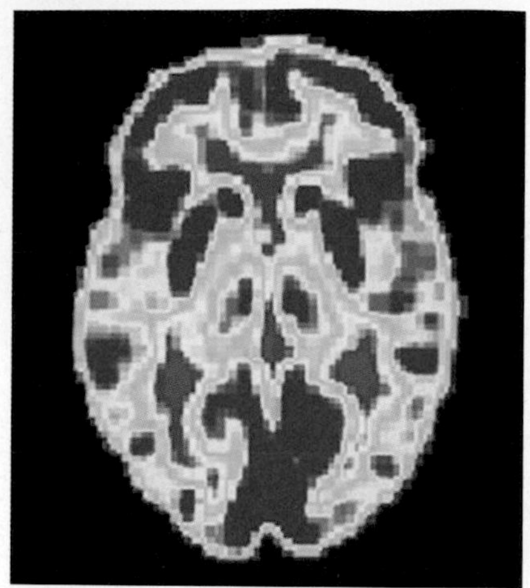

Figure 10 To produce a PET scan, physicians inject a solution into the patient. In the solution, a radioactive isotope is attached to a molecule that will accumulate in the target tissues. When the isotope decays, it produces positrons. These positrons annihilate nearby electrons, producing gamma rays. The PET scanner detects the gamma rays, and the data is used to make a three-dimensional map of the isotope distribution.

Watch a **video about nuclear medicine.**

Artificial Radioactivity

Radioactive isotopes can be formed from stable isotopes by bombardment with protons, neutrons, electrons, photons, or other particles. The resulting unstable nuclei emit radiation until they are converted into stable isotopes. The radioactive nuclei might emit alpha, beta, and gamma radiation, as well as neutrinos, antineutrinos, and positrons. A positron ($^{0}_{+1}e$) is a positively charged particle with the mass of an electron.

Artificially produced radioactive isotopes often are used in medicine. In some medical applications, patients are given radioactive isotopes of elements that are absorbed by specific parts of the body. A physician uses a radiation counter to monitor the activity in the region in question. A radioactive isotope also can be attached to a molecule that will be absorbed in the area of interest, as is done in positron emission tomography, or PET scan. A PET scan of a brain is shown in **Figure 10.**

Radiation is also used to destroy cancer cells. These cells are more sensitive to the damaging effects of radiation because they divide more often than normal cells. Gamma rays from the isotope $^{60}_{27}\text{Co}$ are used to treat cancer patients. Another method is to inject iodine to target thyroid cancer. In a third method, a beam of particles is produced in a particle accelerator and transmitted into tissue in such a way that they decay in the cancerous tissue and destroy the cells.

Nuclear Fission

The possibility of obtaining useful forms of energy from nuclear reactions was first discussed in the 1930s. The most promising results came from bombarding substances with neutrons. In Italy in 1934, Enrico Fermi and Emilio Segrè produced many new radioactive isotopes by bombarding uranium with neutrons.

In 1939 German chemists Otto Hahn and Fritz Strassmann showed that the resulting atoms acted chemically like barium. One week later Lise Meitner and Otto Frisch proposed that the neutrons had caused a division of the uranium into two smaller nuclei, resulting in a large release of energy. The possibility that these reactions could be not only a source of energy but also the basis of explosive weapons was immediately realized.

The division of a nucleus into two or more fragments is called **fission.** For example, the uranium isotope $^{235}_{92}\text{U}$ undergoes fission when it is bombarded with neutrons. The elements barium and krypton are typical results of this fission, as described by the following equation:

$$^{1}_{0}\text{n} + {}^{235}_{92}\text{U} \rightarrow {}^{92}_{36}\text{Kr} + {}^{141}_{56}\text{Ba} + 3({}^{1}_{0}\text{n}) + 173\text{ MeV}$$

The energy released by each fission can be found by calculating the masses of the atoms on each side of the equation. In the reaction above, the total mass on the right side of the equation is 0.186 u smaller than that on the left. The energy equivalent of this mass is 2.78×10^{-11} J, or 173 MeV. This energy appears as the kinetic energy of the products of the fission, as shown in **Figure 11.**

READING CHECK **Confirm** that the fission reaction above releases 173 MeV.

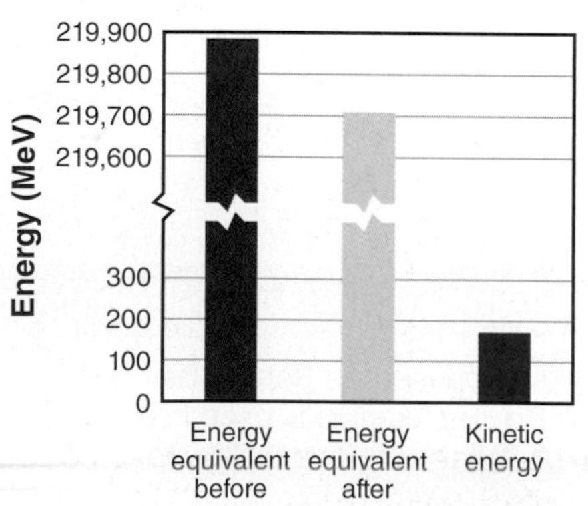

Figure 11 In a fission reaction, the mass of the reactants is equal to the sum of the mass of the products and the mass equivalent of the kinetic energy of the products.

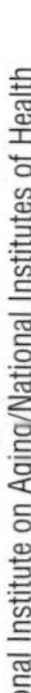

■ **Chain Reaction**

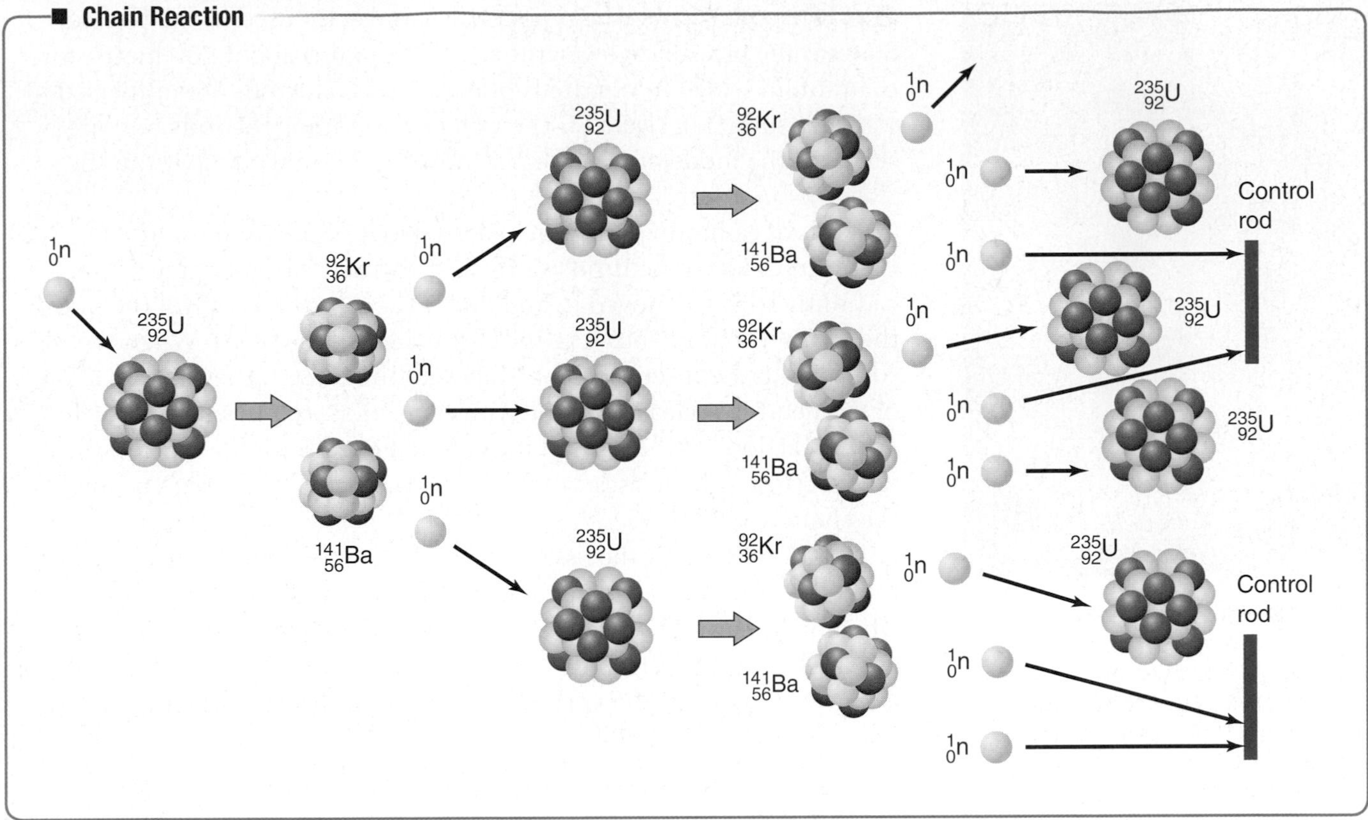

Figure 12 The neutrons produced by the fission of a uranium-235 nucleus can cause more fissions to occur. This can lead to a chain reaction. Controlled chain reactions are required to operate nuclear power plants.

Infer why rods that absorb neutrons might be added to a reactor.

Chain reactions Note that each fission releases three neutrons. These neutrons can collide with additional $^{235}_{92}U$ nuclei, resulting in more fissions. This continual process of repeated fission reactions caused by the release of neutrons from previous reactions is called a **chain reaction.** The process is illustrated in **Figure 12.**

☑ **READING CHECK** **Explain** why the overall release of energy by fission is huge.

Controlling chain reactions To create a controlled and useful chain reaction, the neutrons need to interact with the fissionable uranium at the correct rate. The reaction rate depends on the speed of the neutrons and on the amount of fissionable uranium. Most neutrons released by the fission of $^{235}_{92}U$ atoms have high speeds and are called fast neutrons. Naturally occurring uranium consists of less than 1 percent $^{235}_{92}U$ and more than 99 percent $^{238}_{92}U$. When a $^{238}_{92}U$ nucleus absorbs a fast neutron, it does not undergo fission but becomes a new isotope, $^{239}_{92}U$. The $^{238}_{92}U$ absorbs most of the neutrons. Thus, most neutrons from the original fission do not cause another fission.

To compensate for this, the uranium is broken into small pieces and placed in a moderator, a material that can slow down, or moderate, the fast neutrons. The neutrons collide with relatively light atoms, transfer momentum and energy, and slow down. The moderator slows many fast neutrons to speeds at which they are absorbed more easily by $^{235}_{92}U$ than by $^{238}_{92}U$. This larger number of slow neutrons greatly increases the chances that a neutron released by the fission of a $^{235}_{92}U$ nucleus will cause another $^{235}_{92}U$ nucleus to fission. If there is enough $^{235}_{92}U$ in the sample, a chain reaction can occur. To increase the amount of fissionable uranium, samples are enriched with more $^{235}_{92}U$.

Watch an **animation of a chain reaction.**

Watch a **BrainPOP video about nuclear energy.**

REAL-WORLD PHYSICS

RADIATION TREATMENT Gamma rays destroy both cancerous cells and healthy cells; thus, the beams of radiation must be directed only at cancerous cells.

Figure 13 High-speed particles from fuel rods enter the water faster than the speed of light in water. Photons are emitted, causing the water to glow. This is called the Cerenkov effect and is not the result of radioactivity.

Nuclear reactors The type of nuclear reactor used in the United States is the pressurized water reactor. It contains about 200 metric tons of uranium sealed in hundreds of metal rods. The rods are immersed in water, as shown in **Figure 13.** The water surrounding the rods not only serves as the moderator, but it also absorbs thermal energy from the fission of uranium.

Rods of cadmium metal are placed between the uranium rods. Cadmium absorbs neutrons easily and also acts as a moderator. The cadmium rods are moved in and out of the reactor to control the rate of the chain reaction. Thus, the rods are called control rods. When the control rods are inserted completely into the reactor, they absorb enough of the neutrons released by the fission reactions to prevent any further chain reaction. As the control rods are removed from the reactor, the rate of energy release increases, with more free neutrons available to continue the chain reaction.

Energy released by the fission heats the water surrounding the uranium rods. The water itself doesn't boil because it is under high pressure, which increases its boiling point. As shown in **Figure 14,** this water is pumped to a heat exchanger, where it causes other water to boil, producing steam that turns turbines. The turbines are connected to generators that produce electrical energy.

☑ **READING CHECK** **Explain** why reactors use both water and control rods to control the chain reaction.

Fission of $^{235}_{92}U$ nuclei produces krypton, barium, and other atoms in the fuel rods. Most of these atoms are radioactive. About once a year, some of the uranium fuel rods must be replaced. The old rods no longer have enough uranium to be used in the reactor, but they are still extremely radioactive. Because the radiation is so dangerous to people and the environment, the rods must be stored in a location that can be secured. Methods of permanently storing these radioactive waste products currently are being developed.

REAL-WORLD PHYSICS

MANAGING RISKS Due to health and environmental risks of radioactive materials, scientists and engineers try to construct facilities that resist damage due to natural disasters and devise plans for quick containment of harmful materials. Damage to nuclear power plants in the wake of the March 2011 earthquakes off the east coast of Japan and resulting tsunamis has experts around the world revisiting design specifications and reassessing risks and preparedness.

Nuclear Power Plant

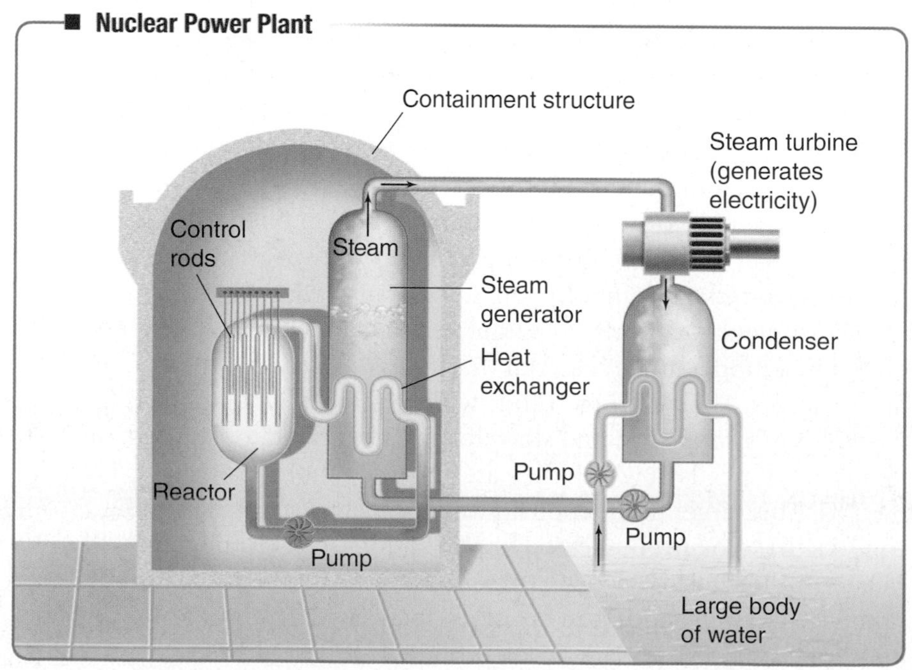

Figure 14 The reactions that occur in nuclear power plants release thermal energy. This energy is ultimately used to turn a turbine, which converts mechanical energy into electrical energy.

Watch a **video of a nuclear power plant.**

Nuclear Fusion

In nuclear **fusion,** nuclei with small masses combine to form a nucleus with a larger mass. In the process, energy is released. You learned earlier in this chapter that the larger nucleus is more tightly bound, so its mass is less than the sum of the masses of the smaller nuclei. This loss of mass corresponds to a release of energy.

An example of fusion is a process that occurs in the Sun. This process, called the proton-proton chain, is shown in **Figure 15.** Four hydrogen nuclei (protons) fuse in several steps to form one helium nucleus. The proton-proton chain can be represented in equation form:

$$^{1}_{1}\text{H} + {}^{1}_{1}\text{H} \rightarrow {}^{2}_{1}\text{H} + {}^{0}_{+1}\text{e} + {}^{0}_{0}\nu \text{ and } {}^{1}_{1}\text{H} + {}^{1}_{1}\text{H} \rightarrow {}^{2}_{1}\text{H} + {}^{0}_{+1}\text{e} + {}^{0}_{0}\nu$$
$$^{1}_{1}\text{H} + {}^{2}_{1}\text{H} \rightarrow {}^{3}_{2}\text{He} + \gamma \text{ and } {}^{1}_{1}\text{H} + {}^{2}_{1}\text{H} \rightarrow {}^{3}_{2}\text{He} + \gamma$$
$$^{3}_{2}\text{He} + {}^{3}_{2}\text{He} \rightarrow {}^{4}_{2}\text{He} + 2({}^{1}_{1}\text{H})$$

The net result (subtracting out the two protons produced in the final step) is that four protons produce one $^{4}_{2}$He, two positrons, and two neutrinos.

The mass of the four protons is greater than the mass of the helium nucleus that is produced. The mass difference is equivalent to sum of the kinetic energy of the resultant particles and the radiant energy of the photons. The energy released by the fusion that produces one helium-4 nucleus is 25 MeV. In comparison, the energy released when one dynamite molecule reacts chemically is about 20 eV, which is about 1 million times smaller.

The repulsive force between the charged nuclei requires the fusing nuclei to have high energies. Thus, fusion reactions take place only when the nuclei have large amounts of thermal energy. The proton-proton chain requires a temperature of about 2×10^{7} K, such as that found in the center of the Sun. Stars that are hot enough can also fuse together nuclei to form larger elements such as carbon, nitrogen, and oxygen. The hottest stars have enough thermal energy to fuse nuclei as large as silicon into iron.

Fusion reactions also occur in a hydrogen bomb, which is also called a thermonuclear bomb. In this device, the high temperature necessary to produce the fusion reaction is produced by exploding a uranium fission bomb, which is often called a nuclear bomb.

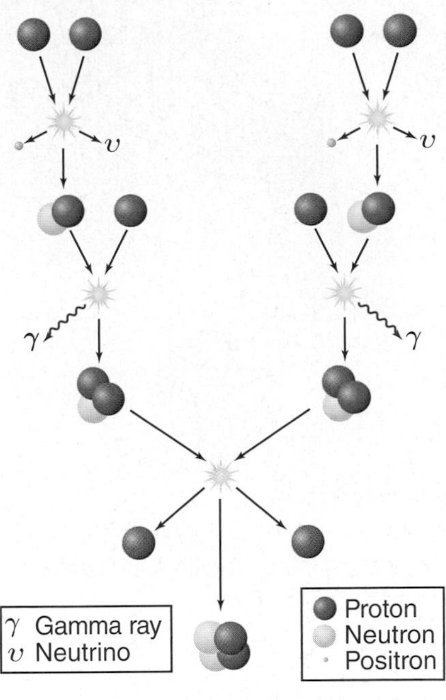

Figure 15 In the Sun, a series of fusion reactions occurs to convert hydrogen into helium.

Confirm that charge is conserved in this reaction.

Watch an **animation about nuclear fusion.**

SECTION 2 REVIEW

Check your understanding.

29. MAINIDEA How can an electron be expelled from a nucleus during beta decay if the nucleus has no electrons?

30. **Nuclear Reactions** The polonium isotope $^{210}_{84}$Po undergoes alpha decay. Write the equation for the reaction.

31. **Half-Life** Use **Figure 9** and **Table 2** to estimate in how many days a sample of $^{131}_{53}$I would have three-eighths its original activity.

32. **Nuclear Reactor** Lead often is used as a radiation shield. Why is it not a good choice for a moderator in a nuclear reactor?

33. **Fusion** Why doesn't helium in a balloon undergo fusion?

34. **Energy** Calculate the energy released in the first fusion reaction in the Sun, $^{1}_{1}\text{H} + {}^{1}_{1}\text{H} \rightarrow {}^{2}_{1}\text{H} + {}^{0}_{+1}\text{e} + {}^{0}_{0}\nu$.

35. **Critical Thinking** Isotopes that experience alpha decay, called alpha emitters, are used in smoke detectors. An emitter is mounted on one plate of a capacitor, and the α particles strike the other plate. As a result, there is a potential difference across the plates. Explain and predict which plate has the more positive potential.

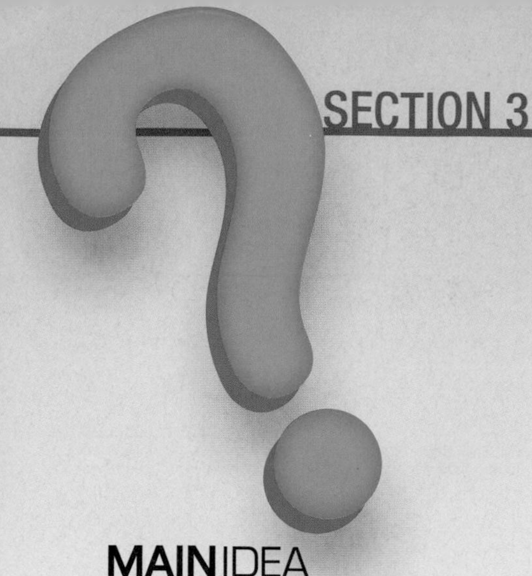

SECTION 3

The Building Blocks of Matter

PHYSICS 4 YOU **You are made of molecules, molecules are made of atoms, and atoms are made of protons, electrons, and neutrons. Are protons, electrons, and neutrons the smallest particles?**

MAIN IDEA

All matter is made of quarks and leptons and interacts with other matter through force carriers.

Essential Questions

- How are particles produced and detected?
- What are the characteristics of the Standard Model of matter?
- How are mass and other forms of energy transformed during pair annihilation and production?
- What is the role of the weak nuclear force?

Review Vocabulary

strong nuclear force an attractive force that binds the nucleus together; is of the same strength between all nucleon pairs

New Vocabulary

Standard Model
quark
lepton
force carrier
pair production
weak nuclear force

Particle Accelerators

The first physicists who studied the nucleus with high-speed particles had to use particles from radioactive sources. Other investigations used cosmic rays produced by not-yet-understood processes in stars. In the early 1930s, scientists developed laboratory devices that could accelerate protons and α particles to energies high enough to penetrate the nucleus. Today, the linear accelerator and the synchrotron are regularly used to accelerate particles to high energies and help scientists make further discoveries about the structure of matter.

Linear accelerators A linear accelerator can be used to accelerate protons, electrons, or their antimatter counterparts (described later in this section). It consists of a series of hollow tubes within a long, evacuated chamber. The tubes are connected to a high-frequency alternating potential difference source, as shown in **Figure 16.** Protons are produced in an ion source similar to those in cathode-ray tubes. When the first tube has a negative potential, protons accelerate into it. They move at a constant velocity because there is no electric field within the tube.

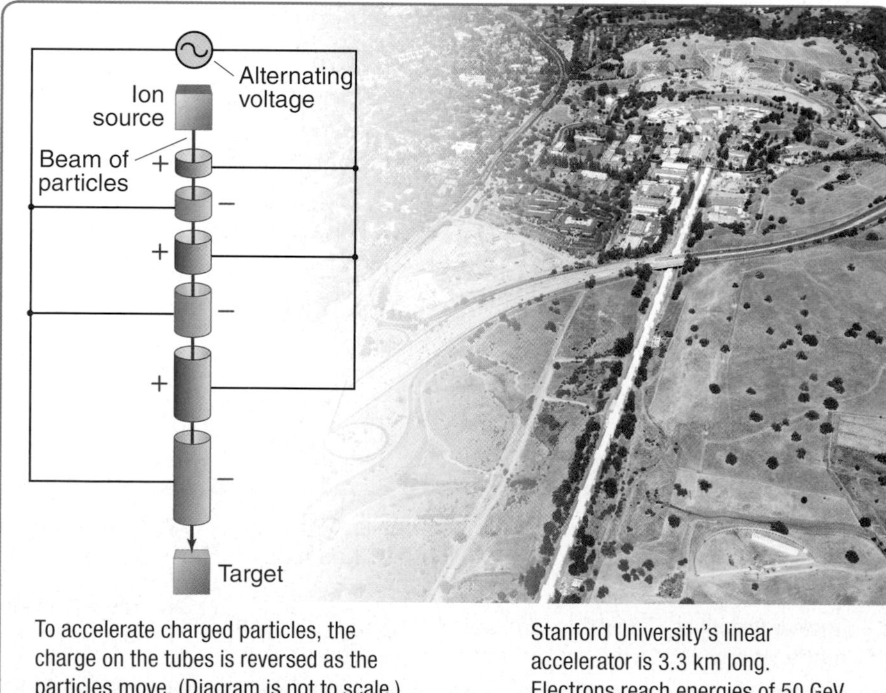

To accelerate charged particles, the charge on the tubes is reversed as the particles move. (Diagram is not to scale.)

Stanford University's linear accelerator is 3.3 km long. Electrons reach energies of 50 GeV.

Figure 16 A linear accelerator accelerates charged particles.

Watch an **animation about a linear accelerator.**

Concepts In Motion

Aerial Archives/Alamy

The length of the tube and the frequency of the potential difference are adjusted so that the potential reverses at the same time the protons reach the far end of the tube. Thus, the second tube has a negative potential in relation to that of the first. The resulting electric field in the gap between the tubes accelerates the protons into the second tube.

This process continues as the protons receive an acceleration between each pair of tubes. Because the particles move faster as they emerge from each successive tube and all the potential differences change at once, they move farther in the same time. Therefore, each tube must be longer than the one before. The protons ride along the crest of an electric field wave, much as a surfboard moves on the ocean. The proton's energy increases by 1×10^5 eV with each acceleration. By the end, the proton's energy might be billions of electron volts. A similar method is used to accelerate electrons. Note that only charged particles can be accelerated in this way.

Because of their higher mass, protons cannot be accelerated in linear accelerators to the extreme energies needed for modern experiments. Instead, linear proton accelerators are used as the initial process to inject protons into circular accelerators where higher energies can be achieved.

☑ **READING CHECK** **Explain** why neutrons cannot be accelerated in particle accelerators.

Synchrotrons Using a magnetic field to direct particles in a circle allows an accelerator to be more compact. A circular accelerator, called a synchrotron, is shown in **Figure 17.** Bending magnets are separated by straight regions where a high-frequency potential difference accelerates the particles. The strength of the magnetic field and the length of the path are chosen so that the particles reach the alternating electric field region precisely when the field's polarity will accelerate them.

One of the largest synchrotrons is the Large Hadron Collider (LHC) at the European Organization for Nuclear Research (CERN) laboratory near Geneva, Switzerland, shown in **Figure 17.** The LHC is so large because it takes a long distance to accelerate particles to the desired energies. The LHC has two proton beams that travel in opposite directions and collide in several interaction regions, and the results are studied. The LHC is designed to accelerate larger nuclei as well.

Figure 17 In a synchrotron, the path and acceleration of the particles are controlled using magnets. The Large Hadron Collider has a circumference of 27 km.

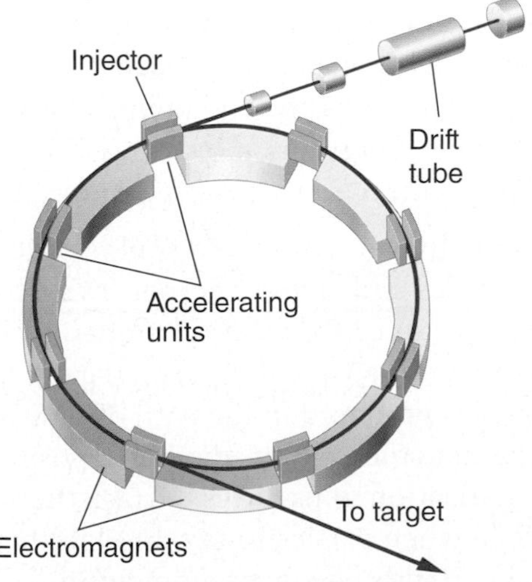

CERN

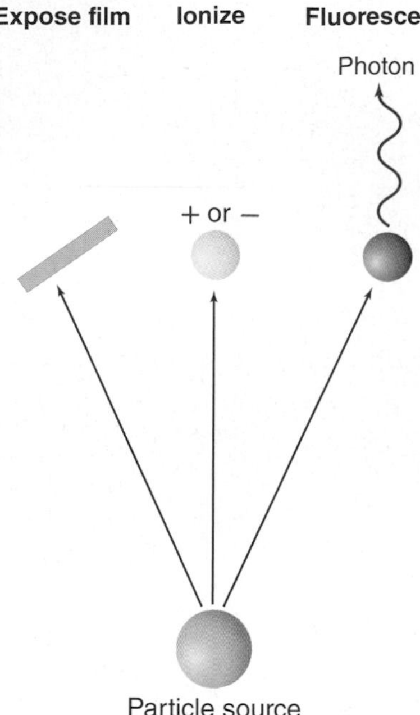

Figure 18 Subatomic particles cannot be seen or felt by humans, but they can be detected when they interact with some types of matter. Particles can expose film, ionize matter, or cause matter to emit photons.

Particle Detectors

The high-speed particles that move through accelerators either collide or crash into targets, producing a spray of subatomic particles. Scientists try to detect these tiny particles to identify and learn about them. To be detected, particles must interact with matter in such a way that people can observe those interactions. Your hand will stop an α particle, yet you will have no idea that the particle struck you. As you read this sentence, billions of neutrinos—many from the Sun—pass undetected through your body. Over the past century, scientists have devised tools to detect and distinguish the products of nuclear reactions.

Recall that radiation from uranium can expose photographic film. Because α particles, β particles, and gamma rays expose photographic film, it is a useful particle detector. Other particle detectors make use of the fact that a high-speed particle can collide with an atom, ejecting one of its electrons. That is, the high-speed particles ionize matter through collisions. In addition, some substances emit photons when exposed to certain types of radiation. These substances also can be used to detect radiation. These three means of detecting radiation are shown in **Figure 18.**

Geiger-Müller tubes A Geiger-Müller tube contains a copper cylinder with a negative charge. Down the center of this cylinder runs a rigid wire with a positive charge. The potential difference across the wire and the cylinder is kept just below the point at which a spontaneous discharge, or spark, occurs.

When a charged particle or gamma ray enters the tube, it ionizes a gas atom between the copper cylinder and the wire. The resulting positive ion is accelerated toward the copper cylinder by the potential difference, and the electron is accelerated toward the positive wire. As these charged particles move toward the electrodes, they create an avalanche of charged particles, and a pulse of current goes through the tube. Geiger counters are useful detectors because they are small enough to transport.

Figure 19 In a bubble chamber, charged particles leave tracks. This false-color photograph shows the paths of several particles.

Explain why the trails bend in different directions.

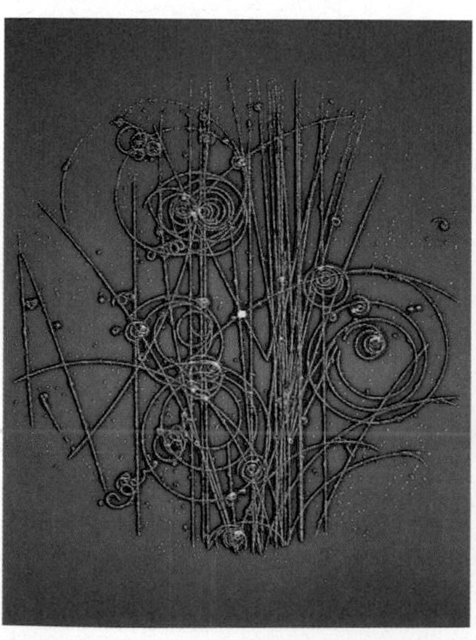

CERN

Condensation trails A device once used to detect particles was the Wilson cloud chamber. The chamber contained an area supersaturated with water vapor or ethanol vapor. Charged particles traveled through the chamber, leaving a trail of ions in their paths. Vapor condensed into small droplets on the ions forming visible trails. In a similar detector, called the bubble chamber, charged particles pass through a liquid held just above the boiling point. The trails of ions cause small vapor bubbles to form, marking the particles' trajectories, as shown in **Figure 19.**

Neutral particles do not leave trails because they do not produce discharges. The laws of conservation of energy and momentum in collisions can be used to infer if any neutral particles were produced.

Modern detectors Recent technology has produced sophisticated detectors to record the paths of particles and analyze properties such as momentum, mass, and energy. A detector called a wire chamber is like a giant Geiger-Müller tube. Huge plates are separated by a small gap filled with a low-pressure gas. A discharge is produced in the path of a particle passing through the chamber. A computer records the discharge position. Solid-state detectors also use the ionization of particles to track their positions. Other detectors emit light when particles pass through them. These are often used to measure a particle's energy or momentum.

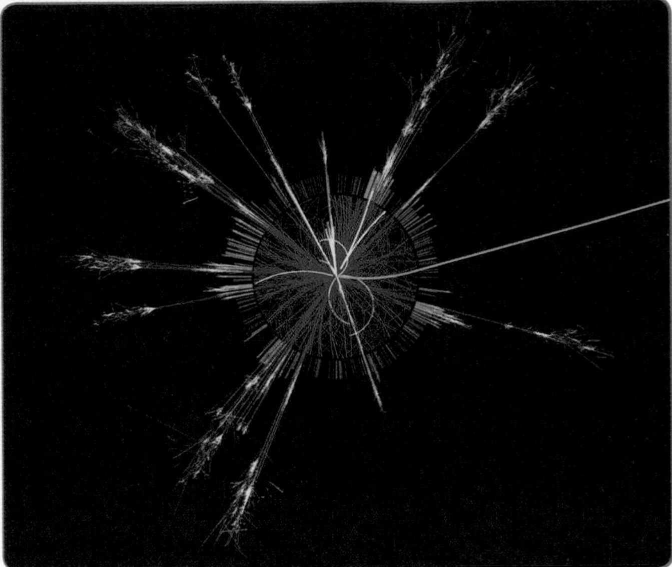

Figure 20 ATLAS (**A** **T**oroidal **LHC** **A**pparatu**S**) monitors tracks from billions of collisions and produces images of some of the most interesting events.

The entire array of detectors used in high-energy accelerator experiments can be huge. For example, the detector known as ATLAS at the Large Hadron Collider (LHC) is 25 m high, as shown in **Figure 20.** ATLAS is actually numerous detectors acting together. It is designed to monitor a billion particle collisions each second and select about 100 of these events that may be of interest to scientists. The recorded data can be used to create a computer picture of the collision events, as shown in **Figure 20.**

Finding New Particles

In 1930 the model of the atom was fairly simple: protons and neutrons surrounded by electrons. More detailed studies of radioactive decay disturbed this simple picture. The α particles and gamma rays emitted by radioactive nuclei have single energies that depend on the decaying nucleus. But β particles are emitted with a wide range of energies that are not always equal to the energy change of the nucleus.

This wide energy range of electrons emitted during beta decay suggested to Niels Bohr that another energy-carrying particle might be involved in nuclear reactions. Wolfgang Pauli in 1931 and Enrico Fermi in 1934 suggested that an unseen neutral particle was emitted with the β particle. Fermi named the particle the neutrino ("little neutral one" in Italian). The particle, which is actually an antineutrino, was not directly observed until 1956.

☑ **READING CHECK** **Explain** why neutrino is an appropriate name.

More and more particles Other studies discovered more particles. The muon, which seemed to be a heavy electron, was discovered in 1937. In 1947 another particle, the pion, was discovered. Experiments with particle accelerators revealed even more particles, with a wide range of masses. Some were much more massive than the proton. Some had a charge, and others were neutral. Some particles had lifetimes as short as 1×10^{-23} s, while others had no detectable decays. Keeping track of the different particles could be difficult. At one point Fermi was asked to identify a particle track and replied, "If I could remember the names of all these particles, I'd be a botanist!"

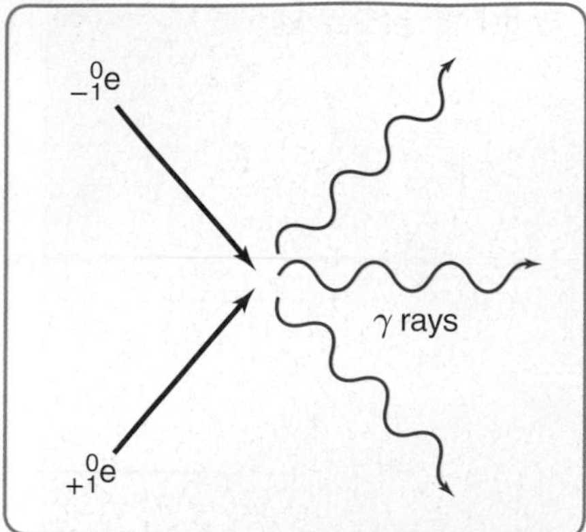

Figure 21 When a particle and its antiparticle collide, they annihilate each other. Here, a positron and an electron collide to produce gamma rays.

Antimatter In the late 1920s, Paul Dirac predicted the existence of an antiparticle associated with each kind of particle. The antiparticle has the same mass and charge magnitude as the particle, but the sign of the charge is opposite. Collectively, these antiparticles are known as antimatter. In some cases, antimatter particles are indicated by a bar over the symbol. In 1932 the electron's positive counterpart, called a positron, was discovered. When a positron and an electron collide, the two can annihilate each other to produce gamma rays, as shown in **Figure 21.** Collisions between other particles and their antiparticles have similar results.

The Standard Model

By the late 1960s, it became clear that protons, neutrons, and pions are not fundamental, or elementary, particles. Physicists divide elementary particles into three families, called quarks, leptons, and force carriers, as shown in **Figure 22.** The model of quarks, leptons, and force carriers as the elementary particles is called the **Standard Model.**

Quarks are elementary particles that form mass and can combine into larger particles such as protons and neutrons. There are six types, called flavors, of quarks. Different combinations of quarks make different particles. For example, particles such as protons and neutrons are made of three quarks and are called baryons. A meson, such as the pion, is made of a quark and an antiquark. **Leptons,** such as electrons and neutrinos, are also elementary particles that have mass but belong to a different family than quarks. There are six flavors of leptons.

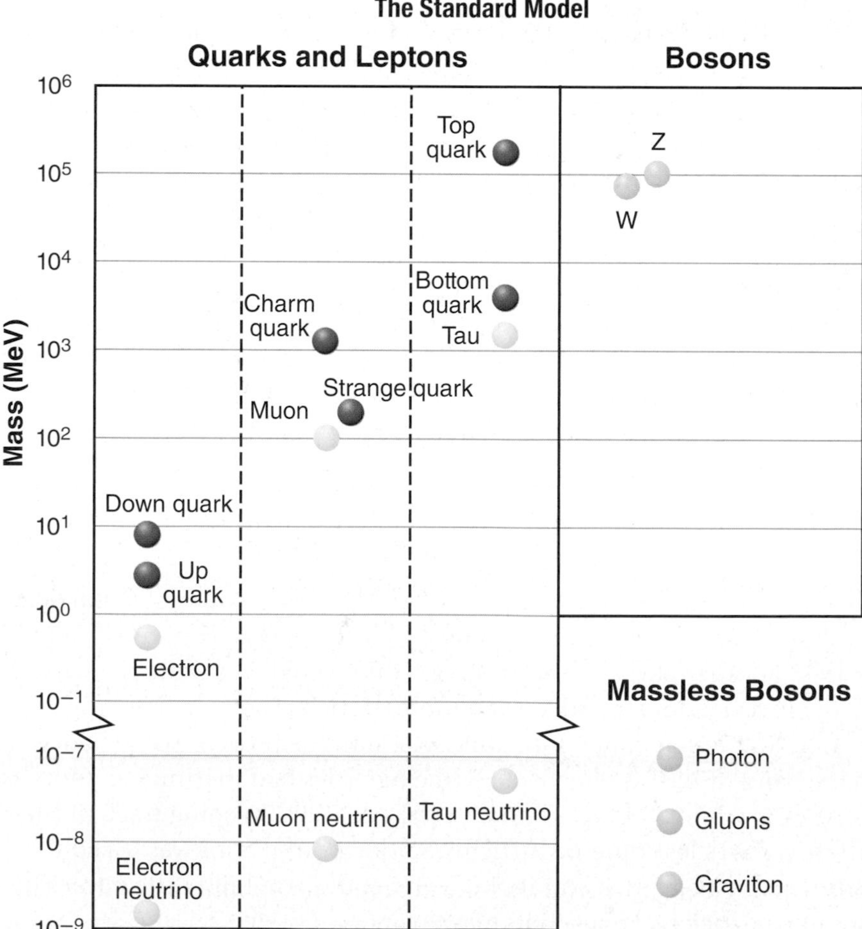

Figure 22 The Standard Model consists of three families. The six known quarks and the six known leptons are divided into three generations, shown as columns in the figure. The everyday world is made from particles in the left-hand column (u, d, e). Particles in the middle column (c, s, μ) are found in cosmic rays and are routinely produced in particle accelerators. Particles in the right-hand column (b, t, τ) are believed to have existed briefly during the earliest moments of the Big Bang and are created in high-energy collisions. The gauge bosons carry the fundamental forces. Masses are stated as energy equivalents, given by Einstein's formula $E = mc^2$.

In 1935, Japanese physicist Hideki Yukawa hypothesized the existence of a particle that could carry the strong nuclear force through space. This idea has developed into the model of **force carriers,** which are particles that transmit forces between objects interacting at a distance. Force carriers are also called gauge bosons. Each force has a corresponding force carrier. Photons carry the electromagnetic force. Gluons carry the strong nuclear force that binds quarks in baryons and mesons. The weak gauge bosons are involved in β decay. Graviton is the name given to the yet-undetected carrier of gravitational force. Experiments are currently underway to find this elusive particle.

Protons and Neutrons

Nucleons—the proton and the neutron—are both made of three quarks, as shown in **Figure 23.** The proton has two up quarks (u) with a charge $+\frac{2}{3}$e and one down quark (d) with a charge $-\frac{1}{3}$e. A proton is described as p = uud. The charge on the proton is the sum of the charges of the three quarks, $\left(\frac{2}{3}+\frac{2}{3}+-\frac{1}{3}\right)e = +e$. The neutron is made up of one up quark and two down quarks: n = udd. The charge of the neutron is zero, $\left(\frac{2}{3}+-\frac{1}{3}+-\frac{1}{3}\right)e = 0$.

Individual free quarks cannot be observed because the strong force that holds them together becomes larger as the quarks are pulled farther apart. In this case, the strong force acts like the force of a spring. It is unlike the electric force, which weakens as charged particles move apart.

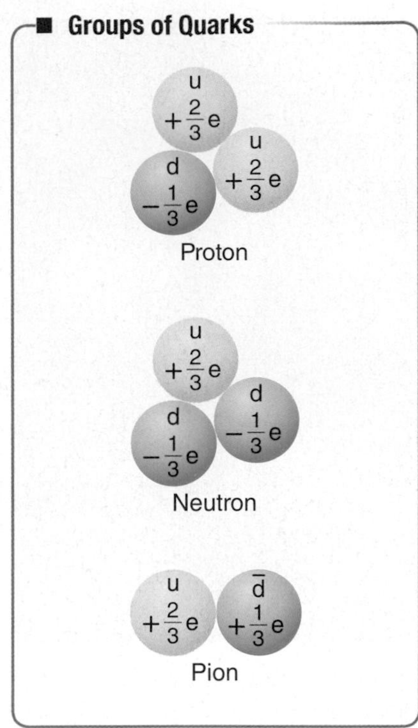

Figure 23 Quarks make up larger particles such as protons, neutrons, and pions. Notice that quarks have fractional charges, but the particles they make have whole-number charges.

Particle Annihilation and Production

Each quark and each lepton has a corresponding antiparticle. A particle and its antiparticle have identical masses. A charged particle and its antiparticle have opposite charges. For example, an up quark (u) has a charge of $+\frac{2}{3}$, while an anti-up quark ($\bar{u}$) has a charge of $-\frac{2}{3}$.

Because a baryon is a group of three quarks, the antiparticle of a baryon is a group of three antiquarks. For instance, a proton (uud) has a charge of +1. The antiproton ($\bar{u}\bar{u}\bar{d}$) has a charge of $-\frac{2}{3}-\frac{2}{3}+\frac{1}{3}=-1$. Likewise, $\bar{u}\bar{d}\bar{d}$ is an antineutron, which has a charge 0. Because a meson is a quark-antiquark pair, its antiparticle is also a quark-antiquark pair.

READING CHECK Determine One type of pion has a u$\bar{d}$ quark composition. What quarks make up its corresponding antipion?

Particle annihilation When a particle and its corresponding antiparticle annihilate each other, the resulting energy can be calculated using $E = mc^2$. An electron and a positron, for example, can collide and produce gamma rays: $e^- + e^+ \rightarrow \gamma + \gamma$. Each has a mass of 9.11×10^{-31} kg. The energy equivalent of the electron-positron pair can be calculated as follows:

$$\begin{aligned} E &= 2(9.11\times10^{-31}\text{ kg})(3.00\times10^{8}\text{ m/s})^2 \\ &= (1.64\times10^{-13}\text{ J})(1\text{ eV} / 1.60\times10^{-19}\text{ J}) \\ &= 1.02\times10^{6}\text{ eV or } 1.02\text{ MeV} \end{aligned}$$

Thus, when an electron and a positron at rest annihilate each other, the sum of the energies of the gamma rays emitted is 1.02 MeV.

Figure 24 Particles are always produced in pairs. If a particle were produced without its antiparticle, charge conservation would be violated. One possibility is for a gamma ray to decay into an electron-positron pair. Two such decays are shown here.

Particle production The inverse of annihilation also can occur. Energy can be converted directly into matter. The conversion of energy into a matter-antimatter pair of particles is called **pair production.**

If a gamma ray with at least 1.02 MeV of energy passes close by a nucleus, for example, a positron and electron pair can be produced: $\gamma \rightarrow e^- + e^+$. This process is shown in a bubble chamber in **Figure 24.** A magnetic field around the bubble chamber caused the oppositely charged particles' tracks to curve in opposite directions. Uncharged gamma rays, entering the detector from the top of the picture, produced no tracks.

If the gamma ray's energy is larger than 1.02 MeV, the excess energy goes into the kinetic energy of the positron and electron. The positron soon collides with another electron and they are both annihilated, resulting in two or three gamma rays with a total energy of no less than 1.02 MeV.

Antiprotons and antineutrons also can be created. Recall that protons and neutrons have approximately 1800 times as much mass as electrons. Thus, the energy needed to create proton-antiproton or neutron-antineutron pairs is larger. The first proton-antiproton pair was produced and observed at Berkeley, California, in 1955.

☑ **READING CHECK** **Estimate** the amount of energy needed to produce a neutron-antineutron pair.

Particle conservation The total number of quarks and the total number of leptons in the universe is constant. That is, quarks and leptons are created or destroyed only in particle-antiparticle pairs. On the other hand, force carriers such as gravitons, photons, gluons, and weak bosons can be created or destroyed if there is enough energy.

Individual reactions, such as $\gamma \rightarrow e^-$ and $\gamma \rightarrow e^+$, however, cannot occur because such events would violate the law of conservation of charge. Reactions such as $\gamma \rightarrow e^- + p$ also do not occur; the pair must be a particle and its corresponding antiparticle.

PRACTICE PROBLEMS

Do additional problems. Online Practice 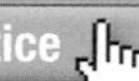

36. For each particle below, determine its pair production partner.

a. positron
b. neutron
c. antiproton
d. muon

37. The mass of a proton is 1.67×10^{-27} kg.

a. Find the energy equivalent of the proton's mass in joules.
b. Convert this value to electron volts.
c. Find the smallest total γ-ray energy that could result in a proton-antiproton pair.

38. A positron and an electron can annihilate and form three gamma rays. Two gamma rays are detected. One has an energy of 225 keV, the other 357 keV. What is the energy of the third gamma ray?

39. The mass of a neutron is 1.008665 u.

a. Find the energy equivalent of the neutron's mass in megaelectron volts.
b. Find the smallest total γ-ray energy that could result in the production of a neutron-antineutron pair.

40. CHALLENGE The mass of a muon is 0.1135 u. It decays into an electron and two neutrinos. What is the energy released in this decay?

Beta Decay and the Weak Interaction

In a stable nucleus, neutrons do not decay, but free neutrons and neutrons in an unstable nucleus can undergo beta decay. Recall that during beta decay a neutron is transformed into a proton by emitting a β particle and an antineutrino ($^{0}_{0}\overline{\nu}$). The antineutrino has a very small mass and is uncharged, but like the photon, it carries momentum and energy. The neutron decay equation is written as follows:

$$^{1}_{0}\text{n} \rightarrow {}^{1}_{1}\text{p} + {}^{0}_{-1}\text{e} + {}^{0}_{0}\overline{\nu}$$

Protons can decay in a similar way. The decay of a free proton has never been observed, but a proton in the nucleus can decay to a neutron with the emission of a positron ($^{0}_{+1}\text{e}$) and a neutrino ($^{0}_{0}\nu$). Recall that this is called positron emission.

$$^{1}_{1}\text{p} \rightarrow {}^{1}_{0}\text{n} + {}^{0}_{+1}\text{e} + {}^{0}_{0}\nu$$

These decays cannot be explained by the strong force. The existence of beta decay indicates that there must be another interaction occurring in the nucleus. The interaction that acts in the nucleus during beta decay is called the **weak nuclear force.** As you might guess, this force is much weaker than the strong nuclear force.

Quark model of beta decay A proton (uud) and a neutron (udd) differ by only one quark. Beta decay occurs in two steps, as shown in **Figure 25.** First, one d quark in the nucleus changes to a u quark with the emission of a W^- boson: $d \rightarrow u + W^-$. The W^- boson is one of the three carriers of the weak nuclear force. In the second step, the W^- boson decays into an electron and an antineutrino. Similarly, a proton in a nucleus can decay into a neutron and a W^+ boson. The W^+ boson then decays into a positron and a neutrino.

The emission of the third weak-force carrier, the Z^0 boson, is not accompanied by a change from one quark to another. The Z^0 boson produces an interaction between the nucleons and the electrons in atoms. This interaction is similar to, but much weaker than, the electromagnetic force holding the atom together and was first detected in 1979. The W^+, W^-, and Z^0 bosons first were observed directly in 1983.

Neutrinos and antineutrinos have long been thought to be massless, but recent experiments that detect neutrinos from the Sun show that neutrinos do have mass. These masses are much smaller than those of any other known particles and have not yet been measured accurately.

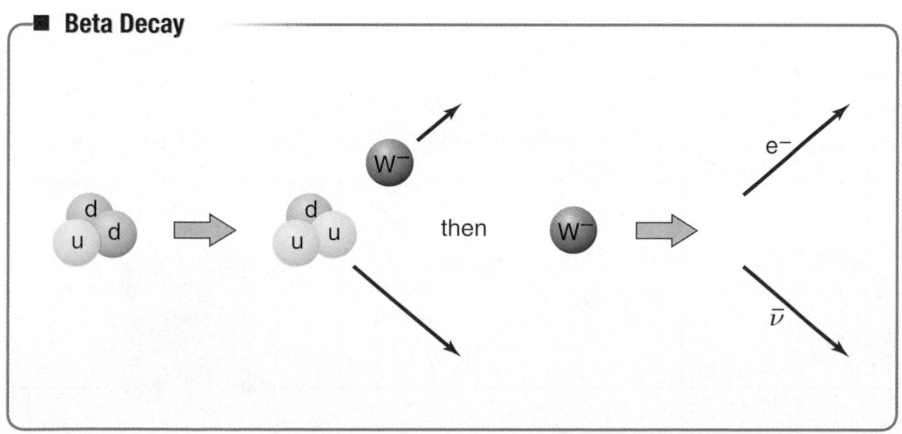

Figure 25 According to the quark model, beta decay of a neutron to a proton occurs in two steps.

Infer Given that charge is conserved during beta decay, what charge must a W^- boson have?

PHYSICS CHALLENGE

NUCLEAR DECAY The uranium isotope $^{238}_{92}U$ decays by α emission and two successive β emissions back into uranium again.

1. Show the three nuclear decay equations.
2. Predict the atomic mass number of the uranium formed.

MiniLAB

WHAT'S IN THE BACKGROUND?

How much background radiation is there?

Testing the Standard Model

You can see in **Figure 22** that the quarks and the leptons are separated into three columns. The everyday world is made from particles in the left-hand columns: protons, neutrons, and electrons. Particles in the middle group are found in cosmic rays and are routinely produced in particle accelerators. Particles in the right-hand family are believed to have existed briefly during the earliest moments of the Big Bang and are created in high-energy collisions.

What determines the masses of the quarks and the leptons? Scientists hypothesize that there is a particle that determines the masses of quarks and leptons, but such a particle has not been observed. The Standard Model is not a theory; it does not explain the masses of particles nor why there are three families of quarks and leptons.

Why are there four forces? The differences among the four fundamental interactions are evident: the forces act on different quantities such as charge or mass, they have different dependencies on distance, and the force carriers have different properties, as shown in **Figure 26.** There are, however, some similarities among the interactions. For instance, the force between charged particles, the electromagnetic force, is carried by photons in much the same way as weak bosons carry the weak interaction. The electromagnetic force acts over a long range because the photon has zero mass, while the weak force acts over short distances because the W and Z bosons are relatively massive. The mathematical structures of the theories of the weak interaction and electromagnetic interaction, however, are similar.

Figure 26 The four fundamental forces are the gravitational force, the electromagnetic force, the strong nuclear force, and the weak nuclear interaction.

The Four Fundamental Forces

Gravitational Force	Electromagnetic Force	Weak Interaction	Strong Force
• affects particles with mass • force carrier: gravitons (not yet found) • infinite range • relative strength: 10^{-36} • always attractive • holds together large-scale structures such as planets and galaxies	• affects electrically charged particles • force carrier: photons • infinite range • relative strength: 1 • can be attractive or repulsive • holds atoms and molecules together	• affects quarks and leptons • force carriers: W^+, W^-, and Z^0 bosons • range: $\sim 10^{-18}$ m • relative strength: 10^{-7} • responsible for beta decay	• affects quarks and hadrons • force carrier: gluon • range: $\sim 10^{-15}$ m • relative strength: 20 • binds quarks into hadrons and holds the nucleus together
	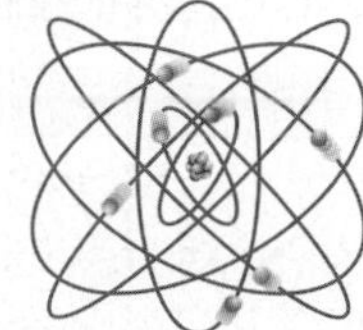	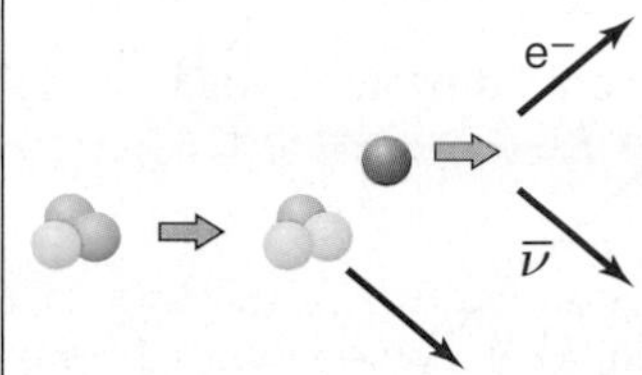	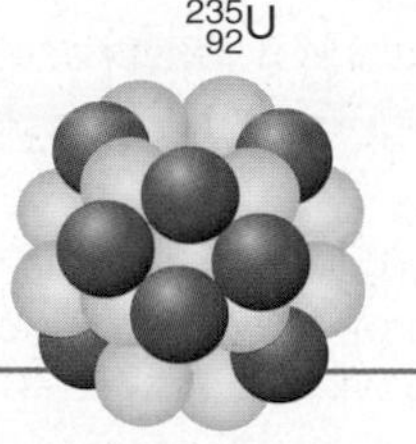

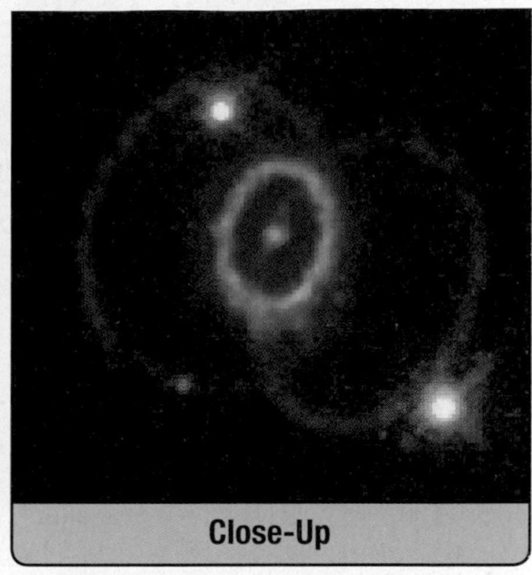

Figure 27 The electromagnetic and weak forces are indistinguishable in a supernova, such as 1987A shown above. When the supernova explodes, it gives off a burst of light and neutrinos. The light and the neutrinos reach Earth at the same time, showing that neutrinos can travel near the speed of light and are produced in supernovas.

Evidence from the cosmos Look at the supernova in **Figure 27.** Astrophysical evidence indicates that during massive stellar explosions, the electromagnetic force and the weak interaction are identical. According to present theories of the origin of the universe, the two forces were identical during the early moments of the cosmos as well. For this reason, the electromagnetic and weak forces are said to be unified into a single force, called the electroweak force.

In the same way that scientists in the 1970s developed a theory of the unified electromagnetic and weak forces, physicists presently are developing theories that include the strong force as well. Work is still incomplete. Hypotheses are being revised and improved, and experiments to test these hypotheses are being planned. A fully unified theory that includes gravitation will require even more effort.

Dark matter Results of studies of galaxies are even more perplexing. These results suggest that matter described by the Standard Model makes up only a small fraction of the mass in the universe. A much larger amount is formed of dark matter, so called because it doesn't interact with photons or ordinary matter except through the gravitational force. In addition, there appears to be dark energy, which is thought to be responsible for accelerating the universe's expansion.

Thus the studies of the tiniest particles that make up nuclei are directly connected to investigations of the largest systems, the galaxies that make up the universe. Elementary particle physicists and cosmologists used to be at opposite ends of the length scale; now they are working together as they try to answer the same question together: "What IS the universe made of?"

SECTION 3 REVIEW

Section Self-Check Check your understanding.

41. **MAIN**IDEA Research the limitations of the Standard Model and possible alternative models.

42. **Nucleus Bombardment** Why would a proton require more energy than a neutron when used to bombard a nucleus?

43. **Particle Accelerator** Protons in the Fermi Laboratory accelerator move counterclockwise when viewed from above. In what direction is the magnetic field of the bending magnets?

44. **Pair Production** **Figure 24** shows the production of two electron-positron pairs. Why does the bottom set of tracks curve less than the top pair of tracks?

45. **Critical Thinking** Consider the following equations:

$$u \rightarrow d + W^+ \text{ and } W^+ \rightarrow e^+ + \nu$$

How could they be used to explain the radioactive decay of a nucleon that results in the emission of a positron and a neutrino? Write the equation in terms of nucleons, rather than quarks.

Wishing On A Star

The Sun and other stars are the most visible energy sources in the universe. What can we learn by considering this stellar source of energy?

A movie opens with a break-in at a nuclear research lab. The top-secret cold fusion plans have been stolen by an enemy spy! You might have seen a movie like this and wondered if anyone is actually researching cold fusion or if it is a made-up technology.

Fusion in the stars As you know, fusion reactions, such as the one shown in **Figure 1**, occur in the Sun and other stars. These reactions have a high-energy yield. Though only a tiny fraction of this energy (in the form of sunlight) reaches Earth, it delivers more energy every second than our civilization uses in an entire year.

Sun on Earth Because of the energy pay-out, it would be useful to recreate fusion reactions on Earth. Deuterium and tritium, the fuel for fusion, are readily available. The product of the reaction is helium gas, a useful material itself. Fusion generators would be clean and could solve many of our energy issues. So what's the problem?

It's hot in here! Nuclear fusion requires immense temperatures. In order to get more energy out than is put in, the fuel must be held at a temperature of at least 100 million K. While not impossible, attaining these temperatures has proven quite challenging.

Cool it! Because it is difficult to obtain the needed temperature conditions, people have searched for a way to fuse nuclei at normal Earth temperatures. Because Earth temperatures are significantly lower than the temperature inside the Sun, this is called cold fusion.

Scientists have found that muons (a flavor of lepton) can be used to catalyze fusion at room temperature. But they have been unable to produce a reaction with a greater energy output than input. For now, cold fusion as a viable power supply remains a thing from the movies.

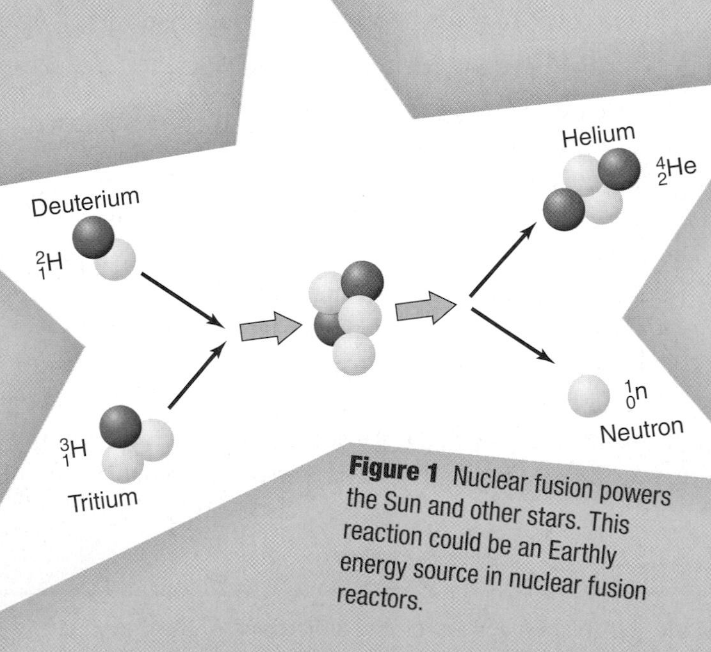

Figure 1 Nuclear fusion powers the Sun and other stars. This reaction could be an Earthly energy source in nuclear fusion reactors.

GOING**FURTHER** >>>

Research the current state of fusion research. Have a class debate about the possibilities of fusion becoming a viable power source.

Tim Ellis/Ikon Images/ Getty Images

CHAPTER 30 STUDY GUIDE

BIGIDEA Splitting and fusing atomic nuclei release energy.

VOCABULARY
- **nucleon** *(p. 802)*
- **atomic number** *(p. 802)*
- **mass number** *(p. 802)*
- **atomic mass unit** *(p. 802)*
- **strong nuclear force** *(p. 804)*
- **mass defect** *(p. 805)*
- **binding energy** *(p. 805)*

SECTION 1 The Nucleus

MAINIDEA The binding energy of the nucleus is equivalent to the mass defect of that nucleus.

- The number of protons in a nucleus is given by the atomic number (Z). The sum of the numbers of protons and neutrons in a nucleus is equal to the mass number (A).
- Atoms with the same number of protons but different numbers of neutrons are called isotopes.
- The strong nuclear force binds the nucleus together.
- The mass defect of a nucleus is the difference between the sum of the masses of its parts and the mass of the assembled nucleus. The energy difference between separated nucleons and the assembled nucleus is the binding energy.
- The mass defect is the energy equivalent of the binding energy of the nucleus.

$$E = mc^2$$

VOCABULARY
- **radioactive** *(p. 808)*
- **alpha decay** *(p. 809)*
- **beta decay** *(p. 809)*
- **gamma decay** *(p. 809)*
- **nuclear reaction** *(p. 810)*
- **half-life** *(p. 812)*
- **activity** *(p. 813)*
- **fission** *(p. 814)*
- **chain reaction** *(p. 815)*
- **fusion** *(p. 817)*

SECTION 2 Nuclear Decay and Reactions

MAINIDEA Radioactive decay releases particles and energy and can change an atom into another element.

- An unstable nucleus decays and can transmute into another element. Radioactive decay produces three kinds of particles. Alpha (α) particles are helium nuclei, beta (β) particles are high-speed electrons, and gamma (γ) rays are high-energy photons. In nuclear reactions, the sums of the mass numbers (A) and the nuclear charges (Z) are not changed.
- The half-life of a radioactive isotope is the time required for half of the nuclei to decay. After t half-lives:

$$\text{remaining} = \text{original}\left(\frac{1}{2}\right)^t$$

  The number of decays of a radioactive sample per second is the activity.
- In nuclear fission, a nucleus is split into two smaller nuclei with a release of neutrons and energy. Nuclear reactors use the energy released in fission to generate electrical energy.
- In nuclear fusion, nuclei with small masses combine to form a nucleus with a larger mass. The loss of mass that occurs in this process corresponds to a release of energy.

VOCABULARY
- **Standard Model** *(p. 822)*
- **quark** *(p. 822)*
- **lepton** *(p. 822)*
- **force carrier** *(p. 823)*
- **pair production** *(p. 824)*
- **weak nuclear force** *(p. 825)*

Games and Multilingual eGlossary

Vocabulary Practice

SECTION 3 The Building Blocks of Matter

MAINIDEA All matter is made of quarks and leptons and interacts with other matter through force carriers.

- Particles are produced by radioactive decay, during high-energy collisions in accelerators, and in stars. They can detected if they expose film, ionize matter, or cause matter to emit photons.
- The Standard Model includes quarks, leptons, and force carriers. Ordinary matter is made of quarks and leptons. Matter interacts with other matter through particles called force carriers.
- When corresponding antimatter and matter particles combine, their mass is converted into gamma rays or lighter matter-antimatter particle pairs. Energy can be transformed into a matter-antimatter pair through pair production.
- The weak nuclear force is a short-range force that allows quarks and leptons to change flavor. It is responsible for beta decay in which a neutron is transformed into a proton or a proton is transformed into a neutron.

CHAPTER 30 ASSESSMENT

Chapter Self-Check

SECTION 1 The Nucleus

Mastering Concepts

46. What force inside a nucleus acts to push the nucleus apart? What force inside the nucleus acts to hold the nucleus together?

47. Define mass defect. To what is it related?

48. Are small or large nuclei generally more unstable?

49. Which isotope has the greater number of protons, uranium-235 or uranium-238?

Mastering Problems

50. What particles and how many of each make up an atom of $^{109}_{47}Ag$?

51. What is the isotopic symbol (the one used in nuclear equations) of a zinc atom composed of 30 protons and 34 neutrons?

52. The isotope $^{32}_{16}S$ has a nuclear mass of 31.97207 u.

a. What is the mass defect of this isotope?

b. What is the binding energy of its nucleus?

c. What is the binding energy per nucleon?

53. The isotope $^{12}_{7}N$ has a nuclear mass of 12.0188 u.

a. What is the binding energy per nucleon?

b. Does it require more energy to separate a nucleon from a $^{14}_{7}N$ nucleus or from a $^{12}_{7}N$ nucleus? The isotope $^{14}_{7}N$ has a mass of 14.00307 u.

54. The two positively charged protons in a helium nucleus are separated by about 2.0×10^{-15} m. Use Coulomb's law to find the electric force of repulsion between the two protons. The result will give you an indication of the strength of the strong nuclear force.

55. The binding energy for $^{4}_{2}He$ is 228.3 MeV. Calculate the mass of a helium isotope in atomic mass units.

SECTION 2 Nuclear Decay and Reactions

Mastering Concepts

56. Define the term *transmutation* as used in nuclear physics, and give an example.

57. Radiation What are the common names for an α particle, a β particle, and a γ particle?

58. What two quantities must always be conserved in any nuclear equation?

59. Nuclear Power What sequence of events must occur for a chain reaction to take place?

60. Nuclear Power What role does a moderator play in a fission reactor?

61. Fission and fusion are opposite processes. How can each release energy?

Mastering Problems

62. Write the complete nuclear equation for the alpha decay of $^{222}_{86}Rn$.

63. Write the complete nuclear equation for the beta decay of $^{89}_{36}Kr$.

64. Complete each nuclear reaction.

a. $^{225}_{89}Ac \rightarrow {}^{4}_{2}He + \_\_\_\_\_\_$

b. $^{227}_{88}Ra \rightarrow {}^{0}_{-1}e + \_\_\_\_\_\_ + \_\_\_\_\_\_$

c. $^{65}_{29}Cu + {}^{1}_{0}n \rightarrow \_\_\_\_\_\_ \rightarrow {}^{1}_{1}p + \_\_\_\_\_\_$

d. $^{235}_{92}U + {}^{1}_{0}n \rightarrow {}^{96}_{40}Zr + 3({}^{1}_{0}n) + \_\_\_\_\_\_$

e. $^{27}_{13}Al + {}^{1}_{0}n \rightarrow \_\_\_\_\_\_ + {}^{4}_{2}He$

65. An isotope has a half-life of 3.0 days. What percentage of the original material will be left after

a. 6.0 days?

b. 9.0 days?

c. 12 days?

66. In an accident in a research laboratory, a radioactive isotope with a half-life of 3 days is spilled. As a result, the radiation is eight times the maximum permissible amount. How long must workers wait before they can enter the room?

67. When the boron isotope $^{11}_{5}B$ is bombarded with protons, it absorbs a proton and emits a neutron.

a. What element is formed?

b. Write the nuclear equation for this reaction.

c. The isotope formed is radioactive and decays by emitting a positron. Write the complete nuclear equation for this reaction.

68. BIGIDEA The first atomic bomb released an energy equivalent of 2.0×10^{1} kilotons of TNT. One kiloton of TNT is equivalent to 5.0×10^{12} J. Uranium-235 releases 3.21×10^{-11} J/atom. What was the mass of the uranium-235 that underwent fission to produce the energy of the bomb?

69. During a fusion reaction, two deuterons ($^{2}_{1}H$) combine to form the helium isotope $^{3}_{2}He$. What other particle is produced?

70. The isotope $^{209}_{84}Po$ has a half-life of 103 years. How long would it take for a 100-g sample to decay so that only 3.1 g of Po-209 was left?

Chapter Self-Check

71. The isotope $^{211}_{86}Rn$ has a half-life of 15 h. What fraction of a sample would be left after 60 h?

SECTION 3 The Building Blocks of Matter

Mastering Concepts

72. **High-Energy Physics** Why would a linear accelerator not work with a neutron?

73. **Forces** In which of the four interactions (strong, weak, electromagnetic, and gravitational) do the following particles take part?

    a. electron

    b. proton

    c. neutrino

74. What happens to the atomic number and mass number of a nucleus that emits a positron?

75. **Antimatter** What would happen if a meteorite made of antiprotons, antineutrons, and positrons landed on Earth?

Mastering Problems

76. What would be the charge of a particle composed of three up quarks?

77. Pions are composed of a quark and an antiquark. The charge of an antiquark is the opposite that of the corresponding quark. Find the charge of a pion made up of the following:

    a. $u\bar{d}$

    b. $d\bar{u}$

78. A synchrotron has a diameter of 2.0 km. Protons circling in it move at approximately the speed of light in a vacuum.

    a. How long does it take a proton to complete one revolution?

    b. The protons enter the ring at an energy of 8.0 GeV. They gain 2.5 MeV each revolution. How many revolutions must they travel before they reach 400.0 GeV of energy?

    c. How long does it take the protons to be accelerated to 400.0 GeV?

    d. How far do the protons travel during this acceleration?

79. Baryons are particles that are made of three quarks. Find the charge on each of the following baryons:

    a. neutron: ddu

    b. antiproton: $\bar{u}\bar{u}\bar{d}$

80. **Figure 28** shows tracks in a bubble chamber. Why might one track curve more than another?

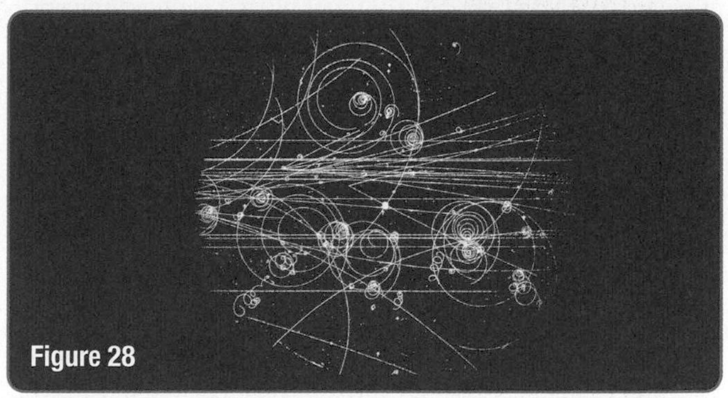

Figure 28

CERN

Applying Concepts

81. **Fission** A Web site claims that scientists have been able to cause iron nuclei to undergo fission. Is the claim likely to be true? Explain.

82. Use the graph of binding energy per nucleon in **Figure 6** to determine whether the reaction $^{2}_{1}H + ^{1}_{1}H \rightarrow ^{3}_{2}He$ is energetically possible.

83. **Isotopes** Explain the difference between naturally and artificially produced radioactive isotopes.

84. **Nuclear Reactor** In a nuclear reactor, water that passes through the core of the reactor flows through one loop, while the water that produces steam for the turbines flows through a second loop. Why are there two loops?

85. The fission of a uranium nucleus and the fusion of four hydrogen nuclei to produce a helium nucleus both produce energy.

    a. Which produces more energy?

    b. Does the fission of a kilogram of uranium nuclei or the fusion of a kilogram of hydrogen produce more energy?

    c. Why are your answers to parts **a** and **b** different?

Mixed Review

86. Each of the following nuclei can absorb an α particle. Assume that no secondary particles are emitted by the nucleus. Complete each equation.

    a. $^{14}_{7}N + ^{4}_{2}He \rightarrow$ ________

    b. $^{27}_{13}Al + ^{4}_{2}He \rightarrow$ ________

87. One of the simplest fusion reactions involves the production of deuterium, $^{2}_{1}H$ (mass = 2.014102 u), from a neutron and a proton. Write the complete fusion reaction and find the energy released.

Chapter Self-Check

88. A $^{232}_{92}$U nucleus (mass = 232.0372 u) decays to $^{228}_{90}$Th (mass = 228.0287 u) by emitting an α particle (mass = 4.0026 u) with a kinetic energy of 5.3 MeV. What must be the kinetic energy of the recoiling thorium nucleus?

89. Problem Posing Complete this problem so that it must be solved using the concept of mass defect: "Radon has 88 protons..."

Thinking Critically

90. Infer The momentum of a gamma ray of energy E is equal to E/c. When an electron-positron pair decays into two gamma rays, both momentum and energy must be conserved. The sum of the energies of the gamma rays is 1.02 MeV. If the positron and electron are initially at rest, what must be the magnitude and direction of the momentum of the two gamma rays?

91. Infer An electron-positron pair, initially at rest, also can decay into three gamma rays. If all three gamma rays have equal energies, what must be their relative directions? Make a sketch.

92. Estimate One fusion reaction in the Sun releases about 25 MeV of energy. Estimate the number of such reactions that occur each second from the luminosity of the Sun, which is the rate at which it releases energy, 4×10^{-6} W.

93. Interpret Data An isotope undergoes radioactive decay. A radiation detector records the number of counts in each 5-min interval. The results are shown in **Table 3.** When the sample is removed, the detector records 20 counts from cosmic rays in 5 min. Find the half-life of the isotope. Note that you should first subtract the 20-count background reading from each result. Then plot the counts as a function of time. From your graph, determine the half-life.

Table 3 Radioactive Decay Measurements

Time (min)	Counts (per 5 min)
0	987
5	375
10	150
15	70
20	40
25	25
30	18

94. Ranking Task Rank the following radioactive samples according to the percentage of original mass remaining, from greatest to least. Specifically indicate any ties.

A. a sample of $^{236}_{94}$Pu that was originally 20 g, after 2.85 years

B. a sample of $^{242}_{94}$Pu that was originally 20 g, after 7.90 years

C. a sample of $^{212}_{82}$Pb that was originally 5.0 g, after 2 days

D. a sample of $^{14}_{6}$C that was originally 5.0 g, after 100 years

E. a sample of $^{194}_{84}$Po that was originally 500 g, after 1 day

95. Reverse Problem Write a physics problem with real objects for which the following equation would be part of the solution:

$$^{230}_{94}\text{Th} \rightarrow ? + ^{4}_{2}\text{He}$$

Writing in Physics

96. Research the present understanding of dark matter in the universe. Why do cosmologists think it must exist? Of what might it be made?

97. Research the hunt for the top quark. Why did physicists hypothesize its existence?

Cumulative Review

98. An EMF of 2.0 mV is induced in a wire that is 0.10 m long when it is moving perpendicularly across a uniform magnetic field at a velocity of 4.0 m/s. What is the magnetic induction of the field?

99. An electron has a de Broglie wavelength of 400.0 nm, the shortest wavelength of visible light.

a. Find the velocity of the electron.

b. Calculate the energy of the electron in eV.

100. A photon with an energy of 14.0 eV enters a hydrogen atom in the ground state and ionizes it. With what kinetic energy will the electron be ejected from the atom?

101. A silicon diode (V = 0.70 V) that is conducting 137 mA is in series with a resistor and a 6.67-V power source.

a. What is the potential difference across the resistor?

b. What is the value of the resistor?

MULTIPLE CHOICE

1. How many protons, neutrons, and electrons are in the isotope nickel-60 ($^{60}_{28}Ni$)?

	Protons	Neutrons	Electrons
A.	28	32	28
B.	28	28	32
C.	32	32	28
D.	32	28	28

2. What has occurred in the following reaction?
$^{212}_{82}Pb \rightarrow ^{212}_{83}Bi + e^- + \overline{\nu}$

A. alpha decay
B. beta decay
C. gamma decay
D. loss of a proton

3. What is the product when pollonium-210 ($^{210}_{84}Po$) undergoes alpha decay?

A. $^{206}_{82}Pb$ **C.** $^{210}_{85}At$
B. $^{208}_{82}Pb$ **D.** $^{210}_{80}Hg$

4. A sample of radioactive iodine-131 emits beta particles at the rate of 2.5×10^8 Bq. The half-life is 8 days. What will be the activity after 16 days?

A. 1.6×10^7 Bq **C.** 1.3×10^8 Bq
B. 6.3×10^7 Bq **D.** 2.5×10^8 Bq

5. Identify the unknown isotope in this reaction:
neutron + $^{14}_{7}N \rightarrow ^{14}_{6}C$ + ?

A. $^{1}_{1}H$ **C.** $^{3}_{1}H$
B. $^{2}_{1}H$ **D.** $^{4}_{2}He$

6. Which type of decay does not change the number of protons or neutrons in the nucleus?

A. positron **C.** beta
B. alpha **D.** gamma

7. Polonium-210 has a half-life of 138 days. How much of a 2.34-kg sample will remain after four years?

A. 0.644 mg **C.** 1.53 g
B. 1.50 mg **D.** 10.6 g

Online Test Practice

8. An electron and a positron collide, annihilate each other, and release their energy as gamma rays. What is the minimum combined energy of the gamma rays? (The energy equivalent of the mass of an electron is 0.51 MeV.)

A. 0.51 MeV **C.** 931.49 MeV
B. 1.02 MeV **D.** 863 MeV

9. The illustration below shows the tracks in a bubble chamber produced when a gamma ray decays into a positron and an electron. Why doesn't the gamma ray leave a track?

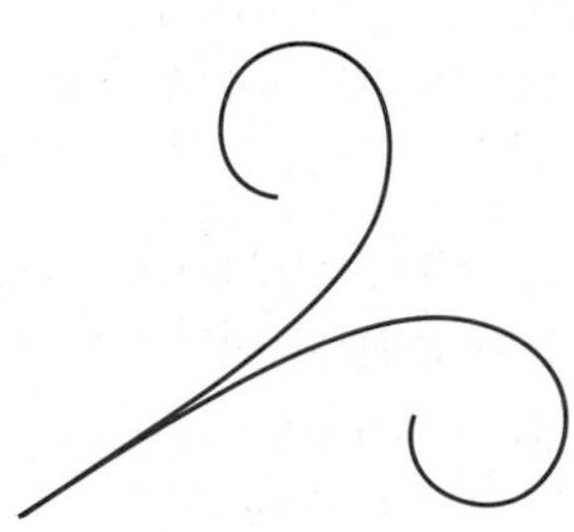

A. Gamma rays move too quickly for their tracks to be detected.
B. Only pairs of particles can leave tracks in a bubble chamber.
C. A particle must have mass to interact with the liquid and leave a track, and the gamma ray is virtually massless.
D. The gamma ray is electrically neutral, so it does not ionize the liquid in the bubble chamber.

FREE RESPONSE

10. The fission of a uranium-235 nucleus releases about 3.2×10^{-11} J. One ton of TNT releases about 4×10^9 J. How many uranium-235 nuclei are in a nuclear fission weapon that releases energy equivalent to 20,000 tons of TNT?

NEED EXTRA HELP?

If You Missed Question	1	2	3	4	5	6	7	8	9	10
Review Section	1	2	2	2	2	2	2	3	3	2

STUDENT RESOURCES

Math Skill Handbook 835

Selected Solutions 860

Reference Tables 885

Glossary ▪ Glosario 894

Index 920

Rubberball/Getty Images

MATH SKILL HANDBOOK

I. Symbols

Symbol	Meaning	Symbol	Meaning
Δ	change in quantity	$\equiv$	is defined as
$\pm$	plus or minus a quantity	$a \times b$ ab $a(b)$	a multiplied by b
$\propto$	is proportional to		
$=$	is equal to	$a \div b$ a/b $\frac{a}{b}$	a divided by b
$\approx$	is approximately equal to		
$\leq$	is less than or equal to		
$\geq$	is greater than or equal to	$\sqrt{a}$	square root of a
$\ll$	is much less than	$\lvert a \rvert$	absolute value of a
$\gg$	is much greater than	$\log_b x$	log to the base b of x

II. Measurements and Significant Figures

Connecting Math to Physics Math is the language of physics. Physicists use mathematical equations to describe relationships among the measurements that they make. Each measurement is associated with a symbol that is used in physics equations. These symbols are called variables.

Significant Figures

All measured quantities are approximate and have significant figures. The number of significant figures indicates the precision of the measurement. Precision is a measure of exactness. The number of significant figures in a measurement depends on the smallest unit on the measuring tool. The digit farthest to the right in a measurement is estimated.

Example: In the figure below, what is the estimated digit for each of the measuring sticks used to measure the length of the rod?

By using the lower measuring tool, the length is between 9 and 10 cm. The measurement would be estimated to the nearest tenth of a centimeter. If the length was exactly on the 9-cm or 10-cm mark, record it as 9.0 cm or 10.0 cm.

By using the upper measuring tool, the length is between 9.5 and 9.6 cm. The measurement would be estimated to the nearest hundredth of a centimeter. If the length was exactly on the 9.5-cm or 9.6-cm mark, record it as 9.50 cm or 9.60 cm.

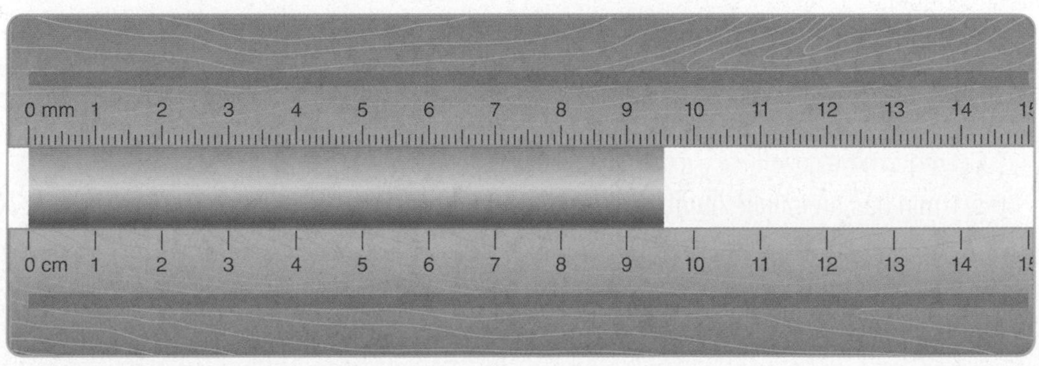

All nonzero digits in a measurement are significant figures. Some zeros are significant, and some are not. All digits between and including the first nonzero digit from the left through the significant figure on the right are significant. Use the following rules when determining the number of significant figures.

1. Nonzero digits are significant.
2. Final zeros after a decimal point are significant.
3. Zeros between two significant figures are significant.
4. Zeros used only as placeholders are not significant.

Example: State the number of significant figures in each measurement.

5.0 g has two significant figures.	Using rules 1 and 2
14.90 g has four significant figures.	Using rules 1 and 2
0.0 has one significant figure.	Using rules 2 and 4
300.00 mm has five significant figures.	Using rules 1, 2, and 3
5.06 s has three significant figures.	Using rules 1 and 3
304 s has three significant figures.	Using rules 1 and 3
0.0060 mm has two significant figures (6 and the last 0).	Using rules 1, 2, and 4
140 mm has two significant figures (just 1 and 4).	Using rules 1 and 4

PRACTICE PROBLEMS

1. State the number of significant figures in each measurement.

a. 1405 m

b. 2.50 km

c. 0.0034 m

d. 12.007 kg

e. 5.8×10^6 kg

f. 3.03×10^{25} mL

There are two cases in which numbers are considered exact and, thus, have an infinite number of significant figures.

1. Counting numbers have an infinite number of significant figures.
2. Conversion factors have an infinite number of significant figures.

Examples:

The factor "2" in 2*mg* has an infinite number of significant figures.

The number 2 is a counting number. It is an exact integer.

The number "4" in 4 electrons has an infinite number of significant figures.

Because you cannot have a partial electron, the number 4, a counting number, is considered to have an infinite number of significant figures.

60 s/1 min has an infinite number of significant figures.

There are exactly 60 seconds in 1 minute, thus there is an infinite number of significant figures in the conversion factor.

Rounding

You can round a number to a specific place value (such as hundreds or tenths) or to a specific number of significant figures. To do this, determine the place to which you are rounding, and then use the following rules.

1. When the leftmost digit to be dropped is less than 5, that digit and any digits that follow are dropped. Then the last digit in the rounded number remains unchanged.
2. When the leftmost digit to be dropped is greater than 5, that digit and any digits that follow are dropped, and the last digit in the rounded number is increased by one.
3. When the leftmost digit to be dropped is 5 followed by a nonzero number, that digit and any digits that follow are dropped. The last digit in the rounded number increases by one.
4. If the digit to the right of the last significant digit is equal to 5 and 5 is followed by a zero or no other digits, look at the last significant digit. If it is odd, increase it by one; if it is even, do not round up.

Examples: Round the following numbers to the stated number of significant figures.

8.7645 rounded to 3 significant figures is 8.76. *Using rule 1*

8.7676 rounded to 3 significant figures is 8.77. *Using rule 2*

8.7519 rounded to 2 significant figures is 8.8. *Using rule 3*

92.350 rounded to 3 significant figures is 92.4. *Using rule 4*

92.25 rounded to 3 significant figures is 92.2. *Using rule 4*

PRACTICE PROBLEMS

2. Round each number to the number of significant figures shown in parentheses.

a. 1405 m (2)

b. 2.50 km (2)

c. 0.0034 m (1)

d. 12.007 kg (3)

Operations with Significant Figures

If using a calculator, do all of the operations with as much precision as the calculator allows, and then round the result to the correct number of significant figures. The number of significant figures in the result depends on the measurements and on the operation.

Addition and subtraction Round the result to the least precise value among the measurements—the smallest number of digits to the right of the decimal points.

Example: Add 1.456 m, 4.1 m, and 20.3 m.

The least precise values are 4.1 m and 20.3 m because they have only one digit to the right of the decimal points.

$$\begin{array}{rl} 1.456 & \text{m} \\ 4.1 & \text{m} \\ +\ 20.3 & \text{m} \\ \hline 25.856 & \text{m} \end{array}$$

Add the numbers.

25.9 m *Round the result to place value of the least precise value.*

Multiplication and division Look at the number of significant figures in each measurement. Perform the calculation. Round the result so that it has the same number of significant figures as the measurement with the least number of significant figures.

Example: Multiply 20.1 m by 3.6 m.

$$(20.1\text{ m})(3.6\text{ m}) = 72.36\text{ m}^2$$

The least precise value is 3.6 m with two significant figures. The product can only have as many digits as the least precise of the multiplied numbers.

72 m *Round the result to two significant figures.*

PRACTICE PROBLEMS

3. Simplify the following expressions using the correct number of significant figures.

a. 5.012 km + 3.4 km + 2.33 km

b. 45 g − 8.3 g

c. 3.40 cm × 7.125 cm

d. 54 m ÷ 6.5 s

Combination When doing a calculation that requires a combination of addition/subtraction and multiplication/division, use the multiplication/division rule.

Examples:

$$x = 19\text{ m} + (25.0\text{ m/s})(2.50\text{ s}) + \frac{1}{2}(-10.0\text{ m/s}^2)(2.50\text{ s})^2$$

$$= 5.0\times10^1\text{ m}$$

19 m only has two significant figures, so the answer should only have two significant figures.

$$\text{slope} = \frac{70.0\text{ m} - 10.0\text{ m}}{29\text{ s} - 11\text{ s}}$$

$$= 3.3\text{ m/s}$$

29 s and 11 s only have two significant figures each, so the answer should only have two significant figures.

Multistep calculations Do not round to significant figures in the middle of a multistep calculation. Instead, round to a reasonable number of decimal places that will not cause you to lose significance in your answer. When you get to your final step where you are solving for the answer asked for in the question, you should then round to the correct number of significant figures.

Example:

$$F = \sqrt{(24\text{ N})^2 + (36\text{ N})^2}$$

$$= \sqrt{576\text{ N}^2 + 1296\text{ N}^2}$$

Do not round to 580 N^2 and 1300 N^2.

$$= \sqrt{1872\text{ N}^2}$$

Do not round to 1800 N^2.

$$= 43\text{ N}$$

Final answer, so it should be rounded to two significant figures.

If you had rounded in each step, you would have obtained an answer of 44 N. This might seem like a small discrepancy, but in more complex calculations, the effects can be large.

For example, if you calculate 5 × 5 × 5 × 5 × 5 without rounding your answer will be 3125, which rounds to 3000. If you round at each step, however, the answer you find is 2500, which rounds to 2000.

III. Fractions, Ratios, Rates, and Proportions

Math Skill Handbook

Fractions

A fraction names a part of a whole or a part of a group. It also can express a ratio (see page 840). A fraction consists of a numerator, a division bar, and a denominator.

$$\frac{\text{numerator}}{\text{denominator}} = \frac{\text{number of parts chosen}}{\text{total number of parts}}$$

Simplification Sometimes, it is easier to simplify an expression before substituting the known values of the variables. Variables often cancel out of the expression.

Example: Simplify $\frac{pn}{pw}$.

Get help **with fractions.**

Personal Tutor

$\frac{pn}{pw} = \left(\frac{p}{p}\right)\left(\frac{n}{w}\right)$ Factor out the *p* in the numerator and the denominator, and break the fraction into the product of two fractions.

$= (1)\left(\frac{n}{w}\right)$ Substitute $(p/p) = 1$.

$= \frac{n}{w}$

Multiplication and division To multiply fractions, multiply the numerators and multiply the denominators.

Example: Multiply the fractions $\frac{s}{a}$ and $\frac{t}{b}$.

$\left(\frac{s}{a}\right)\left(\frac{t}{b}\right) = \frac{st}{ab}$ Multiply the numerators and the denominators.

To divide fractions, multiply the first fraction by the reciprocal of the second fraction. To find the reciprocal of a fraction, invert it—switch the numerator and the denominator.

Example: Divide the fraction $\frac{s}{a}$ by $\frac{t}{b}$.

$\frac{s}{a} \div \frac{t}{b} = \left(\frac{s}{a}\right)\left(\frac{b}{t}\right)$ Multiply the first fraction by the reciprocal of the second fraction.

$= \frac{sb}{at}$ Multiply the numerators and the denominators.

Addition and subtraction To add or subtract two fractions, first write them as fractions with a common denominator. To do this, multiply each fraction by the denominator of the other in the form of a fraction equal to one. Then add or subtract the numerators.

Example: Add the fractions $\frac{1}{a}$ and $\frac{2}{b}$.

$\frac{1}{a} + \frac{2}{b} = \left(\frac{1}{a}\right)\left(\frac{b}{b}\right) + \left(\frac{2}{b}\right)\left(\frac{a}{a}\right)$ Multiply each fraction by a fraction equal to 1.

$= \frac{b}{ab} + \frac{2a}{ab}$ Multiply the numerators and the denominators.

$= \frac{b + 2a}{ab}$ Write a single fraction with the common denominator.

PRACTICE PROBLEMS

4. Perform the indicated operation. Write the answer in simplest form.

a. $\frac{1}{x} + \frac{y}{3}$

b. $\frac{a}{2b} - \frac{3}{b}$

c. $\left(\frac{3}{x}\right)\left(\frac{1}{y}\right)$

d. $\frac{2a}{5} \div \frac{1}{2}$

Ratios

A ratio is a comparison between two numbers by division. Ratios can be written in several different ways. The ratio of 2 and 3 can be written in four different ways:

2 to 3　　2 out of 3　　2:3　　$\frac{2}{3}$

Rates

A rate is a ratio that compares two quantities with different measurement units. A unit rate is a rate that has been simplified so that the denominator is 1.

Example: Write 98 km in 2.0 hours as a unit rate.

98 km in 2.0 hours is a ratio of $\frac{98\text{ km}}{2.0\text{ hours}}$

$\frac{98\text{ km}}{2.0\text{ hours}} = \left(\frac{98}{2.0}\right)\left(\frac{\text{km}}{\text{hour}}\right)$ Split the fraction into the product of a number fraction and a unit fraction.

$= (49)\left(\frac{\text{km}}{\text{hour}}\right)$ Simplify the number fraction.

$= 49$ km per hour or 49 km/h

Example: Write 16 Swedish crowns in 2 U.S. dollars as a unit rate.

16 Swedish crowns in \$2.00 American currency is a ratio of $\frac{16\text{ Swedish crowns}}{2\text{ U.S. dollars}}$

$$\frac{16\text{ Swedish crowns}}{2\text{ U.S. dollars}} = \left(\frac{16}{2}\right)\left(\frac{\text{Swedish crowns}}{\text{U.S. dollars}}\right)$$

$$= (8)\left(\frac{\text{Swedish crowns}}{\text{U.S. dollars}}\right)$$

$= 8$ Swedish crowns per U.S. dollar

or 8 Swedish crowns/U.S. dollar

Proportions

A proportion is an equation that states that two ratios are equal:

$\frac{a}{b} = \frac{c}{d}$, where b and d are not zero.

Proportions used to solve ratio problems often include three numbers and one variable. You can solve the proportion to find the value of the variable. To solve a proportion, use cross multiplication.

Example: Solve the proportion $\frac{a}{b} = \frac{c}{d}$ for b.

Cross multiply: $\frac{a}{b} = \frac{c}{d}$

$bc = ad$ Write the equation resulting from cross multiplying.

$b = \frac{ad}{c}$ Solve for b.

PRACTICE PROBLEMS

5. Solve the following proportions.

a. $\frac{4}{x} = \frac{2}{3}$

b. $\frac{13}{15} = \frac{n}{75}$

c. $\frac{36}{12} = \frac{s}{16}$

d. $\frac{2.5}{5.0} = \frac{7.5}{w}$

IV. Exponents, Powers, Roots, and Absolute Value

Math Skill Handbook

Exponents

An exponent is a number that tells how many times a number (a) is used as a factor. An exponent is written as a superscript. In the term a^n, a is the base and n is the exponent. a^n is called the nth power of a or a raised to the nth power.

Get help with **exponents.**

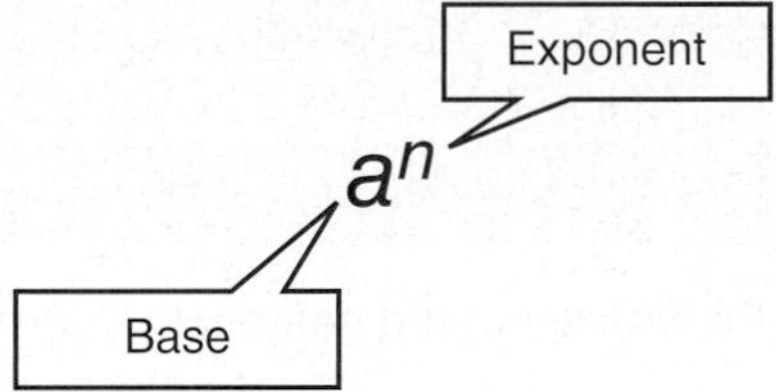

Connecting Math to Physics A subscript is not an exponent. In physics, a subscript is used to further describe the variable. For example, v_0 can be used to represent the velocity at time 0. A subscript is a part of the variable.

Positive exponent For any nonzero number (a) and any integer (n)

$$a^n = (a)(a)(a)\ldots(a) \quad n \text{ times}$$

Examples: Simplify the following exponent terms.

$10^4 = (10)(10)(10)(10) = 10{,}000$

$2^3 = (2)(2)(2) = 8$

Zero exponent For any nonzero number (a)

$$a^0 = 1$$

Examples: Simplify the following zero exponent terms.

$2^0 = 1 \qquad 13^0 = 1$

Negative exponent For any nonzero number (a) and any integer (n)

$$a^{-n} = \frac{1}{a^n}$$

Examples: Write the following negative exponent terms as fractions.

$2^{-1} = \frac{1}{2^1} = \frac{1}{2} \qquad 2^{-2} = \frac{1}{2^2} = \frac{1}{4}$

Square and Cube Roots

A square root of a number is one of its two equal factors. A radical sign ($\sqrt{\ }$) indicates a square root. A square root can be shown as exponent $\frac{1}{2}$, as in $\sqrt{b} = b^{\frac{1}{2}}$. You can use a calculator to find square roots.

Examples: Simplify the following square root terms.

Get help with **square and cube roots.**

Personal Tutor

$\sqrt{a^2} = \sqrt{(a)(a)} = a$

$\sqrt{9} = \sqrt{(3)(3)} = 3$

$\sqrt{64} = \sqrt{(8.0)(8.0)} = 8.0$ Keep two significant figures.

$\sqrt{38.44} = 6.200$ Place two zeros to the right of the calculator answer to keep four significant figures.

$\sqrt{39} = 6.244997\ldots = 6.2$ Round the answer to keep two significant figures.

A cube root of a number is one of its three equal factors. A radical sign with the number 3 ($\sqrt[3]{\ }$) indicates a cube root. A cube root also can be shown as exponent $\frac{1}{3}$, as in $\sqrt[3]{b} = b^{\frac{1}{3}}$.

Example: Simplify the following cube root terms.

$$\sqrt[3]{125} = \sqrt[3]{(5.00)(5.00)(5.00)} = 5.00$$

$$\sqrt[3]{39.304} = 3.4000$$

PRACTICE PROBLEMS

6. Find each root. Round the answer to the nearest hundredth.

a. $\sqrt{22}$

b. $\sqrt[3]{729}$

c. $\sqrt{676}$

d. $\sqrt[3]{46.656}$

7. Simplify by writing without a radical sign.

a. $\sqrt{16a^2b^4}$

b. $\sqrt{9t^6}$

8. Write using exponents.

a. $\sqrt{n^3}$

b. $\frac{1}{\sqrt{a}}$

Operations with Exponents

In the following operations with exponents, *a, b, m,* and *n* can be numbers or variables.

Product of powers To multiply terms with the same base, add the exponents, as in $(a^m)(a^n) = a^{m+n}$.

Quotient of powers To divide terms with the same base, subtract the bottom exponent from the top exponent, as in $a^m/a^n = a^{m-n}$.

Power of a power To calculate the power of a power, use the same base and multiply the exponents, as in $(a^m)^n = a^{mn}$.

***n*th root of a power** To calculate the root of a power, use the same base and divide the power exponent by the root exponent, as in $\sqrt[n]{a^m} = a^{\frac{m}{n}}$.

Power of a product To calculate the power of a product of *a* and *b,* raise both to the power and find their product, as in $(ab)^n = a^nb^n$.

PRACTICE PROBLEMS

9. Write an equivalent form using the properties of exponents.

a. $\frac{x^2t}{x^3}$

b. $\sqrt{t^3}$

c. $(d^2n)^2$

d. $x^2\sqrt{x}$

10. Simplify $\frac{m}{q}\sqrt{\frac{2qv}{m}}$.

Absolute Value

The absolute value of a number (n) is its magnitude, regardless of its sign. The absolute value of n is written as $|n|$. Because magnitudes cannot be less than zero, absolute values always are greater than or equal to zero.

Examples:

$|3| = 3$

$|-3| = 3$

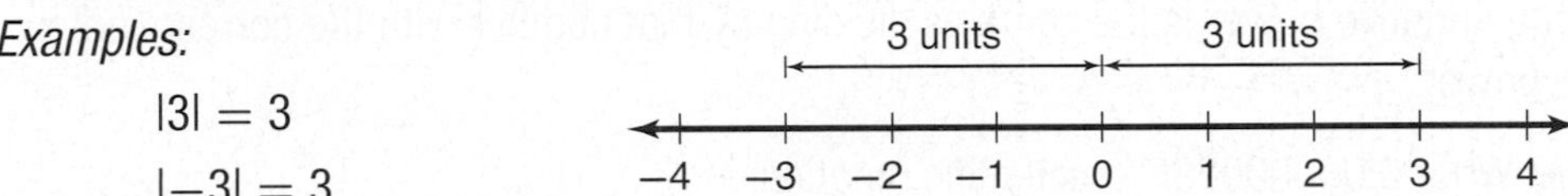

V. Scientific Notation

A number of the form $a \times 10^n$ is written in scientific notation, where $1 \leq a \leq 10$, and n is an integer. The base (10) is raised to a power (n).

Get help with **scientific notation.**

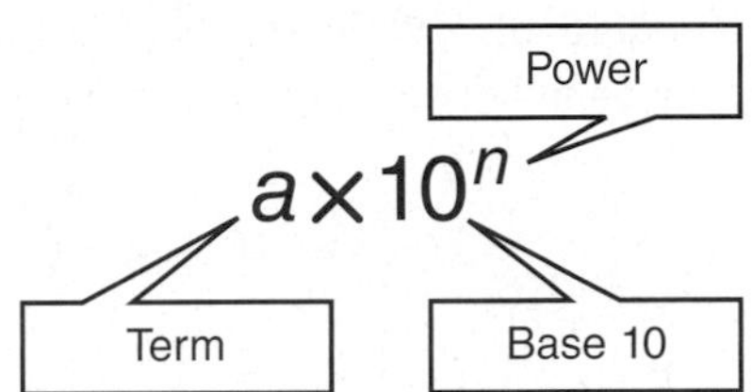

Connecting Math to Physics Physicists commonly use scientific notation to express measurements that are greater than 10 or less than 1. For example, the mass of a proton is written as 6.73×10^{-28} kg. The density of water is written as 1.000×10^3 kg/m^3. This shows, using significant digit rules, that this measurement is exactly 1000 to four significant figures. However, writing the density of water as 1000 kg/m^3 would imply that it has only one significant digit, which is incorrect. Scientific notation helps physicists keep accurate track of significant figures.

Large Numbers—Using Positive Exponents

Multiplying by a power of 10 is like moving the decimal point that same number of places to the right (if the power is positive) or to the left (if the power is negative). To express a large number in scientific notation, first determine the value for a, $1 \leq a \leq 10$. Count the number of decimal places from the decimal point in a to the decimal point in the number. Use that count as the power of 10. A calculator shows scientific notation with e for exponent, as in $2.4\text{e} + 11 = 2.4 \times 10^{11}$. Some calculators use an E to show the exponent, or there is often a place on the display where the calculator can show smaller-sized digits representing the exponent.

Example: Write 7,530,000 in scientific notation.

The value for a is 7.53. (The decimal point is to the right of the first nonzero digit.) So the form will be 7.53×10^n.

$7{,}530{,}000 = 7.53 \times 10^6$ — *There are six decimal places, so the power is 6.*

To write the standard form of a number expressed in scientific notation, write the value of a, and place extra zeros to the right of the number. Use the power and move the decimal point in a that many places to the right.

Example: Write the following number in standard form.

$2.389 \times 10^5 = 2.38900 \times 10^5 = 238{,}900$

Small Numbers—Using Negative Exponents

To express a small number in scientific notation, first determine the value for a, $1 \leq a \leq 10$. Then count the number of decimal places from the decimal point in a to the decimal point in the number. Use that number as the power of 10. Multiplying by a number with a negative power is the same as dividing by that number with the corresponding positive power.

Example: Write 0.000000285 in scientific notation.

The value for a is 2.85. (The decimal point is to the right of the first nonzero digit.) So the form will be 2.85×10^n.

$0.000000285 = 2.85 \times 10^{-7}$ *There are seven decimal places, so the power is −7.*

To express a small number in standard form, write the value for a and place extra zeros to the left of a. Use the power and move the decimal point in a that many places to the left.

Example:

$1.6 \times 10^{-4} = 00001.6 \times 10^{-4} = 0.00016$

PRACTICE PROBLEMS

11. Express each number in scientific notation.

a. 456,000,000 **b.** 0.000020

12. Express each number in standard notation.

a. 3.03×10^{-7} **b.** 9.7×10^{10}

Operations with Scientific Notation

Calculating with numbers written in scientific notation uses the properties of exponents.

Get help with **operations with scientific notation.**

Multiplication Multiply the terms, and add the powers of 10.

Example: Simplify.

$$
\begin{aligned}
(4.0 \times 10^{-8})(1.2 \times 10^{5}) &= (4.0 \times 1.2)(10^{-8} \times 10^{5}) && \text{Group terms and bases of 10.}\\
&= (4.8)(10^{-8+5}) && \text{Multiply terms.}\\
&= (4.8)(10^{-3}) && \text{Add powers of 10.}\\
&= 4.8 \times 10^{-3} && \text{Recombine in scientific notation.}
\end{aligned}
$$

Division Divide the base numbers and subtract the exponents of 10.

Example: Simplify.

$$
\begin{aligned}
\frac{9.60 \times 10^{7}}{1.60 \times 10^{3}} &= \left(\frac{9.60}{1.60}\right) \times \left(\frac{10^{7}}{10^{3}}\right) && \text{Group terms and bases of 10.}\\
&= 6.00 \times 10^{7-3} && \text{Divide terms, and subtract powers of 10.}\\
&= 6.00 \times 10^{4}
\end{aligned}
$$

Addition and subtraction To add and subtract numbers in scientific notation, the powers of 10 must be the same. Thus, you may need to rewrite one of the numbers with a different power of 10. With equal powers of 10, use the distributive property.

Example: Simplify.

$(3.2\times10^5) + (4.8\times10^5) = (3.2 + 4.8)\times10^5$ — *Group terms.*

$= 8.0\times10^5$ — *Add terms.*

Example: Simplify.

$(3.2\times10^5) + (4.8\times10^4) = (3.2\times10^5) + (0.48\times10^5)$ — *Rewrite 4.8×10^4 as 0.48×10^5.*

$= (3.2 + 0.48)\times10^5$ — *Group terms.*

$= 3.68\times10^5$ — *Add terms.*

$= 3.7\times10^5$ — *Round using the addition significant figures rule.*

PRACTICE PROBLEMS

13. Evaluate each expression; express the result in scientific notation.

a. $(5.2\times10^{-4})(4.0\times10^8)$ **b.** $(2.4\times10^3) + (8.0\times10^4)$

VI. Equations

Order of Operations

Scientists and mathematicians have agreed on a set of steps or rules, called the order of operations, so that everyone interprets mathematical symbols in the same way. Follow these steps in order when you evaluate an expression or use a formula.

1. Simplify the expressions inside grouping symbols, such as parentheses ( ), brackets [ ], braces { }, and fraction bars.
2. Evaluate all powers and roots.
3. Do all multiplications and/or divisions from left to right.
4. Do all additions and/or subtractions from left to right.

A useful way to remember these is the phrase *"P*lease *e*xcuse *m*y *d*ear *A*unt *S*ally." The first letter of each word represents an operation: *p*arentheses, *e*xponents, *m*ultiplication, *d*ivision, *a*ddition, *s*ubtraction.

Example: Simplify the following expression.

$4 + 3(4 - 1) - 2^3 = 4 + 3(3) - 2^3$ — *Order of operations step 1.*

$= 4 + 3(3) - 8$ — *Order of operations step 2.*

$= 4 + 9 - 8$ — *Order of operations step 3.*

$= 5$ — *Order of operations step 4.*

Connecting Math to Physics The previous example was shown step-by-step to demonstrate the order of operations. When solving a physics problem, do not round to the correct number of significant figures until after the final calculation.

In calculations involving an expression in a numerator and an expression in a denominator, the numerator and the denominator are separate groups and you should calculate them before dividing the numerator by the denominator. The multiplication/division rule for significant figures is used to determine the final number of significant figures.

Solving Equations

To solve an equation means to find the value of the variable that makes the equation a true statement. To solve equations, apply the distributive property and the properties of equality. Any properties of equalities that you apply on one side of an equation, you also must apply on the other side.

Get help with **using the distributive property.**

Personal Tutor

Distributive property For any numbers a, b, and c,

$$a(b + c) = ab + ac \qquad a(b - c) = ab - ac$$

Example: Use the distributive property to expand the following expression.

$3(x + 2) = 3x + (3)(2) = 3x + 6$

Addition and subtraction properties of equality If two quantities are equal, and the same number is added to (or subtracted from) each, then the resulting quantities also are equal.

$$\text{If } a = b, \text{ then}$$

$$a + c = b + c \quad \text{and} \quad a - c = b - c$$

Example: Solve $x - 3 = 7$ using the addition property.

$$x - 3 = 7$$
$$x - 3 + 3 = 7 + 3$$
$$x = 10$$

Example: Solve $t + 2 = -5$ using the subtraction property.

$$t + 2 = -5$$
$$t + 2 - 2 = -5 - 2$$
$$t = -7$$

Multiplication and division properties of equality If two equal quantities each are multiplied by (or divided by) the same number, then the resulting quantities also are equal.

$$\text{If } a = b, \text{ then}$$

$$ac = bc \text{ and}$$

$$\frac{a}{c} = \frac{b}{c}, \text{ for } c \neq 0$$

Example: Solve $\frac{1}{4}a = 3$ using the multiplication property.

$$\frac{1}{4}a = 3$$
$$\left(\frac{a}{4}\right)(4) = 3(4)$$
$$a = 12$$

Example: Solve $6n = 18$ using the division property.

$$6n = 18$$
$$\frac{6n}{6} = \frac{18}{6}$$
$$n = 3$$

Example: Solve $2t + 8 = 5t - 4$ for t.

$$2t + 8 = 5t - 4$$
$$8 + 4 = 5t - 2t$$
$$12 = 3t$$
$$4 = t$$

Isolating a Variable

Suppose an equation has more than one variable. To isolate a variable—that is, to solve the equation for a variable—write an equivalent equation so that one side contains only that variable with a coefficient of 1.

Connecting Math to Physics Isolate the variable P (pressure) in the ideal gas law equation.

Get help with **isolating a variable.**

$$PV = nRT$$

$$\frac{PV}{V} = \frac{nRT}{V} \quad \text{Divide both sides by } V.$$

$$P\left(\frac{V}{V}\right) = \frac{nRT}{V} \quad \text{Group } \frac{V}{V}.$$

$$P = \frac{nRT}{V} \quad \text{Substitute } \frac{V}{V} = 1.$$

PRACTICE PROBLEMS

14. Solve for x.

a. $2 + 3x = 17$

b. $x - 4 = 2 - 3x$

c. $t - 1 = \frac{x + 4}{3}$

d. $a = \frac{b + x}{c}$

e. $\frac{2x + 3}{x} = 6$

f. $ax + bx + c = d$

Square Root Property

If a and n are real numbers, $n > 0$, and $a^2 = n$, then $a = \pm\sqrt{n}$.

Connecting Math to Physics Solve for v in Newton's second law for a satellite orbiting Earth.

$$\frac{mv^2}{r} = \frac{Gm_Em}{r^2}$$

$$\frac{rmv^2}{r} = \frac{rGm_Em}{r^2} \quad \text{Multiply both sides by } r.$$

$$mv^2 = \frac{Gm_Em}{r} \quad \text{Substitute } \frac{r}{r} = 1.$$

$$\frac{mv^2}{m} = \frac{Gm_Em}{rm} \quad \text{Divide both sides by } m.$$

$$v^2 = \frac{Gm_E}{r} \quad \text{Substitute } \frac{m}{m} = 1.$$

$$\sqrt{v^2} = \pm\sqrt{\frac{Gm_E}{r}} \quad \text{Take the square root.}$$

$$v = \sqrt{\frac{Gm_E}{r}} \quad \text{Use the positive value for speed.}$$

When using the square root property, it is important to consider what you are solving for. Because we solved for speed in the above example, it did not make sense to use the negative value of the square root. Also, you need to consider if the negative or positive value gives you a realistic solution. For example, when using the square root property to solve for time, a negative value might give you a time before the situation even started.

Quadratic Equations

A quadratic equation has the form $ax^2 + bx + c = 0$, where $a \neq 0$. A quadratic equation has one variable with a power (exponent) of 2. It also might include that same variable to the first power. You can estimate the solutions by graphing on a graphing calculator. If $b = 0$, then there is no x-term in the quadratic equation. In this case, you can solve the equation by isolating the squared variable and finding the square root of each side of the equation using the square root property.

Quadratic Formula

You can find the solutions of any quadratic equation by using the quadratic formula. The solutions of $ax^2 + bx + c = 0$, where $a \neq 0$, are given by

$$x = \frac{-b \pm \sqrt{b^2 - 4ac}}{2a}$$

As with the square root property, it is important to consider whether the solutions to the quadratic formula give you a realistic answer to the problem you are solving. Usually, you can throw out one of the solutions because it is unrealistic. Projectile motion often requires the use of the quadratic formula when solving equations, so keep the realism of the solution in mind when solving.

Get help with **using the quadratic formula.**

PRACTICE PROBLEMS

15. Solve for x.

a. $4x^2 - 19 = 17$

b. $12 - 3x^2 = -9$

c. $x^2 - 2x - 24 = 0$

d. $24x^2 - 14x - 6 = 0$

Dimensional Calculations

When doing calculations, you must include the units of each measurement that is written in the calculation. All operations that are performed on the number also are performed on its units.

Connecting Math to Physics The acceleration due to gravity (a) is given by the equation $a = \frac{2\Delta x}{\Delta t^2}$. A free-falling object near the Moon drops 20.5 m in 5.00 s. Find the acceleration (a). Acceleration is measured in meters per second squared.

$$a = \frac{2\Delta x}{\Delta t^2}$$

$$a = \frac{2(20.5\ \text{m})}{(5.00\ \text{s})^2}$$

The number 2 is an exact number, so it does not affect the determination of significant figures.

$$a = \frac{1.64\ \text{m}}{\text{s}^2} \text{ or } 1.64\ \text{m/s}^2$$

Calculate and round to three significant figures.

Unit conversion Use a conversion factor to convert from one measurement unit to another of the same type, such as from minutes to seconds. This is equivalent to multiplying by one.

Connecting Math to Physics Find Δx when $v_i = 67$ m/s and $\Delta t = 5.0$ min. Use the equation $\Delta x = v_i \Delta t$.

$$\frac{60 \text{ seconds}}{1 \text{ minute}} = 1$$

$$\Delta x = v_i \Delta t$$

$$\Delta x = \frac{67 \text{ m}}{\text{s}}\left(\frac{5.0 \text{ min}}{1}\right)\left(\frac{60 \text{ s}}{1 \text{ min}}\right)$$ *Multiply by the conversion factor* $\frac{60 \text{ s}}{1 \text{ min}} = 1$

$$\Delta x = 20{,}100 \text{ m} = 2.0\times10^4 \text{ m}$$ *Calculate and round to two significant figures. The numbers 60 s and 1 min are exact numbers, so they do not affect the determination of significant figures.*

PRACTICE PROBLEMS

16. Simplify $\Delta t = \frac{4.0\times10^2 \text{ m}}{16 \text{ m/s}}$.

17. Find the velocity of a dropped brick after 5.0 s using $v = a\Delta t$ and $a = -9.8$ m/s^2.

18. Calculate the product: $\left(\frac{32 \text{ cm}}{1 \text{ s}}\right)\left(\frac{60 \text{ s}}{1 \text{ min}}\right)\left(\frac{60 \text{ min}}{1 \text{ h}}\right)\left(\frac{1 \text{ m}}{100 \text{ cm}}\right)$.

19. An Olympian ran 100.0 m in 9.87 s. What was the speed in kilometers per hour?

Dimensional Analysis

Dimensional analysis is a method of doing algebra with the units. It often is used to check the validity of the units of a final result and the equation being used, without completely redoing the calculation.

Physics Example Verify that the final answer of $x_f = x_i + v_i t + \frac{1}{2}at^2$ will have the units meters (m).

x_i is measured in meters (m).

t is measured in seconds (s).

v_i is measured in meters per second (m/s).

a is measured in meters per second squared (m/s^2).

Get help with **dimensional analysis.**

Personal Tutor

$$x_f = \text{m} + \left(\frac{\text{m}}{\text{s}}\right)(\text{s}) + \frac{1}{2}\left(\frac{\text{m}}{\text{s}^2}\right)(\text{s})^2$$ *Substitute the units for each variable.*

$$= \text{m} + \text{m}\left(\frac{\text{s}}{\text{s}}\right) + \frac{1}{2}(\text{m})\left(\frac{\text{s}^2}{\text{s}^2}\right)$$ *Simplify the fractions using the distributive property.*

$$= \text{m} + (\text{m})(1) + \frac{1}{2}(\text{m})(1)$$ *Substitute $s/s = 1$, $s^2/s^2 = 1$.*

$$= \text{m} + \text{m} + \frac{1}{2}\text{m}$$ *Everything simplifies to m; thus x_f is in m.*

The factor of $\frac{1}{2}$ in the above does not apply to the units. It applies only to any number values that would be inserted for the variables in the equation. It is easiest to remove number factors such as the $\frac{1}{2}$ when first setting up the dimensional analysis.

VII. Graphs of Relations

The Coordinate Plane

You can locate points on a plane in reference to two perpendicular number lines, called axes. The horizontal number line is called the *x*-axis and represents the independent variable. The vertical number line is called the *y*-axis and represents the dependent variable. A point is represented by two coordinates (*x, y*), which also is called an ordered pair. The value of the independent variable (*x*) always is listed first in the ordered pair. The ordered pair (0, 0) represents the origin.

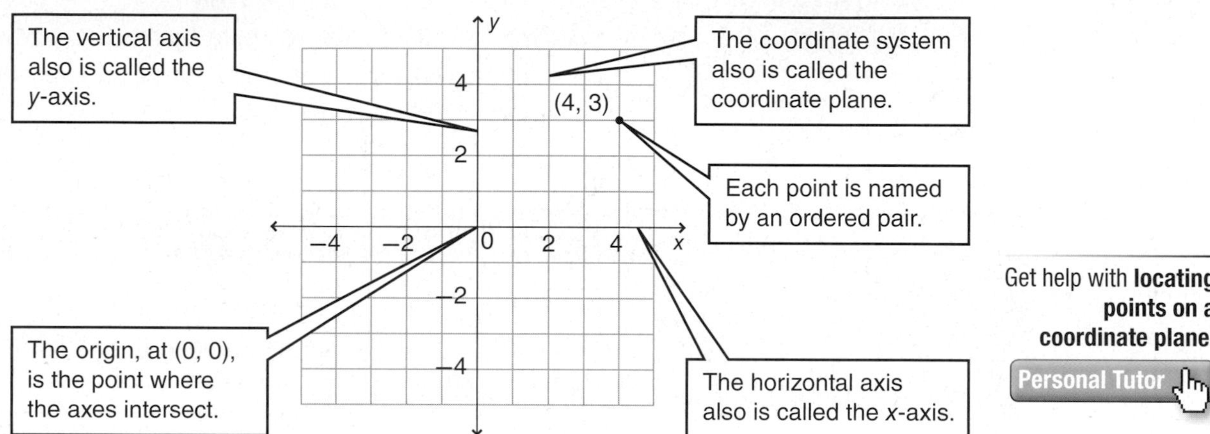

The vertical axis also is called the y-axis.
The coordinate system also is called the coordinate plane.
(4, 3)
Each point is named by an ordered pair.
The origin, at (0, 0), is the point where the axes intersect.
The horizontal axis also is called the x-axis.
y
x
4
2
0
−2
−4

Get help with **locating points on a coordinate plane.**

Personal Tutor

Graphing Data to Determine a Relationship

Use the following steps to graph data.

1. Identify the independent and the dependent variables.
2. Draw two perpendicular axes. Label each axis using the variable names.
3. Determine the range of data for each variable. Use the ranges to decide on a convenient scale for each axis. Mark and number the scales.
4. Plot each data point.
5. If the points seem to lie approximately in a line, draw a best-fit line. The line should be as close to as many points as possible. If the points do not lie in a line, draw a smooth curve through as many points as possible. If there does not appear to be a trend, do not draw any line or curve.
6. Write a title that clearly describes what the graph represents.

Get help with **graphing a line.**

Applied Force (N)	Distance Stretched (cm)
10	0.8
20	2.1
30	2.8
40	4.2
50	4.7

Distance Stretched v. Force Applied

Distance Stretched (cm): 0.0, 0.5, 1.0, 1.5, 2.0, 2.5, 3.0, 3.5, 4.0, 4.5, 5.0

Force Applied (N): 0, 10, 20, 30, 40, 50, 60

Interpolating and Extrapolating

Interpolation is a process used to estimate a value for a relation that lies between two known values. Extrapolation is a process used to estimate a value for a relation that lies beyond the known values. The equation of the best-fit line helps you interpolate and extrapolate.

Example: Using the data and the graph, estimate how far the spring will stretch if a force of 25 N is applied. Use interpolation.

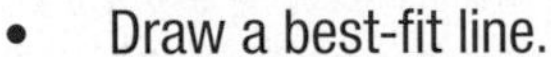

- Draw a best-fit line.
- Draw a line segment from 25 N on the x-axis to the best-fit line.
- Draw a line segment from that intersection point to the y-axis.
- Read the scale on the y-axis. A force of 25 N will stretch the spring 2.4 cm.

Example: Use extrapolation to estimate how far the spring will stretch if a 60-N force is applied.

- Draw a line segment from 60 on the x-axis to the best-fit line. Extend the best-fit line if necessary.
- Read the corresponding value on the y-axis. Extend the axis scale if necessary.
- A force of 60 N will stretch the spring 5.8 cm.

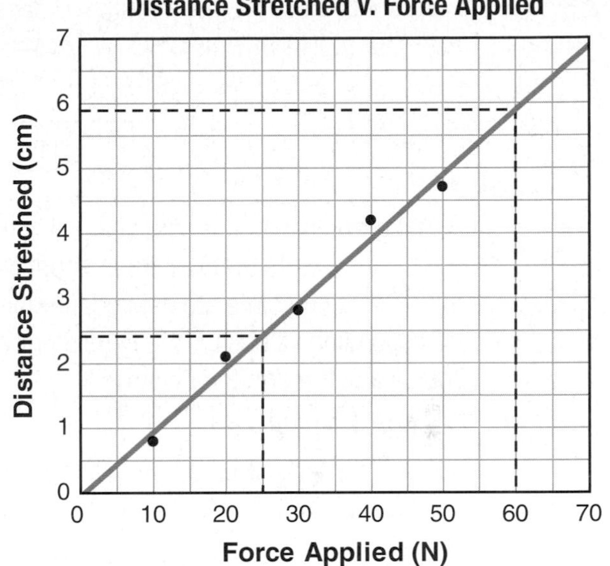

Interpreting Line Graphs

A line graph shows the linear relationship between two variables. Two types of line graphs that describe motion are used frequently in physics.

Connecting Math to Physics

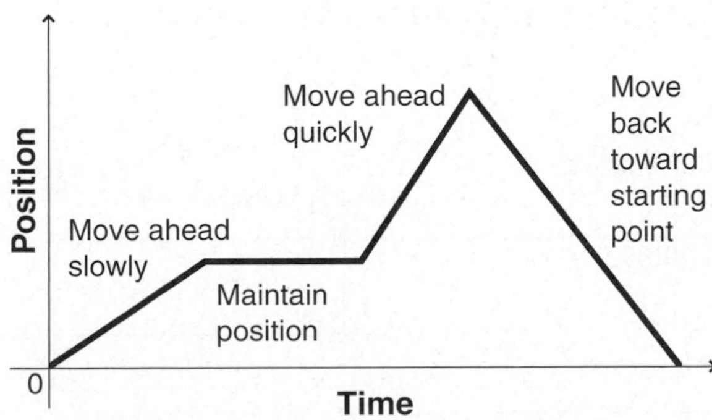

This line graph shows a changing relationship between the two graphed variables.

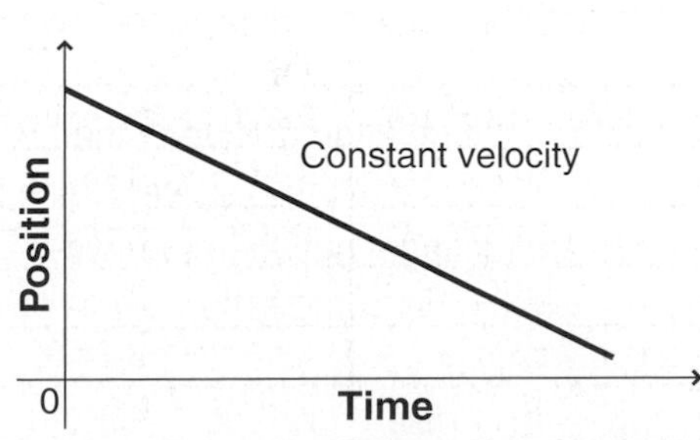

The line graph above shows a constant relationship between the two graphed variables.

Linear Equations

A linear equation can be written as a relation (or a function), $y = mx + b$, where m and b are real numbers, m represents the slope of the line, and b represents the intercept, the point at which the line crosses the y-axis.

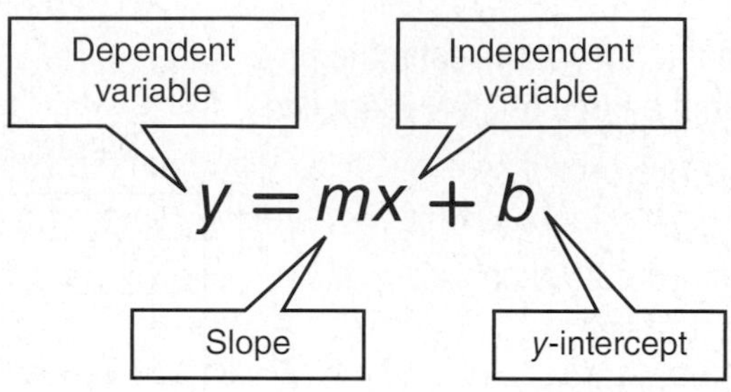

Get help with **identifying equations.**

Personal Tutor

The graph of a linear equation is a line. The line represents all of the solutions of the linear equation. To graph a linear equation, choose three values for the independent variable. (Only two points are needed, but the third point serves as a check.) Calculate the corresponding values for the dependent variable. Plot each ordered pair (x, y) as a point. Draw a line through the points.

Example: Graph $y = -\frac{1}{2}x + 3$

Calculate three ordered pairs to obtain points to plot.

Ordered Pairs

x	y
0	3
2	2
6	0

Ordered Pairs Plot

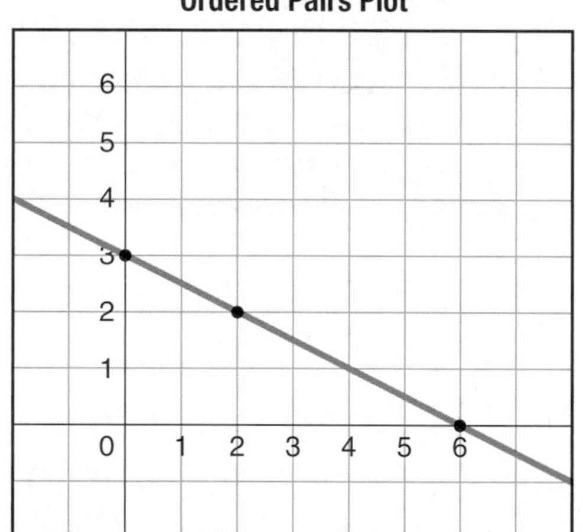

Slope

The slope of a line is the ratio of the change in y-coordinates to the change in x-coordinates. It is the ratio of the vertical change (rise) to the horizontal change (run). This number tells you how steep the line is. It can be a positive number or a negative number.

Get help with **determining slope.**

Personal Tutor

To find the slope of a line, select two points, (x_1, y_1) and (x_2, y_2). Calculate the run, which is the difference (change) between the two x-coordinates, $x_2 - x_1 = \Delta x$. Calculate the rise, which is the difference (change) between the two y-coordinates, $y_2 - y_1 = \Delta y$. Form the ratio.

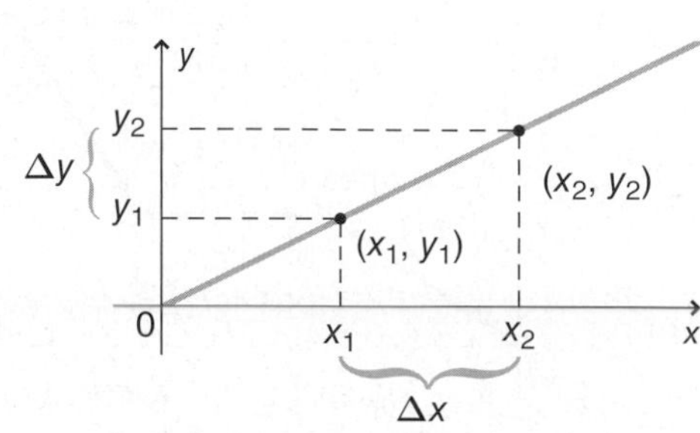

$$\text{Slope } m = \frac{rise}{run} = \frac{y_2 - y_1}{x_2 - x_1} = \frac{\Delta y}{\Delta x}$$

where $x_1 \neq x_2$

Direct Variation

If there is some nonzero constant (m) such that $y = mx$, then y varies directly with x. That means as the independent variable x doubles, the dependent variable y doubles. The variables x and y also are said to be proportional. This is a linear equation of the form $y = mx + b$ in which the value of b is zero. The graph passes through the origin, (0, 0).

Connecting Math to Physics In the force equation for an ideal spring, $F = -kx$, where F is the force on the spring, $-k$ is the spring constant, and x is the spring's displacement, the force on the spring varies directly with (is proportional to) the spring's displacement. That is, the force on the spring increases as the spring's displacement increases.

Inverse Variation

If there is some nonzero constant (m) such that $y = \frac{m}{x}$, then y varies inversely with x.

That means as the independent variable x increases, the dependent variable y decreases. The variables x and y also are said to be inversely proportional. This is not a linear equation because it contains the product of two variables. The graph of an inverse relationship is a hyperbola. This relationship can be written as

$$y = \frac{m}{x}$$
$$y = m\frac{1}{x}$$
$$xy = m$$

Example: Graph the equation $y = \frac{90}{x}$.

Ordered Pairs	
x	***y***
−10	−9
−6	−15
−3	−30
−2	−45
2	45
3	30
6	15
10	9

Inverse Variation

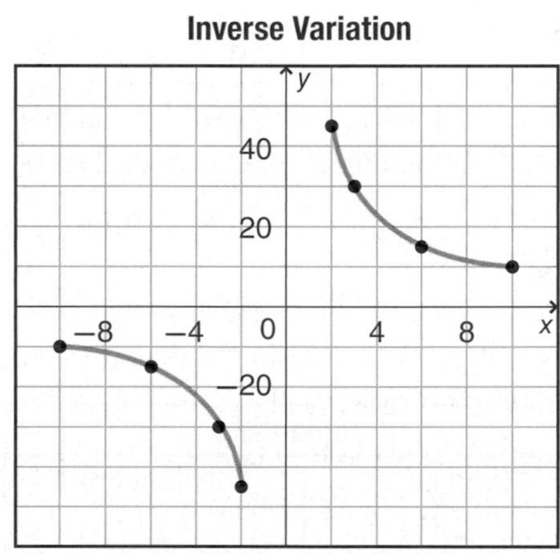

Connecting Math to Physics In the equation for the speed of a wave ($\lambda = \frac{v}{f}$) where λ is wavelength, f is frequency, and v is wave speed, wavelength varies inversely with (is inversely proportional to) frequency. That is, as the frequency of a wave increases, the wavelength decreases. v is constant.

Quadratic Graphs

A quadratic relationship is a relationship of the form

$$y = ax^2 + bx + c, \text{ where } a \neq 0.$$

A quadratic relationship includes the square of the independent variable (x). The graph of a quadratic relationship is a parabola. Whether the parabola opens upward or downward depends on whether the value of the coefficient of the squared term (a) is positive or negative.

Example: Graph the equation $y = -x^2 + 4x - 1$.

Ordered Pairs	
x	y
−1	−6
0	−1
1	2
2	3
3	2
4	−1
5	−6

Quadratic Equation Graph

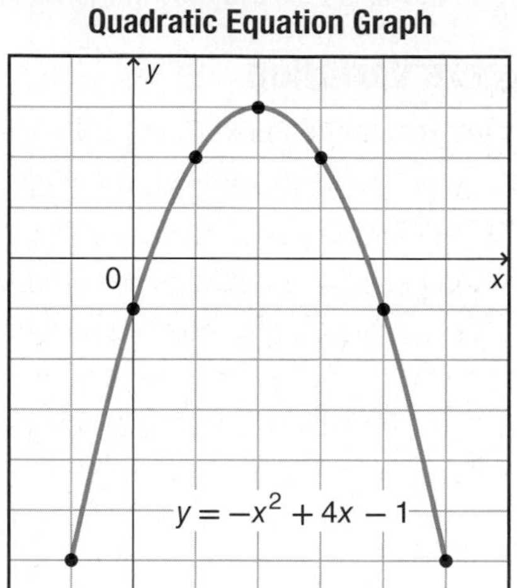

Connecting Math to Physics A position-time graph in the shape of a quadratic relation means that the object is moving at a constant acceleration.

$$x_f = 2\text{ m} + (1\text{ m/s})t + \frac{1}{2}(2\text{ m/s}^2)t^2$$

Ordered Pairs	
Time (s)	Position (m)
0	2
1	4
2	8
3	14
4	22

Position v. Time

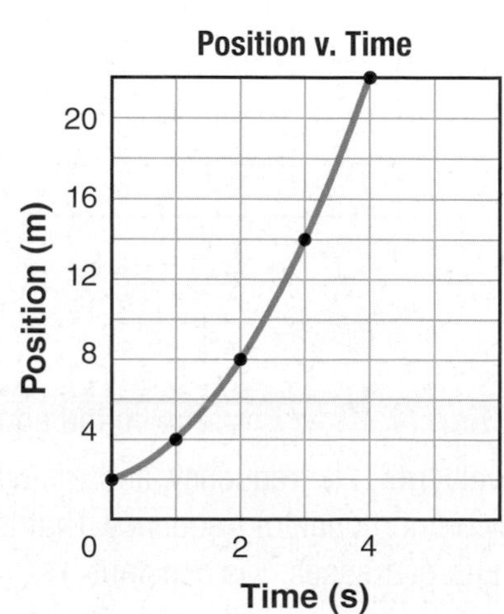

VIII. Geometry and Trigonometry

Perimeter and Area

	Perimeter, Circumference Linear units	**Area** Squared units
Square side a	$P = 4a$	$A = a^2$
Rectangle length l width w	$P = 2l + 2w$	$A = lw$
Triangle base b height h	$P =$ side 1 + side 2 + side 3	$A = \frac{1}{2}bh$
Circle radius r	$C = 2\pi r$	$A = \pi r^2$

Surface Area and Volume

	Surface Area Squared units	**Volume** Cubic units
Cube side a	$SA = 6a^2$	$V = a^3$
Cylinder radius r height h	$SA = 2\pi rh + 2\pi r^2$	$V = \pi r^2 h$
Sphere radius r	$SA = 4\pi r^2$	$V = \frac{4}{3}\pi r^3$

Connecting Math to Physics Look for geometric shapes in your physics problems. They could be in the form of objects or spaces. For example, vectors can sometimes form two-dimensional shapes.

Area Under a Graph

To calculate the approximate area under a graph, cut the area into smaller pieces and find the area of each piece using the formulas shown above. To approximate the area under a line, cut the area into a rectangle and a triangle, as shown below on the left.

To approximate the area under a curve, draw several rectangles from the x-axis to the curve, as shown below on the right. Using more rectangles with a smaller base will provide a closer approximation of the area.

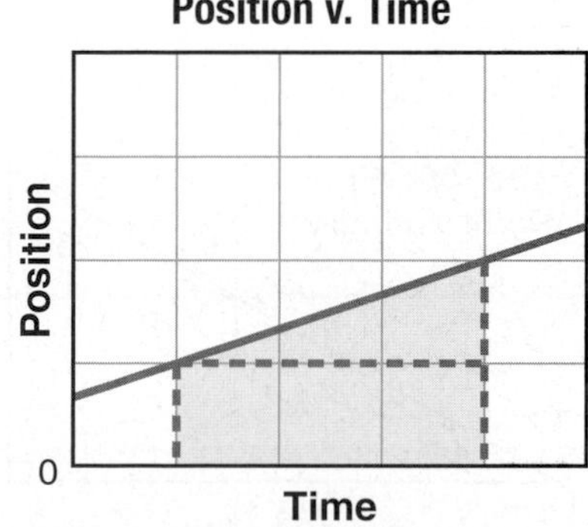

Total area = Area of the rectangle + Area of the triangle

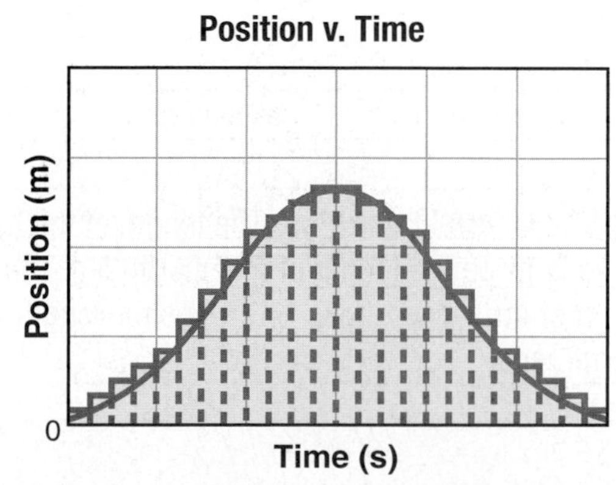

Total area = Area 1 + Area 2 + Area 3 + …

Right Triangles

The Pythagorean theorem states that if a and b are the measures of the legs of a right triangle and c is the measure of the hypotenuse, then

$$c^2 = a^2 + b^2.$$

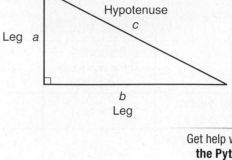

Hypotenuse c
Leg a
b Leg

To determine the length of the hypotenuse, use the square root property. Because distance is positive, the negative value does not have meaning.

$$c = \sqrt{a^2 + b^2}$$

Get help with **using the Pythagorean theorem.**

Example: In the triangle, $a = 4$ cm and $b = 3$ cm. Find c.

$$\begin{aligned} c &= \sqrt{a^2 + b^2} \\ &= \sqrt{(4 \text{ cm})^2 + (3 \text{ cm})^2} \\ &= \sqrt{16 \text{ cm}^2 + 9 \text{ cm}^2} \\ &= \sqrt{25 \text{ cm}^2} \\ &= 5 \text{ cm} \end{aligned}$$

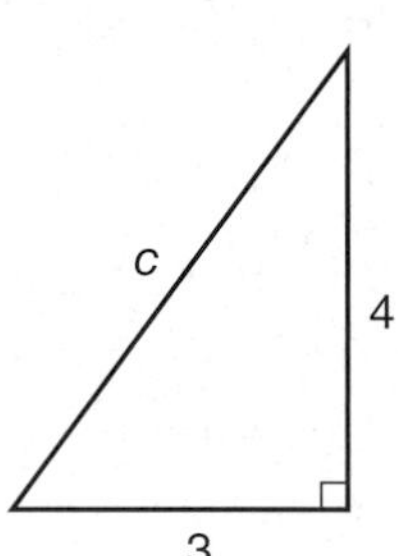

45°-45°-90° triangles The length of the hypotenuse is $\sqrt{2}$ multiplied by the length of a leg.

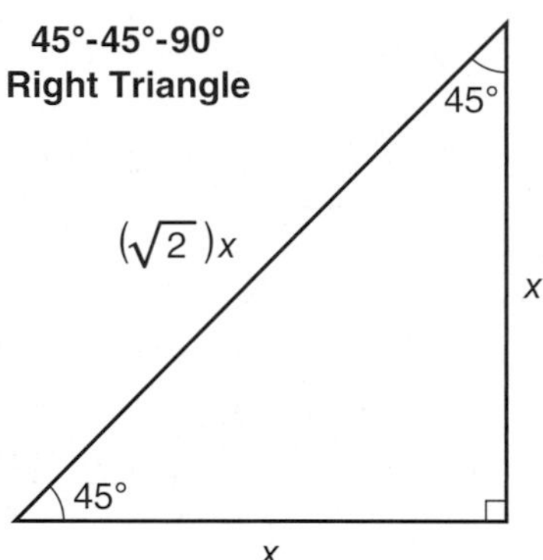

30°-60°-90° triangles The length of the hypotenuse is twice the length of the shorter leg. The length of the longer leg is $\sqrt{3}$ times the length of the shorter leg.

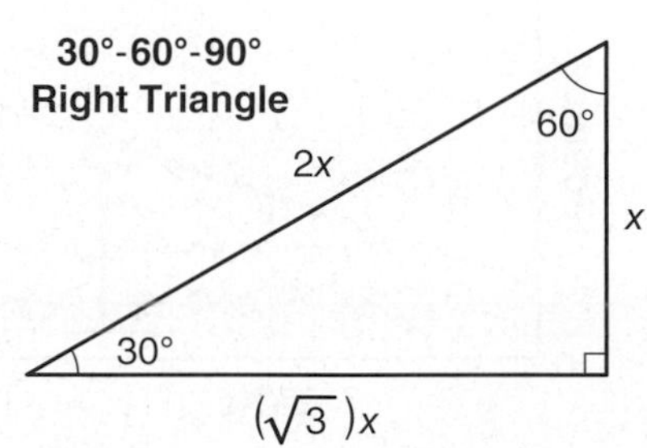

Trigonometric Ratios

A trigonometric ratio is a ratio of the lengths of sides of a right triangle. The most common trigonometric ratios are sine, cosine, and tangent. To memorize these ratios, learn the acronym SOH-CAH-TOA. SOH stands for Sine equals Opposite over Hypotenuse. CAH stands for Cosine equals Adjacent over Hypotenuse. TOA stands for Tangent equals Opposite over Adjacent.

Words	Memory Aid	Symbols
The sine is the ratio of the length of the side opposite to the angle over the length of the hypotenuse.	SOH $\sin\theta = \frac{\text{opposite}}{\text{hypotenuse}}$	$\sin\theta = \frac{a}{c}$
The cosine is the ratio of the length of the side adjacent to the angle over the length of the hypotenuse.	CAH $\cos\theta = \frac{\text{adjacent}}{\text{hypotenuse}}$	$\cos\theta = \frac{b}{c}$
The tangent is the ratio of the length of the side opposite to the angle over the length of the side adjacent to the angle.	TOA $\tan\theta = \frac{\text{opposite}}{\text{adjacent}}$	$\tan\theta = \frac{a}{b}$

Get help with **using trigonometric ratios.**

Personal Tutor

Example: In right triangle *ABC,* if $a = 3$ cm, $b = 4$ cm, and $c = 5$ cm, find $\sin\theta$ and $\cos\theta$.

$$\sin\theta = \frac{3\text{ cm}}{5\text{ cm}} = 0.6$$

$$\cos\theta = \frac{4\text{ cm}}{5\text{ cm}} = 0.8$$

Example: In right triangle *ABC,* if $\theta = 30.0°$ and $c = 20.0$ cm, find *a* and *b.*

$$a = (20.0\text{ cm})(\sin 30.0°) = 10.0\text{ cm}$$

$$b = (20.0\text{ cm})(\cos 30.0°) = 17.3\text{ cm}$$

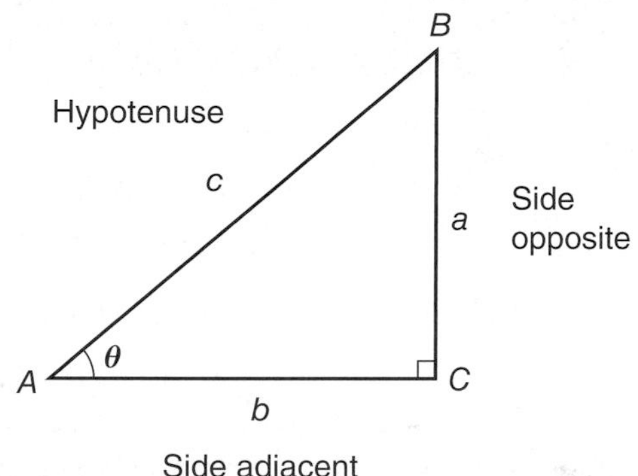

Law of Cosines and Law of Sines

The laws of cosines and sines let you calculate the sides and angles of any triangle.

Law of cosines The law of cosines is the same as the Pythagorean theorem, except for the last term. θ is the angle opposite side *c.* If the angle θ is 90°, the $\cos\theta = 0$ and the last term equals zero. If θ is greater than 90°, its cosine is a negative number.

$$c^2 = a^2 + b^2 - 2ab\cos\theta$$

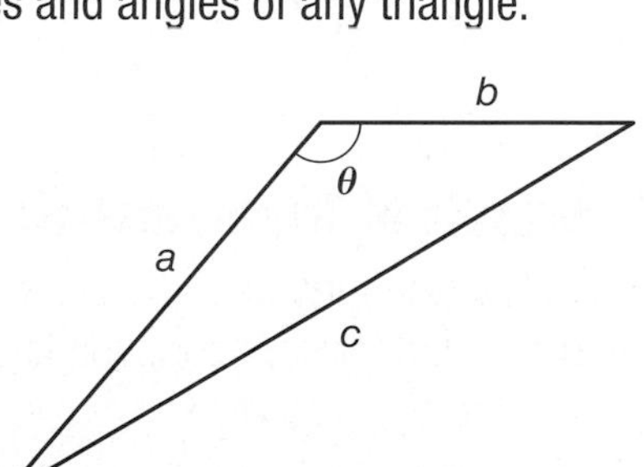

Example: Find the length of the third side of a triangle with $a = 10.0$ cm, $b = 12.0$ cm, $\theta = 110.0°$.

$$c^2 = a^2 + b^2 - 2ab\cos\theta$$

$$c = \sqrt{a^2 + b^2 - 2ab\cos\theta}$$

$$= \sqrt{(10.0\text{ cm})^2 + (12.0\text{ cm})^2 - 2(10.0\text{ cm})(12.0\text{ cm})(\cos 110.0°)}$$

$$= \sqrt{1.00\times10^2\text{ cm}^2 + 144\text{ cm}^2 - (240.0\text{ cm}^2)(\cos 110.0°)}$$

$$= 18.1\text{ cm}$$

Law of sines The law of sines is an equation of three ratios, where *a, b,* and *c* are the sides opposite angles *A, B,* and *C,* respectively. Use the law of sines when you know the measures of two angles and any side of a triangle.

$$\frac{\sin A}{a} = \frac{\sin B}{b} = \frac{\sin C}{c}$$

Example: In a triangle, $C = 60.0°$, $a = 4.0$ cm, $c = 54.6$ cm. Find the measure of angle *A.*

$$\frac{\sin A}{a} = \frac{\sin C}{c}$$

$$\sin A = \frac{a \sin C}{c}$$

$$\sin A = \frac{(4.0\text{ cm})(\sin 60.0°)}{4.6\text{ cm}}$$

$$A = 49°$$

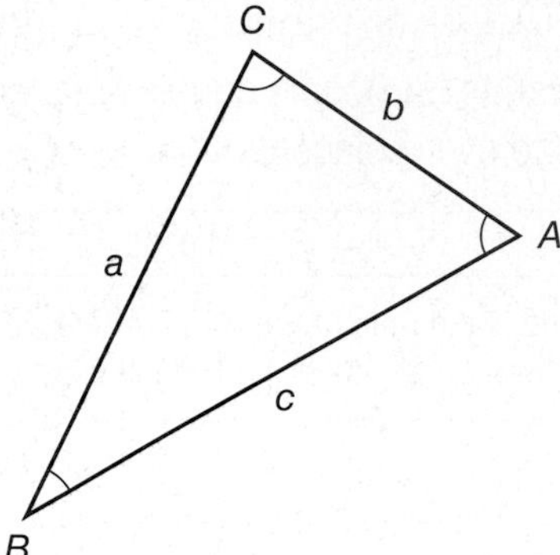

Inverses of Sine, Cosine, and Tangent

The inverses of sine, cosine, and tangent allow you to do the reverse of the sine, cosine, and tangent functions and find the angle. The trigonometric functions and their inverses are as follows:

Trigonometric Function	Inverse
$y = \sin x$	$x = \sin^{-1} y$ or $x = \text{arc} \sin y$
$y = \cos x$	$x = \cos^{-1} y$ or $x = \text{arc} \cos y$
$y = \tan x$	$x = \tan^{-1} y$ or $x = \text{arc} \tan y$

Example: Solve $0.62 = \sin \theta$ for θ.

$$\theta = \sin^{-1} 0.62 = 38°$$

PRACTICE PROBLEMS

20. Solve each of the following for θ.

a. $0.62 = \cos \theta$

b. $0.13 = \cos \theta$

c. $0.53 = \tan \theta$

d. $0.84 = \sin \theta$

Graphs of Trigonometric Functions

The sine function ($y = \sin x$) and the cosine function ($y = \cos x$) are periodic functions. The period for each function is 2π. x can be any real number. y is a real numbers between -1 and 1, inclusive.

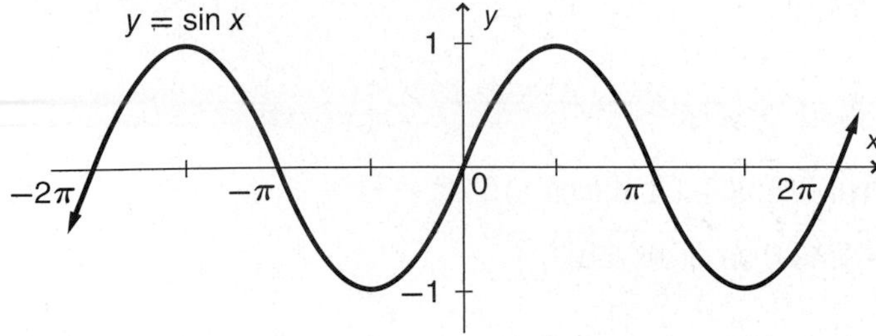

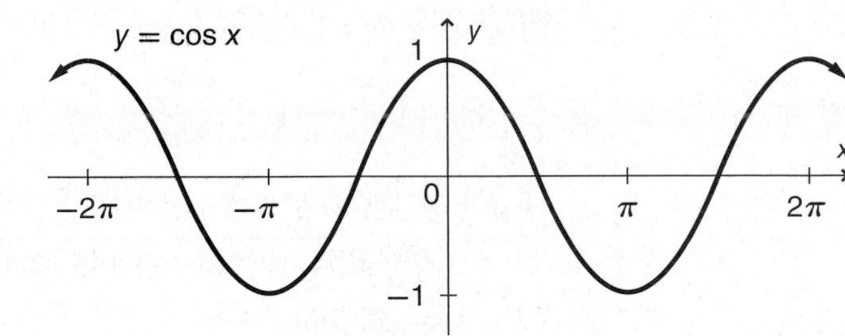

IX. Logarithms

Math Skill Handbook

Logarithms with Base *b*

Let b and x be positive integers with $b \neq 1$. The logarithm of x with base b, written $\log_b x$, equals y, where y is the exponent that makes the equation $b^y = x$ true. The log to the base b of x is the number to which you can raise b to get x.

$$\log_b x = y \text{ if and only if } b^y = x$$

Memory aid: "the log is the exponent"

Get help with **exponents and logarithms.**

Personal Tutor

Examples: Calculate the following logarithms.

$\log_2 \frac{1}{16} = -4$ Because $2^{-4} = \frac{1}{16}$

$\log_{10} 1000 = 3$ Because $10^3 = 1000$

When you want to find the log of a number, you also can use a computer.

Connecting Math to Physics Physicists use logarithms to work with measurements that extend over several orders of magnitude, or powers of 10. Geophysicists use the Richter scale, a logarithmic scale that allows them to rate earthquakes from 0 to 7, or larger, though the power of earthquakes differ by 7 or more powers of 10.

Common Logarithms

Base 10 logarithms are called common logarithms. They often are written without the subscript 10.

$$\log_{10} x = \log x \qquad x > 0$$

Antilogarithms or Inverse Logarithms

An antilogarithm is the inverse of a logarithm. An antilogarithm is the number that has a given logarithm.

Examples: Solve $\log x = 4$ for x.

$\log x = 4$

$x = 10^4$ 10^4 is the antilogarithm of 4.

Connecting Math to Physics An equation for loudness (L) in decibels is $L = 10 \log_{10} R$, where R is the relative intensity of the sound. Calculate R for a fireworks display with a loudness of 130 decibels.

$130 = 10 \log_{10} R$ Divide by 10.

$13 = \log_{10} R$ Use the logarithm rule.

$R = 10^{13}$

When you know the logarithm of a number and want to know the number itself, use a calculator to find the antilogarithm.

PRACTICE PROBLEMS

21. Write the following in exponential form: $\log_3 81 = 4$

22. Write the following in logarithmic form: $10^{-3} = 0.001$

23. Find x. $\log x = 3.125$

CHAPTER 1

21. a.

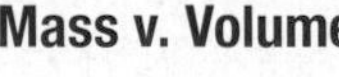

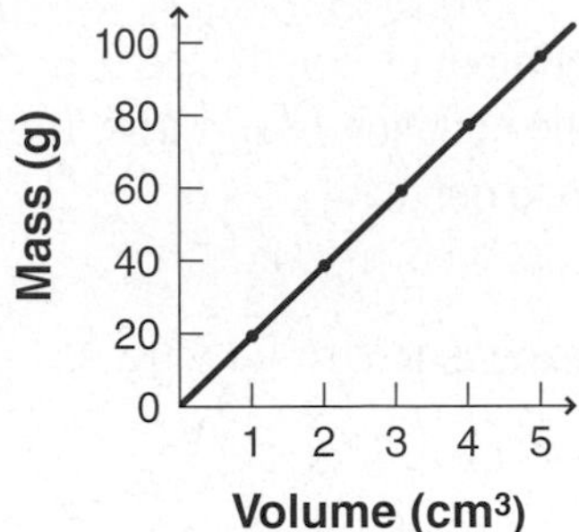

b. a straight line

c. The relationship is linear.

d. slope $= \frac{\Delta y}{\Delta x} = \frac{96.5\text{ g} - 19.4\text{ g}}{5.0\text{ cm}^3 - 1.0\text{ cm}^3}$

$= 19\text{ g/cm}^3$

e. $m = (19\text{ g/cm}^3)V$

f. The mass for each cubic centimeter of pure gold is 19 g.

CHAPTER 2

11. The car begins at a position of 125.0 m and moves toward the origin, arriving at the origin 5.0 s after it begins moving. The car continues beyond the origin.

13. a. at 4.0 s

b. 100.0 m

c. 50.0 m

15. a. 19 s

b. 58 s

c.

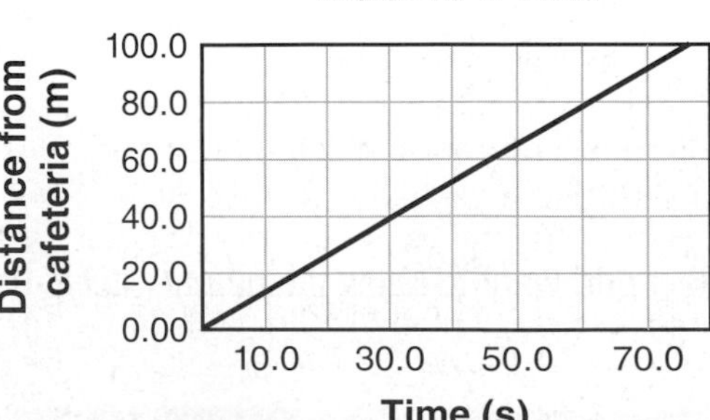

17. runner B

19. approximately 30 m

27. a. Using the points (0 s, 0 m) and (3 s, −1 m)

$\bar{v} = \left|\frac{\Delta x}{\Delta t}\right|$

$= \left|\frac{x_2 - x_1}{t_2 - t_1}\right|$

$= \left|\frac{-1\text{ m} - 0\text{ m}}{3\text{ s} - 0\text{ s}}\right|$

$= |-0.3\text{ m/s}|$

$= 0.3\text{ m/s}$

b. The average velocity is the slope of the line, including the sign, so it is −0.3 m/s or 0.3 m/s north.

29. $\bar{v} = \frac{\Delta x}{\Delta t}$

$= \frac{x_2 - x_1}{t_2 - t_1}$

$= \frac{3.7\text{ cm} - 6.5\text{ cm}}{2.3\text{ s} - 0\text{ s}}$

$= -1.2\text{ cm/s}$

31. The bicycle is moving in the positive direction at a speed of 0.7 km/min.

33. Let east be the positive direction.

$\boldsymbol{x} = \bar{\boldsymbol{v}}t + \boldsymbol{x}_i$

$= (42\text{ km/h})(1.5\text{ h}) + (25\text{ km})$

$= 88\text{ km}$

88 km east

35. Let east be the positive direction.

$\boldsymbol{x} = \bar{\boldsymbol{v}}t + \boldsymbol{x}_i$

$= (-35\text{ km/h})(1.2\text{ h}) + (25\text{ km})$

$= -17\text{ km}$

17 km west

CHAPTER 3

1.

Time (s) 0 1 2 3 4 56 7 8 9 10

3. a. $\bar{a} = \frac{v_2 - v_1}{t_2 - t_1}$

$= \frac{10.0\text{ m/s} - 0.0\text{ m/s}}{5.0\text{ s} - 0.0\text{ s}}$

$= 2.0\text{ m/s}^2$

b. $\bar{a} = \frac{v_2 - v_1}{t_2 - t_1}$

$= \frac{4.0\text{ m/s} - 10.0\text{ m/s}}{20.0\text{ s} - 15.0\text{ s}}$

$= -1.2\text{ m/s}^2$

c. $\bar{a} = \frac{v_2 - v_1}{t_2 - t_1}$

$= \frac{0.0\text{ m/s} - 0.0\text{ m/s}}{40.0\text{ s} - 0.0\text{ s}}$

$= 0.0\text{ m/s}^2$

5. $\bar{a} = \frac{\Delta v}{\Delta t} = \frac{36\text{ m/s} - 4.0\text{ m/s}}{4.0\text{ s}} = 8.0\text{ m/s}^2$

$= 8.0\text{ m/s}^2$ forward

7. a. $\bar{a} = \frac{\Delta v}{\Delta t} = \frac{0.0\text{ m/s} - 25\text{ m/s}}{3.0\text{ s}} = -8.3\text{ m/s}^2$

$= 8.3\text{ m/s}^2$ east

b. half as great (4.2 m/s^2 east)

9. $\bar{a} = \frac{\Delta v}{\Delta t} = \frac{0.75\text{ m/s} - 3.5\text{ m/s}}{10.0\text{ s}} = -0.28\text{ m/s}^2$

$= 0.28\text{ m/s}^2$ west

17. $v_f = v_i + at$

$= 30.0\text{ km/h}$

$+ (1.5\text{ m/s}^2)(6.8\text{ s})\left(\frac{1\text{ km}}{1000\text{ m}}\right)\left(\frac{3600\text{ s}}{1\text{ h}}\right)$

$= 67\text{ km/h}$ east

19. Let the forward direction be positive.

$v_f = v_i + at$

so $t = \frac{v_f - v_i}{a} = \frac{3.0\text{ m/s} - 22\text{ m/s}}{-2.1\text{ m/s}^2} = 9.0\text{ s}$

21. $\Delta x_C = 8.0$ m south; $\Delta x_D = 4.0$ m south

23. $v_f = v_i + at$

$t = \frac{v_f - v_i}{a} = \frac{0.0\text{ m/s} - 1.75\text{ m/s}}{-0.20\text{ m/s}^2} = 8.8\text{ s}$

25. $a = \frac{v_f^2 - v_i^2}{2\Delta x} = \frac{(25\text{ m/s})^2 - (15\text{ m/s})^2}{(2)(125\text{ m})} = 1.6\text{ m/s}^2$

$t = \frac{v_f - v_i}{a} = \frac{25\text{ m/s} - 15\text{ m/s}}{1.6\text{ m/s}^2} = 6.2\text{ s}$

27. $x_f = x_i + v_i t_f + \frac{1}{2}a t_f^2$

$v_i = \frac{(x_f - x_i) - \frac{1}{2}a t_f^2}{t_f}$

$= \frac{350\text{ m} - \frac{1}{2}(0.22\text{ m/s}^2)(18.4\text{ s})^2}{18.4\text{ s}}$

$= 17\text{ m/s}$ west

29. $x_{AB} = v_{AB} t_{AB}$

$= (4.5\text{ m/s})(15.0\text{ min})(60\text{ s/min})$

$= 4050$ m (carrying extra digits)

$x_{BC} = v_i t_f + \frac{1}{2}a t_f^2$

$= (4.5\text{ m/s})(90.0\text{ s})$

$+ \frac{1}{2}(-0.050\text{ m/s}^2)(90.0\text{ s})^2$

$= 203$ m (carrying an extra digit)

$x_{AB} + x_{BC} = 4050\text{ m} + 203\text{ m}$

$= 4.3\times10^2\text{ m}$

31. Let Sunee's initial direction of motion be positive.

Part 1: Constant velocity:

$x = vt$

$= (4.3\text{ m/s})(19\text{ min})(60\text{ s/min})$

$= 4902$ m in the positive direction

Part 2: Constant acceleration:

$x_f = x_i + v_i t + \frac{1}{2}at^2$

$a = \frac{2(x_f - x_i - v_i t)}{t^2}$

$= \frac{(2)(5.0\times10^3\text{ m} - 4902\text{ m} - (4.3\text{ m/s})(19.4\text{ s}))}{(19.4\text{ s})^2}$

$= 0.077\text{ m/s}^2$ in the positive direction

41. a. Let upward be the positive direction.

$v_f = v_i + at,\ a = -9.8\text{ m/s}^2$

$v_f = 0.0\text{ m/s} + (-9.8\text{ m/s}^2)(4.0\text{ s})$

$= -39\text{ m/s} = 39$ m/s downward

b. $x = v_i t + \frac{1}{2}at^2$

$= 0 + \left(\frac{1}{2}\right)(-9.8\text{ m/s}^2)(4.0\text{ s})^2$

$= -78$ m

The brick falls 78 m.

43. $v_f^2 = v_i^2 + 2a\Delta x$, $a = 9.8\text{ m/s}^2$, $v_i = 0$, and $\Delta x = 3.5$ m

$v_f = \sqrt{2a\Delta x}$

so $v_f = \sqrt{(2)(9.8\text{ m/s}^2)(3.5\text{ m})}$

$= 8.3$ m/s

45. a. Choose up as the positive direction
$v_{top} = 0$ m/s; $\boldsymbol{a}_{top} = 9.8$ m/s² downward

b. $v_f^2 = v_i^2 + 2a\Delta x$

$v_i = \sqrt{-v_f^2 + 2a\Delta x}$ where $a = -9.8$ m/s² and $v_f = 0$ at the height of the toss.

$$v_i = \sqrt{-2a\Delta x}$$
$$= \sqrt{-(2)(-9.8\text{ m/s}^2)(0.25\text{ m})}$$
$$= 2.2\text{ m/s}$$

c. $v_f = v_i + at$

$$t = \frac{v_f - v_i}{a} = \frac{-2.2\text{ m/s} - 2.2\text{ m/s}}{-9.8\text{ m/s}^2} = 0.45\text{ s}$$

CHAPTER 4

1.

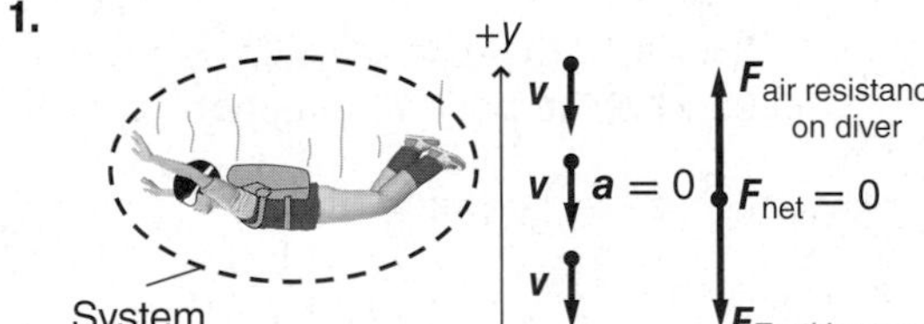

3. Identify the system as the softball.

Motion diagram   Free-body diagram

time $\boldsymbol{a}$ $\boldsymbol{F}_{net}$ $\boldsymbol{F}_{\text{Earth's mass on ball}}$

5. Identify the system as the softball.

Motion diagram   Free-body diagram

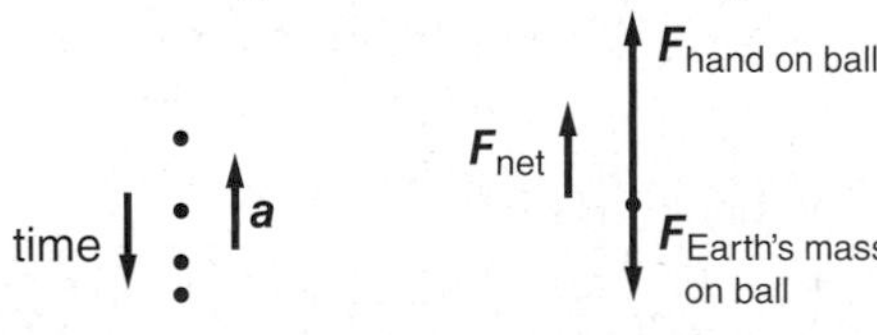

7. $\boldsymbol{F}_{net} = 225\text{ N} - 165\text{ N} = 6.0\times10^{-1}$ N in the direction of the larger force

9. $\boldsymbol{F}_{net} = m\boldsymbol{a}$

$$\boldsymbol{a} = \frac{\boldsymbol{F}_{net}}{m}$$
$$= \frac{2.7\text{ N}}{0.64\text{ kg}}$$

$= 4.2$ m/s² in the direction of the net force

11. a. $\boldsymbol{F}_{\text{left on box}}$ $\boldsymbol{F}_{\text{right on box}}$

b. $\boldsymbol{F}_{net} = \boldsymbol{F}_{right} - \boldsymbol{F}_{left}$
$= 317\text{ N} - 173\text{ N}$
$= 144$ N right

c. First, find the average acceleration.

$$\bar{a} = \frac{\Delta v}{\Delta t}$$
$$= \frac{v_f - v_i}{t_f - t_i}$$
$$= \frac{6.5\text{ m/s} - 0\text{ m/s}}{5\text{ s} - 0\text{ s}}$$
$$= 1.3\text{ m/s}^2$$

Then use the average acceleration to find the mass.

$$F_{net} = ma$$
$$m = \frac{F_{net}}{a}$$
$$= \frac{144\text{ N}}{1.3\text{ m/s}^2}$$
$$= 110\text{ kg}$$

17. $F_g = mg$

$$g = \frac{F_g}{m}$$
$$= \frac{235.2\text{ N}}{22.50\text{ kg}}$$
$$= 10.45\text{ N/kg}$$

19. Use Newton's second law $F_{net} = ma$.
If $F_{\text{bag on groceries}} > 230$ N, then the bag rips.

$F_{net} = F_{\text{bag on groceries}} + F_g$

$F_{\text{bag on groceries}} = F_{net} - F_g$
$= (15\text{ kg})(7.0\text{ m/s}^2)$
$- (15\text{ kg})(-9.8\text{ N/kg})$
$= 105\text{ N} + 147\text{ N}$
$= 252$ N

The bag does not hold.

21. a. Constant speed, so $a = 0$ and $F_{net} = 0$.

$F_{scale} = F_g$
$= mg = (75.0\text{ kg})(9.8\text{ N/kg})$
$= 735$ N

b. $a = 2.00$ m/s²

$F_{scale} = F_{net} + F_g$
$= ma + mg$
$= (75.0\text{ kg})(2.00\text{ m/s}^2)$
$+ (75.0\text{ kg})(9.8\text{ N/kg})$
$= 150\text{ N} + 735\text{ N}$
$= 885$ N

c. $a = -2.00\ m/s^2$

$$F_{scale} = F_{net} + F_g$$
$$= ma + mg$$
$$= (75.0\ kg)(-2.00\ m/s^2) + (75.0\ kg)(9.8\ N/kg)$$
$$= -150\ N + 735\ N$$
$$= 585\ N$$

d. Constant speed, so $a = 0$ and $F_{net} = 0$

$$F_{scale} = F_g = mg$$
$$= (75.0\ kg)(9.8\ N/kg)$$
$$= 735\ N$$

e. The acceleration is upward, so the net force is also upward.

29. The only force acting on the brick is the gravitational attraction of Earth's mass. The brick exerts an equal and opposite force on Earth.

31.

$F_{\text{Earth's mass on ball}}$

The only force acting on the ball is the force of Earth's mass on the ball, when ignoring air resistance. The ball exerts an equal and opposite force on Earth.

33. Identify the bucket as the system and up as positive.

$$F_{net} = F_{\text{rope on bucket}} - F_{\text{Earth's mass on bucket}} = ma$$

$$a = \frac{F_{\text{rope on bucket}} - F_{\text{Earth's mass on bucket}}}{m}$$

$$= \frac{F_{\text{rope on bucket}} - mg}{m}$$

$$= \frac{450\ N - (42\ kg)(9.8\ N/kg)}{42\ kg}$$

$$= 0.91\ m/s^2$$

CHAPTER 5

1.

141 km

65.0 km

125.0 km

$$R^2 = A^2 + B^2$$
$$R = \sqrt{A^2 + B^2}$$
$$= \sqrt{(65.0\ km)^2 + (125.0\ km)^2}$$
$$= 141\ km$$

3. $R^2 = A^2 + B^2 - 2AB\cos\theta$

$$R = \sqrt{A^2 + B^2 - 2AB\cos\theta}$$
$$= \sqrt{(4.5\ km)^2 + (6.4\ km)^2 - (4.5\ km)(6.4\ km)(\cos 135°)}$$
$$= 1.0\times10^1\ km$$

5. Identify north and west as the positive directions.

$$x_{1W} = x_1\sin\theta = (0.40\ km)(\sin 60.0°)$$
$$= 0.35\ km$$
$$x_{1N} = x_1\cos\theta = (0.40\ km)(\cos 60.0°)$$
$$= 0.20\ km$$
$$x_{2W} = 0.50\ km \text{ and } x_{2N} = 0.00\ km$$
$$R_W = x_{1W} + x_{2W} = 0.35\ km + 0.50\ km$$
$$= 0.85\ km$$
$$R_N = x_{1N} + x_{2N} = 0.20\ km + 0.00\ km$$
$$= 0.20\ km$$
$$R = R = \sqrt{R_W^2 + R_N^2}$$
$$= \sqrt{(0.85\ km)^2 + (0.20\ km)^2}$$
$$= 0.87$$
$$\theta = \tan^{-1}\left(\frac{R_W}{R_N}\right)$$
$$= \tan^{-1}\left(\frac{0.85\ km}{0.20\ km}\right)$$
$$= 77°$$

$\boldsymbol{R} = 0.87$ km at 77° west of north

7. The x-component is positive for angles less than 90° and for angles greater than 270°. It's negative for angles greater than 90° but less than 270°.

9. The force will be straight up. Because the angles are equal, the horizontal forces will be equal and opposite and cancel out.

$$F_{combined} = F_{rope1}\cos\theta + F_{rope2}\cos\theta$$
$$= 2F_{rope2}\cos\theta$$
$$= (2)(2.28\ N)(\cos 13.0°)$$
$$= 4.4\ N \text{ upward}$$

19. $F_{\text{Ames on box}} = F_{friction}$

$$= \mu_s F_N$$
$$= \mu_s mg$$
$$= (0.55)(134\ N)$$
$$= 74\ N$$

21. $F_f = \mu_s F_N$

$\mu_s = \frac{F_f}{F_N}$

$= \frac{F_f}{mg}$

$= \frac{403\ \text{N}}{(105\ \text{kg})(9.8\ \text{N/kg})}$

$= 0.39$

23. $F_{net} = F - \mu_k F_N = F - \mu_k mg = ma$

$\mu_k = \frac{F - ma}{mg}$

$= \frac{65\ \text{N} - (41\ \text{kg})(0.12\ \text{m/s}^2)}{(41\ \text{kg})(9.8\ \text{N/kg})}$

$= 0.15$

25. Choose positive direction as direction of car's movement.

$F_{net} = -\mu_k F_N = -\mu_k mg = ma$

$a = -\mu_k g$

Then use the equation $v_f^2 = v_i^2 + 2a(x_f - x_i)$ to find the distance. Let $x_i = 0$ and solve for x_f.

$x_f = \frac{v_f^2 - v_i^2}{2a}$

$= \frac{v_f^2 - v_i^2}{(2)(-\mu_k g)}$

$= \frac{(0.0\ \text{m/s}) - (23\ \text{m/s})^2}{(2)(-0.41)(9.8\ \text{N/kg})}$

$= 66$ m

So he hits the branch before he can stop.

33.

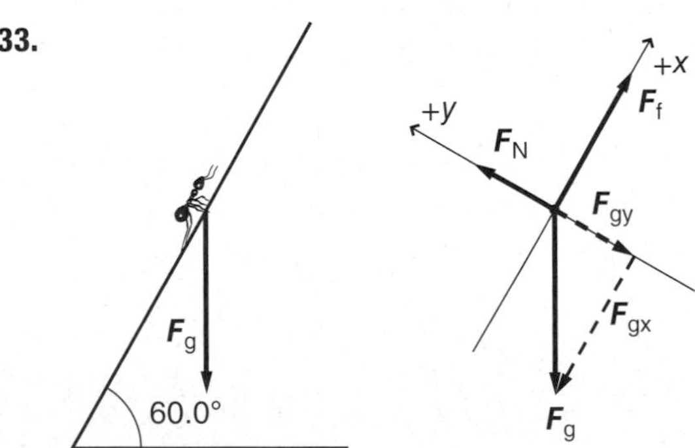

35. $F_N = mg \cos \theta$

$= (43.0\ \text{kg})(9.8\ \text{N/kg})(\cos 35.0°)$

$= 345$ N

37. $a = \frac{F}{m}$

$= \frac{F_g \sin \theta}{m}$

$= \frac{mg \sin \theta}{m}$

$= g \sin \theta$

$= (9.8\ \text{N/kg})(\sin 30.0°)$

$= 4.90\ \text{m/s}^2$

$a = \frac{v_f - v_i}{t_f - t_i}$; let $v_i = t_i = 0$.

Solve for v_f.

$v_f = at_f$

$= (4.90\ \text{m/s}^2)(4.00\ \text{s})$

$= 19.6$ m/s

39. $F_{\text{Stacie's weight parallel with slide}} - F_f = ma$

$a = \frac{F_{\text{Stacie's weight parallel with slide}} - F_f}{m}$

$= \frac{mg \sin \theta - \mu_k F_N}{m}$

$= \frac{mg \sin \theta - \mu_k mg \cos \theta}{m}$

$= g(\sin \theta - \mu_k \cos \theta)$

$= (9.8\ \text{N/kg})((\sin 45°) - (0.25)(\cos 45°))$

$= 5.2\ \text{m/s}^2$

CHAPTER 6

1. a. Since $v_y = 0$,

$\Delta y - v_y t = \frac{1}{2}at^2$

becomes $y = \frac{1}{2}at^2$

or $t = \sqrt{\frac{2\Delta y}{-a}}$

$= \sqrt{\frac{(2)(-78.4\ \text{m})}{-9.8\ \text{m/s}^2}}$

$= 4.0$ s

b. $x = v_x t$

$= (5.0\ \text{m/s})(4.0\ \text{s})$

$= 2.0\times10^1$ m

c. $\mathbf{v}_x = 5.0$ m/s

This is the same as the initial horizontal speed because the acceleration of gravity influences only the vertical motion. For the vertical component, use $v = v_i + at$ with $v = v_y$ and v_i, the initial vertical component of velocity, zero.

At $t = 4.0$ s

$v_y = at$

$= (9.8\ \text{m/s}^2)(4.0\ \text{s})$

$= 39$ m/s

3. $x = v_x t;\ t = \frac{x}{v_x}$

$y = \frac{1}{2}at^2$

$= \frac{1}{2}a\left(\frac{x}{v_x}\right)^2$

$= \frac{1}{2}(-9.8\ \text{m/s}^2)\left(\frac{0.070\ \text{m}}{2.0\ \text{m/s}}\right)^2$

$= -0.0060$ m or -0.60 cm

5. Following the method of the previous practice problem,

Hangtime:

$t = \frac{2v_i \sin\theta}{a}$

$= \frac{(2)(27.0\ \text{m/s})(\sin 60.0°)}{9.8\ \text{m/s}^2}$

$= 4.8$ s

Distance:

$x = v_i(\cos\theta)t$

$= (27.0\ \text{m/s})(\cos 60.0°)(4.8\ \text{s})$

$= 65$ m

Maximum height:

at $t = \frac{1}{2}(4.8\ \text{s}) = 2.4$ s

$y = v_i(\sin\theta)t + \frac{1}{2}at^2$

$= (27.0\ \text{m/s})(\sin 60.0°)(2.4\ \text{s})$

$+ \frac{1}{2}(-9.8\ \text{m/s}^2)(2.4\ \text{s})^2$

$= 28$ m

13. $a_c = \frac{v^2}{r}$,

so $r = \frac{v^2}{a_c} = \frac{(201\ \text{m/s})^2}{5.0\ \text{m/s}^2} = 8.1$ km

15. $a_c = \frac{4\pi^2 r}{T^2} = \frac{4\pi^2(1.4\ \text{m})}{(1.09\ \text{s})^2} = 47\ \text{m/s}^2$

$F_T = ma_c = (0.016\ \text{kg})(47\ \text{m/s}^2)$

$= 0.74$ N

27. $v_{c/g} = v_{w/g} + v_{c/w}$

$= 0.75\ \text{m/s} - 0.02\ \text{m/s} = 0.73\ \text{m/s}$

29. a. $v^2_{p/w} = v^2_{p/b} + v^2_{b/w}$

$v_{p/w} = \sqrt{v^2_{p/b} + v^2_{b/w}}$

$v_{p/w} = \sqrt{(3.8\ \text{m/s})^2 + (1.0\ \text{m/s})^2}$

$= 4.0$ m/s

b. $\theta = \tan^{-1}\left(\frac{v_{p/b}}{v_{b/w}}\right) = \tan^{-1}\left(\frac{1.3\ \text{m/s}}{3.8\ \text{m/s}}\right)$

$= 19°$ south of east

31. a. $v_{p/g} = v_{p/a} + v_{a/g}$

$= 200.0\ \text{km/h} + 50.0\ \text{km/h}$

$= 250.0$ km/h

b. $v_{p/g} = v_{p/a} - v_{a/g}$

$= 200.0\ \text{km/h} - 50.0\ \text{km/h}$

$= 150.0$ km/h

CHAPTER 7

1. $\left(\frac{T_G}{T_I}\right)^2 = \left(\frac{r_G}{r_I}\right)^3$

$r_G = \sqrt[3]{(4.2\ \text{units})^3\left(\frac{32\ \text{days}}{1.8\ \text{days}}\right)^2}$

$= \sqrt[3]{2.34\times10^4\ \text{units}^3} = 29$ units

3. $\left(\frac{T_V}{T_E}\right)^2 = \left(\frac{r_V}{r_E}\right)^3$

$\frac{r_V}{r_E} = \sqrt[3]{\left(\frac{T_V}{T_E}\right)^2}$

$= \sqrt[3]{\left(\frac{225\ \text{days}}{365\ \text{days}}\right)^2}$

$= 0.724$

Thus,

$r_V = 0.724r_E$

5. $\left(\frac{T_M}{T_E}\right)^2 = \left(\frac{r_M}{r_E}\right)^3$ with $r_M = 1.52r_E$

Thus, $T_M = \sqrt{\left(\frac{r_M}{r_E}\right)^3 T_E^2}$

$= \sqrt{\left(\frac{1.52r_E}{r_E}\right)^3(365\ \text{days})^2}$

$= \sqrt{4.68\times10^5\ \text{days}^2}$

$= 684$ days

7. $\left(\frac{T_S}{T_M}\right)^2 = \left(\frac{r_S}{r_M}\right)^3$

$r_S = \sqrt[3]{r_M^3\left(\frac{T_S}{T_M}\right)^2}$

$= \sqrt[3]{(3.9\times10^5\ \text{km})^3\left(\frac{1.00\ \text{days}}{27.3\ \text{days}}\right)^2}$

$= \sqrt[3]{7.96\times10^{13}\ \text{km}^3}$

$= 4.3\times10^4$ km

Selected Solutions

15. $v_M = \sqrt{\frac{Gm_U}{r}}$

$= \sqrt{\frac{(6.67\times10^{-11}\ \text{N·m}^2/\text{kg}^2)(8.68\times10^{25}\ \text{kg})}{(1.29\times10^{8}\ \text{m})}}$

$= 6.70$ km/s

$T_M = \frac{2\pi r}{v}$

$= \frac{2\pi(1.29\times10^{8}\ \text{m})}{(6.70\times10^{3}\ \text{m/s})} = 1.21\times10^{5}\ \text{s}$

$= 1.21\times10^{5}\ \text{s}\left(\frac{1\ \text{h}}{3600\ \text{s}}\right)\left(\frac{1\ \text{Earth day}}{24\ \text{h}}\right)$

$= 1.4$ Earth days

17. a. $v = \sqrt{\frac{Gm_M}{r}}$

$r = r_M + 260$ km

$= 2.44\times10^{6}\ \text{m} + 0.26\times10^{6}\ \text{m}$

$= 2.70\times10^{6}$ m

$v = \sqrt{\frac{(6.67\times10^{-11}\ \text{N·m}^2/\text{kg}^2)(3.30\times10^{23}\ \text{kg})}{2.70\times10^{6}\ \text{m}}}$

$= 2.86\times10^{3}$ m/s

b. $T = 2\pi\sqrt{\frac{r^3}{Gm_M}}$

$= 2\pi\sqrt{\frac{(2.70\times10^{6}\ \text{m})^3}{(6.67\times10^{-11}\ \text{N·m}^2/\text{kg}^2)(3.30\times10^{23}\ \text{kg})}}$

$= 5.94\times10^{3}$ s

$= 1.65$ h

CHAPTER 8

1. a. $\Delta\theta = (60)(-2\pi\ \text{rad})$

$= -120\pi$ rad or -377 rad

b. $\Delta\theta = -2\pi$ rad or -6.28 rad

c. $\Delta\theta = \left(\frac{1}{12}\right)(2\pi\ \text{rad})$

$= \frac{-\pi}{6}$ rad or 0.524 rad

3. $r = \frac{a}{\alpha}$

$= \frac{1.85\ \text{m/s}^2}{5.23\ \text{rad/s}^2}$

$= 0.354$

Thus, the diameter is 0.707 m.

5. Because $\omega = \frac{v}{r}$, if r is increased, ω will decrease. The number of revolutions will also decrease.

11. $\tau = Fr \sin\theta$

So, $F = \frac{\tau}{r \sin\theta}$

$= \frac{35\ \text{N·m}}{(0.25\text{m})(\sin 90.0°)}$

$= 1.4\times10^{2}$ N

13. a. $\tau = Fr \sin\theta$

So, $\theta = \sin^{-1}\left(\frac{\tau}{Fr}\right)$

$= \sin^{-1}\left(\frac{32.4\ \text{N·m}}{(232\ \text{N})(0.234\ \text{m})}\right)$

$= 36.6°$

b. $\tau = Fr \sin\theta$

So, $\theta = \sin^{-1}\left(\frac{\tau}{Fr}\right)$

$= \sin^{-1}\left(\frac{32.4\ \text{N·m}}{(275\ \text{N})(0.234\ \text{m})}\right)$

$= 30.2°$

15. Horizontal: $\theta = 90.0°$

So, $\tau = Fr \sin\theta$

$= mgr \sin\theta$

$= (65\ \text{kg})(9.8\ \text{N/kg})(0.18\ \text{m})(\sin 90.0°)$

$= 1.1\times10^{2}$ N·m

Vertical: $\theta = 0.0°$

So, $\tau = Fr \sin\theta$

$= mgr \sin\theta$

$= (65\ \text{kg})(9.8\ \text{N/kg})(0.18\ \text{m})(\sin 0.0°)$

$= 0.0$ N·m

17. $\tau_{chain} = F_g r$

$= (-35.0\ \text{N})(0.0770\ \text{m})$

$= -2.70$ N·m

Thus, a torque of +2.70 N·m must be exerted to balance this torque.

19. $m_1 = \frac{m_2 r_2}{r_1}$

$= \frac{(0.23\ \text{kg})(1.1\ \text{cm})}{6.0\ \text{cm}}$

$= 0.042$ kg

21. For the two children,
$I = mr^2 + mr^2 = 2mr^2$.
When r is doubled, I is multiplied by a factor of 4.

23. a. $I = mr^2 = (1.0\ \text{kg})(2.0\ \text{m})^2 = 4.0\ \text{kg·m}^2$

b. $I = \left(\frac{1}{2}\right)mr^2 = \left(\frac{1}{2}\right)(1.0\ \text{kg})(2.0\ \text{m})^2$

$= 2\ \text{kg·m}^2$

c. $I = \left(\frac{2}{5}\right)(1.0\ \text{kg})(2.0\ \text{m})^2$

$= 1.6\ \text{kg·m}^2$

25. Torque is now twice as great. The angular acceleration is also twice as great, so the change in angular velocity is twice as great. Thus, the final angular velocity is 32π rad/s, or 16 rev/s.

27. The torque on the wheel comes from either the chain or the string.

$\tau_{chain} = I_{wheel}\alpha_{wheel}$

$\tau_{string} = I_{wheel}\alpha_{wheel}$

Thus,

$\tau_{chain} = \tau_{string}$

$F_{chain}r_{gear} = \tau_{string}$

$F_{chain}r_{gear} = F_{string}r_{wheel}$

$$F_{string} = \frac{F_{chain}r_{gear}}{r_{wheel}} = \frac{(15\ \text{N})(0.14\ \text{m})}{0.38\ \text{m}} = 5.5\ \text{N}$$

29. $$I = \frac{\tau}{\alpha} = \frac{Fr}{\frac{\Delta\omega}{\Delta t}} = \frac{Fr\Delta t}{\Delta\omega}$$

$$= \frac{(16.0\ \text{N})(0.20\ \text{m})(5.00\ \text{s})}{\left(17.0\ \frac{\text{rev}}{\text{min}} - 0.0\ \frac{\text{rev}}{\text{min}}\right)\left(\frac{2\pi\ \text{rad}}{\text{rev}}\right)\left(\frac{\text{min}}{60\ \text{s}}\right)}$$

$= 8.99\ \text{kg·m}^2$

39. a. clockwise:

$\tau_A = F_A r_A$

$= -F_A(0.96\ \text{m} - 0.30\ \text{m})$

$= -(0.66\ \text{m})F_A$

counterclockwise:

$\tau_B = F_B r_B$

$= F_B(0.96\ \text{m} - 0.45\ \text{m})$

$= (0.51\ \text{m})F_B$

b. $\tau_{net} = \tau_A + \tau_B = 0$

so $\tau_B = -\tau_A$

$(0.51\ \text{m})F_B = -(-(0.66\ \text{m})F_A)$

$(0.51\ \text{m})F_B = (0.66\ \text{m})F_A$

c. $F_g = F_A + F_B$

Thus, $F_A = F_g - F_B$

$$= F_g - \frac{(0.66\ \text{m})F_A}{0.51\ \text{m}}$$

$$\text{or } F_A = \frac{F_g}{1 + \frac{0.66\ \text{m}}{0.51\ \text{m}}} = \frac{mg}{1 + \frac{0.66\ \text{m}}{0.51\ \text{m}}} = \frac{(7.3\ \text{kg})(9.8\ \text{N/kg})}{1 + \frac{0.66\ \text{m}}{0.51\ \text{m}}}$$

$= 31\ \text{N}$

d. F_A would become greater, and F_B would be less.

41. Choose the center of mass of the board as the pivot. The force of Earth's gravity on the board is exerted totally on the support under the center of mass.

$\tau_{end} = -\tau_{diver}$

$F_{end}r_{end} = -F_{diver}r_{diver}$

$$\text{Thus, } F_{end} = \frac{-F_{diver}r_{diver}}{r_{end}} = \frac{-m_{diver}gr_{diver}}{r_{end}} = \frac{-(85\ \text{kg})(9.8\ \text{N/kg})(1.75\ \text{m})}{1.75\ \text{m}}$$

$= -8.3\times10^2\ \text{N}$

To find the force on the center support, notice that because the board is not moving,

$F_{end} + F_{center} = F_{diver} + F_g$

Thus, $F_{center} = F_{diver} + F_g - F_{end}$

$= 2F_{diver} + F_g$

$= 2m_{diver}g + m_{board}g$

$= g(2m_{diver} + m_{board})$

$= (9.8\ \text{N/kg})(2(85\ \text{kg}) + 14\ \text{kg})$

$= 1.8\times10^3\ \text{N}$

CHAPTER 9

1. a.

$\boldsymbol{p} = m\boldsymbol{v}$

$$= (725\ \text{kg})(115\ \text{km/h})\left(\frac{1000\ \text{m}}{1\ \text{km}}\right)\left(\frac{1\ \text{h}}{3600\ \text{s}}\right)$$

$= 2.32\times10^4\ \text{kg·m/s eastward}$

b. $\boldsymbol{v} = \frac{\boldsymbol{p}}{m}$

$$= \frac{(2.32\times10^4\ \text{kg·m/s})\left(\frac{3600\ \text{s}}{1\ \text{h}}\right)\left(\frac{1\ \text{km}}{1000\ \text{m}}\right)}{2175\ \text{kg}}$$

$= 38.4\ \text{km/h eastward}$

3. a. $F\Delta t = \boldsymbol{p}_f - \boldsymbol{p}_i = m\boldsymbol{v}_f - m\boldsymbol{v}_i$

$$\boldsymbol{v}_f = \frac{F\Delta t + m\boldsymbol{v}_i}{m} = \frac{(5.0\ \text{N})(1.0\ \text{s}) + (7.0\ \text{kg})(2.0\ \text{m/s})}{7.0\ \text{kg}}$$

$= 2.7$ m/s in the same direction as the original velocity

b. $\boldsymbol{v}_f = \frac{\boldsymbol{F}\Delta t + m\boldsymbol{v}_i}{m}$

$= \frac{(-5.0 \text{ N})(1.0 \text{ s}) + (7.0 \text{ kg})(2.0 \text{ m/s})}{7.0 \text{ kg}}$

$= 1.3$ m/s in the same direction as the original velocity

5.

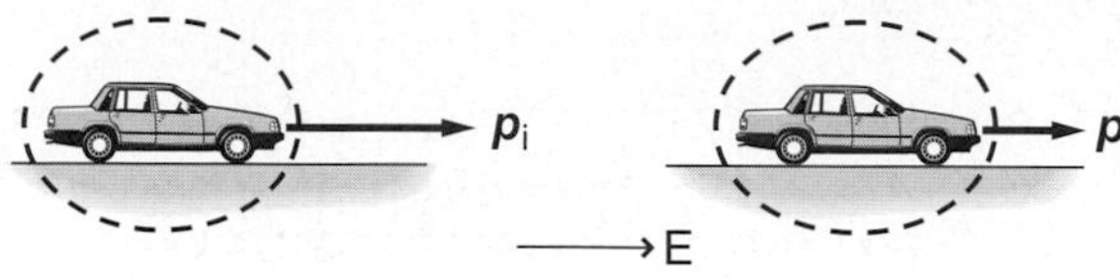

a. $\boldsymbol{F}\Delta t = \Delta \boldsymbol{p} = \boldsymbol{p}_f - \boldsymbol{p}_i$

$\boldsymbol{F} = \frac{\boldsymbol{p}_f - \boldsymbol{p}_i}{\Delta t}$

$\boldsymbol{F} = \frac{\boldsymbol{p}_f - m\boldsymbol{v}_i}{\Delta t}$

$= \frac{(0.0 \text{ kg·km/h}) - (60.0 \text{ kg})(94 \text{ km/h})}{0.20 \text{ s}} \times \left(\frac{1000 \text{ m}}{1 \text{ km}}\right)\left(\frac{1 \text{ h}}{3600 \text{ s}}\right)$

$= 7.8\times10^3$ N opposite to the direction of motion

b. $F_g = mg$

$m = \frac{F_g}{g} = \frac{7.8\times10^3 \text{ N}}{9.8 \text{ N/kg}} = 8.0\times10^2 \text{ kg}$

Such a mass is too heavy to lift. You cannot safely stop yourself with your arms.

7. $\omega_f = 2\pi f_i = 2\pi\left(\frac{6.0\times10^2 \text{ rev}}{\text{min}}\right)\left(\frac{1 \text{ min}}{60 \text{ s}}\right)$

$= 63$ rad/s

(carrying an extra digit)

$\omega_f = \omega_i + \alpha t$

$\omega_f = 0 + \frac{\tau}{I}t$

$\tau = \frac{\omega_f I}{t}$

$= \frac{(62.8 \text{ rad/s})(8.0\times10^{-5} \text{ kg·m}^2)}{0.15 \text{ s}}$

$= 0.033$ N·m

9. $\omega = 2\pi(5.0 \text{ rev/s}) = 31 \text{ rad/s}$

$\tau\Delta t = L_f - L_i = I(\omega_f - \omega_i)$

$= (0.060 \text{ kg·m}^2)(31 \text{ rad/s} - 0 \text{ rad/s})$

$= 1.9$ N·m·s

17. $p_i = p_f$

$mv_{Ai} + mv_{Bi} = 2mv_f$

$v_f = \frac{v_{Ai} + v_{Bi}}{2}$

$= \frac{2.2 \text{ m/s} + 0.0 \text{ m/s}}{2}$

$= 1.1$ m/s

19. $m_b v_{bi} + m_w v_{wi} = (m_b + m_w)v_f$

where v_f is the common final speed of the bullet and piece of lumber.

Because $v_{wi} = 0.0$ m/s,

$v_{bi} = \frac{(m_b + m_w)v_f}{m_b}$

$= \frac{(0.0350 \text{ kg} + 5.0 \text{ kg})(8.6 \text{ m/s})}{0.0350 \text{ kg}}$

$= 1.2\times10^3$ m/s

21. The system is the bullet and the ball.

$m_{bullet}v_{bullet, i} + m_{ball}v_{ball, i} = m_{bullet}v_{bullet, f} + m_{ball}v_{ball, f}$

$v_{ball, i} = 0.0$ m/s and $v_{bullet, f} = -5.0$ m/s

so $v_{ball, f} = \frac{m_{bullet}(v_{bullet, i} - v_{bullet, f})}{m_{ball}}$

$= \frac{(0.0350 \text{ kg})(475 \text{ m/s} - (-5.0 \text{ m/s}))}{2.5 \text{ kg}}$

$= 6.7$ m/s

23. Define the system as the rocket and the fuel.

$p_{r, i} + p_{fuel, i} = p_{rf} + p_{fuel, f}$

where $p_{r, i} + p_{fuel, i} = 0.0$ kg·m/s

If the initial mass of the rocket (including fuel) is $m_r = 4.00$ kg, then the final mass of the rocket is

$m_{r, f} = 4.00 \text{ kg} - 0.0500 \text{ kg} = 3.95 \text{ kg}$

So, $0.0 \text{ kg·m/s} = m_{r, f}v_{r, f} + m_{fuel}v_{fuel, f}$

$v_{r, f} = \frac{-m_{fuel}v_{fuel, f}}{m_{r, f}}$

$= \frac{-(0.0500 \text{ kg})(-625 \text{ m/s})}{3.95 \text{ kg}} = 7.91 \text{ m/s}$

25. Define the system as Carmen, Judi, and the canoe.

$p_{C, i} + p_{J, i} = p_{C, f} + p_{J, f}$

where $p_{C, i} = p_{J, i} = 0.0$ kg·m/s

$m_C v_{C, f} = -m_J v_{J, f}$

So, $v_{J, f} = \frac{-m_C v_{C, f}}{m_J}$

$= \frac{-(80.0 \text{ kg})(4.0 \text{ m/s})}{115 \text{ kg}}$

$= 2.8$ m/s in the opposite direction

27. $p_{i,x} = p_{1,x} + p_{2,x} = 0 + m_2v_{2,i}$

$p_{i,y} = p_{1,y} + p_{2,y} = m_1v_{1,i} + 0$

$p_f = p_i$

$= \sqrt{p_{1,x}{}^2 + p_{i,y}{}^2}$

$= \sqrt{(m_2v_{2,i})^2 + (m_1v_{1,i})^2}$

$v_f = \dfrac{p_f}{m_1 + m_2}$

$= \dfrac{\sqrt{(m_2v_{2,i})^2 + (m_1v_{1,i})^2}}{m_1 + m_2}$

$= \dfrac{\sqrt{((1732\text{ kg})(31.3\text{ m/s}))^2 + ((1383\text{ kg})(-11.2\text{ m/s}))^2}}{1383\text{ kg} + 1732\text{ kg}}$

$= 18.1$ m/s

$\theta = \tan^{-1}\left(\dfrac{p_{i,y}}{p_{i,x}}\right) = \tan^{-1}\left(\dfrac{m_1v_{1i}}{m_2v_{2i}}\right)$

$= \tan^{-1}\left(\dfrac{(1383\text{ kg})(-11.2\text{ m/s})}{(1732\text{ kg})(31.3\text{ m/s})}\right)$

$= 15.9°$ south of east

29. Define the system as the two balls.

$p_{Ci} + p_{Di} = p_{Cf} + p_{Df}$

where $p_{Ci} = 0.0$ kg·m/s

$m_C = m_D = m = 0.17$ kg

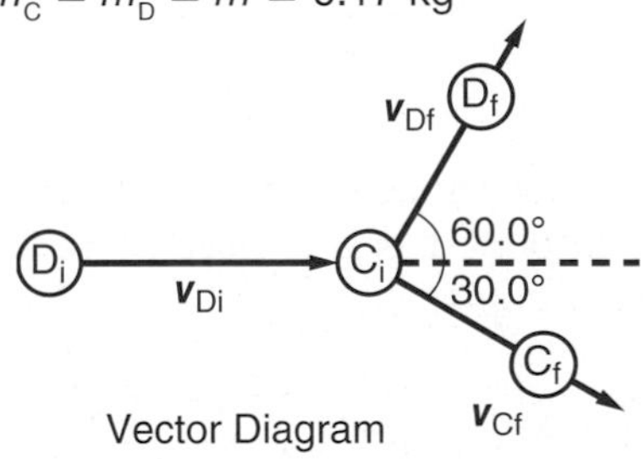

Vector Diagram

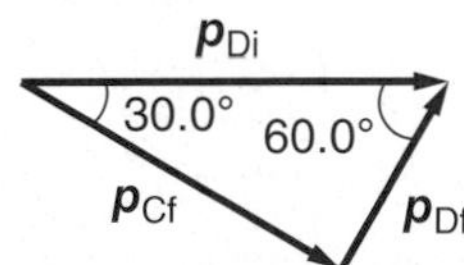

The vector diagram provides final momentum equations for the ball that is initially stationary, C, and the ball that is initially moving, D.

$p_{Cf} = p_{Di} \sin 60.0°$

$p_{Df} = p_{Di} \cos 60.0°$

We can use the momentum equation for the stationary ball to find its final velocity.

$p_{Cf} = p_{Di} \sin 60.0°$

$mv_{Cf} = mv_{Di} \sin 60.0°$

$v_{Cf} = v_{Di} \sin 60.0°$

$= (4.0\text{ m/s})(\sin 60.0°)$

$= 3.5$ m/s, 30.0° to the right

We can use the momentum equation for the moving ball to find its velocity.

$p_{Df} = p_{Di} \cos 60.0°$

$mv_{Df} = mv_{Di} \cos 60.0°$

$v_{Df} = v_{Di} \cos 60.0°$

$= (4.0\text{ m/s})(\cos 60.0°)$

$= 2.0$ m/s, 60.0° to the left

CHAPTER 10

1. **a.** Because $W = Fd$, doubling the force would double the work to 1.35 J.

   b. Because $W = Fd$, halving the distance would cut the work in half to 0.68 J.

3. **a.** $W = Fd = mgd$

   $= (7.5\text{ kg})(9.8\text{ N/kg})(8.2\text{ m})$

   $= 6.0\times10^2$ J

   b. $W = F_gd + 6.0\times10^2$ J

   $= (645\text{ N})(8.2\text{ m}) + 6.0\times10^2$ J

   $= 5.9\times10^3$ J

5. $W = Fd \cos \theta$

   $= (255\text{ N})(30.0\text{ m})(\cos 50.0°)$

   $= 4.92\times10^3$ J

7. **a.** Since gravity acts vertically, only the vertical displacement needs to be considered.

   $W = Fd = (215\text{ N})(4.20\text{ m}) = 903$ J

   b. Force is upward, but vertical displacement is downward, so

   $W = Fd \cos \theta$

   $= (215\text{ N})(4.20\text{ m})(\cos 180.0°)$

   $= -903$ J

9. **a.** Force and displacement are in the same direction.

   $W = Fd$

   $= (25\text{ N})(275\text{ m})$

   $= 6.9\times10^3$ J

   b. The force is downward (−90°), and the displacement is 25° above the horizontal or 115° from the force.

   $W = Fd \cos \theta$

   $= mgd \cos \theta$

   $= (13\text{ kg})(9.8\text{ N/kg})(275\text{ m})(\cos 115°)$

   $= -1.5\times10^4$ J

11. **a.** $P = \dfrac{W}{t} = \dfrac{Fd}{t}$

   $= \dfrac{(145\text{ N})(60.0\text{ m})}{25.0\text{ s}} = 348$ W

   b. t is halved, so P is doubled to 696 W.

Selected Solutions

13. $P = \frac{W}{t} = \frac{Fd}{t}$

$F = \frac{Pt}{d} = \frac{(65\times10^3\text{ W})(35\text{ s})}{17.5\text{ m}}$

$= 1.3\times10^5\text{ N}$

25. $IMA = \frac{r_e}{r_r} = \frac{8.00\text{ cm}}{35.6\text{ cm}} = 0.225$ (doubled)

$MA = \left(\frac{e}{100}\right)IMA = \frac{95.0}{100}(0.225)$

$= 0.214$ (doubled)

$MA = \frac{F_r}{F_e}$ so $F_r = (MA)(F_e)$

$= (0.214)(155\text{ N})$

$= 33.2\text{ N}$ (doubled)

$IMA = \frac{d_e}{d_r}$

So, $d_e = (IMA)(d_r)$

$= (0.225)(14.0\text{ cm})$

$= 3.15\text{ cm}$ (doubled)

27. a. $MA = \frac{F_r}{F_e} = \frac{mg}{F_e}$

$= \frac{(24.0\text{ kg})(9.8\text{ N/kg})}{129\text{ N}}$

$= 1.82$

b. efficiency $= \left(\frac{MA}{IMA}\right) \times 100$

$= \frac{(MA)(100)}{\frac{d_e}{d_r}}$

$= \frac{(MA)(d_r)(100)}{d_e}$

$= \frac{(1.82)(16.5\text{ m})(100)}{33.0\text{ m}}$

$= 91.0\%$

29. efficiency $= \frac{W_o}{W_i}\times100$

$= \frac{F_r d_r}{F_e d_e}\times100$

So, $d_e = \frac{F_r d_r(100)}{F_e(\text{efficiency})}$

$= \frac{(1.25\times10^3\text{ N})(0.13\text{ m})(100)}{(225\text{ N})(88.7)}$

$= 0.81\text{ m}$

CHAPTER 11

1. While stopping:

$E_{transformed} = E_i - E_f = KE_i - 0$

$= \frac{1}{2}mv_i^2$

$= \frac{1}{2}(52.0\text{ kg})(2.5\text{ m/s})^2$

$= 160\text{ J}$

All of her kinetic energy is transformed to other forms of energy when she stops.

To Speed up again:

$W = \Delta E = KE_f - KE_i$

$= \frac{1}{2}mv_f^2 - \frac{1}{2}mv_i^2$

$= \frac{1}{2}(52.0\text{ kg})(2.5\text{ m/s})^2 - \frac{1}{2}(52.0\text{ kg})(0.00\text{ m/s})^2$

$= +160\text{ J}$

3. $KE = \frac{1}{2}mv^2$

$= \frac{1}{2}(7.85\times10^{11}\text{ kg})(2.50\times10^4\text{ m/s})^2$

$= 2.45\times10^{20}\text{ J}$

So $\frac{KE_{comet}}{KE_{bomb}} = \frac{2.45\times10^{20}\text{ J}}{4.2\times10^{15}\text{ J}} = 5.8\times10^4$

5.8×10^4 bombs would be required to produce the same amount of energy used by Earth in stopping the comet.

5. $GPE = mgh$

$= (7.30\text{ kg})(9.8\text{ N/kg})(-0.610\text{ m})$

$= -44\text{ J}$

7. $GPE = mg(h_f - h_i)$

$= (2.2\text{ kg})(9.8\text{ N/kg})(2.10\text{ m} - 0.80\text{ m})$

$= 28\text{ J}$

9. To lift the box to the table:

$W = Fd = mg(h_f - h_i)$
$= \Delta GPE$
$= (10.1\text{ kg})(9.8\text{ N/kg})(1.1\text{ m} - 0.0\text{ m})$
$= 1.1\times10^2\text{ J}$

To slide the box across the table, $W = 0.0$ because the height did not change, the kinetic energy did not change, and we ignored friction.

To lower the box to the floor:

$W = Fd$
$= mg(h_f - h_i)$
$= \Delta GPE$
$= (10.1\text{ kg})(9.8\text{ N/kg})(0.0\text{ m} - 1.1\text{ m})$
$= -1.1\times10^2\text{ J}$

The sum of the three energy changes is $1.1\times10^2\text{ J} + 0.0\text{ J} + (-1.1\times10^2\text{ J}) = 0.0\text{ J}$

17. The system of Earth, bike, and rider remains the same, but now the energy involved is not mechanical energy alone. The rider must be considered as having stored energy, some of which is converted to mechanical energy.

Energy came from the chemical potential energy stored in the rider's body.

19. Using the water as a reference level, the kinetic energy on entry is equal to the potential energy of the diver at the top of his flight. The large diver has

$GPE = mgh$
$= (136\text{ kg})(9.8\text{ N/kg})(3.00\text{ m})$
$= 4.00\times10^3\text{ J}$

To equal this, the smaller diver would have to jump to

$$h = \frac{GPE}{mg}$$
$$= \frac{4.00\times10^3\text{ J}}{(100\text{ kg})(9.8\text{ N/kg})} = 4.1\text{ m}$$

Thus, the smaller diver would have to leap 1.1 m above the platform.

21. Using conservation of momentum:

$mv = (m + M)V$, so

$$v = \frac{(m + M)V}{m}$$
$$= \frac{(0.00800\text{ kg} + 9.00\text{ kg})(0.100\text{ m/s})}{0.00800\text{ kg}}$$
$$= 1.13\times10^2\text{ m/s}$$

23. a.

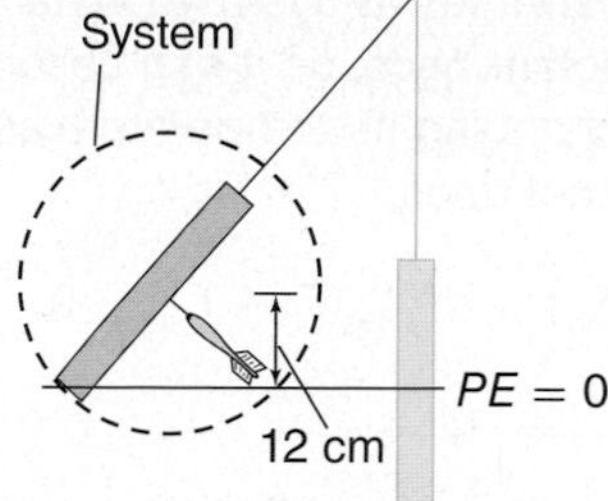

The system includes the suspended target and the dart.

b. Only momentum is conserved in the inelastic dart-target collision, so

$mv_i + MV_i = (m + M)V_f$

where $V_i = 0$ since the target is initially at rest and V_f is the common velocity just after impact. As the dart-target combination swings upward, energy is conserved, so $\Delta GPE = \Delta KE$ or, at the top of the swing,

$$(m + M)gh_f = \frac{1}{2}(m + M)(V_f)^2$$

c. Solve for V_f.

$V_f = \sqrt{2gh_f}$

Substitute V_f into the momentum equation and solve for v_i.

$$v_i = \left(\frac{m + M}{m}\right)\sqrt{2gh_f}$$
$$= \left(\frac{(0.025\text{ kg} + 0.73\text{ kg})}{0.025\text{ kg}}\right)\left(\sqrt{(2)(9.8\text{ N/kg})(0.120\text{ m})}\right)$$
$$= 46\text{ m/s}$$

CHAPTER 12

1. $Q = mC\Delta T$
$= (2.3\text{ kg})(385\text{ J/kg·K})(80.0°\text{C} - 20.0°\text{C})$
$= 5.3\times10^4\text{ J}$

3. a. $Q = mC\Delta T$

$$\Delta T = \frac{Q}{mC}$$
$$= \frac{(8.36\times10^5\text{ J})}{(20.0\text{ kg})(4180\text{ J/kg·K})}$$
$$= 10.0\text{ K}$$

b. The mass of methanol would be 0.80 times the mass of 20.0 L of water, or 16 kg.

$Q = mC\Delta T$

$$\Delta T = \frac{Q}{mC}$$
$$= \frac{8.36\times10^5\text{ J}}{(16\text{ kg})(2450\text{ J/kg·K})}$$
$$= 21\text{ K}$$

c. For temperatures above 0°C, water is the better coolant because it can absorb heat without changing its temperature as much as methanol does.

5. $m_W C_W (T_f - T_{Wi}) + m_M C_M (T_f - T_{Mi}) = 0$

$$C_M = \frac{m_W C_W (T_f - T_{Wi})}{m_M (T_{Mi} - T_f)}$$

$$= \frac{(0.1000\ \text{kg})(4180\ \text{J/kg·°C})(45.0°\text{C} - 35.0°\text{C})}{(0.300\ \text{kg})(100.0°\text{C} - 45.0°\text{C})}$$

$$= 2.53\times10^2\ \text{J/kg·°C}$$

7. $m_W C_W (T_f - T_{wi}) + m_A C_A (T_f - T_{Ai}) = 0$

$$T_f = \frac{m_W C_W T_{Wi} + m_G C_G T_{Gi}}{m_W C_W + m_G C_G}$$

$$= \frac{(0.10\ \text{kg})(4180\ \text{J/kg·K})(16.0°\text{C}) + (0.15\ \text{kg})(840\ \text{J/kg·K})(70.0°\text{C})}{(0.10\ \text{kg})(4180\ \text{J/kg·K}) + (0.15\ \text{kg})(840\ \text{J/kg·K})}$$

$$= 28.5°\text{C}$$

19. $Q = mC\Delta T + mH_f$

$= (0.100\ \text{kg})(2060\ \text{J/kg·°C})(20.0°\text{C})$
$+ (0.100\ \text{kg})(3.34\times10^5\ \text{J/kg})$
$= 3.75\times10^4\ \text{J}$

21. $H_f = 395{,}700\ \text{J} - 61{,}700\ \text{J} = 334{,}000\ \text{J}$
$H_v = 3{,}073{,}000\ \text{J} - 813{,}700\ \text{J}$
$= 2{,}259{,}300\ \text{J}$

23. $Q = mC_{ice}\Delta T + mH_f + mC_{water}\Delta T + mH_v + mC_{vapor}\Delta T$

$= (0.300\ \text{kg})(2060\ \text{J/kg·°C})(0.0°\text{C} - (-30.0°\text{C}))$
$+ (0.300\ \text{kg})(3.34\times10^5\ \text{J/kg})$
$+ (0.300\ \text{kg})(4180\ \text{J/kg·°C})(100.0°\text{C} - 0.0°\text{C})$
$+ (0.300\ \text{kg})(2.26\times10^6\ \text{J/kg})$
$+ (0.300\ \text{kg})(2020\ \text{J/kg·°C})(130.0°\text{C} - 100.0°\text{C})$
$= 9.40\times10^2\ \text{kJ}$

25. $\Delta U = Q - W_{block}$; since $W_{drill} = -W_{block}$ and assuming no heat added to drill:

$\Delta U = 0 + W_{drill} = mC\Delta T$
$W_{drill} = (0.40\ \text{kg})(897\ \text{J/kg·°C})(5.0°\text{C})$
$= 1.8\times10^3\ \text{kJ}$

27. $\Delta U = mC\Delta T$

$= (0.15\ \text{kg})(4180\ \text{J/kg·°C})(2.0°\text{C})$
$= 1.3\times10^3\ \text{J}$

The number of stirs is

$$\frac{1.3\times10^3\ \text{J}}{0.050\ \text{J}} = 2.6\times10^4\ \text{stirs}$$

CHAPTER 13

1. $F = PA = Plw$

$= (1.0\times10^5\ \text{Pa})(1.52\ \text{m})(0.76\ \text{m})$
$= (1.0\times10^5\ \text{N/m}^2)(1.52\ \text{m})(0.76\ \text{m})$
$= 1.2\times10^5\ \text{N}$

3. $P = \frac{F_{g,\ brick}}{A} = \frac{m_{brick}g}{lw}$

$= \frac{\rho Vg}{lw} = \frac{\rho lwhg}{lw} = \rho hg$

$= (11.8\ \text{g/cm}^3)(20.0\ \text{cm})$

$\times (9.8\ \text{N/kg})\left(\frac{1\ \text{kg}}{1000\ \text{g}}\right)\left(\frac{(100\ \text{cm})^2}{(1\ \text{m})^2}\right)$

$= 23\ \text{kPa}$

5. The maximum pressure $P = \frac{F_g}{A} = \frac{mg}{A}$

Therefore, $A = \frac{mg}{P}$

$= \frac{(454\ \text{kg})(9.8\ \text{N/kg})}{5.0\times10^4\ \text{Pa}}$

$= 8.9\times10^{-2}\ \text{m}^2$

7. $PV = nRT$

$$n = \frac{PV}{RT} = \frac{(15.5\times10^6\ \text{Pa})(0.020\ \text{m}^3)}{(8.31\ \text{Pa·m}^3\text{/mol·K})(293\ \text{K})}$$

$= 127.3\ \text{mol}$

$m = (127.3\ \text{mol})(4.00\ \text{g/mol}) = 5.1\times10^2\ \text{g}$

9. $PV = nRT$

$V = \frac{nRT}{P}$

Where $n = \frac{m}{M} = \frac{1.0\times10^3\ \text{g}}{29\ \text{g/mol}}$

and $T = 20.0°\text{C} + 273 = 293\ \text{K}$

$$V = \frac{\left(\frac{1.0\times10^3\ \text{g}}{29\ \text{g/mol}}\right)(8.31\ \text{Pa·m}^3\text{/mol·K})(293\ \text{K})}{1.013\times10^5\ \text{Pa}}$$

$= 0.83\ \text{m}^3$

25. $F_2 = \frac{F_1 A_2}{A_1} = \frac{(55\ \text{N})(2.4\ \text{m}^2)}{(0.015\ \text{m}^2)} = 8.8\times10^3\ \text{N}$

27. $F_{net} = F_g - F_{buoyant}$

$= \rho_{brick}Vg - \rho_{water}Vg$
$= (\rho_{brick} - \rho_{water})Vg$
$= (1.8\times10^3\ \text{kg/m}^3 - 1.00\times10^3\ \text{kg/m}^3)$
$\times (0.20\ \text{m}^3)(9.8\ \text{N/kg})$
$= 1.6\times10^3\ \text{N}$

29. To hold the camera in place, the tension in the wire must equal the net weight of the camera.

$$\begin{aligned} T &= F_{net} \\ &= F_g - F_{buoyant} \\ &= F_g - \rho_{water} V g \\ &= 1250\ \text{N} - (1.00\times10^3\ \text{kg/m}^3) \\ &\quad \times (16.5\times10^{-3}\ \text{m}^3)(9.8\ \text{N/kg}) \\ &= 1.1\times10^3\ \text{N} \end{aligned}$$

31. The buoyant force on the foam must equal 480 N. We are assuming the canoe is made of dense material.

$$F_{buoyant} = \rho_{water} V g$$

$$\begin{aligned} V &= \frac{F_{buoyant}}{\rho_{water} g} \\ &= \frac{480\ \text{N}}{(1.00\times10^3\ \text{kg/m}^3)(9.8\ \text{N/kg})} \\ &= 4.9\times10^{-2}\ \text{m}^3 \end{aligned}$$

39. $L_2 = L_1 + \alpha L_1(T_2 - T_1)$, so

$$\begin{aligned} \Delta L &= \alpha L_1(T_2 - T_1) \\ &= (25\times10^{-6}{}^\circ\text{C}^{-1})(3.66\ \text{m})(39^\circ\text{C} - (-28^\circ\text{C})) \\ &= 6.1\times10^{-3}\ \text{m} = 6.1\ \text{mm} \end{aligned}$$

41. At the beginning, 400 mL of 4.4°C water is in the beaker. Find the change in volume at 30.0°C.

$$\begin{aligned} \Delta V &= \beta V \Delta T \\ &= (210\times10^{-6}{}^\circ\text{C}^{-1})(400\times10^{-6}\ \text{m}^3) \\ &\quad \times (30.0^\circ\text{C} - 4.4^\circ\text{C}) \\ &= 2\times10^{-6}\ \text{m}^3 = 2\ \text{mL} \end{aligned}$$

43. The aluminum shrinks more than the steel. Let L be the diameter of the rod.

$$\begin{aligned} \Delta L_{aluminum} &= \alpha L \Delta T \\ &= (25\times10^{-6}{}^\circ\text{C}^{-1})(0.85\ \text{cm}) \\ &\quad \times (0.0^\circ\text{C} - 30.0^\circ\text{C}) \\ &= -6.38\times10^{-4}\ \text{cm} \end{aligned}$$

For the steel, the diameter of the hole shrinks by

$$\begin{aligned} \Delta L_{steel} &= \alpha L \Delta T \\ &= (12\times10^{-6}{}^\circ\text{C}^{-1})(0.85\ \text{cm})(0.0^\circ\text{C} - 30.0^\circ\text{C}) \\ &= -3.06\times10^{-4}\ \text{cm} \end{aligned}$$

The spacing between the rod and the hole will be

$$\left(\frac{1}{2}\right)(6.38\times10^{-4}\ \text{cm} - 3.06\times10^{-4}\ \text{cm})$$

$$= 1.7\times10^{-4}\ \text{cm}$$

CHAPTER 14

1. $F = kx$

$$\begin{aligned} k &= \frac{F}{x} \\ &= \frac{24\ \text{N}}{0.12\ \text{m}} \\ &= 2.0\times10^2\ \text{N/m} \end{aligned}$$

3. $F = kx$

$$x = \frac{F}{k} = \frac{18\ \text{N}}{56\ \text{N/m}} = 0.32\ \text{m}$$

5. $T = 2\pi\sqrt{\frac{\ell}{g}} = 2\pi\sqrt{\frac{1.0\ \text{m}}{9.8\ \text{N/kg}}} = 2.0\ \text{s}$

7. $T = 2\pi\sqrt{\frac{\ell}{g}}$

$$\begin{aligned} g &= \ell\left(\frac{2\pi}{T}\right)^2 \\ &= (0.75\ \text{m})\left(\frac{2\pi}{1.8\ \text{s}}\right)^2 = 9.1\ \text{N/kg} \end{aligned}$$

15. a. $v = \frac{d}{t}$

$$\begin{aligned} &= \frac{515\ \text{m}}{1.50\ \text{s}} \\ &= 343\ \text{m/s} \end{aligned}$$

b. $T = \frac{1}{f}$

$$\begin{aligned} &= \frac{1}{436\ \text{Hz}} \\ &= 2.29\times10^{-3}\ \text{s} \end{aligned}$$

c. $\lambda = \frac{v}{f}$

$$\begin{aligned} &= \frac{343\ \text{m/s}}{436\ \text{Hz}} \\ &= 0.787\ \text{m} \end{aligned}$$

17. $v = \lambda f$

$$\begin{aligned} &= (0.700\ \text{m})(3.50\ \text{Hz}) \\ &= 2.45\ \text{m/s} \end{aligned}$$

19. $v = \lambda f$, so $\lambda = \frac{v}{f} = \frac{15.0\ \text{m/s}}{6.00\ \text{Hz}} = 2.50\ \text{m}$

21. $v = \lambda f = (0.600\ \text{m})(20.0\ \text{Hz}) = 12.0\ \text{m/s}$

23. The wavelength is one-half of its original value

25. a. $v = \frac{d}{t} = \frac{(2)(465\ \text{m})}{2.75\ \text{s}} = 338\ \text{m/s}$

b. $v = \lambda f$, so $f = \frac{v}{\lambda} = \frac{338\ \text{m/s}}{0.750\ \text{m}} = 451\ \text{Hz}$

c. $T = \frac{1}{f} = \frac{1}{451\ \text{Hz}} = 2.22\times10^{-3}\ \text{s}$

Selected Solutions

CHAPTER 15

1. $v_s = -24.6$ m/s

$$f_d = 523\text{ Hz}\left(\frac{1}{1 - \frac{(-24.6\text{ m/s})}{343\text{ m/s}}}\right)$$

$$= 488\text{ Hz}$$

3. $v = 343$ m/s, $f_s = 475$ Hz,
$v_s = +24.6$ m/s, $v_d = -24.6$ m/s

$$f_d = f_s\left(\frac{v - v_d}{v - v_s}\right)$$

$$= (475\text{ Hz})\left(\frac{343\text{ m/s} + 24.6\text{ m/s}}{343\text{ m/s} - 24.6\text{ m/s}}\right)$$

$$= 548\text{ Hz}$$

5. $v = 343$ m/s, $f_s = 262$ Hz, $f_d = 277$ Hz,
$v_d = 0$ m/s, v_s is unknown

$$f_d = f_s\left(\frac{v - v_d}{v - v_s}\right)$$

Solve this equation for v_s.

$$v_s = v - \frac{f_s}{f_d}(v - v_d)$$

$$= 343\text{ m/s} - \left(\frac{262\text{ Hz}}{277\text{ Hz}}\right)(343\text{ m/s} - 0\text{ m/s})$$

$$= 19\text{ m/s}$$

13. Resonance spacing $= \frac{\lambda}{2} = 1.1$ m

so $\lambda = 2.2$ m
$v = \lambda f = (2.2\text{ m})(440\text{ Hz}) = 970$ m/s

15. Resonance spacing $= \frac{\lambda}{2}$ so using $\lambda = \frac{v}{f}$ the resonance spacing is

$$\frac{\lambda}{2} = \frac{v}{2f} = \frac{343\text{ m/s}}{(2)(440\text{ Hz})} = 0.39\text{ m}$$

CHAPTER 16

1. $$\frac{E_{after}}{E_{before}} = \frac{\left(\frac{P}{4\pi r_{after}^2}\right)}{\left(\frac{P}{4\pi r_{before}^2}\right)} = \frac{r_{before}^2}{r_{after}^2}$$

$$= \frac{(30\text{ cm})^2}{(90\text{ cm})^2} = \frac{1}{9};$$

Therefore, after the lamp is moved the illumination is one-ninth the original illumination.

3. $P = 4\pi(64\text{ cd}) = 256\pi$ lm

So, $E = \frac{P}{4\pi r^2} = \frac{256\pi\text{ lm}}{4\pi(3.0\text{ m})^2} = 7.1$ lx

5. $E = \frac{P}{4\pi r^2} = \frac{2275\text{ lm}}{4\pi(3.0\text{ m})^2} = 2.0\times10^1$ lx

7. $E = \frac{P}{4\pi r^2}$

$$r = \sqrt{\frac{P}{4\pi E}} = \sqrt{\frac{1750\text{ lm}}{4\pi(175\text{ lx})}}$$

$= 0.892$ above the keyboard

$$E = \frac{P}{4\pi r^2} = \frac{1750\text{ lm}}{4\pi(1.5\text{ m})^2}$$

$= 62$ lx, which is far too dim.

To get 175 lx from 1.5 m, you need
$E = \frac{P}{4\pi r^2}$

$P = E(4\pi r^2) = (175\text{ lx})(4\pi)(1.5\text{ m})^2 =$
5.0×10^{-3}-lm bulbs.

17. The relative speed along the axis is much less than the speed of light. Thus, you can use the observed light frequency equation. Because the astronomer and the galaxy are moving away from each other, use the negative form of the observed light frequency equation.

$$f_{obs} = f\left(1 - \frac{v}{c}\right)$$

$$= 6.16\times10^{14}\text{ Hz}\left(1 - \left(\frac{6.55\times10^6\text{ m/s}}{3.00\times10^8\text{ m/s}}\right)\right)$$

$$= 6.03\times10^{14}\text{ Hz}$$

19. Assume that the relative speed along the axis is much less than the speed of light. Thus, you can use the Doppler shift equation.

$$(\lambda_{obs} - \lambda) = \pm\frac{v}{c}\lambda$$

The observed (apparent) wavelength appears to be longer than the known (actual) wavelength of the oxygen spectral line. This means that the astronomer and the galaxy are moving away from each other. So use the positive form of the Doppler shift equation.

$$(\lambda_{obs} - \lambda) = +\frac{v}{c}\lambda$$

Solve for the unknown variable.

$$v = c\frac{(\lambda_{obs} - \lambda)}{\lambda}$$

$$= (3.00\times10^8\text{ m/s})\left(\frac{525\text{ nm} - 513\text{ nm}}{513\text{ nm}}\right)$$

$$= 7.02\times10^6\text{ m/s}$$

CHAPTER 17

1. Water fills in the rough areas and makes the surface smoother. The surface normals are then parallel.

3. a. $\theta_r = \theta_i = 42°$

b. $\theta_{i,\ mirror} = 90° - \theta_i = 90° - 42° = 48°$

c. $\theta_i + \theta_r = 2\theta_i = 84°$

5. $\theta_{r1} = \theta_{i1} = 60°$

$\theta_{i2} = 90° - \theta_{r1}$

$= 90° - 60° = 30°$

13.

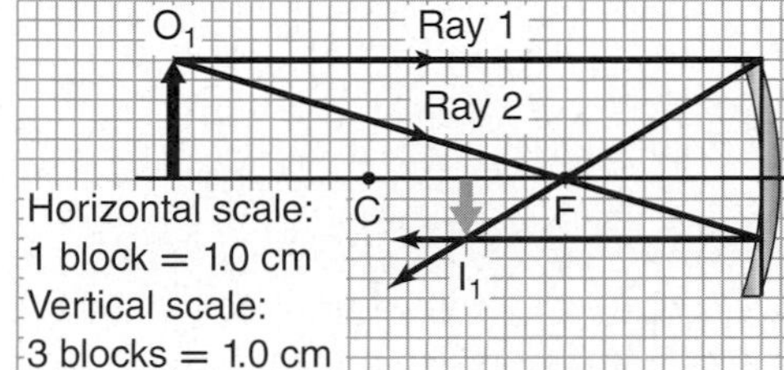

15. $\frac{1}{f} = \frac{1}{x_o} + \frac{1}{x_i}$

$x_i = \frac{x_o f}{x_o - f}$

$= \frac{(20.0\ \text{cm})\left(\frac{16.0\ \text{cm}}{2}\right)}{20.0\ \text{cm} - \left(\frac{16.0\ \text{cm}}{2}\right)} = 13.3\ \text{cm}$

$m \equiv \frac{h_i}{h_o} = \frac{-x_i}{x_o}$

$h_o = \frac{-x_o h_i}{x_i}$

$= \frac{-(13.3\ \text{cm})(3.0\ \text{cm})}{20.0\ \text{cm}}$

$= -2.0\ \text{cm}$

17. $\frac{1}{f} = \frac{1}{x_o} + \frac{1}{x_i}$

$x_o = \frac{x_i f}{x_i - f}$

$= \frac{(16.0\ \text{cm})(10.0\ \text{cm})}{16.0\ \text{cm} - 10.0\ \text{cm}}$

$= 26.7\ \text{cm}$

$m \equiv \frac{h_i}{h_o} = \frac{-x_i}{x_o}$

$h_o = \frac{-x_o h_i}{x_i}$

$= \frac{-(26.7\ \text{cm})(-3.0\ \text{cm})}{16.0\ \text{cm}}$

$= 5.0\ \text{cm}$

19. $\frac{1}{x_o} + \frac{1}{x_i} = \frac{1}{f}$

$x_i = \frac{x_o f}{x_o - f}$

$= \frac{(60.0\ \text{cm})(-13.0\ \text{cm})}{60.0\ \text{cm} - (-13.0\ \text{cm})}$

$= -10.7\ \text{cm}$

$m \equiv \frac{h_i}{h_o} = \frac{-x_i}{x_o}$

$m = \frac{-(-10.7\ \text{cm})}{60.0\ \text{cm}}$

$= +0.178$

$h_i = mh_o = (0.178)(6.0\ \text{cm})$

$= 1.1\ \text{cm}$

21. a. $m \equiv \frac{h_i}{h_o} = \frac{-x_i}{x_o}$

$x_i = \frac{-x_o h_i}{h_o}$

$= \frac{-(2.4\ \text{m})(0.36\ \text{m})}{1.8\ \text{m}}$

$= -0.48\ \text{m}$

b. $\frac{1}{f} = \frac{1}{x_o} + \frac{1}{x_i}$

$f = \frac{x_i x_o}{x_i + x_o}$

$= \frac{(-0.48\ \text{m})(2.4\ \text{m})}{-0.48\ \text{m} + 2.4\ \text{m}}$

$= -0.60\ \text{m}$

CHAPTER 18

1. $n_1 \sin \theta_1 = n_2 \sin \theta_2$

$\theta_2 = \sin^{-1}\left(\frac{n_1 \sin \theta_1}{n_2}\right)$

$= \sin^{-1}\left(\frac{(1.00)(\sin 37.0°)}{1.36}\right)$

$= 26.3°$

3. $n_1 \sin \theta_1 = n_2 \sin \theta_2$

$\theta_2 = \sin^{-1}\left(\frac{n_1 \sin \theta_1}{n_2}\right)$

$= \sin^{-1}\left(\frac{(1.00)(\sin 45.0°)}{2.42}\right)$

$= 17.0°$

5. $n_1 \sin \theta_1 = n_2 \sin \theta_2$

$$n_2 = \frac{n_1 \sin \theta_1}{\sin \theta_2} = \frac{(1.00)(\sin 45.0°)}{\sin 27.7°} = 1.52$$

The medium is float glass.

15. $\frac{1}{f} = \frac{1}{x_o} + \frac{1}{x_i}$

$$x_i = \frac{x_o f}{x_o - f} = \frac{(8.5\text{ cm})(5.5\text{ cm})}{8.5\text{ cm} - 5.5\text{ cm}}$$

$= 15.6$ cm, which rounds to 16 cm

$$m \equiv \frac{h_i}{h_o} = \frac{-x_i}{x_o}$$

$$h_i = \frac{-x_i h_o}{x_o} = \frac{-(15.6\text{ cm})(2.25\text{ cm})}{8.5\text{ cm}} = -4.1\text{ cm}$$

17. $\frac{1}{f} = \frac{1}{x_i} + \frac{1}{x_o}$

with $d_o = d_i$ because

$m \equiv \frac{-x_i}{x_o}$ and $m = -1$

Therefore,

$$\frac{1}{f} = \frac{2}{x_i}$$

$$x_i = 2f = 2(25\text{ mm}) = 50\text{ mm}$$

$$x_o = x_i = 50\text{ mm}$$

19. $x_i = 15$ cm

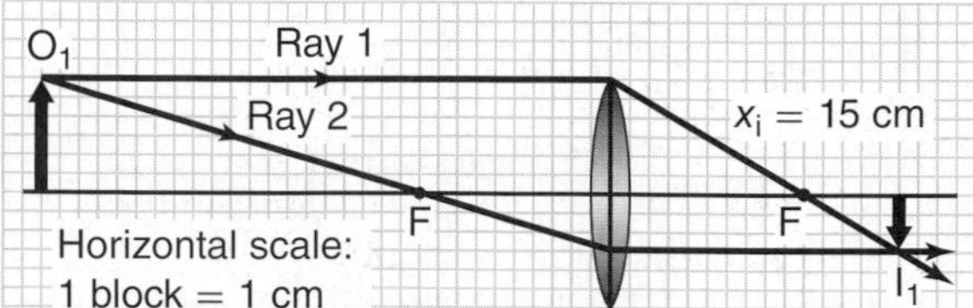

CHAPTER 19

1. $\lambda = \frac{xd}{L}$

$$= \frac{(13.2\times10^{-3}\text{ m})(1.90\times10^{-5}\text{ m})}{0.600\text{ m}} = 418\text{ nm}$$

3. $\lambda = \frac{xd}{L}$

$$d = \frac{\lambda L}{x} = \frac{(632.8\times10^{-9}\text{ m})(1.000\text{ m})}{65.5\times10^{-3}\text{ m}} = 9.66\times10^{-6}\text{ m} = 9.66\ \mu\text{m}$$

5. $2d = \left(m + \frac{1}{2}\right)\frac{\lambda}{n_{oil}}$

For the thinnest film, $m = 0$.

$$d = \left(\frac{1}{4}\right)\frac{\lambda}{n_{oil}} = \frac{635\text{ nm}}{(4)(1.45)} = 109\text{ nm}$$

7. Because $n_{film} > n_{air}$, there is a phase change on the first reflection. Because $n_{air} < n_{film}$, there is no phase change on the second reflection.

For destructive interference to get a black stripe

$$2d = \frac{m\lambda}{n_{film}}$$

For the thinnest film, $m = 1$.

$$d = \frac{\lambda}{2n_{film}} = \frac{521\text{ nm}}{(2)(1.33)} = 196\text{ nm}$$

9. Because $n_{film} > n_{air}$, there is a phase inversion on the first reflection. Because $n_{silicon} > n_{film}$, there is a phase inversion on the second reflection. For destructive interference to keep yellow-green from being reflected:

$$2d = \left(m + \frac{1}{2}\right)\frac{\lambda}{n_{film}}$$

For the thinnest film, $m = 0$.

$$d = \left(\frac{1}{4}\right)\frac{\lambda}{n_{film}} = \frac{555\text{ nm}}{(4)(1.45)} = 95.7\text{ nm}$$

17. $2x_1 = \frac{2\lambda L}{w}$

$L = \frac{(2x_1)w}{2\lambda}$

$= \frac{(2.60\times10^{-2}\text{ m})(0.110\times10^{-3}\text{ m})}{(2)(589\times10^{-9}\text{ m})}$

$= 2.43\text{ m}$

19. $2x_1 = \frac{2\lambda L}{w}$

$\lambda = \frac{(2x_1)w}{2L}$

$= \frac{(24.0\times10^{-3}\text{ m})(0.0295\times10^{-3}\text{ m})}{(2)(60.0\times10^{-2}\text{ m})}$

$= 5.90\times10^{2}\text{ nm}$

21. A full spectrum of color is seen. Because of the variety of wavelengths, dark fringes of one wavelength are filled by bright fringes of another color.

23. $\lambda = d\sin\theta$

$\sin\theta = \frac{\lambda}{d}$

$\tan\theta = \frac{x}{L}$

$x = L\tan\theta$

$= L\tan\left(\sin^{-1}\left(\frac{\lambda}{d}\right)\right)$

$= (0.800\text{ m})\left(\tan\left(\sin^{-1}\left(\frac{421\times10^{-9}\text{ m}}{8.60\times10^{-7}\text{ m}}\right)\right)\right)$

$= 0.449\text{ m}$

25. $\lambda = d\sin\theta$

There is one slit per distance d, so $\frac{1}{d}$ gives slits per centimeter.

$d = \frac{\lambda}{\sin\theta} = \frac{\lambda}{\sin\left(\tan^{-1}\left(\frac{x}{L}\right)\right)}$

$= \frac{632\times10^{-9}\text{ m}}{\sin\left(\tan^{-1}\left(\frac{0.056\text{ m}}{0.55\text{ m}}\right)\right)}$

$= 6.2\times10^{-6}\text{ m} = 6.2\times10^{-4}\text{ cm}$

$\frac{1\text{ slit}}{6.2\times10^{-4}\text{ cm}} = 1.6\times10^{3}\text{ slits/cm}$

CHAPTER 20

9. $F = \frac{Kq_Aq_B}{r_{AB}^{\,2}}$

$= \frac{(9.0\times10^{9}\text{ N·m}^2/\text{C}^2)(-2.0\times10^{-4}\text{ C})(8.0\times10^{-4}\text{ C})}{(0.30\text{ m})^2}$

$= -1.6\times10^{4}\text{ N}$

The forces are attracted to each other.

11. Magnitudes of all forces remain the same. The direction changes to 42° above the negative x-axis, or 138° counterclockwise from the positive x-axis.

13. $F_{\text{B on A}} = K\frac{q_Aq_B}{r_{AB}^{\,2}}$

$= (9.0\times10^{9}\text{ N·m}^2/\text{C}^2)\frac{(2.0\times10^{-6}\text{ C})(3.6\times10^{-6}\text{ C})}{(0.60\text{ m})^2}$

$= 0.18\text{ N}$

direction: toward the right

$F_{\text{C on A}} = K\frac{q_Aq_C}{r_{AC}^{\,2}}$

$= (9.0\times10^{9}\text{ N·m}^2/\text{C}^2)\frac{(2.0\times10^{-6}\text{ C})(4.0\times10^{-6}\text{ C})}{(0.80\text{ m})^2}$

$= 0.1125\text{ N}$

direction: toward the left

$\mathbf{F}_{\text{net}} = \mathbf{F}_{\text{B on A}} - \mathbf{F}_{\text{C on A}}$

$= (0.18\text{ N}) - (0.1125\text{ N})$

$= 0.068\text{ N}$ toward the right

CHAPTER 21

1. $E = \frac{F}{q} = \frac{2.0\times10^{-4}\text{ N}}{5.0\times10^{-6}\text{ C}} = 4.0\times10^{1}\text{ N/C}$

3. The electric force and the gravitational force algebraically sum to zero because the ball is suspended, i.e. not in motion:

$F_g + F_e = 0$, so $F_e = -F_g$

$E = \frac{F_e}{q}$

$q = \frac{F_e}{E} = -\frac{F_g}{E} = -\frac{2.1\times10^{-3}\text{ N}}{6.5\times10^{4}\text{ N/C}}$

$= -3.2\times10^{-8}\text{ C}$

The electric force is upward (opposite the field), so the charge is negative.

5. $\mathbf{E} = \frac{\mathbf{F}}{q}$

$\mathbf{F} = \mathbf{E}q = (27\text{ N/C south})(3.0\times10^{-7}\text{ C})$

$= 8.1\times10^{-6}\text{ N south}$

7. **a.** No. The force on the 2.0-μC charge would be twice that on the 1.0-μC charge.

b. Yes. You would divide the force by the strength of the test charge, so the results would be the same.

9. Because the field strength varies as the square of the distance from the point charge, the new field strength will be one-fourth of the old field strength, or 6.6×10^{3} N/C.

11. $E = \frac{F}{q'} = K\frac{q}{d^2}$

$q = \frac{Ed^2}{K}$

$= \frac{(450\ \text{N/C})(0.25\ \text{m})^2}{(9.0\times10^{9}\ \text{N}\cdot\text{m}^2/\text{C}^2)}$

$= -3.1\times10^{-9}$ C

The charge is negative, because the field is directed toward it.

13. Because the strength of the test charge (q') and the force (F) are directly proportional, $F = (Kqq')/r^2$. Therefore, the ratio of the force to the test charge is constant. $F/q' = Kq/r^2$, assuming that the distance between the charges does not change and the strength of the charged body does not change.

15. $E = K\frac{q}{d^2}$

$= \left(9.0\times10^{9}\ \frac{\text{N}\cdot\text{m}^2}{\text{C}^2}\right)\left(\frac{5.0\times10^{-6}\ \text{C}}{(\sqrt{5^2+5^2}\ \text{m})^2}\right)$

$= 9.0\times10^{2}$ N/C

21. $\Delta V = Ed = (6000\ \text{N/C})(0.05\ \text{m})$
$= 300\ \text{J/C} = 3\times10^{2}$ V

23. $\Delta V = Ed = (2.50\times10^{3}\ \text{N/C})(0.200\ \text{m})$
$= 5.00\times10^{2}$ V

25. $E = \frac{\Delta V}{d} = \frac{275\ \text{V}}{3.5\times10^{-3}\ \text{m}} = 7.9\times10^{4}$ N/C

27. $E = \frac{\Delta V}{d} = \frac{5.00\times10^{2}\ \text{V}}{0.024\ \text{m}} = 2.1\times10^{4}$ N/C

29. $W = q\Delta V = qEd$
$= (1.602\times10^{-19}\ \text{C})(4.5\times10^{5}\ \text{N/C})(0.25\ \text{m})$
$= 1.8\times10^{-14}$ J

31. Gravitational force (weight) downward, friction force of air upward. The two forces are equal in magnitude if the drop falls at constant velocity.

33. $E = \frac{F}{q} = \frac{6.4\times10^{-15}\ \text{N}}{1.602\times10^{-19}\ \text{C}} = 4.0\times10^{4}$ N/C

35. $q = C\Delta V = (27\times10^{-6}\ \text{F})(45\ \text{V})$
$= 1.2\times10^{-3}$ C

37. $\Delta V = \frac{q}{C}$, so the smaller capacitor has the larger potential difference.

$\Delta V = \frac{3.5\times10^{-4}\ \text{C}}{3.5\times10^{-6}\ \text{F}} = 1.1\times10^{2}$ V

39. $q = C\Delta V = (5.3\times10^{-11}\ \text{F})(12\ \text{V})$
$= 6.4\times10^{-10}$ C

CHAPTER 22

1. $P = IV = (2.0\ \text{A})(12\ \text{V}) = 24$ W

3. $P = IV = (0.50\ \text{A})(125\ \text{V}) = 62\ \text{J/s} = 62$ W

5. $I = \frac{P}{V} = \frac{3\times10^{3}\ \text{W}}{75\ \text{V}} = 40$ A

7. Because $P = IV$, the lightbulb's power increases by a factor of $2 \times 3 = 6$.

9. $R = \frac{V}{I} = \frac{3.0\ \text{V}}{2.0\times10^{-4}\ \text{A}} = 1.5\times10^{4}\ \Omega$

11. **a.** $R = \frac{V}{I} = \frac{120\ \text{V}}{0.50\ \text{A}} = 2.4\times10^{2}\ \Omega$

b. $P = IV = (0.50\ \text{A})(120\ \text{V}) = 6.0\times10^{1}$ W

13. **a.** The new value of the current is
$I = \frac{0.60\ \text{A}}{2} = 0.30$ A

$V = IR = (0.30\ \text{A})(2.1\times10^{2}\ \Omega)$
$= 6.3\times10^{1}$ V

b. The total resistance of the circuit is now

$R_{\text{total}} = \frac{V}{I} = \frac{125\ \text{V}}{0.30\ \text{A}} = 4.2\times10^{2}\ \Omega$

Therefore,

$R_{\text{res}} = R_{\text{total}} - R_{\text{lamp}}$
$= 4.2\times10^{2}\ \Omega - 2.1\times10^{2}\ \Omega$
$= 2.1\times10^{2}\ \Omega$

c. $P = IV = (0.30\ \text{A})(6.3\times10^{1}\ \text{V}) = 19$ W

15.

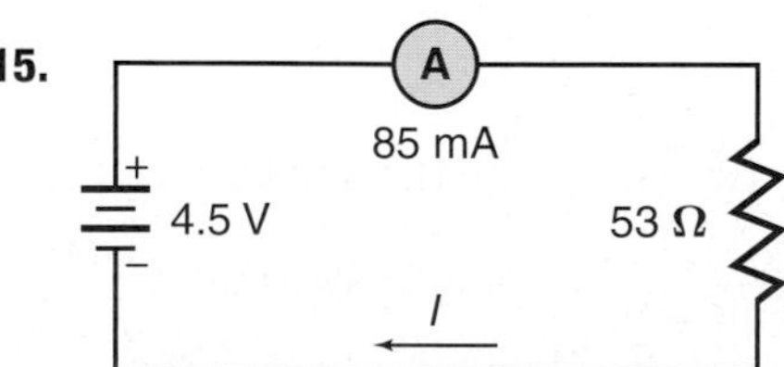

17.

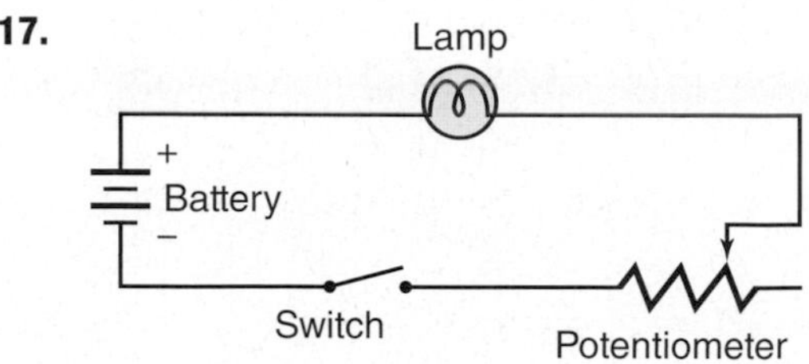

27. a. $I = \frac{V}{R} = \frac{45\ V}{39\ \Omega} = 1.2\ A$

b. $E = \frac{V^2}{R}t$

$= \frac{(45\ V)^2}{(35\ \Omega)}(5.0\ min)(60\ s/min)$

$= 1.7\times10^4\ J$

29. a. $I = \frac{V}{R} = \frac{220\ V}{11\ \Omega} = 2.0\times10^1\ A$

b. $E = I^2Rt = (2.0\times10^1\ A)^2(11\ \Omega)(30.0\ s)$
$= 1.3\times10^5\ J$

c. $Q = mC\Delta T$ with $Q = 0.65E$

$\Delta T = \frac{0.65E}{mC}$

$= \frac{(0.65)(1.3\times10^5\ J)}{(1.20\ kg)(4180\ J/kg\cdot°C)}$

$= 17°C$

31. a. $P = IV = (15.0\ A)(120\ V)$
$= 1800\ W = 1.8\ kW$

b. $E = Pt = (1.8\ kW)(5.0\ h/day)(30\ days)$
$= 270\ kWh$

c. Cost $= (\$0.12/kWh)(270\ kWh)$
$= \$32.40$

33. $E_{charge} = (1.3)IVt$
$= (1.3)(55\ A)(12\ V)(1.0\ h)$
$= 858\ Wh$

$t = \frac{E}{IV} = \frac{858\ Wh}{(7.5\ A)(12\ V)} = 9.5\ h$

CHAPTER 23

1. $R = R_1 + R_2 + R_3$
$= 22\ \Omega + 22\ \Omega + 22\ \Omega$
$= 66\ \Omega$

$I = \frac{\Delta V}{R} = \frac{125\ V}{66\ \Omega} = 1.9\ A$

3. $R = \frac{\Delta V}{I} = \frac{117\ V}{0.06\ A} = 2.0\times10^3\ \Omega$

$R_{bulb} = \frac{R}{10} = \frac{2.0\times10^3\ \Omega}{10} = 2.0\times10^2\ \Omega$

5. $\Delta V_1 = IR_1 = (2.3\ A)(12\ \Omega) = 28\ V$
$\Delta V_2 = IR_2 = (2.3\ A)(15\ \Omega) = 35\ V$
$\Delta V_3 = IR_3 = (2.3\ A)(5\ \Omega) = 12\ V$
$\Delta V_1 + \Delta V_2 + \Delta V_3 = 28\ V + 35\ V + 12\ V$
$= 75\ V$
$=$ voltage of battery

7. a. $I = \frac{\Delta V}{R} = \frac{17.0\ V}{255\ \Omega} = 0.067\ A$

b. First, find the total resistance, then solve for voltage.
$R = R_1 + R_2$
$= 255\ \Omega + 290\ \Omega$
$= 545\ \Omega$
$\Delta V = IR = (0.067\ A)(545\ \Omega) = 37\ V$

c. $P_{total} = I^2R = (0.067\ A)^2(545\ \Omega)$
$= 2.4\ W$
$P_{R_1} = I^2R = (0.067\ A)^2(255\ \Omega)$
$= 1.1\ W$
$P_{R_2} = I^2R = (0.067\ A)^2(290\ \Omega)$
$= 1.3\ W$
$P_{total} = 1.1\ W + 1.3\ W = 2.4\ W$

d. Yes. The law of conservation of energy states that energy cannot be created or destroyed; therefore, the rate at which energy is converted, or power dissipated, will equal the sum of all parts.

9. Because $P = I^2R$, The resistor with the lower resistance will dissipate less power, and thus will be cooler.

11. a. $R = R_1 + R_2 = 22\ \Omega + 33\ \Omega = 55\ \Omega$

b. $I = \frac{\Delta V}{R} = \frac{120\ V}{55\ \Omega} = 2.2\ A$

c. $\Delta V_1 = IR_1$

$= \left(\frac{\Delta V}{R}\right)R_1$

$= \left(\frac{120\ V}{55\ \Omega}\right)(22\ \Omega)$

$= 48\ V$

$\Delta V_2 = IR_2 = \left(\frac{120\ V}{55\ \Omega}\right)(33\ \Omega) = 72\ V$

13. $\Delta V_2 = \frac{\Delta VR_2}{R_1 + R_2}$

$R_1 = \frac{\Delta VR_2}{\Delta V_2} - R_2$

$= \frac{(12.0\ V)(1.2\ k\Omega)}{2.2\ V} - 1.2\ k\Omega$

$= 5.3\ k\Omega$

15. a. The equivalent resistance decreases.

b. Because the equivalent resistance decreases, the total current in the circuit increases.

c. Because the current through each branch is independent, the current through each of the 15-Ω resistors remains the same.

17. A parallel resistor will be required to reduce the resistance.

$$\frac{1}{R} = \frac{1}{R_1} + \frac{1}{R_2}$$

$$\frac{1}{R_1} = \frac{1}{R} - \frac{1}{R_2} = \frac{1}{93\ \Omega} - \frac{1}{150\ \Omega}$$

$R_1 = 2.4\times10^2\ \Omega$

240 Ω in parallel with the 150-Ω resistance

25. By conservation of energy (and power):

$$P_T = P_1 + P_2 + P_3$$
$$= 2.0\text{ W} + 3.0\text{ W} + 1.5\text{ W} = 6.5\text{ W}$$
$$P_T = I\Delta V$$
$$I = \frac{P_T}{\Delta V} = \frac{6.5\text{ W}}{12\text{ V}} = 0.54\text{ A}$$

27. $\frac{1}{R_{par}} = \frac{1}{25} + \frac{1}{22}$

$R_{par} = 12\ \Omega$

$$R = 12\ \Omega + 15\ \Omega = 27\ \Omega$$

$$I = \frac{\Delta V_{source}}{R} = \frac{125\text{ V}}{27\ \Omega} = 4.6\text{ A}$$

$$\Delta V_{toaster} = IR_{toaster} = (4.6\text{ A})(15\ \Omega)$$
$$= 69\text{ V}$$

$$\Delta V_{blender} = \Delta V_{source} - \Delta V_{toaster}$$
$$= 125\text{ V} - 69\text{ V} = 56\text{ V}$$

$$I_{blender} = \frac{\Delta V_{blender}}{R_{blender}} = \frac{56\text{ V}}{22\ \Omega}$$
$$= 2.5\text{ A}$$

CHAPTER 24

1. a. repulsive
b. attractive

3. Earth's geographic north pole is actually its southern magnetic pole and unlike poles attract.

5. a. Because magnetic field strength varies inversely with distance from the wire, the magnetic field at 1 cm will be twice as strong as the magnetic field at 2 cm.
b. Because magnetic field strength varies inversely with distance from the wire, the magnetic field at 1 cm will be three times as strong as the magnetic field at 3 cm.

7. the pointed end

9. Yes. Connect the potentiometer in series with the power supply and the coil. Adjusting the potentiometer for more resistance will decrease the current and the field strength.

19. You would use the right-hand rule for magnetic force on a wire. When you point the fingers of your right hand in the direction of the magnetic field and your thumb in the direction of the wire's conventional (positive) current, the palm of your hand will face in the direction of the force acting on the wire. To use this method, you would need to know the direction of the current and the direction of the field.

21. $F = ILB$

$$B = \frac{F}{IL} = \frac{0.60\text{ N}}{(6.0\text{ A})(0.75\text{ m})} = 0.13\text{ T}$$

23. $F = ILB$

$$I = \frac{F}{BL} = \frac{0.38\text{ N}}{(0.49\text{ T})(0.100\text{ m})} = 7.8\text{ A}$$

25. down

27. $F = qvB$

$$= (2)(1.60\times10^{-19}\text{ C})$$
$$\times (3.0\times10^{4}\text{ m/s})(9.0\times10^{-2}\text{ T})$$
$$= 8.6\times10^{-16}\text{ N}$$

29. $F = qvB$

$$v = \frac{F}{Bq} = \frac{4.1\times10^{-13}\text{ N}}{(0.61\text{ T})(1.60\times10^{-19}\text{ C})}$$
$$= 4.2\times10^{6}\text{ m/s}$$

CHAPTER 25

1. a. $EMF = BLv(\sin\theta)$

$$= (0.4\text{ T})(0.5\text{ m})(20\text{ m/s})(1)$$
$$= 4\text{ V}$$

b. $I = \frac{EMF}{R} = \frac{4\text{ V}}{6.0\ \Omega} = 0.7\text{ A}$

3. a. $EMF = BLv(\sin\theta)$

$$B = \frac{6.0\text{ V}}{(30.0\text{ m})(2.0\text{ m/s})(1)}$$
$$= 0.10\text{ T}$$

b. $I = \frac{EMF}{R} = \frac{6.0\text{ V}}{5.0\ \Omega} = 1.2\text{ A}$

5. a. $V_{eff} = (0.707)V_{max}$

$$= (0.707)(170\text{ V})$$
$$= 120\text{ V}$$

b. $I_{eff} = (0.707)I_{max} = (0.707)(0.70\text{ A})$

$$= 0.49\text{ A}$$

c. $R = \frac{V_{eff}}{I_{eff}} = \frac{\frac{V_{max}}{\sqrt{2}}}{\frac{I_{max}}{\sqrt{2}}} = \frac{V_{max}}{I_{max}} = \frac{170\ \text{V}}{0.70\ \text{A}}$

$= 2.4\times10^2\ \Omega$

7. $P = \frac{1}{2}P_{max}$

$P_{max} = (2)P = (2)(75\text{W}) = 1.5\times10^2\ \text{W}$

17. $V_s = \frac{V_p N_s}{N_p} = \frac{(60.0\ \text{V})(90{,}000)}{300}$

$= 1.80\times10^4\ \text{V}$

$P_s = V_s I_s = (1.80\times10^4\ \text{V})(0.50\ \text{A})$

$= 9.0\ \text{kW}$

$P_P = \frac{9.0\ \text{kW}}{0.95} = 9.47\ \text{kW}$

$I_P = \frac{P_P}{V_P} = \frac{9.47\ \text{kW}}{60.0\ \text{V}} = 1.6\times10^2\ \text{A}$

CHAPTER 26

1. $Bqv = \frac{mv^2}{r}$

$r = \frac{mv}{Bq}$

$= \frac{(1.67\times10^{-27}\ \text{kg})(7.5\times10^4\ \text{m/s})}{(0.080\ \text{T})(1.602\times10^{-19}\ \text{C})}$

$= 0.98\ \text{mm}$

3. $Bqv = Eq$

$v = \frac{E}{B} = \frac{9.0\times10^3\ \text{N/C}}{0.060\ \text{T}} = 1.5\times10^5\ \text{m/s}$

5. $m = \frac{B^2r^2q}{2V_{accel}}$

$= \frac{(7.2\times10^{-2}\ \text{T})^2(0.085\ \text{m})^2(1.602\times10^{-19}\ \text{C})}{(2)(110\ \text{V})}$

$= 2.7\times10^{-26}\ \text{kg}$

7. $\frac{q}{m} = \frac{2V_{accel}}{B^2r^2}$

$r^2 = \frac{2V_{accel}m_{Lithium}}{B^2q}$

$= \frac{2(320\ \text{V})(7)(1.67\times10^{-27}\ \text{kg})}{(1.5\times10^{-2}\ \text{T})^2(1.602\times10^{-19}\ \text{C})}$

$= 2.08\times10^{-1}\ \text{m}^2$

$r = 0.46\ \text{m}$

15. $\lambda = \frac{c}{f} = \frac{3.00\times10^8\ \text{m/s}}{5.70\times10^{14}\ \text{Hz}} = 5.26\times10^{-7}\ \text{m}$

17. $\lambda = \frac{c}{f}$

$f = \frac{c}{\lambda} = \frac{3.00\times10^8\ \text{m/s}}{2.2\times10^{-2}\ \text{m}} = 1.4\times10^{10}\ \text{Hz}$

19. $v = \frac{c}{\sqrt{k}} = \frac{299{,}792{,}458\ \text{m/s}}{\sqrt{1.00054}}$

$= 2.99712\times10^8\ \text{m/s}$

21. $v = \frac{c}{\sqrt{k}}$

$k = \left(\frac{c}{v}\right)^2 = \left(\frac{3.00\times10^8\ \text{m/s}}{2.43\times10^8\ \text{m/s}}\right)^2 = 1.52$

CHAPTER 27

1. $E = \frac{1240\ \text{eV·nm}}{\lambda}$

$= \frac{1240\ \text{eV·nm}}{515\ \text{nm}}$

$= 241\ \text{eV}$

3. $E_b = \frac{1240\ \text{eV·nm}}{320\ \text{nm}}$

$= 3.88\ \text{eV}$

$E_c = \frac{1240\ \text{eV·nm}}{811\ \text{nm}}$

$= 1.53\ \text{eV}$

$c < d < b < a$

5. $(2.3\ \text{eV})\left(\frac{1.60\times10^{-19}\ \text{J}}{1\ \text{eV}}\right) = 3.7\times10^{-19}\ \text{J}$

7. $KE = \frac{1}{2}mv^2$

$= \left(\frac{1}{2}\right)(9.11\times10^{-31}\ \text{kg})(6.2\times10^6\ \text{m/s})^2$

$= (1.75\times10^{-17}\ \text{J})\left(\frac{1\ \text{eV}}{1.602\times10^{-19}\ \text{J}}\right)$

$= 1.1\times10^2\ \text{eV}$

9. $KE = -eV_0$

$= -(-1.602\times10^{-19}\ \text{C})(5.1\ \text{J/C})$

$= 8.2\times10^{-19}\ \text{J}$

11. $KE = -eV_0$

$= -(-1.602\times10^{-19}\ \text{C})(3.2\ \text{J/C})$

$= 5.1\times10^{-19}\ \text{J}$

13. $KE_{max} = \frac{1240\ \text{eV·nm}}{\lambda} - hf_0$

$= \frac{1240\ \text{eV·nm}}{425\ \text{nm}} - 1.95\ \text{eV}$

$= 0.968\ \text{eV}$

15. $W = hf_0 = \frac{hc}{\lambda_0} = \frac{1240\ \text{eV·nm}}{\lambda_0}$

$= \frac{1240\ \text{eV·nm}}{273\ \text{nm}}$

$= 4.54\ \text{eV}$

The metal is most likely silver.

25. $\frac{1}{2}mv^2 = eV$, so

$v = \sqrt{\frac{2eV}{m}}$

$= \sqrt{\frac{(2)(1.602\times10^{-19}\ \text{C})(250\ \text{J/C})}{9.11\times10^{-31}\ \text{kg}}}$

$= 9.4\times10^{6}\ \text{m/s}$

$\lambda = \frac{h}{mv}$

$= \frac{6.63\times10^{-34}\ \text{J·s}}{(9.11\times10^{-31}\ \text{kg})(9.37\times10^{6}\ \text{m/s})}$

$= 7.8\times10^{-11}\ \text{m}$

27. $\lambda = \frac{h}{p}$, so $p = \frac{h}{\lambda}$

$KE = \frac{1}{2}mv^2 = \frac{p^2}{2m} = \frac{\left(\frac{h}{\lambda}\right)^2}{2m}$

$= \frac{\left(\frac{6.63\times10^{-34}\ \text{J·s}}{0.125\times10^{-9}\ \text{m}}\right)^2}{(2)(9.11\times10^{-31}\ \text{kg})}$

$= (1.544\times10^{-17}\ \text{J})\left(\frac{1\ \text{eV}}{1.602\times10^{-19}\ \text{J}}\right)$

$= 96.4\ \text{eV}$

So it would have to be accelerated through 96.4 V.

CHAPTER 28

1. $E_n = \frac{-13.6\ \text{eV}}{n^2}$

$E_2 = \frac{-13.6\ \text{eV}}{(2)^2} = -3.40\ \text{eV}$

$E_3 = \frac{-13.6\ \text{eV}}{(3)^2} = -1.51\ \text{eV}$

$E_4 = \frac{-13.6\ \text{eV}}{(4)^2} = -0.850\ \text{eV}$

3. $\Delta E = E_4 - E_2 = (-13.6\ \text{eV})\left(\frac{1}{4^2} - \frac{1}{2^2}\right)$

$= (-13.6\ \text{eV})\left(\frac{1}{16} - \frac{1}{4}\right) = 2.55\ \text{eV}$

5. $\lambda_{3\text{ to }2} = \frac{hc}{-\Delta E} = \frac{1240\ \text{eV·nm}}{E_3 - E_2}$

$= \frac{1240\ \text{eV·nm}}{-1.51\ \text{eV} - (-3.40\ \text{eV})}$

$= 656\ \text{nm (red line)}$

$\lambda_{4\text{ to }2} = \frac{hc}{-\Delta E} = \frac{1240\ \text{eV·nm}}{E_4 - E_2}$

$= \frac{1240\ \text{eV·nm}}{-0.850\ \text{eV} - (-3.40\ \text{eV})}$

$= 486\ \text{nm (green line)}$

7. $\lambda = \frac{hc}{-\Delta E}$, so

$-\Delta E = \frac{hc}{\lambda} = \frac{1240\ \text{eV·nm}}{304\ \text{nm}} = 4.08\ \text{eV}$

Therefore $E_{excited} = E_{ground} + \Delta E$

$= -54.4\ \text{eV} + 4.08\ \text{eV}$

$= -50.3\ \text{eV}$

CHAPTER 29

1. $\text{free e}^-/\text{cm}^3 = \left(\frac{2\ \text{free e}^-}{\text{atom}}\right)\left(\frac{6.02\times10^{23}\ \text{atoms}}{\text{mol}}\right) \times \left(\frac{1\ \text{mol}}{65.37\ \text{g}}\right)\left(\frac{7.13\ \text{g}}{\text{cm}^3}\right)$

$= 1.31\times10^{23}\ \text{free e}^-/\text{cm}^3$

3. $\text{free e}^-/\text{cm}^3 = \left(\frac{1\ \text{free e}^-}{\text{atom}}\right)\left(\frac{6.02\times10^{23}\ \text{atoms}}{\text{mol}}\right) \times \left(\frac{1\ \text{mol}}{196.97\ \text{g}}\right)\left(\frac{19.31\ \text{g}}{\text{cm}^3}\right)$

$= 5.90\times10^{22}\ \text{free e}^-/\text{cm}^3$

5. $\text{free e}^- = \left(\frac{1.81\times10^{23}\ \text{free e}^-}{\text{cm}^3}\right)\left(\frac{2835\ \text{g}}{2.70\ \text{g/cm}^3}\right)$

$= 1.90\times10^{26}\ \text{free e}^-$ in the tip

7. $\text{free e}^-/\text{atom} = \left(\frac{1\ \text{mol}}{6.02\times10^{23}\ \text{atoms}}\right)\left(\frac{72.6\ \text{g}}{1\ \text{mol}}\right) \times \left(\frac{\text{cm}^3}{5.23\ \text{g}}\right)\left(\frac{1.16\times10^{10}\ \text{free e}^-}{\text{cm}^3}\right)$

$= 2.67\times10^{-13}\ \text{free e}^-/\text{atom}$

9. $\text{free e}^-/\text{atom} = \left(\frac{1\ \text{mol}}{6.02\times10^{23}\ \text{atoms}}\right)\left(\frac{72.6\ \text{g}}{1\ \text{mol}}\right) \times \left(\frac{\text{cm}^3}{5.23\ \text{g}}\right)\left(\frac{4.81\times10^{11}\ \text{free e}^-}{\text{cm}^3}\right)$

$= 1.11\times10^{-11}\ \text{free e}^-/\text{atom}$

11. free e⁻/atom $= \left(\frac{1 \text{ mol}}{6.02\times10^{23}\text{ atoms}}\right)\left(\frac{28.09 \text{ g}}{1 \text{ mol}}\right)$
$\times \left(\frac{\text{cm}^3}{2.33 \text{ g}}\right)\left(\frac{1.89\times10^5 \text{ free e}^-}{\text{cm}^3}\right)$
$= 3.78\times10^{-18}$ free e⁻/atom

$T_K = T_C + 273°$
$T_C = T_K - 273°$
$= 200.0° - 273°$
$= -73°C$

13. From Example Problem 3 you know that there are 4.99×10^{22} Si atoms/cm³,
1.45×10^{10} free e⁻/cm³ in Si, and 1 free e⁻/As atom.

So, e⁻ from As $= (1\times10^4)$(free e⁻ from Si)

However, the ratio of atoms is needed, not electrons.

$$\text{As atoms} = \frac{\text{free e}^- \text{ from As}}{\text{free e}^-/\text{atoms As}} = \frac{(1\times10^4)(\text{free e}^- \text{ from Si})}{\text{free e}^-/\text{atoms As}}$$

$$\text{free e}^- \text{ from Si} = (\text{Si atoms})\left(\frac{\text{free e}^-/\text{cm}^3 \text{ Si}}{\text{Si atoms/cm}^3}\right)$$

Substituting into the expression for As atoms yields

$$\text{As atoms} = \frac{(1\times10^4)(\text{Si atoms})\left(\frac{\text{free e}^-/\text{cm}^3 \text{ Si}}{\text{Si atoms/cm}^3}\right)}{\text{free e}^-/\text{atoms As}}$$

$$\frac{\text{As atoms}}{\text{Si atoms}} = \frac{(1\times10^4)\left(\frac{\text{free e}^-/\text{cm}^3 \text{ Si}}{\text{Si atoms/cm}^3}\right)}{\text{free e}^- \text{ atom As}} = \frac{(1\times10^4)\left(\frac{1.45\times10^{10}}{4.99\times10^{22}}\right)}{1} = 3\times10^{-9}$$

15. $$\frac{\text{As atoms}}{\text{Ge atoms}} = \frac{\left(\frac{\text{doped e}^-}{\text{Ge e}^-}\right)\left(\frac{\text{free e}^-/\text{cm}^3 \text{ Ge}}{\text{Ge atoms/cm}^3}\right)}{\text{free e}^-/\text{atom As}}$$

$$\frac{\text{doped e}^-}{\text{Ge e}^-} = \left(\frac{\text{As atoms}}{\text{Ge atoms}}\right)\left(\frac{\text{Ge atoms/cm}^3}{\text{free e}^-/\text{cm}^3 \text{ Ge}}\right) \times (\text{free e}^-/\text{atom As})$$
$$= \left(\frac{1}{1\times10^6}\right)\left(\frac{4.34\times10^{22}}{1.13\times10^{15}}\right)(1) = 38.4$$

17. $$\frac{\text{doped e}^-}{\text{Si e}^-} = \left(\frac{\text{As atoms}}{\text{Si atoms}}\right)\left(\frac{\text{Si atoms/cm}^3}{\text{free e}^-/\text{cm}^3 \text{ Si}}\right) \times (\text{free e}^-/\text{atom As})$$
$$= \left(\frac{1}{1\times10^6}\right)\left(\frac{4.99\times10^{22}}{4.54\times10^{12}}\right)(1) = 1\times10^4$$

27. $\Delta V_b = IR + \Delta V_d$
$= (0.012 \text{ A})(470\ \Omega) + 0.40 \text{ V}$
$= 6.0 \text{ V}$

29. The anode of one should connect to the cathode of the other, and then the unconnected anode must be connected to the positive side of the circuit.

CHAPTER 30

1. $A - Z =$ neutrons
$234 - 92 = 142$ neutrons
$235 - 92 = 143$ neutrons
$238 - 92 = 146$ neutrons

3. $A - Z = 15 - 8 = 7$ neutrons

5. **a.** Mass defect = (isotope mass)
− (mass of protons and electrons)
− (mass of neutrons)
= 12.000000 u − (6)(1.007825 u)
− (6)(1.008665 u)
= −0.098940 u

b. Binding energy
= (mass defect)(binding energy of 1 u)
= (−0.098940 u)(931.49 MeV/u)
= −92.162 MeV

7. **a.** Isotope mass
= (mass defect)
+ (mass of protons and electrons)
+ (mass of neutrons)
= − 0.113986 u + (7)(1.007825 u)
+ (8)(1.008665 u)
= 15.010109 u

b. Binding energy
= (mass defect)(binding energy of 1 u)
= (−0.113986 u)(931.49 MeV/u)
= −106.18 MeV

15. ${}^{234}_{92}\text{U} \rightarrow {}^{230}_{90}\text{Th} + {}^{4}_{2}\text{He}$

17. ${}^{241}_{95}\text{Am} \rightarrow {}^{237}_{93}\text{Np} + {}^{4}_{2}\text{He}$

19. ${}^{214}_{82}\text{Pb} \rightarrow {}^{214}_{83}\text{Bi} + {}^{0}_{-1}\text{e} + {}^{0}_{0}\bar{\nu}$

Selected Solutions

21. a. $^{33}_{15}P \rightarrow {}^{A}_{Z}X + {}^{0}_{-1}e + {}^{0}_{0}\overline{\nu}$

where $Z = 15 - (-1) - 0 = 16$

$A = 33 - 0 - 0 = 33$

For $Z = 16$, the element must be sulfur. Thus, the isotope is $^{33}_{16}S$.

b. $^{55}_{24}Cr \rightarrow {}^{A}_{Z}X + {}^{0}_{-1}e + {}^{0}_{0}\overline{\nu}$

where $Z = 24 - (-1) - 0 = 25$

$A = 55 - 0 - 0 = 55$

For $Z = 25$, the element must be manganese. Thus, the isotope is $^{55}_{25}Mn$.

23. $^{15}_{7}N + {}^{1}_{1}H \rightarrow {}^{A}_{Z}X + {}^{4}_{2}He$

where $Z = 7 + 1 - 2 = 6$

$A = 15 + 1 - 4 = 12$

For $Z = 6$, the element must be carbon. Thus, the equation is

$^{15}_{7}N + {}^{1}_{1}H \rightarrow {}^{12}_{6}C + {}^{4}_{2}He$

25. 24.6 years = (2)(12.3 years), which is 2 half-lives

$$\text{remaining} = \text{original}\left(\frac{1}{2}\right)^{t}$$

$$= (1.0\ \text{g})\left(\frac{1}{2}\right)^{2}$$

$$= 0.25\ \text{g}$$

27. The half-life of $^{210}_{84}Po$ is 138 days.

There are 273 days or about 2 half-lives between September 1 and June 1. So,

$$\text{activity} = \left(2\times10^{6}\ \frac{\text{decays}}{\text{s}}\right)\left(\frac{1}{2}\right)^{2}$$

$$= 5\times10^{5}\ \text{Bq}$$

37. a. $E = mc^2$

$= (1.67\times10^{-27}\ \text{kg})(3.00\times10^{8}\ \text{m/s})^2$

$= 1.50\times10^{-10}\ \text{J}$

b. $E = \dfrac{1.50\times10^{-10}\ \text{J}}{1.60217\times10^{-19}\ \text{J/eV}}$

$= 9.36\times10^{8}\ \text{eV}$

c. The minimum energy is $(2)(9.36\times10^{8}\ \text{eV}) = 1.87\times10^{9}\ \text{eV}$

39. a. $E = (\text{neutron mass})(931.49\ \text{MeV/u})$

$= (1.008665\ \text{u})(931.49\ \text{MeV/u})$

$= 939.56\ \text{MeV}$

b. The smallest possible γ-ray energy would be twice the neutron energy.

$E_{total} = 2E_{n} = (2)(939.56\ \text{MeV})$

$= 1879.1\ \text{MeV}$

Color Conventions

Displacement vector (***x***)		Negative charge	−
Velocity vector (***v***)		Positive charge	+
Acceleration vector (***a***)		Current direction	
Force vector (***F***)		Electron	
Momentum vector (***p***)		Proton	
Light ray		Neutron	
Object		Coordinate axes	
Image			
Electric field line (***E***)			
Magnetic field line (***B***)			

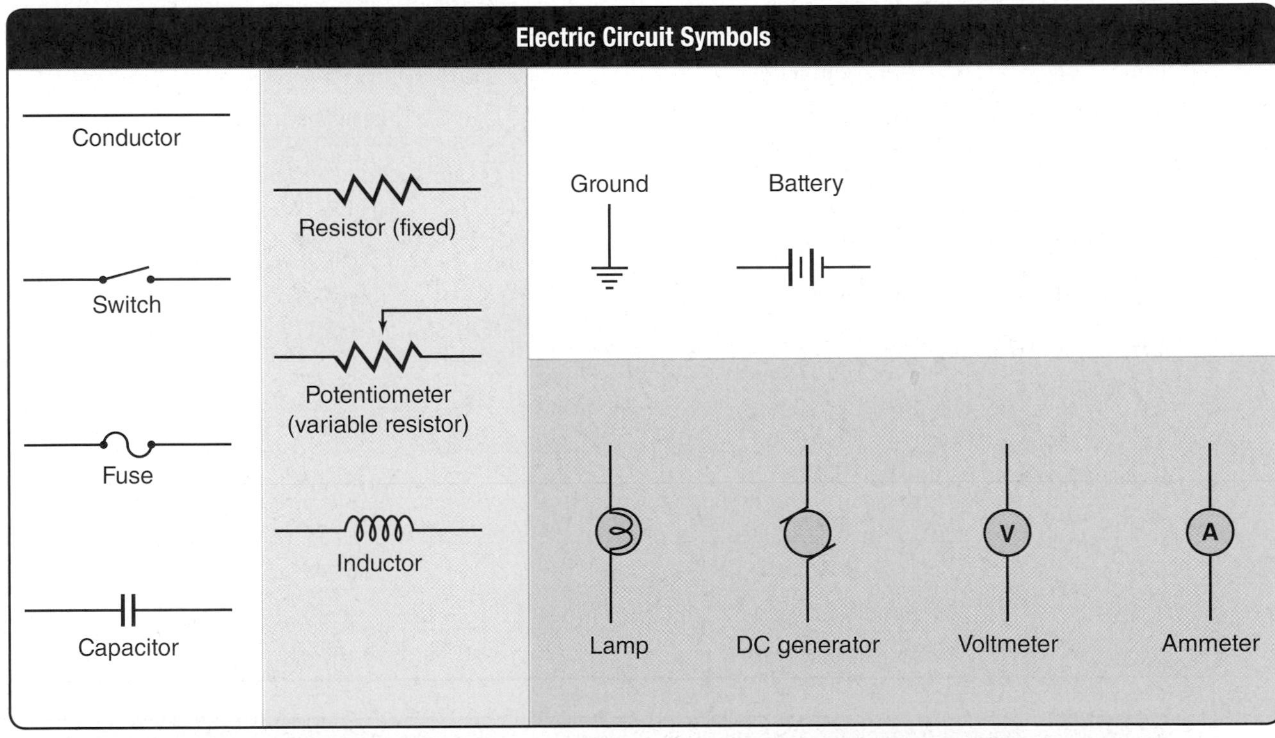

SI Base Units		
Quantity	**Unit**	**Unit Abbreviation**
Length	meter	m
Mass	kilogram	kg
Time	second	s
Temperature	kelvin	K
Amount of a substance	mole	mol
Electric current	ampere	A
Luminous intensity	candela	cd

SI Derived Units				
Quantity	**Unit**	**Unit Symbol**	**Unit Expressed in Base Units**	**Unit Expressed in Other SI Units**
Acceleration	meters per second squared	m/s^2	m/s^2	
Area	meters squared	m^2	m^2	
Capacitance	farad	F	$A^2{\cdot}s^4/(kg{\cdot}m^2)$	
Density	kilograms per meter cubed	kg/m^3	kg/m^3	
Electric charge	coulomb	C	A·s	
Electric field	newtons per coulomb	N/C	$kg{\cdot}m/(A{\cdot}s^3)$	V/m
Electric resistance	ohm	Ω	$kg{\cdot}m^2/(A^2{\cdot}s^3)$	V/A
EMF	volt	V	$kg{\cdot}m^2/(A^2{\cdot}s^3)$	
Energy, work	joule	J	$kg{\cdot}m^2/s^2$	N·m
Force	newton	N	$kg{\cdot}m/s^2$	
Frequency	hertz	Hz	s^{-1}	
Illuminance	lux	lx	cd/m^2	
Magnetic field	tesla	T	$kg/(A{\cdot}s^2)$	N·s/(C·m)
Potential difference	volt	V	$kg{\cdot}m^2/(A{\cdot}s^3)$	W/A or J/C
Power	watt	W	$kg{\cdot}m^2/s^3$	J/s
Pressure	pascal	Pa	$(kg/m)s^2$	N/m^2
Velocity	meters per second	m/s	m/s	
Volume	meters cubed	m^3	m^3	

Useful Conversions		
1 in = 2.54 cm	1 kg = 6.02×10^{26} u	1 atm = 101 kPa
1 mi = 1.61 km	1 oz = 28.4 g	1 cal = 4.184 J
1 mi^2 = 640 acres	1 kg = 2.21 lb	1 eV = 1.60×10^{-19} J
1 gal = 3.79 L	1 lb = 4.45 N	1 kWh = 3.60 MJ
1 m^3 = 264 gal	1 atm = 14.7 lb/in^2	1 hp = 746 W
1 knot = 1.15 mi/h	1 atm = 1.01×10^5 N/m^2	1 mol = 6.02×10^{23} particles

Physical Constants			
Quantity	**Symbol**	**Value**	**Approximate Value**
Atomic mass unit	u	$1.660538782 \times 10^{-27}$ kg	1.66×10^{-27} kg
Avogadro's number	N_A	$6.02214179 \times 10^{23}$ mol^{-1}	6.022×10^{23} mol^{-1}
Boltzmann's constant	k	$1.3806504 \times 10^{-23}$ $Pa \cdot m^3/K$	1.38×10^{-23} $Pa \cdot m^3/K$
Coulomb's constant	K	$8.987551788 \times 10^{9}$ $N \cdot m^2/C^2$	9.0×10^{9} $N \cdot m^2/C^2$
Elementary charge	e	$1.60217653 \times 10^{-19}$ C	1.602×10^{-19} C
Gas constant	R	8.314472 $Pa \cdot m^3/mol \cdot K$	8.31 $Pa \cdot m^3/mol \cdot K$
Gravitational constant	G	6.67428×10^{-11} $N \cdot m^2/kg^2$	6.67×10^{-11} $N \cdot m^2/kg^2$
Mass of an electron	m_e	$9.10938215 \times 10^{-31}$ kg	9.11×10^{-31} kg
Mass of a proton	m_p	$1.672621637 \times 10^{-27}$ kg	1.67×10^{-27} kg
Mass of a neutron	m_n	$1.674927211 \times 10^{-27}$ kg	1.67×10^{-27} kg
Planck's constant	h	$6.62606896 \times 10^{-34}$ $J \cdot s$	6.63×10^{-34} $J \cdot s$
Speed of light in a vacuum	c	2.99792458×10^{8} m/s	3.00×10^{8} m/s

SI Prefix		
Prefix	**Symbol**	**Scientific Notation**
femto	f	10^{-15}
pico	p	10^{-12}
nano	n	10^{-9}
micro	μ	10^{-6}
milli	m	10^{-3}
centi	c	10^{-2}
deci	d	10^{-1}
deka	da	10^{1}
hector	h	10^{2}
kilo	k	10^{3}
mega	M	10^{6}
giga	G	10^{9}
tera	T	10^{12}
peta	P	10^{15}

Moments of Inertia for Various Objects			
Object	**Location of Axis**	**Diagram**	**Moment of Inertia**
Thin hoop of radius r	through central diameter	Axis, r	mr^2
Solid, uniform cylinder of radius r	through center	Axis, r	$\frac{1}{2}mr^2$
Uniform sphere of radius r	through center	Axis, r	$\frac{2}{5}mr^2$
Long, uniform rod of length l	through center	Axis, l	$\frac{1}{12}ml^2$
Long, uniform rod of length l	through end	Axis, l	$\frac{1}{3}ml^2$
Thin, rectangular plate of length l and width w	through center	Axis, l, w	$\frac{1}{12}m(l^2 + w^2)$

Reference Tables

Densities of Some Common Substances	
Substance	**Density (g/cm^3)**
Aluminum	2.70
Cadmium	8.65
Copper	8.92
Germanium	5.32
Gold	19.32
Hydrogen	8.99×10^{-5}
Indium	7.31
Iron	7.87
Lead	11.34
Mercury	13.534
Oxygen	1.429×10^{-3}
Silicon	2.33
Silver	10.5
Water (4°C)	1.000
Zinc	7.14

Melting and Boiling Points		
Substance	**Melting Point (°C)**	**Boiling Point (°C)**
Aluminum	660.32	2519
Copper	1084.62	2562
Germanium	938.25	2833
Gold	1064.18	2856
Indium	156.60	2072
Iron	1538	2861
Lead	327.5	1749
Silicon	1414	3265
Silver	961.78	2162
Water	0.000	100.000
Zinc	419.53	907

Specific Heats			
Material	**Specific Heat, C[J/(kg·K)]**	**Material**	**Specific Heat, C[J/(kg·K)]**
Aluminum	897	Lead	130
Brass	376	Methanol	2450
Carbon	710	Silver	235
Copper	385	Water	4180
Glass	840	Water vapor	2020
Ice	2060	Zinc	388
Iron	450		

Heats of Fusion and Vaporization		
Material	**Heat of Fusion, H_f (J/kg)**	**Heat of Vaporization, H_v (J/kg)**
Copper	2.05×10^{5}	5.07×10^{6}
Gold	6.30×10^{4}	1.64×10^{6}
Iron	2.66×10^{5}	6.29×10^{6}
Lead	2.04×10^{4}	8.64×10^{5}
Mercury	1.15×10^{4}	2.72×10^{5}
Methanol	1.09×10^{5}	8.78×10^{5}
Silver	1.04×10^{5}	2.36×10^{6}
Water (solid)	3.34×10^{5}	2.26×10^{6}

Coefficients of Thermal Expansion at 20°C		
Material	Coefficient of Linear Expansion α (°C^{-1})	Coefficient of Volume Expansion β (°C^{-1})
Solids		
Aluminum	23×10^{-6}	69×10^{-6}
Brass	19×10^{-6}	57×10^{-6}
Concrete	12×10^{-6}	36×10^{-6}
Copper	17×10^{-6}	51×10^{-6}
Glass (soft)	9×10^{-6}	27×10^{-6}
Glass (ovenproof)	3×10^{-6}	9×10^{-6}
Iron, steel	12×10^{-6}	35×10^{-6}
Platinum	9×10^{-6}	27×10^{-6}
Liquids		
Gasoline		950×10^{-6}
Mercury		180×10^{-6}
Methanol		1200×10^{-6}
Water		210×10^{-6}
Gases		
Air (and most other gases)		3400×10^{-6}

Speed of Sound in Various Mediums	
Medium (0°C)	Speed (m/s)
Air (0°C)	331
Air (20°C)	343
Helium (0°C)	972
Hydrogen (27°C)	1310
Water (25°C)	1497
Seawater (25°C)	1533
Rubber	1600
Copper (25°C)	3560
Iron (25°C)	5130
Ovenproof glass	5640
Diamond	12,000

Wavelengths of Visible Light	
Color	Wavelength, λ (nm)
Violet	380–430
Indigo	430–450
Blue	450–500
Cyan	500–520
Green	520–565
Yellow	565–590
Orange	590–625
Red	625–740

Dielectric Constants, *k* (20°C)	
Vacuum	1.0000
Air (1 atm)	1.00059
Neon (1 atm)	1.00013
Glass	4–7
Quartz	4.3
Fused quartz	3.75
Water	80

Solar System Data								
	Mercury	Venus	Earth	Mars	Jupiter	Saturn	Uranus	Neptune
Mass (kg $\times 10^{24}$)	0.330	4.87	5.97	0.642	1899	569	86.8	102
Average radius (m $\times 10^{6}$)	2.44	6.05	6.38	3.40	71.5	60.3	25.6	24.8
Density (kg/m^3)	5427	5243	5515	3933	1326	687	1270	1638
Albedo	0.068	0.90	0.306	0.250	0.343	0.342	0.300	0.290
Average distance from the Sun (m $\times 10^{9}$)	57.91	108.2	149.6	227.9	778.4	1433.5	2872.5	4498.2
Orbital period (Earth days)	88.0	224.7	365.2	687.0	4332	10,759	30,685	60,189
Orbital inclination (degrees)	7.0	3.4	0.0	1.9	1.3	2.5	0.8	1.8
Orbital eccentricity	0.205	0.007	0.017	0.094	0.049	0.057	0.046	0.011
Rotational period (h)	1407.6	5832.5^R	23.9	24.6	9.9	10.7	17.2^R	16.1
Axial tilt (degrees)	0.03	177.4	23.4	25.2	3.1	26.7	97.8	28.3
Average surface temperature (K)	440	737	288	210	163	133	78	73
Gravitational field strength near the surface (N/kg)	3.7	8.9	9.8	3.7	20.9	10.4	8.4	10.7

R indicates retrograde motion.

The Moon	
Mass	0.073×10^{24} kg
Equatorial radius	1738 km
Mean density	3340 kg/m^3
Albedo	0.11
Average distance from Earth	384×10^{3} km
Orbital period	27.3 Earth days
Synodic period (lunar)	29.53 Earth days
Orbital inclination	5.1°
Orbital eccentricity	0.055
Rotational period	655.7 h
Gravitational field strength near the surface	1.6 N/kg

The Sun	
Mass	1.99×10^{30} kg
Equatorial radius	6.96×10^{8} m
Mean density	1408 kg/m^3
Absolute magnitude	+4.83
Luminosity	3.846×10^{26} J/s
Spectral type	G2 V
Rotational period (equatorial)	609.12 h
Mean energy production	0.1937×10^{-3} J/kg
Average surface temperature	5778 K

PERIODIC TABLE OF THE ELEMENTS

Key: Element — Hydrogen; Atomic number — 1; Symbol — H; Atomic mass — 1.008; State of matter

Gas; Liquid; Solid; Synthetic

Metal; Metalloid; Nonmetal; Recently observed

	1	2	3	4	5	6	7	8	9	10	11	12	13	14	15	16	17	18
1	Hydrogen 1 H 1.008																	Helium 2 He 4.003
2	Lithium 3 Li 6.941	Beryllium 4 Be 9.012											Boron 5 B 10.811	Carbon 6 C 12.011	Nitrogen 7 N 14.007	Oxygen 8 O 15.999	Fluorine 9 F 18.998	Neon 10 Ne 20.180
3	Sodium 11 Na 22.990	Magnesium 12 Mg 24.305											Aluminum 13 Al 26.982	Silicon 14 Si 28.086	Phosphorus 15 P 30.974	Sulfur 16 S 32.066	Chlorine 17 Cl 35.453	Argon 18 Ar 39.948
4	Potassium 19 K 39.098	Calcium 20 Ca 40.078	Scandium 21 Sc 44.956	Titanium 22 Ti 47.867	Vanadium 23 V 50.942	Chromium 24 Cr 51.996	Manganese 25 Mn 54.938	Iron 26 Fe 55.847	Cobalt 27 Co 58.933	Nickel 28 Ni 58.693	Copper 29 Cu 63.546	Zinc 30 Zn 65.39	Gallium 31 Ga 69.723	Germanium 32 Ge 72.61	Arsenic 33 As 74.922	Selenium 34 Se 78.96	Bromine 35 Br 79.904	Krypton 36 Kr 83.80
5	Rubidium 37 Rb 85.468	Strontium 38 Sr 87.62	Yttrium 39 Y 88.906	Zirconium 40 Zr 91.224	Niobium 41 Nb 92.906	Molybdenum 42 Mo 95.94	Technetium 43 Tc (98)	Ruthenium 44 Ru 101.07	Rhodium 45 Rh 102.906	Palladium 46 Pd 106.42	Silver 47 Ag 107.868	Cadmium 48 Cd 112.411	Indium 49 In 114.82	Tin 50 Sn 118.710	Antimony 51 Sb 121.757	Tellurium 52 Te 127.60	Iodine 53 I 126.904	Xenon 54 Xe 131.290
6	Cesium 55 Cs 132.905	Barium 56 Ba 137.327	Lanthanum 57 La 138.905	Hafnium 72 Hf 178.49	Tantalum 73 Ta 180.948	Tungsten 74 W 183.84	Rhenium 75 Re 186.207	Osmium 76 Os 190.23	Iridium 77 Ir 192.217	Platinum 78 Pt 195.08	Gold 79 Au 196.967	Mercury 80 Hg 200.59	Thallium 81 Tl 204.383	Lead 82 Pb 207.2	Bismuth 83 Bi 208.980	Polonium 84 Po 208.982	Astatine 85 At 209.987	Radon 86 Rn 222.018
7	Francium 87 Fr (223)	Radium 88 Ra (226)	Actinium 89 Ac (227)	Rutherfordium 104 Rf (261)	Dubnium 105 Db (262)	Seaborgium 106 Sg (266)	Bohrium 107 Bh (264)	Hassium 108 Hs (277)	Meitnerium 109 Mt (268)	Darmstadtium 110 Ds (281)	Roentgenium 111 Rg (272)	Copernicium 112 Cn (285)	Ununtrium * 113 Uut (284)	Ununquadium * 114 Uuq (289)	Ununpentium * 115 Uup (288)	Ununhexium * 116 Uuh (291)		Ununoctium * 118 Uuo (294)

The number in parentheses is the mass number of the longest lived isotope for that element.

* The names and symbols for elements 113, 114, 115, 116, and 118 are temporary. Final names will be selected when the elements' discoveries are verified.

Lanthanide series	Cerium 58 Ce 140.115	Praseodymium 59 Pr 140.908	Neodymium 60 Nd 144.242	Promethium 61 Pm (145)	Samarium 62 Sm 150.36	Europium 63 Eu 151.965	Gadolinium 64 Gd 157.25	Terbium 65 Tb 158.925	Dysprosium 66 Dy 162.50	Holmium 67 Ho 164.930	Erbium 68 Er 167.259	Thulium 69 Tm 168.934	Ytterbium 70 Yb 173.04	Lutetium 71 Lu 174.967
Actinide series	Thorium 90 Th 232.038	Protactinium 91 Pa 231.036	Uranium 92 U 238.029	Neptunium 93 Np (237)	Plutonium 94 Pu (244)	Americium 95 Am (243)	Curium 96 Cm (247)	Berkelium 97 Bk (247)	Californium 98 Cf (251)	Einsteinium 99 Es (252)	Fermium 100 Fm (257)	Mendelevium 101 Md (258)	Nobelium 102 No (259)	Lawrencium 103 Lr (262)

Explore **updates to the periodic table.**

Periodic Table

The Elements

Element	Symbol	Atomic Number	Atomic Mass	Element	Symbol	Atomic Number	Atomic Mass
Actinium	Ac	89	(227)	Molybdenum	Mo	42	95.96
Aluminum	Al	13	26.982	Neodymium	Nd	60	144.24
Americium	Am	95	(243)	Neon	Ne	10	20.180
Antimony	Sb	51	121.760	Neptunium	Np	93	(237)
Argon	Ar	18	39.948	Nickel	Ni	28	58.693
Arsenic	As	33	74.922	Niobium	Nb	41	92.906
Astatine	At	85	(210)	Nitrogen	N	7	14.007
Barium	Ba	56	137.327	Nobelium	No	102	(259)
Berkelium	Bk	97	(247)	Osmium	Os	76	190.23
Beryllium	Be	4	9.012	Oxygen	O	8	15.999
Bismuth	Bi	83	208.980	Palladium	Pd	46	106.42
Bohrium	Bh	107	(272)	Phosphorus	P	15	30.974
Boron	B	5	10.811	Platinum	Pt	78	195.078
Bromine	Br	35	79.904	Plutonium	Pu	94	(244)
Cadmium	Cd	48	112.411	Polonium	Po	84	(209)
Calcium	Ca	20	40.078	Potassium	K	19	39.098
Californium	Cf	98	(251)	Praseodymium	Pr	59	140.908
Carbon	C	6	12.011	Promethium	Pm	61	(145)
Cerium	Ce	58	140.116	Protactinium	Pa	91	231.036
Cesium	Cs	55	132.905	Radium	Ra	88	(226)
Chlorine	Cl	17	35.453	Radon	Rn	86	(222)
Chromium	Cr	24	51.996	Rhenium	Re	75	186.207
Cobalt	Co	27	58.933	Rhodium	Rh	45	102.906
Copernicium	Cn	112	(285)	Roentgenium	Rg	111	(280)
Copper	Cu	29	63.546	Rubidium	Rb	37	85.468
Curium	Cm	96	(247)	Ruthenium	Ru	44	101.07
Darmstadtium	Ds	110	(281)	Rutherfordium	Rf	104	(265)
Dubnium	Db	105	(262)	Samarium	Sm	62	150.36
Dysprosium	Dy	66	162.500	Scandium	Sc	21	44.956
Einsteinium	Es	99	(252)	Seaborgium	Sg	106	(271)
Erbium	Er	68	167.259	Selenium	Se	34	78.96
Europium	Eu	63	151.964	Silicon	Si	14	28.086
Fermium	Fm	100	(257)	Silver	Ag	47	107.868
Fluorine	F	9	18.998	Sodium	Na	11	22.990
Francium	Fr	87	(223)	Strontium	Sr	38	87.62
Gadolinium	Gd	64	157.25	Sulfur	S	16	32.065
Gallium	Ga	31	69.723	Tantalum	Ta	73	180.948
Germanium	Ge	32	72.63	Technetium	Tc	43	(98)
Gold	Au	79	196.967	Tellurium	Te	52	127.60
Hafnium	Hf	72	178.49	Terbium	Tb	65	158.925
Hassium	Hs	108	(270)	Thallium	Tl	81	204.383
Helium	He	2	4.003	Thorium	Th	90	232.038
Holmium	Ho	67	164.930	Thulium	Tm	69	168.934
Hydrogen	H	1	1.008	Tin	Sn	50	118.710
Indium	In	49	114.81	Titanium	Ti	22	47.867
Iodine	I	53	126.904	Tungsten	W	74	183.84
Iridium	Ir	77	192.217	Uranium	U	92	238.029
Iron	Fe	26	55.847	Vanadium	V	23	50.942
Krypton	Kr	36	83.798	Xenon	Xe	54	131.293
Lanthanum	La	57	138.906	Ytterbium	Yb	70	173.04
Lawrencium	Lr	103	(262)	Yttrium	Y	39	88.906
Lead	Pb	82	207.2	Zinc	Zn	30	65.38
Lithium	Li	3	6.941	Zirconium	Zr	40	91.224
Lutetium	Lu	71	174.967	Element 113*	Uut	113	(284)
Magnesium	Mg	12	24.305	Element 114*	Uuq	114	(289)
Manganese	Mn	25	54.938	Element 115*	Uup	115	(288)
Meitnerium	Mt	109	(276)	Element 116*	Uuh	116	(293)
Mendelevium	Md	101	(258)	Element 118*	Uuo	118	(294)
Mercury	Hg	80	200.59				

* Names have not yet been approved by IUPAC.

Safety Symbols	Hazard	Examples	Precaution	Remedy
Disposal	Special disposal procedures need to be followed.	Certain chemicals, living organisms	Do not dispose of these materials in the sink or trash can.	Dispose of wastes as directed by your teacher.
Biological	Organisms or other biological materials that might be harmful to humans	Bacteria, fungi, blood, unpreserved tissues, plant materials	Avoid skin contact with these materials. Wear mask and gloves.	Notify your teacher if you suspect contact with material. Wash hands thoroughly.
Extreme Temperature	Objects that can burn skin by being too cold or too hot	Boiling liquids, hot plates, dry ice, liquid nitrogen	Use proper protection when handling.	Go to your teacher for first aid.
Sharp Object	Use of tools or glassware that can easily puncture or slice skin	Razor blades, pins, scalpels, pointed tools, dissecting probes, broken glass	Practice common-sense behavior and follow guidelines for use of the tool.	Go to your teacher for first aid.
Fume	Possible danger to respiratory tract from fumes	Ammonia, acetone, nail polish remover, heated sulfur, moth balls	Be sure there is good ventilation. Never smell fumes directly. Wear a mask.	Leave foul area and notify your teacher immediately.
Electrical	Possible danger from electrical shock or burn	Improper grounding, liquid spills, short circuits, exposed wires	Double-check setup with teacher. Check condition of wires and apparatus.	Do not attempt to fix electrical problems. Notify your teacher immediately.
Irritant	Substances that can irritate the skin or mucous membranes of the respiratory tract	Pollen, moth balls, steel wool, fiberglass, potassium permanganate	Wear dust mask and gloves. Practice extra care when handling these materials.	Go to your teacher for first aid.
Chemical	Chemicals that can react with and destroy tissue and other materials	Bleaches such as hydrogen peroxide; acids such as sulfuric acid, hydrochloric acid; bases such as ammonia, sodium hydroxide	Wear goggles, gloves, and an apron	Immediately flush the affected area with water and notify your teacher.
Toxic	Substance may be poisonous if touched, inhaled, or swallowed.	Mercury, many metal compounds, iodine, poinsettia plant parts	Follow your teacher's instructions.	Always wash hands thoroughly after use. Go to your teacher for first aid.
Flammable	Flammable chemicals may be ignited by open flame, spark, or exposed heat.	Alcohol, kerosene, potassium permanganate	Avoid open flames and heat when using flammable chemicals.	Notify your teacher immediately. Use fire safety equipment if applicable.
Open Flame	Open flame in use, may cause fire.	Hair, clothing, paper, synthetic materials	Tie back hair and loose clothing. Follow teacher's instruction on lighting and extinguishing flames.	Notify your teacher immediately. Use fire safety equipment if applicable.

Eye Safety

Proper eye protection should be worn at all times by anyone performing or observing science activities.

Clothing Protection

This symbol appears when substances could stain or burn clothing.

Radioactivity

This symbol appears when radioactive materials are used.

Handwashing

After the lab, wash hands with soap and water before removing goggles.

GLOSSARY • GLOSARIO

Multilingual eGlossary

A science multilingual glossary at connectED.mcgraw-hill.com
The glossary includes the following languages: Arabic, Bengali, Chinese, English, Haitian Creole, Hmong, Korean, Portuguese, Russian, Spanish, Tagalog, Urdu, and Vietnamese.

Como usar el glosario en español:
1. Busca el termino en ingles que desees encontrar.
2. El termino en español, junto con la definicion, se encuentran en la columna de la derecha.

Pronunciation Key

Use the following key to help you sound out words in the glossary.

a	back (BAK)	**ew**	food (FEWD)
ay	day (DAY)	**yoo**	pure (PYOOR)
ah	father (FAH thur)	**yew**	few (FYEW)
ow	flower (FLOW ur)	**uh**	comma (CAHM uh)
ar	car (CAR)	**u** (+con)	rub (RUB)
e	less (LES)	**sh**	shelf (SHELF)
ee	leaf (LEEF)	**ch**	nature (NAY chur)
ih	trip (TRIHP)	**g**	gift (GIHFT)
i (i+con+e)	idea, life (i DEE uh, life)	**j**	gem (JEM)
oh	go (GOH)	**ing**	sing (SING)
aw	soft (SAWFT)	**zh**	vision (VIHZH un)
or	orbit (OR but)	**k**	cake (KAYK)
oy	coin (COYN)	**s**	seed, cent (SEED, SENT)
oo	foot (FOOT)	**z**	zone, raise (ZOHN, RAYZ)

ENGLISH — A — ESPAÑOL

absorption spectrum: The characteristic set of wavelengths absorbed by a gas, which can be used to identify that gas. **(p. 755)**

spectrum espectro de absorción: El conjunto característico de longitudes de onda absorbidas por un gas, que puede ser usado para identificarlo. **(pág. 755)**

acceleration: The rate at which the velocity of an object changes. **(p. 61)**

aceleración: La tasa que expresa la variación de la velocidad de un objeto. **(pág. 61)**

accuracy: A characteristic of a measured value that describes how well the results of a measurement agree with the "real" value, which is the accepted value, as measured by competent experimenters. **(p. 16)**

exactitud: La característica de un valor medido que describe cuanto concuerdan los resultados de una medición con el valor "real", aceptado, como ha sido medido por experimentadores competentes. **(pág. 16)**

achromatic (a kroh MA tik) lens: A combination of two or more lenses with different indices of refraction (such as a concave lens with a convex lens) that is used to minimize chromatic aberration. **(p. 506)**

lente acromático: Una combinación de dos o más lentes con índices de refracción diferentes (una lente cóncava y una lente convexa) que es utilizada para minimizar una aberración cromática. **(pág. 506)**

activity: The number of decays per second of a radioactive substance. **(p. 813)**

actividad: El número de desintegraciones por segundo emitidas por una sustancia radiactiva. **(pág. 813)**

adhesive forces: The forces of attraction that particles of different substances exert on one another; responsible for capillary action. **(p. 357)**

fuerzas de adhesión: Fuerzas de atracción que se producen cuando entran en contacto partículas de diferentes sustancias, responsables de la acción capilar. **(pág. 357)**

alpha decay: The radioactive decay process in which an alpha (α) particle is emitted from a nucleus. **(p. 809)**

desintegración alfa: El proceso de desintegración radiactiva donde un núcleo emite una partícula alfa (α). **(pág. 809)**

alpha particle: Massive, positively charged atomic particle that moves at high speed; represented by the symbol α. **(p. 753)**

amorphous (uh MOR fus) solid: A substance having a definite shape and volume but lacking a regular crystal structure. **(p. 367)**

ampere (AM pihr): A flow of electric charge, or electric current, equal to one coulomb per second (1 C/s). **(p. 600)**

amplitude: In any periodic motion, the maximum distance an object moves from the equilibrium position. **(p. 382)**

angular acceleration: The change in angular velocity divided by the time needed to make the change; measured in rad/s^2. **(p. 206)**

angular displacement: The change in the angle as an object rotates. **(p. 204)**

angular impulse–angular momentum theorem: States that the angular impulse on an object is equal to the change in the object's angular momentum. **(p. 240)**

angular momentum: The product of a rotating object's moment of inertia and its angular velocity; measured in $kg \cdot m^2/s$. **(p. 240)**

angular velocity: The angular displacement of an object divided by the time needed to make the angular displacement. **(p. 205)**

antenna: A device that propagates electromagnetic waves through the air. **(p. 714)**

antinode (AN ti nohd): The point with the largest displacement when two wave pulses meet. **(p. 396)**

apparent weight: The support force acting on an object. **(p. 102)**

Archimedes' (ahr kuh MEE deez) principle: States that an object immersed in a fluid has an upward force on it that equals the weight of the fluid displaced by the object. **(p. 361)**

armature (AR muh chur): The wire coil of an electric motor, made up of many loops mounted on an axle or shaft that rotates in a magnetic field; torque on an armature, and the motor's resultant speed, is controlled by varying the current through the motor. **(p. 662)**

atomic mass unit: A unit of mass, u, where u is equal to 1.66×10^{-27} kg; the approximate mass of a proton or neutron. **(p. 802)**

atomic number: The number of protons in an atom's nucleus. **(p. 802)**

average acceleration: The change in an object's velocity during a measurable time interval divided by that specific time interval; measured in m/s^2. **(p. 64)**

partículas alfa: Partículas atómicas masivas, de carga positiva, que se mueven a alta velocidad; representadas por el símbolo α. **(pág. 753)**

sólido amorfo: Una sustancia que tiene forma y volumen definidos, pero que carece de una estructura cristalina regular. **(pág. 367)**

amperios: Un flujo de carga o de corriente eléctrica, igual a un coulomb de carga por segundo (1 C/s). **(pág. 600)**

amplitud: En un movimiento periódico, es la distancia máxima que se desplaza un objeto desde su posición de equilibrio. **(pág. 382)**

aceleración angular: Cambio que experimenta la velocidad angular necesaria para hacer un cambio por unidad de tiempo; se mide en rad/s^2. **(pág. 206)**

desplazamiento angular: El cambio de ángulo hecho por un objeto en rotación. **(pág. 204)**

teorema del momento angular–impulso angular: Establece que el impulso angular sobre un objeto es igual a la variación del momento angular del objeto. **(pág. 240)**

momento angular: Es el producto del momento de inercia de un objeto en rotación por su velocidad angular, se mide en $kg \cdot m^2/s$. **(pág. 240)**

velocidad angular: El desplazamiento angular de un objeto dividido entre el tiempo necesario para realizar tal desplazamiento. **(pág. 205)**

antena: Una estructura que propaga ondas electromagnéticas a través del aire. **(pág. 714)**

antinodo: El punto de máximo de desplazamiento en el encuentro de dos pulsos de onda. **(pág. 396)**

peso aparente: La fuerza de apoyo que actúa sobre un objeto. **(pág. 102)**

principio de Arquímedes: Afirma que un cuerpo sumergido en un fluido recibirá una fuerza vertical ascendente igual al peso del fluido desplazado por dicho cuerpo. **(pág. 361)**

armadura: La bobina de alambre de un motor eléctrico, compuesto de muchos circuitos montados sobre un eje, que rota en un campo magnético; el esfuerzo de torsión en una armadura, y la velocidad resultante del motor, es controlada mediante la variación de corriente a través del motor. **(pág. 662)**

unidad de masa atómica: Unidad de masa, u, donde u es igual a 1.66×10^{-27} kg, es la masa aproximada de un protón o un neutrón. **(pág. 802)**

número atómico: El número de protones en el núcleo de un átomo. **(pág. 802)**

aceleración promedio: El cambio en la velocidad de un objeto durante un intervalo de tiempo mensurable dividido entre el mismo intervalo de tiempo; se mide en m/s^2. **(pág. 64)**

average speed: The distance traveled divided by the time taken to travel that distance; for uniform motion, it is the absolute value of the slope of the object's position-time graph. **(p. 47)**

rapidez promedio: La distancia recorrida por un objeto dividida entre el tiempo que se toma para recorrer esa distancia; para un movimiento uniforme, es el valor absoluto de la pendiente trazada en la gráfica de posición-tiempo de un objeto. **(pág. 47)**

average velocity: The ratio of an object's change in position to the time interval during which the change occurred; for uniform motion, it is the slope of the object's position-time graph. **(p. 47)**

velocidad media: La relación del cambio de posición de un objeto por un intervalo de tiempo durante el cual se produce el cambio; para un cuerpo en movimiento uniforme, es la pendiente de la recta en la gráfica de posición-tiempo de un objeto. **(pág. 47)**

B

band theory: The theory that electric conduction in solids can be better understood in terms of valance and conduction bands separated by forbidden energy gaps. **(p. 779)**

teoría de bandas: La teoría en la que la conducción eléctrica en los sólidos puede ser mejor entendida en términos de bandas de valencias y de conducción que están separadas entre sí por espacios sin energía. **(pág. 779)**

battery: A device made up of several galvanic cells connected together that converts chemical energy to electrical energy. **(p. 598)**

batería: Dispositivo compuesto por varias células galvánicas conectadas entre sí, que convierte la energía química en energía eléctrica. **(pág. 598)**

beat: The oscillation of wave amplitude that results from the superposition of two sound waves with almost identical frequencies. **(p. 426)**

batimiento: La oscilación de amplitud que resulta de la superposición de dos ondas sonoras con frecuencias ligeramente distintas. **(pág. 426)**

Bernoulli's (bur NEW leez) principle: States that as the velocity of a fluid increases, the pressure exerted by that fluid decreases. **(p. 364)**

principio de Bernoulli: Afirma que a medida que incrementa la velocidad de un fluido, disminuye la presión que ejerce. **(pág. 364)**

beta decay: The radioactive decay process in which a neutron is changed to a proton or a proton is changed to a neutron within the nucleus; results in the emission of a beta (β) particle and a neutrino or antineutrino. **(p. 809)**

desintegración beta: El proceso de desintegración radiactiva en la que un neutrón se transforma en un protón o un protón se transforma en un neutrón en el núcleo' resulta en la emisión de una partícula beta (β) y neutrino o un antineutrino. **(pág. 809)**

binding energy: The energy difference between the assembled nucleus and the individual nucleons; it is the energy equivalent of the mass defect and is always negative. **(p. 805)**

energía de enlace nuclear: La diferencia de energía entre el núcleo ensamblado y los nucleones individuales; es el equivalente energético del defecto de masa y tiene siempre carga negativa. **(pág. 805)**

buoyant force: The upward force exerted on an object immersed in a fluid, due to an increase in pressure with increasing depth. **(p. 361)**

empuje hidrostático: La fuerza vertical ascendente ejercida sobre un objeto que está sumergido en un fluido, debido a un aumento de la presión con un aumento de la profundidad. **(pág. 361)**

C

capacitance (ku PA suh tuns): The ratio of the magnitude of charge on one capacitor plate to the electric potential difference between the plates; the slope of the line of a net charge versus potential difference graph. **(p. 585)**

capacitancia: La razón entre la magnitud de una carga en un condensador de placas y la diferencia en el potencial eléctrico entre las placas; la pendiente de la recta en el gráfico de una carga neta en relación con la diferencia potencial. **(pág. 585)**

capacitor: An electrical device used to store electrical energy; made up of two conductors separated by an insulator. **(p. 585)**

condensador: Un dispositivo eléctrico utilizado para almacenar energía eléctrica; se compone de dos conductores separados por un aislante eléctrico. **(pág. 585)**

carrier wave: A specific part of the radio portion of the electromagnetic spectrum assigned to a radio or television station by the Federal Communications Commission. A station broadcasts by varying its carrier wave. **(p. 715)**

center of mass: The point on the object that moves in the same way a point particle would move. **(p. 219)**

centrifugal (sen TRIH fyew gul) "force": The apparent force that seems to pull on a moving object but does not exert a physical outward push on it and is observed only in rotating frames of reference. **(p. 224)**

centripetal (sen TRIH put ul) acceleration: The center-seeking acceleration of an object moving in a circle at a constant speed. **(p. 160)**

centripetal force: The net force exerted toward the center of the circle that causes an object to have a centripetal acceleration. **(p. 161)**

chain reaction: Continual process of repeated fission reactions caused by the release of neutrons from previous fission reactions. **(p. 815)**

charging by conduction: The process of charging a neutral object by touching it with a charged object. **(p. 554)**

charging by induction: The process of charging an object without touching it, which can be accomplished by bringing a charged object close to a neutral object, causing a separation of charges, then separating the object to be charged, trapping opposite but equal charges. **(p. 555)**

chromatic aberration: A lens defect in which light passing through a lens is dispersed, causing an object viewed through a lens to appear ringed with color. **(p. 506)**

circuit breaker: An automatic switch that opens when the current through an electric circuit exceeds a threshold value. **(p. 635)**

closed-pipe resonator: A resonating tube with one end closed to air; its resonant frequencies are odd-numbered multiples of the fundamental. **(p. 245)**

closed system: A system that does not gain or lose mass. **(p. 419)**

coefficient of kinetic friction: The slope of a line on a kinetic friction force v. normal force graph, μ_k. Relates frictional force to normal force and depends on the two surfaces in contact. **(p. 131)**

coefficient of linear expansion: The change in length divided by the original length and the change in temperature. **(p. 369)**

onda portadora: Un parte especifica de la porción del radio del espectro electromagnético que es asignado a una estación de radio o televisión por la Comisión Federal de Comunicaciones. Una estación transmite variando su onda portadora. **(pág. 715)**

centro de masa: El punto en un objeto que se mueve en la misma forma en la que el punto de una partícula se movería. **(pág. 219)**

"fuerza" centrífuga: La fuerza aparenta que parece empujar a un objeto que se aleja del eje de rotación, pero no ejerce un empuje físico real hacia el exterior, y es observable solamente en sistemas de referencia en rotación. **(pág. 224)**

aceleración centrípeta: La aceleración de un objeto que se mueve circularmente a una rapidez constante en búsqueda del centro del eje. **(pág. 160)**

fuerza centrípeta: La fuerza neta ejercida hacia el centro de un círculo y que produce la aceleración centrípeta. **(pág. 161)**

reacción en cadena: Secuencia de reacciones repetidas de fisión causada por la liberación de neutrones a partir de reacciones de fisión anteriores. **(pág. 815)**

carga por conducción: El proceso de cargar un objeto neutro al tocarlo con un objeto cargado. **(pág. 554)**

carga por inducción: Es el proceso de cargar un objeto sin tocarlo, la carga se logra por acercar un objeto cargado al objeto neutro provocando una separación de cargas, posteriormente se separa el objeto que debía cargarse y este captura cargas iguales pero opuestas. **(pág. 555)**

aberración cromática: Un defecto de una lente en el que la luz que la atraviesa se dispersa, causando que el objeto visto parezca rodeado de color. **(pág. 506)**

disyuntor: Un interruptor automático que se abre si la corriente a través de un circuito eléctrico excede el valor de umbral. **(pág. 635)**

resonador de tubo cerrado: Un tubo resonador con una salida de aire cerrada; sus frecuencias de resonancia son de múltiplos impares de la frecuencia fundamental. **(pág. 245)**

sistema cerrado: Un sistema que no gana ni pierde masa. **(pág. 419)**

coeficiente de fricción cinética: La pendiente de la recta en un grafico de fuerza de fricción cinética-fuerza normal, μ_k. Relaciona la fuerza de fricción con la fuerza normal y es dependiente de las dos superficies en contacto. **(pág. 131)**

coeficiente de expansión lineal: El cambio en la longitud dividida entre la longitud original y el cambio de temperatura. **(pág. 369)**

GLOSSARY • GLOSARIO

coefficient of static friction: A dimensionless constant depending on the two surfaces in contact. It is used to calculate the maximum static frictional force that needs to be overcome before motion begins. **(p. 132)**

coefficient of volume expansion: The change in volume divided by the original volume and the change in temperature; is about three times the coefficient of linear expansion because solids expand in three directions. **(p. 369)**

coherent light: Light made up of waves of the same wavelength that are in phase with each other. **(p. 522)**

cohesive forces: The forces of attraction that like particles exert on one another; responsible for surface tension and viscosity. **(p. 356)**

combination series-parallel circuit: A complex electric circuit that includes both series and parallel branches. **(p. 637)**

combined gas law: States that, for a fixed amount of an ideal gas, the pressure times the volume, divided by the Kelvin temperature equals a constant; reduces to Boyle's law if temperature is constant and to Charles's law if pressure is constant. **(p. 351)**

complementary colors: Two colors of light that, when combined, produce white light. **(p. 449)**

components: Projections of a vector parallel to the *x*-axis and another parallel to the *y*-axis. **(p. 125)**

compound machine: A machine consisting of two or more simple machines that are connected so that the resistance force of one machine becomes the effort force of the second machine. **(p. 277)**

Compton effect: The shift in the energy of scattered photons. **(p. 739)**

concave lens: A diverging lens, thinner at its middle than at its edges, that spreads out light rays passing through it when surrounded by material with a lower index of refraction; produces a smaller, virtual, upright image. **(p. 500)**

concave mirror: A mirror that reflects light from its inwardly curving surface, the edges of which curve toward the observer; can produce either an upright, virtual image or an inverted, real image. **(p. 471)**

conductor: A material, such as copper, through which a charge will move easily. **(p. 551)**

consonance: A pleasant set of pitches. **(p. 426)**

coeficiente de fricción estática: Una constante sin dimensión que depende de las dos superficies en contacto. Se utiliza para calcular la fuerza de fricción estática máxima que hay que superar para que comience el movimiento. **(pág. 132)**

coeficiente de dilatación: El cambio del volumen dividido entre el volumen original y el cambio de temperatura; es aproximadamente tres veces el coeficiente de expansión lineal, debido a que los sólidos se amplían en tres direcciones. **(pág. 369)**

luz coherente: Es la luz compuesta por ondas de misma longitud y que guardan una relación de fase constante. **(pág. 522)**

fuerzas de cohesión: Las fuerzas de atracción ejercida entre partículas similares; son responsables por la tensión de superficies y la viscosidad. **(pág. 356)**

circuito mixto en serie y paralelo: Un circuito eléctrico complejo que incluye series y ramas paralelas. **(pág. 637)**

ley general de los gases (ley combinada): Establece que para una cantidad fija de un gas ideal, la presión multiplicada por el volumen, dividida entre la temperatura Kelvin es igual a una constante; se reduce a la ley de Boyle cuando la temperatura es constante o a la ley de Charles cuando la presión es constante. **(pág. 351)**

colores complementarios: Dos colores de la luz, que cuando se combinan, producen luz blanca. **(pág. 449)**

componentes: Las proyecciones de un vector paralela al eje-*x* y al eje-*y*. **(pág. 125)**

máquina compuesta: Es una maquina formada por dos o más máquinas simples que están conectadas de manera que la fuerza de resistencia de una máquina se convierte en la fuerza de apoyo de la segunda máquina. **(pág. 277)**

efecto Compton: El cambio en la energía de fotones dispersos. **(pág. 739)**

lente cóncava: Una lente divergente, más delgada en su parte media que en sus bordes, extiende los rayos de luz que le atraviesan cuando está rodeada por material de menor índice de refracción; produce una imagen más pequeña, virtual y en posición vertical. **(pág. 500)**

espejo cóncavo: Un espejo con un superficie curvada que refleja la luz de su superficie hacia el interior, sus bordes se curvan al observador; puede producir ya sea una imagen virtual en posición vertical o una imagen real invertida. **(pág. 471)**

conductor: Un material a través del cual una carga se mueve con facilidad, como el cobre. **(pág. 551)**

consonancia: Un conjunto de tonos armónicos. **(pág. 426)**

convection: A type of thermal energy transfer that occurs due to the motion of fluid in liquid or gas that is caused by differences in temperature. **(p. 325)**

convección: Un tipo de transferencia de energía térmica que se produce debido al movimiento de fluidos en estado líquido o gaseoso y que es causado por diferencias en la temperatura. **(pág. 325)**

conventional current: the direction in which a positive test charge moves. **(p. 598)**

corriente convencional: La dirección en la cual se mueve una prueba de carga positiva. **(pág. 598)**

convex lens: A converging lens, thicker at its center than at its edges, that refracts parallel light rays so the rays meet at a point when surrounded by a medium with a lower index of refraction; can produce a smaller or larger, inverted, real image or a larger, upright, virtual image. **(p. 500)**

lente convexa: Una lente convergente, más gruesa en su centro que en sus bordes, que refracta los rayos de luz paralelamente hacia un punto central cuando está en un medio con índice menor de refracción; puede producir imágenes reales, invertidas, grandes o pequeñas; o imágenes virtuales, verticales y grandes. **(pág. 500)**

convex mirror: A mirror with the edges curved away from the observer that reflects light from its outwardly curving surface; produces an upright, reduced, virtual image. **(p. 476)**

espejo convexo: Un espejo con bordes curvos alejados del observador que refleja la luz desde su superficie curvada exterior; produce una imagen virtual reducida y en posición vertical. **(pág. 476)**

coordinate system: A system used to describe motion that gives the zero point location of the variable being studied and the direction in which the values of the variable increase. **(p. 37)**

sistema de coordenadas: Un sistema utilizado para describir desde el lugar del punto cero el movimiento de una determinada variable y la dirección en la cual los valores de esa variable incrementan. **(pág. 37)**

Coriolis (kor ee OH lus) "force": The apparent force that seems to deflect a moving object from its path and is observed only in rotating frames of reference. **(p. 224)**

"fuerza" de Coriolis: La supuesta fuerza que parece desviar a un objeto en movimiento de su trayectoria y es observable únicamente en marcos de referencia en rotación. **(pág. 224)**

coulomb (KEW lahm): The SI standard unit of charge; one coulomb (C) is the magnitude of the charge of 6.24×10^{18} electrons or protons. **(p. 558)**

coulomb: El estándar de la unidad de SI de carga; un coulomb (C) es la magnitud de la carga de $6{,}24\times10^{18}$ electrones o protones. **(pág. 558)**

Coulomb's law: States that the force between two point charges varies directly with the product of their charge and inversely with the square of the distance between them. **(p. 557)**

ley de Coulomb: Establece que la fuerza entre dos cargas de puntos es directamente proporcional al producto de sus cargas e inversamente proporcional al cuadrado de la distancia entre ellos. **(pág. 557)**

crest: The high point of a transverse wave. **(p. 390)**

cresta: El punto alto de una onda transversal. **(pág. 390)**

critical angle: The angle of incidence in which a refracted light ray lies along the boundary between two mediums. **(p. 496)**

angulo crítico: El ángulo de incidencia a partir del cual un rayo de luz refractado pasa a lo largo de la frontera entre dos medios. **(pág. 496)**

crystal lattice: A fixed, regular pattern formed when the temperature of a liquid is lowered, the average kinetic energy of its particles decreases and, for many solids, the particles become frozen but do not stop moving and instead, vibrate around their fixed positions. **(p. 367)**

red cristalina: Un patrón fijo y regular que se forma cuando se baja la temperatura de un líquido, disminuye el promedio de la energía cinética de sus partículas y, para muchos sólidos, las partículas se congelan pero no dejan de moverse, en cambio, vibran alrededor de sus posiciones fijas. **(pág. 367)**

D

de Broglie (duh BROY lee) wavelength: The wavelength associated with a moving particle. **(p. 741)**

longitud de onda de De Broglie: La longitud de onda asociada con el movimiento de una partícula. **(pág. 741)**

decibel: The most common unit of measurement for sound level; also can describe the power and intensity of sound waves. Abbreviated dB. **(p. 413)**

decibelios: La unidad de medida del nivel de sonido más utilizado, también describe el poder y la intensidad de las ondas sonoras. **(pág. 413)**

dependent variable: The factor in an investigation that depends on the independent variable. **(p. 18)**

depletion layer: The region around a *pn*-junction diode where there are no charge carriers and electricity is poorly conducted. **(p. 788)**

dielectric: A poor conductor of electric current whose electric charges partially align with an electric field. **(p. 714)**

diffraction: The bending of light around a barrier. **(p. 447)**

diffraction grating: A device consisting of many small, closely-spaced slits that diffract light and form a diffraction pattern that is an overlap of single-slit diffraction patterns; can be used to precisely measure light wavelength or to separate light of different wavelengths. **(p. 534)**

diffraction pattern: A pattern on a screen of constructive and destructive interference of Huygens' wavelets. **(p. 531)**

diffuse reflection: A scattered, fuzzy reflection produced by a rough surface. **(p. 466)**

dimensional analysis: A method of treating units as algebraic quantities that can be cancelled; can be used to check that an answer will be in the correct units. **(p. 11)**

diode: The simplest semiconductor device; conducts charges in one direction only and consists of a sandwich of *p*-type and *n*-type semiconductors. **(p. 788)**

dispersion: The separation of white light into a spectrum of colors by such means as a glass prism or water droplets in the atmosphere. **(p. 498)**

displacement: A change in position having both magnitude and direction; is equal to the final position minus the initial position. **(p. 39)**

dissonance (DIH suh nunts): An unpleasant, jarring set of pitches. **(p. 426)**

distance: The entire length of an object's path, even if the object moves in many directions. **(p. 37)**

domain: A very small group, usually 10–1000 μm on a side, that is formed when the magnetic fields of the electrons in a group of neighboring atoms are aligned in the same direction. **(p. 652)**

dopant (DOH punt): Material with electron donor or acceptor atoms that can be added in low concentration to intrinsic semiconductors, increasing their conductivity by making either extra electrons or holes available. **(p. 783)**

variable dependiente: El factor de una investigación que depende de la variable independiente. **(pág. 18)**

agotamiento de la capa: La región alrededor de un diodo de unión *pn,* donde no hay portadores de carga y la electricidad es débilmente conducida. **(pág. 788)**

dieléctrica: Un mal conductor de corriente eléctrica cuyas cargas eléctricas se alinean parcialmente con un campo eléctrico. **(pág. 714)**

difracción: La curvatura de la luz alrededor de una barrera. **(pág. 447)**

red de difracción: Un dispositivo con varias rendijas cercanas, que difracta la luz, y forma un patrón regular de difracción que es la superposición de los patrones de difracción de cada rendija; se utiliza para medir con precisión la longitud de onda de luz o para separar la luz de diferentes longitudes de onda. **(pág. 534)**

patrón de difracción: Un patrón de interferencia constructiva y destructiva de ondas primarias de Huygens en una pantalla. **(pág. 531)**

reflexión difusa: Una reflexión esparcida y borrosa producida por una superficie rugosa. **(pág. 466)**

análisis dimensional: Un método que trata las unidades como cantidades algebraicas, las cuales pueden ser canceladas; puede ser utilizado para comprobar que la respuesta obtenida estará con las unidades correctas. **(pág. 11)**

diodo: El dispositivo semiconductor más simple, que permite el paso de cargas en una única dirección y consiste en un sándwich de semiconductores tipo-*p* y tipo-*n*. **(pág. 788)**

dispersión: La separación de la luz blanca en un espectro de colores, haciendo uso de medios como un prisma de vidrio o de gotas de agua en la atmósfera. **(pág. 498)**

desplazamiento: Un variación de magnitud y dirección en la posición de un objeto; es igual a la posición final menos la posición inicial. **(pág. 39)**

disonancia: Una serie de tonos desagradable y chocante. **(pág. 426)**

distancia: La longitud total de la trayectoria de un objeto, aun si el objeto se moviera en muchas direcciones. **(pág. 37)**

dominio: Un grupo muy pequeño, generalmente 10-1000 μm de un lado, que se forma cuando los campos magnéticos de los electrones en un grupo de átomos vecinos se alinean en una misma dirección. **(pág. 652)**

dopantes: Material con átomos que son receptores o electrones donantes que se agregan en concentraciones bajas a semiconductores intrínsecos para aumentar su conductividad creando ya sea electrones adicionales o habilitando huecos. **(pág. 783)**

Doppler effect: The change in the frequency of sound or light caused by the movement of either the source, the detector, or both the detector and the source. **(p. 414)**

drag force: The force exerted by a fluid on an object that opposes the object's motion through the fluid; depends on the object's motion and properties and the fluid's properties. **(p. 104)**

efecto Doppler: El cambio en la frecuencia del sonido o de la lux causado ya sea por el movimiento de la fuente, por el detector, o bien por ambos el detector y la fuente. **(pág. 414)**

fuerza de arrastre: La fuerza ejercida por un fluido sobre un objeto para oponerse al movimiento del objeto a través del líquido; depende del movimiento y las propiedades del objeto y de las propiedades del fluido. **(pág. 104)**

E

eddy current: A current generated in a piece of metal that is moving through a changing magnetic field, producing a magnetic field that opposes the motion that caused the current. **(p. 686)**

efficiency: The ratio of output work to input work. **(p. 276)**

effort force: The force a user exerts on a machine. **(p. 274)**

elastic collision: A type of collision in which the kinetic energy before and after the collision remains the same. **(p. 306)**

elastic potential energy: Stored energy due to an object's change in shape. **(p. 298)**

electric circuit: A closed loop or pathway that allows electric charges to flow. **(p. 599)**

electric current: A flow of charged particles. **(p. 598)**

electric field: A property of the space around a charged object that exerts forces on other charged objects. **(p. 570)**

electric field line: A line that indicates the direction of the force due to the electric field on a positive test charge; the spacing between the lines indicates the field's strength; a closer line spacing indicates a stronger field; the lines never cross and they are directed toward negative charges and away from positive charges. **(p. 574)**

electric generator: A device that converts mechanical energy to electrical energy and consists of a number of wire loops placed in a strong magnetic field. **(p. 680)**

electric motor: An apparatus that converts electrical energy into mechanical energy. **(p. 662)**

electric potential difference: The work needed to move a positive test charge from one point to another, divided by the magnitude of the test charge; also called potential difference. **(p. 578)**

electromagnet: A type of magnet whose magnetic field is produced by electric current. **(p. 656)**

corriente de Eddy (o de Foucault): La corriente generada por un pieza de metal que se mueve a lo largo de un campo magnético variable, produciendo un campo magnético que se opone al movimiento que formó la corriente. **(pág. 686)**

eficiencia: La relación entre la energía de salida y energía de entrada. **(pág. 276)**

fuerza aplicada: La fuerza que ejerce un usuario en una máquina simple. **(pág. 274)**

choque elástico: Un tipo de colisión en la cual la energía cinética antes y después de la colisión se mantiene igual. **(pág. 306)**

energía potencial elástica: La energía almacenada de un objeto debido a un cambio de forma. **(pág. 298)**

circuito eléctrico: Un circuito o pasaje cerrado que permite que las cargas eléctricas fluyan. **(pág. 599)**

corriente eléctrica: Un flujo de partículas cargadas. **(pág. 598)**

campo eléctrico: Una propiedad del espacio alrededor de un objeto cargado que ejerce fuerzas hacia otros objetos cargados. **(pág. 570)**

línea de campo eléctrico: una línea que indica la dirección de la fuerza debida al campo eléctrico de una prueba de carga positiva; el espacio entre las líneas indica la fuerza del campo; un espaciamiento angosto indica un campo más fuerte; las líneas nunca se cruzan, y se dirigen hacia las cargas negativas y evadiendo las cargas positivas. **(pág. 574)**

generador eléctrico: Un dispositivo que convierte la energía mecánica en energía eléctrica y consiste en una serie de circuitos de alambres colocados en un campo magnético fuerte. **(pág. 680)**

motor eléctrico: Un aparato que convierte la energía eléctrica en energía mecánica. **(pág. 662)**

diferencia de potencial eléctrico: El trabajo realizado para mover una carga de prueba positiva de un punto a otro, dividido por la magnitud de la carga de prueba; también conocido como la diferencia potencial. **(pág. 578)**

electroimán: Un imán cuyo campo magnético es producido por una corriente eléctrica. **(pág. 656)**

electromagnetic induction: The process of generating a current through a circuit due a changing magnetic field or to the relative motion between a wire and a magnetic field. **(p. 676)**

electromagnetic radiation: Energy that is carried, or radiated, in the form of electromagnetic waves. **(p. 712)**

electromagnetic spectrum: The entire range of frequencies and wavelengths that make up the continuum of electromagnetic waves, including radio waves, microwaves, visible light, and X-rays. **(p. 712)**

electromagnetic wave: Coupled, oscillating electric and magnetic field that travels through space and matter. **(p. 710)**

electron cloud: The region in which there is a high probability of finding an electron. **(p. 766)**

electroscope: A device that is used to detect electric charges and consists of a metal knob connected by a metal stem to two thin metal leaves. **(p. 554)**

electrostatics: The study of electric charges that can be collected and held in one place. **(p. 548)**

elementary charge: The magnitude of the charge of an electron; approximately 1.602×10^{-19} C. **(p. 558)**

emission spectrum: A plot of the intensity of radiation emitted from an object over a range of frequencies. **(p. 729)**

energy: The ability of a system to produce a change in itself or in the world around it; represented by the symbol E. **(p. 270)**

energy level: The quantized amount of energy allowed for electrons in an atom. **(p. 757)**

entropy (EN truh pee): A measure of the energy dispersal in a system. **(p. 338)**

equilibrant: A force that places an object in equilibrium; the same magnitude as the resultant but opposite in direction. **(p. 137)**

equilibrium: The condition in which the net force on an object is zero. **(p. 99)**

equipotential (ee kwuh puh TEN shul): The electric potential difference of zero between two or more positions in an electric field. **(p. 579)**

equivalent resistance: The value of a single resistor that, when it replaces all the resistors in the circuit, results in the same current. **(p. 626)**

excited state: Any energy level of an atom that is higher than its ground state. **(p. 757)**

extrinsic (ek STRIHN zik) semiconductor: Semiconductor with greatly enhanced conductivity resulting from the addition of dopants. **(p. 783)**

inducción electromagnética: El proceso de generar corriente a través de un circuito debido a un campo magnético variable o al movimiento relativo entre el alambre y el campo magnético. **(pág. 676)**

radiación electromagnética: La energía portada o irradiada, en forma de ondas electromagnéticas. **(pág. 712)**

espectro electromagnético: La gama entera de frecuencias y longitudes de onda que componen el continuum de ondas electromagnéticas; incluye a ondas de radio, microondas, luz visible y rayos X. **(pág. 712)**

onda electromagnética: Un campo eléctrico y magnético conectado y variable que viaja a través del espacio y la materia. **(pág. 710)**

nube de electrones: La región en la cual existe una alta probabilidad de encontrar un electrón. **(pág. 766)**

electroscopio: Instrumento que permite determinar las cargas eléctricas, consiste en una perilla metálica conectada por una varilla metálica a dos laminillas de metal. **(pág. 554)**

electrostática: El estudio de cargas eléctricas que pueden ser recogidas y almacenadas en un lugar. **(pág. 548)**

carga elemental: La magnitud de la carga de un electrón, aproximadamente 1.602×10^{-19} C. **(pág. 558)**

espectro de emisión: La intensidad de la radiación emitida por un objeto sobre un conjunto de frecuencias. **(pág. 729)**

energía: La capacidad de un sistema para producir un cambio en sí mismo o en el mundo que le rodea; representado por el símbolo E. **(pág. 270)**

nivel energético: La cantidad de energía cuántica permitida para los electrones de un átomo. **(pág. 757)**

entropía: Es la medición de la energía dispersa en un sistema. **(pág. 338)**

equilibrante: La fuerza que coloca un objeto en equilibrio; tiene la misma magnitud que la fuerza resultante, pero en direcciones opuestas. **(pág. 137)**

equilibrio: La condición en la cual la fuerza neta sobre un objeto es cero. **(pág. 99)**

equipotencial: La diferencia del potencial eléctrico de cero entre dos o más posiciones en un campo eléctrico. **(pág. 579)**

resistencia equivalente: El valor de una sola resistencia que, cuando se reemplaza a todas las resistencias en el circuito, resulta en la misma corriente. **(pág. 626)**

estado excitado: Cualquier nivel de energía de un átomo por encima de su estado fundamental. **(pág. 757)**

semiconductor extrínseco: Semiconductor con gran realce de conductividad producto de la adición de dopantes. **(pág. 783)**

F

farsightedness: A vision defect in which a person cannot see close objects clearly because images are focused behind the retina; also called hyperopia; can be corrected with a convex lens. **(p. 509)**

first law of thermodynamics: States that the change in thermal energy of a system is equal to the heat that is added to the system, minus the work done by the system. **(p. 334)**

fission: The process in which a nucleus is divided into two or more fragments, and neutrons and energy are released. **(p. 814)**

fluid: A material, such as a liquid or gas, that can flow and has no definite shape of its own. **(p. 348)**

focal length: The distance between the focal point and the mirror or lens. **(p. 472)**

focal point: The point where incident light rays that are parallel to the principal axis converge after reflecting from a mirror or refracting through a lens. **(p. 472)**

force: A push or a pull exerted on an object; has both direction and magnitude and may be a contact or a field force. **(p. 90)**

force carrier: A particle that transmits, or carries, forces between objects interacting at a distance; also called a gauge boson. **(p. 823)**

free-body diagram: A physical model that represents the forces acting on a system. **(p. 92)**

free fall: The motion of an object body when air resistance is negligible and the motion can be considered due to the force of gravity alone. **(p. 75)**

free-fall acceleration: The acceleration of an object due only to the effect of gravity. **(p. 75)**

frequency: The number of complete oscillations a point on a wave makes each second; measured in hertz (Hz). **(p. 391)**

fundamental: For a musical instrument, the lowest frequency of sound that will resonate. **(p. 425)**

fuse: A short piece of metal that acts as a safety device by melting and stopping the current when too large a current passes through it. **(p. 635)**

fusion: The process in which nuclei with small masses combine to form a nucleus with a larger mass and energy is released. **(p. 817)**

hipermetropía: Un defecto ocular en la que una persona no puede ver claramente los objetos cercanos porque las imágenes son enfocadas en un punto detrás de la retina; se corrige con una lente convexa. **(pág. 509)**

primera ley de la termodinámica: Afirma que la variación de la energía térmica de un sistema es igual al calor agregado al sistema, menos el trabajo realizado por el sistema. **(pág. 334)**

fisión: El proceso en el cual el núcleo se divide en dos o más fragmentos, liberándose neutrones y energía. **(pág. 814)**

fluido: Un material (como un líquido o un gas) que puede fluir y no tiene forma propia definida. **(pág. 348)**

distancia focal: La distancia entre el punto focal y el espejo o la lente. **(pág. 472)**

punto focal: El punto donde los rayos de luz incidente que son paralelos al eje principal convergen después de reflejarse en un espejo o refractarse a tráves de una lente. **(pág. 472)**

fuerza: Un empuje o jalón ejercido sobre un objeto; posee tanto dirección como magnitud y puede ser de contacto o de un campo de fuerza. **(pág. 90)**

portadoras de fuerza: Una partícula que transmite, o porta, fuerzas entra objetos que interactúan a la distancia, también llamada boson de gauge. **(pág. 823)**

diagrama de cuerpo libre: un modelo físico que representa las fuerzas que actúan en un sistema. **(pág. 92)**

caída libre: El movimiento de un objeto para el cual la resistencia del aire es insignificante y su movimiento puede ser considerado únicamente debido a la fuerza de gravedad. **(pág. 75)**

aceleración de caída libre: La aceleración de un objeto debido únicamente al efecto de la gravedad. **(pág. 75)**

frecuencia: El número de oscilaciones completas que un punto en una onda realiza por segundo; se mide en hercios (Hz). **(pág. 391)**

fundamentales: En un instrumento musical, es la menor frecuencia de resonancia del sonido. **(pág. 425)**

fusible: Un trozo de metal que actúa como un dispositivo de seguridad al derretirse y detener el fluyo de la corriente cuando un mayor flujo de corriente pasa. **(pág. 635)**

fusión: El proceso en el que los núcleos de masa pequeña se combinan para formar un núcleo de mayor masa y para liberar energía. **(pág. 817)**

G

galvanometer (gal vuh NAH muh tur): A device that is used to measure very small currents; can be used as a voltmeter or an ammeter. **(p. 661)**

galvanómetro: Un dispositivo que se utiliza para medir corrientes muy pequeñas; puede ser utilizado como un voltímetro o un amperímetro. **(pág. 661)**

gamma decay: The radioactive decay process in which there is a redistribution of energy within the nucleus but no change in atomic mass or charge, and a gamma ray is emitted. **(p. 809)**

desintegración gamma: El proceso de desintegración radiactiva en el cual hay una redistribución de energía dentro del núcleo, sin ocurrir ningún cambio de masa atómica o de carga eléctrica, y emite un rayo gamma. **(pág. 809)**

gravitational field: A vector quantity that relates the mass of an object to the gravitational force it experiences at a given location; represented by the symbol *g*. **(p. 100)**

campo gravitatorio: Magnitud vectorial que relaciona la masa de un objeto con la fuerza gravitatoria experimentada en un lugar determinado; representada por el símbolo *g*. **(pág. 100)**

gravitational force: The attractive force between two objects that is directly proportional to the mass of the objects. **(p. 182)**

fuerza de la gravedad: La fuerza de atracción entre dos objetos, es directamente proporcional a la masa de los objetos. **(pág. 182)**

gravitational mass: Mass as used in the law of universal gravitation; the quantity that measures an object's response to gravitational force. **(p. 191)**

masa gravitacional: Masa como se define en la ley de la gravedad; cantidad que mide la respuesta de atracción de un objeto ante la fuerza gravitacional. **(pág. 191)**

gravitational potential energy: The stored energy in a system resulting from the gravitational force between objects; represented by the symbol *GPE*. **(p. 295)**

energía potencial gravitatoria: La energía almacenada en un sistema resultado de la fuerza gravitatoria entre objetos; se representa por el símbolo *EPG*. **(pág. 295)**

ground-fault interrupter (GFI): A device that contains an electronic circuit that detects small current differences caused by an extra current path; it opens the circuit, prevents electrocution, and often is required as a safety measure for bathroom, kitchen, and exterior outlets. **(p. 635)**

interruptor de circuito por falla a tierra (ICFT): Un dispositivo que contiene un circuito electrónico capaz de detectar diferencias pequeñas de corrientes causadas por la adición de un trayecto de corriente; abre el circuito, previene la electrocución y usualmente es usado como medida de seguridad en los baños, cocinas y tomas de corriente exteriores. **(pág. 635)**

grounding: The process of removing excess charge by touching an object to Earth. **(p. 556)**

puesta a tierra: El proceso de remover el exceso de carga tocando un objeto en la Tierra. **(pág. 556)**

ground state: State of an atom with the smallest allowable amount of energy. **(p. 757)**

estado fundamental: Estado en el que un átomo posee la cantidad más pequeña de energía permitida. **(pág. 757)**

H

half-life: The time required for half the atoms in a given quantity of a radioactive isotope to decay. **(p. 812)**

vida media: El tiempo requerido para que la mitad de los átomos se desintegren en una determinada cantidad de un isótopo radiactivo. **(pág. 812)**

harmonics: Higher frequencies, which are whole-numbered multiples of the fundamental frequency; give certain musical instruments their own unique timbre. **(p. 425)**

armónicos: Las frecuencias más altas, que son múltiplos de números enteros de la frecuencia fundamental; dota a ciertos instrumentos musicales con su propio y único timbre. **(pág. 425)**

heat: A transfer of thermal energy, which occurs spontaneously from a warmer object to a cooler object; represented by *Q*. **(p. 324)**

calor: Transferencia de energía térmica que ocurre espontáneamente de un objeto más caliente a un objeto más frío; representado por *Q*. **(pág. 324)**

heat engine: A device that continuously converts thermal energy to mechanical energy; requires a high-temperature thermal energy source, a low-temperature receptacle (a sink), and a way to convert the thermal energy into work. **(p. 335)**

heat of fusion: The amount of thermal energy required to change 1 kg of a substance from a solid state to a liquid state at its melting point. **(p. 331)**

heat of vaporization: The amount of thermal energy required to change 1 kg of a substance from a liquid state to a gaseous state at its boiling point. **(p. 331)**

Heisenberg uncertainty principle: States there is a limit to how precisely a particle's position and momentum can simultaneously be measured. **(p. 743)**

Hooke's law: States that the force acting on a spring is directly proportional to the amount the spring is stretched. **(p. 383)**

hypothesis: A possible explanation for a problem using what you know and what you observe. **(p. 6)**

motor térmico: Dispositivo que continuamente convierte energía térmica en energía mecánica; requiere de una fuente de energía térmica a temperatura alta, de un recipiente de temperatura baja (sumidero), y de un mecanismo para convertir la energía térmica en trabajo. **(pág. 335)**

calor de fusión: Es la cantidad de energía térmica necesaria para hacer que un kilogramo de una sustancia pase del estado sólido a líquido en su punto de fusión. **(pág. 331)**

calor de vaporización: La cantidad de energía térmica necesaria para hacer que 1 kg de una sustancia pase de estado liquido a gaseoso en su punto de ebullición. **(pág. 331)**

principio de incertidumbre de Heisenberg: Establece que hay un límite en la precisión en cual se puede medir simultáneamente la posición y el momento de una partícula. **(pág. 743)**

ley de Hooke: Establece que la cantidad de estiramiento de un resorte es directamente proporcional a la fuerza que actúa sobre el resorte. **(pág. 383)**

hipótesis: Una posible explicación de un problema haciendo uso de lo que se sabe y lo que se observa. **(pág. 6)**

I

ideal gas law: For an ideal gas, the pressure times the volume is equal to the number of moles, times the constant (R) and the Kelvin temperature; predicts the behavior of gases remarkably well unless under high-pressure or low-temperature conditions. **(p. 352)**

ideal mechanical advantage: For an ideal machine, is equal to the displacement of the effort force divided by displacement of the resistance force. **(p. 275)**

illuminance (ih LEW muh nunts): The rate at which light strikes a surface, or falls on a unit area; measured in lumens per square meter (lm/m^2) or lux (lx). **(p. 441)**

image: An optical reproduction of an object formed by the combination of light rays reflected or refracted by an optical device. **(p. 468)**

impulse: The product of the average net force on an object and the time interval over which the force acts. **(p. 236)**

impulse-momentum theorem: States that the impulse on an object equals the object's final momentum minus the object's initial momentum. **(p. 237)**

incident wave: A wave pulse that strikes a boundary between two mediums. **(p. 394)**

ley de los gases ideales: Para un gas ideal, es la ecuación de presión por volumen igual al número de moles, por la constante (R) y la temperatura Kelvin; predice aceptablemente la conducta de los gases a menos que estos se encuentren bajo condiciones de presión alta o temperatura baja. **(pág. 352)**

ventaja mecánica ideal: Para una maquina ideal, equivale al desplazamiento de la fuerza de esfuerzo dividido entre la fuerza de resistencia. **(pág. 275)**

iluminancia: La cantidad en la cual la luz incide sobre una superficie por unidad de área; se mide en lúmenes por metro cuadrado (lm/m^2) o lux (lx). **(pág. 441)**

imagen: Una reproducción óptica de un objeto, formado por la combinación de rayos de luz reflejados o refraccionados por un dispositivo óptico. **(pág. 468)**

impulso: Resultado de la fuerza neta promedio en un objeto y el intervalo de tiempo durante el cual actúa la fuerza. **(pág. 236)**

teorema del impulso y momento: Establece que el impulso de un objeto es igual a la resta del impulso final del objeto menos el impulso inicial del mismo. **(pág. 237)**

onda incidente: Un pulso de onda que incide un límite entre dos medios. **(pág. 394)**

incoherent light: Light whose waves are not in phase. **(p. 522)**

independent variable: The factor that is changed or manipulated during an investigation. **(p. 18)**

index of refraction: Determines the angle of refraction of light as it crosses the boundary between mediums; for a given medium, it is the ratio of the speed of light in a vacuum to the speed of light in that medium; represented by the symbol *n*. **(p. 493)**

induced electromotive force: The potential difference across a wire that results from the production of an electric field. **(p. 677)**

inelastic collision: A type of collision in which the kinetic energy after the collision is less than the kinetic energy before the collision. **(p. 306)**

inertia (ihn UR shuh): The tendency of an object to resist changes in velocity. **(p. 98)**

inertial mass: A measure of an object's resistance to any type of force. **(p. 191)**

instantaneous acceleration: The change in an object's velocity at a specific instant. **(p. 64)**

instantaneous position: The position of an object at any particular instant. **(p. 43)**

instantaneous velocity: A measure of motion that tells the speed and direction of an object at a specific instant. **(p. 49)**

insulator: A material, such as glass, through which a charge will not move easily. **(p. 551)**

interaction pair: A pair of forces that are equal in strength but opposite in direction and act on different objects. **(p. 106)**

interference: Results from the superposition of two or more waves; can be constructive or destructive. **(p. 395)**

interference fringes: A pattern of light and dark bands on a screen, resulting from the constructive and destructive interference of light waves passing through two narrow, closely spaced slits in a barrier. **(p. 523)**

intrinsic semiconductor: Pure semiconductor that conducts charge as a result of thermally freed electrons. **(p. 782)**

inverse relationship: A hyperbolic relationship that exists when one variable depends on the inverse of the other variable. **(p. 22)**

isolated system: A closed system on which the net external force is zero. **(p. 245)**

isotope: Each of the differing forms of the same atom that have different masses but have the same chemical properties; atoms with the same number of protons, but different numbers of neutrons. **(p. 706)**

luz incoherente: Luz cuyas ondas no están en fase. **(pág. 522)**

variable independiente: El factor que cambia o se manipula durante una investigación. **(pág. 18)**

índice de refracción: Determina el ángulo de refracción de la luz a su paso frontera través del límite entre dos medios, para un medio específico, es el cociente de la velocidad de la luz en el vacío entre la velocidad de la luz en ese medio; representado por el símbolo *n*. **(pág. 493)**

fuerza electromotiva inducida: La diferencia potencial a lo largo de un alambre que resulta de la producción de un campo eléctrico. **(pág. 677)**

colisión inelástica: Un tipo de choque en el que la energía cinética después del choque es menor que la energía cinética antes del choque. **(pág. 306)**

inercia: La propiedad de un objeto a resistir los cambios en la velocidad. **(pág. 98)**

masa inercial: Es la medida de la resistencia de un objeto a cualquier tipo de fuerza. **(pág. 191)**

aceleración instantánea: El cambio en la velocidad de un objeto en un instante específico. **(pág. 64)**

posición instantánea: La posición de un objeto en cualquier instante en particular. **(pág. 43)**

velocidad instantánea: Una medida del movimiento que indica la velocidad y dirección de un objeto en un instante específico. **(pág. 49)**

aislante: Un material, como el vidrio, a través del cual una carga no se mueva con facilidad. **(pág. 551)**

par de interacción: Un par de fuerzas de igual intensidad, sentidos opuestos, y que actúan sobre objetos diferentes. **(pág. 106)**

interferencia: El resultado de la superposición de dos o más ondas; puede ser constructiva o destructiva. **(pág. 395)**

franjas de interferencia: Un patrón de bandas oscuras e iluminadas sobre una pantalla, resultado de la interferencia constructiva y destructiva de ondas de luz que pasan a través de dos rendijas estrechas y muy próximas entre sí en una barrera. **(pág. 523)**

semiconductor intrínsico: Semiconductor puro que conduce carga como resultado de la liberación térmica de electrones. **(pág. 782)**

relación inversa: Una relación hiperbólica que se da cuando una variable depende de la inversa de la otra variable. **(pág. 22)**

sistema aislado: Un sistema cerrado en el que la fuerza externa neta es cero. **(pág. 245)**

isótopo: Cada una de las diferentes formas de un mismo átomo con masas diferentes, pero con las mismas propiedades químicas; atomos con el mismo numero de protones, pero distinctos números de neutrones. **(pág. 706)**

J

joule (JEWL): The SI unit of work and energy (J); 1 J of work is done when a force of 1 N acts on an object over a displacement of 1 m. **(p. 264)**

julio: La unidad de trabajo y energía del SI (J); 1 J de trabajo equivale al trabajo producido por una fuerza de 1 N cuyo punto de aplicación se desplaza 1 m en la dirección de la fuerza. **(pág. 264)**

K

Kepler's first law: States that the planets move in elliptical paths with the Sun at one focus. **(p. 179)**

primera ley de Kepler: Establece que los planetas se mueven en trayectorias elípticas, con el Sol situado en un foco. **(pág. 179)**

Kepler's second law: States that an imaginary line from the Sun to a planet sweeps out equal areas in equal time intervals. **(p. 179)**

segunda ley de Kepler: Establece que una línea imaginaria desde el Sol a un planeta barre áreas iguales en intervalos iguales de tiempo. **(pág. 179)**

Kepler's third law: States that the square of the ratio of the periods of any two planets is equal to the cube of the ratio of their average distances from the Sun. **(p. 180)**

tercera ley de Kepler: Establece que el cuadrado de la relación de los períodos de dos planetas cualesquiera es igual al cubo de la proporción de sus distancias medias desde el sol. **(pág. 180)**

kilowatt-hour: An energy unit used by electric companies to measure energy sales; 1 kWh is equal to 1000 watts (W) delivered continuously for 3600 s (1 h). **(p. 612)**

kilovatio-hora: Una unidad de energía utilizada por las empresas eléctricas para medir las ventas de energía, 1 kWh es igual a 1000 vatios (W) emitido continuamente durante 3600 s (1 h). **(pág. 612)**

kinetic energy: The energy of an system that is associated with its motion. **(p. 270)**

energía cinética: La energía de un sistema asociada con su movimiento. **(pág. 270)**

kinetic friction: The force exerted on one surface by a second surface when the two surfaces rub against each other because one or both of the surfaces are moving. **(p. 130)**

fricción cinética: La fuerza ejercida sobre una superficie por una segunda superficie, cuando las dos superficies se rozan entre sí porque uno o ambas de las superficies están en movimiento. **(pág. 130)**

L

laser: A device that produces powerful, coherent, directional, monochromatic light that can be used to excite other atoms; the acronym stands for *l*ight *a*mplification by *s*timulated *e*mission of *r*adiation. **(p. 767)**

láser: Un dispositivo que produce luz monocromática poderosa, coherente y direccional que puede ser utilizada para excitar a otros átomos, su acrónimo indica las letras iniciales de su nombre en Ingles que significa *luz* amplificada por emisión de radiación estimulada **(pág. 767)**

law of conservation of angular momentum: States that if there are no net external torques on a closed system, then its angular momentum is conserved. **(p. 251)**

ley de la conservación del momento angular: Establece que, si no hay esfuerzo de torsión neta externa sobre un sistema cerrado, entonces su momento angular se conserva. **(pág. 251)**

law of conservation of energy: States that in a closed, isolated system, energy is not created or destroyed, but rather, is conserved. **(p. 301)**

ley de la conservación de la energía: Establece que, en un sistema cerrado y aislado, la energía no se crea ni se destruye, sino que se conserva. **(pág. 301)**

law of conservation of momentum: States that the momentum of any closed, isolated system does not change. **(p. 245)**

ley de conservación del momento: Establece que el momento de un sistema cerrado y aislado no cambia. **(pág. 245)**

law of reflection: States that the angle of incidence is equal to the angle of reflection. **(p. 398)**

ley de la reflexión: Establece que el ángulo de incidencia es igual al ángulo de reflexión. **(pág. 398)**

law of universal gravitation: States that gravitational force between any two objects is directly proportional to the product of their masses and inversely proportional to the square of the distance between their centers. **(p. 182)**

ley de la gravitación universal: Establece que la fuerza gravitacional entre dos objetos es directamente proporcional al producto de sus masas e inversamente proporcional al cuadrado de la distancia entre sus centros. **(pág. 182)**

lens: A piece of transparent material, such as glass or plastic, that is used to focus light and form an image. **(p. 500)**

lente: Una pieza de material transparente, como el vidrio o plástico, que se utiliza para enfocar la luz y formar una imagen. **(pág. 500)**

Lenz's law: States that an induced current always is produced in a direction such that the magnetic field resulting from the induced current opposes the change in the magnetic field that is causing the induced current. **(p. 684)**

ley de Lenz: Establece que una corriente inducida siempre se produce en una dirección tal que se opone a la variación en el campo magnético que está causando la corriente inducida. **(pág. 684)**

lepton: An elementary particle, such as an electron or a neutrino, that forms mass but belongs to a different family than quarks. **(p. 822)**

lepton: Una partícula elemental, como un electrón o un neutrino, que forma la masa, pero que pertenece a una familia diferente de los quarks. **(pág. 822)**

lever arm: The perpendicular distance from the axis of rotation to the point where force is exerted. **(p. 209)**

brazo de palanca: La distancia perpendicular desde el eje de rotación hasta el punto donde la fuerza se ejerce. **(pág. 209)**

linear relationship: A relationship in which the dependent variable varies linearly with the independent variable. **(p. 18)**

relación lineal: Una relación en la que la variable dependiente varía linealmente con la variable independiente. **(pág. 18)**

line of best fit: A line drawn on a graph as close to all the data points as possible; used to describe data and predict where new data will appear on the graph. **(p. 20)**

línea de mejor ajuste: Una línea trazada sobre un gráfico lo más cercanamente posible a todos los puntos de datos, se utiliza para describir los datos y predecir dónde aparecerán nuevos datos en el gráfico. **(pág. 20)**

longitudinal wave: A mechanical wave in which the disturbance is in the same direction, or parallel to, the direction of wave travel. **(p. 388)**

onda longitudinal: Una onda mecánica en la que la alteración está en la misma dirección, o en una dirección paralela a la dirección de la propagación de la onda. **(pág. 388)**

loudness: Sound intensity as sensed by the ear and interpreted by the brain; depends mainly on the pressure wave's amplitude. **(p. 413)**

sonoridad: La intensidad del sonido detectada por el oído e interpretada por el cerebro; depende principalmente de la presión de la amplitud de la onda. **(pág. 413)**

luminous flux: The rate at which light energy is emitted from a luminous source; measured in lumens (lm). **(p. 440)**

flujo luminoso: La velocidad a la cual se emite la energía luminosa desde una fuente luminosa; se mide en lúmenes (lm). **(pág. 440)**

luminous source: An object, such as the Sun or a lightbulb, that emits light. **(p. 439)**

fuente luminosa: Un objeto, como el Sol o un bombillo incandescente, que emite luz. **(pág. 439)**

M

machine: A device that makes work easier (but does not change the amount of work) by changing the magnitude or the direction of the force exerted to do work. **(p. 274)**

máquina: Un dispositivo que facilita el trabajo (pero no cambia la cantidad), modificando la magnitud o la dirección de la fuerza ejercida para hacer el trabajo. **(pág. 274)**

magnetic field: The area around a magnet, or around any current-carrying wire or coil of wire, where a magnetic force exists. **(p. 653)**

campo magnético: El área alrededor de un imán, o cercano a cualquier cable portador de corriente o rollo de alambre, donde existe una fuerza magnética. **(pág. 653)**

magnetic flux: The number of magnetic field lines that pass through a perpendicular surface. **(p. 654)**

flujo magnético: El número de líneas de un campo magnético que pasan a través de una superficie perpendicular. **(pág. 654)**

magnification: The amount that an image is enlarged or reduced in size, relative to the object. **(p. 478)**

magnificación: La cantidad que una imagen es ampliada o reducida en tamaño, en relación con el objeto. **(pág. 478)**

magnitude: A measure of size. **(p. 38)**

magnitud: Una medida de tamaño. **(pág. 38)**

Malus's law: States that the intensity of light coming out of a second polarizing filter equals the intensity of polarized light coming out of a first polarizing filter times the cosine, squared, of the angle between the polarizing axes of the two filters. **(p. 452)**

mass defect: The difference between the sum of the masses of individual nucleons and the mass of the assembled nucleus. **(p. 805)**

mass number: The number of nucleons in an atom's nucleus. **(p. 802)**

mass spectrometer: Device that uses both electric and magnetic fields to measure the charge-to-mass ratio of positive ions within a material; can be used to determine the masses of ions. **(p. 706)**

measurement: A comparison between an unknown quantity and a standard. **(p. 14)**

mechanical advantage: The ratio of resistance force to effort force. **(p. 274)**

mechanical energy: The sum of kinetic energy and potential energy of the objects in a system. **(p. 301)**

microchip: An integrated circuit consisting of thousands of diodes, transistors, resistors, and conductors. **(p. 792)**

model: A representation of an idea, event, structure, or object to help people better understand it. **(p. 7)**

moment of inertia: The resistance to rotation. **(p. 213)**

momentum: The product of the object's mass and the object's velocity; measured in kg·m/s. **(p. 237)**

monochromatic light: Light having only one wavelength. **(p. 523)**

motion diagram: A series of images showing the positions of a moving object taken at regular (equal) time intervals. **(p. 36)**

mutual inductance: Effect in which a changing current in a coil creates a changing magnetic field that induces a varying *EMF* in a second coil. **(p. 688)**

Ley de Malus: Establece que la intensidad de la luz proveniente de un segundo filtro polarizador equivale a la intensidad de la luz polarizada que proviene del primer filtro polarizador por el coseno al cuadro del ángulo entre los ejes de polarización de los dos filtros. **(pág. 452)**

defecto de masa: La diferencia entre la suma de las masas de los nucleones individuales y la masa del núcleo montado. **(pág. 805)**

número másico: El número de nucleones en el núcleo de un átomo. **(pág. 802)**

espectrómetro de masas: Instrumento que utiliza ambos los campos eléctricos y magnéticos para medir la razón entre carga y masa de los iones en una material; puede ser usado para determinar las masas de los iones. **(pág. 706)**

medición: Una comparación entre una cantidad desconocida y una establecida. **(pág. 14)**

ventaja mecánica: La razón entre la fuerza de resistencia y la fuerza de esfuerzo. **(pág. 274)**

energía mecánica: La suma de energía cinética y energía potenciale de los objetos en un sistema. **(pág. 301)**

microchip: Un circuito integrado que consta de miles de diodos, transistores, resistencias y conductores. **(pág. 792)**

modelo: Representación de una idea, evento, estructura, u objeto para ayudar a la gente a entenderlo mejor. **(pág. 7)**

momento de inercia: La resistencia a la rotación. **(pág. 213)**

momento: El producto de la masa del objeto por la velocidad del objeto; se mide en kg·m/s. **(pág. 237)**

luz monocromática: La luz que tiene sólo una onda longitudinal. **(pág. 523)**

diagrama de movimiento: Una serie de imágenes que muestran las posiciones de un objeto en movimiento tomadas en periodos iguales de tiempo. **(pág. 36)**

inductancia mutua: Efecto en el que una corriente variable en una bobina crea un campo magnético cambiante, que induce una *FEM* variable en una segunda bobina. **(pág. 688)**

N

nearsightedness: A vision defect in which a person cannot see distant objects clearly because images are focused in front of the retina; also called myopia; can be corrected with a concave lens. **(p. 509)**

net force: The vector sum of all the forces on an object. **(p. 93)**

neutral: An object whose positive charges are exactly balanced by its negative charges. **(p. 550)**

miopía: Un defecto ocular en el que una persona no puede ver claramente los objetos distantes, porque las imágenes se enfocan en un punto delante de la retina, se puede corregir con una lente cóncava. **(pág. 509)**

fuerza neta: La suma vectorial de todas las fuerzas sobre un objeto. **(pág. 93)**

neutro: Un objeto cuyas cargas positivas están balanceadas con exactitud por sus cargas negativas. **(pág. 550)**

Newton's first law: States that an object at rest will remain at rest and a moving object will continue moving in a straight line with constant speed, if and only if the net force acting on that object is zero. **(p. 98)**

Newton's second law: States that the acceleration of an object is proportional to the net force and inversely proportional to the mass of the object being accelerated. **(p. 96)**

Newton's second law for rotational motion: States that the angular acceleration of an object is directly proportional to the net torque on it and inversely proportional to its moment of inertia. **(p. 216)**

Newton's third law: States that all forces come in pairs and that the two forces in a pair act on different objects, are equal in strength, and are opposite in direction. **(p. 106)**

node: The stationary point where two equal wave pulses meet and are in the same location, having a displacement of zero. **(p. 396)**

normal: The line in a ray diagram that shows the orientation of the barrier or mirror and is drawn at a right angle, or perpendicular, to the barrier or mirror. **(p. 398)**

normal force: The perpendicular contact force exerted by a surface on an object. **(p. 111)**

nuclear reaction: Occurs when the energy or number of neutrons or protons in a nucleus changes. **(p. 810)**

nucleon: A particle found in the nucleus of an atom; that is, a proton or a neutron. **(p. 802)**

nucleus: The tiny, massive, positively charged central core of an atom. **(p. 753)**

primera ley de Newton: Establece que un objeto en reposo permanecerá en reposo, y un objeto en movimiento continuará moviéndose a una velocidad constante en línea recta, si y sólo si la fuerza neta que actúa sobre ese objeto es igual a cero. **(pág. 98)**

segunda ley de Newton: Establece que la aceleración de un objeto es directamente proporcional a la fuerza neta en el e inversamente proporcional a la masa del objeto que se acelera. **(pág. 96)**

segunda ley de Newton para el movimiento de rotación: Establece que la aceleración angular de un objeto es directamente proporcional a la torsión neta e inversamente proporcional a sul momento de inercia. **(pág. 216)**

tercera ley de Newton: Establece que todas las fuerzas vienen en pares y que las dos fuerzas en un par actúan sobre objetos diferentes, son iguales en intensidad y son opuestas en sentido. **(pág. 106)**

nodo: El punto estacionario en el que dos ondas con igual pulso se encuentran y están en la misma ubicación, teniendo un desplazamiento de cero. **(pág. 396)**

normal: La línea en un diagrama de rayos que muestra la orientación de la barrera o el espejo y esta dibujada en un ángulo recto, o perpendicular, a la barrera o al espejo. **(pág. 398)**

fuerza normal: La fuerza de contacto perpendicular ejercida por una superficie sobre un objeto. **(pág. 111)**

reacción nuclear: Se produce cuando cambia la energía o el número de neutrones o protones en un núcleo. **(pág. 810)**

nucleon: Una partícula encontrada en el núcleo de un átomo; es decir, un protones o un neutron. **(pág. 802)**

núcleo: Un centro pequeño y masiva con carga positiva de un átomo. **(pág. 753)**

O

object: A luminous or illuminated source of light rays. **(p. 468)**

opaque (oh PAYK): A property of a medium that allows that medium to absorb light and reflect some light rather than transmitting it, preventing objects from being seen through it. **(p. 440)**

open-pipe resonator: A resonating tube with both ends open that also will resonate with a sound source; its resonant frequencies are whole-number multiples of the fundamental. **(p. 420)**

origin: The point at which both variables in a coordinate system have the value zero. **(p. 37)**

objeto: Una fuente luminosa o iluminada por los rayos de luz. **(pág. 468)**

opaco: Una propiedad de un medio que permite al medio absorber la luz y reflejar algo de la luz sin transmitirla, previene que los objetos sean vistos a través de él. **(pág. 440)**

resonador de tubo abierto: Un tubo de resonancia con ambos extremos abiertos, que también resonara con una fuente de sonido; sus frecuencias de resonancia son múltiplos de números enteros de la fundamental. **(pág. 420)**

origen: El punto en un sistema de coordenadas en que ambas variables tienen el valor cero. **(pág. 37)**

P

pair production: The conversion of energy into a matter-antimatter pair of particles. **(p. 824)**

parallel circuit: A type of electric circuit in which there are several current paths; its total current is equal to the sum of the currents in the individual branches, and if any branch is opened, the current in the other branches remains unchanged. **(p. 630)**

parallel connection: A type of connection in which there are two or more current paths to follow. **(p. 608)**

particle model: A simplified version of a motion diagram in which the moving object is replaced by a series of single points. **(p. 36)**

pascal: The SI unit of pressure; equal to 1 N/m^2. **(p. 349)**

Pascal's principle: States that any change in pressure applied at any point on a confined fluid is transferred undiminished throughout the fluid. **(p. 359)**

period: In any periodic motion, the amount of time required for an object to repeat one complete cycle of motion. **(p. 382)**

periodic motion: Any motion that repeats in a regular cycle. **(p. 382)**

periodic wave: A wave whose disturbances occur at a constant rate. **(p. 388)**

photoelectric effect: The emission of electrons by certain metals that occurs when they are exposed to electromagnetic radiation. **(p. 731)**

photon: A discrete, quantized bundle of radiation that travels at the speed of light, has zero mass, and has energy and momentum. **(p. 732)**

physics: The branch of science that studies matter and energy and their relationships. **(p. 4)**

piezoelectricity: The property of a crystal that causes it to bend or deform, producing electric vibrations, when a voltage is applied across it. **(p. 717)**

pitch: The highness or lowness of a sound, which depends on the frequency of vibration; can be given a name on the musical scale. **(p. 413)**

plane mirror: A flat, smooth surface from which light is reflected by specular reflection, producing a virtual image that is the same size as the object, has the same orientation, and is the same distance from the mirror as the object. **(p. 468)**

producción de pares: La conversión de energía en un par de materia-antimateria de partículas. **(pág. 824)**

circuito paralelo: Un tipo de circuito eléctrico en el que hay varios trayectos de corrientes, y su corriente total es igual a la suma de las corrientes en las ramas individuales, y si cualquiera de las ramas se abre, la corriente en las otras ramas se mantiene sin cambios. **(pág. 630)**

conexión en paralelo: Un tipo de conexión en la que hay dos o más trayectos de corrientes para seguir. **(pág. 608)**

modelo de partículas: Una versión simplificada de un diagrama de movimiento en el que se sustituye el movimiento del objeto por una serie de puntos individuales. **(pág. 36)**

pascal: Unidad de presión en el SI; igual a 1 N/m^2 **(pág. 349)**

principio de Pascal: Establece que la presión ejercida en cualquier lugar en un fluido encerrado se transfiere por igual en todas las direcciones a través del fluido. **(pág. 359)**

período: En cualquier movimiento periódico, es la cantidad de tiempo necesaria para que un objeto repita un ciclo completo de movimiento. **(pág. 382)**

movimiento periódico: Movimiento que se repite en intervalos regulares de tiempo. **(pág. 382)**

onda periódica: Onda cuyas alteraciones ocurren a un ritmo constante. **(pág. 388)**

efecto fotoeléctrico: Consiste en la emisión de electrones por ciertos metales que ocurre cuando estan expuestos a la radiación electromagnética. **(pág. 731)**

fotón: Paquete cuántico, discreto, nulo en masa, que porta radiación y viaja a la velocidad de la luz, tiene energía e impulso. **(pág. 732)**

física: La rama de la ciencia que estudia la materia y la energía así como sus relaciones. **(pág. 4)**

piezoelectricidad: La propiedad del cristal de doblarse o deformarse, que al aplicársele un voltaje produce vibraciones eléctricas. **(pág. 717)**

tono: Lo más alto o más bajo de un sonido, dependiendo de la frecuencia de vibración; puede dársele un nombre en la escala musical. **(pág. 413)**

espejo plano: Superficie plana y pulida que refleja luz por reflexión especular, produciendo imágenes virtuales del mismo tamaño del objeto, tiene la misma orientación, y tiene la misma distancia del espejo que del objeto. **(pág. 468)**

plasma: A gaslike, fluid state of matter made up of negatively charged electrons and positively charged ions that can conduct electric charge; makes up most of the matter in the universe, such as stars. **(p. 355)**

polarization: Production of light with a specific pattern of oscillation. **(p. 451)**

polarized: The characteristic of magnets that they have two opposite ends called poles. **(p. 650)**

position: The distance and direction from the origin to an object. **(p. 37)**

position-time graph: A graph that can be used to determine an object's velocity and position, as well as where and when two objects meet, made by plotting the time data on a horizontal axis and the position data on a vertical axis. **(p. 41)**

potential energy: Stored energy due to the interactions between objects in a system. **(p. 294)**

power: The rate at which energy is transformed. **(p. 271)**

precision: A characteristic of a measured value describing the degree of exactness of a measurement. **(p. 15)**

pressure: The perpendicular component of a force on a surface divided by the surface's area. **(p. 349)**

primary color: Red, green, and blue, which can be combined to form white light and mixed in pairs to produce the secondary colors yellow, cyan, and magenta. **(p. 449)**

primary pigment: Cyan, magenta, and yellow, each of which absorbs one primary color from white light and reflects two primary colors; can be mixed in pairs to produce the secondary pigments red, green, and blue. **(p. 449)**

principal axis: A straight line perpendicular to the surface of a mirror or lens that divides the mirror or lens in half. **(p. 471)**

principal quantum number: The integer (n) that determines the quantized values of r and E for an electron's orbital radius in hydrogen and the energy of a hydrogen atom. **(p. 760)**

principle of superposition: States that the displacement of a medium caused by two or more waves is the algebraic sum of the displacements of the individual waves. **(p. 395)**

projectile: An object shot through the air, such as a football, that has independent vertical and horizontal motions and, after receiving an initial thrust, travels through the air only under the force of gravity. **(p. 152)**

plasma: Un gas constituido, un estado fluido de la material hecho de carga negativa de electrones y de iones cargados positivamente que pueden conducir carga eléctrica; produce la mayoría de la material del universo, tales como las estrella **(pág. 355)**

polarización: La producción de la luz con un patrón especifico de oscilación . **(pág. 451)**

polarizado: La característica de los imanes de tener dos extremos llamados polos. **(pág. 650)**

posición: La distancia y dirección de un objeto con referencia a su origen. **(pág. 37)**

grafico de posición-tiempo: Una gráfica que puede utilizarse para determinar la velocidad y posición de un objeto, así como dónde y cuándo dos objetos se encuentran, hecho por trazar los datos del tiempo en el eje horizontal y los datos de posición en el eje vertical. **(pág. 41)**

potencial energético: Energía almacenada debido a las interacciones entre los objetos en un sistema. **(pág. 294)**

poder: Tasa a la cual la energía se transforma. **(pág. 271)**

precisión: Una característica de la medida de un valor que describe el grado de exactitud de la medición. **(pág. 15)**

presión: El componente perpendicular de una fuerza sobre una superficie, dividida entre el área de la superficie. **(pág. 349)**

colores primarios: rojo, verde y azul, pueden combinarse para formar luz blanca y mezclarse en parejas para producir los colores secundarios: amarillo, cian y magenta. **(pág. 449)**

pigmento primario: cian, magenta y amarillo, cada uno de ellos absorbe un color primario de la luz blanca y refleja dos colores primarios; se pueden mezclar en parejas para producir los pigmentos secundarios: rojo, verde y azul. **(pág. 449)**

eje principal: Una línea recta perpendicular a la superficie de un espejo o una lente que divide el espejo o la lente por la mitad. **(pág. 471)**

número cuántico principal: El número entero (n) que determina los valores cuánticos de r y E para el radio del orbita de un electrón en el hidrógeno y la energía de un átomo de hidrógeno. **(pág. 760)**

principio de superposición: Establece que el desplazamiento de un medio causado por dos o más ondas es la suma algebraica de los desplazamientos de las ondas individuales. **(pág. 395)**

proyectil: Objeto lanzado en el aire, como un balón de fútbol, que tiene movimientos verticales y horizontales independientes, que después de recibir un impulso inicial, viaja por el aire únicamente por la fuerza de la gravedad. **(pág. 152)**

Q

quadratic relationship: A parabolic relationship that results when one variable depends on the square of another variable. **(p. 21)**

quantized: The property of energy that it exists only in bundles of specific amounts. **(p. 730)**

quantum mechanics: The study of the properties of matter using its wave properties. **(p. 766)**

quantum model: A model of the atom that predicts only the probability that an electron is in a specific region. **(p. 766)**

quark: A tiny elementary particle that forms mass and can combine with other quarks to form larger particle such as protons, neutrons, and pions. **(p. 822)**

relación cuadrática: Una relación parabólica que se produce cuando una variable depende del cuadrado de otra variable. **(pág. 21)**

cuantizada: La propiedad de la energía que existe en paquetes con cantidades específicas. **(pág. 730)**

mecánica cuántica: El estudio de las propiedades de la materia utilizando sus propiedades ondulatorias. **(pág. 766)**

modelo cuántico: Un modelo del atomo que solamente predice la probabilidad de que un electrón se encuentra en una región especifica. **(pág. 766)**

quarks: Partícula elemental diminuta que forma masa y puede combinarse con otros quark a form una partícula mas grande como un protón, neutrón y pión. **(pág. 822)**

R

radian: $\frac{1}{2}\pi$ of a revolution; abbreviated rad. **(p. 204)**

radiation: The thermal transfer of energy by electromagnetic waves. **(p. 325)**

radioactive: Materials with nuclei that emit particles and energy. **(p. 808)**

ray: A line drawn at a right angle to a wavefront; represents the direction of wave travel. **(p. 397)**

Rayleigh criterion: States that if the central bright spot of one image falls on the first dark ring of the second image, the images are at the limit of resolution. **(p. 538)**

ray model of light: A model that represents light as a ray that travels in a straight path, whose direction can be changed only by encountering a boundary. **(p. 439)**

real image: An optical image that is formed by the converging of light rays. **(p. 472)**

receiver: A device that converts oscillating potential differences in an antenna to sound, pictures, or data. **(p. 719)**

reference frame: A coordinate system from which motion is viewed. **(p. 164)**

reference level: The position where gravitational potential energy is defined as zero. **(p. 295)**

reflected wave: A returning wave that results from some of the energy of the incident wave's pulse being reflected backward. **(p. 394)**

refraction: The change in direction of waves at the boundary between two different mediums. **(p. 399)**

radián: $\frac{1}{2}\pi$ de una revolución; se abrevia *rad.* **(pág. 204)**

radiación: La transferencia térmica de energía por ondas electromagnéticas. **(pág. 325)**

radiactivos: Los materiales con núcleos que emiten partículas y energía. **(pág. 808)**

rayo: Una línea trazada en ángulo recto al frente de una onda; representa el sentido del recorrido de la onda. **(pág. 397)**

criterio de Rayleigh: Establece que, si el punto central brillante de una imagen cae en el primer anillo de oscuridad de una segunda, las imágenes estarán en el límite de la resolución. **(pág. 538)**

modelo de rayos de luz: Un modelo que representa la luz como un rayo que recorre una trayectoria recta, cuya dirección solo puede cambiar al encontrar una barrera. **(pág. 439)**

imagen real: Una imagen óptica que se forma por la convergencia de los rayos de luz. **(pág. 472)**

receptor: Un dispositivo que convierte diferencias potenciales oscilantes en una antena al sonido, imágenes, o data. **(pág. 719)**

marco de referencia: Un sistema de coordenadas a través del cual se observa el movimiento. **(pág. 164)**

nivel de referencia: La posición en la el potencial de energía gravitatoria es igual a cero. **(pág. 295)**

onda reflejada: Onda rebotada que proviene del reflejo del pulso de la energía en retroceso de la onda incidente. **(pág. 394)**

refracción: El cambio de dirección que experimenta una onda al pasar de un medio a otro. **(pág. 399)**

Glossary ▪ Glosario

resistance: The measure of how strongly an object or material impedes the flow of electric charge produced by a potential difference; equal to the potential difference divided by the current. **(p. 604)**

resistance force: The force exerted by a machine. **(p. 274)**

resistor: A device with a specific resistance; may be made of long, thin wires, graphite, or semiconductors and often is used to control the current in circuits or parts of circuits. **(p. 605)**

resonance: A special form of periodic motion that occurs when small forces are applied at regular intervals to an oscillating or vibrating object and the amplitude of the vibration increases. **(p. 386)**

resultant: A vector that represents the sum of two other vectors; it always points from the first vector's tail to the last vector's tip. **(p. 40)**

rotational kinetic energy: Kinetic energy of a system due to its rotational motion, proportional to the system's moment of inertia and the square of its angular velocity. **(p. 294)**

resistencia: La medida de la intensidad de oposición de un objeto o material ante un flujo de carga eléctrica producida por una diferencia de potencial; es igual a la diferencia de potencial dividida por la corriente. **(pág. 604)**

fuerza de resistencia: La fuerza ejercida por una máquina. **(pág. 274)**

resistor: Un dispositivo con una resistencia específica, puede ser hecho de alambres largos y delgados, de grafito, o de semiconductores y, a menudo se utiliza para controlar la corriente en circuitos o partes de los circuitos. **(pág. 605)**

resonancia: Una forma especial de movimiento periódico que ocurre cuando se aplican fuerzas pequeñas oscilantes o vibrantes a un objeto en intervalos regulares, aumentando la amplitud de la vibración. **(pág. 386)**

resultante: Un vector que representa la suma de dos vectores; siempre apunta desde la cola del primer vector a la cabeza del último vector. **(pág. 40)**

energía cinética de rotación: La energía cinética de un sistema debido a su movimiento de rotación, es proporcional al momento de inercia del sistema y el cuadrado de su velocidad angular. **(pág. 294)**

S

scalar: A quantity, such as temperature or distance, that is just a number without any direction. **(p. 38)**

scientific law: A statement about what happens in nature and seems to be true all the time. **(p. 8)**

scientific methods: The patterns of investigation procedures. **(p. 5)**

scientific theory: An explanation of things or events based on knowledge gained from many observations and investigations. **(p. 8)**

secondary color: Yellow, cyan, and magenta, each of which is produced by combining two primary colors. **(p. 449)**

secondary pigment: Red, green, and blue, each of which absorbs two primary colors from white light and reflects one primary color; can be produced by mixing pairs of cyan, magenta, and yellow pigments. **(p. 450)**

second law of thermodynamics: States that whenever there is an opportunity for energy dispersal, the energy always spreads out; states that natural processes go in a direction that maintains or increases the total entropy of the universe. **(p. 337)**

escalar: Una cantidad que es meramente un numero sin dirección, como la de la temperatura o la de la distancia. **(pág. 38)**

ley científica: Declaración sobre lo que sucede en la naturaleza y parece ser verdadera todo el tiempo. **(pág. 8)**

métodos científicos: Los patrones de los procedimientos de investigación. **(pág. 5)**

teoría científica: Una explicación de las cosas o eventos basándose en conocimientos adquiridos por varias observaciones e investigaciones. **(pág. 8)**

color secundario: Amarillo, cian y magenta, cada uno de los cuales es producido por la combinación de dos colores primarios. **(pág. 449)**

pigmento secundario: Rojo, verde y azul, cada uno de los cuales absorbe dos colores primarios de la luz blanca y refleja un color primario, puede ser producido por la mezcla de pares de pigmentos de cian, magenta y amarillo. **(pág. 450)**

segunda ley de la termodinámica: Establece que cuando hay una oportunidad para la dispersión de energía, la energía siempre se esparcirá; establece que los procesos naturales van en una dirección que mantienen o aumentan la entropía total del universo. **(pág. 337)**

self-inductance: The property of a wire, either straight or in a coil, to create an induced *EMF* that opposes the change in the potential difference across the wire. **(p. 687)**

semiconductor: Material that behaves as a conductor under certain conditions and as an insulator in others; can be used to make solid-state electronic components. **(p. 778)**

series circuit: A type of electric circuit in which there is only one current path and all current travels through each device; the current is the same everywhere and is equal to the potential difference divided by the equivalent resistance. **(p. 625)**

series connection: A type of connection in which there is only a single current path. **(p. 608)**

short circuit: Occurs when a very low resistance circuit is formed, causing a very large current that could easily start a fire from overheated wires. **(p. 635)**

significant figures: All the valid digits in a measurement, the number of which indicates the measurement's precision. **(p. 12)**

simple harmonic motion: A motion that occurs when the restoring force on an object is directly proportional to the object's displacement from the equilibrium position. **(p. 382)**

simple pendulum: A device that can demonstrate simple harmonic motion when its bob (a massive ball or weight), suspended by a string or light rod, is pulled to one side and released, causing it to swing back and forth. **(p. 386)**

solenoid (SOH luh noyd): A long coil of wire with many spiral loops that is attached to a circuit; fields from each loop add to the fields of the other loops, creating a greater total field strength. **(p. 656)**

sound level: A logarithmic scale that measures sound intensities; depends on the ratio of the intensity of a particular sound wave to the intensity of the most faintly heard sound; unit of measurement is the decibel (dB)**. (p. 413)**

sound wave: A pressure variation transmitted through matter as a longitudinal wave; it reflects and interferes and has frequency, wavelength, speed, and amplitude. **(p. 411)**

autoinducción: La propiedad de un alambre, recto o en forma de una bobina, de crear una fuerza electromotiva inducida opuesta al cambio en la diferencia potencial a lo largo del alambre. **(pág. 687)**

semiconductor: Un material que en determinadas condiciones se comporta como conductor y en otras como aislante; puede ser usado en la fabrica de componentes electrónicos de estado sólido. **(pág. 778)**

circuito en serie: Un tipo de circuito eléctrico en el que hay solamenta una trayectoria por el corriente y toda la corriente viaja a través de cada dispositivo; la corriente es la misma en todas partes y es igual a la diferencia de potencial entre el equivalente de resistencia. **(pág. 625)**

conexión en serie: Tipo de conexión en la que sólo existe un trayecto de corriente. **(pág. 608)**

cortocircuito: Se produce cuando se forma un circuito de muy baja resistencia, causando una corriente muy grande que por el sobrecalentamiento de conductores podría iniciar fácilmente un incendio. **(pág. 635)**

cifras significativas: Todos los dígitos válidos en la medición, indican la precisión de la medición. **(pág. 12)**

movimiento armónico simple: Movimiento periódico que resulta cuando la fuerza restauradora actúa sobre un objeto de manera directamente proporcional a su desplazamiento respecto a su posición de equilibrio. **(pág. 382)**

péndulo simple: Una herramienta que demuestra el movimiento armónico simple, cuando una masa puntual (una bola de masa o peso), suspendida de un hilo o una cuerda sin peso, oscila de un lado a otro. **(pág. 386)**

solenoide: Una bobina larga de cables enrollados en espiral que esta conectagdo a un circuito; los campos de un espiral se añaden a los campos de otras espirales, creando mayor fortaleza al campo total **(pág. 656)**

nivel de sonido: Una escala logarítmica que mide las intensidades de sonido; depende de la relación de la intensidad de una partícula de onda sonora con laintensidad del sonido más débilmente escuchado, su unidad de medida son los decibelios (dB). **(pág. 413)**

onda sonora: Una variación de la presión transmitida a través de la materia como una onda longitudinal reflejando e interfiriendo; tiene frecuencia, longitud de onda, velocidad y amplitud. **(pág. 411)**

specific heat: The amount of energy that must be added to a material to raise the temperature of a unit mass by one temperature unit; measured in J/kg·K. **(p. 325)**

specular reflection: A reflection produced by a smooth surface in which parallel light rays are reflected in parallel. **(p. 466)**

spherical aberration: The image defect of a spherical mirror or lens that does not allow parallel light rays far from the principal axis to converge at the focal point and produces an image that is fuzzy, not sharp. **(p. 474)**

Standard Model: A model of matter in which all elementary particles can be grouped into three families—quarks, leptons, and force carriers. **(p. 822)**

standing wave: A wave that appears to be standing still, produced by the interference of two traveling waves moving in opposite directions. **(p. 397)**

static friction: The force exerted on one surface by a second surface when there is no motion between the two surfaces. **(p. 130)**

step-down transformer: A type of transformer in which the voltage coming out of the transformer is smaller than the voltage put into the transformer. **(p. 688)**

step-up transformer: A type of transformer in which the voltage coming out of the transformer is larger than the voltage put into the transformer. **(p. 688)**

stimulated emission: The process that occurs when an excited atom is struck by a photon having energy equal to the energy difference between the excited state and the ground state—the atom drops to the ground state and emits a photon with energy equal to the energy difference between the two states. **(p. 767)**

streamlines: Lines representing the flow of fluids around objects. **(p. 366)**

strong nuclear force: An attractive force between nucleons that binds the nucleus together; is of the same strength between all nucleon pairs. **(p. 804)**

superconductor: A material with zero resistance that can conduct electricity without thermal energy transformations. **(p. 611)**

surface wave: Disturbance in which the medium's particles follow a circular path that is at times parallel to the direction of wave travel and at other times perpendicular to the direction of wave travel. **(p. 389)**

calor específico: La cantidad de energía que debe añadirse a un material para elevar la temperatura de una unidad de masa por una unidad de temperatura; se mide en J/kg·K. **(pág. 325)**

reflexión especular: Una reflexión producida por una superficie lisa en la que los rayos paralelos de luz se reflejan paralelamente. **(pág. 466)**

aberración esférica: Defecto en la imagen de un espejo o lente esférico que no permite que los rayos paralelos de luz alejados del eje principal converjan en el punto focal, produciendo una imagen borrosa y poco nítida. **(pág. 474)**

Modelo Estándar: Modelo de la materia en el que las partículas elementarías pueden ser agrupadas en tres familias—quarks, leptones, y portadores de fuerza. **(pág. 822)**

onda estacionaria: Una onda que parece permanecer confinada en un espacio, resulta de la interferencia de dos ondas que se mueven en direcciones opuestas. **(pág. 397)**

Fricción estática: La fuerza ejercida por una superficie sobre otra superficie y en donde no existe movimiento entre las dos superficies. **(pág. 130)**

transformador de reducción de voltaje: Un tipo de transformador en el que el voltaje de salida es menor que el voltaje de entrada en el transformador. **(pág. 688)**

transformador de elevación de voltaje: Un tipo de transformador en el que el voltaje saliente del transformador es mayor que el voltaje de entrada del transformador. **(pág. 688)**

emisión estimulada: Proceso que ocurre cuando un átomo excitado recibe un estimulo de un fotón que tiene energía igual a la diferencia de energía entre los dos estados el excitado y el estado fundamental—el átomo baja al estado fundamental y emite un fotón con energía igual a la diferencia de energía entre los dos estados. **(pág. 767)**

líneas de corriente: Líneas que representan el flujo de fluidos alrededor de los objetos. **(pág. 366)**

fuerza nuclear fuerte: Una fuerza de atracción entre nucleones y que se enlaza el nucleo; es de la misma fuerza entre todos los pares de nucleones. **(pág. 804)**

superconductor: Un material con resistencia igual a cero que puede conducir electricidad sin transformaciones de energía térmica. **(pág. 611)**

onda de superficie: Alteración en la que las partículas de un medio siguen un trayecto circular que es a veces paralelo al recorrido de la dirección de la onda y en otras ocasiones perpendicular al recorrido de la dirección de la onda. **(pág. 389)**

system: Object or objects of interest that can interact with each other and the external world. **(p. 91)**

sistema: Objeto u objetos de interés que pueden interactuar uno con el otro y con el mundo exterior. **(pág. 91)**

T

tension: The specific name for the force exerted by a rope or a string. **(p. 109)**

tensión: El nombre específico para la fuerza ejercida por una cuerda o una cadena. **(pág. 109)**

terminal velocity: The constant velocity of an object that is reached when the drag force equals the force of gravity. **(p. 105)**

velocidad terminal: La velocidad constante que un objeto alcanza cuando la fuerza de arrastre iguala la fuerza de gravedad. **(pág. 105)**

thermal conduction: The transfer of thermal energy when particles collide. **(p. 322)**

conducción térmica: La transferencia de energía térmica por la colisión de partículas. **(pág. 322)**

thermal energy: The sum of the kinetic energies and potential energies of the particles in a system. **(p. 300)**

energía térmica: La suma de las energías cinéticas y potenciales de las partículas en un sistema. **(pág. 300)**

thermal equilibrium: The state in which the rates of thermal energy transfer between two objects are equal and the objects are at the same temperature. **(p. 322)**

equilibrio térmico: El estado en cual las tazas de la transferencia de energía térmica entre dos objetos son igual y los objetos se encuentran a la misma temperatura. **(pág. 322)**

thermal expansion: A property of all forms of matter that causes the matter to expand, becoming less dense, when heated. **(p. 354)**

expansión térmica: Una propiedad de todas las formas de la materia que la capacita para extenderse, y volverse menos densa, al calentarse. **(pág. 354)**

thin-film interference: A phenomenon in which light waves reflect from separate boundaries of a thin film and experience constructive and destructive interference. **(p. 527)**

interferencia en película delgada: Un fenómeno en el que las ondas de luz se reflejan en fronteras distinctas de una película delgada y producen interferencia constructiva y destructiva. **(pág. 527)**

thin lens equation: States that the inverse of the focal length of a spherical lens equals the sum of the inverses of the image position and the object position. **(p. 503)**

ecuación de la lente delgada: Establece que la inversa de la distancia focal de una lente esférica es igual a la suma de las inversas de la posición de la imagen y la posición del objeto. **(pág. 503)**

threshold frequency: The certain minimum value at or above which the frequency of incident radiation causes the ejection of electrons from a metal. **(p. 732)**

umbral de frecuencia: El valor mínimo determinado o superior a él, al cual la frecuencia de la radiación incidente provoca la expulsión de electrones de un metal. **(pág. 732)**

time interval: The difference between two times. **(p. 38)**

intervalo de tiempo: La diferencia entre dos tiempos. **(pág. 38)**

torque: The combination of force and lever arm that can cause an object to rotate; the magnitude is equal to the force times the perpendicular lever arm. **(p. 209)**

par de torsión: La combinación de la fuerza y el brazo de palanca que puede causar la rotación de un objeto; la magnitud es igual a la fuerza multiplicada por el brazo de palanca perpendicular. **(pág. 209)**

total internal reflection: An optical phenomenon that occurs when light strikes a boundary between two mediums at an angle of incidence that is greater than the critical angle and all light reflects back into the region of the higher index of refraction. **(p. 496)**

reflexión interna total: Un fenómeno óptico que ocurre cuando la luz golpea a un límite entre dos medios en un ángulo mayor que el ángulo critico, y toda la luz se refleja de nuevo hacia la región con el mayor índice de refracción. **(pág. 496)**

trajectory: The path of a projectile through space. **(p. 152)**

trayectoria: El recorrido de un proyectil a través del espacio. **(pág. 152)**

transformer: A device that can decrease or increase the voltages in AC circuits with relatively little energy loss. **(p. 688)**

transformador: Un dispositivo que puede aumentar o disminuir el voltaje en circuitos CA con una pérdida de energía relativamente pequeña. **(pág. 688)**

transistor: A simple device made of doped semiconducting material that can act as an amplifier, converting a weak signal to a much stronger one. **(p. 791)**

translational kinetic energy: The energy of a system due to the system's change in position. **(p. 270)**

translucent (trans LEW sunt): A property of a medium that allows that medium to transmit light and reflect a fraction of the light, preventing objects from being seen clearly through it. **(p. 440)**

transmitter: a circuit that converts voice, music, pictures, or data to electronic signals, amplifies the signals, and then sends them to an antenna **(p. 714)**

transparent: A property of a medium that allows that medium to transmit light and reflect a fraction of the light, allowing objects to be seen clearly through it. **(p. 440)**

transverse wave: A wave that vibrates perpendicular to the direction of the wave's travel. **(p. 388)**

trough (TROF): The low point of a transverse wave. **(p. 390)**

transmisor: Un dispositivo simple hecho de material semiconductor dopado que puede actuar como amplificador, convirtiendo una señal débil en una señal mucho más fuerte. **(pág. 791)**

energía cinética de traslación: La energía de un sistema causada por un cambio de posición del sistema. **(pág. 270)**

translúcida: Una propiedad de un medio que permite transmitir la luz y reflejar una fracción de la luz, sin permitir que los objetos se vean claramente a través de él. **(pág. 440)**

transmitir: Un circuito que convierte voz, música, fotografías, o data en señales electrónicas, los amplifican, y los manda a una antena. **(pág. 714)**

transparente: Una propiedad de un medio que permite que el medio transmita la luz y reflejar una fracción de la luz, dejando que los objetos sean claramente visibles a través de él. **(pág. 440)**

onda transversal: Una onda que vibra perpendicularmente a la dirección del movimiento de la onda. **(pág. 388)**

depresión: El punto bajo de una onda transversal. **(pág. 390)**

U

uniform circular motion: The movement of an object at a constant speed around a circle with a fixed radius. **(p. 159)**

movimiento circular uniforme: El movimiento de un objeto a una velocidad constante alrededor de un círculo de radio fijo. **(pág. 159)**

V

vector: A quantity, such as position, that has both magnitude and direction. **(p. 38)**

vector resolution: The process of breaking a vector into its components. **(p. 125).**

velocity-time graph: A graph that has velocity plotted on the vertical axis and time plotted on the horizontal axis; its slope is the acceleration of the object whose motion is described by the graph. **(p. 63)**

virtual image: The image formed of diverging light rays. **(p. 468)**

volt: The unit of electric potential difference; equal to one joule per coulomb, 1 J/C. **(p. 578)**

voltage divider: A series circuit that is used to produce a potential difference source of desired magnitude from a battery with a higher potential difference; often is used with sensors such as photoresistors. **(p. 627)**

vector: Una cantidad, como la posición, que tiene tanto magnitud y dirección. **(pág. 38)**

resolución vectorial: El proceso de romper un vector en sus componentes. **(pág. 125)**

gráfica velocidad-tiempo: Un gráfico en la que la velocidad es trazada en el eje vertical y el tiempo en el eje horizontal; su pendiente describe la aceleración del objeto cuyo movimiento está descrito en el gráfico. **(pág. 63)**

imagen virtual: La imagen formada por rayos de luces divergentes. **(pág. 468)**

voltios: La unidad de la diferencia de potencial eléctrico; igual a un julio por culombio, 1 J/C. **(pág. 578)**

divisor de tensión: Un circuito en serie que se utiliza para producir un fuente de diferencia de una magnitud deseada por potencial de una batería con una diferencia potencial mayor; a menudo se usa con sensores como fotoresistores. **(pág. 627)**

W

watt (W): Unit of power; 1 J of energy transferred in 1 s. **(p. 271)**

wave: A disturbance that carries energy through matter or space; transfers energy without transferring matter. **(p. 388)**

wavefront: A line representing the crest of a wave in two dimensions that can show the wavelength, but not the amplitude, of the wave when drawn to scale. **(p. 397)**

wavelength: The shortest distance between points on a wave where the wave pattern repeats itself, such as from crest to crest or from trough to trough. **(p. 390)**

wave pulse: A single disturbance or bump that travels through a medium. **(p. 388)**

weak nuclear force: The interaction that acts in the nucleus during beta (β) decay; much weaker than the strong nuclear force. **(p. 825)**

weight: The gravitational force experienced by an object. **(p. 100)**

weightlessness: An object's apparent weight of zero that results when there are no contact forces supporting the object. **(p. 102)**

work: The transfer of energy that occurs when a force is applied through a distance; equal to the product of the system's displacement and the force applied to the system in the direction of displacement. **(p. 264)**

work-energy theorem: States that when work is done on a system, a change in energy occurs. **(p. 270)**

work function: The energy required to free the most weakly bound electron from a metal; measured by the threshold frequency in the photoelectric effect. **(p. 737)**

watt (W): Unidad de potencia; 1 J de energía transferida en 1 s. **(pág. 271)**

onda: Una perturbación que transporta energía a través de la materia o del espacio; transfiere energía sin transferir materia. **(pág. 388)**

frente de onda: Una línea que representa la cresta de una onda en dos dimensiones mostrando la longitud de onda, pero no la amplitud, cuando es dibujada a escala. **(pág. 397)**

longitud de onda: La distancia más corta entre dos puntos de una onda donde el patrón de la onda se repite, como el de cresta a cresta o de depresión a depresión. **(pág. 390)**

pulso de onda: Una sola o pulso único que viaja a través de un medio. **(pág. 388)**

fuerza nuclear débil: La interacción que durante la desintegración beta (β), actúa en el núcleo, es mucho más débil que la fuerza nuclear intensa. **(pág. 825)**

peso: La fuerza gravitatoria que experimenta un objeto. **(pág. 100)**

ingravidez: El estado de un objeto en el que su peso aparente es igual a cero que resulta cuando no hay fuerzas de contacto que apoya al objeto. **(pág. 102)**

trabajo: La transferencia de energía que se produce cuando se aplica una fuerza por una distancia; es igual al producto del desplazamiento del sistema y de la fuerza aplicada al sistema en la dirección al desplazamiento. **(pág. 264)**

teorema trabajo-energía: Establece que, cuando se realiza trabajo sobre un sistema, se genera un cambio de energía. **(pág. 270)**

función de trabajo: La energía mínima necesaria para liberar el enlace más débil entre los electrones en un metal; se mide por el umbral de frecuencia en el efecto fotoeléctrico. **(pág. 737)**

INDEX

Italic numbers = illustration/photo **Bold numbers = vocabulary term**

A

INDEX

E

INDEX

F

Index

INDEX

Index

N

Index

INDEX

Index

T

Physical Constants			
Quantity	**Symbol**	**Value**	**Approximate Value**
Atomic mass unit	u	$1.660538782 \times 10^{-27}$ kg	1.66×10^{-27} kg
Avogadro's number	N_A	$6.02214179 \times 10^{23}$ mol^{-1}	6.022×10^{23} mol^{-1}
Boltzmann's constant	k	$1.3806504 \times 10^{-23}$ $Pa{\cdot}m^3/K$	1.38×10^{-23} $Pa{\cdot}m^3/K$
Coulomb's constant	K	$8.987551788 \times 10^{9}$ $N{\cdot}m^2/C^2$	9.0×10^{9} $N{\cdot}m^2/C^2$
Elementary charge	e	$1.60217653 \times 10^{-19}$ C	1.602×10^{-19} C
Gas constant	R	8.314472 $Pa{\cdot}m^3/mol{\cdot}K$	8.31 $Pa{\cdot}m^3/mol{\cdot}K$
Gravitational constant	G	6.67428×10^{-11} $N{\cdot}m^2/kg^2$	6.67×10^{-11} $N{\cdot}m^2/kg^2$
Mass of an electron	m_e	$9.10938215 \times 10^{-31}$ kg	9.11×10^{-31} kg
Mass of a proton	m_p	$1.672621637 \times 10^{-27}$ kg	1.67×10^{-27} kg
Mass of a neutron	m_n	$1.674927211 \times 10^{-27}$ kg	1.67×10^{-27} kg
Planck's constant	h	$6.62606896 \times 10^{-34}$ $J{\cdot}s$	6.63×10^{-34} $J{\cdot}s$
Speed of light in a vacuum	c	2.99792458×10^{8} m/s	3.00×10^{8} m/s

SI Derived Units				
Quantity	**Unit**	**Unit Symbol**	**Unit Expressed in Base Units**	**Unit Expressed in Other SI Units**
Acceleration	meters per second squared	m/s^2	m/s^2	
Area	meters squared	m^2	m^2	
Capacitance	farad	F	$A^2{\cdot}s^4/(kg{\cdot}m^2)$	
Density	kilograms per meter cubed	kg/m^3	kg/m^3	
Electric charge	coulomb	C	$A{\cdot}s$	
Electric field	newtons per coulomb	N/C	$kg{\cdot}m/(A{\cdot}s^3)$	V/m
Electric resistance	ohm	Ω	$kg{\cdot}m^2/(A^2{\cdot}s^3)$	V/A
EMF	volt	V	$kg{\cdot}m^2/(A^2{\cdot}s^3)$	
Energy, work	joule	J	$kg{\cdot}m^2/s^2$	$N{\cdot}m$
Force	newton	N	$kg{\cdot}m/s^2$	
Frequency	hertz	Hz	s^{-1}	
Illuminance	lux	lx	cd/m^2	
Magnetic field	tesla	T	$kg/(A{\cdot}s^2)$	$N{\cdot}s/(C{\cdot}m)$
Potential difference	volt	V	$kg{\cdot}m^2/(A{\cdot}s^3)$	W/A or J/C
Power	watt	W	$kg{\cdot}m^2/s^3$	J/s
Pressure	pascal	Pa	$kg/m{\cdot}s^2$	N/m^2
Velocity	meters per second	m/s	m/s	
Volume	meters cubed	m^3	m^3	

PERIODIC TABLE OF THE ELEMENTS

Element — Hydrogen
Atomic number — 1
Symbol — H
Atomic mass — 1.008
State of matter

Gas
Liquid
Solid
Synthetic

Metal
Metalloid
Nonmetal
Recently observed

	1	2	3	4	5	6	7	8	9	10	11	12	13	14	15	16	17	18
1	Hydrogen 1 H 1.008																	Helium 2 He 4.003
2	Lithium 3 Li 6.941	Beryllium 4 Be 9.012											Boron 5 B 10.811	Carbon 6 C 12.011	Nitrogen 7 N 14.007	Oxygen 8 O 15.999	Fluorine 9 F 18.998	Neon 10 Ne 20.180
3	Sodium 11 Na 22.990	Magnesium 12 Mg 24.305											Aluminum 13 Al 26.982	Silicon 14 Si 28.086	Phosphorus 15 P 30.974	Sulfur 16 S 32.066	Chlorine 17 Cl 35.453	Argon 18 Ar 39.948
4	Potassium 19 K 39.098	Calcium 20 Ca 40.078	Scandium 21 Sc 44.956	Titanium 22 Ti 47.867	Vanadium 23 V 50.942	Chromium 24 Cr 51.996	Manganese 25 Mn 54.938	Iron 26 Fe 55.847	Cobalt 27 Co 58.933	Nickel 28 Ni 58.693	Copper 29 Cu 63.546	Zinc 30 Zn 65.39	Gallium 31 Ga 69.723	Germanium 32 Ge 72.61	Arsenic 33 As 74.922	Selenium 34 Se 78.96	Bromine 35 Br 79.904	Krypton 36 Kr 83.80
5	Rubidium 37 Rb 85.468	Strontium 38 Sr 87.62	Yttrium 39 Y 88.906	Zirconium 40 Zr 91.224	Niobium 41 Nb 92.906	Molybdenum 42 Mo 95.94	Technetium 43 Tc (98)	Ruthenium 44 Ru 101.07	Rhodium 45 Rh 102.906	Palladium 46 Pd 106.42	Silver 47 Ag 107.868	Cadmium 48 Cd 112.411	Indium 49 In 114.82	Tin 50 Sn 118.710	Antimony 51 Sb 121.757	Tellurium 52 Te 127.60	Iodine 53 I 126.904	Xenon 54 Xe 131.290
6	Cesium 55 Cs 132.905	Barium 56 Ba 137.327	Lanthanum 57 La 138.905	Hafnium 72 Hf 178.49	Tantalum 73 Ta 180.948	Tungsten 74 W 183.84	Rhenium 75 Re 186.207	Osmium 76 Os 190.23	Iridium 77 Ir 192.217	Platinum 78 Pt 195.08	Gold 79 Au 196.967	Mercury 80 Hg 200.59	Thallium 81 Tl 204.383	Lead 82 Pb 207.2	Bismuth 83 Bi 208.980	Polonium 84 Po 208.982	Astatine 85 At 209.987	Radon 86 Rn 222.018
7	Francium 87 Fr (223)	Radium 88 Ra (226)	Actinium 89 Ac (227)	Rutherfordium 104 Rf (261)	Dubnium 105 Db (262)	Seaborgium 106 Sg (266)	Bohrium 107 Bh (264)	Hassium 108 Hs (277)	Meitnerium 109 Mt (268)	Darmstadtium 110 Ds (281)	Roentgenium 111 Rg (272)	Copernicium 112 Cn (285)	Ununtrium * 113 Uut (284)	Ununquadium * 114 Uuq (289)	Ununpentium * 115 Uup (288)	Ununhexium * 116 Uuh (291)		Ununoctium * 118 Uuo (294)

The number in parentheses is the mass number of the longest lived isotope for that element.

* The names and symbols for elements 113, 114, 115, 116, and 118 are temporary. Final names will be selected when the elements' discoveries are verified.

Lanthanide series	Cerium 58 Ce 140.115	Praseodymium 59 Pr 140.908	Neodymium 60 Nd 144.242	Promethium 61 Pm (145)	Samarium 62 Sm 150.36	Europium 63 Eu 151.965	Gadolinium 64 Gd 157.25	Terbium 65 Tb 158.925	Dysprosium 66 Dy 162.50	Holmium 67 Ho 164.930	Erbium 68 Er 167.259	Thulium 69 Tm 168.934	Ytterbium 70 Yb 173.04	Lutetium 71 Lu 174.967
Actinide series	Thorium 90 Th 232.038	Protactinium 91 Pa 231.036	Uranium 92 U 238.029	Neptunium 93 Np (237)	Plutonium 94 Pu (244)	Americium 95 Am (243)	Curium 96 Cm (247)	Berkelium 97 Bk (247)	Californium 98 Cf (251)	Einsteinium 99 Es (252)	Fermium 100 Fm (257)	Mendelevium 101 Md (258)	Nobelium 102 No (259)	Lawrencium 103 Lr (262)

Z